Wylie and Churchill-Davidson's
A Practice of Anaesthesia

Sixth Edition

Edited by

Thomas E J Healy MSc, MD, FRCA

Professor of Anaesthesia, University of Manchester, and
Honorary Consultant Anaesthetist, Manchester Royal Infirmary,
Withington Hospital and Hope Hospital, Manchester, UK

Peter J Cohen MD JD

Formerly Professor of Anesthesia, University of Pennsylvania Medical Center,
Philadelphia, Pennsylvania, USA

Edward Arnold
A member of the Hodder Headline Group
LONDON   BOSTON   SYDNEY   AUCKLAND

2604
RD81

First published in Great Britain 1960 by Lloyd-Luke (Medical Books) Ltd
Second edition 1966
Third edition 1972
Fourth edition 1978
Fifth edition 1984
Sixth edition 1995 by Edward Arnold, a division of Hodder Headline PLC,
338 Euston Road, London NW1 3BH

Distributed in the Americas by Little, Brown and Company
34 Beacon Street, Boston, MA 02108

Whilst the advice and information in this book is believed to be true and
accurate at the date of going to press, neither the author[s] nor the publisher
can accept any legal responsibility or liability for any errors or omissions
that may be made. In particular (but without limiting the generality of the
preceding disclaimer) every effort has been made to check drug dosages;
however it is still possible that errors have been missed. Furthermore,
dosage schedules are constantly being revised and new side-effects
recognized. For these reasons the reader is strongly urged to consult the
drug companies' printed instructions before administering any of the drugs
recommended in this book.

British Library Cataloguing in Publication Data
A catalogue record for this book is available from the British Library

Library of Congress Cataloging-in-Publication Data
A catalog record for this book is available from the Library of Congress

ISBN 0 340 55309 X (hb)

1 2 3 4 5   95 96 97 98 99

Typeset in Century by
J&L Composition Ltd, Filey, North Yorkshire
Printed and bound in Great Britain by
The Bath Press, Avon

To Cindy – my best friend and wife

P J C

In gratitude to my mother and father who prepared me for life,
and to my wife, Lesley, and family.

T E J

Wylie and Churchill-Davidson's
A Practice of Anaesthesia

Contents

List of Contributors

A. P. Adams MB BS PhD FRCA FANZCA
Professor and Head, Department of Anaesthetics, Guy's Hospital, London, UK

S. P. Allison MD FRCP
Consultant Physician, University Hospital, Nottingham, UK

J. M. Anderton MB ChB DObstRCOG FRCA
Clinical Director/Consultant Anaesthetist, Department of Anaesthesia, Manchester Royal Infirmary, Manchester, UK

Jeffrey M. Baden MB BS FRCA MRCP(UK)
Professor, Department of Anesthesia, Stanford University School of Medicine, Stanford, California, USA; and Chief, Anesthesia Service (112A), Department of Veterans Affairs, Palo Alto, California, USA

Cedric Bainton MD
Professor of Anesthesia and Vice Chairman, Department of Anesthesia, University of California San Francisco; Chief of Anesthesia, San Francisco General Hospital, San Francisco, California, USA

Paul Beatty BSc MSc PhD CEng MBES FIPSM
Senior Lecturer in the Physics of Anaesthesia, Department of Anaesthesia, University of Manchester, Manchester Royal Infirmary, Manchester, UK

M. Bengtsson MD PhD
Professor Emeritus, Linköping University, Department of Anaesthesiology, University Hospital, Linköping, Sweden

D. R. Bevan MB FRCA MRCP
Professor and Head, Department of Anaesthesia, The University of British Columbia and Head, Department of Anaesthesia, Vancouver Hospital and Health Sciences Centre, Vancouver, British Columbia, Canada

Timothy Black MD
Anesthesiologist, Valley Anesthesiology Consultants, Ltd, Attending Staff at Good Samaritan Medical Center and Phoenix Children's Hospital, Phoenix, Arizona, USA

Thomas B. Boulton OBE MB BChir FRCA FDSRCS
Honorary Consulting Anaesthetist, Nuffield Department of Anaesthesia, Oxford, and The Royal Berkshire Hospital, Reading; Honorary Archivist, Association of Anaesthetists of Great Britain and Ireland

J. G. Bovill MD PhD FFARCSI
Professor of Anaesthesia, Department of Anaesthesiology, University Hospital Leiden, Leiden, The Netherlands

T. Andrew Bowdle MD PhD
Associate Professor of Anesthesiology and Pharmaceutics (Adjunct), University of Washington, Seattle, Washington, USA

D. Brennan QC LLB
Recorder of the Crown Court

Jay B. Brodsky MD
Professor of Anesthesiology, Stanford University School of Medicine, Stanford, California, USA

Allan C. D. Brown MB ChB DObst RCOG FRCA
Associate Professor, Departments of Anesthesiology and Otorhinolaryngology, Head and Neck Surgery, University of Michigan; and Director, Difficult Airway Clinic, University of Michigan Medical Center, Ann Arbor, Michigan, USA

Burnell R. Brown, Jr MD PhD, FRCA
Professor and Head, Department of Anesthesiology, University of Arizona College of Medicine, Tucson, Arizona, USA

Joan M. Brown MB ChB FRCA
Consultant Anaesthetist, Queen Victoria Hospital, East Grinstead, Sussex, UK

I. T. Campbell MD FRCA
Senior Lecturer and Honorary Consultant Anaesthetist, University of Manchester, Department of Anaesthesia, Withington Hospital, Manchester, UK

Mark A. Chaney MD
Assistant Professor of Anesthesiology, Loyola University Medical Center, Maywood, Illinois, USA

R. S. J. Clarke BSc MD(Belfast) PhD FRCA FFARCSI
Professor of Anaesthetics, The Queen's University of Belfast (retired); Consultant Anaesthetist, Royal Victoria and Belfast City Hospitals (retired); Past Dean, Faculty of Anaesthetists, Royal College of Surgeons in Ireland

Cynthia B. Cohen PhD JD
Fellow, Kennedy Institute of Ethics, Georgetown University, Washington, DC; Adjunct Associate, The Hastings Center, Briarcliff Manor, New York, USA

Peter J. Cohen MD JD
Formerly Professor of Anesthesia, University of Pennsylvania Medical Center, Philadelphia, Pennsylvania, USA; Graduate, Georgetown University Law Center, Washington, DC, USA

Griselda M. Cooper MB ChB FRCA
Senior Lecturer in Anaesthesia, Department of Anaesthesia and Intensive Care, The University of Birmingham; Honorary Consultant Anaesthetist, Birmingham Maternity Hospital, Birmingham, UK

Michael J. Cousins MD FANZCA FRCA
Professor, Department of Anaesthesia and Pain

Management, Royal North Shore Hospital and University of Sydney, St Leonards, New South Wales, Australia

J. M. Davies MSc MD FRCPC
Professor of Anaesthesia, Foothills Hospital and The University of Calgary, Calgary, Alberta, Canada

James A. DiNardo MD FAAP
Clinical Associate Professor of Anesthesiology, Director, Cardiac Anesthesia, College of Medicine, University of Arizona Health Sciences Center, Tucson, Arizona, USA

F. Donati PhD MD FRCPC
Harold Griffith Professor of Anaesthesia, McGill University; Anaesthetist, Royal Victoria Hospital, Montréal, Québec, Canada

John J. Downes MD
Anesthesiologist-in-Chief and Director, Department of Anesthesiology and Critical Care Medicine, The Children's Hospital of Philadelphia, Philadelphia, Pennsylvania; Professor of Anesthesia and Pediatrics, University of Pennsylvania, Philadelphia, Pennsylvania, USA

Roderic G. Eckenhoff MD
Assistant Professor of Anesthesia and Physiology, Department of Anesthesia and Institute for Environmental Medicine, University of Pennsylvania Medical Center, Philadelphia, Pennsylvania, USA

A. Eleri Edwards MB ChB(Hons) FRCA
Consultant Anaesthetist, Wrexham Maelor Hospital, Wrexham, Clwyd, North Wales, and Institute of Health Studies, Croesnewydd Hall, Croesnewydd Road, Wrexham, Clwyd, Wales

J. Denis Edwards FRCP
Consultant Physician, University Hospital of South Manchester, Manchester, UK

F. R. Ellis MB ChB PhD FRCA
Professor of Anaesthesia, University of Leeds, and Honorary Consultant Anaesthetist, St James's University Hospital, Leeds and Leeds General Infirmary, Leeds, UK

F. H. M. Engbers MD
Staff Anaesthetist, Department of Anaesthesiology, University Hospital Leiden, Leiden, The Netherlands

M. H. Faroqui MBBS FRCA
Consultant Anaesthetist and Honorary Senior Lecturer, University of Birmingham, Queen Elizabeth Hospital, Birmingham, UK

Leonard Firestone MD
Director, Cardiac Anesthesia, Presbyterian Hospital and the University of Pittsburgh, Pittsburgh, Pennsylvania, USA

Susan Firestone MD
Director, Cardiac Anesthesia, Children's Hospital of Pittsburgh and the University of Pittsburgh, Pittsburgh, Pennsylvania, USA

P. Foëx MA DPhil FRCA FANZCA
Nuffield Professor of Anaesthetics, University of Oxford and The Radcliffe Infirmary, Oxford, UK

Clifford B. Franklin MB FRCA MHSM CertMHS
Consultant Anaesthetist, and Chairman of the Medical Audit Committee, Wythenshawe Hospital, Manchester, UK

Edward J. Frink, Jr MD
Associate Professor, Department of Anesthesiology, University of Arizona College of Medicine, Tucson, Arizona, USA

Masahiko Fujinaga MD
Research Associate, Department of Anesthesia, Stanford University School of Medicine, Stanford, California, USA; and Research Associate Teratologist, Department of Anesthesiology (112A), Department of Veterans Affairs, Palo Alto, California, USA

Joseph H. Gaston MB FRCA FRCP(C)
Consultant Anaesthetist, Royal Victoria Hospital, Belfast, UK

E. C. Gordon-Smith MA FRCP FRCPath
Professor of Haematology, St George's Hospital Medical School, London, UK

Jeffrey B. Gross MD BSEE
Professor of Anesthesiology, University of Connecticut School of Medicine, Farmington, Connecticut, USA

Betty L. Grundy MD
Professor of Anesthesiology, Department of Anesthesiology, University of Florida College of Medicine, Gainesville, Florida, USA

C. William Hanson III MD
Assistant Professor of Anesthesia, Surgery and Internal Medicine, Department of Anesthesia, University of Pennsylvania, Philadelphia, Pennsylvania, USA

Stuart A. Hargrave MBChB FRCA
Consultant Anaesthetist, Royal Victoria Infirmary NHS Trust Hospitals, Newcastle upon Tyne; and Honorary Clinical Lecturer, University of Newcastle upon Tyne and the Dental Hospital and School, Newcastle upon Tyne, UK

Thomas E. J. Healy MSc MD FRCA
Professor of Anaesthesia, University of Manchester, and Honorary Consultant Anaesthetist, Manchester Royal Infirmary, Withington Hospital and Hope Hospital, Manchester, UK

John A. Henry MB FRCP
Consultant Physician, Poisons Unit, Guy's Hospital, London, UK

M. Herbert BA PhD CPsychol AFBPsS
Professor of Behavioural Medicine, International Medical College, Kuala Lumpur, Selangor, Malaysia; formerly Senior Lecturer, Behavioural Sciences Section, Nottingham University Medical School, Nottingham, UK

Carol A. Hirshman MD
Professor of Anesthesiology and Critical Care Medicine and Environmental Health Sciences (Physiology) and

Medicine (Pulmonary), The Johns Hopkins Medical Institutions, Baltimore, Maryland, USA

P. M. Hopkins MB BS FRCA
Senior Lecturer in Anaesthesia, University of Leeds, and Honorary Consultant Anaesthetist, St James's University Hospital, Leeds, UK

Christopher J. Hull MB BS DA FRCA
Professor of Anaesthesia, University of Newcastle, and Honorary Consultant Anaesthetist, Royal Victoria Infirmary, Newcastle upon Tyne, UK

Jennifer M. Hunter MB FRCA
Senior Lecturer in Anaesthesia, University Department of Anaesthesia, Royal Liverpool University Hospital, Liverpool, UK

Peter Hutton BSc PhD MB ChB FRCA
Professor of Anaesthesia, University of Birmingham; Honorary Consultant, South Birmingham Health Authority, Birmingham, UK

M. F. M. James MB ChB PhD FRCA
Professor and Chief Anaesthetist, Department of Anaesthetics, University of Cape Town Medical School, Cape Town, South Africa

E. F. Klein Jr MD
Professor and Chairman, Department of Anesthesiology, University of Arkansas for Medical Sciences, Little Rock, Arkansas, USA

Paul Knight MD PhD
Professor and Chairman, Department of Anesthesiology, School of Medicine and Biomedical Sciences, SUNY at Buffalo, New York, USA

Wilhelmina C. Korevaar MD FACPM
Clinical Associate Professor of Anesthesiology, Hahnemann University, Philadelphia, Pennsylvania, USA

Donald A. Kroll MD PhD
Professor of Anesthesiology, Department of Anesthesia, UCLA School of Medicine, Los Angeles, and Acting Chief, Anesthesiology Service, West Los Angeles VAMC, California, USA

J. Lance Lichtor MD
Associate Professor of Anesthesiology and Associate Chairman, Department of Anesthesia and Critical Care, University of Chicago, Chicago, Illinois, USA

M. J. Lindop MA MB FRCA
Consultant Anaesthetist, Department of Anaesthesia, Addenbrooke's Hospital, Cambridge, UK

R. A. Little PhD MRCPath
Honorary Professor of Surgical Science, North Western Injury Research Centre, University of Manchester, Manchester, UK

J. B. Löfström MD PhD
Associate Professor, Linköping University, Department of Anaesthesiology, Linköping University, University Hospital, Linköping, Sweden

Laurence Loh FRCA
Consultant Anaesthetist, Nuffield Department of Anaesthetics, The Radcliffe Infirmary NHS Trust, Oxford, UK

William David Lord BSc(Hons) MBChB FRCA
Consultant Paediatric Anaesthetist, Regional Paediatric Burns Unit, Booth Hall Children's Hospital, Manchester, UK

Bryan E. Marshall MD FRCA FRCP
Horatio C. Wood Professor of Anesthesia, and Director, Center for Anesthesia Research, Department of Anesthesia, University of Pennsylvania, Philadelphia, Pennsylvania, USA

Carol Marshall PhD
Associate Professor, Department of Anesthesia, University of Pennsylvania, Philadelphia, Pennsylvania, USA

I. McConachie MB ChB FRCA
Consultant Anaesthetist, Anaesthetic Department, The Royal Oldham Hospital, Oldham, UK

J. McGeachie MB ChB FRCA
Consultant Anaesthetist, Anaesthetic Department, The Royal Oldham Hospital, Oldham, UK

Alan J. McLintic MRCP(UK) FRCA
Senior Registrar in Anaesthetics, Western Infirmary, Glasgow, UK

H. J. McQuay DM FRCA
Clinical Reader in Pain Relief, Nuffield Department of Anaesthetics, University of Oxford; Honorary Consultant, Oxford Regional Pain Relief Unit, Oxford, UK

George H. Meakin MD FRCA DA
Senior Lecturer in Paediatric Anaesthesia, University of Manchester; Honorary Consultant Anaesthetist, Royal Manchester Children's Hospital, Manchester, UK

Rajinder K. Mirakhur MD PhD FRCA FFARCSI
Senior Lecturer in Anaesthetics, The Queen's University of Belfast; Consultant Anaesthetist, Royal Victoria and Belfast City Hospitals, Belfast, UK

Andrew J. Mortimer BSc MD FRCA
Consultant Anaesthetist, University Hospital of South Manchester, Withington Hospital, Manchester, UK

Stanley Muravchick MD PhD
Professor of Anesthesia at the Hospital of the University of Pennsylvania, Philadelphia, Pennsylvania, USA

Frank L. Murphy MD
Associate Professor of Anesthesia, The Hospital of the University of Pennsylvania, Philadelphia, Pennsylvania, USA

Michael H. Nathanson MB BS MRCP FRCA
Visiting Assistant Professor, Department of Anesthesiology and Pain Management, University of Texas Southwestern Medical Center, Dallas, Texas, USA

Christopher D. Newson MB BS FRCA
Visiting Assistant Professor, Department of

Anesthesiology and Pain Management, University of Texas Southwestern Medical Center, Dallas, Texas, USA

Peter Nightingale FRCA MRCP
Director of Intensive Care and Consultant Anaesthetist, University Hospital of South Manchester, Manchester, UK

Martin L. Norton MSPH MD JD
Emeritus Professor of Anesthesiology, Emeritus Associate Professor of Otolaryngology, Founder, Difficult Airway Clinic, University of Michigan Medical Center, Ann Arbor, Michigan, USA

Charles W. Otto MD FCCM
Department of Anesthesiology/University of Arizona/ College of Medicine, Tucson, Arizona, USA

Barbara J. Pleuvry BPharm MSc PhD MRPharmS
Senior Lecturer in Anaesthesia and Biological Sciences, University of Manchester, Manchester, UK

B. J. Pollard BPharm MD FRCA
Senior Lecturer in Anaesthesia, University of Manchester; Honorary Consultant Anaesthetist, Manchester Royal Infirmary, Manchester, UK

C. J. D. Pomfrett BSc PhD
Lecturer in Neurophysiology applied to Anaesthesia, Department of Anaesthesia, University of Manchester, Manchester, UK

Michael F. Roizen MD
Professor and Chairman, Department of Anesthesia and Critical Care, Professor of Medicine, The University of Chicago, Pritzker School of Medicine, Chicago, Illinois, USA

Timothy W. Rutter MB BS LRCP MRCS BScHon FRCA
Department of Anesthesiology, University of Michigan Medical Center, Ann Arbor, Michigan, USA

John W. Sear MA BSc MBBS PhD FRCA
Reader in Anaesthetics, University of Oxford; Honorary Consultant Anaesthetist, Oxfordshire Health Authority; Nuffield Department of Anaesthetics, John Radcliffe Hospital, Oxford, UK

Hugh L. Seaton MB BS FANZCA
Staff Specialist, Department of Anaesthesia and Pain Management, Royal North Shore Hospital and University of Sydney, St Leonards, New South Wales, Australia

Harry A. Seifert MD
Assistant Professor of Anesthesia, University of Pennsylvania School of Medicine, Philadelphia, Pennsylvania, USA

C. J. Smith BSc PhD MIBiol CBiol MRSC CChem
Head of Multidisciplinary Research and Innovation Centre, North East Wales Institute; presently Principal Scientist, Cortecs Diagnostics, Newtech Square, Deeside Industrial Park, Deeside, Clwyd, North Wales

Theodore C. Smith MD
Professor of Anesthesiology, Loyola University Chicago, Illinois, USA

Kevin Spence MD
Assistant Professor in Anesthesiology, Department of Anesthesiology, University of Arkansas for Medical Sciences, Little Rock, Arkansas, USA

James M. Steven MD
Associate Anesthesiologist, Department of Anesthesiology and Critical Care Medicine, The Children's Hospital of Philadelphia, Philadelphia, Pennsylvania; Assistant Professor of Anesthesia, University of Pennsylvania, Philadelphia, Pennsylvania, USA

L. Strunin MD FRCA FRCPC
BOC Professor of Anaesthesia and Director of the Anaesthetics Unit, The London Hospital Medical College, University of London, London, UK

Alan Tait PhD
Associate Professor, Department of Anesthesiology, University of Michigan Medical Center, Ann Arbor, Michigan, USA

M. J. G. Thomas MA MB DTM&H L/RAMC
Commanding Officer, Army Blood Transfusion Service, Aldershot, Hampshire, UK

J. Gordon Todd FRCA
Consultant Anaesthetist, Western Infirmary, Glasgow, UK

Kevin K. Tremper MD PhD
Professor of Anesthesiology and Chairman, Department of Anesthesiology, University of Michigan Medical Center, Ann Arbor, Michigan, USA

J. M. Turner MBChB FRCA
Consultant Neuroanaesthetist, Addenbrooke's Hospital, Cambridge, UK

Jørgen Viby-Mogensen MD DMSc FRCA
Professor and Chairman, Department of Anaesthesia, National University Hospital, Rigshospitalet, University of Copenhagen, Copenhagen, Denmark

J. Michael Vollers MD
Associate Professor of Anesthesiology, Department of Anesthesiology, University of Arkansas for Medical Sciences, Little Rock, Arkansas, USA

Michael E. Weiss MD
Assistant Clinical Professor of Medicine, University of Washington School of Medicine, Division of Allergy, Seattle, Washington, USA

Paul F. White MD PhD FANZCA
Professor and Holder of the Margaret Milam McDermott Distinguished Chair of Anesthesiology, Department of Anesthesiology and Pain Management, University of Texas Southwestern Medical Center, Dallas, Texas, USA

David J. Wilkinson MB BS FRCA DRCOG
Consultant Anaesthetist, St Bartholomew's Hospital, London; Honorary Curator, Charles King Museum, Association of Anaesthetists of Great Britain and Ireland

Preface

It is over three decades since the First Edition of Wylie and Churchill-Davidson's *A Practice of Anaesthesia* was published. During these years, there have been fundamental changes in medicine, anaesthesia and society. Evolution, with its attendant progress and problems, is clearly evident in this, the Sixth Edition. Manifesting this vast increase in our knowledge, the number of chapters has increased by 50 per cent. No one nation, practice or academic institution has a monopoly of truth; the editors and contributors reflect the significant international and multidisciplinary nature of our specialty.

The book is directed both to the consultant or attending physician and to the registrar or resident in training; it should be of major assistance in giving the senior trainee and examination candidate the opportunity to gain a broad view of anaesthesia as a discipline. As surgery and those allied medical disciplines that require the assistance of anaesthesia have advanced, so has the complexity of the anaesthetic techniques that are required.

Since the appearance of the First Edition, new areas for surgical intervention have been born, ranging from transplantation of major organs to implantation of automatic cardiac defibrillators to surgery performed on the fetus *in utero*. Advances in monitoring techniques have opened new windows on the patient and, combined with ever-present vigilance, have granted us the opportunity to increase our patients' security during anaesthesia.

Our understanding and appreciation of anaesthesia at a molecular level has burgeoned. Knowledge of drug–receptor interactions is increasing rapidly. Use of the fundamental tools of genetic research to examine mechanisms of drug action is flourishing.

The first section introduces the reader to the history and theories of anaesthesia. Since the practice of anaesthesia cannot be separated from certain fundamental scientific principles, these are dealt with in depth. In addition to areas of obvious parochial focus, we have invited contributions dealing with topics of general importance such as operating room transmission of infection, hyperbaric medicine and the role of computers.

In the second section, we move into the theatre and examine the impact of disease and patient status on our practice. The broad areas of the anaesthetist's involvement throughout the perioperative period is apparent.

There is a necessary overlap between the second and third sections; the latter present a comprehensive and up-to-date discussion of the subspecialties of anaesthesia and the application of our practice to specific pathological conditions.

Finally, the last section illustrates the important and exciting ways our specialty interacts with society as a whole. Medicolegal principles (both in the UK and in the USA), quality assurance and medical ethics are thoroughly discussed.

Thomas E. J. Healy
Manchester, UK

Peter J. Cohen
Bethesda, Maryland, USA

Section One

History and Science

The Origins of Modern Anaesthesia

Thomas B. Boulton and David J. Wilkinson

The practice of modern anaesthesia is frequently, and rightly, dated from the demonstration of the inhalation of ether vapour as a means of allaying the pain of surgery by the dentist William Thomas Green Morton (1819–1868) at the Massachusetts General Hospital in Boston USA on 'Ether Day', Friday 16 October 1846.[1,2] Morton did not 'discover' anaesthesia. He was not the first person in history to relieve, or attempt to relieve, surgical pain, nor the first to use ether or the inhalational route for that purpose, but his personal inspiration, his professional conviction, and his successful administration of the right agent, before the right people, in the right place, at the right moment in history, ensured that the news of his success spread rapidly round the world and contributed to a revolution in the practice of surgery;[2–4] as Philip Rhodes has written, 'specific advances attributed to one person . . . do not happen by one person's agency alone, but arise from the climate of opinion and knowledge at the time'.[5]

Morton did not induce unconsciousness in his patient Gilbert Abbott on 16 October 1846;[6] nor probably did he intend to do so, for, like those pioneers who had preceded him and others who followed, the perceived aim was to relieve the pain of surgery rather than to produce oblivion to it. The state which Morton produced was what we would now call 'analgesia' ('a conscious or semiconscious state in which the pain is obtunded by the administration of anaesthetic drugs').[7]

Morton's Boston contemporary, the sage Oliver Wendell Holmes (1809–1894), anatomist, physician and poet, gave the name 'anaesthesia' (in the sense of 'loss of feeling') to Morton's process of 'etherization' in a letter written in the November following Morton's October 1846 demonstration.[1] The word 'anaesthesia' (a combination of two Greek words meaning without feeling) was not new, as Holmes acknowledges, nor does he seem to have been concerned whether or not the patient was conscious.[3] Armstrong Davison[8] and Morch and Major[9] tell us that originally 'anaesthesia' was used by the Greeks including the philosopher Plato (c. 428–c. 348 BC), to designate 'absence of feeling' in the sense of 'boorishness', but it was also employed by Dioscorides in the first century AD to denote the deliberate use of oral and rectal concoctions of mandragora and other herbs to relieve the pain of surgery.[10] The term was also employed by John Elliotson (1791–1868) in the early nineteenth century to describe the process of controlling surgical pain by hypnosis, even though it was more generally used at that time to denote loss of sensation due to neurological disease in the way it is still used by neurologists today.[3,8] If the neurological use is put aside, however, the word 'anaesthesia' employed on its own has come to be used for the scientific discipline devoted to the relief of the pain of surgery (which the Americans now call 'anesthesiology') by employing generally and locally acting drugs, as well as for the state of oblivion produced by some of those drugs.[11] It is therefore preferable to confine the use of the word 'anaesthesia' on its own to denote the scientific discipline, to use the term 'general anaesthesia' for the process of producing unconsciousness during surgery, 'analgesia' ('designating particularly the relief of pain without loss of consciousness') for the relief of pain by systemically acting agents in the conscious or semiconscious patient,[7] and 'local anaesthesia' (or perhaps more correctly 'local analgesia') for the process of abolishing the appreciation of pain in part of the body.[11]

The early anaesthetists in the second half of the nineteenth century were concerned only with relieving the pain of surgery and ensuring that the patient recovered from the effects of the agents required to achieve that end. That was sufficient so far as body surface operations

Table 1.1 The origins and development of modern anaesthesia

c. 3000 BC–c. 1750		Early attempts to relieve the pain of surgery
c. 1750	– 1846	Philosophy and science and the dawn of modern anaesthesia
1846	–c. 1880	Anaesthesia before the acceptance of antiseptic and aseptic surgery
c. 1880	–c. 1925	Anaesthesia and the challenge of major surgery
c. 1925	– 1942	New agents and new concepts
1942	– 1993	The introduction of curare and the revolution in the practice of anaesthesia

similar to those which had been practised before ether anaesthesia was demonstrated by Morton. The introduction of, first antisepsis, and then asepsis, led to a dramatic expansion in the scope of surgery in the 1880s, however, and this ultimately necessitated major changes in anaesthetic techniques and concern for the medical care of the patient undergoing surgery.

The modern physician anaesthetist (USA 'anesthesiologist') in the late twentieth century, in addition to safely securing unconsciousness and pain relief in the face of the very major and traumatic surgical procedures undertaken, now has to manage patients at the extremes of age and those suffering from major systemic disease. It is a compliment to the skills of the modern specialty of anaesthesia that few patients are considered too old, too young or too ill for modern surgical intervention, but it imposes a heavy responsibility.

Today's 'balanced anaesthetic' involves the use of the intravenous and parenteral as well as the inhalational routes of administration, the use of general and local analgesics to inhibit detrimental and somatic autonomic reflexes, and neuromuscular blocking agents to produce profound somatic muscular relaxation including respiratory paralysis, which in turn makes controlled ventilation mandatory. Six phases can be recognized (Table 1.1 and Appendix) but this chapter only deals in detail with the origins of anaesthesia and the events surrounding Morton's demonstration.

Early attempts to relieve the pain of surgery 3000 BC–c. 1750 AD

There is no doubt that from the earliest times attempts were made to relieve the pain of surgical intervention by the oral and rectal administration of herbal preparations.

Physical methods of local anaesthesia, such as the local application of ice or cold water and pressure on nerves, were also employed. It is, however, often difficult to differentiate between genuine contemporary descriptions, myth, and the attributions of subsequent authors sometimes writing many centuries later.[2,3,8–10] It is also important to draw a distinction between the use of drugs and other procedures to induce sleep or to relieve pain as such, and the deliberate employment of these agencies as a means of preventing surgical pain. Many compassionate physicians, including Hippocrates, were culturally ready to relieve the pain of disease and were qualified to use drugs, but contemporary members of the rougher craft of surgery had yet to understand the necessity and advantages of doing so, nor did they have the necessary knowledge or access to pharmacological agents.

Local anaesthesia

A possible use of local anaesthesia in the Old Kingdom of Egypt

It may be that the earliest known illustration of the application of pain relief for surgery is on a panel on the tomb of Ankh-ma-hor, vizier to the Egyptian Pharaoh Teti I, which has been dated around 2300 BC.[12] The panel depicts the preparation of the penis for circumcision as the hieroglyphic characters clearly indicate. The inscription includes the sentences 'The hem-ka priest is circumcising', 'I will make it comfortable', and 'Rub it well that it may be effective' which seem to indicate the application of some form of local anaesthesia. It is possible that the 'Stone of Memphis' (limestone dipped in vinegar releasing carbon dioxide) was being applied, although its effect would be more imagined than actual! It is unlikely that ice was used since it is thought that it was not available in Egypt at that time.[12]

Refrigeration anaesthesia

Armstrong Davison gives an interesting account of the use of refrigeration anaesthesia in his book *The evolution of anaesthesia*. The use of cold to deaden the pain of surgical incisions has been discovered, rediscovered and forgotten several times over the centuries.[8]

Hippocrates (c. 460–357 BC) was aware of the value of the analgesic action of cold water for the relief of swellings and pains in the joints as many physicians down the ages must have been, but there are no known records of the application of this knowledge for the relief of surgical pain until the eleventh century when an unknown Anglo-Saxon monk wrote: 'Again for an eruptive rash let him sit in cold water until it be deadened; then draw him up. Then cut four scarifications around the pocks and let drip as long as he will.'[8]

Johannes Costaeus de Costa of Venice in his *De ineis medicinae praesidiis* referred briefly to the use of cold water, snow and ice in 1595, and the great physician William Harvey (1578–1657), the discoverer of the circulation of blood, treated his own gout by the application of cold compresses after the manner of Hippocrates, but there is no evidence that he or his British colleagues even considered the possibility of the use of cold to lessen surgical pain.[8]

The Danish physician and anatomist Thomas Bartholin (1616–1680), the father of the discoverer of Bartholin's glands, visited and studied the methods of his 'friend and teacher' Marco Aurelio Severeno in 1646. In his treatise *De nivis usu medico* Bartholin describes how Severino cauterized ulcers of the limbs and cut the perineum for the stone painlessly, having first applied snow under the bandages for a quarter of an hour. Severino was crafty enough to dye the snow with 'ultramarine or some other colouring' to conceal the nature of his local anaesthetic agent from the patient.[8]

It will be seen later how refrigeration anaesthesia was discovered once again and received serious attention in the first half of the nineteenth century when clinicians began actively to consider controlling surgical pain.

Nerve compression

The famous French surgeon Ambroise Paré (1510–1590) advocated numbness brought about by the compression of the main nerves to the limb before an amputation.[8] This technique was later revived by Doctor James Moore towards the end of the eighteenth century. Moore invented a screw compressor for the purpose and successfully demonstrated it to the great surgeon John Hunter (1728–1793).[2,4,13] Nerve compression was again employed by Liegard in France in 1837 less than a decade before Morton's demonstration (see p. 15).[2,8]

Centrally acting agents

Oral and rectal preparations

Four principal herbal drugs capable of relieving pain or inducing amnesia or mental detachment were used in the pre-Christian era. These were ethyl alcohol, opium from the poppy *Papaver somniferum*, cannabis from the hemp plant *Cannabis sativa*, and concoctions from the root of the mandrake *Mandragora officinarum*. Mandragora extracts contain various alkaloids including hyoscyamine, hyoscine and mandragorine. The most potent of these from the point of view of the control of the pain of surgery is probably hyoscine with its amnesic effect.[2,8–10,12,14,15]

How far these agents were employed for the relief of the effects of the surgeon's knife in the millennia before the birth of Christ is uncertain, but Hippocrates (c. 460–357 BC) recommended a draught of mandragora juice mixed with wine before surgical intervention, and Herodotus (490–424 BC) recorded the use of both cannabis and mandragora before surgery. Cicero (106–43 BC), the Roman orator, writer and statesman, wrote a diatribe against surgeons who did not employ drugs to allay the pain of surgery, but ignored the undoubted dangers of using the then unstandardized concoctions of the day – an example of not completely informed lay political and journalistic criticism of the medical profession which is not unknown in the mid-twentieth century![10,15]

There were more specific references to the relief of surgical pain in the first century AD. The Roman Pliny (23–79 AD) and the Greek Plutarch (46–125 AD) were both familiar with the concept of the use of drugs to allay the pain of surgery. It is, however, from the writings of the Greek physician and surgeon Pedanius Dioscorides from Anazarba in Asia Minor that we gain the most information about first century anaesthesia.[2,8–10]

Dioscorides was at one time surgeon to the armies of the emperor Nero (37–68 AD) between 54 and 68 AD. He wrote a number of books on materia medica. He describes specifically and interestingly the addition of extracts of *Conium maculatum* (poison hemlock) of the genus *Cicuta* in combination with mandragora to control the initial excitatory spasms produced by the latter herb. *Conium maculatum* extracts contain coniine which has a curare-like action on the neuromuscular junction.[10] This is surely an early example of a 'balanced anaesthetic'! Snow,[2] Armstrong Davison[8] and Moser[10] cite a number of other references to preparations which were used, or allegedly used, for the control of the pain of surgery during the early centuries AD. They include the possible use of the inhalation of the fumes of cannabis by the Chinese surgeon Hoa-Tho in the third century AD,[2] although Armstrong Davison discounts this possibility for lack of contemporary evidence.[8] The references to oral and rectal preparations are mostly from Greek and Roman sources but they include translations of the work of the Arabian philosopher and physician Ibn Sina (979–1037), who is also known by his Latinized name Avicenna.[8,10] Ibn Sina wrote: 'If anyone wishes any of his members cut, let him drink three obols of [mandragora extract] in wine' – another 'balanced' combination of a soporofic without pain controlling properties combined with an analgesic![10]

The soporific sponge (spongia somnifera)

The earliest known description of the mysterious spongia somnifera is in the *Bamberg Antidotarium* of the ninth century, it appears and reappears frequently in the medical literature of the Middle Ages from the ninth to the thirteenth century, and there are occasional mentions up to the seventeenth century[15–18] and an echo in the nineteenth century (p. 19).[19]

A sea-sponge was soaked in various mixtures of solutions of herbal juices including opium, mandragora and hemlock. The sponge was then dried and, when it was needed to produce 'anaesthesia' for surgery, it was reconstituted by dipping it in warm water and applying it to the nostrils.[16,17]

Modern attempts to reproduce the effects of the sponge prepared according to the mediaeval recipes as an inhalational anaesthetic have, not surprisingly, been singularly unsuccessful; but can its alleged efficacy be explained?[17] Could it be that the non-volatile alkaloids dissolved in water were absorbed through the mucosa of the nose[10] or even that the reconstituted solution was squeezed into the mouth as a draught?[2,18] Is it possible that there was an element of hypnosis bolstered by the faith both of the patient and the physician? Yet another alternative suggestion has been made by Infusino *et al.*[17] This is that the sponge was applied as a 'final touch' after a strong 'wound draught' of opium in wine when the patient was already 'too drunk to care' in any case.[17]

The mandrake and witchcraft

Whatever can be said about the value or otherwise of the spongia somnifera, it was certainly less dangerous than the unstandardized use of oral and rectal concoctions of mandragora and other herbs. There were undoubtedly many deaths. Celsus, a lay author, as early as the late first century, and a few years later Galen (129–200), the Greek physician whose teaching dominated orthodox European opinion until well into the sixteenth century, urged caution in the case of soporific herbal concoctions. The use of oral and rectal soporific herbal preparations to control surgical pain by physicians gradually diminished as the centuries went by. This was undoubtedly because of the fatalities occasioned by the lack of standardization and the consequent unreliability. The recipes subsequently fell into the hands of 'wise women' and became associated with witchcraft.

The mandrake in particular came to be linked with black magic. The legends were exacerbated by the resemblance of the mandrake root to the human form. It was believed that the plant screamed as it was torn from the ground and that the screams could be fatal or induce madness if heard by human ears. The operation of uprooting the mandrake was consequently accompanied by magical rituals; a magic circle was drawn and the plant was tied to the tail of a dog to pull it out of the ground while trumpets were sounded to drown out the screams. Some herbalists were, however, less superstitious than others about the dangers associated with the procedure![8,14,15,18]

Mandragora was believed to have other powers in addition to its soporific properties. It was used in love and fertility potions, and it is interesting that, even in modern times, serious statistical evidence has been presented in India purporting to demonstrate that the sex of children can be determined by its administration during pregnancy.[18]

It was dangerous to be associated with the mandragora during the Middle Ages and even later. One of the charges of heresy against St Joan of Arc, who was burnt at the stake in 1431, was that she 'possessed a mandragora',[10] and physicians who were unlucky enough to accidentally cause death by administering concoctions of the root were likely to be arraigned for witchcraft and executed. It is sometimes asserted that the act of relieving pain was considered to be sinful, but it was rather the alleged criminal association with witchcraft which led to the death sentence. The surgeon Ambrose Paré (1510–1590) was a critic of the use of soporific herbal potions to overcome surgical pain[10] but, as has already been noted (see p. 5) he believed in nerve compression to relieve the pain of amputations.[8] After the introduction of chloroform in 1847, by James Young Simpson (1811–1870) of Edinburgh, religious objections were raised about the relief of labour pains by Victorian fundamentalists;[3] however, it is not correct to cite the case of Euphemia MacGalyean of Edinburgh as evidence that, in the sixteenth century, the mere act of seeking to relieve the pains of childbirth was a crime as some authors have done. Euphemia MacGalyean was not put to death by burning in 1590 because she sought to suppress her labour pains. She was condemned as a witch for other reasons; the subsidiary charge relating to her obstetric deliveries arose because of her use of magic powders and other occult measures for the purpose.[8]

The 'discovery' and early medical uses of ether

Claims have been made that various Arabian physicians in the eighth century, or Raymond Lully (1234–1314), a Majorcan who worked at the University of Padua in the thirteenth century, first prepared ether. These claims cannot be upheld.[8,16,20]

The claims of Paracelsus (1493–1541), the Swiss physician, his Italian pupil Valerius Cordus (1515–1544) who published the first 'modern' pharmacopoeia, or perhaps both working together, are more believable,[8,16,20] although the distillate they prepared and called 'sweet oil of vitriol' was probably or possibly a mixture of ether, alcohol and water. This combination was later used orally as an anodyne and known as 'Hoffman's drops'.[20] Paracelsus described how he fed this preparation to hens. This put them into a deep sleep from which they recovered unharmed.

The German chemist Wilhelm Godefroy Froben was the first to put the preparation of ether on a scientific basis in 1735 and to give the substance its modern name of 'aether'; a name derived from the Greek meaning 'burn brilliantly'.[20] Ether thereafter became a popular medicine. Mathew Turner (c. 1725–c. 1790), surgeon and chemist of Liverpool, described some of its uses in his pamphlet *An account of the extraordinary medicinal fluid called aether* published in 1761.[21] It was recommended as a cold surface application for headaches and migraines, as a draught added to water, lemon juice or wine, as an antispasmodic for colic or flatulence and 'snuffed' up the nostrils for 'stubborn' headaches, nervous diseases and fits.[21] The stage was set by the middle of the eighteenth century for the developments in the use of the inhalation of ether, first in pneumatic medicine as a putative cure for phthisis and other lung diseases, then as an intoxicant, and finally as a general anaesthetic in the middle of the next century (see pp. 10, 14–15, 18–26).

Philosophy and science in the eighteenth century

Changes in attitude to pain and disease

The changes in the attitude of man to pain and disease in Western civilization have been succinctly reviewed by both Nicholas Greene[22] and by Donald Caton.[23] There is no doubt that from the earliest times and throughout the Middle Ages pain was regarded as a punishment inflicted by the Almighty for some misdeed. This was especially true of mysterious pathological pain, for which there was no visible aetiology except the imposition of divine will, as opposed to traumatic pain, especially that due to the wounds of battle for which the cause was all too obvious.[22,23] The word pain itself is derived from the Greek poine (a penalty) via the Latin poena (a punishment).[22] On the other hand, in the Judeo-Christian ethic, pain and punishment, having been inflicted by God, could also be reduced by divine intervention as an act of redemption and atonement. The intervention of priests and physicians was consequently often regarded as an aid to the divine intention.[23] It was even considered to be merciful that the worst horrors of judicial execution could be mitigated for repentant criminals, as for example in the case of the offer of analgesics before or after crucifixion[16] or strangulation before burning.[8] The concept that it was believed that pain should not be treated because of its divinely inspired punitive nature has undoubtedly been overemphasized.[22]

The Renaissance in the fourteenth and fifteenth centuries heralded the age of enlightenment and scientific investigation which progressed in the sixteenth and seventeenth centuries. Man was thenceforth regarded as being potentially master of his own environment. Scientific investigation, which had previously been conducted for the glorification of God, began to be regularized and studied for its own sake as an independent discipline; at the same time the active treatment of pain was regarded as a humanitarian act with or without its religious overtones.[23] It was unfortunate that perception had outstripped practicality at the end of the eighteenth century. Standardized morphine had not yet been isolated from the crude drug and there were no effective methods of allaying the pain of surgery. In these circumstances the concept had developed that pain should be nobly borne with discipline, especially by males, and stoicism was regarded as evidence of virility. Those who bore pain with fortitude were highly regarded, but their stoicism was accepted as a matter of necessity rather than something that was desirable. It was certainly not considered to be sinful, or contrary to the will of God, to contemplate the possibility of relieving surgical pain by the beginning of the nineteenth century, but the chances of achieving that objective were generally believed to be impossible. It is also probable that the relief of the pain of surgery had a low priority for physicians faced with the hopeless task of alleviating the suffering of patients dying of the incurable infectious diseases of pneumonia, diphtheria, tuberculosis, malaria, typhoid and cholera.[22,23] By the beginning of the nineteenth century the medical profession had largely abandoned the draconian and ineffective remedies of previous centuries in favour of careful observation, accurate diagnosis and prognosis and preventive measures. Few genuinely effective pharmacological remedies were available and the only major medical advance was the introduction of vaccination against smallpox in 1796 by Edward Jenner (1749–1823) which was a preventive measure. It was not surprising in this atmosphere of 'therapeutic nihilism', that the next great medical advance of surgical anaesthesia should also be preventive in nature.[22–26]

Research and clinical practice in the eighteenth century

Surgical developments

Both the techniques of surgery and the status of surgeons underwent changes in Western Europe during the eighteenth century. Surgery was regarded as a lowly occupation in the early 1700s. Surgeons were despised by the physicians, who considered that the practice of surgery was beneath their dignity, but by 1800 things had changed. Surgery had by then emerged as a scientific discipline taught alongside medicine. This was as the result of the achievements of such men as William Cheselden (1688–1752), Percival Pott (1714–1788), John Hunter (1728–1793) and John Abernethy (1764–1831) in England, and Pierre Joseph Desault (1744–1795) in France. The association of surgeons with barbers was abolished by law in Paris in 1743, and in 1745 in London when the Company (later Royal College) of Surgeons was founded.[24]

The late eighteenth century surgeons were not brutes who gloried in the pain which they inflicted and the blood which they produced. Many of them regarded the performance of operations and the infliction of pain as a cruel necessity ('an humiliating example of the imperfections of science' as John Hunter, the greatest of them all, put it). They dreaded witnessing the distress and agony of their patients, but the only means of alleviating the suffering at their disposal were speed and, to a lesser extent, the manner in which they made their incisions.[24,25] The extraordinary fortitude with which their patients submitted to the agonies of surgery is beyond our comprehension in the late twentieth century.[27]

Cardiopulmonary physiology, chemistry and pneumatic medicine

Cartwright has rightly pointed out that inhalation anaesthesia was the outcome of pneumatic medicine, and that pneumatic medicine was the child of pneumatic chemistry, which in turn had its origins in the comprehension of the respiratory physiology.[28]

Both Leigh[29] and Calverley[30] have given valuable accounts of the long history of the elucidation of the process of respiration and of the function of the lungs and pulmonary circulation. Leonardo da Vinci (1452–1519) depicted and described the mechanical mechanism of inflation and deflation of the lungs, and demonstrated that air consisted of two portions, one of which supported life and combustion while the remainder did not. The eccentric and unpopular Swiss genius Paracelsus (Philippus Aureolus Theophrastus Bombastus von Hohenheim, 1493–1591) observed that the lungs consumed part of the air 'as a stomach concocts meat'. The great Andreas Vesalius of Paris demonstrated in 1542 how the heart of an animal with an open thorax would stop beating effectively unless the lungs were ventilated by blowing into the trachea through a reed, and he gave a description of ventricular fibrillation. The Portuguese scholar Michael Servetus (1511–1551) described the passage of the blood from the right ventricle through the lungs to the left side of the heart, and the change of colour as it did so due to the production of a vital spirit 'engendered . . . by the mingling of the impaired air with the more subtle portion of the blood'. The Italian Andreas Casalpinus (1519–1603) appreciated the principle of the circulation but gave contradictory accounts of the direction taken by the blood round the body, but it was left for William Harvey (1578–1657) of London to provide the definitive description of the circulation in 1616 and to publish it in his *Exercitatio anatomica de motu cordis et sanguinis in animalibus* in 1628. The insufflation experiments of the British experimenters Robert Boyle (1627–1691), Robert Hooke (1635–1703) and Richard Lower (1631–1691) of the Royal Society of London and Oxford University established that movement of the lungs was not necessary for the colour change from venous to arterial blood, and that the process of its passage through the pulmonary circulation blood removed some substance from the air.[29,30]

John Mayow (1645–1679), a medical practitioner in Bath, England, published the results of a remarkable series of experiments in 1674. Many of these were destined to be repeated in the next century by the English scientist Joseph Priestley (1733–1804), Antoine Laurent Lavoisier (1743–1794) of Paris, and others who were unaware of Mayow's work.[30–32] Mayow established that both combustion and respiration consumed a substance ('fire-air') from the atmosphere and that, when a flame died or a mouse expired in a confined atmosphere, a portion of atmospheric air was left which supported neither combustion nor life. He also suggested that something noxious was given out from the blood on expiration, and that the process of respiration was required to sustain muscular function including the action of the heart.[3,8,28–32]

Pneumatic chemistry and physiology and the work of Joseph Priestley

Jean Baptiste Helmont (1577–1644) of Brussels had been the first to recognize the existence of gaseous substances distinct from common air when he observed the effervescent reaction produced by the action of vinegar on calcium carbonate.[3,8,28–32]

Joseph Black (1728–1799), the Scottish physician and chemist, liberated carbon dioxide, which he called 'fixed air', in 1754 by heating calcium carbonate. He also established that carbon dioxide gas was present in expired air.[28–32]

The amazing genius Joseph Priestley (1733–1804), non-conformist theologian and self-taught scientist, turned to the study of chemistry in his thirties. He described a total of ten gases or 'factitious [manufactured or artificial] airs'. Amongst them were oxygen, which he prepared by heating mercuric oxide in 1771 and named 'dephlogisticated air' in 1775, nitrous oxide ('dephlogisticated nitrous air'), which he produced by exposing iron filings and brimstone (sulphur) to nitric oxide ('nitrous air') over water in 1772, and which he identified as a 'new gas' in 1773.[3,8,16,28–32]

There is some doubt as to whether Priestley was actually the first to 'discover' nitrous oxide and oxygen. Joseph Black may have prepared nitrous oxide by heating ammonium nitrate in 1766, but he did not publish his experiment nor identify nitrous oxide as a unique gas. The Swedish scientist Wilhelm Scheele also recorded the existence of oxygen in 1772 but he did not publish his results until 1777,[32] however, as Sir Philip Hartog (1864–1947) once wrote: 'The word "discovery" is often ambiguous in science and leads to misunderstandings'.[32] It will become apparent as the history of anaesthesia unfolds in this chapter that this remark could also be applied to a considerable number of 'discoveries' of special interest to anaesthetists, including inhalation anaesthesia itself (see pp. 18–29), local anaesthesia and the use of curare.

Priestley demonstrated that atmospheric air (common air) supported combustion and the life of small animals, that nitrous oxide also supported combustion but killed small animals, and that nitrogen ('phlogisticated air') and nitric oxide ('nitrous air') were distinct although neither supported combustion or animal life. He also established that air in which animals had died (by exhaustion of the oxygen and exhalation of carbon dioxide) could be made fit for respiration again by growing plants within it.[28–31]

Some of Priestley's earlier experiments were with the plentiful supply of carbon dioxide ('fixed air') generated over fermenting liquor in a brewery in Leeds. He demonstrated *inter alia* that frogs could be made to be apparently dead by exposure to carbon dioxide and yet would recover to normality.

Priestley also noted that carbon dioxide prevented putrefaction and suggested by extension that its inhalation might be used to treat disease. Medicinally he invented apparatus for producing soda water from carbon dioxide, and this was subsequently recommended to the Royal Navy by the College of Physicians as likely to be of use for the prevention of scurvy in the second voyage of exploration (1768–1775) by Captain James Cook (1728–1779).

Priestley's contributions to physics and electricity, inorganic chemistry and physiology were immense and seminal. However, it is apparent from the nomenclature he used that he stuck resolutely to the phlogiston theory. The adherents of this concept postulated that metals gave off 'phlogiston' when heated instead of combining with oxygen to produce an oxide.[31,32]

Priestley was awarded the degree of Doctor of Laws (LLD) by Edinburgh University in 1764 in recognition of his literary and theological publications, and was elected a Fellow of the Royal Society in 1776 in acknowledgement of his work on electricity which preceded his chemical experiments. He had advanced views on civil and religious liberty and toleration, however, and was a supporter of many of the philosophical aims of the French revolution. These beliefs led to his home in Birmingham being burnt down by a mob in 1791. He was rejected by his contemporaries and ultimately emigrated to the United States in 1794 and died in Northumberland, Pennsylvania in 1804.[3,8,16,31,32]

Antoine Laurent Lavoisier (1743–1794), an upper middle class university educated chemist and physiologist, and sometime liberal politician and senior revenue official, was Priestley's contemporary in France.[31,32] It is known that Priestley visited him in Paris and that he communicated with the Swede Scheele, and some have indeed dismissed Lavoisier as a plagiarist who merely modified the work of others. This would be a gross underestimation of his contribution, however. He demolished the phlogiston theory and correctly delineated the process of combustion, and he defined the active function in respiration of Priestley's 'dephlogisticated air' which he renamed 'oxygene', and of Black's 'fixed air' (carbon dioxide), and the passive role of nitrogen ('phlogisticated air'). He also compared respiration to slow combustion, but he was incorrect in postulating that the combustion took place in the lungs and in believing that heat ('caloric') was a transferable substance.[3,8,28–32]

Lavoisier's moderately liberal views were not liberal enough, however, and his early association with tax gathering was the final factor which led him to be condemned to the guillotine in the Terror of 1794.[3,8,16,28–32]

Pneumatic medicine

Priestley's suggestion, made in 1770, that soda water prepared from carbon dioxide might be used to prevent scurvy amongst sailors has already been noted. This concept was based on the theory that, since the disease was supposedly caused by the consumption of putrid stored water, the anti-putrescent effects of carbon dioxide could be utilized to prevent deterioration. Fermenting malt wort had in fact been utilized for the purpose since 1767 on a suggestion made by David MacBride (1726–1778), a physician practising in Dublin.[3,28,31]

Priestley also suggested that fixed air (carbon dioxide) might be used *per rectum* or by diffusion through the skin in the treatment of 'putrid disorders'. Various physicians followed up his suggestion for ailments as diverse as fever, diarrhoea and carcinoma of the breast. A solution of sodium tartrate impregnated with carbon dioxide was also used by mouth to treat gallstones.[32]

Priestley further suggested that the inhalation route might be tried. Thomas Percival (1740–1840) of Manchester demonstrated that carbon dioxide diluted with air could be easily inhaled and began to employ it in the treatment of tuberculosis in 1772, as also did William Withering (1741–1799) who introduced digitalis into medicine. Priestley was happily settled in Birmingham from 1780 to 1791. He was surrounded in the Midlands by a number of progressive physicians and scientists most of whom belonged to the Lunar Society (so called because it met for scientific and literary discussion at the time of the full moon so that its members could return home with greater ease). The membership included the physicians Richard Pearson (1765–1836), William Withering (1741–1799) and Erasmus Darwin (1731–1802), the grandfather of the future discoverer of the theory of evolution, the engineers Mathew Boulton (1728–1809) and James Watt (1736–1819), and the potter Josiah Wedgewood (1730–1795). It is not surprising that Birmingham became the centre of the vogue for pneumatic medicine. Priestley made many suggestions including the possible benefits of the inhalation of oxygen and experimented with the administration of nitrous oxide *per rectum*. Erasmus Darwin administered oxygen, nitrogen and hydrogen by inhalation and had a justifiable faith in the use of oxygen and ether inhalation for some types of asthma.[20,28]

Thomas Beddoes

Thomas Beddoes (1760–1808) was a likeable, compassionate and conscientious physician and no mean poet. He was blessed with an active and imaginative mind, but, unlike Priestley, he based his theories on conceptual intuition rather than experiment, as was commonplace in the period in which he lived.[3,8,28,33–38]

Beddoes was born at Shifnal, Shropshire into a family of landowning farmers. He was an intelligent and industrious schoolboy. He entered Pembroke College of the University of Oxford in 1776 where he diligently learned modern languages and became proficient in French, German, Italian and Spanish. He also studied scientific subjects including botany, geology and the new science of pneumatic chemistry with which he became obsessed for the next 25 years.[33–38] He proceeded to the degree of Bachelor of Arts in 1781. He studied medicine in Edinburgh and London and was awarded his Oxford MD in 1786. His sojourn at Edinburgh included studies under William Cullen (1710–1790), the Professor of Medicine, who certainly believed in the medical value of ether and possibly in its administration by inhalation.[28] This would not be surprising as the work of Turner (1761), to which previous reference has been made (see p. 7), was probably well known by this

time.[21] Beddoes also made good use of his linguistic talents and, amongst other works, translated Scheele's chemical essays which included the latter's work on oxygen.[28,35]

Beddoes took a holiday in Scotland and France after his MD had been conferred and visited Lavoisier in Paris. He was appointed Lecturer in Chemistry to the University of Oxford in 1787. He was an inspired and popular teacher noted for his spectacular experiments and, in 1790 he published an analytical dissertation on the works of John Mayow (see p. 8) who had anticipated the application of pneumatic chemistry 100 years previously. Beddoes also made the acquaintance of Erasmus Darwin during his time at Oxford, and through him Priestley and the other members of the Lunar Society interested in pneumatic medicine in Birmingham (see above).

Beddoes, like Priestley, had revolutionary political ideals. He supported the French Revolution although he was horrified by its excesses; in 1792 he wrote a pamphlet which criticized the younger Pitt's government for declaring war on France. His position in conservative Oxford became untenable and he resigned his lectureship.

Beddoes withdrew to his native Shropshire, where, besides writing a moral tale containing much social and medical advice for the poor entitled *The history of Isaac Jenkins*, his mind turned to the possible hope of utilizing pneumatic medicine as therapy for the then incurable diseases such as tuberculosis. He published an important paper on this subject about this time. This was probably written towards the end of his time at Oxford and was entitled *Observations on the nature and cure of calculus, sea-scurvy, catarrh and fever*.[28]

The origin and foundation of the Pneumatic Institution

Beddoes was persuaded by his friends to pursue his interest and, with their financial help, raised enough money to enable him to set up a laboratory to ascertain the effects of the various factitious airs on various diseases, and to determine the most convenient ways of manufacturing them and enabling patients to inhale them. On the recommendation of Erasmus Darwin and others of his Birmingham friends, Beddoes arrived in the fashionable Bristol, the second city of England, early in 1793 with the objectives of establishing himself as a physician and of experimenting with the gases. He was introduced by another member of the Lunar Society, the Irish author Richard Lovell Edgeworth (1744–1817), to the members of the brilliant intellectual and politically radical circle which existed in Bristol at that time either as residents or frequent visitors. This company included the poets Robert Southey (1774–1843), Samuel Taylor Coleridge (1772–1834) and William Wordsworth (1770–1850), Joseph Cottle, author and publisher, as well as the engineer James Watt (1736–

1819) and the potter Josiah Wedgewood (1730–1795) on their professional or leisure visits to the tin and clay mines of Cornwall and, not least, Davies Giddy (1767–1839) of Penzance (landowner, mathematician and scientist and a former Oxford pupil of Beddoes). Giddy later took his wife's name and ultimately became Sir James Gilbert, President of the Royal Society 1827–1830. All these remarkable men later figured in the events which led up to the introduction of anaesthesia in the 1840s.

Beddoes chose as his residence No. 3 Rodney Place, Clifton, and for his laboratory and clinic, a house in nearby Hope Square close to the medicinal springs of Hotwells, to which many despairing patients suffering from incurable diseases, including cancer and tuberculosis, came to take the waters in the hope of relief.[28]

Beddoes rapidly built up a successful medical practice and between 1793 and 1799 he conducted a series of experiments and clinical trials on the inhalation of gases. He first had the assistance of James Sadler (c. 1750–c. 1810), famous as an early balloonist, to design and manufacture his equipment. Later James Watt (1736–1819) collaborated with him and together they published *Considerations on the medicinal powers of factitious airs* in five parts between 1794 and 1796. James Watt had a very personal interest in pneumatic medicine – his son Gregory (1777–1804) was suffering from consumption (see p. 12).[28,29]

The agents available to Beddoes at that time were, in modern parlance, oxygen, carbon dioxide, hydrogen and impure carbon monoxide, as well as fumes produced by burning feathers or charred meat! All were inhaled in considerable dilution. Beddoes was a Brunonian (a disciple of John Brown (1735–1788), a Scottish physician whose theories he had absorbed while he was a student in Edinburgh). Brown classified diseases as either 'sthenic' or 'asthenic' (stimulatory or depressant) and believed that opposing treatments could restore the balance. He regarded tuberculosis as a stimulatory disease because of its feverish nature, and oxygen as a stimulant gas; his treatment was to dilute the oxygen in the air with nitrogen or hydrogen.[28,33] He not surprisingly achieved some success by treating shortness of breath from any cause (asthenic disease) with oxygen. The production of a pink flush to the face with dilute carbon monoxide in a case of anaemia was less appropriate! Some of his disciples were more enthusiastic about the results than Beddoes himself. 'Cures' of venereal disease and deafness and other diseases were reported at the height of enthusiasm for pneumatic medicine in the last decade of the eighteenth century!ative[28]

It has been said that the reigning Duchess of Devonshire visited Hope Square laboratory and suggested to Beddoes that he should expand his work and found a Pneumatic Medical Institution; be that as it may, Beddoes began to promulgate his ideas and raise further funds in 1794. He was now acknowledged as the leading authority in the country on pneumatic medicine and, by 1797, he had collected sufficient donations and subscriptions (including £1000 from the Wedgewoods) to found an institution.[28]

Beddoes' proposals were ambitious. They included plans for a large outpatient department and a small hospital (which in fact was never built), together with a fully equipped laboratory for the study of pneumatic chemistry and its application to man and animals in health and disease. The staff was to consist of physicians to attend the outpatients, a surgeon, nurses to look after the inpatients in the hospital, an 'operator' (to prepare the apparatus and gases under supervision) and a 'superintendent', who was required to be skilled in chemistry and take charge of the laboratory investigations. Beddoes estimated that 2 years of experimental investigation and clinical trial in such an institution would suffice to evaluate pneumatic medicine and would provide more information than could be obtained in 20 years of private practice.[28,31,33–38]

In October 1798 Beddoes appointed the 20-year-old Penzance surgeon's apprentice Humphry Davy, with whose early contribution to chemistry he was already acquainted (see p. 12), as superintendent on the suggestion of Davies Giddy.[28] Davy began work in the laboratory in Hope Square, but in November 1798 Beddoes took a lease on 6 and 7 Dowry Square. These houses were adapted and the apparatus transferred from Hope Square, and on 21 March 1799 the Pneumatic Medical Institution opened its doors.

The dream of Beddoes to revolutionize medical therapy through the inhalation route came to nothing. He had not even conceived of the notion that the pain of surgery might be relieved, but he had demonstrated that inhalation was a flexible and efficient route for applying medication to the body and, together with James Watt, he had designed and built an apparatus which was capable of delivering inhaled gases to the patient in measured volumes. It was left to Humphry Davy, his young but talented associate, to take the next step towards the introduction of surgical anaesthesia almost by chance, in the course of investigations along the lines which Beddoes had laid down.[28]

Humphry Davy and nitrous oxide 1798–1800

Humphry Davy (1778–1829) came from a family of Cornish farmers and craftsmen and was the protégé of an elderly surgeon from Penzance, John Tonkin. He had a pleasant rural childhood and had developed an amateur interest in natural history and became a gifted poet, but his grammar school education left much to be desired. Davy was apprenticed to another surgeon, John Bingham Borlaise, in 1795 at the age of 16. He apparently showed more interest in the pharmaceutical side of his master's practice than in surgery, and began to study chemistry and to conduct experiments with home-made apparatus on his own account in 1797. Davy met Davies Giddy, the Cornish friend of Thomas Beddoes (see p. 11), in Penzance and the latter gave him free run of the considerable library at his house at Tredrea.[3,16,28] Davy also had another indirect connection with Beddoes. James Watt's consumptive son Gregory, who had been sent to Cornwall in the hope that the climate would benefit him, was a patient of Beddoes and was boarded with Davy's widowed mother. A firm friendship sprang up between the two young men until Gregory's untimely death in 1804.[28,29]

Davy's early research

It was probably from Giddy's library that Davy became acquainted with a copy of Beddoes' and Watt's *Considerations on the medicinal powers of factitious airs* which quoted the extraordinary theory of the American Samuel Latham Mitchill.[28] Mitchill believed that Priestley's 'dephlogisticated nitrous air' (nitrous oxide) was the 'contagion' by which plague was spread, and that if it was inspired by animals the 'most terrible effects' and death would ensue. Davy prepared somewhat impure nitrous oxide by the action of nitrous acid on zinc and showed that small animals could live in it for a longer time than any other gas except air itself, and he also inhaled it himself for short periods without contracting plague. Davy, encouraged by Giddy and young Gregory Watt, sent the results of his experiments with nitrous oxide to Beddoes. Beddoes was impressed by the work of the young self-taught chemist. He was therefore receptive to Giddy's suggestion that he should appoint Davy to the post of superintendent of the new Pneumatic Medical Institution at the age of 20. Davy regarded the appointment as a step towards his goal of formal qualification in medicine which he never reached.[28]

Researches at the Pneumatic Institution

The Institution attracted many patients in its early days. They were encouraged by an advertisement inviting the attendance of persons with many diseases, including consumption, asthma, palsy, dropsy, and venereal complaints, for free treatment. This was not surprising as, such was the enthusiasm of Beddoes, the advertisement confidently predicted that 'a considerable proportion' of patients with consumption 'will be permanently cured', and further that 'none of the methods to be pursued are hazardous or painful'. These high hopes were not, of course, realized and the number of patients attending fell steadily, and later they even had to be paid to attend.[28,34] Pneumatic medicine had, in truth, reached its zenith in the years immediately before the arrival of Davy and the opening of the Institution in 1799; by 1801 Beddoes himself had ceased to believe in its possibilities and was turning his mind to the benefits of preventive medicine. The historical importance of the Pneumatic Institution lies not in its therapeutic achievements, but in the brilliant researches of the young Humphry Davy.[28,31,34-38]

Davy worked in the Institute for just 2 years from March 1799 to March 1801. He experimented with all the gases which were available at the Institute and nearly killed himself by inhaling carbon monoxide, but his main interest lay in the study of nitrous oxide (Priestley's 'dephlogisticated nitrous air') which he introduced to the Institute. He perfected the purity of its preparation by the method of heating ammonium nitrate, originally described by the famous French chemist Count Claude Louis Barthollet (1748–1822). He confirmed or established the specific gravity of the gas, its solubility in water and blood, and the fact that it supported combustion. He noted and measured the rate of uptake of nitrous oxide by his own body when he inhaled the gas using a spirometer designed by William Clayfield (1772–1837), a pupil of James Watt. This advanced piece of equipment enabled the gas to be collected over mercury, thus avoiding the disadvantage of its solubility in water. Davy employed the same apparatus to measure the capacity of his own lungs by inhaling hydrogen.[28,38]

He also studied the effect on animals of breathing as pure nitrous oxide as he could obtain, and observed that the effects 'were very little different from those killed by the privation of atmospheric air'; more importantly he described the progressive reactions of a guinea-pig confined in a bell-jar of nitrous oxide and oxygen which are now known to be stages of anaesthesia – 'struggling' (stage 2, excitation), 'repose' (stage 3, anaesthesia), 'convulsions' (as the oxygen in the limited atmosphere began to be exhausted and carbon dioxide had increased), followed by resuscitation when the animal was removed from the bell-jar and allowed to breathe air.[28,29]

Davy, having established that nitrous oxide was easily inhaled and that it was not lethal to breathe the gas, began to observe its effects on himself and others. The observation of the violent and uncontrolled struggles of the excitement stage unfortunately led Beddoes to believe for a time that nitrous oxide was a stimulant and therefore an antidote to paralysis, and to use it in the treatment of paralytic diseases. Davy's descriptions of his own pleasurable and exhilarating experiences when he inhaled the 'laughing' gas resulted in many professional colleagues and members of the fashionable Bristol circle being eager to experience its unusual inebriating effects; amongst these were the poets Robert Southey (1774–1843) and Samuel Taylor Coleridge (1772–1834), the engineers James Watt (1736–1819) and William Clayfield (1772–1837), Beddoes himself and the physicians Robert Kinglake (1765–1842) and Peter Mark Roget (1777–1869) – of Thesaurus fame.[28,31,34] It is a probable deduction from the length of time that Davy and the other subjects were able to breathe the supposedly 'pure' nitrous oxide that it was, to a greater or lesser extent, diluted with air.

Analgesia and the possible relief of surgical pain

Davy also made the now well known observations relating to the analgesic property of nitrous oxide and its reversibility:

> In one instance while I had a headache from indigestion, it was immediately removed by the effects of a large dose of the gas; though it returned afterwards with much less violence. . . . In cutting one of the unlucky teeth called dentes sapientiae, I experienced an intense inflammation of the gum, accompanied by great pain. . . . I breathed three doses of nitrous oxide. The pain always diminished after the first four or five respirations . . . and uneasiness was for a few minutes swallowed up with pleasure. As the former state of mind, however, returned, the state of the organ returned with it; and I once imagined that the pain was more severe after the experiment than before.[38]

Davy undertook the bulk of his research work on nitrous oxide between December 1799 and April 1800; from then until July 1800 he wrote his book *Researches chemical and philosophical; chiefly concerning nitrous oxide or dephlogisticated nitrous air, and its respiration*.[33,36–39]

The conclusion section of the *Researches* contains the now often quoted paragraph:

> As nitrous oxide in its extensive operation appears capable of destroying physical pain, it may probably be used with advantage during surgical operations in which no great effusion of blood takes place.[38]

The meaning of the last somewhat cryptic claim in this passage has been variously interpreted.[3,28,31] Could it be that Davy accepted Beddoes' Brunonian belief that nitrous oxide was a stimulant and would cause greater bleeding in a major haemorrhagic operation, or, perhaps, that the patient could not survive both the combined excitation due to the gas and the loss of blood, if they occurred at the same time? Cartwright's explanation[28] seems to be simpler and more acceptable – that is, that Davy contemplated a single dose administration for the rapid body surface operations of his day and may have believed that, since nitrous oxide was dissolved in the blood, its effects would be lost if there was a great effusion of blood.[28,31]

The significance of Davy's researches concerning nitrous oxide

There seems little doubt that Beddoes, who had earlier followed Priestley in an interest in electricity, introduced Davy to the study of the subject and, in the months following the publication of the *Researches* Davy was conducting 'galvanic' experiments at the dying Pneumatic Institution.

The Royal Institution in London had been founded in 1799 with the objective of the 'promotion, diffusion and extension of science and useful knowledge'. Davy was head-hunted, on the strength of his publications and growing reputation, for the post of Lecturer in Chemistry at the Institution, by the American born scientist and administrator Count (of the Holy Roman Empire) Benjamin Thomas Rumford (1753–1814). He accepted the position with alacrity and left Bristol in March 1801.[28]

Humphry Davy, of course, subsequently became one of the greatest scientists of his day, President of the Royal Society, and a baronet. Unfortunately the provincial boy of relatively humble origin became increasingly arrogant and jealous of the achievements of others as the years passed. His deprecatory treatment of his assistant and successor Michael Faraday (1791–1867), was inexcusable, more especially as Faraday, like Davy, was self-taught and came from a very similar, rather humble background. Davy gave one course of lectures on pneumatic chemistry after his arrival in London during which he demonstrated the intoxicating effects of nitrous oxide; thereafter he turned his attention to other scientific matters including the discovery of sodium and potassium and the invention of his famous miner's lamp.[28,31]

Davy had, however, sown the seed which was to blossom with the first use of nitrous oxide as an anaesthetic by Horace Wells (1815–1848) in Hartford Connecticut in 1844 (see p. 21). The use of nitrous oxide as an anaesthetic was not because of his reports of its analgesic action, however, or because of his chance remark about its possible use to control surgical pain, which the reviewers of his *Researches* largely ignored. These remarks were really only recalled and brought into prominence after anaesthesia became a universal reality in 1846. The observations on nitrous oxide which appealed to Davy's scientific contemporaries were his descriptions of its peculiar inebriating effects and his dissertations on its chemical properties. Nitrous oxide became an accepted topic in lectures given to medical students of that era and these did not fail to mention the 'curious', 'singular' and 'pleasurable' effects of the 'laughing gas'. One or two descriptions mention that established neuralgic pains were abolished during inhalation of the gas. There were also even very occasional echoes of pneumatic medicine describing its use by inhalation in the treatment of conditions as diverse as asthma, cholera and hydrophobia (rabies) (see pp. 9–11).[31,34]

Nitrous oxide nevertheless required apparatus for its manufacture and immediate storage before inhalation. Its inhalation therefore was largely confined to scientists and medical men for demonstration and experimental purposes, and later by the popular lecturers and showmen who were a feature of the entertainment world in the nineteenth century on both sides of the Atlantic (see pp. 20–21). It was indeed the erstwhile medical student turned itinerant lecturer Gardner Quincy Colton (1814–1898) who ultimately provided the stimulus for the introduction of nitrous oxide as an anaesthetic,[31] and it was to William Herapath (1796–1868), the chemist and popular lecturer on nitrous oxide, that the medical staff of Bristol Royal Infirmary turned on 31 December 1846 to administer the first ether anaesthetic in the city in which Davy had conducted his research over 40 years before.[40]

Davy has been variously chided and excused for not following up his suggestion that nitrous oxide might be used for the relief of the pain of surgical operations.[3,4,28,31,39] His attitude is really not very surprising. Despite his early surgical apprenticeship Davy was primarily a chemist and investigative scientist, the full impact of the new humanitarianism was not apparent in 1800, surgery was a peripheral matter in relation to the overall causes of pain at that time, and clinical surgery was not yet sufficiently developed to be reaching out from being confined to body surface and limbs; moreover, experimental inhalation of 'pure' or concentrated nitrous oxide by medical men and other scientists often resulted in alarming asphyxia, especially in the presence of rebreathing, which was not properly understood.[28,31,41] It may also be that Davy was not sorry to be able to distance himself from the discredited theories of Beddoes in relation to pneumatic medicine, which had been regarded by the medical establishment of the period as an unorthodox product of the lunatic fringe even in its heyday.[28]

The medicinal use and recreational abuse of ether

The oral and surface application of ether as a medicament in the mid-eighteenth century has already been considered (see p. 7) and such usage continued into the nineteenth century.[20–22,42,43]

The story of the suggested use of ether by inhalation by

William Cullen (1710–1790) who lectured to Beddoes when the latter was a student in Edinburgh (see p. 10) cannot be substantiated.[20,28] Ether by inhalation was, however, recommended by the advocates of pneumatic medicine. Erasmus Darwin (1731–1802) prescribed it for the treatment of asthma, and the Birmingham physician Richard Pearson (1765–1836) wrote to Beddoes in 1795 recommending it for catarrh, croup and whooping cough and described its use for the relief of the symptoms of phthisis 2 years later.[20–22,42,43]

Doctor Robert Thornton (1768–1837) of Marylebone, London improved on Pearson's practice of inhaling from a tea cup or wine glass, sometimes with a funnel inverted over them, by placing the ether in a teapot warmed with a candle and having the patient inhale from the spout. William Withering, the eminent physician from Birmingham, used 'a glass vessel resembling what chemists call a tabulated receiver'. This was about 3.5 inches (8.9 cm) in diameter and had two outlets – one for the patient's lips and the other allowing air to enter. Various similar inhalers were devised and used for ether, steam and other substances; their importance to the story of anaesthesia lies in the fact that inhalers were familiar to the early administrators of anaesthesia. Both Morton (see p. 24) and later John Snow (1813–1858), the first specialist anaesthetist in London, based their early apparatus on such devices.[20,22,43]

There are several reports of patients becoming unconscious while inhaling ether for chest disease (for example Lady Hall in 1808) and as a sedative and anodyne. There is also a description of the purposeful use of the agent for its analgesic properties during the examination of sensitive ears by William Wright in 1829.[20,25,33,43] It is interesting that ether was used to relieve distress in 'pulmonary inflammation' early in the nineteenth century by John C. Warren, the surgeon who was to undertake the operation at Morton's demonstration in 1846 (see pp. 21 and 29).[6,25]

A report from the Royal Institution on the properties of ether

An anonymous memorandum in the *Journal of Science and the Arts* (the official organ of the Royal Institution) in 1818 drew attention to the fact that 'when the vapour ether mixed with common air is inhaled, it produces effects very similar to those occasioned by nitrous oxide'.[44] Reference is not made specifically to a pain relieving effect, but the memorandum goes on to warn against the 'imprudent inhalation of ether'. This passage has been attributed to Michael Faraday who was, at that time, Davy's assistant. There does not seem to be any evidence to support this, however, and some authorities believe that the phraseology reads more like a contribution from Davy himself; however, Faraday did detail the precautions which should be taken to avoid the dangerous consequences of the inhalation of impure ether a year later in 1819.[20,28,31]

The abuse of ether

Patients became habituated to the use of ether,[45] and the intoxicating properties of the agent became common knowledge both amongst students and the general public, particularly in Scotland and New England. 'Ether frolics' became a popular pastime in academic circles, particularly in the United States.[3,8,16,20,22,25,43] This was not surprising considering the ease with which ether could be obtained, stored and inhaled compared with nitrous oxide gas. It may also be noted that the parallel practice of drinking ether as a substitute for, and mixed with, alcohol was a problem which persisted in the British Isles and certain continental countries well into the twentieth century.[20]

The relief of surgical pain in the early nineteenth century 1800–1840

The progress in surgical technique and the tradition of caring surgeons deeply distressed by the pain they had necessarily to inflict, which was born in the eighteenth century (see p. 8) continued into the first half of the nineteenth century. The progressive European surgeons included such giants as the remarkable Dominique Jean Larrey (1766–1842), Napoleon's brilliant military surgeon, who, inter alia, successfully employed colostomies for wounds of the abdomen, and Robert Liston (1794–1847) of London who was destined to be the first to perform a major operation under ether in England.[24] They were joined by erudite and compassionate New England surgeons who were the product of the innovative cultural revolution in the United States of America which followed the Declaration of Independence in 1776. John Collins Warren (1778–1856) was such a man. He, himself, endeavoured to find the means of relieving the pain of surgery by the use of 'opium in all its forms' and hypnosis, and it was he who encouraged and supported Morton and invited him to demonstrate the value of etherization while he operated on that fateful day in October 1846 (see pp. 24 and 25).[6,25]

It was developments in the new sciences of chemistry and respiratory physiology from the mid-eighteenth century onwards, and the consequent use of inhalation as a route for the administration of therapeutic agents, which ultimately led to the introduction of universally practical anaesthesia in the 1840s, by which time humanitarian attitudes had made its general acceptance possible; however, during the preceding half century, enlightened individuals attempted to relieve the pain of surgery by techniques based on the physical and oral methods which have already been described.

Physical local anaesthetic techniques

The use of nerve compression by Moore in the 1790s and its revival by Liégard in France in 1837 have already been mentioned (see p. 5).[2,13]

Baron Larrey (1766–1842), Napoleon's military surgeon, noted that soldiers with gangrene extremities due to frostbite suffered little pain at the battle of Eylau during which the temperature was below 14°C. The popular belief that he recorded the observation that pain during amputation was relieved by refrigeration is sadly unfounded, but he did report that bleeding during the operation was reduced;[8,46] however, Arnott (1797–1883), a medical practitioner working in Brighton in England, described a method of surface anaesthesia using pounded ice mixed with salt in 1846 and the method was subsequently employed by a number of London surgeons.[8,47]

Refrigeration anaesthesia was eclipsed by the introduction of inhalation anaesthesia in the same year that Arnott published his results, and by the discovery of pharmacological local anaesthesia in 1884. Refrigeration anaesthesia for amputations in poor risk patients was once again revived in the middle of the twentieth century.[48]

Oral anaesthesia

There are well documented accounts of the use of oral herbal preparations to produce anaesthesia for surgical operations by Seishu Hanaoka (1760–1835), the father of modern Japanese surgery, in the last decade of the eighteenth century and at the beginning of the nineteenth. Seishu Hanaoka practised a form of Western-style surgery which the Japanese had learned from Dutch traders before and during the so-called 'period of isolation' which lasted from 1648 until the intervention of the United States in 1853. His formula included extracts of mandarage (Datura alba). This herb is of Chinese origin but it seems likely that it was imported into Japan from China by the Dutch whose herbalists had noted its close relation to mandragora. Seishu Hanaoka carried out many operations with the aid of his preparation, including operations for cancer of the breast as early as 1805.[49]

Alcohol

The literature includes a number of references to severe wounds being sustained and operated upon, both in military and civilian practice, on patients who were already inebriated prior to the introduction of anaesthesia, but, despite popular belief, there is very little evidence that alcohol was deliberately used routinely; indeed the main use of a small dose of liquor during or after surgery was usually deemed to be for resuscitation.[50]

Hypnosis

The technique of hypnosis is of very ancient origin although its use to relieve the pain of surgery is not well documented until the first half of the nineteenth century in the immediate pre-anaesthetic period.[51] Franz Anton Mesmer (1734–1815) did not use hypnosis to relieve surgical pain. He was concerned in the dubious but lucrative practice of using 'mesmerism' or 'animal magnetism' to cure disease. This resulted in him falling foul of the authorities in both Vienna and Paris.[51] Hypnotism, however, was subsequently applied to surgery by several physicians and surgeons both in France and England. Notable amongst them were Cloquet in France who undertook a mastectomy under its influence in 1829, John Elliotson (1791–1868) of University College Hospital, London, who published a treatise entitled 'Numerous cases of surgical operations without pain' in 1843, and James Esdaile (1808–1859), surgeon to the East India Company, who described a series of cases from Calcutta in 1846.[2,3,8,16,51] John Collins Warren had also experimented with hypnosis as a possible solution to the problem of relief of the pain of surgery before Wells originated nitrous oxide anaesthesia in 1844 and Morton's seminal demonstration in 1846 (see pp. 20–25).[6,25]

Henry Hill Hickman, suspended animation and carbon dioxide

Much has been written about the contribution of Henry Hill Hickman (1800–1830), but the significance accorded to his work depends upon the emphasis which historians place upon originality as opposed to a direct influence upon development.[3,4,8,28,52–55] He never, so far as is known, gave or attempted to give an anaesthetic to a human patient, and yet Sykes awards him first place in his order of merit 'despite the fact that his work did not lead to any practical

results', and was not remembered until after Morton's demonstration in 1846, nor appropriately recognized until the centenary of his death (see p. 18).[4,28,52] Hickman, like his predecessors Beddoes and Davy before him, came from 'one of those old and distinguished families [of] small landowners and farmers'.[28] He was born and christened in Bromfield, Shropshire and he subsequently adopted his mother's maiden name (Hill) as his middle name. He had a surprisingly brief surgeon's training in Edinburgh and qualified as a Member of the Royal College of Surgeons of London (MRCS) in 1820. He practised successively in three towns in his native county – Ludlow (1820–1824), Shifnal (1824–1828) and Tenbury (1828–1829). He died in 1830 at the age of 30 and was buried at Bromfield.[3,4,52–54]

Shifnal was coincidentally the birth place of Thomas Beddoes (see p. 10) but, despite extensive research by several workers (notably Cartwright[28] and Smith[52–54]), no connection has ever been found between pneumatic medicine or Davy's researches into nitrous oxide and Hickman's experiments; moreover Hickman's surviving records refer to experiments conducted at Ludlow before he moved to Shifnal 30 miles away.[34,52–54] Hickman's inspiration sprang from the humanitarian desire of a practising surgeon to relieve the pain of surgery. His proposal was to utilize the already well known state of 'suspended animation'.

Hickman's experiments

'Suspended animation' was regarded as the condition near to death occasioned by asphyxia from such accidents as drowning, hanging without vertebral fracture, strangulation or suffocation, or any other means of depriving the human or animal of atmospheric air, and it was known that resuscitation from this state was possible by restoring the access of the subject to air and by other means, including inflation and the then novel electric (galvanic) stimulation. The evidence suggests that carbon dioxide, which could be easily inhaled, was regarded by Hickman simply as another means of depriving the subject of life-giving air.

Hickman confined small animals (puppies, mice, rabbits and kittens) in bell-jars and rendered them unconscious in three different ways – first by simply allowing them to exhaust the oxygen in the air (asphyxia), secondly by exposure to an atmosphere containing carbon dioxide generated by the action of sulphuric acid on calcium carbonate, and third, by confining them in an atmosphere of carbon dioxide in air provided by his own exhalations. Once the animals were unconscious (in a state of 'suspended animation') he removed them from the container and excised various portions of their anatomies (ears, tails and limbs) before they recovered consciousness. One of the dogs was resuscitated by positive pressure using 'inflatory instruments'. Hickman did make the observation that the animals lost consciousness more rapidly in atmospheres which contained carbon dioxide from the outset of an experiment, but he does not seem to have come to the conclusion that carbon dioxide itself was a narcotic agent.[28,52–54]

Hickman and the Royal Society

Hickman's first attempt to arouse interest in the implications of the experiments which he had conducted at Ludlow was to describe them in a private letter to Thomas Andrew Knight (1759–1838) of Downton Castle on the Herefordshire–Shropshire border close to Hickman's birthplace. Knight was a prominent landowner and a distinguished and acknowledged horticulturist who was a Fellow of the Royal Society. He was also a friend of the (by then Sir) Humphry Davy who frequently visited Downton Castle for the fishing on the River Teme.[52,53] Some authorities have presented Knight as an indifferent patron who, through indolence or failure to grasp the importance of Hickman's experiments, did not consider it worthwhile to discuss them with his friend Sir Humphry Davy or other Fellows of the Royal Society.[3,4,28] Careful research by Denis Smith has provided compelling circumstantial evidence which suggests that it is possible that Knight did discuss Hickman's work with Davy, and that Davy was initially encouraging (probably remembering his own experiments with nitrous oxide) and offered to present an account to the Royal Society; subsequently, however, he may be assumed to have had second thoughts.[52,53] This could have been because he foresaw that there might be danger, or even fatal consequences and he did not wish to be associated with such tragedies, or because there had been criticism of the Royal Society over animal experiments in the not too distant past.[53] Hickman certainly prepared and had printed a pamphlet in the form of an open letter to Knight as 'one of the Presidents of the Royal Society' dated August 1824. The title page included the words 'and read before it [the Royal Society] by Sir Humphry Davy'. These words were neatly obliterated on some copies before public distribution of the pamphlet and there is no evidence that the presentation actually occurred.[52,53] The fact that the introduction to Hickman's pamphlet states that it is written 'at the request of gentlemen of the first rate talent, and who rank high in the scientific world' lends further support to Smith's conjecture.[28,51]

The publication of Hickman's pamphlet may have been delayed because Hickman hoped that indeed his work would be presented to the Royal Society by Sir Humphry Davy. The pamphlet was printed locally at Ironbridge and provoked little interest when it was finally circulated. The only references were factual notes in the *Gentleman's Magazine* and the *Shrewsbury Chronicle* c. 1825 and a diatribe signed with the pseudonym 'Antiquack' in the *Lancet* of 4 February 1826. The latter contains some reasoned if erroneous arguments attempting to refute Hickman's claims, but is marred by words such as 'fool', 'quack', 'humbug' and 'hoax'.[28,52–54]

Hickman, Charles X and the French Royal Academy of Medicine

Hickman had moved to Shifnal in 1824 before the publication of his pamphlet. He remained in practice there until April 1828 when, having been frustrated in England, he set out for Paris, the acknowledged centre of European scientific excellence at that time. He was determined to petition His Most Christian Majesty, the autocratic King Charles X (the last of the legitimate Bourbon kings, who reigned from 1824 to 1830), to permit him to put his case before the French Royal Academy of Medicine.[34,52–54] Denis Smith's researches into Hickman's connections in France, which apparently enabled him to get an accelerated hearing, are fascinating but need not be related here.[28,52–54]

The elegant petition to Charles X survives, but 'the Book' of evidence mentioned in it does not, unless this refers to a copy of Hickman's original pamphlet. It is consequently not known whether additional evidence was submitted based on further experiments by Hickman between 1824 and 1828. The French Royal Academy of Medicine received and considered Hickman's evidence on 28 September 1828. As was customary, a commission was set up to consider the matter. Hickman's proposition was apparently received with incredulity by all except Baron Larrey (see p. 15) who was then aged 62. He was of the opinion that the idea 'deserved the attention of surgeons'.[28,52–54]

The significance of Hickman's work

Hickman returned to England at the end of 1828. He must have been bitterly disappointed. He took up a new practice in Tenbury, but he died 2 years later in 1830 at the age of 30. He was probably suffering from tuberculosis.

We know too little about Hickman. Did he, for instance, conduct further experiments at Shifnal between 1824 and 1828? Is it likely that he did not? Is it possible that, after his indirect (or possibly direct) contact with Sir Humphry Davy, that he used nitrous oxide in his animal experiments? There is some circumstantial evidence that he might have done both of these things and included his results in his presention to the Royal Academy of Medicine in Paris in 1828.[52–54]

Hickman's experiments were forgotten until the dispute

about priority for 'inventing' anaesthesia raged after Morton's seminal demonstration in 1846 (see pp. 27–29). A Doctor Thomas Dudley of Kingswingford near Birmingham attempted to establish Hickman's priority in letters to the *Lancet* in 1847 but failed to trigger support, and Sir James Young Simpson, the discoverer of chloroform, referred to Hickman's experiments in 1870. A definitive article on Hickman by a stepson of one of his daughters, written under the auspices of the Wellcome Historical Foundation, was published in the *British Medical Journal* in 1912, but it was not until the anniversary of his death in 1930 that the Wellcome Foundation succeeded in interesting the Section of Anaesthetics of the Royal Society of Medicine in his contribution. A memorial tablet was unveiled subsequently in Bromsfield Parish Church.[4,28,55]

Hickman's work did not make a direct contribution to the main stream of events which led to the revelation of nitrous oxide or ether anaesthesia in the 1840s. He himself acknowledged that his proposed method of bringing the subject close to death in the state of 'suspended animation' meant that the surgery must be rapid; but he stressed unconsciousness would enable a surgeon to perform operations 'more skilfully', and paradoxically with less haemorrhage. This was because his method involved asphyxia to near cessation of the heart beat rather than the use of the anaesthetic properties of carbon dioxide with its cardiovascular stimulatory action.[28]

Hickman represented the new age of a humanitarian attitude to pain. The historical significance of his contribution is great, but its practical significance is small. Cartwright probably comes nearest to the truth in writing: 'His glory lies in the idea which lies behind his work . . . for he, the first of all men, set out to banish pain by means of experimental investigation.'[28,52–54]

The use of ether by inhalation for the relief of surgical pain before Morton's demonstration in October 1846

Morton was not the first to administer ether deliberately for the relief of surgical pain. There are at least four authenticated but isolated instances of the elective use of ether for the relief of pain of dental or surgical operations before Morton administered it for a dental extraction in his own practice on 30 September 1846,[56] and subsequently for John Collins Warren for a surgical operation on 16 October 1846.[6,25] There is also a fifth report which indicates that it was almost certainly used by mistake before Morton's

administration. It must be said that no accounts of these administrations were made until after Morton's successful demonstration.

William E. Clarke

William E. Clarke (1809–1880) of Rochester, New York who, with his companions, was in the habit of indulging in the inhalations of ether for entertainment while he was a medical student at the Berkshire Medical College in Massachusetts. He successfully administered ether to a Miss Hobbie, who was a young lady friend of his, for the extraction of a tooth by the dentist Elijah Pope in Rochester in January 1842.[8,16,56] He was apparently dissuaded from conducting further experiments by his tutor (Professor E. M. Moore) who believed that the young woman must have been 'in an hysterical freak and feigned unconsciousness'.[57]

Crawford W. Long

Crawford Williamson Long (1815–1879) was the son of a comfortably off Irish Presbyterian plantation owner and politician of Danielsville, Georgia, USA. He had what was, for the period, an exceptionally comprehensive medical education at Transylvania University, Kentucky, the University of Pennsylvania, Philadelphia (from where he graduated MD in 1839) and as a postgraduate student in New York.[57]

Long set up in practice in the small town of Jefferson, Jackson County, Georgia as a general practitioner and surgeon. Long was by all accounts a likeable, hard-working, modest and dedicated physician.[5] His pioneer ether anaesthetics are carefully described in Long's own account[58] and the circumstances which led him to his use of ether to control surgical pain were meticulously considered in an article written in 1877 by the American pioneer gynaecological surgeon James Marion Sims (1813–1883) of New York, after the latter had become a champion of Long's claim for priority.[59]

Long, like Clarke, came to use inhaled ether for the relief of surgical pain through his experience of its exhilarating effects. This was a common practice amongst students and other young people in Georgia at that time. Long had four apprentices in his practice who asked him about the effects of nitrous oxide but Long, not having the means of preparing the gas, told them that its effects were similar to those of ether; thereafter he and his students began to inhale ether frequently, first as a demonstration but very soon as an indulgence.

Long observed that he and his friends frequently bruised themselves unknowingly in their inebriated and excited state after inhaling ether. James Venables, a young student who had previously had experience of inhaling ether for amusement, consulted Long on several occasions about the removal of two 'tumours' (probably sebaceous cysts) on the back of his neck, but was reticent about having them excised because of the prospect of the pain of the surgery. Long, drawing both on his own experience and that of Venables himself, finally suggested to his reluctant patient that it might be possible to remove the cysts painlessly under the influence of ether. Venables agreed and the first cyst was removed under successful etherization on 30 March 1842 and the second on 6 June 1842. Long reported that he carried out one more operation (the amputation of a toe) under ether in 1842 and thereafter, before his paper in 1849, 'one or more surgical operations annually on patients in a state of etherization'. Long was uncertain about the nature of the condition he had produced. He was aware of the use of hypnotism in surgery, but proved to his own satisfaction that his cases were not under its influence by the rather crude method of performing two similar minor operations on the same day on each of two patients, with and without ether.[58,59]

Much has been written about the reasons for Long withholding publication of his results until 1849 after the world had fully realized the benefits of etherization after Morton's demonstration in 1846. Sykes has carefully analysed Long's own reasons for his 'reticence'.[58,60] Long's failure to publicize his discovery does not seem too surprising when one considers his geographical isolation, his modesty, and his determination that he should be absolutely sure of his ground before publicizing his achievement, and because of the rare occasions when cases presented which justified major surgical intervention, in a general practice such as Long's, in those days before antiseptic techniques. Most minor procedures could be undertaken rapidly without prolonged pain.[60] Long subsequently obviously regretted with hindsight that he had not taken the opportunity of testing out etherization for the extraction of teeth, which, no doubt, was a common procedure in his daily practice.[58]

Long's daughter also tells us that some members of the local public considered him to be:

reckless, perhaps mad. It was rumoured that he had a strange medicine by which he could put them to sleep and carve them to pieces without their knowledge. His friends pleaded with him to abandon its use, [lest] in case of a fatality . . . he would be lynched.[57]

Long explained his position as follows:

Had I been engaged in the practice of my profession in a city, where surgical operations are performed daily, the discovery would, no doubt, have been confided to others who would have assisted in the experiments; but occupying a different position I acted differently, whether justifiable or not.[58]

E. E. Marcy and Horace Wells

After Horace Wells had successfully anaesthetized patients with nitrous oxide for dental extraction in Hartford, Connecticut, but had failed in his attempt to convince the medical establishment in Boston of the value of his discovery in 1844 (see pp. 20–22), he returned to Hartford and continued to use nitrous oxide in his practice. He was, however, apparently interested in finding a substitute for nitrous oxide 'which would be attended with less trouble in preparation'. Wells discussed the already well known similarity of the effects of ether and nitrous oxide with the surgeons P. W. Ellsworth and E. E. Marcy. Wells inhaled ether himself and administered it for an extraction on at least one occasion in 1845 but continued to use nitrous oxide because it was easier to inhale. Marcy also gave ether for the removal of a 'wen' (sebaceous cyst) in his office in 1845.[59,61]

E. R. Smilie

The fifth reference to the possible use of ether for the relief of surgical pain before Morton's demonstration at the Massachusetts General Hospital on 16 October 1846 is surprising and is reminiscent of the mediaeval soporific sponge (see p. 6); unlike the reports of Clarke, Long and Marcy, there is no connection with the recreational use of the inebriating properties of ether inhalations.[19]

A letter from a Dr E. R. Smilie of Boston to the *Boston Medical and Surgical Journal* was published on 28 October 1846. It reported the successful use of the inhalation of opium 'to produce insensibility in persons requiring surgical operation'; in retrospect this was not surprising as the vehicle chosen to convey the opium was ether in the form of a warmed 'etherical solution of opium'![19]

Dr Smilie's letter occurs after a vague report in the newspaper, *Boston Journal*, of Morton's extraction of a tooth on 30 September 1846 which records that the patient was put 'into a kind of sleep', and after a report in the *Boston Medical and Surgical Journal* of 21 October 1846 recording 'strange stories' of 'a surgical operation without pain' on 16 October 1846; no details are given of the nature or possible nature of the method or agent in either report.[16,19] Bigelow's leading article (see p. 25) did not appear in the *Boston Medical and Surgical Journal* until 18 November 1846.[4]

It seems clear that Smilie's letter must have been submitted to the *Boston Medical and Surgical Journal* before Morton's demonstration on 16 October and he must have been using his technique for some time before that. Several years later in 1852, when giving evidence to a select committee of the House of Representatives, Smilie stated that he had first administered an etheral solution of opium for the incision of an abscess in the spring of 1844.[16]

New England in the 1840s, dentists and the birth of modern anaesthesia

Both Greene and Vandam carefully examined the factors which made it almost inevitable that the worldwide acceptance of practical inhalation anaesthesia should be the outcome of events which occurred in New England in the fourth decade of the nineteenth century.[22,26]

The progressive changes in philosophical attitudes to pain and its treatment in the first half of the nineteenth century have already been discussed (see p. 15), and the analgesic and hypnotic effects of both nitrous oxide and ether were already well known by 1840. There was also a close interchange of ideas between the scientific intelligentsia of Europe, the United Kingdom and New England but, perhaps, a greater desire to experiment and accept innovations and the ideas of younger men into medical practice in the USA than in European countries, where the medical establishment tended to be traditional and conservative.[22,26]

It was also almost inevitable that New England dentists were to be intimately connected with the introduction of anaesthesia. Dental extraction was probably the most frequent surgical operation practised before the introduction of anaesthesia. Dentists in New England had already attained a quasi-professional and scientific status as an independent profession. Dentistry was less well developed in Europe; a few qualified surgeons in the UK (members of the Royal Colleges of Surgeons) practised dentistry as a specialty in the larger cities[62] and general practitioners undertook extractions as part of their practice, and some learned their trade by apprenticeship, but often extractions were undertaken by artisans and quacks who had not had any formal training.[63]

Greene has also pointed out that, apart from dental extraction, the number of surgical operations undertaken in the early nineteenth century was negligible and, consequently, the frequency with which medical practitioners deliberately inflicted pain was low, whereas dentists had 'a day-to-day incentive for discovering means for the relief of pain'.[22]

Horace Wells and Gardner Quincy Colton and the introduction of nitrous oxide anaesthesia

Horace Wells (1815–1848), the eldest child of a well-to-do landowner, was born at Hartford, Windsor County, Vermont, USA. He had an excellent basic education until he was 19 and then studied dentistry by apprenticeship in Boston before setting up in practice in Hartford, Connecticut in 1836, and quickly becoming the leading member of the profession in that city.[64,65]

Wells had an active and inventive mind and was responsible for several practical inventions including a cement for fixing prosthetic teeth. He was also the author of a small volume on dentistry, and he patented other inventions not connected with dentistry.[3,64,65]

There is evidence that Wells had considered the possibility of nitrous oxide or ether to relieve surgical pain as well as hypnotism before the stage demonstration by the itinerant lecturer Gardner Quincy Colton (1814–1898) which finally convinced him of its efficacy and gave him the means of manufacturing the gas (see p. 21).[31,59,63–66] Colton's assertion in his retrospective pamphlet written in 1886[66] that Wells was specifically 'without knowledge of the suggestion of Humphry Davy concerning the relief of pain' may possibly be true, however, as, as has already been noted, this short passage (see pp. 13–14) received little attention from the reviewers of Davy's *Researches*. There seems to be no doubt, however, that Wells was well aware of the property of both nitrous oxide and ether to cause inebriation followed by insensibility before he came into contact with Colton. It is interesting that the theory of nitrous oxide anaesthesia postulated by Wells was that it had an exhilarating effect that was similar to the observed phenomenon that: 'an individual when much excited from ordinary causes, may receive severe wounds without manifesting the least pain; as, for instance, the man who is engaged in combat . . .' Are these echoes of the interest of Thomas Beddoes in the Brunonian theory in this belief (see p. 11)?[61]

There is also the remote possibility that Wells had already heard of the use of ether by Long for the relief of pain of surgical operations (see p. 29).[16]

Gardner Quincy Colton

Colton (1814–1898) had been a medical student in New York but had had to discontinue his studies because of lack of funds. He took up the role of a travelling scientific lecturer providing entertainment for the public which included hilarious antics in volunteer members of his audience by the inhalation of nitrous oxide.[31] Colton was undoubtedly well aware of Davy's researches.[31,66] Colton, an intelligent man, was destined to play an important part in the development of anaesthesia both in 1844 and some years later.[66]

Colton's expertise, Wells' inspiration and the first use of nitrous oxide for a dental extraction

Colton had arrived in Hartford, Connecticut to give one of his exhibitions on 10 December 1844. There are a number of accounts of the events leading up to the administration of the first nitrous oxide anaesthetic.[3,8,16,31,64–66] All agree on the essential fact that on the morning of 11 December 1844 in his own office Horace Wells himself inhaled nitrous oxide from an oil-silk bag proffered to him by Colton, and that Wells' dental partner John M. Riggs (1810–1885) extracted a troublesome tooth while he was under its influence.[37] Colton adds that on coming round from the nitrous oxide Wells exclaimed: 'It is the greatest discovery ever made. I didn't feel it as much as the prick of a pin.'[66] Other variations state that he also declared that it was 'a new era of tooth pulling'.[64]

Colton's account of the events which preceded the first dental extraction under nitrous oxide, written 40 years later,[66] tells us that:

On the evening of 10 December 1844 I gave an exhibition of the laughing gas in the city of Hartford Connecticut. Amongst those who inhaled it was a young man by the name of [Samuel A] Cooley, who while under its influence, in jumping about, ran against some wooden benches or settees on the stage, bruising his legs badly. After taking his seat, he was astonished to find his legs bloody; and said he did not know he had run against a bench and felt no pain until after the effects of the gas had worn off. Dr Wells – who sat next to him – noticed the circumstance, and . . . asked me why a man could not have a tooth extracted without pain while under the influence of the gas. I replied that I did not know as the idea had never occurred to me. Dr. Wells then said he believed it could be done, and would try it on himself if I would bring a bag of gas to his dental office the next day.[66]

Colton's memory may have been at fault in respect of detail, however. Samuel Cooley himself in a deposition made to the Congress of the United States in 1853 stated that, 'in accordance with the request of several gentlemen', Colton gave a 'private exhibition' on the morning of 11 December 1844 following the 'public exhibition' on the previous evening (10 December), and that it was on this private occasion that he (Cooley) sustained the injuries of which Colton speaks.[3,66] The deposition of Wells' wife confirms that Wells had considered various possible ways of relieving surgical pain, including hypnosis, before Colton's demonstration. She then goes on to describe how she accompanied Wells to the public exhibition on the evening of 10 December and she tells us that Wells himself inhaled the gas, and that she later reproved him for 'making himself ridiculous before a public assembly'. Mrs Wells makes no mention of Cooley's performance or of the injuries he sustained, but she tells us that Wells simply responded that he thought that nitrous oxide 'might be used in extracting teeth'.[31]

An associate of Colton writing in 1894 says that Colton had told him that Wells had inhaled nitrous oxide at a private exhibition before the public demonstration, and that it relieved the pain of an aching tooth from which Wells was suffering.[31] Is it possible that this relief of a painful tooth by inhalation actually occurred either at the public demonstration on 10 December 1844 or at the private demonstration on 11 December 1844, and that this encouraged Wells to use the gas? The story may be too like Humphry Davy's experiment (see p. 13) to be entirely credible, however!

A disastrous demonstration in Boston

After this successful first anaesthetic Colton taught Wells how to make the gas and then went on his way to continue his itinerant exhibitions in other town.[66] Wells successfully used nitrous oxide for extractions for 'twelve or fifteen' of his patients and then felt himself ready to announce his discovery to medical and dental colleagues. To do so he went to Boston early in January 1845 and called upon the leading surgeon John Collins Warren (1778–1856) who (as has already been noted) had been looking for a means of relieving surgical pain, the eminent physician and chemist Charles Thomas Jackson (1805–1880) and William Thomas Green Morton (1819–1868) who had been a pupil of Wells and, although working in Boston, had been in loose partnership with him until October 1844.[31,61,64,65]

Warren invited Wells to lecture on his discovery to the medical students and to demonstrate the use of nitrous oxide for an amputation. Wells addressed the students but the amputation was postponed and he was asked to use nitrous oxide for a dental extraction on one of the students.

Wells tells us that the operation was undertaken before a large number of students and several physicians, but

unfortunately for the experiment the gas bag was removed early by mistake and he was but partially under its influence when the tooth was extracted. He testified that he had experienced some pain, but not as much as usually attends the operation, . . . several [of those present] expressed their opinion that it was a humbug affair (which in fact was all the thanks I got for this gratuitous service). I accordingly left next day.[3,61]

Later accounts suggest that the patient cried out before the extraction was completed and that Wells was actually hissed amidst cries of 'humbug', which may well be true.[64]

Wells had administered nitrous oxide 'twelve or fifteen' times for dental extractions in the relative quiet of his own office using a mouthpiece and an oiled silk bag containing only about 2 litres of nitrous oxide. The small bag was similar to the one used by Colton to produce mere inebriation during his stage exhibitions.[36] These earlier administrations by Wells were doubtless aided by suggestion. It is not surprising that he only produced relatively unsatisfactory analgesia in the tense and challenging atmosphere of a crowded demonstration room.[3]

Wells returned to Hartford immediately after his unfortunate demonstration in Boston and developed a debilitating illness which caused him to 'relinquish entirely' his professional practice as a dental surgeon for several months.[61,65] This episode was almost certainly a forerunner of the mental depression which, after it had been exacerbated by the controversy over the 'discovery' of anaesthesia after Morton's demonstration in 1846, led to his suicide in 1848 at the age of 33 (see pp. 28–29).[61,64,67] Wells continued to be interested in the use of nitrous oxide, however, and surgeons and dentists in Hartford also began to employ successfully the inhalation of the gas to control the pain of their operations, as did Wells himself after he resumed practice in September 1845.[59,64,65] Morton had also become interested in the potential of nitrous oxide and, as will emerge, consulted Wells about its manufacture in July 1845 (see below).[65]

Morton and the events preceding 'Ether Day' 16 October 1846

William Thomas Green Morton (1819–1868) was undoubtedly 'both the prime mover and the immediate agent in the introduction of [inhalational anaesthesia] to the world'.[60,67,68] Accounts of his origins and early life depend on information which he himself supplied in later life; they are to some extent contradictory in detail and may have been embroidered by his biographers.[26] It seems certain, however, that his family was of Scottish origin and his forebears were originally farmers and later small businessmen. Morton was born near Charlton, Massachusetts in 1819. He apparently had an early ambition to study medicine, but his education was cut short as a result of the financial crisis in the United States in the 1830s; after a period as a clerk in a Boston printing firm, and a failure in business on his own account in 1840, he became a student at the newly opened Baltimore College of Dental Surgery (possibly the first such institution in the world). Morton left without obtaining a diploma in 1841 and became an apprentice of the dentist Horace Wells (see above) in Hartford, Connecticut. Morton set up on his own account in nearby Farmington in 1842, but he remained on terms of mutual respect with Wells and continued to study the

latter's advanced techniques.[3,65,67] Morton transferred to Boston and entered into nominal partnership with Wells, who continued to practice in Hartford, in January 1844. The chief objective of the partners was to promote the methods of prosthetic dentistry which Wells had developed in Boston. They were supported by an endorsement of the lack of corrosion produced by Wells' cement from Charles T. Jackson (1805–1880), an eminent Boston physician and chemist who was to play a part in the development of the inhalation of ether as an anaesthetic (see pp. 23 and 28).[3,67]

The partnership did not prove to be profitable and it was terminated by mutual consent, but with expressions of goodwill, in October 1844, 2 months before Wells' first experience with nitrous oxide for dental extraction (see p. 21).[65–68]

Morton's practice in Boston prospered, he extended Wells' ideas, employed a number of assistants and even established a factory for the manufacture of artificial teeth. One great problem remained, however; was it possible to find a way of rendering the extraction of teeth painless prior to the insertion of dentures or the fixing of crowns?

Morton enrolled as a medical student in Boston in March 1844 and married Elizabeth Whitman of Farmington in May of the same year. Charles Jackson, the chemist, was one of Morton's tutors and, as the result of his previous connection, Morton and his young bride began married life boarding with Jackson where they remained until early in 1845.[67]

Morton discussed the problem of dental pain with Jackson during this period and the latter advised the direct application of ether to the gum to produce a local anaesthetic effect. Jackson and Morton also discussed the intoxicating effects of ether at that time.[3,67]

Morton was present at the partial failure of the demonstration of nitrous oxide by Wells in Boston in January 1845 (see p. 21), but he was still attracted by the possibility of pain relief provided by the inhalation of the gas and discussed the practicalities of its manufacture with him in July 1845.[65] This was at a time when Wells and others in Hartford were considering the use of ether as an alternative to nitrous oxide and this possibility may also have been discussed.[59,61,65]

It is not easy to piece together with certainty the exact sequence of events in 1845 and 1846 which led to Morton's public demonstration of the use of ether by inhalation to relieve the pain of surgery on 16 October 1846; nor is it easy to separate conjecture by authors writing dramatically about the events some years later, and some of the anecdotes and attributed quotations seem to be a little too trite! Morton himself composed several depositions during the disputes over priority which followed the demonstration (see pp. 27–29). These include memoranda to the French Academy of Science (1847), to the Congress of the United States (1861 and 1864) and to the scientific and lay public at large in the intervening years and, on the whole and taking into account the evidence of others, there is little reason to doubt that his accounts are reasonably accurate.[3,67]

Morton had certainly become seriously interested in the possibilities of producing stupor with inhaled ether to cover dental procedures by the beginning of 1846 but, though the use of ether by inhalation for short periods for recreational purposes and to relieve asthma and other respiratory conditions was well known and documented, the consensus of opinion was that prolonged inhalation of ether vapour would inevitably result in death. Morton was determined to define the limits and value of ether inhalation and, at the end of June 1846 he engaged a locum dentist (Greville Hayden) to look after his practice with the intention of conducting a series of animal experiments at the farm which he had bought at West Needham (now Wellesley) on the outskirts of Boston.[3,67]

Morton's animal and human experiments

Morton's initial experiments with worms and insects were, not surprisingly, inconclusive, and he turned his attention to his spaniel Nig. The dog was rendered completely unconscious but revived in a few minutes. Some days later Morton himself inhaled ether; he did not lose consciousness but thought that he was stuporous enough for a tooth to be extracted with little pain.

Morton had now exhausted his supply of ether but he obtained a further quantity from a new supplier and was determined to try it on another human being so that he could observe its effect. He set his apprentices to finding a suitable subject who would be willing to have a tooth extracted free of charge if he would first inhale the ether. The apprentices failed in their mission but agreed to inhale ether themselves. The results were unsatisfactory, however, as they did not lose consciousness but became wildly excited. Morton himself inhaled a portion of the newly supplied ether, but he too only became excited and could not reproduce his previous experience.

It is certain that Morton, though frustrated, felt himself on the threshold of a discovery but was loath to share his potentially financially valuable trade secret. He required further advice from someone who was better educated and had a superior knowledge of chemistry. Jackson was the obvious choice. Morton was wary about consulting him, however, because Jackson already had a reputation for attempting to take credit for the inventions and discoveries of others (see p. 27). Morton finally determined to talk to Jackson at the end of September 1846.[3,67]

The interview between Morton and Jackson on 30 September 1846

The interview with Jackson was lengthy. Jackson later maintained that it was on this occasion that he put the idea of using the effect of inhaled ether to lessen the pain of dentistry and surgery into Morton's head. This is manifestly untrue. Morton had certainly experimented with ether before that time and, despite his subsequent written submissions, Jackson contradicted himself verbally (see p. 27).[3,13,67]

Morton's version of what took place is undoubtedly nearer the truth. He purposely did not divulge that he had been experimenting with ether. He first implied that he wished to borrow a gas bag with a view to possibly suggesting to a patient breathing air by hypnosis that he or she was free from pain. Jackson condemned this approach lest Morton be thought to be as big a humbug as Wells (see pp. 21–22). Morton then casually introduced the subject of ether inhalation. Jackson talked about the stupefying effects on students of inhaling ether and led him on to discourse on the purity of samples resulting from various methods of preparation. Jackson also handed Morton a flask with a glass tube through a cork in its neck and suggested that that would be a better way of inhaling ether than putting the fluid in a gas bag.[67]

Morton realized that the ether which he had been using more recently was probably impure and that the impurities might account for the unusual degree of excitement which it had caused. He obtained a specimen of pure rectified ether from his original supplier, shut himself in his office alone and inhaled first from Jackson's flask, which was unsatisfactory, and then from a handkerchief. He became unconscious and when he was recovering, although he was very frightened he demonstrated analgesia by pinching himself.

Morton's first etherization in dental practice and its consequences

That evening (30 September 1846) an emergency patient named Eben Frost called on Morton, requesting extraction under mesmerism. Morton in the presence of his colleague Hayden extracted the tooth under ether inhaled from a handkerchief while Frost was unconscious.

Morton naturally wished to gain the full benefit of immediate commercial exploitation of his discovery, but he was also anxious to conceal its identity until his interests could be protected. The next day (1st October 1846) he therefore both inserted an advertisement recording Eben Frost's extraction in the form of a report in the *Boston Daily News* and consulted Richard Eddy, lawyer and Commissioner for Patents, about the possibility of applying for a patent.[3,67]

Eben Frost's etherization had been an unqualified success but subsequent results were very variable and sometimes alarming; despite this, however, knowledge of Morton's new painless dentistry spread and attracted many patients and Morton experimented with a number of different inhalers.[3,67]

The medical and dental professions, including Jackson, were sceptical and even antagonistic. There was, moreover, a great conceptual difference between provision for the painless extraction of teeth and possible anaesthesia for major cutting surgery.[3,67] Sykes has drawn attention to a statement by Morton's son, Dr W. J. Morton, in the discussion on a paper at a meeting of the American Therapeutic Society in 1911 65 years later. Morton's son told the meeting that, prior to the administration for Warren's operation at the Massachusetts General Hospital on 16 October 1846 'my father, Dr. Morton, had employed [ether] by inhalation for thirty-seven private operations by Henry J. Bigelow'.[4,60,69] This would seem to be very unlikely. The statement also seems to be incompatible with a passage in Bigelow's very important and perceptive first paper on ether anaesthesia (see p. 25), which was written immediately after Warren's operation.[70] This merely records laconically that 'Dr. Morton was understood to have extracted teeth under similar circumstances, without the knowledge of the patient'.[70] Dr W. J. Morton must have been mistaken; after all he was a small child in 1846 and an elderly man by 1911. Given the conventions of the composition of a scientific paper, however, Bigelow's remark is compatible with the probability that the forward looking Henry J. Bigelow had gone out of his way to visit Morton's office and witness dental extractions under ether before 16 October 1846; indeed, although not specifically stated, the dental cases described by Bigelow in his paper could well have been undertaken before the demonstration at the Massachusetts General Hospital.

Henry Bigelow was subsequently a leading champion to Morton's claim to priority and he may have been responsible for encouraging Warren, his senior colleague, to invite Morton to demonstrate the use of his new preparation at the Massachusetts General Hospital as Sykes suggests.[13,68,70]

Morton himself sought and obtained an interview with the formidable John Collins Warren (1778–1856), Senior Surgeon at the Massachusetts General Hospital, who had trained in Edinburgh, London and Paris and who was, without dispute, the leading surgeon in New England. Warren, as has been previously emphasized, was actively interested in methods for the relief of the pain of surgery (see p. 15). He was sympathetic to Morton even though the latter did not divulge the nature of his agent, and on 14 October 1846, Morton received a note from Warren's house surgeon (Dr C. F. Heywood) inviting him to attend the Massachusetts General Hospital at 10 a.m. on Friday 16 October 'to administer to a patient who is then to be operated upon, the preparation you have invented to diminish the sensibility to pain'.[3,67]

The Morton's ether inhaler

Morton must have been painfully aware from his limited dental experience that his technique and apparatus were liable to failure. He was faced with having only 2 days to develop his apparatus further. The idea for an apparatus consisting of a glass globe with inlet and outlet ports into which a sea-sponge soaked in ether was inserted, was originated by the Boston instrument maker Joseph Wightman. The actual inhaler used at the Massachusetts General Hospital on the morning of 'Ether Day' (16 October 1846) was made to Morton's specification by Chamberlain, another instrument maker, and was only ready for use immediately before the demonstration. Bryn Thomas presents convincing evidence, based on a letter which Morton wrote to the *Lancet* in October 1847, that the 125 mm diameter glass inhaler used on Ether Day was fitted with a simple tubular spigot and tap similar to those used on a barrel of beer.[36,71] The patient took the spigot in his mouth and breathed to and fro over the ether sponge. Bryn Thomas determined that, provided the patient breathed wholly through the mouth (the nostrils being pinched between the administrator's fingers), the percentage of ether vapour could be as high as 24 per cent, and sufficient to induce the state of analgesia produced on Ether Day (Bryn Thomas, personal communication, c. 1975). Morton tells us that the inhaler was converted into a one-way valved system after the original demonstration 'late on the night of the 16th [October 1846] at the suggestion of Dr. A. A. Gould, a distinguished physician of Boston' and was used in this form at the second series of operations the next day (17 October).[36,71] It is this refined apparatus which was described by Bigelow and is exhibited at the Massachusetts General Hospital at the present time.[3,36,68,71]

Ether Day Friday 16 October 1846

The operating theatre on the top floor of the Massachusetts General Hospital received maximum light from a domed skylight and is now preserved as the 'Ether Dome'. The tiered seats were crowded by the prominent medical men of Boston by 10 a.m. on Friday 16 October 1846. A number of accounts written by people who were actually present describing the events which took place have come down to us.[6,70,72]

The patient was Gilbert Abbot, a young printer aged 20. He had a vascular tumour on the left side of his neck just below the jaw which 'had probably existed from birth'.[6] He was led in promptly at 10 a.m. and arranged in a sitting posture on an operating chair. Morton had not appeared by 10:15; John Collins Warren raised a 'derisive laugh' by saying 'I presume he [Morton] is otherwise engaged', and then grasped the knife and was about to proceed with the operation.[72] Morton, who had been delayed while Chamberlain put the finishing touches to the inhaler, came hurriedly into the room at that moment accompanied by Eben Frost who had been brought along to authenticate the previous efficacy of the treatment should there be a failure on this occasion.[67]

Warren turned to Morton and 'in a strong voice' said: 'Well, sir, *your* patient is ready.' Morton poured his 'preparation' over the sponge in the inhaler and, having spoken a few words to the patient, persuaded him to inhale the vapour through his mouth. Some accounts say that Morton had attempted to disguise the smell of ether by adding an aromatic such as oil of oranges but, be that as it may, Bigelow tells us that the odour of ether was 'readily recognized in the preparation employed by Dr. Morton'.[67,70]

'After four or five minutes [the patient] appeared to be asleep'.[6] Morton turned to Warren and echoing the latter's earlier remark declared: '*Your* patient is ready, sir.'[72]

The inhalation was discontinued and Warren tells us that:

> I made an incision ... to my great surprise without any starting, crying out, or other indication of pain. ... Then followed the insulation [sic] of the veins, during which he began to move his limbs, cry out and utter extraordinary noises. These phenomena led to some doubt of the success of the application ... until I had, soon after the operation, and on various other occasions asked the question whether he suffered pain. To this he always replied in the negative; adding, however, that he knew of the operation, and comparing the strokes of the knife to that of a blunt instrument passed roughly across his neck.[6]

In his book *Etherization with surgical remarks*, written 2 years later, of the experience of 'about two hundred cases' Warren goes on to say: 'Now that the effects of inhalation are better understood, this [administration] is placed in the class of cases of imperfect etherization'[6] but, despite the imperfection, Ayer tells us that Warren had the vision to turn to those present on Ether Day and say: 'Gentlemen this is no humbug.'[72]

This endorsement of etherization for the relief of the pain of surgery by the internationally renowned senior surgeon of the Massachusetts General Hospital was the first of a succession of events which was to revolutionize the practice of surgery within an amazingly short space of time. It was fortunate that the most recently elected surgeon to the Massachusetts General Hospital, Henry J. Bigelow, was present in the audience. He was certainly profoundly impressed and some accounts say that he actually declared: 'I have seen something today which will go round the world.' He was right, and it was his pen that sent the good news on its way![70]

The importance of Morton's contribution

It is pertinent to consider the significance of Morton's contribution at the moment of his triumph on Ether Day. He had set out to develop a technique that would enable him to extract teeth painlessly before inserting dentures, and ended by making a major contribution to surgery. He might for a time have used etherization as a trade secret to promote his own dental practice, but instead he realized that it might be possible to apply it to cutting surgery, and he brought it to the attention of the medical profession. His motives were undoubtedly mixed; on the one hand, in common with many New Englanders of that entrepreneurial period, he believed that the labourer was worthy of his hire, and was interested in a maximum return on the patent which he had initiated, but, on the other, he was concerned that etherization should be effectively and safely used to the benefit of mankind.

J. Marion Sims, an eminent gynaecological surgeon, writing in 1877 in support of Crawford Long's claim (see p. 18) in an otherwise balanced article, unfairly, and rather patronizingly, deprecates Morton's contribution and emphasizes the importance of the role of the surgeons ('it was not Morton but Warren, Hayward and Bigelow who performed the operations at the Massachusetts Hospital ... that the world owes the immediate and universal use of anaesthesia and surgery').[59] John C. Warren, however, gave the credit to Morton ('The first proposal, for the employment of ether inhalation, for the prevention of pain in surgical operations, was made by Dr. W. T. G. Morton').[6]

It has been said that Morton only applied knowledge which was already known as the direct heir to a process which had started with Beddoes' inhalation of ether, nitrous oxide, and other gases and vapours for therapeutic purposes, followed by the discovery of the inebriating effects of both agents by Humphry Davy and others, and then on to observing the attempt by Wells to press nitrous oxide to the point of insensibility, and learning from him the possibility of using ether for the purpose.[4] This would be unfair, however; it must be recognized that it took research, persistence and courage to press ether inhalation beyond the stage of excitement to analgesia and anaesthesia.

Morton's patent

Morton had applied for a patent for his process in the United States even before he demonstrated its efficacy at the Massachusetts General Hospital on 16 October 1846 (see p. 23). This action has often been condemned, but it must be appreciated that there was a considerable gulf between both the status and the ethics of the established profession of medicine, which altruistically forbade the use of secret remedies and the patenting of technical procedures, and the competitive 'mechanical art of dentistry', which was practised by men like Morton who had no academic qualifications and were largely self-taught. H. J. Bigelow, in his initial paper following Morton's demonstration, commended the concept of reward from public funds for 'discoveries in medical science whose domain approaches to that of philanthropy', but pointed out that one of the greatest fields for the application of etherization would be in dentistry 'many of whose processes are by convention, secret or protected by patent rights'. Bigelow went on to suggest two other possible justifications for a patent which were not connected with the profit motive but would restrict its use: first, that the process of etherization was 'capable of abuse', and could 'readily be applied for nefarious ends', and, second, that the consequences of etherization were not yet thoroughly understood and 'should be restricted to responsible persons'.[70]

The granting of a patent necessarily reveals the nature of a device or process, but restricts its use to the patentee except under licence. The possibility of obtaining a patent for both the inhaler and the fluid was only under consideration by the lawyers at the time of Morton's demonstration on 16 October 1846; Morton therefore endeavoured to conceal the nature of the preparation so that he would not be pre-empted. The lawyers decided that it would be lawful for a patent to be granted, and the application was made on 27 October 1846 and issued on 12 November in the joint names of Morton and Jackson (see p. 27).

The success of the operations on 16 and 17 October 1846 made Warren and his fellow surgeons anxious to extend the benefits of etherization to other patients but, under pressure from the Massachusetts Medical Society and because of the application for the patent and the unknown nature of the fluid, they decided not to use or encourage the use of etherization 'until more liberal arrangements could be made'.[6] Morton ceded the use of his preparation and inhaler to the surgeons of the Massachusetts General Hospital in a letter dated 5 November 1846, together with

'any information, in addition to that which they now possess, which they may think desirable in order to employ it with confidence'. This generous gesture included revealing the nature of the fluid at the first amputation under its influence on 7 November 1846.[3,67]

Morton was generous in his concessions to hospitals in the United States but despite this he continued to try and advertise the process and to enforce his patent, including sending agents to London, Paris and elsewhere. A patent was granted in London on 21 December 1846, the same day as Robert Liston (1794–1847) performed the first major operation in England under ether anaesthesia.

It is sometimes stated that Morton gave his 'preparation' the synonym 'Letheon' in an attempt to disguise its identity before his demonstration on 16 October 1846. This is not the case, however. This name was apparently decided upon at a meeting attended by Morton and H. J. Bigelow and Oliver Wendell Holmes, representing the Massachusetts General Hospital on 21 November 1846, after the identity of the preparation was known.[3,67] The reason for this is not clear, but it may be that the Boston surgeons, having ascertained the identity of the preparation and obtained the freedom to use it themselves, were content to acquiesce in Morton's attempts to enforce his patent elsewhere. Morton certainly used the term extensively in the series of promotional circulars he wrote at the end of 1846 and throughout 1847[16] but his excellent scientific pamphlet, also written in 1847, refers to the agent by its proper chemical name.[73] Other authors used the word 'etherization'[6] but very soon 'anaesthesia' and 'anaesthetic', as suggested by Holmes, became common parlance (see p. 3).[3]

Morton gave up his dentistry and his medical studies but continued to administer ether at the Massachusetts General Hospital and elsewhere and to promote the process in which he passionately believed throughout 1847, both professionally and in an attempt to gain recognition for his patents at home and overseas. By the end of 1847, however, it was apparent that the patent was not enforcable and Morton himself even abandoned the use of his own inhaler in favour of 'a sponge well saturated with ether . . . placed over the nose and mouth of a patient, so that all the air which he breathes must necessarily pass through it'.[3,73] After the end of 1847 Morton retired to the country, and devoted himself to obtaining recognition as the sole discoverer of anaesthesia with the objective of gaining a substantial reward from the United States Congress (see pp. 27–28). Morton probably realized early on that his attempts to extract substantial licence fees by means of his patent might prove futile as he sent his first memorial to Congress claiming a financial reward on 28 December 1846.[3]

The dispute about priority

The claim of Charles Thomas Jackson

No sooner was the use of ether adopted in Boston by the surgeons of the Massachusetts General Hospital than the physician and chemist Charles T. Jackson (see p. 23) laid claim to having given the idea of using ether for the relief of surgical pain to Morton, and even that the latter had acted under his instructions.[2–4,6,13,16,67,74]

Charles T. Jackson (1815–1880) was an internationally acknowledged, eminent, and influential scientist who had had his training at Harvard in Boston and at the Sorbonne in Paris.[67] He had, however, an obsession which ultimately deteriorated into overt insanity (see p. 30). This was that he appears to have believed that, if a great discovery was made it was he (Jackson) who had made it.[4] He had, for example, tried to obtain control of the famous patient Alexis St Martin with the gastric fistula on which the military surgeon William Beaumont performed his well known series of experiments, and also to claim the credit of inventing the electromagnetic telegraph of Samuel F. B. Morse in a blaze of publicity.

Morton acknowledged Jackson's part in the development of his ideas as he saw it, as has already been related (see p. 23). That was that Jackson advised him on the topical use of ether when directly applied to a sensitive nerve, that he discussed the recreational use of ether by students with him, and that he suggested that the best results in such inebriation could be obtained by inhaling the purest ether.

Jackson claimed that he had given Morton the idea of using ether for dental extractions, but all the evidence is that Morton had already experimented with the agent before he approached Jackson (see p. 23). Jackson also refused to give Morton a certificate regarding the safety of the inhalation of ether after the first successful dental extraction (see p. 24), and publicly and in private condemned the use of ether for dental extractions as being both reckless and dangerous.[3,67]

Jackson was not present at Morton's first demonstration (he claimed that he was not notified of Morton's intention) but, following its success, he visited John C. Warren and informed him that he had suggested the use of ether for the relief of surgical pain to Morton. He then pestered Morton to include him in the patent application. Morton was naturally very reluctant at first, but consented after his legal adviser (Richard Eddy, a friend of Jackson's) had emphasized that the support of as eminent a person as Jackson would be invaluable in stifling any criticism. Morton was certainly often concerned, with some justification, that his inferior status as a dentist might weigh against the acceptance of his ideas. Jackson was to receive 10 per cent of the profits.

Jackson was still not certain whether Morton's use of ether for cutting surgery would not lead to fatalities, but he was determined that, if it did prove to be outstandingly successful and generally accepted, he (Jackson) would be internationally recognized as the sole discoverer and Morton would be relegated to the position of an agent who had acted under his instructions.[3,4,13,67] Jackson therefore wrote his version of the events leading to the discovery, placed it in a sealed packet dated 1 December 1846, addressed to Élie de Beaumont, a member of the prestigious French Académie de Sciences in Paris, and dispatched it by mail steamer. This would almost certainly be the 'Acadia' which left Boston on 3 December 1846 also carrying Jacob Bigelow's letter to Boot which alerted London to the use of ether and Edward Warren, Morton's prospective agent in Europe, with samples of the inhaler.[75,76] Beaumont placed the packet in the archives of the French Academy on 28 December 1846. This was the usual practice in the scientific world of the day when an initial discovery was made but further evaluation was required. If the discovery was subsequently confirmed, the originator could claim priority but, if not, the package could be destroyed unopened.

Jackson claimed that he had conceived the idea of using ether to relieve the pain of surgery 'in the winter of 1841–2', after he had rendered himself unconscious by inhaling ether to counteract the respiratory symptoms resulting from the inhalation of chlorine accidentally released in his laboratory (a treatment recommended in the current textbooks of that period including *Pereira's elements of the materia medica* of 1839).[3,13,67] Sykes deduces from the detail in Jackson's description that the incident with the chlorine actually took place, but there is no evidence that the idea of etherization for the relief of surgical pain occurred to Jackson at that time.[13] Jackson's letter to the French Academy did not refer to Morton by name, merely alluding to him as 'a dentist', and he implied that he acted throughout, both in dental practice and at the Massachusetts General Hospital, as Jackson's agent.

The news of Morton's dramatic demonstration on 16 October 1846 reached Paris in the following December and the first successful use of ether for surgery in France was on the 24th of that month.[76–78] The French Academy debated the subject of etherization on 18 January 1847 when de Beaumont dramatically opened and read Jackson's declaration.[8] The result was that for a time, in continental Europe, Jackson, an acknowledged academic scientist, was hailed as the sole discoverer.[3,67] This was in contrast to the United Kingdom and its worldwide sphere of influence which received the news from Henry Bigelow's account of Morton's 16 October 1846 demonstration.[2,3,70] Morton's inhaler was successfully used in Paris on 23 January

1847, but Morton's written rebuttal of Jackson's claims did not reach Paris until 17 May 1847; in 1850 the French Academy divided the Montyon Prize of 5000 francs between Jackson and Morton. Jackson's share was for 'observations and experiments regarding the anaesthetic effect of ether', and Morton's for its 'application'.[3] This decision pleased neither Jackson nor Morton! Morton had his share converted into a suitably inscribed gold medal, much to Jackson's annoyance.

Jackson's claim received less credence in Boston amongst the establishment, but his public vilification of Morton did him and his practice a great deal of harm. Jackson exploited the unpopularity of Morton's initial attempts to defend his patent and, as soon as it had proved worthless, dramatically and publicly renounced his potential share because his 'tender conscience' had come to regard it as 'blood money'. The trustees and medical staff of the Massachusetts General Hospial appointed a commission of inquiry to consider the question of priority. This Committee decided in favour of Morton and described Jackson as a man 'honestly self-deceived' in his assertion that he had given Morton the idea for the use of ether to promote painless surgery. Thereafter Jackson devoted himself to frustrating Morton's claim for a monetary reward from the Congress of the United States (see pp. 28–29).[3,67]

The claim of Horace Wells

Morton, flushed with the success at the successful use of etherization at the Massachusetts on 16 October 1846, wrote to Wells on 19 October. Morton described the successful application of his 'preparation' and invited Wells to go to New York as his agent to 'dispose of rights upon shares'. The letter written by Wells in reply is couched in somewhat cryptic language. It implies that Morton's patent and his concept of licensing the use of his preparation would defeat Morton's objective of making money by selling rights under its terms, but it also suggests that the discovery might make 'a fortune' for Morton; presumably Wells had in mind that the introduction of painless dentistry into Morton's own practice would attract a large number of patients.[3]

Wells visited Morton on 25 October 1846 and Morton demonstrated the use of ether to him. His reaction, according to Mrs Wells, was that Morton's discovery was nothing new and that Morton 'did not know how to use it'. This perhaps confirms that Morton's administrations were by no means perfect at that time![67]

Wells was angered and highly critical when he learned that Morton and Jackson had actually taken out a patent; thereafter, in the remaining months of his life (he committed suicide on 23 January 1848, see p. 30), he pursued his claim to priority with increasing ferocity in a number of

communications to both the scientific and public press and elsewhere.[3,64,65] His main arguments were that 'this discovery does not consist in the use of any one specific vapour, for anything which will cause a certain degree of nervous excitement is all that is required to render the system insensible', and that 'nitrous oxide was safer and less injurious to the system than ether'.[61] The interesting concept that nitrous oxide owes its action to causing 'nervous excitement' has already been considered (see p. 20).

Wells had a perceptive but unstable and quixotic mind. He initiated a number of schemes for making money from inventions and processes unconnected with the profession of dentistry; however, his repeated assertions that he did not seek financial reward for the discovery of the control of pain in surgery, but only recognition and honour as the discoverer of pain relief by inhalation, have the ring of truth.[3,61,64,65]

One impression that Morton and Jackson and other detractors endeavoured to create, that Wells gave up the practice of dentistry entirely after his abortive demonstration in 1844, is palpably false. There is ample evidence that, after his initial mental indisposition (see p. 22), Wells continued to practise dentistry and to experiment and administer nitrous oxide (and later chloroform), first in Hartford and then in New York, up to the time of his death.[3,59,61,64,65] It was, however, purely coincidental that Wells left Boston for Europe at the end of December 1846, having made his first public and private claims for priority over Morton and Jackson. He went to Europe in pursuit of his latest business venture which involved the buying and resale of artistic paintings.[3]

Wells arrived in February during the initial deliberations of the French Academy concerning the counterclaims of Jackson and Morton (see pp. 27–28). He was taken unawares and had no supporting evidence with him, but he prepared a memoir for the consideration of the Academy, and, after his return to Boston he forwarded testimonials and affidavits to the Academy. These submissions resulted in a recollection of Hickman's submission to the Academy (see p. 19).[28,52,64] Wells was elected in 1848 to the status of an Honorary Member of the Faculty of Medicine in recognition of his part in the discovery.

Morton's petitions to Congress

Morton's practice was ruined by the denigration of etherization by rival dentists and the repeated public attacks and innuendos originated by Jackson.[67] He abandoned dentistry and medicine and became a farmer. He exhausted his finances in trying to establish his claim to priority and in petitioning Congress for financial recognition.[3,67]

Morton made three submissions to Congress (in 1846, 1849 and 1851). These were supported by the Trustees and staff of the Massachusetts General Hospital. All three applications were opposed by Jackson and Wells (and following his death in 1848 by the representatives of the latter's widow) and after 1852 by Crawford Long. Jackson visited Long in Athens, Georgia in March 1854 in an attempt to persuade Long to join him in making a joint claim to Congress.[59,60] This Long declined to do, but Jackson none the less supported his case in an attempt to discredit Morton and Wells.[3]

A Bill 'to recompense the discoverer of practical anaesthesia' passed the Senate in April 1854 but was rejected 2 days later by the House of Representatives because of 'the multiplicity of claimants' (a number of others had jumped on the bandwagon in addition to Long, Wells, Morton and Jackson). Morton protested. The protest reached the desk of the President of the United States. A complicated legal procedure involving a staged test case against the Government, whose military and civil medical officers had nominally infringed Morton's patent, was proposed. This was instituted in 1858 but dismissed in 1862. Morton was a broken man both financially and mentally and lived in poverty until his death in 1868 (see p. 30).

The fate of the nineteenth century American originators of anaesthesia

Crawford Williamson Long

Crawford Williamson Long (1815–1879) was a likeable, compassionate and resourceful general practitioner working in an isolated rural practice (see p. 18).[60] His use of ether in 1842 does not appear at first sight to have had any influence on the events which led to Morton's Ether Day, but it is just possible that it did. An authenticated draft letter from Long exists which tells us that:

> an itinerant dentist and surgeon from Boston, Massachusetts were in Jefferson, Jackson County in 1842, and remained for three or four weeks. The dentist practised his profession and the surgeon operated for strabismus – I have always thought it probable that the dentist was Morton or [more plausibly] Wells and a knowledge of my use of ether was obtained at that time.[16]

Long practised all his life in the state of Georgia. He moved briefly to Atlanta in 1850, and to Athens in 1851

where he had a practice with his brother which included the ownership of a drug store. He prospered but, being a Confederate, was almost ruined by the American Civil War of 1860–1865. He looked after the casualties of both sides with dedication and once again prospered in the postwar period after much effort and hard work. He was just 27 when he first used ether for surgery. He died of a cerebral vascular haemorrhage at the age of 62 at a confinement after delivering the baby and, it is said, just managed to ask an attendant to look carefully after the mother and child before he lost consciousness.[59,60]

John Collins Warren

John Collins Warren (1778–1856) is not usually included amongst the pioneers of anaesthesia but he deserves to be. He represents the new breed of skilful and compassionate surgeons of the first half of the nineteenth century who were actively looking for a means of alleviating the excruciating pain of surgery and thereby also extending its scope. He used ether as a respiratory antispasmodic and anodyne, and he experimented with various methods of procuring analgesia, including hypnosis. He encouraged both Wells and Morton in their endeavours, and he recognized the potential of Morton's first etherization for major surgery despite its imperfections. He championed and nurtured anaesthesia in its early days when it was subject to considerable opposition, and he doubtless encouraged his young associate Henry J. Bigelow (1818–1890) to proclaim the great discovery to the world.[6,60] John Collins Warren was the progenitor of a dynasty of New England surgeons.[25] He was 68 years old and still an active surgeon at the time of Morton's demonstration in 1846, and died full of honour 10 years later.

Horace Wells

Horace Wells (1815–1848) was a thoughtful and enigmatic character with an inventive turn of mind. He published a small book in 1838 entitled *An essay on teeth: comprising a brief description of their formation, disease and proper treatment* and, at one time or another, had several business interests outside dentistry. He patented a coal sifter (1839) and a shower bath (1846), he promoted a stage exhibition entitled 'a panorama of nature' in 1845, and he endeavoured to set himself up as an art dealer in 1846 (see p. 28).[64,65]

His final professional act, while engaged in his dispute with Morton and Jackson over the question of priority, was to open a dental office in Chambers Street, New York City in January 1848, and to advertise his services as an extractor of teeth without pain using 'ether and various stimulating gases' and claiming to be the discoverer of the effects of these agents.

Wells had, by this time, become addicted to chloroform while experimenting with that agent which James Young Simpson of Edinburgh had introduced as an anaesthetic in November 1847.[3,4,8,64,65]

Wells died by his own hand on 23 January 1848 after severing his femoral artery with a razor while under arrest for sprinkling acid on the clothes of a prostitute in Broadway. He left a detailed and apparently lucid written confession; in this he stated that he had committed this extraordinary act while under the influence of chloroform, and that he feared that he was becoming a maniac.[64,65]

Wells had strong support from the citizens of Hartford, Connecticut for his claim to be the discoverer of anaesthesia and is commemorated both there and in Paris, France by statues. He was 29 years old when he first inhaled nitrous oxide and had his own tooth extracted (see p. 21), and just 33 when he died.[3,4,59,64,65]

Charles Thomas Jackson

The long and bitter campaign of Charles Thomas Jackson (1805–1880) to supplant Morton as the originator of anaesthesia lasted over 20 years. Jackson's claims were ultimately discredited, but they prevented Morton getting a just financial reward from the Congress of the United States while they were investigated, and it was Jackson's claim that prompted Morton to make his last journey to Washington where he died in 1868 (see below).[3,67]

There now seems to be little doubt that Jackson did administer some anaesthetics himself in the years that followed Ether Day in 1846 and, in 1861, he wrote a *Manual of etherization* primarily intended for military and naval surgeons in the American Civil War. This is a reasonably practical treatise if the familiar diatribe concerning Jackson's part in the discovery is excluded.[13]

Jackson suffered from increasing delusions and paranoia and was finally confined to a mental asylum where he died in 1880.[59] He was 41 years old at the time of Morton's demonstration in 1846 and 75 when he died.

William Thomas Green Morton

The contribution of William Thomas Green Morton (1819–1868) to the development of modern anaesthesia has already been discussed, as have his reasons for attempting to patent his discovery. These were not solely mercenary, and there is no doubt that he had a much more straightforward and likeable character than is sometimes supposed (see pp. 22–26).[16,67] He was not just an opportunist who happened to be in the right place at the right time; before his first public demonstration he had taken the trouble to glean information from those who were more knowledgeable than he was, and had experimented on both animals and himself (see p. 23).

Morton suffered from being a self-educated dentist, a profession then considered in the social scale to be only slightly superior to a tradesman. Some of his contemporaries had difficulty in accepting that a mere dentist could make such a stupendous discovery. If he had been medically qualified he probably would have had less difficulty in establishing his claims for recognition and remuneration, and he might not have made the tactical error of seeking to patent his discovery.

Morton ceased to practise either as a dentist or an anaesthetist after 1848 and devoted his energies to counteracting Jackson's increasingly violent campaign of vilification, and to promoting his justified, but abortive, petitions to Congress for a pecuniary reward. His claims were slowly generally recognized by the profession and public, but his financial position became increasingly desperate. He once more practised anaesthesia and showed his humanitarian instincts by volunteering to administer ether to the wounded close to the field of battle in 1864 during the American Civil War.[79]

Morton died at the age of 48 in 1868 from apoplexy during a heat wave in New York. His objective in making the journey to New York was to seek legal advice in order to bring a suit against Jackson, whose claims had been yet again promoted in another scurrilous article.[3,68]

Jacob Bigelow (1786–1879), Professor of Materia Medica in Boston and father of the surgeon H. J. Bigelow, described Morton in the inscription on his tomb as the 'inventor and revealer' of anaesthetic inhalation;[36] 'inventor', perhaps not, but 'revealer' certainly.

Appendix

A chronological history of anaesthesia and related events

Early references c. BC 2500–c. AD 1750

BC

c.2250   *Babylonian tablet* records a dental filling of henbane to relieve toothache

c. 500   *Hippocrates* described the relief of pain by opium

AD

c. 100   *Dioscorides* of Greece administered a concoction of the root of mandragora to relieve the pain of surgery

c. 150   *Heron of Alexandria* described the first medical piston and barrel syringe

c. 250   *Hua T'o*, a Chinese military surgeon, used Indian hemp (hashish) to render patients unconscious for surgery

c.1200   *Nicolas of Salerno* gave an account of the value of inhalation of fumes from the 'soporific sponge' (soaked in hashish, poppy (opium), mandragora, etc.) for surgical anaesthesia. (Descriptions of the use of the soporific sponge persist throughout the Middle Ages)

c.1540   *Valerius Cordus* synthesized ether ('sweet oil of vitriol') and the Swiss physician *Paracelsus* described the induction of sleep in chickens with its vapour

1596   *Sir Walter Raleigh* (England) described the effects of South American native arrow poison – possibly curare

1616   *William Harvey* (England) described the circulation of blood

1665   *Sir Christopher Wren and Sir Robert Boyle* (England) observed the effect of intravenous injection of opium into a dog

1666   *Richard Lower* (England) described animal to animal blood transfusion

1667   *Jean Denise* (France) attempted therapeutic animal to man transfusion

1730   *August Frobenius* (Germany) gave the name 'ether' to sweet oil of vitriol

1751   *Bailey's English Dictionary* defines anaesthesia as 'a defect in sensation'

The Age of Enlightenment 1750–1846

1754   *Joseph Black* (Scotland) isolated carbon dioxide

1766   *Franz Anton Mesmer* (Germany) publicized 'animal magnetism' (hypnosis)

1771   *Joseph Priestley* (England) discovered oxygen

1772   *Joseph Priestley* (England) discovered nitrous oxide

1794   *Thomas Beddoes* (England) founded the Pneumatic Institute at Bristol with the objective of treating chest diseases by the inhalation of gases and vapours

1796   *James Moore* (England) produced local analgesia for the great John Hunter by compression of peripheral nerves

1800   *Humphry Davy* (England) while acting as Superintendent of Beddoes Institute (see 1794) inhaled nitrous oxide thus reducing the pain of an erupting wisdom tooth. He suggested that the gas might be used for surgical anaesthesia

1807   *Baron Larrey* (France), Napoleon's great military surgeon, observed that refrigerated gangrenous limbs were analgesic at the battle of Preuss Eylan

1818   *Michael Faraday* (England) observed the analgesic effect of inhaling ether

1818   *James Blundell* (England) gave the first human to human blood transfusion

1824   *Henry Hill Hickman* (England) used 'suspended animation' (asphyxia with carbon dioxide) to produce surgical anaesthesia in animal experiments

1825   *Charles Waterton* (England) gave an account of the action of curare which he had brought back from South America

1829   *Cloquet* (France) performed a mastectomy under hypnosis

1843   *John Elliotson* (England) published an account of his use of hypnosis for surgical 'anaesthesia'

1846   *James Esdaile* (England) published an account of the use of hypnosis for major surgery in India

Inhalation anaesthesia before the introduction of antiseptic surgery 1842–1867

1842   *W. E. Clarke and Crawford W. Long* (both of the USA) independently made isolated use of inhaled ether for a dental extraction and the removal of sebaceous cysts respectively

1844   *Horace Wells* (Hartford, Connecticut, USA) a dentist, gave nitrous oxide to himself while his partner removed one of his teeth. He continued to use the gas successfully in his practice but failed at a public demonstration

1844   *E. R. Smilie* (Boston, Massachusetts, USA) opened an abscess from a patient inhaling etheral tincture of opium from a sponge (the soporific sponge again!)

1846   *William Thomas Green Morton* (Boston, Massachusetts, USA) a dentist, first administered ether for dental extractions in his own dental practice and then, on 16 October 1846, for the removal of a tumour of the neck at the Massachusetts General Hospital. *Oliver Wendell Holmes* (Boston, USA) attached the name 'anaesthesia' to Morton's discovery

1846   *Dr Francis Boott* (an American living in London, England) received the news from Boston, USA and then persuaded the dentist *James Robinson* (London, England) to administer ether to a patient and extract a tooth under its influence on 19 December 1846. On the same day in Scotland an amputation was performed under ether at the Dumfries and Galloway Royal Infirmary, Scotland by *Dr William Scott*. Scott had learned of the discovery from *Dr Fraser*, surgeon on the Cunard paddle steamer *Acadia* which had brought the news to England. *Robert Liston*, the famous surgeon, performed the first 'capital' (major) surgical operation – an amputation – under ether in England on 21 December 1846 at the North London (now University College) Hospital. The anaesthetic was administered by a medical student *William Squire*

1847   *James Robinson* (London, England – see 1846) published the first textbook in the world on anaesthesia – *A treatise on the inhalation of the vapour of ether*

1847   *John Snow* (London, England) becomes the first specialist anaesthetist and published his book – *On the inhalation of ether in surgical operations*

1847   *James Young Simpson* (Edinburgh, Scotland) introduced chloroform for the relief of pain in labour in November 1847

1853   *John Snow* (London, England) administered chloroform to Queen Victoria at the birth of Prince Leopold. This effectively silenced theological doubts about the use of anaesthesia

1853   *Alexander Wood* (Edinburgh, Scotland) used a piston and barrel glass syringe and hollow needle to inject morphia subcutaneously in the vicinity of peripheral nerves to relieve neuralgia

1857   *Claude Bernard* (France) demonstrated that curare acts on the myoneural junction

1858   *John Snow's* book on *Chloroform and other anaesthetics* published posthumously

1867   *Joseph Lister* (Glasgow, Scotland) introduced antiseptic surgery

Anaesthesia and the challenge of major surgery 1867– c.1920

1870   *S. S. White* (USA) introduced liquid nitrous oxide cylinders

1872   *Pierre-Cyprien Oré* (Bordeaux, France) gave intravenous anaesthesia with chloral hydrate

1877   *Joseph T. Clover* (London, England) described his portable regulating ether inhaler

1880   *William Macewen* (Glasgow, Scotland) introduced tracheal intubation for operations inside the mouth

1884   *Carl Koller* (Vienna, Austria) introduced topical cocaine anaesthesia for ophthalmology. *W. S. Halsted* (New York, USA) developed the use of cocaine by perineural injection for local anaesthesia

1885   *S. S. White* (USA) introduced compressed oxygen

1887   *Frederic Hewitt* (London, England) described the first nitrous oxide and oxygen apparatus with cylinders

1893   *F. W. Silk* (London, England) founded the London Society of Anaesthetists which became the Anaesthetics Section of the Royal Society of Medicine in 1908

1894   *Schleich* (Germany) popularized infiltration anaesthesia using dilute cocaine

1898   *August Bier* (Kiel, Germany) introduced clinical spinal anaesthesia using cocaine

1901   *Landsteiner* (Vienna, Austria) described blood groups

1902   *Heinrich Braun* (Leipzig, Germany) introduced the use of adrenaline into local anaesthesia

1905   *Heinrich Braun* (Leipzig, Germany) introduced procaine into clinical anaesthesia

1905   *The Long Island Society of Anesthetists* was formed – forerunner of the *American Society of Anesthetists* (1936) and of *Anesthesiologists*

1907   *Jansky* (Prague, Bohemia) developed the concept of blood groups

1909   *George W. Crile* (Ohio, USA) developed anociassociation (combined use of general and local anaesthesia)

1910   *William Moss* (Baltimore, USA) developed practical blood transfusion

1912   *Arthur Läwen* (Germany) injected curare subcutaneously to produce muscular relaxation for surgery

1912   *James T. Gwathmey* (New York, USA) produced his nitrous oxide, oxygen and ether apparatus

1914   *Alber Hustin* (Brussels, Belgium) introduced anticoagulation with citrate

1917   *George W. Crile* (Ohio, USA) introduced anociassociation and blood transfusion to France during the Great War of 1914–1918

1917   *Geoffrey Marshall* (England) in France and *Edmund G. Boyle* (England) introduced machines for use by the British Army based on Gwathmey's

New concepts and new agents c.1920–1942

1920   *Ivan W. Magill* and *E. Stanley Rowbotham* (London, England) developed wide bore tracheal anaesthesia with

nitrous oxide, oxygen and ether for use in plastic surgery. *Arthur E. Guedel* (California, USA) published his first classic paper on the signs of anaesthesia

1923   *Ralph M. Waters* (Wisconsin, USA) introduced carbon dioxide absorption

1924   *P. Fredet* and *R. Perlis* (Paris, France) introduced somnifaine the first intravenous barbiturate anaesthetic

1925   *John Lundy* (Minnesota, USA) developed the theory of 'balanced anaesthesia'

1930   *Brian Sword* (Connecticut, USA) introduced the circle absorption system

1932   Association of Anaesthetists of Great Britain and Ireland founded

1933   *R. J. Minnitt* (Liverpool, England) promoted N_2O and air for obstetric analgesia

1934   *Ralph M. Waters* (Wisconsin, USA) introduced cyclopropane

1934   *J. S. Lundy* (Minnesota, USA) introduced thiopentone

1941   *C. Langton Hewer* (London, England) popularized trichloroethylene

1942   *Harold R. Griffith* and *G. Enid Johnson* (Montreal, Canada) used intravenous curare with spontaneous ventilation in man

The post-curare era 1942 et seq.

1943   *R. R. Macintosh* (Oxford, England) described his curved laryngoscope

1946   *T. Cecil Gray* and *John Halton* (Liverpool, England) began to popularize the use of curare with controlled ventilation

1947   *Torsten Gordh* (Stockholm, Sweden) introduced lignocaine

1948   Faculty of Anaesthetists of the Royal College of Surgeons of England founded. The Faculty became the College of Anaesthetists within the Royal College of Surgeons of England in 1988, and the independent Royal College of Anaesthetists in 1992

1949   *Daniel Bovet* (France) introduced suxamethonium

1956   *H. K. Beecher* (Boston, Massachusetts, USA) published his theories on pain

1956   *Michael Johnstone* (Manchester, England) and *Roger Bryce-Smith* (Oxford, England) introduced halothane

1963   *L. J. Telivuo* (Finland) introduced bupivacaine

1965   *R. Melzack* (USA) and *P. D. Wall* (United Kingdom) described the gate theory of pain

1966   *G. Corssen* and *E. F. Domino* (Michigan, USA) introduced ketamine

1966   *R. W. Virtue* (Colorado, USA) introduced enflurane

1971   *A. B. Dobkin* (New York, USA) introduced isoflurane

1977   *B. Kay* (England) and *G. Rolly* (Belgium) introduced propofol

1984   *A. A. Spence and others* (Glasgow, Scotland) used propofol in a new formulation

1988   *A. I. J. Brain* (England) introduced the laryngeal mask commercially (prototype described 1983)

Bibliography

Atkinson RS, Boulton TB. *History of anaesthesia.* London: Royal Society of Medicine, 1988.

Davison MH. *The evolution of anaesthesia.* Altrincham: John Sherratt, 1965.

Duncum BM. *The development of inhalation anaesthesia.* Oxford: Oxford University Press, 1947.

Fink BR, Morris LE, Stephen CR. *The history of anaesthesia.* Third International Symposium. Chicago: Wood-Library Museum, 1992.

Keys TE. *The history of surgical anaesthesia.* Huntington, New York: Robert E Krieger, 1978.

Rupreht J, Van Lieburg MJ, Lee JA, Erdmann W. *Anaesthesia. Essays on its history.* Heidelberg: Springer-Verlag, 1985.

Sykes WS. *Essays on the first hundred years of anaesthesia.* Vols 1, 2 and 3. Edinburgh: Churchill Livingstone, 1961, 1982.

Thomas KB. *The development of anaesthetic apparatus.* Oxford: Blackwell Scientific Publications, 1975. (Available from the Association of Anaesthetists of Great Britain and Ireland, 9 Bedford Square, London WC1B 3RA, England)

REFERENCES

1. Holmes OW. Letter to EL Snell 1893. In: Keys TH. *The history of anaesthesia.* New York: Dover Publications Inc, 1963: iv–v.
2. Snow J. *On chloroform and other anaesthetics.* London: Churchill, 1858: 1–443.
3. Duncum BM. *The development of inhalation anaesthesia.* Oxford: University Press, 1947: 1–640.
4. Sykes WS. *Essays on the first hundred years of anaesthesia.* Vol 1. Edinburgh: Livingstone Ltd, 1960: 4–171.
5. Rhodes P. *An outline history of medicine.* London: Butterworth, 1985: 3.
6. Warren JC. *Etherization with surgical remarks.* Boston: Ticknor, 1848: 1–100.
7. Hensyl WR, ed. *Stedman's medical dictionary* 25th ed. Baltimore: Williams and Wilkins, 1990: 65.
8. Davison MHA. *The evolution of anaesthesia.* Altrincham: John Sherratt & Son, 1965: 7–236.
9. Morch ET, Major RH. 'Anaesthesia'. Early use of this word. *Current Researches in Anesthesia and Analgesia* 1954; **33**: 64–8.
10. Moser HHP. Early anaesthesia. *Anaesthesia* 1949; **4**: 70–5.
11. Critchley M, ed. *Butterworth's medical dictionary.* 2nd ed. London: Butterworths, 1978: 101–4.
12. Nunn JF. Anaesthesia in ancient times – fact or fable. In: Atkinson RS, Boulton TB, eds. *The history of anaesthesia. International Congress Series No. 134.* London: Royal Society of Medicine, 1989: 21–6.
13. Sykes WS. *Essays on the first hundred years of anaesthesia.* Vol 2. Edinburgh: Livingstone, 1961: 1–187.
14. Bowes JB. Mandrake in the history of anaesthesia. In: Atkinson RS, Boulton TB, eds. *The history of anaesthesia. International Congress Series No. 134.* London: Royal Society of Medicine, 1989: 26–8.
15. Nuland SB. *The origins of anaesthesia.* Birmingham, Alabama: Classics of Medicine Library, 1983: 1–131.
16. Keys TE. *The history of surgical anaesthesia.* New York: Dover Publications, 1963: 3–193.
17. Infusino M, O'Neill YV, Calines S. Hog beans, poppies and mandrake leaves – a test of efficacy of the medieval 'soporific sponge'. In: Atkinson RS, Boulton TB, eds. *The history of anaesthesia. International Congress Series No. 134.* London: Royal Society of Medicine, 1989: 29–33.
18. Couper JL. The mandrake legend. In: Atkinson RS, Boulton TB, eds. *The history of anaesthesia. International Congress Series No. 134.* London: Royal Society of Medicine, 1989: 34–8.
19. Smilie ER. Insensibility produced by the inhalation of the vapor of ethereal solution of opium. *Boston Medical and Surgical Journal* 1846; **35**: 263–4.
20. Cartwright F. The early history of ether. *Anaesthesia* 1960; **15**: 67–9.
21. Slatter EM. The evolution of anaesthesia. 1. Ether in medicine before anaesthesia. *British Journal of Anaesthesia* 1960; **32**: 31–4.
22. Greene NM. Annals of anesthetic history. A consideration of factors in the discovery of anesthesia and their effects on its development. *Anesthesiology* 1971; **35**: 515–22.
23. Caton D. The secularization of pain. *Anesthesiology* 1985; **62**: 485–501.
24. Haeger K. *The illustrated history of surgery.* London: Harold Starke, 1989: 9–288.
25. Warren JC. Address of the President. The influence of anaesthesia. *Transactions of the American Surgical Association* 1897; **15**: 1–25.
26. Vandam LD. The start of modern anaesthesia. In: Atkinson RS, Boulton TB, eds. *The history of anaesthesia. International Congress Series No. 134.* London: Royal Society of Medicine, 1989: 64–9.
27. Clement FW. Surgery before the days of anesthesia. *Anesthesiology* 1953; **14**: 473–89.
28. Cartwright FF. *The English pioneers of anaesthesia (Beddoes, Davy, Hickman).* Bristol: Wright, 1952: 1–338.
29. Leigh JM. Respiration the pulse and animal heat. *Anaesthesia* 1974; **29**: 69–86.
30. Calverley RK. John Snow: 'On asphyxia and on resuscitation of still born children'. *Survey of Anesthesiology* 1992; **36**: 397–404.
31. Smith WDA. *Under the influence. A history of nitrous oxide and oxygen anaesthesia.* London: Macmillan, 1982: 1–188.
32. Hartog PJ. The new views of Priestley and Lavoisier. *Annals of Science* 1941; **5**: 1–56.
33. Bergman NM. Thomas Beddoes (1760–1808). Tuberculosis and the medical pneumatic institution. In: Atkinson RS, Boulton TB, eds. *The history of anaesthesia. International Congress Series No. 134.* London: Royal Society of Medicine, 1989: 477–81.

34. Slatter EM. The evolution of anaesthesia. 4. Pneumatic medicine. *British Journal of Anaesthesia* 1960; **32**: 194–8.

35. Stock JE. *Memoirs of the life of Thomas Beddoes MD.* London: Murray, 1811: 1–415.

36. Thomas KB. *The development of anaesthetic apparatus. A history based on the Charles King collection of the Association of Anaesthetists of Great Britain and Ireland.* Oxford: Blackwell Scientific Publications, 1975: 1–268.

37. Smith WDA. Preface. *Under the influence. A history of nitrous oxide and oxygen anaesthesia.* London: Macmillan, 1982: ix–xxviii.

38. Davy H. *Researches chemical and philosophical chiefly concerning nitrous oxide, or dephlogisticated nitrous air, and its respiration.* London: Johnson, 1800: 1–580. (Facsimilie Edition. London: Butterworth, 1972).

39. Coates DP. Davy's contribution to inhalation anaesthesia. In: Atkinson RS, Boulton TB, eds. *The history of anaesthesia. International Congress Series No. 134.* London: Royal Society of Medicine, 1989: 485–8.

40. Weller RM. Nitrous oxide in Bristol in 1836. A series of lectures by William Herapath (1796–1868). *Anaesthesia* 1983; **38**: 678–82.

41. Coe P. William Allen (1770–1843) the Spitalsfields genius. In: Atkinson RS, Boulton TB, eds. *The history of anaesthesia. International Congress Series No. 134.* London: Royal Society of Medicine, 1989: 481–5.

42. Greene NM. Eli Ives and the medical use of ether prior to 1846. *Journal of the History of Medicine* 1960; **15**: 297–9.

43. Slatter EM. The evolution of anaesthesia. 2. The first English inhalers. *British Journal of Anaesthesia* 1960; **32**: 35–45.

44. Anonymous. Effects of inhaling the vapours of sulphuric ether. *Journal of Science and the Arts* 1818; **4**: 158–9.

45. Editorial. Facts and observations on the inhalation of sulphuric ether vapour as a narcotic and general anodyne with descriptions of the instruments commonly used. *Edinburgh Medical and Surgical Journal* 1847; **67**: 504–5.

46. Wilkinson DJ. History of trauma anesthesia. In: Grande M, ed. *Textbook of trauma anesthesia.* St Louis: Mosby, 1993: 1–34.

47. Bird HM. James Arnott MD (Aberdeen) 1797–1883. A pioneer in refrigeration anaesthesia. *Anaesthesia* 1949; **4**: 10–17.

48. Helliwell PJ. Refrigeration analgesia. *Anaesthesia* 1950; **5**: 58–66.

49. Ogata T. Seishu Hanaoka and his anaesthesiology and surgery. *Anaesthesia* 1973; **28**: 645–52.

50. Whitby JD. Alcohol in anaesthesia and surgical resuscitation. *Anaesthesia* 1980; **35**: 502–3.

51. Gould A. *A history of hypnotism.* Cambridge: University Press, 1992: 1–738.

52. Smith WDA. A history of nitrous oxide and oxygen anaesthesia. Part IV – Hickman and the introduction of certain gases into the lungs. *British Journal of Anaesthesia* 1966; **38**: 58–72.

53. Smith WDA. A history of nitrous oxide and oxygen anaesthesia. Part IV A and B – Further light on Hickman and his times. *British Journal of Anaesthesia* 1970; **42**: 347–53 and 445–58.

54. Smith WDA. A history of nitrous oxide and oxygen anaesthesia. Part IV C, D and E – Henry Hill Hickman in his time. *British Journal of Anaesthesia* 1978; **50**: 519–30, 623–7 and 853–61.

55. CJST. Henry Hill Hickman. A forgotten pioneer of anaesthesia. *British Medical Journal* 1912; **1**: 842–5.

56. Stetson JB. William E Clarke and his 1842 use of ether. In: Fink R, ed. *The history of anesthesia.* Third International Symposium Proceedings, Atlanta, Georgia, 1992. Park Ridge, IL: Wood Library-Museum of Anesthesiology, 1992: 400–7.

57. Stone RF. *Biography of eminent American physicians and surgeons.* Indianapolis: Carlton and Hillenbeck, 1894: 286.

58. Long CW. An account of the first use of sulphuric ether by inhalation as an anaesthetic for surgical operations. *Southern Medical Journal* 1849; **5**: 705–13.

59. Sims JM. The discovery of anaesthesia. *Virginia Medical Monthly* 1877; **4**: 81–99.

60. Sykes WS, Ellis RH, eds. *Essays on the first hundred years of anaesthesia.* Vol 3. Edinburgh: Churchill, 1982: 1–272.

61. Wells H. *A history of the discovery of nitrous oxide gas and other vapours.* Hartford: Gaylord Wells, 1847: 1–14.

62. MacDonald AG. John Henry Hill Lewellin. The first etherist in Glasgow. *British Journal of Anaesthesia* 1993; **70**: 228–34.

63. Ellis RH. Preface. *James Robinson on the inhalation of the vapour of ether.* Eastbourne: Baillière Tindall, 1983: vii–xi.

64. Wells CJ. Horace Wells. *Current Researches in Anaesthesia and Analgesia* 1935; **14**: 176–89.

65. Archer WA. Chronological history of Horace Wells, discoverer of anesthesia. *Bulletin of the History of Medicine* 1939; **7**: 1140–69.

66. Colton CQ. *Anaesthesia. Who made and developed this great discovery? A statement 'delivered during the mellowing of occasion'.* New York: Sherwood, 1886: 3–15.

67 MacQuitty B. *The battle for oblivion.* London: Harrap, 1965: 7–200.

68. Bigelow HT. Etherization. A compendium of its history, surgical use, dangers and discovery. *Boston Medical and Surgical Journal* 1848; **37**: 229–45 and 254–66.

69. Morton WJ. The uses and limitations of general anesthesia as produced by subcutaneous and intravascular injections. *Journal of the American Medical Association* 1911; **56**: 1677.

70. Bigelow HJ. Insensibility during surgical operations produced by inhalation. *Boston Medical and Surgical Journal* 1846; **35**: 309–17.

71. Morton WTG. Letter from Dr. Morton, of Boston U.S. *Lancet* 1847; **2**: 80–1.

72. Ayer W. The discovery of anaesthesia by ether; with an account of the first operation performed under its influence at the Massachusetts General Hospital, and an extract from the record book of the hospital. *Occidental Medical Times* 1896; **10**: 121–9.

73. Morton WTG. *Remarks on the proper mode of administration of sulphuric ether by inhalation.* Boston: Dutton and Wentworth, 1847: 1–44.

74. Gould AB. Charles T Jackson's claim to the discovery of etherization. In: Ruprecht J, Lieburg MJV, Lee JA, Erdman W, eds. *Anaesthesia. Essays on its history.* Berlin: Springer Verlag, 1985: 384–7.

75. Ellis RH. The introduction of ether anaesthesia to Great Britain. 1. How the news was carried from Boston Massachusetts to Gower Street, London. *Anaesthesia* 1976; **31**: 766–77.

76. Trent J. Surgical anaesthesia, 1846–1946. *Journal of the History of Medicine* 1946; **1**: 505–14.

77. Tirer S. Rivalries and controversies during early ether anaesthesia. *Canadian Journal of Anaesthesia* 1988; **35**: 605–11.

78. Secher O. Forty-six 'first anaesthetics in the world'. *Acta Anaesthesiologica Scandinavica* 1990; **34**: 552–6.

79. Morton WTG. The first use of ether an an anesthetic. At the Battle of Wilderness in the Civil War. *Journal of the American Medical Association* 1904; **42**: 1068–72.

Chapter 2

Theories of Narcosis

Peter J. Cohen

What is required for a universal theory of narcosis?
Theories based on correlations
Neurophysiological theories

Biochemical theories
Molecular theories

Although diethyl ether was first administered over 150 years ago, the description of this supreme moment in medicine was not published until 7 years later.[1] Notwithstanding the significant growth in our knowledge of neurophysiology and pharmacology to date, the means by which inhalation anaesthetics produce unconsciousness and freedom from pain is still incompletely understood. This chapter will discuss the various theories that have been proposed to explain the action of these extraordinary drugs.

What is required for a universal theory of narcosis?

There are obvious difficulties which will confront any attempt to formulate a unified theory of narcosis. Perhaps the foremost problem exists on the border between philosophy and science, since the anatomical site of human awareness within the central nervous system (CNS) and, indeed, the molecular events which translate into consciousness are quite uncertain. That being so, how can we explain the actions of drugs which so dramatically *alter* consciousness?

Inhalation anaesthetics possess neither common chemical structure nor similar physical properties. Most evidence suggests that their actions are not mediated through a specific receptor as, for example, are those of the opioids and benzodiazepines. The preponderance of data indicates that the effects of inhalation anaesthetics are not related to stereospecificity. General anaesthesia does not result from formation of strong chemical bonds with structures of the CNS; binding within these areas most likely consists of weak electrostatic forces. Indeed, any theory must account for the action of xenon which, although incapable of

forming any chemical bonds, is a general anaesthetic similar in potency to nitrous oxide.[2]

Inhalation anaesthetics have myriad actions besides the production of analgesia and unconsciousness. While these two effects certainly represent endpoints which must be explained, an acceptable theory must relate to *all* anaesthetic actions. It is clear that myocardial depression, somatic muscle relaxation, uterine atony, production of convulsions by drugs only slightly different in structure from those producing general anaesthesia, interference with cellular growth and replication, and inhibition of mitochondrial respiration are among the many phenomena that must come under the purview of any complete theory. Electroencephalographic effects of the general anaesthetics vary widely which, again, precludes any unitary approach. Differences in analgesic potency among these drugs is a phenomenon requiring objective clarification. The ability of high pressure to reverse some, but not all, effects of anaesthetics is central to any theory of narcosis. The relation between anaesthetic effects and molecular size or 'cutoff effect' is germane: *n*-pentane produces narcosis while *n*-decane is devoid of any such action. The rapid onset and termination of clinical action will obviously be an integral part of any firm theory. Finally, a complete theory of narcosis must be able to predict all actions of yet undiscovered agents. For all the above reasons, as will be stressed throughout this chapter, it is probably naive to attempt an elucidation of a single or unitary mechanism of action.

Before proceeding any further, let us consider what might be meant by a *theory of narcosis*, a question for which the answer is not as obvious as might first appear. For example, is it sufficient simply to demonstrate a *correlation* relating a single physiological attribute of the drug to one of its physical properties? Can we logically extrapolate from knowledge of the *site of action* of inhalation anaesthetics to an all-encompassing theory of anaesthesia? If so, what is the locus at which we should concentrate our examination of anaesthetic action: a macroscopic site such as the reticular activating system or other group of CNS synapses; a cellular or subcellular structure such as the acetylcholine, serotonin, glutamate or α_2-adrenergic receptor; an area responsible for synthesis of an important neurotransmitter; the organelle responsible for aerobic energy production, the mitochondrion; or at a particular molecule such as a specific phospholipid, an ion-conducting channel or a known enzyme whose structure can be altered by an anaesthetic? Is it sufficient only to describe one facet of anaesthetic action such as decreased mitochondrial oxygen uptake, altered CNS electrical activity or perturbation of a certain area of the membrane? At best, even the most sophisticated approach will necessarily give rise to another level of questions, e.g. if descriptive data lead to the conclusion that *membrane expansion* is responsible for narcosis, how does the anaesthetic agent produce membrane expansion in the first place? Furthermore, what is the means by which membrane expansion actually produces anaesthesia? Perhaps the final tale will remain unknown until anaesthetic action in all its ramifications can be explained and predicted by relating them to events at a molecular level. Having said this in introduction, let us now attack the question at hand from a variety of perspectives – perhaps in a manner similar to the blind men examining the elephant.

Theories based on correlations

Lipid solubility

The earliest[3] and still the best[4–7] correlation proposed as an explanation of anaesthetic action related lipid solubility and potency. This correlation extends to agents such as nitrous oxide, xenon and nitrogen which require more than 1 atmosphere (101 kPa) of pressure before general anaesthesia is achieved.[5,7] Originally proposed almost 100 years ago, the *Meyer–Overton hypothesis* stated that narcosis occurs when a critical drug concentration is attained within a crucial lipid of the CNS. Thus, equal anaesthetic doses could be expressed as a constant molar or volume fraction. Lipid solubility is still the property best correlated with both *in vivo* clinical anaesthetic potency[4–7] and anaesthetic potency determined *in vitro* by measuring inhibition of mitochondrial respiration[8,9] (see Figs 2.1 and 2.3, below). In extrapolating from the *correlation* of potency with another anaesthetic property to a general theory dealing with *mechanism of action*, the Meyer–Overton hypothesis suggests that when an anaesthetic dissolves in a lipophilic portion of the membrane, blockade of an essential pore,

perhaps a sodium channel, is produced, thereby preventing depolarization.

If one wishes to develop a unified theory of narcosis, these observations certainly suggest a single underlying principle. Additional evidence for this unitary approach might be parallel dose–response curves of anaesthetics when blockade of painful stimuli is the endpoint, and the *in vivo*[10] and *in vitro*[11] additivity of their potencies. However, correlation with lipid solubility neither defines a specific site of action nor constitutes, in itself, a molecular mechanism of anaesthesia. For example, the ability of general anaesthetics to dissolve in silicon rubber correlates quite well with their *in vivo* potencies.[12] While confirgrational changes produced by anaesthetics in these elastomers may mimic those in a yet unknown site of action, one would be hard-pressed to suggest these data alone can provide an explanation of anaesthetic action. While lipid solubility may be necessary for anaesthetic action, it is not sufficient. Thus, certain highly lipophilic compounds such as *n*-decane demonstrate a size-related 'cutoff effect' and are devoid of either *in vivo* or *in vitro* narcotic effects.[13] Indeed, knowledge of lipid solubility does not allow prediction of whether a drug of the correct size to possess CNS activity will produce anaesthesia or convulsions.[14,15] When an anaesthetic's ability to produce unconsciousness is compared with other endpoints of potency such as analgesia, somatic muscular relaxation or respiratory depression, the same relations do not exist for each agent.[16,17] Finally, while mice may be bred for high or low sensitivity to anaesthetic-induced loss of consciousness (measured by presence or absence of the righting reflex), analysis of CNS constituents demonstrates no convincing relation between lipid composition and anaesthetic requirement.[18] In conclusion, the excellent correlation between the lipid solubility of inhalation anaesthetics and their ability to inhibit the response to noxious stimuli does not, of itself, yield a complete and unified explanation of anaesthetic action.

Action on water molecules

Of considerable historic importance are the concepts of Pauling[19] and Miller[20] who related the mediation of anaesthetic action to changes in an aqueous rather than a lipid site within the CNS. Pauling proposed that a hydrated anaesthetic gas molecule or clathrate (Latin *clathri*, lattice) could stabilize a membrane or occlude essential pores by causing increased order of the molecule structure of water; this aqueous clathrate interfered with depolarization of important pathways within the CNS, thereby producing anaesthesia. In a similar fashion, Miller postulated that physical interaction between water and the anaesthetic resulted in an 'iceberg' whose cage-like structure was able to 'stiffen-up' a membrane, thereby preventing neuronal transmission. Against this theory is the poor correlation of anaesthetic potency with hydrate dissociation pressure shown by compounds such as sulphahexafluoride and diethyl ether.[5,6] Furthermore, while the combination of an anaesthetic that forms a small clathrate with one that forms a large one would be predicted to result in synergism, the effect is usually that of simple addition. Finally, and probably most germane is the fact that neither clathrate nor iceberg can be produced at ambient pressure and body temperature. For these reasons, these theories involving interaction of water and general anaesthetics are no longer considered to be viable.

Binding to specific receptors

In addition to their interaction with lipids and water already described, inhalation anaesthetics can bind to proteins, causing alterations in their configuration.[21] Microtubules, structured with rings of protein, play an important role in maintaining cytoplasmic rigidity and have been proposed to be one of the means by which general anaesthesia affects cellular functions such as mitosis or ciliary beat.[22] Proteins, as well as lipids and water, are building blocks of essential CNS receptors; anaesthetic binding to these structures would be likely to be accompanied by important functional effects.[23] Since ion transport and neurotransmitter release are mediated through such receptors, anaesthetic action at this level must be considered. For example, the membrane receptor protein for acetylcholine interacts with volatile anaesthetics.[24,25] The γ-amino-butyric acid (GABA),[26,27] glutamate,[28,29] and G-protein[30] receptors are also prime candidates for interaction with general anaesthetics.

The exciting concept that inhalation anaesthetics might act on opioid receptors,[31] perhaps through release of exogenous opioids or endorphins, has been supported[32–35] and opposed[35–39] by several studies. As a possible example of similar sites of action of inhalation anaesthetics is the development of tolerance to both the production of *analgesia* and the loss of the *righting reflex* produced by nitrous oxide in rats and mice.[33,35] In contrast, tolerance to cyclopropane's inhibition of the murine *righting reflex* fails to develop.[35] The opioid antagonist naltrexone antagonizes the *analgesia* produced by nitrous oxide in rats and mice, as would be anticipated were general anaesthesia to involve opioid receptors or endorphins.[32] Naloxone, another opioid antagonist, also blocks the *analgesia* which accompanies administration of halothane, enflurane or cyclopropane.[34] However, naltrexone fails to reverse the inhibition of the *righting reflex* produced by nitrous oxide in the rat.[36] The situation is made even more complicated by the observation that neither naloxone nor naltrexone inhibits *analgesia* in the dog when halothane is administered, a finding which led the authors to conclude that endorphins do not play a

role in the action of this drug.[38] Finally, lack of uniformity of anaesthetic–opioid interaction is demonstrated by failure of naloxone to antagonize the *in vitro* effect of halothane on guinea-pig ileum.[39]

These considerations once more suggest that to focus on a single site, to examine only one agent, to attempt to develop a unitary mechanism will be inadequate to explain the multiplicity of actions of all current or future inhalation anaesthetics.

Is it possible that these agents act at non-opioid receptors? Again, there is no single answer. Thus, cross-tolerance between ethanol and nitrous oxide may be demonstrated in the mouse;[40] in contrast, nitrous oxide does *not* demonstrate these findings when barbiturates are used to induce tolerance.[41] This study concluded that these findings, along with the known stereospecificity of barbiturate action, demonstrate unique sites of action for this group of drugs. Conversely, both ethanol and nitrous oxide most likely exert their effects at a non-specific hydrophobic site in the CNS.

An important observation in opposition to the thesis that inhalation anaesthetics may act at specific receptors is the lack of stereospecificity of halothane's action on both synaptic transmission and mobility of fatty-acid chains in lipid bilayers.[42] Moreover, *in vivo* nuclear magnetic resonance imaging fails to demonstrate specific cerebral anaesthetic binding sites in the rabbit brain,[43] while neither halothane nor isoflurane acts directly through rat cortical benzodiazepine receptors *in vivo* or *in vitro*.[44] To complicate this discussion, however, it has been proposed that halothane's negative inotropy results from its dose-dependent reversible binding to voltage-dependent Ca^{2+} channel receptors.[45] Stereospecificity of action by isoflurane on 'particularly sensitive ion channels' in mulluscan CNS neurons is consistent with receptor blockade;[46] however, both isoflurane stereoisomers possess equal efficacy in the ability to disrupt lipid bilayers.[46] The latter finding would again tend to support a non-specific site of action.

At least one study has demonstrated the ability of adenosine to alter anaesthetic potency, an effect reversed by adenosine receptor antagonists.[47] Whether adenosine's effects are secondary to the accompanying hypotension or demonstrate specific interaction with the adenosine receptor remains to be determined.

Recently, the role of the α_2-adrenergic receptor in anaesthetic action has been examined. There is convincing evidence that α_2-receptor agonists appear to increase the analgesic potency of isoflurane and its efficacy in blocking the noxious stimuli of tracheal intubation.[48,49] However, one need not explain these findings by postulating a specific action of the anaesthetic on this receptor since altered central sympathetic tone alone would be sufficient as an explanation of anaesthetic enhancement.

In conclusion, while the story regarding specific receptors is not yet fully told, it appears that at this stage it would be naive to ascribe all anaesthetic actions to a single receptor of defined chemical structure or function at which all inhalation anaesthetics can act.[16,17] It is far more likely that the ubiquitous effects of these drugs are mediated via a combination of proteins, lipids, and water and are most likely not due to a configurational change of a single specific receptor.

Correlations with physical properties

Although numerous studies have dealt with the correlations between anaesthetic potency and a wide variety of physical properties, we have already seen that this approach is not particularly useful. For example, while the ratio of anaesthetizing partial pressure/vapour pressure does not have nearly the wide range as does anaesthetizing partial pressure alone,[4] this does not really inform us about the mode of action of a particular compound. Instead, this ratio is only an expression of the relation of forces acting between anaesthetic molecules existing in the gas phase and those in the liquid phase.[50] This, in turn, is simply an approximation of the tendency of the molecules to attract each other rather than to bind to compounds within the CNS. Similarly, comparison of anaesthetic potency and van der Waals constants[51] again reflects only the interaction of intra- and intermolecular forces, an observation far from a definitive theory of narcosis. Again, the correlation between the potency of some anaesthetics with their effects on the surface tension of a fat–water interface[52] is not necessarily related *a priori* to anaesthetic action, but may again be only an expression of the interaction of intra- and intermolecular

forces. Thus, while correlations may be made between *in vivo* anaesthetic potency and numerous physical properties, with lipid solubility demonstrating the best relation (Fig. 2.1), these data by themselves are not meaningful: they neither directly examine any actual molecular interactions nor explain how these events can produce the overall phenomenon of anaesthesia and its associated events.

Neurophysiological theories

When the effects of anaesthetics on axonal and synaptic transmission are compared, the synapse appears to be the most sensitive.[53,54] This suggests that a likely site of action of these drugs might be at a synaptic level, and that increased numbers of synapses in a neural pathway should be associated with increased anaesthetic sensitivity. These considerations were responsible for the suggestion made 40 years ago that anaesthetics could exert their profound effects by blocking transmission through the multisynaptic reticular activating system (RAS), the pathway felt to be responsible for maintenance of wakefulness.[55]

While this has been deemed an attractive hypothesis of historic interest, it is associated with significant difficulties. Most basic is that the theory merely suggests that anaesthetics act within the CNS, hardly a surprising proposal. Thus, while suggesting a *locus* of anaesthetic action, it does not explain *how* they act. Even if we accept these limitations, we are confronted with additional problems. For example, it is well known that administration of thiopentone (thiopental) is accompanied by decreased cerebral oxygen uptake. If anaesthetics act at

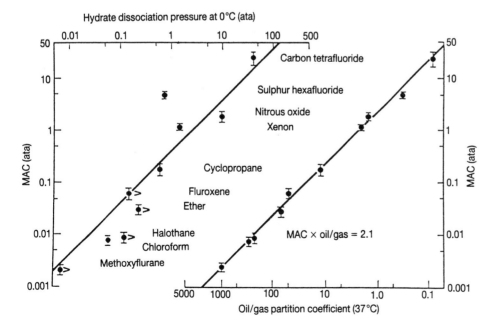

Fig. 2.1 The correlation of MAC (in dogs) with hydrate dissociation pressure (left graph, upper scale) and with lipid solubility (right graph, lower scale). If the data follow the correlation: MAC/hydrate dissociation pressure equals a constant, or MAC × oil/gas partition coefficient equals a constant, then the data should lie along the 45° slopes as indicated. (From Halsey, with permission.[6])

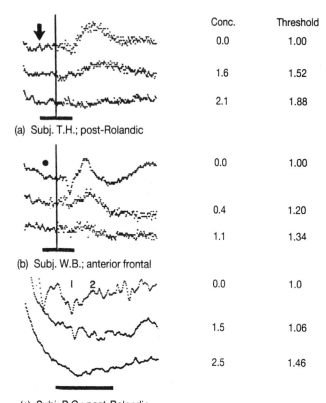

	Conc.	Threshold
	0.0	1.00
	1.6	1.52
	2.1	1.88

(a) Subj. T.H.; post-Rolandic

	0.0	1.00
	0.4	1.20
	1.1	1.34

(b) Subj. W.B.; anterior frontal

	0.0	1.0
	1.5	1.06
	2.5	1.46

(c) Subj. P.C.; post-Rolandic

Fig. 2.2 Effects of cyclopropane on averaged human somatic evoked potentials ($n = 64$). Subject and recording site identified under each set of traces; concentration (conc.) of anaesthetic and absolute thresholds (threshold) as multiples of control thresholds appear at right. Calibrations in (a) and (b): 20 ms to left of vertical line and 100 ms to right; sampling started 5 ms after stimulus. Calibration in (c): 20 ms; sampling started at 4 ms after stimulus. (From Clark, Butler and Rosner, with permission.[58])

the level of the RAS, *surgically* severing this structure would be expected to produce the same effects as would '*chemical*' transection produced by anaesthesia. Yet, surgical ablation of the midbrain RAS of dogs does *not* decrease brain oxygen consumption as does general anaesthesia.[56] Furthermore, thiopentone's effect on brain oxygen uptake is identical in sham-operated and RAS-transected dogs, demonstrating the drug's effect to be independent of that anatomical structure.[56] A further problem in using the neurophysiological approach as the sole means of arriving at a *unitary* theory is that changes in the electroencephalogram and somatosensory evoked potentials in man differ widely among the anaesthetics, suggesting multiplicity of sites of action within the CNS.[57] A fascinating and provocative finding is the observation that while inhalation of subanaesthetic concentrations of cyclopropane has profound effects on somatosensory-evoked potentials compatible with abolition of 'awareness' (Fig. 2.2), the subjects are subjectively cognizant of the sensory stimulus producing this evoked potential.[58] These observations suggest the actual relation of RAS activity to

consciousness is only incompletely elucidated at this time, and that the RAS is not the sole anatomical site of functional or metabolic activation. Thus, it would appear that the effects of anaesthetics on this structure cannot necessarily be the primary explanation of clinical narcosis.

Neurophysiological actions other than abolition of consciousness may be exerted at other sites. For example, anaesthetic-induced muscular relaxation has been related to inhibition of the spinal monosynaptic H-reflex. Again, however, this is only an anatomical *description* rather than a molecular *explanation* of anaesthesia's effects.

Possible mechanisms of neuronal action at *any* site include changing calcium permeability at susceptible synapses or at a pre- or postsynaptic site resulting in decreased release of an important neurotransmitter. Anaesthetics may directly alter neurotransmitter availability, perhaps by increasing the resistance to rupture of neurotransmitter-containing vesicles, an explanation of anaesthetic action analogous to the ability of anaesthetics to protect red blood cells from hypotonic haemolysis. Finally, inhalation anaesthetics may affect membrane structure thereby decreasing the ability of sodium channels to open in response to chemical stimulation. These molecular actions of inhalation anaesthetics as well as the effects of hyperbaric pressure will be further discussed at the conclusion of this chapter.

Biochemical theories

The ability of anaesthetics to exert profound effects on intermediary metabolism has long been known.[59] These agents reversibly reduce oxygen uptake of both the intact organism and of specific organs including the *in vivo* human brain.[60] A prime example of anaesthetic's effects at a cellular level is their interaction with the mitochondrion, the aerobic 'powerhouse' of most cells.[8] All inhalation agents inhibit mitochondrial respiration in a reversible and concentration-dependent fashion,[8,9] a phenomenon even produced by the chemically inert molecule xenon.[61]

There are many similarities between biochemical and clinically important effects of these drugs. For example, their *in vitro* potencies (ED_{50}) are significantly related both to *in vivo* potency (MAC) and lipid solubility (Fig. 2.3).[9] When a homologous series of hydrocarbons is evaluated for ability to produce *in vivo* CNS effects and *in vitro* inhibition of mitochondrial respiration, the cutoff effect for both phenomena appears at the same molecular size.[13] Anaesthetic action is additive when inhibition of mitochondrial respiration is evaluated[11] in a manner analogous to that observed *in vivo*.[10]

These findings, however, should not be extrapolated to

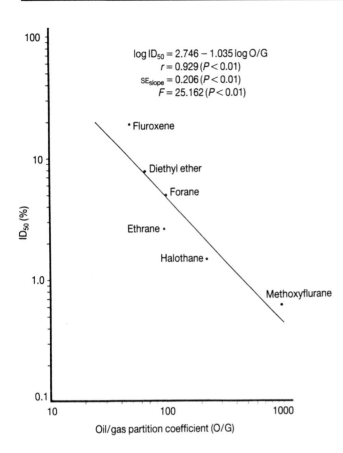

$$\log ID_{50} = 2.746 - 1.035 \log O/G$$
$$r = 0.929 \, (P < 0.01)$$
$$SE_{slope} = 0.206 \, (P < 0.01)$$
$$F = 25.162 \, (P < 0.01)$$

Fig. 2.3 Relation between *in vitro* potency and lipid solubility. ID_{50} is the concentration of anaesthetic which produces 50 per cent inhibition of stage-3 mitochondrial respiration. (From Nahrwold and Cohen, with permission.[9])

the conclusion that inhalation anaesthetics deprive the organism of the ability to conduct aerobic metabolism at a rate sufficient to meet cellular needs. A most important finding regarding this area is that while the rates of synthesis by the CNS of the high-energy phosphates adenosine triphosphate (ATP) and creatine phosphate (CrP) may be significantly decreased by anaesthetics, the rate of CNS utilization is diminished to an *equal* extent.[62,63] As a result of this interaction, the brain's content of ATP and CrP is not lowered by inhalation anaesthetics. Therefore, at this time it would be unwise to conclude that the inhalation anaesthetics act *directly* through interference with cellular metabolic function. Rather, it is likely that the correlations between *in vivo* and *in vitro* effects are fortuitous, suggesting *similar*, but not *identical*, sites of action (perhaps, one relating to control of oxidative metabolism and the other responsible for maintenance of consciousness). Indeed, as will be discussed in the final section of this chapter, that the biochemical and anaesthetic sites of action probably differ is substantially demonstrated by the *inability* of high pressure to reverse either anaesthesia's inhibition of cellular or mitochondrial oxygen uptake, or even analgesia, while yet clearly showing the

ability to antagonize the unconsciousness produced by inhalation anaesthetics.

Anaesthetics produce a variety of other major biochemical effects. Among them is alteration of the flux of calcium ion across mitochondrial and neural membranes.[64] Since this ion is of considerable importance in signal transduction and phenomena such as muscular contraction, neuronal transmission, and heat production, changes in calcium movement might be associated with some of the many actions of anaesthetics. Another biochemical effect is the ability of anaesthetics to increase the concentration of GABA in synaptic areas by inhibiting the rate of its degradation.[26] Since GABA is an important neuroinhibitor, anaesthetic modulation of its action might play a role in its observed effects.[27] As is the case when mitochondrial respiratory depression is evaluated, *in vitro* ability to augment GABA concentration is well correlated with MAC. However, one must question whether or not the comparatively slow (a 20-minute incubation was required) time course of the increase in GABA concentration is consistent with the expected rapid onset of anaesthetic action.[26] In response, it could be postulated that a small increase in GABA in a susceptible area is responsible for profound clinical effects.

In summary, it is evident that anaesthetics act at numerous neural sites, but with significant qualitative and quantitative differences among the many agents and areas studied. Even after complete electrophysiological identification of the anatomical locus or loci of action, the molecular event or events associated with clinical anaesthesia must be understood before a complete theory can be formulated. Let us then complete our examination of the mechanisms of inhalation anaesthesia with a basic exposition of the molecular theory of anaesthesia as currently understood.

Molecular theories

Many investigators have proposed that anaesthetic action is mediated at a susceptible phospholipid membrane and accomplished by altering the physical status of this membrane. For example, clinically effective concentrations of halothane, chloroform, diethyl ether, and methoxyflurane protect erythrocytes from hypotonic haemolysis, a change associated with as little as a 0.4 per cent expansion of the red cell membrane surface area.[65] How might this change in cell membrane surface area be accomplished, and what is its relevance to surgical anaesthesia?

Phospholipid bilayer membranes can coexist as a tightly ordered gel phase and a structurally disordered fluid phase. This is known as *lateral phase separation*.[66] The increased order or compactness of the gel phase allows it to occupy a smaller volume than would the same number of phospholipid molecules existing in the more expanded fluid phase. Movement back and forth between gel and fluid phase, termed *phase transition*, prevents a change in *overall* membrane volume in the face of a *local* alteration in size or configuration, as might occur when an ion channel opens or closes.[66]

Let us now postulate how phase transition may relate to the anaesthetic state. Normally, the membrane immediately surrounding the ion channel exists in the fluid state; the gel phase is located at a greater distance from the site. When the channel opens, perhaps allowing depolarization or chemical transduction, the phospholipid near the channel can convert from a fluid to a gel structure; no net change in membrane volume will then occur. It is clear that anything tending to enhance membrane disorder by increasing the fluid-to-gel ratio will impinge on the ability of the ion channel to open and close freely. Since the ion channels are essential to neuronal or synaptic function, decreased excitability will be produced by anything tending to interfere with their ability to open and close freely. Increased membrane fluidity may prevent additional conformational changes essential for such important events as neurotransmitter binding and release, transport of ions other than those involved in neural depolarization, and important synthetic processes. On the other hand, increased structural order resulting from an augmented gel-to-fluid ratio (as might occur during hyperbaric compression) might be associated with a pathological increase in channel opening and thereby produce heightened excitability. These concepts lead directly to discussion of the interaction of hyperbaric pressure and general anaesthesia.

Pressure reversal of anaesthesia

If narcosis is produced by interference with normal neuronal or synaptic transmission, what may be the molecular events responsible for these profound changes? Might they relate to changes in the ratio of the partitioning of a phospholipid membrane between gel phase and fluid phase?

The phenomenon of *pressure reversal* goes a long way in enabling us to formulate a reasonable answer. The ability of high pressure to antagonize clinical anaesthesia has been well documented for over 50 years.[67] Bioluminescence of the firefly is reversibly inhibited by general anaesthesia,[68] a phenomenon antagonized by hyperbaric pressure.[67] Halothane-anaesthetized tadpoles are awakened when subject to pressures of over 100 atmospheres.[69] In mice, the MAC of nitrous oxide[70] and isoflurane[70,71] is significantly increased by these extreme hyperbaric pressures. This phenomenon, known as *pressure reversal*,

has been demonstrated in a wide variety of species ranging from the slime mould and firefly to higher primates and with both gaseous and intravenous agents.[72] It is therefore not unreasonable to propose that any theory explaining anaesthesia must take into account anaesthesia's reversal under hyperbaric conditions.

We are now in a position to extrapolate from these observations to a possible mechanism of anaesthesia. The *critical volume hypothesis*[73] proposes that narcosis occurs when an inhalation anaesthetic expands the volume of a hydrophobic region by 1.1 per cent, perhaps by interference with gel–fluid transition; restoration to a normal volume ends the anaesthetic state. The critical volume hypothesis also predicts that compression of a normal membrane decreasing its volume by 0.85 per cent produces an effect opposite from that of anaesthesia – increased CNS excitability and convulsions.[74,75]

Let us now summarize the current molecular approach to anaesthetic action using the proposed relation of the lipoprotein bilayer gel-to-fluid transition, membrane expansion and pressure reversal. Because of the excellent correlation between anaesthetic potency and lipid solubility, we may assume that the agents initially dissolve in a lipid component of the CNS, thereby gaining access to a vulnerable structure within a neural or synaptic membrane. Once reaching the lipoprotein bilayer, they increase its fluidity by impeding the fluid-to-gel transition necessary to allow normal ion channel function. The resulting *membrane expansion* produces a significant diminution of neuronal excitability – the result is clinical anaesthesia. In contrast, 'when some hydrophobic region has been compressed beyond a certain critical amount by the application of pressure',[74] the fluid-to-gel transition favours the gel phase and membrane volume is decreased. The clinical correlate in unanaesthetized man exposed to high ambient pressure is development of the *high-pressure neurological syndrome*[74,75] characterized by trembling of the extremities, excitability, respiratory failure, convulsions, unconsciousness, and death.

Although the information given above makes a most attractive story, it still does not account for *all* phenomena associated with anaesthesia, and therefore cannot be presented as an all-encompassing unitary theory. That changes in the gel-to-fluid ratio cannot be the sole explanation of the anaesthetic state is made obvious by considering the effect of temperature on both anaesthetic potency and the physical state of a lipoprotein bilayer. Since hypothermia increases the fraction of a membrane in the ordered gel form, it 'should have some simple effect equivalent to that of pressure' and *antagonize* anaesthesia.[76] However, the opposite pertains: anaesthesia is *potentiated* by hypothermia and unconsciousness is produced by low temperature alone.[77]

That all aspects of general anaesthesia are explained in

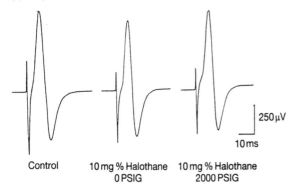

(a) Preganglionic action potential

250 μV

10 ms

Control   10 mg % Halothane   10 mg % Halothane
0 PSIG   2000 PSIG

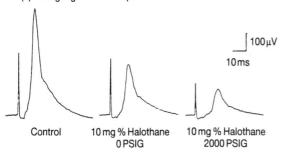

(b) Postganglionic action potential

100 μV

10 ms

Control   10 mg % Halothane   10 mg % Halothane
0 PSIG   2000 PSIG

Fig. 2.4 Interaction between pressure and halothane on neuronal conduction and synaptic transmission in the superior cervical sympathetic ganglion. (a) Preganglionic nerve conducted action potential, showing pressure antagonism of halothane's depressant effect. (b) Response recorded at the postganglionic nerve. Hyperbaric pressure adds to, rather than antagonizes, the depressant effect of halothane. PSIG, pounds per square inch gauge. (From Kendig, with permission.[79])

terms of phase transition, membrane expansion, and pressure reversal may be disputed further. The effect of pressure is non-linear; antagonism produced by a given degree of compression differs among the inhalation anaesthetics.[78] Pressure reversal does not represent 'simply a reciprocal anesthetic expansion–pressure compression phenomenon as predicted by the critical volume hypothesis [but] suggests that anesthetics and pressure act at different sites'.[78]

To argue further, pressure reverses unconsciousness produced by halothane in the intact rat; since halothane anaesthesia is accompanied by inhibition of synaptic transmission in this animal,[53,54] one might logically predict that this *in vitro* effect would also be antagonized by pressure. This is, indeed, the case when the effect of pressure on halothane-inhibited presynaptic neuronal transmission is evaluated. In contrast, however, pressure not only fails to reverse halothane's inhibition of synaptic transmission, but impedes it to an even greater extent than the anaesthetic alone (Fig. 2.4).[79] Similarly, halothane's inhibition of neuronal function in the marine gastropod *Aplysia californica* is not restored by pressure,[80] nor is there

reversal of the decreased ciliary activity produced by halothane in the protozoon *Tetrahymena pyriformis.*[81]

Inhibition of mitochondrial respiration by inhalation anaesthesia, the relationship of MAC, ED_{50}, and lipid solubility, and other examples of the excellent correlation of their *in vivo* and *in vitro* effects have already been discussed. We have also seen that hyperbaric pressure reverses some, but not all, anaesthetic effects. The interaction of pressure and anaesthesia on mitochondrial respiration only adds to the dichotomy inherent in these observations. Thus, while pressure of 50 atmospheres (ata) *reverses* the unconsciousness produced by halothane it *increases* the inhibition of mitochondrial respiration produced by the agent.[82] The same degree of compression also fails to antagonize halothane's depression of respiration in intact monkey kidney cells.[83]

Two major effects of nitrous oxide are analgesia and unconsciousness, the latter requiring nitrous oxide pressures greater than 1 atmosphere. There is again a dichotomy in the effects of hyperbaric pressures on these two anaesthetic actions, indicating significant differences in the drug's several sites of action. While 75 ata of pressure was able to restore the loss of righting reflex produced by 1.5 ata nitrous oxide, the same pressure had absolutely no effect on the analgesia produced by 1.2 ata of the anaesthetic.[84]

These observations were responsible for development of the *multisite expansion hypothesis,*[85] the formulation that different anaesthetic effects are exerted at different cellular sites of action. We have already noted the analogous, but opposite, observation that antagonists of opioids, and presumably endorphins, appear to block the analgesia[32,34] but not the loss of consciousness[36,37] produced by inhalation anaesthetics. These actions of pressure and opioid antagonists again emphasize that *all* inhalation anaesthetics do not exert *all* their effects at the *same* locus and through an *identical* mechanism. The multisite expansion hypothesis appears to be the most viable way of thinking about membrane expansion and pressure reversal at the present time.

As is the case with the many explanations of anaesthetic action we have examined, the exposition of pressure reversal as a means of arriving at a theory of narcosis must also be modified to allow for a variety of different anaesthetic sites, each showing different quantitative and even qualitative effects of anaesthesia. We are forced to acknowledge that the molecular events accompanying clinical anaesthesia are not necessarily identical when the entire panoply of anaesthetic effects including inhibition or aerobic metabolism, depressed myocardial function, neuromuscular blockade, interference with synaptic transmission, binding to certain specific receptors, amnesia, unconsciousness, analgesia and interaction with hyperbaric pressure are considered.

As is the case with so many important drugs, our ability to profit from anaesthetic use does not depend on our intimate knowledge of all aspects of the mechanisms by which they work. Fortunately, burgeoning knowledge of the pharmacology of these agents coupled with appreciation of their actions on normal and pathological physiology has made possible the extraordinary progress of our speciality. The large body of information regarding their actions at a molecular level which we *do* possess will continue to increase as we enter the coming century.

REFERENCES

1. Long CW. An account of the first use of sulphuric ether by inhalation as an anesthetic in surgical operations. *Southern Medical and Surgical Journal* 1849; **5**: 705–13.

2. Cullen SC, Gross EG. Anesthetic properties of xenon in animals and human beings with additional observations on krypton. *Science* 1951; **113**: 580–2.

3. Meyer HH. Zur Theorie der Alkoholnarkose. *Archiv für Experimentelle Pathologie und Pharmakologie* 1899; **42**: 109–18.

4. Miller KW, Paton WDM, Smith EB, Smith RA. Physicochemical approaches to the mode of action of general anesthetics. *Anesthesiology* 1972; **36**: 339–51.

5. Eger EI, Brandstater B, Saidman LJ, Regan MJ, Severinghaus JW, Munson ES. Equipotent alveolar concentrations of methoxyflurane, halothane, diethyl ether, fluroxene, cyclopropane, xenon and nitrous oxide in the dog. *Anesthesiology* 1965; **26**: 771–7.

6. Halsey MJ. Physical chemistry applied to anaesthetic action. *British Journal of Anaesthesia* 1974; **46**: 172–80.

7. Miller KW, Paton WDM, Smith EB. The anaesthetic pressures of certain fluorine-containing gases. *British Journal of Anaesthesia* 1967; **39**: 910–18.

8. Cohen PJ. Effect of anesthetics on mitochondrial function. *Anesthesiology* 1973; **39**: 153–64.

9. Nahrwold ML, Cohen PJ. The effects of forane and fluroxene on mitochondrial respiration: correlation with lipid solubility and *in vivo* potency. *Anesthesiology* 1973; **38**: 437–44.

10. DiFazio CA, Brown RE, Ball CG, Heckel GG, Kennedy SS. Additive effects of anesthetics and theories of anesthesia. *Anesthesiology* 1972; **36**: 57–63.

11. Nahrwold ML, Cohen PJ. Additive effect of nitrous oxide and halothane on mitochondrial function. *Anesthesiology* 1973; **39**: 534–6.

12. White DC, Wardley-Smith B, Halsey MJ. Elastomers as analogues of anaesthetic receptors. *British Journal of Anaesthesia* 1972; **44**: 1020–4.

13. Nahrwold ML, Rapiejko JA, Cohen PJ. The mitochondrion as a model of the anesthetic receptor site. In: Fink BR, ed. *Molecular basis of anesthesia*. New York: Raven Press, 1975: 431–8.

14. Krantz JC, Esquibel A, Truitt EB, Ling ASC, Kurland AA. Hexafluorodiethyl ether (Indoklon) – an inhalant convulsant. *Journal of the American Medical Association* 1958; **166**: 1555–62.

15. Koblin DD, Eger EI, Johnson BH, Collins P, Terrell RC, Speers L. Are convulsant gases also anesthetics? *Anesthesia and Analgesia* 1981; **60**: 464–70.

16. Roth SH. Mechanisms of anaesthesia: a review. *Canadian Anaesthetists' Society Journal* 1980; **27**: 433–9.

17. Deady JE, Koblin DD, Eger EI, Heavner JE, D'Aoust B. Anesthetic potencies and the unitary theory of narcosis. *Anesthesia and Analgesia* 1981; **60**: 380–4.

18. Koblin DD, Dong DE, Deady JE, Eger EI. Selective breeding alters murine resistance to nitrous oxide without alteration in synaptic membrane lipid composition. *Anesthesiology* 1980; **52**: 401–7.

19. Pauling L. The hydrate microcrystal theory of general anesthesia. *Anesthesia and Analgesia Current Research* 1964; **43**: 1–10.

20. Miller SL. A theory of gaseous anesthetics. *Proceedings of the National Academy of Sciences, USA* 1961; **47**: 1515–24.

21. Schoenborn BP. Binding of cyclopropane to sperm whale myoglobin. *Nature* 1967; **214**: 1120–2.

22. Allison AC, Nunn JF. Effects of general anaesthetics on microtubules. A possible mechanism of anaesthesia. *Lancet* 1968; **2**: 1326–9.

23. Tas PWL, Kress HG, Koschel K. General anesthetics can competitively interfere with sensitive membrane proteins. *Proceedings of the National Academy of Sciences, USA* 1987; **84**: 5972–5.

24. Firestone LL, Sauter JF, Braswell LM, Miller KW. Actions of general anesthetics on acetylcholine receptor-rich membranes from *Torpedo californica*. *Anesthesiology* 1986; **64**: 694–702.

25. Brett RS, Dilger JP, Yland KF. Isoflurane causes 'flickering' of the acetylcholine receptor channel: observations using the patch clamp. *Anesthesiology* 1988; **69**: 161–70.

26. Cheng S-C, Brunner EA. Effects of anesthetic agents on synaptosomal GABA disposal. *Anesthesiology* 1981; **55**: 34–40.

27. Moody EJ, Suzdak PD, Paul SM, Skolnick P. Modulation of the benzodiazepine-γ-aminobutyric acid receptor chloride channel complex by inhalation anesthetics. *Journal of Neurochemistry* 1988; **51**: 1386–93.

28. Van den Pol A, Wuarin J-P, Dudek FE. Glutamate, the dominant excitatory transmitter in neuroendocrine regulation. *Science* 1990; **250**: 1276–8.

29. Hirose T, Inoue M, Uchida M, Inagaki C. Enflurane-induced release of an excitatory amino acid, glutamate, from mouse brain synaptosomes. *Anesthesiology* 1992; **77**: 109–13.

30. Gilman AG. G proteins: transducers of receptor-generated signals. *Annual Review of Biochemistry* 1987; **56**: 615–49.

31. Stoelting RK. Opiate receptors and endorphins: their role in anesthesiology. *Anesthesia and Analgesia* 1980; **59**: 874–80.

32. Berkowitz BA, Ngai SH, Finck AD. Nitrous oxide 'analgesia'; resemblance to opiate action. *Science* 1976; **194**: 967–8.

33. Berkowitz BA, Finck AD, Hynes MD, Ngai SH. Tolerance to nitrous oxide analgesia in rats and mice. *Anesthesiology* 1979; **51**: 309–12.

34. Finck AD, Ngai SH, Berkowitz BA. Antagonism of general anesthesia by naloxone in the rat. *Anesthesiology* 1977; **46**: 241–5.

35. Smith RA, Winter PM, Smith M, Eger EI. Rapidly developing tolerance to acute exposures to anesthetic agents. *Anesthesiology* 1979; **50**: 496–500.

36. Smith RA, Wilson M, Miller KW. Naloxone has no effect on nitrous oxide anesthesia. *Anesthesiology* 1978; **49**: 6–8.

37. Bennett PB. Naloxone fails to antagonize the righting reflex in rats anesthetized with halothane. *Anesthesiology* 1978; **49**: 9–11.

38. Pace NL, Wong KC. Failure of naloxone and naltrexone to antagonize halothane anesthesia in the dog. *Anesthesia and Analgesia* 1979; **58**: 36–9.

39. Shiwaku Y, Nagashima H, Duncalf RM, Duncalf D, Foldes F. Naloxone fails to antagonize halothane-induced depression of the longitudinal muscle of the guinea pig ileum. *Anesthesia and Analgesia* 1979; **58**: 93–8.

40. Koblin DD, Deady JE, Dong DE, Eger EI. Mice tolerant to nitrous oxide are also tolerant to alcohol. *Journal of Pharmacology and Experimental Therapeutics* 1980; **213**: 309–12.

41. Koblin DD, Deady JE, Nelson NT, Eger EI, Bainton CR. Mice tolerant to nitrous oxide are not tolerant to barbiturates. *Anesthesia and Analgesia* 1981; **60**: 38–41.

42. Kendig JJ, Trudell JR, Cohen EN. Halothane stereoisomers: lack of stereospecificity in two model systems. *Anesthesiology* 1973; **39**: 518–24.

43. Lockhart SH, Cohen Y, Yasuda N, *et al.* Absence of abundant binding sites for anesthetics in rabbit brain: an in vivo NMR study. *Anesthesiology* 1990; **73**: 455–60.

44. Hanson TD, Warner DS, Todd MM, Baker MT, Jensen NF. The influence of inhalational anesthetics on *in vivo* and *in vitro* benzodiazepine receptor binding in the rat cerebral cortex. *Anesthesiology* 1991; **74**: 97–104.

45. Hoehner PJ, Quigg MC, Blanck TJJ. Halothane depresses D600 binding to bovine heart sarcolemma. *Anesthesiology* 1991; **75**: 1019–24.

46. Franks NP, Lieb WR. Stereospecific effects of inhalational general anesthetic optical isomers on nerve ion channels. *Science* 1991; **254**: 427–30.

47. Seitz PA, Riet M, Rush W, Merrell J. Adenosine decreases the minimum alveolar concentration of halothane in dogs. *Anesthesiology* 1990; **73**: 990–4.

48. Segal IS, Jarvis DJ, Duncan SR, White PF, Maze M. Clinical efficacy of transdermal clonidine during the perioperative period. *Anesthesiology* 1991; **74**: 220–5.

49. Ghignone M, Cavillo O, Quintin L. Anesthesia and hypertension: the effect of clonidine on preoperative hemodynamics and isoflurane requirements. *Anesthesiology* 1987; **67**: 3–10.

50. Miller KW. Intermolecular forces. *British Journal of Anaesthesia* 1974; **46**: 190–5.

51. Wulf RJ, Featherstone RM. A correlation of Van der Waals constants with anesthetic potency. *Anesthesiology* 1957; **18**: 97–105.

52. Clements JA, Wilson KM. The affinity of narcotic agents for interfacial films. *Proceedings of the National Academy of Sciences, USA* 1962; **48**: 1008–14.

53. Larrabee MG, Posternak JM. Selective action of anesthetics on synapses and axons in mammalian sympathetic ganglia. *Journal of Neurosphysiology* 1952; **15**: 91–114.

54. Larrabee MG, Holaday DA. Depression of transmission through sympathetic ganglia during general anesthesia. *Journal of Pharmacology and Experimental Therapeutics* 1952; **105**: 400–8.

55. French JD, Verzeano M, Magoun HW. A neural basis of the anesthetic state. *Archives of Neurology and Psychiatry* 1953; **69**: 519–29.

56. Cucchiara RF, Michenfelder JD. The effect of interruption of the reticular activating system on metabolism in canine cerebral hemispheres before and after thiopental. *Anesthesiology* 1973; **39**: 3–12.

57. Clark DL, Rosner BS. Neurophysiologic effects of general anesthetics: I. The electroencephalogram and sensory evoked responses in man. *Anesthesiology* 1973; **38**: 564–82.

58. Clark DL, Butler RA, Rosner BS. Dissociation of sensation and evoked responses by a general anesthetic in man. *Journal of Comparative and Physiological Psychology* 1969; **68**: 315–19.

59. Cohen PJ, ed. *Metabolic aspects of anesthesia.* Philadelphia: FA Davis, 1975.

60. Cohen PJ, Wollman H, Alexander SC, Chase PE, Behar MG. Cerebral carbohydrate metabolism in man during halothane anesthesia. Effects of $PaCO_2$ on some aspects of carbohydrate utilization. *Anesthesiology* 1964; **25**: 185–91.

61. Bjoraker DG, Cohen PJ. The effect of xenon on mitochondrial respiration: prediction from lipid solubility and *in vivo* potency of known anesthetics. *Federation Proceedings* 1975; **34**: 750.

62. Folbergrova J, Lowry OH, Passonneau JV. Changes in metabolites of the energy reserves in individual layers of mouse cerebral cortex and subjacent white matter during ischaemia and anaesthesia. *Journal of Neurochemistry* 1970; **17**: 1155–62.

63. Brunner EA, Passonneau JV, Molstead C. The effect of volatile anesthetics on levels of metabolites and on metabolic rate in brain. *Journal of Neurochemistry* 1971; **18**: 2301–16.

64. Kress HG, Müller J, Eisert A, Gilge U, Tas PW, Koschel K. Effects of volatile anesthetics on cytoplasmic Ca^{2+} signaling and transmitter release in a neural cell line. *Anesthesiology* 1991; **74**: 309–19.

65. Seeman P, Roth S. General anesthetics expand cell membranes at surgical concentrations. *Biochimica et Biophysica Acta* 1972; **255**: 171–7.

66. Trudell JR. A unitary theory of anesthesia based on lateral phase separations in nerve membranes. *Anesthesiology* 1977; **46**: 5–10.

67. Johnson FH, Brown DES, Marsland DA. Pressure reversal of the action of certain narcotics. *Journal of Cellular and Comparative Physiology* 1942; **20**: 269–76.

68. Ueda I. Effects of diethyl ether and halothane on firefly luciferin bioluminescence. *Anesthesiology* 1965; **26**: 603–6.

69. Halsey MJ, Wardley-Smith B. Pressure reversal of narcosis produced by anaesthetics, narcotics and tranquillisers. *Nature* 1975; **257**: 811–13.

70. Halsey MJ, Eger EI, Kent DW, Warne PJ. High-pressure studies of anesthesia. In: Fink BR, ed. *Molecular basis of anesthesia.* New York: Raven Press, 1975: 353–61.

71. Kent DW, Halsey MJ, Eger EI, Kent B. Isoflurane anesthesia

and pressure antagonism in mice. *Anesthesia and Analgesia, Current Researches* 1977; **56**: 97–101.

72. Miller KW, Wilson MW. The pressure reversal of a variety of anesthetic agents in mice. *Anesthesiology* 1978; **48**: 104–10.

73. Miller KW, Paton WDM, Smith RA, Smith EB. The pressure reversal of general anesthesia and the critical volume hypothesis. *Molecular Pharmacology* 1973; **9**: 131–43.

74. Miller KW. Inert gas narcosis, the high pressure neurological syndrome, and the critical volume hypothesis. *Science* 1974; **185**: 867–9.

75. Miller KW. The opposing physiologic effects of high pressures and inert gases. *Federation Proceedings* 1977; **36**: 1663–7.

76 Hill MW, Hoyland J, Bangham AD. Effects of temperature and n-butanol (a model anaesthetic) on a behavioral function of goldfish. *Journal of Comparative Physiology* 1980; **135**: 327–32.

77. Eger EI, Saidman LJ, Brandstater B. Temperature dependence of halothane and cyclopropane anesthesia in dogs: correlation with some theories of anesthetic action. *Anesthesiology* 1965; **26**: 764–70.

78. Smith RA, Smith M, Eger EI, Halsey MJ, Winter PM. Nonlinear antagonism of anesthesia in mice by pressure. *Anesthesia and Analgesia* 1979; **58**: 19–22.

79. Kendig JJ. Anesthetics and pressure in nerve cells. In: Fink BR, ed. *Molecular mechanisms of anesthesia*. Progress in anesthesiology, Vol. 2. New York: Raven Press, 1980: 59–68.

80. Parmentier JL, Bennett PB. Hydrostatic pressure does not antagonize halothane effects on single neurons of *Aplysia californica*. *Anesthesiology* 1980; **53**: 9–14.

81. Pope WDB, Jones AJ, Halsey MJ, Nunn JF. Pressure enhancement of the depressant effect of halothane on cilial beat. *Canadian Anaesthetists' Society Journal* 1978; **25**: 319–22.

82. Cohen PJ. Effect of hydrostatic pressure on halothane-induced depression of mitochondrial respiration. *Life Sciences* 1983; **32**: 1647–50.

83. Cohen PJ, Bedows E, Brabec MJ, Knight PR. Hyperbaric pressure of 51 atmospheres is without effect on the depression of oxygen uptake in kidney tissue culture produced by halothane. *Life Sciences* 1985; **37**: 1221–4.

84. Cohen PJ. Hyperbaric pressure does not affect the analgesia produced by nitrous oxide in the mouse. *Canadian Journal of Anaesthesia* 1989; **36**: 40–3.

85. Halsey MJ, Wardley-Smith, Green CJ. Pressure reversal of general anesthesia. A multi-site expansion hypothesis. *British Journal of Anaesthesia* 1978; **50**: 1091–7.

Chapter 3

Drug–Receptor Interactions

B. J. Pollard

Much of an anaesthetist's life is spent observing the action of drugs on the body. Applied pharmacology is thus at the centre of every anaesthetic. The anaesthetist's preferred route of administration of drugs is intravenously and therefore the effect is rapid, often immediate. Unwanted effects may also appear with similar rapidity. It is essential therefore for the anaesthetist to be aware of all beneficial and adverse effects of every agent administered.

If one considers only two events, administration of a drug and its subsequent effect, one is immediately led to the logical conclusion that there must be an intermediate stage. The observed effect is not the drug itself but the response of the organism (or certain parts of it) to the presence of the drug, i.e. the drug must exert an effect on cells or organs to bring about the observed response. It is the way in which the drug influences the cell which represents the intermediate stage.

It is appropriate to illustrate this using an example. Take the production of neuromuscular block by tubocurarine. Neuromuscular transmission is present before tubocurarine is added to the system, impulses travel down the nerves and produce contractions of skeletal muscle. Tubocurarine is added and the transmission of the impulse from nerve to muscle ceases. The nerve is still capable of conducting an impulse, transmitter is still being released, and the muscle is still capable of contracting. It is the presence of the tubocurarine which is causing the block in transmission. Remove the tubocurarine and normal transmission is resumed. It would be logical to conclude that the tubocurarine is attaching itself to and reversibly affecting some component which is essential for the transmission process. The result is that transmission is interrupted. But what component of the cell is affected and what is the nature of the interaction?

The historical development of the concept of receptors

At the end of the nineteenth century, the nature of the interaction was a topical subject for discussion. How do drugs work? What is the nature of their action? Is it physical or chemical in nature? Supporters of the hypothesis which favours a physical interaction held the view that drugs exerted their effects by altering certain physical properties of cells, e.g. surface tension or osmotic pressure; supporters of the chemical hypothesis regarded the action of a drug as being mediated by the formation of a chemical union with some structure either on, or within, a cell. The answer came from pioneering work at the beginning of the twentieth century by Langley and Ehrlich.[1–4] This work established the concept of receptors.

Langley's studies centred on the nervous system, particularly with respect to the effects of cholinomimetic compounds on the stimulation of secretion and their antagonism by atropine. These studies led him to suggest that 'there is some substance or substances in the nerve endings or gland cells with which both atropine and pilocarpine are capable of forming compounds'.[1] These he called 'specific receptive substances', and advanced the hypothesis that these substances, present around the nerve ending, were capable of receiving or transmitting stimuli.[3] This theory was later shown to be true. It is interesting to note that during the course of his studies, he examined the alkaloid nicotine and concluded that it was able to form a complex with these specific receptive substances on skeletal muscle.[2]

At around the same time that Langley was busy with his studies on the nervous system, Ehrlich was working with stains and dyes on biological tissues. A number of the early investigations into immunological mechanisms are also attributed to Ehrlich.[4] Ehrlich's work led him to formulate the concept that surface entities existed on cells and that these could exhibit specific binding properties with toxin molecules. These entities he called 'receptors', and it is therefore to Ehrlich that credit for coining the term 'receptor' should be given. Credit for establishing the drug–receptor *theory*, however, should be given to A. J. Clarke.

The early work of Clarke centred around the actions of the glycoside strophanthin. He reached the conclusion that strophanthin acted on the surface of the heart cells without entering them. His initial thoughts were that this action was likely to be of a physical nature.[5] Subsequent work, which centred around the quantitative relation between concentration and action of acetylcholine, however, led him to reconsider this view. In these later studies, he examined the responses of tissue to graded increasing concentrations of acetylcholine. The concentration of acetylcholine and the effect produced by that concentration when plotted in the form of a graph produced a curve which could be expressed mathematically in the following way:

$$x = k \times \frac{y}{(100 - y)}$$

where x = concentration of acetylcholine; y = effect expressed as per cent of maximum possible effect; k = a constant.

The simplest explanation for the relation between x and y would be a reversible monomolecular reaction between two entities, i.e. the drug and a single binding site on the cell. Another crucial piece of information was also provided by Clarke who discovered that there appeared to be no relation between the amount of drug entering a cell and the effect produced. This made it likely that the interaction was taking place on the surface of the cell. Calculating the number of molecules which became attached to each cell then led him to conclude that the interactions could only be taking place on a very small area of the cell surface. This would have been insufficient to exert any significant physical effect.[6,7] Previous views had therefore to be revised, leading to the conclusion that drugs were acting upon a small number of sites on the surface of cells.

Even though the concept of receptors on the surface of cells was soon established, these structures remained a conceptual entity until the 1970s when advanced techniques of electron microscopy allowed them to be actually visualized. Simultaneous advances were taking place in biochemistry which allowed receptors to be isolated and purified. In certain instances it became possible for these receptors to be re-inserted into artificial lipid membranes for further electrophysiological examination.[8] The development of immunological binding techniques at that time was of similar, if not greater importance. Using these techniques

it was possible to raise antibodies to receptors in small animals (rabbits usually).[9,10]

The discovery of various toxins which would attach to receptors also assisted research. The subject of neuromuscular transmission is once again a good example. The toxins from the Taiwanese banded krait, *Bungarus multicinctus* (α-bungarotoxin and β-bungarotoxin) were described. α-Bungarotoxin blocks neuromuscular transmission by binding specifically and irreversibly to acetylcholine receptors.[11] The application of radiolabelled α-bungarotoxin to the neuromuscular junction enables it to be localized almost exclusively to the crests of the folds on the postjunctional membrane.[12] The implication of this observation was that the receptors were located on the shoulders of the folds, a fact which has since been confirmed.

It must be noted that, despite the widespread acceptance of the receptor theory of drug action, and the conclusions that drugs did not act by direct physical mechanisms, there are a small number of substances which probably do act by physical means. One example is the gaseous and volatile general anaesthetic agents which are thought to act by a non-specific action on the membranes of cells (the membrane expansion hypothesis). Certain of the agents which do induce anaesthesia may act through receptor systems (e.g. the barbiturates and the benzodiazepines) but these are probably in the minority.

Definition and classification of receptors

A receptor is an entity which binds to a drug or a transmitter substance and puts into action a chain of events leading to an effect (or part of an effect if it is necessary to exceed a threshold). It must be remembered that there are also non-receptor sites which will bind drugs, e.g. plasma proteins. When a drug binds to a non-receptor site there is no effect. Although these sites possess an affinity for the drug and will bind the drug, they are not receptors in the classic sense. The term 'acceptors' is often used for these. It must be noted that the presence and concentration of inactive acceptor sites may considerably influence the action of a drug by either removing it from the active biophase, thus decreasing the expected effect, or acting as a storage site and potentially prolonging the effect. For greater detail concerning these pharmacokinetic factors the reader is referred to Chapter 4.

The characteristic properties of receptors are those of sensitivity, selectivity and specificity. Most drugs produce marked effects in very low concentrations (sensitivity).

Responses are only produced by a narrow range of chemical substances which possess very similar chemical groupings (selectivity). The response of a cell of a particular type to an agonist drug which is acting on the same set of receptors is always the same because it is determined by the cells themselves (specificity).

Receptors are commonly classified according to the target systems to which they relate, e.g. cholinergic, adrenergic, etc. It must be remembered that the transmitter or mediator may be a hormone, neurotransmitter, or other cellular messenger. Although this classification is logical and is often used, it has a number of inadequacies. Detailed studies with more selective agonists and antagonists have revealed many subdivisions in most classifications, e.g. cholinergic muscarinic M_1, M_2, M_3; adrenergic α_1, α_2, β_1, β_2.

A second method of classification which might be appropriate relates to the anatomical location of the receptors. Most receptors are extracellular (e.g. acetylcholine receptors) although receptors may be intracellular (e.g. the action of local anaesthetics on sodium channels in nerves). The tissue supporting the receptors could also be used to categorize receptors in an anatomical sense, e.g. most cholinergic muscarinic receptors lie on smooth muscle or glandular tissues while a particular subtype of cholinergic nicotinic receptors lies on striated muscle.

There is a third method of classifying receptors which has become available relatively recently following advances in molecular biology. These advances have enabled the isolation and study of single receptors. Subsequent studies have delineated the actual amino acid sequence for some receptor systems. Receptors could therefore be classified in a very detailed manner, according to their individual characteristics and even the gene sequence encoding the receptor. At present such detail is available principally as a research tool.

Chemical bonds involved in the actions of drugs

In view of the existence of receptors to which drugs attach, it is appropriate to consider the forms by which these bond together. These are chemical bonds and may therefore be divided into two principal types, covalent and electrostatic, both of which may exhibit further subdivision.

Covalent bonds

These are strong bonds, formed by the mutual sharing of electrons from the outer electron shells (orbits) surrounding two different atoms. A good example is the bond between carbon and hydrogen (C–H) in which the single electron from the hydrogen atom is shared with an electron from the carbon. This effectively gives the hydrogen two electrons in that shell, which increases its stability. The carbon which started with four electrons now effectively has five. If three more hydrogens (or other suitable groups) each share one electron, then the result will be that the carbon has eight, which is the stable number in that shell. Thus carbon has a valency of four, and readily takes part in covalent bonding. Double bonds are formed by the sharing of two pairs of electrons, and triple by the sharing of three pairs of electrons. If the two atoms involved in the bonding are identical, then the whole entity is electrically neutral. When two dissimilar atoms are involved, however, the electron sharing will not be exactly equal, resulting in one being slightly positively charged and the other having an identical negative charge. This resulting dipole can be amplified by the existence of certain patterns of atoms or groups such that a significant charge may appear at certain points. This has the effect of creating dipoles of sufficient magnitude to enable molecules to create bonds by electrostatic forces. Examples of drugs acting through covalent bonds are the organophosphate anticholinesterases, ecothiopate and dyflos.

Electrostatic bonds

These bonds are formed by attraction of opposing electrical charges. The extent to which molecules may attach to one another (e.g. a drug to a receptor) depends upon the actual charges involved. When the charges are weak, for example when created by unequal distribution of electrons in a covalent bond, then the bonds will be weak. It will only be possible for molecules to attach to each other when they can align themselves closely. It might therefore be expected that such bonds are involved in drug–receptor binding because the shape of most drugs is very specific and complementary to their receptors. The analogy of a lock and key is most appropriate.

Certain molecules may, by gaining an extra proton (hydrogen atom), or losing a proton, become either positively or negatively charged respectively. Such ionization creates more powerful charges and the formation of these charges depends upon the pH of the medium and the pK_a of the molecule. Such strong charges appear to be concerned with non-specific drug binding at non-receptor sites as well as with drug–receptor binding.

There are still weaker ionic bonds which can exist, namely hydrogen bonds. Hydrogen bonding is a specific case of a bond resulting from the existence of an identical dipole or a covalent bond involving a hydrogen atom. Hydrogen bonds are very weak and principally serve as reinforcements for other types of ionic bonds. The importance of hydrogen bonds in biological systems is considerable. The macromolecular structure of DNA and similar proteins depends upon the presence of large numbers of hydrogen bonds.

Van der Waals bonds are weaker and arise from distortions of the outer electron orbits of atoms when they are held in close proximity. They provide very weak electrostatic links which serve to reinforce other bonding and help to stabilize certain structural conformations or interactions.

Structure–activity relations

It has been demonstrated that drugs act at receptors and that these receptors are specific entities on the surface of cells. There are a large number of different receptors, each with its own individual family of related drugs. In general, there is little or no cross-reactivity between drugs in these separate families. For example, tubocurarine has an action at the nicotinic acetylcholine receptor and diazepam at the benzodiazepine receptor. Neither of these agents has any effect on the other receptor. Tubocurarine and diazepam differ considerably in their chemical structure and it might therefore be expected that the characteristics and shape of the molecule are relevant to activity at a receptor. A molecule of a certain shape and/or size is required in order for an effect to appear on a particular receptor system. The whole molecule is not necessarily involved, however. Alcuronium, pancuronium and atracurium, for example, all have the same effect as tubocurarine on the nicotinic acetylcholine receptor whereas midazolam, temazepam and lorazepam all affect the benzodiazepine receptor.

The actual chemical structure of each individual component of each series differs, but there are similarities within the molecules. It would thus appear logical to infer that there must be a component part of each molecule which is common to a series and which possesses the ability to combine with the receptor. That moiety on the benzodiazepine molecule must be different from that on the cholinergic agonists. A particular structure, therefore, has an activity at certain receptors and this is the basis of the study of the relation between structure and activity.

A good analogy for the action of a drug on a receptor as mentioned earlier is that of a lock (receptor) and key (drug) system. The analogy, although crude, does demonstrate some important concepts as follows:

1. The drug is capable of 'unlocking' the response.

2. The lock (receptor) has a certain specificity towards a particular key (drug).
3. The lock may be turned by other keys, which, although not identical to the correct key have certain similarities (master keys or 'skeleton' keys).
4. That key may operate other locks by accident rather than by design.
5. The lock may be 'jammed' by attempting to insert the wrong key. The resulting physical obstruction would prevent the correct key from being used (a pharmacological response may be blocked by an appropriate drug).

The study of the relation between structure and activity has enabled a number of factors with respect to the receptor molecules themselves to be inferred. Information concerning, for example, the number of binding sites and their spatial configuration, the probable nature of the binding forces and the requirements for activation may be deduced. Although the relation between structure and activity continues to be important in the elucidation of receptor structure, newer methods have been developed. Recently introduced computational techniques involving advanced graphics are of particular importance. These enable predictions to be made about the energies between bonds and the conformations of the bound versus unbound states of drug–receptor complexes. Advances in recombinant DNA technology is allowing the primary structures of certain receptors to be elucidated and copied. x-Ray crystallography and nuclear magnetic resonance imaging have also played a vital role. In order to understand structure–activity relationships better, it is appropriate to examine some clearly understood examples.

The various subtypes of acetylcholine receptors are well understood. Although there are a number of acetylcholine receptors they were originally subdivided into two broad categories, nicotinic and muscarinic. This subdivision was determined following observation of the agonist activities of the two alkaloids nicotine and muscarine at acetylcholine receptors. Further subdivisions on the basis of antagonist activity have since been made.

If the structure of acetylcholine is examined it can be seen that there is a cationic centre (positively charged nitrogen) towards one end. Towards the other end of the molecule are two oxygen atoms, one of which exists as a

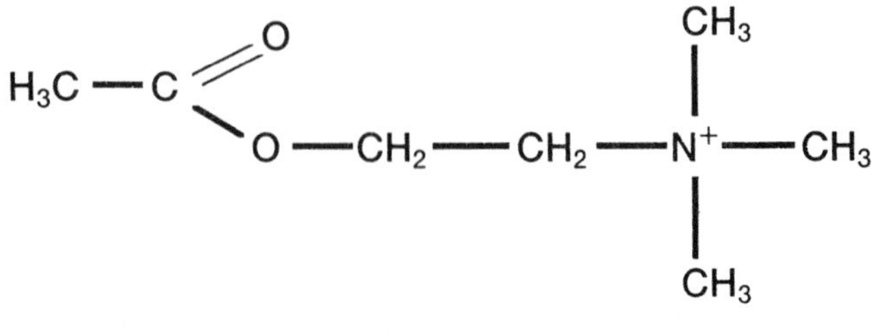

Fig 3.1 The structural formula of acetylcholine.

carbonyl group (C–O) and the other as an ester link (–O–C–) (Fig. 3.1). Both of these oxygens have the potential to take part in hydrogen bonding to other groups (which may be on a receptor). If the structure of acetylcholine is compared with that of nicotine (both are potent agonists at the nicotinic acetylcholine receptor) there are certain structural similarities. The distance between the positive nitrogen and a group capable of hydrogen bonding is 0.59 nm. This second group need not be the same in both molecules, but must behave in a similar fashion. In acetylcholine it is the carbonyl group and in nicotine it is the nitrogen of the pyridine ring.

Acetylcholine is also a potent agonist at the cholinergic muscarinic receptor (it is the transmitter) and muscarine is also a potent agonist at that receptor. A comparison of the structures of acetylcholine and muscarine reveals another similarity. In this case the distance separating the positive nitrogen and a group capable of hydrogen bonding (an oxygen atom in each case) is 0.44 nm. Acetylcholine thus appears to be capable of activating two different receptor systems. A closer inspection would reveal that the acetylcholine molecule is not aligned in exactly the same way in both these situations. It must be remembered that the acetylcholine molecule is a very flexible molecule and can be bent into a number of different shapes (conformations). This factor is very important in the relation between structure and activity. Two molecules with apparent structural differences may both have a potent action at the same receptor because they can both be manipulated into a certain similar shape.

Consider the antagonists at the nicotinic acetylcholine receptor. The original antagonist at the skeletal muscle nicotinic receptor was tubocurarine (the active constituent of curare, the South American arrow poison). It was discovered to contain two positively charged nitrogen groups. The hypothesis was advanced that perhaps other compounds which possessed two positively charged nitrogen atoms might also exhibit neuromuscular blocking activity. A variety of such compounds were then examined for neuromuscular blocking activity and of considerable interest is the polymethylene–bismethonium family (Fig. 3.2).[13,14] An entire series of different compounds were tested, the only difference being variations in the length of the –CH_2– chain. Two distinct phenomena were observed: both ganglionic and neuromuscular blockade could be produced, but the optimum for ganglion blockade was at a chain length of 6 (hexamethonium) and that for neuromuscular blockade at a chain length of 10 (decamethonium). The polymethylene–bismethonium compounds are long flexible molecules. It might therefore be imagined that they could bend themselves into almost any configuration and thus fit almost any receptor system which required the presence of two positively charged nitrogen atoms at a fixed distance. There are, however, other considerations. In a molecule with two positive charges, there will be electrostatic repulsion between the two positive forces which will limit the number of possible conformations.

Care must be taken when inferring action at a receptor system from the simple existence of a certain group, or groups, in a chemical structure. Not every molecule with two positively charged nitrogen atoms is a neuromuscular blocker nor has every neuromuscular blocker two positively charged nitrogens. Furthermore, in vecuronium only one of the nitrogens is positively charged. There may also be more than one mechanism at work. Decamethonium weakly stimulates the receptor before blocking it and in addition has an action at higher concentrations in blocking the receptor channel. This latter action does not occur at the acetylcholine recognition site and is non-competitive in nature.[15] A similar non-competitive action has also been demonstrated for hexamethonium.[16]

Isomerism

The preceding discussions have centred around some fairly simple molecular structures. Few drugs in clinical use have very simple structures, with the possible exception of the volatile and gaseous anaesthetic agents. Although certain simple molecules can demonstrate isomerism, this is more common with complex molecules.

It is important first to consider the terminology. There are a number of important descriptive terms which are used to define molecules and which may also be used to describe relationships between molecules and receptors. The word isomer differentiates two unique molecular entities which are composed of identical numbers of each chemical constituent. Isoflurane and enflurane are isomers. Stereoisomers are isomers in which the constituent atoms or groups differ with respect to their spatial arrangement and fall into two principal groups: geometric isomers and optical isomers. When a molecule contains a double bond or a heterocyclic (non-aromatic) ring system the potential arises for geometric isomerism. A molecule containing a double bond has the potential to exist in two different forms, *cis*

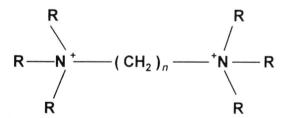

Fig. 3.2 The general structural formula of the methonium series, as examined by Paton and Zaimis.[13] In hexamethonium, $n = 6$; in decamethonium $n = 10$.

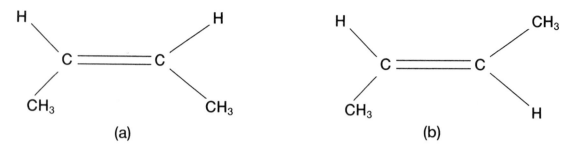

Fig. 3.3   A simple molecule (butene) containing a double bond between two carbon atoms which demonstrates geometric isomerism. The rigidity of the double bond results in two possible structural configurations. (a) Both methyl groups are the same side – the *cis* form. (b) The methyl groups are on opposite sides – the *trans* form.

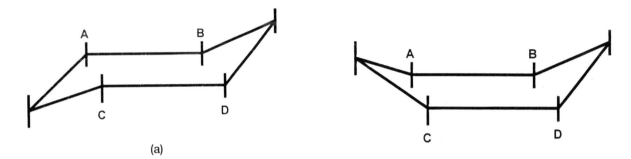

Fig. 3.4   A simple heterocyclic 6-carbon ring showing the (a) chair and (b) boat configurations. Each of the unlabelled bonds would contain a hydrogen atom. The molecules also demonstrate further asymmetry because the additional groups A and B protrude from the upper surface of the molecule and the groups C and D from the lower surface.

and *trans* (Fig. 3.3). A simple heterocyclic ring can exist in 'chair' or 'boat' forms (Fig. 3.4). In addition, because the ring structure is relatively rigid, if certain groups protrude onto the upper surface while others do so onto the lower surface, further asymmetry is introduced (Fig. 3.4). It should thus be clear that two molecules of apparently identical chemical structure can exist in more than one configuration. Furthermore, these may behave as quite different substances with different physicochemical properties.

Returning to the previous discussion about the concept of receptors it was recognized that many receptors are very specific, having an ability to combine with agents in certain conformations only. It is not difficult to see that only one geometric isomer form of a drug might be capable of combining with a particular receptor. The nature of the synthetic process usually means that roughly equal numbers of each possible structure are produced (although certain molecules are capable of conversion from one form to another under certain circumstances). The drug which is injected is therefore usually a mixture of all possible isomers, only one of which may be active.

Further evidence with respect to the relation between structure and activity for certain drugs may be obtained by observation of the activity of individual stereoisomers of a drug. If there is a specific biological receptor involved, then the activity of the various isomers should differ. If there is

no difference, then it is likely that either the action of the drug is not taking place through the medium of a specific receptor system or that the receptor has only one or two loci and thus attachment of a drug is less specific.

Recently, interest has increased in a particular form of isomerism, namely optical isomerism. Optical isomers are a subset of stereoisomers, in which at least two of the isomers are optically active. The term chirality (right- or left-handedness) is often used in this context. Molecules which possess a chiral centre exist in two forms (enantiomers) which cannot be distinguished with respect to the majority of their physical or chemical properties, e.g. melting point, boiling point, lipid solubility, etc. A chiral molecule has at least one isomeric centre, which is usually a carbon atom (Fig. 3.5), but phosphorus, sulphur and nitrogen can also form chiral centres. A mixture of equal portions of each enantiomer is a racemic mixture. If more than one chiral centre is present in the molecule four or more different configurations exist and these are termed diastereoisomers. Diastereoisomers may have different chemical and physical properties.

The usual difference between a pair of chiral isomers relates to their ability to rotate the plane of polarized light – '*l*' or ' ' rotates to the left (anticlockwise) and '*d*' or '+' rotates to the right (clockwise). The capacity to rotate the plane of polarized light is a function of the electronic

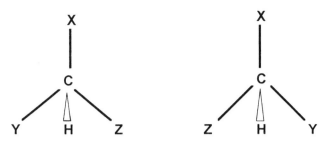

Fig. 3.5 A simple chiral molecule. The two forms can be seen to be mirror images of one another.

properties of the molecule and confusions have arisen in the field of organic chemistry. A second convention was therefore developed, the *R–S* convention. The groups around the chiral centre are sized according to atomic number and depending upon the order, the molecule is assigned the *R* (rectus) or *S* (sinister) designation.[17] Usually, but not always, the particular *R* and *S* rotations correlate with a particular direction of light rotation. In general terms, when one enantiomer in a series shows a certain activity then the equivalent enantiomer of other compounds in that series will also be the active enantiomer, i.e. if the *d*(+) form of one substance is the active one, it will be the *d*(+) form of all others in that series which will also be active.

Biological systems are principally constructed from chiral molecules – most amino acids are in the *l*-form and most sugars are in the *d*-form. It is not surprising therefore that many drugs which possess one or more asymmetric centres exhibit a high degree of stereoselectivity in their interaction with receptors. Stereoselectivity in the action of drugs, not only on receptors but also with respect to ion channels and enzymes, is well established.[18]

Traditionally, physicians and pharmacologists alike have tended to regard any drug as a single agent which would produce a single effect. But is this actually the case when administering a racemic mixture? The answer is not necessarily. The racemic mixture in reality contains a number of 'drugs' each with the potential to behave in a different fashion. Several different pharmacological outcomes are therefore possible depending upon the pharmacological action of each individual enantiomer.

1. All activity could reside in one isomer with the other(s) being completely inactive. It might not be unexpected for all activity to reside in one enantiomer because of the necessity for a particular molecular configuration of drug for attachment to the receptor. It is unusual, however, for the other isomer to be completely inactive. The antihypertensive agent α-methyldopa is one of the few examples of a drug whose desired effect is confined to one optical isomer, the *l*-isomer.[19]

2. All activity could reside in one isomer while the other(s) have alternative pharmacological action(s). Possible alternative actions might include antagonism to the action of the first isomer, totally unrelated actions, or potentially toxic effects. Labetalol is both an α- and β-adrenoceptor antagonist and is a good example of a drug in this category. Labetalol is a mixture of four diastereoisomers by virtue of its two chiral centres. The *RR* isomer has β-blocking activity whereas the *SR* isomer is an α-blocker. The *RS* and *SS* forms are thought to be inactive.[20] L-Dopa is another example. During the development of L-dopa, it was noted that there were a number of serious side effects which included granulocytopenia. Detailed studies revealed that these were confined to the *d*-isomer and not the *l*-isomer. It was the *l*-isomer therefore which was marketed and not the racemic mixture. With respect to anaesthetic drugs, ketamine, which is a chiral molecule, is marketed as a racemic mixture. It might, however, have been advantageous had it not been marketed thus. The *S*(+) enantiomer is approximately three to four times more potent than the *R*( ) enantiomer and the unwanted side effects appear to be principally associated with the *R*( ) enantiomer. Development of the *S*(+) enantiomer alone might therefore have led to a more acceptable anaesthetic agent.

3. Both (all) isomers could have the same action, but with different potencies. It is not uncommon for the various enantiomers to have a similar qualitative action but to differ in potency. Propranolol, verapamil, and warfarin are all examples of drugs which fall into this category. It might be initially supposed that the use of racemic mixtures of these agents would present very few problems. This may not always, however, be the case. Drugs tend not only to be stereoselective in their interactions with receptors, but also in their binding to plasma proteins, their metabolism and their interactions with other drugs. If the pharmacokinetic disposition of the isomers is different, the pharmacodynamic action can vary quite markedly between patients, and even within the same patient at different times.

4. Both (all) isomers could be equal in activity. This situation is uncommon. This should not be unexpected because of the selectivity of receptors for drugs. The two isomers of promethazine have equal antihistamine activity. The isomers of propranolol show an interesting phenomenon, where their effect on insulin secretion is not selective (all of the isomers have an effect) although their β-blocking action is highly selective, being confined to one isomer.

There are a great many considerations with respect to isomers of drugs for which there is not space to consider in this chapter. The different isomers of a drug may differ with respect to absorption, plasma protein binding, distribution, metabolism, renal clearance, drug interactions and pharmacogenetics. The existence of any of these factors is likely to affect the actions of the drugs. It is also possible for certain drugs to undergo conversion between isomers *in vivo* (e.g. ibuprofen).

Pharmacologists are recognizing more and more the advantages of using single isomers rather than a racemic mixture. The separation of one isomer may, however, be complex and not commercially viable in many instances. It is, however, becoming apparent that decisions about safety and efficacy of a compound should extend to understanding the actions and pharmacokinetics of each component enantiomer. It is likely that we will see more drugs presented as single isomers for clinical use in the future.

Quantitative aspects of drug–receptor interactions

This section considers the interpretation of the relationship between drug concentration and observed response. The simplest model is derived from the Law of Mass Action and assumes that two entities (molecules) are combining in a reversible manner to form a third. With respect to drug–receptor interactions, one entity is the drug and the other the receptor.

In its simplest form, the combination of a drug with a receptor may be expressed by the following equation:

$$[\text{Drug}] + [\text{Receptor}] \rightarrow [\text{Drug–receptor complex}]$$

Unless the drug forms a completely irreversible union with the receptor, an uncommon state of affairs, the drug–receptor complex can dissociate again into its two component parts. The equation should really therefore be bidirectional:

$$[\text{D}] + [\text{R}] \underset{K_{21}}{\overset{K_{12}}{\rightleftharpoons}} [\text{DR}] \rightarrow \text{Effect}$$

The rate of reaction is the same as the rate of formation of the drug–receptor complex (K_{12}) and depends upon both the concentration of drug [D] and the concentration of receptors [R]. The rate of dissociation of the drug–receptor complex (K_{21}) depends upon its concentration [DR]. The rates of these two reactions can therefore be summarized as follows:

Rate of formation   (forward reaction) $= K_{12}[\text{D}][\text{R}]$
Rate of dissociation (reverse reaction) $= K_{21}[\text{DR}]$

When steady state is reached,

$$K_{12}[\text{D}][\text{R}] = K_{21}[\text{DR}]$$

therefore

$$\frac{[\text{D}][\text{R}]}{[\text{DR}]} = \frac{K_{21}}{K_{12}} = K_{\text{D}}$$

The ratio K_{21}/K_{12} is usually replaced by the single term K_{D}, the dissociation constant.

Although K_{D} is known as the dissociation constant, theoretically it may be more correct to describe it as the association constant, because it has been defined in terms of the association of the two components, drug and receptor, and not their dissociation. Because of this, some authors use the symbol K for the association constant and it is important to be aware of this potential confusion:

$$K = \frac{1}{K_{\text{D}}}$$

It must be noted that in this scheme, several assumptions have been made. It has been assumed that the reaction is totally reversible, that all receptor sites have an equal affinity for the drug, that binding to some receptor sites does not affect the binding to others, and that there are no other non-specific binding sites present. Clearly, this is an 'ideal' situation and not one which pertains in many clinical circumstances.

A pharmacologist would proceed from here to determine the degree of binding of drugs to receptors using a variety of techniques. These may include the use of radioactively labelled drugs or nuclear magnetic resonance spectroscopy. Complex kinetic analyses can then be used in order to elucidate further the extent of drug–receptor binding, calculate the dissociation constant, etc. Such theoretical considerations are not relevant to the clinical anaesthetist. There are, however, many similarities between some of those more sophisticated analyses and the quantitative evaluation of drug effect. Receptor binding is very important with respect to the characterization of receptors at the molecular level. It is the ability of a receptor to produce an effect which distinguishes the true receptor from any other binding site. The principal means by which many drug receptors are defined is the measurement of a response and the quantification of those responses is central to understanding how receptors do initiate their response.

Agonists

An agonist is a substance which combines with a receptor to produce a response, e.g. acetylcholine.

From the equation of the Law of Mass Action (above) some mathematical rearrangement can be introduced. At equilibrium, the sum of the total number of free receptors [R] and of the total number of bound receptors [DR] must equal the overall total of available receptors which is denoted by the term [R_{TOT}]. Thus,

$$[R_{TOT}] = [R] + [DR]$$

Substituting for [R] in the Law of Mass Action equation gives the following:

$$K_D = \frac{[D][R_{TOT} - DR]}{[DR]}$$

which can be further rearranged to the following equation:

$$\frac{[DR]}{[R_{TOT}]} = \frac{[D]}{K_D + [D]}$$

It should be clear that when all of the receptors [R_{TOT}] are occupied by agonist, the maximum effect (E_{max}) will be produced. When no receptors are occupied, there will be no effect. Between these two extremes, the effect (E) will be proportioned to the concentration of receptor occupied [DR].

Thus

$$E_{max} \text{ is proportional to } [R_{TOT}]$$

and

$$E \text{ is proportional to } [DR]$$

therefore

$$\frac{E}{E_{max}} = \frac{[DR]}{[R_{TOT}]}$$

Substituting this equation in the earlier equation:

$$\frac{E}{E_{max}} = \frac{[D]}{K_D + [D]}$$

Rearranging this equation produces the following result:

$$E = \frac{E_{max}[D]}{K_D + [D]}$$

Although it may not be immediately obvious, this is the equation relating to a rectangular hyperbola (Fig. 3.6). It may be valuable to examine this equation with respect to the different parts of the curve.

Considering the beginning of the curve, when there is no agonist present, there will be no effect, i.e. when [D] = 0, E = 0.

When the effect is at 50 per cent of maximum:

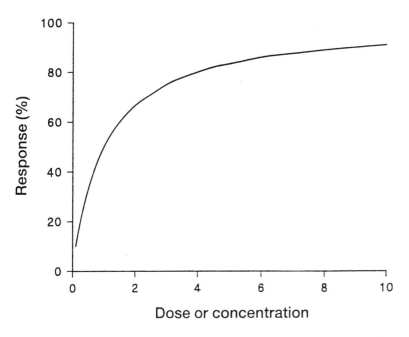

Fig. 3.6 A dose–response relationship.

$$\frac{E}{E_{max}} = 0.5$$

Substituting in the above equation, then

$$\frac{[D]}{K_D + [D]} = 0.5$$

therefore

$$[D] = K_D$$

The dissociation constant K_D is equal to the concentration of drug required to produce an effect 50 per cent of maximum.

When [D] is very high, [D] will be very much greater than K_D and E will be almost equal to E_{max}, i.e. high concentrations of drug produce affects which are close to the maximum effect.

When [D] is very low, [D] is very much less than K_D, and then E is approximately equal to

$$\frac{[D] \times E_{max}}{K_D}$$

or, to put it more simply, at low concentrations of agonist, the effect E becomes almost directly proportional to the agonist concentration. It should now be clear why the relation between agonist combination and effect has the shape of a rectangular hyperbola.

Let us return to the basic equation:

$$\frac{[DR]}{[R_{TOT}]} = \frac{[D]}{K_D + [D]}$$

The proportion of receptors occupied by drug is given by

$$\frac{[DR]}{[R_{TOT}]}$$

Responses measured will depend upon the actual proportion of receptors occupied and the term $[DR]/[R_{TOT}]$ which was earlier replaced by the term E/E_{max}.

If E/E_{max} is denoted instead by 'r', the response, then

$$r = \frac{[D]}{K_D + [D]}$$

Rearranging,

$$[D] = K_D \times \frac{r}{1 - r}$$

This equation is immediately seen to be directly analogous to that derived by A. J. Clarke when he showed that

$$x = \text{constant} \times \frac{y}{100 - y}$$

where x is concentration of drug and y is response as per cent of maximum possible.

It is common to express the relationship between drug concentration and measured effect graphically. When the measured effect is plotted against the concentration of drug, the resulting rectangular hyperbola is particularly difficult to construct and to use. It is therefore common to display the data as the logarithm of the dose (or logarithm of the concentration of drug) against the response, where the response is expressed as a per cent of maximum response. This yields a sigmoid-shaped graph (Fig. 3.7). This transformation is particularly useful because the

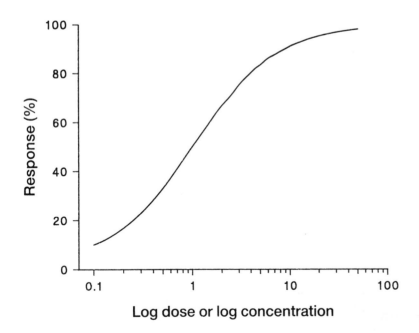

Fig. 3.7 A log dose–response relationship.

section from approximately 20 per cent to 80 per cent of maximal response is converted to a straight line, allowing a more precise estimate of the ED_{50} (or EC_{50}). Additional advantages of this semilogarithmic plot are that it is possible to visualize effects over a wide range of drug doses (concentrations) and also to compare more easily the effects of different agonists and antagonists. Theoretically, if two agonists are acting in exactly the same way on a receptor system, then their log dose–response curves must be parallel and if they are not parallel, then their mechanism of action is unlikely to be the same. The converse does not apply, however, in that if two lines are parallel, then their mechanism of action need not be identical.

The log dose–response graph is difficult to use outside the range 20–80 per cent of maximum response, because it is not linear outside this range. If data can be confined to that central linear part, then statistical analysis becomes much simpler. Clearly, if logarithms are taken of both sides of the equation above:

$$\mathrm{Log}[D] = \mathrm{Log}K_D + \mathrm{Log}\left(\frac{r}{1-r}\right)$$

Thus a graph of log[D] against log $r/(1 - r)$ will produce a straight line, the intercept of which gives the value for $\mathrm{log}K_D$. This relationship is known as the Hill plot (named after A. V. Hill). Strictly speaking, the relationship should be:

$$N \cdot \mathrm{Log}[D] = \mathrm{Log}K_D + \mathrm{Log}\left(\frac{r}{1-r}\right)$$

If the agonist and receptor combine on a 1:1 basis, then N (the slope of the line), should be unity. This is often the case, but not always so.

The term 'Log $r/(1 - r)$' is also referred to as the logit (or logistic) transformation and this is commonly used in biological data handling. The logit transformation has the effect of extending the ends of the scale of 'r' and allows points to be plotted out to approximately 1 per cent and 99 per cent of maximum response. The corresponding logits for values of 'r' of 0, 0.5, 1 are $-\infty$, 0, $+\infty$ respectively.

When considering biological responses, as the concentration of agonist rises, a point will be reached when an effect begins, then increases until a maximum effect is obtained. The two extreme points of zero and 100 per cent must be excluded for two reasons:

1. infinity cannot be plotted on a graph;
2. theoretically, the concentration to just give 100 per cent can be determined but that same value of 100 per cent would also be produced by concentrations of agonist of greater than that to just produce 100 per cent.

There is an alternative approach, namely that of probit analysis. This considers the response as being due to a statistical summation of the probabilities that any individual drug entity and receptor entity will combine to initiate an effect. The response axis is replaced by an axis which represents the proportion (or percentage) of positive receptor interactions for that given dose and response. As the concentration of agonist rises, the proportion (or rate) of receptor interactions increases until the 50 per cent response point is reached. The proportional increase for each unit increase in dose then declines with further increases in dose, although the total number of receptors occupied must be increasing. The frequency distribution of quantal responses to many drugs has been shown to be described by such a normal (Gaussian) distribution. Conversion of a response to its probit value requires complex mathematical methods and it is easier to use either a table of probit values or graph paper which is marked with a probit scale. Corresponding probits for values of r of 0, 0.5, 1 are $-\infty$, 5, $+\infty$, respectively. The effect of probit analysis is almost identical to that of logit analysis in that it expands the ends of the scale of response and converts the sigmoid relation between log concentration and response to a straight line. The same limitations also apply to the points of zero and 100 per cent as to logit analysis.

There is one further technique occasionally applied to biological data, namely arcsine transformation (alternatively known as the angular transformation or inverse sine transformation). The equation $y = \sin^{-1}x$ is applied to the data, where 'x' is the square root of the original variable. As the original variable increases from 0 to 1, y increases from 0 degrees to 90 degrees and equal changes in x correspond to greater changes in y towards the end of the scale. The result is therefore to convert a sigmoid curve (which might be regarded as having the appearance of a section of a sine wave) into a straight line.

Theories underlying the drug–receptor interaction

There are two principal theories underlying the interaction between drugs and receptors, namely the occupation theory and the rate theory. The occupation theory of drug action holds that the observed response is a direct function of the number of receptors occupied by drug. The rate theory recognizes that receptors have to be occupied by drug for the effect to be produced, but holds that the drug is constantly binding with and dissociating from the receptors in a dynamic fashion. The response should then be a function of the rate of occupation of receptors by the agonist.

The occupation theory has its origins in the work of A. J.

Clarke who showed that there were many situations where the effect of a drug was linearly proportional to receptor occupancy. In addition, the maximum response is reached when all of the receptors are occupied by agonist molecules. The occupation theory can readily explain the shape of the dose–response relationships from the equations already described above. There are, however, some discrepancies. When the behaviour of a partial agonist is considered, why does it not result in a maximum response of the same magnitude as a full agonist? The number of receptors is the same and when the drug is in great excess, surely all (or almost all) of the receptors are occupied by drug. It is necessary to postulate that the drug possesses both agonist and antagonist actions, and the original descriptive term of 'dualist' is an apt description. This discrepancy has to be explained by introducing a term 'intrinsic activity'. Full agonists have an intrinsic activity of 1, antagonists 0, and partial agents lie in between. This explanation is regarded with scepticism by some pharmacologists. An additional problem for the occupation theory follows work from Stephenson[21] in which he showed that there were some experimental situations where the response was not exactly proportional to the fraction of receptors occupied.

The rate theory of drug action was advanced by Paton in 1961.[22] This theory can also be shown mathematically to predict the same shapes of graph for the dose–response and log dose–response relationships. The dissociation constant K_D is also equal to the concentration of agonist at half maximal response in this case. The time curve of a response under the rate theory differs from that predicted by the occupation theory. At zero time, as the drug is administered, [DR] is zero. According to the rate theory, the response should immediately rise to a transient peak and then decrease exponentially very rapidly to a plateau (equilibrium response). It is the equilibrium response which is the measured response. Theoretically, the plateau responses are the same as those predicted by the occupation theory. According to the rate theory, however, peak response may considerably exceed the maximum response which would be possible according to the occupation theory. It is hard to reconcile these theoretical considerations and the existence of these instantaneous maximum values is almost impossible to test experimentally.

The points of difference between the two theories of drug action have therefore to be regarded with a certain degree of caution. The only firm conclusion that can be drawn is that one theory holds for certain drug–receptor interactions and the other for different interactions.

Antagonists

An antagonist is a substance which inhibits or blocks the action of an agonist, e.g. tubocurarine. Antagonism of a drug effect can take place in a number of ways. It may be competitive or non-competitive, reversible or irreversible. It may or may not take place at the same receptor and may involve more than one process simultaneously. Each has its own characteristic features.

Competitive antagonism

Competitive antagonists compete with agonists for a common receptor binding site. There are two reactions taking place simultaneously and these are illustrated as follows.

$$[D] \ + \ [R] \ \underset{K_D}{\rightleftharpoons} \ [DR] \ \rightarrow \ \text{Response}$$
$$\updownarrow$$
$$[I] \ + \ [R] \ \underset{K_I}{\rightleftharpoons} \ [IR]$$

The receptor concentration is common to both equations, i.e. both reactions involve the same receptor pool. The extent of the competitive antagonism is dependent upon both the concentration of agonist and its dissociation constant, and also upon the concentration of antagonist and its dissociation constant. In the presence of the competitive inhibitor, the fractional occupation $[DR]/[DR_{max}]$ will decrease. In this equation $[DR_{max}]$ is equal to the maximum available concentration of drug–receptor complexes and is the same as $[R_{TOT}]$.

Because the nature of the interactions is competitive, however, increasing the concentration of agonist [D] can still produce the same maximum response as in the absence of the antagonist. A large excess of agonist drives the equations in the direction that favours a normal response and the combination of antagonist with receptor is reduced to a level at which there is no discernible effect from the agonists. Plots of the log dose of agonist against response produce a family of sigmoid curves, one for each dose (concentration) of antagonist which are all parallel and displaced further to the right as the concentration of competitive antagonist increases (Fig. 3.8).

Non-competitive antagonism

Non-competitive antagonists attach to the receptor, or to a nearby entity, and prevent the receptor from initiating the response whether or not it is activated. The equations

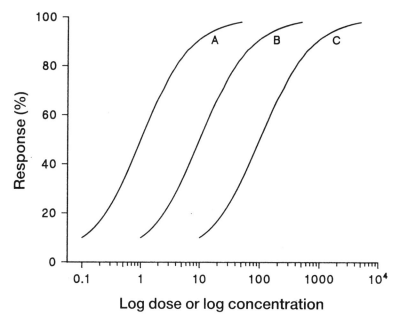

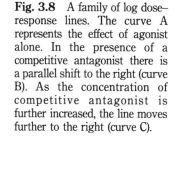

Fig. 3.8 A family of log dose–response lines. The curve A represents the effect of agonist alone. In the presence of a competitive antagonist there is a parallel shift to the right (curve B). As the concentration of competitive antagonist is further increased, the line moves further to the right (curve C).

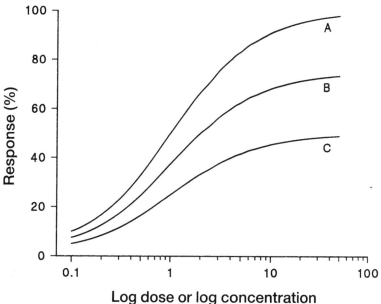

Fig. 3.9 A family of dose-response lines. The curve A represents the effect of agonist alone. In the presence of increasing concentrations of a non-competitive antagonist, the overall shape of the line is retained but the maximum possible response is progressively reduced (curves B and C).

describing the reactions might be similar to those described as follows:

$$[D] + [R] \overset{K_D}{\rightleftharpoons} [DR] \rightarrow \text{Response}$$
$$\uparrow\downarrow$$
$$[I] + [R] \overset{K_I}{\rightleftharpoons} [RI] \quad \text{(inactivated receptor)}$$
$$\uparrow\downarrow$$
$$[D] + [RI] \overset{K_D}{\rightleftharpoons} [DRI] \text{ (no response)}$$

The top two equations are very similar in appearance to the equations for the competitive inhibitor above. A third reaction can, however, take place, when drug [D] can combine with inactivated receptor [RI] with the same K_D as it combines with normal receptors, except that no response is produced. Note that all these equations are reversible and are all interlinked.

The antagonist does not necessarily alter the ability of the drug (agonist) to combine with the receptor which can therefore combine with either the normal receptor or inactivated receptor. The presence of the antagonist has therefore effectively reduced the number of available receptors and because the response depends upon the number of receptors activated, the maximum possible response is reduced. The more non-competitive antagonist that is present, the greater will be the reduction in possible maximum response. The drug should still be combining with the receptor (whether normal or inactivated) with the same affinity (K_D is the same). Other characteristics of the graph, e.g. the ED$_{50}$, will therefore be the same, i.e. the dose to cause 50 per cent of maximum response is the same although it is 50 per cent of a different maximum to the situation pertaining in the absence of antagonist. Increasing the concentration of agonist will have no effect on the maximum response obtainable. Plots of the log dose of agonist against response here produces a family of curves similar to those in Fig. 3.9. It can be seen that graphical analysis can be used to discriminate easily between competitive and non-competitive antagonism.

Irreversible antagonism

It was assumed that the competitive and non-competitive antagonisms described above involved reversible reactions. If the antagonist reacts in an irreversible manner with the receptor, the receptors are likely to be permanently removed from the pool. Under these circumstances it may be necessary to await the manufacture of new receptors. In the meantime, a normal log dose–response relationship will be apparent, but the maximum possible response will be less. A family of curves similar to those in Fig. 3.9 for a non-competitive antagonist will result.

Non-receptor antagonism

It is possible for antagonism to result from a mechanism which is unrelated to block of drug receptors. Examples

might be inhibition of a second messenger system which is activated by the receptor, or alteration in cellular excitability by another means. This type of antagonism is sometimes referred to as physiological antagonism.

It must be remembered that it is possible for some antagonists not to act by one mechanism alone, but by several different mechanisms simultaneously.

Conclusion

The theories and studies of drug–receptor interactions are central to the study of pharmacology. They are central to the action of any drug on any system of the body, and it is essential that every anaesthetist has a good background knowledge of those theories. It is partly for these reasons that details regarding drug–receptor interactions are a very common question during the examinations for the Fellowship Diploma of the Royal College of Anaesthetists.

This chapter has taken a broad view of the subject and only examined some important aspects in much detail. The reader is encouraged to consult any of the common pharmacology texts for further details with respect to this complex, but fascinating, subject.

REFERENCES

1. Langley JN. On the physiology of salivary secretion. Part II. On the mutual antagonism of atropin and pilocarpin, having especial reference to their relations in the sub-maxillary gland of the cat. *Journal of Physiology* 1878; **1**: 339–69.

2. Langley JN. On the contraction of muscle chiefly in relation to the presence of 'receptive' substances. Part 4. The effect of curari and of some other substances on the nicotine response of the sartorius and gastrocnemius muscles of the frog. *Journal of Physiology* 1909; **39**: 235–95.

3. Langley JN. Croonian Lecture 1906 – On nerve endings and on special excitable substances in cells. *Proceedings of the Royal Society of London B* 1906; **78**: 170–94.

4. Ehrlich P. On immunity with special reference to cell life. *Proceedings of the Royal Society of London B* 1900; **66**: 424–48.

5. Clarke AJ. The factors determining tolerance of glucosides of the digitalis series. *Journal of Pharmacology and Experimental Therapeutics* 1913; **4**: 399–424.

6. Clarke AJ. The reaction between acetyl choline and muscle cells. *Journal of Physiology* 1926; **61**: 530–46.

7. Clarke AJ. The reaction between acetyl choline and muscle cells. Part II. *Journal of Physiology* 1927; **64**: 123–43.

8. Boheim G, Hanke W, Barrantes FJ, *et al.* Agonist activated ionic channels in acetylcholine receptor reconstituted into planar lipid bilayers. *Proceedings of the National Academy of Sciences of the USA* 1981; **78**: 3586–90.

9. Conti-Tronconi B, Tzartos S, Lindstrom J. Monoclonal antibodies as probes of acetylcholine receptor structure. 2. Binding to native receptors. *Biochemistry* 1981; **20**: 2181–91.

10. James RW, Kato AC, Rey M-J, Fulpius BW. Monoclonal antibodies directed against the neurotransmitter binding site of nicotinic acetylcholine receptor. *FEBS Letters* 1980; **120**: 145–8.

11. Lee CY. Chemistry and pharmacology of polypeptide toxins in snake venoms. *Annual Review of Pharmacology* 1972; **12**: 265–86.

12. Fertuk HC, Salpeter MM. Localization of acetylcholine receptor by 125 I-labelled alpha-bungarotoxin binding at mouse motor end plate. *Proceedings of the National Academy of Sciences of the USA* 1974; **71**: 1376–80.

13. Paton WDM, Zaimis EJ. The pharmacological actions of polymethylene bis trimethylammonium salts. *Journal of Pharmacology* 1949; **4**: 381.

14. Paton WDM, Zaimis EJ. Curare-like action of polymethylene bis-quaternary ammonium salts. *Nature* 1948; **161**: 718–19.

15. Adams PR, Sakmann B. Decamethonium both blocks and opens endplate channels. *Proceedings of the National Academy of Sciences of the USA* 1978; **75**: 2994–7.

16. Gurney AM., Rang HP. The channel blocking action of methonium compounds on rat submandibular ganglion cells. *British Journal of Pharmacology* 1984; **82**: 623–31.

17. IUPAC. Tentative rules for the nomenclature of organic chemistry, Section E. Fundamental stereochemistry. *Journal of Organic Chemistry* 1970; **35**: 2849–67.

18. Timmermans PBMWM. Steroselectivity in various drug fields. In: Ariens EJ, Sondijn W, Timmermans PBMWM, eds. *Sterochemistry and biological activity of drugs*. Oxford: Blackwells, 1983: 161–80.

19. Gillespie L, Oates JA, Grant JR, Sjoerdsma H. Clinical and chemical studies with α-methyldopa in patients with hypertension. *Circulation* 1962; **25**: 281–91.

20. Brittain RT, Drew GM, Levy GP. The alpha and beta adrenoceptor blocking properties of labetalol and its individual steroisomers. *British Journal of Pharmacology* 1982; **77**: 105–14.

21. Stephenson RP. A modification of receptor theory. *British Journal of Pharmacology* 1956; **11**: 379–93.

22. Paton WDM. A theory of drug action based on the rate of drug–receptor combination. *Proceedings of the Royal Society Series B* 1961; **154**: 21–69.

Chapter 4

The Principles of Pharmacokinetics

C. J. Hull

What has pharmacokinetics to do with anaesthesia?	Factors influencing drug disposition
An apparent volume of distribution	Pharmacokinetic analysis
Drug elimination and clearance	Appendix: exponential functions

What has pharmacokinetics to do with anaesthesia?

Administration of a drug leads to an observable pharmacological effect. Closer observation reveals that the onset of that effect is not instantaneous; indeed it may occur after some considerable delay. Furthermore, the effect may not be constant but may vary greatly in intensity. Indeed, it is commonplace to observe the intensity to reach a peak after some measurable time, and then to decline gradually until eventually it can no longer be detected.

If a further dose is administered the outcome may simply be a repetition of that following the first dose. However, in many cases the intensity profile is quite different. Regular dosing or continuous administration may lead to a continuous pharmacological effect whose intensity may vary or, alternatively, approach a steady state.

Because he is not simply a pharmacological *voyeur* but a practical manipulator who seeks to harness drug action to a specific purpose, the anaesthetist achieves his objectives first by understanding the mechanisms underlying the above phenomena, and then by using models of drug action which can be used to devise optimal dosing strategies. To be effective, such models must encompass both drug disposition within the body and the dynamics of drug action.

Following dosage by a variety of routes (inhalation, transdermal, oral, rectal, etc.) a drug is taken up into the body. The various mechanisms by which this occurs are grouped together under the general term *uptake*. Note that in this context the lumen of the gastrointestinal tract is considered to be part of the *milieu extérieur*, from which uptake takes place just as if it were on the body surface. Of course, many familiar routes of administration (intrathecal, extradural, subcutaneous, intramuscular, intravenous, etc.) bypass this process altogether.

Once in the body, all drugs tend to diffuse passively into adjacent zones and are subject to a variety of transport mechanisms, predominant among which is the circulating blood. Quite often, a drug will be introduced in one part of the body but exert its pharmacological action in another; such transport to the site of action may play a key role in determining its pharmacological profile. The processes by which a drug may be transported about the body are known as *distribution*.

Although some xenobiotics (such as the heavy metals) bind to tissues and become a permanent component of body structure, virtually all the drugs with which anaesthetists are concerned have short effective lifetimes within it. For instance, they may be excreted unchanged in sweat, tears, bile, saliva, milk or urine. Many drugs are poorly excreted but are enzymatically transformed in a variety of body tissues (such as liver, kidney and lung) to water-soluble metabolites (many of which have pharmacological activity) which are readily excreted. Some drugs, such as suxamethonium, atracurium and mivacurium, are metabolized by circulating enzymes in the bloodstream itself. Atracurium is unique in that it has an additional degradation pathway which is independent of enzyme action and requires only a supply of hydroxyl ions. All these mechanisms may be grouped under the term *elimination*.

The above processes are concerned with the movement, or disposition, of a drug about the body. They may be grouped together as *pharmacokinetic processes*. Accordingly, the study of drug disposition is called *pharmacokinetics*.

Drugs act in a great variety of ways. For instance, they may initiate intracellular events by occupying specific recognition sites on cell membranes called receptors. Others work by interaction with enzyme systems, such as by promoting (inducing) or inhibiting enzyme activity, or by behaving as a false substrate which either blocks the enzyme mechanism itself (e.g. neostigmine) or competes with a natural substrate for the enzyme pathway and thereby interferes with the synthesis of a natural hormone (e.g. methyldopa). They may interact directly with cell membrane ion channels or may penetrate cell membranes to interact directly with intracellular elements. All the above mechanisms apply equally to antibiotics and antiviral agents, except that the emphasis is switched from the modification of some natural function to selective lethality.

In each case the drug molecule acts by interaction with some quite specific target structure. The site at which a drug interacts with its target, be it membrane, receptor or enzyme, is known as the *biophase*. This may be a localized tissue zone, as in the case of a focal nerve block, or a very large number of minute but essentially similar sites scattered throughout the body – such as the myriad neuromuscular junctions at which muscle relaxants block neuromuscular transmission. Such a collection of essentially similar sites may be regarded as acting as if it were a single but much larger site whose dynamics may be characterized by relating the time course of drug action with that of biophase drug concentration. As a drug reaches the biophase it interacts with the target structure; this *pharmacodynamic* process may be neither instantaneous nor linear, and adds a further layer of complexity to that already presented by pharmacokinetic processes.

This chapter is concerned with pharmacokinetics. To cover all aspects of the subject within the allotted space would be impossible, so we shall concentrate on some important principles. Readers will find a full and systematic account of the subject in *Pharmacokinetics for anaesthesia.*[1]

Many pharmacokinetic quantities change at rates proportional to their own magnitudes, and therefore must be described in terms of exponential growth and decay. Many readers will already be familiar and comfortable with expressions such as $Y = e^{-kt}$. For those who are not, the essentials are revised in a short appendix at the end of this chapter.

Concept 1. An apparent volume of distribution

When a single dose of a drug is administered intravenously it first undergoes mixing with the plasma and perhaps diffusion into erythrocytes. If no other processes were involved this might be completed within two or three circulation times. However, most drugs diffuse out into the extracellular space and some actually penetrate cells. Therefore the *physical volume of distribution* may range between the volume of the plasma alone (60 ml/kg) to that of total body water (600 ml/kg). However, in most cases drug distribution is infinitely more complex due to ionization and reversible binding to proteins in both plasma and tissues.

As a consequence of pH partition and protein binding, the physical volume of distribution may be a very poor predictor of drug concentrations following administration. In 1934 Dominguez offered an alternative approach.[2] According to his concept a drug may be considered to occupy a certain space which does not necessarily equate with that of the physical volume of distribution. Instead it is defined as that well-stirred, homogeneous space consisting entirely of plasma (or blood) in which the dose of drug *would* have been dispersed in order to explain the observed degree of dilution. That space is known as the *apparent volume of distribution.*

For example, 1 g of a drug is dispersed widely throughout a 70 kg body and at steady state is found in plasma at a concentration of 10 mg/l. The total physical distribution volume could not have exceeded 42 litres, and therefore would have overpredicted the final concentration as 23.8 mg/ml. The apparent volume of distribution is much greater and is calculated to be 1000/10 = 100 litres. Clearly, the apparent volume is not constrained by the physical volume, and extensive tissue binding may result in spectacularly large values (Table 4.1). Similar effects are seen with drugs such as halothane which do not undergo extensive binding but have widely differing solubilities in body tissues.

Table 4.1 Apparent volumes of distribution of some commonly used drugs (typical values taken from the published literature)

Drug	Apparent volume of distribution		
	(litres)		(litres)
Halothane	2530	Thiopentone	120
Imipramine	2100	Lignocaine	110
Chlorpromazine	1400	Midazolam	95
Propofol	1000	Alfentanil	27
Digoxin	750	Vecuronium	12
Fentanyl	330	Warfarin	8
Propranolol	180		

Why should this somewhat illusory concept be preferred to the physical volume of the body? Quite simply, because it provides the essential link between dose, volume and concentration. If a certain plasma concentration C is required, the required dose Xd can be calculated easily as $X\text{d} = C \times V$. Unfortunately, that attractively simple relation is obscured by the fact that drug elimination commences at the instant of administration. Thus by the time a dose has had time to distribute around the body tissues a significant portion is likely to have been eliminated. So, while apparent volume of distribution is a valid, indeed essential concept, it must be considered in the context of simultaneous elimination.

Concept 2. Drug elimination and clearance

All practical drugs must be eliminated from the body in one way or another, otherwise repeated dosing would lead to accumulation and an indefinite duration of action. Drug elimination may take many forms. Drugs may be excreted unchanged by organs such as the kidney, liver or lung, or simply broken down in the blood. The major route is often enzyme-mediated metabolism which can occur at a variety of sites – of which the liver is the most important.

Spontaneous breakdown

Perhaps the most familiar example is the Hoffmann degradation undergone by atracurium. The process is entirely random, so that every molecule of drug has an equal probability of breaking down at any instant. Thus the more molecules of drug there are, the more breakdown events must occur in unit time. It follows that the rate of elimination at any time is proportional to the concentration of unchanged drug. If the body behaved as a single, well-stirred compartment such *first-order kinetics* would cause the plasma concentration C to decline exponentially:

$$\frac{C}{C^0} = e^{-kt}$$

where C^0 is the initial concentration, e the base of the natural logarithm, k the rate constant and t the elapsed time (see Appendix).

Excretion of unchanged drug

Many drugs are, at least in part, excreted unchanged by the liver or kidney. For instance, glomerular filtrate contains the same non-protein-bound drug as plasma, and this is excreted unless some or all is subject to tubular reabsorption. Since the filtrate is a more or less constant fraction of renal blood flow, it follows that the rate of excretion is proportional to the unbound plasma concentration. If reabsorption does occur, the rate of diffusion from renal tubule to capillary is likely to follow similar rules, i.e. proportional to the unbound drug concentration gradient.

In the liver, similar considerations apply. As drug passes through the hepatic sinusoid a small fraction diffuses down an unbound concentration gradient into the hepatocyte whence it is excreted into the bile. Since such processes are essentially concentration-driven, first-order kinetics may be expected to apply.

Enzyme-mediated decomposition

Here, drug molecules must make physical contact with enzyme sites which may be fixed within tissues or

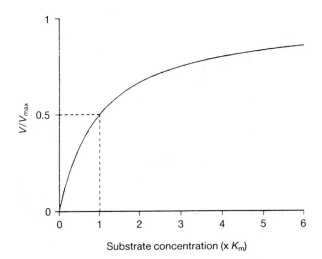

Fig. 4.1 Graphical representation of the Michaelis–Menten equation.

circulating in the blood. Take plasma cholinesterase as an example, which promotes the hydrolysis of a variety of drugs. Each individual enzyme site consumes drug molecules at a rate determined by the substrate concentration and the characteristics of the drug–enzyme interaction itself. The more efficient is the enzyme reaction, the lower need be the substrate concentration to drive it at a given rate. The kinetics of this reaction are expressed by the *Michaelis–Menten* equation:

$$\frac{V}{V_{max}} = \frac{C}{C + K_m}$$

where V is the velocity of reaction (mol of drug consumed per unit time), V_{max} is the maximum possible velocity at infinitely high substrate concentration, C is the substrate concentration (mol/l) and K_m is the Michaelis constant. Inspection of the equation reveals that K_m is numerically equal to the substrate concentration at which the velocity of reaction is $V_{max}/2$. Figure 4.1 shows the equation solved for a wide range of substrate concentrations, normalized as multiples of K_m.

At concentrations much greater than K_m the equation simplifies to:

$$\frac{V}{V_{max}} = \frac{C}{C} = 1$$

Clearly, metabolism proceeds at a fixed rate regardless of any further increase in substrate concentration and a constant mass of drug is eliminated per unit time. This is called *zero-order* metabolism.

At very low concentrations the substrate concentration C

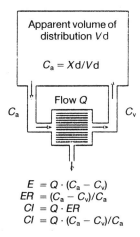

$$E = Q \cdot (C_a - C_v)$$
$$ER = (C_a - C_v)/C_a$$
$$Cl = Q \cdot ER$$
$$Cl = Q \cdot (C_a - C_v)/C_a$$

Fig. 4.2 A simple model for organ-based drug elimination (see text).

is very small compared with K_m so that the equation simplifies to:

$$\frac{V}{V_{max}} = \frac{C}{K_m}$$

This is a linear equation whereby the rate of drug elimination is at all times proportional to the drug concentration at the enzyme site(s). Following *first-order kinetics*, the plasma concentration might be expected to decline exponentially, according to the equation:

$$\frac{C}{C^0} = e^{-kt}$$

Perfusion and compartmental models of drug elimination

The above considerations apply only when drug and enzyme are well mixed in a medium such as blood. In an organ of elimination, such as the liver, other factors become important. In particular, it can no longer be assumed that drug and enzyme sites are well mixed. Consider Fig. 4.2.

Drug in the body is assumed to be dispersed through a single, well-stirred distributional space (corresponding to the apparent volume of distribution Vd). 'Plasma' containing drug at concentration C_a is passed at flow rate Q through an organ of elimination such as the liver. The effluent plasma concentration is C_v. The amount of drug removed from each ml of plasma must be $C_a - C_v$, and the overall rate of drug elimination is $Q(C_a - C_v)$.

Now the extraction ratio (*ER*) can be calculated, as the fraction of drug extracted from each ml of plasma as it passes through the organ:

$$ER = \frac{C_a - C_v}{C_a}$$

While extraction ratio is a most informative measure, it ignores perfusion (*Q*) and therefore is not an index of eliminational performance by the organ as a whole. This is

provided by *clearance* (*Cl*), the product of extraction ratio (*ER*) and organ perfusion (*Q*):

$$Cl = ER \times Q$$

If $ER = 1$, clearance must equal perfusion Q, and therefore represents that volume of plasma entirely cleared of drug per unit time. Since the dimension of Q is volume/time and *ER* is a dimensionless fraction, it follows that the dimension of clearance also must be volume/time. That is, ml/min, l/h etc. If $ER = 0.5$, then $Q/2$ litres per minute *appear* to be cleared of drug per minute, if $ER = 0.1$ $Q/10$ etc.

Now consider the above equation in an expanded form, expressing extraction ratio as above in terms of afferent and efferent concentrations:

$$Cl = \frac{C_a - C_v}{C_a} \times Q$$

But since we have already determined that $Q(C_a - C_v)$ expresses the rate of drug elimination, the above equation can be restated as:

$$\text{Clearance} = \frac{\text{Rate of elimination}}{\text{Concentration}}$$

This is a much more satisfactory expression of efficiency, since it simply expresses the rate of elimination in terms of prevailing drug concentration; the greater the rate of elimination for a given concentration, the more efficient the process. The dimension remains unchanged: volume/time.

But how does this relate to a single well-stirred compartment with first-order elimination? If a small mass of drug is eliminated, the mass of drug remaining decrements accordingly. Therefore, 'rate of elimination' can be restated as 'rate of decrease in mass of drug remaining'. Since mass = concentration × volume, this can be expanded to 'rate of decline in concentration × volume'. Thus the clearance statement becomes:

$$\text{Clearance} = \frac{\text{Rate of concentration decline} \times V}{\text{Concentration}}$$

But since in an earlier section we saw that if a variable Y declines exponentially, the following is true at any time:

$$k = \frac{\text{gradient of } Y}{Y}$$

it follows that:

$$\text{Clearance} = k \times V$$

In this section we have seen how drug elimination can be considered using two quite different models, one based on simple exponential decay and the other on a perfused organ of elimination. So long as our assumptions regarding a well-stirred single compartment remain intact, these models are simply different ways of expressing the same thing. While the perfusion model certainly has the advantage of being more realistic, it also has more parameters; unless these can

Table 4.2 Values of pK$_a$ and percentage unionized at pH 7.4 of some commonly used drugs

Bases	pK$_a$	Per cent unionized	Acids	pK$_a$	Per cent unionized
Pancuronium		0	Frusemide	3.9	0.03
Atropine	9.7	0.5	Warfarin	5	0.4
Propranolol	9.4	1	Chlorothiazide	6.8	20.1
Fentanyl	8.4	9.1	Phenobarbitone	7.4	50
Bupivacaine	8.1	16.6	Thiopentone	7.6	61.3
Morphine	7.9	24	Methohexitone	8	80
Diamorphine	7.6	38.6	Hexobarbitone	8.2	86.3
Ketamine	7.5	44.3	Phenytoin	8.3	88.8
Alfentanil	6.5	89	Isoprenaline	10.1	99.8
Diazepam	3.3	99.99	Propofol	11	99.97

be quantified it is simply a toy having little or no predictive power in the real world.[3]

Factors influencing drug disposition

Many drugs travel to their sites of action and to the tissues and organs from which they are eliminated by means of the bloodstream. Because the blood is itself a complex physiological system it is not surprising to find the disposition of drugs within it equally complex. In the blood a drug is likely to be partitioned between plasma and erythrocytes. In both cases it is likely to be bound to proteins which have a profound influence on its behaviour, and to be further partitioned into ionized and unionized fractions. Finally, it must not be forgotten that some drugs (e.g. suxamethonium, mivacurium, atracurium) are broken down within the blood itself and therefore are not wholly reliant on organ-based elimination mechanisms.

Uptake of drugs into erythrocytes

Erythrocytes are enclosed by a passive lipoprotein membrane having no pores, fenestrae or pinocytotic vesicles by which large or ionized molecules might gain access to the interior. Thus to a great extent the entry of a drug is dependent upon its size and lipid solubility. Highly polar drugs (such as muscle relaxants) or the ionized forms of weak acids and bases cannot enter erythrocytes. The unionized forms of many drugs (such as thiopentone, propofol or fentanyl) are, however, lipid soluble, and therefore enter erythrocytes without difficulty. Once in the erythrocyte a drug may or may not bind to haemoglobin;

this determines the relative concentrations in erythrocyte and plasma.

Ionization of drugs

Some drugs (such as those with a quaternary nitrogen) are ionized under all physiological conditions, while some small molecules (such as ethyl alcohol) remain unionized. However, many drugs are either weak acids or weak bases. At physiological pH all such drugs exist in both ionized and unionized forms, with the relative proportions depending on the pK$_a$ (the pH at which there are equal concentrations of ionized and unionized molecules). See Table 4.2.

The equilibrium is described by the Henderson–Hasselbalch equation. Thus, for a weak acid:

$$pH = pK_a \cdot \log \frac{[\text{ionized}]}{[\text{unionized}]}$$

For a weak base:

$$pH = pK_a \cdot \log \frac{[\text{unionized}]}{[\text{ionized}]}$$

The ratio (r) of ionized to unionized molecules of a weak acid can be determined by:

$$r = 10^{(pH-pK_a)}$$

while that for a weak base is found by:

$$r = 10^{(pK_a-pH)}$$

Then the ionized fraction f_i can be determined using the simple equation:

$$f_i = \frac{r}{r+1}$$

To take a familiar example, thiopentone (pK$_a$ 7.6) in plasma (pH 7.4) yields $r = 0.63$. Then $f_i = 0.39$, or 39 per cent.

This is important because the unionized fraction (i.e. free thiobarbituric acid) can diffuse through cell membranes very much faster than the ionized fraction. Indeed, in the case of specialized capillaries such as those in the brain and placenta where adjacent cells are fused together to form 'tight junctions', and where there are few pores or fenestrae, only the unionized fraction can cross. Drug concentrations on either side of such a membrane approach equilibrium with respect to the unionized fraction alone. Thus, if the tissue pH differs from plasma pH, drug concentrations may differ markedly. For instance, lipid-soluble opioids such as fentanyl diffuse into the lumen of the stomach where, if the pH is low, they become almost wholly ionized. Since the unionized fraction is thereby reduced to a negligible concentration, diffusion of drug continues such that a significant proportion of the administered dose may follow that route. Similar considerations may apply in the renal tubule, where pH influences the extent to which some drugs (such as pethidine) are reabsorbed.

Protein binding of drugs

Many drugs undergo reversible binding to plasma proteins (Table 4.3). A single protein may carry acidic or basic binding sites, sometimes both, and therefore may have differing affinities for individual drugs. Thus acidic drugs, such as penicillins, barbiturates, sulphonamides and salicylates, bind predominantly to albumin. Basic drugs, such as opioids, benzodiazepines and local anaesthetic agents, bind mainly to globulins, lipoproteins and glycoproteins.

Many drugs bind to acute-phase proteins such as α_1 acid glycoprotein, which, because of great variations in concentration, may lead to large differences in protein

Table 4.3 Percentage plasma protein binding of some common drugs

	Per cent bound
Warfarin	99
Diazepam	98
Propofol	98
Bupivacaine	95
Alfentanil	91
Propranolol	90
Fentanyl	82
Thiopentone	80
Lignocaine	65
Atracurium	51
Tubocurarine	45
Morphine	40
Vecuronium	30

Table 4.4 Some drugs which bind to α_1 acid glycoprotein

Alfentanil
Bupivacaine
Chlorpromazine
Imipramine
Lignocaine
Methadone
Propranolol
Quinidine
Verapamil
Warfarin

binding between individual patients (Table 4.4). Readers are referred to a comprehensive review.[4]

The extent to which a drug binds to protein depends on the drug concentration and on the drug–protein affinity:

If one molecule of drug D binds to one site P to yield a reversible drug–protein complex DP:

$$[D] + [P] \underset{k_2}{\overset{k_1}{\rightleftharpoons}} [DP]$$

Affinity between drug and binding site is given by the ratio of the two velocity constants:

$$K_A = \frac{k_1}{k_2}$$

The dissociation constant K_D is the reciprocal:

$$K_D = \frac{1}{K_A} = \frac{k_2}{k_1}$$

If we define occupancy Y as the fraction of all available binding sites occupied by drug molecules:

$$Y = \frac{[DP]}{[DP] + [P]}$$

it can be shown that:

$$Y = \frac{[D]}{[D] + K_D}$$

This hyperbolic function determines that over a range of concentrations well below K_D occupancy Y increases in proportion to [D]. This means that the bound fraction remains approximately constant so long as drug concentration remains $\ll K_D$. Thus for a very potent drug such as fentanyl, where only a tiny fraction of available sites is occupied, protein binding remains approximately 81 per

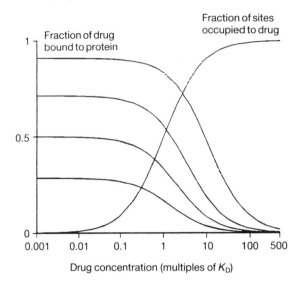

Fig. 4.3 Occupancy of protein binding sites increases with drug concentration. Also drawn is a family of binding plots for four drugs of differing binding affinities.

cent over the whole range of clinically relevant concentrations.

In contrast, very weak drugs with intense protein binding, such as warfarin, may occupy a significant proportion of sites as therapeutic concentrations. As the concentration of such a drug increases the bound fraction diminishes and the unbound, pharmacologically active fraction increases. Thus protein binding approaches *saturation*, at which no more is bound regardless of drug concentration. There is a second, even more important consequence of saturable binding, in that the bound fraction of one drug diminishes the number of binding sites available to another drug with similar binding characteristics (see following section on pharmacokinetic drug interactions).

Since by their very nature plasma proteins are confined to the circulating volume, protein bound drug cannot diffuse out into the extracellular space. Only the unbound fraction is 'diffusible', so the rate of diffusion across a capillary membrane depends upon the concentration gradient of unbound drug (and subject also to pH partition as described above). Once a drug has escaped from the capillary into a tissue, there may be extensive reversible binding with tissue proteins. Clearly the more extensively a drug binds to proteins within a tissue, the more drug will have to enter that tissue before the unbound drug concentration in the tissue approaches that in the plasma of a supplying capillary. Thus the mass of drug in a tissue at equilibrium may depend not only on the physical mass of the tissue but also on its binding affinity for the drug.

The influence of protein binding on drug elimination

As we have seen, glomerular filtrate contains drug at the unbound concentration in plasma. Since, therefore, the unbound concentration in plasma remains unchanged *en passage* through the glomerular capillary, the equilibrium between bound and unbound drug remains undisturbed. It follows that the rate of filtration depends on the degree of binding, since a high free fraction will ensure maximal filtration and *vice versa*.

In the liver it is more complicated. Diffusion of drug from sinusoid into hepatocyte and thus to an enzyme site follows an unbound concentration gradient. Clearly, the concentration achieved at the enzyme site depends on the extent of protein binding in plasma. Now there are several possibilities:

1. If there are a great number of very efficient (i.e. low K_m) enzyme sites, drug is consumed almost as rapidly as it arrives. This disturbs the free:bound equilibrium in plasma to sustain the free concentration. As a single drop of plasma traverses the sinusoid most if not all of the drug is extracted. Protein binding obviously has no influence, and the limiting factor is perfusion.

2. If the enzyme has low efficiency (high K_m) and therefore consumes drug at a low rate, the effect of drug extraction on plasma concentration is minimal. Therefore perfusion is no longer a limiting factor; indeed changes in perfusion may have almost no influence on clearance. The clearance of such a drug is described as *capacity-limited*. Because there is so little extraction the free:bound equilibrium remains undisturbed and bound drug stays bound. Since the enzyme is operating low on its saturation curve (Fig. 4.1), changes in the unbound plasma concentration have a direct influence on the rate of drug elimination. If the drug is poorly protein bound, changes in binding cause proportionate changes in clearance. Thus in the case of a drug whose binding decreases from 35 to 30 per cent, the unbound fraction increases by some 8 per cent as does clearance. However, in the case of drug (such as diazepam) whose protein binding also may decrease by 5 per cent but from 94 to 89 per cent, the unbound fraction (and clearance) increases by 83 per cent. Drugs with capacity-limited clearance and very high protein binding are described as having *restrictive clearance*.

3. If the enzyme has high efficiency (i.e. low K_m) but the reaction consumes drug at a low rate because there are relatively few enzyme sites, extraction is

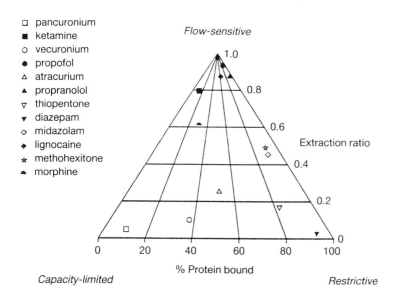

□ pancuronium
■ ketamine
○ vecuronium
● propofol
△ atracurium
▲ propranolol
▽ thiopentone
▼ diazepam
◇ midazolam
◆ lignocaine
★ methohexitone
▲ morphine

Flow-sensitive

Extraction ratio

% Protein bound

Capacity-limited

Restrictive

Fig. 4.4 Perfusion-limited, capacity-limited and restrictive clearance by the liver, illustrated by a range of drugs used in anaesthesia. Based on a diagram by Blaschke[5] and drawn with permission.

low and clearance capacity-limited. Because the enzyme must operate high up on its saturation curve (Fig. 4.1), however, elimination is likely to be non-linear or even zero-order. Because the rate of reaction is relatively insensitive to changes in substrate concentration, the influence of altered protein binding is relatively small.

Understanding the relation between perfusion, extraction and protein binding is facilitated by the diagrammatic form devised by Blaschke,[5] on which the behaviour of some familiar drugs may be plotted (Fig. 4.4).

The importance of first-pass effects

Because of the complexity of the vascular circulation, many drugs undergo quite different patterns of disposition depending on the route of administration. For instance, when a drug is administered by rapid intravenous injection, the whole dose arrives at the pulmonary artery in very high concentration and then must pass through the pulmonary capillary bed before reaching the systemic circulation. While highly ionized drugs (such as muscle relaxants) remain in the bloodstream, unionized molecules are likely to diffuse into the lung tissue. In particular, basic drugs bind reversibly to pulmonary tissue proteins, leading to high pulmonary uptake during the first circulatory pass. Thus more than 80 per cent of a rapidly injected dose of fentanyl may undergo first-pass pulmonary uptake.[6] Although thiopentone is highly lipid soluble it does not bind so avidly to tissue proteins and so only 25 per cent of a dose is taken up in this way.[7] The lung, therefore, has a major influence on the initial volume of distribution. It must be emphasized, however, that much of the drug returns to the circulation very rapidly as plasma concentrations decline behind the initial front; this has the effect of flattening and

spreading the drug pulse. The overall effect is both to delay and to attenuate the peak initial concentration reaching the brain. Other basic drugs such as lignocaine, bupivacaine and propranolol also bind avidly to lung proteins, and it has been noted that pulmonary uptake of fentanyl is reduced from 80 to 20 per cent by pretreatment with propranolol so that drug reaches the systemic circulation more quickly and in much greater concentration.[8]

Many drugs, if administered orally, are absorbed from the gastrointestinal tract into the portal circulation. Those having high hepatic extraction may undergo extensive presystemic metabolism. In contrast to presystemic pulmonary uptake, this phenomenon has the effect of reducing the effective dose. *Bioavailability* is defined as that fraction of an oral dose reaching the systemic circulation following some stated route of administration. It may be estimated by determining serial plasma concentrations following oral and intravenous administration of identical doses in the same subject, and then expressed as the ratio of area under the curves. Thus, if AUC_{iv} and AUC_o are determined following intravenous and oral dosing respectively,

$$\text{Bioavailability} = \frac{AUC_o}{AUC_{iv}}$$

Cumulation

If a drug is administered repeatedly, or by continuous infusion, the mean rate of administration is likely to be greater than the initial rate of elimination. Consequently the mass of drug in the body must increase until, at equilibrium, the plasma concentration reaches a level at which the rate of elimination equals the mean rate of administration. That increase is called *cumulation*. It does not occur if drug clearance is so great that whatever the administration rate there is no recirculation (i.e. clearance $\geq$ cardiac output!). Cumulation, therefore, is a feature of repeated administration or infusion of almost every drug in the anaesthetic armamentarium. It is *particularly* likely to occur when the elimination half-time greatly exceeds the dosing interval (incremental thiopentone is an excellent example).

As defined, cumulation is an inherent feature of almost all drugs. There is, however, a further dimension to this phenomenon. If there is a maximum rate at which a drug can be eliminated from the body, as in the case of thiopentone, whose capacity-limited metabolism saturates at high plasma concentrations,[9] consider what happens if drug is administered at some greater rate. Since dosing rate exceeds the maximum rate of metabolism, the mass of drug in the body continues to rise indefinitely. This is *runaway cumulation*.

The influence of age, sex, body weight and heredity

The disposition of any drug will vary greatly in a normal healthy population. This variation is due in part to differences in body size. Given no other factors, a large body must have a greater volume of distribution than a small one. However, there are strong correlations between body weight and both age and sex, so that identifying the actual sources of variation can be a daunting task. In a large study of adult patients taking a small standard dose of propranolol, sex was found to be the most important pharmacokinetic determinant, with age, smoking habit and racial type also significant.[10] After taking those factors into account, body weight had no significant influence. There are numerous other examples of sex differences, presumably caused by relative differences in the masses of tissues such as muscle and fat. Women also have less efficient clearance of drugs such as temazepam and lignocaine.[11,12] The reasons are not clear, but may be due to sex differences in hepatic enzymes such as the cytochrome P_{450} mixed oxidase system.

In children, body size clearly is a more important factor than in adults. For instance, the distribution volumes of tubocurarine, thiopentone and alfentanil vary directly with body weight, even in quite young infants.[13–15] However, body weight has less influence on clearance, which therefore should be scaled down with some caution.

Drug disposition in neonates often is quite different from that in older infants. This is not surprising, since they have a greater water content and many organs (brain, liver, etc.) are relatively larger than in adults. Plasma protein binding is less intense than in adults, due to lower protein

Table 4.5 Terminal half-times of some common drugs in neonates and adults

	Terminal half-time (hours)	
	Neonates	*Adults*
Bupivacaine	25	1.3
Lignocaine	3	1–2
Diazepam	25–100	15–25
Pethidine	22	3–4
Morphine	7	2–3
Sufentanil	5–19	3–4
Fentanyl	1–7	3–6
Tubocurarine	5.6	2.7

concentrations and to differences in protein composition and affinity for drugs.[16] High plasma concentrations of bilirubin and unconjugated fatty acids may compete for binding sites with acidic drugs such as thiopentone. Immature enzyme systems may lead to reduced clearance, but because fetal cytochrome P_{450} is highly susceptible to induction *in utero*, maternal exposure to inducing agents may result in neonatal mixed oxidase function much in excess of normal.[17] Indeed, there may be dramatic changes in drug metabolizing capacity during the first week of life. All these influences have complex and often unpredictable effects on drug disposition. Thus we find that thiopentone distrubutes to a smaller volume of distribution than expected on the basis of body weight, while that of tubocurarine is greater than expected.[13] Drug elimination may be quite different from the adult pattern; see Table 4.5.

As patients become old their general level of activity diminishes, with a consequent reduction in both muscle and bone mass. Fat, therefore, occupies a greater relative part of the body mass. Consequently water-soluble drugs such as muscle relaxants tend to have smaller apparent volumes of distribution in elderly patients. In contrast, lipid-soluble drugs such as lignocaine, diazepam and pethidine have been shown to have *increased* distribution volumes in the elderly. Since hepatic blood flow is progressively decreased in elderly subjects, highly extracted drugs such as lignocaine and propranolol may exhibit reduced clearance.[18] This effect is compounded by the inevitable decrease in presystemic metabolism, leading to greater bioavailability and even greater plasma concentrations. Less efficiently cleared drugs such as diazepam are influenced by the general reduction in hepatic biomass in elderly subjects, but the effect may be offset by reduced plasma protein binding leading to greater unbound plasma concentrations.[19]

Since ageing is associated with a progressive reduction in renal function (30–50 per cent reduction in creatinine clearance by age 65), it is unsurprising that all drugs dependent on renal excretion are less efficiently cleared in elderly patients.

Elderly patients are also subject to a variety of age-related secondary factors. Thus, they are more prone to debilitating disease, and are exposed to a wider range of drugs, but they smoke and drink less than young adults.

The pharmacokinetic profiles of some drugs are heavily influenced by genetic factors. Inherited variation in the characteristics of an enzyme system is called *genetic polymorphism*. The example best known to anaesthetists must be the genetic variation in cholinesterase characteristics, which has major effects on the rate of hydrolysis of drugs such as suxamethonium and mivacurium. Drugs which are acetylated in the liver (such as hydralazine, procainamide, isoniazid, nitrazepam and phenelzine) show distinct differences in clearance, depending on the acetylator type of the individual concerned. The proportions of 'fast' and 'slow' acetylators vary widely between ethnic groups.[20] The cytochrome P_{450} enzyme system also is subject to genetic polymorphism, influencing the metabolism of drugs such as phenacetin, metoprolol and phenytoin.

Altered disposition in pregnancy

Even in early pregnancy, a number of physiological functions undergo changes such that drug disposition is altered significantly. For instance, gastric uptake of paracetamol is delayed in women who are as little as 12 weeks pregnant.[21] Renal blood flow and glomerular filtration rate are increased from early pregnancy, so that the clearances of drugs such as muscle relaxants and antibiotics are increased. Plasma volume increases by some 50 per cent, total body water by up to 8 litres and body fat by 3–4 kg in the third trimester, leading to large increases in the apparent volumes of distribution of many drugs. These effects are accentuated by reductions in serum albumin and (usually) α_1 acid glycoprotein. These changes are particularly significant in the case of highly bound, poorly extracted drugs such as diazepam and midazolam.[22]

The influence of disease

Many disease conditions are associated with abnormal pharmacokinetics. Changes in drug distribution are often due to loss of body muscle and fat, and to altered protein binding. Both renal and hepatic disease are associated with disturbed protein binding, and a variety of inflammatory and malignant conditions are associated with increased plasma concentrations of α_1 acid glycoprotein.

Hepatic drug elimination may be compromised in many ways; these include loss of hepatic biomass, reduced hepatic blood flow, altered protein binding, and loss of specific enzyme systems both in the liver and in the blood. The effects are highly variable, depending on the nature and severity of the disease process and the drug concerned.

Renal elimination may be compromised by reduced glomerular filtration, complicated by abnormal protein binding. The kidney has a particularly important role in eliminating the water-soluble metabolites of non-polar drugs which do not themselves appear in the urine. Thus renal failure has little effect on the excretion of unchanged morphine, but severely restricts the elimination of morphine 6-glucuronide.[23] It should be noted that some drugs (such as morphine) are *metabolized* in the kidney, and that this may continue despite renal failure so long as organ perfusion is maintained.[24]

Cardiac failure and other low-output states are associated with greatly delayed uptake of drugs administered by any route other than intravenous injection. Even in the latter case, slow venous transit may delay the onset of such drugs as thiopentone and propofol. Poor tissue perfusion may severely restrict the distribution of drugs, so that those organs in which perfusion is maintained (such as the brain) may be exposed to very high drug concentrations. The altered pharmacology of thiopentone in a patient with haemorrhagic shock is an apt example. Both renal and hepatic blood flow tend to decline in proportion to changes in cardiac output; thus the half-time of lignocaine, which is susceptible both to changes in hepatic perfusion and to restricted distribution, may be two to three times greater in cardiac failure patients.

The influence of social poisons; alcohol and tobacco

Several studies have demonstrated pharmacokinetic differences between those who smoke and those who do not. For instance, Walle *et al.*[10] showed clearly that when taking a small standard dose of propranolol, smokers had lower plasma concentrations than non-smokers due to cytochrome P_{450} induction. Regular exposure to marijuana has a similar but even more powerful effect.

Alcohol has a similar effect on the cytochrome P_{450} system, so that regular drinkers may have greater than normal clearance of many drugs. It should be noted, however, that alcohol competes with those other drugs for enzyme sites, so the inebriated drinker may have reduced rather than increased metabolic clearance of other drugs. This principally affects poorly extracted drugs such as tolbutamide, phenytoin and warfarin. As alcoholism progresses towards liver failure the pattern changes to that of diminished hepatic function, involving loss of biomass, reduced perfusion, altered protein binding and loss of both muscle and fat. Thus in patients with alcoholic cirrhosis the half-time of diazepam was found to be doubled.[25]

The influence of other drugs: pharmacokinetic drug interactions

Many drugs influence the uptake of others. For instance, antacids inhibit the absorption of tetracyclines. Drugs such as metoclopramide and cisapride influence uptake of orally administered opioids by modifying the rate of gastric emptying. The uptake of local anaesthetic agents from tissues into which they are injected is delayed by the co-administration of vasoconstrictors such as adrenaline, noradrenaline and octapressin.

Any drug which binds to plasma proteins is likely to inhibit the binding of any other drug with affinity for the same site(s). Thus warfarin competes with salicylates, tolbutamide and sulphonamides for albumin sites.

Drugs which reduce cardiac output are likely to influence the disposition of other drugs; even their own. Halothane, propranolol and propofol are good examples. The effects range from altered distribution to reduced clearance due to the effect on hepatic blood flow.

Many drugs compete for elimination pathways; thus the hydrolysis of suxamethonium is influenced by the co-administration of procaine, cocaine, chloroprocaine, amethocaine, propanidid, diamorphine or mivacurium.

Some drugs influence the metabolism of others by inducing or inhibiting the cytochrome P_{450} mixed oxidase enzyme system. Inducing agents include many common drugs (Table 4.6) as well as chemicals such as benzene, insecticides such as aldrin, dieldrin, lindane, Perthane and DDT, and fungicides such as hexachlorobenzene. Cimetidine is a good example of a P_{450} inhibitor.

Many drugs inhibit other enzymes. Thus monoamine oxidase inhibitors interfere with the metabolism of amphetamines, ephedrine, methoxamine, cocaine, fenfluramine, atropine, phenytoin, L-dopa, phenothiazines, opioids,

Table 4.6 Some drugs which induce hepatic microsomal enzymes

Ethanol	Chlorpromazine
Diphenhydramine	Imipramine
Spironolactone	Nitrazepam
Phenobarbitone	Hydrocortisone
Pethidine	Nitrous oxide
Carbamazepine	Halothane
Tolbutamide	Methoxyflurane

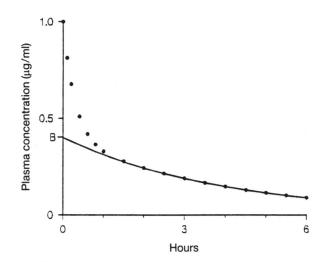

Fig. 4.5 Plasma concentrations following a single dose often follow this form. However, $Y = B \cdot e^{-kt}$ fitted to the later points clearly does not describe the early data.

thiazide diuretics, benzodiazepines and tricyclic antidepressants.

Anaesthetic pharmacology is in fact a maze of complex interactions, many of which involve multiple components. Readers are referred to a comprehensive review.[26]

Pharmacokinetic analysis

Thus far we have maintained the assumption that the space into which a drug distributes may be regarded as equivalent to a single, well-stirred compartment wherein every drug molecule enjoys an equal probability of elimination at any time. Were this to be valid in every case, pharmacokinetic analysis would be simple. The investigator need only fit a simple exponential equation to a set of plasma concentration data in order to determine both distribution volume and rate constant. From those parameters it is both possible and appropriate to express drug disposition in terms of a model comprising but a single compartment (see below).

While this is a useful concept, especially while grappling with pharmacokinetic essentials, it is in the majority of cases a gross oversimplification. Following a single dose, drug concentrations are likely to decline initially much more rapidly than would be predicted by a simple exponential curve fitted to the terminal phase data (Fig. 4.5).

This happens because drug distribution to the various tissues and organs is *not* instantaneous, and the distributional space does not behave as a single, well-stirred compartment. It is evident that in this case the drug has

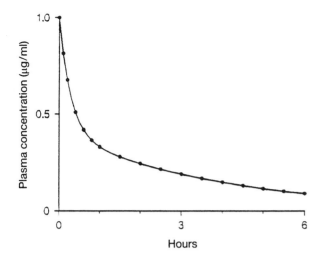

Fig. 4.7 Symbolic representation of the one-compartment open model.

The one-compartment open model

Models of this type usually are represented as a simple box with a descending arrow (Fig. 4.7), the latter indicating the elimination pathway and labelled as a rate constant.

Given the dose $X\mathrm{d}$ and a fitted curve of the type $C\mathrm{p} = B{\cdot}\mathrm{e}^{-\beta t}$, model parameters are determined easily:

$$V\mathrm{d} = \frac{X\mathrm{d}}{B}$$

$$k = \beta$$

The behaviour of the model following a single dose is governed by a simple differential equation:

$$\mathrm{d}X/\mathrm{d}t = -k{\cdot}X$$

where $\mathrm{d}X/\mathrm{d}t$ denotes the rate of change of X, the mass of drug in the model at any time. As expected, the rate of change is determined by X itself and the elimination rate constant k. Integration leads to the familiar equation:

$$\frac{X}{X^0} = \mathrm{e}^{-kt}$$

If the input is not a single dose but a constant infusion, the differential equation becomes:

$$\mathrm{d}x/\mathrm{d}t = k'_{01} - kX$$

Now the behaviour of the model can be predicted for any combination of bolus and infusional dosing:

$$X = \left(X^{\phi} + X\mathrm{d} - \frac{k'_{01}}{k} \right)\mathrm{e}^{-kt} + \frac{k'_{01}}{k}$$

where X^{ϕ} is the mass of drug in the model before dosing (initially zero), $X\mathrm{d}$ is a bolus dose and k'_{01} is the rate at which drug is added as a constant infusion. Concentration can be determined as X/V. In order to model a complex dosing regimen, the procedure is simply divided into a number of segments, in each of which conditions remain unchanged. Thus segment 1 might specify a bolus dose and a constant infusion. Segment 2, starting 1 hour later, might require that the infusion rate be changed to some new value for 2 hours. Segment 1 is calculated for a series of times from 0 to 1 hour, depending on the detail required. Segment 2 takes X^{ϕ} as the last value of X in segment 1, the infusion rate is changed to the new value, and solved for 0–2 hours. Thus each segment is solved quite independently, taking its initial condition as the closing value in the previous

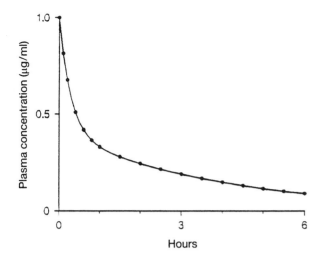

Fig. 4.6 A function comprising the sum of two exponential terms corresponds well with the data points.

initially distributed into a small apparent volume, whence it spread relatively slowly to occupy a much larger volume. In reality there are an almost infinite number of tissue zones, each with a unique profile of volume, perfusion and partition coefficient with respect to plasma. This has encouraged some workers to develop physiologically plausible models of terrifying complexity, but as stated above, realism has little virtue when faced with a long list of parameters demanding precise values!

Perhaps surprisingly, many drugs do behave as if they are in quite simple physical systems, due to the influences of complex factors averaging out into relatively few broad categories. Analysis of plasma concentrations (Cp) by non-linear regression allows the investigator to fit functions comprising one exponential term or the sum of several. Thus the data of Fig. 4.5 were fitted to the equation:

$$C\mathrm{p} = A{\cdot}\mathrm{e}^{-\alpha t} + B{\cdot}\mathrm{e}^{-\beta t}$$

When plotted with the data, this obviously provides a much more satisfactory fit (Fig. 4.6).

In practice the investigator would also fit an equation comprising three terms:

$$C\mathrm{p} = A{\cdot}\mathrm{e}^{-\alpha t} + B{\cdot}\mathrm{e}^{-\beta t} + C{\cdot}\mathrm{e}^{-\gamma t}$$

Then, 'goodness of fit' criteria (such as the maximum likelihood function) for the three solutions can be compared using straightforward statistical tests to determine the most feasible. Depending on the number of terms in the favoured equation, the parameters of a plausible model with one, two or three compartments can be determined. Since compartmental models are widely used as a means of describing the pharmacokinetic characteristics of drugs and also as predictors of response to given dosing regimens, we shall consider the derivation and behaviour of one- and two-compartment models in some detail.

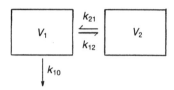

Fig. 4.8 Symbolic representation of the two-compartment open model, assuming that drug is eliminated only from the central compartment.

segment. By this means very complex dosing patterns can be modelled quite easily. Of course, there is a good deal of arithmetic to be done, but that can be performed by a very simple computer program.

The two-compartment open model

Models of this type usually are represented as a double box structure with a descending arrow from the central compartment (Fig. 4.8) indicating the elimination rate constant. It is assumed that both compartments are physically identical to plasma, and that both dosing and elimination are exclusively to and from the central compartment (other configurations can be specified, but are rarely used unless there is a very clear reason for doing so). Drug can move from one compartment to the other, in either direction, at a rate proportional to the concentration difference.

Since multicompartmental models have a number of rate constants, a clearly specified notation is required. Here, in common with the conventions adopted by mainstream anaesthetic journals, rate constants are given two-digit subscripts, the first indicating the source compartment and the second the destination. Thus k_{12} indicates the rate at which drug transfers from compartment 1 to compartment 2. The exterior is denoted by 0, so that elimination from compartment 1 is determined by k_{10}.

The model parameters can be calculated from those specifying the best-fitting equation of the form $C = A \cdot e^{-\alpha t} + B \cdot e^{-\beta t}$, using a sequence of equations:

$$V_1 = \frac{Xd}{A + B}$$

$$k_{21} = \frac{A\beta + B\alpha}{A + B}$$

$$k_{10} = \frac{\alpha\beta}{k_{21}}$$

$$k_{12} = \alpha + \beta - k_{10} - k_{21}$$

$$V_2 = \frac{V_1 k_{12}}{k_{21}}$$

The apparent volume of distribution is simply $V_1 + V_2$ and denoted V^{ss}. The superscript means *steady state*, and

indicates that V^{ss} is the ratio of drug mass in the body to plasma concentration at an infusional steady state. There are other measures of distribution volume, and these will be considered in the following section on non-compartmental methods.

Since drug is assumed to leave the model only via the central (V_1) compartment, clearance is calculated exactly as in the one-compartment model:

$$Cl = V_1 \times k_{10}$$

Once the parameters are established the behaviour of the model in response to any specified input can be determined. Model responses to bolus and infusional dosing can be stated as a set of simultaneous differential equations which are integrated to yield equations for the drug concentrations in both central and deep compartments at any time. To give some flavour of the complexities involved, equations are developed below for the central compartment only; readers seeking more detail should consult a more comprehensive text.

As with the one-compartment model, the method involves solution of dosing successive segments, during each of which dosing conditions remain constant. The effect of a bolus dose is calculated by adding Xd/V_1 to the initial concentration in the central compartment C_1^ϕ.

Then, if:

$$Z = \frac{k'_{01}}{V_1}$$

$$Y = C_1^\phi k_{21} + C_2^\phi k_{12} + Z$$

$$A = \frac{\alpha(C_1^\phi \alpha - Y) + Zk_{21}}{\alpha(\alpha - \beta)}$$

$$B = \frac{\beta(C_1^\phi \beta - Y) + Zk_{21}}{\beta(\beta - \alpha)}$$

$$E = \frac{Z}{k_{10}}$$

the concentration of drug in the central compartment at any time (C_1) can then be calculated:

$$C_1 = Ae^{-\alpha t} + Be^{-\beta t} + E$$

The three-compartment open model

If it can be shown that an equation of the form:

$$C = A \cdot e^{-\alpha t} + B \cdot e^{-\beta t} + C \cdot e^{-\gamma t}$$

is a more likely fit to the data than those comprising fewer terms, a corresponding three-compartment model may be postulated and its parameters calculated. Here, non-linear regression becomes a mathematical nightmare because unless the data follow an almost perfect three-exponential curve there may be a great number of feasible solutions.

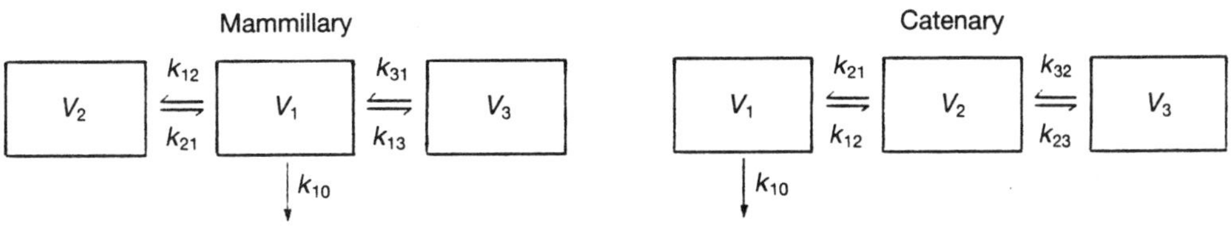

Fig. 4.9 Symbolic representation of mammillary and catenary three-compartment models, assuming that drug is eliminated only from the central compartment.

Furthermore, the investigator must then decide which of the many possible three-compartment configurations is most appropriate. There are two main classes, mammillary and catenary (Fig. 4.9), and each of these can take a variety of forms depending on the number and arrangement of elimination pathways.

Generally, a mammillary form is selected, having one central and two deep compartments, with all drug entering and leaving the model via the central compartment. The response of such a model to bolus and infusional dosing can be calculated in a manner comparable to that shown above. However, the equations are very large indeed and there would be little purpose in reproducing them here. Readers wishing to explore these difficult models should consult a more comprehensive text.

Model-independent approaches to pharmacokinetic analysis

So far it has been assumed that the body can be regarded as a one-, two- or three-compartment system, while all the time recognizing that in truth the number of units behaving as compartments must be very large with an infinitely complex configuration. If it is assumed only that such a large number of compartment-like objects exist, it nevertheless is possible to make surprising progress. The theory of statistical moments considers the times taken for individual drug molecules to pass through the body as a continuous density function which encompasses any number of first-order processes.[27] The mathematical principles are unfamiliar to most clinicians and have not yet made significant impact on clinical pharmacokinetics. However, this approach is finding increasing favour in the analysis of data which can otherwise be analysed only by multicompartmental models whose structures and parameters are difficult to determine and even harder to justify.

Following a single dose of drug, the area under the concentration–time curve (AUC) can be determined as the *zero moment*. If T is the time taken to eliminate all the drug, the mean plasma concentration over that period is AUC/T. Since all the drug was eliminated, the mean rate of elimination must be Xd/T. Since clearance has been stated to be 'rate of elimination divided by concentration', we can write:

$$Cl = \frac{Xd}{T} \div \frac{AUC}{T}$$
$$= \frac{Xd}{AUC}$$

Since we have already seen that in a one-compartment model $Cl = V \times k$, it becomes possible to write:

$$V \cdot k = \frac{Xd}{AUC}$$
$$V = \frac{Xd}{AUC \cdot k}$$

In a multicompartmental model whose final elimination phase has rate constant β, a measure of distribution volume (V^{β}) can be determined which assumes that the body does in fact approximate to a one-compartment system in the late elimination phase:

$$V^{\beta} = \frac{Xd}{AUC \cdot \beta}$$

The limitations are obvious, but because it is not necessary to identify model parameters many investigators have found it useful. V^{β} appears frequently in the published literature.

Now, for each known plasma concentration Cp at time t, the product $Cp \times t$ is calculated and then plotted against t. This generates a second curve called the *moment curve*. The area under this second curve (AUMC) is determined. Now we can calculate the *mean residence time* (MRT), which is the sum of all the residence times for all drug molecules in the body, divided by the number of molecules:

$$MRT = \frac{AUMC}{AUC}$$

The mean residence time is equivalent to the time constant of a one-compartment model, and indicates the time taken for 62.5 per cent of a dose to leave the body. It follows that $T_{\frac{1}{2}} = 0.693 \, MRT$.

It is even possible to calculate V^{ss}, the apparent volume of distribution at steady state:

$$V^{ss} = Cl \cdot MRT$$

Population pharmacokinetics

We have seen that it is not difficult to analyse a set of plasma concentration data in terms of a compartmental model comprising one, two or even three compartments. However, the real difficulties begin when deciding what may legitimately be done with such a model. In truth it does little more than indicate what might be expected to happen if a similar dosing regimen were administered to the same subject again. If applied to other individuals it may be wildly inaccurate because of large differences in pharmacokinetic profiles. To be of predictive value in other individuals it is essential that all possible sources of pharmacokinetic variation are accounted for.

A popular approach has been to take a set of models derived for a group of broadly similar individuals, and then to average each of the parameters to yield an 'average model'. Then, that model is said to represent patients of that age, sex, weight, smoking habits, etc. Unfortunately, this simplistic approach has considerable limitations. First, it should be understood that the confidence with which the constituent models were derived from sets of concentration data will have varied considerably; a near-perfect curve fit yields a robust model while a data set with many 'outliers' may yield a wildly different but highly tenuous model. Yet, all the results are averaged with equal weighting to yield an aggregate model! Clearly, a small proportion of ill-defined models can introduce major errors. Second, such an aggregate model can only hope to emulate the *mean* behaviour of a group of similar subjects; faced with an *individual* the predictive power must be very poor unless he or she has characteristics very close to the mean in every respect.

Even worse, it is commonplace for an investigator to take the model parameters from each subject, scale the distribution volumes by body weight, and *then* average them all together to produce what he calls a 'weight-standardized' model. To apply the model to some other individual, the distribution volumes are scaled up according to his or her body weight. If distribution volume is not proportional to body weight (and in the adult population it is most unusual for it to be so), this procedure may, with the intention of being more scientific, have the opposite effect.

A better alternative is to derive a population model for the drug concerned. In its simplest form this is simply a better way of determining the aggregate model for a group of subjects; the curve-fitting procedure considers the raw data from all the subjects simultaneously, so that the problem of weighting is eliminated. Notwithstanding the limitations, this can be a surprisingly effective procedure.[28]

The most sophisticated approach is to take plasma concentration data from a very large number of subjects, noting also all factors which might possibly contribute to variation in model parameters. Then, a non-linear multiple regression is performed on both the concentration data and the putative influencing factors. This not only identifies those characteristics which influence model parameters but also determines the weighting factors.[29] This is especially important with regard to body weight, which is usually

Table 4.7 A population model for alfentanil (Maitre *et al.*[30])

	Age ≤40	*Age >40*
V_1 (litres)	8.01	8.01
V^{ss} (litres)	34.05	34.05
Cl (ml/min)	356	356 − [0.00264 × (Age - 40)]

significant but with a weighting factor much less than 1. The problem with this method is that many data sets are required from a wide range of patients. Because of such difficulties few such analyses have been reported and inevitably draw on data from a variety of investigators. Table 4.7 offers an example: a population model for alfentanil derived from analysis by the NONMEM computer program.

Readers who remain unconvinced by comments regarding the validity of body weight-scaled models should note that Maitre identified age as an important factor, *but not body weight*.

Models of this type will increasingly become feasible as pharmacokinetic databases become larger and wider ranging. Readers are referred to an authoritative review.[31]

Appendix: Exponential functions

Take a variable Y which decreases from some initial value Y^0 at a rate which is at all times proportional to its current value. Inevitably, with passing time the rate of decline must diminish as Y itself gets smaller (Fig. 4.A1). The equation for such a decay, expressing Y/Y^0 at any time t is:

$$\frac{Y}{Y^0} = \mathrm{e}^{-kt}$$

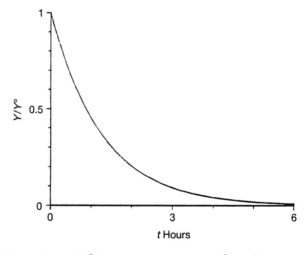

Fig. 4.A1 Y/Y^0 decays exponentially: $Y/Y^0 = \mathrm{e}^{-kt}$.

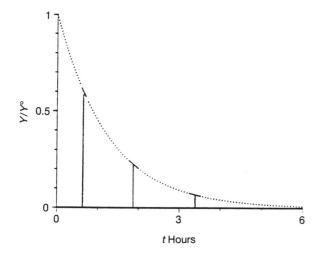

Fig. 4.A2   At any point on the curve, rate constant k is equal to the ratio of gradient to Y/Y^0.

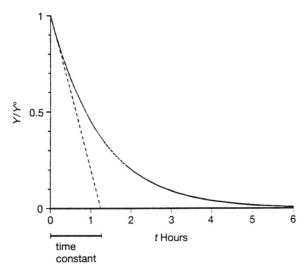

Fig. 4.A3   When an extrapolated line of gradient $-k$ reaches zero, the variable has decayed to 37 per cent of its original value. This period is the *time constant* (τ).

where e is the base of the natural logarithm (2.718 . . .). The exponent is negative because this is a diminishing function.

Although the slope changes constantly, the ratio of slope to Y/Y^0 remains unchanged at all times (Fig. 4.A2). This ratio is called the rate constant and is given the symbol k. Regardless of the direction of change, k is always expressed as a positive number.

The rate at which Y/Y^0 decays can also be characterized as the time taken for it to decline by some fraction of its current value. Consider Fig. 4.A3, in which the initial decay slope is extrapolated to the baseline to create a right triangle of height h and gradient s.

We have seen that $k = s/h$ at all times. This may be rearranged to $s = k/h$. Since $h = 1$, it follows that $s = k$. In any such triangle the gradient is the ratio of height to base,

so that the base must be h/s. As we have determined, $h/s = 1/k$, the base of a triangle created by extrapolation of the initial slope to zero equals $1/k$. The quantity is called the *time constant* and given the symbol τ (tau). The equation for any exponential decay can be expressed in terms of its time constant:

$$\frac{Y}{Y^0} = e^{-\frac{t}{\tau}}$$

How much does Y/Y^0 decay in one time constant? That can be calculated easily by taking $t = \tau$, so that the above equation simplifies to:

$$\frac{Y}{Y^0} = e^{-1} = \frac{1}{e} = \frac{1}{2.718} = 0.368$$

Thus in one time constant (τ), Y/Y^0 decays to 1/e (i.e. 36.8 per cent) of its original value. Although easy to derive, this is not a particularly convenient measure. *Half-time* ($T_{\frac{1}{2}}$), the time taken for an exponentially declining variable to decay by 50 per cent, is widely used, and its relation to the time constant should be clearly understood.

By definition, $Y/Y^0 = 0.5$ at one half-time. Thus:

$$0.5 = e^{-\frac{t}{\tau}}$$

Taking natural logarithms, this reduces to:

$$\ln 0.5 = -\frac{t}{\tau}$$

Since ln 0.5 evaluates to -0.693, the final steps are easy:

$$-0.693 = -\frac{t}{\tau}$$

$$t = \tau \times 0.693$$

The relationship between $T_{\frac{1}{2}}$ and τ is clearly shown in Fig. 4.A4.

As we have seen above, taking the logarithm of e^{-kt}

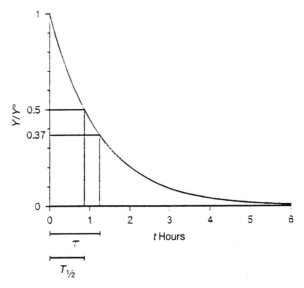

Fig. 4.A4   Time constant (τ) and half-time ($T_{\frac{1}{2}}$).

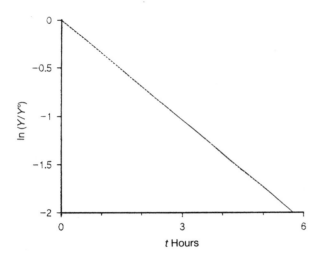

Fig. 4.A5 Plotting $\ln (Y/Y^0)$ against time yields a straight line of gradient $-k$.

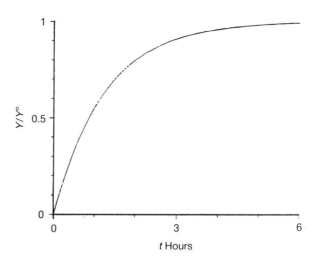

Fig. 4.A6 Y/Y^0 approaches its asymptote (1) by a wash-in exponential function.

yields $-kt$ because that is the power to which the base (in this case e) must be raised to give the number concerned. Applying this method to the equation of the curve we obtain:

$$\ln\left[\frac{Y}{Y^0}\right] = -kt$$

It follows that a plot of Y/Y^0 against time (Fig. 4.A5) will yield a straight line of gradient $-k$.

Exponential processes may go up as well as down. In the simplest case, the equation $Y = e^{kt}$ describes a curve whose gradient *increases* with Y; this might describe the rate of multiplication of bacteria in a culture. Fortunately, such *tearaway* functions do not, as a rule, apply to pharmacology! However, it is not uncommon to find a variable rising towards some final value (known as the *asymptote*) at an ever-reducing rate; this kind of behaviour is typical of drug concentrations during a continuous infusion (Fig. 4.A6).

The equation for this curve is:

$$\frac{Y}{Y^0} = 1 - (e^{-kt})$$

Here the exponential term decreases with time, just as in the washout decay (note e^{-kt}), and represents the diminishing difference between Y/Y^0 and 1.0, the asymptote. Taking logarithms does not resolve this function to a straight line unless it is rearranged slightly:

$$1 - \left[\frac{Y}{Y^0}\right] = e^{-kt}$$

Now, a plot of $\ln(1 - Y/Y^0)$ against time (not illustrated) does yield a straight line of gradient $-k$.

REFERENCES

1. Hull CJ. *Pharmacokinetics for anaesthesia* Oxford: Butterworth-Heinemann Ltd, 1991.
2. Dominguez R. Studies of renal excretion of creatinine. *Proceedings of the Society for Experimental Biology and Medicine* 1934; **34**: 1146–9.
3. Hull CJ. How far can we go with compartmental models? (Editorial). *Anesthesiology* 1990; **72**: 399–402.
4. Kremer JMH, Wilting J, Janssen LHM. Drug binding to human alpha-1-acid glycoprotein in health and disease. *Pharmacological Reviews* 1988; **40**: 1–47.
5. Blaschke TF. Protein binding and kinetics of drugs in liver diseases. *Clinical Pharmacokinetics* 1977; **2**: 32–44.
6. Roerig DL, Kotrly KJ, Vucins EJ, Ahlf SB, Dawson CA, Kampine JP. First pass uptake of fentanyl, meperidine, and morphine in the human lung. *Anesthesiology* 1987; **67**: 466–72.
7. Barratt RL, Graham GG, Torda TA. Kinetics of thiopentone in relation to the site of sampling. *British Journal of Anaesthesia* 1984; **56**: 1385–91.
8. Roerig DL, Kotrly KJ, Ahlf SB, Dawson CA, Kampine JP. Effect of propranolol on the first pass uptake of fentanyl in the human and rat lung. *Anesthesiology* 1989; **71**: 62–8.
9. Stanski DR, Mihm FG, Rosenthal MH, *et al.* Pharmacokinetics of high-dose thiopental used for cerebral resuscitation. *Anesthesiology* 1980; **53**: 169–71.
10. Walle T, Byington RP, Furberg CD, *et al.* Biologic determinants of propranolol disposition: results from 1308 patients in the beta-blocker heart attack trial. *Clinical Pharmacology and Therapeutics* 1985; **38**: 509–18.
11. Divoll M, Greenblatt DJ, Harmatz JS, Shader RI. Effect of age and gender on disposition of temazepam. *Journal of Pharmaceutical Science* 1981; **70**: 1104–7.

12. Wing LMH, Miners JO, Birkett, DJ, Foenander T, Lillywhite K, Wanwimolruk S. Lidocaine disposition – sex differences and effects of cimetidine. *Clinical Pharmacology and Therapeutics* 1984; **35**: 695–701.

13. Matteo RS, McDaniel DD, Lieberman IG, Salanitre E, Diaz J. Pharmacokinetics of *d*-tubocurarine in neonates, infants and children. *Anesthesiology* 1982; **57**: A269 (Abstract).

14. Sorbo S, Hudson RJ, Loomis JC. The pharmacokinetics of thiopental in pediatric surgical patients. *Anesthesiology* 1984; **61**: 666–70.

15. Goresky GV, Koren G, Sabourin MA, *et al.* The pharmacokinetics of alfentanil in children. *Anesthesiology* 1987; **67**: 654–9.

16. Morselli PL, Franco-Morselli R, Bossi L. Clinical pharmacokinetics in newborns and infants, age-related differences and therapeutic implications. *Clinical Pharmacokinetics* 1980; **5**: 485–527.

17. Morselli PL. Clinical pharmacokinetics in neonates. *Clinical Pharmacokinetics* 1976; **1**: 81–98.

18. Castleden CM, Kaye CM, Parsons RL. The effect of age on plasma levels of propranolol and practolol in man. *British Journal of Clinical Pharmacology* 1975; **2**: 203–6.

19. Greenblatt DJ, Allen MD, Harmatz JS, Shader RI. Diazepam disposition determinants. *Clinical Pharmacology and Therapeutics* 1980; **27**: 301–12.

20. McQueen EG. Pharmacological basis for drug reactions. In: Avery GS, ed. *Drug treatment.* New York: Adis Press, 1980.

21. Simpson KH, Stakes AF, Miller M. Pregnancy delays paracetamol absorption and gastric emptying in patients undergoing surgery. *British Journal of Anaesthesia* 1988; **60**: 24–7.

22. Jeffries WS, Bochner F. The effect of pregnancy on drug pharmacokinetics. *Medical Journal of Australia* 1988; **149**: 675–7.

23. Sear JW, Hand CW, Moore RA, McQuay HJ. Studies on morphine disposition: influence of renal failure on the kinetics of morphine and its metabolites. *British Journal of Anaesthesia* 1989; **62**: 28–32.

24. Woolner DF, Winter D, Frendin TJ, *et al.* Renal failure does not impair the metabolism of morphine. *British Journal of Clinical Pharmacology* 1986; **22**: 55–9.

25. Klotz U, Avant GR, Hoyumpa A, *et al.* The effects of age and liver disease on the disposition and elimination of diazepam in adult man. *Journal of Clinical Investigation* 1975; **55**: 347–59.

26. Runciman WB, Mather LE. Effects of anaesthesia on drug disposition. In: Feldman SA, Scurr CF, Paton W, eds. *Drugs in anaesthesia: mechanisms of action.* London: Edward Arnold, 1987: 87–122.

27. Yamaoka K, Nakagawa T, Uno T. Statistical moments in pharmacokinetics. *Journal of Pharmacokinetics and Biopharmaceutics* 1978; **6**: 547–58.

28. Shafer SL, Varvel JR, Aziz N, Scott JC. The pharmacokinetics of fentanyl administered by a computer controlled pump. *Anesthesiology* 1990; **73**: 66–72.

29. Sheiner LB. The population approach to pharmacokinetic data analysis: rationale and standard data analysis methods. *Drug Metabolism Reviews* 1984; **15**: 153–71.

30. Maitre PO, Vozeh S, Heykants J, Thomson DA, Stanski DR. Population pharmacokinetics of alfentanil: the average dose-plasma concentration relationship and inter-individual variability in patients. *Anesthesiology* 1987; **66**: 3–12.

31. Whiting B, Kelman AW, Grevel J. Population kinetics. Theory and clinical application. *Clinical Pharmacokinetics* 1986; **11**: 387–401.

Chapter 5

The Chemical Modulation of Nociceptive Responses and Pain

Barbara J. Pleuvry

Nociceptive responses are those observed when a noxious stimulus is applied and can be observed in all animals. Pain, on the other hand, has an emotional and subjective component and, although it may occur in other animals, can only be expressed by humans. Thus pain is a conscious sensation of distress, suffering or agony associated with, or at least described in terms of, actual or potential tissue damage. Physiological pain occurs when a noxious stimulus activates high threshold sensory receptors (nociceptors). Its effect is to inform the body of potential or actual damage and it is highly correlated with withdrawal reflexes.[1] Pain of this type can be quantified in terms of modality (e.g. heat or pressure), location, duration and intensity. In contrast pathological pain can occur in response to a non-noxious stimulus, as an exaggerated response to a noxious stimulus or even in the absence of a definable stimulus. The result of this hypersensitivity to pain is that further damage is minimized by the avoidance of all stimuli to the affected area, thus promoting healing. However, in its chronic form this type of pain is truly pathological as it has lost any adaptive function.[2]

In the following pages consideration of nociceptive responses, pain and its physiology will focus principally on the chemical neurotransmitters, modulators and mediators involved in these processes. Both peptide and non-peptide molecules are involved and the differences between neuronal transmission utilizing the two classes of molecules are summarized in Table 5.1. Both peptide and non-peptide transmitters may coexist in a single neuron. Some of the peptides which may be involved in the control of nociceptive activity are shown in Table 5.2. It should be noted that the opioid peptides derive from three separate precursors. A single neuron may express only one or, in some cases, two of these precursors and thus their activities may be completely different.

Much of the work concerning the mechanisms involved in the initiation of nociceptive responses has been carried out in the experimental animal, often the rat. Species variation, in the relative proportions of various neuro-transmitters and modulators and the receptors with which they interact, makes extrapolation between species unsafe. For this reason, data have been included where manipulation of a particular transmitter or modulator has been shown to affect pain in humans.

Table 5.1 Differences between peptide and non-peptide neurotransmitters

Peptide transmitter	Non peptide transmitter
Complex molecule, mol. wt 200	Simple molecule, mol. wt <200
Precursor is synthesized by genetic material	Precursor is taken up into neuron
Several cleavage products can be obtained from the same precursor	Single product is synthesized from precursor
Transmitter production is slow	Transmitter production is fast
Action is terminated solely by enzymatic activity	Action is terminated by reuptake into the neuron and by enzymic activity
Termination of action may be slow, e.g. endorphins	Termination of action is fast
Probably involved in long-term changes	Mainly involved in acute changes

Peripheral mechanisms

Nociceptors

The nociceptors triggered by physiological pain are subdivided into two main groups; those associated with slowly conducting (<1 m/s) non-myelinated fibres known as the C fibres and those associated with myelinated Aδ fibres (average conduction 15 m/s). All the nociceptors are

Table 5.2 Principal peptides that may be involved in the control of nociceptive activity

Precursor	Main products	Receptor types
Opioid peptides		
Pro-opiomelanocortin (POMC)	β-Endorphin 1–31*	μ/ε†
Proenkephalin A	[Met] enkephalin	μ/δ
	[Leu] enkephalin	δ
	Peptide E	μ/κ/δ
Prodynorphin	Dynorphin 1–8*	κ
	Dynorphin A 1–17*	μ/κ
	Neoendorphins	κ/δ
Tachykinins		
α-Preprotachykinin	Substance P	NK$_1$
β-Preprotachykinin	Neurokinin A	NK$_2$
	Substance P	NK$_1$
Bradykinin		
Kininogen	Bradykinin	B$_1$/B$_2$
Cholecystokinins (CCK)		
Preprocholecystokinin	CCK-8*	CCK-A/CCK-B
	Caerulin	
Somatostatins		
Preprosomatostatin	Somatostatin	Somatostatin A/B†

* Denotes the number of amino acids in the peptide.
† Existence of receptor type is still open to question.

polymodal as they respond to thermal, mechanical and chemical noxious (or painful) stimuli. However, the nociceptors which are associated with Aδ fibres, respond most strongly to noxious heat and pressure and less strongly to chemical stimulation.[3] Sharp, well localized pain is associated with Aδ fibre activity (first pain) whilst dull burning pain involves C-fibre activation (second pain). In pathological pain, activation of cutaneous touch and pressure afferent fibres (Aβ fibres) may be perceived as pain.

Chemical mediators

The most abundant nociceptors are associated with the free nerve ending of C fibres and these are the receptors activated most strongly by chemical mediators. Tissue injury and inflammation results in the release of many chemicals including bradykinin, histamine, eicosanoids, substance P, 5-hydroxytryptamine, adenosine triphosphate, hydrogen ion and opioid peptides. The role of these substances in the production and modulation of the hyperalgesia associated with tissue damage has been extensively studied.

Bradykinin

Protease activation at the site of tissue injury cleaves plasma and tissue kininogen to the nonapeptide bradykinin, which is one of the most potent pain-producing substances present in inflammatory exudates and damaged tissue. In addition bradykinin is a vasodilator and increases capillary permeability.

In some situations the former activity may be indirect via the mobilization of arachidonic acid leading to the formation of vasodilator prostanoids. Generally kinins have a short-lived action due to kininase activity but, in inflammatory conditions, the acidic environment inhibits kininase and prolongs kinin activity.[4] Two bradykinin receptors are currently recognized: B$_1$ and B$_2$. B$_2$ is the most abundant and appears to be responsible for nociceptor excitation.[5] Currently available B$_2$ antagonists are too short lived to be clinically useful, but a non-selective bradykinin receptor antagonist, D-Arg[4-hydroxy-Pro^3-D-Phe^7]bradykinin, has produced encouraging results in clinical trials for the treatment of pain from burns.[6]

Histamine

Histamine is noted for its involvement in Lewis's triple response to skin injury.[7] However, although it stimulates nociceptors, it produces an itch rather than pain.[8] Although some reports have suggested that high doses may cause pain,[3] it is unlikely that histamine has a role in injury-induced hyperalgesia. The mechanism by which nociceptors can signal both itch and pain is still controversial.[9]

Eicosanoids

The eicosanoids are a group of arachidonic acid metabolites which include the prostaglandins, thromboxanes and leukotrienes. The name is derived from 'eicosa', indicating 20 carbon atoms. Arachidonic acid is a 20-carbon atom fatty acid containing four double bonds, which is usually found esterified in the phospholipids in cell membranes. Arachidonic acid is liberated by activation of phospholipase (A_2 and/or C) which can result from a number of stimuli such as thrombin, bradykinin, antigen–antibody reactions and general cell damage. Once liberated, arachidonic acid is rapidly metabolized by a number of enzyme systems including cyclo-oxygenase and several lipoxygenases.

The prostaglandins (PGs) and thromboxane are products of cyclo-oxygenase activity. PGEs and PGI_2 are most likely to have a role in inflammatory pain. They sensitize nociceptors to other noxious stimuli and they can produce pain in their own right when injected intradermally. Although it is less intense than pain produced by other substances, prostaglandin-induced pain appears to be much longer lasting.[10] Inhibitors of cyclo-oxygenase such as the non-steroidal anti-inflammatory drugs (NSAIDs) are effective in treating inflammatory pain. However, they are also useful in pain which has no obvious inflammatory component such as sudden trauma to otherwise healthy tissue. It is unlikely that peripheral inhibition of cyclo-oxygenase has a role in this type of analgesic activity and there is considerable debate as to whether inhibition of cyclo-oxygenase at any site is relevant.[11] Indeed, part of the antinociceptive effects of diclofenac has been attributed to direct or indirect involvement of central opioid systems.[12]

Nociceptor stimulation has also been observed after leukotriene release, from damaged tissue. Leukotrienes are products of arachidonic acid metabolism via the lipoxygenase pathway, and the D_4 and B_4 variants appear to have a role in nociceptor sensitization.[3] The evidence suggests that their actions are indirect via the release of other noxious mediators such as prostanoids[13] and substance P.[14] Leukotriene B_4 releases hydroxyacids from leucocytes.[15] However, the mechanism by which sensory nerves are stimulated by these acids is unclear.

Substance P

Substance P is one of a group of rapidly acting peptides called the tachykinins, in contrast to bradykinin, which has a slower action. Whilst substance P concentrations increase during nociceptor stimulation, it represents an afferent function of the nociceptor rather than a nociceptive stimulus. Its release from the nerve terminal is associated with the flare response.[9]

5-Hydroxytryptamine

Tissue damage releases 5-hydroxytryptamine (5-HT) from mast cells and blood platelets. In low concentration it sensitizes nociceptors to other noxious substances such as bradykinin, but in higher concentrations it can induce pain directly. It is involved in inflammatory hyperalgesia and a selective $5-HT_3$ receptor antagonist has been demonstrated to reduce hyperalgesia without affecting other signs of inflammation such as oedema formation in the rat.[16]

Adenosine triphosphate (ATP)

Sharp transient pain can be induced by ATP which is released in large quantities by tissue damage. This action is thought to be due to a direct action on ion channels via the ATP-sensitive purine (P_2) receptor and can be blocked by specific P_2 antagonists such as suramin.[3]

Hydrogen ion

As mentioned previously, inflammatory exudates are acidic and it has been reported that protons themselves are capable of inducing a maintained depolarization of a subset of neurons which have the characteristics of nociceptive neurons.[17]

Opioid peptides

There is increasing evidence that opioid peptides can modulate pain transmission at the level of the nociceptors. Inflammatory pain can be relieved by opioid drugs acting at peripheral opioid receptors located on sensory nerves. These receptors have characteristics resembling μ-, δ- and κ-receptors.[18] μ-Receptor activation prevents nociceptor sensitization by switching off the activation of adenylyl cyclase induced by inflammatory mediators such as prostaglandin E_2 (PGE_2).[19] δ- and κ-agonists do not share this property, but they are able to prevent bradykinin-induced hyperalgesia. Bradykinin induces the release of noxious mediators from postganglionic sympathetic nerves. It is likely that δ- and κ-receptors are situated on the sympathetic nerves and prevent the release of noxious mediators.[20] The evidence suggests that the endogenous opioid peptides that activate μ-receptors are released from immunocytes during inflammation. Cells of the immune system contain the genes necessary for the formation of two opioid peptide precursors, pro-opiomelanocortin and proenkephalin. Immunoreactive β-endorphin and metenkephalin increase during inflammation. In contrast dynorphin concentrations remain below the detectable limit.[21] However, antibodies to β-endorphin, but not to metenkephalin, inhibit nociception during inflammation, suggesting that the metenkephalin does not modulate pain transmission at this site. The relevance of these findings to man has been emphasized by the increasing use of peripheral application of opioid drugs to treat inflammatory pain in the clinical situation.[22,23]

Central mechanisms

The nociceptive afferent fibres have cell bodies that lie in the dorsal root ganglion and whose fibres enter the spinal cord and terminate in the dorsal horn. All C fibres and some Aδ fibres synapse in the superficial laminae (I and II), whilst other A fibres penetrate deeper to lamina V. However, not all fibres travel to the dorsal horn via the dorsal root. About 30 per cent of unmyelinated C fibres double back from the dorsal root ganglion and enter the spinal cord via the ventral root, ending up in the same place as the other 70 per cent.[24] The main pain pathways to the thalamus arise from the cells in laminae I and V. Lamina II, also known as the substantia gelatinosa, contain cells which form a network of mainly inhibitory interneurons which may synapse with other neurons either pre- or postsynaptically. It is thought that this network regulates transmission between the nociceptive neurons and those in the spinothalamic tract. It is in this area that Melzack and Wall[25] proposed 'the gate control theory of pain' whereby the passage of impulses giving rise to the sensation of pain can be reduced by activity in other afferent pathways, such as those perceived as touch or heat and by the activity of pathways descending from areas such as the periaqueductal grey (PAG) in the midbrain.

Primary nociceptor afferent transmitters

The dorsal horn neurons show an excitatory response comprising both a fast component and, during repetitive stimulation, a slow component. The principal candidates for the role of sensory neurotransmitter are the excitatory amino acids, adenosine triphosphate and some of the neuropeptides (Fig. 5.1).

Excitatory amino acids

L-Glutamate is the only amino acid which is concentrated in the dorsal horn[26] and iontophoric application causes a depolarization of many spinal neurons which mimics the fast component observed after stimulation of primary afferent fibres. In addition, L-glutamate is released in areas of the spinal cord containing primary afferent terminals.[27] There are a number of receptors for glutamate, namely NMDA (N-methyl-D-aspartate), AMPA (α-amino-3-hydroxy-5-methyl-isoxazole-4-propionate), kainate, LAP4 (L-2-amino-4-phosphonobutyrate) and metabotropic receptors.[28] Studies with selective NMDA antagonists have ruled out activation of this receptor as a mechanism of glutamate's fast activity at this site. This is not to say that NMDA receptors play no role in sensory transmission in the spinal

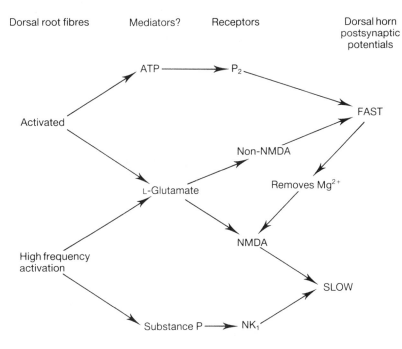

Fig. 5.1 Pain transmission in the dorsal horn of the spinal cord.

cord. The ion channel associated with the NMDA receptors is blocked by magnesium ion at normal resting potentials. However, if the neuronal membrane is depolarized by activation of a non-NMDA receptor or by a neuropeptide receptor, the Mg^{2+} block of the NMDA receptors may be removed. This could result in the synaptic potential being prolonged by NMDA receptor activation, thus giving rise to the slow long duration potentials.[29]

Adenosine triphosphate

ATP can be selectively released from the dorsal horn of the spinal cord[30] and can cause fast depolarization of dorsal horn neurons, an effect not mimicked by other nucleotides. However, the lack of specific receptor antagonists and the difficulties involved in establishing whether a particular neuron uses ATP as a transmitter have precluded any firm conclusions concerning its sensory transmitter function.

Substance P

The tachykinins, substance P and neurokinin A are released within the dorsal horn of the spinal cord in response to noxious stimulation of the primary afferents.[31] Capsaicin, which releases and then depletes substance P from the primary afferent terminals, causes nociceptive deficits. Whilst this is evidence for a role of substance P in pain transmission it is not conclusive since capsaicin also depletes other peptides such as cholecystokinin and somatostatin which are present in primary afferents.[32] Locally applied substance P produces a slow depolarization of dorsal horn neurons and the slow component of the response to nociceptor stimulation is blocked by substance P antagonists.[33] However, the ability of substance P antagonists to provide antinociceptive activity (or analgesia) appears to depend upon the experimental model used and the species of animal studied. In addition the presently available substance P antagonists have low potency and often exhibit partial agonist activity.[34] Thus the case for substance P as a pain transmitter at the primary afferent terminals is still unproven.

Modulators of pain transmission in the spinal cord

Opioid peptides

Both dynorphins and enkephalins are concentrated in the superficial layers of the dorsal horn. Interestingly prodynorphin has been co-localized with substance P in guinea-pig sensory neurons.[35] The distribution of enkephalin containing neurons is more widespread than those containing dynorphin and in some instances they appear to receive only non-nociceptive inputs.[36] The majority of the opioid receptors in the dorsal horn are μ-receptors although δ- and κ-receptors are present. The relative percentage of the various receptors varies considerably from species to species.[37] Whilst the clinical use of opioid receptor agonists prompts us to assume that the endogenous opioid system must be associated with antinociceptive activity and analgesia, the experimental evidence suggests that they may enhance or inhibit nociceptive activity dependent upon the circumstances. Naloxone reduces noxious stimuli evoked responses in the superficial layers of the dorsal horn but enhances the responses of nociceptive neurons in deeper layers.[38] In addition dynorphin A has been reported to have antianalgesic activity.[39]

Biogenic amines

The superficial layers of the dorsal horn contain high concentrations of 5-hydroxytryptamine (5-HT) and noradrenaline derived from cell bodies in the brainstem. These pathways are the descending inhibitory controls which constitute one of the gating mechanisms that control pain transmission in the dorsal horn. Both noradrenaline and 5-HT inhibit the response of the dorsal horn neurons to noxious stimuli.[40] However, the two neuronal systems act independently. Direct application of 5-HT to the spinal cord is analgesic and 5-HT uptake inhibitors have analgesic properties.[41] Prevention of 5-HT reuptake may contribute to the analgesic activity of tramadol[42] which has been successfully marketed in Germany and is currently being launched world wide as an atypical opioid analgesic, which lacks respiratory depressant activity.[43] The 5-HT_2 receptor blocking drug ritanserin antagonizes the antinociceptive activity of tramadol, suggesting that this subtype of 5-HT receptor may be important with respect to 5-HT actions on pain pathways.[42] The contribution of noradrenaline to pain modulation is more controversial as both positive and negative modulation of nociceptive activity have been recorded dependent upon the site studied.[44] Nevertheless local application of agonists at α_2-adrenoceptors produce analgesia and intrathecal and epidural administration of α_2-agonists such as dexmedetomidine and clonidine are receiving considerable attention as potential treatments for pain in man.[45] It has recently been suggested that activation of spinal adrenergic receptors may mediate the antinociceptive activity of dopamine injected into the lumbar subarachnoid space.[46] Neurochemical binding studies suggest the presence of dopaminergic, but not noradrenergic receptors in the lumbar and sacral spinal cord.[47] The antinociceptive activity of dopamine at this site appears to be mediated via D_2 receptors.

Other mediators

Intrathecal administration of many other neurotransmitter substances have been reported to exert antinociceptive effects.[48] Endogenous adenosine release has been reported to be involved in spinally mediated antinociceptive activity and to contribute to the antinociceptive activity of opioid drugs.[49] Furthermore, antinociceptive activity can be induced by inhibition of adenosine kinase.[50] Somatostatin has been localized in interneurons in the superficial layers of the dorsal horn[51] and it has been shown to inhibit nociceptive responses by a mechanism distinct from that exhibited by opioids.[52] There is some controversy concerning somatostatin's antinociceptive activity in the experimental animal. Nevertheless, intrathecal infusion of a stable analogue, octreotide, has been used successfully to treat cancer pain which was refractory to opioids.[53] In many cases, including the antinociceptive activity of intrathecal application of muscarinic cholinergic agonists, agonists at $GABA_B$ receptors (but not $GABA_A$ receptors), calcitonin and vasopressin, the mechanisms and physiological importance of the observed responses are unknown.[48]

Supraspinal modulation of pain transmission

The periaqueductal grey (PAG) in the midbrain is a key centre in the descending inhibitory pathways mentioned above. Electrical stimulation of this area causes a loss of response to noxious stimuli, without any other modality of sensation being affected.[54] It receives connections from the thalamus, hypothalamus and cortex and appears to be the relay station, where the higher centres of the brain can modify nociceptive input. Fibres from the PAG run down to the nucleus raphe magnus which receives an input from spinothalamic neurons via the nucleus reticularis paragigantocellularis. Neurons in these areas, which form part of the rostral ventromedial medulla, respond to noxious stimuli in three ways. One group, known as 'on-cells', show a sudden increase in firing which appears just before and during withdrawal from the noxious stimulus. A second group of 'off-cells' show a decrease in activity during noxious stimulation and the final 'neutral-cells' show no change in firing.[55] Non-noxious stimuli produce only very small changes in the firing rate of these cells and they are believed to have opposing effects on transmission of noxious stimuli: activity in 'on-cells' facilitating nociceptive transmission, whilst activity in 'off-cells' inhibits it (Fig. 5.2). The fibres from the nucleus raphe magnus to the dorsal horn utilize 5-HT or enkephalin as transmitters. A separate descending pathway from the locus ceruleus is noradrenergic.[44]

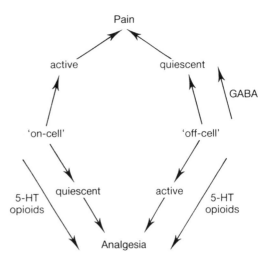

Fig. 5.2 Modulation of the 'on-cells' and 'off-cells' in the rostral ventromedial medulla.

Opioid peptides

High concentrations of opioid receptors are found in the PAG and, nucleus raphe magnus and nucleus reticularis paragigantocellularis and microinjection studies with naloxone have demonstrated that systemic morphine induces analgesia at these sites.[56] Both μ- and δ-receptors[57] have been shown to mediate supraspinal analgesia although μ-receptor activation explains most of the activity of clinically useful opioids.[37] The role of κ-receptor activation in supraspinal analgesia is more debatable.[37,58,59] Recent studies have suggested that the dynorphins, the proposed endogenous ligands of the κ-receptor, actually facilitate pain transmission in the brainstem.[60] A similar role for dynorphins in the spinal cord has been mentioned earlier. It has been suggested that opioids indirectly activate 'off-cells' mentioned above and reduce the activity of 'on-cells',[55] both effects reducing transmission of pain. However, both enhancement and inhibition of descending inhibitory control by opioids has been reported.[61–64] Opioid receptors are found at other sites such as the thalamus and cortex, which may be relevant to nociceptive control. However, concentration in other areas such as the tractus solitarius are more likely to be related to respiratory effects.

Opioid peptides are also found in these midbrain areas. Enkephalin and 5-HT appear to coexist in some neurons, whilst others contain dynorphin and enkephalin.[65] Whilst dynorphin is present in the PAG, it is not involved in modulating pain transmission at this site.[66] Although neurons containing the endorphin precursor pro-opiomelanocortin (POMC) are limited to a few areas of the brain, cell bodies in the hypothalamus project axons to the PAG and locus ceruleus.

In view of the above it is perhaps surprising that opioid

peptide systems do not appear to be active in the day-to-day maintenance of pain threshold.[67] Opioid systems appear to be involved in situations when the pain threshold is increased by factors such as stress.

γ-Aminobutyric acid (GABA)

The nucleus raphe magnus and nucleus reticularis paragigantocellularis are rich in GABAergic neurons[68] and both $GABA_A$ and $GABA_B$ receptors have been demonstrated.[69] However, in contrast with the spinal cord, only $GABA_A$ receptor manipulations influence nociceptive activity. Agonists at $GABA_A$ receptors facilitate responses to noxious stimuli whilst antagonists exhibit antinociceptive activity.[70] Ultrastructural studies have shown that GABA-containing neurons directly contact 'off-cell'[55] and it is likely that they mediate the 'off-cell' pause as iontophoretic application of bicuculline eliminates the 'off-cell' pause.[71] Support for a physiological role of GABA in the supraspinal control of nociception comes from the observation that GABA release into the fourth ventricle is enhanced by noxious stimulation, at least in the rat.[72] The indirect activation of 'off-cells' by opioids described above could be due to inhibition of GABA release, preventing the 'off-cell' pause. However, injection of GABA mimetics (either $GABA_A$ or $GABA_B$) at other supraspinal sites, such as the PAG, appears to be antinociceptive and the overall effect of systemic administration of GABA mimetics in man is analgesia.[48]

Excitatory amino acids

Both glutamate and aspartate containing neurons have been reported in the PAG[73] and there is evidence that projections from these neurons reach the nucleus raphe magnus in the ventromedial medulla.[74] Direct microinjection of glutamate and NMDA into the PAG produces antinociceptive activity which can be blocked by a selective NMDA antagonist.[75] However, in the nucleus raphe magnus non-NMDA excitatory amino acid receptors appear to be involved in the modulation of nociception.[76] These findings contrast with the amplification of noxious stimuli exhibited by excitatory amino acids in the spinal cord.

5-Hydroxytryptamine

In addition to being an important transmitter in the descending pathways, 5-HT has antinociceptive effects when injected directly into the rostral ventromedial medulla. Iontophoretic studies have shown that 5-HT can excite one class of brainstem neurons via an action of 5-HT_2 receptors[77] whilst inhibiting a second class via an action on 5-HT_1-like receptors.[78] Whilst the type of cell stimulated or inhibited was not characterized in these studies, it is

possible that 'off-cells' were stimulated and 'on-cells' inhibited, both effects contributing to a reduction in noxious transmission.[55] A similar mechanism, although indirect, has been suggested for opioid peptides in this region of the brainstem (see above).

Noradrenaline

The actions of noradrenergic systems in the spinal cord have been documented in the greatest detail; however, both α_1 and α_2 binding sites have been detected in the parts of the brainstem controlling pain modulation. Recent evidence suggests that α_1-receptor activation increases responses to noxious stimuli, whilst α_2-receptor activation decreases responses.[55]

Other supraspinal receptor systems

A variety of other transmitters including neurotensin, dopamine, histamine, calcitonin and acetylcholine have been reported to modulate nociceptive activity in the brainstem.[48]

Summary

This summary has concentrated mainly on individual transmitter systems and nociception; however, much of the research has been directed towards examining the effects of transmitter manipulation on the analgesic or antinociceptive activity of established analgesics such as morphine. This also gives us conflicting data. As an example, the cholecystokinin (CCK) group of peptides have been put forward as endogenous antagonists of opioid (i.e. morphine) induced analgesia.[79] This interaction is thought to be via the CCK-B receptor although this conclusion may reflect the relative lack of CCK-A receptors in rat brain, the usual experimental model.[80] Primates have a more generous sprinkling of CCK-A receptors in the CNS. In contrast, CCK in the absence of opioid drugs has analgesic activity[81] and there is evidence that this is mediated via δ-opioid receptors.[82] CCK and opioid peptides have parallel distributions in many brain areas associated with nociception and it may be that the two peptide systems reciprocally modulate CCK-enkephalin release. The end response depends on which type of receptor (μ or δ for the opioids and CCK-A or CCK-B) is involved.[83]

The simplistic view that substance P was the pain transmitter and that opioid peptides inhibited its release and thus prevented transmission of pain must clearly be discarded. Nociception and its control involves a huge repertoire of neurotransmitters and modulators and no

Table 5.3 Chemical modulation of nociceptive activity

Chemical	Site	Effect on noxious transmission	Receptor involved
Bradykinin	Nociceptors	Enhanced	
Histamine	Nociceptors	Enhanced (itch rather than pain)	H_1
Prostaglandins	Nociceptors	Enhanced (sensitize to other noxious chemicals)	$EP_1/EP_2/IP$
Leukotrienes	Nociceptors	Enchanced (release of noxious chemicals)	LTB_4/LTD_4
Substance P	Nociceptors	Probably no role	
	Spinal cord	Enhanced (slow excitation)	NK_1
5-Hydroxytryptamine	Nociceptor	Enhanced	$5\text{-}HT_3$
	Spinal cord	Reduced	$5\text{-}HT_2$
	RVM	Net result reduction (inhibition 'on-cells', excitation 'off-cells')	$5\text{-}HT_1$-like $5\text{-}HT_2$
ATP	Nociceptor	Enhanced	P_2
	Spinal cord	Enhanced	P_2
β-Endophins	Nociceptor	Reduced (prevent sensitization)	μ
	PAG?		
Enkephalins	Nociceptors	Probably no effect	
	Spinal cord	Reduced	μ/δ
	PAG and RVM	Net result reduction (inhibition 'on-cells', excitation 'off-cells')	μ/δ
Dynorphins	Nociceptors?	Reduced (prevents release of noxious mediators)	κ present on sympathetic nerves
	Spinal cord	Enhanced?	κ
	Brainstem	Enhanced?	κ
Excitatory amino acids	Spinal cord	Enhanced	Non-NMDA
		Prolonged	NMDA
	PAG	Reduced	NMDA
	RVM	Reduced	Non-NMDA
Noradrenaline	Spinal cord	Reduced	$α_2$
	Brainstem	Enhanced	$α_1$
		Reduced	$α_2$
Dopamine	Lumber and sacral spinal cord	Reduced	D_2
Somatostatin	Spinal cord	Reduced	Somatostatin A
γ-Aminobutyric acid	Spinal cord	Reduced	$GABA_B$
	PAG	Reduced	$GABA_A/GABA_B$
	RVM	Enhanced (mediate 'off-cell' pause)	$GABA_A$

? Evidence is poor or controversial.
RVM, rostral ventromedial medulla; PAG, periaqueductal grey.

individual chemical group or, indeed, one receptor subtype always facilitates nociception or inhibits it (Table 5.3). Thus our growing understanding of the mechanisms of pain perception and control have so far failed to give direction to research into new treatments for pain in man. Instead the chance observation that a new chemical entity has analgesic activity leads to further unravelling of the complexity of this subject.

REFERENCES

1. Willer JC. Comparative study of perceived pain and nociceptive flexion reflex in man. *Pain* 1979; **3**: 69–80.
2. Woolf CJ. Generation of acute pain: central mechanisms. *British Medical Bulletin* 1991; **47**: 523–33.
3. Rang HP, Bevan S, Dray A. Chemical activation of nociceptive peripheral neurones. *British Medical Bulletin* 1991; **47**: 534–48.
4. Edery H, Lewis GP. Kinin forming activity and histamine in lymph after tissue injury. *Journal of Physiology* 1963; **169**: 568–83.
5. Steranka LR, Manning DC, DeHaas CJ, *et al.* Bradykinin as a pain mediator: receptors are localized to sensory neurons and antagonists have analgesic actions. *Proceedings of the National Academy of Sciences, USA* 1988; **85**: 3245–9.
6. Burch RM, Farmer SG, Steranka LR. Bradykinin receptor antagonists. *Medical Research Review* 1990; **10**: 143–75.
7. Lewis T. *The blood vessels of the human skin and their responses.* London: Shaw & Sons Ltd, 1927.
8. LaMotte RH, Simone DA, Bauman TK, Shain CN, Alreja M. Hypothesis for novel classes of chemoreceptors mediating chemogenic pain and itch. In: Dubner R, Gebhart GF, Bond MR, eds. *Proceedings of the Vth World Congress of Pain.* Amsterdam: Elsevier, 1988: 529–35.
9. Campbell JN, Raja SN, Cohen RH, Manning DC, Khan AA, Meyer RA. Peripheral neural mechanisms of nociception. In: Wall PD, Melzack R, eds. *Textbook of pain.* 2nd ed. Edinburgh: Churchill Livingstone, 1989: 22–45.
10. Campbell WB. Lipid derived autacoids: eicosanoids and platelet activating factor. In: Gilman AG, Rall TW, Nies AS, Taylor P, eds. *Goodman and Gilman's The pharmacological basis of therapeutics.* 8th ed. New York: Pergamon Press, 1990: 600–17.
11. McCormack K, Brune K. Dissociation between the antinociceptive and anti-inflammatory effects of the nonsteroidal anti-inflammatory drugs. A survey of their analgesic efficacy. *Drugs* 1991; **44**: 533–47.
12. Bjorkman RL, Hedner T, Hallman KM, Henning MN, Hedner J. Localization of the central antinociceptive effects of diclofenac in the rat. *Brain Research* 1992; **590**: 66–73.
13. Crook ST, Mattern M, Sarau HM, *et al.* The signal transduction system of the leukotriene D_4 receptor. *Trends in Pharmacological Sciences* 1989; **10**: 103–7.
14. Bloomquist EL, Kream RM. Leukotriene D_4 acts in part to contract guinea pig ileum smooth muscle by releasing substance P. *Journal of Pharmacology and Experimental Therapeutics* 1987; **240**: 523–8.
15. Levine JD, Lam D, Taiwo YO, Donatoni P, Goetzl EJ. Hyperalgesic properties of 15-lipoxygenase products of arachidonic acid. *Proceedings of the National Academy of Sciences, USA* 1986; **83**: 5331–4.
16. Eschalier A, Kayser V, Gilbaud G. Influence of specific $5HT_3$ antagonists on carageenan-induced hyperalgesia in the rat. *Pain* 1989; **36**: 249–55.
17. Bevan S, Yeats J. Protons activate a cation conductance in a subpopulation of rat dorsal root ganglion neurons. *Journal of Physiology* 1991; **433**: 145–61.
18. Stein C, Millan M, Shippenberg TS, Peter K, Herz A. Peripheral opioid receptors mediating antinociception in inflammation. Evidence for involvement of mu, delta and kappa receptors. *Journal of Pharmacology and Experimental Therapeutics* 1989; **248**: 1269–75.
19. Levine JD, Taiwo YO. Involvement of the mu-opiate receptor in peripheral analgesia. *Neuroscience* 1989; **32**: 571–5.
20. Taiwo YO, Levine JD. κ- and δ-opioids block sympathetically dependent hyperalgesia. *Journal of Neuroscience* 1991; **11**: 928–32.
21. Stein C, Hussan AHS, Przewlocki R, Gramsch C, Peter K, Herz A. Opioids from immunocytes interact with receptors on sensory nerves to inhibit nociception in inflammation. *Proceedings of the National Academy of Sciences USA* 1990; **87**: 5935–9.
22. Basbaum AI, Levine JD. Opiate analgesia. How central is a peripheral target? *New England Journal of Medicine* 1991; **325**: 1168–9.
23. Joshi GP, McCarroll SM. Intra-articular morphine for the management of frozen shoulder. *Anaesthesia* 1992; **47**: 627.
24. Coggeshall RE, Applebaum ML, Fazan M, Stubes TB, Sykes MT. Unmyelinated axons in human ventral roots, a possible explanation for the failure of dorsal rhizotomy to relieve pain. *Brain* 1975; **98**: 157–66.
25. Melzack R, Wall PD. Pain mechanisms: a new theory. *Science* 1965; **150**: 971–9.
26. Robert PJ, Keen P, Mitchell JF. The distribution and axonal transport of free amino acids and related compounds in the dorsal sensory neuron of the rat, as determined by the dansyl reaction. *Journal of Neurochemistry* 1973; **21**: 199–209.
27. Takeuchi A, Onodora R, Kawagoe R. The effects of dorsal root stimulation on the release of endogenous glutamate from the frog spinal cord. *Proceedings of the Japanese Academy* 1983; **59**: 88–92.
28. Watson S, Abbott A. *TiPS receptor nomenclature supplement.* Cambridge: Elsevier Science Publishers Ltd, 1991.
29. Mayer ML, Westbrook G, Guthrie PB. Voltage dependent block by Mg^{2+} of NMDA responses in spinal cord neurones. *Nature* 1984; **309**: 261–3.
30. Yoshioka K, Jessell TM. ATP release from the dorsal horn of

rat spinal cord. *Society of Neuroscience Abstracts* 1984; **10**: 993.

31. Duggan AW. Electrophysiology of opioid peptides and sensory systems. *British Medical Bulletin* 1983; **39**: 65–70.

32. Dickenson AH. Pain transmission and analgesia. In: Webster RA, Jordan CC, eds. *Neurotransmitters, drugs and diseases.* Oxford: Blackwell Scientific Publications, 1989: 446–64.

33. Otsuka M, Yanagisawa M. Does substance P act as a pain transmitter? *Trends in Pharmacological Sciences* 1987; **8**: 505–10.

34. Vaught JL. Substance P antagonists and analgesia: a review of the hypothesis. *Life Sciences* 1988; **43**: 1419–31.

35. Przewlocki R. Opioid peptides in relation to antinociception. *Polish Journal of Pharmacology and Pharmacy* 1987; **39**: 609–21.

36. Cruz L, Basbaum AI. Multiple opioid peptides and the modulation of pain: immunohistochemical analysis of dynorphin and enkephalin in the trigeminal nucleus caudalis and spinal cord of the cat. *Journal of Comparative Neurology* 1985; **240**: 331–48.

37. Dickenson AH. Mechanism of the analgesic actions of opiates and opioids. *British Medical Bulletin* 1991; **47**: 690–702.

38. Mokha SS. Differential influence of naloxone on the responses of nociceptive neurons in the superficial versus the deeper dorsal horn of the medulla in the rat. *Pain* 1992; **49**: 405–13.

39. Fujimoto JM, Arts KS, Rady JJ, Tseng LF. Spinal dynorphin A (1–17): possible mediator of antianalgesic action. *Neuropharmacology* 1990; **29**: 609–17.

40. Headley PM, Duggan AW, Griersmith BT. Selective reduction by noradrenaline and 5HT of nociceptive responses of cat dorsal horn neurons. *Brain Research* 1978; **145**: 185–9.

41. Yaksh TL, Wilson PR. Spinal serotonin terminals system mediates antinociception. *Journal of Pharmacology and Experimental Therapeutics* 1979; **208**: 446–53.

42. Driessen B, Reimann W. Interaction of the central analgesic, tramadol, with the uptake and release of 5-hydroxytryptamine in the rat brain *in vitro. British Journal of Pharmacology* 1992; **105**: 147–51.

43. Vickers MD, O'Flaherty D, Szekely SM., Read M, Yoshizumi J. Tramadol: pain relief without depression of respiration. *Anaesthesia* 1992; **47**: 291–6.

44. Fields HL, Basbaum AI. Endogenous pain control mechanisms. In: Wall PD, Melzack R, eds. *Textbook of pain.* 2nd ed. Edinburgh: Churchill Livingstone, 1989: 206–17.

45. Aho M, Erkola O, Korttila K. α_2-Adrenergic agonists in anaesthesia. *Current Opinion in Anaesthesiology* 1992; **5**: 481–7.

46. Liu Q-S, Qiao J-T, Dafny N. D_2 dopamine receptors involvement in spinal dopamine-produced antinociception. *Life Sciences* 1992; **51**: 1485–92.

47. Mouchet P, Manier M, Dietl M, *et al.* Immunochemical study of catecholaminergic cell bodies in the rat spinal cord. *Brain Research Bulletin* 1986; **16**: 342–54.

48. Yaksh TL, Aimone LD. The central pharmacology of pain transmission. In: Wall PD, Melzack R, eds. *Textbook of pain.* 2nd ed. Edinburgh: Churchill Livingstone, 1989: 181–205.

49. Sawynok J, Sweeney MI, White TD. Adenosine release may mediate spinal analgesia by morphine. *Trends in Pharmacological Sciences* 1989; **10**: 186–9.

50. Keil GJ, DeLander GE. Spinally-mediated antinociception is induced in mice by an adenosine kinase, but not an adenosine deaminase, inhibitor. *Life Sciences* 1992; **51**: 171–6.

51. Krukoff TL, Ciriello J, Calaresu FR. Somatostatin-like immunoreactivity in neurons, nerve terminals, and fibres of the cat spinal cord. *Journal of Comparative Neurology* 1986; **243**; 13–22.

52. Sandkuhler J, Fu Q-G, Helmchen C. Spinal somatostatin superfusion *in vivo* affects activity of cat nociceptive dorsal horn neurons: comparison with spinal morphine. *Neuroscience* 1990; **34**: 565–76.

53. Penn RI, Paice JA, Kroin JS. Octreotide: a potent new non-opiate analgesic for intrathecal infusions. *Pain* 1992; **49**: 13–19.

54. Reynolds DV. Surgery in the rat during electrical analgesia induced by focal brain stimulation. *Science* 1969; **164**: 444–5.

55. Fields HL, Heinricher MM, Mason P. Neurotransmitters in nociceptive modulatory circuits. *Annual Review of Neuroscience* 1991; **14**: 219–45.

56. Yaksh TL, AL-Rodhan NRF, Jensen TS. Sites of action of opiates in production of analgesia. *Progress in Brain Research* 1988; **77**: 371–94.

57. Heyman JS, Vaught JL, Raffa RB, Porreca F. Can supraspinal δ-opioid receptors mediate antinociception? *Trends in Pharmacological Sciences* 1988; **9**: 134–8.

58. Leighton GE, Rodriguez RE, Hill RG, Hughes J. κ-Opioid agonists produce antinociception after i.v. and i.c.v. but not intrathecal administration in the rat. *British Journal of Pharmacology* 1988; **93**: 553–60.

59. Millan MJ. κ-Opioid receptors and analgesia. *Trends in Pharmacological Sciences* 1990; **11**: 70–6.

60. Hamann SR, Martin WR. Analgesic actions of dynorphin A (1–13) antiserum in the rat brain stem. *Brain Research Bulletin* 1992; **29**: 605–7.

61. Advokat C. The role of descending inhibition in morphine-induced analgesia. *Trends in Pharmacological Sciences* 1988; **9**: 330–41.

62. Dickenson AH, Le Bars D. Lack of evidence for increased descending inhibition on the dorsal horn of the rat following periaqueductal grey morphine microinjections. *British Journal of Pharmacology* 1987; **92**: 271–80.

63. Gebhart GF, Sandkuhler J, Thalhammer JG, Zimmerman M. Stimulation and morphine in the PAG: inhibition in the spinal cord of nociceptive information by electrical stimulation and morphine microinjection at identical sites in the mid brain of the cat. *Journal of Neurophysiology* 1989; **51**: 75–89.

64. Gogas KR, Presley RW, Levine JD, Basbaum AI. The antinociceptive action of supraspinal opioids results from an increase in descending inhibitory control: correlation of nociceptive behaviour and c-fos expression. *Neuroscience* 1991; **42**: 617–28.

65. Bonica JJ. Biochemistry and modulation of nociception and pain. In: Bonica JJ, ed. *The management of pain.* 2nd ed. Vol. 1. Philadelphia: Lea & Febiger, 1990: 95–132.

66. Basbaum AI, Fields HL. Endogenous pain control systems: brain stem spinal pathways and endorphin circuitary. *Annual Review of Neuroscience* 1984; **7**: 309–38.

67. Frederickson RCA, Chipkin RE. Endogenous opioids and pain: status of human studies and new treatment concepts. In: Fields HL, Besson J-M, eds. *Progress in Brain Research* 1988; **77**: 407–17.
68. Millhorn DE, Hokfelt T, Seroogy K, Verhofstad AAJ. Extent of colocalization of serotonin and GABA in neurons of the ventral medulla oblongata in rat. *Brain Research* 1988; **461**: 159–74.
69. Bowery NG, Hudson AL, Price GW. GABA$_A$ and GABA$_B$ receptor site distribution in the rat central nervous system. *Neuroscience* 1987; **20**: 365–83.
70. Drower EJ, Hammond DL. GABAergic modulation of nociceptive threshold: effects of THIP and bicuculline microinjected in the ventral medulla of the rat. *Brain Research* 1988; **450**: 316–24.
71. Heinricher MM, Haws CM, Field HL. Evidence for GABA-mediated control of putative nociceptor modulating neurons in the rostral ventromedial medulla: iontophoresis of bicuculline eliminates the off cell pause. *Somatosensory and Motor Research* 1991; **8**: 215–25.
72. Ge M, Ueda H, Satoh M. Endogenous GABA release into the fourth ventricle of rat brain *in vivo* is enhanced by noxious stimuli. *Neuroscience Letters* 1988; **92**: 76–81.
73. Clements JR, Madl JE, Johnson RL, Larson AA, Beitz AJ. Localization of glutamate, glutaminase, aspartate and aspartate aminotransferase in the rat mid brain periaqueductal gray. *Experimental Brain Research* 1987; **67**: 594–602.
74. Wiklund L, Behzadi G, Kalen P, *et al.* Autoradiographic and electrophysiological evidence for excitatory aminoacid transmission in the periaqueductal gray projection to the nucleus raphe magnus in the rat. *Neuroscience Letters* 1988; **93**: 158–63.
75. Jacquet YF. The NMDA receptor: central role in pain inhibition in rat periaqueductal gray. *European Journal of Pharmacology* 1988; **154**: 271–6.
76. van Praag H, Frenk H. The role of glutamate in opiate descending inhibition of nociceptive spinal reflexes. *Brain Research* 1990; **524**: 101–5.
77. Davies M, Wilkinson LS, Roberts MHT. Evidence for excitatory 5-HT$_2$-receptors on rat brain stem neurones. *British Journal of Pharmacology* 1988; **94**: 483–91.
78. Davies M, Wilkinson LS, Roberts MHT. Evidence for depressant 5-HT$_1$-like receptors on rat brain stem neurones. *British Journal of Pharmacology* 1988; **94**: 492–9.
79. Wiertelak EP, Maier SF, Watkins LR. Cholecystokinin antianalgesia: safety cues abolish morphine analgesia. *Science* 1992; **256**: 830–3.
80. Hughes J, Woodruff GN. Neuropeptides function and clinical applications. *Arzneimittel-forschung/Drug Research* 1992; **42**: 250–5.
81. Hill RG, Hughes J, Pittaway KM. Antinociceptive action of cholecystokinin octapeptide (CCK8) and related peptides in rats and mice: effect of naloxone and peptidase inhibitors. *Neuropharmacology* 1987; **26**: 289–300.
82. Hong EK, Takemori AE. Indirect involvement of delta opioid receptors in cholecystokinin octapeptide-induced analgesia in mice. *Journal of Pharmacology and Experimental Therapeutics* 1989; **251**: 584–98.
83. Ruiz-Gayo M, Durieux C, Fournie-Zaluski MC, Roques BP. Stimulation of δ opioid receptors reduces the *in vivo* binding of the cholecystokinin (CCK)-B-selective agonist [^{3}H]pBC 264: evidence for a physiological regulation of CCKergic systems by endogenous enkephalins. *Journal of Neurochemistry* 1992; **59**: 1805–11.

Chapter 6

Intravenous Anaesthetic Agents: Induction and Maintenance

R. S. J. Clarke

Thiopentone	Midazolam
Methohexitone	Propofol
Etomidate	Future developments
Ketamine	

Intravenous anaesthesia became possible with drugs available c. 1930 and the concept rapidly became popular with patients and anaesthetists. From the patient's point of view it had the advantage of producing rapid loss of consciousness without excitement, distress or the sensation of smothering often produced by a tightly pressed facemask. For the anaesthetist, there was the advantage of predictable anaesthesia which was (ideally) rapid in onset and without coughing or movements. Unfortunately, up to this period anaesthesia had been produced largely by a single agent and attempts to continue this practice with thiopentone led to the disasters following Pearl Harbor. As a sequel to this the concept of balanced anaesthesia developed, in which each drug had a role best suited to its characteristics. This concept which in practice means one drug for induction of anaesthesia, one for analgesia and one for muscle relaxation, has persisted over the past 40 years.

The early intravenous anaesthetics were either slow in onset of action (pentobarbitone) or had excitatory effects which made them unsatisfactory (hexobarbitone); only thiopentone approached the ideal of being reliable, rapid-acting and smooth. Ralph Waters from Madison, Wisconsin gave the first anaesthetic with thiopentone but John Lundy was the first to publish a report.[1] Its use was not, however, followed by rapid recovery and repeated doses cumulated to produce a really prolonged duration of action. The inhalation agents therefore continued to be regarded as the ideal for maintenance with the intravenous agents used purely for induction. The advent of propofol, with its different pharmacokinetic profile, changed this approach and made it possible to maintain anaesthesia satisfactorily with the same agent that had been used for induction. This concept has required new equipment and a new approach by the anaesthetist but it must be stressed that the concept of balanced anaesthesia and 'nothing in excess' should still hold good.

It is useful to consider the ideal intravenous anaesthetic at this stage and decide how each drug discussed below fits this model. The ideal intravenous anaesthetic should be:

1. available in a stable, non-irritant solution the solvent of which causes no adverse effects in normal usage (ideally water at a pH of 7.4).
2. rapid acting, i.e. in one arm–brain circulation time.
3. short acting, so that recovery is not prolonged after short procedures and so that it can be given repeatedly or by infusion for long procedures without cumulation.
4. devoid of cardiovascular side effects, especially myocardial depression.
5. devoid of respiratory side effects particularly medullary depression.
6. devoid of cerebral excitatory side effects such as rigidity, twitching or convulsions.
7. devoid of cerebral cortical side effects (e.g. dreams or hallucinations) during or at the end of the anaesthesia.
8. devoid of metabolic side effects, such as adrenal suppression or interaction in porphyria.
9. not a histamine liberator.

Thiopentone

Chemistry and physical properties

Thiopentone is a derivative of barbituric acid, substitution of different groupings attached to the ring resulting in compounds of different properties. It differs from its predecessor pentobarbitone in having a sulphur atom, rather than oxygen, attached to one carbon atom resulting in more rapid induction and recovery.

The drug is used in the form of its sodium salt which is

H_5C_2 C O H N $C=S$ $H_3C-CH_2-CH_2-CH$ C N CH_3 O Na

Thiopentone

freely soluble in water, though the solution has only limited stability. Commercial preparations contain sodium carbonate which prevents precipitation of the free acid by atmospheric CO_2 and results in an alkaline solution (pH 10.5–11.0). It is largely non-ionized at body pH, a fact which facilitates its diffusion through membranes.

Dosage and administration

Thiopentone is supplied in vials of 2.5 g or 0.5 g which are made up into a 2.5 per cent solution. The usual induction dose is 4–5 mg/kg.

Pharmacokinetics

When injected intravenously the peak level in vessel-rich tissue is reached in 1–2 minutes. In the plasma 60–80 per cent of the thiopentone is bound to plasma protein and is therefore non-diffusible. Various factors influence binding, such as pH, plasma albumin level and pretreatment with other bound drugs including probenecid and aspirin, but this is probably only clinically important for the immediate effects of very large doses. Because of its high lipid solubility (partition coefficient) it diffuses rapidly into the brain and EEG changes with loss of consciousness can occur in 10 seconds. Levels in the brain begin to fall in a further 60 seconds while the drug is still diffusing into muscle and this rapid redistribution is responsible for the rapid recovery from small doses. Equilibrium with the muscles does not occur for about 15 minutes after administration and thereafter its tissue level parallels that found in plasma. In spite of its high solubility in adipose tissue the concentration here rises only slowly over about 2 hours due to the poor blood supply. Thiopentone then diffuses very slowly out of the fat, maintaining a continuous and clinically significant plasma level for some hours.

Thiopentone is metabolized in the liver and virtually none is excreted unchanged. The hepatic extraction ratio has been estimated as 0.1–0.2.[2] There are three pathways involved:

1. oxidation of the C_5 side chains.

2. oxidative replacement of the sulphur at C_5 to form pentobarbitone.
3. cleavage of the barbiturate ring to form urea and a three-carbon fragment.

The capacity of the liver is such that hepatic dysfunction only causes prolonged recovery from anaesthesia after large doses.

The decline in plasma thiopentone concentration is exponential but it can best be described by three separate curves characterized by three half-lives: $t_{\frac{1}{2}\alpha1}$ of 2–6 minutes corresponding to diffusion into tissues of high blood flow; $t_{\frac{1}{2}\alpha2}$ of 30–60 minutes corresponding to diffusion into adipose tissue and the β half-life ($t_{\frac{1}{2}\beta}$) of 5–10 hours corresponding to the elimination phase.[3]

There are two important practical consequences of the prolonged elimination of thiopentone:

1. The residual effects of even a small single dose will persist for many hours, reducing mental anxiety and having an additive effect with other sedatives.
2. The use of repeated small doses or infusion results in cumulation in adipose tissues and persistent effects for many days.

Factors influencing pharmacokinetics

Pregnancy has little influence on the pharmacokinetics of thiopentone and although the clearance is greater in pregnant patients, the elimination half-life is longer. The drug diffuses freely across the placental barrier but the studies of Morgan and colleagues[4] suggest that the rapid redistribution in the maternal tissues between induction of anaesthesia and delivery is a major factor in producing a more alert baby than the mother's state would suggest. Certainly the umbilical venous plasma concentrations, which should be similar to those entering the fetal brain, were well below the arterial plasma concentrations required to produce anaesthesia in adults.

Renal failure reduces the required induction dose of thiopentone presumably by the resultant hypoproteinaemia and reduced plasma binding. However, clearance and elimination of the drug is unchanged and therefore recovery should not be prolonged.

Liver failure can also lead to hypoproteinaemia and have effects similar to those above. Any reduction in hepatic elimination of thiopentone in cirrhosis will only occur with very high doses.

Increasing age is associated with a lower induction dose of thiopentone probably on the basis of a lower cardiac output and failure to compensate for the effects of the drug on the circulation. The same argument probably applies to other 'poor-risk' patients in whom the induction dose appears to be less than normal and hypovolaemia may also affect the initial redistribution.[5]

Pharmacodynamics

Central nervous system

Thiopentone rapidly diffuses across the blood–brain barrier, causing cortical depression. This is accompanied by characteristic changes in the EEG consisting of high amplitude waves of mixed frequency 10–30 Hz or faster. As anaesthesia deepens, there is progressive suppression of cortical activity interrupted by bursts of high voltage waves at 10 Hz.

The cerebral metabolic rate falls by about 50 per cent with an infusion of thiopentone sufficient to keep the EEG isoelectric. At the same time the cerebral blood flow is reduced by 30–50 per cent. In addition, the reduction in cardiac output and arterial blood pressure lead to secondary effects on cerebral perfusion. A second dose of thiopentone given to neurosurgical patients immediately prior to tracheal intubation prevents the adverse effects on ICP of the hypertension.[6] Similarly, during carotid endarterectomy a bolus dose of 4–5 mg/kg produces advantages from the local metabolic depression if given prior to arterial occlusion.[7] However, when infusions are used for long-term barbiturate coma in the treatment of severe head injury, no beneficial effect can be demonstrated.[8]

Thiopentone is not an analgesic and small doses may increase sensitivity to somatic pain. In addition, the depth of anaesthesia under thiopentone depends as much on the surgical stimulus as on the degree of cortical depression. The patient may be lying quietly relaxed and scarcely breathing until the surgical incision causes a sudden increase in tone with withdrawal movements of the limbs.

Acute tolerance is a term applied to the phenomenon of return of consciousness after a single, large dose occurring at a higher plasma concentration than following smaller doses. A large induction dose usually needs to be followed by larger increments to maintain a constant level of anaesthesia than after smaller induction doses. Since other effects of thiopentone are more directly dose-related, this phenomenon suggests that doses should be kept as low as is compatible with an adequate anaesthetic effect and supported by other drugs in the process of balanced anaesthesia.

Cardiovascular system

Thiopentone has well recognized haemodynamic effects. When given as a bolus it causes a reduction in arterial blood pressure, left ventricular stroke work index and pulmonary wedge pressure, with marked peripheral vasodilatation. However, there is some degree of tachycardia (10–20 per cent) which, with lower doses, contributes to maintenance of the blood pressure and cardiac output.[9,10] The changes appear to be essentially the same in 'healthy' patients and those with ischaemic heart disease but in patients with treated hypertension there is no significant rise in heart rate and a marked fall in cardiac output. Arrhythmias are rare with this as with other i.v. anaesthetics. The ability to compensate is impaired, particularly in hypovolaemic patients and induction is therefore hazardous in patients in compensated shock states. In all such patients resuscitation should, if possible, precede induction of anaesthesia. Patients with fixed cardiac output as with mitral stenosis and cardiac tamponade also require slow administration of the drug with careful assessment of anaesthetic depth, as do patients with chronic hypertension.

Higher doses of thiopentone cause increasing myocardial depression and vasodilatation, but with a normal cardiovascular system the changes are tolerable.[11] In a study in which neurosurgical patients were given high doses over many hours in order to render the EEG isoelectric, it was shown that if one accepts the prolonged recovery period, the only marked disturbances were falls in left and right ventricular stroke work indices (66 and 69 per cent of control).

Thiopentone has been shown to increase the myocardial oxygen consumption in healthy patients but reduces it by as much as 39 per cent in those with ischaemic heart disease. The latter effect is due to a reduced oxygen requirement and even in patients with ischaemic heart disease there is no evidence that it causes adverse effects.[9,12]

Respiratory system

Thiopentone, in common with other barbiturates, is a potent respiratory depressant, that is, it depresses the spontaneous respiratory rate and depth and depresses the sensitivity of the respiratory centre to carbon dioxide. In deep anaesthesia with thiopentone the respiratory drive is maintained largely by the carotid sinus and is related to hypoxia. Carbon dioxide retention may be particularly difficult to detect since the thiopentone may also obtund the usual rise in arterial pressure caused by hypercapnia.

Cough and hiccough are uncommon during induction with thiopentone but there is a heightening of the laryngeal reflexes. This means that minor stimuli frequently cause laryngospasm during light thiopentone anaesthesia. Bronchospasm is rare in normal usage but can occur in asthmatic patients and it is preferable to avoid the drug with them. In many cases, however, of apparent bronchospasm the respiratory difficulties are due to intercostal muscle spasm and coughing and can be reversed by administration of a neuromuscular blocking drug.

Hepatic and renal function

Thiopentone has no effect on either hepatic or renal function in conventional induction doses. This applies even to its use in the presence of pre-existing damage of these organs. There is, however, transient impairment of liver function following large doses and infusions.[13]

Porphyria

This is an inborn error of porphyrin metabolism characterized by cutaneous, central nervous system and alimentary tract disturbances. Porphyrins are pigments composed of four pyrrole rings linked by methane bridges and occur particularly in haemoglobin and cytochrome. They are formed in the liver from aminolaevulinic acid (ALA), a process catalysed by ALA-synthetase. It appears that in some patients this enzyme can be stimulated by certain drugs, increasing porphyrin levels and causing marked excretion of porphyrin in the urine. There are several types of porphyria and not all are adversely affected by thiopentone[14,15] but the dangers are so serious that the drug is best avoided in all types. The classical presentation is with colicky abdominal pain, muscle weakness leading to paralysis, psychiatric manifestations and red coloured urine. When the attack is spontaneous or induced by other drugs it is important to avoid exacerbation by administration of a barbiturate.

Local effects

Pain on injection and evidence of venous irritation are rare with thiopentone even when injected into small veins. The incidence of venous thrombosis is approximately 3–4 per cent. However, thiopentone can cause tissue damage if injected subcutaneously, particularly in large volumes.

The serious danger with thiopentone is intra-arterial injection, though even this has been reduced by the routine use of 2.5 per cent rather than 5 per cent solutions. Injection into an artery leads to precipitation of crystals of insoluble thiopentone in the smaller vessels of the limb. ATP is then released, leading to intimal damage and intravascular thrombosis. Immediately on injection the patient complains of intense pain and the radial pulse may disappear as the arterial system undergoes generalized spasm.[16]

Avoidance is most readily achieved by awareness of the danger and being prepared for aberrant arteries. Treatment includes the use of lignocaine and papaverine into the same vessel, so the needle or cannula should always be left *in situ*.

Other systems

Thiopentone reduces the intraocular pressure, though this is raised to normal or above by subsequent administration of suxamethonium or by tracheal intubation. It has no action on uterine tone, does not interact with myoneural blocking agents and has no clinically important effect on the cortisol response to surgical stress. It appears to have no tendency to cause nausea and vomiting after anaesthesia and its sole use is associated with less sickness than with any of the inhalational anaesthetics.

Methohexitone

Methohexitone

This has many of the same properties as thiopentone and only important differences need be discussed here. It differs chemically in having a CH_3 group attached to an N atom, resulting in more rapid recovery but more marked excitatory phenomena.[15] It is, however, an oxy- rather than a thiobarbiturate which, as has been said, tends to prolong its action.

Dosage and administration

Methohexitone is supplied in vials of 0.5 g which are made up into a 1 per cent solution. This solution is stable in cool conditions for several weeks. It has approximately 2.5 times the potency of thiopentone and the usual induction dose is 2.0–2.5 mg/kg.

Pharmacokinetics

The uptake of methohexitone into the tissues is similar to that of thiopentone as its redistribution half-life ($t_{\frac{1}{2}\alpha 1}=6$ minutes; $t_{\frac{1}{2}\alpha 2}=58$ minutes) but the elimination half-life is significantly shorter ($t_{\frac{1}{2}\beta}=4$ hours).[17] This is due to a higher hepatic excretion ratio and clearance of the drug. Nevertheless, it must be stressed that as with all barbiturates, there is a high fat uptake and recovery to full manual dexterity may take 24 hours.

Pharmacodynamics

Induction with methohexitone is accompanied by spontaneous muscle movements, tremor and hypertonus in approximately 20 per cent of cases (compared with about 4 per cent for thiopentone). The incidence, however, increases with dose and rate of administration and is higher when drugs such as hyoscine, droperidol or the phenothiazines are given for premedication. The incidence is reduced by opioid premedication though this is probably undesirable in the situation where methohexitone is most useful – outpatient anaesthesia.[15]

The cardiovascular effects of methohexitone are similar to those of thiopentone except that the former causes a greater degree of tachycardia and less hypotension.

Pain on injection is a frequent complaint with methohexitone, unlike thiopentone, but venous thrombosis is only minimally more common.

Etomidate

Etomidate

Chemistry and physical properties

Etomidate is water soluble but for reasons of stability it is marketed in a propylene glycol solvent. In this form it is stable at room temperature for 2 years.

Dosage and administration

It is supplied in 10 ml ampoules as a 2 mg/ml solution. The usual induction dose is 0.3 mg/kg.

Pharmacokinetics

Etomidate is about 76 per cent bound to plasma proteins and diffuses readily into the brain. There is a rapid redistribution and elimination phase, the $t_{\frac{1}{2}\alpha1}$ being 2.6 minutes, $t_{\frac{1}{2}\alpha2}$ being 29 minutes and the $t_{\frac{1}{2}\beta}$ being 75 minutes.[18] Etomidate is metabolized mainly in the liver by ester hydrolysis to pharmacologically inactive metabolites. About three-quarters of the administered dose is excreted in the urine in this form, with only about 2 per cent being excreted unchanged.

Pharmacodynamics

Central nervous system

The onset of action is in one arm–brain circulation and the dose–response is reliable and predictable. However, there are frequently severe excitatory effects, unless the drug is preceded by an intravenous opioid or a muscle relaxant is given simultaneously.

Recovery from etomidate is dose related and is faster than from equivalent doses of thiopentone or even methohexitone.

Cardiovascular system

In fit patients 0.3 mg/kg produces a slight increase in cardiac index accompanied by a slight fall in heart rate, arterial pressure and peripheral resistance. Cardiac contractility is enhanced as judged by dP/dt and in general it has been recommended for poor-risk patients by many anaesthetists,[19] provided account is taken of the fall in peripheral resistance by giving adequate fluid replacement. Myocardial oxygen consumption is not increased by etomidate and it appears to have a true but weak coronary vasodilator effect.

Respiratory system

Equipotent doses of etomidate and methohexitone cause a similar shift in the CO_2 response curve[20] but at any given CO_2 tension ventilation is greater after etomidate than after the barbiturate. It may therefore have some stimulant effect on the respiratory centre, though in clinical practice this will be counteracted by any accompanying opioid.

Adrenocortical function

The pharmacokinetic and cardiovascular profile of etomidate made it an ideal drug for long-term infusions and it was used frequently for sedation in intensive care units. It also had a place for neurosurgical operations and other situations suitable for total intravenous anaesthesia. Its safety was, however, questioned by Ledingham and Watt[21] having noted during 1981–1982 a significant increase in mortality in patients surviving more than 5 days from the time of injury, who had received mechanical ventilation. There was no difference in the severity of illness or degree of sepsis between these patients and the lower mortality group treated in the previous 2-year period. The only apparent difference was in the sedation technique, the earlier morphine–benzodiazepine regimen having been replaced by etomidate infusions. They suggested that the causative mechanism might be suppression of adrenocortical function by etomidate by a direct effect on steroid synthesis in the adrenal cortex.

This hypothesis was tested over subsequent years in animal studies which showed that the drug inhibits corticosteroid production whether induced by drugs or ACTH. Subsequent studies by Sear and colleagues[22] confirmed this suppressant effect on adrenocortical function in man. They compared the response to induction followed by infusion of etomidate to that in similar patients anaesthetized with thiopentone, nitrous oxide and halothane. The latter group of patients showed the expected significant increase in plasma cortisol level whereas with the etomidate technique there was a slight decrease. Others have since shown that aldosterone levels are also suppressed by etomidate,[23] suggesting that etomidate inhibits an early stage of steroidogenesis in the adrenal cortex. It now appears that in this respect etomidate behaves quite differently from thiopentone or propofol and that this difference applies even following single induction doses.[24]

Local effects

Since etomidate is normally given in the propylene glycol preparation, it would be expected to cause pain on injection and venous irritation. The incidence of venous sequelae is about 25 per cent and is dose related.[25] More recently haemolysis has been noted as a problem[26] which is not usually severe enough to cause haemoglobinuria since the haemoglobin is taken up by circulating haptoglobin. This problem is avoided if etomidate is dissolved in a fat emulsion.

Other systems

Etomidate, when used for induction, is followed by a higher incidence of emesis than other intravenous anaesthetics.

The drug has been used uneventfully in a known porphyric patient but studies in a rat model have shown that it causes an increase in ALA-synthetase activity and in serum porphyrin levels. It must therefore be regarded as potentially porphyrogenic.[27]

Ketamine

Ketamine

Chemistry and physical properties

Ketamine is a water-soluble drug chemically related to phencyclidine. It forms aqueous solutions of pH 3.5–5.5.

Dosage and administration

It is distributed in 10, 50 and 100 mg/ml strengths and is usually administered in doses of 2 mg/kg i.v. or 10 mg/kg i.m.

Pharmacokinetics

Ketamine has a high lipid solubility and follows a pattern of distribution similar to that of thiopentone. The distribution half-life ($t_{\frac{1}{2}\alpha}$) is around 10 minutes and the elimination half-life ($t_{\frac{1}{2}\beta}$) is approximately 2–3 hours after a single dose. The drug is largely broken down in the liver and excreted in the bile with 20 per cent appearing in the urine in various metabolites. The main metabolite is norketamine and this has hypnotic properties accounting for some of the residual drowsiness in higher plasma levels.[28]

Pharmacodynamics

Central nervous system

Although ketamine has a limited place in clinical anaesthesia, its unusual properties give it a special interest. It produces a condition known as dissociative anaesthesia which is characterized by catalepsy, light sedation, amnesia and marked analgesia. The rate of onset of anaesthesia is much slower than with thiopentone, and it takes usually 20–60 seconds from administration into a fast circulation until a patient stops counting. Even then, the criteria are hard to establish, for often patients appear to gaze sightlessly into space and may not close their eyes for several minutes. Unlike other types of anaesthesia, the eyelash, corneal and laryngeal reflexes are not lost and muscle tone actually increases. This may be accompanied by involuntary muscle movements. A particular advantage of ketamine is that it appears to block awareness in a situation such as caesarean section, where it is desirable to keep concentrations of anaesthetic drugs at a minimum. It has been shown by an isolated arm technique that the incidence of recall is significantly higher using thiopentone–halothane–suxamethonium than using ketamine–halothane–suxamethonium.[29]

Ketamine also produces profound analgesia even in subhypnotic doses and this lasts after recovery of consciousness. The mechanism includes blockade of the spinoreticular pain pathways, depression of the thalamus and depression of the affective–emotional component of pain perception.

Unlike the barbiturates, ketamine increases the cerebral oxygen consumption, cerebral blood flow and cerebrospinal fluid pressure. These changes are independent of any rise in blood pressure and persist even under halothane anaesthesia but can be blocked by thiopentone. They can be dangerous in patients with intracranial pathology in whom they are more marked.

The return of consciousness after ketamine anaesthesia is hard to define because of a phase in which patients seem 'distant' and unaware of their surroundings. In addition, patients often have visual disturbances and may have difficulty in speaking in the recovery phase. At the same time patients have a variety of mental states, ranging from a dreamy, floating feeling to unpleasant dreams and aggressive violence. They can feel isolated, negative, hostile, apathetic, drowsy or drunk. On the whole the severity of symptoms is dose related, severe reactions occurring in 20–50 per cent of patients and made worse by disturbing the patient in this phase. They are commoner in women than men, are blocked by opioid or benzodiazepine premedication and can be treated by a small dose of thiopentone or benzodiazepine.

The emergence upset and dreams are diminished or absent in children and the old and the drug has proved very useful in children undergoing investigative procedures or burns dressings. Even in young adults the severity of sequelae diminishes with repeated use in, for instance, burns dressings.

Cardiovascular system

In contrast to almost all other anaesthetic agents ketamine stimulates the cardiovascular system, causing an increase in heart rate, stroke index, arterial blood pressure and myocardial oxygen consumption. The rise in systolic pressure is usually 20–40 mmHg with a peak after about 5 minutes and a return to normal over the next 10–20 minutes. This sequence is largely due to a direct stimulant effect of the drug on the central nervous system since it acts as a direct myocardial depressant and peripheral vasodilator. These latter effects can lead to a paradoxical circulatory response in critically ill patients which makes it unsuitable for what was once considered a main indication.[30] The drug is also contraindicated in patients with severe coronary artery disease because of the increase in oxygen consumption. The cardiovascular stimulation can be blocked by prior i.v. administration of thiopentone or a benzodiazepine but not by conventional premedication. There is no clear relation between plasma catecholamine concentrations and central haemodynamics although a rise in circulating catechols has been shown.

Ketamine, when given intramuscularly, has a similar action, the anaesthesia and hypertensive response requiring 8–10 minutes to reach a peak. The effect also lasts longer before subsiding.

Respiratory system

Respiratory depression is minimal and transient after clinical doses of ketamine. However, hypoventilation and a fall in PaO_2 can occur after rapid injection of 2 mg/kg in a spontaneously breathing patient on room air.[31] These problems have also been reported after opioid premedication and must certainly be remembered when ketamine is given in disaster situations. Ketamine specifically dilates the bronchial tree and antagonizes the bronchoconstrictor effects of histamine, so that it has theoretical advantages in anaesthesia for the asthmatic.

Secretions in the tracheobronchial tract are stimulated by ketamine and the use of a drying agent before anaesthesia is to be recommended, particularly in children. Coughing, hiccough and laryngospasm are also commoner with ketamine than with thiopentone. It was originally thought that laryngeal protective reflexes were intact under ketamine anaesthesia. However, this is no longer believed and with a typical anaesthetic dose of 2 mg/kg, radiopaque contrast medium can be shown to pass from pharynx to the lung fields. The risks of this are enhanced by sedative premedication.

Isomers of ketamine

The work of White and colleagues[32] showed that the isomers of ketamine had different properties from each other and from the racemic mixture which is normally used. In terms of potency the (+) isomer is 3.4 times more powerful than the (−) isomer and also causes more psychic emergent reactions. It is also a better analgesic as judged by the prevalence of postoperative pain. Despite these differences which give the (+) isomer some advantages, there has been no attempt to market or indeed to study it for clinical safety.

Clinical applications

There are few indications for the use of ketamine as the anaesthetic of choice. While the myocardial stimulation might appear to be valuable in the emergency, hypovolaemic patient, the undesirable cardiovascular effects make it unreliable. The preservation of muscle tone, laryngeal reflexes and respiratory activity could give it a place for anaesthesia at the site of a disaster or battlefield but again the safeguards are not absolute and the ability to protect the airway with certainty and supply additional oxygen is desirable.

The place of ketamine with midazolam and vecuronium in total intravenous anaesthesia for military surgery has been described. It was found to be safe in the typical soldier who is in excellent health and prior to injury requires minimal monitoring.[33] However, any attempt to accelerate the rather slow recovery by the use of flumazenil is accompanied by an increase in hallucination and agitation.[34]

The fact that ketamine can be given intramuscularly with good absorption gives it a place in young children or patients with extensive burns. Its analgesic properties make it particularly useful here and the unpleasant psychic sequelae are rarely encountered in such patients. These points give it a particular advantage in children having neurological investigations, radiotherapy or dentistry. In general, therefore, the place of the drug is assured and it is certainly more widely used in mainland Europe and North America than in the United Kingdom.

Midazolam

Midazolam

Midazolam is the only benzodiazepine which need be considered in the context of an induction agent and it has only a limited place. Its role is largely as a sedative to accompany regional analgesia of all types, ranging from dental extractions to hip replacement.

Chemistry and physical properties

Midazolam belongs to the benzodiazepines but unlike most of this group it is water soluble. This is because its formula includes a ring which opens at pH values below 4.0, imparting water solubility. At the pH of plasma the ring closes and lipid solubility is enhanced.

Pharmacokinetics

Midazolam is highly protein bound (approximately 95 per cent), though not as highly bound as diazepam. The practical implication of this is that patients with a low plasma albumin from any cause will have an enhanced response to it. The drug follows the usual distribution pattern to vessel-rich tissues and later to the poorly perfused fat. Elimination is then dependent on hepatic biotransformation, which converts it into 4-hydroxymidazolam, a metabolite almost devoid of pharmacological activity. The initial redistribution is shorter than with diazepam, contributing to the more rapid recovery from the newer drug. The elimination phase ($t_{\frac{1}{2}\beta}$=2–3 hours) is also more rapid than with diazepam, though slower than with thiopentone or propofol. Elimination is prolonged in elderly patients and following any major surgery ($t_{\frac{1}{2}\beta}$=approximately 5 hours), the latter presumably by interfering with liver blood flow.[35] Placental transmission, as judged by the fetal/maternal plasma ratio in animals, is less for midazolam than for diazepam.

Pharmacodynamics

Central nervous system

This group of drugs acts on specific benzodiazepine receptors which are concentrated in the cerebral cortex, hippocampus and cerebellum. Their action is produced by potentiation of specific depressant interneurons which use γ-aminobutyric acid (GABA) as a transmitter. The release of GABA opens the Cl^- channel, resulting in hyperpolarization of the nerve cell. In this connection it should also be noted that the specific benzodiazepine antagonist, flumazenil, acts by competitive inhibition of these benzodiazepine receptors, thereby blocking the action of midazolam.

The onset of action is slow and the onset of sleep takes 2–5 minutes but with wide interpatient variation. Similarly, the dose required to induce sleep ranges widely around 0.3

mg/kg. However, lower doses (0.05–0.1 mg/kg) will produce drowsiness and amnesia, which is often all that is required in the clinical situation. Amnesia which is an effect common to all benzodiazepines can be undesirable but in dental practice, for instance, may be a valuable adjunct to therapy. Other CNS effects of midazolam which may be required include an anticonvulsant action (e.g. in status epilepticus) and an antihallucinatory action (e.g. after ketamine or in delirium tremens).

Cardiovascular system

Even in large doses the benzodiazepines have little depressant effect on the heart or circulation. Midazolam causes a fall in systemic vascular resistance rather than the rise seen with thiopentone,[36] thus reducing pre- and afterload. While this effect may benefit the patient with a failing heart, it does introduce hazards in hypovolaemic patients. Because of the slow onset of action any cardiovascular depression with the benzodiazepines is often underestimated, though in clinical practice, if used in a full general anaesthetic technique, tracheal intubation may counterbalance any cardiovascular depression.

Respiratory system

Intravenous injection of the benzodiazepines in general can cause respiratory depression, in contrast to the notable safety of this group for oral medication. The depression includes loss of sensitivity to carbon dioxide and both actions are accentuated by the concomitant use of opioids. These effects in turn are more marked in patients with chronic obstructive airway disease. Finally, the use of intravenous benzodiazepines by those not skilled in airway management can lead to unrecognized respiratory obstruction. It is, therefore, highly dangerous to assume that sedation with midazolam is a safe alternative to anaesthesia, permitting the presence of an anaesthetist to be dispensed with.

Local effects

Midazolam, as an aqueous solution, has no irritant effects following intravenous injection. This is seen both in the lack of pain on injection and the absence of venous sequelae.

Propofol

Propofol

Chemistry and physical properties

Propofol is an alkylphenol and is virtually insoluble in water. It was originally solubilized in Cremophor EL but the current formulation is in a 1 per cent emulsion of 10 per cent soya bean oil, 1.2 per cent egg phosphatide and 2.25 per cent glycerol. This is similar to the fat emulsion Intralipid which is used for parenteral nutrition. The pH is 6–8.5. The preparation has a similar viscosity to that of water. Ampoules must not be frozen and should be shaken before use. The emulsion should not normally be mixed with other drugs or infusion fluids.

Dosage and administration

Propofol is supplied in 20 ml vials of the 1 per cent emulsion for induction of anaesthesia. The usual induction dose is 1.5–2.5 mg/kg. It is also available in vials of 50 or 100 ml which are for intravenous infusion and specifically are *not* for multidose use.

Pharmacokinetics

Propofol is 98 per cent protein bound and, being highly lipophilic, is distributed rapidly throughout the blood and vessel-rich tissues. Blood levels then decline with a mean α-phase half-life of 2.5 minutes and a β-phase half-life of 54 minutes.[37] Other workers have identified three phases of 2.3 minutes, 50 minutes and 310 minutes,[38] the slow terminal phase being related to prolonged excretion of the drug from a poorly perfused fat compartment. Propofol, however, is rapidly metabolized in the liver, with a clearance of 1.8–1.9 1/min which accounts for the rapid wakening from anaesthesia. Fentanyl reduces the clearance to approximately 1.3 1/min,[39] probably by reducing propofol distribution and increasing blood level.

The main metabolic product is the glucuronide of propofol produced in the liver, 88 per cent of which is excreted in the urine (with 2 per cent in the faeces).[37] As would be expected, renal disease has little effect on the pharmacokinetics of propofol and because of its high metabolic capacity, moderate degrees of liver damage also have little effect.

Pharmacodynamics

Central nervous system

Propofol in adequate dosage has a rapid onset of action and like thiopentone causes loss of consciousness in 11–15 seconds. Unlike thiopentone, it does not increase the sensitivity to somatic pain[40] and is a satisfactory sedative for patients after major surgery.[41] There is, however, no conclusive evidence of analgesic properties.

The dose required to induce anaesthesia is related to age, being approximately 2 mg/kg under the age of 60 and 1.6 mg/kg above that age.[42] Opioid premedication and pretreatment with i.v. fentanyl reduce the required induction dose. Side effects such as cough, hiccough or muscle movements are comparable to those with thiopentone.

Recovery has been assessed by many different techniques but using reaction times the return to baseline values after propofol 2.5 mg/kg is significantly faster than after thiopentone 5 mg/kg while performance with the latter can be shown to be impaired until the second postoperative day.[43] Another study[44] showed that choice reaction times, although there was a learning effect in the control group, were significantly shorter with propofol than with methohexitone at 20 minutes and much shorter than with thiopentone. The advantage of propofol over methohexitone was absent by 60 minutes but propofol was still significantly better than thiopentone. An additional advantage of propofol is that after short anaesthetics patients wake up, not only free from emetic effects but actually keen to have food. A survey of British anaesthetists in 1988–89 showed that, for outpatients, propofol was the preferred induction agent by 84 per cent.[45]

One of the obvious fields of usefulness for propofol is as an anaesthetic for electroconvulsive therapy (ECT) but there have been suggestions that because the drug is anticonvulsant it is therefore unsuitable. Experimental work has shown that propofol, like the barbiturates, is anticonvulsive,[46] but although it may have an inhibitory action on the induced convulsions, this would not seem to exclude its use in the field. Occasionally convulsions have been reported with or after propofol which further confuses the picture. The subject is usefully reviewed by Sneyd.[47]

Cardiovascular system

The striking clinical feature following the administration is the systemic hypotension without tachycardia and there has been controversy as to the role of central and peripheral causes of this. There is undoubtedly a significant fall in peripheral vascular resistance[48] and this can be demonstrated in patients on cardiopulmonary bypass.[49] The hypotension seen clinically is more marked than with thiopentone[50] and propofol certainly causes a greater fall in peripheral vascular resistance.[51,52] As would be expected the hypotension is greater in older patients and when the drug is injected rapidly.[42]

The other possible causes for the systemic hypotension include a negative inotropic action on the heart and peripheral pooling of blood with a reduction in preload. The latter explanation is supported by a study using gated radionuclide ventriculography[53] which showed no fall in peripheral vascular resistance but the hypotension was exclusively related to a decrease in cardiac index. There was no evidence of impairment of left ventricular performance. However, studies using echocardiographic assessment[54] have demonstrated a negative inotropic action of propofol in contrast to thiopentone which caused minimal changes and etomidate which had least effect. Assessment of the inotropic action of propofol on the myocardium must probably be derived largely from experimental evidence from a number of species. Overall this suggests that increased venous capacitance, decreased vascular resistance and depression of contractility all contribute to the propofol-induced hypotension. It must also be stressed that, while arterial hypotension in the presence of low peripheral resistance may be beneficial in the healthy normovolaemic patient, it may cause a dangerous reduction in myocardial perfusion in the presence of coronary artery narrowing.

Respiratory system

The effect of propofol on respiration is more depressant than that of thiopentone which may be apparent as apnoea of 30–60 seconds. This is not a problem in clinical anaesthesia and patients easily tolerate assisted ventilation. However, it does delay the uptake of volatile agents if spontaneous ventilation is left unaided. Respiratory reflexes appear to be depressed, making tracheal intubation and insertion of a laryngeal mask easier than with thiopentone.

Local effects

Pain on injection has been reported since the first studies of propofol in the Cremophor formulation. It is less with the emulsion but still is a feature, rising to 40 per cent when the drug is injected into small veins on the back of the hand. The most successful remedies to minimize injection pain seem to involve giving lignocaine 20–40 mg either mixed into the propofol immediately before administration or its injection alone into the vein before the propofol.[55] The former is the simpler technique and the addition does not appear to affect the stability of the emulsion.

Venous thrombosis and phlebitis are not clinical problems with propofol, the incidence of sequelae being similar to that with most aqueous solutions.

Other systems

Studies of adrenocortical function, renal function and liver function have not shown any significant abnormality in the postoperative period. In addition, there is no evidence of histamine liberation with the current formulation of propofol, nor does the frequency of hypersensitivity reactions appear to be any higher than with other intravenous solutions.

No problems have been attributed to the particular preparation of propofol administered for induction of anaesthesia. Cases of metabolic acidosis and fatal myocardial failure have been reported in children receiving high dose infusions for prolonged sedation[56] and this technique

is no longer acceptable until more is understood about the effects of such infusions.

Future developments

Steroid anaesthetics have been used intermittently over 50 years[57] and in terms of therapeutic index the actual drugs have proved to be superior to those in use at present. However, most of the products have been fat soluble and in the case of Althesin (alphaxalone) the vehicle, Cremophor EL, caused hypersensitivity reactions. The water-soluble alternatives, hydroxydione and minaxalone, have had other disadvantages. Pregnanolone has been given clinically in a fat emulsion and has proved satisfactory in early trials[58] and may well prove to have a long-term place. Newer water-soluble steroids are also being studied and one of them may also prove to be a safe alternative to the present agents.

REFERENCES

1. Lundy JS, Tovell RM. Some of the newer local and general anaesthetic agents. Methods of their administration. *Northwest Medicine (Seattle)* 1934; **33**: 308–11.

2. Stanski DR, Watkins WD. Drug disposition in anaesthesia. In: Scurr CF, Feldman S, eds. *The scientific basis of clinical anesthesia*. New York: Grune and Stratton, 1982: 148.

3. Morgan DJ, Blackman GL, Paull JD, Wolf LJ. Pharmacokinetics and plasma binding of thiopental. II: Studies at cesarean section. *Anesthesiology* 1981; **54**: 474–80.

4. Morgan DJ, Blackman GL, Paull JD, Wolf LJ. Pharmacokinetics and plasma binding of thiopental. I: Studies in surgical patients. *Anesthesiology* 1981; **54**: 468–73.

5. Christensen JH, Andreasen F, Janssen JA. Increased thiopental sensitivity in cardiac patients. *Acta Anaesthesiologica Scandinavica* 1985; **29**: 702–5.

6. Unni VKN, Johnston RA, Young HSA, McBride RJ. Prevention of intracranial hypertension during laryngoscopy and endotracheal intubation. Use of a second dose of thiopentone. *British Journal of Anaesthesia* 1984; **56**: 1219–23.

7. Bendsten AO, Cold GE, Astrup J, Rosenorn J. Thiopental loading during controlled hypotension for intracranial aneurysm surgery. *Acta Anaesthesiologica Scandinavica* 1984; **28**: 473–7.

8. Ward JD, Becker DP, Miller JD, *et al.* Failure of prophylactic barbiturate coma in the treatment of severe head injury. *Journal of Neurosurgery* 1985; **62**: 383–8.

9. Reiz S, Balfors E, Friedman A, Haggmark S, Peter T. Effects of thiopentone on cardiac performance, coronary hemodynamics and myocardial oxygen consumption in chronic ischemic heart disease. *Acta Anaesthesiologica Scandinavica* 1981; **25**: 103–10.

10. Filner BE, Karliner JS. Alterations of normal left ventricular performance by general anesthesia. *Anesthesiology* 1976; **45**: 610–21.

11. Todd MM, Drummond JC, Hoi Sang U. The hemodynamic consequences of high dose thiopental anesthesia. *Anesthesia and Analgesia* 1985; **64**: 681–7.

12. Sonntag M, Hellberg K, Schenk H-D, *et al.* Effects of thiopental (Trapanal) on coronary blood flow and myocardial metabolism in man. *Acta Anaesthesiologica Scandinavica* 1975; **19**: 69–78.

13. Kawar P, Briggs LP, Bahar M, *et al.* Liver enzyme studies with disoprofol (ICI 35,868) and midazolam. *Anaesthesia* 1982; **37**: 305–8.

14. Leading Article. Latent acute hepatic porphyria. *Lancet* 1985; **1**: 197–8.

15. Dundee JW, Wyant GM. *Intravenous anaesthesia*, 2nd ed. Edinburgh: Livingstone, 1988.

16. Dundee JW. Intra-arterial thiopental. *Anesthesiology* 1983; **59**: 154–5.

17. Hudson RJ, Stanski DR, Burch PG. Pharmacokinetics of methohexital and thiopental in surgical patients. *Anesthesiology* 1983; **59**: 215–19.

18. Van Hamme MJ, Ghoneim MM, Ambre JJ. Pharmacokinetics of etomidate, a new intravenous anesthetic. *Anesthesiology* 1978; **49**: 274–7.

19. Lindeburg T, Spotoff H, Pregard-Sorensen M, Skopsted T. Cardiovascular effects of etomidate used for induction and in combination with fentanyl-pancuronium for maintenance of anaesthesia in patients with valvular heart disease. *Acta Anaesthesiologica Scandinavica* 1982; **26**: 205–8.

20. Choi SD, Spaulding BC, Gross JB, Apfelbaum JL. Comparison

of the ventilatory effects of etomidate and methohexital. *Anesthesiology* 1985; **62**: 442–7.

21. Ledingham IMcA, Watt I. Influence of sedation on mortality in critically ill multiple trauma patients. *Lancet* 1983; **1**: 1270.

22. Sear JW, Allen MC, Gales M, *et al.* Suppression by etomidate of normal cortisol response to anaesthesia and surgery. *Lancet* 1983; **2**: 1028.

23. Fragen RJ, Shanks CA, Molpeni A, Avram MJ. Effects of etomidate on hormonal responses to surgical stress. *Anesthesiology* 1984; **61**: 652–6.

24. Wanscher M, Tonnesen E, Huttel M, Larsen K. Etomidate infusion and adrenocortical function. A study in elective surgery. *Acta Anaesthesiologica Scandinavica* 1985; **29**: 483–5.

25. Zacharias M, Clarke RSJ, Dundee JW, Johnston SB. Venous sequelae following etomidate. *British Journal of Anaesthesia* 1979; **51**: 779–82.

26. Nebauer AE, Doenicke A, Hoernecke R, Angster R, Mayer M. Does etomidate cause haemolysis? *British Journal of Anaesthesia* 1992; **69**: 58–60.

27. Harrison GG, Moore MR, Meissner PN. Porphyrinogenicity of etomidate and ketamine as continuous infusions. *British Journal of Anaesthesia* 1985; **57**: 420–3.

28. Domino EF, Domino FE, Smith RE, *et al.* Ketamine kinetics in unpremedicated and diazepam-premedicated subjects. *Clinical Pharmacology and Therapeutics* 1984; **36**: 645–53.

29. Baraka A, Louis F, Noueihid R, Diab M, Dabbous A, Sibai A. Awareness following different techniques of general anaesthesia for caesarean section. *British Journal of Anaesthesia* 1989; **62**: 645–8.

30. Waxman K, Shoemaker WC, Lippmann M. Cardiovascular effects of anesthetic induction with ketamine. *Anesthesia and Analgesia* 1980; **59**: 355–8.

31. Zsigmond EK, Matsuki A, Kothary SP, Jallad M. Arterial hypoxemia caused by intravenous ketamine. *Anesthesia and Analgesia* 1976; **55**: 311–14.

32. White PF, Ham J, Way WL, Trevor AJ. Pharmacology of ketamine isomers in surgical patients. *Anesthesiology* 1980; **52**: 231–9.

33. Restall J, Tully AM, Ward PJ, Kidd AG. Total intravenous anaesthesia for military surgery. A technique using ketamine, midazolam and vecuronium. *Anaesthesia* 1988; **43**: 46–9.

34. Restall J, Johnston IG, Robinson DN. Flumazenil in ketamine and midazolam anaesthesia. *Anaesthesia* 1990; **45**: 938–40.

35. Harper KW, Collier PS, Dundee JW, Elliott P, Halliday NJ, Lowry KG. Age and nature of operation influence the pharmacokinetics of midazolam. *British Journal of Anaesthesia* 1985; **57**: 866–71.

36. Al-Khudhairi D, Whitwam JG, Chakrabarti MK, Askitopoulou H, Grundy EM, Powrie S. Haemodynamic effects of midazolam and thiopentone during induction of anaesthesia for coronary artery surgery. *British Journal of Anaesthesia* 1982; **54**: 831–5.

37. Cockshott ID, Propofol (Diprivan) pharmacokinetics and metabolism – an overview. *Postgraduate Medical Journal* 1985; **61**: 45–50.

38. Kay NH, Sear JW, Uppington J, Cockshott ID, Douglas EJ. Disposition of propofol in patients undergoing surgery. A comparison in men and women. *British Journal of Anaesthesia* 1986; **58**: 1075–9.

39. Briggs LP, White M, Cockshott ID, Douglas EJ. The pharmacokinetics of propofol (Diprivan) in female patients. *Postgraduate Medical Journal* 1985; **61** (Suppl 3): 58–9.

40. Briggs LP, Dundee JW, Bahar M, Clarke RSJ. Comparison of the effect of di-isopropyl phenol (ICI 35 868) and thiopentone on response to somatic pain. *British Journal of Anaesthesia* 1982; **54**: 307–11.

41. Grounds RM, Lalor JM, Lumley J, Royston D, Morgan M. Propofol infusion for sedation in the intensive care unit: preliminary report. *British Medical Journal* 1987; **294**: 397–400.

42. Dundee JW, Robinson, FP, McCollum JSC, Patterson CC. Sensitivity to propofol in the elderly. *Anaesthesia* 1986; **41**: 482–5.

43. Herbert M, Makin SW, Bourke JB, Hart EA. Recovery of mental abilities following general anaesthesia induced by propofol (Diprivan) or thiopentone. *Postgraduate Medical Journal* 1985; **61** (Suppl 3): 132.

44. O'Toole DP, Milligan KR, Howe JP, McCollum JSC, Dundee JW. A comparison of propofol and methohexitone as induction agents for day case isoflurane anaesthesia. *Anaesthesia* 1987; **42**: 373–6.

45. Mirakhur RK. Drugs used by anaesthetists. *Anaesthesia* 1990; **45**: 500–1.

46. Lowson S, Gent JP, Goodchild CS. Anticonvulsant properties of propofol and thiopentone: comparison using two tests in laboratory mice. *British Journal of Anaesthesia* 1990; **64**: 59–63.

47. Sneyd JR. Excitatory events associated with propofol anaesthesia: a review. *Journal of the Royal Society of Medicine* 1992; **85**: 288–91.

48. Prys-Roberts C, Davis JR, Calverley RK, Goodwin NW. Haemodynamic effects of infusion of di-isopropyl phenol (ICI 35 868) during nitrous oxide anaesthesia. *British Journal of Anaesthesia* 1983; **55**: 105–11.

49. Boer F, Ros P, Bovill JG, Van Brummelen P, Van der Krogt J. Effect of propofol on peripheral vascular resistance during cardiopulmonary bypass. *British Journal of Anaesthesia* 1990; **65**: 184–9.

50. McCollum JSC, Dundee JW. Comparison of induction characteristics of four intravenous anaesthetic agents. *Anaesthesia* 1986; **41**: 995–1000.

51. Lippmann M, Paicius R, Gingerich S, *et al.* A controlled study of the haemodynamic effects of propofol versus thiopental during anesthesia induction. *Anesthesia and Analgesia* 1986; **65**: S89.

52. Vohra A, Thomas AN, Harper NJN, Pollard BJ. Non-invasive measurement of cardiac output during induction of anaesthesia and tracheal intubation: thiopentone and propofol compared. *British Journal of Anaesthesia* 1991; **67**: 64–8.

53. Lepage JY, Pinaud ML, Helias JH, *et al.* Left ventricular function during propofol and fentanyl anesthesia in patients with coronary artery disease: assessment with a radionuclide approach. *Anesthesia and Analgesia* 1988; **67**: 949–55.

54. Gauss A, Heinrich H, Wilder-Smith OHG. Electrocardiographic assessment of the haemodynamic effects of propofol: a comparison with etomidate and thiopentone. *Anaesthesia* 1991; **46**: 99–105.

55. Johnson RA, Harper NJN, Chadwick S, Vohra A. Pain on injection of propofol. *Anaesthesia* 1990; **45**: 439–42.

56. Parke TJ, Stevens, JE, Rice ASC, *et al.* Metabolic acidosis and fatal myocardial failure after propofol infusion in children: five case reports. *British Medical Journal* 1992; **305**: 613–16.

57. Clarke RSJ. Steroid anaesthesia (editorial). *Anaesthesia* 1992; **47**: 285–6.

58. Powell H, Morgan M, Sear J. Pregnanolone: a new steroid intravenous anaesthetic. Dose finding study. *Anaesthesia* 1992; **47**: 287–90.

Volatile Anaesthetic Agents and their Delivery Systems

Michael Cousins and Hugh Seaton

Modern volatile anaesthetic agents
Occupational exposure to volatile anaesthetics

Anaesthetic vaporizers

The evolution of modern volatile anaesthetic agents has been dictated by increasing concerns for safety and ease of use. The prediction in 1932[1] of the 'ideal' properties of current fluorinated inhalation agents, is a reminder of the substantial recent progress in the era 1970–1990. Table 7.1 illustrates eight prototypic volatile agents, and their

Table 7.1 Prototypic agents

Agent (year introduced)	Problems
Diethyl ether (1842)	Flammable and explosive Bronchial secretions Nausea and vomiting Hepatic damage
Chloroform (1847)	Arrhythmias Hepatic necrosis
Ethyl chloride (1848)	Flammable and explosive Cardiovascular depression Narrow therapeutic range Hydrolysed by soda lime
Divinyl ether (1933)	Flammable and explosive CNS stimulation, seizures Salivation Hepatic, renal impairment ?Mutagenic
Trichloroethylene (1934)	Hydrolysed by soda lime to toxin C_2Cl_2 Bradyarrhythmias ?Mutagenic
Ethyl vinyl ether (1947)	Flammable and explosive CNS excitation ?Mutagenic
Fluroxene (1953)	Flammable in clinical range Vomiting common Rare hepatic damage
Methoxyflurane (1960)	Fluoride nephrotoxicity Hepatic damage

Compiled from Secher O. Physical and chemical data on anaesthetics. *Anaesthesiologica Scandinavica Supplementum* 1971; **42.**

shortcomings, that have shaped our requirements for safety today. The desirable properties of the ideal volatile anaesthetic agent have been outlined, and in many ways represent an identification of the problems of agents of the past and attributes of those currently in use[1,2] (Table 7.2). Development of large series of investigational compounds has led to understanding of structure–activity relation as they relate to physical properties (Table 7.3). Saturated vapour pressure or volatility, molecular stability and solubilities are measured or predicted and in turn allow extrapolation to pharmacokinetic attributes.[3] For example, among structurally similar compounds, an increase in molecular weight is associated with increased anaesthetic potency: isoflurane is four times more potent than

Table 7.2 Ideal properties of a volatile agent[1]

Characteristic	Implications
Molecular stability	Stable to light, alkali or soda lime Safe in closed circle system Not corrosive Long shelf-life in all storage conditions No requirement for preservatives Non-flammable or explosive Not metabolized
Potent	Allows use in high concentration of oxygen Allows use as sole agent
Low solubility	Rapid rise in alveolar concentration Rapid induction and recovery Ready adjustment of depth of anaesthesia
Not pungent	Permits smooth induction by inhalation
Anaesthetic specific effects	Analgesia, amnesia, hypnosis No CNS excitation No cardiovascular or respiratory effects No adverse drug interactions No organ-specific toxicity

Table 7.3 Generalizations on the relations between structure and activity by Rudo and Krantz (cited in reference 3)

Development	Result
Increasing halogenation of hydrocarbons and ethers	Increasing potency Increasing arrhythmogenesis: F < Cl < Br < 1 Full halogenation usually convulsant
Increasing fluorination of ethers	Partial fluorination caused convulsant action Full fluorination decreased potency Increasing stability, less flammable
Choice of halogenated ether	Methyl ethyl ethers vs diethyl ethers more potent, stable and better anaesthetics Vinyl ethers tend to be unstable and toxic Thioethers: unpleasant odour, potent and toxic
Mode of halogenation	Ethers containing an assymmetric carbon atom are good anaesthetics (i.e. –CHFCl; –CHFBr; –CHClBr and –CFClBr) Produce mixed depression and stimulation if the rest of the molecule is heavily halogenated

Modern volatile anaesthetic agents

Halothane

Halothane is a halogenated hydrocarbon, $CF_3CHClBr$. Synthesized by Suckling of Imperial Chemical Industries in 1951, Raventos first published studies indicating its efficacy and safety in 1956. Clinical evaluation by Johnstone began in 1956 establishing halothane as a versatile, potent anaesthetic agent. Available for the first time was an agent that was non-combustible, apparently non-toxic and allowed a relatively rapid, clear-headed recovery compared to existing agents.[1]

desflurane as a result of replacing one fluorine atom in desflurane by a chlorine atom.[1] In addition, a relation exists between lipid solubility and potency; that is, the product of the oil/gas partition coefficient and MAC is similar for many agents. Table 7.4 lists the physicochemical properties of agents of historical and current interest.

Physical properties

Halothane is a colourless liquid with a sweet ethereal odour, and is thus favoured for inhalational induction in children. It is non-flammable in clinical concentrations. It is presented in amber-coloured bottles as exposure to light causes decomposition, but it does not react with soda lime at 22°C (degrades at 2.2 per cent per hour at 54°C);[1] thymol (0.01 per cent) is added as a stabilizer. Aluminium, brass and lead are corroded by halothane in the presence of water vapour, but copper and chromium are unaffected. Although halothane is quite soluble in rubber, it is dissolved very little in polyethylene.

It has a molecular weight of 197 Da, a boiling point of 50°C and a saturated vapour pressure of 32.2 kPa (242 mmHg) at 20°C, which makes it suitable for delivery through a temperature- and flow-controlled vaporizer.[4]

Pharmacokinetics

Halothane has a blood/gas solubility of 2.3. This relative insolubility permits rapid attainment of alveolar–arterial partial pressure consistent with rapid induction and recovery. The blood/gas solubility is lower in neonates and infants than adults, due to differences in blood constituents, particularly protein and lipid ratios. This contributes to a more rapid rise of the alveolar to inspired anaesthetic partial pressure, and shorter induction time.[5]

Elimination and metabolism

Based on the recovery of metabolites, up to 20 per cent of halothane is metabolized and 80 per cent is cleared from the body by exhalation. Mass balance studies performed more recently indicate that metabolism may be as high as 40–50 per cent, particularly as alveolar concentration decreases: in swine, at an alveolar concentration of 0.0026 per cent, halothane is cleared entirely by hepatic metabolism.[4] Metabolism accelerates the terminal elimination of halothane to a level similar to that of isoflurane.[6]

Halothane is metabolized both oxidatively and reductively by cytochrome P_{450}. Both pathways can be induced by compounds such as phenobarbitone and polychlorinated biphenyls. Cimetidine, and possibly isoflurane, inhibit oxidative metabolism of halothane. The oxidative metabolism products are inorganic bromide, chloride and trifluoracetic acid, while fluoride is one of the reductive metabolites. The concentration of metabolites peaks 24 hours postoperatively, and they are eliminated by renal excretion during the following week.[4] Prolonged anaesthesia with halothane does not result in serum fluoride level exceeding 20 µmol/l.[7] Bromide may reach concentrations assumed to produce drowsiness in elderly patients. Trifluoroacetic acid is not toxic but the trifluoroacetyl intermediate is thought to play a role in the major liver toxicity of halothane.[4] The reductive pathway of halothane

Table 7.4 Physicochemical properties of volatile anaesthetic agents

Agent (Units)	Empirical formula	Molecular weight (Da)	Boiling point (°C)	SVP at 20°C (kPa)	SVP at 20°C (mmHg)	Blood gas	Oil gas	Flammable or explosive	MAC anaesthetic conc* (%)	Stabilizers	Impurities	Stable in soda lime
Chloroform	$CHCl_3$	119.38	61.2	21.3	160	10.3	265	No	1–2*	Ethanol 0.6–1%	Phosgene, chlorine, chlorides, aldehydes, tetrachloromethane	Yes
Desflurane	$C_3H_2F_6O$	168.04	23.5	88.3	664	0.42	19.0	No	6–7	None	–	Yes
Diethyl ether	$C_4H_{10}O$	74.12	34.6	58.8	442	12.1	65	Yes	1.92	Diphenylamine 0.001%	Peroxides, ethanol, acetaldehyde	Yes
Enflurane	$C_3H_2CF_5O$	184.5	56.5	22.9	172	1.9	96.0	No	1.68	None	–	Yes
Ethyl chloride	C_2H_5Cl	64.52	13.1	131.4	988	3	–	Yes	3–4.5*	Ethanol & phenylnaphthyl-amine 0.01%	HCl, acetic acid, aldehydes, ethylene	No – slowly hydrolysed to HCl & ethanol
Fluroxene	$C_2H_5F_3O$	126.04	43.2	38.0	286	1.37	47.7	Yes	3.4	Phenylnapthyl-amine 0.01%	–	Yes
Halothane	$C_2HBrClF_3$	197.39	50.2	32.4	243.3	2.3	224	No	0.75	Thymol 0.01%	HCl, HBr, phosgene, others	Yes
Isoflurane	$C_3H_2CF_5O$	184.5	48.5	31.9	240	1.4	91.0	No	1.15	None	–	Yes
Methoxyflurane	$C_3H_4Cl_2F_2O$	164.97	104.8	3.1	23	13	825	Flammable	0.16	Toluol 0.1%	–	Yes
Sevoflurane	$C_4H_3F_7O$	200.5	58.5	21.3	160	0.60	53.0	No	2	None	–	No
Trichloroethylene	C_2HCl_3	131.4	86.7	8.6	64.5	9.15	960	No	0.17	Thymol 0.01% waxolene blue 0.001%	Dichloroacetylene, phosgene, CO, CO_2	No – produces dichloroacetylene (neurotoxic)
Vinyl ether	C_4H_6O	70.09	28.3	73.5	553	2.8	54.6	Yes	2–4*	Ethanol 4% & phenylnaphthyl-amine 0.01%	Formaldehyde, acetaldehyde, acetic acid, peroxides	Yes

Compiled from Secher O. Physical and chemical data on anaesthetics. *Acta Anaesthesiologica Scandinavica Supplementum* 1971; **42** and Dobkin AB. Development of new volatile inhalation anaesthetics. *Monographs in Anaesthesiology*, Vol. 6. Amsterdam: Elsevier/North Holland Biomedical Press, 1979.
* Anaesthetic inhaled concentrations, not MAC value.

biotransformation was cited in 1976 as being capable of producing reactive metabolites,[8] difluorochloroethylene, trifluorochloroethane, free radicals, and possibly difluoro-bromochloroethylene, and efforts were made to reproduce conditions responsible for 'halothane hepatitis'. Present consensus holds that this pathway could be responsible for the common but usually transient liver damage seen after halothane anaesthesia[9] (see below).

Toxicity

Halothane hepatitis   Following the enthusiastic adoption of halothane into clinical practice in 1956, reports gathered of postoperative jaundice and death, with massive hepatic necrosis first reported in 1958. As a response to this, the American National Halothane Study was commissioned in 1964 – a retrospective, non-randomized, multicentre survey of approximately 856 500 patients over the years 1959–1962. Published in 1969, it found that massive hepatic necrosis occurred in 1 in 10 000 operations, and that 1 in 22 000–35 000 was associated with halothane. This and subsequent studies identified obese, middle-aged patients, particularly females, as being at higher risk, and previous halothane anaesthetics had been administered in up to 82 per cent of cases.[10] The mortality rate approached 90 per cent in some series.[11]

In 1964 it was found that halothane was metabolized, and this gave rise to two theories as to the aetiology of halothane hepatitis. The first held that a reactive intermediate metabolite of biotransformation, probably of the reductive pathway, was hepatotoxic. Subsequent animal studies utilizing hypoxic, metabolism-induced models variably reproduced liver damage under these conditions. The second hypothesis was that an immunological mechanism was responsible, noting the association of repeated exposure, delayed onset, eosinophilia, rash and arthralgia. Based on these two theories, guidelines for the use of halothane were published in 1986 by the Committee on the Safety of Medicines in the United Kingdom. These stated that exposure to halothane within 3 months, previous adverse reactions to halothane or other halogenated hydrocarbon anaesthetic, family history of this, or the presence of pre-existing liver disease contraindicated the use of halothane.[11,12] Another effect of these perceptions was that biotransformation of volatile anaesthetic agents became an issue for anaesthetists world wide, and the success of later, more stable agents was assured.

Recent work has isolated antibodies to the trifluoroacetyl (TFA) halide metabolite of halothane's oxidative pathway.[13] The antibodies found in the serum of patients with halothane hepatitis are anti-endoplasmic reticulum antibodies directed against a trifluoroacetylated carboxylesterase.[14] An enzyme-linked immunosorbent assay (ELISA) has been developed and shown to be a diagnostic marker against antibodies to halothane-altered hepatocyte anti-

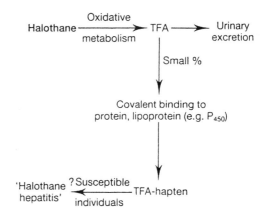

Fig. 7.1 Proposed mechanism of severe liver injury following halothane anaesthesia.

gens. Anti-TFA is found in high titre in patients suspected of having 'halothane hepatitis', and is absent in patients with cirrhosis or viral hepatitis and in patients without liver necrosis after halothane administration; hence it is considered diagnostic. The only interfering entity is primary biliary cirrhosis which gives a 15 per cent false-positive result. The current concept of 'halothane hepatitis' is that a small percentage of incompletely metabolized TFA entity acylates proteins of cytochrome P_{450}, forming a hapten, which, in certain individuals (1 in 5000–10 000) provokes a severe, and frequently fatal, inflammatory

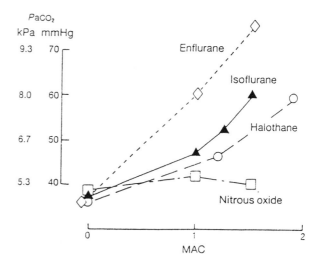

Fig. 7.2 Effect of volatile anaesthetic agents on respiratory function. Healthy male volunteers were given one of four anaesthetics in oxygen. Increasing levels of enflurane, isoflurane and halothane, but not nitrous oxide, increased Pa_{CO_2}. The order of their ability to depress respiratory function (highest to lowest) was as follows: enflurane (Calverley *et al.*[74]), isoflurane (Cromwell *et al.*[75]), halothane (Bahlman *et al.*[76]) and nitrous oxide (Winter *et al.*[77]). Nitrous oxide was given in a pressure chamber at 1.1 and 1.55 atmospheres (total pressure in both cases was 1.9 atmospheres). (Reproduced with permission from Ohio Medical Products, Airco Inc, PO Box 7550, Madison, Wisconsin 53707, USA.)

heptatis over 6–14 days, with production of detectable plasma anti-TFA antibodies (Fig. 7.1). The reason for this idiosyncratic immune response is unknown;[9] however, there is evidence in animal and human studies of an inherited susceptibility. Avoiding re-exposure in patients who have had a previous adverse reaction to halothane demonstrated by unexplained pyrexia or jaundice, or a family history of sensitization to the drug, may reduce the incidence of halothane hepatitis. In these cases, halothane-free equipment should be used, and exposure to other volatile halogenated anaesthetics should be avoided.[11]

Malignant hyperthermia   Maligant hyperthermia (MH) susceptibility is rare but the true incidence is uncertain. It has been suggested that fulminant MH occurs in 1:250 000 anaesthetics of all types but with an incidence of 1:50 000 with inhalational agents in conjunction with suxamethonium. Abortive MH may occur in as many as 1:4300 cases when inhalational agents are used after suxamethonium. Among the inhalational agents halothane has been the one most often involved in triggering the syndrome in susceptible individuals.

Pharmacodynamics

Potency

The MAC value for halothane is 0.75 per cent in oxygen and is reduced to 0.29 per cent when administered with 70 per cent nitrous oxide. MAC is reduced for all volatile anaesthetics by concomitant administration of opioids or sedatives, age, pregnancy, hypothermia (by 5 per cent per °C), hyponatraemia, and profound metabolic acidosis, hypotension or hypoxia.[4,15]

Respiratory effects

Halothane depresses ventilation, resulting in carbon dioxide retention. The ventilatory response to carbon dioxide and hypoxaemia is reduced, and the slope of the carbon dioxide response curve is decreased with increasing depth[4,16] (Fig. 7.2). The concentration producing apnoea is estimated to be 2.3 MAC; this is greater than for other agents in current use.[17] Tidal volume decrease is greater than dead space decrease and consequently dead space to tidal volume ratio increases by about 15 per cent at 1.1 MAC.[16] Halothane inhibits bronchoconstriction and induces bronchodilatation. In an isolated preparation halothane inhibits hypoxic pulmonary vasoconstriction, but the significance of this in the development of intraoperative hypoxaemia is controversial.[18] Pulmonary vascular resistance is little changed.[2]

Cardiovascular effects

Halothane induces a dose-dependent reduction in systemic arterial pressure due to myocardial depression, reducing cardiac output by 20–50 per cent (Fig. 7.3). Baroreceptor

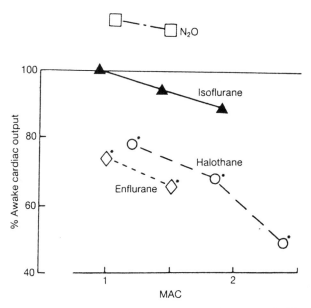

Fig. 7.3 Effect of volatile anaesthetics on cardiac output. Neither isoflurane (Stevens *et al.*[78]) nor nitrous oxide (Winter *et al.*[77]) depressed cardiac output below awake levels in volunteers. In contrast, halothane (Eger *et al.*[79]) and enflurane (Calverley *et al.*[80]) decreased output significantly (asterisks) and did so to a greater extent at deeper levels of anaesthesia. (Reproduced with permission from Ohio Medical Products, Airco Inc, PO Box 7550, Madison, Wisconsin 53707, USA.)

reflexes are blunted, so reflex tachycardia is absent. Total peripheral resistance is slightly reduced. After 2–5 hours of halothane anaesthesia, circulation returns to normal, possibly mediated through sympathetic activation.[4] Patients with coronary artery disease undergoing coronary or peripheral revascularization demonstrate reduction in myocardial oxygen consumption due to coronary artery vasodilatation and decreased perfusion pressure, rather than impairment of coronary autoregulation.[19] The circulatory effects of carefully administered halothane are considered benign in healthy subjects. Halothane causes increased automaticity, slows conduction through A-V nodal and His–Purkinje fibres.[20] More arrhythmias, exacerbated by both endogenous and exogenous catecholamines, occur with this than other agents. Patients on long-term β-antagonist agents, and those with pre-existing low-grade A-V block may develop significant bradyarrhythmias when anaesthetized with halothane. As halothane decreases sarcoplasmic Ca^{2+} release, additive effects with calcium antagonists such as verapamil and diltiazem can be expected in patients with cardiac failure or conduction disturbances.[21,22] It may induce arrhythmias in patients treated with aminophylline.

CNS effects

Some halogenated anaesthetic agents cause a dose-related reduction in cerebral oxygen consumption until neuronal function is abolished and the EEG is isoelectric.[23] This phenomenon is not seen with halothane. The halogenated agents 'uncouple' cerebral metabolism and cerebral blood flow at 1 MAC. As a result, in normotensive normocapnic subjects, 1.6 MAC of halothane increases CBF almost fourfold as a result of vasodilation while cerebral metabolic rate is decreased. These changes may lead to increased intracranial pressure, which is ameliorated by hyperventilating the patients prior to, but not after, introducing halothane. Cerebrospinal fluid formation is reduced by halothane.[23] Halothane alone has not been reported to cause convulsions in humans; isolated reports of halothane-related seizures were associated with the use of nitrous oxide. Halothane has been used in children to terminate status epilepticus resistant to conventional anticonvulsants.[24] At 1 MAC halothane anaesthesia, sensory evoked potentials (SEPs) display reduced amplitude with latency prolongation, but do not significantly detract from intraoperative monitoring in spinal surgery.[25]

Neuromuscular effects

Halothane is a generally poor muscle relaxant, and potentiates neuromuscular blockade by non-depolarizing agents and suxamethonium (succinylcholine) to a degree less than other agents.[4] The ED_{50} of non-depolarizing agents are reduced by only 10 per cent at 1 MAC halothane and the ED_{50} of suxamethonium during halothane anaesthesia is 1.5 times that during isoflurane anaesthesia.[26] Halothane relaxes uterine muscles, allowing instrumentation, and may increase bleeding after curettage, but no adverse effects are seen when low concentrations are used to avoid awareness during caesarean section.[4]

Renal effects

All volatile anaesthetic agents transiently depress renal function with urine output, glomerular filtration rate (GFR), renal blood flow (RBF), and electrolyte excretion being reduced, whilst ADH levels rise. Halothane is associated with a moderate increase in renal vascular resistance and a reduction in renal blood flow in response to reduced cardiac output and increased circulating catecholamine levels. Conflicting evidence exists for the effect of halothane and other volatile agents on the preservation of renal autoregulation to changes in mean arterial pressure.[27]

Hepatic effects

Intestinal and hepatic blood flow is reduced in proportion to the reduction in cardiac output; portal venous oxygen content is markedly reduced and dependence on hepatic arterial oxygen supply increases. Autoregulation of flow in the preportal and hepatic arterial circulations is not apparent during halothane anaesthesia in dogs, and hepatic circulation is more compromised during anaesthesia with halothane than enflurane.[28] As stated above, halothane anaesthesia is associated with minor hepatic

dysfunction, as evidenced by a rise in transaminases. This occurs in 8–40 per cent of individuals anaesthetized with halothane in the absence of clinical or other liver-related laboratory abnormalities, and incidence is increased after multiple exposures.[9,29] A more sensitive index of drug-induced hepatocellular damage which correlates better with hepatic histology is serum glutathione-S-transferase (GST). Halothane anaesthesia produces a significant increase in plasma GST concentration 3 hours after anaesthesia, with a secondary peak occurring at 24 hours. It has been speculated that the peak at 3 hours is a result of alterations in the hepatic blood flow and oxygenation. The secondary peak is thought to be a result of the toxic effect of a metabolite. Cimetidine, an inhibitor of oxidative halothane metabolism, does not influence the magnitude of this increase in GST concentration.[14] Manipulation of reductive metabolism, and associated effects on GST in man, has not been reported.

Other effects

Increases in white cell count, neutrophil count and blood sugar level are recorded after anaesthesia with halothane in volunteers.[30]

Enflurane

Enflurane is a methyl ethyl ether, CHF_2OCF_2CHFCl, an isomer of isoflurane. It was the 347th compound developed by Dr Ross Terrell and his co-workers at Ohio Medical Products in 1963, and after initial screening by Krantz in 1964 it was investigated, and released in the USA in 1972.

Physical properties

Enflurane is a colourless liquid with a slightly ethereal smell. It is non-flammable in clinical concentrations, does not corrode metals and is resistant to decomposition by physical (e.g. ultraviolet light, heat) or chemical (soda lime or strong base) means – hence no stabilizers are added. It is less soluble than halothane in rubber and plastics.

It has a molecular weight of 185.4 Da, a boiling point of 56.5°C and a saturated vapour pressure of 22.9 kPa (172 mmHg) at 20°C, which makes it suitable for delivery through a temperature- and flow-controlled vaporizer.[4]

Pharmacokinetics

The blood/gas partition coefficient of enflurane, 1.9, is less than that of previously developed volatile agents, but higher than those of isoflurane (1.40), desflurane (0.42) and sevoflurane (0.6–0.7) which allows alveolar concentration to rise to inspired concentration more rapidly than halothane, but less so than the newer agents. This permits moderately rapid induction and elimination phases, and facilitates control of depth via inspired partial pressure. Enflurane has a much lower solubility in fat than halothane which reduces the total amount of enflurane accumulated during long anaesthetic procedures.[4]

Elimination and metabolism

Enflurane is more resistant to metabolism than halothane as only 2–8.5 per cent is excreted following hepatic oxidative metabolism. Attack on the enflurane molecule occurs at the β carbon, yielding difluoromethoxydifluoro-acetic acid, and chloride and fluoride ions. The extent of biotransformation to fluoride ion is insufficient to result in clinically significant nephrotoxicity, unlike methoxyflur-ane.[29] The defluorination of enflurane in humans and animals is not inducible with either phenobarbitone or methylcholanthrene; however, increased defluorination rates have been documented in rare cases. Patients who are 'fast acetylators' treated with isoniazid produce high concentrations of hydrazine, which increases the fluorina-tion pathway by inducing the formation of cytochrome P_{451}.[4] It is possible that these patients, anaesthetized with enflurane, could develop mild fluoride nephropathy, if concurrently exposed to isoniazid, or other hydrazine-containing drugs, such as hydralazine.[4,8,29]

Toxicity

Reactive intermediate metabolites of enflurane form covalently bound acetylated hepatic protein adducts. Although the hapten produced from enflurane is chemi-cally distinct from that of halothane, it may act as an immunogen in a manner similar to TFA in halothane metabolism, and may be recognized by anti-TFA ELISA. 'Enflurane hepatitis' has been reported, and may follow previous exposure to halothane or enflurane, but is unlike halothane hepatitis in that female sex and obesity are not associated and mortality rate is much lower.[29] The possibility exists of cross-sensitization by enflurane in a patient with a past history of halothane hepatitis, hence this substitution is not recommended.[9] Abnormal liver function tests occur in a quarter of patients following enflurane anaesthesia, but not more frequently following repeated exposure.[29]

Nephrotoxic concentrations of inorganic fluoride (greater than 50 μmol/l) leading to high output renal failure are rare. High but not toxic concentrations of inorganic fluoride have been reported in obese patients and after prolonged enflurane anaesthesia, and associated defects in renal concentrating ability resolve after 24 hours.[4,27] This is because inorganic fluoride peaks at about 6 hours and is rapidly cleared, since the small 'depot' of enflurane in blood and fat dissipates quickly. Enflurane has been implicated in contributing to postoperative deterioration in renal function in two patients with pre-existing renal disease.[29]

Pharmacodynamics

Potency

The MAC of enflurane, 1.68 per cent, is greater than that of halothane (0.75 per cent) and isoflurane (1.15 per cent) in oxygen, and the ability to deliver higher MAC levels determines inhalational induction rate with this agent. Whilst this potency allows simultaneous administration of nearly 100 per cent oxygen, the addition of 70 per cent nitrous oxide reduces enflurane MAC to 0.57. The MAC is also reduced by pharmacological and physiological factors outlined in the section on halothane.

Respiratory effects

Enflurane, in clinical concentrations, is non-irritant to the airways. It depresses ventilation in a dose-related fashion but to a degree greater than halothane or isoflurane, making it less suitable if spontaneous respiration is required.[4] In clinical use, this respiratory depression is antagonized by surgical stimulation. The concentration producing apnoea is estimated to be 1.6 MAC.[17] Enflurane reduces bronchomotor tone in vitro. Isolated case reports exist of enflurane causing bronchoconstriction in asthmatic patients, but in contrast this agent has been used successfully in the treatment of status asthmaticus.[18]

Cardiovascular effects

In vitro studies indicate that enflurane decreases myocar-dial contractility to a greater degree than halothane. In humans, however, less myocardial depression is found with enflurane than halothane. As in the case of other volatile agents, enflurane decreases systemic arterial pressure dose dependently. This is caused by both myocardial depression and reduced total peripheral resistance. Recovery from cardiovascular depression may occur during long-term enflurane anaesthesia, as for halothane.[4] Enflurane causes greater coronary vasodilatation than halothane in dogs. In patients during coronary or peripheral revascularization, 1 MAC enflurane is associated with a normal adjustment in the distribution of coronary blood flow in response to a decrease in regional myocardial blood flow.[19] Enflurane may increase heart rate but the incidence of arrhythmias is low. Catecholamine increases are well tolerated, allowing enflurane anaesthesia in patients with phaeochromocy-toma.[4] However, the dose–response curve for ventricular arrhythimas and dose of exogenous adrenaline is not parallel to that for halothane, even though the enflurane curve is favourably shifted to the right; thus arrhythmias may occur unpredictably in association with enflurane and adrenaline (Fig. 7.4). The haemodynamic effects of enflurane are marked following administration of propano-lol in dogs, and a withdrawal of 20 per cent of blood volume is poorly tolerated.[4] Experimentally, enflurane combined

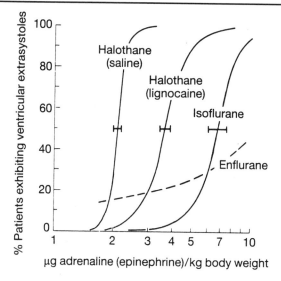

Fig. 7.4 Interaction of volatile anaesthetic agents and adrenaline (epinephrine) on ventricular extrasystole production. Groups of patients were anaesthetized with 1.25 MAC of halothane, isoflurane or enflurane in oxygen. Their tracheas were intubated and ventilation was controlled to produce normocapnia. Within each group, patients received various doses of adrenaline (one dose per patient) injected submucosally, and the occurrence of three or more extrasystoles (a positive response) was noted. The data thus obtained were analysed to give the graphs in this figure. The doses estimated to produce extrasystoles in 50 per cent of patients were as follows: halothane (adrenaline injected in saline), 2.1 µg/kg; halothane (adrenaline injected with 0.5 per cent lignocaine), 3.7 µg/kg; isoflurane (adrenaline in saline), 6.7 µg/kg; enflurane (adrenaline in saline), 10.9 µg/kg. (Data from Johnston et al.[81] Reproduced with permission from Ohio Medical Products, Airco Inc. PO Box 7550, Madison, Wisconsin 53707, USA.)

with calcium antagonists causes more myocardial depression than halothane or isoflurane.[4,22]

CNS effects

Enflurane causes a dose-related depression of CMRO₂, and an increase in cerebral blood flow (CBF) with loss of autoregulation similar to halothane at 1 MAC in normotensive, normocapnic subjects, but, as with isoflurane, at 1.6 MAC, CBF is only doubled. Glucose metabolism by the choroid plexus is enhanced and cerebrospinal fluid production increased by enflurane.[23] These events lead to an increase in intracranial pressure, which may be attenuated by hyperventilation or the use of thiopentone. In both normal and epileptic patients, increasing depth of enflurane anaesthesia is characterized by the appearance of high voltage spikes, then spike and dome activity with intermittent periods of burst suppression at concentrations higher than 1.5 MAC. These burst suppression patterns are thought to be excitatory events, unlike those of other volatile agents.[24,31] Seizure or EEG spiking activity is more readily produced in the presence of hypocapnia and auditory stimulation.[24] Enflurane may be contraindicated in patients with a history of epilepsy or EEG epileptogenic foci; enflurane anaesthesia and hyperventilation have been

used to activate silent epileptogenic foci intraoperatively, to delineate the site of seizure activity before discrete surgical excision. Postoperative seizure activity related to enflurane has been reported up to 9 days later, and prolonged EEG changes documented.[24] At 0.5 MAC enflurane anaesthesia, intraoperative SEPs display reduced amplitude with latency prolongation.[25]

Neuromuscular effects

Enflurane depresses skeletal muscle activity to a degree similar to isoflurane and greater than halothane. Like isoflurane it markedly potentiates non-depolarizing neuromuscular blocking agents.

Renal effects

Renal blood flow, glomerular filtration rate and urinary flow decrease in enflurane anaesthesia as with other volatile anaesthetics. Inorganic fluoride is produced by hepatic metabolism of enflurane and nephrotoxicity, although rare, has been described as outlined above.

Hepatic effects

In common with other volatile anaesthetics, enflurane depresses splanchnic blood flow due to systemic haemodynamic depression whilst splanchnic bed oxygen consumption remains constant. Hepatic blood flow is reduced, as portal vein flow decreases significantly, with hepatic arterial flow essentially unchanged. Autoregulation of preportal and hepatic arterial vascular beds is maintained during enflurane anaesthesia.[28] Enflurane is similar to isoflurane in producing less depression of hepatic oxygenation than halothane in conditions of hypotension.

Miscellaneous effects

Enflurane may be better at preventing a rise in cortisol and ACTH prior to surgical incision than halothane.[32] Enflurane is probably less suitable for anaesthetizing neonates than halothane.[4] The effect on uterine smooth muscle is similar to halothane. Enflurane is capable of triggering malignant hyperthermia in susceptible individuals.

Isoflurane

Isoflurane is a methyl ethyl ether, CF₃CHClOCHF₂, an isomer of enflurane and was first synthesized in 1965. It was the 469th compound developed by Dr Ross Terrell and his co-workers at Ohio Medical Products, and was recognized to have some advantages over halothane and enflurane. From 1971, it had been used clinically in 3000 patients when it was reported in poorly controlled studies (Corbett, 1976) to cause hepatic neoplasia in mice (cited in Eger[33]). This delayed FDA approval, pending reinvestigation, until

1979, and from 1981 it was the most widely used potent inhaled agent in North America.[33]

Physical properties

Isoflurane is a slightly pungent colourless liquid. It is non-flammable in clinical concentrations, does not corrode metals and is resistant to decomposition by physical (e.g. ultraviolet light, heat) or chemical (soda lime, 0.15 per cent per hour at 60°C[34] or strong base) means – hence no stabilizers are added.

It has a molecular weight of 184.5 Da, a boiling point of 48.5°C and a saturated vapour pressure of 31.9 kPa (240 mmHg) at 20°C, which makes it suitable for delivery through a temperature- and flow-controlled vaporizer. As the SVP of isoflurane is similar to that of halothane, 32.4 kPa (243.3 mmHg), a vaporizer calibrated for halothane can be used to deliver isoflurane relatively accurately, although interchanging agents is not recommended.[33]

Pharmacokinetics

The blood/gas partition coefficient of isoflurane, 1.4, is less than that of previously developed volatile agents, but higher than those of desflurane (0.42) and sevoflurane (0.6–0.7) which allows alveolar concentration to rise to inspired concentration more rapidly than halothane or enflurane, but less so than the newer agents (Fig. 7.5). This permits relatively rapid induction with isoflurane (somewhat limited by its pungency), good control of depth via inspired partial pressure, and rapid elimination of the agent which promotes recovery and reduces the quantity available for hepatic metabolism. Like halothane, isoflurane has a lower blood/gas partition coefficient in neonates and infants

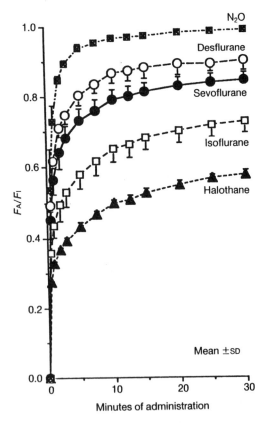

Fig. 7.5 Wash-in curves for volatile anaesthetic agents. The pharmacokinetics of sevoflurane and isoflurane during their administration are defined as the ratio of end tidal anaesthetic concentration (F_A) to inspired anaesthetic concentration F_I (i.e. F_A/F_I) (mean ± SD). Consistent with its relative blood/gas partition coefficient, the F_A/F_I of sevoflurane increases more rapidly than that of isoflurane or halothane but slower than that of N_2O or desflurane. (Reproduced with permission from Yasuda N, Lockhart SH, Eger II EI, *et al. Anesthesia and Analgesia* 1991; **72**: 316–24.[47])

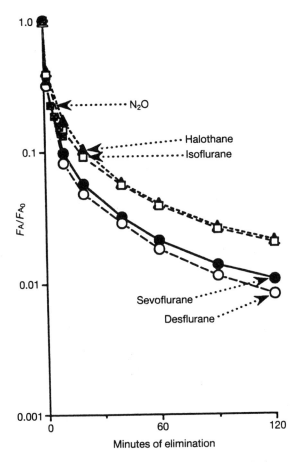

Fig. 7.6 Early wash-out curves for volatile anaesthetic agents. Elimination of each anaesthetic is defined as the ratio of end tidal anaesthetic concentration (F_A) to the F_A immediately before the beginning of elimination (F_{AO}) (i.e. F_A/F_{AO}). Over the 120-minute period portrayed, the elimination of sevoflurane is faster than that of isoflurane by a factor of about 1.6 (note the logarithmic scale of the ordinate). The elimination rate of sevoflurane is slightly lower than that of desflurane, whereas the elimination rates of isoflurane and halothane are indistinguishable. Elimination of N_2O is slower (*n* only equals 2) than with sevoflurane because equilibration with N_2O was 8–10 times longer. (Reproduced with permission from Yasuda N, Lockhart SH, Eger II EI, *et al. Anesthesia and Analgesia* 1991; **72**: 316–24.[47])

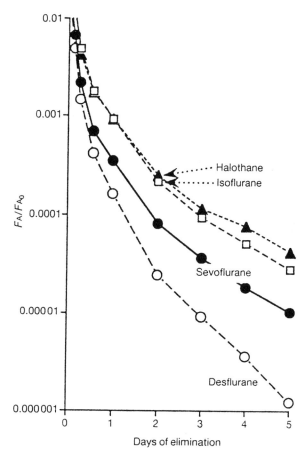

Fig. 7.7 Late wash-out curves for volatile anaesthetic agents. The terminal elimination of sevoflurane is faster than that of isoflurane but is slower than that of desflurane. After 2–5 days there is a two- to threefold difference between the values for sevoflurane and isoflurane. The difference between the values for desflurane and sevoflurane progressively increase after day 1 and approach eightfold by day 5. (Reproduced with permission from Yasuda N, Lockhart SH, Eger II EI, *et al. Anesthesia and Analgesia* 1991; **72**: 316–24.[47])

compared with adults, due to differences in blood composition; this increases induction rate in these age groups.[5]

Elimination and metabolism

Due to the lower solubility of isoflurane in blood and fat, elimination faster than for halothane would be expected; however, in the immediate postoperative period halothane is eliminated at a similar rate to isoflurane, since halothane's metabolism significantly contributes to its clearance, whereas for isoflurane this is not the case (Fig. 7.6). From 24 to 48 hours postoperatively, isoflurane clearance is faster than halothane (Fig. 7.7). Isoflurane is 0.2–0.3 per cent metabolized (this figure may be higher[9]), via an oxidative pathway on cytochrome P_{450}. The main pathway of isoflurane metabolism is believed to begin when an active oxygen atom is inserted into the ethyl α-C–H bond. The resulting unstable compound is hydrolysed to difluoromethanol and trifluoroacetic acid. Both compounds,

particularly the latter, are a major source of inorganic fluoride,[7] although in man, trifluoroacetic acid predominates. Isoflurane metabolism is inducible with phenobarbitone.[8]

Toxicity

The trifluoroacetyl oxidative metabolite is the same as that identified as part of the immunogenic hapten producing massive hepatic necrosis with halothane, and therefore an individual who has had a documented hepatic reaction to halothane should not be subsequently exposed to isoflurane.[9] In human hepatocyte cell cultures, isoflurane has been shown to be virtually non-toxic.[14] Isoflurane administration for 20 MAC-hours leads to sufficient metabolism to inorganic fluoride that subclinical renal dysfunction is possible, although not reported.[7]

Pharmacodynamics

Potency

The MAC of isoflurane is 1.15 per cent, which lies between that of halothane (0.75 per cent) and enflurane (1.68 per cent) in oxygen. Whilst this potency allows simultaneous administration of nearly 100 per cent oxygen, the addition of nitrous oxide reduces isoflurane MAC by 0.1 per cent per 10 per cent nitrous oxide.[20] Other drugs and physiological factors reduce MAC as they do for halothane.

Respiratory effects

Isoflurane, like other potent inhalation agents, depresses ventilation in a dose-related fashion. Arterial carbon dioxide partial pressure increases with alveolar concentration of isoflurane in subjects breathing spontaneously, to an extent intermediate between halothane and enflurane. The ventilatory response to an increase in arterial carbon dioxide partial pressure is depressed by 50–70 per cent at 1 MAC and at 2 MAC it is abolished, with higher carbon dioxide partial pressure further depressing respiration.[33] The ventilatory response to hypoxia is more profoundly depressed; even at subanaesthetic levels, 0.1 MAC produces a 50–70 per cent depression, and at 1.1 MAC it is abolished. The concentration producing apnoea is estimated to be 1.7 MAC.[17] Respiratory mechanics are altered such that functional residual capacity and pulmonary compliance decrease slightly at 1 MAC, and airways resistance increases, probably due to decreased lung volume. The changes in functional residual capacity and compliance return toward normal at 2 MAC. Like halothane, isoflurane at 1.5 MAC attenuates bronchoconstriction due to methacholine or antigens in animal studies and is well tolerated by patients with chronic airflow limitation and asthmas.[33] The risk of ventricular arrhythmias in patients receiving aminophylline and halothane makes isoflurane a safer alternative.[21] Hypoxic pulmonary vasoconstriction is inhibited in studies on isolated canine lung, but although pulmonary vascular resistance is little affected in man, it has been used to treat pulmonary hypertension.[2,3]

Cardiovascular effects

Isoflurane produces less depression of myocardial contractility *in vivo* than halothane or enflurane, although *in vitro* it demonstrates a dose-dependent reduction in maximal velocity of shortening and mean maximal developed force.[33] In human subjects, 1–1.8 MAC isoflurane does not affect dP/dt or cardiac output (see Fig. 7.3), although stroke volume falls. This may be due to increased sympathetic activity, particularly beta, as seen from the sustained myocardial contractility, increased heart rate and peripheral vasodilatation, particularly muscle blood flow. Whatever the mechanism, isoflurane demonstrates a wider margin of safety before myocardial depression occurs. Dose-dependent decrease in systemic arterial pressure is caused by a decrease in total peripheral vascular resistance; in association with these changes ventricular work and myocardial oxygen consumption are decreased.[4]

Tachycardia occurring with isoflurane anaesthesia may be due to increased beta activity or a baroreceptor-mediated reflex invoked by hypotension; this reflex may be better preserved in isoflurane anaesthesia than with halothane or enflurane.[33] Tachycardia increases myocardial oxygen demand and reduces supply time and, particularly in the presence of vagolytic drugs, hypotension or coronary artery disease, may be undesirable. Isoflurane does not affect the incidence of atrial, nodal or ventricular arrhythmias, and does not slow conduction through the His–Purkinje fibres.[20] The dose of adrenaline required to produce extrasystoles in 50 per cent of subjects is three times that for equianaesthetic halothane exposure (6.7 vs 2.1 μg/kg at 1.25 MAC) (see Fig. 7.4).[20] Isoflurane is expected to have less marked interactions with β-blockers than halothane but similar additive effects with calcium antagonists have been found experimentally.[4,22]

Isoflurane dilates epicardial resistance vessels. Its coronary vasodilatory action is weaker than adenosine or dipyridamole, but considerably stronger than halothane or enflurane.[18] There is evidence, both clinical and experimental, that isoflurane can cause redistribution of myocardial blood flow, intercoronary and transmural 'steal' which may lead to myocardial ischaemia.[35] Steal-prone coronary anatomy is described as one or more total occlusions of a major coronary artery, and a concomitant haemodynamically significant stenosis of the collateral-supplying vessel, and is evident on 12 per cent of angiograms from the Coronary Artery Surgery Study registry.[35] Coronary vasodilatation is dose dependent and is not observed in concentrations less than 0.75 per cent. Where isoflurane has been used as an adjunct to high dose opioids in coronary revascularization an increased inci-

dence of ischaemia has not been noted. If isoflurane is used in doses likely to produce coronary vasodilatation it is suggested that close control of systemic haemodynamic variables be maintained.[18] Prospective outcome studies have not demonstrated an increase in perioperative ischaemic events in patients in whom isoflurane was used.[35]

CNS effects

Isoflurane causes a dose-related depression of the central nervous system with progressive electroencephalographic (EEG) changes that parallel those produced by other agents. Concentrations less than 0.5 MAC may increase alpha frequency from 8–12 Hz to more than 15 Hz with associated increase in voltage. Amnesia occurs at about 0.4 MAC, and is associated with shift of higher voltage activity from posterior to anterior regions of the brain. At 1 MAC, EEG frequency decreases and voltage increases further; 1.5 MAC causes burst suppression with decrease in voltage to isoelectric levels at 2 MAC. Seizure activity is not induced, even in association with hypocapnia, repetitive auditory stimulation or nitrous oxide.[23] Isoflurane has been used to treat refractory status epilepticus.[24] Latency of auditory evoked brainstem potentials increases with 1–2 per cent isoflurane and 0.5 per cent isoflurane reduces amplitude with latency prolongation in intraoperative SEPs, an effect similar to halothane and enflurane.[25] Cerebral blood flow is not changed in normocapnic, normotensive volunteers given 0.6 or 1.1 MAC but doubles at 1.6 MAC; cerebral oxygen consumption is reduced in a dose-related manner to 2 MAC when the EEG becomes isoelectric. Autoregulation of cerebral blood flow response to arterial pressure changes is retained to 1.5 MAC, and a decrease in $PaCO_2$ from normal produces a nearly proportional decrease in cerebral blood flow. These characteristics allow greater indirect control of intracranial pressure than available with earlier agents, and protection of the brain against ischaemia produced by profound hypotension, properties desirable in neurosurgical anaesthesia.[23,33] Stable intracranial pressure can be maintained on induction with simultaneous introduction of hyperventilation and isoflurane vapour. Cerebrospinal fluid formation and absorption is not affected by isoflurane.[23] Isoflurane does not increase, and probably decreases, intraocular pressure. Intellectual function in volunteers given 6 hours of isoflurane anaesthesia returns to normal within 2–4 days.

Neuromuscular effects

Isoflurane depresses skeletal muscle activity to a degree similar to enflurane, greater than halothane. Like enflurane it markedly potentiates non-depolarizing neuromuscular blocking agents, and also enhances the effects of suxamethonium. Recovery from these effects is rapid because of the low solubility and increased muscle blood flow of isoflurane. Rigidity and shivering are seen in recovery in some 10 per cent of patients.[36] The effect on uterine smooth muscle is similar to halothane. Isoflurane is capable of triggering malignant hyperthermia in susceptible individuals. Isoflurane has been associated with three reported cases, up to 1991, which were confirmed by muscle biopsy and contracture testing, and five cases which have not been substantiated.[14]

Renal effects
Renal blood flow, glomerular filtration rate and urinary flow decrease in isoflurane anaesthesia as with other volatile anaesthetics. Negligible amounts of inorganic fluoride are produced by hepatic metabolism of isoflurane and nephrotoxicity has not been described.

Hepatic effects
Isoflurane reduces hepatic arterial resistance and maintains total hepatic perfusion better than either halothane or enflurane. Isoflurane may also be advantageous for perfusion of the preportal organs as it appears to reduce vascular resistances in this area slightly more than enflurane or halothane.

Other effects
Increases in white cell count and neutrophil count have been noted 24 hours after anaesthesia with isoflurane as with halothane. Blood sugar level also rises.[30]

Desflurane

Between 1959 and 1966, Terrell and associates at Ohio Medical Products (now Anaquest, a division of British Oxygen Company) synthesized more than 700 compounds in order to develop a better volatile anaesthetic agent. The 653rd, $CF_2HOCFHCF_3$, desflurane, was discarded at the time as its synthesis involved a potentially explosive step with elemental fluorine and because its vapour pressure, close to atmospheric pressure, precluded its use in contemporary vaporizers. Reinvestigated in 1987, its stability and low solubility suggest that it may be advantageous in ambulatory surgery with its emphasis on short post-anaesthetic recovery time.[1]

Physical properties

Desflurane is a clear, slightly pungent volatile liquid. Structurally it differs from isoflurane only in the substitution of fluorine for chlorine at the α-ethyl carbon. It has a molecular weight of 168 Da, a boiling point of 23.5°C and a saturated vapour pressure of 88.3 kPa at 20°C (664 mmHg), which makes it unsuitable for delivery through conventional vaporizers. Direct metering vaporization is required;

an electrically heated vaporizer is maintained at 23–25°C and combined with a flowmeter calibrated for vapour output.[1] It is stable to light and in contact with soda lime (does not degrade significantly below 80°C),[34] requiring no stabilizers. It is non-corrosive, and is non-flammable in clinical conditions.

Pharmacokinetics

Desflurane has a blood/gas partition coefficient of 0.42, comparable to nitrous oxide or cyclopropane, and one-third that of isoflurane. This relative insolubility causes rapid achievement of alveolar–arterial tension consistent with rapid induction and recovery (see Figs 7.5–7.7). Desflurane's oil/gas partition coefficient of 18.7 predicts diminished potency compared to that of halothane or isoflurane, according to Meyer–Overton kinetics.[37] Induction of anaesthesia with desflurane may be more rapid than with other agents if pungency does not limit the rate at which inspired concentration can be increased. In volunteers, the ratio of alveolar to inspired partial pressure of the agent (F_A/F_I) is 0.82 after 10 minutes of exposure; comparable values for halothane and enflurane are 0.42 and 0.52 respectively.[38] Despite the similarity to the blood solubility of nitrous oxide (0.47), F_A/F_I of nitrous oxide at anaesthetic concentrations increases more rapidly due to the concentration effect and lower tissue solubility (see Fig. 7.5).[6] However, when nitrous oxide is administered in trace concentrations the rate of uptake is similar to that of desflurane, as the acceleration in uptake caused by the concentration effect is absent.[38] Desflurane has a lower solubility in all tissues studied than other volatile agents. This suggests that altering the depth of anaesthesia during surgery is accomplished most readily with desflurane.[1] Reawakening time is approximately twice as fast with desflurane compared to isoflurane, and is related to depth of anaesthesia rather than exposure time.[39] Long-term recovery is more rapid and complete with desflurane compared to isoflurane as by day 2 post-administration, F_A/F_{A0} exceeds by a factor of 10 that of isoflurane and halothane, by day 5 a factor of 20 (see Fig. 7.7).[6]

Elimination and metabolism
Desflurane, being halogenated solely with fluorine, resists biodegradation.[6] In human volunteers exposed to 7.35 MAC-hours, less than 0.02 per cent desflurane was metabolized and no changes in serum fluoride level were detected.[30] Studies in rats pretreated with phenobarbitone to cause enzyme induction demonstrate thousand-fold increases in plasma and urinary inorganic fluoride ion and urinary organic fluoride concentrations in rats exposed to halothane and methoxyflurane; isoflurane caused a measurable but smaller increase, but the concentrations of fluoride after desflurane were virtually indistinguishable

from those of control animals.[40] Similarly, in swine exposed to approximately 5.5 MAC-hours of isoflurane or desflurane, plasma concentrations of fluoride ion were three times greater than control values immediately after and at 4 hours after exposure to isoflurane, whereas there was no detectable increase immediately after anaesthesia with desflurane and only a 17 per cent increase after 4 hours.[1,40]

Toxicity

The physical stability of desflurane suggests that organ-specific toxicity with this agent will prove to be minimal or absent. Preliminary work has supported this contention. In volunteer studies no alterations in liver or renal function have been determined.[1,40] A 7.35 MAC-hour exposure in volunteers demonstrated no change in tests of hepatocellular integrity (plasma alanine transferase activity), synthetic function (serum albumin, prothrombin time, partial thromboplastin time) or renal function (serum creatinine, blood urea nitrogen).[30] This lack of toxicity extended to an enzyme-induced hypoxic rat model. Current evidence, therefore, indicates that desflurane is metabolized significantly less than other anaesthetics and that this is associated with a marked absence of organ-specific toxic effects.[1,41]

Pharmacodynamics

Potency

The MAC value for desflurane is 4.6–7.25 per cent in oxygen. The lower value was obtained from a study utilizing tetanic stimulus.[38] MAC is reduced by 50 per cent when administered with 60 per cent nitrous oxide. MAC decreases with age of subjects.[37] MAC-awake, the concentration midway between the value permitting response to command and that just preventing response, is 2.42 per cent.[38]

Respiratory effects

Desflurane is a mild airway irritant; inhalational induction may be accompanied by marked secretion, coughing, and occasionally laryngospasm, particularly as the inspired concentration rises through 6–7 per cent.[37] Airway support may be required at inspired concentrations over 2–3 per cent, oral airways are tolerated at levels of 4 per cent in oxygen.[42] Desflurane causes a dose-dependent depression of respiration. At concentrations up to 1.24 MAC, this is comparable to isoflurane. At deeper levels, up to 1.66 MAC, depression is greater, to a degree similar to enflurane. The concentration producing apnoea is estimated to be 1.8 MAC. Desflurane depresses ventilation primarily by reducing tidal volume; increased ventilatory frequency does not compensate and arterial carbon dioxide tension increases.[17] Effects on bronchomotor tone are likely to be similar to other agents.[1]

Cardiovascular effects

In healthy human volunteers, desflurane, in common with isoflurane, decreases systemic vascular resistance and mean arterial blood pressure. Unlike other inhaled halogenated anaesthetics, desflurane does not alter cardiac index, even with deep levels of anaesthesia (1.66 MAC). A sustained cardiac output and resistance to depression of contractility may result from better sustained autonomic activity.[43] Unlike isoflurane, the maintenance of cardiac index at light levels of desflurane anaesthesia is not dependent on increased heart rate, but at deeper levels of desflurane anaesthesia, increased heart rate may contribute to the maintenance of cardiac index. The unchanged cardiac output in the presence of increased right-heart filling pressure, decreased systemic vascular resistance, and increased heart rate, suggests that desflurane decreases myocardial contractility, although probably less than that produced by other halogenated anaesthetics. Decreased myocardial compliance may play a role in the decreasing myocardial function[44] or, like isoflurane, desflurane may prolong isovolaemic relaxation without altering ventricular compliance.[43] Prolonged desflurane anaesthesia produces a return of cardiovascular parameters towards pre-anaesthetic values, changes similar to those seen with all other agents except isoflurane. The aetiology of this is uncertain but appears unlikely to be due to increased β-sympathetic activity or products of metabolism as has been suggested for earlier agents.[44] Tachycardia occurs at deeper levels (> 1.66 MAC) than with isoflurane, and induction with desflurane may be associated with transient tachycardia and hypertension.[43] Stability of cardiac rhythm is notable during desflurane anaesthesia, and it does not sensitize the myocardium to catecholamines.[42]

CNS effects

Desflurane produces a pattern of increasing cortical depression, comparable to equipotent doses of isoflurane. Desflurane significantly suppresses EEG activity; prominent burst suppression is seen at 1.24 MAC.[31] Hypocapnic ($PaCO_2$ 3.3 kPa; 24 mmHg) patients with intracranial tumours have cerebral blood flows at 1 and 1.5 MAC that do not differ between desflurane and isoflurane. The increases in cerebral blood flow with imposed increases in $PaCO_2$ in these patients tends to be less with desflurane than with isoflurane. At 0.5 MAC, desflurane increases intracranial pressure in humans with supratentorial mass lesions to the same extent as isoflurane.[43] Cerebral metabolic rate is depressed in dogs in a dose-related manner similar to the result obtained with isoflurane.[31] Somatosensory evoked potentials are decreased by desflurane in a dose-related manner, 1.5 MAC producing a 45 per cent depression.[43] No epileptiform activity is seen, even in the presence of hypocapnia and repetitive auditory stimulus in volunteers.[31] Recovery, as discussed above, is rapid with desflurane, with less cognitive function impairment, pain, drowsiness, fatigue, clumsiness and confusion in the early postoperative period in comparison with isoflurane. Rapid emergence has, however, been associated with delirium with this agent, and concomitant administration of opioids has been advocated.[45]

Neuromuscular effects

Desflurane significantly depresses neuromuscular function in the absence of neuromuscular-blocking drugs, and augments the action of pancuronium and suxamethonium to a degree similar to isoflurane. It has depressant actions at sites in addition to the postjunctional membrane, perhaps at muscle peripheral to the junction or a prejunctional action, resulting in tetanic fade.[26] Desflurane can produce malignant hyperthermia in swine.[43]

Renal effects

After approximately 1 MAC-hour exposure to desflurane, mean inorganic fluoride concentration did not exceed the control value (0.79 µmol/l) at any time during the following week.[1] Nephrotoxicity has not been described.[43]

Hepatic effects

Metabolism of desflurane is minimal; however, small amounts of trifluoroacetate are found in blood and urine after anaesthesia with desflurane, hence caution should be exercised in patients at risk for 'halothane hepatitis'.[43] Hepatotoxicity has not been described with desflurane to date.

Other effects

Increases in white cell neutrophil count have been observed after 90 minutes of subanaesthetic levels of desflurane, and persisted for 4 days after several MAC-hours of anaesthesia. Blood glucose concentration rises during anaesthesia with desflurane in normal volunteers but returns to normal within 24 hours.[30] Desflurane does not produce mutagenic changes in the Ames test.[43]

Sevoflurane

In the early 1970s Regan, a research pharmacologist at Travenol Laboratories, Morton Grove, Illinois, identified a fluorinated isopropyl ether, sevoflurane (fluoromethyl-2,2,2-trifluoro-1-[trifluoromethyl] ethyl ether), as a potent anaesthetic agent with a low solubility in blood.

Physical properties

Sevoflurane is a clear colourless volatile liquid, notable for its low pungency. It has a molecular weight of 200.053 Da, a boiling point of 58.5°C and a saturated vapour pressure of 21.3 kPa (160 mmHg) at 20°C, characteristics that are similar to halothane, enflurane and isoflurane, making it suitable for delivery through standard vaporizers. Sevoflurane is less chemically stable than other currently employed agents. It is degraded in contact with soda lime, in increasing amounts with increasing temperature: at 22°C it is degraded at 6.5 per cent per hour, increasing by 1.6 per cent per hour per degree increase in temperature reaching 57.4 per cent degradation per hour at 54°C.[1] Baralyme degrades sevoflurane at an even greater rate at higher temperatures. The two main degradation products have been studied intensively in rats and humans with no evidence of toxicity under normal clinical conditions. High concentrations of one of the degradation products (three to four times 'maximal' clinical doses) caused reversible renal dysfunction in rats. Sevoflurane is not flammable in concentrations used in anaesthesia.

Pharmacokinetics

Sevoflurane has a blood/gas partition coefficient of 0.60–0.68 and an olive oil/gas partition coefficient of 47–53.0.[46] The low blood/gas solubility causes rapid achievement of alveolar–arterial tension consistent with rapid induction and recovery (see Figs 7.5 and 7.6). The lean tissue/blood partition coefficient of sevoflurane, greater than desflurane and isoflurane but less than halothane, suggests that cerebral concentration is moderately well correlated with inspired concentration using this agent. Unlike halothane and isoflurane there is no age effect on the blood/gas partition coefficient, probably due to its very low value.[5]

Elimination and metabolism

In addition to degradation by soda lime, sevoflurane also undergoes biodegradation to a similar extent to enflurane, which decreases the terminal elimination half-life of sevoflurane (see Fig. 7.7).[47] The primary products of metabolism are hexafluoroisopropyl alcohol, carbon dioxide and the fluoride ion.[8] Pretreatment with ethanol or phenobarbitone to induce hepatic enzymes increases biodegradation.[48]

Toxicity

Sevoflurane's toxic potential has now been clarified. In rats pretreated with phenobarbitone, exposure to 3 per cent sevoflurane in oxygen for 2 hours for 4 consecutive days, or 2.5 per cent sevoflurane in oxygen for 12 hours did not cause morphological hepatic changes, nor did it increase serum alanine aminotransferase or aspartate aminotransferase. Sevoflurane administered to hypoxic rats with induced hepatic microsomal enzymes, both in the presence and absence of soda lime, results in an incidence and extent of hepatotoxicity similar to that seen with isoflurane.[48]

The fluoride concentrations seen after sevoflurane anaesthesia appear to be similar to those seen after enflurane anaesthesia in terms of MAC-hours exposure. After about 1 MAC-hour exposure to sevoflurane in one study a plasma fluoride concentration of 22.1 μmol/l was found.[1] Prolonged anaesthesia with 1–4 per cent sevoflurane and nitrous oxide in patients with normal hepatorenal function is associated with a rise in serum inorganic fluoride that may exceed 50 μmol/l. In contrast with methoxyflurane, which produces high fluoride levels for several days and is associated with high output renal failure, serum and urinary inorganic fluoride declines to almost half the maximum level within 48 hours after cessation of sevoflurane, due to its lower blood/gas and tissue solubilities. This combined with the minimal metabolism of sevoflurane in the kidney (in contrast to methoxyflurane), reduces the risk of nephrotoxicity.[49]

Pharmacodynamics

Potency
The MAC value for sevoflurane is approximately 2 per cent (1.71–2.05 per cent) in oxygen and is reduced to 0.66 per cent when administered with 63.5 per cent nitrous oxide.[50]

Respiratory effects
Sevoflurane does not cause breath-holding or coughing, and induction is rapid and pleasant. Sevoflurane's respiratory effects are similar to those seen with the other agents, although a comparative study of the effects on medullary respiratory neurons in cats found sevoflurane to be twice as depressant as halothane in a dose-related manner.[48] In humans, 1.1 MAC produced almost the same degree of respiratory depression as halothane, and at 1.4 MAC sevoflurane produced more profound respiratory depression than halothane. Depression of respiration is evidenced by a rise in arterial carbon dioxide tension (about 20 per cent at 1 MAC) and decrease in the slope of the carbon dioxide response curve. This is usually accompanied by an increase in respiratory rate, but insufficient to maintain minute volume.[16] However, the rapid elimination of sevoflurane results in less postoperative respiratory depression than seen with halothane.[43] Sevoflurane's effects on dead space to tidal volume ratio, dead space and alveolar ventilation are the same as those of halothane.[16] Bronchodilatation occurs and sevoflurane will relax bronchiolar smooth muscle constricted by histamine or acetylcholine, but may be slightly less effective than halothane. Hypoxic pulmonary vasoconstriction is not inhibited by 1 MAC sevoflurane in dogs.[43]

Cardiovascular effects
The cardiovascular effects of sevoflurane are similar to, or more desirable than, those of isoflurane. It causes a lesser decrease in systemic arterial pressure and diastolic pressure decreases more than systolic.[48] Heart rate may be slower with sevoflurane, resulting from depression of sympathetic activity without alteration of parasympathetic activity, and a blunting or abolition of sympathetic responses to painful stimuli. Although both agents cause depression, myocardial contractility may be better preserved with sevoflurane than with halothane. In dogs, sevoflurane decreases myocardial oxygen consumption without decreasing myocardial blood flow. Sevoflurane appears to be a coronary artery dilator. Myocardial oxygen extraction ratio and myocardial lactate extraction decrease with sevoflurane, and sevoflurane decreases coronary flow reserve, but to a lesser extent than isoflurane or halothane. Myocardial blood flow is greater and resistance to flow less with either sevoflurane or isoflurane than with halothane; however, sevoflurane appears to preserve a normal ratio of endocardial to epicardial flow in the face of low coronary perfusion pressures. Sevoflurane does not appear to be arrhythmogenic, and in dogs the arrhythmogenic dose of adrenaline (epinephrine) equals or exceeds that found during anaesthesia with halothane or isoflurane. Sevoflurane has been used to manage the anaesthesia for resection of phaeochromocytoma in humans.[43]

CNS effects
The effects of sevoflurane on the brain appear to be similar to other potent inhaled halogenated anaesthetics. Sevoflurane has effects on cerebral blood flow, oxygen consumption and intracranial pressure similar to equipotent concentrations of isoflurane.[48] The EEG effects of sevoflurane differ from other volatile anaesthetics. In light planes, an increase in both frequency (10–14 Hz) and amplitude is associated with unconsciousness; and at deep levels, slower, low-voltage, 5–8 Hz activity also appears. However, the predominant 10–14 Hz activity persists with a further increase in amplitude.[24] No EEG or motor evidence of seizure activity has been noted, and sevoflurane modestly increases the threshold for convulsions from administration of lignocaine. However, its effects on epileptic patients have not been reported. Sevoflurane does not appear to increase cerebral blood flow in some animal studies, but this may result from a decrease in systemic arterial blood pressure that masks a cerebral vasodilatory effect. Rats anaesthetized with sevoflurane show a dose-related increase in the diameter of cerebral vessels and increase in cerebral blood flow. In rabbits, isoflurane and sevoflurane are comparable in their capacities to increase intracranial pressure and decrease cerebral metabolic rate. Relative to a nitrous oxide–fentanyl anaesthetic in rats, sevoflurane decreases the neurological deficit that results from hypotension in the presence of ligation of a carotid artery.[43]

Neuromuscular effects
Sevoflurane is a good muscle relaxant and appears to have a stronger potentiating effect for pancuronium or

vecuronium in humans when compared to previous studies with halothane or enflurane.[48] The degree of relaxation produced is sufficient to permit tracheal intubation without facilitation by muscle relaxants.[43] Like other halogenated anaesthetics sevoflurane can induce malignant hyperthermia in swine.[48]

Renal effects

Although there is some evidence that sevoflurane decreases renal blood flow and that prolonged exposure results in plasma inorganic fluoride concentrations which may exceed nephrotoxic values, there has been no evidence of gross changes in renal function in either animals or man. However, in these studies renal function was assessed by relatively simple tests, such as plasma creatinine, blood urea nitrogen and uric acid concentration, urinary volume and concentrating ability; more sensitive tests of tubular function, such as measurement of urinary retinol binding protein (RBP) or β–N-acetylglucosaminidase (NAG) were not performed.[1,49] (See toxicity section.)

Hepatic effects

No reports of clinical hepatotoxicity in humans have been made following the administration of over 2 million anaesthetics prior to October 1994. Conflicting evidence from animal studies, however, have shown biochemical and histological evidence of sevoflurane-induced hepatotoxicity, and it has been suggested that hepatotoxicity was related to a decrease in hepatic blood flow. The precise role of altered hepatic blood flow in the genesis of hepatotoxicity remains unclear, as hepatic blood flow has been shown to increase in a canine model. Sevoflurane decreased protein synthesis in rats at clinically relevant concentrations (albumin 68.3 per cent; fibrinogen 67.7 per cent; transferrin 62.3 per cent) and, with the exception of the effect of enflurane on albumin synthesis, this was greater with sevoflurane than with any other agent tested.[1]

Other effects

The Ames test indicates that sevoflurane is not mutagenic.[43]

Occupational exposure to volatile anaesthetics

Mutagenicity and carcinogenicity

Volatile anaesthetics reversibly inhibit cell multiplication, but not at clinically used concentrations. The ED_{50} for inhibition of multiplication of mammalian cell lines *in vitro* occurs at around 3 MAC, and effects are noted on all phases of the cell cycle. However, except in the case of some anaesthetics containing a vinyl group (fluroxene and divinyl either), there is no evidence that chromosome breaks, sister chromatid exchanges or mutagenesis are produced. The Ames test, which detects mutagenicity and carcinogenicity, has reported negative results with halothane, enflurane and isoflurane. Animal studies and human epidemiological surveys also suggest that modern volatile anaesthetics are not carcinogenic.[51,52]

Teratogenicity

The first epidemiological study of spontaneous abortion and malformations was published in 1967 when Vaisman reported a Russian study on 303 (193 men and 110 women) anaesthetists. The author noted that among 31 pregnant anaesthetists, one pregnancy in three had ended in spontaneous abortion, two women gave birth prematurely, and in one case a child had congenital malformations. The anaesthetic in use during the period of exposure was most commonly ether, but nitrous oxide and halothane had been used. Further epidemiological studies appeared to support the hypothesis that exposure of female personnel to trace anaesthetic vapours increased the risk of spontaneous abortion or major congenital anomaly. None of the studies was rigorous enough to establish a causal relationship, and the presence of multiple confounding variables (such as stress, viral exposure, radiation and chemicals such as methylmethacrylate monomer, hexachlorophene and ethylene oxide) makes risk assessment difficult. Most authors conclude that female theatre personnel have an abortion rate approximately double the normal incidence, but that nitrous oxide is more likely contributory than volatile agents.[51–53] With respect to development of neurological deficits after *in utero* exposure to volatile anaesthetic agents, at either trace or anaesthetic levels, there is insufficient evidence to permit quantification of risk.[54]

Behavioural impairment

No consistent decrements in test performance have been demonstrated in laboratories studying trace exposure (up to 200 parts per million halothane) to volatile anaesthetics. Consistent and reproducible decrements have only been associated with subanaesthetic concentrations of approximately 0.1–0.15 MAC (greater than 500 parts per million halothane or 1500 parts per million enflurane). Reaction times and error rates are increased in a dose-related manner; however, at 0.2 per cent halothane, the agent can be detected by smell and subjective feelings of drowsiness are experienced. Immediately following exposure to anaesthetic concentrations of volatile agents, cognitive and psychomotor function is impaired. At 1–4 days there is subjective residual impairment, but there are no objective criteria substantiating long-term decrements in mental acuity or psychomotor function as being due to volatile anaesthetic agents.[54]

Hepatic and renal toxicity

Evidence for development of hepatic injury (manifest as enzyme abnormalities, fatty changes or focal hepatic necrosis) and nephrotoxicity has accumulated in animal models of trace exposure to some halogenated agents. Despite these experimental models and epidemiological reports suggesting an increased risk of non-viral hepatic or renal disease in exposed personnel, no conclusive evidence exists for hepatic or renal impairment occurring in man as a result of occupational exposure to halogenated anaesthetic agents.[29]

Anaesthetic vaporizers

The purpose of an anaesthetic vaporizer is to deliver, accurately and safely, a clinically useful concentration of volatile anaesthetic agent. Each agent evaporates in ambient conditions to a degree which is dictated by saturated vapour pressure and ambient temperature. Safe anaesthesia requires that the fully saturated vapour be delivered to the patient predictably and reliably diluted in a carrier gas, over a range of concentrations.

The ether-soaked sponge in a bottle used in Morton's initial demonstration of 1846, and masks, such as the Schimmelbusch, used in the ensuing period, relied on the skill of the anaesthetist to titrate dose and effect. Fortunately ether's high blood solubility left a wide margin for error, but introduction of modern agents, vaporizers and monitoring standards have made predictable anaesthetic depth a reality. The evolution of the modern vaporizer has seen many devices incorporating a series of important design improvements, and any classification must address these. Table 7.5 is based on a previous classification by Dorsch and Dorsch.[55]

Method for regulating output concentration

Variable bypass vaporizers

Variable bypass vaporizers split the total fresh gas flow into two streams, each flow depending on the ratio of resistances. The carrier gas stream is diverted through the vaporizing chamber to become saturated with vapour before rejoining the bypass flow; these mix and pass to the common gas outlet. A single control knob, which must lock in the 'off' position and turn counterclockwise to open, is linked to allow adjustment of a variable orifice, which in modern vaporizers is in the inlet side.

Table 7.5 Classification of vaporizers[55]

Method for regulating output concentration	Variable bypass Measured flow
Method of vaporization	Flow-over with/without wick Bubble-through Flow-over or bubble-through

Location	Outside the breathing system
	Inside the breathing system
	    In circle
	    Inhaler or 'draw-over' type

Temperature compensation	None By supplied heat By flow alteration
Agent specificity	Agent specific Multiple agent

Measured flow vaporizers

Measured flow vaporizers of the Copper Kettle type utilize a flowmeter to provide a measured amount of oxygen to the vaporizer, all of which passes through the chamber, and becomes fully saturated. It is then diluted by addition to the balance of the fresh gas flow through the anaesthetic machine. These devices consist of three parts: the vaporizer with inbuilt thermometer, accurate flowmeter, and a circuit control valve to switch the vaporizer output into the circuit. In order to set a concentration, one needs to know: the agent's saturated vapour pressure (SVP), atmospheric pressure (P_{atm}), fresh gas flow (F_{fg}) and vaporizer flowmeter flow (F_{vap}):

$$\text{per cent delivered} = \frac{E_{vap} \times SVP}{P_{atm} \times F_{fg} - SVP \times F_{vap}}$$

It is important not to turn on the flow to the vaporizer flowmeter before the other flowmeters or else a lethal concentration of volatile agent may be delivered.

Method of vaporization

The accuracy of modern vaporizers relies largely on the attainment of vapour-saturated carrier gas for dilution in the fresh gas flow. Saturation is ensured by bringing carrier gas into close proximity with the liquid agent. In a flow-over vaporizer the carrier gas stream is directed over the surface of a volatile liquid to remove vapour. The efficiency of vaporization will depend on three factors: the surface area of liquid agent, the velocity of carrier gas flow and the height of gas flow above the liquid. In 1905 Levy presented the 'regulating chloroform inhaler', a draw-over vaporizer in which the concept of splitting the fresh gas flow was first introduced as a means to attain a predictably high output from a vaporizing chamber.[56] Devices of this period utilized baffles or spiral partitions to increase the transit time; addition of wicks through which the agent moves by capillary action has proved to be a solution retained in most current vaporizers. Wicks have the added advantage of reducing liquid movement when vaporizers are transported.

The familiar Boyle bottle employed the bubble-through method to increase the gas–liquid interface. It can be converted in use from flow-over to bubble-through by depressing a plunger attached to a hood over the carrier gas flow, which directs the gas below the surface, and trebles the output concentration of ether.[57] Efficiency increases as the gas–liquid interface increases; the Copper Kettle vaporizer contains a sintered bronze disc at its base which produces a very fine stream of bubbles with a long transit time through the liquid agent and thus achieves complete vaporization within the vaporizer. Knowledge of the vapour pressure, gas flow through the kettle, and bypass flow permits accurate calculation of the anaesthetic concentration. A problem with this approach is that foaming may occur within the vaporizer.

Location

Continuous flow-type vaporizers are mounted outside the patient circuit, most commonly between flowmeters and common gas outlet. This provides a stable mounting and heat transfer. All measured-flow vaporizers are located here on a dedicated circuit, and it is the recommended position for variable bypass devices. A vaporizer added between the common gas outlet and the patient circuit is convenient, but poses certain hazards (see below). Vaporizers designed for out-of-circuit use may not incorporate 22 mm fittings, and must be marked with their designated flow direction. Details of some vaporizers currently manufactured for out-of-circuit use are found in Table 7.6.

In contrast, vaporizers designed for use inside the breathing system must have standard 22 mm fittings or threaded weight-bearing fittings. Inlet and outlet as well as flow direction must be marked, and low internal resistance is the most important feature.

In-circle vaporizers such as the Goldman are simple, inefficient devices but may be very effective in a circle system due to the recirculation of vapour-laden gas. They may be located at the inspiratory or expiratory side; agent monitoring is recommended as lethal concentrations may build up, and moisture condenses in the vaporizer, affecting output.

Early vaporizers, such as that of Snow, were of the inhaler type in which carrier gas (usually air) is drawn through the device by a patient's respiration. Modern draw-over vaporizers for use with limited or no compressed gas supplies are listed in Table 7.7. These are supplied with self-inflating bags, tubing and spill valves to enable assisted or controlled ventilation and oxygen supplementation.

Temperature compensation

As volatile agent is vaporized, energy is lost in the form of heat. As the temperature of the liquid decreases, so does the vapour pressure. In order to maintain a constant vapour output in the face of this heat loss, two alternative means have been employed. One is to supply heat and the other is to alter the flow of carrier gas to compensate for heat loss.

Supplied heat was the earliest method used to ensure a constant vaporizer temperature. Warmed water baths were most commonly used, electrically heated elements were also incorporated. As the specific heat and thermal conductivity of copper is high this metal has been widely used in the manufacture of vaporizers. Heat can be conducted from supporting structures as a form of thermal buffering. Interestingly the development of a pressurized vaporizer for desflurane has seen the reintroduction of supplied heat to ensure saturated vapour output.

Most variable bypass vaporizers alter the flow ratio to the vaporizing chamber and the bypass in response to temperature change. Flow is automatically adjusted by a thermostatic mechanism which opens and closes the outflow of the vaporizing chamber. As vapour pressure varies non-linearly with temperature, and compensation is usually linear, operational temperature range must be limited to manufacturer's specifications. In other vaporizers temperature compensation must be made manually by adjusting the flow, depending on chamber temperature and accompanying nomograms. Note that sudden changes in ambient temperature of about 20°C, such as may occur on moving a vaporizer from storage into use, require 1–2 hours for temperature compensation in order to maintain concentrations within stated tolerances.[58]

Effects of carrier gas compensation

Vaporizer output varies with changes in the composition of the carrier gases.[59] Large variations in vaporizer output lasting up to 30 minutes after introducing or discontinuing nitrous oxide have been documented and attributed to the high solubility of nitrous oxide in the volatile agent.[60] Nitrous oxide thus absorbed or liberated from the liquid anaesthetic decreases or increases respectively the total gas flow through the vaporizing chamber. If carrier gas composition is markedly changed in a non-rebreathing system, the aberrant output may be transmitted rapidly to the patient, resulting in inappropriate change in depth of anaesthesia. Circle systems will tend to buffer any such changes; however, low flow systems may also exhibit unexpected gas composition change.[61]

The carrier gas used for calibration of vaporizers may be air or oxygen. Gas mixes with differing viscosities alter delivered agent concentration in many variable bypass vaporizers, an effect that is usually clinically insignificant, particularly if agent monitoring is in use. Air and helium

Table 7.6 Out-of-circuit vaporizers

	Dräger-Vapor-19.2,19.3	Engström Elsa	Penlon PPV Sigma	Siemens 950 series	Tec 4 series	Tec 5 series	Tec 6
Agent models	Halothane (H) Enflurane (E) Isoflurane (I)	Halothane Enflurane Isoflurane	Halothane Enflurane Isoflurane	Halothane Enflurane Isoflurane	Halothane Enflurane Isoflurane	Halothane Enflurane Isoflurane	Desflurane
Delivered range	H 0–4, 0–5% E 0–5, 0–7% I 0–5% ± 15%	Accuracy ±5–25%	H 0–4 & 5% E 0–5 & 7% I 0–5% ± 20–25%	H 0.2–4% E 0.2–5% I 0.2–5% ± 10%	H 0–5% E 0–7% I 0–5%	H 0–5% E 0–7% I 0–5%	1–18%
Control mechanism	Variable bypass, control cone in chamber outlet path	Gas driven pulsed injection	Variable bypass, needle valve in chamber outlet path	Valve directs driving gas, pressure injects agent into fresh gas flow	Variable bypass, radial groove	Variable bypass, radial groove	Agent metered as gas
Temperature compensation	Expansion rod, valve in chamber inlet 15–35°C ±10%	Heated vaporizing chamber to 70°C	Liquid-filled expansion bellows, valve in bypass	No: 10°C rise causes 10% output rise	Bimetallic strip, valve in outflow 18–35°C	Bimetallic strip, valve in outflow 18–35°C	Electrically heated, pressurized chamber
Capacity (ml)	200 wick 60	N/A	165 ± 10 wick 7	125 wick 20	125 wick 35	300 wick 75	450
Tipping	45° limit	N/A	Avoid 45° limit	Avoid 45° limit	Non-spillable to 180°	Non-spillable to 180°	
Back-pressure compensation	Yes to ± 20%	?	Yes	N/A	Yes	Improved over Tec 4	Yes
Other features	Interlock device, Selectatec available	Agent level sensor; one agent deliverable; monitor feedback; self-calibrates	Selectatec compatible	For use with Servo 900 series & gas blender. High inlet pressure (3–500 kPa)	Selectatec interlock	Selectatec interlock, rear dial release, rapid filling, 3 year service	Selectatec interlock, status alarm and shutdown
Weight (kg)	7.5	Integral	5.7–6.1	1.4	7.2	7.0	
Filler type	Keyed filler or spout	Direct from bottles	Keyed filler, screw cap	Proprietary keyed filler	Keyed filler or funnel	Keyed filler or funnel	Specific for desflurane
Flow range (l/min)	0.5–15	0.4–10 esp. low flow	1–10	> 35	0.25–15	0.2–15	
Manufacturer	Drägerwerk Ag Lübeck, Germany	Gambro Engström AB, Sweden	Penlon Ltd, UK	Seimens-Elema AB, Sweden	Ohmeda, BOC Health Care Ltd	Ohmeda BOC Health Care Ltd	Ohmeda, BOC Health Care Ltd

Table 7.7 Draw-over vaporizers

	PAC series (Drawover Tec) Ohmeda	Ether Pac (Ethertec) Ohmeda	EMO Epstein, Macintosh, Oxford	OMV Oxford Miniature Vaporizer Penlon	Afya Dräger
Agent models	Trichloroethylene (T) Halothane Enflurane Isoflurane	Diethyl ether	Diethyl ether	Halothane Trichloroethylene Isoflurane	Diethyl ether
Delivered range	T 0–1.5% H 0–5% E 0–5% I 0–5%	0–15%	0–20% ± 0.75%	H 0–4% T 0–1% I 0–5%	0–20%
Control mechanism	Variable bypass, radial groove	Variable bypass, radial groove	Variable bypass	Variable bypass	Variable bypass, control cone in chamber outlet
Operational temperature & compensation	18–35°C Bimetallic strip, valve in outflow	18–35°C Bimetallic strip, valve in outflow	13–23°C Ether-filled bellows, 1.2 litre water bath, visual indicator	15–30°C Water jacket and flow design only, no compensation	16–28°C Copper/water heat sink & manual adjustment
Capacity (ml)	85 wick 13	150 wick 35	450 cloth wicks	50	
Tipping	Yes	Yes	Avoid	Unsafe	Yes
Back-pressure compensation	Non-return valve, circuit valves	Non-return valve, circuit valves	Circuit valves	Circuit valves	Non-Return valve, circuit valves
Other features	Non-spill, unaffected by motion, fits oxygen supplementation	Non-spill, unaffected by motion	Combines with Oxford Inflating Bellows for controlled ventilation	Often used in series with EMO or other OMV (as part of Triservice apparatus)	Based on Vapor, fits OMV, oxygen supplement
Filler type	Keyed filler (not trichloroethylene) or screw cap	Screw cap	Screw cap	Screw cap	Screw cap
Resistance	0.06 kPa/l per min	0.06 kPa/l per min	0.125 kPa at 40 l/min	0.08 kPa at 40 l/min	

reduce, whereas nitrous oxide increases, delivered concentration, the changes being in the order of 2–15 per cent from that obtained with oxygen.[58,62]

Effects of altered barometric pressure

Variable bypass vaporizers deliver the same partial pressure with changes in barometric pressure, but concentration as volumes per cent delivered will change:

$$D = \frac{C \times 760}{P}$$

where D is the output concentration in volumes per cent at barometric pressure, P is in mmHg and C is the dial setting in volumes per cent. However, since it is the partial pressure which dictates depth of anaesthesia, the clinical effect of a given setting will be independent of atmospheric pressure.

Effects of intermittent back-pressure

Pressure fluctuations encountered during ventilation, or as a result of activation of the oxygen flush, can result in delivery of a concentration greater than that set on the dial. Back-pressure disturbs the normal ratio of flows to the vaporizing chamber and bypass, increasing the gas flow in the former, via a pulse of retrograde flow. This pumping effect is greater if the vaporizing chamber is large or nearly empty, if the set concentration is low, if large amplitude swings in pressure occur, and if expiratory pressure drop is sudden.[58] First documented in the Fluotec Mark II,[55] incorporation of smaller vaporizing chambers with longer chamber inlet paths in contemporary models has reduced the magnitude of the discrepancy. At high flow rates the same vaporizer delivered reduced concentrations, a phenomenon ascribed to a pressurizing effect of pressure fluctuations during ventilation. Likewise, modern vaporizer design has rendered this problem clinically insignificant.

Hazards

Incorrect agent

The 'worst case', of setting a halothane-filled enflurane vaporizer at 2 per cent, gives four times the intended MAC-dose.[57] Keyed filling systems minimize the risk of accidental vaporizer filling.

Multiple agent

Combinations of volatile agents have been popular in the past, either in series (trilene or methoxyflurane in combination with other agents) or as azeotropic mixtures. Modern volatile agents eliminate one hazard of mixing: halothane, enflurane and isoflurane do not react chemically or form azeotropes.[57] Currently, however, vaporizer design seeks to minimize the risk of multiple exposure. Interlock devices, such as the Selectatec system, prevent use of adjacent vaporizers and most isolate the fresh gas flow by bypassing deactivated vaporizers. Physical disconnection is a simpler method, whereby a single pair of male–female slip-fit connections are used to connect only the intended vaporizer. Where these systems are not employed for vaporizers in series, cross-contamination is possible. To minimize risk to the patient, agents of equal potency should be arranged so that the one with the lower vapour pressure is upstream; likewise agents of equal vapour pressure should be arranged with the less potent upstream. If explosive agents or drugs incompatible with soda lime, such as trilene, are used they should be downstream.[57]

Overdosage

Tipping, foaming, overfilling or reflux admitting liquid agent into the patient circuit or vaporizer flowmeter are all documented causes of volatile agent overdosage. Reversed flow of carrier gas in copper kettle-type devices has caused major morbidity as liquid agent is forced into the stem of the loving cup, and reflux into the patient circuit. During testing and use of copper kettle devices, frequent observation of the sight glass to check the liquid anaesthetic level is recommended as rapid decreases may indicate misconnection of the unit.[63] Improperly calibrated vaporizers can cause delivery of the wrong concentration of volatile agent, resulting in either insufficient or excessive anaesthesia.

Contamination

The discoloration of anaesthetic agents in vaporizers indicates ageing or contamination of the fluid and should alert the user to the need for corrective action. In no instance should this signal be discredited or ignored. Prompt drainage and replacement of discoloured agent is recommended, as failure to do so may mask later problems or conditions requiring immediate attention.[64] Contaminants causing discoloration may be oils or plasticizers from seals or wicks, particularly if these have been recently purchased or replaced,[65,66] or thymol, present as a stabilizer in halothane. Thymol may crystallize in valve mechanisms of older-design vaporizers and causes sticking.[67] Corrosion of the bypass valve mechanism by aqueous solution has been documented as a cause of fatal overdosage of anaesthetic agent.[68] Water may enter vaporizers during cleaning or by condensation during filling, and can not be removed without disassembling the unit.

Leaks

Loss of fresh gas flow may cause low patient circuit flows with rebreathing and has been documented as responsible for cardiac arrest.[69] This complication can be detected by pre-use leak test and intraoperative capnography. When a leak is detected and corrected, the test should be repeated to exclude the presence of a second leak.[70]

Recommendations

1. Examine all vaporizers before each use: check fluid levels, fill the correct agent to the proper level, and seal the filling port. Turn the vaporizer off before leak testing. Use a low-flow leak test (e.g. 30 ml/min) with a pressure gauge at the common gas outlet, and turn each vaporizer on and off sequentially to exclude fresh gas leakage. Turn all vaporizers off.[71]

2. Keep all vaporizers connected to the anaesthetic machine and use a selector valve to engage the appropriate unit; do not use an arrangement requiring frequent removal or interchange of vaporizers.[68]

3. Avoid placing vaporizers between common gas outlet and patient circuit.[72] High internal resistance reduces oxygen flush (vaporizers in this position should permit 70 l/min flow) and does not bypass the vaporizer, so that a high concentration of volatile agent may be delivered to the patient when high flow oxygen is required. In addition, high resistance at common gas outlet trips the upstream high pressure relief valve and causes loss of fresh gas flow, and may deliver high back-pressure to vaporizers on the back-bar. A vaporizer at the common gas outlet in addition to back-bar vaporizers adds the risk of delivering multiple

agents simultaneously. A device in this position is also potentially unstable, allowing tipping; potential for misconnection and subsequent flow reversal is greater.

4. After shipping, servicing, repairing or modifying vaporizers or connections, visually check connections, inspect the unit for damage and test vaporizer output with a vapour analyser.[73] Document any work performed on vaporizers to show repairs or modifications performed and the testing performed to ensure the unit is ready for use.

5. Monitor inspired and expired anaesthetic concentration during anaesthesia by the use of an agent-specific monitor.

Servicing

Halothane vaporizers should be drained at least monthly, the old anaesthetic discarded, and the unit refilled with fresh halothane. No other cleaning procedure is necessary between services performed by qualified service personnel. Enflurane and isoflurane vaporizers do not need such frequent maintenance because these are pure substances which do not contain thymol. Drainage to eliminate products of decomposition or inadvertently added impurities, such as those from wicks, should be performed periodically. The output of vaporizers should be verified twice a year and the manufacturer's recommendations for overhaul and calibration intervals, which vary from 1 to 3 years, should be followed.[72]

REFERENCES

1. Jones RM. Desflurane and sevoflurane: inhalation anaesthetics for this decade? *British Journal of Anaesthesia* 1990; **65**: 527–36.

2. Heijke S, Smith G. Editorial II. Quest for the ideal inhalation anaesthetic agent. *British Journal of Anaesthesia* 1990; **64**: 3–6.

3. Halsey MJ. Reassessment of the molecular structure-functional relationships of the inhaled general anaesthetics. *British Journal of Anaesthesia* 1984; **56**: 9S.

4. Dale O, Brown Jr BR. Clinical pharmacokinetics of the inhalational anaesthetics. *Clinical Pharmacokinetics* 1987; **12**: 145–67.

5. Malviya S, Lerman J. The blood/gas solubilities of sevoflurane, isoflurane, halothane, and serum constituent concentrations in neonates and adults. *Anesthesiology* 1990; **72**: 793–6.

6. Yasuda N, Lockhart SH, Eger II EI, *et al.* Kinetics of desflurane, isoflurane, and halothane in humans. *Anesthesiology* 1991; **74**: 489-98.

7. Murray JB, Trinick TR. Plasma fluoride concentrations during and after prolonged anaesthesia: a comparison of halothane and isoflurane. *Anesthesia and Analgesia* 1992; **74**: 236–40.

8. Mazze RI. Metabolism of inhaled anaesthetics: implications of enzyme induction. *British Journal of Anaesthesia* 1984; **56**: 27S.

9. Brown BR. Hepatotoxicity of inhaled anaesthetics. *Current Opinion in Anesthesiology* 1989; **2**: 414-17.

10. Dundee JW. Problems of multiple inhalation anaesthetics. *British Journal of Anaesthesia* 1984; **56**: 63S.

11. Neuberger JM. Halothane and hepatitis. Incidence, predisposing factors and exposure guidelines. *Drug Safety* 1990; **5**: 28–38.

12. Brown BR, Gandolfi AJ. Adverse effects of volatile anaesthetics. *British Journal of Anaesthesia* 1987; **59**: 14–23.

13. Pohl LR, Kenna JG, Satoh H, *et al.* Neoantigens associated with halothane hepatitis. *Drug Metabolism Reviews* 1989; **20**: 203–17.

14. Rosenberg PH. Modern aspects of the toxicity of inhaled anaesthetics. *Current Opinion in Anesthesiology* 1991; **4**: 518–21.

15. Steinbereithner K. Introduction to quantitative anaesthesia. In: Torri G, Damia G, eds. *Update in inhalational anaesthetics.* New York: Worldwide Medical Communications, 1989; Chapter 6: 145–57.

16. Doi M, Ikeda K. Respiratory effects of sevoflurane. *Anesthesia and Analgesia* 1987; **66**: 241–4.

17. Lockhart SH, Rampil IJ, Yasuda N, Eger II EI, Weiskopf RB. Depression of ventilation by desflurane in humans. *Anesthesiology* 1991; **74**: 484–8.

18. Pasch, T, Kamp H-D, Petermann H. Inhalation anaesthesia for the patient with respiratory disease. In: Torri G, Damia G, eds. *Update in inhalational anaesthetics.* New York: Worldwide Medical Communications, 1989; Chapter 2: 52–6.

19. Reiz S. Myocardial ischaemia associated with general anaesthesia. A review of clinical studies. *British Journal of Anaesthesia* 1988; **61**: 68-84.

20. Eger II EI. Isoflurane: a review. *Anesthesiology* 1981; **55**: 559–76.

21. Jones RM. Clinical comparison of inhalation anaesthetic agents. *British Journal of Anaesthesia* 1984; **56**: 57S.

22. Durand P-G, Lehot J-J, Foex P. Calcium-channel blockers and anaesthesia. *Canadian Anaesthetists' Society Journal* 1991; **38**: 75–89.

23. Frost EAM. Inhalation anaesthetic agents in neurosurgery. *British Journal of Anaesthesia* 1984; **56**: 47S.

24. Modica PA, Tempelhoff R, White PF. Pro- and anticonvulsant effects of anesthetics (Part I). *Anesthesia and Analgesia* 1990; **70**: 303–15.

25. Schwartz DM, Bloom MJ, Pratt Jr RE, Costello JA. Anesthetic effects on neuroelectric events. *Seminars in Hearing* 1988; **9**: 99–112.

26. Caldwell JE, Laster MJ, Magorian T, *et al.* The neuromuscular effects of desflurane, alone and combined with pancuronium or succinylcholine in humans. *Anesthesiology* 1991; **74**: 412–18.

27. Cousins MJ, Skowronski G, Plummer JL. Anaesthesia and the kidney. *Anaesthesia and Intensive Care* 1983; **11**: 292–320.

28. Andreen M. Inhalation versus intravenous anaesthesia. Effects of the hepatic and splanchnic circulation. *Acta Anaesthesiologica Scandinavica* 1982; **26**: (Suppl 75): 25–31.

29. Cousins MJ, Plummer JL, Hall P De la M. Toxicity of volatile anaesthetic agents. *Clinics in Anaesthesiology* 1984; **2**: 551–75.

30. Weiskopf RB, Eger EI, Ionescu P, *et al.* Desflurane does not produce hepatic or renal injury in human volunteers. *Anesthesia and Analgesia* 1992; **74**: 570–4.

31. Rampil IJ, Lockhart SH, Eger II EI, Yasuda N, Weiskopf RB, Cahalan MK. The electroencephalographic effects of desflurane in humans. *Anesthesiology* 1991; **74**: 434–9.

32. Weissman C, Hollinger I. Modifying systemic responses with anesthetic techniques. *Anesthesiology Clinics of North America* 1988; **6**: 221–35.

33. Eger II EI. The pharmacology of isoflurane. *British Journal of Anaesthesia* 1984; **56** (Suppl 1): 71S–99S.

34. Eger EI. Stability of I-653 in soda lime. *Anesthesia and Analgesia* 1987; **66**: 983–5.

35. Priebe HJ. Isoflurane and coronary hemodynamics. *Anesthesiology* 1989; **71**: 960–76.

36. Levy WJ. Clinical anaesthesia with isoflurane. A review of the multicentre study. *British Journal of Anaesthesia* 1984; **56** (Suppl 1): 101S–112S.

37. Rampil IJ, Lockhart SH, Zwass MS, *et al.* Clinical characteristics of desflurane in surgical patients: minimum alveolar concentration. *Anesthesiology* 1991; **74**: 429–33.

38. Jones RM, Cashman JN, Eger II EI, Damask MC, Johnson BH. Kinetics and potency of desflurane (I-653) in volunteers. *Anesthesia and Analgesia* 1990; **70**: 3–7.

39. Smiley RM, Ornstein E, Matteo RS, Pantuck EJ, Pantuck CB. Desflurane and isoflurane in surgical patients: comparison of emergence time. *Anesthesiology* 1991; **74**: 425-8.

40. Jones RM, Koblin DD, Cashman JN, *et al.* Biotransformation and hepatorenal function in volunteers after exposure to desflurane (I-653). *British Journal of Anaesthesia* 1990; **64**: 482–7.

41. Holmes MA, Weiskopf RB, Eger II EI, Johnson BH, Rampil IJ. Hepatocellular integrity in swine after prolonged desflurane (I-653) and isoflurane anesthesia: evaluation of plasma alanine aminotransferase activity. *Anesthesia and Analgesia* 1990; **71**: 249–53.

42. Jones RM, Cashman JN, Mant TGK. Clinical impressions and cardiorespiratory effects of a new fluorinated inhalation anaesthetic, desflurane (I-653), in volunteers. *British Journal of Anaesthesia* 1990; **64**: 11–15.

43. Eger EI. Current status of sevoflurane and desflurane. *Anesthesia and Analgesia* 1992; **74** (Suppl): 81–7.

44. Weiskopf RB, Cahalan MK, Eger II EI, *et al.* Cardiovascular actions of desflurane in normocarbic volunteers. *Anesthesia and Analgesia* 1991; **73**: 143–56.

45. Ghouri AF, Bodner M, White PF. Recovery profile after desflurane–nitrous oxide versus isoflurane–nitrous oxide in outpatients. *Anesthesiology* 1991; **74**: 419–24.

46. Strum DP, Eger II EI. Partition coefficients for sevoflurane in human blood, saline and olive oil. *Anesthesia and Analgesia* 1987; **66**: 654–6.

47. Yasuda N, Lockhart SH, Eger II EI, *et al.* Comparison of kinetics and sevoflurane and isoflurane in humans. *Anesthesia and Analgesia* 1991; **72**: 316–24.

48. Weiskopf RB. New inhaled anesthetics. *Current Opinion in Anesthesiology* 1989; **2**: 421-4.

49. Kobayashi Y, Ochiai R, Takeda J, *et al.* Serum and urinary inorganic fluoride concentrations after prolonged inhalation of sevoflurane in humans. *Anesthesia and Analgesia* 1992; **74**: 753–7.

50. Katoh T, Ikeda K. The minimum alveolar concentration (MAC) of sevoflurane in humans. *Anesthesiology* 1987; **66**: 301–3.

51. Nunn JF. Faulty cell replication: abortion, congenital abnormalities. *International Anesthesiology Clinics* 1981; **19**: 77–97.

52. Baden JM. Chronic toxicity of inhalation anaesthetics. *Clinics in Anaesthesiology* 1983; **1**: 441–54.

53. Edling C. Anesthetic gases as an occupational hazard. A review. *Scandinavian Journal of Work, Environment and Health* 1980; **6**: 85–93.

54. Rice SA. Behavioural toxicity of inhalation anaesthetic agents. *Clinics in Anaesthesiology* 1983; **1**: 507–19.

55. Dorsch, JA, Dorsch SE. Classification of vaporizers. In: Dorsch JA, Dorsch SE, eds. *Understanding anesthesia equipment.* Baltimore: Williams & Wilkins, 1984; Chapter 4: 81.

56. Zuck D. The development of the anaesthetic vaporizer. The contribution of A.G. Levy. *Anaesthesia* 1988; **43**: 773–5.

57. Ohmeda product information, Omeda, Australia.

58. Drägerwerk Ag product information: Dräger-Vapor® 19.n anaesthetic vaporizer, Lübeck, Germany.

59. Prins L, Strupat J, Clement J, Knill RL. An evaluation of gas density dependence of anaesthetic vaporizers. *Canadian Anaesthetists' Society Journal* 1980; **27**: 106–10.

60. Gould DB, Lampert BA, MacKrell TN. Effect of nitrous oxide solubility on vaporizer aberrance. *Anesthesia and Analgesia* 1982; **61**: 938–40.

61. Scheller MS, Drummond JC. Solubility of N_2O in volatile anesthetics contributes to vaporizer aberrancy when changing carrier gases. *Anesthesia and Analgesia* 1986; **65**: 88–90.

62. Penlon Ltd Document no. PT 192 UI. *PPV Sigma vaporizer. User instruction manual.* Oxford: Oxonian Rewley Press, January 1992.

63. Crossed copper kettle vaporizer connections in foregger anesthesia machines. *Health Devices* 1984; **13**: 322–4.

64. North American Drager anesthesia vaporizers. *Health Devices* 1986; **15**: 332.

65. Weldon ST, Williams-Van Alstyne SI, Gandolfi AJ, Blitt CD. Production and characterization of impurities in isoflurane vaporizers. *Anesthesia and Analgesia* 1985; **64**: 634-9.

66. Gandolfi AJ, Blitt CD, Weldon ST. Discoloration and impurities in isoflurane vaporizer. *Anesthesia and Analgesia* 1983; **62**: 366.

67. Carter KB, Gray WM, Railton R, Richardson W. Long-term performance of Tec vaporizers. Basis for a rational maintenance policy. *Anaesthesia* 1988; **43**: 1042–6.

68. Water in halothane vaporizers. *Health Devices* 1985; **14**: 326–7.

69. Vaporizer leak with Mapleson breathing circuits. *Health Devices* 1986; **15**: 344.

70. Pre-use anesthesia check fails to find faults. *Health Devices* 1988; **17**: 274.

71. Pre-use checklist for anesthesia units (machines and accessories). *Health Devices* 1988; **17**: 275.

72. High risk equipment. *Health Devices* 1988; **17**: 48–50.
73. Concentration calibrated vaporizers. *Health Devices* 1987; **16**: 112.
74. Calverley RK, Smith NT, Jones CW, *et al*. Ventilatory and cardiovascular effects of enflurane anesthesia during spontaneous ventilation in man. *Anesthesia and Analgesia* 1978; **57**: 610–18.
75. Cromwell TH, Eger II EI, Stevens WC, *et al*. Forane uptake, excretion and blood solubility in man. *Anesthesiology* 1971; **35**: 401–8.
76. Bahlman SH, Eger II EI, Halsey MJ, *et al*. The cardiovascular effects of halothane in man during spontaneous ventilation. *Anesthesiology* 1972; **36**: 494–502.

77. Winter *et al*. ASA abstracts 1972; 103–4.
78. Stevens WC, Cromwell TH, Halsey MJ, *et al*. The cardiovascular effects of a new inhalation anesthetic, Forane, in human volunteers at constant arterial carbon dioxide tension. *Anesthesiology* 1971; **35**: 8–16.
79. Eger II EI, Smith NT, Stoelting RK, *et al*. Cardiovascular effects of halothane in man. *Anesthesiology* 1970; **32**: 396–409.
80. Calverley RK, Smith NT, Prys-Roberts C, *et al*. Cardiovascular effects of enflurane anesthesia during controlled ventilation in man. *Anesthesia and Analgesia* 1978; **57**: 619–28.
81. Johnston RR, Eger II EI, Wilson C. A comparative interaction of epinephrine with enflurane, isoflurane and halothane in man. *Anesthesia and Analgesia* 1976; **55**: 709–12.

Chapter 8

Neuromuscular Transmission and Neuromuscular Disease

Jørgen Viby-Mogensen

Anatomy of the neuromuscular junction
Physiology of the neuromuscular transmission
Neuromuscular block

Factors that may affect neuromuscular transmission and neuromuscular block
Neuromuscular blocking agents in patients with neuromuscular disorders

The arrow poison used for hunting by the native people of South America has been known for centuries. Shortly after the first Spaniards arrived in the New World in the sixteenth century accounts of the mysterious poison began to appear. Among the more spectacular personalities reporting on the poison was Sir Walther Raleigh (1552–1618), who at that time was a well known pirate, poet, adventurer and sailor. He described the poison in 1596, and it was one of his captains who named the poison 'Ourari'.

Among others, the French scientist Charles-Marie de la Condamine (1701–1774) and the English scientist Edward Bancroft (1744–1821) brought back to Europe samples of the curare poison. For many years these samples were the basis for experiments in different parts of Europe. Some of the more famous results of these experiments were the findings of Benjamin Brodie (1783–1862) and his assistant Edward Nathaniel Bancroft (1772–1842; son of Edward Bancroft). They showed that the poison paralysed the

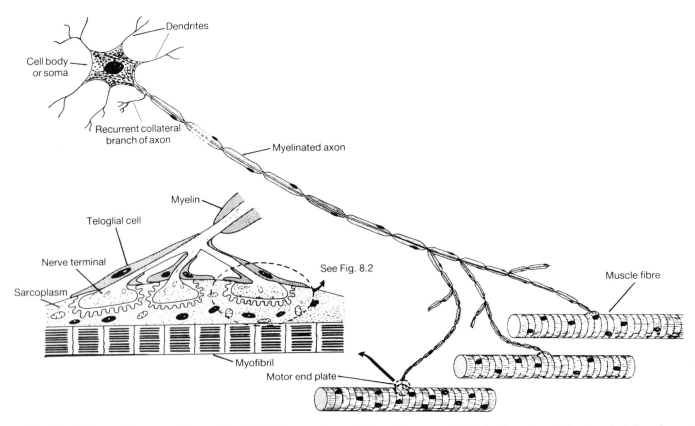

Fig. 8.1 Diagram of a motor unit containing focally innervated muscle fibres. A motor end plate is enlarged as the inset on the left, and a unit of this is further enlarged in Fig. 8.2. (Reproduced with permission from Bowman WC, Rand MJ. *Textbook of pharmacology.* 2nd ed. Oxford: Blackwell Scientific Publications, 1980.)

respiratory muscles, and that an animal given curare could be kept alive if ventilated.[1,2] However, it was not until 1856, when Claude Bernard (1818–1878) published his classic experiments on frogs, that it became clear that curare acted peripherally, causing paralysis at the junction of the nerve and the muscle.[3] Finally, in the 1930s H. H. Dale, W. Feldberg and M. Vogt convincingly showed that acetylcholine, the transmitter at the neuromuscular junction, acted on striated muscle, while the effect of curare was exerted at the acetylcholine receptor site.[4] Since then the neuromuscular junction has undoubtedly been the most studied junction in the body. The basic concept of acetylcholine as the chemical transmitter, being synthesized in the nerve endings and acting on postsynaptic receptors, has not been changed over the years. However, in recent years important advances in modern technology, not least in electron microscopy, electrophysiology, immunology, and DNA technology, have much increased our knowledge of the transmission process.

Anatomy of the neuromuscular junction

Motor units

Striated muscles are innervated by the lower motor neurons which pass uninterrupted from the spinal cord to the muscles. Each motor neuron innervates many muscle fibres (Fig. 8.1) and the neuron together with its muscle fibres form a motor unit. The number of muscle fibres in a unit depends on the function of the muscle in question, varying from less than 10 (the extraocular muscles) to more than 1700 (postural muscles). The more delicate the movements, the fewer muscle fibres per motor neuron.[5] Most muscle fibres in man are focally innervated, i.e. they receive their innervation at one focal point (the motor end plate), often at the muscle's mid portion (*en plaque* ending). However, tonic muscles, such as the extraocular muscles, the facial muscles and the intrinsic laryngeal muscles, are multiply innervated with several motor end plates distributed over the muscle fibres (*en grappe* endings). The reaction of a motor unit and of the muscle fibre itself is an all-or-none process; i.e. there is a maximum contraction of all muscle fibres in the unit or no contraction at all. The force of a muscle contraction is graded by and proportional to the number of motor units activated.

The neuromuscular junction

The motor neurons lose their myelin sheaths when they reach the muscle fibre and divide into a number of branches ending in small swellings embedded in the muscle fibre membrane and forming the neuromuscular junction (Fig. 8.1). The nerve endings contain vesicles for the chemical transmitter acetylcholine (see later). There is no direct contact between the nerve terminal and the muscle fibre. They are separated by a gap of 20–50 nm: the synaptic or junctional cleft. The terminal Schwann cells extend to the muscle fibre (Fig. 8.2), enclosing the neuromuscular junction and separating it from the extracellular fluid. The junctional cleft is filled with a collagen structure named the basement membrane. To this membrane, which extends into the clefts of the junctional folds, is attached most of the acetylcholinesterase present at the neuromuscular junction.

The muscle membrane adjacent to the nerve ending is heavily folded, forming the secondary clefts. The acetylcholine receptors are concentrated at the shoulders of these clefts, in such a way that the receptors lie adjacent to the active zones at the nerve endings (see later). Each neuromuscular end plate has 10^6–10^7 nicotinic receptors.[6] The receptors have recently been isolated and purified, and the amino acid sequence of the different polypeptides of the receptor is now known. The receptor consists of a pentamer of four different proteins (subunits), arranged in the form of a rosette around a central ion channel (Fig. 8.3). The receptor protrudes on both sides of the cell membrane, but more so on the synaptic side, where it reaches about 6 nm into the synaptic gap. The subunits of the adult receptor are called α, β, ϵ and δ. Two of the subunits (α) are identical, and each contains one binding site for acetylcholine. The molecular weight of the whole receptor is around 250 000 daltons, the α-units each having a molecular weight around 40 000 daltons, and the other subunits between 49 000 and 67 000 daltons.[7] The receptor is synthesized in the muscle cell and within a few hours inserted in the cell membrane.

Two types of postsynaptic cholinergic receptors exist. In fetal muscle and in denervated muscles the receptors are inserted over the whole muscle surface and are more or less free floating in the lipid layers of the cell membrane. These so-called extrajunctional nicotinic receptors, instead of an ϵ-subunit, contain a γ-subunit with fewer basic amino acids and with a lifetime of only 17–20 hours. On the other hand, the postsynaptic nicotinic receptors of innervated muscles are concentrated at the neuromuscular junction and they have a lifetime measured in days to weeks, i.e. much longer than the 'fetal' extrajunctional nicotinic receptors. The two types of postsynaptic nicotinic receptors react differently to agonists as well as antagonists (see later).

The muscle

The contractile elements of muscle cells are the myofilaments: the thick myosin filaments and the thin actin filaments with attached troponin and tropomysin. These filaments interdigitate and slide over each other when the muscle cell contracts. The myofilaments are grouped into larger cylindrical structures called myofibrils. Surrounding the myofibrils is the sarcoplasmic reticulum, which constitutes a network of tubules and sacs and acts as a reservoir for calcium. The sarcoplasmic reticulum comes into close proximity with the invaginations of the sarcolemma, the transverse tubules (T-tubules), and there are indications for the existence of a special receptor-like protein structure between the sarcoplasmic reticulum and the T-tubules.[7] These tubules convey the electrical impulses (the action potential) from the muscle surface into the sarcoplasmic reticulum, thereby triggering the liberation of calcium and the contractions of the myofilaments.

Sometimes muscles are classified into *tonic fibres* (with more than one end plate) and *twitch fibres* (with only one end plate). Only a few human muscles are tonic muscles as defined above (i.e. extraocular muscles and laryngeal muscles). Most muscles are of the 'twitch' type, which on the basis of their histological appearance and speed of contractions can be divided into two types: slow-twitch, red fibres, and fast-twitch, white fibres. Predominantly tonic muscles, such as the soleus, have a high percentage of slow-twitch, red fibres. Predominantly phasic muscles, such as the tibialis anterior and the diaphragm, have a high proportion of fast-twitch, white fibres. However, the composition of fibre types in human muscles varies tremendously, both between muscles and individuals, and most human muscles show no striking preponderance of either fibre type.

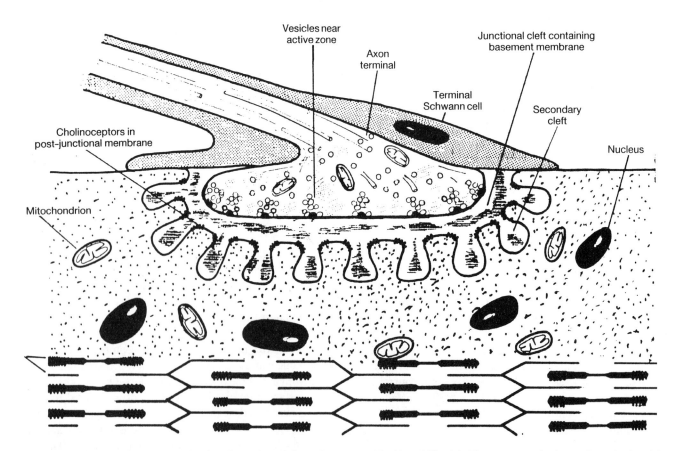

Fig. 8.2 Diagram of neuromuscular junction enlarged from the motor end plate of Fig. 8.1. The axon terminal contains mitochondria, microtubules and acetylcholine-containing vesicles. (Reproduced with permission from Bowman WC, Rand MJ. *Textbook of pharmacology.* 2nd ed. Oxford: Blackwell Scientific Publications, 1980.)

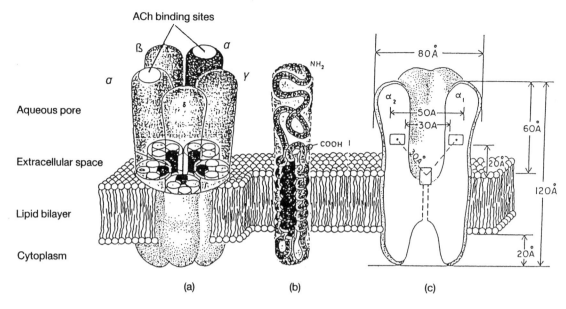

Fig. 8.3 A schematic model showing the nicotinic ACh receptor localized in the lipid bilayer (Reproduced with permission from Sastry BVR. Nicotinic receptor. *Anaesthetic Pharmacology Review* 1993; **1**: 6–19.) (a) Five homologous subunits (α, α, β, γ, δ) of the ACh receptor may combine to form a transmembrane aqueous pore. Both α-subunits contain ACh binding sites. Most of the receptor is localized in the extracellular space. Each subunit has four membrane spanning domains which are evident from the cross-section of the subunits. The dark areas of each subunit face the aqueous pore. (b) One of the α-subunits is shown separately. The polypeptide chains of each subunit are postulated to cross the lipid bilayer as α-helices. (c) Approximate size of ACh receptor.

Physiology of the neuromuscular transmission

The transmission of impulses from nerve to muscle is mediated by acetylcholine, released by depolarization of the motor nerve terminal. Acetylcholine diffuses across the junctional cleft and acts upon the postjunctional nicotinic receptors. Stimulation of these receptors triggers the excitation–contraction coupling sequence of the muscle. Acetylcholine is hydrolysed in the synaptic cleft by the enzyme acetylcholinesterase.

The motor nerve terminal

Acetylcholine synthesis and storage

The acetylcholine in the motor nerve terminal is synthesized in the axoplasma from choline and acetylcoenzyme A (acetyl-CoA) by a process facilitated by the enzyme choline acetyltransferase. Most of the choline necessary for this process is derived from the extracellular fluid, from which it is transported into the nerve terminal by a carrier-mediated transport system. The extracellular choline is derived partly from hydrolysed acetylcholine and partly from the diet. Also,

some is synthesized in the liver. Acetyl-CoA is synthesized in the nerve terminal and choline acetyltransferase in the central cell body.

About 20 per cent of the acetylcholine in the nerve terminal is present as free acetylcholine in the axoplasma, and 80 per cent is contained within the vesicles, each containing about 4–5×10^5 molecules of acetylcholine. Vesicles are spherical bodies present in large numbers in the nerve terminal. The vesicles, which are synthesized in the cell body and later transported down to the nerve terminal, have an external diameter of about 45 nm and a wall of a bilayer lipid membrane. They are found all over the nerve terminal, though they tend to concentrate in certain areas, opposite the crests of the postsynaptic membrane ('active zones') (Fig. 8.4). Experts do not agree as to the significance of the vesicles. There are two theories: the *vesicular exocytosis* and the *membrane gate* hypotheses.[7] According to the former and prevailing theory, a nerve impulse causes a calcium-mediated exocytosis of large quantities of vesicular acetylcholine. According to this theory the content of a vesicle represents the minimal functioning unit of evoked acetylcholine release. Contrary to this, the membrane gate theory holds that the acetylcholine within the vesicles is a reserve pool, and that the released acetylcholine in response to a nerve impulse represents quantal release of acetylcholine dissolved in the axoplasma.[8] For more information on these two hypotheses the interested reader is referred to reference

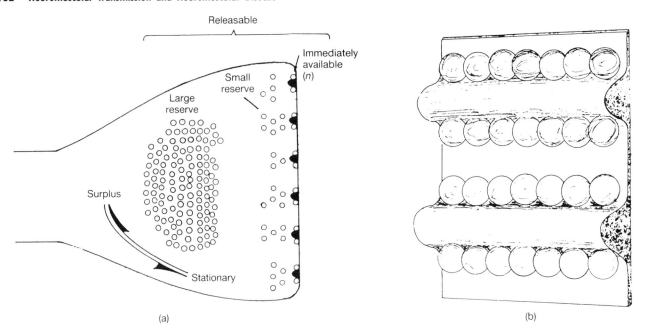

Fig. 8.4   The separate pools of acetylcholine stored within a nerve terminal. (a) Available acetylcholine is represented as being held within vesicles. The immediately available store is that aligned with the active zones. The reserve store is composed of large and small substores. All of the vesicular acetylcholine is thought to be eventually releasable. Stationary and surplus acetylcholine (the latter being present only after cholinesterase inhibition) are dissolved in the axoplasm. (b) Portion of terminal membrane showing two active zones (Reproduced with permission from Bowman WC, Rand MJ. *Textbook of pharmacology.* 2nd ed. Oxford: Blackwell Scientific Publications, 1980.)

7. The following description of the presynaptic events relating to neuromuscular transmission is based upon the vesicular hypothesis.

Separate pools or stores of acetylcholine exist within the nerve terminal. Most of the acetylcholine (about 80 per cent) can be released by nerve impulses (the releasable pool), but some can not (the non-releasable pool or stationary pool). The releasable pool consists of the acetylcholine contained within the vesicles, whereas the non-releasable pool is the acetylcholine of the axoplasma. The releasable pool is often divided into the immediately available and the reserve pool (Fig. 8.4). The 'active zones' of the terminal membrane belongs to the immediately available stores of acetylcholine.[7]

Acetylcholine release

Acetylcholine is released from the nerve terminal both spontaneously and as a result of depolarization of the nerve terminal. Small changes (0.5–1.0 mV) occur regularly, about every second, in the resting potential of the postsynaptic membrane in inactive muscles. These *miniature end-plate potentials (MEPPs)* are thought to be caused by random release of a quantum or packet of acetylcholine, and most experts believe that the quantum is the content of one vesicle. As the MEPPs are so small, they do not generate action potentials. The vast majority (probably > 99 per cent) of spontaneously released acetylcholine is, however, non-quantal. The mechanism underlying non-quantal release is unknown as is the reason. It has, however, been hypothesized that the acetylcholine exerts a trophic effect on the muscle fibre.

In contradistinction to these two types of spontaneous acetylcholine release the evoked release is caused by the arrival of an impulse to the nerve terminal. Depolarization of the nerve terminal leads to the release of several hundred (100–400) quanta of acetylcholine, the number of quanta depending on the type of muscle. According to the vesicular hypothesis the release of the many quanta of acetylcholine is caused by simultaneous opening of a large number of acetylcholine containing vesicles at the active zones adjacent to the crests of the muscle end plate.

Calcium is important in the acetylcholine release process. Following depolarization of the nerve terminal extracellular Ca^{2+} passes through opened calcium channels into the axoplasma. Inside the nerve terminal calcium binds to specific binding proteins (calmodulin and calcitonin gene-related peptide) important for the activation of enzymes necessary for the acetylcholine release. The exact mechanism underlying the fusion of the vesicles with the terminal membrane is not known. It is thought, however, that the protein synaptin I, present in the axon terminals and fixed to the surface of the vesicles, binds the vesicles to the cytoskeleton in the unstimulated nerve. Following nerve stimulation and opening of the calcium channels a $Ca^{2+}/$calmodulin-dependent protein kinase dissociates synaptin I or part of it from the vesicles. In this way the vesicles can move more freely in the axoplasma and make contact with the terminal membrane in order to empty their content of acetylcholine into the synaptic cleft.

Feedback control of acetylcholine release

Some experts believe that there is a positive as well as a negative feedback mechanism involved in the release of acetylcholine, and there is pharmacological evidence for the existence of at least two types of nicotinic and two types of muscarinic receptors at motor nerve endings.[7] The role of these receptors could be to enhance transmitter mobilization and release during especially high frequency stimulation (> 2 Hz). The mechanism by which these receptors exert their action is unknown, and it is still controversial whether or not these receptors have any physiological role in regulating the acetylcholine release.

The synaptic cleft

Junctional acetylcholinesterase is the enzyme responsible for the hydrolysis of acetylcholine in the synaptic cleft. Initially, following evoked release of the transmitter, most of the acetylcholine 'avoids' the acetylcholinesterase to reach and react with the postsynaptic nicotinic receptors. However, ultimately following release from the receptor, all molecules of acetylcholine are hydrolysed by acetylcholinesterase to inactive choline and acetate.

Acetylcholinesterase is a protein attached to the basement membrane, and probably also to membranes of the motor end plates and the nerve terminals. Each molecule of the enzyme is able to bind and hydrolyse several molecules of acetylcholine. The total number of active sites per motor end plate is close to or equal to the total number of acetylcholine receptors. It has been estimated that for each molecule of acetylcholine released by a nerve impulse, there are at least 10 active enzyme sites available. This arrangement ensures that each acetylcholine molecule only reacts once with the receptor, after which it is rapidly (< 1 ms) hydrolysed. Accordingly, under normal physiological conditions there is no accumulation of acetylcholine from one nerve stimulation to the next.

The end plate

At rest the transmembrane potential across the postsynaptic membrane is about 90 mV, with the inside of the cell being negative relative to the outside (i.e. −90 mV). This potential difference results from an excess of positively charged ions outside the cell (see Electrolyte imbalance, later), and is similar to that which is seen in other excitable

membranes of the body. The binding of two molecules of acetylcholine to the two α-subunits of the cholinergic receptor induces a conformational change in the proteins of the receptor. This results in an opening of a channel in the receptor complex, which allows cations to flow through the membrane in accordance with their concentration and electrical gradients. Though all cations (Na^+, K^+ and Ca^{2+}) may pass the channel, the most important change is a net inward flow of Na^+, constituting the *end-plate current*. As a result of this current flow, the *membrane potential* falls (the membrane partly depolarizes) and the *end-plate potential* (EPP) is produced. At a certain threshold value of the EPP (around −50 mV) the fall in the potential opens specific sodium channels in the end-plate membrane, allowing sodium ions to enter the cell. Hereby an action potential (AP) is generated, making the external surface of the end-plate membrane negative in relation to the inside. The AP draws current from the surrounding muscle fibre membrane and this results in the opening of specific voltage operated sodium channels in the muscle fibre membrane, triggering a propagated AP in the muscle fibre. By way of the T-tubules, the AP reaches the sarcoplasmic reticulum from which calcium is released and the contraction of the muscle initiated.

Margin of safety

As for many other physiological processes in the body, neuromuscular transmission has a margin of safety: the

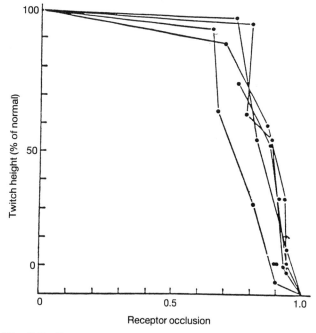

Fig. 8.5 Receptor occupancy required to produce depression of twitch height. There is no depression of twitch response until over 70% (0.7) receptors are occupied by the non-depolarizing neuromuscular agent. (Reproduced with permission from reference 9.)

number of postsynaptic cholinergic receptors by far exceeds the number required to trigger a muscle action potential under normal conditions. Thus 70–80 per cent of the receptors have to be occupied by an antagonist such as a non-depolarizing neuromuscular blocking agent before the response to nerve stimulation is affected (Fig. 8.5).[9] Accordingly, during recovery from neuromuscular block even with normal inspiratory force, vital capacity and sustained head lift for 5 seconds, etc., 70–80 per cent of all receptors can still be occupied by the neuromuscular blocker.[10] An understanding of the concept, margin of safety, is of great importance to the anaesthetist, not least during recovery from anaesthesia and in the recovery ward. At a time when both clinical criteria and response to nerve stimulation indicate sufficient recovery of neuromuscular function, about 70 per cent of receptors may still be occupied by the neuromuscular blocking agent.

Neuromuscular block

Non-depolarizing neuromuscular block: competitive antagonism

The classical description of the effect of the non-depolarizing neuromuscular blocking agents is that of competitive antagonism: the drug competes with the agonist acetylcholine for the binding sites on the postsynaptic receptors and the proportion of receptors occupied by acetylcholine and neuromuscular blocking drug is determined by their respective concentrations and affinities for the receptor. The higher the concentration of drug in the synaptic cleft, relative to the concentration of acetylcholine, the more receptor sites are occupied by the drug and the deeper is the neuromuscular block. It follows that recovery from the block can be achieved either by lowering the concentration of blocking drug in the synaptic cleft or by increasing the concentration of acetylcholine, for instance by inhibiting the enzyme that normally rapidly hydrolyses acetylcholine (acetylcholinesterase). However, the competition between drug and acetylcholine is biased in favour of the drug. Whereas the blocking drug only has to bind to one α-subunit of the receptor to block it, both α-subunits must be occupied by acetylcholine to open up the ion channel. This may explain why it is very difficult or even impossible by injection of an anticholinesterase drug to reverse an intense neuromuscular block caused by a large dose of a non-depolarizing drug. However, channel block (see later) may also play a role in this situation.

Depolarizing neuromuscular block

Phase I block

The depolarizing neuromuscular blocking agent suxamethonium (succinylcholine) acts as the name implies by depolarizing the neuromuscular end plate. Unlike the transmitter acetylcholine, suxamethonium is not hydrolysed by acetylcholinesterase in the synaptic cleft. Instead it is hydrolysed by cholinesterase in plasma (pseudocholinesterase, acylcholine-acylhydrolase EC 3.1.1.8). As a consequence, the clearance of suxamethonium from the synaptic cleft depends on diffusion from the cleft into the plasma and is much slower than the clearance of acetylcholine. Therefore suxamethonium is able to react repeatedly with the receptor, causing a longer lasting depolarization of the end plate. In contrast, acetylcholine normally only reacts once with the receptor before it is hydrolysed. This continuous depolarization of the end plate inactivates the voltage-dependent sodium channels of the muscle membrane adjacent to the end plate, thereby preventing depolarization of the muscle membrane and initiation of an action potential. This inactivation of the voltage-dependent sodium channels (sometimes called accommodation block) lasts until suxamethonium has diffused away from the synaptic cleft and the end plate is repolarized.

Phase II block

If suxamethonium remains at the neuromuscular junction for an extended period of time – either because of a prolonged infusion in a patient with normal plasma cholinesterase activity or because a relative overdose has been given to a patient with genotypically abnormal plasma cholinesterase – the characteristics of the block changes.[11,12] The membrane potential gradually recovers to normal though the neuromuscular transmission is still blocked. The original depolarizing block changes to a non-depolarizing-like block, characterized by tetanic fade, post-tetanic facilitation and fade in the train-of-four response. This change in block characteristics has been described as a change from a phase I to a phase II block. Other less fortunate designations used are dual block, mixed block or desensitization block. The term 'desensitization block', especially, should not be used synonymous with phase II block (though it has been suggested that desensitization of receptors is the basis for the development of the clinical phenomenon of a phase II block, this has never been convincingly shown – see later). In the following only the terms phase I and phase II block will be used.

Numerous theories have been proposed for the mechanism behind the development of phase II block, but none has gained general acceptance. Some researchers are convinced that phase II block is caused by desensitization of the receptors, some believe that it is caused by conformational changes in the receptor proteins, some that the reason is abnormal electrolyte balances over the end plate caused by the prolonged initial depolarization, and others again ascribe it to channel blockade (see later). The explanation for the change in the nature of the block is, however, very complex, and a phase II block most probably is an expression of both pre- and postsynaptic changes.

The management of a patient with a phase II block depends on the cholinesterase activity and genotype.[11,12] In genotypically normal patients a phase II block can be antagonized with a cholinesterase inhibitor a few minutes after discontinuation of the suxamethonium administration. In patients with abnormal genotypes, the effect of i.v. injection of a cholinesterase inhibitor is unpredictable. It can potentiate the block, produce an initial improvement in neuromuscular transmission followed by potentiation of the block, or produce a partial reversal, all depending on the time elapsed since the administration of suxamethonium and the dose of the cholinesterase inhibitor. The explanation for this difference in response between genotypically normal and abnormal patients is that in patients with abnormal genotypes, the quantity as well as the quality of the plasma cholinesterase is changed, and suxamethonium is very slowly or not hydrolysed at all in plasma. High concentrations of free suxamethonium will therefore persist in plasma and at the neuromuscular junction. The resulting neuromuscular block is a 'mixed' block with both a depolarizing (phase I) and a non-depolarizing-like (phase II) element. The depolarizing part dominates initially after the administration of suxamethonium, but later the non-depolarizing-like part of the block becomes more important. It is safe practice not to try to reverse the block, but rather keep the patient anaesthetized and ventilated until the usual clinical criteria indicate that the patient has fully recovered from the block.

Fresh frozen plasma and blood have been used to treat prolonged apnoea. However, the risk of transfusion of blood or plasma in a patient presenting no other indication for this treatment exceeds the slight risk of maintaining anaesthesia and ventilation until spontaneous breathing is sufficient.

Desensitization block

More than 35 years ago Thesleff showed that neuromuscular block caused by acetylcholine, suxamethonium and decamethonium applied to end plates for prolonged periods was not due to a persistent depolarization, but to a decrease in receptor sensitivity.[13] The initial depolarization subsided, and the membrane potential recovered to the normal level. The neuromuscular block, however, remained

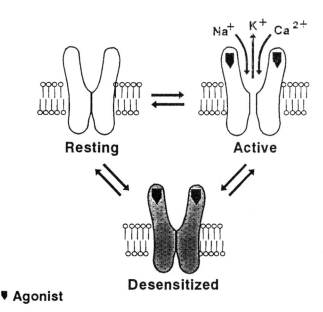

Fig. 8.6 Different normal stages of the acetylcholine receptor.

as long as the drugs were present. In some way or another, the receptors had turned refractory to the effect of the drugs.

Evidence available today indicates that receptor desensitization is a physiological phenomenon occurring even when no external agonists (or antagonists) are applied. Thus under normal conditions a receptor on the end plate may exist in three different states: resting (ion channel closed), active (ion channel open), or desensitized (ion channel closed). The receptor appears to be constantly shifting between these three states (Fig. 8.6). Several factors may promote receptor desensitization. First of all high concentrations of both agonists and antagonists speed up the process of desensitization, because agonists and antagonists bind to the desensitized receptors and tend to keep them in this state. Many other drugs, such as local anaesthetics, volatile anaesthetics, some intravenous anaesthetics, and some calcium channel blockers, may also hasten the rate of desensitization.

The mechanism by which receptor desensitization occurs is unknown. It is generally believed, however, that the basis is a conformational change in the receptor proteins, caused by phosphorylation of one or more of the amino acids constituting the receptor proteins.

The clinical significance of desensitization has been the subject of discussion over the years. Some consider it to be the basis for suxamethonium block, even following normal intubation doses (1 mg/kg), others that it constitutes the basis for the development of phase II block following prolonged infusion of suxamethonium, and the issue has not yet been settled. Whether or not receptor desensitization plays any role in connection with a competitive neuromuscular block, or its reversal, is still not known.

Channel block

Many drugs may produce postjunctional block without causing depolarization of the end plate or competing with acetylcholine for the receptor (competitive block). They act with the receptors at sites that are different from the acetylcholine recognition sites, preventing the ion channel from opening.[14] Three different mechanisms have been proposed (Fig. 8.7): open-channel block, closed-channel block, and alteration of the lipid environment of the receptor. In *open-channel block* the channel-blocking drugs act by occluding the channels in their open state. Open-channel blocking drugs only exert their action if the channel is open, and their effect increases with receptor use; also, the blockade is more effective when the blocking drug concentration is high. Increased receptor use is associated with an increased concentration of acetylcholine in the synaptic cleft, and open-channel block can not therefore be reversed by injection of an anticholinesterase drug. On the contrary, anticholinesterases (and calcium) often potentiate this type of block. Local anaesthetics, barbiturates, and some antibiotics and both depolarizing and non-depolarizing neuromuscular blocking agents are examples of drugs that can cause open-channel block of the nicotinic cholinergic receptor. The clinical significance of this is disputed. Though some researchers have assigned a significant role to this mode of action, even in connection with normal use of non-depolarizing muscle relaxants, most researchers believe that only a very small fraction of the receptors are normally blocked in the open position. The possibility of open-channel block should, however, be borne in mind when very high doses of neuromuscular blocking agents have been used. A difficulty in reversing a block caused by very high doses of a non-depolarizing drug may be due to the presence of open-channel blockade.

In *closed-channel block* the drug binds to the ion channel in its closed position, near the external opening, preventing the opening of the channel (Fig. 8.7). Some antibiotics, tricyclic antidepressant drugs, and quinidine may cause closed-channel block.

Some lipid-soluble drugs, like the inhalation anaesthetics and some alcohols, may dissolve in the membrane lipids, thereby changing the channel properties.[15]

Factors that may affect neuromuscular transmission and neuromuscular block

Temperature

Effect of temperature on neuromuscular transmission

The mechanism by which temperature alters the neuromuscular transmission is complex. Temperature may

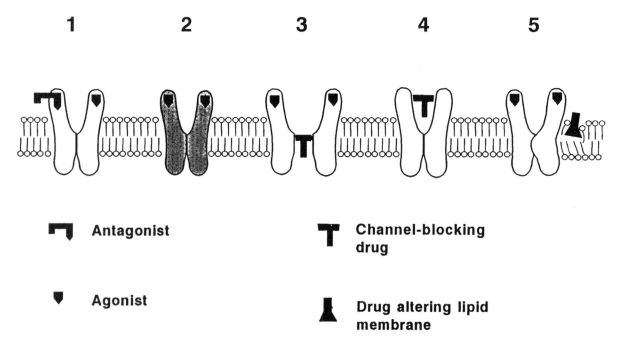

Fig. 8.7 Possible modes of action of muscle relaxants on acetylcholine receptors: 1, competitive inhibition; 2, desensitization; 3, blockade of open ion channels; 4, blockade of closed ion channels; 5, alteration of lipid environment. (Modified from reference 14.)

influence events taking place in the motor nerve, in the synaptic cleft, at the end plate and in the muscle.[7,16] A decrease in temperature causes a decrease in conduction along the nerve, but an increase in evoked release of acetylcholine from the motor nerve terminal. In the synaptic cleft a fall in temperature reduces the activity of acetylcholinesterase, thereby increasing the concentration of acetylcholine. Postsynaptic receptor sensitivity increases with decreasing temperature. A fall in temperature may cause depolarization of the end plate and a prolongation of the repolarization phase. These changes are, however, normally of little clinical significance, because of the marked margin of safety of the neuromuscular junction (see above, Margin of safety). More important is probably the effect of temperature on the muscle contraction itself. The type of muscle fibres in the muscle influences the response: the response of the predominantly *slow twitch* (red) and *fast twitch* (white) muscles, decrease and increase respectively, in response to cooling.[17]

In clinical practice, the twitch tension of the adductor pollicis muscle decreases with a fall in muscle temperature. Eriksson *et al.*[16] found that peripheral cooling of the skin to

27°C was associated with a decrease in the first response in TOF of 3–4 per cent per °C and in the TOF ratio of 1 per cent per °C (Fig. 8.8). In connection with central cooling the twitch tension may decrease even more from 14 to 20 per cent per °C.[18] It is therefore essential to keep both core and peripheral temperature as near normal as possible when monitoring neuromuscular function.

Effect of temperature on the effect of neuromuscular blocking agents

For the anaesthetist knowledge of the effect of temperature on the action of muscle relaxants is at least as important as knowledge of the effect of temperature on neuromuscular transmission *per se*. The topic is, however, controversial. Both animal and human studies have shown conflicting results: hypothermia has been found to increase or decrease the effect of *d*-tubocurarine, to increase or to have no effect on the effect of pancuronium, to increase the paralysis produced by suxamethonium, and finally to prolong the effect of both atracurium and vecuronium. This variability in findings is perhaps not surprising, considering all the different investigational techniques used, and that changes in temperature, besides affecting virtually all processes involved in neuromuscular transmission and muscle contractions, may also affect, for instance, the pharmacokinetics of the muscle relaxants, blood flow, potassium concentration and catecholamine levels. The informed anaesthetist should expect a prolonged effect of all commonly used neuromuscular blocking agents following a significant drop in core and/or muscle temperature.

Electrolyte imbalance

Except for calcium and magnesium, which are dealt with elsewhere, only changes in the plasma concentration of potassium may have a significant influence on the neuromuscular transmission and hence the action of muscle relaxants.

In the resting state the potassium gradient across the end plate is the most important factor in keeping the resting membrane potential around −90 mV (the Nernst equation: E_m (in mV) = 61 log $[K^+]_o/[K^+]_i$, where E_m is the potential difference across the membrane, $[K^+]_o$ and $[K^+]_i$ the potassium concentration outside and inside the cell, respectively). An acute decrease in extracellular potassium, unaccompanied by a similar change in intracellular potassium, hyperpolarizes the membrane, making it more resistant to depolarization by acetylcholine and thus more sensitive to the effect of the non-depolarizing muscle relaxants. In agreement with this, animal studies have demonstrated an increased sensitivity to *d*-tubocurarine and pancuronium in chronically potassium depleted

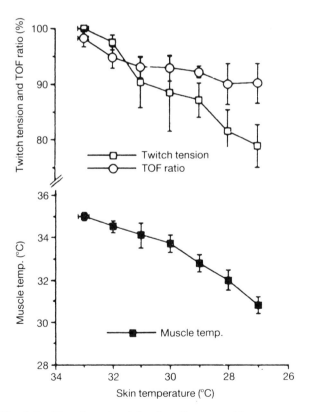

Fig. 8.8 The effect of peripheral cooling upon skin temperature (over the adductor pollicis muscle) and corresponding changes in muscle temperature (hypothenar muscle), twitch tension (%) and train-of-four (TOF) ratio (given in %) following continuous train-of-four nerve stimulation in eight patients receiving an opioid-based nitrous oxide–oxygen anaesthesia. Data are presented as mean ± SD. (Reproduced with permission from reference 16.)

animals,[19,20] and Waud and Waud[21] in *in vitro* experiments found a decreased sensitivity to the same drugs with (acute) increases in extracellular potassium.

Though no good clinical investigations are available, the evidence indicates that patients with very low plasma potassium concentrations (< 2.5 mmol/l), whether acute or chronic, would require a decreased dose of non-depolarizing muscle relaxant.

Acid–base changes

Though changes in hydrogen ion concentrations (pH) have little effect on the neuromuscular transmission *per se*,[22,23] acid–base disturbances may influence the effect of muscle relaxants in many ways. Changes in pH may influence membrane conduction, muscle contractility, ratio of intracellular to extracellular potassium, binding properties of the acetylcholine receptor, and affinity of the muscle relaxant for the receptor, to mention just a few of the possibilities. It should be no surprise, therefore, that the results of studies of the effect of acid–base changes on the effect of muscle relaxants are contradictory, and that often the results obtained *in vitro* contrast with those from clinical studies. Recently, however, Ono *et al.*,[24,25] using rat phrenic nerve-hemidiaphragm preparations, convincingly showed that the monoquaternary neuromuscular blocking agents (*d*-tubocurarine and vecuronium) reacted to acid–base changes differently from the bisquaternary agents (metocurine, pancuronium and alcuronium), and this was independent of how the pH was modified. A decrease in pH potentiated *d*-tubocurarine and vecuronium block, whereas it antagonized metocurine, pancuronium and alcuronium block. An increase in pH antagonized *d*-tubocurarine and vecuronium block, but potentiated metocurine, pancuronium and alcuronium. This opposite effect of the two types of agent was explained on the basis of an alteration in the binding properties of the receptor sites, and – for the monoquaternary drugs – from a greater affinity and specificity for the receptors during acidosis caused by changes in the ionization of the molecule. These findings are in accordance with results of some previous studies,[26,27] but in contrast to others.[28,29] The available evidence does, however, indicate that the effect of *d*-tubocurarine and possibly vecuronium are especially influenced by acid–base changes, whereas the other non-depolarizing drugs are less influenced.

The antagonism of both pancuronium[19,30] and *d*-tubocurarine[31] by anticholinesterases has been shown to be impaired by respiratory acidosis and metabolic alkalosis. This effect might be the result of depressed muscle contractility rather than a failure of neuromuscular transmission. Thus acidosis in itself is known to decrease muscle contractility (twitch tension), whereas alkalosis has the opposite effect.[29]

The effect of the depolarizing neuromuscular blocking agent suxamethonium is antagonized by metabolic as well as respiratory acidosis.[27]

Drug interactions at the neuromuscular junction

Drug interactions at the neuromuscular junction may, in principle, take place at at least three different sites: at the

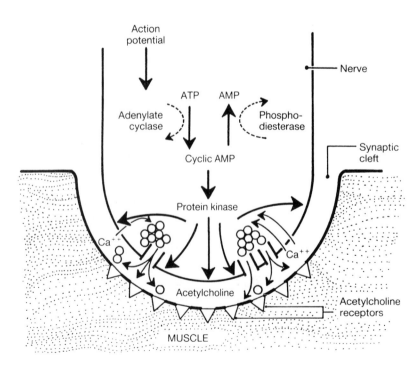

Fig. 8.9 Diagrammatic representation of the enzyme cascade involved in the release of acetylcholine at the nerve terminal. ATP, adenoside triphosphate; AMP, adenosine monophosphate.

nerve terminal, in the synaptic cleft and at the postsynaptic membrane – or at all three sites at the same time.

Some drugs such as the local anaesthetics may interfere with the propagation of the nerve terminal action potential. Others may modify calcium flow into the nerve terminal either by interacting with the calcium channels (aminoglycosides) or by interfering with one or more of the enzymes involved in the control of calcium flow into the nerve terminal (theophylline and azathioprine) and thus the release of transmitter (Fig. 8.9). Certain drugs (some antibiotics and lithium) act by inhibiting the synthesis of acetylcholine.

In the synaptic cleft any drug with anticholinesterase activity may interfere with the enzymatic hydrolysis of acetylcholine.

At the postsynaptic membrane some drugs (certain antibiotics) have an effect similar to the non-depolarizing muscle relaxants (competitive inhibition), others exert their effect by reducing open-channel lifetime, thereby impairing ion channel conductance. Other drugs again may enter and occlude open ion channels or promote desensitization.

A brief description of some of the most important clinical interactions follows. The list is in no way exhaustive. For more information the reader is referred to previous surveys.[32–35]

Drugs causing increased sensitivity to muscle relaxants

Antibiotics
Some antibiotics can produce neuromuscular blockade on their own, although this is seldom seen in patients without neuromuscular disease. However, the use during anaesthesia of certain antibiotics (i.e. aminoglycosides, polypeptides, tetracyclines, clindamycin and lincomycin) may increase the sensitivity to muscle relaxants, and postoperative recurarization has been seen following injection of antibiotics.[33]

The mechanism of action is complex and not fully understood. Suggested mechanisms include a reduction in the evoked release of acetylcholine, a decrease in sensitivity of the nicotinic receptor, and channel blockade. The channel block may explain why it may be difficult to reverse a neuromuscular block when either polymyxin, clindamycin or lincomycin has been administered.[35] Channel block is not competitive, but depends on the channel being open. The greater the activation of the channel the deeper the block. Therefore, anticholinesterases and calcium may at least in theory increase channel block.

Anticholinesterases
Only a relatively small fraction of an administered dose of suxamethonium or mivacurium actually reaches the neuromuscular junction, because the drugs are rapidly hydrolysed in plasma by plasma cholinesterase. Drugs that inhibit plasma cholinesterase may therefore cause a prolonged response to these two muscle relaxants. However, in patients with genotypically normal plasma cholinesterase, a decrease in enzyme activity does not cause a very prolonged response (Fig. 8.10).[36] Organophosphate pesticides, cyclophosphamide, ecothiopate eyedrops and bambuterol (carbamylated terbutaline used in the treatment of bronchial asthma) are the drugs most likely to cause clinically significant depression of plasma cholinesterase activity.[12,36]

Inhalational anaesthetic agents
Inhalational anaesthetics depress neuromuscular function in a dose-dependent way. The mechanism of action is complex and only partly understood.[7,32] The inhalational anaesthetic agents decrease the release of acetylcholine presynaptically,[37] but the main effect seems to be exerted on the ion channels of the postsynaptic membrane. However, anaesthetic agents do not bind to the receptor at the acetylcholine binding site. Rather, they seem to

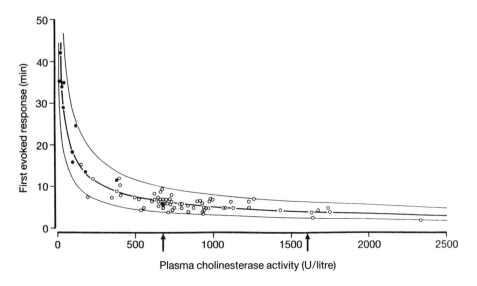

Fig. 8.10 Relation between enzyme activity and time to first evoked response to train-of-four nerve stimulation following administration of suxamethonium 1 mg/kg i.v. in patients with normal plasma cholinesterase genotype. The closed circles represent 10 patients given bambuterol 30 mg orally 2 h before the suxamethonium administration. The open circles represent 41 patients not given bambuterol. Mean curve and 95% prediction interval are given. Arrows indicate normal values of plasma cholinesterase activity. (Reproduced with permission from reference 36.)

dissolve in the lipid of the membrane, thereby influencing channel function (see Fig. 8.7). It may also be that they bind to the receptor proteins at a site different from the acetylcholine binding site.[7] In agreement with this, inhalational anaesthetics potentiate muscle relaxants, the longer acting non-depolarizing agents such as pancuronium and d-tubocurarine being more affected than the shorter acting drugs atracurium and vecuronium.[38] Enflurane and isoflurane are more powerful than halothane in potentiating neuromuscular blocking agents.

When using potent inhalational anaesthetic agents, the maintenance dose of the muscle relaxant may be reduced 25–30 per cent.

Intravenous anaesthetic agents
Intravenous anaesthetic agents increase acetylcholine release presynaptically and at the same time decrease the sensitivity of the post-synaptic membrane to acetylcholine, and these two effects normally balance out each other.[35] Therefore, clinically significant interactions are not seen with the commonly used intravenous anaesthetic agents. Ketamine, however, has been found to potentiate the effect of d-tubocurarine but not pancuronium.[32]

β-Adrenergic blocking agents
β-Adrenergic blocking agents have been reported to aggravate or unmask myasthenia gravis[39] and to induce myasthenic syndrome.[40] Although two reports indicated prolongation of the action of non-depolarizing muscle relaxants,[41,42] the evidence for interaction between these drugs and neuromuscular function is inconclusive. It still remains to be proved that an interaction occurs in man with the drug concentrations used clinically. The mechanism underlying a possible interaction is not known.

Calcium channel blockers
Calcium channel blocking drugs (verapamil, nifedipine, diltiazem, and flumarizine) do not block neuromuscular transmission themselves when used in clinically relevant doses. However, it is now well established that they do potentiate the effect of non-depolarizing muscle relaxants and that they occasionally may cause difficulty in reversing neuromuscular block with neostigmine.[43–45] Probably, however, this effect is relatively unimportant, except in situations in which the margin of safety of the neuromuscular junction is reduced. The site of action of the calcium channel blockers is both pre- and post-junctional,[46,47] but the different drugs may act at different sites. For instance, verapamil blocks both the fast sodium channels and the slow calcium channels, while nifedipine only blocks the latter.[45]

Local anaesthetics
Local anaesthetics are 'fast channel' blockers, thus depressing the propagation of nerve impulses, the release of acetylcholine, the sensitivity of the postsynaptic membrane to acetylcholine and the excitability of the muscle cell membrane. No wonder therefore, that local anaesthetics enhance the neuromuscular block due to both depolarizing and non-depolarizing muscle relaxants.[33,48] Recently, it has been shown that even normal doses of epidurally injected local anaesthetics may potentiate the effect of normal doses of non-depolarizing muscle relaxants.[49]

Magnesium sulphate
Magnesium decreases acetylcholine release from the nerve terminal, reduces the sensitivity of the post-junctional membrane to acetylcholine and depresses the excitability of the muscle cell membrane.[32,35,50] Accordingly magnesium has been shown to enhance neuromuscular block from all muscle relaxants. Several cases have been reported of prolonged neuromuscular blockade in patients with pre-eclamptic toxaemia treated with magnesium sulphate. Calcium antagonizes the effect of magnesium.

Drugs causing decreased sensitivity to muscle relaxants

Some antiepileptic drugs (phenytoin and carbamazepine), azathioprine (used as an immunosuppressive agent for organ transplantation), corticosteroids, and methylxanthines (aminophylline and theophylline) have been reported to cause resistance to the effect of non-depolarizing muscle relaxants.[33,34] The mechanisms underlying this effect are uncertain. However, aminophylline and azathioprine are phosphodiesterase inhibitors. An inhibition of phosphodiesterase in the nerve terminal would result in an increased level of cyclic adenosine monophosphate (cAMP) and possibly also of acetylcholine (compare Fig. 8.9).

Neuromuscular blocking agents in patients with neuromuscular disorders

Many very different diseases are classified as neuromuscular diseases, and the majority of them are rare and difficult to differentiate, unless one is dealing with them on a daily basis. However, patients with neuromuscular disorders often share the same three problems in relation to anaesthesia, independent of the type of disorder: they may have a cardiomyopathy, a restricted respiratory capacity and an abnormal response to muscle relaxants. The response of patients with the more common

Table 8.1 Response of patients with neuromuscular disease to neuromuscular blocking agents

Neuromuscular disorder	Response to Non-depolarizing agents	Suxamethonium
Intracranial lesions		
Hemiplegia	Resistance	Hyperkalaemia Cardiac arrest
Multiple sclerosis	Normal	Hyperkalaemia? Contracture?
Diffuse intracranial lesions	Normal	Hyperkalaemia
Spinal cord lesions		
Paraplegia	Increased?	Hyperkalaemia Cardiac arrest
Amyotrophic lateral sclerosis	Increased	Contracture Hyperkalaemia?
Peripheral nerve lesions		
Peripheral neuropathies	Increased?	Hyperkalaemia?
Muscular denervation	Resistance	Contracture Hyperkalaemia
Disuse atrophy	Resistance	Hyperkalaemia
Neuromuscular junction lesions		
Myasthenic syndrome	Increased (no fade in tetanic response (50 Hz))	Increased or normal
Myasthenia gravis	Increased (+ fade in tetanic response)	Variable
Muscular lesions		
Myotonias	Normal or increased	Contracture Cardiac arrest
Muscular dystrophies	Normal	Contracture Hyperkalaemia Rhabdomyolysis Myoglobinuria

neuromuscular disorders to muscle relaxants is reviewed here. For more in-depth information about neuromuscular disorders and for a discussion of malignant hyperthermia, the reader is referred to Chapter 46.

For the purpose of clarity in this chapter the neuromuscular disorders are divided into five main groups according to the site of the primary lesion:[51] intracranial lesions, spinal cord lesions, peripheral nerve lesions, neuromuscular junction lesions, and muscular lesions (Table 8.1).

Intracranial lesions

Hemiplegia

Patients with hemiplegia may react abnormally to both non-depolarizing and depolarizing neuromuscular blocking

agents. Affected muscles of hemiplegic patients are resistant to non-depolarizing blocking drugs and react abnormally to nerve stimulation. Thus the response to nerve stimulation may vary greatly between an affected and a non-affected muscle.[52] The administration of non-depolarizing drugs should therefore always be guided by the use of a nerve stimulator on the non-affected side. Otherwise the degree of neuromuscular block in for instance the respiratory muscles may be underestimated.

After administration of suxamethonium severe hyperkalaemia with ventricular fibrillation and cardiovascular collapse has been described several times in patients with hemiplegia. Hyperkalaemia may be seen from 1 week to 6 months after the cerebrovascular catastrophy.

The common mechanism behind the abnormal responses to the two types of neuromuscular blocking agents is probably a spread of receptors beyond the neuromuscular transmission caused by a lack of innervation, as described in more detail under the section 'Muscular denervation'.

Multiple sclerosis

Patients with multiple sclerosis react normally to non-depolarizing neuromuscular blocking agents. Case reports exist, however, of hyperkalaemia[53] and a myotonic-like syndrome[54] after the use of suxamethonium.

Diffuse intracranial lesions

Many reports have shown hyperkalaemia and cardiac arrest in patients with intracranial lesions following suxamethonium. The mechanism is not clear. It has been suggested but not proved that the abnormal response is due to muscle atrophy caused by immobilization (disuse atrophy).

Spinal cord lesions

Paraplegia

Contrary to another upper motor neuron disorder, hemiplegia, paraplegic patients in one study have been found to have an increased sensitivity to non-depolarizing neuromuscular blocking drugs.[51] This finding which is difficult to explain has not been substantiated by other studies. The problem obviously needs further investigation.

Hyperkalaemia followed by cardiac arrest has repeatedly been shown to occur in paraplegic patients following administration of suxamethonium. Probably the increased sensitivity to suxamethonium occurs within a week of the trauma. It may persist for months.

Amyotrophic lateral sclerosis

Patients with amyotrophic lateral sclerosis have increased sensitivity to non-depolarizing agents. Most probably this is due to a decreased synthesis of acetylcholine in the nerve terminal, caused by degeneration of the motor ganglia of the anterior horn cells. Contracture in response to suxamethonium injection has been described.

Peripheral nerve lesions

Peripheral neuropathies

Peripheral neuropathy may be caused by a variety of systemic disorders, such as diabetes mellitus, vitamin B_{12} deficiency and uraemia, by drugs, such as nitrofurantoin and vincristine, and by heavy metals. Patients with peripheral neuropathy have been found to be sensitive to non-depolarizing neuromuscular blocking drugs. Most patients seem to react normally to suxamethonium, but hyperkalaemia has been reported.[55]

Muscular denervation

Within a few weeks of denervation of a muscle extrajunctional chemosensitivity develops, and the response to both depolarizing and non-depolarizing drugs changes. Suxamethonium may cause hyperkalaemia and contracture, and the response to non-depolarizing drugs is decreased. There is no general agreement explaining the mechanism by which this altered response occurs, but most probably the explanation is as follows: in the human body two types of nicotinic acetylcholine receptors exist. In addition to the one in the neuromuscular junction of normal adults (the junctional nicotinic receptor), a different receptor type exists in denervated muscles and in the fetus (the extrajunctional nicotinic receptor). At the molecular level the two receptors differ only in one way: in the extrajunctional receptor the ε-subunit has been replaced by a γ-subunit. Though these two subunits are not very different the change in receptor protein does influence the function of the receptor. Extrajunctional receptors are more sensitive to agonists (acetylcholine and suxamethonium) and more resistant to antagonist (non-depolarizing drugs).

The fetal muscles before they are innervated contain only extrajunctional receptors dispersed all over the muscle cell membrane, but along with the development of innervation, muscle cells start to produce junctional receptors. In neonates and infants the two receptor types coexist, but in the active, normal adult person only junctional receptors exist. However, if the nerve activity to a muscle is reduced or totally abolished, as may be the case following peripheral nerve lesions (or disuse atrophy,

strokes and spinal cord disorders), the muscle starts producing extrajunctional receptors. These are inserted all over the cell membrane and in the neuromuscular end plate, a phenomenon called up-regulation of the receptor. As a consequence, the sensitivity to agonists as well as antagonists changes: the muscle gets more sensitive to the injection of suxamethonium, which may cause hyperkalaemia and contracture, and more resistant to the effect of non-depolarizing drugs. The hyperkalaemia following suxamethonium is normally attributed partly to the changed sensitivity of the receptor *per se* and partly to the greatly increased number of receptors spread all over the muscle cell. The contracture has been explained on the basis of a direct effect of calcium upon the contractile elements of the muscles, brought about by an increase in Ca^{2+} movements across the muscle membrane caused by suxamethonium.

Disuse atrophy

Bedridden and immobilized patients sometimes require two to three times greater doses of non-depolarizing relaxants for paralysis,[56] and they may respond with an increased potassium efflux after the administration of suxamethonium.[57] These changes appear to be caused by atrophy of the nerve and muscle (disuse atrophy), leading to the development of extrajunctional nicotinic receptors (receptor up-grading) as described in the above section, 'Muscular denervation'. The resistance to non-depolarizing relaxants seems to occur earlier (within 2 weeks) than the disuse-related potassium efflux (14 days or longer). The hyperkalaemia seen following suxamethonium in patients with disuse atrophy is less pronounced than the hyperkalaemia seen in patients after total muscle denervation.

Neuromuscular junction lesions

Myasthenic syndrome (Eaton–Lambert syndrome)

The myasthenic syndrome is a rare disorder of impaired acetylcholine release from the motor nerve terminal due to an autoimmune based destruction of the sites in the prejunctional membrane responsible for the acetylcholine release (the active sites). The syndrome is most often associated with small cell carcinoma of the lung, but has also been found in patients without any malignancy.[58] The symptoms are proximal muscle weakness, sometimes accompanied by paraesthesiae and aches. The patients are extremely sensitive to non-depolarizing agents and long-lasting paralysis has been described following normal doses of these drugs. Both normal and increased sensitivity to suxamethonium has been described.

Patients with a myasthenic syndrome may respond

Table 8.2 Characteristics of myasthenia gravis and myasthenic syndrome[59,65]

	Myasthenia gravis	Myasthenic syndrome
Sex	More common in women than in men	Almost entirely men
Age of onset	Commonly 20–40 years	Commonly 50–70 years
Presenting signs	Weakness of external ocular, bulbar, and facial muscles Fatigue on activity Muscle pains uncommon Tendon reflexes normal	Weakness and fatiguability of proximal limb muscles Transient increase in strength on activity precedes fatigue Muscle pains common Tendon reflexes reduced or absent
Pathological state	The thymus gland most often abnormal, and 15–20% of patients have thymomas	Small cell bronchogenic carcinoma usually present
Electromyographic response (non-treated patients)	Initial muscle action potential relatively normal Voltage decrement to repeated stimulation Good response to anticholinesterases	Initial action potential abnormally small Voltage increment to repeated stimulation Poor response to anticholinesterases
Response to muscle relaxants	Increased sensitivity to non-depolarizing blockers Response to suxamethonium variable	Increased sensitivity to non-depolarizing blockers Normal or increased response to suxamethonium

poorly to anticholinesterases. However, 3,4-diaminopyridine, which increases the release of acetylcholine from the motor nerve terminals by a selective block of potassium channels, may improve neuromuscular function following a non-depolarizing neuromuscular block.[59]

Table 8.2 summarizes and compares characteristics of myasthenic syndrome and myasthenia gravis.

Myasthenia gravis

Myasthenia gravis is an autoimmune disease of the neuromuscular junction resulting from a production of IgG antibodies against the nicotinic receptors. This leads to a reduction in the number of active postsynaptic (and probably also presynaptic) nicotinic receptors, caused partially by a receptor breakdown and partially by channel block by the antibodies. Often myasthenia gravis occurs with other autoimmune diseases such as thyroid hypofunction, rheumatoid arthritis and systemic lupus erythematosus. Most patients with myasthenia gravis have IgG antibodies against nicotinic receptors in plasma.

The characteristic symptoms of myasthenia gravis are muscle weakness and fatiguability, most often affecting facial, bulbar, arm and neck muscles; but other muscles may be involved as well. The disease may progress slowly over years, or rapidly over less than 6 months, leading to generalized muscle weakness with wasting, and respiratory insufficiency.

The diagnosis is made from the characteristic clinical and electromyographic findings (Table 8.2), from the demonstration of IgG antibodies against the acetylcholine receptor in plasma, or from the result of a systemic or regional curare test.[60]

The four cornerstones in the treatment of myasthenia gravis are anticholinesterases, plasmapheresis, immunosuppressive drugs (corticosteroids, azathioprine and cyclosporin A), and thymectomy. The cholinesterase inhibitors (neostigmine and pyridostigmine) of course only represent symptomatic therapy. Plasmapheresis with removal of circulating antibodies from the blood often leads to short-lived (days to weeks) improvement in the condition of the patient, and is therefore often used as a preparation for thymectomy or for controlling acute exacerbations. Corticosteroids are successful in most patients, whereas the effect of azathioprine and cyclosporins is less pronounced. Cyclosporin A is only used in severe cases because of the serious side effects (nephro- and hepatotoxicity). Thymectomy helps 80–90 per cent of patients, and today most patients with myasthenia gravis have this operation.

The response to suxamethonium is variable and depends on the patient's medical treatment. In non-treated patients, the decreased number of functional receptors makes the end plates 'resistant' to the effect of suxamethonium (Fig. 8.11), and the dose necessary for smooth tracheal intubation therefore has to be increased. In patients treated with anticholinesterases, the plasma cholinesterase activity is decreased. More suxamethonium therefore reaches the end plate, with the result that the neuromuscular block is potentiated and sometimes prolonged. It appears that in myasthenia gravis phase II block develops rapidly.

Patients with myasthenia gravis are extremely sensitive to non-depolarizing muscle relaxants, but again the response depends on the severity of the disease and the therapeutic regimen (Fig. 8.12). Some authorities therefore advocate that non-depolarizing muscle relaxants should not be used at all for myasthenia gravis. Instead they rely on deep inhalational anaesthesia, both for tracheal intubation

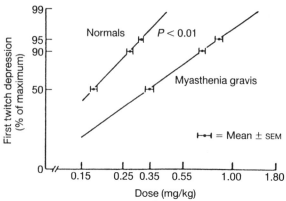

Fig. 8.11 Mean suxamethonium dose–response in normal and myasthenic patients. Horizontal axis shows suxamethonium cumulative dose (log scale). Vertical axis shows response (logit transformation). (Reproduced with permission from reference 63.)

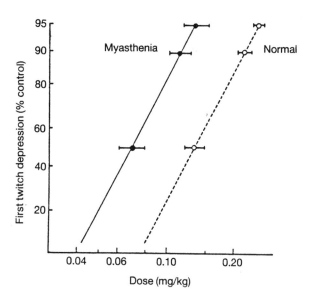

Fig. 8.12 Dose–response relations for atracurium in patients with myasthenia gravis ($n = 5$, cumulative dose technique) and in normal individuals ($n = 10$, cumulative dose technique). The logit transformation of first twitch depression is plotted as a function of the logarithm of the dose. Error bars represent SEM. (Reproduced with permission from reference 64.)

and maintenance of anaesthesia. However, practically all the non-depolarizing muscle relaxants have been used with success in myasthenia gravis with neuromuscular monitoring. The intermediate duration muscle relaxants atracurium and vecuronium seem particularly suited because of their relatively short duration of action. There are many reports of safe titrated use of both atracurium and vecuronium. Mivacurium may be less suitable at least in anticholinesterase treated patients, as it is hydrolysed by plasma cholinesterase. Prolonged response to mivacurium has been found in patients with low or abnormal plasma cholinesterase activity.[61,62]

Whether anticholinesterase therapy should or should not be discontinued before anaesthesia is a matter for debate. Some argue that the anticholinesterase therapy should be discontinued the day before surgery for thymectomy, the argument being that thymectomy might alter the response to the anticholinesterases with the danger of a myasthenic crisis. Others are in favour of continuing the medication until the time of surgery in order to ensure maximal muscle power.

Muscular lesions

The muscular dystrophies

The muscular dystrophies (MD) are dealt with in Chapter 46.

Patients with MD seem to react normally to non-

depolarizing neuromuscular blocking agents, though prolonged duration of action has been found with the regional curare test.

Several reports exist of cardiac arrest, hyperkalaemia, muscle rigidity and masseter spasm following the use of suxamethonium in patients with MD and a connection between MD and malignant hyperthermia has been proposed. This has, however, not been proved. Though suxamethonium has been used in patients with MD without complications, it is probably safest to avoid suxamethonium.

Myotonic disorders

Myotonia is dealt with in Chapter 46.

Patients with a myotonic disorder often react normally to non-depolarizing neuromuscular blocking agents, but increased sensitivity has been seen. Suxamethonium may provoke a generalized sustained myotonic contracture that may last from minutes to many hours. Suxamethonium is therefore best avoided in these patients. Neostigmine and cold (shivering) also may provoke myotonic contracture.

━━━━━━━━━━━━━━━━━━━━━━━━━━ REFERENCES ━━━━━━━━━━━━━━━━━━━━━━━━━━

1. Brodie BC. Experiments and observations on the different modes in which death is produced by certain vegetable poisons. *Philosophical Transactions of the Royal Society of London* 1811: 178–208.

2. Brodie BC. Further experiments with South American arrow poison. *Philosophical Transactions of the Royal Society of London* 1812: 205–27.

3. Bernard MC. 3° action du curare de la nicotine sur le système

musculaire. *Compte Rendu des Séances de la Société de Biologie* 1850; **2**: 195–6.

4. Dale HH, Feldberg W, Vogt M. Release of acetylcholine at voluntary motor nerve endings. *Journal of Physiology (London)* 1936; **86**: 353–80.

5. Buchtal F. The general concept of the motor unit. *Research Publications. Association for Research in Nervous and Mental Diseases* 1960; **38**: 3–30.

6. Peper K, Bradley RJ, Dreyer F. The acetylcholine receptor at the neuromuscular junction. *Physiological Reviews* 1982; **62**: 1271–340.

7. Bowman WC. *Pharmacology of neuromuscular function.* 2nd ed. London: John Wright and Sons Ltd, 1990.

8. Oorschot DE, Jones DG. The vesicle hypothesis and its alternatives. A critical assessment. *Current Topics in Research on Synapses* 1987; **4**: 85–153.

9. Paton WDM, Waud DR. The margin of safety of neuromuscular transmission. *Journal of Physiology* 1967; **191**: 59–90.

10. Viby-Mogensen J. Clinical assessment of neuromuscular transmission. *British Journal of Anaesthesia* 1982; **54**: 209–23.

11. Østergaard D, Viby-Mogensen J. Prolonged apnea after succinylcholine. Etiology, diagnosis and management. *Problems in Anesthesia* 1989; **3**: 455–64.

12. Viby-Mogensen J. Cholinesterase and succinylcholine. *Danish Medical Bulletin* 1983; **30**: 129–50.

13. Thesleff S. The mode of neuromuscular block caused by acetylcholine, nicotine, decamethonium and succinylcholine. *Acta Physiologica Scandinavica* 1955; **34**: 218–31.

14. Dreyer F. Acetylcholine receptor. *British Journal of Anaesthesia* 1982; **54**: 115–30.

15. Kenedy RD, Galindo AD. Comparative site of action of various anaesthetic agents at the mammalian myoneural junction. *British Journal of Anaesthesia* 1975; **47**: 533–40.

16. Eriksson LI, Lennmarken C, Jensen E, Viby-Mogensen J. Twitch tension and train-of-four ratio during prolonged neuromuscular monitoring at different peripheral temperatures. *Acta Anaesthesiologica Scandinavica* 1991; **35**: 247–52.

17. Close R, Hoh JFY. Influence of temperature on isometric contractions of rat skeletal muscles. *Nature* 1968; **217**: 1179–80.

18. Heier T, Caldwell JE, Sessler DI, Kitts JB, Miller RD. The relationship between adductor pollicis twitch tension and core, skin, and muscle temperature during nitrous oxide-isoflurane anesthesia in humans. *Anesthesiology* 1989; **71**: 381–4.

19. Miller RD, Roderick LL. Diuretic-induced hypokalaemia, pancuronium neuromuscular blockade and its antagonism by neostigmine. *British Journal of Anaesthesia* 1978; **50**: 541–4.

20. Waud BE, Mookerjee A, Waud DR. Chronic potassium depletion and sensitivity to tubocurarine. *Anesthesiology* 1982; **57**: 111–15.

21. Waud BE, Waud DR. Interaction of calcium and potassium with neuromuscular blocking agents. *British Journal of Anaesthesia* 1980; **52**: 863–6.

22. Gessell R, Mason A, Brassfield CR. Acid-humoral intermediation in the rectus abdominus of the frog. *American Journal of Physiology* 1944; **142**: 131–9.

23. Sokoll MD, Thesleff S. Effects of pH and uranyl ions on action potential generation and acetylcholine sensitivity of skeletal muscle. *European Journal of Pharmacology* 1968; **4**: 71–6.

24. Ono K, Ohta Y, Morita K, Kosaka F. The influence of respiratory-induced acid–base changes on the action of non-depolarizing muscle relaxants in rats. *Anesthesiology* 1988; **68**: 357–62.

25. Ono K, Nagano O, Ohta Y, Kosaka F. Neuromuscular effects of respiratory and metabolic acid–base changes in vitro with and without nondepolarizing muscle relaxants. *Anesthesiology* 1990; **73**: 710–16.

26. Baraka A. The influence of carbon dioxide on the neuromuscular block caused by tubocurarine chloride in the human subject. *British Journal of Anaesthesia* 1964; **36**: 272–8.

27. Crul-Sluijter EG, Crul JF. Acidosis and neuromuscular blockade. *Acta Anaesthesiologica Scandinavica* 1974; **18**: 224–36.

28. Payne JP. The influence of changes in blood pH and the neuromuscular blocking properties of tubocurarine and dimethyl tubocurarine in the cat. *Acta Anaesthesiologica Scandinavica* 1960; **4**: 85–90.

29. Gencarelli PJ, Swen J, Koot HWJ, Miller RD. The effects of hypercarbia and hypocarbia on pancuronium and vecuronium neuromuscular blockades in anesthetized humans. *Anesthesiology* 1983; **59**: 376–80.

30. Wirtavuori K, Salmenpera M, Tammisto T. The effect of hypocarbia and hypercarbia on the antagonism of pancuronium-induced neuromuscular blockade with neostigmine in man. *British Journal of Anaesthesia* 1982; **54**: 57–61.

31. Miller RD, van Nyhuis LS, Eger EI, Way WL. The effect of acid–base balance on neostigmine antagonism of d-tubocurarine-induced neuromuscular blockade. *Anesthesiology* 1975; **42**: 377–82.

32. Miller RD. Neuromuscular blocking agents. In: Smith NT, Miller RD, Corbascio AN, eds. *Drug interactions in anesthesia.* Philadelphia: Lea & Febiger, 1981: 249–69.

33. Viby-Mogensen J. Interaction of other drugs with muscle relaxants. *Seminars in Anaesthesia* 1985; **4**: 52–64.

34. Østergaard D. Drug interactions. *Current Opinion in Anaesthesiology* 1989; **2**: 497–500.

35. Østergaard D, Engbæk J, Viby-Mogensen J. Adverse reactions and interactions of the neuromuscular blocking drugs. *Medical Toxicology and Adverse Drug Experience* 1989; **4**: 351–68.

36. Bang U, Viby-Mogensen J, Wirén JE, Theil-Skovgaard L. The effect of bambuterol (carbamylated terbutaline) on plasma cholinesterase activity and succinylcholine-induced neuromuscular blockade in genotypically normal patients. *Acta Anaesthesiologica Scandinavica* 1990; **34**: 596–9.

37. Hughes R, Payne JP. Interaction of halothane with non-depolarizing neuromuscular blocking drugs in man. *British Journal of Clinical Pharmacology* 1979; **7**: 485–90.

38. Rupp SM, Miller RD, Gencarelli PJ. Vecuronium-induced neuromuscular blockade during enflurane, isoflurane, and halothane anesthesia in humans. *Anesthesiology* 1984; **60**: 102–5.

39. Weisman SJ. Masked myasthenia gravis. *Journal of the American Medical Association* 1949; **141**: 917–18.

40. Heriskomi Y, Rosenberg P. β-Blockers and myasthenia gravis. *Annals of Internal Medicine* 1975; **83**: 834–5.

41. Harrak MD, Way WL, Katzing BG. The interaction of d-tubocurarine with antiarrhythmic drugs. *Anesthesiology* 1970; **33**: 406–10.

42. Rozen MS, Whan FM. Prolonged curarization associated with propranolol. *The Medical Journal of Australia* 1972; **1**: 467–9.

43. van Poorten JF, Dhasmana KM, Kuypers RSM, Erdmann W. Verapamil and reversal of vecuronium neuromuscular blockade. *Anesthesia and Analgesia* 1984; **63**: 155–7.

44. Jones RM, Cashman JN, Casson WR, Broadbent MP. Verapamil potentiation of neuromuscular blockade: failure of reversal with neostigmine but prompt reversal with edrophonium. *Anesthesia and Analgesia* 1985; **64**: 1021–5.

45. Bikhazi GB, Leung I, Flores C, Mikati HMJ, Foldes FF. Potentiation of neuromuscular blocking agents by calcium channel blockers in rats. *Anesthesia and Analgesia* 1988; **67**: 1–8.

46. Anderson KA, Marshall RJ. Interactions between calcium entry blockers and vecuronium bromide in anaesthetized cats. *British Journal of Anaesthesia* 1985; **57**: 775–81.

47. Durant NN, Nguyen N, Katz RL. Potentiation of neuromuscular blockade by verapamil. *Anesthesiology* 1984; **60**: 298–303.

48. Matsuo S, Rao DBS, Chaudry I, Foldes FF. Interaction of muscle relaxants and local anesthetics at the neuromuscular junction. *Anesthesia and Analgesia* 1978; **57**: 580–7.

49. Toft P, Kirkegaard Nielsen H, Severinsen I, Helbo-Hansen HS. Effect of epidurally administered bupivacaine on atracurium-induced neuromuscular blockade. *Acta Anaesthesiologica Scandinavica* 1990; **34**: 649–52.

50. Baraka A, Yazigi A. Neuromuscular interaction of magnesium with succinylcholine–vecuronium sequence in the eclamptic parturient. *Anesthesiology* 1987; **67**: 806–8.

51. Azar I. Muscle relaxants in patients with neuromuscular disorders. In: *Clinical pharmacology. Vol. 7. Muscle relaxants, side effects and a rational approach to selection.* New York and Basel: Marcel Dekker, Inc, 1987.

52. Moorthy SS, Hilgenberg JC. Resistance to non-depolarizing muscle relaxants in paretic upper extremities of patients with residual hemiplegia. *Anesthesia and Analgesia* 1980; **59**: 624–7.

53. Cooperman LH. Succinylcholine induced hyperkalemia in neuromuscular disease. *Journal of the American Medical Association,* 1970; **213**: 1867–71.

54. Weintraud MI, Megaled MS, Smith BH. Myotonic-like syndrome in multiple sclerosis. *New York State Journal of Medicine* 1970; **70**: 677–9.

55. Fergusson RJ, Wright DJ, Willey RF, Crompton GK, Grant IWB. Suxamethonium is dangerous in polyneuropathy. *British Medical Journal* 1981; **282**: 298–9.

56. Gronert GA. Disuse atrophy with resistance to pancuronium. *Anesthesiology* 1981; **55**: 547–9.

57. Fung DL, White DA, Jones BR, Gronert GA. The onset of disuse-related potassium efflux to succinylcholine. *Anesthesiology* 1991; **75**: 650–3.

58. O'Neill JH, Murray NMF, Newsom-Davis J. The Lambert–Eaton myasthenic syndrome: a review of 50 cases. *Brain* 1988; **111**: 577–96.

59. Telford RJ, Hollway TE. The myasthenic syndrome: Anaesthesia in a patient treated with 3.4 diaminopyridine. *British Journal of Anaesthesia* 1990; **64**: 363–6.

60. Brown JC, Charlton JE, White DJK. A regional technique for the study of sensitivity to curare in human muscle. *Journal of Neurology, Neurosurgery and Psychiatry* 1975; **38**: 18–26.

61. Østergaard D, Jensen FS, Jensen E, Skovgaard LT, Viby-Mogensen J. Influence of plasma cholinesterase activity on recovery from mivacurium-induced neuromuscular blockade in phenotypically normal patients. *Acta Anaesthesiologica Scandinavica* 1992; **36**: 702–6.

62. Østergaard D, Jensen FS, Jensen E, Skovgaard LT, Viby-Mogensen J. Mivacurium-induced neuromuscular blockade in patients with atypical plasma cholinesterase. *Acta Anaesthesiologica Scandinavica* 1993; **37**: 314–18.

63. Eisenkraft JB, Book WJ, Mann SM, Papatesta AE, Hubbard M. Resistance to succinylcholine in myasthenia gravis: a dose–response study. *Anesthesiology* 1988; **69**: 760–3.

64. Smith CE, Donati F, Bevan DR. Cumulative dose–response curves for atracurium in patients with myasthenia gravis. *Canadian Journal of Anaesthesia* 1989; **36**: 402–6.

65. Wise RP, Wylie WD. The thymus gland. Its implications in clinical anesthetic practice. In: Jenkins MT, ed. *Clinical anesthesia: anesthesia for patients with endocrine disease.* Philadelphia: Davies & Co, 1964: Chapter II.

RECOMMENDED READING

1. Bowman WC. *Pharmacology of neuromuscular function.* 2nd ed. London: John Wright and Sons Ltd, 1990.

2. Azar I. Muscle relaxants in patients with neuromuscular disorders. In: Azar I. *Clinical pharmacology. Vol. 7. Muscle relaxants, side effects and a rational approach to selection.* New York and Basel: Marcel Dekker, Inc, 1987.

3. Østergaard D, Engbæk J, Viby-Mogensen J. Adverse reactions and interactions of the neuromuscular blocking drugs. *Medical Toxicology and Adverse Drug Experience* 1989; **4**: 351–68.

Chapter 9

Muscle Relaxants and Clinical Monitoring

David R. Bevan and François Donati

It is now more than 50 years since the introduction of neuromuscular blocking drugs into clinical anaesthesia. Their appearance changed the philosophy of anaesthetic practice. The triad of hypnosis, analgesia and muscle relaxation could now be provided by separate agents without the cardiovascular risks of providing all three with the same technique – deep inhalational anaesthesia or extensive regional anaesthesia. However, new problems appeared, including postoperative ventilatory inadequacy with residual neuromuscular block.

Depolarizing agents

Neuromuscular blockade can be achieved with drugs that depolarize the end plate. The initial action of these agents is characterized by an acetylcholine-like effect, which is the source of most of their undesired actions. This category of drugs includes decamethonium and suxamethonium but only suxamethonium is used in clinical practice, because of its unique neuromuscular profile. Suxamethonium is the only neuromuscular blocking drug with a short onset (1–1.5 minutes) and a short duration of action (5–10 minutes).

Neuromuscular pharmacology

Depolarizing neuromuscular relaxants probably act by producing desensitization of postsynaptic receptors.[1] When an agonist is present at the neuromuscular junction for an extended period of time, the threshold required to trigger an action potential and a contraction in muscle increases to a level greater than that of the depolarization, and neuromuscular blockade follows. This phenomenon also occurs with acetylcholine if the neurotransmitter concentration is maintained elevated at the neuromuscular junction

Table 9.1 Complications of suxamethonium

Cardiovascular	Hyperkalaemia
Fasciculations	Allergic reactions
Muscle pains	Masseter spasm
Increased IOP	Malignant hyperthermia
Increased IGP	Atypical cholinesterase

by pharmacological or experimental manipulations. Under normal physiological conditions, however, acetylcholine is broken down very rapidly by acetylcholinesterase, and desensitization does not occur. Although suxamethonium is a short-acting agent by pharmacological standards, the drug remains at the neuromuscular junction for a sufficiently extended time (minutes) for desensitization, a phenomenon which takes place within milliseconds, to occur.

The side effects of suxamethonium (Table 9.1) are related to its ability to depolarize the end plate, that is to keep receptors open for a relatively long time. This produces a flow of potassium ions from the inside to the outside of the cell. Because this process is generalized in all muscle cells of the body, an increase in extracellular potassium concentration occurs. The magnitude of this hyperkalaemia depends of course on the number of receptors activated, and may be very high in pathological states which involve proliferation of receptors.[2] In normal patients, the increase in K^+ concentration is 0.5–1.0 mmol/l, is transient, and is of little practical consequence.

Fasciculations are manifestations of suxamethonium's acetylcholine-like, agonist activity, but they probably have a presynaptic origin.[3] Small doses of non-depolarizing relaxants have enough presynaptic activity to block fasciculations, but have little postsynaptic, neuromuscular blocking effect of their own. However, defasciculating doses of non-depolarizing relaxants block a sufficient number of postsynaptic receptors to decrease significantly the number of postsynaptic receptors available for binding with suxamethonium. As a result, the dose of suxamethonium

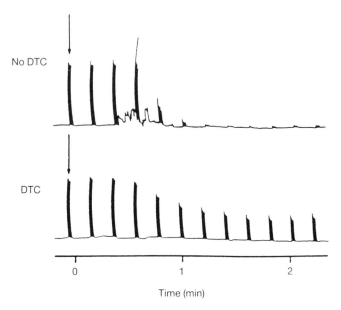

Fig. 9.1 Fasciculations and twitch augmentation after suxamethonium 0.5 mg/kg with (bottom trace) and without pretreatment with tubocurarine (DTC), 0.05 mg/kg. (Reproduced, with permission, from Szalados JE, Donati F, Bevan DR. Effect of d-tubocurarine pretreatment on succinylcholine twitch augmentation and neuromuscular blockade. *Anesthesia and Analgesia* **71**: 55–9. © International Anesthesia Research Society 1990.)

Characteristics of blockade

Initially, depolarizing blockade is characterized by (1) absence of fade with train-of-four (TOF) or tetanic stimulation (Fig. 9.2), (2) absence of post-tetanic facilitation, and (3) increased blockade with anticholinesterase drugs such as neostigmine and edrophonium. However, prolonged administration of suxamethonium produces a type of blockade which has non-depolarizing characteristics, such as (1) fade with TOF or tetanic stimulation, (2) post-tetanic facilitation, and (3) antagonism of blockade with anticholinesterase agents.[6] This type of blockade has been termed 'non-depolarizing', 'dual', and 'phase II'. The last expression is preferred because it does not imply any specific mode of action, and probably reflects our ignorance of the exact mechanism. The transition from phase I, or depolarizing, to phase II block is gradual, and is characterized by increased TOF fade. It normally occurs after 7–10 mg/kg have been given, which normally corresponds to 30–60 minutes of paralysis. Recovery after phase II is longer than after phase I block. Onset of phase II block is also associated with tachyphylaxis, that is an increased requirement for the same degree of blockade. Tachyphylaxis is thought of as a manifestation of the antagonism between phase I and phase II blocks.

must be increased to produce the same effects (Fig. 9.1).[3,4] The same presynaptic action of depolarizing agents produces repetitive muscle stimulation and an increase in twitch height (Fig. 9.1). It is unclear at present whether muscle pains and fasciculations are related, and the mechanism for the former is unknown.

The other neuromuscular effect of suxamethonium is its ability to produce an increase in muscle tension in certain muscle groups, in the absence of nerve stimulation. This is most apparent and clinically important at the masseter, but it can also be observed at the adductor pollicis. This increase in tension is highly variable from patient to patient, and its magnitude (a few hundred grams or less) is usually of no clinical importance. However, masseter spasm is probably an exaggerated form of this response[5] but its relation to malignant hyperthermia is uncertain.

Pharmacokinetics of suxamethonium

Suxamethonium was introduced into clinical practice decades before pharmacokinetic methods and principles became generalized. Furthermore, the application of kinetic analysis to suxamethonium is limited by its rapid *ex vivo* degradation and by the lack of adequate models for rapidly metabolized drugs which never achieve a steady state. Nevertheless, it is known that suxamethonium is hydrolysed rapidly by plasma cholinesterase. Indirect estimates of the half-life of the drug yield values of 2–4 minutes.[7]

The rapid breakdown of suxamethonium is dependent on the activity of the enzyme. Decreases or increases in plasma cholinesterase activity due to drugs or pathological processes do not normally induce clinically significant changes in onset and duration of neuromuscular blockade. However, qualitative, genetically produced alterations in the

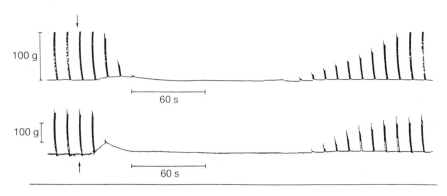

Fig. 9.2 Increases in resting tension of adductor pollicis (top) and masseter (bottom) after suxamethonium 1 mg/kg. (Reproduced, with permission, from Plumley *et al.*, 1990.[5])

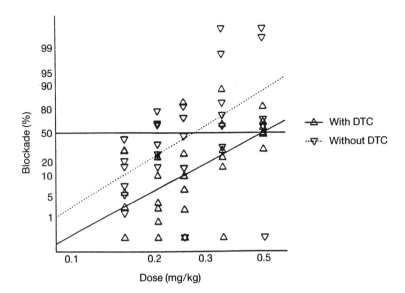

Fig. 9.3 Dose–response curves for suxamethonium with and without tubocurarine pretreatment. (Reproduced with permission, from Szalados JE, Donati F, Bevan DR. Nitrous oxide potentiates succinylcholine neuromuscular blockade in humans. *Anesthesia and Analgesia* **72**: 18–21. © International Anesthesia Research Society 1991.)

enzyme, present in approximately one in 2000 individuals, produce a considerable increase in the duration of action of the drug.[8]

Pharmacology

Dose–response

The dose required to produce 95 per cent blockade (ED_{95}) at the adductor pollicis is in the range 0.3–0.5 mg/kg in patients anaesthetized without inhalational agents. The ED_{50} is 0.2–0.3 mg/kg (Fig. 9.3).[4,9] It appears that in the presence of nitrous oxide, suxamethonium blockade is more intense, and the ED_{50} and ED_{95} are at the lower end of the range mentioned above.[10] If given immediately after induction of anaesthesia, as is normal in clinical practice, the ED_{50} and ED_{95} are 0.3 and 0.5 mg/kg, respectively.[4] Like non-depolarizing agents, suxamethonium spares the diaphragm, and doses required for diaphragmatic paralysis are greater than those mentioned above. However, contrary to the action of non-depolarizing drugs, suxamethonium has a special affinity for laryngeal adductor muscles, and doses required for vocal cord paralysis are less than for comparable blockade at the adductor pollicis.[11].

Onset and duration

Suxamethonium is characterized by a rapid onset of action. With subparalysing doses (less than 0.3–0.5 mg/kg), maximum blockade is achieved within 1.5–2 minutes at the adductor pollicis.[4,11] Larger doses (1–1.5 mg/kg) lead to the abolition of neuromuscular activity within approximately one minute. Onset is even more rapid at the diaphragm or the larynx.[4] Onset times are shorter in children than in adults.

The duration of action of the drug is less dose-dependent than for longer acting agents. This is because, in general, doubling the dose of a drug produces a prolongation of effect equal to one half-life.[7] In the case of suxamethonium, doubling the dose increases the half-life by only 2–4 minutes. Doses of 1 mg/kg have a duration of action of 10–12 minutes.[9]

Drug interactions

Depolarizing drugs interact with non-depolarizing drugs. Usually, depolarizing blockade is antagonized by non-depolarizing agents and potentiated by anticholinesterase drugs. On the other hand, phase II blockade is potentiated by non-depolarizing blocking drugs and antagonized by anticholinesterase agents.[6] For example, after small defasciculating doses of non-depolarizing agents, suxamethonium requirements are increased.[3,4] *d*-Tubocurarine, 0.05 mg/kg, given 3 minutes before suxamethonium, increases the requirement of the latter by a factor two.[4] The situation is more difficult to predict when the non-depolarizing relaxant has been given in a dose sufficient to produce a block of its own, as may occur when additional relaxation may be required at the end of anaesthesia. In general, suxamethonium intensifies a shallow block but antagonizes more intense blockade. Neostigmine and pyridostigmine have two separate effects on suxamethonium blockade. They deepen phase I block by virtue of their effects at the neuromuscular junction and they delay suxamethonium metabolism by their inhibitory effect on plasma cholinesterase.[12] The net effect is potentiation and prolongation of suxamethonium blockade, and this may be encountered when suxamethonium is given after reversal agents are administered. Edrophonium probably has much less effect on plasma cholinesterase,[12] but may potentiate phase I blockade because of its effects at the neuromuscular junction.[6] Paradoxically, the dose requirements for vecuronium, atracurium and pancuronium administered after full recovery from suxamethonium blockade are reduced, compared with when any of these non-depolarizing drugs is given alone.[13]

The ED_{50} and ED_{95} of suxamethonium appear to be reduced in the presence of anaesthetic vapours. With prolonged infusions (2–3 hours), suxamethonium requirements decrease when isoflurane or enflurane are used.

Side effects

The use of suxamethonium is limited considerably by its many side effects. In some patients such as those with dystrophia myotonica, the drug is absolutely contraindicated.

Cardiovascular

Suxamethonium has somewhat unpredictable cardiovascular effects, because at least three separate mechanisms of action may play a role. Because of its structural similarity to acetylcholine, vagal effects such as bradycardia and even asystole may be observed, especially in children. For an unknown reason, these deleterious effects are seen more commonly after a second dose of suxamethonium. Pretreatment with an anticholinergic, atropine or glycopyrronium (glycopyrrolate), is usually effective in abolishing these vagal effects. However, suxamethonium is associated with an increase in circulating catecholamines. This effect might counteract the vagal influences and might be the reason why bradycardia is less frequent than would be anticipated from the drug's structure alone.[14] Finally, suxamethonium produces hyperkalaemia,[2] which may in turn cause cardiovascular effects of its own, especially if the patient is at risk for large serum potassium increases.

Fasciculations

Rapid administration of a bolus of suxamethonium is usually followed by uncoordinated muscle twitching which appears in some patients as violent contractions, especially in muscular adults. Fasciculations are infrequent in children and infants. The mechanism of this side effect is most likely a depolarization of nerve terminals through suxamethonium's action at the presynaptic receptors.[3] This produces antidromic (backward) firing in the nerve, with propagation of the action potential to all branches supplying a motor unit. Fasciculations are blocked effectively by doses of non-depolarizing relaxants sufficient to block presynaptic receptors.[3] These doses are usually much smaller than those required for depression of twitch height. For example, d-tubocurarine, 0.05 mg/kg, which does not produce any detectable blockade of its own, is effective in preventing fasciculations if given at

least 3 minutes before suxamethonium (Fig. 9.1).[4] Gallamine, 0.2 mg/kg, is also effective. Atracurium, 0.03 mg/kg, pancuronium, 0.01 mg/kg or vecuronium, 0.007 mg/kg may also be used as defasciculants, but these drugs are not as effective as d-tubocurarine or gallamine for this purpose. Many other drugs, such as lignocaine (lidocaine), diazepam, thiopentone, fentanyl, and enflurane, have been proposed as defasciculants, but they are probably less effective than non-depolarizing drugs and they may have some actions of their own.[14] The common mechanism of action is most likely a membrane stabilizing effect which tends to inhibit propagation of antidromic action potentials. 'Self-taming' doses of suxamethonium have also been used. This involves giving a small dose (10 mg or so), followed 1 minute later by the intubating dose. This practice has been abandoned because the self-taming dose sometimes produces blockade of its own,[14] an event which might be emotionally disastrous should it occur in an awake patient.

Muscle pains

Patients receiving suxamethonium often complain of muscle pains 24–48 hours after the surgical procedure. The pains are similar to those following intense exercise and are often severe. Because of the subjective nature of the complaint, it is difficult to determine the effectiveness of any prophylaxis. Nevertheless, a recent meta-analysis has demonstrated the beneficial effect of pretreatment with defasciculating doses of non-depolarizing relaxants.[15]

Intraocular pressure (IOP)

Suxamethonium is associated with a small increase in intraocular pressure, which is of little clinical importance except in cases of eye injury. Pretreatment with defasciculating doses of non-depolarizing relaxants is ineffective in preventing this increase.[14] The importance of the increase in IOP must be weighed against the much greater increases which may occur with inadequate anaesthesia, hypertension, mechanical pressure on the eye globe, bucking, coughing, and the consequences of inadequate airway control when the anaesthetic technique does not include suxamethonium.

Intracranial pressure (ICP)

Although there have been conflicting reports, suxamethonium probably increases ICP, and pretreatment with non-depolarizing drugs probably has some protective effect. At least some of this effect on ICP might be due to the increase in blood flow associated with the fasciculation-induced increase in CO_2.

Intragastric pressure (IGP)

Suxamethonium is associated with an increase in IGP. However, this increase is of no consequence, even in patients with full stomachs, because there is a corresponding increase in 'barrier pressure', that is the pressure inside the cardiac sphincter.[14]

Hyperkalaemia

The small, transient increase in serum potassium concentration associated with suxamethonium administration is clinically unimportant in most patients. However, severe hyperkalaemia may occur in individuals with spinal cord transection or extensive burns.[2] This phenomenon may also be observed in patients with a space-occupying CNS lesion, stroke, immobility, trauma and a large number of CNS or skeletal muscle diseases.[14]

Allergic reactions

Suxamethonium has probably been incriminated in allergic and anaphylactic reactions more often than any other drug used in anaesthesia. The true incidence of the complication is unknown, because of incomplete reporting and the concurrent use of other drugs, which makes causal relations uncertain. With more and more patients exposed to other relaxants, the incidence of anaphylactic reactions to non-depolarizing agents is expected to increase.

Masseter spasm

In certain muscles of the body, the normal effect of suxamethonium is to produce an increase in muscle tension, even in the absence of nerve stimulation. This phenomenon has been observed at the adductor pollicis, but is certainly most prominent at the masseter. Interestingly, laryngeal muscles do not show this response.[11] The magnitude of the increase in masseter muscle tension is usually of the order of a few hundred grams, and can be overcome easily by the laryngoscopist. However, a small number of patients show an abnormally large increase in masseter tension which makes intubation impossible. This is probably an exaggerated form of the normal response,[5] although this interpretation is not accepted by all experts in the field. Defasciculating doses of non-depolarizing relaxants do not alter this response significantly. However, paralysing doses of atracurium abolish masseter tension increases,[5] suggesting that the effect is mediated through postsynaptic acetylcholine receptors.

Malignant hyperthermia

Suxamethonium has been associated, together with the potent vapours, with malignant hyperthermia. There is an association between masseter spasm and malignant hyperthermia, but the former is usually not followed by the latter. The nature of this association is unknown. Furthermore, not all workers agree about the safety of continuing an anaesthetic during which isolated masseter rigidity was found.

Abnormal plasma cholinesterase

Suxamethonium is broken down by an enzyme which may be quantitatively or qualitatively abnormal. Certain conditions, such as pregnancy, liver disease, malnutrition, are associated with decreased activity of plasma cholinesterase.[8] Although the duration of action of suxamethonium is increased in these conditions, the prolongation is usually small and of no clinical significance. However, certain individuals have a genetically determined deficiency in plasma cholinesterase activity. The synthesis of the enzyme is controlled by a gene which is absent or abnormal in approximately 1:2000 people.[8] These patients have paralysis of a few hours' duration following usual doses of suxamethonium. Treatment for this condition is by mechanical ventilation until neuromuscular function returns. Plasma cholinesterase is preserved in plasma, but the risk of transfusions most likely outweighs that of short-term ventilation. Heterozygous individuals, i.e. those with one normal and one abnormal gene, have a clinically unimportant prolongation of blockade.

Clinical use

The main use of suxamethonium is to facilitate tracheal intubation. It is given after administration of the anaesthetic induction agent in doses of 1–1.5 mg/kg (2–3 times ED_{95} obtained without nitrous oxide). The reasons for giving doses greater than the ED_{95} is to provide relaxation to certain muscles, such as the diaphragm, which exhibit resistance to suxamethonium, and to allow for an increased requirement for the relaxant in certain individuals. The dose should be increased to 2–2.5 mg/kg if a defasciculating dose of non-depolarizing relaxant has been administered. The dose required in children is greater than that in adults (approximately 2 mg/kg), and defasciculating doses are not needed. Duration of paralysis is in the 8–12 minute range,[9] although some patients may have paralysis for 15–20 minutes. Facilitation of intubation with suxamethonium offers the advantage of a rapid onset, and the possibility to terminate the anaesthetic quickly should an unanticipated airway management problem occur. Suxamethonium is indicated especially for emergency patients with a full stomach, or if a difficult intubation is anticipated provided that difficulty in ventilation with a mask is not anticipated. If mask ventilation and intubation are believed to be difficult, paralysing agents should be avoided.

Muscle relaxation can be maintained with suxamethonium, but its use for this purpose is declining because of the availability of intermediate and short duration non-depolarizing alternatives. However, the use of suxamethonium is indicated for 20–30 minute procedures, a time which may be too short for atracurium or vecuronium. If suxamethonium is administered for a longer time, the infusion rate must be increased because of tachyphylaxis, from a mean of 50 µg/kg per minute to 80–100 µg/kg per minute. Recovery for durations of less than 1 hour is generally rapid.

Non-depolarizing agents

Non-depolarizing neuromuscular blocking drugs (NDA) are quaternary ammonium compounds which compete with acetylcholine (ACh) and bind with at least one or two α-subunits of the ACh receptor (AChR) at the postsynaptic membrane. At least 75 per cent of receptors must be occupied – 'margin of safety' – before transmission is impaired and if more than 90 per cent of receptors are occupied, transmission fails.[16] Non-depolarizing neuromuscular blocking drugs may also enter and block the ion channel when it is open but the 'channel block' is of little clinical importance. The NDAs may act at a presynaptic receptor to modify the release of ACh[17] and this is probably responsible for the run-down in ACh release and fade in muscle tension in response to repeated stimulation during partial neuromuscular block. Several non-depolarizing relaxants produce cardiovascular effects by actions at the sympathetic ganglia and nerve endings, and at the vagus. Some stimulate histamine release and all may, rarely, produce anaphylaxis. During the last decade attempts have been made to develop more potent neuromuscular relaxants in the anticipation that their circulatory effects will be reduced.

Up-/down-regulation of the acetylcholine receptor

Sensitivity to NDAs may vary due to an increase or decrease in the number of ACh receptors.[18] Sensitivity is reduced in severely burned patients, or in denervation syndromes, immobility, sepsis, trauma, and after chronic exposure to muscle relaxants (Table 9.2) as a result of receptor multiplication, mainly extrajunctional – *up-regulation*. Such 'new' receptors revert to the fetal [α(2)β, γ, δ subunits] rather than the adult [α(2), β, δ, ε] receptor type. Resistance to NDAs is often associated with hyperkalaemia in response to suxamethonium. In other situations, e.g. myasthenia gravis, sensitivity to NDAs is increased as a result of reduction in ACh receptors – *down-regulation*.

Table 9.2 Conditions associated with up-/down-regulation of skeletal muscle acetylcholine receptors[4]

Resistance to NDAs, hyperkalaemia to suxamethonium
  Increased AChR
    Burns
    Denervation syndromes; upper motor neuron, lower motor neuron, multiple sclerosis
    Immobilization
    Sepsis
    Chronic NDA exposure
Resistance to NDAs, no hyperkalaemia
    Cerebral palsy
    Anticonvulsant use
Sensitivity to NDAs, resistance to suxamethonium
  Decreased AChR
    Myasthenia gravis
    Exercise

Characteristics of non-depolarizing block

Non-depolarizing block is characterized by a slow onset to maximal effect (5–12 minutes) and slow recovery, in comparison with suxamethonium. Central muscles (diaphragm, larynx, masseter, orbicularis oculi) tend to be affected earlier and recover from the block sooner than those of the periphery (adductor pollicis), probably as a result of preferential perfusion.

The characteristic feature of non-depolarizing block is failure to maintain muscle tension in response to repeated stimulation. Tetanic (50 or 100 Hz) or train-of-four stimulation (2 Hz) results in fade of tetanic or TOF responses which is more obvious during recovery from the block than during its onset (Fig. 9.4).[19] Tetanic stimulation is followed by augmentation of the response to single stimulation, post-tetanic facilitation, and this has been made use of in determining the *post-tetanic count* (see below). Although NDAs produce flaccid paralysis, the muscles are still able to respond to direct stimulation.

The neuromuscular block of NDAs is reversed by the anticholinesterases, edrophonium, neostigmine or pyridostigmine: some recovery can be induced with small doses of suxamethonium.

Pharmacological principles

Dose–response

Dose–responses for NDAs are established by giving single or incremental small doses of relaxant during stable anaesthesia. The cumulative dose–response curves produced by repeated doses underestimate the potency of short-acting compounds unless precautions are taken to

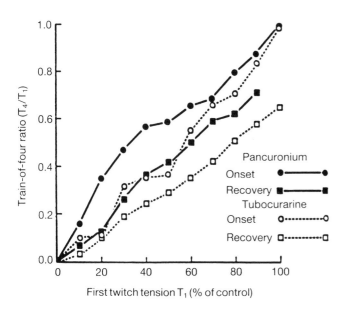

Fig. 9.4 Train-of-four fade during onset and recovery of neuromuscular block after pancuronium or tubocurarine. (Reproduced, with permission, from Robbins et al., 1984.[19])

Table 9.3 Sensitivities of different muscle groups to non-depolarizing relaxants in descending order

Geniohyoid
Masseter
Adductor pollicis
Abdominal muscles
Diaphragm
Orbicularis oculi
Laryngeal adductors

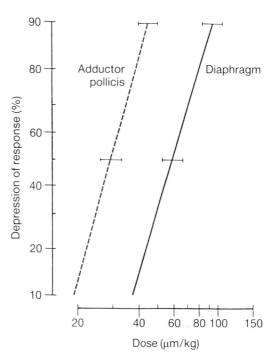

Fig. 9.5 Mean cumulative dose–response curve for pancuronium for the adductor pollicis twitch tension and diaphragm electromyogram. (Reproduced, with permission, from Donati et al., 1986.[21])

Table 9.4 Factors which influence speed of onset of action of muscle relaxants

Perfusion
  Cardiac output
  Muscle blood flow
Clearance
Potency
Dose
Priming

prevent disappearance of the drug.[20] Responses of the thumb are the most frequently observed after stimulation of the ulnar nerve with single supramaximal stimuli or from the first response, T_1, to TOF stimulation. Mean doses producing 50, 90 or 95 per cent reduction of twitch height, acceleration, or EMG are known as the $ED_{50,90 \text{ or } 95}$ respectively.

Although, because of its convenient access, the adductor pollicis is most frequently assessed, other muscles have different sensitivities (Table 9.3). The diaphragm requires approximately twice the dose of relaxant to produce the same effect as in the adductor pollicis (Fig. 9.5).[21]

Onset and duration

Perfusion

The duration of neuromusclar block is determined by the time for plasma relaxant concentration to decrease below that necessary to produce block. However, there is a delay between peak arterial plasma concentrations and maximal block which has been modelled mathematically using the

concept of an 'effect compartment' to which access is controlled by a rate constant k_{eo}[22] which is determined by the factors which influence access to and removal of drug from the receptor (Table 9.4). Rapid delivery of drug is the probable reason for the more rapid onset of block in small children, with their more dynamic circulation than in the elderly.[23] Onset is not the same in all muscles; it is more rapid at the diaphragm and larynx than at the adductor pollicis (Fig. 9.6).[24] Repeated muscle contraction by repeated nerve stimulation before administration of the relaxant accelerates onset, probably as a result of preferential perfusion. The onset of gallamine block in dogs was delayed when the flood flow to the limb was reduced.

Clearance

If metabolism or redistribution is very rapid, onset time is reduced, but this is probably only important for

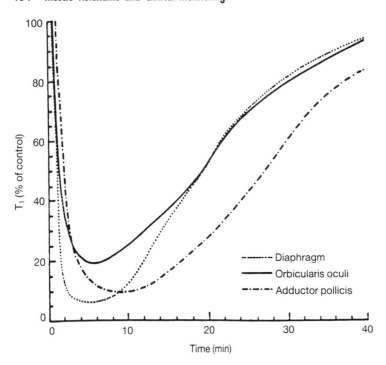

Fig. 9.6 Course of action of vecuronium, 0.07 mg/kg, at the orbicularis oculi, diaphragm and adductor pollicis muscles in humans. (Reproduced, with permission, from Donati *et al.*, 1990.[24])

suxamethonium and, perhaps, for mivacurium. The depolarizing or non-depolarizing nature of the block does not, of itself, affect onset time. Thus when suxamethonium was administered to patients with abnormal plasma cholinesterase (reduced metabolism), its onset was delayed and was no more rapid than that of atracurium in the same patients.

Potency

Low potency drugs have a more rapid onset of action.[25] Transfer of the NDA from the circulation to the receptor occurs rapidly and does not appear to be a critical factor in determining onset. However, the neuromuscular junction has a very high concentration of receptors, and a large proportion of them must be bound to relaxant before neuromuscular blockade can be detected. Potent drugs are expected to have a slower onset of action, because fewer molecules are injected and the concentration gradient between the synaptic cleft and the receptor is less than with low potency compounds. In man, the most potent NDA, doxacurium, has the slowest onset (12–15 minutes for subparalysing doses). The speed of onset of equipotent doses of gallamine, *d*-tubocurarine, and pancuronium, is inversely related to potency[26] (Fig. 9.7). Potent relaxants have high affinity for the receptor and their tendency to wear off slowly can be attributed to 'buffered diffusion' involving repetitive binding and unbinding from the receptor. Less potent drugs have less repetitive binding and a higher concentration gradient encourages a faster onset of action. The iontophoretic administration of NDAs, *in vitro*, confirmed that more potent drugs have a slower onset and recovery.[27]

Dose

The time to maximal block is independent of dose if less than 100 per cent block is attained. After large doses total paralysis is achieved before peak concentrations are achieved at the junction and the time to disappearance of twitch decreases with increasing dose. For most drugs, rapid onset is achieved only at the expense of delayed recovery.

Priming

Of the many attempts to speed the onset of non-depolarizing NDAs, 'priming' is the most popular. An initial small dose occupies the receptor 'sink' and, when followed by the main dose, a more rapid onset occurs[23] but the effect is small and carries the disadvantage of symptoms in sensitive muscles, such as the eyes and the upper airway muscles. Regurgitation and aspiration of gastric contents have been described after 'priming' and it has been shown that priming doses of atracurium and vecuronium depress swallowing in humans.

Classification

The NDAs are classified according to the time from injection to 90 per cent recovery of twitch tension of the adductor pollicis: short-acting, 10–20 minutes; intermediate-acting, 20–30 minutes; and long-acting, 30–60 minutes (Table 9.5).

Since the introduction of *d*-tubocurarine into clinical practice 50 years ago, more than 50 muscle relaxants have been introduced. Few (suxamethonium, tubocurarine,

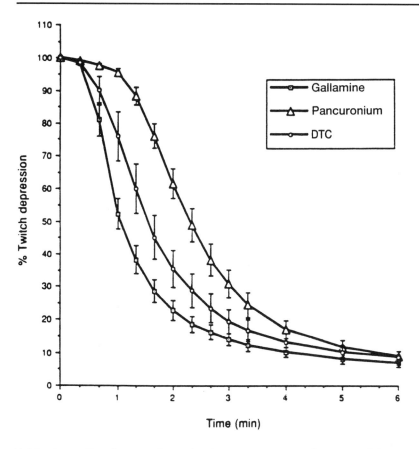

Fig. 9.7 Onset of neuromuscular blockade of the adductor pollicis in patients given equipotent doses of gallamine, tubocurarine (DTC) and pancuronium. (Reproduced with permission, from Kopman, 1989.[26])

Table 9.5 Classification of non-depolarizing neuromuscular blocking drugs (NDAs)

Short-acting
  Mivacurium
Intermediate-acting
  Atracurium
  Rocuronium
  Vecuronium
Long-acting
  Alcuronium
  Doxacurium
  Gallamine
  Pancuronium
  Pipecuronium
  Tubocurarine

Table 9.6 Typical neuromuscular activity of non-depolarizing blocking drugs

	ED_{50} (µg/kg)	ED_{95} (µg/kg)	Onset (min)	RI (min)	T_{90} (min)
Tubocurarine	250	500	6	25–35	70–90
Alcuronium	70	145	6	30	60
Atracurium	120	200	5–6	10–15	20–25
Doxacurium	15	25	10–14	N/A	80–100
Gallamine	750	2000	5–7	25–40	70–80
Metocurine	130	280	5	30–40	80–90
Mivacurium	50	80	2–3	10–15	20 .
Pancuronium	35	60	4–5	25	60
Pipecuronium	30	60	5–6	30–40	80–90
Rocuronium	150	300	2–3	10–15	20
Vecuronium	25	80	5–6	10–15	20

Potency (ED_{50}, ED_{95}), onset to maximum effect, 25–75 per cent recovery index (RI), and time to 90 per cent T_1 recovery (T_{90}) after and ED_{95} dose in adults

gallamine, metocurine, pancuronium, atracurium and vecuronium) have succeeded. Discussion will be restricted to these and to the newly introduced agents mivacurium, rocuronium, doxacurium and pipecuronium (Table 9.6).

Short-acting agents

Mivacurium

Mivacurium is a recently introduced benzylisoquinoline compound (Fig. 9.8). It is short acting because of metabolism by plasma cholinesterase.

The ED_{95} in adults is 70–80 µg/kg and 90–110 µg/kg in children. Drug degradation occurs before it reaches the neuromuscular junction where concentrations of mivacurium are small. Recovery occurs slightly more rapidly after reversal with neostigmine or edrophonium. The rapid recovery makes mivacurium an ideal agent for outpatient surgery where it may be given by repeated bolus injection or by infusion (6–8 µg/kg per minute). In the presence of atypical cholinesterase, the duration of action is considerably prolonged ($2 \times ED_{95}$ doses last for 6–8 hours) and is

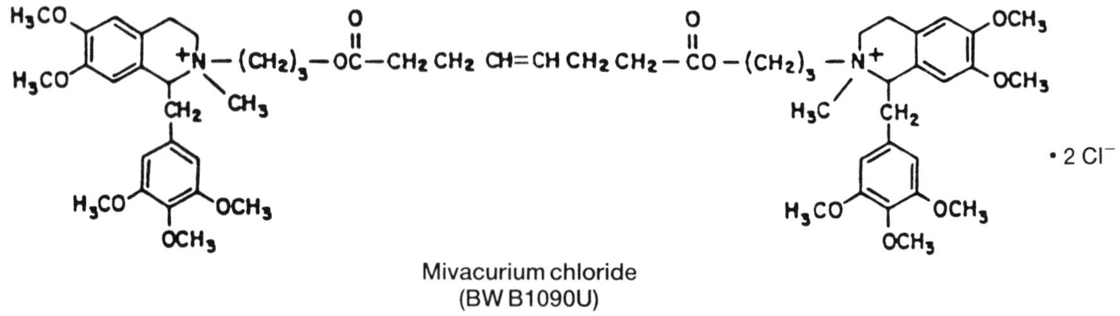

Mivacurium chloride
(BW B1090U)

Fig. 9.8 Mivacurium – structure.

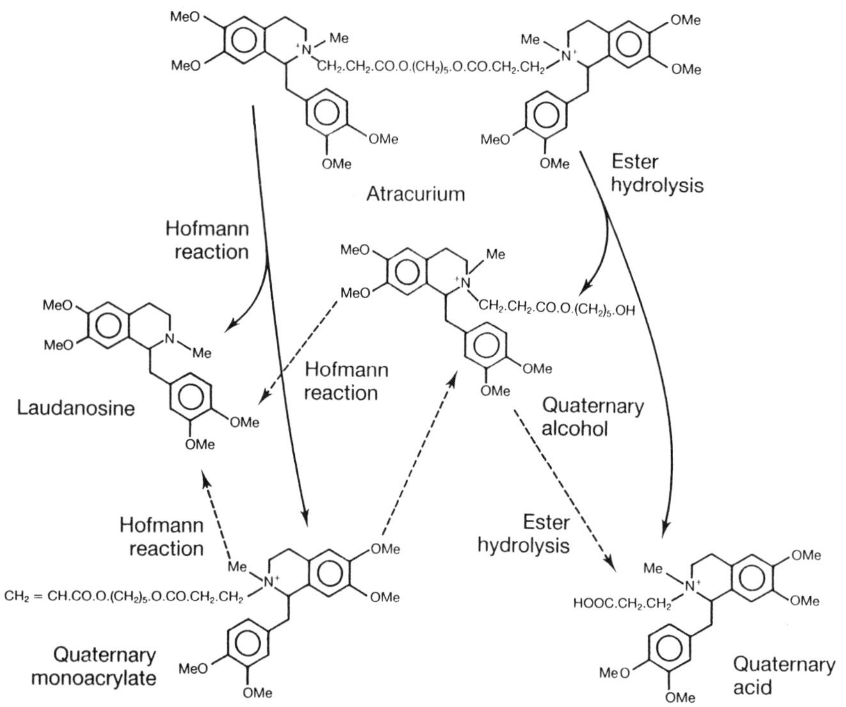

Fig. 9.9 Atracurium structure and metabolism.

treated with sedation and mechanical ventilation until muscle power returns. Mivacurium releases histamine when administered rapidly in doses of 2–3 × ED95, and this is probably responsible for its haemodynamic profile, brief 10–20 per cent decrease in blood pressure and increase in heart rate, which is similar to atracurium.

Intermediate-acting agents

Atracurium

Atracurium is a bisquaternary ammonium benzylisoquinoline compound. It was introduced because of its minimal cardiovascular effects and because spontaneous Hofmann degradation (Fig. 9.9), at body pH and temperature, produced an agent whose intermediate duration of action was independent of the liver and kidney.[28] Also, non-

specific plasma esterases may be responsible for up to two-thirds of the metabolism.

In humans the ED95 of atracurium is 200 µg/kg, maximal block occurs at 5-6 minutes, similar to most other NDAs and slower than suxamethonium, but intubation can usually be performed about 2 minutes after 0.5 mg/kg. Spontaneous recovery is rapid and the recovery index (25–75 per cent recovery of T_1) of 10–15 minutes remains constant during prolonged administration by either intermittent boluses or infusion.

It has been suspected that the metabolites of atracurium are toxic. Laudanosine, at high doses, causes convulsions in animals but they have not been seen in humans. Similarly the potential cellular toxicity of the acrylates have not been confirmed.

Large doses (> 2 × ED95) administered rapidly do produce some histamine release which may lead to cutaneous flushing and a brief, small decrease in blood pressure and an increase in heart rate. However, the commonest cardiovascular effect is bradycardia from the unopposed vagal stimulation by high doses of opioids or surgical stimulation. Histamine release seldom causes difficulty and may be avoided by slow administration or prevented by preoperative H_1- and H_2- receptor blockade. In animals, atracurium produces dose-related bronchospasm which is prevented by H_1-receptor block but bronchospasm is rare in clinical practice. Anaphylaxis has been described.

Atracurium may be given as a bolus at the start of anaesthesia to facilitate tracheal intubation. Neuromuscular blockade may be maintained either by intermittent boluses ($ED_{95} \times 0.75$) or by continuous infusion (4–10 µg/kg per minute).[29] Requirements seem to be similar to adults in the very young and in the elderly. No cumulation is seen although the requirements may decrease by more than 50 per cent in the presence of inhalational anaesthetic agents – especially enflurane.

Vecuronium

Vecuronium, produced by demethylation of pancuronium at the 2-piperidino position, is an intermediate duration aminosteroid which is devoid of cardiovascular activity.[30] Demethylation decreased ACh-like activity, reducing the cardiovascular effects, and increasing lipophilicity which increased plasma clearance by the liver compared with pancuronium.

The ED_{95} of vecuronium is 50 µg/kg and is similar at all ages. The onset of action of subparalysing doses is 5–6 minutes but good intubating conditions are produced within about 2 minutes after 100 µg/kg. Onset is enhanced by 'priming' with vecuronium 10 µg/kg although this may be associated with respiratory impairment including reductions in FEV_1, FVC, FRC and forced expiratory flow. Recovery to 25 per cent T_1 occurs in 30-40 minutes but more rapidly in children and more slowly in infants and in the elderly – the latter probably a result of decreased plasma clearance in the very young and very old. The shorter duration of action of vecuronium than pancuronium is not a result of increased clearance but of enhanced distribution. Consequently, when administered for long periods or in high dosage, peripheral storage sites become saturated and the duration of action is dependent upon metabolism and excretion.

Vecuronium is deacetylated to 3-OH, 17-OH and 3,17-$(OH)_2$ metabolites. The 3-OH derivative has about 60 per cent of the neuromuscular blocking potency of vecuronium, is excreted by the kidney and has been blamed for prolonged block in ICU patients with multisystem disease given vecuronium (usually in large doses by infusion) to facilitate mechanical ventilation.[31]

Vecuronium has no intrinsic cardiovascular actions. Bradycardia may occur due to exaggerated vagal stimulation by narcotics or surgical procedures and this may be due to amplification of the usual response. Vecuronium excretion is decreased slightly in renal failure and considerably in severe hepatic disease resulting in a prolonged duration of action. The duration is also prolonged and recovery was delayed in the obese but, as pharmacokinetic variables were not affected, the prolongation is probably due to a relative overdose. Dosage, in the obese, should be calculated on ideal body weight.

Vecuronium is used to facilitate tracheal intubation and the dose can be increased to 2–3 × ED_{95} without producing cardiovascular effects. Maintenance of relaxation is with repeated boluses, 1–2 mg, or infusion, 1–2 µg/kg per minute. The doses required to maintain a constant effect will decrease by approximately 50 per cent over 2–3 hours. Prolonged administration is associated with slower spontaneous recovery than after atracurium but the block can be reversed very easily after either agent even though TOF fade for the same T_1 depression is less for vecuronium than for atracurium during recovery from neuromuscular block. The administration of vecuronium by infusion to facilitate artificial ventilation has become very popular in ICUs because of the drug's innocuous haemodynamic profile. Such administration is seldom with adequate monitoring and may be followed by weakness lasting for several weeks or months.[31] It is uncertain whether such complications are specific to the drug or are related to drug interactions or multi-organ failure. There is little doubt that long-term administration should be performed only with adequate neuromuscular monitoring.

Rocuronium

Rocuronium is a recently introduced desacetoxy derivative of vecuronium which is currently undergoing clinical evaluation. It is an intermediate-acting NDA without cardiovascular effects but with only about one-seventh the potency of vecuronium (ED_{95} 300 µg/kg).

The low potency results in more rapid onset of action than vecuronium. Maximal blockade at the adductor pollicis occurs after about 3 minutes (vecuronium 5–6 minutes) and at the laryngeal adductors in about 1.5 minutes (vecuronium 3 minutes, suxamethonium 1 minute). Intubating conditions after rocuronium, 600 µg/kg, were found to be excellent at 90 seconds compared with 60 seconds after suxamethonium, 1 mg/kg. However, 90 per cent twitch recovery was more rapid after suxamethonium, 11.3 minutes compared with 36.1 minutes. Nevertheless, rocuronium is the first NDA with a speed of onset which is comparable with that of suxamethonium.

Preliminary studies in humans suggest that, as in animals, it is devoid of cardiovascular effects. Similarly its pharmacokinetic behaviour resembles vecuronium.

Until now, only a few experimental studies have been reported using rocuronium. If the early evidence is supported, it will no doubt replace vecuronium as an intermediate-acting agent and, in many situations, be used in place of suxamethonium.

Long-acting drugs: older agents

Tubocurarine

The first neuromuscular blocking agent to be given under controlled conditions during anaesthesia was a crude extract of curare – 'Intocostrin' – by Griffith and Johnson in 1942.[32] Supplies in England were short so that Gray and Halton used an authenticated curare extract,[33] manufactured by Burroughs Wellcome, following Harold King's isolation of *d*-tubocurarine chloride from a sample of 'curare' in the British museum. Curare is still used, but the frequent cardiovascular complications encouraged the search for new, more specific agents.

The ED_{95} for tubocurarine is 500 μg/kg. The onset of action, 5–6 minutes, and recovery, time to 90 per cent return of thumb twitch (60–90 minutes), are slow. Initially, it was used in small doses in North America, 5–15 mg, to provide surgical relaxation. In Europe the larger doses, 30–45 mg, formed part of a balanced anaesthetic technique with intense neuromuscular blockade. Small doses, 3 mg, have been found to be superior to other NDAs in preventing suxamethonium myalgias.[15]

Tubocurarine is a monoquaternary compound but the second nitrogen atom undergoes protonation at body pH. Little metabolism occurs and it is excreted in the bile and in the urine. Its duration of action is increased in renal failure. The potency of tubocurarine (mg/kg) is similar at all ages but the increased volume of distribution in infants conceals a real increase in sensitivity of the neuromuscular junction, probably to all muscle relaxants, in this age group. The prolonged duration in the very young and very old results from decreased glomerular filtration in these age groups.

Tubocurarine produces dose-dependent hypotension which, in humans, is due to histamine release, although it also produces ganglion and vagal blockade.[34] Nevertheless, bradycardia is more common than tachycardia. Skin flushing is common and equivalent histamine release requires at least three times the dose of atracurium or mivacurium and twice the equivalent of metocurine.

Tubocurarine has two advantages: it is cheap and it prevents suxamethonium myalgias. Thus, its principal use during anaesthesia is in those countries and institutions where the cost of drugs is of paramount importance. Given slowly and in low doses it is a safe and effective muscle relaxant, particularly in those patients in whom some hypotension is not hazardous. In North America, probably its most frequent use is as 'precurarizing' agent before suxamethonium.

Metocurine

Methylation of two of the hydroxy groups of tubocurarine produced a compound that was more potent and produced fewer cardiovascular effects. It enjoyed brief popularity until the introduction of the haemodynamically neutral atracurium and vecuronium.

The ED_{95} of metocurine is 280 μg/kg. Its onset of action is slow and it has a slow rate of recovery. Metocurine undergoes little metabolism and the kidney is a more important route of excretion than for tubocurarine. There is a wider separation of the neuromuscular and cardiovascular effects for metocurine than for tubocurarine[34] and it results in less histamine release.

The combination of metocurine and pancuronium has been recommended to maintain cardiovascular stability for patients undergoing cardiac surgery. The sympathomimetic activity of pancuronium was counteracted by the histamine releasing properties of metocurine. Now that NDAs are available without cardiovascular effects, it has no therapeutic role.

Gallamine

Gallamine was the first synthetic NDA to be introduced into clinical practice but it has been superseded by agents with fewer side effects.

It has low potency, ED_{95} 2.5 mg/kg, which ensures that its onset is more rapid than other NDAs (Fig. 9.7). Otherwise its course of action resembles other long-acting NDAs, except that from the same level of block it is less easy to reverse than tubocurarine and pancuronium.

Gallamine is excreted, unchanged, almost entirely by the kidney so that its duration of action may be prolonged considerably in renal failure. It is associated with a dose-related tachycardia largely a result of vagal blockade but there is also some sympathetic stimulation. The side effects of gallamine have made it mainly of historical interest, although some favour its use, in small (10–20 mg) doses, as a precurarizing agent before suxamethonium.

Pancuronium

Pancuronium is a synthetic NDA synthesized by attaching two quaternary ammonium groups to a rigid steroid nucleus with the aim of producing a compound with a fixed inter-ammonium distance.[35] Pancuronium is a long-acting, slow onset NDA. The ED_{95} is 60 μg/kg. As with all NDAs, the onset occurs more rapidly in infants and children.

Pancuronium is excreted mainly in the urine but also via the liver. Consequently, renal failure is associated with an increase in terminal half-life and prolonged action. However, once recovery has been demonstrated reparalysis does not occur. Similarly, the slower recovery from block in the elderly is probably a result of decreased glomerular filtration. Pancuronium is metabolized by deacetylation and one of the metabolites, 3-OH pancuronium, has one-half the neuromuscular blocking potency of the parent compound.

In large doses, $2 \times ED_{95}$, pancuronium is associated with modest increases in heart rate, blood pressure and cardiac output. These may be hazardous in the patient with severe coronary artery disease in whom tachycardia may produce myocardial ischaemia. The cause of the cardiovascular changes is not clear. Some are due to vagolytic effects but sympathomimetic activity may occur from block of muscarinic receptors, which normally provide some brake on ganglionic transmission, and from catecholamine release. Anaphylactic reactions after pancuronium appear to be less frequent than after any other neuromuscular blocking drug used in anaesthesia.

The use of pancuronium has diminished considerably since the introduction of the intermediate NDAs. However, it retains some popularity as an inexpensive relaxant for use during long surgical procedures. The onset of action is too slow to recommend its use to facilitate tracheal intubation even though the onset can be accelerated, slightly, using the 'priming' technique. Some advocate its use in the cardiac patient to counteract the vagal bradycardia induced by high doses of opioids.

Alcuronium

Alcuronium is a semi-synthetic bisquaternary NDA, introduced in 1961, in the unproven anticipation that it might have a shorter duration of action than tubocurarine. It resembles other long-acting relaxants, such as pancuronium and tubocurarine, in onset, duration, and recovery of block.

Alcuronium is about twice as potent as tubocurarine, ED_{95} is 150 µg/kg. In common with other NDAs, it is potentiated by inhalational anaesthetic agents. Its pharmacokinetic and pharmacodynamic activities are very similar to pancuronium and tubocurarine. It is not metabolized in man and is excreted unchanged mainly in the urine. Thus, its action is prolonged in renal failure.

The cardiovascular effects of alcuronium are less than those of pancuronium and tubocurarine. Heart rate increases by about 10 per cent due to its vagolytic action but blood pressure is usually unchanged or decreases slightly, probably from a ganglionic effect. Alcuronium does not release histamine, although anaphylaxis has been described.

Alcuronium seems to have no particular advantage as a neuromuscular blocking agent. Although its cardiovascular actions are less than those of pancuronium or tubocurarine, this does not compare with the intermediate agents or the new long-acting compounds. Its slow onset of action has ensured that it is not used to facilitate intubation even after 'priming'. Its most common use was to maintain neuromuscular block with modest doses, 0.2–0.3 mg/kg, and it was probably these small doses which suggested that its

duration of action was shorter than other long-acting NDAs.

Long-acting drugs: newer agents

Twenty years ago, Kitz and Saverese recommended the development of three groups of neuromuscular blocking drugs:[36] a short-acting replacement for suxamethonium; an agent with an intermediate duration of action, and replacements for the long-acting tubocurarine and pancuronium. All should be free of cardiovascular effects. The first to become available were the intermediate agents atracurium and vecuronium; there is still no adequate suxamethonium replacement, but doxacurium and pipecuronium are the recently introduced long-acting compounds.

Doxacurium

Doxacurium is a benzylisoquinoline compound and is the most potent NDA yet described. It is devoid of cardiovascular effects in clinical doses.

During balanced anaesthesia the ED_{95} of doxacurium is 25–30 µg/kg. It has a very slow onset to maximum effect (10–13 minutes at subparalysing doses) and the duration of clinical relaxation after $2 \times ED_{95}$ doses is 90–120 minutes which is similar to pancuronium, as is the recovery index. Recovery is slower in the elderly. The onset of action is slightly more rapid in children than in adults and recovery is quicker – duration is reduced by half. The block can be reversed with neostigmine and, at more superficial levels, with edrophonium.

Doxacurium is excreted mainly unchanged in the urine. It is a weak substrate for plasma cholinesterase, but less than 10 per cent is metabolized. There have been no reports of its use in renal failure.

Doxacurium is free of cardiovascular effects and does not release histamine at least at doses up to $3 \times ED_{95}$. The major inconvenience is its slow onset of action, which mitigates against its use to facilitate tracheal intubation. Inevitably, reports will occur of prolonged recovery and persistant curarization after anaesthesia as for all long-acting NDAs.[37]

The use of doxacurium is restricted to prolonged operations in patients for whom cardiovascular stability is mandatory. Consequently, it has been used during cardiac surgery although opioid-induced bradycardia should be predicted. The exact place of doxacurium in anaesthetic practice has not yet been determined, but will probably be limited.

Pipecuronium

Pipecuronium may be considered the aminosteroid equivalent of doxacurium. It is a long-acting NDA without haemodynamic effect. Pipecuronium has similar pharmacokinetic and pharmacodynamic properties to pancuronium and these are not affected in the elderly. The ED_{95} is 50–60 µg/kg. The onset time of $2 \times ED_{95}$ doses is 3–4 minutes and the duration of clinical anaesthesia is 90–120 minutes. In common with other NDAs, pipecuronium is potentiated by inhalational anaesthetic agents. The metabolic pathways have yet to be defined but are likely to be similar to pancuronium and vecuronium. It is excreted, mainly unchanged, in the urine.

Pipecuronium has no effect on the vagus, autonomic ganglia or sympathetic system and does not release histamine when used in clinical doses. The principal problem that is likely to be associated with its use is persistent curarization after anaesthesia.

The principal use for pipecuronium is likely to be during prolonged operations in patients with myocardial ischaemia. Again, it will give no protection against opioid bradycardia. Some doubt whether it holds any advantage over large doses of vecuronium. The duration of action of pipecuronium, 60 µg/kg, is similar to vecuronium 200 µg/kg, but the onset time of the latter is more rapid.

Reversal of neuromuscular blockade

The commonest cause of death due to anaesthesia is respiratory failure and, in at least one survey, residual neuromuscular blockade was the principal contributing factor.[38] Since Viby-Mogensen and his colleagues showed that 30–40 per cent of patients exhibited residual weakness after anaesthesia using long-acting NDAs[37] it is difficult to believe that the two are unrelated. Also, the high mortality after the use of curare, described by Beecher and Todd in the USA in 1954,[39] occurred before reversal of neuromuscular blockade was common practice.

The action of muscle relaxants is usually reversed and controlled with neuromuscular monitoring. The goal is to restore neuromuscular activity so that the patient can maintain adequate pulmonary ventilation and a patent airway. Specifically, using train-of-four stimulation of the ulnar nerve, the aim is a train-of-four ratio of at least 0.7. This is achieved with the anticholinesterases (AChE) edrophonium, neostigmine, or pyridostigmine. Two other agents have been tried unsuccessfully. Germine monoacetate restores neuromuscular transmission in myasthenia gravis and in the Eaton–Lambert syndrome, probably by producing repetitive stimulation, and 4-aminopyridine increases ACh release but has only a weak and slow anticurare effect and it produces central stimulation after crossing the blood–brain barrier. Neither has been used in anaesthesia.

Table 9.7   Anticholinesterase actions

Enzyme inhibition
Agonist activity
Repetitive stimulation
Neostigmine block
Vagal stimulation

Anticholinesterase pharmacology

Mechanism of action

The inhibition of AChE may be only one of the several mechanisms involved in reversing neuromuscular blockade (Table 9.7). After its release from the nerve terminal ACh passes through an AChE barrier in the basement membrane. Thus enzyme inhibition increases the amount of ACh delivered to the postsynaptic membrane. Neostigmine and pyridostigmine are hydrolysed by AChE and, in the process, the enzyme is carbamylated. Edrophonium is not broken down by AChE, but the different mechanisms of action are probably unimportant in clinical practice. They do not explain the very rapid action of edrophonium.

Anticholinesterases have presynaptic effects which, like suxamethonium, may result in fasciculations and can be prevented with small doses of NDAs. They also have direct effects at the end plate, but usually only at high concentrations. The anticholinesterases have a ceiling effect[40] *in vitro* and this may explain why intense neuromuscular block cannot be reversed. Their anticurare effect may be augmented by atropine and adrenaline.

Anticholinesterases have been suspected of causing neuromuscular block but very large doses are required. Increased fade in response to tetanic stimulation has been observed after administration of neostigmine, but not edrophonium, when used to reverse neuromuscular block. Clinically, this effect appears to be unimportant.

Dose–response

Dose–response curves have been constructed for each of the anticholinesterases in reversing various levels of block either during a continuous infusion or after a bolus dose of several relaxants. Potency estimates among the reversal agents differ depending upon the circumstance and method of stimulation. For example, reversal of an intense (99 per cent) atracurium block produced curves, for both edrophonium and neostigmine, to the right of those obtained in reversing 90 per cent block. However, the curve for edrophonium was flatter for reversing the intense block[41] (Fig. 9.10). During reversal of 90 per cent pancuronium block, 12 times as much edrophonium as neostigmine is required to achieve 80 per cent T_1, but 25 times as much is required to achieve T_4 of 0.5.

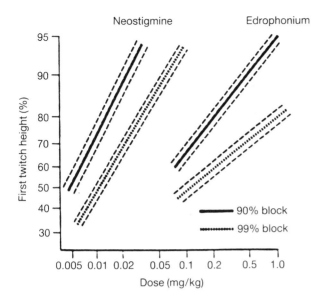

Fig. 9.10 Dose–response curves for neostigmine and edrophonium in antagonizing intense (99%) and modest (90%) atracurium-induced block. (Reproduced, with permission, from Donati F, Smith CE, Bevan DR. Dose-response relationships for edrophonium and neostigmine as antagonists of moderate and profound atracurium blockade. *Anesthesia and Analgesia* **68**: 13–19. © International Anesthesia Research Society 1989.)

Onset–duration

The durations of action of edrophonium, neostigmine, and pyridostigmine are similar. The effect depends upon both the spontaneous rate of recovery from the NDA and the acceleration produced by the reversal agent. Recurarization will only occur if the duration of the latter is much less than that of the former. However, the onset times are very different: the peak effect of edrophonium occurs at 1–2 minutes, neostigmine at 7–11 minutes, and pyridostigmine at greater than 16 minutes. The rapid effect of edrophonium allows estimation of recovery and the need for additional reversal agent very soon after the end of anaesthesia. Conversely, the long delay of pyridostigmine is too slow for clinical purposes.

Muscarinic effects

The bradycardic action of anticholinesterases can be attenuated with an anticholinergic agent such as atropine or glycopyrronium which is usually mixed with the anticholinesterase. The time course of the bradycardia is similar to the neuromuscular effect – rapid for edrophonium, slower for neostigmine and pyridostigmine. Edrophonium requires less atropine than an equivalent dose of neostigmine to avoid bradycardia.[42] Atropine, 0.6 mg, should be added to each 25 mg edrophonium and each 1.25 mg neostigmine. Because of the slow onset of glycopyrronium, atropine is preferred with edrophonium. Similarly, to

avoid an initial tachycardia, glycopyrronium may be indicated with neostigmine.

Anticholinesterases may produce salivation and increase bowel activity. Whilst anticholinergic drugs may inhibit the former, increased peristalsis may persist and be of particular concern to the day-care patient. Some studies have suggested an increase in bowel anastomotic leakage after the use of anticholinesterases but this may depend upon surgical practice and is not a reason to avoid reversal of neuromuscular blockade.

Anticholinesterases have been used effectively and in low dosage to reverse the phase II block produced by suxamethonium. However, in patients with atypical plasma cholinesterase the effect of anticholinesterases is unpredictable and they should be avoided.

Factors affecting reversal

Intensity of block

Following a standard dosage of reversal agent the time to recovery to any specific point, e.g. TOF 0.7, is dependent upon the intensity of block when the reversal agent is given[41] (Fig. 9.10). Intense blocks are reversed better with neostigmine than with edrophonium. There is a general consensus that in the presence of complete block, reversal should not be attempted.

Choice of relaxant

Recovery from neuromuscular block is dependent upon spontaneous recovery and its acceleration by an anticholinesterase. Thus, the intermediate relaxants appear to be 'easier to reverse' because a specified point of recovery is achieved earlier (Table 9.8). The use of long-acting NDAs is associated with residual curarization in the post-anaesthetic recovery room[43] which probably reflects the difficulty of detecting clinically small degrees of block. After atracurium and vecuronium recovery is so rapid that the period at risk is much shorter. Recovery of neuromuscular function after reversal of a block produced by an infusion of NDA is

Table 9.8 Approximate doses of neostigmine and edrophonium required to produce 80 per cent (ED$_{80}$) and 50 per cent (ED$_{50}$) recovery of T$_1$ 10 minutes after reversal of 90 per cent neuromuscular block

	Tubocurarine	Pancuronium	Atracurium	Vecuronium
Neostigmine (mg/kg)				
ED$_{50}$	0.017	0.013	0.010	0.010
ED$_{80}$	0.045	0.045	0.022	0.024
Edrophonium (mg/kg)				
ED$_{50}$	0.270	0.170	0.110	0.180
ED$_{80}$	0.880	0.680	0.440	0.460

slower than after bolus injections and neostigmine is more effective than edrophonium in this situation.

Dose of reversal agent

Reversal of neuromuscular block is dose-dependent (Table 9.8) and the ceiling effect seen *in vitro* does not appear to occur in clinical practice.

Age

Reversal of neuromuscular block appears to be easier in small infants[44] and more difficult, except after atracurium, in the elderly[45] than in young adults. These responses probably reflect the rates of spontaneous recovery in these age groups.

Acid–base balance

There is a suspicion that neuromuscular block is more difficult to reverse in the cachectic patient and that this is due to a more pronounced block in the presence of acidosis. However, recent studies have failed to demonstrate such a relation.

Drug interactions

Although there are reports of the potentiation of NDAs with several drugs, e.g. antibiotics, anaesthetic agents, anticonvulsants, anticancer drugs, etc., there are few reports of impaired reversal of block. Recovery of function is slower when the block is reversed during the course of enflurane anaesthesia but if the anaesthetic administration is terminated at the time of anticholinesterase administration, recovery is impaired only slightly. It seems unlikely that drug interactions lead to serious clinical sequelae as long as NDAs and their reversal agents are administered with appropriate monitoring.

Clinical recommendations

Reversal regimens should be based upon a consideration of the factors which modify reversal (Table 9.9). Only in exceptional circumstances should reversal be omitted after the use of NDAs. Full return of function must be

Table 9.9 Recommended doses of neostigmine or edrophonium according to response to ulnar nerve TOF stimulation

TOF visible twitches	Fade	Agent	Dose (mg/kg)
None	Avoid reversal until some evoked response		
≤ 2	++++	Neostigmine	0.07
3–4	+++	Neostigmine	0.04
4	++	Edrophonium	0.5
4	+/−	Edrophonium	0.25

demonstrated. Unfortunately, small degrees of block are impossible to detect clinically. However, the recent introduction of mivacurium which has a very rapid rate of recovery may offer the possibility of avoiding reversal. This may be of considerable benefit to the patient undergoing outpatient surgery, in whom the alimentary symptoms of reversal may be troublesome. For all other agents reversal should be attempted and confirmed.

Monitoring

The first nerve stimulators were introduced into clinical anaesthesia to help in the diagnosis of infrequent suxamethonium prolonged apnoea. However, it has become apparent that close neuromuscular monitoring has other advantages and should be applied to all patients receiving muscle relaxants. This recommendation is based first on the variable individual response to muscle relaxants. For example, after *d*-tubocurarine, 0.1 mg/kg, blockade has been found to vary between 0 and 100 per cent.[46] The same phenomenon has been observed with atracurium, and is probably true of all muscle relaxants. Thus, there is a need to assess the response in each individual patient. The second compelling reason to use nerve stimulators to monitor the effect of relaxants is their narrow 'therapeutic window'. Neuromuscular blockade occurs over a relatively narrow range of receptor occupancy. Animal experiments suggest that detectable blockade does not occur unless 75–80 per cent of receptors are occupied and paralysis is complete at 90–95 per cent receptor occupancy.[16] Although these numbers may change with species and from muscle to muscle within the same species,[16] the gap between too little and too much paralysis might be small. Furthermore, adequate muscle relaxation corresponds to a still narrow range, i.e. between 75 and 95 per cent twitch depression, which corresponds to an even narrower receptor occupancy range, perhaps 85–90 per cent.

In addition, neuromuscular monitoring is performed for different purposes depending upon the patient, the surgical procedure and especially the time during the procedure. Requirements for relaxation are different during onset and maintenance of paralysis and the situation is different during recovery.

Physiology of nerve stimulation

Nerve membranes are electrically excitable, and an electrical current supplied in the vicinity of the nerve can generate an action potential which propagates to the synapse, and induces the release of acetylcholine. If neurotransmission is not interrupted, the process leads to depolarization of the end plate, propagation of an action potential in muscle and contraction of the muscle fibre. If there is a sufficient concentration of neuromuscular relaxant at the end plate, acetylcholine release does not result in sufficient depolarization for the subsequent events in muscle to occur. The process is all-or-none at each end plate, i.e. each muscle fibre either contracts fully or not at all. Partial paralysis is the result of the sum of all-or-none responses of individual fibres, and not on the sum of individual partial responses in each fibre. An estimate of the degree of paralysis can be made by either measuring the force of contraction in the muscle tested, or its electrical activity (the EMG).

A nerve responds to the density of current passing through it. In any given individual with a fixed arrangement of electrodes, the current density is proportional to the current delivered. It also depends on the distance between the electrodes, the electrical properties of the tissue near the electrodes, and the distance from the electrodes to the nerve. The voltage across the electrodes is the current delivered multiplied by the electrical impedance. This impedance depends on skin thickness, shape and size of the electrodes, sweat production, etc. Thus, it is preferable to use stimulators which deliver a constant current rather than a constant voltage because of the possibly changing skin impedance with time.

To generate an action potential in a nerve axon, the potential difference across the membrane must reach a certain threshold, which is a value more positive than the normal negative resting value. If current is delivered to a whole nerve trunk, and if its intensity is too low for threshold to be reached in any of the fibres, no action potential will be propagated. As current is increased, certain nerve fibres reach threshold and propagate an action potential. If the stimulating current is increased further, all nerve fibres are sufficiently depolarized to initiate an action potential. Increasing the stimulus intensity would not further increase the number of action potentials and supramaximal current is then said to have been applied. In the absence of neuromuscular relaxants, all nerve action potentials reach the synapse, and trigger the release of enough acetylcholine to initiate muscle contraction. Thus, in this setting, the shape of the intensity of the muscle response as a function of current applied to the nerve is sigmoid, and little, if any, increase in response occurs if current is increased beyond the supramaximal value (Fig. 9.11).[47]

Once an action potential has been generated, the nerve is insensitive to further stimulation for a certain interval, the refractory period, which is usually 1–2 ms in duration. If the duration of the electrical stimulus is greater than the refractory period, another action potential may be triggered

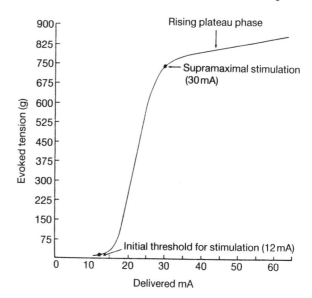

Fig. 9.11 Representative milliamp-evoked tension response curve. (Reproduced, with permission, from Kopman and Lawson, 1984.[47])

and neuromuscular monitoring may be biased. If, however, the stimulus duration is too short, there might be insufficient time for the nerve membrane to reach threshold. This can be compensated for by a larger intensity stimulus, but the voltage required may be very high. This phenomenon occurs with durations of less than 0.1–0.2 ms. Most modern stimulators have adopted a stimulus duration which is a compromise between these two extremes. A stimulus duration of 0.2 ms is a feature of most stimulators, although 0.1–0.3 ms is the usual range found.

Stimulation patterns

Many modes of stimulation are available on stimulators, to meet various requirements depending on the clinical circumstances, and relying chiefly on the fade which occurs when relatively high frequency stimulation is applied during non-depolarizing blockade. At frequencies greater than 0.1 Hz (one stimulus every 10 seconds), the height of the second and subsequent twitches is decreased compared with the first. Using electromyography, the decrement is found to decrease with increasing frequency up to 2 Hz, and then fade stays constant in the range 2–50 Hz. When measurement of force is used, the same pattern is found except at frequencies greater than 7–10 Hz, when the muscle has insufficient time to relax between stimulations. A tetanic contraction, or fusion of individual twitches, occurs when the frequency is greater than 20–30 Hz. Thus, the peak tetanic tension is the sum of the responses to the first few stimuli, and fade represents the decrease in response as stimulation is maintained. As a

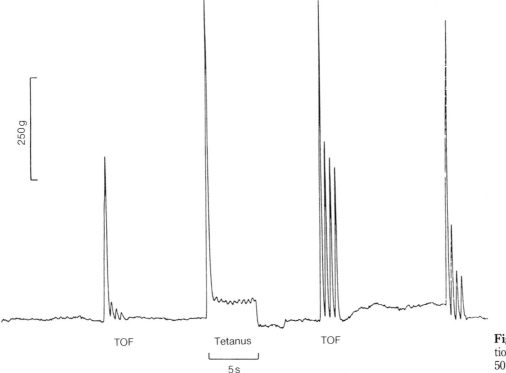

TOF   Tetanus   TOF

5 s

Fig. 9.12 Post-tetanic facilitation of TOF by tetanic stimulus, 50 Hz × 5 seconds.

consequence, tetanic fade measured by mechanomyography (force) is less than that by electromyography.

If low frequency stimulation is used after tetanic stimulation is applied, the response will be greater than it was before the tetanic response (Fig. 9.12). The magnitude of this 'post-tetanic potentiation' depends on the frequency and on the duration of the tetanus. Fade is almost certainly explained by a decrease in acetylcholine release caused by blockade of the presynaptic receptors. The origin of post-tetanic potentiation is less clear, but might be due to displacement of relaxant molecules by heavy bombardment by acetylcholine molecules during the tetanus.

Single twitch

If the nerve is stimulated by single electrical impulses at intervals of greater than 10 seconds, 'single twitch' stimulation is said to have been applied. Unless 100 per cent blockade is present, the magnitude of the response cannot be interpreted unless it is compared with a pre-relaxant, control value. Although this requirement limits the clinical applicability of single twitch stimulation, this mode is used frequently in research and the degree of single twitch depression is a useful yardstick by which to assess the potency of drugs or other modes of stimulation. For example, the ED_{95} of a drug, defined as the dose producing a mean of 95 per cent single twitch depression, and the disappearance of a fourth response after TOF stimulation corresponds to 70–75 per cent single twitch depression.

Tetanus

When stimulating frequencies are greater than 15–20 Hz, fusion of the individual twitch responses occurs and results in a prolonged contraction. During non-depolarizing blockade, this tetanic response is not sustained, but displays fade, and the degree of fade is directly related to the degree of single twitch depression. This implies that, contrary to single twitch stimulation, a reasonably accurate assessment of the degree of neuromuscular blockade can be made with tetanic stimulation without a previous, pre-relaxant control. As a test, tetanic fade is more sensitive than single twitch stimulation because some degree of fade is still present when twitch height recovers to 100 per cent of its pre-relaxant value. Most stimulators deliver tetanic stimulations at frequencies of either 50 or 100 Hz. Fade with frequencies of 100 Hz or greater may occur in the absence of muscle relaxants. For this reason, 50-Hz frequency is most commonly used. Traditionally, a 5-second duration has been recommended for tetanic stimulation, but there are few data for such a recommendation. The major drawback of tetanic stimulation is the presence of post-tetanic facilitation, which implies that after a tetanus has been applied, the response to further stimulation becomes greater than it would have been without such an intervention. The duration of post-tetanic stimulation depends on the frequency and the duration of stimulation. For a 50-Hz, 5-second tetanus, post-tetanic potentiation persists for at least 1–2 minutes.

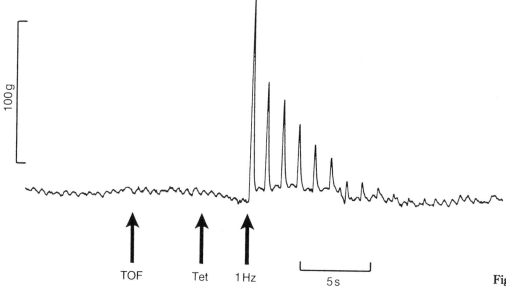

TOF   Tet   1 Hz   5 s

Fig. 9.13 Post-tetanic count.

Train-of-four (TOF)

The most popular mode of stimulation for the clinical monitoring of neuromuscular function consists of four pulses delivered every 0.5 seconds (frequency of 2 Hz). With such a small number of stimuli delivered, there is no post-tetanic facilitation, and the neuromuscular junction recovers rapidly, within 10–12 seconds. After this interval, TOF stimulation can be reapplied. The frequency is low enough so that the muscle has time to relax completely between each stimulus. In addition, the human senses have enough time to estimate each response separately. Fortunately, the degree of fade at 2 Hz is approximately the same as at 50 Hz,[48] and the sensitivities of TOF and 50-Hz tetanus are comparable. In addition, there is a direct relationship between first twitch height and TOF ratio, defined as the height of the fourth response divided by the height of the first. Thus, TOF monitoring can be interpreted without control, pre-relaxant values.

Post-tetanic count (PTC)

During intense neuromuscular blockade, the response to single twitch, TOF and tetanic stimulations are all zero. Thus, it is impossible to estimate the degree of blockade, or to predict for how long blockade will persist. The post-tetanic count (PTC) was designed to be used in this circumstance, and it applies the post-tetanic facilitation principle. It consists of a 5-second, 50-Hz tetanus followed, 3 seconds later, by at least 20 stimulations delivered at 1 Hz.[49] During intense blockade, there is no detectable response to the tetanic stimulation. However, the 1-Hz twitches become facilitated, and may be detectable (Fig. 9.13). Moreover, the number of detectable twitches is inversely related to the degree of blockade. If 12–15 post-tetanic twitches are

observed, that is if the PTC is 12–15, reappearance of a response to TOF stimulation is imminent.[49] A PTC of one indicates deeper blockade, and the time to reappearance of a response to TOF stimulation depends on the pharmacology of the drug. For example, this interval is 30–45 minutes for pancuronium and 10–15 minutes for atracurium or vecuronium. Intermediate values represent intermediate intervals. Because tetanic stimulation may facilitate further tetanic responses, it is recommended not to apply PTC more often than every 5 minutes.

Double-burst stimulation (DBS)

The visual and tactile assessment of residual blockade is often difficult (see below). Trained observers are often unable to detect fade when the actual TOF ratio is as low as 0.2–0.3. To improve the detection rate, a new mode of stimulation, which consists of two short tetani separated by an interval long enough to allow the muscle to relax, has been proposed. Thus, double-burst stimulation (DBS) is made up of three stimuli 20 ms apart (i.e. delivered at a frequency of 50 Hz), followed by an identical burst of three stimuli delivered at the same frequency.[50,51] Each burst leads to fusion of the contractile responses, so that two contractions are observed after DBS. During partial non-depolarizing blockade, the response to the second burst is less than that of the first, so fade is present. The magnitude of this fade is similar to TOF fade. In other words, the height of the second burst divided by the height of the first is similar to the TOF ratio. However, human senses can detect DBS fade better. Most studies suggest that DBS fade is usually detected when TOF ratio is less than 0.5–0.6 which is an improvement over TOF fade (Fig. 9.14).[50]

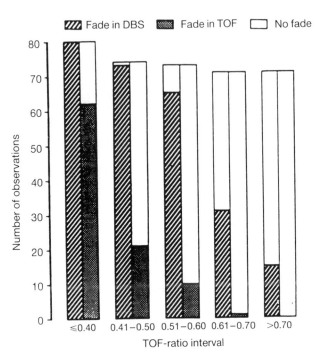

Fade in DBS   Fade in TOF   No fade

Fig. 9.14 Manually detectable fade in response to DBS and TOF stimulation. (Reproduced, with permission, from Drenck *et al.*, 1989.[50])

Measuring the response

Choice of muscle

The response to neuromuscular relaxants is different in different muscles, both in terms of onset and intensity of blockade. Thus, it would appear logical to choose a clinically relevant muscle as the site of monitoring. For instance, laryngeal muscles are important for tracheal intubation, abdominal relaxation is desirable during many surgical procedures, and the function of respiratory and upper airway muscles must be assessed during recovery. In practice, however, it is inconvenient to switch sites and more importantly, most of the above muscles are not clinically accessible. Therefore, it is better to choose one site of stimulation and recording which is convenient and accessible, and be aware of the possible differences between the monitored response and the response of other important and clinically relevant muscle groups.

Choosing the ulnar nerve and assessing the response of the adductor pollicis has many advantages. One of the hands is usually accessible during most surgical procedures, the response can be measured relatively easily with various methods, and a lot of information is available regarding the differences between adductor pollicis and other key muscles. The ulnar nerve may be stimulated by applying surface electrodes to the overlying skin either at the wrist or at the elbow. For this purpose, silver–silver chloride ECG

electrodes are adequate in most, if not all, circumstances. Although effective, needle electrodes are not recommended because of the possibility of bleeding, infection, nerve damage and accidental dislodgement. Stainless steel, ball-shaped electrodes attached rigidly to a nerve stimulator are less reliable than silver–silver chloride surface electrodes. Usually, a lower threshold for supramaximal stimulation is achieved if the negative electrode (cathode) is placed closer to the nerve. Thus, it is best to apply the negative electrode over the part of the nerve which is most superficial.[52] For the ulnar nerve, these locations are at the anteromedial border of the wrist or the olecranon fossa at the elbow. The positive electrode can be positioned along the path of the nerve, approximately 7–10 cm away from the cathode.

Stimulation of the ulnar nerve at the wrist results in contraction of the adductor pollicis, the interossei muscles, and the flexors and adductors of the little finger which make up the hypothenar eminence. The first dorsal interosseous and the adductor pollicis have similar sensitivities to non-depolarizing relaxants, but the muscles of the hypothenar eminence are somewhat more resistant than the adductor pollicis. Thus, the degree of paralysis is usually less when observing the movement of the little finger. Stimulating the ulnar nerve at the elbow also produces a contraction of the flexor carpi ulnaris, which flexes and adducts at the wrist, and of the medial half of the flexor digitorum profundus, which results in flexion of the fourth and fifth digits. No study has dealt with the comparative sensitivity of these muscles, but clinical experience suggests that they probably are more resistant than the adductor pollicis to the effect of neuromuscular relaxants. Because of the varying sensitivities of the hand and forearm muscles and because most data on relaxants have been obtained with the adductor pollicis, it is suggested to observe the response of the thumb.

It is sometimes necessary to use other locations than the hand for the purpose of neuromuscular monitoring. The peroneal nerve may be stimulated at the head of the fibula, producing dorsiflexion of the foot. Plantar flexion may be produced by stimulation of the posterior tibial nerve, behind the medial malleolus.[52] Current evidence suggests that the sensitivity of the foot muscles is approximately the same as that of the hand muscles. During many surgical procedures, the facial nerve is also accessible. Observation of the contraction of the orbicularis oculi muscle is then possible. It is, however, important to realize that non-depolarizing neuromuscular relaxants produce less blockade at the orbicularis oculi than the adductor pollicis. Paradoxically, onset of blockade is faster at the orbicularis oculi.[24]

Visual and tactile means

Once a site is chosen for stimulation, one must record the response. The simplest and cheapest way to do this is to look and feel for it. With ulnar nerve stimulation, this is best accomplished by holding the patient's thumb with one's finger and feel for the response. Twitch count can be accomplished reliably with this method, i.e. the number of observed twitches following TOF stimulation during surgery corresponds to the measured number of responses. However, inaccuracies in the visual and tactile assessment are frequent during recovery. Train-of-four fade is difficult to detect, even by experienced investigators, when a considerable degree of residual blockade is still present, corresponding to a TOF ratio as small as 0.3 (Fig. 9.14). When DBS is used, detection of fade is possible when the TOF ratio is 0.5–0.6 or less. Evidence of respiratory impairment can be found even at TOF ratios exceeding the usually recognized value of 0.7. Recent studies have indicated residual weakness in some patients at values in the range 0.85–0.9.[41,51]

Force measurement

An obvious solution to the lack of accuracy of tactile and visual assessment is to measure contractile force with a force transducer. Such a device can provide a pre-relaxant control and an accurate evaluation of TOF ratio. Unfortunately, such a system is bulky, requires some skill to set up, is expensive, and can usually be adapted to only one muscle. For these reasons, force transducers are not commonly used in clinical practice. If such a system is used, however, there is no need for DBS, because the DBS ratio is numerically equivalent to the TOF ratio, and DBS was designed only to improve visual and tactile detection of residual blockade.

Electromyography (EMG)

The electrical activity in the muscle can also be quantified by recording the electromyographic (EMG) signal. The EMG is the electrical activity which spreads within muscle cells and precedes mechanical contraction. When recorded from hand muscles, the signal is typically 5–20 mV, and its duration is 10–15 ms. Most EMG units process the signal to derive peak amplitude, or peak-to-peak amplitude or, more commonly, the area under the curve. The differences between these measurements are not clinically important. Provided that the EMG and force are recorded from the same muscle, a good agreement is usually found between force and EMG measurements. As with force measurement, EMG recording of DBS is unnecessary. In addition, EMG recordings are not necessary after tetanic stimulation, because EMG fade is unchanged in the range 2–50 Hz.[48] Most EMG units are less bulky and easier to set up than force transducers, but

cost remains a problem. An advantage of the EMG technique is the possibility of applying it to various muscles. Ideally, one of the three recording electrodes should be placed on the belly of the muscle, a second electrode over the insertion, and the indifferent electrode should be positioned near the other two. In the case of the adductor pollicis, two electrodes should be placed on the palmar surface of the hypothenar eminence and near the metacarpophalangeal joint of the thumb, respectively. Unfortunately, these locations are subject to movement artefacts and a better recording may be obtained with dorsal interosseous recording, which is done by applying electrodes on the dorsal surface of the hand, one over the muscular portion of the web between the first two fingers, and the other over the metacarpophalangeal joint of the second finger. Recordings from the first dorsal interosseous are similar to those of the adductor pollicis. The hypothenar eminence is also a convenient location for EMG recordings, but the intensity of blockade is usually less than at the adductor pollicis.

Accelerography

Recently, relatively inexpensive devices which depend on the measurement of acceleration have been introduced. The principle of operation is Newton's law, which states that force is equal to mass times acceleration. Thus, assuming that mass remains constant, acceleration is proportional to force. The device consists of a small acceleration transducer, which must be taped to the thumb and connected to a control unit. The system is very sensitive to changes in the position of the hand and movement of the thumb must be unhindered. Good correlations were found between force and accelerographic measurements.[53]

Clinical uses

Neuromuscular monitoring is used in anaesthesia and in the intensive therapy unit to (1) determine the optimum time for tracheal intubation, (2) assess the intensity of blockade during a procedure, and (3) evaluate the degree of recovery for extubation.

Onset

Relaxation during tracheal intubation depends on the depth of anaesthesia and neuromuscular blockade in the muscles of the jaw, upper airway, larynx and respiratory system. Experimental data are limited to the masseter, vocal cord adductors and diaphragm, which all exhibit a faster onset

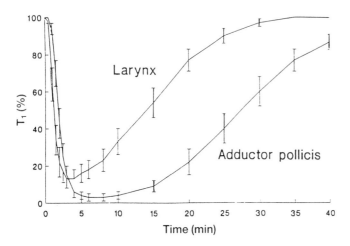

Fig. 9.15 First twitch height (T_1) for laryngeal adductors and adductor pollicis after vecuronium 0.07 mg/kg. (Reproduced, with permission, from Donati et al., 1991.[24])

of blockade than peripheral muscle, both for depolarizing and non-depolarizing drugs (Fig. 9.15).[11,24,54] For example, with subparalysing doses of vecuronium or pancuronium maximal blockade is reached within 3–4 minutes at the diaphragm, larynx or masseter, compared with 5–7 minutes at the adductor pollicis. This difference in onset times is probably related to the differences in muscle blood flow and distance to the central circulation between centrally located and peripheral muscle. In addition, there is a wide range of sensitivity among the muscles which must be relaxed for adequate intubating conditions. Although the masseter requires approximately the same dose of non-depolarizing or depolarizing agent for identical degrees of blockade, the diaphragm needs considerably more. With non-depolarizing relaxants, the adductor muscles of the larynx appear even more resistant than those of the diaphragm.[54] For example, 90 per cent blockade is achieved at the adductor pollicis with a mean vecuronium dose of 0.04 mg/kg, while the same degree of blockade requires 0.07 mg/kg for the adductor muscles of the larynx.[54] Similar results were obtained with other non-depolarizing relaxants, such as atracurium, mivacurium and rocuronium. Therefore, monitoring the adductor pollicis during onset of blockade is of limited value. In practice, three situations might be encountered: if the dose is insufficient to block the adductor pollicis, then one must assume that it is also insufficient for vocal cord paralysis and more relaxant is needed. However, this will be known, for non-depolarizing drugs, only after 5–7 minutes. If, however, the dose given blocks the adductor pollicis but is insufficient for laryngeal blockade, intubating conditions are likely to be less than adequate. Finally, if the dose is large enough for paralysis of vocal cord muscles and adductor pollicis, intubating conditions will be adequate, but the clinician might miss the best time for intubation because laryngeal blockade occurs

before the adductor pollicis is paralysed. The problem can be overcome by (1) always giving an 'adequate' dose, i.e. that which will paralyse all muscle in all patients and probably is equal to at least $3 \times ED_{95}$ at the adductor pollicis, and wait 3 minutes; or (2) expect inadequate intubating conditions in some patients; or (3) use facial nerve–orbicularis oculi monitoring, because the intensity and time course of orbicularis oculi blockade, at least for atracurium and vecuronium, is close to that of the diaphragm and the vocal cords.[24,54]

Surgical relaxation

The management of relaxation during surgery should follow two principles: (1) an adequate degree of relaxation should be provided, and (2) reversibility should be possible at the end of the procedure. As the effectiveness of the anticholinesterase agents edrophonium, neostigmine and pyridostigmine decrease dramatically if blockade is greater than 85–90 per cent,[51] it has been recommended in the past never to paralyse patients completely. The relaxant dosage was adjusted to leave at least one twitch visible, as assessed at the adductor pollicis. With shorter-acting and more predictable drugs, complete paralysis is possible and sometimes desirable, because the diaphragm[21] and abdominal muscles may not be adequately relaxed when some residual activity is seen at the adductor pollicis. The degree of blockade can be evaluated by the PTC method applied to the adductor pollicis,[49] or by monitoring the orbicularis oculi, the response of which follows the diaphragm closely.[24] However, proper knowledge of the pharmacology of the relaxant drug used is important to ensure that sufficient recovery occurs before reversal is attempted.

Monitoring recovery

Ulnar nerve–adductor pollicis monitoring should be performed before administration of the reversal agent. At least one, and preferably two palpable twitches should be present in response to TOF stimulation. If blockade is more intense, mechanical ventilation should be continued until recovery is observed before giving a reversal agent.[51] If four twitches are seen and felt by visual and manual examination, it should not be assumed that complete neuromuscular recovery is present[50,51] (Fig. 9.14). Reversal agents may be omitted only if confirmatory evidence that complete recovery has occurred is present. The absence of fade with DBS must be interpreted with caution. Full neuromuscular recovery may normally be assumed to be present if head lift, hand grip, tongue protrusion and eye opening are sustained.[39] Similarly, it is unnecessary to administer reversal agents if recovery has been present for a time which exceeds the normal recovery time for the drug. If, for example, four twitches have been felt to be equal on manual evaluation for the last 45 minutes of surgery in a patient receiving atracurium, full recovery may be assumed to be present. The same assumption could not be made if pancuronium had been used.

Neuromuscular monitoring should also be used in the same manner to assess the effect of the reversal agent. Although fade in response to TOF stimulation should be interpreted as inadequate reversal, failure to detect fade with one's senses should not be taken as proof of adequate recovery. The same can be said of DBS, although the degree of confidence could be greater with this mode of stimulation. These simple tests should be completed by close observation of the patient and use of clinical tests.

Conclusion

The introduction of neuromuscular blocking drugs into clinical anaesthesia practice allowed the development of the concept of using different drugs to produce different effects. All are potent and, potentially, dangerous. However, by administering the appropriate drug in the appropriate situation in the correct dose, by monitoring the intensity of relaxation to avoid complete block, and to confirm adequate return of function, the dangers can be reduced. Nevertheless, one should remember that the commonest cause of anaesthetic-related mortality is ventilatory failure and that muscle relaxants are important contributing agents.

REFERENCES

1. Kimura I, Tsuneki H, Kondoh T, Kimura M. Enhancing effect by nicotinic acetylcholine receptor channel blockers, including beta-eudesmol, on succinylcholine-induced inhibition of twitch tension and intracellular Ca^{++} in mouse diaphragm muscle. *Journal of Pharmacology and Experimental Therapeutics* 1991; **256**: 24–32.

2. Gronert GA, Theye RA. Pathophysiology of hyperkalemia induced by succinylcholine. *Anesthesiology* 1975; **43**: 89–99.

3. Hartman GS, Flamengo SA, Riker WF. Succinylcholine: mechanism of fasciculations and their prevention by d-tubocurarine or diphenylhydantoin. *Anesthesiology* 1986; **65**: 405–13.

4. Szalados JE, Donati F, Bevan DR. Effect of d-tubocurarine pretreatment on succinylcholine twitch augmentation and neuromuscular blockade. *Anesthesia and Analgesia* 1990; **71**: 55–9.

5. Plumley MP, Bevan JC, Saddler JM, Donati F, Bevan DR. Dose related effects of succinylcholine on adductor pollicis and masseter muscles in children. *Canadian Journal of Anaesthesia* 1990; **37**: 15–20.

6. Lee C. Train-of-four fade and edrophonium antagonism of neuromuscular block by succinylcholine in man. *Anesthesia and Analgesia* 1976; **55**: 663–7.

7. Cook DR, Wingard LB, Taylor FH. Pharmacokinetics of succinylcholine in infants, children, and adults. *Clinical Pharmacology and Therapeutics* 1976; **20**: 493–8.

8. Whittaker M. Plasma cholinesterase variants and the anaesthetist. *Anaesthesia* 1980; **35**: 174–97.

9. Vanlinthout LEH, Van Egmond J, De Boo T, Lerou JGC, Wevers RA, Booij LHDJ. Factors affecting magnitude and time course of neuromuscular block produced by suxamethonium. *British Journal of Anaesthesia* 1992; **69**: 29–35.

10. Szalados JE, Donati F, Bevan DR. Nitrous oxide potentiates succinylcholine neuromuscular blockade in humans. *Anesthesia and Analgesia* 1991; **72**: 18–21.

11. Meistelman C, Plaud B, Donati F. Neuromuscular effects of succinylcholine on the vocal cords and adductor pollicis muscles. *Anesthesia and Analgesia* 1991; **73**: 278–82.

12. Mirakhur RK. Edrophonium and plasma cholinesterase activity. *Canadian Anaesthetists' Society Journal* 1986; **33**: 588–90.

13. Ono K, Manabe N, Ohta Y, Morita K, Kosaka F. Influence of suxamethonium on the action of subsequently administered vecuronium or pancuronium. *British Journal of Anaesthesia* 1989; **62**: 324–6.

14. Donati F, Bevan DR. Suxamethonium – current status. In: Norman J, ed. *Clinics in Anaesthesiology Volume 3/Number 2 Neuromuscular Blockade*. London: WB Saunders Company, 1985: 371–84.

15. Pace NL. Prevention of succinylcholine myalgias: a meta-analysis. *Anesthesia and Analgesia* 1990; **70**: 477–83.

16. Paton WD, Waud DR. The margin of safety of neuromuscular transmission. *Journal of Physiology (London)* 1967; **191**: 59–90.

17. Bowman WC. Prejunctional and postjunctional cholinoceptors at the neuromuscular junction. *Anesthesia and Analgesia* 1980; **59**: 935–43.

18. Martyn JAJ, White DA, Gronert GA, Jaffe RS, Ward JM. Up-and-down regulation of skeletal muscle acetylcholine receptors. *Anesthesiology* 1992; **76**: 822–43.

19. Robbins R, Donati F, Bevan DR, Bevan JC. Differential effects of myoneural blocking drugs on neuromuscular transmission. *British Journal of Anaesthesia* 1984; **56**: 1095–9.

20. Smith CE, Donati F, Bevan DR. Cumulative dose-response with infusion: a technique to determine neuromuscular blocking potency of atracurium and vecuronium. *Clinical Pharmacology and Therapeutics* 1988; **44**: 56–64.

21. Donati F, Antzaka C, Bevan DR. Potency of pancuronium at the diaphragm and the adductor pollicis muscle in humans. *Anesthesiology* 1986; **65**: 1–5.

22. Sheiner LB, Stanski DR, Vozeh S, *et al.* Simultaneous modeling of pharmacokinetics and pharmacodynamics: application of

d-tubocurarine. *Clinical Pharmacology and Therapeutics* 1979; **25**: 358–71.

23. Bevan JC, Donati F, Bevan DR. Attempted acceleration of the onset of action of pancuronium: effects of divided doses in infants and children. *British Journal of Anaesthesia* 1985; **57**: 1204–8.

24. Donati F, Meistelman C, Plaud B. Vecuronium neuromuscular block at the diaphragm, the orbicularis oculi, and adductor pollicis muscles. *Anesthesiology* 1990; **73**: 870–5.

25. Bowman WC, Rodger IW, Houston J, Marshall RJ, McIndewar LI. Structure:action relationships among some desacetoxy analogs of pancuronium and vecuronium in the anesthetized cat. *Anesthesiology* 1988; **69**: 57–62.

26. Kopman AF. Pancuronium, gallamine, and d-tubocurarine compared: is speed of onset inversely related to drug potency? *Anesthesiology* 1989; **70**: 915–20.

27. Law Min JC, Bekavac I, Glavinovic MI, Donati F, Bevan DR. Iontophoretic study of speed of action of various muscle relaxants. *Anesthesiology* 1992; **77**: 351–6.

28. Savarese JJ, Basta SJ, Ali HH, Sunder N, Moss J. Neuromuscular and cardiovascular effects of BW 33A (atracurium) in patients under halothane anesthesia. *Anesthesiology* 1982; **57**: A262.

29. Martineau RJ, St-Jean B, Kitts JB, *et al.* Cumulation and reversal with prolonged infusions of atracurium and vecuronium. *Canadian Journal of Anaesthesia* 1992; **39**: 670–6.

30. Savage DS, Sleigh T, Carlyle I. The emergence of ORG NC 45, 1-[(2 beta, 3 alpha, 5, 16 beta, 17 beta)-3, 17-bis(acetoxy)-2-(piperidinyl)-androstan-16-yl]-1-methyl-piperidinium bromide, from the pancuronium series. *British Journal of Anaesthesia* 1980; **52**: 3S–9S.

31. Segredo V, Caldwell JE, Matthay MA, Sharma ML, Gruenke LD, Miller RD. Persistent paralysis in critically ill patients after long-term administration of vecuronium. *New England Journal of Medicine* 1992; **327**: 524–8.

32. Griffith HG, Johnson GE. The use of curare in general anesthesia. *Anesthesiology* 1942; **3**: 418–20.

33. Gray TC, Halton J. A milestone in anaesthesia? (d-tubocurarine chloride). *Proceedings of the Royal Society of Medicine* 1946; **39**: 400–10.

34. Hughes R, Chapple DJ. Effect of non-depolarizing neuromuscular blocking agents on peripheral autonomic mechanisms in cats. *British Journal of Anaesthesia* 1976; **48**: 59–68.

35. Buckett WR, Hewitt CL, Savage DS. Pancuronium bromide and other steroidal neuromuscular blocking agents containing acetylcholine fragments. *Journal of Medicinal Chemistry* 1973; **16**: 1116–24.

36. Savarese JJ, Kitz R. Does clinical anesthesia need new neuromuscular blocking agents? Editorial. *Anesthesiology* 1975; **42**: 236–9.

37. Viby-Mogensen J, Jørgensen BC, Ørding H. Residual curarization in the recovery room. *Anesthesiology* 1979; **50**: 539–41.

38. Lunn JN, Hunter AR, Scott DB. Anaesthesia-related surgical mortality. *Anaesthesia* 1983; **38**: 1090–6.

39. Beecher HK, Todd DP. A study of the deaths associated with anesthesia and surgery. Based on a study of 599,548 anesthesias in ten institutions 1948–1952 inclusive. *Annals of Surgery* 1954; **140**: 2–34.

40. Bartkowski RR. Incomplete reversal of pancuronium neuromuscular blockade by neostigmine, pyridostigmine and edrophonium. *Anesthesia and Analgesia* 1987; **66**: 594–8.

41. Donati F, Smith CE, Bevan DR. Dose-response relationships for edrophonium and neostigmine as antagonists of moderate and profound atracurium blockade. *Anesthesia and Analgesia* 1989; **68**: 13–19.

42. Cronnelly R, Morris RB, Miller RD. Edrophonium: duration of action and atropine requirement in humans during halothane anesthesia. *Anesthesiology* 1982; **57**: 261–6.

43. Bevan DR, Smith CE, Donati F. Postoperative neuromuscular blockade: a comparison between atracurium, vecuronium and pancuronium. *Anesthesiology* 1988; **69**: 272–6.

44. Meakin G, Sweet PT, Bevan JC, Bevan DR. Neostigmine and edrophonium as antagonists of pancuronium in infants and children. *Anesthesiology* 1983; **59**: 316–21.

45. McCarthy GJ, Cooper R, Stanley JC, Mirakhur RK. Dose-response relationships for neostigmine antagonism-induced neuromuscular block in adults and the elderly. *British Journal of Anaesthesia* 1992; **69**: 281–3.

46. Katz RL. Neuromuscular effects of d-tubocurarine, edrophonium and neostigmine in man. *Anesthesiology* 1967; **28**: 327–36.

47. Kopman AF, Lawson D. Milliamperage requirements for supramaximal stimulation of the ulnar nerve with surface electrodes. *Anesthesiology* 1984; **61**: 83–5.

48. Lee C, Katz RL. Fade of neurally evoked compound electromyogram during neuromuscular block by d-tubocurarine. *Anesthesia and Analgesia* 1977; **56**: 271–5.

49. Viby-Mogensen J, Howardy-Hansen P, Chraemmer-Jorgensen B, Ording H, Engbaek J, Nielsen A. Posttetanic count (PTC): a new method of evaluating an intense nondepolarizing neuromuscular blockade. *Anesthesiology* 1981; **55**: 458–61.

50. Drenck NE, Ueda N, Olsen NV, *et al.* Manual evaluation of residual curarization using double burst stimulation: a comparison with train-of-four. *Anesthesiology* 1989; **70**: 578–81.

51. Bevan DR, Donati F, Kopman AF. Reversal of neuromuscular blockade. *Anesthesiology* 1992; **77**: 785–805.

52. Hudes E, Lee KC. Clinical use of peripheral nerve stimulators in anaesthesia. *Canadian Journal of Anaesthesia* 1987; **34**: 525–34.

53. Meretoja OA, Werner MU, Wirtavuori K, Luosto T. Comparison of thumb acceleration and thenar EMG in a pharmacodynamic study of alcuronium. *Acta Anaesthesiologica Scandinavica* 1989; **33**: 545–8.

54. Donati F, Meistelman C, Plaud B. Vecuronium neuromuscular blockade at the adductors muscles of the larynx and adductor pollicis. *Anesthesiology* 1991; **74**: 833–7.

RECOMMENDED READING

General

1. Katz RL, ed. *Muscle relaxants: basic and clinical aspects.* Orlando: Grune & Stratton, 1984.
2. Bowman WC, ed. *Pharmacology of neuromuscular function.* 2nd ed. London: Wright, 1990.
3. Bevan DR, Bevan JC, Donati F. *Muscle relaxants in clinical anaesthesia.* Chicago: Year Book Medical Publishers, 1988.

New drugs

Doxacurium

1. Koscielniak-Nielsen ZJ, Law-Min JC, Donati F, Bevan DR, Clement P, Wise R. Dose-response relations of doxacurium and its reversal with neostigmine in young adults and healthy elderly patients. *Anesthesia and Analgesia* 1992; **74**: 845–50.
2. Basta SJ, Savarese JJ, Ali HH, *et al.* Clinical pharmacology of doxacurium chloride: a new long-acting nondepolarizing relaxant. *Anesthesiology* 1988; **69**: 478–86.

Pipecuronium

1. Pittet J-F, Tassonyi E, Morel DR, Gemperle G, Richter M, Rouge J-C. Pipecuronium-induced neuromuscular blockade during nitrous oxide-fentanyl, isoflurane and halothane anesthesia in adults and children. *Anesthesiology* 1989; **71**: 210–13.

2. Larijiani GE, Bartkowski RR, Azad SS, *et al.* Clinical pharmacology of pipecuronium bromide. *Anesthesia and Analgesia* 1989; **68**: 734-9.

Mivacurium

1. Savarese JJ, Ali HH, Basta SJ, *et al.* The clinical neuromuscular pharmacology of mivacurium (BW 1090). *Anesthesiology* 1988; **68**: 723-32.
2. Goudsouzian NG, Alifirmoff JK, Eberly C, *et al.* Neuromuscular and cardiovascular effects of mivacurium in children. *Anesthesiology* 1989; **70**: 327-42.

Rocuronium

1. Cooper R, Mirakhur RK, Clarke RSJ, Boules Z. Comparison of intubating conditions after administration of ORG 9426 (rocuronium) and suxamethonium. *British Journal of Anaesthesia* 1992; **69**: 269–73.
2. Wierda JMKH, de Wit APM, Kuizenga K, Agoston S. Clinical observations on the neuromuscular blocking action of ORG 9426, a new steroidal agent. *British Journal of Anaesthesia* 1990; **64**: 521–3.

Reversal

1. Bevan DR, Donati F, Kopman AF. Reversal of neuromuscular blockade. *Anesthesiology* 1992; **77**: 785–805.

Chapter 10

Physiology of Nerve Conduction and Local Anaesthetic Drugs

J. B. Löfström and M. Bengtsson

Nerve conduction

The peripheral nerve contains, apart from different connective tissue barriers, nerve fibres of different sizes and with varying degrees of myelinization (Fig. 10.1). The thick, heavily myelinated motor or sensory fibres transmit signals to striated muscles or conduct signals from peripheral sensory receptors (*A fibres*, Aα fibres are related to motor function, proprioception and reflex activity, Aβ fibres are also related to motor function and in addition transmit touch and pressure sensations, Aγ fibres control muscle spindle tone, Aδ fibres transmit pain and temperature sensations as well as signals of tissue damage). The smallest myelinated fibres are preganglionic autonomic *B fibres*. The postganglionic autonomic fibres are unmyelinated *C fibres*. Similar fibres also transmit information regarding pain and temperature. The speed of conduction is very fast in heavily myelinated fibres (up to 100 m/s), less in thinner fibres and slow in C fibres (down to 0.5 m/s).

The surface membrane of the axon of the peripheral A and B fibres is wrapped in several layers of the Schwann cell containing myelin (sphingomyelin). This almost completely isolates the axon surface membrane (Fig. 10.2).

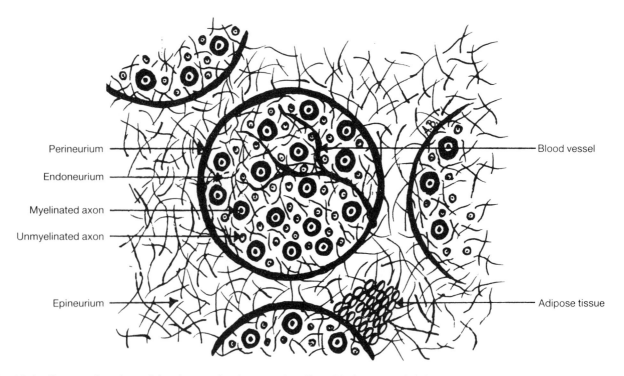

Fig. 10.1   Cross-section of a peripheral nerve showing nerve bundles with the connective tissue sheath barriers surrounding the axons. There are about twice as many unmyelinated fibres as myelinated fibres.

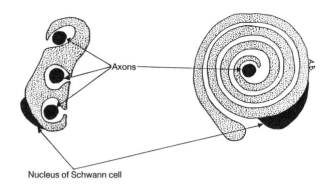

Fig. 10.2 Cross-sectional drawing of unmyelinated and myelinated nerve fibres.

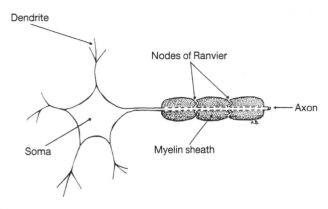

Fig. 10.3 Schematic drawing of a neuron with a medullated nerve fibre showing the microvillous interdigitations of adjacent Schwann cells at the nodes of Ranvier.

Unmyelinated nerve fibres are evaginated into the membrane and cytoplasm of a Schwann cell and conduct changes in potential along the whole axon membrane. There are, at regular intervals, open gaps between different Schwann cells, the nodes of Ranvier. At these gaps, the axon membrane is in contact with the interstitial fluid and its electrolytes (Fig. 10.3).

The ionic content of the interstitial fluid and the axon are different. Outside the axon membrane the sodium (Na^+) content is high (142 mmol/l) and inside low (14 mmol/l). Potassium (K^+), on the other hand, is low on the outside (4 mmol/l) and high on the inside (140 mmol/l).

The cell membrane is composed of a palisade of lipid molecules having a lipophilic end and a hydrophilic end. Large protein molecules are located in this palisade which contains the major ionic channels. A small leakage of sodium and potassium occurs through potassium–sodium 'leak' channels. Such a channel is far more permeable to potassium than to sodium.

The interior of the axon with its axoplasm is negatively charged (−90 mV), with respect to the outside (Fig. 10.4a). This resting potential is maintained by the sodium–potassium pump. More positive charges are pumped to the outside of the axon than to the inside (three Na^+ to the outside for every two K^+ to the inside). The main channel through the membrane is the voltage-gated sodium channel, closed at rest, which when activated, allows an influx of Na^+ into the axon. The influx of Na^+ results in a reversal of the membrane potential which results in depolarization. This may be displayed as an action potential (Fig. 10.4).

There are also potassium channels which when opened allow an outflow of K^+ and repolarization of the cell membrane. In addition there are Ca^{2+} channels which are very slow to be activated (slow channels) in contrast with the Na^+ channels (fast channels). Furthermore, if the concentration of Ca^{2+} in the interstitial fluid is low the Na^+ channels are activated by a small increase in the membrane

potential above the resting level. The nerve fibre becomes highly excitable and spontaneous discharge may occur.

The initial event in an action potential is a rapid influx of Na^+ through the opened activation gate. Within a few ten thousandths of a second after the activation gate opens, the inactivation gate closes, preventing further influx of Na^+ through the cell membrane (Fig. 10.4c).

When the resting membrane potential is caused to rise markedly, the sodium channels open, allowing an influx of Na^+. This further increases the membrane potential, stimulating more sodium channels to open. There is thus a positive feedback mechanism opening the sodium channels.

An action potential will not occur until an increase in resting potential of the order of 30 mV has been provoked by, for example, a mechanical or chemical disturbance of the cell membrane, or by electrical stimulation. In larger nerves a rise from the resting potential to −65 mV usually results in the explosive development of an action potential, an all or nothing event. This potential is said to be the threshold for stimulation. The membrane potential then begins to recover. The inactivation gate will not open until the membrane potential has almost returned to the resting level (refractory period).

Repolarization of the axon membrane starts as the voltage-gated potassium channels are slowly activated, increasing the exit of potassium through the membrane. Full recovery of the resting membrane potential is seen within a few 10^{-4} seconds (Fig. 10.4c). Depolarization and repolarization are related to movements of ions, though the total number of these moving is extremely low. Thus millions of action potentials can be provoked without a marked change in ionic content on the two sides of the cell membrane. With time, however, it becomes necessary to re-establish the sodium–potassium membrane concentration differences by the action of the Na^+–K^+ pump. Sodium is to be returned to the outside and potassium to the inside of the axon membrane. This requires energy, provided by

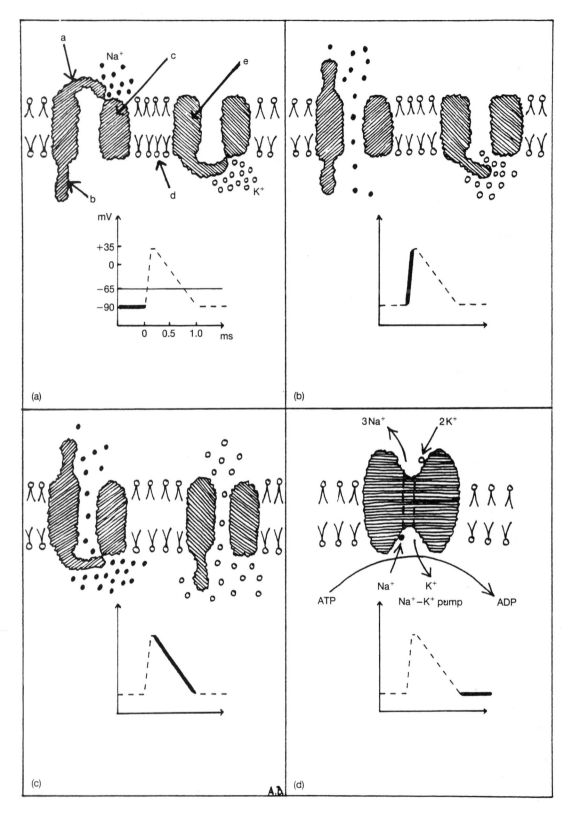

Fig. 10.4   Characteristics of the voltage-gated sodium and potassium channels, the Na^+-K^+ pump and the nerve membrane action potential. (a) Resting (closed) state: a, activation gate; b, inactivation gate; c, sodium channel protein; d, phospholipid bilayer; e, potassium channel protein. The horizontal line denotes threshold potential (-65 mV). (b) Sodium channel in the activated, open state (depolarization). Potassium channel slowly opens, allowing increased potassium diffusion outward through the channel. (c) Sodium channel closes (inactivated state), potassium channel in the open state with potassium ions flowing down the electrochemical gradient out of the cell (repolarization). (d) Upon completion of the action potential, sodium and potassium ions are actively transported back to the cell exterior and interior, respectively, by the electrogenic sodium–potassium pump (three Na^+ ions to the outside for two K^+ ions to the inside).

adenosine triphosphate (Fig. 10.4d). The activity of this pump is greatly enhanced by the accumulation of sodium within the cell membrane.

An action potential is propagated to the adjacent parts of a nerve and thus spreads in both directions away from the stimulus. In the case of myelinated fibres, the cell membrane can only be depolarized at the nodes of Ranvier where the axon membrane comes into direct contact with the interstitial fluid. The action potential thus jumps from one node of Ranvier to the next. This is known as 'saltatory conduction'. Saltatory conduction allows the depolarization process to hop along the nerve, thereby increasing the speed of conduction. Furthermore, energy is conserved as only the nodes are depolarized.

Membrane stabilization factors, such as a high extra-cellular calcium concentration, decrease membrane permeability and reduce its excitability. It is, however, local anaesthetic drugs in particular that we associate with membrane stabilization. These prevent depolarization and thereby impulse transmission.

Chemistry, pharmacology and physiology of local anaesthetic drugs

Historical notes on local anaesthetic agents

When the Spaniards penetrated the realm of the Incas they came across chasquis (the 'postman') and the habit of chewing coca leaves. The leaves were kept in a wooden box and lime ash in a small gourd. The coca leaves were chewed with the lime, the latter increasing the uptake of the agent we today call cocaine. Central nervous system stimulation was enjoyed and the mucous membrane in the mouth was anaesthetized. The saliva was not swallowed but spat out as, once swallowed, the substance becomes ineffective as a CNS stimulant. The lime stained the mouth white. This short historical note summarizes the main characteristics of local anaesthetic agents: local anaesthesia, uptake and effect are improved by a raised pH, CNS stimulation and degradation by the liver before it enters the general circulation.

Cocaine was first used to anaesthetize the cornea (Koller, 1884) and later used in attempts to 'freeze' nerves (Halsted, Hall). It was not until procaine came into use that surgeons began to explore the possibility of blocking tissue pain on a wider scale, e.g. by subcutaneous infiltration by injection close to nerves and nerve plexuses. Procaine was

introduced into clinical practice in 1905; rapid metabolism contributes to its safety. Its duration of action is, however, brief but may be extended if the injection into the surgical field is made more or less continuous. 2-Chloroprocaine (1952) is a 'modern version' of procaine. Amethocaine (1928) is a potent but also a rather toxic drug and is now mainly used for intrathecal anaesthesia.

A new type of local anaesthetic agent was introduced during the middle years of the twentieth century, the amide. Lignocaine (1943) was followed by mepivacaine (1956), prilocaine (1960), bupivacaine (1957), and by etidocaine (1971). Ropivacaine, an S-isomer, is undergoing clinical trials. The amides are more toxic than the esters but have a favourable therapeutic ratio. Over the past 30 years, these local anaesthetics, including ropivacaine, have been extensively studied with respect to their blocking characteristics, pharmacokinetics, general effects and toxicity. The clinical use of regional blockade has increased markedly during this time.

The agents to be discussed are shown in Table 10.1.

Chemistry

The local anaesthetic agents used today all have a similar chemical configuration, i.e. they have one aromatic lipophilic part (benzene ring) and one hydrophilic part

Table 10.1 Local anaesthetic agents discussed in this chapter

Amethocaine (tetracaine, Pantocaine®, Decicain®, Pontocaine®)
Bupivacaine (Carbostesin®, Duracaine®, Marcaine®, Sensorcaine®)
2-Chloroprocaine (Nesacaine®)
Emla®
Etidocaine (Duranest®)
Lignocaine (lidocaine, Duncaine®, Xylocaine®)
Mepivacaine (Carbocaine®, Polocaine®, Scandicaine®)
Prilocaine (Citanest®, Distanest®, Xylonest®)
Procaine (Ethocaine®, Novocaine®)
Ropivacaine (currently undergoing clinical trials)

Generic names and some trade names (®), in alphabetical order.

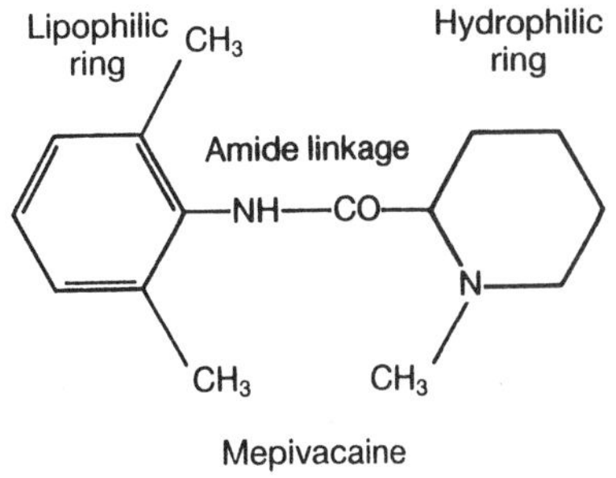

Fig. 10.5 A schematic drawing of the chemical structure of an amide local anaesthetic, mepivacaine.

(quaternary amine), connected by an intermediate link, either an *ester (–COO–)* or an *amide (–NHCO–)* (Fig. 10.5).

Esters such as chloroprocaine, procaine and tetracaine are broken down rapidly by plasma cholinesterase. *para*-Aminobenzoic acid is a degradation product which in rare cases may cause an allergic reaction. Most of the local anaesthetics used today are amides such as bupivacaine, etidocaine, lignocaine, mepivacaine or prilocaine. They are used in the racemic form which contrasts with the new drug, ropivacaine, which is a stereoisomeric compound.

The amides are slowly metabolized in the liver and the metabolites as well as the parent drug (to some extent) are excreted via the urine and bile. Some reabsorption from the intestines occurs. Prilocaine is the agent most rapidly degraded, which accounts for its low toxicity. *ortho*-Toluidine is, however, formed during this process, producing methaemoglobin, giving the patient, who has received a large dose of this agent, a bluish-pale skin colour several hours after the introduction of the block. Up to 25 per cent of the haemoglobin may be converted to methaemoglobin and cease to carry oxygen. Methylene blue injected intravenously will reverse the process or prevent its occurrence.[1] Nowadays prilocaine is mainly used in intravenous regional anaesthesia (Bier's block) because of its low toxicity. The most commonly used dose, 200 mg (40 ml of 0.5 per cent solution) does not cause clinically significant methaemoglobinaemia in the adult. There are indications, however, that infants and neonates might develop methemoglobin more easily than adults. Metabolites, formed during the degradation of the other amides such as bupivacaine, seem not to reach a toxic level.[2]

Pharmacokinetics

Local anaesthetic agents exist in two forms. The lipophilic form is poorly water soluble, unionized, exists at a higher pH and is freely diffusible across cell membranes. The hydrophilic form is highly water soluble, ionized, exists at a lower pH and penetrates membranes poorly. At a certain pH, called pK_a, 50 per cent of the drug is unionized and 50 per cent is ionized. It is the lipophilic basic form which penetrates cell membranes. The pK_a of local anaesthetic agents varies from 8.9 (procaine) to 7.7 (etidocaine).

Figure 10.6 illustrates the sequence of events when a local anaesthetic agent such as mepivacaine, dissolved in water, penetrates the body tissues after subcutaneous injection.[3,4] The pK_a in this example is 7.74. The pH of extracellular fluid is 7.4 and at this pH mepivacaine is predominantly hydrolysed (68 per cent) only 32 per cent is in the unionized lipophilic form which can penetrate into the cell. Inside the cell the pH is 7.10 and the drug dissociates into an ionized part (81 per cent) and a lipophilic basic form (19 per cent). In this example mepivacaine is

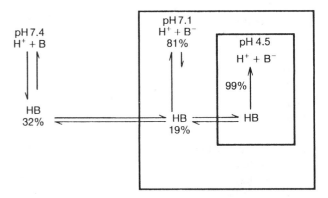

Fig. 10.6 Hypothetical model of the tissue distribution of a basic amine. See text. (Modified with kind permission of the author[3])

supposed to penetrate into the lysosomes which have a pH of only 4.5. The drug becomes 99 per cent ionized and is bound intralysosomally. There will be almost no lipophilic basic form available to allow the drug to pass back into the subcutaneous tissue, where it can be taken up into the blood and eliminated.

This figure also illustrates what happens when a fetus becomes severely acidotic. The local anaesthetic agent will be trapped in the fetus and the most important route of elimination, i.e. via the mother, is no longer available. Furthermore, drug metabolism in the newborn is less effective than in the adult.

A local anaesthetic agent given by epidural injection is taken up rapidly by the circulation[5] (Table 10.2). It is also fairly rapidly distributed to tissues with a high blood flow, i.e. the $\alpha t_{\frac{1}{2}}$ is brief. This is particularly so for mepivacaine and prilocaine. The differences between agents is more marked when one considers the effect of uptake in less well perfused tissues, i.e. the $\beta t_{\frac{1}{2}}$ is no longer for bupivacaine and etidocaine (high lipid solubility and strong protein binding) compared with the other local anaesthetic agents.

Table 10.2 Pharmacokinetic properties of amide local anaesthetics

Agent	α $t_{\frac{1}{2}}$ (min)	β $t_{\frac{1}{2}}$ (min)	γ $t_{\frac{1}{2}}$ (hours)	VD_{SS} (l)	*Clearance* (l/min)
Bupivacaine	2.7	28.0	3.5	72	0.47
Etidocaine	2.2	19.0	2.6	133	1.22
Lignocaine	1.0	9.6	1.6	91	0.95
Mepivacaine	0.7	7.2	1.9	84	0.78
Prilocaine	0.5	5.0	1.5	261	2.84
Ropivacaine	–	20	1.8	59	0.73

$t_{\frac{1}{2}}\alpha$-phase for uptake by the rapidly equilibrating tissues, β-phase illustrates the distribution to slowly perfused tissues and the γ-phase metabolism and excretion. VD_{SS}, volume of distribution at steady state. (Adapted from Covino[5] with kind permission Rational Drug Therapy and Astra Pain Control 1990.)

The $\gamma t_{\frac{1}{2}}$ phase represents metabolism and excretion. This phase is longest for bupivacaine and etidocaine. Prilocaine undergoes very rapid redistribution which is indicated by the large volume of distribution at steady state (VD_{SS}). Prilocaine is also rapidly metabolized by the liver, illustrated by a high clearance rate.

Local anaesthetics are, to varying degrees, bound to plasma protein, mainly α_1-acid glycoprotein (AAG) (acute phase and stress protein) and to a minor extent to albumin. The extent of binding may be altered by different pathologies; as a result, the free fraction or unbound plasma drug level can vary greatly from patient to patient. Patients with an inflammatory disease, cancer, myocardial infarct, or those in the postoperative period have a high level of protein binding (mainly AAG), and thus a low free fraction in the plasma. In contrast, neonates and patients with the nephrotic syndrome have a low level of binding protein and as a consequence a high free plasma level. Theoretically, there ought to be a closer relation between the unbound plasma drug level than between the total plasma drug level and its general effect. A high plasma level of bupivacaine in the postoperative period when AAG levels are high might, for example, be associated with a non-toxic level of free unbound bupivacaine. The significantly lower cord/maternal plasma concentration ratios of bupivacaine and etidocaine, compared with the more short-acting agents, does not result in bupivacaine and etidocaine being less toxic because the neonate has a very low level of AAG and the free drug concentration is very similar in both maternal and fetal concentrations. In a steady state situation, however, any change in plasma binding will be buffered by redistribution and clearance. The amount of drug bound in the plasma is very low in comparison with the total amount bound to all body tissues. In the event of an inadvertent intravascular injection the binding capacity of blood proteins will be saturated, and differences in binding capacity will be of no consequence.[6,7]

The physiological disposition of local anaesthetic agents

The term physiological disposition describes the chain of events by which circulating drugs such as local anaesthetics are taken up by body tissues and eventually eliminated.

Uptake from the site of injection

The uptake of a local anaesthetic agent from the site of injection depends on several factors: the concentration of the solution, volume injected, partition coefficient of the drug injected, blood supply to the area of injection, etc. *Adrenaline* in a concentration of 5 μg/ml (1:200 000) is often used to reduce blood flow and slow the rate of absorption of the local anaesthetic agent, thus reducing the plasma concentration and prolonging the duration of action.[8] Adrenaline is most effective in combination with lignocaine, less effective with mepivacaine or prilocaine solutions as these agents seem to have an anti-adrenaline-like effect. These agents are, therefore, preferred in a plain solution for patients with cardiovascular disease. It should be mentioned that there are reports that an adrenaline concentration of 1:600 000 suffices in a lignocaine block.[9] The addition of adrenaline has little effect on blockade due to etidocaine and virtually none with bupivacaine.

The use of adrenaline can be followed by restlessness, cold sweating, unstable circulation and arrhythmias in those with heart disease or uraemia. Even those with normal cardiac states may feel palpitations after injection of adrenaline. Postoperative bleeding may occur. When used in the finger, tip of the nose or ear or penis, necrosis has been reported.

Adrenaline added to local anaesthetics can be used in a total dose of 15 μg to indicate an accidental intravenous injection of local anaesthetic (see review on the test dose[10]).

Octapressin has been used as an alternative to produce pronounced vasoconstriction without having the cardiac and mental effects of adrenaline. (Note, octapressin is a coronary vasoconstrictor.) Other drugs such as noradrenaline and phenylephrine have also been used but adrenaline seems to produce the best vasoconstriction when used with a local anaesthetic agent.

Maximum non-toxic blood and plasma concentrations

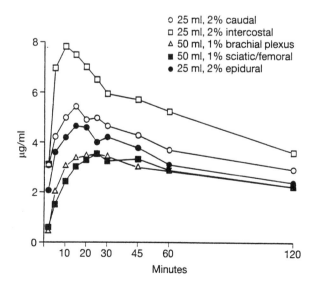

Fig. 10.7 Plasma concentrations of mepivacaine after a single injection of 500 mg. (Adapted from Tucker *et al.*[11] With kind permission from the authors and *Anesthesiology.*)

vary with the local anaesthetic agent. These are achieved quickly with plain lignocaine, but far more slowly when bupivacaine is used. The site of injection into a highly perfused tissue is also important, being followed by a rapid absorption and an early, high blood concentration. The high blood levels seen in connection with intercostal nerve block[11] are of special interest (Fig. 10.7). The uptake into the blood of amethocaine administered via the upper airways is also very rapid, behaving as though the drug has been injected intravenously. In contrast, the blood level following an intrathecal injection is low.

The uptake of local anaesthetic agents into tissues is related to lipid solubility, binding to proteins, the pK_a of the agent, tissue pH, etc. (Table 10.3). The highly perfused brain and heart have a tissue concentration similar to that of blood. The uptake of local anaesthetic agents or their degradation products is also high in the liver and kidney. They may be found in the intestine following excretion in the bile and some reabsorption takes place.

The uptake of local anaesthetic agents (lignocaine and bupivacaine have been studied) into lung tissue (*in vivo*) is high.[3,12] The lungs act as a buffer, most of the drug

Table 10.3 Chemical structure, physicochemical and pharmacological properties of the commonly used local anaesthetics

Agent	Chemical configuration			Physicochemical properties			
	Aromatic lipophilic	Intermediate chain	Amine hydrophilic	Molecular weight (base)	pK_a (25°C)	Partition coefficient	Protein binding (%)
Esters							
Procaine				236	8.9	0.02	6
Amethocaine				264	8.5	4.1	76
Chloroprocaine				271	8.7	0.14	–
Amides							
Prilocaine				220	7.9	0.9	55
Lignocaine				234	7.9	2.9	64
Mepivacaine				246	7.6	0.8	78
Bupivacaine				288	8.1	27.5	96
Etidocaine				276	7.7	141	94
Ropivacaine				275	8.1	6.1	94

Adapted from Covino[4] with kind permission *British Journal of Anaesthesia* and *Pain Control*, Astra, Södertälje, Sweden.

remaining in the lungs for a brief period (10–15 seconds). It has been calculated that in such a case the maximal arterial blood concentration reaches only one-third of its possible maximum without the buffering effect of the lung. Propranolol has been found to significantly depress removal by the lung of low doses of bupivacaine.[12]

Elimination

Local anaesthetic agents (esters) such as procaine, chloroprocaine and tetracaine undergo rapid biodegradation. The half-life of procaine in the blood is so brief that a toxic blood level is hard to achieve even when injected in 1 per cent concentration more or less continuously. Similarly, chloroprocaine has a very short half-life (minutes), even in the newborn, and has therefore been mainly used for obstetrics.

Biodegradation of the amides takes place in the liver and depends therefore on blood flow and enzyme activity in that organ.[13,14] Elimination is prolonged by a reduced splanchnic blood flow, a severely diseased liver and age.[15] The β-blocker propranolol causes reduction in splanchnic blood flow and thereby the speed of degradation of local anaesthetics. Drugs affecting liver enzyme activity, such as cimetidine and ranitidine, do not seem, however, to affect significantly the pharmacokinetics of bupivacaine or lignocaine.[16–18]

The biodegradation of prilocaine is very rapid.

Inhibition of nerve conduction by local anaesthetic agents

The lipophilic form is poorly water soluble, not ionized and has a high pH. The hydrophilic form is highly water soluble, ionized and has a low pH. Local anaesthetic drugs are available for clinical use in ampoules containing the water-soluble hydrochloride salt, and thus have a low pH (the pH of a plain solution is about 6). Chloroprocaine, which has a high pK_a, can only be dissolved in water at a low pH (about 3.5). The oxidation of adrenaline is reduced in adrenaline-containing solutions by reducing the pH (pH 4–5) and by the addition of an antioxidant such as sodium metabisulphite.

The pK_a of local anaesthetic agents varies from 8.9 (procaine) to 7.6 (etidocaine). A drug with a high pK_a such as bupivacaine (pK_a 8.1) is mostly ionized and only 15 per cent is in the unionized form at a pH of 7.40 and, therefore, available to penetrate a nerve, as it is the lipophilic unionized form that penetrates cell membranes. This

penetration of the axon cell membrane occurs at the nodes of Ranvier in the myelinated nerve fibres.

Inside the axon the local anaesthetic agent is again separated into non-ionized and ionized forms. It is probably the ionized form that blocks the sodium channels from the inside and thus prevents depolarization by preventing the fast inflow of sodium through the axon cell membrane. The sodium channels have specific receptors for local anaesthetic molecules. When two to four nodes of Ranvier are blocked the propagation of a nerve impulse through saltatory (jumping) transmission is prevented. The 'natural' local anaesthetic agents tetrodotoxin and saxitoxin appear to block the sodium channels from the outside.

The minimum blocking concentration is defined as the lowest concentration which will block a nerve *in vitro*. The minimum blocking concentration of a local anaesthetic agent is correlated with the partition coefficient and the potency of the drug which gives a clue to the minimum concentration required in clinical practice. The minimum blocking concentration depends on nerve fibre size, temperature, pH, amount of surrounding connective tissue, stimulation frequency, etc.

The blocking characteristics of the various local anaesthetic agents are related to their physicochemical properties. Thus a high lipid solubility (i.e. partition coefficient) means a high uptake into the nerves and thus a high potency. Agents such as procaine and chloroprocaine have a low partition coefficient (0.02 and 0.14, respectively) and are, relatively speaking, not potent drugs. One may compensate for low lipid solubility by increasing the concentration of the local anaesthetic solution injected. Thus, chloroprocaine is used in a 3 per cent solution. Chloroprocaine is very rapidly hydrolysed so this concentration will not increase the risk of a toxic reaction. Lignocaine, mepivacaine and prilocaine have a higher partition coefficient (0.8–2.9) and are thus more potent. The most potent local anaesthetic agents are bupivacaine and etidocaine (partition coefficient 27.5 and 141, respectively). These agents are also the most potent local anaesthetic drugs in common use (amethocaine is also a very potent drug, partition coefficient 4.1, but is rarely used today except for intrathecal anaesthesia). The greater the potency the greater is the toxicity, so drugs such as bupivacaine are used in low concentrations (0.12–0.5 (0.75) per cent). Etidocaine has the highest lipid solubility but is required in a concentration of 1.5 per cent to produce a deep extradural blockade. Etidocaine is supposedly taken up in the epidural fat to such an extent that only a limited amount of the injected drug is free to penetrate into the neural tissues. Etidocaine has also a large volume of distribution and a high clearance rate (see Table 10.3), which significantly reduces the toxicity of this agent.

There is an inverse correlation between pK_a and onset time. The onset time for lignocaine, mepivacaine, prilocaine and etidocaine, pK_a being 7.7 or slightly higher, is similar and short, being longer for bupivacaine (pK_a 8.1) and much longer in the case of amethocaine, chloroprocaine (in a low concentration) and procaine (pK_a 8.5, 8.7 and 8.9, respectively).

Protein binding also affects the blocking characteristics of a local anaesthetic agent (see Table 10.3). High protein binding is associated with a prolonged duration of action. Procaine has a low protein binding (6 per cent) and has a very short duration of action in contrast with amethocaine (protein binding 76 per cent), etidocaine and bupivacaine (94 and 95 per cent, respectively) which have a prolonged action. Lignocaine, mepivacaine and prilocaine are intermediate in protein binding and have an intermediate duration of action.

The duration of action depends not only on the properties of the anaesthetic agent. Blood circulation in the area of injection is of great importance. The duration of action is more prolonged when the drug is injected around a peripheral nerve or brachial plexus than when injected into the lumbar extradural space. The effect of the local anaesthetic agent on the local circulation is also of importance. Lignocaine seems to promote a significant increase in blood flow, thereby limiting the duration of its action. Mepivacaine and prilocaine seem not to have this effect. The main reason for the prolonged action of etidocaine and bupivacaine seems to be that these drugs firmly adhere to the neural tissue. The addition of adrenaline (5 µg per ml or 1:200 000) will markedly prolong the duration of lignocaine, but has less effect on mepivacaine, prilocaine and etidocaine and virtually no effect on the action of bupivacaine.

Carbonated solutions have been tried, in an attempt to shorten onset time and to obtain a deeper block. Carbon dioxide diffuses rapidly through the nerve, lowers the intracellular pH and possibly increases the intracellular concentration of the ionized form of the local anaesthetic. Several studies reporting the influence of carbonization have been reported but no convincing proof of its efficacy has been presented. The alkalinization of local anaesthetic solutions by adding a small amount of sodium bicarbonate does seem, however, to improve the blocking ability of bupivacaine, mepivacaine and prilocaine (more rapid onset, longer duration of action) (positive effect,[19,20] no effect[21]). One problem with adding bicarbonate to a local anaesthetic solution, to increase the pH, is that it may cause precipitation of the local anaesthetic agent.[22,23]

Pain sensation can be evaluated by the pinprick test or laser skin stimulation. The degree of motor blockade is usually evaluated by using the Bromage scale or measuring objectively the changes in motor strength.[24] The Bromage scale is used to evaluate the degree of muscle relaxation in the lower extremities when more than 60 per cent of the muscle power has been eliminated by an intrathecal or epidural blockade. Surprisingly from the clinical point of view, such an objective study indicates that 0.5 per cent bupivacaine produces an almost complete motor blockade, though the duration is far shorter than after 1.5 per cent etidocaine.[25]

Differential blocks

The impression gained from the clinical use of local anaesthetics is that a weak solution will block mainly the finer nerve fibres such as pain fibres. A higher concentration is required to block motor fibres and the sensation of touch. This phenomenon is called differential block. It has also been assumed that if a good motor block exists that the pain fibres will also be blocked. It is quite obvious that bupivacaine in a weak concentration (0.125–0.25 per cent) can effectively block pain perception while preserving fairly good motor function. This may be used to great advantage, for example, in obstetrics. If an effective motor block is needed, 0.5–0.75 per cent has to be used. Etidocaine may produce a deep motor blockade but still leave some pain fibres unblocked. Thus, patients may complain of pain during a knee operation under epidural anaesthesia with 1 per cent etidocaine despite complete paralysis of that extremity; to avoid this eventuality, 1.5 per cent etidocaine solution should be used.

Results from four types of study suggest that the mechanism of differential blockade is complicated. Contrary to the general view, reports from *in vitro* studies suggest that the concentration of local anaesthetic needed to block heavily myelinated A fibres and less myelinated B fibres is less than that required to block non-myelinated C fibres.[26–28] Several studies indicate, for example, that subarachnoid and extradural blockade of moderate extension as well as stellate ganglion block may only produce a partial sympathetic block.[29] These findings may explain why high epidural anaesthesia may fail to block surgical stress response.[30] It has also been shown using sensory evoked potentials (SEPs) that in spite of clinically good surgical anaesthesia, impulses from the blocked area can still reach the brain.[29,31,32]

Clinical implications

The doses used clinically and the 'highest' recommended or 'maximum' dose for the various local anaesthetic agents are shown in Table 10.4 and the onset times and durations of action of these agents and the effect of adrenaline are shown in Table 10.5.

Table 10.4 Highest recommended dose of different local anaesthetic agents

Agent	Concentrations used clinically	Highest recommended dose
Amethocaine	1% intrathecally	20 mg
Bupivacaine	0.125–0.5% 0.75% (not to be used in epidurals for obstetrics)	2 mg/kg body wt max. 150 mg over 4 h, max during 24 h 400 mg
	0.5% intrathecally plain or with glucose	20 mg
Chloroprocaine	3%	800 mg 1000 mg with adrenaline
Emla®	Intact skin 1.5–2 g/10 cm^2	
Etidocaine	1–1.5%	4 mg/kg body wt 300 mg
Lignocaine	0.5–2%	200 mg 500 mg with adrenaline
	5% in 7.5% glucose intrathecally, topical anaesthesia	100 mg
	4%	200 mg
	Viscous® 2%	300 mg
Mepivacaine	0.5–2%	5 mg/kg body wt 350 mg
Prilocaine	0.5–2%	400 mg in UK 600 mg in Sweden > 8 mg/kg body wt risk of methaemoglobin formation

In alphabetical order.

When evaluating the correct dose for a block the anaesthetist has to consider the volume and concentrations required to obtain a good block as well as the safety of the dose. The highest dose recommended by the pharmaceutical company manufacturing the drug is related to a fit, healthy patient, less than 50–60 years of age, weighing 70–80 kg and given as a single injection. Other factors to consider are allergy to local anaesthetic agents, age, reduced cardiac output and poor liver function. Special consideration must also be given to the site of injection and to the half-life of the drug to be used.

Potency (relative toxicity): lignocaine, mepivacaine, prilocaine 2, etidocaine 6 and bupivacaine 8.

Physiological effects

Effect on the circulation

The effects of local anaesthetic agents on the circulation were originally evaluated using anaesthetized animals. The

Table 10.5 Onset time and duration of action of the local anaesthetics commonly used for peripheral nerve blocks

Agent	Onset time (min)	Duration of action	Effect of adrenaline
Bupivacaine	20–30	8–9 hours	× 1
Chloroprocaine	Fast in 3% solution	30–60 min	
Etidocaine	15–20	6–8 hours	× 1.5
Lignocaine	20	60 min	× 4
Mepivacaine	20	2 hours	× 2
Prilocaine	20	2 hours	× 2
(Ropivacaine	20	8–9 hours	× 1)

Adrenaline reduces the blood circulation in the area into which the local anaesthetic solution is injected, thereby decreasing the absorption rate of the drug and prolonging the duration of action of, for example, lignocaine (duration prolonged 4 times). Bupivacaine, on the other hand, is strongly fixed to the tissue and thus adrenaline has little or no effect on the drug absorption or duration of action. The duration of action of epidural blockade is only 50–60 per cent of that seen following peripheral nerve blocks.

results obtained by these studies were not always those expected from clinical experience – namely that the systemic circulation is generally stable even when a large dose is given as, for example, during a plexus blockade.

Infusions of local anaesthetics in awake dogs and in volunteers have shown a stimulatory effect on the circulation. Cardiac output increases as does heart rate, blood pressure has a tendency to rise and the peripheral resistance to fall.[13,33] A relation between increasing blood levels of lignocaine and an increase in cardiac output has been demonstrated when lignocaine was infused postoperatively into the epidural space, to maintain a constant level of analgesia.[34] The stimulation of the general circulation is partly due to CNS stimulation. Local anaesthetic agents also stimulate contraction of the smooth muscles in the capacitance vessels and thus promote an increase in venous return and an increase in cardiac output.[33] Intra-arterial infusion results in a local increase in smooth muscle tonus.[35] The central stimulatory effect provoking a relaxation of the peripheral vasculature seems to override a local contraction of the smooth muscle. It is apparent that when one evaluates the circulatory effects of an epidural blockade one has to consider both the specific circulatory effects of the local anaesthetic agent used and the partial sympathetic block produced. There are reports of a decreased cardiac output in combination with a marked increase in peripheral resistance when bupivacaine was infused intravenously.[36,37] This finding might be related to the fall in cardiac output due to a direct cardiac effect (see below).

In moderate clinical doses local anaesthetic agents have no or little effect on the myocardium; their effect is on the conduction system of the heart. Lignocaine is used extensively to combat ventricular arrhythmias. Varying degrees of block of cardiac conduction can be produced by local anaesthetics. These agents should therefore not be used without precautions (for example, pacemaker, isoprenaline infusion) in a patient with, for example, Stokes–Adams disease.

This general view on the cardiac effect of local anaesthetics has been challenged by reports that several cardiac arrests have been provoked by bupivacaine and etidocaine.[38] The Food and Drug Administration in the USA considered some 50 deaths, allegedly caused by these two local anaesthetics, bupivacaine being mainly used in obstetrics. An obvious overdose was given in a few cases. In others, a rather modest dose was used. Deaths due to bupivacaine appear to be 'an American disease', possibly due to the use of 0.75 per cent solution for epidural blockade in obstetrics. Animal studies have shown that a bupivacaine infusion can provoke severe cardiac arrhythmias, mainly ventricular, not seen during lignocaine infusion. Lignocaine and bupivacaine differ in respect of their effect on the heart muscle. Both drugs penetrate into the myocardium rapidly. Lignocaine leaves the muscle rapidly, where bupivacaine leaves the heart muscle slowly. The slow release of bupivacaine from the myocardium explains why it can be very difficult to treat bupivacaine arrhythmias successfully.[39–42] There are indications that the myocardium of a woman at term might be more sensitive to bupivacaine than that of a non-pregnant woman. The new local anaesthetic agent ropivacaine has similar blocking characteristics but seems to be about 25 per cent less toxic[43] than bupivacaine.[44–48]

During regional anaesthesia a depression of the electrocardiograph T wave is often seen. This is, as a rule, not a sign of myocardial depression but of an increase in heart rate,[33] or the hypokalaemic effect of adrenaline provoked by a β_2-stimulation and its metabolic effects.[49,50]

Local anaesthetic agents have specific effects on organ blood flow.[51] Thus, umbilical and placental blood flow are markedly reduced when local anaesthetics are infused intra-arterially. The decrease in flow is most marked for bupivacaine, less for lignocaine, which explains the bradycardia and fetal distress often reported when bupivacaine is used for paracervical blockade. Calcium blockers have been reported to prevent this reduction in placental blood flow.

Local anaesthetic agents (lignocaine, mepivacaine and ropivacaine) cause an increase in pulmonary vascular resistance.[33,39] It is reasonable to assume that this increase is due to a direct stimulation of smooth muscle in the pulmonary vessel wall. The clinical significance of this increase in pulmonary artery pressure and increased load on the right heart in patients with chronic lung disease, who already have a considerably raised pulmonary artery pressure, has not as yet been studied.

Splanchnic vascular resistance is decreased and an increase in blood flow occurs when local anaesthetic agents are infused intravenously. This increase may in fact be explained by the rise in cardiac output, described above, but the increase in liver metabolism caused by the biodegradation of the local anaesthetic agent may also contribute to this effect. A decrease in splanchnic venous blood volume has been demonstrated during epidural anaesthesia, possibly caused by a direct stimulatory effect by the local anaesthetic agent on the portal vessels.

Effect on respiration

During severe convulsions produced by infusion of local anaesthetics in awake dogs, Pa_{CO_2} decreased, a change which compensated for a metabolic acidosis resulting from intense muscle activity.[33]

In studies of human volunteers, blood gas tensions were not influenced by the intravenous infusion of a large dose of bupivacaine, lignocaine or mepivacaine.[33] There are some indications that local anaesthetics increase the response to hypercapnia. Results of these studies seem to indicate that local anaesthetics are not respiratory depressants. However, unexpected and unexplained deaths reported in the literature might be the result of a poor response of the respiratory centre and the chemoreceptors to hypercapnia and in particular to hypoxaemia,[52] suggesting that local anaesthetics depress the hypoxic response.

In another perhaps more conclusive volunteer study hypoxic stimulation during lignocaine, mepivacaine or bupivacaine infusion (plasma levels typical of those reported during major regional anaesthesia) resulted in a slight but significant increase in response.[52]

There is therefore no absolute evidence that local anaesthetics depress the ventilatory response to hypercapnia or hypoxia.

Miscellaneous effects

Besides the blocking effects of local anaesthetic drugs on neural conduction (effect on sodium channels within the nerve cell membrane) these compounds have several other properties and uses. It has been known for some time that lignocaine has the ability to suppress irritable foci in the heart and brain because of its stabilizing effect on cell membranes. This constitutes the basis for lignocaine treatment of ventricular arrhythmias and epileptic seizures. It has also been known for quite some time that the intravenous administration of local anaesthetics in non-toxic doses can block a variety of airway reflexes such as laryngobronchospasm, the cough response and the increase in blood pressure elicited by various airway manipulations such as laryngoscopy, tracheal intubation, etc. Some other properties which have attracted increasing interest over recent years might confer new important indications for local anaesthetics.

Antithrombotic effects

Local anaesthetic agents inhibit platelet aggregation *in vitro*, but only at concentrations greater than the peak plasma concentrations found during clinical use.[54] A suggested mechanism is that either calcium influx is blocked or the intracellular calcium stores are mobilized. The inhibition of platelet aggregation is unlikely to be due to effects of the metabolites of local anaesthetics. Furthermore, raised antithrombin III levels associated with epidural anaesthesia may be one of several reasons (for example improved rheological conditions as well as an improvement in fibrinolysis function)[55,56] for the beneficial effect of this technique in the prevention of thrombosis formation. Epidural anaesthesia itself might also be associated with an inhibitory effect on platelet aggregation, probably through a mechanism unrelated to a direct local anaesthetic inhibition.[54] Topical application of lignocaine, in particular, has provided evidence for an antithrombotic effect in laser-induced vessel wall injury.[57] An increase in local production of prostacyclin from the endothelial cells when challenged with a local anaesthetic might be of importance in explaining the antithrombotic effect of local anaesthetics.

Effects on smooth muscle

In lower concentrations local anaesthetic agents have a direct stimulatory effect on the smooth muscles in blood vessels. Higher concentrations, however, may provoke vasodilatation. Likewise a stimulatory effect (increased peristaltic reflex) is seen on gastrointestinal smooth muscle with low concentrations of local anaesthetics, while high concentrations have an inhibitory effect. The cellular mechanism suggested proposed interference with Ca^{2+}-dependent and Ca^{2+}-independent K^+ channels. The clinical implication of the effect on the smooth musculature in the gut is that local anaesthetics given intraperitoneally or as an intravenous infusion may induce a faster return of propulsive motility in the colon in the postoperative period.[58,59]

Anti-inflammatory effects

Potent anti-inflammatory properties shown by amide local anaesthetics such as lignocaine may contribute to the stimulatory effects of amide local anaesthetics on the paralysed gut. The intraperitoneal administration of local anaesthetics has been shown to induce potent inhibition of peritonitis.[60] These anti-inflammatory effects of lignocaine involve the inhibition of prostaglandin synthesis, inhibition

of the migration of granulocytes into the inflammatory area, and the inhibition of granulocyte release of lysosomal enzymes and their production of tissue-toxic oxygen free radicals.[61] This dose-dependent inhibition of the activity of neutrophils, monocytes and lymphocytes usually occurs at higher concentrations and longer exposure than those occurring during clinical regional anaesthesia. Modulation of the immune response by regional anaesthesia is therefore not clinically feasible to the same extent as it is possible to modulate metabolic and endocrine responses.[62,63]

Both bupivacaine and lignocaine (without preservatives) exert bactericidal activity at clinical concentrations.

Analgesic effects

Systemically administered local anaesthetic drugs were reported to produce analgesia more than four decades ago.[64] In some chronic pain states a continuation of the analgesic response has been reported weeks after termination of an infusion. Thus, intravenous lignocaine infusion has been found effective in chronic painful diabetic neuropathy,[65] in adiposis dolorosa (Dercum disease) and chronic pain of the deafferentation type as well as in acute pain states, for example postoperative pain[66] and burn pain.[67]

Subcutaneous lignocaine infusion has also been used with success in malignant pain states. The mechanism underlying the analgesic effect is not clear but results obtained in decerebrated rats favour an action at spinal cord monosynaptic and polysynaptic reflexes. It is suggested that this may result from a central modulatory mechanism in the dorsal horn by activation of the endogenous opioid system by systemic lignocaine, and not from conduction blockade of pain fibres.

Ectopic discharge originating in neuromas can also be stopped by lignocaine in doses substantially lower than those required to block the propagation of nerve impulses already generated.[68,69] The analgesic response may also be partially due to a sedative effect.

Local toxicity

Experimental studies have shown toxic effects of local anaesthetics on various tissues. Local tissue reactions to local anaesthetics may depend on the type of local anaesthetic, concentration of the drug, site and technique of injection and the physical and chemical characteristics (pH, osmolality, additives such as preservatives, vasoconstrictors, etc.).

Neurotoxicity

Amino esters seem to be more neurotoxic than amino amides. Thus tetracaine and chloroprocaine are more neurotoxic than lignocaine and bupivacaine.[70] The extra-fascicular application of 3 per cent chloroprocaine or 1 per cent tetracaine in rats caused endoneural oedema in addition to Schwann cell injury and axonal dystrophy. In animals bupivacaine seems to be irritant to tissue in a concentration $\geq$ 0.6 per cent and to nerves in concentrations of 0.75–1 per cent. Reversible vacuolization and disruption of myelin sheaths in rat sciatic nerves following repeated exposure to bupivacaine 0.5 per cent has been seen.[71] Significant histopathological changes (lymphocyte accumulation) in spinal cord sections of rabbits have been reported following hyperbaric 0.75 per cent bupivacaine. Other researchers have not identified any local neurotoxicity using 0.75 per cent bupivacaine. An increase in neurotoxicity may be seen after the addition of adrenaline. In an experimental study,[72] however, no evidence of histological nerve damage was found following extrafascicular injection of local anaesthetic agents in concentrations provided routinely. This included short- and long-acting agents with and without adrenaline. Intrafascicular injection, on the other hand, is associated with an increase in toxicity, especially when carbonated lignocaine, tetracaine or procaine is used.[72] This increase in toxicity after intraneural injection may in part be due to ischaemia of the nerve caused by an increase in endoneural pressure to greater than the endoneural capillary perfusion pressure, thereby rendering the segment ischaemic.[70]

A multifactorial aetiology is probably responsible for the neuropathies reported following regional anaesthesia. In addition to mechanical factors, bleeding and infection, which may cause neurological damage, other factors may be involved. At least four factors seem to be involved in the cases with spinal complications following the subarachnoid injection of a large dose of chloroprocaine: spinal hypoperfusion (low blood pressure due to high blockade), prolonged exposure to chloroprocaine, low pH of the preparation (3–3.5) and exposure to sulphite[73,74] (sodium bisulphite, an antioxidant added to the commercial preparation).

Myotoxicity

Bupivacaine, tetracaine and mepivacaine have myotoxic effects.[71] The myotoxicity of bupivacaine is so well recognized that this drug has been routinely used in research laboratories to produce degeneration of skeletal muscle fibres.

Adverse reactions[75]

These include toxic reaction, reactions not related to the drug (for example, a vasovagal syncope), hypotension provoked by a high spinal anaesthesia, an idiosyncratic reaction (for example, methaemoglobin formation following the degradation of prilocaine and allergic reactions).

Toxic reaction

The 'new' local anaesthetics, the amides, are very effective nerve blockers which means that there is by and large a favourable relation between the dose required for block and the 'maximum dose'.

Prevention
Toxic reactions can be reduced by the implementation of aspiration tests, slow injection of incremental doses, and observation of and verbal contact with the patient. The dose to be given must be calculated with great care and in selected cases the use of a test dose with a solution containing adrenaline (15 µg). Equipment necessary for the adequate treatment of a toxic reaction should be available. (See Treatment, below.)

Symptoms
An overdose, relative or absolute, or an accidental intravenous injection might provoke a toxic reaction. The initial symptoms or the mild symptoms are *circumoral numbness*, *tinnitus*, *light headedness*, *confusion* ('something is wrong'). Small *muscle twitches* are often seen in the early stages of a major regional block. If the symptoms progress sudden *major convulsions*, similar to epileptic seizures, follow. Acidosis and/or hypoxia makes the convulsions worse.[76,77] As the toxic process progresses, these symptoms of CNS excitation are replaced by signs of general depression. Breathing will become more and more ineffective largely due to airway obstruction and the convulsions. The patient becomes cyanotic, the pulse irregular and *respiratory* and *circulatory arrest* ensue.

There is no strict correlation between blood level of a local anaesthetic agent and 'toxic' symptoms. Vasovagal syncope plays a role and this explains the grave hypotension sometimes seen with the injection of a small dose of a local anaesthetic agent or just a needle prick.

Clinical experience confirms that the first serious symptoms of a toxic reaction involves the stimulation of the brain. Signs of cardiac depression is a later event, seen at blood concentrations higher than those provoking CNS symptoms. There seems to be a 1:2–4 relation between the cumulative convulsive dose and the cumulative lethal dose[39] (Table 10.6). In this respect bupivacaine (and possibly etidocaine) are exceptions as fatal arrhythmias can be provoked by even moderate amounts of the drug. Arrhythmias seen in connection with bupivacaine are very difficult to treat.[78]

In very rare cases, 1 per cent of the adverse reactions, amide local anaesthetic agents have provoked an *allergic reaction* (IgE-mediated reactions such as skin rash,

Table 10.6 Comparative CNS and CVS (cardiovascular) toxicity of local anaesthetics (mean ± SEM), intravenous bolus dose (mg/kg) causing convulsions and death in unanaesthetized dogs

Agent	Convulsive dose	Lethal dose
Bupivacaine	3.6 ± 0.1	7.7 ± 0.8
Etidocaine	4.6 ± 0.0	9.2 ± 0.0
Lignocaine	11.8 ± 0.6	47.5 ± 5.6
Mepivacaine	15.2 ± 1.4	39.0
Ropivacaine	3.6 ± 0.2	13.8 ± 2.1

Reproduced with kind permission from Feldman.[39]

bronchospasm, Quincke's oedema, hypotension).[79] A sudden loss of consciousness and a severe fall in blood pressure is not necessarily an anaphylactic reaction. A vasovagal syncope is more likely if no true allergic symptoms are seen and serological tests are negative. In suspected cases of allergy to local anaesthetic agents a provocation test is useful. Very small doses from 0.01 mg in increasing amounts up to 10 mg are given at 15-minute intervals, first intracutaneously, then subcutaneously and finally, if no reaction occurs, intravenously. To exclude a local vasomotor response to the intradermal injection, a small amount of saline should precede the intradermal injection of the local anaesthetic agent. If there are indications of a late reaction 24 hours should elapse after the first injection before the following ones are given. All tests should be carried out with preservative-free solutions. There are indications that methylparaben, in particular, used as a preservative in some local anaesthetic solutions, and possibly sodium bisulphite can induce an allergic reaction.

Cross-sensitivity does not exist between esters and amides.

Treatment
Muscle twitches are quite often seen 5–10 minutes after the injection of a local anaesthetic for a major regional block. Gentle reassurance and perhaps some thiopentone (15–50 mg) or possibly diazepam (1–2 mg takes effect after more than 2–3 minutes but is a respiratory depressant and may decrease the margin of safety[80]) intravenously will reverse this unpleasant symptom as well as other weak toxic effects. Slight confusion can also be reversed in this way, the patient becoming lucid.

Severe excitatory events require immediate treatment if severe hypoxia is to be avoided. Oxygen by mask with gentle assisted ventilation might stop the convulsions.[76] If a large dose of the local anaesthetic agent has been injected then it is likely that the convulsions will continue for a considerable period of time. Tracheal intubation should be carried out after the injection of suxamethonium i.v. (1 mg/kg body wt). If an intravenous access has been lost because of the convulsions, the injection of suxamethonium 3 mg/kg

body wt into the pectoralis major muscle will be followed by complete muscle relaxation within 90 seconds, allowing easy intubation.

It has been suggested that excessive cortical activity might damage the cortex. Thiopentone infusion should, therefore, be started, to prevent the convulsions when the patient is no longer paralysed, without causing marked cardiovascular depression. Acidosis must also be corrected.

If a marked cardiovascular depression (hypotension, signs of myocardial depression, bradycardia, conduction block) is encountered, adrenaline or isoprenaline (isoproterenol)[81] should be infused. Arrhythmias seen in association with bupivacaine are very difficult to treat. Lignocaine infusion has been tried with some success.

Summary of the clinical characteristics of some local anaesthetic agents

The esters

Procaine is a poor local anaesthetic, with a slow onset, weak potency and short duration of action. The main advantage of this drug is that it has low toxicity and is rapidly broken down in the plasma. Procaine is cheap and is still used in developing countries.

Chloroprocaine has the same low toxicity as procaine. Onset time of the 3 per cent solution is short. Biodegradation is rapid and therefore the duration of action is short. It has mainly been used in obstetrics and then often in combination with other agents such as bupivacaine to obtain a fast onset with the more prolonged duration. Such a use has not been documented in a well controlled study. Serious nerve damage following accidental intrathecal injection has been reported (an effect of the low pH of the solution and of bisulphite used to stabilize the local anaesthetic solution). Today EDTA is used instead of bisulphite. Back pain has been associated with the use of this drug for epidural analgesia.

Amethocaine has a high potency and is also more toxic than lignocaine, mepivacaine and prilocaine. Amethocaine is mainly used for intrathecal analgesia.

All three drugs mentioned above (procaine, chloroprocaine and amethocaine) are metabolized by plasma cholinesterase. A degradation product is *para*-aminobenzoic acid, known sometimes to provoke an allergic reaction.

The amides

The amides are very stable and tolerate brief autoclaving. They are metabolized by the liver, the $\gamma t_{\frac{1}{2}}$ being 1.5–3.5 hours. Accumulation is likely if repeated doses via a catheter are not carefully chosen.

Lignocaine, mepivacaine and prilocaine are very good

agents with a fast onset and with a useful safety margin. Lignocaine is rather short-lasting unless adrenaline is included in the solution. Mepivacaine and prilocaine provide a good, moderately long-lasting block without the need for adrenaline to be added to the local anaesthetic solution. These drugs are thus to be preferred in patients with cardiovascular disease. Prilocaine is the least toxic agent and is in particular suited for intravenous regional analgesia. ortho-Toluidine is, however, a degradation product which provokes significant methaemoglobin formation when a large dose of this drug has been given. Many anaesthetists are reluctant to use this drug in obstetrics because of methaemoglobin formation in the fetus or infant.

Bupivacaine has a moderately prolonged onset of action and a long-lasting effect, mostly affecting pain fibres in contrast to etidocaine. Etidocaine 1.5 per cent has a rapid onset and a prolonged duration of action. The effect of etidocaine but not of bupivacaine is prolonged by adrenaline. Both drugs are more toxic than the first three amides mentioned in this section. There is today great concern about the safety of 0.75 per cent bupivacaine in higher doses. Severe refractory cardiac arrhythmias have been reported. It should not be used in this concentration for an epidural blockade in obstetrics. Ropivacaine is a recently introduced local anaesthetic agent at present undergoing clinical trials. Its local anaesthetic effect is similar to that of bupivacaine, but with a more potent block of pain fibres. The reason for the interest in this drug is that it is less cardiotoxic than bupivacaine. In lower concentrations (0.5 per cent or less) it provokes vasoconstriction, similar to the effect of adrenaline. In higher concentrations (for example 1 per cent) an anti-adrenaline-like effect is seen.[51]

Emla[82] is a eutectic mixture (oil/water emulsion) of lignocaine 2.5 per cent and prilocaine 2.5 per cent which produces good analgesia when applied on the skin. Indications for its use are skin biopsy, skin grafting, venepuncture (children), arterial puncture, removal of excessive granulation such as genital warts, surgical debridement of leg ulcer, circumcision, postherpetic neuralgia, decreasing the dermal response to radiation, otitis externa (note, if Emla reaches the middle ear an ototoxic effect can result.) Application time on intact skin should be 1–3 hours and on mucous membranes 5–10 minutes. Only low plasma concentrations (< 0.2 μg/ml of both drugs) have been reported. Emla produces some vasoconstriction in the skin vessels, sometimes making venepuncture difficult. It should not be used in small children (< 3 months) and on children 3–12 months who undergo medical treatment with drugs inducing methaemoglobin formation (e.g. sulpha drugs).

REFERENCES

A fairly complete list of references should have contained several hundred titles, which space does not permit. Therefore, only the most relevant and recent publications are included.

Cousins MJ, Bridenbaugh PO, eds. Neural blockade in clinical anesthesia and management of pain. 2nd ed. Philadelphia: JB Lippincott, 1988: Chapters 2–4.

Current Opinion in Anaesthesiology Reviews 1989; **2**: 606–11. 1990; **3**: 727–30. 1991; **4**: 665–9. 1992; **5**: 672–5.

Guyton AC. Membrane potentials and action potentials. In: Textbook of medical physiology. 8th ed. Philadelphia: WB Saunders, 1991: 51–66.

Löfström JB, Sjöstrand U, eds. Local anaesthesia and regional blockade. Pharmacology, physiology and clinical effects. Monographs in anaesthesiology. Vol 15. Amsterdam: Elsevier, 1988.

1. Hjelm M, H:son Holmdahl M. Biochemical effects of aromatic amines. Cyanosis, methaemoglobinaemia and Heinz-body formation induced by a local anaesthetic agent (prilocaine). Acta Anaesthesiologica Scandinavica 1965; **2**: 99–120.

2. Pere P, Tuominen M, Rosenberg PH. Cumulation of bupivacaine, desbutylbupivacaine and 4-hydroxybupivacaine during and after continuous interscalene brachial plexus block. Acta Anaesthesiologica Scandinavica 1991; **35**: 647–50.

3. Post C. Studies on the pharmacokinetic function of the lung with special reference to lidocaine. Medical dissertations, Linköping University, Sweden, 1979: No. 73.

4. Covino BG. Pharmacology of local anaesthetic agents. British Journal of Anaesthesia 1986; **58**: 701–16.

5. Covino BG. Pharmacology of local anesthetic agents. Rational Drug Therapy 1987; **21**: 1–9.

6. Tucker GT. Pharmacokinetics of local anaesthetics. British Journal of Anaesthesia 1986; **58**:717–31.

7. Tucker GT. Is plasma binding of local anesthetics important? Acta Anaesthesiologica Belgica 1988; **39**: 147–50.

8. Scott B. Adrenaline in local anesthetic solutions. Acta Anaesthesiologica Belgica 1988; **39**: 159–61.

9. Ohno H, Watanabe M, Saitoh J, Saegusa Y, Hasegawa Y, Yonezawa T. Effect of epinephrine concentration on lidocaine disposition during epidural anesthesia. Anesthesiology 1988; **68**: 625–8.

10. Blomberg RG, Löfström JB. The test dose in regional anaesthesia. Acta Anaesthesiologica Scandinavica 1991; **35**: 465–8.

11. Tucker GT, Moore DC, Bridenbaugh PO, Bridenbaugh LD, Thompson GE. Systemic absorption of mepivacaine in commonly used regional block procedures. Anesthesiology 1972; **37**: 277–87.

12. Rothstein P, Cole JS, Pitt BR. Pulmonary extraction of [^{3}H]bupivacaine: modification by dose, propranolol and

interaction with [^{14}C]5-hydroxtryptamine. *Journal of Pharmacology and Experimental Therapeutics* 1986; **240**: 410–14.

13. Wiklund L. Human hepatic blood flow and its relation to systemic circulation during intravenous infusion of bupivacaine and etidocaine. *Acta Anaesthesiologica Scandinavica* 1977; **21**: 189–99.

14. Wiklund L, Jorfeldt L. Splanchnic turn-over of some energy metabolites and acid–base balance during intravenous infusion of lidocaine, bupivacaine or etidocaine. *Acta Anaesthesiologica Scandinavica* 1981; **25**: 200–8.

15. Veering BT, Burm AG, Vletter AA, van den Hoeven RAM, Spierdijk J. The effect of age on systemic absorption and systemic disposition of bupivacaine after subarachnoid administration. *Anesthesiology* 1991; **74**: 250–7.

16. Pihlajamäki KK, Lindberg RLP, Jantunen ME. Lack of effect of cimetidine on the pharmacokinetics of bupivacaine in healthy subjects. *British Journal of Clinical Pharmacology* 1988; **26**: 403–6.

17. Flynn RJ, Moore J, Collier PS, McClean E. Does pretreatment with cimetidine and ranitidine affect the disposition of bupivacaine? *British Journal of Anaesthesia* 1989; **62**: 87–91.

18. Brashear WT, Zuspan KJ, Lazebnik N, Kuhnert BR, Mann LI. Effect of ranitidine on bupivacaine disposition. *Anesthesia and Analgesia* 1991; **72**: 369–76.

19. Solak M, Aktürk G, Eriyes I, Özen M, Çolak M, Duman E. The addition of sodium bicarbonate to prilocaine solution during i.v. regional anesthesia. *Acta Anaesthesiologica Scandinavica* 1991; **35**: 572–4.

20. Quinlan JJ, Oleksey K, Murphy FL. Alkalinization of mepivacaine for axillary block. *Anesthesia and Analgesia* 1992; **74**: 371–4.

21. Verborgh C, Claeys M-A, Camu F. Onset of epidural blockade after plain or alkalinized 0.5% bupivacaine. *Anesthesia and Analgesia* 1991; **73**: 401–4.

22. Peterfreund R, Datta S, Ostheimer GW. pH adjustment of local anesthetic solutions with sodium bicarbonate: laboratory evaluation of alkalinization and precipitation. *Regional Anesthesia* 1989; **14**: 265–70.

23. Ikuta PT, Vasireddy AR, Raza SM, Winnie AP, Durrani Z, Masters RW. pH adjustment schedule for the amide local anesthetics. *Regional Anesthesia* 1989; **14**: 229–35.

24. Axelsson K, Widman GB. A comparison of bupivacaine and tetracaine in spinal anaesthesia with special reference to motor block. *Acta Anaesthesiologica Scandinavica* 1985; **29**: 79–86.

25. Axelsson K, Nydahl P-A, Philipson L, Larsson P. Motor and sensory blockade after epidural injection of mepivacaine, bupivacaine, and etidocaine – a double-blind study. *Anesthesia and Analgesia* 1989; **69**: 739–47.

26. Gissen AJ, Covino BG, Gregus J. Differential sensitivity of mammalian nerve fibers to local anesthesia agents. *Anesthesiology* 1980; **53**: 467–74.

27. Rosenberg PH, Heinonen E, Jansson S-E, Gripenberg J. Differential nerve block by bupivacaine and 2-chloroprocaine. *British Journal of Anaesthesia* 1980; **52**: 1183–9.

28. Wildsmith JAW, Gissen AJ, Takman B, Covino BG. Differential nerve blockade: esters v. amides and the influence of pK_a. *British Journal of Anaesthesia* 1987; **59**: 379–84.

29. Malmqvist EL-Å. Sympathetic neural blockade during regional analgesia. Clinical investigations in man. Medical Dissertations, Linköping University, Sweden, 1992: No. 366.

30. Rutberg H, Håkanson E, Kehlet H. Trauma and stress – the effect of neural blockade. In: Löfström JB, Sjöstrand U, eds. *Local anaesthesia and regional blockade. Monographs in anaesthesiology.* Vol 15. Amsterdam: Elsevier, 1988: 259–72.

31. Lund C, Selmar P, Hansen OB, Kehlet H. Effect of intrathecal bupivacaine on somatosensory evoked potentials following dermatomal stimulation. *Anesthesia and Analgesia* 1987; **66**: 809–13.

32. Lund C, Hansen OB, Kehlet H. Effect of epidural 0.25% bupivacaine on somatosensory evoked potentials to dermatomal stimulation. *Regional Anesthesia* 1989; **14**: 72–7.

33. Jorfeldt L, Löfström B, Pernow B, Persson B, Wahren J, Widman B. The effect of local anaesthetics on the central circulation and respiration in man and dog. *Acta Anaesthesiologica Scandinavica* 1968; **12**: 153–69.

34. Sjögren S, Wright B. Blood concentration of lidocaine during continuous epidural blockade. *Acta Anaesthesiologica Scandinavica* 1972; **16**: 51–6.

35. Jorfeldt L, Löfström B, Pernow B, Wahren J. The effect of mepivacaine and lidocaine on forearm resistance and capacitance vessels in man. *Acta Anaesthesiologica Scandinavica* 1970; **14**: 183–201.

36. Hasselström LJ, Mogensen T, Kehlet H, Christensen NJ. Effects of intravenous bupivacaine on cardiovascular function and plasma catecholamine levels in humans. *Anesthesia and Analgesia* 1984; **63**: 1053–8.

37. Beal JL, Freysz M, Timour Q, Bertrix L, Lang J. Haemodynamic effects of high plasma concentrations of bupivacaine in the dog. *European Journal of Anaesthesiology* 1988; **5**: 251–60.

38. Albright GA. Cardiac arrest following regional anesthesia with etidocaine or bupivacaine. Editorial views. *Anesthesiology* 1979; **51**: 285–7.

39. Feldman HS. The relative acute systemic toxicity of selected local anesthetic agents. Medical Dissertations, Uppsala University, Sweden, 1989: No. 226.

40. Moller RA, Covino BG. Cardiac electrophysiologic effects of lidocaine and bupivacaine. *Anesthesia and Analgesia* 1988; **67**: 107–14.

41. Pitkanen M, Feldman HS, Arthur GR, Covino BG. Chronotropic and inotropic effects of ropivacaine, bupivacaine, and lidocaine in the spontaneously beating and electrically paced isolated, perfused rabbit heart. *Regional Anesthesia* 1992; **17**: 183–92.

42. Reiz S, Häggmark S, Johansson G, Nath S. Cardiotoxicity of ropivacaine – a new amide local anaesthetic agent. *Acta Anaesthesiologica Scandinavica* 1989; **33**: 93–8.

43. Scott DB, Lee A, Fagan D, Bowler GMR, Bloomfield P, Lundh R. Acute toxicity of ropivacaine compared with that of bupivacaine. *Anesthesia and Analgesia* 1989; **69**: 563–9.

44. Arthur GR, Feldman HS, Covino BG. Comparative pharmacokinetics of bupivacaine and ropivacaine, a new amide local anesthetic. *Anesthesia and Analgesia* 1988; **67**: 1053–8.

45. Concepcion M, Arthur GR, Steele SM, Bader AM, Covino BG. A new local anesthetic, ropivacaine. Its epidural effects in humans. *Anesthesia and Analgesia* 1990; **70**: 80–5.

46. Katz JA, Knarr D, Bridenbaugh PO. A double-blind comparison of 0.5% bupivacaine and 0.75% ropivacaine

administered epidurally in humans. *Regional Anesthesia* 1990; **15**: 250–2.

47. Kerkkamp HEM, Gielen MJM, Edström HH. Comparison of 0.75% ropivacaine with epinephrine and 0.75% bupivacaine with epinephrine in lumbar epidural anesthesia. *Regional Anesthesia* 1990; **15**: 204–7.

48. Brockway MS, Bannister J, McClure JH, McKeown D, Wildsmith JA. Comparison of extradural ropivacaine and bupivacaine. *British Journal of Anaesthesia* 1991; **66**: 31–7.

49. Hahn RG. Decrease in serum potassium concentration during epidural anaesthesia. *Acta Anaesthesiologica Scandinavica* 1987; **31**: 680–3.

50. Toyoda Y, Kubota Y, Kubota H, *et al*. Prevention of hypokalemia during axillary nerve block with 1% lidocaine and epinephrine 1:100,000. *Anesthesiology* 1988; **69**: 109–12.

51. Löfström JB. The effect of local anesthetics on the peripheral vasculature. *Regional Anesthesia* 1992; **17**: 1–11.

52. Caplan RA, Ward RJ, Posner K, Cheney FW. Unexpected cardiac arrest during spinal anesthesia: a closed claims analysis of predisposing factors. *Anesthesiology* 1988; **68**: 5–11.

53. Johnson A, Löfström JB. Influence of local anaesthetics on ventilation. *Regional Anesthesia* 1991; **16**: 7–12.

54. Gibbs NM. The effect of anaesthetic agents on platelet function. *Anaesthesia and Intensive Care* 1991; **19**: 495–520.

55. Modig J. Studies on the significance of lumbar epidural anaesthesia and of local anaesthetics as antithrombotic prophylaxis. In: Löfström JB, Sjöstrand U, eds. *Local anaesthesia and regional blockade. Monographs in anaesthesiology.* Vol 15. Amsterdam: Elsevier, 1988: 199–208.

56. Tuman KJ, McCarthy RJ, March RJ, DeLaria GA, Patel RV, Ivankovich AD. Effects of epidural anesthesia and analgesia on coagulation and outcome after major vascular surgery. *Anesthesia and Analgesia* 1991; **73**: 696–704.

57. Luostarinen V, Evers H, Lyytikäinen M-T, Scheinin A, Wahlén A. Antithrombotic effects of lidocaine and related compounds on laser-induced microvascular injury. *Acta Anaesthesiologica Scandinavica* 1981; **25**: 9–11.

58. Rimbäck G, Cassuto J, Faxén A, Högström S, Wallin G, Tollesson PO. Effect of intra-abdominal bupivacaine instillation on postoperative colonic motility. *Gut* 1986; **27**: 170–5.

59. Rimbäck G, Cassuto J, Tollesson P-O. Treatment of postoperative paralytic ileus by intravenous lidocaine infusion. *Anesthesia and Analgesia* 1990; **70**: 414–19.

60. Rimbäck G, Cassuto J, Wallin G, Westlander G. Inhibition of peritonitis by amide local anesthetics. *Anesthesiology* 1988; **69**: 881–6.

61. Eriksson AS, Sinclair R, Cassuto J, Thomsen P. Influence of lidocaine on leukocyte function in the surgical wound. *Anesthesiology* 1992; **77**: 74–8.

62. Hole A, Breivik H. Local anaesthetics, regional blocks, and immune defence. In: Löfström JB, Sjöstrand U, eds. *Local anaesthesia and regional blockade. Monographs in anaesthesiology.* Vol 15. Amsterdam: Elsevier, 1988: 209–20.

63. Salo M. Effects of anaesthesia and surgery on the immune response. *Acta Anaesthesiologica Scandinavica* 1992; **36**: 201–20.

64. Glazer S, Portenoy RK. Systemic local anesthetics in pain control. *Journal of Pain and Symptom Management* 1991; **6**: 30–9.

65. Bach FW, Jensen TS, Kastrup J, Stigsby B, Dejgård A. The effect of intravenous lidocaine on nociceptive processing in diabetic neuropathy. *Pain* 1990; **40**: 29–34.

66. Cassuto J, Wallin G, Högström S, Faxén A, Rimbäck G. Inhibition of postoperative pain by continuous low-dose intravenous infusion of lidocaine. *Anesthesia and Analgesia* 1985; **64**: 971–4.

67. Jönsson A, Cassuto J, Hanson B. Inhibition of burn pain by intravenous lignocaine infusion. *Lancet* 1991; **338**: 151–2.

68. Chabal C, Russel LC, Burchiel KJ. The effect of intravenous lidocaine, tocainide, and mexiletine on spontaneously active fibers originating in rat sciatic neuromas. *Pain* 1988; **38**: 333–8.

69. Devor M, Wall PD, Catalan N. Systemic lidocaine silences ectopic neuroma and DRG discharge without blocking nerve conduction. *Pain* 1992; **48**: 261–8.

70. Selander D. Nerve toxicity of local anaesthetics. In: Löfström JB, Sjöstrand U, eds. *Local anaesthesia and regional blockade. Monographs in anaesthesiology.* Vol 15. Amsterdam: Elsevier, 1988: 77–97.

71. Kyttä J, Heinonen E, Rosenberg PH, Wahlström T, Gripenberg J, Huopaniemi T. Effects on repeated bupivacaine administration on sciatic nerve and surrounding muscle tissue in rats. *Acta Anaesthesiologica Scandinavica* 1986; **30**: 625–9.

72. Gentili F, Hudson AR, Hunter D, Kline DG. Nerve injection injury with local anesthetic agents: a light and electron microscopic, fluorescent microscopic, and horseradish peroxidase study. *Neurosurgery* 1980; **6**: 263–72.

73. Covino BG, Marx GF, Finster M, Zsigmond EK. Prolonged sensory/motor deficits following inadvertent spinal anesthesia. *Anesthesia and Analgesia* 1980; **6**: 399–400.

74. Wang BC, Hillman DE, Spielholz NI, Turndorf H. Chronic neurological deficits and nesacaine-CE – an effect of the anesthetic, 2-chloroprocaine, or the antioxidant, sodium bisulfite? *Anesthesia and Analgesia* 1984; **63**: 445–7.

75. Reynolds F. Adverse effects of local anaesthetics. *British Journal of Anaesthesia* 1987; **59**: 78–95.

76. Moore DC, Crawford RD, Scurlock JE. Severe hypoxia and acidosis following local anesthetic-induced convulsions. *Anesthesiology* 1980; **53**: 259–60.

77. Englesson S, Matousek M. Central nervous system effects of local anaesthetic agents. *British Journal of Anaesthesia.* 1975; **47**: 241–6.

78. Scott DB. Toxicity caused by local anaesthetic drugs. Editorial. *British Journal of Anaesthesia* 1981; **53**: 553–4.

79. Glinert RJ, Zachary CB. Local anesthetic allergy. Its recognition and avoidance. *Journal of Dermatologic Surgery and Oncology* 1991; **17**: 491–6.

80. Gerard J-L, Edouard A, Berdeaux A, Duranteau J, Ahmad R. Interaction of intravenous diazepam and bupivacaine in conscious dogs. *Regional Anesthesia* 1989; **14**: 298–303.

81. Clermont G, Lacombe P, Couture P, Garand M, Savard D, Blaise G. Reversal of bupivacaine cardiotoxicity by isoproterenol in vivo. *Regional Anesthesia* 1992; **17**: 166.

82. Juhlin L, Evers H. EMLA: a new topical anesthetic. *Advances in Dermatology* 1990; **5**: 75–92.

Chapter 11

Drugs Acting on the Cardiovascular System

Pierre Foëx

Adrenergic agonists and antagonists	Calcium antagonists
Cholinergic agonists	ACE inhibitors
Ganglionic blocking agents	Vasodilators
Antiarrhythmic drugs	Antihypertensive therapy
Inotropes	Hypertensive emergencies

Adrenergic agonists and antagonists

Adrenergic receptors

Catecholamines produce a wide range of effects mediated by the activation of adrenoceptors. These receptors have been subclassified into α- and β-receptors, themselves subdivided into α_1- and α_2-, β_1- and β_2-receptors.[1–4] Dopaminergic receptors have also been subdivided into two subtypes, the presynaptic D_2- and the postsynaptic D_1-receptors.

α- and β-Adrenoceptors are membrane receptors with seven membrane-spanning domains linked to guanine nucleotide binding regulatory (G) proteins.[5] The G-proteins are involved in the activation of one or more second messenger-effector systems such as adenylate cyclase, phospholipases, potassium and calcium ions channels, and the sodium/proton (Na^+/H^+) antiport. α-Adrenoceptors should be subdivided into at least four subtypes (α_{1A}, α_{1B}, α_{2A}, α_{2B}). This subdivision may become clinically relevant as α_2-adrenoceptor agonists (clonidine, azepexole, dexmedetomidine) may play a role in anaesthesia, pain relief, and ischaemia prevention.

Receptors have two main characteristics: (1) affinity for a specific molecule (the transmitter) and (2) the triggering of a chain of reactions leading to a physiological response. Chemicals other than the natural transmitters may bind to the receptor and cause either activation (agonists) or inactivation (antagonists). Receptors may be located on the cell membrane, in the cytoplasm, or at the surface of intracellular organelles. The wide variety of locations of adrenergic receptors (Table 11.1) explains the multiplicity of the effects of adrenergic stimulation.

Adrenergic receptors are subject to regulation. Desensitization causes a reduction of the efficacy of an agonist for the receptor. It may occur because of uncoupling of the receptor G-protein, sequestration of the receptors into intracellular vesicles, or destruction of the receptors. The latter is termed down-regulation. With low catecholamine background activity, the number of receptors increases (up-regulation). Conversely, exposure to high catecholamine concentrations (chronic cardiac failure, phaeochromocytoma) decreases the number of receptors (down-regulation). This also occurs after administration of exogenous catecholamines and their derivatives.[6] Chronic β-adrenoceptor blockade causes an increase in β-adrenoceptor density. This may contribute to the rebound hypertension and worsening of myocardial ischaemia which are observed after abrupt withdrawal of β-adrenoceptor blockade.[7]

Postsynaptic α_1- and β_1-receptors are sensitive to noradrenaline, while presynaptic receptors are sensitive to adrenaline. Because of these differences in sensitivity, α_1- and β_1-receptors behave as transmitter receptors, while presynaptic α_2- and β_2-receptors behave as hormonal receptors.

Table 11.1 Location of adrenergic receptors

Location	Subtype of receptor
Heart	β_1, β_2, α_1
Blood vessels	α_1, α_2, β_1, β_2
Lungs (bronchi)	β_2, β_1, α_1
Gut	α_1, β_1
Uterus	α_1, β_2
Eye	α_1, β_2
Skin	α_1
Liver	α_1
Kidney	β_1, α_2
Pancreas	β_2, α_2
Adipose tissue	β_1
Platelets	α_2

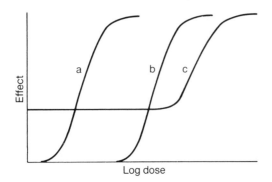

Fig. 11.1 Dose–response curves: effects of pure antagonists and partial agonists. Curve a, dose–response to a pure agonist; curve b, dose–response in the presence of an antagonist; curve c, dose–response in the presence of a partial agonist.

In usual situations, cardiovascular regulation is influenced mostly by activation of sympathetic nerve endings (α_1- and β_1-receptors). However, in stress situations, adrenaline is liberated and there is intense stimulation at α_2- and β_2-receptors.

Three types of drugs interact with the receptors: the agonists, partial agonists and antagonists. Most of the interactions are competitive. In the presence of an antagonist, the dose–response curve to the agonist is displaced to the right (Fig. 11.1). Partial agonists exert a stimulating effect on the receptors and prevent the association of agonists with the receptors.

Adrenergic agonists

Noradrenaline, adrenaline, and dopamine are naturally occurring catecholamines which, with their structurally related synthetic derivatives, appear to act on at least five types of receptors (Table 11.2). Stimulation of the postsynaptic α_1-receptors causes peripheral arteriolar constriction and venoconstriction. Stimulation of the presynaptic α_2-receptors decreases the release of the neurotransmitter. β_1-Receptor stimulation is responsible for most of the cardiac effects and β_2-receptor stimulation for most of the peripheral effects of β-adrenergic stimulation. However, β_2-receptors are also found in the myocardium. They constitute 15 per cent of the population on β-adrenoceptor in ventricular muscle and 25 per cent in atrial muscle. Finally, specific dopaminergic receptors in the renal, mesenteric, coronary, and cerebral vascular beds are the mediators of dopamine-induced vasodilatation. This response is most pronounced in the mesenteric and renal arteriolar territories. The haemodynamic effects of adrenergic receptor stimulation are summarized in Table 11.3.

With the exception of salbutamol and low dose dopamine that are used to obtain either peripheral or splanchnic and renal vasodilatation, catecholamines are predominantly used to increase the inotropic state of the myocardium and to increase arterial pressure. Unless the drug causes vasodilatation (e.g. isoprenaline, dobutamine, and dopexamine) arterial pressure increases because of the improved performance of the cardiac pump and/or peripheral vasoconstriction.

Catecholamines and their derivatives are used in the treatment of cardiogenic shock following myocardial infarction, cardiac failure following cardiac surgery, circulatory failure complicating septicaemia, to improve cardiac performance during artificial ventilation with positive end-expiratory pressure, and to increase cardiac output above the normal range to improve oxygen delivery and oxygen consumption. As increases in inotropy are frequently associated with increases in heart rate, myocardial oxygen consumption may increase markedly. In patients with coronary heart disease, this may cause myocardial ischaemia.

Haemodynamic profile of the catecholamines

The haemodynamic profile of some of the catecholamines used in clinical practice is summarized in Fig. 11.2, while the usual doses are listed in Table 11.4. For most of the catecholamines, administration by continuous infusion is necessary because of their very short duration of action.

Table 11.2 Major sites of action of adrenergic agonists

Drug		α_1	α_2	β_1	β_2	Dopaminergic
Noradrenaline		+++	+++	+	+	
Adrenaline		+	+	+	+	
Isoprenaline				+	+	
Dopamine	ld					+
	md			+		
	hd	+		+		
Dobutamine				+	(+)	
Salbutamol					+	

ld, low dose (<5 µg/kg per minute); md, doses between 5 and 15 µg/kg per minute; hd, doses in excess of 15 µg/kg per minute.
(+), Weak effect; +, moderate effect; +++, very strong effect.

Table 11.3 Cardiovascular effects of adrenergic receptor stimulation

Receptor	Effects
α_1	Peripheral vasoconstriction
	Venoconstriction
	Increased inotropy
α_2	Presynaptic sympathetic inhibition
	Vasoconstriction
β_1	Positive chronotropy
	Positive inotropy
	Increased atrioventricular conduction
	Increased myocardial excitability
β_2	Peripheral vasodilatation
	Presynaptic sympathetic stimulation
	Positive chronotropy
	Increased atrioventricular conduction
DA_1	Renal vasodilatation
	Mesenteric vasodilatation
DA_2	Presynaptic dopaminergic inhibition

DA_1, DA_2, dopaminergic.

Table 11.4 Usual doses of catecholamines

Drug	Infusion rate (μg/kg per minute)
Noradrenaline	0.01–0.07
Adrenaline	0.06–0.18
Isoprenaline	0.02–0.18
Dopamine	
Dopaminergic	1–5
β_1 effect	5–15
α and β effects	>15
Dobutamine	2–40
Salbutamol	0.2–0.5
Dopexamine	2.5–10.0

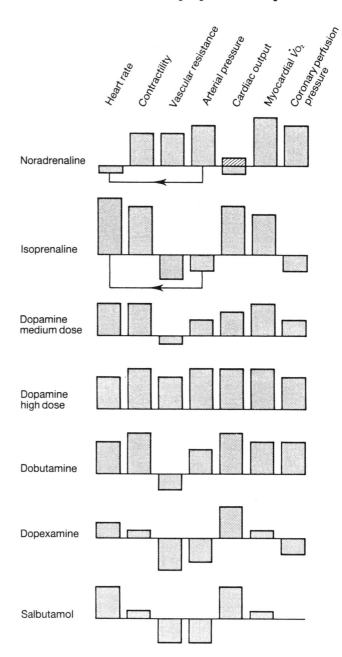

Fig. 11.2 Haemodynamic profile of some catecholamines and their derivatives.

Noradrenaline

The main haemodynamic effects of noradrenaline are α-adrenoceptor-mediated arteriolar vasoconstriction, venoconstriction and β_1-adrenoceptor-mediated positive inotropy resulting in increases in arterial pressure while cardiac output may increase or decrease. In response to the increase in arterial pressure, there is a baroreceptor-mediated bradycardia. Renal, hepatic, and muscle blood flows are all substantially reduced. However, in septic shock, normalization of haemodynamics by noradrenaline improves renal function.[8] The effects of noradrenaline on the myocardium are a combination of increased afterload (vasoconstriction), increased preload (venoconstriction) and increased contractility. The latter is caused by increased

myoplasmic calcium and by an increase in the sensitivity of the contractile proteins to calcium. When the coronary arteries are narrowed, coronary blood flow may increase because of the increase in aortic diastolic pressure, thus improving myocardial metabolism and myocardial performance. However, oxygen demand may still exceed supply. Noradrenaline is used increasingly frequently in septic shock when peripheral vascular resistance is greatly reduced. Noradrenaline is also used to prevent hypotension during venodilator therapy with glyceryl trinitrate. In congestive heart failure, the inotropic response is reduced because of β-receptor 'down-regulation', and cardiac output may decrease.

Adrenaline

Adrenaline acts on the peripheral α- and β-adrenoceptors and on the cardiac β-receptors. The increase in myocardial contractility is accompanied by an increase in heart rate, cardiac automaticity, and atrioventricular node conduction. Because adrenaline causes less vasoconstriction and hypertension than noradrenaline the direct chronotropic effect (β_1-adrenoceptor stimulation) is less inhibited by baroceptor reflexes. At low doses, cardiac stimulation is predominant and systemic vascular resistance may decrease, while at high doses α_1-mediated vasoconstriction becomes important. Renal blood flow and glomerular filtration rate are reduced by adrenaline. The development of adrenaline-induced tachycardia and arrhythmias may compromise coronary perfusion.

Isoprenaline

Isoprenaline acts on the β_1- and β_2-adrenoceptors. Its effects include increases in inotropy, chronotropy, atrioventricular conduction, automaticity, and peripheral and pulmonary vasodilatation. Large increases in cardiac output are obtained, facilitated by the unbridled chronotropic response, contributed to by a reflex increase secondary to arterial hypotension. Because of the tachycardia and reduced aortic diastolic pressure, coronary perfusion may be adversely affected and, in the presence of coronary artery disease, myocardial ischaemia can develop. Indeed, experimentally, isoprenaline has been shown to increase the size of myocardial infarction and to increase the extent of ST segment elevation. Thus, isoprenaline is not indicated in the treatment of circulatory failure caused by myocardial ischaemia or infarction. However, when circulatory failure is associated with bradycardia, isoprenaline is useful, particularly in the emergency treatment of heart block. In addition, isoprenaline may be advantageous when patients exhibit exaggerated responses to the α-adrenoceptor mediated effects of adrenaline, noradrenaline and dopamine.[9]

Dopamine

The effects of dopamine on the circulation are dose-dependent (see Table 11.2). In low doses, dopamine acts predominantly on receptors in the renal and mesenteric vasculature, increasing blood flow, glomerular filtration, and natriuresis. With higher doses of dopamine, β_1-adrenoceptor stimulation becomes predominant, myocardial contractility is enhanced, and both cardiac output and arterial pressure increase. The elevation of arterial pressure tends to lessen the chronotropic response to β_1-receptor stimulation. As dopamine maintains or improves the coronary perfusion pressure the risk of myocardial ischaemia is less than with isoprenaline. At the highest doses, α_1-adrenoceptor-mediated peripheral vasoconstriction occurs.

Dobutamine

This β_2-adrenoceptor agonist causes less β_2-receptor stimulation than isoprenaline, and much less α-adrenoceptor stimulation than noradrenaline. Dobutamine does not exert any effect on the dopaminergic receptors. Dobutamine, for an equivalent enhancement of contractility, causes less tachycardia than isoprenaline. Enhancement of left ventricular performance is reflected in reductions of left ventricular end-diastolic pressure and volume which enhance coronary perfusion. However, oxygen demand may still outstrip oxygen supply and dobutamine may worsen myocardial ischaemia. Dobutamine is an effective inotropic agent in the treatment of circulatory failure after cardiac surgery and is also very useful for patients with low cardiac output syndromes. As dobutamine causes an inotropic response associated with vasodilatation, it represents the model of an 'inodilator'. Thus other drugs offering the same therapeutic profile, i.e. the phosphodiesterase inhibitors, are often evaluated with reference to dobutamine.[10]

Dopexamine

Dopexamine hydrochloride is a potent β_2-adrenoceptor agonist with activity at dopaminergic DA_1- and DA_2-receptors. As it is a potent inhibitor of the uptake-1 process, it potentiates the actions of endogenous catecholamines. Dopexamine possesses one-third of the potency of dopamine on DA_1-receptors. Dopexamine causes renal and splanchnic vasodilatation and increases natriuresis. Experimental studies suggest that dopexamine increases the likelihood of recovery of renal function following haemorrhagic shock.[11] The reduced resistance contributes to moderate increases in cardiac output and heart rate.[12]

Table 11.5 Comparison of dopexamine, dopamine and dobutamine

Receptors	α	β_1	β_2	DA_1	DA_2	Uptake$_1$ inhibition
Dopexamine	0	(+)	+++	++	+	+++
Dopamine	+++	++	(+)	+++	++	++
Dobutamine	++	+++	++	0	0	+

0, no effect; (+), weak effect; +, moderate effect; ++, strong effect; +++, very strong effect.

The inotropic activity of dopexamine is probably due to β_2-adrenoceptor stimulation since its activity on the β_1-adrenoceptor is very weak. In chronic heart failure, β_1- and not β_2-adrenoceptors are down-regulated. This reduces the efficacy of β_1-adrenoceptor agonists but does not prevent the inotropic effect of dopexamine.

The place of dopexamine in intensive care appears to be in patients with low output circulatory failure associated with an elevated vascular resistance. Improvement of oxygen delivery to the splanchnic region contributes to a reduction of the risk of multiple organ failure. A comparison of dopexamine, dopamine, and dobutamine is shown in Table 11.5.

Salbutamol

Salbutamol is a β_2-adrenoceptor agonist with selectivity for bronchial and vascular smooth muscle. The major effect of salbutamol on the cardiovascular system is to cause peripheral vasodilatation. This facilitates left ventricular ejection and reduces ventricular wall stress. Salbutamol causes increases in heart rate. This may result from stimulation of the cardiac β_2-adrenoceptors and reflexes elicited by systolic hypotension. As salbutamol causes more increase in heart rate than other peripheral vasodilators its usefulness in the treatment of circulatory failure associated with ischaemic heart disease is limited. However, salbutamol is useful in the face of protracted vasoconstriction.

Phenylephrine

Phenylephrine is a relatively selective α_1-adrenoceptor agonist. It has weak α_2-agonist activity and some activity at β-adrenoceptors. As phenylephrine causes dose-dependent vasoconstriction, arterial pressure rises and there is reflex bradycardia. Cardiac output is slightly decreased, renal and splanchnic blood flows are markedly reduced. Phenylephrine may be used to counteract the hypotension of spinal or epidural anaesthesia.

Ephedrine

Ephedrine increases the release of noradrenaline and has weak α- and β-adrenoceptor agonist properties. Ephedrine increases blood pressure by peripheral vasoconstriction and cardiac stimulation. The effects of ephedrine are variable. Ephedrine continues to be used routinely in the prevention or treatment of hypotension resulting from regional anaesthesia.

Clonidine

Clonidine is an α_2-agonist that causes presynaptic α_2-mediated inhibition of noradrenaline release from nerve terminals, reduces renin secretion and acts on medullary pathways involved in vasomotor control. This may be the most important mechanism for clonidine in the control of hypertension. Clonidine has been used successfully in the prevention of the large variations in arterial pressure observed during anaesthesia and surgery,[13] and has been shown to minimize the release of adrenaline and noradrenaline.[14] Clonidine causes sedation, decreases the minimum alveolar concentration of inhalational anaesthetics, and reduces opioid requirements.[15] These effects suggest that α_2-adrenoceptors may play a role in the mechanisms of anaesthesia and pain relief, possibly as a result of altered ion fluxes.

Dexmetedomidine

Dexmetedomidine is a more selective α_2-agonist than clonidine; it is a full agonist known to reduce the requirements for opioids, thiopentone, and inhalation anaesthetics. In doses of 1 and 2 µg/kg, dexmetedomidine causes a decrease in plasma catecholamines in healthy volunteers, associated with reductions in mean arterial pressure and cardiac output.[16] Such doses cause sedation and result in loss of responsiveness in most subjects. This is associated with minimal ventilatory depression.[17]

Adrenoceptor antagonists: β-adrenoceptor antagonists

As a group, β-adrenoceptor antagonists exhibit five main characteristics:

1. they are competitive antagonists at the β-receptors;
2. they may exhibit receptor subtype selectivity;
3. they may exhibit partial agonist activity;
4. they possess, at very high doses, 'quinidine-like activity';
5. they may cause an increase in the number of β-receptors, when administered over a prolonged period of time.

Competitive antagonism

β-Blockers displace the dose–response curve to isoprenaline or dobutamine to the right. Because they are competitive, their effects can be reversed by the administration of large doses of β-adrenoceptor agonists.

Receptor subtype selectivity

The structure of the β-adrenoceptor antagonists is relatively similar to that of the sympathomimetic amines. Some β-adrenoceptor antagonists have greater affinity for the cardiac (β_1) than for the peripheral (β_2) adrenoceptors. Propranolol is considered non-selective, since it is almost equally as effective in blocking β_1- and β_2-adrenoceptors. Similarly, nadolol, oxprenolol, pindolol and timolol are non-

Table 11.6 Classification of the β-adrenoceptor blockers

	Cardioselective (β₁)	Non-selective (β₁ and β₂)
Pure agonists	Atenolol Metoprolol	Nadolol Propanolol Sotalol Timolol Esmolol Labetalol*
Partial agonists	Acebutolol Practolol	Alprenolol Oxprenolol Pindolol

* Labetalol is also an α-adrenoceptor blocker.

selective β-adrenoceptor antagonists. Cardioselective antagonists (acebutolol, atenolol, metoprolol) act predominantly on the β₁-adrenoceptors, reducing the positive inotropic and chronotropic responses to isoprenaline without inhibiting peripheral vasodilatation and bronchodilatation. Nevertheless, when large doses of β₁-adrenoceptor blockers are administered, they eventually block both β₁- and β₂-receptors.

Partial agonist activity

Partial agonists stimulate β₁- or β₁- and β₂-adrenoceptors, an effect previously called intrinsic sympathomimetic activity. However, their presence at receptor sites prevents the effect of other agonists.[18] Partial agonists increase heart rate and contractility. Typical examples include acebutolol, oxprenolol, pindolol and xamoterol. They cause less resting bradycardia and, in the case of borderline heart failure, may be better tolerated than pure antagonists (Table 11.6).

Membrane stabilization

Membrane stabilization is caused by inhibition of sodium transport. With the exception of atenolol and nadolol, the β-adrenoceptor antagonists cause some membrane stabilization in the heart and, therefore, resemble quinidine but only at concentrations that are considerably greater than those required to obtain adequate clinical β-adrenoceptor blockade.

Up-regulation of β-adrenoceptors

Prolonged administration of β-adrenoceptor blockers causes an increase in the number of β-adrenoceptors. This up-regulation may be partly responsible for the adverse effects of the abrupt withdrawal of β-adrenoceptor blockers.

Antiarrhythmic action

The major antiarrhythmic effect of β-adrenoceptor blockade is the prevention of the arrhythmogenic effect of the endogenous and exogenous catecholamines. Membrane stabilization is probably of little importance. However, increased duration of the action potential contributes to the antidysrhythmic efficacy of sotalol.[19] As β-adrenoceptor blockers decrease heart rate and contractility, thereby reducing myocardial oxygen consumption, they are also effective in the treatment of ischaemia-related arrhythmias.

Absorption and elimination

Lipophilic β-blockers (metoprolol, pindolol, propranolol, timolol) are almost completely absorbed from the gastro-intestinal tract. The more hydrophilic β-blockers (atenolol, nadolol) are absorbed to a much lesser extent (25–50 per cent). Because of degradation in the gut wall and liver (first-pass effect), bioavailability may be as low as 25 per cent (nadolol) or between 50 and 100 per cent (pindolol). The most lipophilic drugs have the highest bioavailability.[20,21]

Though data on bioavailability are very important when oral administration is to be replaced by intravenous administration, in practice it is necessary to titrate the intravenous administration to obtain the desired effect.

Haemodynamic effects of β-adrenoceptor blockade

All β-adrenoceptor blockers reduce heart rate and cardiac output because they reduce the effect of β₁-receptor stimulation on the heart. Their effects are more pronounced when sympathetic activity is exaggerated and less pronounced when it is depressed. β₂-Adrenoceptor blockade increases peripheral vascular resistance as it leaves α-adrenoceptor stimulation unopposed. Changes in heart rate and cardiac output are of smaller magnitude with partial agonists. During exercise, the heart rate response is less and cardiac output increases less than in the absence of β-blockade.

Effect on respiratory function

β-Adrenoceptor blockade may worsen airway obstruction especially when non-selective antagonists are used. In addition, they may decrease the ventilatory response to CO_2.

Reversal of β-adrenoceptor blockade

Reversing the effect of β-adrenoceptor blockade is seldom warranted. However, bradycardia may cause concern and this may be treated with atropine. If it becomes necessary to increase the inotropic state of the myocardium, β-adrenoceptor agonists are effective, but the doses required may be 5–20 times the usual doses. When non-selective β-adrenoceptor antagonists have been administered, large doses of the non-selective agonist isoprenaline are effective. When cardioselective β_1-adrenoceptor antagonists have been administered, a β_1-agonist, such as dobutamine, should be administered.

Withdrawal syndrome

Sudden withdrawal of adrenergic blockade may be dangerous. The syndrome of withdrawal consists of development of ventricular arrhythmias, worsening of angina, myocardial infarction and even sudden death. This may reflect β-adrenoceptor up-regulation.

Cardiac protection by β-adrenoceptor blockers

β-Receptor blockers improve the stability of the cardiovascular system and protect the myocardium in patients with coronary heart disease,[22,23] hypertensive heart disease, dysrhythmias, or obstructive cardiomyopathies.

β-Adrenoceptor antagonists in anaesthesia

It has been demonstrated that β-adrenoceptor blockade causes only modest reductions in cardiac output under anaesthesia in hypertensive patients, while decreasing the incidence of arrhythmias and myocardial ischaemia after laryngoscopy and intubation. It is recommended that their administration be continued until the day of surgery,[24,25] and postoperatively (if necessary by the intravenous route).

β-Adrenoceptor antagonists may be used deliberately in order to prevent anaesthetic-related dysrhythmias (laryngoscopy, tracheal intubation, bronchoscopy, dental surgery, cardiac and vascular surgery, neurosurgery, surgery of the thyroid gland, phaeochromocytoma, administration of catecholamines). They are also used to prevent or limit the tachycardia that is associated with induced hypotension, and to prevent or treat anaesthesia- and surgery-related hypertensive crises caused by sympathetic overactivity. Even a single dose of a β-blocker given as premedication decreases the incidence of episodes of myocardial ischaemia.[26]

It is beyond the scope of this review to discuss the small differences that exist between different β-adrenoceptor blockers. However, two drugs deserve special consideration, one, esmolol, because of its very short duration of action, the other, xamoterol, because it is a partial agonist used in the treatment of heart failure.

Esmolol

Esmolol is a relatively selective β_1-receptor antagonist that undergoes rapid hydrolysis by esterases.[27,28] Hydrolysis by erythrocyte esterases reduces the terminal half-life of esmolol to 9.2 minutes. Onset of blockade occurs in about 5 seconds with a maximum reached after about 1 minute. Esmolol is 50-fold less potent than propranolol and sixfold less potent than metoprolol.[29,30] Infusions of esmolol 50–200 μg/kg per minute cause a degree of blockade similar to that of 3–6 mg propranolol.

Perioperative tachycardia and hypertension are the main indications for esmolol.[31] In coronary artery surgery patients, esmolol given throughout the early phases of surgery decreases the cardiocirculatory effects of induction of anaesthesia, intubation, sternotomy and aortic dissection.[32] Esmolol is also effective in the control of postoperative hypertension following coronary artery bypass surgery. A possible adverse effect of esmolol is hypotension. It is dose-related and resolves within 30 minutes after the reduction of the dosage.

Ultrashort-acting β-adrenoceptor blockade is a useful therapeutic modality when sympathetic overactivity is limited in time. Moreover, the rapid reversibility of β-blockade may make esmolol safer than longer acting β-blockers in the perioperative period.

Xamoterol

Xamoterol is a partial agonist at the β_1-adrenoceptors with little effect on vascular resistance. Its partial agonist activity amounts to 45 per cent of the maximum activity of the full agonist isoprenaline.[33] In patients with poor left ventricular function xamoterol causes significant increases in cardiac index associated with reductions in pulmonary wedge pressure; it improves exercise tolerance and attenuates the exercise-induced increase in heart rate.[34] In chronic heart failure, xamoterol significantly increases exercise duration (bicycle ergometer) and improves both breathlessness and tiredness.[35]

Adrenoceptor antagonists: α-adrenoceptor antagonists

α-Adrenoceptor antagonists are used in the treatment of hypertensive emergencies, and in the long-term treatment of arterial hypertension and chronic cardiac failure.

With the exception of phenoxybenzamine, the α-adrenoceptor antagonists are competitive blockers. They can exhibit selectivity for the postsynaptic α_1-receptors (prazosin) or the presynaptic α_2-receptors (yohimbine). Phenoxybenzamine and phentolamine are non-selective antagonists while labetalol blocks α_1-, β_1- and probably β_2-receptors. The administration of α-adrenoceptor antagonists causes peripheral vasodilatation in addition to some venodilatation. The resulting arterial hypotension causes baroreceptor-mediated increases in sympathetic activity.

Phenoxybenzamine

Phenoxybenzamine is a non-selective, non-competitive α-adrenoceptor antagonist with a slow onset of action. The marked vasodilatation produced by phenoxybenzamine is useful in the short-term preservation of human kidneys for transplantation. The major place for phenoxybenzamine is in the treatment and preoperative preparation of patients suffering from phaeochromocytoma, usually in association with β-adrenoceptor blockers.[36]

Phentolamine

Phentolamine is a weak, non-selective, competitive α-adrenoceptor antagonist. Less potent than phenoxybenzamine, phentolamine causes vasodilation mostly by a direct effect on vascular smooth muscle. It has a rapid onset of action and its effects last for approximately 15–30 minutes. In patients suffering from myocardial infarction and chronic heart failure phentolamine increases stroke volume and decreases the pulmonary capillary wedge pressure. The occurrence of tachycardia may be a significant disadvantage. Phentolamine can abolish α-adrenoceptor-mediated coronary artery spasm. In patients with phaeochromocytoma, phentolamine is effective in the treatment of paroxysmal hypertension during operative manipulations of the tumour. Phentolamine, administered as a bolus dose or by continuous infusion, may be used to control intra- and postoperative hypertension and also to control the perfusion pressure during extracorporeal circulation for cardiac surgery.

Prazosin

Prazosin is a selective, competitive, α_1-adrenoceptor blocker which reduces arterial and venous tone. Because α_2-receptors are not blocked, the negative feedback effect of noradrenaline on its own release is not inhibited. This explains why tachycardia and renin stimulation do not occur. The major side effects of prazosin are postural hypotension and occasionally a precipitous decrease in arterial pressure after the first dose. Prazosin is used in the treatment of congestive cardiac failure and has been shown to increase cardiac output and decrease the pulmonary capillary wedge pressure.

Labetalol

Labetalol differs from other adrenergic receptor antagonists in that it blocks both α- and β-adrenoceptors. Less potent an α-blocker than phentolamine, it is also much less potent a β-blocker than propranolol. Labetalol does not block the presynaptic α_2-receptors, but blocks both β_1- and the β_2-receptors. Labetalol produces a decrease in arterial pressure because of the simultaneous reductions in systemic vascular resistance and cardiac output. Labetalol is effective and provides smooth control of arterial pressure in hypertensive emergencies.

Thymoxamine

Thymoxamine is an α_1-selective antagonist. Although it may be effective in the treatment of circulatory failure after cardiac surgery, most of the published work relates to the treatment of peripheral vascular disease. Thymoxamine improves regional peripheral blood flow after oral, intravenous, or intra-arterial administration.

Cholinergic agonists and antagonists

Cholinergic agonists

These drugs may either bind to the cholinergic receptors (direct action) or inhibit the action of the enzyme cholinesterase, thus increasing the concentration of endogenous acetylcholine.

Muscarinic agonists appear to stimulate the enzyme guanylate cyclase, increase potassium flux across the cell membrane, and increase the turnover of inositol phospholipids in cell membranes.

Nicotinic agonists modify the conformation of nicotinic receptors so that sodium and potassium ions are allowed to diffuse down their concentration gradients: depolarization of the nerve cell or neuromuscular end plate occurs.

By increasing the membrane permeability to potassium in atrial muscle and in sinoatrial and atrioventricular nodes, and by decreasing the slow calcium current, cholinergic agonists reduce the pacemaker rate and decrease the conduction velocity. The mechanisms of peripheral arteriolar and venous dilatation by acetylcholine involves the release of nitric oxide by the vascular endothelium.

Edrophonium is the only cholinergic stimulant used for its cardiovascular effects, as it may convert supraventricular tachyarrhythmias into sinus rhythm. The vagolytic effect of edrophonium has been used to convert atrial tachycardia to sinus rhythm. It is used as an intravenous bolus dose at 10–20 mg.

Cholinergic antagonists

These can be subdivided into antimuscarinic and antinicotinic agents. The latter comprise ganglion-blocking and neuromuscular-blocking drugs.

Muscarinic receptor blocking agents cause tachycardia in moderate and high doses, while in low doses they may cause bradycardia because of central vagal stimulation. Atrioventricular conduction increases and any parasympathetically mediated depression of cardiac performance is abolished. Sympathetic cholinergic vasodilatation is blocked by muscarinic antagonists.

Atropine

Atropine is effective in preventing or abolishing the bradycardia associated with vagal discharge, irrespective of its cause. It is also effective in preventing the bradycardia associated with the administration of anticholinesterases to reverse neuromuscular blockade, and in preventing or abolishing the bradycardia associated with direct carotid sinus stimulation.

Glycopyrronium

Glycopyrronium (glycopyrrolate) has a longer duration of action than atropine and provides prolonged cardiac protection against the muscarinic effects of neostigmine. The onset of action of both glycopyrronium and neostigmine are similar and heart rate remains stable when they are administered together. Glycopyrronium does not penetrate the blood–brain barrier and this minimizes the risk of central anticholinergic syndrome.

Ganglionic blocking agents

Drugs that block this subtype of nicotinic receptor exert powerful cardiovascular effects because they block both sympathetic and parasympathetic transmissions. Ganglion blockade may be caused by depolarizing drugs which stimulate before blocking transmission (nicotine), or by non-depolarizing drugs which block transmission without causing any stimulation. Hexamethonium, trimetaphan and pentolinium are competitive non-depolarizing ganglion blockers. Their effects on the circulation depend on the level of sympathetic activity. Usually the capacitance vessels are more sensitive to ganglion blockers than the resistance vessels and blood pooling on assuming the erect position is a major determinant of orthostatic hypotension. Myocardial contractility is reduced because of reduced sympathetic activity while heart rate increases because of reduced vagal tone. Cerebral blood flow is maintained but renal blood flow decreases.

Hexamethonium

Hexamethonium is a selective nicotinic receptor antagonist. Tachyphylaxis develops rapidly because of the dominance of the muscarinic pathway when the nicotinic pathway is blocked.

Pentolinium

Pentolinium is more potent and longer acting than hexamethonium. As well as reducing sympathetic activity it appears to reduce plasma renin activity.

Trimetaphan

Trimetaphan is a competitive ganglion-blocking drug which exhibits some direct vasodilatation. Interruption of sympathetic outflow produces vasodilatation. The latter is relatively slow both in onset and recovery. The association of parasympathetic blockade results in mydriasis and tachycardia. Mydriasis makes neurological assessment difficult; this is an important drawback in neurosurgery. Tachyphylaxis makes it difficult to achieve a stable blood pressure reduction.

Antiarrhythmic drugs

Electrophysiology

The electrophysiological classification of the antiarrhythmic drugs introduced by Vaughan Williams subdivides antiarrhythmics into four or five classes (see Table 11.7).

Class I

Drugs in this class interfere with the fast sodium inward current. They are subclassified into three groups. Class 1a drugs depress conduction of both sinus beats and premature beats and prolong the duration of the action potential. Class 1b drugs depress conduction of premature beats and conduction through ischaemic tissue; they shorten the duration of the action potential. Class 1c drugs cause a marked depression in conduction of both normal and premature beats, without altering the duration of the action potential.

Class II

Cardiac arrhythmias can be initiated or exacerbated by stress or emotions; thus drugs with antisympathetic effects, such as β-adrenoceptor blockers, may be regarded as antiarrhythmics. In addition, some β-blockers have a class I action on cardiac muscle and some prolong the duration of the action potential.[37]

Table 11.7 Vaughan Williams' classification of antiarrhythmic drugs

	Class I	Class II	Class III	Class IV
	Na+ channel blockers	β-Blockers	K+ channel blockers	Ca2+ channel blockers
1a	Quinidine Procainamide Disopyramide	β-Blockers	Amiodarone Bretylium tosylate Sotalol	Verapamil Diltiazem
1b	Lignocaine Phenytoin Mexiletine Tocainide			
1c	Flecainide Encainide Propafenone			

Class III

Class III drugs prolong the duration of the action potential by selectively decreasing potassium ion conductance.[38] They are very effective in prolonging atrial refractoriness and abolishing an experimentally induced circus movement.[39] It is still unclear whether the antiarrhythmic effect is solely due to the increased duration of the action potential since most agents possess other actions such as antiadrenergic activity.

Class IV

The antiarrhythmic effects of class IV drugs result from blockade of the second inward current carried by calcium ions.[40] Blockade of the calcium channels produces a slowing of conduction and prolongs the refractoriness within sinus and atrioventricular nodes. The sinus rate decreases, the PR interval is lengthened, and there is reduced ventricular response to atrial arrhythmias. Calcium antagonists (especially verapamil and diltiazem but not nifedipine and its analogues) are effective in terminating episodes of paroxysmal supraventricular tachycardia. They slow the ventricular response in atrial flutter and fibrillation and, less reliably, control multifocal atrial tachycardia.

In the ischaemic myocardium, the second inward current could take over the depolarizing function, so that a slow action potential may conduct impulses at low velocity, encouraging the development of re-entry.

The main side effect of calcium antagonists is hypotension. The association of calcium channel blockers with β-blockers or disopyramide may cause severe cardiac depression in patients with poor left ventricular function.

Class V

This class of drugs includes specific bradycardic agents.[41] Alinidine has a selective bradycardic action on the sinoatrial node because it reduces the slope of slow diastolic depolarization. It does not block the sodium channels, does not have a negative inotropic action and does not block the positive inotropic effect of the β-adrenoceptors.[42] The effect of alinidine may be mediated by a reduction in the chloride current.

Antiarrhythmic drugs within the Vaughan Williams classification

Class Ia

Quinidine
Quinidine slows down conduction in the bundle of His, decreases the rate of diastolic depolarization, and prolongs the effective refractory period. It is effective in the prophylaxis of supraventricular dysrhythmias, particularly those involving re-entry, and in the treatment of ventricular tachycardia. It facilitates cardioversion and maintenance of sinus rhythm following cardioversion.[43]

Procainamide
Procainamide is particularly effective in the treatment of ventricular arrhythmias. The negative inotropic action of procainamide may cause cardiac failure. Procainamide is also used to maintain sinus rhythm after DC conversion.

Disopyramide
Disopyramide is effective in patients with atrial and ventricular arrhythmias. Negative inotropy is more

pronounced in patients with pre-existing heart failure. Disopyramide is successful in maintaining patients in sinus rhythm after conversion of atrial fibrillation[44] and in the treatment of supraventricular tachycardia associated with the Wolff–Parkinson–White syndrome.

All three drugs increase sinus rate because of their anticholinergic properties and a reflexly mediated increase in sympathetic tone. However, caution should be exercised as slowing of heart rate may occur in patients with pre-existing sinus node disease and in the transplanted hearts.

Class Ib

Lignocaine

Lignocaine is the first-line drug for the treatment of ventricular arrhythmias associated with acute myocardial infarction, digitalis toxicity and, together with cardioversion, for the treatment of ventricular fibrillation. Neurological side effects have been well documented, ranging from paraesthesiae, drowsiness, nausea, confusion to convulsion and respiratory arrest,[45] especially in the elderly. In patients with poor left ventricular function circulatory failure may occur, and in those with conduction disorders there is the risk of heart block.

Mexiletine

Mexiletine is available for oral and intravenous administration, and is effective in the treatment of ventricular arrhythmias especially after myocardial infarction. However, overall mortality may be increased.[46] Arrhythmias (sinus bradycardia, atrial fibrillation and heart blocks) may occur.

Tocainide

Tocainide is available in oral and intravenous forms. Its therapeutic indications are the same as those of mexiletine.[47] However, the relatively high incidence of neutropenia and of interstitial pneumonitis precludes its use as first-choice drug.

Class Ic

Flecainide

Flecainide is an oral antiarrhythmic with an elimination half-life averaging 14 hours. Flecainide is effective in the control of supraventricular tachycardias. However, the drug is arrhythmogenic and raises the threshold for both defibrillation and ventricular pacing.

Encainide

The overall drug action is that of its metabolites O-demethyl encainide and 3-methoxy-O-demethyl encainide.[48] Encainide is effective in the control of ventricular

dysrhythmias but may not be effective in the prevention of sustained ventricular tachycardia.

Propafenone

Propafenone is effective in the treatment of ventricular and supraventricular arrhythmias. It has some β-blocking activity in addition to class Ic effects.

Class II agents

β-Adrenoceptor blockers are effective in the treatment of arrhythmias caused by increased sympathetic activity. They are also effective in the management of arrhythmias associated with myocardial infarction. These arrhythmias are due to a number of factors including increases in plasma catecholamines, leakage of potassium due to hypoxia, rise in free fatty acids, and alterations in action potential duration (shortened by hypoxia; lengthened by acidosis), all of which enhance the heterogeneity of repolarization.

β-Blockers cause significant reductions in sinus rate and decrease the number of both supraventricular and ventricular arrhythmias.[49] This is associated with a reduction in the incidence of ventricular fibrillation, and a reduction in mortality.[50] β-Blockers may also reduce the risk of cardiac rupture.[51]

Class III agents

Amiodarone

Amiodarone is a potent antiarrhythmic agent in the control and prophylaxis of most supraventricular and ventricular arrhythmias.[52] Amiodarone has non-competitive antiadrenergic actions, resulting from the inhibition of the coupling of β-receptors with the regulatory unit of the adenylate cyclase complex.[53] In addition, amiodarone reduces the density of β-adrenoceptors.

Amiodarone is unique in its slow hepatic metabolism resulting in an elimination half-life of 26–107 days.[54]

The haemodynamic effects of amiodarone include reductions in heart rate, blood pressure, systemic and coronary vascular resistances. In patients with normal ventricular function or only mild left ventricular dysfunction, intravenous amiodarone exerts a mild negative inotropic effect associated with peripheral vasodilatation.[55] The negative inotropy is more prominent in patients with moderate or severe left ventricular dysfunction.[56] However, continuous infusion of doses up to 1000 mg per day is well tolerated even by patients with severe left ventricular dysfunction.

Oral amiodarone is highly effective in paroxysmal and persistent atrial fibrillation.[57] It is also effective in the management of drug-resistant paroxysmal supraventricular tachycardias and in the prevention of the recurrence of tachycardia in the Wolff–Parkinson–White syndrome.[58]

Oral amiodarone is highly effective in the treatment of ventricular arrhythmias even when other antiarrhythmic drugs have failed.[54] These advantages must be balanced against the potentially serious side effects. The most serious complication is pulmonary toxicity resulting in diffuse bilateral interstitial changes accompanied by patchy alveolar infiltrates, and mild hypoxaemia.[59]

As amiodarone contains iodine (75 mg for each 250 mg tablet) and is deiodinated *in vivo*, excess iodine may cause hypothyroidism. Microdeposits of the drug or its metabolites in the cornea occur during long-term therapy with amiodarone.

Amiodarone reduces both the renal and non-renal elimination of digoxin.[60] As amiodarone prevents the development of ventricular ectopic activity, digoxin toxicity is characterized by sinus arrest and gastrointestinal symptoms.

Serious arrhythmias may occur when class I antiarrhythmics are used together with amiodarone; therefore, their dosage should be reduced. Additive suppressive effects on sinus and atrioventricular nodes may occur with β-adrenoceptor blockers and calcium antagonists.

Sotalol

Sotalol reduces the time-dependent outward K^+ current activated during the plateau of the action potential resulting in a dose-dependent increase in the action potential duration of atrial, ventricular and Purkinje fibres.[61] It is also a non-selective β-receptor antagonist.

Sotalol reduces heart rate and cardiac output. The class III effects of sotalol are not unique. Other β-adrenoceptor blockers have a 'chronic class III effect' that may reduce the incidence of sudden death after myocardial infarction.[62]

Clofilium

This quaternary ammonium is a highly selective and potent class III agent. It prolongs the duration of cardiac action potential, and increases ventricular and atrial refractoriness without altering the conduction velocity in normal or ischaemic myocardium.

Clofilium does not act on α- and β-adrenoceptors. Its electrophysiological effects are enhanced by halothane; this facilitates the development of ventricular arrhythmias, particularly when bradycardia is present.[63]

Bretylium

Bretylium causes an immediate release of noradrenaline from adrenergic terminals, and subsequently prevents the further release of noradrenaline. In addition, bretylium prolongs the duration of the cardiac action potential, and in very high doses blocks both sodium and potassium channels. The antiarrhythmic and antifibrillatory activity of bretylium is such that it has been termed a 'pharmacological defibrillator'. The main indication is the treatment of ventricular arrhythmias resistant to conventional treatment.[64]

Class IV agents

Verapamil

Verapamil reduces vascular smooth muscle tone, decreases myocardial contractility and is effective in the control of supraventricular arrhythmias. Intravenous boluses (2.5 mg repeated if necessary after 30 seconds) of verapamil (under ECG monitoring) are effective in the control of the ventricular rate in atrial fibrillation but may cause hypotension. Conduction defects are a contraindication.

Diltiazem

Diltiazem acts as an arteriolar dilator with less negative inotropy than verapamil. It is used more in the treatment of myocardial ischaemia than in the treatment of supraventricular arrhythmias.

Bepridil

Bepridil is a calcium antagonist that slows intramyocardial conduction velocity and increases the Q–Tc interval and refractoriness.[65] These effects suggest class III activity, though studies have shown bepridil to block sodium and calcium but not potassium channels.[66]

Antiarrhythmics outside the Vaughan Williams classification

Adenosine

Adenosine is an endogenous nucleoside with an extremely short half-life. The main antiarrhythmic effects of adenosine are mediated by A_1-receptors linked to an inhibitory guanine-binding protein and are characterized by a negative chronotropic effect on the sinus node and negative dromotropic effect on atrioventricular conduction.[67] Adenosine is effective in terminating paroxysmal supraventricular tachycardias involving the atrioventricular node as it blocks the anterograde limb of re-entrant circuits. As the duration of action is short, adverse reactions are also short-lived. This is an advantage over other drugs as the success rate in terminating paroxysmal supraventricular tachycardias is high.[68]

Cardiac glycosides

Cardiac glycosides cause an increase in K^+ conductance associated with a slowly developing decrease in the action potential duration, while resting membrane potential and action potential amplitude are essentially unchanged.[69] Shortening of the action potential duration by toxic doses of cardiac glycosides contributes to their arrhythmogenicity, as the refractory period is reduced.

By contrast, antiarrhythmic effects of cardiac glycosides are essentially mediated by the autonomic nervous system.[70] Vagomimetic effects of cardiac glycosides may be due to an increased sensitivity of the baroreceptors, activation of the vagal centres, inhibition of the peripheral sympathetic effects on the heart, or increased sensitivity to acetylcholine.

Atrial fibrillation and atrial flutter are the most frequent indications for cardiac glycosides. Atrial flutter is frequently converted into atrial fibrillation, and recent atrial fibrillation may revert to sinus rhythm over a few hours. This effect may be facilitated by the addition of disopyramide.[71] The main advantage of digoxin is its weak positive inotropic action. This may play an important role when the myocardium is compromised and any negative inotropic drug would worsen cardiac function.

There is no evidence that prophylactic preoperative digitalization of patients with marginal cardiac failure is of any benefit, but appropriate digoxin therapy to maintain normal heart rates (55–70 beats/min) in patients with atrial fibrillation is very important. Patients in atrial fibrillation who present for urgent surgery with heart rates over 100/ min should not be treated with intravenous digoxin, but should have their heart rates *cautiously* slowed under ECG control by intravenous cardioselective β-adrenoceptor antagonists.[72]

Perioperative management of arrhythmias

The first step in the treatment of arrhythmias occurring during anaesthesia and surgery is *not* the administration of antiarrhythmic drugs but the correction of factors which may have contributed to their development such as hypoxia, hypercapnia, hypocapnia and metabolic acidosis, or inadequate depth of anaesthesia. Once these factors have been corrected, most arrhythmias disappear. If they persist, treatment is indicated when the circulation is impaired, or when the arrhythmia is likely to develop into a life-threatening type.

Hypokalaemia poses difficult problems. Chronic hypokalaemia is usually accompanied by intracellular potassium deficiency so that the ratio of intracellular to extracellular potassium concentration may remain normal or near normal. Rapid potassium replacement may worsen the electrophysiology of the myocardium. If chronic hypokalaemia is accompanied by arrhythmias, surgery should be postponed and correction of potassium depletion should be achieved over several days and not attempted over a few

hours.[73] Acute hypokalaemia, however, should be corrected before anaesthesia and surgery.

With hyperkalaemia, the risks of further increases in potassium concentration causing life-threatening arrhythmias are such that correction should be attempted. In case of chronic renal failure haemodialysis may be necessary before surgery. During anaesthesia, arrhythmias may occur when large volumes of stored blood are given rapidly, causing sudden and severe hyperkalaemia. The intravenous administration of calcium may be necessary to suppress these arrhythmias. Other measures such as the administration of sodium bicarbonate may be necessary to cause a shift of potassium into the cells. In addition, glucose and insulin may be needed.

Supraventricular arrhythmias

Sinus bradycardia caused by vagal stimulation is frequently accompanied by hypotension and responds well to intravenous atropine or glycopyrronium. If atropine is ineffective, it may be necessary to administer a β_1-receptor agonist (dobutamine or isoprenaline).

Supraventricular tachyarrhythmias can be controlled by β-adrenoceptor blockers, calcium channel blockers and digitalis. Verapamil and diltiazem are effective in the treatment of supraventricular tachycardias. However, they reduce the inotropic state of the myocardium; their administration during anaesthesia with halogenated anaesthetics may cause exaggerated myocardial depression. However, they can be used successfully because the beneficial effects of bringing the heart rate to within the normal range may outweigh the detrimental effect on the circulation of myocardial depression. If calcium influx blockers are used, they should be given in small doses to achieve the desired effect by titration. In the absence of hypokalaemia digitalis may be preferred to calcium blockers or β-adrenoceptor blockers when the quality of the myocardium is poor.

Ventricular arrhythmias

Lignocaine is the drug of choice for the intraoperative treatment of acute ventricular arrhythmias (ventricular premature beats and ventricular tachycardia). The usual dose is 1 mg/kg body weight followed by an infusion to maintain blood levels between 2 and 5 μg/ml. High concentrations may cause myocardial and central nervous system depression. Isolated ventricular premature beats may not require treatment. However, treatment is necessary if they are frequent, assume the R on T configuration, or appear in salvoes.

When ventricular arrhythmias are obviously caused by sympathetic overactivity, β-adrenoceptor blockade is the treatment of choice. A small initial dose should be followed by further small incremental doses to achieve control of the arrhythmia.

Ventricular tachycardia is probably best treated with cardioversion when it impairs the circulation. In some patients ventricular fibrillation may occur and prove resistant to electroconversion. In this situation, bretylium tosylate may either restore a normal rhythm (acting as 'pharmacological defibrillator') or facilitate the conversion of ventricular fibrillation into a normal rhythm.

Inotropes

The first step in the management of low output states is to optimize the ventricular filling. If this fails to restore cardiac output, inotropic drugs and vasodilators may be necessary. As catecholamines and their derivatives have already been discussed (see Adrenergic agonists), this section will deal with cardiac glycosides, phosphodiesterase inhibitors, calcium entry promoters and calcium sensitizers.

Cardiac glycosides

Cardiac glycosides are used in the management of congestive heart failure, particularly when associated with atrial fibrillation. Cardiac glycosides decrease the activity of the membrane-bound enzyme Na^+-K^+/(Mg^{2+}-dependent) ATPase. This increases the cytosolic sodium ion concentration.[74] The uninhibited fraction of the sodium pump is stimulated, increasing the rate of exchange of calcium. Increases in both cytosolic free Ca^{2+}, and the cellular calcium transients suggest increased exchange and releasability of calcium from cellular stores.[75] The combined inotropic-bradycardic action of digitalis is unique and the weak inotropy of digoxin is useful in the treatment of congestive cardiac failure, improving survival when added to vasodilators and diuretics. In acute left ventricular failure, however, diuretics and more powerful inotropes such as dobutamine, dopamine, amrinone or enoximone have superseded digitalis, unless failure is associated with valvular disease and atrial fibrillation.

Cardiac glycosides may accumulate in patients with poor renal function. This may be accentuated following anaesthesia and surgery because of the alterations in renal function associated with the stress response to surgery. Digitalis toxicity may be precipitated during anaesthesia if the patient is hyperventilated and made hypocapnic as acute hypocapnia causes hypokalaemia. There is a particular risk in patients taking diuretics, whose plasma K^+ concentration is already at the lower limit of normal (3 mmol/l).

Phosphodiesterase inhibitors

Phosphodiesterase (PDE) is the enzyme responsible for the breakdown of cyclic AMP into 5-AMP. Inhibition of this enzyme results in an increase in the concentration of cyclic AMP and, in turn, in the activation of protein kinases. This increases calcium entry through calcium L-channels, resulting in an increase in myoplasmic Ca^{2+}. Phosphodiesterase inhibitors increase cardiac contractility and reduce peripheral vascular resistance,[76,77] hence they can be regarded as inodilators.

Until 1978 the only oral inotropic agents for use in patients with chronic heart failure were the digitalis glycosides. In 1978 the first clinical report on the inotropic effect of amrinone, a drug selective for the myocardial phosphodiesterase (PDE III or IV) was published and amrinone was hailed as a 'new digitalis'.

Bipyridine derivatives

Amrinone exerts haemodynamic effects similar to dobutamine. Given intravenously, it increases cardiac output by 30–70 per cent in patients with heart failure without changing heart rate and blood pressure. The lack of effect on heart rate and blood pressure is an advantage over the sympathomimetic amines. Amrinone has no significant effect on cardiac conduction but decreases the atrial and atrioventricular functional refractory periods. Amrinone is effective in patients with poor left ventricular function after cardiopulmonary bypass, patients with borderline ventricular function undergoing non-cardiac surgery, and after myocardial infarction. In addition, amrinone reverses the depressant effect of halothane in isolated heart papillary muscle.[78] Long-term administration of amrinone is complicated by a high incidence of side effects, particularly thrombocytopenia and ventricular arrhythmias; in addition the therapeutic results appear to be variable.

Milrinone causes more vasodilatation than amrinone, resulting in marked reduction of left ventricular end-diastolic pressure and arterial hypotension. Beneficial in patients with systemic or pulmonary congestion, milrinone's inotropic potential is limited to patients with high filling pressures. Milrinone may be particularly useful in patients who are relatively insensitive to catecholamines because of β-adrenoceptor down-regulation.[79]

In small clinical studies, milrinone was found to be much less toxic than amrinone. The recent Prospective Randomized Milrinone Survival Evaluation has shown milrinone to increase mortality by approximately 30 per cent.[80] While it was widely accepted that depression of cyclic AMP was responsible for the reduced performance of the chronically failing heart, the disappointing effects of the phosphodiesterase inhibitors and of the partial β-adrenoceptor agonist xamoterol suggest that reduced cAMP may be an adaptive response to protect myocardial cells from further injury. Thus the short time gains of enhanced inotropy do not necessarily predict better outcome.[81]

Imidazoline derivatives

Enoximone and piroximone are imidazolines with known PDE inhibitory properties. Their effects are similar to those of dobutamine but they cause greater reductions in pulmonary wedge pressure.

Enoximone is active both intravenously and orally. It has a very favourable inotropic to chronotropic dose ratio (less than 0.1) and a wide margin of safety. The positive inotropic effect is accompanied by vaso- and venodilatation. Enoximone is effective in the treatment of low output states following cardiac surgery,[82] and reverses the negative inotropy of verapamil.[83]

Long-term treatment with enoximone increases exercise tolerance, left ventricular function and exercise time.[84] This long-term effect, however, is not always observed and the place of enoximone is mainly in the short-term treatment of exacerbations of cardiac failure, or during the waiting period for cardiac transplantation.[85]

Piroximone increases cardiac output, reduces pulmonary and systemic vascular resistance and improves contractility. Though increases in contractility may increase the risk of myocardial ischaemia, piroximone has been shown to increase the anginal threshold in patients with coronary heart disease.

Calcium sensitizers

Sensitivity of troponin-C to calcium is decreased by ischaemia, hypoxia and acidosis, resulting in depression of contractility. By contrast, sensitivity is increased by α-adrenoceptor stimulation,[86] and by some phosphodiesterase inhibitors. The advantage of drugs increasing the sensitivity of troponin-C to calcium is that increased myoplastic calcium (an effect of all the other inotropic drugs) facilitates arrhythmias and increases the energy requirements of the cardiac cells as more outward transport of calcium is needed.

Pimobendan

Pimobendan, a benzimidazole derivative, is a weak phosphodiesterase inhibitor (20–30 per cent of the inhibition obtained by milrinone) that sensitizes troponin-C to calcium. The addition of pimobendan to standard therapy of heart failure increases exercise tolerance and peak oxygen uptake. Long-term the effects of pimobendan compare favourably with those of enalapril.[87]

Calcium entry promoters

While many calcium antagonists (nifedipine, nimodipine, nicardipine) are antagonists at the nitrendipine receptors, thereby reducing calcium entry into the cells, calcium entry promoters are agonists at these receptors. New drugs with a selective effect on cardiac cells are being developed.

Calcium antagonists

Calcium ions play a decisive role in the electrical activity of the heart, in the excitation–contraction coupling of skeletal, cardiac and vascular smooth muscle, and in the release of neurotransmitters at the presynaptic junctions.

Calcium current

The slow (calcium) inward current (I_{si}) is distinct from the sodium current, underlies impulse conduction in nodal tissue and is responsible for the plateau phase of the action potential. It is increased by adrenergic agonists, ischaemia and hypoxia. The slow inward current plays an important role in the development of arrhythmias. Ischaemia inactivates sodium channels so that electrical activation is initiated by the calcium current. This causes delays in impulse propagation which facilitate re-entry.

The calcium channels have two separate mechanisms. One, on the extracellular side of the sarcolemmal membrane, is voltage dependent. The other, on the cytoplasmic side of the channel is less voltage dependent and appears to be regulated by cyclic nucleotides such as cyclic AMP and cyclic GMP. Voltage dependence means that when the cell is depolarized the gate is open, and when the cell is repolarized, the gate is closed. Nucleotide dependence means that a critical phosphorylation reaction at the inner gate of the Ca^{2+} channel determines the state of the gate.[88]

Voltage-operated channels can be divided into three subtypes, L, N and T.[89] Transient calcium currents are carried by channels with a short mean lifetime and a low conductance (T channels). Other channels have a long mean lifetime and a large conductance (L channels), while some channels (N channels) have intermediate characteristics. In cardiac muscle, the major pathway for calcium entry is through L channels.[90]

Table 11.8 Classification of calcium antagonists

Selective for slow Ca^{2+} channels

I	Verapamil and derivatives (verapamil, gallopamil, anipamil)
II	Dihydropyridines (nifedipine, nicardipine, nimodipine, nisoldipine, nitrendipine)
III	Diltiazem

Non-selective for slow Ca^{2+} channels

IV	Diphenylpiperazines (cinnarizine, flunarizine)
V	Prenylamine derivatives (prenylamine, fendiline)
VI	Others (bepridil, caroverine, perhexiline)

In smooth muscle, particularly arterial muscle cells, L and T channels,[91] as well as receptor-operated and stretch-operated channels, have been described.[92] Calcium binds with calmodulin. The calcium–calmodulin complex activates myosin light chain kinase which catalyses the phosphorylation of myosin light chain. This initiates shortening of vascular smooth muscle.

Specific blockade of the slow channels

The chemical heterogeneity of the calcium antagonists suggests that they act at different receptor sites functionally linked with the calcium channels (Table 11.8).

Cardiovascular effects of the calcium antagonists

The selective inhibition of the transmembrane influx of Ca^{2+} is responsible for the depression of sinus automaticity, atrioventricular conduction, and vascular tone.

Calcium antagonists protect the ischaemic myocardium by reducing heart rate and contractility while inducing vasodilation, all of which decrease oxygen consumption. At the same time, they decrease coronary vascular resistance and thus facilitate oxygen supply. Moreover, calcium blockers prevent coronary spasm. As calcium overload plays an important role and contributes to permanent damage in totally ischaemic muscle, calcium channel blockers minimize the cellular damage caused by ischaemia.[93]

Therapeutic indications

Myocardial ischaemia
The potency of the calcium antagonists as coronary dilators is of greatest benefit in the treatment of vasospastic and unstable angina. Calcium antagonists are also effective in the treatment of effort-induced angina. However, in patients with severe left ventricular dysfunction or congestive cardiac failure their negative inotropy may be exaggerated.

Arterial hypertension
Calcium antagonists reduce arterial pressure in direct proportion to the degree of hypertension. Nifedipine is effective in the treatment of malignant hypertension complicated by encephalopathy and left ventricular failure.

Congestive heart failure
Afterload reduction by nifedipine and nicardipine is associated with improved subendocardial perfusion and left ventricular relaxation.[94]

Hypertrophic cardiomyopathy
The major effect of calcium antagonists is to improve relaxation, so that the relation between pressure and volume during diastole becomes more normal.

Cerebral artery vasospasm
Vasospasm is commonly associated with subarachnoid haemorrhage and trauma. Nifedipine and nimodipine are effective in blocking experimental vasospasm. Nimodipine improves the functional prognosis after rupture of intracranial aneurysm.[95] However, increases in intracranial pressure may occur in head injured patients.

Interactions between calcium antagonists and anaesthesia

Interactions with verapamil and diltiazem
Halothane, enflurane, isoflurane, and propofol cause dose-dependent depression of contractile performance, may modify atrioventricular conduction, and may cause peripheral vasodilatation. These effects reflect a reduction of Ca^{2+} fluxes across the cell membrane and within the cells.[96,97] Thus, potentiation of cardiac depression, vasodilatation and slowing of conduction may be expected in the presence of calcium antagonists.[98,99]

Intriguing and, as yet unexplained, is the observation of left ventricular apical dysfunction with the combined administration of verapamil and isoflurane or halothane,[100,101] probably caused by exaggerated disruption of calcium fluxes.

Interactions with nifedipine
Nifedipine causes peripheral vasodilatation and may increase cardiac output.[102] However, during high dose fentanyl anaesthesia, the increase in cardiac output may be associated with hypotension.[103] By contrast, during anaesthesia with halothane or enflurane left ventricular performance decreases with the administration of nifedipine and cardiac output may be reduced.

Protection against perioperative ischaemia

Diltiazem has been used successfully as an intravenous infusion to reduce the frequency of ischaemic ST segment depression in patients with ischaemic heart disease undergoing non-cardiac surgery and anaesthetized with fentanyl and nitrous oxide.[104] In contrast, oral medication with calcium antagonists alone did not seem to prevent the development of ischaemia in patients undergoing cardiac surgery, while β-adrenoceptor blockade was effective.[105]

Need for inotropic support

When exaggerated cardiovascular depression results from the association of calcium antagonists and inhalational anaesthetics, an i.v. bolus dose of calcium increases the inotropic state of the myocardium and restores cardiac output and arterial pressure for approximately 5 minutes. However, intravenous calcium is ineffective in reversing atrioventricular conduction disorders.[106]

Catecholamines are not always effective in the treatment of severe calcium antagonist overdose, probably because calcium fluxes are too heavily blocked and thus, myoplasmic calcium cannot be elevated. Glucagon, a polypeptide hormone which exerts positive inotropic and chronotropic actions, independent from β-adrenoceptors or noradrenaline depletion, may offer a useful alternative to calcium and catecholamines.[107] Amrinone has also been shown in several experimental models to reverse the effects of verapamil.[108]

ACE inhibitors

Renin, angiotensin, and aldosterone play a role in the development of hypertension as well as in fluids and electrolytes regulation. Conversion of angiotensin I into angiotensin II involves the angiotensin converting enzyme (ACE) which can be inhibited by ACE inhibitors. The same enzyme is responsible for the breakdown of bradykinin into inactive metabolites. Thus ACE inhibitors prevent the formation of a potent vasoconstrictor (angiotensin II) and facilitate the accumulation of a potent vasodilator (bradykinin). ACE inhibitors may exert some of their effects by preventing the interaction between angiotensin and sympathetic activity, thus reducing sympathetic tone and noradrenaline release. Interactions with other vasoactive substances may also contribute to the vaso- and venodilatation of the ACE inhibitors and explain their efficacy in patients with normal renin levels.

Captopril is a stable inhibitor of ACE.[109] Its sulphhydryl moiety has been implicated in various side effects (skin rash, loss of sense of taste).

Enalapril, an *N*-carboxyalkyl dipeptide, is devoid of the sulphydryl group and is metabolized into a highly potent ACE inhibitor: enalaprilat.[110] The latter is more potent than captopril.

Lisinopril is the lysine derivative of enalaprilat and is also a non-sulphur containing compound. Lisinopril is a potent ACE inhibitor which is effective both in reducing mean arterial pressure in hypertensive patients, and in the treatment of congestive heart failure. In patients with heart failure lisinopril appears to cause a greater improvement in exercise performance than captopril.[111]

Cardiovascular effects

ACE inhibitors cause a decrease in blood pressure principally due to a decrease in systemic vascular resistance, while cardiac output remains unchanged or increases. The absence of reflex tachycardia has been attributed to a resetting of the baroreflex. Prolonged treatment of hypertension with ACE inhibitors can lead, after a few months, to a regression in left ventricular hypertrophy.[112]

ACE inhibitors in congestive heart failure (CHF)

More than 2 000 000 people in the United States are thought to suffer from CHF and new cases develop at a rate of 2 per 1000 population per year.[113] The overall mortality exceeds 50 per cent in 5 years from the initial diagnosis and, for patients with refractory symptoms, it is about 20 per cent per year.

The clinical syndrome of CHF is caused by the inability of cardiac output to meet the demands of peripheral tissues. This activates neural and humoral factors including sympathetic tone, circulating catecholamines, and plasma renin activity. This results in vasoconstriction that further reduces stroke volume.

Conventional treatment of heart failure with cardiac glycosides, and diuretics, may aggravate vasoconstriction while vasodilators improve cardiac performance by reducing afterload. ACE inhibitors cause significant reductions in systemic and pulmonary pressures with a marked decrease in pulmonary wedge pressure. Heart rate tends to fall while stroke volume and cardiac output increase markedly.[114] Vasodilatation is more marked in the renal circulation, the rise in renal blood flow accounting for 50 per cent of the increase in cardiac output.[115] Arteriolar dilatation is accompanied by relaxation of venous tone.

Exercise capacity is improved by about 20 per cent and the mortality of patients with CHF is reduced.

ACE inhibitors in hypertension

Although the greatest antihypertensive efficacy may be expected in high-renin hypertension, many low-renin patients respond to ACE inhibitors. At variance with other vasodilators, ACE inhibitors do not cause sodium and fluid retention. A first-dose hypotensive effect has been noted especially in patients with high plasma renin activity. Low dose captopril and enalapril are effective in a large proportion of patients with uncomplicated hypertension and have few side effects.

Renovascular hypertension

Because of the increased renin from kidneys with a significant renal artery stenosis, renovascular hypertension is very responsive to ACE inhibitors. Abrupt, profound hypotension is likely to occur in response to the first dose of an ACE inhibitor. This extreme responsiveness can be used as a test of the surgical curability of renovascular hypertension. However, there is a risk of deterioration of renal function when lesions are bilateral, when renal function is marginal or when atherosclerotic lesions are extensive. The concurrent use of a loop diuretic increases the risk of renal failure.

Chronic renal failure

In hypertensive patients with chronic renal failure, ACE inhibitors may be effective and may improve renal function. However, acute or insidious renal failure may also develop especially when they are administered together with a loop diuretic.

The ultimate aim in reducing blood pressure is the prevention of vascular disease, especially in the coronary and cerebral arteries, and to prevent the reduction in arterial compliance which increases the workload of the heart and increases the rate of development of hypertrophy.[116] Pressure reduction with converting enzyme inhibitors (and calcium antagonists) is associated with an increase in vascular compliance and a reduction in cardiac hypertrophy. Moreover, the reduction in both angiotensin II and sympathetic activity improves coronary blood flow.[117]

Side effects of ACE inhibitors

Severe hypotension may occur in patients with high plasma renin activity especially after the first dose of captopril and enalapril. Drug-induced renal insufficiency occurs mostly in patients with renal artery stenosis and is usually reversible after discontinuation of ACE inhibitors. Hyperkalaemia may occur because of the inhibition of aldosterone and in patients with renal insufficiency, hyperkalaemia may reach life-threatening levels. Moreover, potassium-sparing diuretics and prostaglandin synthetase inhibitors can induce uraemia and excessive potassium retention when associated with ACE inhibitors.

Relatively little is known of the modification of the effects of anaesthesia by ACE inhibitors. Some studies suggest that chronic treatment of arterial hypertension with enalapril accentuates the hypotensive effect of anaesthesia.[118] Hypotension is attributed to a reduction in preload (corrected by moderate volume loading) and peripheral vascular resistance (corrected by phenylephrine). In patients undergoing cardiac surgery, vascular responsiveness has been found to be profoundly reduced by preoperative ACE inhibitor treatment.

Vasodilators

Several groups of vasodilators have already been discussed: in the sections on adrenergic agonists and antagonists (α-adrenoceptor blockers, α_2-adrenoceptor agonists, β_2-adrenoceptor agonists, ganglion-blocking agents), calcium channel blockers, and ACE inhibitors.

Important to the understanding of modern concepts of the control of vascular tone is the role of the endothelium. The endothelial cells of the vascular system modulate vascular tone by releasing vasodilator and vasoconstrictor substances such as endothelium-derived relaxing factor (EDRF), endothelin (a potent vasoconstrictor), prostacyclin (a vasodilator) and thromboxane (a vasoconstrictor).

Nitric oxide

Endogenous nitric oxide

The discovery of nitric oxide as a vasodilator stems from observations showing that the vasodilator effect of acetylcholine requires the integrity of the vascular endothelium. This suggested to Furchgott and Zawadzki[119] that the effect of acetylcholine was mediated by an endothelium-derived relaxing factor, now known to be nitric oxide (NO) or an NO-containing species such as nitrosothiol.

Glyceryl trinitrate, sodium nitroprusside, nitrosothiols and molsidomine release NO and can be considered as prodrugs. They do not need an intact endothelium to cause vasodilatation, hence their efficacy in diseases in which endothelial function is impaired (hypertension, athero-sclerosis, vasospastic disorders).

Exogenous nitric oxide

Because NO binds rapidly with haemoglobin, inhaled NO exerts its effects on the pulmonary but not the systemic vasculature.[120] This makes it unique as all other pulmonary vasodilators cause some systemic vasodilatation. In addition, within the lungs NO dilates vessels only in the ventilated areas while other vasodilators dilate vessels throughout the lungs, including the poorly ventilated areas, thus increasing venous admixture.[121] Beneficial effects of NO have been reported in the acute respiratory distress syndrome, after mitral valve surgery in patients known to have long-standing pulmonary hypertension, and after cardiac transplantation. Because of possible toxicity, concentrations should be kept as low as possible, in the range of 40–180 p.p.m.

Endothelins antagonists

The endothelial cells generate vasoconstrictor substances known as endothelins.[122] Three isoforms have been identified, all having 21 amino acids. Their vasoconstricting power is ten times that of angiotensin II, and their duration of action is much longer. As the systemic, renal, coronary and pulmonary vasoconstrictor effects of endothelin are enhanced by the NO inhibitor L-NMMA, it is suggested that endothelin and NO are in balance. Imbalance may be responsible for the development of atherosclerosis, congestive heart failure, essential hypertension and pulmonary hypertension. The inhibition of the effects of endothelin by calcium channel blockers such as nifidepine suggests that endothelin may be an agonist at nitrendipine receptors.[123]

Flosequinan is an antagonist of endothelin-1 that inhibits the endothelin-induced formation of inositol triphosphate and prevents protein kinase C activation. Flosequinan is more effective than enalapril in the treatment of chronic heart failure.[124]

Nitrates

Nitroglycerin (glyceryl trinitrate) was first used in angina pectoris by William Murrel in 1879. More recently nitrates have gained popularity in the management of congestive heart failure. Nitrates improve left ventricular dynamics, decreasing preload and increasing cardiac output.[125,126] Nitrate tolerance has been documented in animal models and to a lesser extent in humans.[127] This is particularly important as far as transdermal nitroglycerin patches are concerned.

Nitroglycerin and organic nitrates are relaxants of vascular smooth muscle in virtually all vascular beds. Their effect involves NO and other nitrosothiols. The latter activate guanylate cyclase (cGMP), the common pathway for smooth muscle relaxation, via a decrease in intracellular free calcium possibly caused by enhanced calcium extrusion from the cytosol.[128]

Vascular effects of nitrates

At low concentrations nitrates are potent venodilators. Blood flow to the extremities is increased and there is venous pooling in the splanchnic circulation whereas blood volume is reduced in the lungs, heart and liver. Reductions in right and left ventricular filling pressures improve right and left ventricular failure. However, in the presence of a relatively low preload, nitrates may decrease cardiac output.

At relatively low concentrations, nitrates increase arterial conductance and compliance. The diameter of vessels increases, and the impedance of the vascular tree is reduced. The increase in pulse pressure may cause pounding headaches. High concentrations of nitrates cause arteriolar dilatation and hypotension. This, in turn, may elicit reflex sympathetic activation.

Nitrates cause dilatation of epicardial conductance coronary arteries. They enhance collateral flow,[129] and have the ability to dilate eccentric stenoses, particularly the eccentric stenoses which possess sufficient smooth muscle.[130] Moreover, they reduce the myocardial compressive forces in patients with high left ventricular end-diastolic pressure. Nitrates are effective in reversing and preventing vasospasm.

Ischaemic heart disease

Nitrates remain the gold standard for antianginal activity. Few patients with exercise angina do not obtain relief from nitroglycerin because of the combined systemic and coronary actions of the drug. For patients with relatively frequent anginal attacks, long-acting, as opposed to short-acting nitrates are of proven value, the most convincing evidence being for isosorbide dinitrate.[131] However, long-acting nitrates are not recommended for patients who experience fewer than two or three episodes of angina per week, as tolerance develops.

Nitrate therapy is particularly effective in stable angina associated with impaired systolic function, left ventricular enlargement, or history of left ventricular failure. In contrast, nitrate therapy should not be used as mono-therapy in patients with stable angina associated with arterial hypertension.

Patients with variable angina threshold benefit from nitrates as excessive coronary vascular tone is a determinant of the anginal syndrome. Intravenous nitroglycerin is very effective,[132] later replaced by oral long-acting nitrates (isosorbide dinititrate) or nitroglycerin ointment. Nitrates are effective probably because adrenergically increased vascular tone plays a major role in unstable angina. For the same reason, nitrates are particularly effective in vasospastic angina.[133]

Congestive heart failure

Nitrates are unique in the treatment of congestive cardiac failure because of their efficacy in reducing ventricular preload. Though venodilatation occurs with low doses in the normal heart, higher doses may be necessary to achieve adequate preload reduction in the face of cardiac failure. Nitrate therapy results in long-term improvement of exercise capacity.[134] In patients with cardiac failure, nitrates do not appear to cause hypotension. The greatest efficacy is in patients with large hearts and ejection fractions less than 40 per cent, also in those with mitral regurgitation. The combination of nitrates and vasodilators (isosorbide dinitrate plus hydralazine) has been shown to decrease mortality when compared with placebo or a vasodilator alone.[135]

Short-acting drugs

Sublingual nitroglycerin and isosorbide dinitrate, as aerosol sprays, are the major formulations for acute episodes of angina. Isosorbide dinitrate has a slower onset of action and lasts longer than nitroglycerin. Buccal or transmucosal nitroglycerin provides immediate and sustained release for several hours.

Intravenous nitroglycerin is adsorbed onto plastic tubing. However, little difference in efficacy has been found between polyvinylchloride (conventional) and special delivery sets made of polyethylene.

Long-acting drugs

Buccal or transmucosal nitroglycerin may ensure delivery of the drug for up to 6 hours. Oral isosorbide dinitrate has a half-life of 1–2 hours and is metabolized into two active compounds, 2- and 5-isosorbide mononitrate; 5-isosorbide mononitrate is almost completely bioavailable, is not metabolized by the liver and has a half-life of 3–4 hours.

Topical or dermal nitrates have been shown to be effective in angina and congestive cardiac failure. Nitroglycerin ointment may exert beneficial effects for up to 3–6 hours, as it delivers therapeutic plasma nitroglycerin concentrations.[136] In order to avoid tolerance, nitrate-free intervals of 6–8 hours are necessary.

Transdermal discs or patches of nitroglycerin, which consist of nitroglycerin bound to a silicone matrix, allow for the very slow release of the drug across the skin. There is considerable controversy about this formulation, both in terms of doses to be given and risk of tolerance. The latter may be avoided by nitroglycerin-free intervals of 10–12 hours.

Prostanoids (prostacyclin and prostaglandin E_1)

These prostanoids are powerful vasodilators. While prostacyclin exerts its effects on both the systemic and pulmonary vasculatures, prostaglandin E_1 exerts its effects predominantly on the pulmonary vessels. Prostacyclin can be used in patients with acute peripheral ischaemia and in patients with pulmonary hypertension in order to decrease the right ventricular afterload. Because of its systemic effects, there is usually a need for noradrenaline to prevent a substantial reduction in systemic arterial pressure. This has also been found to be necessary during administration of prostaglandin E_1.[137]

Antihypertensive therapy

The Medical Research Council trial has shown that in mild hypertension stroke rate and incidence of all cardiovascular events are decreased by treatment.[138] Diuretics reduce stroke rate more than β-blockers, but they are ineffective in reducing coronary events.[139]

The question of the level of diastolic blood pressure at which treatment is justified has been answered by the Hypertension Detection and Follow-up Program Cooperative Group.[140] After 5 years of intensive stepped care, all-cause mortality was 16.9 per cent lower in well controlled than in less rigorously treated patients; the highest difference (20.3 per cent reduction in mortality) was in the group with diastolic pressure between 90 and 104 mmHg. Such data support the view that antihypertensive therapy should be initiated when diastolic pressure exceeds 90 mmHg.

Diuretics

Diuretics are inexpensive, easy to administer and effective in a high proportion of patients and over prolonged periods of time.

Thiazide-type diuretics

The antihypertensive effect reaches a plateau at relatively low doses with which hypokalaemia is usually limited to a reduction of 0.4–0.6 mmol/l. Magnesium depletion may also occur and may contribute to arrhythmias. Impaired glucose tolerance and hyperuricaemia are observed; diabetes and gout may develop.

Potassium sparing diuretics

These are usually given in association with thiazide-type diuretics to patients with thiazide-induced hypokalaemia. There is a risk of hyperkalaemia, particularly in patients with renal failure, diabetes or hypoaldosteronism. This risk is enhanced by the concomitant administration of non-steroidal anti-inflammatory drugs, ACE inhibitors, and potassium supplements. The most commonly used diuretics in this group are amiloride, spironolactone and triamterene.

Loop diuretics

These are not used for treatment of hypertension except in the presence of chronic renal insufficiency, or of an oedematous state (chronic heart failure, nephrotic syndrome, cirrhosis). The antihypertensive effect is dose dependent without the ceiling effect of the thiazides.

Adrenergic receptor antagonists

β-Adrenoceptor antagonists

β-Adrenoceptor antagonists are effective in the long-term treatment of hypertension as monotherapy and in the management of cardiac conditions associated with arterial hypertension (supraventricular tachyarrhythmias, hypertrophic cardiomypathy, angina). β-Blockers reduce blood pressure by decreasing cardiac output. This may explain the reduced exercise tolerance and the cold extremities. Fatigue, depression, insomnia, vivid dreams, bronchospasm, Raynaud's phenomenon, and cold extremities are side effects of β-blockade. Hydrophilic β-blockers (acebutolol, atenolol, nadolol) are less likely to cause CNS side effects that lipophilic ones. Selective β-blockers do not prolong insulin-induced hypoglycaemia as much as non-selective blockers.[141]

β-Blockers appear to be less effective in the elderly than in the younger age group. Young patients with a hyperkinetic circulation respond particularly well to β-blocker monotherapy. Patients with angina, hypertrophic cardiomyopathy, or supraventricular arrhythmias also benefit most.

α-Adrenoceptor antagonists

Non-selective α-adrenoceptor antagonists such as phenoxybenzamine and phentolamine have been superseded by selective α_1-adrenoceptor blockers, such as prazosin, terazosin and doxazosin. They reduce peripheral vascular resistance and increase vascular capacitance; they do not interfere with the feedback control on noradrenaline release. Diastolic pressure is well controlled by α_1-blockers.[142] Syncopal episodes are rare (less than 1 per cent). The risk of first-dose effect (orthostatic hypotension leading to syncope), particularly in the elderly, can be reduced by giving a small initial dose at bedtime. In patients with diabetes, asthma, and peripheral vascular disease, α_1-blockers may offer advantages over thiazides and β-blockers.

Prazosin

Prazosin, a selective α_1-receptor antagonist, is effective singly or in association with other drugs. There is little increase in sympathetic activity or plasma renin activity. Sodium and water retention do occur and the efficacy of prazosin is enhanced by the addition of a diuretic.[143] Prazosin may be given in association with β-blockers, α-methyldopa, and direct vasodilators. As prazosin causes little change in glomerular filtration rate, it can be used safely in renal hypertension.[144]

Terazosin

Terazosin, a long acting selective α_1-blocker with a half-life of 12 hours, is less potent than prazosin and has a slower onset of action.[145] Terazosin is effective in the control of hypertension as monotherapy. It is also used as a complement to β-blockers and diuretics.[146] The first dose phenomenon (severe orthostatic hypotension) is less pronounced than with prazosin.

α- and β-Adrenoceptor antagonists

Labetalol is useful in the treatment of all grades of hypertension. It reduces vascular resistance. Heart rate and cardiac output do not increase because of β-blockade.[147] Orthostatic hypotension may occur. Though afterload is reduced by labetalol, great care should be exercised in patients with cardiac failure.

Calcium antagonists

Diltiazem, nifedipine, nicardipine, and verapamil decrease blood pressure in hypertensive patients through vasodilatation. Nifedipine is the most potent vasodilator. Its administration causes an increase in heart rate unless a β-blocker is added. This association is well tolerated while that of verapamil or diltiazem with a β-blocker may result in disorders of conduction and marked cardiac depression.

Nifedipine is effective in the treatment of hypertensive crises (encephalopathy, intracranial haemorrhage, acute left ventricular failure, and aortic dissection), associated with blood pressures exceeding 200/120 mmHg. Small doses or oral or sublingual nifedipine may be effective in 5–60 minutes as it reduces systemic and pulmonary pressures and improves cardiac output.[148] The advantages of nifedipine are: rapid oral absorption (sublingual administration is not faster than bite-swallow capsules); relatively long duration of action (3–5 hours); blood pressure reduction function of the initial level of hypertension, maintenance of cardiac output, and coronary vasodilatation.

Verapamil

Verapamil reduces vascular resistance. Activation of the sympathetic nervous system occurs but its effects on the myocardium are minimized by the negative inotropy and chronotropy of verapamil.

Angiotensin converting enzyme inhibitors

Captopril and enalapril are particularly useful in patients with hypertension and left ventricular dysfunction. Low dose diuretics associated with an ACE inhibitor achieve control in 75 per cent of patients with mild hypertension.[149]

The acute antihypertensive response to an ACE inhibitor is a function of the pretreatment plasma renin activity, but this relation is no longer obvious during long-term therapy.[150] As the frequency of renal artery disease increases with age, the possibility of renovascular hypertension should be kept in mind when ACE inhibitors are given to elderly patients as their administration may precipitate acute renal failure.

Hypotension may occur in patients with renovascular hypertension, high renin essential hypertension, diuretic pretreatment, and congestive cardiac failure. Low doses of captopril (6.25–12.5 mg) or enalapril (2.5 mg) should be given in the early phase of treatment.

Centrally acting sympatholytic agents

Clonidine and methydopa are α_2-receptor agonists that reduce blood pressure by reducing sympathetic efferent discharge. This is often accompanied by lethargy and fatigue. As there is an association between sympathetic activity and the development of left ventricular hypertrophy, centrally acting sympatholytic agents may be

expected to prevent or cause regression of hypertrophy.[151] Such a regression is desirable since left ventricular hypertrophy increases cardiovascular mortality.[152]

Other antihypertensive drugs

Direct vasodilators

The most commonly used are hydralazine and minoxidil. Long-term administration of hydralazine may cause a systemic lupus erythematoid (SLE) syndrome. Minoxidil is the most powerful vasodilator and a loop diuretic is needed to prevent vasodilator-induced oedema. A major side effect of minoxidil is hirsutism.

Serotonin antagonists

The effects of serotonin on the circulation are mediated by two subtypes of receptors S_1 and S_2. The diversity of vascular responses is influenced by the distribution of receptor subtypes. Vasoconstriction is largely mediated by S_2-receptors on platelets and endothelial cells. Moreover, activation of S_2-receptors amplifies the effects of histamine, angiotensin, prostaglandin $F_{2\alpha}$ and noradrenaline.[153] Direct effects of serotonin on the heart and the circulation include increases in cardiac contractility, cardiac output, and pulmonary vascular resistance. However, S_1-receptor activation in the endothelium causes the release of endothelium-derived relaxing factor, prostacyclin and vasoactive intestinal peptide (VIP). Inhibition of the release of noradrenaline results in systemic vasodilatation.[154,155]

In the elderly and in hypertensive patients, serotonin activity is increased. This could cause an increase in the tone of the capacitance vessels. More importantly, damaged vessels (a common feature of atherosclerosis), lose their protective endothelial barrier and their S_1-receptor-mediated vasodilatation. In addition serotonin could contribute to hypertension by an increase in aldosterone secretion.[156]

Ketanserin

Ketanserin is a highly selective S_2 antagonist. Its administration decreases peripheral vascular resistance by antagonizing the effects of S_2-receptor activation on vessels and platelets. It also unmasks the effect of S_1-receptor activation on blood vessels and on the release of noradrenaline (unopposed inhibition). Some of the effects of ketanserin may be due to α-adrenergic blockade.

Ketanserin lowers systolic and diastolic pressures. The main mechanism is systemic vasodilatation, resulting in increases in renal, splanchnic, cerebral and cutaneous blood flow, but not cardiac output. It is especially effective in the elderly where serotonergic stimulation plays an important role.[157] The effects of ketanserin are augmented by the association with β-adrenergic blockers. Side effects include orthostatic hypotension, sleep disturbances and anxiety, fatigue and sedation.[158] Ketanserin can be administered orally and intravenously.

Hypertensive emergencies

Perioperative hypertension

β-Adrenoceptor blockers blunt the hypertensive responses to some of the stimuli encountered during anaesthesia and surgery. Calcium channel blockers (nifedipine, nimodipine, nicardipine) also minimize the risk of hypertensive responses to anaesthetic and surgical maneouvres mostly by reducing vascular resistance. In addition they may reduce the release of catecholamines.

The adrenergic responses may be decreased by the administration of an α_2-adrenoceptor agonist used as a premedicant drug. Premedication with clonidine has been shown to minimize the increase in plasma noradrenaline during surgery of the abdominal aorta, and to decrease the blood pressure and heart rate variability in treated hypertensive patients undergoing anaesthesia and surgery.[15,159,160]

Postcardiac surgical hypertension

Hypertension occurs in 30–60 per cent of patients after cardiac surgery.[161] The risks of postoperative hypertension include cerebrovascular accidents, disruption of suture lines, dissection of the aorta, and myocardial ischaemia.

Neurohormonal changes include elevated catecholamines, renin, angiotensin, and vasopressin. They persist for several days after surgery especially if tracheal intubation is maintained, or sedation and pain relief are inadequate. Postoperative hypertension is more common in hypertensive patients, and in those with a high ejection fraction, indicative of good left ventricular function.[162]

The main haemodynamic disturbance is an increase in systemic vascular resistance, hence the efficacy of a rapidly acting vasodilator such as sodium nitroprusside with which arterial pressure is easily titrated to the desired level. Excessive reductions in diastolic arterial pressure and coronary steal may cause myocardial ischaemia. Because of these risks, nitroglycerin is often used rather than nitroprusside, and it is particularly effective when filling pressures are abnormally high.

In some patients hypertension is associated with a hyperdynamic circulation and responds well to β-adrenoceptor blockers. In this respect, the ultrashort-acting β-blocker esmolol given as a continuous infusion is particularly useful. Blood pressure may be titrated, cardiac output is usually only minimally reduced, diastolic pressure is well maintained, heart rate is reduced, and pulmonary shunting is not increased.[163]

Combined α- and β-blockade with labetalol or nicardipine are effective. Once blood pressure control has been achieved buccal or sublingual nifedipine medication may be practical.

Associations of drugs may prove more effective than single drugs. Ketanserin has been shown to reduce the sodium nitroprusside requirements and to improve the quality of blood pressure control.[164]

Hypertensive emergencies in patients with intracranial disease

Marked haemodynamic fluctuations occur during neurosurgical procedures including excessive responses to laryngoscopy, intubation, skin incision, handling of brain structures and extubation. As autoregulation of cerebral blood flow is disturbed both before and after surgery in the neurosurgical patient, hypertensive episodes may be accompanied by marked increases in cerebral blood flow and in intracranial pressure. Prevention of hypertensive episodes is a prerequisite for a safe perioperative outcome.[165]

As hypertensive episodes during surgery and during emergence from anaesthesia may occur in a very high percentage of patients, antihypertensive drugs are often required. The choice of the drug must take into consideration its effect on the cerebral circulation.

Most conventional antihypertensive agents such as sodium nitroprusside, nitroglycerin, hydralazine, and trimetaphan impair cerebral autoregulation and may increase cerebral blood flow and cerebral blood volume. Of the β-adrenoceptor blockers, esmolol may be more effective and more easily controlled than others.[31] Labetalol does not alter intracranial pressure and compliance. Urapidil, an α_1-adrenoceptor blocker with some β_1-blocking efficacy and a central hypotensive activity,[166] is effective in the control of hypertensive episodes and does not increase the intracranial pressure even in patients with brain tumours.[167]

REFERENCES

1. Ahlquist RP. A study of the adrenotropic receptors. *American Journal of Physiology* 1948; **153**: 586–600.

2. Arnold A, McAuliff JP, Ludena FP, Brown TG Jr, Lands AM. Lipolysis and sympathomimetic amines. *Federation Proceedings* 1966; **25**: 500.

3. Lands AM, Arnold A, McAuliff JP, Luduena FP, Brown TG. Differentiation of receptor systems activated by sympathomimetic amines. *Nature* 1967; **215**: 597–8.

4. Langer SZ. Presynaptic regulation of catecholamine release. *Biochemical Pharmacology* 1974; **23**: 1793–800.

5. Schwinn DA. Adrenoceptors as models for G protein-coupled receptors: structure, function and regulation. *British Journal of Anaesthesia* 1993; **71**: 77–85.

6. Van Tits LJH, Michel MC, Grosse-Wilde H, *et al.* Catecholamines increase lymphocyte β2-adrenergic receptors via a β2-adrenergic, spleen-dependent process. *American Journal of Physiology* 1990; **258**: E191–202.

7. Aarons RD, Nies AS, Gal J, Hegstrand LR, Molinoff PB. Elevation of beta-adrenergic receptor density in human lymphocytes afer propranolol treatment. *Journal of Clinical Investigation* 1980; **65**: 949–57.

8. Martin C, Eon B, Saux P, Aknin P, Gouin F. Renal effects of norepinephrine used to treat septic shock patients. *Critical Care Medicine* 1990; **18**: 282–5.

9. Lesch M. Inotropic agents and infarct size. Theoretical and practical considerations. *American Journal of Cardiology* 1976; **37**: 508–13.

10. Mager G, Klocke RK, Hopp H-W, Hilger HH. Phosphodiesterase III inhibition or adrenoreceptor stimulation: milrinone as an alternative to dobutamine in the treatment of severe heart failure. *American Heart Journal* 1991; **121**: 1974–82.

11. Chintala MS, Lockandwala MF, Jandhyala BS. Protective effect of dopexamine hydrochloride in renal failure after acute haemorrhage in anaesthetized dogs. *Journal of Autonomic Pharmacology* 1990; **10** (Suppl 1): 95–102.

12. Smith GW, O'Connor SE. An introduction to the pharmacologic properties of Dopacard (dopexamine hydrochloride). *American Journal of Cardiology* 1988; **62**: 9C–17C.

13. Quintin L, Viale JP, Hoen JP, *et al.* Oxygen uptake after major abdominal surgery: effect of clonidine. *Anesthesiology* 1991; **74**: 236–41.

14. Hayashi Y, Maze M. Alpha2adrenoceptor agonists and anaesthesia. *British Journal of Anaesthesia* 1993; **71**: 108–18.

15. Ghignone M, Calvillo O, Quintin L. Anesthesia and hypertension: the effect of clonidine on perioperative hemodynamics and isoflurane requirements. *Anesthesiology* 1987; **67**: 3–10.

16. Bloor BC, Ward BS, Belleville JP, Maze M. Effects of intravenous dexmetedomidine in humans. *Anesthesiology* 1992; **77**: 1134–42.

17. Belleville JP, Ward DS, Bloor BC, Maze M. Effects of intravenous dexmetedomidine in humans. *Anesthesiology* 1992; **77**: 1125–33.

18. Ablad B, Brogard M, Ek L. Pharmacologic properties of H56/28 a beta-adrenergic receptor antagonist. *Acta Pharmacologica Toxicologica (Kopenhaven)* 1967; **25** (Suppl 2): 9–40.

19. Bennett DH. Acute prolongation of myocardial refractoriness by sotalol. *British Heart Journal* 1982; **47**: 521–6.

20. Regardh C-G. Pharmacokinetics of β-adrenoceptor antagonists. In: Poppers PJ, van Dijk B, van Elzakker AHM, eds. β-Blockade and anaesthesia. Rijswijk, The Netherlands: Astra Pharmaceutica, 1982: 29–45.

21. Feely J, de Vane PJ, Maclean D. Beta-blockers and sympathomimetics. *British Medical Journal* 1983; **286**: 1043–7.

22. Yusuf S, Ramsdale D, Peto R, *et al.* Early intravenous atenolol treatment in suspected acute myocardial infarction. *Lancet* 1980; **2**: 273–6.

23. Hjalmarson A. International beta-blocker review in acute and postmyocardial infarction. *American Journal of Cardiology* 1988; **61**: 26B–9B.

24. Prys-Roberts C, Foëx P, Biro GP, Roberts JG. Studies of anaesthesia in relation to hypertension V. Adrenergic beta-receptor blockade. *British Journal of Anaesthesia* 1973; **45**: 671–81.

25. Foëx P. Alpha- and beta-adrenoceptor antagonists. *British Journal of Anaesthesia* 1984; **56**: 751–65.

26. Stone JG, Foëx P, Sear J, Johnson LL, Khambatta HJ, Triner L. Myocardial ischemia in untreated hypertensive patients: effect of a single small oral dose of a beta-blocker. *Anesthesiology* 1988; **68**: 495–500.

27. Gorcyski RJ. Basic pharmacology of esmolol. *American Journal of Cardiology* 1985; **56**: 3F–13F.

28. Quon CY, Stampfli HF. Biochemical properties of blood esmolol esterase. *Drug Metabolism and Drug Disposition* 1989; **13**: 420–4.

29. Reynolds RD, Gorczynski RJ, Quon CY. Pharmacology and pharmacokinetics of esmolol. *Journal of Clinical Pharmacology* 1986; **26** (Suppl A): A3–A14.

30. Zaroslinski J, Borgman RJ, O'Donnell JP, *et al.* Ultra-short acting beta-blockers: a proposal for the treatment of the critically ill patient. *Life Sciences* 1982; **31**: 899–907.

31. Cucchiara RF, Benefiel DJ, Matteo RS, De Wood M, Albin MS. Evaluation of esmolol in controlling increases in heart rate and blood pressure during endotracheal intubation in patients undergoing carotid endarterectomy. *Anesthesiology* 1986; **65**: 528–31.

32. Newsome LR, Roth JV, Hug CC, Nagle D. Esmolol attenuates hemodynamic responses during fentanyl-pancuronium anesthesia for aortocoronary bypass surgery. *Anesthesia and Analgesia* 1986; **65**: 451–6.

33. Nuttall A, Snow HM. The cardiovascular effects of ICI 118587: a beta1-adrenoceptor partial agonist. *British Journal of Pharmacology* 1982; **77**: 381–8.

34. Molajo AO, Bennett DH. Effect of xamoterol (ICI 118587) a new beta1-adrenoceptor partial agonist, on resting haemodynamic variables and exercise tolerance in patients with left ventricular dysfunction. *British Heart Journal* 1985; **54**: 17–21.

35. The German and Austrian Xamoterol Study Group. Double-blind placebo-controlled comparison of digoxin and xamoterol in chronic heart failure. *Lancet* 1988; **1**: 489–93.

36. Hull CJ. Phaeochromocytoma. Diagnosis, preoperative

preparation and anaesthetic management. *British Journal of Anaesthesia* 1986; **58**: 1453–68.

37. Vaughan Williams EM, Dennis PD, Garnham C. Circadian rhythm of heart rate in the rabbit; prolongation of action potential duration by sustained beta-adrenoceptor blockade is not due to associated bradycardia. *Cardiovascular Research* 1986; **20**: 528–35.

38. Snyders DJ, Katzung BG. Clofilium reduces the plateau potassium current in isolated cardiac myocytes. *Circulation* 1985; **72-III**: 233.

39. Feld GK, Vankatesh N, Singh BN. Pharmacologic conversion and suppression of experimental canine atrial flutter: differing effects of *d*-sotalol, quinidine and lidocaine and significance of changes in refractoriness. *Circulation* 1986; **74**: 197–204.

40. Singh BN, Vaughan Williams EM. A fourth class of antiarrhythmic agents? Effects of verapamil on ouabain toxicity, on atrial and ventricular intracellular potentials, and on other features of cardiac function. *Cardiovascular Research* 1972; **6**: 109–19.

41. Kobinger W. Specific bradycardic agents. In: Vaughan Williams EM, ed. *Antiarrhythmic drugs*. Berlin: Springer-Verlag, 1989: 423–52.

42. Millar JS, Vaughan Williams EM. Pacemaker selectivity. Effects on rabbit atria of ionic environments and of alinidine, a possible anion channel antagonist. *Cardiovascular Research* 1981; **15**: 335–50.

43. Grande P, Sonne B, Pedersen A. A controlled study of digoxin and quinidine in patients DC reverted from atrial fibrillation to sinus rhythm. *Circulation* 1986; **74** (Suppl II): 101.

44. Karlson BW, Throstensson I, Abjorn C. Disopyramide in the maintenance of sinus rhythm after electroconversion of atrial fibrillation – a placebo controlled one year follow-up study. *Circulation* 1986; **74** (Suppl II): 101.

45. Turner WM. Lidocaine and psychotic reactions. *Annals of Internal Medicine* 1982; **97**: 149–50.

46. Impact Research Group. International mexiletine and placebo anti-arrhythmic coronary trial: 1. Report on arrhythmia and other findings. *Journal of the American College of Cardiology* 1984; **4**: 1184–93.

47. Pottage A. Clinical profiles of newer class I antiarrhythmic agents – tocainide, mexiletine, encainide, flecainide and lorcainide. *American Journal of Cardiology* 1983; **52**: 24c–31c.

48. Elharrar V, Zipes D. Effects of encainide and metabolites (MJ14030 and MJ9444) on canine cardiac Purkinje and ventricular fibres. *Journal of Pharmacology and Experimental Therapeutics* 1982; **220**: 440–7.

49. Rossi PRF, Yusuf S, Ramsdale D, Furze L, Sleight P. Reduction of ventricular arrhythmias by early intravenous atenolol in suspected acute myocardial infarction. *British Medical Journal* 1983; **286**: 506–10.

50. Yusuf S, Peto R, Lewis J, Collins R, Sleight P. Beta blockade during and after myocardial infarction: an overview of the randomized trials. *Progress in Cardiovascular Disease* 1985; **27**: 335–71.

51. Julian D, Chamberlain D, Sandoe E, *et al.* Mechanisms for the early mortality reduction produced by beta blockade started early in acute myocardial infarction. *Lancet* 1988; **1**: 921–3.

52. Singh BN, Collett JT, Chew CYC. New perspectives in the pharmacologic therapy of cardiac arrhythmias. *Progress in Cardiovascular Disease* 1980; **22**: 243–301.

53. Gagnol JP, Devos C, Clinet M, Nokin P. Amiodarone: biochemical aspects and hemodynamic effects. *Drugs* 1985; **29** (Suppl 5): 1–10.

54. Rotmensch HH, Belhassen B. Amiodarone in the management of cardiac arrhythmias: current concepts. *Medical Clinics of North America* 1988; **72**(2): 321–58.

55. Kosinski EJ, Albin JB, Young E, Lewis SM, Leland OS. Hemodynamic effects of intravenous amiodarone. *Journal of the American College of Cardiology* 1984; **4**: 565–70.

56. Bellotti G, Silva LA, Esteves Filho A, *et al.* Hemodynamic effects of intravenous administration of amiodarone in congestive heart failure from chronic Chagas' disease. *American Journal of Cardiology* 1983; **52**: 1046–9.

57. Gold MI, Brown M, Coverman S, Herrington C. Heart rate and blood pressure effects of esmolol after ketamine induction and intubation. *Anesthesiology* 1986; **64**: 718–23.

58. Feld GK, Nademanee K, Weiss J, Stevenson W, Singh BN. Electrophysiologic basis for the suppression by amiodarone of orthodromic supraventicular tachycardias complicating pre-excitation syndromes. *Journal of the American College of Cardiology* 1984; **3**: 1298–307.

59. Rakita L, Sobol SM, Mostow N, Wrobel T. Amiodarone pulmonary toxicity. *American Heart Journal* 1983; **106**: 906–15.

60. Fenster PE, White NW Jr, Hanson CD. Pharmacokinetics evaluation of the digoxin-amiodarone interaction. *Journal of the American College of Cardiology* 1985; **5**: 108–12.

61. Carmeliet E. An electrophysiologic and voltage-clamp analysis of the effects of sotalol on isolated cardiac muscle and Purkinje fibers. *Journal of Pharmacology and Experimental Therapeutics* 1985; **232**: 817–25.

62. Norwegian Multicentre Study Group. Timolol-induced reduction in mortality and reinfarction in patients surviving acute myocardial infarction. *New England Journal of Medicine* 1981; **304**: 801–7.

63. Steinberg MI, Smallwood JK. Clofilium and other class III agents. In: Vaughan Williams EM, ed. *Antiarrhythmic drugs*. Berlin: Springer-Verlag, 1989: 399–412.

64. Frame VB, Wang HH. Importance of interaction with adrenergic neurons for antifibrillatory action of bretylium in the dog. *Journal of Cardiovascular Pharmacology* 1986; **8**: 336–45.

65. Lynch JJ, Rahwan RG, Lucchesi BR. Antifibrillatory actions of bepridil and butyl-MDI, two intracellular calcium antagonists. *European Journal of Pharmacology* 1985; **111**: 9–15.

66. Yatani A, Brown AM, Schwartz A. Bepridil block of cardiac calcium and sodium channels. *Journal of Pharmacology and Experimental Therapeutics* 1986; **237**: 9–17.

67. Rankin AC, Brooks R, Ruskin JN, McGovern BA. Adenosine and the treatment of supraventicular tachycardia. *American Journal of Medicine* 1992; **92**: 655–64.

68. Hood MA, Smith WM. Adenosine versus verapamil in the treatment of supraventicular tachycardia: a randomized double-crossover trial. *American Heart Journal* 1992; **123**: 1543–9.

69. Herzig S, Lüllmann H. Effects of cardiac glycosides at the

cellular level. In: Vaughan Williams EM, ed. *Antiarrhythmic drugs.* Berlin: Springer-Verlag, 1989: 545–63.

70. Gillis RA, Quest JA. The role of the nervous system in the cardiovascular effects of digitalis. *Pharmacological Reviews* 1980; **31**: 19–97.

71. Gavaghan TP, Feneley MP, Campbell TJ, Morgan JJ. Atrial tachyarrhythmias after cardiac surgery: results of disopyramide therapy. *Australian and New Zealand Journal of Medicine* 1985; **15**: 27–32.

72. Gorven AM, Cooper GM, Prys-Roberts C. Haemodynamic disturbances during anaesthesia in a patient receiving calcium channel blockers. *British Journal of Anaesthesia* 1986; **58**: 357–60.

73. Wong KC, Kawamura R, Hodges MR, Sullivan SP. Acute intravenous administration of potassium chloride to furosemide pretreated dogs. *Canadian Anaesthetists' Society Journal* 1977; **55**: 203–11.

74. Lee CO, Abete P, Pecker M, Sonn JK, Vassalle M. Strophantidin inotropy: role of intracellular sodium ion activity and sodium-calcium exchange. *Journal of Molecular and Cellular Cardiology* 1985; **17**: 315–30.

75. Morgan JP. The effects of digitalis on intracellular calcium transients in mammalian working myocardium as detected with aequorin. *Journal of Molecular and Cellular Cardiology* 1985; **17**: 1065–75.

76. Evans DB. Overview of cardiovascular physiologic and pharmacologic aspects of selective phosphodiesterase peak III inhibitors. *American Journal of Cardiology* 1989; **63**: 9A–11A.

77. Naccarelli GV, Goldstein RA. Electrophysiology of phosphodiesterase inhibitors. *American Journal of Cardiology* 1989; **63**: 34A–40A.

78. Komai M, Rusy BF. Inotropic effect of amrinone in rabbit papillary muscle: reversal of myocardial depressant effect of halothane. *Canadian Journal of Physiology and Pharmacology* 1984; **62**: 1382–6.

79. Colucci WS, Wright RF, Jaski BE, Fifer MA, Braunwald E. Milrinone and dobutamine in severe heart failure: differing hemodynamic effects and individual patient's responsiveness. *Circulation* 1986; **73** (Suppl 3): 175–83.

80. Packer M, Carver JR, Rodeheffer RJ, *et al.* Effect of oral milrinone on mortality in severe chronic heart failure. *New England Journal of Medicine* 1991; **325**: 1468–75.

81. Curfman GD. Inotropic therapy for heart failure – and unfulfilled promise. *New England Journal of Medicine* 1991; **325**: 1509–10.

82. Gonzalez M, Desager J-P, Jacquemart J-L, Chenu P, Muller T, Installe E. Efficacy of enoximone in the management of refractory low-output states following cardiac surgery. *Journal of Cardiothoracic Anesthesia* 1988; **2**: 409–18.

83. Dage RC, Kariya T, Hsieh CH, *et al.* Pharmacology of enoximone. *American Journal of Cardiology* 1987; **60**: 10C–14C.

84. Treese N, Erbel R, Pilcher J, *et al.* Long-term treatment with oral enoximone for chronic congestive heart failure: the European experience. *American Journal of Cardiology* 1987; **60**: 85C–90C.

85. Loisance D, Sailly JC. Cost-effectiveness in patients awaiting transplantation receiving intravenous inotropic support. *European Journal of Anaesthesia* 1993; **10** (Suppl 8); 9–13.

86. Endoh M, Hiramoto T, Ischihata A, Takanashi M, Inui J. Myocardial α_1-adrenoceptors mediate positive inotropic effect and changes in phosphatidylinositol metabolism. *Circulation Research* 1991; **68**: 1179–90.

87. Katz SD, Kubo SH, Jessup M, *et al.* A multicentre randomized, double-blind placebo-controlled trial of pimobendan, a new cardiotonic and vasodilator agent in patients with severe congestive heart failure. *American Heart Journal* 1992; **123**: 95–102.

88. Tsien RW. Calcium channels in excitable membranes. *Annual Review of Physiology* 1984; **45**: 341–58.

89. Spedding M. Three types of Ca^{2+} channel explain discrepancies. *Trends in Pharmacological Sciences* 1987; **8**: 115–16.

90. Morad M, Cleeman L. Role of Ca^{2+} channel in development of tension in heart muscle. *Journal of Molecular and Cellular Cardiology* 1987; **19**: 527–53.

91. Benham CD, Hess P, Tsien RW. Two types of calcium channels in single smooth muscle cells from rabbit ear artery studied with whole-cell and single-channel recordings. *Circulation Research* 1987; **61** (Suppl I): 10–16.

92. Lansman JB, Hallam TJ, Rink TJ. Single stretch-activated ion channels in vascular endothelial cells as mechanotransducers? *Nature* 1987; **325**: 811–13.

93. Nayler WG, Panagiotopoulos S, Elz JS, Sturrock WJ. Fundamental mechanism of action of calcium antagonists in myocardial ischemia. *American Journal of Cardiology* 1987; **59**: 75B–83B.

94. Burlew BS, Gheorghiade M, Jafri SM, Goldberg AD, Goldstein S. Acute and chronic hemodynamic effects of nicardipine hydrochloride in patients with heart failure. *American Heart Journal* 1987; **114**: 793–804.

95. Petruk KC, West M, Mohr G, *et al.* Nimodipine treatment in poor-grade aneurysm patients. Results of a multicenter double-blind placebo-controlled trial. *Journal of Neurosurgery* 1988; **66**: 505–17.

96. Terrar DA, Victory JGG. Effects of halothane on membrane currents associated with contraction in single myocytes isolated from guinea-pig ventricle. *British Journal of Pharmacology* 1988; **94**: 500–8.

97. Puttick RM, Terrar DA. Effects of propofol and enflurane on action potentials, membrane currents and contraction of guinea-pig isolated ventricular myocytes. *British Journal of Pharmacology* 1992; **107**: 559–65.

98. Reves JG, Kissin I, Lell WA, Tosone S. Calcium entry blockers: uses and implications for anesthesiologists. *Anesthesiology* 1982; **57**: 504–18.

99. Atlee JL, Hamann SR, Brownlee SW, Kreigh C. Conscious state comparisons of the effects of the inhalation anesthetics and diltiazem, nifedipine, or verapamil on specialized atrioventricular conduction times in spontaneously beating dog hearts. *Anesthesiology* 1988; **68**: 519–28.

100. Videcoq M, Arvieux CC, Ramsay JG, *et al.* The association isoflurane-verapamil causes regional myocardial dysfunction. *Anesthesiology* 1987; **67**: 635–41.

101. Ramsay JG, Cutfield GR, Francis CM, Devlin WH, Foëx P.

Halothane-verapamil causes regional myocardial dysfunction in the dog. *British Journal of Anaesthesia* 1986; **58**: 321–6.

102. Sodoyama O, Ikeda K, Matsuda I, Fukunaga AF, Bishay EG. Nifedipine for control of postoperative hypertension. *Anesthesiology* 1983; **59**: A18.

103. Griffin RM, Dimich I, Jurado R, Kaplan JA. Haemodynamic effects of diltiazem during fentanyl-nitrous oxide anaesthesia. *British Journal of Anaesthesia* 1988; **60**: 655–9.

104. Godet G, Coriat P, Baron JF, *et al.* Prevention of intraoperative myocardial ischemia during noncardiac surgery with intravenous diltiazem: a randomized trial versus placebo. *Anesthesiology* 1987; **66**: 241–5.

105. Slogoff S, Keats AS. Does chronic treatment with calcium entry blocking drugs reduce perioperative myocardial ischemia? *Anesthesiology* 1988; **68**: 676–80.

106. Lehot JJ, Leone B, Foëx P. Calcium reverses global and regional myocardial dysfunction caused by the combination of verapamil and isoflurane. *Acta Anaesthesiologica Scandinavica* 1987; **31**: 441–7.

107. Zaritsky AL, Horowitz M, Chernow B. Glucagon antagonism of calcium channel blocker-induced myocardial dysfunction. *Critical Care Medicine* 1988; **16**: 246–51.

108. Makela VHM, Kapur PA. Amrinone and verapamil-propranolol induced cardiac depression during isoflurane anaesthesia in dogs. *Anesthesiology* 1987; **66**: 792–7.

109. Cushman DW, Ondetti MA. Inhibitors of angiotensin converting enzyme for treatment of hypertension. *Biochemical Pharmacology* 1980; **29**: 1871–7.

110 Vlasses PH, Larijani GE, Conner DP, Ferguson RH. Enalapril, a nonsulfhydryl angiotensin-converting enzyme inhibitor. *Clinical Pharmacology* 1985; **4**: 27–40.

111. Power ER, Chiaramida A, DeMaria AN, *et al.* A double-blind comparison of lisinopril with captopril in patients with symptomatic congestive heart failure. *Journal of Cardiovascular Pharmacology* 1987; **9** (Suppl 3): S82–8.

112. Dunn FG, Oigman W, Ventura HO, Messerli FM, Kobrin I, Frohlich ED. Enalapril improves systemic and renal hemodynamics and allows regression of left ventricular mass in essential hypertension. *American Journal of Cardiology* 1984; **53**: 105–8.

113. Franciosa JA. Epidemiologic patterns, clinical evaluation, and long-term prognosis in chronic congestive heart failure. *American Journal of Medicine* 1986; **80** (Suppl 2B): 14–21.

114. Romankiewicz JA, Brogden RN, Heel RC, Speight TM, Avery GS. Captopril: an update review of its pharmacological properties and therapeutic efficacy in congestive heart failure. *Drugs* 1983; **25**: 6–40.

115. Packer M, Medina N, Yushak M, Meller J. Hemodynamic patterns of response during long-term captopril therapy for severe chronic heart failure. *Circulation* 1983; **68**: 803–12.

116. Safar ME. Focus on the large arteries in hypertension. *Journal of Cardiovascular Pharmacology* 1985; **7** (Suppl 2): S1–S4.

117. Magrini F, Shimizu M, Roberts N, Fouad FM, Tarazi RC, Zanchetti A. Converting enzyme inhibition and coronary blood flow. *Circulation* 1987; **75** (Suppl I): 168–74.

118. Colson P, Saussine M, Seguin JR, Cuchet D, Chaptal PA, Roquefeuil B. Hemodynamic effects of anesthesia in patients chronically treated with angiotensin converting enzyme inhibitors. *Anesthesia and Analgesia* 1992; **74**: 805–8.

119. Furchgott RF, Zawadzki JV. The obligatory role of endothelial cells in the relaxation of arterial smooth muscle. *Nature* 1980; **288**: 373–6.

120. Moncada S, Palmer PMJ, Higgs EA. Nitric oxide: physiology, pathophysiology and pharmacology. *Pharmacological Reviews* 1991; **43**: 108–42.

121. Frostell C, Fratacci MD, Wain JC, Jones R, Zapol WM. Inhaled nitric oxide. A selective pulmonary vasodilator reversing hypoxic pulmonary vasoconstriction. *Circulation* 1991; **83**: 2038–47.

122. Yanagisawa M, Kurihara H, Kimura S, *et al.* A novel potent vasoconstrictor peptide produced by vascular endothelial cells. *Nature* 1988; **332**: 411–15.

123. Kramer BK, Nishida M, Kelly RA, Smith TW. Endothelins. *Circulation* 1991; **85**: 350–6.

124. Silke B, Tennet H, Fischer-Hansen J, Keller N, Heikkila J, Salminen K. A double-blind, parallel-group comparison of flosequinan and enalapril in the treatment of chronic heart failure. *European Heart Journal* 1992; **13**: 1092–100.

125. Cohn JN, Franciosa JA. Vasodilator therapy of cardiac failure. (First of two parts). *New England Journal of Medicine* 1977; **297**: 27–31.

126. Cohn JN, Franciosa JA. Vasodilator therapy of cardiac failure. (Second of two parts). *New England Journal of Medicine* 1977; **297**: 254–8.

127. Cowan JC. Nitrate tolerance. *International Journal of Cardiology* 1986; **12**: 1–19.

128. Popescu LM, Foril CP, Hinescu M, Panoiu C, Cinteza M, Gherasim L. Nitroglycerin stimulates the sarcolemma Ca^{2+} extrusion of coronary smooth muscle cells. *Biochemical Pharmacology* 1985; **34**: 1857–60.

129. Cohen MV, Downey JM, Sonnenblick EH, Kirk ES. The effects of nitroglycerin on coronary collaterals and myocardial contractility. *Journal of Clinical Investigation* 1973; **52**: 2836–47.

130. Brown G, Bolson E, Petersen RB, Pierce CD, Dodge HT. The mechanisms of nitroglycerin action. Stenosis vasodilatation as a major component of drug response. *Circulation* 1981; **64**: 1089–97.

131. Dalal JJ, Yao L, Parker JO. Nitrate tolerance: influence of isosorbide dinitrate on the haemodynamic and antianginal effects of nitroglycerin. *Journal of the American College of Cardiology* 1983; **2**: 115–20.

132. Conti CR, Hill JA, Feldman RL, Mehta JL, Pepine CJ. Nitrates for treatment of unstable angina pectoris and coronary vasospasm. *American Journal of Medicine* 1983; **74** (Suppl 6B): 28–32.

133. Ginsburg R, Lamb I, Schroeder JS, Hu M, Harrison DC. Randomized double-blind comparison of nifedipine and isosorbide dinitrate therapy in variant angina pectoris due to coronary arterial spasm. *American Heart Journal* 1982; **103**: 44–9.

134. Leier CV, Huss P, Magorien RD, Unverferth DV. Improved exercise capacity and differing arterial and venous tolerance during chronic isosorbide dinitrate therapy for congestive heart failure. *Circulation* 1983; **67**: 817–22.

135. Cohn JN, Archibald DG, Ziesche S, *et al.* Effect of

vasodilator therapy on mortality in chronic congestive heart failure. Results of a Veterans Administration Cooperative Study. *New England Journal of Medicine* 1986; **374**: 1547–52.

136. Armstrong PW, Armstrong JA, Marks LS. Pharmacokinetic-hemodynamic studies of nitroglycerin ointment in congestive heart failure. *American Journal of Cardiology* 1980; **40**: 670–6.

137. Vincent JL, Leon M, Berre J, Melot C, Kahn RJ. Addition of enoximone to adrenergic agents in the management of severe heart failure. *Critical Care Medicine* 1992; **20**: 1102–6.

138. Medical Research Council Working Party. MRC trial of treatment of mild hypertension: principal results. *British Medical Journal* 1985; **291**: 97–102.

139. Greenberg G, Thompson SG, Brennan PJ. The relationship between smoking and the response to antihypertensive treatment in mild hypertensives in the Medical Research Council's trial treatment. *International Journal of Epidemiology* 1987; **16**: 25–30.

140. Hypertension Detection and Follow-up Program Cooperative Group. Persistence of reduction in blood pressure and mortality in the Hypertension Detection and Follow-up Program. *Journal of the American Medical Association* 1988; **259**: 2113–22.

141. Wright AD, Barber SC, Kendall MJ, Poole PH. Beta-adrenoceptor blocking drugs and blood sugar control in diabetes mellitus. *British Medical Journal* 1979; **1**: 159–61.

142. Grimm RH. Alpha$_1$-antagonists in the treatment of hypertension. *Hypertension* 1989; **13** (Suppl I): I-131–6.

143. Colucci WS. New developments in alpha-adrenergic receptor pharmacology: implications for the initial treatment of hypertension. *American Journal of Cardiology* 1983; **51**: 639–43.

144. Gunnells JC Jr. Treating the patient with mild hypertension and renal insufficiency. *American Journal of Cardiology* 1983; **51**: 651–6.

145. Kynch J. Pharmacology of terazocin. *American Journal of Medicine* 1986; **80** (Suppl 5B): 12–19.

146. Chrysant SG. Experience with terazosin administered in combination with other antihypertensive agents. *American Journal of Medicine* 1986; **80** (Suppl 5B): 55–61.

147. MacCarthy EP, Bloomfield SS. Labetalol: a review of its pharmacology, pharmacokinetics, clinical uses, and adverse effects. *Pharmacotherapy* 1983; **3**: 193–217.

148. Takekoshi N, Murakami E, Murakami H, *et al*. Treatment of severe hypertension and hypertensive emergency with nifedipine, a calcium antagonistic agent. *Japanese Circulation Journal* 1981; **45**: 852–60.

149. Weinberger MH. Comparison of captopril and hydrochlorothiazide alone and in combination in mild to moderate essential hypertension. *British Journal of Clinical Pharmacology* 1982; **14**: 127S–31S.

150. Hodson GP, Isles CG, Murray GD, Usherwood TP, Webb DJ, Robertson JI. Factors related to first dose hypotensive effect of captopril; prediction and treatment. *British Medical Journal* 1983; **286**: 832–4.

151. Fouad FM, Nakashima Y, Tarazi RC, Salcedo EE. Reversal of left ventricular hypertrophy in hypertensive patients treated with methyldopa. *American Journal of Cardiology* 1982; **49**: 795–801.

152. Pooling Project Research Group. Relationship of blood pressure, serum cholesterol, smoking habit, relative weight, and ECG abnormalities to incidence of major coronary events; final report of the Pooling Project. *Journal of Chronic Diseases* 1978; **31**: 201–306.

153. Houston D, Vanhoutte P. Serotonin and the vascular system: role in health and disease and implications for therapy. *Drugs* 1986; **31**: 149–63.

154. Vanhoutte PM, Luscher TF. Serotonin and blood vessel wall. *Journal of Hypertension* 1986; **4** (Suppl 1): S29–S35.

155. Breuer J, Meschig R, Breuer H, Arnold G. Effects of serotonin on the cardiopulmonary circulatory system with and without 5-HT$_2$-receptor blockade by ketanserin. *Journal of Cardiovascular Pharmacology* 1985; **7** (Suppl 7): s23–s25.

156. Mantero F, Rocco S, Opocher G, Armanini D, Boscaro M, D'Agostino D. Effect of ketanserin in primary aldosteronism. *Journal of Cardiovascular Pharmacology* 1985; **7** (Suppl 7): s172–5.

157. De Cree J, Hoing M, De Ryck M, Symoens J. The acute antihypertensive effect of ketanserin increases with age. *Journal of Cardiovascular Pharmacology* 1985; **7** (Suppl 7): s126–7.

158. Breckenridge A. Ketanserin – a new antihypertensive agent. *Journal of Hypertension* 1986; **4** (Suppl 1): s13–16.

159. Engleman RM, Haag B, Lemeshow S, Angelo A, Rousou JH. Mechanism of plasma catecholamine increases during coronary artery bypass and valve procedures. *Journal of Thoracic and Cardiovascular Surgery* 1983; **86**: 608–15.

160. Ghignone M, Noe C, Calvillo O, Quintin L. Anesthesia for ophthalmic surgery in the elderly: the effects of clonidine on intraocular pressure, perioperative hemodynamics, and anesthetic requirements. *Anesthesiology* 1988; **68**: 707–16.

161. Gray RJ. Postcardiac surgical hypertension. *Journal of Cardiothoracic Anesthesia* 1988; **2**: 678–82.

162. Roberts AJ, Niarchos AP, Subramanian VA, *et al*. Systemic hypertension associated with coronary artery bypass surgery. Predisposing factors, hemodynamic characteristics, humoral profile, and treatment. *Journal of Thoracic and Cardiovascular Surgery* 1977; **74**: 846–59.

163. Gray RJ, Bateman TB, Czer LSC, Conklin C, Matloff JM. Comparison of esmolol and nitroprusside for acute post-cardiac surgical hypertension. *American Journal of Cardiology* 1987; **59**: 887–92.

164. Hodsman NBA, Colvin JR, Kenny GNC. Effect of ketanserin on sodium nitroprusside requirements, arterial pressure control and heart rate following coronary artery bypass surgery. *British Journal of Anaesthesia* 1989; **52**: 527–31.

165. Van Aken H, Cottrell JE, Anger C, Puchstein C. Treatment of intraoperative hypertensive emergencies in patients with intracranial disease. *American Journal of Cardiology* 1989; **63**: 43C–7C.

166. Van Zwieten PA, De Jonge A, Wilffert B, Timmermans PB, Beckeringh JJ, Thoolen MJMC. Cardiovascular effects and interaction with adrenoceptor of urapidil. *Archives Internationales de Pharmacodynamie et de Thérapie* 1985; **276**: 180–1.

167. Van Aken H, Puchstein C, Sicking K, Koning H. Antihypertensive treatment with urapidil does not increase ICP in patients. *Intensive Care Medicine* 1986; **12**: 106–7.

Chapter 12

Physiology and Pathophysiology of the Cardiovascular System

Pierre Foëx

Cardiovascular physiology	Pathophysiology

Cardiovascular physiology

The circulatory system delivers oxygen and fuels to, and removes carbon dioxide and metabolic byproducts from, the tissues. The circulation also transports hormones, vasoactive and immune substances, and serves to maintain body temperature. Its efficacy depends on the integrity of the myocardium, the characteristics of the vascular beds, the adequacy of coronary blood flow, and the integrity of the autonomic nervous system. In addition, the requirements of the specialized circulations must be satisfied. Many patients presenting for surgery suffer from diseases of the cardiovascular system.

The cardiac cell

Ultrastructure

Cardiac cells are branched filament-like structures, 10–20 μm in diameter and 50–100 μm long, attached to one another by intercalated discs. Approximately every 2 μm in their longitudinal axis, transverse (T) tubules penetrate the cells and facilitate their activation.[1,2] The sarcolemma and the membranes of the sarcoplasmic reticulum and mitochondria control ion fluxes. Sodium ion fluxes are essential to the depolarization of the cells, while calcium ion fluxes are central to the excitation–contraction coupling. The basic unit is the sarcomere. It contains myofilaments of actin and myosin (Fig. 12.1). The thick filaments (myosin) exhibit lateral projections disposed in a helix pattern. The thin filaments (actin) form an hexagonal array around the myosin filaments from which they are separated by a distance of approximately 15 μm.

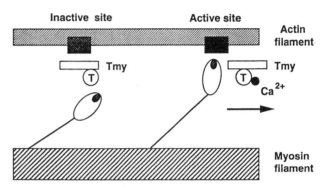

Fig. 12.1 According to the cross-bridge theory of contraction, lateral projection of the thick filaments with their hinged arm and rotating head are attached for a very short time to active binding sites on the thin filaments. They slide and move to the next active binding site. Tropomyosin (Tmy) prevents the formation of cross-bridges under resting conditions. Calcium ions (Ca^{2+}) alter the tropomyosin–troponin complex to allow cross-bridge formation. T, troponin.

Contraction

Following depolarization, the thick and thin filaments slide past each other.[3,4] This is facilitated by the lateral projections of the myosin filaments which are mobile and possess an articulated head attached to an arm which is capable of sustaining tension (Fig. 12.1). When the arm abducts and the head comes closer to the actin filaments, cross-bridges are formed. Head rotation pulls on the arm and causes sliding.[2,5] At rest, the active sites are blocked by tropomyosin. This block is removed by the binding of calcium ions to the troponin complex. The energy for cross-bridge cycling is provided by adenosine triphosphate (ATP).

For isometric contractions (contractions without short-ening), developed tension depends upon the initial length of the sarcomere (Fig. 12.2). Tension reaches its maximum at the optimal initial sarcomere length for cross-bridge formation.[6] Beyond the optimal initial length, the overlap of actin and myosin is reduced: tension is submaximal. Below the optimal initial length, the overlap of actin and

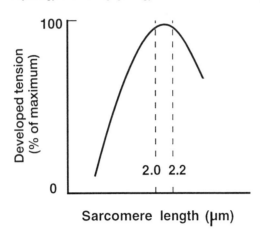

Fig. 12.2 Initial fibre lengths of between 2.0 and 2.2 μm are associated with the maximum tension development by isolated sarcomeres.

myosin filaments is too extensive and fewer cross-bridges can develop.

Activation of cardiac fibres

A potential difference (−60 mV to −90 mV) exists between the interior and exterior of the myocyte. This results from the high concentration of potassium ions (K^+) in the intracellular fluid and the high permeability of the cell to potassium ions. In response to changes in the permeability for sodium, calcium and potassium ions, action potentials are generated and transmitted.

During the plateau phase of the action potential calcium ions enter the cardiac cells. This triggers the release of calcium from the sarcoplasmic reticulum, and raises the concentration of myoplasmic calcium from 10^{-7} to 10^{-5} mmol/l. As calcium associates with troponin, the troponin-mediated inhibition of the actin–myosin interaction is suppressed. Relaxation occurs when calcium is taken up by the sarcoplasmic reticulum (active transport), making it possible for troponin to inhibit the actin–myosin interaction.

Calcium ions penetrate the sarcomere through voltage-dependent and receptor-dependent channels, and by passive diffusion across the sarcolemma. The voltage-dependent channels are subdivided into three types, L (long-lasting, large current), T (transient, tiny current) and N (neurally located). The calcium channels seem to have two separate gating mechanisms. One, on the extracellular side of the sarcolemmal membrane, is voltage dependent. The other, located on the cytoplasmic side, is regulated by nucleotides (cyclic AMP, cyclic GMP). An energy dependent Ca^{2+}-Na^+ pump may facilitate either influx or egress of calcium in exchange for sodium ions. Finally an ATP-dependent calcium pump is responsible for the greater part of the flux of calcium out of the cell.

Cardiac performance: from fibre to intact heart

Isolated heart fibre

The relation between force or tension and initial sarcomere length forms the cellular basis for Starling's law of the heart.[7] The relation between force or tension and velocity of shortening forms the cellular basis for the study of contractility.[8,9] An increase in preload increases the maximum force the muscle may develop but does not influence the maximum velocity at zero load. Conversely, an increase in contractility increases both the maximum developed force and the maximum velocity at zero load (Fig. 12.3). The models used to describe cardiac muscle usually imply three elements, i.e. contractile element, series elastic element, and parallel elastic element (Fig. 12.4). The arrangement of these elastic elements explains the resting tension which is much higher than in skeletal muscle and increases as a non-linear function of sarcomere length.

Intact heart

The external work developed during each contraction varies with alterations in preload, contractility and afterload.

Relations between ventricular work and filling pressure (ventricular function curve), pressure and output (pump function curve) and tension and dimensions (dynamic pressure dimension loops), can be studied.

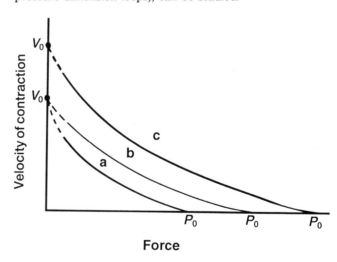

Fig. 12.3 Changes in preload and in contractility have characteristic effects on the relation between force and velocity of contraction of cardiac muscle. An increase in preload increases the maximum load (P_0) and shifts the relation upwards (a to b) without change in velocity at zero load (V_0). An increase in contractility, however, shifts the relation upwards (a to c) with increases in both P_0 and V_0. Note that velocity at zero load (V_0) is usually termed V_{max}.

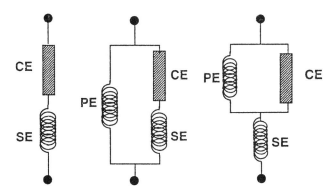

Fig. 12.4 Models of contraction for skeletal (two-element model), and cardiac muscle (three-element models). Contractile element (CE) and series elastic element (SE) account for skeletal muscle contraction. For cardiac muscle, resting tension is accounted for by the parallel elastic element (PE).

Indices derived from V_{max}

Force–velocity curves have been constructed for the intact heart and V_{max} extrapolated for zero load. However, the first derivative of left ventricular pressure, which is easily measured in animal experiments and can be obtained in man during cardiac catheterization,[10] is more commonly used and can be shown to relate to the velocity of contraction of the contractile element. The maximum rate of change of left ventricular pressure, LVdP/dt_{max}, is little influenced by changes in aortic pressure and heart rate but is relatively sensitive to changes in preload.

Ventricular function curves

Ventricular function curves express the mechanical activity of the ventricle as a function of the preload.[11] Ventricular stroke work, stroke volume or cardiac index can be plotted against right or left ventricular end-diastolic pressure, the latter represented by the pulmonary capillary wedge pressure. A shift of the function curve upwards and to the left indicates improved contractility and a shift downwards and to the right, myocardial depression (Fig. 12.5). However, similar displacements may be caused by changes in vascular resistance.[12] Alterations in ventricular compliance further complicate the interpretation of ventricular function curves: the end-diastolic ventricular pressure does not relate linearly to fibre length and ventricular volume. Thus, when changes in ventricular compliance occur, the same end-diastolic pressure may represent different end-diastolic volumes. With their limitations ventricular function curves allow some estimation of the function of the cardiac pump and make it easier to determine how to optimize cardiac output (Fig. 12.5). Recently, the relation between stroke work and end-diastolic volume has been re-examined, found to be linear, and termed recruitable stroke work.[13] Increases in inotropy cause an upward shift in this relation (Fig. 12.6).

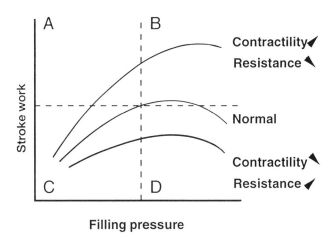

Filling pressure

Fig. 12.5 Ventricular function curves are obtained by plotting stroke work (or stroke volume) against ventricular filling pressure (or volume). Positive inotropic interventions and reductions in vascular resistance shift the curve upwards. Conversely, negative inotropy and increases in vascular resistance shift the curve downwards. Therapeutic decisions are facilitated by consideration of these relations. Values in area B indicate the need for vasodilators or diuretics; area C, volume load; area D, inotropes, vasodilators, or aortic balloon counterpulsation. Area A corresponds to a normal or hyperdynamic circulation.

Pressure–volume relations

In 1895 Frank described the pressure–volume relation for the frog heart.[14] Pressure–volume curves can be constructed for the intact ventricle, and they form a loop. Each loop has an ascending segment which represents the isovolumic contraction, a horizontal segment representing the ejection phase, a vertical segment corresponding to the isovolumic relaxation and a final horizontal segment representing ventricular filling (Fig. 12.7). At end systole, depending on the resistance to ejection, the pressure–volume loops fall on a straight line, the end-systolic pressure–volume line (Fig. 12.7), the slope of which is

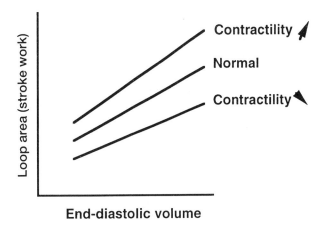

End-diastolic volume

Fig. 12.6 The relation between stroke work and preload (end-diastolic volume), the preload recruitable stroke work, is linear over a relatively large range of preload. The slope is an index on contractility.

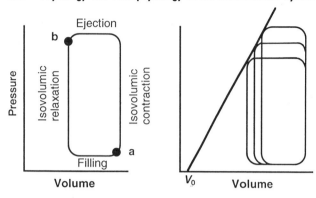

Fig. 12.7 When ventricular pressure and volume are measured simultaneously and are plotted against one another, the resulting pressure–volume loop shows the four phases of the cardiac cycle (left panel). In the face of changes in resistance to ejection, the end-systolic points of the loops fall on the end-systolic pressure–volume line (right panel). The slope of this line is an index of contractility.

characteristic of the inotropic state of the myocardium and is termed maximum elastance.[15] The area of the pressure–volume loop corresponds to the external work developed during each systole. The area delineated by the pressure–volume line at end systole and the loop corresponds to the potential energy for the cardiac cycle (Fig. 12.8).[16]

Ventricular relaxation

Isovolumic relaxation

In many pathological conditions (myocardial ischaemia, hypertrophic cardiomyopathies, left ventricular failure) abnormalities of diastolic function occur before systolic function is depressed. This may be explained by the active, energy-requiring nature of relaxation (about 15 per cent of the total energy of the cardiac cycle). Relaxation of the heart fibre is controlled by total load and inactivation.[17] Sensitivity to total load can be explained by the balance between number of cross-bridges and load; the greater the load, the faster fibres relax. This is further facilitated by the rapid filling of the coronary arteries which increases wall thickness, intramyocardial pressure and the total load.

Inactivation depends upon the active transport of calcium ions (ATP-dependent calcium pump) from the troponin to the sarcoplasmic reticulum. The cytosolic calcium concentration may remain abnormally high (and relaxation abnormally slow or incomplete) when (1) calcium pump fails; (2) calcium leaks from the sarcoplasmic reticulum; (3) the myocytes are unable to extrude calcium; or (4) when the affinity of the troponin for calcium is exaggerated. With myocardial ischaemia, too little ATP is available, calcium transport to the sarcoplasmic reticulum is diminished and the inhibition of the actin–myosin interaction is delayed and incomplete. This results in a

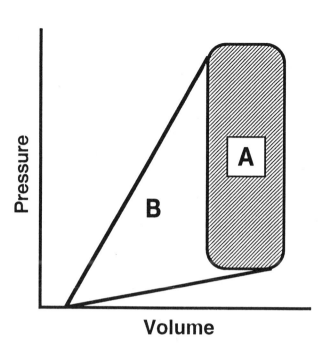

Fig. 12.8 The area of the pressure–volume loop represents the stroke work (A), while the area delineated by the end-systolic pressure–volume line, the left side of the loop, and the line joining V_0 and the end-diastolic point (B) represents the potential energy of the ventricle.

slower relaxation of the ventricle. Sympathomimetic amines enhance relaxation of normal myocardium and delay that of ischaemic myocardium, reflecting catecholamine-induced asynchrony of contraction in compromised myocardium. Conversely, β-adrenoceptor blockade depresses relaxation in the normal myocardium but, because of the reduction in left ventricular asynchrony, improves that of the compromised myocardium.

Ventricular filling

While isovolumic relaxation is caused by the dissociation of the cross-bridges between actin and myosin, ventricular filling is due to the elongation of the sarcomeres that follows. It has three phases: rapid filling, diastasis and atrial contribution to filling (Fig. 12.9). The efficiency of filling is decreased when the end-systolic volume is increased, when cross-bridges do not dissociate, and when fibrosis develops in the myocardium.

Ventricular static compliance

This is the relation between pressure and volume at end diastole (Fig. 12.10). Myocardial ischaemia and hypertrophic cardiomyopathies increase the stiffness of the ventricle so that the relation between pressure and volume

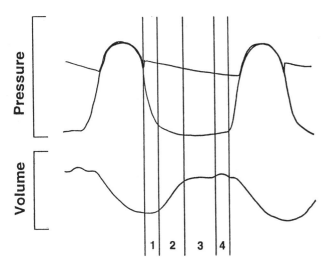

Fig. 12.9 Ventricular diastole can be divided into four phases: isovolumic relaxation (1), rapid filling (2), diastasis (3), and atrial contribution to filling (4); in addition to the relation between end-diastolic pressure and volume (see Fig. 12.10).

becomes much steeper (Fig. 12.10). Inverventions that improve myocardial oxygenation and nitrates improve left ventricular compliance.

The cardiac cycle

The normal human heart weighs between 250 and 300 g, contracts 70–75 times per minute at rest, pumping 5 litres of blood per minute.

Each cardiac cycle (Fig. 12.11) consists of a period of contraction (systole) followed by a period of relaxation (diastole). Atrial systole is initiated by the spontaneous generation of an action potential in the sinoatrial node. After a delay due to the slow conduction velocity of the atrioventricular node, the ventricles are activated. Tension develops rapidly without change in intracavitary volume (isovolumic contraction). Once the intraventricular pressures exceed the pulmonary artery and aortic pressures, ejection starts and most of the volume is ejected during the first third of systole. The ejection fraction, the ratio of stroke volume and end-diastolic volume, is approximately 66 per cent. As left and right ventricular pressures fall at the end of the active phase of contraction, the aortic and pulmonary valves close. The elastic recoil of the valves is responsible for the dicrotic notch. Initially, the ventricles relax without change in volume (isovolumic relaxation). As the rapid fall in intracavitary pressure allows the mitral and tricuspid valves to open, filling starts and is rapid in the early phase and slower during the later phase of diastole. Atrial contraction may contribute as much as 25 per cent of total ventricular filling. The normal values for intracavitary pressures are given in Table 12.1.

For a heart rate of 75 beats/min, the duration of each cardiac cycle is approximately 800 ms (systole: 300 ms; diastole: 500 ms). With heavy exercise heart rate may attain 180 beats/min. The duration of systole is reduced by about 33 per cent to 200 ms and that of diastole by about 75 per cent to 125 ms. This ultimately limits ventricular filling.

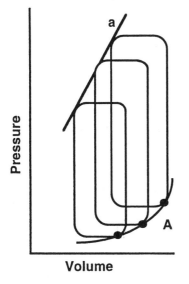
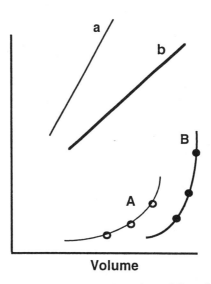

Fig. 12.10 With alterations in ventricular filling, the end-diastolic points of pressure–volume loops fall on the end-diastolic pressure–volume curve (denoted A on the left-hand panel), while the end-systolic points of the pressure–volume loops fall on the end-systolic pressure volume line (a). Note that the end-diastolic pressure–volume relationship is exponential (A). The right-hand panel shows that in cardiac failure and with myocardial ischaemia, the ventricle dilates and becomes stiffer so that at end diastole the pressure–volume loops fall on a new curve (B) that is shifted to the right, and is steeper than the normal curve (A). In addition, cardiac failure depresses the end-systolic pressure–volume relationship (b instead of a).

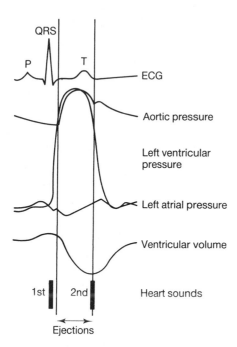

Fig. 12.11 Schematic representation of the events of the cardiac cycle, including aortic, ventricular and left atrial pressure, ventricular volume, ECG and heart sounds. Beginning and end of ejection are denoted by vertical lines.

The vascular beds

The systemic and pulmonary vascular circuits consist of series-coupled sections. The elasticity of the large arteries and the resistance to flow in the peripheral vessels reduce the changes in arterial and pulmonary pressures in respect of the changes in ventricular pressures.

The *arteries* offer little resistance to flow, are distensible, and damp the pulsatile output of the ventricles. Their elasticity (the media contains an internal elastic membrane) allows systolic storage and diastolic propulsion of blood flow between contractions. When blood reaches the smaller arteries flow is relatively steady throughout the cardiac cycle. However, in the pulmonary circulation flow remains pulsatile.

As large arteries subdivide into much smaller arteries, the proportion of elastic fibres decreases, the walls become thinner and wall tension decreases. In the arterioles, the media consists almost entirely of smooth muscle with a rich

Table 12.1 Normal pressures in the circulation

	kPa	mmHg
Right atrium	0.6	4
Right ventricle	3.8/0.6	28/4
Pulmonary artery	3.8/1.8	28/12
Left atrium	1.2	8
Left ventricle	17.3/1.2	130/8
Aorta	17.3/9.3	130/70

nerve supply. Their calibre varies and this determines flow distribution throughout the body and allows arterial pressure to be maintained in the face of changes in cardiac output.

The *precapillary resistance vessels* (small arteries and arterioles) offer the greater part of the total resistance to flow. Small changes in their radius cause large changes in resistance. At each branching, the combined cross-sectional area of the branches exceeds that of the stem (usually by a factor of 1.2–1.3). However, the radius of each branch is smaller than that of the stem and *resistance* increases with each branching. The precapillary sphincters determine the size of the capillary exchange area by altering the number of open capillaries.

The *capillary exchange vessels*, limited by a single layer of endothelial cells, form a dense network with large cross-sectional area, large surface area and short length. Systemic capillaries have an average radius of 3 μm, a length of 750 μm, a cross-sectional area of 30 μm^2, and a surface area of 15 000 μm^2. Under resting conditions, 25–35 per cent of the capillaries are open and the effective total exchange surface is of the order of 250–350 m^2.

The pulmonary capillaries are wider (4 μm) and shorter (350 μm) than the systemic capillaries; their effective surface area is approximately 60 m^2. In heavy exercise it can increase to 90 m^2. The average transit time through lung capillaries is about 1 second (0.35 second in heavy exercise).

Transcapillary exchange

Filtration/absorption

Exchanges depend on the hydrostatic pressure and the colloid–osmotic pressure (Starling's hypothesis). The net rate and direction of fluid movement (J_v) is a function of the net filtration pressure ($P_{capillary} - P_{interstitium}$) and the net colloid osmotic pressure ($\pi_{plasma} - \pi_{interstitium}$) corrected for the characteristics of the capillary membrane: the reflection coefficient σ. In addition, filtration depends on the surface area (S) and the hydraulic conductance of the wall (L_p). Thus, capillary fluid exchange may be expressed as follows:

$$J_v = L_p S[(P_{capillary} - P_{interstitium}) - \sigma(\pi_{plasma} - \pi_{interstitium})]$$

Filtration remains within acceptable limits when pressure at the arteriolar end of the capillary segment is about 4 kPa (30 mmHg) and venous pressure about 2 kPa (15 mmHg). Exceptions are the glomeruler capillaries (pressure of the order of 9 kPa [70 mmHg]) and the pulmonary capillaries (pressure of about 1.3 kPa [10 mmHg]). The plasma colloid–osmotic pressure, mostly due to albumin, is approximately 3 kPa (25 mmHg). Bradykinin, kallidin and histamine increase capillary permeability and increase the leakage of proteins into the interstitial space.

Excess fluid filtration towards the interstitial space causes oedema. The mechanisms involved include excessive capillary pressure (hydrostatic oedema), insufficient colloid osmotic pressure (hypoproteinaemia), increased permeability caused by inflammatory agents (inflammatory oedema), and finally obstruction of lymphatic drainage.

Diffusion
For uncharged soluble particles diffusion depends on the diffusion constant and the concentration difference per unit distance. Lipid-soluble substances (including oxygen and carbon dioxide) diffuse almost freely, while water-soluble molecules pass slowly through the endothelial cells. For water-soluble small molecules smaller than the pore size, diffusion is free only across the pore area. For molecules with a radius close to that of the pores, diffusion is restricted and depends on the geometry of the pore and friction between molecule and pore wall. In tissue injury differences in diffusibility of water and water-soluble solutes cause osmotic gradients resulting in the accumulation of fluid in intracellular and interstitial spaces. For large molecules diffusion is extremely restricted and additional passage is possible through large pores or capillary leaks which represent only a very small proportion of the pore except in fenestrated capillaries and discontinuous capillaries.[18,19]

The *postcapillary resistance vessels* (venules and small veins) determine the ratio between pre- and postcapillary resistance. This controls the hydrostatic pressure in the capillaries and contributes to the regulation of fluid transfer. The large cross-sectional area, large volume and low resistance of the postcapillary resistance vessels allow venous return with a low pressure gradient.

Veins have a small amount of elastin and smooth muscle in their walls. Pressure at the venular end of capillaries is approximately 1.3 kPa (10 mmHg) and at the entrance of the vena cava in the heart approximately 0.5 kPa (4 mmHg). The venous system contains about two-thirds of the total blood volume and may take up 90 per cent of fluid loads.

Resistance in the circulation

The pressure difference $(P_1 - P_2)$ across the vascular beds and the resistance (R) which opposes blood flow determines flow $(\dot{Q})$ through blood vessels:

$$\dot{Q} = \frac{P_1 - P_2}{R} \qquad (1)$$

Resistance (R), can be calculated as:

$$R = \frac{8\eta L}{\pi r^4} \qquad (2)$$

where r is the radius of the vessel, η the viscosity of the liquid and L the length of the vessel. Equations 1 and 2 can be rewritten as Poiseuille's Law (3):

$$\dot{Q} = \frac{(P_1 - P_2)\ \pi r^4}{8\eta L} \qquad (3)$$

Flow depends upon the cross-sectional area of the vessel (πr^2) and the velocity of the fluid which is proportional to the square of the radius.

Vascular resistance can be calculated by dividing the pressure gradient by the flow. Peripheral (or systemic) vascular resistance is approximately eight times larger than pulmonary vascular resistance.

Viscosity
Viscosity is the resistance to flow due to the friction of molecules in the moving stream and depends mainly on the concentration of the red cells. Blood viscosity *in vivo* is less than *in vitro* because of axial streaming of the cells. Using water as reference, blood viscosity at normal haematocrit is about 3.6, increasing to 5.6 at a haematocrit of 60 per cent.

Impedance
Because the arterial bed offers elasticity, resistance and inertia to the pulsatile flow, blood flow should be compared with a.c. and not d.c. current.[20] The dynamic relation between pulsatile pressure and flow (input impedance) is more meaningful than that of mean pressure and mean flow (vascular resistance). This is particularly true of the pulmonary circulation because of the pulsatile nature of blood flow throughout the pulmonary vasculature.

Pressure waveforms
When blood is forced into distensible vessels, some blood is stored and the intermittent input is transformed into a more even outflow. Delays in transmission, creep, hysteresis and wave reflection from branching sites explain the complex shape of the pressure waveforms. Autonomic stimulation and arteriosclerosis reduce both vascular distensibility and pressure 'storing' so that peak pressures become higher.

Ageing is associated with an increasing rigidity of the vascular tree. This contributes to an increase in systolic pressure. In addition, wave reflection (from major branching sites) increases the load presented to the left ventricle, resulting in left ventricular hypertrophy.

Interactions in the cardiovascular system

Interactions between heart and vascular beds

Vascular impedance is a major determinant of stroke volume[20] and of ventricular wall stress. The latter is often considered the true ventricular afterload. Increases in

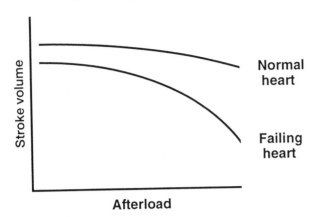

Fig. 12.12 The relation between stroke volume and afterload (resistance) is altered as a function of the quality of the ventricular muscle. With depressed inotropy, any increase in resistance is associated with a marked reduction in stroke volume.

afterload decrease output unless they are caused by interventions which also increase contractility. When an increase in afterload is applied to the depressed heart, stroke volume decreases and the pressure–volume relations are altered (Fig. 12.12).

Interventricular interactions

The pericardium is a common compliance for both ventricles. Once a critical level of ventricular dilatation has been reached, pericardial compliance is less than ventricular compliance. Beyond this point, increases in the volume of one ventricle decrease the space available to the other. Indeed, in patients with raised pulmonary vascular impedance the right ventricle dilates, its end-diastolic pressure may exceed that of the left ventricle and the interventricular septum bulges into the left ventricle, reducing its compliance. Thus, left ventricular performance is no longer solely dependent upon its own inotropic state and afterload but also upon the state of the right ventricle.[21,22]

Electrophysiology

The conduction system

The action potential spreads through a specialized conduction sytem consisting of the atrioventricular (AV) node, the bundle of His, its branches, and the Purkinje network. The sinoatrial (SA) node, the AV node and the remaining conduction tissue can initiate action potentials and serve as pacemakers. The SA node has the fastest discharge rate. The transmission velocity of the action potential is slow (0.03–0.05 m/s) in the SA and AV nodes, fast in the bundle of His (0.8–1.0 m/s) and very fast in the

Purkinje tissue (5 m/s). Depolarization is propagated from fibre to fibre by the intercalated discs.

Action potentials

The resting potential (−60 to −90 mV) is determined essentially by the concentration gradient of potassium ions across the sarcolemma and is maintained by chemical concentration gradients across the membrane, internal and external electrostatic forces and the membrane itself.[23–26] The magnitude of the action potential can be calculated using the Nernst equation:

$$E_K = \frac{RT}{FZ_K} \ \ln \ \frac{[K_O^+]}{[K_i^+]}$$

where E_K is the equilibrium potential for K^+; Z_K the valence for K^+; $[K_0^+]$ the K^+ concentration outside and $[K_i^+]$ inside the cell, R the gas constant, T the absolute temperature, and F the Faraday number.

The cell membrane, through the opening and closure of ion channels and the operation of ion pumps, maintains the resting potential, generates and propagates the action potential. At rest, the permeability of the cell membrane to potassium ions is much greater than that for sodium and calcium ions. The ratio of permeabilities $P_{K+}:P_{Na+}$ is approximately 100:1. The high extracellular sodium concentration would cause the resting potential to fall rapidly (sodium leak) in the absence of active outward transport of sodium ions. The enzyme Na^+,K^+-ATPase supplies the energy for this active process which is impeded by ischaemia (inadequate substrate availability) and by digitalis.

The sinus node

In diastole, the sinus node pacemaker cells are polarized (−50 to −60 mV inside). The Na^+,K^+-ATPase of the membrane ensures that the sodium ion concentration remains low inside the cell. However, a slow inflow of positively charged ions (sodium, calcium or both) causes spontaneous depolarization. This phenomenon suddenly becomes much faster because of increased permeability to calcium ions. This results in an action potential. Cyclical changes in membrane conductance may be caused by the action of catecholamines on the adenylate cyclase system. Under the effect of an inward calcium current, catecholamines could be released and then bind to the outside of the cell membrane. This, in turn, would favour the inward calcium current. A positive feedback mechanism would explain the exponential nature of the depolarization of the pacemaker cells.

Atrial, Purkinje and ventricular muscle fibres

These muscle fibres have a resting membrane potential of −80 to −90 mV. Depolarization is fast (100–500 V/s)

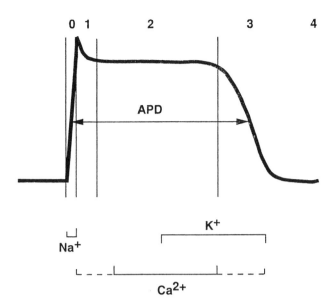

Fig. 12.13 Diagrammatic representation of an action potential of a contractile ventricular cell, with the main ion fluxes responsible for depolarization and repolarization.

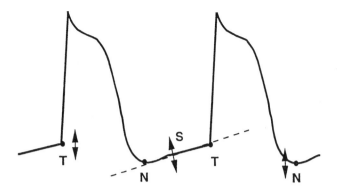

Fig. 12.14 Alterations in autonomic activity modify heart rate by means of their effect on the spontaneous depolarization (S), the threshold for rapid depolarization (T) and the level or repolarization (N).

because of an explosive increase in sodium ion conductance (phase 0) followed by a rapid increase in chloride ion conductance. The cells become positively charged because sodium ions enter the cells faster than potassium ions leave.

Repolarization of cardiac cells is slow. A brief rapid phase brings the potential to zero volt (phase 1). This is followed by a plateau phase at about zero volt (phase 2) and a slow phase (phase 3) which brings the membrane potential back to its resting value (Fig. 12.13). The plateau phase of the action potential results from a slow flux of calcium and sodium ions. The slow inward current of calcium ions opposes the repolarizing effect of sodium ion extrusion, thus increasing the action potential duration. Moreover, the inward migration of calcium ions ensures the triggered release that is necessary for the excitation–contraction coupling. Eventually, potassium fluxes bring the membrane potential back to its original level.

After each depolarization, the cell will not respond to another stimulation irrespective of its intensity (the absolute refractory period). Later, the cell will respond only to an intense stimulation (relative refractory period). This is followed by the vulnerable period associated with the peak of the T wave of the electrocardiogram.

The differences in ion fluxes between sinus node cells and contractile cells (respectively slow and fast response cells) may relate to the gating of sodium ion channels. The gates are probably charged electrical fields in the matrix of the membrane. If gates remain closed, inward flux of sodium is minimal and the action potential has a much slower upslope. In ischaemia, fast cells may be converted into slow cells exhibiting less negative resting potential and slower phase 0 depolarization. This transformation is

accompanied by a tendency to spontaneous depolarization leading to ectopic rhythms.

The control of heart rate by the autonomic nervous system is due to alterations in ionic permeability especially in sinus node cells (Fig. 12.14). Three major types of alterations in the action potential contribute to changes in heart rate: (1) modifications in the slope of diastolic depolarization; (2) changes in the threshold potential; (3) changes in the magnitude of the resting potential.

Vagal stimulation increases the resting potential (hyperpolarization), decreases the rate of spontaneous depolarization, and increases the threshold above which explosive increases in sodium and calcium conductance occur. These combined mechanisms decrease the overall rate of sinus node discharge. Acetylcholine reduces the duration of the action potential. Adrenergic stimulation causes hyperpolarization, increases the rate of spontaneous depolarization, and reduces the duration of the action potential. This facilitates the development of ectopic rhythms.

The coronary circulation

Anatomy

The left main coronary artery divides into two major branches, the left anterior descending and the circumflex artery. The former supplies the septum and a larger mass of the wall of the left ventricle than the circumflex or the right coronary arteries. The right coronary artery perfuses the inferior wall of the left ventricle in 85 per cent of adult humans.

The perfusion fields of the large epicardial coronary vessels interdigitate with each other. Capillary loops from adjacent beds also provide some anastomoses. The intramural coronary vessels penetrate into the myocardium from the epicardial vessels, form a dense vascular network leading to a subendocardial plexus. The capillary density is of the order of 2–$5 \times 10^5/mm^2$ with a capillary:fibre ratio of 1:1 and an intercapillary distance of 14–19 μm.

Epicardial veins drain the effluent blood from the left ventricle, converge into the coronary sinus and empty into the right atrium. Anterior superficial veins drain the right ventricle into the right atrium. Thebesian veins enter directly into the cardiac cavities.

The total volume of blood contained in the coronary circulation is of the order of 6–15 ml/100 g of myocardium. Blood present in the coronary vessels contributes to the 'architecture' of the myocardium (the stiffness of the wall) and its volume is about 6–10 times that 'pumped' into the coronary vasculature for each heart beat.

Physiology

Metabolic regulation

The myocardium is almost completely dependent on aerobic metabolism. Because of the high oxygen extraction a close linkage exists between oxygen demand and coronary blood flow.

Oxygen delivery to the myocardium can be augmented by an increase in coronary blood flow, a small increase in oxygen extraction, and by a decrease in the diffusion distance between capillaries and myocytes caused by recruitment of capillaries. In the normal coronary circulation, coronary blood flow can increase by five- to sixfold, and coronary oxygen extraction by approximately 20 per cent.[27,28]

The major determinants of myocardial oxygen demand are heart rate, contractility, and wall stress. The latter is the product of ventricular pressure multiplied by the radius of the ventricular cavity.[29] Other factors play a lesser role (Table 12.2).[30]

Heart rate   Increases in heart rate cause an almost linear increase in oxygen consumption.

Contractility   Inotropic interventions increase myocardial oxygen consumption, even when heart rate and wall stress are kept constant.

Table 12.2 Major determinants of myocardial oxygen consumption

Heart rate
Contractility
Wall tension (systole and diastole)
Shortening (ejection)
Electrical activation
Integrity of cell membranes

Wall stress   As wall stress increases myocardial oxygen consumption increases steeply.

Flow generation   It is usually stated that flow work contributes little to oxygen consumption. However, recent studies have shown that the oxygen cost of flow work has been underestimated.

Activation   Electrical activation of cardiac membranes requires only a very small amount of oxygen.

Basal metabolic requirements   The energy needed to maintain the metabolic integrity of the cardiac cells can be supplied by about 25 per cent of the resting coronary blood flow.

Mediators of the metabolic vasodilatation

A number of mediators have been suggested. To qualify as a metabolic mediator of the coupling of vascular resistance and myocardial metabolism, the substance must be a potent vasodilator, have an adequate endogenous source, be able to gain access to vascular smooth muscle, and reach concentrations which are sufficient to produce maximum vasodilatation under physiological conditions.[27]

Oxygen   Arterial hypoxaemia causes profound coronary vasodilatation. However, similar degrees of dilatation occur in severe experimental anaemia when arterial Po_2 is normal. In both hypoxia and anaemia, the arterial oxygen content is decreased. However, a low arterial oxygen content is not a prerequisite for maximal coronary vasodilatation. Indeed, after brief periods of coronary occlusion vasodilatation persists at a time arterial Po_2, arterial oxygen content, myocardial Po_2 and coronary sinus Po_2 are either normal or higher than normal. This suggests that low intramyocardial Po_2 (caused by hypoxia, anaemia or ischaemia) is only the trigger for the release of mediators.

Potassium ions   Potassium ion concentration is an important determinant of skeletal muscle blood flow during exercise. However, in cardiac muscle increases in metabolism cause smaller increases in potassium ion concentration, and vasodilatation may persist after potassium levels have returned to normal.

Carbon dioxide and pH   Alterations in arterial Pco_2 (and the resulting changes in pH) increase (hypercapnia) or decrease (hypocapnia) coronary blood flow. However, in many situations (post-ischaemic reactive hyperaemia, anaemia) coronary blood flow is elevated at a time when arterial, myocardial and coronary sinus Pco_2 are either normal or decreased. Changes in local Pco_2 may modify the effectiveness of other chemical mediators or act as triggers without carbon dioxide being a mediator.

Adenosine   Adenosine is directly involved with the synthesis and breakdown of high energy phosphates and is capable of causing maximal vasodilatation, particularly of the subendocardial vessels. A close correlation has been demonstrated between tissue adenosine, rate of release of adenosine and coronary blood flow.[31,32] Recently, several types of receptors for adenosine have been described. Stimulation of A_2-receptors is responsible for vasodilatation.[33,34] However, the lack of modification of vasodilator responses by adenosine deaminase, and the striking dissociation between coronary sinus levels of adenosine and coronary vascular resistance during prolonged infusion of noradrenaline cast some doubts on the central role of adenosine.

Arachidonic acid derivatives   Prostaglandin I_2 (PGI_2) is a powerful coronary vasodilator. Conversely, thromboxane A_2 and leukotrienes cause coronary vasoconstriction. Myocardial ischaemia, hypoxaemia, adrenergic nerve stimulation, alterations in myocardial metabolism are associated with the release of prostaglandins from the myocardium. Thus prostaglandins may contribute to the 'tuning' of the local coronary vasoregulation through a balance between vasodilatation (prostaglandin I_2) and vasoconstriction (thromboxane A_2, leukotrienes). In the vicinity of atheromatous lesions, damage to the endothelium may alter the balance of prostaglandin synthesis, facilitating vasoconstriction.

Endothelial relaxing factors (EDRFs)   Arterial and venous endothelium can synthesize a dilator substance which diffuses from the endothelium into the underlying vascular smooth muscle.[35] This substance is now identified as nitric oxide (NO) whose half-life is very short (a few seconds). EDRF activates the enzyme guanylate cyclase, thereby increasing the concentration of cyclic guanosine monophosphate (cGMP). This causes vasorelaxation. The role of EDRF in the control of normal coronary vascular tone is controversial. However, in the presence of atheroma, destruction of the endothelium may prevent the local release of EDRF and promote exaggerated vasoconstriction.

Endothelins   Several types of endothelins have been identified as well as two types of endothelin receptors.[36] The cardiac effects of endothelins include increased inotropy, mediated by an increase in myoplasmic Ca^{2+}, and coronary vasoconstriction. In the intact heart coronary vasoconstriction is of such magnitude that cardiac function is depressed by ischaemia, secondary to the reduction in coronary blood flow, rather than enhanced by the direct effect of endothelin on the myocytes.

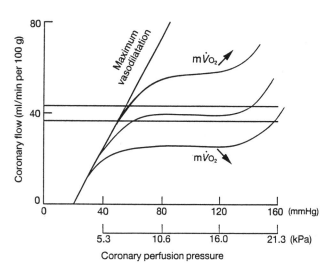

Fig. 12.15 The perfusion pressure between which coronary autoregulation exists is a function of the level of oxygen consumption. As oxygen consumption increases, the range of autoregulated flow becomes narrower. With maximal coronary vasodilatation, autoregulation ceases, and flow becomes pressure dependent.

Autoregulation

Autoregulation of coronary blood flow means that flow is not pressure dependent provided metabolic activity remains constant (Fig. 12.15). The lower limit of autoregulation is 6–9 kPa (50–70 mmHg) while the upper limit is 17–20 kPa (130–150 mmHg). Autoregulation is more effective in epi- than in endocardial vessels. Profound vasodilatation and the presence of coronary artery stenoses impair autoregulation, so that coronary blood flow becomes pressure dependent. Collateral blood flow to ischaemic areas is also pressure dependent and not autoregulated. Brief periods of ischaemia can also temporarily impair autoregulation making flow pressure dependent. The same mediator or mediators involved in local regulation of coronary blood flow are likely to be responsible for autoregulation.

Reactive hyperaemia

Intense myocardial ischaemia produces maximal coronary dilatation. This is exemplified by the marked increase in coronary blood flow that follows short periods of coronary occlusion. Reactive hyperaemia may be used to assess the coronary reserve and to evaluate the extent of vasodilatation caused by drugs.

Coronary perfusion pressure

The coronary perfusion pressure is often calculated as mean arterial minus coronary sinus pressure, or as diastolic

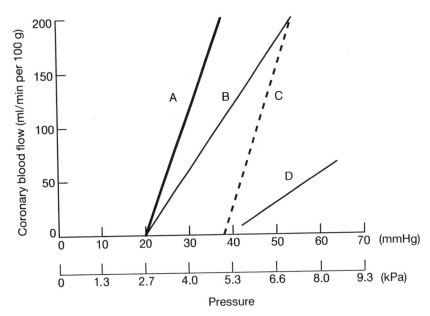

Fig. 12.16 Diagrammatic representation of the relations between pressure and flow in the coronary circulation. The lines represent several situations during which coronary vascular tone is altered. Line A represents maximal coronary vasodilatation (i.e. for any perfusion pressure, coronary flow is higher than in any other situation). Line B represents less than maximal coronary vasodilatation. Note that the lines A and B have a common intercept, the coronary closing pressure, the pressure at which flow is zero (here 2.7 kPa; 20 mmHg). Line C represents maximal vasodilation associated with altered intra-myocardial forces such that the coronary closing pressure is high (here 4.9 kPa; 37 mmHg). Line D represents resting coronary tone. It is apparent that there is scope for vasodilatation (D to C) and for a reduction in coronary closing pressure.

arterial minus ventricular end-diastolic pressure. However, studies of the dynamic pressure–flow relations show that zero flow occurs at inflow pressures (critical closing pressures) substantially higher than diastolic intracavity pressure (Fig. 12.16). The pressure at which flow ceases has been termed the critical closing pressure. It is of the order of 1.5–2.5 kPa (12–18 mmHg) when maximal vasodilatation is present and it is much higher (over 4 kPa; 30 mmHg) when vascular tone is normal.[37] This suggests that coronary resistance depends on compressive forces in the myocardium itself rather than on diastolic ventricular pressure. The critical closing pressure may be modified by drugs: it is decreased by halothane, nitrous oxide, and isoflurane.[38]

In the normal heart, systolic compressive forces are substantially higher in the left than in the right ventricle and consequently, systolic flow to the left ventricular myocardium is much less than flow to the right ventricular myocardium. The ratio of diastolic to systolic flow is 6:1 in the left and 2:1 in the right ventricle.

Transmural myocardial perfusion

Capillary density is greatest at the subendocardium and the responsiveness of the coronary vessels to stimuli is non-homogenous so that vascular resistance across the wall is not uniform. The greater capillary density and lower

vascular resistance in the subendocardium are essential for the adequate perfusion of an area of high metabolic activity.

Neural control

The sympathetic nerves to the heart provide the coronary arteries and veins with a relatively dense innervation. The vagus nerves give abundant cholinergic innervation to the conduction system and the coronary vessels. α- and β-receptors are present as well as muscarinic–cholinergic receptors.

The effects of sympathetic and parasympathetic stimulation are often difficult to ascertain because their direct action on the coronary vasculature may be overshadowed by the indirect effects resulting from changes in myocardial oxygen consumption: local and autonomic regulation may act in opposite directions (Fig. 12.17).

In isolated perfused heart preparations, α-adrenoceptor stimulation causes marked increases in coronary vascular resistance, but does not modify the transmural distribution of coronary blood flow. β₂-Adrenoceptor stimulation causes some coronary vasodilatation, while intracoronary acetylcholine causes near maximal coronary dilatation.

In the intact circulation sympathetic nerve stimulation increases heart rate, myocardial contractility, arterial pressure and oxygen consumption. This is associated with metabolically mediated coronary vasodilatation. Vasodilatation is more pronounced in the presence of

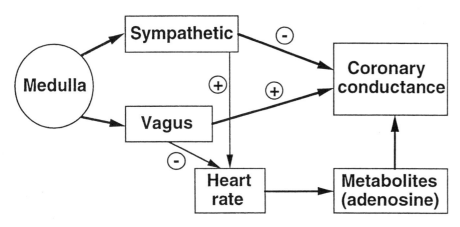

Fig. 12.17 Diagrammatic representation of the influence of sympathetic and parasympathetic systems on the coronary circulation. The direct neurogenic effects may be antagonized by metabolic regulation.

α-adrenoceptor blockade, even in territories supplied by narrowed coronary arteries. This indicates that α-adrenoceptor stimulation prevents the full effect of the local metabolic regulation of coronary blood flow. By contrast, vagal stimulation causes global haemodynamic changes (bradycardia, hypotension) within one cardiac cycle, so that myocardial oxygen consumption is immediately reduced and coronary vasoconstriction ensues. However, the reduction in coronary blood flow in response to vagal stimulation is less than justified by the reduction in metabolic demand.

Anaesthesia

Most anaesthetic agents do not interfere with the local metabolic regulation of coronary blood flow. When coronary blood flow decreases during anaesthesia with halothane and enflurane it is because of the reduction in myocardial oxygen consumption. When coronary blood flow increases with intravenous agents, it reflects the increased oxygen demand caused by tachycardia. These changes are therefore entirely consistent with unaltered local regulation. The lack of effect of anaesthesia on the regulation of coronary blood flow is confirmed by the absence of significant modifications of oxygen extraction. Isoflurane, desflurane and sevoflurane differ from the other inhalational anaesthetics. During their administration, coronary blood flow remains relatively unchanged even though oxygen demand is decreased.[39] This indicates that isoflurane, desflurane and sevoflurane interfere with the local regulation of coronary flow. While relative vasodilatation is unlikely to have an adverse effect on the normal myocardium, dilatation of normal vessels may compromise flow to neighbouring areas supplied by narrowed vessels. Coronary steal may occur when arteriolar vasodilators are administered in the presence of coronary artery stenoses, especially when patients with a 'steal prone' configuration of their coronary circulation. A steal phenomenon has been suggested to explain the development of ischaemia during isoflurane anaesthesia.[40]

The lack of effect of anaesthesia on the control of normal coronary flow does not mean that vasomotor control is totally spared. Under conditions of maximal vasodilatation anaesthesia may modify the coronary closing pressure and alter the relation between perfusion pressure and coronary blood flow (see Fig. 12.16, C to A).

Alterations in carbon dioxide tension exert marked effects on coronary blood flow. Hypocapnia decreases coronary flow more than oxygen consumption so that the arterial to coronary sinus oxygen content difference widens. As a result myocardial $P\text{CO}_2$ decreases. Conversely, hypercapnia increases coronary blood flow more than oxygen consumption (the arterial to coronary sinus oxygen content difference narrows and myocardial $P\text{O}_2$ increases.

Cardiac metabolism

The heart requires a continuous high level supply of energy. The most efficient supply derives from the tricarboxylic acid cycle which leads to the electron transport chain (Fig. 12.18). The rate-limiting enzymes are stimulated by decreasing concentrations of ATP and creatine phosphate (CP), so that more substrate can be handled and more high energy phosphates produced when energy requirements increase. The main rate-limiting enzymes are citric synthetase and isocitric dehydrogenase.

Carbohydrates

At rest, glucose is not the preferred fuel source. However, in stress and hypoxic conditions, the insulin-dependent sarcolemmal transport of glucose is stimulated. Phosphofructokinase controls the glycolytic pathway. Its activity is inhibited by excess citrate, and enhanced by adenosine diphosphate (ADP), adenosine monophosphate (AMP) and inorganic phosphates. Glycolysis provides a rapid source of energy during sudden bursts of heart work, as it can start within 5 seconds of increased demand.

Lactate is taken up by the heart during normal aerobic metabolism. Conversion of lactate into pyruvate is controlled by lactic dehydrogenase (LDH). This enzyme is stimulated by a high NAD (nicotinamide-adenine nucleotide):NADH (reduced NAD) ratio. In hypoxic conditions,

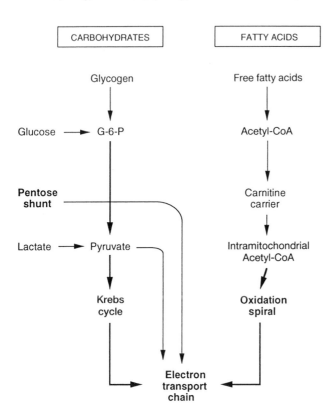

Fig. 12.18 Diagrammatic representation of the main metabolic pathways in the myocardium.

regeneration of NAD is inadequate, the tricarboxylic acid cycle is less active and lactate cannot be used. In such circumstances, myocardial lactate uptake is replaced by lactate production.

Lipid metabolism

Fatty acids traverse the cell membrane by passive diffusion and by a protein transfer mechanism which passes them from albumin to sarcolemmal proteins. Fatty acids are esterified and the glyceroesters are oxidized. Plasma triglycerides cannot pass the cell membrane unless they have been hydrolysed by a lipoprotein lipase. The availability of fatty acids for oxidation in the tricarboxylic cycle depends on the activity of sarcoplasmic lipase (enhanced by catecholamines and inhibited by high levels of fatty acids).

Carnitine

Acetyl-coenzyme A (acetyl-CoA) must be transported across the mitochondrial membrane to enter the tricarboxylic acid cycle. Carnitine (an amino acid), and the enzyme acetyl carnitine transferase are necessary. A number of processes involve acetyl-CoA and carnitine. These include the extramitochondrial formation of acetyl-CoA and of acetyl carnitine from acetyl-CoA, the translocation of acetyl carnitine into the mitochondrial space, the transformation (within the mitochondrial space) of acetyl

carnitine into carnitine with formation of acetyl-CoA. In addition, acetyl-CoA enters the fatty acid oxidation spiral, and there is transfer of carnitine out of the mitochondrial space. Note that carnitine is not synthesized in the heart but has to be supplied by the blood.

Hypoxia

Hypoxia causes marked increases in coronary blood flow so that glycolytic products do not accumulate in the myocytes but are removed from the myocardium. Glucose transport across the sarcolemma is enhanced and glycogen stores are rapidly depleted. However, anaerobic metabolism cannot meet the energy demand so that hydrogen ions accumulate. Eventually, the intracellular acidosis blocks the glycolytic process. Fatty-acid metabolism is decreased by hypoxia. The lack of oxygen interrupts the electron transport chain. In addition, increases in inorganic phosphates (P_i), AMP, and ADP inhibit a number of reactions in the fatty-acid pathway. As fatty acids accumulate, their uptake decreases and there may be exaggerated formation of triglycerides. The accumulation of fatty acids may have an adverse effect on the cell membrane and on the electrophysiology of the cardiac cell.

Ischaemia

In ischaemia, the changes caused by oxygen lack are compounded by the effect of lack of blood flow. Metabolic products and carbon dioxide cannot be removed and therefore tissue acidosis develops very rapidly. The appearance of lactate in coronary sinus blood is a marker of ischaemia. High coronary sinus concentrations of inosine and hypoxanthine are also found in ischaemia. When cellular damage has occurred, leakage of the enzymes creatine kinase (CK) and lactic dehydrogenase (LDH) occurs. The relatively specific CK-MB fraction and LDH2 fraction can be used as indicators of cell damage. Recent evidence suggests that cardiac troponin I is more specific and avoids false diagnoses associated with CK-MB.[102]

Right ventricle and the pulmonary circulation

Right ventricular function

In patients with acute respiratory failure, right ventricular dysfunction is an important determinant of circulatory failure. Pulmonary arterial hypertension and intermittent positive pressure ventilation, especially when associated with positive end-expiratory pressure, increase the afterload of the right ventricle. As in an unstressed state, the right ventricle resembles a passive conduit more than a pump, its physiology has been studied less than that of the left ventricle,[41] and its role in acute circulatory failure has been neglected.

The right ventricle consists of two functionally distinct regions, the inflow and the outflow tracts. They form a crescent-shaped, thin-walled cavity which is bordered by the concave free wall and the convex interventricular septum. The right ventricular free wall is attached to both the wall of the left ventricle and the septum and this facilitates right ventricular emptying.

Contraction of the right ventricle includes a downward motion of the tricuspid valve and an inward motion of the free wall. As the left ventricle contracts, the inward motion of the septum contributes to right ventricular emptying. For the free wall, contraction starts in the inflow tract and progresses towards the outflow tract over at least 25 ms. This delay results from slow electrical activation. Functionally, in the early phase of systole, the inflow tract contracts and the outflow tract dilates. The outflow tract appears to act as a buffer to protect the pulmonary vasculature against acute rises in pressure. The regional differences in right ventricular contraction and relaxation are enhanced by interventions such as positive end-expiratory pressure (PEEP) and pulmonary hypertension.

Determinants of right ventricular performance
Because of its thin wall, the right ventricle is more sensitive to increases in afterload than the thick-walled left venticle. Increases in resistance to ejection cause a greater decline in right ventricular than in left ventricular stroke volume. Increases in preload, and inotropic interventions shift the relation between stoke volume and resistance.[41]

Right ventricular compliance is greater than left ventricular compliance. However, because both ventricles are contained within the pericardium, right ventricular

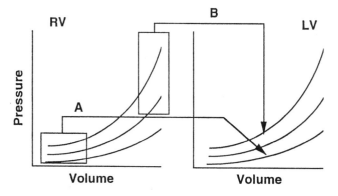

Fig. 12.19 As presures are low in both ventricles during diastole, and the curvature of the septum may vary, alterations in the filling pressure of the right ventricle may modify left ventricular compliance and *vice versa*. High filling pressures (B) in the right ventricle result in a reduction of left ventricular compliance, while low filling pressures (A) are associated with increased compliance.

compliance is influenced by the filling pressure of the left ventricle (Fig. 12.19). Thus, increases in left ventricular end-diastolic pressure decrease right ventricular compliance. The interactions between right and left ventricles are further accentuated by the presence of the interventricular septum. Normally, the gradient in systolic and diastolic pressure between the ventricles is such that the septum bows into the right ventricle. However, when acute or chronic increases in right ventricular volume or pressure occur, the septum may be shifted toward the left ventricle. The latter becomes distorted and less distensible. This explains why in many patients suffering from the adult respiratory distress syndrome, the pulmonary capillary wedge pressure is elevated even though the end-diastolic volume of the left ventricle is not increased. This indicates that right ventricular dilatation (produced by pulmonary hypertension) has reduced left ventricular volume and compliance. Pharmacologically mediated reductions in pulmonary vascular resistance unload the right ventricle and increase left ventricular compliance, presumably by reducing the degree of leftward shift of the septum.

Coronary blood flow
Coronary flow in the right ventricle occurs during both systole and diastole because the coronary driving pressure (the difference between aortic pressure and right ventricular pressure) is high throughout the cardiac cycle. However, in the presence of pulmonary hypertension, right ventricular pressure increases and the difference between aortic pressure and right ventricular pressure is narrowed. Systolic coronary flow is maintained only as long as aortic pressure remains high. If it decreases, coronary blood flow decreases dramatically and the wall of the right ventricle becomes ischaemic. It is now recognized that an adequate gradient between aortic and right ventricular pressure is essential if right ventricular function is to be maintained in the face of pulmonary hypertension. Otherwise right ventricular coronary flow is insufficient to meet oxygen demand and the right ventricle may fail. The pressure gradient can be restored by pulmonary vasodilatation or peripheral vasoconstriction.[42] However, most drugs that dilate the pulmonary circulation also decrease systemic vascular resistance and therefore reduce both systemic and pulmonary pressures. This may not be the case with very low inhaled concentrations of nitric oxide.

Pulmonary circulation

The distribution of pulmonary blood flow is regulated by the alveolar P_{O_2}. Hypoxic pulmonary vasoconstriction diverts blood flow away from the poorly ventilated areas of the lung, making it possible for oxygenation to be maintained under a wide range of conditions.

The alveolar capillaries provide the air–blood interface

necessary for the transport of oxygen and carbon dioxide. This interface is expandable to allow for oxygen transfer to increase when required, and thin to make it efficient. However, the thin interface allows fluid leakage which would impair oxygenation without efficient lymphatic drainage.

While the systemic circulation must meet the variable oxygen requirements of many different organs and cope with postural variations, the pulmonary circulation distributes flow through a single organ. The hydrostatic gradients in the lungs are relatively small (2.9 kPa; 30 cmH_2O) in the erect subject so that a low perfusion pressure is adequate. Because of the hydrostatic gradient, the dependent regions of the lungs are better perfused than the upper regions. Capillary flow is pulsatile throughout the cardiac cycle and flow velocity is 200 ml/s during systole and 50 ml/s during diastole. Red cells cross the capillary network in about 700 ms. With exercise, pulmonary vascular resistance decreases, capillary pulsatility increases and flow velocity reaches 460 ml/s during systole and 100 ml/s during diastole.

A unique feature of the pulmonary circulation is that the dimensions of its vessels depend on three pressures, namely the pulmonary artery, *alveolar* and venous.[43] Lung capillaries can expand into the alveolar space and therefore the transpleural pressure gradient (the pressure difference between alveoli and pleural cavity) is a major determinant of their diameter. The balance of pressures across the pulmonary capillaries can be illustrated in a three-zone model (West model). In the upper lung zones (zone 1), pulmonary arterial pressure may be lower than alveolar pressure, the capillary may be collapsed, and there is little or no blood flow, especially with hypovolaemia and increased alveolar pressure. In the intermediate zone (zone 2), pulmonary arterial pressure exceeds alveolar pressure and blood flow is a function of the pressure gradient between pulmonary arterial and alveolar pressure (waterfall effect). In the lower zones (zone 3), both arterial and venous pressure exceed the alveolar pressure and the capillaries remain distended.

Pulmonary vascular resistance is low and can be further decreased by recruitment and distension. Recruitment occurs when a critical opening pressure is reached. This mechanism could explain changes in blood flow through the upper zones of the lungs. Distension is the only mechansim by which blood flow may increase through the lower zones of the lungs because the capillaries are already open.

The effects of airway pressure are not limited to the capillaries. The calibre of the large pulmonary arteries and veins depends to some extent on the intrapleural pressure; the calibre of pulmonary arteries and veins inside the parenchyma depends on changes in lung volume. These vessels dilate with lung inflation because of traction on their thin walls, and are less influenced by alveolar pressure. Conversely, capillaries are strongly influenced by alveolar pressure and less by lung distension alone. Note that in a low pressure system small variations in interstitial pressure may substantially modify the transmural pressure. The normal interstitial pressure of the lung is estimated at − 1 kPa (− 8 mmHg).

Hypoxic pulmonary vasoconstriction

Graded decreases in airway oxygen tension below 13 kPa (97 mmHg) produce graded increases in pulmonary vascular resistance within seconds of the onset of hypoxia. This response occurs in isolated perfused lungs as well as in the intact lung. Thus, the basic mechanism must be intrinsic to the lung. A direct effect of hypoxia on the vascular smooth muscle or hypoxia-induced inactivation of a vasodilator substance have been postulated. No single mediator has been identified and it may be that several vasoactive substances are involved, including catecholamines, histamine, serotonin, angiotensin II, bradykinin, prostaglandin, acetylcholine and ATP.[44]

Cytochrome *c* oxidase has been proposed as the oxygen sensor for the pulmonary hypoxic vasoconstriction. The transduction involves an increase in cytoplasmic calcium. This is indirectly confirmed by the inhibition of the hypoxic vasoconstriction by calcium influx blockers. The autonomic nervous system may play a role as pulmonary vessels contain both α- and β-adrenoceptors and sympathectomy (and α-adrenoceptor blockade) blunts the hypoxic pulmonary vasoconstriction.

The lungs can produce large quantities of arachidonic acid metabolites including PGI_2, thromboxane and leukotrienes. Some of these metabolites cause vasodilatation (PGI_2) while others (thromboxane, leukotrienes) cause vasoconstriction. These substances may play a role in the hypoxic pulmonary vasoconstriction but evidence is still controversial.

Hypoxic pulmonary vasoconstriction, enhanced by hypercapnia and acidaemia, constitutes an almost ideal intrinsic regulatory system based on oxygen itself. In poorly ventilated areas of the lung, hypoxia (and the associated elevation of carbon dioxide tension) causes vasoconstriction so that blood flow is not 'wasted' to an area where gas exchange would be ineffective.

Inhibition of hypoxic vasoconstriction by halothane has been well documented in a variety of animal models but is modest. Other halogenated anaesthetics and diethyl ether decrease the hypoxic vasoconstriction. Nitrous oxide causes only a small reduction in its magnitude. The intravenous anaesthetic agents do not appear to modify pulmonary hypoxic vasoconstriction.[45]

Effects of ventilation on the circulation

Phasic alterations in stroke volume and asynchrony in the output of the ventricles occur during spontaneous breathing and artificial ventilation. Intermittent positive pressure ventilation, particularly when the inspiratory phase is prolonged, increases the mean intrathoracic pressure and decreases the cardiac output. This reduction results from reduced venous return, increased pulmonary vascular resistance, and reduced pulmonary blood volume. Because hypovolaemia increases the effects of artificial ventilation on the circulation, great emphasis has been put on decreased venous return.

With the introduction of PEEP in order to improve oxygenation in patients with severe lung damage, the effects of artificial ventilation have become even more obvious. Several mechanisms are responsible for the reduction in cardiac output caused by PEEP. The high mean intrathoracic pressure impedes venous return. This explains the correction of the low output syndrome by fluid loading. However, the beneficial effects of inotropic agents suggest that PEEP causes myocardial depression. Experimentally, the plasma of animals ventilated with PEEP depresses isolated heart muscle preparations. Stretching the lungs by PEEP may cause the release of a negative inotropic substance. Metabolites of arachidonic acid may be involved.

In a number of studies, evidence for myocardial depression is lacking and an alternative explanation, for the reduced cardiac output, is a decrease in ventricular compliance. Indeed, when the pericardium is intact PEEP increases the afterload of the right ventricle. This results in an acute dilatation of the right ventricle and an increase in the radius of curvature of the septum. In turn, this causes a reduction in the cross-sectional area of the left ventricle both at end diastole and end systole.[46] In such circumstances, volume loading may further increase the radius of curvature of the septum and as the left ventricle becomes even less able to fill, cause circulatory failure.

Pathophysiology

The major risk factors for cardiovascular complications of anaesthesia and surgery are cardiac failure, coronary heart disease, hypertensive heart disease and major rhythm disorders.

Cardiac failure

Causes of ventricular failure

Cardiac failure is the inability of either or both ventricles to maintain cardiac output in the face of increased impedance to ejection (ejection failure) or impaired ventricular filling (filling failure). The pathological processes include ischaemia, cardiomyopathy, myocarditis, ageing and valvular disease.

Myocardial ischaemia

Ischaemia is the commonest pathological process which diminishes myocardial function. Decreased coronary blood flow results in an imbalance between myocardial oxygen demand and supply, with consequent impairment in both contraction and relaxation of cardiac muscle. Impaired contraction leads to failure of either or both ventricles to eject against a normal or increased hydraulic load, whereas impaired relaxation leads to decreased diastolic filling and increased stiffness of the ventricles. Regional ischaemia may lead to ventricular dyskinesia with uncoordinated contraction. In addition, ischaemia of the conduction system and the atria can lead to dysrhythmias or partial heart block with consequent failure in coordinated ventricular function.

Cardiomyopathy

There are three distinct types of cardiomyopathy: *hypertrophic*, *dilated* (congestive), and *restrictive*.[47]

Hypertrophic (obstructive) cardiomyopathy is characterized by massive ventricular hypertrophy. The ventricles relax irregularly and fill more slowly than normal. Systolic function is powerful and blood is rapidly ejected from the ventricles, often resulting in pressure gradients within the left ventricular cavity, especially under conditions of enhanced inotropic activity. Obstruction can be alleviated by drugs which impair contractile performance, e.g. non-selective β-adrenoceptor antagonists or calcium channel blockers.

Dilated cardiomyopathy is characterized by marked dilatation of both ventricles and impaired contractile performance. The main conditioning factors are excessive alcohol consumption and previous viral myocarditis. Characteristically left ventricular ejection fraction is extremely low (10–30 per cent), leading to pulmonary venous engorgement, pulmonary hypertension and subsequently right ventricular failure.

Restrictive cardiomyopathy is associated with deposition of organic material in the endomyocardium causing a restriction of inflow into the ventricles. The haemodynamic changes of restrictive cardiomyopathy resemble those of constrictive pericarditis, in that there is rapid filling of the ventricles in early diastole, but an almost complete cessation of filling during late diastole. Amyloid disease of the heart is the main cause of restrictive cardiomyopathy in temperate climates.

Ageing
Arterial walls stiffen with age. Arteriosclerotic changes in the aorta and larger arteries cause an increase in systolic and a decrease in diastolic arterial pressure. These age-related changes result in a secondary increase in left ventricular work which, if associated with atheromatous coronary vascular disease, may lead to ventricular failure.[48]

Valvular heart disease
Pathology which results in stenosis or incompetence of the cardiac valves is likely to result in failure of the ventricle adjacent to that valve.

Aortic stenosis is a major cause of left ventricular failure. Myocardial oxygen consumption is high (raised left ventricular systolic pressure and ventricular hypertrophy); and myocardial blood flow is low (reduced aortic diastolic perfusion pressure). Increases in heart rate reduce the duration of diastole, decrease coronary blood flow and precipitate ischaemic ventricular failure. Similarly, reductions in diastolic arterial pressure (vasodilators) decrease coronary flow and must be avoided. Anaemia associated with aortic stenosis may also result in myocardial ischaemia as a result of impaired O_2 transport.

In *aortic incompetence* forward flow is facilitated by a relatively low vascular resistance, while regurgitant flow is increased when vascular resistance is high. Because of the influence of diastolic time, regurgitant flow is also enhanced during bradycardia. Left ventricle dilatation is more marked than hypertrophy, and mechanical activity is wasted by the regurgitant flow.

Mitral stenosis makes flow through the narrowed mitral valve depend on left atrial pressure and on the contribution of atrial systole (late diastolic flow). Duration of diastole is an important factor, and tachycardia decreases the time for ventricular filling. Atrial fibrillation compromises late diastolic filling, and any increase in heart rate is poorly tolerated. Thus, digitalis therapy is important because of its negative chronotropy.

The operational range for left atrial pressure (LAP) is decreased, and as mean LAP is usually above 15 mmHg, any factors which encourage overfilling of the atria will lead to pulmonary oedema early in the history of mitral stenosis. Later, pulmonary hypertension develops and leads to right ventricular failure.

Mitral regurgitation results in dilatation of both left atrium and left ventricle. The regurgitant flow causes pressure to be high in the atrium during systole. Any increase in systemic vascular resistance will limit left ventricular forward ejection and thus encourage retrograde flow into the atrium.

Haemodynamics of ventricular failure

Understanding ventricular failure is facilitated by consideration of the dynamic relations between pressure and volume (see Fig. 12.7). For a normal heart the end-diastolic pressure–volume relation is relatively steep, while for the failing heart it is flatter. In addition, the end-diastolic pressure–volume relation is flatter for the normal heart than for the failing heart (Fig. 12.20). Thus the failing heart responds to changes in resistance by smaller changes in systolic pressure, larger increases in diastolic pressure, and greater decreases in stroke volume (and cardiac output) than the normal heart.[49] In addition, drugs which impair the inotropic state of the myocardium have more effect in the failing than in the normal heart. The failing heart also has a greater potential for self-perpetuating myocardial

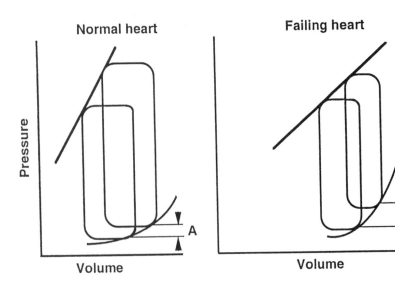

Normal heart   **Failing heart**

Pressure | Volume (axes)

Fig. 12.20 During each cardiac contraction, the instantaneous relation between pressure and volume describes a loop. The loops are constrained between the end-diastolic and end-systolic pressure–volume lines. In the failing heart the ventricle is stiffer (the end-diastolic line is steeper) and the poorer contractility is associated with a flatter end-systolic pressure–volume line (right panel). Increases in resistance cause a greater increase in end-diastolic pressure (B vs A), a smaller increase in pressure and a reduction in stroke volume (the width of the loop decreases).

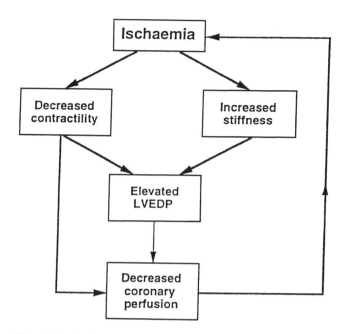

Fig. 12.21 Ischaemia causes a vicious circle leading to increased functional damage.

depression in the presence of poor coronary perfusion (Fig. 12.21).

The anaesthetic management of patients with a failing heart must therefore minimize any negative inotropic effects, prevent increases in vascular resistance, and maintain ventricular filling. The monitoring of these variables can be improved by the use of a pulmonary artery balloon catheter, to determine the pulmonary occluded pressure, an indirect estimate of left ventricular end-diastolic pressure. Right atrial pressure (which can be determined simultaneously with a double lumen pulmonary artery catheter) is elevated when right ventricular failure is present.

Right ventricular failure (RVF)

Right ventricular failure associates enlargement of the liver, peripheral oedema, and pleural effusion. It must be corrected before surgery. Right ventricular failure may be caused by right ventricular infarction, pulmonary valvular stenosis or, most commonly, by pulmonary hypertension secondary to chronic obstructive airway disease. Pulmonary hypertension may also be the result of repeated pulmonary thromboembolism, mitral valve disease, or chronic left ventricular failure. Acute right ventricular failure may be precipitated by hypoxia and/or hypercapnia. Indeed, short episodes of hypoxia associated with difficult intubation, or inadequate ventilation during recovery from anaesthesia, can precipitate acute pulmonary arteriolar constriction. As the pericardium is relatively non-compliant, acute right ventricular dilatation caused by pulmonary hypertension may precipitate a leftward shift of the interventricular septum, so that failure of left ventricular filling complicates the impairment of right ventricular ejection.

In patients with severe pulmonary hypertension with right ventricular hypertrophy (cor pulmonale), right ventricular function may be acutely disturbed if systemic arterial pressure decreases, causing coronary perfusion pressure to the right ventricle to be diminished. This in turn causes right ventricular ischaemia and worsens failure.

Where right ventricular failure is associated with a ventricular septal defect (VSD) and pulmonary hypertension (Eisenmenger's complex) it is important to avoid anaesthetics which will cause systemic arteriolar dilatation (enflurane or isoflurane). A decrease in the systemic vascular resistance relative to that in the pulmonary arteries will result in a reversal of the shunt through the VSD, decreasing pulmonary blood flow, causing hypoxia, and acutely precipitating right ventricular failure. Systemic vascular resistance must be maintained at a level which will optimize the interventricular shunt in favour of maximal pulmonary blood flow. The use of nitrous oxide (50–60 per cent) which causes systemic α-adrenoceptor mediated arteriolar constriction[50,51] will provide such conditions in association with low concentrations of halothane, and to a lesser extent, enflurane.

Left ventricular failure (LVF)

Pulmonary oedema is the most severe manifestation of LVF, but in most patients the impairment of LV function is less obvious. Basal râles, persistent unexplained tachycardia, and a third heart sound (gallop rhythm) are more common indicators of LVF. The presence of a third heart sound (S3) in diastole indicates a decreased diastolic left ventricular compliance.[52] Giant v waves in the pulmonary artery wedge pressure trace may appear in patients who develop an acute reduction of diastolic compliance because of myocardial ischaemia.[53] In the presence of reduced diastolic compliance, left ventricular filling pressures must be kept within very narrow limits. Reduced filling results in decreased cardiac output, and overfilling causes pulmonary oedema. Thus, careful monitoring of the pulmonary occlusion pressure is an essential part of the management of anaesthesia in patients with LVF.

The failing left ventricle is exquisitely sensitive to the myocardial depressant effects of inhalational anaesthetics.[54] Even nitrous oxide causes more depression of the failing heart than the normal. Increases in systemic vascular resistance caused by sympathoadrenal activation, laryngoscopy, or cross-clamping of the aorta, cause acute dilatation of the failing heart with marked reductions in stroke volume. Once the filling pressure has been optimized by judicious fluid administration, decreasing the hydraulic load of the left ventricle by arteriolar dilators makes it

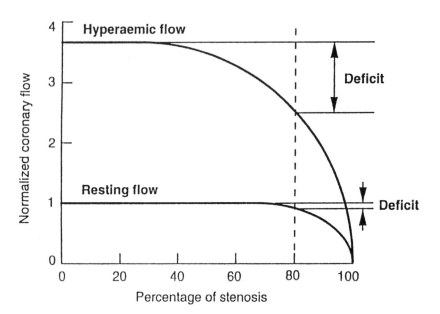

Fig. 12.22 Depending on the metabolic requirements, the effects of coronary stenosis on coronary flow and flow deficit are small (resting state), or large (maximum oxygen consumption). The flow deficit is the difference between the flow that would exist without stenosis and the flow through that stenosis. It is clear that the effect of the same stenosis is much greater when metabolic requirements are high.

possible to maintain optimal cardiac output and arterial pressure.

Ischaemic heart disease

Pathophysiology of the coronary circulation

Effects of coronary stenosis
Atherosclerotic obstruction develops as plaques within the vessel wall, most commonly in proximal segments and major bifurcations of the epicardial vessels. Progressive coronary obstruction decreases the coronary flow reserve (the difference between maximum and resting flow) as soon as the stenosis exceeds 45 per cent of the diameter of the vessel (Fig. 12.22). Resting flow decreases when the stenosis reaches 80–90 per cent of the diameter.[55]

The diameter of stenotic coronary artery segments is not fixed but changes with the distending pressure and under the influence of neural and humoral factors. Stenotic segments may dilate in response to nitroglycerin and other vasodilator stimuli, and collapse when the distending pressure decreases.

Where the endothelium is disrupted, local regulation of coronary blood flow may be altered by the absence of EDRF or by the excessive release of endothelin. In addition, microthrombosis and microlysis may occur, resulting in transient ischaemia.

Coronary spasm
Spasm occurs usually just distal to a fixed obstructive stenosis, and may cause complete occlusion. The exact cause of spontaneous spasm is unknown, but hypercholesterolaemia may increase the responsiveness to vasoconstrictors, probably by increasing the access of fat-soluble substances to the vascular smooth muscle. Spasm is thought to be important not only in variant (Prinzmetal) angina but also in exercise-induced angina, acute myocardial infarction and sudden cardiac death.[56]

Collateral circulation
In man, native coronary collaterals are small in diameter and number. In response to myocardial ischaemia, they widen and the large increase in diameter overstretches the vessel wall. Thickening occurs, so that the collateral vessels eventually resemble normal coronary arteries. In some patients normal ventricular function may be maintained in the face of total occlusion of a major coronary artery. This suggests that collateral channels provide substantial coronary flow. Moreover, perfusion scintigrams during exercise show that perfusion may be well maintained in areas supplied by an occluded coronary artery in the presence of angiographically demonstrated collaterals. The development of collateral vessels may result in the 'steal-prone' configuration of the coronary circulation.[57] This occurs when a coronary artery is occluded, another artery is narrowed and there are collateral channels joining two arterial territories. This occurs in approximately 25 per cent of patients with coronary artery disease. Presence of a steal-prone configuration of the vessels increases the risk of ischaemia when coronary vasodilators are administered.

Myocardial stunning and hibernation
Short episodes of myocardial ischaemia, lasting as little as 10–15 minutes, may result in prolonged reductions in high energy phosphates, alteration in the ultrastructure of the myocytes, and depression of mechanical function which may last for 4–5 days[58] or even longer. This has been

termed myocardial stunning. During this period, the heart may be unable to tolerate stresses.

Recently, it has been recognized that myocardium with poor coronary blood flow may fail to contract, while the integrity of the cells is maintained. Should blood flow be restored by coronary artery surgery or angioplasty, function may improve substantially. This phenomenon has been termed myocardial hibernation.[59]

Impaired coronary blood flow caused by atherosclerosis is responsible for an imbalance between myocardial oxygen demand and myocardial oxygen supply resulting in the impairment of both systolic and diastolic function. The impairment in systolic function causes reduced systolic contraction and paradoxical wall motion so that ischaemic regions of the ventricle may behave like an aneurysm. Often the earliest alteration caused by ischaemia is an increase in the stiffness of the ventricle.[60] Ischaemia of the conduction system can cause heart block. This further reduces the capability for left ventricular filling as the atrial contribution to ventricular filling is absent. In addition, ischaemia decreases the conduction velocity of the myocardium so that delayed impulses may cause re-entrant dysrhythmias.

Causes of ischaemia

Conventionally ischaemia is attributed mainly to haemodynamic aberrations that either increase oxygen demand in the face of limited supply (i.e. tachycardia and hypertension) or decrease supply more than demand (diastolic hypotension, elevated end-diastolic wall tension). In the presence of atheroma, distal vessels have lost their dilatory reserve and, therefore, local regulation of coronary blood flow is unable to maintain the matching of oxygen supply and oxygen demand. Indeed, in the anaesthetic setting, many episodes of myocardial ischaemia can be ascribed to haemodynamic aberrations. Pre-eminence of the local control of coronary blood flow must be questioned in the light of the increasingly large number of ischaemic episodes that are not associated with (or caused by) haemodynamic aberrations.

It is now clear that coronary vascular tone is modulated by alterations in autonomic activity so that α-adrenoceptor stimulation causes a relative reduction in coronary flow or prevents coronary flow from increasing as much as the increased metabolism needs. Conversely, vagal stimulation causes a relative increase in coronary blood flow or prevents coronary flow from decreasing in parallel with the oxygen requirements. Stimulation of the baroreceptors may cause changes in coronary blood flow that do not match the changes in haemodynamics. Moreover, especially in the presence of atheroma, local disruption of the endothelium is likely to be associated with a profound dysregulation in the release of vasoactive substances. Release of endothelium-derived relaxant factor (nitric oxide) causes vasodilatation whereas release of endothelin causes vasoconstriction. Similarly, release of prostacyclin causes vasodilatation whereas release of thromboxane causes vasoconstriction. Imbalance in these vasoregulator systems may be responsible for myocardial ischaemia in the absence of haemodynamic aberrations.

Recent myocardial infarction

The relation between previous myocardial infarction and subsequent postoperative reinfarction has been demonstrated in sequential studies at the Mayo Clinic,[61,62] showing that patients with recent myocardial infarction (up to 6 months) were at greater risk of perioperative reinfarction than normal patients of the same age and sex. Ten years ago, Rao, Jacobs and El-Etr[63] reported that aggressive monitoring and therapy during and after surgery could decrease the incidence of perioperative myocardial infarction to less than 2 per cent even in patients who had suffered previous myocardial infarction within 3 months of surgery. They advocated the use of a balloon-tipped pulmonary artery catheter and direct intra-arterial pressure monitoring for a period of 4 days starting on the evening before surgery, with stringent control of heart rate, arterial pressure and both right and left ventricular filling pressures. This is probably the optimal approach to the management of patients who require urgent or emergency surgery shortly after myocardial infarction. More recently, Shah and colleagues[64] have shown the incidence of reinfarction to be much lower than previously reported. This is not surprising as the general level of monitoring has improved over the years, anaesthetic techniques tend to provide more stable haemodynamics, and patients with acute myocardial infarction benefit from treatment (β-blockade; thrombolysis), investigations (coronary angiography), and often early coronary revascularization (angioplasty or coronary bypass surgery). This must have contributed to the reduction in the risk of reinfarction after anaesthesia and surgery.

Coronary artery grafts

Patients who have had successful coronary artery bypass grafts have a lower incidence of cardiovascular complications during anaesthesia and surgery than patients with coronary artery disease who have not been grafted.[65] However, the recently grafted patient may be as much at risk during subsequent surgery as the patient with a recent myocardial infarction. Apart from the general considerations outlined above, arterial hypotension must be specifically avoided in such patients to minimize the possibility of thrombosis in the coronary artery grafts during periods of decreased coronary blood flow. Recent evidence suggests that coronary artery graft procedures and coronary angioplasty may increase the left ventricular ejection fraction, thereby reducing the risk of non-cardiac surgery.[66]

Silent myocardial ischaemia

Silent myocardial ischaemia has been recognized for over 30 years[66] and is similar to symptomatic myocardial ischaemia in all respects except pain. Silent myocardial ischaemia (SMI) occurs in totally asymptomatic patients (type 1 SMI); in patients with previous myocardial infarction (type 2 SMI); and in patients with angina (type 3 SMI). In the latter group, silent ischaemia may represent up to 75–90 per cent of all the episodes of myocardial ischaemia.[67,68] In all groups, silent ischaemia is associated with an adverse prognosis. This is not surprising as patients are unaware of silent ischaemia and therefore continue with their activities. It is possible that freedom from pain may result from enhanced release of β-endorphins.[69]

A number of recent studies in groups of surgical patients suggest that silent myocardial ischaemia is a major determinant of adverse outcome including cardiac death, myocardial infarction, new angina, unstable angina, acute left ventricular failure, and life-threatening dysrhythmias.[70-74] While the relation between acute ischaemia, new angina, unstable angina, myocardial infarction and cardiac death is obvious, the association with life-threatening dysrhythmias and acute left ventricular failure needs to be discussed. Ischaemia causes a slowing of intraventricular conduction so that activation progresses more slowly in the ischaemic region and, thus, reaches normal distal myocardium during the vulnerable phase of repolarization, setting off re-entry dysrhythmias. Ischaemia causes an immediate reduction (or suppression) in systolic wall motion. This leads to a reduction in stroke volume and increases both end-diastolic volume and wall stress. This, in turn, further compromises coronary blood flow. In addition, regional ischaemia is associated with a generalized increase in wall stiffness both in the ischaemic and normal myocardium.[75] This results in a marked reduction in global left ventricular compliance, which facilitates the development of acute left ventricular failure.

Recent studies have shown that silent ischaemia is accompanied by increased deterioration after anaesthesia and surgery, such that both the number and duration of episodes of ischaemia increase. This results in a large increase in the 'total ischaemic burden'.[76] Postoperative silent ischaemia is likely to be multifactorial. Haemodynamic aberrations, especially tachycardia, play a role. Indeed, there is good evidence that heart rate shifts towards higher values after surgery.[77] Alterations in coagulation facilitating microthrombosis followed by thrombolysis are also likely to play a role. Indeed, the fibrinolytic activity

reaches its nadir 3–5 days after surgery, while hyper-coagulability reaches its peak after 7–15 days.[78,79] Finally, postoperative hypoxaemia, down to levels of less than 80 per cent saturation, is frequently observed particularly during the night and may be associated with silent myocardial ischaemia.[80,81]

The long-term importance of silent ischaemia has been emphasized by Mangano and colleagues, as their studies have shown that the prognosis at 2 years is worse in patients with postoperative silent ischaemia, and dramatically worse in those who have suffered postoperative myocardial infarction.[82] Whether silent ischaemia and postoperative infarction are simply markers of more severe coronary heart disease, or contribute to a poor long-term prognosis is unclear. However, better protection of the myocardium in the postoperative period may influence outcome.

Hypertensive heart disease

During hypertensive episodes, systolic arterial pressure usually increases more than diastolic arterial pressure (DAP), so that the increase in myocardial oxygen consumption is likely to exceed the improvement in oxygen supply provided by the higher coronary perfusion pressure. When coronary arteries are stenosed, such hypertensive episodes cause acute ischaemia which may precipitate ventricular dilatation and failure, creating a vicious circle to further impair myocardial blood flow.

While mild hypertension (DAP < 100 mmHg) does not constitute a serious threat,[83] moderate hypertension (DAP 100–115 mmHg) accompanied by involvement of target organs such as brain (cerebrovascular disease), heart (coronary artery disease) and kidneys (atheromatous occlusions accompanied by impaired renal function), and severe uncontrolled hypertension (DAP > 115 mmHg) are serious risk factors in the surgical patient.

Changes in vascular smooth muscle tone cause greater changes in systemic vascular resistance in hypertensive patients[84] than in their normotensive counterparts. These exaggerated haemodynamic responses of the poorly controlled hypertensive patient can lead to increased morbidity. Uncontrolled hypertension has been consistently identified as a pre-existing factor which increases the risk of perioperative myocardial infarction[62] and both transient and permanent neurological deficits after carotid surgery.[85]

Left ventricular hypertrophy is a consistent accompaniment of severe uncontrolled hypertension, and constitutes a specific problem in relation to anaesthesia. Patients with hypertrophic left ventricles require higher filling pressures (8–15 mmHg) than normal, and are more sensitive to the effects of junctional rhythms which remove the atrial contribution to ventricular filling.

The management of general anaesthesia in the well controlled hypertensive patient has been reviewed extensively and the responses of these patients to regional blockade has also been studied. While lumbar or thoracic epidural block, either alone or in combination with light general anaesthesia, are well tolerated by the controlled hypertensive patient,[86] profound hypotension has been observed in 50 per cent of uncontrolled hypertensives following lumbar epidural block. Thus, all severely hypertensive patients and those with moderate hypertension and target organ involvement presenting for surgery should have their arterial pressure well controlled by previous drug therapy. Recently, poor control of hypertension has been shown to be associated with a very high incidence of silent myocardial ischaemia.[87,88]

Dysrhythmias and heart blocks

Cardiac arrhythmias occur frequently during anaesthesia and surgery; most are of short duration, compromise the circulation only minimally and do not require pharmacological interventions. However, in poor risk patients, cardiac arrhythmias may seriously compromise cardiac output. Occasionally they herald life-threatening complications and require active treatment.

Dysrhythmias result from abnormalities of impulse initiation, impulse conduction or a combination of both.[89,90] Abnormal impulse initiation is caused by localized changes in ionic currents and may be expressed as automaticity or triggered activity. Abnormal impulse conduction and re-entry result from propagating impulses which do not cease after complete activation but may re-excite the atria or the ventricles after the end of the refractory period. Re-entry can be promoted by slowing the conduction velocity or shortening the refractory period. These abnormalities may result from reduced negativity of the resting potentials and reduced sodium inward current. In addition, at all times conduction is faster along the long axis of the fibres than in the direction perpendicular to their orientation.

Dysrhythmias are frequently an indication of myocardial ischaemia as the latter impairs the repolarization of the Purkinje fibres and enhance re-entry.[91]

Junctional rhythm is a common feature of anaesthesia with either volatile or intravenous agents, and results in the loss of the atrial contribution to ventricular filling. This is particularly disruptive to the patient with left ventricular hypertrophy who needs higher filling pressures to maintain cardiac output. No specific and reliable methods are available to reverse junctional rhythms, other than decreasing the concentration of the relevant anaesthetic.

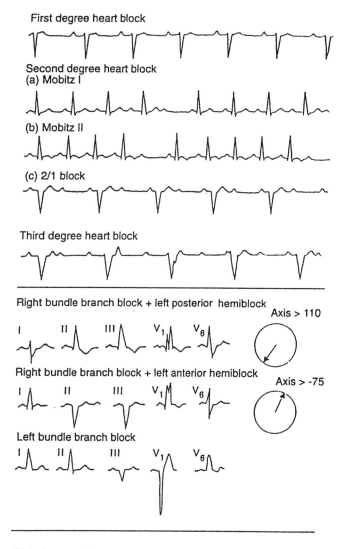

Fig. 12.23 Examples of electrocardiograms of conduction disorders.

Some patients with major conduction disorders are likely to develop complete heart block during anaesthesia. Figure 12.23 shows electrocardiographs of various degrees of heart block. First degree heart block is recognized by a prolongation of the PR interval >0.2 second. In itself it is not harmful, but is an indication of a potential for progression to a more severe degree of block, through the additive effects of drugs which impair atrioventricular conduction. These include halogenated anaesthetics, propofol, neostigmine, calcium channel blockers and β-blockers

Second degree block can be recognized by three patterns: Mobitz type I (more commonly known as the Wenckebach phenomenon), Mobitz type II and 2/1 block. In the former the PR interval becomes progressively longer until finally an atrial beat is blocked and the subsequent ventricular beat is dropped. Mobitz type II is the result of prolongation of the absolute refractory period and is a much more serious condition. The PR interval is fixed, but the ventricles do not respond to every atrial stimulation. In some patients, the ventricules respond to every second atrial contraction (2/1 block). Drugs which depress conduction should be avoided in these patients, and a cardiologist should be consulted about the insertion of a temporary pacemaker before anaesthesia.[92]

Third degree heart block implies that all supraventricular impulses are blocked. Patients with this severity of block will have a permanent pacemaker fitted, and are therefore less of a problem to manage during anaesthesia.[93]

Right bundle branch block (RBBB) is characterized by the rSR configuration in the right ventricular leads (V_1, V_2) with widening of the QRS complex. This is a common condition in patients with ischaemic heart disease, and in itself is not particularly dangerous. However, it is occasionally associated with partial block of either the anterior (commonly) or posterior (uncommonly) fascicles of the left bundle. The resulting bifascicular block is potentially dangerous, although clinical studies during anaesthesia in patients with bifascicular block[94] or in cardiological patients[95] suggest that elective transient pacing is not required.

Left bundle branch block (LBBB) is characterized by widening of the QRS (>0.12 second) associated with an rSR pattern in the left venticular leads. This condition, although not dangerous in itself, is also an indicator of potential problems in patients with ischaemic heart disease or hypertension.

Conditions other than these conventional blocks must also be considered (Table 12.3). The management of patients with sick sinus syndrome can be made safer with previous insertion of a transvenous pacemaker, allowing drug therapy to control episodes of tachycardia to be safely implemented.[96] Sick sinus syndrome should be considered as a preoperative diagnosis in older patients who have slow resting heart rates which do not increase in response to Valsalva's manoeuvre. Carotid sinus massage is of particular value in the demonstration of sick sinus syndrome or sinoatrial node disease, in the absence of electrocardiographic changes.[97]

It is often assumed that regional or local anaesthesia are

Table 12.3 Indications for pacemakers

Symptomatic first degree heart block
Symptomatic second degree (Mobitz I) heart block
Second degree (Mobitz II) heart block
Third degree heart block
Symptomatic bifascicular block
LBBB + first degree heart block
Sick sinus syndrome
Slow rates unresponsive to drugs

potentially safer than general anaesthesia in patients with conduction disorders. Substantial proportions of administered doses of local anaesthetics are absorbed systemically and may further impair cardiac conduction,[98,99] leading to severe heart block or cardiac arrest.

It is essential in the presence of atrioventricular and/or intraventricular conduction disorders to establish whether episodes of circulation inefficacy have occurred. A history of dizzy spells, blackouts, funny turns is likely to indicate intermittent major blockade of impulse transmission justifying the insertion of a transvenous pacemaker irrespective of the type of block. Anaesthetists are often more concerned by atrioventricular and intraventricular blocks than cardiologists because anaesthesia may precipitate the development of complete heart block in patients with moderately compromised atrioventricular conduction. Anaesthetic agents may reduce AV conduction; anaesthesia often causes an arrhythmia and premature beats are known to facilitate the development of complete heart block. In addition, alterations in potassium concentration (hypocapnic IPPV, blood transfusion) may occur and modify cardiac excitability. This is why the insertion of a temporary pacemaker is often necessary before elective or emergency surgery, even though permanent pacing may not be indicated.[100] The indication for temporary pacing is imperative if patients with atrioventricular conduction disorders and with bifascicular blocks have presented Stokes–Adams attacks or unexplained fainting or blackouts (Table 12.3). Even in the absence of symptoms, the indication for temporary pacing is almost imperative for third degree heart block, and Mobitz type II blocks. Temporary pacing is also indicated in the case of severe bradycardia accompanied by cardiac failure, angina or syncope, in the case of atrial fibrillation with very slow heart rate and in the case of sick sinus syndrome. Failure to use a pacemaker when indicated may cause fatalities.[101]

REFERENCES

1. Nayler WG, Gevers W, Opie LH. Heart cells and organelles. In: Opie LH, ed. *The heart*. London: Grune & Stratton, 1984: 15–29.

2. Katz AM. Congestive heart failure. Role of altered myocardial cellular control. *New England Journal of Medicine* 1975; **293**: 1184–91.

3. Huxley AF, Niedergerke R. Structural changes in muscle during contraction. *Nature* 1954; **173**: 971–3.

4. Huxley HE, Hanson J. Changes in cross-striation of muscle during contraction and stretch and their structural interpretation. *Nature* 1954; **173**: 973–6.

5. Pollack GG, Kreuger JW. Sarcomere dynamics in intact cardiac muscle. *European Journal of Cardiology* 1976; **4** (suppl): 53–65.

6. Kreuger JW, Pollack GH. Myocardial sarcomere dynamics during isometric contraction. *Journal of Physiology (London)* 1975; **251**: 627–43.

7. Starling EK. *The Linacre lecture on the law of the heart, given at Cambridge, 1915*. London: Longmans, 1918.

8. Hill AV. Heat of shortening and dynamic constants of muscle. *Proceedings of the Royal Society of London (Biology)* 1938; **126**: 136–95.

9. Abbott BC, Mommaerts WFHM. A study of inotropic mechanisms in the papillary muscle preparation. *Journal of General Physiology* 1959; **42**: 533–51.

10. Mason DT, Spann Jr JF, Zelis R. Quantification of the contractile state of the intact human heart. *American Journal of Cardiology* 1970; **26**: 248–57.

11. Sarnoff SJ, Case RB, Berglund E, Sarnoff L. Ventricular function: I. Starling's law of the heart studied by means of simultaneous right and left ventricular function curves in the dog. *Circulation* 1954; **9**: 706–18.

12. Chatterjee K, Parmely WW. Vasodilator therapy for chronic heart failure. *Annual Review of Pharmacology and Toxicology* 1980; **20**: 475–512.

13. Glower DD, Spratt JA, Snow ND, *et al*. Linearity of the Frank–Starling relationship in the intact heart: the concept of preload recruitable stroke work. *Circulation* 1985; **71**: 994–1009.

14. Frank O. Zur Dynamik des Herzmuskels. *Zeitschrift für Biologie* 1895; **32**: 370–437.

15. Sagawa K. The ventricular pressure–volume diagram revisited. *Circulation Research* 1978; **43**: 677–87.

16. Suga H, Hisano R, Goto Y, Yamada O, Igarashi Y. Effect of positive inotropic agents on the relation between oxygen consumption and systolic pressure volume area in canine left ventricle. *Circulation Research* 1983; **53**: 306–18.

17. Brutsaert DL, Rademakers FE, Sys SU. Triple control of relaxation: implications in cardiac diseases. *Circulation* 1984; **69**: 190–6.

18. Renkin EM. Multiple pathways of capillary permeability. *Circulation Research* 1975; **41**: 735–43.

19. Michel CC, Clough G. Capillary permeability and transvascular fluid balance. In: Sleight P, Vann Jones J, eds. *Scientific foundations of cardiology*. London: Heinemann Medical Books Ltd, 1983: 25–30.

20. Milnor WR. Arterial impedance as ventricular afterload. *Circulation Research* 1975; **36**: 565–70.

21. Glantz SA, Misbach GA, Moores WY, *et al*. The pericardium substantially affects the left ventricular diastolic pressure–volume relationship in the dog. *Circulation Research* 1978; **42**: 433–41.

22. Laver MB, Strauss HW, Pohost GM. Right and left ventricular geometry: adjustments during acute respiratory failure. *Critical Care Medicine* 1979; **7**: 509–19.

23. Draper MH, Weidmann S. Cardiac resting and action potentials recorded with an intracellular electrode. *Journal of Physiology (London)* 1951; **115**: 74–94.

24. Noble D. A modification of the Hodgkin–Huxley equations to Purkinje-fibre action and pace-maker potentials. *Journal of Physiology (London)* 1962; **160**: 317–52.

25. Beeler Jr GW, Reuter H. The relation between membrane potential, membrane currents and activation of the contraction in ventricular myocardial fibres. *Journal of Physiology (London)* 1970; **207**: 211–29.

26. Beeler Jr GW, Reuter H. Membrane calcium current in ventricular myocardial fibres. *Journal of Physiology (London)* 1970; **207**: 191–209.

27. Rubio R, Berne RM. Regulation of coronary blood flow. *Progress in Cardiovascular Disease* 1975; **18**: 105–22.

28. Markus M, Wright C, Doty D, *et al*. Measurement of coronary flow velocity and reactive hyperaemia in the coronary circulation of humans. *Circulation Research* 1981; **49**: 877–91.

29. Braunwald E. Control of myocardial oxygen consumption. *New England Journal of Medicine* 1971; **307**: 1618–27.

30. Markus ML. Metabolic regulation of coronary blood flow. In: *The coronary circulation in health and disease*. New York: McGraw-Hill, 1983: 65–92.

31. Rubio R, Wiedmeier WT, Berne RM. Relationship between coronary blood flow and adenosine production and release. *Journal of Molecular and Cellular Cardiology* 1974; **6**: 561–7.

32. Sparks HV, Bardenheuser H. Regulation of adenosine formation by the heart. *Circulation Research* 1986; **58**: 193–201.

33. Sabouni MH, Cushing DJ, Makujina SR, Mustafa SJ. Inhibition of adenylate cyclase attenuates adenosine receptor mediated relaxation in coronary artery. *Journal of Pharmacology and Experimental Therapeutics* 1991; **259**: 508–11.

34. Freilich A, Tepper D. Adenosine and its cardiovascular effects. *American Heart Journal* 1992; **123**: 1324–8.

35. Moncada S, Palmer PMJ, Higgs, EA. Nitric oxide: physiology, pathophysiology and pharmacology. *Pharmacological Reviews* 1991; **43**: 108–42.

36. Kramer BK, Nishida M, Kelly RA, Smith TW. Endothelins. *Circulation* 1991; **85**: 350–6.

37. Bellamy RF. Diastolic coronary pressure–flow relation in the dog. *Circulation Research* 1978; **43**: 92–101.

38. Verrier ED, Edelist G, Macke C, Robinson S, Hoffman JIE. Greater coronary vascular reserve in dogs anesthetized with halothane. *Anesthesiology* 1980; **53**: 445–59.

39. Merin RG, Bernard J-M, Doursot M-F, Cohen M, Chelly JE. Comparison of the effects of isoflurane and desflurane on cardiovascular dynamics in the chronically instrumented dog. *Anesthesiology* 1991; **74**: 568–74.

40. Reiz S, Balfors E, Sorensen MB, Ariola S, Friedman A, Truedson H. Isoflurane: a powerful coronary vasodilator in patients with coronary artery disease. *Anesthesiology* 1983; **59**: 91–7.

41. Weber KT, Janicki JS, Schroff SG, Likoff MJ, St John Sutton MG. The right ventricle, physiologic and pathophysiologic considerations. *Critical Care Medicine* 1983; **11**: 323–8.

42. Ghignone M, Girling L, Prewitt RM. Volume expansion versus norepinephrine in treatment of a low cardiac output complicating an acute increase in right ventricular afterload in dogs. *Anesthesiology* 1984; **60**: 132–5.

43. West JB. Pulmonary gas exchange. *International Review of Physiology* 1977; **14**: 83–106.

44. Fishman AP. Vasomotor regulation of the pulmonary circulation. *Annual Review of Physiology* 1980; **42**: 211–20.

45. Sykes MK. Effect of anaesthetics and drugs used during anaesthesia on the pulmonary circulation. In: Altura BM, Halevy S, eds. *Cardiovascular actions of anesthetics and drugs used in anesthesia.* vol 2. Basel: Karger, 1987: 92–125.

46. Jardin D, Farcot JC, Boisante L, Curien N, Margariaz A, Bourdarias JP. Influence of positive end-expiratory pressure on left ventricular performance. *New England Journal of Medicine* 1981; **304**: 387–92.

47. Goodwin JF. The frontiers of cardiomyopathy. *British Heart Journal* 1982; **48**: 1–18.

48. Lakatta EG, Fleg JL. Aging of the adult cardiovascular system. In: Stephen CR, Assaf RAE, eds. *Geriatric anesthesia, principles and practice.* Boston: Butterworths, 1986: 1–26.

49. Laskey WK, Kussmaul WG, Martin JL, Kleaveland JP, Hirschfield JW, Shroff S. Characteristics of vascular hydraulic load in patients with heart failure. *Circulation* 1985; **72**: 61–71.

50. Smith NT, Eger EI II, Stoelting RK, Whayne TF, Culklken DJ, Kadis LB. The cardiovascular and sympathomimetic responses to the addition of nitrous oxide on the circulation during enflurane anesthesia in man. *Anesthesiology* 1970; **32**: 345–9.

51. Smith NT, Calverly RK, Prys-Roberts C, Eger EI II, Jones CW. Impact of nitrous oxide on the circulation during enflurane anesthesia in man. *Anesthesiology* 1979; **48**: 345–9.

52. De Werf F, Boel A, Geboers J, et al. Diastolic properties of the left ventricle in normal adults and in patients with third heart sounds. *Circulation* 1984; **69**: 1070–80.

53. Waller JL, Kaplan J. Anaesthesia for patients with coronary artery disease. *British Journal of Anaesthesia* 1981; **53**: 757–65.

54. Filner BE, Karliner JS. Alterations of normal left ventricular performance by general anesthesia. *Anesthesiology* 1976; **45**: 610–21.

55. Gould KL, Lipscomb K, Hamilton GW. Physiologic basis for assessing critical coronary stenosis. *American Journal of Cardiology* 1974; **33**: 87–94.

56. Maseri A, L'Abbate A, Chierchia S, et al. Significance of spasm in the pathogenesis of ischemic heart disease. *American Journal of Cardiology* 1979; **44**: 778–82.

57. Buffington CW, Davis KB, Gillispie S, Pettinger M. The prevalence of steal-prone coronary anatomy in patients with coronary artery disease: an analysis of the coronary artery surgery study registry. *Anesthesiology* 1988; **69**: 721–7.

58. Braunwald E, Kloner RA. The stunned myocardium:

prolonged post-ischemic ventricular dysfunction. *Circulation* 1982; **66**: 1146–9.

59. Kloner RA, Przyklenk K, Patel B. The stunned and hibernating myocardium. *American Journal of Medicine* 1989; **86**: 14–22.

60. Nogoni H, Hess OM, Bortone AS, Ritter M, Caroll JD, Krayenbuhl HP. Left ventricular pressure-length relation during exercise-induced ischemia. *Journal of the American College of Cardiology* 1989; **13**: 1062–70.

61. Tarhan S, Moffitt, EA, Taylor WF, Guiliani ER. Myocardial infarction after general anesthesia. *Journal of the American Medical Association* 1973; **220**: 1451–4.

62. Steen PA, Tinker JH, Tarhan S. Myocardial re-infarction after anesthesia and surgery. *Journal of the American Medical Association* 1978; **239**: 2566–70.

63. Rao TLK, Jacobs KH, El-Etr AA. Reinfarction following anesthesia in patients with infarction. *Anesthesiology* 1983; **59**: 499–505.

64. Shah KB, Kleinman BS, Sami H, et al. Re-evaluation of perioperative myocardial infarction in patients with prior myocardial infarction undergoing noncardiac operations. *Anesthesia and Analgesia* 1990; **70**: 240–7.

65. Foster ED, Davis KB, Carpenter JA, Abele S, Fray D. Risk of noncardiac operations in patients with defined coronary disease: the Coronary Artery Surgery Study (CASS) registry experience. *Annals of Thoracic Surgery* 1986; **41**: 42–50.

66. Rankin JS, Muhlbaier LH, Behar VS, Fedor JM, Sobiston JC. The effect of coronary revascularization on left ventricular function in ischemic heart disease. *Journal of Thoracic and Cardiovascular Surgery* 1989; **90**: 818–32.

67. Holter NJ. New method for heart studies. *Science* 1961; **134**: 1214–20.

68. Deanfield JE, Selwyn AP, Chierchia S, et al. Myocardial ischaemia during daily life in patients with stable angina: its relation to symptoms and heart rate changes. *Lancet* 1983; **2**: 753–8.

69. Droste C, Roskamm H. Silent myocardial ischemia. *American Heart Journal* 1989; **118**: 1087–92.

70. Raby KE, Goldman L, Creager MA, et al. Correlation between preoperative ischemia and major cardiac events after peripheral vascular surgery. *New England Journal of Medicine* 1989; **321**: 1296–300.

71. Ouyang P, Gerstenblith G, Furman WR, et al. Coronary artery disease: frequency and significance of early postoperative silent myocardial ischemia in patients having peripheral vascular surgery. *American Journal of Cardiology* 1989; **64**: 1113–16.

72. Mangano D, Browner W, Hollenberg M, London M, Tubau J, Tateo I. Association of perioperative myocardial ischemia with cardiac morbidity and mortality in men undergoing non-cardiac surgery. *New England Journal of Medicine* 1990; **323**: 1781–8.

73. McCann RL, Clements FM. Silent myocardial ischemia in patients undergoing peripheral vascular surgery: incidence and association with perioperative morbidity and mortality. *Journal of Vascular Surgery* 1991; **9**: 583–7.

74. Raby KE, Barry J, Creager MA, Crook F, Weisberg MC, Goldman L. Detection and significance of intraoperative and postoperative myocardial ischemia in peripheral vascular

surgery. *Journal of the American Medical Association* 1992; **268**: 222–7.

75. Marsch SCU, Wanigasekera VA, Ryder WA, Wong LSS, Foëx P. Graded myocardial ischemia is associated with a decrease in diastolic distensibility of remote non-ischemic myocardium in anesthetized dog. *Journal of the American College of Cardiology* 1993; **22**: 899–906.

76. Landesberg G, Luria MH, Cotev S, *et al.* Importance of long-duration post-operative ST-segment depression in cardiac morbidity after vascular surgery. *Lancet* 1993; **341**: 715–19.

77. Knight AA, Hollenberg M, London MJ, *et al.* Perioperative myocardial ischemia: importance of the preoperative ischemic pattern. *Anesthesiology* 1988; **68**: 681–8.

78. Ichinose Y, Hara N, Ohta M, Hayasi S, Yagawa K. Appearance of thrombosis-inducing activity in the plasma of patients undergoing pulmonary resection. *Chest* 1991; **100**: 693–7.

79. Tuman KJ, McCarthy RJ, March RJ, DeLaria GA, Patel RV, Ivankovich AD. Effects of epidural anesthesia and analgesia on coagulation and outcome after major vascular surgery. *Anesthesia and Analgesia* 1991; **73**: 696–704.

80. Rosenberg J, Rasmussen V, van Jensen F, Ullstad T, Kehlet H. Late postoperative episodic and constant hypoxaemia and associated ECG abnormalities. *British Journal of Anaesthesia* 1990; **65**: 684–91.

81. Reeder MK, Muir AD, Foëx P, Goldman MD, Loh L, Smart D. Postoperative myocardial ischaemia: temporal association with nocturnal hypoxaemia. *British Journal of Anaesthesia* 1991; **67**: 626–31.

82. Mangano DT, Browner WS, Hollenberg M, Tateo IM. Long-term cardiac prognosis following noncardiac surgery. *Journal of the American Medical Association* 1992; **268**: 233–9.

83. Goldman L, Caldera DL. Risks of general anesthesia and elective operation in the hypertensive patient. *Anesthesiology* 1979; **50**: 285–92.

84. Prys-Roberts C. Anaesthesia and hypertension. *British Journal of Anaesthesia* 1984; **56**: 711–24.

85. Asiddao CB, Donegan JH, Whitesell RC, Kalbfleisch JH. Factors associated with perioperative complications during carotid endarterectomy. *Anesthesia and Analgesia* 1982; **61**: 631–7.

86. Dagnino J, Prys-Roberts C. Anaesthesia in relation to hypertension. VI. Cardiovascular responses to extradural blockade of treated and untreated hypertensive patients. *British Journal of Anaesthesia* 1984; **56**: 1065–73.

87. Muir AD, Reeder MK, Sear JW, Foëx P. Preoperative silent ischaemia has a relationship with hypertension. *British Journal of Anaesthesia* 1992; **69**: 540P.

88. Hemming AE, Howell S, Sear JW, Foëx P. Admission blood pressure is predictive of silent myocardial ischaemia in treated hypertension. *British Journal of Anaesthesia* 1993; **71**: 760P.

89. Hoffman BF, Cranefield PF. The physiological basis of cardiac arrhythmias. *American Journal of Medicine* 1964; **37**: 670–84.

90. Wit AL, Rosen MR. Pathophysiologic mechanisms of cardiac arrhythmias. *American Heart Journal* 1983; **106**: 798–811.

91. Atlee JR III. Anaesthesia and cardiac electrophysiology. *European Journal of Anaesthesia* 1985; **2**: 215–56.

92. Santini M, Carrara P, Benhar M, *et al.* Possible risks of general anesthesia in patients with intraventricular conduction disturbances. *Pace* 1980; **3**: 130–7.

93. Zaidan JR. Pacemakers. *Anesthesiology* 1984; **60**: 319–34.

94. Relationship of right bundle-branch block and marked left axis deviation to complete heart block during general anesthesia. *Anesthesiology* 1976; **44**: 64–6.

95. McAnulty JH, Shahbudin HR, Murphy ES, *et al.* A prospective study of sudden death in 'high-risk' bundle-branch block. *New England of Journal of Medicine* 1978; **299**: 209–15.

96. Reid DS. Disorders of cardiac conduction: sick sinus syndrome. *British Journal of Hospital Medicine* 1984; **31**: 341–52.

97. McConachie I. Value of preoperative carotid sinus massage. *Anaesthesia* 1987; **42**: 636–8.

98. Gupta PK, Lichstein E, Chadda KD. Lidocaine-induced heart block in patients with bundle branch block. *American Journal of Cardiology* 1974; **33**: 487–92.

99. Coriat P, Harari A, Ducardonnet A, Tarot JP, Viars P. Risk of advanced heart block during extradural anaesthesia in patients with right bundle branch block and left anterior hemi-block. *British Journal of Anaesthesia* 1981; **53**: 545–8.

100. Wynands JE. Anesthesia for patients with heart block and artificial cardiac pacemaker. *Anesthesia and Analgesia* 1976; **55**: 626–32.

101. Ponka JL. Arteriosclerotic heart disease and surgical risk. *American Heart Journal* 1977; **93**: 1–2.

102. Adams JE, Sicard GA, Allen BT, *et al.* Diagnosis of perioperative myocardial infarction with measurement of cardiac troponin I. *New England Journal of Medicine* 1994; **330**: 670–4.

Chapter 13

Effects on Respiratory Control of Anaesthetic Agents and Adjuvants

Mark A. Chaney and Theodore C. Smith

Control of breathing has both conscious and unconscious components. The central nervous system (CNS) is composed of voluntary and involuntary (autonomic) sections which control the actions and functions of the body. Thus we run from danger (voluntary) and alter the perfusion of organs with blood (involuntary), both under the control of the CNS. Some body systems are almost entirely controlled by one or the other. There is little we can do to increase urine output by mental effort for example, but we can chew and swallow as much food as is available, or nearly so. The respiratory system can be affected in major fashion by both divisions of the CNS. Thus we have beautiful arias from divas, dull lectures from professors, sneezing, eating, defaecating, lifting, all of which alter respiratory rate and rhythm by voluntary effort. Yet all the while there is a certain average alveolar ventilation matched to current metabolic needs, which continues awake, asleep, anaesthetized, in coma, albeit with varied matching.

The transition between normal awake and anaesthesia-altered breathing occurs twice for each anaesthetization. General anaethesia and many anaesthetic adjuvant drugs produce alteration in respiration which the anaesthetist must understand and plan to deal with. Some of the effects are due to actions on muscles, on blood vessels and circulation, and on the geometry (size) of the lung. Some involve interaction with pre-existing disease. But most involve the central nervous system.

Physiology of respiration

Oxygen is moved from the ambient atmosphere to the mitochondrial electron transport system by a series of alternately convective and diffusive steps. *Respiration* is used by biochemists to mean the intracellular use of oxygen, and perhaps anaesthetists should use the term *ventilation* to describe the mass movement of air into and out of the lung. In this chapter, ventilation and respiration are used interchangeably. Moreover, they are restricted to mean the average breathing over a period of minutes that persists despite temporary and voluntary interruption for deglutition, vocalization, elimination, etc. These voluntary acts do alter the pattern of breathing (rate, tidal volume, *I/E* ratio, etc.) but not the matching of averaged ventilation to concurrent metabolism, nor the proportionality factor in the matching.

The organs and tissues involved in breathing

The act of breathing is a recurrent cycle. Muscles contract, and by exerting force on bones, cartilage, ligaments and organs, enlarge the thorax. This creates a lower than atmospheric pressure in alveoli, and air moves (inspiration) through the airway into the alveoli. Muscles relax, causing positive alveolar pressure and outflow of air (expiration). The timing (duration) and degree of contraction (tidal volume) depend on the action of neurons in the medulla. These neurons, in the respiratory centre, stimulate motor neurons in the cervical and thoracic cord, and are themselves modulated by neurons in the pons, by reflex input from receptors in the lung, aorta, carotid and elsewhere, and by the cortex for voluntary efforts (Fig. 13.1). The average volume of gas moved into and out of the lungs over time affects the composition of blood leaving the lung. This is in turn detected by the various reflex receptors and the amount of brain activity altered to maintain a particular homoeostatic level of the blood components and other inputs (largely oxygen partial

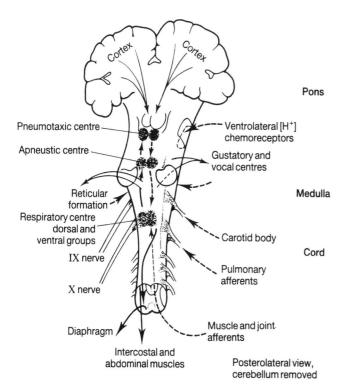

Fig. 13.1 Schematic view of the central nervous system elements of respiratory control, from a posterolateral view. The respiratory centre is composed of bilateral groups of dorsal (mainly inspiratory) and ventral (mainly expiratory cells) which fire cyclicly to stimulate motor neurons innervating diaphragm, intercostal, and other respiratory muscle. This cyclic activity is modulated by input from IX and X cranial nerves, by hydrogen ion concentration on the ventolateral medullary surface, by groups of cells in the pons, and by cells in the reticular formation, all resulting in an average minute volume that is matched to contemporary metabolism so that oxygen tension and hydrogen ion concentration at vital areas are kept constant. For short periods and other, non-respiratory purposes, the actual frequency and tidal volume can be altered voluntarily, by input to the medullary centres from the cortex and from other brainstem centres.

pressure in blood and hydrogen ion concentration in brain or cerebrospinal fluid).[1]

The brainstem centres

Concentrations of neurons that affect breathing were found in the brainstem in the early nineteenth century. In the medulla are bilateral groups that fire on inspiration, and on expiration, stimulating motor neurons in the spinal cord. This medullary respiratory centre has connection with two pontine centres: the lower is called the apneustic centre, and the higher the pneumotaxic centre. Anatomically the medullary cells are part of the nucleus tractus solitarii (NTS) and nucleus ambiguus (NA). The pneumotaxic centre is part of the nucleus parabrachialis media (NPBM). The apneustic 'centre' is diffuse. The centres were named by the effect of brainstem section on breathing (Table 13.1). Note that, with some sections, blocking the vagus alters the ventilation in pattern and volume.

Connection of the centres

Neuroanatomists have traced neural input from pulmonary receptors (stretch, irritant, and juxtacapillary receptors) via the vagus to the NTS, wherein the respiratory centres are found. There are additionally neurons on the lateral surfaces of the upper medulla that connect to the NTS. There is also input from other cranial nerves (trigeminal, facial and glossopharyngeal) as well as from the cerebral cortex. Afferents from the NTS go to the NPBM, to the NA, and to the contralateral motor neurons in C_{3-5} and T_{1-12} via the cord's lateral funiculus.

Table 13.1 Results of brain section on pattern and volume of breathing

Level of section	Vagi	Breathing
Above pneumotaxic centre	Intact	Unchanged
	Blocked	Very slow deep breaths
Between pneumotaxic and apneustic centre	Intact	Very slow deep breaths
	Blocked	Inspiratory holds with short expirations (apneusia)
Between apneustic centre and medullary centre	Intact	Deep breaths slowly
	Blocked	Deep breaths slowly
Below medullary centre	Intact	No respiratory effort
	Blocked	No respiratory effort

Neural patterns of firing

Inspiration begins with activity (called centrally initiated activity, CIA) of some of the more dorsal medullary respiratory neurons, the inspiratory group. The number of cells firing and their frequency depend on the integrated activity of various afferents, notably from chemoreceptors in the carotid body (largely oxygen sensitive) and on the medullary surface (largely $[H^+]$ sensitive). They may be initiated or inhibited by voluntary effort through afferents from the cortex for brief periods but over a time scale of minutes they are principally modulated by the chemoreceptors.

When the CIA reaches some level, neurons progressively stop firing, decreasing the muscle force and initiating exhalation. This CIA level is set in part by pulmonary afferents (stretch reflex) and in part by other CNS activity, acting functionally like an 'off switch'. As the CIA activity decays, and perhaps as exhalation triggers deflation receptors, a point is reached like a threshold, when a new inspiration is started ('on switch'). The triggering of the 'off' and 'on' switches is modulated by voluntary effort, gustatory sequences, cardiovascular reflexes and input from the reticular formation presumably related to level of consciousness.

The sum of these actions serves to maintain a level of overall alveolar ventilation that provides homoeostasis, compensating for pulmonary, circulatory, and metabolic diseases, and providing for the voluntary efforts of alimentation, elimination, communication and transportation.

The maintenance of homoeostasis – the ventilation–carbon dioxide diagram

The major function of the complex control system is to regulate hydrogen ion concentration in body fluids, and to provide sufficient oxygenation to pulmonary capillary blood. In normal man carbon dioxide tension is controlled by hydrogen ion regulation through the Henderson relation (the product of hydrogen ion concentration and carbon dioxide tension is constant). This normally assures almost complete oxygenation of arterial blood. The role of normal oxygen tension accounts for less than 10 per cent of resting breathing in healthy man. At high altitude (10 000 feet above sea level or more) or when disease affects oxygenation, the second function becomes pre-eminent.

The carbon dioxide excretion hyperbola

Practically all the carbon dioxide that the body produces is eliminated in the expired gas. Since only trace amounts are eliminated from the mucosal lining of the airways, one can assume that only alveolar expirate carries CO_2 out of the body. The actual amount of CO_2 excreted ($\dot{V}CO_2$ is the symbol for CO_2 excretion) can be calculated by collecting all the gas expired for a minute ($\dot{V}E$) and analysing it for the fraction that is CO_2 ($F\bar{E}CO_2$). Algebraically:

$$\dot{V}CO_2 = \dot{V}E \times F\bar{E}CO_2 \qquad (1)$$

Since all the $\dot{V}CO_2$ is assumed to come from the alveolar ventilation, the same answer($\dot{V}CO_2$) should be obtained by multiplying the alveolar ventilation ($\dot{V}A$) by the fractional concentration of CO_2 in alveolar gas ($FACO_2$):

$$\dot{V}CO_2 = \dot{V}A \times FACO_2 \qquad (2)$$

Equation (2) is in fact the defining equation for alveolar ventilation and depends on what estimate one uses for alveolar CO_2 concentration. If one uses the ventilation-weighted average obtained by end tidal gas analysis, the calculated alveolar ventilation differs from $\dot{V}E$ by the *anatomical dead space* ventilation, normal about one-third of tidal volume. If one assumes that $FACO_2$ is best estimated from arterial gas analysis (a perfusion weighted average that uses the relation: $FACO_2 = PaCO_2/(PB - PH_2O)$), the alveolar ventilation calculated from equation (2) differs from $\dot{V}E$ by the *physiological dead space* ventilation. Neither is 'correct': they differ depending on the convention accepted to estimate 'average' or 'ideal' alveolar gas. In healthy man they are very nearly the same, but when disease, mechanical ventilation, pulmonary hypotension or other conditions increase alveolar dead space, the difference increases. Since the analysis based on arterial blood reflects the conditions experienced by the rest of the body, that is perhaps the preferable choice.

The form of equation (2) is that of a rectangular hyperbola and has the general equation:

$$\begin{aligned} \text{Constant} &= x \text{ times } y \\ \dot{V}CO_2 &= \dot{V}A \text{ times } FACO_2 \end{aligned}$$

A graph of this equation in the first Cartesian quadrant is a curve asymptotic to the x and y axes.[1] By convention we usually plot $FACO_2$ on the horizontal axis and $\dot{V}A$ on the vertical axis (see Fig. 13.2a). The normal point (defined as the $PACO_2$ and $\dot{V}A$ at rest) can be connected to the axes to form a rectangle with an area proportional to $\dot{V}CO_2$. This area is calculated by multiplying the ventilation (a flow) times a dimensionless fraction proportional to $PACO_2$ (from equation (3) and yields a flow – the CO_2 excretion rate. But a given amount of CO_2 can be excreted by different pairs of $\dot{V}A$ and $FACO_2$. For example, in Fig. 13.2(a) the other two dots define rectangles with the same area, also proportional to $\dot{V}CO_2$. In fact any point lying on the hyperbola of the graph of equation (2) represents such a pair (Fig. 13.2b), and represents a possible steady state value for a body. How does the central control decide where on the hyperbola to

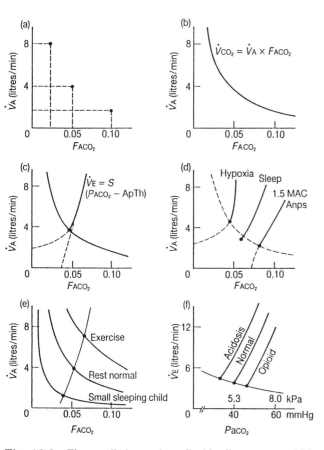

Fig. 13.2 The ventilation–carbon dioxide diagram, on which can be plotted the CO_2 excretion hyperbola for steady state ventilation and F_{CO_2}, and the ventilatory response to CO_2 ($\dot{V}ER_{CO_2}$). (a) Three of the infinite number of pairs of alveolar ventilation and alveolar F_{CO_2} which eliminate a constant CO_2 load. (b) The hyperbola on which all such points lie, and its equation. (c) A typical response to elevating alveolar, with linear extrapolation to the apnoeic threshold (ApTh), and a curvilinear response to artificially lowering CO_2. Various subjects respond between these two dashed lines at lower F_{ACO_2}. (d) Typical alterations to the $\dot{V}ER_{CO_2}$ in which the slope is steeper (e.g. hypoxia) or flatter (e.g. anaesthesia) than normal. (e) The effect of different metabolic rates for CO_2 on the resting point, given a normal $\dot{V}ER_{CO_2}$. (f) The common presentation of this diagram: the CO_2 axis is truncated to common values, the CO_2 stimulus expressed as arterial tension (Pa_{CO_2}) rather than alveolar fraction (F_{ACO_2}), and the ventilation expressed as a directly measured minute volume ($\dot{V}E$) rather than a calculated or theoretical alveolar ventilation ($\dot{V}A$). A normal $\dot{V}ER_{CO_2}$, and examples of responses in metabolic acidosis and opioid premedication are shown.

stay? That follows from the second kind of line that can be graphed on the same coordinate system. While the CO_2 excretion hyperbola is a basic truth, following from its definition, the second line must be found by experimentation. It is called the ventilatory response to CO_2 ($\dot{V}ER_{CO_2}$).

The ventilatory response to CO_2

While there is no physical fact that dictates it, experiments show that a straight line describes the increasing ventilation that results from inspiratory CO_2, either exogenous CO_2 or rebreathing expired CO_2 (Fig. 13.2c). This line is described by the following equation:

$$\dot{V}E = S\,(Pa_{CO_2}) + \text{a constant}$$
$$y = mx + b$$

It may be slightly manipulated by both multiplying and dividing the constant by S, and factored to yield:

$$\dot{V}E = S\,(Pa_{CO_2}) + \text{ApTh} \qquad (3)$$

S is the slope of the line, normally 15–23 l/min per kPa (2–3 l/min per mmHg). The constant (abscissal intercept where ventilation is zero) is called the apnoeic threshold (ApTh). It is found by *extrapolating* the ventilatory response to zero ventilation and is normally about 4.8 kPa (36 mmHg). Apnoeic threshold is a name for a point, and not a fact. Normal humans do not usually become apnoeic after hyperventilation lowers P_{CO_2},[2] while spontaneously breathing anaesthetized patients are apnoeic when CO_2 is lowered even a little below their current resting level.[3]

The *normal homoeostatic point*, i.e. $\dot{V}A$ and F_{ACO_2} and thus the body fluid pH (through action of the Henderson equation relating [H$^+$] to P_{CO_2}), is set by the *intersection of the two lines* on the ventilation–CO_2 diagram. Circumstances alter the two lines. Characteristically the $\dot{V}ER_{CO_2}$ can increase or decrease in slope (Fig. 13.2d) as level of consciousness changes. It can also be shifted left or right with constant slope, which is seen with premedicant opioid drugs, certain CNS stimulants, and acid–base alterations. The CO_2 excretion hyperbola can also change as a result of a higher or lower CO_2 production. Figure 13.2(e) shows two such examples of a large and a small metabolic rate for CO_2. Finally, the graph usually appears as shown in Figure 13.2(f): the abscissa is truncated and enlarged, and the ordinate is usually given as total ventilation $\dot{V}E$, measured directly. Expired ventilation is usually 50 per cent larger than alveolar ventilation, representing a dead space:tidal volume fraction of $\frac{1}{3}$:

$$\dot{V}E = \dot{V}A\,(1 - V_D/V_T)$$

Tests of ventilatory control

Since it is the current state of the brain that provides the set point for homoeostasis, several tests have been developed to assess the state.[4] Perhaps more important for anaesthetists, these tests can be used to examine drug effects. Several general approaches are in use: (1) the ventilatory response to CO_2 may be studied in several ways, usually by steady state or rebreathing (Read) methods; (2) the ventilatory

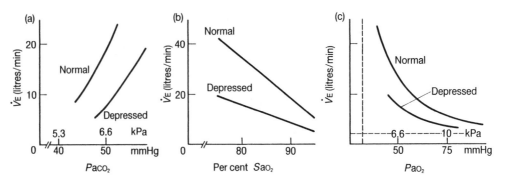

Fig. 13.3 Three common displays of the ventilatory response to hypoxia ($\dot{V}_E R \downarrow O_2$). (a) A ventilation–$CO_2$ diagram: hypoxia causes an increased slope, which is flattened by conditions which depress the hypoxic response, e.g. sub-MAC anaesthesia. (b) A plot of the ventilatory response to hypoxia in which the stimulus is arterial haemoglobin saturation. A straight line with negative slope of 2 l/min per percentage saturation fit the data. Depressants flatten this slope. (c) The same $\dot{V}_E R \downarrow O_2$ but plotted as a rectangular hyperbolic function of arterial P_{O_2}. Depression is represented by a curve closer to asymptotes, and therefore a quantitatively smaller increase in breathing with decrease in oxygen tension. The latter graph is sometimes plotted with the P_{O_2} axis reversed.

response to hypoxia may be observed in several ways; (3) arterial blood analysis; or (4) a variety of other, less frequently used tests may be incorporated in a drug trial, such as the airway pressure 100 ms (one-tenth second) after occluding inspiration ('P point one'). In general all give satisfactory measure of drug effects, and the differences within one of the types, e.g. CO_2 or hypoxic response, are important only in designing the drug trial. For example, a short-acting drug's time course cannot be studied with a steady state method requiring 10 minutes or more per point. Sometimes more than one observation is necessary to distinguish similar appearing states. For example, during spinal anaesthesia, a dose of 15 mg of morphine will elevate P_{CO_2} into the 50s, as will 1.5 MAC anaesthesia with isoflurane. But should an additional stress such as steep Trendelenburg position be requested, we would expect an awake patient under spinal block to maintain his (elevated) P_{CO_2}. The unresponsive anaesthetized patient will not. The explanation is that the two have different slopes to their CO_2 response curves.

Ventilatory response to CO_2

Two different varieties of this test are common. One, the steady state technique, employs two or three increased but constant levels of inspired CO_2, each for long enough to reach a steady state. Together with a resting point, this test requires half an hour or more and hence is not suitable for following rapid changes. Very rapid drug changes, e.g. i.v. induction of anaesthesia, may be followed accurately with Lambertsen's isohypercapnoeic steady state method.[5] Normally there is a characteristic slope of 15–23 l/min per kPa (2–3 l/min per mmHg) CO_2 and an apnoeic threshold (extrapolated) in the mid 30s. As a good generalization, pharmacological effects which do *not* cause loss of consciousness, shift the position of the response left (stimulation) or right (depression) with no slope change.

Pharmacological effects which do alter consciousness change the slope. Fear, arousal and in fact adrenaline (epinephrine) infusion steepen the response, as does hypoxia! Sleep, anaesthesia and coma depress the response. If the slope is altered, we cannot draw any conclusion about displacement of the curve, largely because we do not know the point about which the response curve rotates, when slope changes. Early workers assumed it was around the apnoeic threshold but more careful work extrapolating drug response curves generally show it is often in the fourth Cartesian quadrant (i.e. a negative ventilation with a positive CO_2.) The errors of extrapolation are such that no information results from such attempts. However, in some REM sleep states the resting point is found at a low CO_2 and elevated ventilation, but the slope is half of normal. This response intersects the normal response, giving this confusing result: the resting ventilation is higher than normal (stimulation) but the slope is flatter (depression). Similar paradoxes have been described during evolving drug effects (with ketamine and etomidate for example[6,7]).

The second test involves total rebreathing of expired gas without CO_2 absorption. The size of the reservoir determines the rate of rise of CO_2. Commonly rises of 2.0–2.7 kPa (15–20 mmHg) are observed over 5–15 minutes. Drug effects evolving over many minutes are quantifiable. The results are well represented by straight lines but the slope is slightly less than steady state values, and the apnoeic threshold is shifted to the right by 0.9–1.3 kPa (7–10 mmHg). Read's modification uses a bag containing enough CO_2 to elevate alveolar CO_2 above mixed venous tension. Thereafter CO_2 rises at the same rate as during apnoea, typically 0.4–0.5 kPa (3–4 mmHg)/min. The slope is usually 1.5–2 l/min and apnoeic threshold approximately 6 kPa (45 mmHg). This test has been used to follow drug effects changing over 10–15 minute periods. Drug effects evolving over many minutes may be studied using Read's modifications of the rebreathing methods, which require only 4 or 5 minutes but have a slightly higher variability and a slightly lower slope than steady state points.

Ventilatory response to hypoxia

Lowering oxygen concentrations below normal ambient levels increases ventilation and the slope of the ventilatory response to CO_2 (Fig. 13.3a). The latter, while real, is not as easy to measure accurately enough to construct dose–response curves for drugs. The increase in ventilation can be more easily measured as a function of increasing hypoxia at constant PCO_2 (decreasing oxygenation). Two different displays of the results follow from two different choices to represent the stimulus (hypoxia) – arterial oxyhaemoglobin saturation or arterial tension (Fig. 13.3b and c). The latter is the older technique, from which the most data have been collected. It makes three assumptions however, which have *not* been examined *vis-à-vis* their implications in pharmacological analysis. The first assumption is that the response is described by a hyperbola. While there is no physiological or physical rationale for this choice, it does seem to fit the data, and hence is as valid as fitting a straight line to CO_2 responses. Second, the horizontal asymptote is believed to be the ventilation in the absence of carotid body stimulation, easily achieved by inhaling 35 per cent oxygen or more for a few minutes. This at least gives a determinable asymptote. The third is critical; the PO_2 at which ventilation becomes infinite. This vertical asymptote is neither experimentally determinable nor physiologically based. It is guessed! The value $PO_2 = 32$ (mmHg) is commonly chosen. After having guessed two variables (asymptotes) the response is then 'quantified' by fitting a hyperbola to the data and calculating its curvature, a parameter known as 'A'. The value of this approach is that it gives a single number to characterize the response. A large body of data has accumulated from use of this method.

The alternative method for hypoxic responsiveness ($\dot{V}ER{\downarrow}O_2$) uses arterial saturation to estimate the stimulus. When this is done, the resulting data seem characterized by a straight line with negative slope. The slope varies, flattening, with increasing drug depression. No assumptions about asymptote are required. The data can be fitted by straightforward regression methods and an unambiguous slope derived. The intercept is often above 100 per cent and is ignored.

There is a connection between the two methods. The shape of the oxyhaemoglobin dissociation can be closely described as a hyperbolic function of desaturation, at least over the ranges of PO_2 that can be used in a hypoxic challenge. Thus one could, in theory, mathematically link parameter A with the slope of the response to desaturation. The choice between the two is largely a matter of

inclination. Both are called the ventilatory response to hypoxia ($\dot{V}ERCO_2$).

One other presentation of hypoxic responsiveness needs identification. In the illustration Fig. 13.3(c) the oxygen tension *increases* to the right, representing *decreasing* stimulus. This seems backward to some so they present the data as a mirror image of Fig. 13.3(c) with oxygen tension decreasing (to 32 or to zero) to the right. At first glimpse this looks different, but should present no real problem. The value calculated for parameter A is, of course, invariant.

Other tests of responsiveness

A variety of estimates of brain activity in respiratory regulation have been devised, but few have been used in drug studies and none extensively. One is worth noting: the airway pressure generated 100 ms (one-tenth second) after occlusion of airway in inspiration known as the *P point one* ($P_{0.1}$). Ideally the output of the brain should be studied as the neuronal traffic in efferent nerves. For respiration this would be the activity in respiratory centres, or at least in motor neurons involved. Even in animals stable chronic studies of this nature are difficult and invasive, and out-of-the-question in humans. Breathing observations, that is to say observation of respiratory rate, volume, pattern, etc., are used as surrogates of neuronal activity. But the quantitative nature of these observations is modulated by the physical characteristics of the thorax and lungs – e.g. compliance and resistance – and any change in these will alter the effect of the brain which will itself respond, reflexly, by a compensating alteration. The $P_{0.1}$ test, however, assumes that in the first 100 ms of inspiration against a totally occluded inspiratory circuit, the negative pressure generated in the airway is proportional to respiratory drive. One-tenth of a second is too short a time for a reflex arc – perception, transmission, CNS, change and afferent action require a longer period. The chest does not really change shape and muscle length is essentially constant in the interrupted breath. The occlusion is transient and usually applied only once in a series of breaths; thus, no major effect on overall breathing occurs. Conditions which alter rib cage–diaphragm interaction (diseases, usually) may invalidate the assumption that $P_{0.1}$ represents neural drive.[8] In general, the value for $P_{0.1}$ falls with the same depressants that depress $\dot{V}ERCO_2$ and when incorporated in drug studies, confirm that the results represent CNS effect. Moreover, similar dose–response curves have been obtained when the $\dot{V}ERCO_2$ and $P_{0.1}$ are used simultaneously.

Simple observations are often used to characterize drug effect: tidal volume, respiratory rate, inspiration:expiratory period, resting and end tidal CO_2, resting minute volume and arterial blood analysis at rest breathing air. The changes may be small and may, because of the many interactions and compensation for these and other changes, not be proportional to drug dose, and may be otherwise questioned. In general the use of stress is more informative as to mechanism and degree of effect.

Many factors, other than the drug being studied, have significant effects on respiratory control mechanisms and must be considered when interpreting results of investigations. In addition to primary stimuli of $[H^+]$ and PO_2, respiration is affected by age, sex, blood pressure, metabolic rate, temperature, time of day, genetic factors, exercise, mechanical loads, instrumentation apparatus, interindividual variability, posture, psychological profile, and concomitant use of other CNS or muscle affecting drugs. The mere awareness that an investigator is studying one's breathing significantly affects respiratory rate and tidal volume. The level of consciousness, in and of itself, exerts profound influence on respiratory control mechanisms. Drugs, therefore, may alter respiratory control via specific receptor mechanisms or by simply decreasing the level of consciousness in a non-specific manner.

Significant differences exist between species. A 70 kg man may need 2 mg/kg of a drug for premedication. A 7 kg dog requires 20 mg/kg for the same effect while a cat given 2 mg/kg exhibits gross hyperactivity and sham rage. The discussion below concerns data obtained in humans.

Depression may be variously defined: an increase in the arterial CO_2 tension at rest, absent alkalosis, and a decrease in the slope of the ventilatory response to CO_2 are the most common interpretations of depression. Even if an investigation proves beyond doubt that an anaesthetic drug causes 'respiratory depression', this does not necessarily translate into problems when the drug is used in the clinical setting. A statistically significant depression may be clinically acceptable or even desirable.

Inhalation agents

All inhalation agents cause dose-dependent depression of the slope of the $\dot{V}ERCO_2$. Interestingly, the least depressing (fluroxene, cyclopropane, ether and methoxyflurane) are no longer utilized today, whereas the most potent respiratory depressants (halothane, enflurane, isoflurane) are in common clinical use. The former group had in common activation of the sympathetic nervous system with measurable increases in circulating adrenaline and noradrenaline (epinephrine and norepinephrine). The latter do not cause such activation. Nitrous oxide is intermediate in both its slope and catecholamine effects.

Mechanism of action

The mechanism underling the respiratory depressant effects of the anaesthetic agents remains to be elucidated. Central and peripheral mechanisms have been extensively explored in human as well as animal models. The obvious CNS effects favour a central site. Major effects may be mediated via preferential suppression of intercostal muscle function with relative sparing of the diaphragm.[8] This could have direct effects on muscle, or decreased motor neuron activity. Other investigators have discovered a correlation between the results obtained from preoperative pulmonary function tests and the magnitude of respiratory depression induced by the halogenated agents, suggesting that altered pulmonary mechanics may also play a significant role.

Common halogenated agents

Halothane, enflurane, isoflurane and (to the extent that present studies are available) the newest agents sevoflurane and desflurane, share many similarities in respiratory effects and but few differences, largely attributed to smell and pungency.

Effects on resting ventilation

The halogenated agents cause a dose-dependent decrease in tidal volume and increase in respiratory rate with moderate differences between the agents at equipotent concentrations (Figs 13.4 and 13.5). Trichlorethylene produced the most tachypnoea and fluroxene the least. For most agents except isoflurane the tachypnoea increases with depth. The increase in respiratory rate does not compensate for a simultaneous decrease in tidal volume: a dose-dependent decrease in alveolar ventilation occurs with corresponding increases in arterial carbon dioxide tension (Fig. 13.6). At low levels (0.1 MAC), little change is observed in alveolar ventilation although the $\dot{V}ERCO_2$ may be depressed. With increasing inspired levels, however, consciousness is altered and alveolar ventilation begins to decrease. Enflurane has been shown to be the most ventilatory depressant of the halogenated agents. During spontaneous ventilation at 1.0 MAC enflurane, isoflurane, and halothane, the corresponding arterial carbon dioxide tensions were approximately 8.6, 6.6 and 6.0 kPa (65, 50, and 45 mmHg), respectively. At enflurane levels above 1.5 MAC most became apnoeic, making this agent difficult to use during spontaneous ventilation. Others have found somewhat lower CO_2 tensions, especially during surgical stimulation, in the 6.6–7.3 kPa (50–55 mmHg) range for enflurane.[9]

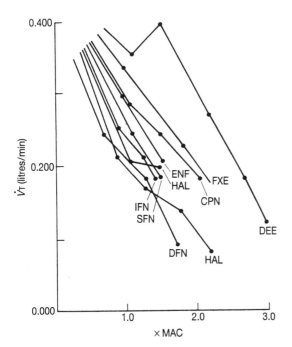

Fig. 13.4 Effects of various anaesthetic agents and concentrations on the tidal volume of spontaneously breathing humans. The three agents which increase catecholamine levels in blood (DEE, diethyl ether; FXE, fluroxene; CPN, cyclopropane) lie to the right and above modern halogenated agents (HAL, halothane; SFN, sevoflurane; IFN, isoflurane) but have largely similar slopes of decreased tidal volume with increased concentration when measured on an equipotential scale of multiples of minimum alveolar concentration (MAC). DFN, desflurane; ENF, enflurane.

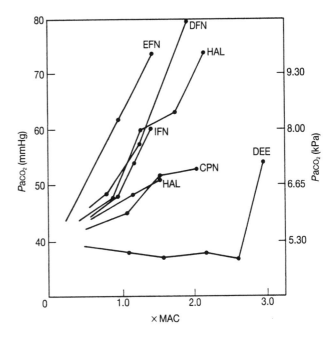

Fig. 13.6 Effects of various anaesthetic agents and concentrations on the arterial CO_2 tension of spontaneously breathing humans. Except for diethyl ether (DEE), there is a clearly incremental increase in arterial P_{CO_2} for each increment of anaesthetic concentration. Different reports have somewhat different absolute values (see for examples two studies of halothane, HAL), but all agents (even ether at more than 2.5 MAC) depress respiration. CPN, cyclopropane; DFN, desflurane; EFN, enflurane; IFN, isoflurane.

Effects on the ventilatory response to carbon dioxide

Much as with their effects on resting ventilation, the halogenated agents cause a dose-dependent depression of the $\dot{V}_{E}RCO_2$ with little difference between the agents at equipotent concentrations (Fig. 13.7). At high levels of halogenated agent (2.5 MAC), essentially no increase in ventilation is observed following a carbon dioxide challenge.

Effects of apnoeic threshold

Awake man may become apnoeic if his CO_2 tension is lowered.[2] For those that do, breathing resumes approximately when the CO_2 rises to the extrapolated apnoeic threshold, or even sooner. By contrast, there is near uniform apnoea in anaesthetized man when the CO_2 tension is lowered. Spontaneous breathing does not resume when the CO_2 rises to the extrapolated apnoeic threshold, but instead apnoea persists to a higher CO_2. The ventilatory response between the apnoeic threshold and the unstimulated steady state resting point, during inhalation anaesthesia, is steeper than the $\dot{V}_{E}RCO_2$ at higher CO_2. In awake man it is as steep or less steep.

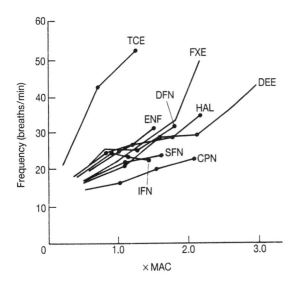

Fig.13.5 Effects of various anaesthetic agents and concentration on the respiratory frequency of spontaneously breathing humans. In general, anaesthesia raises the frequency, and increasing depth raises the frequency further. This is exaggerated with trichlorethylene (TCE) and less marked with cyclopropane (CPN) and least marked with isoflurane (IFN). DEE, diethyl ether; DFN, desflurane; ENF, enflurane; FXE, fluroxene; HAL, halothane; SFN, sevoflurane.

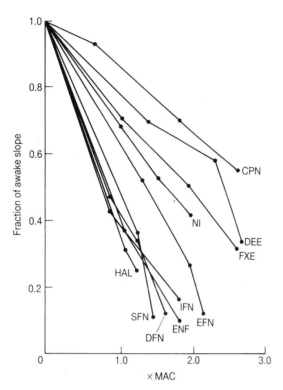

Fig.13.7 The slope of the ventilatory response to CO_2 as affected by various agents and depths. NI refers to nitrous oxide and Thalamonal (Innovar): all other symbols are as previously noted. Those agents which activate the adrenergic autonomic nervous system produce somewhat less flattening of the response to CO_2. There is a similarity in the effects of most halogenated noradrenergic stimulating agents, such that the ability to maintain PCO_2 and hence [H^+] when a stress impairs CO_2 elimination is seriously blunted.

Surgical stimulation

Investigators have described ventilatory stimulation induced by surgical stimulation. In patients anaesthetized with 1.1 MAC enflurane, minute ventilation increases approximately 30 per cent and arterial carbon dioxide tension decreases approximately 10 per cent.[10,11] Similar ventilatory changes have been associated with halothane and isoflurane anaesthesia. Although surgical stimulation increases minute ventilation, it does not appear to affect chemoresponsiveness, and the slope of the $\dot{V}ERCO_2$ remains unchanged (Fig. 13.8). No mechanism for these effects is known, but the rapidity of the ventilatory changes suggests a neurogenic mechanism involving peripheral nerve fibres. Other factors associated with surgical stimulation which may contribute to the ventilatory effects include increases in catecholamine levels, changes in the metabolic rate and carbon dioxide production, and fluctuations in cardiac output, all of which have been reported to be ventilatory stimulants in certain settings.

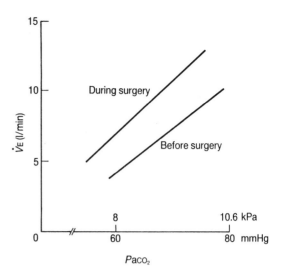

Fig. 13.8 The ventilatory response to CO_2 during clinical anaesthesia before and during surgery. In this example, the depth of enflurane anaesthesia was kept constant. Initiation of operation caused ventilation to increase and arterial PCO_2 to decrease, indicating that the stimulation partially reversed depression. But the slope of the $\dot{V}ERCO_2$ was not significantly increased, remaining less than 25 per cent of normal. Such a flattened response would not help a spontaneously breathing patient defend homoeostasis.

Time adaptation

The magnitude of ventilatory depression induced by the halogenated agents is affected by time.[12] The ventilatory depression induced by constant depth halothane anaesthesia is relatively stable for up to 3 hours' duration, but thereafter, is reduced (Fig. 13.9). Time adaptation has also been demonstrated with enflurane and isoflurane. Enflurane-induced ventilatory depression appears to undergo the most significant adaptation associated with time, being reduced maximally at 7 hours. It is unclear as to why time adaptation occurs, but changes in chemical or hormonal levels of substances known to stimulate ventilation (noradrenaline, corticosteroids) have been suggested as possible aetiological agents. Adaptation has little clinical impact.

Chronic obstructive pulmonary disease

Ventilatory depression induced by halothane is significantly greater in patients with emphysema than in normal patients at equipotent concentrations. The magnitude appears to correlate with the patient's preoperative FEV_1 but not the arterial blood gas tension. It is assumed that enflurane and isoflurane act in a similar manner. Normal preoperative arterial blood tension does not preclude significant intraoperative ventilatory depression. Controlled or assisted ventilation is strongly recommended when using halogenated agents in patients

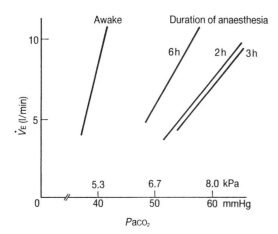

Fig. 13.9 The change in $\dot{V}_E$RCO$_2$ during prolonged anaesthesia. No such change occurs in the first several hours, but with continued anaesthesia there is a progressive but partial reversal of the response.

with emphysema because of their greater susceptibility to ventilatory depression. The close relation between the magnitude of ventilatory depression induced by the halogenated agents and the preoperative FEV$_1$ points to altered pulmonary mechanics as the explanation.

Effects on the ventilatory response to hypoxia

It has been known for half a century that anaesthetics cause depression of the $\dot{V}_E$RCO$_2$. Only since 1978, however, has it been appreciated that they also cause depression of the $\dot{V}_E$R$\downarrow$O$_2$. There are major differences in the magnitude of depression by the halogenated agents at equipotent concentrations of the responses to carbon dioxide and hypoxia. While low levels of halogenated agent, 0.1 MAC, will cause little depression of the $\dot{V}_E$RCO$_2$, this same level will cause significant depression of the $\dot{V}_E$R$\downarrow$O$_2$ (Fig. 13.10).[13] Higher levels of halogenated agent, 1.1 MAC, will totally abolish the hypoxic response.

The halogenated agents also attenuate hypercapnia-induced augmentation of the $\dot{V}_E$R$\downarrow$O$_2$.[13] Sedating or subanaesthetic levels (0.1 MAC) of the halogenated agents markedly impair peripheral chemoreflexes originating in the carotid bodies that initiate the hypoxic response. This impairment in a normal patient with a normal and intact hypercapnic response does not necessarily pose a problem. However, if this impairment occurs in patients who depend on their hypoxic drive to breathe, they will be in danger. This includes those with chronic obstructive pulmonary disease, with surgical or disease-induced ablation of the carotid reflex, with acute metabolic acidaemia, and those who have taken other drugs known to depress the $\dot{V}_E$RCO$_2$ (opioids, benzodiazepines, etc.). Sedating or subanaesthetic levels of the halogenated agents are often present in patients upon arrival in the recovery area following

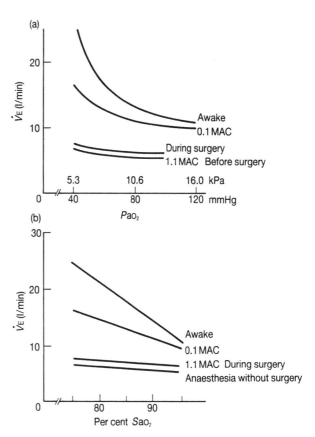

Fig. 13.10 The ventilatory response to hypoxia, as affected by subanaesthetic and anaesthetic doses of halothane, and by subsequent surgery. Even 10 per cent of a minimum anaesthetic concentration blunts the response, and 1.1 MAC nearly abolishes it. Surgery during light anaesthesia produces no significant change.

administration of general anaesthesia. These patients may also receive other drugs that depress the $\dot{V}_E$RCO$_2$. Therefore, patients on arrival in the recovery area, following administration of general anaesthesia with the halogenated agents, are at risk of developing hypoxaemia. Appropriate monitoring and supplemental oxygen should be utilized as dictated by the clinical situation.

The effects of surgical stimulation during anaesthesia on the hypoxic response are comparable to its effects on the hypercapnic response.[10] Minute ventilation increases and arterial carbon dioxide tension decreases, yet no effect on chemoresponsiveness is seen (Fig. 13.10).

Sevoflurane and desflurane

Sevoflurane and desflurane are the two most recently introduced anaesthetic agents. Few studies of their ventilatory depressant effects are available but the investigations performed indicate that they behave much like the other halogenated agents.[14,15] Both agents cause a dose-dependent decrease in tidal volume and in alveolar ventilation despite an increase in respiratory rate. Arterial carbon dioxide tension increases. Both agents cause dose-dependent depression of the slope of $\dot{V}ER\text{CO}_2$.

Nitrous oxide

Nitrous oxide has low potency. Its effects can only be investigated in subanaesthetic doses or under hyperbaric conditions. Winter *et al.* exposed nine normal volunteers in a hyperbaric chamber to 1.55 atm nitrous oxide and 0.32 atm oxygen.[16] The volunteers displayed a dramatic increase in respiratory rate (15 breaths/min baseline to 47 breaths/min) and a dramatic decrease in tidal volume (0.72 litres baseline to 0.33 litres). Minute ventilation increased significantly (9.95 l/min baseline to 15.4 l/min) yet $Pa\text{CO}_2$ remained unchanged from baseline 5–5.3 kPa (38–40 mmHg). The slope of the $\dot{V}ER\text{CO}_2$ curve was significantly decreased when compared to baseline from 24.1 to 5.6 l/min per kPa (3.2 to 0.75 l/min per mmHg). This is comparable to that of equipotent concentrations of halothane and isoflurane. Anaesthesia with 0.6 MAC N_2O and enough Thalamonal (Innovar) to produce unresponsiveness reduced the slope more than the older sympathomimetic agents, but less than some of the halogenated agents. Nitrous oxide, when substituted for a portion of a halogenated agent, results in less ventilatory depression and a lower arterial CO_2 tension than is caused by equipotent concentrations of the halogenated agents alone. Simply stated, a mixture of 0.5 MAC halogenated agent and 0.5 MAC nitrous oxide will cause less ventilatory depression than 1.0 MAC halogenated agent. This 'ventilatory sparing' effect of nitrous oxide has been well documented with halothane, enflurane and isoflurane. The underlying mechanism may involve sympathetic nervous system stimulation. The cardiorespiratory effects suggest similar stimulation. Winter *et al.* observed sweating, dilated and divergent pupils, increased muscle tone, increased blood glucose, catecholamines, corticosteroids, and elevated white blood cell counts in his volunteers.[53]

Nitrous oxide also affects the $\dot{V}ER\downarrow O_2$. Even at low concentrations, 0.1 MAC, nitrous oxide decreases the hypoxic response to approximately 60 per cent of awake control values.[17]

Opioids systemically administered

Morphine-like drugs were once called narcotic analgesics, i.e. pain relieving drugs which facilite sleep (narcosis). Now known as opioids, they are perhaps the oldest and best studied drugs known to man. Opium use, for its euphoric effects, can be traced back over 4000 years. Since the respiratory effects were first noted (approximately 600 years ago) innumerable investigations documenting respiratory depression have appeared. This does not translate into a complete understanding of opioid-induced respiratory depression and the indirect effects of changes in conscious level which occur.

Mechanism of action

Opioid receptors were first isolated in the human brain in 1973, and have subsequently been found to reside in high concentrations in areas thought to play significant roles in the control of respiration. These include the medulla, pons, fourth ventricle, nucleus tractus solitarius, and nucleus ambiguus. Very minute amounts of opioids, when injected into the cerebrospinal fluid bathing these structures, result in profound respiratory depression. Although opioid receptors exist in the spinal cord and are clearly involved in mediating analgesia, their effects on respiration are minimal. Opioid receptors have also been isolated in lung parenchyma but their role at this time is unknown.

Opioid receptor types and their physiological effects are recent knowledge.[18] in 1978, morphine's analgesic and respiratory depressant effects were found to be mediated via at least three different opioid receptor mechanisms. The μ_1-receptor subtype mediates analgesia whereas the μ_2-receptor subtype mediates respiratory depression. κ-Receptors appear to mediate analgesia with little effect on respiration while contributions from δ- and σ-receptors are unclear (σ-receptors may, in fact, stimulate respiration).

Endogenous opioids possessing μ-receptor agonist properties are found in the central nervous system. Their role in normal respiratory control is uncertain. They may, however, be important in states characterized by abnormal respiratory control: the subtype μ_2 may mature earlier than μ_1 in some newborns. Stimulation of a relatively increased population of μ_2-receptors may, theoretically, play a role in the development of infant apnoea syndromes or sudden infant death syndrome. In some patients with chronic obstructive pulmonary disease the administration of naloxone (a non-specific μ-receptor antagonist) is associated with clinical improvement.

Respiratory effects

The many drugs possessing μ-receptor agonist properties induce qualitatively similar dose-dependent depression in respiration. Quantitative differences, however, do exist and are dictated by the dose, route of administration, pharmacokinetic profile, and pharmacodynamic profile of the specific drug given. Not all opioids are capable of producing a maximal μ effect. These drugs are called partial agonists. A few drugs bind to receptors with no effect: these are called antagonists. There is a characteristic decrease in respiratory rate after opioids. A decrease in alveolar ventilation and a corresponding increase in arterial carbon dioxide tension results, with a normal to slightly increased tidal volume. When used clinically, opioids may produce only subtle changes in alveolar ventilation by reduction of the respiratory rate, tidal volume, or both.

Opioids with μ-receptor agonist properties significantly affect the $\dot{V}E \rm{R} co_2$ and $\dot{V}E \rm{R} \downarrow o_2$. In patients administered opioids in doses that retain consciousness, only displacement to the right typically occurs. In some patients very high doses of opioids (> 3 mg/kg morphine or > 25 μg/kg fentanyl) may be administered that will stop spontaneous ventilation without producing complete unresponsiveness. They are well below their apnoeic threshold, as there has not been time for CO_2 to increase markedly. These narcotized patients will breathe if directed to do so, although their attention span is measured in seconds. When unconsciousness does occur, a decreased slope will occur. These different effects of the opioids on the $\dot{V}E \rm{R} co_2$ at increasing doses are probably due to the changing level of consciousness.[19] During normal sleep (without morphine), the slope of the $\dot{V}E \rm{R} co_2$ decreases. Morphine facilitated sleep gives an even flatter response.

Opioids also decrease the $\dot{V}E \rm{R} \downarrow o_2$. As with the inhalation agents, this hypoxic response is more sensitive than the response to carbon dioxide, but since the hypoxic drive accounts for only 10 per cent of normal respiratory drive, it is not clinically apparent in most patients.

The augmentation of opioid-induced ventilatory depression by a decreased level of consciousness has important clinical implications. Opioids may be confidently administered in relatively large amounts without much concern for ventilatory depression so long as the dose is titrated to partial relief of pain. If pain that stimulates consciousness persists, it prevents augmented depression. Patients who exhibit significant opioid-induced ventilatory depression can be partially 'reversed' by arousal or stimulation to increase the level of consciousness. Any concomitant decreases in consciousness (fatigue, benzodiazepines, anaesthesia, coma, etc.) will augment opioid-induced ventilatory depression.

Investigators have long searched for opioids to provide analgesia without the risk of ventilatory depression. Early 'successes' are faulted for many reasons, including differing definitions and measurements of 'ventiltory depression', the subjective quality of pain reports, concomitant use of other known respiratory depressants, changing levels of consciousness, and the unknown relation between opioid blood concentration and opioid brainstem concentration (the presumed site of action). Drugs with a limited potential to depress breathing were found to have limited analgesia as well (partial agonists). All opioids possessing μ-receptor agonist properties will cause ventilatory depression, but new drugs may have non-μ analgesic activity. Hope for an analgesic without respiratory depression now concentrates on finding either a κ active agent with acceptable side effects or by anatomical localization to the spinal cord by epidural or intrathecal administration.

Fully effective classical opioid agonists

The many pain relieving drugs of the opioid class differ largely in potency (the dose required to produce an effect), minor side effects, and pharmacokinetics. All relieve severe pain and all depress ventilation.

Morphine

Morphine produces a dose-dependent depression of ventilation which is characteristic of μ-receptor agonists. With increasing dose, morphine decreases alveolar ventilation and increases arterial carbon dioxide tension, but changes in respiratory rate and tidal volume are variable. The respiratory rate, except at extremes, is not a reliable clinical indicator of the magnitude of morphine-induced ventilatory depression. Bradypnoea may not be present even when arterial carbon dioxide tensions exceed 12 kPa (90 mmHg). Morphine decreases the hypercapnic and hypoxic responses in doses as low as 7.5 mg administered subcutaneously.[20] The hypoxic response appears to be more sensitive to morphine-induced depression than the former.

Peak ventilatory depressant effects of morphine occur within 10 minutes after i.v. injection and nearly an hour after subcutaneous injection. Intramuscular effect is intermediate and depends on muscle blood flow. The slight delay after i.v. usage is due to morphine's relative lipophobic nature, compared with the newer opioids. Plasma concentrations of morphine may be an unreliable predictor of ventilatory depression.[21–23] Morphine, 10 mg intravenously, may measurably reduce the $\dot{V}E \rm{R} co_2$ for up to 10 hours even when plasma levels are negligible.

The lipophobic properties of morphine may also explain why the newborn appears to be more susceptible to developing ventilatory depression following its use.[24] Newborns, who have an immature blood–brain barrier, may allow a higher fraction of a given dose of morphine to cross into the cerebrospinal fluid than older patients with mature blood–brain barriers. Highly lipophilic opioids are believed to cross immature and mature blood–brain barriers equally well. Morphine, therefore, should be administered in smaller doses (μg/kg basis) to neonates in order to avoid significant ventilatory depression. Highly lipophilic opioids may be administered in the doses (μg/kg basis) used for older patients, with similar degrees of ventilatory depression to be expected. The age at which the blood–brain barrier becomes 'mature' is unknown.

Fentanyl

Fentanyl causes ventilatory depression (Fig. 13.11) that is qualitatively similar but quicker in onset and shorter in duration than equianalgesic doses of morphine. Fentanyl is approximately 580 times more lipophilic than morphine and its quick onset is likely to result from the rapid entry of the drug into the cerebrospinal fluid. Maximum ventilatory depression from an intravenous dose of fentanyl occurs within 5 minutes. Its action is briefer largely due to more rapid detoxification in the liver. Unlike morphine, plasma concentrations of fentanyl do correlate with the magnitude of ventilatory depression.

Even though fentanyl is a relatively short-acting drug,

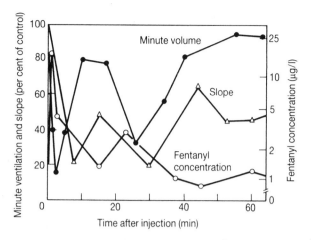

Fig. 13.12 Minute volume, slope (S) of $\dot{V}ERCO_2$ and blood concentrations of fentanyl after administration of 500 μg i.m. to one volunteer. Note the parallel recurrence of reduced $\dot{V}E$ and S of $\dot{V}ERCO_2$ in the time between 20 and 40 minutes after injection, and the opposite increase in fentanyl concentration just preceding it. This is attributed to enteric recycling of the fentanyl. In other subjects the recycling was demonstrated somewhat later.

there are reports of delayed ventilatory depression following its use. Fentanyl, in small doses, may cause significant depression of the $\dot{V}ERCO_2$ lasting 4 hours. Secondary increases in plasma fentanyl concentrations are reported to have occurred in four of seven patients 30–90 minutes post-injection.[25] A second ventilatory depression correlated with the secondary increases in plasma fentanyl concentration (Fig. 13.12) and two patients developed ventilatory insufficiency requiring treatment. The mechanism underlying the secondary increase in plasma fentanyl concentration is possibly enterosystemic recirculation. Fentanyl is transferred from plasma across the stomach lining into gastric juice. The drug then travels to the small bowel where it may be reabsorbed, causing the secondary increase in plasma fentanyl concentration. In man, 16 per cent of an intravenous fentanyl dose is recovered from gastric juice within 10 minutes of injection. Other possible causes of secondary increases in plasma fentanyl concentration include sequestration of the drug in lung or muscle with subsequent release. Whatever the cause, one must remember that even though fentanyl is a relatively short-acting drug, delayed ventilatory depression is possible with its use.

Sufentanil and alfentanil

Sufentanil and alfentanil both cause ventilatory depression similar to equianalgesic doses of fentanyl. In common with fentanyl, sufentanil and alfentanil are highly lipophilic drugs, being approximately 1270 and 104 times more lipophilic than morphine, respectively. Ventilatory depression following an intravenous dose is quick in onset. Pharmacokinetic data predict, and studies have demonstrated, that alfentanil produces depression shorter in duration than equianalgesic doses of fentanyl which is

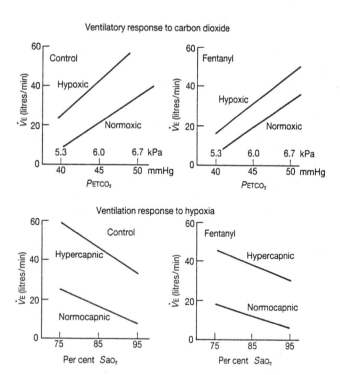

Fig. 13.11 Typical ventilatory response to carbon dioxide and hypoxia before (left) and after (right) premedication with an opioid. In this example, 100 μg of fentanyl is shown, but similar results are found with 10 mg morphine or 80 μg pethidine. Fentanyl flattens the normal hypoxic response, whether expressed as slope of the $\dot{V}ERCO_2$ during hypoxia or the slope of the $\dot{V}ER\downarrow O_2$ itself. Fentanyl has relatively little effect on the $\dot{V}ERCO_2$ except to shift it a few mmHg to the right.

attributable to more rapid metabolism as well as redistribution. Sufentanil is intermediate. Plasma concentrations do correlate with ventilatory depression.

Several recent case reports have described severe delayed postoperative ventilatory depression, responsive to naloxone, occurring following use of an alfentanil continuous intravenous infusion. The delayed ventilatory depression associated with an alfentanil continuous infusion occurred as late as 70 minutes following termination of the infusion. Secondary increases in plasma alfentanil concentrations do occur, but have not yet been linked to episodes of delayed ventilatory depression. It thus appears that alfentanil, much like fentanyl, possesses the ability to cause delayed ventilatory depression, especially when given in a continuous intravenous infusion.

Pethidine

Pethidine (meperidine) causes ventilatory depression similar to morphine in equianalgesic doses. Pethidine is approximately 28 times more lipophilic than morphine, thus its ventilatory effects will have a somewhat quicker onset. Maximum ventilatory depression from an intramuscular injection of pethidine varies widely, from 15 to 110 minutes. Sequestration of pethidine in gastric juice has been reported, raising the possibility of secondary increases in the plasma concentration of the drug.

Methadone

Methadone, a potent μ-receptor agonist with little κ or σ effect, is administered to opioid addicts to prevent withdrawal symptoms. The drug has a very long elimination half-life (15–25 hours), thus maintenance requires only one dose per day. A single dose of methadone, 30 mg i.v., can cause measurable ventilatory depression for up to 8 days. Patients on methadone provide a unique model for examining whether or not tolerance develops to the ventilatory depressant effects of μ-receptor agonists. Patients administered methadone exhibit initially an increase in resting $P\text{a}CO_2$ and a decrease in hypercapnic responses following their daily dose.[26] However, when methadone treatment persists past 5 months, the resting arterial carbon dioxide tension is gradually normalized and the $\dot{V}ERCO_2$ is unchanged following their daily dose. The $\dot{V}ER\downarrow O_2$ also improves, yet still exhibits depression following the daily dose. Thus, the $\dot{V}ERCO_2$ developed full tolerance while the $\dot{V}ER\downarrow O_2$ developed only partial tolerance.

Opioid partial agonist-antagonists

First n-allyl norcodeine and then n-allyl normorphine (nalorphine) were found to stimulate ventilation in patients deeply narcotized with morphine. Puzzlingly, given alone in small doses they depressed breathing much like morphine. This was explained by Bellville and Fleischli in 1968:[27] both the opioid and the n-allyl-nor-compounds act by binding to an opioid receptor. However, the latter is not fully effective. As receptor occupancy increases, the effect reaches a ceiling. Further receptor occupancy by these partial agonists produces no more effect. If some receptors are occupied by effective agonists, the partial agonist may bind to others, increasing depression. However, if many of the receptors are already occupied by a fully efficacious agonist (e.g. morphine) administration of a partial agonist may replace (competitively) some of the more depressing morphine with less depressing drug. Thus these partial agonists may increase depression (given in the presence of small doses of a more efficacious depressant) or reverse depression (given in the presence of an effect above the ceiling).

Not all partial agonists are n-allyl-compounds. Codeine is a typical partial agonist, as is probably pethidine. Not all partial agonists have the same ceiling. Pethidine reaches its ceiling at near-anaesthetic doses, failing to relieve severe pain. Codeine reaches its ceiling but with a lower effect, relieving only mild to moderate pain, at a cost of moderate elevation of $P\text{a}CO_2$. Pentazocine has a similar intermediate ceiling. Levallorphan is like nalorphine. Some compounds may be thought of as having a zero ceiling: they do not produce either analgesia or respiratory depression but do bind to opioid receptors. These compounds may be properly called pure opioid antagonists. Naloxone and naltrexone are examples.

The ability to produce analgesia is not sufficient criteria for clinical use of full or partial agonists. Severity of its side effects was responsible for nalcodeine not being used clinically, for nalorphine being abandoned, and for limited popularity of pentazocine. These side effects are probably attributable to σ- and possibly κ-opioid receptor effects, and non-specific actions as well.

Nalorphine

Nalorphine was the first clinically useful mixed opioid agonist-antagonist. It is a μ-receptor and κ-receptor partial agonist, and σ-receptor full agonist. Nalorphine's ceiling effect on ventilatory depression (a maximal rise of about 0.7 kPa [5 mmHg] $P\text{a}CO_2$) was discovered in 1966 and provided the stimulus for further research into opioid receptor mechanisms. Today, it is not utilized clinically because of its psychotomimetic effects, short half-life, and the availability of more efficacious antagonists.

Pentazocine

Pentazocine was the first partial agonist to be widely used clinically. It is a μ-receptor partial agonist, and κ- and σ-receptor agonist. Pentazocine is one-half to one-fourth as potent as morphine in producing analgesia. Characteristic ventilatory depression occurs at low doses and a ceiling effect about 1.33–2 kPa (10–15 mmHg) $PaCO_2$. Pentazocine is difficult to study at higher doses because of its psychotomimetic effects.

Butorphanol

Butorphanol is a weak μ-receptor partial agonist, and κ- and σ-receptor agonist. It is five to eight times as potent as morphine in producing analgesia. Butorphanol, 2 mg i.m., is equivalent to morphine, 10 mg i.m., in analgesic and ventilatory depressant effects. A ceiling effect on ventilatory depression exists at doses of 0.03–0.06 mg/kg. Increasing the butorphanol dose above this level will not increase $PaCO_2$ above about 6.6 kPa (50 mmHg), but will increase the duration of depression. Butorphanol may be useful in decreasing marked fentanyl-induced ventilatory depression without reversing analgesia.

Buprenorphine

Buprenorphine is a μ-receptor partial agonist, as well as partial agonist for κ- and σ-receptors with very high μ affinity, but the κ- and σ-receptor agonist activity is relatively insignificant. Despite buprenorphine's very high μ-receptor affinity, receptor association and dissociation is very slow. Peak effects, therefore, may not occur until 3 hours after a dose is administered and effects may persist for up to 10 hours. Buprenorphine is 30 times as potent as morphine in producing analgesia. A ceiling effect on ventilatory depression, with $PaCO_2$ in the 7 kPa (mid-50 mmHg) range, exists at doses of 0.15–1.2 mg, intravenously in adults. Buprenorphine, like butorphanol, may be useful in reversing fentanyl-induced ventilatory depression without reversing analgesia.

Nalbuphine

Nalbuphine is a μ-receptor antagonist, κ-receptor partial agonist, and σ-receptor full agonist. In lower doses it is equipotent with morphine in producing analgesia but a ceiling effect is found at about 6.65 kPa (50 mmHg) $PaCO_2$. Maximum analgesia is attained with nalbuphine, 20–30 mg intravenously in adults, but is not equivalent to the same amount of morphine[28] (Fig. 13.13). Maximum ventilatory depression induced by nalbuphine is equivalent to that induced by morphine, 15 mg.

Intrathecal and epidural opioids

Administering opioids directly to the spinal cord should in theory give excellent analgesia while limiting side effects

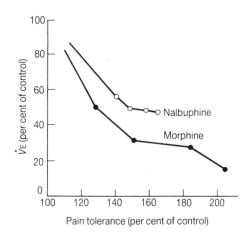

Fig. 13.13 Comparison of increasing doses of nalbuphine and morphine on their ventilatory and analgesic effects. Note that progressive increments in morphine produce increases in both depression and analgesia while nalbuphine, nearing its ceiling effect, gives little increment at increasing dose.

(no respiratory depressant receptors are found in the cord). The use of intrathecal and epidural opioids in man, first reported in 1979, has enjoyed widespread popularity, because of excellent analgesia without motor, sensory, or autonomic deficits typically induced by intrathecal and epidural local anaesthetics. The major disadvantage of the technique, however, is the onset of respiratory depression, occasionally severe to the point of apnoea, with early or delayed onset.

Respiratory depression

Early depression, 30–120 minutes following intrathecal or epidural administration, is likely to be due to vascular uptake of the drug. It is typically associated with epidural use (larger doses utilized, and rapid vascular uptake from the epidural space), but has been reported following at least one intrathecal use. Although vascular uptake is important, it does not seem to account for all the episodes of early respiratory depression following administration of intrathecal or extradural opioids.

Delayed respiratory depression, occurring 4–18 hours following administration of intrathecal or epidural opioids, is thought to be due to cephalad spread of the drug in the cerebrospinal fluid. It can occur with intrathecal or epidural use of opioids. Dural penetration of epidurally administered opioids is related to the physicochemical properties of the drug. Highly lipophilic agents (fentanyl, sufentanil) traverse the dura rapidly when compared to less lipophilic agents (morphine). Once the opioid is present in cerebrospinal fluid, whether by intrathecal or epidural administration, it can travel cephalad to the brainstem, producing respiratory depression. Highly lipophilic opioids, however, dissolve readily in neural tissue, limiting the amount of drug available for cephalad spread. Fentanyl and sufentanil, therefore, are not associated with a high incidence of delayed respiratory depression. Conversely, morphine, being less lipophilic, and more hydrophilic, is allowed to travel cephalad to a greater extent and is associated with a higher incidence of delayed respiratory depression. Most case reports describing delayed respiratory depression following intrathecal or epidural use of opioids involve morphine.

Intrathecal morphine

Numerous reports describe delayed ventilatory depression following use of intrathecal morphine. Most are due to a relative overdose (3–20 mg) either by design, drug error, or intrathecal migration of an epidural catheter. Ventilatory depression typically occurs 4–12 hours post-injection and is responsive to naloxone. Lower doses of morphine, 0.1–1.0 mg, may be given to decrease ventilatory depression yet still obtain excellent analgesia. The most important factor influencing development of delayed ventilatory depression following intrathecal morphine is the dose of drug administered. The $\dot{V}ERCO_2$ is unchanged up to 24 hours following 0.1 mg or 0.25 mg intrathecal morphine.[29] Excellent analgesia may be obtained with both doses. Larger doses are clearly associated with a higher incidence of delayed ventilatory depression. However, significant interindividual variability exists; doses of morphine as low as 0.3 mg injected intrathecally have initiated delayed ventilatory depression.

Epidural morphine

Early or delayed ventilatory depression may result from epidural morphine. Following lumbar epidural morphine, 10 mg, peak serum concentrations of the drug from vascular uptake occur at 8 minutes post-injection, and peak cervical cerebrospinal fluid concentrations of the drug from cephalad spread occur at 120 minutes post-injection.[30] Therefore, initial vascular uptake of the drug may cause early ventilatory depression and late cephalad spread of the drug in cerebrospinal fluid may cause delayed ventilatory depression.

Numerous studies have explored the incidence of ventilatory depression following epidural morphine. Some document significant ventilatory depression while others do not. However, sensitive tests (ventilatory response to carbon dioxide) indicate that epidural morphine, 5 mg, may induce ventilatory depression lasting 24 hours.[31]

Recent large studies suggest that the incidence of clinical ventilatory depression following epidural morphine is less than 1 per cent, and is extremely rare in healthy patients when use of additional parenteral opioids is avoided. These studies also stress the insensitivity of respiratory rate used alone to diagnose ventilatory depression. A global assessment focusing on level of consciousness is more useful. Although the incidence of ventilatory depression following epidural morphine is low, it is not zero, and may occur in the absence of known or suspected risk factors. Life-threatening, delayed ventilatory depression following extradural morphine has been reported with doses as low as 2.5 mg and up to 22 hours post-injection.

Intrathecal fentanyl

Highly lipophilic fentanyl has an advantage over morphine because less drug is allowed to travel cephalad in the cerebrospinal fluid to cause delayed ventilatory depression. Clinical experience appears to validate a lesser incidence of delayed depression when fentanyl is compared with morphine, and perhaps a shorter 'delay'. Prudence dictates that patients receiving intrathecal fentanyl be as closely monitored as with morphine to detect early as well as delayed ventilatory depression.

Epidural fentanyl

Clinically significant ventilatory depression following epidural fentanyl is very rare. Following lumbar epidural fentanyl, 1 μg/kg, the serum concentration of the drug from vascular uptake is negligible, but peak lumbar cerebrospinal fluid concentration of the drug occurs at 5–30 minutes post-injection.[32] Cervical cerebrospinal fluid concentration of fentanyl, resulting from cephalad spread, peaks at 10–45 minutes post-injection, but peak cervical concentrations are only 10 per cent of peak lumbar concentrations.[32] This indicates that ventilatory depression resulting from cephalad spread of the drug in cerebrospinal fluid, though possible, is unlikely following epidural fentanyl.

Other centrally administered opioids

Sufentanil

Several reports describe uncommon but life-threatening ventilatory depression following epidural sufentanil. The ventilatory depression occurs approximately 30 minutes post-injection and has been reported with doses of sufentanil as low as 20 μg. Very little information exists about the ventilatory effects of intrathecal sufentanil.

Pethidine

Pethidine possesses local anaesthetic properties and has been used as an intrathecal anaesthetic agent. Intrathecal administration of pethidine, 100 mg, does not result in significant delayed ventilatory depression. Pain relief may be inferior to that after other opioids.

Several reports describe early ventilatory depression following epidural pethidine. The ventilatory depression occurs approximately 30 minutes post-injection and has been reported with doses as low as 50 mg. There are no reports of delayed ventilatory depression following epidural pethidine, and no studies examining hypoxic or hypercapnic responsiveness.

Other opioids

Numerous other opioids have been given intrathecally and epidurally including alfentanil, methadone, pentazocine, butorphanol, buprenorphine, and nalbuphine. Insufficient data exist at the present time to draw conclusions regarding their ventilatory depressant effects. One report describes two patients who developed early ventilatory depression following lumbar epidural buprenorphine, 0.3 mg, resistant to naloxone. The resistance to naloxone may be secondary to buprenorphine's high affinity for opioid receptors.

Summary and recommendations

The incidence of clinically significant ventilatory depression from the central use of opioids is probably much less than 1 per cent. Certain techniques do increase the risk: the use of hydrophilic opioids (morphine), large doses, repeated doses, intrathecal as opposed to epidural administration, concomitant use of intravenous opioids or other central nervous system depressants, and the use of a thoracic catheter as opposed to a lumbar catheter. Certain patients are also at increased risk of developing ventilatory depression: the elderly, the debilitated, the opioid 'virgin', those with coexisting pulmonary disease, and possibly those with increased intrathoracic pressure from coughing, vomiting, or mechanical ventilation. However, even in the absence of any identifiable risk factors, some patients will develop life-threatening ventilatory depression following intrathecal or extradural administration of opioids.[33]

Sedatives/hypnotics

Many drugs are administered to patients with the goals of decreasing anxiety (sedation) and the level of consciousness (hypnosis). All possess the ability to initiate significant, life-threatening respiratory depression in patients concomi-

tantly receiving other drugs known to induce respiratory depression. The barbiturates were almost the only drugs available to provide sedation (calming) before 1946, and they were not very effective in subhypnotic (sleep-inducing) doses. When first meprobamate, then chlorpromazine, and finally a flood of efficacious drugs followed, their manufacturers called them anxiolytics, ataractics, antipsychotics, antidepressants, even analeptics to distance them from (poorly) sedative barbiturates.[34] Anaesthetists now used barbiturates largely for induction of anaesthesia, a hypnotic use. Several benzodiazepines are used for premedication, a few for deliberate loss of awareness or with the hope of producing amnesia. Several other agents with a variety of properties and uses are also available. All, without exception, exert major depressive effects on ventilatory control.

Barbiturates

The thiobarbiturates thiopentone (thiopental) and thiamylal along with the oxybarbiturate methohexitone (methohexital) are ultrashort-acting drugs used for induction of anaesthesia. The barbiturates may be considered together when discussing their respiratory effects, for they all exert clinical effects via the same specific receptor mechanism.

Mechanism of action

Barbiturates produce clinical effects though a specific receptor protein in the brain, part of the γ-aminobutyric acid (GABA) modulated chloride ion channels that exert CNS inhibitory effects (hyperpolarizing synapses). The receptor is adjacent to the receptor for GABA, as is the receptor protein for benzodiazepines noted below.

Respiratory effects (Fig. 13.14)

Barbiturates produce respiratory effects in common with all hypnotics in that respiratory depression tends to parallel the decrease in the level of consciousness. Small sedative doses of barbiturate have little clinical effect on the $\dot{V}ER_{CO_2}$. Larger doses that decrease the level of consciousness decrease alveolar ventilation, increase Pa_{CO_2} and decrease the slope of the $\dot{V}ER_{CO_2}$. Large doses of barbiturates may also cause apnoea.

The $\dot{V}ER\downarrow_{O_2}$ is significantly decreased by the barbiturates, more than the $\dot{V}ER_{CO_2}$. In the small doses used for premedication, the $\dot{V}ER\downarrow_{O_2}$ is significantly decreased for up to 90 minutes.

All ultrashort-acting barbiturates are highly lipid soluble, and thus exert their respiratory effects rapidly. Thiopentone, 4.0 mg/kg i.v., decreases the $\dot{V}ER_{CO_2}$ maximally within one arm vein to brain circulation time.[35] Redistribution,

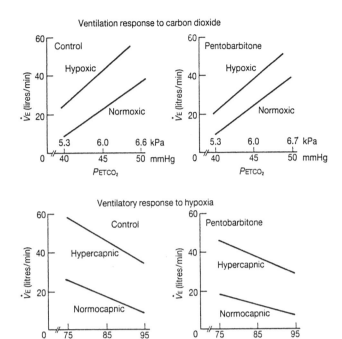

Fig. 13.14 Typical ventilatory responses to carbon dioxide and hypoxia before (left) and after (right) premedication with a barbiturate. In this example, 100 mg pentobarbitone is shown but similar results are seen with quinalbarbitone, amylobarbitone, etc. The ventilatory response to normoxic hypercapnia is not significantly altered, but the hypoxic response is clearly blunted.

however, terminates the action of a single dose of thiopentone quickly, and the respiratory response is almost restored 6 minutes following the dose. Methohexitone, 1.5 mg/kg i.v., acts similarly. Apnoea is a common clinical occurrence following an induction dose of a barbiturate, but rarely lasts longer than 30 seconds. The β half-life (elimination) of the barbiturates is not short (thiopentone 6–18 hours, methohexitone 2–6 hours). If large doses or multiple doses are administered, prolonged respiratory effects may occur.

Patients with chronic obstructive pulmonary disease appear to be more sensitive to the respiratory depressant effects of the barbiturates. In common with all sedatives, concomitant administration of other drugs known to induce hypnosis or respiratory depression may initiate life-threatening respiratory depression.

Benzodiazepines

The benzodiazepines are a class of drugs used widely for their sedative (amnesic, anticonvulsant, anxiolytic) and hypnotic (sleep-inducing) properties. Respiratory depres-

sion, however, poses a constant threat with their use, especially for the elderly.

Mechanisms of action

Benzodiazepines exert effects by occupying a specific benzodiazepine receptor that modulates the chloride activated by GABA, a major inhibitory mechanism in the brain. It is still unknown, however, how the spectrum of benzodiazepine's clinical effects (amnesia, anticonvulsant activity, anxiolysis, and unresponsiveness) are mediated. It could be that they are simply a result of fractional receptor occupancy: benzodiazepine receptor occupancy below 20 per cent may be sufficient to produce anxiolysis, a 30–50 per cent occupancy for hypnosis, and more than 60 per cent occupancy unconsciousness. Benzodiazepine receptors are found in areas of the brain thought to play significant roles in respiratory control (medulla, pons, etc.). However, other areas of brain, such as the cerebral cortex, possess far greater numbers of benzodiazepine receptors. Benzodiazepines are likely to exert respiratory depressant effects simply by decreasing the level of consciousness and not via specific benzodiazepine receptor interaction in areas of the brain thought to be involved in control of respiration. However, some reports suggest that γ-aminobutyric acid may be intimately involved in respiratory control.[36]

Respiratory effects (Fig. 13.15)

Since the benzodiazepines' clinical effects are mediated via a specific receptor mechanism, the many drugs possessing benzodiazepine receptor agonist properties induce qualitatively similar changes. They do not differ in the magnitude of the depression induced when administered in clinically equivalent amounts. With a dose-dependent depression of alveolar ventilation there is a corresponding increase in $PaCO_2$. Their effects on respiratory rate and tidal volume are variable.

Benzodiazepines, in sedative doses, have little effect on the slope of the ventilatory response to carbon dioxide in normal man. The incidence of respiratory depression induced by benzodiazepines is increased in the elderly, patients with chronic obstructive pulmonary disease, patients with some severe hepatic or renal disorders, and in patients concomitantly receiving other drugs known to induce respiratory depression paralleling the decrease in level of consciousness. Larger (hypnotic) doses decrease the slope progressively. Benzodiazepines also decrease the ventilatory response to hypoxia.

Diazepam

Diazepam has been studied often, with conflicting results. A sleep dose of diazepam i.v., in elderly males, produced minimal changes ($PaCO_2$ 5.3–5.6 kPa, 40–42 mmHg) in one

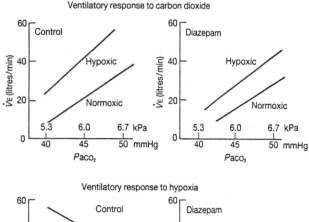

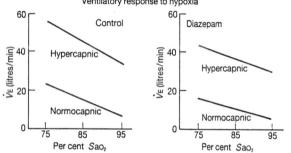

Fig. 13.15 Typical ventilatory responses to carbon dioxide and hypoxia before (left) and after (right) premedication with a benzodiazepine. In this example, 10 mg diazepam were given by mouth 90 minutes earlier, but similar results are seen with 2 mg lorazepam and other benzodiazepines. There is little effect on the normoxic hypercapnic response, minor blunting of the normocapnic hypoxic response and significant blunting of the hypoxic hypercapnic response.

study[37] but a smaller dose raised $PaCO_2$ 0.9 kPa (7 mmHg) in another.[38] Life-threatening apnoea and Cheyne–Stokes breathing have results from as little as 2.5 mg i.v. The majority of investigations document a decrease in the slope of the ventilatory response to carbon dioxide curve, without displacement to the right, which parallels the decrease in level of consciousness. A minority of patients, however, will demonstrate an increase in the slope of the ventilatory response to carbon dioxide.[39] Failure to account for this small subgroup of patients who demonstrate respiratory stimulation when given diazepam may explain the many conflicting investigations. Other benzodiazepines may exhibit variability in respiratory effects but these have not been studied.

Diazepam, in small doses that do not affect the ventilatory response to carbon dioxide, significantly decreases the ventilatory response to hypoxia.

Respiratory effects begin within 1 minute after i.v. injection, peak at 3 minutes, and are minimal at 30 minutes. The quick onset and short duration of respiratory effects following a *single* dose of diazepam are due to the drug's high lipid solubility, and rapid

redistribution. The β half-life is long, 20–90 hours, and there is an active metabolite, desmethyldiazepam. Therefore larger doses or multiple doses may cause significantly prolonged respiratory effects.

Lorazepam

Lorazepam exhibits respiratory effects that are similar to diazepam when administered in clinically equivalent amounts. It is approximately five times as potent, in its clinical effects, as diazepam. Lorazepam causes a decrease in the slope of the $\dot{V}ERCO_2$, which parallels the decrease in level of consciousness.

Lorazepam appears to be a relatively safe drug, with few case reports describing life-threatening respiratory depression following its use. In normal patients, doses as high as 7.5 mg orally, 6.0 mg intramuscularly, and 5.0 mg intravenously, have produced minimal clinical respiratory effects. With high doses of lorazepam, however, it is difficult for the patient to remain awake and prolonged sedation and distressing amnesia are likely.

Lorazepam is not as lipophilic as diazepam and therefore penetrates the brain only slowly. It is better absorbed after i.m. injection. Respiratory effects are generally not prolonged, but lorazepam's β half-life of 11–22 hours suggests that if large doses or multiple doses are administered, prolonged respiratory effects may occur.

Midazolam

Midazolam, the first clinically useful water-soluble benzodiazepine, exhibits respiratory effects that are similar to diazepam when administered in clinically equivalent amounts. It is approximately three to four times as potent in its clinical effects as diazepam. Midazolam exhibits pH-dependent solubility. It is water soluble in a buffered acidic medium (pH 3.5), yet upon entering the bloodstream (pH 7.4) it becomes more lipid soluble than diazepam. Respiratory effects of midazolam, 0.2 mg/kg i.v., begin within 1 minute, peak at 3 minutes, and are minimal at 15 minutes.[40] Its β half-life of 1.7–2.6 hours makes it the most attractive benzodiazepine for continuous intravenous infusion.

Midazolam causes a decrease in the slope of the $\dot{V}ERCO_2$, which parallels the decrease in level of consciousness. The decrease begins at intravenous doses as low as 0.05–0.075 mg/kg, yet clinical respiratory depression is usually minimal following intravenous doses up to 0.1 mg/kg in normal patients. Midazolam, in small doses that do not effect the $\dot{V}ERCO_2$, significantly decreases the hypoxic response.

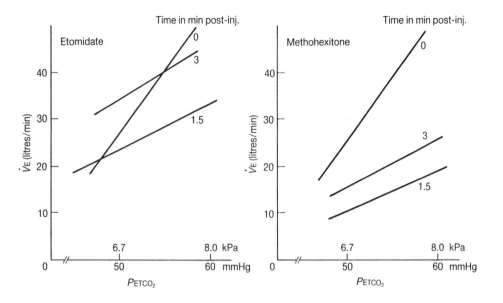

Fig. 13.16 The ventilatory response to carbon dioxide just before, 1.5 minutes and 3 minutes after induction of anaesthesia with etomidate and methohexitone. The two different agents have almost the same effect on the slope of the $\dot{V}E R CO_2$, but the resting (unstimulated) ventilation is consistently higher after etomidate than methohexitone. To some extent, ketamine resembles etomidate, and both thiopentone and midazolam resemble methohexitone.

Other agents usually used intravenously

Etomidate

Etomidate, an imidazole-derived hypnotic agent, is an alternative for induction of anaesthesia. Its mechanism of action may involve modulation of GABA activity. Etomidate produces respiratory effects that are similar to all hypnotics in that respiratory depression tends to parallel the decrease in the level of consciousness.

Etomidate may cause less respiratory depression than other commonly utilized induction agents, and may even be associated with brief periods of hyperventilation. Choi *et al.* compared respiratory effects of induction doses of etomidate, 0.3 mg/kg, with methohexitone, 1.5 mg/kg, both given intravenously.[7] Both drugs decreased the slope of the $\dot{V}E R CO_2$ with maximum depression occurring 30–120 seconds post-injection. However, the two agents differed in their effects on displacement of the curve. Methohexitone caused a downward shift of the carbon dioxide ventilatory response curve compared with etomidate (Fig. 13.16). Induction of anaesthesia with etomidate induced a significant increase in minute ventilation that was maximal 3.5 minutes post-injection. Methohexitone induced a decrease in minute ventilation that was maximal within 60 seconds post-injection. The increase in minute ventilation associated with etomidate was achieved by an increase in the respiratory rate: tidal volume remained unchanged.

The demonstration by Choi *et al.*[7] that both etomidate and methohexitone decrease the slope of the ventilatory response to carbon dioxide curve suggests a similar effect at the medullary centre, which modifies ventilatory drive in response to changing carbon dioxide tension. However, the differing effects on the displacement of the curve indicates

that etomidate alone may directly stimulate ventilation independent of the carbon dioxide tension. Such an upward shift of the ventilatory response to carbon dioxide is uncommon although such a curve may be seen with ketamine[6] and in some REM sleep states. The increase in minute ventilation associated with etomidate makes it useful during induction of anaesthesia, when maintenance of spontaneous ventilation is desired. Apnoea may be associated with the use of etomidate, and the incidence is influenced significantly by concomitant use of other drugs known to induce respiratory depression. Oddly, premedication with an opioid has been shown to decrease the incidence of apnoea associated with etomidate.

The respiratory effects of a single intravenous induction dose of etomidate, 0.2–0.3 mg/kg, are transient and terminated by redistribution. The β half-life of etomidate is 2.9–5.3 hours; therefore, prolonged respiratory effects may occur following large doses or multiple doses of the drug. The hypoxic respiratory drive has not been studied under etomidate.

Propofol

Propofol, an alkylphenol, has achieved widespread popularity for induction and maintenance of anaesthesia, as well as for sedation. The mechanism of its respiratory effects are believed to result from a decrease in the level of consciousness. Induction doses of propofol, 1.0–2.5 mg/kg i.v., decrease tidal volume and mildly increase respiratory rate. Changes in alveolar ventilation are variable. The hypercapnic response decreases maximally within 90 seconds: depression is still significant 20 minutes later,[33] despite recovery of consciousness.

The incidence of prolonged apnoea associated with propofol is greater than after other intravenous agents. Induction doses may initiate apnoeic episodes lasting more than 3 minutes.

Recent evidence from animal studies suggest that propofol has powerful depressant effects on peripheral chemoreceptors, and thus, the ventilatory response to hypoxia.

Ketamine

Ketamine, a phencyclidine derivative, has unique respiratory effects that clearly separate it from other anaesthetics, intravenous or inhalational. Selective depression of certain areas of the cerebral cortex and thalamus along with stimulation of the limbic system probably play key roles. Ketamine produces dose-related unconsciousness and profound analgesia, but also possesses many undesirable side effects, most importantly emergence delirium. Characteristic respiratory effects of ketamine include an increased tidal volume with a variable effect on respiratory rate. Most investigations report minute ventilation unchanged while others document increased minute ventilation for up to 20 minutes following a single dose.

A consistent respiratory effect of ketamine is a significantly increased inspiratory time/expiratory time ratio which leads to either preservation of, or an actual increase in, the functional residual capacity. In contrast with the halogenated inhalation anaesthetics, ketamine may preserve intercostal muscle function, which also has a beneficial effect on maintaining functional residual capacity. Ketamine appears to be unique in that it is the only anaesthetic, intravenous or inhalational, shown to preserve or actually increase the functional residual capacity.

Ketamine, 1.1–3.0 mg/kg given slowly i.v., causes a dose-dependent displacement to the right of the $\dot{V}ERCO_2$, yet even in doses large enough to produce unconsciousness reliably (> 1.8 mg/kg) the slope of the curve is unchanged.[11] However, when administered to children as a rapid intravenous bolus, 2 mg/kg, ketamine did decrease the slope of the ventilatory response to carbon dioxide curve, although it returned to normal during a continuous infusion maintaining unconsciousness (40 µg/kg per minute).[41] Ketamine's potential ability to maintain a normal slope of the $\dot{V}ERCO_2$ despite producing unconsciousness is also unique among anaesthetics, intravenous or inhalational.

The incidence of apnoea associated with ketamine is high, 60 per cent in some studies, and may be prolonged. A child given 10 mg/kg i.m., followed by 1.6 mg/kg i.v., became apnoeic for 2 hours post-injection. Paediatric patients with increased intracranial pressure appear to be at significant risk for developing apnoea following ketamine.

Ketamine, with only limited respiratory depressant properties, beneficial respiratory effects (maintenance of functional residual capacity and slope of the $\dot{V}ERCO_2$ at doses producing unconsciousness) is useful if one is willing to accept the undesirable side effect.

Droperidol

Droperidol, a butyrophenone, produces marked tranquillity and cataleptic immobility. Its mechanism may involve activity within the central nervous system at sites where dopamine, noradrenaline and serotonin exert their effects. The respiratory effects of droperidol are minor. At 0.3 mg/kg i.v. there are usually no effects on the $\dot{V}ERCO_2$. Doses as high as 0.44 mg/kg intravenously produce no clinical respiratory effects. An occasional patient may exhibit either respiratory depression or respiratory stimulation. The variability is probably explained by the variable incidence and nature of the undesirable side effects of large doses of droperidol (unusual psychological reactions and extrapyramidal reactions).

Droperidol measurably *increases* the ventilatory response to hypoxia. A single intravenous dose of 2.5 mg almost doubles the response, probably related to block of endogenously released dopamine.[42] Dopamine appears to play an integral role in carotid body chemoregulation and has been shown to decrease the $\dot{V}ER\downarrow O_2$. In fact, droperidol can reverse existing depression of the $\dot{V}ER\downarrow O_2$ after an intravenous infusion of dopamine, 3.0 µg/kg per minute. Droperidol, with its ability to increase or maintain the hypoxic responsiveness, may be useful in patients who either depend on their hypoxic drive to breathe (severe pulmonary disease) or have had their hypoxic drive to breathe altered (post-carotid endarterectomy, halogenated inhalation agents, etc.) but with the customary use of elevated oxygen concentration it would be dangerous to rely on this effect to maintain ventilation.

The use of droperidol significantly increased with the introduction of Thalamonal (Innovar), a fixed combination of fentanyl, 50 µg/ml, and droperidol, 2.5 mg/ml. Fentanyl provides analgesia while droperidol provides sedation, hypnosis, and profound antiemetic effects. Droperidol does not augment fentanyl-induced respiratory depression. Respiratory effects of Thalamonal, therefore, are a result of fentanyl alone.

Lignocaine

Lignocaine (lidocaine), the most widely used local anaesthetic, possesses both respiratory depressant as well as respiratory stimulant properties. Its major use as an anaesthetic adjuvant lies in its antitussive action, although, like procaine, it adds to the analgesia of N_2O, which requires near-convulsive doses. A bolus of lignocaine, 1.5 mg/kg i.v., decreases the slope of the $\dot{V}ERCO_2$ within 90 seconds. The effects are transient, however, and last only 2.5 minutes. In contrast, a continuous infusion of lignocaine, 60 µg/kg per minute i.v., increases the slope of the $\dot{V}ERCO_2$. The contrasting effects of lignocaine probably result from rapidly changing central nervous system lignocaine levels, with transient high levels following bolus administration inducing respiratory depression and lower levels, as seen with a continuous infusion, inducing respiratory stimulation. The increase in the slope of the $\dot{V}E$RCO$_2$ does correlate with plasma levels of lignocaine.

Lignocaine has variable effects on the ventilatory response to hypoxia but most subjects experience depression of the ventilatory response to hypoxia with clinically relevant doses.[44] When lignocaine is used for arrhythmia control, antitussive effects, or regional anaesthesia (epidural), plasma concentrations of the drug are commonly in the range that enhance the $\dot{V}E$RCO$_2$ and may depress the ventilatory response to hypoxia.

Other anaesthetic adjuvants

Rarely in modern practice are one or two agents employed for total anaesthesia. A variety of classes of pharmaceuticals offer specific effects that enhance surgical conditions, safety, patient acceptance, etc. Some of these also have respiratory effects. Three classes of drugs will be considered: agents used to reverse effects, agents used to stimulate breathing, and neuromuscular blockers.

Reversal agents

Many agents have been used in attempts to reverse drug-induced respiratory depression. Some reverse respiratory depression via a specific receptor mechanism, such as naloxone and flumazenil. Others, such as physostigmine, attempt to reverse respiratory depression via non-specific central nervous system arousal.

Opioid antagonists

At the turn of the century n-allyl-norcodeine was known to alter the effects of morphine in rabbits, but they became vicious and unpredictable. At mid-century Eckenhoff introduced nalorphine for morphine poisoning. It had far fewer psychomimetic effects. The last quarter of the century has seen several more such drugs approved. The first, naloxone (n-allyl-noroxymorphone) is widely used and has not been improved upon, except for slower kinetics, as with naltrexone and nalmefene.

Naloxone
Naloxone is a competitive antagonist at μ, κ, σ and δ opioid receptors. The drug's ability to antagonize μ-receptors has led to its widespread clinical application in reversing the opioid respiratory depression.[45]

Intravenous naloxone, 3.66 µg/kg loading dose followed

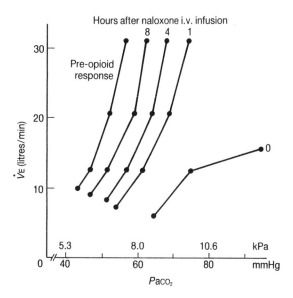

Fig. 13.17 The ventilatory response to carbon dioxide before and after 1 mg/kg morphine, and its reversal with naloxone. Morphine results in loss of spontaneous activity, lowered or absent response to spoken word, and a markedly flattened response to CO$_2$. Naloxone almost immediately restored awareness, and the slope of the $\dot{V}E$RCO$_2$ although at first this was displaced to the right (higher PCO$_2$). With time the $\dot{V}E$RCO$_2$ reverted toward the control level. The improvement with time represents a constant naloxone level supported by infusion and progressive hepatic metabolism of the initial dose of morphine.

by 3.66 µg/kg per hour continuous infusion, promptly reverses morphine-induced respiratory depression.[22] With this regimen, the slope normalizes with return of responsiveness (Fig. 13.17). With further doses the $\dot{V}E$RCO$_2$ is shifted to the left, minute ventilation increases, and resting end tidal carbon dioxide tension decreases. Bladder distension and vomiting, however, were common and unwanted side effects with this regimen.

It may be possible to manipulate the dose of naloxone administered in order to reverse respiratory depression and maintain some analgesia. Naloxone's affinity for each opioid receptor population is unequal. Animal studies suggest that at low concentrations, binding takes place preferentially at high affinity μ-receptors, which mediate respiratory depression. With increasing concentrations, naloxone binding takes place at μ as well as other opioid receptor populations. In humans, low-dose intravenous naloxone infusion (5 µg/kg per hour) appears to prevent epidural morphine-induced respiratory depression without affecting analgesia whereas higher doses (10 µg/kg per hour) appear not only to prevent respiratory depression but also to decrease analgesia.

It is important to understand naloxone's pharmacokinetic properties when the drug is used clinically. It is a highly lipid-soluble drug, with an α half-life of 4 minutes and a β half-life of 1–2 hours. Peak clinical effects following i.v., i.m. and s.c. administration of naloxone are at 3, 15, 20 minutes, respectively. Clinical effects of a single injection last only 30–90 minutes, depending on the dose administered. Naloxone's transient clinical effects following a single injection may be secondary to the drug's rapid egress out of the brain (redistribution and rapid hepatic metabolism) as well as its differential affinity for opioid receptor populations at different drug concentrations. A high interindividual variation in responses also appears to exist. Repeated intravenous injection or a continuous intravenous infusion of naloxone, therefore, is often required to maintain reversal of opioid-induced respiratory depression. Multiple case reports describe the reappearance of opioid-induced respiratory depression following single intravenous injections of naloxone.

Patients developing significant respiratory depression following administration of partial agonists with high receptor affinity may be resistant to reversal by naloxone. A report describes two patients who developed early naloxone-resistant respiratory depression following lumbar extradural buprenorphine, 0.3 mg.

Administration of very large doses of naloxone to normal conscious humans does not appear to affect respiratory control. No discernible physiological changes are observed with daily doses of naloxone, 90 mg s.c., for 2 weeks or upon its abrupt withdrawal. Naloxone does, however, appear to influence respiratory control in patients with chronic obstructive pulmonary disease. Naloxone restores load-compensating responses and clinically improves episodes of acute respiratory failure in certain patients with chronic obstructive pulmonary disease. This suggests that endogenous opioids may play a significant role in respiratory control in these patients.

There are scattered reports of naloxone activity in patients given non-opioid depressants.[45] Similarly it has been used in very large doses in endotoxic and hypovolaemic shock, brain and spinal cord ischaemia, schizophrenia, constipation, pruritus and thalamic pain syndrome.

Significant complications from naloxone reversal of opioid effect are rare. Sudden severe pain with overenthusiastic dosage is treated with more opioid. Rare but more serious is acute pulmonary oedema in young healthy subjects, arrhythmia and even rupture of a cerebral aneurysm.

Benzodiazepine antagonists

Only one is clinically available.

Flumazenil
Flumazenil, an imidazobenzodiazepine derivative, is the first clinically available benzodiazepine receptor antagonist. It demonstrates high affinity and great specificity for benzodiazepine receptors. Flumazenil is not a pure antagonist and does possess very weak agonist proper-

ties. However, large doses of flumazenil, 0.1 mg/kg i.v., cause no discernible clinical effects in patients not receiving benzodiazepine agonists.

Flumazenil reverses benzodiazepine-induced hypnosis, but its ability to reverse benzodiazepine-induced respiratory depression is inconsistent. Flumazenil, in doses as low as 0.3 mg i.v., has been shown by some to completely reverse the benzodiazepine-induced decrease in the slope of the $\dot{V}_E$RCO$_2$. Others have revealed only partial improvement in the slope with 1.0 mg i.v. Flumazenil, 1.0 mg i.v., administered after midazolam, 0.13 mg/kg i.v., rapidly reverses hypnosis and the displacement of $\dot{V}_E$RCO$_2$, yet has no effect on the decreased slope.[46] Some investigators have actually demonstrated an augmentation of the benzodiazepine-induced decrease in the slope of the $\dot{V}_E$RCO$_2$.

Sedation and respiratory depression induced by benzodiazepines may reappear following initial reversal by a single dose of flumazenil. Flumazenil's clearance is higher, and its half-life is shorter, than all the clinically used benzodiazepines. Clinical effects of a single dose of flumazenil, 1.0 mg i.v., peak within 5 minutes of administration and last 45–60 minutes. Repeated intravenous injections or a continuous intravenous infusion of flumazenil, therefore, may be required to maintain reversal of benzodiazepine-induced clinical effects, including respiratory depression.

In summary, flumazenil reliably reverses benzodiazepine-induced hypnosis. Whether or not it reliably reverses all clinical components of benzodiazepine-induced respiratory depression is unclear. Clinically, this means a patient may 'wake up' after receiving flumazenil, yet benzodiazepine-induced respiratory depression may still be present.

Anticholinesterases

Drugs which antagonize cholinesterase and which cross the blood–brain barrier, produce generalized arousal from increased acetylcholine activity.

Physostigmine
Physostigmine, a naturally occurring acetylcholinesterase inhibitor, may reverse the CNS effects of a variety of diverse compounds including benzodiazepines, opioids, butyrophenones, tricyclic antidepressants, anticholinergics and antihistamines. Prompt reversal of diazepam-induced coma and respiratory arrest with physostigmine, 0.4–2.0 mg i.v., has been reported. Animal studies reveal that coma induced by diazepam, 80–120 mg/kg intraperitoneally, is fully reversed within 10 minutes of administration of physostigmine. Other investigations and case reports, however, fail to demonstrate improvement in sedation or respiratory depression following this drug. Some investigators find reliably reversed benzodiazepine-induced sedation but with

no effect on benzodiazepine-induced depression of hypercapnic or hypoxic responses. Clinically, this means a patient may 'wake up' after receiving physostigmine but with persistent depression of the ventilatory response. Physostigmine reversal of opioid-induced respiratory depression with maintenance of analgesia has also been reported within 5–10 minutes, but other investigations and case reports have failed to demonstrate improvement.

The highly unpredictable reversal of sedation and respiratory effects relate to its non-specific mechanism of action. Physostigmine rapidly crosses the blood–brain barrier and inhibits acetylcholinesterase, increasing acetylcholine levels, thus causing generalized central nervous system arousal. The clinical effects of a single dose of physostigmine, 1–2 mg i.v., peak within 10 minutes of administration and last approximately 60 minutes. Recurrence of sedation and respiratory depression have been reported following a single dose of physostigmine.

With the development of more specific receptor antagonists such as naloxone and flumazenil, clinical use of physostigmine in reversal of drug-induced respiratory depression is decreasing. However, when significant respiratory depression occurs following administration of a drug with no clinically useful specific receptor antagonist, physostigmine remains an option in treatment.

Respiratory stimulants

Respiratory stimulant properties have been demonstrated for many drugs.[47,48] Only doxapram, progesterone, and the catecholamines retain clinical applicability.

Analeptics

The introduction of doxapram in the early 1960s resulted from a continuing search for analeptic drugs to treat barbiturate overdose. Those in use prior to doxapram (picrotoxin, pentaethylenetetrazole, nikethamide) have very low margins of safety. Doses stimulating respiration were very close to doses inducing convulsions. However, doxapram, an acknowledged central analeptic, produces selective respiratory stimulation of carotid chemoreceptors with a convulsive dose/respiratory stimulant dose ratio of 20 or more.

Doxapram, 1 mg/kg i.v., increases respiratory rate, tidal volume, and minute ventilation. Doxapram, 2 mg/kg i.m., induces displacement to the left of $\dot{V}E R CO_2$. Some investigators have utilized doxapram ('pharmacological sighs') postoperatively to improve respiratory function, blood gas tensions and pH. Doxapram may also reverse respiratory depression induced by pethidine, 1 mg/kg i.m., without altering analgesia.

Doxapram's mechanism of action makes it attractive for use in patients with chronic obstructive pulmonary disease. Patients with carbon dioxide retention often depend on their hypoxic drive to maintain ventilation. If supplemental oxygen is administered to these patients a decrease in ventilation may occur, along with increasing hypercapnia and acidosis. Doxapram, by stimulating carotid chemoreceptors, may prevent this decrease in ventilation. In fact, Moser et al. showed that in patients with chronic obstructive pulmonary disease and carbon dioxide retention, pretreatment with doxapram significantly decreased the incidence of subsequent hypercapnia or acidosis after supplemental oxygen.[48] However, side effects have included tachycardia, hypertension, restlessness, agitation, frank psychosis, and two unexplained deaths.

The use of doxapram to stimulate ventilation in the immediate postoperative period following general anaesthesia with the halogenated inhalation agents may be unproductive. Doxapram's stimulant properties are profoundly depressed by very low levels of halogenated inhalation agents, much in the same way as the $\dot{V}E R CO_2$.

Respiratory stimulation induced by a single injection of doxapram is transient. The half-life is a few minutes only. A continuous intravenous infusion should be used when administering this drug. Inadvertent discontinuation of an intravenous doxapram infusion may have disastrous results, as respiratory depression can occur rapidly. Continuous intravenous infusion of doxapram has been used for over 48 hours without any evidence of tachyphylaxis or dose-related toxicity.

In summary, doxapram is a drug with very limited application. Its primary clinical use appears to be when brief respiratory stimulation is desired. This time period for respiratory stimulation is beneficial in that other modes of therapy (oxygen, bronchodilators, pulmonary toilet, etc.) can be used that may prevent the need for tracheal intubation and mechanical ventilatory support.

Catecholamines

It has always been recognized that fear, excitement and muscular effort cause obvious increased breathing, accompanied by increased blood concentrations of adrenaline and noradrenaline. Relatively early in the history of respiratory control investigations these hormones were studied, indeed before present methods including $\dot{V}E R CO_2$ and $\dot{V}E R{\downarrow}O_2$ measurements were used or at least before they were standardized. Consequently, although exogenous adrenergics do stimulate breathing, the mechanism is not fully understood.

Noradrenaline has no effect on breathing in the absence of a hypoxic drive. It increases the drive for any given hypoxic stress, by further increasing the slope of the $\dot{V}E R CO_2$, and by increasing the curvature (parameter A) of the $\dot{V}E R{\downarrow}O_2$.[49] In contrast, other evidence variously suggested both a normoxic increase in slope and non-CO_2-mediated increased breathing.

Adrenaline acts as does noradrenaline at the carotid body (an effect which is abolished by hyperoxia or section of the nerve circuit involved in the Hering–Breuer reflex). In addition, its β-adrenergic activity stimulates metabolism, increases metabolic rate for CO_2 and thus increases breathing. Additionally, any pressor drug may be expected to increase carotid body blood flow and indirectly decrease the afferent activity of the chemoreceptors.

Patients with high adrenergic activity and corresponding lower PaO_2 values would seem to be at risk of apnoea when induction of anaesthesia by rapid i.v. medication 'turns off' the arousal, at least until CO_2 rises.

Dopamine significantly decreases the ventilatory response to hypoxia, decreasing a portion of the $\dot{V}ER_{CO_2}$ mediated via the peripheral chemoreceptors. Unlike the halogenated inhalation agents, however, augmentation of $\dot{V}ER{\downarrow}O_2$ by hypercapnia is preserved with dopamine.[50] Evidence strongly suggests that dopamine exerts its respiratory depressant effects at the peripheral chemoreceptors located in the carotid bodies via specific dopaminergic receptor stimulation. Depression of the ventilatory response to hypoxia and hypercapnia occur at dopamine doses that predominantly stimulate dopaminergic receptors (> 5 µg/kg per minute i.v.) and depressant effects are completely reversed by administration of a drug possessing dopaminergic receptor antagonistic properties, such as droperidol. In fact, when droperidol is administered alone, it increases (almost doubles) the ventilatory response to hypoxia. Blockade of α- and β-receptors has no effect on dopamine's respiratory depressant effects.

Endogenous dopamine is present in human carotid bodies and may play a central role in peripheral chemoreceptor neurotransmission. Release of dopamine in the carotid body is associated with reduced carotid sinus nerve activity. The functional role, however, that dopamine plays in respiratory control is far from certain. In fact, when dopamine is administered in doses that stimulate α- and β-receptors as well as dopaminergic receptors (> 10 µg/kg per minute i.v.) an increase in ventilation may be seen.

Catecholamines other than dopamine have also been isolated from carotid bodies, including noradrenaline and possibly adrenaline, both of which may effect respiration through changes in sympathetic nervous system activity. β-Receptor agonists, such as noradrenaline and isoprenaline (isoproterenol), significantly increase resting ventilation while α-receptor agonists, such as phenylephrine, moderately decrease resting ventilation.[51] A β-receptor agonist-induced increase in ventilation is prevented by pretreatment with a β-receptor antagonist and an α-receptor agonist-induced decrease in ventilation is prevented by pretreatment with an α-receptor antagonist. Suppression of peripheral chemoreceptor activity by the administration of supple-

mental oxygen blocks the increase in resting ventilation associated with β-receptor agonists but does not block the decrease in resting ventilation associated with α-receptor agonists.

In summary, dopaminergic receptor stimulation decreases the ventilatory response to hypoxia and carbon dioxide, β-receptor stimulation increases resting ventilation, and α-receptor stimulation decreases resting ventilation. Dopaminergic and β-receptor effects appear to be mediated via peripheral chemoreceptors located in the carotid bodies, whereas α-receptor effects do not.

Hormones other than catecholamines

Progesterone is a potent stimulant increasing the ventilatory response to both hypercapnia and hypoxia. Increased levels in pregnancy and during the luteal phase of the menstrual cycle are accompanied by increases in minute ventilation and decreases in arterial carbon dioxide tension. In normal patients, male or female, exogenous progesterone increases tidal volume and minute ventilation and decreases arterial carbon dioxide tension. Little or no effect on respiratory rate is seen.

Progesterone has been used clinically for its respiratory stimulant properties. In the Pickwickian syndrome (obesity, hypercapnia, hypoxaemia, polycythaemia, and cor pulmonale), sublingual medroxyprogesterone acetate improves arterial carbon dioxide and oxygen tension, and decreases haematocrit and mean pulmonary artery pressures. The therapeutic effects disappear when the drug is discontinued. Medroxyprogesterone acetate has similar beneficial effects when administered to patients chronically hypoxaemic and polycythaemic secondary to living at high altitudes. Progesterone therapy stimulates respiration in patients with sleep apnoea syndromes without an obstructive component. Patients with chronic obstructive pulmonary disease and with carbon dioxide retention exhibit highly variable respiratory changes following progesterone, and reliable improvements in blood gas tensions and pH have yet to be demonstrated.

The mechanism underlying progesterone's ability to stimulate respiration probably involves the CNS. Results from animal studies indicate that peripheral chemoreceptors are not involved. Progesterone readily crosses the blood–brain barrier: cerebrospinal fluid concentrations of the hormone in pregnant patients are approximately 10 per cent of corresponding plasma concentrations. The beneficial respiratory effects of progesterone may also relate to its ability to increase glomerular filtration rate and inhibit aldosterone, thus inducing a diuresis. The long-term effects of progesterone therapy are unknown yet may be significant, especially in males.

Hormones other than progesterone may also influence

respiration by little understood mechanisms.[52] These include other gonadal hormones, thyroid hormones, adrenocortical hormones, and adrenocorticotrophic hormones.

Muscle relaxants

The muscle relaxants have no consistent CNS effects. Their respiratory effects originate solely from their ability to paralyse muscles. Partial paralysis reveals no consistent change in the ventilatory response to carbon dioxide.[53] Some investigations document an increased respiratory drive in patients partially paralysed without sedation, but this increase is likely to be secondary to anxiety.

Characteristic respiratory changes observed in patients with partial paralysis include a decreased tidal volume and an increased respiratory rate. Eventually the increase in respiratory rate is unable to compensate for the decrease in tidal volume. The decrease in alveolar ventilation then allows the arterial carbon dioxide tension to increase. The magnitude of the decrease in alveolar ventilation depends on the degree of respiratory muscle paralysis.

Respiratory muscles exhibit significant differences in sensitivity to muscle relaxants. The pharyngeal muscles of the upper airway and the accessory muscles of respiration appear to be the most sensitive to neuromuscular blockade. Abdominal muscles and expiratory intercostals are more sensitive than are the inspiratory intercostals. The diaphragm is the least sensitive.

In patients with partial paralysis, inspiratory flow rates are lower than expiratory flow rates. This is probably due to the high sensitivity to neuromuscular blockade exhibited by the pharyngeal muscles of the upper airway creating upper airway obstruction. In fact, with increasing degrees of partial paralysis, upper airway resistance will increase and inspiratory flow rates will decrease prior to significant reductions in vital capacity or chest wall muscle strength. Adequate ventilation can be maintained with significant reductions in vital capacity or chest wall muscle strength. Adequate ventilation can be maintained with significant amounts of partial paralysis (hand grip strength 6 per cent of control) if a patent airway is maintained.[53] At these levels of partial paralysis, however, significant impairment of vital respiratory functions such as coughing, swallowing, and deep breathing are severely impaired.

Muscle relaxants may affect respiratory control mechanisms in other ways. As an example, near paralysis (0–5 per cent grip strength but with 40 per cent vital capacity) prolongs breath-holding times 300–400 per cent in volunteers and also decreases sensation of breathlessness. Differential sensitivity of intercostals and diaphragm may affect diaphragm–rib cage interaction. If paralysis is followed by a decrease in functional residual capacity, the lung falls to a lower, stiffer portion on its pressure–volume curve. The same neural output would produce less minute ventilation, a non-CNS ventilatory depression.

Summary

Breathing continues without voluntary effort to match gas exchange with metabolic rate such that breathing is proportional to demands. The study of the control of breathing is the study of the proportionality factor in large part. As long as the level of consciousness remains normal an increase in carbon dioxide tension in the central chemoreceptor environment increases respiratory minute volume by about 22.5 l/min per kPa (3 l/min per mmHg). Drugs may shift the resting point (the pair of values for P_{CO_2} and $\dot{V}_E$ found at rest) to the left and up (stimulation) or to the right and down (depression), without important change in the response to CO_2. Normally the response to decreasing P_{O_2} plays little role in resting breathing, but can become predominant, at altitude, asphyxia, or with some diseases.

When the level of consciousness is altered, there are marked effects on breathing. The response to CO_2 is altered by changing the proportionality constant called S (which is the slope of the line describing the increase in ventilation with an increase in CO_2). It falls to a small fraction of its normal value, and may even become negative, implying that CO_2 narcosis is adding to the depression. The hypoxic response is not only blunted with small changes in consciousness (e.g. one-tenth of the anaesthetic concentration producing anaesthesia) but essentially ablated at moderate anaesthesia depth.

The interaction of the many agents and adjuvants used in current practice is of importance to the practitioner. They are best understood by understanding their direct (receptor-specific) and indirect (generalized) mechanisms of action.

REFERENCES

1. Berger AJ, Mitchell RA, Severinghaus JW. Regulation of respiration. *New England Journal of Medicine* 1977; **297**: 92–7 (Part One), 138–43 (Part Two), 194–201 (Part Three).
2. Bainton C, Mitchell RA. Posthyperventilation apnea in awake man. *Journal of Applied Physiology* 1966; **21**: 411–15.
3. Hickey RF, Fourcade HE, Eger EI, *et al.* The effects of ether,

halothane, and forane on apneic thresholds in man. *Anesthesiology* 1971; **35**: 32–7.

4. Lourenco RV, ed. Clinical methods for the study of regulation of ventilation (report of symposium). *Chest* 1976; **70** (supplement): 109–95.

5. Gross JB, Smith L, Smith TC. Time course of ventilatory response to carbon dioxide after intravenous diazepam. *Anesthesiology* 1982; **57**: 18–21.

6. Hamza J, Ecoffey C, Gross JB. Ventilatory response to CO_2 following intravenous ketamine in children. *Anesthesiology* 1989; **70**: 422–5.

7. Choi SD, Spaulding BC, Gross JB, Apfelbaum JL. Comparison of the ventilatory effects of etomidate and methohexital. *Anesthesiology* 1985; **62**: 442–7.

8. Tusiewicz K, Bryan AC, Froese AB. Contributions of changing rib cage–diaphragm interactions to the ventilatory depression of halothane anesthesia. *Anesthesiology* 1977; **47**: 327–37.

9. Rosenberg M, Tobias R, Bourke D, Kamat V. Respiratory response to surgical stimulation during enflurane anesthesia. *Anesthesiology* 1980; **52**: 163–5.

10. Lam AM, Clement JL, Knill RL. Surgical stimulation does not enhance ventilatory chemoreflexes during enflurane anaesthesia in man. *Canadian Journal of Anaesthesia* 1980; **27**: 22–8.

11. Bourke DL, Malit LA, Smith TC. Respiratory interactions of ketamine and morphine. *Anesthesiology* 1987; **66**: 153–6.

12. Fourcade HE, Larson CP, Hickey RF, Bahlman SH, Eger EI. Effects of time on ventilation during halothane and cyclopropane anesthesia. *Anesthesiology* 1972; **36**: 83–8.

13. Knill RL, Gelb AW. Ventilatory responses to hypoxia and hypercapnia during halothane sedation and anesthesia in man. *Anesthesiology* 1978; **49**: 244–51.

14. Doi M, Ikeda K. Respiratory effects of sevoflurane. *Anesthesia and Analgesia* 1987; **66**: 241–4.

15. Warltier DC, Pagel PS. Cardiovascular and respiratory actions of desflurane: is desflurane different from isoflurane? *Anesthesia and Analgesia* 1992; **75**: 517–31.

16. Winter PM, Hornbein TF, Smith G, Sullivan D, Smith KH. Hyperbaric nitrous oxide anesthesia in man: determination of anesthetic potency (MAC) and cardiorespiratory effects. *Abstracts of Scientific Papers – American Society of Anesthesiologists Annual Meeting* 1972; 103–4.

17. Knill RL, Clement JL. Variable effects of anaesthetics on the ventilatory response to hypoxaemia in man. *Canadian Journal of Anaesthesia* 1982; **29**: 93–9.

18. Shook JE, Watkins WD, Camporesi EM. Differential roles of opioid receptors in respiration, respiratory disease, and opiate-induced respiratory depression. *American Review of Respiratory Disease* 1990; **142**: 895–909.

19. Forrest WH, Bellville JW. The effect of sleep plus morphine on the respiratory response to carbon dioxide. *Anesthesiology* 1964; **25**: 137–41.

20. Weil JV, McCullough RE, Kline JS, Sodal IE. Diminished ventilatory response to hypoxia and hypercapnia after morphine in normal man. *New England Journal of Medicine* 1975; **292**: 1103–6.

21. Camporesi EM, Nielsen CH, Bromage PR, Durant PAC. Ventilatory CO_2 sensitivity after intravenous and epidural

morphine in volunteers. *Anesthesia and Analgesia* 1983; **62**: 633–40.

22. Johnstone RE, Jobes DR, Kennell EM, Behar MG, Smith TC. Reversal of morphine anesthesia with naloxone. *Anesthesiology* 1974; **41**: 361–7.

23. Rigg JRA. Ventilatory effects and plasma concentration of morphine in man. *British Journal of Anaesthesia* 1978; **50**: 759–65.

24. Way WL, Costley EC, Way EL. Respiratory sensitivity of the newborn infant to meperidine and morphine. *Clinical Pharmacology and Therapeutics* 1965; **6**: 454–61.

25. Stoeckel H, Schuttler J, Magnussen H, Hengstmann JH. Plasma fentanyl concentrations and the occurrence of respiratory depression in volunteers. *British Journal of Anaesthesia* 1982; **54**: 1087–95.

26. Santiago TV Pugliese AC, Edelman NH. Control of breathing during methadone addiction. *American Journal of Medicine* 1977; **62**: 347–54.

27. Bellville JW, Fleischli G. The interaction of morphine and nalorphine on respiration. *Clinical Pharmacology and Therapeutics* 1968; **9**: 152–61.

28. Gal TJ, DiFazio CA, Moscicki J. Analgesic and respiratory depressant activity of nalbuphine: a comparison with morphine. *Anesthesiology* 1982; **57**: 367–74.

29. Abboud TK, Dror A, Mosaad P, *et al*. Mini-dose intrathecal morphine for the relief of post-cesarean section pain: safety, efficacy, and ventilatory responses to carbon dioxide. *Anesthesia and Analgesia* 1968; **67**: 137–43.

30. Gourlay GK, Cherry DA, Plummer JL, Armstrong PJ. Cousins MJ. The influence of drug polarity on the absorption of opioid drugs into CSF and subsequent cephalad migration following lumbar epidural administration: application to morphine and pethidine. *Pain* 1987; **31**: 297–305.

31. Abboud TK, Moore M, Zhu J, *et al*. Epidural butorphanol or morphine for the relief of post-cesarean section pain: ventilatory responses to carbon dioxide. *Anesthesia and Analgesia* 1987; **66**: 887–93.

32. Gourlay GK, Murphy TM, Plummer JL, Kowalski SR, Cherry DA, Cousins MJ. Pharmacokinetics of fentanyl in lumbar and cervical CSF following lumbar epidural and intravenous administration. *Pain* 1989; **38**: 253–9.

33. Etches RC, Sandler AN, Daley MD. Respiratory depression and spinal opioids. *Canadian Journal of Anaesthesia* 1989; **36**: 165–85.

34. Steen SN. The effects of psychotropic drugs on respiration. *Pharmacology and Therapeutics Bulletin* 1976; **2**: 717–41.

35. Blouin RT, Conrad PF, Gross JB. Time course of ventilatory depression following induction doses of propofol and thiopental. *Anesthesiology* 1991; **75**: 940–4.

36. DeFeudis FV. Minireview: GABA and respiratory function. *General Pharmacology* 1984; **15**: 441–4.

37. Pearce C. The respiratory effects of diazepam supplementation of spinal anaesthesia in elderly males. *British Journal of Anaesthesia* 1974; **46**: 439–41.

38. Catchlove RFH, Kafer ER. The effects of diazepam on the ventilatory response to carbon dioxide and on steady state gas exchange. *Anesthesiology* 1971; **34**: 9–13.

39. Bailey PL, Andriano KP, Goldman M, Stanley TH, Pace NL.

Variability of the respiratory response to diazepam. *Anesthesiology* 1986; **64**: 460–5.

40. Gross JB, Zebrowski ME, Carol WD, Gardner S, Smith TC. Time course of ventilatory depression after thiopental and midazolam in normal subjects and in patients with chronic obstructive pulmonary disease. *Anesthesiology* 1983; **58**: 540–4.

41. Ward DS. Stimulation of hypoxic ventilatory drive by droperidol. *Anesthesia and Analgesia* 1984; **63**: 106–10.

42. Gross JB, Caldwell CB, Shaw LM, Laucks SO. The effects of lidocaine on the ventilatory response to carbon dioxide. *Anesthesiology* 1983; **59**: 521–5.

43. Gross JB, Caldwell CB, Shaw LM, Apfelbaum JL. The effect of lidocaine infusion on the ventilatory response in hypoxia. *Anesthesiology* 1984; **61**: 662–5.

44. Sawynok J, Pinsky C, LaBella FS. Minireview on the specificity of naloxone as an opiate antagonist. *Life Sciences* 1979; **25**: 1621–31.

45. Gross JB, Weller RS, Conard P. Flumazenil antagonism of midazolam-induced ventilatory depression. *Anesthesiology* 1991; **75**: 179–85.

46. Hickey RF, Severinghaus JW. Regulation of breathing: drug effects. In: Hornbein TF, ed. *Regulation of breathing. Part II.* New York: Marcel Dekker (Publishers), Inc, 1981: 1251–308.

47. Pavlin EG, Hornbein TF. Anesthesia and the control of ventilation In: Geiger SR, exec. ed. *Handbook of physiology, Section 3: The respiratory system, Volume II: Control of breathing. Part 2.* Bethesda; American Physiological Society, 1986: 793–813.

48. Moser, KM, Luchsinger PC, Adamson JS, *et al.* Respiratory stimulation with intravenous doxapram in respiratory failure: a double-blind co-operative study. *New England Journal of Medicine* 1973; **288**: 427–31.

49. Cunningham DJC, Heg EN, Patrick JM, Lloyd BB. The effect of noradrenaline infusion on the relation between pulmonary ventilation and the alveolar P_{O_2} and P_{CO_2} in man. *Annals of the New York Academy of Sciences* 1963; **109**: 756–71.

50. Sabol SJ, Ward DS. Effect of dopamine on hypoxic–hypercapnic interaction in humans. *Anesthesia and Analgesia* 1987; **66**: 619–24.

51. Heistad DD, Wheeler RC, Mark AL, Schmid PG, Abboud M. Effects of adrenergic stimulation on ventilation in man. *Journal of Clinical Investigation* 1972; **51**: 1469–75.

52. Lyons HA. Centrally acting hormones and respiration. *Pharmacology and Therapeutics Bulletin* 1976; **2**: 743–51.

53. Gal TJ, Smith TC. Partial paralysis with d-tubocurarine and the ventilatory response to CO_2: an example of respiratory sparing? *Anesthesiology* 1976; **45**: 22–8.

Chapter 14

Clinical Physiology and Pathophysiology of the Respiratory System

Bryan E. Marshall, C. William Hanson III and Carol Marshall

Preoperative evaluation	**Postoperative period**
Intraoperative management	

We will discuss here aspects of perioperative pulmonary function that may influence the management of and eventual outcome for, patients presenting for anaesthesia for surgical procedures. The development of a plan[1] is of primary importance and it is a truism that the anaesthetist who runs through the entire procedure in his mind beforehand, including the most likely complications, is the one whose practice is the least punctuated with alarms. Problems are thereby anticipated, and measures that are introduced early in a pathophysiological sequence are often effective in preventing or reversing a deterioration in function.

This process of planning begins as soon as the surgical procedure is identified. The nature of the disease and/or the surgical approach will indicate whether respiratory function is likely to be impaired. With experience much more detail is contained in knowledge about the surgeons involved and the postoperative care available but in this context the discussions will focus on the objective facts upon which planning decisions, diagnoses and therapeutic approaches can be based. These facts constitute the clinical physiology and pathophysiology of the respiratory system.

Clearly the process from admission of a patient to discharge from the hospital is a continuum but it is convenient and conceptually useful to divide this discussion of perioperative respiratory function into four phases as follows:

1. Preoperative evaluation
2. Intraoperative management
3. Early postoperative period
4. Late postoperative period

While the basic assessment of pulmonary function is the same at all times, each of the phases identified above is associated with some special opportunities or potential complications that deserve emphasis.

Preoperative evaluation

The tasks for the preoperative period are to determine whether the patients' respiratory system is impaired and if so to evaluate the functional reserve; to decide what preparative or therapeutic measures to institute prior to surgery and to weigh the risks of the proposed surgical procedure.[2]

Interview and physical examination

An anaesthetist will approach the patient usually knowing only the initial diagnosis, proposed surgical procedure and the general needs of the specific surgeon. By the end of the chart review, interview and examination of the patient, a preliminary plan should have evolved in which the mechanical, gas exchange and respiratory control aspects of lung function have been assessed, the risks to the respiratory system of the surgical procedure and anaesthetic drugs and techniques evaluated, the likely times and patterns of acute pulmonary deterioration identified and the course of action explained to the patient. The purpose of this section is to provide the facts and concepts on which this process is based. It should be emphasized that this is not a simple process like a decision tree in a computer programme but rather one in which the experienced anaesthetist always has several alternative plans in mind simultaneously. Each plan is tested against the known facts as they are elicited and the components of the evolving plan are arrived at by mental jumps. These jumps represent integrations of knowledge and probabilities that become faster and more comprehensive only with experience.

To evaluate the respiratory system the interview should include direct questions about whether the patient has had respiratory illness previously. Specific mention of asthma, emphysema, chronic bronchitis, pneumonia, pleurisy and heart trouble may elicit a positive answer when respiratory illness in general has been denied. In every patient the past or present history of smoking, wheezing and coughing should be established and the current exercise and sleeping habits and limits described. For most patients this is sufficient to exclude respiratory disease and the physical examination can be confined to auscultation of the chest and examination of the airway to establish ease of intubation. A history of respiratory pathophysiology, even far in the past, and especially if associated with previous anaesthesia, should be regarded seriously. Thus a patient free of respiratory signs or symptoms now, reporting spontaneous pneumothorax in the past is more likely to have bullae capable of rupture than the normal population. At the other extreme a patient reporting frequent hospitalization with chronic cardiopulmonary disease is a signal perhaps for a cardiac or pulmonary consultation before arriving at final plans.

The most judgement is required for the in-between patient, who describes some signs and symptoms of respiratory involvement. If the episodes occurred in the past the nature, extent and precipitating factors should be established with particular emphasis as to whether any sequelae such as cough or exercise limitation persist to the present and whether they are events likely to be repeated during the present admission. For example, less concern would be elicited by a history of atelectasis resulting from inhalation of a foreign body than one of aspiration pneumonitis following a previous general anaesthetic.

If questions reveal current symptoms of respiratory disorders they should be pursued in order to decide how much impairment of respiratory function has occurred and whether further treatment is warranted prior to surgery.

It is more important to examine closely the severity of current symptoms than it is to list the names of previous diagnoses. If there is a history of cigarette smoking, how long and how much should be recorded as pack-years and the patient should be encouraged to stop before surgery. If a cough is reported, the time of occurrence (i.e. first thing in the morning or at other times) should be established and if productive, the colour and consistency of the sputum should be noted, preferably by inspection. A purulent sputum should be treated before elective surgery and active respiratory toilet instituted for 2–3 days prior to surgery.

Wheezing may be reported particularly in association with asthma, chronic bronchitis and heart disease. While it is important to establish the differential diagnosis it is equally important for the anaesthetist to note that in such patients the bronchi are sensitive to stimuli precipitating further bronchospasm and especially the intubation of the trachea during light levels of anaesthesia. Furthermore the response to bronchodilators should be noted both by history and perhaps by testing (see below).

Sleeping habits are useful indicators particularly if they have changed. A preference for sleeping propped up by several pillows is familiar as a symptom of left heart failure but less familiar perhaps is the diagnosis of sleep apnoea syndrome which would be supported by diurnal somnolence and morning headache in an obese patient.

The history of tolerance to exercise is of key significance to the anaesthetist attempting to judge the extent of the respiratory reserve. Usually this can be judged by the exertion that coincides with shortness of breath. A patient incapable of conversation while walking on the flat is severely handicapped while one who reports interference with conversation only after ascending three flights of steps is normal. Such commonplace examples avoid the semantic difficulties that come with the use of the term dyspnoea but sometimes it may be necessary to establish the difference. Shortness of breath is defined as breathing that reaches the level of consciousness while in dyspnoea there is a sensation that life is in jeopardy.

Most of the information essential for the anaesthetist can be acquired readily through this history and chart review but it is also essential to examine the face, mouth, teeth and chest. The purpose of this examination is to identify problems that pertain to techniques of anaesthesia or that might influence the outcome of surgery. Intubation of the trachea may be difficult[3] if the patient is unable to open the jaw sufficiently, if soft tissue lesions obstruct the pharynx or larynx, or if the trachea is displaced by mediastinal or cervical masses. Intubation may be more difficult if neck extension is prevented because of suspected cervical fracture or impossible as a result of fibrosis. Kyphoscoliosis or ankylosing spondylitis may not only complicate airway access but also may compromise respiratory function by restricting chest wall movement and be a contraindication for spinal or epidural anaesthesia.

These symptoms and signs provide a general indication as to whether a patient has, or may be prone to develop, respiratory impairment and also permit a crude estimate of the nature and extent of the abnormality. For more objective evidence of the latter a battery of pulmonary function tests is available.

Tests of pulmonary function

Lung diseases may be grouped into restrictive, obstructive and interstitial diseases. This distinction is achieved by the use of the pulmonary function tests discussed below and is useful as the basis for perioperative management. Restrictive lung impairments include all those diseases and circumstances that result in reducing lung volumes. These may result from destructive or space-occupying pathology that has reduced the available lung tissue or they may result from previous lung surgery. Similarly kyphoscoliosis, a body cast, skeletal muscle weakness, pain or increased intra-abdominal pressure may be the cause of restrictive lung impairment because chest wall movement is diminished. Obstructive lung diseases include not only the diseases intrinsic to the airway wall such as cystic fibrosis and asthma but also all of those patients diagnosed with chronic bronchitis, where the problem is one of hypersecretion and repeated infection; and emphysema where loss of parenchymal tissue is characteristic. Frequently chronic bronchitis and emphysema are both present and thus the designation of all these conditions as chronic obstructive lung disease is a practical expedient. But the pathophysiological distinction between the extreme dyspnoea characterizing the 'pink puffer' of late emphysema and the hypoxaemic patient with cor pulmonale characterized as the 'blue bloater' of severe chronic bronchitis obviously present somewhat different challenges to the anaesthetist and illustrate well the concept that the functional pathophysiology is more critical than the diagnostic label.

Interstitial diseases include all those pathological processes that are initiated in the tissues of the alveolar wall bounded by the alveolar epithelium on one side and the vascular endothelial cells on the other. A list of non-malignant causes of interstitial diseases includes diseases resulting from inhalation of dust, gases and vapours, secondary effects of drugs, radiation inflammation, repeated aspiration and a wide variety of diseases of unknown origin such as Goodpasture's syndrome and amyloidosis. The destruction, distortion and fibrosis of the critically thin gas exchanging surface of the alveoli is the cause of functional changes and the effects vary depending on the pattern and extent of damage from each of these pathological processes.

For the pulmonary physician there is now a broad spectrum of diagnostic tests available but the anaesthetist needs to be familiar only with those that provide direct information as to the functional state of the respiratory system. These tests can be conveniently divided into four categories: mechanics of breathing, gas exchange, control of ventilation, and pulmonary circulation.

Mechanics of breathing

These tests measure all those aspects of breathing that influence the volumes and convective movement, or flow, of respiratory gases in the lungs. The classic pulmonary function tests are readily performed with a spirometer, combined with a helium or nitrogen analyser to measure functional residual capacity (FRC).

With this instrumentation the lung can be divided into a series of static volumes: residual volume (RV), expiratory reserve volume (ERV), tidal volume (V_T) and inspiratory reserve volume (IRV). Two or more volumes are combined into convenient composite volumes called capacities, namely total lung capacity (TLC), functional residual capacity (FRC), inspiratory capacity (IC) and vital capacity (VC). In healthy, young, seated adults, the normal proportions are shown in Fig. 14.1; and in Table 14.1 the predicted normal range of volumes for the age, sex and weight are shown.

Lung volumes change with advancing age. The chest wall stiffens and muscle tone declines so that VC decreases and RV increases. At some reduced lung volume small

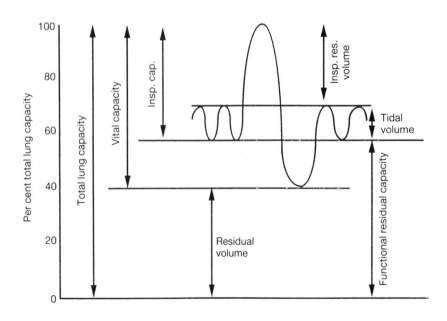

Fig. 14.1 Lung volumes and capacities as per cent of total lung capacity in an adult subject.

Table 14.1 Pulmonary function tests: regression relations for predicting lung function from age (years), height (metres) and weight (kg) in healthy adults

Test	Sex	Regression coefficients Height	Age	Weight	Constant
Total lung capacity (litres)	M	7.8			−7.3
	F	7.9	−0.008		−7.49
Vital capacity (litres)	M	5.2	−0.002		−3.60
	F	5.2	−0.018		−4.36
Functional residual capacity (litres)	M	3.2			−2.94
	F	5.3		0.02	−4.74
FEV_1 (litres)	M	3.7	−0.028		−1.59
	F	2.67	−0.027		−0.54
FEV_1/FVC (%)	M		−0.373		+91.8
	F		−0.261		+92.1
Closing volume/VC (%)	M		+0.357		+0.562
	F		+0.293		+2.812
Arterial Po_2 (kPa)			−0.032		+13.96
Arterial Pco_2 (kPa)					37–43
Diffusing capacity (CO) (ml/min per kPa)	M	4.34	−0.026		−2.34
	F	2.83	−0.021		−0.35
Airway resistance (cmH$_2$O/l per second)					0.5–2.0
Lung compliance (l/cmH$_2$O)					0.09–0.40

Adapted from reference 4.

airways tend to close (the closing volume) but with advancing age this tends to occur at greater volumes. Total lung capacity remains constant but the ratio of RV to TLC increases with age.

Of these volumes the most useful in the present context is the vital capacity. For restrictive or interstitial lung diseases a reduction to less than 50 per cent of the predicted VC is indicative of moderate to severe disease. The VC is not altered by moderate obstructive lung disease.

Dynamic tests such as maximum ventilatory volume (MVV), forced expiratory volume in one second (FEV_1) or forced vital capacity (FVC) are measures of the ability of the respiratory system to develop high gas flow rates. Normal values for a young male adult are: MVV = 170 l/min and $FEV_1/FVC \geq 83$ per cent. The presence of moderately severe obstructive lung disease is characterized by MVV or FEV_1/FVC of 50 per cent or less of predicted while these values remain in the normal range with moderately severe restrictive or interstitial lung diseases. Functional tests such as these require cooperation of the patient and

maximum effort. It is essential to perform three repetitions of each test because reproducibility is a good sign that these prerequisites are being met. For the present discussion it is assumed that neuromuscular disease and cardiac failure are not present but either one may be associated with impaired VC and FEV_1.

In many patients with chronic obstructive pulmonary disease there is active bronchoconstriction. The response to a bronchodilator is therefore a quantitative measure of the reversibility of airway obstruction. A greater than 15 per cent increase in the FVC and FEV_1 is regarded as a significant response.

With moderately severe interstitial disease the pattern of pulmonary function test results is similar to that of restrictive disease with normal MVV and reduction to about 50 per cent of predicted values for VC and FEV_1 and normal FEV_1/FVC. It should be noted that none of the tests measures small airway disease and no routine test is available for this purpose.

The diffusing capacity for carbon monoxide (D_{LCO}) is a

complex test that depends on diffusion of carbon monoxide across the alveolar capillary membrane, the volume of capillary blood and the reaction rate with haemoglobin. The test is valued because it is a sensitive indicator of restrictive and interstitial lung disease; a reduction of 50 per cent represents moderate severity.

Gas exchange

In healthy subjects the respiratory system can satisfy a wide range of demands. Thus the resting oxygen consumption ($\dot{V}O_2$ = 250 ml/min) and carbon dioxide clearance ($\dot{V}CO_2$ = 200 ml/min) of a normal adult is met with a tidal volume of 500 ml at a frequency of 12 breaths/min for a total expiratory minute volume ($\dot{V}E$) of 6 l/min. With exercise the oxygen and carbon dioxide exchange and the expired minute volume may increase three- or fourfold and oxygen consumption in a trained athlete is capable of increasing 40-fold.

With advancing age and a sedentary life style this large functional reserve is reduced but it is even more curtailed by the presence of disease. While the previous section was concerned with objective quantification of these limits, this section is concerned with the changes that influence gas exchange.

For each breath (V_T) only about 70 per cent of the respired gas reaches the gas exchanging alveoli while the rest washes in and out of the airways and is designated dead space (V_D).

The effective ventilation each minute that results in gas exchange is called the alveolar ventilation ($\dot{V}A$) and the relation with the other volumes is expressed as:

$$\dot{V}_A = \dot{V}_E - (\dot{V}_D \times f) \qquad (1)$$

where f is the frequency of respiration.

Ventilation is almost entirely determined by the requirement for carbon dioxide clearance and a useful relation is:

$$\dot{V}_A = 863 \times \dot{V}CO_2/PaCO_2 \qquad (2)$$

The constant 863 converts the gas volumes from body temperature and pressure to standard conditions and also corrects the carbon dioxide from partial pressure to per cent concentration. This equation states that in a steady state the alveolar ventilation is inversely proportional to the $PaCO_2$. Thus doubling the alveolar ventilation will halve the $PaCO_2$ and the equation is the basis for adjusting the rates and volumes of mechanical ventilators.

The dead space is normally about one-third of the tidal volume but this may increase in many pulmonary diseases to 50 per cent or more. Excessive minute ventilation in the absence of increased metabolic carbon dioxide production and with normal or only slightly increased $PaCO_2$ is due to increased dead space and the fractional volume of the dead space can be calculated:

$$V_D/V_T = (PaCO_2 - PECO_2)/PaCO_2 \qquad (3)$$

where $PECO_2$ is the partial pressure of carbon dioxide measured in a collection of the entire mixed expired gases. This dead space volume is called the physiological dead space because it includes both the anatomical or airways dead space volume and the gas that reaches non-perfused alveoli or the alveolar dead space. The anatomical dead space is altered only to a limited extent by most drugs and diseases. In contrast the alveolar dead space is not only altered by disease but by any manoeuvre that reduces the effective pulmonary artery blood pressure (Fig. 14.2); for example, mechanical ventilation or hypovolaemic hypotension.

Until now the discussion has emphasized the bellows functions of the lungs and these gas volume and flows are easy to measure. However, the critical function of the lung is gas exchange in which blood flow is equally important although less accessible.

For oxygen exchange the linear relation described for carbon dioxide in equation (2) does not hold. However, just as the arterial carbon dioxide tension is the best measure of the effectiveness of ventilation, so the arterial oxygen tension is the best overall estimate of the efficiency of oxygen exchange. The partial pressure of oxygen in air (PIO_2) is 20 kPa (149 mmHg) when warmed and humidified in the nose. In the alveolar gas the oxygen tension is reduced by uptake of oxygen and diluted by excretion of carbon dioxide, so that the alveolar oxygen tension (PAO_2) is 14 kPa (105 mmHg). In normal subjects the distribution of ventilation/perfusion ratios results in some inefficiency of exchange so that the mixed end capillary blood oxygen tension (PcO_2) of perfused alveoli may be about 0.5 kPa less than the mixed alveolar gas. Furthermore, the admixture of mixed venous blood from non-ventilated alveoli (pulmonary shunt) and from thebesian and bronchial veins results in a further reduction in arterial oxygen tension by about 1 kPa. The normal arterial oxygen tension of a healthy young adult is therefore about 12.7 kPa (95 mmHg) and the alveolar to arterial oxygen difference is approximately 2 kPa. These relations are summarized in Fig. 14.3. With advancing age the arterial oxygen tension decreases:

$$PaO_2 = 13.6 - 0.04 \times (\text{age in years}) \qquad (4)$$

In all severe respiratory diseases states and earlier in the course of interstitial diseases the alveolar to arterial oxygen tension difference is increased as a result of increased venous admixture and increased distribution of ventilation/perfusion ratios and the PaO_2 decreases while $PaCO_2$ remains normal. If ventilatory failure occurs so that $PaCO_2$ begins to increase then the alveolar and the arterial oxygen tensions will decrease.

The total inefficiency of oxygen exchange expressed as the alveolar to arterial oxygen tension difference can also be

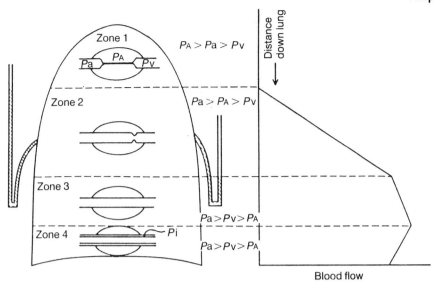

Fig. 14.2 Gravitationally determined changes in pulmonary blood flow distribution in the erect posture. For all zones, the intravascular pressures vary directly with their vertical distance down the lung. The alveolar pressure is the same everywhere in the lung; therefore, zones 1, 2, and 3 are differentiated by the relative magnitudes of the alveolar (P_A), pulmonary arterial (P_a), and pulmonary venous (P_v) pressures. In zone 4, the flow rate decreases again, perhaps because the vessels are compressed by increasing interstitial tissue pressure (P_i). (Redrawn from West JB. *Ventilation/blood flow and gas exchange*. 3rd ed. Philadelphia: FA Davis, 1982, with permission.)

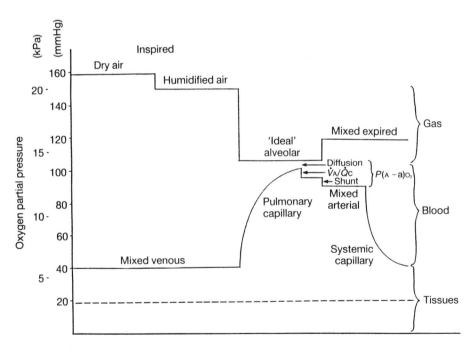

Fig. 14.3 The changes in oxygen tension in the gas and blood phases during transfer between the air and the cells; see text for discussion (from reference 16 with permission).

expressed as the equivalent amount of venous blood that would have to be mixed with the end capillary blood to result in the same arterial oxygen tension. This quantity is calculated as a percentage of the cardiac output and is designated the venous admixture per cent. If air is replaced by 60–100 per cent oxygen as the inspired gas, the differences in concentrations of oxygen, due to ventilation/perfusion differences between different alveoli, are eliminated. The inefficiency of oxygen that is now measured can again be calculated in terms of a percentage of the cardiac

output, containing mixed venous oxygen concentrations, that would have the same effect. This latter quantity is designated the pulmonary shunt per cent. Comparison of the venous admixture per cent and the pulmonary shunt per cent provides a useful measure of how much of the inefficiency is due to atelectasis (pulmonary shunt per cent) and how much is due to maldistribution of ventilation/perfusion ratios (venous admixtures per cent – pulmonary shunt per cent).

In most hypoxaemic lung diseases abnormality of the

ventilation/perfusion ratio distribution is prominent but in the perioperative period atelectasis is common and therefore this distinction is an important one.

Control of ventilation

The rhythmic sequence of efferent signals that causes activation of the respiratory muscles is generated by control centres – chiefly in the pons and medulla of the brain. Chemoreceptors on the ventral surface of the medulla are sensitive to pH and P_{CO_2} and others in the carotid body are sensitive to P_{O_2}. Both provide feedback control to the neuronal centres. Still other receptors respond to irritants in the airway, stretch in the smooth muscle and distension in the alveolar septa that also influence ventilation and are often evident clinically. Thus many individuals develop wheezing (bronchospasm) when exercising in cold air where the irritant receptors are stimulated in the airways; and tachypnoea is common in left heart failure as oedema fluid distends the lung tissue and so forth.

The fact that ventilation is primarily adjusted to satisfy clearance of carbon dioxide was alluded to in the previous section. The arterial carbon dioxide tension is normally 5.3 kPa (40 mmHg) and if this tension is increased artificially by 1 mmHg (where 1 kPa = 7.5 mmHg), the ventilation is increased by 3 l/min. The change in minute ventilation per mmHg change of carbon dioxide ($\Delta \dot{V}_E/\Delta P_{CO_2}$) is a useful index of both the neural drive and the ability of the chest wall and diaphragm to respond. An even simpler index of the same properties is the ratio of tidal volume to the time for inspiration (V_T/T_I). Further information about the neural drive can be obtained from the airway occlusion test which measures the pressure generated in the first 100 ms after occlusion $P_{0.1}$. These tests are not usually performed as part of the presurgical preparation or assessment of patients with pulmonary disease but they have their counterpart during recovery from anaesthesia (see below). It is helpful to keep in mind that respiratory control is commonly altered in association with respiratory disease.

When resting ventilation approaches the maximum attainable in a patient, there is a tendency for the drive to be reduced so that the actual P_{CO_2} increases and the systemic and central pH returns towards normal with a partially compensated respiratory acidosis. Restoring the Pa_{CO_2} to normal in these patients by assisting ventilation will result in apnoea when ventilation is discontinued. Similarly patients with severe chronic obstructive pulmonary disease often develop sufficient hypoxaemia that a significant part of their respiratory drive is due to carotid body stimulation. Administering increased oxygen to such patients will result in acute respiratory depression, respiratory acidoses, mental confusion and even coma as the arterial P_{CO_2} increases.

Pulmonary circulation

Although the pulmonary circulation is less well characterized than the systemic, nevertheless advances in radiography, the introduction of the flow directed catheter and the development of radionuclide scanning techniques have provided details of regional and overall haemodynamic performance. Cardiomegaly, pulmonary embolism, pulmonary hypertension and pulmonary oedema can be diagnosed and quantified with these techniques.

Work of several decades ago established that the pulmonary circulation was a low pressure, low resistance system dominated by gravity in which regional flow distribution could be understood as dependent on differences between intravascular blood pressures and intra-alveolar gas pressures.[4] These concepts have proved very useful for understanding normal pulmonary haemodynamics and even the dependent pattern of occurrence of oedema and so forth in pathophysiological states (see Fig. 14.2).

However, the overall concept of the pulmonary circulation as a passive system of tubes and of flow distribution dominated by gravity has required modification recently. The recognition that blood flow tends to decrease from the hilum out in an onion-skin pattern as well as vertically even in the normal lung and the active regulation of pulmonary vascular tone by hypoxia (hypoxic pulmonary vasoconstriction, HPV) provide anatomical and physiological evidence of non-gravitational inhomogeneity for blood flow distribution. Furthermore, in disease states active changes in tone and non-gravitational alterations of flow distribution usually dominate over the gravitational effects so as to make the latter of less importance. Many humoral, neuronal, pathological and pharmacological factors can influence pulmonary vascular tone but only HPV can be considered an intrinsic feedback control mechanism constantly in operation.

Another improvement in understanding concerns the shape of the pressure–flow relations. As perfusion pressure increases, flow generally increases slowly at first and then to a greater extent. At increased pressures the flow–pressure relation becomes almost linear until finally pulmonary oedema supervenes. Thus the flow–pressure relation is a curvilinear one with the pulmonary vascular resistance decreasing, as flow increases. Some years ago it was proposed that at zero flow all the passive pulmonary vascular tubes were closed and that the curvature resulted when pressure was increased because different tubes opened at different pressures. That suggestion explained the curve and has been widely accepted, even up to the present, despite the fact that no physical basis for, or evidence of, vascular closure in this fashion has been presented. More recently the recognition that the pulmonary vasculature has elastic walls and that the curved flow–pressure relation follows from this property of distensibility has become the preferred concept and one that is well founded on experimental observation.[5] In the current literature the flow–pressure relation is often described as a straight line with a slope and an intercept extrapolated to the pressure axis. In the older interpretation the intercept was called an opening pressure and was identified with some postulated site in the vascular bed at which that pressure was restricting flow. This often resulted in vague and confusing conclusions because the site of flow restriction seemed to have no identifiable anatomical location. With the newer concept of a distensible circuit such inconsistencies do not appear and this discussion will be confined to interpretation in this framework.

Pulmonary hypertension[6] is defined as a pressure exceeding 35 cmH$_2$O in the absence of increased cardiac output. In association with pulmonary disease the causes are loss of vascular bed (by fibrosis, hypertrophy, embolism and tissue destruction) and vasoconstriction, particularly by hypoxia and hypoxaemia during acute infections but also on occasion by other intrinsic and extrinsic vasoconstrictors. Because the pulmonary circulation can normally accommodate such large changes of flow with moderate changes in pressure, the occurrence of pulmonary hypertension is evidence of an extensive abnormality.

If the pulmonary artery pressure is increased and cannot be accounted for by increased left atrial pressure (as in left heart failure) or increased cardiac output, then it is due to an increased vascular resistance. This in turn is attributable to loss of vascular bed by disease or surgery or to removal of the vasculature by hypertrophy, oedema or to active constriction. These relations are summarized by:

$$PAP = (PVR \times \dot{Q}T) + P_{LA} \qquad (5)$$

Where PAP is pulmonary arterial pressure; PVR is pulmonary vascular resistance; $\dot{Q}T$ is cardiac output and P_{LA} is left atrial or pulmonary artery occlusion pressure. A particularly revealing test in this regard is to observe the changes in pulmonary artery pressure and pulmonary artery occlusion (wedge) pressure when the cardiac output is increased during exercise. The loss of reserve, for example, after pneumonectomy in the presence of chronic lung disease is clearly apparent during exercise (Fig. 14.4) although values may be normal at rest. Such invasive and complex tests are not normally performed but prior to resection of lung cancers the outcome of pneumonectomy should be considered and the function of the remaining lung may be specifically measured by occluding the pulmonary artery of the diseased lung or by differential spirometry. If in these circumstances the pulmonary artery pressure exceeds 35 cmH$_2$O, or the function of the remaining lung does not exceed the minimum normal, then the procedure is associated with a poor prognosis.

Considerable information concerning the distribution of both perfusion and ventilation can be obtained from radioisotopic scans of the lungs. These tests have the advantage of being essentially non-invasive and only moderately expensive; however, they should be used for specific diagnostic purposes and not as part of a routine battery of tests. Uncertainty about the state of the pulmonary circulation is a sufficient justification for the insertion of a flow-directed pulmonary artery catheter for monitoring during anaesthesia. This will allow measurement of pulmonary artery pressure, pulmonary artery occlusion pressure and cardiac output, from which the pulmonary vascular resistance can be calculated and all the

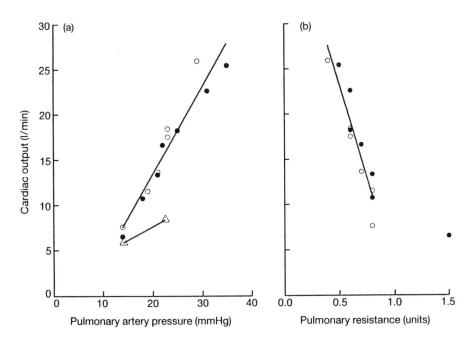

Fig. 14.4 Relation between pulmonary artery pressure and cardiac output during exercise in normal subjects (open and closed circles) and patients following pneumonectomy (open triangles). While both groups start at similar points the reduction in vascular bed following pneumonectomy is revealed by the reduced slope of the flow–pressure line.

components of equation (5) can then be followed during anaesthesia, surgery and the early postoperative period.

Lung water balance and pulmonary oedema

The blood–gas interface consists of the alveolar epithelium and the capillary endothelium lining the 300 million alveoli. The gas exchanging surface has an area of 70 m^2 and is exquisitely thin ($\approx$0.5 µm). The gas pressure of the alveolar side oscillates a few cm of H$_2$O pressure around zero during normal breathing while the hydrostatic pressure of the capillary blood side has to be approximately 12 cmH$_2$O positive for flow to continue. Pulmonary oedema does not occur only because several mechanisms exist in the normal lung to prevent it; when one or more of these fail, water begins to accumulate first in the perivascular and the peribronchial tissue where it may be visualized on a radiograph as increased vascular markings, bronchial cuffing and Curly's lines. When the tissue spaces are filled, alveoli begin to fill, at first in a scattered manner and finally with generalized flooding and the appearance of frothy fluid at the airway.

The principal factors governing water movement in the lung are summarized by the Starling equation:[7]

$$\dot{Q}w = Kw[(Pc - Pi) - \sigma(\pi c - \pi i)] \qquad (6)$$

In this equation $\dot{Q}w$ is the net flow of water and is usually positive, so that net flow is from the capillary to the interstitial space. The equation states that the net flow is equal to the product of the hydraulic conductivity (Kw) of water and the pressure difference resulting when all the terms inside the square brackets are evaluated.

The hydraulic conductivity is a complex function that includes the size of the lung and all the factors hindering the bulk movement of water into the tissues. It is important to note that there is normally no hindrance to exchange of water across the capillary membrane and the Kw is not a measure of capillary permeability for water. It is more dependent on the compliance of the tissue and bulk movement of water through the gel of the interstitial tissue than it is on the state of the capillary wall. In oedema states this Kw term increases because the interstitial tissues change from gel to liquid state and the capillary membrane permeability may remain unimpaired.

The hydrostatic pressure in the capillary (Pc) and interstitial (Pi) spaces are normally believed to be in the region of 12 and -5 cmH$_2$O respectively while the oncotic pressures in the capillary (πc) and interstitial (πi) compartments are -35 and -20 cmH$_2$O respectively. For these values the pressure difference therefore amounts to $[(12 - -5) - \sigma(-35 - -20)] = +2$ cmH$_2$O where $\sigma = 1$ as it is normally. This means that a small net excess of

water is continually flowing through the interstitial tissues into the perivascular tissue spaces where it is removed by the lymphatic drainage.

The sigma (σ) is called the reflection coefficient and is the measure of the ability of the capillary wall to maintain an oncotic pressure difference between the capillary and interstitial compartments. The normal reflection coefficient of between 0.9 and 1.0 means that almost every plasma protein molecule that contacts openings or channels through the capillary membrane is reflected back or prevented from leaving the capillary. Water may therefore leave the capillary freely but the plasma proteins only very slowly and therefore an oncotic pressure gradient is established that opposes the bulk flow of water.

Furthermore if the capillary hydrostatic pressure increases as a result of left heart failure or overtransfusion there will be an increase in water flow across the membrane and three mechanisms provide compensation to prevent oedema. First and most important there is an increase in lymph flow to carry away the excess water. Second, and simultaneously, because the plasma protein molecules are prevented from crossing the membrane, the water entering the lymphatics carries with it some of the interstitial protein so that the interstitial oncotic pressure decreases. Starling's equation reveals that the increased hydrostatic pressure can be directly offset by such a compensation. Finally, the interstitial pressure can change from negative to zero, and because of the nature of the ground substance that constitutes the gel this can represent a considerable change in volume with little change in pressure. All of these factors occur simultaneously and are capable of preventing oedema over a wide range of physiological stresses.

In disease states any of these components may be impaired. Not only may capillary hydrostatic pressure be increased, as referred to above, but plasma oncotic pressure may be reduced by malnutrition, defective synthesis of protein, excessive loss of protein through the kidneys or at burn surfaces or simply by haemodilution. Furthermore, a multitude of factors may increase permeability by damaging the capillary wall. Such factors include sepsis, allergy, toxins, inflammation and stretching as by increased hydrostatic pressure or even hyperinflation of the lung.

It is important to recognize that the factors influencing capillary permeability and those causing changes in capillary hydrostatic or oncotic pressure are different although both types of cause often coexist in disease. Failure to recognize both these facts have led some to disregard the clinical utility of the Starling relation. There is, for example, a tendency currently to conclude that the selection of colloid-containing versus pure crystalloid solutions for resuscitation is unimportant because of contradictory experimental results. There are good economic reasons to reduce unnecessary use of colloid and clear experimental evidence that the compensatory resources of the lung are sufficient to permit significant variation in oncotic pressure without adverse outcome. Yet to disregard the advantages to be obtained by working within the rationale set out by the Starling equation is to disregard the scientific basis of our modern understanding of lung water balance.

A normal plasma oncotic pressure is one of the principal defences against pulmonary oedema. If this oncotic pressure is low, then the reserve against oedema is directly and proportionally reduced, but only so long as the relative impermeability of the capillary membrane to plasma protein is maintained. If membrane permeability increases, the entire Starling equation collapses and if colloid infusion does not improve pulmonary oedema in a patient with a low plasma oncotic pressure, then in the absence of fluid overload, this is evidence of impaired capillary function.

It is worth emphasizing in this context that 35 cmH_2O is an apparently limiting design feature of the lung. Plasma oncotic pressure usually has this value, pulmonary artery pressures exceeding 35 cmH_2O are defined as pulmonary hypertension and a vital capacity manoeuvre requires a transpulmonary pressure of 35 cmH_2O. This is roughly true for all animals and it appears that fluid balance is the most fundamental underlying consideration.

While it is seldom that detailed information will be available concerning the pulmonary circulation except in the special circumstances of some cardiac and thoracic surgical procedures, the likelihood of pulmonary oedema should always be considered before any major surgery. Total plasma protein and albumin/globulin composition allow calculation of oncotic pressure; the history of resuscitation, of cardiac failure and of proposed surgery together with the physical findings provide an index of suspicion. Whether to place a flow-directed catheter, whether to use colloid or crystallized replacement solutions and a specific plan of action should oedema occur are necessary components of the anaesthetic plan to minimize pulmonary functional changes.

Risks and outcome

The preoperative evaluation is predicated in the belief that by so doing the risks and outcome of the proposed procedure can be favourably influenced. For the pulmonary system in particular the simplicity and widespread availability of pulmonary function testing and blood gas measurements has encouraged the search for specific numbers on which decisions as to the risks and long-term outcome of anaesthesia and surgery can be used.

Several excellent and realistic studies[8–11] have provided data but it must be emphasized that these are guidelines to be used mainly as indices as to where the most and least

successful outcomes lie and not as authority to rule out anaesthesia or surgery. Thus all the studies quoted indicate that FEV_1 of less than 0.5 litres is associated with a very poor prognosis and yet others have shown[12] that patients can survive with such values. A recent review[13] of all studies of the past 25 years attempting to identify preoperatively patients at risk for postoperative complications concluded that the only successes were in assessing patients prior to lung resection and, in severely affected patients, split lung screening and exercise testing were found useful. But, overall, spirometry and blood gas studies had little correlation with risk in most patients for thoracic, abdominal or cardiac surgery.

No long-term difference in risk or outcome has been established between the choices of general or regional anaesthesia that relate to the pulmonary system. Even in asthmatics, what is gained by avoidance of stimulating the airway by the use of regional anaesthesia is often lost by increased bronchial sensitivity as a result of interference with the autonomic nervous system. Furthermore, following all surgical procedures, even with normal lung function, there is a 6–10 per cent incidence of postoperative complications. Most of these are pulmonary complications but the incidence is greatest following intrathoracic or intra-abdominal procedures and least following surgery at peripheral sites. The conclusion from numerous studies is that there is no one pulmonary function test that allows reliable prediction of complication; rather the results of a battery of tests considered with the other available data allow some general conclusions about the threat to and from the pulmonary system.

The incidence of pulmonary complications increases from 10 per cent in patients free of pulmonary disease to 70–100 per cent in those with abnormal pulmonary function tests.[8] As mentioned above, the nature and site of surgery are important influences. There is some agreement that the FVC and FEV_1 are the best predictors and Stein et al.[11] showed that the institution of preoperative antibiotics, bronchial dilation and chest physiotherapy decreased the postoperative complications in patients with abnormal pulmonary function to 20 per cent.

Cardiac surgery is associated with a 20 per cent incidence of pulmonary complications but this is independent of preoperative pulmonary function.[14] Perhaps this is due to the more complete preparation and selection of patients for cardiac surgery. Curiously the median sternotomy incision is associated with less impairment than are intercostal thoracotomy incisions or even an upper abdominal incision. Cardiac surgery is one area in which the choice of anaesthesia does influence the postoperative course. The incidence of pulmonary complications has been shown to be increased in patients managed with high dose narcotics and ventilated postoperatively compared to others managed with inhalational agents and extubated in the first 3 hours postoperatively.[15] Certainly in this instance the choice of anaesthesia determines the time of extubation but the numbers are too small to tell if outcome was influenced.

Perhaps the most direct statement can be made concerning the risks and outcome of thoracic surgery because the state of the remaining lung is so critical. The methods of assessing this state have been discussed previously and the conclusions may be summarized as follows. If the remaining lung is normal, the risks of the procedure are not increased and although there may be exercise limitation following pneumonectomy even this is not usually marked in the long term. In contrast, if the pulmonary function tests are less than 50 per cent of the predicted normal for the remaining one lung, the patient is in the high risk category with a mortality rate approaching 40 per cent if pneumonectomy is performed. If FEV_1 is less than 0.5 litres or the PAP greater than 60 cmH_2O, survival is unlikely.

Spirometric tests, diffusing capacity and blood gas values ($PaO_2 \leq 6$ kPa; 45 mmHg) are therefore effective indices for identifying the normal and the severely impaired but also provide a less definite but nevertheless useful idea of the relative degree of risk for those in between. Pneumonectomy is the most demanding procedure but all other intrathoracic lung restrictions and upper abdominal surgical procedures impose a severe stress. For pneumonectomy the most precise information about the postoperative performance is obtained by split lung studies, differential lung scans or differential pulmonary artery occlusion. If the patient cannot tolerate these, in the pulmonary physiological sense, it is unlikely that the postoperative course will be successful. Even after a wedge resection or lobectomy, the remaining lung on that side is often non-functioning effectively for some hours or even days postoperatively and the result is similar to a pneumonectomy. Such preoperative differential studies can be of great assistance in complex situations.

Conclusions

Impairment of the respiratory system occurs to some extent during and after all anaesthesia and surgery. Usually the impairment is transient but atelectasis, oedema, pneumonia and hypoxaemia are more serious complications that are particularly likely to occur in the presence of preoperative cardiopulmonary disease, obesity and old age and following abdominal and thoracic surgical procedures.

Reduction in the incidence of these complications is the purpose of preoperative evaluation. Preparation of patients with pulmonary impairment prior to surgery include antibiotic and physiotherapy treatment to eliminate active infection, digoxin, diuretics and low flow oxygen therapy to control cor pulmonale; education to ensure optimal compliance, and training to perform incentive spirometry; and the control of bronchospasm with aerosol, oral or intravenous bronchodilators.

Intraoperative management

Arterial hypoxaemia

With regard to pulmonary function, the principal concern during anaesthesia and surgery is the avoidance of arterial hypoxaemia.[16] Normally, the arterial oxygen tension (PaO_2) is greater than 12 kPa (90 mmHg) and the saturation of haemoglobin is greater than 95 per cent but these values decrease with age.[17] It is desirable to maintain the oxygen saturation at or greater than 95 per cent ($PaO_2 \leq 13$ kPa; 97.5 mmHg) at all times intraoperatively. At lower values the steepness of the haemoglobin dissociation curve limits the oxygen availability for delivery to the tissue and reduces the reserve against tissue hypoxaemia should some other factor, such as blood loss or cardiac depression, supervene.

The transport of oxygen from the air to the tissue cell occurs by alternating steps of convection and diffusion (see Fig. 14.3). Convection carries the oxygen into the lungs during ventilation; diffusion directs the flow of oxygen down the partial pressure (tension) gradient from the alveolar gas to the haemoglobin in the red cell; convection then carries the oxygen out to the tissues in the blood and diffusion down the partial pressure gradient delivers the oxygen to the mitochondria inside the cells where it is consumed and carbon dioxide generated. Anaesthesia and surgery interfere chiefly with the convective processes and there is no evidence that diffusion of oxygen is directly impaired by anaesthetic drugs or surgical procedures. The effectiveness of diffusion may be impaired by increased path length as occurs in oedema (pulmonary or tissue) or by failure to administer sufficient oxygen in the inspired gas mixture but those are indirect results.

Arterial hypoxaemia during general anaesthesia is defined as unexpectedly reduced oxygen tension or saturation. The use of pulse oximetry has revealed that episodes of hypoxaemia are more common and severe than was previously believed. One such survey of all patients undergoing anaesthesia for a variety of surgical procedures reveals mild hypoxaemia (SpO_2 86–90 per cent) in 53 per cent of patients, moderate hypoxaemia (SpO_2 <81–85 per cent) in 34 per cent and severe hypoxaemia (SpO_2 <81 per cent) in 20 per cent. These episodes occurred at all stages of the anaesthetic management (i.e. induction, maintenance, and recovery) but were much more frequent during general

Table 14.2 Causes of reduced arterial oxygen tension

Diminished P_{IO_2}
  Flow meter error
  Cessation of oxygen flow
  Anaesthetic vaporized in air
  Inert gas excretion

Diminished alveolar ventilation
  Respiratory depression
    Central
      Neurological disease
      Premedication
      Anaesthetic agent
    Peripheral
      Neuromuscular
      Relaxant drugs
  Increased respiratory dead space
    Drugs (e.g. atropine)
    Mechanical ventilation
    Hypotension
    Cardiopulmonary disease
    Anaesthesia equipment
  Mechanical impairment
    Airway obstruction
      Oropharyngeal soft tissue
      Vomit or regurgitation
      Uncleared secretions
      Foreign body
      Respiratory disease
    Thoracic wall
      Bone and joint disease
      Posture or immobilization
      Surgical retraction, etc.

Causes of increased $P_{(A-a)O_2}$
  Impaired diffusion
  Increased pulmonary shunt
    Anatomical
    Alveolar
  Increased scatter of ventilation/perfusion ratios
  Increased inspired oxygen tension
  Decreased mixed venous P_{O_2}
    Reduced cardiac output
    Increased oxygen consumption
    Anaemia
  Shift of haemoglobin-dissociation curve
    pH and/or P_{CO_2} effect
    Temperature effect
    Altered erythrocyte 2,3-DPG

than regional anaesthesia.[18] All of the circumstances that can lead to this result can be classified into three categories, reduced inspired oxygen tension (P_{IO_2}), reduced alveolar oxygen tension (P_{AO_2}) or increased alveolar to atrial oxygen tension difference $P_{(A-a)O_2}$. The nature of a disturbance can therefore be readily identified (Table 14.2) with the monitoring equipment in routine use during modern anaesthesia. Thus the inspired oxygen tension is measured with an oxygen sensor in the anaesthetic circuit, saturation of haemoglobin with a pulse oximeter and the end tidal carbon dioxide with a capnometer at the airway.

The diagnosis and treatment of the causes of decreased P_{IO_2} and decreased P_{AO_2} needs to be undertaken rapidly and is usually easy. There are occasions when failure to intubate the trachea seems unlikely because, for example, the cords were visualized, but if carbon dioxide is not detected with the capnograph then any delay in re-intubation is unwarranted. Sometimes airway obstruction, chest wall spasm or bronchial spasm may prevent adequate ventilation; if repositioning of the airway does not improve matters, then a short-acting muscle relaxant will usually restore control in the first two cases and parenteral or aerosolized bronchodilator will improve gas exchange in bronchospasm. Of the bronchodilators, atropine intramuscularly or intravenously, terbutaline, isoprenaline or ipratropium aerosol or glycopyrronium (glycopyrrolate) parentally or instilled into the airway may be effective. It is interesting to note that atropine preoperatively does not usually reduce the quantity of sputum produced.[19] In the presence of bronchospasm thiopentone is not contra-indicated for induction[20] and ketamine[21] and the inhalational anaesthetics do reduce airway resistance.[22]

Because all anaesthetics depress respiration, the arterial carbon dioxide tension will be increased in patients breathing spontaneously.[17] Surgical incision and manipulation may stimulate respiration but only transiently. Short procedures can safely be undertaken while the arterial carbon dioxide tension is allowed to increase to 6.7–8.0 kPa (50–60 mmHg) but the prevailing opinion is that for longer and/or more demanding procedures the body pH should be maintained at approximately 7.4 when the body temperature is 37°C and the P_{aCO_2} is 5.3 kPa.[23] Large departures from these parameters are thought to influence adversely the general functioning of enzymes and the control and physiological functions. This view is not supported by overwhelming definitive evidence nor by the failure to observe adverse outcomes with techniques such as deliberate hyperventilation, or maintenance of pH at 7.4 during deliberate hypothermia. Nevertheless, with so complex a state the prevailing view expressed above provides a rational basis for management decisions.

Assuming that the inspired and alveolar oxygen tensions are normal and that the end tidal carbon dioxide tension is not increased then a persistent desaturation (or decreased P_{AO_2}) is due to increased $P_{(A-a)O_2}$; the causes of which may be intra- or extrapulmonary.

The extrapulmonary causes, which are sometimes significant, include: shifts in the dissociation curve as a result of variations in body temperature, pH or the level of 2,3-diphosphoglycerate which causes alterations in the oxygen tension at which haemoglobin is 50 per cent saturated (P_{50}). More commonly the extrapulmonary causes of an increased $P(A-a)O_2$ are those leading to decreased mixed venous oxygen tension $P\bar{v}O_2$. The mixed venous oxygen content $C\bar{v}O_2$ is defined by:

$$C\bar{v}O_2 = CaO_2 - (\dot{V}O_2/\dot{Q}T) \qquad (7)$$

Therefore, the $C\bar{v}O_2$, and hence the $P\bar{v}O_2$, is decreased when CaO_2 is decreased or when $\dot{V}O_2$ increases and/or $\dot{Q}T$ decreases. During general anaesthesia $\dot{V}O_2$ usually decreases to about 80 per cent of the normal resting value but it is not uncommon for $\dot{Q}T$ to be disproportionately decreased, at least transiently during induction (e.g. with thiopentone) or as a result of increasing depth of anaesthesia or surgical manipulation. If the body temperature has declined during anaesthesia, the restoration of temperature is achieved during the early postoperative period by generalized shivering which is accompanied by large increases (two- to fourfold) in oxygen consumption. If the cardiac output is not similarly increased the $C\bar{v}O_2$ (and $P\bar{v}O_2$) decreases.

The reason that decreased $P\bar{v}O_2$ results in a decreased PaO_2 is because oxygen exchange in the lung is not perfect. Some blood crosses the lung as if not exposed to oxygen exchange (see below); this mixed venous blood (venous admixture) therefore mixes with the oxygenated alveolar capillary blood and reduces the final oxygen tension of blood reaching the aorta. The lower the $P\bar{v}O_2$ the lower the final PaO_2, and, since the PaO_2 is not directly altered by this relation, the greater the $P(A-a)O_2$ difference.

Three-compartment model

The intrapulmonary causes of increased $P(A-a)O_2$ are of particular interest because these are the causes of hypoxaemia with pulmonary disease. In addition the state of general anaesthesia also results in increased $P(A-a)O_2$. The $P(A-a)O_2$ is an effective measure of the efficiency of gas exchange: the greater the value the more oxygen tension is being lost in the exchange process. Traditionally three causes of inefficiency[24] are identified; alveolar dead space ($VDALV$) ventilation/perfusion inequality ($\dot{V}/\dot{Q}$) and pulmonary shunt ($\dot{Q}s/\dot{Q}t$) (Fig. 14.5).

Dead space

The concept of the gas exchange as the result of three compartments is particularly useful clinically because different factors influence each compartment and they can be distinguished with simple techniques. When gas is inspired during ventilation the first gas that enters the alveolar gas is the gas filling the airways. This gas was the last of the alveolar gas from the previous breath. Similarly the last gas entering the airways at the end of inspiration does not take part in the gas exchange and is breathed out

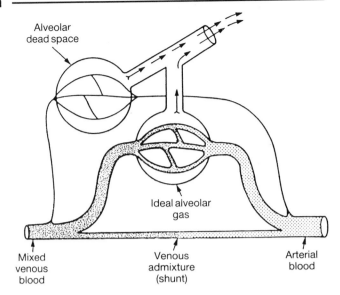

Fig. 14.5 The three-compartment model of gas exchange in the lungs. The upper compartment represents unperfused alveoli the ventilation of which is alveolar dead space. The lower compartment represents non-ventilated alveoli the perfusion of which is venous admixture or shunt. The middle compartment is the only line in which gas exchange is occurring and it is assumed that ventilation and perfusion are homogeneously matched or 'ideal'. In reality there is a spread of ventilation/perfusion mismatching in the gas exchange compartment but this is calculated in this model as contributing both dead space to the upper compartment and venous admixture to the lower one. These concepts permit the complexity of the gas exchange to be reduced to measurable quantities. (From reference 24 with permission.)

again at the beginning of expiration. This to-and-fro ventilation of the airways is wasted ventilation or anatomical dead space. The alveolar dead space, as its name suggests, is the result of ventilation of alveoli that are not perfused and therefore not taking part in gas exchange. Together the alveolar and anatomical dead spaces compose the physiological dead space which is normally about one-third of the tidal volume and can be calculated using equation (1).

The alveolar dead space is quantified by substituting the end tidal carbon dioxide tension $PETCO_2$ for $PECO_2$ in equation (1). The anatomical dead space is obtained as the difference between the physiological and the alveolar dead space.

The anatomical dead space may be increased by disease and by some drugs (i.e. bronchodilators) but generally not to an important extent. In contrast, the alveolar dead space is not only increased in association with many pulmonary diseases but is also increased whenever the distribution of blood flow is impaired. Institution of mechanical ventilation, positive end-expiratory pressure (PEEP) and hypotension all increase alveolar dead space, often dramatically. During deliberate hypotension in subjects with normal lungs, the dead space may increase from the normal 30 per cent to 70 per cent of tidal volume. In such circumstances the minute ventilation must be increased if hypercapnia is to be avoided.

How to achieve such an increase in ventilation in patients with chronic lung disease has long been of concern. Patients with restrictive disease often exhibit limited compliance so that excessive pressures accompany ventilation, and 'trapping' of gas occurs with obstructive diseases when all the inspired gas is not exhaled and FRC increases. In patients with restrictive lung disease increased ventilation is best achieved by increasing the rate, while in those with obstructive lung disease the volume for each breath can be increased by prolonging the breaths at slower flow rates. There is no one formula to guide the selection of mechanical ventilation settings but the process is one of arriving at a compromise that permits effective oxygen and carbon dioxide exchange.[25]

During mechanical ventilation the use of a tracheal tube results in a small reduction in dead space; however, an inappropriate selection of equipment (i.e. adult circuit for an infant) connecting the ventilator to the tracheal tube may impose an additional source of wasted ventilation in the form of an apparatus dead space.

Ventilaton/perfusion ratio distribution and shunt

Normally the alveolar ventilation in an adult is about 5 l/min and the perfusion is about 6 l/min so that the mean ventilation/perfusion ratio overall is about 0.8. If all alveoli received this same ratio of ventilation to perfusion the gas exchange would be optimized. However, there is normally a small gravitationally determined vertical gradient such that the ratio decreases from the apex to the base of the lungs. As a result the alveolar oxygen tension increases in passing from the dependent to the non-dependent regions of the lung. Because of the shape of the oxygen dissociation curve for haemoglobin, the final oxygen tension obtained by mixing equal volumes of blood, one of which has a saturation of 100 per cent (P_{O_2} = 20 kPa; 150 mmHg) and the other a saturation of 50 per cent (P_{O_2} = 3.3 kPa; 24.7 mmHg), will be a saturation of 75 per cent (i.e. mean of 100 and 50) but the P_{O_2} of the mixture will be approximately 5.3 kPa (39.7 mmHg), much less than the mean of 20 and 3.3 kPa. For oxygen the greater the spread of $\dot{V}/\dot{Q}$ ratios the greater the difference between the mean alveolar and the mean arterial oxygen tension and the greater the $P(A-a)O_2$. This effect is exaggerated with increased blood flow from the low $\dot{V}/\dot{Q}$ regions (i.e. those with the lowest oxygen tension).

The physiological basis for this gravitationally determined variation in $\dot{V}/\dot{Q}$ has been nicely described by the difference between the alveolar gas pressures and the pulmonary artery and pulmonary venous pressures. The zones (see Fig. 14.2) described by West[4] provide a conceptual basis for understanding how the gas pressure and vascular pressures interact to alter dead space and gas exchange. However, it is now recognized that even in normal lungs there is a non-gravitationally determined increase in flow from the hilum to the periphery and in essentially all chronic lung diseases the distribution of $\dot{V}/\dot{Q}$ that is responsible for the severe hypoxaemia is not gravitationally determined.

As emphasized above the inefficiency of gas exchange that results is attributable not to the mean $\dot{V}/\dot{Q}$ but to the distribution around the mean. A convenient way to summarize the abnormality for the blood flow is as the standard deviation (SD) of the log of the $\dot{V}/\dot{Q}$ ratio (ln SD(Q)) and for the ventilation as ln SD(V). A disease state or the effects of a drug can then be characterized as changes in the mean of the log ln SD(Q) or ln SD(V). Another advantage of this approach is that the distribution of the $\dot{V}/\dot{Q}$ ratio can be estimated independently by infusing a mixture of inert gases (usually six) of widely differing solubility and calculating what distribution would result in the same retention and excretion of each gas. While this technique depends on a number of assumptions, nevertheless it yields predictions that match both the gas exchange and more recently, the haemodynamic changes that are observed as a result of disease, anaesthetic drugs and respiratory manoeuvres.

It can be appreciated that dead space corresponds to a $\dot{V}/\dot{Q}$ ratio that is infinitely large. At the other extreme a $\dot{V}/\dot{Q}$ ratio of 0 corresponds to regions of the lung that are perfused but not ventilated.

The extent of this latter abnormality can be calculated by estimating what percentage of the cardiac output would have to be mixed venous blood in order for the final arterial mixture to have the observed $P(A-a)O_2$. This so-called venous admixture is therefore calculated by:

$$\dot{Q}s/\dot{Q}_T\% = (Ca_{O_2} - C\bar{v}_{O_2})/(Cc'_{O_2} - C\bar{v}_{O_2}) \qquad (8)$$

Although perfusion of atelectatic lung is accurately represented by this calculation there are two other sources of blood flow that are also included. One is the blood flow that returns from the bronchial circulation into the pulmonary veins and the other the thebesian veins that return venous blood from the ventricular wall. This blood flow, called the anatomical shunt, is a small fraction of the cardiac output, normally less than 3 per cent, but it is important to note that although the blood involved is venous, it does not necessarily have the same oxygen tension as the mixed venous blood that is assumed in the venous admixture calculation. In addition in some congenital cardiac abnormalities or pulmonary vascular obstructions the bronchial circulation may become greatly expanded.

The second additional source of venous admixture is the abnormality attributable to the spread of $\dot{V}/\dot{Q}$ ratios. The venous admixture is calculated with the assumption that the end capillary oxygen tension is equal to the alveolar oxygen tension. All the sources of $P(A-a)O_2$ are therefore included. It is also assumed that the admixture is of blood with mixed venous oxygen composition but clearly that will not be true in this instance because all $\dot{V}/\dot{Q}$ ratios greater than zero will deliver blood with oxygen tension greater than mixed venous blood.

Despite these qualifications the calculation of venous admixture is very useful because it can be measured readily and in most clinical circumstances allows a distinction between atelectasis and increased $\dot{V}/\dot{Q}$ ratio distribution. Venous admixture measured while breathing 100 per cent oxygen is due only to atelectatic lung regions and the usually very small anatomical shunt, while that measured when breathing air contains all three components. If the venous admixture is constant on air and oxygen then atelectasis is the cause, if substantially less when breathing oxygen than air then an increased $\dot{V}/\dot{Q}$ distribution is the cause.

General anaesthesia

Hypoxaemia resulting from respiratory depression, airway obstruction, circulatory collapse or inappropriate inspired gas mixtures is a potential complication of any anaesthetic technique whether local, regional or general. These events often develop rapidly and may become life threatening but the pathophysiology is well understood (see Table 14.2) and the conduct of safe anaesthesia is to a large extent designed to prevent, diagnose and treat these situations as discussed previously.

However, it is also recognized that during general anaesthesia there is an additional cause of hypoxaemia that is due to a deterioration in the pulmonary exchange of oxygen.[16,26] This phenomenon has received much attention from investigators and a substantial body of data is therefore available and yet this source of hypoxaemia is widely misunderstood and its fundamental causes remain essentially unknown. A current description of the pathophysiology is summarized as follows.

The induction and maintenance of general anaesthesia results in a decreased resting lung volume so that the functional residual capacity (FRC) and the compliance of

Table 14.3 Changes from the awake state in human subjects for shunt, venous admixture and low $\dot{V}A\dot{Q}$ perfusion with type of general anaesthetic and method of ventilation[†]

	Injectable anaesthetics		Reference	Inhalational anaesthetics		Reference
Spontaneous respiration						
From oxygen measurement				Sh = + 3*	[6.1 ± 8.3]	27
				Sh = 9.5	[14.8 ± 17.7]	28
				Sh = 1.7		29
				VA = 13	[18.2 ± 20]	27
From inert gas measurement	Sh = + 0.4	[0.9 ± 3]	31	Sh = + 2.5	[10.5 ± 5.1]	29
	Sh = + L = + 2.8	[3.9 ± 7]	31			
Mechanical ventilation						
From oxygen measurement	VA = + 5	[8 ± 3]	30	Sh = + 8	[11.1 ± 17]	27
				VA = + 9.2	[15.1 ± 25]	27
				VA = + 3.7	[9.2 ± 10]	32
From inert gas measurement	Sh = + 2.6	[3.1 ± 4]	31	Sh = + 5.3	[6.5 ± 11]	33
	Sh + L = + 5.7	[6.8 ± 6]	31	Sh = + 4.5	[4.8 ± 8.2]	32
	Sh = − 1	[1.0 ± 1.6]	35	Sh = + 2	[2.6 ± 8]	34
	Sh + L = − 3.5	[9.9 ± 5.2]	35	Sh = + 13.8	[15.1 ± 10.2]	36
				Sh + L = + 10.2	[11.9 ± 16]	33
				Sh + L = + 7.7	[9.8 ± 91]	32
				Sh + L + 6.2	[12.2 ± 8]	34
				Sh + L = + 32.2	[37.8 ± 12.3]	36

* For example: this entry +3 is the mean change in per cent shunt from the awake state; [6.1 ± 8.3] is the actual mean ± one SD for the per cent shunt during anaesthesia.

† Sh, shunt from oxygen or inert gas technique; VA, venous admixture from oxygen measurements; Sh + L, shunt plus low $\dot{V}A/\dot{Q}$ regions measured by inert gas technique.

the lung and chest wall are decreased. These changes lead to the development of atelectasis in dependent regions of the lung and pulmonary shunting and perfusion of low $\dot{V}/\dot{Q}$ regions increases. Hypoxic pulmonary vasoconstriction (HPV) reduces the blood flow to both atelectatic and low $\dot{V}/\dot{Q}$ regions so as to reduce the effect of the abnormalities on gas exchange; but anaesthetic agents inhibit HPV and the deterioration in oxygenation is exacerbated. The result of all these changes is that pulmonary shunt and the ln SD(Q) are increased during general anaesthesia; thus alveolar to arterial oxygen tension difference (P(A−a)O_2) is increased and at any inspired oxygen tension the arterial oxygen tension is less than it would be in the awake state. While each statement in this sequence can be supported by published data none is without some controversy and these qualifications are discussed in more detail below.

The occurrence of an unexpectedly low arterial oxygen tension during general anaesthesia has been reported in hundreds of papers over three or four decades. Because of the normal variability in the efficiency of oxygen exchange, the rigorous examination of the question requires that measurements be obtained before as well as during anaesthesia. Studies that meet these criteria for human subjects are collected in Table 14.3 both for direct measurements of oxygen exchange by shunt and/or venous admixture and the indirect measurements that correlate with oxygen exchange using the multiple inert gas technique to evaluate shunt and/or shunt plus low $\dot{V}/\dot{Q}$ regions. The table demonstrates that for inhalational anaesthesia, however measured, this change is always positive from the awake to the anaesthetized state, demonstrating that a deterioration in lung function occurs. With injectable anaesthesia the values change less and one study accorded improvement in gas exchange during anaesthesia with mechanical ventilation. The variability is large and the values of the mean ± 1 SD demonstrate that in some individuals significant arterial hypoxaemia will occur. In general there is no difference

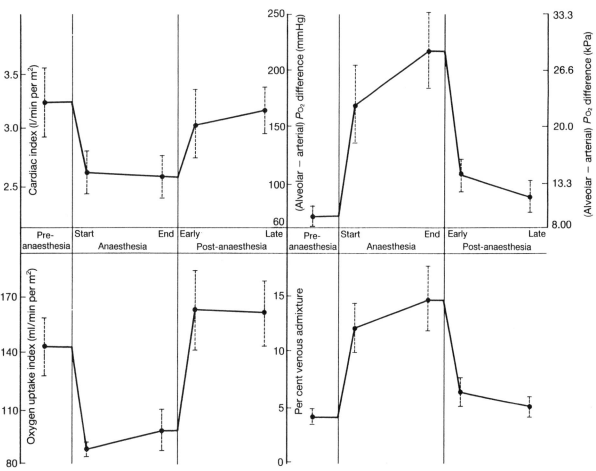

Fig. 14.6 Changes in oxygen tension and venous admixture before, during and after halothane anaesthesia with mechanical ventilation. Note that the P(A−a)O_2 increases as soon as anaesthesia is induced (30 min) and persists while anaesthesia is maintained (2.5 h) but reverses towards normal as the anaesthetic is eliminated (early = 40 min and late = 3.5 h). These changes are not due to extrapulmonary causes because the cardiac output and oxygen consumption decrease approximately proportionately. The intrapulmonary abnormality associated with anaesthesia is calculated as venous admixture and because the inspired gas approximates 100 per cent oxygen, the cause is increased pulmonary shunt. (From reference 28 with permission.)

Table 14.4 Per cent change in functional residual capacity with general anaesthesia in human subjects

Anaesthetic	Intravenous*	Reference	Inhalational†	Reference
Spontaneous respiration	−33	40	−30	45
	−23	41	−20	46
	−14	42	−16	47
	−28	43	−16	48
	$\bar{x} = -24.5$		$\bar{x} = -21$	
Mechanical ventilation	−33	40	−18	49
	−11	44	−15	50
	−25	41	−17	51
	−8	42	−15	52
	−26	43	−15	53
	$\bar{x} = -21$		$\bar{x} = -18$	

FRC measured with a variety of techniques.
* Intravenous agents include barbiturates, narcotics and benzodiazepines.
† Inhalational agents include halothane, cyclopropane, methoxyflurane and nitrous oxide.

between spontaneous or mechanical ventilation but the paucity of studies with intravenous agents is striking.

Some studies not listed in this table have attempted to compare intravenous and inhalational agents by changing from one agent to the other during the course of the anaesthetic but there is strong evidence that the events that lead to the impairment of gas exchange occur rapidly with the induction and are reversed almost as rapidly during emergence (Fig. 14.6) and that awakening is required ·to reverse the effects of the induction agent.[37,38]

Most of the research into the mechanisms responsible for the deterioration in oxygen exchange have focused on alteration of pulmonary mechanics.[39] Numerous sophisticated studies have established unequivocally that the resting lung volume or functional residual capacity, and pulmonary compliance are reduced. In Table 14.4 are collected the reports of changes in functional residual capacity with onset of general anaesthesia. The table shows a remarkable homogeneity in the approximately 20 per cent reduction of FRC that is observed whenever general anaesthesia is induced whether by injectable or inhalational agent and whether ventilation is spontaneous or mechanical. That a decreased resting lung volume might lead to impairment of gas exchange appears straightforward. First the normal distribution of ventilation/perfusion ratios is in part determined by gravity in the normal lung and both blood flow and ventilation decrease on vertically ascending the lung. When the lung volume is reduced each region moves down the pressure–volume curve for the lung, so that, in contrast to the awake state, the more compliant lung regions are now in non-dependent regions. Now the ventilation and perfusion distribution is altered so that, particularly with mechanical ventilation, the decrease in blood flow remains but ventilation increases on ascending

the lung and the spread of ventilation/perfusion ratios is increased. Venous admixture and perfusion of low $\dot{V}/\dot{Q}$ regions will therefore increase but this does not explain the increased shunting which is a prominent change observed with anaesthesia. In patients with lung disease where oxygenation is already impaired by increased spread of $\dot{V}/\dot{Q}$ ratios general anaesthesia exacerbates the problem but in these patients the abnormal distribution in ventilation and perfusion is seldom dependent on gravity and the effects of the anaesthetic are due to a combination of actions.

In healthy young subjects at some lung volume between the functional residual capacity and the residual volume, some small airways close. This volume is the closing capacity (CC). With increasing age the closing capacity increases at a greater rate than the functional residual capacity and in supine subjects over the age of about 45 years they become equal. Consequently in older subjects, at some point in tidal expiration, small airways close and ventilation of those regions becomes decreased. With general anaesthesia the FRC decreases but CC usually does not, so that airway closure during tidal ventilation occurs routinely in subjects over the age of 35 years (Fig. 14.7). These changes also contribute to the development of low $\dot{V}/\dot{Q}$ regions in the lung but again they do not by themselves account for pulmonary shunting.

Some of the earliest papers reporting the occurrence of hypoxaemia during anaesthesia recognized that atelectasis was probably occurring but it was not until the application of computerized tomography to this question that atelectasis was confirmed and its site shown to be in the dependent regions of the lung. In a series of convincing studies Hedenstierna[26] and his colleagues have demonstrated that general anaesthesia is accompanied by a very rapid development in dependent atelectasis, which is

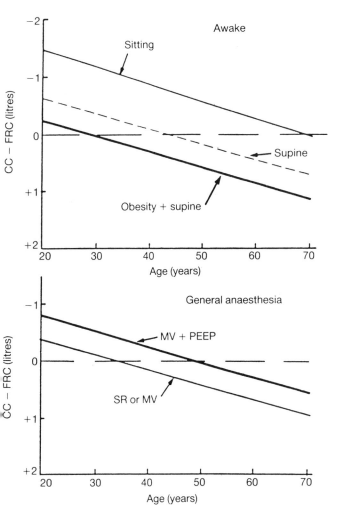

Fig. 14.7 Changes in the relation between closing capacity (CC) and functional residual capacity (FRC) with age. When (CC−FRC) is less than zero some small airways will close during part or all of the respiratory cycle with consequent loss of efficiency of gas exchange. Upper panel: in awake subjects the cephalic movement of the diaphragm when posture changes from the sitting to the supine reduces FRC but not CC so that the difference line crosses zero at a younger age and this is further exaggerated with obesity. Lower panel: during general anaesthesia the FRC is further reduced with little change in CC so that some airways will be closed even in healthy subjects under the age of 35 and is only partially reversed by mechanical ventilation (MV) with PEEP. SR, spontaneous respiration. (Data from references 54, 55 and 56.)

partially reversed by positive end-expiratory pressure, occurs with spontaneous and mechanical ventilation and the extent of which correlates approximately with the increases in shunt and shunt plus low $\dot{V}/\dot{Q}$ regions measured with the multiple inert gas techniques.

The coincidence of a decreased FRC and atelectasis seems natural and this is further reinforced by the observations that the injectable drugs thiopentone (pentobarbital) and ketamine appear to be exceptions. Ketamine does not reduce FRC or cause impaired oxygen exchange in children.[57] In adults ketamine with spontaneous respiration is not associated with the development of atelectasis (by

tomography) nor with significant increase in shunt.[31] And finally in sheep[58] and dogs[59] thiopentone (pentobarbital) is not associated with decreased FRC, atelectasis or abnormal oxygen exchange.

Despite this, the relation between FRC and atelectasis is not a simple one. Decreasing the FRC, to the same extent as during anaesthesia, by binding the chest wall and abdomen, does not induce atelectasis or abnormal gas exchange either in the sitting[60] or supine[61] position. Application of PEEP while increasing the FRC as noted above has an unpredictable effect on the arterial oxygen tension. Furthermore, the explanation cannot simply be that absorption atelectasis is occurring because there is no difference in the changes observed[36] with nitrogen or nitrous oxide, although the latter is 25 times more soluble and diffusible and would be absorbed proportionately faster.

The basic causes for the observed changes in lung mechanics therefore remain unknown although it seems likely that both central neurogenic and local factors contribute. Changes in the relative position of the chest wall and diaphragm[43,62,63] appear to be critical and they are the result of a complex interaction involving central depression in respiratory drive, reduction of muscular tone, changes of lung and chest wall compliance, cranial movement of diaphragm and abdominal contents, changes in tissue and airway resistance and thoracic blood volume. Details are important for this analysis and clearly more attention needs to be paid to the influence of the induction agent, the use of relaxants and intubation of the trachea.

The extent to which a particular degree of atelectasis or low $\dot{V}/\dot{Q}$ region causes a decrease in arterial oxygen tension depends first on the activity of hypoxic pulmonary vasoconstriction (HPV). When a region of the lung becomes hypoxic the small pulmonary arteries serving that region constrict and their blood flow is diverted to less hypoxic regions. The result is an increased efficiency in oxygen exchange but with an increased pulmonary vascular resistance. The simplest example is atelectasis for here there is no alveolar gas and the oxygen tension $P_{\mathrm{s}}O_2$ that stimulates HPV is that of the mixed venous blood. HPV is maximally stimulated when $P_{\mathrm{s}}O_2$ or in the case of atelectasis $P_{\bar{v}}O_2$, is 4 kPa (30 mmHg) so even the normal $P_{\bar{v}}O_2$ of 5 kPa (37.5 mmHg) provides a near maximal and highly efficient reduction in the blood flow to atelectic lung regions.[5] The effectiveness of the response depends on the magnitude of the atelectasis. If 50 per cent of the lung is atelectatic HPV can reduce the blood flow to about half the normal flow and a shunt of 25 per cent would be the result; but if only a small subsegment, say 10 per cent of the lung, is atelectatic the flow may be reduced 70 or 80 per cent and a shunt of 2 or 3 per cent would be observed. Some concept of the degree of abnormality developing during anaesthesia can be obtained by looking again at the values for shunt

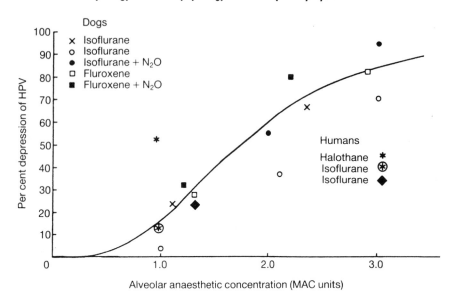

Fig. 14.8 Depression of HPV by inhalational agents demonstrates a sigmoid dose–response relation in both animal and human subjects. (Data from references 65, 66, 67 and 68.)

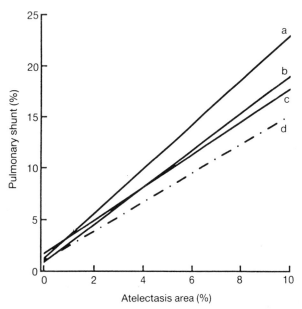

Fig. 14.9 Correlation of atelectasis area per cent and pulmonary shunt per cent. Transverse computed tomography scans of the lower thorax permitted the identification of dependent densities; these were quantified by comparing this area with total intrathoracic area and expressed as the atelectasis area per cent. Shunt was measured by the multiple inert gas technique. All data are for human subjects during general anaesthesia with inhalational agents halothane or enflurane for lines a, b and c, and the injectable agent ketamine for line d. Note the smaller slope of the line from ketamine suggests that hypoxic pulmonary vasoconstriction with the injectable anaesthetic preserves oxygenation better than with the inhalational agents. Data for lines a, b, c and d are from references 32, 33, 69 and 31, respectively.

that have been observed during anaesthesia in Table 14.3. From oxygen measurements mean values of 10 per cent with a standard deviation of 10 per cent are a reasonable estimate. In the presence of active HPV, 10–20 per cent shunt would be expected from an atelectasis involving 20–

40 per cent of the lung. Such changes are consistent with the densities observed with computerized tomography.

Of particular importance to anaesthetists is the recognition that a number of factors can reduce the activity of HPV.[64] Trauma, sepsis and vasodilator drugs (e.g. nitroprusside, nitroglycerin, nitric oxide and isoprenaline) may reduce or abolish HPV, while in the presence of drugs that vasoconstrict the pulmonary vasculature (e.g. noradrenaline, dopamine and histamine) the effectiveness of HPV may be reduced. Furthermore, all the inhalational anaesthetics inhibit HPV in approximately the same dose-related manner (Fig. 14.8) while injectable anaesthetics do not. There have been no definitive studies comparing the effects of injectable and inhalational agents but Fig. 14.8 suggests that on average, HPV is inhibited by 20–40 per cent with inhalational agents in the clinical range of 1–1.5 MAC alveolar concentrations. It is therefore interesting to note in Fig. 14.9 that the correlation between the degree of atelectasis on computerized tomography and the shunt, measured by multiple inert gas techniques, has a steeper slope with the inhaled anaesthetics and that this difference from the injectable anaesthetics amounts to a shunt that is 16–50 per cent greater in the three studies with inhalational agents; such a difference is accounted for by inhibition of HPV.

Conclusions

To summarize this discussion two points deserve emphasis. The first is that most patients do not develop severe hypoxaemia during anaesthesia, the impairment is moderate and an adequate arterial oxygen tension is maintained with inspired oxygen concentrations of 25–30 per cent. However, it is those few who do develop hypoxaemia that are the focus of attention and therefore in analyisng results in the literature it is not the mean values that are so important for the individual but the variability around the mean or the standard deviation. This essential point has been emphasized previously.[36] It is evident that in any one individual the normal deterioration may be exacerbated by trauma, drugs, disease, or obesity and that even in individuals otherwise free of cardiopulmonary disease any of the pathophysiological events, increased spread of $\dot{V}/\dot{Q}$ ratios, occurrence of atelectasis or inhibition of HPV, may be increased.

The second point concerns the debate about whether the occurrence of atelectasis or inhibition of HPV is more important in the aetiology of hypoxaemia during general anaesthesia. The controversy is a spurious one for both are necessary. Without atelectasis there would not be an increased shunt, but without HPV, hypoxaemia would be a problem in almost every patient and one-lung anaesthesia would always be untenable without additional measures to reduce blood flow to the collapsed lung. Of the individual variables mentioned above the extent of atelectasis is increased with obesity and some disease states but most of the variables, including drugs, trauma, sepsis and disease, exert their adverse influence principally by inhibition of HPV.

The corollary of this conclusion for the clinical management of patients is to recognize that increased pulmonary shunt and increased ventilation/perfusion abnormality will occur with general anaesthesia, but if hypoxaemia also occurs, and the extrapulmonary causes (hypoventilation, hypoxia or circulatory failure) can be ruled out, then it is most probable that the cause is impairment of HPV.

Postoperative period

For practical reasons it is useful to divide the pulmonary changes in the postoperative period into two phases, the early and the late. The early phase is associated with reversing the effects of anaesthesia and the surgical manipulations while the late phase concerns complications that arise during the postoperative period. It should be appreciated that this division into early and late phases was introduced[16] to emphasize the concepts and, as will be apparent below, very often no actual separation will be detectable.

Early postoperative

With most modern anaesthetic techniques the recovery of consciousness and physiological function is rapid with the termination of the surgical procedure. An exception to this is following the use of high dose narcotic techniques for cardiac surgery and, as discussed previously, mechanical ventilation for 12 hours or so postoperatively is usually necessary in such patients.

Following general anaesthesia the early postoperative period is often associated with fluctuations in arterial saturation even when pulmonary function is normal. Some of the causes, such as increased oxygen consumption with shivering, residual ventilatory depression with injectable anaesthetics or fluid overload with pulmonary oedema have already been referred to previously. However, it is important whenever ventilation is inadequate to consider also whether the reversal of muscle relaxant or narcotic agonist have been inadequate. Simple bedside tests of muscle strength and ventilation measure the ability of the patient to lift his/her head off the pillow and the inspiratory pressure or expiratory flow rate.

Restoration of normal spontaneous ventilation may be delayed following periods of prolonged hypercapnia. During hypercapnia the pH of the cerebrospinal fluid (CSF) is at first acid and respiration is stimulated at the medullary chemocentres. With time CSF bicarbonate is retained and CSF pH returns toward normal. If at this time, the anaesthetic drugs reducing the respiratory drive are reduced an initial decrease in arterial carbon dioxide tension is reflected in a marked CSF alkalosis and central ventilatory drive is reduced until the CSF bicarbonate is excreted.

In some patients with chronic lung disease, part of the respiratory drive results from arterial hypoxaemia, and typically there is some degree of hypercapnia in the preoperative period. While this is not apparent during the intraoperative period, it may complicate the restoration of spontaneous ventilation if undue hypoxaemia is to be avoided in the postoperative period. It is also worth noting that the hypoxic ventilatory drive is abolished by even subanaesthetic concentrations of inhalational anaesthetics. The rules for trial of extubation and spontaneous ventilation in such patients are the same as for any patient following respiratory intensive care (Table 14.5) with the recognition that patients who retain carbon dioxide chronically may be extubated with some degree of hypercapnia.

Additional special causes of arterial desaturation follow the use of 70 per cent nitrous oxide in the anaesthetic gas mixture. If this anaesthetic is abruptly terminated the large volume of nitrous oxide that is excreted into the lung so dilutes the inspired gas that 'diffusion hypoxia' occurs. This effect is transient and easily avoided by ventilating with

Table 14.5 Criteria indicating continued need for, or successful weaning from mechanical ventilation

	Need for ventilation	Successful weaning from ventilator
Vital capacity (ml/kg)	$\leq$ 15	> 15
Inspiratory force (cmH$_2$O)	$\leq$ −25	> −25
Respiratory rate	$\geq$ 30	< 30
F/V$_T$ (respiratory rate ÷ tidal volume)	$\geq$100	< 100
Minute ventilation (l/min)	> 10	$\geq$ 10
PaO$_2$ (PEEP < 5 cmH$_2$O; FIO$_2$ < 0.50)	< 60	$\geq$ 60

oxygen for a few minutes after disconnecting nitrous oxide. The other special abnormalities of lung function that are associated with the intraoperative period diminish as the anaesthetic agents are cleared. Following inhalational and short-acting injectable agents are largely reversed in the first postoperative hour. In particular the dependent atelectasis, increased low $\dot{V}/\dot{Q}$ regions and increased $P(A-a)O_2$ are restored to normal. However, in some circumstances this does not occur so that the signs and symptoms of the early and late complications of the postoperative period are merged.[70]

Late postoperative phase

If the early postoperative phase is dominated by the reverse of the effect of anaesthesia, the late postoperative period reflects the onset of pulmonary abnormalities resulting from the effects of surgery. The predominant complications are atelectasis, pneumonia and thromboembolism.

Atelectasis and pneumonia result primarily from reluctance to move the lungs so that sighing, coughing and changing position are reduced. It can be readily appreciated that old age, chronic lung disease, obesity, abdominal and thoracic incisions and pain are the exacerbating factors. Although the atelectasis is typically in dependent lung regions, these complications are distinct from those that occur intraoperatively. This conclusion is supported by the observation that atelectasis in the late postoperative period has the same incidence whether general or regional anaesthesia is employed intraoperatively. In these patients atelectasis and impaired gas exchange are established by the first to third postoperative

days. If a regional anaesthetic was employed during surgery the deterioration in pulmonary function only becomes evident postoperatively. In contrast, following a general anaesthetic, atelectasis established during anaesthesia may seem to persist for a prolonged period or merge with the new causes promoting atelectasis postoperatively.

Atelectasis occurs when those forces which tend to preserve alveolar patency are overcome by those forces which tend towards alveolar collapse. Normal respiration, sighs and coughs clearly have a role in maintaining alveolar patency and functional residual capacity, while splinting and impaired diaphragmatic motility invariably lead to atelectasis. Haldane identified the physiological effects of splinting in 1919.[71] Since then, researchers have shown that the postoperative or post-injury period is associated with significant changes in pulmonary mechanics[72–74] including reduction in vital capacity (VC) and functional residual capacity (FRC). Beecher[72] described the FRC or 'subtidal volume' as 'the most significant of all lung volumes' and noted its decrease following laparotomy in 1932. Several subsequent studies have shown that the vital capacity is reduced to about 40 per cent of preoperative values after upper abdominal surgery, and the vital capacity declines to 70 per cent of preoperative values.[75–77] While various mechanisms are involved in these changes in mechanics, pain, due to the incision, plus the effects of rapid shallow breathing, without periodic sighs, lead to the fall in FRC. Impairment of the cough mechanism, because of pain, leads to retained pulmonary secretions, which may become inspissated and produce obstruction, or infected and produce pneumonia.

Alleviation of pain in the injured or postsurgical patient has both humanitarian and therapeutic consequences. A variety of different studies have shown that different approaches to pain management can shorten stay, lessen morbidity[78] and modify the pulmonary sequelae of surgery. Intermittent injection of opiates was the standard treatment for many years, and it was only in the mid-1980s that studies compared the utility of this approach with alternatives. Modern pain management options include patient-controlled analgesia (PCA), epidural analgesia with local anaesthetics, opiates or combinations of the two, and regional blockade with local anaesthetics, as with intercostal nerve blocks, intrapleural instillation of local anaesthetic, or continuous axillary block. Intravenous non-steroidal anti-inflammatory agents are a new addition to the arsenal of approaches available for treating pain in the postoperative period.

Each of the aforementioned approaches has advantages and disadvantages (Table 14.6). Traditional bolus administration of opiates by the intramuscular or subcutaneous route suffers from several drawbacks. There is generally a significant delay between perception of pain by the patient

Table 14.6 Advantages and disadvantages of a variety of approaches to acute, postoperative or post-injury pain

Technique	Analgesia	Sedation	Respiratory depression	Respiratory mechanics	Early mobilization	Risks
PRN injection	Intermittent	Intermittent	Intermittent	—	—	Minimal
Patient-controlled analgesia	Good	Minimal	Minimal	—	—	Minimal
Epidural narcotic	Excellent	Minimal	Can be profound	↑VC, FRC vs PRN	++	Infection, neural damage
Epidural narcotic with local anaesthetic	Excellent	Minimal	Can be profound	↑VC, FRC vs PRN	++	Infection, neural damage
Regional	Excellent	Absent	Minimal	↑VC vs PRN	—	Risks of local anaesthetic
Non-steroidal anti-inflammatory	Comparable to opiates	Absent	None	—	—	GI, renal, ↓platelet fxn

++ indicates studies where early mobilization was achieved, and a dash indicates a lack of experimental data.

and attainment of adequate opiate blood level. Intermittent injections are required, which may be uncomfortable for the patient and require interventions by busy nursing staff. Most importantly, blood levels cycle between inadequate and supratherapeutic with intermittent bolus administration. The use of PCA generally results in more consistent patient comfort and lower total narcotic use, but requires a cooperative patient who has been given some instruction in its use.

Epidural narcotics may cause delayed and profound respiratory depression if administered or monitored inappropriately;[79] and the use of even low concentrations of local anaesthetics in the epidural space may be accompanied by hypotension and motor blockade. Regional blocks are often technically difficult and unsuited for sustained administration. PCA programmes require a significant investment in equipment and education both for patients and nursing staff. These impediments might prohibit the implementation of new approaches to pain management were it not for the growing recognition that effective pain management can have demonstrable effects on patient length of stay, morbidity and mortality.

Following thoracotomy, patients receiving epidural narcotics are reported to have both better pain control and significantly better FVC and FEV_1 than patients receiving an intravenous regimen. Ullman[74] compared two groups of patients with rib fractures and additional studies have demonstrated that patients treated with epidural morphine were ventilated for a shorter period, had a lower incidence of tracheostomy and spent less time in the ICU and the hospital than controls treated with intravenous morphine. Epidural analgesia significantly improves ventilatory function with flail chest injuries.[80,81] Morbidly

obese patients who underwent gastroplasty with postoperative epidural analgesia had fewer pulmonary complications and left the hospital sooner than controls,[82] which was attributed to earlier ambulation. Transcutaneous electrical nerve stimulator (TENS) units[83] and intercostal nerve blocks[84,85] can partially restore vital capacity towards normal, when compared to standard perioperative pain management. The improvement in postoperative vital capacity is seen both after thoracotomies and upper abdominal operations, and in a variety of techniques, when contrasted to standard management (Fig. 14.10).

While chronic pain management is a well established discipline, acute perioperative pain management has only recently gained acceptance. Practitioners have a number of alternatives for the management of postoperative or post-injury pain, and can tailor their approach based on the risk/benefit ratio for an individual patient.

The most important concern of management during this period is the provision of adequate pain relief while at the same time encouraging early mobilization and chest physiotherapy. The traditional treatment of pain by intermittent and uncontrolled bolus injections of narcotics has been supplemented with a variety of other techniques including patient-controlled narcotic administration, epidural narcotics, epidural and local analgesia and intercostal block or infiltration of incision site with local anaesthetics. Each of these special procedures have strong advocates and there is a general impression that with any of them the patient's well-being is improved. However, it is not clear that respiratory function is specifically benefited and each of these techniques is associated with some additional potential complications. For example, narcotics, whether patient controlled or epidural, may interfere with

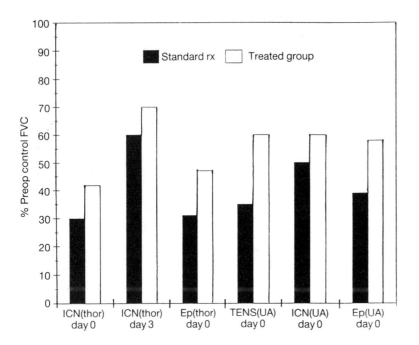

Fig. 14.10 Summarizing several studies comparing standard (intravenous or intramuscular opiates) pain management to alternatives. Improved vital capacity is seen in thoracic (thor) and upper abdominal (UA) operations with epidural analgesia (Ep), transcutaneous electrical nerve stimulation (TENS), and intercostal nerve blocks (ICN). Enhanced VC is evident immediately postop (day 0) and extends well into the recovery period (day 3). See text for references.

respiratory drive and respiratory failure may be precipitated occasionally but unpredictably. Epidural local analgesia may impair movement and autonomic function and with local infiltration, anaesthetic toxicity and increased infection rate at the wound site have been reported.

An additional interesting correlation is that between the occurrence of apnoea, the administration of narcotics, and disturbance of sleep patterns. It has been reported that patients receiving infusions of morphine demonstrate a far greater number of periods of desaturation than those receiving regional analgesia for the first 2 hours postoperatively.[86] These hypoxaemic episodes occurred during sleep and were related to obstructive apnoea. Administration of oxygen did not alter the frequency of apnoea and therefore the morphine is probably exerting an effect on the central oscillatory drive. That some change of an even more general nature is occurring in the postoperative period is suggested by the observation that sleep apnoea is common during the rapid eye movement (REM) stage of sleep. Sleep patterns are grossly disturbed after anaesthesia and surgery and when REM sleep reappears, about 24 hours after surgery, it occurs to a greater extent and hypoxaemic episodes coincide with its appearance. The significance of these observations for the postoperative care of patients remains to be established.

The principal preventive measure against atelectasis is therefore a 'stir up' regimen with incentive spirometry, respiratory therapy, early mobilization from bed rest to chair sitting and walking. If atelectasis and/or pneumonia is suspected, early and aggressive treatment is essential. Measurements of lung volume demonstrate severe impair-

ment of vital capacity postoperatively, and even for such apparently minor procedures as haemorrhoidectomy a 30 per cent decrease is usual. The duration of these deficiencies varies with vital capacity which is only partially restored 6 months after pneumonectomy compared with full recovery 2 or 3 days after lower abdominal or peripheral procedures. Arterial hypoxaemia and postoperative pulmonary function generally reach their nadir at 24–48 hours following thoracic and abdominal surgery and this is the period to be especially watchful if complications are becoming established. Complications occurring beyond this time are not usually considered the direct result of the anaesthesia and surgery.

Venous thrombosis, particularly of pelvic or lower abdominal veins, is not uncommon with prolonged procedures, immobile posture and any additional cause of thrombogenicity. Included in the latter are any cause of increased coagulability or more often intraoperative positioning of supporting pillows, straps or posts in such a position as to impair venous return. The conditions encouraging venous thrombosis are those occurring during anaesthesia and surgery although the pulmonary embolism that results may not become manifest for several days. The use of elastic bandages and other compressive measures for the lower limbs, attention to positioning and the institution of low level anticoagulants prior to surgery has greatly reduced the instance of this most serious complication but only with constant vigilance will it remain so.

Conclusions

The postoperative period is one of rapidly changing pulmonary pathophysiology. The immediate concerns are those related to the lingering effects of the anaesthetic agents and techniques but beyond about 3 hours the predominant concerns are those due to complications rising from the site, extent and incisional pain of the surgery. This period is therefore a particularly challenging one in which the interaction of pharmacology, physiology, medicine and surgery necessitate a high degree of diagnostic skill and therapeutic decisiveness.

REFERENCES

1. Muravchick S. *The anesthetic plan*. Philadelphia: Mosby Year Book, 1991: 1–82.
2. Grippi MA, Metzgar LF, Krupinski AV, Fishman AP. Pulmonary function testing. In: Fishman AP, ed. *Pulmonary diseases and disorder*. 2nd ed. New York: McGraw-Hill, 1988: 2469–522.
3. Fincuane BT, Santora AH. *Principles of airway management*. Philadelphia: FA Davis, 1988.
4. West JB. *Respiratory physiology. The essentials*. 4th ed. Baltimore: Williams & Wilkins, 1990.
5. Marshall BE, Marshall C. A model for hypoxic pulmonary vasoconstriction. *Journal of Applied Physiology* 1988; **64**: 68–77.
6. Marshall BE, Marshall C. Pulmonary hypertension. In: Crystal RG, West JB, Cherniak N, Weibel R, eds. *The lung: scientific foundations*. New York: Raven Press, 1991: Chapter 5.2.6.
7. Staub NC, Taylor AE. *Edema*. New York: Raven Press, 1984.
8. Boysen PG. *Risk and outcome in anesthesia*. 2nd ed. In: Brown DL, ed. Philadelphia: Lippincott, 1992: 77–102.
9. Jackson CV. Preoperative pulmonary evaluation. *Archives of Internal Medicine* 1988; **148**: 2120–8.
10. Tisi GM. Preoperative evaluation of pulmonary function. *American Review of Respiratory Disease* 1979; **119**: 293–310.
11. Stein M, Cassara EL. Reoperative pulmonary evaluation and therapy for surgery patients. *Journal of the American Medical Association* 1970; **211**: 787–90.
12. Milledge JS, Nunn JF. Criteria for fitness for anaesthesia patients with chronic obstructive lung disease. *British Medical Journal* 1975; **3**: 670–3.
13. Zibrak JD, O'Donnell CR, Marton K. Indication for pulmonary function testing. *Annals of Internal Medicine* 1990; **112**: 763–71.
14. Cann HD, Stevens PM, Adaniya R. Preoperative pulmonary function and complications after cardiothoracic surgery. *Chest* 1979; **76**: 130–8.
15. Lichtenthal RR, Wade LD, Niemyski PR. Respiratory management after cardiac surgery with inhalational anesthesia. *Critical Care Medicine* 1983; **11**: 603–9.
16. Marshall BE, Wyche MQ. Hypoxemia during and after anesthesia. *Anesthesiology* 1970; **37**: 516–17.
17. Hickey RF, Severinghaus JW. Regulation of breathing: drug effects. In: Hornbein TF, ed. *Regulation of breathing*. New York: Marcel Dekker, 1981; 17: Part II, 1251–98.
18. Moller JT, Johannessen NW, Berg H, Espersen K, Larsen LE. Hypoxemia during anaesthesia: an observer study. *British Journal of Anaesthesia* 1991; **66**: 437–44.
19. Lopez-Vidriero MT, Costello J, Clark TJH, Das I, Keal EE, Reid L. Effect of atropine on sputum production. *Thorax* 1975; **30**: 543–6.
20. Grunberg G, Cohen JD, Keslin J, Gassner S. Facilitation of mechanical ventilation in status asthmaticus with continuous intravenous thiopental. *Chest* 1991; **99**: 1216–19.
21. Huber FC, Reves JG, Gutierrez J, Corssen G. Ketamine: the effect on airway resistance in man. *Southern Medical Journal* 1972; **65**: 1776–80.
22. Brichant JF, Gunst SJ, Warner DO, Rehder K. Halothane, enflurane, and isoflurane depress the peripheral vagal motor pathway in isolated canine tracheal smooth muscle. *Anesthesiology* 1991; **74**: 325–32.
23. Rahn H, Prakash O. *Acid–base regulations and body temperature*. Boston: Martinus Nijhoff, 1985.
24. Nunn JF. *Applied respiratory physiology*. 3rd ed. London: Butterworth, 1987: 350–78.
25. Pietak S, Weenig CS, Hickey RF, Fairley HB. Anesthetic effects on ventilation in patients with chronic obstructive pulmonary disease. *Anesthesiology* 1975; **42**: 160–6.
26. Hedenstierna G. Gas exchange during anaesthesia. *British Journal of Anaesthesia* 1990; **64**: 507–14.
27. Price HL, Cooperman LH, Warden JC, Morris JJ, Smith TC. Pulmonary hemodynamics during general anesthesia in man. *Anesthesiology* 1969; **30**: 629–36.
28. Marshall BE, Cohen PJ, Klingenmaier CH, Aukberg S. Pulmonary venous admixture before, during, and after halothane: oxygen anesthesia in man. *Journal of Applied Physiology* 1969; **27**: 653–7.
29. Colgan FJ, Whang TB. Anesthesia and atelectasis. *Anesthesiology* 1968; **29**: 917–22.
30. Campbell FJM, Nunn JF, Peckett BW. A comparison of artificial ventilation and spontaneous respiration with particular reference to ventilation–blood flow relationships. *British Journal of Anaesthesia* 1958; **30**: 166–75.
31. Tokics L, Strandberg A, Brismar B, Lundquist H, Hedenstierna G. Computerized tomography of the chest and gas exchange measurements during ketamine anaesthesia. *Acta Anaesthesiologica Scandinavica* 1987; **31**: 684–92.
32. Gunnarsson L, Tokics L, Gustavsson H, Hedenstierna G. Influence of age on atelectasis formation and gas exchange impairment during general anaesthesia. *British Journal of Anaesthesia* 1991; **66**: 423–32.
33. Tokics L, Hedenstierna G, Strandberg A, Brismar B, Lundquist H. Lung collapse and gas exchange during general anesthesia: effects of spontaneous breathing, muscle paralysis,

and positive end-expiratory pressure. *Anesthesiology* 1987; **66**: 157–67.

34. Gunnarsson L, Tokics L, Lundquist H, *et al.* Chronic obstructive pulmonary disease and anaesthesia: formation of atelectasis and gas exchange impairment. *European Respiratory Journal* 1991; **4**: 1106–16.

35. Anjou-Lindskog E, Broman L, Broma M, Holmgren A, Settergren G, Öhqvist G. Effects of intravenous anesthesia on $\bar{v}_A/Q$ distribution. *Anesthesiology* 1985; **62**: 485–92.

36. Dueck R, Young I, Clausen J, Wagner PD. Altered distribution of pulmonary ventilation and blood flow following induction of anesthesia. *Anesthesiology* 1980; **52**: 113–25.

37. Marshall BE, Hoffman DA, Neufeld GR, Goltone ET. Influence of induction agent on pulmonary venous admixture. *Canadian Anaesthetists' Society Journal* 1974; **21**: 461–6.

38. Benumoff J, Scott AD, Gibbons JA. Halothane and isoflurane only slightly impair arterial oxygenation during one-lung ventilation in patients. *Anesthesiology* 1987; **67**: 910–15.

39. Rehder K, Sessler AD, Mass HM. General anesthesia and the lung. In: Murray JF, ed. *Lung disease: state of the art.* New York: American Lung Association, 1976: 367–89.

40. Bergman NA. Distribution of inspired gas during anesthesia and artificial ventilation. *Journal of Applied Physiology* 1963; **18**: 1085–9.

41. Westerbrook PR, Stubbs SE, Sessler AD, Rehder K, Hyatt RE. Effects of anesthesia and muscle paralysis on respiratory mechanics in normal man. *Journal of Applied Physiology* 1973; **34**: 81–6.

42. Rehder K, Hatch DJ, Sessler AD, Marsh HM, Fowler WS. Effects of general anesthesia, muscle paralysis, and mechanical ventilation on pulmonary nitrogen clearance. *Anesthesiology* 1971; **35**: 591–601.

43. Hedenstierna G, Löfström B, Lundh R. Thoracic gas volume and chest-abdomen dimensions during anesthesia and muscle paralysis. *Anesthesiology* 1981; **55**: 499–506.

44. Laws AK. Effects of induction of anaesthesia and muscle paralysis on functional residual capacity of the lungs. *Canadian Anaesthetists' Society Journal* 1968; **15**: 325–31.

45. Don HF, Wahba M, Cuadrado L, Kelkar K. The effects of anesthesia and 100 percent oxygen on the functional residual capacity of the lungs. *Anesthesiology* 1970; **32**: 521–9.

46. Don HF, Wahba WM, Craig DB. Airway closure, gas trapping, and the functional residual capacity during anesthesia. *Anesthesiology* 1972; **36**: 533–9.

47. Hewlett AM, Hulands GH, Nunn JF, Heath JR. Functional residual capacity during anaesthesia. *British Journal of Anaesthesia* 1974; **46**: 486–94.

48. Hickey RF. Effects of halothane anesthesia on functional residual capacity and alveolar–arterial oxygen tension difference. *Anesthesiology* 1973; **35**: 20–4.

49. Hewlett AM, Hulands GH, Nunn JF, Milledge JS. Functional residual capacity during anaesthesia. *British Journal of Anaesthesia* 1974; **46**: 495–503.

50. Dueck R, Prutow RJ, Davies JH, Clausen JL, Davidson TM. The lung volume at which shunting occurs with inhalation anesthesia. *Anesthesiology* 1988; **69**: 854–61.

51. Hedenstierna G, Strandberg A, Brismar B, Lundquist H, Svensson L, Tokics L. Functional residual capacity, thoracoabdominal dimensions, and central blood volume during

general anesthesia with muscle paralysis and mechanical ventilation. *Anesthesiology* 1985; **62**: 247–54.

52. Dobbinson TL, Nisbet HIA, Pelton DA, Levison H. Functional residual capacity (FRC) and compliance in anaesthetized paralysed children. *Canadian Anaesthetists' Society Journal* 1973; **20**: 322–33.

53. Hedenstierna G, McCarthy G. Mechanics of breathing, gas distribution and functional residual capacity at different frequencies of respiration during spontaneous and artificial ventilation. *British Journal of Anaesthesia* 1975; **47**: 706–11.

54. LeBlanc P, Ruff F, Milic-Emil J. Effects of age and body position on 'airway closure' in man. *Journal of Applied Physiology* 1970; **28**: 448–51.

55. Craig DB, Wahba WM, Don HF, Couture JG, Becklake MR. Closing volume and its relationship to gas exchange in seated and supine posture. *Journal of Applied Physiology* 1971; **31**: 717–21.

56. Gilmure I, Burnham M, Craig DB. Closing capacity measurements during general anesthesia. *Anesthesiology* 1976; **45**: 477–82.

57. Shulman D, Beardsmore CS, Aronson HB, Godfrey S. The effect of ketamine on the functional residual capacity in young children. *Anesthesiology* 1985; **62**: 551–6.

58. Dueck R, Rathbun M, Greenburg AG. Lung volume and VA/Q distribution response to intravenous versus inhalation anesthesia in sheep. *Anesthesiology* 1984; **61**: 55–65.

59. Lundquist H, Hedenstierna G, Ringertz H. Barbiturate anaesthesia does not cause pulmonary densities in dogs: a study using computerized axial tomography. *Acta Anaesthesiologica Scandinavica* 1988; **32**: 162–5.

60. Klineberg PL, Rehder K, Hyatt RE. Pulmonary mechanics and gas exchange in seated normal man with chest restrictions. *Journal of Applied Physiology* 1981; **51**: 26–32.

61. Tokics L, Hedenstierna G, Brismar B, Strandberg Å, Lundquist H. Thoracoabdominal restriction in supine men: CT and lung function measurements. *Journal of Applied Physiology* 1988; **64**: 599–604.

62. Froese AB, Bryan AC. Effects of anesthesia and paralysis on diaphragmatic mechanics in man. *Anesthesiology* 1974; **41**: 242–55.

63. Rehder K, Knopp TJ, Sessler AD. Regional intrapulmonary gas distribution in awake and anesthetized-paralyzed prone man. *Journal of Applied Physiology* 1978; **45**: 528–35.

64. Marshall BE. Hypoxic pulmonary vasoconstriction. *Acta Anaesthesiologica Scandinavica Supplementum* 1990; **34**: 37–41.

65. Benumof JL, Wahrenbrock EA. Local effects of anesthesia as regional hypoxic pulmonary vasoconstriction. *Anesthesiology* 1975; **43**: 525–32.

66. Domino KB, Borowec L, Alexander CM, Marshall C, Marshall BE. Influence of isoflurane on hypoxic pulmonary vasoconstriction in dogs. *Anesthesiology* 1986; **64**: 423–9.

67. Rogers SN, Benumoff JL. Halothane and isoflurane do not decrease PaO_2 during one lung ventilation in intravenously anesthetized humans. *Anesthesia and Analgesia* 1985; **64**: 946–54.

68. Bindslev L, Hedenstierna G, Santesson J, Gottliieb I, Caevallhas A. Ventilation–perfusion distribution during

inhalation anaesthesia. *Acta Anaesthesiologica Scandinavica* 1981; **25**: 360–71.

69. Hedenstierna G, Tokics L, Strandberg A, Lundquist H, Brismar B. Correlation of gas exchange impairment to development of atelectasis during anaesthesia and muscle paralysis. *Acta Anaesthesiologica Scandinavica* 1986; **30**: 183–91.

70. Lindberg P, Gunnarsson L, Tokics L, Secher E, Lundquist H, Brismar B, Hedenstierna G. Atelectasis and lung function in the postoperative period. *Acta Anaesthesiologica Scandinavica* 1992; **36**: 546–53.

71. Haldane JS, Meakins JL, Priestley JG. The effect of shallow breathing. *Journal of Physiology* 1919; **52**: 433–53.

72. Beecher HK. Effect of laparotomy on lung volume: demonstration of a new type of pulmonary collapse. *Journal of Clinical Investigation* 1932; **12**: 651–8.

73. Craig DB. Postoperative recovery of pulmonary function. *Anesthesia and Analgesia* 1981; **60**: 46–52.

74. Ullman DA, Wimpy RE, Fortune JB, *et al.* The treatment of patients with rib fractures using continuous thoracic epidural narcotic infusion. *Regional Anesthesia* 1989; **14**: 43–7.

75. Ali J, Weisel RD, Layug AB, *et al.* Consequences of postoperative alterations in respiratory mechanics. *American Journal of Surgery* 1974; **128**: 376–82.

76. Myers JR, Lembeck L, O'Kane H, *et al.* Changes in functional residual capacity of the lung after operation. *Archives of Surgery* 1975; **110**: 576–82.

77. Alexander JL, Spence AA, Parikh RK, *et al.* The role of airway closure in postoperative hypoxaemia. *British Journal of Anaesthesia* 1973; **5**: 34–40.

78. Yeager MP, Glass DD, Neff RK. Epidural anesthesia and analgesia in high-risk surgical patients. *Anesthesiology* 1987; **66**: 729–36.

79. Reiz S, Wetsberg M. Side effects of epidural morphine. *Lancet* 1980; **2**: 203–4.

80. Mackersie RC, Shackford DR, Hoyt DB, Karagianes TG. Continuous epidural fentanyl analgesia: ventilatory function, improvement with routine use in therapy of blunt chest trauma. *Journal of Trauma* 1987; **27**: 1207–12.

81. Dittman M, Wolff G. A rationale for epidural analgesia in the treatment of multiple rib fractures. *Intensive Care Medicine* 1978; **4**: 193–201.

82. Rawal N, Sjöstrand U, Christofferson E, *et al.* Comparison of intramuscular and epidural morphine for postoperative analgesia in the grossly obese: influence on postoperative ambulation and pulmonary function. *Anesthesia and Analgesia* 1984; **63**: 583–92.

83. Ali J, Yaffe C, Serrette C. The effect of transcutaneous electric nerve stimulation on postoperative pain and pulmonary function. *Surgery* 1981; **89**: 507–12.

84. Engberg G. Relief of postoperative pain with intercostal blockade compared with the use of narcotic drugs. *Acta Anaesthesiologica Scandinavica Supplementum* 1978; **47**: 284–8.

85. Toledo-Pereyra LH, DeMeester TR. Prospective randomized evaluation of intrathoracic intercostal nerve block with bupivacaine on postoperative ventilatory function. *Annals of Thoracic Surgery* 1979; **27**: 203–5.

86. Catley DM, Thornton C, Jordan C, Lehane JR, Rydston D, Jones JG. Pronounced episodic oxygen desaturation in the postoperative period: its association with ventilatory patterns and analgesic regimen. *Anesthesiology* 1985; **63**: 20–8.

Chapter 15

Clinical Physiology and Pathophysiology of Acid–Base Balance

J. Michael Vollers, Kevin Spence and E. F. Klein Jr

Definitions	Methods of assessment of acid–base imbalance
Production of acids	Clinical effects of acid–base derangements
Acid–base homoeostasis	Therapy for acid–base disorders

Evaluation of the state of acid–base balance in the body is of vital importance to the clinician, as it reflects the function or dysfunction of multiple organ systems. Not only is acid–base status reflective of pathology, but acid–base imbalance itself will affect the function of almost every enzyme system (and therefore every organ system) in the body.

In spite of the importance of hydrogen ion (H^+) balance, many clinicians find acid–base analysis confusing at best. Comprehension of acid–base physiology has been hindered by the use of a non-linear measurement system (a negative logarithmic scale), complex and clinically irrelevant phraseology ('standard bicarbonate', 'alkali reserve'), and confusing double negatives ('negative base excess').

Definitions

Acids and bases

The method which helps most to understand acid and base balance relies on the Bronsted–Lowry definition of acids as proton (H^+) donors, and bases as proton acceptors. In solution, an acid (HA) will dissociate to form H^+ and a base (A^-), described by the equation:

$$HA \; (\overset{k_1}{\underset{k_2}{\rightleftarrows}}) \; H^+ + A^- \qquad (1)$$

The reaction follows the principle of mass action, which states that the product of reactants on one side of the equation is proportional to the product of reactants on the other side of the equation. The proportions in the relation are determined by the dissociation constants; if the dissociation constant k_1 is larger than k_2, then $[H^+]$ and $[A^-]$ will represent a larger fraction of the total reactants.

Henderson, in 1909, described this principle in the following formula:

$$[H^+] = K \frac{[HA]}{[A^-]}, \text{ where } K = \frac{k_1}{k_2} \qquad (2)$$

This demonstrates that the concentration of hydrogen ions $[H^+]$ in the solution depends on the dissociation constant and the *ratio* of the concentrations of buffer pairs A^- and HA. In 1916, Hasselbalch transformed Henderson's equation into the negative logarithm, solving for pH:

$$pH = pK_a + \log \frac{[A^-]}{[HA]}, \text{ where } p = -\log \qquad (3)$$

pK_a is the negative logarithm of the dissociation constant of the acid, and equals the pH at which the acid is 50 per cent dissociated (equal concentrations of A^- and HA). pK_a describes the strength of dissociation, known as the strength of the acid; due to the negative log terminology, however, a lower pK_a denotes a stronger acid. Lactic acid ($pK_a = 3.86$) is a stronger acid than carbonic acid ($pK_a = 6.1$) because at any given pH lactic acid will be more dissociated and release more H^+ than carbonic acid.

Acids may be positively or negatively charged, or neutral. Acids and bases are not defined on the basis of their charge but on their ability to donate or accept a proton. Whether a substance acts as an acid (i.e. donating a proton) or a base also depends on the concentration of H^+ in solution (the pH), and the degree of dissociation of the substance (the pK).

The pH system

Although the concentrations of most ions are expressed in terms of moles or equivalents, the concentration of hydrogen ions is conventionally expressed as pH. pH describes the *negative* logarithm (to the base 10) of the concentration of H^+. Conversion from ion concentration to a logarithmic number is a source of confusion for most clinicians. First, the non-linearity of the logarithmic system (in which a change of one pH unit reflects a tenfold change in H^+ concentration) impairs the visualization of the respective change in concentration. Note that over the range of clinically viable pH values, the H^+ concentration changes 1000 per cent (pH 6.8 = $[H^+]$ 160 nmol/l, pH 7.8 = $[H^+]$ 16 nmol/l), but the numerical value of the pH units changes by only 13 per cent. Second, the terminology employs a negative or inverse value assignment, such that an *increase* in pH describes a *decrease* in the H^+ concentration. Although H^+ concentration would be easily understood using the measurement of nanomoles (nmol, or 10^{-9} moles), the use of the pH system is so engrained as to require conformity within the medical community.

It is also important to note that, due to the logarithmic scale, equal changes in the pH values do not correlate with equal changes in the H^+ concentration. As an example, a change in pH from 7.4 to 7.0 (40 nmol/l $[H^+]$ to 100 nmol/l $[H^+]$) represents a change of 60 nmol/l in $[H^+]$, whereas an equal change in pH from 7.4 to 7.8 (40 nmol/l $[H^+]$ to 16 nmol/l $[H^+]$) represents a 24 nmol/l change in $[H^+]$.

Measurement of acid–base status

Clinical acid–base balance depends in large part on the chemistry of buffers. A buffer is a solution containing a poorly dissociated or weak acid (HA) and its base (A^-), with the ability to resist a change in pH upon addition of a strong acid or base. The most important buffer pair in the extracellular fluid is carbonic acid (H_2CO_3) and bicarbonate (HCO_3^-). By convention, the interaction of this buffer pair serves as the basis for measurement of acid–base balance.

When CO_2 is added to water, the following reaction takes place:

$$H_2O + CO_2 \rightleftarrows H_2CO_3 \rightleftarrows H^+ + HCO_3^- \tag{4}$$

The initial reaction hydrating CO_2 to produce carbonic acid is slow, requiring minutes to achieve equilibrium, and would be inadequate to enable gas exchange in the pulmonary capillaries. This reaction is greatly accelerated by the catalytic enzyme carbonic anhydrase, found in erythrocytes and other tissues, but not in plasma. In the second phase of the reaction, carbonic acid dissociates freely to bicarbonate and H^+ (approximately 96 per cent of the carbonic acid is dissociated at physiological pH).

If the components of the carbonic acid–bicarbonate reaction are inserted into the Henderson–Hasselbalch equation (equation 15.3):

$$pH = pK_a + \log\frac{[HCO_3^-]}{[H_2CO_3]} \tag{5}$$

The concentration of carbonic acid is difficult to measure physiologically, because of its continuous and almost complete dissociation to H^+ and HCO_3^-. It is therefore convenient (and reasonably accurate) to substitute the amount of dissolved CO_2 in solution, described as αP_{CO_2} (α is the solubility coefficient for CO_2, or 0.23 mM/l per kPa (0.03 mM/l per mmHg) at 37°C). Under physiological conditions, the pK_a for this reaction is 6.1; this is variable with temperature and pH. The working form of the equation for plasma at 37°C is:

$$pH = 6.1 + \log\frac{[HCO_3^-]}{0.23\ P_{CO_2}} \tag{6}$$

This formula is clinically useful in the measurement of acid–base disturbances for several reasons. First, the components of this equation are used to describe abnormalities in metabolic disorders (HCO_3^-) as well as respiratory disorders (Pa_{CO_2}). Although neither component *in vivo* uniquely describes a specific type of acid–base disturbance (i.e. alterations in Pa_{CO_2} from respiratory disorders will have a direct effect on $[HCO_3^-]$), the distinction between the components is useful in characterizing most primary acid–base imbalances. Second, the value of any component (pH, P_{CO_2}, or $[HCO_3^-]$) of the Henderson–Hasselbalch equation may be derived from direct measurement of the other two components. In clinical practice, the value of serum bicarbonate is typically derived from the measured values of pH and Pa_{CO_2}.

The third and most important clinical value of this equation is the fact that pH is not determined by the absolute values of HCO_3^- or P_{CO_2}, but by the *ratio* of the two. Under normal conditions, assuming HCO_3^- to be 24 mmol/l and Pa_{CO_2} to be 5.3 kPa (40 mmHg), the ratio can be described. After multiplying the Pa_{CO_2} by the solubility coefficient of CO_2 in plasma (0.23 mmol/l per kPa or 0.03 mmol/l per mmHg), the ratio is 24/1.2 or 20:1. Because the pH equals pK (6.1) plus log20 (1.3), under normal conditions the pH is 7.4. Alterations in the ratio of HCO_3^- to Pa_{CO_2} may describe a primary acid–base imbalance. An increase in the ratio (through an increase in $[HCO_3^-]$ or a decrease in Pa_{CO_2}) results in an increased pH (alkalosis); a decrease in the ratio (through loss of HCO_3^- or increase in the Pa_{CO_2}) produces a decreased pH (acidosis). It is important to note that the pH will remain normal despite a change in the levels of either HCO_3^- or Pa_{CO_2} *if the ratio remains the same* (i.e. if Pa_{CO_2} is increased and $[HCO_3^-]$ is increased proportionally). This is the basis for maintenance of acid–base homoeostasis through normal compensatory mechanisms, as discussed below.

Production of acids

Essentially all acids in the body are the products of metabolic processes, as a result of both normal and abnormal metabolism. Carbon dioxide, produced in the mitochondria as an end product of carbohydrate metabolism, constitutes the major acid load produced in the body. Assuming an average basal CO_2 production of 200 ml/min (or 288 l/day), the net daily production of acid in the form of CO_2 is 12 960 mEq (the term mmol/l is not usually used for a gas). Because the body is able to eliminate this acid via the lungs, CO_2 is commonly referred to as a 'volatile' or 'respiratory' acid.

The majority of the metabolic or 'non-volatile' acids are the metabolites of protein degradation. Sulphuric acid results from metabolism of cysteine and methionine. Hydrochloric acid (HCl) is produced from the degradation of lysine, arginine and histidine. Phosphoric acid ($H_2PO_4^-$), is an additional non-volatile acid derived from the normal dietary consumption of phosphate. The metabolic acid load is offset in part by the production of HCO_3^- from the metabolism of aspartate, glutamate and some organic anions (i.e. citrate). The balance of metabolic acid and HCO_3^- production results in the net creation of approximately 70 mmol/day of acid, or 1 mmol/kg per day.

Lactic acid is usually considered a marker of anaerobic metabolism, and as such it is often seen as an indicator of hypoxia, poor perfusion, or some disturbance in the delivery of oxygen to the tissues. It is important to realize that approximately 1400 mmol/day of lactic acid are produced in the course of normal metabolism, mostly by skeletal muscle, skin, and red blood cells. This lactate is converted to sodium lactate in the plasma which is then oxidized by the liver to regenerate sodium bicarbonate. A small portion of the lactic acid is cleared by the kidney, through renal excretion, oxidation, or conversion by gluconeogenesis. Lactic acid, pyruvic acid, acetoacetic acid, and β-hydroxybutyric acid are all normal metabolic products; abnormalities of production or elimination may be seen in pathological states (diabetic ketoacidosis, anaerobic metabolism, etc.). These metabolic acids have the advantage of being further degraded to CO_2 and H_2O, and eliminated by the respiratory system. Inorganic acids and other acids which are not converted to CO_2 are excreted by the kidney, which is normally responsible for the loss of approximately 1 mEq/l per day of H^+.

Acid–base homoeostasis

Acid–base homoeostasis requires both the elimination of acids produced by metabolism, and the production or recovery of sufficient base to balance any remaining acid load. While it is classically thought that the concentration of H^+ in the body is carefully regulated, H^+ concentrations acutely compatible with life may range tenfold from 16 to 160 nmol/l (pH 6.8–7.8): it is remarkable that no other ion in the body demonstrates such a wide range of variability. The intrinsic defence against pH change maintains the normal range of H^+ between 35 and 45 nmol/l (pH 7.35–7.44) through three mechanisms: buffering, compensation, and correction. The chemical buffer systems of the body act immediately to minimize the change in pH on addition of an acid or a base. Compensation describes the physiological attempt to restore the ratio of $[HCO_3^-]/P\text{CO}_2$ to normal, by alteration of whichever variable was not primarily deranged. If, for example, the primary disorder was metabolic and resulted in a change in $[HCO_3^-]$, physiological compensation attempts to adjust the $P\text{CO}_2$ in a similar direction by altering ventilation, endeavouring to restore the ratio $[HCO_3^-]/P\text{CO}_2$ to normal. Finally, the ultimate homoeostatic response to acid–base imbalance is through correction of the primary aetiology, if possible.

Buffering systems

The ability to maintain a narrow range of $[H^+]$ in normal and pathological conditions is largely due to the buffering systems of the body. Buffers are solutions which typically consist of a weak acid (HBuffer) and its salt (NaBuffer) that resist or minimize the change in $[H^+]$ with addition of a stronger acid or base. Strong acids (such as hydrochloric acid) are buffered by the salts of a weak acid:

$$\text{HCl} + \text{NaBuffer} \rightleftarrows H^+ + Cl^- + Na^+ + \text{Buffer}^- \rightleftarrows \text{HBuffer} + \text{NaCl} \qquad (7)$$

HCl, a strong acid, is largely dissociated and the H^+ released from its dissociation combines with the dissociated Buffer^- to produce the weak acid HBuffer. Because HBuffer is a weaker acid than HCl, it is less dissociated, and less H^+ is released by its dissociation than if the HCl dissociated in the absence of a buffer. Similarly, a strong base such as sodium hydroxide (NaOH) is buffered by a weak acid:

$$\text{NaOH} + \text{HBuffer} \rightleftarrows Na^+ + OH^- + H^+ + \text{Buffer}^- \rightleftarrows \text{NaBuffer} + H_2O \qquad (8)$$

In the absence of a buffer, the dissociated OH^- from the strong base NaOH would combine with H^+, decreasing the concentration of H^+ and raising pH. Because the NaBuffer is a weaker base than NaOH, it is less dissociated and the $[H^+]$ changes less.

The most effective buffer systems in the body are those in large concentration (available to buffer more H^+), and those with a pK_a close to the initial pH of the solution to be buffered (weaker acids and their salts are more effective buffers against strong acids and bases). Of the many buffer systems in the body, the most important are the carbonic acid–bicarbonate, haemoglobin, protein, and phosphate buffers. The carbonic acid–bicarbonate buffer is predominant in the extracellular fluid, while the remaining non-carbonic buffers are important intracellularly.

Carbonic acid–bicarbonate system

Carbonic acid and its salt (usually $NaHCO_3$ in the extracellular fluid and $KHCO_3$ or $Mg(HCO_3)_2$ intracellularly) is a major buffer of *metabolic* acid or alkali as represented:

$$HCl + NaHCO_3 \rightleftarrows NaCl + H_2CO_3 \rightleftarrows NaCl + H_2O + CO_2 \tag{9}$$

The net result of acid addition to the bicarbonate buffer is the conversion of the strong acid to CO_2, which may then be eliminated by ventilation. Because the body acts to maintain a relatively constant P_aCO_2, essentially all the carbonic acid formed from the buffering of a strong acid is eliminated by the lungs as CO_2. As CO_2 is removed, HCO_3^- is consumed in equal parts (because carbonic acid is in equilibrium with both CO_2 and HCO_3^-). It is therefore important to note that, for a bicarbonate solution without other buffers, the change in $[HCO_3^-]$ reasonably reflects the amount of H^+ added to or removed from the system.

A strong base, such as NaOH, may also be buffered by carbonic acid:

$$NaOH + H_2CO_3 \rightleftarrows NaHCO_3 + H_2O \tag{10}$$

Continuous production of CO_2 from metabolic processes maintains the supply of carbonic acid for the buffering of a strong base.

It is necessary to remember that *the carbonic acid–bicarbonate system functions to buffer metabolic acid and base only*, and is unable to buffer respiratory acid. If CO_2 is added to a bicarbonate buffer *in vitro*, H^+ and HCO_3^- are produced in equal amounts:

$$CO_2 + H_2O + NaHCO_3 \rightleftarrows H^+ + HCO_3^- + NaHCO_3 \tag{11}$$

If enough CO_2 were added to produce a 40 nmol/l increase in $[H^+]$, the effect on pH would be large (assuming a starting pH of 7.4, or an $[H^+]$ of 40 nmol/l, the resultant $[H^+]$ would be 80 nmol/l, or a pH of 7.1). Because $[HCO_3^-]$ would increase equally, 40 nmol/l $[HCO_3^-]$ is added to a normal $[HCO_3^-]$ of 24 nmol/l, yielding a total of 24.000 040 mmol/l: a clinically unmeasurable increase in $[HCO_3^-]$. For practical purposes, it may be assumed that the $[HCO_3^-]$ does not change on addition of CO_2 to the system *in vitro*.

Because the bicarbonate buffer has a low pK_a relative to physiological pH, and a small concentration of its components, it would be expected to play only a minor role in the maintenance of acid–base balance in the body. The efficacy of this buffer system *in vivo*, however, is equivalent to that of all the other buffers combined; this is due to the respiratory and metabolic regulation of the buffer

components, CO_2 and HCO_3^-. This important distinction between the *in vitro* and *in vivo* capacity of the buffer will be discussed in greater detail later in the chapter.

Haemoglobin system

Quantitatively, haemoglobin is the predominant non-carbonic buffer in the extracellular fluid, and it functions to buffer both respiratory and metabolic acids. Haemoglobin acts to buffer and transport CO_2 through two methods. First, CO_2 can combine directly with the terminal amino acids of the haemoglobin molecule to form carbamino compounds. Carbamino carriage accounts for 15–25 per cent of the total CO_2 in the blood; a much smaller amount of CO_2 is carried by plasma proteins. Second, CO_2 is catalysed within the erythrocyte by carbonic anhydrase, to produce H^+ and HCO_3^-. The H^+ produced by carbamino formation and by carbonic acid degradation (as well as from dissociation of a metabolic acid) is buffered by haemoglobin, which exists within the red blood cell as a weak acid (HHb) and its potassium salt (KHb). Because haemoglobin is a weaker acid (pK_a 6.8) than carbonic acid (pK_a 6.1), H^+ is buffered by haemoglobin and HCO_3^- is increased proportionally:

$$H^+ + HCO_3^- + KHb \rightleftarrows HHb + K^+ HCO_3^- \tag{12}$$

As H^+ is buffered within the erythrocyte, the concomitant increase in $[HCO_3^-]$ causes the diffusion of HCO_3^- into plasma along a concentration gradient. As HCO_3^- diffuses out of the cell, chloride (Cl^-) moves intracellularly to maintain electrical neutrality: this is referred to as the 'chloride shift'. Therefore, while most of the H^+ is buffered within the cell, most of the change in $[HCO_3^-]$ is observed in the plasma.

The buffering capacity of haemoglobin is increased in the reduced (deoxygenated) state, due to the Haldane effect, which, simply stated, notes that reduction of haemoglobin makes it a weaker acid. Haemoglobin in the reduced state has a greater capacity to carry CO_2 as carbamino compounds and provides increased availability of buffer sites for H^+. With oxygenation in the lung, haemoglobin becomes a stronger acid and the converse is true: CO_2 is released from the carbamino groups, and H^+ is released from the haemoglobin to react with $KHCO_3$:

$$HHb + O_2 + K^+ + HCO_3^- \rightleftarrows KHbO_2 + H_2O + CO_2 \tag{13}$$

CO_2 diffuses out of the erythrocyte and is eliminated by the lung; this creates a gradient for diffusion of HCO_3^- back into the cell, in exchange for Cl^-.

Protein buffers

In addition to haemoglobin, other plasma proteins serve as effective buffers, both because of the large total concentration of proteins in the body, and because the pK_a of some of the proteins approximates 7.4. Although most amino acids are bound by peptide linkages, some have free acidic radicals in the form of ·COOH, which can liberate H^+ to buffer excess base:

$$\cdot COOH + OH^- \rightleftarrows -COO^- + H_2O \qquad (14)$$

Other amino acids have free basic radicals, commonly in the form of ·NH_3OH, to buffer acid:

$$\cdot NH_3OH + H^+ \rightleftarrows -NH_3^+ + H_2O \qquad (15)$$

The protein buffer system is predominantly intracellular, where the concentration of proteins are comparatively greater than in the extracellular fluid. It is estimated that three-quarters of the body's buffering capacity is intracellular. It is the ability of CO_2 (and, to a lesser degree, H^+ and HCO_3^-) to diffuse across cell membranes that enables the intracellular protein buffer system to participate also in the buffering of extracellular fluid.

Phosphate buffer

The phosphate buffer system is the largest inorganic buffer, and is of most importance intracellularly. Its function is similar in nature to the bicarbonate buffer, and with a pK_a of 6.8, it is effective at physiological pH.

$$HCl + Na_2HPO_4 \rightleftarrows NaH_2PO_4 + NaCl \qquad (16)$$
$$NaOH + NaH_2PO_4 \rightleftarrows Na_2HPO_4 + H_2O \qquad (17)$$

Equations (16) and (17) represent the buffering of a strong acid and a strong base, respectively. The conversion of a strong acid or base to a weak acid or base results in the liberation of less H^+, and therefore a smaller change in pH. The efficacy of the phosphate buffer in the extracellular space is limited because the concentrations of the buffer are less than 10 per cent of the components of the bicarbonate buffer system. Phosphate buffer concentrations are highest within the cell, and because the intracellular pH (estimated at 6.9) is about the same as the pK_a of the buffer, the buffering potential is even greater. For similar reasons, the buffer is particularly important in renal tubular fluid: phosphate is greatly concentrated in the tubules, and the renal tubular fluid is much more acidic than plasma, more closely approximating to the pK_a of the phosphate buffer system.

Interrelation of buffers

It is not only the total concentration of buffers in the body but their interrelation which results in the remarkable buffering capacity of the body. The buffering capacity of the intact organism is greater than the sum of the various intracellular and extracellular buffers, due to the common link of the hydrogen ion. When a condition results in a change in $[H^+]$, it causes the balance of all of the buffer systems to change at the same time, because the buffer systems, in essence, buffer one another. This interrelation between the carbonic and non-carbonic buffers will also distort the relation of $[HCO_3^-]$ and $PaCO_2$ as 'markers' of respective metabolic and respiratory imbalances. For example, in a pure bicarbonate buffer, the change in $[HCO_3^-]$ is equivalent to the amount of metabolic acid or base added, but with concurrent buffering by non-carbonic buffers the change in $[HCO_3^-]$ is not as great as expected. Similarly, recall that alterations in PCO_2 do not change the $[HCO_3^-]$ in an isolated bicarbonate buffer. The non-carbonic buffers will buffer CO_2 and result in a change in $[HCO_3^-]$. The greater the concentration of non-carbonic buffers, the greater will be the buffering of H^+ for a given change in CO_2, and the greater the change in $[HCO_3^-]$. As a result, in a mixed buffer solution as occurs physiologically: (1) $[HCO_3^-]$ serves as a marker of metabolic acid change, but not in a linear fashion, and (2) the change in $[HCO_3^-]$ is not independent of respiratory acid–base changes.

Compensation

Acid–base disturbances are not only minimized by the multiple buffer systems, but by compensatory regulation of the elimination of acids and base through ventilation and renal excretion. An alteration in pH stimulates homoeostatic mechanisms to restore the ratio of $[HCO_3^-]/PaCO_2$, in an attempt to normalize pH.

Ventilation

Respiratory compensation for any given metabolic acid–base disorder is limited to the change in ventilatory excretion of CO_2. Respiratory drive is controlled by the chemosensitive area of the medulla, and, to a lesser degree, by chemoreceptors in the carotid bodies. Chemically, these regulatory centres are responsive to both H^+ and CO_2 ions. The hydrogen ion appears to be the prime mediator affecting the medullary centre. Changes in blood $PaCO_2$ are rapidly reflected in the cerebrospinal fluid (CSF) PCO_2; the CO_2 is then rapidly converted to carbonic acid, which is dissociated to H^+. Because of the limited buffering capacity of the CSF, a given change in the PCO_2 of CSF results in a greater change in $[H^+]$ than is seen in interstitial fluid.

In addition to the homoeostatic control of $PaCO_2$, alteration in ventilation serves to compensate for metabolic acid–base disturbances. A decrease in $[HCO_3^-]$ and consequent decrease in pH will stimulate an increase in ventilation, which changes the ratio of $[HCO_3^-]/PaCO_2$ toward normal by lowering $PaCO_2$. An increase in $[HCO_3^-]$ causes a reduction in alveolar ventilation and results in elevation of $PaCO_2$. This compensatory response may be limited by hypoxaemia resulting from hypoventilation, which produces stimulation of the ventilatory centre via stimulation of the carotid bodies.

Renal elimination

Quantitatively, the major function of the kidney is to recover the HCO_3^- produced by the titration of acid against the buffer systems of the body. The amount of HCO_3^- filtered by the kidney is approximately 4320 mmol/day (24 mmol/l times 180 l/day); virtually all of it is reabsorbed. H^+ is titrated into the tubular fluid, and carbonic anhydrase catalyses the production of H_2O and CO_2 (which is rapidly reabsorbed). Carbonic anhydrase within the cell converts the CO_2 back into HCO_3^-, which exits into the peritubular blood. Additionally, the kidneys generate new HCO_3^- by the metabolism of glutamine in the proximal tubules, producing CO_2 and NH_3. The NH_3 is secreted into the tubular fluid as a buffer, and combines with H^+ to form NH_4^+, the CO_2 combines with H_2O to form HCO_3^-, which is then absorbed. If the kidney is unable to excrete the NH_4^+, absorption of the NH_4^+ essentially titrates the HCO_3^-, and negates the process of new HCO_3^- generation.

In addition to recovering HCO_3^-, the kidneys must excrete an amount of acid equal to the daily production of metabolic acid, or approximately 70 mmol/l per day. These free acids cannot be excreted directly, but are first buffered by the weaker acids ammonia (NH_3/NH_4^+) and phosphate ($HPO_4^{2-}/H_2PO_4^-$), and, to a lesser degree, other buffer species in the renal tubular fluid. Both the excretion of acid and resorption of HCO_3^- depend on the active secretion of H^+ into the tubular fluid. The amount of H^+ secreted is regulated by changes in the plasma $[HCO_3^-]$, plasma pH, and PCO_2. In the condition of systemic acidosis (from the loss of HCO_3^- or an increased $PaCO_2$), the reduction in intracellular pH stimulates release of H^+ into the tubular fluid, thereby increasing recovery of HCO_3^-. The metabolism of glutamine is also stimulated by intracellular acidosis; the increased production of new HCO_3^- and NH_4^+ results from synthesis of new enzymes and may take several days for complete adaptation. Conversely, metabolic and respiratory alkalosis inhibit the resorption of HCO_3^- by the kidney, through reduction of H^+ secretion. Because the secretion of H^+ in the kidney is tied to the concomitant reabsorption of Na^+ for electrical balance, the process is also impacted by aldosterone. An increase in aldosterone secretion stimulates the reabsorption of Na^+ and the excretion of H^+ into tubular fluid, resulting in the increased reabsorption of HCO_3^-. The opposite effect is seen with reduced aldosterone levels.

Methods of assessment of acid–base imbalance

Despite the efficiency and abundance of the body's buffer systems and compensatory mechanisms, acid–base disturbances occur. Measured values of pH, $PaCO_2$, and the calculated value of HCO_3^- are required for the determination of the primary disorder, the degree of compensation, and the prescribed therapy for any given disturbance. Because the mechanisms of acid–base homoeostasis are complex, many systems have been proposed for characterization and quantification of the disturbance. The majority of these evaluation systems are flawed because simplification of a complex process is inherently difficult.

In vitro methods

The most convenient method of measurement of acid–base disorders employs the Henderson–Hasselbalch equation for carbonic acid–bicarbonate; therefore, several methods have been proposed to correct for the presence of other buffers in order to evaluate the entire system as if it were a bicarbonate buffer.

Alkali reserve method

Prior to the development of pH and CO_2 electrodes, one attempt at simplification involved the titration of $PaCO_2$ of a plasma sample back to 5.3 kPa (40 mmHg); this was believed to correct any respiratory disorder by establishing a normal $PaCO_2$. The total content of CO_2 ($ctCO_2$) was determined, and called the combining power or alkali reserve. The HCO_3^- was measured, then calculation of the pH was made using either a graph or the Henderson–Hasselbalch equation. Because plasma or serum was used in this determination, the buffering power of haemoglobin was not accounted for, and this system is no longer used in clinical practice.

Standard bicarbonate system

The standard bicarbonate system attempted to correct the deficiencies of the alkali reserve system by using whole blood. Furthermore, this method assured that the haemoglobin was fully oxygenated, and the measurement was performed at 38°C, rather than at room temperature (as was done for alkali reserve). The blood sample was equilibrated against gas with a $PaCO_2$ of 5.3 kPa (40 mmHg), and the HCO_3^- was measured. Normal standard bicarbonate is 24 (22–26) mmol/l. This system, like the less accurate alkali reserve system, fails to account for the buffering systems outside the blood.

Astrup method

The Astrup method is another variation of the above systems. After an initial pH measurement, the blood sample is equilibrated against two different CO_2 concentrations, and the pH at each of these CO_2 values is noted. The values of these two measurements are used to plot pH against the logarithm of $Pa CO_2$, and a line is drawn between the two points; this line describes all possible pH and corresponding $Pa CO_2$ values for that sample. The $Pa CO_2$ of the sample is determined from the intersection of the initial pH value with the line. Other values, including standard bicarbonate, actual bicarbonate, total CO_2 content, buffer base concentration, and base excess, may also be determined graphically from the Astrup method.

Buffer base and base excess systems

The buffer base and base excess system take different approaches to the description and *in vitro* analysis of acid–base disorders. Compared to the previously described techniques, these systems attempt to compensate for the buffering capacity of non-carbonic buffer systems. To do so, the dual buffer system is converted to the equivalent of a carbonic acid–bicarbonate buffer, in which the change in $[HCO_3^-]$ directly reflects a change in H^+ of the system.

Conversion to a single buffer system *in vitro* is performed through the process of acid titration. Rather than equilibrating the blood sample to a known $P CO_2$ value, the sample is altered (by acid titration using CO_2) to a pH value of 7.4. Recall that at pH 7.4 the components of all the buffer systems (except for the bicarbonate buffer) not only exist in a defined ratio (based on pK_a), but in fixed total concentrations; that is, the total concentrations of haemoglobin or plasma proteins do not change with alterations in pH. Therefore, as the pH is restored to 7.4 by the manipulation of $P CO_2$, the non-carbonic buffer pairs exist in the same concentrations and ratios as in the absence of an excess acid or base load. The magnitude of buffering is then assumed by the bicarbonate buffer system, and the change in the value of HCO_3^- reflects the quantity of acid or base added to the system. Note that the bicarbonate buffer system is essentially unaffected by respiratory acidosis because the addition of CO_2 results in equal production of H^+ and buffer (HCO_3^-) from carbonic acid. The buffer base and base excess systems are therefore considered to reflect metabolic disturbances only.

Both the buffer base and base excess systems utilize the Siggaard–Andersen nomogram, which relates $Pa CO_2$, pH, and HCO_3^- based on *in vitro* titration curves. This allows the determination of any variable by plotting the other two

measured values. Calculated values of the $ctCO_2$ and base excess are also available from the nomogram. Calculation of the base excess value requires measurement of the haemoglobin value, as the nomogram attempts to compensate for the non-carbonic buffer component.

Buffer base is a description of the amount of buffering capacity in whole blood, with a normal value of 45–50 mEq/l. (Buffer base is a term which involves the buffering capacity of a variety of buffers, including bicarbonate and proteins. This term was defined in units of mEq/l and as such may not easily or appropriately be converted to SI units.) Conceptually, the addition of H^+ must result in either free H^+ (resulting in a change in pH) or binding to a buffer (with a consequent reduction in the remaining buffer capacity). Evaluation of the pH and the buffer capacity should therefore more clearly define any acid–base abnormality than determination of pH and $ctCO_2$.

Base excess does not describe the total buffering capacity, but instead reports the difference in buffering ability from normal. Normal buffer base is subtracted from the calculated value of the sample, and the difference is reported as base excess. An increase in buffering capacity (buffer base greater than 50 mEq/l, or base excess) may result from a decrease in metabolic acid production or an increase in buffer content (increased bicarbonate, haemoglobin, proteins, or phosphates). A decreased buffer base (less than 45 mEq/l) may result from the production of excess metabolic acids or a decreased buffer content. This is also described as a negative base excess, or base deficit. The simplicity of the base excess system (normal values +2.5 to −2.5 mEq/l) and logical association (base excess correlates with metabolic alkalosis, base deficit corresponds to metabolic acidosis) accounts for its widespread clinical usage. Although convenient, deficiencies exist with the base excess and buffer base systems, as with all the *in vitro* analyses of acid–base disorders. The buffer base system only predicts acute disturbances, and is unable to evaluate adequately any degree of compensation to the acute disorders. In addition, the base excess system will indicate the presence of a metabolic acidosis which does not exist in the condition of severely elevated levels of CO_2; this is due to the action of extravascular buffers which are not evaluated by the base excess method.

In vivo measurements

All the *in vitro* methods of acid–base analysis, including the base excess system, are compromised in the evaluation of acid–base disturbances of the body as a whole. There are three major reasons for this limitation: (1) the size of the intravascular buffer system evaluated *in vitro* is much smaller than the total buffering capacity of the body; this will lead to an overestimation of the severity of an acid–base disturbance; (2) the *in vitro* systems fail to account for the interaction of the acid–base balance of the blood with buffer systems of the interstitial and intracellular fluid; (3) the *in vitro* analysis of acid–base imbalance is unable to evaluate the degree of physiological compensation for the primary disorder.

In vitro systems analyse blood to evaluate the change in acid–base balance, but because of the size of the interstitial space relative to blood volume, the *in vitro* buffer capacity is about one-third to one-fifth the *in vivo* value. Estimations of the magnitude of an acid–base disturbance using *in vitro* buffering values may grossly overestimate the effect on the whole organism. One method to compensate for the difference between *in vitro* and *in vivo* buffering values, if using the Siggaard–Andersen nomogram, is to use a haemoglobin value of one-third that reported for the patient.

The magnitude of the *in vivo* buffering capacity is due to the ability of H^+, CO_2 and HCO_3^- to diffuse between the blood and the extravascular fluids and tissues. When the $PaCO_2$ of blood is increased, there is a large increase in $[HCO_3^-]$ and a small decrease in pH, due to the large concentration of non-carbonic buffers (haemoglobin and plasma proteins) in the blood. In the extracellular fluid, however, an equivalent increase in PCO_2 produces a much smaller increase in $[HCO_3^-]$, because the extracellular fluid has both a lower total buffer concentration than blood, and a much smaller concentration of non-carbonic buffers. *In vivo*, blood is in equilibrium with the extracellular fluid, and HCO_3^- will diffuse across most capillaries into the interstitial fluid until equilibration is reached; this results in a fall in plasma $[HCO_3^-]$ and an increase in interstitial $[HCO_3^-]$. It is by this mechanism that the *in vitro* base excess system identifies a metabolic acidosis in the circumstances of a severe hypercapnic acidosis.

Because the *in vitro* evaluation systems are designed to evaluate acute metabolic disorders, they are unable to characterize or evaluate the normal compensatory mechanisms of acid–base homoeostasis. Normal renal compensation for a respiratory acidosis will be identified by the base excess system as a metabolic alkalosis complicating a respiratory acidosis; similarly, a metabolic acidosis will also be diagnosed in the condition of a compensated respiratory alkalosis. No data are provided by the base excess system by which to judge the degree of compensation in a chronic acid–base disturbance. The ability of the base excess system to define the extent of a metabolic abnormality is negated in the compensated chronic acid–base disorder.

In summary, the *in vitro* analytical methods suffer from inaccuracy in the estimation of the magnitude of an acid–base disorder, and are unable to account for either the acute interaction between buffer systems or the changes related to compensatory mechanisms. It is therefore more reasonable and accurate to evaluate acid–base disturbances

utilizing information from *in vivo* titration curves. These are data derived from the collation of normal human values of pH, $PaCO_2$, and HCO_3^- in acute and chronic disorders. Based on these *in vivo* values, sample values may be compared to the normal response, and the deviation from predicted may be characterized and quantified for both acute and chronic disorders.

The merit of the *in vivo* analytical method is in the recognition that changes occur in both $[HCO_3^-]$ and $PaCO_2$ in both acute and chronic acid–base disturbances.[1] In the situation of an acute or chronic respiratory acidosis, for example, there will be changes in $[HCO_3^-]$ and pH *in vivo* that are not predicted by *in vitro* methods of analysis. This is due to both the interaction between buffers and secondary to normal compensatory mechanisms. While the primary aetiology can be determined using the markers of respiratory disorders ($PaCO_2$) and metabolic disturbances (HCO_3^-), changes in the other marker and pH can be quantified and compared to normal values for the acute or chronic condition.

In vivo *evaluation of respiratory acidosis*

With an acute increase in $PaCO_2$, the increase in carbonic acid results in an increase in both $[H^+]$ and $[HCO_3^-]$. Acutely, each 1.3 kPa (10 mmHg) rise in $PaCO_2$ results in an increase in $[HCO_3^-]$ of 0.08 mmol/l, and an increase in $[H^+]$ of 8 nmol/l. Although the relation between $PaCO_2$ and $[H^+]$ is linear, conversion to the logarithmic pH scale distorts this association. As an approximate conversion, each 1.3 kPa (10 mmHg) rise in $PaCO_2$ produces a pH decrease of about 0.07 units, within the $PaCO_2$ range of 4–8 kPa (30–60 mmHg). This generalization is less accurate as the $PaCO_2$ deviates further from normal.

Renal compensation for the respiratory acidosis results in excretion of H^+ and reabsorption of HCO_3^-, raising the pH but not completely normalizing it. As the ratio of $[HCO_3^-]$ to $PaCO_2$ approaches normal, the change in $[HCO_3^-]$ is greater and the change in pH is less than for an acute disturbance. Therefore, under conditions of chronic hypercapnia with renal compensation, each 1.3 kPa (10 mmHg) increase in $PaCO_2$ results in a HCO_3^- rise of 4 mmol/l, and an increase in $[H^+]$ of 3.2 nmol/l, or an approximate 0.03 unit decrease in pH.

In vivo *evaluation of respiratory alkalosis*

Under conditions of acute hyperventilation and hypocapnia, the levels of carbonic acid are reduced and $[H^+]$ and $[HCO_3^-]$ are decreased. An acute decrease of 1.3 kPa (10 mmHg) in the $PaCO_2$ results in a decrease in $[HCO_3^-]$ of 2 mmol/l. For each 1.3 kPa (10 mmHg) decrease in $PaCO_2$, $[H^+]$ decreases by 8 mmol/l (an equivalent magnitude of change as in respiratory acidosis, but opposite in direction), and the pH increases by approximately 0.08 units.

Renal compensation for chronic hypocapnia causes a decrease in tubular H^+ secretion and diminished HCO_3^- reabsorption. As plasma $[HCO_3^-]$ decreases, the ratio of $[HCO_3^-]$ to $PaCO_2$ approaches normal, and the change in pH is less in comparison with an acute respiratory alkalosis. In the chronic compensated respiratory alkalosis, each 1.3 kPa (10 mmHg) decrease in $PaCO_2$ is associated with a 6 mmol/l decrease in $[HCO_3^-]$, and a decrease in $[H^+]$ of 1.7 nmol/l, corresponding to a pH increase of approximately 0.03 units.

In vivo *evaluation of metabolic acidosis*

Excess metabolic acid in the system produces an almost immediate increase in alveolar ventilation. Peripheral chemoreceptors in the carotid bodies stimulate ventilatory drive, resulting in a decrease in $PaCO_2$ and attenuation of the fall in pH. Although reduction in $PaCO_2$ is rapid, the respiratory response is only capable of a 50–75 per cent compensation for the metabolic acidosis (i.e. an uncorrected metabolic acidosis which would reduce the pH to 7.0 is normally compensated to a pH of 7.2–7.3). *In vivo* measurements of the normal response to metabolic acidosis show that the $PaCO_2$ is decreased by 0.15 kPa (1.1 mmHg) for each 1 mmol/l reduction in $[HCO_3^-]$. An alternate method of predicting $PaCO_2$ associated with a pure metabolic acidosis uses the formula:

$$\text{Predicted } PaCO_2 \text{ (kPa)} = 0.2 [HCO_3^-] + 1.06 (\pm 0.27), \text{ or}$$
$$\text{Predicted } PaCO_2 \text{ (mmHg)} = 1.5 [HCO_3^-] + 8 (\pm 2)$$
(18)

Because the respiratory compensation ameliorates the change in pH, *in vivo* measurements show that for each 4 mmol/l decrease in $[HCO_3^-]$, the pH decreases by 0.08 units. Due to the logarithmic nature of the pH system, this generalization becomes less accurate as the $[HCO_3^-]$ is further from normal.

In vivo *evaluation of metabolic alkalosis*

The normal compensatory response to a pure metabolic alkalosis is a rapid depression of respiratory drive, resulting in an increase in $PaCO_2$ as an attempt to restore the ratio of $[HCO_3^-]/PaCO_2$ to normal. This compensation may be limited if hypoxaemia results from hypoventilation, particularly in the patient breathing room air. It appears that the respiratory stimulus from hypoxaemia is stronger than the compensatory depression due to alkalosis, and oxygenation is maintained despite a metabolic alkalosis. Although less well compensated than other acid–base disturbances, hypercapnia can occur in the presence of metabolic alkalosis, and the following formula is used to predict the resultant $PaCO_2$:

$$\text{Predicted } PaCO_2 \text{ (kPa)} = 0.9 [HCO_3^-] + 2.67 (\pm 0.27), \text{ or}$$
$$\text{Predicted } PaCO_2 \text{ (mmHg)} = 0.7 [HCO_3^-] + 20 (\pm 2)$$
(19)

Utilizing *in vivo* titration curves

There are three basic steps in the evaluation of acid–base disorders using *in vivo* titration curve data.

Evaluate the arterial pH. Normal arterial pH values range from 7.35 to 7.45; acidaemia is defined as an arterial pH of less than 7.35, and alkalaemia is defined as an arterial pH of greater than 7.45. The primary disturbance of acid–base balance will shift the pH in the direction of that disorder (i.e. a metabolic acidosis will produce an acidaemia); while compensatory mechanisms will minimize the alteration in pH, they are unable to restore it to normal values. Note that a normal pH does not preclude the presence of an acid–base disturbance: two concurrent disorders of equal magnitude but opposite direction (as a metabolic acidosis and respiratory alkalosis) may result in a normal pH.

Evaluate the markers for the aetiology of the primary disturbance. In general, $PaCO_2$ is used as a descriptor of respiratory disorders, and HCO_3^- serves as an indicator of metabolic disturbances. As an example, consider the blood gas sample: pH = 7.30, $PaCO_2$ = 4.53 kPa (34 mmHg), HCO_3^- = 19 mmol/l. The low $PaCO_2$ describes a respiratory alkalosis, and the reduction in HCO_3^- concentration suggests a metabolic acidosis. Because the pH is acidaemic, the primary aetiology will usually be the metabolic acidosis. The reduction in $PaCO_2$ may be a compensatory measure, and should be evaluated as follows.

Compare the observed values with those predicted. In the previous example of a metabolic acidosis, the HCO_3^- value is reduced by 5 mmol/l. The predicted respiratory response is a 0.15 kPa (1.1 mmHg) reduction in $PaCO_2$ for each 1 mmol/l decrease in HCO_3^-. This should reduce the $PaCO_2$ by 0.75 kPa (5.5 mmHg), to a level of 4.55 kPa (34.1 mmHg). This example describes an appropriate respiratory compensation for a metabolic acidosis.

Clinical effects of acid–base derangements

Much has been written regarding the aetiology and pathophysiology of acid–base disturbances, but definitive

Table 15.1   Predicted changes in acid–base abnormalities *in vivo*

	Primary alteration $Paco_2$ kPa (mmHg)	*Associated with a change in* HCO_3^- mmol/l	*pH* units
Respiratory acidosis			
Acute	↑ 1.33 (10)	↑ 0.8	↓ 0.07
Chronic	↑ 1.33 (10)	↑ 4	↓ 0.03
Respiratory alkalosis			
Acute	↓ 1.33 (10)	↓ 2	↑ 0.08
Chronic	↓ 1.33 (10)	↓ 6	↑ 0.03

	Primary alteration HCO_3^- mmol/l	*Associated with a change in* $Paco_2$ kPa (mmHg)	*pH* units
Metabolic acidosis	↓ 4	↓ 0.6 (4.4)	↓ 0.08
	Alternatively: Predicted $Paco_2$ (kPa) = 0.2 [HCO_3^-] +1.06 (±0.27) Predicted $Paco_2$ (mmHg) = 1.5 [HCO_3^-] + 8 (±2)		
Metabolic alkalosis	Predicted $Paco_2$ (kPa) = 0.9 [HCO_3^-] + 2.67 (±0.27) Predicted $Paco_2$ (mmHg) = 0.7 [HCO_3^-] + 20 (±2)		

Table 15.2   Summary: clinical effects of acid–base imbalance

	Decreased pH			*Increased pH*		
	Direct	*Indirect*	*Clinical*	*Direct*	*Indirect*	*Clinical*
Cerebral blood flow	↑	↑	↑	↓	Ø	↓
Heart rate	↓	↑	↑	Ø	Ø	Ø
Cardiac inotropy	↓	↑	↔	Ø	↔	↔
Systemic arterial tone	VD	VC	VD	VC	Ø	VC
Systemic venous tone	VC	VC	VC	Ø	Ø	Ø
Pulmonary arterial tone	VC	VC	VC	Ø	VD	VD
Airway tone	↓	↑	↑	↑	↓	↑
Uterine blood flow	↑	↓	↔	↓	Ø	↓
Renal blood flow	↑	↓	↓	Ø	Ø	Ø
Ionized Ca^{2+}	↑	Ø	↑	↓	Ø	↓
Serum potassium	↑	Ø	↑	↓	Ø	↓

↓, decreases; ↑, increases; ↔, variable response; VC, vasoconstriction; VD, vasodilation; Ø, minimal change or no effect. See text for details.

descriptions of the clinical effects of these derangements are limited. Characterization of the clinical effect of acid–base disturbance is difficult for many reasons. In some instances, the effects have not been effectively studied, and in many studies the available data are often conflicting. Clinical disturbances of acid–base balance usually are a mixture of metabolic (H^+ and HCO_3^-) and respiratory ($Paco_2$) disorders; distinguishing the exact aetiology for the clinical effect may be difficult. Variability of the clinical response is not only dependent on the aetiology of the disturbance, but may also depend on the rate of change of the acid–base imbalance. Additionally, the direct effect of

the disturbance may be masked by the indirect response of the sympathetic nervous system, or by concomitant changes in electrolyte concentrations (such as potassium or calcium), each of which may alter the end organ response. Despite this disclaimer, it can be broadly stated that organ function tends to be optimal within the normal range of pH, and that outside the normal pH range one usually sees depression of organ or system activity. In the following discussion of the effects of acid–base disturbance on various organ systems, the focus will be mostly on acidaemia, which is both more common clinically, and better studied. This is not meant to minimize the significance of alkalosis, which in many cases may be more harmful than acidosis to organ function (Table 15.2).

Cardiovascular effects

Cardiac rate

Tachycardia frequently results from a reduction in pH from 7.4 to 7.1. The aetiology of the response is felt to be indirect, caused by adrenaline (epinephrine) release from the adrenal medulla in response to the acidaemia.[2] The direct effect of acidaemia on the isolated heart preparation, or in the patient whose sympathetic nervous system is blocked, is bradycardia. This effect is more dramatic at a pH below 7.1. The reduction in heart rate may be the result of an increased vagal tone and an accumulation of acetylcholine (ACh) due to decreased metabolism of ACh in the acid environment.[3,4] Clinically, with either respiratory or metabolic acidosis, one would expect to see an initial increase in heart rate (assuming normal sympathetic function), followed by a decreasing heart rate with extremes of pH, as the direct depressant effects become more predominant.

Cardiac rhythm

Dysrhythmias, of both atrial and ventricular origin, are more prevalent with acid–base derangements. It is difficult to ascertain whether it is the pH alone or the extracellular potassium levels, which are closely tied to pH, that serves as the mechanism for the dysrhythmias. A change of 0.1 units in pH will generally cause the serum potassium level to change 0.5–1.5 mmol/l in the opposite direction; in the acidotic state, high extracellular $[H^+]$ causes a reduction in the intracellular $[K^+]$. This change in intracellular potassium concentration alters the resting membrane potential of the cardiac conduction system. Arrhythmogenicity may also be due to pH related alterations in calcium or magnesium levels, or the levels of circulating catecholamines. Although the data are conflicting, most studies have demonstrated that the threshold for ventricular fibrillation is lowered by decreasing the pH, while an increase in the pH raises the fibrillatory threshold.[5–7]

Under anaesthesia with a volatile anaesthetic, the spontaneously breathing patient will develop a respiratory acidosis (resting $PaCO_2$ with halothane at 1.25 MAC: 6.13 kPa (46 mmHg); with isoflurane: 6.93 kPa (52 mmHg); with enflurane: 8.26 kPa (62 mmHg)). Because of the increase in circulating catecholamines and the decrease in pH one may see an increase in ventricular ectopy. Of the volatile anaesthetics, halothane lowers the arrhythmogenic threshold the most, probably by the additional myocardial sensitization to catecholamine effects. Because the pH change in healthy patients is minimal, this is usually not a problem of clinical significance. When ventilatory depression due to volatile anaesthetics is combined with other factors, such as obstructive lung disease, poor neuromuscular function, or other drugs which depress ventilation, the $PaCO_2$ may rise excessively and cause significant dysrhythmias.

Cardiac contractility

In the acidotic state, an increase in inotropy is commonly seen, secondary to the indirect effects of increased circulating catecholamines. This is in contrast to the direct effect of acidosis on the isolated heart, which causes a reduction in contractility.[3,8] The mechanism is thought to be impairment of calcium entry into the cells and/or the decreased release of calcium from intracellular stores.[9,10] In the normally innervated heart, negative inotropy is usually not appreciated clinically until the pH falls below 7.2, due to the offsetting effects of the concurrent catecholamine release.[3] The overall clinical effect will depend on which stimulus is greater, the direct effect of the pH change or the indirect effect of catecholamine release. Variability of the response is expected, based on the patient's cardiac reserve and autonomic status; patients with sympathetic blockade or those taking calcium channel blockers may demonstrate a pronounced negative inotropy with relatively mild changes in pH.

Alkalosis may produce a mild increase in contractility, secondary to the increased responsiveness of the myocardium to circulating catecholamines in an alkaline environment. This increase in contractility may be at the expense of a proportionally greater coronary oxygen extraction, as alkalosis also increases coronary vascular resistance and causes a left shift in the oxyhaemoglobin dissociation curve, resulting in a decreased delivery of oxygen to the myocardium. It has been shown that, at equal cardiac workloads, coronary sinus lactate levels are higher in the alkalotic patient than in the patient at normal pH; this suggests a greater cardiac extraction of oxygen in alkalosis and a smaller myocardial oxygen reserve. Alkalosis will also reduce the concentration of unbound plasma calcium, which may also reduce contractility.

Cardiac output

In situations of mild acidaemia, increased levels of circulating catecholamines may cause an increase in both cardiac rate and contractility, consequently increasing cardiac output. Additionally, acidaemia decreases arterial vascular tone and increases venous tone; the combination of increased preload, reduced afterload, increased heart rate and contractility will augment cardiac output up to a point. As the pH nears 7.0, the direct effects of the acidaemia overtake the indirect stimulation, and cardiac output will fall secondary to a reduction in both contractility and heart rate. Such a decrease in cardiac output coupled with an increase in venous return may predispose the patient with limited cardiac reserve to congestive heart failure.

Systemic vascular effects

Generally, the direct effect of a reduction in pH on arterial vasculature is vasodilation[11] (the exception to this rule being its effect on the pulmonary vasculature, see below). This effect is much more pronounced in respiratory acidosis than in metabolic acidosis, because CO_2 can cross cell membranes more readily than can bicarbonate. In a respiratory acidaemia, CO_2 enters the cell and is converted to carbonic acid which then dissociates to H^+ and HCO_3^-. The resultant decrease in intracellular pH is considered to be the mechanism directly responsible for relaxation of the vascular smooth muscle. In a metabolic acidaemia, the limited diffusibility of the acid (compared to CO_2) alters the intracellular pH to a lesser degree. With mild degrees of metabolic acidosis, the indirect effects from increased levels of circulating catecholamines may result in vasoconstriction, and only with further decreases in pH (usually below 7.2) would vasodilation be expected.[12] Alkalosis tends to produce vasoconstriction (again, with the exception of the pulmonary vasculature). Systemic blood pressure may or may not change despite alterations in vascular tone, dependent on concurrent changes in cardiac output.

The overall response of individual vascular beds may differ significantly, due to the contribution of direct and indirect effects. The direct effect of metabolic acidosis is dilation of the arterial vessels of the skin and muscle, in the renal and splanchnic beds, the uterine arteries, and the coronary arteries; the direct effect of a comparable respiratory acidosis produces a greater vasodilation of the vascular beds of the kidney, gastrointestinal tract, and uterus. Because of the indirect effects of an intact sympathoadrenal axis, however, both respiratory and metabolic acidosis result in clinical vasoconstriction of the renal and splanchnic beds; changes are less consistent in the other vasculature. In contrast, the direct response to metabolic or respiratory acidosis in the pulmonary bed is vasoconstriction; this is augmented greatly by an intact sympathoadrenal system or in the presence of hypoxia.

Coronary vascular resistance is closely tied to myocardial oxygen demand, which in turn is related to factors such as heart rate, contractility, preload, and afterload. Because all of these variables respond differently to acid–base imbalance, it is difficult to simply define the coronary response to acidosis or alkalosis in the intact animal. In the isolated heart, however, the coronary vasculature responds to acidosis by dilation, and to alkalosis by constriction.[13,14] With hyperventilation,

coronary blood flow can be reduced significantly, resulting in the elevation of coronary sinus lactate levels and myocardial ischaemia.

The venous system is one of the few vascular beds in which the direct and indirect effects of acidosis produce the same results. Venoconstriction results from an increase in hydrogen ion concentration as well as from an increase in catecholamine levels.[15] Acidosis therefore results in an increase in venous return to the heart, which may increase cardiac work particularly in an acid environment which may impair myocardial contractility.

Respiratory effects

The respiratory response to acidosis is increased minute ventilation ($\dot{V}E$). The increase is manifest by an increase in tidal volume and a slight increase in respiratory rate (Kussmaul respirations). $\dot{V}E$ is most markedly increased by a CO_2 challenge; the ventilatory response to comparable metabolic acidosis is only half as great. The medullary chemoreceptors are responsible for 80 per cent of the total respiratory response to hypercapnia; a lesser percentage of the respiratory stimulus originates from peripheral chemoreceptors in the carotid bodies.[16,17] The carotid receptors are stimulated by changes both in pH and in $PaCO_2$, while the central receptors are stimulated by the passage of CO_2 across the blood–brain barrier and the subsequent increase in $[H^+]$ of the CSF. The reduction in pH at the medullary level, from either a respiratory or metabolic acidosis, stimulates efferent respiratory drive to increase ventilation. Respiratory acidosis produces a greater ventilatory response than does a comparable metabolic acidosis because the blood–brain barrier is more permeable to CO_2 than metabolic acids. The normal CO_2 response curve demonstrates that for every kPa increase in $PaCO_2$, $\dot{V}E$ increases by 15–23 l/min (2–3 l/min per mmHg). Ventilatory stimulus peaks at a $PaCO_2$ of approximately 13.3 kPa (100 mmHg); further increases in $PaCO_2$ result in respiratory depression. Volatile anaesthetics also depress the slope of the ventilatory response to CO_2 in a dose-dependent fashion.

Airway resistance is affected by changes in acid–base equilibrium; changes in airway tone are mediated by both the local effect of CO_2 and the indirect effect of sympathoadrenal stimulation. An increase in $PaCO_2$ produces a locally mediated bronchodilation, due to the intracellular diffusion of CO_2 and subsequent production of $[H^+]$, resulting in smooth muscle relaxation. In the intact animal, however, the indirect effect of bronchoconstriction predominates, and results in increased airway resistance, decreased tracheal volume, and increased resistive work of breathing.[18] Bronchoconstriction in response to hypercapnia is mediated through vagal stimulation, and is minimized or abolished by atropine or vagotomy.[19] Conversely, a reduction in $PaCO_2$ produces a locally mediated bronchoconstriction which may be partially offset by centrally mediated bronchodilation. The broncho-constriction associated with hypocapnia is clinically predominant, and is useful in the matching of ventilation to blood flow in normal lungs and in pathological conditions such as pulmonary embolism.

While acid–base imbalance does not directly alter pulmonary oxygenation of blood, oxygen transport may be affected by both acidosis and alkalosis. Oxygen transport depends on both cardiac output and oxygen carrying capacity; acid–base imbalance has an impact on each of these variables. Cardiac output may, as previously mentioned, be preserved in acidosis with a pH as low as approximately 7.2. With a more severe acidosis, cardiac output may fall, proportionally reducing oxygen transport. The effect of acid–base imbalance on the affinity of haemoglobin for oxygen is described by the Bohr effect, which states that an increase in $[H^+]$ reduces the affinity of haemoglobin for oxygen. This shifts the oxygen dissociation curve to the right and results in an increased availability of oxygen to the tissues. Conversely, alkalosis increases the affinity for oxygen, and reduces tissue oxygenation. The rightward shift of the oxyhaemoglobin dissociation curve in acidosis occurs immediately, but within 12–36 hours the concentration of 2,3-diphosphoglycerate (2,3-DPG) falls and restores the oxygen affinity, shifting the oxygen dissociation curve back to the left.[20] 2,3-DPG concentration decreases because glycolysis is impaired in an acid environment and the glycolytic intermediates in the 2,3-DPG production pathway become depleted.

The response of the pulmonary vasculature to pH change is the opposite of that in the other vascular beds, in that acidosis increases pulmonary vascular tone, and alkalosis reduces the tone of the vascular bed.[21–23] While there is some evidence that an increase in $PaCO_2$ results in direct vasodilation as in other tissues, this effect is overridden in an acid environment by vasoconstriction of the pre- and postcapillary sphincters of the pulmonary vascular bed.[24] The degree of vasoconstriction caused by acidosis is small when compared to hypoxic pulmonary vasoconstriction. The hypoxic vasoconstrictor response is, however, greatly enhanced in the presence of acidosis.[25] With acidosis, one can expect to see an increase in pulmonary artery pressures and an increase in the calculated pulmonary vascular resistance. This is due not only to capillary sphincter constriction, but to an increase in venous return secondary to acidosis-induced venoconstriction. Additionally, with mild degrees of acidaemia, the catecholamine-mediated increase in cardiac output produces an increase in pulmonary blood flow, and results in increased pulmonary artery pressures.

Gastrointestinal effects

Studies of the response of the splanchnic vascular bed to acidosis yield conflicting data. In some animal studies, respiratory acidosis increases splanchnic blood flow. This effect was not seen in metabolic acidosis, and the increase in blood flow persisted after sympathetic blockade. This suggests that CO_2 directly reduces the intracellular pH of the vessel endothelium to produce vasodilation.[26] Epstein studied the effects of respiratory acidosis in human subjects and reported an increase in splanchnic vascular resistance.[27] The discrepancy between these studies is probably due to the vasoconstrictor effect of sympathetic stimulation, overwhelming local vasodilation.

There has also been little research on the effect of pH imbalance on gastric motility. Isolated strips of the oesophagus and stomach are shown to have both a decreased spontaneous rate and amplitude of intrinsic neural discharge when bathed in an acid medium.[28] These effects may be magnified by concurrent derangements of potassium, magnesium, and calcium. Clinically, this may correlate with vague abdominal pain, nausea, and vomiting often seen with acidosis.

Renal effects

The renal vascular response to acidosis depends on the balance of direct and systemic factors, but vasoconstriction is observed clinically. In situations of mild respiratory acidosis, renal vascular resistance may be unchanged or decreased, but as acidosis worsens, the renal vascular resistance increases and renal blood flow decreases. Metabolic acidosis of a given pH appears to produce a much greater increase in vascular resistance than does a comparable respiratory acidosis.[29,30] This is probably due to the ability of CO_2 in a respiratory acidosis to cross the cell membrane and reduce the intracellular pH; the local vasodilation produced ameliorates the catecholamine-induced vasoconstriction. Direct vasodilation is not a feature of metabolic acidosis, and catecholamine-mediated vasoconstriction is therefore unopposed.

Uteroplacental effects

Acid–base imbalance affects the fetus directly through placental transmission, and indirectly by alteration in placental perfusion. Diffusion of CO_2 across the placenta is rapid, and respiratory alterations in $PaCO_2$ result in a rapid change in fetal $PaCO_2$ in the same direction. Passage of HCO_3^- and H^+ appears to be much slower, such that a metabolic acidosis or alkalosis will cause little direct change in fetal pH over several hours. Alterations in fetal pH will produce similar effects in fetal organ function as is described for the adult.

The direct result of acidosis on the uterine circulation is vasodilatation; the effect is slightly greater in respiratory acidosis than from a metabolic acidosis. With severe hypercapnia, the vasodilation is opposed by vasoconstriction mediated by sympathoadrenal stimulation, so that in the intact acidotic animal there is little change in uterine blood flow.[31] The direct effect of alkalosis is vasoconstriction of the uterine vessels; respiratory alkalosis produces a significantly greater reduction in uteroplacental blood flow than does a metabolic alkalosis. Alkalosis also causes a leftward shift of the maternal oxyhaemoglobin dissociation curve and increases the affinity of haemoglobin for oxygen; this reduces the amount of oxygen available to the fetus. The combination of vasoconstriction and reduced oxygen delivery seen in severe alkalosis (pH > 7.65) will cause fetal hypoxaemia and metabolic acidosis.[32,33] In a milder respiratory alkalosis (such as in active labour), placental perfusion is maintained, and the low maternal $PaCO_2$ equilibrates across the placenta. This increases fetal pH and shifts the fetal oxyhaemoglobin dissociation curve to the left, enhancing oxygen uptake by fetal haemoglobin, and results in a higher fetal oxygen saturation than in a metabolic alkalosis of comparable degree.

Neuroendocrine effects

The cerebrovascular response to $PaCO_2$ is similar to that of other vascular beds. An increase in $PaCO_2$ results in vasodilation and an increase in cerebral blood flow. An increase in $PaCO_2$ to 10.6 kPa (80 mmHg) will approximately double cerebral blood flow; reducing the $PaCO_2$ to 2.7 kPa (20 mmHg) will halve the blood flow. There is no further reduction in blood flow at a $PaCO_2$ of less than 2.7 kPa; this is thought to be secondary to the accumulation of lactic acid in the periarteriolar tissue, producing local vasodilation which limits cerebral vasoconstriction.[34]

Neurological changes from acid–base imbalance stem from the ability of CO_2 to cross the blood–brain barrier. The resultant change in pH of the CSF impairs neuronal function, and may lead to mental status change and coma.[35] The clinical effects are more prominent in a respiratory acidosis or alkalosis because the blood–brain barrier is more permeable to CO_2 than to metabolic acids.

Respiratory acidosis may also result in a decrease in body temperature due in part to an impairment in central temperature regulation,[36] but also due to a reduction in cellular metabolism and an increased heat loss from acidosis-induced cutaneous vasodilatation.

The neuroendocrine response to acidosis is an increase in the level of circulating catecholamines. Adrenaline release from the adrenal glands is increased, and the sympathetic nervous system is stimulated to directly release noradrenaline from nerve terminals.[37,38] Noradrenaline synthesis is also increased to match the increased rate of catecholamine release. With mild degrees of acidosis, the elevated catecholamine levels tend to counteract the depressant effects of acidaemia on the organ systems. As the pH falls, cells become less responsive to catecholamine stimulation, and depression of function becomes more apparent despite increased catecholamine levels.[39]

Electrolyte effects

Calcium exists in three forms in the plasma. Approximately 50 per cent is non-ionized and bound to plasma proteins, and therefore unavailable for diffusion into the tissues. Five per cent of the calcium is diffusible but bound to other plasma components (such as citrate), and 45 per cent of the plasma calcium exists in the ionized (chemically active) form. In conditions of acidosis, hydrogen ions compete for the negatively charged binding sites on albumin, displacing calcium and increasing the proportion of serum ionized calcium. Conversely, alkalosis results in an increase in available protein binding sites, a reduction in ionized calcium concentrations, and clinical hypocalcaemia.[40] Tetany and/or mild disturbances of cardiac contractility are occasionally seen as a result of such hypocalcaemia. Although nomograms are available to estimate ionized calcium concentrations from measurements of total calcium, albumin, and pH, direct measurement of ionized calcium is the only accurate method of assessing free unbound calcium.

It is well known that pH and serum potassium levels [K$^+$] are usually reciprocal: when pH falls, serum [K$^+$] increases. In acidaemic conditions, hydrogen ions enter the cell along a concentration gradient, and potassium ions move into the extracellular fluid to maintain intracellular electrical neutrality. It is commonly quoted that a 0.6 mmol/l change in [K$^+$] occurs for each 0.1 unit change in pH, but in several studies the relations between [K$^+$] and pH have been shown to be non-linear.[41] These variations may be due to other factors controlling potassium homoeostasis including catecholamines, aldosterone, insulin, and the non-linearity of the pH system itself.

Therapy for acid–base disorders

An understanding of the diagnosis, magnitude, and physiological impact of acid–base disorders is certainly important in clinical practice. Indeed, one of the goals of the many evaluation systems (buffer base, base excess, etc.) was to define and quantify the magnitude of a metabolic acidosis so that appropriate amounts of buffer or base could be administered. Logically, therapy to correct such an imbalance should be beneficial to the patient, but 'historically appropriate' therapy is now met with increasing scrutiny and disfavour. The question remains: when is therapy indicated, and how (or should) acid–base disorders be treated?

Alkalosis

Clinically significant alkalosis, while troublesome, is generally a less common and less ominous disturbance than acidosis. Metabolic alkalosis can be classified as either chloride responsive or chloride resistant in nature, as distinguished by the measurement of urinary chloride levels. Urinary chloride values of less than 10 mmol/l are usually consistent with an alkalosis associated with intravascular volume depletion, as from diuretic use, hyperemesis, nasogastric suctioning, and diarrhoea. Both the alkalosis and hypovolaemia will respond to or correct with the administration of a salt containing solution. Sodium chloride will commonly restore intravascular volume and promote bicarbonate excretion, correcting the alkalosis. Administration of potassium chloride is also necessary to correct the alkalosis and restore concomitant hypokalaemia, which may be exacerbated or unmasked as the pH falls. Hypokalaemia must be corrected to effectively treat the alkalosis because, in the hypokalaemic state, the kidneys will alter the usual sodium/potassium ion exchange to a sodium/hydrogen ion exchange to promote potassium conservation. This will result in acidification of the urine which maintains the alkalosis.

A chloride-resistant alkalosis, as determined by urinary chloride levels greater than 20 mmol/l, is less common, and is usually related to mineralocorticoid excess, thiazide or loop diuretics, or potassium supplementation. These patients characteristically are normovolaemic or hypervolaemic, have normal urine sodium and chloride levels, and do not respond to the administration of chloride salts. Therapy is aimed at resolution of the underlying disorder.

Carbonic anhydrase inhibitors such as acetazolamide may be effective in the therapy of metabolic alkalosis by decreasing the renal reabsorption of filtered sodium and bicarbonate, producing a net loss of bicarbonate. In rare cases (as with a pH of > 7.6), hydrochloric acid (HCl) may be infused to reduce the pH to more physiological levels. The dose of HCl equal to the calculated base excess (equation 20) may be infused via a central venous line over 24 hours as a 0.1 N (100 mmol/l) or 0.2 N (200 mmol/l) solution, with frequent monitoring of arterial blood gases and electrolyte

values. Ammonium chloride is also used as an acidifying agent (given parenterally or enterally), as is arginine hydrochloride (although it is not approved by the Federal Drug Administration for this usage).

$$\text{Base excess} = \text{weight (kg)} \times (\text{actual bicarbonate} - \text{normal bicarbonate}) \times 0.4 \quad (20)$$

Respiratory alkalosis will result only if alveolar ventilation exceeds CO_2 production. Hypoxaemia is one of the more common causes of hypocapnia, as ventilatory drive is stimulated by a $Pa\text{CO}_2$ of less than approximately 8 kPa (60 mmHg). Head injuries, neurological disorders, and hysteria are other aetiologies for respiratory alkalosis. In general, therapy is aimed at determination and correction of the underlying disease.

Acidosis

Respiratory acidosis is the result of ventilatory failure. The aetiologies of respiratory failure are legion, and may include any drug or disease depressing central or peripheral neural control of ventilation (trauma, tumours, myasthenia gravis, narcotics, etc.), or interfering with the mechanics of respiration (trauma, asthma, emphysema, neuromuscular relaxants, etc.). Therapy for the acidosis should include ventilatory support, but focus on the diagnosis and treatment of the underlying disease process.

The presence of a metabolic acidosis usually indicates a more severe underlying disorder than does alkalosis, and much more attention has been given to therapy. The archetypal clinical condition producing a metabolic (lactic) acidosis is cardiac arrest. Recommendations regarding alkali therapy for this disorder have been debated for two decades. In the early 1970s common clinical practice included the administration of sodium bicarbonate ($NaHCO_3$) at the initiation of cardiopulmonary resuscitation, and the use of additional doses at regular timed intervals. By the 1980s, these recommendations were modified so that most alkali therapy would be guided by arterial blood gases and pH. Current advisories warn against routine bicarbonate administration as the value of sodium bicarbonate is questionable during cardiac arrest.[42]

There are several reasons for the current reassessment of routine therapy using $NaHCO_3$ for cardiac arrest or low flow conditions[43] (see Chapter 71).

1. In the cardiac arrest state, acidosis is the product of an accumulation of CO_2 and lactic acid. Hypercapnia results from aerobic metabolism and inadequate CO_2 elimination, while lactic acidosis is the product of anaerobic metabolism. Subsequent therapy with $NaHCO_3$ may in fact exacerbate the acidosis. Approximately 10 per cent of an administered 50 ml ampoule of $NaHCO_3$ is converted to CO_2 (see

equation 21): this is the equivalent of 200 ml of CO_2, or the same amount of CO_2 produced in 1 minute of basal aerobic metabolism in the average adult. If the CO_2 cannot be cleared with appropriate pulmonary perfusion and ventilation, the additional hypercapnia may compound the existing lactic acidosis.[44]

$$NaHCO_3 + H^+ \rightleftarrows H_2CO_3 \rightleftarrows CO_2 + H_2O \qquad (21)$$

2. The additional hypercapnia from $NaHCO_3$ administration creates a gradient for CO_2 to enter the cells, and because the cell membrane is more permeable to CO_2 than H_2CO_3, the intracellular pH is reduced.[45] The situation is similar in the central nervous system, where permeability of the blood–brain barrier to CO_2 results in disproportionate acidosis of the CSF in a respiratory acidosis.[46]

3. Administration of several ampoules of 8.4 per cent bicarbonate solution (20 000 mmol/l) can result in a significant osmolar load. Each 50 ml ampoule contains approximately 100 mmol, and will raise the serum osmolality by 3 mmol/l in the average adult.[47]

4. $NaHCO_3$ administration (and the elimination of produced CO_2) may lead to an extracellular alkalosis, a shift of the oxygen dissociation curve to the left, and a reduction in available oxygen at the tissue level. The alkalosis will also decrease serum ionized calcium levels (by as much as 25 per cent), and may lead to impairment of myocardial contractility.[48,49]

5. Because of the variable effects of $NaHCO_3$ administration on acid–base balance, the haemodynamic benefits may also be unpredictable. Cardiac output and blood pressure may increase, decrease, or remain the same. A reduction in ionized calcium may impair cardiac output, or the output may increase simply as a result of volume loading rather than correction of pH. In one study comparing bicarbonate to saline infusion, cardiac output was comparably increased in both groups, and thought to be the result of volume loading alone. In another prospective, randomized, crossover study, patients with lactic acidosis were treated with saline or $NaHCO_3$, and no differences were seen in cardiac output, blood pressure, or systemic or pulmonary vascular resistance. Of the 14 patients studied, 13 were receiving catecholamine infusions; despite concerns regarding the reduced responsiveness to catecholamines in acidotic conditions, pH correction did not result in an increased efficacy of the infusions.[50]

6. The administration of $NaHCO_3$ has been shown in animal models to reduce the hepatic clearance of lactic acid. This is postulated to be the result of hepatic cellular dysfunction secondary to increased intracellular CO_2 and acidosis.[57]

These same questions have challenged the utility of sodium bicarbonate in therapy for other acidotic states as well. While respiratory clearance of CO_2 produced by $NaHCO_3$ may not be of significance as it is in the cardiac arrest or low flow state, many of the aforementioned problems are of concern.

Considering the current unknowns in the therapy of acid–base disturbances, clinical decisions may be difficult. The following guidelines may assist the decision and method of therapy:

1. Treat the underlying disorder, if at all possible.
2. Adjust controlled ventilation to assist in compensation for the pH imbalance.
3. Recognize the potential side effects of therapy for acidoses, particularly in the low flow state.
4. At the extremes of pH, beware that the buffering capacity is limited, and that small changes in the hydrogen ion concentration may have a profound effect on the pH. In these situations (i.e. with serum bicarbonate concentrations of 10–12 mmol/l), the judicious administration of a buffer may be helpful in avoiding the physiological impairment of extreme acidosis.

Summary

Because of the multiple factors contributing to the clinical effects of acid–base imbalance, a clear understanding of the pathophysiology is difficult. In general, the direct effects of a mild alkalaemia or acidaemia are overshadowed by the indirect effects of activation of the sympathoadrenal axis. In conditions of a more profound acid–base imbalance, the direct depressant effects begin to predominate. Because of the relative diffusibility of CO_2 and the resultant alteration of intracellular pH, respiratory acidoses or alkaloses typically produce a more significant clinical effect than metabolic disturbances of a similar pH.

Therapy for acid–base disorders is controversial: the perceived benefits are now in question, and the risks are both theoretical and actual. At the present time, therapy should best be approached with the intent to identify and correct the underlying disorder, with an understanding of the potential clinical impact of both the pH imbalance and its treatment.

REFERENCES

1. Brackett NC, Cohen JJ, Schwartz WB. Carbon dioxide titration curve of normal man. The effect of increasing degrees of acute hypercapnia on acid–base equilibrium. *New England Journal of Medicine* 1965; **272**: 6–12.

2. Wildenthal K, Mierzwiak DS, Myers RW, Mitchell JH. Effects of acute lactic acidosis on left ventricular performance. *American Journal of Physiology* 1968; **214**: 1352–9.

3. Mitchell JH, Wildenthal K, Johnson Jr RL. The effects of acid–base disturbances on cardiovascular and pulmonary function. *Kidney International* 1972; **1**: 375–89.

4. Clowes GA, Hopkins A, Simeone F. Comparison of physiological effects of hypercapnia and hypoxia in production of cardiac arrest. *Annals of Surgery* 1955; **142**: 446–51.

5. Gerst PH, Fleming WH, Malm JR. A quantitative evaluation of the effects of acidosis and alkalosis upon the ventricular fibrillation threshold. *Surgery* 1966; **59**: 1050–60.

6. Rogers RM, Spear JF, Moore EN, Horowitz LH, Sonne JE. Vulnerability of canine ventricle to fibrillation during hypoxia and respiratory acidosis. *Chest* 1973; **63**: 986–94.

7. Kerber RE, Pandian NG, Hoyt R, *et al*. Effect of ischemia, hypertrophy, hypoxia, acidosis, and alkalosis on canine defibrillation. *American Journal of Physiology* 1983; **244**: H825–31.

8. Pannier JL, Brutsaert DL. Contractility of isolated cat papillary muscle and acid–base changes. *Archives of International Pharmacodynamics and Therapeutics* 1968; **172**: 244–6.

9. Katz AM. Contractile proteins of the heart. *Physiological Reviews* 1970; **50**: 63–158.

10. Williamson JR, Safer B, Rich T, Schaffer S, Kobayashi K. Effects of acidosis on myocardial contractility and metabolism. *Acta Medica Scandinavica* 1976; **587** (suppl): 95–112.

11. Kontos HA, Richardson DW, Patterson Jr JL. Vasodilator effect of hypercapnic acidosis on human forearm blood vessels. *American Journal of Physiology* 1968; **215**: 1403–5.

12. Downing SE, Talner NS, Gardner TH. Cardiovascular responses to metabolic acidosis. *American Journal of Physiology* 1965; **208**: 237–42.

13. McElroy WT, Gerdes AJ, Brown Jr EB. Effects of CO_2, bicarbonate, and pH on the performance of isolated perfused guinea pig hearts. *American Journal of Physiology* 1958; **195**: 412–16.

14. Daugherty Jr RM, Scott JB, Dabney JM, Haddy FJ. Local effects of O_2 and CO_2 on limb, renal, and coronary vascular resistances. *American Journal of Physiology* 1967; **213**: 1102–10.

15. Sharpey-Schafer EP, Semple SJG, Halls RW, Howarm S. Venous constriction after exercise and its relation to acid–base changes in venous blood. *Clinical Science* 1965; **29**: 397–406.

16. Biscoe TJ. Carotid body: structure and function. *Physiological Reviews* 1971; **51**: 437–95.

17. Wollman H, Smith TC, Stephen GW, Colton 3d ET, Gleaton HE, Alexander SC. Effects of extremes of respiratory and metabolic alkalosis on cerebral blood flow in man. *Journal of Applied Physiology* 1968; **24**: 60–5.

18. Daly MdeB, Lambertsen CJ, Schweitzer A. The effects upon bronchial musculature of altering the oxygen and carbon dioxide tensions of blood perfusing the brain. *Journal of Physiology* 1953; **119**: 292–341.

19. Nadel JA, Widdicombe JG. Effect of changes in blood gas tensions and carotid sinus pressure on tracheal volume and total lung resistance to airflow. *Journal of Physiology* 1962; **163**: 13–33.

20. Chanutin A, Curnish RR. Effect of organic and inorganic phosphates on the oxygen equilibrium of human erythrocytes. *Archives of Biochemistry and Biophysiology* 1967; **121**: 96–102.

21. Duke HN, Killick EM, Marchant JV. Changes in pH of the perfusate during hypoxia in isolated perfused cat lungs. *Journal of Physiology* 1960; **153**: 413–22.

22. Bergofsky EH, Lehr DE, Fishman AP. The effect of changes in hydrogen ion concentration on the pulmonary circulation. *Journal of Clinical Investigation* 1962; **41**: 1492–502.

23. Tenny SM. Sympatho-adrenal stimulation by carbon dioxide and the inhibitory effect of carbonic acid on epinephrine response. *American Journal of Physiology* 1956; **187**: 341–6.

24. Barer GR, Shaw JW. Pulmonary vasodilator and vasoconstrictor actions of carbon dioxide. *Journal of Physiology (London)* 1971; **213**: 633–45.

25. Shubrooks Jr SJ, Schneider B, Dubin H, Turino GM. Acidosis and pulmonary hemodynamics in hemorrhagic shock. *American Journal of Physiology* 1973; **225**: 225–9.

26. McGinn FP, Mendel D, Perry PM. The effects of alterations of CO_2 and pH on intestinal blood flow in the cat. *Journal of Physiology (London)* 1967; **192**: 669–80.

27. Epstein RM, Wheeler HO, Frumin MH, Hubif DV, Papper EM, Bradley SE. The effect of hypercapnia on estimated hepatic blood flow, circulating splanchnic blood volume, and hepatic sulfobromophthalein clearance during general anesthesia in man. *Journal of Clinical Investigation* 1961; **40**: 592–8.

28. Schulze-Delrieu K, Lepsien G. Depression of mechanical and electrical activity in muscle strips of opossum stomach and esophagus by acidosis. *Gastroenterology* 1982; **82**: 720–4.

29. Stone JE, Wells J, Draper WB, Whitehead RW. Changes in renal blood flow in dogs during the inhalation of 30 percent CO_2. *American Journal of Physiology* 1958; **194**: 115–19.

30. Norman JN, MacIntyre J, Shearer JR, Craigen IM, Smith G. Effect of carbon dioxide on renal blood flow. *American Journal of Physiology* 1970; **219**: 672–6.

31. Blechner JN, Stenger VG, Eitzman DV, Prystowski H. Effects of maternal metabolic acidosis on the human fetus and newborn infant. *American Journal of Obstetrics and Gynecology* 1967; **99**: 46–54.

32. Moya F, Morishima HO, Shnider JM, James LS. Influence of maternal hyperventilation on the newborn infant. *American Journal of Obstetrics and Gynecology* 1965; **91**: 76–84.

33. Motoyama EK, Rivard G, Acheson F, Cook CD. The effects of changes of maternal pH and PCO_2 on the PO_2 of fetal lambs. *Anesthesiology* 1967; **28**: 891–903.

34. Berne RM, Winn HR, Rubio R. The local regulation of cerebral blood flow. *Progress in Cardiovascular Disease* 1981; **24**: 243–60.

35. Posner JB, Plum F. Spinal fluid pH and neurologic symptoms

in systemic acidosis. *New England Journal of Medicine* 1967; **277**: 605–13.

36. Schaefer KE, Messier AA, Morgan C, Baker 3d GT. Effect of chronic hypercapnia on body temperature regulation. *Journal of Applied Physiology* 1975; **38**: 900–6.

37. Nahas GG, Steinsland OS. Increased rate of catecholamine synthesis during respiratory acidosis. *Respiratory Physiology* 1968; **5**: 108–17.

38. Nahas GG, Ligou JC, Mehlman B. Effects of pH changes on O_2 uptake and plasma catecholamine levels in the dog. *American Journal of Physiology* 1960; **198**: 60–6.

39. Hornbein TF, Griffo ZJ, Roos A. Quantitation of chemoreceptor activity: interrelation of hypoxia and hypercapnia. *Journal of Neurophysiology* 1961; **24**: 561–8.

40. Moore EW. Ionized calcium in normal serum, ultrafiltrates, and whole blood determined by ion-exchange electrodes. *Journal of Clinical Investigation* 1970; **49**: 318–34.

41. Goodkin DA, Narins RG. Quantitation of serum potassium changes during acute acid–base disturbances. In: Whelton PK, Whelton A, eds. *Potassium in cardiovascular and renal medicine*. New York: Marcel Dekker, 1986: 67–78.

42. Albarran-Sotelo R, Atkins JM, Bloom RS, *et al*. Cardiovascular pharmacology I. In: *Textbook of advanced cardiac life support*. American Heart Association, 1987: 97–113. (not produced by a known publishing company).

43. Young GP. Reservations and recommendations regarding sodium bicarbonate administration in cardiac arrest. *Journal of Emergency Medicine* 1988; **6**: 321–3.

44. Weil MH, Rackow EC, Trevino R, Grundler W, Falk JL, Griffel MI. Difference in acid-base state between venous and arterial blood during cardiopulmonary resuscitation. *New England Journal of Medicine* 1986; **315**: 153–6.

45. Bishop RL, Weisfeldt ML. Sodium bicarbonate administration during cardiac arrest. Effect on arterial pH, P_{CO_2}, and osmolality. *Journal of the American Medical Association* 1976; **235**: 506–9.

46. Berenyi K, Wolk M, Killip T. Cerebrospinal fluid acidosis complicating therapy of experimental cardiopulmonary arrest. *Circulation* 1975; **52**: 319–24.

47. Mattar JA, Weil MH, Shubin H, Stein L. Cardiac arrest in the critically ill. II. Hyperosmolal states following cardiac arrest. *American Journal of Medicine* 1974; **56**: 162–8.

48. Pedersen KO. Binding of calcium to serum albumin. II. Effect of pH via competitive hydrogen and calcium ion binding to the imidazole groups of albumin. *Scandinavian Journal of Clinical and Laboratory Investigation* 1972; **29**: 75–83.

49. Lang RM, Fellner SK, Neumann A, Bushinsky DA, Borow KM. Left ventricular contractility varies directly with blood ionized calcium. *Annals of Internal Medicine* 1988; **108**: 524–9.

50. Cooper DJ, Walley KR, Wiggs BR, Russell JA. Bicarbonate does not improve hemodynamics in critically ill patients who have lactic acidosis. A prospective, controlled study. *Annals of Internal Medicine* 1990; **112**: 492–8.

51. Arieff AI, Leach W, Park R, Lazarowitz VC. The systemic effects of $NaHCO_3$ in experimental lactic acidosis in dogs. *American Journal of Physiology* 1982; **242**: F586–91.

Chapter 16

Calcium and Magnesium in Anaesthesia

M. F. M. James

Calcium	Magnesium

Calcium and magnesium represent the major divalent cations in the body. Calcium, like sodium, is the predominant extracellular ion, and has complex control mechanisms which regulate the extracellular content of calcium within fairly narrow limits. Magnesium, on the other hand, is, like potassium, predominantly an intracellular ion, and the regulatory mechanisms which control the extracellular concentrations of magnesium are relatively poorly developed. Calcium is the major structural mineral in the body, but of more physiological relevance is the growing appreciation of its predominant role as an intracellular regulator.

Calcium

Regulation of serum calcium concentration

Calcium is present in the body in greater amounts than any other mineral. The adult human body contains about 1100 g of calcium or 1.5 per cent of total body weight, the majority of which (99 per cent) is in the skeleton. The plasma concentration is approximately 2.5 mmol/l and consists of three fractions (Table 16.1): protein bound, complexed with anions and freely diffusible ionized calcium (Ca^{2+}). It is the ionized fraction which is responsible for the physiological effects of calcium. Calcium is the principal component of bone, but calcium ions are crucial for blood coagulation, neurotransmitter release, muscle contraction and intracellular regulation.[1]

The plasma concentration of calcium is normally maintained within a very narrow range, varying by less than 10 per cent despite large movements of calcium across gut, bone, kidney and other tissues. Several hormones including parathyroid hormone (PTH), calcitriol (1,25-dihydroxyvitamin D_3) and calcitonin regulate the ionized fraction of plasma calcium (approximately 50 per cent of total plasma calcium) by modulating calcium fluxes to and from the extracellular fluid. The secretion rates for these hormones are, to varying degrees, inversely related to the extracellular calcium concentration, thus completing a negative feedback loop.

Calcium is absorbed by active transport from the small intestine and this active transport is increased by calcitriol which is a metabolite of vitamin D produced primarily in the kidney. Synthesis of calcitriol is partly regulated in a negative feedback fashion by plasma calcium and phosphate and partly regulated by parathyroid hormone. Parathyroid hormone has a small and probably clinically insignificant effect on gut absorption of calcium, its major action on calcium uptake being via the regulation of calcitriol production.

Parathyroid hormone is produced by the parathyroid glands and its role is to regulate the concentration of plasma-ionized calcium (Ca^{2+}). PTH regulates Ca^{2+} by direct effects on the bone and the kidney and indirect actions (via calcitriol production) on gut (Fig. 16.1). In the kidney PTH acts on the proximal and distal tubules to increase the renal tubular reabsorption of calcium and to depress the reabsorption of phosphate, resulting in a rise in plasma calcium and a fall in plasma phosphate. PTH also decreases renal tubular reabsorption of bicarbonate, resulting in a hyperchloraemic metabolic acidosis, which further elevates Ca^{2+}. In bone PTH increases the rate of bone resorption by osteoplasts but may also increase bone formation by stimulating osteoblastic activity. The major effect of PTH on the skeleton may, therefore, be to increase bone turnover, although the net effect is to increase circulating Ca^{2+}. Parathyroid hormone secretion is

Table 16.1 Components of total calcium content of human plasma (mmol/l)

Ionized calcium (Ca^{2+})	1.18
Complexed to HCO_3^-, phosphate, citrate	0.16
Protein bound	1.16
Total	2.50

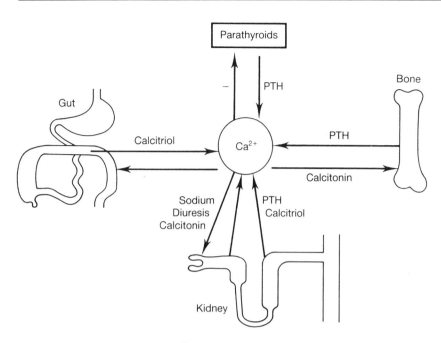

Fig. 16.1 Outline of the control of calcium metabolism. Parathyroid hormone (PTH) is released in inverse proportion to the ionized calcium concentration (see text for details).

stimulated by a low plasma Ca^{2+} and inhibited by a high Ca^{2+} in a complex mechanism which is incompletely understood. As most hormonal exocytotic processes are dependent on increases in intracellular Ca^{2+} this inverse relation with Ca^{2+} in the parathyroid acinar cells has been called the parathyroid paradox. Possible mechanisms involved in this process have recently been reviewed.[2] Magnesium has a similar effect on PTH secretion with a potency of approximately one-third that of calcium.

Calcitonin is a peptide hormone which is released mainly from the parafollicular (C cells) of the thyroid gland in humans. Total thyroidectomy does not result in a fall of calcitonin levels to zero, suggesting that there are some non-thyroidal sites of origin probably including the thymus, the adrenal and pituitary glands. Stimulus to the release of calcitonin is an increase in plasma Ca^{2+} concentration although its release is also affected by a glucagon, gastrin and β-adrenergic agonists. Calcitonin has many actions that are opposite to those of parathyroid hormone with respect to the maintenance of plasma Ca^{2+}. Although calcitonin inhibits bone resorption and decreases renal tubular reabsorption of calcium, sodium, phosphate, magnesium and potassium, major alterations in thyroid gland function result in only minor disturbances in skeletal or plasma calcium homoeostasis, and the role of calcitonin in human physiology is dubious.

Physiological actions of calcium

Calcium functions as an intracellular messenger in virtually all eukaryotic cells, mediating a wide range of responses from brief single events to repetitive and prolonged actions. There is considerable diversity in the manner in which calcium functions as a messenger in different types of

tissue. The details of the calcium messenger system are extremely complex and this review only offers a general framework within which to discuss some of these events.

Cellular regulation

Since calcium plays such a crucial role in cellular regulation, it is vital that intracellular calcium levels are very tightly regulated. Intracellular calcium levels are very low. Free cytosolic calcium is in the range of 100–200 nmol/l (10^{-7} molar) giving a transcellular gradient of approximately 10^4 M. This gradient is maintained partly because the cell membrane has a low natural permeability to divalent cations and partly by at least two ATP-dependent pumps: the ATP-driven calcium pump and the sodium/calcium exchange pump. The first of these pumps is a low-capacity, high-affinity system directly dependent on ATP which fine tunes the intracellular calcium concentration; the sodium–calcium exchange mechanism is a high-capacity, low-affinity system which moves large quantities of calcium out of the cell, and is driven by the transmembrane sodium gradient ($[Na^+]_o/[Na^+]_i$).[3] In this latter pump, three sodium ions are exchanged for each calcium ion and the rate of removal of calcium is directly proportional to the $[Na^+]_o/[Na^+]_i$ ratio. This mechanism is indirectly ATP dependent as the capacity of this pump depends on the sodium gradient across the cell membrane which is maintained by the sodium/potassium pump. Both of these pump mechanisms have their capacity increased by increases in intracellular calcium $[Ca^{2+}]_i$, thus achieving a compensatory increase in calcium efflux for any increased calcium influx and maintaining a dynamic balance within the cytosol. As the intracellular calcium rises the calcium binds to calcium effector proteins such as calmodulin, and the resultant calcium–calmodulin complexes associate with and activate a number of protein kinases and proteases which in turn govern other aspects of cell function. The ATPases which govern the efficiency of the calcium pumps are activated by the calcium–calmodulin complex and the increase in calmodulin activity which follows an increase in $[Ca^{2+}]_i$ enhances the maximal capacity of these pumps for extruding calcium.[4]

The intracellular organelles also play a vital role in the maintenance of calcium homoeostasis. In the mitochondria an efficient calcium pump drives calcium from the cytosol into the mitochondrial matrix where calcium is stored in a non-ionic calcium–phosphate complex. Calcium leaks back from the mitochondria on a non-stimulus-dependent base. Mitochondria have a high capacity for calcium and can act as a buffer helping to stabilize cytosolic calcium concentration by increasing the rate of net uptake in periods of cytosolic calcium accumulation, particularly in the case of calcium overload in hypoxia. However, excess accumulation of calcium by the mitochondria inhibits ATP production, and thus this mechanism is at best a short-term safety valve and not part of the normal homoeostatic mechanisms for calcium regulation.[4] The other major intracellular organelle involved in calcium regulation is the endoplasmic or, in the case of muscle, sarcoplasmic reticulum. In many cells this organelle provides the chief source of calcium for rapid release in response to transient stimulation. A rise in intracellular calcium stimulates ATPases via, among other mechanisms, calcium–calmodulin complexes and this, in turn, activates an ATP-dependent pump which drives the calcium into the endoplasmic reticulum. Unlike the situation in the mitochondria, calcium is stored in the endoplasmic reticulum in the ionic state and is released from there in response to a variety of stimuli which will be discussed later.

Under normal circumstances, calcium enters the cell as a result of conformational changes in the proteins that constitute the calcium channels. There is debate as to the nature of these calcium channels, but broadly they can be categorized into two groups: those channels which admit calcium in response to depolarization, termed potential-dependent or voltage-operated channels (VOC), and those channels which admit calcium as a result of changes produced by the stimulation of one or other receptors, referred to as receptor-operated channels (ROC). It is important to appreciate that the cell does not tolerate the accumulation of intracellular calcium. Where there is a short-term, rapid calcium signal, transient increases in calcium occur as a result of increased calcium influx through the various calcium channels or via a release of calcium from intracellular calcium pools or both. Where there is a sustained response, continued accumulation of calcium does not occur. After the initial transient increase

in intracellular calcium concentration, the influx of calcium may be maintained but an increased efflux of calcium from the cell occurs, thus balancing the new influx rate. It is thought that sustained responses are maintained by the rate of calcium cycling rather than by a sustained increase in intracellular calcium concentration.

The calcium signal

Alterations in cytosolic calcium produce a variety of effects depending on the tissue involved. In general terms the influx of calcium produces a signal which is transmitted into an intracellular response via a family of calcium-modulated proteins including troponin-C, parvalbumin, S-100 proteins, calbindins and calmodulin, all of which have structural features in common.[5] Calmodulin is present in all eukaryotic cells, suggesting that it has a general function, whereas the other calcium-binding proteins have specific functions associated with certain differentiated cells.[5] The mechanism of generating and responding to the calcium signal are significantly different in different types of tissue.

Skeletal muscle

When a wave of depolarization reaches the nerve terminal VOCs are activated which result in calcium influx into the motor nerve terminal. This influx of calcium combines with calmodulin and activates ATP-dependent transporters which discharge acetylcholine into the synaptic cleft. Some form of this mechanism is common to all forms of neurosecretion including the release of neurotransmitters in the brain and of acetylcholine and catecholamines peripherally, although the process is highly complex with many variations on this basic scheme occurring, especially within the central nervous system.[6] As depolarization spreads over the sarcolemma the T-tubule system carries the wave of depolarization deep into the substance of the muscle and into close apposition with the cisternae of the sarcoplasmic reticulum (SR). In skeletal muscle, the information flow from the T-tubules to the SR occurs within a triad consisting of the T-tubule and two terminal cisternae of the SR. Within this triad, the cisternal SR is linked to the T-tubule by a specific protein with junctional processes or 'feet'. This area of the triad binds ryanodine which in low concentrations releases calcium from the SR and at higher concentrations closes the SR calcium channel[7] and has been termed the ryanodine receptor. Although the exact mechanism is unclear, it appears that depolarization, itself, stimulates the release of calcium directly from the SR in large quantities, which binds with troponin-C and triggers off the contraction. Simultaneously the rise in calcium causes the formation of calcium–calmodulin complexes which in turn activate ATPases

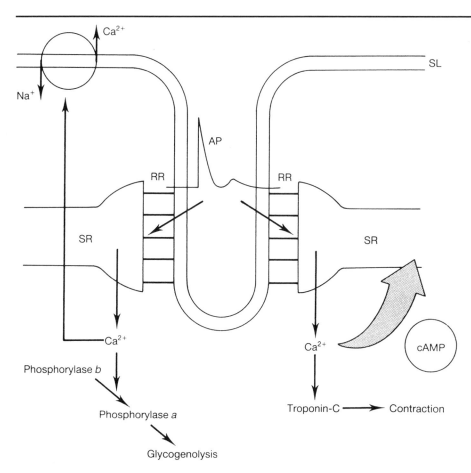

Fig. 16.2 Regulation of calcium and muscular contraction in skeletal muscle. Calcium is released directly from the sarcoplasmic reticulum (SR) by the action potential (AP), through the ryanodine receptor (RR). The sodium–calcium exchanger on the sarcolemma (SL) and the cyclic AMP regulated pump on the SR remove calcium from the myofibril, terminating contraction. Calcium simultaneously activates glycogenolysis to provide energy for contraction (see text for details).

which rapidly pump the calcium back into the sarcoplasmic reticulum thus terminating the contraction. The rise in calcium also increases cellular metabolic rate. It stimulates the mitochondria increasing the rate of respiration, and activates phosphorylase kinase which in turn converts phosphorylase *b* to the active phosphorylase *a* which then catalyses glycogenolysis, thus providing more energy for muscle activity. Calcium influx triggering further calcium release is not a major mechanism in skeletal muscle, and skeletal muscle contraction is relatively independent of extracellular calcium concentration. In malignant hyperthermia (MH), the defect is probably at the ryanodine receptor, which, in response to a given stimulus, allows a continued leak of calcium from the sarcoplasmic reticulum resulting in sustained contracture, increase in the metabolic rate and an increase in lactate production as a result of the increased rate of glycogenolysis outstripping the oxygen supply.

Myocardium

The myocardium depends on both extracellular and intracellular sources of calcium for the production of myocardial contraction. Depolarization of the myocardium opens voltage-operated (slow) channels which allow the influx of calcium. This is reflected in the different shape of the myocardial action potential in which the plateau phase represents the inward calcium current. The influx of calcium then stimulates the SR to release further calcium – the so-called calcium-triggered calcium release (Fig. 16.3). Within the sarcomere of skeletal muscle, the relation of the contractile response of the myofibril to calcium concentration is considerably less steep than that in skeletal muscle. In the former, the response curve is so steep that muscular contraction is virtually an all-or-none response, whereas in cardiac muscle a much more graded response is possible, with differing quantities of calcium release provoking differing contractile power. Under normal physiological circumstances, maximal inotropic stimulation produces no more than 50 per cent of the full contractile response as compared to that obtained in skinned muscle cells exposed to high calcium concentrations.[8] This reserve capacity allows for a range of responsiveness of the sarcomere to a given level of stimulation, thus allowing for variations in contractile power, which can be produced by modifying the quantity of calcium released into the sarcomere at any given time. Variations in contractility can be produced through variations in intracellular calcium availability caused by either physiological agents or drugs. It is also worth noting that the calcium-induced calcium release is graded, with the peak $[Ca^{2+}]_i$ being proportional to the calcium influx.[9]

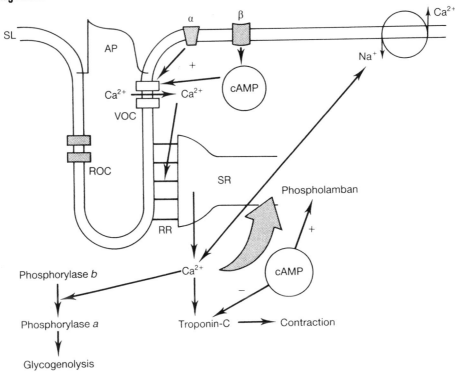

Fig. 16.3 Regulation of calcium and muscular contraction in cardiac muscle. Calcium enters the cell through voltage-operated channels (VOC) which are opened by the action potential (AP), which exhibits a plateau representing the calcium current. The calcium influx stimulates the ryanodine receptor (RR) to release calcium from the sarcoplasmic reticulum (SR). α-Adrenergic stimulation (α) directly stimulates the opening of the VOC, and β-adrenergic stimulation (β) increases the probability of channel opening through cyclic AMP. cAMP also increases the rate of dissociation of calcium from troponin-C and stimulates phospholamban which increases uptake of calcium by the SR. The other major route for calcium removal is through the sodium–calcium exchanger on the sarcolemma (SL). Receptor-operated channels will open and admit additional calcium, when appropriately stimulated (see text for details).

Alterations in contractility

α-Adrenergic stimulation affects the probability of opening of the sarcolemmal calcium channel, allowing an increased influx of calcium with each contraction, possibly by an action mediated by diacylglycerol. β-Stimulation, on the other hand, similarly increases the probability of channel opening through a cAMP-dependent process thus increasing contractility. Uptake of calcium by the SR in the myocardium is accelerated by a special regulatory protein called phospholamban. Cyclic AMP-dependent protein kinase phosphorylates phospholamban, increasing the rate of re-uptake of calcium by the SR. Thus an increase in cAMP facilitates relaxation, as well as contraction. Adrenergic stimulation may also increase the rate of phosphorylation of troponin-I which in turn increases the rate of dissociation of calcium from troponin-C, further enhancing the speed of relaxation. However, the increase in calcium flux will also increase the metabolic rate of the myocardial cells, thus increasing oxygen demand, and may also predispose the ischaemic myocardium to calcium overload.

Phosphodiesterase inhibitors, which decrease the rate of cAMP destruction, will exert an effect on calcium flux and on myofibrillar responses similar to that of the catechola-mines.[10] The cardiac glycosides are primarily inhibitors of the Na^+/K^+ ATPase pump of the sarcolemma, and thus increase the intracellular sodium concentration. As a consequence of the rise in $[Na^+]_i$, the efficiency of the Na^+/Ca^{2+} pump is reduced, thus allowing the concentration of intracellular free calcium to rise. This results in an increase in SR calcium stores, with increased release at stimulation, but also increases the resting $[Ca^{2+}]_i$ which will decrease cardiac relaxation. This may be beneficial in the dilated heart, but will increase oxygen consumption, and may induce diastolic dysfunction by allowing accumulation of cytosolic calcium, particularly if heart rate is allowed to increase.[11] At high concentrations of catecholamines, an increase in cAMP appears to desensitize troponin-C to further increases in $[Ca^{2+}]_i$, thus limiting the maximal force development in response to catecholamine stimulation, which may act as a protective mechanism, partially defending the myocardium against excessive catechola-mine stimulation. The sensitivity of myofilaments to calcium is also impaired by hypoxia, acidic pH – which appears to reduce the affinity of troponin-C for troponin-I – and by increased inorganic phosphate.[12] The sarcoplasmic reticulum of the myocardium does not appear to be susceptible to the MH disorder, possibly because of differences in the mechanism of calcium signalling.

Smooth muscle

The muscular structure in vascular smooth muscle is quite different to that in striated muscle, with a far greater actin to myosin ratio, and an absence of striations. The structure of smooth muscle allows for the generation of greater tension, at lower energy expenditure than occurs in striated muscle, but at the cost of a much slower shortening speed. The situation regarding calcium is also quite different. Direct transfer of calcium across the membrane and calcium-triggered calcium release both occur as they do in cardiac muscle, but depolarization of the plasma membrane need not take place as calcium may enter the smooth muscle cell through either the VOCs or ROCs. Smooth muscle has the capacity for sustained response and this invokes a quite different messenger system. There are two principal mechanisms involved, both of which involve the action of phospholipase C on a membrane-bound polyphosphoinositide – phosphatidylinositol 4,5–bisphosphate (PIP$_2$). The first is the inositol phosphate system which maintains the release of calcium from the sarcoplasmic reticulum for a shorter period of time and the second is the protein C–kinase mechanism which maintains contraction, either by a direct effect through phosphorylation of myosin light chain kinase (MLCK) or via sustained calcium cycling across the cell membrane. These will be considered in more detail under cellular responses. α-Adrenergic stimulation both increases transmembrane calcium flux, and directly stimulates phospholipase C, thus increasing [Ca^{2+}]$_i$ by direct channel opening and through inositol triphosphate (IP$_3$)-mediated SR calcium release. There is also evidence that the reduction in membrane-bound PIP$_2$ may result in inhibition of the ATP-dependent calcium pump, thus decreasing calcium extrusion, and effectively increasing intracellular calcium.[13] The action of calcium within smooth muscle is also quite different. It is hypothesized that calcium binds to calmodulin, and this complex then activates MLCK. The activated MLCK then catalyses phosphorylation of the regulatory myosin light chain, resulting in actin–myosin interaction and hydrolysis of ATP. It is worth noting that the VOCs are more sensitive to the effects of calcium channel blocking agents than are ROCs.

Cellular responses

This field is complex and has been extensively reviewed recently.[2,14] When a hormone receptor which stimulates calcium-mediated processes is activated two events occur. There is an increase in calcium influx across the cell membrane and an activation of the hormone phospholipase C. This catalyses the hydrolysis of phosphatidylinositol 4,5-bisphosphate (PIP$_2$). This molecule is broken down to form diacylglycerol (DG) and IP$_3$. Both of these products of PIP$_2$ cleavage can act as intracellular messengers. Diacylglycerol

activates protein kinase C which can, in turn, activate a G-protein to maintain calcium flux across the cell membrane or can stimulate cell responses in its own right through phosphorylation of proteins. IP$_3$ directly affects the endoplasmic reticulum, causing calcium release and a transient rise in the calcium concentration in the cytosol. The production of diacylglycerol also activates the breakdown of membrane fats to produce arachidonate, giving rise to prostaglandins and leukotrienes which are local hormones for intercellular communication and regulation of local cell responses. It is thought that the continuous cycling of calcium across the cell membrane results in continuous production of diacylglycerol and protein kinase C which are also direct mediators of the cellular response. The resultant activation of intracellular proteins such as calmodulin produces the appropriate cellular responses, including glycogenolysis, lipolysis and various secretory processes including neurosecretion, insulin release, TSH release and degranulation of mast cells. It is probable that an array of other cell functions yet to be delineated are similarly stimulated.

Calcium-related hypoxic cell damage

Recent evidence suggests that calcium accumulation within hypoxic cells may contribute to loss of viability. The failure of the ATP-dependent Na$^+$/K$^+$ ATPase pump will result in the influx of sodium in excess of any efflux of potassium with consequent depolarization and water accumulation. The rise in [Na$^+$]$_i$ impairs the sodium–calcium exchange mechanism, thus limiting the ability of the cell to extrude calcium, reversing the normal direction of the Na$^+$/Ca^{2+} exchanger and resulting in a rise in [Ca^{2+}]$_i$.[15] Depolarization will open voltage-operated calcium channels, allowing a further influx of calcium. The influx of calcium will be further enhanced by increased permeability of the cell due to membrane damage, osmotic swelling, failure of ATP-dependent mechanisms which extrude calcium, and – especially in the brain – the opening of receptor-operated calcium channels by excitatory chemicals, many of which may be amino acids.[16] Glutamate, which is thought to be the major excitatory amino acid, acts as a mixed agonist at three receptor subtypes – kainite, quisqualate and N-methyl-D-aspartate (NMDA). The first two non-specifically increase cell permeability to monovalent ions, and thus increase sodium influx into the cell, producing depolarization and a rise in intracellular sodium and water content. The NMDA receptor opens ROCs, increasing calcium influx. It has also been suggested that the calcium-release channel of the SR is degraded during ischaemia, with the consequence that the increased permeability of the SR membrane would contribute to the cytosolic calcium overload seen in the ischaemic myocardium.[7] The intracellular accumulation of calcium plays a vital role in the pathogenesis of ischaemic injury. The calcium-mediated activation of phospholipases results in breakdown of membrane phospholipids with accumulation of free fatty acids, and disruption of intracellular organelles. Calcium may also be involved in the conversion of xanthine dehydrogenase to xanthine oxidase, which is a major source of the superoxide radical released during reperfusion.[17] Furthermore, calcium accumulation will trigger the activation of phosphorylase, initiating glycogenolysis with subsequent accumulation of lactate. This may be of particular relevance in neuronal ischaemia, the mechanisms of which have been recently reviewed.[16]

The accumulation of calcium by ischaemic cells will adversely affect energy production. Not only will oxidative phosphorylation be impaired by hypoxia, but the accumulation of calcium by the mitochondria further impairs mitochondrial function, so that restoration of oxidative phosphorylation is inhibited. Hypoperfusion appears to increase $[Ca^{2+}]_i$ in vascular smooth muscle, particularly on reperfusion[18] with consequent smooth muscle contraction leading to the 'no-reflow' phenomenon.

In summary, calcium is involved in many of the processes which may lead to cell death. The loss of ATP allows accumulation of $[Ca^{2+}]_i$ through the opening of calcium channels, through the loss of integrity of the cell membrane and through calcium-induced calcium release from the endoplasmic reticulum membranes. The uptake of calcium by the mitochondria impairs energy production further. The increase in $[Ca^{2+}]_i$ activates phospholipases with consequent breakdown of membrane phospholipid. This in turn induces free fatty acid accumulation which results in increased prostaglandin and leukotriene production. This progression of self-sustaining events will eventually lead to cell death.

Therapeutic use

Calcium solutions for intravenous administration are generally available in two forms, the chloride and the gluconate, which is metabolized to citrate. Calcium chloride contains 27 per cent by weight free calcium whereas the gluconate contains only 9 per cent. Both solutions provide almost immediately available calcium ions in solution. The chloride solution is very irritating to veins, and should not be administered through a peripheral line. Excessively rapid administration of either solution can produce acute elevations of plasma Ca^{2+} which may induce arrhythmias.[19] Neither solution should be given through a line through which blood or blood products are being administered, and calcium-containing solutions (such as Ringer's lactate) should be flushed out of an intravenous line before blood products are administered. The major use

of injectable calcium solutions is the correction of hypocalcaemia. The most common cause of hypocalcaemia is a reduced plasma albumin, and other causes include acute pancreatitis, hypoparathyroidism, malabsorption syndromes and vitamin D deficiency. In anaesthetic practice, acute hypocalcaemia can arise from rapid haemodilution and citrate loading as a result of rapid infusion of blood products. Alkalosis reduces the proportion of ionized calcium in the plasma and produces the typical symptoms of hypocalcaemia including tetany, neuromuscular irritability, circumoral paraesthesia, laryngospasm and seizures.

The use of calcium infusions in clinical practice has declined in recent years with the appreciation of the role which calcium may play in cell damage consequent on hypoxia. It is now widely recommended that calcium be excluded from most cardiopulmonary resuscitation regimens as administering excess calcium may worsen reperfusion injury. Calcium is indicated in circumstances such as citrate toxicity where calcium deficiency is the presumed cause of cardiac arrest, where potassium excess is suspected, or in cases of calcium antagonist toxicity.[20] The use of calcium as a myocardial stimulant following cardiopulmonary bypass is controversial, with many authorities now arguing that calcium should not be given until the myocardium is fully reperfused, and others advocating its complete avoidance. Robertie et al. concluded that calcium should not be administered routinely upon emergence from cardiopulmonary bypass, but that correction of significant hypocalcaemia was indicated, in which case treatment should be titrated against frequent estimations of ionized calcium in blood.[21]

The role of calcium supplementation during blood transfusion is also the matter of some debate. Although the citrate in transfused blood will remove calcium from the circulation, this effect is dependent on the rate of infusion and is negligible unless the rate of infusion is at least 100 ml/min in the average 70 kg adult.[22] Calcium stores in the body are vast and the removal of citrate is normally sufficiently rapid to prevent any significant depletion of calcium concentration in the plasma, unless the infusion of blood is prolonged. However, during massive blood transfusion, the administration of calcium may have significant beneficial effects on myocardial performance and vascular tone. This effect depends on the rapid accumulation of citrate in excess of the capacity of the liver to metabolize this chelator, with consequent binding of ionized calcium in the circulation. This is particularly important when liver perfusion is impaired by hypotension or liver damage, and is, of course, particularly important in liver transplantation. Again, calcium administration should be titrated against plasma Ca^{2+} wherever possible. Hypocalcaemia is not a significant cause of coagulation difficulties, as the calcium concentrations required to impair

Table 16.2 Causes of hypercalcaemia

Common
  Artefactual: hyperalbuminaemia, hypergammaglobulinaemia
  Malignancy with skeletal metastases (e.g. breast, lung)
  Haematological malignancy (myeloma, lymphoma)
  Primary hyperparathyroidism

Uncommon
  Tertiary hyperparathyroidism – renal transplantation, chronic renal failure
  Vitamin D toxicity
  Increased responsiveness to vitamin D (e.g. sarcoidosis)
  Immobility – Paget's disease
  Milk alkali syndrome
  Thyrotoxicosis
  Thiazide diuretics
  Adrenal failure
  Phaeochromocytoma
  Familial hypocalciuric hypercalcaemia
  Haemodialysis: high dialysate calcium

coagulation are far below those compatible with myocardial function.

Hypercalcaemia

Hypercalcaemia has many causes which are summarized in Table 16.2. Patients should be evaluated for abnormalities of renal, cardiac or central nervous systems. Mild hypercalcaemia is seldom symptomatic and rarely warrants special treatment. Emergency treatment of hypercalcaemia is indicated when the serum Ca^{2+} exceeds 3.5 mmol/l. Rarely emergency parathyroidectomy may be precipitated by very high levels of serum calcium (> 5 mmol/l). In the preoperative management of hypercalcaemia, the most crucial factor is the re-establishment of cardiovascular homoeostasis and adequate renal function, rather than the reduction of the plasma Ca^{2+} to any predetermined value. Fluid replacement should be with saline solutions and once the plasma volume is re-expanded, the use of diuretics such as frusemide can assist in the reduction of serum calcium levels. The use of diuretics prior to full rehydration can exacerbate the hypercalcaemia. Correction of the concomitant hypophosphataemia will inhibit the absorption of calcium from the gut, increase bone uptake of calcium as well as correct the cardiac and skeletal muscle weakness which accompanies phosphate deficits. The use of sodium phosphate can no longer be recommended due to the risk of calcium phosphate deposition in the tissues. Other agents which will lower the plasma Ca^{2+} include mithramycin, calcitonin and glucocorticoids. Mithramycin, a cytotoxic, has been employed for the reduction of serum calcium levels as it impedes bone resorption, but it may have renal, hepatic and haematological toxic effects and must be used with caution.

Recently there has been a great deal of interest in the use of the diphosphonates which inhibit osteoplastic activity and also reduce the serum calcium. Both of these agents will take several days to be effective. The diuresis used to lower the serum calcium may produce hypokalaemia and hypomagnesaemia with a resultant increase in risk of cardiac dysrhythmias. Once adequate fluid balance has been established, the management is that of the underlying condition. There is no literature to support any specific plasma calcium concentration as a cutoff value above which anaesthesia becomes unacceptably hazardous.

Magnesium

Magnesium is the fourth most abundant cation in the body and the second most plentiful intracellular cation after potassium and is one of the most important regulators of intracellular biochemistry. The ion is responsible for the activation of over 300 enzyme systems including most of the enzymes involved in energy metabolism and is an essential cofactor in oxidative phosphorylation and the production and functioning of ATP. Other processes dependent on magnesium include the production of DNA and RNA and protein synthesis. Magnesium is an important regulator of the behaviour of calcium in the body, and has been described as the physiological calcium antagonist.[23]

Magnesium metabolism

The average human body contains approximately 24 g (1000 mmol) of magnesium. Of this about half is present in bone and about 20 per cent in skeletal muscle. The remainder is found in other body tissues especially the liver and the heart. Only 1 per cent of whole body magnesium is in the extracellular fluid compartment including the plasma. In the plasma, approximately one-third of the total plasma magnesium is bound to protein and approximately 5.5 per cent complexed to phosphate, citrate and other compounds. The major regulatory system for absorbed magnesium is the kidney which excretes between 120 and 150 mg of magnesium per 24 hours in an individual on a normal diet, although during periods of parenteral magnesium administration very much higher excretion rates of magnesium can be achieved. Filtered magnesium is absorbed in the renal tubules by active transport, which appears to have a transport maximum of only slightly above the normal plasma magnesium concentration (1.1 mmol/l). The regulatory mechanisms governing renal reabsorption and excretion of magnesium are poorly understood and numerous hormones have been suggested as being involved in this process. However, no single mechanism seems adequate to explain the regulation of magnesium excretion at a renal level. There is evidence that fluctuations in calcium, total inorganic phosphorous, potassium, parathyroid hormone, calcitonin and aldosterone can influence magnesium turnover. However, since magnesium itself can influence the secretion or metabolism of most of these factors, the relation between them and control of total body magnesium is unclear.[24]

Distribution and exchange

Like calcium, magnesium crosses the cell membrane only with difficulty. Intracellular magnesium stores are, however, very much higher than those of calcium. Total intracellular magnesium varies between 10 and 28 mmol/l, the vast majority of which is stored in the nucleus, in the mitochondria and in the microsomes, whilst some is complexed with various lipid components of the cell. Free magnesium inside the cell is at a level of 1–3 mmol/l. Entry of magnesium into the cell occurs only during cell activity and requires active transport. It is stimulated by anabolic processes and insulin. Loss of magnesium from the cell occurs in response to catecholamine stimulation, potassium loss, catabolic processes, hypoxia and a fall in serum magnesium. Within the cell magnesium competes with calcium for binding sites on the calcium effector proteins, and activates mechanisms involved in the control of $[Ca^{2+}]_i$. The uptake of calcium by the sarcoplasmic reticulum is highly magnesium dependent and magnesium deficiency may thus predispose to intracellular calcium overload. Magnesium activates the Na^+/K^+ ATPase pump and blocks the efflux of potassium from the cell, possibly by closing a specific potassium channel.[25] This may well be the mechanism by which magnesium counteracts hypokalaemia.

Skeletal muscle

In skeletal muscle, magnesium is an essential regulator of calcium release at the nerve ending and in excess can compete with calcium for the calcium channel.[26] Within skeletal muscle magnesium is necessary for normal energy metabolism and in its absence muscle weakness occurs. Although it is a major regulator of calcium release from the sarcoplasmic reticulum, it does not readily penetrate the intracellular space and for this reason magnesium is of limited value in the management of malignant hyperthermia.

Myocardium

In the myocardium, magnesium is essential for the regulation of calcium as well as for the maintenance of normal energy metabolism and hence myocardial contractility. In magnesium-deficient states, the myocardium is extremely sensitive to all types of arrhythmias, especially those associated with digitalis and catecholamines, probably through its actions in determining Na^+, K^+ and Ca^{2+} fluxes. Magnesium limits the access of calcium into the myocardium in response to catecholamine stimulation and also minimizes the efflux of potassium and increase in lactate production that normally accompanies catecholamines stimulation.

Smooth muscle

In smooth muscle, magnesium inhibits calcium influx and also acts as a direct α-blocker. It is, therefore, a major regulator of vascular tone and there is some evidence that it may be important in certain forms of hypertension. In excess, magnesium is a powerful vasodilator, which is particularly effective in the vascular beds of the brain, kidney, coronary and splanchnic vessels and the uterus.

Metabolism

Magnesium inhibits many calcium-mediated processes within the cells regulating calcium stimulation. It is important in the release of certain neurotransmitters and hormones, notably TSH and insulin, and in the degranulation of mast cells. It is crucial in protein synthesis and regulates, either directly or indirectly, most of the enzymes involved in the glycolytic pathway. Magnesium regulates $[Ca^{2+}]_i$ by competing with calcium for transmembrane channels and for binding sites on protein kinases.

Magnesium deficiency

The generally accepted normal range for serum magnesium is 0.75–1.0 mmol/l and there is general agreement that hypomagnesaemia exists when the serum concentration is less than 0.7 mmol/l, although some authors use a serum concentration of less than 0.75 mmol/l. It should be appreciated, however, that as magnesium is primarily an intracellular ion whole body magnesium depletion may exist in the presence of normal or even elevated serum magnesium concentration.

Causes of magnesium deficiency

Magnesium deficiency develops frequently in a wide variety of conditions and these are summarized in Table

Table 16.3 Causes of magnesium deficiency

Decreased intake and absorption
  Poor diet
  Malnutrition
  Prolonged i.v. therapy
  Bowel dysfunction
  Pancreatic failure

Increased requirements
  Pregnancy
  Rapid growth and development

Gastrointestinal losses
  Prolonged diarrhoea
  Small bowel fistulas
  Billiary fistulas

Renal losses
  Recovering ATN
  Drugs
    Thiazide diuretics
    Mannitol
    Aminoglycosides
    Cisplatin
    Digoxin
    Ethanol
  Stress
    Aldosterone
    Catecholamines
  Hypokalaemia
  Hypophosphataemia
  Hypercalciuria
  Diabetes

16.3. Despite this large array of causes for hypomagnesaemia, magnesium deficiency is seldom sought, and hypomagnesaemia is frequently missed. Several drugs may contribute significantly to magnesium deficiency, particularly diuretics, and this is compounded by the concomitant use of cardiac glycosides. Infusions of adrenaline and dopamine have been shown to produce significant reductions in serum magnesium concentration although dobutamine and salbutamol apparently do not.

Diagnosis

The diagnosis of magnesium deficiency is complicated.[27] Determination of serum magnesium concentration is easy to perform, but it is only of value in assessing acute changes in magnesium status. In evaluating the serum magnesium concentration, severe hypoalbuminaemia should be excluded, as 30 per cent of the serum magnesium content is protein bound; a normal free magnesium could thus be misinterpreted as a magnesium deficit. Magnesium concentration is better determined in serum rather than plasma, as anticoagulants may interfere

with the estimation; haemolysis of the specimen must be avoided. Where the serum magnesium is low, a total body deficit is highly likely, unless some acute event, such as the correction of diabetic ketoacidosis immediately precedes the performance of the estimation. However, since 99 per cent of total body magnesium is intracellular, considerable total body deficits can occur in the presence of a normal serum magnesium. The magnesium content of various cellular elements has been advocated for the assessment of whole-body magnesium status, including red blood cells, muscle and bone, but none of these is entirely satisfactory in the clinical setting. Reasonable estimates of the magnesium balance of the patient can be obtained by a combined evaluation of serum concentration and renal excretion. Decreased renal excretion of < 0.5 mmol per 24 hours should raise suspicion of magnesium deficiency; an elevated renal magnesium excretion (> 5 mmol per 24 hours) in the presence of low serum concentration suggests renal magnesium wasting. In the absence of magnesium wasting, a suspicion of magnesium deficiency in the presence of a normal serum magnesium concentration can be substantiated by the magnesium load test in which 30 mmol $MgSO_4$ is administered as an infusion over 8 hours, and the urine collected for the 24 hours starting at the commencement of the infusion. A retention of more than 60 per cent of the load is highly suggestive of magnesium deficiency.

Potassium and calcium abnormalities frequently accompany magnesium deficiency, but neither can be used as an accurate pointer to the possibility of magnesium deficiency. Nevertheless, the occurrence of persistent hypokalaemia or hypocalcaemia should alert the clinician to the possibility of magnesium deficiency.

Consequences of magnesium deficiency

Despite the wide range of biochemical processes in which magnesium is involved, there is no clearly defined syndrome attributable to magnesium lack, but rather a spectrum of non-specific manifestations affecting various organ systems which may be due to many factors of which magnesium deficiency is but one. This is probably because there is great variation in the individual response to magnesium deficiency, and hypomagnesaemia rarely occurs in isolation. However, magnesium deficiency almost certainly contributes to the impaired function of a number of organ systems in critically ill patients.

Cardiovascular manifestations
Electrocardiographic changes produced by magnesium deficiency are non-specific and include widening of the QRS complex and peaked T waves. Severe magnesium deficiency results in prolongation of the PR and QT intervals, ST segment depression and T-wave flattening.

The most important cardiovascular pathology associated with magnesium deficiency is the development of arrhythmias, and the relevance of hypomagnesaemia to the occurrence of both supraventricular and ventricular arrhythmias in magnesium-deficient patients has been reviewed recently.[28] Digoxin-induced arrhythmias may be refractory to conventional anti-arrhythmic agents but respond well to intravenous magnesium sulphate.[29] It is debatable whether magnesium deficiency itself is arrhythmogenic, as the evidence for this is inconclusive. Magnesium appears to act as a pharmacological agent in its anti-arrhythmic action rather than as a simple nutrient correcting a deficit.

Magnesium deficiency appears to increase the risk of sudden death in patients suffering from myocardial infarction and several studies have documented a loss of myocardial magnesium content following ischaemic injury.[30] Hypomagnesaemia may predispose to myocardial infarction, and may increase the risk of sudden death following an ischaemic episode.

Central nervous system manifestations
Personality changes and psychiatric manifestations have been described in association with magnesium deficiency; these may include depression, agitation, confusion, anxiety and delirium. Seizures have been described in association with magnesium deficiency but this is an infrequent occurrence. Magnesium deficiency may contribute to the seizures of the alcohol withdrawal syndrome; it also interferes with thiamine utilization in the brain and may aggravate the Wernicke–Korsakoff encephalopathy.

Neuromuscular manifestations
Magnesium plays an important role in skeletal muscle contraction and interacts with calcium at several steps in the excitation–contraction coupling process. Magnesium deficiency may produce frank tetany and Chvostek's and Trousseau's signs have both been associated with magnesium deficiency. These changes are, however, uncommon. Of more importance in the intensive care unit is that magnesium deficiency may result in muscle weakness, particularly of the respiratory muscles, and thus interfere with weaning of patients from mechanical support. Correction of magnesium deficiency has been shown to improve respiratory muscle power.[31]

Magnesium replacement

Magnesium replenishment should be considered in all patients who are at risk (see Table 16.3), particularly those who are persistently hypokalaemic, hypocalcaemic or who have significant cardiac arrhythmias. The measurement of serum magnesium concentration, although limited, can act as a first-line guide to those needing magnesium replacement. Gambling *et al.* considered that replacement of magnesium deficits should precede elective surgery in all patients at risk.[32]

Where magnesium replacement is considered necessary, a typical replacement regimen in a patient with normal renal function would consist of the intravenous infusion of 40 mmol (10 g $MgSO_4$) over the first 24 hours, followed by 24 mmol (6 g $MgSO_4$) for 2–5 days. In patients with abnormal renal function hypomagnesaemia is less likely to occur and magnesium represents a significantly greater therapeutic risk when given by the intravenous route in such patients. Daily oral or intravenous administration of 4–8 mmol of magnesium can prevent further magnesium depletion once replacement is complete.

Pharmacology of hypermagnesaemia

Raised serum magnesium levels are rare in clinical medicine as absorption of the ion from the gastrointestinal tract is limited and the renal elimination of any excess magnesium is extremely rapid, excretion of a magnesium load by the kidney being complete within 4–8 hours. Consequently, parenteral administration is the only way to achieve a significant increase in serum magnesium levels for therapeutic purposes. The volume of distribution of injected magnesium approximates to the volume of the extracellular space and thus a loading dose of approximately 16 mmol (4 g $MgSO_4$) is required to raise the plasma magnesium concentration by 1 mmol/l. Continuous infusions (at an hourly rate of 15–30 mg/kg), or intramuscular injections are necessary to maintain serum magnesium levels in the generally accepted therapeutic range of 1.5–4 mmol/l.

An intravenous loading dose given by slow intravenous injection or infusion produces a short-lived sensation of cutaneous warmth and flushing. Peripheral resistance is reduced in proportion to the serum magnesium and declines by about 20–30 per cent in response to a doubling of the serum magnesium concentration. This is accompanied by a compensatory increase in cardiac output – which prevents hypotension from occurring – and a 22 per cent increase in renal blood flow.[33] Thus infusion of magnesium produces a decrease in afterload and increase in cardiac output without a significant increase in stroke work.

Central nervous system

When applied directly to neural tissue, magnesium has a marked depressant effect, and perfusion of the cerebral ventricles with magnesium-rich solutions can cause unconsciousness. Magnesium acts as a central calcium antagonist, producing cerebral vasodilatation, and it is also an NMDA antagonist. However, there is little evidence that parenterally administered magnesium salts have any significant central nervous system depressant effects. The original description of magnesium-induced anaesthesia probably owed more to muscle relaxation and hypoxia than to anaesthesia. Provided ventilation is maintained, even very high levels of serum magnesium produce virtually no CNS depression.[34] Thompson et al.[35] described a 60 per cent reduction in minimum alveolar concentration of halothane in magnesium-treated rats which they ascribed to a central effect of the ion, although this has not been substantiated. The reason for the absence of major CNS depression is that magnesium penetrates the blood–brain barrier poorly and that its level in the CSF is extremely tightly controlled, probably by an active transport mechanism. Even in cases where high levels of serum magnesium have been maintained for several days CSF magnesium has been found to be only marginally above normal. The question therefore arises as to whether or not magnesium has major anticonvulsant properties which would justify its use in obstetrics for the management of pre-eclampsia. Although magnesium penetrates the blood barrier poorly, small but significant increases in CSF magnesium concentrations following magnesium treatment have been demonstrated. Conflicting evidence as to whether or not this is sufficient to explain the anticonvulsant action on an electrophysiological basis exists and has been summarized by Cotton et al.,[36] and the dispute currently remains unresolved. An alternative explanation for the anticonvulsant action of magnesium in gestational proteinuric hypertension is its cerebral vasodilator action which reverses the cerebral vasospasm now thought to be a principal cause of the convulsions.[37]

Peripheral nervous system

Magnesium appears to have little axonal action on peripheral nerves, although it has been suggested that it may potentiate the action of local anaesthetics. The major peripheral action of magnesium is by interfering with the release of neurotransmitter substances at all synaptic junctions.

Motor end plate

Magnesium produces a dose-dependent presynaptic inhibition of neurotransmitter release in peripheral nerves due to competition with calcium for membrane channels on the presynaptic terminal. At the neuromuscular junction, magnesium concentrations of 5 mmol/l and above produce significant neuromuscular blockade, which is characterized by being reversed by increasing stimulus frequency, and which does not appear to demonstrate fade. Magnesium will potentiate the action and prolong the duration of even the shorter-acting non-depolarizing muscle relaxants,[38] and may precipitate severe muscular weakness in patients with the Lambert–Eaton syndrome or myasthenia gravis. Non-depolarizing relaxants must be used in reduced doses and at increased dosage intervals in patients who are significantly hypermagnesaemic. Nerve stimulators should be used to titrate the dose of the relaxant to effect. Where difficulties in reversal of neuromuscular blockade occur, calcium may be of use in re-establishing normal motor function. Magnesium does not shorten the onset time of non-depolarizing relaxants. It is generally assumed that magnesium also prolongs the action of depolarizing relaxants, but there is no evidence that this is of any clinical significance. Acute hypermagnesaemia does not affect the duration of a single dose of suxamethonium, either in normal subjects or in magnesium-treated pre-eclamptic mothers.[39] Recent data suggest that pre-eclamptic patients may have reduced cholinesterase levels which may produce prolongation of the action of suxamethonium, but that magnesium has no effect on this phenomenon. Patients treated with $MgSO_4$ do not demonstrate fasciculations and acute administration of $MgSO_4$ prior to the use of suxamethonium appears to prevent the release of potassium provoked by the relaxant.[40] The possibility exists that magnesium may reduce the incidence and severity of suxamethonium-induced muscle pains, and may make suxamethonium usable in those circumstances in which the risks of excessive potassium release currently make the relaxant contraindicated, but, to the present time, no clinical use has been made of these interactions.

Autonomic nerve terminals

Magnesium exerts a similar, although less well known, effect on autonomic ganglia and on vagal nerve terminals to that demonstrated at the motor end plate. Serum magnesium concentrations exceeding 2.5 mmol/l produce progressive inhibition of the release of catecholamines from both adrenergic nerve terminals and from the adrenal medulla. Magnesium at higher levels produces ganglionic blockade.[40]

Cardiovascular system

Vessels

Hypermagnesaemia decreases peripheral vascular tone by both a direct action on the blood vessels and by interference with the action of a wide range of vasoconstrictor substances. Magnesium reduces peripheral resistance in a dose-dependent manner by a combination of mechanisms which include calcium antagonism at calcium leak channels, VOCs and ROCs, antagonism of hormonal-induced vasoconstriction and possible interactions with prostacyclin and endothelium-derived relaxing factor.[33] The interaction of magnesium with prostacyclin may inhibit platelet function. In addition to its direct effects on the vessel wall, raised serum magnesium levels may also reduce peripheral vascular tone by a number of other mechanisms including sympathetic blockade and inhibition of catecholamine release.

Myocardium

The action of magnesium ions on cardiac function and, in particular, its interaction with other relevant ions, is complex and has been reviewed recently.[25] Since magnesium is a calcium antagonist, hypermagnesaemia should, theoretically, produce a reduction in myocardial contractile force, and this has been demonstrated in the isolated heart. In intact animals and humans, the evidence for cardiac depression by clinically useful levels of hypermagnesaemia is minimal, and it appears that the reduction in peripheral resistance more than offsets any decrease in contractility.[40] Of particular importance is the fact that magnesium does not appear to impair the cardiotonic action of adrenaline, even at plasma concentrations which reverse catecholamine-induced increases in blood pressure.[41]

Cardiac rhythm

Magnesium infusions have a variable and unpredictable action on heart rate. On the isolated heart, hypermagnesaemia, as might be expected, reduces heart rate. However, in the intact subject, the inhibition of vagal acetylcholine release produced by magnesium often overrides the intrinsic slowing, and a mild tachycardia commonly occurs. At mild levels of hypermagnesaemia little electrophysiological effect is seen, and the only consistent finding at higher concentrations is a slowing of conduction through the AV node.[33] Magnesium has been shown to be effective in treating a variety of forms of serious arrhythmias including those associated with digitalis toxicity, hypokalaemia, alcoholism and myocardial infarction. It has also been shown to be as effective as propranolol, and more effective than verapamil, in controlling the arrhythmias associated with adrenaline administration,[42] and may protect against bupivacaine-induced arrhythmias. It has also been recommended for intractable ventricular tachycardia and fibrillation, *torsade de pointes* and multifocal atrial tachycardia. A bolus dose of 2 g MgSO$_4$ administered over 5 minutes is generally recommended followed by an infusion of the agent at a rate determined by the desired serum magnesium level, but usually in the range of 1–2 g/h.

No firm guidelines for the most appropriate serum magnesium concentration in these circumstances have been established, but a concentration of between 1.5 and 2 mmol/l has been reported as both safe and effective.[40]

Respiratory system

Magnesium has no effect on central respiratory drive, and its only respiratory depressant effect is due to the neuromuscular block which it produces. Because of its smooth muscle relaxant properties, magnesium is an effective bronchodilator, and it has been used successfully in severe asthma, although the evidence is not, at present, strong enough to recommend its use in this area. The ability of magnesium to inhibit catecholamine-induced arrhythmias without worsening the bronchospasm may make magnesium a useful agent in those patients in whom β-agonist therapy is causing disturbances of cardiac rhythm, but this possibility remains to be investigated.

Genitourinary system

Magnesium has been widely used in obstetric practice, particularly in the United States, both in the management of gestational proteinuric hypertension – where it is used primarily as an anticonvulsant – and as a tocolytic for the management of premature labour. It is also an effective renal vasodilator, and exerts a significant diuretic effect. This mechanism is dose dependent, and increases the renal clearance of magnesium in proportion to the rate of rise of the plasma magnesium concentration, thus inducing elimination of the ion during periods of rapid administration.

Musculoskeletal system

Magnesium exerts its major effect on muscles at the motor end plate, although it does have minor calcium antagonist properties on the muscle itself. This may have some benefit in the management of malignant hyperpyrexia but the effect is small and probably not of great clinical significance.[40]

Therapeutic use of magnesium infusions

Clinical usage of magnesium infusions has, up until recently, been almost entirely limited to the control of convulsions in gestational proteinuric hypertension. Recently, however, appreciation of the multiple actions of the ion which may be of use in various clinical circumstances has seen a marked increase in interest in the use of magnesium infusions in a number of conditions. In particular, it has become apparent that magnesium should be regarded first and foremost as a cardiovascular drug with the actions of both a calcium antagonist – with vasodilator and antiarrhythmic effects – and an adrenergic antagonist with principally α-antagonistic actions. It is these actions of the ion which have led to considerable interest in the field of cardiology and cardiac anaesthesia and surgery.

Cardiac applications

The value of hypermagnesaemia in the treatment of arrhythmias is well established.[29] It is now clear that the major role of magnesium in the management of these rhythm disturbances is not primarily related to correction of magnesium deficiency but is due to pharmacological actions of the ion.[33] A number of recent studies have suggested that magnesium infusions may have a beneficial role in the management of acute myocardial infarctions. These studies have been summarized in a meta-analysis by Teo et al.[43] and supported by another large prospective study.[44] In this latter study, a loading dose of 8 mmol was given over 10–15 minutes and a further 65 mmol infused over the next 24 hours. This regimen produced a doubling of the serum magnesium concentration in their patients. The conclusion of these studies is that magnesium intervention appears to reduce the risk of death following myocardial infarction by between 36 and 55 per cent. Confirmation of this apparently beneficial effect will await the outcome of larger multicentre studies.

Cardiovascular anaesthesia

Magnesium may be of benefit during anaesthesia for cardiopulmonary bypass. The use of magnesium-containing cardioplegia is well established and widely used, but by no means universal. Magnesium has been shown to be of benefit in the early phase of ischaemic contracture of the myocardium and may be of benefit to the recovering ischaemic heart muscle. It will not interfere with the inotropic action of adrenaline, even in the postbypass patient. There is a serious risk of hypomagnesaemia following cardiac surgery and hypomagnesaemic patients have an increased risk of postoperative dysrhythmias.[45] The role of prophylactically administered magnesium in these circumstances remains to be established but there is a case to be made for and against the use of therapeutic injections of magnesium in the postcardiac patient. Whatever the merits of this argument, it would seem logical at the very least to avoid magnesium depletion scrupulously in the cardiac surgical patient, particularly in

view of the adverse effects hypomagnesaemia may have in the patient with ischaemic heart disease.

The vasodilator and antiarrhythmic properties of the agent suggest that it may also be of use during aortic cross-clamping for major vascular surgery, and the agent has been used successfully where other, more conventional forms of treatment have been inadequate. The action of magnesium as an NMDA antagonist suggests that magnesium may offer some protection against ischaemia to the spinal cord during the surgical repair of suprarenal aortic aneurysms.

Whilst magnesium may be considered for the control of the hypertensive response to tracheal intubation the tendency of magnesium to produce a moderate tachycardia limits the usefulness of this agent in patients with ischaemic heart disease. In those patients in whom a mild increase in heart rate is not of great importance, $MgSO_4$ (40–60 mg/kg) may be a very useful method of controlling the response as it not only inhibits the rise in blood pressure, but also prevents catecholamine release, and is antiarrhythmic.

Obstetric use

The use of magnesium in obstetrics raises a number of considerations of importance to anaesthetists. The interaction of magnesium with muscle relaxants has been discussed, but it must be emphasized that the presence of significant hypermagnesaemia in the pregnant patient makes the use of defasciculating doses of non-depolarizing relaxants prior to induction of anaesthesia not only unnecessary, but potentially hazardous. Of more interest to the obstetric anaesthetist is the ability of magnesium to alter the pressor response to intubation in patients with gestational proteinuric hypertension. In a recent study $MgSO_4$ (40 mg/kg) was found to be superior to either lignocaine (1.5 mg/kg) or alfentanil (10 μg/kg) for the control of the hypertensive response to intubation in terms of the numbers of hypertensive episodes found, and also produced less fetal depression than did alfentanil. A combination of 30 mg/kg $MgSO_4$ with 7.5 μg/kg alfentanil has been shown to be superior to the larger dose of either agent alone, with no increase in side effects. For the severe pre-eclamptic, a combination of $MgSO_4$ and alfentanil may be superior to magnesium alone.[46]

Of additional interest in this field are recent studies which have shown that magnesium will increase the likelihood of modest hypotension during epidural anaesthesia in normotensive ewes. This action of magnesium did not, however, appear to be associated with increased risk to the fetus, probably because cardiac output was sustained. Whilst these actions may be beneficial in controlling pressor responses to events such as intubation or straining during labour, they may also worsen hypotension in the presence of haemorrhage. Whether or not magnesium will increase the likelihood or severity of hypotension in gestational patients with proteinuric hypertension receiving epidural analgesia has not yet been studied. However, present evidence would tend to indicate that the sustained cardiac output and uterine blood flow will minimize the hazards associated with hypotension to both mother and fetus.[47]

Phaeochromocytoma

The antiadrenergic, antihypertensive and antiarrhythmic actions of magnesium, together with its ability to inhibit the release of catecholamines, have been advanced as a rationale for the use of $MgSO_4$ infusions in the anaesthetic management of patients undergoing resection of a phaeochromocytoma. In a series of 17 anaesthetics[48] and 20 subsequent cases it has been shown that magnesium effectively controls the release of catecholamines and hypertensive responses at induction of anaesthesia, intubation of the trachea and during surgical stimulation. A loading dose of 40–60 mg/kg followed by an infusion of 2 g/h together with 2 g incremental bolus doses has been effective in controlling the cardiovascular responses. Very large doses are necessary to maintain therapeutic serum concentrations (which appear to be in the range of 2.5–4 mmol/l) and totals of up to 18 g of magnesium sulphate have been administered over a 2-hour operation. Additional sodium nitroprusside infusions may be needed on rare occasions during periods of tumour handling. As the safety of magnesium in pregnant patients is so well established $MgSO_4$ has obvious appeal in the pregnant patient who also happens to have a phaeochromocytoma.[40]

Although magnesium does not inhibit the β_1-adrenergic actions of adrenaline, nor diminish the effect of adrenaline on cardiac output, nevertheless, on present knowledge, caution must be advised when this agent is used in a patient with a phaeochromocytoma complicated by impaired myocardial contractility. Calcium is useful to antagonize the effects of magnesium where hypotension following tumour excision, which does not respond readily to fluid loading, occurs.

Vecuronium is the muscle relaxant agent of choice, and because of its short duration of action, no problems of reversal of muscular relaxation have been experienced in association with the large doses of $MgSO_4$ administered.

Intensive care

The deleterious effects of magnesium deficiency are now well recognized in the intensive care unit, and most units now monitor magnesium levels in an effort to ensure that hypomagnesaemia does not develop. Magnesium infusions have been shown to reduce cardiovascular instability and to inhibit the release of catecholamines in patients with very severe tetanus, either on its own or in conjunction with clonidine.[49] However, because magnesium lacks significant central sedative properties, it is inadequate as a sole agent, and should be used in conjunction with deep sedation regimens.

Ischaemic protection

The calcium antagonist role of magnesium has led many researchers to consider the possibility that magnesium may have a valuable role to play in the protection and resuscitation of ischaemic tissue. In acute myocardial ischaemia, magnesium appears to be of value in decreasing mortality in the immediate post-infarction period, an action which does not appear to be related to its antiarrhythmic properties.[33] The possibility exists that magnesium may be of benefit in cerebral resuscitation, as it not only acts as a calcium antagonist, but also is an effective NMDA antagonist and a potent cerebral vasodilator. These actions have been shown to be of benefit in laboratory experiments, but there is no clinical evidence of significant benefit at the present time.

Conclusions

Magnesium is a major electrolyte in the body, and it has significant metabolic effects. Consequently, magnesium deficiency may have serious consequences, which may be subtle and hard to pinpoint, but which may still adversely affect patient outcome. As a therapeutic tool, magnesium has many potential applications, particularly in the fields of cardiovascular control and ischaemic protection. Future studies in this area will establish more clearly those circumstances in which the use of this ion is of clinical value.

REFERENCES

1. Landers DF, Becker GL, Wong KC. Calcium, calmodulin, and anesthesiology. *Anesthesia and Analgesia* 1989; **69**: 100–12.

2. Pocotte SL, Ehrenstein G, Fitzpatrick LA. Regulation of parathyroid secretion. *Endocrine Reviews* 1991; **12**: 291–301.

3. Rassmussen H. The calcium messenger system (first of two parts). *New England Journal of Medicine* 1986; **314**: 1094–101.

4. Carafoli E. Intracellular calcium regulation, with special attention to the role of the plasma membrane calcium pump. *Journal of Cardiovascular Pharmacology* 1988; **12 (Suppl 3)**: S77–S84.

5. Heizman CW, Hunziker W. Intracellular calcium binding molecules. In: Bronner, P., ed. *Intracellular calcium regulation.* New York: Wiley-Liss, 1990: 212–48.

6. Anwyl R. Modulation of vertebrate neuronal calcium channels by transmitters. *Brain Research Reviews* 1991; **16**: 265–81.

7. Holmberg SRM, Williams AJ. The calcium-release channel from cardiac sarcoplasmic reticulum: function in the failing and acutely ischaemic heart. *Basic Research in Cardiology* 1992; **82 (Suppl 7)**: 255–68.

8. Rüegg JC. The vertebrate heart: modulation of calcium control. In: Rüegg JC. *Calcium in muscle activation. A comparative approach.* Berlin: Springer-Verlag, 1988: 165–99.

9. Tibbits GF, Hamman BN. Regulation of myocardial contractility. *Medicine and Science in Sports and Exercise* 1991; **23**: 1140–4.

10. Skoyles JR, Sherry KM. Pharmacology, mechanisms of action and uses of selective phosphodiesterase inhibitors. *British Journal of Anaesthesia* 1992; **68**: 293–302.

11. Morgan JP. Abnormal intracellular modulation of calcium as a major cause of cardiac contractile dysfunction. *New England Journal of Medicine* 1991; **325**: 625–31.

12. Solaro RJ. Regulation of Ca^{2+}-signalling in cardiac myofilaments. *Medicine and Science in Sports and Exercise* 1991; **23**: 1145–8.

13. Casteels R, Wuytack F, Raeymakers L, Himpens B. Ca^{2+}-transport ATPases and Ca^{2+}-compartments smooth muscle cells. *Zeitschrift für Kardiologie* 1991; **80 (Suppl 7)**: 65–8.

14. Rassmussen H. The calcium messenger system (second of two parts). *New England Journal of Medicine* 1986; **314**: 1164–70.

15. Haigney MCP, Miyati H, Lakatta EG, Stern MD, Silverman HS. Dependence of hypoxic cellular calcium loading on Na^+–Ca^{2+} exchange. *Circulation Research* 1992; **71**: 547–57.

16. Siesjö BK. Pathophysiology and treatment of focal cerebral ischaemia. Part I: pathophysiology. *Journal of Neurosurgery* 1992; **77**: 169–84.

17. Murdoch R, Hall J. Brain protection: physiological and pharmacological considerations. Part I: the physiology of brain injury. *Canadian Journal of Anaesthesia* 1990; **37**: 663–71.

18. Siesjö BK. Cerebral circulation and metabolism. *Journal of Neurosurgery* 1984; **60**: 883–908.

19. Drop LJ. Ionized calcium, the heart and hemodynamic function. *Anesthesia and Analgesia* 1986; **64**: 432–51.

20. Baskett PJF. Advances in cardiopulmonary resuscitation. *British Journal of Anaesthesia* 1992; **69**: 182–93.

21. Robertie PG, Butterworth JF, Royster RL, *et al.* Normal parathyroid hormone responses to hypocalcemia during cardiopulmonary bypass. *Anesthesiology* 1991; **75**: 43–8.

22. Denlinger JK, Nahrwold ML, Gibbs PS, Lecky JH. Hypocalcaemia during rapid blood transfusion in anaesthetized man. *British Journal of Anaesthesia* 1976; **48**: 995–1000.

23. Iseri LT, French JH. Magnesium: nature's physiological calcium blocker. *American Heart Journal* 1984; **108**: 188–93.

24. Littledike ET, Goff J. Hormonal control of magnesium metabolism. In: Altura BM, Durlach J, Seelig MS, eds. *Magnesium in cellular processes and medicine.* Basel: Karger, 1987: 164–82.

25. White RE, Hartzell HC. Magnesium ions in cardiac function. *Biochemical Pharmacology* 1989; **38**: 859–67.

26. Krendel DA. Hypermagnesemia and neuromuscular transmission. *Seminars in Neurology* 1990; **10**: 42–5.

27. Elin RJ. Assessment of magnesium status. *Clinical Chemistry* 1987; **33**: 1965–70.

28. Iseri LT, Allen BJ, Brodsky MA. Magnesium therapy of cardiac arrhythmias in critical-care medicine. *Magnesium* 1989; **8**: 299–306.

29. Tzivoni D, Keren A. Suppression of ventricular arrhythmias by magnesium. *American Journal of Cardiology* 1990; **65**: 1397–9.

30. Abraham AS. Potassium and magnesium status in ischaemic heart disease. *Magnesium Research* 1988; **1**: 53–7.

31. Molloy DW, Dhingra S, Sloven F, Wilson A, McCarthy DS. Hypomagnesemia and respiratory muscle power. *American Review of Respiratory Disease* 1984; **129**: 497–8.

32. Gambling DR, Birmingham CL, Jenkins LC. Magnesium and the anaesthetist. *Canadian Journal of Anaesthesia* 1988; **35**: 644–54.

33. Woods KL. Possible pharmacological actions of magnesium in acute myocardial infarction. *British Journal of Pharmacology* 1991; **32**: 3–10.

34. Somjen G, Hilmy M, Stephen CR. Failure to anesthetize human subjects by intravenous administration of magnesium sulfate. *Journal of Pharmacology* 1966; **154**: 652–9.

35. Thompson SW, Moscicki JC, Difazio CA. The anesthetic contribution of magnesium sulphate and ritodrine hydrochloride in rats. *Anesthesia and Analgesia* 1988; **67**: 31–4.

36. Cotton DB, Janusz A, Berman RF. Anticonvulsant effects of magnesium sulfate on hippocampal seizures: therapeutic implications in preeclampsia–eclampsia. *American Journal of Obstetrics and Gynecology* 1992; **166**: 1127–36.

37. James MFM. The role of the anaesthetist in the management of pre-eclamptic toxaemia of pregnancy. In: Atkinson RS, Adams AP, eds. *Recent advances in anaesthesia and analgesia.* London: Churchill Livingstone, 1991: 137–55.

38. Sinatra RS, Philip BK, Naulty JS, Ostheimer GW. Prolonged neuromuscular blockade with vecuronium in a patient treated with magnesium sulfate. *Anesthesia and Analgesia* 1985; **64**: 1220–2.

39. Baraka A, Yazigi A. Neuromuscular interaction of magnesium with succinylcholine-vecuronium sequence in the eclamptic parturient. *Anesthesiology* 1987; **67**: 806–8.

40. James MFM. Clinical use of magnesium infusion in anesthesia. *Anesthesia and Analgesia* 1992; **74**: 129–36.
41. Prielipp RC, Zaloga GP, Butterworth JF, *et al.* Magnesium inhibits the hypertensive but not the cardiotonic actions of low-dose epinephrine. *Anesthesiology* 1991; **74**: 973–9.
42. Mayer DB, Miletich DJ, Feld JM, Albrecht RF. The effects of magnesium salts on the duration of epinephrine-induced ventricular tachyarrhythmias in anesthetized rats. *Anesthesiology* 1989; **71**: 923–8.
43. Teo KK, Yusuf S, Collins R, Held PH, Peto R. Effects of intravenous magnesium in suspected acute myocardial infarction: overview of randomized trials. *British Medical Journal* 1991; **303**: 1499–503.
44. Woods KL, Fletcher S, Roffe C, Haider Y. Intravenous magnesium sulphate in suspected acute myocardial infarction: results of the second Leicester intravenous magnesium intervention trial (LIMIT–2). *Lancet* 1992; **339**: 1553–8.
45. Aglio LS, Stanford GG, Maddi R, Boyd JL, Nussbaum S, Chernow B. Hypomagnesemia is common following cardiac surgery. *Journal of Cardiothoracic and Vascular Anesthesia* 1991; **5**: 201–8.
46. Ashton WA, James MFM, Janicki PK, Uys PC. Attenuation of the pressor response to intubation in hypertensive proteinuric pregnant patients undergoing caesarean section by magnesium sulphate with and without alfentanil. *British Journal of Anaesthesia* 1991; **67**: 741–7.
47. Vincent RD Jr, Chestnut DH, Sipes SL, Weiner CP, DeBruyn CS, Bleuer SA. Magnesium sulfate decreases maternal blood pressure but not uterine blood flow during epidural anesthesia in gravid ewes. *Anesthesiology* 1991; **74**: 77–82.
48. James MFM. The role of magnesium sulphate in the anaesthetic management of phaeochromocytoma – a report of 17 anaesthetics. *British Journal of Anaesthesia* 1989; **61**: 616–23.
49. Sutton DN, Tremlett MR, Woodcock TE, Nielsen MS. Management of autonomic dysfunction in severe tetanus: the use of magnesium sulphate and clonidine. *Intensive Care Medicine* 1990; **16**: 75–80.

ADDITIONAL READING

Altura BM, Durlach J, Seelig MS, eds. *Magnesium in cellular processes and medicine.* Basel: Karger, 1987.

Bronner P, ed. *Intracellular calcium regulation.* New York: Wiley-Liss, 1990.

Lassere B, Durlach J. *Magnesium: a relevant ion.* London: John Libbey and Company Ltd, 1991.

Rüegg JC. *Calcium in muscle activation. A comparative approach.* Berlin: Springer-Verlag, 1988.

Chapter 17

Metabolism and the Metabolic Response to Injury

I. T. Campbell and R. A. Little

Energy requirements	Body composition
Whole body metabolism	Techniques for measuring body mass and composition
Thermoregulation	Metabolic response to trauma and sepsis
Energy expenditure and nutrient oxidation	Provision of nutritional support in trauma, sepsis and multiple organ
Starvation	failure

The normal metabolic relations between supply and demand for energy and protein within the body and the homoeostatic adjustments which take place when these are disturbed by an episode of trauma and/or sepsis are discussed along with the methods used to measure these changes. The details of the metabolic (nutritional) support for such an episode are given in Chapter 43.

Energy requirements

Energy is defined as the ability to do work. The unit of energy is the joule which is the work done when a force of 1 Newton moves through a distance of 1 metre. In the living organism energy is derived from the oxidation of foodstuffs, principally carbohydrate, fat and protein. The summary equations for these processes are given in Table 17.1.[1] Oxygen is consumed, carbon dioxide and water are produced and energy is released, usually via a number of intermediary steps in metabolism with the liberation of discrete packages of energy, principally in the form of adenosine triphosphate (ATP).

The energy in the equations in Table 17.1 is given as units of heat – kilocalories (kcal) – defined as 1000 calories, the calorie being the amount of heat required to raise the temperature of 1 g of water by 1°C. This is the traditional

Table 17.1 Oxidation of carbohydrate (glucose), fat (palmitate) and amino acids (AA)

$$1 \text{ glucose} + 6O_2 \longrightarrow 6CO_2 + 6H_2O + 673 \text{ kcal}$$

$$1 \text{ palmitate} + 23O_2 \longrightarrow 16CO_2 + 16H_2O + 2398 \text{ kcal}$$

$$1 \text{ AA} + 5.1O_2 \longrightarrow 4.1CO_2 + 0.7 \text{ urea} + 2.8H_2O + 475 \text{ kcal}$$

unit of measurement in human energetics. Heat is a form of energy. The relation between mechanical and heat energy is given by the mechanical equivalent of heat:

$$1 \text{ kcal} = 4.18 \text{ joules}$$

The kcal has been the traditional unit because in the early studies of energetics in humans heat output was technically the easiest parameter to measure. With the institution of the SI system of units the kcal was supposedly replaced by the joule and its multiples, the kilojoule (kJ) and the megajoule (MJ), but this move appears to have been confounded by user resistance and it is still common practice to report units in kcals; many journals report results in both units.

Whole body metabolism

In Table 17.1 it can be seen that oxygen is consumed and, in the case of carbohydrate and fat, carbon dioxide, water and energy (heat) produced. Energy expenditure can be assessed by measuring the energy produced in the equation in Table 17.1 as heat by means of direct calorimetry and as external work. Alternatively, oxygen consumed (or carbon dioxide produced) can be measured and assumptions made about the amount of energy production that that level of oxygen consumption (or CO_2 production) implies; this is the technique of indirect calorimetry.

Energy in man is required for basal metabolism (maintenance of ionic gradients, work of breathing, heart work, etc.), thermoregulation and the performance of external work. Energy is also expended in the digestion and assimilation of food consumed via the gastrointestinal tract or given intravenously; this is referred to as specific dynamic action, or more recently as dietary induced thermogenesis or the thermic effect of food.

Thermoregulation

In the hibernating rodent and in the newborn human (possibly extending for the first 6 months of life) energy expenditure can be increased by sympathetic activation of brown adipose tissue – so-called non-shivering thermogenesis. This mechanism of heat production is said not to occur in adult man. The only way the adult human can increase heat output and, for example, defend his body temperature against a cold stress is by muscular contraction, either involuntarily by shivering, or by increasing voluntary activity. The other main defence against cold in man is to minimize heat loss by peripheral vasoconstriction and with chronic cold exposure there is evidence that the efficiency of this mechanism increases. The thermoneutral range for adult (naked) man starts at about 27–28°C, i.e. at temperatures less than this heat output increases, in man by shivering or by an increase in muscle tone, in the hibernating rodent or the neonate by non-shivering thermogenesis. This response is modified by the differing degrees of insulation provided by varying thicknesses of subcutaneous adipose tissue. In normal circumstances reaction to environmental temperature is affected by clothing, and in hospital patients, by bed coverings.

Energy expenditure and nutrient oxidation

Energy expenditure is most easily measured by measurement of oxygen consumption for which consumption of 1 litre of oxygen represents an energy expenditure of 4.6–5.0 kcal (depending on the relative proportions of the different nutrients being metabolized). It can also be predicted from various equations. Basal energy expenditure can be predicted from age, height, weight and sex. Factors (5, 10, 20 per cent, etc.) can be added to allow for activity, temperature (classically energy expenditure rises by 13 per cent for every °C rise in body temperature) and pathology; this latter point will be dealt with later.

It can be seen from Table 17.1 that when carbohydrate (glucose) is oxidized the ratio of carbon dioxide produced to oxygen consumed is 1. The corresponding ratio when fat is oxidized is 0.7. At the tissue level this ratio is termed the respiratory quotient; when measured by respiratory gas exchange it is termed the respiratory exchange ratio. When protein metabolism has been accounted for (see below) the non-protein respiratory quotient (or respiratory exchange ratio) gives an indication of the relative proportions of carbohydrate and fat being oxidized.

There are, however, practical limitations to the use of respiratory gas exchange to assess the metabolism of fat and carbohydrate, particularly in relation to factors disturbing respiratory patterns. In naive subjects any instrumentation on or around the face, such as facemasks and mouthpieces, induces disturbances in ventilatory patterns, usually hyperventilation, and this obviously increases the rate of apparent CO_2 production and thus respiratory exchange ratio in a way that does not reflect metabolism. Hypoventilation such as that produced by opiates has the reverse effect. The other major metabolic abnormalities that disturb respiratory exchange ratio are those of hydrogen ion production; a metabolic acidosis elevates respiratory exchange ratio and a metabolic alkalosis depresses it.

Oxidation of protein produces carbon dioxide and water as well as energy and nitrogen-containing products of metabolism excreted in the urine (Table 17.1) principally in the form of urea, but other products of protein metabolism include uric acid, creatinine and ammonia. The quantity of protein metabolized over a period can be assessed from the urinary nitrogen content. Classically the presence of 1 g of nitrogen in the urine denotes the oxidation of 6.25 g of protein. Thus the total quantity of protein oxidized over a period is obtained from the total urinary nitrogen excreted over that period with a correction made, if necessary, for alterations in blood urea levels. It is not normal clinical practice to measure urinary nitrogen, but urinary urea can be measured and a reasonable assumption for clinical purposes is that 80 per cent of the nitrogen in the urine is present in the form of urea. Urinary nitrogen can be calculated accordingly.[2]

Starvation

Trauma and sepsis are often accompanied by varying degrees of undernutrition. Starvation[3] is defined as a lack of exogenous energy substrate and may be relative or absolute. Following injury, appetite is depressed and spontaneous intake is often inadequate to meet protein and energy requirements. The normal metabolic responses to starvation or semistarvation are modified by trauma and sepsis in ways that will be described, but in order to understand these modifications a knowledge of the response to simple starvation is desirable.

Energy and protein requirements are normally derived from the dietary intake. Ten to twenty per cent of energy comes from oxidation of protein, with fat and carbohydrate providing the remaining 80–90 per cent in roughly equal proportions. Food consumed in excess of immediate requirements is stored, carbohydrate as glycogen in the liver and muscle, fat in adipose tissue and amino acids go to protein mostly in the liver and muscle. Amino acids consumed in excess of requirements for protein synthesis are deaminated and the amino group metabolized – principally to urea. The carbohydrate consumed in excess of requirements and the carbohydrate residues from amino acid deamination are used as energy, or go to form new glucose via hepatic gluconeogenesis, or are synthesized into fat.

Most tissues normally derive their energy requirements from a mixture of carbohydrate, fat and protein as discussed above, but some, such as nervous tissue and blood cells and haemopoietic tissue, as well as wound tissue, are obligatory consumers of glucose. During starvation adaptive processes are all directed towards the conservation of energy and protein. Glycogen stores are small and are used up in 24–48 hours. Only liver glycogen contributes to the maintenance of blood glucose, the glycogen within muscle being used only by the muscle tissue itself. However, glycolysis in some circumstances, such as severe exercise or 'shock', results in the formation of lactate which diffuses into the bloodstream and may then be converted to glucose by the liver.

When glycogen stores are exhausted glucose requirements are met by breakdown of protein and urinary nitrogen excretion increases accordingly. Most of the body's tissues such as muscle, including cardiac muscle, derive their energy from fat oxidation. The liver starts to synthesize ketones – acetoacetate and β-hydroxybutyrate – from fatty acids and these ketones are metabolized by the central nervous system and to a certain extent replace glucose as a metabolic fuel. As this keto-adaptation takes place, so the amount of protein broken down and contributing to gluconeogenesis diminishes. Urinary nitrogen excretion which rises slightly during the first week of starvation from 8–12 g/day to 14–16 g/day now diminishes to 4–5 g/day. The amount of protein broken down decreases with keto-adaptation from around 75 g/day during the first few days to 20 g/day by week 3. Total glucose consumption by the brain decreases from 140 to 40 g/day.

Carbohydrate stores are used up very rapidly. Fat and protein stores in an individual of normal build last about 60–70 days, although there are cases on record of very obese individuals surviving without food for over a year. Fat in the form of adipose tissue represents the energy store. Protein, however, as well as being a 'store' of gluconeogenic precursors, also constitutes the structure and function of the body, so any diminution in protein mass represents a diminution in function.

Energy is conserved by a decrease in energy expenditure brought about by a decrease in resting metabolism and by a decrease in spontaneous activity. Over time there is an absolute decrease in resting metabolism of about 30 per cent but also a loss in body weight; when the decrease in energy expenditure is corrected for the loss in weight the decrement in energy expenditure is nearer 15 per cent. The fall in resting metabolism is due to a decrease in mass of some of the metabolically more active tissues such as the liver, kidney and gastrointestinal tract, particularly the gastrointestinal mucosa which tends to become atrophic; this can cause problems when a normal intake is resumed. There is a loss of skeletal and heart muscle mass and a decrease in heart rate.

The hormonal responses to fasting are well characterized. Low insulin levels, caused by a decrease in blood glucose, result in an increase in glucagon which provokes hepatic glycogenolysis, the initiation of gluconeogenesis and the breakdown of muscle protein. Glucagon increases over the first 24–48 hours then decreases and by day 10 has reverted to prefasting values. Growth hormone in the absence of an energy intake is a catabolic and lipolytic hormone and is seen to rise in the first 24–48 hours then decreases over the ensuing 10–20 days. There is an initial rise in catecholamines, perhaps provoked by the 'stress' of starvation, then an overall decrease. Tri-iodothyronine levels diminish and this probably contributes slightly to the fall in energy expenditure.

Body composition

Energy intake, in the form of food (ingested, or infused intravenously) in excess of requirements, results in an increase in body mass. In the healthy individual this is manifest principally as an increase in body fat or adipose tissue. If energy intake is less than expenditure, either because expenditure is increased and/or intake is decreased, body mass is lost. The relative change in adipose and lean tissue depends on the circumstances in which the disturbance of intake and expenditure has occurred and are due to factors as yet poorly defined. In health the relevant factors include voluntary dieting, perhaps accompanied by varying degrees of exercise. In sepsis, trauma and malignant disease, appetite is depressed and a diminution in voluntary intake occurs as well as a decrease in voluntary activity. If artificial nutritional support is instituted the intention is that the rate of loss of body mass is diminished. If the degree of nutritional support is greater than the level of energy expenditure there may be an increase in adipose tissue deposition.

The human body is made up principally of fat, protein and minerals such as the calcium salts in bone. It has become traditional to think of the body as a two-compartment model, variously called fat mass and fat-free mass, or adipose tissue and lean body mass. The two sets of terms are not synonymous. They do approximate to each other but lean body mass contains structural fat such as in the nervous system and in cell membranes. The lean tissue element is said to represent the body's metabolically more active tissue and the fat or adipose tissue represents an energy store for the active tissue. Such a model has obvious limitations; it implies a uniform and static structure and activity of the lean tissue element, which is obviously not true, and implies that adipose tissue is metabolically inactive, and that too is demonstrably untrue. Despite this there is an obvious need to differentiate degrees of leanness and fatness between individuals, as well as changes in body mass and lean tissue within individuals and this type of two-compartmental model, in the absence of a more satisfactory and practical alternative, is the one that is most widely used.

Relatively few human bodies have been analysed and the details published. The total in all comes to little more than 50, including those dissected in the nineteenth century.[4,5] Muscle contributes on average about 50 per cent of the lean body mass and bone about 21 per cent. The skin constitutes about 8 per cent and the remainder – internal organs etc. – about 21 per cent.[4]

Techniques for measuring body mass and composition

All techniques for measuring body composition *in vivo* are of necessity indirect and make assumptions about the distribution of certain elements within the body, or the density of the two compartments referred to earlier, and most assume a normal state of hydration. Body weight is the most usual index of body mass and composition but of course takes no account of differences in composition in terms of fat and lean tissue. Body weight normally varies by up to 0.5 kg from day to day and can change, in healthy individuals, by up to 1 kg due to fluctuations in body water.[6] In disease states these figures are obviously much greater.

An index of body mass which takes height into account is the body mass index (BMI). This is given by:

$$BMI = wt\ (kg)/height\ (m)^2$$

The normal range is 18–25. Above 25 the individual is said to be 'overweight' and over 30 to be 'obese'.

Densitometry and skinfold thickness

The relation between percentage fat and body density is described by the equation

$$Per\ cent\ fat = \left(\frac{a}{body\ density} - b\right) \times 100$$

This relation was originally derived empirically,[7] but can be derived from first principles.[8] The constants a and b depend on the density of lean tissue and fat which in turn depend on the temperatures at which the measurements are made. At 37°C the density of fat-free mass is assumed to be 1.1 kg/l and fat to be 0.9 kg/l. With these assumptions the values of a and b are 4.95 and 4.5 respectively.[9]

Prediction equations have been derived experimentally relating skinfold thickness measurements to body density. Body density is derived from underwater weighing, allowing for or measuring residual gas in the gastrointestinal tract and in the lungs. The equations most widely used are those of Durnin and Womersley.[9] The skinfold sites most commonly measured are over the biceps and triceps muscles, just below the inferior angle of the scapula and just above the iliac crest in the midline.

In clinical practice these sorts of measurements are rarely made. In normal clinical circumstances mid upper arm circumference is the most useful index of lean body mass or muscle mass and triceps skinfold thickness alone an index of body fat. From these two measurements two other indices of lean body mass – arm muscle area and arm muscle circumference – can be derived using conventional formulae.[10]

Total body water and isotope dilution[11]

The use of total body water to measure lean body mass assumes that (1) all water in the body is contained within lean body mass (i.e. adipose tissue is anhydrous) and (2) constitutes 73 per cent of the lean body mass. Total body water is measured using isotope dilution, either with the radioactive isotope ^3H_2O or the stable isotope deuterium oxide (^2H_2O) which requires mass spectrometric techniques for analysis. The tracer can be administered either orally or intravenously and the aqueous concentration in blood or urine measured after the isotope has come to equilibrium with total body water. This normally takes 4–6 hours and all losses (urinary, wound drainage and gastrointestinal) over that time are measured and taken into account in the ensuing calculation.

More complex techniques of isotope dilution have been devised involving injection of multiple isotopes of sodium, potassium and labelled red cells.[11] Detailed discussion of these, however, are beyond the scope of this chapter.

Metabolic response to trauma and sepsis

Classically, the metabolic response to injury has been classified into two phases, the 'ebb' phase and the 'flow' phase. In the former, a period immediately following injury, metabolic activity is reduced, body temperature may be depressed and spontaneous activity diminished. There is an increase in substrate availability, but a decreased ability to utilize it, and insulin secretion is inhibited. The 'flow' phase is a period of increased metabolic activity, of regeneration and repair and an increase in body temperature. Appetite is diminished and the body derives its energy requirements, at least in part, from endogenous stores of fat and protein, but there is an ongoing need for glucose by tissues such as the central nervous system and wound tissue. This ongoing requirement is met by the process of gluconeogenesis. Body protein is broken down and there is a rise in the excretion of nitrogen-containing products of metabolism, principally urea. Any use of endogenous protein as a metabolic fuel results in a loss of structure and a diminution in function. As the organism recovers it enters the 'anabolic' phase, food intake increases, lean tissue is laid down and activity and function return towards normal but with a reduced body mass.[12]

Study of the response to trauma has the advantage that there is a clearly defined starting point and the factors responsible for its initiation can be investigated. Metabolic responses to sepsis are broadly similar to those seen in trauma, but it is often difficult to decide exactly when a patient becomes septic. If the traumatic or septic insult is overwhelming or blood loss massive and treatment not instituted, or homoeostatic mechanisms are unable to cope because of, for example, a preceding period of illness or malnutrition, organ systems fail and the organism dies, by a process that has been termed necrobiosis.[13]

The ebb phase

A common misconception is that the 'ebb' phase immediately after injury represents the period before resuscitation and thus implies that the metabolic responses at this time are secondary to circulatory hypovolaemia and a failure of oxygen delivery. This is not so; a failure of oxygen transport is the hallmark of necrobiosis whereas the ebb phase is a complex neuroendocrine response[14,15] characterized by a mobilization of energy reserves and changes in the central control of a number of homoeostatic reflexes.

The ebb phase is the early stage after injury during which oxygen transport may remain adequate. The ebb phase can, therefore, be transient or persist for 24 hours or more. It includes the pattern of physiological and metabolic changes associated with the preparation for 'fight or flight' (the defence or alerting reaction) on which are superimposed the responses to fluid loss from the circulation and/or tissue damage and pain associated with injury.

Thermoregulation

In the experimental animal major changes in thermoregulation occur after injury. The first systematic studies were done by Rosenthal and his colleagues in the 1940s. They noted that following haemorrhage mice (treated with saline) were unable to maintain body temperature when transferred from an ambient temperature of 26–29°C to 18°C. The fall in body temperature could be prevented or reversed by either the administration of 100 per cent oxygen or the intravenous injection of whole blood.[16] The conclusion was that the effect of haemorrhage on body temperature was secondary to changes in oxygen delivery. However, a very different picture was seen following tissue injury when the fall in body temperature (and oxygen consumption) on transfer to the lower ambient temperature could not be corrected by increasing tissue oxygen delivery.[16] These studies suggested that tissue injury in some way, which could not be related to a change in oxygen delivery, impaired heat production on cold exposure.

A more complete description of these changes in body temperature and heat production during the ebb phase showed, in the rat, an inhibition of thermoregulatory heat production resulting from an activation of noradrenergic neurons in the hindbrain, from which axons descend in the ventral noradrenergic bundle to liberate noradrenaline in the region of the dorsomedial nucleus of the hypothalamus. The changes can be initiated by nociceptive stimuli triggered by the application of tourniquets to the hindlimb and they are exacerbated by fluid loss from the circulation. However, thermoregulatory heat production can be stimulated in such injured animals if the ambient temperature is lowered sufficiently.[17]

The inhibition of thermoregulatory heat production by limb ischaemia in the rat involves both skeletal muscle and brown adipose tissue. Lower temperatures have to be applied to the skin or the hypothalamus to induce shivering in skeletal muscle and the ambient temperature at which heat transfer from interscapular brown adipose tissue commences is reduced for injured animals. The thermogenic activity of brown adipose tissue is also decreased shortly after scalding injury in the rat.[18] However, after both types of injury brown adipose tissue thermogenesis could be activated by the injection of exogenous noradrenaline, strongly supporting the suggestion that the reduction in heat production in the 'ebb' phase of the response to injury in experimental animals is due to a change in central control rather than to an impairment of peripheral effector mechanisms. These inhibitory effects of injury affect the mechanisms of heat loss as well as of heat production.

The evidence for a similar pattern of 'ebb' phase in man is not nearly so convincing. Core and whole body (calculated from the core and mean skin values) temperatures are reduced acutely after accidental injury and the reduction is directly related to the severity of the injury.[19] It was also noted following accidental injury that patients did not shiver when their body temperatures fell below the normal threshold for the onset of shivering. It is difficult to be sure that the changes in man are central in origin because in the most severely injured, plasma lactate concentrations are elevated and impairment of oxygen transport cannot be excluded. There is, however, some evidence for a change in thermoregulatory control at this time in that in man the normal relation between core temperature and the selection of an ambient temperature which maximizes thermal comfort is lost and all subjects select a high temperature irrespective of body temperature.[20]

There is no convincing evidence for a reduction in metabolic rate in man shortly after accidental trauma, although values as low as 55 per cent of predicted basal have been measured after severe injuries.[21] Oxygen consumption calculated from the data provided by Cournand and his colleagues[22] suggested that heat production is maintained in critically injured patients, even before resuscitation, at higher than normal values.[23] Oxygen consumption calculated from a modification of the Fick equation showed that it was above normal in 10 and reduced in only 2 of 16 patients vigorously resuscitated shortly after severe multiple injuries,[23] a trend that was supported by direct measurements of oxygen consumption in a group of critically ill patients in the emergency room.[24] There is also no evidence for a reduction in metabolic rate immediately following surgery (data reviewed by Barton et al.[25]). In fact there is often a rise in metabolic rate at this time which cannot be simply explained as a response to intraoperative cooling.[26]

A recent study has shown that after burn injury children rapidly develop a pyrexia and not a fall in body temperature as would be expected from both experimental work[27] and the clinical studies in adults described above. The increase in body temperature and heat content is maximal within the first 12 hours following injury and is due to an upward resetting of metabolic control rather than to the use of occlusive dressings and a high ambient temperature.[28,29] In such children body temperature was positively related to plasma levels of interleukin-6 (IL-6). Interleukin-1 (IL-1) was not detectable in plasma at this time but was found, together with IL-6, in high concentrations in blister fluid.[30] Thus it is possible that IL-6 produced by macrophages and endothelial cells activated by burning injury may have a role in the mediation of the acute hypermetabolic response to such injuries in children, perhaps involving the central release of prostaglandins which in turn influence hypothalamic thermoregulatory control. This suggestion is supported by the reduction in core temperature elicited in burned children by paracetamol (acetaminophen), a cyclo-oxygenase inhibitor.[31] IL-6 has also been detected in the plasma of adult burned patients.[32]

In summary, there does seem to be good experimental evidence for a reduction in body temperature and heat production during the ebb phase response to injury which is not due to failure of oxygen transport but to a change in central metabolic control. However, the clinical evidence for such a response is not nearly so convincing and what data there are suggest that metabolic rate at this time is increased not decreased. This apparent discrepancy may be explained by differences between species in the relative importance of heat loss and heat production for thermal homoeostasis. As discussed above, injury inhibits the efferent pathways for both heat loss and heat production and so whilst an animal such as the rat which depends mainly on heat production might be expected to show an inhibition of heat production after injury, an animal such as man, in whom heat loss predominates, might be expected to show an inhibition of heat loss. Indeed there is some evidence that heat loss is inhibited acutely following burn injury in children[33] and lower limb fractures in adult man.[20]

Hormonal changes

The hypothalamic–pituitary–adrenal axis is perhaps the most thoroughly investigated response to injury. Nociceptive stimuli increase, via a complex series of neuronal interactions involving 5-HT, GABA and the opioids, the release of corticotrophin releasing factor (CRF) by the parvocellular nuclei of the posterior hypothalamus. CRF is secreted into the capillary plexus of the hypophyseal portal system and is then carried to the adenohypophysis where it stimulates adrenocorticotrophic hormone (ACTH) secretion. CRF is the main stimulus for ACTH production but vasopressin released concomitantly from the neurohypophysis following activation of the magnocellular (supraoptic) nuclei of the hypothalamus[34] is also involved.[35] ACTH stimulates the secretion of cortisol from the adrenal cortex although the relation between plasma ACTH and cortisol concentrations acutely after injury is complex.[36] Plasma cortisol, both free and bound, concentrations are higher after injuries of moderate severity than after minor trauma but more severe injuries are associated with lower concentrations. This cannot be attributed to low ACTH levels; after severe injury plasma ACTH concentrations are raised to around the concentration needed for maximal stimulation of the adrenal cortex and reduced perfusion of the adrenal cortex may be important.

Of the other anterior pituitary hormones growth hormone, the endorphins and prolactin are released acutely after injury and once again the relations with severity are complex.[36] The plasma concentrations of thyrotrophic hormone (TSH) appear to be normal immediately after injury and although tri-iodothyronine (T_3) may start to fall at this time changes in the control of thyroid hormone concentrations are a feature of the flow phase.[14] The release of vasopressin from the posterior pituitary is increased in the ebb phase and its plasma concentration is directly related to the severity of the injury.[37]

The other major component of the neuroendocrine response to injury is a consequence of the increased activity of the sympathetic nervous system. As mentioned above many of the afferent inputs associated with the appreciation of danger, fluid loss from the circulation and tissue damage are integrated within the hypothalamus which in turn modifies the activity of the preganglionic sympathetic neurons in the intermediolateral columns of the spinal cord. Increased activity of the sympathetic nervous system leads to the release of noradrenaline (NA) from postganglionic nerve fibres and adrenaline (A) from adrenal medullary cells which are analogous to postganglionic neurons. Acutely after injury there are rapid increases in the plasma concentrations of NA, A and dopamine which are directly related to the severity of injury.[38,39] The increases are sufficient to influence the secretion of other hormones and, as will be discussed below, the mobilization of energy substrates. Plasma insulin concentrations are often low acutely after severe injuries despite a marked hyperglycaemia and this is a result of a suppression of insulin secretion by adrenaline acting on pancreatic α-adrenergic receptors.[40] In contrast, the secretion of glucagon is stimulated by raised catecholamine concentrations, this time by a β-adrenergic receptor mechanism[41] and as expected plasma glucagon levels are raised after injury.[14]

Metabolic changes

The increased activity of the sympathetic nervous system acutely after injury leads to mobilization of energy substrates, stimulating glycogenolysis and lipolysis. The main stimulus for breakdown of glycogen in both skeletal muscle and liver is adrenaline, although glucagon and vasopressin may also have a role in the liver. This glycogenolysis leads to hyperglycaemia either directly due to liberation of glucose from the liver or indirectly, via the Cori cycle, from lactate released from skeletal muscle. The hyperglycaemia, which is directly related to the severity of injury,[42] is potentiated after severe injuries by the reduction in glucose utilization in skeletal muscle following the inhibition of insulin secretion by raised adrenaline levels and by the development of intracellular insulin resistance. The mechanism of this early insulin resistance is still unclear although both glucocorticoids and cytokines may be involved.

The changes in carbohydrate metabolism in the 'ebb' phase can be interpreted as defensive. In addition to providing fuel for fight or flight, the hyperglycaemia may also play a role in the compensation of post-traumatic fluid loss both through the mobilization of water associated with glycogen and through its osmotic effects. The decrease in glucose clearance associated with insulin resistance can also be considered protective in that it prevents the wasteful use of the mobilized glucose, which is an essential fuel for the brain and the wound, at a time when a supply of nutrients may be limited.

Plasma concentrations of non-esterified fatty acids (NEFA) and glycerol are also raised after injury, reflecting the mobilization of triacylglycerol stores in adipose tissue. The relation with severity, however, is also complex; for example, NEFA concentrations are lower after severe than after moderate injuries. This may be related to either metabolic (e.g. stimulation of re-esterification within adipose tissue by the raised lactate levels associated with severe injuries) or circulatory (e.g. poor perfusion of fat depots) factors.

Although the major changes in protein metabolism following injury are associated with the flow phase (see below) the acute phase plasma protein response[43] is initiated during the ebb phase. A number of plasma proteins increase in concentration (e.g. C-reactive protein (CRP) and fibrinogen) although there is always a lag of approximately 6 hours before changes are seen. The cytokine IL-6 released from activated macrophages, etc. after injury may be responsible for inducing the hepatic synthesis of such acute phase proteins.[44] After surgery it has been shown that the rise in IL-6 precedes that of CRP and a weak but significant relation has been demonstrated between their serum levels.[45] Such a delay or lag is not seen for the proteins that show an acute phase decrease in concentration after injury. The rapid fall in, for example, albumin concentration cannot be attributed to a reduction in its rate of synthesis but is due to changes in its distribution between intra- and extravascular compartments secondary to an increase in microvascular permeability.[44,46] There is also a reduction in plasma amino acid concentration after, for example, major surgery and a marked negative correlation has been found between the plasma concentrations of IL-6 and glutamine.[47]

The flow phase

Thermoregulation

Increases in metabolic rate and core temperature are characteristic features of the 'flow' phase. The increase in metabolic rate is said to be directly related to the severity of injury.[9,48] The largest increases have been reported after major burns (>40 per cent total body surface area) treated by exposure,[49] although the introduction of early excision and grafting of such wounds has reduced the hypermetabolism. The increases in oxygen consumption are associated with improved survival in the critically ill.[50] Any increase in metabolic rate has to be met by an increase in tissue oxygen delivery secondary to an increase in minute ventilation, cardiac output and tissue blood flow. If a patient is unable to achieve these increases unaided then assistance with, for example, mechanical ventilation and the administration of inotropes is indicated. An increase in the extraction of the delivered oxygen by the tissue is also important, although it has been calculated that if the oxygen saturation of mixed venous blood falls below 30 per cent then reoxygenation in the lungs may not be completed and this leads to a progressive arterial hypoxaemia.[51] This problem would be exacerbated if pulmonary gas exchange were impaired – a common problem in the critically ill.

When metabolic rates are measured in patients who would be expected to show a marked flow phase response (e.g. 1 week after a major head injury or with sepsis complicating intra-abdominal surgery) they are often close to predicted basal. This apparent discrepancy is related to problems with predicting energy expenditure in patients who would have been bedridden for a period and/or have lost weight (especially lean body mass) because of their catabolic state and inadequate nutritional support. Thus a hypermetabolic state is superimposed on conditions which tend to lower energy expenditure.

The pathogenesis of the hypermetabolism is unclear although a number of factors have been proposed and it is quite possible that they may all contribute although their relative roles may depend on the underlying clinical condition. An upward central, resetting of metabolic control is strongly suggested by the studies of Wilmore and colleagues.[52] They have shown an increase in energy

Table 17.2 Phases of the metabolic response to trauma

Defence reaction		
Changes in homoeostatic reflex activity (thermoregulatory & cardiovascular)	Increase in metabolic rate	Restoration of organ/whole body structure and function
Reduction in appetite, changes in gut motility	Wound 'organ'	
Activation of hypothalamopituitary adrenal axis	Insulin resistance	
Increase in plasma hormone levels (e.g. catecholamines)	Increase in skeletal muscle breakdown – loss of lean body mass	
Mobilization of energy reserves	Increase in urinary nitrogen loss	
Changes in fuel utilization	Organ dysfunction	
Initiation of acute phase plasma protein response		

expenditure in burned patients even at ambient temperatures within the thermoneutral range. This was not secondary to an increased afferent cold input stimulating metabolism; in fact, the burned patient at this time is warm both peripherally and centrally. It seems most likely that the burn wound triggers a cytokine/prostanoid 'cascade' which elicits an increase in sympathetic nervous system activity. Plasma catecholamine concentrations are raised during the 'flow' phase following burns and combined α- and β-adrenergic receptor blockade reduces the hypermetabolic response to such injuries.[52] It should be noted, however, that although plasma catecholamine levels are raised during the hypermetabolic phase of the response to burning injury and also to a lesser extent following severe head injuries, they are close to normal at a comparable time after major non-head injuries.[12]

It has been suggested that one mechanism by which catecholamines may stimulate energy expenditure is by an increase in substrate cycling. This is a process in which although there is no change in the amount of either substrate or its metabolic products, ATP is used and energy expenditure has to be increased for its resynthesis. Two such cycles, glycolysis–gluconeogenesis and triglyceride–fatty acid, have been shown to be increased during the period 9–48 days following severe 60–95 per cent burns. The increase in the triglyceride–fatty acid cycle is due to β-adrenergic stimulation.[53] The contribution that such 'cycling' makes to total energy expenditure is difficult to assess, but an estimate of at least 15 per cent has been made.[53]

An increase in sympathetic activity will also stimulate grown adipose tissue thermogenesis in the small mammal and its activity is increased concomitantly with an increase in whole body oxygen consumption following a scald injury in the rat.[18] Its role in energy balance in adult man is controversial. There is little evidence for its importance under normal conditions although it has been reported to become so in a number of situations where circulating catecholamine levels are increased for some time, for example during chronic cold exposure or after the development of a phaeochromocytoma,[54] but the situation after severe injury or sepsis is unknown.

As alluded to previously, the severely injured are frequently not fed for a number of days following their accident. This may, perhaps surprisingly, contribute to the increase in heat production during the flow phase as a period of starvation enhances the thermogenic response to glucose.[55]

A feature of the flow phase is increased catabolism of lean tissue and there has been considerable debate about the contribution this makes to the increase in heat production.[56] Early predictions that the hypermetabolism could be fully explained by the thermic effect of protein metabolism[57] have not been substantiated. It now seems that the contribution of protein oxidation to total heat (energy) production is no more than 20–30 per cent, even in patients with multiple injuries or in those following burns of up to 60 per cent of body surface area.[58] The situation may be different after head injuries when values of up to 34 per cent have been calculated,[59–61] although some of these data have to be interpreted with caution as a number of the patients were treated with dexamethasone which augments the catabolic response to head injury.[62]

The evaporation of water from the surface of a burned wound or an area of granulation tissue will increase energy expenditure. For example, the energy cost of the latent heat of evaporation of a transcutaneous water loss of 3 l/day, which is not unusual after severe burns,[63] will increase energy expenditure by over 1500 kcal/day. The impact of this evaporative water loss on energy balance can be reduced by the application of impermeable dressings or by raising ambient temperature to reduce dry heat losses. However, although such manoeuvres may ameliorate the hypermetabolism associated with a burn injury, they do not abolish it; indeed, Wilmore has agreed that the evaporative water loss is not the cause of the hypermetabolism, but instead acts as a convenient route for the dissipation of the excess calories generated by an upward setting of metabolic control.[48]

The wound itself may be the most important influence on heat production in the 'flow' phase. It has been suggested that the wound (e.g. burn, fracture site, abscess, etc.) should be considered as an extra organ 'grafted' on to the body by injury/infection.[64] The wound is a heterogeneous tissue consisting of polymorphonuclear leucocytes, monocytes, fibroblasts, endothelial and epithelial cells, all of which are metabolically very active. The wound has a large blood supply which is not under neural control, and indeed much of the increase in cardiac output during the 'flow' phase may be directed to the wound.[64] One feature of the wound very relevant to a discussion of metabolic rate is the ability of monocytes and endothelial cells to produce cytokines.[65]

The burn wound, involving a limb, has been studied extensively and has been shown to be a consumer of large amounts of glucose and to a lesser extent of oxygen.[66] The glucose is converted to lactate in the presence of oxygen (i.e. aerobic glycolysis) but, as less ATP is generated by glycolysis than by oxidative phosphorylation, a large amount of glucose has to be taken up and converted to lactate to meet the energy demands of the wound. The lactate produced is carried to the liver where it is reconverted to glucose. This is an energy consuming process which is reflected in an increase in hepatic oxygen consumption.[67] Thus the wound becomes a user of glucose at a time when other tissues such as skeletal muscle reduce their uptake of glucose secondary to insulin resistance.[68] Thus it can be seen that the wound may increase whole body heat production in two ways. First there is the increase in the peripheral glucose/lactate cycle and second the wound acts as a focus for triggering the cytokine/prostanoid 'cascade' which, as discussed above, may lead to a central upward resetting of metabolic control.

Metabolic changes

Whole body protein turnover is increased after injury with the balance between synthesis and breakdown being modified by the severity of injury and the influence of nutritional intake on synthesis.[69] Thus increasing severities of injury cause increasing rates of both synthesis and breakdown whilst undernutrition reduces synthesis. However, after the most severe injuries the increase in breakdown predominates and cannot be counteracted by even the most aggressive nutritional support.[70]

The most obvious site for the net increase in protein breakdown is skeletal muscle although it is likely that, just as in starvation, muscle in the diaphragm, the wall of the gut and the heart is also affected. The breakdown of myofibrillar protein is reflected by the increase in urinary excretion of 3-methylhistidine which is directly related to the amount of damaged muscle rather than to the injury severity score (ISS).[71] However, the use of 3-methylhistidine as a specific marker of skeletal muscle breakdown is complicated by its liberation from other organs such as the gut.[72] The increase in urinary creatinine after injury is, however, directly related to ISS although much of the creatinine comes from muscle distant from the site of injury, emphasizing the general nature of the catabolic flow phase.[73]

The increase in proteolysis provides amino acids as precursors for hepatic gluconeogenesis. Although the plasma levels of a number of amino acids, such as alanine, fall at this time their hepatic extraction is increased because of increases in hepatic blood flow.[25] The increase in hepatic gluconeogenesis, at a time when plasma concentrations of glucose and insulin are increased, is one of the facets of insulin resistance discussed in more detail below. One amino acid of particular interest is glutamine, the intracellular concentration of which falls from its normally high levels after injury/sepsis.[74,75] Glutamine released from muscle is an important fuel for the lymphocytes and macrophages activated by injury. Also it has recently been implicated in the maintenance of the gut mucosa, the integrity of which is compromised after injury.[76]

Hyperglycaemia and inappropriately high plasma insulin concentrations are features of the flow phase although an exception to this pattern may be seen after very severe injuries, such as burns, when the prolonged rise in plasma catecholamine concentration maintains adrenergic suppression of insulin secretion. There is also an exaggerated pancreatic insulin response to glucose which may be related to the increased plasma concentration of arginine, an insulin secretagogue.[77,78] The concomitant elevations in plasma glucose and plasma insulin concentration are the hallmarks of insulin resistance which involve both liver and muscle. Also hepatic glucose production is not inhibited as expected by hyperglycaemia and hyperinsulinaemia during the flow phase. There is an increase in glucose turnover at this time although because of the prevailing insulin resistance the peripheral utilization of glucose is less than expected from the raised glucose and insulin concentrations (see review by Barton et al.[25]) This impairment in glucose disposal has been demonstrated using glucose/insulin clamp techniques after thermal and non-thermal injuries and also in septic surgical patients.[68-81] As expected, part of insulin resistance is found in uninjured skeletal muscle (Table 17.2). The relation between glucose uptake and plasma insulin concentration shows a marked reduction in both the maximum response and in sensitivity to insulin,[68] suggesting changes in both the receptor binding of insulin and in the intracellular pathways.[82]

An important role has been suggested for the counter-regulatory hormones (e.g. glucagon, adrenaline, cortisol and growth hormone) in the pathogenesis of insulin resistance. The plasma concentrations of all these hormones are elevated at some time during the response to injury.[14] Infusion of glucagon, adrenaline and cortisol over a 3-day period mimic some of the features of the flow phase – peripheral insulin resistance and increases in metabolic rate and urinary nitrogen excretion.[83,84] These responses were enhanced if the volunteers were also injected intramuscularly with the inflammatory agent etiocholanolone which induces IL-1 production.[85] However, the plasma concentrations of the counter-regulatory hormones needed to elicit this pattern of response are closer to those found in the acute 'ebb' phase rather than in the 'flow' phase when the endogenous levels of these hormones are falling, except after the most severe injuries, to or close to normal.[12] The suggestion is that other humoral factors may have a role and the cytokines are likely candidates. They are able to reproduce, by central and peripheral mechanisms, many of the acute and flow phase responses to injury such as changes in the plasma concentration of stress hormones,[86] acute phase protein synthesis,[44] central resetting of metabolic activity,[87] stimulation of pituitary–adrenal cortex axis[88] and changes in glucose homoeostasis.[89,90] A role for IL-1, acting via prostaglandin E_2, in the stimulation of muscle proteolysis has also been proposed[91] although this has not been supported by more recent studies using recombinant IL-1.[92] Although it is perhaps an oversimplification to assume that any single cytokine is predominant they may, by acting collectively in a coordinated way, both locally (e.g. IL-1) and systemically (e.g. IL-6 and TNF), be very important.

Once glucose has been taken up during the 'flow' phase it seems that in those receiving low dose glucose infusions or those receiving enteral carbohydrate its rate of oxidation is increased.[78,93] However in 'clamp' studies when large amounts of glucose or insulin are given either to injured or septic patients there is evidence that glucose utilization is impaired (Table 17.3).[68,81] Indeed, in such patients fat oxidation, which should be suppressed, persists.

As the flow phase progresses plasma NEFA concentrations fall as the sympathetic drive to lipolysis wanes, although after major injuries such as severe burns they may remain high.[78,94] Fatty-acid oxidation is, however, greater than expected from the plasma NEFA concentration.[78,95] Turnover which is normally directly proportional to concentration is also disproportionately increased[96] although there is no clear relation between NEFA turnover and oxidation. The turnover of endogenous and of infused triacylglycerol is also enhanced in the hypermetabolic state.[96–98] Injury causes similar changes in the relation between the turnover and the plasma concentration of glycerol. Thus, in patients with burns glycerol turnover is increased in relation to its concentration and also to the turnover of NEFAs, implying increased re-esterification within adipose tissue.[53]

In the fasted uninjured subject fat oxidation is suppressed by insulin released by the intravenous administration of large amounts of glucose. However, in hypermetabolic patients this suppression is incomplete and fat oxidation continues.[25] The reason for this continuing preferential oxidation of fat is not known but it is an important factor to be considered when planning nutritional support for the injured/septic patient.

Provision of nutritional support in trauma, sepsis and multiple organ failure

It is self-evident that unless nutritional support is provided the injured body will derive its energy and protein requirements entirely from its endogenous reserves of fat and (structural) protein.

Energy expenditure can be assessed either by measuring it or by estimating basal requirements from weight, age, height and sex using one of a number of standard formulae and adding correction factors for injury, sepsis, burns, etc. as discussed earlier (see also Chapter 54). Expenditure can be measured using indirect calorimetry. Formerly a research tool, there are now instruments available that

Table 17.3 Glucose uptake by forearm at end of 'clamp' in septic and injured patients

		Uptake (μmol/min per 100 ml)
Hyperglycaemic	Control	5.8±2.0
	Septic	1.1±0.9
Euglycaemic – hyperinsulinaemic (200 m-units/min per m^2)	Control	6.3±1.0
	Injured	3.2±0.9

Data taken from references 68 and 81.

enable this technique to be used in routine clinical practice.[99,100] A major potential problem, however, is in extrapolating from a brief period of measurement to a 'true' 24-hour expenditure. Most hospital patients are relatively inactive, so if several discrete measurements are made over a period it is a reasonable assumption, in this population, that measured expenditure will reflect true 24-hour expenditure.

Problems arise also in measuring energy expenditure in patients requiring high inspired oxygen concentrations. In measuring oxygen consumption inspired and expired gas concentrations are measured; inspired and expired volumes can also be measured but it is more usual to measure only one and calculate the other from the inspired and expired nitrogen concentrations (N_I and N_E respectively), i.e. the non-O_2 non-CO_2 concentrations. Thus $V_E = V_I \times N_I/N_E$. This assumes a steady state, and that net nitrogen exchange is zero. As the inspired oxygen concentration (F_IO_2) increases it is obvious that the amount of nitrogen in the system decreases[101] and in practice with most commercial instruments the system ceases to function at an F_IO_2 of about 0.7,[99] above which it is more accurate to measure inspired and expired volumes separately.[102]

The conventional recommendations for nutritional support have in the past been to cover basal requirements, an added factor for stress requirements 'plus 1000 kcal for anabolism and weight gain'.[103] Relative requirements for protein (nitrogen) and energy can be worked out on the basis of 1 g N/200 kcal non-protein energy for the unstressed patient, 1 g N/150 kcal for the uncomplicated surgical patient and a figure of 1 g N/100 kcal for the septic patient or the patient in multiple organ failure. These recommendations have on the whole been based on first principles and relatively little work has been done to determine what contribution nutritional support makes to the maintenance, or gain, of lean (or even fat) tissue. This has been due in part to the fact that most of the techniques used to assess body composition depend on a normal state of hydration as discussed earlier. In the acutely ill patient fluid is retained, capillary walls become permeable, fluid passes into the extracellular tissue, and excessive quantities of the fluid given during resuscitation are retained.

In the septic or traumatized patient rates of gluconeogenesis are about twice those seen in normal controls.[104] Provision of energy substrate suppresses gluconeogenesis completely in the normal individual, but in the septic patient it only decreases by 50 per cent, so that theoretically the most one could expect in providing nutritional support for the severely stressed patient would be to decrease gluconeogenesis and the rate of lean tissue breakdown, not stop it completely.[104] Only two studies have attempted to assess rates of tissue wasting in relation to this type of illness. One fed around 42 kcal/kg lean body mass per day over 2 weeks with amino acids equivalent to 20 g N/day to patients in multiple organ failure and showed a decrease in total body protein stores of about 10 per cent.[70] Body composition measurements were made by *in vivo* neutron activation analysis. Energy expenditure was not measured but body fat increased. The other[105] showed that lean tissue, as denoted by mid arm circumference, wasted away regardless of the level of energy intake relative to expenditure.

Surprisingly it is only recently that any effort has been made to examine systematically changes in body composition after elective major surgery. After uncomplicated major surgery Hill[106] showed that without any artificial nutritional support, 3 kg of weight was lost over 2 weeks, then regained over 3 months. Those with pre-existing deficits went on to gain weight over 6 months and many patients overshot their pre-illness weight by 2–3 kg due to accumulation of fat. The 3 kg body was composed of 1.4 kg fat, 0.6 kg protein and 1 kg water. Many patients felt unduly fatigued postoperatively. Most had recovered by 1 month and those with severe preoperative weight loss were generally better by 3 months, although facets other than pure protein loss appeared to be involved.[107]

Measurements of energy expenditure in patients who have suffered trauma, elective surgery or burns appear to show an elevation in metabolic rate proportional to the severity of the pathological insult. In the patient suffering multiple organ failure this does not appear to be the case. Patients on average have a measured expenditure generally about 120–130 per cent of predicted basal, but with a very wide spread of individual values that are not related to normal clinical scores or other markers of severity of illness.[108–110] The only way to assess energy expenditure in these patients is to measure it, either using indirect calorimetry as discussed earlier or by calculating $\dot{V}o_2$ using cardiac output measurements and measurements of mixed venous and arterial oxygen content – the so-called reverse Fick technique.[23] Not surprisingly energy expenditure in many instances is related to the degree of disturbance/sedation/paralysis/spontaneous activity,[104] but no studies as yet have examined the relation between level of sedation/muscular activity and measured 24-hour energy expenditure.

Gluconeogenesis persists in the septic state and also in the patients with multiple organ failure despite apparently adequate levels of nutrient intake. The largest source of endogenous amino acids is obviously muscle and the amino acids released in the greatest quantities from muscle are alanine and glutamine. Glutamine is a primary fuel for the gut and for the immune system and macrophages and is the most abundant intramuscular free amino acid, accounting for 60 per cent of the free amino acid pool. The size of the pool diminishes under conditions of trauma and sepsis. Protein breakdown within the muscle releases the branch chain amino acids. Their amino groups transfer to pyruvate to form alanine and to α-ketoglutarate to form glutamine. Alanine is a major gluconeogenic precursor and glutamine, in addition to being a metabolic fuel for the gut and the immune system, also releases ammonia in the kidney which buffers fixed acids excreted by the kidney.

It has come to be recognized in recent years that the gut has a major role in the maintenance of gluconeogenesis in the critically ill. Gut mucosal permeability is often increased in critical illness due to reduced splanchnic blood flow and hypoperfusion as well as local injury/surgery/sepsis. The subsequent release of endotoxin from the gut lumen, and the release of inflammatory mediators from elsewhere, again increases gut permeability and results in persistence of protein breakdown, mediated probably by tumour necrosis factor. The future for the treatment of this sort of condition, in addition to surgical drainage of septic foci, probably lies in blockade of the various hormonal and cytokine mediators of the catabolic state and the provision of specific substrates such as glutamine and α-ketoglutarate.[106]

REFERENCES

1. Flatt JP. Energetics of intermediary metabolism. In: *Assessment of energy metabolism in health and disease.* Columbus, Ohio: Ross Laboratories, 1980: 77–87.
2. Lee HA, Hartley TF. A method of determining daily nitrogen requirements. *Postgraduate Medical Journal* 1975; **51**: 441–5.
3. Cahill GF. Starvation in man. *New England Journal of Medicine* 1970; **282**: 668–75.

4. Clarys JP, Martin AD. The concept of adipose tissue-free mass. In: Norgan NG, ed. *Human body composition and fat distribution.* Wageningen: Euro Nut, 1985: 49–63.

5. Knight GS, Beddoe AH, Streat SJ, Hill GL. Body composition of two human cadavers by neutron activation and chemical analysis. *American Journal of Physiology* 1986; **250**: E179–85.

6. Adam JM, Best TW, Edholm OG. Weight changes in young men. *Journal of Physiology* 1961; **156**: 38P.

7. Rathbun EN. Pace studies on body composition. The determination of total body fat by means of body specific gravity. *Journal of Biological Chemistry* 1945; **158**: 667–72.

8. Keys A, Brozek J. Body fat in adult man. *Physiological Reviews* 1953; **33**: 245–325.

9. Durnin JVGA, Womersley J. Body fat assessed from total body density and its estimation from skinfold thickness: measurements on 481 men and women aged 16 to 72 years. *British Journal of Nutrition* 1974; **32**: 77–97.

10. Grant JP, Custer PB, Thurlow J. Current techniques of nutritional assessment. *Surgical Clinics of North America* 1981; **61**: 437–63.

11. Shizgal HM. Body composition and nutritional support. *Surgical Clinics of North America* 1981; **61**: 729–41.

12. Frayn KN. Hormonal control of metabolism in trauma and sepsis. *Clinical Endocrinology* 1986; **24**: 577–99.

13. Stoner HB. Metabolism after trauma and in sepsis. *Circulatory Shock* 1986; **19**: 75–87.

14. Barton RN. The neuroendocrinology of physical injury. *Baillière's Clinical Endocrinology and Metabolism* 1987; **1**: 355–74.

15. Gann DS, Amaral JF. Endocrine and metabolic responses to injury. In: Schwartz SI, Shires GT, Spence FT, eds. *Principles of surgery.* New York: McGraw-Hill, 1988: 1–68.

16. Tabor H, Rosenthal SM. Body temperature and oxygen consumption in traumatic shock and haemorrhage in mice. *American Journal of Physiology* 1947; **149**: 459–64.

17. Stoner HB. Studies on the mechanism of shock: the impairment of thermoregulation by trauma. *British Journal of Experimental Pathology* 1969; **50**: 125–38.

18. Rothwell NJ, Little RA, Rose JG. Brown adipose tissue activity and oxygen consumption after scald injury in the rat. *Circulatory Shock* 1991; **33**: 33–6.

19. Little RA, Stoner HB. Body temperature after accidental injury. *British Journal of Surgery* 1981; **68**: 221–4.

20. Little RA, Stoner HB, Randall P, Carlson G. An effect of injury on thermoregulation in man. *Quarterly Journal of Experimental Physiology* 1986; **71**: 295–306.

21. Little RA, Stoner HB, Frayn KN. Substrate oxidation shortly after accidental injury in man. *Clinical Science* 1981; **61**: 789–91.

22. Cournand A, Riley RA, Bradley SE, *et al.* Studies of the circulation in clinical shock. *Surgery* 1943; **13**: 964–95.

23. Edwards JD, Redmond AD, Nightingale P, Wilkins G. Oxygen consumption following trauma – a reappraisal in severely injured patients requiring mechanical ventilation. *British Journal of Surgery* 1988; **75**: 690–2.

24. Skootsky SA, Abraham E. Continuous oxygen consumption measurement during initial emergency department resuscitation of critically ill patients. *Critical Care Medicine* 1988; **16**: 706–9.

25. Barton RN, Frayn KN, Little RA. Trauma, burns and surgery, In: Cohen RD, Lewis N, Alberti KGMM, Denman AM, eds. *The metabolic and molecular basis of acquired disease.* Vol. I. London: Baillière Tindall, 1990: 684–717.

26. Carli F, Aber VR. Thermogenesis after major elective surgical procedures. *British Journal of Surgery* 1987; **74**: 1041–5.

27. Little RA. The impairment of thermoregulation by trauma during the first days of life of the rabbit. *Biology of the Neonate* 1974; **24**: 363–74.

28. Childs C. Fever in burned children. *Burns* 1988; **14**: 1–6.

29. Childs C, Stoner HB, Little RA, Davenporte PJ. A comparison of some thermoregulatory responses in healthy children and in children with burn injury. *Clinical Science* 1989; **77**: 425–9.

30. Childs C, Ratcliffe RJ, Holt I, Little RA, Hopkins SJ. The relationship between interleukin-1, interleukin-6 and pyrexia in burned children. In: Dinarello CA, Kluger M, Powanda M, Oppenheim J, eds. *Physiological and pathological effects of cytokines – progress in leukocyte biology.* New York: Alan R Liss, 1990: 295–300.

31. Childs C, Little RA. Acetaminophen (paracetamol) in the management of burned children with fever. *Burns* 1988; **14**: 343–8.

32. Guo Y, Dickerson C, Chrest FJ, Adler WH, Munster AM, Whitchurch RA. Increased levels of circulating interleukin-6 in burn patients. *Clinical Immunology and Immunopathology* 1990; **54**: 361–71.

33. Childs C, Stoner HB, Little RA. Cutaneous heat loss shortly after burn injury in children. *Clinical Science* 1992; **83**: 117–26.

34. Gann DS, Lilly MP. The endocrine response to injury. *Progress in Critical Care Medicine* 1984; **1**: 15–47.

35. Buckingham JC. Hypothalamic-pituitary responses to trauma. *British Medical Bulletin* 1985; **41**: 203–11.

36. Barton RN, Stoner HB, Watson SM. Relationships among plasma cortisol, adrenocorticotrophin and severity of injury in recently injured patients. *Journal of Trauma* 1987; **27**: 384–92.

37. Anderson ID, Forsling ML, Little RA, Pyman JA. Acute injury is a potent stimulus for vasopressin release. *Journal of Physiology* 1989; **416**: 28P.

38. Davies CL, Newman RJ, Molyneux SG, Grahame-Smith DG. The relationship between plasma catecholamines and severity of injury in man. *Journal of Trauma* 1984; **24**: 99–105.

39. Frayn KN, Little RA, Maycock PF, Stoner HB. The relationships of plasma catecholamines to acute metabolic and hormonal responses to injury in man. *Circulatory Shock* 1985; **16**: 229.

40. Frayn KN, Maycock PF, Little RA, *et al.* Factors affecting the plasma insulin concentration shortly after accidental injury in man. *Archives of Emergency Medicine* 1987; **4**: 91–9.

41. Porte D, Robertson RP. Control of insulin secretion by catecholamines, stress and the sympathetic nervous system. *Federation Proceedings* 1973; **32**: 1792–6.

42. Stoner HB, Frayn KN, Barton RN, *et al.* The relationships between plasma substrates and hormones and the severity of injury in 277 recently injured patients. *Clinical Science* 1979; **56**: 563–73.

43. Fleck A. Nutrition, protein metabolism and fluid balance. In: Kox W, Gamble J, eds. *Baillière's Clinical Anaesthesiology – Fluid resuscitation 2.* London: Baillière Tindall, 1989: 625–48.

44. Gauldie J, Richards C, Harnish G, *et al.* Interferon beta 2/B-cell

stimulatory factor type 2 shares identity with monocyte-derived hepatocyte-stimulating factor and regulates the major acute plasma protein response in liver cells. *Proceedings of the National Academy of Sciences of the USA* 1987; **84**: 7251–5.

45. Cruickshank AM, Fraser WD, Burns HJG, *et al.* Response of serum interleukin-6 in patients undergoing elective surgery of varying severity. *Clinical Science* 1990; **79**: 161–5.

46. Fleck A, Colley CM, Myers MA. Liver export proteins and trauma. *British Medical Bulletin* 1985; **41**: 265–73.

47. Parry Billings M, Newsholme EA, Baigrie R, Lamont PM, Morris PJ. Effects of major and minor surgery on plasma glutamine and cytokine concentrations. *Proceedings of the Nutrition Society* 1992; **51**: 107A.

48. Wilmore DW. *The metabolic management of the critically ill.* New York: Plenum Press, 1977.

49. Matsuda T, Clarke N, Hariyani GD, Bryant RS, Hanumadass ML, Kagan RJ. The effect of burn wound size on resting energy expenditure. *Journal of Trauma* 1987; **27**: 115–18.

50. Shoemaker WC. Hemodynamic and oxygen transport patterns in septic shock: physiologic mechanisms and therapeutic implications. In: Sibbald WJ, Sprung CL, eds. *Perspectives on sepsis and septic shock.* Fullerton, CA: Society of Critical Care Medicine, 1986: 203–34.

51. Dantzker DR. The influence of mixed venous PO_2 on arterial oxygenation. In: Vincent JL, ed. *Uptake in intensive care and emergency medicine – update 1990.* Berlin: Springer-Verlag, 1990: 131–7.

52. Wilmore DW, Long JM, Mason AD, Skreen RW, Pruitt BA. Catecholamines: mediator of the hypermetabolic response to thermal injury. *Annals of Surgery* 1974; **180**: 653–68.

53. Wolfe RR, Herndon DN, Jahoor F, Miyoshi H, Wolfe M. Effect of severe burn injury on substrate cycling by glucose and fatty acids. *New England Journal of Medicine* 1987; **317**: 403–8.

54. Lean MEJ, James WPT. Brown adipose tissue in man. In: Trayhurn P, Nicholls DG, eds. *Brown adipose tissue.* London: Edward Arnold, 1986: 339–65.

55. Mansell PI, Fellows IW, MacDonald IA. Enhanced thermogenic response to epinephrine after 48-h starvation in humans. *American Journal of Physiology* 1990; **258** *(Regulatory, Integrative and Comparative Physiology* 27): R87–93.

56. Little RA. Heat production after injury. *British Medical Bulletin* 1985; **41**: 226–31.

57. Cuthbertson DP. Alterations in metabolism following injury. Part 1. *Injury* 1980; **11**: 175–89.

58. Duke JH, Jorgensen SB, Broell JR, Long CL, Kinney JM. Contribution of protein to caloric expenditure following injury. *Surgery* 1970; **68**: 168–74.

59. Clifton GL, Robertson CS, Grossman RG, Hodge S, Foltz R, Garza A. The metabolic response to severe head injury. *Journal of Neurosurgery* 1984; **60**: 686–96.

60. Dickerson RN, Guenter PA, Gennarelli TA, Dempsey DT, Nullen JL. Increased contribution of protein oxidation to energy expenditure in head-injured patients. *Journal of the American College of Nutrition* 1990; **9**: 96–8.

61. Hadfield JM, Little RA. Substrate oxidation and the contribution of protein oxidation to energy expenditure after severe head injury. *Injury* 1992; **23**: 183–6.

62. Greenblatt SH, Long CL, Blakemore WS, Dennis RS, Rayport

M, Geiger JW. Catabolic effect of dexamethasone in patients with major head injuries. *Journal of Parenteral and Enteral Nutrition* 1989; **1**: 372–6.

63. Davies JWL. *Physiological responses to burning injury.* London: Academic Press, 1982.

64. Wilmore DW. The wound as an organ. In: Little RA, Frayn KN, eds. *The scientific basis for the care of the critically ill.* Manchester: Manchester University Press, 1986: 45–59.

65. Fong U, Moldawer LL, Shires GT, Lowry SF. The biologic characteristics of cytokines and their implications in surgical injury. *Surgery, Gynecology and Obstetrics* 1990; **170**: 363–78.

66. Wilmore DW, Aulick LH, Mason AD, Pruitt BA. Influence of the burn wound on local and systemic responses to injury. *Annals of Surgery* 1977; **186**: 444–58.

67. Wilmore DW, Goodwin CW, Aulick LH, Powanda MC, Mason AD, Pruitt BA. Effect of injury and infection on visceral metabolism and circulation. *Annals of Surgery* 1980; **192**: 491–502.

68. Henderson AA, Frayn KN, Galasko CSB, Little RA. Dose-response relationships for the effects of insulin on glucose and fat metabolism in injured patients and control subjects. *Clinical Science* 1991; **80**: 25–32.

69. Clague MB, Keir MJ, Wright PD, Johnston IDA. The effects of nutrition and trauma on whole-body protein metabolism in man. *Clinical Science* 1983; **65**: 165–75.

70. Streat SJ, Beddoe AH, Hill GL. Aggressive nutritional support does not prevent protein loss despite fat gain in septic intensive care patients. *Journal of Trauma* 1987; **27**: 262–6.

71. Threlfall CJ, Stoner HB, Galasko CSB. Patterns in the excretion of muscle markers after trauma and orthopedic surgery. *Journal of Trauma* 1981; **21**: 140–7.

72. Rennie MJ, Millward DJ. 3-Methylhistidine excretion and the urinary 3-methylhistidine/creatinine ratio are poor indicators of skeletal muscle protein breakdown. *Clinical Science* 1983; **65**: 217–25.

73. Threlfall CJ, Maxwell AR, Stoner HB. Post-traumatic creatinuria. *Journal of Trauma* 1984; **24**: 516–23.

74. Vinnars E, Bergstrom J, Furst P. Influence of the postoperative state on the intracellular free amino acids in human muscle tissue. *Annals of Surgery* 1975; **182**: 665–71.

75. Milewski PJ, Threlfall CJ, Heath DF, Holbrook IB, Wilford K, Irving MH. Intracellular free amino acids in undernourished patients with or without sepsis. *Clinical Science* 1982; **62**: 83–91.

76. O'Dwyer ST, Smith RJ, Hwang TL, Wilmore DW. Maintenance of small bowel mucosa with glutamine enriched parenteral nutrition. *Journal of Parenteral and Enteral Nutrition* 1989; **13**: 579–85.

77. Fajans SS, Floyd JC, Knopf RF, Conn JW. Effect of amino acids and proteins on insulin secretion in man. *Recent Progress in Hormone Research* 1967; **23**: 617–62.

78. Frayn KN, Little RA, Stoner HB, Galasko CSB. Metabolic control in non-septic patients with musculoskeletal injuries. *Injury* 1984; **16**: 73–9.

79. Black PR, Brooks DC, Bessey PQ, *et al.* Mechanisms of insulin resistance following injury. *Annals of Surgery* 1982; **196**: 420–35.

80. Brookes DC, Bessey PQ, Black PR, *et al.* Post-traumatic insulin

resistance in uninjured forearm tissue. *Journal of Surgical Research* 1984; **37**: 100–7.

81. White RH, Frayn KN, Galasko CSB, Little RA. Hormonal and metabolic responses to glucose infusion in sepsis studied by the hyperglycemic glucose clamp technique. *Journal of Parenteral and Enteral Nutrition* 1987; **11**: 345–53.

82. Kahn CR. Insulin resistance, insulin sensitivity, and insulin unresponsiveness: a necessary distinction. *Metabolism* 1978; **27** (Suppl 2): 1893.

83. Bessey PQ, Watters JM, Aoki TT, Wilmore DW. Combined hormonal infusion simulates the metabolic response to injury. *Annals of Surgery* 1984; **200**: 264–80.

84. Gelfand RA, Matthews DE, Bier DM, Sherwin RS. Role of counterregulatory hormones in the catabolic response to stress. *Journal of Clinical Investment* 1984; **74**: 2238–48.

85. Watters JM, Bessey PQ, Dinarello CA, Wolfe SM, Wilmore DW. Induction of interleukin-1 in humans and its metabolic effects. *Surgery* 1985; **98**: 298–306.

86. Van der Poll T, Romijn JA, Endert R, Borm JJJ, Buller HR, Sauerwein HP. Tumor necrosis factor mimics the metabolic response to acute infection in healthy humans. *American Journal of Physiology* 1991; **261**: E457–65.

87. Dinarello C. Interleukin-1. *Reviews of Infectious Diseases* 1984; **6**: 51–6.

88. Roh MS, Drazenovich KA, Barbose JJ, Dinarello CA, Cobb CF. Direct stimulation of the adrenal cortex by interleukin-1. *Surgery* 1987; **102**: 140–6.

89. Del Rey A, Besedovsky H. Interleukin-1 affects glucose homeostasis. *American Journal of Physiology* 1987; **253**: 794–8.

90. Tredget EE, Yong Ming UY, Zhong S, *et al.* Role of interleukin-1 and tumor necrosis factor on energy metabolism in rabbits. *American Journal of Physiology* 1988; **255**: E760–8.

91. Clowes GHA, George BC, Villee CA, Saravis CA. Muscle proteolysis induced by a circulating peptide in patients with sepsis or trauma. *New England Journal of Medicine* 1983; **308**: 545–52.

92. Moldawer LL, Svaninger G, Gelin J, Lundholm KG. Interleukin-1 and tumor necrosis factor do not regulate protein balance in skeletal muscle. *American Journal of Physiology* 1987; **253**: C766–73.

93. Long CL, Spencer JL, Kinney JM, Geiger JW. Carbohydrate metabolism in man: effect of elective operations and major injury. *Journal of Applied Physiology* 1971; **31**: 110–16.

94. Batstone GF, Alberti KGMM, Hinks L, *et al.* Metabolic studies in subjects following thermal injury. Intermediary metabolites, hormones and tissue oxygenation. *Burns* 1976; **2**: 207–25.

95. Birkhahn RH, Long CL, Fitkin DL, *et al.* A comparison of the effects of skeletal trauma and surgery on the ketosis of starvation in man. *Journal of Trauma* 1981; **21**: 513–18.

96. Nordenstrom J, Carpentier YA, Askanazi J, *et al.* Free fatty acid

mobilization and oxidation during total parenteral nutrition in trauma and infection. *Annals of Surgery* 1983; **198**: 725–35.

97. Wilmore DW, Moylan JA, Helmkamp GM, Pruitt BA. Clinical evaluation of 10 per cent intravenous fat emulsion for parenteral nutrition in thermally injured patients. *Annals of Surgery* 1973; **178**: 503–13.

98. Wolfe RR, Shaw JHF, Durkot MJ. Effects of sepsis on VLDL kinetics: responses in basal state and during glucose infusion. *American Journal of Physiology* 1985; **248**: E732–40.

99. Regan CJ, Snowdon SL, Campbell IT. Laboratory evaluation and use of the Engstrom Metabolic Computer in the clinical setting. *Critical Care Medicine* 1990; **18**: 871–7.

100. Takala J, Keinanen O, Vaisanen P, Kari A. Measurement of gas exchange in intensive care: laboratory and clinical validation of a new device. *Critical Care Medicine* 1989; **17**: 1041–7.

101. Ultman JS, Bursztein S. Analysis of error in the determination of respiratory gas exchange at varying FIO_2. *Journal of Applied Physiology (Respiratory, Environmental, Exercise Physiology)* 1981; **50**: 210–16.

102. Svensson KL, Sonander HG, Stenqvist O. Validation of a system for measurement of metabolic gas exchange during anaesthesia with controlled ventilation in an oxygen consuming lung model. *British Journal of Anaesthesia* 1990; **64**: 311–19.

103. MacBurney M, Wilmore DW. Rational decision-making in nutritional care. *Surgical Clinics of North America* 1981; **61**: 571–82.

104. Shaw JHF, Klein S, Wolfe RR. Assessment of alanine, urea and glucose interrelationships in normal subjects and in patients with sepsis with stable isotope tracers. *Surgery* 1985; **97**: 557–62.

105. Green CJ, Helliwell TR, McClelland P, *et al.* Arm circumference and energy balance in acute illness. *Proceedings of the Nutrition Society* 1990; **49**: 17A.

106. Hill GL. *Disorders of nutrition and metabolism in clinical surgery.* Edinburgh: Churchill Livingstone, 1992: 19–32.

107. Editorial. Postoperative fatigue. *Lancet* 1979; **1**: 84–5.

108. Dickerson RN, Vehe KL, Mullen JL, Feurer ID. Resting energy expenditure in patients with pancreatitis. *Critical Care Medicine* 1991; **19**: 484–90.

109. Weissman C, Kemper M, Hyman AI. Variation in the resting metabolic rate of mechanically ventilated critically ill patients. *Anesthesia and Analgesia* 1989; **68**: 457–61.

110. Weissman C, Kemper M, Elwyn DH, Askanazi J, Hyman AJ, Kinney JM. The energy expenditure of the mechanically ventilated critically ill patient. An analysis. *Chest* 1986; **89**: 254–9.

Chapter 18

Autologous Blood Transfusion

M. J. G. Thomas

Homologous blood transfusion
Autologous transfusion
Predeposit donation
Acute normovolaemic haemodilution

Intraoperative cell salvage
Postoperative collection
The future

The term autologous transfusion describes the administration of any blood component which has been donated by the intended recipient. This chapter will first consider why autologous transfusion should be employed, will then discuss the various methods, with the benefits and problems associated with each, and finally will speculate on the possible advances and necessary responses during the next decade. By the end of the chapter, hopefully the reader will be in a position to decide whether the claim made in the technical manual of the American Association of Blood Banks,[1] that 'a recipient who serves as his or her own donor receives the safest possible transfusion', is in fact true.

When reviewing the requirements of a patient who has lost blood, or may in the future be subjected to a procedure which involves such a loss, the first question that needs to be asked is not whether autologous transfusion is indicated, but whether any blood transfusion is necessary.

Homologous blood transfusion

Indications

The main indications for transfusion are the replacement of circulating volume, the correction of haemostatic defects and the replacement of oxygen carrying capacity.

In a fit individual, when blood loss is rapid as may happen during trauma or surgery, the initial problem is not reduced oxygen carrying capacity, but hypovolaemia. If this can be corrected by the use of crystalloids or colloids, the patient's normal physiological compensatory mechanisms will lead to recovery without the transfusion of blood. Once the circulatory volume has been adequately restored,

oxygen perfusion of the tissues and organs returns to normal.

It is generally accepted that patients can tolerate limited dilutional anaemia [haematocrit (packed cell volume) 25–30 per cent] remarkably well, as long as the circulating blood volume is maintained and adequate arterial oxygen saturation is preserved.[2] A fall in haematocrit activates mechanisms which provide adequate oxygenation of the tissues.

The first is triggered by enhanced blood fluidity leading to increased cardiac output. Fluidity depends on haematocrit, plasma viscosity and red cell deformability. The dilutional effect of the crystalloid or colloid used in resuscitation after blood loss results, if the red cell deformability is normal, in a lower red cell concentration and a decrease in plasma viscosity. A linear decrease in haematocrit results in an exponential increase in the blood fluidity. The increase in fluidity promotes ventricular emptying and decreases the afterload, thereby increasing venous return which in turn leads to an increase in stroke volume and cardiac output, while the heart rate remains constant. This holds true as long as the haematocrit does not fall below 25 per cent.[3] The increased cardiac output provides a higher blood flow to the organs and increased coronary blood flow results from the combined effect of increased fluidity and coronary dilatation.

The second compensatory mechanism depends on increased oxygen extraction from the arterial blood. Hint[4] calculated that systemic oxygen transport capacity increases as the haematocrit falls, until a level of 30 per cent is reached, after which the capacity decreases.

Postoperatively the decision to transfuse must take into account the level and duration of anaemia, the complexity of the operation, the possibility of further blood loss and the presence of coexisting conditions which might impair oxygen transport, such as inadequate cardiac, pulmonary or renal function, or cerebrovascular disease. However, current experience suggests that it is not necessary to transfuse homologous blood when the haematocrit is greater than 30 per cent.

Risks

Disease transmission

Although HIV has been perceived by the public as the most dangerous disease transmitted by blood transfusion, there are numerous other viral diseases, such as hepatitis B, hepatitis C, CMV and HTLV 1 and 2, which are bloodborne. In some countries hepatitis C is a thousand times more common than HIV and 50 per cent of those infected progress to chronic aggressive hepatitis or chronic persistent hepatitis, 10 per cent of whom will develop cirrhosis. In addition, organisms such as *Treponema pallidum*, microfilaria and the protozoa responsible for malaria and trypanosomiasis are carried in blood. The prevalence of these organisms varies from country to country. In general the transmission of syphilis and viral diseases has been greatly reduced by the introduction of microbiological screening of homologous blood donations.

Immune and allergic reactions

These, which are listed in Table 18.1, are probably the most frequent adverse reactions to blood transfusion. An acute intravascular haemolytic reaction, which is frequently fatal, is the most important, but fortunately the most infrequent.

Alterations in host immune function

Although in some cases these alterations may be beneficial, for example increased renal allograft survival and decreased recurrence of Crohn's disease, in the main they are detrimental. The latter effects cause more serious viral infections, increased postoperative bacterial infection and possibly an increased rate of solid tumour recurrence. In experimental animals, mortality due to burn wound sepsis, intraperitoneal injection of *Escherichia coli*, or caecal ligation and puncture, is doubled if the animal is

Table 18.1 Immunological and allergic reactions

Red cells
    Acute intravascular haemolytic
    Delayed extravascular
White cells
    Febrile reactions
    Graft-vs-host disease
Platelets
    Post-transfusion purpura
    Refractiveness to platelet transfusion
Miscellaneous anaphylactic, urticarial or pyrexial

Table 18.2 Animal *in vitro* indicators

Decreased cell-mediated immunity indicated by:
    Decreased skin test reactivity
    Decreased contact sensitivity
    Inhibition of lymphocyte response to antigen and mitogen
Macrophage migration inhibition
Increased production of immunosuppressive substances such as prostaglandin E, thromboxane and prostacyclin
Decreased generation of interleukin-2

Table 18.3 Human *in vitro* indicators

Decreased circulating T cells, B cells, NK cells and monocytes
Decreased lymphocyte response to antigen and mitogen
Suppression of lymphocyte reactivity
Inhibition of cell-mediated lympholysis
Increased suppressor cell activity and decreased T-4:T-8 ratio
Decreased interleukin-2 production

transfused with allogenic blood.[5] The *in vitro* indicators of altered immune function in animals are listed in Table 18.2 and those in humans in Table 18.3.

Fluid overload

The transfusion of blood, as with any other fluid, can lead to circulatory overload.

Bacterial contamination

Blood is an ideal bacterial culture medium. Donations which have been collected improperly, or stored at the wrong temperature, may become contaminated. Transfusion of such blood may be fatal. One of the more worrying aspects of this problem is the emergence of *Yersinia enterocolitica* as a contaminating organism.

Clerical errors

Clerical errors can lead to blood intended for one patient being transfused to another with potentially catastrophic results.

Risk avoidance

Exposure to the above risks can be minimized by adopting a policy of blood conservation. Such a policy should include a maximum surgical blood order schedule (MSBOS),[6] the use of group and screening procedures, measures to reduce intraoperative blood loss[7] and the use of alternatives to homologous blood.

Maximum surgical blood order schedule

The Hospital Transfusion Committee (HTC) should review the blood usage for all surgical procedures within the hospital, with a view to agreeing a tariff for each operation. It is suggested that the number of units of blood allocated for any procedure should be the average number of units actually used in that type of operation. If this is less than two, then no blood should be cross-matched, but the patient should be blood grouped and screened for atypical serological antibodies. Single unit transfusions should be avoided except in paediatric cases.

Measures to reduce blood loss

These include induced hypotension, the use of topical fibrin glue and pharmacological haemostatic agents such as DDAVP[8,9] and aprotinin.[10] Patients taking aspirin or non-steroidal anti-inflammatory drugs should have these stopped at least 1 week prior to surgery if possible.

Alternatives to homologous blood

Haematinics should be employed to raise the haemoglobin level when anaemia is chronic or where there is no pressing urgency to return the level to normal. The loss of circulatory volume can often be corrected by using crystalloids or colloids rather than blood. Where there is, in addition, a requirement for increased oxygen carrying capacity, alternatives to blood such as perfluorochemical emulsions and haemoglobin solutions, have yet to find clinical acceptance. However, the use of autologous blood may be an acceptable alternative to homologous transfusion.

Autologous transfusion

During the past decade autologous transfusion has been encouraged, to avoid transmission of viral disease. Now that an increasing number of these diseases can be detected by laboratory screening tests, the rationale for autologous transfusion is moving towards the advantages to be gained by avoiding immunosuppression. Triulzi[11] calculated the 'in hospital' infection rates in 109 orthopaedic patients. The 24 who received homologous transfusions had an infection rate of 20.8 per cent, whereas the rates were 3.3 per cent in the 60 receiving autologous blood and 4 per cent in the 25 who received no blood. Those receiving homologous blood had an average hospital stay of 12.3 days, whereas those receiving autologous or no blood had an average stay of 9.7 days. There was also a difference in the number of days that the two groups were pyrexial and number of days on

Table 18.4 Advantages of autologous transfusion

Eliminates the risk of transfusion reactions
Eliminates the risk of disease transmission
Eliminates the risk of alloimmunization to red cells, white cells, platelets or plasma proteins
Eliminates the risk of transfusion transmitted graft-vs-host disease
Allows safe transfusions in patients with multiple alloantibodies or with rare blood groups
Predeposit donation 'stimulates' erythropoiesis prior to surgery
Haemodilution improves tissue oxygen perfusion by lowering the blood viscosity
Provides blood cover for some Jehovah's Witnesses
Provides readily available blood in cases of major haemorrhage
Reduces the demand on homologous blood supply in remote areas or developing countries
Gives patients the psychological benefit of actively participating in their treatment
Boosts the donor pool, as many autologous donors subsequently become homologous donors

antibiotics. The two groups differed in the duration of the surgical procedure and the amount of blood transfused, but multivariate analysis confirmed that, whilst surgical factors strongly correlated with homologous transfusion, only homologous transfusion itself was related to the predisposition to infection.

Evidence that early tumour recurrence is associated with homologous blood transfusion is equivocal.[12] Although it is almost certainly untrue that transfusion acts as a surrogate marker for clinical severity, it may well be that the type of blood component transfused is a relevant factor when considering tumour recurrence. Blumberg and Heal[13] state that, although there is no direct evidence to show that autologous transfusion is beneficial, there is certainly a theoretical advantage.

The major benefits of autologous transfusion are listed in Table 18.4 and the disadvantages in Table 18.5.

Physicians in California are required to inform patients

Table 18.5 Disadvantages of autologous transfusion

Complex logistics for collection, storage and transfusion of the correct unit to the appropriate patient
Only suitable for certain operative procedures
Tendency to overtransfuse because 'it is there'
If the surgical procedure is delayed, the blood may become outdated
Bacterial contamination
Patient is made grossly anaemic by either too frequent predeposit donation or overhaemodilution
Coagulation defects
Incorrect techniques can cause red cell haemolysis
Equipment for intraoperative blood salvage is expensive and requires trained staff
Tends to give a misguided impression that normal homologous donations are unsafe

of all transfusion options including autologous transfusion and to schedule non-urgent surgical procedures at such a time as will allow adequate predeposit donation. In 11 further states of the USA, where appropriate, patients must be informed of the option for making autologous predeposit donations.

Setting up a programme

Once the decision has been made to have an autologous transfusion programme, the first task is to identify the physician who is to be in charge. This could be a surgeon, anaesthetist or the haematologist in charge of the blood bank. He must draw up protocols for the various procedures, the training of the operators, the servicing and maintenance of the equipment and its cleaning. The latter should be arranged in consultation with the Hospital Control of Infection Committee. These protocols should then be submitted to the HTC for approval, who in turn will review the safety precautions and appropriateness of the service. The physician in charge must keep a register of the adverse events related to the techniques and regularly report these to the HTC.

Methods

There are four methods of autologous transfusion: predeposit donation alternatively known as storage for defined need (SFDN), acute normovolaemic haemodilution, intraoperative cell salvage and postoperative collection. The most effective programme is to use these methods in combination as appropriate. Ferrari and his colleagues[14]

use a system in which patients predeposit red cells and plasma. One to three units of erythrocytes suspended in saline together with an average of 1.6×10^{11} platelets and 460 ml plasma are collected preoperatively, i.e. normovolaemic haemodilution. At the end of the procedure, the tubing is disconnected from the central collection line and connected to the standard intraoperative blood salvage (IBS) suction cannula. In addition the red cells that remain in the blood oxygenator are concentrated and returned to the patient. It can be seen in Table 18.6 that 85 per cent of those given autologous transfusion required no homologous blood or blood components, the remaining 15 per cent having a dramatically reduced exposure.

Predeposit donation

In suitable cases, approximately 70 per cent of patients can have their total surgical blood requirement satisfied using predeposit donation. Not all the risks of homologous transfusion are obviated by this procedure, as it is still subject to the dangers of bacterial contamination, documentation errors, especially if performed by personnel unused to blood donation procedures, and anaemia in patients because of the frequent repeated donations. The amount of blood to be donated for any given procedure should be determined by the HTC, who should publish a schedule for optimal preoperative collection of autologous blood (SOPCAB),[15] similar to the MSBOS. Only procedures in which homologous transfusion would be indicated under similar clinical circumstances should be scheduled for predeposit donation.

Table 18.6 Homologous blood requirements at San Martino Hospital, 1987

	Group A*	Group B	Group C
No. of patients	152	105	123
Predeposited units of erythrocytes	2.4	2.3	0
Haematocrit (%) at operation/7 days later	36.3/33.7	37.0/34.1	41.3/35.4
Homologous blood needs at operation			
Erythrocyte units (patients)	6(3)	9(4)	424(all)
Plasma and platelet units (patients)	24(3)	91(18)	430(all)
Requirements of homologous blood components in the week of operation			
Erythrocyte units (patients)	22(9)	38(15)	64(25)
Plasma and platelet units (patients)	6(2)	89(19)	161(25)
Total no. of patients exposed to homologous blood components	12(8%)	28(27%)	123(100%)
Non-A, non-B hepatitis. No. of patients (per % of total patients)			
Raised alanine aminotransferase	2(1.3%)	4(3.8%)	15(12%)
Clinical disease	0	1(1%)	7(5.7%)

* Group A, predeposit autodonation, IBS and autologous platelet support; group B, predeposit autodonation and IBS; group C, conventional transfusion support with homologous blood components.

Patient selection

Initial suitability for predeposit donation should be assessed by the physician with clinical responsibility for the patient. In the United Kingdom a standard letter is used by the referring physician and a standard consent form signed by the donor. The final responsibility for ensuring that a patient's health is satisfactory lies with the doctor who undertakes the predeposit procedure.

Criteria

The criteria for predeposit donations are less strict than those for normal homologous blood donors. The precise criteria for patient eligibility vary from country to country and national guidelines should be consulted.[16–18] In the UK the haemoglobin concentration should normally be greater than 11 g/dl and never below 10 g/dl. Paediatric patients (see below) are suitable for predonation as long as they are able to comprehend what is being asked of them and are willing to cooperate. De Palma[19] reports that he has drawn blood from patients as young as 7 years (23 kg). Parental consent is mandatory.

Contraindications

Active bacterial infection

Patients who have active infections may be bacteraemic. If such blood is drawn, the bacteria may proliferate during storage, leading to fatal reactions. With *Yersinia enterocolitica*, patients have been infected without exhibiting any gastrointestinal symptoms.[20]

Patients with cardiac disease

These patients should be assessed by a cardiologist before being allowed to predeposit. Spiess *et al*.[21] have reported their experience with a group of patients suffering from cardiovascular disease who were bled weekly until 4 days prior to operation. The donations were taken in the postanaesthesia care unit and non-invasive haemodynamic monitoring used to ensure adequate fluid replacement. All the patients fell within the American Society of Anesthetists risks classification 2 or 3, including those with cardiac dysrhythmias, poorly controlled hypertension requiring two or more medicaments, congestive heart failure, valvular or congenital heart disease, demonstrable cerebral vascular insufficiency or a history of myocardial infarction, angina, or cerebrovascular accident. Patients with unstable angina, aortic stenosis or who had had a myocardial infarction within the previous 6 months were excluded. In more than 20 per cent of the patients systolic hypotension occurred during donation. Orthostatic hypotension occurred in 16 per cent and serious dysrhythmias in 3 per cent. However, only 2 per cent of patients actually suffered syncope. From these data, Spiess *et al.* postulate that it is unsafe to rely solely on subjective complaints to assess the safety of the process, and that patients with a history of cardiovascular disease should only be bled under haemodynamic monitoring. At present the absolute contraindications to predeposit are:

1. Significant aortic stenosis.
2. Prolonged and/or frequent unstable angina.
3. Significant narrowing of the left main coronary artery.
4. Cyanotic heart disease.
5. Uncontrolled hypertension.

Loss of consciousness

Patients who have previously been blood donors and have had a prolonged fainting attack should not be accepted.

Impaired placental flow

Pregnant patients should not predonate if they are suffering from a disease, such as hypertension, pre-eclamptic toxaemia or diabetes mellitus, which is associated with impaired placental flow and/or intrauterine growth retardation.

Method

Patients can donate 450 ml of blood at weekly intervals, the last donation being at least 4 days and preferably 1 week prior to surgery. They should commence oral iron prior to their first donation and continue until the day of surgery. Adults weighing less than 50 kg and paediatric patients require special consideration. The volume withdrawn at any one time should not exceed 12 per cent of the patient's estimated blood volume. They should have blood drawn into Pedipacks containing 35 ml of anticoagulant and which are suitable for collection of up to 250 ml of blood. The physiological response in patients taking β-blockers and/or ACE inhibitors is compromised by their treatment; they should therefore be given isovolaemic crystalloid replacement to minimize the hazardous sequelae, which may follow a sudden reduction in blood volume. When donating late in pregnancy, patients should lie in the lateral position because the weight of the uterus impedes the venous return when the patient is lying on her back.

Table 18.7 Details to be entered on autologous blood pack label

BLOOD FOR AUTOLOGOUS TRANSFUSION ONLY
Surname:
First names:
Date of Birth:
Hospital no.:
Date of collection:
Date of expiry:
ABO and Rh(D) types:
Lab reference no.:
Patient's signature:

Labelling

The patient, or in the case of a child the parent, should sign the pack label to confirm that the details on the label (apart from the ABO and (D) type which may not have been entered when the first unit is drawn) are correct. This should be done while the patient is actually on the donation couch. Details of the information which should be on the label are shown in Table 18.7. The signature can later be compared, as part of a pretransfusion checking procedure, with the one on the consent form, which by then will be in the patient's notes. The label should be affixed to the blood pack during the donation, with an adhesive suitable for refrigerated storage. This label should not occlude the information given on the manufacturer's standard pack label. Care must be taken when two patients are being bled at the same time to prevent crossover of labels and/or samples.

Place of collection

The reaction rate in predeposit donors who meet normal donor criteria is the same as that for homologous donors; therefore it is perfectly safe to collect blood from such predeposit donors in a non-hospital setting. Reactions are especially likely in patients less than 17 years old, females weighing less than 50 kg, first time donors and those with a previous history of a reaction. Those least likely to have a reaction are male donors aged 66 and over.[22] AuBuchon and Popovsky[23] showed that even when donors did not meet the above criteria there were only four severe reactions (one TIA, three cases of angina) in a series of 886 donations. From their studies, they calculated that only seven patients in every 10 000 would require any form of medical intervention and 99.5 per cent of predeposit donors, not meeting homologous criteria, would have no reaction or one that would require only simple first aid procedures.

Storage

Whole blood units collected into CPDA-1 may be stored for 35 days and those where the red cells are separated and resuspended in an optimal additive solution, such as SAG-M, may be stored for 42 days. Blood must be stored in a proper blood bank at a temperature of $4° \pm 2°C$. The blood need not be stored in a separate blood bank but must be kept in an area segregated from homologous blood for routine issue.

Testing

The degree to which autologous units should be tested is a matter of debate.[24] As predeposited blood is to be returned to the patient from whom it has been taken, some people question the need for grouping and microbiologically screening such donations. The argument against screening is that if the donation is infected, the patient will already be carrying that disease, so reinfusion will have no deleterious effect. Those who advocate normal processing put forward the following arguments:

1. The blood should be cross-matched to ensure that a clerical error has not occurred and an incorrect donation assigned to the patient.
2. To process a small number of units differently from the normal laboratory routine is likely to lead to mistakes.
3. In some centres, quality control checks prevent the release of blood which has not been fully screened.
4. If blood is found to be infected, this will alert the staff to the inherent dangers.

In the United Kingdom, guidelines state that all packs should be ABO and rhesus D blood grouped. The first donation should be screened for atypical serological antibodies, in case the patient needs additional homologous blood, and the first and last donations should be screened for the normal microbiological markers.[18] If the blood is found to be positive most people argue that all such blood should be discarded as there is a danger of mistakenly transfusing it to another patient. However, Dzik and Devarajan[25] suggest that each case should be treated as a distinct entity and the decision as to whether to retain or discard the unit depends on a number of variables, including the likelihood of the patient requiring transfusion, the possible adverse effects on the patient if given a routine homologous unit and the possibility of harm to an unintended recipient who is mistakenly transfused with the unit. A useful decision analysis formula is described.

When the patient is admitted to hospital, a sample of blood should be taken for compatibility testing. This should at least consist of an ABO group and rhesus check on the patient and all the donations, but many laboratories feel that it is better to do a full cross-match, as this is less likely to disrupt the normal laboratory procedure and lead to clerical errors. Each unit should have a standard compatibility label fixed to it prior to issue.

Disposal of unused blood

Whether unused units of autologous blood should be issued to other patients is again a matter of debate.[24] Proponents of the practice feel that a valuable resource should not be wasted, whilst others point out that autologous donors frequently do not meet the criteria laid down for homologous donors. In the United Kingdom, predeposited blood, if not transfused must not be issued to other patients, and plasma must not be put into routine plasma pools.[18]

Quality control

At least 1 per cent of packs collected should, when time expired or not required for use, be cultured both aerobically and anaerobically.

Erythropoietin

Erythropoietin has been advocated by some physicians[26] as an adjunct to the predeposit programme to allow a greater number of units to be collected. Goodnough et al.[27] reported the result of a double blind trial in which patients were either given erythropoietin, 600 iu/kg body weight intravenously twice a week, or a placebo. Both groups were taking 325 mg of ferrous sulphate three times a day orally. During a 3-week period Goodnough et al. attempted to collect 6 units of blood from each patient. Patients were not allowed to donate if their haematocrit was less than 34 per cent. The erythropoietin group donated an average of 5.4 units with a mean red cell volume of 961 ml, as opposed to the placebo group who donated an average of 4.1 units with a mean red cell volume of 683 ml. The mean usage rate in the study was 3.7 units, so the fact that the placebo group were able to donate an average of 4.1 units is highly significant,[28] as there were very few cases when more than 4 units were used and this amount is sufficient to cover most operations. Erythropoietin is extremely expensive and in addition is not without problems. It needs to be given 2 weeks before the commencement of the donation programme as there is a delay of 2–3 weeks prior to a noticeable rise in haematocrit. There is also a theoretical disadvantage, in that exogenous erythropoietin suppresses endogenous production for up to a week after stopping

exogenous administration. Far more importantly, the use of erythropoietin is associated with an increased risk of thrombosis.[29] Goodnough et al.[27] recorded one case of peripheral arterial thrombosis in their series. It is not clear whether this effect is due to the raised haematocrit and increased blood viscosity, or whether erythropoietin has a more specific effect on the coagulation mechanism. There is therefore an obvious potential danger to patients with cardiovascular disease or atherosclerosis.[30]

Pregnancy

Although the number of transfusions in pregnancy are relatively few, any complications can have a major impact on this young healthy population. In contrast to most recipients, these patients still have a long life ahead of them with ample time for the long-term sequelae to appear. There are three major concerns that have been raised with regard to predeposit during pregnancy.

Safety

The increased maternal blood volume and the physiological sparing of uteroplacental perfusion in the face of changes in maternal cardiac output and blood pressure minimize the dangers to the fetus which occur following the reduction in maternal blood volume after donation. Kruskall et al.[31] report no fetal problems or untoward sequelae at delivery for mother or infant following 61 donations from 39 patients. However, Tabor[32] points out that mothers of small babies and non-surviving fetuses have very little increase in plasma volume, and feels that further studies should be established to evaluate the effects on the fetus and, more importantly, the developmental milestones during the first few years of life.

Need for transfusion

This technique should only be used for patients with a high risk of requiring blood transfusion. Examples of such patients are those with placenta praevia or placenta accreta, a history of a previous postpartum haemorrhage or multiple pregnancies. Two or three units will cover the majority of these cases.

Cost

As the cost of predeposit is not inconsiderable this technique should be confined to high risk cases.

Problems

Phlebotomy

Skettino *et al.*[33] report that bleeding 'high risk' patients can cause anxiety amongst phlebotomists. This can be reduced by ensuring that dependable medical support is closely available and that the phlebotomists employed have had recent experience in patient care and a knowledge of, and a commitment to, the benefits of autologous transfusion.

Overtransfusion

A major problem that has recently been identified with predeposit donation is that there is a tendency to overtransfuse. The practice of 'giving it because it is there' should be resisted and patients should only be transfused if the clinical circumstances would normally indicate a homologous blood transfusion. This will inevitably lead to wastage of autologous blood and this possibility must be explained to the patient prior to commencement on the programme.

Delays in procedure

As the maximum shelf life of predeposited blood stored at 4°C is 42 days, delays in surgical procedures are bound to cause problems. Such delays may be due to illness of the patient or surgeon, hospital bed non-availability or domestic or social causes. One method that has been attempted to overcome this is the so-called 'leapfrogging' of units,[24] whereby the oldest unit is tranfused back to the patient and a fresh unit is taken in its place.

An alternative strategy is to freeze the blood immediately after donation and store it until required. This is especially useful for patients who have multiple alloantibodies, who are able to donate when they are healthy in case there is a future requirement for surgical procedures, or early in pregnancy, thus avoiding the risks of third trimester donation. The storage of autologous blood from such patients suffering from sickle cell anaemia has been investigated.

At present the only methods available for cryopreservation utilize glycerol as a cryoprotectant agent. After thawing, the glycerol must be removed by washing in a variety of solutions, which makes the method expensive, both in equipment and the use of trained personnel, time consuming and the resulting product is composed of red cells suspended in saline with a 24-hour shelf life.

An alternative method has been developed[34] and is at present under clinical trials. This utilizes hydroxyethyl-starch (HES), a widely used artificial plasma expander, as the cryoprotectant agent. As HES is a pharmaceutical plasma expander there is no need to remove it prior to

transfusion, so that predeposited blood can be made available for transfusion within 20 minutes of request. The resulting mixture of red cells and HES has the same volume expansion properties as whole blood. In addition, further trials are being performed to determine whether the cryopreserved red cells can be resuspended in the plasma which had been removed from them prior to freezing.

Acute normovolaemic haemodilution

As described earlier, patients can tolerate an anaemia with a haematocrit of 25–30 per cent, as long as normovolaemia is maintained. This is the basis for acute normovolaemic haemodilution (ANH). In the operating suite, 1–2 litres of blood are removed with simultaneous infusion of crystalloid and/or colloid to maintain the blood volume.

Surgical blood loss during the procedure will consist of blood with a lowered haematocrit because of the haemodilution, and so will represent proportionally less of the patient's red cell mass.

This technique should be considered for paediatric patients where predeposit donation had been impossible due to age and/or poor venous access. In such cases a formula, based on starting haematological values and body weight, should be used to allow the patient to be bled down to a specific haematocrit.

Careful assessment of patients with cardiovascular disease and/or decreased respiratory reserve is required and these people should only be subjected to haemodilution when invasive monitoring can be used to ensure adequate tissue oxygenation. ANH should not be used in those patients listed in Table 18.8.

The blood collected is normally divided into red cells and platelet-rich plasma (PRP), the red cells being available during the operation and the plasma retained to correct microvascular bleeding after haemostasis has been attained. If only PRP is required, as in some cardiac cases, machines are available for collecting this specific product.

The red cells and PRP should normally be stored at room

Table 18.8 Contraindications to acute normovolaemic dilution

Haematocrit less than 24%
Patients with limited ability to increase cardiac output such as valvular heart disease and intra- or extracardiac shunting
Patients with respiratory insufficiency needing mechanical ventilation
Patients with haemostatic defects

temperature in the operating theatre, the red cells only being stored at 4°C if a delay in transfusion of more than 4 hours is expected. When surgical haemostasis has been achieved, the red cells, if not already transfused, and PRP are reinfused, simultaneously giving a diuretic to mobilize the crystalloid or colloid given at the beginning of the procedure. In ANH crystalloid is preferred, because it is more easily mobilized, there is less risk of volume overload and it is cheaper. Ferrari's method[14] has already been described, Takaori[35] suggests the use of ANH when a blood loss of more than 400 ml is expected. His protocol commences with the patient receiving 500 ml of Ringer's lactate after which 600 ml of blood is drawn, which is replaced by 600 ml of dextran 70. If blood loss of more than 1 litre is expected, a further 600 ml of blood is taken and replaced with an equal volume of dextran 70. He considers that this protocol is sufficient to cover blood loss of up to 2 litres, but if a greater loss is anticipated ANH should be combined with intraoperative cell salvage.

Intraoperative cell salvage

Introduction

Predeposit donation is limited by the time that the blood can be stored and by the ability of patients to donate at frequent intervals. ANH is limited by the patient's total blood volume and by haemodynamic considerations. Postoperative salvage is limited by mechanical problems and possible bacterial contamination. Intraoperative cell salvage (ICS), in contrast, can be used throughout the surgical procedure and is able to replace blood in proportion to the amount lost.

History

The history of ICS dates back to the Cornish obstetrician, James Blundell, who commenced animal experiments after attending a patient who had exsanguinated from a postpartum haemorrhage. The first autologous transfusion was performed by John Duncan of Edinburgh in 1886. Interesting histories of ICS are given by Pineda and Valbonesi[36] and Williamson and Taswell.[37]

Methods and equipment

The basis of ICS is to collect shed blood from the operative field, into a sterile container where the blood may or may not be further processed, prior to returning it to the patient. To prevent the blood clotting, either the patient has to be anticoagulated prior to the operation, or anticoagulant has to be added to the blood at the sucker tip. Anticoagulation cannot be delayed until the blood has arrived in the container because, unlike the situation in normal blood donation, the coagulation factors have been activated in the operative field.

There are three basic methods for ICS: semicontinuous flow centrifugation, canister collection and single-use disposable reservoirs.

Disposable reservoirs

Shed blood is aspirated using a vacuum pressure of less than 150 mmHg, into a single-use self-contained disposable reservoir (Fig. 18.1). Unless the patient has been anticoagulated prior to surgery, the blood should be anticoagulated using citrate in a ratio recommended by the manufacturer. When the reservoir is full, the blood can either be immediately reinfused by gravity using a standard giving set and microaggregate filter, or stored at room temperature until required. This method has the advantage of being cheap and requiring no specialized equipment or specially trained personnel, but is slow, only suitable for small volumes and the product is unwashed and has a low haematocrit.

Canister collection

The blood is aspirated, via a double lumen sucker that allows anticoagulant to be mixed at the sucker tip, through a 170 μm filter into a rigid reservoir containing a disposable liner. When the canister is full, the liner is removed and the blood reinfused through a standard transfusion set and microaggregate filter. Prior to reinfusion the blood can be washed using a standard cell washer which is usually situated in the blood bank. If this is to be done, the blood must be labelled, prior to being taken from the operating theatre, with sufficient identification details to ensure that there can be no danger of the blood being returned to the wrong patient.

Semicontinuous flow centrifugation

Blood is aspirated, using a double lumen sucker, into a reservoir. The blood is washed with saline prior to being reinfused. There are two varieties of equipment available, the so-called slow flow machines which produce one unit every 7–10 minutes and the fast flow systems which produce a unit in less than 3 minutes.

Slow flow system
The basic layout for all these systems is similar (Fig. 18.2). Roller pumps transfer the blood, from the reservoir, and the

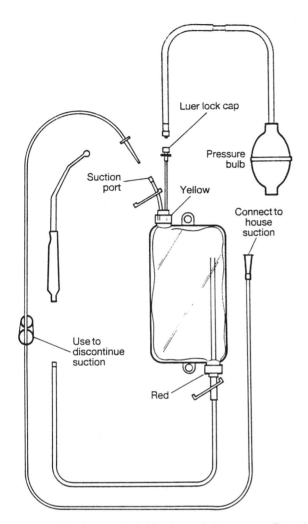

Fig. 18.1 Autologous whole blood transfusion system. (Summit Medical Ltd, Gloucestershire, England)

wash solutions into a plastic disposable centrifuge bowl. Tubing clamps control the fluid pathways and the centrifuge washes, separates and concentrates the red cells. A separate disposable plastic harness, composed of the reinfusion and waste bags, interconnecting tubing, centrifuge bowl and a double lumen aspiration assembly is used for each patient. Typically the blood is collected and held in a reservoir until this is full. The blood is then pumped into the centrifuge bowl where it is washed and concentrated prior to being pumped into the reinfusion bag.

Rapid flow system

The basic configuration is as for the slow flow systems, but the use of microprocessor technology, larger bore tubing and higher capacity pumps gives a much faster turn round.

Both these systems have the advantage of producing large quantities of blood with a high haematocrit and free from debris and other contaminants. They have the disadvantage of being expensive, requiring trained

personnel to operate and are not suitable for all paediatric patients as they have not been optimized for processing volumes of less than 125 ml blood.

Technique

Technical faults causing red cell destruction should be avoided because, in spite of the majority of haemolysed red cells being removed during the wash procedure, some free haemoglobin escapes over into the reinfused blood and the more cells that are haemolysed the less are available to be returned to the patient. To avoid haemolysis the following precautions should be taken:

1. Blood should be aspirated by placing the tip below the surface of the blood. The surgeon must avoid skimming, as aspiration of air with the blood will lead to turbulence and haemolysis.
2. The sucker should have a plastic tip with multiple holes. This is especially important during orthopaedic surgery as aspirated debris may occlude some of the holes, thereby increasing turbulence.
3. The equipment should be used in conjunction with a dedicated suction regulator to prevent the vacuum rising above 150–250 mmHg.

Anticoagulant

The manufacturer's instructions must be followed, but most advise either citrate in a ratio of 1:5 to 1:1, or heparin at a concentration of 30 000 units per litre of saline, with 15 ml of heparin/saline being added to each 100 ml of blood (ratio heparin to blood 1:7).

Wash solution

The volume of saline wash solution should be three to four times that of the volume of the bowl, although for orthopaedic cases this may be increased to six to seven times with a minimum volume of 1500 ml. There are, however, no firm data to suggest that a better product is obtained by using the larger volume of wash solution.

Reinfusion

The blood must be reinfused using a standard blood filter with or without a microaggregate filter. As there may well be residual air in the bag, the blood should not be reinfused using a pressure system unless an air dectector is incorporated in the instrument.

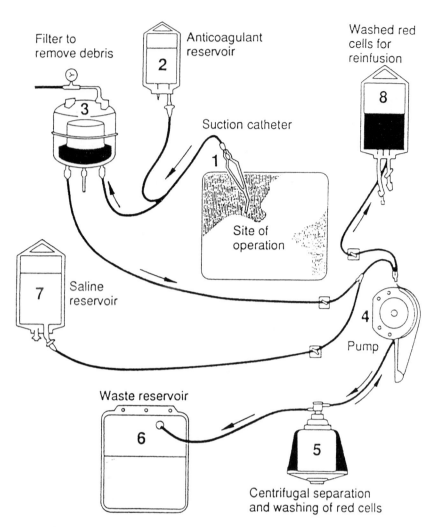

Fig. 18.2 The component parts of the Dideco cell saver blood salvage system (Dideco Ltd, Midhurst, Sussex).

Quality assurance

It is vital, in all instances, to have a detailed protocol covering the operation of the machine and its maintenance. In the United States, protocols should conform to the AABB standards[38] and similar guidelines are at present being produced for the United Kingdom. Records must be kept of each procedure both in a log, the standard details of which are in Table 18.9, and in the patient's case notes. In addition, the director of the blood programme must ensure that each machine is regularly checked for product sterility, that the concentration and wash cycles are regularly validated and that regular preventive maintenance is performed and recorded.

Indications

Theoretically ICS can be used whenever intraoperative blood loss will require either intra- or postoperative transfusion and the blood is shed into a clean wound

from which it can be retrieved at a rate permitting aspiration without due haemolysis.[18] However, this is too vague a concept for adequate planning as equipment has to be prepositioned and an operator available. A list of the more basic indications is given in Table 18.10. Using such guidelines, the Mayo Clinic provides 17 per cent of its total blood requirement from ICS.[39]

Cardiovascular

ICS was originally used mainly for cardiovascular surgery. The canister collection method should be used when the anticipated blood loss is less than 1500 ml. A semicontinuous flow centrifugal method is preferred when clinical experience indicates a greater loss.[39]

Vascular surgery

Both in the reconstruction of an aortic aneurysm and in the treatment of aortic rupture, there is a requirement for rapid salvage and return of blood. In these cases fast flow instruments are usually best. Using ICS, Hallett et al.[40]

Table 18.9 Details to be entered in intraoperative cell salvage log

Date of procedure
Operator's name
Patient's identification details
Nature of operation and name of surgeon and anaesthetist
Duration of operation
Amount of blood salvaged and reinfused
Type of anticoagulant and amount used
Machine number
Lot no. and expiry date of disposables
Any complications arising during the procedure

Table 18.10 Indications for use of intraoperative cell salvage

Anticipated blood loss greater than 20% of the patient's estimated blood volume
Homologous blood would ordinarily be cross-matched for the operation
More than 10% of patients undergoing this procedure require transfusion
The mean transfusion for the procedure exceeds 1 unit

reported that in elective aortic surgery 78 per cent of patients required no homologous blood.

Orthopaedic

In hip and spinal surgery, the use of ICS, especially when combined with predeposit donation, can usually avoid the need for homologous blood. In knee joint replacements the bleeding is usually insufficient to justify ICS; however, there may be an indication for postoperative collection.

Liver transplantation

Although early work suggested that ICS could only provide about 50 per cent of the blood requirement, recent improvements in technique mean that for many patients the total requirement for blood and blood components can be met by autologous transfusion.

Neurosurgery

Autologous transfusion has not been widely used in neurosurgery, probably because surgeons are worried about the thromboplastic potential of brain tissue or neurological tumours. There is no evidence to suggest that this anxiety is well founded but increasingly ICS is being successfully used during the refashioning of arteriovenous malformations.

Trauma

The American Association of Blood Banks have laid down guidelines[41] for the use of ICS in trauma. The main difficulties are the unpredictability of the requirement and the fact that many wounds are contaminated. Other difficulties are posed by the need to collect blood from multiple trauma sites and the need to predict, at an early stage, which patients will benefit from ICS. In spite of these difficulties ICS is being used increasingly following trauma.

Ectopic pregnancy

Thorough washing of the blood is required to prevent the reinfusion of amniotic fluid leading to emboli and disseminated intravascular coagulopathy. Twycross[42] has pointed out that ICS is especially useful in developing countries where blood stocks are scarce.

Plastic surgery

ICS is widely used in operations dealing with vascular malformations.

Jehovah's Witnesses

Although many refuse all types of transfusion, some will accept intraoperative cell salvage as long as the blood does not 'break contact' with the patient. In such cases the use of semicontinuous centrifugal machines can be life saving.

Serological antibodies

Patients with rare and/or multiple serological antibodies and/or other cross-match problems can be candidates for ICS. The most effective use is in conjunction with predeposit donations.

Contraindications

The presence of malignancy or infection were initially considered absolute contraindications to the use of ICS; however, these are now considered to be only relative contraindications.

Malignancy

Malignant cells shed into the wound can be aspirated and are not all removed or destroyed by the washing procedure. The addition of chemotherapeutic agents, such as Mitomycin-C, to the wash solutions, has failed to show any benefit. However, where the malignancy is remote from the operative site, for example a patient with breast malignancy having an abdominal aneurysm repaired, or when the malignancy is already disseminated, there is no contraindication to ICS. In addition, the possible immuno-suppressive effects of homologous transfusion and the prognosis, that is whether the surgery is designed to be curative or palliative, must be taken into account when considering the use of ICS. Hart et al.[43] describe 33 patients who underwent radical cystectomy for transitional cell carcinoma of the bladder. During the operation, ICS was used and the mean volume of blood reinfused was 1500 ml. The patients were followed up for a mean of 2 years during which time there was no evidence of tumour dissemination.

Infection

To return blood containing live bacteria to a patient is obviously hazardous, especially so in immunocompromised patients such as those undergoing liver transplantation. Experimentally it has been shown that washing, while markedly reducing the number of organisms, does not remove them all. The addition of broad-spectrum anti-biotics to the irrigation and wash solutions, kills faecal cocci and bacteroids but not other anaerobic species. However, when performing cardiac surgery on patients with endocarditis, the organism is already widely dissemi-nated in the bloodstream, and so no additional problems should be introduced when using ICS, as this is not different from recirculating blood through the cardiopul-monary bypass apparatus. Timberlake and McSwain[44] successfully used ICS in 11 trauma patients with extensive penetrating thoracoabdominal injuries. The salvaged blood, which was contaminated with abdominal contents, was washed and reinfused in conjunction with broad-spectrum antibiotics. Although, in elective procedures for which adequate homologous blood is available, it is not acceptable at present to salvage blood from an infected site, in an emergency, where there is massive blood loss, even when potentially contaminated, ICS with antibiotic cover may be appropriate, and because of the immunosuppressive effect of homologous transfusion, even preferred. It could be that the immunosuppressive effect of homologous transfusion would be more likely to lead to postoperative infection than the use of ICS. The author has contended for many years that 'it is better to have a live patient who is septicaemic, than a dead patient who is not'.

Patients suffering from viral diseases

Patients who are HIV, hepatitis B or hepatitis C positive are not harmed by ICS, but there are possible hazards for the operator and other staff within the operating theatre, as blood salvaging equipment can implode, causing the infected blood to be dispersed as an aerosol.

Product

The product is immediately available at body temperature. Blood collected by ICS has a high 2,3-diphosphoglycerate content so that the haemoglobin can easily offload oxygen. The oxygen dissociation curve is normal or even slightly right-shifted, in contrast to that of homologous banked blood which is markedly left-shifted and normal oxygen release is not achieved for 6–12 hours post-transfusion. ICS red cells have increased osmotic resistance with an excellent 24-hour post-transfusion survival, and a normal cell life ($T_{\frac{1}{2}}$ 24 days). This may be because only the younger and fitter cells survive the collection and washing procedure.

Washed saline suspended red cells

Typically 10 per cent of the red cells are haemolysed and lost in the washing procedure. The level of free haemoglobin in the salvaged blood will be between 200 and 500 mg/dl although washing removes 50–70 per cent of this. The mean cell volume, mean cell haemoglobin and mean cell haemoglobin concentration are normal and the majority of the cells, when viewed microscopically, are morphologically normal. The haematocrit of the reinfused product is between 45 and 65 per cent.[37] Platelet numbers are reduced and their function is grossly impaired, probably due to the release of β-thromboglobulin. Washing removes all the plasma proteins, including most of the clotting factors, as well as most of the anticoagulant. Although complement activation occurs during cell salvage, the washed product is complement free.

Unprocessed salvaged blood

As this blood is not concentrated, the haemoglobin level is between 7 and 9 g/dl and is sometimes as low as 4 g/dl.[37] The plasma haemoglobin is normally in the range of 60–250 mg/dl but may be up to 2000 mg/dl. There is a marked increase in fibrin degradation products and D-dimers. Anticoagulant is added to the shed blood as it is aspirated from the operative field and this is not removed during processing. Concentrations of heparin of 3 iu/ml and higher have been found in the reinfused blood.

Complications

Air embolism

Although formerly associated with the old Bentley apparatus, which is no longer manufactured, two cases of fatal embolism have been reported as recently as 1986.[37]

The newer instruments have air detectors to prevent the reinfusion of air and draining the reinfusion bag into a separate transfer pack prior to transfusion effectively eliminates this risk.

Reinfusion of haemolysed red cells

If the vacuum pressure on the sucker is too high or excess turbulence has been caused by improper aspiration technique, haemolysed red cells can be reinfused into the patient. Haemolysis is not a major problem with the centrifugal machines, where the blood has been washed, but with other methods reinfusion can lead to renal damage. The level required to cause such damage is not known.

Coagulation disorders

Prolongation of the clotting mechanism can be caused by reinfusion of residual anticoagulant or unrecognized dilution of the coagulation factors.

Thrombocytopenia

If ICS is used to process large volumes of blood which are returned to the patient, quite marked thrombocytopenia can occur. The platelet count usually returns to normal within 48–96 hours postoperatively. Thrombocytopenia can be avoided by transfusing, at the end of the operation, platelet-rich plasma (PRP) previously collected by ANH.

Microfibrillar collagen haemostatic material

These products promote platelet adherence and aggregation leading to local haemostasis. Deleterious effects could occur if such material is salvaged with the red cells and retranfused.

Hypocalcaemia

This is a rare complication when citrate is used as an anticoagulant and large volumes of blood are reinfused, particularly in patients with liver disease whose ability to metabolize citrate is depressed.

Fat embolism

Theoretically this can occur in trauma patients but is avoided by the use of microaggregate filters.

Toxic antibiotic effects

Topical antibiotics, such as neomycin, used in irrigation solutions are aspirated and could be reinfused. Although this is a potential hazard, no problems have actually been reported.

Cost-effectiveness

The equipment required for ICS, especially the rapid flow semicontinuous centrifugal machines, is extremely expensive to purchase and requires dedicated staff who must be properly trained. It is obvious that, if the case load is inadequate, such instrumentation cannot be cost-effective. When assessing cost-effectiveness, capital costs, depreciation, costs of cross-matching, staff salaries and training costs must be taken into account. Statistics on offsetting costs such as length of hospital stay, use of antibiotics, rates of postoperative infection leading to increased morbidity and mortality, not forgetting transfusion transmitted diseases, are yet to be collected. The number of units of banked blood that must be replaced to make ICS break even has been variously estimated between 2 and 8.3 units, though most estimates are between 3 and 4 units. In hospitals where it is not cost-effective to utilize centrifugal machines, savings can still be made using a canister-type blood salvage system and, if the blood bank has cell washers, complementing this by washing the blood before reinfusion.

It is probable that some form of ICS will be cost-effective in any procedure associated with a loss of more than 2 units of blood, especially when combined with predeposit donation or acute normovolaemic haemodilution.

Postoperative collection

This method, which should always be part of an overall autologous blood programme, is mainly used after cardiothoracic surgery and sometimes after joint replacement. The suction drain is connected either to a disposable collection device or to a cardiotomy reservoir. Anticoagulation is seldom required as the blood is defibrinated within the mediastinum, but if the bleeding is brisk, citrate should be added to the container. The vacuum pressure should be between 0 and 40 mmHg. Because of the dangers of bacterial contamination, the time between the start of collection and reinfusion should be less than 6 hours. The product contains no clotting factors, including fibrinogen, has a haematocrit of 15–20 per cent and contains considerable free haemoglobin and products of clot lysis. When volumes greater than 800 ml of unwashed blood are reinfused, it has been shown that platelet function is deranged and the patient develops a mild coagulopathy.

The thrombin time increases in relation to the amount of blood reinfused. A number of other complications following the reinfusion of unwashed blood have been described including upper airway oedema requiring intubation[45] and non-cardiogenic pulmonary oedema, which is believed to be due to platelet and/or complement activation leading to the capillary leak phenomenon. Roberts et al.[46] describe a series in which the use of unwashed mediastinal shed blood did not decrease the amount of homologous blood transfused, although these findings are challenged by other authors[47] who believe that any reduction in exposure to homologous blood is worthwhile. Washed blood, as used in ICS, has been shown to be safe. Williamson and Taswell[37] make the point that, as the safety of washed blood is proved, the burden of proof that unwashed is better now lies with its advocates.

The future

How and by what means can autologous transfusion progress over the next decade? The first problem that must be addressed is that of diversity. At present, autologous blood programmes conform to no set protocol. In the main they are based on a single hospital. This impedes the collection of standardized national statistical data. The feasibility of an integrated community blood programme has been shown by Giordano et al.[48] in Southern Arizona. They have increased the amount of autologous blood to 20 per cent of the centre's total collection; however, this was not done without a great deal of effort and extensive marketing skills. Nicholls[7] points out that, although requests for a national programme in Australia have been made over many years, no such programme is yet available. Little is known about the overall morbidity and mortality due to autologous transfusion. National registers must be established to allow problems to be identified and evaluated. This will ensure that any risks inherent in autologous transfusion do not outweigh the benefits of avoiding the risks of homologous blood.

At present autologous transfusion is underused[49] and the demand is patient led. If this technique is to be more widely practised an extensive education programme[50] will be required with committed blood bank staff. As autologous transfusion is more widely practised and physicians become aware of its benefits, the number of referrals into the programme will increase and a rapid rise in its usage can be expected. The Mayo Clinic collected 175 units of autologous blood from 285 patients in 1983 rising to 7850 units from 2725 patients in 1987.

At present, even in very experienced hands, at least 30 per cent of blood shed into the operative field is lost in sponges and swabs. Trials to improve this and other techniques must be initiated.

Whether white cell depletion will remove the immuno-suppressive effect of homologous blood, thereby removing an advantage of autologous blood, must be thoroughly explored. Jensen et al.[51] have recently reported data that suggest that the modulation of the acute immunological response to surgery, compromising the natural antibacterial defences, is much less if white cell depleted homologous blood, rather than whole blood, is transfused. Their results showed little difference in postoperative infection rates between those who had autologous blood and those given white cell depleted homologous blood.

On the other hand, as more and more microbiological screening tests are introduced in an effort to make homologous blood safer, the cost of each unit will rise and autologous blood will thus become more cost-effective. In addition, cheap hollow fibre blood concentration devices are being developed which could well supplant centrifugation. Initial studies suggest that these devices will be able to produce 500 ml of concentrated red cells, plasma and platelets per minute, while eliminating all particles less than 80 000 Da.

Will autologous transfusion continue to be cost-effective? Will a genetically engineered non-antigenic sterilizable artificial red cell ever be produced? As posed at the beginning of this chapter, 'does a recipient who serves as his or her own donor receive the safest possible transfusion'? Only time will tell.

REFERENCES

1. Autologous transfusion. In: Walker RH, ed. *Technical manual.* 10th ed. Arlington, Virginia: American Association of Blood Banks, 1990: 433.
2. Messmer K. Hemodilution – possibilities and safety aspects. *Acta Anaesthesiologica Scandinavica* 1988; **32** (Suppl 89): 49–53.
3. Messmer K, Sunder-Plassman L, Klovekorn W, Holper K. Circulatory significance of hemodilution: rheological changes and limitations. *Advances in Microcirculation* 1972; **4**: 1–77.
4. Hint H. The pharmacology of dextran and physiological background of the clinical use of rheomacrodex. *Acta Anaesthesia Belgica* 1986; **19**: 119–38.
5. Mezrow CK, Bergstein I, Tartter PI. Postoperative infections following autologous and homologous blood transfusions. *Transfusion* 1992; **32** (1): 27–30.
6. Friedman BA, Oberman HA, Chadwick AR, Kingdon K. The

maximum surgical blood order schedule and surgical blood use in United States. *Transfusion* 1976; **17**: 163–8.

7. Nicholls MD. Autologous blood transfusion – 1989. *Medical Journal of Australia* 1988; **149**: 515–17.

8. Kobrinsky NL, Letts M, Patel LR, *et al.* 1-Desamino-8-D-arginine vasopressin (desmopressin) decreases operative blood loss in patients having Harrington rod spinal fusion surgery: a randomized, double-blinded control trial. *Annals of Internal Medicine* 1987; **107**: 446–50.

9. Salzman EW, Weinstein MJ, Weintraub RM, *et al.* Treatment with desmopressin acetate to reduce blood loss after cardiac surgery: a double-blind randomized trial. *New England Journal of Medicine* 1986; **314**: 1402–6.

10. Bidstrup BP, Royston D, Taylor KM. Reduction in blood loss and blood use after cardiopulmonary by-pass with high dose aprotinin (Trasylol). *Journal of Thoracic and Cardiovascular Surgery* 1989; **97**: 364–72.

11. Triulzi DJ, Vanek K, Ryan DH, Blumberg N. A clinical and immunologic study of blood transfusion and postoperative bacterial infection in spinal surgery. *Transfusion* 1992; **32** (6): 517–24.

12. Blumberg N, Chuang-Stein C, Heal JM. The relationship of blood transfusion, tumor staging, and cancer recurrence. *Transfusion* 1990; **30** (4): 291–4.

13. Blumberg N, Heal JM. Transfusion and host defenses against cancer recurrence and infection. *Transfusion* 1989; **29** (3): 236–45.

14. Ferrari M, Zia S, Valbonesi M, *et al.* A new technique for hemodilution, preparation of autologous platelet-rich plasma and intraoperative blood salvage in cardiac surgery. *International Journal of Artificial Organs* 1987; **10**: 47–50.

15. Axelrod FB, Pepkowitz SH, Goldfinger D. Establishment of a schedule of optimal preoperative collection of autologous blood. *Transfusion* 1989; **29** (8): 677–80.

16. Widman FK, ed. *Standards for blood banks and transfusion services*. 15th ed. Arlington, Virginia: American Association of Blood Banks, 1993: L1.100–L1.500.

17. Guidelines for autologous blood transfusion. *Vox Sanguinis* 1989; **57**: 278–80.

18. BCSH Transfusion Task Force. Guidelines for autologous transfusion. Pt 1 – Preoperative autologous donation. *Transfusion Medicine* 1993; **3**: 307–16.

19. DePalma L, Luban NLC. Autologous blood transfusion in pediatrics. *Pediatrics* 1990; **85** (1): 125–8.

20. Richards C, Kolins J, Trindade CD. Autologous transfusion-transmitted *Yersinia enterocolitica. Journal of the American Medical Association* 1992; **268** (12): 1541–2.

21. Spiess BD, Sassetti R, McCarthy RJ, *et al.* Autologous blood donation: hemodynamics in a high risk patient population. *Transfusion* 1992; **32** (1): 17–22.

22. McVay PA, Andrews A, Kaplan EB, *et al.* Donation reactions among autologous donors. *Transfusion* 1990; **30** (3): 249–52.

23. AuBuchon JP, Popovsky MA. The safety of preoperative autologous blood donation in the nonhospital setting. *Transfusion* 1991; **31** (6): 513–17.

24. Zuck TF, Carey PM. Autologous transfusion practice. Controversies about current fashions and real needs. *Vox Sanguinis* 1990; **58**: 234–53.

25. Dzik WH, Devarajan S. Should autologous blood that tests positive for infectious diseases be used or discarded? *Transfusion* 1989; **29** (8): 743–5.

26. Zanjani ED, Ascensao JL. Erythropoietin. *Transfusion* 1989; **29** (1): 46–57.

27. Goodnough LT, Rudnick S, Price TH, *et al.* Increased preoperative collection of autologous blood with recombinant human erythropoietin therapy. *New England Journal of Medicine* 1989; **321**: 1163–8.

28. Bell K, Gillon J. Erythropoietin and preoperative blood donation. *New England Journal of Medicine* 1990; **322**: 1158.

29. Canadian Erythropoietin Study Group. Association between human recombinant erythropoietin and quality of life and exercise capacity of patients receiving haemodialysis. *British Medical Journal* 1990; **300**: 573–8.

30. Gillon J, Bell K, Prescott RJ. Autologous transfusion – too far, too soon? *Vox Sanguinis* 1991; **61**: 81–3.

31. Kruskall MS, Leonard S, Klapholz H. Autologous blood donations during pregnancy: analysis of safety and blood use. *Obstetrics and Gynecology* 1987; **70**: 938–41.

32. Tabor E. Potential risks of blood donation during pregnancy for autologous transfusion. *Transfusion* 1990; **30** (1): 76.

33. Skettino S, Ferguson K, Andrews A, *et al.* Donor room personnel attitudes toward autologous donors. *Transfusion* 1991; **31** (3): 249–53.

34. Thomas MJG. Military experience in blood supply applicable to disaster medicine. In: Castelli D, Genetet B, Habibi B, Nydegger U, eds. *Transfusion in Europe*. Paris: Arnette SA, 1990: 237–8.

35. Takaori M. Perioperative autotransfusion: haemodilution and red cell salvaging. Symposium Report *Canadian Journal of Anaesthesia* 1991; **38** (5): 604–7.

36. Pineda AA, Valbonesi M. Intraoperative blood salvage. *Baillière's Clinical Haematology* 1990; **3** (2): 385–403.

37. Williamson KR, Taswell HF. Intraoperative blood salvage: a review. *Transfusion* 1991; **31** (7): 662–75.

38. Widman FK, ed. *Standards for blood banks and transfusion services*. 15th ed. Arlington, Virginia: American Association of Blood Banks, 1993: L.2000–L.2900.

39. Williamson KR, Taswell HF. Indications for intraoperative blood salvage. *Journal of Clinical Apheresis* 1990; **5**: 100–3.

40. Hallett, JW Jr, Popovsky MA, Ilstrup DM. Minimizing blood transfusions during abdominal aortic surgery: recent advances in rapid autotransfusion. *Journal of Vascular Surgery* 1987; **5**: 601–6.

41. Stehling LC, ed. *Guidelines for blood salvage and reinfusion in surgery and trauma*. Arlington, Virginia: American Association of Blood Banks, 1990.

42. Twycross WR. Autologous blood transfusion for ruptured ectopic pregnancy. *Medical Journal of Australia* 1989; **150**: 165–6.

43. Hart OJ 3rd, Klimberg IW, Wajsman Z, Baker, J. Intraoperative autotransfusion in radical cystectomy for carcinoma of the bladder. *Surgery, Gynecology and Obstetrics* 1989; **165**: 302–6.

44. Timberlake GA, McSwain NE Jr. Autotransfusion of blood contaminated by enteric contents: a potentially life-saving measure in the massively hemorrhaging trauma patient? *Journal of Trauma* 1988; **28**: 855–7.

45. Woda R, Tetzlaff JE. Upper airway oedema following autologous blood transfusion from a wound drainage system. *Canadian Journal of Anaesthesia* 1992; **39** (3): 290–2.

46. Roberts SR, Early GL, Brown B, *et al.* Autotransfusion of unwashed mediastinal shed blood fails to decrease banked blood requirements in patients undergoing aortocoronary bypass surgery. *American Journal of Surgery* 1991; **162**: 477–80.

47. Fuller JA, Buxton BF, Picken J, *et al.* Haematological effects of reinfused mediastinal blood after cardiac surgery. *Medical Journal of Australia* 1991; **154**: 737–40.

48. Giordano GF, Dockery J, Wallace BA, *et al.* An autologous blood program coordinated by a regional centre: a 5-year experience. *Transfusion* 1991; **31** (6): 509–12.

49. Autologus transfusion. In: Walker RH, ed. *Technical manual.* 11th ed. Bethesda, Maryland: American Association of Blood Banks, 1993: 491.

50. Toy PTCY, McVay PA, Stauss RG, *et al.* Improvement in appropriate autologous donations with local education: 1987 to 1989. *Transfusion* 1992; **32** (6): 562–4.

51. Jenson LS, Anderson AJ, Christiansen PM, *et al.* Post-operative infection and natural killer cell function following blood transfusion in patients undergoing elective colorectal surgery. *British Journal of Surgery* 1992; **79**: 513–16.

Chapter 19

Haemorrhage, Blood Volume and the Physiology of Oxygen Transport

Timothy W. Rutter and Kevin K. Tremper

The physiological determinants of oxygen transport	Autologous transfusions
Minimizing blood loss	Red cell substitutes

It has long been recognized that the maintenance of optimal oxygen delivery is the core concern in anaesthesia and critical care medicine.[1] The delivery of oxygen to the cell mitochondria although a continuous process may be regarded as occurring in four phases.[2,3] During the first phase, inspired gas is transplanted to the alveoli by the mechanism of breathing, gas exchange then takes place between the oxygen in the alveoli and the bloodstream, the oxygenated blood is then transported to the tissues, and finally, diffusion between the bloodstream tissue capillaries and the cell mitochondria occurs. The complicated mechanism of respiration is necessary in order to maintain the adequate end-capillary oxygen tension necessary for aerobic metabolism. This process has evolved to deal with the limitations of oxygen delivery to cells living in a normal aqueous environment with an enormous oxygen consumption. These limitations arise from the slow process of molecular diffusion of gases in liquids and the low solubility of oxygen in water. For multicellular organisms the diffusion gradient deteriorates as the diffusion distance increases. Henry's law states that the amount of oxygen dissolved in solution is proportional to its partial pressure

(P_{O_2}). Therefore arterial blood at a P_{O_2} of 100 mmHg (13.3 kPa) contains only 0.3 ml O_2/dl. It is apparent that a human would be unable to maintain aerobic metabolism at the mitochondria by employing pure molecular diffusion over long distances in a water-based environment. The solution has been the development of a far more efficient oxygen transport system, the cardiovascular system and the blood, using haemoglobin to increase the oxygen-carrying capacity of this aqueous solution (Fig. 19.1).

The need for this efficient oxygen transplant system is reinforced by the small stores of oxygen in the body, approximately 20 ml/kg.[4–6] The functional residual capacity of adult lungs contains approximately 400 ml of oxygen which can be increased by approximately 2.35 litres by inspiring 100 per cent oxygen. In addition, arterial blood contains an oxygen reserve of approximately 1000 ml. The tissues, interstitial fluid and myoglobin probably account for further storage of 200 ml of oxygen. Oxygen stores in the blood, from the perspective of oxygen utilization, are of limited value because they have to maintain the pressure gradients between the alveoli and the mitochondria and in an emergency they are rapidly consumed in the absence of

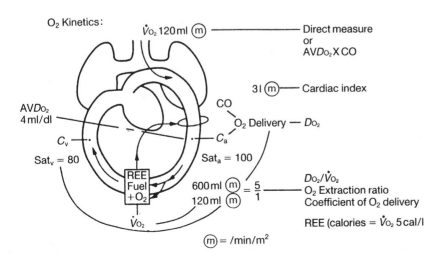

Fig. 19.1 Schematic representation of oxygen transport physiology depicting oxygen uptake in the lungs and oxygen delivery to the cardiovascular system.

an oxygen supply.[7,8] Failure of oxygen transport results in hypoxia which is defined as an inadequate tissue oxygenation due to either inadequate blood flow or low arterial oxygen content (Table 19.1).

Hypoxaemic hypoxia due to low arterial oxygen content may be the result of a decrease in haemoglobin, PO_2, or saturation of the haemoglobin molecule (SaO_2). The amount of oxygen, therefore, that is delivered to the tissues depends on the blood flow (cardiac output) and on the quantity of oxygen that is carried by the blood (O_2 content). These two variables determine systemic oxygen transport, the physiology of which will be the focus of this chapter. Within this context, how variations in blood volume and specifically haemorrhage, alter this vital function of oxygen delivery to the tissues will be discussed.

The physiology of O_2 transport is significant from another point of view. Anaesthetists transfuse more than half of the 20 million units of the blood and blood components administered each year in the United States. Although this life-saving treatment is accomplished with very little acute mortality, the risk of transmitting infectious agents with significant morbidity and mortality is present with every unit. The risk of transmitting AIDS has been estimated to be between 1:40 000 and 1:100 000 per unit.[9] The more commonly transmitted virus is non-A, non-B hepatitis, now known as hepatitis C. This virus has been associated with disease transmission at a rate of approximately 1 per cent/unit, with nearly half of these patients developing clinical hepatitis and up to one-third going on to chronic disease. Even with the recent introduction of an assay for the antibody to hepatitis C, due to the latency of seroconversion there will still be a significant rate of disease transmission.[10] Six months after clinical hepatitis as many as 40 per cent of patients have not seroconverted.[10] These factors must be part of the decision process in choosing how and when to replace blood. The importance of maintaining oxygen transport while minimizing risk to the patient requires careful consideration by the practising anaesthetist.

The physiological determinants of oxygen transport

The importance placed upon the physiology of oxygen transport is reflected in the number of variables used to measure the efficiency of the oxygen delivery system and the number of monitoring systems that have been developed to assess oxygen transport.[11,12]

Table 19.1 Classification of hypoxia

Ischaemic hypoxia	↓Blood flow, normal CaO_2
Hypoxaemic hypoxia	↓CaO_2
Hypoxic hypoxaemia	↓PaO_2, ↓SaO_2, normal Hb
Anaemic hypoxaemia	↓Hb, normal PaO_2, SaO_2
Toxic hypoxaemia	↓SaO_2, normal PaO_2 (i.e. MetHb or COHb)

Oxygen tension: inspired, alveolar and arterial

One of the simplest ways of quantifying pulmonary function with respect to oxygenation is to compare the alveolar−arterial oxygen tension difference (A−a gradient). To do this, we must first estimate the alveolar oxygen tension using the 'alveolar gas equation'.

$$PAO_2 = FIO_2 (Pb - 6.3) - PaCO_2 \times F$$
$$(PAO_2 = FIO_2 (Pb - 47) - PaCO_2 \times F)$$

where PAO_2 is alveolar oxygen tension, kPa (mmHg); FIO_2 is the fraction inspired oxygen; Pb is the barometric pressure, kPa or (mmHg); 6.3 kPa (47 mmHg) is the vapour pressure of water at body temperature (37°C); $PaCO_2$ is arterial PCO_2, estimate of alveolar PCO_2, kPa (mmHg); F is 1/respiratory exchange ratio.

The normal value for PAO_2 is 13.26 kPa (102 mmHg) at sea level. Note that the normal PAO_2 will decrease with increasing altitude (decreasing Pb), increasing $PaCO_2$ and obviously with decreasing FIO_2. Consequently, a hypercapnic patient at high altitude would have a substantially lowered PAO_2. A patient in Denver with a $PaCO_2$ of 7.8 kPa (60 mmHg) would be expected to have a PAO_2 of only 6.76 kPa (52 mmHg). Once the PAO_2 is estimated, pulmonary function can be roughly assessed by obtaining PaO_2 from an arterial blood sample and calculating an A−a gradient (normal value 0.65–1.3 kPa; 5–10 mmHg). Unfortunately, the A−a gradient itself is a function of FIO_2, increasing to approximately 13 kPa (100 mmHg) at an FIO_2 of 1.0. For this reason, the arterial/alveolar ratio (PaO_2/PAO_2) = 0.7. This a/A ratio is less dependent on FIO_2 than the A−a gradient. As with many other physiological variables, the expected 'normal' value of PaO_2 changes with the patient's age. The following expression has been used to predict an expected PaO_2 as a function of age:

$$\text{'Normal' } PaO_2 \simeq 13 - 0.04 \times \text{age(years) kPa}$$
$$= (100 - 0.3 \times \text{age (years) mmHg)}$$

This expression assumes that the patient is breathing room air at sea level and is normocapnic. Arterial hypoxaemia (hypoxic hypoxaemia) can be produced by three mechanisms: decreased PAO_2, right-to-left pulmonary shunt, or pulmonary diffusion barrier. If PAO_2 is adequate, the most common cause of hypoxaemia in perioperative patients is an increased pulmonary shunt ($\dot{Q}_{sp}/\dot{Q}_t$). Since the calculation of pulmonary shunt is based on oxygen content, it will be discussed after oxygen content and oxygen transport variables have been defined.

Oxygen content

The oxygen content of the blood is that quantity of oxygen contained in the red cell added to the quantity dissolved in the plasma and is defined as the volume of oxygen in ml carried in one dl of blood. It is normally calculated from the equation:

$$CaO_2 = Hb \times 1.37 \times SaO_2 + 0.023 \times PaO_2 \text{ kPa}$$
$$(CaO_2 = Hb \times 1.37 \times SaO_2 + 0.0034 \times PaO_2) \text{ mmHg}$$

where the letter a signifies an arterial blood sample. CaO_2 equals the arterial oxygen content in ml/dl of blood, Hb equals the haemoglobin concentration in g/dl of blood, 1.37 is equal to the volume of oxygen in ml carried by 1 g of fully saturated haemoglobin, SaO_2 is equal to the fractional haemoglobin saturation, and 0.023 and 0.0034 are equal to the solubility coefficient of oxygen in plasma (ml of oxygen/dl plasma, kPa or mmHg respectively) and PaO_2 is equal to the arterial oxygen tension measured in kPa or mmHg. The red cells' contribution to oxygen content is, therefore, governed by the concentration of active haemoglobin and

by the ability of haemoglobin to combine with oxygen, i.e. the saturation. The saturation depends on the partial pressure of oxygen and the position of the oxygen dissociation curve (Fig. 19.2).

Fractional haemoglobin saturation is defined as the ratio of oxygen haemoglobin to total haemoglobin in per cent:

$$\frac{SaO_2}{HbO_2 + Hb + metHb + COHb}$$

Multiwavelength co-oximeters are required to measure the saturation. Another haemoglobin saturation that is frequently used is referred to as functional saturation which ignores the presence of methaemoglobin and carboxyhaemoglobin:

$$SaO_2 = \frac{HbO_2}{HbO_2 + Hb}$$

This functional haemoglobin saturation is the one provided by blood gas analysis and estimates saturation from PaO_2 and the oxyhaemoglobin dissociation curve (Fig. 19.2). Since methaemoglobin and carboxyhaemoglobin do not carry oxygen only fractional saturation should be used when accurately calculating oxygen content.

With a haemoglobin of 15 g and normal values of PaO_2 and SaO_2, the arterial oxygen content is 20 ml/dl blood, 20 volumes per cent, or 9 mmol/l. Normal oxygen content of blood is very similar to that of room air at sea level.

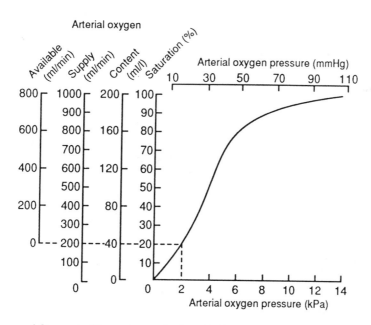

Fig. 19.2 O_2 dissociation curve and O_2 supply. The content scale gives the O_2 content (ml/l blood) assuming a normal Hb concentration of 14.5 g/dl. The 'supply' scale gives the systemic O_2 flow per minute assuming a cardiac output of 5 l/min. The 'available' scale allows for the fact that many vital tissues cannot extract the last 20 per cent of oxygen from haemoglobin because they cannot tolerate a capillary PO_2 below about 2–2.6 kPa (15–20 mmHg). Note that when the arterial PO_2 lies on the steep portion of the curve, a small increase in tension (e.g. from PO_2 of 3.3 to 5.3 kPa (25–40 mmHg) causes a large increase in saturation (from 40 to 70 per cent) and a large increase in available O_2 (from 200 to 500 ml/min).

Oxygen delivery

The total quantity of oxygen delivered to the tissues per minute is a function of the oxygen content and the cardiac output (Fig. 19.2). This overall flow rate of oxygen to the tissues is called the oxygen delivery (O_2 del).[13–15] With a normal cardiac output of 5 l/min for a 70 kg adult this will result in a normal (O_2 del) of 44.5 mmol oxygen per minute or 1000 ml oxygen per minute. Since the normal cardiac output depends on the size of the patient, cardiac output is commonly indexed to body surface area (CI = CO/BSA l/m^2 per minute) using a normal cardiac index (CI) of 3 l/m^2 per minute oxygen delivery as shown in the following equation:

$$O_2 \text{ delivery} = CaO_2 \times CI \text{ in ml } O_2/m^2 \text{ per minute}$$

In an adult patient with a normal CI of 3 l/m^2 per minute, the oxygen delivery index would be 600 ml O_2/m^2 per minute or approximately 27 mmol O_2/m^2 per minute.

This measure of the oxygen delivery index is a value of total oxygen transport but may be misleading as it does not take into consideration the regional variations in oxygen transport to specific organ systems.

Available oxygen (see Fig. 19.2)

The concept of available oxygen implies that a proportion of the oxygen supply cannot be extracted. This is determined by the need to maintain the diffusion gradients and requires a working pressure at the capillary level of approximately 2 kPa (15 mmHg). Thus, 150–200 ml of the total O_2 delivery represents the core tissue requirement below which aerobic metabolism cannot occur. The level of oxygen supply above this lower limit necessary to preclude anaerobic metabolism is of considerable importance in the management of the critically ill.[16–18]

Oxygen consumption ($\dot{V}O_2$)

Approximately one-fifth to one-quarter of all the oxygen delivered to the periphery is consumed in maintaining the metabolic functions of the body. This represents an oxygen consumption of 150 ml/m^2 per minute or 6.7 mmol O_2/m^2 per minute. Human tissue consumes approximately 5 ml of oxygen for every dl of blood flow, and since the normal arterial oxygen content is 20 ml/dl of blood, 75 per cent of the oxygen remains in the venous blood returning to the heart. The mixed venous oxygen saturation is therefore 75 per cent. The oxygen consumption ($\dot{V}O_2$) can be calculated by multiplying the arterial venous oxygen content difference of approximately 5 volumes per cent (2.2 mmol/l) by the cardiac output:

$$\text{Oxygen consumption} = (CaO_2 - C\bar{v}O_2) \times \text{cardiac output}$$

As with the oxygen delivery, the consumption may be indexed to adult body size so that the equation becomes:

$$\text{Oxygen consumption index} = (CaO_2 - C\bar{v}O_2) \times CI$$
$$\dot{V}O_2 = (20 \text{ ml/dl} - 15 \text{ ml/dl}) \times 3 \text{ l/m}^2 \text{ per min} \times 10 \text{ dl/l}$$
$$\dot{V}O_2 = 150 \text{ ml/m}^2 \text{ per min}$$

This relation between oxygen consumption, the arterial venous oxygen content difference and the cardiac output was first described by Fick and predicts that any increase in oxygen demand requires an increase in cardiac output if the arterial–venous difference is to remain constant.[19]

Mixed venous oxygen tension, saturation and content

Since we are unable to measure the mitochondrial or working tissue PO_2 in clinical practice the mixed venous PO_2 can be a basis on which to judge the adequacy of tissue oxygenation.[20,21] This has been measured by sampling expired air during rebreathing[22] or as is practice today, by sampling mixed venous blood via a pulmonary artery catheter.[23,24] The normal value in man is 5–6 kPa (39 mmHg). The normal mixed venous oxygen content of $C\bar{v}O_2$ of 12–15 vol per cent corresponds to a mixed venous haemoglobin saturation $S\bar{v}O_2$ of 72–78 per cent. The oxygen extraction ratio (O_2 ext) is the relation between the consumption of oxygen to its delivery as shown in the following equation:

$$O_2 \text{ ext} = \frac{(CaO_2 - C\bar{v}O_2) \times CI}{CaO_2 \times CI}$$

The normal extraction ratio of 25 per cent indicates a wide margin of safety built into the oxygen delivery system. If the O_2 delivery decreases, the extraction ratio from arterial blood will increase, thus lowering the $C\bar{v}O_2$. An increase in $\dot{V}O_2$ will produce the same result if there is not a compensating increase in O_2 delivery. From Fick's equation it can be seen that if CaO_2 and $\dot{V}O_2$ are constants, $C\bar{v}O_2$ (also $P\bar{v}O_2$ and $S\bar{v}O_2$) will follow changes in cardiac index. However, in acute care settings with haemorrhage and rapid fluid resuscitation there are dynamic changes in oxygen content and $\dot{V}O_2$ which negate the usefulness of $S\bar{v}O_2$ monitoring as an early indicator of low cardiac output states.[25–27] In addition, as it is a mean value for all blood returning to the heart it does not represent regional variations in flow which may distort its true significance.[28–30] When tissue perfusion limits oxygen supply there is a redistribution of blood flow from tissues with low oxygen extraction to tissues with high oxygen extraction ensuring oxygen supply to the heart and brain.

Small contributions from these sources to venous return

Table 19.2 Distribution of blood flow

Basal tissue oxygen exchange			Blood flow		
Tissue	Arteriovenous oxygen vol %	$P\bar{v}O_2$ kPa (mmHg)	Basal ml/min	Exercise ml/min	Cardiac failure ml/min
Heart	11.4 (11)*	3.0 (23)	250 (4)†	1000	300
Muscle	8.4 (30)	4.4 (34)	1200 (21)	22 000	1200
Brain	6.3 (20)	4.3 (33)	750 (13)	750	500
Splanchnic (liver)	4.1 (25)	5.6 (43)	1400 (24)	300	800
Kidney	1.3 (7)	7.3 (56)	1100 (19)	250	350
Skin	1.0 (2)	7.8 (60)	500 (9)	600	50
Other			600	100	200
Total flow			5800	25 000	34 000
Oxygen uptake			240	2000	300

* % of oxygen delivery.
† % of total cardiac output.

may lead to normal or falsely elevated readings for $S\bar{v}O_2$. Normal or elevated $S\bar{v}O_2$ values may also occur in high output states, large left-to-right shunts and in cases of cyanide poisoning. The $S\bar{v}O_2$ does not measure oxygen consumption, cardiac output or systemic oxygen delivery but is a valuable guide to the ratio of oxygen delivery to consumption[31,32] and with the above caveats $S\bar{v}O_2$ is considered by many to be a reliable monitor in acute care settings.[33,34]

Mixed venous oxygen pulmonary shunt and hypoxaemia

The concept of 'absolute' intrapulmonary shunt assumes that blood flowing through the lung is divided into two streams: one with 'ideal' oxygenation producing a PO_2 equal to PAO_2 and another which picks up no oxygen, and has a PO_2 equal to $P\bar{v}O_2$. Given this assumption, an oxygen flow balance can be calculated for the left heart as follows:

$$\text{Oxygen flow in} = \text{oxygen flow out}$$
$$\text{Ideal } O_2 + \text{shunt } O_2 = O_2 \text{ del}$$
$$CAO_2 \times \dot{Q}\text{ideal} + C\bar{v}O_2 \times \dot{Q}\text{sp} = CaO_2 \times \dot{Q}\text{t}$$

In the above equation, CAO_2 is calculated from the oxygen content equation, assuming the PO_2 equals the PAO_2. 'Ideal' represents the pulmonary blood flow of that ideally oxygenated blood. $\dot{Q}\text{sp}$ is the shunt blood flow and $\dot{Q}\text{t}$ equals the total cardiac output. Since $\dot{Q}\text{t}$ equals the sum of $\dot{Q}\text{ideal} + \dot{Q}\text{sp}$, $\dot{Q}\text{t} - \dot{Q}\text{sp}$ can be substituted into the above equation for $\dot{Q}\text{ideal}$. After algebraic rearrangement, the ratio of shunt blood flow to total blood flow ($\dot{Q}\text{sp}/\dot{Q}\text{t}$ can be calculated:

$$\frac{\dot{Q}\text{sp}}{\dot{Q}\text{t}} = \frac{\dot{C}AO_2 - \dot{C}aO_2}{\dot{C}AO_2 - \dot{C}\bar{v}O_2} \times 100 \text{ per cent}$$

In actuality, pulmonary shunt is not usually produced by two distinct blood flows. Pulmonary blood perfuses lung units with various ventilation to perfusion ratios, resulting in less than ideal oxygen uptake. Pulmonary shunt is a useful concept and a useful method of quantifying decreased pulmonary function with respect to oxygenation. It is often referred to as venous admixture because the shunt blood is venous blood being admixed to the alveolar oxygenated blood. This terminology emphasizes the importance of venous oxygenation on the ultimate arterial oxygenation in the presence of a pulmonary shunt (Fig. 19.3).

During anaesthesia some causes of hypoxaemia are obvious (ventilator disconnect, airway obstruction, etc.) while others are not (low $P\bar{v}O_2$ in the presence of a pulmonary shunt). The larger the $\dot{Q}\text{sp}/\dot{Q}\text{t}$ the greater the effect $P\bar{v}O_2$ has on PaO_2, e.g. at 100 per cent shunt $P\bar{v}O_2$ is PaO_2, and the less effect increasing FIO_2 will have on increasing PaO_2. Consequently, any lowering of $P\bar{v}O_2$ will lower PaO_2 in the presence of a shunt. As discussed earlier, $P\bar{v}O_2$ ($C\bar{v}O_2$, $S\bar{v}O_2$) decreases if there is a decrease in O_2 del (i.e. a decrease in CaO_2 or CO) or increase in $\dot{V}O_2$. Since $\dot{Q}\text{sp}/\dot{Q}\text{t}$ invariably increases under anaesthesia, blood loss is common (decreasing Hct, i.e. CaO_2) and most anaesthetic agents depress myocardial function (decreasing CO), the stage is set for hypoxaemia even in the presence of adequate ventilation with high concentrations of oxygen. Because of the potentially severe consequences of unrecognized hypoxaemia, continuous oxygen monitoring has become a standard of care.[12]

Haemoglobin and oxygen transport

The tetrameric protein haemoglobin, a respiratory pigment, is of crucial importance in the transport of oxygen and

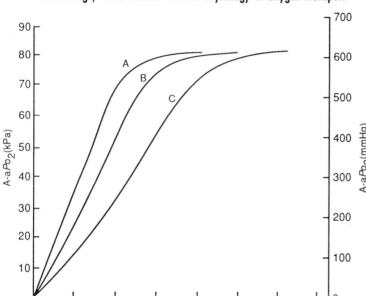

Fig. 19.3 Relation between A−a P_{O_2} and percentage right-to-left shunt when breathing oxygen. Hb = 15 g/dl, P_{CO_2} = 5.3 kPa (40 mmHg). Curves are drawn for three different arteriovenous content differences: A, 3.13 mmol/l (7 vol per cent); B, 2.23 mmol/l (5 vol per cent); C, 1.34 mmol/l (3 vol per cent).

carbon dioxide by the blood. Haem is a tetrapyrrole ring with a central iron atom and globin, an unbranched chain of amino acid units. The haemoglobin molecule is formed in the cytoplasm and the principal species in the normal adult is HbA. This denotes that the structural formula of the four globin chains is $a_2 b_2$. Other haemoglobin species may be present but usually in smaller amounts. Haemoglobin F ($a_2 g_2$) which is found in the neonate, has the property of binding oxygen more tightly than adult haemoglobin, thus its suitability for transport of oxygen in the fetus, i.e. higher saturation or lower P_{O_2}. Haemoglobin F is gradually replaced by adult haemoglobin over the first few months of life. The ability of haemoglobin to function as an oxygen transporter depends on the haem moiety or tetrapyrrole ring with a central iron atom that is able to reversibly bind oxygen. This function is used in a variety of subcellular mechanisms including the myoglobin and cytochrome systems. The haemoglobin molecule combines with four molecules of oxygen which, in theory, equals 1.39 ml/g of haemoglobin or 0.062 mmol/g Hb. There are small amounts of methaemoglobin and carboxyhaemoglobin normally present, and a more usual figure for the oxygen combining power would be 0.059 mmol/l per g or 1.34 ml/g Hb.[35,36] Blood therefore contains approximately 200 ml O_2/l or 9 mmol/l. The oxygen equilibrium curve for haemoglobin under normal physiological conditions is sigmoidal (Fig. 19.2). This results from the specific characteristics of the haemoglobin molecule due to a haem–haem interaction.[37] This interaction allows the oxygenation of one subunit of the haemoglobin molecule to alter the oxygen affinity of the other subunits. This interaction also depends on the configuration of the globin chains. The further recruitment of subunits within the haemoglobin molecule for oxygen combination will occur at a lower partial pressure of oxygen than otherwise would be required, e.g. the rapidity with which the fourth subunit of haemoglobin combines with oxygen occurs is approximately 300 times the rate for the first subunit.[38] Perhaps more importantly oxygen release from the haemoglobin will occur in the same way, i.e. the release of oxygen from the first haemoglobin subunit facilitates release of further oxygen.[39] This enhancement of oxygen uptake and release is the cause of the sigmoidal shape of the oxygen dissociation curve and defines the percentage saturation of haemoglobin at any partial pressure of oxygen.

The importance of the shape of the curve is several-fold. The rapid descent allows a large fraction of oxygen to be

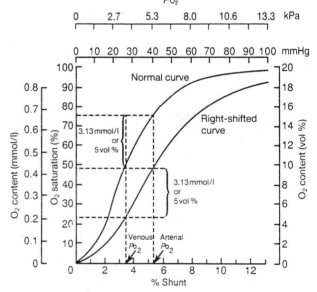

Fig. 19.4 The effect of a rightward shift of the dissociation curve when the arterial P_{O_2} is on the steep part of the curve. There is no increase in oxygen delivery when P_{50} is increased.

released to the tissues, as blood flows through them, with a modest drop in the partial pressure of oxygen. As the partial pressure of oxygen falls from an arterial level of 13.0 kPa (100 mmHg) to a venous level of 5.2 kPa (40 mmHg), 25 per cent of the oxygen is released. However, as the pressure further decreases to 2.6 kPa (20 mmHg), an additional 25 per cent is released, approximately doubling the original amount (Fig. 19.4). The relatively high 'head of oxygen pressure' at the tissue level in microvascular beds provides sufficient oxygen partial gradient pressure to ensure diffusion of oxygen to cells at the periphery.

The position of the curve is defined by the P_{50} (see Fig. 19.5), that being the partial pressure of oxygen at which the haemoglobin is 50 per cent saturated. The P_{50} at pH 7.4 is approximately 3.5 kPa or 26.6 mmHg. When the curve shifts to the right the P_{50} will increase and decreases when the curve shifts to the left. These alterations in the position of the curve greatly affect the rate at which oxygen can be delivered to the cells from capillaries. These factors are particularly important in the region of the venous P_{O_2}, on the rapidly descending portion of the sigmoidal curve, where a right shift leads to a much lower venous saturation, and thus for any given arterial venous partial pressure of oxygen difference there is much greater oxygen content difference.

The important factors determining the position of the curve are shown in Table 19.3.

Carbon dioxide

The change in oxygen affinity resulting from an alteration in carbon dioxide concentration was first described by

Table 19.3 Some causes of shifts of the O_2 haemoglobin dissociation curve

P_{50} decreased	P_{50} increased
By direct action	By direct action
Decreased [H$^+$]	Increased [H$^+$]
Decreased P_{CO_2}	Increased P_{CO_2}
Decreased temperature	Increased temperature
Abnormal Hb	Abnormal Hb
Fetal Hb	
Carboxyhaemoglobin	
Methaemoglobin	
By decreasing DPG	By increasing DPG
Increased [H$^+$]	Decreased [H$^+$]
Decreased thyroid hormone	Thyroid hormone
Hyperoxia	Hypoxaemia
Panhypopituitarism	Congestive heart failure
Blood storage	Hepatic cirrhosis

Bohr, and has come to be known as the Bohr effect (Fig. 19.6). This effect, which is independent of the effect due to the change in the pH brought about by the presence of carbon dioxide, is due to an interaction between carbon dioxide and the end terminal amino acid groups of the β-chain of the haemoglobin molecule.[40] When carbon dioxide binds, affinity for oxygen by the molecule is reduced. The converse is also true. Haemoglobin is therefore intimately involved not only in the transport of oxygen to the periphery but also in the transfer of carbon dioxide to the lungs where it is excreted. However, while the alteration in the P_{50} brought about by an increase in the partial pressure of carbon dioxide in the bloodstream is of significant proportions under normal physiological conditions, it may

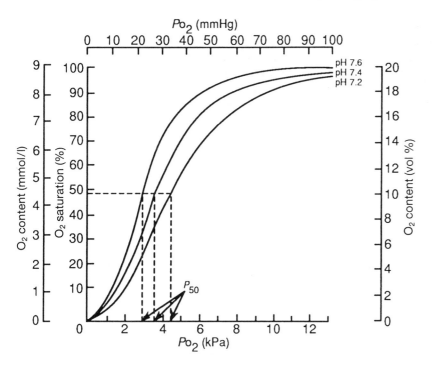

Fig. 19.5 Normal dissociation curves (after Severinghaus, 1966). The position of the curve is determined by the P_{50}.

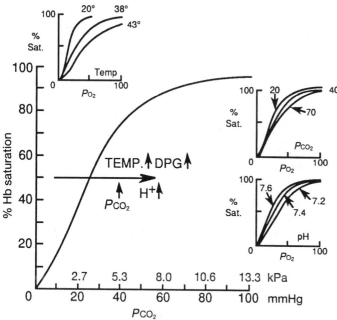

Fig. 19.6 Rightward shift of the O_2 dissociation curve by increase of H^+, P_{CO_2}, temperature, and 2,3-diphosphoglycerate (DPG).

be important in situations where oxygen saturation is low or in conditions that lead to a low level of diphosphoglycerate.[41]

Hydrogen ion

The change in the position of the oxygen saturation curve, or more specifically, the alteration of oxygen affinity as a result of an increasing hydrogen ion concentration is also known as the Bohr effect but should probably be more specifically called the hydrogen ion Bohr effect (Fig. 19.6). A reduction in the pH from 7.4 to 7.3, an increase in the hydrogen ion concentration of approximately 10 nmol/l will increase P_{50} by approximately 0.3 kPa (3 mmHg). The molecular mechanism for the Bohr effect is the interaction between the hydrogen ions and specific amino acid residues on the globin moiety, causing an alteration in their tertiary structure, such that the affinity for oxygen is changed.[40] There is, therefore, a reciprocal relation between the hydrogen ion and oxygen binding of haemoglobin.

Temperature

In addition to the effect of carbon dioxide and hydrogen ion concentration on the affinity of haemoglobin for oxygen, alterations in body temperature will also cause an alteration in the position of the oxygen dissociation curve according to the equation:

$$\frac{\Delta \log P_{50}}{\Delta T} = 0.023$$

Under these circumstances an increase in the temperature of approximately 5°C will increase the P_{50} by approximately 0.9 kPa (7 mmHg).[40]

These effects of carbon dioxide, hydrogen ion concentration and temperature have almost instantaneous effects on the position of the dissociation curve shown in Fig. 19.6. From the point of view of maintaining homoeostasis, these effects are beneficial to the organism. They enable haemoglobin to transport carbon dioxide to the lungs for excretion, and increase the buffering power of haemoglobin for the hydrogen ion. Under certain circumstances the amount of acid produced may be extremely large. In actively exercising muscle, for example, this buffering capacity is able to soak up the hydrogen ion concentration with very little change in the erythrocyte or plasma pH and allow it to be transported to the kidneys for excretion. In addition, the relation between carbon dioxide, hydrogen ion concentration and temperature allows oxygen to be delivered more effectively at the periphery to tissues with raised metabolic activity, thus facilitating aerobic metabolism.

2,3-Diphosphoglycerate

Further changes in the position of the oxygen dissociation curve result from the products of metabolism within the red cell, namely ATP and 2,3-diphosphoglycerate (2,3-DPG) (Fig. 19.6). Changes in 2,3-DPG concentration are the most important and are often seen in chronic hypoxaemia, heart disease and anaemia. Patients who have sickle cell anaemia have a chronic right shift of the oxygen dissociation curve, associated with changes in 2,3-DPG concentration, which probably allows them to withstand the effects of anaemia relatively well. 2,3-DPG is a metabolic product of anaerobic metabolism with a normal intraerythrocyte concentration of 15 μmol/g Hb. It is able to bind with the β-chains when haemoglobin is in its deoxygenated form and increases oxygen availability.[42] Increasing the 2,3-DPG concentration to twice basal erythrocyte concentrations leads to an increase in the P_{50} from 3.3 to 4.9 kPa (25 to 37 mmHg) and favours oxygen release.[39] The mechanism of action by which the levels of 2,3-DPG increase seems to depend on the level of intracellular hydrogen ion. This may result from increased intracellular or intraerythrocyte levels of reduced haemoglobin in these conditions or by an overall increase in the hydrogen ion concentration of whole blood. However, in acute situations the pH and 2,3-DPG interact to determine oxygen affinity.[43] Any change in the hydrogen ion concentration produces an alteration in oxygen affinity and a delayed and opposite effect via 2,3-DPG. Any increase in the hydrogen ion concentration will move the curve to the right which results in an intracellular diminution of the 2,3-DPG concentration, restoring the oxygen dissociation curve to its original position. The converse effect is true in alkalosis. As the effects on hydrogen ion concentration are more immediate and those of 2,3-DPG more long acting, the presence of an acidosis or alkalosis, and its attendant alteration in the oxygen dissociation curve, may alter oxygen affinity for some time until 2,3-DPG compensation occurs. In patients who have an otherwise normal physiological environment such a change may not be of great consequence. However, in patients with altered blood flow the consequences may be more severe. An example of this would be the vasoconstrictor effect of alkalosis on specific organ systems such as the brain and heart, diminishing the oxygen available to those tissues.[41]

Changes in the levels of 2,3-DPG in the erythrocyte can be of importance in transfusion physiology. During the storage of red cells there is a reduction in the levels of 2,3-DPG to extremely low levels at approximately 1–2 weeks.[3] This reduction will occur at a slower rate in citric phosphate dextrose solution than in acid citrate dextrose blood. Frozen cells retain 2,3-DPG to a greater extent than normally stored cells. Blood that has been stored in liquid form and subsequently transfused into the patient has an alteration in its oxygen affinity as shown by a decrease in the P_{50} which returns to normal over a period of 24–48 hours. The transfusion of small amounts of stored blood may be of minimal consequence, physiologically, to a patient; however, massive transfusions of blood to patients who have altered tissue blood flow may produce marked changes in oxygen transport and oxygen affinity such that oxygen delivery may be impaired.[3]

Oxygen transport from the blood to the mitochondrium

The regulation of oxygen transport between the systemic capillary blood and the tissue cell is brought about by simple diffusion in the same manner that gas exchange occurs in the lungs. Fick's law of diffusion describes the process of the molecular transport down a concentration gradient:

$$J = -D\frac{dc}{dx}$$

where J equals mass flux or rate of transport of the diffusing gas (oxygen) through a plane. D, diffusivity, is the diffusion constant in units of cm^2/s; dc/dx is the concentration gradient where c is the concentration of oxygen as a function of distance, x. To determine the volume of oxygen being transported we multiplied this equation times the surface area through which transport takes place. Additionally, to use oxygen partial pressure the solubility constant (α) of oxygen in water must be included. The above equation changes to:

$$\frac{mlO_2}{min} = AD\alpha\frac{(p_1-p_2)}{\Delta x}$$

where Δx is the diffusion distance and p_1 and p_2 the partial pressure of oxygen at the capillary wall and the mitochondria respectively. A represents the surface area through which transport takes place, i.e. the surface areas of the capillary walls. Because the diffusion distances and the area of transport are often difficult to measure, the term permeability has been substituted for the combination of area (A), diffusivity (D), solubility (α) and diffusion distance (x):

$$p = \frac{AD\alpha}{\Delta x}$$

The diffusion constant, D for all gases diffusing in liquids is in the range of 10^{-5} cm^2/s and is inversely proportional to the square of the molecular weight of the gas. The diffusion constant therefore depends on the type of tissue through which it is diffusing, and also the gas being transferred. In the lung, the size of the tissue bed and thickness of the blood gas barrier are favourable to diffusion. In the tissues, the size of the tissue bed varies significantly in terms of density of capillaries per unit tissue size and therefore, the distance for gas diffusion will vary proportionately. Oxygen delivery is therefore flow-dependent and will increase as recruitment of capillaries occurs as in actively exercising muscle.

The composition of the interstitial fluid spaces is important. The presence of oedema will increase the diffusion distance, diminish the rate of oxygen delivery and produce an unfavourable balance between supply and demand for oxygen. As the oxygen diffuses away from the capillary, it is consumed, and the partial pressure of oxygen within the tissue will fall. At a critical point the PO_2 reaches a level at which all the delivered oxygen is utilized by the tissues, and oxygen delivery is unable to meet demands. This appears to occur when PO_2 is approximately 0.4 kPa (3 mmHg), which would appear to be the working pressure for the diffusion of oxygen intracellularly to the mitochondria. The presence of a significantly higher partial pressure of oxygen at the arterial end of the capillary ensures that these diffusion gradients remain intact, that aerobic metabolism can continue and that tissue anoxia will not occur.

Blood volume

The central role of the circulating blood volume for the maintenance of oxygen delivery has been stressed. Therefore, perioperative fluid management is of critical importance in the case of the surgical patient requiring detailed evaluation by the anaesthetist. The blood makes up a unique proportion of the extracellular fluid volume being composed of extracellular water contained in the plasma and intracellular water within the erythrocytes. In the normal adult the blood volume approximates 5000 ml, of which 60 per cent is made up of the plasma and 40 per cent contained within the formed elements and constitutes the haematocrit. Compartmental analysis to measure these two fractions by 'tagged red cells' or dye dilution allows calculation of total blood volume – although of limited practical clinical value, e.g.

$$\text{Blood volume} = \text{plasma volume} \times \frac{100}{100 - 0.87 \times \text{haematocrit}}$$

The blood volume (more specifically the plasma volume) represents a small fraction (5 per cent) of the total body

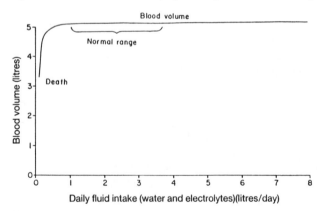

Fig 19.7 Effect on blood volume of marked changes in daily fluid intake. Note the precision of blood volume control in the normal range.

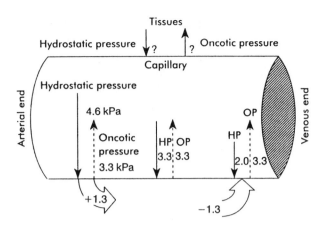

Fig. 19.8 Schematic representation of a normal capillary depicting the forces involved in the Starling equilibrium. At the arterial end, hydrostatic pressure (HP) exceeds oncotic pressure (OP) and there is a net efflux of capillary fluid. At the venous end hydrostatic pressure is reduced and the oncotic pressure causes a net reabsorption of interstitial fluid. Pressure values are in kPa. (Adapted from Maxwell MH. The nephrotic syndrome in adults. *Postgraduate Medicine* 1958; **23**: 427.)

water but it is the precise regulation of this key portion that determines the human's well-being (Fig. 19.7).

The system is volume driven such that the loss of red cells will be compensated for by an increase in the plasma volume until such time as the red cell mass can be reconstituted; likewise any alteration in plasma proteins. The reconstitution of the blood volume from interstitial and intracellular fluid reserves depends on the transcapillary forces that make up the Starling equilibrium (Fig. 19.8).

The Starling equilibrium (Fig. 19.8) is in fact a disequilibrium whereby approximately 2 ml/min of fluid is lost from the intravascular space to the interstitium due to a positive filtration coefficient of 0.87 ml/min per kPa (6.67 ml/min per mmHg) for the whole body. This sensitive homoeostatic mechanism allows major changes in transcapillary fluid exchange for relatively minor alterations in applied pressure in order to react rapidly to any challenges to loss of blood volume. However, in the case of acute haemorrhage, this may take many hours to complete, requiring that circulating blood volume be maintained, as opposed to absolute blood volume by the regulatory mechanisms shown in Fig. 19.9, particularly the need to increase cardiac index.

The regulation of oxygen transport

A number of variables therefore determine oxygen delivery in the adult patient. These are the concentration of haemoglobin, the oxygen saturation, the cardiac output and the ability to onload oxygen in the lung and offload oxygen at the periphery. The PaO_2 which is one of the

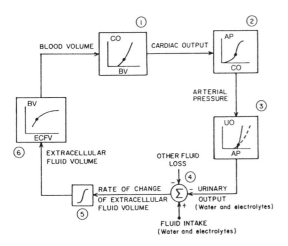

Fig. 19.9 Basic feedback mechanism for control of blood volume and extracellular fluid volume (the points on each of the curves represent normal values) where CO is the cardiac output, BV is the blood volume, AP is arterial pressure, UO is urinary output, and ECFV is the extracellular fluid volume.

factors that governs haemoglobin saturation may be followed as a measure of adequate oxygen delivery to the tissues and used to guide therapy. However, a patient can have serious physiological derangements in O_2 delivery and yet retain a normal Pa_{O_2}: haemorrhagic anaemia which may be acute or chronic is a prime example. In either case the oxygen content of the blood will be reduced but the oxygen delivery may be markedly different depending on the ability of the organism to mount a satisfactory response. This may be determined by several factors including age. Haemoglobin production is under the control of the kidneys where an unfavourable balance in the supply and demand for oxygen by renal tissue leads to the production of erythropoietin. These changes occur slowly and can approximately double the rate of red cell production from 1 to 2 per cent per day. Further, in chronic anaemia the oxyhaemoglobin curve is shifted to the right, the degree depending on the cause, and proportional to the severity of the anaemia.[44] This is due to an increase in the concentration of intraerythrocyte 2,3-DPG.[45,46] The normal physiological response to anaemia is to increase cardiac output, thereby tending to normalize O_2 delivery.[47,48] The concentration of haemoglobin at which the cardiac index begins to rise varies depending on the physiological state and age but is usually in the range of 8–9 g Hb/dl blood.[38] At these levels the fall in oxygen content can be compensated for by a moderate increase in the P_{50} as shown in Fig. 19.10. In this figure, arteriovenous content difference is an indirect measure of cardiac output, that is, cardiac output increases as arteriovenous content difference decreases.

Below these levels oxygen demand must be met by either increasing cardiac output, or by increasing oxygen extraction and allowing the mixed venous oxygen content

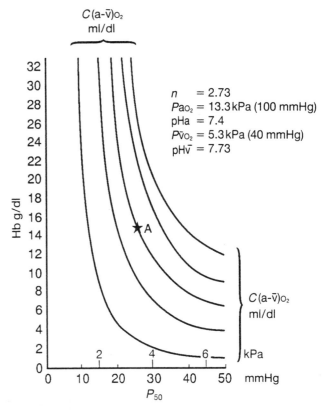

Fig. 19.10 The relation between haemoglobin concentration (Hb) and P_{50} at different levels of cardiac output, expressed as arteriovenous content differences, $C(a-\bar{v})O_2$. Point A represents normal values for these three variables. Normal arterial and mixed venous PO_2 and pH and an oxygen–haemoglobin equivalent of 1.34 ml/g have been assumed.

to fall (Fig. 19.10). Conversely, if the cardiac output is low as here measured by the $C(a-\bar{v})O_2$ difference the ability to tolerate even moderate anaemia is compromised and further reductions in Hb will lead to a further increase in the $C(a-\bar{v})O_2$.

After acute surgical haemorrhage, a patient may be anaemic, will have received a large amount of stored blood with an increased affinity for oxygen and simultaneously have a diminished circulating blood volume with a low cardiac output. This patient may have a significant reduction in oxygen delivery (DO_2), an increase in oxygen extraction and an extremely tenuous balance between supply of oxygen and its demand by the tissues for oxidative phosphorylation (Fig. 19.11).

The importance of this $\dot{V}O_2/DO_2$ ratio can be further appreciated in hypermetabolic states as found in thermal injuries when oxygen demand may outstrip oxygen delivery and oxygen delivery reserves (Fig. 19.12).

The kinetics of oxygen transport require that there be complex and as yet poorly understood compensatory mechanisms to autoregulate DO_2 and $\dot{V}O_2$. Figure 19.13 indicates that in acute onset anaemia DO_2 can only be met

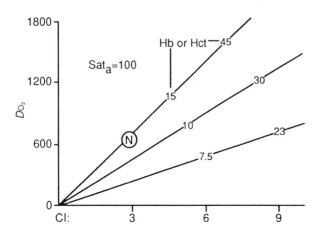

Fig. 19.11 The relationship of oxygen delivery (D_{O_2}) to changes in haemoglobin concentrations (Hb) or haematocrit (Hct) at normal saturation and changes in the cardiac index (CI).

by an increase in cardiac index; likewise, any fall in oxygen content due to acute hypoxia (Fig. 19.14). The increase in cardiac index maintains the normal ratio of D_{O_2} to $\dot{V}_{O_2}$ of 4:1 and ensures that aerobic metabolism can continue. However, an acute reduction in cardiac index cannot induce any beneficial response to facilitate oxygen delivery. Instead, oxygen demand must be met by an increase in oxygen extraction with a consequent decrease in the mixed venous oxygen content and a decrease in the $D_{O_2}/\dot{V}_{O_2}$ ratio (Fig. 19.15).

The actual $D_{O_2}/\dot{V}_{O_2}$ ratio at which oxygen consumption becomes supply dependent is a matter of conjecture but a 2:1 ratio is often regarded as a critical point.[16,17,49] Several studies indicate that below this level anaerobic metabolism with the attendant production of lactic acid occurs.[50–52]

However, the significance of oxygen delivery values may differ depending on how they are produced, i.e. variations in content versus variations in flow and distribution of flow.

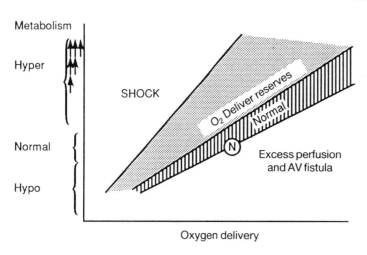

Fig. 19.12 The relationship between oxygen delivery (D_{O_2}) and oxygen consumption ($\dot{V}_{O_2}$) with increasing oxygen demand. N represents the balance between delivery of oxygen to consumption with maintenance of a normal $D_{O_2}/\dot{V}_{O_2}$ ratio. Hyperbolic states deplete oxygen delivery reserves and the normal $D_{O_2}/\dot{V}_{O_2}$ ratio to a point beyond which compensatory mechanisms can no longer provide adequate oxygenation and a state of shock is said to exist.

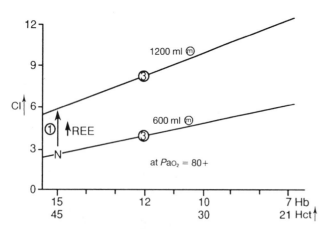

Fig. 19.13 Relation between the cardiac index (CI) and the haemoglobin curve (Hb) as haematocrit (Hct) at normal saturation. N represents the 'normal' change in cardiac index with increasing anaemia. 3 Increasing metabolic expenditure (REE) shifts the curve to a higher level of cardiac index 1 in order to match oxygen delivery to increasing demand. This effect is accentuated by increasing anaemia. Pa_{O_2} is measured in mmHg.

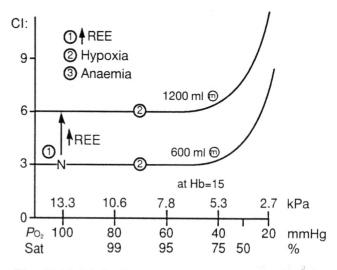

Fig. 19.14 Relation between the cardiac index (CI) and the partial pressure of oxygen (P_{O_2}) and the saturation of haemoglobin (Sat) at a normal haemoglobin concentration (Hb = 15), at two different levels of metabolic expenditure (REE) 1. ⓜ, /min per m².

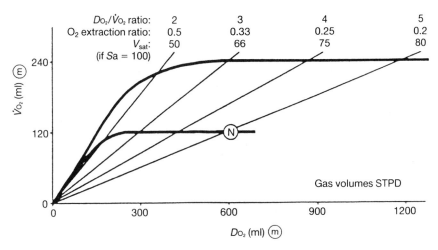

Fig. 19.15 Relation between oxygen consumption ($\dot{V}O_2$) and oxygen delivery (DO_2) with changes in oxygen delivery to oxygen consumption ratios ($DO_2/\dot{V}O_2$), oxygen extraction ratios and changes in mixed venous saturation. N represents the point on a normal curve with a $DO_2/\dot{V}O_2$ of 5. In both normal and increased metabolic states as the $DO_2/\dot{V}O_2$ approaches 2, the oxygen consumption $\dot{V}O_2$ becomes supply dependent. ⓜ, /min per m².

Several studies have demonstrated lactate acidosis without mixed venous desaturation.[53,54] Lactate acidosis and lactate ratio has also been demonstrated to be relatively insensitive to changes in haemoglobin until there is less than 6.9 g/dl and PaO_2 until it is less than 5.3 kPa (40 mmHg).[55,56]

In clinical practice when there is increased oxygen consumption it remains independent of oxygen delivery as long as there is a proportionate increase in oxygen delivery and the ratio of DO_2 to $\dot{V}O_2$ is greater than 2 (Fig. 19.15). This is seen in shock, trauma, infection and sepsis where oxygen consumption may be increased to as much as 20 per cent. In the severely traumatized septic or burned patient this may clearly exceed 100 per cent.[17,57] If the $DO_2/\dot{V}O_2$ falls below the threshold level a state of supply dependency is considered to exist. In Fig. 19.15 this threshold, the knee on the curve, occurs at approximately 330 ml/mm² per minute of DO_2 if the $\dot{V}O_2$ is 160 ml/mm² per minute.[58]

Below these levels of DO_2, increasing oxygen extraction cannot compensate for the decrease in oxygen supply and oxygen debt is incurred. Over time, organ system failure occurs and a state of circulatory shock is considered to exist. The biphasic response to the challenge of anaemia, hypoxia or falling cardiac output has been questioned.[18] It has been the view of some authors that in the severely ill, such as patients with adult respiratory distress syndrome (ARDS), that a condition of supply dependency exists at all levels of oxygen delivery DO_2.[59,60] The methodology of these experiments has been challenged on the basis of a lack of independency of the measurements made.[61] Specifically, the need to measure oxygen consumption and oxygen delivery separately and not from the same variables, i.e. CaO_2, $C\bar{v}O_2$ and cardiac output.[49] In the studies in which the measurements have been independently confirmed, the biphasic response for varying $DO_2/\dot{V}O_2$ ratios has been demonstrated.[32,50,62]

Oxygen transport transfusion therapy

Taking these matters into account, what is an acceptable level of haemoglobin in clinical practice? As the haematocrit

decreases the tissues may either extract more oxygen, if blood flow remains constant, increase blood flow if the volume of oxygen extracted is to remain constant, increase capillary recruitment or decrease oxygen consumption.[63] Blood flow usually increases due to an increase in stroke volume in normovolaemic anaemia; however, in hypovolaemic anaemia there is an increase in heart rate and possibly stroke volume. The increase in stroke volume reflects the decrease in end systolic volume as the myocardium operates against a diminished impedance.[64] Since blood viscosity decreases with haematocrit and decreases with increasing flow rate (blood is a non-Newtonian fluid) normovolaemic anaemia decreases cardiac afterload (systemic vascular resistance). This assumes intravascular volume is maintained and there is ample cardiac reserve. The effect of viscosity on flow is predicted by Poiseuille's law:

$$F = \frac{(P_1 - P_2)\pi r^4}{8\mu L}$$

where F is the flow, $P_1 - P_2$ is the pressure drop, μ is the viscosity and L is the length of the tube. From the above equation, it can be seen that flow is very sensitive to vessel radius (r) as well as changes in length, pressure drop and viscosity. What is not obvious in the above equation is that viscosity is not a constant but decreases with increasing flow, thereby giving it a positive feedback to increasing flow during progressive anaemia. The change in flow with viscosity will also reflect the total cross-sectional area of the vascular bed such that the profound effect of viscosity on flow is greatest in the postcapillary venules.[65]

However, the $P\bar{v}O_2$ at which DO_2 becomes critical in normovolaemic anaemia is considerably higher than in hypoxic hypoxaemia or ischaemic hypoxia which would indicate that the $P\bar{v}O_2$ does not accurately reflect tissue oxygenation in this circumstance. This has been interpreted as a limitation on oxygen extraction at lower levels of haematocrit due to significant reductions in the erythrocyte capillary transit time offsetting beneficial rheological effects of haemodilution.[66,67]

In clinical practice there are no prospective controlled studies which evaluate the risk or benefits of maintaining a specific preoperative haematocrit. Historically, a haematocrit of <30 per cent has been an indication for a preoperative or intraoperative blood transfusion even in a young healthy patient.[68] Although oxygen carrying capacity decreases linearly with haematocrit, physiological oxygen transport may be optimal at a haematocrit in the low 30s because of the opposing effects of decreasing carrying capacity and decreasing blood viscosity. Retrospective data have demonstrated little benefit in transfusing patients to a haematocrit of >30 per cent[69]. Non-controlled studies and case reports have demonstrated patients can survive haematocrits of 10 per cent or less if carefully managed.[70,71] Fortune et al.[72] conducted a prospective study in severely injured trauma patients maintaining haematocrits near 30 per cent in one group and near 40 per cent in a control group. They not only found no improvement in cardiopulmonary function in the patients with the higher haematocrit but in fact found an increase in pulmonary shunt associated with increased transfusions. Singbartl et al.[73] recently presented a study of ASA I–III patients undergoing elective orthopaedic procedures and requested the avoidance of homologous blood transfusions. They were monitored continuously with ST segment analysis. The 77 ASA I and II patients demonstrated no ST segment changes at haematocrits of 30 per cent, while 11 of the 73 ASA III patients showed ST segment changes at this haematocrit. When the haematocrit was reduced to 24 per cent, five of the ASA I and II patients were noted to have ST segment changes and eight additional ASA III patients developed ST segment changes. The ASA III patients' lowest haematocrit averaged 19 per cent, at which point seven additional patients developed ST segment changes (26 of 73 patients with ST segment changes). The ASA I and II patients' lowest haematocrit averaged 14 per cent which resulted in eight additional (13 of 77) patients developing ST segment changes. None of the patients developed myocardial infarction perioperatively. Three of the ASA III patients complained of chest pain and were successfully treated with blood transfusions and nitroglycerin.[73]

A recent report describes a case of haemodilutional anaemia in a patient refusing transfusion.[74] An opportunity arose to study oxygen delivery to profound levels of normovolaemic haemodilution. This critically ill patient was able to tolerate a haemoglobin value as low as 4.9/dl and these authors suggested that the critical oxygen delivery rate is 184 ml/m^2 per minute, that is, the level at which $\dot{V}O_2$ becomes supply dependent during haemodilution in humans.

Reviewing the animal literature may shed light on reasonable limits to target for transfusion therapy. Wilkerson et al. produced progressive normovolaemic anaemia in baboons to determine the limits of cardiac decompensation.[75] They found baboons tolerated anaemia to a haematocrit of 10 per cent, below which there was significant cardiac decompensation. Six of the seven baboons survived to a haematocrit of 4 per cent. Cardiac output and left ventricular coronary blood flow progressively increased as haematocrit decreased to 10 per cent. Below a haematocrit of 10 per cent left ventricular blood flow decreased and cardiac decompensation ensued. Left ventricular oxygen extraction remained relatively constant.[75] Levine et al. evaluated recovery of baboons from an experimental exploratory laparotomy with haemodilution to a haematocrit of 15 per cent.[76] All 19 baboons recovered without complication during a 2-month period of postoperative observation.

These animal data are consistent with clinical studies investigating haemodilution. Normovolaemic anaemia is well tolerated when there is adequate cardiac reserve to compensate for decreasing oxygen carrying capacity by increasing blood flow. The primary compensation involves increased stroke volume down to a haematocrit of ≈20 per cent and increased heart rate at lower haematocrits. When cardiac reserve is limited, one would assume anaemia would be less tolerated. Two studies have evaluated the haemodynamic consequences of haemodilution in rats with myocardial infarction or chemically induced myocardial depression.[77,78] In the infarction model a 25 per cent reduction in haematocrit was tolerated but the compensating increase in cardiac output with haemodilution was restricted. Any further haemodilution resulted in progressive cardiac decompensation.[77] In a model of myocardial depression induced by disopyramide, rats were haemodiluted to a haematocrit of 20 per cent.[78] As might be expected, haemodilution was progressively less well tolerated in these rats.

The following conclusions may be drawn regarding perioperative transfusion from the available animal and clinical data. Healthy patients with good cardiac function tolerate haematocrit of 20 per cent or below if adequately volume resuscitated. In patients with impaired myocardial function a haematocrit of 30 per cent may be required. Any signs of inadequate cardiac compensation must be followed closely to assess the need for transfusion.

What alternative solutions to the avoidance of homologous blood transfusion exist? They may be classified as minimizing blood loss, autologous transfusion, intraoperative blood salvage and red cell substitutes.

Minimizing blood loss

Intraoperative blood loss is primarily determined by the procedure to be performed and the surgical technique. It has been recently reported in a group of patients who had significant preoperative anaemia and who refused blood transfusion, that postoperative mortality was most closely associated with intraoperative blood loss and not necessarily preoperative haematocrit.[79] In this group of 29 patients the 12 patients who lost <500 ml had no postoperative mortality whereas those who lost >500 ml had significant postoperative mortality.[79] Although surgical blood loss is not under the primary control of the anaesthetist there are methods that may be used to reduce intraoperative blood loss: induced hypotension and treatment with drugs that effect coagulation.

Induced hypotension during a surgical procedure may accomplish two goals. First, it may decrease blood loss during the surgery and may also produce a drier surgical field. Although the use of induced hypotension to decrease blood loss was first reported in 1950, it was not until 1966 that Eckenhoff and Rich performed the first controlled study.[80] They demonstrated blood loss could be decreased by 50 per cent or more if the mean blood pressure was lowered to between 50 and 65 mmHg. Earlier studies not only had no control groups, most did not even measure blood loss but used visual estimates. In considering the use of induced hypotension the following questions might be asked: What procedures are appropriate for the technique? Where should the blood pressure be maintained? What agents are most effective in inducing controlled hypotension? What are the complications and contraindications to its use? Fortunately over the past 30 years a substantial amount of experimental and clinical data have been presented to answer these questions. The benefits of induced hypotension for minimizing blood loss have been demonstrated most dramatically in major orthopaedic procedures, total hip arthroplasty and spine surgery. The technique has also been used effectively in procedures involving the head and neck and major cancer surgery. Although some studies have shown variable effects on blood loss, the majority of studies have demonstrated at least a 50 per cent decrease in blood loss. Induced hypotension has been shown to be more effective than intraoperative haemodilution in decreasing the need for blood transfusion.[81] The target mean arterial pressure is usually between 60 and 65 mmHg measured relative to the surgical site, although this blood pressure may be titrated to a visual observation of bleeding at the surgical site. It is also important, if possible, to elevate the surgical site to improve venous drainage. The target blood pressure of 50–65 mmHg is chosen because it approaches the lower limit of cerebral autoregulation. In patients with poorly treated hypertensive disease the autoregulatory curve is shifted to the right and the lower limit for induced hypotension should be elevated accordingly. It has been demonstrated that it is indeed the lowering of blood pressure and not cardiac output that affects blood loss.[82] A variety of methods have been used to lower blood pressure including infusion of sodium nitroprusside, nitroglycerin, trimetaphan, adenosine, α- and β-adrenergic blockade, calcium channel blockade, all the inhalation agents, spinal and epidural anaesthesia. Combinations of the above methods have also been used. The most commonly used agents are sodium nitroprusside, trimetaphan and high dose isoflurane. Nitroprusside has the advantage of rapid onset and short duration but has the disadvantage of the potential toxicity of its breakdown product, cyanide. If doses are maintained at 8 µg/kg per minute or less, cyanide toxicity does not appear to be a risk. Nitroprusside has other disadvantages in that it induces stimulation of the sympathetic nervous system and the renin–angiotensin system. This activation contributes to apparent resistance which necessitates increasing the infusion rate of nitroprusside. Pretreating patients with propranolol has been shown to reduce the amount of nitroprusside required. More recently esmolol has been used to accomplish the same effect. Captopril pretreatment (3 mg/kg) has also been used in an analogous fashion to block effects of the stimulation of the renin–angiotensin system. Ganglionic blockade with infusion of trimetaphan is also effective in producing hypotension without reflex stimulation of the sympathetic nervous system. It does, however, produce its hypotensive effect by decreasing cardiac output along with peripheral vascular resistance. Nitroprusside has the advantage of maintaining cardiac output. Isoflurane would appear to be the inhalation agent of choice for it produces a dose-dependent reduction in blood pressure while maintaining cardiac output and decreasing cerebral metabolic oxygen consumption. The myocardial depression of isoflurane is more pronounced in the elderly.

Regardless of the agent used, because of the importance of tight control of blood pressure, invasive arterial blood pressure monitoring is recommended, unless only very mild degrees of hypotension are planned. This will also allow for the frequent measurement of blood gases and haematocrit. Although induced hypotension to some degree may be appropriate for all patients undergoing procedures with significant anticipated blood loss, the technique is not recommended for patients with cerebral or coronary artery disease. It is also relatively contraindicated in anaemic patients especially when using agents which block the physiological compensations for anaemia. (Recent animal studies investigating the combination of haemodilution and adenosine-induced hypotension will be discussed below.[83,84]) Consequently, the degree of hypotension and the method by which it is induced should be titrated to the specific patient and procedure.

Drug therapy: aprotinin, ε-aminocaproic acid, tranexamic acid, desmopressin and erythropoietin

The fibrinolytic enzyme system functions as a basic defence mechanism to control the excess deposition of fibrin. The fibrinolytic system acts in balance with the coagulation system through a series of activators and inhibitors. The key substance in the fibrinolytic system is plasminogen which when activated releases the proteinase plasmin which cleaves peptide bonds in fibrin, producing fibrinolysis. Over the past decade research has been directed toward purifying substances which activate fibrinolysis (streptokinase, urokinase, TPA, etc.) and other substances which inhibit the fibrinolytic system. Synthetic inhibitors of fibrinolysis include ε-aminocaproic acid and tranexamic acid. Aprotinin is a naturally occurring proteinase inhibitor with inhibitory effects on human plasmin, trypsin, and tissue kallikrein. In recent years, these drugs have been investigated for use in the treatment of postoperative bleeding disorders, e.g. bleeding associated with a cardiac bypass and prostatic surgery. ε-Aminocaproic acid and tranexamic acid, the two synthetic agents, act at the same binding site on plasminogen and exert their inhibitory effect. Tranexamic acid is approximately seven times as potent as ε-aminocaproic acid and therefore can be used in lower doses with fewer side effects. The usual dose of tranexamic acid is 0.5–1 g (10–15 mg/kg) intravenously two to three times daily immediately following urological surgery for 2–3 days. An ε-aminocaproic acid dose of 2.5 g pre- and postbypass has been demonstrated to reduce blood loss following cardiopulmonary bypass by 10–20 per cent.[85,86] Although one might be concerned about an increased risk of thrombotic complications, none was noted in these studies. These drugs are contraindicated in patients with suspected disseminated intravascular coagulation (DIC) or in patients in whom fibrinolysis is necessary to prevent potentially fatal thrombosis.

Aprotinin has been used in recent studies to increase clotting and decrease fibrinolytic activity following cardiopulmonary bypass for cardiac surgical patients. It has also been speculated that high dose aprotinin may have a platelet protective effect during cardiac bypass. Royston et al. reported that high dose aprotinin decreased blood loss in patients undergoing repeat open heart surgery from 1500 ml in the control group to 280 ml in the aprotinin group.[87] The dose of aprotinin was 280 mg prior to sternotomy followed by a 70 mg/h infusion with an additional 280 mg added to the pump prime solution. Although the exact dose and timing of aprotinin treatment is yet to be determined, it is evident that a portion of the dose must be given prior to initiation of cardiac bypass. The reduction in blood loss has been confirmed by multiple other controlled blinded studies.[88] Aprotinin also substantially increased the activated clotting time (ACT) when the usual celite activator is used.[89] It has been recommended that when the celite activator is used in the ACT device to assess the adequacy of heparinization that the ACT be greater than 750 seconds (instead of greater than 400 seconds) before extracorporeal circulation is initiated.[88] It has been recently demonstrated that if kaolin is used as the activator in determining the ACT normal, ACT values should be used. The prolonged ACT is an interaction between celite, heparin and aprotinin and falsely increases the ACT value.

Desmopressin acetate (DDAVP) is a synthetic analogue of L-arginine vasopressin (antidiuretic hormone). It has been demonstrated to increase von Willebrand's factor, increase factor VIII activity, and have some non-specific enhancement of platelet function. It also has been reported to decrease postcardiac bypass blood loss in a dose of 0.3 µg/kg.[90] In spite of initial enthusiasm for its use, follow-up studies have not found a significant reduction in blood loss with DDAVP treatment. DDAVP is indicated in the prophylactic treatment of patients with haemophilia A and type I von Willebrand's disease. It is not effective in types II or III.[91]

Epogen is human recombinant erythropoietin which is identical to the human erythropoietin produced in the kidney in response to anaemia (tissue hypoxia) which in turn stimulates the bone marrow to produce red blood cells. This drug is now available and is currently being used to treat patients with renal failure and other diseases that result in chronic anaemia. Patients we see in the operating room with chronic renal failure may already be on treatment with erythropoietin to maintain their haematocrit in the normal range. The major side effects seen in 50 per cent of these patients is hypertension. Erythropoietin treatment has also been used to increase the volume of red cells obtained during preoperative autologous donation before elective surgery.[92] The efficacy of preoperative erythropoietin treatment in general is being evaluated for patients who are to undergo procedures with significant anticipated blood loss. Since it usually requires 5–7 days after the onset of acute anaemia before a significant reticulocyte count is generated, it would appear reasonable that preoperative treatment with erythropoietin may be of use to accelerate RBC production. Results of these studies are pending.

Autologous transfusions

Autologous transfusion is discussed in Chapter 18 and therefore is merely touched on in this chapter for completeness.

There are three methods of acquiring autologous blood for transfusion: preoperative donation and storage, acute preoperative phlebotomy and haemodilution, and perioperative blood salvage from the surgical site. Preoperative donation has become routine at most medical centres for elective procedures which may require blood transfusion. In 1991 there will be an estimated 500 000 units predonated in the US. It has the advantages of avoiding transfusion reactions (excluding clerical errors), disease transmission (bacteraemia is an absolute contraindication), and it stimulates red blood cell production (because it stimulates endogenous erythropoietin production). Long-term storage in the form of frozen red cells is particularly useful for patients with unusual antibodies. For an in-depth discussion refer to the American Assoication of Blood Banks manual.[93] It should be kept in mind that autologous blood donation is not an inexpensive alternative. At the University of Michigan the patient charge is $260/unit of autologous or homologous blood including the cost of testing. A few aspects to consider regarding autologous transfusion are that it requires approximately 72 hours to normalize plasma proteins so the last donation should be completed at least 3 days prior to surgery. All patients should receive iron supplements, 325 mg three times a day. 'High risk' patients may be able to donate. Mann et al. reviewed the files of 342 high risk patients who requested autologous transfusion and found a 4 per cent rate of mild adverse reactions, which is the same rate as normal donors.[94] Finally, it is currently recommended that even an autologously donated unit of blood should not be transfused unless required. This is because the most likely cause of a lethal haemolytic transfusion reaction is due to clerical errors and this may occur with autologous as well as homologous blood transfusions.[95]

Acute preoperative phlebotomy and isovolaemic haemodilution is accomplished by removal of blood and simultaneous resuscitation with crystalloid or colloid. The purpose is to lower the haematocrit but not the vascular volume prior to surgical blood loss and then have the ability to transfuse several units of the patient's fresh whole blood after the major blood loss has been completed. It may be necessary to retransfuse sooner if the haematocrit is approaching an unacceptable level. There are several advantages to this technique. First, it is fast, easy, inexpensive (≈$15/unit), and takes less planning than preoperative autologous donation. It has the obvious limitations of being able to provide a limited number of units of blood with decreasing haematocrit in each, and can only be contemplated in patients who are not anaemic preoperatively. Since it is in effect producing acute anaemia, it has all the same relative contraindications, e.g. significant cardiac and cerebral vascular disease, although it has been used successfully in patients undergoing coronary artery surgery. Since it will dilute coagulation factors as well as platelets and red blood cells, it would be contraindicated in patients with a coagulopathy. Finally, since it will also involve diuresis of the haemodilution fluid, patients with significant renal disease would also not be good candidates. The limits to which a patient should be haemodiluted are similar to those limits of intraoperative anaemia in general, with the additional concern that patients are going to have ongoing blood loss with a surgical procedure. It would be of little use to have to immediately reinfuse the blood which has been withdrawn to maintain an acceptable haematocrit.

Crystalloid (with a 3:1 volume replacement) and colloid (with a 1:1 volume replacement) have been used as the haemodilution fluid. Crystalloid has the advantage of being inexpensive and easily diuresed. Colloid has the advantage of requiring less volume because it maintains intravascular volume. In addition, haemodilution with dextran has been shown to decrease the blood viscosity and thereby improve the fluidity of the blood.[96] If hetastarch is used it has been recommended that the dose be maintained below 20 ml/kg because of potential effects on factor VIII activity which appear to exceed the effects of haemodilution.[97] This effect may not be associated with pentastarch.[98] The amount of blood which may be withdrawn can easily be estimated from the following equation if the patient's current blood volume (BV), the patient's initial haematocrit (Hi), and the desired endpoint haematocrit (He) are known.

$$\text{blood volume withdrawn} = BV \times \underline{Hi - He}$$

$$\text{where } Hav = \frac{Hi + He}{2}$$

For an adult, 70 kg patient starting with a haematocrit of 40 per cent and ending with a haematocrit of 30 per cent, approximately 1500 ml may be withdrawn. Blood is collected in standard blood bags containing anticoagulant. It is most convenient to have a scale to weigh the blood as it is being withdrawn to keep a continuous estimate of the amount of resuscitation fluid required. The procedure may be accomplished in healthy patients using two large bore intravenous lines. It is convenient to have an arterial catheter for continuous monitoring of blood pressure and frequent sampling of blood for haematocrit. It is also convenient but not necessary to have a central venous catheter to assess the adequacy of resuscitation. A pulmonary artery catheter may or may not be required depending on the preoperative condition of the patient. A urinary catheter is also necessary but this would usually be standard for any patient undergoing a large blood loss procedure. After the blood is withdrawn it is labelled and kept in the operating room at room temperature. It may be kept safely at room temperature for up to 6 hours. If it is anticipated that the blood will not be transfused within 6 hours it should be refrigerated and used within 24 hours. As the blood is withdrawn it should be agitated slightly to withdrawal the initiation of tachycardia may indicate inadequate volume resuscitation.

Anaemia is usually considered a relative contraindication to the application of induced hypotension but haemodilution has been used in conjunction with induced hypotension for blood conservation. The combination of acute haemodilution and induced hypotension has been investigated in animals. Crystal *et al.* induced hypotension to 50 per cent of the control blood pressure with adenosine after the dog haematocrit had been haemodiluted to half of the control value.[84] The authors found that adenosine preserved adequate myocardial oxygen supply to demand ratio due to its vasodilatory effect on the coronary blood vessels.[85] There was a reduction in oxygen delivery to other organs including the brain and kidney. One would conclude from these data that the combination of haemodilution and induced hypotension might only be appropriate in healthy patients and only undertaken to a mild degree, i.e. only moderate hypotension during moderate haemodilution.

Intraoperative blood salvage has become routine in many institutions, initially for cardiac surgery but now for a wide variety of procedures in which blood can be retrieved from the surgical site. An estimated 360 000 patients per year use this technique. It is relatively contraindicated in cases where there may be bacterial or malignant cell contamination. There are three basic techniques of intraoperative salvage and reinfusion: semicontinuous flow centrifugation, canister collection with disposable liner, and single-use self-contained revision. The first type is the most common and results in washed red cells with a haematocrit of ≈60 per cent. It costs ≈$200 for the disposables but multiple units of blood may be processed. The canister and self-contained type reinfuse unwashed blood but few data are available on the quality of this product. The blood from the canister type can be sent to the blood bank for washing. None of these transfused products will have functioning platelets or coagulation factors.

Red cell substitutes

Two types of artificial oxygen transporting fluids (red cell substitutes) have been under investigation: stroma-free haemoglobin (SFH), and perfluorochemical emulsions (PFC). Both have potential clinical applications but neither has developed to the point of clinical utility. Perfluorochemicals are inert, immiscible liquids which have an oxygen solubility of approximately 20 times that of normal plasma. These inert fluids must be emulsified into aqueous electrolyte solutions, forming a suspension of small (0.1 μm) particles. There has been difficulty producing stable emulsions and to date the suspensions that have been available for clinical use only contain 10 per cent perfluorochemical on a volume basis. Therefore the emulsions themselves are 'anaemic'. Additionally, since they transport oxygen by direct solubility, to be effective they require very high PaO_2 values. In the clinical studies that have been conducted patients with fluorocrits (volume per cent of perfluorochemical in the blood) of approximately 2 per cent have been able to transport an additional 1.5 volume per cent of oxygen at a PaO_2 of nearly 66.5 kPa (500 mmHg).[71] This is approximately the same contribution as 1 g/dl of saturated haemoglobin.

Haemoglobin solutions have been studied since the turn of the century. Unfortunately, when haemoglobin is removed from the red blood cell and purified its P_{50} is reduced to the range of 1.6–1.9 kPa (12–14 mmHg) (normal 3.6 kPa; 26.7 mmHg) so though it takes up oxygen it cannot readily release it. Also, due to the small size, the dissociated α and β haemoglobin chains are quickly diuresed. Recently, human recombinant haemoglobin has been developed as a blood substitute.[99] The α-chains of this recombinant haemoglobin are modified. The addition of one amino acid links the two α-chains together which maintains the tetrameric structure of haemoglobin when it is in solution. This prevents the haemoglobin from being diuresed. The current product has a P_{50} of 4.3 kPa (32 mmHg) and is being tested in a solution of 7 g/dl with a normal oncotic pressure of 3.3 kPa (25 mmHg). This recombinant haemoglobin solution could be a major breakthrough. Both of these substitutes are cleared from the vascular space by the reticuloendothelial system within an approximate 24-hour half-life. Therefore neither will provide oxygen transport capability for very long. Although the progress to date looks promising, there is still significant work required before a haemoglobin solution is available for clinical use.

Anaesthetists' attitudes to anaemia and transfusion will probably require re-evaluation in the 1990s. Overall, it would appear that the most cost-effective ways of avoiding homologous blood transfusion are to avoid blood loss and to employ acute haemodilution. These techniques also eliminate the risk of disease transmission. In the future there may be more aggressive use of drugs and ultimately of blood substitutes, not only to avoid iatrogenic disease but to ensure that oxygen transport requirements are achieved.

REFERENCES

1. Barcroft J. Presidential address on anoxaemia. *Lancet* 1920; **2**: 485–9.

2. Nunn JF. *Applied respiratory physiology*. London: Butterworth, 1987.

3. Sykes, MK, McNicot MW, Campbell EJM. *Respiratory failure*. Oxford: Blackwell Scientific, 1976.

4. Cross CE, Packer BS, Altman M, *et al.* The determination of total body exchangeable oxygen stores. *Journal of Clinical Investigation* 1968; **47**: 2402–10.

5. Harris P. Lactic acid and phlogiston debt. *Cardiovascular Research* 1969; **3**: 381–90.

6. Finch CA, Lenfant MD. Oxygen transport in man. *New England Journal of Medicine* 1972; **286**(8): 407–15.

7. deHaan EJ. Cell respiration. *Folia Medica Neerlandica* 1970; **13**: 90–100.

8. Baue AE, Soyeed MM. Alterations in the functional capacity of mitochondria in hemorrhagic shock. *Surgery* 1970; **68**: 40–7.

9. Alter MJ, Pascell RM, Shih JW, *et al.* Detection of antibody C virus in prospectively followed transfusion recipients with acute and chronic non-A, non-B hepatitis. *New England Journal of Medicine* 1989; **321**: 1494.

10. Alter MJ, Hadler SL, Judson FN, *et al.* Risk factors for acute non-A, non-B hepatitis in the United States and association with hepatitis C virus infection. *Journal of the American Medical Association* 1990; **264**: 2231.

11. Tremper KK, Barker SJ. Monitoring of oxygen. In: Lake C, ed. *Clinical monitoring*. London, Philadelphia: WB Saunders, 1990.

12. Eichhorn JM, Cooper JB, Cullen DJ, *et al.* Standards for patient monitoring during anesthesia at Harvard Medical School. *Journal of the American Medical Association* 1986; **1256**: 1017–20.

13. Richards DW. *The circulation in traumatic shock in man*. *Harvey Lectures* 1943–44; **39**: 217.

14. Nunn JF, Freeman J. Problems of oxygenation and oxygen transport during hemorrhage. *Anaesthesia* 1964: **19**: 206.

15. Sullivan SF. Oxygen transport. *Anesthesiology* 1972; **37**: 140.

16. Bartlett RH. Critical care. In: Greenfield LJ, Mulholland MW, Oldham KT, Zelenock GB. *Surgery, scientific principles and practice*. Chapter 8. Philadelphia: JB Lippincott, 1993: 195–222.

17. Bartlett RH. Critical care and metabolism. *Bulletin of the American College of Surgeons* 1989; **74**: 10–15.

18. Cain SM. Acute lung injury. Assessment of tissue oxygenation. *Critical Care Clinics* 1986; **2**(3): 537–50.

19. Wade OC, Bishop JM. *Cardiac output and regional blood flow*. Oxford: Blackwell Scientific Publications Ltd, 1962.

20. Cain SM. Oxygen transport and consumption. In: Tinker J, Zapel WM, eds. *Care of the critically ill patient*, 2nd ed. Berlin: Springer Verlag, 1992: Chapter 5.

21. Barcroft J. *Features in the architecture of physiological function*. Cambridge: Cambridge University Press, 1938: 71.

22. Cerretelli P, Cruiz JC, Fashi LE, *et al.* Determination of mixed venous O_2 and CO_2 tensions and cardiac output by rebreathing method. *Respiratory Physiology* 1966; **1**: 258–64.

23. Scheinmann MM, Brown MA, Rapaport E. Critical assessment of use of central venous oxygen saturation as a mirror of mixed venous oxygen in severely ill cardiac patients. *Circulation* 1969; **40**: 165–72.

24. Muir AL, Kirby BJ, King AJ, *et al.* Mixed venous oxygen saturation in relation to cardiac output in myocardial infarctions. *British Medical Journal* 1970; **4**: 276–8.

25. Norfleet EA, Watson CB. Continuous mixed venous oxygen saturation measurement: a significant advance in hemo-dynamic monitoring? *Journal of Clinical Monitoring* 1985; **1**: 245–58.

26. Divertie MB, McMichan JL. Continuous monitoring of mixed venous oxygen saturation. *Anesthesiology* 1984; **85**: 423–8.

27. Birman H, Hag, A, Hew E, *et al.* Continuous monitoring of mixed venous oxygen saturation in hemodynamically unstable patients. *Anesthesiology* 1984; **86**: 753–6.

28. Carron CC, Snyder JV. Hyperdynamic severe intravascular sepsis depends on fluid resuscitation in cynomolgus monkey. *American Journal of Physiology* 1982; **243**: R-131–41.

29. Miller MJ. Tissue oxygenation in clinical medicine: an historical review. *Anesthesia and Analgesia* 1982; **61**(6): 527–35.

30. Lee MD, Wright F, Barber R, Stanely L. Central venous oxygen saturation in shock. *Anesthesiology* 1972; **36**(5): 472–8.

31. Zwischenberger JB, Colley RE, Kirsh MM, Decker RE, Bartlett RH. Does continuous monitoring of mixed venous oxygen saturation accurately reflect oxygen delivery and oxygen consumption following coronary artery bypass grafting? *Surgical Forum* 1986; **37**: 66–8.

32. Cilley RE, Polley TZ, Zwischenberger JB, Toomasian JM, Bartlett RH. Independent measurement of oxygen consumption and oxygen delivery. *Journal of Surgical Research* 1988; **47**: 242–7.

33. Nelson LD. Continuous venous oscimetry in surgical patients. *Annals of Surgery* 1986; **203**: 329–33.

34. Shenaq SA, Casar G, Ott H, Crawford ES. Continuous monitoring of mixed venous oxygen saturation during aortic surgery. *Chest* 1987; **92**: 796–9.

35. Gregory IC. The oxygen and carbon monoxide capacities of foetal and adult blood. *Journal of Physiology (London)* 1974; **236**: 625.

36. Von Hufner CG. Neue Versuche zur Bestimmung der Sauerst-off Capacitah des Blut Forbstoffs. *Archiv für Anatomie und Physiologie Physiol. Abt*, 1984; 130.

37. Adair GS. The hemoglobin system vs. the oxygen dissociation curve of hemoglobin. *Journal of Biological Chemistry* 1925; **63**: 529.

38. Gillies IDS. Anaemia and anaesthesia. *British Journal of Anaesthesia* 1974; **46**: 589.

39. Perutz MF. Stereochemistry of coopertive efforts in hemo-globin. *Nature (London)* 1970; **228**: 726.

40. Woodson RD. Hemoglobin structure and oxygen transport. *Principles of transfusion medicine*. Baltimore: Williams and Wilkins, 1991.

41. Bellingham AJ, Deto JC, Lenfant C. Regulatory mechanisms of oxygen affinity in acidosis and alkalosis. *Journal of Clinical Investigation* 1971; **50**: 700.

42. Hamasaki N, Asakura T, Minakami S. Effect of oxygen tension on glycolysis in human erythrocytes. *Journal of Biochemistry (Tokyo)* 1970; **68**: 157–61.

43. Rapaport S. The regulation of glycolysis in mammalian erythrocytes. *Essays in Biochemistry* 1965; **1**: 69–103.

44. Torrance J, Jacobs P, Restrepo A, Eschbach J, Lenfant C, Finch CA. Intraerythrocytic adaptation to anemia. *New England Journal of Medicine* 1970; **283**: 165.

45. Greenwald I. A new type of phosphoric acid compound isolated from blood, with some remarks on the effect of substitution on the rotation of L-glyceric acid. *Journal of Biological Chemistry* 1925; **63**: 339.

46. Bluesch R, Bluesch RE. The effect of organic phosphates from the human erythrocyte on the allosteric properties of hemoglobin. *Biochemical and Biophysical Research Communications* 1967; **26**: 162.

47. Sharpey Schafer EP. Cardiac output in severe anemia. *Clinical Science* 1944; **5**: 125.

48. Murray JF. Anemia and cardiac function. *UCLA Forum Medical Science* 1970; **10**: 309.

49. Bartlett RH, Dechert RE. Oxygen kinetics: pitfalls in clinical research. *Journal of Critical Care* 1990; **5**(2): 77–80.

50. Cilley RE, Scharenberg AM, Bongiorno PF, Guire, KE, Bartlett RH. Low oxygen delivery produced by anemia, hypoxia and low cardiac output. *Journal of Surgical Research* 1991; **51**: 425–33.

51. Simmons DH, Alpas AP, Tashkin DP, Coulson A. Hyperlactemia due to arterial hypoxemia or reduced cardiac output or both. *Journal of Applied Physiology* 1978; **45**: 195–202.

52. Kasnitz P, Druger GL, Yorra F, Simmons DH. Mixed venous oxygen tension and hyperlactemia. *Journal of the American Medical Association* 1976; **236**: 570–4.

53. daLuz PL, Cavanilles JM, Michaels S, Weil MH, Shubin H. Oxygen delivery, anoxic metabolism and hemoglobin oxygen affinity (P_{50}) in patients with acute myocardial infarction and shock. *American Journal of Cardiology* 1975; **36**: 148–54.

54. Cohen RD, Simpson R. Lactate metabolism. *Anesthesiology* 1975; **43**: 661–73.

55. Seibert DJ, Ebaugh FG Jr. Assessment of tissue anoxemia in chronic anemia by the arterial lactate pyruvate ratio and excess lactate formation. *Journal of Laboratory and Clinical Medicine* 1967; **69**: 177–82.

56. Green NM, Talner NS. Blood lactate pyruvate and lactate pyruvate ratios in congenital heart disease. *New England Journal of Medicine* 1964; **270**: 1331–6.

57. Wolf GT, Coter S, Perel A, Manny J. Dependence of oxygen consumption on cardiac output in sepsis. *Critical Care Medicine* 1987; **15**(3): 198–203.

58. Shibutani K, Kamatsu T, Kubal K, Sanchala V, Kumar V, Bizzarri DV. Critical level of oxygen delivery in anesthetized man. *Critical Care Medicine* 1983; **11**(8): 640–3.

59. Danek SJ, Lynch JP, Weg JG, Dantzker DR. The dependence of oxygen uptake on oxygen delivery in the adult respiratory distress syndrome. *American Review of Respiratory Diseases* 1980; **122**: 387–95.

60. Gutierez G, Pohil RJ. Oxygen consumption is linearly related to oxygen supply in critically ill patients. *Journal of Critical Care* 1986; **1**(1): 45–53.

61. Archie JP. Relationship of oxygen delivery and oxygen consumption. Letter to Editor. *Journal of Critical Care Medicine* 1986; **12**(8): 695.

62. Hirschl RB, Heiss KF, Cilley RE, Hultquist KA, Housner BS, Bartlett RH. Oxygen kinetics in experimental sepsis. *Surgery* 1992; **112**(1): 37–44.

63. Krogh A. The supply of oxygen to the tissues and the regulation of capillary circulation. *Journal of Physiology* 1919; **53**: 457.

64. Fowler NO, Holmes JC. Blood viscosity and cardiac output in acute experimental anemia. *Journal of Applied Physiology* 1975; **39**: 453.

65. Mirhasheni S, Breit GA, Chavez Chavez RH, Inlaglietta M. Effects of hemodilution on skin microcirculation. *American Journal of Physiology* 1988; **254**: 411.

66. Gutierrez G. The role of oxygen release and its effect on oxygen tension: a mathematical analysis. *Respiratory Physiology* 1986; **63**: 79.

67. Messmer K. Hemodilution. *Surgical Clinics of North America* 1975; **55**: 659.

68. Kowalyshyn TJ, Prager D, Young J. A review of the present status of preoperative hemoglobin requirements. *Anesthesia and Analgesia* 1972; **51**: 75.

69. Cezer LA, Shoemaker WC. Optimal hematocrit value in critically ill postoperative patients. *Surgery, Gynecology and Obstetrics* 1978; **147**: 363.

70. Tremper KK, Freedman AE, Levine EM, *et al.* The preoperative treatment of severely anemic patients with perfluorochemical emulsion oxygen transporting fluid, Fluosol-DA. *New England Journal of Medicine* 1982; **307**: 277.

71. Lichtenstein A, Echhart WF, Swanson KJ, *et al.* Unplanned intraoperative and postoperative hemodilution: oxygen transport and consumption during severe anemia. *Anesthesiology* 1988; **69**: 119.

72. Fortune JB, Feustrel PJ, Saifi J, *et al.* Influence of hematocrit on cardiopulmonary function after acute hemorrhage. *Journal of Trauma* 1987; **27**: 243.

73. Singbartl G, Becker M, Frankenberger C, Maleszka H, Schleinzer W. Intraoperative on-line ST segment analysis with extreme normovolemic hemodilution. *Anesthesia and Analgesia* 1992; **74**: S295.

74. van Woerkens ECSM, Trouwborst A, van Lanschot JJB. Profound hemodilution: what is the critical level of hemodilution at which oxygen delivery-dependent oxygen consumption starts in an anesthetized human? *Anesthesia and Analgesia* 1992; **75**: 818.

75. Wilkerson DK, Rosen AL, Lakshman R, *et al.* Limits of cardiac compensation in anemic baboons. *Surgery* 1988; **103**: 665.

76. Levine E, Rosen A, Sehgal L, *et al.* Physiologic effects of acute anemia: implications for a reduced transfusion trigger. *Transfusion* 1990; **30**: 11.

77. Kobayashi H, Estafanous FG, Fetnat M, *et al.* Effects of myocardial infarction on hemodynammic response to variable degree of hemodilution. *Anesthesia and Analgesia* 1988; **67**: S117.

78. Estafanous FG, Wafaie S, Tarazi RC. Effects of cardiac

depression on hemodynamic response to hemodilution. *Anesthesiology* 1985; **63**: A38.

79. Spence RK, Carson JA, Poses R, *et al.* Elective surgery without transfusion: influence of preoperative hemoglobin level and blood loss on mortality. *American Journal of Surgery* 1990; **159**(3): 320–4.

80. Eckenhoff JE, Rich JC. Clinical experience with deliberate hypotension. *Anesthesia and Analgesia* 1966; **45**: 21.

81. Barbier-Bohm G, Desmonts JM, Couderc E, *et al.* Comparative effects of induced hypotension and normovolemic hemodilution on blood loss in total hip arthroplasty. *British Journal of Anaesthesia* 1980; **52**: 1039.

82. Sivarajan N, Amory DW, Everett GB, *et al.* Blood pressure not cardiac output, determines blood loss during induced hypotension. *Anesthesia and Analgesia* 1989; **59**: 203.

83. Crystal GJ, Rooney MW, Salem MR. Regional hemodynamics and oxygen supply during isovolemic hemodilution alone and in combination with adenosine induced controlled hypotension. *Anesthesia and Analgesia* 1988; **67**: 211.

84. Crystal GJ, Rooney MW, Salem MR. Myocardial blood flow and oxygen consumption during isovolemic hemodilution alone and in combination with adenosine induced controlled hypotension. *Anesthesia and Analgesia* 1988; **67**: 539.

85. Del Rossi AI, Cernaiann AC, Botros S, *et al.* Prophylactic treatment of postperfusion bleeding using EACA. *Chest* 1989; **96**: 27.

86. Harrow JC, Hlavacek J, Strong MD, *et al.* Prophylactic tranexamic acid decreases bleeding after cardiac operations. *Journal of Thoracic and Cardiovascular Surgery* 1990; **99**: 70.

87. Royston D, Bidstrup BP, Taylor KM, Sapsford RN. Effects of aprotinin on need for blood transfusion after repeat open heart surgery. *Lancet* 1987; **2**: 1289.

88. Royston D. High-dose aprotinin therapy: a review of the first five years' experience. *Journal of Cardiothoracic and Vascular Anesthesia* 1992; **6**: 76.

89. Wang JS, Lynn CY, Hung WT, Carp RB. Monitoring of heparin induced anticoagulation with kaolin-activated clotting time in cardiac surgical patients treated with aprotinin. *Anesthesiology* 1992; **77**: 1080–4.

90. Salzman EW, Weinstein MJ, Weintraub RM, *et al.* Treatment with desmopressin acetate to reduce blood loss after cardiac surgery. *New England Journal of Medicine* 1986; **314**: 1402.

91. Mannucci PM. Desmopressin: a nontransfusional form of treatment for congenital and acquired bleeding disorders. *Blood* 1988; **72**: 1449.

92. Goodnough TL, Rednick S, Price RH, *et al.* Increase preoperative collection of autologous blood with recombinant human erythropoietin therapy. *New England Journal of Medicine* 1989; **321**: 1163.

93. Garner RJ, Silverglrid AJ. *Autologous and directed blood programs.* Arlington, Virginia: American Association of Blood Banks, 1987.

94. Mann S, Sacks HJ, Goldfinger D. Safety of autologous blood donation prior to elective surgery for a variety of potentially high risk patients. *Transfusion* 1983; **23**: 229–32.

95. Myhre DA. Fatalities from blood transfusions. *Journal of the American Medical Association* 1980; **244**: 1333.

96. Messner K, Kessler M, Sunder-Plassmann L. Hemorheologic effects of intentional hemodilution. In: Shoemaker WC, Tavares BM. *Current topics in critical care medicine.* Basel, Switzerland: S Karger, 1976: 130.

97. Stump DC. Effects of hydroxyethylstarch on blood coagulation, particularly factor 8. *Transfusion* 1985; **25**: 249.

98. Strauss RG, Sanfield C, Henriksen RA. Pentastarch may cause fewer effects on coagulation than hetastarch. *Transfusion* 1988; **28**: 257.

99. Looker D, Abbott-Brown D, Cozart P, *et al.* A human recombinant hemoglobin designed for use as a blood substitute. *Nature* 1992; **356**: 258.

Chapter 20

Haemostasis, Haemoglobinopathies and Anaesthesia

E. C. Gordon-Smith

Two topics in which haematological practice and anaesthetics interact are addressed here. These are the problems of bleeding and thrombosis on the one hand and the effects of changes in the structure of haemoglobin on function on the other. The chapter is therefore divided into two separate sections.

Haemostasis, thrombosis and anaesthesia

Haemorrhage and thrombosis are two major problems which affect the outcome of surgery. Thrombosis with its attendant risk of embolism and infarction is, of the two, the more dangerous and thus many measures are taken to reduce the risk of thrombosis during and after surgery. These measures inevitably interfere with normal haemostatic mechanisms. Many patients receive preoperative treatment or prophylaxis with drugs which alter haemostatic homoeostasis. Therefore, a knowledge of the action and interaction of such drugs, with practice based upon that knowledge, will help to reduce or anticipate problems with surgery, anaesthesia and above all postoperative care.

Haemostatic mechanisms

Formation of a stable haemostatic plug at the site of injury

To some extent an understanding of the way in which blood loss is controlled at the site of vessel injury has been hindered by the historical experimental processes by which the components of thrombus formation have been discovered and explained. Thus it has been conventional to consider platelet function, clotting factors and blood vessels as activating separate or at least separable

haemostatic pathways, whereas in reality, or perhaps more correctly *in vivo*, the three components are inextricably linked and act in an elegant harmony. The endpoint of normal control of blood loss following injury is a plug or thrombus which consists of fused platelets, adherent to damaged tissue surfaces and to vessel endothelium around the site of injury, enmeshed in strands of stabilized, cross-linked fibrin. Such a plug has already activated the pathways of its own destruction by the time it is formed.

An explanation of the haemostatic mechanisms is made easier, however, by describing the individual components separately and bringing the whole together at the end. Most of the laboratory tests of haemostatic function also measure the integrity or otherwise of individual pathways or components.

Platelets

Platelets are produced from megakaryocytes in the bone marrow. The normal platelet's life span is 7–10 days and the normal count 150–400 $\times$ 10^9/litre. Newly produced platelets are larger than old platelets and are more responsive to activation, an observation which influences the action taken in the face of thrombocytopenia, depending on whether increased destruction or failure of production is the main cause. The platelet membrane plays a key role in haemostasis, not only because it carries receptors which respond to factors exposed or released by tissue injury which lead to platelet adhesion and aggregation, but also because the phospholipid of the platelet provides structure on which activated coagulation factors are brought together in the optimum stereospecific fashion for rapid conversion of prothrombin to thrombin. Specific glycoproteins (GP) on the platelet surface are receptors for aggregating agents, coagulation factors and inhibitors.

Within the platelet are a variety of granules which package highly active substances. Activation of the platelet membrane by collagen or thrombin leads to the *release reaction* whereby the contents of these granules are discharged into an open canicular system in the platelet and thence into the surrounding medium. The release

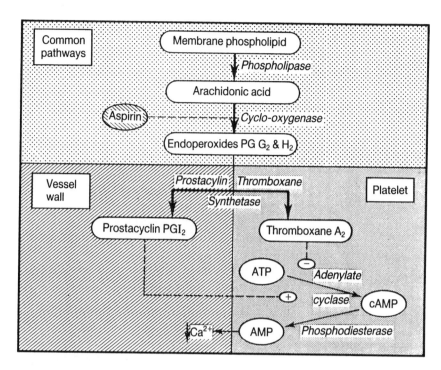

Fig. 20.1 Pathways involved in platelet aggregation. Intracellular levels of free Ca^{2+} are controlled by cyclic AMP (cAMP). Low levels of Ca^{2+} inhibit aggregations, high levels promote. The end products of the cyclo-oxygenase pathways, thromboxane A_2 and prostacyclin have opposing effects on adenylate cyclase, the enzyme which controls cAMP levels, inhibiting or increasing the activity respectively. Aspirin inactivates cyclo-oxygenase through acetylation, the platelet enzyme being more readily inhibited than the endothelial enzyme.

reaction also liberates arachidonate from the membrane, a substrate which increases the formation of thromboxane A_2. Thromboxane A_2 is formed via the cyclo-oxygenase pathway and inhibits the enzyme adenylate cyclase necessary for the conversion of ATP to cyclic AMP and prevention of calcium release (see Fig. 20.1). The opposite effect, activation of adenylate cyclase, inhibition of the release reaction and prevention of platelet adhesion and aggregation is provided by prostacyclin (PGI_2) released by undamaged vascular endothelial cells, also through the cyclo-oxygenase pathway. Aspirin is an inhibitor of cyclo-oxygenase and hence leads to a deficiency of thromboxane A_2 production and defective platelet plug formation.

The release of ADP and thromboxane A_2 from the activated platelets causes additional platelets to aggregate at the site of injury, so that a platelet mass, large enough to plug the injured vessels, builds up.

Thus the platelet plays a central role in the initial achievement of haemostasis, helps to localize the formation of a stable clot at the site of injury and, perhaps, initiates healing. Deficiency in platelet numbers has a modest effect on haemostasis; abnormalities of platelet function have more profound results.

Coagulation – the formation of fibrin

The process of coagulation depends on the activation of various procoagulant factors following tissue injury or contact with activated platelets with the final result of converting fibrinogen, a soluble precursor, into fibrin.

Fibrin is a monomer which will form an insoluble complex of polymers, subsequently stabilized through covalent linking. Fibrinogen is converted to fibrin by the action of thrombin, a serine protease derived from its inactive precursor prothrombin (factor II). Most of the activated factors in the coagulation pathways are serine protease enzymes, two exceptions being factor V and factor VIII (antihaemophilic factor) which are enzyme cofactors. Even these two cofactors exist in an inactive form in the undisturbed circulation.

It has been conventional to divide the pathways which lead to the formation of thrombin into the intrinsic and extrinsic pathways. The intrinsic pathway comprises those reactions which lead to the formation of a clot simply on the exposure of blood to negatively charged surfaces (e.g. kaolin, glass), all the factors necessary for coagulation being present in the circulating blood (in inactive form). The extrinsic system is activated by factors released from damaged tissue, thromboplastins. The terms only have meaning in relation to the coagulation screening tests, the activated partial thromboplastin time (APTT) also known as partial thromboplastin time with kaolin (PTTK) for the intrinsic system and the prothrombin time (PT) for the extrinsic system.

Most significant steps in the production of fibrin *in vivo* are shown in Fig. 20.2. The complex of activated factor X(Xa), cofactor V and platelet membrane rapidly converts prothrombin to thrombin and initiates clotting. Antithrombin III is an inhibitor of factor X activation. Its action is greatly enhanced by heparin. Tests are available for measuring the level of factor Xa.

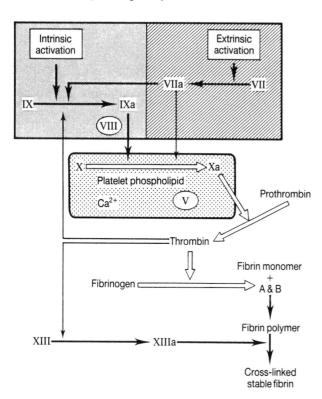

Fig. 20.2 Production of stable fibrin clot. Intrinsic activation initiates coagulation following exposure of subendothelial collagen and other negatively charged compounds. Extrinsic activation requires the release of tissue factor from perivascular tissues. Both pathways are activated by injury. Factor Xa activity is increased by several orders of magnitude through the complex with phospholipid and cofactor, factor V.

There are many other factors which may be involved in the formation of a fibrin clot under special circumstances, disease, or laboratory conditions, but which in general are unimportant for the production of the normal haemostatic plug.

Vitamin K dependent proteins

Vitamin K is essential for the γ-carboxylation of certain clotting factors and some anticoagulant factors. In the absence of the carboxylation the coagulant and anti-coagulant proteins cannot bind to phospholipids or platelets so that the rate of coagulation or fibrinolysis

becomes too slow for meaningful control. The major vitamin K factors are listed in Table 20.1.

Deficiency of vitamin K occurs in neonates, particularly premature infants with jaundice and breast-fed infants. In older children and adults vitamin K deficiency occurs in the presence of obstructive jaundice, malabsorption syndromes such as sprue and coeliac disease in elderly patients with poor diet, and in patients requiring intensive care. The major inhibition of vitamin K activation comes from the use of oral anticoagulants to block carboxylation. Liver disease will amplify the effect of vitamin K deficiency on factors II (prothrombin), VII, IX, and X which are synthesized in the liver. More severe liver failure exerts its effect on coagulation mainly through failure of production of factors II, VII, IX, and X and also factor V. The administration of vitamin K intravenously will correct the coagulation defects produced by that deficiency, providing liver function is adequate. Vitamin K will also reverse the action of oral anticoagulants.

Fibrinolysis and other regulators of haemostasis

It should be clear from the foregoing section that there are a wide variety of positive and negative pathways which control the activation and deactivation of the coagulation system and which prevent the inappropriate formation of thrombus in the intact circulation but permit the rapid development of a haemostatic plug at the site of injury.

Fibrinolytic system

Dissolution of fibrin occurs through the activity of a powerful protease, plasmin, derived from the circulating proenzyme plasminogen. Plasminogen is activated by a number of substances, particularly tissue plasminogen activator (tPA). tPA is released from endothelial cells following activation and binds strongly to fibrin, thus localizing plasmin activity to the site of thrombus formation. Urokinase is another naturally occurring plasminogen activator which is commercially available as

Table 20.1 The major coagulation factors

Vitamin K dependent group	Thrombin dependent group
Prothrombin (II) VII, IX, X	Fibrinogen (I), V, VIII, XIII
Require Ca^{2+} for activation	All acted on by thrombin
Active forms are serine proteases	Fibrin substrate or cofactors
VII, IX, X are not consumed in coagulation (present in serum)	Consumed in coagulation process (not present in serum)
Stable, preserved in fresh frozen plasma	V and VIII lose activity in stored plasma. Fibrinogen stable

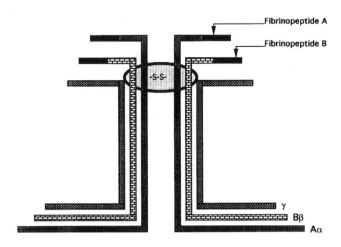

Fig. 20.3 Scheme of fibrinogen structure. Fibrinogen consists of a dimer of three dissimilar chains, Aα, Bβ and γ held together by disulphide bridges. Thrombin splits off fibrinopeptides A and B which exposes binding sites which allow the fibrin monomers to combine covalently on a lattice framework which is stabilized by further disulphide links catalysed by factor XIIIa.

a fibrinolytic agent, as is recombinant human tPA. Streptokinase is a plasminogen activator produced by haemolytic streptococci.

Fibrin degradation

The formation and dissolution of fibrin leads to the production of various fibrin peptides which have regulatory, diagnostic and therapeutic implications. Fibrinogen consists of two identical subunits each consisting of three dissimilar peptides, Aα, Bβ, and γ (see Fig. 20.3). Thrombin splits off fibrinopeptides A and B in the conversion of fibrinogen to fibrin. Plasmin degrades fibrin into a number of fragments which collectively are the fibrin degradation products (FDP). FDPs inhibit the conversion of fibrin to fibrinogen and have anticoagulant properties themselves (Fig. 20.4).

Plasma inhibitors of activated coagulation factors

Antithrombin III inhibits the protease activity of thrombin and other serine proteases by forming a covalently bound, stable complex. Heparin greatly potentiates the formation of this complex. Heparin loses much of its activity in the absence of antithrombin III, deficiency of which may be seen in disseminated intravascular coagulation. Congenital deficiency of antithrombin III is associated with an increased risk of pathological thrombosis.

Protein C and protein S are two vitamin K dependent factors, deficiency of either of which leads to an increased risk of spontaneous thrombosis. Protein C is a protease which destroys activated factors V and VIII, the cofactors necessary for thrombin production. Protein C activity is enhanced by binding to the platelet surface in the presence

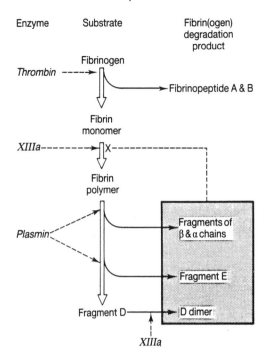

Fig. 20.4 Fibrinolysis and formation of fibrin degradation products.

of its cofactor protein S. Protein C is activated at the endothelial cell following the formation of a thrombin/thrombomodulin complex at the cell surface.

Detection of abnormalities in the haemostatic mechanisms

The detection of potential problems of haemostasis in surgical patients depends more on detailed enquiry of the patient than reliance on laboratory tests. The patient known to have a coagulation defect only presents a problem in emergency situations and if colleagues who are expert in haematological and blood transfusion therapy have not been involved in management from the earliest time. Patient enquiry needs to include a careful drug history as well as eliciting a history of previous bleeding or bruising problems. Family history is also essential – carriers of factor VIII deficiency and other coagulation defects may have severe surgical bleeding even though they may have no spontaneous problems. Particular questions should be asked about the use of aspirin or non-steroidal anti-inflammatory drugs and the use of the contraceptive pill (Table 20.2).

Screening tests for coagulation pathways should be requested preoperatively if there is a worry concerning haemostasis from the history, if the patient is known to have a disease which may interfere with haemostasis, if the patient is to receive peroperative heparin or if the patient is to have emergency surgery and the history is not available.

Table 20.2   Enquiries relevant to haemostasis to be made preoperatively

Enquiry	Action to be taken if response is positive
Past or recent history of bleeding or bruising	Blood count, coagulation screen. Advice of haemostasis team
Family history of bleeding	Blood count, coagulation screen. Advice of haemostasis team
Drug history – warfarin	Discontinue 72 h preoperatively. Use heparin if necessary. In emergency reverse with FFP ± vitamin K
aspirin	Discontinue 72 h preoperatively. Give platelets if bleeding excessive
non-steroidal anti-inflammatory drug	Discontinue 24 h preoperatively. Watch for GI haemorrhage if anticoagulants used postoperatively
Past history or family history of thrombosis	If positive or doubtful discuss thrombophilia screening with haematology laboratory. LMW heparin should be used in all high risk situations
Oestrogen-containing contraceptive pill	Stop 4 weeks prior to elective surgery. For emergency surgery use LMW heparin prophylaxis
Hormone replacement therapy	Adhere scrupulously to normal prophylactic measures

Screening tests for coagulation disorders

Abnormalities of haemostatic pathways may be measured using appropriate screening tests, that is tests which measure the results of interference at any one or more of a number of steps along a defined pathway, rather than specific reactions or factor concentrations.

Prothrombin time (PT)

The prothrombin time measures the integrity of the so-called extrinsic pathway, including factors VII and the common pathway factors V, X, prothrombin and fibrinogen. Thromboplastin is added in the form of brain extract to initiate the reaction. Variation in thromboplastins has led to the formulation of the International Normalisation Ratio (INR) for the control of oral anticoagulant therapy. The normal PT is 10–14 seconds.

Partial thromboplastin time with kaolin (PTTK)

This measures the integrity of the intrinsic pathway including factors VIII, IX (XI, XII) as well as V, X, prothrombin and fibrinogen. Normal PTTK times are 30–40 seconds before a clot first forms.

Thrombin time (TT)

The thrombin time is sensitive to deficiency or functional abnormality of fibrinogen and to inhibitors of thrombin activity including the presence of FDPs. The normal range is 15–16 seconds. Heparins prolong the TT and the presence of heparin may complicate the management of patients with disseminated intravascular coagulation (DIC).

Bleeding time

The bleeding time is the time it takes for a clot to form after cuts have been made on the forearm under controlled conditions. Normal bleeding time is 3–8 minutes. Prolongation of the bleeding time indicates abnormalities of platelet function or number. The bleeding time is difficult to perform because of problems in standardizing the conditions and should be carried out by the haemostasis staff rather than by inexperienced medical staff.

The screening tests used in the diagnosis of coagulation disorders are summarized in Table 20.3.

The screening tests and, if needed, factor assays, may indicate which factors are deficient, but usually an induced coagulopathy perioperatively is produced by multiple deficiencies. The special case of drug-induced coagulopathies is considered later. Patients coming to surgery with coagulation problems should be as fully assessed as possible so that appropriate replacement materials may be available.

The major blood products which are available and their content of blood coagulation factors are shown in Table 20.4. The main concern, and one for which there is no definitive answer, is when to correct laboratory abnormalities in coagulation tests in the presence of known causes of coagulopathy. Decisions should be taken on the basis of knowledge of the risks of severe haemorrhage in any situation, the potential for rapid correction of haemostatic problems should the need arise and a balance between haemorrhagic risk and fluid overload as well as the possibility of giving expensive and not entirely safe drugs unnecessarily. The platelet count as well as the screening tests will determine replacement strategy.

Medical and surgical acquired coagulopathies

Liver disease

Most patients with liver disease have some evidence of coagulation disorder as shown by prolongation of the prothrombin time. There is some debate over the level of increase in the INR which should prompt preoperative

Table 20.3 Screening test for coagulation disorders

Test	Normal range*	Abnormalities indicated by prolongation	Common causes of disorder
Thrombin time (TT)	14–16 s	Deficiency or abnormalities of fibrinogen. Inhibition of thrombin by heparin or EDPs	Heparin therapy. Disseminated intravascular coagulation (DIC)
Prothrombin time (PT)	10–14 s	Deficiency or inhibition of V, VII, X. prothrombin, fibrinogen	Warfarin therapy. Liver disease. Heparin, DIC, vitamin K deficiency
Partial thromboplastin time with kaolin (PTTK)	30–40 s	Deficiency or inhibition of V, VII, IX, X, prothrombin, fibrinogen	Heparin therapy. DIC. Haemophilia. Christmas disease, von Willebrand's disease, liver disease
Bleeding time	3–8 min	Abnormal platelets formation. Reduced platelet number	Inherited disorders of platelet function. Drug-induced renal failure, von Willebrand's disease

* Normal range must be established by local laboratories and regular quality control exerted.

Table 20.4 Blood products used for haemostasis

Blood product	Main coagulation factors	Uses	Comments
Fresh frozen plasma (FFP)	Fibrinogen, prothrombin, factors V, X, XI, some factor VIII, IX, VII	1 Correction of coagulation disorders in liver disease 2 Replacement of factors in DIC or following major trauma 3 Replacement during massive transfusion	ABO compatible only should be given (contains anti-A, B antibodies) Stored at < −30°C Potential source of HIV, hepatitis viruses Should be used within 30 min of rapid thawing
Cryoprecipitate	Fibrinogen (50–80 g/dl) Factor VIII (2–5 u/ml)	1 Additional source of fibrinogen and factor VIII in massively transfused patients 2 Source of factor VIII for mild haemophiliacs or heterozygotes at risk 3 To correct bleeding time and permit surgery in patients with platelet storage disease	No ABO antibodies present Single donor but potential source of virus transmission. Useful addition to FFP if volume overload is a problem Stored at −20°C Stable for 6 months
Albumin solution	No coagulation factors	Volume replacement	Small risk of virus transfusion
Platelet concentrates	Platelets	1 In disorders of platelet production 2 In platelet storage disease 3 In massively transfused patients	ABO compatible should be used Can transmit viruses including CMV (unless white cell depleted) Not effective in immune thrombocytopenia

correction. If the platelet count is greater than 80×10^9/litre and the bleeding time is normal, administration of fresh frozen plasma (FFP) preoperatively is unnecessary if the INR is < 1.6. Patients with liver disease should receive vitamin K intravenously preoperatively, 10 mg i.v. daily for 3 days, but this will not correct the coagulation disorder in the presence of severe liver failure. ABO compatible fresh frozen plasma (FFP) should be available peroperatively in case needed. Above an INR of 1.8, or in the presence of thrombocytopenia or platelet dysfunction, FFP should be given within 6 hours of the start of the operation. It may require 1–2 litres of FFP to correct coagulation defects so fluid overload is a very real problem. The smaller volume of cryoprecipitate for fibrinogen replacement makes this product an attractive adjunct to FFP. The effectiveness of coagulation control is monitored using both the PT and the PTTK. The platelet count should be determined at the same time.

Cardiopulmonary bypass

The need for coagulation replacement therapy during cardiopulmonary bypass surgery has greatly diminished. This has been achieved by a number of measures including more effective bypass machines causing less platelet consumption, less blood loss at surgery, the use of blood saving equipment and the use of aprotinin to reduce bleeding. On the other hand, the widespread use of aspirin in patients undergoing bypass surgery has led to a significant increase in blood loss. If excessive haemorrhage occurs during cardiopulmonary bypass surgery, the first emergency measure, given before laboratory results are available, is the administration of ABO compatible blood or platelets followed by FFP.

Trauma and the critically ill patient

The approach to the critically ill, bleeding and massively transfused patient also depends on individual circum-

stances. A platelet count and coagulation screen are early prerequisites for management. Packed red cells and whole blood stored for more than 24 hours do not contain platelets or significant amounts of factor VIII. Patients transfused 10 or more units over 24 hours usually show evidence of increased haemorrhage. It is prudent to replace these factors in the massively transfused patient, giving 5×10^9 platelets/10 kg body weight and 2 units (1 unit = approximately 200 ml) of FFP for each 8–10 units of blood transfused.

A degree of platelet consumption, with or without evidence of disseminated intravascular coagulation (DIC), is common in these patients and a platelet count below 100 $\times 10^9$/litre warrants platelet transfusion. A prothrombin ratio (INR) greater than 1.8 is suggestive of acute liver dysfunction and warrants coagulation factor replacement by FFP and cryoprecipitate. If there is evidence of DIC, indicated by a prolongation of the TT and the presence of FDPs, and/or the patient is likely to require postoperative heparin, a measure of plasma antithrombin III is helpful since it may be consumed along with other factors in DIC and heparin will not exert its anticipated anticoagulant activity. In most, if not all such cases restoration of blood volume, blood pressure and essential organ perfusion is

Table 20.5 Disorders of platelets

Disorder	Mechanisms	Causes
Reduced number (thrombocytopenia)	Failure of production	Leukaemias Cytotoxic drugs Radiation Aplastic anaemia Myelofibrosis Myelodysplasia HIV infection

	Increased destruction	Immune destruction
		  ITP
		  Drug induced
		Intravascular coagulation (DIC)
		Hypersplenism

Abnormalities of function	Acquired	Drugs
		  aspirin
		  NSAIDS
		  heparin
		Myeloproliferative diseases
		  essential thrombocythaemia
		  PRV
		  myelofibrosis
		Myelodysplastic syndromes

	Constitutional	Glanzmann's syndrome Bernard–Soulier syndrome von Willebrand's disease (platelet type)

more important than early correction of the coagulation defects.

Platelet disorders

Platelet disorders at the time of surgery and anaesthesia may arise from a deficiency or abnormal function. The major causes are shown in Table 20.5. When platelet numbers are reduced it is important to know whether this is because of reduced production or increased destruction. When there is a reduced production the effect may be overcome by giving platelet transfusions. The prolongation of the bleeding time is the function of the reduction in number of platelets but it is not useful in predicting who will suffer excessive surgical bleeding; indeed, preoperative assessment of the thrombocytopenic patient to decide on prophylactic platelets is generally unsatisfactory.[1] Patients with platelet counts above 50×10^9/litre with no evidence of impaired platelet function do not need prophylactic platelets for surgery which would not normally require red blood cell transfusions but platelets for that patient should be available in case of mishaps. Transfusions of red blood cells lowers the platelet count, probably by increasing the consumption of platelets through complement activation, so platelets should be given to the thrombocytopenic patient who receives a blood transfusion. If the patient has a platelet count of less than 50×10^9/litre prophylactic platelets may be advisable. The dose, type of platelets received (CMV negative, white cell depleted, HLA matched, etc.) should be decided in conjunction with the transfusion laboratory and the timing of the administration determined. Usually it is advisable to have raised the count to 30×10^9/litre presurgery and then to give platelets in theatre according to bleeding tendency. Further platelets should be given postoperatively to maintain the count above 30×10^9/litre.

A different situation arises when the thrombocytopenia is due to destruction of platelets. The most common cause is the presence of antiplatelet antibodies in immune thrombocytopenic purpura (ITP). The platelets in this condition have a very short life span and the majority are very 'young' platelets. These respond vigorously to activation and the bleeding time is much less affected than would be expected from their number. The usual dilemma is whether to give platelets if an urgent splenectomy is contemplated. Fortunately, this arises much less frequently now that intravenous immunoglobulin is available to raise the platelet count within 24 hours. In general there is little role for platelet transfusions in these patients before the spleen is isolated during surgery and indeed the need for such transfusions at all is now uncommon.

Platelet transfusions

The 'dose' of platelets to be given should be determined in conjunction with the transfusion service since different methods of preparation will lead to different numbers of platelets in the local units. In general about 5×10^9 platelets/10 kg body weight is a suitable prophylactic dose post-surgery if indicated.

Anticoagulant and antiplatelet therapy

An increasing number of indications for giving anticoagulant or antiplatelet drugs to ward off thrombotic complications have been identified. Many patients, particularly in the older age group, will be receiving this treatment when they come to surgery. For the anaesthetist there are a number of questions to be addressed, namely what effect will the anticoagulants have on blood loss, should they be continued during the period of surgery and does the presence of anticoagulant or antiplatelet drugs influence the route of anaesthetic?

Heparin

The use of subcutaneous heparin for prophylaxis against venous thrombosis in patients at high risk during surgery has become standard practice and has greatly reduced the incidence of this complication[2] and the mortality and morbidity that goes with it.[3] Heparin is not a single drug substance but a range of mucopolysaccharides extracted from animal material. The molecular weight of such material ranges from 15 to 18 kDa. Hydrolysis of natural heparin yields a variety of low molecular weight (LMW) heparins with molecular weights 4–6 kDa depending on the process used. Low molecular weight heparins have better bioavailability than high molecular weight when given subcutaneously and have a longer half-life so that they need to be given less frequently. Subcutaneous LMW heparin given once a day seems to be as efficacious as standard heparin given by continuous intravenous infusion in treating proximal vein thrombosis.[4] The main complication of heparin prophylaxis throughout surgery is increased blood loss though this appears to be less with the single dose LMW compared with twice daily subcutaneous unhydrolysed heparin. In general the low dose of subcutaneous heparin does not produce meaningful changes in the laboratory parameters of coagulation. Therefore routine monitoring of heparin levels or effects are not worthwhile and management of excessive haemorrhage depends on clinical skill.

Subcutaneous heparin and regional anaesthesia

Although readily detectable changes in coagulation parameters cannot be shown, subcutaneous heparins are effective anticoagulants and may be a hazard when epidural or spinal anaesthesia is used. The danger is the increased risk of spinal haematoma and cord or corda equina compression with the development of paraplegia. Although trauma to a blood vessel during central nerve blockade is relatively common,[5] spinal haematoma is very rare and most cases occur in patients with a congenital or acquired coagulation disorder.[6] It would seem prudent to avoid central nerve blockade in patients who are fully anticoagulated or receiving active fibrinolytic therapy at the time of anaesthesia.[7] In practice such occurrences will be rare since surgical or obstetric procedures are not carried out under such circumstances and there is usually time to reverse or modify the anticoagulant regimen before surgery commences.

With low dose heparin the situation is less clear. Administration of the block should be delayed for at least 4 hours after the initial dose of subcutaneous heparin to avoid the peak systemic levels which occur at about 2 hours. It may be acceptable in the context of the particular surgery involved to give the first dose of subcutaneous heparin after the central block has been established but the rarity of the complication means that such advice is based on theoretical considerations rather than clinical trial data.

If anticoagulant therapy (full dose) is started during operation or postoperatively in a patient receiving epidural anaesthesia, the catheter should be in place well before surgery starts and should only be removed when there has been a period of 2 hours without heparin postoperatively, the timing of which will require discussion with the surgical team. Some surgical teams recommend that epidural anaesthesia be abandoned if there is a traumatic introduction of the catheter in patients who will receive full anticoagulants[8] but there are no cases to support such an obviously ultra-safe practice.

Heparin induced thrombocytopenia and thromboembolism

A particularly devastating complication of heparin therapy is the occurrence of thrombocytopenia and thromboembolism. Thrombocytopenia due to heparin dependent antiplatelet antibodies is a well recognized complication, usually occurring after 5–7 days of heparin therapy. In the majority of cases the thrombocytopenia develops over 2 or 3 days with a moderate decline in numbers. In a small proportion of patients the fall in platelets is more rapid and is associated with massive arterial thrombosis including femoral artery occlusion, myocardial infarction or cerebral artery thrombosis. Immediate withdrawal of heparin is essential and may prevent permanent damage but the withdrawal must be total including exclusion of hepflushes for central venous lines and of heparin added to intravenous fluids. Minute amounts of heparin are all that is necessary to perpetuate or extend the thrombosis. The mechanism of this devastating complication is not entirely clear but it seems likely that heparin dependent antiplatelet antibodies can react with the endothelium to promote platelet plugging – the thrombi are noticeable for their very white appearance. If heparin induced antiplatelet antibodies are suspected, an *in vitro* agglutination test should be carried out in the haemostasis laboratory to demonstrate their presence, using unmodified and LMW heparins. In some instances LMW heparins do not cross-react with the high molecular weight forms and can be used to continue anticoagulation, but they must not be used until the agglutination test is shown to be negative with LMW. The incidence of heparin induced thrombocytopenia is probably less with the LMW heparins. Alternative anticoagulants which do not cross-react are under clinical investigation, including a heparinoid (mixture of heparan sulphate and dermatan sulphate), and hirudin, main anticoagulant of the leech, *Hirudo medicinalis*. The heparinoid (lomoparin) has been used to treat heparin induced thrombocytopenia.[9] All patients who receive heparin for more than 7 days should have the platelet count checked and alternative anticoagulants substituted if the platelet count is falling.

The main relevance of this complication to anaesthetic practice is in the Intensive Care Unit. The staff should be aware that once the syndrome is suspected, it must be confirmed and all heparin must be avoided.

Antiplatelet drugs

Aspirin is widely used to reduce the risk of coronary occlusion, not only by patients with vascular disease but also by a substantial proportion of the normal adult population over 40 years. Aspirin inhibits cyclo-oxygenase and reduces platelet aggregation and adhesion. The inhibition is irreversible and the effect of aspirin on bleeding time may take 2–3 days to become normal as unaffected platelets are produced to dilute out the aspirin-affected circulating platelets.

There are very few data concerning the use of regional anaesthesia in patients receiving aspirin (or other antiplatelet drugs such as non-steroidal anti-inflammatory drugs, or dipyridamole). One study from the Mayo Clinic failed to find any adverse effects of aspirin or other antiplatelet drugs on the outcome of spinal or epidural anaesthesia.[10] It would seem reasonable not to withhold such anaesthesia if the indications are clear but to be aware of potential heightened risk. If there are doubts in an individual case a transfusion of normal platelets will dilute the effect of the aspirin-treated platelets.

Exposure to antiplatelet drugs will increase the risk of surgical haemorrhage whatever the route of anaesthesia.

Aspirin should be stopped 1 week before surgery if possible, non-steroidal anti-inflammatory drugs 2 days presurgery. This is particularly important for patients who take antiplatelet drugs regularly preoperatively and who are anticoagulated peroperatively. This combination may be common in patients undergoing orthopaedic surgery.

Aspirin and antiplatelet drugs not only interfere with platelet function but also increase the likelihood of gastrointestinal haemorrhage due to gastric erosions. The introduction of intravenous non-steroidal anti-inflammatory agents (for example, ketorolac) for pain control in day-case patients, while avoiding the problems of opiate sedation in these patients, has increased the risk of postoperative gastrointestinal haemorrhage. These agents should be avoided in patients who receive peroperative and/or postoperative subcutaneous heparin.

Vitamin K antagonists

Warfarin and the coumarin derivatives are widely used anticoagulants for patients with previous thromboembolic disease, patients with cardiovascular prostheses, atrial fibrillation (to prevent stroke) and many other indications. If possible patients on warfarin should be converted to heparin preoperatively because bleeding complications due to heparin are easier to control. Warfarin should be stopped at least 72 hours preoperatively and full anticoagulant doses of heparin instituted, or subcutaneous low dose heparin begun if the reason for anticoagulation warrants the lower dose. If the substitution by heparin proves impracticable, the effects of warfarin may be reversed by giving vitamin K, 10 mg i.v., and checking the INR (or prothrombin time). Occasionally larger doses may be required. Once the vitamin K has been given the patient will be resistant to the reintroduction of warfarin for several days. If it is proposed to reintroduce oral anticoagulants in the early postoperative period, fresh frozen plasma, 2×250 ml, may reduce the INR to 1.0–1.5, at which level most surgery will be as safe as it ever is. Such manoeuvres are best carried out with the prior warning of and assistance from the haemostasis laboratory.

Risk factors for venous thrombosis and pulmonary emboli

Patients who undergo prolonged abdominal surgery or surgery to the legs, particularly if they require a period of reduced mobility postoperatively, are at high risk from venous thrombosis and embolism. Patients with cancer may also be at increased risk. It is these patients who require anticoagulant measures during and/or after surgery with subcutaneous heparin.[3] There are, however, risk factors which will increase the likelihood of venous

thrombosis even more and which may be reduced preoperatively (Table 20.6).

Thrombophilia

The term thrombophilia has been coined to describe patients who have an increased risk of venous thrombosis, mostly on the basis of a congenital deficiency of fibrinolytic factors. It is not relevant to discuss the diagnosis and pathogenesis of these disorders except to point out that deficiencies of antithrombin III, protein C and protein S have each been found to raise the risk of thrombosis. Potent changes in these factors as well as coagulation factors occur during pregnancy, postpartum and at the menopause. That like changes may be induced by oestrogen-containing contraceptive pills and by hormone replacement therapy (HRT) has led to worries that they may increase the surgical thrombotic risk.

Oestrogen-containing contraceptive pill

Ethinyl oestradiol is the most commonly used oestrogen in the combined contraceptive pill. Synthetic oestrogens increase the activity of various procoagulant factors and also reduce the activities of antithrombin III and protein C. It might, therefore, be expected that oestrogen-containing contraceptive pills would increase the risk of thrombotic complications post-surgery. In 1985 it was suggested that this was the case and that oestrogen-containing contraceptive pills should be discontinued at least 4 weeks before elective surgery when delay in ambulation was anticipated or operations on the leg were proposed.[11] Formal studies to test this clinical suggestion have produced conflicting results[12] but in view of the biological effects upon the coagulation and fibrinolytic systems it would seem advisable to adopt the recommendation that the oestrogen-containing pill be stopped 4 weeks before surgery.

Table 20.6 Risk factors for thromboembolism

Related to venous stasis	Thrombophilic states
Immobility	Hereditary
Dehydration	Antithrombin III deficiency
Cardiac failure	Protein C deficiency
Stroke	Protein S deficiency
Pelvic obstruction	Acquired
Nephrotic syndrome	Lupus anticoagulant
Varicose veins	Paroxysmal nocturnal
Hyperviscosity	haemoblobinuria
Sickle cell disease	Pregnancy and puerperium
	Oestrogens
Multifactorial	Hormone replacement therapy(?)
Age	Surgery – abdominal and hip
Obesity	Malignant disease
Sepsis	Major trauma

Progesterone containing 'mini pills' do not produce this effect.

When emergency surgery is required or the oestrogen-containing pill has been continued anyway, the patient and the surgical team should be aware of the potentially increased risk and anticoagulant measures adopted.

Hormone replacement therapy (HRT)

HRT is given to an increasing proportion of postmenopausal women to reverse the undesirable effects of the menopause including osteoporosis and perhaps the increasing risk of coronary artery disease. HRT contains oestrogens and so might be expected to increase the risk of venous thrombosis. However, the biological evidence that there is an increased thrombotic risk in postmenopausal women given HRT is much less strong than for premenopausal women given the contraceptive pill. There is no clinical study which has been designed to show the effect of HRT on operative thromboembolism. There are even differences in the effects of the same oestrogens given orally or non-orally (by patches) which make general rules even more difficult to define.[13,14] There is thus no indication that HRT should be discontinued 4 weeks before surgery but the usual prophylactic measures against thrombo-embolism, anticoagulants and mobilization should be meticulously applied to patients on HRT.

Summary

Modification to the coagulation system in one way or another has become an important part of prophylaxis against disease, particularly in an ageing population. The anaesthetist needs to be aware of these trends so that appropriate enquiry can be made in the assessment of patients for surgery and modification to the route of anaesthesia and surgical practice instituted if indicated. Table 20.2 summarizes the main enquiries to be made.

Haemoglobinopathies and anaesthesia

The normal adult haemoglobin molecule (HbA) consists of a tetramer made up of globin chains, two α and two β, with an iron containing haem group inserted into a 'haem pocket' in each chain. Abnormalities associated with most of the 141 α-chain amino acids or 146 β-chain amino acids have been recorded as well as gene deletions which give rise to α- or β-thalassaemia syndromes. Most of the

substitutions are relatively rare and are of interest mainly to the patient and haematologists. The occurrence of the substitution of the normal glutamic acid at the β6 position by valine is, on the other hand, very common and of interest and importance to many branches of medicine. It is the abnormality which gives rise to HbS, sickle haemoglobin.

Sickle haemoglobin, HbS

The presence of sickle haemoglobin in the heterozygous state (HbA/S) protects to some extent against the lethal effects of *Plasmodium falciparum* infection (falciparum malaria). This is why a gene which produces lethal effects in the homozygous state nevertheless developed and spread in various parts of the world where falciparum malaria was common. The substitution in the β6 position, glutamic acid replaced by valine, determines that deoxy-HbS molecules tend to stack together in rod-like crystals which may precipitate within the red blood cell. This tendency to form insoluble complexes increases with the concentration of HbS molecules and is influenced by the intracellular and extracellular environment of the red cell. Repeated precipitation and resolubilization of sickle haemoglobin eventually leads to damage to the red cell which becomes rigid, irreversibly sickled and liable to block small blood vessels, producing the sickle 'crisis'. The likelihood of crisis is determined by the total expression of haemoglobin genes within the red cell. The various sickle cell syndromes are summarized in Table 20.7.

Sickle cell anaemia

Sickle cell anaemia (HbS/S) occurs in the homozygous state when two HbS genes are inherited. It is the most severe of the sickle cell syndromes. In the homozygous state HbS has a low O_2 affinity which means that a higher proportion than with HbA will be in the deoxy state at mixed venous PO_2 and that the patient has a haemoglobin level usually about 7–9 g/dl. The increased efficiency of the low affinity HbS in delivering O_2 to tissues means that this is the physiological level of haemoglobin and does not indicate functional anaemia. Patients with sickle cell anaemia should not receive transfusions just because the level of Hb is lower than normal. It will be seen that patients with HbA/S or HbS/C have normal or near normal haemoglobin levels (Table 20.7). This is because combinations of HbS with HbA or HbC has a near normal O_2 affinity.

The clinical effects of sickle cell anaemia are widespread and many, but may be classified into a few distinct syndromes (Table 20.7). Painful vascular occlusion crises are the most common. The acute chest crisis is the most

Table 20.7 Sickle cell syndromes

Name	Haemoglobins	Clinical features
Sickle cell disease	HbS/HbS	Haemoglobin 7–9 g/dl Painful crises Chest syndrome Girdle syndrome Thrombotic risks
Sickle/β thalassaemia	HbS/-	Haemoglobin 7–9 g/dl Variable but usually as severe as sickle cell disease
Sickle/C disease (S–C disease)	HbS/HbC	Haemoglobin 10–12 g/dl Painful crises uncommon Thrombotic risk Crises in pregnancy Renal papillary necrosis Retinopathy
Sickle cell trait	HbS/HbA	Haemoglobin normal (11–14 g/dl) Clinically normal Risk only with prolonged vascular stasis and hypoxia

serious and commonest cause of death. The chest crisis is particularly devastating because of the tendency for symptoms and signs to improve with primary treatment, only to relapse acutely and often fatally, 24–48 hours later unless the HbS has been replaced by exchange transfusion. Other syndromes may also occur due to sequestration of red cells in the liver or spleen, haemolytic crises or aplastic crises associated with infection.

In addition to the sickle cell crises patients usually exhibit hyposthenuria (inability to concentrate urine) and functional asplenia. The former means that they readily become dehydrated and the latter that they have an increased susceptibility to infection, particularly from encapsulated organisms.

Factors which aggravate the propensity to crisis are hypoxia, acidosis, fever, dehydration, cold or chilling, and circulatory stasis. From an anaesthetic point of view two main queries have to be addressed: should the patient have preoperative exchange transfusion to remove the sickle cell haemoglobin and what special precautions should be taken with patients with sickle cell anaemia whether exchanged or not? The first problem, however, is to determine whether a patient has sickle cell anaemia, particularly in the emergency situation.

Diagnosis of sickle cell anaemia

The major population carrying the sickle haemoglobin gene is derived from Central and West Africa and hence it is common in the Afro/Caribbean/American black population. Additional, separate populations are found in the Mediterranean basin, especially northern Greece, and in the Middle East, particularly around the Gulf. The sickle gene

also developed in the indigenous population of the Indian subcontinent. The possibility of sickle cell anaemia or a sickle cell syndrome should be suspected and sought in these populations before surgery is carried out. The most reliable way to establish the diagnosis is through haemoglobin electrophoresis, particularly as this will provide a complete picture of the haemoglobin profile of the individual. In the urgent situation such information may not be available and the following steps should be taken:

1. *Blood count and film.* A haemoglobin value above 11.0 g/dl makes HbS/S unlikely, but does not distinguish HbA/A, HbA/S, HbS/C. A haemoglobin of 8 g/dl or less raises the possibility of HbS/S and this may become obvious if the blood film is examined for sickled cells. Intermediate haemoglobin levels are less helpful. The blood film in HbA/S or HbS/C does not show sickled cells.

2. *Sickle test.* This is a rapid screening test which relies on the gelation of HbS in an acid environment and is used to identify patients who carry the HbS gene. It does not distinguish between patients with HbA/S (trait), HbS/C or HbS/S. A negative test excludes sickle cell anaemia or HbS/C disease. The blood film will again help to confirm or deny HbS/S but still leaves the problem of HbS/C disease.

3. *Haemoglobin electrophoresis.* This should be carried out at the earliest opportunity if records are not available. In any event if there is any doubt the assistance of the paediatric/haematology teams should be sought as appropriate.

Anaesthesia in sickle cell anaemia

For major elective surgery most centres, where appropriate blood transfusion services exist, perform exchange transfusion preoperatively to lower the percentage of HbS in the blood to below 30 per cent. However, blood transfusion carries additional risk in these patients unless specifically genotyped blood is available and major delayed transfusion reactions are not uncommon. In parts of the world where the sickle gene is common but blood transfusion less readily available, major surgery may be successfully carried out using precautions based on understanding the sickle process. The following guidelines apply to patients with HbS/S or HbS/C underoing surgery with anaesthesia, general or regional, without exchange transfusion, but are also valid even for exchanged transfused elective patients.

Preoperative care

Surgery should be carried out whilst the patient is in a steady state without evidence of ongoing crisis. Dehydration is a major hazard for these patients. They must be given intravenous fluid to cover the period of preoperative starvation. This is always important but especially so if the surgery is likely to be delayed.

If the patient has HbS/S with a preoperative haemoglobin level of 7–9 g/dl and exchange transfusion is not planned, transfusion should not be performed except to cover blood loss. Transfusion of patients with HbS/S without removing some of the HbS increases the risk of crisis and of venous thrombosis. The shifted O_2 dissociation curve allows the patients to tolerate this degree of 'anaemia' well.

Induction of general anaesthesia
It would seem reasonable to preoxygenate the patient with 100 per cent O_2 for 5 minutes before induction to make sure that as much HbS as possible is in the oxy form in case there is delay in intubation. Once the airway is established 30–50 per cent inspired oxygen is usually preferred. There are no clinical trials to demonstrate the basis for such common sense approaches.

Although acidosis is a major precipitating factor for sickle cell crisis, the use of alkalinization measures has not helped in avoiding crises or improving the outcome of surgery and it is not recommended except to correct a metabolic acidosis.

Peroperative care

There are no data to suggest which type of anaesthetic agents are best tolerated. Hypotensive agents should be avoided in patients with HbS/S or HbS/C disease. The patient should be protected from chilling during surgery. Regional circulatory stasis should be avoided; if it is part of the surgical procedure exchange transfusion should precede surgery.

Postoperative care

This is the most hazardous period for patients with sickle disease (both HbS/S and HbS/C). Great care is required to make sure that the patient does not suffer from hypoxia, dehydration or cold during recovery. They should be given O_2 (100 per cent) by mask and i.v. fluids continued. They should be carefully monitored for evidence of hypotension or circulatory slowing. Once the patient has fully recovered from the anaesthesia mobilization and other antithrombotic measures should be started as early as possible because sickle cell anaemia and HbS/C disease are each risk factors for thrombosis.

Sickle cell haemoglobin C, HbS/C disease

The HbC gene is common in West Africa and the combination HbS/C is seen in populations who take their origin from that part of the world. HbS/C differs from HbS/S in that crises are less frequent though equally severe when they do occur, the haemoglobin level is near normal (usually > 11.0 g/dl) and the patient may be unaware that he/she has a sickle cell syndrome. Thrombotic complications are at least as frequent as with HbS/S disease. If exchange transfusion is required preoperatively the final proportion of HbA should be greater than 50 per cent and the total haemoglobin level maintained above 50 per cent.

━━━━━━━━━━━━━━━━━━━━━━━━ **REFERENCES** ━━━━━━━━━━━━━━━━━━━━━━━━

1. Rapaport SI. Preoperative hemostatic evaluation: which tests if any? *Blood* 1983; **61**: 229–31.
2. Colditz GA, Tuden RL, Oster G. Rates of venous thrombosis after general surgery: combined results of randomised clinical trials. *Lancet* 1986; **2**: 143–6.
3. Collins R, Scrimgeour A, Yusuf S, Peto R. Reduction in fatal pulmonary embolism and venous thrombosis by perioperative administration of subcutaneous heparin. *New England Journal of Medicine* 1988; **318**: 1162–73.
4. Hull RD, Raskob GE, Pineo GF, *et al.* Subcutaneous low

molecular weight heparin compared with continuous intravenous heparin in the treatment of proximal vein thrombosis. *New England Journal of Medicine* 1992; **326**: 975–85.

5. McNeill MJ, Thorburn J. Cannulation of the epidural space: a comparison of the 18- and 16-gauge needles. *Anaesthesia* 1988; **43**: 154–5.

6. Sage DJ. Epidurals, spinal and bleeding disorders in pregnancy: a review. *Anaesthesia and Intensive Care* 1990; **18**: 319–26.

7. Wildsmith JAW, McClure JH. Editorial. Anticoagulant drugs and central nerve blockade. *Anaesthesia* 1991; **46**: 613–14.

8. Stanley TH. Anticoagulants and continuous epidural anesthesia. *Anesthesia and Analgesia* 1980; **59**: 394–5.

9. Chong BH, Ismail F, Cade J, Gallus AS, Gordon S, Chesterman CR. Heparin-induced thrombocytopenia: studies with a new low molecular weight heparinoid Org 10172. *Blood* 1989; **73**: 1592–6.

10. Horlocker TT, Wedel DJ, Offord KP. Does preoperative antiplatelet therapy increase the risk of hemorrhagic complications associated with regional anesthesia? *Anesthesia and Analgesia* 1990; **70**: 631–4.

11. Guillebaud J. Editorial. Surgery and the pill. *British Medical Journal* 1985; **291**: 498–9.

12. Whitehead EM, Whitehead MI. Editorial. The pill, HRT and postoperative thromboembolism: cause for concern? *Anaesthesia* 1991; **46**: 521–2.

13. Lindberg MB, Crona N, Stigendal L, *et al.* A comparison between effects of oestradiol valerate and low dose ethinyl oestradiol on haemostasis parameters. *Thrombosis and Haemostasis* 1989; **61**: 65–9.

14. De Lingieres B, Basdevant A, Thomas G, *et al.* Biological effects of estradiol-17β in postmenopausal women: oral versus percutaneous administration. *Journal of Clinical Endocrinology and Metabolism* 1986; **62**: 536–41.

Chapter 21

Maternal and Fetal Effects of Anaesthesia

Masahiko Fujinaga and Jeffrey M. Baden

Maternal considerations
Placental considerations
Fetal considerations

Anaesthetic implications
Related topics

This chapter describes factors involved in the anaesthetic management of women during pregnancy other than during delivery. The basic objective is to promote maternal and fetal safety, especially the avoidance of teratogenic effects of anaesthesia and surgery. First, the basic biology of pregnancy and human development is described so that known and potential adverse effects of anaesthesia can be better assessed.

Notes in terminology

An important but confusing issue when discussing pregnancy and human development is the terminology used. First, one should remember that embryologists define life as beginning at fertilization, whereas obstetricians define it as beginning at the last menstrual period about 2 weeks earlier (Fig. 21.1). The term embryology generally refers to the study of the whole prenatal period. However, the embryonic stage refers only to the limited period after the bilaminar embryonic disc is formed (second week) and before the end of organogenesis when most of the major organs have been established (the end of eighth week by embryological definition; Fig. 21.1). Both embryologists and obstetricians define the term fetus as the period from the end of organogenesis to birth. For clinical convenience, obstetricians also divide the nine calendar months of pregnancy into three 3-month periods or trimesters. In this chapter, we will use trimesters as much as possible but we will also often use embryological definitions when gestational age is expressed in weeks. Another term used in obstetrics is fetal death which is defined as death occurring when the fetus weighs more than 500 g. This is approximately at the 20th week of pregnancy by the obstetrical definition (Fig. 21.1), and death before this time is called an abortion. To be consistent with the definition of the fetal period, embryologists regard fetal death as occurring any time after the eighth week of gestation. In this chapter, we generally use the terms fetus or fetal for the whole period of pregnancy to avoid these complexities, although other terms are also used as appropriate. In addition, when the term conceptus is used, it includes both fetal and extraembryonic tissues, such as the placenta.

Maternal considerations

Changes in physiology during pregnancy

During pregnancy, changes in physiology occur due to changes in anatomy and biochemistry, and they must be understood for appropriate management of anaesthesia. Because many of them become most significant at term, they are the major concerns of obstetric anaesthesia and are discussed in Chapter 64. In this chapter, we briefly summarize these changes, focusing on their time course during pregnancy. For further information, readers should consult more detailed reviews.[1–5]

The underlying aetiology of many physiological changes that occur during pregnancy are only partly understood. Hormonal effects are believed responsible for most changes that are prominent in the first trimester whereas pressure from the enlarging uterus on adjacent structures is believed responsible for most changes in the second half of pregnancy. Additional factors are increased metabolic demands from the growing uterus, fetus, placenta and breasts, and development of the uteroplacental circulation which acts as a haemodynamically significant low pressure arteriovenous shunt.

Hormone levels
Human chorionic gonadotropin (HCG) is produced almost as soon as the trophoblast is formed and can be detected in the serum by the end of the first week and in the urine by

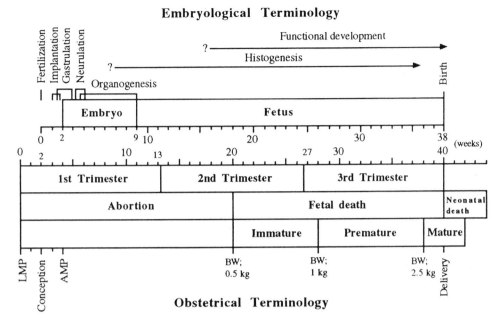

Fig. 21.1 Terminology used by embryologists and obstetricians for the stages of pregnancy. AMP, anticipated menstrual period; BW, body weight; LMP, last menstrual period.

the end of the second week. The serum concentration of HCG peaks during the 10th or 11th week, and then rapidly falls and remains low until term (Fig. 21.2). In contrast, other hormones, such as human chorionic somatomammotropin (HCS or human placental lactogen), progesterone and oestrogens, continuously increase during pregnancy and reach peak levels at full term (Fig. 21.2).

Cardiovascular function

Blood volume begins to increase in the first trimester and plateaus at approximately 50 per cent above the non-pregnant level in the third trimester (Fig. 21.3). Although red blood cell mass also rises steadily throughout

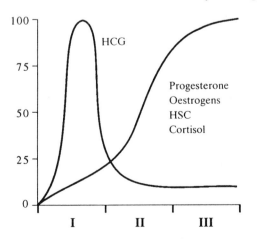

Fig. 21.2 Changes in some hormone levels during pregnancy. x-axis indicates the trimester and y-axis indicates the per cent of peak level during pregnancy.

pregnancy, it increases relatively less than total blood volume; thus, red blood cell count, haemoglobin, and haematocrit decline. Cardiac output increases steadily throughout pregnancy and reaches a peak of about 40 per cent above normal at term. Much of the increase is due to increased uterine blood flow which is 3 per cent of the total cardiac output during the first trimester but 10–15 per cent during the third trimester. Because systemic vascular resistance decreases throughout pregnancy, blood pressure remains unchanged or only slightly increased.

Respiratory function

Changes in respiratory function during pregnancy are due mainly to the respiratory stimulant effects of progesterone, increased total body oxygen consumption and mechanical effects of the enlarging uterus. Vasodilatation throughout the respiratory tract occurs early and is seen as increased vascular markings in the lungs by chest radiograph. Airway resistance decreases during pregnancy due to the relaxant effect of progesterone on bronchial smooth muscle. Hyperventilation usually begins in the first trimester, probably because of the stimulant effects of progesterone. Changes in lung volumes and capacities also begin early (Fig. 21.3). In particular, functional residual capacity (FRC) decreases significantly, thus increasing the potential for impaired oxygenation. Although the enlarging uterus produces an elevation of the diaphragm, total lung capacity and vital capacity remain unchanged because of a compensatory increase in the anterior–posterior and transverse diameters of the chest.

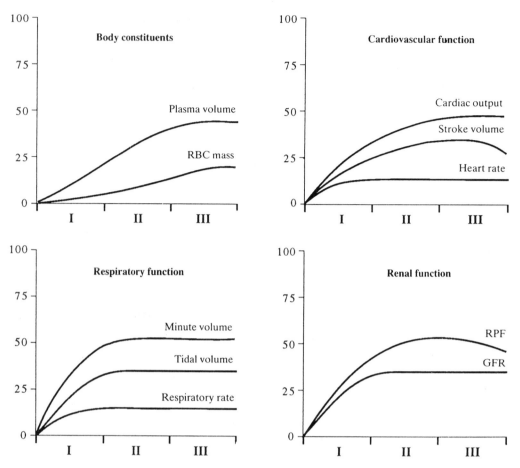

Fig. 21.3 Changes in some physiological parameters during pregnancy. *x*-axis indicates the trimester and *y*-axis indicates the per cent increase above non-pregnant values.

Renal function

Changes in renal function are due mainly to increased levels of ACTH, ADH, aldosterone, cortisol, HCS and thyroid hormone. Glomerular filtration rate (GFR) starts increasing early and plateaus at about 40 per cent above non-pregnant levels by midpregnancy (Fig. 21.3). Renal plasma flow (RPF) also increases by as much as 50 per cent and peaks by the end of the second trimester, remaining high until term (Fig. 21.3).

Gastrointestinal function

Gastrointestinal motility decreases probably because of a direct effect of progesterone, but possibly because of an indirect, inhibitory effect of progesterone on plasma motilin, a recently identified hormone which stimulates gastro-intestinal activity. Diminished lower oesophageal sphincter tone is apparent as early as the first trimester. After the second trimester, the stomach and intestines are gradually pushed upward by the enlarging uterus.

Coagulation and fibrinolytic functions

Plasma levels of factors VII, X, and XII and fibrinogen (factor I) increase during pregnancy, leading to a hypercoagulable state; prothrombin (factor II) is only slightly affected, if at all. Plasma fibrinogen concentrations begin to increase by the end of the first trimester and almost double by term. In contrast, fibrinolytic activity is considerably depressed during pregnancy. These changes protect the mother against catastrophic haemorrhage at the time of delivery, but they also increase her susceptibility to thrombosis.

Hepatic function

Although there are no gross morphological changes in the liver, there are functional changes which are reflected in laboratory data.[6] One of the concerns for anaesthetists is the plasma cholinesterase level which decreases signifi-cantly as early as the first trimester and remains low until delivery. Proposed aetiologies include haemodilution, hepatic dysfunction, hypoalbuminaemia, and elevated oestrogen levels.

Changes in sensitivity for anaesthetic agents

During pregnancy, the dose requirement for many anaesthetic agents is reduced. MACs for many inhalational anaesthetics have been reported to decrease by as much as 40 per cent in pregnant ewes probably because of the sedative effects of progesterone and/or the increased levels of endogenous opiates. Local anaesthetic requirement also decreases by as much as 30 per cent from the first trimester. Again, progesterone has been implicated as the underlying cause. Recent animal studies have shown that there is a greater sensitivity of nerve fibres to local anaesthetics. Epidural venous engorgement reduces the volume of both the cerebrospinal fluid and the epidural space and leads to additional decreases in drug requirement for spinal and epidural anaesthesia as pregnancy advances.

Placental considerations

The placenta starts to form soon after implantation and establishes some uteroplacental circulation between mother and fetus by the beginning of the fourth week. Major functions and activities of the placenta are metabolism, endocrine secretion and molecular transfer. One metabolic function that is especially noticeable during early pregnancy is synthesis of glycogen, cholesterol, and fatty acids which serve as sources of nutrients and energy for the embryo. The placenta secretes various kinds of hormones including HCG, HCS, human chorionic thyrotropin, human chorionic corticotropin, progesterone and oestrogens. Transfer of various molecules is a vital function of the placenta and is an important consideration in anaesthetic management. It is discussed below in detail.

Placental transfer of drugs and respiratory gases

The placental membrane is often called the 'placental barrier', a generally inappropriate term because only a few very large and/or charged molecules are unable to cross it. Almost all molecules can cross the placental membrane by one of the following four mechanisms: simple diffusion, facilitated diffusion, active transport, and pinocytosis.[7] Most drugs and metabolites cross the placenta freely by simple diffusion except for those structurally similar to amino acids. The rate of transfer by simple diffusion depends on permeability, tension gradient, relative affinities of maternal and fetal blood, maternal and fetal blood flows, and the area available for placental exchange.[8] In general, the driving force for placental drug transfer is the concentration gradient of free drug between maternal and fetal blood. Compounds with molecular weights less than 500, like most anaesthetic agents, cross the placenta unimpeded. Local anaesthetics and opioids are weak bases with a relatively low degree of ionization and considerable lipid solubility; thus, they easily cross the placenta. In contrast, only about 10–20 per cent of muscle relaxants cross the placenta because they are less lipophilic and more ionized. The effects of maternal plasma protein binding on transfer of drugs to the fetus are complicated, but in general, drugs that are highly bound are slower to cross to the fetus and achieve lower levels than less bound drugs.

Transport of O_2 and CO_2 in the placenta is by simple diffusion and is almost as efficient as in the lungs. If O_2 transport is interrupted for even a few minutes, the fetus is at great risk because there is essentially no storage of O_2. Since the amount of O_2 reaching the fetus is primarily flow limited, the most common cause of fetal hypoxia is a decrease in either uterine or fetal blood flow. The actual partial pressure and content of O_2 of fetal blood is a complex function of many factors including the shape of the O_2 dissociation curve for fetal and maternal haemoglobin, the haemoglobin content of fetal and maternal blood and the Bohr effect.[8] The partial pressure and content of CO_2 in fetal blood is less complicated; the dissociation curve for fetal and maternal haemoglobin is nearly linear and maternal and fetal CO_2 tensions are closely correlated over a wide range of values.

Placental blood flow

Placental blood flow, the most important factor in proper placental function, is largely passive. As aptly stated by Reynolds, it depends on many factors including maternal cardiac output and blood pressure, vasomotor tone of the uterine vessels, state of uterine contraction and pathological changes in the placenta.[8] Effects of anaesthetic-related agents on uterine blood flow are extremely complicated and not always easy to predict. For example, blood flow is generally well maintained or even increased at 1–1.5 MAC of halothane and isoflurane but can fall precipitously at 2.0 MAC. A detailed review of such effects is found elsewhere.[8,9]

Fetal considerations

Notes on embryology and teratology

Usual outcome of pregnancy

Spontaneous abortion is a biological process that is believed to play an essential role in human natural selection. Almost 80 per cent of human conceptions are estimated to be lost eventually, and 40 per cent are lost even before pregnancy is discovered.[10] Half of the early abortions have chromosomal abnormalities.[11] Despite the elimination of most abnormal fetuses, approximately 3 per cent of newborns have congenital malformations,[12] and of these, about one-third are considered life threatening.[13] Table 21.1 summarizes the estimated causes of developmental defects in humans.[12] About 25 per cent are thought to be associated with known genetic defects or chromosomal aberrations, whereas most of the rest are still listed as of unknown origin. Drug and environmental chemicals account for only a small percentage of abnormalities with known causes.

Brief summary of human development

The processes of human development and their susceptibility to teratogenic effects have been described often[7,14–16] and are briefly summarized here.

Gametogenesis

Gametogenesis is the formation and maturation of gametes (sperm and ova) which are the specialized generative cells involved in fertilization. The main event of gametogenesis is the halving of chromosome number from 46 to 23. Although there is increasing evidence that teratogenic effects may occur by changes in gametes induced by

Table 21.1 Known causes of developmental defects in humans

Known genetic transmission	20%
Chromosomal aberration	3–5%
Environmental causes	
Radiations	<1%
Infections	2–3%
Maternal metabolic imbalance	1–2%
Drug and environmental chemicals	2–3%
Unknown	65–70%
Total	100%

Modified from reference 12.

environmental factors including drugs, it is very difficult to identify such causal relations.

Fertilization and blastocyst formation

Human development begins at fertilization when a sperm unites with an ovum to form a single cell zygote. The essential feature of this complex cytological process is the recombination of hereditary material from two different sources and their subsequent redistribution in the genome of the offspring. Approximately 30 hours after fertilization, the zygote begins a series of rapid mitotic divisions, called cleavage, resulting in many daughter cells called blastomeres. Approximately 3 days after fertilization, a solid ball of 12–16 blastomeres, the morula, moves into the uterus. The following day, spaces start to appear between the central cells of the morula. These soon fill with fluid and coalesce to form a cavity which divides the cells into an inner cell mass (embryoblast) which gives rise to the embryo and the trophoblast which gives rise to part of the placenta. At this stage, the conceptus is called a blastocyst. Before implantation, the conceptus lies free within the fallopian tube and uterus, relying on their secretions for its nutrition. Many experimental studies have shown that exposure to teratogens during the preimplantation stages rarely causes the embryo to develop abnormalities. The reasons are probably twofold; first, the blastocyst does not receive effective levels of the teratogens until it actually comes into direct contact with the endometrium, and second, the embryonic cells at this stage are totipotential. The few malformations that are induced during this period can be explained by direct genetic damage to the embryo or by effects on the mother, such as maternal toxicity, endometrial changes and changes in oviductal fluid.

Implantation

Implantation of the blastocyst is a continuous process that begins at the end of the first week and is completed during the second week. As implantation progresses, morphological changes occur in the inner cell mass, resulting in formation of the embryonic disc which consists of two layers, the epiblast and hypoblast. Concurrently, lacunae (isolated spaces) appear in the syncytiotrophoblast which surrounds the blastocyst and they become filled with a mixture of maternal blood and secretions from the endometrial glands. The opening of uterine vessels into the lacunae is the beginning of the uteroplacental circulation. Successful implantation depends on both trophoblastic and endometrial factors and failure of this process results in early miscarriage. For example, large doses of oestrogens are known to interfere with these processes and are used for contraception. Women who miscarry during this period often do not know they are pregnant.

Embryonic period

From the time that the bilaminar embryonic disc is formed during the second week until the end of the eighth week when all major organs have appeared, the developing human is called an embryo (Fig. 21.1). This period is characterized by intense cellular activity that is required for important developmental processes including gastrulation, neurulation and organogenesis. Gastrulation is the process by which the inner cell mass is converted into two layers, epiblast and hypoblast, and eventually into three layers, ectoderm, mesoderm and endoderm. This is a crucial process of early development when organized migrations of cells and tissues take place to establish the basic body plan. Gastrulation begins during the second week and is completed by the end of the third week, after which most internal and external structures gradually begin to appear. Neurulation is another major event involving formation of the neural plate and closure of the neural tube; it begins during the third week and is completed by the end of the fourth week. Damage during neurulation causes neural tube defects, which may result in embryonic/fetal death, if severe, or congenital malformations such as exencephaly and spina bifida, if less severe. Organogenesis is defined as the period when all major organs are formed. Although it is difficult to determine precisely the beginning and ending of organogenesis, it is generally assumed to be between the fourth and eighth weeks,[16] after which the embryo has unquestionably human characteristics. The embryonic stage is by far the most susceptible period to teratogens. Pregnancy is often first noticed during the embryonic period since the third week of development coincides with the week following the first missed menstrual period. However, bleeding at the expected time of menstruation does not rule out pregnancy because sometimes there may be implantation bleeding.

Fetal period

Although transition from the embryonic to the fetal period is not marked by any abrupt change and therefore is somewhat arbitrary, the fetal period is generally considered to start at the beginning of the ninth week when differentiation of most organs is complete (see Fig. 21.1). It is characterized by marked histogenic changes, rapid growth, and functional maturation. The rate of body growth during the fetal period is very rapid, especially starting in the second trimester, whereas the rate of head growth is relatively slower. Overall weight gain is phenomenal during the final weeks of pregnancy. Although there is a general decrease in susceptibility to major teratogenic insults as differentiation progresses, several organs including the central nervous system, external genitalia, palate and teeth are not yet fully established and are still subject to structural damage. Minor congenital defects and tissue and organ damage may

occur at any time during the fetal period, and functional disturbances may develop even after birth.

Basic principles in teratology

Most teratologists accept the principle that teratogenicity is the end result of three basic interdependent factors, namely, genetic background (genotype), stage of development when exposure occurs and intensity of exposure. Furthermore, most believe Karnofsky's principle which states that any agent can be shown to be teratogenic providing enough is given at the right time.[17] The genotype of an individual influences not only the sensitivity to a particular teratogen but also the teratogen's pharmacological profile including its pharmacokinetics, pharmacodynamics and metabolic pathways. The stage of development when exposure occurs determines the severity of damage, the target tissues/ organs and the types of defects. Figure 21.4 presents critical phases of development when different organs show susceptibility to teratogens. As with other toxicological phenomena, the effects of teratogens are dose dependent.[12] For further information on teratology, readers should consult two excellent introductory textbooks[12,17] and two more extensive reviews.[18,19]

Manifestation of teratogenic outcomes

Death, structural (morphological) abnormality, growth retardation, and functional deficiency have been regarded as the four possible outcomes following a teratogenic insult.[12] Many now use the term developmental toxicity to cover these outcomes rather than the classic term teratogenicity that we will mainly use and that is sometimes regarded as only covering structural abnormality. Death may occur at any time during pregnancy and depending on the stage is called abortion, fetal death or stillbirth in humans (Fig. 21.1) and resorption or dead-in-utero in animals. The causes of intrauterine death are difficult to investigate and poorly understood. They may involve not only the fetus but also the placenta and mother. Structural abnormalities, often called malformations or congenital anomalies (defects), were the major subject of classic teratology but are now regarded as only one aspect of teratogenicity. They are usually categorized into malformation, deformation, disruption or dysplasia, depending on their pathogenesis and aetiology.[20] Severe structural abnormalities often lead to death, although death may occur without structural abnormalities being present. Growth retardation has received much less attention. However, it is now increasingly accepted as another manifestation of teratogenicity and is thought related to such diverse factors as placental insufficiency and genetic and environmental factors. Functional deficiency includes such variables as motor ability, sociability, emotionality and

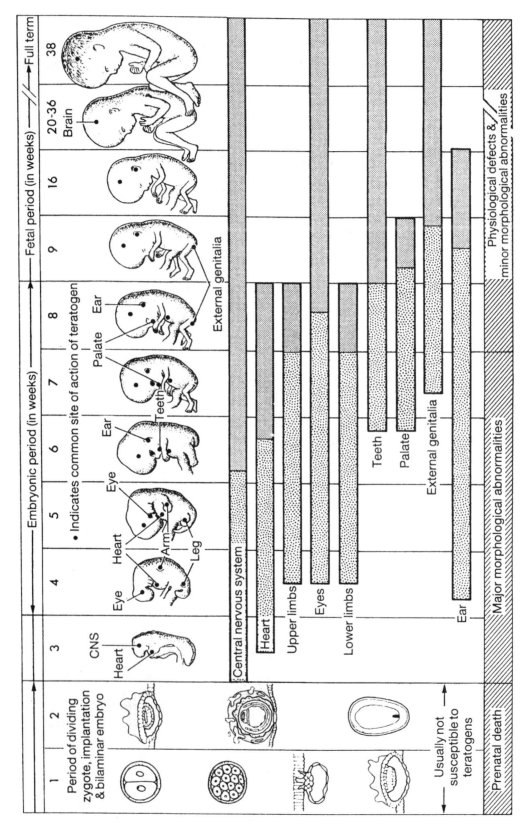

Fig. 21.4 Schematic illustration of human development and critical periods when different organs are most susceptible to teratogens. Periods of high susceptibility are indicated by stippling, whereas, periods of less susceptibility are indicated by shading. (Reproduced with kind permission of Dr K. L. Moore and W. B. Saunders Company from reference 7.)

learning capacity, and is associated with its own area of research called behavioural teratology or neurobehavioural toxicology.[21] In contrast to structural abnormalities which are mainly induced during organogenesis, functional deficiency is induced during late pregnancy or even after birth when the central nervous system is still developing (Fig. 21.1).

Teratogenic risk assessment

Since the thalidomide tragedy in the early 1960s, regulatory agencies in many countries have initiated requirements for animal testing of drugs aimed at detecting teratogenic risk to humans. For example, in the United States, three segment studies are now requested by the Food and Drug Administration (FDA) in at least two mammalian species, and at two or three dosage levels. Segment I studies involve toxicology tests addressing general fertility and reproductive performance. Segment II studies involve classic teratology tests addressing embryotoxicity, including birth defects. Segment III studies involve tests on peri- and postnatal development, addressing such problems as labour, delivery, lactation and neonatal viability. Other countries have similar systems but some have additional guidelines to evaluate behavioural teratogenicity. The FDA has developed a rating system for drugs based on those unpublished animal studies reported by manufacturers and available epidemiology studies. Drugs are assigned to one of the following categories: A (controlled studies show no risk), B (no evidence of risk in humans), C (risk cannot be ruled out), D (positive evidence of risk), or X (contra-indicated in pregnancy). The category for a particular drug is included in its package labelling or in such publications as *Physician's Desk Reference*.[22] However, this rating system has been widely criticized as an inadequate assessment of teratogenic risk in humans.[23,24] One of many reasons is that segment II tests are usually designed so that the highest dose produces maternal toxicity, and at this dose it is common for embryotoxicity to occur. In the absence of human data, a compound which causes such toxicity is usually put in category C and in fact, most drugs have been given the C designation.[24] Unfortunately, the category C designation does not give any indication of what risk potential there may be for humans.[24]

Most known human teratogens were first recognized in clinical situations rather than animal studies. Subsequently, animal studies have been used to confirm causal relations between suspected agents and the defects they induce, and to determine underlying mechanisms. There are several lists of human teratogens proposed by various investigators or organizations (Table 21.2). The lists differ somewhat in their details because assessment of teratogenic risk is not always straightforward.[13,24–27] They also include only drugs about which there are adequate data, a small percentage of drugs in common use. One reason is that epidemiological studies which should be the basis for risk assessment in humans are difficult and expensive to perform and have not been considered justified for most drugs or chemicals. Furthermore, among the 60 000 synthetic chemicals which are now in use, only about 4000 have been tested in animals.[27] In Table 21.2, anaesthetics are listed as unlikely teratogens on the basis that they have neither been definitely linked to birth defects nor shown to be completely safe. No consideration is given

Table 21.2 Definite, possible and unlikely teratogens in humans

Human teratogens
Radiation
  Atomic weapons, radioiodine, therapeutic
Infections
  Cytomegalovirus, herpesvirus hominis ? I and II, parvovirus B-19 (erythema infectiosum), rubella virus, syphilis, toxoplasmosis, Venezuelan equine encephalitis virus
Maternal metabolic imbalance
  Alcoholism, cretinism (endemic), diabetes, folic acid deficiency, hyperthermia, phenylketonuria, rheumatic disease and congenital heart block, virilizing tumours
Drugs and environmental chemicals
  Aminopterin and methylaminopterin, androgenic hormones, busulphan, captopril, chlorobiphenyls, cocaine, coumarin anticoagulants, cyclophosphamide, diethylstilboestrol, diphenylhydantoin, enalapril, etretinate, iodides and goitre, lithium, mercury (organic), methimazole and scalp defects, penicillamine, 13-*cis*-retinoic acid, tetracyclines, thalidomide, trimethadione, valproic acid
Possible human teratogens
  Binge drinking, carbamazepine, chronic villus sampling, cigarette smoking, disulfiram, ergotamine, high vitamin A, lead, primidone, streptomycin, toluene abuse, varicella virus, zinc deficiency
Unlikely human teratogens
  Agent Orange, *anaesthetics*, aspartame, aspirin (but aspirin in the 2nd half of pregnancy may increase cerebral haemorrhage during delivery), Bendectin (antinauseants), birth control pills, illicit drugs (marijuana, LSD), metronidazole (Flagyl), oral contraceptives, rubella vaccine, spermicides, video display screens

Modified from reference 25 with kind permission of Dr T. H. Shepard and The John Hopkins University Press.

Table 21.3 Summary of epidemiological studies on fetal outcome after non-obstetrical operation

	Mazze and Källén (1989)[32]	Duncan et al. (1986)[31]	Crawford and Lewis (1986)[30,a]	Brodsky et al. (1980)[29]	Shnider and Webster (1965)[34,a]	Smith (1963)[35]
Source	Swedish records	Manitoba records	Hospitals records	Dental personnel postal survey	Hospital records	Hospital records
Period	1973–81	1973–81[b]	1968–84	1968–78[c]	1959–64	1957–61
No. of pregnancies examined	720 000	?	?	12 929	9073	18 248
No. of operations (% of total)	5405 (0.75)	2565	59	287 (2.22)	129 (1.42)	67 (0.36)
During 1st trimester	2252	?	35	187	45	10
During 2nd trimester	1881	?	?	100	42	45
During 3rd trimester	1272	?	?	0	42	12
Categories of anaesthesia						
General anaesthesia	2929	911	35	?	45	24
Regional anaesthesia	369	46	0	?	15	43
Local anaesthesia	375	337	0	?	50	0
No anaesthesia	0	503	0	?	19	0
Unknown	1732	768	0	?	0	0
Manifestation of teratogenicity (1st/2nd/3rd trimester)						
Death						
Spontaneous abortion	Not examined	Neg[d]	??	Pos/Pos/?	Not examined	??
Stillbirth	Neg/Neg/Neg	?				
Postnatal death	Neg/Pos/Pos	Not examined			Neg	
Structural abnormality	Neg/Neg/Neg	Neg	??	Neg/Neg/?	Neg	??
Growth retardation		Not examined	??	Not examined		
Very low birth weight (< 500 g)	Pos/Pos/Neg				Neg	??
Low birth weight (< 2500 g)	Pos/Pos/Pos				Neg	??
Premature delivery	Pos/Pos/Pos				??	
Functional deficiency	Not examined	Not examined	Not examined	Not examined	Not examined	Not examined

[a] Data from patients having cervical cerclage were included in the original reports but have been omitted from this table.
[b] Entire population of Manitoba, Canada, of approximately 1 million.
[c] Dentists, dental assistants and their spouses in the United States. Response rate was 70%.
[d] There was no difference between control and operated groups. However, there was a significant increase for the subgroup having general anaesthesia.
?, No data presented; ??, no statistical analysis was made.

to the fact that, if inappropriately used, they may produce severe changes in maternal physiology and can cause teratogenic effects (discussed later).

Evaluation of animal studies

For most drugs including many used in anaesthetic practice, animal studies provide the only data available to assess teratogenic risk in humans. However, evaluation of *in vivo* animal studies is not straightforward.[24,26,27] For example, as mentioned earlier, any agent can be shown to be teratogenic in animals providing enough is given at the right time.[17] Indeed, most animal studies are designed to maximize the likelihood of obtaining positive results. Thus, the presence of dose–response data is very important in assessing teratogenic risk. A dose of test agent at which no adverse effects occur is the no observed adverse effects level (NOAEL). If the NOAEL (dose/kg) divided by the human exposure level (dose/kg) is 100 or greater, most scientists consider the test agent relatively safe for humans although obviously there is no absolute division between safe and unsafe.[24,26] Selecting which species of animal to use for hazard assessment is also somewhat arbitrary. Ideally, selected animals should metabolize and distribute the test drug in ways similar to humans. In addition to classic *in vivo* tests, various kinds of *in vitro* models have been developed, some aimed at screening chemicals and drugs of interest and others aimed at elucidating mechanisms of teratogenicity. They have proved to be very successful for the latter purpose, but there is less certainty about their role as screening tests. The current consensus is that *in vitro* studies and non-mammalian *in vivo* studies should not be used for human risk assessment.

Several reference books listed in various compendiums are available which provide reviews of animal studies and epidemiology studies of drugs and other chemicals.[24,28] In the United States, at least two online systems are also available. These are REPROTOX (Reproductive Toxicology Center, Columbia Hospital for Women Medical Center, 2440 M Street, N.W., Suite 217, Washington, DC 20037–1404; phone, within the USA, 1–800–535–4100; international, 202–293–5946/5947), and REPRORISK which includes REPROTEXT (Reproductive Hazard Reference), Shepard's Catalog of Teratogenic Agents and TERIS (The Teratogen Information System) (Micromedex, Inc., 600 Grant Street, Denver, Colorado 80203–3527; Phone, within the USA, 1–800–525–9083; international, 1–303–831–1929).

Teratogenic risk of anaesthesia

Epidemiological studies

Epidemiological studies of teratogenicity from anaesthesia are very difficult to perform for several reasons. First, there is an extremely high background incidence of spontaneous abortion, and thus, large numbers of patients are required to detect statistical increases. Second, the overall incidence of birth defects is very low as is the incidence of operations during pregnancy; again, studies must include large numbers of patients. Third, most patients are prescribed many drugs during pregnancy and it is difficult to separate their effects.[27] Finally, it is difficult to separate effects of the anaesthetic drug or procedure of interest from effects of the surgery or the original condition that necessitated it. Despite these difficulties, several reports have been published on fetal outcome following anaesthesia and these are summarized in Table 21.3[29–35] Reports on obstetrical operations, mainly Shirodkar's procedure, were not included in the table because the risk of a poor outcome is so high following these procedures.[30,34,36]

Several conclusions can be drawn from these studies with some degree of certainty. First, structural abnormalities are an unlikely consequence of anaesthesia and surgery no matter when in pregnancy they occur. Second, the incidence of low birth weight and premature delivery is increased if patients are subjected to anaesthesia and surgery at any time during pregnancy. Finally, the incidence of postnatal death is increased if patients are subjected to anaesthesia and surgery during the second and third trimesters. Much less certain is whether anaesthesia and surgery increase the incidence of spontaneous abortion. At first sight the studies of both Duncan and colleagues[31] and Brodsky and colleagues[29] give credence to there being an increased incidence. However, studies of spontaneous abortion are notoriously difficult to interpret. One difficulty in their results is a reported background incidence of spontaneous abortion below 10 per cent. Such a low number emphasizes the dangers of responder bias and failure to medically confirm spontaneous abortion since it is well known than the normal incidence is many-fold greater. It should be noted that results from almost all the epidemiological studies relate to the combination of anaesthesia and surgery and no conclusion can be made about whether one, the other or both is causal. Also, no conclusions about postnatal functional deficiency after maternal anaesthesia and surgery can be made since relevant studies have not been performed.

Teratogenic effects of abnormal physiological conditions

If anaesthesia eventually proves to be a contributing factor to poor fetal outcome following surgery, changes in maternal physiology are even more likely to be an aetiology than direct toxicity by anaesthetic drugs. The ultimate result of most physiological insults is fetal hypoxia which can certainly cause developmental toxicity to various degrees. Unfortunately, designing well controlled, *in vivo* experiments to study teratogenic effects of a specific physiological change is very difficult because most physiological parameters are interdependent and change continually and concurrently. Indeed, although numerous reports of such topics as hypoxia and acidosis during pregnancy have been written, almost no conclusions have been drawn. For example, most experiments of hypoxia have been performed by exposing animals to low O_2 concentrations for various durations, but since arterial O_2 tensions have not been reported, interpretation of results is almost impossible. In contrast to the uncertain results from *in vivo* studies, results from many *in vitro* studies have shown teratogenic effects of hypoxia under relatively well defined conditions. Nevertheless, it is difficult to extrapolate such results to clinical situations. All that can be stated at present is that maternal physiological parameters should be kept within normal ranges to avoid potential adverse fetal outcome.

Interestingly, mammalian embryos are under hypoxic conditions before the placenta starts to function fully. During this period, they use the anaerobic glycolysis pathway to provide most of their energy requirements rather than the Krebs cycle and electron transport system. Many studies have shown that rodent embryos during this period require a low O_2 tension to grow normally *in vitro*, and a high or even normal O_2 tension can cause teratogenic effects. Whether there is any clinical implication is not known. It is commonly believed, however, that maternal hyperoxia near term does not cause fetal retrolental fibroplasia or premature closure of the ductus arteriosus, whereas both these conditions can occur if high O_2 tensions are delivered to the neonate.

Increasing concern has been expressed in recent years that changes in body temperature can cause teratogenic effects, and there is now reasonable evidence that hyperthermia does so in humans.[37] Although hyperthermia, in general, rarely occurs in modern anaesthetic practice, cases of malignant hyperpyrexia do occur and probably are associated with adverse fetal outcome. Hypothermia is also known to cause teratogenicity in animals,[38] although there are many case reports of uneventful birth after hypothermia used during cardiac surgery.[9]

Teratological studies of anaesthetic agents in animals

Many animal studies aimed at investigating adverse reproductive effects of drugs used in anaesthetic practice have been performed. It is beyond the scope of this chapter to describe these studies in detail, but a few salient comments about various classes of anaesthetic drugs are made below. Two general comments are in order. First, as already noted, physiological effects produced by the drug under test conditions must be taken into account when interpreting results. It is surprising how infrequently this is done and, hence, how often conclusions about a drug's direct toxicity are invalid. The second comment is that many drugs used in anaesthesia have such potent physiological effects that they are often not tested in animals at the same high dosages used clinically. This is because a large number of animals, usually rodents, must be used in reproductive studies, and it is not practical to control the many physiological changes that occur in each animal. In contrast anaesthetists carefully control such changes in each patient to prevent morbidity. The situation with many anaesthetic drugs is quite the reverse of normal toxicological screening when concentrations of drugs far above those used clinically are tested.

Inhaled anaesthetics

A number of studies have been conducted at trace, subanaesthetic and anaesthetic concentrations of inhaled anaesthetics using various species of animals. Among currently used agents, N_2O is the only one that has been consistently shown to be teratogenic in mammals (discussed later). It is generally regarded as a weak teratogen because exposure to at least 50 per cent for 24 hours is required to cause teratogenicity in rats, although its behavioural teratogenicity remains to be fully investigated. Although there is not an exact equivalence, 24 hours of exposure from gestational day 8 in rats is equivalent to approximately 2–3 days of exposure during the third week of pregnancy in humans. One should remember that although the exposure required to produce teratogenic effects in rats is long, the difference in sensitivity between rats and humans is unknown.

Opioids

Many investigators have reported that the opioids are teratogenic when given in high doses to animals. In most studies, however, severe respiratory depression must have occurred and was the probable cause of the teratogenicity. In contrast, many well designed studies aimed at minimizing respiratory depression indicate that currently used opioids are not teratogenic at concentrations clinically achieved.[39]

Table 21.4 Types of non-obstetric operations performed on Swedish pregnant patients from 1973 to 1981

	1st trimester	2nd trimester	3rd timester	Total
No. of cases	2252	1881	1272	5405
(% of total cases)	(41.6)	(34.8)	(23.5)	(100.0)
Type of operation (%)	100.0	100.0	100.0	100.0
Central nervous system	6.7	5.4	5.6	6.0
Eye, ear, nose and throat	7.6	6.4	9.5	7.8
Heart, lung	0.7	0.8	0.6	0.7
Abdominal	19.9	30.1	22.6	24.6
Genitourinary, gynaecological	10.6	23.3	24.3	18.6
Laparoscopy	34.1	1.5	5.6	16.1
Orthopaedic	8.9	9.3	13.7	10.3
Endoscopy	3.6	11.0	8.6	7.5
Skin	3.8	3.2	4.1	3.7
Miscellaneous	4.0	4.2	5.5	4.6

Modified from reference 33.

Muscle relaxants

Muscle relaxants are especially difficult to test in animals because of the respiratory paralysis that they cause in the mother, and only few published reports are available. The few studies performed in rodents have indicated that none of the clinically used muscle relaxants is teratogenic.[40] There are many studies showing that prolonged disruption of muscle activity causes limb anomalies in chicks. However, these effects are seldom seen in mammals, probably because the placental barrier prevents complete paralysis of the fetus.

Local anaesthetics

Commonly used local anaesthetics have not been found to cause teratogenic effects at clinical concentrations, with the exception of cocaine if it is classified as a local anaesthetic. Lignocaine (lidocaine) has been reported to cause neural tube defects in embryos of mice and chicks *in vitro*. However, we recently demonstrated that it does not produce similar lesions in rat embryos, suggesting that the effect is species specific. Although several studies have reported behavioural deficits following lignocaine, the results are inconsistent and further work is necessary to clarify the issue. Cocaine is the only local anaesthetic that has been shown to be a human teratogen.[25] Fortunately, it is hardly used today in anaesthetic practice, although patients addicted to it often require anaesthesia.

Intravenous induction agents

None of intravenous induction agents, including barbiturates, benzodiazepines, and ketamine, have been shown to be teratogenic at clinical concentrations. Although there are some studies reporting teratogenic effects of diazepam at higher than clinical concentrations, these reports have been widely criticized because of inappropriate experimental design and conclusions. The consensus at present is that diazepam is not a human teratogen.[25]

Anaesthetic implications

Non-obstetrical surgery

Type of operations

According to the Swedish registry which contains data on the whole Swedish population of approximately 8 million, 5405 non-obstetric operations were performed among the women who delivered 720 000 babies between 1973 and 1981.[33] Most operations occurred in the first trimester (41.6 per cent) with lesser numbers occurring in the second (34.8 per cent) and third (23.5 per cent) trimesters (Table 21.4). Laparoscopy (34.1 per cent) was the most common operation performed during the first trimester, whereas this procedure made up only 1.5 per cent and 5.6 per cent during the second and third trimesters, respectively. Appendectomy was the most frequently performed procedure in the second and third trimesters.

Consideration before operation

All women of childbearing age scheduled for surgery should be carefully questioned regarding the possibility of pregnancy. Even if pregnancy is identified immediately after a missed menstrual period, the embryo is in the third week of pregnancy when gastrulation has already begun (Fig. 21.1). At this time and later, the embryo is susceptible to teratogenic insult. Nevertheless, the most frequent and serious error is unnecessary delay of an urgently required surgery. The reason is that delay in treating such a patient leads to an increase in both maternal and fetal morbidity and mortality that far outweighs any potential risk to the fetus. The common reasons for delay are innate fear of anaesthesia based on old assumptions, failure to perform indicated diagnostic procedures such as x-rays for fear of affecting the fetus and difficulty in arriving at a correct diagnosis.[41] In general, if the surgical problem is not an acute one, operation should be postponed until after delivery. If that is not possible, the operation may be performed anytime during pregnancy as discussed earlier. The patient and her family should be informed that currently available data indicate that there is increased risk of low birth weight, premature delivery, and postnatal death after surgery at any stage of pregnancy, no increased risk of congenital defects even after surgery during the first trimester, and unknown risks for abortion and behavioural deficiency.

Choice of anaesthesia

To date, there is no evidence that any anaesthetic drug or technique is safer than another. Thus, the type of anaesthesia should be that considered appropriate for the particular case and should be one that minimizes changes in maternal physiology. Similarly, appropriate premedication of any kind should be given if necessary. Some have recommended that N_2O should not be administered to pregnant women or, if it is used, that it be given with folinic acid to bypass the metabolic block created by methionine synthase inactivation. However, the scientific evidence at the present time does not support this recommendation (discussed later). Based on current information, we recommend that N_2O be used if it is necessary for the appropriate management of anaesthesia, especially as techniques that compensate for N_2O's absence from an anaesthetic regimen have not been shown to be safer.

Some considerations during anaesthesia

Choosing the appropriate dosages of anaesthetic agents requires extra caution because sensitivity to many drugs is increased from as early as the first trimester. Doses of local anaesthetics used for epidural anaesthesia should be further decreased later in pregnancy because of the reduced volume of the epidural space. Inhalational anaesthesia is rapidly induced in the pregnant patient because decreased FRC results in less dilution of inspired gases and quicker achievement of desired alveolar concentrations. Although still debated, most clinicians believe that the dosage of suxamethonium (succinylcholine) should be reduced because of low cholinesterase levels during pregnancy. Prevention of pulmonary aspiration also needs special attention during pregnancy because of decreased gastric motility, increased gastric content and

difficulty of tracheal intubation because of weight gain. Increased vascularity of the respiratory tract mucosa can result in profuse bleeding during and after placement of a tracheal tube.

After the first trimester, the enlarging uterus may compress the aorta and inferior vena cava and thereby decrease venous return and cardiac output. A substantial decrease in uteroplacental perfusion may follow even if overt maternal hypotension is absent. Thus, great care should be taken to provide adequate uterine displacement during both regional and general anaesthesia if progressive fetal deterioration is to be avoided. Maternal blood pressure is often high during either regional or general anaesthesia in the supine position because of elevated catecholamine levels secondary to light anaesthesia. Again, one cannot assume that placental perfusion is adequate. Blood loss is generally well tolerated by the pregnant patient because of increased blood volume and red cell mass. Nevertheless, catastrophic blood loss is not uncommon and must be anticipated.

Hypoxia and hypercapnia occur more rapidly in the pregnant patient than in the non-pregnant patient because of decreased FRC which results in reduced O_2 storage in the lungs, increased O_2 consumption, early airway closure, and decreased cardiac output in the supine position. Thus, preoxygenation is recommended before tracheal intubation and supplemental O_2 during regional anaesthesia. Although its clinical significance is still controversial, maternal hyperventilation may result in respiratory alkalosis and hypocapnia and may cause adverse effects in the fetus.

Fetal monitoring

Fetal heart rate can reliably be monitored transabdominally by Doppler techniques as early as the end of the first trimester and may be useful in evaluating fetal status during surgery. Although great store is placed on heart rate variability, one should remember that many drugs used for anaesthesia result in a reversible decrease in variability without serious consequences. Later in pregnancy when the uterine fundus is well above the pelvic rim, uterine contractions can be monitored using a tocodynamometer. This technique may be useful during non-abdominal surgery and after any surgery for determining whether tocolytic therapy should be employed to suppress preterm labour.

After the operation

Although not usually the role of the anaesthetist post-operative care of the pregnant patient and newborn requires special attention to factors such as maternal bleeding and infection, premature delivery, perinatal development and early detection of congenital defects.[42]

In vitro fertilization

In vitro fertilization in humans began in the late 1970s and is becoming increasingly common. Evidence so far indicates that the procedure is not associated with an increased incidence of congenital defects or spontaneous abortion once pregnancy is established, although data are limited. For many years, laparoscopy was the standard method for oocyte retrieval and is still used. However, ultrasound guided follicular aspiration, either transvaginal or percutaneous–transvesicular, is rapidly replacing laparoscopy and does not require general anaesthesia.

Premedication

Premedication of any kind may be administered as appropriate since no one regimen has been found to be safer than another. Generally, long-acting sedatives or opioids should be avoided if oocyte retrieval is to be done as an outpatient procedure. Because adult outpatients are considered by many to be at greater risk of aspiration than inpatients, they are often prophylactically administered antacids, histamine blockers and metoclopramide. Whether such treatment is effective in this setting is still being debated.

Local anaesthesia

For ultrasound guided oocyte retrieval, local anaesthesia has proved to be entirely satisfactory in most patients. The local anaesthetic is infiltrated into the suprapubic area for the transabdominal approach or vaginal wall for the transvaginal approach. When laparoscopy is used, the need for the Trendelenburg position and pneumoperitoneum make local anaesthesia unsuitable for most patients.

Regional anaesthesia

Regional anaesthesia is rarely used for oocyte retrieval. For ultrasound guided oocyte retrieval in the outpatient setting, it is usually unnecessary. Furthermore, such a technique as spinal anaesthesia is associated with an unacceptably high incidence of headaches in young women. For laparoscopy, regional anaesthesia has the same disadvantages as local anaesthesia.

General anaesthesia

Laparoscopic follicular aspiration is usually performed under general anaesthesia which provides excellent muscle relaxation, the ability to control ventilation, elimination of patient anxiety, complete analgesia and amnesia, and a still surgical field.[43] To date, no single anaesthetic regimen has been shown to be safer than another. There are some theoretical concerns about the use of N_2O because of its ability to reduce cell division in certain tissues such as bone marrow. However, in at least one study, no difference in fertilization or pregnancy rate was found whether N_2O was used or not.[44]

Laparoscopy requires a Trendelenburg (head-down) position and pneumoperitoneum. Both conditions reduce vital capacity by increasing cephalad pressure on the diaphragm and causing the mediastinum to shift cephalad. A similar mechanism may cause the tracheal tube to migrate into a main-stem bronchus. A head-down position increases pulmonary blood volume, leading to decreased pulmonary compliance and diminished functional residual capacity, factors that predispose the patient to postoperative atelectasis. Gas embolism and pneumothorax are rare but potentially serious complications from pneumoperitoneum. Hypercapnia can also occur from reabsorption of the CO_2 used for producing the pneumoperitoneum. Far more common, however, is haemorrhage from blind needle placement which accounts for almost half of all complications of laparoscopy.

During laparoscopy, special care should be taken to avoid brachial plexus injury which is more common than normal because of the head-down position, especially if shoulder braces are used to prevent the patient slipping on the operating table. These devices can compress and stretch brachial plexus nerves by depressing the clavicle into the retroclavicular space. The incidence may be decreased by avoiding unnecessarily steep head down and not hyperextending the arms if they must be abducted.

At the conclusion of surgery, the patient should be returned slowly to the horizontal position to allow gradual shunting of blood from the thorax and upper extremities to the lower extremities so that severe hypotension does not occur.

Induced abortion

Over 1.5 million pregnancies are terminated voluntarily each year in the United States alone. The procedure is not without risk. For example, although the mortality rate had fallen eightfold since 1972, the Center for Disease Control in the United States found there were still 0.5 deaths per 100 000 abortions in 1981. In one report of 193 deaths investigated between 1972 and 1985, 29.4 per cent followed complications from general anaesthesia.[45]

Fetal therapy/surgery

Fetal therapy/surgery is a relatively new area of medicine that continues to expand rapidly. Some procedures such as intrauterine blood transfusion for erythroblastosis fetalis and placement of vesicoamniotic shunt catheter are considered minor since they can be performed percutaneously. Others such as repair of diaphragmatic hernia and excision of cystic adenomatoid malformation are more major and are performed via a hysterotomy. No doubt as more experience is gained and technology advances, some procedures will move from the major to the minor category. A recent excellent review by Rosen describes the experiences of a fetal treatment programme at the University of California, San Francisco.[46]

Minor procedures

Local anaesthetic infiltration of the maternal abdomen is usually sufficient for percutaneous placement of needles and catheters. Adequate fetal sedation may be achieved via placental transfer of drugs such as opioids and benzodiazepines administered to the mother. However, occasionally maternal comfort may be inadequate when multiple attempts are needed for needle placement or when dealing with overly anxious patients. Furthermore, fetal movement may not be entirely prevented by fetal sedation and may make the procedure very difficult to complete; it also increases the incidence of bleeding, trauma, and compromise to the umbilical circulation. When both maternal discomfort and excessive fetal movement occur, general anaesthesia should be considered. If control of fetal movement is all that is needed, it can be safely achieved by direct intramuscular or intravascular administration of a neuromuscular blocking agent to the fetus. Prophylactic use of tocolytic agents for control of uterine irritability may be appropriate when contemplating more extensive procedures.

Major procedures

General anaesthesia is usually used for procedures involving hysterotomy. Placement of a lumbar epidural catheter for postoperative analgesia with opioids is often performed before induction of anaesthesia.[46] Volatile anaesthetics are particularly useful for maintenance of anaesthesia because they reliably anaesthetize both mother and fetus, although some delay in producing adequate levels of anaesthesia in the fetus should be anticipated because of the fetal circulation. Volatile anaesthetics also provide satisfactory uterine relaxation for surgery and decrease uterine irritability. However, deep anaesthesia with volatile agents should be avoided since it can produce an unacceptable depression of the fetal cardiovascular system and progressive fetal acidosis. Maternal considerations for anaesthesia during intrauterine fetal surgery are similar to those already described for non-obstetrical operations. Neuromuscular blocking agents may be needed for fetal immobilization, but because the placental transfer of these agents is limited, they should preferably be injected into the fetus either intramuscularly or into the umbilical vein under

ultrasonic guidance. Preterm delivery is a major complication of fetal surgery. Thus, tocolytic agents including ritodrine, terbutaline or magnesium sulphate are routinely administered postoperatively.[46]

Related topics

Mechanisms of N_2O-induced teratogenicity in animals

Nitrous oxide (N_2O) is the only inhalational anaesthetic that has definitely been shown to be teratogenic in mammals. Fink and his colleagues at the University of Washington, Seattle performed the seminal work in this area. They reported in 1967 that pregnant rats exposed continuously to 50 per cent N_2O for 2, 4 or 6 days starting on gestational day 8 had increased incidences of resorptions and skeletal abnormalities; gestational day 0 was defined as the day when a copulatory plug was observed.[47] Subsequently, they established that exposure to 70 per cent N_2O for 24 hours on gestational day 8 consistently caused teratogenic effects, and since then, this model has been used extensively to investigate the mechanisms of N_2O-induced teratogenicity. Later, the threshold concentration for effect was found to be about 50 per cent; the threshold exposure time for effect is yet to be accurately determined.

The adverse reproductive effects that follow N_2O exposure in rats are now clearly established. They are resorptions (embryonic/fetal loss), skeletal abnormalities including major and minor rib and vertebral defects, visceral abnormalities including situs inversus and growth retardation. The effects of N_2O on postnatal function have not been studied using the standard model although preliminary data from other models suggest that functional deficiencies can occur. The days of gestation when embryos are most susceptible to the effects of N_2O are day 8 (resorption, situs inversus, minor skeletal anomalies such as extra cervical rib), day 9 (major skeletal malformations) and day 11 (resorptions).[48] Although the aetiology of the resorptions is unclear, it has been demonstrated that after exposure on day 8, the highest incidence of fetal death occurs between days 13 and 14.[49] This is the time when the liver takes over from the yolk sac as the dominant organ for haemopoiesis, and it has been suggested that N_2O exposure on day 8 might damage primordial liver cells. The aetiology of the resorptions after exposure on day 11 may be due to the failure of the embryo to switch from dependence on the yolk sac to dependence on chorioallantoic placenta for vital nutritional and metabolic functions which occurs on this day.[48]

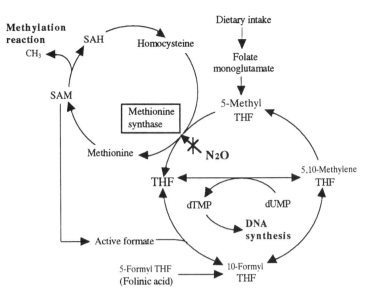

Fig. 21.5 Metabolic pathways affected by the inhibition of methionine synthase by N_2O. SAH, S-adenosyl homocysteine; SAM, S-adenosyl methionine; THF, tetrahydrofolate; dTMP, deoxythymidine; dUMP, deoxyuridine.

Until recently, the sole cause of N_2O-induced teratogenicity had been thought to be the oxidation of vitamin B_{12} which cannot then function as a coenzyme for methionine synthase. This enzyme catalyses the transmethylation from methyltetrahydrofolate to homocysteine to produce tetrahydrofolate and methionine (Fig. 21.5). The expected result of its inhibition is decreased tetrahydrofolate which may lead to impaired DNA synthesis and decreased methionine which may lead to impaired methylation reactions. Inactivation of methionine synthase by N_2O occurs rapidly in animals and humans and is known to cause a pernicious anaemia-like syndrome consisting of subacute combined degeneration of the spinal cord, megaloblastic anaemia and pancytopenia in humans. Because the haematological changes in humans are prevented by folinic acid (5-formyl tetrahydrofolate) administered with N_2O, presumably because DNA synthesis is restored to normal, impaired DNA synthesis had been proposed to account for N_2O teratogenicity.[50]

However, the following evidence indicates that lack of methionine rather than tetrahydrofolate is the main cause of N_2O-induced teratogenicity, and that sympathetic stimulation rather than methionine synthase inhibition accounts for the situs inversus. First, maximum reduction of methionine synthase activity and a decrease in DNA synthesis occur at dosages of N_2O that are well below those that produce teratogenic effects.[51] Second, supplementation with folinic acid which should restore DNA synthesis to normal partially reduces the high incidence of only one type of malformation, minor skeletal anomalies.[52] The third line of evidence comes from an *in vitro* rat whole embryo culture system that has recently been used to elucidate mechanisms of N_2O-induced teratogenicity.[53] Studies using this system have demonstrated that supplementation with methionine, but not with folinic acid, almost completely prevent N_2O-induced growth retardation and all malforma-

tions other than situs inversus.[54] They have also shown that α_1-adrenergic stimulation mimics N_2O-induced situs inversus.[55] Clearly, N_2O-induced teratogenicity is multifactorial. The mechanisms involved in the production of teratogenicity by lack of methionine and of situs inversus by sympathetic stimulation are undergoing intense study.

Teratogenic effects of waste anaesthetic gases in the workplace

Many epidemiological studies have been conducted to evaluate the reproductive performance of personnel who work in operating rooms or dental offices, and their spouses. These studies have been extensively reviewed in numerous articles and book chapters and will not be described again here. Most investigators agree with the report from the epidemiologists who were commissioned in 1980 by the American Society of Anesthesiologists to conduct an independent review of studies existing at the time.[56] They found that there were small increases in the rate of spontaneous abortion (30 per cent) and congenital defects of offspring (20 per cent) among women directly exposed to waste anaesthetic gases. The increases were within the range that normally occurs with studies of the type reviewed and that are often due to various biases and uncontrolled confounding factors. No consistent increase of adverse reproductive effects were found for wives of exposed males. The epidemiologists pointed out that even if the increases were real, many factors such as x-rays and viruses could have accounted for them. Based on this review and subsequent registry-based studies[57,58] that are generally regarded as more credible than the original studies, most now believe that hazards from low levels of waste anaesthetic gases are minimal or absent. Nevertheless, this issue is still under active investigation and an occasional report adds fuel to the concept that hazards may exist under certain circumstances. For example, in a recent study, the National Institute of Environmental Health Sciences of the United States reported that high levels of N_2O exposure in dental assistants adversely affect a woman's fertility.[59]

REFERENCES

1. Hytten FE, Leitch I. *The physiology of human pregnancy.* 2nd ed. Oxford: Blackwell Scientific Publications, 1971.

2. Cole PL, St John Sutton M. Normal cardiopulmonary adjustments to pregnancy: cardiovascular evaluation. *Cardiovascular Clinics* 1989; **19**: 37–56.

3. Dafnis E, Sabatini S. The effect of pregnancy on renal function: physiology and pathophysiology. *American Journal of Medical Science* 1992; **303**: 184–205.

4. Kase NG, Reyniak JV, Bergh PA. Endocrinology of pregnancy. In: Cherry SH, Merkatz IR, eds. *Complications of pregnancy: medical, surgical, gynecologic, psychosocial, and perinatal.* 4th ed. Baltimore: Williams & Wilkins, 1991: 916–39.

5. Bonnar J, Daly L, Sheppard BL. Changes in the fibrinolytic system during pregnancy. *Seminars in Thrombosis and Hemostasis* 1990; **16**: 221–9.

6. Chitkara U, Bowers C. Laboratory assessment and values of importance in pregnancy. In: Cherry SH, Merkatz IR, eds. *Complications of pregnancy: medical, surgical, gynecologic, psychosocial, and perinatal.* 4th ed. Baltimore: Williams & Wilkins, 1991: 1226–64.

7. Moore KL, Persaud TVN. *The developing human. Clinically oriented embryology.* 5th ed. Philadelphia: WB Saunders Company, 1993.

8. Reynolds FJM. The fetus and placenta. In: Churchill-Davidson HC, ed. *Wylie & Churchill-Davidson's A practice of anaesthesia.* 5th ed. Chicago: Year Book Medical Publishers, Inc, 1984: 1069–89.

9. Strickland RA, Oliver WC, Chantigian RC, Ney JA, Danielson GK. Anesthesia, cardiopulmonary bypass, and the pregnant patient. *Mayo Clinic Proceedings* 1991; **66**: 411–29.

10. Roberts CJ, Lowe CR. Where have all the conceptions gone? *Lancet* 1975; **1**: 498–9.

11. Rushton DI. The nature and causes of spontaneous abortions with normal karyotypes. In: Kalter H, ed. *Issues and reviews in teratology.* Vol. 3. New York: Plenum Press, 1985: 21–63.

12. Wilson JG. *Environment and birth defects.* New York: Academic Press, 1973.

13. Shepard TH. Human teratogens: how can we sort them out? *Annals of the New York Academy of Sciences* 1986; **477**: 105–15.

14. Arey LB. *Developmental anatomy. A textbook and laboratory manual of embryology.* 7th ed. Philadelphia: WB Saunders, 1974.

15. O'Rahilly R, Müller F. *Human embryology and teratology.* New York: Wiley-Liss, 1992.

16. Sadler TW. *Langman's medical embryology.* 6th ed. Baltimore: Williams & Wilkins, 1992.

17. Wilson JG, Warkany J, eds. *Teratology. Principles and techniques.* Chicago: The University of Chicago Press, 1965.

18. Kalter H, ed. *Issues and reviews in teratology.* Vols 1–6. New York: Plenum Press, 1983–1993.

19. Wilson JG, Fraser FC, eds. *Handbook of teratology.* Vols 1–4. New York: Plenum Press, 1977.

20. Spranger J, Benirschke K, Hall JG, *et al.* Errors of morphogenesis: concepts and terms. *Journal of Pediatrics* 1982; **100**: 160–5.

21. Riley EP, Vorhees CV, eds. *Handbook of behavioural teratology.* New York: Plenum Press, 1986.

22. *Physician's desk reference.* 47th ed. Montvale, NJ: Medical Economics Co, Inc, 1993.

23. Friedman JM, Little BB, Brent RL, Cordero JF, Hanson JW, Shepard TH. Potential human teratogenicity of frequently prescribed drugs. *Obstetrics and Gynecology* 1990; **75**: 594–9.

24. Scialli AR. *A clinical guide to reproductive and developmental toxicology.* Boca Raton: CRC Press, 1992.

25. Shepard TH. *Catalog of teratogenic agents.* 7th ed. Baltimore: The John Hopkins University Press, 1992.

26. Jelovsek FR, Mattison DR, Young JF. Eliciting principles of hazard identification from experts. *Teratology* 1990; **42**: 521–33.

27. Schardein JL. *Chemically induced birth defects.* New York: Marcel Dekker, Inc, 1985.

28. Scialli AR. The proliferation of reference books in developmental toxicology. *Reproductive Toxicology* 1991; **5**: 459–61.

29. Brodsky JB, Cohen EN, Brown BW, Wu ML, Whitcher C. Surgery during pregnancy and fetal outcome. *American Journal of Obstetrics and Gynecology* 1980; **138**: 1165–7.

30. Crawford JS, Lewis M. Nitrous oxide in early human pregnancy. *Anaesthesia.* 1986; **41**: 900–5.

31. Duncan PG, Pope WDB, Cohen MM, Greer N. The safety of anesthesia and surgery during pregnancy. *Anesthesiology* 1986; **64**: 790–4.

32. Källén B, Mazze RI. Neural tube defects and first trimester operations. *Teratology* 1990; **41**: 717–20.

33. Mazze RI, Källén B. Reproductive outcome after anesthesia and operation during pregnancy: a registry study of 5405 cases. *American Journal of Obstetrics and Gynecology* 1989; **161**: 1178–85.

34. Shnider SM, Webster GM. Maternal and fetal hazards of surgery during pregnancy. *American Journal of Obstetrics and Gynecology* 1965; **92**: 891–900.

35. Smith BE. Fetal prognosis after anesthesia during gestation. *Anesthesia and Analgesia* 1963; **42**: 521–6.

36. Aldridge LM, Tunstall ME. Nitrous oxide and the fetus. A review and the results of a retrospective study of 175 cases of anaesthesia for insertion of Shirodkar suture. *British Journal of Anaesthesia* 1986; **59**: 1348–56.

37. Edwards MJ. Hyperthermia as a teratogen: a review of

experimental studies and their clinical significance. *Teratogenicity, Carcinogenicity and Mutagenicity* 1988; **6**: 563–82.

38. Smoak IW, Sadler TW. Hypothermia: teratogenic and protective effects on the development of mouse embryos *in vitro*. *Teratology* 1991; **43**: 635–41.

39. Fujinaga M, Mazze RI. Teratogenic and postnatal developmental studies of morphine in Sprague–Dawley rats. *Teratology* 1988; **38**: 401–10.

40. Fujinaga M, Baden JM, Mazze RI. Developmental toxicity of nondepolarizing muscle relaxants in cultured rat embryos. *Anesthesiology* 1992; **76**: 999–1003.

41. Slater GI, Aufses AH. Complications of pregnancy: surgical aspects. In: Cherry SH, Merkatz IR, eds. *Complications of pregnancy: medical, surgical, gynecologic, psychosocial, and perinatal*. 4th ed. Baltimore: Williams & Wilkins, 1991: 722–31.

42. Fabro S, Scialli AR. The role of the obstetrician in the prevention and treatment of birth defects. In: Kalter H, ed. *Issues and reviews in teratology*. Vol. 3. New York: Plenum Press, 1985: 1–20.

43. Spielman FJ. *In vitro* fertilization. In: James FM, Wheeler AS, Dewan DM, eds. *Obstetric anesthesia: the complicated patient*. 2nd ed. Philadelphia: FA Davis Company, 1988: 545–54.

44. Rosen MA, Roizen MF, Eger EI, *et al*. The effect of nitrous oxide on in vitro fertilization success rate. *Anesthesiology* 1987; **67**: 42–4.

45. Atrash HK, Cheek TG, Hogue CR. Legal abortion mortality and general anesthesia. *American Journal of Obstetrics and Gynecology* 1988; **158**: 420–4.

46. Rosen MA. Anesthesia for procedures involving the fetus. *Seminars in Perinatology* 1991; **15**: 410–17.

47. Fink BR, Shepard TH, Blandau RJ. Teratogenic activity of nitrous oxide. *Nature* 1967; **214**: 146–8.

48. Fujinaga M, Baden JM, Mazze RI. Susceptible period of nitrous oxide teratogenicity in Sprague–Dawley rats. *Teratology* 1989; **40**: 439–44.

49. Fujinaga M, Baden JM, Shepard TH, Mazze RI. Nitrous oxide alters body laterality in rats. *Teratology* 1990; **41**: 131–5.

50. Nunn JF, Chanarin I. Nitrous oxide inactivates methionine synthase. In: Eger EI, ed. *Nitrous oxide/N_2O*. New York: Elsevier Science, 1985: 211–33.

51. Baden JM, Serra M, Mazze RI. Inhibition of fetal methionine synthetase by nitrous oxide. *British Journal of Anaesthesia* 1984; **56**: 523–6.

52. Mazze RI, Fujinaga M, Baden JM. Halothane prevents nitrous oxide teratogenicity in rats, folinic acid does not. *Teratology* 1988; **38**: 121–7.

53. Baden JM, Fujinaga M. Effects of nitrous oxide on day 9 rat embryos grown in culture. *British Journal of Anaesthesia* 1991; **66**: 500–3.

54. Fujinaga M, Baden JM. Mechanisms of nitrous oxide (N_2O)-induced reproductive toxicity in rat embryos grown in culture. *Teratology* 1992; **45**: 476. (Abstract)

55. Fujinaga M, Maze M, Hoffman BB, Baden JM. Activation of α-1 adrenergic receptors modulates the control of left/right sidedness in rat embryos. *Developmental Biology* 1992; **150**: 419–21.

56. Buring JE, Hennekens CH, Mayrent SL Rosner B, Greenberg ER, Colton T. Health experiences of operating room personnel. *Anesthesiology* 1985; **62**: 325–30.

57. Ericson HA, Källén AJB. Hospitalization for miscarriage and delivery outcome among Swedish nurses working in operating rooms 1973–1978. *Anesthesia and Analgesia* 1985; **64**: 981–8.

58. Hemminki K, Kyyrönen P, Lindbohm ML. Spontaneous abortions and malformations in the offspring of nurses exposed to anesthetic gases, cytostatic drugs and other potential hazards in hospitals, based in registered information of outcome. *Journal of Epidemiology and Community Health* 1985; **39**: 141–7.

59. Rowland AS, Baird DD, Weinberg CR, Shore DL, Shy CM, Wilcox AJ. Reduced fertility among women employed as dental assistants exposed to high levels of nitrous oxide. *New England Journal of Medicine* 1992; **327**: 993–7.

Chapter 22

Ventilators and Anaesthetic Breathing Systems

Paul Beatty

Physiological considerations in the use of breathing systems and ventilators	Breathing systems
Artificial ventilators	

The first steps in artificial ventilation were made in the sixteenth and seventeenth centuries when Vesalius and Hook used a bellows to support the ventilation of a dog.[1, 2] Ventilatory support during surgery became necessary as thoracic techniques advanced but were severely limited by the 'pneumothorax problem'. In 1900 Matas[3] demonstrated intermittent positive pressure ventilation (IPPV) which was followed in 1905 by the first practical apparatus for surgical IPPV.[4] In 1904 Sauerbruch demonstrated negative pressure ventilation (NPV). The polio epidemics of the 1940s and 1950s accelerated the development of ventilators.[5] The concept of a physiological ventilation waveform has now been challenged by high-frequency jet ventilation.

In contrast with the increasing complexity of ventilators, anaesthetic breathing systems have remained simple, comprising bags, tubing and simple valves. Initially, the administration of volatile agents (chloroform, ether) and nitrous oxide used separate breathing systems, but by 1876 Clover had combined nitrous oxide and ether administration into one apparatus.[6] Oxygen and nitrous oxide were combined (1886) by Hillischer[7] but all three elements of the modern anaesthetic gas machine were not incorporated in one machine until the twentieth century with machines such as Gwathmey's nitrous oxide, oxygen and ether apparatus (1914)[8] and Wellesley's 'Boyle' machine (1917).[9] About 1930 Magill invented the Magill attachment. The use of soda lime to absorb carbon dioxide was exploited in 1926 by Waters[10] in the 'to-and-fro' system and by Sword in 1930 who used flap valves to create a circle system.[11]

Physiological considerations in the use of breathing systems and ventilators

Respiratory mechanics

The movement of gas in and out of the respiratory system is achieved during spontaneous ventilation by the combined action of the diaphragm and the muscles of the chest which increase the volume of, and reduce the pressure within, the thoracic cavity, creating a positive pressure difference from the atmosphere to the alveoli. Whether spontaneous or controlled ventilation is used, a pressure gradient is required.

Three mechanical properties of the respiratory system will determine the rate and efficiency with which gas will flow. These are:

- *Resistance*: the frictional effect that opposes the flow of gas in the airways and includes the viscous resistance of the tissues.
- *Compliance*: the relation between pressure and volume of gas flow that expresses the energy dissipated in stretching components of the respiratory system.
- *Inertance*: the component that expresses the storage of kinetic energy in the tissues and gases in the respiratory system during ventilation.

In practice the effect of inertance is relatively small and can be ignored. The respiratory system may be considered to be a compliance fed by a resistance.

Resistance

Tissue (viscous) resistance is largely a product of friction between tissues in the chest wall. It can be increased in some pathological conditions but in general it remains relatively constant even during anaesthesia.

Airway resistance is the result of resistance to the flow of gas itself. It may be expressed as a pressure drop per unit volume flow. For laminar gas flow in a tube it is given by:

$$R = 8\eta \; \frac{L}{\pi r^4}$$

where η is the viscosity, r the radius of the pipe and L the length of the pipe.

Thus increased length or decreased radius will increase resistance. There are further factors. Some parts of the

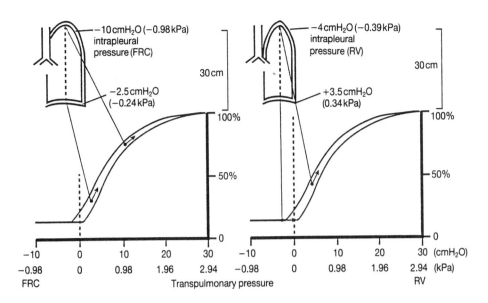

Fig. 22.1 Regional differences in the static transpulmonary pressure–lung volume curve. Ventilation at residual volume (RV) leads to regions at the base of the lung operating at volumes near to closing volumes. At a total lung volume close to functional residual capacity (FRC) there is still a distribution of compliance across the regions of the lung due to the weight of the lung but even the regions at the base of the lung do not approach closing volume. (Reproduced with permission from West JB. *Ventilation/blood flow and gas exchange.* Oxford: Blackwell Scientific Publications, 1965.)

respiratory tract, such as the larynx, contain sharp changes in radius which behave like orifices, rather than tubes and tend to generate turbulent flow. In turbulent flow, resistance is given by:

$$R = \rho \, \frac{L}{4\pi^2 r^5}$$

where ρ is the density.

Resistance in the airway is increased during turbulent flow and is influenced by changes in the mucosal surface such as inflammation. However, since resistance depends on air flow velocity, it tends to be confined to the proximal part of the bronchial tree, particularly within the larynx. In normal patients 20–30 per cent of the airway resistance is incurred proximal to the cricoid cartilage in the region of turbulent flow.

The sum of airway and tissue resistance is the total respiratory resistance. Total respiratory resistance has been found to double during anaesthesia from 0.3 to 0.6 kPa/l per second. This may be due to the fall in functional residual capacity (FRC) and an associated reduction in airway radius. Connection to any breathing system or ventilator further changes the total respiratory resistance by substituting a variety of tracheal, laryngeal or nasal devices for all or parts of the upper respiratory tract.

Compliance

The relation between volume and the pressure of the gas in the alveoli is termed compliance. It is defined as the volume increase per unit increase in pressure (l/kPa) and has two components associated with the lung and the chest wall. The sum of these is the total compliance:

$$\frac{1}{\text{Total compliance}} = \frac{1}{\text{Lung compliance}} + \frac{1}{\text{Thoracic cage compliance}}$$

There is interpatient variability in compliance but much of this variability can be related to differences in FRC.

Compliance is a measure of elastic properties. The relation between pressure and volume for a balloon made of an elastic material, such as rubber, is non-linear. At low volumes the pressure above atmospheric pressure required to inflate the balloon is high because the radius of the balloon is small and its compliance is therefore low. This effect is true even for an ideal material, i.e. one obeying Hooke's law, but with real materials the effect is enhanced by non-linear characteristics at low extensions. As volume increases, the pressure required to produce a unit increase in volume falls and compliance is higher. As the material stretches towards its elastic limit the pressure per unit increase in volume rises again and compliance is reduced. This effect is true for all the regions of the lung and gives rise to the pressure–volume curves shown in Fig. 22.1. For a healthy lung near FRC, most regions will be in the middle of the pressure–volume curve and their compliances will be low. However, at lung volumes close to residual volumes (RV) or in cases of reduced FRC, some units will be in the low compliance region at low positive pressures, i.e. just above atmospheric pressure, and thus overall lung compliance will be reduced.

The term specific compliance has been used to facilitate a comparison between the elastic properties of the lungs of different patients.

$$\text{Specific compliance} = \frac{\text{Total compliance}}{\text{FRC}}$$

The normal range of specific compliance is 0.004–0.007 kPa.

Anaesthesia reduces FRC and thus reduces compliance.[12,13] Pathological responses to anaesthesia, such as anaphylaxis, can induce dramatic and critical reductions in compliance. However, after the general reduction in FRC, the most important effect reducing compliance will be that of posture. Total compliance has been found to fall by 38 per cent when going from the lithotomy position to a prone position with the head down and FRC is reduced when supine.[14,15]

Whereas resistance can be thought of as a relatively fixed linear quantity, compliance is distinctly non-linear. Measurements of static compliance in spontaneously ventilating subjects can be made by inspiring a measured volume from a spirometer and then breath-holding with the glottis open. The pressure at the mouth can be considered to be the same as that in the alveoli, as no gas is flowing, and the compliance can be calculated. However, at the moment the breath is held pressure does not immediately become constant. It reaches a peak from which it then subsides to a plateau. This effect is due to redistribution of gas from 'fast' to 'slow' lung units and stress relaxation in the tissues of the chest. It is also an indicator of non-linear mechanical effects and hysteresis in the tissues.

A simple resistance–compliance model of lung mechanics

It remains instructive to use a simple resistance–compliance model when considering ventilator action.

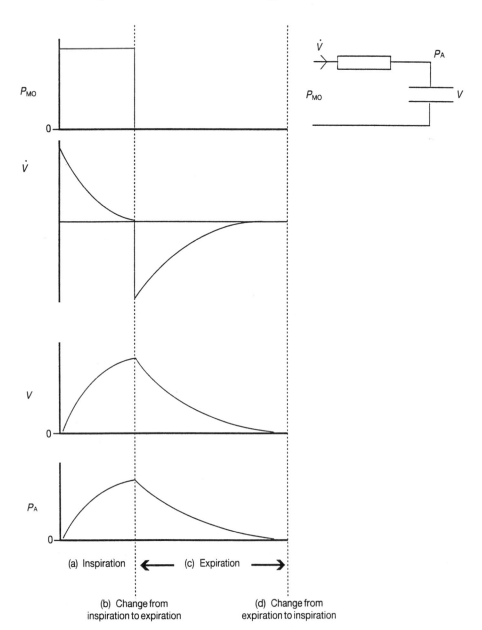

Fig. 22.2 The pressure, flow and volume curves for a simple resistance–compliance model of ventilated lung under constant pressure ventilation. P_{MO}, pressure at the mouth; $\dot{V}$, the flow; V, lung volume; P_A, alveolar pressure.

A resistance–compliance model with the pressure, flow and volume curves for IPPV using a constant inspiratory pressure is shown in Fig. 22.2. Since pressure at the mouth is constant, as the lung fills the flow rate falls as the alveolar pressure approaches the pressure at the mouth asymptotically. In fact the rate at which the lung fills is directly proportional to the pressure difference between the airway and the alveoli, the lung therefore fills exponentially and flow rate falls exponentially. Exponential changes are characterized by their time constants. In this case the time constant of the lung (T) is given by:

$$\text{time constant } (T) = \text{total compliance } (C) \times \\ \text{pulmonary resistance } (R)$$

The model shows that a given tidal volume is a balance between the transpulmonary pressure difference, inspiratory time and resistance; furthermore there may be a considerable difference between the peak airway pressure and the peak alveolar pressure.

The lung empties passively and exponentially during expiration and another effect becomes clear: for a given expiratory time the pressure in the lung at end expiration is determined by the resistance.

Dead space

Dead space is the most important physiological factor affected by breathing systems. In the case of the patient it is physiological dead space that is important but, when breathing system efficiency is considered, it is anatomical dead space that dominates.

Physiological dead space can be defined as that part of tidal volume which does not participate in gas exchange.[16] Its derivation uses the Bohr mixing equation and arterial $P\text{CO}_2$ values obtained using blood gas analysis. It thus includes the effects of $\dot{V}_A/\dot{Q}$ mismatch and A−a differences as well as the effects of apparatus and anatomical dead space.

Anatomical dead space can be defined in two ways:

- *Morphological*: the volume of the conducting passages which are not covered in respiratory epithelium. Usually measured volumetrically.
- *Functional or series*: the fraction of inspired tidal volume which is expired unchanged at the beginning of expiration. Quantified using simultaneous measurements of expired gas composition and volume (Fowler's method[17]), in which the boundary between dead space gas and alveolar gas is determined by graphical display of the changes in expired volume and carbon dioxide.

The first definition of anatomical dead space is not clinically accessible for measurement but the second can be measured during anaesthesia. It will not be equivalent to the morphological dead space, however, by definition, and it will be affected by connection to the breathing system which replaces those parts of the upper respiratory tract which contribute to anatomical dead space. In studies using artificial lung models with the anatomical dead space simulated with a tube of known volume, the measurements of functional anatomical dead space are invariably lower than the volumes of the tubes used to simulate it.[18]

Artificial ventilators

The functional classification of ventilators

The functionality of a ventilator may be considered using three major approaches:

- The method of producing a positive pulmonary pressure gradient.
- The method of generating the gas flow in the ventilator.
- The method of separating inspiration from expiration (cycling).

These approaches are not independent of each other. For instance, an IPPV ventilator will generate its positive airway pressure from a source of gas, with properties which determine the characteristics of the airway pressure and predispose the designer to use certain methods of cycling. However, there is a distinct operational separation between them. While the selection of a method of pressure gradient generation is chosen on clinical grounds, the second two approaches are concerned with ventilator control. It is appropriate to consider methods of negative pulmonary pressure gradient production separately from the issue of flow generator characteristics and cycling.

Types of artificial ventilation and other methods of respiratory support

Intermittent positive pressure ventilation (IPPV)
IPPV is the common form of artificial ventilation. It mimics the physiological pattern of spontaneous ventilation with rates of the order of 10 breaths/min and tidal volumes in excess of patient dead space. It creates a positive transpulmonary gradient by raising airway pressure above atmospheric pressure.

Intermittent negative pressure ventilation (INPV)
The use of INPV is often associated with the polio epidemics of the 1930s and 1950s though it was originally

conceived for use during surgery. It has been superseded by IPPV for the support of critically ill patients in hospital, but its use has increased for home ventilation of neurologically impaired patients or those requiring ventilatory support at night.

The working principles require the thorax to be surrounded with a rigid container inside which a lower than atmospheric pressure is created. This in turn creates the required transpulmonary pressure gradient.

Initially, INPV as used for surgical purposes was a complete room with the head of the patient protruding through a wall. In the 1950s this concept was replaced by the tank ventilator or 'iron lung'. In this the patient was surrounded completely by a tank approximately the size of a hospital bed. The patient had a window of some kind to look through. Access was clearly difficult and movement highly restricted.

A far more convenient design is that used for the cuirass in which only the thorax is surrounded by the rigid container. The system is lightweight and affords better access and freedom of movement and is more suitable for home use.

High-frequency ventilation (HFV)

HFV is provided by rapid breathing rates and tidal volumes of less than the dead space. It is in general a positive airway pressure technique but there is one version of the system that employs a cuirass and negative pressure.

HFV has been defined as being ventilation at rates greater than four times those of the normal subject[19] but a wide variety of techniques have been used with a range of tidal volumes and breathing rates. The systems are classified in a somewhat arbitrary fashion using respiration rate but three categories have been defined.[20]

High-frequency positive pressure ventilation (HFPPV) through a normal tracheal tube, using breathing rates of 60–100 breaths/min (1.0–1.7 Hz) and a tidal volume of 2–3 ml/kg of body weight, was the first practical form of HFV.[21]

High-frequency jet ventilation (HFJV) is the most widely used form of HFV.[22] Inspired gases are delivered to the airway at five times atmospheric pressure through a catheter in the trachea. Breathing rates are between 100 and 400 breaths/min (1.7–6.7 Hz). At these frequencies, tidal volume has little meaning and it is more logical to think in terms of maximum inspiratory gas flow rates, which are of the order of 100 l/min, and overall gas flow, 400 ml/min.

High-frequency oscillation ventilation (HFOV), breathing rate of 400–2400 breaths/min (6.7–40 Hz), is superimposed by a loudspeaker or piston pump[23] on a continuous flow of oxygen into the subject. At such high frequencies the acoustic properties of the bronchial tree play a significant if not dominant role.

External high-frequency oscillation (EHFO) uses a semi-rigid cuirass to generate high-frequency pressure oscilla-tions at greater and less than atmospheric pressure. Tidal volume is between 1.5 and 6.5 ml/kg and breathing rate, 60–720 breaths/min (1–12 Hz), placing it in the HFOV band for classification. Maximum CO_2 elimination occurs at 30 breaths/min.[24]

With such a range of frequencies and methods it is not surprising that the mechanisms that allow HFV to achieve gas exchange are not well understood. It is difficult to measure gas concentrations, pressures and flows at these frequencies and it is likely that several different physical factors combine to achieve gas exchange for any particular method. It is difficult to determine which effect dominates at any one time. At the alveolar level it is likely that gas exchange is by diffusion between the alveolar and the pulmonary capillaries as in IPPV. However, above this level there are several possible mechanisms, which are summarized below.

Convective streaming During HFV the velocity of inspiratory gas flow is much higher than the expiratory flow. This results in a greater gas velocity at the centre of the airway during inspiration than in expiration. A volume of gas in the middle of the airway will travel further down the airway during inspiration than it will travel back on the subsequent expiration. Thus in the centre of the airway there will be a net flow of fresh gas into the lung. At the periphery of the airway, where the velocity difference will be small due to viscosity effects, there will be a net flow of gas out of the lung.[25]

Direct alveolar ventilation It has been suggested that for some alveoli the dead space between them and the ventilator is low enough for the small tidal volumes of HFV to be in excess of that dead space. This is due to distance or to gas flow dynamics that reduce the effective dead space.[26]

Turbulent dispersion Normally gas velocities are not high enough to cause turbulence in small airways. However, this may not be the case during HFV; mixing caused by turbulence will abolish the axial distribution of gas concentration in the small airways and improve gas exchange.

Acoustic resonance Acoustic resonance in the bronchial tree is well known[27] and this, too, would be likely to cause improved mixing of gas in the small airways.

The advantages of HFV during anaesthesia are associated with lower peak airway pressures, lower pressure and volume excursions and its capacity to deliver ventilation in difficult circumstances or with a disrupted airway. HFPPV gives lower chest excursions that have advantages for some operative fields. HFJV can be used during bronchoscopy. EHFO avoids the need for intubation. The reduced airway pressures can be of advantage for neonatal use when the risk of barotrauma is reduced. However, the reduced pressures do not improve cardiac index, and may increase the complications of artificial ventilation without eliminating those of IPPV. The high pressures involved pose new problems for safety and these ventilators tend to be noisy. HFV assists weaning from mechanical ventilation during intensive therapy but it may reduce spontaneous respiratory effort.

Other methods of respiratory support

The rocking bed It is known that the abdominal contents increase intrapleural pressure when a patient moves from the erect to the supine position. In 1932 it was shown that the weight of the abdominal contents can generate a tidal volume of 300 ml during positional change.[28] This is the basis of the rocking bed in which a patient is tipped backwards and forwards through an angle of 30° about the horizontal. The technique is used primarily for those requiring intermittent or nocturnal ventilatory support who have a marginal impairment of function or neurological disease.

Extracorporeal and peritoneal oxygenation The primary function of artificial ventilation is to maintain gas exchange. The use of substitute gas exchange systems should therefore be mentioned when artificial ventilation is considered. Two methods have been adopted.

Extracorporeal membrane oxygenators (ECMO), using venoarterial bypass, have been used to treat severe and acute respiratory failure. The results of oxygenation alone are not encouraging, though they are improved when carbon dioxide is removed.

The value of using oxygen carrying perfluorocarbon emulsions has been demonstrated in animals and experimented with in human subjects. The technique is easier than ECMO and uses equipment similar to that used for peritoneal dialysis but is not a routine treatment at this stage of development.

The classification of artificial ventilators by control strategy and flow generator type

Despite the variety of ventilation methods, IPPV remains the most common form. It is therefore logical to describe the characteristics and control of ventilators from the point of view of IPPV.

A flow diagram for constant inspiratory pressure ventilation from the beginning of inspiration to the end of expiration is shown in Fig. 22.2. There are four distinct phases:

(a) Inspiration.
(b) The change from inspiration to expiration.
(c) Expiration.
(d) The change from expiration to inspiration.

Phases (a) and (c) are clear since they are seen during normal healthy ventilation. Phases (b) and (c) are less obvious, and at first sight seem to be only related to the ventilator and its control. These phases are indeed the key control phases of the ventilator but they also have physiological counterparts.

Phase (a) is dominated by the characteristics of the flow generator in the ventilator and the resistance and compliance of the ventilated lung. In practice neither of these two factors is under the direct control of the anaesthetist. The former was set by the ventilator manufacturer and the latter by the patient. Their interaction determines whether an appropriate tidal volume can be delivered in the available time. The anaesthetist has control over phase (b).

In phase (b) monitored parameters of flow, volume, pressure or time reach thresholds set using the ventilator controls. When the threshold is reached the ventilator cycles itself to expiratory mode. In some ventilators the controls relate directly to the monitored variables of pressure, time, volume or flow. In others they relate to physiological variables but assume characteristics for the patient that may not be true for a given individual. In the most sophisticated ventilators the controls relate to physiological variables calculated using the monitored values for the individual patient. Such ventilators will control the characteristics and settings of the flow generator during inspiration so as to obtain the desired tidal volume within the constraints set by the controls of the ventilator.

Phase (c) follows the general pattern of phase (a). It is dominated by the characteristics of the patient and the ventilator or, to be specific, since expiration tends to be a passive function, the characteristics of valves and tubes in the ventilator breathing system. Phase (d) is a second control phase and its function is determined by threshold controls.

The implication of this four-phase analysis of ventilator action is that a ventilator's functional classification requires four terms:

(a) The inspiratory flow generator type (e.g. constant flow, constant pressure).
(b) The type of parameter used as a threshold for cycling in phase (b) (time, pressure, flow, volume).
(c) The expiratory flow regimen (e.g. constant flow, constant pressure).
(d) The type of parameter used as a threshold for cycling in phase (d) (time, pressure, flow, volume).

The four terms may be contracted into three or two if the methods used in phase (a) and (c) or phase (b) and (d) are the same. A ventilator that uses different phase (b) and (d) parameters may be described as a mixed cycle ventilator.

These interactions have been summarized in separate

ways. Mushin et al.[29] have described all the interactions and the variables that affect ventilation in a complex butterfly diagram. Mapleson described ventilator classification in a way that mirrors the four-phase description presented here.[30]

Pressure and flow characteristics during IPPV

Pressure generators

The action of a pressure generator utilizes control of the airway pressure to produce gas flow into the lung. The ventilator controls the pressure at the airway and ignores the effect of resistance on flow. For a constant pressure generator at three times the time constant for the model, 95 per cent of the terminal tidal volume will have been achieved. (Time constant is time for the change to be completed assuming the initial rate of change is maintained throughout the change.)

The rate of flow of gas into the lung is related not only to the time constant for the model but also to the magnitude of the constant pressure, since it is proportional to the pressure difference across the resistive part of the model. At higher constant pressure, more volume will be delivered in a given time for the same time constant. In these circumstances the time required to produce a given tidal volume falls as the maximum attainable tidal volume rises. At high constant pressure the rate of change of flow and volume is so swift that they can be approximated to straight lines over the time required to produce realistic tidal volumes.

At low airway pressures the exponential nature of the changes has to be taken into account. This is especially true when, for a low constant pressure generator, compliance or resistance changes. The effect of halving compliance or doubling resistance, while the time constant is altered is shown in Fig. 22.3. If compliance is halved, maximum tidal volume falls as does the time required to reach a fixed proportion of it, but assuming that the desired tidal volume is still less than the maximum achievable, the period of time to achieve the desired tidal volume will increase. Doubling resistance has a similar effect on the time required to achieve a desired tidal volume, in this case because the time constant increases.

When a low constant pressure generator is combined with time cycling, an attractively simple method of control, variations in compliance and resistance can markedly affect delivered tidal volume.

The archetypal pressure generator is a weighted concertina bellows, in which the pressure generated is determined by the weight. Such an arrangement is exploited in the Manley ventilator. A gas flow derived from the pipeline supply to the anaesthetic machine can also act as a constant pressure source as in the Ohmeda 7000 series. Devices such as blowers or circulators have been used inside circle breathing systems to assist spontaneous ventilation, or as the ventilator flow generators. These can be regarded as constant low pressure generators though the pressure generated need not be constant.

If an elastic compliance is charged with gas and allowed to discharge, the pressure at its outlet will be determined by the distension of the elastic material and will change with volume. It will thus be a variable pressure generator. An interesting side effect of this method of flow generation is that without fresh gas flow into the system the alveolar pressure generated and the bellows pressure converge at some value of PEEP and the volume originally in the bag redistributes itself between the bellows and the lung, achieving an equilibrium. With fresh gas entering the system no convergence pressure is achieved, as at some point the fresh gas flow forces pressure in both the bag and lung to rise. In ventilators with this type of flow generator, desired tidal volumes are usually achieved before this secondary effect of fresh gas is seen, but the same situation can arise in anaesthetic breathing systems, in which inspiratory gas flow from a bag and fresh gas flow contribute towards the end of expiration in a way that increases airway pressure. The interaction between ventilators and breathing systems is not constant, each has characteristics of the other.

The advantages of pressure generators can be summarized as follows:

- *Inherent pressure regulation*: this reduces the risk of barotrauma.
- *Mechanically simpler*: therefore they should be more reliable.
- *Low internal and source impedance*: reduces the effect of shunting impedances in the ventilators and improves efficiency (see later, 'The effects of ventilator internal resistance and compliance').

The main disadvantage is that the volume delivered is neither inherently related to the ventilator characteristics nor control but to the characteristics of the patient.

Flow generators

A discussion of the characteristics of flow generators follows the general structure of the discussion for pressure generators.

The relation between pressure and flow for a constant flow generator is shown in Fig. 22.3. Lung volume and alveolar pressure increase linearly. Since pressure drop across the resistance of the model must fall, while flow is constant, airway pressure must fall. The relation between

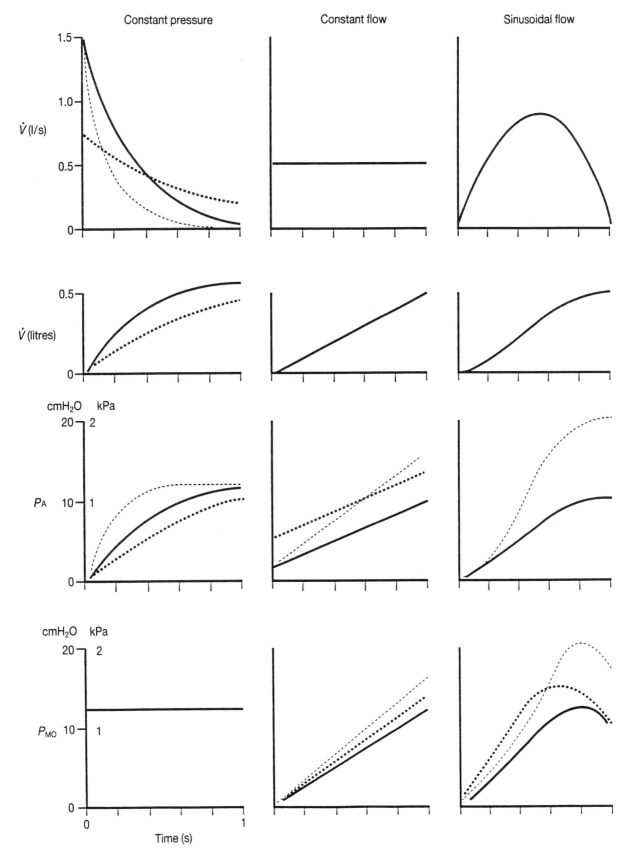

Fig. 22.3   The effect on different types of flow generator of doubling airway resistance (- - - -) or halving lung compliance (- - - -) from the normal adult values (————).

pressure and flow is of little direct clinical interest, however, since it gives little insight into the performance of the ventilator in the face of changing compliance and resistance.

A flow generator lends itself to control by flow or volume. Flow generators, such as an electrically driven displacement pump as in the Kontron ABT ventilators or a ventilator using a simple reciprocating piston, invite a variable flow configuration. The relations between flow and volume for a sinusoidal variable flow generator are shown in Fig. 22.3 at normal compliance, halved compliance and doubled resistance for the compliance–resistance model. Tidal volume remains constant but alveolar pressure rises dramatically for reduced compliance and airway pressure rises to achieve the same tidal volume when resistance is doubled. This reflects the way in which flow generators deliver tidal volumes regardless of the mechanical characteristics of the patient's chest. The advantages and disadvantages of the generator follow from this basic property.

The advantages are:

- *Controllable, defined tidal volume*: regardless of the condition of the respiratory system.
- *A ventilator capable of being more reactive to the patient*: ventilation manoeuvres such as IMV or MMV are easier to provide and control.

The disadvantages are:

- *High airway pressures*: higher risk of barotrauma.
- *More monitoring is required*: this is difficult at small tidal volumes, e.g. for neonates.
- *More susceptible to internal compliance and resistances*: shunting compliances and resistance affect performance and efficiency (see following section).

The effects of ventilator internal resistance and compliance

Not all pressure or flow generators can be categorized in a simple way. In a Venturi injector, a high-pressure, high-velocity gas stream through a jet is used to entrain gas from a low pressure source using the Venturi effect. This functions as a pressure source but with a high internal resistance of its own. Similarly, a bellows driven by a piston or a high-pressure gas supply, acts as a pressure transformer, and is in theory a constant pressure source. However, the bellows has an internal compliance and the friction in the piston will affect the pressure if not its constancy. In both these cases the internal resistances or compliances of the flow generator will affect performance. Even if the ventilator itself is solely flow or pressure, the placing of a breathing system between it and the patient may alter its performance in a similar way. The internal

resistance or compliance of a ventilator is an important consideration when optimizing a ventilator's performance.

Any internal compliance in a flow generator will shunt the resistance and compliance of the patient's lung and add to the effective compliance. Thus the flow of gas will be divided between the lung and the internal compliance. Any gas flowing into this compliance will be effectively lost for the purposes of ventilation (indeed this is referred to as the 'lost gas effect'). At a typical value for internal compliance of 0.5 ml/kPa, which is small compared to the typical compliance of a breathing system, 6 ml/kPa, the compliance similarly shunts the lung compliance. However, the significance of the internal compliance rises with increasing airway resistance and with respiratory frequency.

Internal source resistance will, in the case of pressure generators, be in series with the resistance of the breathing system. The pressure generated will be divided between the internal resistance and the respiratory resistance. This is not likely to be a problem in the case of internal compliance and flow generators for normal adult patients but if the source resistance approaches the respiratory resistance, as is the case with some humidifiers, then small changes in the airway resistance or in the breathing system may markedly affect the delivered tidal volume. In the case of bronchospasm, the pressure attained at the airway may altogether cease to be sufficient to deliver an adequate tidal volume.

Control systems in practical ventilators

Methods of cycling

In the four-phase description of ventilator action and control presented in the section on classification (see p. 423), it was noted that, in the control phases (b and d), four mechanisms could be used to trigger the initiation of the inspiratory or expiratory phases and to cycle the ventilator; these were time, pressure, flow and volume.

In the early stages of ventilator development the mechanisms that could be used to achieve cycling were very few and it was easy to identify examples of purely time, pressure, flow or volume cycled ventilators. However, a wide variety of transducers and methods are now used to measure and change the required trigger. The increased level of sophistication now being used for this task is reflected in the design of ventilators. Ventilators that are likely to be encountered in modern clinical use are often difficult to categorize by cycling method. The examples given below are used to illustrate the technology in use.

Time cycling

The Kontron ABT-4100 ventilator is a time cycled microprocessor controlled ventilator. Ventilation is provided by a motor driven bellows connected to a breathing system from which gas is drawn during expiration. The external controls set analogue voltages that correspond to minute ventilation, breathing rate and I:E ratio. The required volume, expiratory and inspiratory times are then calculated by the microprocessor, which controls the speed and position of the motor to give a physiological flow waveform. Apart from stopping and starting the motor, the cycling of the ventilator is controlled by the operation of two solenoid valves which allow expired gas to pass out directly to the atmosphere during expiration and seal the expiratory limb of the breathing system for inspiration as soon as the calculated expiratory time has elapsed.

Pressure cycling

Pressure cycling is probably the commonest form of ventilator cycling since it is cheap and technically easy to implement with electronic or mechanical pressure-sensitive transducers or valves and is compatible with simple pressure generator flow sources. The techniques used for detecting the airway pressures are, however, varied. Ventilators such as the Puritan Bennet 7200 series use electronic detection and microprocessor pressure control. They achieve accurate control, adaptable performance and ventilation in the face of the most adverse clinical conditions. In contrast, purely fluidic or pneumatic ventilators such as the Medishield Fuidor or the Ohio Fluidic 'Anaesthesia' ventilator use fluidic or pneumatic components to control pressure cycling. These are simple, small and reliable ventilators. However, the simplicity and flexibility of pressure cycling can best be illustrated not by a sophisticated ventilator but by the 'humble' Minivent.[31]

The Minivent is a small ventilator the size of an anaesthetic 'T'-piece or non-rebreathing valve. It is normally used in a non-rebreathing system, fitted to the patient side of the reservoir bag. Inside the ventilator there is a bobbin made of a magnetic material, that can switch backwards and forwards to obstruct either the port connected to the reservoir bag (inspiratory port) or the port connected to atmosphere (expiratory port). In the inspiratory port there is a permanent magnet which attracts the bobbin so that at rest it occludes the inspiratory port. On the expiratory port, there is a knurled knob that controls the distance the bobbin must move to seal the expiratory port. The third port is connected to the patient. With the bobbin sealing the inspiratory port the patient can expire to atmosphere via the expiratory port. The fresh gas flow entering the reservoir bag increases inspiratory port pressure until the force on the bobbin is sufficient to overcome the magnetic force on the bobbin. The inspiratory port opens, the bobbin flies back to seal the expiratory port, a distance set by the control knob, and inspiration from the reservoir bag occurs. The bag acts as a

discharging compliance. At first sight this seems like pressure cycling but the pressure at which the expiration ceases depends on the volume in the reservoir bag and hence, to a first approximation it is volume cycling. When the difference between the pressures in the reservoir bag and airway has dropped sufficiently, the magnet moves the bobbin back to close the inspiratory port. The force required to do this depends on the distance from the bobbin to the magnet and thus the control knob position controls the I:E ratio. Expiration is thus pressure cycled. Other examples of mechanical pressure cycling are found in the Bird series of ventilators.

Flow cycling

The concept of flow cycling has never been popular, and at the time of writing, no intentionally designed flow cycler is available. There is no particular reason for this, and many volume cycling ventilators derive their volume measurements from flow transducers. However, in the case of some ventilators certain modes may result in flow cycling.

The Ohmeda Manley 'Servovent' is a mixed cycle ventilator. Gas for inspiration is delivered to the patient from a bellows by a spring connected to a piston and cylinder mechanism. As the bellows empties, the gas in the cylinder is forced out via several valves, and an adjustable resistor which controls the rate of emptying of the bellows and hence inspiratory flow. The gas being displaced from the cylinder also holds the expiratory valve of the breathing system closed. When the cylinder is empty the expiratory valve opens and expiration starts. Thus cycling is primarily volume controlled, depending on the volume of gas in the cylinder at the beginning of inspiration. However, if on inspiration the flow of gas from the cylinder drops below a level that is sufficient to hold the expiratory gas valve shut, then expiration is initiated. This flow rate will be set by the inspiratory flow rate controller. If this occurs before the cylinder empties, the ventilator is flow cycled.

Expiratory time is controlled by a second adjustable resistor, that controls the time required to charge a cylinder on a pressure-controlled valve connected to a relay valve. When pressure in the chamber reaches a threshold value inspiration is triggered. During expiration the cylinder and piston are recharged with driving gas. Expiration is time cycled.

Volume cycling

Several ventilators used for both intensive therapy and in the operating theatre are volume cycled. These tend to make use of Venturi respirometer (Engstrom Erica series) or pneumotachograph (Siemens Servovent 900 series) flow measurement devices and integrate the signals obtained to control tidal volumes. Discharging compliance ventilators often incorporate volume cycling.

Good examples of modern volume cycled ventilators are

the Ohmeda 7000 series. These are mixed cycle ventilators, using volume cycling on inspiration and time cycling on expiration. The high pressure gas supply to the ventilator is used to squeeze a 'bag in a box' mechanism using a fixed flow rate determined by a microprocessor. The gas is supplied to the bellows housing through five tuned orifices which allow five flow rates of 2, 4, 8, 16 and 32 controlled by solenoid valves. These allow for step increases in flow of 2/l min, from 2 to 62 l/min. During inspiration the valves are opened at the flow rate calculated to give a tidal volume with an inspiratory time determined from control settings of minute ventilation, respiration rate and I:E ratio. The valves remain open until a turbine respirometer has registered the intended tidal volume, and then they close. They remain closed for a calculated expiratory time, during which period expiration is via the breathing system attached to the ventilator. The return of the bellows to the top of its housing is regulated by a passive exhaust valve connected to the bellows.

PEEP and specialized ventilation features

Positive end-expiratory pressure (PEEP)
The application of PEEP increases transpulmonary pressure and moves the working point of basal lung units upwards on the volume transpulmonary pressure curve, increasing FRC and decreasing the likelihood of reaching closing volume during expiration or of absorption atelectasis. Collapsed alveoli are expanded and the compliance may be increased. Shunt is decreased and oxygenation improved.[32–34]

PEEP can be achieved in a variety of ways. Expiration is predominantly a passive process, with the patient expiring to atmosphere. The resistance–compliance model shows that if an additional resistance is placed in the expiratory path the expiratory time constant will be increased and in a given time the lung will discharge less gas. If inspiration is triggered before the alveolar volume and pressure have approached the resting volumes, a PEEP will exist. PEEP is often controlled in this manner with an adjustable resistor placed in the expiratory path. The disadvantage of this system is that the level of PEEP is flow dependent. An alternative method is to place a threshold resistor such as a column of water or a spring loaded valve in the expiratory path. In this case no gas will flow until the threshold pressure of the valve is reached. It is arguable whether either of these systems is really a PEEP system. The threshold resistor increases pressures throughout the expiratory phase and not just at end expiration. However, the resistor type of control also increases mean expiratory pressure. Valves can be designed to be pressure controlled using suitable servomechanisms. These in principle could be used to deliver a true PEEP, strictly limited to the

moment when expiratory flow falls to end-expiratory values.

PEEP is used when hypoxaemia is severe and when the patient is at risk from pulmonary oxygen toxicity. It is useful for maintaining oxygenation in the face of pulmonary oedema, probably because it redistributes intra-alveolar fluid. It does nothing to reverse underlying disease processes. Optimal PEEP is the lowest level which appreciably reduces shunt and improves oxygenation without increasing cardiovascular depression and the risk of barotrauma. Typical values are 5–15 cmH$_2$O (0.49–1.47 kPa), though higher levels have been suggested.[35]

PEEP can also be applied during spontaneous ventilation and is denoted by the letters sPEEP. It can be delivered using a facemask, thus avoiding tracheal intubation. Valving comparable with that used to achieve PEEP during full IPPV can be employed. It is very similar to continuous positive airway pressure (CPAP).

Continuous positive airway pressure (CPAP)
CPAP is a continuous positive airway pressure throughout the breathing cycle. It may be used during weaning from IPPV or may be used in the care of patients whose ventilatory performance does not indicate the need for IPPV. It avoids some of the complications of IPPV but still carries the risks of barotrauma and cardiovascular depression.

CPAP cannot be achieved simply by the use of valves or resistors. A source of gas must be included. During expiration the unidirectional valve prevents expired gas flowing from the patient back into the reservoir bag. This ensures that throughout expiration the pressure in the breathing system is fixed by the threshold pressure of the distal positive pressure valve. During inspiration, provided the fresh gas flow rate is sufficiently high to ensure a continuous spillage of gas from the positive pressure valve, inspired pressure will still be raised to the operating threshold pressure of the valve.

Other methods of mechanical ventilation

Inverse ratio ventilation This mode of ventilation reverses the normal inspiratory to expiratory time ratio (I:E ratio) so that inspiration is prolonged and expiratory time is reduced. The result in terms of pressure is very similar to PEEP, since it raises end-expiratory alveolar pressure by reducing the volume of gas discharged from the lung. Also, it may improve ventilation of low lung compartments. It is often combined with an inspiratory hold or pause. Typical ratios between 1.1:1 and 1.7:1 are used to achieve optimal oxygen delivery, but higher ratios have also been recommended.

Differential lung ventilation Differential and single lung ventilation are essential in certain surgical procedures (thoracotomy) but differential lung ventilation has also been suggested in the presence of unilateral lung disease. Two ventilators with a double lumen tracheal tube are used. Asynchrony of the lungs does not seem to be a problem and coordination of the separate lung inspirations is not required. In principle such a system could be tuned to maximize ventilation–perfusion matching.

Patient-triggered ventilation modes

During pure IPPV the cycling of the ventilator is only affected by the respiratory properties of the patient's lungs or the ventilator settings. In certain modes of ventilation the patient, by voluntary respiratory effort, controls the pattern of respiration which may be spontaneous or artificial.

Intermittent mandatory ventilation (IMV) IMV allows the patient to breathe spontaneously between a preset number of mandatory breaths. It was originally introduced to facilitate weaning from full IPPV but is now accepted as a mode of ventilation for the non-paralysed patient. A modification to the basic system is synchronized IMV (SIMV) in which mandatory breaths are delivered in synchrony with spontaneous breaths, triggered by detection of subatmospheric pressure in the airway due to attempted inspiration. If no spontaneous efforts are detected by the ventilator within a given time, then a mandatory breath is supplied. IMV and SIMV can be adjusted to gradually increase the effort required by the patient and the proportion of spontaneous to mandatory breaths as weaning progresses. IMV often uses two breathing systems, a CPAP breathing system to support the spontaneous breaths and a second ventilator system to provide the mandatory breaths. The two are connected by two one-way valves.

Mandatory minute volume ventilation (MMV) In this mode the ventilator is set to ensure a preset minute ventilation by detecting and assisting spontaneous ventilation or attempted ventilation. The patient is thus assured of a constant ventilation despite any variation, minute to minute, in his or her ability to breathe. Hyperventilation is, theoretically, prevented and, unlike IMV, self-weaning is possible.

Pressure support or assisted ventilation Pressure support or assisted ventilation is initiated by the patient. Once initiated the ventilator applies a constant, preset airway pressure, e.g. 10 cmH$_2$O (1 kPa), until a given time has elapsed or there has been a fall in inspiratory flow, to predetermined fractions of respectively inspiratory time or inspiratory flow. Pressure support lowers oxygen consumption, decreases patient inspiratory work and shortens weaning times.[36]

Sigh

Many ventilators provide a 'sigh'. A sigh is a deep breath approaching vital capacity. It occurs spontaneously about once per minute, and is followed by a reduction in the work of breathing and an increase in oxygenation. Its adoption as a technique for improving IPPV rested on a belief that it could improve FRC and help prevent miliary atelectasis. However, studies using an artificial sigh have not demonstrated these benefits, due possibly to the change in ventilation–perfusion match in the artificially ventilated lung compared with the normal, upright, spontaneously ventilating lung.

Additional monitors and features

Ventilator alarms

Ventilators may fail and therefore must include threshold alarms, audible and visual. The tendency is to integrate the alarms into the overall functions of the anaesthetic machine. In the 7000 series ventilator (Ohmeda CD anaesthesia machines) the alarm functions are integrated into a single display. However, individual ventilator alarms still have a place. They have the advantage that they can be independent of the power supply to the main machine (e.g. battery powered) or remotely sited; they ensure a uniform standard of alarm level. Any independent alarm must detect disconnection (low airway pressure) and high airway pressure.[37] The alarm must be reliable. Troublesome alarms tend to be switched off or ignored.

Ventilator alarms may be divided into four types:

- *Pressure*: which respond to pressure changes in the breathing system.
- *Flow*: that respond to inspiratory or expiratory flows.
- *Spirometric*: which respond to changes in patient inspired or expired volumes.
- *Capnographic*: that respond to changes in the patient's carbon dioxide waveform.

Pressure alarms may be either high or low. The high-pressure alarm protects against obstruction alarming when a safe airway pressure is exceeded. Low-pressure alarms provide a warning of disconnection. When inflation pressure does not exceed a minimum threshold in a fixed time (usually 15–30 s), a warning is provided. An adjustable low-pressure alarm, set to just below the peak airway pressure, may be activated either by increased leak, increased compliance or reduced patient resistance. An alarm for negative pressure or a fall below a threshold pressure may be required for CPAP systems. Sophisticated alarms may analyse and detect failure in the pressure waveform. The beginning of expiration is detected by a rapid fall in the airway pressure (Drager ventilator monitor in IPPV/IMV mode). Pressure alarms have the advantage of being simple to operate and will detect incomplete disconnection. Care has to be taken with the siting of the tube connecting the pressure sensors to the airway system to prevent blockage with water or other material.

Spirometric alarms detect the adequacy of ventilation and thus can be adjusted to measure the target settings of tidal volume or minute ventilation. A volumetric alarm consisting of a disposable rotating vane anemometer around which clips a photodetector rotation sensor is used in the Ohmeda 5420. The sensor can be mounted to detect expiratory or inspiratory volumes. The battery powered unit displays flow, respiration rate and either minute or tidal volume on a small LCD screen. Upper and lower alarm thresholds for tidal or minute ventilation can be set using thumb-wheel switches. The anemometer used is quite bulky, adding additional dead space to the breathing system, and the sensor, with its associated cable, is quite heavy. The fresh gas flow may be sufficient to suppress the alarm if the sensor remains partially attached to the breathing system.

The capnograph is an effective ventilation alarm. The detection of expiratory carbon dioxide in the breathing system is a clear indication of connection to the patient and is not easy to confuse except when carbon dioxide is added to breathing gases at the end of anaesthesia to stimulate spontaneous ventilation. Capnograph sampling tubes are small diameter and can be used not only as simple threshold detectors but also, since the capnograph signal takes on unique features for different fault conditions, for highly sophisticated fault detection.[38]

The online calculation of P–V loops

Increasing microprocessor power has meant that computers may not only control the ventilator, but also the information used for control, such as pressure, flow and volume may be displayed. Online displays of these parameters are available, on many ventilators designed primarily for intensive therapy, using integrated graphical displays to show both the variation of the parameter with time and may also display pressure–volume and flow–volume loops.

A simplified pressure–volume loop is shown in Fig. 22.4. The work done in moving from the origin to any point on the curve is given by:

$$W = \int_0^v P \mathrm{d}V$$

If the whole curve is traced then the difference between the expiratory and the inspiratory work is given by the area of the loop. The pressure–volume loop thus quantifies the work done in overcoming pulmonary resistance. A line joining the end tidal points on this curve gives the overall change in volume with pressure and thus the dynamic

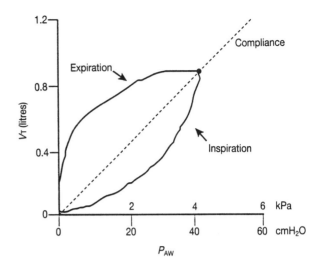

Fig. 22.4 An idealized volume–pressure loop for a ventilated subject.

compliance of the patient's lungs. In ventilators with specialized ventilation features such as IMV, MMV, etc., online P–V loops allow assessment of the work required to trigger a breath and the effects of pressure support. However, though in quantitative physiological terms the P–V loop is interesting, in terms of ventilator control, its real use is qualitative rather than quantitative. The shape of the loop changes characteristically with the type of breath and conditions such as lung overdistension are easily identified. The flow–volume loop finds similar qualitative application, indicating in an easily accessible fashion the effects of changes such as those induced by bronchodilators.

Indirect calorimetry

During intensive care, indirect calorimetry (see Chapter 17) is of increasing clinical interest and the calculation of oxygen uptake during anaesthesia has been used to monitor the state of the cardiovascular system. Indirect calorimetry enables the mean differences in the volumes of inspired and expired oxygen and carbon dioxide to be measured. The respiratory quotient (RQ) and other derived parameters such as resting energy expenditure can then be calculated. Indirect calorimetry is made available at the bedside by using integrated modules connected to ventilators. Many of the flow and volume measurements that are required are readily available as a result of the operation of the ventilator itself. The main difficulties occur during IPPV and at high oxygen concentrations which limit the applicability and accuracy of systems.

The testing of ventilators

Such a wide range of ventilator flow generator systems, methods of cycling and additional features, all of which

Table 22.1   Test parameters for ventilators

Parameter	Units	Test apparatus accuracies	Test target
		Max zero error (sensitivity error as % reading)	
Compliance	ml/kPa	0 (5%)	Record
Resistance	kPa or l/s	0 (10%)	Record
Time (inspiratory or expiratory)	s	0 (1%)	0 (5%)
I:E ratio (inspiratory time %)	% vent. pref.	1 (0%)	2 (0%)
Frequency	breaths/min	0 (1%)	0 (5%)
Ventilation	l/min	0 (5%)	0 (5%)
Mouth pressures	kPa	0.02 (1%)	0.05 (2%)
Alveolar presures	kPa	0.02 (1%)	0.05 (2%)
Supply pressure	kPa	0 (1%)	0 (2%)
Steady or mean flow	l/min	0 (2%)	Record
Instantaneous flows	l/min	Adult 1.0 Paed. 0.3 Neon. 0.1 All (5%)	0 (5%)
Delivered volume (tidal or minute)	l or ml	0 (5%)	0 (5%)
Instantaneous volumes	l or ml	Adult. 10 ml Paed. 3 ml Neon. 1 ml All (5%)	0 (5%)

interact with the patient lungs and upon whose perform-ance the safe ventilation of the patient depends, requires that ventilators should be maintained and tested. Standard-ized testing and performance verification methods are essential for objective comparison of ventilators.

Table 22.1 shows a range of test parameters for testing ventilator performance.[39, 40]

The accuracy required of the measurements and also the accuracy required by the measurement systems used for testing are shown in Table 22.1.

The measurements recommended when testing different ventilators include, in addition to those suggested in Table 22.1, tests for trigger performance in specialized modes of ventilation, tests of IMV and MMV performance and tests of rebreathing and dead space with associated breathing systems.

The test apparatus simulates the resistance–compliance model used to describe lung mechanics and ventilator action during IPPV. Resistances are simulated using sintered glass of various sizes. Compliance is either simulated using the compliance of a volume of gas held in a large volume or using a spring-loaded bellows system. In the case of the former, great care must be taken to ensure isothermal conditions. The model lung should be packed with conductive copper wool and made of copper to conduct away any heat generated by the repeated compression of the gas. When simulated spontaneous breathing is required a piston pump with a specified flow waveform can be connected to the compliance.

The standard apparatus may be unwieldy and imprac-tical for routine ventilator testing. In these circumstances proprietary systems, calibrated and certified as secondary standards, may be used. These devices range from simple spring loaded compliance–resistance models to fully transduced, microprocessor-controlled instruments capable of performing automatic assessment, producing a written report and logging the results for computer record keeping.

Breathing systems

The ideal anaesthetic breathing system

The primary requirement of any breathing system is to maintain the delivery of oxygen and removal of carbon dioxide and during anaesthesia, the delivery of nitrous oxide and volatile agents.

Most patients, at least at induction or at recovery, breathe spontaneously. No accidental additional resistance should be imposed on the patient's breathing. During artificial ventilation the system should not add to the internal resistance and compliance of the ventilator, so that the volume set on the ventilator is what is delivered to the patient.

The breathing system must be capable of delivering anaesthesia to all sizes and shapes of patient from the smallest baby to the largest adult, despite variations in patient characteristics and lung mechanics.

The breathing system should limit the waste of expensive volatile drugs but when gases are spilled to the atmosphere, health and safety considerations require that they do not pollute the working environment. Thus, the breathing system should be suitable for scavenging without affecting its function.

Ease of access, patient visibility and the least pressure on the patient's airway are desirable. The breathing system should therefore have a low weight, and use as few tubes and components as possible. These components should be interchangeable and able to be reconfigured to other systems if required. Their design should allow the fine control of the breathing system characteristics. They should be sterilizable or disposable to minimize cross-infection risk.

The breathing system should be reliable, easy to use and fail safe, since its manipulation is often required in critical situations and failure may go undetected in some circumstances. It should facilitate the use of monitoring apparatus.

This checklist is clearly unattainable for any particular system but it serves to test the overall performance of any practical breathing system.

The classification of breathing systems

Systems of classification

Systems of classification have been codified.[41] The first is based on a notional boundary to the breathing system that restricts the flow of expired gas and fresh gas flow. This system is summarized below and leads to the terms open, semi-open, closed and semi-closed:

- *Open*: a breathing system with infinite boundaries and no restrictions on the entry of fresh gas.
- *Semi-open*: a partially bounded breathing system with some restriction on fresh gas entry.
- *Closed*: a fully bounded breathing system with no provision for gas overflow.
- *Semi-closed*: a fully bounded breathing system with provision for venting fresh gas. These can be further classified into;
  (a) semi-closed rebreathing systems
  (b) semi-closed absorption systems
  (c) non-rebreathing systems.

In 1985 Conway[42] rejected this system and terms such as open and closed because of the way the terms had taken on a wide variety of meanings. In its place he suggested a 'more rational approach' with two broad categories:

- *Absorber systems*: those using chemical carbon dioxide absorption.
- *Non-absorber systems*: those not using carbon dioxide absorption.

This simpler classification has the advantage of clarity with a clear functional difference between the two categories. While some functional properties, relevant in clinical practice, are better described using the semi-open/semi-closed classification, the second classification is the more widely used.

Performance criteria

Resistance to breathing

Intuitively it is clear that resistance to breathing will have physiological effects during both spontaneous and controlled ventilation. In spontaneous ventilation, extra inspiratory resistance will increase the effort the patient has to make to breathe in. It would be expected that excessive inspiratory resistance would reduce ventilation. Excessive expiratory resistance is unlikely under anaesthesia to evoke an active expiration on the part of the patient but even with passive expiration the airway pressure will rise, imposing PEEP, affecting FRC and limiting tidal volumes. A breathing system can impose two sorts of resistance.

All components will impose resistances analogous to airway resistance. The resistance will increase with increasing flow rate and with transition from laminar to turbulent flow. However, breathing systems include valves and other components which are pressure operated. When the pressures required to open a valve are generated by the patient, there will be a type of threshold resistance for which resistance is extremely high until the threshold pressure is achieved. These two sorts of resistance would be expected to have different physiological effects. This intuitive view was examined by Nunn and Ezi-Ashi in 1961.[43] In a study of the effect of threshold and flow resistances on a group of patients spontaneously ventilating through an experimental breathing system they found that both forms of resistance did indeed reduce ventilation, even to the point of apnoea. They recommended that the maximum threshold pressure for a spontaneously ventilating patient should be 4 cmH_2O with a maximum flow resistance of 1 l/cmH_2O. This argument leads to the resistance of a breathing system being defined in terms of threshold pressures and flow resistances measured under steady state continuous flows into the system or its components. Many comparisons of breathing systems and performance standards have been written based on this view. However, ventilation is not a steady unidirectional process. It is reciprocal and variable at rates that suggest that fully developed flow regimens are unlikely ever to establish themselves in the breathing system.

The alternative method of testing breathing systems accepts that the mechanical performance of the system is dynamic and needs to be assessed in a dynamic way.[44] In this view the breathing system characteristics are assessed by examining the energy or work expended by a simulated patient breathing through the system. Changes in the energy expended or additional work of breathing (additional because it is added to the work of breathing associated with the patient's own airways and chest), are measured using the same principles as are used to measure work of breathing from pressure–volume loops. The advantage of this approach is that it takes into account all the sources of energy loss in the breathing system, without assuming types of resistance or flow conditions. The disadvantage is that the method views the performance of the whole system and does not lead easily to testing individual components. There is also some debate about the physiological significance of the measurements. Several authors have adopted this approach but as yet there is no minimum standard for resistance measured in this fashion.

Rebreathing characteristics and fresh gas efficiency
The objective in defining the rebreathing characteristics and fresh gas efficiency of a breathing system is to determine the fresh gas flow rates for clinical use. This has been approached in a number of ways, but there is no generally recognized laboratory or clinical measurement of rebreathing or method of comparison. However, three methods of approach have proved more useful than others.

Clinical studies have tended to examine breathing system performance from the point of view of capnography. This approach has the advantage of providing a measurement that is easily applicable in the operating theatre. In spontaneously ventilating patients, a minimum inspired carbon dioxide concentration[45] or rises in end tidal carbon dioxide concentration indicate significant rebreathing. However, the most successful method is that of Kain and Nunn[46] in which the physiological responses of patients are measured as fresh gas flows are reduced. This method has been shown to be the most reliable.[47] During controlled ventilation capnography has tended to be applied in some breathing systems to eliminate particular features of the capnograph trace.[48]

Fresh gas flows have also been determined by extrapolation from laboratory experiments. These have concentrated on simulating clinical studies using model lungs,[18, 49] applying modified clinical criteria to determine rebreathing,[50, 51] making absolute measures of volumes of carbon dioxide rebreathed, or determining increases in functional dead space. In some cases laboratory studies have been confirmed by clinical studies. These studies have enabled the fresh gas flow values required to achieve normocapnia for specific breathing systems to be obtained. Two methods appear to be applicable to all breathing systems.

Miller used a model lung to simulate spontaneous ventilation and to measure rebreathing in terms of the increases in functional dead space to tidal volume ratio.[18] Using this system he was able to compare several different systems. His study led him to propose a structural and functional classification for breathing systems.

Two terms, fractional utilization and fractional delivery, have been used in several guises by different authors to measure breathing system efficiency.[52] They have been defined[53] as, respectively, the volume of fresh gas involved in gas exchange and the volume of fresh gas delivered to the alveoli for gas exchange by the breathing system. They have the advantage that they can be measured both in the operating theatre and in the laboratory but have only been used to define the rebreathing characteristics of a limited range of breathing systems.

Theoretical studies have played little role in clinical guidance except to underpin experimental measurements. They have been used to develop models based on assumptions about gas mixing and valve operation, treating the systems as black boxes into which gas flows and from which gas is withdrawn, or have made other assumptions about flow and dead space. The most important and extensive have been those of Conway.[54, 55]

Non-absorber breathing systems

The Mapleson alphabet

The most familiar nomenclature for non-absorber breathing systems is that of Mapleson. First used in 1954,[56] it was not proposed as a classification system as such, but has been used in this way. It sees breathing systems as being made up of three components, an expiratory valve, a length of tubing and a reservoir bag connected to a steady source of fresh gas flow and the patient. Initially five systems were described (A, B, C, D, E) with the sixth (F) added by Willis *et al.* in 1975.[57] The configuration with Mapleson alphabet is shown in Fig. 22.5.

A (Magill, Lack, enclosed afferent reservoir, preferential flow, Humphrey ADE, Carden Ventmaster)

The behaviour of the Mapleson A during spontaneous ventilation is shown in Fig. 22.6. At the start of expiration the reservoir bag is empty due to the previous inspiration and the expiratory valve closed. At the start of expiration gas from the airway dead space enters the breathing system. The reservoir bag begins to fill with the expired

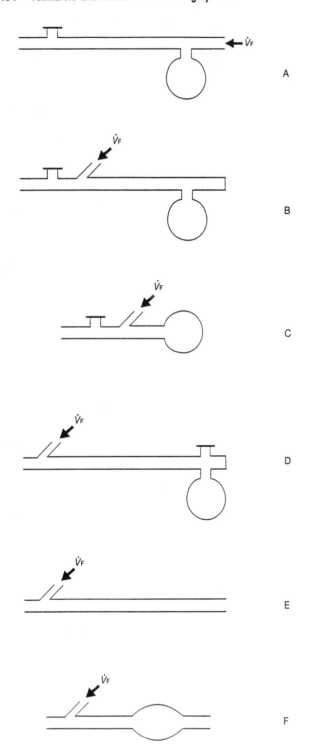

Fig. 22.5 Schematic of the Mapleson alphabet of non-absorber (semi-open) breathing systems.

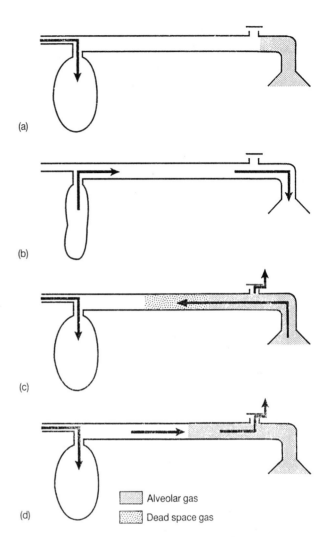

Fig. 22.6 The action of the Mapleson A (Magill) breathing system during spontaneous ventilation. (a) At the end of expiration the breathing system is essentially free of gas contaminated with carbon dioxide and the reservoir bag is being filled by fresh gas flow. (b) As inspiration begins the patient draws gas from the reservoir bag and the fresh gas flow. (c) When expiration starts, at first uncontaminated gas enters the breathing system, followed by gas contaminated with carbon dioxide. This gas passes the expiratory valve. The reservoir bag is filled by the fresh gas flow. (d) When the reservoir bag is sufficiently full it raises the internal pressure in the system and the valve opens. All subsequent contaminated gas passes out through the valve and the system tubing is swept clean by the fresh gas flow.

gas and the fresh gas flow. This continues until the reservoir bag is filled sufficiently to raise the pressure in the whole system to the opening pressure of the expiratory valve. This is achieved at a volume sufficient to supply the next inspiration since the reservoir bag has a non-linear pressure–volume curve whereby pressure remains low

initially. When the valve opening pressure is reached further expiratory gas is deflected out of the expiratory valve. Thus, for a given setting of valve opening pressure and fresh gas flow, the valve can be made to open at the point where expired dead space gas, uncontaminated by carbon dioxide, gives way to alveolar gas which needs to be removed from the breathing system. The theoretical limit for the Mapleson A efficiency is therefore equal to alveolar ventilation. In fact this maximum efficiency is obtained only if the valve and fresh gas flow are tuned so that the valve opens just after the alveolar gas has started to pass up the main tubing of the system. When the valve opens, the fresh gas flow no longer fills the reservoir bag but flushes the gas at the valve end of the breathing system out through the expiratory valve. Thus if the valve opens too early fresh gas flow is wasted and efficiency falls.

In the classic Magill, the valve is mounted at the patient 'Y'-piece close to the mouth. This can be inconvenient. The Lack system[58] removes the valve back to a mounting on the anaesthetic machine, linking it to the same place in the breathing system via a separate tube that can be coaxial or parallel to the original. The analysis presented above assumes that little mixing of fresh, dead space and alveolar gas occurs in the system. The extra tube has sufficient volume to store gas that would otherwise leave by the expiratory valve. Gas will thus be mixed thoroughly in this tube. However, this mixing has no effect on performance and it seems that once the gas has passed into the extra tubing it has effectively left the breathing system as effectively as if it had been vented immediately to atmosphere.

During controlled ventilation using IPPV, pressure at the mouth must rise above atmospheric pressure to drive inspiration, in contrast with spontaneous ventilation when pressure remains below or equal to atmospheric pressure. If the reservoir bag is converted into a 'bag in a bottle' and the valve is set to the opening pressure used for spontaneous ventilation, as inspiration starts the valve will open and most of the gas intended for inspiration will spill through the valve. To ensure an adequate tidal volume the valve opening pressure will have to be greatly increased, which will result in very late opening during expiration and most of the alveolar gas will remain in the breathing system and will be rebreathed. The unmodified Mapleson A is thus unsuitable for IPPV.

In response, several authors have presented modified Mapleson A systems in which the expiratory valve is essentially held shut during inspiration by a secondary mechanism which is released on expiration, allowing an appropriate expiratory valve opening pressure for maximum efficiency. The enclosed afferent reservoir systems that have been studied are essentially modified Mapleson A systems in which the valve has been removed from the head of the patient in the manner of a parallel Lack.[59, 60] The Humphrey ADE[61, 62] and Carden Ventmaster avoid the problem of the Mapleson A, during IPPV, by arranging for an easy changeover to D or B systems. The Carden system is in any case an enclosed afferent reservoir in its own right.

The Mapleson A is a mechanically simple system. The resistance to inspiratory flow is mostly caused by inertia in the reservoir bag, provided it does not collapse during inspiration. Excess fresh gas flow will tend to slightly increase system pressure towards the end of inspiration, thus providing some positive pressure assist. During expiration the resistance depends on the valve resistance, the valve opening pressure, bag inertia and fresh gas flow. Excessive fresh gas flow can increase expiratory resistance but the predominant component will be the valve. Dynamic testing of Mapleson A systems indicates generally low resistances. The resistance of the anaesthetic tubing does not play a significant part in the resistance of adult breathing systems at normal respiratory volumes and rates. However, the use of a coaxial system could, by reducing the effective area of the inspiratory tube, increase resistance. A parallel Lack system avoids this risk and the increased radius of the outer inspiratory tube reduces the resistance in the coaxial system. Modified systems tend to have higher resistances at equivalent flows because of the raised inertia of the enclosed reservoir and the higher resistances of the expiratory valve which tend to have higher open resistances than a simple Heidbrink type valve.

B and C systems

The performance of the B and C systems has been studied less, probably because their use is largely limited to resuscitation. The difference between the systems is the length of the tubing between the valve and the reservoir. If alveolar gas passes into this tube or the reservoir bag then it will be mixed and rebreathed. A high flow is necessary to prevent this occurring. The systems are therefore very inefficient and the resistance to breathing is raised.

D, E and F systems (T-piece, Bain, Humphrey ADE, Carden Ventmaster)

In terms of rebreathing, performance of the D, E and F systems can be considered together. The E and F differ from the D in that there is no expiratory valve and the reservoirs (the tube in the E and the open-ended bag in the F) are open to the atmosphere. However, essentially, the reservoir and valve in the D serve the same function as the reservoirs in the E and F in that they store, mix and return to the patient proportions of the expired alveolar gas. The performance is dominated by the fresh gas entry flow.

The action of the group can be described by reference to the D. On expiration, dead space and alveolar gas will be mixed and will flow back into the reservoir. The valve opens as the reservoir fills and pressure rises and the mixed gases are vented. Thus alveolar and dead space gas are always mixed and there can be no selective venting of alveolar gas. The efficiency of a D must always be less than that of an A. On inspiration, provided the fresh gas flow is in excess of the peak inspiratory flow rate, no rebreathing can occur, but it is possible to prevent rebreathing at flows less than this peak inspiratory flow rate. At the end of expiration or in any expiratory to inspiratory pause, the fresh gas flow rate will greatly exceed the expiratory flow rate. In this case the fresh gas will tend to fill the distal part of the breathing system with gas, free of carbon dioxide.

This will be breathed in during inspiration when the inspiratory flow rate exceeds the fresh gas flow. If the fresh gas flow rate is too low to ensure sufficient stored fresh gas to prevent rebreathing, then at the end of inspiration a mixed gas will be rebreathed. As this gas passes a capnograph placed at the patient's mouth, a small blip of carbon dioxide will be seen just prior to the expiratory part of the wave. Thus, titrating fresh gas flow to just remove this blip is a way of ensuring the lowest fresh gas flow with no rebreathing in the Mapleson D.

Fresh gas efficiency during controlled ventilation with the D is essentially unchanged. During spontaneous ventilation the opening pressure of the expiratory valve would normally be set at a level near to that required to deliver adequate tidal volumes if the reservoir is used as a 'bag in a box'. The gas will be vented at the start of inspiration rather than at the end of expiration but the composition of this gas will be the same. In the E and F the open reservoirs require closing during inspiration using IPPV. Several devices have been designed for this purpose but since the destination of the gas in the reservoir is immaterial for the breathing system efficiency, any ventilator that can create a positive pressure in the tubing and then vent to atmospheric pressure for expiration is sufficient. In the case of the 'T'-piece for paediatric use, a thumb placed over the reservoir end is sufficient to generate positive pressure due to fresh gas flow.

More clearly than any other breathing system in the Mapleson alphabet, the D illustrates the nature of this type of non-rebreathing system. Viewed from one point of view, the D is not so much a non-rebreathing system as controlled rebreathing system. The fresh gas flow in the D has to be matched to the minute ventilation of the patient for normocapnia to be achieved. End tidal carbon dioxide depends on both minute ventilation and fresh gas flow. Rose and Froese[52] recognized this fact for the T-piece, by plotting achievable end tidal carbon dioxide values as isopleths on a minute ventilation versus fresh gas flow diagram.

In terms of resistance, it is the fresh gas flow that determines the system behaviour. The presence of the inlet near to the patient's mouth raises the pressure in this region when compared to the A. During inspiration this pressure ensures a positive pressure assist, even during spontaneous ventilation. At high fresh gas flows it increases expiratory resistance. This effect is most marked in the coaxial version of the D, the Bain, in which the fresh gas flow is delivered through a central narrow pipe that directly faces the Y-piece. Since the fresh gas is supplied via a thin pipe, the Bain, unlike the Lack, does not have to have increased outer radius to reduce the resistance in the breathing system tubing.

The circle system as a non-absorber breathing system

It is common to think of circle systems as absorber systems only. This point of view ignores three factors: that circles are commonly used at fresh gas flow rates that exceed basal requirements and therefore spill gas to the atmosphere; that in some countries it is common practice to switch the absorber out of system to create controlled rebreathing; that circle systems are used as non-absorber systems in some ITU ventilators and their fresh gas flow efficiency is still relevant to economy if not pollution. Thus in certain circumstances, circle systems, even with absorbers, share features in common with rebreathing non-absorber systems.

A feature of a circle used as a non-absorber system when compared with the systems of the Mapleson nomenclature is that the one-way valves that direct expiratory and inspiratory flow ensure that if gas is rebreathed it will be thoroughly mixed with fresh gas. However, this does not make the positioning of the fresh gas inlet or the expiratory valve any the less important. Snowden et al.[63] showed that moving the fresh gas inlet in a circle system from a position distal to the inspiratory valve to a position proximal to it improved the efficiency of the system by 10 per cent. Similarly Eger and Ethans[64] showed changes in fresh gas flow efficiency with component position. Zbinden et al. have shown variations in the efficiency of circles incorporated in sophisticated ventilators.[65] The efficiencies of some of the models they tested showed lower efficiencies than either Mapleson A or D configurations.

Resistance in the circle system is less susceptible to increases in fresh gas flow but the presence of the inspiratory and expiratory valves invariably increases the resistance significantly, during both inspiration and expiration, to levels in excess of other non-rebreathing systems except at very high fresh gas flow rates.

Absorber breathing systems

Methods of absorption

Carbon dioxide absorption was first used experimentally by John Snow in 1850 and by Alfred Coleman for dental anaesthesia in 1869. However, there was no widespread use of the system until soda lime became easily available after World War I.

Soda lime is a mixture of calcium hydroxide (94 per cent), sodium hydroxide (5 per cent) and, optionally, potassium hydroxide (1 per cent). Small amounts of silica may be added to react with calcium hydroxide to form calcium silicate which is very hard. The quantity of silica added determines the overall hardness of the soda lime and the amount of dust it produces. The components are fused and then pulverized to granules. The material may be sintered into blocks of particular shape and porosity which are smaller and more efficient than the granule form but more expensive. Soda lime's natural colour is white but indicators may be added to the granules which change colour as the soda lime expires by detecting the acids formed as part of the absorption reaction.

The bulk of the carbon dioxide is absorbed by the chemical reactions shown below:

$$CO_2 + 2NaOH \rightarrow H_2O + Na_2CO_3 + heat$$
$$Na_2CO_3 + Ca(OH)_2 \rightarrow 2NaOH + CaCO_3$$

and to a lesser extent by:

$$CO_2 + Ca(OH)_2 \rightarrow H_2O + CaCO_3 + heat$$

The byproduct of both processes is heat, sometimes referred to as the heat of neutralization. Water temperatures in excess of 50°C have been recorded in absorber canisters and warmth of the absorber canister is a practical indicator of correct operation. The raised temperature results in the water vaporizing and precipitating either in the absorber canister or elsewhere in the breathing system. This effect is known as 'rain out'. It has the advantage of humidifying the rebreathed gases, and maintaining the humidity of the soda lime itself which is required for efficient absorption near to the optimum 12 per cent, but drainage of the water needs to be allowed for in absorber design.

Soda lime reacts with certain volatile agents to form carbon monoxide. In most cases the reaction is very slow and only becomes marginally significant if an atmosphere of the highly volatile agent remains sealed in an absorber for a long period and the absorber is then connected to a patient without being flushed with fresh gas. Soda lime reacts strongly with sevoflurane and must not be used with this agent. An alternative carbon dioxide absorber substance is baralyme, which uses barium hydroxide in place of the sodium hydroxide used in soda lime. Baralyme is less efficient as an absorber and more expensive than soda lime.

To-and-fro systems

In 1926 Waters introduced the first absorber breathing system which is shown in Fig. 22.7. The Waters system has the basic configuration of a Mapleson C. In addition to the tubing connecting the valve and fresh gas inlet to the reservoir bag, an absorber canister is included. The fresh gas is mixed during expiration with the expired gas and passes into the reservoir bag via the canister. During inspiration, as in a Mapleson A, the inspired gas is provided

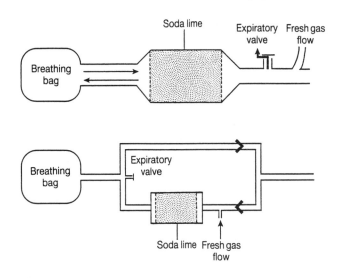

Fig. 22.7 Schematic layouts of the two general designs of absorber breathing system: To-and-fro (top), circle system (bottom).

by the reservoir and the fresh gas flow, but the soda lime removes carbon dioxide from the reservoir. The fresh gas flows required to prevent carbon dioxide rebreathing are less than the alveolar ventilation.

The to-and-fro systems have the advantage of reduced gas use and spillage combined with the low resistance to breathing of a mechanically simple system. Unfortunately, four problems have prevented their general use:

- *Dead space progressively increases as the soda lime reacts with carbon dioxide*: since this occurs initially from the patient end.
- *Channelling*: the soda lime granules settle so that a channel forms between the uppermost part of the canister and the soda lime through which gas can flow without being purged of carbon dioxide.
- *Soda lime dust inhalation*: from the canister which is close to the patient airways.
- *Weight and bulk of the apparatus*: the canister at the patient's head creates more traction on the tracheal tube than a circle or non-absorber system.

Circle systems

The main reasons for using a circle system are those of economy and pollution control.

It has been calculated that the cost of gases used in an oxygen, nitrous oxide and halothane anaesthetic for an operation lasting 100 minutes using a closed circle system is 13.5 per cent of the cost of the same anaesthetic using a non-absorber system with a fresh gas flow of 7 l/min. This cost saving, however, has to be offset against the higher capital cost of the breathing system itself and the associated monitoring equipment.

Operating theatre pollution was considered to be extremely important in the late 1970s and early 1980s, with anaesthetic agents implicated in spontaneous abortion and other effects on staff. However, when scavenging systems are used, operating theatre pollution levels are greatly reduced. However, medical gases, notably volatile agents, many of which are CFCs, and nitrous oxide, which is a greenhouse gas, are general atmospheric pollutants. In the case of volatile agents, if present targets are achieved, medical uses of CFCs, which are exempt from control at present, will be the largest contributor to CFC use in Europe by 2100. The wider use of circle systems at low fresh gas flow rates is a worthwhile environmental exercise.

The components of a circle system are shown in Fig. 22.7. They consist of two unidirectional valves, that direct the flow round the system, a carbon dioxide absorber, a reservoir bag, an expiratory valve and an inlet port for fresh gas flow. The relative positions of these components around the circle can have marked effects on performance. Three of the more useful properties of the configurations are:

- *Fresh gas inlet in the inspiratory limb*: fresh gas is supplied directly to the patient, increasing the speed of response to changes in inspired gas concentrations and minimizing fresh gas spillage from the valve.
- *The valve spills gas upstream of the absorber*: this conserves soda lime.
- *The absorber is placed between the valve and the fresh gas inlet*: this ensures that the fresh gas flow does not dry the soda lime, reducing its absorption efficiency.

The safety of the system at low flows depends on the correct functioning of the unidirectional valves. Failure can be difficult to detect, even with a capnograph. The standard design has been a horizontal flat disc valve of low weight sitting on a vertical valve seat under a transparent dome. This arrangement allows the anaesthetist to make a visual inspection of valve operation. The design, though reliable and safe, precludes the valves being placed close to the patient Y-piece, the optimum position for fresh gas efficiency.

The valves in a circle system may be replaced with a fan circulator driven by a gas turbine, electric motor or a Venturi device.[66] The circulator is set to maintain the inspiratory limb pressure above that in the expiratory limb to ensure a directed flow. In practice such devices do not replace valves but provide a supplementary source of power to generate the gas flows. The resistance to breathing in circle systems is generally higher than in non-absorber systems during similar clinical applications. This is due to the energy dissipated in overcoming the resistance of the unidirectional valves. A circulator may reduce the valve opening pressures and reduce resistance to breathing. Circulators have been used most in paediatric anaesthesia but have to be used with caution.

The main performance criterion, vaporizer use apart, is fresh gas flow rate. Circle systems can be used with a wide range of fresh gas flow rates. At flow rates greater than alveolar ventilation they behave as non-absorber systems, with end tidal concentrations independent of minute volume.

The logical design is the completely closed circle system. Since carbon dioxide is absorbed, the fresh gas flow into a circle system can be reduced to the basal metabolic oxygen requirement (approximately 225 ml/min) plus the anaesthetic gases. Circle systems that can achieve this by control of fresh gas flow have been reported.[67] The total fresh gas flow required varies from approximately 700 ml/min at the start of anaesthesia when nitrous oxide absorption is high to approximately 350 ml/min at 100 minutes into the operation. A useful rule of thumb for the decrease in the nitrous oxide requirement with time is that it is equal to a constant multiplied by the square root of time. The value of the constant varies from study to study but a safe estimate by Severinghaus[68] is 1000. It has been suggested that the form of this rule indicates that nitrous oxide use will never fall to zero during an operation, possibly because of loss from the skin or wound, depending on surgical procedure.[69]

A completely closed circle system highlights a further difficulty with low- or very low-flow anaesthesia. The body is saturated with nitrogen at atmospheric concentrations. This nitrogen is present in the lungs but it is also dissolved in all the body tissues. Removal of lung nitrogen can be quickly achieved at the induction of anaesthesia by preoxygenation, but the dissolved nitrogen will start to be expelled from the body through the lungs as soon as a denitrogenated mixture is breathed. This nitrogen builds up in a closed circle system and has to be removed. It may be argued that the presence of nitrogen in a breathing system may reduce absorption atelectasis, but it also reduces the inhaled anaesthetic concentration and even with advanced monitoring is difficult to measure. Mass spectrometry or Raman spectrometry is used to measure it but in most cases it has to be estimated using the discrepancy between the inspired partial pressures of the measurable gases and atmospheric pressure.

A further concern at low-flow rates, in circle systems, is the potential buildup of nitrous oxide to produce hypoxic mixtures if nitrous oxide flow rates are incorrectly adjusted. Spence[70] showed that for fresh gas flow rates of 2 l/min or more there was no significant difference between fresh gas concentrations and inspired concentrations in a circle system. He suggested that a 50:50 mixture of nitrous oxide and oxygen at 1 l/min was a safe fresh gas flow rate after lung denitrogenation.

Circle systems, even in the hands of the most enthusiastic practitioner, are likely to be used as low-flow systems with a fresh gas flow in the range of 1–2 l/min. Oxygen concentrations must be monitored at these flows.

Vaporizers and circle systems

Three types of vaporization have been used in conjuction with circle breathing systems: 'plenum', 'drawover' and direct injection.

Plenum vaporizers are high resistance and are suitable for use out of circuit (VOC). The flow of gas into a plenum device is split, part passing into the vaporization chamber to be saturated with agent vapour. The gas streams are mixed before leaving the vaporizer. The exit concentration can be controlled between the saturated vapour pressure and zero by varying the ratio of gas flowing into the bypass and vaporization chamber. Plenum vaporizers require temperature compensation.

A VOC is placed in the fresh gas flow line and is usually mounted on the anaesthetic machine. With VOC the lower the fresh gas flow the larger the difference between the fresh gas vapour concentration and the inspired concentration. This is due to dilution of the incoming agent by the gas circulating in the breathing system, which has had much of its vapour content removed by the patient. In a Mapleson A, with a VOC, the fresh gas flow is such that the mass of agent supplied to the breathing system is always well in excess of that required by the patient. The inspired concentration is thus independent of the patient's uptake.

The overpressure technique in which high concentrations are applied briefly at the start of anaesthesia in order to accelerate the increase in brain agent levels, is also compromised. Apart from increasing fresh gas flow, to raise the mass of agent being delivered, the only solution is to prime the circle with a high concentration of agent before the start of anaesthesia.

This discrepancy also affects the maintenance of concentrations during the middle part of anaesthesia such that there is a discrepancy between the concentration set on the vaporizer and that inspired. The use of a sidestream agent monitor is indicated.

The other problem of the VOC at low flow is that any change in delivered gas concentration is buffered by the volume of the circle itself and thus an effect is delayed. The capacity of a typical circle breathing system is 5–7 litres, which provides a safety margin for oxygen failure.

Drawover vaporizers are low-resistance vaporizers and can be used in the circuit (VIC). The flow of gas is not split as in a plenum vaporizer; all of it flows through a low-resistance vaporization chamber. The calibration of drawover vaporizers can only approximate to their output which depends on the level of agent in the chamber as well as flow rate of gas. The latter varies with the breathing cycle. Thus output concentration will vary and as the gas recirculates it is possible for the concentration of gas in the system to climb towards the saturated vapour pressure of the agent at the ambient temperature. In fact the agent concentration is limited by the patient uptake and the fresh gas flow, which controls the amount of recirculation and gas lost from the system.

During controlled ventilation at fresh gas flow rates of 1 l/min, inspired agent concentration will still rise steadily. Inspired agent monitoring is mandatory in these circumstances. During spontaneous ventilation a complex interaction between the patient and the breathing system occurs. As concentration rises the patient's respiratory drive is depressed, circulatory flow is reduced and less agent is vaporized but the necessity for an inspired agent monitor is obvious.

Direct injection of inhalational agent into the circle avoids the disadvantages of both VOC and VIC. The method has been used since 1847. In its modern form bolus injection is replaced by infusion of agent using a syringe pump. The coupling of a syringe pump control to an inhalational agent monitor leads logically to a simple and effective method for servo-controlled anaesthesia but such systems are as yet not fully safe, especially when computer control is used.

Techniques for fuel injection, pressurized vaporization and above ambient vaporization have led to new VOC and VIC design. Further advances are likely.[71]

REFERENCES

1. Vesalius A. De humani corporis fabrica libri septum. Basel, 1543.
2. Hook R. An account of an experiment made by Mr. Hook, of preserving animals alive by blowing through their lungs with bellows. *Philosophical Transactions of the Royal Society* 1667; **2**: 539–40.
3. Matas R. Intralaryngeal insufflation for the relief of acute surgical pneumothorax: its history and methods with a description of the latest devices for this purpose. *Journal of the American Medical Association* 1900; **34**: 1371–5.
4. Morch ET. History of mechanical ventilation. In: Kirby RR, Banner MJ, Downs JB, eds. *Clinical applications of ventilatory support*. Edinburgh: Churchill Livingstone, 1990: 1–61.
5. Lassen HCA. A preliminary report on the 1952 epidemic of poliomyelitis in Copenhagen. *Lancet* 1953; **1**: 37–41.
6. Clover JT. An apparatus for administering nitrous oxide gas and ether singly or combined. *British Medical Journal* 1876; **2**: 74–5.
7. Bryn-Thomas K. *The development of anaesthetic apparatus.* London: Association of Anaesthetists/Blackwell Scientific, 1975: 116.
8. Gwathmey JT. *Anaestheisa.* New York: Appleton, 1914: 174.

9. Bryn-Thomas K. The development of anaesthetic apparatus. London: Association of Anaesthetists/Blackwell Scientific, 1975: 148.

10. Waters RM. Clinical scope and utility of carbon dioxide filtration in inhalational anaesthesia. *Anesthesia and Analgesia* 1924; **3**: 20–4.

11. Sword BC. The closed circle method of administration of gas anaesthesia. *Anesthesia and Analgesia* 1930; **9**: 198–224.

12. Froese AB, Bryan AC. Effects of anaesthesia and diaphragmatic paralysis in man. *Anesthesiology* 1974; **41**: 242–55.

13. Rehder K, Sessler AD, Marsh HM. General anaesthesia and the lung. *American Review of Respiratory Disease* 1975; **112**: 541–4.

14. Agostini E, Mead J. Statics of the respiratory system. In: Fenn WO, Rahn H, eds. *Handbook of physiology (Section 3): Respiration.* Vol 1, Chapter 13. Washington: American Physiological Society, 1964: 387–409.

15. Nunn JF. Respiratory aspects of anaesthesia. *Applied respiratory physiology.* 3rd ed. London: Butterworths, 1987: 358.

16. Nunn JF. *Applied respiratory physiology.* 2nd ed. London: Butterworths, 1977: 227–8.

17. Fowler WS. Lung function studies: II The respiratory deadspace. *American Journal of Physiology* 1948; **154**: 405–10.

18. Miller DM. Breathing systems for use in anaesthesia. *British Journal of Anaesthesia* 1988; **60**: 555–64.

19. Hill S. Clinical applications of body ventilators. *Chest* 1986; **90**: 897–905.

20. Mortimer AJ. High frequency ventilation. *Current Anaesthesia and Critical Care* 1989; **1**: 11–18.

21. Oberg PA, Sjostrand U. Studies of blood pressure regulation I: Common carotid artery clamping in studies of carotid sinus baroreceptor control of the systemic blood pressure. *Acta Physiologica Scandinavica* 1969; **75**: 276–86.

22. Klain M, Smith RB. High Frequency percutaneous transtrachael jet ventilation. *Critical Care Medicine* 1977; **5**: 280–7.

23. Lukenheimer PP, Rafflenbeul W, Keller H, Frank I, Dickhut HH, Fuhrman C. Application of transtracheal pressure oscillation as a modification of diffusion respiration. *British Journal of Anaesthesia* 1972; **44**: 627.

24. Hayek Z, Sohar E. External high frequency oscillation – concept and practice. *Intensive Care World* 1993; **10**: 36–40.

25. Scherer PW, Haselton FR. Convective exchange in oscillatory flow through bronchial tree models. *Journal of Applied Physiology* 1982; **53**: 1023–33.

26. Ross BB. Influence of bronchial tree structure on ventilation in the dog's lung as inferred in measurements of a plaster cast. *Journal of Applied Physiology* 1957; **10**: 1–4.

27. Lin ES, Smith BE. An acoustic model for the patient undergoing artificial ventilation. *British Journal of Anaesthesia* 1987; **59**: 256–64.

28. Eve FC. Actuation of the inert diaphragm by a gravity method. *Lancet* 1932; **2**: 995–8.

29. Mushin WW, Rendell-Baker L, Thompson PW, Mapleson WW, eds. The control of automatic ventilators. *Automatic ventilation of lung.* 3rd ed. Oxford: Blackwell Scientific, 1980: 152–3.

30. Mushin WW, Rendell-Baker L, Thompson PW, Mapleson WW, eds. *Automatic ventilation of lung.* 3rd ed. Oxford: Blackwell Scientific Publications, 1980: 62–131.

31. Beatty PCW. Respiratory systems, artificial. In: Payne PA, ed. *Concise encyclopaedia of biological and biomedical measurement systems.* Oxford: Pergamon Press, 1991: 330–5.

32. Barach AL, Martin J, Eckman M. Positive pressure respiration and its application to the treatment of acute pulmonary oedema. *Annals of Internal Medicine* 1938; **12**: 754–9.

33. Ashbaugh DG, Bigelow DB, Petty TL, Levine BE. Acute respiratory distress in adults. *Lancet* 1967; **2**: 319–21.

34. Grummitt RM, Jones JG. The physiology of artificial ventilation. *Current Anaesthesia and Critical Care* 1989; **1**: 3–10.

35. Qvist J, Andersen JB, Pemberton M, Bennike KA. High level PEEP in severe asthma. *New England Journal of Medicine* 1982; **307**: 1347–8.

36. Brochard L, Pluskwa F, Lemaire F. Improved efficacy of spontaneous breathing with inspiratory pressure support. *American Review of Respiratory Disease* 1987; **136**: 411–15.

37. Turner AE. Evaluation of ventilator alarms. *Journal of Medical Engineering and Technology* 1984; **8**: 270–6.

38. Orr JA, Westenskow DR, Farrell BS. Information content of three breathing circuit monitors: a neural network analysis. *Anesthesiology* 1992; **37**: A517.

39. *BS 5724 Medical Electrical Equipment: Section 3.12 Method of declaring parameters for lung ventilators.* London: BSI, 1991.

40. *BS 5724 Medical Electrical Equipment: Section 2.13 Specification of anaesthetic machines.* London: BSI, 1990.

41. Conway CM. Anaesthetic breathing system. In: Scurr FC, Feldman SA, eds. *Scientific foundations of anaesthesia.* 3rd ed. London: Heinemann, 1973: 399–405.

42. Conway CM. Anaesthetic breathing systems. *British Journal of Anaesthesia* 1985; **57**: 649–57.

43. Nunn JF, Ezi-Ashi TI. The respiratory effects of resistance to breathing in anesthetized man. *Anesthesiology* 1961; **22**: 174–85.

44. Kay B, Beatty PCW, Healy TEJ, Accoush MEA, Calpin M. Change in the work of breathing imposed by five anaesthetic breathing systems. *British Journal of Anaesthesia* 1983; **55**: 1239–47.

45. Humphrey D. The Lack, Magill and Bain anaesthetic breathing systems: a comparison in spontaneously breathing anaesthetized adults. *Journal of the Royal Society of Medicine* 1982; **75**: 513–24.

46. Kain ML, Nunn JF. Fresh gas economics of the Magill circuit. *Anesthesiology* 1968; **29**: 964–74.

47. Barrie JR, Beatty PCW, Campbell IT, Meakin G. Determination of the onset of rebreathing in an enclosed afferent reservoir breathing system in anaesthetised, spontaneously ventilating adults: a comparison of three methods. *European Journal of Anaesthesiology* 1994; **11**: 187–91.

48. Meakin G, Coates AL. An evaluation of rebreathing with the Bain system during anaesthesia with spontaneous ventilation. *British Journal of Anaesthesia* 1983; **55**: 487–96.

49. Tyler CKG, Barnes PK, Rafferty MP. Controlled ventilation with a Mapleson A (Magill) breathing system: reassessment using a lung model. *British Journal of Anaesthesia* 1989; **62**: 462–6.

50. Chan ASH, Bruce WE, Soni N. A comparison of anaesthetic breathing systems during spontaneous ventilation. *Anaesthesia* 1989; **44**: 194–9.

51. Ooi R, Lack JA, Soni N, Whittle J, Pattison J. The parallel Lack anaesthetic breathing system. *Anaesthesia* 1993; **48**: 409–14.

52. Rose DK, Froese AB. The regulation of Pa_{CO_2} during controlled ventilation of children with a T-piece. *Canadian Anaesthetists' Society Journal* 1979; **26**: 104–13.

53. Beatty PCW, Meakin G, Healy TEJ. Fractional delivery of fresh gas: a new index of the efficiency of semi-closed breathing systems. *British Journal of Anaesthesia* 1992; **69**: 474–7.

54. Conway CM. A theoretical study of homeostasis in the Magill circuit. *British Journal of Anaesthesia* 1976; **48**: 441–5.

55. Conway CM. Aleveolar gas relationships during the use of semi-closed rebreathing anaesthetic systems. *British Journal of Anaesthesia* 1976; **48**: 865–9.

56. Mapleson WW. The elimination of rebreathing in various semi-closed anaesthetic systems. *British Journal of Anaesthesia* 1954; **26**: 323–32.

57. Willis BA, Pender JW, Mapleson WW. Rebreathing in a T-piece: volunteer and theoretical studies of the Jackson-Rees modification of the Ayre's T-piece during spontaneous ventilation. *British Journal of Anaesthesia* 1975; **47**: 1239–45.

58. Lack JA. Theatre pollution control. *Anaesthesia* 1976; **31**: 259–62.

59. Miller DM, Miller JC. Enclosed afferent reservoir breathing systems – description and clinical evaluation. *British Journal of Anaesthesia* 1988; **60**: 469–75.

60. Jennings AJ, Michell BC, Beatty PCW, Meakin G, Healy TEJ. The physical characteristics of an enclosed afferent reservoir breathing system. *British Journal of Anaesthesia* 1992; **68**: 625–9.

61. Humphrey D, Brock-Utne JG, Downing JW. Single lever Humphrey ADE low flow universal anaesthetic system I: Comparison with dual lever ADE Magill and Bain systems in anaesthetized spontaneously breathing adults. *Canadian Anaesthetists' Society Journal* 1986; **33**: 698–709.

62. Humphrey D, Brock-Utne JG, Downing JW. Single lever Humphrey ADE low flow universal anaesthetic system II: Comparison with the Bain system in anaesthetized adults during controlled ventilation. *Canadian Anaesthetists' Society Journal* 1986; **33**: 710–18.

63. Snowden SL, Powell, Fadl ET, Utting JE. The circle system without absorber. *Anaesthesia* 1975; **30**: 323–32.

64. Eger EI, Ethans CT. The effects of inflow, overflow and valve placement on the economy of the circle system. *Anesthesiology* 1968; **29**: 93–100.

65. Zbinden AM, Feigenwinter P, Hutmacher M. Fresh gas utilization of eight circle systems. *British Journal of Anaesthesia* 1991; **67**: 492–9.

66. Jorgensen S, Hansen LK. The Venturi anaesthesia circuit I: An all purpose breathing circuit for inhalational anaesthesia. *Acta Anaesthesiologica Scandinavica* 1985; **29**: 269–72.

67. Beatty PCW, Wheeler MFS, Kay B, Cohen AT, Parkhouse J. A versatile closed circuit. *British Journal of Anaesthesia* 1982; **54**: 689–97.

68. Severinghaus JW. The rate of nitrous oxide uptake in man. *Journal of Clinical Investigation* 1954; **33**: 1183–6.

69. Beatty PCW, Kay B, Healy TEJ. Measurement of the rates of nitrous oxide uptake and nitrogen excretion in man. *British Journal of Anaesthesia* 1984; **56**: 223–32.

70. Spence AA. A guided approach to low flow anaesthetic systems. In: Zorab J, ed. *Lectures in anaesthesiology.* Vol 2. Oxford: Blackwell Scientific Publications, 1986: 1–12.

71. White DC. Vaporization and vaporizers. *British Journal of Anaesthesia* 1985; **57**: 658–71.

Chapter 23

Essential Monitoring

Anthony P. Adams

Closed claim studies in the USA	Volumetric measurements of respiration
Monitoring standards	Capnography
Alarms	Photoacoustic spectroscopy
Invasive monitoring	Raman scattering
Arterial (blood) pressure (ABP)	Oximetry
Central venous pressure (CVP)	Temperature
Electrocardiogram (ECG)	

Classically, information about a patient's progress was obtained by very simple clinical observations – a finger on the pulse, an ear to listen to breath and heart sounds (stethoscope) and an eye to see the chest movements and the patient's colour. The value today of these kinds of clinical observations cannot be overemphasized despite the wonders of modern technology. It was always rightly emphasized that the finger needed to be glued metaphorically to the patient so that observations would be continuous and thus prevent any untoward events from being missed. It is common in the USA for a monaural oesophageal stethoscope to be used for every patient: a safety pin, to fix the tubing to the anaesthetist's shirt had the advantage of fixing him to the patient. Advances in anaesthesia and surgery have demanded a better flow of continuous information (on line) from the patient, and from the anaesthetic and other equipment used, to the anaesthetist. The anaesthetist or the intensive care clinician has many tasks to perform simultaneously and modern equipment can provide the necessary information continuously and immediately so problems may be identified very early and corrected appropriately. Monitors are the 'unsleeping eyes' that allow us to 'see the invisible', especially early hypoxia. However, monitors are not likely to be of value unless the user can recognize and act appropriately on what is seen.

Hazards of equipment result not only from mechanical failure[1-4] but also from improper interaction between operator and machines. The most obvious dangers have largely long been eliminated from anaesthetic machines and associated equipment in developed countries but hazards of varying subtlety remain. Nevertheless, after human error (see below), equipment problems are the second most common cause of preventable critical incidents in anaesthesia. A critical incident is a human error or equipment failure that could have led (if not discovered or corrected in time), or did lead, to an undesirable outcome, ranging from increased length of hospital stay to death.[5] Included within this range of undesirable outcomes is increased stay in a recovery room or intensive care unit. Disconnection of some part of the breathing system is very common: so common in fact that Cooper identified the general problem of disconnection (involving the breathing system, i.v. lines, monitoring apparatus) separately in his analysis of major errors and equipment failures in anaesthesia management. A study commissioned jointly by the US and Canadian governments found that disconnections are so common that many anaesthetists consider them routine.[6] Although this particular problem has been discussed by experts from the Department of Health (DoH),[7] the message appears not to be getting across; indeed, important information, e.g. that male–female components of the breathing system should be mated by a 'push and twist' technique (rather than by simply making the connection a push fit), so as to force a good connection,[8] appears often ignored possibly because of a failure to convey and publicize the information sufficiently.

Monitoring systems can never reduce the importance of training, continuing education, clinical skill, and vigilance in providing safe anaesthesia and high quality care.[9] Indeed, intense training is required in the correct use and interpretation of monitors. At what point do advanced safety features become positive hazards? When manufacturers are trying to reduce cockpit clutter, why are we prepared to put up with a plethora of different monitors or multifunction push buttons on the monitors? How can we prevent the brain being overloaded with too many good intentions? Litigation in the USA has forced hospitals into extensive monitoring of patients to protect themselves from multimillion dollar malpractice claims. In the UK, sums awarded in damages for medical negligence are rising:

payments above £500,000 are becoming commonplace, and there are cases where brain damaged patients are being awarded £2–4 million.

Monitoring in the operating room or in the intensive care unit may be considered from the viewpoint of (1) monitoring the performance of the equipment (i.e. the supply of oxygen and the breathing system), and (2) monitoring the patient. The analysis of gases from the anaesthetic machine must be established when the equipment is checked at the start. The Association of Anaesthetists' anaesthetic machine check-out[10] is novel since it requires the use of an oxygen monitor, attached at the common gas outlet, to verify that when the oxygen flowmeter and emergency oxygen flush are opened the gas issuing is indeed oxygen and that no 'crossover' or contamination has occurred up to this point. This will also verify the correct operation of any antihypoxic device incorporated into the anaesthetic machine.

The patient is monitored by clinical observations, continuous monitoring devices and by intermittent monitoring devices (Table 23.1). Additional specialized monitoring is required in certain instances: (1) where there is pre-existing medical disease; (2) when special techniques such as controlled hypotension are being used; (3) when surgery involves the lungs, cardiovascular or central nervous systems; or (4) when major blood loss is

expected. Furthermore, the same standards of monitoring should be applied when sedation or anaesthesia is given for operations or procedures of brief duration including those performed outside the conventional operating theatre. Monitoring should be continued during transfer of the patient and in the recovery room: a full range of monitoring devices, as employed during anaesthesia, should be available. International standards for a safe practice of anaesthesia – developed by The International Task Force on Anaesthesia Safety – have recently been published[11] and are recommended to anaesthetists throughout the world. Their purpose is to provide guidance and assistance to anaesthetists, professional societies, hospital administrators, and governments to improve the quality and safety of anaesthesia. These standards, which include standards of monitoring during anaesthesia, were adopted by the World Federation of Societies of Anaesthesia in 1992.

Closed claim studies in the USA

A 'closed claim'[12] for an adverse anaesthetic outcome typically consists of relevant hospital and medical records,

Table 23.1 Summary of standards for basic intraoperative monitoring of the American Society of Anesthesiologists, October 21, 1992 and the recommendation for standards of monitoring of the Association of Anaesthetists of Great Britain and Ireland, 1988

UK: Association of Anaesthetists	USA: American Society of Anesthesiologists
Continuous presence of an adequately trained anaesthetist and clinical observations always strongly emphasized	
• Regular arterial pressure and heart rate measurements (recorded) • Continuous display of ECG throughout anaesthesia (can provide warning of impending circulatory failure due to myocardial ischaemia, conduction defects and arrhythmias) • Continuous analysis of gas mixture oxygen content (with audible alarm) • Oxygen supply failure alarm • Tidal/minute volume measurement • Ventilator disconnection alarm • Pulse oximeter • Capnography with moving trace • Temperature measurement available • Neuromuscular monitoring available	• Arterial pressure and heart rate measurements every 5 min at least* (recorded) • Continuous display of the ECG throughout anaesthesia (from beginning of anaesthesia until preparing to leave the anaesthetizing location)* • Continuous measurement of gas mixture oxygen content* (with low limit alarm) • Ventilator disconnect alarm (audible) • Pulse oximeter* • Capnography (must be used to verify correct position of endotracheal tube;* thereafter use is strongly encouraged) • Temperature measurement available (when changes intended, anticipated or unexpected *shall* be measured) • Neuromuscular monitoring available • Expired volume measurement • Adequate illumination and exposure of the patient is necessary to assess colour*
Monitoring to be continued throughout the period of induction, anaesthesia and recovery	ASA differentiates between *continual* meaning 'repeated regularly and frequently in steady rapid succession' whereas *continuous* means 'prolonged without any interruption at any time' * Under extenuating circumstances those items marked with an asterisk may be waived and reason recorded

narrative statements from involved health care personnel, expert and peer reviews, deposition summaries, outcome reports, and the cost of settlement or jury award.[13] Each claim is reviewed by an anaesthetist who has been specially trained for participation in the Closed Claim Study. Information from 'closed claims' has been acquired from 20 insurance organizations involving a total of 1541 cases (dental damage – the most frequent injury in any survey – is not considered). The most common cause of injury is difficulty in management of the respiratory system (34 per cent of cases). This also represented the source of the most common untoward result, namely death and brain damage. The three most common respiratory-related critical incidents are inadequate ventilation of the lungs, oesophageal intubation and difficult intubation. In an in-depth analysis of a sample of cases of inadequate ventilation it was found that the majority of cases were due to poor monitoring of the respiratory system (42 per cent) and personnel problems (13 per cent) such as impaired vigilance of the anaesthetist. Patient pathophysiology, such as obesity and impaired lung function, were noted in only 10 per cent of cases. Oesophageal intubation was the most frequent critical incident in the study – brain damage or death was the outcome in 94 such cases. It is notable that in 59 cases the person administering the anaesthetic claimed to have auscultated both sides of the chest and thought the tube was properly placed; this points out the fallacy of this sign alone.[14] The next most common respiratory emergency was difficult intubation comprising 6 per cent of the total cases. Detailed accounts of the closed claim studies have been published as the information has been gathered.[15–17]

Monitoring standards

The Dutch Health Council, which is the main Advisory Council to the Ministry of Health, produced a pioneering advisory report on monitoring standards in 1978.[18] This was accepted and sent out to all hospitals with the requirement that they had to comply within 3 years; a team of inspectors under the Chief Inspector of Medical Affairs checks compliance with this report. In Holland there has now been a reduction in the number of operating rooms covered by one anaesthetist supervising nurse anaesthetists (a 1:2 ratio).

In 1986 the American Society of Anesthesiologists (ASA) instituted minimal monitoring standards[19] following the example of those promulgated by the Boston hospitals attached to Harvard Medical School.[20] These standards have since been upgraded: from January 1989 there is a provision that a quantitative method of assessing oxygenation such as pulse oximetry shall be employed in the operating room, and from January 1992, in the recovery room. A further modification of the standards became effective from January 1991 (and modified October 1992): 'when an endotracheal tube is inserted, its correct positioning in the trachea must be verified by clinical assessment and by identification of carbon dioxide in the inspired gas. End-tidal CO_2 analysis, in use from the time of endotracheal tube placement, is strongly encouraged'.[21] The ASA did not agree to a suggestion to make *continuous* capnography a formal rule because it was felt that to do so would require a number of caveats and exceptions, particularly regarding paediatric patients. Nevertheless, in the USA there is acceptance of end tidal carbon dioxide and pulse oximetry monitoring as a general standard, both in the operating theatre and the recovery room. Another conclusion from the ASA Closed Claims Study for data from 1079 analysed claims 'provides solid evidence' that the *combination* of pulse oximetry and capnometry can expect to help prevent anaesthetic-related morbidity and mortality.[22] The benefits of insisting upon oximetry and capnography are demonstrated by the fact that by June 1989 Massachusetts had experienced no hypoxic deaths since 1988.[23]

In January 1989 the adoption of standards became law in the state of New York. Every patient receiving a general anaesthetic via a tracheal tube is required to be monitored with an end tidal carbon dioxide monitor, an inline oxygen monitor, and some kind of disconnection alarm. The state of New Jersey adopted (effective 21 February 1989) hospital licensing standards for anaesthesia care.[23–25] These standards are now required by law, enforced by state inspectors, and a hospital can be closed if the law is broken. The standard covers the governance of the Department of Anesthesia (must be a board certified anesthesiologist), the anaesthesia machine (i.e. diameter index and pin index safety systems, vaporizer interlock system), pre- and postanaesthesia care, equipment maintenance, patient monitoring (i.e. pulse oximetry for all patients including conscious sedation, body temperature monitoring on every patient), education and quality assurance. Already the effect of standards promulgated and laws enacted consequent upon the adoption of the Harvard Standards by the American Society of Anesthesiologists and other bodies has been to reduce the rate of rise of premiums and now even to reduce premiums for malpractice insurance; it is becoming evident that this is a consequence of proper monitoring and not entirely due to greater awareness of the problem.[26,27] Advances in technology have played a very important part: although, for example, ten different measurements were originally required in Holland, this number has now been reduced by combination into fewer monitors, or even one. Technological advances have reduced the bulk and improved the reliability of monitors.[28]

In 1987 the Australasian Faculty of Anaesthetists and

the Canadian Anaesthetists' Society published minimal monitoring standards[29] and *Guidelines to the practice of anaesthesia*.[30] Use of integrated monitoring has improved the safety record of Canadian anaesthetists. In 1988 they were taken out of the most expensive membership category of the Canadian Medical Protective Association, which provides professional insurance for doctors. The Association of Anaesthetists of Great Britain and Ireland (AAGBI) *Recommendations for standards of monitoring in anaesthesia and recovery* was published in 1988.[31] This stresses the importance of the *constant presence* of the anaesthetist and the absolutely fundamental tenets of clinical observation backed up by constant online information from monitors about the condition of the patient, the functioning of the apparatus and the interactions between all of these. The document also addresses regional anaesthesia, and the recovery and transport of patients. Fifty per cent of successful claims in the USA are due to failure to monitor, poor preoperative assessment, absence from theatre or inadequate recovery and follow-up. The first UK CEPOD report[32] showed that knowledge was not applied in 75 per cent of anaesthetic-related deaths but it is a fallacy to think that because equipment doesn't feature largely in the report that it isn't important.

Alarms

Importance of setting appropriate alarm limits

When a monitor is switched on defaults for the thresholds of the various variables are automatically set. If any such threshold is violated an alarm will be triggered. The default is a value(s) set up by the manufacturer in the expectation that it is of use to the majority of users. Sometimes these settings are appropriate and sometimes not. For example, the default low heart rate alarm limit on one popular monitor is 40 b.p.m. – hardly appropriate for an anaesthetic conducted in a child when the anaesthetist would want to be alerted well before such a serious bradycardia developed. The alarm limits on anaesthetic monitoring equipment rarely seem to be set up appropriately by the anaesthetist before the anaesthetic is started. There is a tendency to set such limits later on in the course of the anaesthetic. There is too much reliance on the default values of individual monitors and an amazing lack of knowledge of the actual values involved. The presence of many similar and over-obtrusive alarms in the operating

theatre continue to confuse and irritate[33] and there is a need to rethink completely the characteristics of the different audible warnings. Intelligent alarm systems are being developed to organize these alarms, on the assumption that they will shorten the time that anaesthetists need to detect and correct faults.[34] False positives often occur because the user does not fully understand the operation of the equipment and does not set appropriate alarm limits. False-positive alarms generate a sense of disregard and disbelief, yet modern equipment correctly set up is indicating that there really is something wrong. Errors of judgement by the anaesthetist are a major problem[35-37] and this is usually a reflection of the anaesthetist's inexperience, ignorance or lack of skill. A judgmental error occurs when the action taken is indeed the action intended, i.e. a bad judgement; these are errors based on faulty decisions, e.g. choice of facemask anaesthesia for a patient with a full stomach.

Invasive monitoring

Invasive monitoring, used in the appropriate circumstances, provides vital information about a patient's current status and response to therapy. However, such monitoring techniques are not without risk. One need only consider the vast list of problems which may follow cannulation of the internal jugular vein. Other techniques which pose risks are arterial and particularly pulmonary artery catheterization. Thus it is essential that those who practise such techniques are correctly taught and remain exposed sufficiently to them subsequently that skills are maintained. The ASA issued a guidance on invasive monitoring in 1984 which aims to reduce the risks to the patient by members of the Anesthesia Care Team by stating that (1) the decision to use invasive monitoring is a medical judgement and should, therefore, be made only by a qualified physician; (2) invasive monitoring techniques should be prescribed by a physician and depending on its risk should be applied only by a competent and trained physician, or under the immediate medical attention of such a person. The ASA statement provides for training and accrediting of non-physician members of the team who may perform invasive monitoring (CVP catheterization via the upper extremity and arterial cannulation) if approved at local medical staff level by the anaesthetic department and the active medical staff. Insertion of pulmonary artery catheters is a relatively hazardous procedure and should only be performed by a properly trained physician.

Arterial (blood) pressure (ABP)

When the American neurosurgeon Harvey Cushing was visiting Bern in Switzerland in 1901 he heard about Scipione Riva Rocci's method of measuring blood pressure which had been described in 1896. Accordingly he went to see him to learn more about this method which was in continuous daily use by the bedside of every patient in the Ospedale di Matteo in Pavia near Milan. Cushing was so impressed that when he returned to the Johns Hopkins Hospital in the United States he insisted that his anaesthetists used the technique during his operations. The method was later enhanced by listening to the Korotkoff sounds – described in 1905 – by auscultation over the brachial artery with a stethoscope. This is still a superb method for determining arterial pressure.

Indirect methods – intermittent non-invasive blood pressure

There have been many attempts to automate the measurement of arterial pressure using indirect methods. Most of the methods work satisfactorily in normal patients but are unreliable in hypotensive or restless patients. Systolic pressure can be detected more reliably than diastolic pressure. A number of physical principles have been used such as electrical impedance changes, double and triple cuff pneumatic methods; a microphone has been used for detecting the Korotkoff sounds and movements of the arterial wall have been sensed using ultrasound.

The technique of oscillotonometry was made popular by von Recklinghausen who developed two instruments, the scala alternans and the scala altera. The technique was one which used two (double) cuffs which overlapped one another slightly. Both cuffs are inflated to above systolic pressure. The upper cuff is slowly deflated stepwise, thus eventually allowing jets of blood to pass under the upper cuff once its pressure had fallen below the systolic pressure. The lower cuff which is still inflated forms the sensing cuff and picks up the signals (systolic pressure) as the jets of blood impinge on it. At this time the lower cuff is slowly deflated, and the diastolic pressure is taken as the point when the oscillations detected by this cuff die away.

The automated oscillotonometric technique uses a single cuff and a pressure transducer measures both the pressure and the oscillations. The onset of the oscillations and the point of maximum oscillation are sensed to give the systolic and mean pressures respectively.

Continuous non-invasive blood pressure

A well known instrument is the Finapress. A small cuff is positioned over a finger and is attached to a transducer and an air pump. A light emitting diode (LED) and a photocell are included in the cuff so the system is in effect a photoplethysmograph. The volume of blood in the arteries varies with the degree of distension during systole. The microprocessor senses this volume at a set point according to the mean arterial pressure and a servocontrol constantly adjusts the amount of air and hence pressure in the cuff to maintain the photoplethysmograph output constant. A continuous tracing of arterial pressure is obtained. However, despite extensive trials the method has not been generally thought to be sufficiently reliable.

A new development is a method based upon the relation between pulse wave velocity (PWV) and blood pressure which was initially described over 60 years ago. In any individual, changes in pulse wave velocity in the short term are due primarily to changes in blood pressure. Studies using two pressure transducers placed a known distance apart on a major artery identified a linear relation between PWV and mean blood pressure. Development of these concepts means that blood pressure may now be measured continuously and non-invasively. One pair of sensors measures blood pressure, heart rate and arterial oxygen saturation by pulse oximetry. The system is based on photometric measurements which do not rely on compression or occlusion of arteries to track blood pressure (Artrac™ system). The blood pressure cuff provides a calibration, or baseline, value. The sensors are placed on the forehead and a digit, and are used to measure changes in pulse transit time, microvascular volume and other physiological values related to the blood pressure. Changes in these values are used in an empirically derived equation that calculates changes in blood pressure from the initial calibration value.

Invasive method

In low cardiac output states pulses may be poorly palpable and Korotkoff sounds difficult to hear while the intra-arterial pressure may be only moderately reduced. This method is justified for difficult or major cases as the method is more reliable on a beat-to-beat basis compared with intermittent methods. In addition, blood samples can be obtained for analysis. Monitoring of intra-arterial pressure is easily accomplished by percutaneous insertion of a 20 or 22 gauge cannula into the radial or dorsalis pedis artery. The brachial artery may be used but as this is an end artery it ought, if possible, to be avoided. Occasionally the femoral artery may be used but is best avoided in case of problems resulting from any resulting infection or thrombosis. Long-term patency is assisted by intermittent or continuous flushing devices using heparinized saline (approximately 1 iu heparin per ml 0.9 per cent NaCl solution). Miniature transducers can be attached to the arterial cannula thus eliminating long fluid-filled catheters. Disposable transducers are now inexpensive.

Central venous pressure (CVP)

Traditionally CVP has been estimated using a catheter passed into the right atrium and connected to a saline manometer. Care must be taken that the reference zero of the system is in the phlebostatic axis of the patient. When the patient is supine the zero reference may be taken as level with the suprasternal notch. The small error should not pose a problem since the anaesthetist is interested in trends and responses to therapy. Changes in CVP are also used to assist in the diagnosis of hypovolaemia, cardiac failure and tamponade. Sykes has pointed out that the correct use of the CVP is to assess the response to repeated small (200 ml) transfusions of fluid. In a patient with hypovolaemia, the CVP rises slightly after each such bolus and then rapidly falls to the preinfusion level.[38] In a normovolaemic patient the postinfusion fall in pressure is slower, taking about 3 minutes to return to the control value, while in a patient with fluid overload the CVP will be high and the fall in pressure after infusion will be extremely slow.

Electrocardiogram (ECG)

It is usual to use a three-lead configuration during anaesthesia. The best configuration is the CM5 lead where the right electrode (negative) is positioned in the second or third intercostal space on the right side of the manubrium. The positive lead (left) is positioned in the fifth intercostal space in the anterior axillary line. The indifferent third lead may be positioned anywhere convenient. Since the mass of the heart lies between the positive and negative leads a good signal is obtained. Although information about heart rate and rhythm may be obtained by other leads the value of the CM5 configuration for anaesthetists is that signs of subendocardial ischaemia are more readily detected. The name CM5 (for central manubrium, 5th intercostal space) comes from the original position of the right electrode in the sternal notch which is an uncomfortable place in an awake patient. It is important for the anaesthetist to hear the sound of each heart beat, whether from the ECG signal or from the pulse oximeter. The audible 'bleep' of the monitor should therefore be switched on; even small changes in heart rate will be instantly apparent.

Volumetric measurements of respiration

Monitoring respiration is often difficult unless the patient is receiving artificial ventilation. Leaks in the breathing system during spontaneous ventilation pose problems when a conventional anaesthetic facemask is used. Measurement of minute volume and tidal volume may be made using simple devices such as the Wright respirometer (anemometer). However, delicate instruments like this are easily damaged. Dry displacement gas meters are rugged, reliable and quite accurate.[39] They are resistant to the effects of water vapour and can therefore be left in circuit continuously. Pneumotachographs have also been used but are difficult to use clinically, especially for long periods, because of condensation of water vapour in the pressure lines and the calibration of the instrument changes during inspiration and expiration. Changes in the diameter of the chest and abdomen can be sensed in various ways. Impedence pneumography can be used to record changes in the electric impedance of the chest during respiration and a temperature sensor placed in the airway will record changes in air temperature during the phases of respiration and count the number of breaths per minute. Temperature probes may become displaced or covered with secretions and the impedance pneumogram may continue to record respiratory efforts made by the patient even though complete respiratory obstruction has occurred.

Capnography

Capnography is the study of the shape or design of the changing concentrations of CO_2 in respired gas. A high speed capnogram gives detailed information about each breath while overall changes in CO_2 may be followed at a slower paper speed. The capnograph is an excellent early warning system. The association of certain patterns with specific circumstances is now recognized and the curves are often diagnostic; indeed, the effect of different breathing systems and malfunctions produce their own individual 'signature' capnograms.

Luft developed the principle of capnography in 1943 using the fact that CO_2 is one of the gases which absorbs infrared (IR) radiation of a particular wavelength. Infrared radiation is absorbed by all gases with more than two atoms in the molecule. If there are only two atoms, absorption only occurs if the two atoms are dissimilar. A capnometer is an instrument that measures the concentration of CO_2 numerically. A device which continuously records and displays CO_2 concentration in the form of a tracing or waveform is called a capnograph and the tracing on recording paper is called a capnogram. The introduction of capnography into routine clinical practice was pioneered by Smalhout and Kalenda in The Netherlands.[40,41] The presentation of the CO_2 waveform obtained from breathing systems used in anaesthesia and intensive care in the analogue (waveform) format, i.e. the capnograph or capnogram, is vastly to be preferred to a meter or even a fast digital display. Indeed, both the latter are useless in anaesthetic practice where the breath-by-breath waveform needs to be displayed to permit continuous monitoring and analysis and is essential where fractional rebreathing techniques are employed, such as with the Mapleson D/E/ F and Bain breathing systems, because a meter or digital display cannot indicate the CO_2 concentration of the end tidal CO_2 plateau in the face of CO_2 also appearing in the inspiratory part of the respiratory cycle. The measurement of changing CO_2 concentrations is of value in recognizing abnormalities of metabolism, ventilation and circulation. In the normal state alveolar CO_2 concentration is maintained within rather narrow limits independent of the metabolic state or the size of the physiological dead space. Thus alveolar CO_2 concentration can serve as a valuable guide to CO_2 homoeostasis during prolonged periods of mechanical ventilation of the lungs required in anaesthesia, intensive care or other conditions associated with altered breathing.

Carbon dioxide in respired gas may be continuously measured by mass spectrometry, IR analysis or photo-acoustic spectrometry. Infrared rays are given off by all warm objects and are absorbed by non-elementary gases (i.e. those composed of dissimilar atoms); certain gases absorb particular wavelengths, producing absorption bands on the IR electromagnetic spectrum. The intensity of IR radiation projected through a gas mixture containing CO_2 is diminished by absorption which allows the CO_2 absorption band to be identified and is proportional to the amount of CO_2 in the mixture.

Infrared rays have a wavelength greater than 1 μm and thus lie beyond the visible spectrum (0.4–8 μm); CO_2 shows strong absorption in the far IR at 4.3 μm and so this wavelength in the far IR range is used. A narrow band IR filter prevents the passage of light which would otherwise be absorbed by gases other than CO_2. There is some overlap in the absorption bands of other gases (e.g. N_2O distorts the absorption bands for CO_2). Hence, due allowance must be made for any interfering gases; N_2O molecules also interact with CO_2 molecules to produce a collision broadening effect which affects the sensitivity of the IR CO_2 analyser (see below).

An IR analyser, put simply, consists of a source of IR radiation, an analysis cell, a reference cell and a detection cell. In the Luft system, rays of light from the source are filtered so as to obtain the required wavelengths and then pass through the analysis cell to fall on the detector which contains pure CO_2. Any IR radiation which is not absorbed by the gases in the analysis cell is absorbed in the detector and heats the CO_2. The pressure in the detector (which is in effect a differential micromanometer as opposite sides are subjected to the light transmitted through the measuring and reference cells respectively) will vary according to the heating effect from the IR radiation. These alinear changes are suitably detected, amplified and displayed; modern instruments use a linearizing circuit. Drift is reduced by interrupting or chopping the IR beam with a rotating shutter at 25–100 Hz; the pulses of IR radiation thus produce pulses of pressure in the detector cell.

The modern alternative to the classical Luft system uses an LED to produce light of the required wavelength together with a solid state photodetector (instead of relying on a micromanometer) to measure the amount of light reaching it alternately via the measuring and reference cells. The beam is chopped 4000 times a minute. In some designs the chopper is omitted and the infrared LED is switched on and off by a microprocessor. Some instruments dispense with the need for a reference cell and instead obtain a CO_2 zero for reference from the sample cell itself at a time when the cell is known to contain CO_2. Thus, in both types of capnograph the electrical output consists of a series of pulses the height of which varies with the CO_2 concentration in the analysis cell. The IR cell is the most critical part of the system and must be protected from contamination by liquids or particulate matter as these invariably cause high erroneous readings because of their high IR absorbance.

It has been common for capnographs to be provided with an automatic zeroing device which returns the trace to the baseline just as the next inspiration is sensed. This was of help in the past because of the drift experienced with electrical components. However, this facility is nowadays somewhat limiting because it does not permit the instrument to be used in the presence of breathing systems where it may be normal to have some CO_2 in the inspired mixture, such as controlled fractional rebreathing during intermittent positive pressure ventilation of the lungs with circuits such as the Bain breathing system.

Analysis of inspired oxygen, expired CO_2 and arterial oxygen saturation should be standard during every

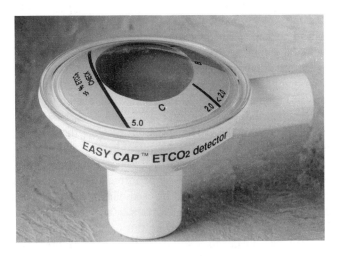

Fig. 23.1 Nellcor Easy Cap end tidal CO$_2$ detector.

produce a capnogram where the curves are raised above the baseline and are sinusoidal in form without a plateau. A poor connection at some point along the sampling tube can allow the entrainment of air thus producing an abnormal capnogram. Gas which has been sampled by the analyser is usually discarded and not returned to the breathing circuit because of the inconvenience involved as well as the possibility of interfering with the gaseous homogeneity of the breathing system. However, the loss of gas through sampling may pose a problem in paediatric practice or with the use of closed or low-flow circle breathing systems; in such circumstances gas may be returned to the breathing system or an extra equivalent fresh gas inflow provided to compensate.

anaesthetic. Certainly all intubated patients should be monitored by capnography to detect problems with tracheal intubation: systems which display the moving waveform of the capnogram are the best[42] although the end tidal carbon dioxide detector (Fig. 23.1; Nellcor Easy Cap end tidal CO$_2$ detector) which uses a pH chemical indicator is useful in many situations.[43]

Sidestream analysers

These draw gas continuously from the sampling site via a small bore tube of about 2 m length to the measuring cell for analysis and display. The bore should be not greater than 1–2 mm to avoid mixing of gas, to resist the introduction of foreign matter and to facilitate the withdrawal of small sample volumes. The length of the sampling tube should be kept as short as possible so as to obtain a fast response time for greatest accuracy. The sampling tube must be impermeable to CO$_2$ and so polyvinylchloride (PVC) is often used but because halogen hydrocarbons react with this material a tube made of Teflon^{TM} is preferred in anaesthetic practice. Several manufacturers provide sampling tubes and T-pieces which are hydrophobic, i.e. treated with a material which resists the entry of water. The sampling rate is of the order of 50–500 ml/min and can be varied either continuously or by preset steps; however, the flow control is often inconveniently positioned at the rear of some instruments so that the user may be unaware of the sampling rate. This may pose a hazard, for instance when excessive sampling rates are accidentally applied to children's breathing systems. Excessive sampling rates should generally be avoided in case loss of gas upsets the homoeostasis of the breathing system. On the other hand, too low a sampling rate may

Sampling sites

It used to be common for the tip of the sampling catheter to be positioned just above the carina so as to obtain a good sample of alveolar gas. However, this is inconvenient and the risk of aspiration of secretions and water into the apparatus is considerable. It is now usual to position the end of the sampling tube at the proximal end of the tracheal tube using a T-piece adapter; this is also analogous with the position of the IR measuring cell in mainstream analysers (see below). Care must be taken in obtaining the correct sampling position when T-piece breathing systems on the Ayre principle are used lest fresh gas from the anaesthetic machine is drawn into the sampling tube along with the expired gas sample to produce, because of the dilution, an erroneous low value for the end tidal sample. The same problem occurs with the use of lung ventilators which produce a constant flow. The problem can be prevented by interposing a right angle adapter (such as that used with a facepiece) between the breathing system (Ayre's T-piece or Bain system) and the tracheal tube and interposing the sampling tube on the patient's side of the angle piece. When conventional twin breathing hoses are used with a Y-piece, the sampling site should be as close to the patient's mouth as possible; if a catheter mount is used, care must be taken to ensure that the sampling site is not at the junction of the Y-piece with the catheter mount because of the extra dead space introduced by the mount. The same consideration obtains with the use of filters and condenser humidifiers despite the increased risk of water getting into the sampling tube. Manufacturers produce a wide variety of adapters for sampling and these are recommended; special adapters to cope with the special problems of small children are also available; the use of a hypodermic needle inserted through non-disposable breathing tubes to achieve access to the respired gas is to be deplored.

Mainstream analysers

These do not draw gas but incorporate the analysis cell with IR source, detector and associated electronics into a specially designed airway adapter which is interposed into the breathing system. There is no specific reference cell. This form of 'no loss' system offers the advantages of a very fast response. The possibility of condensation of water vapour is prevented by heating the measuring chamber to about 40°C but there remains the possibility of contamination from secretions which absorb IR radiation and lead to a spurious high value for CO_2 concentration. The added dead space is a disadvantage in infants. In clinical practice the sidestream system is often to be preferred purely because of the expense involved in the repair of accidental damage sustained to the delicate components in the mainstream adapter.

Calibration and interference

Unfortunately, N_2O absorbs some light at the most convenient CO_2 wavelength of 4.3 μm. Furthermore, the absorbance properties of CO_2 molecules are affected by the presence of N_2O molecules which collide with them. The effect is to cause the absorption spectrum to become broader with the result that the degree of overlap in the absorption bands of different gases varies according to the gas concentrations. This collision broadening effect (sometimes also called pressure broadening) results from the fact that the IR absorption of CO_2 is based on the vibrational motion of the molecules. When the CO_2 molecule vibrates in a crowd of other molecules the collisions affect their vibrational energy states and thus the absorption of IR light. The degree of interaction depends on the ambient pressure and on the mass and nature of the neighbouring molecules. The overall effect when CO_2 is measured in gas mixtures containing N_2O is an overestimate by an amount of about 10 per cent for a mixture containing 50 per cent N_2O, 45 per cent O_2 and 5 per cent CO_2. This is overcome commercially by introducing an electronic bias to the results through compensation buttons for N_2O and often for O_2 as well; sophisticated instruments also monitor N_2O (by another infrared LED) and O_2 concentrations simultaneously to provide a continuous and varying compensation according to the changing concentration of the interfering gases.

Other considerations include the influence of water vapour, atmospheric pressure and calibration procedures. Erroneous results will occur if water or particulate matter which have high IR absorbance enter the cell. An effective water separation system is required for continuous use but a filter may produce an undesirable sinusoidal curve because of a mixing effect. A water trap is used in sidestream analysers to remove water in particulate form before it can enter the analysis cell. The design of the trap relies on gravitational forces to separate drops of water from the gas stream and the trap must be frequently dried out and attention directed to prevent water accumulating.

Because the principle of capnography is based on measurement of the partial pressure of CO_2, the method is somewhat affected by changes in the barometric pressure. It is for this reason that calibration procedures with gaseous mixtures must be performed using the same type of sampling tube as will be used when the analyser is connected to the patient system; omission of the standard long narrow 2 m sampling tube during such calibration procedures will fail to take account of the large pressure drop across the ends of the tube and a measuring error will therefore result during subsequent clinical use. Equipment suppliers now provide canisters of gas mixtures containing known amounts of CO_2 in a mixture of anaesthetic gases and vapours. Although modern instruments provide means for electronic calibration, regular checks using such gaseous calibration are recommended. Some modern capnometers sense changes in barometric pressure and automatically correct the CO_2 reading. In cases of doubt a rough check may be made by a healthy individual making a forced vital capacity into the sampling tube and observing the peak CO_2 reading which should be about 5.0–5.5 per cent (mean 5.3 kPa; 40 mmHg). The presence of water vapour also affects the reading since the temperature of the patient is usually 37°C whilst that of the instrument cell is, say, 25°C, i.e. a difference of 3 kPa (23 mmHg) in P_{H_2O}, resulting in an overestimate of P_{CO_2} by 0.15 per cent (0.15 kPa; 1.13 mmHg).

The normal capnogram

In conditions of cardiovascular stability, the end tidal CO_2 concentration ($P_{ET}CO_2$) bears a constant relation to P_aCO_2 and the normal $P_{ET}CO_2 - P_aCO_2$ difference is 0.7 kPa (5 mmHg). If the alveoli from all areas of the lung are emptying synchronously, $P_{ET}CO_2$ will be synonymous with P_aCO_2. The normal capnogram is shown in Fig. 23.2. The fast speed (12.5 mm/s) is essential to detect changes in individual respiratory cycles. When expiration begins the first part of the gas passing out of the patient's mouth is composed of gas from the mechanical and anatomical dead space and since this normally contains no CO_2 the capnograph registers zero. Next a sharply rising front is seen which represents the mixing of dead space gas with alveolar gas. It is important to note that the end-expiratory plateau which follows is not an isocapnic trace but that there is a very slight and steady increase in the end tidal CO_2 concentration as the alveolar fraction is expelled from

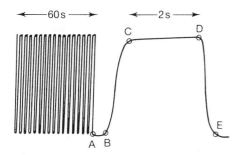

Fig. 23.2 The normal capnogram at slow and fast recording speeds. A–B denotes the expiration of gas from the anatomical dead space (containing zero CO_2); B–C denotes the exhalation of mixed expired gas; C–D is the alveolar or end tidal 'plateau'; D denotes the onset of inspiration and D–E denotes the return of the CO_2 tracing to the baseline as fresh gas is inspired.

the lungs. This effect is exaggerated in patients with chronic bronchitis and emphysema. There then follows a sharp downward return of the trace towards zero as inspiration begins. It is common to see a ripple effect superimposed on this downward part of the trace, the so-called cardiac oscillations, although this is now thought to be due to small gas movements created largely by the pulsations of the aorta. These oscillations are especially noticeable at slow respiratory rates and when opioids such as fentanyl have been given. It is claimed that individual drugs produce their own characteristic shapes or signatures and observation of this part of the capnogram is helpful in this respect.

The normal capnogram seen at low (trend) speed recording (25 mm/s) is rather like looking at an open box of matches and observing the striped effect of the heads of the matches all lined up next to one another. Trends are best observed at this recording speed. Reference has already been made to the often distinctive 'fingerprint' or 'signature' shapes produced in the capnogram in special circumstances. Such pattern recognition (as with the electrocardiogram) enables instant detection of problems with the patient or breathing apparatus. Figure 23.3 shows the normal characteristic capnogram 'signature' of a healthy patient undergoing uneventful IPPV of the lungs with a Bain breathing system. The small amount of CO_2 appearing during inspiration (the rebreathing wave)

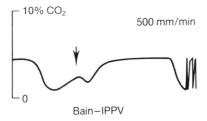

Fig. 23.3 The normal 'signature' capnogram of a patient anaesthetized using IPPV with the Bain breathing system. Note the rebreathing wave (arrow) during inspiration.

represents the controlled fractional rebreathing which enables a predetermined $PaCO_2$ to be achieved in the face of large tidal volumes (thus avoiding excessive hypocapnia with all its attendant disadvantages) and hence the maintenance of good oxygenation. The exact position and shape of this rebreathing wave depends on the magnitude of the fresh gas flow from the anaesthetic machine.

The abnormal capnogram

Abnormalities should be sought by analysing the various phases of the capnogram for individual breaths as well as observing the trends occurring over a period of time. Five characteristics should be inspected, viz. height, frequency, rhythm, baseline and shape. Rebreathing may be observed by the appearance of carbon dioxide in the inspired gas; this may or may not be abnormal depending on the type of breathing system used. Blunting of the normal sharp upward deflection as expiration proceeds (mixed alveolar gas) occurs with atelectasis or with a leak in the breathing circuit. A second alveolar plateau superimposed on the first (the so-called 'camel' capnogram) represents one lung emptying slightly before the other and may be seen in patients in the lateral position. Small indentations or dips in the final portion of the alveolar plateau may represent gas movement due to someone leaning on the chest, or the patient making a very small movement of the diaphragm as a muscle relaxant is wearing off (the so-called 'curare cleft'). Complete disconnection or total obstruction of the breathing system shows a sudden return to a zero trace, but torsion or kinking of the tracheal tube or breathing circuit, with some gas passage remaining, shows a peaked effect in place of the normal end tidal plateau. Disconnection alarms are not infallible: the commonest types monitor the pressure in the airway and the thresholds for 'high' and 'low' need to be set carefully according to the prevailing circumstances. Under some circumstances partial or even complete disconnections of some part of the breathing system may go undetected should cyclic pressure changes prevent the alarm from operating. Hence, continuous capnography is valuable.

Unrecognized intubation of the oesophagus instead of the trachea is a common problem in anaesthetic practice and accounts for many unfortunate outcomes. It is often thought that intubation of the oesophagus by mistake for the trachea is associated with a zero CO_2 signal. However, this is not necessarily the case because previous attempts to inflate the lungs by facemask drive some expired gas into the oesophagus and stomach so that when oesophageal intubation unknowingly occurs some of this CO_2 returns to the breathing system to appear on the capnograph when IPPV is attempted. Such a tracing has a peaked effect and does not resemble the normal end tidal plateau; the concentration of CO_2 is low and gastric CO_2 decreases to zero after about a minute (five 'breaths'). The golden rule to remember with capnography is that any trace other than the normal pattern with a smooth distinct end tidal plateau should immediately suggest that something is wrong somewhere.

In a few countries CO_2 is still used by some anaesthetists as part of the anaesthetic technique to stimulate breathing, to aid nasal intubation, or to prevent hypocapnia in paediatric anaesthesia; in such circumstances the capnograph should be mandatory to warn of an excess of PCO_2 or its accidental administration. In the United Kingdom such incidents of inadvertent CO_2 administration continue to occur with tragic results. In The Netherlands capnography in anaesthetic practice has been mandatory by law since 1978. During spontaneous breathing it should be borne in mind that return of the CO_2 tracing to the zero level between each breath does not necessarily imply that rebreathing is not occurring because the retention of CO_2 stimulates breathing; simultaneous measurement of expired volume should indicate whether or not this is the case. However, failure of the CO_2 trace to return to zero during spontaneous breathing with any breathing system indicates the presence of gross rebreathing of carbon dioxide.

The measurement of CO_2 fraction in mixed expired gas ($FECO_2$), to permit the calculation of physiological dead space and the rate of elimination of CO_2 ($\dot{V}CO_2$), is usually performed by analysing mixed expired gas over a period of time by collection into a Douglas bag. However, a more convenient method for obtaining $FECO_2$ is to measure the instantaneous expiratory flow ($\dot{V}E$) and expiratory carbon dioxide fraction ($FE'CO_2$ or $ETCO_2$). The product of these two signals gives carbon dioxide flux and integration with time yields expiratory carbon dioxide volume. Processing of these signals is achieved online electronically and provides particularly useful information for patients in intensive care who are receiving IPPV. It is not practicable for commercial systems to measure both flow and $ETCO_2$ at the same point in the system; therefore the CO_2 production is overestimated. The error occurs in the measurement of the instantaneous flow of CO_2 to and from the patient as a result of rebreathing. This corresponds to about 24 ml of end-expiratory gas per breath using the standard Y-piece and tubing of the Siemens-Elema Servo ventilator with the Siemens-Elema CO_2 analyser. This problem can be much decreased by the use of non-return valves in the Y-piece.

Continuous monitoring of $ETCO_2$ is of value for the early detection of the malignant hyperthermia syndrome (MHS) and dramatic rises in PCO_2 occur before body temperature has much increased. Continuous CO_2 monitoring is also useful in other conditions associated with increases in metabolism such as shivering, pain and seizures.

Reductions in cardiac output are accompanied by reductions in carbon dioxide output and $ETCO_2$ monitoring is quite a useful adjunct in monitoring cardiogenic or hypovolaemic shock. A gradually diminishing $ETCO_2$ concentration should alert the anaesthetist to the possibility that blood loss has been more than is realized. When a sudden interruption in pulmonary perfusion occurs such as with cardiac arrest (or cardiac failure or gross arrhythmias), the transport of CO_2 to the lungs ceases and the capnograph tracing quickly and exponentially decays towards zero. The capnograph is also a very sensitive monitor of venous air embolism and is second only in sensitivity to the Doppler ultrasonic technique without the disadvantage of interference by surgical diathermy. The blockage in the lung capillaries caused by emboli disturbs the existing ventilation/perfusion ratio so that gas exchange in affected units is either grossly impaired or ceases. The overall effect is thus one of an increased physiological dead space with gas returning from affected alveolar units (having the same composition of fresh gas, i.e. not containing CO_2) hence diluting the mixed expired gas containing CO_2 from unaffected regions of the lung to produce a lowering of the $ETCO_2$ concentration. This monitor is of particular value in neurosurgical operations and continuous monitoring of the carbon dioxide tracing is important. In such circumstances even the slightest reduction in the $ETCO_2$ concentration should raise the possibility that an air embolus has occurred. However, the more expensive nitrogen meter is a better monitor for this purpose because, once denitrogenation has been achieved as part of the anaesthetic process, any ingress of air (whether as an air embolus or as an air leak into the breathing system) will produce a strong positive signal.

Carbon dioxide measurements are not accurate, especially in children, if the response time of the CO_2 analyser is too slow and its output fails to reach the actual carbon dioxide concentration at the end of each breath. An accurate high-frequency response is essential when end tidal PCO_2 is monitored during paediatric anaesthesia. Six infrared capnometers and one multiplexed mass spectrometer were assessed in a laboratory study in the face of increasing respiratory rates from 8 to 101 cycles/min.[44] At or below frequencies of 31 cycles/min, four capnometers overestimated and three underestimated the true $PETCO_2$. At frequencies above 31 cycles/min, six capnometers underestimated and one overestimated $PETCO_2$. The differences in displayed CO_2 from known CO_2 over the entire range of frequencies studied was between -16.4 mmHg and $+6.6$ mmHg although if two suspect values obtained from one capnograph are removed from the reported results the range narrows to between $+6.6$ mmHg and -11.4 mmHg and is $+6.6$ mmHg to -7.4 mmHg if the mass spectrometer results are also discarded (see below). The cause of the underestimation in $PETCO_2$ is thought to be the mixing of adjacent breaths during transport down the sampling catheters and in the analysis chamber; the long sampling line (50 m) of the multiplexed mass spectrometer system presumably contributes to this error. The Hewlett-Packard 47210A capnometer was the instrument least affected, presumably because its mainstream analysis cell requires no sampling tube.

The rise time of a capnograph is the time taken for the analyser output to respond to a sudden step change in CO_2 concentration, i.e. the time (T_{90}) it takes for the analyser to change from 10 per cent of the final value to 90 per cent of the final value. Alternatively, the response time, T_{70}, is used in place of T_{90} because the 70 per cent point is a steeper part of the response curve; for all practical purposes, T_{90} is twice the value of T_{70}. The rise times of capnographs for clinical use have values which range from 50 ms to 600 ms. The distortion in the CO_2 waveform is a function of the rise time of the analyser. The rise times of 11 commercially available CO_2 analysers were measured in a laboratory study: only six instruments responded quickly enough to be accurate for rates up to 100 breaths/min. All 11 responded rapidly enough to measure end tidal CO_2 concentration with 5 per cent accuracy when ventilatory rates were less than 30 breaths/min. To measure CO_2 output ($\dot{V}CO_2$) with 5 per cent accuracy an analyser should have a rise time of 20 ms; the analyser rise time (for analysers with

T_{70} rise times <200 ms) can be estimated clinically to within 10 ms ($\pm$8 ms SD) by a simple breath hold and forced exhalation, thus providing an estimate of the accuracy of CO_2 measurements in adults or children. Hence, the limitations of capnography should be appreciated in special situations such as in paediatric anaesthesia. Some capnographs show a better performance than others. It has been recommended that a CO_2 analyser should have a T_{90} of less than 100 ms to measure ETCO$_2$ accurately in adults; when the ventilatory rate is high, as in children, even faster rise times are recommended. A sidestream capnometer should only be considered reliable if the total delay time is less than the respiratory cycle time; total delay time can be reduced by increasing the rate of gas sampling, or by reducing the length of the sampling tube, or both, although other factors are involved.

Capnography is not a measurement solely of respiratory function, and capnograms must be interpreted in conjunction with other clinical findings. The capnogram, like the electrocardiogram (ECG), requires systematic analysis to obtain the best information (viz. baseline, height, frequency, rhythm, shape). Various monographs detail numerous such examples.

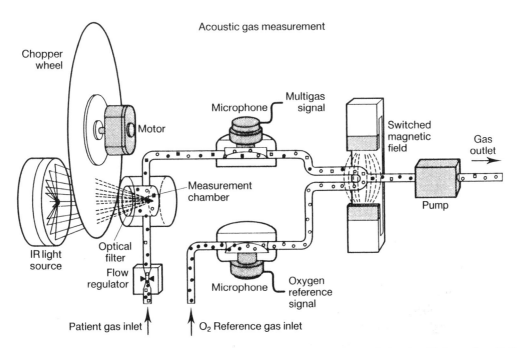

Fig. 23.4 Principle of photoacoustic spectroscopy. The chopper consists of three concentric bands of holes and so divides the incident infrared (IR) beam into three different parts which pulsate at different frequencies. The varied frequencies allow for differentiation between the three signals caused by the carbon dioxide, nitrous oxide and anaesthetic agent. The light is filtered before entering the measurement chamber. The frequencies and wavelengths of the incident light are thus optimized to match the IR absorption spectra of the three gases to be measured. In the measurement chamber the gases are excited and the absorption of the IR light causes each gas to expand and contract at a frequency equal to the pulsation frequency of the appropriate IR beam. This periodic pressure fluctuation is detected by the condenser microphone. (Reproduced by courtesy of Brüel & Kjaer Ltd.)

Photoacoustic spectroscopy

Monitors for analysis of anaesthetic vapours are becoming more widespread. They have a special value in low-flow and closed circle techniques since overdosage and insufficient anaesthesia (awareness) should be apparent at an early stage. The eavesdropping microphones used in World War II and even today in some businesses as bugging devices, can be put to good use in medicine. Such instruments (Brüel & Kjær) use the principle of photo-acoustic spectroscopy (PAS):[45] with the exception of oxygen, all gas and vapour concentrations are determined by this technique (Fig. 23.4). Much the same basic principles are used in the infrared (IR) monitor, namely the ability of CO_2 and N_2O and anaesthetic agents to absorb IR light. However, the measurement techniques differ. IR spectroscopy uses the optical method whereas PAS uses the acoustic technique. This is based on the fact that if IR energy is applied to a gas the gas will expand and lead to an increase in pressure. If the applied energy is delivered in pulses the pressure will be intermittent, thus resulting in a fluctuating pressure. If the pulsation lies within the audible range an acoustic signal is produced which can be detected by a microphone. The advantages claimed for PAS over the basic IR method are high accuracy, increased reliability, reduced preventive maintenance and the need for less frequent calibration. Moreover, as the amount of IR light absorbed is measured directly without use of a reference cell, zero drift is non-existent with PAS. If there is no gas in the chamber there can be no acoustic signal. A magnetoacoustic technique is used to measure gaseous oxygen concentrations. In some instruments the same microphone is used for both measurement methods. This means that the rise time is the same for all gases and gives a true real time relation between all the gas waveforms.

Raman scattering

Raman scattering, a powerful and widely used laboratory spectroscopic technique, was discovered in 1928. Accurate high speed analysis of gases requires an intense monochromatic light source such as a laser. Raman scattering of light provides continuous and trend measurement of anaesthetic and respiratory gases. A precisely regulated laser beam (helium–neon laser) is recirculated hundreds of times through a gas cell. During each pass through the cell the laser beam interacts with a flowing sample of gas being continuously withdrawn from the anaesthesia breathing system. When a photon from the laser beam collides with a gas molecule in the cell, its energy excites the vibrational and rotatational modes of the molecule. As the molecule loses energy after the collision, it re-emits scattered light at a lower energy and consequently at greater wavelength. The amount of wavelength 'shift' corresponds to characteristic vibrational and rotational energies of the gas molecule. Because these shifts are different for each gas, the wavelength components present in Raman scattered light provide specific chemical identification of the gases present. The quantities of scattered photons produced at each wavelength are directly proportional to the concentration of a particular gas within the cell. Proprietary filters and photon counting circuits are used to measure simultaneously the scattered light at multiple wavelengths. The molecules of each respiratory and anaesthetic gas in the sample give out distinct Raman scattered light with characteristic wavelengths when irradiated by the laser beam. The number of scattered photons for each wavelength is directly proportional to the concentration of a particular gas present in the mixture. The RASCAL (**RA**man **SC**attering Ana**L**ysis) is one such instrument (Ohmeda Ltd).

Oximetry

Transient hypoxaemia is common during anaesthesia, often resulting from hypoventilation during induction and recovery or from minor degrees of obstruction. Measures can be taken to avoid its occurrence when hypoxia may be anticipated such as during endoscopy, one-lung anaesthesia or possible difficult intubation. At other times it is unexpected, may be unrecognized and lethal. For decades attempts have been made to find a convenient and reliable means of monitoring the delivery of oxygen to the tissues. Matthes published at least 20 papers on oximetry between 1934 and 1944 and may safely be regarded as the father of oximetry. In the mid-1930s Professor Robert Brinkman used the newly invented barrier layer photocell to measure oxygen saturation of blood. During World War II non-invasive ear oximeters were developed for use in aviation research; this led to the introduction of the classic Atlas and Cyclops oximeters, but although the benefits of continuous oximetry were appreciated, technical problems prevented their routine clinical use. In 1948 Brinkman substituted the conventional technique of light transmission for a reflection measurement. In the late 1960s Shaw developed a self-calibrating eight wavelength ear oximeter which eventually became the Hewlett-Packard ear oximeter. However, the introduction of the Clark oxygen electrode in 1956 directed clinical thinking and technology for the ensuing 30 years away from the concept of saturation and focused instead upon the tension of oxygen in the blood as the driving gradient for oxygen between the inspired air and the mitochondria of tissue cells.

General principles of pulse oximetry

Many devices have been used as pulse meters, e.g. mercury-in-rubber strain gauges, microphones, piezoelectric crystals, Doppler devices and photoelectric cells. Although Hertzman reported the use of photoelectric finger plethysmography in 1937, it was not until 1975 that the concept of pulse oximetry was reported from Japan, developed by Minolta, and tested by other Japanese researchers. The introduction of the Nellcor pulse oximeter into clinical practice by Yelderman and New in 1983 to measure arterial oxygen saturation easily by non-invasive means has opened up the prospect of reliable and continuous monitoring of oxygen saturation in every patient.[46] Pulse oximetry has now emerged as the most useful method as it is simple, non-invasive and accurate under most circumstances.[42,47] Cyanosis is notoriously difficult to detect clinically due to lighting conditions and variability amongst individual observers; it is even more difficult to detect where the epidermis is thickened, the skin is pigmented or there is pigment associated with jaundice or Addison's disease. Cyanosis and bradycardia are late signs of hypoxaemia and pulse oximetry represents a very significant advance in patient safety because even astute clinicians do fail to detect cases of severe arterial desaturation. It is worth remembering that SaO_2 will not decrease until the PaO_2 is below 11.3 kPa (85 mmHg) because of the shape of the oxyhaemoglobin dissociation curve. A useful guide in the clinical range of oxygen saturation between 90 per cent and 75 per cent is that the relation is roughly $PaO_2 \simeq SaO_2 - 30$. Furthermore, when the low threshold alarm sounds as saturation falls below the default setting of 90 per cent the corresponding PO_2 is 7.7 kPa (57.8 mmHg) on the standard oxygen dissociation curve.

The basis of oximetry is to shine light of known intensity and given wavelength through a substance in solution and to measure the amount of light which is transmitted through it. The chosen wavelength depends on the absorption spectrum of the substance under investigation. The fundamental law (Lambert–Beer law) governing the transmission, or the absorption, of the light is $I_t = I_0 e^{-Ecd}$ (where I_0 is the intensity of the incident light, and I_t is the intensity of the light after transmission through a solution of a substance of concentration, c; d is the distance that the light has to travel through the substance, e is the base of natural logarithms, and E is a proportionality constant known as the extinction coefficient). The term Ecd is called

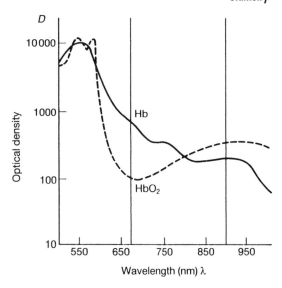

Fig. 23.5 Absorption spectrum of reduced (Hb) haemoglobin (solid line) and oxygenated (HbO$_2$) haemoglobin (dashed line). D is the optical density (an index of the opaqueness of the medium) and λ is the wavelength in nanometres. 810 nm is one of the isobestic points where the absorbance of the two forms of haemoglobin is the same. The two vertical lines denote wavelengths in the red and infrared parts of the spectrum used by the light emitting diodes of pulse oximeters.

the absorbance or optical density, D, of the solution. It can be shown that $D = 2.303 \log_{10}(I_0/I_t)$. In the red region of the spectrum, at a wavelength of 650 nm, there is a large difference in optical absorption between reduced haemoglobin and oxyhaemoglobin (Fig. 23.5). When haemoglobin is oxygenated the transmission of light is increased. In the near infrared region of the spectrum, at 805 nm, there is an isobestic point. There are several of these points and they represent wavelengths at which the optical absorption of fully reduced and fully oxygenated haemoglobin are equal. Hence, a measurement at this wavelength determines the total amount of haemoglobin present, and the difference in output between the measurements at the two wavelengths (650 nm and 805 nm) is an index of the oxygen saturation of the blood. However, when light is shone through a substance or tissue it is reflected and scattered as well as being transmitted and the Lambert–Beer law is to be regarded as entirely empirical.

A pulse oximeter analyses the changes in the transmission of light through any pulsating arterial vascular bed. The amount of light transmitted, such as through the nailbed of the finger or the lobe or pinna of the ear, depends on the amount absorbed by the various structures present, such as skin, muscle, bone, venous and capillary blood, etc. The path length that the light has to travel through the finger is constant until it is changed (increased and decreased) due to expansion and relaxation from the entry and exit of pulsing arterial blood into the system to produce the familiar plethysmographic waveform. The amount of light absorbed and transmitted will then alter. This 'pulse

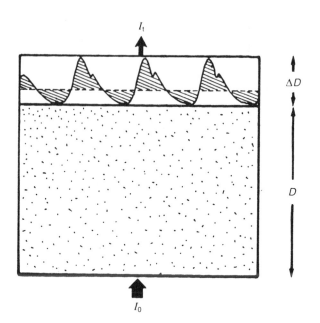

Fig. 23.6 Principle of pulse oximetry. D is the fixed absorbance of light shone through the non-pulsating structures (e.g. bone, muscle, skin, veins) of a tissue; $\triangle D$ is the 'pulse added' component, or variable absorbance, of light as a result of the pulsating arterial system in the tissue. I_o is the intensity of the incident light and I_t is the intensity of the transmitted light emerging from the tissue.

added' signal (Fig. 23.6) is subtracted from the background transmission signal by a microprocessor-controlled empirical algorithm computed hundreds of times per second; the saturation is averaged over a short time interval of 3–6 seconds so as to produce a rapid response. The algorithm is created by measuring pulse-added absorbances in healthy, awake volunteers breathing various gas mixtures. These absorbances are then correlated with actual oxygen saturations obtained by arterial sampling and a CO oximeter. The differing light absorption characteristics of oxygenated and non-oxygenated blood are thus used to compute continuously the oxygen saturation of the arterial blood.

Two narrow band light emitting diodes (LEDs) are used at wavelengths of 660 nm in the red and 910 nm in the infrared (IR) part of the spectrum; some manufacturers use an LED at 940 nm instead of the latter wavelength but as either value corresponds to a plateau on the curve the isobestic point of 810 nm is in effect mimicked (it is not technically possible to produce a reliable LED for 810 nm). Because only two wavelengths are used the pulse oximeter cannot distinguish between more than two forms of pigment, i.e. reduced and oxyhaemoglobin. If a third form is present such as carboxyhaemoglobin (COHb) it is not clear what the pulse oximeter will actually measure. Each LED switches on and off at about 720 Hz (red on, red off and IR on, both off, etc.) and a single broad band photodiode detects the amount of light transmitted. The photodetector produces outputs for the transmitted light from each of the diodes and also for any ambient light detected during the off periods in the 720 Hz cycle. The LEDs and photodetector are housed in a probe which is attached to a digit, the pinna or lobe of the ear, or across the foot, heel or hand in the case of infants. Skin, flesh, bone and venous blood reflect and absorb a constant amount of light; arterial blood, by contrast, absorbs varying amounts of light because of the pulsatile blood flow. Until recently, all commercially available LEDs were homostructures, i.e. they consisted of pn junctions formed in one type of material. Over the past few years new types of LEDs (hetereostructures) such as those based on the aluminium gallium arsenide (AlGaAs) system have appeared with significant improvements in light output efficiency and even further advancements in LED technology are due to appear.

Calibration and performance

The calibration of the instrument is preset during manufacture and cannot be altered by the user; it is unfortunate that no external standards have so far been developed. Most instruments are programmed to check their internal calibration and sensor function when switched on and intermittently thereafter. The majority of pulse oximeters are tested down to saturation values of 70 per cent, the lower limit of tolerable hypoxaemia; the error above this value is about ±2 per cent, and ±3 per cent in the range 70–50 per cent saturation. The PaO_2 corresponding to a saturation of 70 per cent on the standard oxygen dissociation curve is 5 kPa (37 mmHg). Below this point on the dissociation curve the calibration is generally achieved by extrapolation and in this region the detection of very small optical signals together with the need to reject artefacts requires the averaging of the data over several seconds, and this procedure can affect the response to rapid changes in saturation. In a study of six different makes of pulse oximeter it was found that some instruments continued to show nearly normal saturation when the true saturation was down to 40–70 per cent, and some routinely failed to indicate any saturation at all during the critical period of falling saturation. Response times tend to be longer for desaturation than for increasing saturation, perhaps as a result of an increase in finger and ear blood flow when hypoxic blood reaches the tissue. Other studies have explored the range of saturation 95.5 per cent down to 56 per cent and have found good correlation with arterial blood samples. Studies of oxygen saturation in volunteers submitted to hypoxic gas mixtures have shown a number of differences between the various instruments available; however, improved models are continually appearing. The arterial oxygen saturation is usually displayed digitally with the provision of an audible tone accompanying each pulse so that the pitch falls as saturation falls. The magnitude of the pulse waveform is indicated either by a bar graph or else by display of the pulse waveform; however, in some instruments this form of waveform presentation can be rather misleading because decreases in signal produced, e.g. by vasoconstriction, are compensated by an increase in gain of the amplifier.

The pulse oximeter is designed to be insensitive to changes in haemodynamics although situations where the pulse is diminished, such as in peripheral vascular disease, hypothermia, extreme hypotension or vasoconstriction, may produce too weak a signal. However, from Lawson and his colleagues' study[48] it appears that a pulse oximeter is able to estimate saturation even in states of severely reduced peripheral blood flow. They found that when the arm was progressively occluded using a blood pressure cuff the flow rate, which was measured with a laser Doppler flow probe, decreased to approximately 9 per cent of the control value before the pulse oximeter (Nellcor N-100) was unable to estimate the saturation. When the cuff was released from total occlusion the pulse oximeter regained a measured pulse and an estimation of saturation at only 4 per cent of the original blood flow. No attempt was made in this study to compare the saturations obtained with an independent method, but another study addressed the problem that venous congestion may cause low readings of saturation. In this study the effect of venous congestion,[49] on the average value of SaO_2, measured by pulse oximetry in the finger tip and finger base, was found to be between 1.5 and 7.8 per cent lower respectively than when the digit was elevated. This effect is thought to be a result of venous pulsations effected through shunting of arterial blood through open arteriovenous anastomoses in the cutaneous circulation. Desaturation and resaturation are best determined by centrally placed sensors, e.g. tongue, cheek or ear in preference to fingers or toes. Movement artefacts are a nuisance but only a problem in non-anaesthetized patients. The Nellcor N-200 model has the facility to synchronize the determinations of arterial saturation with the patient's ECG and thus is particularly useful in monitoring awake patients.

Forms of haemoglobin such as methaemoglobin (MetHb) and carboxyhaemoglobin (COHb) cannot be distinguished from oxyhaemoglobin by pulse oximetry. Because COHb is red it is 'seen' by the pulse oximeter as oxyhaemoglobin, therefore in the presence of COHb the oximeter reads the sum of the O_2Hb and the COHb. A heavy smoker with 10 per cent of his haemoglobin combined as COHb shows a 2–3 per cent overestimate of haemoglobin saturation when 90 per cent of the available haemoglobin is in fact saturated. The effective acute 'anaemia' of carbon monoxide poisoning does not alter the oximeter reading and thus the instrument does not alert to the low arterial oxygen content under these circumstances. It appears that a pulse oximeter 'sees' COHb as 90 per cent O_2Hb and 10 per cent reduced Hb.

Methaemoglobinaemia has a more complex effect because the ability of the pulse oximeter to register desaturation is impaired and no indication of the functional haemoglobin is given. Methaemoglobin is seen as a large absorbance in both the red and infrared wavelengths of the pulse oximeter; this large pulsatile absorbance seen in both wavelengths forces the ratio of the added pulse absorbance towards unity which is interpreted by the pulse oximeter as a saturation of 85 per cent O_2Hb and 15 per cent reduced Hb. Therefore, in the presence of significant amounts of methaemoglobin the pulse oximeter will tend towards a reading of 85 per cent regardless of the saturation. The Radiometer 'OXImeter' has provision for adjusting the read-out from the instrument according to known amounts of abnormal forms of haemoglobin determined separately from the Radiometer OSM3 haemoximeter. This approach has obvious value when dealing with victims of smoke inhalation when high concentrations of COHb are common.

Dyes which absorb light in the region of 660 nm, such as indocyanine green, indigo carmine and especially methylene blue, may also cause sudden false low readings for a time after intravenous injection. The time course for these changes is about one to two circulation times and recovery to baseline values occurs within 3 minutes of bolus dye injection in young healthy subjects breathing room air.

Artefacts may result from interference from radio-frequency diathermy apparatus, light of high intensity and some infrared heat lamps. Because the photodetector can measure weak signals, oximeters are designed to reject ambient light. When the intensity of ambient light is high the photodetector cannot sense light transmitted through tissue or calculate SaO_2. Certain operating theatre lights may cause interference and cause erroneous high saturation readings. It is recommended that the probe be shielded by wrapping in a thin opaque metal foil. One study considered the effects of nail polishes of different colours: black, blue and green nail polish lower pulse oximeter readings of SaO_2. Blue and green produced greater decreases than purple and red; black produced an intermediate decrease. Nail polish should therefore be routinely removed before pulse oximetry. The probe should always be carefully placed so that no undue pressure or torque is applied and careful inspection made of both site of application and probe. Other problems appear to be few, such as minor burns, abrasions and a sun tanning effect.

Clinical applications of pulse oximetry

Pulse oximetry is now a valuable routine monitor in anaesthesia and is especially valuable in patients with skin pigmentation, and those with poor physical status and a low arterial oxygen tension, i.e. below 13 kPa (95 mmHg). Pulse oximetry is also of value for patients undergoing regional anaesthesia, sedation, endoscopies, and procedures of brief duration, as severe hypoxaemia is far commoner than hitherto realized. The clinical use of pulse oximetry has led to many interesting observations such as the high incidence of hypoxaemia during manual ventilation (compared with ventilator controlled or spontaneous ventilation), and in patients in the lithotomy position. It also reveals the hypoxaemia of inadvertent bronchial intubation. Although pulse oximetry provides rapid non-invasive evaluation of total blood oxygenation and displays information about pulse amplitude, the conclusion that pulse oximeters monitor tissue perfusion and tissue oxygen delivery must be approached with caution.

Pulse oximetry is a useful monitor in the preparation of patients before anaesthesia (e.g. while intravascular lines are being inserted), for one-lung ventilation and as a valuable tool during surgical manipulation of the pulmonary artery such as banding or clamping when pulmonary artery flow is reduced. It is a valuable monitor of oxygenation in the recovery room, intensive care unit and ward and also during transfer of patients. Hypoxaemia (defined as a SaO_2 of 90 per cent or less (=PaO_2 7.7 kPa, 58 mmHg)) occurred in 35 per cent, and severe hypoxaemia (defined as a SaO_2 of 85 per cent or less (=PaO_2 6.7 kPa, 50 mmHg)) occurred in 12 per cent, of ASA class 1 or 2 patients breathing room air during their transfer from the operating theatre to the recovery room.[50] Postoperative hypoxaemia did not correlate with anaesthetic agent, age, duration of anaesthesia or level of consciousness. Smith and Crul have demonstrated early profound decreases in SaO_2 in healthy patients during transfer to the recovery room in spite of the administration of 100 per cent oxygen for 5 minutes at the end of the anaesthetic.[51] They suggest that ventilation/perfusion mismatch is most likely to be responsible; no patient given 2 l/min by nasopharyngeal catheter during transport had a SaO_2 below 90 per cent. Children are potentially more susceptible than adults to airway closure and disturbances in pulmonary gas exchange. In children 21 per cent oxygen (room air) is a potentially hypoxic mixture in the early recovery period. Substantial desaturation has been observed in patients receiving either traditional parenteral analgesia or epidural morphine following caesarean section. Because oxygen is innocuous under most circumstances the safe course is to provide air enriched with oxygen during postoperative transfer and in the early recovery period. In the intensive care unit pulse oximetry is also valuable as an online non-invasive indicator of the benefits or otherwise of therapy; it is a particularly useful monitor during chest physiotherapy and suctioning procedures.

A Danish group have recently published a large multicentre prospective controlled study of the value of pulse oximetry on the outcome of anaesthetic care.[52,53] Eighty per cent of anaesthetists in this study felt more secure with pulse oximetry. Despite the enormous effort involved and over 20 000 patients studied, the sample was too small to conclude that pulse oximetry had any effect on patient outcome. However, intuitively it is believed that brief exposure to mild levels of hypoxaemia may be harmful to severely ill patients, whereas healthy patients probably tolerate longer periods of more severe hypoxia. The technology provides for early warning and enables earlier intervention; additional time is available to resolve a problem.[54]

Temperature

Temperature measurement is useful in the prevention of hypothermia during long operations and as a monitor of methods used to keep patients warm. In some institutions a single disposable oesophageal probe, which incorporates a unipolar ECG, a thermistor and a hollow tube connecting to a monaural stethoscope, is routine for every case. The nasopharynx is another popular site for temperature measurement. For major operations, and especially in children, the use of a second more peripheral temperature sensor (e.g. on a digit) is used to allow measurement of the temperature gradient and hence provides an index of cardiac output and peripheral perfusion. In 1992 the *ASA standards for postoperative care*[55] were supplemented by a new reference to the monitoring of temperature. The relevant section now reads: '. . . particular attention should be given to monitoring oxygenation, ventilation, circulation and temperature.'

Intracranial pressure (ICP)

The measurement of ICP is a useful adjunct in the management of neurological patients in the intensive care unit. ICP measurement is valuable in cases of trauma to indicate the onset of cerebral oedema or haemorrhage. The most accurate method relies on ventricular puncture. Extradural sensors are less reliable but may be left *in situ* for longer periods. The technique demands electronic systems for the best results. Absolute sterility and an aseptic technique are vital. Monitoring of the EEG or its derivatives, and the problem of monitoring the depth of anaesthesia are discussed in Chapter 24.

Computer-assisted monitoring

The advent of the microprocessor has resulted in the automation of clinical monitoring and the collection of a lot of data which can be used as a permanent record and for purposes of audit and research. However, early enthusiasm often produced instruments which were difficult to use and presented far too much information without much attempt to present it in order of priority. A common problem is the requirement for the anaesthetist to have to scroll through pages of menus on the screen in order to get the information wanted. Automatically generated records of the anaesthetic process are now being introduced.

Conclusions

Continuous observation of anaesthetized patients has been considered vital for patient safety since the earliest days of anaesthesia. There are now sufficient reports which indicate that dedicated monitoring instruments can benefit the patient as an extension of the anaesthetist's human senses. Indeed, in several countries standards of monitoring have now been laid down by professional anaesthetic organizations or even by law. To a great extent this has come about because of the escalating rise in medical malpractice insurance premiums. Monitoring instruments are not a substitute for careful clinical observation.

However, good monitoring benefits not only the patient by constant and close presentation of many vital signs and other parameters but also the anaesthetist by freeing him from many manual tasks, thus helping to relieve fatigue and tension and so concentrating his attention better on the overall care of his patient. Capnography and pulse oximetry can non-invasively provide continuous information that would not otherwise be available and the use of reliable, audible and visual alarms increases patient safety particularly when access is restricted or room lighting is reduced. There is a strong analogy between the anaesthetist and the civil airline pilot. Both must rely heavily on instruments for information and the fallacies of human error are well known to both professions. In some operations the anaesthetist cannot get near to his patient – rather like flying in cloud; however, with the use of adequate instruments (monitors) everything should be perfectly safe. Cooper and his colleagues have shown that human error, rather than equipment failure, is overwhelmingly responsible for anaesthetic mishaps. Attention must nevertheless be directed towards combining the various monitors used in anaesthesia into a common system so as to simplify and reduce the number of controls and switches. It is imperative that sufficient reliable equipment is purchased and great attention must be paid to the training of anaesthetists in its use and to proper maintenance and checking. There is a powerful argument for the use of routine capnography and pulse oximetry in all patients in addition to traditional monitors such as the ECG, blood pressure, etc., which may be expected to make a considerable contribution to patient safety.

REFERENCES

1. Eger EI II, Epstein RM. Hazards of anesthesia equipment. *Anesthesiology* 1964; **25**: 490–504.
2. Rendell-Baker L. Problems with anaesthetic gas machines and their solutions. In: Rendell-Baker L, ed. Problems with anaesthetic and respiratory equipment. *International Anesthesiology Clinics* 1982; **20**(3): 1–82.
3. Schreiber P. *Anaesthesia equipment. Performance, classification and safety.* Berlin: Springer-Verlag, 1972.
4. Schreiber P. *Safety guidelines for anesthesia systems.* Pennsylvania: The North American Dräger Company, 1985.
5. Cooper JB, Newbower RS, Kitz RJ. An analysis of major errors and equipment failures in anesthesia management: considerations for prevention and detection. *Anesthesiology* 1984; **60**: 34–42.
6. Interim report available from study of accidental breathing system disconnections. *Medical Devices Bulletin* 1984; **2**: 1–2.
7. Spurring PW, Small LFG. Breathing system disconnexions and misconnexions. A review of some common causes and some suggestions for safety. *Anaesthesia* 1983; **38**: 683–8.
8. Health Equipment Information, No. 150. *Evaluation of breathing attachments for anaesthetic apparatus to BS 3849*: 1965. London: Department of Health & Social Security, 1965.
9. Wood MD. Monitoring equipment and loss reduction: an insurer's view. In: Gravenstein JS, Holtzer JF, eds. *Safety and cost containment in anesthesia.* London: Butterworths, 1988: 47–54.
10. *Checklist for anaesthetic machines.* A recommended procedure based on the use of an oxygen analyzer. London: Association of Anaesthetists of Great Britain & Ireland, 1990.
11. International Standards for a Safe Practice of Anaesthesia. *European Journal of Anaesthesiology* 1993; **10** (Suppl 7): 12–15.

12. Cheney FW. ASA closed claims study. *American Society of Anesthesiologists Newsletter* 1989; **53**(11): 8–9.

13. Caplan RA, Posner KL, Ward RJ, Cheney FW. Adverse respiratory events in anesthesia: a closed claims analysis. *Anesthesiology* 1990; **72**: 828–33.

14. Birmingham PK, Cheney FW, Ward RJ. Esophageal intubation. A review of detection techniques. *Anesthesia and Analgesia* 1986; **65**: 886–91.

15. Caplan RA, Ward RJ, Posner KL, Cheney FW. Unexpected cardiac arrest during spinal anesthesia: a closed claims analysis of predisposing factors. *Anesthesiology* 1988; **68**: 5–11.

16. Caplan RA, Posner KL, Ward RJ, Cheney FW. Peer review agreement for major anesthetic mishaps. *Quality Review Bulletin* 1989; **14**: 363–8.

17. Cheney FW, Posner KL, Caplan RA, Ward RJ. Standard of care and anesthesia liability. *Journal of the American Medical Association* 1989; **261**: 1599–603.

18. Committee of the Health Council of The Netherlands. *Advisory report on anaesthesiology. Part I: Recent developments in anaesthesiology.* The Hague, The Netherlands: Government Publishing Office, 1978.

19. American Society of Anesthesiologists. Standards of basic intraoperative monitoring. *American Society of Anesthesiologists Newsletter* 1986; **50**: 9.

20. Eichhorn JH, Cooper JB, Cullen DJ, Maier WR, Philip JH, Seeman RG. Standards of patient monitoring during anaesthesia at Harvard medical school. *Journal of the American Medical Association* 1986; **256**: 1017–20.

21. American Society of Anesthesiologists. Standards for basic intra-operative monitoring (Amended by House of Delegates on October 18, 1989 to become effective January 1, 1990). *American Society of Anesthesiologists Newsletter* 1990; **54**(5): 17–22.

22. Dull DL, Tinker JH, Caplan RA, Ward FW. ASA closed claims study: can pulse oximetry and capnography prevent anesthesia mishaps? *Anesthesia and Analgesia* 1989; **68**: S74.

23. Moss E. New Jersey enacts anesthesia standards. *American Patient Safety Foundation Newsletter* 1989; **4**(2): 13–18.

24. Anesthesia Monitoring: meeting the new standards of care. *ERCI Technology for Anesthesia* 1989; **9**(8): 1–20.

25. Health Facilities Evaluation and Licensing. *New Jersey Register*, CITE 21 N.J.R., 21 February 1989.

26. Moss E. New Jersey continues as a center of anesthesia regulatory activity. *Anesthesia Patient Safety Foundation Newsletter* 1990; **5**(1): 4–5.

27. Moss E. Insurance premiums cut in New Jersey. *Anesthesia Patient Safety Foundation Newsletter* 1990; **5**(1): 1–2.

28. Sykes MK. Essential monitoring. *British Journal of Anaesthesia* 1987; **59**: 901–12.

29. Cass NM, Crosby WM, Holland RB. Minimal monitoring standards. *Anaesthesia and Intensive Care* 1988; **16**: 110–13.

30. Canadian Anaesthetists' Society. *Guidelines to the practice of anaesthesia.* 1987.

31. *Recommendations for standards of monitoring during anaesthesia and recovery.* Issued 1988. London: Association of Anaesthetists of Great Britain & Ireland.

32. Buck N, Devlin H, Lunn JN. *The report of a confidential enquiry into perioperative deaths.* London: The Nuffield Provincial Hospitals Trust, 1987.

33. Hedley-Whyte J. Monitoring and alarms – philosophy and practice. In: Dinnick OP, Thompson PW, eds. Some aspects of anaesthetic safety. *Baillière's Clinical Anaesthesiology* 1988; **2**(2): 379–89.

34. Westenskow DR, Orr JA, Simon FH, Ing D, Bender H-J, Frankenberger H. Intelligent alarms reduce anesthesiologist's response time to critical faults. *Anesthesiology* 1992; **77**: 1074–9.

35. Cooper JB, Newbower RS, Kitz RJ. An analysis of major errors and equipment failures in anesthesia management: considerations for prevention and detection. *Anesthesiology* 1984; **60**: 34–42.

36. Cooper JB, Newbower RS, Long CD, McPeek B. Preventable anesthesia mishaps: a study of human factors. *Anesthesiology* 1978; **49**: 399–406.

37. Gaba DM, Maxwell M, DeAnda A. Anesthesia mishaps: breaking the chain of accident evolution. *Anesthesiology* 1987; **66**: 670–6.

38. Sykes MK. Central venous pressure as a guide to the adequacy of transfusion. *Annals of the Royal College of Surgeons of England* 1963; **33**: 185.

39. Adams AP, Vickers MDA, Munroe JP, Parker CW. Dry displacement gas meters. *British Journal of Anaesthesia* 1967; **39**: 174–83.

40. Smalhout B, Kalenda Z. *An atlas of capnography.* 2nd ed. Zeist, The Netherlands: Kerckebosch, 1981.

41. Swedlow DB. Capnometry and capnography: the anesthesia early disaster warning system. In: Katz RL, ed. *Seminars in Anesthesia* 1986; **5**: 194–205.

42. Adams AP. Capnography and pulse oximetry. In: Atkinson RS, Adams AP, eds. *Recent advances in anaesthesia*, Vol 16. London: Churchill Livingstone, 1989: 155–75.

43. O'Flaherty D, Adams AP. The end-tidal carbon dioxide detector. *Anaesthesia* 1990; **45**: 653–5.

44. From RP, Scamman FL. Ventilatory frequency influences accuracy of end-tidal CO_2 measurements: analysis of seven capnometers. *Anesthesia and Analgesia* 1988; **67**: 884–6.

45. Møllgaard K. Acoustic gas measurement. *Biological Instrumentation and Technology* 1989; **23**: 495–7.

46. Yelderman M, New W Jr. Evaluation of pulse oximetry. *Anesthesiology* 1983; **59**: 349–52.

47. Payne JP, Severinghaus JW. *Pulse oximetry.* London: Springer-Verlag, 1986.

48. Lawson D, Norley I, Korbon G, Loeb R, Ellis J. Blood flow limits and pulse oximeter signal detection. *Anesthesiology* 1987; **67**: 599–603.

49. Kim J-M, Arakawa K, Benson KT, Fox DK. Pulse oximetry and circulatory kinetics associated with pulse volume amplitude measured by photoelectric plethysmography. *Anesthesia and Analgesia* 1986; **65**: 1333–9.

50. Hanning CD. 'He looks a little blue down this end'. Monitoring oxygenation during anaesthesia. *British Journal of Anaesthesia* 1985; **57**: 359–60.

51. Smith DC, Crul J. Early postoperative hypoxaemia during transport. *British Journal of Anaesthesia* 1988; **61**: 625–7.

52. Moller JT, Pedersen T, Rasmussen LS, *et al.* Randomized evaluation of pulse oximetry in 20,802 patients: I. Design,

demography, pulse oximetry failure rate and overall complication rate. *Anesthesiology* 1993; **78**: 436–44.

53. Moller JT, Johannessen NW, Espersen K, *et al*. Randomized evaluation of pulse oximetry in 20,802 patients: II.

Perioperative events and postoperative complications. *Anesthesiology* 1993; **78**: 445–53.

54. Orkin FK, Cohen MM, Duncan PG. The quest for meaningful outcomes. *Anesthesiology* 1993; **78**: 417–22.

Chapter 24

Monitoring the Central Nervous System

Betty L. Grundy

Monitoring the central nervous system is an essential part of every anaesthetic. Effects of anaesthetic agents on the brain and spinal cord are usually evaluated to determine whether anaesthesia is adequate for proposed diagnostic or therapeutic interventions. Constant vigilance is required to prevent rare but devastating complications that may result in lasting damage to the central nervous system. Most of this monitoring is effected by direct clinical assessment with no special equipment or techniques. Additional monitoring may be needed in several situations, however, particularly when general anaesthesia or coma limits full clinical evaluation of neurological function. This chapter addresses monitoring methods and techniques useful in the operating room or critical care unit when the functional integrity of structures and pathways in the central nervous system may be at risk. Monitors of function, pressure, flow, and metabolism are discussed.

Clinical neurological assessment

When feasible and appropriate, clinical assessment of neurological function in the awake patient is highly cost-effective. This is commonly used in high-risk patients when procedures can be satisfactorily completed under local or regional anaesthesia.

Local or regional anaesthesia

Clinical evaluation of the alert patient is particularly important during certain procedures. In some of these,

management of systemic medication may be critical. For example, narcotic administration during radiofrequency lesioning for trigeminal neuralgia may prevent appropriate localization if the patient's typical pain cannot be reproduced by stimulation. Benzodiazepines can hamper or preclude evaluation of parkinsonian tremor and its improvement during stereotactic thalamotomy. Similarly, benzodiazepines may interfere with cortical mapping of language areas during resection of nearby tumours.[1]

Techniques for minimizing discomfort are therefore tailored to facilitate each procedure. Radiofrequency lesioning of the fifth cranial nerve is usually done with intermittent small doses of methohexitone (methohexital) from which the patient can rapidly recover. This allows testing for the blink reflex, typical symptoms of neuralgia, or possible sensory loss. Vasoactive drugs may be needed during this procedure to control cardiovascular responses to painful stimulation. Stereotactic procedures for movement disorders are usually performed with either no medication or minimal doses of a short-acting narcotic. Benzodiazepines are specifically avoided during functional neurosurgery for movement disorders. For intraoperative mapping of cortical language areas, craniotomy may be done with local anaesthesia and small doses of droperidol, dimenhydrinate, and a small bolus followed by a very low dose infusion of a narcotic such as fentanyl, sufentanil, or alfentanil.[2] In young or very apprehensive patients, sedation may be accomplished during the craniotomy with an infusion of propofol. Patients are usually sufficiently alert for cortical mapping within 20–25 minutes after the infusion is discontinued.[3]

Clinical evaluation of the alert patient under regional anaesthesia is a highly cost-effective method of monitoring brain function during carotid endarterectomy (CEA). Sedation must be minimal so that the patient is indeed alert, and often only very small doses of narcotic agents are given. This approach allows observation of the precise time of onset of neurological deficits and, secondarily, may be useful in determining the probable cause of each deficit.[4,5] In a series of 359 patients having CEA with selective shunting under cervical plexus block, only 14 (4 per cent) had neurological deterioration during test occlusion of the carotid artery and had shunts inserted.[5] Only one patient in this series had a postoperative deficit that could be attributed to cerebral hypoperfusion during carotid occlusion. The other 20 postoperative deficits were related to arterial hypotension, thromboembolism, or reperfusion. Only six of the 359 patients had deficits lasting longer than 24 hours. In alert patients needing a shunt, the usual manifestation of severe ischaemia is complete loss of consciousness within seconds of test occlusion of the carotid artery. When ischaemia is less severe and the onset of symptoms less abrupt, symptoms can often be completely alleviated by raising the arterial blood pressure. This manoeuvre may avoid use of a shunt, which is itself associated with an increased risk of intraoperative embolism.

Wake-up test

The 'wake-up test' is another clinical monitoring technique that requires no special equipment.[6,7] It provides definite evidence of intact motor function, whereas the most frequently used electrophysiological monitoring techniques reflect primarily dorsal column function. A major disadvantage of the wake-up test is that it can be used only intermittently and infrequently, though it can be used more than once during a single operation. Furthermore, it may present some risks of its own. Anecdotal reports describe uncontrolled movements of awakened patients, with dislocation of orthopaedic instrumentation, or of monitoring or life support devices. Spontaneous inspiration could produce air embolism. Controlled ventilation and adequate analgesia help prevent complications during the wake-up test. Many clinicians having access to evoked potential (EP) monitoring omit the wake-up test during spinal surgery so long as EPs are monitored constantly and do not deteriorate. Any important change in EPs would be cause for performing a wake-up test. Appropriate emotional support should be given to the patient preoperatively, intraoperatively, and postoperatively to minimize the possibility of subsequent psychological problems.

The Glasgow coma scale[8] and classifications of patients with aneurysms and subarachnoid haemorrhage[9,10] are used virtually universally in critical care units. These clinical assessments are discussed elsewhere in this volume.

Electrophysiological monitoring – general considerations

The electroencephalogram (EEG) and evoked potentials (EPs) are valuable tools for assessing the functional integrity of the nervous system when clinical neurological examination is severely hampered by altered states of consciousness. They can be used to monitor neural structures and pathways at risk; to identify specific parts of the brain during stereotactic procedures or open craniotomy; to identify cranial and peripheral nerves and evaluate their functional integrity; and to monitor responses to interventions and treatments that affect neurological function.[11,12] Prompt recognition of changes in the functional integrity of the nervous system may allow therapeutic interventions by the anaesthetist or surgeon, or both, to decrease the risk of neurological injury.

Electrophysiological monitoring of the nervous system may be useful in the operating room or critical care unit when four conditions are met.[11] First, a part of the nervous system amenable to monitoring must be at risk or otherwise require identification or assessment that cannot be done clinically. Second, equipment and personnel must be available to record and interpret waveforms. Third, appropriate sites must be available for recording (and for stimulation for EPs). Finally, if the monitoring is to serve any practical purpose for the individual patient, there must be some possibility of intervening to improve function if deterioration is detected.

When EEG or EP monitoring is undertaken in the operating room, the active cooperation of the anaesthetist is needed for best results. Close communication between the anaesthetist and the monitoring team is essential, so that normal and reversible changes due to anaesthetic drugs and manipulations can be distinguished from changes that should prompt corrective action to preserve neurological function. Some important therapeutic interventions, such as manipulation of arterial blood pressure, are the direct responsibility of the anaesthetist. Monitoring of EPs which are particularly sensitive to anaesthetics, such as motor EPs or those sensory EPs emanating from the cerebral cortex, may be impossible without the active cooperation of the anaesthetist.

In many settings the anaesthetist is primarily responsible for performing EEG or EP monitoring in the operating room. An expert technologist is needed for reliable and continuous monitoring, but a physician must take responsibility for interpreting changes in the context of all that is happening to a patient physiologically and pathophysiologically. Only thus can appropriate interventions be carried out in a timely fashion. Serious errors in interpretation of EP changes, with resulting inappropriate management of patients and potentially devastating outcomes, can occur even in experienced hands if the interpretation of EEG or EPs is done by a physician who is outside the operating room and who is not fully aware of all concurrent events.

Although no large-scale prospective randomized clinical trials have shown that electrophysiological monitoring (or any other monitoring) changes patient outcome, numerous controlled experiments in animals have demonstrated an association between EEG or EP changes and permanent neurological injury. Documented risks of injury to the nervous system during certain neurosurgical, orthopaedic, and vascular operations, combined with knowledge of pathophysiological aspects of these injuries, support the concept of intraoperative monitoring. The accumulating collective body of knowledge regarding associations between intraoperative EEG or EP changes and neurological outcome in patients continues to strengthen the case for intraoperative electrophysiological monitoring during procedures associated with known risks to the nervous system.

Reports of so-called 'false-positive' and 'false-negative' results of intraoperative electrophysiological monitoring are of obvious concern to physicians considering use of these techniques. At times these terms may be inappropriate. For example, how would one define a 'false-negative' change in blood pressure? 'False-negative' and 'false-positive' may more suitably describe recordings made in the diagnostic laboratory, where the goals of testing are diagnosis and prediction, than in the operating room, where the goal is continual optimization of function during periods of recognized risk to minimize the possibility of lasting damage. Untoward intraoperative events that can be detected with EEG or EP monitoring, although occurring only occasionally, may be of devastating proportions. Analysis of published reports suggests that most reports of so-called 'false-negative' results in intraoperative EP monitoring may be due to inappropriate application of techniques, technical difficulties, and erroneous interpretation of EP changes in the operating room.[13]

Electroencephalography

The electroencephalogram (EEG) represents the on-going spontaneous electrical activity of the brain. Signals recorded from the scalp consist of shifting dipole fields, which are generated by the graded summation of excitatory and inhibitory postsynaptic potentials in cortical neurons.[14] The only cells in a position to generate these dipole fields at the cortical surface or scalp are the pyramidal cells found in the layers 2, 3, and 5 of the cerebral cortex. These cells and their long dendritic trees are oriented perpendicularly to the cortical surface. A depth electrode inserted into this region shows a phase reversal midway through the granular layer.

EEG activity is often described in terms of nominal frequencies: δ (less than 4 Hz); θ (4 to less than 8 Hz); α (8–13 Hz); and β (greater than 13 Hz). Widely differing conditions can produce similar frequencies. Delta-activity is typical of metabolic encephalopathy or cerebral ischaemia, but it is also seen during deep stages of sleep and with deep anaesthesia. Theta activity is common in children and adolescents and during anaesthesia. Alpha-rhythm is the characteristic EEG pattern in the alert but relaxed adult with eyes closed and is most prominent in occipital leads. Normal α-rhythm breaks up upon opening of the eyes or during mental activity such as mental arithmetic. Not all α-frequencies constitute normal α-rhythms. Frequencies in this range that spread over the entire head and do not react to stimulation such as eye-opening are seen during halothane anaesthesia and in α-coma. Alpha-coma, typically seen after an injury of the brainstem, may be irreversible. Frequencies in the β-range are typical of intense mental activity but are also produced by low doses of barbiturates or benzodiazepines. For example, activity in this range may be seen after an initial small dose of thiopentone (thiopental) but may also persist, particularly in frontal areas, up to 2 weeks after discontinuation of ambulatory doses of diazepam. Thus, the frequency ranges of EEG activity are never diagnostic of a particular state of the brain but require interpretation within the relevant clinical context.

Methods

Recording

An electrode placed on the scalp records primarily that activity arising within a 2–2.5 cm radius. Correct localization and careful application of electrodes are important, even though standard locations are often modified to meet the needs of a specific situation. Scalp electrodes are positioned according to designations of the International Ten Twenty System[15] (Fig. 24.1). They must be securely attached to minimize artefact from movement. Although initial application is simpler with electrode paste, collodion and gauze provide far more secure fixation. We prefer silver/silver chloride cup electrodes, 9 or 10 mm in diameter, applied with collodion and gauze, and filled with conductive gel. For prolonged recording, cup electrodes may be sealed with plastic tape or collodion to prevent drying of the electrode gel. Disposable silver/silver chloride

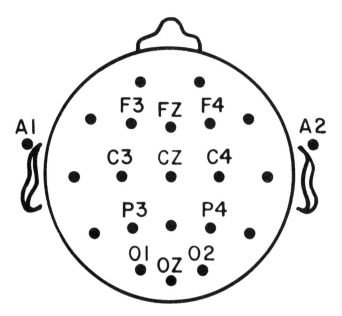

Fig. 24.1 Electrode positions designated by the International Ten Twenty System.[15] Even numbers are to the right of the midline, odd numbers to the left. The letters designate anatomical areas of the brain (F, frontal; C, central; P, parietal; O, occipital, ear). The locations not labelled in this drawing also have standard letter and number designation. (From Grundy,[353] with permission.)

adhesive electrodes manufactured for electrocardiographic recording are often suitable for locations without hair. Those with large conductive areas and resistant to drying are best.

Skin is prepared by rubbing with alcohol, then with dry gauze. An abrasive salt gel on a cotton-tipped applicator is used to lightly abrade the skin, reducing impedance to transmission of the small electrophysiological signals. With cup electrodes, a blunt-tipped needle is used to place gel inside the cup and may also be used for additional light abrasion of the skin if necessary to obtain sufficiently low impedances.

Subdermal needle electrodes, usually platinum, are easily inserted and can be used within the sterile surgical field. Because the surface of needle electrodes is small, impedances tend to be higher than with cup or disc electrodes. Wire spiral subdermal electrodes, similar to fetal scalp electrodes, are more secure than the usual needle electrodes but are also more uncomfortable. They may be inserted with local refrigeration anaesthesia obtained with a volatile spray such as ethyl chloride. For intraoperative corticography, arrays of electrodes embedded in a silastic or Teflon sheet are more convenient and less cumbersome than the traditional saline-soaked wick electrodes.

Regardless of the type of electrode, impedances must always be measured and should be low and matched. We prefer impedances of 1000–3000 ohms, although 5000–10 000 ohms may be acceptable for EEG if the impedances between recording electrodes are matched. These higher

impedances often prove unsatisfactory for EP monitoring in the operating room or critical care unit. More attention is required for EEG or EP electrodes than for electrocardiographic (ECG) electrodes because the signals are much smaller. Note that the standardization mark on a clinical ECG machine is 1 *millivolt*, while that on a clinical EEG machine may be 2–100 *microvolts*. The calibration mark for far-field EP may be a few hundred *nanovolts*.

Signal processing, display, and measurement

Once acquired, EEG signals are amplified and filtered to minimize frequencies outside the range of interest (usually 0.5–70 Hz in the diagnostic laboratory, 1–30 Hz in the operating room or critical care unit). The traditional diagnostic EEG recording consists of a paper strip chart showing 16–32 channels as a plot of voltage against time (Fig. 24.2). This representation is in the time domain. Multiple channels are required to show regional activity of the brain.

Numerous techniques for signal processing have been developed to facilitate data compression and interpretation. When these are used, the unprocessed analogue signal

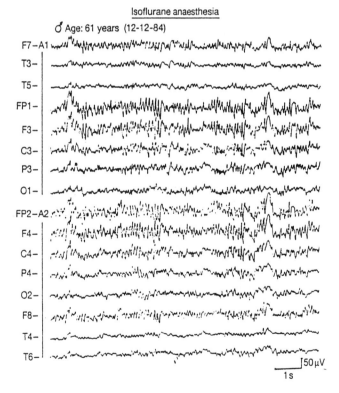

Fig. 24.2 This 16-channel EEG shows characteristic patterns under anaesthesia. These include widespread anteriorly maximum rhythmic (WAR) activity which is usually in the lower β- or α-frequency range. Anteriorly maximum intermittent slow waves (AIS) are also well demonstrated. In addition, more widespread persistent slow waves (WPS) are seen, which are of somewhat lower amplitude than the AIS waves. (From Blume and Sharbrough,[29] with permission.)

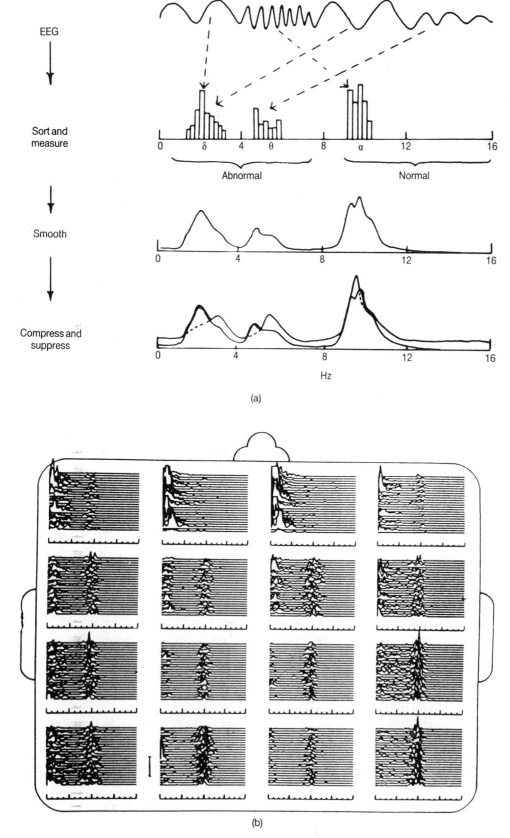

EEG

Sort and
measure

Abnormal

Normal

Smooth

Compress and
suppress

Hz

(a)

(b)

Fig. 24.3 The compressed spectral array (CSA), as popularized by Bickford. (a) Signal processing. (b) A 16-channel CSA recorded from a patient with a right occipital tumour. Note the attenuation of activity in the right occipital leads. The slow activity in frontal leads actually represents eye blinks rather than δ-frequency EEG activity. (From Bickford,[354] with permission.)

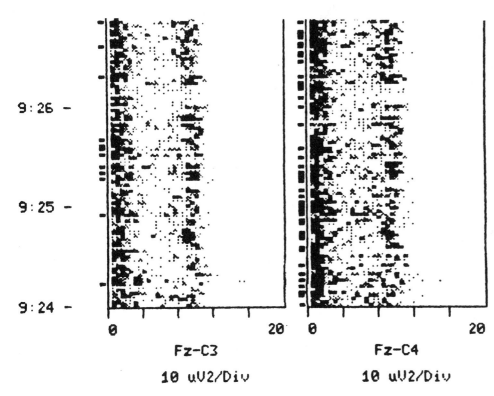

Fig. 24.4 The density modulated spectral array (DSA). This display shows information also derived by power spectral analysis. It uses varying shades of grey rather than 'mountains and valleys' to indicate the intensity of EEG activity at various frequencies. This two-channel tracing depicts bimodal activity, with predominant amplitudes in the δ- and α-frequency ranges, somewhat more intense on the right. Clock times are shown to the left of the tracings. (From Grundy,[353] with permission.)

must still be monitored to exclude artefact. The most popular method of signal processing in commercially available monitors is power spectral analysis,[16] which mathematically converts the analogue EEG signal from the time domain to the frequency domain (Fig. 24.3). This technique forms the basis for display of the compressed spectral array (CSA) and the density-modulated spectral array (DSA) (Fig. 24.4). Quantitative measurements derived from the power spectrum, such as the peak power frequency (PPF), the median power frequency (MPF), and the spectral edge frequency (SEF), give a simple quantitative estimate of EEG activity, but much of the information contained in the original signal is lost.[17,18] Aperiodic analysis (Fig. 24.5), which depends on measurements of amplitude and frequency in the time domain,[19] is relatively simple and is better than spectral analysis at preserving information on varying EEG patterns such as burst suppression or epileptic activity.

Most 'brain mapping' is based on power spectral analysis. Diagrams of the head with different colours representing types and amounts of activity change dynamically with the electrical activity of the brain. Although this sounds intriguing, it has a number of drawbacks. Inclusion of artefact can be a major problem if the raw signal is not monitored. We still know too little about what should be mapped. Dozens of maps, according to frequencies, amplitudes, and specific measurements, can be generated from a single brief epoch of EEG. The quantitative analyses from which the maps are constructed

do permit interesting statistical comparisons, and this area invites further investigation.

Power spectral analysis assumes that component frequencies in the EEG are independent and do not interact. This is not strictly true. Bispectral analysis[20–22]

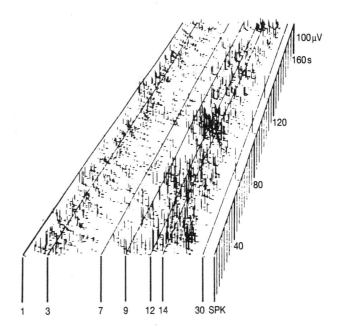

Fig. 24.5 A display derived from aperiodic analysis as developed by Demetrescu.[19] Each vertical line represents an EEG wave in the time domain. A track is added to the right of the display for spike activity, which is detected much more effectively by aperiodic analysis than by power spectral analysis. Elapsed time is shown to the right.

analyses coherence between any two components of the spectral array and thus reflects some non-linear characteristics of the EEG. Like the weather, the EEG is actually a deterministic chaotic process sensitively dependent on initial conditions, even though a given sample may appear to be random or stochastic.[23–25] Commercially available monitors do not yet have sufficient computational power for online analyses of chaotic systems, but analyses of deterministic chaos can be expected to gain importance as monitors become more powerful.

Clinical applications

Both traditional multichannel EEG recording and automated EEG signal analysis are used in the operating room and the critical care unit. Historically most well established are monitoring during carotid endarterectomy and EEG recording as an adjunct to other tests in the diagnosis of brain death.

EEG monitoring during carotid endarterectomy

Monitoring of brain activity during carotid endarterectomy (CEA) is surrounded by controversy.[26] Some surgeons always use a shunt; some never use a shunt; and some use selective shunting based on monitoring of neurological function in the awake patient, distal carotid stump pressures, EEG, somatosensory evoked potentials (SEPs), cerebral blood flow, or cerebral blood flow velocity. Some of this controversy may be engendered by the misconception that the only available intervention is insertion of a shunt. Impaired function can often be resolved by other measures such as raising arterial pressure or adjusting retractors

within the surgical field. In most large reported series, serious morbidity and mortality are in the range of 2 per cent or less; yet reports continue to appear describing neurological deficits at far greater rates. Although a few teams with reasonably low rates of serious neurological morbidity after CEA do not use brain monitoring of any kind, it is clear that some patients who have marked changes in brain function intraoperatively go on to suffer severe and permanent neurological deficits.[27–29]

Changes in the electrical activity of the brain during CEA do not invariably indicate permanent injury. Once electrical activity is affected, the EEG cannot be used to determine whether cells are dying or are merely in the 'ischaemic penumbra', with sufficient flow to maintain cellular integrity of neurons but insufficient flow to support electrical activity.[30,31] When EEG monitoring is used to determine which patients will have a shunt inserted, some patients do receive shunts who would do well without them. In this situation, however, patients suffering severe and permanent brain injury cannot be distinguished from those whose ischaemic cerebral segments are merely in the 'ischaemic penumbra'. The risk of shunting is clearly less than the risk of omitting a shunt when marked EEG changes are seen.[28,29]

A report of 176 CEAs done with EEG monitoring without shunting describes EEG changes in 55 patients. Of these 55, five (9 per cent) had related neurological deficits upon awakening from anaesthesia. Two of these patients, who showed major EEG changes within 12 and 20 seconds of carotid occlusion, respectively, had severe and permanent brain damage.[28] These two patients were among the 22 with major EEG changes. During episodes of major EEG change, activity in the 8–15 Hz range was markedly attenuated or absent, and δ activity at 1 Hz or less was at

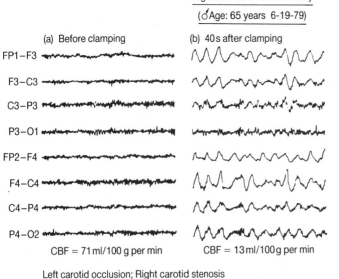

Right carotid endarterectomy

(♂ Age: 65 years 6-19-79)

(a) Before clamping   (b) 40 s after clamping   (c) After shunting

FP1–F3

F3–C3

C3–P3

P3–O1

FP2–F4

F4–C4

C4–P4

P4–O2

CBF = 71 ml/100 g per min   CBF = 13 ml/100 g per min   CBF = 46 ml/100 g per min

Left carotid occlusion; Right carotid stenosis

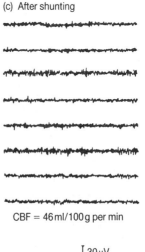

30 µV

2 s

Fig. 24.6 EEG changes during carotid endarterectomy. The paper speed is 15/s, half that shown in Fig. 24.2. The typical changes of cerebral ischaemia are seen with carotid occlusion and correlated with the dramatic fall in cerebral blood flow. Placement of a temporary shunt restores a satisfactory level of cerebral blood flow and returns the EEG to near its preocclusion pattern. (From Blume and Sharbrough,[29] with permission.)

least doubled (Fig. 24.6). Thus, 9 per cent of the patients with marked EEG changes had related neurological injuries. By contrast, among the 121 patients with no intraoperative EEG changes, there were no new deficits on awakening from anaesthesia.

Optimal EEG monitoring during CEA requires multiple channels, usually 16, to detect regional changes in brain activity. Further, it requires constant and knowledgeable observation of the EEG. An unassisted anaesthetist has no hope of accomplishing this. Such an individual can, however, monitor two or four symmetrical EEG channels with electrodes placed over the watershed areas of the middle cerebral arteries. Although focal changes will be missed, major catastrophic ischaemia can be detected.

Two-channel EEG monitoring during 103 operations prompted shunt insertion in 14 patients.[32] None of these 14 suffered new ischaemic injuries. Six of the unshunted patients awoke postoperatively with new neurological deficits. The two-channel monitoring had indicated a late ischaemic event, when it was thought no longer possible to insert a shunt, in five. No ischaemic changes were seen on the EEG from the sixth patient with a new deficit. Seventy other patients having CEA under cervical plexus block anaesthesia also had two-channel EEG monitoring. Aperiodic analysis of the two channels was less reliable than clinical neurological assessment in detecting intraoperative ischaemic events.[33] In two other series, neither the four-channel density-modulated spectral array[34] nor a set of 18 quantitative descriptors of the EEG power spectrum[35] adequately substituted for the 16-channel EEG interpreted by a neurologist. Elsewhere, EEG power changes were found to be more sensitive than variations in spectral edge frequency as indicators of intraoperative cerebral ischaemia.[36]

The largest series of CEAs monitored with 16-channel EEG is that at the Mayo Clinic,[27] where several thousand patients have been operated upon under general anaesthesia with selective shunting based on continuous EEG monitoring and intermittent measurements of regional cerebral blood flow. In this series, no patient has awakened with a major new neurological deficit that was not foretold by intraoperative EEG changes.[29] These patients were continuously monitored from before induction of anaesthesia until awakening from anaesthesia permitted clinical neurological assessment.

Such continuous monitoring is important for reliable results. Monitoring limited to the period of carotid occlusion and reperfusion will miss a large proportion of insults. Studies of patients operated upon under regional anaesthesia show that only a small percentage of perioperative neurological injuries occur during carotid occlusion when selective shunting is used.[4,5]

EEG monitoring during cardiac surgery

Few of the reported studies in this controversial area describe interventions made on the basis of EEG data; most were attempts to predict outcome and were therefore of very limited use in patient management. The value of any monitoring device in the care of an individual patient depends on at least the possibility of intervention to prevent or ameliorate deterioration. In an early series of 75 patients having EEG monitoring during cardiac surgery, 15 had clinically important episodes of hypotension during cardiopulmonary bypass.[37] In all 15, EEG changes were seen which consisted of slowing and/or loss of amplitude. No postoperative neurological deficits were seen in the seven patients whose EEG changes were transient, but the eight patients whose abnormalities persisted postoperatively developed neurological deficits that correlated with the EEG changes.

One-fourth of another 25 patients having cardiac surgery with EEG analysis had burst-suppression EEG patterns during cardiopulmonary bypass.[38] Spectral analysis with averaging over time destroyed this typical pattern. Moreover, the power spectra were multimodal in most of the patients. It was concluded that, in a large proportion of patients, univariate descriptors of the EEG spectrum are inadequate to describe the complexity of the EEG during anaesthesia.

Despite the recognized limitations of spectral analysis, particularly of a limited number of channels, these techniques can provide valuable information about global cerebral function. Such limited monitoring should not be expected to demonstrate subtle changes,[39] but the information provided can be valuable. Timely warning may lead to prompt restoration of adequate cerebral perfusion or institution of other protective measures. EEG has also been used to determine safe levels of hypothermia for circulatory arrest[40] and to detect burst-suppression produced by thiopentone when this drug is given before aortic declamping in an attempt to protect the brain.[41]

Two reports of intraoperative interventions during cardiac surgery, which were based on multichannel quantitative EEG monitoring, described dramatic differences in rates of postoperative neuropsychological dysfunction when physicians intervened on the basis of EEG changes. Although both studies used sequential groups of patients rather than randomization, the differences attributed to EEG monitoring are impressive. The rates of postoperative deterioration in neuropsychological function fell from 44 per cent without interventions to 5 per cent with interventions[42] and from 29 per cent without interventions to 4 per cent with interventions,[43] respectively. These reports invite additional investigation.

EEG monitoring during intracranial operations

Recordings directly from the surface of the brain or from depth electrodes are essential for determining the extent of resection when epileptogenic foci are removed.

During intracranial vascular neurosurgery, simultaneous monitoring of EEG and EPs may be particularly helpful. Thiopentone may be given as a protective measure before temporary occlusion of key intracranial arteries, and EEG recording is used to detect burst-suppression at a time when EP can still be recorded.[44] High-dose barbiturate anaesthesia, with or without mild hypothermia, may also be used to prevent or treat serious brain swelling after operations for arteriovenous malformations. Titration of effective doses in these settings is greatly facilitated by EEG monitoring. Both EEG and EPs can be recorded from the same electrodes.

Electroencephalography in the critical care unit

The most important application of EEG recording in critical care is for detection of epileptic activity. Alterations in consciousness may be due to continual epileptic activity without motor manifestations, and in some cases this has not been previously suspected.[45] A predictable sequence of EEG changes is seen during generalized status epilepticus when it is not treated.[46] Patterns include discrete electrographic seizures, followed by waxing and waning seizure activity, continuous seizure, continuous seizure with flat periods, and, finally, periodic epileptiform discharges on a relatively flat background. A patient who is either having seizures or comatose and shows any of these EEG patterns should be considered to be in generalized status epilepticus and should be aggressively treated to stop all clinical and electrical seizure activity. This is to prevent further neurological injury.

In coma due to head injury, EEG patterns reflect the degree of rostrocaudal neurological deterioration. Loss of such EEG characteristics as sleep-like activity, alternating patterns, and reactivity to stimulation indicates an unfavourable prognosis.[47] The EEG may be more sensitive than computerized tomography (CT) in detecting the extent of brain pathology in eclamptic women.[48] In neonates treated with extracorporeal membrane oxygenation for severe respiratory failure, EEG seizure activity indicates a significantly increased rate of adverse outcomes.[49] Bursts of slow waves seen in patients with subarachnoid haemorrhage from ruptured aneurysms are highly predictive of angiographic vasospasm.[50]

EEG recording, like other tests such as measurement of cerebral blood flow, is used more commonly in the United States than in Great Britain to help confirm the diagnosis of brain death. Technical guidelines for EEG recording in suspected brain death, recently revised, were published by the American Electroencephalographic Society in 1994.[51]

Rigorous technical standards must be met during both recording and interpretation, and repeated recordings may be indicated. Electrocerebral inactivity can be produced by drug overdose and/or hypothermia even in a normal brain, and particular caution must be used in diagnosing brain death in infants and young children. Tests for the absence of cerebral blood flow are more reliable and in many centres have largely replaced EEG recordings for the purpose of confirming brain death. In more than 25 000 radioisotope angiograms performed in one centre, no condition other than brain death was associated with the demonstration of absent cerebral perfusion.[52]

Evoked potentials (EPs)

Evoked potentials (EPs) are the electrophysiological responses of the nervous system to sensory, electrical, magnetic, or cognitive stimulation. They reflect the functional integrity of structures from which potentials arise and of pathways traversed between the site of stimulation and the neural generators of the evoked electrophysiological activity.[11,12]

Classifications of evoked potentials

EPs are primarily classified according to the *type of stimulation* used to elicit electrophysiological signals. Although EPs are most often elicited by sensory or electrical stimulation, event-related potentials can also be evoked by movement, thought processes, or other identifiable events. Different types of stimulation activate different parts of the nervous system. Additional useful classifications of EPs are according to the *post-stimulus latency* of an evoked peak or complex, the *distance separating the neural generator from the recording electrode*, and the *neural structures from which EP waveforms are thought to arise*. A framework for the classification of EPs useful to the anaesthetist is outlined in Table 24.1.

Type of stimulation

The modes of stimulation to be considered here include *somatosensory, auditory, visual, trigeminal, cognitive,* and *motor*. All these except cognitive EPs have been used in the operating room, but only somatosensory and auditory stimulation are in routine clinical use for intraoperative monitoring.

Post-stimulus latency

The time between application of a stimulus and the occurrence of a peak or complex in the EP waveform, the

Table 24.1 Classifications of EP

Mode of stimulation (for intraoperative EP monitoring)
 Somatosensory: electrical current
 Auditory: clicks or tones, delivered by ear-insert transducers
 Visual: flash, delivered by light-emitting diodes mounted in
  opaque goggles over closed eyes
 Trigeminal: electrical current
 Motor: electrical current or magnetic transients
Post-stimulus latency
 Short (less than 10 to less than 40 ms after stimulus)
 Intermediate (approximately 20–120 ms)
 Long (approximately 120–500 ms)
Distance from neural generator to recording electrode
 Near-field evoked potentials
 Far-field evoked potentials
Purported neural generators
 Cerebral cortex
 Subcortical structures of the brain
 Spinal cord
 Cranial nerve, peripheral nerve, or nerve plexus
 Sensory receptor

Reproduced with permission from Grundy BL. The electroencephalogram and evoked potential monitoring. In: Blitt CD, Hines R, eds. *Monitoring in anesthesia and critical care medicine*. Churchill Livingstone, New York, 1994.[355]

post-stimulus latency, is measured in milliseconds (ms). This latency is characterized as *short* (less than 10–15 ms for brainstem auditory EPs or less than 40 ms for somatosensory EPs), *intermediate* (usually 20–120 ms), or *long* (usually 120–500 ms).

Short-latency sensory EPs are subcortical in origin or represent only initial cortical activity, are less variable than later potentials, and are less affected by anaesthetic agents. When recorded from scalp electrodes, short-latency subcortical EPs are far-field potentials and of smaller amplitude than the on-going EEG. Short-latency EPs may be near-field potentials, with greater amplitudes and better signal-to-noise ratios, when recorded directly over a peripheral source. For example, an electrode placed on the skin at Erb's point records a near-field short-latency potential from the brachial plexus. In recording short-latency potentials, rapid stimulus rates can be used with relatively little loss of signal. This is because recovery from each individual stimulus is faster than for potentials of intermediate or long latency, presumably due to the fact that fewer synapses are involved.

Sensory EPs of intermediate latency are near-field potentials when recorded from the scalp and are therefore of greater amplitude than short-latency potentials. They arise from the primary sensory areas of the cerebral cortex and associated areas.[53] Recording can be done with fewer repetitions of the stimulus than needed for far-field short-latency potentials, but slower stimulus rates must be used. These EPs are affected by anaesthetics to a greater extent than short-latency EPs, but to a lesser extent than long-

latency potentials. Reliable monitoring can be done during general anaesthesia.

Long-latency EPs such as the P300 are thought to reflect cognitive function or physiological responses to pain. Their current usefulness is in research rather than in clinical monitoring.

The *central conduction time (CCT)* and *conduction velocity (CV)* are measurements derived from post-stimulus latencies of EP components. They facilitate quantitative comparisons of EPs over time, between sites, and among populations of normal and abnormal patients. CCT is calculated by measuring the post-stimulus time intervals between EP peaks generated in peripheral or cranial nerve, cervical spinal cord, or brainstem, and later peaks generated in midbrain, thalamus, or primary sensory cortex. Right-to-left asymmetries in CCT are particularly helpful diagnostically as well as in the operating room and critical care unit. Conduction velocity for a peripheral or cranial nerve can be estimated from the post-stimulus latency of evoked electrical activity and the measured distance between the stimulus site and the recording electrode. Spinal cord CV can be estimated from the time interval between EPs detected by different pairs of electrodes placed at intervals along the spine, or, intraoperatively, from electrodes on the cord or in the epidural space.

Distances separating neural generators from recording electrodes

Near-field potentials are those recorded from electrodes within a few centimetres (perhaps 2–3 cm) of their sites of origin. Near-field EPs recorded from the scalp are of cortical origin and have voltages similar to those of the on-going EEG. Only 32–128 individual responses need be averaged to make the near-field intermediate-latency cortical EPs apparent. Near-field potentials can also be recorded from electrodes placed over peripheral nerves or from invasive electrodes placed intraoperatively directly on or near structures such as the spinal cord or auditory nerve. In near-field recording, the EP waveform is markedly affected by electrode location. Polarity inversion between bipolar channels localizes the signal of interest to the electrode that is common to both channels. This kind of localization is key for cortical mapping to identify the sensorimotor strip during craniotomy. Another practical application of this phenomenon lies in the fact that inappropriate placement of electrodes can make EP recording difficult or impossible.

By contrast, *far-field EPs* are relatively little affected by small changes in electrode position. Signals travel by volume conduction to electrodes far removed from their anatomical sites of origin. Because signal strength decreases with distance, far-field potentials are much smaller than near-field potentials. They are far smaller than the background EEG, so that hundreds or thousands of individual responses must be averaged to extract the EP signal from the 'noise' of the on-going EEG.

Neural structures from which EPs arise

Cortical EPs, like the spontaneous EEG, arise in the pyramidal cells of the cerebral cortex. Several types of cortical EPs are known.[54] Those recorded over the primary sensory receiving areas are *primary specific responses*. These potentials are complex but consistent, can be produced by either meaningful or non-meaningful stimuli, and show little habituation. Waveforms arising in regions adjoining the primary sensory areas may have secondary waves not seen in the primary specific complex, habituate quickly, and are called *secondary specific responses. Non-specific responses*, also called *late* or simply *event-related responses*, can be recorded widely from the frontal and temporal regions regardless of the mode of sensory stimulation. They habituate rapidly with monotonous stimulation but are enhanced when the stimulus is meaningful or the subject focuses attention on the stimulus. Shifts in the surface negative potential of the cerebral cortex that depend on some relation between a signal and a subsequent action or decision by the subject are *contingent responses*. Of these, the best known is the *contingent negative variation* (CNV), a gradual increase in the surface negativity seen over frontal areas of the brain for 15–30 seconds between a conditional warning stimulus and an unconditional signal that requires decision or action by the subject. *Antecedent and imaginary responses* can be recorded over frontal cortex for about a second before voluntary movement and can be produced by intent alone when the movement never occurs. Of these cortical

responses, only the primary specific response is of interest for intraoperative monitoring.

Subcortical potentials reflect some combination of ascending volleys in sensory axons and firing or potential shifts of neurons in subcortical nuclei. Both auditory and somatosensory EPs reflect activity arising in the medulla, pons and thalamus, as well as thalamocortical radiations.

Potentials generated by the spinal cord and spinal nerve roots can be recorded invasively or non-invasively, but non-invasive recordings are small and thousands of repetitions must usually be averaged. EPs can be recorded from the *brachial plexus* by using a surface electrode at Erb's point, and similar potentials can be recorded from electrodes overlying the *lumbar plexus*.

Nerve action potentials (NAPs) and the evoked *electromyogram* (EMG) are used in nerve conduction studies to evaluate the function of peripheral nerves.[55] An electrode placed on or near a peripheral nerve detects a triphasic potential when the nerve is depolarized. An initial positive deflection is recorded as the potential approaches the electrode. With the net flow of positive ions into the nerve, depolarization and a dominant negative potential occur, followed by a small positive deflection as the area of depolarization moves away from the electrode. An electrode placed on the proximal cut end of a peripheral nerve detects only a large positivity.

The *M wave* is the directly evoked muscle action potential produced by stimulation of a peripheral motor or mixed nerve. The latency of its onset is the time required for action potentials in the fastest conducting fibres that are stimulated to reach nerve terminals and activate the muscle. When the recording electrode is directly over the muscle end plate, the M wave is biphasic (negative/positive) and latency measurement is most accurate. If recording is

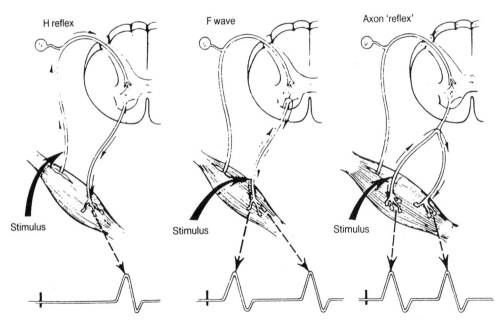

Fig. 24.7 Late muscle responses following peripheral nerve stimulation. See text for explanation. (Reproduced with permission from Daube JR. Nerve conduction studies. In: Aminoff MJ, ed. *Electrodiagnosis in clinical neurology.* Churchill Livingstone, New York, 1992: 292.)[55]

not directly over the end plate, the M wave is triphasic (positive/negative/positive) and onset latency is less precise. With supramaximal stimulation, the area of the negative phase of the M wave is directly proportional to the number of muscle fibres depolarized, but this area is also influenced by the distance separating the muscle from the recording electrode.[55]

Late muscle responses to stimulation of peripheral nerves include the H reflex, the F wave, and the axon 'reflex'[55,56] (Fig. 24.7). These responses are particularly useful in diagnosing abnormalities of the proximal segments of peripheral nerves, such as Guillain–Barré syndrome, thoracic outlet syndrome, and lumbar disc disease with radiculopathy. The *H reflex* is the electrophysiological equivalent of the clinical deep tendon reflex. To record the H reflex, the posterior tibial nerve is stimulated in the popliteal fossa with the cathode positioned proximally to avoid anodal block. The active recording electrode is placed over the soleus muscle and the indifferent electrode over the Achilles tendon. The H reflex is a monosynaptic response served by the same pathways traversed by deep tendon reflexes. Because the stimulus is electrical rather than a muscle stretch, the response is not affected by alterations in muscle spindle sensitivity. Afferent Ia fibres carry the impulse through the dorsal horn of the lumbar spinal cord and synapse with motor neurons that innervate the soleus-gastrocnemius muscle complex. An efferent NAP in the motor nerve generates a compound muscle action potential (CMAP) which is recorded as the H reflex. The H reflex can be consistently recorded only from the soleus muscle in adults, and the amplitude and latency fluctuate widely with the level of excitability of the central nervous system. During the first 24 hours after spinal cord injury, both the H reflex and deep tendon reflexes are depressed. The tendon reflexes remain depressed for weeks. By contrast, the H reflex rapidly recovers[57] and, after recovery from spinal shock, shows a marked increase in amplitude.[58] This reflects the recognized increase in excitability of central reflexes in chronic spinal cord injury.

The *F wave* (Fig. 24.7) can be easily recorded from most distal skeletal muscles. When a motor nerve is stimulated, impulses are transmitted not only distally (to activate muscle, producing the M wave), but also antidromically. If the cathode of the stimulating pair is placed proximally to the anode along the nerve, to avoid proximal anodal block, a few anterior horn cells will be activated and produce orthodromic potentials along the same nerve. These late potentials produce a small muscle response 20–50 ms after stimulation, actual latencies depending on the distance between the site of simulation and the spinal cord. Because the F wave provides a rough measure of conduction in central segments of motor fibres, it is particularly useful in diagnosing Guillain–Barré syndrome; central nerve

segments may be predominantly involved in the early stages of this disease.[59,60]

The *axon 'reflex'* is an abnormal response seen in pathological states when an axon branches in the peripheral nerve. Electrical stimulation distal to the site of branching produces a late muscle response through the abnormal axon branch (see Fig. 24.7). Other delayed responses can be seen when some axons are conducting more slowly than others or with irritability of peripheral nerves and repetitive firing after each stimulation.

Potentials generated by sensory receptors include the electrocochleogram and electroretinogram, which are of greater interest to otologists and ophthalmologists, respectively, than to anaesthetists. In clinical monitoring of auditory and visual EPs these potentials from sensory receptors can be used to document satisfactory stimulation when later waveforms are abnormal or cannot be recorded.

Methods for monitoring evoked potentials

Equipment

A system for recording EPs consists of stimulators and their transducers; electrodes, connectors and cables; filters and signal amplifiers; a computer to control stimulation and recording, process the acquired electrophysiological signals, and display them for measurement and evaluation; and recorders for paper and electromagnetic data storage. Guidelines for EP recording[61] and for intraoperative monitoring of EPs[62] have been suggested by the American Electroencephalographic Society.

For adequate quality control, it is necessary to be able to observe the unprocessed signal (after initial filtering and amplification) and suspend recording when the signal contains excessive noise. The available options for automatic artefact rejection, although they are very helpful, depend on amplitude criteria and may allow artefact not exceeding certain limits to be included in the signal. A common offender is the bipolar cautery. Similarly, the automatic artefact rejection may exclude physiological EEG or EP signals under certain conditions and must be reset or turned off to permit recording.

Most workers consider at least two channels necessary for effective EP monitoring, and at least four channels highly desirable. Other desirable features for EP systems used to monitor in the operating room or critical care unit include the capacity to display and store several waveforms, automatic counting and display of the number of repetitions accepted and rejected in an average, and automatic storage of stimulus and recording parameters with each waveform. Accurate documentation is greatly facilitated and human errors in data collection are minimized by placing as much of the monitoring as possible under programme control.

Stimulation

For several reasons, methods used for stimulation in the operating room differ in some respects from those used in the diagnostic electrophysiology laboratory. The unconscious patient is neither able to warn of impending injury nor able to cooperate with testing procedures. Special precautions are necessary for electrical safety and to avoid interfering with or contaminating the sterile surgical field. Some operations provide opportunities for invasive stimulation and recording that are not available in the diagnostic laboratory. Specific techniques for stimulation to elicit EP are detailed below.

Signal processing

Early workers used *superimposition* and *summation* of signals to demonstrate EPs. Widespread use of EP recording awaited the development of digital computers. Today, virtually all clinical EP recording uses *averaging* of digitized EEG to extract the EP 'signal' from the 'noise' of the background EEG and other electrical activity.[63] Because EEG activity is to some extent random, while EP activity is time-locked to a stimulus, averaging of multiple epochs or 'sweeps' of EEG activity immediately following the stimulus increases the signal-to-noise ratio by the square root of the number of such epochs averaged. The averaged digital signal is then converted to an analogue waveform for display and measurement.

Many other methods of signal analysis have been applied to EP signals. If the EP waveform cannot be seen in the usual average, however, more complex signal processing may yield spurious results. Particular caution must be used in applying advanced signal processing techniques when the signal-to-noise ratio is less than optimal. Even traditional filtering and smoothing, when carried to excess with noisy waveforms, can lead to erroneous interpretation of EPs and possible mismanagement of patients. Highly processed noise may resemble EP waveforms while bearing no relation to electrophysiological activity. Excessive filtering and smoothing must be avoided. Comparisons of EP waveforms to averages done with no stimulation and to waveforms recorded from other sites often aid interpretation. Familiarity with the usual averaged signals is essential, so that newly processed signals can be compared with customary averaged waveforms to avoid erroneous interpretations.[64] A broad range of signal processing techniques have been applied to EP data[65] but only averaging is in widespread clinical use.

Evoked potentials, such as the EEG, are customarily recorded in the time domain as plots of voltage against time. They are then described in terms of the post-stimulus latencies and peak-to-peak amplitudes in the averaged waveform. Identification of peaks is by visual pattern recognition by a trained observer. Peaks should not be automatically identified by the maximum deflection within a given time window, because neurological abnormalities and other factors can markedly prolong latencies. In such cases, a peak might still be identified within the time window but would not be the correct peak. In some cases, prolonged post-stimulus latencies may be normal. For example, cortical SEPs recorded after stimulation of the posterior tibial nerve at the ankle in a very tall individual are later than those recorded in a person of average height, simply because a greater distance must be traversed during signal transmission. Barring such explanations, however, prolonged absolute or interpeak latencies indicate delayed conduction. Particular patterns of latency change can help localize abnormalities.

EP amplitudes are usually measured from peak to peak rather than from baseline to peak, in order to minimize the effects of underlying low frequency activity or baseline drift. Amplitudes are more variable than latencies and have therefore been less often examined in diagnostic testing or monitoring. Amplitude changes, however, may be more sensitive than latency changes as indicators of abnormal function.[66,67] Differences in amplitudes of EPs elicited by stimulation at different sites may be diagnostically important.[68]

Numerous methods have been suggested for quantitative characterization of EPs including, among others, measurement of component wave areas,[69] power spectral analysis,[70] noise estimation,[71] correlation analysis,[72] stepwise discriminant analysis,[73] significance probability mapping,[74] and parametric predictive modelling.[75] Other statistical techniques have also been employed. Colour or grey-scale mapping of EPs ('brain mapping') has been recently popularized, but the place this will hold in clinical EP monitoring is not yet clear. Computer programs that extrapolate from a few electrodes to give a relatively detailed map can be misleading, and the value of the colour map as compared to the EP waveforms is not yet fully understood.

Steady state EPs are produced by continuous repetitive stimulation at rates of 6–8/s or greater, with a constant interstimulus interval. The brain is assumed to approach rapidly a relatively stable state of responsiveness, and one records a train of overlapping responses with constant frequency and only minimally varying amplitude. When steady state EPs are recorded, analysis of specific components similar to that done with transient EPs is no longer possible. Instead, the sinusoidal waveform produced is characterized by measuring the amplitude and phase angle of the response relative to the train of stimuli.[76] The steady state EPs can also be analysed by performing

Fourier analysis of the train of responses and computing the power spectra for trains of responses evoked at each of several stimulus frequencies. Ratios between the power spectra recorded over symmetrical areas of the right and left hemispheres of the brain should be less than 2.0 in normal subjects.[77] Although steady state EPs seem inherently attractive for monitoring because they can be rapidly recorded and quantitatively measured, little experience has been gained with this technique in the operating room or intensive care unit.

Interpreting evoked potentials

Once EP peaks are identified by pattern recognition, precise interpretation depends on reproducible quantitative representation of data and statistical characterization of appropriate databases. EPs vary with gender and age, so that normal data should be established separately for males and females in each age group. Once a normal data set is established, and the 95 per cent and 99 per cent tolerance limits for quantitative measurements are determined, each EP recording can be classified as normal, abnormal, or technically inadequate. Normal data sets have been published, but it is important if these published normal values are used to show that the techniques and values in the literature can be reproduced in the individual laboratory.[61]

In the operating room, each patient serves as his own control. Successful monitoring can be performed even when baseline EPs are abnormal, so long as reproducible signals can be obtained. The degree and duration of EP change consistent with preservation of function in the monitored pathway are not precisely known. EP values consistent with recovery vary according to the pathway being monitored, the nature of the underlying pathophysiology, and other factors that are not fully defined. We do know that even complete obliteration of brainstem auditory evoked potential (BAEP) signals may be compatible with preservation of neurological function when the insult is retraction of the auditory nerve and the EP changes are reversible in the operating room.[78,79] By contrast, obliteration of SEPs during operations on the aorta for more than 15–30 minutes may be associated with recovery of dorsal column function but permanent loss of motor function.[80]

Confounding variables such as changes in body temperature, anaesthetic agents, and arterial tensions of respiratory gases must be considered in interpreting intraoperative EP changes, and monitoring is greatly facilitated when the anaesthetist can maintain a steady state both pharmacologically and physiologically during critical monitoring periods.

Somatosensory evoked potentials (SEPs)

The most widely used EPs in the operating room, as in the diagnostic laboratory, are the SEPs. These potentials are used to monitor brain, spinal cord, and peripheral nerves.

Neurophysiology

Electrical stimulation of a somatosensory nerve preferentially activates the largest sensory fibres in the peripheral nerve (group I). These fibres subserve vibratory and proprioceptive sensation, and NAPs can easily be recorded over proximal segments of the nerve. According to a somewhat simplified classical description, the first-order neurons of the SEPs lie in the dorsal root ganglion, with distal axons in the peripheral nerve and proximal axons coursing through the dorsal nerve root, entering the dorsal root entry zone of the cord at its medial division, and continuing cephalad in the posterior funiculus, or dorsal columns, of the cord. The fasciculus cuneatus conducts impulses from the upper extremity and thorax; it is lateral to the fasciculus gracilis, which lies next to the dorsal median septum and carries fibres from the lower extremity. The first-order neurons of the somatosensory system synapse uncrossed in the nucleus gracilis and nucleus cuneatus of the lower medulla. Second-order neurons pass through the decussation of the medial lemniscus in the lower medulla and ascend as the medial lemniscus to the thalamus, where they synapse with third-order neurons in the ventral posterolateral nucleus of the thalamus. Axons of the third-order neurons lie in the thalamocortical radiations and project through the posterior limb of the internal capsule to the postcentral gyrus of the parietal lobe. This is the pathway classically considered to be traversed by the SEPs. In actual fact, however, only a small number of the lumbosacral dorsal root fibres in the fasciculus gracilis reach the upper cervical cord, and even fewer fibres in the fasciculus cuneatus reach the medulla. The pathway as classically described, with direct conduction of the first-order neuron to the dorsal column nuclei, is greatly oversimplified. Interneuronal relays along this pathway are common.[81–83]

Even with intervening synapses, EPs produced by stimulation of the upper extremity seem to be conducted predominantly ipsilaterally in the dorsal columns. By contrast, EPs elicited by stimulation of the lower extremity may be carried to a substantial extent in the lateral funiculus.[84] Stimulation of the posterior tibial nerve with intensity set at a subject's motor threshold activates group I fibres that synapse and ascend in the dorsal spinocerebellar tract. The fibres synapse in nucleus Z at the spinomedullary junction. The impulse is then transmitted to the ventral posterolateral thalamic nucleus and through the thalamocortical radiations to the primary sensory cortex.[85] Studies of spinal cord lesions in the monkey, cat and dog are consistent with SEP conduction in all quadrants of the spinal cord but predominantly in the ipsilateral dorsal lateral funiculus.[86]

A number of reproducible waveforms are generated as the SEP impulse is transmitted from the peripheral nerve to the brain. The NAP shows a triphasic form, as described above. Additional potentials, probably composed of NAP and excitatory postsynaptic potentials, are generated where the nerve roots enter the cord.[87] Recordings from the skin over the spine or directly from the spinal cord sometimes show an initial triphasic spike, probably produced by primary afferent activity. Subsequent slow activity, the N wave, has been attributed to cellular activity in Rexed's laminae III and IV of the dorsal horn, with an early deflection which is also presynaptic and is thought to reflect activity in afferent terminals.[88] A subsequent slow positive wave is related to primary afferent depolarization, a process by which axoaxonic synapses are thought to produce presynaptic inhibition.[83]

Figure 24.8 shows an intermediate-latency SEP after median nerve stimulation, as recorded in a four-channel array for intraoperative monitoring. Purported neural generators of the short-latency SEPs are outlined in Table 24.2. A negative peak recorded over the contralateral sensory cortex approximately 20 ms after stimulation of the median nerve at the wrist seems to be the initial cortical component of the SEPs. This peak is called the N20. (Peaks in the SEPs are labelled N for negative or P for positive, with a following number designating the nominal post-stimulus latency. Other systems are also used to label SEP peaks, but they sometimes lead to needless confusion about which peak is being discussed.) Although some observers believe that the N20 arises in the thalamus or thalamocortical radiations, all agree that the following positivity, about

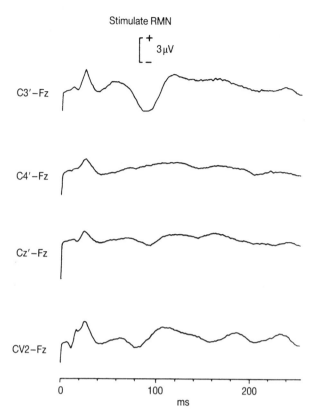

Fig. 24.8 Normal intermediate-latency SEPs recorded during anaesthesia. The same electrode montage can be used to record activity elicited by stimulation of either median nerve or the posterior tibial nerve. The last tracing, recorded from the skin overlying the spine of the second cervical vertebra, is often easier to record during anaesthesia than in the awake patient because muscle activity is suppressed. If this site is not available, does not give satisfactory responses, or is rostral to the likely points of injury intraoperatively, NAPs should be recorded during stimulation at each site to document signal input. (From Grundy,[11] with permission.)

23–25 ms after the stimulation of the median nerve of the wrist, arises in the somatosensory cortex.

In far-field recordings, new deflections are generated whenever the impulse enters a new and different volume conductor (base of finger or thumb; wrist; shoulder; spinal cord; foramen magnum). The moment the impulse reaches a boundary, current density changes suddenly in the two adjacent conducting media, giving rise to an apparent standing potential.[89] This mechanism probably plays an important role in the generation of early positive waves recorded from the scalp with a non-cephalic reference. These early positive waves are obscured in frontally referenced recordings and are not of concern during intraoperative monitoring of SEPs.

A popular technique for measuring central conduction time (CCT)[90] uses the negativity recorded from the skin overlying the spine of the second cervical vertebra, referenced to FZ or FPZ, as a benchmark for subcortical

Table 24.2 Purported generators of short-latency somatosensory evoked potentials

Peak	Generator
N9	Brachial plexus
N11	Spinal roots or dorsal columns
N13, 14	Spinal cord grey matter or dorsal columns
N14, 15	Brainstem and/or thalamus
N20	Primary somatosensory cortex

Stimulation: median nerve; recording: clavicle, mastoid process, 2nd cervical vertebra, primary somatosensory cortical area; reference electrode: FZ or non-cephalic. (Early positive waves are obscured in frontally referenced recordings.) N means negative peak. See text for references. From Grundy[11] with permission.

activity. The N20 is used as the indicator of cortical activity, and the CCT is the time required for impulse transmission between these two points. If the cervical potential is indistinct or unavailable, and the initial cortical positivity is clearer than the N20, it is often practical to substitute a measurement of elapsed time between the Erb's point potential and the initial cortical positivity.

In clinical neurology, abnormal SEPs are seen when demyelinating diseases, tumours, infarcts, or other structural abnormalities disrupt the conducting pathway.[91,92] Cortical SEPs are altered in patients with focal lesions of the somatosensory system at any point along the neuraxis, whether lesions are in the peripheral nerve, nerve plexus, spinal cord, brainstem, midbrain, or cortex. SEPs are particularly helpful in the diagnosis of multifocal or diffuse lesions, such as multiple sclerosis.

Methods

Somatosensory stimulation

Somatosensory EPs can be elicited by stimuli that are electrical, mechanical, thermal, or magnetic. *Electrical stimulation* is most commonly used in both the diagnostic laboratory and the operating room. Electrical current is easy to apply, measure and control. It is delivered to a sensory or mixed nerve via surface or subdermal electrodes. The median and ulnar nerves at the wrist, the posterior tibial nerve at the ankle, and the common peroneal nerve at the knee are mixed nerves commonly stimulated for diagnostic or monitoring purposes. The sural nerve is sometimes tested as well. When peripheral nerves are stimulated, it is best to stimulate each extremity singly in turn. Stimulation of pathways not at risk from the surgical procedure helps distinguish operative effects on EPs from effects of systemic factors such as changes in temperature, anaesthetic agents, or arterial tensions of respiratory gases. Simultaneous stimulation of nerves in both lower extremities (bilateral stimulation) gives larger responses than stimulation of a single nerve, but unilateral lesions of the cord can be missed.[93]

For most clinical SEP monitoring, electrode pairs are placed on the skin over peripheral nerves. Electrical current flowing between the two electrodes depolarizes the nerve at the cathode (−). For sensory stimulation, the cathode is placed proximal to the anode. This is because some axons may be hyperpolarized at the anode (+) and, at least theoretically, this hyperpolarization might block proximal transmission of the impulse ('anodal block'). Stimulators delivering constant currents are preferable to those giving constant voltages. It is actually the current that stimulates the nerve, and constant-current stimulators automatically adjust voltage to compensate for changes in electrode impedance so that the stimulus is kept constant. Stimulation of a mixed nerve that is sufficient to produce a motor twitch is called supramaximal, and it is often stated that further increases in current produce no change in the SEPs. In actual practice, however, it is clear that additional current usually produces augmentation of SEP amplitude. For clinical testing, some practitioners use the lowest current that produces a visible motor twitch; others use the highest current easily tolerated by the patient. A reasonable compromise that provides a level of reproducibility is use of a current equal to the sum of sensory and motor thresholds to electrical stimulation. In any case, the actual stimulus intensity used should be recorded, and it should be kept constant during critical monitoring periods. During anaesthesia, current is often increased to some arbitrary level such as 20 mA.

When a mixed nerve is stimulated, the signal ultimately reaching the central nervous system is complex and somewhat dispersed over time.[94] It includes not only orthodromic sensory impulses but also antidromic motor impulses and subsequent sensory impulses due to the muscle contraction produced by orthodromic motor impulses. Pure sensory stimulation can be performed with ring electrodes on the fingers, and cutaneous receptors can be stimulated with surface electrodes to test sensory function in specific dermatomes. Dermatomal SEPs can be used to localize radiculopathies diagnostically or in the operating room. Pure somatosensory stimulation produces SEPs that are smaller and less reproducible than those produced by electrical stimulation of mixed nerve trunks, and this technique is used only for special indications in the operating room or critical care unit.

The spinal cord can be directly stimulated by epidural or subarachnoid electrodes.[95–98] When recording is also from within the spinal canal, very rapid rates of stimulation can be used (30–50 Hz). This is because few if any synapses are involved and recovery from each individual stimulus is rapid. Very large amplitudes are achieved (100–150 μV), so only 50–100 repetitions need be averaged. Direct spinal stimulation is non-specific and involves multiple pathways. Non-specific stimulation of the cauda equina has also been described.[99] Because of the non-specific stimulation of multiple pathways, unilateral cord injury could be missed with either of these methods.[93] Invasiveness is also a relative disadvantage. *Mechanical stimulation* can be used to evoke SEPs,[100] but this technique is less reproducible than electrical stimulation and is essentially experimental. Late cortical EPs produced by *laser heat stimulation* are thought to represent a psychophysiological pain response. Each painful stimulus, however, produces a second-degree burn.[101] The potentials are large enough for single responses to be seen with special filtering techniques, so repeated stimulation and averaging of the response are not necessary. Thermal stimulation is valuable for pain research but not practical for clinical application.

Magnetic stimulation does not require electrode contact. Nerve trunks can be stimulated by time-varying magnetic fields.[102,103] Magnetic stimulation is more comfortable than electrical, but techniques for adequately reproducible localization are not yet fully developed.

Recording

Selection and arrangement of electrode locations for recording depend on the pathways to be monitored and the number of channels available. Signals are recorded over peripheral nerves, plexi, spine and scalp. Various invasive techniques have been described, with electrodes placed intraoperatively in bone,[104] interspinous ligament,[105,106] epidural space[107,108] and subarachnoid space.[97] Of the invasive techniques, only epidural and ligamentous recordings have been widely used in the United States. SEPs recorded from electrodes placed stereotactically in the depths of the brain can be used to localize subcortical structures for diagnostic or therapeutic purposes,[109,110] but these applications have been largely displaced by modern stereotactic localization based on computerized tomographic radiological techniques or magnetic resonance imaging.

Clinical experience with SEP monitoring

Operations for scoliosis

The largest collective experience with intraoperative EP recording is with SEPs, the first EPs made practical for routine intraoperative monitoring.[111] Members of the Scoliosis Research Society drew attention to the incidence of cord injury during operative treatment for scoliosis in 1975.[112] Numerous investigators have since described their experiences with SEP monitoring during orthopaedic procedures on the spine.[97,104,113–127] Orthopaedic surgeons were the most enthusiastic early users of SEP monitoring in the operating room, even though the incidence of spinal cord injury during orthopaedic surgery is much lower than the incidence of damage to the cord during neurosurgery to the cord itself. The spinal cord is damaged in less than 1 per cent of operations for scoliosis with Harrington rod instrumentation,[112] but procedures that involve sublaminar wiring involve a greater risk of injury.

All observers report that the majority of patients monitored during spine surgery show either no clinically important changes or changes that are only transient. In one series of 300 spinal procedures, four patients had transient SEP changes that improved after specific surgical interventions.[118] All four were neurologically intact post-operatively. By contrast, three patients had SEP changes that did not resolve intraoperatively, and all three had neurological deficits after surgery. In the entire series, there were no new neurological deficits unpredicted by the intraoperative SEP monitoring.

In a series of 137 patients having segmental spinal instrumentation with SEP monitoring, 41 per cent of the patients who had sublaminar wiring procedures had at least transient changes in the SEPs, and nine patients (6.6 per cent) were left with some neurological deficit.[116] Several patients had postoperative paraesthesias not foretold by SEP monitoring, but the three patients in this series with major motor deficits (2.2 per cent) all had marked intraoperative SEP changes. Amplitude changes of the nominal P40 peak were seen in most of the 61 patients in another series.[119] Wake-up tests were performed in those patients who showed notable SEP changes. The intra-operative tests for voluntary motor function were abnormal, and subsequent reduction of spinal distraction was associated with postoperative recovery of function. One patient was found to be paraplegic when examined 2 hours postoperatively, even though intraoperative SEPs had been normal. Repeat SEP recording was performed when the deficit was discovered, and SEPs were then absent. It seems likely that the onset of this deficit occurred after SEP monitoring was stopped, and this case emphasizes the importance of continuous neurological monitoring whenever possible. Continuation of SEP monitoring until the patient is sufficiently alert for clinical neurological examination can be expected to help avert such unfortunate outcomes.

In another 45 patients monitored during surgery for scoliosis, notable alterations in SEPs were seen in only two.[56] In one patient, SEPs elicited by stimulation of the lower extremity disappeared during exposure of the spine, before any instrumentation. A wake-up test confirmed paraplegia. Emergency myelography was normal, and the patient later regained antigravity strength in the lower limbs. In the second case, SEPs to stimulation of the lower extremities disappeared after passage of a sublaminar wire. The patient was paraplegic upon intraoperative testing of voluntary motor function, and the paraplegia persisted postoperatively. The SEPs seen after stimulation of the median nerve remained normal in both of these patients. The other 43 patients in this series had stable SEPs intraoperatively and no neurological deficits postoperatively.

A published case report describes a patient who was paraplegic after the second operation in a staged procedure for scoliosis even though intraoperative monitoring detected 'no change in the SEP'.[128] The waveforms published in this report, however, show that SEP latencies after induction of anaesthesia for the second operation were 5 ms later than those recorded during the first operation.

Perhaps there actually was a significant change in the SEPs in this case. When there is any doubt about SEPs intraoperatively, the wake-up test[6,7] should be used.[129]

A 1991 report of 1168 operations on the spine described decreases of SEP amplitude greater than 50 per cent in 119 patients (10.2 per cent).[122] Thirty-two of these 119 had clinically detectable neurological changes after surgery. In 35 operations, SEP amplitude was restored rapidly intraoperatively either spontaneously or by repositioning a recording electrode. None of these 35 patients had postoperative neurological deficits. Persistent minor SEP changes were seen in 52 patients but were not followed by neurological problems. No neurological injuries occurred in patients who had amplitude decreases of less than 50 per cent. Several factors were identified with an increased risk of postoperative neurological deficit: neuromuscular scoliosis, sublaminar wiring, magnitude of SEP decrement, and limited or absent intraoperative recovery of SEP amplitude.

Reported interventions made on the basis of SEP changes during spinal surgery have included readjustment of spinal instrumentation, reversal of induced hypotension, decreases in the amount of surgical correction of the deformity, and removal of bone grafts. When sudden changes are seen in SEPs during spinal surgery, an immediate wake-up test should be performed if at all possible. When SEPs are monitored continuously and are stable, many teams omit wake-up tests.

Simultaneous stimulation of both lower extremities can lead to false-negative results.[130] SEPs elicited by stimulation of either lower extremity can be easily recorded from the same midline electrodes on the scalp. If injury occurs to one side of the cord while the other side continues to function, bilateral stimulation can produce unaltered SEPs, whereas stimulation of each lower extremity separately would clearly show the loss of signal during stimulation of the extremity ipsilateral to the injury.[93]

Operations for spinal injury

The wake-up test may be dangerous for the patient with an acutely injured and severely unstable spine, because uncontrolled movement might produce additional serious injury to the spinal cord. SEP monitoring therefore seems attractive as an aid to intraoperative management of these patients.[131] SEPs were monitored in 11 patients who had operations for decompression and stabilization of acute injuries.[132] SEP amplitudes increased upon decompression of the cord in four patients, but clinical outcome did not seem related to whether SEPs improved or stayed the same. One patient who had marked intraoperative improvement of SEPs regained posterior column function but remained paraplegic and never recovered the ability to sense pain. In a separate report, the same workers described SEPs in 10 patients with incomplete cord lesions due to trauma.[133] All

10 showed improvement in SEPs upon decompression, and all but one subsequently showed considerable functional return. Perhaps this conflict between reports from the same group is more apparent than real; it might be due to differences in emphasis rather than to differences in actual outcomes. Paralysis despite preservation of dorsal column function and SEPs may be more likely in the non-operative setting.

During 76 operations for spinal fracture-dislocation[56] changes were seen in two. In one patient, SEPs were lost for 5 minutes after the passage of a Luque wire and then recovered. The patient had no neurological deficit postoperatively. In the second patient, SEPs to stimulation of the right leg disappeared during spinal instrumentation while SEPs to stimulation of the left leg were unchanged. Later during the same operation the patient had a myocardial infarction and could not be resuscitated. At autopsy, the right side of the cord showed acute haematomyelia with mechanical disruption while the left side was essentially intact.[93] Monitoring with only bilateral stimulation of the lower extremity would not be expected to detect this kind of injury.

Other operations within the spinal canal

SEP monitoring has been used during resection of spinal cord tumours and other lesions impinging on the spinal canal.[107,108,134–139] SEPs are often abnormal or absent preoperatively in symptomatic patients, and if reproducible waveforms cannot be obtained during anaesthesia, monitoring cannot be done. Interactions among pathophysiological factors, anaesthetic effects, and stimulus and recording techniques are incompletely understood, but some clinical guidelines may prove useful. Slowing the rate of stimulation to the range of 0.9–3.1 Hz may be helpful, as may increasing the stimulus current and/or duration. Omission of inhaled anaesthetics may also be useful. Etomidate amplifies cortical SEPs[140] and may be an appropriate anaesthetic agent in this setting. If reproducible waveforms can be recorded, even though these may be abnormal, useful monitoring can be done. Several reports even describe improvement in SEPs with decompression of the spinal cord. SEP monitoring has been used by several investigators during resection of spinal arteriovenous malformations.[141,142] This seems particularly useful when large feeding vessels are available for test occlusion or when resection can be stopped if marked SEP changes occur.

Many neurosurgeons consider SEP recording indispensable during operations to relieve intractable pain by placing deliberate lesions in the dorsal route entry zone of the cord.[56,83] The appropriate segment for lesioning is located by using stimulation of peripheral nerves at multiple levels with monopolar recording from the cord. Unwanted numbness in the lower extremities, formerly the most frequent complication of this procedure, can usually be avoided by monitoring cortical SEP elicited by stimulation of the posterior tibial nerve at the ankle or the peroneal nerve at the knee.

Good results have been reported in monitoring evoked spinal potentials during operations for scoliosis.[97] This technique proved less satisfactory during operations for spinal cord tumours. During resection of six extramedullary and 14 intramedullary spinal cord tumours, an epidural electrode caudal to the level of the tumour was used to stimulate the cord and potentials were recorded from an epidural electrode rostral to the tumour.[143] Increases in the amplitude of the evoked spinal potentials were seen after removal of the tumour in four extramedullary and three intramedullary cases. Five of the six patients who had decreased amplitudes of this potential after removal of the tumour developed motor deficits after surgery. Four patients, however, developed postoperative deficits even though spinal cord EPs were stable throughout the operation. Of these four, two had intramedullary tumours and experienced postoperative symptoms consistent with unilateral spinal cord damage. One patient had changes in EP amplitude but no postoperative deficit. Despite the high amplitude of the evoked spinal cord potential, the reproducible waveforms, and the robust character of this potential during anaesthesia, the investigators concluded that, because 20 per cent of the intraoperative spinal cord injuries were not detected, this technique was insufficiently sensitive to monitor patients with spinal cord tumours.

Operations on the aorta

The spinal cord may suffer ischaemic injury during operations on the thoracic or abdominal aorta, or even during angiography. The blood supply to the anterior two-thirds of the cord consists of a single artery that traverses the entire anterior aspect of the cord. The anterior spinal artery is supplied by branches of the vertebral arteries and numerous tributaries, including the thyrocervical trunk and several radicular branches in the thoracic and lumbar regions. Usually one radicular branch is larger than the others and more critical for cord perfusion. This vessel arises as a left intercostal branch, between T9 and L2 in most patients, and is known as the greater medullary artery or the artery of Adamkiewicz. Two posterior spinal arteries, also arising from the vertebral arteries, supply the posterior third of the cord. Occlusion of the anterior spinal artery causes infarction of the anterior part of the cord, with loss of pain, temperature and crude touch below the lesion. If the lesion extends far enough posteriorly, bilateral spastic paralysis results. When the posterior spinal arteries continue to function, a patient with anterior spinal artery

occlusion may be paraplegic while position and vibration sense, as well as SEPs, are preserved.

The recognized anterior spinal artery syndrome naturally poses questions about the possible usefulness of SEP monitoring during operations on the aorta, and several reports address this problem. If useful, SEP monitoring would seem highly desirable during operations on the thoracic aorta; as many as 24 per cent of patients may be paraplegic postoperatively. The risk of cord injury is even greater during operations for large, complex thoraco-abdominal abnormalities of the aorta.

Abdominal aortic surgery poses far less risk to the cord, but injuries do occur. The risk of paraplegia is lower in patients with coarctation, probably because the collateral blood supply to the cord is better. A report describing eight cases of damage to the cord following operations on the abdominal aorta also reviewed 36 additional cases from the literature.[144] Fifteen of the 44 patients discussed had intact proprioception, and three more had unspecified partial sensory function preserved. SEPs were not recorded, but presumably the patients with proprioceptive function intact would also have had preserved SEPs.

Several experimental studies of compromised aortic flow have included SEP monitoring.[145–150] Microsphere studies of cord perfusion in dogs[147–149] showed that maintenance of distal aortic pressure at or above 70 mmHg kept spinal cord blood flow and SEPs within an acceptable range, while distal aortic pressures below 40 mmHg abolished SEPs.[151] Occlusion of a critical intracostal artery in other animals markedly decreased cord perfusion and abolished SEPs.[152] When cord perfusion was restored 5 minutes after SEP loss, no paraplegia resulted. By contrast, when aortic occlusion was maintained for 10 minutes after SEP loss, two-thirds of the animals were paraplegic postoperatively.[153]

During operations on the aorta in patients, the cases described in the literature to date suggest that marked change or loss of SEPs lasting longer than some critical period, probably somewhere between 15 and 45 minutes, may be associated with paraplegia. If SEPs recover intraoperatively during this type of surgery but alteration or obliteration has lasted beyond the critical period (which probably differs somewhat from patient to patient), the patient can be expected to have paraplegia with preservation of both dorsal column function and SEPs.

In one reported series, no SEP changes were seen in patients with distal pressures greater than 60 mmHg during aortic occlusion.[148,154] Two patients had distal pressures less than 40 mmHg, associated with SEP loss that occurred between 15 and 25 minutes after aortic occlusion. In one of these patients the SEPs could not be elicited for 60 minutes before aortic flow was restored, and this patient was subsequently paraplegic. In a series of 25 patients with coarctation, SEPs were lost for more than 30 minutes in six patients; five of these were paraplegic postoperatively.[155] Ten of 13 patients with thoracoabdominal aortic aneurysms lost SEPs at intervals of 17–40 minutes after aortic occlusion. The patient with the longest period of SEP obliteration, 59 minutes, was subsequently paraplegic.[156]

Methodological issues cloud the conclusions reached in one report of an attempt to assess the impact of distal aortic perfusion and SEP monitoring on outcomes of 198 operations for thoracoabdominal aneurysms.[157] Patients were not randomized to have operations with or without these adjuncts. The patients who did not have distal aortic perfusion and SEP monitoring were those in whom these measures could not be used for either technical or administrative reasons. Furthermore, 11 of the 20 neurological deficits began between 12 hours and 21 days after the initial postoperative examination. This latter group included six of the nine patients who died in the postoperative period, and in all six of these cord dysfunction progressed to complete paraplegia before death. Failure to monitor SEPs until completion of the operation is another complicating factor in this series. Monitoring was usually discontinued 30 minutes after aortic reconstruction if SEPs were unchanged, and up to 2 hours after reconstruction if they were normal. It is not unlikely that SEPs may have deteriorated after monitoring was stopped in the three patients of these 198 who had immediate postoperative neurological deficits despite lack of intraoperative SEP change. Interestingly, 12 patients in this series had normal SEPs during aortic reconstruction, which deteriorated upon restoration of aortic flow. One of these 12 had a neurological deficit immediately postoperatively and three others developed delayed deficits. Despite the methodological limitations in this report, there was an association between level of intraoperative SEP change and final neurological outcome. Thirteen per cent of patients with no SEP change had neurological complications, compared to 32 per cent of those who had SEP changes without recovery during the monitoring period.

A summary report of 1509 patients having thoracoabdominal aortic operations by a single surgeon between 1960 and 1991[158] included 234 patients with paraplegia or paraparesis (16 per cent). Factors associated with spinal cord injury were total aortic clamp time, extent of aortic repair, aortic rupture, patient age, proximal aortic aneurysm, and history of renal dysfunction. During the more recent years survival rates improved but rates of paraplegia changed little. Late onset of paraplegia was related to postoperative complications, particularly respiratory failure and hypotension. A separate report describing variables predictive of outcome in 832 of these same patients[159] showed that the use of atriofemoral bypass eliminated the increased risk associated with prolonged aortic occlusion times. The authors of this analysis concluded that the two most important factors in patients

who were paraplegic or paraparetic postoperatively were the degree and duration of spinal cord ischaemia and failure to successfully reimplant critical intercostal arteries.

In a series of 33 patients having operations on the descending thoracic or a thoracoabdominal aorta,[160] SEPs were stable and spinal cord function was intact postoperatively in 17 patients with distal aortic perfusion pressures greater than 60 mmHg. In these 17 patients, aortic occlusion times ranged from 23 to 105 minutes and distal perfusion was maintained by either shunt or bypass. In the other 16 patients in this series, aortic occlusion times ranged from 16 to 124 minutes and SEP loss was associated with inadequate maintenance of distal perfusion pressure or interruption of flow in intercostal arteries. Loss of SEPs for more than 30 minutes was followed by a 71 per cent incidence of paraplegia, but when SEP loss did not occur or was limited to less than 30 minutes spinal cord function was invariably found to be intact postoperatively.

Simultaneous recordings from scalp electrodes and electrodes placed over the third lumbar vertebra showed that during ischaemia of a peripheral nerve produced by occlusion of the abdominal aorta or femoral artery, the lumbar spinogram could disappear rapidly while cortical potentials were preserved.[161] Presumably a desynchronized afferent volley from the partially ischaemic peripheral nerve was not adequate to produce a lumbar potential but was sufficient for amplification at the cortical level.

Recent reports describing aggressive approaches to preservation of spinal cord function during surgery on the thoracoabdominal aorta suggest that progress is being made. A multimodality protocol used in 42 patients included complete intercostal reimplantation whenever possible, cerebrospinal fluid drainage to improve spinal cord perfusion pressure, maintenance of proximal hypertension during aortic occlusion, moderate hypothermia, high-dose barbiturates, and avoidance of hyperglycaemia.[162] In addition, mannitol, steroids, and calcium channel blockers were used to minimize reperfusion injury. None of the 42 patients managed with this protocol had postoperative spinal cord injuries, compared to 6 per cent of 108 historical control patients. More recently, the same group used atriofemoral bypass with local cooling of intercostal and visceral arteries during segmental resection of high-risk thoracoabdominal aneurysms.[163] Each segment of the aorta was perfused with cold crystalloid prior to occlusion and attempts were made to reimplant all pairs of intercostal arteries from T8 to L2. Warm ischaemia of each segment was thus limited to 30 minutes. Seven of the 23 patients described in this report had dissections or ruptures that necessitated emergency surgery and one of these seven was paraplegic postoperatively – a remarkably low incidence in these very high-risk patients. Another approach to regional hypothermia of the spinal cord during aortic surgery was tested in rabbits.[164] With normothermia, 40 per cent of the animals were paraplegic after 20 minutes of ischaemia, 75 per cent after 40 minutes, and 100 per cent after 60 minutes. Perfusion of the epidural space around the lumbar segments of the cord with isotonic saline at 5°C prevented any spinal cord injury even after 60 minutes of ischaemia.

Motor EPs (MEPs), are less sensitive than SEPs for reliably predicting motor function after aortic occlusion in animals.[165] Spinal EPs may also be less sensitive than cortical SEPs for detecting clinically important spinal cord ischaemia. Postoperative paraparesis has been described postoperatively in a patient who had well maintained spinal EPs throughout aortic cross-clamping.[166]

Ten of 22 patients, monitored during operations for coarctation,[80] had SEP changes. Loss of SEPs for 30 minutes in one patient was followed by paraplegia. SEP obliteration for 14 minutes in another patient was associated postoperatively with transient paraesthesias in the lower extremity. A third patient in this group had abrupt and reproducible loss of SEPs upon test occlusion of the aorta. Because of the SEP changes, this patient had subclavian-aortic bypass rather than resection of the coarctation (Fig. 24.9). SEPs recovered after full restoration of flow, and the patient suffered no neurological injury.

SEP monitoring may be helpful during spinal angiography and embolization. In a series of 41 patients, injection of contrast material rapidly reduced SEP amplitudes.[167] Improved SEPs after embolization correlated with improved clinical function. One patient in this group suffered irreversible SEP loss and a new neurological deficit.

SEP monitoring during carotid endarterectomy

The middle cerebral artery is the artery most often embolized in patients having carotid endarterectomy (CEA), and its watershed areas are the most at risk from cerebral hypoperfusion during carotid occlusion. The primary cortical response to stimulation of the median nerve at the wrist arises in the area at risk. A number of reports describe SEP monitoring during CEA.

SEPs,[168] like EEGs,[169] have detected otherwise unsuspected shunt occlusions. Complete loss of SEPs was seen during 17 of 400 operations.[170] Seven of these patients had no shunt inserted, and five of the seven had postoperative neurological deficits. Three of the ten patients, who did have shunts inserted, had neurological deficits after surgery. Thus, patients who lost SEPs during carotid occlusion had a much higher incidence of postoperative neurological deficit when no shunt was used (71 per cent) than when a shunt was inserted (30 per cent). In a series of 675 CEAs, SEPs had a diagnostic sensitivity of 60 per cent in predicting neurological outcome, and specificity of 100 per cent.[171]

Some of the apparent discrepancies among reports of EEG and SEP monitoring during CEA may be related to

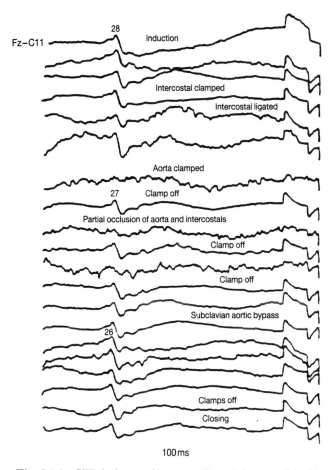

Fig. 24.9 SEP during aortic surgery. Potentials were elicited by stimulation of the posterior tibial nerve during thoracotomy for repair of an aortic coarctation. The cervical spinal evoked potential disappeared during test occlusion of the aorta and returned promptly when the clamp was removed. The same phenomenon was seen during two subsequent attempts at aortic clamping. The surgeon elected to perform a subclavian-aortic bypass rather than the usual patch-graft repair, which requires complete aortic occlusion. SEP were normal during the rest of the procedure. There was no neurological deficit postoperatively. (A 1 μV, 9 ms calibration pulse is shown at the end of each waveform.) (Reproduced with permission from Friedman WA, Theisen GJ, Grundy BL. Electrophysiological monitoring of the nervous system. In: Stoelting RK, Barash PG, Gallagher TJ, eds. *Advances in anesthesia*. Chicago: Year Book Medical, 1989: 231–77.)[13]

the fact that there is no general consensus on criteria for abnormality of either. This problem is clearly illustrated by a report of 193 CEAs under general anaesthesia that compared SEPs according to different criteria.[172] An increase in central conduction time of more than 20 per cent from the anaesthetic baseline control recording was 100 per cent sensitive and 89 per cent specific for early postoperative neurological morbidity. When the criterion of amplitude reduction exceeding 50 per cent was used, sensitivity was 86 per cent and specificity 96 per cent. For complete loss of the cortical response, sensitivity was 71 per cent and specificity 99 per cent. In attempting to prevent neurological injury, a high level of sensitivity in detecting possible ischaemic injury is far more important than specificity. Most clinicians would prefer a criterion with a sensitivity as close to 100 per cent as possible (no undetected brain injuries) even at the cost of a lower level of specificity.

A few studies have directly compared SEP and EEG monitoring during CEA under general anaesthesia. Results were not uniform. One study concluded that conventional EEG and SEP monitoring had similar sensitivity and specificity.[173] Evaluation of this result is difficult, however, in light of the fact that SEPs were monitored only before, during, and for 15 minutes after carotid occlusion rather than continuously from before induction of anaesthesia until awakening. Also, no shunt was used in any patient regardless of EEG or SEP changes.

Selective shunting was used in a series of 53 CEAs.[174] In these patients, SEP monitoring was less sensitive than 16-channel EEG recording for detecting intraoperative cerebral ischaemia. EEG evidence of ischaemia was seen in 23 patients during carotid occlusion, while only 10 of these had an increase in SEP latency of 0.1 ms or greater and only one of the 23 had a decrement in SEP amplitude of 50 per cent or greater.

The third comparison of EEGs and SEPs during CEA included 151 operations.[175] No change was seen in either EEG or SEPs upon carotid occlusion in 120 cases. In 16 cases, cortical SEPs were severely depressed; these were the only cases in which a bypass shunt was used. Fifteen patients with more moderate SEP changes had no shunts, and none of these 15 had new neurological deficits postoperatively. Two deficits were seen in patients who had marked SEP abnormalities as well as EEG changes, and one occurred in a patient with no reported change in EEG or SEPs.

Most intraoperative SEP monitoring examines only subcortical and initial cortical activity, usually through the initial cortical positivity. This peak is normally seen 20–25 ms after stimulation of the median nerve at the wrist. SEPs of intermediate latency, thought to arise in immediately adjacent areas of the cortex, may be more sensitive to cerebral ischaemia than the short-latency SEPs most often used for monitoring during CEA.[176]

Taken together, the most reliable methods of brain monitoring during CEA seem to be clinical assessment of the alert patient operated upon under regional anaesthesia and 16-channel analogue EEG, continuously monitored by a knowledgeable observer from before induction of anaesthesia until the patient is sufficiently awake for postoperative neurological assessment. When neither of these techniques is available, SEP or more limited EEG monitoring can provide valuable information to guide intraoperative management.

SEP monitoring during cardiac surgery

Less work has been done with EP monitoring during cardiac surgery than during CEA. Hypothermia increases latencies of all major components of the SEPs.[177] Later components are more markedly affected than early components. Brachial plexus injury related to self-retaining sternal retractors can be detected with SEPs elicited by stimulation of the median or ulnar nerve.[178]

All cortical SEP activity was lost at 18°C in nine profoundly hypothermic infants. SEPs recovered at an average time of 30 minutes after rewarming.[179] One infant had a prolonged time to recovery of SEPs after rewarming and went on postoperatively to develop seizures. Two babies with normal recovery of SEPs had cortical blindness. This suggests that EEGs or multimodal EPs may be preferable for monitoring patients during cardiopulmonary bypass. SEPs would not be expected to reflect function of the occipital cortex.

SEP and nerve conduction studies in peripheral nerve surgery

Management of injuries to peripheral nerves or the brachial plexus is greatly facilitated by intraoperative recording of EPs and nerve action potentials (NAPs).[56,180–182] The most important contribution of electrophysiological monitoring in this setting is determination of the type of injury to the peripheral nerve: neurapraxia, axonotmesis, or neurotmesis. Neurapraxia is a reversible injury; the variable loss of distal function usually improves within days or weeks without surgical intervention. Intraoperatively, the most important distinction is between axonotmesis, in which axons and myelin are interrupted but the surrounding connective tissue is preserved, and neurotmesis, when not only axons and myelin but also the surrounding connective tissue elements are disrupted. At surgery 2–3 months after injury, the appearance of either lesion is most often that of a 'neuroma-in-continuity'. The distinction between the two types of injury in this case cannot normally be made by either visual inspection or observation of the muscle response to proximal nerve stimulation. No muscle contraction will be produced in either case. It is the distinction between neurotmesis and axonotmesis, however, that determines the appropriate operative treatment. With neurotmesis, no natural pathway exists for regeneration of the nerve. The intervening neuroma should in this case be resected and the cut ends of the nerve reanastomosed, using a nerve graft if necessary. By contrast, when the connective tissue is intact, as in axonotmesis, the natural pathway already provided for regeneration of the nerve is far superior to the pathway provided by reanastomosis. In this situation, only neurolysis should be done.

This dilemma can be resolved electrophysiologically. In the axonotmetic injury, sufficient nerve regrowth (about 1 mm/day) will have occurred for conduction of the NAPs across the neuroma-in-continuity. Thus, with stimulation proximal to the lesion and recording from the exposed nerve distal to the lesion, transmission of the NAPs indicates an axonotmetic lesion and failure of transmission (no distal NAPs) determines that the lesion is neurotmetic. Sensory fibres can be tested by stimulation just distal to the neuroma and recording of cortical SEPs. The decision between neurolysis only and excision with reanastomosis or grafting then becomes straightforward.

SEP monitoring during intracranial surgery

Recordings of SEPs from multiple electrodes placed directly on the surface of the cerebral cortex can be used to locate precisely the rolandic fissure during resection of lesions in adjacent areas.[183,184]

More frequently, SEPs recorded from scalp electrodes are used to monitor the adequacy of brain perfusion during operations for intracranial aneurysms or arteriovenous malformations (AVMs). Studies in baboons have shown that SEPs elicited by stimulation of the median nerve begin to deteriorate at cerebral blood flow levels of approximately 15 ml/100 g per minute.[185] Neuronal injury, however, does not occur until flows fall to 10 ml/100 g per minute or lower.[186] Thus, SEP monitoring can warn of cerebral ischaemia before irreversible damage occurs.[187] When induced hypotension is planned, SEPs can be monitored during a trial of hypotension prior to placement of brain retractors so that pressures below those that alter the SEPs can subsequently be avoided.[188] Reduction in SEP amplitude, increasing central conduction time, or loss of the cortical response can be used as guides for manipulating arterial blood pressure, timing temporary vascular occlusion, or repositioning imperfectly placed aneurysm clips.[189,190] If EPs are lost during temporary occlusion of an intracranial artery, postoperative neurological deficit will be less likely if flow can be established within 10–12 minutes.[191,192]

Ischaemia in areas not monitored by the EP recordings, such as aphasia, may not be detected.[189] In areas of the cerebral cortex adjacent to but not within the monitored pathway, ischaemia will not be detected unless it is so severe that it extends to areas within the monitored pathway. SEP monitoring is most valuable during surgery for aneurysms of the middle cerebral artery[190] or internal carotid artery.[192] By contrast, undetected injury is relatively frequent during surgery for aneurysms at the tip of the basilar artery. Here ischaemia of motor pathways can occur while somatosensory pathways are unaffected. The ischaemia is sometimes sufficiently severe to extend into the somatosensory pathway and alter SEPs, but unpredictable motor deficits are particularly likely in patients with aneurysms at the basilar bifurcation. Unfortunately, this is the most common location of aneurysms in the posterior circulation.

During 16 operations for basilar tip aneurysms, ten patients were monitored with both SEPs and BAEP, four with BAEP only, and two with SEPs only.[193] Only two patients had EP changes that were considered significant. In one, transient changes in BAEP were related to retraction of the auditory nerve and pons, while in the other progressive attenuation and transient loss of SEPs were seen during basilar artery occlusion. Four patients in this series had new ischaemic brainstem deficits postoperatively that were not predicted by intraoperative monitoring.

Most SEP monitoring for intracranial vascular surgery has been done with median nerve stimulation. For anterior communicating and anterior cerebral artery aneurysms, however, posterior tibial nerve stimulation is more appropriate. The leg and foot areas of the cortex are at greatest risk during these procedures. SEPs elicited by stimulation of the posterior tibial nerve have been used to determine the safety of sacrificing an enlarged anterior cerebral artery that was the main feeding vessel of a large AVM.[194]

SEP changes with positioning

SEP changes related to positioning have been seen as incidental findings in patients who had median nerve SEPs done for control purposes while lower extremity SEPs were being monitored during scoliosis fusion. During vertebral periosteal elevation and decortication, a patient positioned on a four-poster frame can be unintentionally moved so that an upper poster supports the patient at the axilla rather than at the chest wall. With appropriate repositioning, the SEP to median nerve stimulation quickly recovers. It seems likely that such a patient may be spared an injury of the brachial plexus.

Transient loss of cortical SEPs, while cervical potentials were preserved, has been seen when positioning a patient with a large posterior fossa tumour that displaced the brainstem.[195] The reversible SEP changes were attributed to transient brainstem ischaemia. The patient was eventually operated upon in a somewhat modified position that did not obliterate the cortical SEPs, but SEP loss occurred after about 2 hours of surgery and the patient was left with a profound hemiparesis. In two other patients,[196] peripheral and cortical SEP components were lost when patients were placed in the park-bench position for neurosurgical procedures. Because peripheral components as well as cortical waveforms were affected, it was possible to determine that the problem was peripheral in nature. SEPs were restored by repositioning an axillary roll under one patient and repositioning the arm of the other. This manoeuvre restored the SEPs so that the neurosurgical procedure could be monitored and it almost certainly prevented compression injuries of the brachial plexus.

SEP monitoring in the critical care unit

SEPs can provide information of diagnostic and prognostic value in patients with head trauma.[197–201] The central conduction time is particularly useful in comatose patients,[199] and the extent of CCT prolongation is usually correlated with the extent of injury. CCT decreases with clinical recovery. Only in patients with supratentorial injuries, however, are nearly normal SEPs correlated with good outcomes.[202] When the primary injury is in the brainstem, prolonged CCT can also be seen in patients who make good recoveries. Patients who die or have severe disability may initially have normal CCT on one side but absent cortical SEPs on the other. In one series of patients with severe head injuries (Glasgow coma scale 7 or less) EP measurements predicted outcome more accurately than did motor findings, intracranial pressure measurements, or pupillary light reaction.[203] In children with mild to moderately severe closed head injuries, long-latency SEP findings were correlated with long-term deficits in school performance.[204] A comparison of SEPs and motor EPs (MEPs) in 60 patients with head injury and 35 with non-traumatic coma showed that SEPs were better predictors of outcome than were MEPs.[205] All the patients in this series with bilaterally absent cortical SEPs died, whereas all with bilaterally preserved SEPs and CCT less than or equal to 6.5 ms survived. Approximately one-third of the patients in this series had bilaterally preserved electromyographic responses to transcranial stimulation but died. Asymmetries in CCT between the two hemispheres are usual in head-injured patients and are the most important criterion for recognizing the onset of cerebral ischaemia.

CCT is also prolonged in patients comatose from non-traumatic causes,[206] although experience is less in these patients than in patients with head injuries. Central hypothermia prolongs CCT,[199,207] but hypothermia of the limbs with central normothermia may not.[208] During barbiturate coma with EEG silence or marked burst suppression, CCT can be recorded but is prolonged.[44,199]

In patients who have suffered cardiac arrest, SEP measurements correlate well with outcomes.[209–211] Absence of cortical or thalamocortical activity after a cardiac arrest, while Erb's point or cervical recordings are intact, portends a poor outcome. If these findings persist for several days after cardiopulmonary arrest, it is very likely that the patient will die or remain in a persistent vegetative state.

Intermediate-latency SEPs elicited by stimulation of the median nerve (erroneously called long-latency potentials in the report) were 100 per cent accurate in predicting good versus poor outcomes among 66 patients who were successfully resuscitated from cardiac arrest but were still unconscious and mechanically ventilated.[211] The N70 peak was found between 74 and 116 ms in all 17 patients with good outcomes. In the 49 patients with poor outcomes, the N70 was either absent (*n*=35) or delayed to a latency between 121 and 171 ms (*n*=14). Among 57 asphyxiated infants, normal SEPs recorded within 3 days of birth were highly associated with normal outcomes at 18–24 months.[212]

SEP latencies and regional cerebral blood flow measurements in grade IV aneurysm patients differed statistically from those of patients in all other grades.[213,214] CCT measurements can be used in the critical care unit to detect the onset of vascular spasm and monitor its course.[215,216] In baboons subjected to ligation of the middle cerebral artery, CCT and regional cerebral blood flow showed a highly significant correlation.[185]

SEP monitoring may be particularly helpful in infants or comatose patients with known or suspected injuries of the spinal cord, because clinical assessment of cord function is severely hampered by inability of the patient to cooperate. The most important prognostic sign for spinal cord recovery is whether any function exists shortly after injury. Considerable recovery may occur over many months. Electrophysiological tests are also important for detecting peripheral nerve injuries in patients who have injuries of the spinal cord or who are comatose. Electrophysiological evaluation of the patient with spinal cord injury has been reviewed.[217]

Brainstem auditory evoked potentials (BAEPs)

Basic neurophysiology

Brainstem auditory evoked potentials (BAEPs) consist of a series of positive and negative waves that can be recorded in the far field from the vertex and ear, or from the vertex and a non-cephalic reference (Fig. 24.10). The BAEP is

Table 24.3 Purported generators of brainstem auditory evoked potentials

Peak	Generator
I	Acoustic nerve
II	Intracranial acoustic nerve and/or cochlear nucleus (medulla)
III	Superior olive (pons)
IV	Lateral lemniscus (pons)
V	Inferior colliculus (midbrain)
VI	Medial geniculate (thalamus)
VII	Thalamocortical radiations

Listed peaks are positive at the vertex. See text for references.
From Grundy[11] with permission.

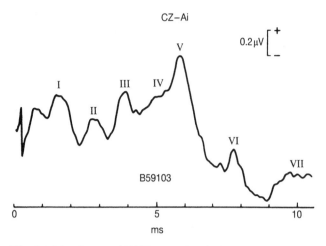

Fig. 24.10   A normal BAEP recorded during anaesthesia, with the vertex positive. That is, the grid-1 electrode is placed on the vertex and the grid-2 electrode on the earlobe. Activity which is positive (+) at the vertex is shown as an upward deflection. Purported generators of the labelled peaks are listed in Table 24.3.

subcortical in origin and provides no information about the functional integrity of the cerebral cortex (Table 24.3).

When sound waves strike the tympanic membrane, vibrations are transmitted through the middle ear by way of the ossicular chain to reach the inner ear. Pressure is applied to the cochlear fluid by means of stapes footplate displacement at the oval window. The fluid-filled chambers of the snail-shaped cochlea are separated by the basilar membrane, which supports the inner and outer hair cells in rows. When it is stimulated by sound waves, the basilar membrane moves like a travelling wave that starts at the base of the cochlea and moves toward the apex. The distance traversed from the base of the cochlea toward the apex is a function of the frequency of the stimulating sound, with low-frequency sounds travelling a greater distance than high-frequency sounds and therefore reaching farther toward the apex. The basilar membrane at the base of the cochlea responds mainly to very high frequencies while that at the apex responds to low frequencies. The mechanical stimulus of membrane motion is transduced by the receptor cells (inner and outer hair cells) into a form that can excite the auditory neurons that innervate them. Each inner hair cell connects to several nerve fibres, while one nerve fibre supplies several outer hair cells. The electrocochleogram (ECochG) consists of the responses of the cochlea and auditory nerve to sound. It includes the cochlear microphonic and summating potential, which originate in the cochlear receptor cells, as well as the compound action potential of the extracranial portion of the auditory nerve.

The first-order afferents of the auditory system are the spiral ganglion cells in the cochlea. Their central processes form the auditory portion of the eighth cranial nerve. These units discharge spontaneously but increase their discharges

above the baseline rate in response to sounds. They synapse in the cochlear nuclear complex, where most of the afferents are ipsilateral but some units are also affected by inputs from the contralateral ear. Wave I of the BAEP arises from the extracranial portion of the acoustic nerve. Wave II arises from the intracranial portion of the acoustic nerve and/or the cochlear nucleus (medulla).

The central portion of the auditory pathway is complex. The neurons of the cochlear nuclear complex receive considerable non-primary input from other neurons in the cochlear nucleus as well as from the superior olivary complex and the lateral lemniscus. The neurons of the cochlear nucleus connect to several nuclei on both sides, including the nuclei of the superior olivary complex through the nucleus of the trapezoidal body, and the inferior colliculus through the lateral lemniscus. The inferior colliculus transmits impulses to the medial geniculate body and thence to the auditory cortex. Wave III of the BAEP is thought to arise primarily in the superior olivary complex (pons); wave IV from the lateral lemniscus (pons); wave V from the inferior colliculus (midbrain); wave VI from the medial geniculate (thalamus); and wave VII from the thalamocortical radiations. These designations are clinically useful, although they represent considerable simplification of the actual situation. For example, it seems clear that each component of the BAEP as recorded from the vertex and ear represents some combination of afferent volleys in nerve fibres with nerve action potentials that may arise in more than one site.

Methods

Auditory stimulation

Ideally, baseline EPs will be recorded preoperatively with the same techniques and stimulators to be used intraoperatively. Pure-tone audiometry and testing of speech discrimination before anaesthesia and operation are thought to be essential by a few authors. At the very least, the ear canals and tympanic membranes should be examined with an otoscope.

Because many patients have some hearing loss preoperatively, broadband filtered clicks are preferable to tones unless pure-tone audiograms have been done preoperatively to test hearing at the frequencies to be used for monitoring. Clicks with predominant frequencies near 2000 Hz minimize the effects of high-frequency hearing loss on EPs. This is particularly important in elderly patients.

Most clinicians prefer to set the intensity of auditory stimulation in relation to the individual patient's hearing threshold. Transducers should be regularly calibrated, so that actual stimulus intensity is known in 'decibels peak equivalent sound pressure level' (dB pe SPL). The patient's auditory threshold ('sensation level', SL) for the stimulus to be used should be tested, and stimuli 60–70 dB greater are used for intraoperative EP monitoring (60–70 dBSL). 'Hearing level' (HL) refers to the mean sensation level established in a group of normal subjects with the equipment to be used for monitoring. A stimulus intensity set according to HL or dBSPL might be inadequate for a patient with partial hearing loss. Conductive hearing loss, if compensated for by increased stimulus intensity, produces little change in BAEP. By contrast, abnormalities of the eighth nerve or of central auditory pathways typically produce marked changes in the BAEP.

In the operating room, stereo headphones are usually impractical because they interfere with access to the surgical field during operations that pose risks of injury to auditory pathways. Transducers attached to moulded ear inserts are ideal, because the acoustic signal is delivered directly to the ear canal and cannot be obstructed by compression of soft tubing or earpieces. This transducer produces artefact in the nearby ear electrode, however, which can obscure as much as 2 ms of the initial portion of the BAEP when single-polarity stimuli are used. The stimulus artefact can be satisfactorily reduced by using clicks or tones of alternating polarity. That is, the initial sound pressure alternates between positive (compression) and negative (rarefaction), so that the stimulus artefact in the averaged EPs cancels out.

To avoid the stimulus artefact when single polarity clicks are used, foam ear inserts can be connected with plastic tubing to speakers a few inches from the ear. Rarefaction clicks have been most widely used, but compression clicks are also satisfactory. In a proportion of normal subjects, rarefaction clicks are converted to compression clicks while traversing the ear canal. Click polarity is less important than keeping the stimulus constant throughout monitoring.

When speakers are removed from the ear by even a few inches of tubing, allowance must be made in measuring EP peaks for the time elapsed during transmission of the click or tone through the tube connecting the speaker to the earpiece. Manufacturers should specify the duration of this time delay; it can also be directly measured in an audiology laboratory. Because changes from the individual patient's baseline values are the parameters of interest for monitoring, it is best to record baseline BAEP with the same transducers to be used intraoperatively.

Recording

Recording in at least two channels, with at least periodic stimulation of the ear contralateral to the site of surgery, is helpful not only for quality control but also for help in identifying peaks when waveforms are abnormal or unusual. If central auditory conduction time is prolonged, and wave V is not well seen, sweep times should be lengthened to 15 or even 20 ms so that a delayed but recordable peak will not be missed. BAEP waveforms shown in this chapter show vertex positivities (or ear negativities) as upward deflections.

Intraoperatively, it is possible to record the compound nerve action potential directly from the acoustic nerve in the surgical field. With the recording electrode directly on the nerve, the action potential is in the near field and of greater amplitude than when recorded from surface electrodes. Although this technique is popular with some electrophysiologists, others find that it adds little to information gained non-invasively. Physical movement of the electrode causes noise that interferes with recording, but many fewer sweeps must be averaged to demonstrate the action potential recorded directly from the nerve.

Clinical experience with BAEP monitoring

BAEPs are used to monitor the functional integrity of the acoustic nerve and brainstem auditory pathways in patients having neurosurgical procedures in the posterior cranial fossa, and in comatose patients in critical care units.

Intraoperative BAEP monitoring

In a series of 54 neurosurgical operations in the cerebellopontine angle,[79] intraoperative alterations in these potentials were reported in 37 cases. An increase in wave V latency of 1.5 ms was considered abnormal, and more than half the patients in this series had changes of this magnitude or greater. In 22 of the 54 operations, BAEP changes were related to retraction of the cerebellum or brainstem. These changes progressed to virtual obliteration in six cases. Changes were also seen with the combination of relative hypotension and hypocapnia and with positioning for retromastoid craniectomy.[79,218] In six cases, the BAEP was virtually obliterated with positioning. Except for two patients in whom the acoustic nerve was deliberately sacrificed, the BAEP returned towards normal when the head was returned to a neutral position at the end of the procedure and hearing was preserved.

In 32 of the 37 cases in this series during which BAEP changes were reported intraoperatively, waveforms recovered well toward baseline in the operating room.[79] In the remaining five patients hearing was diminished preoperatively and BAEPs recorded before the induction of anaesthesia were abnormal but reproducible. Two of these patients lost the BAEP prior to surgical incision, perhaps because of local ischaemia related to anaesthesia and positioning in already compressed nerves. One patient underwent sectioning of the eighth nerve as treatment for tinnitus, and in the four others the eighth nerve had to be cut during resection of large acoustic neuromas. All five of the patients with irreversible loss of the BAEP were deaf in the affected ear after operation. In all 54 cases, the presence or absence of BAEP at the end of anaesthesia correctly predicted the presence or absence of auditory function postoperatively.

Deafness is a recognized risk of surgery for microvascular decompression of cranial nerves. Before the introduction of BAEP monitoring during these operations,[78,79] ipsilateral hearing loss was reported in 1–4 per cent of patients having decompression of the fifth cranial nerve for trigeminal neuralgia, and partial hearing loss occurred in as many as 20 per cent of these patients.[219] The risk of damaging the auditory nerve is greatest during operations to decompress the seventh cranial nerve for treatment of hemifacial spasm, because of the proximity of the seventh and eighth nerves.[220]

In 21 patients monitored during microvascular decompression of cranial nerves,[221] prolongations of wave I, III, and V absolute latencies as well as I–III and I–V interpeak latencies were predictable with retraction of the cerebellum. Transient obliteration of BAEP occurred in four cases, and the surgeons were warned only for disappearance of the BAEP. The BAEP is an exquisitely sensitive monitor of auditory function. When cranial nerve function is primarily at risk, warning can perhaps be withheld until the BAEP is lost. In contrast, because the brainstem is far more sensitive to ischaemia and hypoxia than the cranial nerve, earlier warning might be chosen when the brainstem or its perfusion is primarily at risk.

In another report, intraoperative BAEP recordings correctly predicted postoperative function of the auditory nerve after surgery in the cerebellopontine angle.[222] Ten of 66 patients developed delays greater than 1.5 ms in 'BAEP latencies' (peak or peaks not specified). These patients had diminished hearing postoperatively that recovered within 30 days. Six patients had irreversible loss of the BAEP past wave I. Of these, five had an apparently intact eighth nerve; but all six had profound hearing loss postoperatively. One patient in this series had diffuse brainstem injury related to uncontrollable cerebellar oedema and lost the BAEP past wave II intraoperatively. This patient failed to awaken from anaesthesia and died 1 month later. Another patient, who underwent resection of a meningioma from the lateral ventricle, had extensive cortical damage despite normal BAEP intraoperatively. In light of the fact that BAEPs are entirely subcortical in origin, this is not surprising.

Somewhat more disconcerting is the report of a brainstem stroke during microvascular decompression while BAEPs were normal intraoperatively. Postoperatively, however, auditory function was intact.[223] Clearly, BAEPs do not necessarily reflect function in all areas of the brainstem. They can be expected to show only those changes that affect the auditory pathways.

A number of reports describe the use of BAEP monitoring in attempts to preserve hearing during removal of acoustic neuromas.[224–228] A comparison of 90 patients who had BAEP monitored with 90 unmonitored historical controls, matched for tumour size and preoperative hearing, showed that in patients with tumours less than 2 cm in size monitoring was associated with a greater likelihood that hearing would be preserved and a greater likelihood that preserved hearing would be useful.[228]

BAEP in critical care

Brain-injured and comatose patients have been monitored in the intensive care unit with BAEP recordings. In 30 patients with severe head injuries, all those patients who recovered had normal BAEP throughout the clinical course, while patients who were severely disabled or vegetative had at least transient BAEP abnormalities.[229] In another 85 patients comatose after head trauma, BAEP findings also correlated with outcome status.[202] Both primary and secondary brainstem lesions affected BAEP. Most of the patients who died or survived with severe disabilities had abnormal BAEPs and abnormal SEPs. In 53 comatose patients who had simultaneous BAEP and intracranial pressure recordings, acute increases in intracranial pressure to levels greater than 40 mmHg were not always followed by changes in BAEP.[230] In these patients, prolongation of BAEP brainstem transmission time in response to elevations in intracranial pressure was almost always followed by brain death, while lack of BAEP change in association with increases in intracranial pressure was associated with a high probability of survival.

When no BAEP can be recorded from a comatose patient, it is necessary to demonstrate that the peripheral hearing mechanism is functional. This can be done by showing presence of wave I of the BAEP or by recording the electrocochleogram or cochlear microphonic. If none of these can be recorded, the patient might simply be deaf. Peripheral injuries to the auditory system are common in head injury.

Visual evoked potentials (VEPs)

Relatively little experience has been gained with monitoring VEPs in the operating room and intensive care unit. This may be partly explained by the remarkable complexity of the visual system and by logistic considerations.

Neurophysiology

Parallel processing in multiple channels is a key characteristic of the visual system.[231] At least seven parallel channels of ganglion cells are thought to process visual information in the primate retina. Different classes of neurons specialize in recognizing particular features of visual stimuli such as luminance, contrast, and colour. Specific groups of neurons in the retina project to separate specialized brain regions, beginning with different laminae of the lateral geniculate body (LGB). Impulses are transmitted from the LGB to area 17 in the occipital cortex and thence along multiple pathways to areas 18, 19, and the mid-temporal areas. The inferior temporal cortex is another important area for visual processing, and multiple links exist between different areas. It has been estimated that more than half the cerebral cortex is devoted primarily to visual function in primates and presumably in man, with at least nine visiotopically organized areas. Thus, visual responses involve large areas of the temporal and parietal lobes as well as the occipital lobe.

Different kinds of stimuli can be used to preferentially activate particular structures in the retina and associated visual pathways. For example, in a patient with impaired

night vision, full-field flash stimulation would be used to evaluate the rod retinal receptor system, while a pattern-reversal VEP, the most sensitive electrophysiological test for function of the optic nerve, would be appropriate in a patient with suspected multiple sclerosis. In retrochiasmatic disorders, VEPs elicited by full-field pattern stimulation may be normal even in the presence of dense homonymous hemianopia. Attempts to detect unilateral retrochiasmatic lesions with hemifield VEPs have met with varying success.

Methods

VEPs are widely distributed over the occipital and parietal areas. They can be recorded from multiple locations between the vertex and inion, with a reference anterior to the vertex or off the head.

In the diagnostic laboratory, the visual system is most often stimulated with a reversing chessboard pattern on a television screen. The reversing pattern activates edge receptors in the retina, and the potentials produced are more sharply defined, with better reproducibility within and among subjects, than the VEPs produced by flash stimulation (Figs 24.11 and 24.12). Designation of the type of stimulus is important for interpretation.[61] For pattern-reversal VEPs (P-VEPs or PREPs), various check sizes are used and are described as the number of degrees of visual arc subtended by a single square. This arc is a function of the check size on the television screen and the distance between the screen and the subject's eye. The subject must wear corrective lenses if these are needed and must

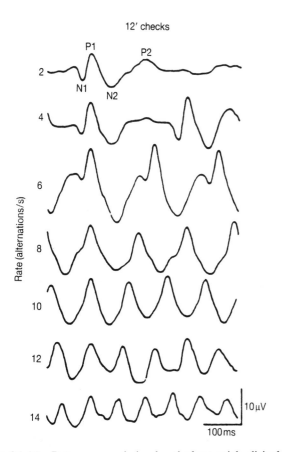

Fig. 24.11 Pattern-reversal visual evoked potentials elicited by stimulation at different alternation rates. The alternation rate per second is shown to the left of the records. Note the change from transient to steady state VEPs at 8 alternations per second. (Reproduced with permission from Sokol S. Visual evoked potentials. In: Aminoff MJ, ed. *Electrodiagnosis in clinical neurology.* Churchill Livingstone, New York, 1992: 441–66.)[356]

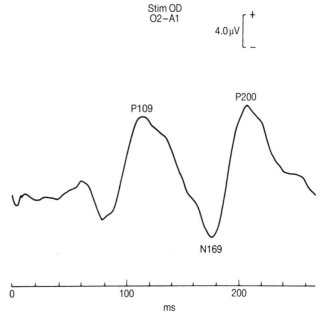

Fig. 24.12 Flash-evoked potential elicited by stimulation through closed eyelids, as for intraoperative VEP monitoring. (From Grundy,[353] with permission.)

cooperate by keeping the eyes open and focusing on a spot in the centre of the screen. Because the test requires the subject's cooperation, P-VEPs cannot be recorded in anaesthetized or comatose patients.

Flashes of light can be used to evoke VEPs (Fig. 24.12). Stroboscopic flash is sometimes used in the laboratory or intensive care unit, but, because of limited access to the patient's eyes, stroboscopic flash is rarely used in the operating room. Arrays of light-emitting diodes (LEDs) can also be used to elicit VEPs (LED-VEPs). The colour of the diodes should be noted, although it seems that any colour flash through closed eyelids will be perceived as red.

In the operating room, LED arrays mounted in swimmer's goggles are most often used for visual stimulation. A bland ophthalmic ointment and clear tape should be used to protect the eyes under the goggles, and great care must be taken to assure that the goggle housing does not slip off the bony ridge of the orbit to compress the globe of the eye. Pressure on the eye is a particular risk, because the operations during which VEPs are of greatest interest (those at the optic chiasm, in the anterior cranial fossa, and in the retrobulbar area) place the surgical field and the visual stimulators in immediate proximity. In such cases, any change in the VEPs that does not resolve immediately should prompt a careful check of the position of the visual stimulator. Prolonged pressure on the eye can produce central retinal artery thrombosis and permanent untreatable blindness in that eye. Some investigators have used LEDs mounted on scleral contact lenses of the sort that are commonly used to protect the eye during major craniofacial operations. This stimulator might also pose some risk of injury to the eye, particularly when the eye (and contact lens) are not accessible for inspection intraoperatively. Better visual stimulators are needed.

For best results with VEP monitoring in the operating room or critical care unit, the pupils should be maximally dilated. The visual response is directly related to 'retinal illuminance', the amount of light reaching the retina. This cannot be completely standardized when eyes are closed, but the pupillary constriction seen with narcotics can severely limit the quality of VEP recording. When mydriatic agents are used to dilate the pupil artificially, the ocular signs of neurological deterioration are temporarily lost. If the patient is awakened immediately after surgery for clinical neurological evaluation and is doing well, the dilated pupils may not be a serious problem. Otherwise, confusion about the patient's status could result. Individualized decisions must be made in consultation with the surgeon and anaesthetist.

Clinical applications of VEP monitoring

The neurophysiological complexities and logistic constraints described above undoubtedly explain the limited intraoperative use of VEP monitoring to date. Applications are essentially limited to procedures that might affect the anterior visual pathways or optic chiasm, such as operations in the anterior cranial fossa or on the pituitary gland. Because of the exquisite sensitivity of VEPs to anaesthetic agents, meaningful monitoring is useful only with careful maintenance of a pharmacological steady state and avoidance of high concentrations of volatile inhaled anaesthetics. Several interesting observations have been reported, including intraoperative improvement in abnormal VEPs followed by improved visual fields postoperatively.[12]

Somewhat more work has been done with VEPs in the critical care unit. In a series of 57 asphyxiated term infants who had SEPs and VEPs recorded during the first 3 days of life,[212] abnormal VEPs were always followed by abnormal neurological outcomes. In the same patients, normal SEPs virtually guaranteed a normal outcome. Outcome prediction was better with both VEPs and SEPs than with either alone. Abnormal VEPs were also reliable predictors of unfavourable outcomes after head injury.[203] Steady state VEPs recorded during neurosurgical procedures or in the intensive care unit showed changes within seconds of surgical and/or medical decompression of intracranial hypertension.[232]

Motor evoked potentials (MEPs)

Motor pathways can be stimulated with electrical current or magnetic transients applied to peripheral or cranial nerves or nerve roots, to the spinal cord, or to the motor strip of the cerebral cortex. Motor nerve conduction studies are important in both the diagnostic laboratory and the operating room. Intraoperatively, direct stimulation is used to identify branches of the facial nerve or other motor nerves.

Monitoring of motor pathways in the spinal cord is of particular interest during operations that pose risks of injury to anterior segments of the cord with little effect on the posterior cord. SEPs are conducted predominantly in the dorsal columns, and in chronic injury or with isolated microsurgical or vascular insults to the anterior cord, SEPs and posterior cord function may remain intact while motor function is lost. Intraoperative insults to the anterior cord, however, are almost always accompanied by at least temporary alterations to posterior column function and SEPs. The 'wake-up test'[6,7] is a reliable and simple method for testing motor function, but during some operations this test is not feasible.

Neurophysiology

The full length of a central motor pathway can be monitored electrophysiologically by stimulating the motor cortex transcranially either electrically or magnetically and recording nerve action potentials (NAPs) in peripheral nerves or evoked compound muscle action potentials (CMAPs) from muscles. For stimulation of the spinal cord, electrodes can be placed into the epidural or subarachnoid space blindly, with fluoroscopic control, or under direct vision.

When stimulation of motor pathways is done at the level of the cervical spinal cord or lower, recording of peripheral NAPs may be misleading. Antidromic sensory impulses recorded from peripheral nerves might be mistaken for motor NAPs, because some sensory nerve fibres traverse these pathways from the periphery to the cervical cord with no intervening synapses. If antidromic sensory potentials were mistaken for motor potentials, serious misinterpretations might be made and injury to the patient could result. This is not a problem with cortical motor stimulation, because no sensory nerves pass between points of stimulation and recording without intervening synapses. Recordings from peripheral nerves following rostral stimulation of the spinal cord (the so-called 'neurogenic motor evoked potential') were initially thought by some observers to reflect the function of motor pathways.[233] Subsequent studies in animals, however, showed that the sciatic nerve response to spinal cord stimulation was abolished by dorsal column transection.[234]

Methods

Traditional stimulation of motor nerves in the surgical field is done by placing the cathode in nearby subcutaneous tissue while the exploring anode is held in the surgeon's hand. Non-invasive stimulation of peripheral nerves is similar to that used for eliciting SEPs, except that the cathode is distal to prevent anodal block. The spinal cord can be stimulated by subarachnoid or epidural electrodes or by electrodes placed directly on the spinal cord. Evoked muscle activity can be monitored by direct observation or by recording the CMAPs. Inadvertent mechanical stimulation of the facial nerve by instruments in the surgical field produces electromyographic activity that can be used to provide audible feedback to the surgeon. This is similar to detection of unintended mechanical stimulation of spinal nerve roots during laminectomy by observation of the resulting muscle contractions.

Monitoring of the muscle activity evoked by any of these methods requires avoidance of complete neuromuscular blockade. Whether partial blockade lessens the reliability of MEP monitoring is not known. Complete resolution of this question would require extensive studies that considered, among other factors, the differing susceptibilities of different muscles to neuromuscular blocking agents. Recording of peripheral NAPs avoids the requirement for preservation of neuromuscular transmission. Only with cortical stimulation, however, can one be sure that the recorded NAP is motor rather than sensory.

Direct electrical stimulation of the motor cortex during open craniotomy has been extensively studied and is in widespread clinical use. Current levels of 0.25 to a few mA are sufficient. Five to 15 seconds of stimulation with pulses of 0.3 ms duration at 50–60 Hz have not been associated with adverse effects in patients. Similar techniques are used with epidurally implanted electrodes prior to seizure surgery.

Initial reports of transcranial electrical stimulation described an anodal plate 3 by 5.5 cm attached to the scalp over the motor cortex, with a curved 4 cm disc electrode on the hard palate as the cathode.[235] Biphasic pulses of about 20 mA and 17–28 Hz were applied for 0.75–1 ms. Averaged EPs were recorded directly from the spinal cord after 250–500 stimuli. Movement was seen only at the higher levels of stimulation. Intraoperative monitoring was done by using currents of 20–40 mA and 5–25 Hz, applied for several milliseconds. Multiple cathodal plates or a headband around the base of the skull permit stimulation with lower current levels. Although no seizure activity was described after this type of stimulation, hypertension and tachycardia were produced in some patients. These could usually be minimized by reducing the rate and duration of stimulation.

High voltages are required to transmit the necessary current across the skin and skull to the motor cortex. Single shocks of 500–700 V and 1000 mA have been used to demonstrate slowed conduction in patients with multiple sclerosis, and no apparent adverse effects have been reported.[236] The risk of cortical injury is reduced by minimizing current density. This can be done by using lower currents and larger electrodes.

Repeated electrical stimulation of the cerebral cortex with current densities sufficient to elicit after-discharges can produce an epileptic focus that continues to fire in the absence of further stimulation. In experimental animals, electrical stimulation, continued beyond the production of after-discharges, is used to produce a new epileptic focus in previously normal cortex. This phenomenon is called 'kindling'. During clinical electrocorticography, the EEG is monitored for after-discharges following direct cortical stimulation. If after-discharges are seen, subsequent stimulation can be reduced or eliminated so that kindling is prevented. Although seizures can be produced in epileptics, there are no reports of kindling due to direct electrical cortical stimulation in the clinical setting. Calculations of current densities at the surface of the cerebral cortex during transcranial electrical stimulation suggest that these are near those current densities achieved during direct cortical stimulation.

Transcranial magnetic stimulation is less well localized than electrical stimulation. In addition to exciting neurons that generate descending motor potentials, it activates excitatory intracortical axon collaterals to neighbouring pyramidal cells.[237] Inhibitory neurons projecting to the same pyramidal cells are also activated.[238] At low rates of stimulation (0.3–1 Hz) these connections balance each other and no horizontal spread of excitation is seen. With stimulation at higher frequencies, of which recently available stimulators are capable, the faster conduction in myelinated monosynaptic excitatory collaterals outstrips the slower conduction of inhibitory potentials. Temporal summation of excitatory postsynaptic potentials thus proceeds more rapidly than temporal summation of inhibitory postsynaptic potentials. The imbalance between excitatory and inhibitory corticocortical connections can produce widespread horizontal activation of the cortex sufficient that re-entering self-generating activity causes a seizure, even in subjects with no predisposing factors.

Studies in normal volunteers showed that the number of transcranial magnetic stimuli required to produce excitatory spread fell as stimulus frequency and intensity increased.[239] A healthy 35-year-old woman had a generalized convulsion lasting 57 seconds following the third train of 10 Hz stimuli, having had a total of 30 seconds of stimulation. The seizure was followed by post-ictal confusion for 20 minutes and flaccid paralysis of the right side of the body which resolved over 35 minutes. The only potential risk factors in this subject were in the family history. Uncomplicated febrile convulsions had occurred in a sister and a daughter, and another sister had a drug-induced convulsion (drug not specified). These investigators derived parameters estimating a maximum number of stimuli that could be given at various rates and intensities. According to their estimates, magnetic stimulation at 25 Hz and 200 per cent threshold can be safely applied for only four impulses. Further investigation is warranted, but for the present it appears that electrical stimulation may be safer than magnetic stimulation for producing MEPs.

It should also be noted that many commercially available magnetic stimulators produce noise sufficient to impair hearing in both patient and physician.[240,241] Susceptibility to noise-induced hearing loss may be increased by the effects of muscle relaxants on the protective middle ear acoustic reflex.[242] Three of nine volunteers in one study[239] had decreased hearing after only 240 seconds of rapid transcranial magnetic stimulation. This had resolved 4 hours later. Violent propulsion of ferromagnetic objects

within the field of the magnetic stimulator is an ever-present possibility, even though the field is relatively small.

Precautions for transcranial magnetic stimulation at rates of 1 Hz or faster have been suggested, based on studies in normal conscious volunteers.[239] Both patient and physician should wear earplugs. Precautions must be taken to prevent burns from metal electrodes that can be overheated by magnetic pulses.[243] Subjects should be informed of the possibility of seizures, particularly when there is a family history of seizure activity. As EEG after-discharges are not sufficiently reliable to predict a risk of seizures occurring, spreading muscle activity that increases with continued stimulation should be considered the first sign of epileptogenic activity. Data are available to estimate the risk at particular stimulus rates and intensities.[239]

Clinical applications of MEP monitoring

Monitoring of motor activity elicited by stimulation of cranial or motor nerves is well accepted and straightforward. Such monitoring is considered indispensable during operations on peripheral nerves and during mastoidectomy or procedures in the posterior cranial fossa. Auditory monitoring of electromyographic activity inadvertently produced by surgical instruments can rapidly alert the surgeon to possible impending injury of the facial nerve.[244,245] Six of ten patients so monitored during partial neuromuscular blockade had marked decreases in CMAPs intraoperatively and postoperatively had severe impairment of facial nerve function.[246] The cranial nerves that supply the extraocular muscles can be monitored during cranial base surgery by stimulating within the surgical field and recording from subdermal electrodes placed near the extraocular muscles.[247]

Studies of MEPs and spinal cord potentials produced by transcranial stimulation of the motor cortex have provided some disappointing information. These potentials are exquisitely sensitive to anaesthetic agents[248] (Fig. 24.13). Attempts to provide adequate anaesthesia without abolishing these potentials suggest that intravenous anaesthesia with nitrous oxide in concentrations no greater than 50–60 per cent may be the most suitable.[249–251] An initial report of MEP monitoring in 98 cases indicated that responses to stimulation of the motor cortex that were recorded from the spinal cord were relatively resistant to anaesthetics but changed so little in amplitude and latency with injury conditions that their use was compromised.[252]

MEPs elicited by direct electrical stimulation of the motor cortex and recorded over the lumbar spinal cord were compared to SEPs elicited by stimulation of the sciatic nerve and recorded over the thoracic spinal cord in 27 dogs subjected to 60 minutes of occlusion of the proximal descending aorta. SEPs were more sensitive to ischaemic spinal cord injury than MEPs, although MEPs were more

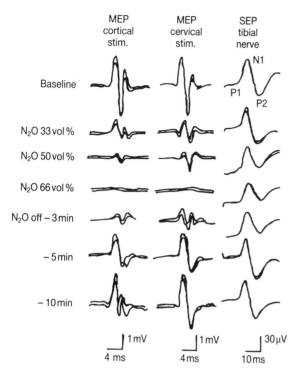

Fig. 24.13 Changes in rodent motor and somatosensory EP produced by nitrous oxide. MEPs represent the electromyographic activity elicited by cortical and cervical stimulation respectively. SEPs were recorded from cortical electrodes after stimulation of the tibial nerve. Note that MEPs are far more sensitive than SEPs to the anaesthetic. (From Zentner and Ebner,[248] with permission.)

specific. MEPs were lost in only nine of 20 dogs with cord injury and were not lost in any of the seven dogs that were neurologically normal. SEPs, by contrast, were lost in 19 of 20 paraplegic or paraparetic dogs but also in three of seven normal dogs. After reperfusion, MEPs in all the neurologically injured dogs returned to normal and the recovered MEPs were still present at 24 hours despite paraparesis or paraplegia. Thus, MEPs were highly specific (100 per cent) but had such a low sensitivity (46 per cent) that they were not reliable predictors of neurological injury.[165] This study indicates that SEPs are more sensitive than MEPs to loss of both motor and sensory function in the cord during aortic occlusion. SEPs should be more specific with the use of partial cardiopulmonary bypass or a bypass shunt to provide distal aortic perfusion and prevent ischaemia of the peripheral nerve.

MEP monitoring deserves investigation during operations for aneurysms at the bifurcation of the basilar artery. Unpredictable motor deficits are particularly likely in these patients, as motor areas are much closer than sensory pathways to the artery in question and may become ischaemic while sensory pathways are spared and SEPs unchanged.

SEPs also perform better than MEPs in predicting outcomes of coma. MEPs and SEPs were recorded in 60 patients with traumatic coma and 35 patients with non-traumatic coma.[205] All patients in both groups with bilaterally absent cortical SEPs died. When SEPs were preserved bilaterally and central conduction times were 6.5 ms or less, all patients survived. By contrast, 31 per cent of patients with traumatic coma and 39 per cent with non-traumatic coma had bilaterally preserved MEPs to transcranial stimulation but died. Only the bilateral absence of MEPs was associated with certain death. In 22 other patients who had coma of diverse aetiology, MEPs were closely related to outcomes and appeared to be better predictors than the Glasgow coma scale.[253] Of the six patients in this group who had bilaterally absent MEPs, however, one made a good recovery. Outcome prediction in this series was improved by using both SEPs and MEPs.

Six patients with severe chronic tetraparesis caused by head injury (three cases), basilar thrombosis (two cases), or global hypoxia (one case) had normal MEP thresholds and latencies following transcranial magnetic stimulation. MEP amplitudes were greater than normal. These findings did not correlate with the clinical severity of paresis, particularly the severe paresis of the muscles from which MEPs were recorded, the thenar and abductor hallucis muscles. Despite the different underlying causes of the tetraparesis in these patients, all were shown using CT scans to have diffuse atrophy of cerebral white matter. Thus, intact conduction of the fastest corticospinal efferents is not sufficient for normal voluntary muscle strength.[254]

In summary, exploration of the uses of MEPs for monitoring in the operating room and critical care unit is still at an early stage. Although these techniques seem intuitively appealing, early reports describe a number of disappointing results.

Intracranial pressure (ICP) monitoring

Neurons and glia require perfusion with oxygenated blood in order to function and survive. The cranial vault is a relatively closed space and therefore subject to increases in pressure if the volume of its contents rises beyond the limit of compensatory mechanisms. Monitoring of intracranial pressure in patients with mass lesions or intracranial catastrophes may facilitate appropriate clinical management when intracranial pressure levels are so high that they threaten cerebral perfusion.

Physiology

Cerebrospinal fluid (CSF) makes up only about 150 ml of the 1650 ml volume of the central nervous system contained within the cranial vault and the spinal column. Normally, about 500 ml of CSF is produced daily by the choroid plexus within the ventricles of the brain. This fluid circulates to the cisterna magna and the subarachnoid space before being absorbed by arachnoid villi in the sagittal venous sinus. Normal intracranial pressure (ICP) is usually less than 15 mmHg. Small changes in volume can be compensated for by movement of CSF out of the cranial vault into the spinal canal, but larger changes cannot. With an expanding supratentorial lesion, rostrocaudal deterioration in neurological function may be seen. If this process continues unabated, cerebral perfusion pressure falls, the brainstem and cerebellar tonsils herniate through the foramen magnum, and death results. Pressure differences between compartments of the intracranial vault can lead to other herniation syndromes, such as upward transtentorial herniation due to a rapidly expanding haematoma in the posterior fossa or uncal herniation beneath the falx with rapid formation of an intracerebral haematoma. Thus, pressure is not necessarily uniform throughout the cranial vault.

In the normal situation, small increases in intracranial volume produce only small and transient changes in intracranial pressure. As the limits of compensation are approached, however, intracranial compliance is much less and a small change in volume can cause a marked change in intracranial pressure (Fig. 24.14). Moment-to-moment changes in intracranial volume are often explained by changes in cerebral blood flow and blood volume. Autoregulation maintains cerebral blood flow constant between mean arterial pressures of approximately 50 and 150 mmHg, while cerebral blood flow is quite sensitive to either hypoxia or changes in the arterial partial pressure of carbon dioxide. ICP normally shows fluctuations with moment-to-moment changes in arterial blood pressure and respiration. Changes in ICP with hypercapnia or hypoxia can be dangerous in patients with decreased intracranial compliance. Patients with intracranial hypertension may show intermittent plateau waves, during which ICP is sustained for several minutes at levels of 50–100 mmHg.[255] This is a dire prognostic sign.

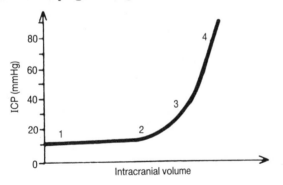

Fig. 24.14 Idealized pressure–volume curve for the intracranial vault. (From Shapiro,[357] with permission.)

Methods

Devices for monitoring ICP can be classified according to whether the transducer is located inside the cranial vault or externally.[256] Three devices using external fluid-coupled transducers will be discussed: the ventricular catheter, the subarachnoid screw or bolt, and the subdural catheter.

Measurement of ventricular fluid pressure by means of a catheter inserted into the lateral ventricle through a burr hole, fluid-coupled to an external transducer, is the standard to which other methods are compared.[257] More than three decades of experience have shown this technique to be relatively safe, although it is the most invasive of the methods currently in use for continuous monitoring of ICP. The catheter can be used to withdraw small amounts of CSF to reduce intracranial hypertension, or small amounts of preservative-free saline can be injected to test intracranial compliance. After subgaleal or subcutaneous tunnelling to help reduce the risk of infection, the catheter is attached to a strain gauge transducer and collecting system. Alternatively, the ventricular catheter may be attached to a subcutaneous reservoir which is then tapped with a small needle for pressure measurement. The reservoir can be used not only to measure pressures but also to inject drugs or withdraw CSF. Continuous monitoring of ICP is not practical with the reservoir technique. When intracranial hypertension has produced small slit-like ventricles or when mass lesions have distorted the normal anatomy of the lateral ventricles, placement of a ventricular catheter may be difficult. Also, use of this technique is associated with some small risk of epidural, subdural, intracerebral, or ventricular haemorrhage.

Another fluid-coupled ICP monitor with an external transducer system is the subarachnoid screw or bolt.[258] A threaded hollow screw is inserted through a burr hole just far enough for its unthreaded tip to protrude about a millimetre beyond the inner surface of the dura. The screw is then linked to the external transducer through a saline-filled tubing. Small leaks and frequent obstruction are the most common problems with this device.

A third type of external fluid-coupled transducer uses a catheter in the subdural space. The so-called 'throat cup catheter' is a ribbon-like catheter usually placed in the subdural space at the time of craniotomy.[259] A concavity on one side of the catheter near its tip is placed facing the brain surface beyond the limits of bony exposure. This 'cup' communicates with the saline-filled lumen of the catheter which, in turn, is linked to an external transducer. The most commonly encountered problems with this device are small leaks and obstructions of the catheter lumen. Attempts to place the cup catheter through a burr hole rather than at the conclusion of craniotomy may result in penetration into abnormally swollen or soft brain tissue.

Devices that place the transducer within the cranial vault

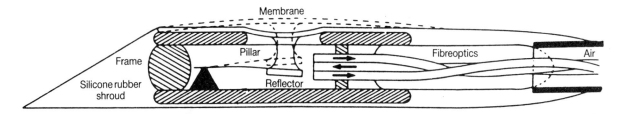

Fig. 24.15 Cross-section of the original Ladd intracranial pressure transducer. Deflection of mirror by ICP diminishes the amount of light transmitted along the fibreoptic columns. Counterbalancing air pressure maximizes light transmission when equal to ICP. (From Levin,[260] with permission.)

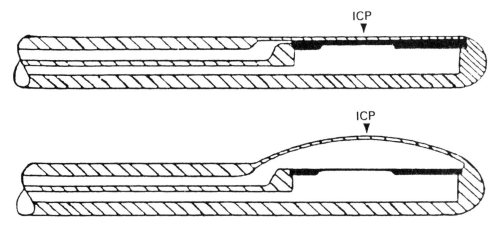

Fig. 24.16 Cross-section of Gaeltec intracranial pressure device. The top view shows how the transducer at the tip measures the difference between pressure transmitted across the membrane and atmospheric pressure. In the bottom view, zero determination is made *in vivo* by injecting 0.3 ml air to equalize pressure across the transducer. (From Barnett and Chapman[256].)

minimize problems with obstruction, leaks, and possibly infection. The Ladd, Gaeltec, and Camino monitors merit description.

The Ladd intracranial pressure monitor (Fig. 24.15) detects changes in ICP by deflection of a mirror which in turn alters return of fibreoptically transmitted light.[260,261] An external monitor detects the change in returned light and activates a servo system which increases the air pressure in the intracranial transducer just enough to return the mirror to its original position. The ICP is equal to the counterbalance in air pressure generated by the monitor. This transducer can be placed in either the epidural or the subdural space. It has the relative disadvantage that it cannot be calibrated or zeroed *in vivo*. Also, it provides no waveforms for visual monitoring. Erroneous readings may be due either to air leaks or to fractures in the fibreoptic cable. With epidural placement, falsely high readings can be produced if the sensor is not exactly parallel with the dura or if prolonged use causes a change in the compliance of the dura.

The Gaeltec ICP monitor[262,263] contains a strain gauge transducer at its tip, one side being coupled to atmospheric pressure by a rigid-walled tubing and the other reflecting ICP through a thin membrane (Fig. 24.16). The transducer continually generates an electrical signal that can be easily visualized on an external monitor. This transducer can be zeroed, *in vivo*, by injecting 0.3 ml of air to open the potential space between the transducer and the overlying membrane, equalizing pressure on the two sides of the strain gauge. The Gaeltec transducer is most often used in the epidural position, but its design permits subdural placement as well. Although this transducer can fail by either breakage of electrical connections or ruptures in the intracranial membrane, these problems are easily identified by failure to zero properly. Pneumocephalus has been detected following repeated attempts to zero this device by injecting incremental boluses of air when the membrane was ruptured.[264]

A comparison of the Gaeltec transducer in the subdural space to a fluid-filled catheter in the subdural space showed that 72 per cent of the Gaeltec subdural pressure readings corresponded to simultaneously measured intraventricular pressure, while only 44 per cent of the fluid-filled subdural catheter readings corresponded.[263] Erroneous readings from the Gaeltec system have been related to radio waves from a personal beeper system.[265]

The Camino miniaturized fibreoptic intracranial pressure monitor determines intracranial pressure directly from the amount of light reflected off a pressure-sensitive diaphragm at the tip of a fibreoptic catheter[266] (Fig. 24.17). This system does not require counterpressure or pneumatics. It can be inserted into the subdural space, into a ventricular catheter or into brain parenchyma. The transducer can be calibrated but not zeroed *in vivo*.[267] The Camino system is safe, accurate and reliable. Comparisons of readings from this monitor with values from an intraventricular catheter have for the most part shown very close correspondence.[266,268,269] A comparison of intraparenchymal

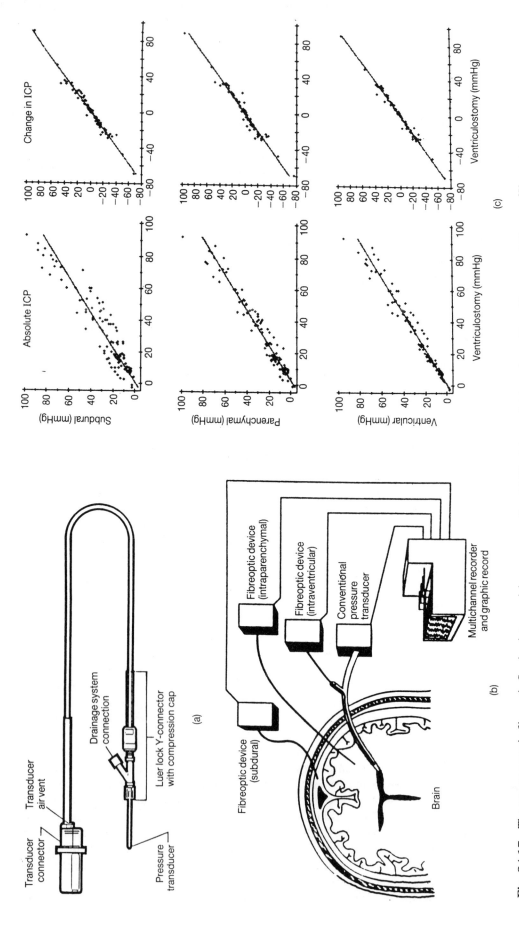

Fig. 24.17 The non-pneumatic fibreoptic Camino intracranial pressure monitor. (a) The intracranial transducer. (From Barnett and Chapman[256].) (b) Setup of Camino system for monitoring at different intracranial locations in dogs. (c) Comparison of data from Camino ICP monitor at three different sites to simultaneous measurements from an intraventricular catheter. ((b) and (c) from Crutchfield *et al.*[267] with permission.)

pressures measured with the Camino fibreoptic system and ventricular fluid pressures measured from a ventricular catheter in 10 patients showed that the fibreoptic intraparenchymal pressures exceeded ventricular fluid pressure in 66 per cent of paired measurements.[270] The mean difference was 9.2 ± 7.8 mmHg. Tissue pressures may not uniformly equal ventricular fluid pressures, even in the supratentorial region. The Camino fibreoptic device can be re-zeroed after insertion only by temporarily removing it from the patient. The sensor can fail due to breakage of optical fibres, but this is easily detectable.

Clinical applications of ICP monitoring

Monitoring of ICP is recommended for all patients who are comatose as a result of head injuries, even when initial computerized tomography shows no signs of intracranial hypertension.[271] Multimodality monitoring, including such measurements as blood flow velocity in the middle cerebral artery and jugular bulb venous oxygen saturation, discussed below, add valuable information. Still, cerebral perfusion pressure, derived from ICP and intra-arterial pressure measurements, is thought to be the most important monitoring parameter in patients with severe head injuries.[272]

Monitoring of ICP is also valuable in non-traumatic coma. It has been considered essential in managing patients having liver transplantation for fulminant hepatic failure.[273] Epidural transducers may be safest in these patients, even though they are less precise than intraventricular catheters or intraparenchymal fibreoptic pressure transducers. In one study of such patients, fatal haemorrhage occurred in 1 per cent of patients who had epidural ICP monitoring, whereas subdural and intraparenchymal devices were associated with fatal haemorrhage in 5 per cent and 4 per cent of patients respectively.[274]

Cerebral blood flow

Only a few of the numerous methods available for measuring or estimating cerebral blood flow (CBF) are practical in the operating room and the critical care unit. Positron emission tomography (PET), single photon emission computerized tomography (SPECT), stable xenon computerized tomography, and magnetic resonance techniques, with few exceptions, require that the patient be transported to another part of the hospital. Although they can be used repeatedly, they cannot be used continuously at the bedside.

CBF studies using inhaled, intravenous, or intracarotid ^{133}Xe have been used in both the operating room and the critical care unit.[275–278] The number of measurements that can be made in a short period of time is limited. This technique also requires handling of radioactive materials in the acute care setting and is cumbersome and expensive. CBF studies using radioactive xenon provide much valuable clinical information but are available in only a few centres.

Thermal diffusion[279,280] and laser Doppler flow-metry[281,282] can be used during open craniotomy, when probes can be placed directly on the brain, or in the critical care unit with probes inserted through burr holes. Laser Doppler flowmetry provides information about such a small volume of brain tissue (about 1 mm^3)[283] that its clinical relevance is not always clear. CBF monitoring by thermal diffusion also evaluates only a small part of the brain. Both methods, however, do permit continuous monitoring at the bedside. Also feasible at the bedside is ocular pneumoplethysmography, which can be used to assess ocular perfusion.[284] This may be helpful in patients with cerebrovascular disorders. All three of these methods are essentially experimental at present.

Transcranial Doppler (TCD)

Methods

Although Doppler ultrasound recording of blood flow velocity has long been done in the extracranial arteries, in the intracranial arteries during craniotomy, and through open fontanelles in infants, transcranial recordings awaited use of instruments with lower frequencies.[285] The skull is an important obstacle to penetration of ultrasound from conventional instruments, which use frequencies in the range of 5–10 MHz. Use of frequencies in the range of 1–2 MHz, and testing through the thin bone of the temporal region, made TCD ultrasound recording of flow velocity in the arteries at the base of the brain feasible. A range-gating system is used to acquire data at a distance from the probe which is selected by the operator. An acoustic lens focuses the ultrasonic beam to a width of approximately 4 mm at the focal length of the lens, which is approximately 5 cm (Fig. 24.18). Positioning of the probe is critical in most subjects and sometimes several minutes of searching for the maximum amplitude of the Doppler signals may be required. Once the maximum amplitude signal is identified, its spectrum is displayed on the instrument and measured by an operator-controlled cursor. The instrument then calculates the flow velocity at that level by using the

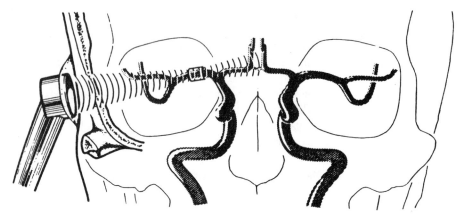

Fig. 24.18 Frontal view of the TCD probe directed toward the middle cerebral artery (MCA). The cylinder around the MCA indicates the observation region (sampling volume) for the Doppler recording. The distance from the middle of the cylinder to the probe corresponds to the depth setting. (From Aaslid *et al.*,[285] with permission.)

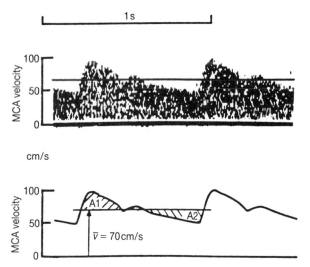

Fig. 24.19 Spectral display of the Doppler signal from the middle cerebral artery (MCA). The horizontal line through the spectra represents a cursor that can be moved up or down the display by the operator. The spectra are shown above and an outline of them below. The cursor was placed so that the areas A1 and A2 were judged equal, and the corresponding velocity was calculated by using the Doppler equation. (From Aaslid *et al.*,[285] with permission.)

Doppler equation. Figure 24.19 shows the cursor setting used to measure the mean flow velocity. The peak flow velocity is measured by simply moving the cursor to the highest point on the spectral display and the diastolic (minimum) flow velocity by moving the cursor to the lowest point of the curve describing height of the spectral display.

Although TCD techniques have the great advantages of being non-invasive and usable in the operating room or critical care unit at the bedside, they have important limitations. First, the signal cannot be acquired in all subjects. The incidence of inability to record the signal is greater with advancing age and in women. Thus, lack of a signal does not necessarily indicate abnormality. The test is much more important when positive results are obtained.

The second important limitation of this method is that velocity does not directly measure flow. The use of velocity to estimate flow depends on the critical assumption that the diameter of the vessel remains constant. Flow varies according to the square of the radius of the vessel. In some applications (see below), the opposite assumption is made, that flow remains constant or decreases. Under this assumption, an increase in flow velocity indicates narrowing (spasm) of the artery rather than increased cerebral blood flow. Results can be confusing when these assumptions are not met. Both assumptions obviously represent an oversimplification of the actual situation, and the degree of oversimplification varies both from setting to setting and from patient to patient.

A study of 1194 middle cerebral arteries in 597 Japanese patients showed that blood flow velocities could be recorded in only 77 per cent.[286] In women of age 70 years or older, successful bilateral recording of middle cerebral artery flow velocity was possible in only 17 per cent. A subset of 18 elderly patients were re-examined after increasing the standard transducer power of 100 mW/cm^2 to 400 mW/cm^2. Signals were then detected in two-thirds of those patients with recording barriers at the standard power.

Intraoperative measurements of cerebral arterial diameters were made through the operating microscope in 12 patients subjected to moderate alterations in blood pressure (30 ± 16 mmHg) and end tidal CO_2 (14 ± 6 mmHg).[287] The mean diameter change in the carotid, middle cerebral, and vertebral arteries was less than 4 per cent, but smaller arteries showed much larger changes in diameter. For example, diameters of the anterior cerebral artery and M2 segment of the middle cerebral artery changed in diameter as much as 29 per cent and 21 per cent, respectively. This suggests that blood flow velocities in the carotid, middle cerebral, and vertebral arteries more faithfully reflect flow than do changes in blood flow velocity in the anterior cerebral artery and the distal segments of the middle cerebral artery.

Blood flow velocities were recorded in 97 normal subjects 2–84 years of age.[288] Measurements were made at three

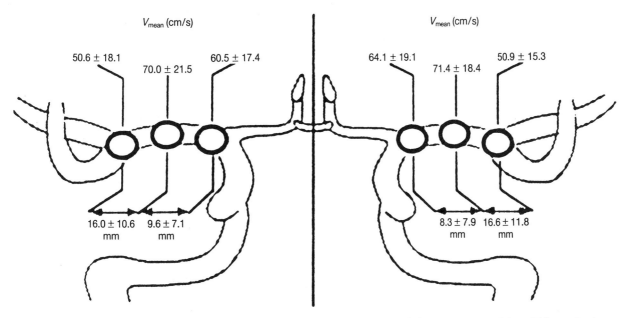

Fig. 24.20 Variability of flow velocities recorded at different sites along the course of the M1 segment of the middle cerebral artery in normal subjects. (From Laumer *et al.*,[288] with permission.)

different points among the M1 segment of the middle cerebral artery, with positioning based on the ultrasonic arteriogram obtained by three-dimensional Doppler imaging. Marked variability from point to point was seen, with maximum flow velocity of about 70 cm/s approximately 10 mm distal to the carotid bifurcation (Fig. 24.20).

A number of reports indicate good correlation between CBF and CBF velocity measured by TCD. Measurements of CBF velocity after injection of acetazolamide in normal subjects showed changes in flow velocity corresponding to changes that would be expected in CBF.[289] Flow velocity in the basilar artery fell significantly after rotation of the head to the side of a hypoplastic vertebral artery (artery asymmetry of greater than 75 per cent), but no such changes were seen in subjects with symmetrical vertebral arteries.[290] Middle cerebral artery flow velocity in patients on chronic dialysis varied with age and arterial oxygen content, as would be expected for CBF.[291] Direct comparisons of CBF and CBF velocity in animals showed a high degree of correlation under different anaesthetic conditions[292,293] and during angiotensin-induced arterial hypertension,[294] but the findings indicated that the amount of change in CBF with hypertension may have been underestimated.

Intravenous injection of stable air microbubbles bound to galactose microparticles has been used to enhance the TCD signal in patients. Significant enhancement was achieved in approximately 85 per cent of 20 subjects.[295] Minimal side effects occurred after 12.4 per cent of all injections (97 injections in 20 patients) and were reversible. The authors considered this a safe contrast medium, but many anaesthetists would have concerns about such injections.

Intraoperative applications of TCD monitoring

In the operating room, transcranial Doppler (TCD) monitoring has been used most frequently during CEA and during open heart surgery. Preoperative measurements of middle cerebral artery flow velocity in 90 patients scheduled for CEA showed that all patients who had major TCD changes during intraoperative carotid occlusion had flow velocity reduced preoperatively to zero by digital compression of the carotid artery.[296] In 114 cases of carotid surgery under regional anaesthesia, TCD and distal carotid artery stump pressure each detected 97 per cent of symptomatic ischaemic episodes.[297] The two methods together detected 100 per cent, and no intraoperative strokes occurred in this series. In another 238 operations, back pressure was correlated with Doppler flow velocity but TCD was more helpful in detecting ischaemia.[298] When there were no intraoperative ischaemic changes on TCD, stroke was significantly less likely without a shunt (2/175 no shunt vs 2/12 shunt). Flow velocities were always in the 'no ischaemia' category during shunting. In patients who had marked ischaemic changes on the EEG, those who had no shunt were much more likely to have strokes (4/9) than patients who had a shunt inserted (0/13). In 29 patients who were classified as having only mild ischaemia, the only stroke occurred in one of 20 patients who were not given a shunt. In another series of 44 patients, the criterion of a fall in mean middle cerebral artery flow velocity to less than 40 per cent of the baseline value detected the ischaemia associated with EEG flattening in all three patients in whom this was seen.[299] In these, flow velocity measurements were considerably more accurate than stump pressures for detecting ischaemia.

TCD monitoring is useful during CEA for detecting not only ischaemia but also such events as embolism or thrombosis.[300] Embolism is particularly frequent following shunt insertion and upon final restoration of flow.[301] A study of 130 consecutive CEAs with TCD and EEG monitoring detected 75 episodes of embolization in 55 patients.[302] EEG changes occurred during only two of the 75 embolic episodes. Another comparison of EEG and TCD monitoring in 23 patients permitted verification of shunt function in all of the 11 patients who had shunts.[303] Middle cerebral artery flow velocity during carotid occlusion was 14.7 ± 8.5 cm/s in the four patients who had EEG changes. By contrast, flow velocity averaged 24.1 ± 15.5 cm/s in patients whose EEG did not change.

A comparison of SEP monitoring with TCD monitoring in 97 patients undergoing 103 carotid operations showed that SEPs could always be recorded while TCD recording was possible in only 78 of the 97 patients (80 per cent).[304] In five patients mean flow velocity in the middle cerebral artery was reduced by more than 60 per cent without relevant SEP changes, but one patient with critical SEP findings had only 33 per cent reduction in flow velocity. A retrospective study of 1495 CEAs performed in 11 centres with TCD monitoring[305] found severe ischaemia in 7.2 per cent of operations, which cleared spontaneously in about half the cases. If the ischaemia persisted, the rate of severe stroke was very high, but shunting protected against stroke in such cases. If no ischaemia was detected, the stroke rate was higher with a shunt present but not so great as in patients with severe ischaemia who had no shunts.

TCD monitoring of CBF velocity can be performed during non-pulsatile normothermic or hypothermic cardiopulmonary bypass.[306,307] During normothermic bypass, autoregulation with variations in arterial pressure is preserved, but during hypothermia flow velocity becomes pressure-passive. The partial pressure of CO_2, however, continues to exert control over middle cerebral artery flow velocity even during hypothermia. A decline in neuropsychological function following cardiopulmonary bypass has been related to the number of emboli detected intraoperatively by TCD monitoring.[308]

Measurement of anterior fontanelle pressure in 25 neonates and infants 3–210 days of age permitted calculation of cerebral perfusion pressure.[309] In this patient population, CBF could not be detected by TCD when perfusion pressure fell to 9 (± 2 SD) mmHg. It was detected, however, when cerebral perfusion pressure was increased to 12 (± 1) mmHg.

Middle cerebral artery flow velocities were measured in 22 small children who had cardiac operations under profound hypothermia with or without cardiac arrest.[310]

Ten of the 12 children who had cardiac arrest had absent diastolic blood flow velocity immediately following cardiopulmonary bypass, whereas only one of the 10 patients who had low flow maintained during profound hypothermia showed this pattern. Diastolic flow velocity was restored 54–328 minutes after resumption of flow. The temporary loss of diastolic flow velocity was attributed to brain oedema and increased intracranial pressure in the patients who had circulatory arrest.

TCD applications in the critical care unit

One of the earliest applications of TCD monitoring is probably the most frequent application in the critical care unit still: detection of intracranial vasospasm following subarachnoid haemorrhage.[285] In this setting, one assumes that blood flow either remains constant or decreases so that an increase in velocity indicates a narrowing of blood vessels rather than increased flow. Serial TCD examinations, neurological evaluations, and CBF determinations were carried out in 21 patients with subarachnoid haemorrhage due to ruptured intracranial aneurysms.[311] Flow velocities were significantly elevated for the group of eight patients with spasm on the fourth through twelfth days after haemorrhage. Furthermore, the increases in flow velocities preceded the clinical signs of ischaemia. Correlations between flow velocities and angiographically measured diameters of the anterior cerebral artery and middle cerebral artery in 12 of the 21 patients, however, were not well correlated. Flow velocities in patients Hunt and Hess grade V were relatively low on admission.

A cautionary note is raised by more recent reports.[288,312] In a prospective study of 66 patients treated with nimodipine, those in grades IV and V had the lowest peak flow velocities.[288] A significant increase in peak flow velocity in the days preceding the onset of delayed ischaemic neurological deficit was found in only three of 11 cases. The other eight patients who developed spasm had constant high or constant or constant low velocities or decreasing velocities over time before onset of the deficit.

Flow velocities seem to be less helpful in detecting spasm of the anterior cerebral artery than spasm of the middle cerebral artery.[312] In 41 patients who had transcranial Doppler recording and cerebral angiography within the same 24-hour period, sensitivity for detection of spasm in the middle cerebral artery was 86 per cent and specificity 98 per cent. For the anterior cerebral artery, however, sensitivity was only 13 per cent while specificity was 100 per cent. Among patients with aneurysms of the anterior communicating artery, sensitivity was 14 per cent and specificity 90 per cent. The authors of this report emphasize the pitfalls of making treatment decisions on the basis of normal flow velocities in the anterior communicating artery. Reliability of flow velocities from the internal carotid artery and the circle of Willis may be problematic due to the unfavourable angle of insonation of ultrasound from the temporal area. Thus, flow velocities recorded from the middle cerebral artery are much more likely to be useful than those recorded from other arteries at the base of the brain.

Transcranial Doppler recordings made daily or every second day after subarachnoid haemorrhage for a 14-day period showed that the highest recorded velocity was greater in the 47 patients who developed delayed ischaemic neurological deficits, but this velocity was often recorded only after the onset of the deficit.[313] There was no significant difference between the groups in the peak velocities recorded before the onset of neurological deficit. Serial transcranial Doppler recordings could still be used to predict the delayed neurological deficit, however, by examining the rate of increase in flow velocity over time. A maximum velocity increase of 65 ± 5 cm/s per 24-hour period was seen in patients who developed subsequent deficits, compared to 47 ± 3 cm/s in those who did not. Serial recordings thus seem to be of much greater value than isolated recordings in attempts to predict the development of vasospasm after subarachnoid haemorrhage. Flow velocity recordings may also be valuable if done before and after interventional neuroradiological treatment of vasospasm. In two of four patients who had intracranial angioplasty, flow velocities remained elevated after the procedure despite initial anatomical correction of the spasm.[314] Repeat angiography showed new areas of vasospasm which were treated by papaverine infusion or repeat angioplasty. These treatments improved neurological status and decreased spasm as seen on arteriography. Persistence of elevated flow velocities following intracranial angioplasty suggests a need for repeat angiographic evaluation and possibly further therapy.

Various pulsatility indices have been derived in attempts to quantify aspects of the transcranial Doppler waveform other than simple velocity measurements. These include:

- the Gosling pulsatility index (GI):[315,316]
  $(V_{systolic} - V_{diastolic})/V_{mean}$;
- the first Fourier pulsatility index (FPI):[318] first Fourier coefficient/V_{mean};
- the Pourcelot resistance index (RI):[317]
  $(V_{systolic} - V_{diastolic})/V_{systolic}$.

In a recent report, however, the pulsatility indices of patients with delayed ischaemic deficits were completely within the range of those in patients without delayed ischaemic deficits.[318] The complexity of multiple factors affecting pulsatility indices may explain this unexpected finding. Pulsatility indices measured proximal to vasospastic segments reflect not only the resistance induced by vasospasm but also the unknown peripheral resistance. Distal resistances may well have been lowered to compensate for decreased flow in a narrowed proximal segment. TCD flow velocity recordings can detect neither significant narrowing of second- or third-order vessels nor functional anastomoses between different arterial territories.

Pulsatility indices, although they may indicate global changes in cerebrovascular resistance due to subarachnoid haemorrhage, cannot give sufficient information about local vascular changes to predict delayed ischaemic deficits accurately. The unknown variable is the volume flow in the short or penetrating branches that supply deep structures from which the first signs of ischaemia are usually noted. Concurrent and fluctuating abnormalities in arterial blood pressure, intracranial pressure, diameters of proximal and distal vessels, interstitial oedema, and operative trauma confuse the interpretation of pulsatility indices in these patients.

TCD recordings may be helpful in evaluating patients following acute strokes. In 48 patients seen within 5 hours of the onset of an ischaemic stroke, ultrasound evaluation by duplex scanning of the carotid artery, TCD recordings, cerebral angiography, and computerized tomography were performed.[319] Doppler signals were unobtainable transcranially in 6.25 per cent of these patients. In the patients who had successful TCD recording, this technique had a sensitivity of 80 per cent and a specificity of 90 per cent in detecting intracranial arterial obstructions seen on angiography. For obstructions of the internal carotid artery and the M1 segment of the middle cerebral artery, sensitivity was 95 per cent and specificity 92 per cent. Twenty-five patients had comparisons of TCD recordings and single-photon emission computerized tomography (SPECT) within 6 hours of onset of acute hemispheric brain infarctions and again 1 week later.[320] Middle cerebral artery flow velocity was abnormal in only 15 of the 25 cases, in comparison with abnormal cerebral blood flow on SPECT of the region supplied by the middle cerebral artery. Although there was a highly significant correlation between flow velocity in the middle cerebral artery and perfusion in the middle cerebral artery region ($r = 0.63$, $P < 0.001$) and between flow velocity and hemispheric perfusion ($r = 0.59$, $P < 0.002$), flow velocities did not reflect the state of cerebral perfusion in every case.

In acute stroke of undetermined cause, TCD techniques may be useful for diagnosing paradoxical embolization through a patent foramen ovale. A 35-year-old man with no obvious aetiology for an acute stroke had microbubbles in the middle cerebral artery detected with TCD ultrasound examination.[321] Gaseous emboli have been demonstrated during deep-sea diving, and their presence correlates with the occurrence of decompression sickness.[322,323] Although gaseous emboli produce the sharpest Doppler signals, solid emboli can also be detected. A computer algorithm has been developed that distinguishes between artefact and embolic signals with a sensitivity of 97 per cent and a specificity of 97 per cent.[324]

Patients with head trauma may have several abnormalities that can be monitored non-invasively with TCD recordings. These include increased ICP,[325–327] ischaemia,[328] intracranial haematomas,[326] and subarachnoid haemorrhage with attendant cerebral vasospasm.[329–332] Severe brain oedema accompanied by rapid increases in intracranial pressure and decreases in blood flow velocity with increased pulsatility indices may also be seen.[333] Thus, either increased or decreased flow velocities may be abnormal and each recording must be interpreted in its clinical context. In pyogenic meningitis in children, a close correlation between resistance index and mean arterial pressure may suggest a loss of cerebrovascular autoregulation.[334] In survivors of acute bacterial meningitis, the final resistance index is lower than that measured upon admission. In contrast to the marked increases in initial blood flow velocities seen in patients with bacterial infections, the flow velocities in the basal cerebral arteries are not elevated in patients with uncomplicated viral infections.[335]

Serial TCD recordings in five patients with increased intracranial pressures who went on to brain death showed a progressive reduction in flow velocities as intracranial pressure increased and cerebral perfusion pressure decreased. Diastolic flow velocities were diminished to a greater extent than systolic flow velocities, until finally pendular flow occurred with flow away from the probe during diastole.[336]

Cerebral metabolism

Jugular venous oxygen saturation (SjO₂)

Intermittent measurements of oxygen tension or saturation in blood drawn from the jugular bulb by the surgeon were used in the past to assess the adequacy of cerebral perfusion during carotid occlusion. The method fell into disuse after it became clear that regional ischaemia, even when severe, was often missed.[337] The continuous monitoring of SjO_2 made possible by currently available oximetric fibreoptic catheters, however, has rekindled interest. General principles of monitoring venous oxygen saturation continuously are addressed elsewhere in this text. SjO_2 monitoring can be used to detect widespread acute changes in cerebral oxygenation.[338] Falls in SjO_2 suggest hypoxia or ischaemia;[339–342] elevations suggest clinically important hyperperfusion.[343] When taken alone, however, an increase or decrease in SjO_2 may be either good or bad.

Simultaneous determinations of oxygen saturations from both left and right internal jugular veins were made in 32 patients with head injuries.[344] Repeat measurements showed a high degree of accuracy and reliability, but considerable biological variation between the two sides was seen in a significant number of patients. A 95 per cent probability was found that between 30 and 64 per cent of patients could have had, at some point during monitoring, a difference greater than 15 per cent between saturations in the two internal jugular veins. This difference might have affected treatment decisions. For example, a venous saturation of 60 per cent is within normal limits, but when SjO_2 falls below 45 per cent intervention is urgently indicated. It was not possible in these patients to predict which jugular vein would show the greater degree of desaturation. In two of the 32 patients, the side with higher saturation levels initially had the lower saturations later in the clinical course.

As placement of fibreoptic catheters in both internal jugular veins cannot be recommended as a routine, two rules of thumb have been suggested for determining the side of insertion. By the first, an attempt is made to catheterize the jugular vein thought to have greatest flow; or, when there seems to be no difference, the right jugular vein is chosen because cannulation is both simpler and safer. Alternatively, the sizes of the jugular foramina can be assessed on the computerized tomographic images that will already be available in these patients. The side with the larger foramen is chosen. A functional test has also been suggested, assessing the relative effect on intracranial pressure produced by compressing each internal jugular vein.[341,345] The location of brain injury (right vs left, superficial vs deep) is not a reliable guide to selection of the most appropriate jugular vein for catheterization. Regional cerebral blood flow differs by at least 25 per cent from the global value in almost half of patients with severe head injuries.[346]

Despite these recognized limitations of SjO_2 monitoring, it seems that multivariate optimization of global cerebral oxygenation, based on global cerebral delivery and extraction, may be clinically important for a variety of intracranial disorders that are predominantly diffuse and acute.[347] Trends in jugular venous oxygen saturation values are more useful than isolated single readings, so that continuous monitoring with fibreoptic oximetry represents an important advance in monitoring techniques for patients with global changes in cerebral oxygenation. Care must be taken, however, to interpret data in the appropriate clinical context and to harbour a healthy scepticism when values do not fit the clinical picture.

Near-infrared spectroscopy (NIRS)

Another non-invasive method for monitoring brain oxygenation, which so far has been relatively little used, is near-infrared spectroscopy. Multiple narrow-band wavelengths of light, generated either by refraction and filtering light from an incandescent source or by multiple lasers, are delivered by fibreoptic strands to the skin surface and directed through a segment of the brain. The lasers most often used are not hazardous to the skin. In neonates and infants, the scalp and skull are thin enough and the head small enough that the delivering fibreoptic bundle (optode) can be placed over the temporal area on one side of the head and the collecting fibreoptic bundle over the opposite temple. The intervening tissues scatter the light considerably, but most of the absorption of light in this pathway is by oxygenated haemoglobin, deoxygenated haemoglobin, and oxidized cytochrome aa_3. The absorption spectra of these three substances were determined in experiments done in animals. After signal processing and mathematical analyses, data are displayed on four channels that represent deoxyhaemoglobin, oxyhaemoglobin, total volume of haemoglobin (taken as a surrogate measurement for blood volume), and cytochrome aa_3.[348]

Relative rather than absolute units are displayed on some instruments. An approximate quantification, however, can be obtained by calculating the length of the differential light path. The mean distance traversed by the photons is determined by measuring the time of flight of an optical pulse traversing the head from one fibreoptic bundle to the other. The differential path length (mean distance travelled by the photons divided by the interoptode distance) can then be used to estimate quantitative values for oxyhaemoglobin, deoxyhaemoglobin, cerebral blood volume, and cerebral blood flow. For valid use of this estimate, the instrument should be able to separate precisely the light absorption of cytochrome aa_3 from oxyhaemoglobin. Determinations based on these measurements are valid only for that part of the tissue actually traversed by the light beam. Presently available NIRS instruments, particularly those for use in adults, have a number of limitations. Artefact is common, particularly with movements of the head.

Preterm neonates are ideal candidates for NIRS monitoring. Term infants, older children, and adults have thicker skulls and larger heads that limit transmission of the photons. When the distance between the temples is greater than 7–9 cm, the light transmitter and receiver must be placed closer together. In this case, light is reflected rather than transilluminated.[349]

Special problems hamper interpretation of reflectance NIRS. Because the thicknesses of scalp and skull vary considerably, appropriate focusing to facilitate separation of the superficial tissue signals from signals reflecting brain tissue is quite uncertain. Also, only a small wedge of cortex can be monitored, compared to a much greater segment of the frontal lobes in preterm neonates. When values are derived from reflected rather than transilluminated light, it

Table 24.4 Factors causing near-infrared spectroscopy changes

Parameter	Factors causing a decrease
Oxyhaemoglobin	A decrease in oxygen saturation
	Decreased cerebral blood flow
	Decreased concentration of oxyhemoglobin
Deoxyhaemoglobin	An increase in oxygen saturation
	Relief of obstruction to cerebral venous return
	Decreased inflow of desaturated blood
	Decreased concentration of deoxyhaemoglobin
Blood volume	A decrease in cerebral blood flow
	Relief of obstruction to cerebral venous return
	Decreased amount of haemoglobin
Cytochrome aa_3 oxidation	Decreased oxygen availability
	Decreased metabolic activity
	Increased electrons on respiratory chain

Increases in oxyhaemoglobin, deoxyhaemoglobin, blood volume, and cytochrome aa_3 oxidation occur with the opposite conditions. Reprinted with permission from *Journal of Clinical Monitoring*, **7**, 330, 1991.[358]

is not possible to determine with certainty how much light reached the brain. This complicates estimates based on absorption, which are derived from measurements of light returned to the collecting optode. A recent study in adults of a commercially available instrument showed no change in intracerebral oxygenation while middle cerebral artery flow was doubled with deliberate hypercapnia.[350] This suggests that this particular device does not provide data on intracerebral oxygenation. It may be that placing the optodes farther apart would help.

Even in preterm neonates, several different factors may be responsible for similar changes in NIRS data (Table 24.4). The information provided at present by near-infrared spectroscopy can only be interpreted within a specific clinical context, and appropriate interpretation may be difficult.

Other methods

Other methods that have been suggested for monitoring cerebral metabolism include simultaneous optical monitoring of localized cerebrocortical NADH and brain microvessel haemoglobin oxygen saturation,[351] and reflectance pulse oximetry measurements made from the retinal fundus by using a specially constructed pulse oximeter probe mounted in a contact lens.[352] These methods are still experimental. Metabolic studies using SPECT, PET and magnetic resonance spectroscopy (MRS) are of great interest but cannot be used for continuous monitoring in the acute care setting.

Conclusions

New developments in monitoring technology, in concert with our rapidly growing understanding of the function and malfunction of the central nervous system, offer exciting new ways of monitoring the brain and spinal cord. Many of these methods are still experimental, but some provide important information that may well have a dramatic effect on the outcome of patient care in the operating room and critical care unit. Nevertheless, clinical assessment of the awake patient, when this is possible and reasonable, is still the most important technique available for monitoring the central nervous system. The additional methods described in this chapter become important primarily, although not solely, when clinical assessment is severely constrained by anaesthesia or coma.

REFERENCES

1. Haglund MM, Berger MS, Shamseldin M, Lettich E, Ojemann GA. Cortical localization of temporal lobe language sites in patients with gliomas. *Neurosurgery* 1994; **34**: 567–76.
2. Gignac E, Manninen PH, Gelb AW. Comparison of fentanyl, sufentanil and alfentanil during awake craniotomy for epilepsy. *Canadian Journal of Anaesthesia* 1993; **40**: 421–4.
3. Silbergeld DL, Mueller WM, Colley PS, Ojemann GA, Lettich E. Use of propofol (Diprivan) for awake craniotomies: technical note. *Surgical Neurology* 1992; **38**: 271–2.
4. Peitzman AB, Webster MW, Loubeau JM, Grundy BL, Bahnson HT. Carotid endarterectomy under regional (conducive) anaesthesia. *Annals of Surgery* 1982; **196**: 59–64.
5. Steed DL, Peitzman AB, Grundy BL, Webster MW. Causes of stroke in carotid endarterectomy. *Surgery* 1982; **92**: 634–41.
6. Vauzelle C, Stagnara P, Jouxvinroux P. Functional monitoring of spinal cord activity during spinal surgery. *Clinical Orthopaedics and Related Research* 1973; **93**: 173–8.
7. Sudhir KG, Smith RM, Hall JE, Hansen DD. Intraoperative awakening for early recognition of possible neurologic sequelae during Harrington-rod spinal fusion. *Anesthesia and Analgesia* 1976; **55**: 526–8.
8. Jennett B, Teasdale G. Aspects of coma after severe head injury. *Lancet* 1977; **1**: 878–81.
9. Botterell EH, Lougheed WM, Scott JW, Vandewater SL. Hypothermia, and interruption of carotid, or carotid and vertebral circulation, in the surgical management of intracranial aneurysms. *Journal of Neurosurgery* 1956; **13**: 1–42.
10. Hunt WE, Hess RM. Surgical risk as related to time of intervention in the repair of intracranial aneurysms. *Journal of Neurosurgery* 1968; **28**: 14–20.
11. Grundy BL. Monitoring of sensory evoked potentials during neurosurgical operations: methods and applications. *Neurosurgery* 1982; **11**: 556–75.
12. Grundy BL. Intraoperative monitoring of sensory-evoked potentials. *Anesthesiology* 1983; **58**: 72–87.
13. Friedman WA, Theisen GJ, Grundy BL. Electrophysiologic monitoring of the nervous system. In: Stoelting RK, Barash PG, Gallagher TJ, eds. *Advances in anesthesia.* Chicago: Year Book Medical Publishers, 1988; 231–89.
14. Speckmann E-J, Elger CE. Introduction to the neurophysiological basis of the EEG and DC potentials. In: Niedermeyer E, Lopes Da Silva F, eds. *Electroencephalography: basic principles, clinical applications, and related fields.* Baltimore: Williams & Wilkins, 1993: 15–26.
15. Jasper HH. The ten-twenty electrode system of the International Federation. *Electroencephalography and Clinical Neurophysiology* 1958; **10**: 371–3.
16. Lopes da Silva F. EEG analysis: theory and practice. In: Niedermeyer E, Lopes Da Silva F, eds. *Electroencephalography: basic principles, clinical applications, and related fields.* Baltimore: Williams & Wilkins, 1993: 1097–123.
17. Nuwer MR. Quantitative EEG: I. techniques and problems of frequency analysis and topographic mapping. *Journal of Clinical Neurophysiology* 1988; **5**: 1–43.
18. Nuwer MR. Quantitative EEG: II. frequency analysis and topographic mapping in clinical settings. *Journal of Clinical Neurophysiology* 1988; **5**: 45–85.
19. Demetrescu M. The aperiodic character of the EEG: a new approach to data analysis and condensa. *Physiologist* 1975; **18**: 189.
20. Dumermuth G, Huber PJ, Leiner B, Gasser T. Analysis of the interrelations between frequency bands of the EEG by means of the bispectrum: a preliminary study. *Electroencephalography and Clinical Neurophysiology* 1971; **31**: 137–48.
21. Vernon J, Bowles, S, Sebel PS, Chamoun N. EEG bispectrum predicts movement at incision during isoflurane or propofol anesthesia. *Anesthesiology* 1992; **77**: A502.
22. Nakamura M. Waveform estimation from noisy signals with variable signal delay using bispectrum averaging. *IEEE Transactions on Biomedical Engineering* 1993; **40**: 118–27.
23. Pradhan N, Dutt DN. A nonlinear perspective in understandig the neurodynamics of EEG. *Computers in Biology and Medicine* 1993; **23**: 425–42.
24. Law SK, Nunez PL, Wijesinghe RS. High-resolution EEG using spline generated surface laplacians on spherical and ellipsoidal surfaces. *IEEE Transactions on Biomedical Engineering* 1993; **40**: 145–53.
25. Fell J, Roschke J, Beckmann P. Deterministic chaos and the 1st positive Lyapunov exponent: a nonlinear analysis of the human electroencephalogram during sleep. *Biological Cybernetics* 1993; **69**: 139–46.
26. Grundy BL, Heros R. Ischemic cerebrovascular disease. In: Matjasko J, Katz J, eds. *Clinical controversies in neuroanesthesia and neurosurgery.* Orlando: Grune & Stratton, 1986: 1–76.
27. Sundt TM, Sharbrough FW, Piepgrass DG. Correlation of cerebral blood flow and electroencephalographic changes during carotid endarterectomy: with results of surgery and

haemodynamics of cerebral ischaemia. *Mayo Clinic Proceedings* 1981; **56**: 533–43.

28. Blume WT, Ferguson GG, McNeill DK. Significance of EEG changes at carotid endarterectomy. *Stroke* 1986; **17**: 891–7.

29. Blume WT, Sharbrough FW. EEG monitoring during carotid endarterectomy and open heart surgery. In: Niedermeyer E, Lopes Da Silva F, eds. *Electroencephalography: basic principles, clinical applications, and related fields*. Baltimore: Williams and Wilkins, 1993: 747–56.

30. Symon L, Lassen NA, Astrup J, Branston NM. Thresholds of ischaemia in brain cortex. *Advances in Experimental Medicine and Biology* 1977; **94**: 775–82.

31. Astrup J, Symon L, Branston NM, Lassen NA. Cortical evoked potential and extracellular K^+ and H^+ at critical levels of brain ischemia. *Stroke* 1977; **8**: 51–7.

32. Tempelhoff R, Modica PA, Grubb Jr RL, Rich KM, Holtmann B. Selective shunting during carotid endarterectomy based on two-channel computerized electroencephalographic/compressed spectral array analysis. *Neurosurgery* 1989; **24**: 339–44.

33. Silbert BS, Koumoundouros E, Davies MJ, Cronin KD. Comparison of the processed electroencephalogram and awake neurological assessment during carotid endarterectomy. *Anaesthesia and Intensive Care* 1989; **17**: 298–304.

34. Kearse LA, Martin D, McPeck K, Lopezbresnahan M. Computer-derived density spectral array in detection of mild analog electroencephalographic ischemic pattern changes during carotid endarterectomy. *Journal of Neurosurgery* 1993; **78**: 884–90.

35. Young WL, Moberg RS, Ornstein E, *et al.* Electroencephalographic monitoring for ischemia during carotid endarterectomy: visual versus computer analysis. *Journal of Clinical Monitoring* 1988; **4**: 78–85.

36. Hanowell LH, Soriano S, Bennett HL. EEG power changes are more sensitive than spectral edge frequency variation for detection of cerebral ischemia during carotid artery surgery: a prospective assessment of processed EEG monitoring. *Journal of Cardiothoracic and Vascular Anesthesia* 1992; **6**: 292–4.

37. Stockard JJ, Bickford RG, Myers RR, Aung MH, Dilley RB, Schauble JF. Hypotension-induced changes in cerebral function during cardiac surgery. *Stroke* 1974; **5**: 730–46.

38. Levy WJ. Intraoperative EEG patterns: implications for EEG monitoring. *Anesthesiology* 1984; **60**: 430–4.

39. Bashein G, Nessly ML, Bledsoe SW, *et al.* Electroencephalography during surgery with cardiopulmonary bypass and hypothermia. *Anesthesiology* 1992; **76**: 878–91.

40. Coselli JS, Crawford ES, Beall Jr AC, Mizrahi EM, Hess KR, Patel VM. Determination of brain temperatures for safe circulatory arrest during cardiovascular operation. *Annals of Thoracic Surgery* 1988; **45**: 638–42.

41. Metz S, Slogoff S. Thiopental sodium by single bolus dose compared to infusion for cerebral protection during cardiopulmonary bypass. *Journal of Clinical Anesthesia* 1990; **2**: 226–31.

42. Arom KV, Cohen DE, Strobl FT. Effect of intraoperative intervention on neurological outcome based on electroencephalographic monitoring during cardiopulmonary bypass. *Annals of Thoracic Surgery* 1989; **48**: 476–83.

43. Edmonds HL, Griffiths LK, Vanderlaken J, Slater AD, Shields CB. Quantitative electroencephalographic monitoring during myocardial revascularization predicts postoperative disorientation and improves outcome. *Journal of Thoracic and Cardiovascular Surgery* 1992; **103**: 555–63.

44. Drummond JC, Todd MM,. Schubert A, Sang H. Effect of the acute administration of high dose pentobarbital on human brain stem auditory and median nerve somatosensory evoked responses. *Neurosurgery* 1987; **20**: 830–5.

45. Brenner RP. The electroencephalogram in altered states of consciousness. *Neurologic Clinics* 1985; **3**: 615–31.

46. Treiman DM. Generalized convulsive status epilepticus in the adult. *Epilepsia* 1993; **34**: S2–S11.

47. Rumpl E. [Electro-neurological correlations in early stages of posttraumatic comatose states. I. The EEG at different stages of acute traumatic secondary midbrain and bulbar brain syndrome (author's transl.)] EEG/EMG 1979; **10**: 148–57.

48. Moodley J, Bobat SM, Hoffman M, Bill PLA. Electroencephalogram and computerised cerebral tomography findings in eclampsia. *British Journal of Obstetrics and Gynaecology* 1993; **100**: 984–8.

49. Streletz LJ, Bej MD, Graziani LJ, *et al.* Utility of serial EEGs in neonates during extracorporeal membrane oxygenation. *Pediatric Neurology* 1992; **8**: 190–6.

50. Rivierez M, Landau-Ferey J, Grob R, Grosskopf D, Philippon J. Value of electroencephalogram in prediction and diagnosis of vasospasm after intracranial aneurysm rupture. *Acta Neurochirurgica* 1991; **110**: 17–23.

51. American Electroencephalographic Society. Minimum technical standards for EEG recording in suspected cerebral death. *Journal of Clinical Neurophysiology* 1994; **11**: 10–13.

52. Goodman JM, Heck LL. Confirmation of brain death at bedside by isotope angiography. *Journal of the American Medical Association* 1977; **238**: 966–8.

53. Desmedt JE, Brunko E. Functional organization of far-field and cortical components of somatosensory evoked potentials in normal adults. *Progress in Clinical Neurophysiology* 1980; **7**: 27–50.

54. Walter WG. Evoked response general. In: van Leeuwen WS, Lopes da Silva FH, Kamp H. *Handbook of electroencephalography and clinical neurophysiology: evoked responses*. Amsterdam: Elsevier Scientific Publishing Co, 1975: 20–32.

55. Daube JR. Nerve conduction studies. In: Aminoff MJ, ed. *Electrodiagnosis in clinical neurology*. New York: Churchill Livingstone, 1992: 283–326.

56. Friedman WA, Theisen GJ, Grundy BL. Electrophysiologic monitoring of the nervous system. In: Stoelting RK, Barash PG, Gallagher TJ, eds. *Advances in anesthesia*. Chicago: Year Book Medical Publishers, 1988: 231–89.

57. Diamantopoulos E, Zander OP. Excitability of motor neurons in spinal shock in man. *Journal of Neurology, Neurosurgery and Psychiatry* 1967; **30**: 427–31.

58. Little JW, Halar EM. H-reflex changes following spinal cord injury. *Archives of Physical Medicine and Rehabilitation* 1985; **66**: 19–22.

59. Kimura J, Butzer JF. F-wave conduction velocity in Guillain–Barré syndrome. *Archives of Neurology* 1975; **32**: 524–9.

60. King D, Ashby P. Conduction velocity in the proximal segments of a motor nerve in the Guillain–Barré syndrome.

Journal of Neurology, Neurosurgery and Psychiatry 1976; **39**: 538–44.

61. American Electroencephalographic Society. Guidelines on evoked potentials. *Journal of Clinical Neurophysiology* 1994; **11**: 40–73.

62. American Electroencephalographic Society. Guidelines for intraoperative monitoring of sensory evoked potentials. *Journal of Clinical Neurophysiology* 1994; **11**: 77–87.

63. Cooper R, Osselton JW, Shaw JC. *EEG technology.* London: Butterworths, 1980: 1–344.

64. Nagelkerke NJD, deWeerd JPC, Strackee J. Some criteria for the estimation of evoked potentials. *Biological Cybernetics* 1983; **48**: 27–33.

65. Grundy BL. Sensory-evoked potentials. In: Boulton AA, Baker GB, Boisvert DPJ, eds. *Neuromethods-8.* Clifton, New Jersey: Humana Press, 1988: 375–433.

66. Achor LJ, Starr A. Auditory brain stem responses in the cat. I. Intracranial and extracranial recordings. *Electroencephalography and Clinical Neurophysiology* 1980; **48**: 154–73.

67. Achor J, Starr A. Auditory brain stem responses in the cat. II. Effects of lesions. *Electroencephalography and Clinical Neurophysiology* 1980; **48**: 174–90.

68. Desmedt JE, Noel P. Average cerebral evoked potentials in the evaluation of lesions of the sensory nerves and of the central somatosensory pathway. In: Desmedt JE, ed. *New developments in electromyography and clinical neurophysiology.* Basel: S. Karger, 1973: 352–71.

69. Anthony PF, Durrett R, Pulec JL, Hartstone JL. A new parameter in brain stem evoked response: component wave areas. *Laryngoscope* 1979; **89**: 1569–78.

70. Boston JR. Spectra of auditory brainstem responses and spontaneous EEG. *IEEE Transactions on Biomedical Engineering* 1981; **28**: 334–41.

71. Wong PK, Bickford RG. Brainstem auditory evoked potentials: the use of noise estimate. *Electroencephalography and Clinical Neurophysiology* 1980; **50**: 25–34.

72. Keller I, Madler C, Schwender D, Poppel E. Analysis of oscillatory components in perioperative AEP-recordings: a nonparametric procedure for frequency measurement. *Clinical Electroencephalography* 1990; **21**: 88–92.

73. Donchin E, Herning RI. A simulation study of the efficacy of stepwise discriminant analysis in the detection and comparison of event related potentials. *Electroencephalography and Clinical Neurophysiology* 1975; **38**: 51–68.

74. Duffy FH, Bartels PH, Burchfiel JL. Significance probability mapping: an aid in the topographic analysis of brain electrical activity. *Electroencephalography and Clinical Neurophysiology* 1981; **51**: 455–62.

75. Cerutti S, Chiarenza G, Liberati D, Mascellani P, Pavesi G. A parametric method of identification of single-trial event-related potentials in the brain. *IEEE Transactions on Biomedical Engineering* 1988; **35**: 701–11.

76. Tang Y, Norcia AM. Improved processing of the steady-state evoked potential. *Electroencephalography and Clinical Neurophysiology* 1993; **88**: 323–34.

77. Celesia GG. Steady-state and transient visual evoked potentials in clinical practice. *Annals of the New York Academy of Sciences* 1982; **388**: 290–305.

78. Grundy BL, Lina A, Procopio PT, Jannetta PJ. Reversible evoked potential changes with retraction of the eighth cranial nerve. *Anesthesia and Analgesia* 1981; **60**: 835–8.

79. Grundy BL, Jannetta PJ, Procopio PT, Lina A, Boston JR, Doyle E. Intraoperative monitoring of brainstem auditory evoked potentials. *Journal of Neurosurgery* 1982; **57**: 674–81.

80. Kaplan BJ, Friedman WA, Alexander JA, Hampson SR. Somatosensory evoked potential monitoring of spinal cord ischemia during aortic operations. *Neurosurgery* 1986; **19**: 82–90.

81. Cracco RQ, Evans B. Spinal evoked potential in the cat: effects of asphyxia, strychnine, cord section and compression. *Electroencephalography and Clinical Neurophysiology* 1978; **44**: 187–201.

82. Higgins AC, Pearlstein RD, Mullen JB, Nashold Jr BS. Effects of hyperbaric oxygen therapy on long-tract neuronal conduction in the acute phase of spinal cord injury. *Journal of Neurosurgery* 1981; **55**: 501–10.

83. Nashold Jr BS, Ovelmen-Levitt J. Sharpe R, Higgins AC. Intraoperative evoked potentials recorded in man directly from dorsal roots and spinal cord. *Journal of Neurosurgery* 1985; **62**: 680–93.

84. Cohen AR, Young W, Ransohoff J. Intraspinal localization of the somatosensory evoked potential. *Neurosurgery* 1981; **9**: 157–62.

85. York DH. Somatosensory evoked potentials in man: differentiation of spinal pathways responsible for conduction from the forelimb vs hindlimb. *Progress in Neurobiology* 1985; **25**: 1–25.

86. Snyder BG, Holliday TA. Pathways of ascending evoked spinal cord potentials of dogs. *Electroencephalography and Clinical Neurophysiology* 1984; **58**: 140–54.

87. Gasser HS, Graham HT. Potentials produced in the spinal cord by stimulation of dorsal roots. *American Journal of Physiology* 1933; **103**: 303–20.

88. Gelfan S, Tarlov IM. Differential vulnerability of spinal cord structures to anoxia. *Journal of Neurophysiology* 1955; **18**: 170–88.

89. Kimura J, Mitsudome A, Yamada T, Dickens QS. Stationary peaks from a moving source in far-field recording. *Electroencephalography and Clinical Neurophysiology* 1984; **58**: 351–61.

90. Symon L, Hargadine J, Zawirski M, Branston N. Central conduction time as an index of ischaemia in subarachnoid haemorrhage. *Journal of the Neurological Sciences* 1979; **44**: 95–103.

91. Chiappa KH, Ropper AH. Evoked potentials in clinical medicine: part I. *New England Journal of Medicine* 1982; **306**: 1140–50.

92. Chiappa KH, Ropper AH. Evoked potentials in clinical medicine: part 2. *New England Journal of Medicine* 1982; **306**: 1205–11.

93. Friedman WA, Richards R. Somatosensory evoked potential monitoring accurately predicts hemi-spinal cord damage: a case report. *Neurosurgery* 1988; **22**: 140–2.

94. Burke D, Skuse NF, Lethlean AK. Cutaneous and muscle afferent components of the cerebral potential evoked by electrical stimulation of human peripheral nerves. *Electroencephalography and Clinical Neurophysiology* 1980; **51**: 579–88.

95. Shimoji K, Kano T, Higashi H. Epidural recording of spinal electrogram in man. *Electroencephalography and Clinical Neurophysiology* 1971; **30**: 236–9.

96. Shimoji K, Kano T. Evoked electrospinogram: interpretation of origin and effects of anesthetics. *International Anesthesiology Clinics* 1975; **13**: 171–89.

97. Tamaki T, Tsuji H, Inoué S, Kobayashi H. The prevention of iatrogenic spinal cord injury utilizing the evoked spinal cord potential. *International Orthopaedics* 1981; **4**: 313–17.

98. Tamaki T, Noguchi T, Takana H. Spinal cord monitoring as a clinical utilization of the spinal evoked potential. *Clinical Orthopaedics and Related Research* 1984; **184**: 58–64.

99. Lueders H, Hahn J, Gurd A, *et al.* Surgical monitoring of spinal cord function: cauda equina stimulation technique. *Neurosurgery* 1982; **11**: 482–5.

100. Pratt H, Starr A. Mechanically and electrically evoked somatosensory potentials in humans: scalp and neck distributions of short-latency components. *Electroencephalography and Clinical Neurophysiology* 1981; **51**: 138–47.

101. Carmon A, Mor J, Goldberg J. Evoked cerebral responses to noxious thermal stimuli in humans. *Experimental Brain Research* 1976; **25**: 103–7.

102. Zarola F, Rossini PM. Nerve, spinal cord and brain somatosensory evoked responses: a comparative study during electrical and magnetic peripheral nerve stimulation. *Electroencephalography and Clinical Neurophysiology* 1991; **80**: 372–7.

103. Kunesch E, Knecht S, Classen J, Roick H, Tyercha C, Benecke R. Somatosensory evoked potentials (seps) elicited by magnetic nerve stimulation. *Electroencephalography and Clinical Neurophysiology* 1993; **88**: 459–67.

104. Brown RH, Nash Jr CL. Current status of spinal cord monitoring. *Spine* 1979; **4**: 466–70.

105. Hahn JF, Lesser R, Klem G, Lueders H. Simple technique for monitoring intraoperative spinal cord function. *Neurosurgery* 1981; **9**: 692–5.

106. Lueders H, Gurd A, Hahn J, Andrish J, Weiker G, Klem G. A new technique for intraoperative monitoring of spinal cord function: multichannel recording of spinal cord and subcortical evoked potentials. *Spine* 1982; **7**: 110–15.

107. Macon JB, Poletti CE. Conducted somatosensory evoked potentials during spinal surgery. Part 1: control conduction velocity measurements. *Journal of Neurosurgery* 1982; **57**: 349–53.

108. Macon JB, Poletti CE, Sweet WH, Ojemann RG, Zervas NT. Conducted somatosensory evoked potentials during spinal surgery. Part 2: clinical applications. *Journal of Neurosurgery* 1982; **57**: 354–9.

109. Bertrand C, Martinez SN, Hardy J, Molina-Negro P, Velasco F. Stereotactic surgery for Parkinsonism: microelectrode recording, stimulation, and oriented sections with a leucotome. *Progress in Neurological Surgery* 1973; **5**: 79–112.

110. Haider M, Ganglberger JA, Groll-Knapp E, Schmid H. Averaged cortical and subcortical potentials during stereotactic operations in humans. In: Speckmann EJ, Caspers H, eds. *Origin of cerebral fluid potentials.* Stuttgart: Georg Thieme, 1979: 141–50.

111. Nash Jr CL, Lorig RA, Schatzinger LA, Brown RH. Spinal cord monitoring during operative treatment of the spine. *Clinical Orthopaedics and Related Research* 1977; **126**: 100–5.

112. MacEwen GD, Bunnell WP, Sviram K. Acute neurologic complications in the treatment of scoliosis: a report of the Scoliosis Research Society. *Journal of Bone and Joint Surgery [Am]* 1975; **57**: 404–8.

113. Engler GL, Spielholz NI, Bernhard WN, Danziger F, Merkin H, Wolff T. Somatosensory evoked potentials during Harrington instrumentation for scoliosis. *Journal of Bone and Joint Surgery [Am]* 1978; **60**: 528–32.

114. Grundy BL, Nash Jr CL, Brown RH. Arterial pressure manipulation alters spinal cord function during correction of scoliosis. *Anesthesiology* 1981; **54**: 249–53.

115. Grundy BL, Nash Jr CL, Brown RH. Deliberate hypotension for spinal fusion: prospective randomized study with evoked potential monitoring. *Canadian Anaesthetists' Society Journal* 1982; **29**: 452–62.

116. Wilber RG, Thompson G, Shaffer JW, Brown RH, Nash CL. Postoperative neurological deficits in segmental spinal instrumentation: a study using spinal cord monitoring. *Journal of Bone and Joint Surgery [Am]* 1984; **66**: 1178–87.

117. Bradshaw K, Webb JK, Fraser AM. Clinical evaluation of spinal cord monitoring in scoliosis surgery. *Spine* 1984; **9**: 636.

118. Brown RH, Nash CL, Berilla JA, Amaddio MD. Cortical evoked potential monitoring: a system for intra-operative monitoring of spinal cord function. *Spine* 1984; **9**: 256–61.

119. Mostegl A, Bauer R. The application of somatosensory-evoked potentials in orthopaedic spine surgery. *Archives of Orthopaedic and Trauma Surgery* 1984; **103**: 179–84.

120. Veilleux M, Daube JR, Cucchiara RF. Monitoring of cortical evoked potentials during surgical procedures on the cervical spine. *Mayo Clinic Proceedings* 1987; **62**: 256–64.

121. Sloan TB, Ronai A, Koht A. Reversible loss of somatosensory evoked potentials during anterior cervical spinal fusion. *Anesthesia and Analgesia* 1986; **65**: 96–9.

122. Forbes HJ, Allen PW, Waller CS, *et al.* Spinal cord monitoring in scoliosis surgery – experience with 1168 cases. *Journal of Bone and Joint Surgery [Br]* 1991; **73**: 487–91.

123. Kalkman CJ, ten Brink SA, Been HD, Bovill JG. Variability of somatosensory cortical evoked potentials during spinal surgery. Effects of anesthetic technique and high-pass digital filtering. *Spine* 1991; **16**: 924–9.

124. Dunne JW, Field CM. The value of non-invasive spinal cord monitoring during spinal surgery and interventional angiography. *Clinical and Experimental Neurology* 1991; **28**: 199–209.

125. Williamson JB, Galasko CS. Spinal cord monitoring during operative correction of neuromuscular scoliosis. *Journal of Bone and Joint Surgery [Br]* 1992; **74**: 870–2.

126. Hicks RG, Burke DJ, Stephen JP. Monitoring spinal cord function during scoliosis with Cotrel–Dubousset instrumentation. *Medical Journal of Australia* 1991; **154**: 82–6.

127. Ashkenaze D, Mudiyam R, Boachie-Adjei O, Gilbert C. Efficacy of spinal cord monitoring in neuromuscular scoliosis. *Spine* 1993; **18**: 1627–33.

128. Ginsburg HH, Shetter AG, Raudzens PA. Postoperative paraplegia with preserved intraoperative somatosensory

evoked potentials: case report. *Journal of Neurosurgery* 1985; **63**: 296–300.

129. Friedman WA, Grundy BL. Monitoring of sensory evoked potentials is highly reliable and helpful in the operating room. *Journal of Clinical Monitoring* 1987; **3**: 38–44.

130. Molale M. False negative intraoperative somatosensory evoked potentials with simultaneous bilateral stimulation. *Clinical Electroencephalography* 1986; **17**: 6–9.

131. Nash CL, Schatzinger LH, Brown RH, Brodkey J. The unstable thoracic compression fracture: its problems and the use of spinal cord monitoring in the evaluation of treatment. *Spine* 1977; **2**: 261–5.

132. Spielholz NI, Benjamin MV, Engler GL, Ransohoff J. Somatosensory evoked potentials during decompression and stabilization of the spine: methods and findings. *Spine* 1979; **4**: 500–5.

133. Spielholz NI, Benjamin MV, Engler G, Ransohoff J. Somatosensory evoked potentials and clinical outcome in spinal cord injury. In: Popp AJ, Bourke RS, Nelson LR, Kimelberg HK, eds. *Neural trauma.* New York: Raven Press, 1979: 217–22.

134. McCallum JE, Bennett MH. Electrophysiologic monitoring of spinal cord function during intraspinal surgery. *Surgical Forum* 1975; **26**: 469–71.

135. McPherson RW, North RB, Udvarhelyi GB, Rosenbaum AE. Migrating disc complicating spinal decompression in an achondroplastic dwarf: intraoperative demonstration of spinal cord compression by somatosensory evoked potentials. *Anesthesiology* 1984; **61**: 764–7.

136. Nuwer MR. Use of somatosensory evoked potentials for intraoperative monitoring of cerebral and spinal cord function. *Neurologic Clinics* 1988; **6**: 881–97.

137. Schramm J. Spinal cord monitoring: current status and new developments. *Central Nervous System Trauma* 1985; **2**: 207–27.

138. Schramm J, Kurthen M. Recent developments in neurosurgical spinal cord monitoring. *Paraplegia* 1992; **30**: 609–16.

139. Cioni B, Meglio M, Moles A, Tirendi M, Visocchi M. Spinal somatosensory evoked potential monitoring during microsurgery for syringomyelia: case reports. *Stereotactic and Functional Neurosurgery* 1991; **57**: 123–9.

140. McPherson RW, Sell B, Traystman RJ. Effects of thiopental, fentanyl, and etomidate on upper extremity somatosensory evoked potentials in humans. *Anesthesiology* 1986; **65**: 584–9.

141. Owen MP, Brown RH, Spetzler RF. Excision of intramedullary arteriovenous malformation using intraoperative spinal cord monitoring. *Surgical Neurology* 1979; **12**: 271–6.

142. Grundy BL, Nelson PB, Doyle E, Procopio PT. Intraoperative loss of somatosensory-evoked potentials predicts loss of spinal cord function. *Anesthesiology* 1982; **57**: 321–2.

143. Koyanagi I, Iwasaki Y, Isu T, *et al.* Spinal cord evoked potential monitoring after spinal cord stimulation during surgery of spinal cord tumors. *Neurosurgery* 1993; **33**: 451–60.

144. Szilagyi DE, Hageman JH, Smith RF, Elliott JP. Spinal cord damage in surgery of the abdominal aorta. *Surgery* 1978; **83**: 38–56.

145. Hitchon PW, Loboski JM, Wilkinson TT, Yamada T, Torner JC, Gant PR. Direct spinal cord stimulation and recording in hemorrhagic shock. *Neurosurgery* 1985; **16**: 796–800.

146. Coles JG, Wilson GJ, Sima AF, Klement P, Tait GA. Intraoperative detection of spinal cord ischemia using somatosensory cortical evoked potentials during thoracic aortic occlusion. *Annals of Thoracic Surgery* 1982; **34**: 299–306.

147. Laschinger JC, Cunningham JN, Isom OW, Nathan IM, Spencer FC. Definition of the safe lower limits of aortic resection during surgical procedures on the thoracoabdominal aorta: use of somatosensory evoked potentials. *Journal of the American College of Cardiology* 1983; **2**: 959–65.

148. Laschinger JC, Cunningham JN, Nathan IM, Knopp EA, Cooper MM, Spencer FC. Experimental and clinical assessment of the adequacy of partial bypass in maintenance of spinal cord blood flow during operations on the thoracic aorta. *Annals of Thoracic Surgery* 1983; **36**: 417–26.

149. Laschinger JC, Cunningham JN, Nathan IM, Krieger K, Isom OW, Spencer FC. Intraoperative identification of vessels critical to spinal cord blood supply: use of somatosensory evoked potentials. *Current Surgery* 1984; **41**: 107–9.

150. Laschinger JC, Cunningham JN, Cooper MM, Krieger KH, Nathan IM, Spencer FC. Prevention of ischemic spinal cord injury following aortic cross-clamping: use of corticosteroids. *Annals of Thoracic Surgery* 1984; **38**: 500–7.

151. Laschinger JC, Cunningham Jr JN, Baumann FG, Isom OW, Spencer FC. Monitoring of somatosensory evoked potentials during surgical procedures on the thoracoabdominal aorta: II. Use of somatosensory evoked potentials to assess adequacy of distal aortic bypass and perfusion after thoracic aortic cross-clamping. *Journal of Thoracic and Cardiovascular Surgery* 1987; **94**: 266–70.

152. Laschinger JC, Cunningham Jr JN, Baumann FG, Cooper MM, Krieger KH, Spencer FC. Monitoring of somatosensory evoked potentials during surgical procedures on the thoracoabdominal aorta: III. Intraoperative identification of vessels critical to spinal cord blood supply. *Journal of Thoracic and Cardiovascular Surgery* 1987; **94**: 271–4.

153. Laschinger JC, Cunningham Jr JN, Cooper MM, Baumann FG, Spencer FC. Monitoring of somatosensory evoked potentials during surgical procedures on the thoracoabdominal aorta: I. Relationship of aortic cross-clamp duration, changes in somatosensory evoked potentials, and incidence of neurologic dysfunction. *Journal of Thoracic and Cardiovascular Surgery* 1987; **94**: 260–5.

154. Cunningham JN, Laschinger JC, Merkin HA, *et al.* Measurement of spinal cord ischemia during operation upon the thoracic aorta: initial clinical experience. *Annals of Surgery* 1982; **196**: 285–96.

155. Krieger KH, Spencer FC. Is paraplegia after repair of coarctation of the aorta due principally to distal hypotension during aortic cross-clamping? *Surgery* 1985; **97**: 2–6.

156. Mizrahi EM, Crawford ES. Somatosensory evoked potentials during reversible spinal cord ischemia in man. *Electroencephalography and Clinical Neurophysiology* 1984; **58**: 120–6.

157. Crawford ES, Mizrahi EM, Hess KR, Coselli JS, Safi HJ, Patel VM. The impact of distal aortic perfusion and somatosensory evoked potential monitoring on prevention of paraplegia

after aortic aneurysm operation. *Journal of Thoracic and Cardiovascular Surgery* 1988; **95**: 357–67.

158. Svensson LG, Crawford ES, Hess KR, Coselli JS, Safi HJ. Experience with 1509 patients undergoing thoracoabdominal aortic operations. *Journal of Vascular Surgery* 1993; **17**: 357–70.

159. Svensson LG, Crawford ES, Hess KR, Coselli JS, Safi HJ. Variables predictive of outcome in 832 patients undergoing repairs of the descending thoracic aorta. *Chest* 1993; **104**: 1248–53.

160. Cunningham Jr JN, Laschinger JC, Spencer FC. Monitoring of somatosensory evoked potentials during surgical procedures on the thoracoabdominal aorta: IV. Clinical observations and results. *Journal of Thoracic and Cardiovascular Surgery* 1987; **94**: 275–85.

161. Fava E, Bortolani EM, Ducati A, Ruberti U. Evaluation of spinal cord function by means of lower limb somatosensory evoked potentials in reparative aortic surgery. *Journal of Cardiovascular Surgery* 1988; **29**: 421–7.

162. Hollier LH, Money SR, Naslund TC, *et al.* Risk of spinal cord dysfunction in patients undergoing thoracoabdominal aortic replacement. *American Journal of Surgery* 1992; **164**: 210–14.

163. Fehrenbacher JW, McCready RA, Hormuth DA, *et al.* One-stage segmental resection of extensive thoracoabdominal aneurysms with left-sided heart bypass. *Journal of Vascular Surgery* 1993; **18**: 366–71.

164. Vanicky I, Marsala M, Galik J, Marsala J. Epidural perfusion cooling protection against protracted spinal cord ischemia in rabbits. *Journal of Neurosurgery* 1993; **79**: 736–41.

165. Elmore JR, Gloviczki P, Harper CM, *et al.* Failure of motor evoked potentials to predict neurologic outcome in experimental thoracic aortic occlusion. *Journal of Vascular Surgery* 1991; **14**: 131–9.

166. Ihaya A, Morioka K, Noguchi H, *et al.* [A case report of descending thoracic aortic aneurysm associated with anterior spinal artery syndrome despite no marked ESP changes]. *Kyobu Geka* 1990; **43**: 843–6.

167. Berenstein A, Young W, Ransohoff J, Benjamin V, Merkin H. Somatosensory evoked potentials during spinal angiography and therapeutic transvascular embolization. *Journal of Neurosurgery* 1984; **60**: 777–85.

168. Sasaki T, Takeda R, Ogasawara T, *et al.* [Monitoring of somatosensory evoked potentials during extracranial revascularization]. *Neurologia Medico-chirurgica (Tokyo)* 1989; **29**: 280–4.

169. Artru AA, Strandness Jr DE. Delayed carotid shunt occlusion detected by electroencephalographic monitoring. *Journal of Clinical Monitoring* 1989; **5**: 119–22.

170. Schweiger H, Kamp HD, Dinkel M. Somatosensory-evoked potentials during carotid artery surgery: experience in 400 operations. *Surgery* 1991; **109**: 602–9.

171. Horsch S, De Vleeschauwer P, Ktenidis K. Intraoperative assessment of cerebral ischemia during carotid surgery. *Journal of Cardiovascular Surgery* 1990; **31**: 599–602.

172. Russ W, Thiel A, Moosdorf R, Hempelmann G. [Somatosensory evoked potentials in obliterating interventions of the carotid bifurcation]. *Klinische Wochenschrift* 1988; **66** (supplement 14): 35–40.

173. Lam AM, Manninen PH, Ferguson GG, Nantau W. Monitor-

ing electrophysiologic function during carotid endarterectomy: a comparison of somatosensory evoked potentials and conventional electroencephalogram. *Anesthesiology* 1991; **75**: 15–21.

174. Kearse Jr LA, Brown EN, McPeck K. Somatosensory evoked potentials sensitivity relative to electroencephalography for cerebral ischemia during carotid endarterectomy. *Stroke* 1992; **23**: 498–505.

175. Fava E, Bortolani E, Ducati A, Schieppati M. Role of SEP in identifying patients requiring temporary shunt during carotid endarterectomy. *Electroencephalography and Clinical Neurophysiology* 1992; **84**: 426–32.

176. Markand ON, Dilley RS, Moorthy SS, Warren Jr C. Monitoring of somatosensory evoked responses during carotid endarterectomy. *Archives of Neurology* 1984; **41**: 375–8.

177. Markand ON, Warren CH, Moorthy SS, Stoelting RK, King RD. Monitoring of multimodality evoked potentials during open heart surgery under hypothermia. *Electroencephalography and Clinical Neurophysiology* 1984; **59**: 432–40.

178. Hickey C, Gugino LD, Aglio LS, Mark JB, Son SL, Maddi R. Intraoperative somatosensory evoked potential monitoring predicts peripheral nerve injury during cardiac surgery. *Anesthesiology* 1993; **78**: 29–35.

179. Coles JG, Taylor MJ, Pearce JM, *et al.* Cerebral monitoring of somatosensory evoked potentials during profoundly hypothermic circulatory arrest. *Circulation* 1984; **70** (supplement 1): 96–102.

180. Kline DG, Judice DJ. Operative management of selected brachial plexus lesions. *Journal of Neurosurgery* 1983; **58**: 631–49.

181. Kaplan BJ, Gravenstein D, Friedman WA. Intraoperative electrophysiology in treatment of peripheral nerve injuries. *Journal of the Florida Medical Association* 1984; **71**: 400–3.

182. Murase T, Kawai H, Masatomi T, Kawabata H, Ono K. Evoked spinal cord potentials for diagnosis during brachial plexus surgery. *Journal of Bone and Joint Surgery [Br]* 1993; **75**: 775–81.

183. Allison T. Localization of sensorimotor cortex in neurosurgery by recording of somatosensory evoked potentials. *Yale Journal of Biology and Medicine* 1987; **60**: 143–50.

184. Suzuki A, Yasui N. Intraoperative localization of the central sulcus by cortical somatosensory evoked potentials in brain tumor. Case report. *Journal of Neurosurgery* 1992; **76**: 867–70.

185. Hargadine JR, Branston NM, Symon L. Central conduction time in primate brain ischemia: a study in baboons. *Stroke* 1980; **6**: 637–42.

186. Branston NM, Strong AJ, Symon L. Extracellular potassium activity, evoked potential and tissue blood flow. Relationships during progressive ischaemia in baboon cerebral cortex. *Journal of the Neurological Sciences* 1977; **32**: 305–21.

187. Strong AJ, Goodhardt MJ, Branston NM, Symon L. A comparison of the effects of ischaemia on tissue flow, electrical activity and extracellular potassium ion concentration in cerebral cortex of baboons. *Biochemical Society Transactions* 1977; **5**: 158–60.

188. Hargadine JR. Intraoperative monitoring of sensory evoked

potentials. In: Rand RW, ed. *Microneurosurgery*. St Louis: CV Mosby, 1985: 92–110.

189. Friedman WA, Kaplan BL, Day AL, Sypert GW, Curran MT. Evoked potential monitoring during aneurysm operation: observations after fifty cases. *Neurosurgery* 1987; **20**: 678–87.

190. Friedman WA, Chadwick GM, Verhoeven FJ, Mahla M, Day AL. Monitoring of somatosensory evoked potentials during surgery for middle cerebral artery aneurysms. *Neurosurgery* 1991; **29**: 83–8.

191. Symon L, Momma F, Murota T. Assessment of reversible cerebral ischaemia in man: intraoperative monitoring of the somatosensory evoked response. *Acta Neurochirurgica, Supplementum (Wien)* 1988; **42**: 3–7.

192. Mizoi K, Yoshimoto T, Piepgras DG, Schramm J. Permissible temporary occlusion time in aneurysm surgery as evaluated by evoked potential monitoring. *Neurosurgery* 1993; **33**: 434–40.

193. Little JR, Lesser RP, Luders H. Electrophysiological monitoring during basilar aneurysm operation. *Neurosurgery* 1987; **20**: 421–7.

194. Grundy BL, Nelson PB, Lina A, Heros RC. Monitoring of cortical somatosensory evoked potentials to determine the safety of sacrificing the anterior cerebral artery. *Neurosurgery* 1982; **11**: 64–7.

195. McPherson RW, Szymanski J, Rogers MC. Somatosensory evoked potential changes in position-related brain stem ischemia. *Anesthesiology* 1984; **61**: 88–90.

196. Mahla ME, Long DM, McKennett J, Green C, McPherson RW. Detection of brachial plexus dysfunction by somatosensory evoked potential monitoring: a report of two cases. *Anesthesiology* 1984; **60**: 248–52.

197. Greenberg RP, Mayer DJ, Becker DP, Miller JD. Evaluation of brain function in severe human head trauma with multimodality evoked potentials: part 1. Evoked brain-injury potentials, methods, and analysis. *Journal of Neurosurgery* 1977; **47**: 150–62.

198. Greenberg RP, Becker DP, Miller JD, Mayer DJ. Evaluation of brain function in severe human head trauma with multimodality evoked potentials: part 2. Localization of brain dysfunction and correlation with posttraumatic neurological conditions. *Journal of Neurosurgery* 1977; **47**: 163–77.

199. Hume AL, Cant BR. Central somatosensory conduction after head injury. *Annals of Neurology* 1981; **10**: 411–19.

200. Kawahara N, Sasaki M, Mii K, Takakura K. Reversibility of cerebral function assessed by somatosensory evoked potentials and its relation to intracranial pressure – report of six cases with severe head injury. *Neurologia Medico-chirurgica (Tokyo)* 1991; **31**: 264–71.

201. Firsching R, Frowein RA. Multimodality evoked potentials and early prognosis in comatose patients. *Neurosurgical Review* 1990; **13**: 141–6.

202. Rumpl E, Prugger M, Gerstenbrand F, Brunhuber W, Badry F, Hackl JM. Central somatosensory conduction time and acoustic brainstem transmission in post-traumatic coma. *Journal of Clinical Neurophysiology* 1988; **5**: 237–60.

203. Anderson DC, Bundlie S, Rockswold GL. Multimodality evoked potentials in closed head trauma. *Archives of Neurology* 1984; **41**: 369–74.

204. Ruijs MBM, Keyser A, Gabreels FJM, Notermans SLH.

Somatosensory evoked potentials and cognitive sequelae in children with closed head-injury. *Neuropediatrics* 1993; **24**: 307–12.

205. Zentner J, Ebner A. [Somatosensory and motor evoked potentials in the prognostic assessment of traumatic and non-traumatic comatose patients]. *EEG/EMG* 1988; **19**: 267–71.

206. Hume AL, Cant BR, Shaw NA. Central somatosensory conduction time in comatose patients. *Annals of Neurology* 1979; **5**: 379–84.

207. Lutschg J, Pfenninger J, Ludin HP, Vassella F. Brainstem auditory evoked potentials and early somatosensory evoked potentials in neurointensively treated comatose children. *American Journal of Diseases of Children* 1983; **137**: 421–6.

208. Cant BR, Shaw NA. Central somatosensory conduction time: method and clinical applications. In: Cracco RQ, Bodis-Wollner I, eds. *Evoked potentials*. New York: Alan R Liss, 1986: 58–67.

209. Brunko E, Zegers de Beyl D. Prognostic value of early cortical somatosensory evoked potentials after resuscitation from cardiac arrest. *Electroencephalography and Clinical Neurophysiology* 1987; **66**: 15–24.

210. Rothstein TL, Thomas EM, Sumi SM. Predicting outcome in hypoxic-ischemic coma. A prospective clinical and electrophysiologic study. *Electroencephalography and Clinical Neurophysiology* 1991; **79**: 101–7.

211. Madl C, Grimm G, Kramer L, *et al.* Early prediction of individual outcome after cardiopulmonary resuscitation. *Lancet* 1993; **341**: 855–8.

212. Taylor MJ, Murphy WJ, Whyte HE. Prognostic reliability of somatosensory and visual evoked potentials of asphyxiated term infants. *Developmental Medicine and Child Neurology* 1992; **34**: 507–15.

213. Symon L, Wang AD, Costa E, Silva IE, Gentili F. Perioperative use of somatosensory evoked responses in aneurysm surgery. *Journal of Neurosurgery* 1984; **70**: 269–75.

214. Rosenstein J, Wang ADJ, Symon L, Suzuki M. Relationship between hemispheric cerebral blood flow, central conduction time, and clinical grade in aneurysmal subarachnoid hemorrhage. *Journal of Neurosurgery* 1985; **62**: 25–30.

215. Symon L, Hargadine J, Zawirski M, Branston NM. Central conduction time as an index of ischaemia in subarachnoid haemorrhage. *Journal of the Neurological Sciences* 1979; **44**: 95–103.

216. Wang AD, Cone J, Symon L, Costa da Silva IE. Somatosensory evoked potential monitoring during the management of aneurysmal subarachnoid hemorrhage. *Journal of Neurosurgery* 1984; **60**: 264–8.

217. Grundy BL, Friedman W. Electrophysiological evaluation of the patient with acute spinal cord injury. *Critical Care Clinics* 1987; **3**: 519–48.

218. Grundy BL, Procopio PT, Jannetta PJ, Lina A, Doyle E. Evoked potential changes produced by positioning for retromastoid craniectomy. *Neurosurgery* 1982; **10**: 766–70.

219. van Loveren H, Tew JM, Keller JT, Nurre MA. A 10-year experience in the treatment of trigeminal neuralgia. *Neurosurgery* 1982; **57**: 757–64.

220. Jannetta PJ. Neurovascular compression in cranial nerve and systemic disease. *Annals of Surgery* 1980; **192**: 518–25.

221. Friedman WA, Kaplan BJ, Gravenstein D, Rhoton Jr AL. Intraoperative brain-stem auditory evoked potentials during posterior fossa microvascular decompression. *Journal of Neurosurgery* 1985; **62**: 552–7.

222. Raudzens PA, Shetter AG. Intraoperative monitoring of brainstem auditory evoked potentials. *Journal of Neurosurgery* 1982; **57**: 341–8.

223. Piatt JH, Radtke RA, Erwin CW. Limitations of brain stem auditory evoked potentials for intraoperative monitoring during a posterior fossa operation: case report and technical note. *Neurosurgery* 1985; **16**: 818–21.

224. Hardy Jr RW, Kinney SE, Lueders H, Lesser RP. Preservation of cochlear nerve function with the aid of brain stem auditory evoked potentials. *Neurosurgery* 1982; **11**: 16–19.

225. Ojemann RG, Levine RA, Montgomery WM, McGaffigan P. Use of intraoperative auditory evoked potentials to preserve hearing in unilateral acoustic neuroma removal. *Journal of Neurosurgery* 1984; **61**: 938–48.

226. Glasscock ME, Hays JW, Minor LB, Haynes DS, Carrasco VN. Preservation of hearing in surgery for acoustic neuromas. *Journal of Neurosurgery* 1993; **78**: 864–70.

227. Levine RA, Montgomery WW, Ojemann RG, McGaffigan PM. Monitoring auditory evoked potentials during acoustic neuroma surgery. *Annals of Otology, Rhinology and Laryngology* 1984; **93**: 116–23.

228. Harper CM, Harner SG, Slavit DH, *et al.* Effect of BAEP monitoring on hearing preservation during acoustic neuroma resection. *Neurology* 1992; **42**: 1551–3.

229. Facco E, Munari M, Casartelli Liviero M, *et al.* Serial recordings of auditory brainstem responses in severe head injury: relationship between test timing and prognostic power. *Intensive Care Medicine* 1988; **14**: 422–8.

230. Garcia-Larrea L, Artru F, Bertrand O, Pernier J, Mauguière F. The combined monitoring of brain stem auditory evoked potentials and intracranial pressure in coma: a study of 57 patients. *Journal of Neurology, Neurosurgery and Psychiatry* 1992; **55**: 792–8.

231. Celesia GG. Visual evoked potentials in clinical neurology. In: Aminoff MJ, ed. *Electrodiagnosis in clinical neurology.* New York: Churchill Livingstone, 1992: 467–90.

232. Zaaroor M, Pratt H, Feinsod M, Schacham SE. Real-time monitoring of visual evoked potentials. *Israel Journal of Medical Sciences* 1993; **29**: 17–22.

233. Owen JH, Bridwell KH, Grubb R, *et al.* The clinical application of neurogenic motor evoked potentials to monitor spinal cord function during surgery. *Spine* 1991; **16**: S385–90.

234. Su CF, Haghighi SS, Oro JJ, Gaines RW. 'Backfiring' in spinal cord monitoring. High thoracic spinal cord stimulation evokes sciatic response by antidromic sensory pathway conduction, not motor tract conduction. *Spine* 1992; **17**: 504–8.

235. Levy WJ, York DH, McCaffrey M, Tanzer F. Motor evoked potentials from transcranial stimulation of the motor cortex in humans. *Neurosurgery* 1984; **15**: 287–302.

236. Cowan JM, Rothwell JC, Dick JP, Thompson PD, Day BL, Marsden CD. Abnormalities in central motor pathway conduction in multiple sclerosis. *Lancet* 1984; **2**: 304–7.

237. DeFelipe J, Conley M, Jones EG. Long-range focal collateralization of axons arising from cortico-cortical cells in monkey sensory-motor cortex. *Journal of Neuroscience* 1986; **6**: 3749–66.

238. Stefanis C, Jasper H. Recurrent collateral inhibition in pyramidal tract neurons. *Journal of Neurophysiology* 1964; **27**: 855–77.

239. Pascual-Leone A, Houser CM, Reese K, *et al.* Safety of rapid-rate transcranial magnetic stimulation in normal volunteers. *Electroencephalography and Clinical Neurophysiology* 1993; **89**: 120–30.

240. Counter SA, Borg E, Lofqvist L, Brismar T. Hearing loss from the acoustic artifact of the coil used in extracranial magnetic stimulation. *Neurology* 1990; **40**: 1159–62.

241. Counter SA, Borg E, Lofqvist L. Acoustic trauma in extracranial magnetic brain stimulation. *Electroencephalography and Clinical Neurophysiology* 1991; **78**: 173–84.

242. Counter SA, Borg E. Acoustic middle ear muscle reflex protection against magnetic coil impulse noise. *Acta Otolaryngologica (Stockholm)* 1993; **113**: 483–8.

243. Roth BJ, Pascual-Leone A, Cohen LG, Hallett M. The heating of metal electrodes during rapid rate transcranial magnetic stimulation: a possible safety hazard. *Electroencephalography and Clinical Neurophysiology* 1992; **85**: 116–23.

244. Harner SG, Daube JR, Ebersold MJ. Electrophysiologic monitoring of facial nerve during temporal bone surgery. *Laryngoscope* 1986; **96**: 65–9.

245. Harner SG, Daube JR, Ebersold MJ, Beatty CW. Improved preservation of facial nerve function with use of electrical monitoring during removal of acoustic neuromas. *Mayo Clinic Proceedings* 1987; **62**: 92–102.

246. Lennon RL, Hosking MP, Daube JR, Welna JO. Effect of partial neuromuscular blockade on intraoperative electromyography in patients undergoing resection of acoustic neuromas. *Anesthesia and Analgesia* 1992; **75**: 729–33.

247. Sekhar LN, Moller AR. Operative management of tumours involving the cavernous sinus. *Journal of Neurosurgery* 1986; **64**: 879–89.

248. Zentner J, Ebner A. Nitrous oxide suppresses the electromyographic response evoked by electrical stimulation of the motor cortex. *Neurosurgery* 1989; **24**: 60–2.

249. Zentner J. Noninvasive motor evoked potential monitoring during neurosurgical operations on the spinal cord. *Neurosurgery* 1989; **24**: 709–12.

250. Jellinek D, Platt M, Jewkes D, Symon L. Effects of nitrous oxide on motor evoked potentials recorded from skeletal muscle in patients under total anesthesia with intravenously administered propofol. *Neurosurgery* 1991; **29**: 558–62.

251. Jellinek D, Jewkes D, Symon L. Noninvasive intraoperative monitoring of motor evoked potentials under propofol anesthesia: effects of spinal surgery on the amplitude and latency of motor evoked potentials. *Neurosurgery* 1991; **29**: 551–7.

252. Levy Jr WJ. Clinical experience with motor and cerebellar evoked potential monitoring. *Neurosurgery* 1987; **20**: 169–82.

253. Facco E, Baratto F, Munari M, *et al.* Sensorimotor central conduction time in comatose patients. *Electroencephalography and Clinical Neurophysiology* 1991; **80**: 469–76.

254. Netz J, Homberg V. Intact conduction of fastest corticospinal

efferents is not sufficient for normal voluntary muscle strength: transcranial motor cortex stimulation in patients with tetraplegia. *Neuroscience Letters* 1992; **146**: 29–32.

255. Risberg J, Lundberg N, Ingvar DH. Regional cerebral blood volume during acute transient rises of the intracranial pressure (plateau waves). *Journal of Neuroscience* 1969; **31**: 303–10.

256. Barnett GH, Chapman PH. Insertion and care of intracranial pressure monitoring devices. In: Ropper AH, Kennedy SF, eds. *Neurological and neurosurgical intensive care*. Rockville, Maryland: Aspen Publishers, 1988: 43–55.

257. Lundberg N. Continuous recording and control of ventricular fluid pressure in neurosurgical practice. *Acta Psychiatrica et Neurologica Scandinavica* 1960; **149** (supplement 36): 1–193.

258. Vries JK, Becker DP, Young HR. A subarachnoid screw for monitoring intracranial pressure: technical note. *Journal of Neurosurgery* 1973; **39**: 416–19.

259. Wilkinson HA. The intracranial pressure-monitoring cup catheter: technical note. *Neurosurgery* 1977; **1**: 139–41.

260. Levin AB. The use of a fibreoptic intracranial pressure monitor in clinical practice. *Neurosurgery* 1977; **1**: 266–71.

261. Marcotty SF, Levin AB. A new approach in epidural intracranial pressure monitoring. *Journal of Neurosurgical Nursing* 1984; **16**: 54–9.

262. Roberts PA, Fullenwider C, Stevens FA, Pollay M. Experimental and clinical experiences with a new solid state intracranial pressure monitor with in vivo zero capability. In: Ishii S, Nagai H, Brock M, eds. *Intracranial pressure V.* Berlin: Springer-Verlag, 1983: 104–5.

263. Barlow P, Mendelow AD, Lawrence AE, Barlow M, Rowan JO. Clinical evaluation of two methods of subdural pressure monitoring. *Journal of Neurosurgery* 1985; **63**: 578–82.

264. Gentleman D, Mendelow AD. Intracranial rupture of a pressure monitoring transducer: technical note. *Neurosurgery* 1986; **19**: 91–2.

265. Betsch HM, Aschoff A. [Measurement artifacts in Gaeltec intracranial pressure monitors due to radio waves from personal beeper systems]. *Anasthesiologie, Intensivmedizin, Notfallmedizin, Schmerztherapie* 1992; **27**: 51–2.

266. Ostrup RC, Luerssen TG, Marshall LF, Zornow MH. Continuous monitoring of intracranial pressure with a miniaturized fibreoptic device. *Journal of Neurosurgery* 1987; **67**: 206–9.

267. Crutchfield JS, Narayan RK, Robertson CS, Michael LH. Evaluation of a fibreoptic intracranial pressure monitor. *Journal of Neurosurgery* 1990; **72**: 482–7.

268. Weinstabl C, Richling B, Plainer B, Czech T, Spiss CK. Comparative analysis between epidural (Gaeltec) and subdural (Camino) intracranial pressure probes. *Journal of Clinical Monitoring* 1992; **8**: 116–20.

269. Yablon JS, Lantner HJ, McCormack TM, Nair S, Barker E, Black P. Clinical experience with a fibreoptic intracranial pressure monitor. *Journal of Clinical Monitoring* 1993; **9**: 171–5.

270. Schickner DJ, Young RF. Intracranial pressure monitoring: fibreoptic monitor compared with the ventricular catheter. *Surgical Neurology* 1992; **37**: 251–4.

271. Osullivan MG, Statham PF, Jones PA, et al. Role of intracranial pressure monitoring in severely head-injured patients without signs of intracranial hypertension on initial computerized tomography. *Journal of Neurosurgery* 1994; **80**: 46–50.

272. Chan KH, Dearden NM, Miller JD, Andrews PJ, Midgley S. Multimodality monitoring as a guide to treatment of intracranial hypertension after severe brain injury. *Neurosurgery* 1993; **32**: 547–53.

273. Ascher NL, Lake JR, Emond JC, Roberts JP. Liver transplantation for fulminant hepatic failure. *Archives of Surgery* 1993; **128**: 677–82.

274. Blei AT, Olafsson S, Webster S, Levy R. Complications of intracranial pressure monitoring in fulminant hepatic failure. *Lancet* 1993; **341**: 157–8.

275. Hoedt-Rasmussen K, Sveinsdottir E, Lassen NA. Regional cerebral blood flow in man determined by intra-arterial injection of radioactive inert gas. *Circulation Research* 1966; **18**: 237–47.

276. Olesen J, Paulson OB, Lassen NA. Regional cerebral blood flow in man determined by the initial slope of the clearance of intra-arterially injected ^{133}Xe: theory of the method, normal values, error of measurement, correction for remaining radioactivity, relation to other flow parameters and response to $PaCO_2$ changes. *Stroke* 1971; **2**: 519–40.

277. Lassen NA, Henriksen L, Paulson OB. Regional cerebral blood flow in stroke by 133-xenon inhalation using emission tomography. *Stroke* 1981; **12**: 284–8.

278. Prough DS, Rogers AT. What are the normal levels of cerebral blood flow and cerebral oxygen consumption during cardiopulmonary bypass in humans? *Anesthesia and Analgesia* 1993; **76**: 690–3.

279. Dickman CA, Carter LP, Baldwin HZ, Harrington T, Tallman D. Continuous regional cerebral blood flow monitoring in acute craniocerebral trauma. *Neurosurgery* 1991; **28**: 467–72.

280. Carter LP, Weinand ME, Oommen KJ. Cerebral blood flow (CBF) monitoring in intensive care by thermal diffusion. *Acta Neurochirurgica Supplementum (Wien)* 1993; **59**: 43–6.

281. Tamaki N, Ehara K, Fujita K, Shirakuni T, Asada M, Yamashita H. Cerebral hyperfusion during surgical resection of high-flow arteriovenous malformations. *Surgical Neurology* 1993; **40**: 10–15.

282. Steinmeier R, Bondar I, Bauhuf C. Assessment of cerebral haemodynamics in comatose patients by laser Doppler flowmetry – preliminary observations. *Acta Neurochirurgica Supplementum (Wien)* 1993; **59**: 69–73.

283. Haberl RL, Villringer A, Dirnagl U. Applicability of laser-Doppler flowmetry for cerebral blood flow monitoring in neurological intensive care. *Acta Neurochirurgica Supplementum (Wien)* 1993; **59**: 64–8.

284. Nicholas GG, Hashemi H, Gee W, Reed JF. The cerebral hyperfusion syndrome – diagnostic value of ocular pneumoplethysmography. *Journal of Vascular Surgery* 1993; **17**: 690–5.

285. Aaslid R, Markwalder TM, Nornes H. Noninvasive transcranial Doppler ultrasound recording of flow velocity in basal cerebral arteries. *Journal of Neurosurgery* 1982; **57**: 769–74.

286. Itoh T, Matsumoto M, Handa N, et al. Rate of successful recording of blood flow signals in the middle cerebral artery using transcranial doppler sonography. *Stroke* 1993; **24**: 1192–5.

287. Giller CA, Bowman G, Dyer H, *et al.* Cerebral arterial diameters during changes in blood pressure and carbon dioxide during craniotomy. *Neurosurgery* 1993; **32**: 737–42.

288. Laumer R, Steinmeier R, Gonner F, Vogtmann T, Priem R, Fahlbusch R. Cerebral hemodynamics in subarachnoid hemorrhage evaluated by transcranial doppler sonography. 1. Reliability of flow velocities in clinical management. *Neurosurgery* 1993; **33**: 1–9.

289. Mancini M, Dechiara S, Postiglione A, Ferrara LA. Transcranial doppler evaluation of cerebrovascular reactivity to acetazolamide in normal subjects. *Artery* 1993; **20**: 231–41.

290. Hedera P, Bujdakova J, Traubner P. Blood flow velocities in basilar artery during rotation of the head. *Acta Neurologica Scandinavica* 1993; **88**: 229–33.

291. Macko RF, Ameriso SF, Akmal M, *et al.* Arterial oxygen content and age are determinants of middle cerebral artery blood flow velocity. *Stroke* 1993; **24**: 1025–8.

292. Werner C, Hoffman WE, Baughman VL, Albrecht RF, Shulte J. Effects of sufentanil on cerebral blood flow velocity, and metabolism in dogs. *Anesthesia and Analgesia* 1991; **72**: 177–81.

293. Kochs E, Hoffman WE, Werner C, Albrecht RF, Schulte J. Cerebral blood flow velocity in relation to cerebral blood flow, cerebral metabolic rate for oxygen, and electroencephalogram analysis during isoflurane anesthesia in dogs. *Anesthesia and Analgesia* 1993; **76**: 1222–6.

294. Werner C, Kochs E, Hoffman WE, Blanc IF, Esch JSA. Cerebral blood flow and cerebral blood flow velocity during angiotensin-induced arterial hypertension in dogs. *Canadian Journal of Anaesthesia* 1993; **40**: 755–60.

295. Ries F, Honisch C, Lambertz M, Schlief R. A transpulmonary contrast medium enhances the transcranial doppler signal in humans. *Stroke* 1993; **24**: 1903–9.

296. Chiesa R, Minicucci F, Melissano G, *et al.* The role of transcranial Doppler in carotid artery surgery. *European Journal of Vascular Surgery* 1992; **6**: 211–16.

297. Bergeron P, Benichou H, Rudondy P, Jausseran JM, Ferdani M, Courbier R. Stroke prevention during carotid surgery in high risk patients (value of transcranial Doppler and local anesthesia). *Journal of Cardiovascular Surgery (Torino)* 1991; **32**: 713–19.

298. McDowell Jr HA, Gross GM, Halsey JH. Carotid endarterectomy monitored with transcranial Doppler. *Annals of Surgery* 1992; **215**: 514–18.

299. Jørgensen LG, Schroeder TV. Transcranial Doppler for detection of cerebral ischaemia during carotid endarterectomy. *European Journal of Vascular Surgery* 1992; **6**: 142–7.

300. Romner B, Bergqvist D, Lindblad B. Blood flow velocity in the middle cerebral artery and carotid artery stump pressure during carotid endarterectomy. *Acta Neurochirurgica (Wien)* 1993; **121**: 130–4.

301. Naylor AR, Wildsmith JAW, McClure J, Jenkins AM, Ruckley CV. Transcranial doppler monitoring during carotid endarterectomy. *British Journal of Surgery* 1991; **78**: 1264–8.

302. Jansen C, Vriens EM, Eikelboom BC, Vermeulen FEE, Vangijn J, Ackerstaff RGA. Carotid endarterectomy with transcranial doppler and electroencephalographic monitoring

303. Schneider PA, Rossman ME, Torem S, Otis SM, Dillery RB, Bernstein EF. Transcranial Doppler in the management of extracranial cerebrovascular disease: implications in diagnosis and monitoring. *Journal of Vascular Surgery* 1988; **7**: 223–31.

304. Thiel A, Russ W, Zeiler D, Dapper F, Hempelmann G. Transcranial Doppler sonography and somatosensory evoked potential monitoring in carotid surgery. *European Journal of Vascular Surgery* 1990; **4**: 597–602.

305. Halsey Jr JH. The International Transcranial Doppler Collaborators: risks and benefits of shunting in carotid endarterectomy. *Stroke* 1992; **23**: 1583–7.

306. Lundar T, Lindegaard KF, Frøysaker T, Aaslid R, Grip A, Nornes H. Dissociation between cerebral autoregulation and carbon dioxide reactivity during nonpulsatile cardiopulmonary bypass. *Annals of Thoracic Surgery* 1985; **40**: 582–7.

307. Endoh H, Shimoji K. Changes in blood flow velocity in the middle cerebral artery during nonpulsatile hypothermic cardiopulmonary bypass. *Stroke* 1994; **25**: 403–7.

308. Markus H. Transcranial doppler detection of circulating cerebral emboli – a review. *Stroke* 1993; **24**: 1246–50.

309. Taylor RH, Burrows FA, Bissonnette B. Cerebral pressure–flow velocity relationship during hypothermic cardiopulmonary bypass in neonates and infants. *Anesthesia and Analgesia* 1992; **74**: 636–42.

310. Astudillo R, Vanderlinden J, Ekroth R, *et al.* Absent diastolic cerebral blood flow velocity after circulatory arrest but not after low flow in infants. *Annals of Thoracic Surgery* 1993; **56**: 515–19.

311. Sekhar LN, Wechsler LR, Yonas H, Luyckx K, Obrist W. Value of transcranial Doppler examination in the diagnosis of cerebral vasospasm after subarachnoid hemorrhage. *Neurosurgery* 1988; **22**: 813–21.

312. Lennihan L, Petty GW, Fink ME, Solomon RA, Mohr JP. Transcranial doppler detection of anterior cerebral artery vasospasm. *Journal of Neurology, Neurosurgery and Psychiatry* 1993; **56**: 906–9.

313. Grosset DG, Straiton J, Mcdonald I, Cockburn M, Bullock R. Use of transcranial doppler sonography to predict development of a delayed ischemic deficit after subarachnoid hemorrhage. *Journal of Neurosurgery* 1993; **78**: 183–7.

314. Hurst RW, Schnee C, Raps EC, Farber R, Flamm ES. Role of transcranial doppler in neuroradiological treatment of intracranial vasospasm. *Stroke* 1993; **24**: 299–303.

315. Gosling RG, Dunbar G, King DH. The quantitative analysis of occlusive peripheral arterial diseases by a non-intrusive ultrasonic technique. *Angiology* 1971; **22**: 52–5.

316. Gosling RG, King KH. Arterial assessment by Doppler-shift ultrasound. *Proceedings of the Royal Society of Medicine* 1974; **67**: 447–9.

317. Pourcelot L. Diagnostic ultrasound for cerebral vascular disease. In: Donald J, Levis S, eds. *Present and future of diagnostic ultrasound*. Rotterdam: Kooyker Scientific Publications, 1976: 141–7.

318. Steinmeier R, Laumer R, Bondar I, Priem R, Fahlbusch R. Cerebral hemodynamics in subarachnoid hemorrhage evaluated by transcranial doppler sonography. 2. Pulsatility

– a prospective study in 130 operations. *Stroke* 1993; **24**: 665–9.

indices – normal reference values and characteristics in subarachnoid hemorrhage. *Neurosurgery* 1993; **33**: 10–19.

319. Camerlingo M, Casto L, Censori B, Ferraro B, Gazzaniga GC, Mamoli A. Transcranial doppler in acute ischemic stroke of the middle cerebral artery territories. *Acta Neurologica Scandinavica* 1993; **88**: 108–11.

320. Zanette EM, Pozzilli C, Roberti C, Toni D, Lenzi GL, Fieschi C. Transcranial doppler ultrasonography and single photon emission tomography following cerebral infarction. *Cerebrovascular Diseases* 1993; **3**: 370–4.

321. Massaro AR, Hoffmann M, Sacco RL, Ditullio M, Homma S, Mohr JP. Detection of paradoxical cerebral embolism using transcranial doppler in a patient with infarct of undetermined cause. *Cerebrovascular Diseases* 1993; **3**: 116–19.

322. Spencer MP. Decompression limits for compressed air determined by ultrasonically detected blood bubbles. *Journal of Applied Physiology* 1976; **2**: 229–35.

323. Butler BD, Robinson R, Fife C, Sutton T. Doppler detection of decompression bubbles with computer assisted digitization of ultrasonic signals. *Aviation Space and Environmental Medicine* 1991; **62**: 997–1004.

324. Markus H, Loh A, Brown MM. Computerized detection of cerebral emoli and discrimination from artifact using doppler ultrasound. *Stroke* 1993; **24**: 1667–72.

325. Homburg AM, Jakobsen M, Enevoldsen E. Transcranial doppler recordings in raised intracranial pressure. *Acta Neurologica Scandinavica* 1993; **87**: 488–93.

326. Cardoso ER, Kupchak JA. Evaluation of intracranial pressure gradients by means of transcranial Doppler sonography. *Acta Neurochirurgica Supplementum (Wien)* 1992; **55**: 1–5.

327. Shigemori M, Kikuchi N, Tokutomi T, *et al.* Monitoring of severe head-injured patients with transcranial Doppler (TCD) ultrasonography. *Acta Neurochirurgica Supplementum (Wien)* 1992; **55**: 6–7.

328. Chan KH, Dearden NM, Miller JD. Transcranial Doppler-sonography in severe head injury. *Acta Neurochirurgica Supplementum (Wien)* 1993; **59**: 81–5.

329. Sander D, Klingelhofer J. Cerebral vasospasm following post-traumatic subarachnoid hemorrhage evaluated by transcranial doppler ultrasonography. *Journal of the Neurological Sciences* 1993; **119**: 1–7.

330. Chan KH, Dearden NM, Miller JD. The significance of posttraumatic increase in cerebral blood flow velocity: a transcranial Doppler ultrasound study. *Neurosurgery* 1992; **30**: 697–700.

331. Goraj B, Rifkinsonmann S, Leslie DR, Kasoff SS, Tenner MS. Cerebral blood flow velocity after head injury – transcranial doppler evaluation – work in progress. *Radiology* 1993; **188**: 137–41.

332. Martin NA, Doberstein C, Zane C, Caron MJ, Thomas K, Becker DP. Posttraumatic cerebral arterial spasm: transcranial Doppler ultrasound, cerebral blood flow, and angiographic findings. *Journal of Neurosurgery* 1992; **77**: 575–83.

333. Sanker P, Richard KE, Weigl HC, Klug N, van Leyen K. Transcranial Doppler sonography and intracranial pressure monitoring in children and juveniles with acute brain injuries or hydrocephalus. *Childs Nervous System* 1991; **7**: 391–3.

334. Goh D, Minns RA. Cerebral blood flow velocity monitoring in

pyogenic meningitis. *Archives of Disease in Childhood* 1993; **68**: 111–19.

335. Haring HP, Rotzer HK, Reindl H, *et al.* Time course of cerebral blood flow velocity in central nervous system infections – a transcranial doppler sonography study. *Archives of Neurology* 1993; **50**: 98–101.

336. Klingelhöfer J, Conrad B, Benecke R, Sander D. Intracranial flow patterns at increasing intracranial pressure. *Klinische Wochenschrift* 1987; **65**: 542–5.

337. Larson CP, Ehrenfeld WK, Wade JG, Wylie EJ. Jugular venous oxygen saturation as an index of adequacy of cerebral oxygenation. *Surgery* 1967; **62**: 31–9.

338. Cruz J, Miner ME, Allen SJ, Alves WM, Gennarelli TA. Continuous monitoring of cerebral oxygenation in acute brain injury: injection of mannitol during hyperventilation. *Journal of Neurosurgery* 1990; **73**: 725–30.

339. Raggueneau JL, Bellec C, Paoletti C, *et al.* [Evaluation of cerebral blood flow in severe brain injuries by SVO_2 measurement in the internal jugular vein]. *Agressologie* 1991; **32**: 369–74.

340. Kuwabara M, Nakajima N, Yamamoto F, *et al.* Continuous monitoring of blood oxygen saturation of internal jugular vein as a useful indicator for selective cerebral perfusion during aortic arch replacement. *Journal of Thoracic and Cardiovascular Surgery* 1992; **103**: 355–62.

341. Dearden NM, Midgley S. Technical considerations in continuous jugular venous oxygen saturation measurement. *Acta Neurochirurgica Supplementum (Wien)* 1993; **59**: 91–7.

342. Cruz J. Cerebral oxygenation. Monitoring and management. *Acta Neurochirurgica Supplementum (Wien)* 1993; **59**: 86–90.

343. Yamada M, Kawaguchi M, Furuya H, *et al.* [Perioperative monitoring of cerebral circulation in a patient with Takayasu disease]. *Masui* 1991; **40**: 1546–50.

344. Stocchetti N, Paparella A, Bridelli F, *et al.* Cerebral venous oxygen saturation studied with bilateral samples in the internal jugular veins. *Neurosurgery* 1994; **34**: 38–44.

345. Dearden NM. Jugular bulb venous oxygen saturation in the management of severe head injury. *Current Opinion in Anaesthesia* 1991; **4**: 279–86.

346. Marion DW, Darby J, Yonas H. Acute regional cerebral blood flow changes caused by severe head injuries. *Journal of Neurosurgery* 1991; **74**: 407–14.

347. Cruz J, Raps EC, Hoffstad OJ, Jaggi JL, Gennarelli TA. Cerebral oxygenation monitoring *Critical Care Medicine* 1993; **21**: 1242–6.

348. Brazy JE. Cerebral oxygen monitoring with near infrared spectroscopy: clinical application to neonates. *Journal of Clinical Monitoring* 1991; **7**: 325–34.

349. McCormick PW, Stewart M, Goetting MG, Dujovny M, Lewis G, Ausman JI. Noninvasive cerebral optical spectroscopy for monitoring cerebral oxygen delivery and hemodynamics. *Critical Care Medicine* 1991; **19**: 89–97.

350. Harris DN, Bailey SM. Near infrared spectroscopy in adults. Does the Invos 3100 really measure intracerebral oxygenation? *Anaesthesia* 1993; **48**: 694–6.

351. Rampil IJ, Litt L, Mayevsky A. Correlated, simultaneous, multiple-wavelength optical monitoring in vivo of localized cerebrocortical NADH and brain microvessel hemoglobin

oxygen saturation. *Journal of Clinical Monitoring* 1992; **8**: 216–25.

352. Dekock JP, Tarassenko L, Glynn CJ, Hill AR. Reflectance pulse oximetry measurements from the retinal fundus. *IEEE Transactions on Biomedical Engineering* 1993; **40**: 817–23.

353. Grundy BL. Electrophysiologic monitoring: electroencephalography and evoked potentials. In: Newfield P, Cottrell JE, eds. *Neuroanesthesia: Handbook of clinical and physiologic essentials.* Boston: Little, Brown and Company, 1991: 30–58.

354. Bickford RG. Computer analysis of background activity. In: Remond A, ed. *EEG informatics: a didactic review of methods and applications of EEG data processing.* Amsterdam: Elsevier, 1977: 215–32.

355. Grundy BL. The EEG and evoked potential monitoring. In: Blitt CD, Hines R, eds. *Monitoring in anesthesia and critical care medicine* New York: Churchill Livingstone, 1994; 423–89.

356. Sokol S. Visual evoked potentials. In: Aminoff MJ, ed. *Electrodiagnosis in clinical neurology.* New York: Churchill Livingstone, 1992: 441–66.

357. Shapiro HM. Intracranial hypertension: therapeutic and anesthetic considerations. *Anesthesiology* 1975; **43**: 445–71.

358. Brazy JE. Cerebral oxygen monitoring with near infrared spectroscopy: clinical application to neonates. *Journal of Clinical Monitoring* 1991; **7**: 325–34.

Chapter 25

Allergic Drug Reactions in Anaesthesia, Pathophysiology and Management

Michael E. Weiss and Carol A. Hirshman

Mechanisms of allergic drug reactions
Classification of allergic reactions
Anaphylaxis/mast cell activation
Clinical manifestations of anaphylaxis

Treatment of anaphylaxis
Determining the cause of allergic reactions
Specific allergic reactions often seen by the anaesthetist

Fatal drug reactions have been estimated to occur in 0.1 per cent of medical inpatients and 0.01 per cent of surgical inpatients[1]. Anaphylactic reactions occur between 1:5000 and 1:25 000 anaesthetic cases in Australia, with a 3.4 per cent mortality.[2] Drug sensitization has likewise been implicated in 4.3 per cent of deaths and 5.6 per cent of cerebral damage in cases of anaesthetic mishaps reported in the United Kingdom.[2] The incidence of anaphylaxis during anaesthesia in the United States is unknown.

Mechanisms of allergic drug reactions

Allergic reactions to drugs are caused by the interaction between drugs or their metabolites and various effector cells of the immune system. Since most drugs are small organic molecules (molecular weights less than 1000 daltons), they are not readily recognized by immune effector cells. Therefore, it is usually necessary for the drug or the drug metabolite to combine with host proteins to form a drug–protein (or drug metabolite–protein) complex which is capable of stimulating a host immune response. This drug–protein complex must have multiple antigenic sites to stimulate both initial sensitization and subsequent allergic reactions. Initial exposure and sensitization may induce production of drug-specific antibodies or T-lymphocytes with drug-specific receptors on the cell surface. Readministration of the drug at a later date may induce an allergic drug reaction in a sensitized individual (Fig. 25.1).

Classification of allergic reactions

Gell and Coombs initially classified four types of immunopathological reactions (Table 25.1).[3]

Type I reactions – immediate hypersensitivity

These reactions result from the interaction of antigens with preformed antigen-specific IgE antibodies that are bound to tissue mast cells and/or circulating basophils via high affinity IgE receptors. Cross-linking two or more IgE receptors by multivalent drug (antigen) leads to the release of both preformed and newly generated mediators. Release of these mediators can lead to urticaria, laryngeal oedema, and bronchospasm with or without cardiovascular collapse.

Type II reactions – cytotoxic antibodies

These reactions result when IgG or IgM antibody reacts with a cell-bound antigen (i.e. blood group antigens, penicillin determinants bound to RBCs). The antigen–antibody interaction activates the complement system resulting in cell lysis. Type II reactions may also be complement independent. IgG or IgM antibody may bind to cell membrane bound antigen, resulting in neutrophil or macrophage attachment and activation via an IgG or IgM Fc receptor. This opsonization can result in injury to the

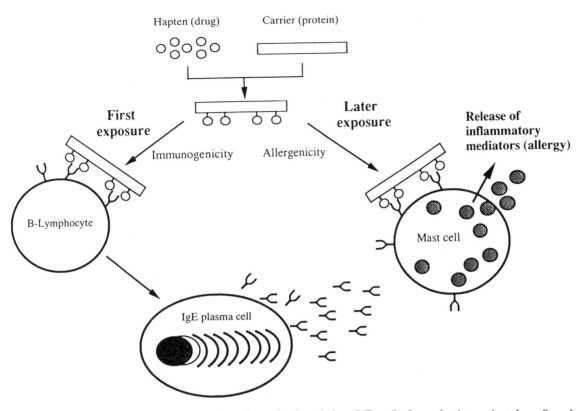

Fig. 25.1 Illustration of drug hapten combining with carrier molecule to induce IgE antibody production against drug. On subsequent exposure, drug combines with IgE antibody on mast cell surface, leading to anaphylactic reaction. (Reprinted with permission from Weiss ME, Hirshman C. Anaphylactic reactions and anesthesia. In: Rogers M, Tinker T, Covino B, eds. *Principles and practice of anesthesiology.* St Louis: CV Mosby Co, 1992: 2457–81.)

Table 25.1 Classification of immunopathological reactions according to the scheme of Gell and Coombs

Type of reaction	Description	Antibody	Cells	Other	Clinical reactions
I	Anaphylactic (reagenic) Immediate hypersensitivity	IgE	Basophils, mast cells		Anaphylaxis, urticaria
II	Cytotoxic or cytolytic	IgG, IgM	Any cell with isoantigen	C', RES	Coombs + haemolytic anaemia; drug induced nephritis; transfusion reaction; Rh disease
III	Immune complex disease	Soluble immune complexes (Ag–Ab)	None directly	C'	Serum sickness; drug fever; glomerulonephritis
IV	'Delayed' or cell-mediated hypersensitivity	None known	Sensitized T-lymphocytes		Contact dermatitis
V	Ideopathic	? ? ? ?	? ? ? ?	Maculopapular eruptions Eosinophilia Stevens–Johnson syndrome Exfoliative dermatitis

C', complement; RES, reticuloendothelial system; Ag–Ab, antigen–antibody; (?) indicate that the immunopathological mechanism is in doubt.
Reprinted with permission from Weiss ME, Adkinson Jr NF. *Clinical Allergy* 1988; **18**: 515–40.

antigen laden cell. Examples of type II reactions include ABO-incompatible transfusion reactions and drug-induced haemolytic anaemia, leucopenia, or thrombocytopenia.

Type III reactions – immune complexes (Arthus reaction)

Antigen-specific IgG or IgM antibodies form circulating complexes with antigens. The complexes lodge in tissue sites, fix complement, and attract polymorphonuclear leucocytes which attempt to phagocytize the immune complexes. The release of potent enzymes from the phagocytic cells results in tissue damage. Immune complex reactions typically appear 7–14 days after the continual exposure of antigen. Examples include serum sickness and possibly drug fever.

Type IV reactions – cell-mediated hypersensitivity

These reactions are not mediated by an antibody, but rather by T-lymphocytes that are specifically sensitized to recognize a particular antigen. After being modified by antigen-processing cells (i.e. macrophages, Langerhan cells), the modified antigen is presented in association with MHC class II molecules to the T-lymphocyte. The sensitized T-lymphocyte recognizes the processed antigen through an antigen-specific T-cell receptor. This triggers the T-cell to release cytokines which orchestrate the immune response by recruiting and stimulating proliferation of other lymphocytes and mononuclear cells which ultimately cause tissue inflammation and injury. Examples of cell-mediated immune reactions are tuberculin skin testing, contact dermatitis (i.e. poison ivy), and graft vs host disease.

Idiopathic reactions

Some reactions have obscure immunological pathogenesis and are not included in Gell and Coombs' classification system. These reactions include various skin eruptions such as pruritus, maculopapular (morbilliform) exanthemas, erythema multiforme, erythema nodosum, photosensitivity reactions, fixed drug reactions and exfoliative dermatitis.

Other drug reactions caused by unknown mechanisms include the Stevens–Johnson syndrome involving rash (usually erythema multiforme) plus the involvement of two or more mucous membranes and Lyell's syndrome, also known as toxic epidermal necrolysis (epithelial bullae and desquamation). Both Stevens–Johnson and Lyell's syndromes are often associated with the administration of a medication, but they can occur in situations in which no medications have been recently given. The medications that are most frequently associated with Stevens–Johnson syndrome and Lyell's syndromes include phenobarbitone (phenobarbital), nitrofurantoin, trimethoprim–sulphamethoxazole, penicillins (especially ampicillin and amoxycillin), phenytoin and oxyphenbutazone.[4,5] Patients who have had Stevens–Johnson or Lyell's syndromes associated with a specific medication should not receive that medication in the future. Attempts to desensitize individuals to medications causing these reactions are also not recommended.

Interstitial and/or alveolar pneumonitis, pulmonary oedema and fibrosis have occurred secondary to drug administration. The mechanism involved in these pulmonary hypersensitivity reactions is unclear. Cholestatic and hepatocellular inflammatory liver changes have been associated with certain medications. Fortunately the hepatic inflammation usually subsides when the offending drug is withdrawn. Interstitial nephritis, glomerulonephritis and the nephrotic syndrome have been induced secondary to medication administration. Generalized lymphadenopathy has occasionally been described in patients receiving phenytoin, sulphonamide, and penicillin medications. Lists of medications most frequently associated with the different types of above reactions can be found in references.[6,7]

Levine proposed classifying adverse reactions to an antigen such as penicillin according to their time of onset (Table 25.2).[8] Immediate reactions occur within the first hour after antigen administration; it may involve anaphylaxis, laryngeal oedema, urticaria, or wheezing. Accelerated reactions occur 1–72 hours after antigen presentation; they most commonly involve urticaria. Late reactions begin more than 72 hours after the onset of therapy: anaphylaxis does not occur in the course of continuous antigen therapy; maculopapular eruptions are most common, but type II–IV reactions also occur in this time frame. Allergic reactions may also be classified according to their predominant clinical manifestations as seen in Table 25.3.

Anaphylaxis/mast cell activation

Anaphylaxis is characterized by a sudden and often unexpected onset; even with treatment, catastrophe may ensue. For this reason, it is considered to be the most devastating form of drug allergy. Release of vasoactive and bronchoconstrictive mediators which accompanies activation of mast cells and basophils is the unifying aetiology of

Table 25.2 Classification of allergic reactions based on their time of onset

Reaction type	Onset	Clinical reactions
Immediate	0–1 h	Anaphylaxis
		Hypotension
		Laryngeal oedema
		Urticaria/angioedema
		Wheezing
Accelerated	1–72 h	Urticaria/angioedema
		Laryngeal oedema
		Wheezing
Late	> 72 h	Morbilliform rash
		Interstitial nephritis
		Haemolytic anaemia
		Neutropenia
		Thrombocytopenia
		Serum sickness
		Drug fever
		Stevens–Johnson syndrome
		Exfoliative dermatitis

Table 25.3 Classification of allergic reactions according to their predominant clinical manifestations

Anaphylaxis	Laryngeal oedema
	Hypotension
	Bronchospasm
Cutaneous reactions	Urticaria/angioedema
	Vasculitis
	Stevens–Johnson syndrome
	Exfoliative dermatitis
	Contact sensitivity
	Fixed drug eruption
	Toxic epidermal necrolysis
	Pruritus
	Maculopapular (morbilliform) rash
	Erythema multiforme
	Erythema nodosum
	Photosensitivity reactions
Destruction of formed elements of blood	Haemolytic anaemia
	Neutropenia
	Thrombocytopenia
Renal reactions	Interstitial nephritis
	Glomerulonephritis
	Nephrotic syndrome
Serum sickness	
Drug fever	
Sytemic vasculitis	
Lymphadenopathy	

the anaphylactic syndromes. Overwhelming and precipitous physiological changes associated with the anaphylactic syndrome result directly from release of these mediators, and are characterized by urticaria, laryngeal

oedema, nausea, vomiting, abdominal pain, diarrhoea, bronchospasm and cardiovascular collapse. 'Anaphylaxis' is the term classically applied to IgE antibody-mediated allergic reactions. Similar signs and symptoms may occur without release of IgE globulins, a phenomenon previously called anaphylactoid reactions. Throughout this chapter, we will refer to the clinical syndrome as 'anaphylaxis'; the term will not necessarily suggest a specific mechanism.

IgE-mediated anaphylaxis

Anaphylaxis typically results from initial stimulation with an antigen, usually a protein, capable of inducing production of IgE globulins; subsequent re-exposure triggers production and release of IgE antibodies; these, in turn, may form a complex with both tissue mast cells and circulating basophils, both of which contain high affinity IgE receptors.[9] This attachment, occurring via the Fc region of the IgE molecule, allows the Fab locus of the IgE antibody to bind to the antigen. Re-exposure to either antigen or hapten requires antigens that are functionally multivalent, i.e. have two or more antigenic sites. This results in direct bridging of cell surface IgE receptor molecules and subsequent cross-linking with the IgE antibody. In turn, this is followed by activation of membrane-associated enzymes and a complex biochemical cascade producing an influx of extracellular calcium. The cascade continues with mobilization of intracellular calcium which activates the release of preformed granule-associated mediators as well as synthesis of new mediators from cell membrane phospholipids (see Fig. 25.1)[10,11]

This chapter will examine several clinically significant examples of antigens associated with IgE antibody-mediated allergic reactions: insulin, chymopapain, muscle relaxants, latex, and β-lactam antibiotics.

Complement-mediated reactions

Activation of the complement system produces the membrane attack unit (C5b through to 9) and synthesis of anaphylatoxins, low molecular weight peptides C3a, C4a, and C5a (Fig. 25.2).[12] Anaphylatoxins are responsible for major pathology by releasing mediators from mast cells and basophils. In addition, they directly increase vascular permeability, produce a smooth muscle contraction, aggregate platelets, and stimulate macrophages to produce thromboxane.[12,13] The complement cascade may be activated through two pathways: classical and alternative. The classical pathway may be triggered when IgG or IgM antibodies bind to antigens. An important example of this is haemolytic ABO-incompatible blood transfusion reactions. Another example of classical pathway complement

Pathway	Activation	Component proteins	Humorally active fragments

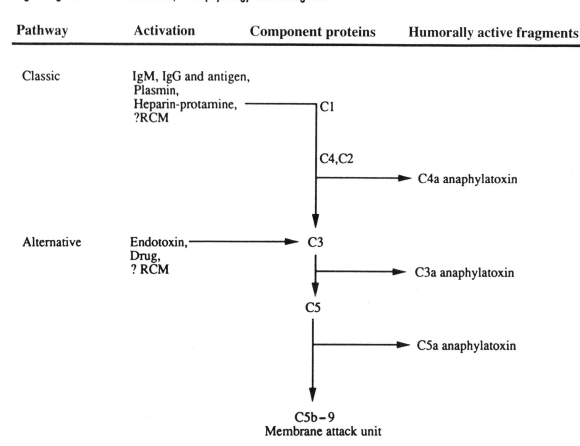

Fig. 25.2 Schematic diagram outlining pathways, stimuli, and humorally active fragments produced from complement activation. (Modified with permission from Levy JH. *Anaphylactic reactions in anesthesia and intensive care*. Boston: Butterworth, 1986.)

activation has been demonstrated for heparin–protamine complexes *in vitro*[14] and *in vivo*.[15,16] Clinical anaphylaxis may occur when preformed immune complexes or IgG aggregates are injected as these are capable of activating complement.[17] Following multiple transfusions, patients deficient in IgA antibodies may develop IgG antibodies to IgA. These antibodies, not unexpectedly, can activate complement and develop anaphylaxis.[18]

The alternative pathway may be stimulated to activate complement by lipopolysaccharides (endotoxins),[19] Althesin,[20] radiocontrast media,[21] and membranes used for cardiopulmonary bypass and dialysis.[22]

Pharmacological (non-immunological) mast cell activators

Mast cell mediator release may be triggered by certain agents through mechanisms whose aetiology is non-immunological. The exact mechanism of this form of mediator release is poorly understood, but release is non-cytotoxic. Agents producing mediator release by mechanisms other than the classical or alternative immune pathways include opiates (especially morphine and

codeine)[23,24] and neuromuscular blocking agents such as atracurium, and *d*-tubocurarine.[25]

Presumed abnormalities of arachidonic acid metabolism

Anaphylactic responses may be precipitated by aspirin and structurally unrelated non-steroidal anti-inflammatory drugs (NSAIDs), a phenomenon occurring in approximately 1 per cent of the population.[26] Adults with asthma are at greater risk; aspirin sensitivity or intolerance occurs in 10–20 per cent, most commonly in middle age.[26] The triad of aspirin-sensitive asthma consists of rhinitis, sinusitis, and nasal polyps. Aspirin intolerance is far from benign, for bronchospasm may be extremely severe.[26] Production of anaphylaxis by aspirin and NSAIDs may be related by their inhibition of cyclo-oxygenase, the enzyme responsible for the conversion of arachidonic acid to prostaglandins and thromboxanes (Fig. 25.3). This inhibition could decrease levels of bronchodilating prostaglandins or, by shunting arachidonic acid to the 5-lipoxygenase pathway, might produce synthesis of vasoactive and bronchoconstrictive leukotrienes.[26] In support of this theory is the finding that

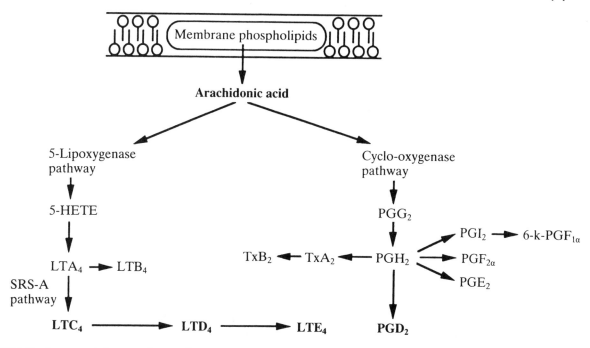

Fig. 25.3 Newly generated vasoactive mediators produced by metabolism of arachidonic acid. Mediators in bold illustrate the predominant products produced by human mast cells. (Reprinted with permission from Weiss ME, Hirshman C. Anaphylactic reactions and anesthesia. In: Rogers M, Tinker T, Covino B, eds. *Principles and practice of anesthesiology.* St Louis: CV Mosby Co, 1992: 2457–81.)

aspirin-sensitive asthmatics are 1000-fold more sensitive to the bronchoconstrictive effects of LTE_4 than are non-sensitive asthmatics, a sensitivity explained by the greater production of LTE_4 when arachidonic acid is shunted through the 5-lipoxygenase pathways.[27]

Mediators of anaphylactic reactions

Any of the mechanisms already discussed may bring about release of mediator from mast cell and/or basophils. These mediators include both preformed compounds stored in granules and substances which are newly synthesized in response to appropriate stimulus. Their release is capable of causing a number of pathophysiological responses, each of which can produce life-threatening anaphylaxis. Preformed and newly synthesized mediators released from mast cells and/or basophils and their physiological and biological actions are seen in Table 25.4.

Clinical manifestations of anaphylaxis

The time from exposure to a triggering agent to the onset of anaphylaxis as well as its manifestations vary from individual to individual. Symptoms usually begin within minutes after exposure, but may not occur for as long as 1–2 hours. Among the many symptoms described by conscious patients is a sense of impending doom. In humans, the primary target organs in humans are skin, gut, and the cardiorespiratory system, the latter being of major importance (Table 25.5). When anaphylaxis occurs outside the operating rooms, 70 per cent of fatalities resulted from respiratory and 24 per cent from cardiovascular complications.[28] During anaesthesia, cardiovascular signs predominate.[29]

Cutaneous manifestations

The first signs and symptoms may include erythema, flushing and pruritus, especially of the palms, soles, and groin. These often progress to urticaria and angioedema.

Gastrointestinal manifestations

In the conscious patient, gastrointestinal signs and symptoms may include nausea, vomiting crampy abdominal pain, and intense, often bloody, diarrhoea.

Table 25.4 Biological actions and physiological manifestations of mast cell mediators

Mediators	Biological action	Physiological manifestations
Preformed		
Histamine	Smooth muscle contraction	Bronchospasm, abdominal pain, diarrhoea, nausea, vomiting
	Vasodilation	Tachycardia, hypotension
	Increases vasopermeability	Oedema, urticaria/angioedema, influx of inflammatory cells
	Stimulates mucus secretion	Excessive respiratory/gastrointestinal secretions, inspissation
ECF-A	Eosinophil chemotaxis	Inflammation
NCA	Neutrophil chemotaxis	Inflammation
Neutral proteases	Cleave amino acids from proteins	?
	Stimulate mucus secretions	?
Proteoglycans (heparin and chondroitin sulphate)	Anticoagulant	?
Newly generated		
PGD$_2$	Smooth muscle contraction	Bronchospasm, abdominal pain, diarrhoea, nausea, vomiting
	Vasodilation	Tachycardia, hypotension
	Stimulates mucus secretion	Excessive respiratory/gastrointestinal secretions, inspissation
	Enhances basophil mediator release	Potentiates reactions
	Inhibits platelet aggregation	?
LTC$_4$/D$_4$/E$_4$ (SRS-A)	Smooth muscle contraction	Bronchospasm, abdominal pain, diarrhoea, nausea, vomiting
	Vasodilation	Tachycardia, hypotension
	Increases vasopermeability	Oedema, urticaria/angioedema, influx of inflammatory cells
	Stimulates mucus secretion	Excessive respiratory/gastrointestinal secretions, inspissation
PAF	Smooth muscle contraction	Bronchospasm, abdominal pain, diarrhoea, nausea, vomiting
	Vasodilation	Tachycardia, hypotension
	Increases vasopermeability	Oedema, urticaria/angioedema, hypotension
	Decreases inotropy of heart	Hypotension
	Neutrophil stimulation	?
	Platelet aggregation	?

Reprinted with permission from Weiss ME, Hirshman C. Anaphylactic reactions and anesthesia. In: Rogers M, Tinker T, Covino B, eds. *Principles and practice of anesthesiology.* St Louis: CV Mosby Co, 1992: 2457–81.

Respiratory manifestations

A life-threatening effect of anaphylaxis is asphyxia resulting from laryngeal oedema; this must always be anticipated and carefully assessed. Early manifestations of upper respiratory involvement in the conscious patient include hoarseness, dysphonia, and increasing respiratory distress. Anaphylaxis involving the lower respiratory tract of a conscious patient will present as shortness of breath, tightness of the chest, cough and wheezing. In the anaesthetized patient, difficulty in ventilating and increased airway pressure will be obvious.

Cardiovascular manifestations

Hypotension and tachycardia are frequent cardiovascular signs of anaphylaxis. Symptoms range from lightheadedness and faintness to overt syncope. Cardiovascular complications are legion – dysrhythmias, myocardial infarction and cardiovascular collapse.

There are numerous other manifestations of anaphylaxis including nasal, ocular and palatal pruritus, sneezing, diaphoresis, disorientation and incontinence. Spontaneous recurrence of anaphylaxis has been reported to occur after a delay of 8–24 hours, a phenomenon equivalent to late-phase reaction.

Intraoperative/perioperative anaphylaxis

Anaphylaxis during surgery is a sudden and frightening challenge, even for the skilled and experienced physician. Yet, if evaluation and treatment are not prompt tragedy will be certain. In this setting, frequent administration of a variety of medications is often the rule, making a specific diagnosis of the offending agent difficult. Early signs and symptoms of anaphylaxis will not be obvious in the unconscious and draped patient. Most important, anaesthetics themselves alter both mediator release and effect, thus delaying prompt recognition of the occurrence of anaphylaxis.[30] As a result, the anaesthetist may suddenly be confronted with cardiovascular collapse, a relatively late

Table 25.5 Clinical manifestations of anaphylaxis

System	Signs and symptoms
Cutaneous	Pruritis
	Flushing
	Erythema
	Urticaria/angioedema
Gastrointestinal	Nausea
	Abdominal pain
	Diarrhoea
	Vomiting
Respiratory	Laryngeal oedema
	Hoarseness
	Dysphonia
	'Lump in throat'
	Chest tightness
	Dyspnoea
	Cough
	Wheezing
	Cyanosis
	Increase in peak airway pressure
Cardiovascular	Light-headedness
	Faintness
	Syncopy
	Tachycardia
	Hypotension
	Dysrhythmias

Signs in bold print are most likely to occur in patients during anaesthesia.
Reprinted with permission from Weiss ME, Hirshman C. Anaphylactic reactions and anesthesia. In: Rogers M, Tinker T, Covino B, eds. *Principles and practice of anesthesiology.* St Louis: CV Mosby Co, 1992: 2457–81.

Table 25.6 Differential diagnosis of anaphylaxis

Vasovagal reaction
Dysrhythmia
Myocardial infarction
Overdose of medication or illicit drugs
Pulmonary embolism
Seizure disorder
Cerebral vascular accident
Aspiration
Globus hystericus
Fictitious asthma
Hereditary angioedema
Physical or idiopathic urticaria
Serum sickness
Carcinoid tumours
Systemic mastocytosis
Irritant bronchospasm
Pulmonary oedema

Reprinted with permission from Weiss ME, Hirshman C. Anaphylactic reactions and anesthesia. In: Rogers M, Tinker T, Covino B, eds. *Principles and practice of anethesiology.* St Louis: CV Mosby Co, 1992: 2457–81.

event in the syndrome, during what appeared to be an uneventful anaesthetic.[31] If continuous invasive haemodynamic monitoring is available during an anaphylactic reaction, one will observe decreased systolic, diastolic, and mean arterial pressure with a reduction in systemic vascular resistance; cardiac output and stroke volume may increase or decrease depending on severity. In the ventilated patient, a sudden increase in airway pressure will herald anaphylaxis-induced decreased pulmonary compliance. Decreased oxygen saturation and cyanosis may be noted.[29]

Differential diagnosis

All 'reactions' are not anaphylaxis. In conscious patients, vasovagal reactions often mimic the more serious response to drug injection (Table 25.6). A vasovagal faint usually is associated with a frightening and painful injection or situation. Although the patient may complain of nausea and lightheadedness and demonstrate syncope with either anaphylaxis or a vasovagal reaction, the latter is accompanied by pallor rather than cyanosis. Vasovagal

reactions are not associated with pruritus or respiratory difficulty; furthermore, symptoms are almost immediately relieved once the patient is supine. Vasovagal reactions are often accompanied by profuse diaphoresis and bradycardia, without flushing, urticaria, angioedema, pruritus, or wheezing. A supine position, reassurance, and, if required, intravenous atropine will usually suffice.

Obvious additions to the differential diagnosis of sudden collapse also include dysrhythmia, myocardial infarction, aspiration of food and foreign body, pulmonary embolism, seizure disorder, hypoglycaemia, and stroke.

Laryngeal oedema, especially when accompanied by abdominal pain, should make one think of hereditary angioedema. Globus hystericus and fictitious asthma need to be considered when respiratory symptoms are present.

While serum sickness and urticaria may bring anaphylaxis to mind, serum sickness generally occurs 6–21 days after the antigenic stimulation and is frequently associated with fever, lymphadenopathy, arthralgias, arthritis, nephritis and neuritis. There are many far less serious conditions that can mimic anaphylaxis, such as cold urticaria (especially if generalized), idiopathic urticaria, carcinoid tumours, systemic mastocytosis and, perhaps, an overdose of medication.

In anaesthetized patients, the differential diagnosis of symptoms associated with anaphylaxis includes any cause of cardiovascular collapse, severe hypotension and bronchospasm, any of which must receive prompt diagnosis and treatment. Things to consider include myocardial infarction or dysrhythmia, drug overdose, pulmonary embolism, irritant-induced bronchospasm, pulmonary oedema, and aspiration of gastric contents (Table 25.6).

Table 25.7 Management of anaphylaxis

Initial therapy

1. Stop administration or reduce absorption of offending agent
   If antigen given s.c.:
   Venous tourniquet proximal to site
   Adrenaline (1:1000) into antigen site
2. Maintain airway and administer 100% O_2
   Aerosolized adrenaline
   Intubation or cycothyrotomy
3. Rapid intravascular volume expansion
   25–50 ml/kg (2–4 litres) of crystalloid or colloid for hypotension
4. Administer adrenaline
   0.01 ml/kg of 1:1000 i.v.
   Titrate as needed
   10 ml of 1:10 000 – tracheal administration
5. Discontinue all anaesthetic agents
6. Consider use of MAST

Secondary therapy

1. Administer antihistamine
   Diphenhydramine 1 mg/kg i.v. or i.m. (maximal dose 50 mg)
   Ranitidine 1 mg/kg i.v. (maximal dose 50 mg)
2. Administer glucocorticoids
   Hydrocortisone – 5 mg/kg initially and then 2.5 mg/kg repeat 4–6 h
   Methylprednisolone – 1 mg/kg initially and repeat 4–6 h
3. Administer aminophylline
   Loading dose – 5–6 mg/kg
   Continuous infusion – 0.4–0.9 mg/kg per hour (check blood level)
4. Administer inhaled β_2-agonists
5. Continuous catecholamine infusion
   Adrenaline – 0.02–0.05 µg/kg per min (2–4 µg/min)
   Noradrenaline – 0.05 µg/kg per min (2–4 µg/min)
   Dopamine – 5–20 µg/kg per min
6. Administer sodium bicarbonate
   0.5–1 mg/kg initially, titrate using ABGs

Reprinted with permission from Weiss ME, Hirshman C. Anaphylactic reactions and anesthesia. In: Rogers M, Tinker T, Covino B, eds. *Principles and practice of anesthesiology*. St Louis: CV Mosby Co, 1992: 2457–81.

Treatment of anaphylaxis

If an anaphylactic reaction is not recognized promptly, death may occur within minutes.[32] As the initiation of therapy is delayed, the expectation of fatality increases.[28] This is not surprising, since the major pathological consequences of anaphylaxis relate to compromised cardiovascular and respiratory function. Meticulous monitoring of vital signs, specifically of blood pressure, airway patency and adequacy of ventilation is crucial in assessing the severity of the reaction and response to therapy. Treatment of anaphylactic reactions can be divided into initial and secondary therapies (Table 25.7).

Initial therapy

It is essential that administration of the antigen be halted and its further absorption prevented. Intravenous infusions can be discontinued promptly, while other routes of administration demand clear thinking and creativity. If the responsible antigen was administered subcutaneously (i.e. insulin or 'desensitization' therapy) a venous tourniquet may be placed proximal to the site and 0.01 mg/kg of aqueous adrenaline (epinephrine) 1:1000 (maximal dose 0.3–0.5 ml) injected directly into the antigen 'depot' in order to reduce the local circulation and absorption of antigen.

One hundred per cent oxygen should be administered through a patent airway; adequacy of oxygenation may be monitored with a pulse oximeter or, if appropriate, arterial blood gases. If the trachea is not already intubated and laryngeal oedema is impinging on the airway, prompt intubation is essential. Oedema of the airway may be treated with aerosolized adrenaline (three inhalations of 0.16–0.20 mg adrenaline per inhalation from a metered dose inhaler) or by a nebulizer (8–15 drops of 2.25 per cent adrenaline in 2 ml normal saline). If upper airway compromise is refractory to these measures or is progressing too rapidly, a needle catheter cricothyrotomy or emergency surgical cricothyrotomy may be necessary.

As soon as anaphylaxis is diagnosed, intravenous access must be established (if not already in place) and blood pressure sustained with isotonic crystalloid or colloid. Rapid administration of 25–50 ml/kg (2–4 litres in an adult) or normal saline or lactated Ringer's solution is an essential part of initial therapy. Military antishock trousers (MAST suit) has been shown to be useful in combating anaphylaxis-induced hypotension.[33,34] By compressing the lower part of the body, the MAST suit aids perfusion to vital organs and may also improve conditions for gaining peripheral venous access in the upper extremities.[34]

Adrenaline is the mainstay of initial pharmacological treatment. The dose is 0.01 mg/kg, 1:1000 (maximal dose 0.3–0.5 ml) s.c. or i.m. Treatment may be repeated every 15–20 minutes as required. If hypotension or laryngospasm is severe, 0.1 mg/kg of 1:10 000 adrenaline (100 µg/ml) should be given intravenously (maximal dose, 1–5 ml). If warranted by specific conditions, this dosage may be increased. Obvious risks of intravenous adrenaline include cardiac arrhythmias, myocardial infarction, or stroke. If an intravenous line is not immediately available, 0.5 ml of 1:1000 adrenaline should be given intramuscularly, or 10 ml of 1:10 000 adrenaline administered through the tracheal tube. It should be remembered that in the presence of impaired circulation accompanying anaphylaxis, absorption of intramuscular or subcutaneous adrenaline is most unreliable.

General anaesthesia must be discontinued; the inhalation anaesthetics are negative inotropes, decreasing systemic vascular resistance and interfering with reflexes which normally compensate for hypotension. Since comparatively high doses of catecholamines will be used to treat many of the manifestations of anaphylaxis, drugs which sensitize the heart to these agents, e.g. halothane, may be hazardous.

Secondary treatment

After the most emergent symptoms have been stabilized, use of other pharmacological agents may be warranted.

Antihistamines, e.g. diphenhydramine 1 mg/kg (up to 50 mg), administered intravenously or intramuscularly, are useful in relieving symptoms of pruritus. Evidence that H_2-receptor antagonists are useful in treating anaphylaxis is lacking; however, some workers believe that ranitidine, 1 mg/kg i.v., is reasonable when hypotension is persistent since peripheral vasodilation may be exacerbated by the effects of histamine on endothelial H_2-receptors. Because intravenous cimetidine has been reported to cause short-term hypotension, it is wise to avoid its use.[35]

Glucocorticoids are considered to be useful as prophylaxis of possible late-phase reactions, and in treating persistent bronchospasm. This family of drugs is not appropriate in the initial treatment of anaphylaxis since they have no immediate effects. Hydrocortisone, 5 mg/kg (up to 200 mg initial dose) and then 2.5 mg/kg every 6 hours or methylprednisolone 1 mg/kg initially and every 6 hours as indicated may be given.

If bronchospasm persists, intravenous aminophylline is a useful therapy. A loading dose of 5–6 mg/kg should be followed by a continuous infusion of 0.4–0.9 mg/kg per hour. Aminophylline produces an acute release of catecholamines;[36] this may be responsible for its amelioration of airway constriction. It may be more prudent to use β_2-agonists by aerosol.

Infusions of catechols are useful in treating persistent hypotension. Adrenaline has two potential benefits when treating combined hypotension and bronchospasm. The suggested starting dose of adrenaline is 0.02–0.05 µg/kg per minute (2–4 µg/min). Evaluation of tissue perfusion is a rational endpoint of titration. Because tachycardia may be bothersome when more than 8–10 µg/min of adrenaline are required, noradrenaline may be an effective substitute. Its suggested starting dose is 0.05 µg/kg per minute (2–4 µg/min); again, this should be titrated to maintain tissue perfusion. Blood pressure may also be maintained with dopamine. A dose of 5–20 µg/kg per minute will maintain

cardiac output, and may very well improve coronary, cerebral, renal and mesenteric blood flow.

If acidosis is diagnosed, intravenous sodium bicarbonate in an initial dose of 0.5–1 mg/kg is appropriate. Acid–base status must be monitored using arterial blood gas levels to guide further therapeutic interventions.

Response to therapy is usually prompt; however, in spite of prompt and rational therapy, not all patients respond quickly. Thus, even when physicians were absolutely prepared to treat anaphylaxis resulting from deliberate sting challenge (while assessing the efficacy of immunotherapy), severe, protracted and difficult to treat hypotension was observed.[37]

A not-unexpected difficulty in therapy is the increased use of β-adrenergic blocking agents (e.g. propranolol).[38] This is a major problem when prompt treatment of anaphylaxis with β-agonists is desirable. The solution is not totally clear. A limited number of studies have suggested that the MAST suit and the administration of atropine or glucagon may be of benefit.[17]

Determining the cause of allergic reactions

When anaphylaxis occurs during surgery, the patient's future health depends on being able to identify the responsible antigen.

A comprehensive history including anaesthetic experience, previous evidence of allergy and any concurrent illness is essential (Table 25.8). Identification of the

Table 25.8 Techniques used to evaluate the cause of allergic reactions

Detailed history
In vivo tests
Skin tests for immediate hypersensitivity reactions
DTH skin tests
Patch tests
Incremental provocative challenge
In vitro tests
Total serum IgE
Assays to measure complement activation
Basophil histamine release
Measurements of mediators – serum/urine
Lymphocyte blast transformation
Antigen-specific IgG, M antibody (ELISA, RIA)
Antigen-specific IgE antibody (RAST)

Bold print denotes the techniques most often used to evaluate prior reactions.
Reprinted with permission from Weiss ME, Hirshman C. Anaphylactic reactions and anesthesia. In: Rogers M, Tinker T, Covino B, eds. *Principles and practice of anesthesiology.* St Louis: CV Mosby Co, 1992: 2457–81.

offending agent will be facilitated by developing a record of the temporal events: times, drugs administered (dosage and duration), surgical events, clinical manifestations of allergy as well as the response to therapy. Previous use of either the suspected medication or a structural analogue must be investigated. Earlier diagnostic testing or drug challenge should be noted. All drugs used by the patient should be evaluated as potential triggering agents. Agents used recently are more likely suspects than are those taken continuously without adverse effects. Unfortunately, the perioperative period is often associated with use of several new medications; as a result, history alone may be insufficient for specific diagnosis.

Immunodiagnostic tests

Skin testing for immediate hypersensitivity reactions

Testing should be done by individuals experienced in interpreting the data; facilities and skilled personnel for resuscitation are essential. While reagents are available for investigating more common hypersensitivity reactions (e.g. pollen or bee-sting), suspected anaphylaxis triggered by perioperative drugs is more difficult because of lack of availability of relevant drug metabolites or appropriate multivalent testing reagents. None the less, intradermal skin testing is still an important and useful tool.

Skin testing has long been used to examine IgE-mediated penicillin allergy[39] and is a mainstay in evaluating allergies to muscle relaxants,[40,41] barbiturates,[41,42] chymopapain,[43] streptokinase,[44] insulin,[45] latex[46,47] and a variety of other drugs. Specific protocols for skin testing are available[48,49] and need not be detailed in this chapter.

To minimize untoward events, a scratch or puncture test should be performed prior to the more definitive intradermal test.[49] Appropriate skin testing concentrations of medications commonly used in anaesthetic practice have been published.[40,50] If there is no published experience with skin testing for a suspected drug, one must consider the possibility of false-positive diagnosis resulting from non-specific irritation. Therefore, any positive response should be validated by testing the allergen in five normal controls. Similarly, in the absence of experience with the skin-testing reagent, negative results should be disregarded until their reliability is established with separate studies. Care must be taken to ensure absence of any medication capable of intefering with positive skin tests, especially H$_1$-antihistamines, tricyclic antidepressants, and sympathomimetic agents). Obviously, all testing should be accompanied by appropriate positive (histamine) and negative (diluent) controls.

Other in vivo tests

Delayed (tuberculin-like) skin testing is of little, if any, use in testing patients who have experienced life-threatening anaphylaxis. Patch tests may be of value in cases of contact dermatitis even if the eruption was provoked by systemic administration of a drug.

In vitro tests

Total serum IgE levels

Elevated levels of total serum IgE have been reported after allergic reactions;[51] however, this determination is rarely, if ever, useful in diagnosing anaphylaxis during anaesthesia and surgery.

Assays to measure complement activation

Complement activation may be involved in specific allergic pathology. This activation is accompanied by decreases in complement components, i.e. C4, C3 or total haemolytic complement (C$_H$50). Alternatively, generation of products of complement activation (C3a, C4a, C5a, etc.) may be a hallmark of its activation.

Release of histamine and other mediators by basophils

There is always a danger in re-exposing a patient at risk to a suspected allergen. For this reason, *in vitro* testing may be desirable. When incubated with a precipitating antigen, washed leucocytes containing basophils with IgE antibody on their cell surface will release histamine and other mediators.[52] The response is correlated with those produced by direct immediate skin testing.[53,54] Although this *in vitro* basophil histamine release assay avoids the danger of repeat anaphylaxis, the test is time consuming, requires whole blood drawn immediately before the test, and currently limited to research laboratories. This assay has been used to demonstrate sensitivity to thiopentone (thiopental),[55] muscle relaxants[40,56] and penicillins.[57] Appropriate controls are mandatory since non-specific non-immunological release of histamine may follow *in vitro* exposure to some agents.

Measurements of mediators

Blood obtained during or shortly after the allergic episode may be analysed for evidence of mediator release. Histamine, PGD_2, or high molecular weight neutrophil chemotactic factor are amenable to such measurement.[37,58] Urine may also be tested for metabolites of histamine or PGD_2. Unfortunately, changes in plasma concentrations of histamine and PGD_2 are evanescent, thereby limiting clinical utility of this technique. Bioassays to measure serum neutrophil chemotactic factor are difficult and fraught with considerable variability. Assaying serum tryptase (a protease released specifically from mast cells) is a recent promising development in studying mast cell-mediated allergic reactions since this substance may be elevated for hours following release.[59]

Lymphocyte blast transformation

Lymphocyte blast transformation measures blastogenesis by measuring the uptake of the DNA precursor, tritiated thymidine, after the patient's lymphocytes have been exposed to antigens *in vitro*. However, there is lack of agreement about the test's ability to diagnose drug allergy.[60] Furthermore, the determination is complex, takes considerable time, while there is considerable variability and lack of reproducibility when different laboratories are compared.[60]

RAST testing

A solid-phase radioimmunoassay, termed the radioallergosorbent test (RAST), was first introduced in 1967. This technique measures circulating allergen-specific IgE antibody. The allergen is first attached to a solid phase such as a carbohydrate particle, paper disc, polystyrene wall of a test tube or a plastic microtitre well. The resulting complex is then incubated with the serum being evaluated; specific antibody of all immunoglobin classes becomes bound. The particles are then washed, and a second incubation carried out with a radiolabelled, highly specific anti-IgE antibody, after which results are compared to a positive reference serum and a negative control serum. Bound radioactivity is directly related to the allergen-specific IgE antibody content in the original serum. This *in vitro* test is of considerable potential utility, since RAST correlates well with skin test endpoint titration, basophil histamine release, and provocation tests.[53,61,62] However, application of RAST to diagnose drug hypersensitivity due to IgE antibody is limited due to insufficient knowledge of drug metabolites which may act as either antigen or hapten. In 1971 a RAST was developed to measure IgE antibody to the major determinant of penicillin.[63] More recently, RASTs capable of measuring IgE antibody to insulin,[45] chymopapain,[43] muscle relaxants,[64,65] thiopentone,[66] trimethoprim,[67] protamine[68] and latex[47,69,70] have become available. As in all testing, the user must be careful; false-positive tests result from high non-specific binding, high total serum IgE levels or poor technique.[71] False-negative tests may occur due to interference of high levels of IgG 'blocking antibodies' or inability to maximize assay sensitivity.[72]

Unfortunately, commercialization of this technique has led to abuse in recent years. Some commercial laboratories have sold RAST for the diagnosis of numerous drug allergies when an IgE-mediated aetiology has not been demonstrated (i.e. local anaesthetics and radiocontrast media). The clinical consequences of relying upon information from such inappropriate RAST are a cause for anxiety.

Measurement of specific IgG or IgM antibodies

Other than drug-induced thrombocytopenia, haemolytic anaemia and agranulocytosis, there is often poor correlation between *in vitro* measurement of antigen-specific IgG and IgM antibodies and *in vivo* allergic drug reactions. Recent evidence suggests that certain protamine reactions are mediated through protamine-specific IgG antibody.[68,73]

Requirement for same drug in the future

The physician's resourcefulness and creativity will be challenged when confronted by a history of serious allergy to a drug which is now required. Risks and benefits must be analysed clearly. If equally effective drugs without cross-reactivity are available, the problem is solved. If these substitutes fail, induce unacceptable side effects or are clearly less effective, then cautious administration of the suspected drug following, if possible, a desensitization protocol or premedication regimen may have to be considered.

These prophylactic techniques have mostly been tested and validated when confronting a previous reaction to radiocontrast medium in a patient who requires repeat administration.[21] These reactions are not IgE mediated. Furthermore, even with IgE-mediated anaphylaxis, there is little to support the use of antihistamines or steroids as primary prophylaxis.[74,75] Acute desensitization protocols are available for patients with a history of allergic reactions to penicillin,[39,76,77] insulin,[78] sulphonamides,[79] heterologous antisera[78] and aspirin.[26] The details of each protocol are not described here since they may be found in the above references.

Specific allergic reactions often seen by the anaesthetist

Muscle relaxants

Cardiovascular collapse, tachycardia, urticaria and broncho-spasm may result from muscle relaxant-induced anaphylaxis. Vervloet *et al.* and Baldo, Fisher and Harle[40,56,64,65] have demonstrated that these reactions are mediated through IgE. In support of this hypothesis are positive Prausnitz–Küstner tests, basophil histamine release, inhibition of basophil histamine release after desensitization to anti-IgE and the finding of drug-specific IgE antibodies in sera from patients with documented adverse responses to muscle relaxants.[40,56,64,65] It has been proposed that the IgE antibodies complex with the quaternary or tertiary ammonium ions of the muscle relaxant.[64] Extensive *in vitro* cross-reactivity has been reported between muscle relaxants and other compounds with quaternary and tertiary ammonium ions.[64] The latter are not only present in many drugs, but are found in foods, cosmetics, disinfectants and industrial materials. The clinical significance of *in vitro* cross-reactivity is unclear but some workers have proposed that sensitization may result from environmental exposure.[64] Muscle relaxants which contain two ammonium ions seem to be functionally divalent; thus, they can cross-link cell surface IgE and precipitate mediator release from mast cells and basophils without haptenizing to carrier molecules.[56] When the ammonium ions are less than 0.4 nm apart, the drug does not appear to induce histamine release; the optimal length for cross-linking cell surface IgE is $\geqslant$ 0.6 nm.[80] Muscle relaxants possessing a rigid backbone between the two ammonium ions (such as pancuronium and vecuronium) are less able to initiate anaphylaxis than more flexible molecules.[80]

Atopy is not a significant predictor of anaphylactic reactions to muscle relaxants.[81] It is unclear why 90–95 per cent of muscle relaxant-induced anaphylactic reactions occur in females.[82] Sensitization to ammonium ion epitopes in cosmetics has been postulated to explain this phenomenon.[64]

Although the precise incidence of allergy to muscle relaxants is unknown, its occurrence is less common in the United States than in France or Australia (personal observation). Although skin testing and RAST have been used to evaluate the presence of IgE antibodies to muscle relaxants,[41,64,79] additional studies are needed to determine the predictive value of these techniques.

Barbiturates

Use of thiobarbiturates has been followed by acute allergic episodes, most of them associated with thiopentone.[31] Both non-immune and IgE-mediated mechanisms may be responsible and are supported by recent evidence (personal observation).[31,83] Patients in whom anaphylaxis followed induction of general anaesthesia with thiopentone have demonstrated positive immediate skin tests to the thiobarbiturate.[41,84] RAST testing for thiopentone has been reported recently[66] and release of histamine by mast cells in response to *in vitro* thiopentone has been described.[55,85] It should be stressed that the ability of skin testing and RAST to predict thiopentone-induced anaphylaxis is currently uncertain and requires further study.

Local anaesthetics

It is not uncommon for a patient to describe adverse reactions to local anaesthetics; as a result, they may be warned of 'allergy' to these agents. However, a true allergic response to injected local anaesthetics is exceedingly rare. Events masquerading as allergic reactions to local anaesthetics most likely represent a vasovagal 'faint', toxicity produced by high blood levels (often due to inadvertent intravascular administration) or the normal physiological effects of adrenaline either injected with the local anaesthetic or secreted secondary to anxiety. These sympathetic effects include tremor, diaphoresis, tachycardia, palpitations and hypertension. Psychomotor responses such as hyperventilation should be considered. True toxicity (overdose), as opposed to allergy, involves both the central nervous and cardiovascular systems, and is manifested by slurred speech, euphoria, dizziness, excitement, nausea, emesis, disorientation, or convulsions.[86] Vasovagal reactions are characterized by bradycardia, sweating, and pallor; rapid improvement ensues when the patient is placed supine.

True allergy to local anaesthetics is rare, and will produce symptoms consistent with IgE-mediated reactions such as urticaria, bronchospasm and anaphylactic shock. When anaphylaxis is suspected, acceptable documentation of an IgE-mediated response will usually be lacking, although there have been a few exceptions.[87] IgE-mediated sensitivity has also been reported for parabens, preservatives used in local anaesthetics, although this, too, is rare.[88]

Local anaesthetics are divided into two chemical groups (Table 25.9). Group I contains benzoate esters; these anaesthetics may cross-react with each other, but not with group II drugs. Group II agents include mostly amides and do not demonstrate substantial cross-reactivity.

When a patient presents with a history of suspected

Table 25.9 Protocols for evaluation of local anaesthetic allergy

Step*	Route	Volume (ml)	Dilution
(a)			
1	Intradermal	0.02	1:1000†
2	Intradermal	0.02	1:100
3	Intradermal	0.02	1:10
4	Intradermal	0.02	Undiluted
5	Subcutaneous	0.3	Undiluted
(b)			
1	Puncture		Undiluted
2	Subcutaneous	0.1	Undiluted
3	Subcutaneous	0.5	Undiluted
4	Subcutaneous	1.0	Undiluted
5	Subcutaneous	2.0	Undiluted

* Administer at 15-minute intervals.
† If history is strongly suggestive of IgE mediated reaction, start with puncture at 1:1000 dilution.
Reprinted with permission from Weiss ME. Drug allergy. In: Bush R, ed. *Medical clinics of North America – clinical allergy*. Philadelphia: WB Saunders Co, 1992; **76**: 857–82.

allergy to a local anaesthetic, detailed evaluation is necessary not just for current planning but for possible future needs. A complete history of the purported allergic response should be obtained. The likelihood of non-allergic responses (discussed above) should be entertained. If the index of suspicion is sufficiently high, skin testing and incremental drug challenge may be deemed appropriate. The Johns Hopkins University protocol appears in Table 25.9(a). The more aggressive protocol employed at Northwestern University is presented in Table 25.9(b).

The local anaesthetic tested should be the most suitable for the planned surgery. In order to avoid potential cross-reactivity, it should be in a group different from that of the alleged allergen. If the suspected local anaesthetic allergen is unknown, a group II anaesthetic (frequently lignocaine) should be selected. If the history is suggestive of an IgE-mediated reaction or there is a question of sensitivity to paraben, a paraben-free formulation should be used for testing, challenge and actual administration for anaesthesia. Adrenaline should be avoided in any solution used for skin testing as it may mask a positive skin test[1] or produce untoward effects of its own.

Narcotics

Narcotics cause non-immunologically mediated histamine release from skin mast cells, producing changes which might be confused with an allergic reaction. *In vitro* data suggest that mast cells derived from skin are quite sensitive to narcotic-induced histamine release; in contrast, mast cells originating in the lung and gastrointestinal tract, as well as

circulating basophils, do not release histamine when exposed to narcotics.[23,24,89] The important clinical message is that most opiate-induced abnormalities are self-limiting cutaneous reactions, restricted to hives, pruritus or mild hypotension, the latter of which is easily treated by fluid administration. IgE antibodies which bind epitopes contained in opiate narcotics may be induced;[90–92] none the less, the pharmacological release of non-immune mediators induced by opiates is far more common than the rare event produced by morphine-specific IgE antibodies.[93] Since codeine, morphine and pethidine (Demerol) will frequently produce a 'positive' skin test due to non-immunological skin mast cell histamine release, these tests must be interpreted cautiously.

Radiocontrast media

Use of conventional high-osmolar, radiocontrast media (HORCM) injections may be associated with adverse reactions. The usual response begins 1–3 minutes after intravascular injection. Vasomotor reactions including nausea, vomiting, flushing, or warmth occur in 5–8 per cent of patients.[21] More serious anaphylactoid reactions follow HORCM infusions in 2–3 per cent of patients; its symptoms are urticaria, angioedema, wheezing, dyspnoea, hypotension, or even death.[78] Fatalities produced by radiocontrast media occur in about 1:50 000 intravenous procedures,[78] resulting in as many as 500 deaths annually.

Atopic patients are reported to be twice as susceptible to adverse reactions as are non-atopic individuals.[94] It is essential to obtain a detailed history in all patients who are to receive HORCM since those with a prior reaction have approximately a 33 per cent (range 17–60 per cent) chance of repeating the adverse response if re-exposed.[78]

While the aetiology of adverse reactions to contrast media is not completely understood at present, liberation of histamine seems to accompany most reactions.[21] It is interesting that elevations in plasma histamine have been described in the absence of haemodynamic changes or anaphylaxis.[21] Activation of serum complement, either through the classical or alternative pathway, is observed after intravascular injection of HORCM. The contrast medium is also able to induce histamine release from mast cells and basophils without evidence of complement activation.[21] *In vitro* studies suggest that the hypertonicity of conventional HORCM (seven times the osmality of plasma) results in non-immunological mediator release from mast cells and basophils,[21] and there is no evidence that IgE-mediated mechanisms play a role in radiocontrast media reactions.

As already mentioned, repeat exposure to radiocontrast media in a patient with a history of adverse response poses an increased (17–60 per cent) risk.[21] Pretreatment with

Table 25.10 Reaction rate to contrast media in patients with previous radiocontrast media reactions

HORCM and no premedication	33% (range of 17–60%)
HORCM and premedication*	9–3.1%
LORCM and no premedication	2.7%
LORCM and premedication*	0.7%

* Premedication = prednisone (50 mg p.o.) 13, 7, and 1 h before, diphenhydramine (50 mg i.m. or p.o.) 1 h before, and ephedrine (25 mg p.o.) 1 h before if no contraindication to its use. (? Ranitidine as well).
HORCM, high osmolality radiocontrast media; LORCM, low osmolality radiocontrast media.
Reprinted with permission from Weiss ME. Drug allergy. In: Bush R, ed. *Medical clinics of North America – clinical allergy.* Philadelphia: WB Saunders Co, 1992; **76**: 857–82.

prednisone (50 mg) 13, 7 and 1 hour before contrast media administration along with diphenhydramine (50 mg) 1 hour before RCM administration reduces this risk to 9 per cent.[95] Almost all reactions in pretreated patients are mild.[21] The addition of ephedrine (25 mg) 1 hour before RCM administration (in patients without angina, arrhythmia, or other contraindications for ephedrine administration) resulted in a reaction rate of 3.1 per cent in one study.[21] It might be expected that the addition of an H_2-receptor antagonist,[96] such as cimetidine or ranitidine, might further decrease the incidence of RCM reactions, although studies to verify such benefit have not been done.

Recently, newer low osmolality radiocontrast media (LORCM) have been developed. The LORCM have only about twice the osmolality of plasma and appear to induce fewer adverse reactions. In patients with a history of a prior reaction to conventional HORCM, the incidence of an adverse reaction when contrast media is readministered is lower with LORCM (2.7 per cent) than with conventional HORCM (33 per cent) (Table 25.10).[97] Unfortunately, LORCM cost 20 times more than conventional HORCM, and replacing LORCM for all procedures in which HORCM are now used would increase the cost from US $100 million annually to US $1.5 billion in the United States alone.[97] A recent study, using historical controls, showed that combining premedication with LORCM is most beneficial (a reaction rate of only 0.7 per cent) in preventing reactions in high risk individuals (prior reaction to HORCM) (Table 25.10).[98] It would seem prudent (and cost-effective) to limit the use of LORCM (with premedication) to patients who have had prior reactions to HORCM.

Protamine

Protamine sulphate is a polycationic, strongly basic small protein with a molecular weight of 4300. Protamine is used to reverse anticoagulation produced by heparin, and to retard the absorption of some types of insulin. With the frequent use of intravenous heparin during cardiopulmonary bypass, cardiac catheterization, haemodialysis and phoresis, increasing reports of life-threatening adverse reactions are not surprising.

Adverse reactions to this important medication include rash, urticaria, bronchospasm, pulmonary vasoconstriction, and systemic hypotension leading at times to cardiovascular collapse and death.[99,100]

There are several groups of patients in whom adverse reactions to protamine should be anticipated. Diabetics receiving daily subcutaneous injections of insulin containing protamine (NPH and PZI) have a 40- to 50-fold increased risk of life-threatening reactions if protamine is administered intravenously.[101,102] Another group putatively at increased risk for protamine reactions are men who have undergone vasectomies. Studies have shown that disruption of the blood–testis barrier accompanying vasectomy causes a 20–33 per cent incidence of haemagglutinating autoantibodies against protamine-like compounds.[103] Protamine reactions in vasectomized men have been described[104] and we have recently found that 34.5 per cent of vasectomized men have antiprotamine IgG antibodies in their sera, compared to 1 per cent in age-matched controls.[105] No anti-protamine IgE antibodies were found in either vasectomized or control sera.

Since protamine is produced from the matured testis of salmon or related species of fish belonging to the family Salmonidae or Clupeidae, it has been suggested that individuals allergic to fish may have serum antibodies directed against protamine. Conversely, commercial protamine preparations may be contaminated with fish proteins to which fish-allergic patients may react. In spite of this theoretical possibility, evidence demonstrating an increased risk of protamine reactions in fish-allergic patients is lacking and is limited to rare case reports.[104]

Finally, previous exposure to intravenous protamine given for reversal of heparin anticoagulation may increase the risk for a reaction on subsequent protamine administration.[99] No systematic study has been reported in which the human immune response to protamine following intravenous administration has been evaluated.

The exact mechanisms of acute protamine reactions are incompletely understood. *In vitro* studies have suggested that protamine causes direct non-immunological release of histamine from hamster and rat peritoneal mast cells.[106] However, exposure of human basophils and mast cells to protamine concentrations up to 100 μg/ml is not followed by significant histamine release.[107,108] Tobin *et al.*[67] proposed that protamine does not cause non-immunological basophil histamine release in humans, but may potentiate IgE-mediated release of histamine from human basophils *in vitro* through a polycationic-recognition site. Some protamine reactions may be associated with complement activation either through protamine–heparin complexes[14–16,109] or through a protamine- and complement-fixing antiprotamine IgG antibody interaction.[110] These reactions are associated with the generation of thromboxane, a pulmonary vasoconstrictor,[111,112] and may lead to pulmonary artery pressure elevation. Lakin[110] demonstrated that protamine-specific IgG antibodies caused protamine reactions by activating complement, while others[113,114] have also reported the presence of protamine-specific IgG antibodies in small numbers of protamine reactors. We have demonstrated that whether or not diabetic patients were treated with protamine insulin, the finding of serum antiprotamine IgG antibody is associated with a significant risk of acute protamine reactions.[68] Kurtz et al.[115] used a double antibody radioimmunoprecipitation assay to demonstrate that 38–91 per cent of diabetics dependent on protamine insulin had serum IgG antibodies to protamine. Nell and Thomas employed ELISA to show that 38 per cent of 319 NPH insulin-treated diabetic patients had IgG antibodies to protamine in their sera, while only 2.5 per cent of 202 normal controls possessed the same antibody.[116] Recent data suggest that protamine can inhibit the ability of plasma carboxypeptidase N to convert the anaphylatoxins and bradykinins to less active *des arg* metabolites, thus allowing these vasoactive compounds to produce their pathological haemodynamic effects.[117]

We have developed an agarose-based RAST to measure antiprotamine IgE antibodies; this enabled us to show that when diabetic patients receive protamine-insulin injections, the presence of serum antiprotamine IgE antibody is a significant risk factor for acute protamine reactions.[68] It seems likely that an antibody-mediated mechanism is the cause of the increased numbers of protamine reactions seen in diabetics who are dependent on protamine insulin. One should consider prescreening these high risk patients for the presence of antiprotamine antibodies prior to undertaking elective procedures that might require intravenous protamine. If such antibodies are present, special precautions could be taken or alternative heparin antagonists, such as hexadimethrine (which in the USA must be obtained from the FDA by compassionate use IND) could be substituted (personal observations).[118]

Skin testing with protamine is not useful in distinguish-

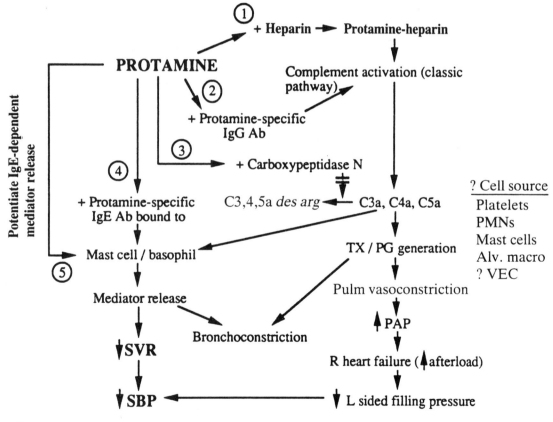

Fig. 25.4 Possible mechanism for acute protamine reactions. **1** Protamine combines with heparin to activate complement via the classic pathway. The generation of anaphylatoxins may cause mast cell or basophil mediator release or may stimulate thromboxane and prostaglandin generation, leading to pulmonary vasoconstriction. If severe, this could lead to acute right-sided heart failure and systemic hypotension. **2** Protamine combines with antiprotamine IgG antibody, leading to complement activation with the same cascade as noted in number 1. **3** Protamine may inhibit the ability of carboxypeptidase N to convert the anaphylatoxins to their less active *des arg* metabolites. **4** Protamine cross-links cell surface antiprotamine IgE antibodies on mast cells and basophils leading to mediator release and systemic vascular dilatation and hypotension and/or bronchoconstriction. **5** Protamine may combine with a polycationic activation site on mast cells and basophils and potentiate IgE-mediated histamine release. (Reprinted with permission from Weiss ME, Adkinson NF. Allergy to protamine. In: Vervloet D, ed. *Clinical reviews in allergy: anesthesiology and allergy.* New York: Humana Press, 1991: 339–55.)

ing between subjects with significant serum antiprotamine IgE antibody and controls.[100,119,120] Protamine may be an incomplete or univalent antigen that first must combine with a tissue macromolecule or possibly heparin to become a complete, multivariant antigen capable of eliciting mediator release.[119] Lowenstein and colleagues[112,121] found that pulmonary artery pressure elevation following rapid protamine injection in three patients was associated with elevations of thromboxane B_2 and C5a, while plasma histamine levels did not change. In contrast, a patient who developed a decrease in systemic vascular resistance and profound systemic hypotension without pulmonary artery hypertension had a tenfold elevation in plasma histamine without change in thromboxane B_2 or C5a. These data suggest that more than one mechanism may be responsible for adverse reactions associated with protamine (Fig. 25.4).

Rubber (latex) allergy

Natural rubber (latex) is a milky juice obtained from the sap tree, *Hevea brasiliensis*, that grows in the Amazon region. During the manufacturing process, various small molecular weight chemicals may be added before the latex is vulcanized (cured with heat and sulphur – 130°C for 5–30 minutes). These small molecular weight chemicals (accelerators, curing agents, antioxidants, stabilizers, dyes, bleaches, retarders, and blowing products) help give natural rubber its desired final properties, such as colour, texture and elasticity. Proteins make up 2–3 per cent of the final latex product. These proteins are water soluble and heat and chemically stable. Latex is presently used in over 40 000 products and is a component of many medical devices, including gloves, urinary catheters, condoms, intubation tubes, anaesthesia masks, dental cofferdams, rebreathing masks, barium enema cuffs (recently recalled), ostomy bags and medication stoppers.

It has long been known that latex exposure could cause contact dermatitis. Contact dermatitis is caused by a type 4, T-cell-mediated, delayed hypersensitivity allergic reaction in which one of the small molecular weight chemicals (accelerators or antioxidants), added during the manufacturing process, acts as a hapten and triggers sensitized T-lymphocytes to recruit and stimulate the proliferation of other lymphocytes and mononuclear cells. This ultimately leads to tissue inflammation and dermatitis.[122]

It was not until 1979 that the first case of contact urticaria due to latex was reported.[123] Since then, latex has been shown to cause not only contact urticaria, but also, generalized urticaria, asthma, and anaphylaxis, which was first reported in 1987.[124] Over 10 deaths have been caused by latex anaphylaxis.[125] The cause of these reactions is an IgE antibody directed against one or more proteins in latex.[46] In patients with this antilatex IgE antibody, latex

exposure may lead to a type 1, immediate hypersensitivity allergic reaction resulting in mast cell and basophil degranulation and potentially life-threatening anaphylaxis.

Certain population groups are at increased risk of developing anaphylactic reactions to latex. These include health care workers, especially those with frequent exposure to latex gloves. Studies have shown that 6–10 per cent of operating room medical personnel have IgE antibodies directed against latex proteins.[126,127] Another group at very high risk are children with myelomeningocele or congenital urological abnormalities.[128–130] In fact, this group of patients may be at greatest risk of severe latex reactions. In one study involving five mylodysplasia clinics, 187 families were questioned and 25 per cent of the children with myelomeningocele had a positive history of allergic reactions to latex products.[131] Investigators have found that 71/127 (56 per cent) patients with myelomeningocele had IgE antibodies to latex.[132–134] It is thought that repetitive exposures to latex products (gloves, catheters, etc.) at an early age from frequent urinary catheterization and multiple operations and procedures leads to latex sensitization.

Patients with atopy (allergic rhinitis, asthma, or atopic dermatitis) are also more likely to develop latex reactions.[70,128,129] In addition, individuals with hand eczema who are exposed to latex gloves have a high risk of latex anaphylaxis.[70,135] Whether eczema, by disrupting the natural skin barrier, allows latex antigens to better reach immunocompetent cells below the epidermis and therefore increase the chance for latex sensitization, or whether eczema, iteslf, is an early manifestation of latex allergy, is presently unknown. Patients with banana allergy appear to be at increased risk for latex reactions and one study showed that IgE antibody to latex protein also bound to a protein in banana.[136]

What then can be done to minimize the risk of latex reactions during anaesthesia? First, information on latex allergy and the population groups at increased risk for such reactions needs to be disseminated. During the anaesthetist's preoperative evaluation, specific questions concerning prior reactions to latex or rubber products should be asked.[93] Patients who have suffered systemic, severe intraoperative reactions to latex often have a prior history of contact urticaria or angioedema when exposed to rubber products such as gloves, rubber balloons, latex condoms, diaphragms, or pacifiers. Latex allergy should be suspected in patients who have had prior, unexplained intraoperative anaphylaxis. In a recent study, 5/43 (10 per cent) cases of intraoperative anaphylaxis at one hosptial were found to be caused by latex allergy.[137] The temporal pattern of intraoperative anaphylaxis due to latex is different from anaphylaxis secondary to anaesthetic agents. Intraoperative anaphylaxis due to latex allergy often occurs 20–150 minutes after induction, when latex-gloved hands contact mucosal and peritoneal surfaces, leading to systemic absorption of latex proteins.[138] This is in contrast to reactions related to anaesthetic agents, which typically occur at the time of induction.

Confirmation of latex allergy should be obtained from latex epicutaneous skin tests using either natural latex obtained from latex manufacturing companies, an extract prepared by incubating latex gloves in phosphate buffered saline, or by skin testing through a moistened latex glove.[47,135,139] A commercially available skin test reagent (natural latex) has recently become available in Canada (Bencard Lab., Mississauga, Ontario). Unless one is quite familiar with all aspects of allergy skin testing, the patient should be referred to an allergist for such tests and their interpretation. Epicutaneous skin testing using serial dilutions of latex has a high safety profile and allows one to observe a dose-dependent sensitivity.

Intradermal tests are generally not needed and have provoked systemic reactions in some patients. A provocative challenge (use test) where the individual wears a moistened latex glove or glove finger has also resulted in systemic reactions and generally is not recommended. A commercial latex RAST (Latex disc k 82, Pharmacia Diagnostics, Piscataway, NJ) is available but is not as sensitive as epicutaneous skin testing. The latex RAST may be negative between 20 and 45 per cent of the time in patients with positive skin tests to latex.[69,70] Recently, there have been a few case reports of patients who have had systemic reactions to epicutaneous skin tests. Therefore, it may be prudent to first do a latex RAST and if negative, proceed to epicutaneous skin tests.

If patients require surgery and have confirmed latex sensitivity, or have convincing histories of latex reactions, then avoidance of latex products, especially latex gloves, is mandatory.[140] Strong consideration should also be given to latex avoidance in all patients with myelomeningocele or congenital urological abnormalities, since the prevalence of latex sensitization appears to be so high in this population.

Avoidance of latex gloves can be achieved using synthetic rubber gloves made from material such as Neolon (neoprene) [Deseret Medical Inc.] or Elastyren (styrene butadiene) [Hermal Pharmaceutical Labs Inc.]. Since synthetic gloves are expensive, may have properties which make their use less desirable to some surgeons, and are not always readily available, medical personnel caring for latex allergic patients can also use vinyl gloves taped in place at the wrist over washed latex gloves. One study suggested that thorough washing of latex gloves prior to their use was sufficient to remove the allergenic antigen,[47] but until confirmatory studies are done, this is not recommended in latex allergic patients.

Preoperative treatment of latex allergic patients with steroids and antihistamines has been suggested without any evidence of efficacy. Since pretreatment has never been demonstrated to prevent IgE-mediated allergic reactions and since pretreatment may mask early signs of anaphylaxis, its use is not generally recommended. Obviously, the anaesthetist needs to monitor for early signs of anaphylaxis and be ready to treat it aggressively, should it occur.

Since we have become a very mobile society, the patient with latex allergy must be instructed to mention this condition to all future health care providers. Latex allergic patients should wear medic-alert information, should be given an Epi Pen and instructed in its use, and should for safety carry various sizes of non-latex gloves for potential emergency use.

While latex and rubber products have been used for many years, it is only in the last decade that reports of contact urticaria and systemic anaphylactic reactions to latex have been described. It is possible that in the past, a connection to latex exposure and these reactions was not made, but this is unlikely. It is also possible that the recent increased need for latex products, especially gloves and condoms, because of the HIV epidemic, has resulted in increased exposure and sensitization. Some have suggested that a change in the manufacturing process has led to an increased antigenicity of latex products. These possible explanations for the recent development of IgE-mediated allergy to latex await future scientific validation.

Chymopapain

Intradiscal injection of chymopapain (chemonucleolysis) has been used to treat herniated lumbar intervertebral discs. Anaphylaxis is a significant problem, with an incidence of 1 per cent; fatalities occur with a frequency of 0.14 per cent.[43] Development of anaphylaxis is three times more likely in women than in men.[1] Chymopapain is derived from a crude fraction (papain) from the papaya tree. Meat tenderizers, cosmetics, beer, or soft contact lenses may be responsible for prior exposure to chymopapain.[22] Reactions to chymopapain may be IgE mediated.[43,78] IgE antibodies to chymopapain have been demonstrated using *in vivo* skin testing and *in vitro* immunoassays;[43] additional studies will be needed to demonstrate the predictive value of these tests. Prescreening and decreased use of chemonucleolysis has resulted in a decreased frequency of chymopapain reactions.

Streptokinase

Streptokinase, a protein derived from group C β-haemolytic streptococci, is a thrombolytic agent used to dissolve acute obstructing thrombi in coronaries and retinal veins. The incidence of allergic reactions is unknown; reports range from 1.7 to 18 per cent.[141] *In vivo* skin tests and *in vivo* assays can detect IgE streptokinase antibodies. While protocols for skin testing have been advised prior to instituting thrombolytic therapy with streptokinase,[44] relatively few patients have been evaluated and the tests' predictive abilities are not yet known. Use of recombinant tissue plasminogen activator (rTPA) for thrombolysis may decrease the need for streptokinase; however, a recent randomized, prospective study was unable to show any advantages for rTPA over streptokinase in preserving myocardial function after an acute myocardial infarction.[142]

Mannitol

Mannitol or other hyperosmotic agents may cause direct, non-immune mediated histamine release from circulating basophils and mast cells.[31] Evidence for immunologically mediated reactions is lacking. Slow infusion of these hyperosmolar agents helps to prevent the problem.

Methylmethacrylate

Methylmethacrylate bone cement is used to attach prosthetic joints to raw bone during orthopaedic surgery. Cardiopulmonary complications of methylmethacrylate include hypotension, hypoxaemia, non-cardiogenic pulmonary oedema, and even cardiac arrest. While many aetiologies have been proposed, none relies on immune mechanisms.[31]

Penicillin antibiotics

Penicillin antibiotics are the most common drugs responsible for allergic reactions in or out of the operating theatre. Their exact frequency is not known, but a range of incidence from 0.7 to 8 per cent has been published.[143] Anaphylactic reactions occur in 0.004–0.015 per cent of patients treated with penicillin, with fatalities occurring approximately once in every 50 000–100 000 exposures.[143] Anaphylaxis to penicillin accounts for 500–800 deaths each year.[144] All varieties of immunopathological reactions described by Gell and Coombs have been seen with penicillin (see Table 25.1).[3] The pathogenesis of some reactions to penicillin are obscure; these have been labelled idiopathic. Among them are the common maculopapular rash, eosinophilia, Stevens–Johnson syndrome, exfoliative dermatitis, and toxic epidermal necrolysis (see Table 25.1). Ampicillin induces rashes with much greater frequency than penicillin, although the reasons for this are not known.[39,145] Pseudoanaphylactic reactions have been observed after i.m. or accidental i.v. injection of procaine penicillin, and are most likely due to a combination of a toxic and embolic phenomenon caused by procaine.[146] Because of the risk of life-threatening anaphylaxis, IgE-mediated reactions are the most important manifestation of penicillin allergy.

Because of penicillin's low molecular weight (only 356), its potential as an allergen depends on an initial covalent combination with tissue macromolecules, presumably proteins. This produces the multivalent hapten–protein complex required for both the induction of an immune response and the elicitation of an allergic reaction.[147] Thirty years ago, Levine and Parker showed that under physiological conditions, the β-lactam ring in penicillin spontaneously opens to form the penicilloyl group.[148] Recent evidence suggests that this reaction may be facilitated by low molecular weight molecules in serum.[149,150]

The penicilloyl group has been designated the *major determinant* because about 95 per cent of the penicillin molecules that irreversibly combine with proteins form penicilloyl moieties,[8] a reaction which occurs with the prototype benzylpenicillin as well as virtually all semisynthetic penicillins. Benzylpenicillin can also be degraded by other metabolic pathways to form additional antigenic determinants.[151] These derivatives are formed in small quantities, stimulate a variable immune response, and are therefore termed *minor determinants*. Thus, exposure to penicillin and other β-lactams will result in the production of IgE antibodies against a variety of substances: haptenic derivatives and the major and minor determinants. Anaphylactic reactions to penicillin are usually mediated by IgE antibodies directed against minor determinants, although some anaphylactic reactions have occurred in patients with only penicilloyl-specific IgE antibodies.[8,151,152] Accelerated and late urticarial reactions are generally mediated by penicilloyl-specific IgE antibodies, a major determinant.[8]

Allergic responses are more frequent following parenteral than oral penicillin.[76] However, this may be more related to dose than route of administration. Thus, the incidence of allergic reactions to oral penicillin are similar to those following i.m. procaine penicillin when equivalent doses are administered.[153] Individuals with a history of previous penicillin reactions have a four- to sixfold increased risk of subsequent reactions with penicillin compared to those with a negative history.[154] Although a careful history should always be obtained, this will not always prevent an untoward response to penicillin since most serious and fatal allergic reactions to penicillin and β-lactam antibiotics occur in patients without any evidence of prior sensitivity. These patients may have become sensitized only during their last penicillin therapy, or (less likely) by occult environmental exposures.

Prevalence of skin test reactivity and adverse reactions to penicillin

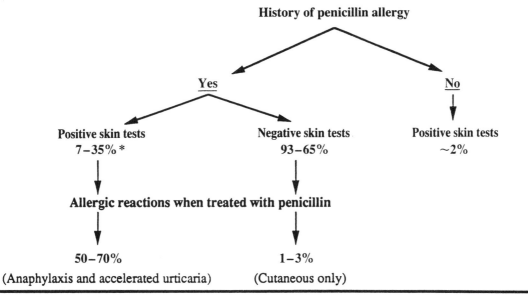

Fig. 25.5 Prevalence of positive and negative skin tests, and subsequent allergic reactions in patients treated with penicillin (based on studies using both PPL and MDM as skin test reagents). (Reprinted with permission from Weiss ME, Adkinson Jr NF. Beta-lactam allergy. In: Mandell LG, Douglas Jr RG, Bennett JE, eds. *Principles and practice of infectious disease.* 3rd ed. New York: Churchill Livingstone Inc, 1989: 264–9.)

Approximately 10–20 per cent of hospitalized patients report a history of allergy to penicillin. However, many of these patients have been either incorrectly labelled as allergic to penicillin or have lost their sensitivity. The most useful single predictor of an individual's likelihood of an acute IgE-mediated reaction is to be found in the skin test response to major and minor penicillin determinants (Fig. 25.5). Multiple penicilloyl determinants have been linked to a weakly immunogenic polylysine carrier molecule to form penicilloyl-polylysine (PPL). This has been used successfully as a skin test reagent to detect *major determinant* IgE antibodies.[155] PPL reagent (Pre-Pen; Kremers-Urban) is available commercially in the United States.

Minor determinants are labile and cannot be readily synthesized in multivalent form. As a result, skin testing for minor determinants usually employs a mixture of benzylpenicillin, its alkaline hydrolysis product (benzyl-penicilloate), and its acid hydrolysis product (benzylpenil-loate); this combination is called the minor determinant mixture (MDM). A complete MDM reagent is not commercially available in the United States at this time. Benzylpenicillin diluted to a concentration of 10 000 u/ml [10^{-2} M] may serve as the sole minor determinant reagent, but will fail to diagnose approximately 5–10 per cent of skin test reactive patients.[156,157] Unfortunately, some of these patients with false-negative MDM testing may be at risk of serious anaphylactic reactions.[158] None the less, benzylpenicillin is a reasonable alternative to MDM for use

in patients *without* impressive histories of IgE-mediated reactions. For patients whose histories are more suggestive of an IgE-mediated anaphylactic reaction, we advise referral to a centre where testing with complete MDM is available. The actual technique of performing penicillin skin tests has been described previously[49] and will not be discussed in detail here.

RASTs have been developed to detect IgE antibodies against the penicilloyl determinant.[63] At the present there is no *in vitro* RAST for minor determinant antibodies. Therefore, RAST and other *in vitro* analogues have limited clinical utility.

When appropriate tests are administered by experienced workers, negative responses can be most assuring. Therapeutic doses of penicillin administered to patients with histories of penicillin allergy but with negative skin tests to PPL and MDM are rarely followed by IgE-mediated reactions; reactions which do occur are almost always mild and self-limited. About 1 per cent of skin test negative patients will develop accelerated urticarial reactions and approximately 3 per cent will develop other milk reactions.[39] Penicillin anaphylaxis has not been reported in patients with negative skin tests which indicate that this important antibiotic may be safely given. Recently Blanca and associates in Spain suggested that skin testing with major and minor determinants of benzylpenicillin may not predict all individuals with ampicillin/amoxycillin specific allergy.[159] Mendelson *et al.* studied 443 patients with a history of penicillin allergy who were skin test negative to PPL and a MDM of penicillin. The 443 patients tolerated a 10-day penicillin challenge, but when given 10 days of amoxycillin, 18.7 per cent developed cutaneous reactions (Mendelson, personal communication). Thus amoxycillin/ampicillin seems to have a greater propensity for causing cutaneous allergic reactions. It has been postulated that the diamino acyl side chain contained in ampicillin/amoxycillin more readily allows for the formation of linear polymers of varying lengths, which may explain the higher cutaneous reaction rate seen with ampicillin/amoxycillin. Studies in the 1960s showed that patients with late-occurring maculopapular rashes from ampicillin/amoxycillin could safely be retreated with the antibiotics without any increased risk of acute allergic reactions. Skin testing in these individuals with PPL and MDM is useful to rule out concomitant type I, IgE-mediated sensitivity.

A few patients with positive skin tests have been treated with therapeutic doses of penicillin. Their risk of an anaphylactic or accelerated allergic reaction ranged from 50 to 70 per cent.[39] Therefore, if skin tests are positive, equally effective, non-cross-reacting antibiotics should be substituted when available.

If alternative antibiotics fail, produce unacceptable side effects, or clearly are less effective, a desensitization protocol should be considered in order to allow administration of penicillin. Such protocol markedly reduces the risk of anaphylaxis in patients with a positive skin test. Infections in which this situation is most common are subacute bacterial endocarditis due to enterococci, brain abscess, bacterial meningitis, overwhelming infections with staphylococci or *Pseudomonas* organisms such as osteomyelitis or sepsis, *Listeria* infections, neurosyphilis or syphilis during pregnancy.

Acute penicillin desensitization must be carried out in an intensive care setting. Premedication with antihistamines or steroids is not recommended – these drugs are not effective prophylactically and may mask early signs of reactivity which might otherwise trigger modification of the protocol.[74,75] Protocols are available for penicillin desensitization using both the oral and parenteral routes.[76,77] Approximately one-third of patients undergoing desensitization will manifest a transient allergic reaction during either desensitization or subsequent treatment.[76] These reactions are usually mild and self-limited in nature, but may be severe. Once desensitized, the patient's treatment with penicillin must not be delayed or the risk of an allergic reaction will increase.

Cephalosporins

Like penicillins, cephalosporins possess a β-lactam ring, but the five-membered thiazolidine ring is replaced by the six-membered dihydrothiazine ring. It was not long after the introduction of cephalosporins into clinical use that allergic reactions, including anaphylaxis, were reported. The possibility of cross-reactivity between cephalosporins and penicillins was considered.[160] Immuno- and bioassays evaluating IgG, IgM and IgE antibodies have clearly demonstrated cross-reactivity between penicillins and cephalosporins in animals and man.[161–163] Primary cephalosporin allergy, in patients without penicillin allergy, has also been reported, but the incidence has not been determined precisely.[164,165] Studies have been limited as the haptenic determinants involved in cephalosporin allergy are unknown. Similarly, the precise incidence of clinically relevant cross-reactivity between the penicillins and the cephalosporins is unknown; while it is probably small, it should not be discounted on statistical grounds since life-threatening anaphylactic cross-reactivity has been described. Therefore, patients with positive skin tests to any penicillin reagent should probably not receive cephalosporin antibiotics unless substitutes are clearly less efficacious. If cephalosporin drugs are necessary, they must be given cautiously, possibly with the aid of a modified desensitization protocol. In children, serum sickness is 15 times more likely to occur secondary to administration of cefaclor compared with amoxycillin.[166]

New β-lactam antibiotics

Two new classes of β-lactam antibiotics are the carbapenems (imipenem) and monobactams (aztreonam). Initial studies suggest significant cross-reactivity between penicillin determinants and imipenem, suggesting that it might be prudent to withhold carbapenems from penicillin skin test positive patients.[167] In contrast, preliminary investigations suggest only weak cross-reactivity between aztreonam and other β-lactam antibiotics; this suggests the possibility that aztreonam may be administered safely to most if not all penicillin allergic subjects.[168,169]

Vancomycin

Hypotension is the most serious immediate adverse effect of vancomycin. Direct myocardial depression[170] and non-immunologically mediated histamine release[171,172] are proposed aetiologies. Hypotension is most common during rapid infusion of vancomycin, or when a concentrated solution is used.[170] Vancomycin-associated hypotension is most common during surgery and anaesthesia; this may be related to the concomitant use of other drugs which produce vasodilation and/or have a negative inotropic effect.[170] Vasodilation which is responsible for hypotension is often associated with the 'red neck' or 'red man's syndrome'; this consists of intense erythematous discoloration of the upper trunk, arms and neck and may be associated with pruritus. Vancomycin has also been associated with sudden development of throbbing pain or spasm in the chest or parasternal muscles without evidence of myocardial ischaemia. To minimize these sometimes frightening reactions, vancomycin should be infused slowly, allowing at least 60 minutes, and in a dilute solution (500 mg/dl). Reactions should be treated by stopping the infusion, administering an antihistamine, and, if necessary, using medications to counteract the hypotension.

Sulphonamides

Sulphonamides are frequently responsible for drug-induced skin eruptions and drug fever which often appear during the seventh to tenth days of treatment. Less common manifestations include vasculitis, pulmonary reactions, the Stevens–Johnson syndrome, and urticaria. A resurgence of sulphonamide use resulted from the introduction of trimethoprim–sulphamethoxazole which is effective in treating a variety of infections. Hospitalized patients show an incidence of reactions to trimethoprim–sulphamethoxazole of 3–6 per cent.[173] For reasons that are unknown, reactions occur approximately 10–15 times more frequently in patients with the acquired immunodeficiency syndrome (AIDS).[174] The immunochemistry of sulphonamide allergy is not completely understood, although recent evidence suggests that some reactions are mediated through IgE antibodies.[175,176] Hepatic metabolism is necessary for the conversion of native sulphonamide into its immunogenic metabolite, N^4-sulphonamidoyl.[177] Preliminary studies using multiple N^4-sulphonamidoyl residues attached to a polytyrosine carrier as a reagent for skin testing reagent have been completed;[178] obviously, additional work is required to evaluate its clinical utility. Isolated examples of desensitization to sulphonamides have been reported using different desensitization protocols,[79,179] although severe reactions have been precipitated by desensitization to trimethoprim–sulphamethoxazole.[180]

Aspirin

Reactions to aspirin (ASA) are confined to the skin (urticaria and/or angioedema) and respiratory system (bronchospasm, rhinitis and sinusitis). Aspirin-induced bronchospasm is rare in the asthma-free patient. Bronchospasm resulting from ASA is seen in approximately 10 per cent of asthmatics over 10 years old, in 30–40 per cent of asthmatics with nasal polyps, rhinitis and sinusitis, and in 60–85 per cent of asthmatics with a history of other aspirin-induced reactions.[26] Reactions to aspirin can be severe, difficult to treat, and may even result in death.

The mechanism of ASA reactions is not currently understood, although it does not appear to be IgE mediated. Aspirin and other NSAIDs which inhibit cyclo-oxygenase may shunt arachidonic acid metabolism through the 5-lipoxygenase pathway, producing increased amounts of leukotrienes, C_4, D_4, and E_4 (see Fig. 25.3). Recent evidence suggests that aspirin-sensitive asthmatics are 1000-fold more susceptible to the bronchoconstrictive effects of LTE_4 than are asthmatics without aspirin sensitivity.[27] Since essentially all NSAIDs cross-react with ASA, they should be avoided by ASA-sensitive asthmatics.[26] If ASA or other NSAIDs are required for treatment of a disease in a patient with a history of aspirin sensitivity, aspirin desensitization may be indicated.[181]

Insulin

Insulin consists of two polypeptide chains joined by disulphide bonds. The most common commercial preparations contain both bovine and porcine insulin. Bovine insulin differs from human insulin in three of its amino acids; only one amino acid is different when porcine and human insulins are compared. These small differences may be responsible for the immunogenicity of commercial insulins. Another cause of immunogenicity may be changes in tertiary structure. Altered tertiary structure is felt to be one of the causes of allergy to recombinant human insulin.[182]

Most patients receiving daily injections of insulin demonstrate antibodies to insulin during the first few weeks of therapy.[183] However, the incidence of clinical insulin allergy in these patients is 5–10 per cent. In the majority of cases, the reactions are localized.

Local reactions are usually mild with erythema, induration, burning, and pruritus at the injection site.[1] They usually occur within the first 1–4 weeks after insulin therapy has begun. Reactions almost always disappear in 3–4 weeks even though insulin administration is continued. Occasionally they will persist, a phenomenon which may precede a systemic reaction.[1] Discontinuing insulin solely because of local reaction may actually increase the risk of systemic reactions should insulin therapy be resumed at a later time.

Most local allergic reactions can be treated with H_1-receptor antagonists. There is evidence that local allergic reactions coincide with production of IgE antibodies against insulin, while symptoms disappear with the concomitant production of IgG antibodies directed at insulin.[1]

Systemic allergic reactions to insulin with generalized urticaria, angioedema, bronchospasm and hypotension are very rare. If a systemic allergic reaction to insulin occurs, insulin should not be discontinued. Instead, the dose should be reduced by approximately two-thirds, and then slowly increased (by 2–5 units per treatment) until an adequate dose can be administered safely. In patients where insulin has been discontinued because of a systemic allergic reaction, desensitization may be cautiously attempted if insulin therapy is again required. The least allergenic insulin may be selected by comparative cutaneous testing of bovine, porcine and human insulin. Desensitization protocol schedules have been published.[78] The incidence of insulin allergy has decreased, but has not been eliminated, with increased use of recombinant human insulin.

REFERENCES

1. DeSwarte RD. Drug allergy. In: Patterson R, ed. *Allergic diseases: diagnosis and management.* Philadelphia: JB Lippincott Co, 1989: 505–661.

2. Fisher MMcD, More DG. The epidemiology and clinical features of anaphylactic reactions in anaesthesia. *Anaesthesia and Intensive Care* 1981; **2**: 226–34.

3. Gell PGH, Coombs RRA. Classification of allergic reactions responsible for clinical hypersensitivity and disease. In: Gell PGH, Coombs RRA, Hachmann PJ, eds. *Clinical aspects of immunology.* Oxford: Blackwell Scientific Publications, 1975: 761–81.

4. Chan H, Ster RS, Arndt KA, *et al*. The incidence of erythema multiforme, Stevens–Johnson syndrome, and toxic epidermal necrolysis. *Archives of Dermatology* 1990; **126**: 43–7.

5. Roujeau J, Guillaume J, Fabre J, Penso D, Flechet M, Girre J. Toxic epidermal necrolysis (Lyell syndrome). *Archives of Dermatology* 1990; **126**: 37–42.

6. Anderson JA, Adkinson Jr NF. Allergic reactions to drugs and biologic agents. *Journal of the American Medical Association* 1987; **258**: 2891–9.

7. Van Arsdel PP. Adverse drug reactions. In: Middleton E, Reed C, Ellis EF, eds. *Allergy: principles and practice.* St Louis: CV Mosby Company, 1983: 1389–414.

8. Levine BB. Immunologic mechanisms of penicillin allergy. A haptenic model system for the study of allergic diseases of man. *New England Journal of Medicine* 1966; **275**: 1115–25.

9. Metzger H, Alcaraz G, Hohman R, Kinet J-P, Pribluda V, Quarto R. The receptor with high affinity for immunoglobin E. *Annual Review of Immunology* 1986; **4**: 419–70.

10. Ishizaka T. Mechanisms of IgE-mediated hypersensitivity. In: Middleton E Jr, Reed CE, Ellis EF, Adkinson NF Jr, Yunginger JW, eds. *Allergy: principles and practice.* St Louis: CV Mosby Co, 1988: 71–93.

11. Siraganian RP. Histamine secretion from mast cells and basophils. *Trends in Pharmacological Sciences* 1983; **4**: 432–7.

12. Ghebrehiwet B. The complement system: mechanisms of activation, regulation, and biological functions. In: Kaplan AP, ed. *Allergy.* New York: Churchill Livingstone, 1985: 131–52.

13. Yancey KB, Hammer CH, Harvath L, Renfer L, Frank MM, Lawley TJ. Studies of human C5a as a mediator of inflammation in normal human skin. *Journal of Clinical Investigation* 1985; **75**: 486–95.

14. Rent R, Ertel N, Eisenstein R, Gewurz H. Complement activation by interaction of polyanions and polycations. I. Heparin-protamine induced consumption of complement. *Journal of Immunology* 1975; **114**: 120–4.

15. Kirklin JK, Chenoweth DE, Naftel DC, *et al*. Effects of protamine administration after cardiopulmonary bypass on complement, blood elements, and the hemodynamic state. *Annals of Thoracic Surgery* 1986; **41**: 193–9.

16. Best N, Sinosich MJ, Teisner B, Grudzinskas JG, Fisher MM. Complement activation during cardiopulmonary bypass by heparin–protamine interaction. *British Journal of Anaesthesia* 1984; **56**: 339–43.

17. Wasserman SI, Marquardt DL. Anaphylaxis. In: Middleton E Jr, Reed CE, Ellis EF, Adkinson NF Jr, Yunginger JW, eds. *Allergy: principles and practice.* St Louis: CV Mosby Co, 1988: 1365–76.

18. Vyas GN, Perkins HA, Fudenberg HH. Anaphylactoid transfusion reactions associated with anti-IgA. *Lancet* 1968; **2**: 312–15.

19. Fearon DT, Ruddy S, Schur PH, McCabe WR. Activation of the properdin pathway of complement in patients with gram-negative bacteremia. *New England Journal of Medicine* 1975; **292**: 937–40.

20. Watkins J, Clark A, Appleyard TN, Padfield A. Immune medicated reactions to althesin (alphaxalone). *British Journal of Anaesthesia* 1976; **48**: 881–6.

21. Greenberger PA. Contrast media reactions. *Journal of Allergy and Clinical Immunology* 1984; **74**: 600–5.

22. Craddock PR, Fehr J, Brigham KL, Kronenberg RS, Jacob HS. Complement and leukocyte-mediated pulmonary dysfunction in hemodialysis. *New England Journal of Medicine* 1977; **296**: 769–74.

23. Hermens JM, Ebertz JM, Hanifin JM, Hirshman CA. Comparison of histamine release in human mast cells by morphine, fentanyl, and oxymorphone. *Anesthesiology* 1985; **62**: 124–9.

24. Ebertz JM, Hermens JM, McMillan JC, Uno H, Hirshman CA, Hanifin JM. Functional differences between human cutaneous mast cells and basophils: a comparison of morphine-induced histamine release. *Agents and Actions* 1986; **18**: 455–62.

25. North FC, Kettelkamp N, Hirshman CA. Comparison of cutaneous and *in vitro* histamine release by muscle relaxants. *Anesthesiology* 1987; **66**: 543–6.

26. Stevenson DD. Adverse reactions to aspirin and nonsteroidal anti-inflammatory drugs. Presented at the *12th International Congress on Clinical Immunology* 1985: 20–5.

27. Arm JP, O'Hickey SP, Spur BW, Lee TH. Airway responsiveness to histamine and leukotriene in subjects with aspirin-induced asthma. *American Review of Respiratory Disease* 1989; **140**: 148–53.

28. Barnard JH. Studies of 400 Hymenoptera sting deaths in the United States. *Journal of Allergy and Clinical Immunology* 1973; **52**: 259–64.

29. Fisher M, Hirshman CA. Hypersensitivity to drugs and other substances. In: Scurr C, Feldman S, Soni N, eds. *Scientific foundations of anaesthesia: the basis of intensive care.* Chicago: Year Book Medical Publishers, Inc, 1990: 642–9.

30. Kettelkamp NS, Austin DR, Cheek DBC, Doiwnes H, Hirshman C. Inhibition of *d*-tubocurarine-induced histamine release by halothane. *Anesthesiology* 1987; **66**: 666–9.

31. Levy JH. *Anaphylactic reactions in anesthesia and intensive care.* Boston: Butterworths, 1986: 19–27.

32. James LP, Austen KF. Fatal systemic anaphylaxis in man. *New England Journal of Medicine* 1964; **270**: 597–603.

33. Bickell WH, Dice WH. Military antishock trousers in a patient with adrenergic-resistant anaphylaxis. *Annals of Emergency Medicine* 1984; **13**: 189–90.

34. Loehr MM. Suit up against anaphylaxis. *Emergency Medicine* 1985; **April**: 127–8.

35. Smith CL, Bardgett DM, Hunter JM. Haemodynamic effects of the i.v. administration of cimetidine or ranitidine in the critically ill patient: a double-blind prospective study. *British Journal of Anaesthesia* 1987; **59**: 1397–402.

36. Tobias JD, Kubos KL, Hirshman CA. Aminophylline does not attenuate histamine-induced airway constriction during halothane anesthesia. *Anesthesiology* 1989; **71**: 723–9.

37. Smith PL, Kagey-Sobotka A, Bleecker ER, *et al*. Physiologic manifestations of human anaphylaxis. *Journal of Clinical Investigation* 1980; **66**: 1072–80.

38. Jacobs RL, Geoffrey WR Jr, Fournier DC, Chilton RJ, Culver WG, Beckmann CH. Potentiated anaphylaxis in patients with drug-induced beta-adrenergic blockade. *Journal of Allergy and Clinical Immunology* 1981; **68**: 125–7.

39. Weiss ME, Adkinson NF Jr. Immediate hypersensitivity reactions to penicillin and related antibiotics. *Clinical Allergy* 1988; **18**: 515–40.

40. Vervloet D, Nizankowska E, Arnaud A, Senft M, Alazia M, Charpin J. Adverse reactions to suxamethonium and other muscle relaxants under general anesthesia. *Journal of Allergy and Clinical Immunology* 1983; **71**: 552–9.

41. Fisher MM. Intradermal testing in the diagnosis of acute anaphylaxis during anaesthesia – results in five years experience. *Anaesthesia and Intensive Care* 1979; **7**: 58–61.

42. Moscicki RA, Sockin SM, Corsello BF, Ostro MG, Bloch KJ. Assessing the predictive value of skin testing for general anesthetic agent hypersensitivity *Journal of Allergy and Clinical Immunology* 1989; **83** (abstr): 270.

43. Grammer LC, Patterson R. Proteins: chymopapain and insulin. *Journal of Allergy and Clinical Immunology* 1984; **74**: 635–40.

44. Dykewicz MS, McGrath KG, Davison R, Kaplan KJ, Patterson R. Identification of patients at risk for anaphylaxis due to streptokinase. *Archives of Internal Medicine* 1986; **146**: 305–7.

45. Hamilton RG, Rendell M, Adkinson NF Jr. Serological analysis of human IgG and IgE anti-insulin antibodies by solid-phase radioimmunoassays. *Journal of Laboratory and Clinical Medicine* 1980; **96**: 1022–36.

46. Slater JE. Rubber anaphylaxis. *New England Journal of Medicine* 1989; **320**: 1126–30.

47. Morales C, Basomba A, Carreira J, Sastre A. Anaphylaxis produced by rubber glove contact. Case reports and immunological identification of the antigens involved. *Clinical and Experimental Allergy* 1989; **19**: 425–30.

48. Norman PS. Skin testing. In: Rose NR, Friedman H, Fahey JL, eds. *Manual of clinical laboratory immunology.* Washington, DC: American Society for Microbiology, 1986: 660.

49. Adkinson NF Jr. Tests for immunological drug reactions. In: Rose NF, Friedman H, eds. *Manual of clinical immunology.* Washington, DC: American Society for Microbiology, 1986: 692–7.

50. Fisher M. Intradermal testing after anaphylactoid reaction to anaesthetic drugs: practical aspects of performance and interpretation. *Anaesthesia and Intensive Care* 1984; **12**: 115–20.

51. Etter MS, Helrich M, Mackenzie CF. Immunoglobulin E fluctuation in thiopental anaphylaxis. *Anesthesiology* 1980; **52**: 181–3.

52. Lichtenstein LM, Osler AG. Studies on the mechanisms of hypersensitivity phenomena. IX. Histamine release from leukocytes by ragweed pollen antigen. *Journal of Experimental Medicine* 1964; **120**: 507–30.

53. Norman PS, Lichtenstein LM, Ishizaka K. Diagnostic tests in ragweed hay fever. A comparison of direct skin tests, IgE antibody measurements, and basophil histamine release. *Journal of Allergy and Clinical Immunology* 1973; **52**: 210–24.

54. Bruce CA, Rosenthal RR, Lichtenstein LM, Norman PS. Diagnostic tests in ragwood-allergic asthma. A comparison of direct skin tests, leukocyte histamine release, and quantitative bronchial challenge. *Journal of Allergy and Clinical Immunology* 1974; **53**: 230–9.

55. Hirshman CA, Peters J, Cartwright-Lee I. Leukocyte histamine release to thiopental. *Anesthesiology* 1982; **56**: 64–7.

56. Vervloet D, Arnaud A, Senft M, *et al*. Leukocyte histamine release to suxamethonium in patients with adverse reactions to muscle relaxants. *Journal of Allergy and Clinical Immunology* 1985; **75**: 338–42.

57. Pienkowski MM, Kazmier WJ, Adkinson NF Jr. Basophil histamine release remains unaffected by clinical desensitization to penicillin. *Journal of Allergy and Clinical Immunology* 1988; **82**: 171–8.

58. Atkins PC, Norman M, Weiner H, Zweiman B. Release of neutrophil chemotactic activity during immediate hypersensitivity reactions in humans. *Annals of Internal Medicine* 1977; **86**: 415–18.

59. Schwartz LB, Metcalfe DD, Miller JS, Earl H, Sullivan T. Tryptase levels as an indicator of mast-cell activation in systemic anaphylaxis and mastocytosis. *New England Journal of Medicine* 1987; **316**: 1622–6.

60. Anonymous. Council on Scientific Affairs. *In vitro* testing for allergy. Report II of the Allergy Panel. *Journal of the American Medical Association* 1987; **258**: 1639–43.

61. Plaut M, Lichtenstein LM, Henney CS. Properties of a subpopulation of T cells bearing histamine receptors. *Journal of Clinical Investigation* 1975: **55**: 856–74.

62. Hagedorn HC, Jensen BN, Kraup NB, Woodstrup I. Protamine insulate. *Journal of the American Medical Association* 1936; **106**: 177–80.

63. Wide L, Juhlin L. Detection of penicillin allergy of the

immediate type by radioimmunoassay of reagins (IgE) to penicilloyl conjugates. *Clinical Allergy* 1971; **1**: 171–7.

64. Baldo BA, Fisher MM. Substituted ammonium ions as allergenic determinants in drug allergy. *Nature* 1983; **306**: 262–4.

65. Harle DG, Baldo BA, Fisher MM. Detection of IgE antibodies to suxamethonium after anaphylactoid reactions during anaesthesia. *Lancet* 1984; **1**: 930–2.

66. Harle DG, Baldo BA, Smal MA, Wajon P, Fisher MM. Detection of thiopentone-reactive IgE antibodies following anaphylactoid reactions during anaesthesia. *Clinical Allergy* 1986; **16**: 493–8.

67. Tobin MC, Karns BK, Anselmino LM, Thomas LL. Potentiation of human basophil histamine release by protamine: a new role for a polycation recognition site. *Molecular Immunology* 1986; **23**: 245–53.

68. Weiss ME, Nyhan D, Peng Z, *et al*. Association of protamine IgE and IgG antibodies with life-threatening reactions to intravenous protamine. *New England Journal of Medicine* 1989; **320**: 886–92.

69. Turjanmaa K, Reunala T, Rasanen L. Comparison of diagnostic methods in latex surgical glove contact urticaria. *Contact Dermatitis* 1988; **19**: 241–7.

70. Wrangsjo K, Wahlbert JE, Axelsson GK. IgE-mediated allergy to natural rubber in 30 patients with contact urticaria. *Contact Dermatitis* 1988; **19**: 264–71.

71. Hamilton RG, Adkinson NF Jr. Serological methods in the diagnosis and management of human allergic disease. *Critical Reviews in Clinical Laboratory Sciences* 1984; **21**: 1–18.

72. Zeiss CR, Grammer LC, Levitz D. Comparison of the radioallergosorbent test and a quantitative solid-phase radio-immunoassay for the detection of ragweed-specific immuno-globulin E antibody in patients undergoing immunotherapy. *Journal of Allergy and Clinical Immunology* 1981; **67**: 105–10.

73. Lichtenstein LM, Normal PS, Kagey-Sobotka A, Adkinson NF Jr, Golden DBK. The immunologic basis for the efficacy of immunotherapy. In: Kerr JW, Ganderton MA, eds. *XI International Congress of Allergology and Clinical Immunology*. London: The Macmillan Press Ltd, 1983: 285–9.

74. Mathews KP, Hemphill FM, Lovell RG, Forsythe WE, Sheldon JM. A controlled study on the use of parenteral and oral antihistamines in prevailing penicillin reactions. *Journal of Allergy* 1956; **27**: 1–15.

75. Sciple GW, Knox JM, Montgomery CH. Incidence of penicillin reactions after an antihistaminic simultaneously administered parenterally. *New England Journal of Medicine* 1959; **261**: 1123–5.

76. Sullivan TJ, Yecies LD, Shatz GS, Parker CW, Wedner JH. Desensitization of patients allergic to penicillin using orally administered beta-lactam antibiotics. *Journal of Allergy and Clinical Immunology* 1982; **69**: 275–82.

77. Adkinson NF Jr. Penicillin allergy. In: Lichtenstein LM, Fauci A, eds. *Current therapy in allergy, immunology and rheumatology*. Ontario, Canada: BS Decker, 1983: 57–62.

78. Patterson R, DeSwarte RD, Greenberger PA, Grammer LC. Drug allergy and protocols for management of drug allergies. *New England Regional Allergy Proceedings* 1986; **4**: 325–42.

79. Smith RM, Iwamoto GK, Richerson HB, Flaherty JP. Trimethoprim–sulfamethoxazole desensitization in the acquired immunodeficiency syndrome. *Annals of Internal Medicine* 1987; **106**: 335.

80. Didier A, Cador D, Bongrand P, *et al*. Role of the quaternary ammonium ion determinants in allergy to muscle relaxants. *Journal of Allergy and Clinical Immunology* 1987; **79**: 578–84.

81. Charpin D, Benzarti M, Hemon Y, *et al*. Atopy and anaphylactic reactions to suxamethonium. *Journal of Allergy and Clinical Immunology* 1988; **82**: 356–60.

82. Charpin J, Vervloet D, Nizankovska E. Adverse reactions due to muscle relaxants. *Journal of Allergy and Clinical Immunology* 1988; **82**: 443–8.

83. Binkley K, Cheema A, Sussman G, *et al*. Generalized allergic reactions during anaesthesia: 28 consecutive cases. *Journal of Allergy and Clinical Immunology* 1990; **85**: 230(Abstract).

84. Dolovich J, Evans S, Rosenbloom D, Goodacre R, Rafajac FO. Anaphylaxis due to thiopental sodium anesthesia. *Canadian Medical Association Journal* 1980; **123**: 292–4.

85. Hirshman CA, Edelstein RA, Ebertz JM, Hanifin JM. Thiobarbiturate-induced histamine release in human skin mast cells. *Anesthesiology* 1985; **63**: 353–6.

86. Schatz M. Skin testing and incremental challenge in the evaluation of adverse reactions to local anesthetics. *Journal of Allergy and Clinical Immunology* 1984; **74**: 606–16.

87. deShazo RD, Nelson HS. An approach to the patient with a history of local anesthetic hypersensitivity: experience with 90 patients. *Journal of Clinical Immunology* 1979; **63**: 387–94.

88. Nagel JE, Fuscaldo JT, Fireman PL. Paraben allergy. *Journal of the American Medical Association* 1977; **237**: 1594–5.

89. Lawrence ID, Warner JA, Cohan VL, Hubbard WC, Kagey-Sobotka A, Lichtenstein LM. Purification and characterization of human skin mast cells. Evidence for human mast cell heterogeneity. *Journal of Immunology* 1987; **139**: 3062–9.

90. Harle DG, Baldo BA, Coroneos NJ, Fisher MM. Anaphylaxis following administration of papaveretum. Case report, implication of IgE antibodies that react with morphine and codeine and identification of an allergenic determinant. *Anesthesiology* 1989; **71**: 489–94.

91. Zucker-Pinchoff B, Ramanathan S. Anaphylactic reaction to epidural fentanyl. *Anesthesiology* 1989; **71**: 599–601.

92. Bennett MJ, Anderson LK, McMillan JC, Ebertz JM, Hanifin JM, Hirshman CA. Anaphylactic reaction during anaesthesia associated with positive intradermal skin test to fentanyl. *Canadian Anaesthetists' Society Journal* 1986; **33**: 71–4.

93. Weiss ME, Adkinson Jr NF, Hirshman CA. Evaluation of allergic drug reactions in the perioperative period. *Anesthesiology* 1989; **71**: 483–6.

94. Enright T, Chua-Lim A, Duda E, Lim DT. The role of a documented allergic profile as a risk factor for radiographic contrast media reaction. *Annals of Allergy* 1989; **62**: 302–5.

95. Kelly JF, Patterson R, Lieberman P, Mathison DA, Stevenson DD. Radiographic contrast media studies in high risk patients. *Journal of Allergy and Clinical Immunology* 1978; **62**: 181–4.

96. Kaliner M, Schelhamer J, Ottesen EA. Effects of infused histamine: correlation of plasma histamine levels and symptoms. *Journal of Allergy and Clinical Immunology* 1982; **69**: 283–9.

97. King BF, Hartman GW, Williamson Jr B, Leroy AJ, Hattery RR. Low-osmolality contrast media: a current perspective. *Mayo Clinic Proceedings* 1989; **64**: 976–85.

98. Greenberger P, Patterson R. The prevention of immediate generalized reactions to radiocontrast media in high-risk patients. *Journal of Allergy and Clinical Immunology* 1991; **87**: 867–72.

99. Sharath MD, Metzger WJ, Richerson HB, *et al*. Protamine-induced fatal anaphylaxis. *Journal of Thoracic and Cardiovascular Surgery* 1985; **90**: 86–90.

100. Weiss ME, Adkinson NF Jr. Allergy to protamine. In: Vervloet D, ed. *Clinical reviews in allergy: anesthesiology and allergy.* New York: Humana Press, 1991: 339–55.

101. Stewart WJ, McSweeney SM, Kellett MA, Faxon DP, Ryan TJ. Increased risk of severe protamine reactions in NPH insulin-dependent diabetics undergoing cardiac catheterization. *Circulation* 1984; **70**: 788–92.

102. Gottschlich GM, Gravlee GP, Georgitis JW. Adverse reactions to protamine sulfate during cardiac surgery in diabetic and non-diabetic patients. *Annals of Allergy* 1988; **61**: 277–81.

103. Samuel T. Antibodies reacting with salmon in human protamines in sera from infertile men and from vasectomized men and monkeys. *Clinical and Experimental Immunology* 1977; **30**: 181–7.

104. Knape JTA, Schuller JL, De Haan P, De Jong AP, Bovill JG. An anaphylactic reaction to protamine in a patient allergic to fish. *Anesthesiology* 1981; **55**: 324–5.

105. Adourian U, Fuchs E, Adkinson NF Jr, Hirshman CA. Incidence of anti-protamine antibodies in vasectomized males. *Anesthesiology* 1990; **73**: A1257 (Abstr).

106. Keller R. Interrelations between different types of cells. *International Archives of Allergy* 1968; **34**: 139–44.

107. Foreman JC, Lichtenstein LM. Induction of histamine secretion by polycations. *Biochimica et Biophysica Acta* 1980; **629**: 587–603.

108. Sauder RA, Hirshman CA. Protamine-induced histamine release in human skin mast cells. *Anesthesiology* 1990; **73**: 165–7.

109. Cavarocchi NG, Schaff HV, Orszulak TA, Homburger HA, Schnell WA, Pluth JR. Evidence for complement activation by protamine–heparin interaction after cardiopulmonary bypass. *Surgery* 1985; **98**: 525–31.

110. Lakin JD, Blocker TJ, Strong DM, Yocum MW. Anaphylaxis to protamine sulfate mediated by a complement-dependent IgG antibody. *Journal of Allergy and Clinical Immunology* 1977; **61**: 102–7.

111. Degges RD, Foster ME, Dang AQ, Read RC. Pulmonary hypertensive effect of heparin and protamine interaction: evidence for thromboxane B2 release from the lung. *American Journal of Surgery* 1987; **154**: 696–9.

112. Morel DR, Zapol WM, Thomas SJ, *et al*. C5a and thromboxane generation associated with pulmonary vaso- and bronchoconstriction during protamine reversal of heparin. *Anesthesiology* 1987; **66**: 597–604.

113. Grant JA, Cooper JR, Albyn KC, Buhner D, Farnam J, Yunginger JW. Anaphylactic reactions to protamine in insulin-dependent diabetics after cardiovascular procedures. *Journal of Allergy and Clinical Immunology* 1984; **73**: 180 (Abstr).

114. Gottschlich GM, Georgitis JW. Protamine-specific antibodies in protamine anaphylaxis. *Annals of Allergy* 1987: 180 (Abstr).

115. Kurtz AB, Gray RS, Markanday S, Nabarro JDN. Circulating IgG antibody to protamine in patients treated with protamine-insulins. *Diabetologia* 1983; **25**: 322–4.

116. Nell LJ, Thomas JW. Frequency and specificity of protamine antibodies in diabetic and control subjects. *Diabetes* 1988; **37**: 172–6.

117. Skidgel RA, Tan F, Jackman H, Zsigmond EK, Erdos EG. Portamine inhibits plasma carboxypeptidase N (CPN), the inactivator of anaphylatoxins and kinins. *Federation Proceedings* 1988; **2**: A1382 (Abstr).

118. Doolan L, McKenzie I, Krafchek J, Parson B, Buxton B. Protamine sulphate hypersensitivity. *Anaesthesia and Intensive Care* 1981; **9**: 147–9.

119. Weiss ME, Chatham F, Kagey-Sobotka A, Adkinson Jr NF. Serial immunological investigations in a patient who had a life-threatening reaction to intravenous protamine. *Clinical and Experimental Allergy* 1990; **20**: 713–20.

120. Weiler JM, Gelhaus MA, Carter JG, *et al*. A prospective study of the risk of an immediate adverse reaction to protamine sulfate during cardiopulmonary bypass surgery. *Journal of Allergy and Clinical Immunology* 1990; **85**: 713–19.

121. Lowenstein E. Lessions from studying an infrequent event: adverse hemodynamic response associated with protamine reversal of heparin. *Journal of Cardiothoracic Anesthesia* 1989; **3**: 99–107.

122. Adams RM. Diagnosis of allergic contact dermatitis of occupational origin. *Clinical Reviews of Allergy* 1986; **4**: 323–38.

123. Nutter AF. Contact urticaria to rubber. *British Journal of Dermatology* 1979; **101**: 597–8.

124. Axelsson JGK, Johansson SGO, Wrangsjo K. IgE-mediated anaphylactoid reactions to rubber. *Allergy* 1987; **42**: 46–50.

125. Gelfand DW. Barium enemas, latex balloons, and anaphylactic reactions. *American Journal of Roentgenology* 1991; **156**: 1–2.

126. Turjanmaa K, Reunala T. Contact urticara from rubber gloves. *Dermatology Clinics* 1988; **6**: 47–51.

127. Turjanmaa K. Incidence of immediate allergy to latex gloves in hospital personnel. *Contact Dermatitis* 1987; **17**: 270–5.

128. Kelly K, Selock M, Davis JP. Anaphylactic reactions during general anesthesia among pediatric patients – U.S., January 1990 – January 1991. *Morbidity and Mortality Weekly Report* 1991; **40**: 437–43.

129. Sussman GL, Tarlo S, Dolovich J. The spectrum of IgE-mediated responses to latex. *Journal of the American Medical Association* 1991; **265**: 2844–7.

130. Nguyen DH, Burns MW, Shapiro GG, Mayo ME, Murrey M, Mitchell ME. Intraoperative cardiovascular collapse secondary to latex allergy. *Journal of Urology* 1991; **146**: 571–4.

131. Meeropol E, Kelleher R, Bell S, Leger R. Allergic reactions to rubber in patients with myelodysplasia. *New England Journal of Medicine* 1990; **323**: 1072.

132. Sandberg ET, Slater JE, Roth DR, Abramson SL. Rubber-specific IgE in children in a spina bifida clinic. *Journal of Allergy and Clinical Immunology* 1992; **89**: 223 (Abstr).

133. Yassin MS, Sanyurah S, Lierl MB, *et al*. Evaluation of latex allergy in patients with meningomyelocele. *Annals of Allergy* 1992; **69**: 207–11.

134. Mathew SN, Melton A, Wagner W, Battisto JR. Latex

hypersensitivity: prevalence among children with spina bifida and immunoblotting identification of latex proteins. *Journal of Allergy and Clinical Immunology* 1992; **89**: 225 (Abstr).

135. Turjanmaa K, Laurila K, Makinen-Kiljunen S, Reunala T. Rubber contact urticaria. Allergenic properties of 19 brands of latex gloves. *Contact Dermatitis* 1988; **19**: 362–7.

136. M'Raihi L, Carpin D, Bongrand P. Cross-reactivity between latex and banana. *Journal of Allergy and Clinical Immunology* 1991; **87**: 129–30.

137. Leynadier F, Pecquet C, Dry J. Anaphylaxis to latex during surgery. *Anaesthesia* 1989; **44**: 547–50.

138. Gold M, Swartz JS, Braude BM, Dolovich J, Shandling B, Gilmour RF. Interoperative anaphylaxis: an association with latex sensitivity. *Journal of Allergy and Clinical Immunology* 1991; **87**: 662–6.

139. Ownby DR, Tomlanovich M, Sammons N, McCullough J. Anaphylaxis associated with latex allergy during barium enema examinations. *American Journal of Roentgenology* 1991; **156**: 903–8.

140. Slater JE. Latex allergy – what we know. *Journal of Allergy and Clinical Immunology* 1992; **90**: 279–81.

141. McGrath KG, Patterson R. Anaphylactic reactivity to streptokinase. *Journal of the American Medical Association* 1984; **252**: 1314–17.

142. White HD, Rivers JT, Maslowski AH, *et al*. Effect of intravenous streptokinase as compared with that of tissue plasminogen activator on left ventricular function after first myocardial infarction. *New England Journal of Medicine* 1989; **320**: 817–21.

143. Idsoe O, Guthe T, Willcox RR, De Weck AL. Nature and extent of penicillin side-reactions, with particular reference to fatalities from anaphylactic shock. *Bulletin of the World Health Organization* 1968; **38**: 159–88.

144. Sheffer AL. Anaphylaxis. *Journal of Allergy and Clinical Immunology* 1985; **75**: 227–33.

145. Shapiro S, Siskin V, Slone D, Lewis GP, Jick H. Drug rash with ampicillin and other penicillins. *Lancet* 1969; **2**: 969–72.

146. Galpin JE, Chow AW, Yoshikawa TT, Guze LB. 'Pseudoana-phylactic' reactions from inadvertent infusion of procaine penicillin G. *Annals of Internal Medicine* 1974; **81**: 358–9.

147. Eisen HN. Hypersensitivity to simple chemicals. In: Lawrence HS, ed. *Cellular and humoral aspects of the hypersensitive states.* New York: PB Hoeber, 1959: 111–26.

148. Levine BB. Immunochemical mechanisms involved in penicillin hypersensitivity in experimental animals and in human beings. *Federation Proceedings* 1965; **24**: 46–50.

149. Sullivan TJ. Facilitated haptenation of human proteins by penicillin. *Journal of Allergy and Clinical Immunology* 1989; **83**: 255 (Abstr).

150. DiPiro JT, Hamilton RG, Adkinson NF Jr. Facilitation of penicilloation of proteins by serum cofactors. *Journal of Allergy and Clinical Immunology* 1990; **85**: 192(Abstr).

151. Levine BB, Redmond AP. Minor haptenic determinant-specific reagins of penicillin hypersensitivity in man. *International Archives of Allergy and Applied Immunology* 1969; **35**: 445–55.

152. Levine BB, Redmond AP, Fellner MJ, Voss HE, Levytska V. Penicillin allergy and the heterogeneous immune responses of man to benzylpenicillin. *Journal of Clinical Investigation* 1966; **45**: 1895–906.

153. Adkinson NF Jr, Wheeler B. Risk factors for IgE-dependent reactions to penicillin. In: Kerr JW, Ganderton MA, eds. *XI International Congress of Allergology and Clinical Immunology.* London: Macmillan Press Ltd, 1983; 55–9.

154. Sogn DD. Prevention of allergic reactions to penicillin. *Journal of Allergy and Clinical Immunology* 1987; **78**: 1051–2.

155. Parker CW. The immunochemical basis for penicillin allergy. *Postgraduate Medical Journal* 1964; **40**: 141–55.

156. Sullivan TJ, Wedner HJ, Shatz GS, Yecies LD, Parker CW. Skin testing to detect penicillin allergy. *Journal of Allergy and Clinical Immunology* 1981; **68**: 171–80.

157. Parker CW. Drug therapy (first of three parts). *New England Journal of Medicine* 1975; **292**: 511–14.

158. Gorevic PD, Levine BB. Desensitization of anaphylactic hypersensitivity specific for the penicilloate minor determinant of penicillin and carbenicillin. *Journal of Allergy and Clinical Immunology* 1981; **68**: 267–72.

159. Blanca M, Vega JM, Garcia J, *et al* . Allergy to penicillin with good tolerance to other penicillins: study of the incidence in subjects allergic to betalactams. *Clinical and Experimental Allergy* 1990; **20**: 475–81.

160. Grieco MH. Cross-allergenicity of the penicillins and the cephalosporins. *Archives of Internal Medicine* 1967; **119**: 141–6.

161. Petz L. Immunologic cross-reactivity between penicillins and cephalosporins: a review. *Journal of Infectious Diseases* 1978; **137**: S74–9.

162. Shibata K, Atsumi T, Itorivchi Y, Mashimo K. Immunological cross-reactivities of cephalothin and its related compounds with benzylpenicillin (penicillin G). *Nature* 1966; **212**: 419–20.

163. Abraham GN, Petz LD, Fudenberg HH. Immunohaemato-logical cross-allergenicity between penicillin and cephalothin in humans. *Clinical and Experimental Immunology* 1968; **3**: 343–57.

164. Abraham GN, Petz LD, Fudenberg HH, Cephalothin hypersensitivity associated with anticephalothin antibodies. *International Archives of Allergy and Applied Immunology* 1968; **34**: 65–74.

165. Ong R, Sullivan T. Detection and characterization of human IgE to cephalosporin determinants. *Journal of Allergy and Clinical Immunology* 1988; **81**: 222.

166. Heckbert SR, Stryker WS, Coltin KL, Manson JE, Platt R. Serum sickness in children after antibiotic exposure: estimates of occurrence and morbidity in a health maintenance organization population. *American Journal of Epidemiology* 1990; **132**: 336–42.

167. Saxon A, Beall GN, Rohr AS, Adelman DC. Immediate hypersensitivity reactions to beta-lactam antibiotics. *Annals of Internal Medicine* 1987; **107**: 204–15.

168. Adkinson NF Jr, Swabb EA, Sugerman AA. Immunology of the monobactam aztreonam. *Antimicrobial Agents and Chemotherapy* 1984; **25**: 93–7.

169. Adkinson NF Jr, Wheeler B, Swabb EA. Clinical tolerance of the monobactam aztreonam in penicillin allergic subjects. Abstract (WS–26–4) presented at the *14th International Congress of Chemotherapy, June 23–28, Kyoto, Japan.*

170. Southorn PA, Plevak DJ, Wright AJ, Wilson WR. Adverse

effects of vancomycin administered in the perioperative period. *Mayo Clinic Proceedings* 1986; **61**: 721–4.

171. Verburg KM, Bowsher RR, Israel KS, Black HR, Henry DP. Histamine release by vancomycin in humans. *Federation Proceedings* 1985; **44**: 1247.

172. Levy JH, Kettlekamp BA, Goertz P, Hermans J, Hirshman CA. Histamine release by vancomycin: a mechanism for hypotension in man. *Anesthesiology* 1987; **67**: 122–5.

173. Jick H. Adverse reactions to trimethoprim-sulfamethoxazole in hospitalized patients. *Reviews of Infectious Diseases* 1982; **4**: 426–8.

174. Gorden FM, Simon GL, Wofsy CB. Adverse reactions to trimethoprim-sulfamethoxazole in patients with the acquired immunodeficiency syndrome. *Annals of Internal Medicine* 1984; **100**: 495–9.

175. Carrington DM, Earl HS, Sullivan TJ. Studies of human IgE to a sulfonamide determinant. *Journal of Allergy and Clinical Immunology* 1987; **79**: 442–7.

176. Gruchalla RS, Sullivan TJ. *In vitro* and *in vivo* studies of immunologic reactivity to sulfamethoxazole. *Journal of Allergy and Clinical Immunology* 1990; **85**: 157(Abstr).

177. Rieder MJ, Uetrecht J, Shear NH, Cannon M, Miller M, Spielberg SP. Diagnosis of sulfonamide hypersensitivity reactions by *in vitro* 'rechallenge' with hydroxylamine metabolites. *Annals of Internal Medicine* 1989; **110**: 286–9.

178. Gruchalla RS, Sullivan TJ. Detection of IgE to sulfamethoxazole by skin testing. *Journal of Allergy and Clinical Immunology* 1991; **87**: 231(Abstr).

179. Finegold I. Oral desensitization to trimethoprim-sulfamethoxazole in a patient with acquired immunodeficiency syndrome. *Journal of Allergy and Clinical Immunology* 1986; **78**: 905–8.

180. Sher MR, Suchar C, Lockey RF. Anaphylactic shock induced by oral desensitization to trimethoprim/sulfamethoxazole (TMP-SMZ). *Journal of Allergy and Clinical Immunology* 1986; **77**: 133.

181. Stevenson DD. Diagnosis, prevention, and treatment of adverse reactions to aspirin (ASA) and nonsteroidal anti-inflammatory drugs (NSAID). *Journal of Allergy and Clinical Immunology* 1984; **74**: 617–22.

182. Patterson R, Lucerna G, Metz R, *et al.* Reaginic antibody against insulin: demonstration of antigenic distinction between native and extracted insulin. *Journal of Immunology* 1969; **103**: 1061–71.

183. Yalow RS, Berson SA. Immunologic aspects of insulin. *American Journal of Medicine* 1961; **31**: 882–91.

Chapter 26

Computers in Anaesthesia

J. G. Bovill and F. H. M. Engbers

Computer basics	Closed-loop control in anaesthesia
Networks	Simulators and training devices
Computer-controlled infusions	Automated anaesthetic records

As in virtually every other field of modern society, computers and computer technology have made inroads into anaesthetic practice. Every new physiological monitor and most of the newer generation of ventilators rely on this technology. The generation of graphical trends, simultaneous graphical and alphanumerical displays, arrhythmia detection and ST-segment analysis would not be possible without the dedicated processors present in these devices. Often, of course, the presence of computers in our equipment is not obvious to the user. However, in other fields of anaesthesia computers are beginning to make a more obvious impact. And of course many anaesthetists, in common with most other professionals, will use the enormous processing power of the modern personal computer for word processing, administrative and statistical purposes, to name but a few. Thus the anaesthetist of the future will need to be computer literate.

Computer basics

Computers can be divided into two broad types, analogue and digital. Analogue computers solve problems by operating on continuous variables whereas digital computers operate on discontinuous or discrete values. Today analogue computers are only used for highly specialized tasks, e.g. by engineers for complex simulations. In general, when speaking about computers one is referring to digital computers.

Mechanical or electromechanical counting machines or computers were devised in the seventeenth century. The French scientist and philosopher Blaise Pascal invented a digital adding machine in 1642, using a mechanical gear system, to help him in computations for his father's business accounts. The first automatic digital computer was conceived in the 1830s by the English inventor, Charles

Babbage. Called the Analytical Engine, it was never constructed, largely because the precision engineering required was not available at that time. Currently a group is constructing a working example of Babbage's Analytical Engine, and the first stage has recently been completed. The first all-purpose modern electronic computer was the ENIAC (Electronic Numerical Integrator and Calculator), completed in 1946 by J.P. Eckert and J.W. Mauchly at the University of Pennsylvania. ENIAC was 1000 times faster than its electromechanical predecessors and could execute up to 5000 basic arithmetical operations per second (a modern computer can execute several million instructions per second). ENIAC was an externally programmed machine that utilized plugboards on which the sequence of operations required to solve any particular problem were arranged by hand. Each new problem required the arrangement to be manually altered. In later versions the instructions were preprogrammed into the computer. These first generation computers were massive machines and, since they used valve technology, had a power consumption equivalent to that of a small town.

The real breakthrough in computer technology came with the discovery of the transistor in the 1950s. Its small size, much greater reliability and its relatively low power consumption made it far superior to the valve. During the 1960s and 1970s electronic components were further miniaturized, resulting in dramatic improvements in computer technology. Today's computers use very large scale integration (VLSI) technology, in which hundreds of thousands of electronic components are contained on a silicon chip (integrated circuit) less than 5 mm^2.

The 'brain' of a computer is the central processing unit (CPU), an integrated circuit that contains all the arithmetic, logic and control circuitry needed to allow the computer to function. It is the CPU that interprets and executes instructions. Every digital computer performs five basic functions: input, storage, control, processing and output. The most common input sources are the keyboard and magnetic disk units, but could equally well be from an ECG monitor. Input data are stored in various internal storage devices, referred to as memory, until needed. Also held in memory are the instructions that govern the functioning of the computer and the program that will operate on the input data. The primary internal storage device is a high-speed random-access memory (RAM). The CPU can both read from and write to RAM, and external programs and the data that they operate on must be read into RAM, e.g. from the hard disk, before execution. RAM is, however, volatile and all stored information is lost when power is turned off. The second type of internal memory is ROM (read-only memory), which in contrast to RAM is permanent. ROM stores the programs the computer needs to function, such as the BIOS (basic input/output system). BIOS is a piece of operating system software that handles the input and output functions and various routines that test and report errors in the components of the computer when power is switched on. The control function involves retrieval of instructions from storage in the appropriate sequence and relaying the proper command to the arithmetic/logic unit where the actual processing takes place. The output device that receives processed data from the computer can take many forms, such as the computer screen, printer or magnetic disk. When extremely large amounts of information must be stored mass-storage systems such as optical disks, developed from compact disc and videodisc technology, may be used. Information is encoded by a laser beam on the disc platter as a pattern of pits. The more sophisticated of these systems can store in excess of one hundred thousand million bytes, far more than the highest capacity hard disk. Unlike the hard disk, however, data stored on the optical disk is permanent. They are therefore referred to as 'write once read many' devices or WORMs.

Information in a digital computer is stored as groups of binary digits called bits. Each bit can be in one of only two possible states, 'on' or 'off'. These states are more often represented logically by the symbols '1' and '0', and internally as +5 volts and 0 volts respectively. A pattern of eight bits is called a byte, and a number of bytes, the actual number varying between computers, is referred to as a word. Although modern computers work with 2-byte words, 4-byte words or 8-byte words, the byte remains the basic unit in computer technology. In computer terminology one kilobyte (1 KB) represents 1024 (2^{10}) bytes, not 1000 bytes. Although computers actually work in the binary system (i.e. base 2), a binary number of even modest size is very long and difficult for humans to manipulate easily (365 decimal = 101101101 binary). Therefore, computer programmers often use another means of representation, the hexadecimal number system (base 16) in which a group of 4 bits (a nibble) is encoded as one hexadecimal digit. Using 4 bits it is possible to count from zero up to 15 ($2^4 - 1$). In order to count up to 15 using

Table 26.1 Computer number systems

Decimal	Binary	Octal	Hexadecimal
0	0	0	0
1	1	1	1
2	10	2	2
3	11	3	3
4	100	4	4
5	101	5	5
6	110	6	6
7	111	7	7
8	1000	10	8
9	1001	11	9
10	1010	12	A
11	1011	13	B
12	1100	14	C
13	1101	15	D
14	1110	16	E
15	1111	17	F

single digits the decimal number system is extended using the letters A, B, C, D, E, F (Table 26.1). The decimal number 365 is represented in hexadecimal as 16D. Occasionally the octal system is used. This has its origin in the time when some mainframe computers had an internal organization based on 6-bit units rather than 8-bit bytes. The octal digit requires 3 bits ($2^3 = 8$, hence octal) and a 6-bit unit could be made up of two octal digits.

The processor consists of two basic types of components, registers and operational circuits. Registers are storage elements that temporarily hold binary digits representing the instruction or the data to be processed. The operational circuits include binary logic elements that perform the arithmetic and logical operations of the computer. Communication between the various parts of the microprocessor, and with external devices, is by means of three special communication lines called buses (Fig. 26.1). The address

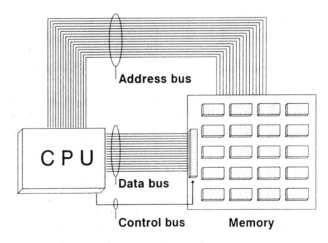

Fig. 26.1 Schematic representation of the connections between memory and the central processing unit (CPU) of a microprocessor.

bus has the job of enabling the path for the communication between the CPU and ROM or RAM. The data bus passes the actual information to and from the CPU, and the control bus defines the type of communication, e.g. read or write. For reliable data transfer to be possible all the bus signals must be accurately synchronized by means of an internal clock. Depending on the speed of the CPU and the memory access time, the clock can operate at frequencies up to 60 MHz. The speed of the computer is a function of the clock speed because each instruction takes at least one clock tick.

Digital information can be transmitted between the computer and external devices in either serial or parallel format. In the case of serial transmission digital data is sent in sequence on a single wire, whereas in the case of parallel transmission all eight bits of a byte are sent simultaneously on eight separate wires. Parallel transmission is generally much faster than serial, but for electrical reasons is restricted to cable distances of about 3 metres. For longer distances serial transmission must be used. There are many different types of serial interfaces, the most popular being the RS-232C and the newer RS-422. Transmission speed is indicated in bauds (one baud equals one bit per second). Because each transmitted byte requires at least 2 extra bits to indicate the beginning and end of the byte, a baud rate of 9600 baud represents the transmission of only about 960 rather than 1200 bytes or characters per second. Networks use a special kind of serial transmission in which the information is modulated so that transmission rates of 10 megabit per second are possible.

The power of the digital computer rests not only on its speed and capacity for manipulating large quantities of data but also on its ability to perform complicated sequences of such actions under control of a stored program. The first step in programming should be to express the solution of the problem to be solved as a series of step-by-step procedures – such a solution is called an algorithm. The next step is to translate the algorithm into a language that is understood by the computer. Early generation computers required that the individual steps of the algorithm were entered, in the correct sequence, in binary or hexadecimal format, a task that was both tedious and very prone to errors. Subsequently a programming language close to machine language, called assembly language, was created. Each step of the algorithm was represented by a mnemonic code designed to assist the memory of the programmer. For example, the assembly language instruction 'ADD AX, ADR' (add the contents of memory address 'ADR' to register AX) is much easier to remember than the hexadecimal equivalent '132 7D3'. Having coded the complete algorithm in assembly language, a translator program or assembler is then used to translate the mnemonic into a form that can be loaded into the computer memory ready for execution. Assembly language programming is still used when high speed

processing is needed. However, most programming today is done using high-level languages such as Fortran, Pascal or C which are much easier to use than assembly language.

Networks

The enormous increase in the computing power and storage capacity of the modern PC in recent years has been paralleled by an even greater increase in the size and complexity of the programs that are available, and in the data files that sometimes are generated. The original IBM PC had as little as 16 KB memory and one or two 128 KB floppy disks. Today many commercial programs occupy up to 10 MB storage capacity and require 1 MB (1024 KB) of internal memory. These large programs can often be most appropriately installed on a central computer in a local area network where they can be shared by several users concurrently. In simplest terms a local area network (LAN) is a high-speed communication link, allowing personal computers to communicate with each other, with minicomputer or mainframes, and with other electronic devices. One of the major reasons for using a network includes sharing programs, data files and resources such as printers and mass-storage systems that need to be accessed by many people.

The computers and devices in a LAN are connected using a system of cables, network interfaces and appropriate communication software. Most early LANs used coaxial cables but nowadays other types are also widely used, such as twisted-pair or fibreoptic cables. The network interface is the physical connection between the computers or other devices and the network cabling system. Its function is to organize and control the transmission of data. Most PC-based LANs operate in a client/server environment in which some or all of the application processing is done on the client computer, while a dedicated computer called a file server is used to provide controlled access to files and access to network printers or mass-storage facilities. The particular computer that is used as a file server might be a standard PC or a specialized machine designed specifically for that purpose. In any case, it is the software running in that machine that defines it as a file server.

Computer-controlled infusions

The increasing interest in total intravenous anaesthesia (TIVA) has focused attention on the role of computers for this application. TIVA makes use of a continuous intravenous infusion of hypnotics, analgesics and muscle relaxants at a variable rate determined by the patient's response to surgical or anaesthetic stimuli. While manually adjusted infusions do provide stable anaesthesia, very frequent adjustments to the infusion rate are usually required, often in conjunction with additional boluses of the anaesthetic drugs, in order to achieve relatively stable plasma concentrations. In order to achieve rapidly and maintain a constant plasma concentration over a period of time requires a combination of three processes: (1) a bolus dose calculated to fill the central compartment (which includes blood) to the required concentration: (2) a constant rate infusion to replace drug lost by elimination; and (3) an exponentially decreasing infusion that will replace drug lost from the plasma by transfer or distribution to peripheral tissues. While the administration of a bolus and a constant rate infusion poses no great technical problems, the administration of an infusion at a rate which must constantly change according to a predetermined exponential function is obviously much more demanding. This is only possible with the use of an infusion pump that can be controlled by a computer, suitably programmed with the appropriate pharmacokinetic data of the drug to be infused. This combination of a bolus, fixed rate and exponentially declining infusion rate is the basis of the BET (bolus, elimination, transfer) scheme described by Schwilden.[1]

When a combination of a hypnotic and an opioid, e.g. propofol and alfentanil, are used for TIVA, it is usually satisfactory to maintain relatively stable plasma or blood concentrations of one component, for example the hypnotic, while varying the concentration of the second component according to the response of the patient to changing surgical stimuli. While the BET scheme would be suitable for administration of the constant concentration component, it is too inflexible for administering the variable component. Fortunately it is relatively simple to program a computer to solve continuously the pharmacokinetic equations and derive the infusion rate required to achieve any desired plasma concentrations, or to move from one concentration to a higher one. When moving from a higher to a lower concentration, the infusion pump is stopped, and only restarted when the computer has calculated that the (predicted) plasma concentration has decreased to the level requested by the anaesthetist.[2]

Several commercially available syringe infusion pumps incorporate a serial, usually RS-232, interface allowing two-way communication with a computer. This bidirectional communication is essential since not only must the computer send instructions about infusion rate to the pump, it must also be able to read information from the pump, e.g. to confirm the actual rate, detect errors in the communication process and the presence of alarm

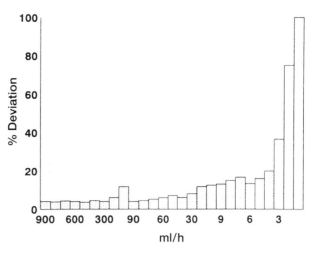

Fig. 26.2 Percentage deviation between the selected flow rate and the measured flow rate at different selected rates, for an Ohmeda 9000 syringe pump.

situations such as empty syringe, an obstruction in the tube, etc. Additionally, the pump must be capable of providing a wider range of infusion rates (up to 1200 ml/h) than conventional syringe pumps, and should have minimal 'backlash', i.e. delayed engagement of the driving mechanism. This is particularly important when starting up at low flow rates, e.g. after refilling the syringe. This delay cannot be detected by the driving computer. Another problem that we have found to be associated with low infusion rates is that delivery of the fluid may be considerably less than anticipated (Fig. 26.2). This is caused by the requirement for the pump motor to have advanced by a certain amount before receiving the next command to set the pump infusion rate. Kenny and White[3] have also reported this problem in their system, an Ohmeda 9000 syringe pump interfaced to a Psion POS200 hand-held microcomputer, when slow infusion rates were frequently transmitted to the pump. The fault was corrected by transmitting an infusion rate command to the pump only if it was different from the rate previously set.

Several computer-controlled infusion devices, implementing compartmental pharmacokinetic models, have been described.[2-4] These systems allow the pump rate to be updated at short, e.g. 5-second, intervals and provide the experienced user with a flexibility in manipulating plasma concentration comparable to that offered by calibrated vaporizers for the volatile anaesthetic agents. Furthermore, it is possible for one computer to control two pumps, e.g. delivering propofol and alfentanil simultaneously but independently. This obviously increases the flexibility of the system, but does require that fail-safe error checking is incorporated in the software to prevent unintentional changes in the concentration of the wrong drug. At the present time none of these systems has been made commercially available; there remain considerable legal and regulatory hurdles to be overcome before this can happen. It is likely that a commercial product will consist of a single module incorporating both the syringe pump mechanism and the computer hardware and software required to drive the syringe. Ideally such a module should contain a library of pharmacokinetic data sets covering the full range of potential TIVA drugs.

For many anaesthetists a stumbling block to the more widespread use of TIVA, including the use of computer-controlled infusions, is the difficulty sometimes encountered in judging the depth of anaesthesia. Clinical signs used to indicate light anaesthesia during balanced anaesthesia, e.g. haemodynamic changes or increased autonomic activity, often prove unreliable during TIVA, especially when moderate to high concentrations of an opioid are used. Unfortunately there is no universally reliable monitor of the depth of anaesthesia, although several have been investigated, with varying success. The most successful have used either a parameter derived from the electroencephalogram (EEG) or sensory evoked responses. Both EEG analysis and evoked response monitoring rely on computer technology.

Most anaesthetic drugs cause slowing of the EEG in a dose-related fashion, the exception being the benzodiazepines that shift the EEG frequency distribution towards higher frequencies. The most widely used technique for EEG frequency analysis uses the fast Fourier transform (FFT), a mathematical technique that transforms successive epochs of a signal in the time domain, i.e. a voltage that varies with time, into one in the frequency domain in which the signal is represented in terms of amplitude, phase and frequency. Often amplitude is replaced by power which is equal to the square of the amplitude. For FFT analysis to be carried out in real time requires the processing power of modern computers. Changes in EEG frequency are frequently presented as a single parameter derived from the frequency spectrum such as the median frequency or spectral edge frequency. The median frequency is that frequency which equally divides the power in the EEG spectrum – in other words it is the 50th percentile of the cumulative power distribution. In a similar manner the spectral edge frequency represents the 95th percentile, i.e. the frequency below which is found 95 per cent of the power in the spectrum.

Whereas the EEG represents the global summation of electrical activity of the brain, an evoked response is the electroencephalographic change in response to a specific stimulus. The auditory evoked response (AER) is the EEG response to an auditory stimulus, such as a click presented via headphones. The amplitude of the response to a single click is of very low amplitude, a few microvolts, whereas the EEG, EMG and other extraneous electrical activity is of the order of 50 to several hundred microvolts. Fortunately the evoked response is time-locked to the originating stimulus whereas the EEG and other electrical signals are essentially random, so that if repeated stimuli at 5–10 Hz are presented to the patient, the auditory response can be extracted using computer averaging techniques. Of the three commonly used stimulus modalities, the AER appears to be the most promising as a possible predictor of anaesthetic depth. In particular, changes in the middle latency or early cortical responses, that is waves with latencies of 8–45 ms, are the most consistent in their behaviour, with a close correlation between concentration of inhalational and intravenous anaesthetics and changes in latency and amplitude.[5] Even more relevant, however, is that the AER shows an arousal pattern in response to surgical stimulation, in some patients almost reverting to the pattern seen in the awake patient.[6,7] Depth of anaesthesia represents a balance between the depressant effect of the anaesthetic and the stimulating effect of surgery. It is the ability of AER to reflect that balance that makes it a potentially useful monitor.

A disadvantage of the AER is that as many as 1024 averages may be required to reliably acquire a valid response; at a stimulus frequency of 6 Hz this would take 2.8 minutes, a response time much too slow to provide a reliable real-time index of anaesthetic depth. But here again computer technology has come to the rescue. Using a moving average of the incoming responses it is possible to obtain a reasonable approximation to a real-time display of the AER. It is also possible, by doubly differentiating the AER wave, to derive a numerical index giving an indication of anaesthetic depth.[7] Kenny et al.[8] have described the successful use of closed-loop control of anaesthesia with propofol in spontaneously breathing patients undergoing body surface surgery. A level of arousal score was derived from analysis of a 3-second moving average of the AER and used as the input to a proportional-integral controller. The controller altered the predicted blood propofol concentration to minimize the difference between the measured and target level of the arousal score.

Closed-loop control in anaesthesia

Closed-loop control systems are widely used in industry and system control theory has been extensively studied by engineers. Closed-loop feedback control systems accept an input control variable, compare its value with a predetermined set-point and use the difference (the error parameter) in a computer algorithm to determine the amount by which the controlled process must be modified to achieve the set-point. In other words, the output feeds back its information

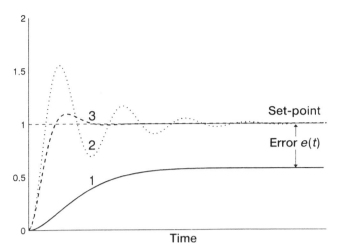

Fig. 26.3 Response of a closed-loop feedback control system to a unit step input. Curve 1, the response of a proportional controller with a control constant (K_P) of low value. The response is sluggish and the set-point is never reached. Curve 2, the response of a proportional controller with a control constant (K_P) of high value. There is a rapid response but the system exhibits instability with marked oscillation about the set-point. Curve 3, the response of a PID controller. The set-point is quickly reached with minimum overshoot.

to modify the input to the system. As an example, if blood pressure is the controlled variable, the difference between the measured and preset pressure, i.e. the error signal, could be used to control the increase or decrease in the infusion rate of sodium nitroprusside so as to reduce the error signal to zero. A natural example of closed-loop feedback control is the baroreceptor system. Changes in blood pressure alter the firing frequency of the baroreceptors. This information is processed in the CNS which then adjusts heart rate and peripheral resistance so as to return the blood pressure to its previous level.

Two broad classes of control systems are available, open-loop control and closed-loop control. In an open-loop system a prediction is made about the input needed to give the desired output, based on knowledge about the system and past experience, and the input adjusted accordingly. An example, already discussed, is a computer-controlled infusion in which adjustments to the infusion rate to achieve a desired concentration of drug are based on predictions obtained from a pharmacokinetic model. Because sensors are not yet available to monitor online drug concentrations, a closed-loop system cannot be used to control blood concentration. To close the loop it is necessary to be able to measure the system output. This measurement can be direct, e.g. blood pressure, or indirect. Because there is no direct method for monitoring anaesthesia, closed-loop systems designed to maintain constant depth of anaesthesia make use of indirect variables such as the EEG or auditory evoked potentials. Closing the loop can make the system more accurate but it can also make it very oscillatory or even unstable. This

problem arises when delays occur within the system which causes corrective action to be applied too late, leading to alternative overcorrection and undercorrection. In practice a choice has to be made between the conflicting requirements of accuracy and stability.

The simplest form of closed-loop control is one in which the error signal is multiplied by a constant, K_P, to yield a signal which is the input to the process, e.g. pump infusion rate. This arrangement is called proportional control. The numerical value of the constant K_P determines the amount of corrective effort which is applied for a given magnitude of error. By varying the value of K_P the dynamic behaviour of the system can be altered. For low value of K_P, the corrective effort is small and the response likely to be sluggish, and the set-point may never be reached (Fig. 26.3). If K_P is very large, significant oscillation or instability is likely to result. For this reason one or more additional control actions are usually added to proportional control. Proportional plus integral (or PI) control involves adding to the proportional term a signal proportional to the time integral of the error. Thus, the controller output (C_s) may be represented by the equation:

$$C_s(t) = K_P \times e(t) + K_I \int e(t) \ dt$$

where $e(t)$ is the error signal at time t and K_I is the integral control constant. PI controllers have the advantage that even the smallest error eventually produces a corrective signal of sufficient magnitude to ensure that the error signal is reduced to zero, i.e. the set-point is always reached. In addition, by careful selection of the control constants, K_P and K_I, it is possible to achieve a rapid response without excessive overshoot or oscillation. If, in practice, damping is a real problem, the damping of the system can be further increased by adding a derivative feedback loop to the control action. The derivative signal responds to the rate of change, or derivative, of the error signal. Controllers that incorporate all three control actions are known as proportional-integral-derivative (PID) controllers (Fig. 26.4). The equation for a PID controller is:

$$C_s(t) = K_P \times e(t) + K_I \int e(t) \ dt + K_D \frac{d(e(t))}{dt}$$

In general, controllers can be tuned to provide the best possible combination of accuracy, response time, and damping by adjusting the value of the control constants. Increasing the integral constant increases the accuracy of the system and increasing the derivative component reduces oscillations and improves the response time.[9] A common example of the use of a PID control system is the computer disk drive. When the head is required to read information from a distant track it initially moves towards that track at maximum speed, slowing as it approaches the designated track under the influence of the PID algorithm, eventually stopping without overshooting.

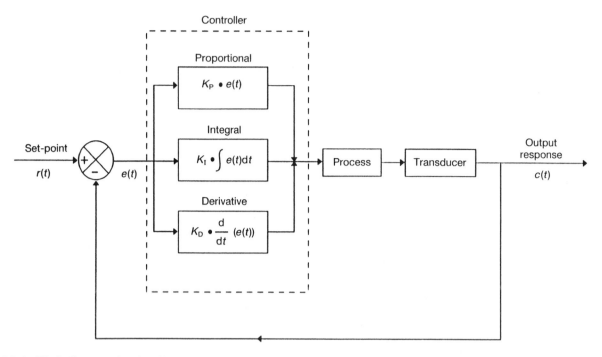

Fig. 26.4   Block diagram of a closed-loop feedback system with PID control.

In recent years engineers have increasingly moved from PID control to fuzzy logic control. Fuzzy logic is based on traditional logic theory and is widely used in artificial intelligence (AI) applications. One of the advantages of fuzzy logic control is that it can be applied to linear and non-linear processes. Most processes, both in engineering and biology, are non-linear. To use PID control for a non-linear process requires special techniques such as adaptive control. Fuzzy logic control is also much less sensitive to interference in the control circuit than PID control. An additional advantage is that it does not require the detailed mathematical description of the controlled process needed for PID control. Fuzzy logic control has been applied to the automatic control of blood pressure.[10]

Closed-loop feedback control systems have been used in medicine for many years. Attempts to control blood glucose accurately within normal limits by means of feedback-driven insulin pumps were made in the 1970s, with moderate success.[11] Feedback control of the delivery of drugs is, in principle, an efficient and safe method of administration in that the amount delivered is automatically adjusted to the requirements of the individual patient, bypassing the need to make any assumptions about the pharmacodynamics or pharmacokinetics of the drug in that individual. The technique has a long history in anaesthesia. In 1937 Gibbs *et al.*[12] reported the first systematic study into the effects of anaesthetic agents on the EEG, and suggested the possibility of employing the EEG for controlling the depth of anaesthesia. In the 1950s and early 1960s numerous reports appeared in the anaesthetic and neurophysiological literature documenting the EEG effects of inhalational and intravenous anaesthetic agents. Several investigators developed systems whereby the integrated output from EEG amplifiers was used to drive electromechanical devices controlling the delivery of cyclopropane[13] or syringe pumps delivering a barbiturate.[14] The compressed spectrum of the EEG was used in 1950 by Bickford[15] for feedback control of anaesthesia with ether, and subsequently for the control of thiopentone anaesthesia.[16] More recently Schwilden and his colleagues have applied these techniques to control automatically the depth of anaesthesia with methohexitone[17] and propofol.[18] They used the median frequency of the EEG as the control variable. This group has also demonstrated the reliability, in clinical practice, of closed-loop feedback control of intravenous anaesthesia with propofol, in combination with an open-loop computer-controlled delivery system for alfentanil.[19] Computer-based techniques to control the delivery of halothane have also been described, using the mean arterial pressure as the response variable.[20] More recently Monk *et al.*[21] described an adaptive feedback, or self-tuning, controller to regulate the inspired concentration of isoflurane to achieve controlled hypotension in patients undergoing ENT surgery. Adaptive control systems are those in which the control constants do not remain fixed but are automatically updated according to the response of the patient. With the availability of monitors to measure accurately the concentration of inhalational agents, these devices have been used to provide feedback input (for complete review see O'Hara *et al.*[9]).

During anaesthesia and in the immediate postoperative period maintenance of haemodynamic stability often

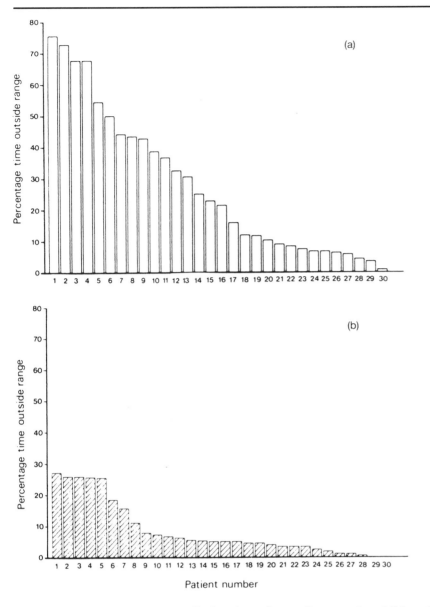

Fig. 26.5 (a) Percentage time spent outside the prescribed systolic pressure range for each of 30 patients whose arterial pressure was controlled by an infusion of sodium nitroprusside regulated by the nursing staff. (b) Percentage time spent outside the prescribed systolic pressure range for each of 30 patients whose arterial pressure was controlled by a computerized closed-loop infusion of nitroprusside. (From Reid and Kenny[26] with permission.)

requires intensive input from medical and nursing staff to regulate administration of fluids and vasoactive drugs. Computer-controlled feedback techniques have been applied to control the infusion of blood and blood substitutes for the expansion of blood volume in cardiac surgical patients.[22] End-expiratory left atrial pressure was used as the feedback signal. Subsequently Sheppard[23] reported the successful use of closed-loop feedback control of arterial blood pressure in hypertensive cardiac surgical patients with an infusion of sodium nitroprusside. Since then there have been many reports of this method for controlling postoperative hypertension, and its superiority over manual control has been demonstrated in several studies[24–26] (Fig. 26.5). The sensitivity of these systems is illustrated in Fig. 26.6.

Muscle relaxation during anaesthesia is one of the few clinical examples in which quantitative measurement of the therapeutic endpoint can be easily and frequently obtained.

In addition the dose–response characteristics of muscle relaxants are well known and, in skilled hands, the consequences of overdosage or underdosage need not be serious for the patient. It is, therefore, not surprising that feedback control of neuromuscular block has been extensively investigated. The systems have varied from simple mechanical or electronically controlled devices to computer-based adaptive feedback control systems.[9,27,28] To avoid undesirable oscillations in the neuromuscular block, the control algorithm needs to incorporate a pharmacodynamic model to account for the hysteresis effect, i.e. the significant time delay between plasma concentrations and the pharmacological effect, that occurs with all muscle relaxants. For this reason, the more rapidly acting relaxants, such as atracurium and vecuronium, that have less hysteresis, have been favoured.

Closed-loop control also has a role in closed-circuit anaesthesia. Both for economic reasons, in particular the

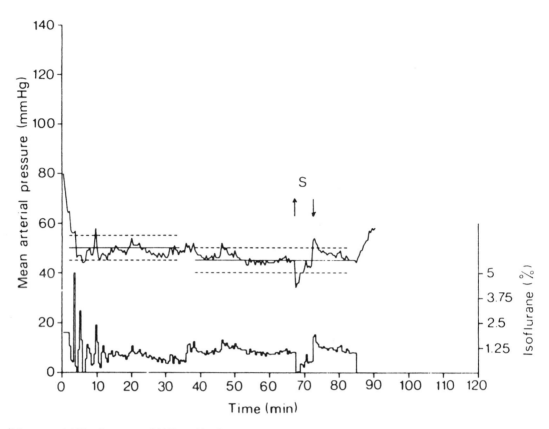

Fig. 26.6   Mean arterial blood pressure (MAP) and isoflurane inspired concentration in a patient during controlled hypotension in whom regulation of blood pressure was achieved by means of automatic closed-loop control of isoflurane inspired concentration. Initial control to 50 mmHg target pressure (——) is followed by a period of 45 mmHg. (— — — = ± 5 mmHg control band.) Two perturbations in MAP, mimicked by transducer shifts at the points indicated by S, resulted in changes in the isoflurane inspired concentration to regain regulation. The controller was switched off at the end of surgery to allow the MAP to return towards normal. (From Monk *et al.*[21] with permission.)

high cost of the newer volatile anaesthetic agents, and to reduce the risk of polluting the environment, there has been a movement towards the use of low-flow or totally closed-circuit anaesthesia. It has been estimated that decreasing fresh gas flow from 5 l/min to 2.5 l/min while maintaining the anaesthetic concentration of the inhalational agent could save 225 million US dollars per year world wide.[29] The use of a closed-circuit system – that is one in which the fresh gas flow is less than $25 \times (kg)^{3/4}$ ml/min – requires that several fundamental problems be overcome.[30] The uptake of nitrous oxide and the volatile agents constantly changes, so that their administration needs to be precisely controlled in order to maintain a constant alveolar partial pressure. This requires the facility to continuously monitor gas concentrations within the circuit. Fortunately, today there are several devices capable of doing this, often to a high degree of accuracy. Another, but very much related problem, is that at very low flow rates the current generation of vaporizers are both inaccurate and sluggish in their response. For example, when using isoflurane in a closed circuit, it takes 5 minutes to achieve an alveolar concentration of only 0.5 per cent, even with the vaporizer

on its maximum setting of 5 per cent. The solution is to incorporate the monitoring of the anaesthetic concentration into an automatic closed-loop control system and, second, to replace the conventional vaporizer by a system whereby the liquid anaesthetic is injected directly into the circuit – the exact amount to be injected being calculated and delivered under computer control. Several such systems have been described and validated.[31,32] A commercially available closed-loop anaesthesia machine (Physioflex) has been developed that incorporates PID control of the delivery of gaseous and volatile anaesthetics, in addition to ventilator control using feedback of end tidal carbon dioxide.[33] These systems accurately control the amount of oxygen fed into the system to maintain the required O_2 concentration in the fully closed circuit; therefore the anaesthetist can be presented with continuous information about the patient's O_2 consumption. None the less, despite the theoretical advantages of totally closed-circuit anaesthetic devices, utilizing highly sophisticated technology, it remains to be verified whether in practice they offer substantial clinical or economical advantages over the simpler minimum flow techniques, with fresh gas flows between 600 and 1500 ml/min. Minimum flow anaesthesia can be performed reliably and safely using equipment that is already available.

Simulators and training devices

In recent years there has been a marked awareness of and interest in the topic of quality assurance in anaesthesia. As a result attention has been focused on the role that simulators and training devices can play in improving and maintaining standards of anaesthetic care. Simulators are widely used in industry and in aviation. During their early training airline pilots spend many hours in a simulator learning how to cope with the complex cockpit systems and the many problems with which they may be confronted during actual flight. This will include exposure to many emergency situations, for example, engine failure at critical periods such as takeoff, engine fire, etc., which are fortunately uncommon in real life. Indeed some airlines now train their aircrew entirely in dynamic simulators. Even after full certification the commercial pilot is still required to undergo regular compulsory evaluation checks several times each year on a flight simulator. These simulators are very high fidelity systems that can accurately replicate the aircraft and its environment, and are fully capable of making a trainee aware of the consequences of his action. If a wrong decision leads to a catastrophe, it is allowed to happen – without in any way exposing passengers or other crew to any actual physical risk.

The analogy between aviation and anaesthesia has frequently been drawn. In anaesthesia, as in aviation, simulation and training devices offer many advantages. Traditionally anaesthetic skills have been acquired by anaesthetists-in-training by hands-on experience with real patients, hopefully under the supervision of a senior colleague. Quite rightly, that colleague will intervene when a situation threatens the well-being of the patient but, of course, the learning process for the trainee is compromised. In a simulated situation the instructor can let a potentially critical situation develop to its natural conclusion. We all learn from our mistakes – in a simulator trainees can even be encouraged to make mistakes and by observing the consequences the learning process is enhanced. As in most other walks in life, anaesthetists learn best by active rather than by passive learning. Simulators are also ideally suited to maintaining the competence and skills of qualified anaesthetists, however long they have been in the specialty, and are particularly useful for upgrading competence in handling those

uncommon but potentially fatal problems that require a rapid and correct response.

The first anaesthetic simulator, Sim I, was developed in 1969 by Denson and Abrahamson.[34] Since then several such devices have been developed. Some of these, such as the CASE device described by Gaba and DeAnda[35] and the Leiden Anaesthesia Simulator developed in the authors' department, attempt to recreate the reality of the operating theatre environment. The trainee is required to provide anaesthesia using conventional apparatus and obtains information about the status of his 'patient' from conventional monitors. Like its aviation counterpart, the modern anaesthetic simulator is a sophisticated computer-based device, incorporating complex physiological, pharmacological and pathological models. In addition, in the true simulator as opposed to a training device (see below), simulated signals such as ECG, pressure curves, etc. are generated and presented in the correct format and with accurate timing to the monitors. Further, data must be read from external sources, for example from the ventilators, gas monitors and keyboard input from the instructor, and integrated into the control algorithm and the control models, thereby enabling the responses to intervention by the trainee to be simulated realistically.

A distinction is usually made between simulators and training devices. A simulator is a machine that attempts to reproduce or represent the exact or nearly exact phenomena likely to occur in the real world, e.g. a complex operating theatre environment. Training devices often concentrate on specific aspects of anaesthetic knowledge such as pharmacology or physiology, using computer-screen-based simulation rather than reproducing the real operating theatre environment. Training devices are very useful educational tools. There are several that simulate the uptake and distribution of inhalational anaesthetics[36,37] or neuromuscular function and the administration of neuromuscular blocking drugs.[38] Others are available that simulate intravenous anaesthesia (IVA-SIM, produced by the Department of Anaesthesia in Bonn, Germany). A much more sophisticated training device is the Anesthesia Simulator Consultant (ASC, University of Washington) which creates the workspace of the anaesthetist on the screen of a personal computer and simulates the patient's responses to a wide variety of interventions. Its case library contains a wide variety of critical incidents, emergency situations and difficult clinical problems. ASC is an expanded version of a system described by Schwid and O'Donnell.[39] Both the full scale simulation such as that developed by Gaba and DeAnda[35] and training devices such as the ASC have proved their value in the education of trainees and for improving the performance of qualified anaesthetists.[40–42] A heart–lung bypass simulator has been described that can be used to teach perfusionists, anaesthetists and surgeons the principles of cardiopulmonary bypass and allow them to gain experience with conducting bypass both in routine and complex cases.[43]

In the Department of Anaesthesia in Leiden all new trainees spend several hours on the simulator, gaining experience with basic techniques and apparatus, before they are allowed to participate in anaesthesia in the operating theatre. Trainees are also required to demonstrate basic competence on the simulator before being allowed to take part in the on-call rota. The simulator is also used to allow members of staff to gain experience with new apparatus before it is introduced into routine clinical use. The Leiden simulator has also proved to be invaluable in evaluating new anaesthetic monitoring equipment at an early stage in its industrial development.

Automated anaesthetic records

Anaesthetic records fulfil several functions. A properly prepared record, accurately documenting the progress of anaesthesia, is of value not only to the anaesthetist in detecting untoward trends during the case but also is important in assisting staff involved in the postoperative care of the patient. It is an essential source of information for those who may be responsible for providing subsequent anaesthetics to that patient. Anaesthetic records are also a valuable source of information for education and research purposes, and are required for medicolegal purposes in most countries. In a court of law the medical record is considered the primary source of facts, and juries tend to believe what is written over what is said – 'if it isn't documented, it wasn't done.' A full and accurate record may make the vital difference between the defensible and the indefensible in a claim for damages in which professional negligence is alleged. Unfortunately it is probably true that the majority of anaesthetic records do not reflect a 'full and accurate record' of what occurred to the patient during anaesthesia. Even with the best will in the world it is virtually impossible to record accurately all events, vital signs, drug administration, etc., as they occur, especially during induction, recovery or during critical incidents. At these times the anaesthetist is, quite rightly, concerned with more important aspects of patient care and data must be recorded retrospectively. Even during times of relative stability, it is probably not possible to record data manually more often than every 5 minutes while adequately performing the other clinical tasks and demands on the anaesthetist's time.[44] The need to enter data more frequently than this requires some form of automated record keeping if the anaesthetist is to cope with the growing complexity of pharmacological intervention and physiological monitoring.

An automated anaesthetic record system is a computer-based device that acquires data from physiological monitors, ventilators, etc., as well as data entered directly by the anaesthetist, and displays these, in a suitable format, online to the anaesthetist. In addition the acquired data will be stored to enable the automatic generation and printing of the anaesthetic record at the end of the case as well as allowing subsequent retrieval for clinical, administrative or research purposes. The core of such a system is data acquisition. Most modern equipment used in anaesthesia and intensive care has the facility to output data in digital form, allowing relatively straightforward connection to a computer. It is this ability to exchange information between medical devices that forms the foundation of modern data management in anaesthesia. Digital output from these devices is almost universally in serial form using a RS-232 protocol. Unfortunately there are many variations within this so-called serial communication standard. In the future it is to be hoped that all manufacturers will agree on an international standard for data transfer between medical equipment. A potential standard is that proposed by the American Institute of Electrical and Electronic Engineers (IEEE), the IEEE P1703 medical information bus.

Even when a monitor has a digital output, there may be circumstances where one might want to capture a signal in analogue form, e.g. to store a run of arrhythmias in the ECG. The digital output of most ECG monitors only contains basic information such as heart rate or possibly RR interval. The analogue signal, normally available for output to a printer, must first be converted to digital form using an analogue-to-digital convertor (ADC) in order to capture and store electronically an analogue ECG signal in digital format. When the output of a device is only available in analogue form, it must, of course, first be digitized by an ADC before it can be input to the automated record system. When carrying out analogue-to-digital conversion it is important that the correct sampling frequency is chosen. In order to allow complete recovery of an analogue signal without distortion, the original signal must be sampled at a rate greater than twice that of the highest frequency component contained in that signal. This minimum sampling frequency, often known as the Nyquist frequency (after H. Nyquist of Bell Laboratories), is a restriction placed on sampling by the *sampling theorem*. Note that the sampling rate must be *greater* than twice the highest frequency in the sampled signal and not equal to the latter. Sometimes the sampling theorem is erroneously stated on this point. Failure to observe this condition of the sampling theorem will result in distortion when the analogue signal is reconstructed from the digital data, since the reconstructed signal will contain frequencies not present in the original signal. This phenomenon is known as aliasing (Fig. 26.7). A

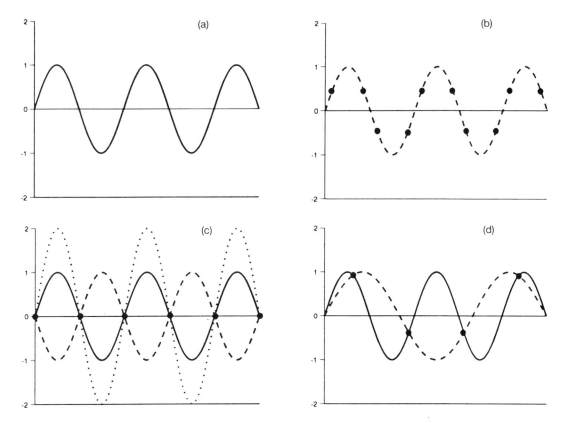

Fig. 26.7 Illustration of the importance of the correct sampling rate when digitizing analogue signals. (a) Sinusoidal input signal. (b) When the sampling interval is less than half that of the period of the input signal, i.e. the sampling frequency is greater than twice that of the input signal, a unique curve can be fitted through the sampled data points (●). (c) Here the sampling frequency is equal to that of the input signal, and an infinite number of curves with amplitudes that are integer multiples of the original can be fitted through the sampled data points. (d) When the sampling interval is greater than half that of the period of the input signal, the reconstructed curve is not unique.

sampling frequency of 10 Hz would be adequate to acquire a capnogram but would result in excessive distortion of an arterial blood pressure curve, for which the minimum sampling frequency should be 150 Hz.[45] For rapidly changing biological signals, e.g. evoked responses, the Nyquist frequency lies above 1 kHz. ECG information above 100 Hz is not required for most applications; however, if the ECG is sampled at 200 Hz, it must first be filtered to remove information above 100 Hz to prevent aliasing, and it is important to note that only a portion of the original signal remains in the digitally reconstructed signal.

The correct choice of sampling frequency is related to the resolution of the ADC. In general the resolution of an ADC is one part in 2^N, where N is the number of bits in the binary representation. High resolution is essential to prevent distortion when sampling complex, rapidly changing signals. With 8-bit resolution the signal will only be sampled at intervals corresponding to 1/256 of full scale; with 12-bit resolution the sampling interval will be 1/4096 of full scale. For an 8-bit converter having a range of 10 V, the absolute error is $\pm$ 20 mV. This is 0.39 per cent of full scale; at smaller input voltages the relative error will be higher. Thus with a signal having a range of 10 V, a signal

value lying between 4.98 V and 5.02 V will be seen by an 8-bit converter as 5 V. A 12-bit ADC can give a full scale resolution of 0.024 per cent and only signal values lying between 4.999 V and 5.001 V will be seen as 5 V.

In addition to automatic data capture, there is also a need for some form of manual input to the record system, for example to enter drug or i.v. fluid administration, events, etc. The most common method of input is via the computer keyboard but other possibilities include touch-screen, mouse or trackerball driven menus, light pen or voice recognition. Murchie and Kenny[46] compared input by keyboard, light pen or voice recognition for speed and efficacy among intensive care nurses. The average speed of data entry was significantly faster when using the standard 'QWERTY' keyboard compared to the other methods, being three times faster than with voice recognition input. The latter was also associated with significantly more errors. In the future other technologies will be developed, such as the use of bar codes on ampoules, prefilled syringes and infusion bags, which will not only simplify input about drug and fluid administration but also reduce the risk of input errors.

An important aspect of automated record systems is

validation of data input. Concern is often expressed that erroneous or artefactual values of physiological parameters might be incorporated into an automated record with possible medicolegal consequences. The automated record system will, and should, faithfully record the information presented to it. In contrast the anaesthetist keeping a handwritten record is unlikely to record readings that are considered artefactual or unexpectedly out of range. There is also a tendency to record high values a bit lower and low values a bit higher, a phenomenon known as 'smoothing'.[47] It is possible to incorporate artefact detection algorithms in the automated record system software that would allow 'flags' to be attached to those stored items considered as artefacts. In many cases the artefactual data may be physiologically unrealistic or even impossible. Blood pressures of 300/300 mmHg, negative blood pressures or a zero pressure recorded within 1 second of a normal blood pressure, such as might occur during a catheter flush or transducer zeroing, are obvious artefacts that would be of no medicolegal significance.

Another aspect that must be considered is the security of the database containing the automated records. It must not be possible to edit data once it is stored in the database, although errors may be corrected at the time of entry. While an alteration to a handwritten record can easily be detected this is much more difficult, if not impossible, to detect with an electronic record. The security of the database becomes even more important if it is maintained on a central computer serving a local area network. It is then vitally important that adequate safeguards are built into the system to prevent unauthorized access, and to preserve patient confidentiality.

The use of automated record systems in a local area network (LAN) rather than as stand alone devices offers many advantages. The ability to link into the hospital information computer system can give immediate, online access to laboratory data and other information of importance for the clinical management of the patient. Storage of data centrally on a file server is also more economical, and improves data security. It is possible for the complete anaesthetic records of upwards of 15 000 patients to be stored on one optical WORM storage device, and details of any patient recalled within a few seconds. One of the fears concerning automated records is loss of information due to computer failure. This risk can be reduced to a minimum when a LAN is used. During a case, as well as all data being transmitted as they are entered down the network for storage on the central storage facility, they can also be stored on hard disk in the local computer in the operating theatre. Should the local computer fail, then all information up to that moment is safely stored centrally. Conversely, should either the network or the file server fail, the anaesthetist can confidently continue with local storage.

The chance of both the network and the local computer going down simultaneously is extremely remote.

With a network, intraoperative data can also be immediately available to medical and nursing staff responsible for the patient in the recovery room or intensive care. And of course using a network offers administrative advantages. Having virtually instantaneous access to a database containing detailed information about all patients anaesthetized by the department during the past year or years makes evaluation of trends in, e.g., drug usage, and preparation of statistical reports very easy for heads of departments. In the future we are likely to see automated record systems incorporated into anaesthetic workstations, that will integrate ventilator and monitors with intelligent alarms in one apparatus. Such a workstation could also offer the possibility of decision support to the anaesthetist based on some form of expert system.[48]

Several investigators have compared handwritten and automated anaesthetic records. Lerou et al.[49] quantified the differences in the documented values of eight physiological variables recorded online by a commercially available automated system with the values from the same patients recorded on handwritten anaesthetic records. The incidence of erroneous data in the handwritten record was between 1 and 11 per cent while the incidence of missing data varied from 8 to 31 per cent, depending on the parameter. Most missing or erroneous data occurred during the period of induction and at the end of the case, when 'after-the-fact' data are often recorded by the anaesthetist from memory. Erroneous data were more frequently observed for the recorded blood pressures than for any other variables, with 57 per cent more erroneous data observed for diastolic than for systolic pressure. This is probably due to anaesthetists preferentially memorizing systolic blood pressure and entering a preconceived diastolic value, since records are often constructed some minutes after the event they describe, frequently during the middle of the case.

Similar discrepancies between handwritten and automated blood pressure records were noted by Cook et al.[50] and explained by faulty reconstruction of handwritten records from memory and bias in favour of less controversial values. No handwritten record contained a diastolic pressure above 110 mmHg whereas the automated system recorded diastolic pressure over 110 mmHg 33 times in 15 of the 46 patients, including eight measurements greater than 125 mmHg. Thus it would appear that those factors that may cause significant inaccuracy in the handwritten record, including observer bias, missed readings and errors of memory, may be avoided by using automated data management systems.[51] It is often claimed that these systems, by accurately recording what is happening to the patient, and by freeing the anaesthetist from the drudgery of maintaining a handwritten record, allows him to devote more time to direct patient care, and that they thereby contribute to improvement in patient safety. However, this claim is mainly based on theoretical speculation, and the evidence is largely anecdotal. There is a need for a well conducted study in this field. An argument used by opponents of these systems is that they might reduce, rather than increase, the attention the anaesthetist pays to the patient. The discipline of having to note mentally and record manually a value for physiological parameters at regular intervals may, in fact, enhance patient contact and increase the vigilance of the anaesthetist. To quote Saunders:[52] 'I do believe that automated record systems will be implemented and will be extremely useful, but we must exercise great care in their design and implementation, lest they wind up doing more harm than good'. This sentiment could, perhaps, apply to all uses of computers in anaesthesia.

REFERENCES

1. Schwilden H. A general method for calculating the dosage scheme in linear pharmacokinetics. *European Journal of Clinical Pharmacology* 1981; **20**: 379–86.
2. Jacobs JR. Algorithm for optimal linear model-based control with application to pharmacokinetic model-driven drug delivery. *IEEE Transactions on Biomedical Engineering* 1990; **37**: 107–9.
3. Kenny GN, White M. A portable target controlled propofol infusion system. *International Journal of Clinical Monitoring and Computing* 1992; **9**: 179–82.
4. Alvis JM, Reves JG, Govier AV, *et al.* Computer assisted continuous infusion of fentanyl during cardiac anesthesia: comparison with a manual method. *Anesthesiology* 1985; **61**: 41–9.
5. Jones JG. Use of the auditory evoked response to evaluate depth of anaesthesia. In: Bonke B, Fitch W, Millar K, eds. *Memory and awareness in anaesthesia.* Amsterdam: Swets and Zeitlinger, 1990: 303–15.
6. Thornton C, Konieczko, K, Jones JG, Jordon C, Dore CJ, Heneghan CPH. Effect of surgical stimulation on the auditory evoked response. *British Journal of Anaesthesia* 1988; **60**: 372–8.
7. Thornton C, Barrowcliffe MP, Konieczko K, *et al.* The auditory evoked response as an indicator of awareness. *British Journal of Anaesthesia* 1989; **63**: 113–15.
8. Kenny GNC, McFadzean W, Mantzaridis H, Fisher AC. Closed-loop control of anesthesia. *Anesthesiology* 1992; **77**: A328.
9. O'Hara DA, Bogen DK, Noordergraaf A. The use of computers for controlling the delivery of anesthesia. *Anesthesiology* 1992; **77**: 563–81.
10. Ying H, Sheppard L, Tucker D. Expert-system-based fuzzy control of arterial pressure by drug infusion. *Medical Progress and Technology* 1988; **13**: 203–15.
11. Kadish AH. Automatic control of blood sugar. A servomechanism for glucose monitoring and control. *American Journal of Medical Electronics* 1964; **3**: 82–6.
12. Gibbs FA, Gibbs EL, Lennox WG. Effect on the electroencephalogram of certain drugs which influence nervous activity. *Archives of Internal Medicine* 1937; **60**: 154–66.
13. Bellville JW, Artusio JF, Bulmeer MW. Continuous servo-motor integration of the electrical activity of the brain and its application to the control of cyclopropane anesthesia. *Electroencephalography and Clinical Neurophysiology* 1954; **6**: 317–20.
14. Bickford RG. Use of frequency discrimination in the automatic electroencephalographic control of anesthesia (servo-anesthesia). *Electroencephalography and Clinical Neurophysiology* 1951; **3**: 83–6.
15. Bickford RG. Automatic electroencephalographic control of general anaesthesia. *Electroencephalography and Clinical Neurophysiology* 1950; **2**: 93–6.
16. Kiersey DK, Faulconer A, Bickford RG. Automatic electroencephalographic control of thiopental anesthesia. *Anesthesiology* 1954; **15**: 356–64.
17. Schwilden H, Schüttler J, Stoeckel H. Closed-loop feedback control of methohexitone anesthesia by quantative EEG analysis in humans. *Anesthesiology* 1987; **67**: 341–7.
18. Schwilden H, Stoeckel H, Schüttler J. Closed-loop feedback control of propofol anaesthesia by quantative EEG analysis in humans. *British Journal of Anaesthesia* 1989; **62**: 290–6.
19. Schüttler J, Kloos S, Ihmsen H, Schwilden H. Clinical evaluation of a closed-loop dosing device for total intravenous anesthesia based on EEG depth of anesthesia monitoring. *Anesthesiology* 1992; **77**: A501.
20. Fukui Y, Smith NT, Fleming RA. Digital and sampled-data control of arterial blood pressure during halothane anesthesia. *Anesthesia and Analgesia* 1982; **61**: 1010–15.
21. Monk CR, Millard RK, Hutton P, Prys-Roberts C. Automatic arterial pressure regulation using isoflurane: comparison with manual control. *British Journal of Anaesthesia* 1989; **63**: 22–30.
22. Sheppard LC, Kouchoukos NT, Kurtts MA. Automated treatment of critically ill patients following operations. *Annals of Surgery* 1968; **168**: 596–604.
23. Sheppard LC. Computer control of the infusion of vasoactive drugs. *Annals of Biomedical Engineering* 1980; **8**: 431–44.
24. Murchie CJ, Kenny GN. Comparison among manual, computer-assisted, and closed-loop control of blood pressure after cardiac surgery. *Journal of Cardiothoracic Anesthesiology* 1989; **3**: 16–19.
25. Keogh BE, Jacobs J, Royston D, Taylor KM. Microprocessor-controlled hemodynamics: a step towards improved efficiency and safety. *Journal of Cardiothoracic Anesthesia* 1989; **3**: 4–9.
26. Reid JA, Kenny GNC. Evaluation of closed-loop control of arterial pressure after cardiopulmonary bypass. *British Journal of Anaesthesia* 1987; **59**: 247–55.
27. Jaklitsch RR, Westenskow DR, Pace NL, Streisand JB, East KA. A comparison of computer-controlled versus manual administration of vecuronium in humans. *Journal of Clinical Monitoring* 1987; **3**: 269–76.

28. Uys PC, Morrell DF, Bradlow HS, Rametti LB. Self-tuning, microprocessor-based closed-loop control of atracurium-induced neuromuscular blockade. *British Journal of Anaesthesia* 1988; **61**: 685–92.

29. Lampotang S, Nyland ME, Gravenstein N. The cost of wasted anesthetic gases. *Anesthesia and Analgesia* 1991; **71**: S151.

30. Couto da Silva JM, Aldrete JA. A proposal for a new classification of anaesthetic gas flows. *Acta Anaesthesiologica Belgica* 1990; **41**: 253–8.

31. Westenskow DR, Zbinden AM, Thomson DA, Kohler B. Control of end-tidal halothane concentration. Part A: Anaesthesia breathing system and feedback control of gas delivery. *British Journal of Anaesthesia* 1986; **58**: 555–62.

32. Zbinden AM, Frei F, Westenskow DR, Thomson DA. Control of end-tidal halothane concentration. Part B: Verification in dogs. *British Journal of Anaesthesia* 1986; **58**: 563–71.

33. Verkaaik APK, Erdmann W. Respiratory diagnostic possibilities during closed circuit anaesthesia. *Acta Anaesthesiologica Belgica* 1990; **41**: 177–88.

34. Denson JS, Abrahamson S. A computer-controlled patient simulator. *Journal of the American Medical Association* 1969; **208**: 504–8.

35. Gaba DM, DeAnda A. A comprehensive anesthesia simulation environment: re-creating the operating room for research and training. *Anesthesiology* 1988; **69**: 387–94.

36. Philip JH. Gas man – an example of goal oriented computer-assisted teaching which results in learning. *International Journal of Clinical Monitoring and Computing* 1986; **3**: 165–73.

37. Heffernan PB, Gibbs JM, McKinnon AE. Teaching the uptake and distribution of halothane. A computer simulation program. *Anaesthesia* 1982; **37**: 9–17.

38. Jaklitsch RR, Westenskow DR. A simulation of neuromuscular function and heart rate during induction, maintenance, and reversal of neuromuscular blockade. *Journal of Clinical Monitoring* 1990; **6**: 24–38.

39. Schwid HA, O'Donnell D. The Anesthesia Simulator-Recorder: a device to train and evaluate anesthesiologists' responses to critical incidents. *Anesthesiology* 1990; **72**: 191–7.

40. Gaba DM. Improving anesthesiologists' performance by simulating reality. *Anesthesiology* 1992; **76**: 491–4.

41. Schwid HA, O'Donnell D. Anesthesiologists' management of simulated critical incidents. *Anesthesiology* 1992; **76**: 495–501.

42. Schwid HA O'Donnell D. Educational computer simulation of malignant hyperthermia. *Journal of Clinical Monitoring* 1992; **8**: 201–8.

43. Leonard RJ. A total heart/lung bypass simulator. *ASAIO Transactions* 1988; **34**: 739–42.

44. Gravenstein JS, de Vries A, Beneken JEW. Sampling intervals for clinical monitoring of variables during anesthesia. *Journal of Clinical Monitoring* 1989; **5**: 17–21.

45. Siegel LC, Pearl RC. Pressure measurement artifact with analog-to-digital conversion. *Journal of Clinical Monitoring* 1990; **6**: 318–21.

46. Murchie CJ, Kenny GN. Comparison of keyboard, light pen and voice recognition as methods of data input. *International Journal of Clinical Monitoring and Computing* 1988; **5**: 243–6.

47. Block Jr FE. Normal fluctuation of physiologic cardiovascular variables during anesthesia and the phenomenon of 'smoothing'. *Journal of Clinical Monitoring* 1991; **7**: 141–5.

48. Prakash O, Meiyappan S. Anesthesia support systems. *International Journal of Clinical Monitoring and Computing* 1992; **9**: 131–9.

49. Lerou JGC, Dirksen R, van Daele M, Nijhuis GMM, Crul JF. Automated charting of physiological variables in anesthesia: a quantitative comparison of automated versus handwritten anesthesia records. *Journal of Clinical Monitoring* 1988; **4**: 37–47.

50. Cook RI, McDonald JS, Nunziata E. Differences between handwritten and automatic blood pressure records. *Anesthesiology* 1989; **71**: 385–90.

51. Thrush DN. Are automated anesthesia records better? *Journal of Clinical Anesthesia* 1992; **4**: 386–9.

52. Saunders RJ. The automated anesthetic record will not automatically solve problems in record keeping. *Journal of Clinical Monitoring* 1990; **6**: 334–7.

Chapter 27

Operating Theatre Transmission of Infection

Paul R. Knight and Alan R. Tait

Historically, anaesthetists have shown only passing interest in the field of infectious disease, allowing nursing and other operating room personnel to deal with the challenge of infection control and surveillance. For most anaesthetists, the issue of infection only becomes important when it is necessary to cancel an elective case when a patient presents with an acute infectious process. However, with the increasing number of immunocompromised and/or infected patients presenting for surgery there has been greater interest in the relation between anaesthesia, infection and the immune response. Indeed, numerous studies have demonstrated suppression of the immune response in patients undergoing anaesthesia and surgery. Although the clinical significance of anaesthetic and surgical-induced immunosuppression is unclear at this time, the anaesthetist needs to be aware of the potential consequences of this occurring. In addition, anaesthetists by virtue of belonging to a high risk speciality have become more aware of their ability to contract occupationally transmitted diseases. In particular, with the increasing exposure to patients with acquired immune deficiency syndrome (AIDS) or hepatitis B virus (HBV) the anaesthetist has developed a heightened awareness of infection control and prevention in the perioperative period.

A major clinical challenge facing anaesthetists, now and in the future, is to develop ongoing strategies to improve the quality of care of surgical patients and favourably influence their medical outcome. Traditionally, anaesthetists accept responsibility for anaesthetic related morbidity and mortality occurring within 48 hours following surgery. However, perioperative anaesthetic practices and procedures may have an impact on complications that do not become manifest until well after the 48-hour post-anaesthetic period. Indeed, occupational transmission of infections with HIV or HBV may not produce a disease until months or years after exposure.

The appearance of an infectious diathesis following anaesthesia and surgery is a major cause of morbidity and mortality and as such is an important indicator of the quality of health care in the perioperative period. Successful control of infection during this period not only requires application of the technical knowledge to ensure sterility and cleanliness, but also requires a thorough understanding of the host–parasite relationship. This understanding is important since many perioperative anaesthetic practices and procedures can have an impact upon the incidence of postoperative infectious complications by altering mechanisms which control the equilibrium of the host–parasite relationship. For these reasons, not only should the anaesthetist be able to apply the principles of infection control but he/she should also have a basic familiarity with how an infection becomes established, what factors influence this process, how this infection produces a disease (pathogenesis) and how the disease alters the normal physiology of the patient (pathophysiology). An understanding of these processes can increase the ability of the anaesthetist to develop strategies for the care of patients presenting with acute infections or for those who are at an increased risk of developing an infectious illness postoperatively.

The anaesthetist may either directly transmit an infectious agent to the patient or increase the susceptibility of the patient to the risk of subsequent postoperative nosocomial infections by practices and procedures performed in the operating theatre. In addition, the anaesthetist is also at risk of contracting an infection from the patient. In particular, the risk of HIV and HBV infections and their dire consequences have served to heighten the awareness of the physician to the possibility of patient to physician transmission and have led to the institution of, and compliance with, protocols designed to decrease the risk of transmission of these types of blood-borne infections.

This chapter examines the area of infectious disease as it applies to the perioperative period.

Epidemiology

Approximately 70 per cent of all acute illness can be attributed to infectious agents.[1] Community-based infections are principally viral in nature, whereas nosocomial infections are primarily bacterial and are particularly prevalent in patients with decreased immune function. Patients receiving allografts, immunosuppressive drugs or treatments, or those with AIDS are particularly prone to a number of opportunistic infections. Viral infections are estimated to cause only 5 per cent of nosocomial infections and although they can be associated with immunosuppression they more typically reflect the patterns of viral infections seen in the community at that time.

Transmission of hospital-acquired infections typically occurs by direct inoculation via one of the following routes:

1. The respiratory tract, following contact with an infected individual, from air-conditioning units, contaminated respiratory therapy and/or anaesthesia equipment, or from aspiration of gastric contents.
2. The urinary tract from indwelling catheter placement.
3. The skin and mucous membranes, from direct contact with infectious lesions or material or as a result of surgery, trauma or burns.
4. The blood, via the placement of indwelling catheters or from transfusions of blood or blood products.
5. Via the placement of grafts and prostheses.

In the otherwise healthy individual, exposure to microorganisms in the hospital usually results in a mild or aborted infection. However, many of these infections may become life-threatening in patients with lowered resistance. Factors that may predispose the patient to infection include age, pre-existing disease, diabetes or other non-infectious illnesses, smoking, malnutrition, cancer and transplant

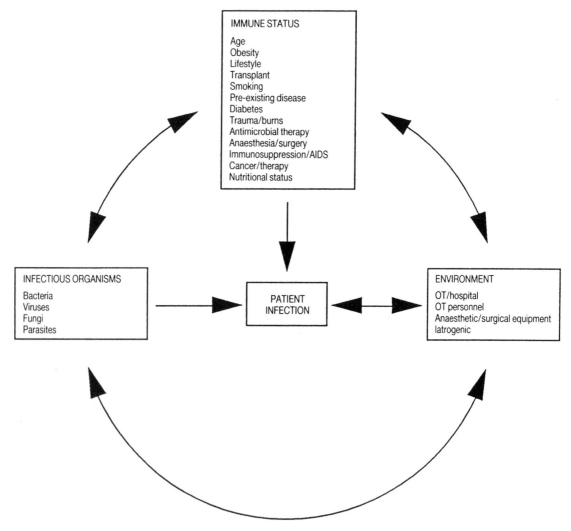

Fig. 27.1 Schematic representation of the interrelation of factors that predispose the patient to infection in the operating theatre (OT). Note that infection may be bidirectional between patient and personnel.

patients receiving immunosuppressive drugs, burn or trauma patients, and patients with acquired or congenital defects in immune responsiveness. The interrelation of factors that predispose the patient to infection are outlined in Fig. 27.1.

Control of nosocomial infections should be directed at interrupting the infectious cycle. Denying a portal of entry for the micro-organism into the body can be attained by the use of barrier techniques, i.e. gloves, masks, eyewear and attention to aseptic technique in placing indwelling catheters. Universal precautions are detailed elsewhere in this chapter but are critical in reducing the potential for occupational transmission of hepatitis B virus and HIV. In addition, the simple hygienic practice of washing hands will go a long way to prevent dissemination of infectious material. These techniques are not only important in protecting the patient from infection but are critical in protecting the anaesthetist from patient-borne pathogens. Adequate sterilization or disinfection of anaesthetic and surgical equipment must be maintained and air-conditioning units should be cultured routinely for growth of potential pathogens. Awareness of the patient's disease status is paramount. Sophisticated laboratory tests are now available to assist in the rapid diagnosis of infectious agents. These tests, together with clinical and epidemiological information, can be used to identify sources of infection and develop long-term strategies for effective control and prevention.

In addition to the responsibility of employing appropriate practices for prevention of occupational transmission of disease, anaesthetists must be aware of how anaesthetics *per se* affect the patient's immune response and ability to resist infection. Recognition of the interactions between anaesthesia, immunology and infection may be critical in optimizing patient management and subsequent outcome.

Nosocomial transmission of infectious disease

Resistance to infection

Hospital patients usually become infected as a result of inoculation of pathogenic organisms following a breakdown in the natural barriers to infection and/or a concomitant decrease in the patient's ability to mount a sufficient immune response. The natural barriers include the skin and mucous membranes of the gastrointestinal tract, the conjunctiva, the respiratory tract, and the urinary tract. The presence of local tissue damage, extremes of age, severe debilitating illness, drug immunosuppression, or

biological properties of the infectious agent, are factors which are associated with limitation of the patient's ability to respond adequately to an invading micro-organism.

Initial barriers to invasion

The patient's initial defence is to deny the pathogenic micro-organism a portal of entry. The skin is the largest organ of the body and provides both a mechanical barrier to infection by virtue of its low water content and constant desquamation and a chemical barrier due to the secretion of long chain fatty acids and the presence of resident non-pathogenic microbial flora. The conjunctiva possesses special defences in the form of tears which contain bacteriostatic agents. Violation of the epidermal layer by tissue destruction following either surgery, trauma or burns, or iatrogenic invasion from intravenous or intra-arterial catheters, permits access of the infectious agent to areas of the body that are less resistant to its invasiveness and more conducive to its replication.

The endothelial linings of the genitourinary and gastrointestinal tracts also provide barriers to invading micro-organisms. These linings secrete mucus which contains various proteins that are inhibitory to microbial growth, bactericidal enzymes such as lysozyme and specific secretory immunoglobulins that directly destroy or inhibit invading microbial agents. Damage to the endothelial linings of the gastrointestinal and urinary tract as a result of disease, e.g. renal calculi, or abrasions by mechanical devices such as urinary catheters can facilitate the entry of pathogenic agents.

The respiratory tract is of special concern to the anaesthetist due to the possibility of mechanical trauma associated with airway instrumentation. Tracheal tubes, bronchoscopes, and oral airways can all abrade the endothelial lining as well as bypass upper respiratory tract defences and interfere with the normal function of the mucociliary clearing system. Barrier protection of the upper airway involves the nasal hair and the labyrinthine effect of the nasal anatomy. The mucociliary clearing system protects the lower respiratory tree by ciliary propulsion of a constantly moving mucus blanket cephalad. The mucus blanket, which incidentally contains IgA antibodies and other bacteriostatic substances, such as lysozyme and lactoferrin, is capable of removing unwanted particles that inadvertently enter the lower pulmonary tract. IgA antibodies prevent micro-organisms from adhering to respiratory epithelial surfaces. The use of non-humidified air may dry the mucus secretions and interfere with the normal ciliary clearing system. Also, obtundation of reflexes during anaesthesia may allow for the regurgitation of acidic and/or particulate gastric contents which may chemically or mechanically damage the respiratory epithelium and ciliary blanket. Aspiration of both gastrointestinal and upper respiratory tract material, in conjunction with inhibition of lower respiratory tract clearing mechanisms, may also serve to seed pathogenic organisms into the lower respiratory tract.

Host defences beyond the natural barriers

Once pathogenic micro-organisms successfully pass through the initial boundaries and gain a portal of entry, they are subjected to a number of non-adaptive and adaptive host responses. Three major factors determine their success in establishing an infection. The first is the micro-organism's ability to overcome the patient's defence mechanisms and the second is the viability of the patient's immune response. These two factors determine the invasiveness of the microbial agent or how well a given micro-organism will spread systemically or to contiguous regions. The third factor is the micro-organism's virulence or ability to produce pathology in the patient. Certain agents may be extremely virulent but not invasive (e.g. *Clostridium tetani*). Conversely, some agents, particularly viruses, are very invasive but not always virulent. Furthermore, the ability of a microbial agent to be invasive and/or virulent depends on where it successfully enters the body, the number of organisms in the inoculum, and whether the environment at the point of inoculation is healthy or damaged. Finally, the patient's own defences may contribute to the virulence of a microbial illness. Acute and chronic inflammatory reactions contribute to both local and systemic tissue destruction and may be responsible for much of the symptomatology experienced by the patient during the course of an infectious illness.

Once the micro-organisms have penetrated the initial external barriers, the next response is the mobilization of non-specific defences. These are mobilized in response to chemoattractants that are produced by microbial activity at the site(s) of inoculation. These chemoattractants originate primarily from two sources: (1) from areas of tissue damage, and (2) from foreign antigens found on the surface of, or released from, the micro-organisms themselves. Chemoattractants such as cytokines and other paracrine substances are produced by injured tissues and macrophages in response to microbes, microbial products, or products of tissue injury and can stimulate the recruitment of other phagocytes (macrophages, neutrophils, and eosinophils) as well as natural killer (NK) cells to form the first line of non-adaptive cytocidal activity.

The phagocytes, or leucocytes, which are found circulating freely in the blood, in secretions, or in the interstitial tissue spaces, kill and digest micro-organisms using a combination of low pH, digestive enzymes, and by the production of highly reactive oxygen free radicals. NK cells attach to, and kill, cells containing foreign markers. The acute inflammatory response, which is initiated by this

sequence of events, promotes further recruitment of inflammatory cells and enhances the amount of tissue damage. In addition to the phagocytic and NK cells there are a number of systems such as the complement system and prostaglandins which augment phagocytosis and activate and enhance other defence mechanisms to increase the overall intensity of the inflammatory response. Although most micro-organisms are destroyed by this acute non-adaptive response, several pathogenic organisms have strategies to circumvent it.

The final adaptive line of defence includes both humoral and cellular components of the immune system as well as cells and mechanisms (complement) of the acute non-adaptive response. The adaptive and non-adaptive responses are highly integrated. For example, antibodies increase the efficacy of neutrophils and macrophages. Macrophages engulf foreign antigens which are processed and presented to lymphocytes for recognition and elaboration of a response specific to that antigen. The stimulated macrophage also releases specific cytokines (monokines) which in addition to being chemotactic are responsible for the recruitment and coordination of various cells of the immune response.

The humoral arm of the adaptive response involves the production and release of immunoglobulins. These compounds are proteins which are produced by plasma cells following differentiation from a subset of B-lymphocytes. Their principal function is the recognition of specific antigens. Antibodies are classified by the protein subunits that make up the individual molecules and include IgA, IgG, IgE, IgM, and IgD species. These compounds can be further differentiated according to function and site of action. The development of the immunoglobulin-producing plasma cells involves a complex series of events involving the trapping and processing of antigen by macrophages and the interaction of the macrophages with the B-cells. T-lymphocytes, which are discussed below, are also involved in this initial recognition. These two lymphocyte subtypes interact extensively in the development of the adaptive immune response and augment each other's activity. Recognition of antigen by the B-cells triggers the development of a population of 'memory cells' which rapidly identify and respond to subsequent exposures to the same antigen. This phenomenon forms the basis for vaccination against infection.

Immunoglobulins destroy microbial agents in many ways. They may directly damage or inhibit the ability of the micro-organism to be invasive or pathogenic by binding to key proteins on the micro-organism's surface and/or to substances released by the micro-organism which are responsible for the agent's virulence (i.e. endotoxin). Antibodies also enhance the antimicrobial activity of the non-adaptive response. They can facilitate lysis of a foreign cell by helping bind complement to it. Attachment of antibody to microbial agents with or without agglutination enhances phagocytosis by macrophages and neutrophils.

The cellular response to infection is mediated by a subset of specific lymphocytes that recognize foreign proteins on the surface of micro-organisms or infected cells, attach to these sites and initiate cytolysis. T-lymphocytes are so named because, although they also originate in the bone marrow, they require 'conditioning' by the thymus. There are several subsets of T-lymphocytes. T-helper lymphocytes and T-suppressor lymphocytes are responsible for control and modulation of the adaptive immune response while the cytotoxic T-lymphocytes are responsible for the direct attack and destruction of microbial agents. Like the B-cell, the cytotoxic T-lymphocytes require interaction with macrophages to recognize specific antigens on foreign cells. Antibody-like proteins on the surface of the lymphocyte are responsible for the initial cell-to-cell attachment. Free circulating antibodies can also facilitate this recognition.

Microbial agents possess a number of interesting defences against the adaptive response. HIV, for example, uses a docking protein that is specific for the T-helper lymphocyte and replicates selectively in these cells. As the number of T-helper lymphocytes decreases over time, the body's ability to control the virus infection becomes limited. Many viruses, e.g. herpes simplex virus, can enter cells and become quiescent and activate only when changes in the host parasite equilibrium, i.e. immune suppression, favour its replication and concomitant pathology.

With the explosion of medical knowledge and the increasing numbers of immunosuppressed patients presenting for surgery, it is inevitable that anaesthetists will be faced with problems that were less prevalent in the past. Infection control strategies for these types of patients must be directed at preventing the establishment of an infection by carefully maintaining the patient's natural barriers. This entails providing strict sterile control techniques when the barriers are violated to avoid introducing micro-organisms into locations that the patient can no longer adequately protect.

Effects of anaesthesia and surgery on the resistance to infection

Anaesthesia and surgery alter the response to a number of the components of the immune system. While these changes alter the immunological status of the patient, they may or may not affect the outcome of an infectious disease. Since much of the pathology and symptomatology of an infectious disease is mediated by the negative side of the immune response, an inhibition of this response may be beneficial to some patients. Evidence for this scenario has been presented. For these reasons, anaesthetists should be aware of the interaction of anaesthesia and surgery with the immune system and how this may alter the eventual outcome of an infectious process.

Anaesthetics have been reported to decrease both the numbers and response of key effectors of the body's non-adaptive defence system. Leucopenia, bone marrow suppression, impaired phagocytosis, and decreased leucocyte motility have all been demonstrated following exposure to a number of different anaesthetic agents and/or surgical stimulation.[2,3] Natural killer cells demonstrate an insensitivity to cytokine stimulation for up to 14 days following exposure of experimental animals to one of several different anaesthetic agents, although basal cytotoxic activity appears unaffected.[4] Paradoxically, while the production of at least one of the chemical effectors of the inflammatory response, the superoxide anion, is inhibited, the sensitivity of target tissues, i.e. endothelial cells, to oxidants during exposure to volatile anaesthetics is increased.[5,6] This example serves to highlight that the interaction of anaesthesia and surgery with the immune response is multifactorial and that blanket statements regarding outcome as a result of this interaction may be difficult to make.

Anaesthetic agents can also affect the function of the adaptive immune system.[2,3] Although preformed antibodies are not affected by anaesthesia and surgery there is a decrease in antibody production. Lymphocyte transformation, a step important in 'turning on' the adaptive immune function of these cells, is also decreased. Like NK cells the cytotoxicity of lymphocytes that require sensitization to antigen is also decreased following exposure to anaesthetic agents.

The effect of other non-anaesthetic agents or stimuli on the immune response must also be considered in assessing a patient's ability to respond to an infectious disease during the perioperative period. For example, blood and blood components may decrease the cell-mediated immune response to microbial invasion.[7,8] The stress of surgery is also a potent inhibitor of the immune system. Antibody production, phagocytosis, and T-lymphocyte activity are all decreased as a result of this stimulus. The effects of surgical stress on the immune response appear to be related to the increased release of catecholamines and adrenocorticoids which may occur during periods of light anaesthesia and/or hypoxia. It would appear prudent, therefore, to avoid situations that may encourage release of the substances, particularly if the patient is immunosuppressed.

The effect of all these factors on the outcome of an infectious process is currently being examined and will be discussed in greater depth later in the chapter. However, many studies of anaesthesia, infection and outcome are considered controversial. At clinically relevant concentrations volatile anaesthetics inhibit virus replication but have little or no effect on bacterial biology in vitro.[9,10] In vivo experiments have been equivocal in that some suggest that anaesthesia exacerbates an infectious disease whereas others suggest that the pathogenesis of infection is less severe following anaesthesia.[11–13] The reason for these apparent differences is due, in part, to differences between the experimental models, the host–parasite interaction, the pathogen under study, and the physical status of the subjects being examined. When all these factors are taken into account these authors believe that anaesthesia and surgery do not predispose an otherwise healthy individual to an increased risk of infection, but feel that patients who are immunodeficient are probably at a greater risk of developing an infection due to the added suppressive effect of anaesthesia and surgery. Patients presenting with a bacterial respiratory tract infection or a viral hepatitis are likely to have an increase in the severity of the disease as a result of anaesthesia and surgery, while patients presenting with a mild viral upper respiratory tract infection may actually see a decrease in symptoms despite a slight decrease in the immune response. The data required to predict the effect of anaesthesia and surgery on the morbidity and mortality of patients presenting with other types of infections are currently not available.

Specific infections

Cutaneous infections

Cutaneous infections arise following the subcutaneous inoculation of pathogenic micro-organisms into areas where there is suppression of the local defence system. This frequently arises following surgical trauma, pressure injury or placement of catheters. The oedema associated with these events limits blood flow and prevents adequate activation of the acute non-adaptive immune response. Unrelieved pressure can lead to tissue necrosis which damages the skin barrier and inhibits the development of the appropriate leucocyte response. In addition, once certain bacteria become established, they can release compounds that inhibit neutrophil and macrophage functions such as phagocytosis. Local trauma to the skin predisposes the injured dermis to a cellulitis and subsequent deep bacterial tissue invasion. Subcutaneous infections, abscesses, myositis and haematogenous spread may result from this process. Infectious agents may also be introduced directly into a wound. Intravascular dissemination of cutaneous infections can also lead to septic thrombophlebitis, bacteraemia with and without sepsis, bacterial endocarditis and deep tissue abscess formation. The common micro-organisms associated with this type of infection process are group A β-haemolytic *Streptococcus*, and *Staphylococcus aureus*. Mixed Gram-negative infections and *Candida* may be isolated from similar infections in debilitated patients. Prevention of cutaneous infections in the perioperative

period should focus on the use of good sterile technique and limiting the amount of local trauma when positioning the patient or placing an intravascular catheter. Pressure-induced breakdown of the skin in susceptible parts of the body is prevented by proper body placement and padding of bony prominences and other areas of focused pressure. This is particularly important in the elderly patient and the patient with a suppressed immune system.

In the United States more than 25 000 patients per year develop a bacteraemia as a result of the use of an intravascular catheter. Since many of these devices are placed in the perioperative period, the anaesthetist must understand the risks of infectious complications and be aware of the methods that can be used to decrease the likelihood of infection following placement of peripheral intravenous lines, central venous lines, Swan–Ganz pulmonary artery catheters, and arterial lines.

Several sources of bacterial infection exist following intravascular line placement. These include endogenous bacterial flora on the skin at the site of entry of the catheter, contamination of the infusate, contamination at hub junctions, and contamination of accessory devices such as transducers, manometers, and stopcocks. Factors which predispose the patient to intravascular infections include: (1) suppressed immune states (extremes of age, chemotherapy, severe illness, blood dyscrasias); (2) changes in normal skin function (burns, psoriasis, alteration of normal microbial flora); (3) hygiene of medical personnel (handwashing); (4) spread from other infected loci (local as well as haematogenous seeding); (5) catheter factors (composition {plastic vs steel, flexible vs stiff, thrombogenic}, size, use {infusion vs haemodynamic monitoring}, number of lumina); (6) placement (site {central vs peripheral; femoral vs subclavian}, duration of placement {>72 hours vs <72 hours}, type {cutdown vs percutaneous, emergent vs elective}, and the skill of person inserting the catheter).

Micro-organisms most associated with intravascular line infections include: (1) *Staphylococcus* spp. (most common), *Pseudomonas* spp., *Corynebacterium* spp. from colonization of the skin; (2) *Candida, Klebsiella* spp., *Enterobacter* spp., *Serratia* spp. and *Citrobacter* spp. or *Pseudomonas* spp. from contaminated infusate; and (3) *Streptococcus* spp., *Staphylococcus* spp., *Escherichia* spp. from other infected loci.

Intravascular infections may result in the seeding of micro-organisms to different organs. Catheter-initiated suppurative thrombophlebitis may serve as a potential source for infectious emboli. Infectious thrombophlebitis is probably a result of bacterial entrapment and growth in a local thrombus or fibrin deposit. Perivascular inflammation with endothelial damage results and infectious purulent material may fill up the lumen of the vein and serve as a source of haematogenous spread of the bacteria. Alternatively, bacteria may not disseminate but grow in the infusate, catheter site, or at a nidus anywhere along the administration set. The release of toxic substances, such as endotoxin, from these microbial colonies may subsequently produce fever and sepsis.

A high index of suspicion that catheter sepsis is present is suggested by a fever of unknown origin, a positive blood culture for *Staphylococcus*, *Candida* or unusual micro-organisms, and bacteraemia or septicaemia unresponsive to antimicrobial therapy. A presumptive diagnosis may be made based on the presence of bacteraemia or sepsis in a patient with an indwelling intravascular line and no other source of infection, cellulitis or phlebitis at the catheter site, >15 microbial colonies from the cultured catheter tip, resolution of fever following removal of the catheter, or evidence of septic thrombi distal to the indwelling intravascular line.

Since the anaesthetist must administer many intravenous agents during anaesthesia, care must be undertaken to avoid exposing the patient unnecessarily to infectious micro-organisms. Several measures can be undertaken to prevent the introduction of micro-organisms or their products into the lines used for intravenous access. These will be outlined in greater detail elsewhere in this chapter but, in general, are aimed at maintaining sterility and adhering to sound hygienic practice.

Nosocomial respiratory tract infections

Fifty per cent of all hospital-acquired infections of the pulmonary tract occur in the postoperative period or intensive care unit and are, therefore, of major concern to the anaesthetist.[14] In particular, pneumonia is associated with mortality ranging from 20 to 50 per cent in university tertiary care centres.[15] Reducing the incidence of this particular type of infection would, therefore, have a major impact on the quality of perioperative care.

Many factors have been implicated in increasing the risk of hospital-acquired respiratory tract infections. First and foremost is the association of pneumonia with the placement of a tracheal tube for short-term procedures and long-term respiratory support. Placement of a tracheal tube increases the risk of pulmonary infection fourfold over that experienced by patients undergoing surgical procedures that do not require intubation. The incidence of respiratory tract infection in patients requiring tracheal tube placement, in one report, was as high as 17–20 per cent.[16]

Major risk factors associated with tracheal tube placement include the presence of intracranial pressure monitors, the use of cimetidine, and the presence of a tracheostomy.[17] A number of other factors may also account for the increase in pneumonia associated with tracheal intubation. These include: (1) the increased use of tracheal intubation for longer more debilitating surgery and in sicker patients; (2) bypassing the filtration functions of

the upper airways and interfering with the mucociliary 'moving blanket' to produce pooling of secretions around the cuff of the tracheal tube; (3) mechanical abrasion and irritation of the respiratory mucosa by the tracheal tube; (4) impairment of normal laryngeal function following tracheal intubation can lead to an increased likelihood of aspiration of upper respiratory tract secretions and bacterial flora and/or gastric contents. Obesity, extremes of age, severity of the presenting illness, and the perioperative use of antibiotics have all been associated with an increased risk of pneumonia following surgery. Interestingly, in one report the mortality from nosocomial pneumonia in patients with chronic respiratory disease was no different from that seen in patients without chronic respiratory disease.[18] Respiratory equipment itself may also serve as a source of nosocomial pulmonary contamination and pneumonia. The use of good cleaning and sterilization procedures and the use of disposable tubing have significantly reduced this as a major factor; however, respiratory equipment still serves as a source for bacterial contamination of the respiratory tract, particularly in the ICU patient.

The major factor influencing the outcome of a nosocomial pneumonia infection is the aetiological agent responsible for causing the disease. The prognosis associated with Gram-negative bacteria, particularly *Pseudomonas*, is considerably worse than that associated with Gram-positive bacteria which, in turn, is worse than that associated with a viral pneumonia.[15] Gram-negative bacilli produce a more severe pneumonia than Gram-positive bacteria due to several factors, including the increased amounts of antibiotics required to treat them and, more importantly, the severe necrotizing haemorrhage associated with their pathology. Unfortunately, while the aetiological agent in community-acquired pneumonia is usually *Streptococcus pneumoniae*, a Gram-positive bacteria, 60 per cent of the nosocomial pneumonias are caused by Gram-negative bacilli. While viral nosocomial pneumonias are usually not fatal, death can occur, particularly in children requiring surgery for repair of congenital heart defects. The most common aetiological agents responsible for nosocomial pneumonia are outlined in Table 27.1. Interestingly, although *Legionella* accounts for less than 5 per cent of nosocomial pneumonias overall, in some hospitals it may cause over 30 per cent of lower respiratory infections.[19] Respiratory syncytial virus (RSV) and influenza A are two important aetiological agents for nosocomial respiratory tract infections and have received some attention in this regard. Nosocomial fungal pulmonary infections are rare and usually found in immune suppressed patients. Less common agents responsible for less than 1 per cent each of nosocomial pneumonias include *Candida*, *Aspergillus*, *Cryptococcus*, *Histoplasma*, *Mycoplasma pneumoniae*, *Branhamella catarrhalis*, *Haemophilus influenzae*, *Pneumocystis pneumoniae* and cytomegalovirus. It is

Table 27.1 Causes of nosocomial pneumonia

Agent	Incidence	Source of contamination	Associated findings
Common aetiological agents			
Gram-positive bacteria			
Staphylococcus aureus	12.9%	Upper respiratory tract flora	Seen in elderly and debilitated patients or in patients after a viral infection
Streptococcus pneumoniae	<3%	Upper respiratory tract flora	Most common cause of community-acquired pneumonias; also seen in patients with chronic obstructive pulmonary disease and following influenza virus infections
Gram-negative bacteria			
Pseudomonas aeruginosa	16.9%	Colonization of respiratory tract from environmental sources	Primarily a nosocomial pneumonia seen in patients with cystic fibrosis, burn patients, patients on long-term antibiotics, steroids, immunosuppressive drugs and patients with long-term hospital stays
Klebsiella spp.	11.6%	Colonization of respiratory tract from environmental sources	Primarily a nosocomial pneumonia associated with altered states of consciousness, aspiration pneumonitis, patients on antibiotics, steroids, or antimetabolites
Enterobacteriaceae spp.	9.4%	Colonization of upper respiratory tract from gastrointestinal tract with aspiration into the lower respiratory tract	As above
Escherichia coli	6.4%	As above	As above
Serratia marcescens	5.8%	Environmental sources	As above
Proteus spp.	4.2%	Environmental sources	As above
Respiratory syncytial virus (RSV)	Hospital endemic; may account for up to 40% of infant nosocomial pneumonias in isolated settings	Exposure of infectious droplets from an infected person	Can be a major source of morbidity and mortality in hospitalized children in isolated locations
Legionella pneumophila	Hospital endemics; may account for up to 30% of nosocomial pneumonias in isolated settings	Contamination of potable water. Air-conditioning units	Associated with an underlying illness, older males who smoke or patients on immunosuppressive drugs. Tends to occur in isolated outbreaks
Less common aetiological agents	<1% each: *Haemophilus influenzae, Branhamella catarrhalis, Mycoplasma pneumoniae, Candida, Aspergillus,* influenza A, cytomegalovirus, *Pneumocystis carinii*		
Unknown aetiology	~40%		

important to note at this point that determining the causative agent of a nosocomial pneumonia is not always possible. An infectious agent can not be identified as the aetiological agent in up to 40 per cent of hospital-acquired pneumonias.

Nosocomial pneumonias usually develop as a result of either bacterial colonization of the upper airway or as a consequence of bacteraemia. Aspiration of pathogenic bacteria that have colonized the upper airway is the major cause of nosocomial pulmonary infection and the most important to the anaesthetist. Colonization of the upper airway with Gram-negative bacilli is dependent on several factors which include: (1) the length of time spent in the hospital; (2) the use of antibiotics which depress normal endogenous flora; (3) the existence of an underlying pulmonary disease; (4) the severity of illness; (5) the presence of azotaemia; (6) tracheal intubation; and (7) alkalinization of gastric contents. The most important way to control environmental colonization of the upper airway is by hand-washing. Colonization of the upper airways with the patient's endogenous gastrointestinal flora is, however, more difficult. For example, alkalinization of gastric contents with drugs such as cimetidine promote overgrowth of Enterobacteriaceae spp. in the gastrointestinal tract and establishment of the organism in the oral airways.

The mechanism by which endogenous bacterial flora of the upper airways are replaced with Gram-negative bacteria depends primarily on bacterial factors, i.e. the presence of pili, epithelial cell receptors, fibronectin on the mucosal surface, and the protease content in the oral cavity. Fibronectin coating of mucosal cells supports the adherence and growth of Gram-positive organisms. Proteases in the saliva destroy the fibronectin coating and permit the adherence of Gram-negative bacteria to the buccal cells and subsequent replication.[20] Since the presence of Gram-negative bacteria colonization of the upper airways increases the risk of pneumonia (23 per cent versus 3.3 per cent incidence when the upper airway still has its endogenous, predominantly Gram-positive flora), the anaesthetist should evaluate any cultures that have been obtained and preoperatively assess the likelihood of Gram-negative colonization of the upper airway. This is particularly true if the patient has been institutionalized for an extended period.[21]

Pre-existing respiratory tract infections

Preoperative respiratory tract infections have long been thought to predispose the patient to an increased rate of intraoperative and postoperative complications. The increased secretions and airway reactivity that often accompany these infections are thought to increase the risk of laryngospasm and bronchospasm intraoperatively. Furthermore, the potential for the development of post-operative pneumonia in patients presenting with an upper respiratory tract infection is of major concern. Bacterial infections of the upper airways are primarily due to the establishment of normal flora in locations in which they do not usually grow. Examples of these types of infections include otitis media, gum and other maxillofacial abscesses, acute or chronic sinusitis, and bacterial pharyngitis. These infections may serve to seed lower respiratory tract infections.

Viral nasopharyngitis (the common cold) accounts for approximately 95 per cent of upper respiratory tract infections and may be caused by a representative from the myxovirus, paramyxovirus, adenovirus, picornavirus, coronavirus, or rhinovirus groups. These types of infection tend to be self-limiting and virus replication usually is localized to the nasopharyngeal mucosa. Other infections of the respiratory tract that the anaesthetist should be familiar with include acute laryngitis, laryngotracheal bronchitis (croup), and epiglottitis. Laryngitis is a result of hyperaemia and oedema of the larynx and may be of viral or bacterial origin. Further mechanical irritation may make the oedema worse. The presence of an exudate suggests a streptococcal infection. This infection should be treated with cold mist and antibiotics, if appropriate. Croup occurs primarily in young children and presents, characteristically, as stridor on inspiration. A number of viral agents may cause this illness and secondary bacterial infection is not uncommon.

Epiglottitis is a rapidly progressive cellulitis of the epiglottis caused by *Haemophilus influenzae*. This is a true anaesthetic emergency. When epiglottitis is suspected, a definitive diagnosis should be confirmed only in a location in which tracheal intubation and emergency tracheostomy can be immediately performed by skilled personnel. A cherry red epiglottis is pathopneumonic in establishing the diagnosis. Bacterial tracheitis may mimic epiglottitis but usually occurs in older children.

If a patient presents for surgery with a lower respiratory tract infection, the procedure should be cancelled if at all possible. Bacterial infections may be disseminated throughout the lung due to inhibition of the normal bronchial toilet and positive pressure ventilation. Viral lower respiratory tract infections, unlike viral upper respiratory tract infections, are frequently associated with secondary bacterial infections. Lower respiratory tract infections include viral or bacterial pneumonia, viral or mycoplasmal bronchitis, tracheobronchitis, and bronchiolitis. *Haemophilus influenzae*, *Streptococcus pneumoniae*, and other normal oropharyngeal bacterial flora are also involved in the pathogenesis of the chronic bronchitis of smokers, and play a role in producing acute exacerbations of the disease.

Only 18 per cent of patients presenting with an acute pneumonia will be free of an underlying disease. This is important to recognize when evaluating the patient preoperatively. The pneumonia associated with influenza virus can be used to illustrate this importance. This pneumonia appears in two forms and usually appears several days after the symptoms of the primary infection have abated. Primary influenza virus pneumonia is almost always associated with underlying heart disease. Secondary bacterial pneumonia following an influenza virus infection is usually found in the very young, the elderly, in patients with pre-existing pulmonary disease, or in patients who are pregnant. Since these patients are at risk of contracting a secondary bacterial lower respiratory tract infection, it would seem prudent to forego elective anaesthesia and surgery if there is a history of an influenza-like infection in the previous 4–6 weeks.

Bacteraemia, endocarditis and septicaemia

Nosocomial infections of the blood occur primarily as a result of iatrogenic procedures used to treat the patient. As discussed, placement of a percutaneous intravascular device accounts for 75 per cent of all nosocomial bacteraemias. However, bacteraemia may also result when mucosal surfaces having a high concentration of microflora, such as those lining the oropharyngeal cavity and the large intestine, are traumatized or instrumented. Bacteraemia may also occur following instrumentation of the urinary tract, from wounds or surgical trauma to the gastro-intestinal or urogenital systems, or from pre-existing infections of the gastrointestinal, urinary, or respiratory tract. Oral intake should be instituted as soon as possible in patients who have been on long-term total parenteral nutrition to prevent overwhelming translocation of normal gut flora such as *Escherichia coli* into the bloodstream. For the anaesthetist, the most important source of bacteraemia is that which can follow instrumentation of the airway. This risk is increased in the patient who is difficult to intubate, requires a double lumen tracheal tube, or undergoes rigid bronchoscopy. In addition, nasal placement of a tracheal tube in patients with sinusitis may serve to seed the infection in the lower respiratory tract.

As with nosocomial pneumonia, a diagnosis of Gram-negative bacteraemia and septicaemia carries with it very high mortality. Patients who are debilitated or have been in the hospital for an extended period of time are the most likely to have Gram-negative bacterial colonization of the oropharynx. Care must be taken during laryngoscopy to minimize any trauma to the patient's oropharyngeal mucosa. In addition, proper handling and cleaning of the laryngoscope blades following use in a patient with Gram-negative colonization of the oropharynx is critical in preventing transmission of infection to subsequent patients.

Endotoxins, which are constituents of the bacterial cell wall, are responsible for the lethal pathogenesis of Gram-negative septicaemia. Lipopolysaccharide (LPS), extracted from the cell wall of *E. coli*, is capable of producing death in experimental animals by induction of an acute phase reaction. Tumour necrosis factor (TNF), interleukin-1 and interleukin-6 (IL-1 and IL-6) are elevated systemically in LPS treated animals and patients and have been implicated in the pathogenesis of the disease. Evidence for this hypothesis is based on the observation that monoclonal antibodies directed specifically against TNF decrease mortality in experimental animals.[22] Experiments such as these suggest that treatments directed at inhibiting the biological effects of TNF, IL-1 or IL-6 may offer hope for decreasing the high mortality from Gram-negative septi-caemia. However, until such therapies are available haemodynamic support, antibiotics and surgery are the mainstays of treatment.[23]

Although *E. coli* septicaemias are the most common, other agents such as *Klebsiella, Enterobacter, Serratia* and *Pseudomonas* are important aetiological agents. *Klebsiella, Proteus* and *Serratia* septicaemias originate primarily from colonization of the oropharynx. *Pseudomonas* septicaemia has the highest mortality rate of all the Gram-negative bacteraemias and is seen primarily in the immunosuppressed patient or burn victim.

Gram-positive bacteraemias do not carry as high a mortality as the Gram-negative types although infective endocarditis is a feared complication in patients with valvular heart disease. *Staphylococcus* and group A β-haemolytic *Streptococcus* are the most common Gram-positive pathogens to be isolated from blood cultures and generally result from the placement of intravascular catheters.

Candida, the fungal agent most frequently isolated from blood cultures, is found primarily in debilitated patients. Isolation of *Candida* from the blood of relatively healthy patients usually follows seeding from a point source such as an intravascular catheter and can be easily treated by removing the nidus of infection.

Endocarditis and antibiotic prophylaxis

Infectious endocarditis is an infection involving the internal lining of the heart (endocardium). While this infection may occur anywhere on the endocardium the primary locus is on the heart valves. The pathogenesis of endocarditis results in fibrin deposition and platelet aggregation at a site of endothelial damage or trauma. Alterations in the endo-cardial surface provide a nidus for microbial seeding and adherence and may result in vegetations which can subsequently release as emboli. Classically, the organisms associated with infectious endocarditis are *Staphylococcus, Streptococcus* and *Neisseria*; however, Gram-negative bacteria, *Candida, Chlamydia, Rickettsia* and viruses have been isolated.

Since alteration of the endocardial surface appears to be important in the pathogenesis of the infection, certain patient groups are at increased risk for developing this infection during the perioperative period. These include patients with congenital heart disease (particularly those with turbulent flow due to shunting) or valvular heart disease, patients with pacemaker wires and those with intravascular shunts. Intravenous drug abusers, especially cocaine addicts, are at significant risk for infectious endocarditis involving the tricuspid valve. Patients who have an increased risk for contracting infectious endo-carditis should be given prophylactic antibiotic coverage prior to surgery. Treatment should not begin until immediately before the start of the procedure.[24] Recommended therapy is based on the type of procedure and the most likely associated organism, e.g. *Streptococcus* from the mouth, *Enterococcus* from the gastrointestinal tract and *Staphylococcus* from the skin.

Urinary tract infections

Although the anaesthetist is not directly involved with urinary tract infections intraoperatively, many surgical patients have an urinary catheter placed to monitor urine output. Urinary tract infections are the most common type of nosocomial infections and 80 per cent are associated with the use of an urethral catheter. The majority of urinary tract infections are caused by the patient's own endogenous intestinal flora that migrate across the perineum and colonize the periurethral area. These organisms enter the bladder by insertion of the catheter through the infected meatus or by migration up the catheter lumen or catheter–mucosal interface. The most commonly isolated organism is *E. coli* followed by *Pseudomonas, Klebsiella, Proteus, Staphylococcus* and *Enterobacter*. The presence of the catheter in the bladder appears to promote the growth of bacteria which are not normally seen in a non-catheterized bladder. Factors which increase the risk of developing a catheter-related urinary tract infection include duration of catheterization, absence of an urimeter, diabetes, female gender, concurrent renal disease, lack of antibiotic prophylaxis, interrupting the closed urinary catheter system, and equipment contamination. Long-term bladder catheterization presents further problems that need not be discussed here.

Most urinary tract infections are asymptomatic and resolve after the catheter is removed. While damage may be done to the bladder itself, the major concern with bacteriuria is the development of acute pyelonephritis, renal stones, or a perinephritic abscess. Urinary tract infections are also responsible for approximately 15 per cent of nosocomial bacteraemias and are a source of significant mortality.

Maintenance of a closed system and avoiding contamination of the reservoir's drain tube are protocols which decrease the incidence of catheter-related urinary tract infections. Removing catheter obstructions and maintaining an adequate flow of urine also decrease the likelihood of a urinary tract infection in the perioperative period. The anaesthetist must not contribute to the incidence of urinary tract infection by committing errors in the care of the urinary catheter.

Herpes virus

Herpes simplex viruses (HSV) are among the most ubiquitous agents found in man. Two distinct serotypes of this DNA virus are evident, distinguished clinically by different modes of transmission and age distributions. HSV-1 is extemely common in childhood and is transmitted primarily via the oral route. Once infected, the virus remains dormant throughout life, residing in the sensory ganglia that innervate the primary site of infection. Activation of the infection may occur following a number of stimuli, including stress and immunosuppression. Active lesions generally occur around the mouth, lips and skin above the waist. Culture of the lesions produces high HSV-1 titres; therefore, anaesthetists should be particularly careful to wear gloves when in contact with the oral mucosa of infected patients.[25] Health care workers including dentists, respiratory therapists and anaesthetists may be particularly vulnerable to herpetic whitlow following unprotected contact with open lesions from an infected patient.[26] HSV-2 is most often transmitted venereally or following vaginal delivery of the newborn. In general, HSV-2 infection occurs in adolescence or early adulthood via sexual contact. Although HSV infections are sporadic they may be particularly painful and psychologically damaging. Despite the fact that there is no vaccine or cure for the disease, oral acyclovir has proved efficacious in reducing the frequency and severity of the infection. For the anaesthetist, universal precautions are appropriate for managing patients with open lesions. Gloves are particularly important during procedures such as tracheal intubation, suctioning and bronchoscopy. Such precautions are bidirectional since patients may also be at risk of acquiring HSV from an infected anaesthetist. In a recent survey 16 per cent of anaesthetists provided anaesthetic care while infected with HSV.[27] Immunosuppressed patients or those with extensive burns are also at increased risk of HSV infection. Infected personnel should be appropriately protected or should avoid treating these patients during exacerbations. Fortunately, since the virus is particularly susceptible to changes in environment, hand-washing and drying are generally sufficient to prevent transmission of the agent.

Varicella-zoster virus (VZV) is a highly contagious member of the herpes virus family. In addition to transmission via direct contact, air-borne and droplet spread are the primary mechanisms. In general, most adults have acquired immunity to this virus in childhood and nosocomial outbreaks of infection are typically limited to children and immunocompromised patients. When managing the patient with shingles or varicella the anaesthetist should wear a mask in addition to gloves and if non-immune should avoid contact with the patient altogether.

Cytomegalovirus (CMV) is another member of the family Herpetoviridae. It is estimated that 60–80 per cent of individuals over the age of 30 possess CMV antibody. Occupational transmission of CMV appears to be limited;[28] however, anaesthetists who are pregnant should avoid contact with infected patients. CMV transmission has been associated with blood transfusions with an estimated risk of 2.5 per cent per unit of blood given. CMV is particularly serious in immunosuppressed patients or those undergoing organ transplantation. It is commonly associated with allograft rejection or loss and may predispose these patients to fatal opportunistic infections. In addition, CMV has been linked to Kaposi's sarcoma, and carcinomas of the prostate, cervix and colon.

Epstein–Barr virus is most commonly associated with infectious mononucleosis but has also been implicated with Burkitt's lymphoma and chronic fatigue syndrome. Transmission is predominantly via intimate oral contact, i.e. kissing, although transmission via blood transfusion has been documented.

Hepatitis virus

In terms of prevalence and risk, hepatitis is perhaps the most important virus to which the anaesthetist is exposed on an occupational basis. There are four types of hepatitis that are important to the anaesthetist. Of greatest importance is hepatitis B virus. It is estimated that there are 8700 HBV infections among health care workers exposed to blood, in the United States annually. Approximately 400–440 of these individuals require hospitalization and 200 die from acute or chronic HBV infection.[29] Occupational transmission from infected patients usually occurs following an accidental needlestick injury or contact with contaminated blood or blood products. HBV seroconversion among anaesthetists has been estimated at between 12.7 and 48.6 per cent.[29] Of those infected with HBV, 5–10 per cent become chronic carriers with continued presence of hepatitis B surface antigen for up to 6 months in the blood. The high risk of acquiring occupational HBV is based primarily on its ease of transmission. Studies in chimpanzees and humans show that dilutions as low as 10^{-7}–10^{-8} of the virus are sufficient for transmission. Also, rates of transmission via needlestick injury are estimated between 10 and 35 per cent. A vaccine for HBV is available and has proved efficacious. However, immune globulin is still the treatment of choice when the vaccine is unavailable or contraindicated. Other methods of protecting

the anaesthetist from occupational HBV are discussed later in this chapter.

Delta hepatitis is another important type of hepatitis. The delta agent is an incomplete virus that requires simultaneous infection with HBV to replicate. Frequently this occurs when a chronic carrier of HBV is exposed to the delta agent.[30] It is estimated that approximately 30 per cent of all fulminant hepatitis is attributed to the delta agent.

Non-A, non-B hepatitis (NANBH) is the most important cause of post-transfusion hepatitis today. Recently, hepatitis C virus (HCV) has been identified as the aetiological agent in the majority of cases of NANBH post-transfusion hepatitis. It is estimated that 150 000–170 000 new HCV infections occur annually in the United States.[31] Not all patients with NANBH test positive for anti-HCV assays; therefore, diagnosis of NANBH is often by exclusion of all other types of hepatitis producing agents.

Hepatitis A virus (HAV) is the most common type of hepatitis. In Europe and the United States the prevalence in young adults varies between 30 and 70 per cent but is less than 10 per cent in small children.[32] HAV produces an acute disease with a mortality rate below 0.5 per cent. Transmission is via the faecal–oral route and, therefore, anaesthetists are at no greater risk of acquiring this disease than are the general public. Other viruses are also capable of producing clinical hepatitis and should be excluded serologically when making a diagnosis. Such viruses include EBV, CMV, HSV, rubella and the yellow fever virus.

Human immunodeficiency virus

During the period from June 1981 to May 1991, 179 136 cases of acquired immune deficiency syndrome (AIDS) were reported to the Centers for Disease Control (CDC). Furhermore, present estimates indicate that 1–1.5 million individuals in the United States and 10 million world wide may be infected with the human immunodeficiency virus.[33] HIV, the causative agent of AIDS, is an RNA retrovirus that causes depletion of the helper subset of T-lymphocytes (T4 or $CD4^+$ T-cells) and profound immunosuppression. Other cells including dendritic cells and macrophages may also be infected. HIV-1 is the principal causative agent in the United States and Europe; however, in parts of Africa where the disease is endemic, a subtype HIV-2 is prevalent and is a major causative agent in heterosexual transmission of the disease.

Since the formal discovery of AIDS in 1981, a number of risk factors have been established for the disease. Reports issued by the CDC suggest that 57 per cent of the cases occur via male homosexual/bisexual contact. Nineteen per cent of cases are due to intravenous drug abuse, 5 per cent from heterosexual contact, and 2 per cent recipients of blood or blood products. Approximately 13 per cent of cases have multiple modes of exposure and 4 per cent are of undetermined aetiology. Occupational transmission of HIV accounts for approximately 0.1 per cent of AIDS cases. As of September 1992 the CDC had received reports of 32 health care workers in the United States with documented occupational acquisition of HIV and 69 with possible transmission.[34] Of those with documented occupationally acquired HIV, 84 per cent had percutaneous exposure, 13 per cent mucocutaneous exposure and 3 per cent had both types of exposure. The populations at greatest risk of occupational transmission appear to be clinical laboratory technicians and nurses. The total number of individuals with occupationally acquired HIV is probably higher than that shown, since many cases are not reported and some may not seek HIV testing following exposure. Although there are no reports of occupational transmission among anaesthetists, a recent survey showed that only 46.1 per cent sought treatment following a contaminated needlestick and of those just over half sought HIV testing.[27]

Clinically, the continuum of the disease can be divided into three phases. The initial or window phase follows exposure, during which the patient is infectious but may not have seroconverted. Typically, seroconversion occurs within 2–6 weeks of exposure; however, in some cases seroconversion may not occur for 6 months or longer. It is imperative, therefore, that individuals with risk factors for HIV infection do not donate blood. Usually within a week of the initial exposure, the individual may experience an acute flu-like illness characterized by fever, lymphadenopathy, rash and fatigue. Symptoms usually subside within a week or two. Following the initial phase there is an asymptomatic phase which may last for 5–10 years. As this phase progresses subtle changes in immune function occur. T4-lymphocyte counts decrease and once below $200/mm^3$ of blood (normal = $1000/mm^3$) the patient becomes particularly prone to numerous opportunistic infections. The late or symptomatic stage is characterized, therefore, by numerous recurrent infections, lymphadenopathy, weight loss and in some cases neurological manifestations such as dementia and peripheral neuropathy. Although a number of drugs are now available to slow the progression of the disease and improve quality of life, death usually occurs within 2–3 years of late stage onset.

The infected patient

Patients presenting at surgery with an acute infection pose a considerable challenge to the anaesthetist. Not only must the anaesthetist be aware of the potential for occupational acquisition of the infection but there must be an understanding of how anaesthesia *per se* may effect pathogenesis and outcome.

Occupational transmission of infection

It is only recently, with the advent of the AIDS epidemic, that the potential for occupational transmission of disease has been addressed in the operating theatre and among health care works in general. Given the increasing incidence

Table 27.2 Occupational risk of important infections among anaesthetists

Organism	Source	Transmission	Prevention	Risk
Hepatitis B virus (HBV)	Infected blood, blood products or bodily fluids	Needlesticks, contact with mucous membranes, open cuts or sores	Avoid needlesticks. Universal precautions. Immunization	Significant in the unvaccinated individual
Delta agent hepatitis	See HBV	Requires a concurrent HBV infection	Universal precautions. No vaccine available	Unknown
Non-A, non-B hepatitis (NANBH, HCV)	See HBV	See HBV	Universal precautions. No vaccine available	Unknown
Human immuno-deficiency virus (HIV)	See HBV	See HBV	Avoid needlesticks. Universal precautions	Low; however, contraction is probably 100% fatal
Herpes simplex virus (HSV)	Infected fluid from open lesions	Contact with mucous membrane, open cuts or sores	Gloves, hand-washing	Minimal, herpetic whitlow important
Varicella-zoster virus (VZV)	Respiratory secretions/ droplets	Respiratory tract, direct contact	Masks, gloves	Minimal
Cytomegalovirus (CMV)	Secretions, blood	Direct contact. Associated with transfusions and transplants	Masks, gloves, avoid contact if pregnant	Minimal; however, infection during pregnancy may pass to infant

of this disease, it is inevitable that anaesthetists today will be presented with an increasing number of patients who are at risk for HIV and HBV or who are involved with procedures that involve large replacement volumes of blood. In addition to the risk of acquiring HIV and HBV in the workplace there are other occupational infections that may be acquired by the anaesthetist. These infections are outlined in Table 27.2.

As the modes of HIV transmission became better defined the CDC issued guidelines aimed at reducing the potential for occupational transmission of this disease.[35] These guidelines have recently been updated to include occupational HBV transmission.[36] Although these guidelines pertain in general to all personnel involved with high risk patients or material they have specific applications to the practice of anaesthesia. Included in the CDC guidelines are recommendations to adopt infection control protocols involving the concept of 'universal precautions'. Under a universal precautions policy, health care workers are to assume that all patients are infected and that barrier techniques be employed to avoid contact with the patient's blood and bodily fluids. This applies to any exposure to blood and blood products, semen, vaginal fluids, cerebrospinal fluid, synovial, pericardial, peritoneal and amniotic fluids and any bodily fluid that may contain blood. In addition, all laboratory specimens containing the virus (e.g. suspensions of concentrated virus) should be considered as part of these precautions. Universal precautions are not only important in protecting the health care worker from HIV and HBV but are also important in protection against the myriad of opportunistic infections that are often associated with these patients. This is particularly important in light of the increasing incidence of multidrug-resistant tuberculosis among AIDS patients. Under these conditions the wearing of gloves and regular hand-washing should be routine. Masks, goggles and gowns are recommended for procedures in which the splattering of blood or contact with bodily fluids is anticipated. For the anaesthetist this may involve procedures such as tracheal intubation, bronchoscopy and endoscopy. Undoubtedly, the greatest concern for occupational exposure to HIV or HBV is the contaminated needlestick. According to the CDC, 80 per cent of all occupational exposures occur in this way. In a recent survey of anaesthetists in the United States, 36.7 per cent had received at least one contaminated needlestick within the previous 12 months.[27] Patently, this is of concern given that seroconversion from a contaminated needlestick is estimated to be approximately 0.5 per cent for HIV, and may be as high as 35 per cent for HBV. Indeed, in one study, anti-HBs and anti-HBc (hepatitis B surface and core antibodies respectively) were detected in 18.8 per cent of anaesthetists.[37] The importance of the hepatitis vaccine cannot, therefore, be understated. In an attempt to reduce the incidence of needlesticks among health care workers, the CDC recommends that needles should never be bent, broken or recapped, and should be placed in puncture-proof containers after use. A number of manufacturers have developed disposable syringe and needle systems that avoid the need to recap. These employ a standard design with a cylindrical sheath that slides over the contaminated needle after use. Other methods of transmission have been postulated and deserve mention. These include dissemination of papilloma virus in the smoke from laser surgery and the aerosolization of infected blood and tissue during invasive orthopaedic procedures.

As with all guidelines, efficacy is related to the degree of compliance. For example, in a survey of emergency nurses in Columbus, Ohio, USA, universal precautions were followed only 16.5 per cent of the time.[38] Indeed, the CDC suggests that 40 per cent of all occupational exposures to HIV could have been avoided if universal guidelines had been employed. However, despite the fact that a number of studies[39,40] suggest that health care workers in general are not adequately following universal precautions, two recent studies suggest that compliance among anaesthetists is improving. Rosenberg et al.[41] showed that there was a significant improvement in self-protection among anaesthetists surveyed between 1990 and 1991 including the wearing of gloves for arterial line placement and the prevalence of hepatitis vaccination. Also, Tait showed that compliance with CDC regulations increased as the risk from the patient increased. For example, gloves were used by 46.1 per cent of anaesthetists if the patient was low risk and 86.3 per cent and 94.7 per cent if the patients were high risk (antibody status unknown) and HIV seropositive respectively.[27] However, in this study, 69.3 per cent of anaesthetists admitted recapping needles on a frequent basis. Although it is encouraging to note that anaesthetists as a group appear to be altering their practices in light of the AIDS epidemic, many of the documented needlestick injuries involving transmission of the virus have occurred during emergency situations. It is important, therefore, that many of these preventive practices become automatic and that extra vigilance is taken at these times.

Anaesthetic considerations for the infected patient

General dogma indicates that anaesthesia is a contra-indication for patients who present at surgery with an acute infection. Nowhere is this more evident than for the patient who presents with an upper respiratory tract infection (URTI). The practice of rescheduling these patients until asymptomatic is based on the premise that anaesthesia increases the risk of perioperative complications and exacerbates their postoperative course. For example, in a recent study by Cohen and Cameron[42] children with URTIs were found to be two to seven times more likely to experience perioperative respiratory-related adverse effects than asymptomatic children. Studies by Tait and Knight,[43] however, demonstrated no increase in perioperative complications for children with acute uncomplicated URTIs who underwent minor surgery. Furthermore, these studies demonstrated fewer respiratory symptoms in the infected children following anaesthesia compared to a matched group of infected children who had not received anaesthesia.[43] Unfortunately, comparison of these types of studies is difficult due to differences in population demographics, surgical procedures, URTI criteria and study design. Generally, therefore, decisions to proceed with anaesthesia in the face of an URTI are based largely on personal preference and experience. However, if the anaesthetist does decide to cancel a case due to the presence of an URTI then the patient should not be rescheduled for a minimum of 4 weeks. This is based on studies that clearly demonstrate increased airway reactivity in patients for 4–6 weeks following an URTI.[44] Changes in airway reactivity may predispose the patient to intraoperative coughing or laryngospasm, particularly if the airway requires instrumentation. Based on the authors' research, it is their opinion that otherwise healthy patients presenting with acute uncomplicated URTIs in which the secretions are clear and the patient is afebrile may be considered candidates for surgery. Similarly, patients with allergic or vasomotor rhinitis or patients with chronic rhinitis appear to be at no greater risk of perioperative complications. Patients who present with severe nasopharyngitis, fever greater than 38°C, productive cough or flu- or croup-like symptoms should be rescheduled. Potential intraoperative complications in these patients include laryngospasm, bronchospasm, apnoea, hyperthermia, stridor and blockage of the tracheal tube. Postoperatively patients may experience oxygen desaturation, atelectasis, postintubation croup and pneumonia. Recommendations for the management of these patients include regional techniques, avoidance of tracheal intubation, humidified gases and antisialagogues.

Anaesthesia for the patient with AIDS poses a challenge to the anaesthetist since these patients frequently present with multi-organ disease. In addition to the numerous opportunistic infections, many AIDS patients have decreased pulmonary function due to *Pneumocystis carinii* or tuberculosis. Also, patients on AZT therapy may present with anaemia and clotting abnormalities. Neurological function may also be affected leading to dementia, seizures and peripheral neuropathies. These findings may be important in decisions to use regional techniques. AIDS patients who present with wasting syndrome due to persistent diarrhoea may be hypovolaemic and have electrolyte imbalances. Furthermore, autopsy studies of AIDS patients describe multiple cardiac lesions including pericardial effusion, fibrinous pericarditis, right ventricular hypertrophy and evidence of Kaposi's sarcoma in the pericardium and myocardium.[45]

Anaesthesia and infection

The finding that anaesthesia may reduce the severity of a viral respiratory infection is intriguing and may be related to the effect of anaesthesia on the immune response. As discussed earlier, studies show that anaesthetics affect a number of individual components of the immune response. These findings are important as they lend further support to the hypothesis that volatile anaesthetic agents inhibit the body's ability to mount an inflammatory response.

That anaesthetic suppression of the immune response may not always be deleterious is suggested by a number of animal studies. For example, in one early study diethyl ether increased the survival of puppies inoculated with distemper virus.[46] More recently, studies have shown that volatile anaesthetic agents decrease the pathogenesis of influenza infection in mice by suppressing the recruitment of intra-alveolar immune cells and decreasing the inflammatory response in the lung.[12,13] This is further supported by work that demonstrated that mice infected with influenza virus and subsequently treated with low dose cyclosporin (a T-cell suppressant) showed less pulmonary injury and lower cumulative mortality than a group of mice that were infected but not treated with the drug.[47] These studies suggest that although there was suppression of the T-cell response the animals were able to contain the virus and suggests that immunosuppression of certain T-cells may be beneficial in reducing pathogenesis in this model.

These findings are, however, not without contradiction. In one study, for example, mice infected with murine hepatitis and anaesthetized with halothane had greater mortality than similarly infected unanaesthetized mice.[11] However, these differences may be related to the virus and/or the effect of anaesthesia on liver function since patients who receive anaesthesia while incubating HBV, or who have acute hepatitis infections, show further deterioration in liver function postoperatively. Although many of these experiments have been carried out in animals the findings suggest important interactions between the volatile anaesthetic agents and infection (particularly viral) both in terms of the immune response and outcome. These observations, therefore, may have important ramifications for the infected or immunocompromised patient undergoing anaesthesia and surgery.

The immunocompromised patient

It is well known that certain patient groups are readily predisposed to infection. Patients who are elderly, obese, diabetic or have poor nutritional status (e.g. alcoholics, drug abusers) are all prone to infection. In addition, patients with extensive burns have higher rates of infection. Today, the anaesthetist is also faced with an increasing number of patients who are immunocompromised by congenital or acquired disease, or who are receiving immunosuppressive drugs for transplantation or malignant disease. In addition, many of these patients present with unusual infections that have become opportunistic in the face of decreased immune function. Indeed, infection is one of the leading causes of death in these patients. Pneumonia, for example, accounts for up to 40 per cent of deaths in children and adults with cancer.[48,49] Furthermore, 35 per cent of renal transplant recipients develop pneumonia in the first year following transplantation.[50]

At some time, patients with defects in one or more of the components of the immune response will present for surgery. It is important, therefore, that anaesthetists understand the relation between anaesthesia, infection and the immune response and how their practice may influence outcome. Defects in phagocytosis and granulocytopenia, for example, are frequently seen in patients with aplastic anaemia, acute leukaemia or those on myelosuppressive therapy. These patients frequently present with bacterial infections of the lungs, oropharynx, perianal region and the skin. Infectious organisms in these patients typically include most Gram-negative bacteria and fungi such as *Candida* and *Aspergillus*. Pneumonia with *Haemophilus influenzae* or *Streptococcus* is particularly prevalent in patients with humoral dysfunction, which is associated with agammaglobulinaemia, multiple myeloma and chronic lymphocytic leukaemia. Patients with Hodgkin's disease, those undergoing radiotherapy and those with AIDS have defects in cell-mediated cytotoxicity and are particularly prone to fungal and herpes virus infections. AIDS patients are frequently associated with HBV and other sexually transmitted diseases such as syphilis and gonorrhoea. Opportunistic infections in this group of patients include pneumocystis pneumonia, herpes simplex and zoster viruses, cytomegalovirus, tuberculosis, oral candidiasis, Epstein–Barr virus and a particularly virulent form of Kaposi's sarcoma. Patients with severe combined immunodeficiency syndrome (SCID) have an absence of both humoral and cell-mediated immunity. These patients have marked lymphopenia, thymic abnormalities and a lack of immunoglobulin and are particularly prone to severe bacterial and viral respiratory disease.

As discussed earlier, a number of *in vitro* and *in vivo* studies have shown that volatile anaesthetics suppress immune function in both normal and infected models. The effect of anaesthesia on the immunocompromised individual is less well understood, although one might hypothesize that the immunosuppressive effect of general anaesthesia and/or surgery might add to the degree of immunosuppression and affect outcome in these patients. This may be particularly important for the cancer patient undergoing anaesthesia and surgery. Animal studies suggest that anaesthesia increases pulmonary metastases following anaesthesia with either thiopentone (thiopental),[51] chloroform or diethyl ether.[52] In a study by Lundy *et al.*[53] halothane alone had little effect on the development of pulmonary metastases in mice; however, halothane together with surgery decreased cell-mediated cytotoxicity and increased metastatic spread postoperatively. Clinically, these findings are important as it has been shown that in the immediate postoperative period the ability of lymphocytes to kill tumour cells is diminished.[54] Several studies point to suppression of NK activity as a factor in the development of postoperative metastases and infection. In one of these studies, high doses of either morphine (30 mg/kg), fentanyl (0.3 mg/kg) or sufentanil (0.06 mg/kg) significantly suppressed NK activity in rats for up to 12 hours following administration.[55] This effect was blocked by administration of the opiate antagonist naltrexone. This study also showed that NK cytotoxicity was increased by pretreatment of the animals with polyinosinic:polycytidylic acid (poly I:C), an interferon inducer, but was subsequently reduced to baseline activity by fentanyl. Similar findings by Markovic *et al.*[4] showed that anaesthesia with halothane or isoflurane inhibited interferon-stimulated cytotoxicity of NK cells in mice for 10–14 days following administration. Interestingly, if interferon-stimulated NK cytotoxicity occurred prior to anaesthesia, then there was no decrease in activity in the postsurgical period. These studies suggest that pretreatment of immunocompromised surgical patients with immune adjuvant may result in decreased infectious morbidity and mortality from compromised NK cell defences.

Prevention of infection

Preoperative blood screening

In the period between 1981 and January 1990, 3027 cases of transfusion-associated AIDS were reported to the CDC. Over 1000 of these cases were associated with the administration of clotting factors to haemophiliacs. Recipients of blood products now comprise approximately 2 per cent of the total AIDS cases. The safety of the reserve blood supply has, however, improved dramatically since the advent of the AIDS epidemic and with the introduction of the ELISA test for HIV antibodies in 1985. Although the risk of contaminated blood or blood products has been reduced, 5 per cent of post-transfusion hepatitis cases are due to HBV despite negative HBsAg titres in the donor blood.[56] Similarly, blood from an HIV infected patient may show negative on test if sampling occurred during the window between infection and seroconversion. Current evidence suggests that the risk of contracting AIDS from a blood transfusion is approximately 1 in 250 000.[57] Non-A, non-B hepatitis (NANBH) is estimated to cause 90 per cent of post-transfusion hepatitis. There is no specific test for NANBH other than by exclusion and by the presence of elevated serum transaminases.

Prophylaxis for perioperative infections

In general, prophylactic immunization is not required preoperatively; however, the importance of HBV vaccination cannot be overemphasized for anaesthesia personnel and patients from high risk groups. Recent surveys in the United States and the UK showed that the prevalence of hepatitis vaccination among anaesthetists was approximately 80 per cent.[27,41,58] Hepatitis vaccines are both safe and efficacious. Approximately 90 per cent of vaccinated individuals develop adequate protective antibodies (anti-HBs). Influenza A vaccines may also be appropriate for elderly or high risk patients during influenza epidemics. Although anaesthetists are not intimately involved with antibiotic therapy, it is important that they are aware of the potential for infection, particularly in the immunocompromised patient, and are careful to optimize the patient's management. In general, preoperative prophylactic antibiotic therapy is only required for patients who are at risk of developing postoperative infections. For example, instillation of aerosolized gentamicin has been found efficacious for patients requiring long-term tracheal intubation in units with Gram-negative outbreaks. Care should be taken with immunosuppressed patients, however, in that aggressive treatment of infections with antibiotics may alter normal bacterial flora and increase their susceptibility to nosocomial pathogens.

A number of antiviral agents have been developed and are proving efficacious in controlling viral infections.[59] Acyclovir is a purine nucleoside analogue with an acyclic side chain which works by inhibiting viral DNA synthesis. Acyclovir is particularly important in use against herpes virus types 1 and 2, varicella-zoster virus and Epstein-Barr virus. Many immunocompromised patients presenting with these infections may benefit from this drug perioperatively.

Vidarabine (adenosine arabinoside) has also proved useful in herpes virus infections. Treatment with this drug has reduced HSV encephalitis mortality from 70 to 28 per cent. Amantadine hydrochloride is a symmetrical tricyclic amine that specifically inhibits the replication of influenza A viruses. Prophylactic amantadine has been shown to be efficacious both in the community and hospital setting. Its use is recommended for the elderly or chronically ill patient in whom vaccination is contraindicated or unavailable. Tribavirin (ribavirin) is a purine nucleotide analogue which has proved effective in the treatment of respiratory syncytial virus in infants and children.

Hygienic practices in anaesthesia

It has been known for many years that the practice of anaesthesia has the potential for transmitting a number of infectious agents to the patient. Indeed, Skinner in 1873 decribed the practice of using the same inhaler on multiple patients without any type of cleaning. Nowadays, much of the anaesthesia equipment comes as sterile, disposable, single-use units designed for convenience and as a means to reduce cross-contamination and transmission of infection. However, despite the fact that disposable anaesthetic equipment is a multimillion dollar industry, there are still questions as to its efficacy in reducing perioperative transmission of disease.[60,61] For example, one investigator evaluated the incidence of postoperative pulmonary infections in patients who had used sterile disposable anaesthesia equipment and bacterial filters and found that the incidence of infection was no different from those patients in which clean re-usable equipment was used.[61] Recent evidence also suggests that contamination of disposable and re-usable anaesthesia equipment is quite prevalent. In one study, 25 per cent of circle systems were found to be contaminated with *Staphylococcus* and *Pseudomonas* species. Forty-four per cent of ventilators were also contaminated. Pathogens identified included *Klebsiella pneumoniae, Pseudomonas aeruginosa, Aspergillus* spp. and *Xanthomonas maltophila*. Stopcocks and injection ports also provide a potential source of contamination. In a recent study by Dryden and Brickler,[62] 46 per cent of stopcocks used during anaesthesia were contaminated. Care should be taken, therefore, to keep them free of blood and to cover the connectors with a sterile cap.

In addition to the potential of infection from contaminated anaesthesia equipment the practice of anaesthesia may promote infection since it often requires violation of the body's mechanical barriers. For example, placement of intravenous and intra-arterial catheters, instrumentation of the airway and mechanical ventilation all provide potential vehicles for transmission of infectious agents. Devices that penetrate the body or come in contact with sterile areas must be sterile at the time of use. Stringent aseptic technique should be employed whenever these devices are being used. Other equipment such as laryngoscope blades, masks, breathing circuits and oesophageal stethoscopes that do not penetrate body surfaces, yet may come in contact with mucous membranes, must be free of contamination, yet need not be sterile. Pulse oximeter probes, stethoscopes and blood pressure cuffs that do not normally come into contact with the patient or only touch unbroken skin should be cleaned with a detergent or disinfectant at the end of the day or when obviously contaminated. In general, the routine sterilization or disinfection of the insides of the anaesthesia machine is not required; however, valves and soda-lime canisters should be cleaned and disinfected with each change of absorbent. The working surfaces of anaesthesia machines are, however, prone to contamination and should be disinfected after each patient use. This is particularly important in light of a recent study that showed that 51 per cent of anaesthesia records for coronary and vascular procedures were soiled and 36 per cent tested positive for blood contamination. Given the minute amounts of HBV needed for transmission and the unusual ways in which this virus can be transmitted, this type of contamination may be hazardous to both anaesthesia personnel and medical record keepers.[63]

Sterilization implies the killing of all micro-organisms including viruses and spores, whereas disinfection, depending on the level, destroys only vegetative organisms but not necessarily all bacterial and fungal spores or viruses. For the most part, sterilization involves the application of heat and pressure or gases to kill micro-organisms. Steam sterilization is the most popular technique, employing saturated steam under pressure. This technique is recommended for a wide range of articles, including metal, cloth, glass and some thermo-resistant plastics. Resterilization or repeated disinfection of disposable equipment is not recommended, however, since these products invariably contain plastics that degrade with these processes. Ethylene oxide sterilization is another popular technique. Gaseous ethylene oxide at a humidity of 30–40 per cent acts as an alkylating agent to kill any micro-organisms. This technique is particularly suited to the sterilization of anaesthesia machines and electronic equipment. A drawback of this method, however, is that it takes much longer than steam sterilization due to the aeration period that is required to remove toxic residues following the sterilization process. Other methods include radio-sterilization and formaldehyde sterilization. Adequate disinfection can be achieved using a number of commercially available agents including glutaraldehyde preparations, sodium hypochlorite solutions, 70 per cent alcohol, phenolic compounds and hexachlorophene. Two of the more important viruses, HIV and HBV, can be easily destroyed by appropriate sterilization/disinfection techniques. Hepatitis B

virus can survive at 25°C and drying for 1 week but is easily killed by steam sterilization at 121°C and 793.5 kPa (115 p.s.i.) for 20 minutes. Two per cent activated glutaraldehyde may be used when steam sterilization is unsuitable. HIV, on the other hand, is fairly labile and is susceptible to many commonly used disinfectants, including 70 per cent ethyl alcohol, 1 per cent lysol, and 1:10 dilution of 5.25 per cent sodium hypochlorite. The virus is quickly inactivated at 56°C.

The use of a common syringe for the administration of drugs to more than one patient is a potentially dangerous anaesthetic practice even if a new needle is used for each individual. In one study, approximately 40 per cent of anaesthetists admitted using a common syringe on multiple patients at some time.[41] It is known, for example, that minute amounts of blood may be inadvertently aspirated into the syringe with each patient use.[64] This may be sufficient to transmit HBV. Recently Froggatt et al.[65] reported six patients who contracted acute HBV infections from a re-used syringe that had been in contact with a contaminated stopcock from a hepatitis B carrier. Similarly, single-dose ampoules should be used as such. Many of the single-dose ampoules do not contain preservatives and, therefore, may become contaminated with bacteria if allowed to remain open for long periods of time. Recently, surveys from the CDC have implicated extrinsic contamination of propofol with cases of postoperative infection.[66] Drugs should be drawn up as near to the time of use as possible and discarded if not completely used. This practice avoids the potential for drug contamination and is cost-effective should the drug not be required. Infusion solutions should also be used on a one-time, one-patient basis. Studies show that contaminated infusions have resulted in bacterial and fungal infections when used on multiple patients.[66] Multidose vials generally contain preservatives and, therefore, may be used on more than one patient. However, care must be taken to use a new sterile needle and syringe for each withdrawal and to take care to clean the rubber septum with alcohol prior to use. Vials that are obviously contaminated should be discarded. Although transmission of infection by this route is uncommon there have been reports of viral and bacterial infections linked to contaminated multidose vials.[67,68]

Hand-washing is a much overlooked yet important means of reducing perioperative transmission of infection. For example, Gwaltney et al.[69] showed that during experimental rhinovirus infections, transmission of the virus occurred in 73 per cent of cases via the hands and only 8 per cent via sneezing and coughing. Anaesthesia personnel who have open sores from psoriasis, dermatitis or have herpetic whitlow should wear gloves at all times or avoid patient contact during exacerbations. In addition, anaesthesia personnel who have active respiratory infections should wear gloves and a mask. In a recent survey of hygienic practices among anaesthetists 92 per cent of respondents admitted administering anaesthesia while harbouring a respiratory infection, 16 per cent with herpetic lesions and 10 per cent with either psoriasis or dermatitis.[26] The wearing of protective clothing such as masks and gloves is not only important in protecting the patient from infection but is essential in protecting anaesthesia personnel from potential occupational transmission of infection from an infected patient. This issue, particularly with reference to HIV and HBV infections, has already been addressed; however, it highlights the need for a greater awareness of infection control and hygienic practices in the operating theatre.

Although a cause–effect relation between anaesthesia practice and postoperative infection is not easy to establish, the fact that many postopertive infections result from the types of pathogens identified on anaesthetic equipment makes a strong argument for greater attention to hygienic practice. It is important, therefore, particularly in these days of extended case loads and rapid turnaround times, that standards of hygiene are maintained by anaesthesia personnel both in terms of the protection of their patients and, indeed, of themselves.

Treatment for occupational exposure to infectious agents

For the most part, prevention of occupational exposure to infectious agents can be achieved through attention to universal precautions. If occupational exposure to blood, blood products or bodily fluids should occur, the site should be washed immediately and the date, time and nature of the exposure documented. Source individuals should be sought and evaluated for the epidemiological and clinical likelihood of HIV and/or HBV infection. Most exposures via cuts or needlesticks from low risk patients can be treated by cleaning the wound with an antiseptic solution or ointment. However, exposures to the blood or bodily fluids of high risk patients requires greater attention to detail. Anaesthetists not vaccinated against HBV should be counselled to do so. Patients who are considered high risk, e.g. homosexual or bisexual men and i.v. drug abusers, should undergo serological testing immediately; however, if consent is denied or cannot be obtained the anaesthetist should undergo testing as soon as possible. If the anaesthetist initially tests negative for HIV, retesting should be instituted every 6 weeks for a minimum of 6 months. If positive for the virus, then post-exposure prophylaxis should be discussed together with counselling to restrict further transmission. Recently, post-exposure prophylaxis with zidovudine (AZT, Retrovir) has been suggested for health care workers.[70] Zidovudine is a thymidine analogue that inhibits replication of most retroviruses. In patients with advanced AIDS, zidovudine

has been shown to increase both the length and quality of life and delay the progress of the disease in patients with early HIV infections. Results of two studies suggest that the efficacy of post-exposure zidovudine prophylaxis for health care workers is inconclusive; however, many physicians advocate its use. Didanosine (ddl) has recently been licensed to treat patients who may be intolerant or unresponsive to zidovudine. The side effects of this drug are pancreatitis and peripheral neuropathy.

Epidemiological investigation of an infectious outbreak in the operating theatre

Clusters of cases involving infection among surgical patients or operating theatre personnel should be evaluated to determine if an outbreak or epidemic exists. The

Table 27.3 Epidemiological investigation of an infectious outbreak in the operating theatre (OT)

Define the problem
    Establish and verify the diagnosis for reported cases
    Is the appearance of cases unusual?
    Is there an epidemic?

Characteristics of the outbreak
    Person: age, sex, type of surgical procedure, etc.
    Time: time and date of exposure and onset. Construct an
        epidemic curve
    Place: geographic distribution of the cases, e.g. OT, ICU
    Attack rates: numbers of individuals infected, deaths

Preliminary analysis
    Possible source of infection: single exposure? Vehicles and/or
        vectors, i.e. person-to-person, iatrogenic?
    Seek a common denominator and unusual or secondary cases

Formulate a causal hypothesis
    Base hypothesis on available information, e.g. suspected
        aetiological agent, source of infection, mode of transmission
        and periods of exposure

Test the hypothesis
    Are there significant differences in attack rates between those
        exposed and those not exposed?
    Was the source of infection common to all cases?
    Examine the environment of exposure, e.g. anaesthesia/surgical
        equipment, aseptic technique, OT personnel, etc.

Conclusions
    Base conclusions on all the pertinent evidence and not on a
        single circumstance
    Identify the population at risk
    Plan strategies for long-term prevention and control

primary objective of the investigation is to determine a diagnosis and common source and to identify ways to prevent further transmission of the pathogen. Table 27.3 describes the steps that should be followed when investigating a disease outbreak.

In general, successful investigations of a disease outbreak involve good detective work. It is imperative that a differential diagnosis be made initially and a final diagnosis made using accurate clinical and laboratory information. Epidemic curves should be generated to identify cases in terms of times of onset and exposure. Attack rates, i.e. the number of persons exposed to the suspected aetiological agent who become infected, should be calculated to correlate exposure with outcome. If there appears to be a geographical or specific distribution of cases, spot maps should be generated to track them.

Successful verification of this type of hypothesis involves demonstration that: (1) there is a significant difference in the attack rates between individuals exposed to the suspected aetiological agent and those not exposed; and (2) either there was a common source to all infected cases or that no other mode of transmission could account for the person and place distribution of the cases. Failure to confirm the hypothesis usually results from poor execution of the tests or from an incorrect statement of the initial hypothesis. Under these circumstances the hypothesis should be restated and tested again.

Conclusion

Control of infection in the operating theatre provides an interesting challenge to our profession in the future. Over the last two decades, medicine has made striking advances in patient care; however, this in itself has also created new problems in the field of infection control. As medical knowledge and technology progress, anaesthetists will take care of an increasing number of patients who are at some risk for infection. Knowledge of how anaesthetic agents, blood transfusions and surgery affect the immune response to infection will become more important in the care of certain groups of surgical patients. The days of the empirically derived clinical doctrine and the technician approach to the infected patient will not serve if we are to make any impact on improving the quality of care in these patients in both the operating theatre and the intensive care setting.

The anaesthetist must be familiar with the pathogenesis and pathophysiology of a presenting infectious disease and with those that may occur as a complication of anaesthesia and surgery. The anaesthetist must understand the concept of the host–parasite equilibrium, how anaesthesia and surgery may affect this relationship, and which groups of patients are most likely to suffer untoward problems as a result of a shift in equilibrium. Knowledge of these relationships together with sound hygienic practices will better prepare the anaesthetist for the management of these patients both in terms of limiting the potential for transmission of infection to the patient and indeed to themselves.

AIDS, cancer, transplant, debilitated and geriatric patients are some of the more interesting high risk patients to which the anaesthetist will be increasingly exposed. We must be prepared to provide optimal care for these patients in the perioperative period. In order to do this we simply must have sufficient working knowledge in the field of infectious disease and its relation to anaesthesia and surgery. The medical care of the patient in the operating room theatre is our responsibility. Anaesthetists can make a difference in the outcome of these patients both in terms of limiting transmission and modifying the infectious process and its sequelae.

REFERENCES

1. Hall CB. Nosocomial viral respiratory infections: perennial weeds on pediatric wards. *American Journal of Medicine* 1981; **70**: 670–6.
2. Knight PR, Tait AR. Immunological aspects of anesthesia. In: Nunn JF, Utting J, Brown B. *General anaesthesia*. 5th ed. Boston: Butterworth Publishing Co, 1988; 283–93.
3. Stevenson GW, Hall SC, Rudnick S, Seleny FL, Stevenson HC. The effect of anesthetic agents on the human immune response. *Anesthesiology* 1990; **72**: 542–52.
4. Markovic SN, Knight PR, Murasko DM. Inhibition of interferon stimulation of natural killer cell activity in mice anesthetized with halothane or isoflurane. *Anesthesiology* 1993; **78**: 700–6.
5. Nakagawara M, Takeshige K, Takamatsu J, Takahashi S, Yoshitake J, Minakami S. Inhibition of superoxide production and Ca^{2+} mobilization in human neutrophils by halothane, enflurane and isoflurane. *Anesthesiology* 1986; **64**: 4–12.
6. Shayevitz JR, Varani J, Ward PA, Knight PR. Halothane and isoflurane increase pulmonary artery endothelial cell sensitivity to oxidant-mediated injury. *Anesthesiology* 1991; **74**: 1067–77.
7. Waymack JP, Miskell P, Gonce S. Alterations in host defense associated with inhalation anesthesia and blood transfusion. *Anesthesia and Analgesia* 1989; **69**: 163–8.
8. George CD, Morell PJ. Immunologic effects of blood transfusion upon renal transplantation, tumor operations and

bacterial infections. *American Journal of Surgery* 1986; **152**: 329–37.

9. Bedows E, Davidson BA, Knight PR. Effect of halothane on the replication of animal viruses. *Antimicrobial Agents and Chemotherapy* 1984; **25**: 719–24.

10. Johnson BH, Eger EI. Bactericidal effects of anesthetics. *Anesthesia and Analgesia* 1979; **58**: 136–8.

11. Moudgil GC. Influence of halothane on mortality from murine hepatitis (MHV$_3$). *British Journal of Anaesthesia* 1973; **45**: 1236.

12. Penna AM, Johnson KJ, Camilleri J, Knight PR. Alterations in influenza A virus specific immune injury in mice anesthetized with halothane or ketamine. *Intervirology* 1990; **31**: 188–96.

13. Tait AR, Heckenkamp LJ, Knight PR. Influenza virus-induced injury in the mouse: comparison of the effects of halothane, enflurane and isoflurane. *Anesthesiology* 1992; **77**: A376.

14. Eickhoff JC. Pulmonary infections in surgical patients. *Surgery Clinics of North America* 1980; **60**: 175–83.

15. Pennington JE. Nosocomial respiratory infection. In: Mandell GL, Douglas, RG, Bennett JE. *Principles and practice of infectious disease.* 3rd ed. New York: Churchill Livingstone, 1990: 2199–205.

16. Garibaldi RA, Britt MR, Coleman ML, Reading JC, Pace NL. Risk factors for postoperative pneumonia. *American Journal of Medicine* 1981; **70**: 677–80.

17. Craven DE, Kunches LM, Kilinsky V, Lichtenberg DA, Make BJ, McCabe WR. Risk factors for pneumonia and fatality in patients receiving continuous mechanical ventilation. *American Review of Respiratory Disease* 1986; **133**: 792–6.

18. Stevens RM, Teres D, Skillman JJ, Feingold DS. Pneumonia in an intensive care unit. *Archives of Internal Medicine* 1974; **134**: 106–11.

19. Yu VL, Kroboth FJ, Shonnard J, Brown A, McDearman S, Magnussen M. Legionnaire's disease: new clinical perspective from a prospective pneumonia study. *American Journal of Medicine* 1982; **73**: 357–61.

20. Abraham SN, Beachey EH, Simpson WA. Adherence of *Streptococcus pyogenes, Escherichia coli,* and *Pseudomonas aeruginosa* to fibronectin-coated and uncoated epithelial cells. *Infection and Immunity* 1983; **41**: 1261–8.

21. Johanson Jr WG, Pierce AK, Sanford JP, Thomas GD. Nosocomial respiratory infections with gram-negative bacilli. *Annals of Internal Medicine* 1972; **77**: 701–6.

22. Pennington JE. TNF: therapeutic target in patients with sepsis. *American Society of Medicine News* 1992; **58**: 479–82.

23. Murray MJ, Kumur M. Sepsis and septic shock. *Postgraduate Medicine* 1991; **90**: 199–208.

24. Prevention of bacterial endocarditis *Medical Letter* 1984; **26**: 3–4.

25. Daniels CA, LeGoff SG. Shedding of infectious virus/antibody complexes from vesicular lesions of patients with recurrent herpes labialis. *Lancet* 1975; **2**: 524–8.

26. Orkin FK. Herpetic whillow – occupational hazard to the anesthesiologist. *Anesthesiology* 1970; **33**: 671–3.

27. Tait AR. Prevention of HIV and HBV transmission among anesthesiologists: a survey of anesthesiology practice. *Anesthesiology* 1992; **77**: A1083.

28. Dworsky ME, Welch K, Cassady G, Stagno S. Occupational risk for primary cytomegalovirus infection among pediatric caseworkers. *New England Journal of Medicine* 1983; **309**: 950–3.

29. Berry AJ, Greene ES. The risk of needlestick injuries and needlestick-transmitted diseases in the practice of anesthesiology. *Anesthesiology* 1992; **77**: 1007–21.

30. Rizzetto M. The delta agent. *Hepatology* 1983; **3**: 729–37.

31. Centers for Disease Control. Public health service inter-agency guidelines for screening donors of blood, plasma, organs, tissues, and semen for evidence of hepatitis B and hepatitis C. *Morbidity and Mortality Weekly Report* 1991; **40**: 1–17.

32. Hadler SC, Margoli HS. Viral hepatitis. In: Evans AS, ed. *Viral infections of humans.* 3rd ed. New York: Plenum Medical, 1989: 351–91.

33. Centers for Disease Control. The HIV/AIDS epidemic: the first 10 years. *Morbidity and Mortality Weekly Report* 1991; **40**: 357–69.

34. Centers for Disease Control. Surveillance for occupationally acquired HIV infection. *Morbidity and Mortality Weekly Report* 1992; **41**: 823–5.

35. Centers for Disease Control. Recommendations for the prevention of HIV in the health care setting. *Morbidity and Mortality Weekly Report* 1987; **36**: 2S–18S.

36. Centers for Disease Control. Guidelines for the prevention of human immunodeficiency virus and hepatitis B virus to health-care and public safety workers. *Morbidity and Mortality Weekly Report* 1989; **38** (S6): 1–37.

37. Berry AJ, Isaacson IJ, Kane MA, *et al.* A multicenter study of the prevalence of hepatitis B viral serologic markers in anesthesia personnel. *Anesthesia and Analgesia* 1984; **63**: 738–42.

38. McCray E, Martone WJ. Preventing HIV exposure among patients and staff. *AIDS Patient Care* 1987; **Sept**: 32–4.

39. Becker MH, Janz NK, Band J, Bartley J, Snyder MB, Gaynes RP. Noncompliance with universal precautions policy: why do physicians and nurses recap needles? *American Journal of Infection Control* 1990; **18**: 232–9.

40. Courington KR, Patterson SL, Howard RJ. Universal precautions are not universally followed. *Archives of Surgery* 1991; **126**: 93–6.

41. Rosenberg AD, Bernstein D, Skovron ML, Ramanathan S, Turndorf H. Changing practice habits of anesthesiologists: accidental needlesticks. *Anesthesiology* 1992; **77**: A1084.

42. Cohen MM, Cameron CB. Should you cancel the operation when a child has an upper respiratory tract infection? *Anesthesia and Analgesia* 1991; **72**: 282–8.

43. Tait AR, Knight PR. The effects of general anesthesia on upper respiratory tract infections in children. *Anesthesiology* 1987; **67**: 930–5.

44. Empey DW, Laitinen LA, Jacobs L, Gold WM, Nadel JA. Mechanisms of bronchial hyperreactivity in normal subjects after upper respiratory tract infection. *American Review of Respiratory Disease* 1976; **113**: 131–9.

45. Lewis W. AIDS: cardiac findings from 115 autopsies. *Progress in Cardiovascular Disease* 1989; **32**: 207–15.

46. Donovan CA. The influence of diethyl ether inhalation in canine distemper. *Veterinary Medicine, Small Animal Clinics* 1968; **63**: 345–7.

47. Shiltknecht E, Ada GL. In vivo effects of cyclosporine on influenza A virus-infected mice. *Cellular Immunology* 1985; **91**: 227–39.

48. Hughes WT. Fatal infections in childhood leukemia. *American Journal of Diseases of Children* 1971; **122**: 283–7.

49. Levine AS, Graw Jr RG, Young RC. Management of infections in patients with leukemia and lymphoma: current concepts and experimental approaches. *Seminars in Hematology* 1972; **9**: 141–79.

50. Murphy JF, McDonald FD, Dawson M, *et al.* Factors affecting the frequency of infection in renal transplant recipients. *Archives of Internal Medicine* 1976; **136**: 670–7.

51. Lundy J, Lovett EJ, Conran P. Pulmonary metastases, a potential biologic consequence of anesthetic-induced immuno-suppression by thiopental. *Surgery* 1977; **82**: 254–6.

52. Agostino D, Cliffton EE. Anesthetic effect on pulmonary metastases in rats. *Archives of Surgery* 1964; **88**: 735–9.

53. Lundy J, Lovett EJ, Hamilton S, Conran P. Halothane, surgery, immunosuppression, and artificial pulmonary metastases. *Cancer* 1978; **41**: 827–30.

54. Vose BM, Moudgil GC. Effect of surgery on tumour-directed leukocyte respones. *British Medical Journal* 1975; **1**: 56–8.

55. Beilin B, Martin FC, Shavit Y, Gale RP, Liebeskind JC. Suppression of natural killer cell activity by high-dose narcotic anesthesia in rats. *Brain, Behavior, and Immunity* 1989; **3**: 129–37.

56. Browne RA, Chernesky MA. Infectious disease and the anaesthetist. *Canadian Journal of Anaesthesia* 1988; **35**: 655–65.

57. Dodd RY. The risk of transfusion-transmitted infection. *New England Journal of Medicine* 1992; **327**: 419–21.

58. Maz S, Lyons G. Needlestick injuries in anaesthetists. *Anaesthesia* 1990; **45**: 677–8.

59. Hayden FG, Douglas RG. Antiviral agents. In: Mandell GL, Douglas RG, Bennett JE, eds. *Principles and practice of infectious diseases.* 2nd ed. New York: John Wiley & Sons, 1979; 270–86.

60. du Moulin GC, Hedley-Whyte J. Bacterial interactions between anesthesiologists, their patients, and equipment. *Anesthesiology* 1982; **57**: 37–41.

61. Feeley TW, Hamilton WK, Xavier B, Moyers J, Eger EI. Sterile anesthesia breathing circuits do not prevent postoperative pulmonary infection. *Anesthesiology* 1981; **54**: 369–72.

62. Dryden GE, Brickler J. Stopcock contamination. *Anesthesia and Analgesia* 1979; **58**: 141–2.

63. Merritt WT, Zuckerberg AL. Contamination of the anesthetic record. *Anesthesiology* 1992; **77**: A1102.

64. Lutz CT, Bell CE, Wedner HJ, Krogstad DJ. Allergy testing of multiple patients should no longer be performed with a common syringe. *New England Journal of Medicine* 1984; **310**: 1335–7.

65. Froggatt JW, Dwyer DM, Stephens MA. Hospital outbreak of hepatitis B in patients undergoing electroconvulsive therapy. *Interscience Conference on Antimicrobial Agents and Chemotherapy* 1991; **347**: 157.

66. Centers for Disease Control. Postsurgical infections associated with an extrinsically contaminated intravenous anesthetic agent – California, Illinois, Maine, and Michigan. *Morbidity and Mortality Weekly Report* 1990; **39**: 426–33.

67. Alter MJ, Ahtone J, Maynard JE. Hepatitis B virus transmission associated with a multiple-dose vial in hemodialysis unit. *Annals of Internal Medicine* 1983; **99**: 330–3.

68. Bawden JC, Jacobsen JA, Jackson JC, *et al.* Sterility and use patterns of multiple-dose vials. *American Journal of Hospital Pharmacology* 1982; **39**: 294–7.

69. Gwaltney JM, Moskalski PB, Hendley JO. Hand-to-hand transmission of rhinovirus colds. *Annals of Internal Medicine* 1978; **88**: 463–7.

70. Centers for Disease Control. Public health service statement on management of occupational exposure to human immuno-deficiency virus, including considerations regarding zido-vudine post-exposure use. *Morbidity and Mortality Weekly Report* 1990; **39**: 1–4.

Chapter 28

Anaesthesia and the Immune System

A. Eleri Edwards and C. J. Smith

Despite the fact that the complexities of the immune system are only poorly understood, clinicians have long accepted the concept that the immune system is primarily a protective mechanism against infection. George Bernard Shaw stated in the *Doctor's Dilemma* (1906): 'There is at bottom only one genuinely scientific treatment of all diseases and that is to stimulate the phagocytes.' He also wrote that: 'Chloroform has done a lot of mischief. It has enabled every fool to become a surgeon.' Whilst neither statement is strictly true, their appearance in a popular play at the turn of the century reflects the public interest in both immunology and anaesthesia at that time.

Undoubtedly, developments in anaesthesia have resulted in a growing number of older and more ill patients undergoing surgery of increasing complexity whilst developments in intensive care have ensured the survival of patients who previously would have died. Developments in the management of cardiorespiratory, renal, hepatic and metabolic function have revealed an increasing need to re-examine critically the effects of clinical anaesthesia and intensive therapy on aspects of the immune system and to re-evaluate the relations between trauma, stress and anaesthesia, particularly with regard to their implications for sepsis and malignancy.[1, 2]

It is inappropriate in this chapter to examine the total structure and function of the immune system and for this, readers are referred to standard textbooks of immunology.[3–8] However, in considering the interface between anaesthesia and immunology, it is important to recognize that these are two complex areas which are as yet only poorly understood. The effects of anaesthesia have been mainly studied in terms of the physiological changes in patients to whom anaesthetic and adjuvant agents have been administered. The detailed comprehension of molecular events responsible for the anaesthetic state is only a recent development. At the molecular or even cellular level, the influence of anaesthesia and individual anaesthetic agents is still without a clear description. The immune system is essentially a system which functions at the molecular and cellular level with a multiplicity of cell types, cell functions, molecular interactions dependent on intercellular signalling, which is specifically organized to interact with, or react to, extraneous materials, chemicals or cells, which invade the patient. These substances include anaesthetic agents. Thus it is necessary to have an understanding of immunological mechanisms against which the process of administering anaesthesia can be set.

Simplistically, the role of the immune system in combating infection is to recognize organisms and other materials which are 'non-self'. This mobilizes the body's defences and develops memory which results in immunity against further episodes of the same infection (acquired immunity). A more difficult concept is that of autoimmunity when the body fails to recognize 'self' and mobilizes an immune response against its own tissues, e.g. Graves' disease, Hashimoto's disease and systemic lupus erythematosus (SLE). In this context, it is important to appreciate that the immune system recognizes 'self' as a result of the presence of cell surface markers. In fact, the opposite view is probably more accurate; in order to recognize 'non-self', T-lymphocytes must identify an antigen plus major histocompatibility complex class I molecules (MHC-I) on the cell's surface (MHC restriction). During maturation of T-lymphocytes in the thymus, those which are capable of recognizing 'self' antigens are eliminated. Thus, these cells do not enter the system, i.e. a censoring mechanism destroys these cells, and so 'self' is, for practical purposes, ignored.

A secondary but major function of the immune system is to scavenge ageing and damaged cells[9] and whilst the immune system has self-tolerance to cell surface molecules, intracellular molecules may still be potential antigens and as such can, if released, elicit an immune response. This is demonstrated in patients with certain autoimmune diseases, e.g. SLE, mixed connective tissue disease (MCTD) and rheumatoid arthritis, who develop antinuclear antibodies. Moreover, this self-tolerance may be significantly altered by disease or trauma. The lowering of immunity

1 Involvement in autoimmunity

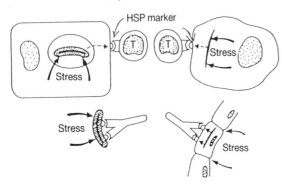

2 Involvement in tumour rejection

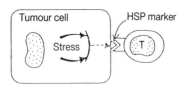

3 Immune response to pathogens

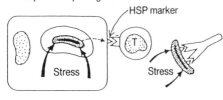

4 Assembly of immunologically important molecules

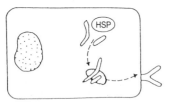

Fig. 28.1 Roles of chaperonins in cellular functions.
1–3: HSP (heat shock protein) is expressed on the surface of 'stressed' cells allowing interaction with T-lymphocytes and consequently antibody-mediated action against the cause of the stress or the stressed cell itself.
4:     HSP assists in the assembly of antibody mollecules by maintaining peptide interactions.

that accompanies surgery, burns and other stress is considered to be part of a protective mechanism which preserves slightly damaged cells from being rejected and scavenged.[1,2] There are indications that strenuous exercise may decrease the overall response of the immune system, possibly a self-protective mechanism to prevent rejection of transiently stressed cells. It has been observed that athletes under intensive training appear to be prone to infection.[10]

Thus, stimulation of the mechanism that protects the body from external invaders (for example, bacteria) may increase tissue damage following injury and conversely, natural depression of the immune system following injury, as a protective mechanism, may increase the vulnerability of the body to invasion by infective agents.

It has recently been suggested that a possible explanation for these apparently conflicting roles for the immune system may be the result of the common evolutionary development of all cells and organisms – phylogenesis. At a cellular level, the response to a variety of stressful stimuli such as increased temperature is characterized by an increased synthesis of specific stress proteins, known as 'heat shock proteins' or 'chaperonins'.[11–19] These appear to have an important protective function in protecting proteins responsible for cellular function (Fig. 28.1). Heat shock proteins are present in all known eukaryotic cells (cells with nuclei), including bacteria.[20] Human immunological memory for heat shock proteins is probably established early in life at the time when gut and skin flora are established and then boosted by repeated exposure to minor infections throughout life.

In critically ill patients, hypoperfusion leads to the expression of heat shock proteins on cell surfaces whilst at the same time hypoperfusion at mucosal surfaces, for example, gastrointestinal and respiratory, predisposes to increased uptake of bacteria which will themselves express heat shock proteins.[14] In bacteria, the heat shock proteins are a natural surface marker which may be used by the immune system as a universal recognition signal, whereas in humans the heat shock proteins are normally intracellular and not available to the immune system.[14] However, these proteins have a high degree of structural homology such that antibodies to bacterial heat shock proteins will also react with human heat shock proteins. Thus, the same mechanisms can be employed by the immune system to remove bacteria and stressed (damaged, virally infected, abnormal) hosts cells.[14,16,18,21,22] This can be a disadvantage in that once activated by invading bacteria, the mechanism may also remove slightly stressed cells. Conversely, depression of the immune system following trauma, whilst advantageous in conserving damaged cells, may result in an increased susceptibility to infection. Much experimental work is directed to explain the clinical significance of these findings.[23,24]

This brief and, of necessity, selective introduction identifies the problem for anaesthetists, not only in considering how their activities affect the immune system, but also in identifying those clinical situations where stimulation may be beneficial or conversely where depression would be advantageous.

Further, the anaesthetist needs to be aware of the immune system as a functional, integrated system and to give it the same consideration as is given to any other vital system. As stated, its prime function is defence and it is organized on multiple levels to resist the incursions of 'foreign' organisms and materials, to isolate and destroy those that penetrate the outer defences, to counteract massively those that successfully invade the body, and to mount continual surveillance, removing both 'foreign' and unwanted host materials and cells.

The immune system can be divided into two parts – the innate and the adaptive – which combine to prevent the invasion of foreign organisms. The innate system includes the skin, body hair, mucous membranes and mucus, respiratory tract cilia, gastric acid and serum opsonins. Inhalational and intravenous anaesthetics and analgesics depress ciliary activity and the cough reflex.[25] The innate defences are easily circumvented during anaesthesia. Tracheal and tracheotomy tubes bypass the barriers represented by the nasal mucosa, the lysozymal activity of the saliva, the cough reflex, the upper respiratory tract ciliary activity, and the specific activity of secreted IgA in binding foreign antigens. Also, direct pressure on local tissues may compromise cell viability and lymph flow. All invasive procedures such as central lines and urinary catheters are also direct portals of entry. In addition, it must be remembered that the lines themselves may initiate an immune response as may their placement, e.g. latex rubber.[26]

The adaptive immune system consists of those parts of the immune system which can respond (adapt) to invasion. This includes both the cellular (responses mounted by cells such as natural killer cells and cytotoxic T-cells) and humoral (antibody production) arms of the systems. In its entirety, the immune system presents a comprehensive defence; however, the experimental and scientific data have led to its division into separate sections, interactions which often fail to be recognized. Only in recent years have the consequences of divorcing the activities of the subunits of this complex and interactive system been recognized. An example of this is the importance of an integral, functioning gastrointestinal tract in the innate immune system and as part of the adaptive system.

The essential role of the gastrointestinal associated lymphoid tissues (GALT) and the mucosal associated lymphoid tissues (MALT) in the immune process have probably been underestimated. The MALT and GALT comprise accumulations of subepithelial lymphoid tissue guarding the respiratory, alimentary and genitourinary tracts. These accumulations occur as diffuse collections of lymphocytes, plasma cells and phagocytes throughout the lamina propria of the intestinal wall, or as more clearly organized tissues with well formed follicles. These latter include the lingual, palatine and pharyngeal tonsils, the small intestinal Peyer's patches and the appendix. These form a separate interconnected system within which cells committed to IgA or IgE synthesis circulate. Thus, the gut

can mount a local immune response independent of a systemic response; for example, immunization against poliomyelitis with oral live poliomyelitis vaccine (Sabin Vaccine) gives better protection than the injected inactivated poliomyelitis vaccine (Salk Vaccine) although both induce serum IgG and IgA antibodies. The local IgA produced in response to the oral vaccine partly blocks the uptake of pathogenic bacteria into the circulation. Splanchnic ischaemia compromises mucosal and subepithelial lymphoid function, decreasing the effectiveness of this protective barrier, which may be of fundamental importance in the development of sepsis.[27]

The adaptive immune response is centred upon the lymphatic system, the basic organization of which is to protect the central core from invasion. However, the scale of activity of the lymphatic system is not immediately obvious. Only 3 per cent of neutrophils are found in the circulating blood pool and another 3 per cent in the marginal pool where they remain for approximately 10 hours before entering the tissues. Similarly, only 2 per cent of the total lymphocyte pool is to be found in the blood circulation of an adult. However, massive numbers of white cells are known to transit through the peripheral blood en route to and from tissues, for example 5×10^{11} lymphocytes travel through the blood each day with a mean transit time of 30 minutes and the daily output of neutrophils from the bone marrow into the circulation is 10^{11}.[1] Hence, the major proportion of the leucocyte population is in the lymphatic tissues and ducts or in transit through tissues to the lymphatic system.

Lymphoid tissues and nodes are located in such a way that invaders and antigens are screened rapidly and an immune response mounted at the earliest opportunity with immune complexes being processed within the lymphatic system. Impairment of lymphatic function and flow can have significant clinical consequences.

Animal experiments have demonstrated that increased intrathoracic pressure reduces thoracic lymph flow[28,29] and high positive end-expiratory airway pressure during controlled ventilation depresses bacterial clearance from the peritoneal cavity.[30] Interference with lymph flow through lymph ducts and glands compromises the normal process by which antigens are transported to lymph nodes for presentation to B-lymphocytes, initiating antibody production. The further consequences of this will be to allow invading bacteria to reproduce and potentially leads to septicaemia. Maintaining spontaneous respiration, when possible, and hence normal thoracic and abdominal lymph flow in both the respiratory and gastrointestinal tract may be of more than theoretical importance particularly in intensive care patients, and is therefore worthy of further study.

A further consequence of the impairment of lymphatic function and flow will be increased antigen concentrations entering the bloodstream, as opposed to the lymphatic system, resulting in the formation of immune complexes confined to the vascular system. The removal of such complexes can be by several routes. Macrophages may ingest complexes and then present fragments as antigenic stimuli for T-lymphocytes which will then activate the B-lymphocytes to produce antibodies. Extracellular enzymes released by leucocytes can degrade complexes. Erythrocytes can bind soluble complexes to their surface and transport them to the spleen where they are stripped and the complexes destroyed. The first two of these processes depend on recirculation of the white blood cells through the lymphatic system. The third process depends only on a functional peripheral blood circulation and spleen. Damage or bypass of the lymphatic system will impair the first two processes, resulting in higher concentrations of immune complexes within the circulatory system. The consequences of such high levels are deposition of insoluble complexes within tissues. This deposition occurs principally within capillary beds. Complexes delivered from the venous system will be deposited primarily in the lungs and this may be one explanation for the common lung complications arising after trauma, sepsis, transfusions etc., all of which may contribute to the adult respiratory distress syndrome (ARDS).[31] Similarly, deposition in the kidneys may contribute to renal failure, the deposition of complexes being exacerbated by a diminution of blood flow in the capillaries. In other tissues, particularly the liver, there is the independent capability to destroy immune complexes by enzymatic action and, additionally, the liver is guarded by fixed macrophages (Kupffer cells) which can remove immune complexes.

Clinical use of the intravenous route by clinicians for the administration of foreign material, for example, drugs, complex fluids, and nutrients, inevitably bypasses the normal initial immune processes which may, therefore, give rise to the direct formation of immune complexes in the circulation at concentrations which can overwhelm the removal processes and cause deposition in tissues with consequent damage. Anaesthetic practice involves the intravenous administration of numerous drugs. It has been speculated that whilst individual agents may not stimulate an immune response, or themselves be antigenic, combinations of drugs given in rapid succession may act synergistically to produce significant effects in the immune system. It is postulated that the injection of different drugs in rapid response during anaesthesia may generate colloidal precipitates leading to deposition of colloid in the microvasculature of the lungs, initiating a local inflammatory response, and may give rise to aggregate anaphylaxis.[32]

Anaesthetic agents

As early as 1875 Bernhard[33] stated:

> Let us remember that chloroform does not act solely on the nerve tissues, far from that it has actions on all the tissues and attacks each one at a time, which is a function of their susceptibility. An anaesthetic is not a special poison for the nervous system, it anaesthetises all cells, benumbing all tissues and stopping their irritability.

Paradoxically, while trying to protect the patient from the effects of stress and trauma, the anaesthetist deliberately chooses to use drugs that interfere with the activities of a variety of chemical messengers which mediate the immune response. In attempting to control the nervous, respiratory and circulatory systems, the anaesthetist's many drugs must be assumed to have the potential of blocking, modifying, changing and interacting with both the chemical messengers and cell receptors of the immune system. Although the exact mode of action of anaesthetic agents is unknown, there is growing evidence that there is cell membrane receptor involvement.[34,35]

There are few studies of the effects of prolonged human exposure to anaesthetic agents for purely experimental purposes, as this is not ethically acceptable and animal findings cannot be invariably extrapolated to assess the effects of anaesthesia on human immune function. Thus, most studies have been undertaken either by measuring the effects of *in vitro* anaesthetic exposure of lymphocytes and leucocytes isolated from the blood of normal volunteers, or by examination of the distribution and function of lymphocytes and leucocytes isolated from the blood of patients undergoing dissimilar anaesthetic and surgical procedures. In such studies, it must be recognized that the cells in the blood samples may not be representative of the cells within the whole body and the function and numbers of peripheral blood lymphocytes and leucocytes may not reflect the activity and capacity of the lymphocytes and leucocytes within the lymphatic system. Additionally, removal of the cells from the body isolates these cells from the normal immunoregulatory system of the body and the removal process itself may affect the responsiveness of the cells. For example, separation of lymphocytes from whole blood usually involves the use of a density gradient and centrifugation; cells passing through the gradient may become coated with polymer which would interfere with the efficiency of cell surface receptors and markers. In addition, blood concentrations of anaesthetic agents obtained during clinical anaesthesia may be considerably greater than the concentrations in the tissues and lymphoid organs where many of the cellular responses to antigenic stimuli occur.

Nevertheless, despite the difficulties in testing, there is an extensive literature, related to the influence of individual drugs used in anaesthetic practice, on aspects of the immune system.[36] Numerous *in vitro* tests have been employed including phagocytosis, migration, neutrophil polarization studies and tests which depend on the production of excited oxygen radicals. The value of those studies based on lectin stimulation tests is limited as lymphocyte responses which are reproducible in controlled *in vitro* studies have proved unreliable as indicators of responses in human clinical studies[37] and commercial phytohaemagglutinin (PHA) preparations are known to contain different mitogenic components.[38] Conclusions from these studies should therefore be viewed with caution. However, valuable information has been obtained dealing with the dose–response effects of differing anaesthetic agents on the functions of isolated leucocytes, and most studies demonstrate that increasing concentrations of inhalation, intravenous and local anaesthetics result in dose-dependent inhibition of leucocyte activity.

Moudgil[39] has shown that thiopentone, ketamine, pethidine, morphine, chlorpromazine, diazepam and local anaesthetics significantly depress phagocytic activity but that in general drugs that did not depress the central nervous system did not directly depress phagocytic activity, e.g. muscle relaxants, prednisolone, aspirin. Studies by Smith and Edwards[40] demonstrated significant depression of random leucocyte and lymphocyte migration by thiopentone at serum levels normally achieved following induction of anaesthesia. Depression by morphine only occurred at concentrations 1000-fold greater than normal serum analgesic levels. However, such levels are observed in opioid addicts in whom there is evidence of immunosuppression and increased susceptibility to infections.[41] These effects are reversible and time dependent. Lorazepam, pancuronium and atropine resulted in no depression at levels 1000-fold above normal clinical serum levels.[40] It is interesting to note that adrenaline and hydrocortisone had no effect on random migration of leucocytes and lymphocytes at clinical concentrations *in vitro*[40] despite demonstration of enhanced leucocyte and lymphocyte migration in patients receiving 'light balanced' anaesthesia associated with an increase in plasma adrenaline.[42] This was not observed in surgical patients following epidural anaesthesia and in whom there was no increase in plasma adrenaline. Thus it appears that adrenaline may indirectly affect leucocyte and lymphocyte responses *in vivo*.

This finding that adrenaline is associated with enhanced function is of significance and will be discussed later with respect to the effects of trauma and stress. The presence of adrenergic receptors on peripheral blood lymphocytes is also of significance as in the *in vitro* study, the presence of adrenaline at 1 in 200 000 lowers the concentrations at which lignocaine and bupivacaine affect cell activity.[40] Rarely are combinations of drugs tested during *in vitro*

studies although this might better approximate to the clinical situation. Lignocaine and bupivacaine have been demonstrated to be toxic to both leucocytes and lymphocytes at concentrations used in wound infiltration, concentrations which are bactericidal.[43,44]

Increasing use of intravenous sedation in intensive therapy together with the development of novel tests in leucocyte function continue to stimulate further studies. O'Donnell *et al.*[45] demonstrated that thiopentone and propofol produced significant (approximately 50 per cent) inhibition of neutrophil polarization at clinically relevant concentrations. In contrast, they demonstrated that intralipid 10 per cent augmented polarization and reduced the degree of inhibition produced by propofol. However, Jensen *et al.*[46] concluded that propofol at clinically relevant concentrations may adversely affect leucocyte locomotion *in vitro*. In their studies, propofol, emulsified propofol (Diprivan) and intralipid 10 per cent all produced significant depression. Studies by Weiss *et al.*[47] measured production of oxygen radicals by polymorphonuclear cells and demonstrated significant depression by both diazepam and midazolam at clinically relevant concentrations. Undoubtedly, further *in vitro* and *in vivo* studies will help to clarify the situation. However, the inhibitory effect of etomidate on adrenal steroid genesis by inhibition of P_{450}-mediated hydroxylations was only demonstrated after its introduction in intensive care was noted to be associated with an increased mortality.[48,49] Its subsequent withdrawal from use for intensive care sedation demonstrates the crucial role of careful clinical audit.

In vitro studies of inhalation agents by Moudgil *et al.*[50] demonstrate that both halothane and nitrous oxide significantly depress both leucocyte and lymphocyte migration at clinical concentrations whereas isoflurane and enflurane have little effect. In contrast, Nunn[51,52] reported that while halothane produces reversible inhibition of lymphocyte activity, it has no effect on neutrophil chemotaxis. *In vitro* studies by Lippa *et al.*[53] demonstrated significant depression of human granulocyte chemiluminescence by both halothane and ethrane whereas Welch and Zaccari[54] found that while enflurane at clinical concentrations significantly inhibits chemiluminescence, isoflurane does not. Erskine and James demonstrated that isoflurane, but not halothane, stimulates neutrophil chemotaxis.[55] The fact that isoflurane, a chemical isomer of enflurane, produces responses which are at variance with those of enflurane demonstrates the sophistication of the recognition mechanisms of the immune system. There are many examples of this type of structural isomerism. The classic experiments demonstrating the sensitivity of antibody recognition to chemical structure were carried out by Landsteiner and van der Scheer.[56] They demonstrated that antibodies raised to *meta*-aminobenzene sulphonate were less reactive to the *ortho* and *para* isomers and to arsonate and carboxylate derivatives of aminobenzene. Therefore, it must always be remembered that these systems are highly agent specific and can differentiate between chemical isomers and the possibility remains that an isomer which is an anaesthetic without being an immunotoxin may yet be identified.

Studies on the effects of anaesthetic agents on the immune function of animals were undertaken as long ago as 1904 when Rubin[57] demonstrated that anaesthetic inhalation agents, i.e. ether and chloroform, decreased the resistance of rabbits to infection. Most studies in animals have shown some depressant effects on the immune response. However, it is noted that homoeostatic control in animals is often difficult to achieve and results cannot be extrapolated readily to humans. Dose–responses in animals and in humans bear little relation to one another.

There have been few studies in healthy volunteers but in one such study Doenicke et al.[58] noted marked depression of PHA stimulated lymphocyte mitosis when healthy volunteers were exposed to halothane for 3 hours. Unfortunately, these volunteers also received thiopentone, pancuronium and nitrous oxide which logically precludes the clear implication of any one of these agents as the sole depressing agent. Furthermore, the assay system, mitogen stimulation of human lymphocytes, is one whose applicability has been questioned. The validity of the test in comparative studies depends on the assumption that individual samples from different patients will all respond in the same way and to the same extent to what is in reality a poorly defined and poorly understood stimulus.[37]

However, not all studies in humans have produced ill-defined or inconclusive results. The observation that patients with tetanus who were exposed to nitrous oxide anaesthesia for several days suffered from depression of bone marrow activity led to further studies which demonstrated that nitrous oxide specifically interferes with vitamin B_{12} metabolism.[59] This depressive effect is normally first observed 8–12 hours after exposure but in patients under intensive care the effect is evident after only 2 hours. No adverse immunological responses were reported after the administration of isoflurane for 34 days in a patient with tetanus[60] and studies on its use in ventilated patients requiring intensive therapy[61] demonstrated a more satisfactory degree of sedation and catecholamine response than with midazolam over a 24-hour period. These findings along with reports of stimulation of neutrophil function[55] suggest that isoflurane is worthy of further study for long-term use in intensive therapy. However, potentially toxic concentrations of serum inorganic fluoride were observed in the patient with tetanus although no clinical effect on renal function was found.[60]

Following 35 years of controversy over the existence of 'halothane hepatitis' it has recently been recognized that halothane is metabolized oxidatively to give the trifluoro-acetyl halide (TFH), a highly reactive compound, which is incorporated onto the surface of hepatocytes, thus altering the microsomal liver protein. This altered microsomal protein is recognized as non-self (i.e. a neo-antigen) and in susceptible patients the ensuing immune response may lead to fulminating hepatitis. A review by Elliott and Strunin[62] considers the implications for the continuing use of halothane, enflurane, isoflurane and desflurane.

Having considered the experimental evidence, there is little to suggest that individual anaesthetic agents significantly affect the immune response at concentrations and exposure times obtained during routine surgical operations. The body is able to correct even severe alterations in immune response provided vital functions are maintained and the underlying disease treated.[1] However, in intensive care, elucidation of the effects on the immune system of long-term sedation awaits further studies.

Intravenous fluids and nutrition

Anaesthetists increasingly use complex intravenous solutions and there is growing interest in their immunological effects. While crystalloid solutions have not been demonstrated to affect the immune response, dextrans, gelatin and hydroxethyl starch have all been shown to produce adverse reactions. Following infusion of gelatin in healthy volunteers, Brodin and Hesseluik demonstrated a significant decrease in fibronectin concentration after 48–72 hours.[63] Hydroxethyl starch, with its high molecular weight and long half-life in vivo, has been shown to block the reticuloendothelial system reversibly – 30 per cent of the infused material is taken up by the reticuloendothelial system, mostly in the liver and spleen, and final elimination is very slow. Low concentrations are still apparent in the plasma 4 months after infusion.[64]

Severe malnutrition depresses immune responses which, in some patients, can be reversed by nutritional support.[65] Studies in which the effectiveness of enteral nutrition has been compared with total parenteral nutrition (TPN) have shown a higher incidence of septic complications in patients receiving TPN.[66] This may be due to the need to use central venous cannulae but intravenous lipid emulsions have been shown to be highly immunosuppressive. Immunoglobulin synthesis is inhibited, there is suppression of the entire reticuloendothelial system particularly neutrophil function and a decrease in the T-helper/T-suppressor ratio.[67] Lipid free parenteral nutrition in contrast with standard parenteral nutrition given preoperatively has been associated with improved outcomes.[68] In neonates given intralipid, fat accumulation in the pulmonary capillaries has been reported; this has been associated with a sixfold increase in the incidence of staphylococcal bacteraemia.[69] Reports of decreased perfusion of the gut, muscle, skin and other organs following intravenous intralipid combined with reports of serum agglutination of intralipid in acutely ill patients indicates a need to reassess its clinical use.[70] Furthermore, these considerations also apply to the use of propofol in intralipid especially for long-term sedation in critically ill patients in intensive care.[71]

In addition, total parenteral nutrition leads to considerable mucosal atrophy by starvation of the gut mucosa. Enteral feeding has been shown to be important in maintaining normal gut integrity and blood flow. Luminal nutrients are an essential source of nutrients for intestinal cells and for maintaining intestinal secretory IgA levels, thus limiting bacterial translocation. The gut is one of the most important structures in the immune defence system – with its local immune tissues, it acts as a barrier between the intestinal contents and the blood circulation with the cells of the GALT censoring all material crossing the barrier.[72] Traditionally, following surgery, surgeons have awaited the return of bowel sounds or the passage of flatus before commencing enteral feeding. However, 'normal post-op ileus' has been shown to affect only stomach and colon; small intestinal motility and function are maintained and early feeding via a nasojejunal tube or a feeding jejunostomy is recommended.[73] Further, hepatobiliary abnormalities have also been reported following TPN.[74]

The use of selective decontamination of the digestive tract (SDD), by administering non-absorbable antibiotics which are confined to the gastrointestinal tract, has not been proved to improve survival significantly in patients receiving intensive care although the incidence of respiratory tract infections was reduced by 60 per cent.[75] In contrast, the use of live yoghurt to recolonize the gastrointestinal tract in patients who have received extensive antibiotic therapy does appear advantageous. Interestingly, it should be recognized that certain amino acids, particularly glutamine and arginine, are important substrates for the maintenance of gastrointestinal integrity and function. Further, these amino acids also play an important role in maintaining leucocyte activity and function.[76,77]

Meta-analysis of clinical studies[78] showed little if any clinically important benefit in surgical patients having received TPN. The incidence of septic complications is substantially raised and only in severely malnourished patients did the benefits outweigh the risks. This is further considered in Chapter 43.

Blood transfusion

Blood transfusion is the commonest form of transplant; whole blood transfusion results in the transfer of immunocompetent cells and is a significant stimulus to the recipient's immune system. There is growing evidence that perioperative blood transfusion in surgical patients has an immunosuppressive effect and is associated with an increased risk of cancer recurrence and decreased patient survival. There is an increased susceptibility to postoperative infection but renal graft survival is enhanced following transplantation.[79]

Burrows and Tartter[80] first reported a positive correlation between blood transfusion and the recurrence of colorectal carcinoma in 1982. Their report stimulated further studies and a similar correlation was observed in patients with malignancies in other organs, e.g. breast[81] and lung.[82] Due to difficulties in standardizing clinical studies with regard to the status of the patient, the extent and nature of the malignancy and the indications used for blood transfusion, conclusions based on retrospective studies have been viewed with caution and results are awaited from prospective randomized studies on the effects of homologous blood transfusion on the recurrence of cancer. However, meta-analysis of results presented in studies between 1982 and 1990 supports the hypothesis that perioperative blood transfusion is associated with an increased risk of recurrence of colorectal carcinoma and death from this malignancy.[83] The estimated cumulative odds ratio of a negative outcome in the presence of transfusion was 1:6.9; this represents an increase in the odds of suffering a negative outcome for transfused patients of 69 per cent. The calculated odds ratio for recurrence was 80 per cent. In addition, clinical studies have demonstrated an increased risk of postoperative infection in patients following blood transfusion for surgery for colorectal cancer, for Crohn's disease and following intestinal injury. It is postulated that immunosuppression may be a factor in the disproportionately higher risk of developing AIDS following blood transfusion.[84] The finding in the 1970s that renal allograft survival was improved in patients who had previously received a blood transfusion[85] altered transfusion practice in transplant surgery, even though there was recognition that some patients might form lymphocytotoxic antibodies which might eliminate them as recipients. The introduction of cyclosporine A is now further altering practice by improving survival in non-transfused patients and eliminating the problem of sensitizing patients.[86]

Experiments using animals have produced conflicting results when the effects of transfusion on tumour growth have been studied. However, Marquet et al.[87] and Singh et al.[88] have demonstrated that tumour growth increased in animals receiving allogeneic blood compared with animals receiving syngeneic blood and Singh et al.[89] also demonstrated that the growth of lung metastases was enhanced by allogeneic whole blood, red blood cells and leucocytes but not plasma. Increased mortality from wound burn sepsis has been shown in rats receiving allogeneic blood compared with rats receiving syngeneic blood or saline. Those receiving allogeneic blood showed decreased macrophage migration in response to inflammatory stimuli and a 150 per cent increased rate of production of prostaglandin E, which is known to be immunosuppressive, and an 80 per cent increase in the level of corticosterone.[90] Other clinical studies have shown a decrease in cell-mediated immunity (CMI), decreased helper/suppressor T-cell ratios, reduced natural killer (NK) cell activity and decreased generation of interleukin-2 (IL-2) by splenocytes.[91]

Thus, a strong association, probably a causal relation, exists between perioperative blood transfusion and a negative outcome in patients following surgery for malignant disease and, in addition, there is an associated increase in infection. Changing surgical and anaesthetic practice to minimize blood loss and hence the use of homologous blood is recommended. There is increasing interest in the use of predeposited autologous blood, the use of preoperative normovolaemic haemodilution and perioperative blood salvage. When homologous blood transfusion is unavoidable, leucocyte-free red blood cell concentrates would appear to be the method of choice to minimize immunological complications, but further studies are awaited before this is recommended for routine use.[92] Thus, on clinical grounds homologous blood transfusion, however closely cross-matched, should probably be confined to those patients in whom it is considered that additional red cells are of critical importance for tissue oxygenation.

When considering transfusion it is important to note in making such a decision that experimental studies have shown that neutrophil activity is diminished in poorly perfused wounds and that granulocytes lose their killing capacity at $PO_2 < 2.6$ kPa (19.5 mmHg). In normal circumstances the PO_2 in a cleanly incised and closed wound is 4.7–6.0 kPa (35.2–45 mmHg) whereas this decreases significantly in dead space wounds.[93] SAG-M red blood cell concentrates, i.e. red blood cells in a solution of saline, adenine, glucose and mannitol, is recommended where tissue hypoxia due to inadequate blood oxygen transport might otherwise occur. The reported survival of a patient whose haemoglobin fell to 14 g/1 following emergency caesarean section, who would not consent to the use of blood products is worthy of note.[94]

A review of the role of recombinant growth factors in transfusion medicine[95] describes the history of the recognition, purification and identification of erythropoietin and its subsequent cloning, expression and characterization as a recombinant product. This occurred over a relatively brief period (1953–1985) while its introduction into clinical work has been even more rapid. Clinical studies are being undertaken in patients preoperatively to correct anaemia and for Jehovah's witnesses underoing cardiac surgery. Trials are being undertaken to assess its effectiveness in patients during the procurement of autologous blood. In addition, clinical trials are being conducted on colony stimulating factors (CSF) – glycoprotein hormones that regulate the production and function of myeloid blood cells, e.g. granulocyte-macrophage colony stimulating factor (GM-CSF) and IL-3. The use of these haematopoietic growth factors may be a further method of reducing requirements for extraneous red blood cells and platelets.

Trauma, stress and sepsis

The interrelations between trauma, stress and sepsis and the immune response are an area of considerable interest to the anaesthetist. A review by Salo[1] states:

> Basically the immune response to anaesthesia and surgery is a beneficial reaction, needed in local defence and wound healing, and in preventing the body from making autoantibodies against its own tissues. The response may, however, contribute to the development of post-operative infection and spread of malignant disease.

As stated previously, possible explanations lie in the dual and at times conflicting roles of the immune system to guard against infection and to scavenge damaged and stressed host cells. The concept that the immune system may be divided into two major complementary branches, one specialized in distinguishing 'self' from 'non-self', and the other 'stressed' from 'non-stressed', is now being considered and the role of γδ T-lymphocytes bearing receptors which recognize stress proteins is being studied.[96–100]

Clinically, the metabolic effects of surgery, including adrenocortical stimulation, hyperglycaemia, acute phase protein synthesis and an increased leucocyte count, have been considered to be a response to a combination of somatic and autonomic afferent neuronal stimuli and the release of cytokines from damaged tissues with anaesthesia as a compounding factor. In attempting to establish the significance of these factors, it is noted that complete afferent neuronal blockade, achieved by regional anaesthesia and analgesia for surgery of the limbs and pelvic organs, leads to suppression of the classical hormonal response to injury, while the acute phase response is unaffected.[101] Specific receptors for many neuroendocrine factors have been identified on lymphocytes, monocytes and other immunocompetent cells,[2] including receptors for catecholamines, acetylcholine and endogenous opioids (for example, endorphins). Further, activated lymphocytes and monocytes are capable of synthesizing and secreting the same hormones and neuroendocrine messengers for which they bear receptors. Thus whilst a relation between the effects of differing types of anaesthesia on the autonomic and hormonal response to injury can be demonstrated, effects on the immune response are still unclear.

Whilst there is a general assumption that suppression of the autonomic response is advantageous, studies by Edwards et al.[42] demonstrated that in standardized clinical conditions there was an increase in leucocyte and lymphocyte migration associated with an increase in adrenaline production in patients receiving a light balanced general anaesthesia as opposed to those receiving an epidural technique or a combination of both techniques. This study questioned the assumption that 'stress-free' anaesthesia is necessarily advantageous particularly in relation to the immune response. Tissue trauma induces cell migration to the site of injury, a process which is not inhibited and in fact may be enhanced by light balanced general anaesthesia. A study of patients undergoing thyroid surgery under acupuncture anaesthesia demonstrated significant increases in the levels of catecholamines

associated with an increase in β-endorphin levels and in many ways mimicked the effects of light balanced general anaesthesia on the immune response with possible stimulation of lymphocyte function.[102]

In addition, studies by Kehlet demonstrated that inhibition of the catabolic hormonal response did not inhibit increases in cytokine release following surgery.[101] A study by Moore et al.[103] demonstrated that extradural blockade from T4–S5 dermatomes for pelvic surgery had no effect on circulating IL-6 concentrations although anterior pituitary hormone secretion was prevented and catecholamine-mediated metabolic changes blocked. The release of cytokines from damaged tissues and their effects on a wide variety of other tissues, particularly immunological cells, is an area of rapidly expanding interest.[104] The cytokines are polypeptides including the interleukins (IL-1–IL-12 to date), tumour necrosing factor (TNFα and TNFβ) and the interferons (Table 28.1). Interleukin-6 is the main cytokine released after routine surgery and increases with increasing severity of surgery (Fig. 28.2).[105] The increase in IL-6 levels correlates well with observed increases in serum C-reactive protein (CRP) following trauma. CRP is an acute phase protein synthesized and secreted by hepatocytes, and it is interesting to note that IL-6 was previously known as 'hepatocyte stimulating factor'. The complexity and confusion of the system is reflected by other synonyms for IL-6 including B-cell stimulatory factor-2, interferon-β2, 26 kDa protein, hybridoma-plasmacytoma growth factor

Table 28.1 Summary of cytokine effects and actions

Interleukin-1α and 1β [IL-1α, IL-1β]	Activate T-cells and induce lymphokine release
	Act as cofactors stimulating B-cells to proliferate and differentiate
	Induce GM-CSF and IL-6 production in bone marrow stromal cells leading to induction of haemopoiesis
	Act on non-lymphoid cells: fibroblasts proliferate and release prostaglandins and collagenase: hepatocytes release acute phase proteins: osteocytes and chondrocytes resorb bone and break down cartilage respectively
	Act on tissues: synovium releases prostaglandins and collagenase, muscle undergoes proteolysis: induce fever and sleep by action on the CNS
Interleukin-2 IL-2	Induces T, B, NK cells and oligodendrocytes to proliferate
	Causes B-lymphocytes to differentiate
	Augments NK non-MHC restricted killing and increases macrophage cytotoxicity
Interleukin-3 IL-3	Acts on pluripotent stem cells of the bone marrow supporting the growth of several lineages, eosinophils, granulocytes and macrophages
	Activates eosinophils
	Supports basophil growth
	Enhances macrophage cytotoxicity
Interleukin-4 IL-4	In B-lymphocytes induces class II MHC, enhances IgG and IgE production and stimulates growth and differentiation
	Stimulates growth of T-cells and mast cells
	Activates macrophages
Interleukin-5 IL-5	In B-cells induces growth, enhances IgA and IgM production and IL-4 induced IgE production
	Promotes growth and differentiation of eosinophil precursors in bone marrow
Interleukin-6 IL-6	In B-cells stimulates Ig synthesis, in T-cells promotes IL-2 production and differentiation of cytotoxic cells
	Promotes differentiation in pluripotent stem cells
	Induces acute phase protein release by hepatocytes
Interleukin-7 IL-7	Induces proliferation in T-cells, thymocytes and pre-B-cells

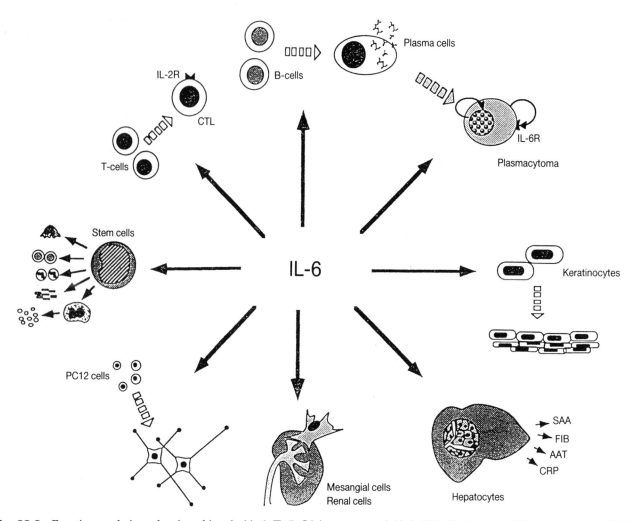

Fig. 28.2 Functions and sites of action of interleukin-6 (IL-6). SAA, serum amyloid A; FIB, fibrinogen; AAT, α_1-antitrypsin; CRP, C-reactive protein.

and myeloid blood cell differentiation-inducing protein, denoting previously identified functions. The growing use of monoclonal antibodies to identify specific cytokines and acute phase proteins will enable their precise functions and interrelations to be studied (Fig. 28.3). The physiological functions of the acute phase response is unclear but it is known to occur in all warm blooded animals in response to serious tissue damage.

However, the development of minimally invasive techniques for surgical operations has clearly demonstrated the clinical advantages of minimizing tissue trauma. A study comparing the hormonal and metabolic responses to cholecystectomy conducted by either conventional laparotomy or laparoscopy demonstrated that circulating IL-6 levels, concentrations of circulating CRP and white blood cell counts were significantly decreased in patients undergoing laparoscopy compared with laparotomy. Plasma cortisol and catecholamine concentrations were not significantly different in the two groups. Both groups received similar amounts of volatile agent. During

the first and second postoperative days, analgesic consumption was less in the laparoscopy group whilst vital capacity, forced expiratory volume in 1 second (FEV_1), and PaO_2 were significantly greater. This resulted in a reduction in hospital stay of nearly 50 per cent in the laparoscopy group, i.e. 2.75 days versus 5.0 days. Thus it appears that decreased tissue trauma was the main factor in the improved outcome in the laparoscopy group.[106] Further studies need to be undertaken to confirm these findings and to unravel the precise role of cytokines as mediators of immunological responses.

While IL-6 is the main cytokine released after trauma, and circulating IL-1 and TNFα levels do not normally increase, both IL-1 and TNFα have been implicated as mediators of the 'sepsis syndrome' – also termed the 'mediator syndrome'.[107,108] Initially, TNRα concentrations rise sharply, peaking at 1 hour, with a period of activity of approximately 30–90 minutes. TNFα induces IL-1 expression which peaks at 2 hours with a period of clinical activity of 2–12 hours. Both TNFα and IL-1 can stimulate

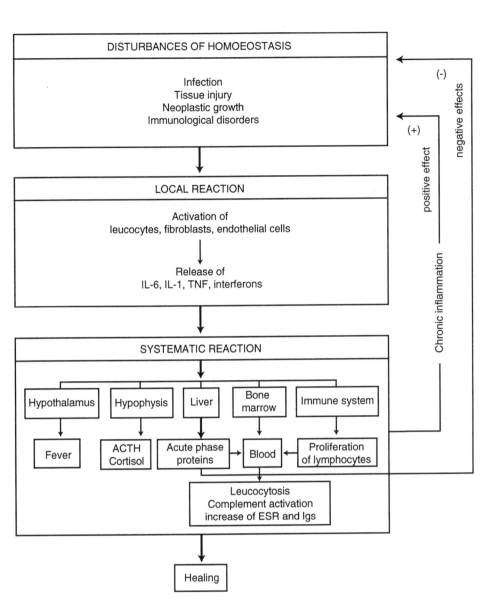

Fig. 28.3 Feedback mechanisms showing positive and negative influences of cyotkines.

IL-6 expression which peaks at 6 hours and has a considerably longer period of clinical activity lasting 2–4 hours. High TNF concentrations are associated with high mortality rates but the association between IL-1 plasma levels and mortality rates remains unclear.

The development of antibodies which interact specifically with certain cytokines, e.g. antibodies to TNFα and TNFβ, raises the possibility of selectively blocking the systemic effects of sepsis. However, cytokines are clinically important intercellular messengers for maintaining homoeostatic balance and only appear harmful when released systemically in high concentrations; thus selective blockade of an individual cytokine may have disadvantages. Studies using monoclonal antibodies to TNF have produced encouraging results in animals and appear safe when

tested in phase 1 human clinical studies and await further clinical trials.[109]

Bacterial endotoxin is a potent stimulator of the sepsis syndrome. It is a lipopolysaccharide present in the outer cell membrane of Gram-negative bacteria which is released into the circulation when the cells replicate or die.[110] Endotoxin may enter the circulation from sites infected with Gram-negative aerobic bacteria (GNAB) or from the bowel, where GNAB are present as part of the normal gut flora. Normally, as already stated, the functioning gut immune system prevents this occurring but direct trauma and surgery of the gastrointestinal tract, ischaemic injury during episodes of circulatory insufficiency, and gut malnutrition during fasting and prolonged parenteral nutrition are all examples of clinical situations which give rise to Gram-negative septicaemia and endotoxaemia. The development of human and murine monoclonal IgM antibodies to the terminal lipid A fraction of bacterial endotoxin raises the possibility of selectively blocking their systemic effects by antibodies to prevent stimulation of macrophages, and the release of cytokines, including TNF and prostaglandins. As a result of difficulties in standardizing clinical trials in critically ill patients and the complexities of the immune system, trials on the use of the HA-1A human monoclonal antibody have proved inconclusive. While mortality appeared to be decreased in those patients proved to have a Gram-negative bacteraemia, it appeared to be increased in those where the bacteraemia was not apparent.[111] A study using murine IgM monoclonal antibody in septic patients reduced mortality in patients not in refractory shock, an effect that was not influenced by the presence or absence of Gram-negative bacteraemia.[112] Further studies will be needed before its value as a clinical tool can be assessed.

Evidence of interest to anaesthetists has been accumulating in support of the view that overproduction of L-arginine-derived nitric oxide (NO) is a major cause of endotoxin- and cytokine-induced circulatory shock.[113] Endotoxin and the cytokines TNF, IL-1 and IL-2 induce the enzyme nitric oxide synthase (NOS) which results in an increase in the production of nitric oxide. The concentration of endothelium-derived nitric oxide maintains the vasculature in a permanent state of active vasodilatation and is thus of paramount importance in the regulation of blood pressure and organ blood flow distribution.[114] However, endotoxin and cytokines also induce an isomer of nitric oxide synthase, present in phagocytic cells, which is important in the nitric oxide mediated killing of bacteria and tumour cells. Thus, an immune response designed to act against bacterial pathogens and tumour cells is also implicated in the circulatory failure associated with septic shock.[115]

Inhibitors of nitric oxide synthases such as N^G-monomethyl-L-arginine (L-NAME), which ameliorate the hypotension induced by bacterial lipopolysaccharide in the anaesthetized rat,[116] are now being tested against septic shock in humans. However, inhibition of basal NO release by non-enzyme selective NOS inhibitors (such as L-NAME) may decrease the blood flow and oxygen supply to vital organs and further investigations are needed which address the role of NOS inhibition under conditions of multiple organ failure. A study using anaesthetized rabbits demonstrated that the formation of NO by the constitutive NOS opposes hypoxic pulmonary vasoconstriction and supports blood flow to hypoxic alveoli resulting in a reduction in arterial oxygen tension. Inhibition of NO synthesis by L-NAME reduced the blood flow to the hypoxic lung areas, resulting in an increase of arterial oxygen tension.[117] The role of NOS inhibitors and inhalation of nitric oxide in conditions associated with a mismatching of ventilation to perfusion such as adult respiratory distress syndrome (ARDS) may represent a novel therapeutic approach.[118]

The finding that glucocorticoid agonists block endotoxin-induced biosynthesis of TNF and that pretreatment with dexamethasone in anaesthetized rats having received endotoxin prevents increased NO synthesis is of interest to anaesthetists.[114] However, the use of high dose steroids in patients with severe sepsis and septic shock did not result in an improved outcome.[119] A controlled clinical trial of high dose methylprednisolone demonstrated in contrast that general suppression of the immune system rendered the patients more susceptible to secondary infections.[120] Again of interest to the anaesthetist is that endotoxin stimulated IL-6 production by hepatic Kupffer cells is decreased by the simultaneous production of prostaglandin E_2 (PGE$_2$). Indomethacin, by blocking synthesis of PGE$_2$, increases the amount of IL-6 released. The clinical significance of this finding in patients with trauma and sepsis treated with non-steroidal anti-inflammatory drugs awaits clarification.

While the interface between trauma, stress and sepsis is clearly intricate and elucidation awaits further studies, an additional compromising factor is the relation between the immune system, the endocrine system and the central nervous system. Stress, distress and a variety of psychiatric illnesses, notably the affective disorders, are increasingly reported to be associated with immunosuppression. Interdisciplinary research in psychoneuroimmunology and neuroendocrinoimmunology suggests that stress may have a direct influence on the immune system via hypothalamic pituitary peptides and the sympathetic branch of the autonomic nervous system. Feedback regulatory loops between the CNS, the endocrine system and the immune system have been proposed.[121]

It is not possible to consider here all the evidence for factors of importance in the interactions of trauma, stress and sepsis and many have been omitted, e.g. prostaglandins, thromboxanes, platelet activating factors, to name but a few (see Chapter 66).

At present, minimizing trauma, maintaining tissue perfusion and oxygenation and maintaining the integrity of the gastrointestinal tract are of prime clinical importance, combined with the avoidance of factors which impair lymphatic flow and function. We are now entering a phase of development in clinical medicine during which it is anticipated that there will be major changes in clinical practice resulting from the physicians' ability to modulate the immune system and to reduce adverse factors associated with the surgical insult.

Patient factors

Patients at the extremes of life have impaired immune systems and appear prone to infection. Preterm infants, especially those of low birth weight, are hypogammaglobulinaemic because of inadequate placental transfer of IgG. In addition, neonates cannot produce antibodies to polysaccharide antigens and antibodies to protein antigens are of low affinity.[122] As a general rule, organ function, including cardiac, respiratory, renal and hepatic function, decreases by 1 per cent for each year of life after 30. Such changes will inevitably compromise immune responses. Indeed, the immune system itself apparently declines with age. Specific changes in T-lymphocyte effector/suppressor ratios have been demonstrated in the elderly (Morgan, 1993, personal communication), specifically a significant increase in suppressor cells. It is postulated that this may again be a protective mechanism to preserve stressed, slightly damaged and ischaemic tissues from destruction but conversely will increase susceptibility to infection.

Pregnancy may be regarded as a prime example of the modulation of the immune status to accommodate foreign, non-self, tissue. During pregnancy, the maternal immune system will tolerate the growth of the fetus, primarily as a result of the presence of the functioning placenta. However, there is little evidence that anaesthesia has specific immunological consequences at present for either the mother or fetus during pregnancy.

Patients with primary immune deficiencies, e.g. hypogammaglobulinaemia, need replacement therapy before surgery and anaesthesia[123] and there is increasing use of intravenous immunoglobulin therapy in patients with secondary immune deficient states, e.g. lymphoproliferative diseases.[124] A case report in which three different types of general anaesthesia were given on three separate occasions to a 14-month-old infant with congenital agranulocytosis, Kostmann's syndrome, demonstrated a profound reduction in neutrophil numbers irrespective of the agents used. These cases are rare but it would seem prudent to use regional anaesthesia, if appropriate.[125]

Patients with acquired immunodeficiency syndrome (AIDS) are themselves prone to infections but in addition there is concern at the possible transmission of infection not only from patients to medical and nursing personnel but also from AIDS infected personnel to patients. Advice on this subject is contained in a publication by the Association of Anaesthetists of Great Britain and Ireland 1993 – *HIV and other blood borne viruses*.[126] However, there is as yet no indication as to preferred anaesthetic practice.

Severe malnutrition is associated with immunosuppression and poor diabetic control interferes with leucocyte chemotaxis, opsonization, phagocytosis and cell-mediated immune responses.[127] Severe alcoholics and opioid addicts are prone to infections and smoking affects mucosal immunity.[128] Patients undergoing therapy with corticosteroids, immune suppressive drugs and cytotoxic drugs are prone to infection and, as such, are in a special category. The effects of antibiotics are uncertain but may also reduce immune activity.[129,130] Patients with known allergies and atopic patients are susceptible to anaphylactic and anaphylactoid reactions. The response of such patients to trauma and sepsis is less certain but it is of interest to note that adrenaline remains the first-line treatment for severe reactions.

These examples merely serve to indicate that it is necessary to assess each patient individually with regard to immune status prior to the administration of anaesthesia. The anaesthetist may be enabled, with increasing knowledge, to provide clinical treatments more appropriate to such special cases.

Conclusion

The prime function of the anaesthetist is still to relieve the pain and anxiety of patients undergoing surgical operations with general or regional anaesthesia while retaining physiological homoeostasis. During routine anaesthetic and surgical practice, the immune response may pass unnoticed, and greater understanding of the immune system may not appear necessary. However, there is still an unacceptably high incidence of postoperative morbidity and mortality resulting from infection and malignancy and, while intensive care has aided postoperative recovery rates in all ages, attention should now be directed to aspects of the immune response and function to provide further improvements in survival rates. Evidence suggests that in future, specific control and modification of immune responses may improve outcomes but at present within our current knowledge it is now incumbent on the anaesthetist to consider how to avoid inducing unnecessary immune-mediated complications.

Undoubtedly, a knowledge of immunology is now essential for all anaesthetists and of particular importance in intensive care. The immunological consequences of present practices need re-examination and both scientific and clinical studies should be undertaken with the introduction of any new drugs and techniques to determine the consequences of their application on the immune system.

REFERENCES

1. Salo M. Effects of anaesthesia and surgery on the immune response. *Acta Anaesthesiologica Scandinavica* 1992; **36**: 201–20.

2. Abraham E. Physiological stress and cellular ischaemia: relationship to immunosuppression and susceptibility to sepsis. *Critical Care Medicine* 1991; **19**: 613–18.

3. Roitt IM. *Essential immunology.* 8th ed. Oxford: Blackwell Scientific Publications, 1993.

4. Male D, Brostoff B, Cooke A. *Advanced immunology.* London: Gower Medical Publishing, 1987.

5. Paul W. *Fundamental immunology.* New York: Raven Press, 1984.

6. Wilson JD, Simpson SI. *Diagnostic immunology and serology: a clinician's guide.* Lancaster: MTP Press, 1980.

7. Sheehan C. *Clinical immunology: principles and laboratory diagnosis.* Philadelphia: JB Lippincott, 1990.

8. Chapel H, Heaney M. *Essentials of clinical immunology.* 3rd ed. Oxford: Blackwell Scientific Publications, 1993.

9. Khansari DN, Murgo AJ, Faith RE. Effects of stress on the immune system. *Immunology Today* 1990; **11**: 170–5.

10. Sharp C, Parry Billings M. Can exercise damage your health? *New Scientist* 1992; No. 1834, 33–7.

11. Craig EA. The heat shock response. *Critical Reviews in Biochemistry* 1985; **18**: 239–80.

12. Lindquist S, Craig EA. The heat shock proteins. *Annual Review of Genetics* 1988; **22**: 631–77.

13. Pelham H. Heat shock proteins coming in from the cold. *Nature* 1988; **332**: 776–7.

14. Jindal S, Dudani AK, Singh B, Harley CB, Gupta RS. Primary structure of a human mitochondrial protein homologous to the bacterial and plant chaperonins and to the 65-kilodalton mycobacterial antigen. *Molecular and Cellular Biology* 1989; **9**: 2279–83.

15. Ellis RJ. Molecular chaperonins. *Seminars in Cell Biology* 1990; **1**: 1–72.

16. Jones DB, Hunter NR, Duff GW. Heat shock protein 65 as a B cell antigen of insulin dependent diabetes. *Lancet* 1990; **336**: 583–5.

17. Young RA. Stress proteins and immunology. *Annual Review of Immunology* 1990; **8**: 401–20.

18. Kaufman SHE. Heat shock protein and the immune response. *Immunology Today* 1990; **11**: 129–36.

19. Van Eden W. Heat shock protein in auto-immune arthritis: a critical contribution based on the adjuvant arthritis model. *APMIS* 1990; **98**: 383–94.

20. Lindquist S. The heat-shock response. *Annual Review of Biochemistry* 1986; **55**: 1151–91.

21. Grossman Z, Herbermann RB. 'Immune surveillance' without immunogenicity. *Immunology Today* 1986; **7**: 128–31.

22. Lappe M. Immune recognition of preneoplastic and neoplastic cell surfaces. *Journal of the National Cancer Institute* 1972; **35**: 49–55.

23. Child DF, Smith CJ, Williams CP. Heat shock protein and the double insult theory for the development of insulin dependent diabetes. *Journal of the Royal Society of Medicine* 1993; **86**: 217–18.

24. Latchman DS. Heat shock proteins and human disease. *Journal of the Royal College of Physicians of London* 1991; **5**: 295–9.

25. Lee KS, Park SS. Effect of halothane, enflurane and nitrous oxide on tracheal ciliary activity in vitro. *Anesthesia and Analgesia* 1980; **59**: 426–30.

26. Garred P, Olsen J, Mollues TE, Bilde T, Glahn BE. Biocompatibility of urinary catheters: effect on complement activation. *British Journal of Urology* 1989; **63**: 367–71.

27. Marston A, Bulkley DB, Fiddian-Green RG, Haghund UH, eds. *Splanchnic ischemia and multiple organ failure.* London: Edward Arnold, 1989.

28. Frostell C, Blomquist H, Hedenstierma G, Halbig I, Pieper R. Thoracic and abdominal lymph drainage in relation to mechanical ventilation and PEEP. *Acta Anaesthesiologica Scandinavica* 1987; **31**: 405–12.

29. Zabner J, Angeli LS, Martinez RR, Sanchez de Leon R. The effect of graded administration of positive end expiratory pressure on the fluid filtration rate in isolated rabbit lungs, using normal lungs, hydrostatic oedema lungs and oleic acid induced oedema. *Intensive Care Medicine* 1990; **16**: 89–94.

30. Last M, Kurtz K, Stein TA, Wise L. Effect of PEEP on the rate of thoracic duct lymph flow and clearance of bacteria from peritoneal cavity. *American Journal of Surgery* 1983; **145**: 126–30.

31. Wiener-Kronish JP, Gropper MA, Matthay MA. The adult respiratory distress syndrome: definition and prognosis, pathogenesis and treatment. *British Journal of Anaesthesia* 1990; **65**: 107–29.

32. Watkins J. Intravenous therapy and 'immunological' disasters. *Theoretical Surgery* 1986; **1**: 103–12.

33. Bernhard C. *Anesthésiques et l'asphyxie.* Paris: Baillière, 1875.

34. Pockock G, Richards CD. Cellular mechanisms in general anaesthesia. *British Journal of Anaesthesia* 1991; **66**: 116–28.

35. Halsey MJ, Prys Roberts C, Strunin L. Receptors and transmembrane signalling: cellular and molecular aspects of anaesthesia. *British Journal of Anaesthesia* 1993; **71**: 1.

36. Smith CJ, Edwards AE, Hunter A. Immunology, anaesthesia and surgery: an historical survey biased towards adverse reactions and clinical interactions. *Theoretical Surgery* 1989; **3**: 210–17.

37. Smith CJ, Edwards AE, Gemmell LW, Ferguson BJM, Gower DE, Gough JD. Lectin-stimulated lymphocyte responses: use and misuse in clinical research. *European Journal of Anaesthesiology* 1988; **5**: 269–78.

38. Smith CJ, Allen JC. Commercial PHA preparations contain different mitogenic components. *Cell and Tissue Kinetics* 1978; **11**: 659–63.

39. Moudgil GC. Effect of premedicants, intravenous anaesthetic agents and local anaesthetics on phagocytosis in vitro. *Canadian Anaesthetists' Society Journal* 1981; **28**: 597–602.

40. Smith CJ, Edwards AE, Gower DE, Ferguson BJM, Williams CP. Leucocyte migration: effects of in vitro response to anaesthetic agents: possible potentiation of effects by adrenaline. *European Journal of Anaesthesiology* 1992; **9**: 463–72.

41. Brown SM, Stimmel B, Taub RN, *et al.* Immunologic

dysfunction in heroin addicts. *Archives of Internal Medicine* 1974; **134**: 1000–6.

42. Edwards AE, Smith CJ, Gower DE. Anaesthesia, trauma, stress and leucocyte migration: influence of general anaesthesia and surgery. *European Journal of Anaesthesiology* 1990; **7**: 185–96.

43. Schmidt RM, Rosenkranz HS. Antimicrobial activity of local anaesthetics: lidocaine and procaine. *Journal of Infectious Diseases* 1970; **121**: 597–607.

44. Rosenberg PH, Renkonen OV. Antimicrobial activity of bupivacaine and morphine. *Anesthesiology* 1985; **62**: 178–9.

45. O'Donnell NG, McSharry CP, Wilkinson PC, Asbury AJ. Comparison of the inhibitory effect of propofol, thiopentone and midazolam on neutrophil polarization in vitro in the presence or absence of human serum albumin. *British Journal of Anaesthesia* 1992; **69**: 70–4.

46. Jensen AG, Dahlgren C, Eintrei C. Propofol decreases random and chemostatic stimulated locomotion of human neutrophils in vitro. *British Journal of Anaesthesia* 1993; **70**: 99–100.

47. Weiss M, Mirow N, Birkhahn A, Schneider M, Werner P. Benzodiazepines and their solvents influence neutrophil granulocyte function. *British Journal of Anaesthesia* 1993; **70**: 317–21.

48. Wagner RL, White PF. Etomidate inhibits adrenocortical function in surgical patients. *Anesthesiology* 1984; **61**: 647–51.

49. Wagner RL, White PF, Kan PB, Rosenthal MH, Feldman D. Inhibition of adrenal steroid genesis by the anaesthetic etomidate. *New England Journal of Medicine* 1984; **310**: 1415–21.

50. Moudgil GC, Gordon J, Forrest JB. Comparative effects of volatile anaesthetic agents and nitrous oxide on human leucocyte chemotaxis in vitro. *Canadian Anaesthetists' Society Journal* 1984; **31**: 631–7.

51. Nunn JF, Sharp JA, Kimball KL. Reversible effect of an inhalational anaesthetic on lymphocyte motility. *Nature* 1970; **226**: 85–6.

52. Nunn JF, Sturrock JE, Jones AJ, *et al.* Halothane does not inhibit human neutrophil function in vitro. *British Journal of Anaesthesia* 1979; **51**: 1101–8.

53. Lippa S, DeSole P, Meucci E, Littarru GP, DeFrancisci G, Magalini SI. Effect of general anaesthetics on human granulocyte chemiluminescence. *Experientia* 1983; **39**: 1386–8.

54. Welch WD, Zaccari BS. Enflurane but not isoflurane inhibits neutrophils. *Anesthesiology* 1981; **55**: A293.

55. Erskine R, James MFM. Isoflurane but not halothane stimulates neutrophil chemotaxis. *British Journal of Anaesthesia* 1990; **64**: 723–7.

56. Landsteiner K, Van der Scheer J. Cross reactions of immune sera to azoproteins. *Journal of Experimental Medicine* 1936; **63**: 325–39.

57. Rubin G. The influence of alcohol, ether and chloroform on natural immunity in relation to leucocytosis and phagocytosis. *Journal of Infectious Diseases* 1904; **1**: 425–44.

58. Doenicke A, Grote B, Suttman H. Effects of halothane on the immunological system in healthy volunteers. *Clinical Research Reviews* (Janssen Research Foundation) 1981; **1**: 23–5.

59. Lassen HCA, Henriksen E, Neukirch R. Treatment of tetanus: severe bone marrow depression after prolonged nitrous oxide anaesthesia. *Lancet* 1956; **1**: 527–30.

60. Stevens JJWM, Griffin RM, Stow PJ. Prolonged use of isoflurane in a patient with tetanus. *British Journal of Anaesthesia* 1993; **70**: 107–9.

61. Kong KL, Willatts SM, Prys-Roberts C, Harvey JT, Gorman S. Plasma catecholamine concentration during sedation in ventilated patients requiring intensive therapy. *Intensive Care Medicine* 1990; **16**: 171–4.

62. Elliott RH, Strunin L. Hepatotoxicity of volatile anaesthetics. *British Journal of Anaesthesia* 1993; **70**: 339–48.

63. Brodin B, Hesseluik JF. Reticuloendothelial response to administration of plasma substitutes. *Baillière's Clinical Anaesthesia* 1988; **2**: 691–706.

64. Mishler JM. *The pharmacology of hydroxyethyl starch*. Oxford: OUP, 1982.

65. Daly JM, Dudrick SJ, Copeland EM. Intravenous hyperalimentation – effect on delayed cutaneous hypersensitivity in cancer patients. *Annals of Surgery* 1980; **192**: 587–92.

66. Moore F, Feliciano D, Andrassy R, *et al.* Enteral feeding reduces post-operative septic complications. *Journal of Parenteral and Enteral Nutrition* 1991; **15** (suppl): 225.

67. Seidner DL, Mascioli EA, Istfan NW, Porter KA, Sellock K, Blackburn JA. Effects of long-chain triglyceride emulsions on reticuloendothelial system function in humans. *Journal of Parenteral and Enteral Nutrition* 1989; **10**: 53–63.

68. Muller JM, Keller HW, Brenner V, Walter M, Holzmuller W. Indications and effects of preoperative parenteral nutrition. *World Journal of Surgery* 1986; **10**: 53–63.

69. Freeman J, Goldmann DA, Smith NE, Sidebottom DG, Epstein MF, Platt R. Association of intravenous lipid emulsion and coagulase-negative staphylococcal bacteraemia in neonatal intensive care units. *New England Journal of Medicine* 1990; **323**: 301–8.

70. Bulow J, Madsen J, Hajgaard L. Reversibility of the effects on local circulation of high lipid concentrations in blood. *Scandinavian Journal of Clinical and Laboratory Investigation* 1990; **50**: 291–6.

71. Parke TJ, Stevens JE, Rice ASC, *et al.* Metabolic acidosis and fatal myocardial failure after propofol infusion in children: five case reports. *British Medical Journal* 1992; **305**: 613–16.

72. Alverdy JC, Arys E, Moss GS. Total parenteral nutrition promotes bacterial translocation from the gut. *Surgery* 1988; **104**: 185–9.

73. Maynard ND, Bihari DJ. Postoperative feeding. *British Medical Journal* 1991; **303**: 1007–8.

74. Clarke PJ, Ball MJ, Kettlewell MGW. Liver function tests in patients receiving parenteral nutrition. *Journal of Parenteral and Enteral Nutrition* 1991; **15**: 54–9.

75. Meta-analysis of the digestive tract trialists' collaborative group. Selective decontamination of the digestive tract. Paper. *British Medical Journal* 1993; **307**: 525–32.

76. Newsholme EA, Parry-Billings M. Properties of glutamine release from muscle and its importance for the immune system. *Journal of Parenteral and Enteral Nutrition* 1990; **14**: 635–75.

77. Kirk SJ, Barbul A. Role of arginine in trauma, sepsis and immunity. *Journal of Parenteral and Enteral Nutrition* 1990; **14**: 2265–95.

78. Detsky AS, Baker JP, O'Rourke K, Goel V. Perioperative

parenteral nutrition: a meta-analysis. *Annals of Internal Medicine* 1987; **107**: 195–203.

79. Salo M. Immunosuppressive effects of blood transfusion in anaesthesia and surgery. *Acta Anaesthesiologica Scandinavica* 1988; **32** (suppl 89): 26–34.

80. Burrows L, Tartter PI. Effect of blood transfusion on colonic malignancy recurrence rate. *Lancet* 1982; **2**: 662 (Letter).

81. Tartter PI, Burrows L, Papatestas AE, Lesiuck G, Aufses Jr AH. Perioperative blood transfusion has prognostic significance for breast cancer. *Surgery* 1985; **97**: 225–30.

82. Hyman NH, Fostr Jr RS, DeMeules JE, Constanza MC. Blood transfusions and survival after lung cancer resection. *American Journal of Surgery* 1985; **149**: 502–7.

83. Chung M, Steinmetz OK, Gordon PH. Perioperative blood transfusion and outcome after resection for colorectal carcinoma. *British Journal of Surgery* 1993; **80**: 427–32.

84. Hardy AM, Allen JR, Morgan WM, Curran JW. The incidence rate of acquired immunodeficiency syndrome in selected populations. *Journal of the American Medical Association* 1985; **253**: 215–20.

85. Opelz G. Current relevance of the transfusion effect in renal transplantation. *Transplantation Proceedings* 1985; **17**: 1015–22.

86. Groth CG. There is no need to give blood transfusions as pre-treatment for renal transplantation in the cyclosporine era. *Transplantation Proceedings* 1987; **19**: 153–4.

87. Marquet RL, de Bruin RWF, Dallinga RJ, Singh SK, Jeekel J. Modulation of tumor growth by allogeneic blood transfusion. *Journal of Cancer Research and Clinical Oncology* 1986; **111**: 50–63.

88. Singh SK, Margreet RL, de Bruin RWF, Westbrock DN, Jeetel J. Promotion of tumour growth by blood transfusions. *Transplantation Proceedings* 1987; **19**: 1473–4.

89. Singh SK, Margreet RL, Westbrock DN, Jeetel J. Enhanced growth of artificial tumour metastases following blood transfusion: the effect of erythrocytes, leukocytes and plasma transfusion. *European Journal of Cancer and Clinical Oncology* 1987; **23**: 1537–40.

90. Waymack JP, Gallou L, Barcelli V, Trocki O, Alexander JW. Effect of transfusions on immune function. III alterations in macrophage arachidonic acid metabolism. *Archives of Surgery* 1987; **122**: 56–60.

91. Brunson ME, Alexander JW. Mechanisms of transfusion-induced immunosuppression. *Transfusion* 1990; **30**: 651–8.

92. Smith R. Filtering white cells from blood for transfusion. *British Medical Journal* 1993; **306**: 810.

93. Hunt TK. Surgical wound infections: an overview. *American Journal of Medicine* 1981; **70**: 712–18.

94. Brimacombe J, Skippen P, Talbutt P. Acute anaemia to a haemoglobin of 14 g L^{-1} with survival. *Anaesthesia and Intensive Care* 1991; **19**: 581–3.

95. Goodnough LT. The role of recombinant growth factors in transfusion medicine. *British Journal of Anaesthesia* 1993; **70**: 80–6.

96. Holoshitz J, Koning F, Coligan JE, De Bruyn J, Strober S. Isolation of CD4-CD8 myobacteria-reactive T lymphocyte clones from rheumatoid arthritis synovial fluid. *Nature* 1989; **339**: 226–9.

97. Modlin RL, Pirmez C, Hofman FM, *et al.* Lymphocytes bearing antigen-specific gamma delta T-cell receptors accumulate in human infectious disease lesions. *Nature* 1989; **339**: 544–8.

98. Boru W, Happ MP, Dallas A, *et al.* Recognition of heat shock proteins and gamma delta cell function. *Immunology Today* 1990; **11**: 40–3.

99. Asarnow DM, Kuziel WA, Bonyhadi M, Tigelaar RE, Tucker PW, Allison JP. Limited diversity of gamma delta antigen receptor genes of Thyo 1+ dendritic epidermal cells. *Cell* 1988; **55**: 837–47.

100. Janeway Jr CA, Jones B, Hayday A. Specificity and function of T-cells bearing gamma delta receptors. *Immunology Today* 1988; **9**: 73–6.

101. Kehlet H. The modifying effect of general and regional anaesthesia on the endocrine-metabolic response to surgery. *Regional Anaesthesia* 1982; **7**: S38–S48.

102. Kho HG, van Egmond J, Zhuang CF, Zhang GL, Lin GF. The patterns of stress response in patients undergoing thyroid surgery under acupuncture anaesthesia inclusion. *Acta Anaesthesiologica Scandinavica* 1990; **34**: 563–71.

103. Moore CM, Desborough JP, Burrin JM, Hall GM. IL-6 and the pituitary hormone response to surgery. *Journal of Endocrinology* 1992; **132** (suppl): 207.

104. Hall GM, Desborough JP. Interleukin 6 and the metabolic response to surgery. *British Journal of Anaesthesia* 1992; **69**: 337–8.

105. Cruickshank AM, Fraser WD, Burus HJG, VanDamme J, Shenkin A. Response of interleukin-6 in patients undergoing elective surgery of varying severity. *Clinical Science* 1990; **79**: 161–5.

106. Joris J, Cigarini I, Legrand M, *et al.* Metabolic and respiratory changes after cholecystectomy performed in a laparotomy or laparoscopy. *British Journal of Anaesthesia* 1992; **60**: 341–5.

107. Bone RC. The pathogenesis of sepsis. *Annals of Internal Medicine* 1991; **115**: 457–69.

108. Glauser MP, Zanetti G, Baumgartner JD, Cohen J. Septic shock: pathogenesis. *Lancet* 1991; **338**: 732–6.

109. Exley AR, Cohen J, Buurman W, *et al.* Monoclonal antibody to TNF in severe septic shock. *Lancet* 1990; **335**: 1275–7.

110. Reitschel ET, Brade L, Braudenburg K. Chemical structure and biological activity of bacterial and synthetic lipid-A. *Reviews of Infectious Diseases* 1987; **9**: 5527–36.

111. Ziegler EJ, Fisher CJ, Sprung CL, Straube RC, Sadoff JC, Foulke GE. Treatment of gram negative bacteraemia and septic shock with HA-1A human monoclonal antibody against endotoxin. *New England Journal of Medicine* 1991; **324**: 429–36.

112. Greenman RL, Schein RMH, Markin MA. A controlled clinical trial of ES murine monoclonal IgM antibody to endotoxin in the treatment of Gram-negative sepsis. *Journal of the American Medical Association* 1991; **266**: 1097–102.

113. Kilbourn RG, Griffith OW. Overproduction of nitric oxide in cytokine-mediated and septic shock. *Journal of the National Cancer Institute* 1992; **84**: 827–31.

114. Moucade S, Palmer RMJ, Higgs EA. Nitric oxide: physiology, pathophysiology and pharmacology. *Pharmacological Reviews* 1991; **43**: 109–42.

115. Hibbs JB, Taintor RR, Vavrin J. Macrophage cytotoxicity: a

role for L-arginine deaminase and immune nitrogen oxidation of nitrite. *Science* 1987; **235**: 473–6.

116. Thiemermann C, Vane JR. Inhibition of nitric oxide synthesis reduces the hypotension induced by bacterial lipopolysaccharide in the anaesthetised rat. *European Journal of Pharmacology* 1990; **182**: 591–5.

117. Sprague RS, Thiemermann C, Vane JR. Endogenous EDRF opposes hypoxic pulmonary vasoconstriction and supports blood flow to hypoxic alveoli in anaesthetized rabbits. *Proceedings of the National Academy of Sciences of the USA* 1992; **89**: 8711–15.

118. Rossaint R, Gerlach H, Falke KJ. Inhalation of nitric oxide – a new approach in severe ARDS. *European Journal of Anaesthesiology* 1994; **11**: 43–51.

119. Ledingham I, McArdle C. Prospective study of the treatment of septic shock. 1978; **1**: 1194–7.

120. Bone RC, Fisher CJ, Clemmer TP, Slotman GJ, Metz CA, Balk RA and the Methyl-prednisolone Severe Sepsis Study Group. A controlled clinical trial of high-dose methyl-prednisolone in the treatment of severe sepsis and septic shock. *New England Journal of Medicine* 1987; **317**: 653–8.

121. Khansari DN, Murgo AJ, Faith RE. Effects of stress on the immune system. *Immunology Today* 1990; **11**: 170–5.

122. Baker CJ, Melish ME, Hall RT, Casto DT, Vasau U, Givner LB. Intravenous immunoglobulin for the prevention of nosocomial infection in low birth weight neonates. *New England Journal of Medicine* 1992; **327**: 213–19.

123. Medical Research Council Working Party on hypogammaglobulinaemia in the United Kingdom. *Lancet* 1969; **1**: 163–8.

124. Cooperative Group for the study of immunoglobulin in chronic lymphatic leukemia IVIG, for the prevention of infection in chronic lymphatic leukemia. *New England Journal of Medicine* 1988; **319**: 902–7.

125. Fenner SG, Cashman JN. Anaesthesia and congenital agranulocytosis: influence of anaesthetic agent on neutrophil numbers in a patient with Kostmann's syndrome. *British Journal of Anaesthesia* 1991; **66**: 620–4.

126. *HIV and other blood borne viruses.* London: Association of Anaesthetists of Great Britain and Ireland.

127. Rayfield EJ, Ault MJ, Keusch GT, Brother MJ, Nechemias C, Smith H. Infections and diabetes: the case for glucose control. *American Journal of Medicine* 1982; **72**: 439–50.

128. Editorial – smoking and immunity. *Lancet* 1990; **335**: 1561–3.

129. Hanser WE, Remington JS. Effect of antibiotics on the immune response. *American Journal of Medicine* 1982; **72**: 711–16.

130. Shalit I. Immunological aspects of new quinolones. *European Journal of Clinical Microbiology and Infectious Diseases* 1991; **10**: 262–6.

Section Two

General Systems

Chapter 29

Preoperative Assessment and Premedication for Adults

Michael F. Roizen* and J. Lance Lichtor

The goals of preoperative medical assessments are to improve patient outcome; to reduce patients' anxiety by acquainting them with their doctors and by explaining procedures and protocols; and to obtain informed consent. Medical assessments enable physicians to reduce morbidity by optimizing health status and planning perioperative management. Because perioperative morbidity and mortality increase with the severity of pre-existing disease,[1-7] careful evaluation and treatment should reduce their occurrence. Consequently, patients would benefit from a reliable method of preoperative assessment by which laboratory tests could be selected.

From the 1940s to the 1960s, preoperative medical assessment relied primarily on accurate history-taking and physical examination. Then, in the late 1960s, multiphasic screening laboratory tests were introduced. The ease of ordering and low cost of obtaining many tests made this new mode of evaluation attractive. As a result, many hospitals, anaesthetic departments and outpatient surgery centres made rather arbitrary rules – recommendations that became requirements – regarding the tests that should be performed before elective surgery. When, with good intentions, anaesthetists and surgeons tried to follow those rules, problems arose. Physicians believed that, by ordering inexpensive batteries of tests, they could efficiently screen for disease. Instead, because anaesthetists were often still trying to determine which tests to order before surgery, what to do about an unexpectedly abnormal result on the morning of surgery, or how abnormal a result had to be before requesting a consultant opinion, the preoperative multiphasic screening with multiple tests became imprac-

tical. Non-selective testing produced so many false-positive and false-negative results that the subsequent harm vastly outweighed any possible benefit.

Although laboratory screening tests can aid in optimizing a patient's preoperative condition once a disease is suspected or diagnosed, they have several shortcomings: they frequently fail to uncover pathological conditions; the abnormalities they detect do not necessarily affect patient care or outcome; and they are inefficient in screening for asymptomatic diseases. Finally, most abnormalities discovered on preoperative screening, or even on admission screening for non-surgical purposes, are not recorded (other than in the laboratory report) or appropriately pursued.

Domoto et al.[8] examined the yield and effectiveness of a battery of 19 screening laboratory tests performed routinely in 70 functionally intact elderly patients (average age, 82.6 years) who resided at a chronic care facility. The 70 patients underwent 3903 screening tests. 'New abnormal' results occurred primarily in five of the 19 screening tests; most of these 'new abnormalities' were only minimally outside the normal range. Only 4 (0.1 per cent of all tests ordered) led to change in patient management, none of which, Domoto and colleagues concluded, benefited any patient in an important way.

Wolf-Klein and colleagues[9] retrospectively studied the results of annual laboratory screening on a population of 500 institutionalized and ambulatory elderly patients (average age, 80 years). From the 15 000 tests performed, 756 new abnormalities were discovered, 690 of which were ignored. Of those new abnormalities, 66 were evaluated; 20 new diagnoses resulted, 12 of which were treated. Two patients of the 500 ultimately may have benefited from eradication of asymptomatic bacteriuria (although eradication of asymptomatic bacteriuria has not

* The section of this chapter written by Dr Roizen is modified from Roizen MF and Hurd M: Preoperative patient evaluation in Rogers MC, ed. *Current practice in anesthesiology*, 2nd ed. St Louis: Mosby-Year Book, with the authors' permission.

been shown to improve the quality of life or extend the life span).

Studies show that the history and physical examination are the best measures for screening for disease. Delahunt and Turnbull[10] evaluated patients who were assessed preoperatively for varicose vein stripping or inguinal herniorrhaphy. For 803 patients who underwent 1972 tests, only 63 abnormalities were uncovered in those patients whose history or physical findings had not indicated the need for tests; but in no instance did the discovery of these abnormalities influence patient management. Rosselló and associates[11] retrospectively evaluated 690 admissions for elective paediatric surgical procedures. The history and physical examination indicated the probability of abnormalities in all 12 patients in whom an abnormality was found through laboratory testing. Clinical diagnosis, and not laboratory testing, was the apparent basis for any change in operative plans.

Narr and colleagues[12] discovered no benefit from screening tests for the 3782 patients of ASA physical status I in their study. Macpherson et al.[13] found that, of 3096 results of laboratory tests that had been normal within one year of surgery, upon repeat testing of patients, only 13 results had changed to a range unacceptable for surgery. These 13 were predictable from a change in patient history. Patient history was crucial in allowing 26 new diagnoses to be made by Mitchell et al.[14] for 550 patients on the general internal medicine ward of a university hospital.

Even in a referral population, history and physical examination determine more than 90 per cent of the clinical course when a patient is referred for consultation for cardiovascular, neurological or respiratory diseases.[15] Other studies also have demonstrated that the history and physical examination accurately indicate all areas in which subsequent laboratory testing proves beneficial to patients. For example, Rabkin and Horne[16,17] examined the records of 165 patients with 'new' (i.e. a change from a previous tracing) abnormalities on the electrocardiogram (ECG) that were potentially 'surgically significant' (i.e. that might affect perioperative management or outcome). In only two instances were the anaesthetic or surgical plans altered by the discovery of new abnormalities on an ECG that were not indicated by history. Thus, for these 165 patients, for whom the benefits of a laboratory test should have been maximal because a new abnormality was detected preoperatively, the history or physical examination determined case management most of the time. Even in one of the two instances of altered case management – a patient with atrial fibrillation – the physical examination should have indicated that an ECG needed to be performed. A history or physical examination was not available for the other patient.

In summary, the studies cited above point to the adequacy of patient history as an independent means of assessing patients preoperatively. Laboratory tests are useful after assessment by history has indicated a likelihood of disease or of high risk, but screening laboratory tests seem to be largely superfluous to the management of patient care. History and physical examination are the most effective ways to screen for disease.

Screening laboratory tests may even present extra risk to the patient. Unnecessary testing may lead physicians to pursue and treat abnormalities based on borderline positive and false-positive results. Few studies examine whether increased tests and the follow-up of false-positive tests adversely affect patients. In a retrospective examination of the adverse effects of chest radiographs on a population of 606 patients, 386 extra chest radiographs were ordered without being indicated.[18] In those 386 patients, one elevated hemidiaphragm and probable phrenic nerve palsy was found that may have resulted in improved care for that patient. In addition, three lung shadows were found that resulted in three sets of invasive tests, including one thoracotomy, without discovery of disease. These procedures caused considerable morbidity, including one pneumothorax and four months of disability.

In another study, Turnbull and Buck[19] examined the charts of 2570 patients undergoing cholecystectomy, to determine the value of preoperative tests. The history and physical examinations successfully indicated all tests that ultimately benefited the patients, with four possible exceptions. But again, in those four patients it is doubtful if any benefit actually occurred. Among them was one patient whose emphysema was detected only by chest x-ray; he underwent preoperative physiotherapy without subsequent postoperative complications. Two patients had unsuspected hypokalaemia (3.2 and 3.4 mmol/l or mEq/l, respectively) and received potassium treatment before surgery. No harm occurs to patients undergoing an operation with this degree of hypokalaemia, and severe potential harm may be caused by treating such patients with oral or intravenous potassium. The fourth patient in whom possible benefit occurred received a blood transfusion before cholecystectomy for an asymptomatic haemoglobin concentration of 9.9 g/dl. Since cholecystectomy is not normally associated with major blood loss, it is concluded that this patient also received no benefit and only the risk of transfusion from that preoperative laboratory test. Thus, it is not clear that any patient in this study benefited from preoperative screening tests that were not indicated.

In another study, only two patients at most (whose asymptomatic bacteriuria was eradicated) benefited from the 9270 screening tests that were obtained.[20] At least one patient was seriously harmed from the pursuit of abnormalities on screening tests and their treatment; this woman developed atrial fibrillation and congestive heart failure after thyroid therapy was instituted for borderline low thyroxine and free thyroxine index tests. It is unclear if these investigators examined other patients for potential harm arising similarly from the pursuit and treatment of abnormalities.

When batteries of laboratory tests yield abnormal results that are neither pursued nor noted, medicolegal risk for physicians increases. Extra testing – testing that is not warranted by findings on a medical history – does not serve as medicolegal protection against liability. Roizen[21] reviewed a series of studies showing that 30–95 per cent of all unexpected abnormalities found on preoperative laboratory tests are not noted on the chart preoperatively. Many reports of preoperative radiographs, for example, are not on the chart before anaesthesia is administered. This lack of notation occurs not only at university medical centres but in community hospitals also. Failure to pursue an abnormality that has been detected poses a greater risk to medicolegal liability than does failure to detect that abnormality. In this way, extra testing can result in extra medicolegal risk to physicians.

Random preoperative testing is inefficient for operating room schedules. Surgeons say that they order preoperative tests to satisfy the anaesthetist: they find it easier just to order all the tests and let the anaesthetist sort them out. Surgeons also believe that it is much more efficient to order batteries of tests than for the anaesthetist, who sees the patient the night before or the morning of surgery, to try to get the tests on an emergency basis. These surgeons apparently do not realize that the abnormalities detected are not discovered until the night before or the morning of surgery, if at all. Then abnormal results on these tests delay or postpone schedules, as extra effort and time are wasted in obtaining consultant reviews of false-positive or slightly abnormal results.

Implementation of preoperative evaluation

There are at least three methods for organizing preoperative evaluation efficiently. First, the surgeon or anaesthetist who sees the patient before a scheduled procedure can obtain the history and perform the physical examination. Second, a clinic can be set up in an outpatient facility so that these two tasks are performed early enough to ensure that laboratory tests or consultations can be obtained without delaying schedules. Third, a questionnaire answered by the patient can be used to indicate likely disease processes and appropriate laboratory tests.

Of the first method one might ask: Can the appropriate testing be easily generated from the surgeon's preoperative visit? One study found that it could. At the University of California, San Francisco, Kaplan et al.[22] found that even a partial history conveyed enough information to indicate correctly all but 22 abnormalities (none of which affected patient outcome) in more than 2785 preoperative blood tests obtained (counting the complete blood count and simultaneous multichannel analysis of six variables (SMA 6) as one test). Knowing only the admission diagnosis, previous discharge diagnoses and scheduled operation, and using previously determined indications for laboratory testing, enabled detection of virtually all abnormalities that would have been detected by routine screening.

As regards the patient questionnaire, several groups have tested the effects and sensitivity of orally administered or written questionnaires as a means of linking the selection of laboratory tests with a patient's medical history.[21,23] In 1987, McKee and Scott[24] used an orally administered set of 17 questions and patient demographics to select preoperative tests for 400 patients. They found that age was the best predictor of abnormalities on preoperative tests. Complications occurred most commonly in patients who reported positive symptoms on the questionnaire and who were older.

A recent study determined that the responses of patients to written questions can be used to predict the laboratory tests that will yield abnormal results for those patients.

Table 29.1 Laboratory test recommendations for asymptomatic healthy patients scheduled to undergo non-blood-loss peripheral surgical procedures

Age (years)	Tests indicated	
	For men	For women
≤ 40	None	?Pregnancy test*
40–49	Electrocardiogram	Haemoglobin or haematocrit ?Pregnancy test*
50–64	Electrocardiogram	Haemoglobin or haematocrit Electrocardiogram
65–74	Haemoglobin or haematocrit Electrocardiogram BUN/glucose ?Chest x-ray=	Haemoglobin or haematocrit Electrocardiogram BUN/glucose ?Chest x-ray=
≥ 75	Haemoglobin or haematocrit Electrocardiogram BUN/glucose ?Chest x-ray=	Haemoglobin or haematocrit Electrocardiogram BUN/glucose ?Chest x-ray=

* If patient cannot definitely rule out pregnancy.
= Benefit/risk ratio of chest x-ray for asymptomatic individuals over 60 years of age is not clear, but it seems that risk exceeds benefit until the patient is older than 74 years.

Table 29.2 Simplified strategy for preoperative testing

Preoperative condition	HGB		WBC	PT/PTT	PLT/BT	Elect	Creat/BUN	Blood Gluc	SGOT/Alk PTAse	X-ray	ECG	Preg.	T/S
	M	F											
Procedure with blood loss	X	X											X
Procedure without blood loss													
Neonates	X	X											
Age < 40		X											
Age 40–49		X									m		
Age 50–64		X									X		
Age ≥ 65	X	X					X	X		±	X		
Cardiovascular disease							X			X	X		
Pulmonary disease										X	X		
Malignancy	X	X	*		*					X			
Radiation therapy			X							X	X		
Hepatic disease				X					X				
Exposure to hepatitis									X				
Renal disease	X	X				X	X						
Bleeding disorder				X	X								
Diabetes						X	X	X			X		
Smoking ≥ 20 pack-years	X	X								X			
Possible pregnancy												X	
Diuretic use						X	X						
Digoxin use						X	X				X		
Steroid use						X		X					
Anticoagulant use	X	X		X									
CNS disease			X			X	X	X			X		

Not all diseases are included in this Table. The anaesthetist's own judgement is needed regarding patients with diseases not listed.
Symbols: ±, perhaps obtain; *, obtain for leukaemias only; X, obtain; m, obtain for men only
Abbreviations: BT, bleeding time; Creat/BUN, creatinine or blood urea nitrogen; Elect, electrolytes (i.e. sodium, potassium, chloride, carbon dioxide and proteins); HGB, haemoglobin; PLT, platelet count; PT, prothrombin time; PTT, partial thromboplastin time; SGOT/Alk PTAse, serum glutamic oxaloacetic transaminase and alkaline phosphatase; T/S, blood typing and screen for unexpected antibodies; WBC, white blood cell count.
Modified from references 21, 22 and 26.

After the patient answers the questionnaire, a plastic overlay reveals what tests are indicated. If the patient cannot answer the questions, a standard group of tests is ordered. Even in a tertiary care (specialist) hospital that admits very sick patients, more than 60 per cent of those laboratory tests now routinely obtained could be eliminated.

The protocols outlined in Tables 29.1 and 29.2 are guidelines for using clinical judgement in ordering laboratory tests. A careful history and physical examination of the patient are required, with special attention to testing whenever indicators of disease entities listed in the Tables are discovered. The protocol clearly places the burden on whoever takes the history to do so accurately.

Obtaining a medical history

A medical history can be provided by a number of sources because patients undergoing surgery move through a continuum of medical care to which a primary care physician, an internist, an anaesthetist and a surgeon contribute. No aspect of medicine requires a greater degree of cooperation among physicians than does the performance of a surgical operation and the perioperative care of a patient. The importance of integrating practice is even greater in the context of the increasing life span of our population. As the number of the elderly and the very old (those older than 85 years of age) grows, so does the incidence among surgical patients of co-morbidities and multiple drug regimens, knowledge of which is crucial to successful patient management. At a time when medical information is encyclopaedic and medical subspecialization proliferates, it is difficult for even the most conscientious physician to keep abreast of the medical issues relevant to perioperative patient management. Thus it is useful to coordinate care with the patient's general practitioner (primary care physician). Even when a good disease history is available, a screening review of these systems is beneficial: airway, chest, cardiovascular, renal, central nervous system, peripheral nervous system, gastrointestinal, musculoskeletal and genitourinary. Questions should

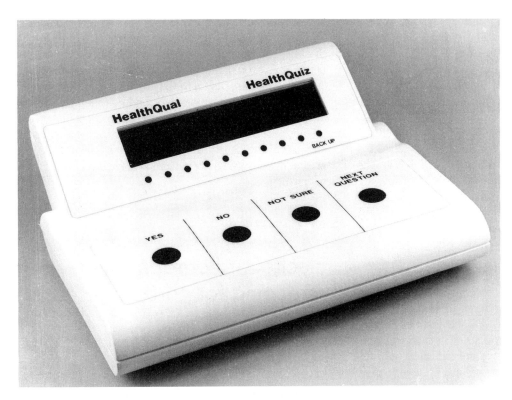

Fig. 29.1 The HealthQuiz, a hand-held computer that enables a patient to answer questions about his or her medical history by pressing one of three buttons.

also be asked regarding medications, allergy, use of alcohol and drugs, smoking, family history of disease, activities and exercise tolerance. Our patients take 10–15 minutes to respond to such questions by pressing a button on the HealthQuiz (Fig. 29.1), an automated device for taking a health history. Positive responses on the HealthQuiz to questions about an organ system elicit more questions about the system. Based on a patient's responses to the questions, certain laboratory tests are indicated. The rationale for the choice of tests relative to response is beyond the scope of this chapter, but some specific tests deserve comment.

outside these ranges, we recommend seeking alternative diagnoses before anaesthesia or surgery. Haematocrit or haemoglobin levels should be determined preoperatively for all surgical patients older than 60 years of age. Haematocrit or haemoglobin level and red blood cell antigen screenings (cross-matched blood) is warranted for all patients undergoing procedures involving possible loss of more than 2 units per 70 kg of body weight. WBCs appear to be rarely indicated in asymptomatic individuals but might be considered when a prosthesis is to be inserted.

Haemoglobin, haematocrit and white blood cell counts

How many asymptomatic patients have a degree of abnormality on haematocrit (packed cell volume) or the white blood cell count (WBC) that alters perioperative management? The following values do not merit intervention perioperatively: a haematocrit of 27–54 for patients undergoing operations during which major blood loss is unlikely, and a WBC of 2400–16 000/mm^3 for individuals not undergoing insertion of a prosthesis. When values fall

Blood chemistry and urinalysis

What blood chemistry would have to be abnormal, and how abnormal would the results have to be, to justify changing perioperative management? Abnormal hepatic or renal function might change the choice and dose of anaesthetic or adjuvant drugs. About 1 in 700 supposedly healthy patients is actually harbouring hepatitis, and 1 in 3 of these will become jaundiced. However, in our prospective study of the HealthQuiz involving 3500 patients, no asymptomatic patient who denied exposure to hepatitis became jaundiced after uneventful surgery.[25] These data imply that either the screening history suffices or the incidence of asymptomatic hepatitis is decreasing. Unexpected abnormalities are reported for 2–10 per cent of patients with multiphasic screening, and these abnormalities lead to many additional tests that usually (in approximately 80 per cent of cases) have no significance for the patient. Unexpected abnormalities that are significant arise in 2–5 per cent of patients studied. Of these abnormalities, approximately 70 per cent are related to blood glucose and blood urea nitrogen (BUN) levels. The four to 19 additional tests on the screening SMAs 6–21 panels lead to very few important discoveries affecting anaesthesia. In fact, the false-positive rate is so high (96.5 per cent for the test for calcium) that more harm than benefit is likely to result for asymptomatic patients who receive these tests.

When a screening test for hepatitis is desired because the incidence of hepatitis is 0.14 per cent or because one wishes to avoid the potential legal problems of postanaesthetic jaundice, only three tests appear to be justified: serum glutamic oxaloacetic transaminase (SGOT), blood glucose and BUN. Even then, the last two are indicated only for patients more than 64 years of age. In fact, if the data from our group on asymptomatic liver disease can be generalized, no blood chemistry tests are warranted for patients younger than 64 years of age. Furthermore, if the antibody test that detects non-A, non-B hepatitis (now called hepatitis C) proves as useful after infection has occurred, the medicolegal risk posed by postanaesthetic jaundice will be even less.

Abnormalities are commonly found on urinalysis.[21] The quality of urinalysis results obtained by dipstick technique has been variable at best. In addition, these abnormal results usually do not lead to beneficial changes in management. Most of the results that do lead to beneficial changes could have been obtained by history or determination of BUN and glucose levels, tests that are already recommended for all patients over 64 years of age. Thus, urinalysis, although initially inexpensive, becomes an expensive test to justify on a cost–benefit or benefit–risk basis.

Chest radiographs

What abnormalities on chest radiographs would influence management of anaesthesia? Certainly, it may be important to know about the existence of tracheal deviation; mediastinal masses; pulmonary nodules; a solitary lung mass; aortic aneurysm; pulmonary oedema; pneumonia; atelectasis; new fractures of the vertebrae, ribs or clavicles; dextrocardia; or cardiomegaly before proceeding to

anaesthesia and surgery. However, a chest radiograph probably would not detect the degree of chronic lung disease requiring a change in anaesthetic technique any better than would the history and physical examination. Abnormalities are rare in the asymptomatic individual.[21] In fact, if a patient is asymptomatic and less than 74 years of age, the risks of chest x-ray examination probably exceed the possible benefits. This analysis is, of course, predicated on maximizing benefit to society in general, as one cannot predict in advance which patients will benefit or which will be harmed.

Electrocardiography and screening for cardiac disease

Although individual episodes of myocardial ischaemia may not produce symptoms, almost all patients without diabetes or autonomic insufficiency have symptoms that lead one to screen for myocardial disease. An algorithm is presented in Fig. 29.2 for pursuing laboratory testing for cardiovascular disease.[21]

Pulmonary function testing

These expensive tests are rarely needed but are useful for determining those who may benefit from bronchodilator therapy and for predicting the risk of thoracic operations before obtaining informed consent. The indications for pulmonary function tests in Table 29.3 seem justified by the data.[21]

Clotting function studies

Virtually no asymptomatic patient in the literature has had unequivocal benefit from clotting function studies preoperatively. Most show symptoms or have a medication history suggesting that clotting function tests may be necessary. Aspirin at 60 mg/70 kg of body weight per day does not seem to pose a risk for bleeding, but the data are not available for 300 mg or more within 12 hours of surgery. Because the pharmacology of aspirin changes

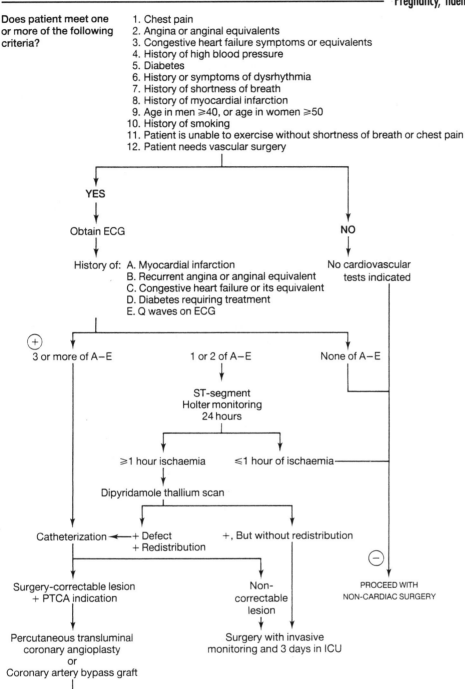

Does patient meet one or more of the following criteria?

1. Chest pain
2. Angina or anginal equivalents
3. Congestive heart failure symptoms or equivalents
4. History of high blood pressure
5. Diabetes
6. History or symptoms of dysrhythmia
7. History of shortness of breath
8. History of myocardial infarction
9. Age in men ≥40, or age in women ≥50
10. History of smoking
11. Patient is unable to exercise without shortness of breath or chest pain
12. Patient needs vascular surgery

YES → Obtain ECG → History of:
A. Myocardial infarction
B. Recurrent angina or anginal equivalent
C. Congestive heart failure or its equivalent
D. Diabetes requiring treatment
E. Q waves on ECG

NO → No cardiovascular tests indicated

(+) 3 or more of A–E

1 or 2 of A–E → ST-segment Holter monitoring 24 hours → ≥1 hour ischaemia → Dipyridamole thallium scan; ≤1 hour of ischaemia

None of A–E

Catheterization ← + Defect + Redistribution; +, But without redistribution

Surgery-correctable lesion + PTCA indication → Percutaneous transluminal coronary angioplasty or Coronary artery bypass graft → Other surgery

Non-correctable lesion → Surgery with invasive monitoring and 3 days in ICU

(−) PROCEED WITH NON-CARDIAC SURGERY

Fig. 29.2 How to decide which cardiovascular laboratory tests to obtain: use the history to segregate patients into groups to test and/or monitor invasively.

when more than 2 g/70 kg is consumed per day, patients should be evaluated if they cannot stop aspirin early enough to have no appreciable level of acetylsalicylic acid for 24 hours before surgery (the period without acetylsalicylic acid necessary to generate the approximately 50 000 new platelets/mm^3 needed for normal platelet aggregation) or if surgical haemostasis cannot be assured or a regional anaesthetic involving a closed space is planned.

Pregnancy, haemoglobinopathy and AIDS screening

Tests for the acquired immunodeficiency syndrome (AIDS) and pregnancy and screening for haemoglobinopathy and malignant hyperthermia raise ethical issues that may

Table 29.3 Indications for pulmonary function tests

History of at least one of the following:
1. Chronic obstructive pulmonary disease
2. Shortness of breath
3. Orthopnoea

Also at least one of the following needs to be determined:
1. Reversibility of bronchospasm
2. Baseline condition if mechanical ventilation is expected preoperatively
3. Risk for lung resection (maximal mid-expiratory flow rate, maximum breathing capacity, diffusing capacity for carbon monoxide)

require close attention to institutional policy and the immediate availability of counselling services. Moreover, all of these tests have associated risks. The physician may therefore decide to limit testing to only at-risk populations (e.g. for pregnancy testing, only female patients who believe they may possibly be pregnant).

Testing of the asymptomatic population for AIDS is not likely to be the most effective way of uncovering the disease unless one or more risk factors are present. Few of the more than 180 000 people in the USA who have had AIDS have not been gay, had sex with a prostitute or multiple partners, had unprotected sex (homo- or heterosexual), used intravenously administered drugs and shared needles, been stuck with a needle, cared for a family member with AIDS, been born of a woman with AIDS, or received a blood transfusion after 1979. One programme, screening for human immunodeficiency virus (HIV) in asymptomatic individuals, was able to produce an 'acceptably low false-positive rate' by diagnosing HIV infection only after one sample of blood produced positive results on four different tests and after a second sample of blood had been used for verification. Thus, for pregnancy, haemoglobinopathies and AIDS, the history is still the best at identifying those at risk for a condition.

Towards a technology for efficient, less expensive, quality care

Even when anaesthetists attempt to choose tests selectively based on the history and physical examination of a patient, errors are made in ordering tests. When surgeons and anaesthetists agreed on indications for testing, 30–40 per cent of patients who should have had tests did not get them, and 20–40 per cent of patients who should not have had tests got them. For instance, Blery and co-workers[26] used a protocol based on suspected disease to order

preoperative tests selectively for 3866 surgical patients in France, and found that, even after clinicians had been educated with regard to indications, 30 per cent of tests ordered were not indicated and another 22 per cent of tests that were indicated were not obtained. Thus, by ordering tests in the usual way, surgeons and anaesthetists both increased costs and failed to obtain possibly valuable information. Blery subsequently questioned anaesthetists to assess whether management of the patient suffered from omission of one or more preoperative tests. Only 0.2 per cent of omitted tests would have possibly been useful. Blery did not examine how many times such data might have led to pursuits that harmed patients.

Charpak *et al.*[27] examined the value of preoperative screening chest radiographs in 3849 patients. Surgeons and anaesthetists agreed that any lung or cardiovascular disease, malignant disease, current smoking history in patients over 50 years of age, major surgical emergencies, immunodepression or lack of prior health examination in immigrants were indications for ordering a chest radiograph. The surgeons ordered or did not order the chest radiograph after seeing the patient. Even with this agreement on indications, of a total of 1426 chest radiographs that should have been ordered in this group of 3849 patients, 271 were ordered although they were not recommended, and 596 were not ordered although they were recommended. Clinical judgements may account for some of these decisions, but it is presumed that many extra chest radiographs ordered or not ordered were simply errors. If so many errors occur in a single test trial, more laboratory tests are likely to generate more errors. The problem is how to order tests that are appropriate for each patient without decreasing efficiency. It appears that selective ordering strategies are better than non-selective methods for ordering tests, but an even easier, more efficient method exists.

At the University of Chicago and at least 100 other institutions, a preoperative health quiz is given to patients on a four-button computer machine. Improvements in technology, graphics and voice properties make it relatively inexpensive to display (or read) clearly questions about a patient's health on a portable $20 \times 15 \times 2.5$ cm ($8 \times 6 \times 1$ in) hand-held box.* The surgeon or anaesthetist can have this box in the office and, after both have agreed on the indications, it can be used by the patient to suggest the tests needed. The box simply asks patients yes/no

---

* Dr Roizen has been involved in the development of three systems designed to facilitate preoperative ordering of indicated tests. The University of Chicago is developing one of these methods, a lap-top video preoperative health questionnaire, into a commercial product. If the product is successful, Dr Roizen will benefit financially, as the University distributes a royalty and/or partial ownership right in such commercialized inventions to its faculty. Nellcor has obtained the rights to commercialize this product.

questions about their health. It then generates a printout of the answers to the questions and a symptom summary, as well as suggested laboratory tests based on the agreed indications and the patient's answers. It also gives reminders about items in history important to anaesthetic care, such as allergies and capped teeth. HealthQuiz suggests tests that would be appropriate for each patient based on that patient's medical history and accepted indications for testing. The physician, surgeon or anaesthetist can then override or add to the suggested tests to be ordered preoperatively. With the information from such a questionnaire, the anaesthetist can spend 'quality time' when meeting a patient preoperatively and have the time to emphasize those issues pertinent to his condition.

Premedication

The main reason for premedicating patients before surgery is to decrease their anxiety. Other valid reasons are to promote haemodynamic stability intraoperatively; to reduce the possibility of aspiration of stomach contents; and to reduce postoperative nausea, vomiting and pain. Certain premedicants cause drowsiness, which, although beneficial preoperatively, may not be desirable for the day-stay (ambulatory) surgical patient. In addition to causing drowsiness, some premedicants may also increase postoperative nausea or reduce postoperative nausea and pain.

Incidence of preoperative anxiety

As has been suggested, one major benefit of a preoperative assessment clinic may be to reduce patient anxiety. When we see a patient for the first time in the preoperative 'holding' area, we may sense that the patient is anxious. Around him preparations are being made for what to us or to other hospital workers is a routine procedure; to the patient the event may seem catastrophic. Indeed, the patient may have felt anxious from the time he learned that surgery was necessary, and this feeling of anxiety may last up to several days after surgery.[28] Furthermore, his level of anxiety is not necessarily highest on the day of surgery (Fig. 29.3). There are many reasons for preoperative anxiety: fear of the unknown or of postoperative nausea or pain; fear of the loss of control during anaesthesia; and fear, based on previous experience or the experiences of others, of not being asleep during surgery.[29]

It is sometimes difficult to know how anxious a patient is. For some, even the need to remove dentures may be associated with increased preoperative stress. Therefore, it is good to ask whether a patient is feeling anxious. Every individual is unique, but some patterns of anxiety have been distinguished. Increased anxiety has been noted in women, in patients accompanied by a friend or relative to the preoperative holding area, in those undergoing mutilating surgery or surgery for malignancy, in younger patients or in those undergoing surgery for the first time, and in patients requiring a second anaesthetic after a bad first experience with anaesthesia.

Reducing anxiety without medication

Not all anxiety needs to be treated with medication. In fact, certain non-pharmacological techniques have been shown to reduce anxiety.[30] Preoperative visits, informational booklets or audiovisual materials, and even hypnosis can reduce anxiety. Evoking a relaxation response, in which attention is focused on a positive or neutral theme as distracting thoughts are passively ignored, has also had limited success in reducing pre- and postoperative anxiety, pain sensation and postoperative narcotic requirement.[31] Techniques such as the relaxation response are best started at least 24 hours before surgery, but the ideal time to conduct a preoperative interview for the purpose of reducing perioperative stress is unclear.

Benzodiazepines

When medication is the treatment choice to reduce anxiety, the benzodiazepines – namely midazolam, diazepam,

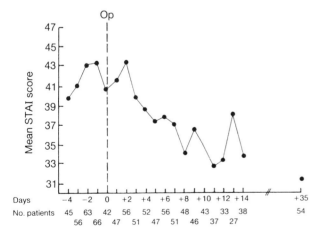

Fig. 29.3 Patient anxiety from 4 days before surgery to 35 days after surgery (when most patients had been discharged), measured using the State-Trait Anxiety Inventory (STAI). Anxiety was high before surgery and also 2 days after surgery. 'Op' = day of surgery. (Reproduced, with permission, from Johnston M. Anxiety in surgical patients. *Psychol Med* 1980; **10**: 145–52.)

lorazepam and triazolam – are the drugs routinely used. Midazolam is the benzodiazepine most commonly used perioperatively. It is water-soluble, has a short half-life (initial distribution half-life, 7.2 minutes; elimination half-life, 2.5 hours (range, 2.1 to 3.4 hours)). The short half-life of midazolam makes it suitable for patient-controlled administration for pre- and postoperative control of anxiety. Elimination half-life is 5.6 hours in elderly patients and up to 8.4 hours in obese patients.[32] Midazolam is highly lipophilic at physiological pH. In Europe, this benzodiazepine is available in both intravenous and tablet form, but in the USA it is available only in intravenous form. Midazolam is unique in that approximately 50 per cent of an orally administered dose undergoes first-pass metabolism in the liver.[33] In patients with chronic liver disease, therefore, systemic availability of the drug is greater when it has been taken orally. Midazolam also may be metabolized extra-hepatically.[34] An *in vitro* study showed that cyclosporin A may inhibit midazolam metabolism, but this inhibition has not been observed *in vivo*.[35] Midazolam does not produce thrombosis or thrombophlebitis, and its metabolites do not have sedative effects.

Diazepam, which is available in tablet and intravenous form in the USA and in Europe, is a popular drug for reduction of preoperative anxiety, especially when patients can be treated earlier than one day before surgery. The distribution half-life of diazepam is 1 hour, and excretion half-life is 32.9 ± 8.8 hours.[36] As with midazolam, elimination half-life is longer in older individuals, primarily because of an increased volume of distribution (Fig. 29.4). Diazepam requires microsomal non-conjugative pathways for elimination and therefore should not be used for

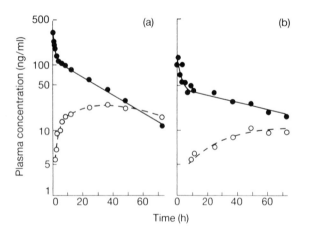

Fig. 29.4   Diazepam (solid line) and desmethyldiazepam (broken line) plasma concentration versus time after intravenous injection of diazepam (0.1 mg/kg) in two healthy adults with no laboratory evidence of disease: a 20-year-old (a) and a 67-year-old (b). The elimination half-life of diazepam was 21.6 hours for the younger person and 51.9 hours for the older person. (Reproduced, with permission, from Klotz U, Avant GR, Hoyumpa A, *et al*. The effects of age and liver disease on the disposition and elimination of diazepam in adult man. *J Clin Invest* 1975; **55**: 347–59.)

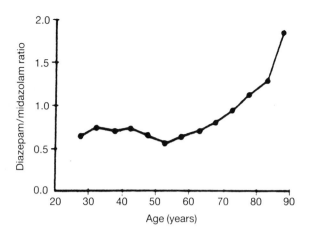

Fig. 29.5   Above 60 years of age, subjects who need sedation for upper gastrointestinal endoscopy become progressively more sensitive to midazolam, compared to diazepam. (Reproduced, with permission, from Scholer SG, Schafer DF, Potter JF. The effect of age on the relative potency of midazolam and diazepam for sedation in upper gastrointestinal endoscopy. *J Clin Gastroenterol* 1990; **12**: 145–7.)

patients with acute hepatitis. Two hours after administration of diazepam, the presence of desmethyldiazepam, a metabolite with similar pharmacological properties, can be detected; clearance of desmethyldiazepam takes 36 hours. The prolonged action of diazepam is thus attributable not only to its extended excretion half-life but also to the long half-life of its metabolite. In patients up to 60 years of age, diazepam is one-half to one-fourth as potent as midazolam. In patients older than 60 years who become progressively more sensitive to midazolam, the relative potency of diazepam increases (Fig. 29.5).[37] Propylene glycol, which is used to promote solubility of the injectable form of diazepam, produces pain on injection and thrombophlebitis. Venous thrombosis is more frequent in older patients (Fig. 29.6) and when diazepam is injected into smaller veins. Orally administered diazepam is well absorbed from the intestine, and plasma levels peak after 60 minutes. Injection of diazepam into the gluteal region, particularly in areas of fat, produces low plasma levels of the drug. If a narcotic or atropine is co-administered with oral diazepam, the plasma concentration of diazepam decreases, although this effect can be reversed by metoclopramide. Serum levels of diazepam may increase if food is ingested 5 hours after intravenous administration, possibly because diazepam is excreted in the bile and reabsorbed from the intestine.

Lorazepam is approximately four times as potent as diazepam: 2.5 mg of orally administered lorazepam is equivalent to 10 mg of diazepam.[38] Onset of action of lorazepam is slower; peak drug effect does not occur for 40 minutes after intravenous injection.[39] Duration of action is approximately three or four times greater than that of diazepam; mean elimination half-life of lorazepam is 14–15 hours. The metabolites of lorazepam are not active and fewer venous sequelae result from lorazepam than with

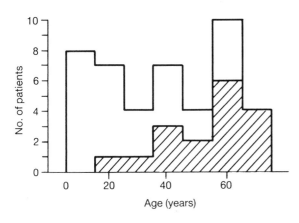

Fig. 29.6 The incidence of venous thrombosis increases with age. The clear outline represents the total number of patients given diazepam, 10 mg. The shaded outline represents the number of patients having venous thrombosis 7–10 days later. (Reproduced, with permission, from Hegarty JE, Dundee JW. Sequelae after the intravenous injection of three benzodiazepines – diazepam, lorazepam and flunitrazepam. *Br Med J* 1977; **2**: 1384–5.)

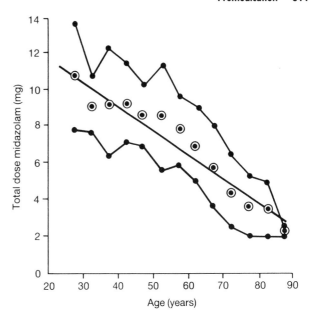

Fig. 29.7 As age increases, the dose of midazolam required to provide sedation adequate for upper gastrointestinal endoscopy decreases. (Reproduced, with permission, from Scholer SG, Schafer DF, Potter JF. The effect of age on the relative potency of midazolam and diazepam for sedation in upper gastrointestinal endoscopy. *J Clin Gastroenterol* 1990; **12**: 145–7.)

diazepam (15 per cent vs 39 per cent). Absence of recall and drowsiness last longer after lorazepam than after diazepam. In one study, 50 per cent of patients given lorazepam did not recall events from most of the day.[40]

Adverse reactions can occur with any of the benzodiazepines, and the large number reported after treatment with triazolam has curtailed its use. Confusion, bizarre behaviour, amnesia, agitation and hallucinations were reported in the elderly after treatment with triazolam.

The benzodiazepines reduce anxiety and produce amnesia and drowsiness. Generally these effects are inversely related to patient age. For example, the dose of midazolam required to produce sedation decreases approximately 15 per cent per decade of life (Fig. 29.7).[41] The dose of midazolam should therefore be adjusted for the elderly, who may be more sensitive than young patients to its effects.

The effect of the benzodiazepines on the incidence of seizures has not been determined. In animals, diazepam reduced seizure threshold and protected against seizures caused by local anaesthetics. However, no study of patients undergoing regional anaesthesia has demonstrated that benzodiazepines reduce the incidence of seizures. Diazepam 0.14 mg/kg in humans did not blunt the degree of severity of the number of CNS symptoms after lignocaine (lidocaine) 1 mg/kg.[42] Also, convulsions have been associated with the use of midazolam in a few cases.

Adverse effects attributed to treatment with benzodiazepines include respiratory and cardiac depression; psychomotor impairment, which may delay discharge from hospital or return to normal acitivities; and amnesia. Thrombocytopenia and pancytopenia have occurred after short-term use of lorazepam. Allergic reactions associated with the benzodiazepines are extremely rare. More common

are perception distortions such as fantasies or hallucinations. Some patients who are treated with benzodiazepines experience sexual fantasies and subsequently allege sexual assault. To forestall such potential allegations, it has been suggested that another person be present when sedating patients.

Benzodiazepines can cause hypotension. When midazolam was first used for the elderly, it received adverse publicity in the lay press because of deaths attributed to hypotension and respiratory depression. This publicity led to an awareness of the need to reduce the dose of midazolam for older patients. Recent data suggest that, compared with diazepam, midazolam places patients at no additional risk.[43] The decrease in blood pressure observed in healthy individuals given benzodiazepines is trivial (10 per cent) but in older patients, particularly those with heart disease, blood pressure may decrease from 20 to 35 per cent and may be associated with apnoea. The decreases in blood pressure are greater when other drugs (e.g. alfentanil, methohexitone (methohexital), morphine, thiopentone (thiopental) and the inhaled agents) are used with benzodiazepines during induction or maintenance of anaesthesia. Decrease in blood pressure is accentuated if the patient has cardiac disease. Loss of baroreflex control of heart rate occurs after benzodiazepines, although loss is less than that with the inhalational agents.

Because the benzodiazepines also depress respiratory function, oxygen supplementation greatly improves oxygenation during sedation with benzodiazepines. After a dose

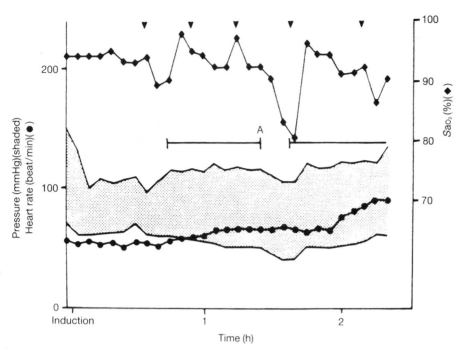

Fig. 29.8 A patient is given incremental doses of midazolam at various times (arrows) and the patient's oxygen saturation decreases (point A) when the oxygen catheter is removed. Blood pressure, heart rate and arterial oxygen saturation (Sao_2) are measured. The horizontal lines indicate administration of oxygen, 2 l/min. (Reproduced, with permission, from Smith DC, Crul JF. Oxygen desaturation following sedation for regional analgesia. *Br J Anaesth* 1989; **62**: 206–9.)

of midazolam, oxygenation can decrease to the point of cardiorespiratory arrest and death. This decrease can be noted during endoscopic procedures and during regional conduction block performed without the use of supplemental oxygen (Fig. 29.8).[44] Because oxygen saturation may decrease whenever benzodiazepines are given, routine oxygen supplementation, with or without continuous monitoring of arterial oxygenation, is recommended. The respiratory response to carbon dioxide is also blunted by the benzodiazepines. Respiratory depression after narcotics is further potentiated by the benzodiazepines. This fact is important because more than 50 per cent of patients who died of respiratory failure associated with midazolam also received narcotics. Decreased response to carbon dioxide after midazolam is much more profound in patients with

chronic obstructive pulmonary disease.[45] Risk of respiratory failure also may be increased when intravenous benzodiazepines are used to sedate patients with neurological injuries.

Psychomotor impairment lasted for up to 7 hours after diazepam and up to 3 hours after midazolam in volunteers.[46] In patients, however, for whom factors such as pain and nausea also affect recovery, the differences recorded in volunteers may not be important. As expected, psychomotor function after lorazepam is impaired longer than after diazepam or midazolam, and so lorazepam is probably not appropriate for day-stay (ambulatory) surgery.

Amnesia that occurs after premedication with benzodiazepines can be a problem. In one report, patients who received midazolam did not remember seeing their anaesthetist or surgeon before surgery. Amnesia can last for up to 4 hours after lorazepam and is more intense than after diazepam (Figs. 29.9 and 29.10). In controlled studies, retrograde amnesia has not been demonstrated in patients who received benzodiazepines.

Recommendations

For the patient seen at least 24 hours before surgery who seems anxious about the procedure despite adequate reassurance and explanation, oral diazepam 10 mg/70 kg is prescribed for 06:00 on the day of surgery, or for 08:00 if the surgery is later than noon. This dose may be adjusted according to the patient's age, with a reduction of 5–10 per cent per decade for patients more than 20 years of age. For example, a 70-year-old patient would receive 25–50 per cent

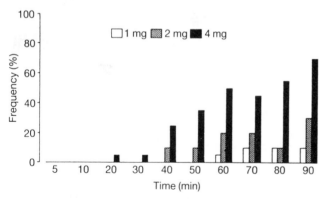

Fig. 29.9 Percentage failure to recognize a picture (assessed 6 hours after surgery), shown at the times indicated, after oral administration of 1, 2 or 4 mg of lorazepam. (Reproduced, with permission, from McKay AC, Dundee JW. Effect of oral benzodiazepines on memory. *Br J Anaesth* 1980; **52**: 1247–57.)

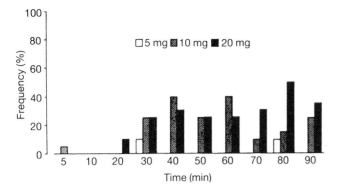

Fig. 29.10 Percentage failure to recognize a picture (assessed 6 hours after surgery), shown at the times indicated, after oral administration of 5, 10 or 20 mg of diazepam. (Reproduced, with permission, from McKay AC, Dundee JW. Effect of oral benzodiazepines on memory. *Br J Anaesth* 1980; **52**: 1247–57.)

less diazepam than a 20-year-old. For the patient seen in the preoperative waiting area who seems anxious despite reassurance, midazolam, 0.03 mg/kg intravenously, with a similar downward adjustment for increasing age, is recommended.

β-Adrenergic blocking drugs

β-Adrenergic receptor blocking drugs have been shown to reduce anxiety in stressful situations such as test-taking, public speaking or performance, and participation in athletics. They can even improve performance. β-Adrenergic blocking drugs are as effective as benzodiazepines in reducing perioperative anxiety without the disadvantage of impairing function.[47] It is unclear if patients regularly taking β-adrenergic blocking drugs have less anxiety. Nevertheless, premedication with these drugs before surgery deserves further investigation.

α₂-Adrenergic agonists and haemodynamic stability

Premedication of patients with α₂-adrenergic agonists such as clonidine has reduced extremes of blood pressure intraoperatively. Clonidine depresses sympathetic activity by activating central medullary α₂-adrenergic receptors that then inhibit sympathetic outflow. Preoperative administration of clonidine reduces anaesthetic requirement and decreases extremes in arterial blood pressure during anaesthesia (Fig. 29.11). Minimum alveolar concentration (MAC) is not reduced, however, simply because central release of catecholamines is reduced. Dexmedetomidine, a more potent α₂-agonist, still reduced MAC in rats depleted of noradrenaline (norepinephrine).[48] Hypnotic anaesthetic action is therefore probably mediated via the activation of central α₂-adrenoreceptors. Indeed, part of the mechanism by which clonidine decreases extremes in blood pressure during anaesthesia may be by reducing anaesthetic requirement. Decreases in variability in blood pressure with the α₂-adrenergic agonists may be more apparent in hypertensive than in normotensive patients.

That haemodynamic variability decreases with clonidine has been noted during intubation; during abdominal, head and neck, and orthopaedic surgery; and in the elderly (Fig. 29.12). Haemodynamic stability has not been reported, though, in all studies with clonidine.

One of the effects of clonidine is increased sedation, an increase that does not always decrease anxiety. To maintain stable levels of the drug throughout surgery, it also may be useful to apply a transdermal clonidine patch when administering clonidine orally. Recovery may or may not be prolonged. Certainly, less anaesthesia is required with clonidine, which may lessen the need for analgesia. Another beneficial effect of the α₂-agonists may be the prevention of

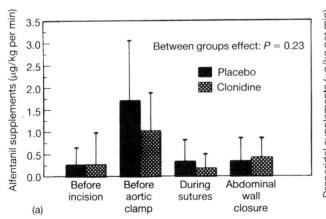

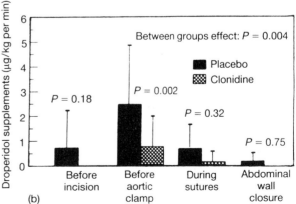

Fig. 29.11 (a) Mean ± SD doses of alfentanil, after either placebo or clonidine, injected to control blood pressure during each of the four periods. (b) If alfentanil proved insufficient, mean ± SD doses of droperidol, injected to control blood pressure. (Reproduced, with permission, from Engelman E, Lipszyc M, Gilbart E, Van der Linden P, Bellens B, Van Romphey A, de Rood M. Effects of clonidine on anesthetic drug requirements and hemodynamic response during aortic surgery. *Anesthesiology* 1989; **71**: 178–87.)

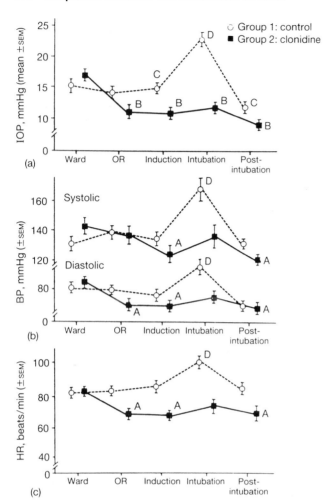

Fig. 29.12 Clonidine decreases (a) intraocular pressure (IOP), (b) blood pressure (BP) and (c) heart rate (HR) responses to stimulation during elective ophthalmic surgery under general anaesthesia. Variables were measured before premedication ('Ward'), on arrival at the operating room ('OR'), after induction ('Induction') within 3 minutes of tracheal intubation ('Intubation') and 8 minutes after tracheal intubation ('Post-intubation'). Differences were significant (A, $P < 0.05$; B, $P < 0.01$) when values were compared with values on the ward. Differences were significant (C, $P < 0.05$; D, $P < 0.001$) when values for a given measurement were compared between groups. (Reproduced, with permission, from Ghignone M, Noe C, Calvillo O, Quintin L. Anesthesia for ophthalmic surgery in the elderly: the effects of clonidine on intraocular pressure, perioperative hemodynamics, and anesthetic requirement. *Anesthesiology* 1988; **68**: 707–16.)

muscle rigidity resulting with narcotics. Whether they shorten recovery time has not been established.

Adverse effects of clonidine include dry mouth, which depends, like sedation, on plasma concentration of drug. Hypotension, particularly with doses of 0.3 mg or more, and bradycardia also may occur. In awake patients, clonidine, 5 μg/kg, can attenuate the effect of intravenous atropine (Fig. 29.13). Oral clonidine augments the pressor responses to intravenous ephedrine before and during general anaesthesia. In anaesthetized patients, the pressor effects of

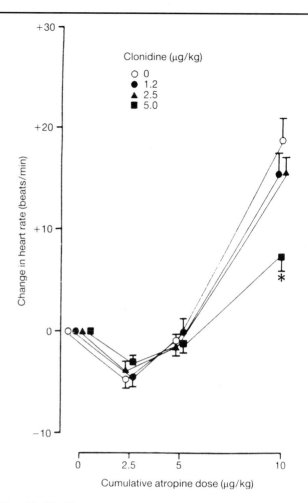

Fig. 29.13 Heart rate responses to a cumulative intravenous dose of atropine, 2.5, 5 and 10 μg/kg, in patients also given oral doses of clonidine 0 ($n = 20$), 1.2 ($n = 20$), 2.5 ($n = 20$) and 5.0 μg/kg. *Significantly different from no clonidine ($P < 0.01$); clonidine, 1.2 μg/kg ($P < 0.05$); and clonidine, 2.5 μg/kg ($P < 0.01$). (Reproduced, with permission, from Nishikawa T, Dohi S. Oral clonidine blunts the heart rate response to intravenous atropine in humans. *Anesthesiology* 1991; **75**: 217–22.)

ephedrine are potentiated by α-adrenoreceptor-mediated vasoconstriction, whereas, in awake patients, the effect of ephedrine is potentiated predominantly by a β-adrenergic effect.

Dexmedetomidine, a more selective agonist than clonidine, has been studied in a few clinical trials. Dexmedetomidine reduced thiopentone requirement, decreased blood pressure lability during laryngoscopy and intubation, reduced postoperative narcotic requirement, increased fatigue and caused dry mouth. Probably because of the reduced anaesthetic requirement, patients had less nausea and recovery was rapid. α₂-Adrenergic agonists seem to be attractive alternatives to the benzodiazepines, especially for the treatment of individuals who are or are likely to be hypertensive in surgery. For a more comprehensive review of α₂-adrenoceptor agonists in anaesthesia, see Maze and Tranquilli.[49]

Restriction of oral intake before surgery

Vomiting and aspiration of gastric contents during induction of anaesthesia can cause pulmonary damage if the volume of aspirate reaches 25 ml/kg and the pH is ≤ 2.5, although fluids of lower pH and smaller volume have also caused damage. The amount of fluid in the stomach necessary to produce a corresponding volume of aspirate is unknown. Patients who are pregnant or morbidly obese, who have a hiatus hernia or who smoke (smokers tend to be more anxious than non-smokers) are at greater risk of aspiration. Other predisposing factors for aspiration include neurological disorder, oesophageal disease, extremes of age and emergency or night surgery. Drugs that reduce anxiety should theoretically decrease the risk of aspiration. In one study, the benzodiazepine diazepam was found to reduce gastric secretion and total acid output.[50]

Anaesthetists usually instruct patients to take nothing by mouth for 6–8 hours before surgery to reduce the risk of aspiration of gastric contents. However, liquids taken 2–3 hours before surgery may actually reduce gastric volume and increase gastric pH. When patients who drank tea or coffee with milk and ate one slice of buttered toast on the morning of surgery were compared with patients who had fasted, gastric volume and pH did not differ between groups.[51] In another comparison, the volume of aspirated gastric contents was greater after patients drank 100 ml of water than after 50 ml of water.[52] Approximately 20 per cent of patients fast for more than 12 hours, which may increase the risk of perioperative dehydration, particularly in older patients. One advantage of allowing coffee drinkers their morning coffee is that the incidence of headaches after surgery is reduced.[53] The Canadian Anaesthetists Society recommends that unrestricted clear fluids be allowed until 3 hours before the scheduled time of surgery. With the liberalization of 'nil by mouth' policy, it is unclear if the incidence of vomiting, possibly with aspiration, will increase because, during induction of anaesthesia, some patients who claimed to have fasted according to instructions vomit a partially digested meal.

It is difficult to know if all patients should be asked to fast for only 3 hours before surgery or for 6 hours. We allow healthy patients without other risk factors to drink small amounts of clear liquids before elective surgery. If patients normally drink coffee in the morning, they may drink one cup of black coffee 2–3 hours before surgery. Fasting restrictions for patients at greater risk may be different. Patients at increased risk may remain at risk even if they fast for 6 hours. These patients may benefit from pharmacological control of the risks of aspiration with drugs that decrease gastric volume and increase gastric pH.

Pharmacological control of the risks of aspiration

Histamine H_2-receptor antagonists, such as cimetidine, ranitidine, famotidine and nizatidine, increase pH in the stomach by blocking the action of histamine on histamine H_2-receptors and thus inhibiting gastric stimulation in response to acetylcholine, histamine or gastrin. The antagonist effects of cimetidine begin approximately 60–90 minutes after administration and last for about 3 hours. Other histamine H_2-receptor antagonists are several times more potent than cimetidine: nizatidine, three times; ranitidine, four to six times; and famotidine, 20 times.

Cimetidine inhibits cytochrome P_{450} more than the other commonly prescribed H_2-receptor antagonists and therefore potentiates drugs that depend on certain cytochrome P_{450} microsomal enzyme systems in the liver for metabolism. Potentiation can occur after only 1 day of therapy. Thus metabolism of diazepam and midazolam is affected by cimetidine, but metabolism of lorazepam is not.[54] Ranitidine has a much weaker effect on cytochrome P_{450} systems so

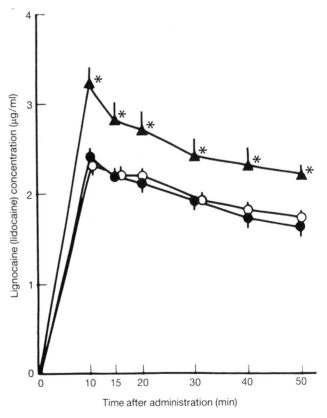

Fig. 29.14 Mean plasma concentration of lignocaine administered epidurally after famotidine (●), cimetidine (▲) or control (○). Bars represent SEM; * $P < 0.01$ compared to control or famotidine. (Reproduced, with permission, from Kishikawa K, Namiki A, Miyashita K, Saitoh K. Effects of famotidine and cimetidine on plasma levels of epidurally administered lignocaine. *Anaesthesia* 1990; **45**: 719–21.)

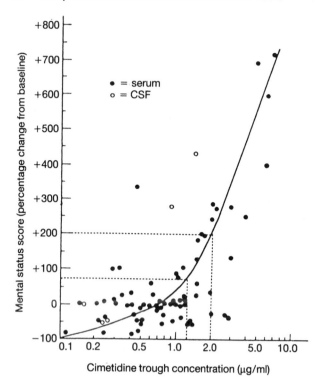

Fig. 29.15 Cimetidine sometimes causes mental confusion. Scores for mental status (percentage change from baseline) vs trough concentrations of cimetidine in serum and cerebrospinal fluid (CSF) are plotted. A 200 per cent increase over baseline was judged to constitute significant mental impairment. A normal trough concentration of cimetidine is less than 1.25 μg/ml. (Reproduced, with permission, from Schentag JJ, Cerra FB, Calleri G, DeGlopper E, Rose JQ, Bernhard H. Pharmacokinetic and clinical studies in patients with cimetidine-associated mental confusion. *Lancet* 1979; **1**: 177–81.)

the interaction of ranitidine is not as great as the interaction of cimetidine with other drugs. Famotidine and nizatidine do not appreciably bind cytochrome P$_{450}$.

Blood flow in the liver also decreases by 25 per cent after acute administration of cimetidine to the patient who has fasted, and by 33 per cent after chronic administration. Therefore, drugs such as propranolol, for which elimination depends on blood flow in the liver, are eliminated more slowly if cimetidine is also administered.[55] Other drugs that interact with cimetidine, with recommendations for therapy in brackets, include: oral anticoagulants, e.g. warfarin [monitor coagulation variables]; theophylline [decrease maintenance dose by 50 per cent but do not change loading dose]; phenytoin and carbamazepine [reduce maintenance dose by 33 per cent]; and lignocaine (lidocaine) [reduce maintenance dose by 50 per cent]. Plasma levels of bupivacaine and lignocaine will be higher during regional blocks in patients treated with cimetidine (Fig. 29.14).

Cimetidine has been associated with mental confusion within 48 hours of the first dose, especially in older patients (Fig. 29.15).[56] This deterioration occasionally follows kidney

or liver pathology. Symptoms include restlessness, confusion, disorientation, agitation, hallucinations, focal twitching, seizures and unresponsiveness; apnoea is associated with trough concentrations of cimetidine of more than 2 μg/ml. Confusion is much less common with ranitidine.

Omeprazole is a substituted benzimidazole that inhibits the gastric enzyme (hydrogen–potassium–adenosine triphosphatase (ATPase)) mediating the production of hydrochloric acid in the parietal cell. The half-life of omeprazole ranges from 0.3 to 2.5 hours. Its metabolite is active and binds irreversibly to gastric hydrogen–potassium–ATPase. There are few studies of omeprazole in the perioperative setting, and they report mixed results. Like cimetidine, omeprazole inhibits certain cytochrome P$_{450}$ systems and therefore reduces metabolism of drugs depending on these systems for metabolism.

Sodium citrate and Bicitra (a commercial preparation of sodium citrate and citric acid) are soluble (absorbable) antacids useful in buffering acidic gastric contents. These soluble antacids increase gastric pH but, unfortunately, may also increase gastric volume.

Metoclopramide is a dopamine antagonist. It increases lower oesophageal sphincter pressure, speeds gastric emptying and prevents or alleviates nausea and vomiting. It is especially effective when used with other drugs such as sodium citrate (Bicitra) or cimetidine. Although metoclopramide is an effective antiemetic, this action is not uniformly present in patients undergoing surgery. Anaesthetists generally administer a dose of 0.10–0.15 mg/kg; to prevent emesis after cisplatin therapy, oncologists administer doses as high as 2 mg/kg. Side effects of drowsiness, agitation, restlessness and dystonia occur most frequently in the young, within 36 hours after the start of treatment, and only rarely after a single dose. Metoclopramide inhibits plasma cholinesterase and so recovery time after suxamethonium (succinylcholine) muscle blockade may be prolonged.[57]

Recommendations

Routine treatment of patients for prevention of acid aspiration is unwarranted. However, for patients at risk, including the morbidly obese, those who have hiatus hernia and those who are pregnant, metoclopramide and an H$_2$-receptor antagonist (but not cimetidine in the older patient) may be preferred.

Other premedicants

Narcotics are administered to decrease anxiety and to sedate the patient before surgery; to control hypertension intraoperatively, primarily during tracheal intubation; and to decrease pain postoperatively. Various studies of

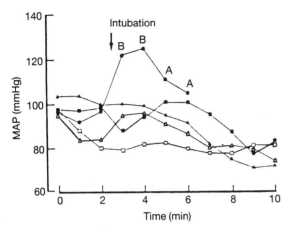

Fig. 29.16 Narcotic administration controls hypertension associated with tracheal intubation. Mean arterial blood pressure (MAP) is plotted against time after laryngoscopy and intubation. Control subjects were given saline (●). Study subjects were given either: alfentanil, 15 µg/kg (∧) or 30 µg/kg (□), intravenously after administration of thiopentone; or fentanyl, 5 µg/kg, before (■) or after (▲) administration of thiopentone. A and B indicate significant increases (A, $P < 0.05$; B, $P < 0.01$). (Reproduced, with permision, from Black TE, Kay B, Healy TEJ. Reducing the haemodynamic responses to laryngoscopy and intubation. A comparison of alfentanil with fentanyl. *Anaesthesia* 1984; **39**: 883–7.)

fentanyl, however, have shown that it relieves anxiety, is useful with benzodiazepines and is not more effective than placebo. Many studies purporting to examine relief of anxiety with narcotics do not use control subjects or test only a single dose of narcotic. Narcotics can control intraoperative hypertension, particularly during tracheal intubation (Fig. 29.16). After intubation, hypotension can recur.

Premedication with alfentanil, butorphanol, fentanyl, morphine, nalbuphine or pethidine (meperidine) to prevent anxiety results in nausea, although administration of atropine with narcotics can decrease nausea and vomiting.[58] Narcotics also prolong gastric emptying time. Metoclopramide 10 mg i.v. antagonizes this effect of morphine.[59]

Oral transmucosal fentanyl, primarily for premedication of children, also may be useful for adults. Whether fentanyl is administered orally or parenterally, the side effects are similar.

We give narcotics preoperatively for relief of pain, to supplement block in patients undergoing regional anaesthesia, and to attenuate the cardiovascular reflexes in response to tracheal intubation in patients with a history of hypertension.

Anticholinergic drugs are administered as premedication to decrease secretions and bronchospasm and to control hypotension or bradycardia. The resulting dryness of the mouth is uncomfortable for the patient but is helpful for procedures on the mouth or bronchus. A patient's anxiety

can naturally lead to dry mouth, which is exacerbated by deprivation of fluids. All anticholinergics are similar in that they decrease salivation, although, compared to atropine, glycopyrrolate produces less tachycardia, pyrexia and blurred vision. The anticholinergics may be useful to control hypotension or bradycardia during induction or maintenance of anaesthesia. Anticholinergic drugs do not prevent conditions that lead to acid aspiration.

Preoperative control of postoperative nausea

Nausea and vomiting, the most common complications after surgery, are usually the reason for hospital admission of day-stay (ambulatory) surgery patients. Nausea occurs more frequently after certain operations such as eye surgery. It also seems to occur more frequently in younger patients and in women more than in men (Fig. 29.17).[60] Certain premedicants, such as narcotics, worsen postoperative nausea and vomiting. Other drugs that control perioperative nausea and vomiting include droperidol, ondansetron and transdermally administered hyoscine (scopolamine). Lower doses of droperidol (0.25–0.5 mg i.v.) seem to be more effective than higher doses and do not delay recovery.[61]

Ondansetron, a serotonin antagonist, is used primarily for treatment of chemotherapy-induced nausea and vomiting. Ondansetron is equally or more effective than metoclopramide, which has been the standard treatment

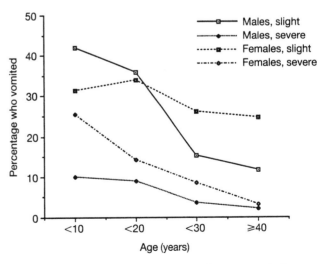

Fig. 29.17 Female patients are more likely to vomit than are male patients after surgery. The incidence decreases with age for both sexes. (Reproduced, with permission, from Burtles R, Peckett BW. Postoperative vomiting: some factors affecting its incidence. *Br J Anaesth* 1957; **29**: 114–23.)

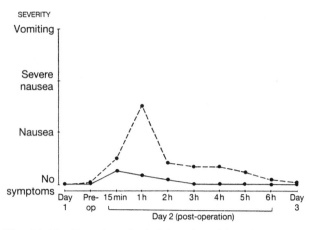

Fig. 29.18 Transdermal administration of hyoscine (scopolamine) (solid line) given the evening before surgery controls postoperative nausea and vomiting better than a placebo (broken line). (Reproduced, with permission, from Tolksdorf W, Meisel R, Müller P, Bender H-J. Transdermales scopolamin (TTS-Scopolamin) zur prophylaxe postoperativer übelkeit and erbrechen. *Anaesthesist* 1985; **34**: 656–62.)

for chemotherapy-induced nausea. The half-life of ondansetron is approximately 3.5 hours in adults, shorter in children and longer in the elderly (mean time, 7.9 hours). The primary adverse effect is headache, although patients may also have diarrhoea, constipation, sedation and transient minor elevations in liver function tests. The drug is not associated with extrapyramidal reactions. The antiemetic effect of ondansetron is enhanced with dexamethasone. Ondansetron, 4–8 mg, proved to be an effective antiemetic for patients undergoing anaesthesia. In one study in which ondansetron significantly reduced postoperative nausea, patients given placebo and then a 'rescue' antiemetic of metoclopramide (20 mg i.v.) and hydroxyzine (25 mg i.v.) had the same incidence of nausea as patients given ondansetron.[62]

Transdermally administered hyoscine, popular in the prevention of motion sickness, shows promise for the prevention of postoperative nausea and vomiting. For hyoscine levels to peak, the patch must remain on the skin for 8 hours. The patch is effective for up to 72 hours, and placement of the patch on patients at least 8 hours before surgery is recommended (Fig. 29.18).[63] A patch placed 1–2 hours before surgery is not effective. Effectiveness may also depend on the other drugs used for surgery: with higher doses of narcotics during an anaesthetic, nausea and vomiting may be greater even if the patch is used. The

patch is effective in older patients, although some older patients develop transient delirium or psychosis. The hyoscine patch should be avoided in patients over 60 years of age. Anisocoria results if patients scratch the patch and then touch their eyes.

Nausea should be treated with drugs when patients have a history of severe nausea and vomiting after anaesthesia; when they report the tendency to become nauseated easily (e.g. car sickness); and when they are to undergo a procedure, such as eye surgery, associated with a high incidence of nausea. If the need for treatment can be determined at least 12 hours before surgery, a hyoscine patch can be placed the night before surgery. For patients who are not seen 12 hours preoperatively or who are over 60 years of age, a small dose of droperidol is given before or immediately after induction of anaesthesia.

Summary

The preoperative visit is essential for preparing patients for anaesthesia and surgery. During this time, a screening history can be obtained using a questionnaire. The patient's responses can be used to select laboratory tests and to disclose disease states or potential high risk. A physical examination can be performed at this time, with attention to the airway, chest and cardiovascular and central nervous systems. Non-pharmacological or pharmacological methods may be used to reduce perioperative anxiety. For pharmacological intervention to control anxiety, the benzodiazepines are useful preoperatively: oral diazepam for treatment in advance of surgery and midazolam immediately before surgery.

Administering drugs that may increase nausea is undesirable. Possible therapy for postoperative nausea can be considered during the preoperative visit, and prophylactic therapy for acid aspiration of gastric contents is advisable for patients known to be at risk.

Finally, it is important to note that, in the studies cited in this chapter, doses were controlled and times were measured accurately. Apart from a study environment, it may not be possible to follow such strict protocols, and the effects of drugs may be greater or less than expected or desired.

REFERENCES

---

1. Vacanti CJ, VanHouten RJ, Hill RC. A statistical analysis of the relationship of physical status to postoperative mortality in 68,388 cases. *Anesthesia and Analgesia* 1970; **49**: 564–6.

2. Lewin I, Lerner AG, Green SH, Del Guercio LRM, Siegel JH. Physical class and physiologic status in the prediction of operative mortality in the aged sick. *Annals of Surgery* 1971; **174**: 17–31.

3. Goldman L, Caldera DL, Nussbaum SR, *et al.* Multifactorial index of cardiac risk in noncardiac surgical procedures. *New England Journal of Medicine* 1977; **297**: 845–50.

4. Keats AS. The ASA classification of physical status – a recapitulation. *Anesthesiology* 1978; **49**: 233–6.

5. Rehder K. Clinical evaluation of isoflurane: complications during and after anaesthesia. *Canadian Anaesthetists' Society Journal* 1982; **29** (suppl): S44–8.

6. Cohen MM, Duncan PG. Physical status score and trends in anaesthetic complications. *Journal of Clinical Epidemiology* 1988; **41**: 83–90.

7. Ziffren SE, Hartford CE. Comparative mortality for various surgical operations in older versus younger age groups. *Journal of the American Geriatrics Society* 1972; **20**: 485–9.

8. Domoto K, Ben R, Wei JY, Pass TM, Komaroff AL. Yield of routine annual laboratory screening in the institutionalized elderly. *American Journal of Public Health* 1985; **75**: 243–5.

9. Wolf-Klein GP, Holt T, Silverstone FA, Foley CJ, Spatz M. Efficacy of routine annual studies in the care of elderly patients. *Journal of the American Geriatrics Society* 1985; **33**: 325–9.

10. Delahunt B, Turnbull PRG. How cost effective are routine preoperative investigations? *New Zealand Medical Journal* 1980; **92**: 431–2.

11. Rosselló PJ, Cruz AR, Mayol PM. Routine laboratory tests for elective surgery in pediatric patients: are they necessary? *Boletin – Asociacion Medica de Puerto Rico* 1980; **72**: 614–23.

12. Narr BJ, Hansen TR, Warner MA. Preoperative laboratory screening in healthy Mayo patients: cost-effective elimination of tests and unchanged outcomes. *Mayo Clinic Proceedings* 1991; **66**: 155–9.

13. Macpherson DS, Snow R, Lofgren RP. Preoperative screening: value of previous tests. *Annals of Internal Medicine* 1990; **113**: 969–73.

14. Mitchell TL, Tornelli JL, Fisher TD, Blackwell TA, Moorman JR. Yield of the screening review of systems: a study on a general medical service. *Journal of General Internal Medicine* 1992; **7**: 393–7.

15. Sandler G. Costs of unnecessary tests. *British Medical Journal* 1979; **2**: 21–4.

16. Rabkin SW, Horne JM. Preoperative electrocardiography: its cost-effectiveness in detecting abnormalities when a previous tracing exists. *Canadian Medical Association Journal* 1979; **121**: 301–6.

17. Rabkin SW, Horne JM. Preoperative electrocardiography: effect of new abnormalities on clinical decisions. *Canadian Medical Association Journal* 1983; **128**: 146–7.

18. Roizen MF, Kaplan EB, Schreider BD, Lichtor JL, Orkin FK. The relative roles of the history and physical examination, and laboratory testing in preoperative evaluation for outpatient surgery: the 'Starling' curve of preoperative laboratory testing. *Anesthesiology Clinics of North America* 1987; **5** (1): 15–34.

19. Turnbull JM, Buck C. The value of preoperative screening investigations in otherwise healthy individuals. *Archives of Internal Medicine* 1987; **147**: 1101–5.

20. Levinstein MR, Ouslander JG, Rubenstein LZ, Forsythe SB. Yield of routine annual laboratory tests in a skilled nursing home population. *Journal of the American Medical Association* 1987; **258**: 1909–15.

21. Roizen MF. Preoperative evaluation. In: Miller RD, ed. *Anesthesia*, 3rd ed., vol. 1. New York: Churchill Livingstone, 1990: 743–72.

22. Kaplan EB, Sheiner LB, Boeckmann AJ, *et al.* The usefulness of preoperative laboratory screening. *Journal of the American Medical Association* 1985; **253**: 3576–81.

23. Lutner RE, Roizen MF, Stocking CB, *et al.* The automated versus the personal interview. Do patient responses to preoperative health questions differ? *Anesthesiology* 1991; **75**: 394–400.

24. McKee RF, Scott EM. The value of routine preoperative investigations. *Annals of the Royal College of Surgeons of England* 1987; **69**: 160–2.

25. Apfelbaum J, Roizen MF, Robinson D, *et al.* How frequently do asymptomatic patients benefit from the pursuit of abnormalities in their preoperative test results? *Anesthesiology* 1990; **73**: A1254.

26. Blery C, Szatan M, Fourgeaux B, *et al.* Evaluation of a protocol for selective ordering of preoperative tests. *Lancet* 1986; **1**: 139–41.

27. Charpak Y, Blery C, Chastang C, Szatan M, Fourgeaux B. Prospective assessment of a protocol for selective ordering of preoperative chest x-rays. *Canadian Journal of Anaesthesia* 1988; **35**: 259–64.

28. Johnston M. Anxiety in surgical patients. *Psychological Medicine* 1980; **10**: 145–52.

29. McCleane GJ, Cooper R. Forum. The nature of pre-operative anxiety. *Anaesthesia* 1990; **45**: 153–5.

30. Wallace LM. Psychological preparation as a method of reducing the stress of surgery. *Journal of Human Stress* 1984; **10**: 62–77.

31. Lawlis GF, Selby D, Hinnant D, McCoy CE. Reduction of postoperative pain parameters by presurgical relaxation instructions for spinal pain patients. *Spine* 1985; **10**: 649–51.

32. Greenblatt DJ, Abernethy DR, Locniskar A, Harmatz JS, Limjuco RA, Shader RI. Effect of age, gender, and obesity on midazolam kinetics. *Anesthesiology* 1984; **61**: 27–35.

33. Greenblatt DJ, Divoll M, Abernethy DR, Ochs HR, Shader RI. Clinical pharmacokinetics of the newer benzodiazepines. *Clinical Pharmacokinetics* 1983; **8**: 233–52.

34. Park GR, Manara AR, Dawling S. Extra-hepatic metabolism of midazolam. *British Journal of Clinical Pharmacology* 1989; **27**: 634–7.

35. Li G, Treiber G, Meinshausen J, Wolf J, Werringloer J, Klotz U. Is cyclosporin A an inhibitor of drug metabolism? *British Journal of Clinical Pharmacology* 1990; **30**: 71–7.

36. Klotz U, Antonin K-H, Bieck PR. Pharmacokinetics and plasma binding of diazepam in man, dog, rabbit, guinea pig and rat. *Journal of Pharmacology and Experimental Therapeutics* 1976; **199**: 67–73.

37. Scholer SG, Schafer DF, Potter JF. The effect of age on the relative potency of midazolam and diazepam for sedation in upper gastrointestinal endoscopy. *Journal of Clinical Gastroenterology* 1990; **12**: 145–7.

38. Dundee JW, McGowan WAW, Lilburn JK, McKay AC, Hegarty JE. Comparison of the actions of diazepam and lorazepam. *British Journal of Anaesthesia* 1979; **51**: 439–46.

39. Dundee JW, Lilburn JK, Nair SG, George KA. Studies of drugs given before anaesthesia. XXVI: lorazepam. *British Journal of Anaesthesia* 1977; **49**: 1047–56.

40. Fragen RJ, Caldwell N. Lorazepam premedication: lack of recall and relief of anxiety. *Anesthesia and Analgesia* 1976; **55**: 792–9.

41. Bell GD, Spickett GP, Reeve PA, Morden A, Logan RF. Intravenous midazolam for upper gastrointestinal endoscopy: a study of 800 consecutive cases relating dose to age and sex of patient. *British Journal of Clinical Pharmacology* 1987; **23**: 241–3.

42. Haasio J, Hekali R, Rosenberg PH. Influence of premedication on lignocaine-induced acute toxicity and plasma concentrations of lignocaine. *British Journal of Anaesthesia* 1988; **61**: 131–4.

43. Arrowsmith JB, Gerstman BB, Fleischer DE, Benjamin SB. Results from the American Society for Gastrointestinal Endoscopy/US Food and Drug Administration collaborative study on complication rates and drug use during gastro-intestinal endoscopy. *Gastrointestinal Endoscopy* 1991; **37**: 421–7.

44. Smith DC, Crul JF. Oxygen desaturation following sedation for regional analgesia. *British Journal of Anaesthesia* 1989; **62**: 206–9.

45. Gross JB, Zebrowski ME, Carel WD, Gardner S, Smith TC. Time course of ventilatory depression after thiopental and midazolam in normal subjects and in patients with chronic obstructive pulmonary disease. *Anesthesiology* 1983; **58**: 540–4.

46. Galletly D, Forrest P, Purdie G. Comparison of the recovery characteristics of diazepam and midazolam. *British Journal of Anaesthesia* 1988 **60**: 520–4.

47. Dyck JB, Chung F. A comparison of propranolol and diazepam for preoperative anxiolysis. *Canadian Journal of Anaesthesia* 1991; **38**: 704–9.

48. Segal IS, Vickery RG, Walton JK, Doze VA, Maze M. Dexmedetomidine diminishes halothane anesthetic require-ments in rats through a postsynaptic alpha$_2$-adrenergic receptor. *Anesthesiology* 1988; **69**: 818–23.

49. Maze M, Tranquilli W. Alpha-2 adrenoceptor agonists: defining the role in clinical anesthesia. *Anesthesiology* 1991; **74**: 581–605.

50. Birnbaum D, Karmeli F, Tefera M. The effect of diazepam on human gastric secretion. *Gut* 1971; **12**: 616–18.

51. Miller M, Wishart HY, Nimmo WS. Gastric contents at induction of anaesthesia: is a 4-hour fast necessary? *British Journal of Anaesthesia* 1983; **55**: 1185–8.

52. Brocks K, Jensen JS, Schmidt JF, Jørgensen BC. Gastric contents and pH after oral premedication. *Acta Anaesthesio-logica Scandinavica* 1987; **31**: 448–9.

53. Fennelly M, Galletly DC, Purdie GI. Is caffeine withdrawal the mechanism of postoperative headache? *Anesthesia and Analgesia* 1991; **72**: 449–53.

54. Klotz U, Reimann I. Delayed clearance of diazepam due to cimetidine. *New England Journal of Medicine* 1980; **302**: 1012–14.

55. Feely J, Wilkinson GR, Wood AJJ. Reduction of liver blood flow and propranolol metabolism by cimetidine. *New England Journal of Medicine* 1981; **304**: 692–5.

56. Schentag JJ, Cerra FB, Calleri G, DeGlopper E, Rose JQ, Bernhard H. Pharmacokinetic and clinical studies in patients with cimetidine-associated mental confusion. *Lancet* 1979; **1**: 177–81.

57. Kao YJ, Turner DR. Prolongation of succinylcholine block by metoclopramide. *Anesthesiology* 1989; **70**: 905–8.

58. Riding JE. Post-operative vomiting. *Proceedings of the Royal Society of Medicine* 1960; **53**: 671–7.

59. McNeil MJ, Ho ET, Kenny GNC. Effect of i.v. metoclopramide on gastric emptying after opioid premedication. *British Journal of Anaesthesia* 1990; **64**: 450–2.

60. Burtles R, Peckett BW. Postoperative vomiting: some factors affecting its incidence. *British Journal of Anaesthesia* 1957; **29**: 114–23.

61. Millar JM, Hall PJ. Nausea and vomiting after prostaglandins in day case termination of pregnancy: the efficacy of low dose droperidol. *Anaesthesia* 1987; **42**: 613–18.

62. Bodner M, White PF. Antiemetic efficacy of ondansetron after outpatient laparoscopy. *Anesthesia and Analgesia* 1991; **73**: 250–4.

63. Tolksdorf W, Meisel R, Müller P, Bender H-J. Transdermales scopolamin (TTS-scopolamin) zur prophylaxe postoperativer übelkeit und erbrechen *Anaesthesist* 1985; **34**: 656–62.

Chapter 30

Neonatal Anaesthesia

James M. Steven and John J. Downes

Neonatal physiology
Neonatal clinical pharmacology
Neonatal anaesthetic equipment
Preanaesthetic evaluation and preparation
Intraoperative anaesthetic management

Regional anaesthesia
Postanaesthetic care
Complications of anaesthesia
Conclusion

The neonate, an infant from birth to age 28 days, who requires anaesthesia presents unique and formidable challenges to the general anaesthetist. These patients face significantly greater risks of anaesthesia-related cardiac arrest, complications and death than do older children and adults.

An understanding both of the adaptive physiology of the newborn to extrauterine existence and of the responses of the neonate to drugs forms the basis for safe neonatal anaesthetic practice. Not only the paediatric anaesthetist but also the general anaesthetist need to understand the fundamentals of applied neonatal physiology and pharmacology, and how neonates differ from older children and adults. The general anaesthetist may be called upon to assist in stabilizing the neonate with a surgical condition prior to transport to a specialist paediatric centre. In more rural areas with limited transport facilities, or in less developed countries, the general anaesthetist must be capable of providing full anaesthetic care to a neonate. However, the neonate requiring anaesthesia for any type of procedure, but especially for major operations, should be transferred whenever feasible to a specialist paediatric centre staffed by fully qualified paediatric anaesthetists and surgeons.

Neonatal physiology

The newborn's physiological adaptations to extrauterine life of concern to the anaesthetist involve the pulmonary and cardiovascular systems, the central and autonomic nervous systems, metabolism, thermal homoeostasis, and fluid and electrolyte balance.

Pulmonary and cardiovascular systems

Because of the close functional interactions between the pulmonary and cardiovascular systems in the neonate we will consider the physiology of these two anatomically different systems together. We focus on the transfer of oxygen from the external environment (placenta or atmospheric air) to the tissues, and on the removal of carbon dioxide from the tissues to the external environment.

Fetal cardiopulmonary development

The lungs develop in the 4-week-old embryo as a ventral pouch extending into the mesenchyme from the primitive pharynx.[1,2] Concurrently, a vascular tube forms in the mesenchyme ventral to these lung buds, to begin the development of the heart and great vessels. By 8 weeks the epithelial cords, which will become the trachea and bronchi with segmental branches, have pushed caudad and ventrally into the mesenchyme accompanied by a pulmonary arterial trunk and branching vessels. The four-chambered heart and great vessels have already developed, reflecting the disparity in cardiovascular development that precedes pulmonary maturation throughout fetal, neonatal and infant life.

The tracheobronchial tree, by 16 weeks of gestation, achieves 16 branching generations ending in the terminal bronchiole, the distal growth centre for development of the gas exchange unit, the acinus.[2] If the lungs fail to develop all 16 branches by this stage, through disease, genetic defect or the pressure of a space-occupying lesion (e.g. diaphragmatic hernia), they never will and pulmonary hypoplasia will result.

The pulmonary capillaries begin to approximate the primitive alveolar saccules at 24 weeks, and the type 2 alveolar epithelial cells initiate secretion of an immature surfactant at the alveolar gas–liquid interface. This establishes the potential for alveolar gas exchange with pulmonary capillary blood. In most fetuses, however, the diffusion barrier for oxygen, the matching of pulmonary capillary flow with alveolar gas, the stability of the alveoli and the central control of breathing remain inadequate for effective spontaneous and unsupported ventilation until at least 26 weeks' gestation.[1,3]

As the fetus approaches 34–36 weeks, the proliferating acini develop multiple branching saccules and a few true alveoli.[3] The pulmonary capillaries envelop these terminal units, the alveolar capillary membrane reduces its thickness and the type 2 epithelial cells secrete mature surfactants (phosphatidyl choline and phosphatidyl glycerol).[3,4]

Total and lung dynamic compliance as well as airways conductance, when related to lung volume (functional residual capacity), differ little in the neonate, child and adult.[5] The neonate, however, exhibits a high thoracic compliance which reflects the exceptional compressibility of the thoracic cage that prevents rib fractures during vaginal delivery.[5] By term, the diaphragm has increased its muscle fibre content and potential power such that it can exert a transpleural subatmospheric pressure up to 80 cmH$_2$O in initiating the first breath of air.

Total lung volume, crying vital capacity and functional residual capacity (FRC) of the neonate are all significantly less per unit body mass compared to the child over 7 years.[6] This discrepancy is even more striking when compared to the infant's metabolic needs and alveolar ventilation (V̇A). The average FRC of a normal newborn (30 ml/kg) is only slightly less than the mean adult value (34 ml/kg),[6] but the FRC/V̇A ratios differ greatly (newborn 0.23, adult 0.56). This indicates a much greater reserve volume of gas in the lung of the adult in relation to alveolar ventilation and oxygen consumption. The normal fetal lung and bellows system at term can, nevertheless, assume the burden of gas exchange to meet the total metabolic needs of extrauterine existence. Only when cardiopulmonary disorders afflict the neonate does this diminished pulmonary reserve volume become clinically problematic.

Cardiopulmonary adaptation to extrauterine life

The fetal cardiovascular system performs admirably with the placenta as the organ of fetal external gas exchange. The parallel circuit flow patterns in the fetus must convert

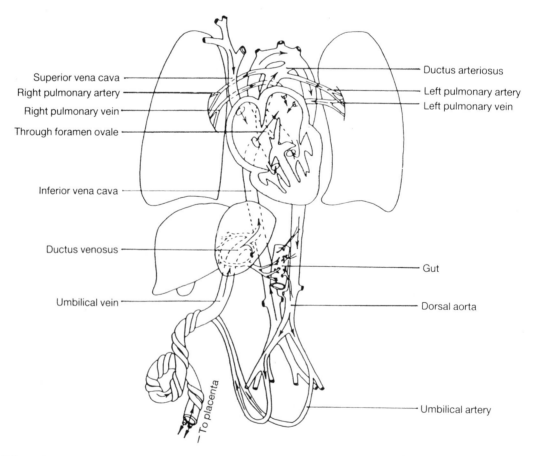

Fig. 30.1   Schematic representation of the fetal circulation (not to scale). (Reproduced, with permission, from Avery ME. *The lung and its disorders in the newborn*, Philadelphia: WB Saunders, 1964.)

to a series circuit at birth (Fig. 30.1). This is accomplished through a dramatic increase in pulmonary blood flow accompanied by closure of the ductus arteriosus and foramen ovale. Central to this conversion is an abrupt and substantial reduction in the high pulmonary vascular resistance of the fetus, which allows only 10 per cent of the right ventricular output to flow through the lungs.[7,8] In the term fetus, pulmonary and systemic vascular resistances and mean pressures, as well as right and left ventricular volumes and wall thicknesses, are roughly equal. Following the first few breaths and lung inflation at birth, sufficient pulmonary blood flow for adequate gas exchange normally occurs, yet the cardiovascular adaptations to a low pressure pulmonary system and a higher pressure systemic system, comparable to that of the adult, evolve only gradually over the first 6 postnatal months.

The myocardium during development undergoes many changes in membrane proteins, contractile myofilaments and extracellular matrix.[9] The myofilament content increases and contractile force improves due to maturation of cross-bridge attachments, myofibril orientation and strengthening of the cytoskeleton and extracellular matrix.

In the early days following birth, left ventricular wall thickness increases considerably while only minimal thickening of the right ventricular wall occurs.[9] These changes reflect the altered impedance of the systemic circuit (increased) and pulmonary circuit (decreased). Electrophysiological conduction matures in parallel with these changes, with the gradual emergence of left ventricular predominance in the ECG during the first year.

Although the neonatal heart responds to some degree as predicted by the Frank–Starling principle, the neonate adjusts cardiac output to meet rapidly changing metabolic demands primarily by altering heart rate rather than stroke volume. This occurs because the relatively incompliant ventricular walls tend to limit the range of stroke volume.[9] A heart rate that declines 20–30 per cent below normal levels for a neonate (e.g. less than 100 beats per minute) may be associated with a reduction in cardiac output.[10] In response to acute hypoxaemia, the neonate initially increases cardiac output by raising heart rate, but as hypercapnia or lactic acidaemia ensues, stroke volume decreases.[10,11] Changes in regional vascular resistance allow an increased blood flow to the heart, brain and liver compared to other regions; however, pulmonary blood flow will decrease if there is concomitant acidosis. In the later phases of hypoxaemia associated with apnoea, bradycardia and eventually a fall in cerebral blood flow ensue.[12]

Failure of the newborn to establish adequate alveolar ventilation and pulmonary blood flow will be followed quickly by failure to recover from the birth asphyxia that results from virtually every vaginal and most caesarean deliveries.[13] Persistent acidaemia and hypoxaemia, in turn, cause relaxation of ductal smooth muscle as well as pulmonary vasoconstriction[8] which induces diminished right ventricular compliance. This leads to right-to-left shunting across the foramen ovale. Opening of ductus and foramen ovale, combined with raised pulmonary vascular resistance, reduces pulmonary blood flow by diverting right ventricular output to the systemic circulation. The infant remains in, or relapses to, a condition of perfusion by a parallel fetal circulatory system designed for placental rather than pulmonary gas exchange. A spiral into fatal asphyxia ensues unless interrupted immediately by externally controlled ventilation with oxygen, and elevation of blood pH.

The principal differences in fetal and neonatal red blood cells, compared with those of the older infant, are the content of fetal haemoglobin and the relatively low concentrations of 2,3-diphosphoglycerate (2,3-DPG). Fetal haemoglobin (HbF) binds oxygen more readily than adult haemoglobin (HbA), and, combined with lower 2,3-DPG levels, results in a left-shifted oxyhaemoglobin dissociation

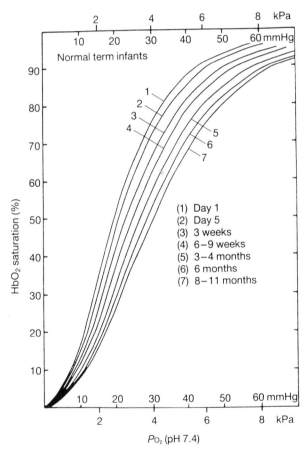

Fig. 30.2 The P_{50} on day 1 of 2.7 kPa (20 mmHg) shifts to 4 kPa (30 mmHg) by 12 months of age due to increasing red cell concentrations of 2,3-diphosphoglycerate; this and the replacement of fetal by adult haemoglobin result in a rightward shift from the fetal to adult positions of the oxyhaemoglobin dissociation curve. (Reproduced, with permission, from Oski F, Delivoria-Papadopoulos M. The red cell 2,3-diphosphoglycerate and tissue oxygen release. *Journal of Pediatrics* 1970; **77**: 941–56.)

curve and a lower P_{50} (PaO_2 at SaO_2 50 per cent) of 2.7 kPa (20 mmHg) versus 4 kPa (30 mmHg) at 1 year of age (Fig. 30.2).[14] This favours uptake of oxygen in the placenta or lung, but also decreases the volume of oxygen unloaded in the peripheral tissues. Nevertheless, provided capillary flow persists at normally high levels, and arterial haemoglobin content and saturation remain in the normal range, tissue oxygen delivery readily meets metabolic demands of the healthy neonate.

The newborn (age 0–7 days) must make five major cardiopulmonary adaptations to survive after birth.

1. Reflexly initiate breathing and expand the lungs with gas. Air must permeate the lung liquid with minute gas bubbles during the first breath, creating a transient foam which reaches the alveoli and begins alveolar gas exchange with blood. This establishes a functional residual capacity with concomitant rapid absorption of fetal lung liquid by the pulmonary lymphatics. During this process, the phospholipid surfactant material suspended in the lung liquid must spread over the newly formed alveolar gas–liquid–epithelial membrane interface, thereby stabilizing the expanded alveoli and permitting gas exchange with pulmonary capillary blood throughout the ventilatory cycle.

2. Convert the fetal parallel circulatory pattern with high pulmonary vascular resistance and minimal pulmonary blood flow *to an adult series circulatory pattern* with low pulmonary vascular resistance that allows the entire output of the right ventricle to flow through the lungs. This conversion entails three major steps: (a) reduction in pulmonary vascular resistance that occurs with expansion of the lungs with gas, elevation of alveolar oxygen tension, and decrease in both blood and alveolar carbon dioxide tension with a concomitant increase in blood pH[13] (Fig. 30.3; Table 30.1); (b) functional closure of the overlapping flaps of the foramen ovale which occurs as pulmonary blood flow and left atrial volume increase, thereby raising left atrial pressure above that of the right atrium; and (c) functional closure of the ductus arteriosus due to intense contraction

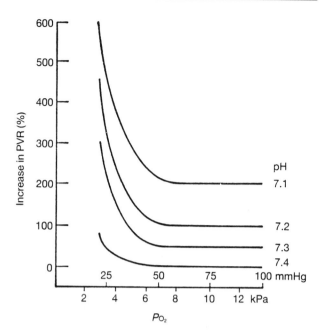

Fig. 30.3 Relationship of pulmonary vascular resistance (PVR) to systemic arterial pH and PaO_2 in the newborn calf. (Reproduced, with permission, from Rudolph AM. *Congenital diseases of the heart.* Chicago: Year Book Medical, 1974.)

of ductal smooth muscle induced by increasing PaO_2 above 5.3 kPa (40 mmHg), adrenergic stimulation, and withdrawal of prostaglandin E_2 which maintained ductal patency *in utero*.[11]

Various autonomic and endocrine factors, especially other endoperoxides such as prostacyclin, play important roles in reducing pulmonary vascular resistance, closing the ductus, raising systemic vascular resistance, and integrating the timing of the cardiovascular events with pulmonary events to favour optimal cardiopulmonary adaptation. Anatomical closure (fibrosis) of the ductus has usually been completed by a postnatal age of 4 weeks, whereas anatomical sealing of the foramen ovale will take more than 6 weeks with probe-patent interatrial communication in 50 per cent of children at age 5 years and in 25

Table 30.1 Normal arterial pH, blood gas tensions, haematocrit (mean ± SD)

	Age		
	1 h[13]	*24 h*[13]	*1–24 mo*[15,16]
Arterial pH	7.33 (0.03)	7.37 (0.03)	7.40 (0.03)
$PaCO_2$ (mmHg)	36 (4)	33 (3)	34 (4)
(kPa)	4.8 (0.5)	4.4 (0.4)	4.5 (0.5)
Base excess (mEq/l)	−6.0 (1)	−5.0 (1)	−3.0 (3)
PaO_2 (mmHg) (FIO_2 0.21)	63 (11)	73 (10)	83 (11)
(kPa)	8.4 (1.5)	9.7 (1.3)	11.1 (1.5)
Haematocrit (vol %)	53 (53)	55 (7)	35 (2.5)

per cent of young adults, although the volume of shunted blood has no physiological significance.

3. *Recover from birth asphyxia*, a normal consequence of vaginal and often of caesarean section delivery, in which umbilical arterial pH, $PaCO_2$, bicarbonate concentration and PaO_2 reflect a mixed respiratory and non-respiratory (metabolic) acidaemia and hypoxaemia. The healthy term newborn achieves recovery from birth asphyxia in 1–6 hours through rapid uptake of oxygen and increased haemoglobin saturation following the onset of ventilation, excretion of excess carbon dioxide stores by ventilation, and metabolism of tissue lactic acid by the liver (Table 30.1).

4. *Establish adequate spontaneous ventilation.* The centrally integrated control of breathing, even in the full-term newborn (37–42 weeks' gestation), has not matured and differs from that of older infants and children in that:

(a) the CO_2–ventilatory response curve, whilst of normal slope, has a lower resting alveolar CO_2 tension and thus a leftward displacement comparable to that of the pregnant mother; the lower $PaCO_2$ persists until at least 24 months of age[15] (Table 30.1; Fig. 30.4a);

(b) the ventilatory response to inhalation of a hypoxic gas mixture, rather than being hyperventilation as in the adult, consists of a transient hyperventilation (peripheral chemoreceptor response) followed by significant reduction in minute volume which persists after inhalation of air has been restored; this response is corroborated by a depressed ventilatory response to increased inspired carbon dioxide in 15 per cent oxygen, the opposite of that seen in adults (Fig. 30.4b). This reduced response normally ceases by age 3 weeks, but the hyperventilatory adult response does not develop until an unknown later age;[17] and

(c) periodic breathing, consisting of intermittent episodes of short apnoea (less than 5 seconds) interspersed with normal or tachypnoeic respiratory frequencies, occurs commonly in preterm infants and has been observed up to 52 weeks post-conception in term infants.[17] Periodic breathing causes no clinically important impairment of gas exchange.

Thus, certain ventilatory control responses fundamental to survival, such as the hyperventilatory response to hypoxia, are not well developed in the term newborn, and are even less developed in preterm infants. More than 70 per cent of preterm newborns incur periodic breathing, nearly half of these suffering prolonged apnoeic episodes (over 15 seconds) accompanied by bradycardia and arterial oxyhaemoglobin desaturation below 90 per cent. Ventilatory control matures in relation to post-conceptual age from 26 weeks, when protracted apnoea with severe hypoxaemia usually occurs postnatally, through to 37 weeks when apnoea causing hypoxaemia rarely occurs.

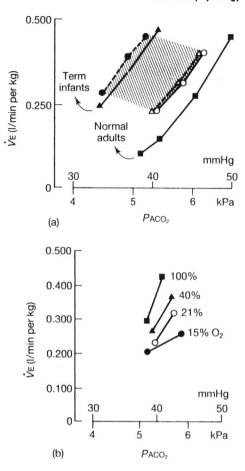

Fig. 30.4 (a) Steady state CO_2 response curves in preterm infants breathing periodically (○), regularly (●) or semiperiodically (intermediate) (△) with curves for term infants (▲) and normal adults (■). (b) Steady state CO_2 response curves at different inspired O_2 concentrations: the lower the inspired O_2, the lower the slope of CO_2 response. (Reproduced, with permission, from Rigatto H. Apnea. *Pediatric Clinics of North America* 1982; **29**: 1105–16.)

5. *Maintain consistent ventilation and lung volumes.* This presumes integrity of ventilatory control and stability of alveoli as well as chest wall and diaphragm, and adequate pulmonary capillary perfusion. The newborn's pulmonary function, compared to that of the adult, is presented in Table 30.2.

Central and autonomic nervous systems

Developmental aspects of the neonatal nervous system relevant to anaesthetic care are the following.

1. The neonatal brain receives approximately one-third of the cardiac output, in contrast to the adult brain's one-seventh of the cardiac output. Autoregulation of global cerebral flow over the full range of normal arterial pressures exists even in unstressed preterm infants.[18] However, this autoregulation appears to be

Table 30.2 Normal pulmonary function (mean values)[6]

	Newborn	Adult
Oxygen consumption (ml/kg per min)	6.4	3.5
Carbon dioxide production (ml/kg per min)	6.0	3.0
Exhaled minute volume (ml/kg per min)	210	90
Alveolar ventilation (ml/kg per min)	130	60
Respiratory rate (breaths/min)	35	15
Tidal volume (ml/kg)	6	6
Vital capacity (ml/kg)	35	70
Functional residual capacity (ml/kg)	30	34
FRC/V_A ratio	0.23	0.57
Total lung capacity (ml/kg)	63	86
Anatomical dead space (ml/kg)	2.5	2.0
Physiological dead space/tidal volume ratio	0.3	0.3
Tracheal diameter (mm)	4	16
Tracheal length (mm)	57	120

more effective in limiting severe surges during acute systemic hypertension than in maintaining flow during systemic hypotension.[18] The cerebrovascular reactivity of the human neonate to $Pa\text{CO}_2$ has been documented, but the response to hypoxaemia (vasodilation) remains uncertain.[18]

2. The blood–brain barrier consists of capillary endothelial walls without fenestrations, with tight junctions and an outer sheath of astrocytic foot processes.[19] Specialized transport mechanisms transfer nutrients and electrolytes across the barrier. Numerous animal studies and a few observations in the human neonate indicate that the mammalian blood–brain barrier remains immature at birth, but develops rapidly in postnatal life.[19] The less mature barrier permits more readily the passage of larger, lipid-soluble molecules such as unbound bilirubin (leading to kernicterus and brain damage) and various drugs.

3. The neonate has fairly advanced development of innervation in the cardiovascular and respiratory systems; parasympathetic and sympathetic functions exist, but do not mature until later in infancy.[20] The newborn's bradycardic response to hypoxaemia and its attenuation by atropine illustrate the predominant parasympathetic response to this and other major autonomic stimuli (such as laryngoscopy).

4. The fetus and newborn, by the time of viability (24–25 weeks' gestation), respond to noxious stimuli with facial grimaces as well as cardiovascular and metabolic stress responses, suggesting perception of pain. Interruption of those responses during painful procedures in the preterm neonate by anaesthetics or narcotics appears to improve the infant's physiological recovery from the procedure,[21] provided the agents and techniques employed did not themselves cause severe physiological derangements (e.g. systemic arterial hypotension).

The physiological stress response observed in preterm neonates undergoing thoracotomy for ligation of a patent ductus arteriosus with 50 per cent nitrous oxide–curare anaesthesia consists of exceptionally high plasma levels of catecholamines, adrenocorticosteroids, glucose, insulin and gluconeogenic substrates. The addition of fentanyl (10 μg/kg) almost completely attenuates these metabolic responses.[21,22] Other data indicate that, under similar anaesthetic circumstances, fentanyl in doses up to 20 μg/kg may be needed to prevent tachycardia and hypertension in newborns.

Whether elicitation of the metabolic and cardiovascular stress response can be equated with the perception of pain (and therefore suffering) by the infant remains conjectural. The issue seems moot because, with only nitrous oxide analgesia, the ensuing systemic arterial hypertension may place the infant at an increased risk of intracranial haemorrhage. Thus, the anaesthetist will usually seek to attenuate the stress response as well as to prevent the perception of pain by providing sufficient anaesthesia.

Renal system and fluid balance

By 36 weeks of gestation, all nephrons seen in the mature kidney have formed, although cortical development continues for months. However, the newborn's kidneys are not functionally mature. Increased renal blood flow causes glomerular filtration rate to rise abruptly in the first few hours after birth.[23] The maximum concentrating ability of the neonate's kidneys during water deprivation or excess loss reaches only 900 mmol/1, 35 per cent less than adult levels (see Table 30.3). On the other hand, the term newborn's capacity to dilute urine following an excessive water load is quite mature and approximates adult levels by 6 weeks of age.[23] Tubular maximal threshold for many solutes, including drugs and p-aminohippuric acid, and maximal concentrating ability increase with advancing post-conceptual age (Table 30.3). The healthy infant's

Table 30.3 Normal basal fluid volumes (range or mean ± SD)

	Age	
	1–30 days	6–24 mo
Urine volume (ml/kg per hour)[24]	1–5	2–4
Urine osmolality (mmol/l)[24]	40–900	50–1400
Extracellular volume (% body wt)[25]	42 (6)	34 (4)
Blood volume (ml/kg)[26,27]	82–107	70–75

kidney at 12–24 months of age approaches an adult level of performance.

The preterm newborn's renal function is even less effective, but normally exhibits striking postnatal improvement by the age of 6 months. The preterm neonate, however, remains ill equipped to conserve sodium if subjected to acute sodium deprivation. On the other hand, sodium restriction seems beneficial in the first postnatal week during routine fluid management of preterm newborns weighing less than 1000 g.[28] The preterm infant's ability to excrete an excessive water load is suboptimal due to lower glomerular filtration. Indeed, infants born at less than 1500 g take 18 months or more to achieve a level of function comparable to that of the adult.[29]

The newborn's total body water constitutes nearly 75 per cent of its body mass compared with 50–60 per cent in the child and adult. Extracellular fluid (ECF) comprises over half of the total body water, or roughly 40 per cent of body mass, in the newborn[25] (Table 30.3; Fig. 30.5). This ECF decreases rapidly over the first few postnatal weeks to approximately 30 per cent of body mass (Fig. 30.5). The contraction of the ECF will not be associated with clinically significant weight loss during the first postnatal week, even in preterm infants with provision of increasing exogenous energy (272 J or 65 cal/kg per day on day 1 to 418 J or 100 cal/kg per day on day 4).[30] Blood volume varies with post-conceptual age in relation to body mass, from approximately 80–110 ml/kg in the neonate, decreasing to about 70–75 ml/kg in the older infant (Table 30.3).[31–33] Serum levels of electrolytes and osmolality differ little with age, whereas levels of energy substrates (glucose and total protein) and metabolic products (urea nitrogen) increase with age (Table 30.4).

Table 30.4 Normal blood chemistries (range)

	Age	
	1–30 days[34]	*1–24 months*[35]
Sodium (mmol/l or mEq/l)	134–152	139–146
Potassium (mmol/l or mEq/l)	5.0–7.7	3.0–6.0
Chloride (mmol/l or mEq/l)	92–114	98–106
Calcium ion (mmol/l or mEq/l)	2.2–2.5	2.2–2.5
Calcium, total (mg/dl)	9.0–11.0	8.8–11.0
Glucose (fasting) (mg/dl)	40–90	60–100
Total protein (g/dl)	5.9–8.5	6.1–7.9
Blood urea nitrogen (mg/dl)	3–12	5–18
Osmolality (mmol/kg)	290–310	290–310

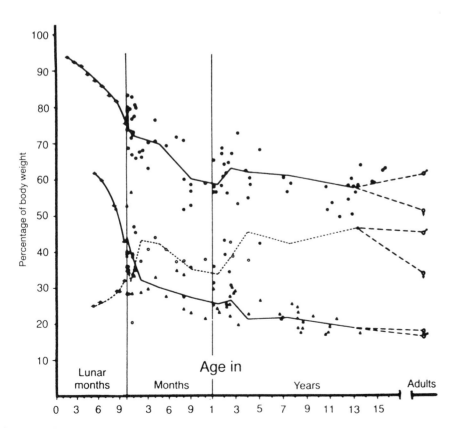

Fig. 30.5 Body water compartments as percentages of body weight from early fetal life to adult life. ●, total body water; ▲, extracellular water; ○, intracellular water. (Adapted from Friis-Hansen GJ[25] by Hill LL. Body composition, normal electrolyte concentrations, and maintenance of normal volume, tonicity, and acid–base metabolism. *Pediatric Clinics of North America* 1990; **37**: 241–56, with permission.)

Water requirements in the preterm neonate differ considerably from those in term neonates. The preterm infant, especially one less than 1500 g, has an immature skin and an exceptionally high ratio of surface area to body mass. This results in unusually large evaporative water losses in relation to metabolic rate, which become even larger during exposure to ambient warmers and photo-therapy units.[23] Accurate daily weights and hourly urine output, coupled with precise records of fluid intake, have proved to be necessary for optimal fluid management. The same issues confront larger preterm and full-term neonates, but their evaporative losses are more predictable.[36]

Thus, the healthy newborn successfully adjusts its fluid balance and extracellular electrolytes to the needs of extrauterine life quite well if birth weight exceeds 1500 g, if given exogenous fluids and calories, and if not subjected to major thermal stress or surgical operations. Outside those limits, the kidneys, like the lungs, cannot be expected to compensate for adverse events nearly as well in the neonate as they will in the 6-month-old infant, larger child or adult.

Metabolism and thermal homoeostasis

Oxygen consumption in the warm, resting newborn averages 5 ml/kg per minute on day 1, increasing to 7–8 ml/kg per minute by day 7.[37] This relatively high oxygen consumption persists throughout infancy, declining gradually in later childhood and adolescence to adult levels of 3–4 ml/kg per minute. The output of carbon dioxide parallels these trends. The respiratory quotient ($\dot{V}CO_2/\dot{V}O_2$) varies from 0.7 to 1.0 in the neonate, depending on diet and stress responses.

Because metabolism drives minute ventilation and cardiac output, in unstressed neonates both are double the levels observed in adults in relation to body mass (see Table 30.2). These increases in ventilation and cardiac output are achieved primarily through acceleration of their rate functions (breaths or beats per minute). Thus, a resting minute volume in the neonate of about 200 ml/kg per minute is associated with a tidal volume of 6 ml/kg and a frequency of 35–40 breaths/min.[38] This contrasts with the adult values of 90 ml/kg per minute, a tidal volume also of 6 ml/kg and a frequency of 15–20 breaths/min. Cardiac output averages 200 ml/kg per minute at rates of 100–180 beats/min in the neonate,[39] compared with 80–100 ml/kg per minute and rates of 55–95 in the adult. Similarly, basic resting energy requirements of the newborn are more than double those of the older child and young adult (580 vs 146 J or 120 vs 35–50 kcal/kg per day).[36]

Glucose is the primary energy substrate for the neonatal brain and myocardium. Even with appropriate nutrition, liver glycogen stores in the neonate, especially the preterm infant,[40] remain limited compared to those of the older infant. Blood glucose levels therefore need to be maintained near optimal levels (40–90 mg/dl)[41] in the sick neonate. Similarly, calcium reserves and their control are less well developed in the neonate than in older infants. Serum ionized calcium should be evaluated throughout the course of neonatal illness and surgery,[42] particularly if large volumes of citrated blood products are given; optimal serum levels are 2.2–2.5 mEq/l.

All humans are homoeothermic, from the smallest preterm newborn to the elderly adult. As such, they attempt to maintain a stable core temperature in the presence of changing environmental conditions. Newborns possess increasing capabilities to defend core temperature, in direct relation to advancing post-conceptual age, through the maturation of central and autonomic thermal controls and the acquisition of body fat. The naked neonate less than 1000 g, for example, lies virtually defenceless against an environmental temperature below 32°C, whereas a healthy term newborn of 3000 g can successfully maintain its core temperature, at least temporarily and at great metabolic cost, in an environmental temperature as low as 22°C.[43,44]

To maintain a core temperature of 37°C despite a cool environment (less than 32°C), the neonate increases metabolism, especially that of highly perfused interscapular and retroperitoneal fat depots ('brown fat'), and thereby raises endogenous heat production as well as oxygen consumption.[43] This non-shivering thermogenesis, the major neonatal thermal defence mode, depends on: (1) intact surface and central responses to the difference between the skin temperature and that of the environment; (2) intact sympathetic nerve endings to release the catecholamines that stimulate the catabolism of fat; (3) the cardiopulmonary capacity to increase oxygen uptake and delivery; and (4) the availability of fat stores to provide free fatty acids which undergo complete metabolism locally or in distant sites and generate calories. Oxygen consumption rises in the term newborn in direct relation to the increasing difference between the abdominal skin temperature over the liver (mean 36°C) and the environmental temperature as that difference exceeds 4°C[43] (Fig. 30.6). The newborn's minimal oxygen requirement occurs when the difference between the skin and environmental temperatures is less than 2–4°C. This commonly prevails at an environmental temperature of 32–34°C with a neonate's abdominal skin temperature at 36°C; such a condition has been termed the 'neutral thermal state'. The preterm infant or any sick neonate should be maintained under these thermal conditions whether in a delivery room, intensive care unit, operating room or during transport.

The environmental temperature of the usual operating room is between 20 and 22°C, creating a temperature difference that causes a threefold increase in oxygen consumption in healthy awake term neonates (Fig. 30.6).[43] This lies at the margin of the infant's ability to defend its core temperature. Thermal stess also elevates plasma catecholamine levels, causing pulmonary and systemic vasoconstriction and metabolic acidaemia; this can precipitate a return to a fetal circulatory pattern, especially if the neonate cannot increase cardiopulmonary function to meet the raised metabolic demand.

Shivering develops in the neonate but appears to become a major source of heat production only after 6 months postnatal age. Anaesthetics and neuromuscular blockade interfere with both non-shivering and shivering thermogenesis. Preservation of core temperature with minimal metabolic stress in an infant exposed to the usual temperatures of an operating room requires application of external sources of heat (see below).[44]

Neonatal clinical pharmacology

The commonly used anaesthetics, muscle relaxants, narcotics and other adjuvant drugs differ in their pharmacokinetics and pharmacodynamics in the neonate as compared to the older child and adult. These differences can be attributed in part to those age-related variations described above in lung volumes, alveolar ventilation, cardiac function and distribution of blood flow, the relative sizes of fluid compartments and metabolism. In addition, other factors in the neonate such as decreased plasma protein binding and hepatic enzyme immaturity affect pharmacokinetics.[45] The major variations which the anaesthetist needs to consider are as follows.

1. The greater volumes of body water and the extracellular fluid compartment in the neonate result in a substantially larger volume of distribution; this leads to the need for larger doses per unit body mass for certain drugs (e.g. suxamethonium (succinylcholine)) to obtain a specific response in the neonate compared with the older child.

2. Body composition differs considerably between the neonate and the adult; the solid vital organs (brain, heart, liver and kidneys) constitute 18 per cent of the body weight of the neonate but only 5 per cent in the adult, whilst muscle mass represents 25 per cent of the newborn's and 40 per cent of the adult's weight. Fat comprises from 3 per cent at 1000 g to 15–28 per cent of the term neonate's weight, and ordinarily

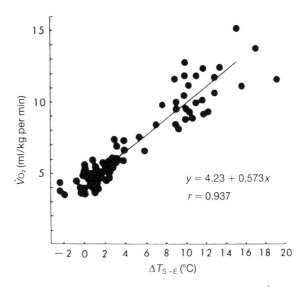

Fig. 30.6 Relation between oxygen consumption ($\dot{V}o_2$) and temperature gradient between skin temperature and environment (ΔT_{S-E}) in full-term human newborns with varying deep-body and skin temperatures. (Reproduced, with permission, from Adamson K, Gandy GM, James LS. The influence of thermal factors upon oxygen consumption of the newborn human infant. *Journal of Pediatrics* 1965; **66**: 495–508.)

The graph shows $\dot{V}o_2$ (ml/kg per min) on the y-axis and ΔT_{S-E} (°C) on the x-axis, with fitted equation $y = 4.23 + 0.573x$, $r = 0.937$.

about 15–18 per cent of the young adult's weight.[46] In the neonate, therefore, a much larger fraction of a dose of an inhaled or injected drug will be distributed to the highly perfused vital organs rather than to muscle or fat.

3. Neonates, especially if born preterm, have significantly lower levels of albumin and other plasma proteins than do older children. Intravenously injected drugs ordinarily bind in part to protein; therefore, they will more readily diffuse into tissues to produce their effects. Fentanyl, sufentanil and ketamine are among the drugs so affected. Metabolic acidosis, decreased total serum protein concentration and abnormally elevated serum unconjugated bilirubin further impair protein binding of drugs.[46]

4. Drugs with high lipid solubility, which include the thiobarbiturates and narcotics, diffuse more readily into the neonatal brain than into that of the older child or adult because of the higher permeability of the neonatal blood–brain barrier.[19] Furthermore, virtually all drugs that penetrate the barrier will be taken up more rapidly by the neonatal brain because of the high proportion of the cardiac output dedicated to cerebral blood flow (34 per cent vs 14 per cent in the adult).[18]

5. Most hepatic enzyme systems in the neonate that contribute to microsomal drug metabolism are either inactive or immature.[46,47] However, neonates activate or develop these enzyme systems as a function of postnatal age (as opposed to postconceptual age), or because exposure to a drug induces a certain enzyme

to operate. The rate of development, however, varies widely among individual infants. Oxidation reactions are particularly underdeveloped in the newborn; these affect the metabolism of thiobarbiturates, fentanyl, morphine, pethidine (meperidine) and ketamine. Hydrolysis reactions may approach adult functional levels, affecting the metabolism of the ester and amide local anaesthetics and suxamethonium. Oxidation reactions tend to reach mature functional levels within the first postnatal week, whereas conjugation reactions (with acetate, glucuronic acid and amino acids) develop over the first 3 postnatal months.[47]

6. The lower glomerular filtration rate of the neonatal kidney results in slower elimination of most drugs and their metabolites, and does not approach levels of adult function until later in infancy. Elimination of conjugated metabolites requires maturity of proximal tubular function which does not occur until approximately 6 months postnatal age in term infants,[23] and up to 18 months in preterm infants.[29]

Anaesthetic and adjuvant drugs

Inhalation agents

The uptake of nitrous oxide and volatile agents and their distribution to vital organs occur more rapidly in the neonate than in older children and adults.[48] A higher alveolar ventilation, smaller functional residual capacity and proportionately greater blood flow to vital organs, combined with comparatively lower solubility in the blood of neonates,[49] account for this phenomenon.

The average minimum alveolar concentration (MAC) of halothane increases from 0.87 per cent in the term newborn to 1.1 per cent at 1 month; it remains at that level until 6 months and then declines to approximately 0.95 per cent at 12 months and 0.9 per cent at 3 years of age[50] (Fig. 30.7). The reasons for these alterations in potency with age remain unclear. The healthy neonate and young infant, however, clearly suffer a greater susceptibility to cardiovascular depression from halothane than do older children or adults,[51] as evidenced by a 33 per cent incidence of systemic arterial hypotension at 1.0 MAC in infants.[50] Clinically significant arterial hypotension (decreases below control of more than 25 per cent) with a concomitant slowing of heart rate[52] also occurs in up to 50 per cent of preterm infants anaesthetized with halothane. Whether these findings represent a hazard to the healthy neonate because of impaired blood flow to vital organs remains debatable, but systemic hypotension in sick infants clearly leads to harmful effects.

Isoflurane exhibits similar age-dependent variations in

MAC (Fig. 30.8).[53] In anaesthetic concentrations, isoflurane lowers systemic arterial pressure in healthy term neonates by decreasing systemic vascular resistance as well as by myocardial depression, and, despite maintenance of preinduction heart rate, cardiac output decreases.[54] Earlier studies[55] found cardiac output sustained in some infants with increased heart rates. However, in preterm infants, systemic arterial pressure decreased even more with isoflurane than with halothane at comparable anaesthetic concentrations (30 per cent below preinduction pressure).[52] Infants 6 months and older undergoing isoflurane anaesthesia for elective repair of congenital cardiac defects exhibited clinically insignificant impairment of left ventricular ejection fraction and systemic arterial pressure.[56] Despite these conflicting results in clinical studies comparing agents in infants, our clinical experience, and that of other clinicians, suggests that both halothane and isoflurane in effective anaesthetic concentrations in infants cause clinically important systemic arterial hypotension.[57,58] The bradycardia associated with halothane can be reversed with atropine.[59] Both agents in concentrations below 1 MAC, if supplemented by nitrous oxide and/or narcotics, can provide adequate anaesthesia and circulatory stability in all but the sickest and smallest neonates.

Enflurane offers only one theoretical advantage over halothane or isoflurane for neonatal anaesthesia, namely a considerably higher threshold for tachydysrhythmias as demonstrated in children.[60] Its major drawbacks are an unacceptable odour on induction, cardiovascular depressant effects (comparable to halothane) and electroencephalographic patterns indicating cortical irritability with hypocapnia and higher anaesthetic concentrations.[61]

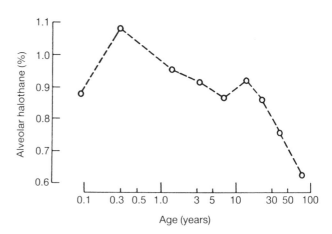

Fig. 30.7 Minimum alveolar concentration (MAC) of halothane vs patient age. Note the higher requirements in infants and teenagers. Neonates have lower requirements than older infants. (Reproduced, with permission, from Coté CJ. Practical pharmacology of anesthetic agents, narcotics and sedatives. In: Coté CJ, Ryan JF, Todres ID, Goudsouzian NG. *A practice of anesthesia for infants and children*, 2nd ed. Philadelphia: WB Saunders, 1993: 113–14.)

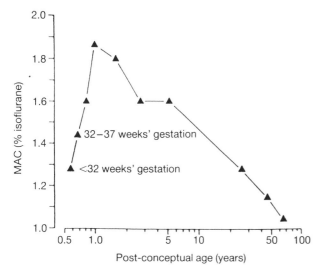

Fig. 30.8 The MAC of isoflurane and postconceptual age. Values for postconceptual age were obtained by adding 40 weeks to the mean postnatal age for each age group. The MAC of isoflurane in preterm neonates is significantly less than in full-term neonates and older infants 1–6 months of postnatal age. (*P*<0.005). (Reproduced, with permission, from LeDez KM, Lerman J. The minimum alveolar concentration (MAC) of isoflurane in preterm neonates. *Anesthesiology* 1987; **67**: 301–7.)

Enflurane seems to have no practical application at this time in neonatal anaesthesia.

Desflurane has been studied to a very limited extent in neonates. The MAC at ages 1–6 months appears to be 9.4 per cent,[62] which is decreased to 7.5 per cent by the addition of 60 per cent nitrous oxide.[63] Paediatric clinical trials in patients from 1 month to 12 years of age have described an unacceptably high incidence of breath holding, stridor and laryngospasm during inhalational induction.[64] Cardiovascular responses in infants resemble those seen with light halothane anaesthesia. The only potential advantage of the agent in neonatal anaesthesia evident at this time, like that observed in older patients, is rapid excretion and awakening. This effect, attributable to a blood/gas partition ratio similar to that of nitrous oxide,[65] rarely offers important advantages to the neonate. In infants anaesthetized with this agent, postanaesthetic pain should be prevented by regional block or narcotics administered in time to be effective prior to termination of the anaesthetic.

Narcotics

Narcotic anaesthesia, in conjunction with pancuronium or vecuronium, offers advantages to the sick or very small (under 1500 g) neonate whose cardiovascular system may be severely impaired by full anaesthetic concentrations of potent volatile agents. The addition of nitrous oxide and/or low dosage halothane or isoflurane reduces the dosage of fentanyl required to attenuate the neonatal stress response (evidenced by increases in systemic arterial pressure and heart rate 20–30 per cent above preanaesthetic levels). Fentanyl in total doses of 10–15 µg/kg, with air or oxygen, causes minimal systemic arterial hypotension in term neonates undergoing a variety of major operative procedures.[66,67] Nevertheless, fentanyl in anaesthetic doses of 10–20 µg/kg can decrease systemic arterial pressure in preterm infants,[52] and depresses significantly the baroreceptor reflex control of the newborn's heart rate.[67] This drug therefore cannot be regarded as completely free of potential adverse haemodynamic effects, and those infants who do experience hypotension will usually respond to a bolus infusion of intravenous fluid (e.g. Ringer's lactate). In comparison to older infants and children, fentanyl has a greatly prolonged yet highly variable elimination half-life in preterm neonates (mean ± SD of 18 ± 9 hours)[68] whereas the term neonate, similar to the adult, has a mean (± SD) elimination half-life of 5.3 ± 1.2 hours.[69] This half-life can be prolonged three- to fourfold in neonates with markedly increased intra-abdominal pressure.[69] Sufentanil, an analogue of fentanyl, has been used in neonatal anaesthesia but offers no advantages over fentanyl. Morphine (in doses of 0.05–0.2 mg/kg) has been employed as an adjunct to general anaesthesia in neonates for over 30 years, but rarely as the primary agent. In infusion doses of 20–100 µg/kg per hour in mechanically ventilated sick term newborns, the elimination half-life of morphine averaged 7 hours compared with 4 hours in older infants (age 3–9 weeks).[70] Unfortunately, neither morphine nor pethidine has undergone careful pharmacodynamic study in the neonate, although both drugs clearly depress ventilation, as does fentanyl.

Postoperative management includes the need for short-term mechanical ventilation in many neonates (4–24 hours), with intensive care nursing observation and monitoring for at least 24 hours to ensure safe recovery from the narcotic effects. Our experience indicates that narcotic antagonists, though effective immediately, do not have sufficient duration of action after a single dose, nor reliability of effect with constant infusion, to forego mechanical support of ventilation. Thus, we reserve narcotic anaesthesia for infants undergoing major intracavitary procedures who will receive intensive care for up to 24 hours post-anaesthesia. These conditions do not, of course, apply to the use of narcotics to supplement inhalation agents, such as fentanyl (in doses of 1 µg/kg to a total of 3 µg/kg) or morphine (0.05–0.2 µg/kg). The prolonged elimination half-life of fentanyl and morphine must be considered in neonates who will be expected to maintain unassisted ventilation following anaesthesia.

Relaxants

In healthy neonates, the commonly used non-depolarizing muscle relaxants (pancuronium and vecuronium) are effective and safe in doses similar to those used in children and adults.[71] Individual variability in duration of effect, however, seems to be much greater, and onset of action appears to be more rapid. The similarity in effective dosage over the age range[72,73] occurs because the greater sensitivity of the immature myoneural junction of the neonate to non-depolarizing relaxants is offset by their larger volume of distribution.[74] Pancuronium increases heart rate and systemic arterial pressure in infants as well as in older patients[71] whereas vecuronium, atracurium and metocurine seem to be free of significant haemodynamic effects in neonates. The duration of blockade due to pancuronium and vecuronium varies considerably from infant to infant, but on the average lasts longer than in older children and adults, probably because of diminished renal clearance.[71] Indeed, the supposedly shorter acting vecuronium has a threefold greater duration of action in the neonate than in the child (59 min vs 18 min).[75] Atracurium's relatively shorter duration of effect offers little advantage for most procedures in neonates. As in older patients, halothane and isoflurane reduce the effective dose requirement and prolong the duration of non-depolarizing blockade in the neonate.[71]

Infants 1–10 weeks of age require substantially larger doses (in mg/kg) of suxamethonium to achieve effective blockade than observed in children and adults.[76] This increased dosage requirement in neonates appears to be caused, in part, by a larger volume of distribution. Neonates develop tachyphylaxis and a phase 2 block similar to that seen in adults at doses over 5 mg/kg.[71] Neonates require intravenous doses (in mg/kg) of atropine and neostigmine similar to older children and adults for reversal of neuromuscular blockade.

Intravenous induction agents

Thiopentone (thiopental) and ketamine offer practical means of achieving rapid induction of anaesthesia for the purpose of tracheal intubation in the neonate, with the same secondary effects as found in older patients. Thiopentone dose requirements for induction are increased in the infant under 12 months as compared to the adult (ED_{50} 6–7 mg/kg vs 4.0–4.5 mg/kg) (Fig. 30.9),[77] but the onset of unconsciousness appears more rapidly in the infant due to their higher brain blood flow. As in the older patient, myocardial depression[78] and systemic vasodilatation can result in systemic arterial hypotension following intravenous thiopentone in the hypovolaemic or seriously ill infant.

Ketamine in neonates offers the major advantages of minimal circulatory depression, no direct elevation of

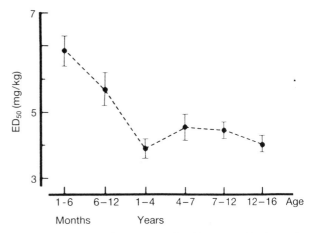

Fig. 30.9 Estimated $ED_{50} \pm$ SE for thiopentone in the various age groups. (Reproduced, with permission, from Jonmarker C, Westrin P, Larsson S, Werner D. Thiopental requirements for induction of anesthesia in children. *Anesthesiology* 1987; **67**: 104–7.)

pulmonary vascular resistance and negligible respiratory depression. The total dose requirements for anaesthesia vary widely among individuals but are considerably higher in infants under 6 months of age than in older children.[79] The intravenous induction dose requirement, however, seems to be similar at all ages (2 mg/kg).[79] Intramuscular ketamine (4–6 mg/kg) is effective for induction prior to achieving difficult vascular access, but time to onset of unconsciousness varies among infants. Elimination is prolonged due to slower metabolism and renal excretion.[47] The drug seems to be most suitable for induction of anaesthesia in some neonates with complex cardiovascular lesions, and in infants whose blood volume status is uncertain. Ketamine transiently increases intracranial pressure but this effect usually poses minimal concern in a neonate with an open fontanelle.

Neonatal anaesthetic equipment

The two anaesthetic gas delivery systems employed widely today for neonates are the various modifications of the Mapelson-D system and the circle absorber system with small-bore tubing. The original Mapelson-D modification, a simple, safe, partial rebreathing system first developed by Jackson Rees, has been widely employed throughout Europe, Canada and the USA from 1960 through 1980. The Mapelson modification most commonly used over the past decade has been the coaxial system of Bain in which the fresh gas flows through an inner tube while warmed expired gas flows through the larger outer tube.[80] Unfortunately, it wastes larger volumes of gases and vapours because of the need for higher total flows.[81] The desire to conserve costly new volatile agents and limit operating room pollution has led to a return to the circle absorber systems, modifications including small-bore delivery system tubing, a 0.5 litre rebreathing bag and neonatal bellows for the anaesthetic machine's mechanical ventilator. The machine must have an accurate airway pressure manometer to display the peak and end-expired inflating pressures imposed on the infant's airway by both manual and mechanical ventilation. As with adult patients, continuous assessment of inspired oxygen concentration and of expired concentrations of carbon dioxide and anaesthetic gases and vapours is a requisite for safe practice.

Currently recommended masks for infants of all sizes are transparent, and include soft, easily conforming cushioned rims which enable the anaesthetist to achieve a gas-tight fit without excessive pressure on the skin, eyes or nose (which can be compressed easily, causing nasal airway obstruction). Infant sizes of Guedel plastic oropharyngeal airways (sizes 000 through 0) should be available to relieve supraglottic soft-tissue airway obstruction.

The laryngeal mask airway (LMA) affords benefits for neonates in a few specific circumstances.[82] As with any anaesthetic by mask, this device is designed to deliver inhaled anaesthetics to spontaneously ventilating patients, although it will permit controlled ventilation if airway resistance and lung compliance are near normal. Because neonates exhibit a propensity for hypotension with inhaled volatile anaesthetics, the use of an LMA should be limited to brief procedures in healthy neonates. It enables access to the face for some superficial procedures (e.g. ophthalmological examinations) which the placement of the facemask will impede. In addition, the LMA can bypass soft-tissue obstruction in neonates with congenital malformations that compromise the airway (e.g. Pierre Robin syndrome). Placed after topical anaesthesia in such a neonate, the LMA permits a smooth inhalation induction or controlled ventilation to facilitate tracheal intubation with any of several techniques.[83]

The neonate should have the trachea intubated to ensure airway patency throughout the anaesthetic and early emergence. Infant laryngoscope blade configurations vary but the Miller series (sizes 0 and 1), preferably equipped with a cannula and proximal nipple attachment for continuous oxygen flow (2 LPM) during laryngoscopy, is widely used; this simple device (Oxyscope, Puritan-Bennet Corp., Foregger Medical Div., Langhorne, PA) considerably reduces the likelihood of hypoxaemia in infants undergoing tracheal intubation.[84]

Uncuffed Magill and Murphy polyvinylchloride tracheal tubes constitute the most commonly employed tubes for

Table 30.5 Dimensions of tracheal tubes

Age	ID* (mm)	Connector† (tube end)(mm)	Length‡ (OT, NT)(cm)
Preterm	2.5	3.0	10–11
Term to 3 months	3.0	3.0	11–12

* ID = internal diameter (minimum). Tubes should conform to the international standard ISO 5361:1993
† Connectors should have machine ends with tapered 15 mm male fittings conforming to the international standard ISO 7228:1993. In the UK, the machine end may have a tapered 8 mm male fitting conforming to the British national standard.
‡OT = orotracheal (may be cut to desired shorter length); NT = nasotracheal

infants today (Table 30.5). Such tubes should meet national and international standards, including tissue compatibility tests. Special tubes, such as those with internal spirally wound wire ('armoured tubes') to resist compression or kinking, are also available in neonatal sizes. The Cole tracheal tube widens in diameter at the predicted level of the cords to preclude bronchial intubation. Paediatric anaesthetists no longer use this tube because the large proximal portion impairs visualization of the larynx during intubation, and the impingement of the wider shoulder of the proximal tube on the larynx can inflict pressure damage on the cords and adjacent structures.[85] Cuffed tracheal tubes are rarely indicated or necessary in neonatal anaesthesia because an adequate tracheal seal can usually be achieved at the level of the relatively narrowed subglottic tracheal diameter.[86]

The tracheal tube connector with a 15 mm tapered male fitting at the machine end has been for 30 years the only approved international standard for all anaesthetic and critical care applications, including those in the neonate. In the UK, however, there exists a national standard for an 8 mm fitting for use in neonates. The patient end of such fittings, which insert into the tube, must be as large or slightly larger than the internal diameter of the tracheal tube (2.5 or 3.0 mm) to prevent imposition of additional resistance and difficulty with passage of a suction catheter.

Overhead radiant warming devices, combined with a calibrated heated mattress at a maximum temperature of 39°C[87] and heated humidification in the breathing system[88] have proved necessary to preserve the core temperature of preterm infants, and reduce thermal stress in all infants exposed to ambient temperatures of operating rooms that are comfortable for personnel (20–22°C).

Calibrated infusion pumps provide precise and controlled injection of maintenance fluids and continuous drug infusions with far greater reliability than gravity-dependent drip systems. With maintenance fluid volumes of 4 ml/kg per hour, such precision seems warranted for all but the briefest and simplest procedures. When infused rapidly, fluids such as blood products and 5 per cent albumin or Ringer's lactate

Table 30.6 Surgical lesions and their anaesthetic consideration

Airway abnormalities:
   Choanal atresia
   Pierre Robin syndrome
   Neoplasm
   Laryngeal stenosis
	Airway obstruction
	Difficult intubation
	Asphyxia
	Aspiration
	Pneumothorax

Diaphragmatic hernia and eventration	Asphyxia Hypoplastic lungs Volatile pulmonary vascular resistance Gastrointestinal distension Congenital heart disease Pneumothorax Small abdominal cavity
Oesophageal atresia and tracheo-oesophageal fistula (TOF, TEF)	Gastric distension (TOF) Pooled pharyngeal secretions Congenital heart disease Pneumonitis Diminished ability to ventilate after gastrostomy (TOF)
Congenital heart disease	Asphyxia Cardiac failure Pulmonary or systemic hypoperfusion Pulmonary oedema Arrhythmia Systemic emboli
Omphalocele and gastroschisis	Asphyxia Hypovolaemia Hypothermia Compression of abdominal and thoracic contents Congenital heart disease (omphalocele) Shock

Gastrointestinal obstruction without ischaemia:
   Intestinal atresia
   Imperforate anus
   Pyloric stenosis
   Incarcerated hernia
   Hirschsprung's disease
   Meconium ileus
	Regurgitation
	Aspiration
	Fluid/electrolyte imbalance

Gastrointestinal obstruction with ischaemia:
   Necrotizing enterocolitis
   Volvulus
   Intussusception
   Strangulated hernia
	Regurgitation
	Aspiration
	Fluid/electrolyte imbalance
	Sepsis
	Occult blood loss
	Shock

Myelomeningocele	Positioning for induction Prone during operation Hydrocephalus
Sacrococcygeal teratoma	Massive blood loss Prone during operation Hypothermia

solutions should be warmed to 32–37°C in a temperature-calibrated blood warming device.

Preanaesthetic evaluation and preparation

Anaesthetic management of a newborn or young infant facing operation requires consideration of the surgical lesion and its systemic effects, of the potential for operative blood loss, and of special requirements such as the patient's position in relation to monitoring devices and vascular catheters. The anaesthetist must also be concerned about the possibility of associated anomalies. Congenital cardiovascular malformations occur in 23 per cent of infants with diaphragmatic hernia,[89] 20 per cent with omphalocele,[90] 15 per cent with oesophageal atresia[91] and 12 per cent with imperforate anus.[92] The more common lesions threatening the life of a neonate, and their major associated clinical problems, are outlined in Table 30.6. Controversy exists regarding the optimum time to undertake anaesthesia and operation for lesions requiring early correction in the neonate. Babies with lesions devastating in their physiological consequences, such as gastroschisis, require immediate operation, whereas those with a lesion such as imperforate anus may easily tolerate a postponement of 24 hours or longer. Before deciding when to operate, one must consider the consequences of delay in repair of the lesion, the degree of birth asphyxia or other stresses on the infant, and the possible presence of associated anomalies or other disorders.

Preparation of a sick infant for transfer to a referral (specialist) centre or to an operating room aims at stabilizing the cardiopulmonary system, body temperature and metabolic functions and at providing energy substrates to meet immediate metabolic needs. A protocol for this is outlined in Table 30.7 and appropriate drug dosages are given in Table 30.8. Proper preparation plays a major role in the perioperative results of neonatal anaesthesia and surgery. The rush to transfer an infant or to operate at the expense of the time required for appropriate stabilization can only increase morbidity and mortality.

The pharynx must be cleared of secretions and any obstruction by soft tissues. In the infant who is obtunded or who has an anomaly of the upper airway, this may be achieved most readily by orotracheal intubation. Patients with a $PaCO_2$ over 6.7–8 kPa (50–60 mmHg) will usually require preoperative tracheal intubation and controlled ventilation. Prior to tracheal intubation, all infants should receive intravenous atropine to prevent bradycardia during laryngoscopy. Decompressing the stomach reduces the likelihood of regurgitation and aspiration, and should

therefore precede airway manipulation in any infant likely to have a substantial gastric liquid volume. The inspired oxygen concentration should be adjusted to maintain the PaO_2 within the normal neonatal range.

A secure intravenous line must be established to provide fluid, drugs and metabolic therapy. This can usually be accomplished by placing a plastic catheter (20–24 gauge) percutaneously or by cutdown in a peripheral vein. An umbilical vein catheter passed into the inferior vena cava at its junction with the right atrium, with its location confirmed by x-ray, can serve as a central venous route, but poses the hazard of periportal cirrhosis and other sequelae, particularly if the distal orifice does not cross the ductus venosus and empties into the portal circulation.[93]

The prevention and correction of metabolic imbalance requires a comprehensive approach. Ambient temperatures of 32–34°C around the infant minimize oxygen consumption and metabolic stress. Infusion rates of 8–12 ml/kg per hour for 2–3 hours may be required to replace insensible and urinary fluid losses before and during anaesthesia and to maintain a normal volume of intravascular fluid. Patients with hypovolaemia associated with sequestration of proteins and fluid into the intestinal lumen or other body compartments require infusion of 5 per cent albumin (10–20 ml/kg) to restore an adequate circulating blood volume. Fresh-frozen plasma provides comparable colloid volume replacement, but the risk of transmitting component-borne infectious pathogens has generally limited its use to neonates with coagulopathy. The infant with a blood

Table 30.7 Preoperative stabilization of vital systems

Clear airway
Oxygen to achieve PaO_2 6.7–9.3 kPa (50–70 mmHg)
Decompress stomach
Warm to 37°C (colon), 36°C (skin over liver)
Establish intravenous route:
    Plastic cannula (24 gauge or larger)
    Percutaneous, cutdown, umbilical vein
Correct acidosis if arterial pH < 7.30:*
    Respiratory: ventilate if $PaCO_2$ > 6.7 kPa (> 50 mmHg)
    Metabolic: sodium bicarbonate* if arterial base excess > −6
    mEq/l or HCO_3^- < 18 mEq/l
Correct dehydration (crystalloid solutions):
    Insensible losses
    Gastrointestinal losses
    Other losses
Correct hypovolaemia: Ringer's lactate, albumin, plasma, erythrocytes, whole blood
Correct hypoglycaemia (<40 mg/dl)†

---

* Correct with sodium bicarbonate 1–2 mEq/kg by slow infusion (10–30 minutes). When arterial base excess >−10 mEq/l, calculate actual base deficit (BE × 0.3 × body weight) to estimate the dose of bicarbonate necessary to achieve a complete repair of the acidosis.
† Use 10 per cent dextrose in 0.2 per cent saline. For blood glucose 20–40 mg/dl, administer 1 ml/kg i.v. bolus, while a blood glucose of < 20 mg/dl should be treated with 2 ml/kg i.v. bolus. Follow the bolus dose with an infusion of 3–4 ml/kg per hour and frequently recheck blood glucose.

Table 30.8   Intravenous drug dosages in infants

Adjuvant	Newborn	Infant
Atropine	0.04 mg/kg (<4 kg)	0.02 mg/kg (>4 kg), minimum 0.16 mg i.v.
Tubocurarine:		
Initial	0.3 mg/kg increments to effect	0.6 mg/kg
Maintenance	25% initial total dose	25% initial dose
Pancuronium:		
Initial	0.1 mg/kg	0.1 mg/kg
Maintenance	20% initial dose	20% initial dose
Metocurine	0.4 mg/kg initial dose	0.4 mg/kg initial dose
Atracurium	0.4 mg/kg initial dose	0.4 mg/kg initial dose
Vecuronium	0.1 mg/kg initial dose	0.1 mg/kg initial dose
Suxamethonium	2 mg/kg	1 mg/kg
Neostigmine	0.07 mg/kg	0.07 mg/kg
Thiopentone (2.5%)	3–4 mg/kg	4–7 mg/kg
Ketamine	2 mg/kg	2 mg/kg
Fentanyl:		
Supplementary	1–5 µg/kg	1–5 µg/kg
Primary anaesthetic	10–50 µg/kg	10–50 µg/kg
Morphine	0.05–0.2 mg/kg	0.05–0.2 mg/kg

Cardiovascular drugs	Dosages
Adrenaline (epinephrine):	
Initial	1–10 µg/kg
Maintenance	0.1–2.0 µg/kg per min
Isoprenaline (isoproterenol)	0.1–2.0 µg/kg per min
Dopamine	1–20 µg/kg per min
Dobutamine	2–20 µg/kg per min
Calcium gluconate (10%)	30–60 mg/kg
Frusemide (furosemide)	0.5–1.0 mg/kg
Lignocaine (lidocaine) (1%)	1.0 mg/kg

glucose value of 20–40 mg/dl should initially receive a rapid infusion of 10 per cent dextrose 1 ml/kg, followed by an infusion of 10 per cent dextrose in lactated Ringer's solution at 3–4 ml/kg per hour to maintain normal blood glucose levels.[40] The infant whose blood glucose is below 20 mg/dl should receive a bolus infusion of 2 ml/kg 10 per cent dextrose. When, despite restoration of metabolic homoeostasis, the arterial pH remains below 7.30 due to a base deficit exceeding 6 mEq/l, a slow correction (over 10–30 minutes) with sodium bicarbonate is indicated (see Table 30.7).

During this period of stabilization an arterial line should be inserted in any critically ill infant. The arterial line permits not only safe, convenient blood sampling for measurement of ventilatory and metabolic variables but also continuous direct monitoring of systemic arterial pressure (SAP). We prefer the right radial artery, although the umbilical artery is technically easier and relatively safe to use.[93,94] Because of the risk of intracranial emboli when temporal artery catheters are flushed, they are rarely employed.

In conjunction with heart rate, SAP provides the most readily available continuous estimate of cardiac output and blood flow in the neonate; collection of urine output in an external bag also provides an assessment of renal perfusion. The electrocardiogram, respiratory rate, and skin and colon temperatures should be monitored continuously throughout the perioperative period.

Intraoperative anaesthetic management

Selection of anaesthetic agents and adjuvant drugs

The neonate, especially the term newborn, reacts reflexly to all sudden noxious stimuli, including pain, with a mass startle reaction followed by crying. Tachycardia, systemic hypertension and increased skeletal muscle tone accompany this reaction. Recent studies indicate that this sympathetically mediated response impairs normal metabolism and, if sustained with minimal or no relief, may increase postanaesthetic morbidity and mortality in sick neonates.[95] We therefore recommend the prevention, or at least the attenuation, of pain perception in all infants whose vital system stability permits the administration of potent analgesics, local or regional blocks, or general anaesthesia as required by the operative procedure.

All gaseous and volatile anaesthetics currently available for clinical use have been administered to neonates. Halothane is unique among the potent agents in the limited irritation it causes to the airway, and is the agent of choice for inhalation induction for neonates with suitable cardiovascular function. Isoflurane differs from halothane primarily in its tendency to maintain or increase heart rate, thereby in some infants providing a more satisfactory cardiac output during the maintenance of anaesthesia. Nevertheless, the sensitivity to myocardial depression that neonates exhibit to all volatile anaesthetics usually precludes the administration of sufficient concentrations, even with the addition of nitrous oxide to provide satisfactory operating conditions. Non-depolarizing muscle relaxants assure both adequate haemodynamics and good operating conditions, and usually are employed for all but very brief procedures in healthy neonates. Nitrous oxide, none the less, provides a useful adjunct to either potent volatile anaesthetics or narcotics except when the infant requires an inspired oxygen concentration greater than 50 per cent, when bowel obstruction exists or during an extensive abdominal operation.

Opioid analgesics, particularly fentanyl and sufentanil, possess valuable attributes in their ability to blunt haemodynamic and endocrinological evidence of operative stress. Opioid dose must be limited to minimal amounts to prevent clinically significant ventilatory depression in neonates who will not receive postoperative surveillance in an intensive care unit. Despite the haemodynamic stability that neonates usually exhibit with anaesthetic levels of opioids, those who present with life-threatening conditions (e.g. septic shock, myocardial failure) often suffer substantial hypotension following small doses of these analgesics. In this limited group of infants, who presumably are sustained on the basis of substantial release of 'stress' hormones, the administration of any drugs that blunt that release must be undertaken with great caution. Ketorolac, a non-steroidal anti-inflammatory drug, appears to provide substantial analgesia without the risk of respiratory depression. Experience with this drug in neonates, however, remains quite limited. The doses of narcotics, muscle relaxants and other drugs used in the conduct of anaesthesia are presented in Table 30.8.

Induction of anaesthesia

The postnatal and postconceptual age of the neonate, physical status, relative risk of pulmonary aspiration of regurgitated gastric contents and nature of the surgical lesion affect the selection of induction technique. Regardless of these considerations, the immediate goals are: (1) to attenuate or, cardiopulmonary stability permitting, eliminate pain perception and the major stress response; (2) to secure a tracheal airway and control ventilation of the lungs as quickly as feasible; and (3) to attach special monitoring devices and insert the intravascular lines necessary during and following the operation. This should be accomplished with minimal disturbance of the infant's cardiopulmonary or metabolic status, including thermal stress. In virtually all instances, induction of anaesthesia should be preceded by administration of atropine to prevent bradycardia.

The pulmonary aspiration of gastric contents poses the same serious risk of chemical pneumonitis in the neonate as in the adult because the gastric pH, within 6 hours after birth, decreases from over 6 to mean of 2.5 units.[96] Controversy exists among paediatric anaesthetists as to whether awake tracheal intubation or a rapid intravenous induction technique with tracheal intubation offers greater safety. The weight of opinion sides with the awake technique in the hands of the anaesthetist who only occasionally anaesthetizes neonates. Laryngoscopy and intubation can be difficult in a vigorous infant, and leads to a maximal, though transient, stress response which could be deleterious in a preterm neonate at risk of intracranial haemorrhage. In critically ill infants, especially those with cardiopulmonary disease, ventilation of the lungs with 100 per cent oxygen, analgesia and sedation with intravenous fentanyl or ketamine, neuromuscular blockade with suxamethonium and gentle cricoid pressure facilitate tracheal intubation and cause minimal physiological disturbance. Suxamethonium does not cause muscle fasciculations nor increased gastric pressure in neonates, eliminating the need for an anti-fasciculating dose of a non-depolarizing relaxant. Compared with older children, the neonate's smaller functional residual capacity in relation to oxygen consumption will result in much earlier onset of hypoxaemia. Regardless of the technique, therefore, ventilation of the infant's lungs with an inspired oxygen concentration of 100 per cent for 2 minutes before and after the intubation should prevent, or at least reduce, any hypoxaemia associated with the apnoea or breath holding during intubation. An audible leak (with a stethoscope) around the tracheal tube at 20–40 cmH$_2$O airway pressure provides some assurance that the fit is not excessively tight. Maintaining the neonate's head in the correct position during mask ventilation and laryngoscopy reduces airway soft-tissue obstruction and enhances visualization of the larynx. The optimal position involves moderate flexion of the cervical spine with extension of the atlanto-occipital joint (the 'sniffing position'). In the supine position with the shoulders flat on the operating table, the neonate's prominent occiput causes the appropriate cervical flexion, and elevation of the symphysis menti will result in the desired extension of the atlanto-occipital joint. A small sponge rubber 'doughnut' under the occiput will help to hold the head steady in the midline. A trained assistant should be available to hold the infant in desired positions and to hand equipment to the anaesthetist whenever necessary so that the anaesthetist can focus on the infant at all times.

Maintenance of anaesthesia

Ensuring adequate alveolar ventilation and oxygen delivery to vital organs coupled with prevention of pain constitute the essence of anaesthetic maintenance in the neonate as it does in patients of all ages. The high level of technology available in anaesthetic locations today should maximize that assurance, but at times can distract the anaesthetist and imperil the infant. Virtually continuous observation of the neonate's breath sounds, chest expansion, skin and blood colour, heart rate and systemic arterial pressure should guide the adjustment of anaesthetic agents to meet the patient's and the surgeon's needs. We prefer, in most neonates, to use a combination of a narcotic (usually fentanyl by intermittent injection), minimal concentrations of isoflurane and a non-depolarizing muscle relaxant such as pancuronium or vecuronium (for anaesthesia lasting less

than 30 minutes). Nitrous oxide should be reserved for normoxic infants who do not have intestinal obstruction, congenital diaphragmatic hernia, omphalocele or gastroschisis.

Monitoring

Close observation of the infant and the operative field combined with appropriate monitoring devices greatly enhances the safety of anaesthesia in infants (Table 30.9). The new minimum monitoring guidelines call for the same monitors in the infant as in the adult: precordial stethoscope (preferably amplified and electronically filtered), blood pressure (preferably automated) with appropriate cuff sizes, pulse oximetry, electrocardiogram, temperature, and end-tidal carbon dioxide and an anaesthetic record. Accurate non-invasive blood pressure determination requires a sufficiently wide cuff (bladder width more than 50 per cent the circumference of the mid-upper arm).[97] Manufacturers of automated blood pressure devices recommend a slightly larger cuff extending from the axilla to the antecubital fossa. In neonates and young infants, normal blood pressure varies with age (Table 30.10). Rectal or oesophageal temperature probes readily detect hypothermia in infants.

Most neonates and young infants undergoing intracavity procedures, as well as any critically ill neonate, need an intra-arterial catheter for precise and timely monitoring of blood pressure, as well as sampling of blood for pH, gas tensions, haematocrit, and fluid and electrolyte changes. The radial artery of the full-term newborn will accept a 22 gauge catheter inserted either percutaneously or by cutdown. Rapid flushing of a radial arterial catheter must be avoided because this could cause cerebral emboli.[99] Less accessible, but perhaps safer, are the dorsalis pedis and posterior tibial arteries in the foot. Umbilical arteries can be used, but carry an associated risk of aortic thrombosis, renal hypertension, and splanchnic and lower extremity ischaemia.[93] Continuous infusion of dilute heparin (1 unit/ml) at 1–3 ml/h through a flush system inhibits clot formation in any arterial catheter while permitting continuous monitoring of pressure.[94]

A central venous pressure line located at the vena cava–right atrial junction can be useful in evaluating adequacy of fluid replacement. Central venous pressure (CVP) monitoring in neonates is generally reserved for infants undergoing operations that entail massive blood loss, infants with myocardial dysfunction, or circumstances in which alternative means of evaluating tissue perfusion (e.g. urine flow) are unavailable. In neonates, the CVP is neither a sensitive nor a specific indicator of hypovolaemia and should not be used as a 'target' variable. The determination of CVP may help guide therapy in the hypotensive, tachycardic neonate

Table 30.9 Monitoring

Clinical information	Device
Breath sounds and heart sounds	Precordial or oesophageal stethoscope
Heart rhythm	Electrocardiograph, lead 2

Systemic arterial pressure	Infant cuff
	Oscillometer
	Doppler transducer
	Intra-arterial catheter (22–24 gauge) and transducer:
	    Radial artery
	    Dorsalis pedis artery
	    Posterior tibial artery
	    Umbilical artery (3.5–5 Fr.)

Central venous pressure	Central venous catheter:
	    Umbilical vein (3.5–5 Fr.)
	    Internal jugular

Temperature	Thermistor probe:
	    Rectal
	    Oesophageal
	    Nasopharyngeal

Ventilation:
  Inspiratory pressure
  Breath sounds
  Inspired O_2
  End-tidal CO_2
	Airway pressure manometer
	Stethoscope
	Oxygen analyser
	Mass spectrometer or infrared capnometer

Blood gases:
  Sao_2
  $Paco_2$
	Pulse oximeter
	End-tidal CO_2

Blood loss	Small-volume suction traps Weighed sponges Serial haematocrits
Urine volume	Urinary catheter or bag Small-volume urinometer

Table 30.10 Normal cardiovascular function (mean and range or 2 SD)

	1–30 days	1–24 months
Heart rate (beats/min)[35]	125 (70–190)	120 (80–160)
Systemic arterial pressure (mmHg)	82 (11) 51 (10)	94 (12) 55 (9)
Cardiac index (ml/kg per min)[20]	200–250	170–200

when reasonable fluid administration has failed to restore haemodynamic responses. Pulmonary artery catheters are seldom indicated because right and left heart function are rarely discordant in neonates undergoing non-cardiac surgery. Urine output in the infant can be measured accurately with a small urinometer. Similarly, operative field suction lines used in infants should be short and flow into calibrated low volume traps.

Temperature control

Maintaining the neonatal body temperature, even for short superficial procedures, requires a multifaceted approach. An overhead radiant warmer during induction and emergence, placing the intubated patient's head and limbs in plastic sandwich bags[100] and covering the remainder of the patient with an adhesive plastic drape or reflective blanket, minimize loss of heat to the environment. This applies even when room temperature is raised to nearly the maximum tolerated by the surgical team personnel (approximately 28°C or 82°F). Once the infant has been covered by plastic adherent drapes the ambient temperature can be lowered to the more comfortable 20–22°C. As mentioned above, a servo-controlled, circulating water or air blanket,[87] in conjunction with a device to heat and humidify the anaesthetic gases[88] (and heat the inspiratory tubing to prevent condensation), will provide the energy (caloric) input usually needed to maintain an infant's core temperature, even when a body cavity is open.

Intraoperative fluid therapy

The daily maintenance energy and total fluid requirements of neonates, established in 1957[36] (Table 30.11), indicate that 4.184 J (1 cal) require 1 ml of water for metabolism. Of the commercially available intravenous fluids 0.2 per cent saline most closely resembles the obligatory fluid losses of the neonate, and usually is infused with 5 per cent glucose. The infused volumes are small (4 ml/kg per hour) in the neonate and young infant, tempting anaesthetists to use this hypotonic maintenance solution for third space replacement as well. In addition, surgeons frequently continue infusing this intraoperative fluid in the post-operative period. When used in this manner, 0.2 per cent saline can produce hyponatraemia. Neonates readily handle the added sodium and therefore most paediatric anaesthetists now use glucose in 0.45 per cent saline or lactated Ringer's solution for maintenance and lactated Ringer's solution for third space fluid replacement. In some infants, 5 per cent glucose solutions may produce hyperglycaemia. Hyperglycaemia in stressed newborns has been associated with intracranial bleeding, thought to be due to diuresis, cellular dehydration and serum hyperosmolality. The volume of 5 per cent glucose solutions should be limited, therefore, to 15–20 ml/kg during anaesthesia. In the light of the propensity of neonates for both hypo- and hyperglycaemia, their blood glucose levels should be determined regularly during a surgical procedure.

Third space losses (Table 30.11) in neonates with necrotizing enterocolitis, omphalocele or gastroschisis can exceed a blood volume.[101] Third space replacement therapy should be re-evaluated when the total replacement reaches 40 ml/kg to verify the need for continued replacement. On the other hand, preterm infants or neonates with a diaphragmatic hernia or congenital heart disease may not tolerate large volumes of third space replacement without developing interstitial pulmonary oedema. Serial determinations of haematocrit, serum osmolality, total serum protein, electrolytes including ionized calcium, and urinary

Table 30.11 Fluid and blood management

Maintenance fluids[36] 5–10% dextrose in 0.45% saline	<10 kg: 100 ml/kg per 24 hours
'Third-space' replacement[101] Lactated Ringer's	3–10 ml/kg per hour during operation depending on operative site; re-evaluate when total dose reaches 40 ml/kg
Colloid replacement (in lieu of blood if haematocrit >35) Fresh-frozen plasma (in coagulopathy) 5% albumin/lactated Ringer's	10–20 ml/kg 10–20 ml/kg

Blood replacement[101,102]
Estimated blood volume (EBV)
Acceptable haematocrit (Hct$_a$) = 35 vol%
Allowable blood loss (ABL) calculation:
$$ABL = \frac{(\text{starting haematocrit} - \text{Hct}_a)}{\text{current haematocrit}} \times EBV$$
Replace ABL <10 ml/kg with lactated Ringer's
Replace ABL >10 ml/kg with 5% albumin in lactated Ringer's (or fresh-frozen plasma when coagulopathy occurs)

Techniques:
  Warm fluids and blood
  Calibrated syringe pumps
  Manual syringe injection for colloid and blood

output assist in the accurate management of fluid therapy. Serial determinations of PaO_2 at a constant inspired oxygen concentration will detect the onset of hypoxaemia due to developing interstitial pulmonary oedema caused by excess extracellular fluid volume.

The normal neonatal cardiovascular system compensates adequately for acute variations of up to 10 per cent in blood volume. This permits 'priming' neonates with up to 10 ml/kg of blood just prior to a bloody surgical procedure, such as resection of a sacrococcygeal teratoma, when the blood loss will be difficult to measure.[102] A single 500 ml unit of whole blood or even 250 ml of packed erythrocytes will equal or exceed one blood volume in a preterm or young neonate. All the coagulation, temperature and metabolic problems of multiple blood volume transfusions occur earlier in neonates because the rates of infusion (in ml/kg per minute) tend to be higher. Hyperkalaemia and hypocalcaemia often develop during high volume, rapid transfusion in infants. Whilst dilutional coagulopathies occur less frequently than electrolyte disturbances, the normally diminished levels of many coagulation factors (e.g. II, IX, X, XI, XII)[103] render neonates more vulnerable to this phenomenon as well.

Glucose requirements vary widely in neonates and should be evaluated periodically by determination of blood glucose levels. We advise the use of blood glucose screening devices at least hourly, supplemented by laboratory analysis prior to anaesthesia, every 1–2 hours during an operation, and after the infant's arrival in the neonatal intensive care unit. Preterm infants tend to have less stable glucose homoeostasis. A blood level between 20 and 40 mg/dl should be treated as described in Table 30.7. A blood glucose below 20 mg/dl should receive a bolus infusion of 200 mg/kg glucose (2 ml/kg of a 10 per cent glucose solution).

Recovery from anaesthesia

Awakening and recovery from anaesthesia usually occurs more rapidly in neonates than in adults, for the same reasons that induction is more rapid. Tracheal extubation prior to awakening may result in laryngospasm. The initial treatment of laryngospasm consists of constant positive airway pressure with forward mandibular thrust and extension of the atlanto-occipital joint to relieve residual supraglottic soft-tissue obstruction. If haemoglobin oxygen saturation falls below 85 per cent, we advocate intravenous suxamethonium (succinylcholine) (2.0 mg/kg) *before* the onset of cyanosis or bradycardia. Failure to prevent strong inspiratory efforts against a closed glottis can cause severe pulmonary oedema in an otherwise healthy patient of any age,[104] although because of their highly compliant, soft thoracic cage, this occurs less often in the neonate.[105]

Neuromuscular blockade must always be reversed with appropriate drugs if extubation of the trachea is planned at the termination of the anaesthetic. Because hypothermia delays recovery of neuromuscular function following non-depolarizing muscle relaxants, reversal should not be attempted unless the patient's core temperature is at least 35°C.[71] The suitability for, and then the adequacy of, neuromuscular blockade reversal should be assessed by nerve stimulation prior to awakening, and prior to extubation, by determining the presence and vigour of flexion of the hips with the raising of one or both legs off the operating table. This latter clinical sign corresponds to a negative inspiratory force of -35 cmH$_2$O.[106]

Perioperative considerations for selected malformations

A summary of perioperative considerations for common neonatal conditions requiring surgery, as noted previously, appears in Table 30.6. An expanded discussion of the most common congenital and acquired disorders that exert a major impact on the conduct of anaesthesia follows. Further detailed information regarding these and other disorders can be found in the paediatric anaesthesia texts listed under 'Further Reading'.

The difficult airway

This group of neonates poses a significant challenge to even the most skilled paediatric anaesthetist. Because it incorporates a diverse population of neonates with different airway abnormalities and associated malformations, there is no single formula for successful management. Close cooperation with an otolaryngologist, whether or not the operating surgeon, frequently proves valuable during induction of anaesthesia and tracheal intubation. An awake intubation affords the safety of spontaneous ventilation should the attempt fail. The placement of the laryngeal mask after topical oropharyngeal anaesthesia permits safe mask induction, and even controlled ventilation following the administration of muscle relaxants, provided the airway resistance and lung compliance are nearly normal.[83] This technique facilitates direct laryngoscopy with conventional or modified laryngoscopes such as the anterior commissure scope, a cylindrical laryngoscope that prevents soft-tissue obstruction of the endoscopist's view.[107] Whilst fibreoptic bronchoscopes can be employed in neonates, the diameter of all bronchoscopes large enough to incorporate a suction port (3.6 mm outer diameter) precludes the placement of neonatal tracheal tubes over them. Modified techniques are necessary in which angiographic guidewire, placed in the trachea via the suction port under bronchoscopic visualization, serves as a guide for the tracheal tube.[108] Alternatively, blind methods facilitated by a lighted stylette[109] or finger palpation of the glottis[110] have been reported to have impressive success in experienced hands.

Congenital diaphragmatic hernia

The most important determinant of outcome with congenital diaphragmatic hernia (CDH) resides in the degree of pulmonary hypoplasia. Several physiological indices have been proposed that may provide prognostic information, the most widely accepted of which is the ventilatory index advocated by Bohn:

$$\text{Ventilatory index} =$$
$$\text{mean airway pressure} \times \text{respiratory rate}$$

Neonates with CDH who have a $PaCO_2$ under 5.3 kPa (40 mmHg) with an index under 1000 nearly always survive; those with $PaCO_2$ levels over 6.7 kPa (50 mmHg), or those whose $PaCO_2$ could be lowered to 5.3 (40) only by very high levels of ventilation (index >1000), usually die.[111,112]

The anaesthetic management centres around preventing distension of the bowel that occupies the left hemithorax (minimal positive pressure mask ventilation, evacuation of the stomach, avoidance of nitrous oxide), and a strategy designed to lower pulmonary vascular resistance (hyperventilation, blunting neuroendocrine response to noxious stimuli with opioids, alkalinization, and possibly pulmonary vasodilating medications). The postoperative course of infants with congenital diaphragmatic hernia depends on the degree of lung development as well as the unstable pulmonary vascular resistance changes that these infants exhibit. Whilst urgent repair of the defect represents the most common strategy, some centres currently advocate delayed repair in the belief that the unstable nature of the pulmonary vascular bed will diminish in the first few days following transition from fetal circulation, at which time the pulmonary abnormalities that follow the operation will be better tolerated.

Extracorporeal membrane oxygenation (ECMO) has gained widespread acceptance for the management of selected infants with this malformation both before and after surgery, although the impact of this therapy on outcome remains controversial.[113] Whilst one might expect that this intervention addresses the pulmonary vascular component of the pathophysiology, 40–50 per cent will still succumb from inadequate pulmonary parenchymal development. The advent of specific pulmonary vasodilators such as nitric oxide may obviate the need for ECMO in the neonates in whom it is likely to be efficacious.[114]

Oesophageal atresia and tracheo-oesophageal fistula

There are five major classifications of oesophageal atresia and tracheo-oesophageal fistula (TOF), yet 85 per cent have oesophageal atresia with a fistula from the trachea to the distal oesophagus.[115] Neonates with TOF require particular attention to airway and ventilatory management in order to prevent massive gastric dilatation. To minimize positive pressure mask ventilation and inflation of the stomach, these infants should undergo tracheal intubation while breathing spontaneously, either awake or after inhalational induction. The tracheal tube should be positioned just above the carina in order to limit gas flow into the fistula, which typically arises 0.5–1 cm more proximally in the trachea. Muscle relaxants provide satisfactory operating conditions without excessive cardiovascular depression, but controlled ventilation must be undertaken with careful attention to gastric distension because massive gastric dilatation can have a catastrophic impact. Whilst a staged approach to repair that includes initial gastrostomy will avoid the problem of gastric distension, it provides a route for the egress of gas administered with positive pressure ventilation,[116] especially in the infant with severe pulmonary disease.[117] When lung compliance becomes substantially diminished, gas delivered to the trachea preferentially escapes via the gastrostomy. Placing the tip of the tracheal tube distal to the fistula minimizes gastric distension or escape of gas.

Congenital heart disease

A comprehensive consideration of specific congenital heart malformations is beyond the scope of this chapter. In general, the anaesthetist is best served by a physiological approach to these myriad lesions. Developing an understanding of the physiological derangement imposed by a given malformation in terms of hypoxaemia, volume overload or pressure overload should serve as a foundation upon which desirable manipulations of haemodynamic variables are planned. We advocate analysing each infant's cardiovascular physiology with respect to the optimal manipulation of preload, heart rate, contractility, and pulmonary and systemic vascular resistance. The goal is to promote satisfactory systemic and myocardial perfusion while minimizing the magnitude of the physiological derangement imposed by a particular lesion. Such a scheme incorporates perioperative fluid management, anaesthetic agents, strategy of ventilation, and preferable resuscitation drugs based upon their predicted haemodynamic effects.

Omphalocele and gastroschisis

Although these abdominal wall defects have different embryological origins that convey implications for associated malformations, the intraoperative management is virtually identical. Yaster and colleagues[118] have proposed a management strategy whereby primary repair is undertaken only if it results in an increase in intragastric and central venous pressures under 2.7 and 0.5 kPa (20 and 4 mmHg) respectively. Excessive pressure elevations not only risk dehiscence but also impair respiratory mechanics and abdominal organ perfusion. When excessive intra-abdominal pressure will occur with primary closure, a staged reduction is carried out in the intensive care unit over several days by gradual constriction of a silicone elastomeric abdominal silo. Mechanical ventilation, occasionally in conjunction with muscle relaxants, usually proves necessary to maintain ventilation and facilitate reduction. The elevated intra-abdominal pressure that accompanies these repairs can also result in a protracted elimination of drugs metabolized by the liver, such as fentanyl.[69,119]

Gastrointestinal ischaemia and peritonitis

Neonates with congenital or acquired lesions that result in gastrointestinal ischaemia, perforation or peritonitis are among the most physiologically fragile infants confronting the anaesthetist. The primary derangement, systemic hypotension, results from a massive capillary leak as well as other cardiovascular effects of sepsis. In their most fulminant form, these lesions necessitate third space fluid replacement up to 100 ml/kg per hour. The administration of massive volumes of crystalloid solutions should be accompanied by albumin to maintain normal serum levels (see Table 30.11). It often becomes necessary to administer blood products to correct the anaemia, thrombocytopenia and coagulopathy that accompany operations for these conditions. When elevated central venous pressure or lack of a positive arterial pressure response to fluid administration suggest myocardial depression, administration of inotropes is indicated (see Table 30.8).

Regional anaesthesia

Regional anaesthesia in neonates provides the theoretical advantage of reduced requirements for other anaesthetic agents and excellent postoperative analgesia.[120] It also exposes these infants to the risks of both a general anaesthetic and a regional technique, because a substantial proportion of neonates will continue to move if awake during an operation under regional block. Nevertheless, some forms of regional anaesthesia now play an increasing role in neonatal anaesthesia care.

Ilioinguinal–iliohypogastric block

For the amelioration of postoperative pain, we advocate that the surgeon infiltrate the ilioinguinal and iliohypogastric nerves under direct vision in neonates undergoing inguinal herniorrhaphy under general anaesthesia. This prevents the distortion of tissues in the operative field that may accompany percutaneous infiltration. Bupivacaine 1 mg/kg (0.25 per cent solution) is suitable for this purpose.

Spinal anaesthesia

Spinal anaesthesia offers an alternative for herniorrhaphy in the preterm infant[121,122] who is less than 60 weeks postconception and therefore at risk for postoperative apnoea.[123] The neonatal spinal cord ends at L3, and the dural sac ends at S3; lumbar puncture is therefore performed at L4/5 or L5/S1. Placing the patient in the sitting position makes the block technically easier to perform and decreases the incidence of dry taps. Extreme flexion of the neonate's neck produces airway obstruction, so the head must be supported while the block is being introduced.[124] Bupivacaine (0.5 per cent) 0.5 mg/kg or amethocaine (tetracaine) 0.4–0.5 mg/kg in 10 per cent dextrose is suitable, but the use of amethocaine up to 0.8 mg/kg has been reported.[122] Adrenaline (epinephrine) increases the duration of the block from approximately 60 minutes to over 100 minutes.[122] The volumes are small and therefore the dead space volume of the spinal needle must be measured and that volume (0.05–0.10 ml) added to the proposed injectate volume. Spinal anaesthesia in infants can result in an excessively high level of block with apnoea and systemic hypotension if the legs and hips are lifted to place an electrocautery ground pad[125] while the infant lies supine. Sedation with ketamine during spinal anaesthesia resulted in prolonged apnoea with bradycardia in eight of nine infants in one series,[126] suggesting that the benefits of this technique are lost, with concomitant sedative administration, in preterm infants.

Penile nerve block

This block provides intraoperative and postoperative pain relief for circumcision.[127] There are several techniques, but 0.5 per cent bupivacaine, 1 mg/kg, without adrenaline, injected in the midline just caudad to the symphysis pubis is appropriate. The needle is inserted in the midline at the base of the penis and walked off the symphysis, similar to walking off a rib in an intracostal nerve block.

Caudal anaesthesia

The caudal approach to the epidural space has experienced a resurgence of interest among paediatric anaesthetists.[128] Whilst that resurgence was initially driven by the desire to administer local anaesthetics and opioids for postoperative analgesia, some employ this approach as the primary anaesthetic for high-risk neonates and former preterm infants. The principal advantage of caudal epidural over spinal anaesthesia in neonates is its longer duration of action. In infants, the prominence of the sacral cornua that delineate the sacral hiatus serves to make this approach to the epidural space significantly easier than in adults.

Many formulae for calculating caudal doses have been advanced; 0.25 per cent bupivacaine 1 ml/kg with 1 : 200 adrenaline added is suitable. The haemodynamic consequences of the associated sympathetic block, as with spinal anaesthetics, are rarely detectable in infants. Total spinal block has been reported, presumably from inadvertent subarachnoid injection of the local anaesthetic.[129] A few centres have reported adding opioid analgesics to the caudal mixture in infants as young as 1 day of age. Given the propensity of neonates for disorders of respiratory drive, caudal narcotics would only seem reasonable following extremely painful procedures when very close surveillance is available. Indeed, postoperative apnoea with bradycardia can follow caudal anaesthesia without opioids in the former preterm infant.[130]

Postanaesthetic care

In the immediate post-anaesthesia period, vigilance needs to be directed at maintaining the infant's airway, alveolar ventilation and oxygenation. Close attention to the rate and pattern of breathing will detect airway obstruction, breathing irregularities and apnoea in the neonate whose trachea has been extubated. Any former preterm infant (born at less than 37 weeks' gestation), who is less than 60 weeks postconceptual age and has undergone a general anaesthetic faces an increased risk of prolonged apnoea (over 15 seconds).[123,131] This risk increases substantially in former preterm infants less than 44 weeks post-conception,[123] and can result in severe hypoxaemia and bradycardia.[132] In our view, all infants born preterm and less than 60 weeks post-conception, and term infants less

than 44 weeks post-conception, should have cardiorespiratory monitoring (ECG and impedance pneumograph) or pulse oximetry for a minimum of 24 hours following general anaesthesia.

Intermittent arterial pH and gas tensions, as well as continuous pulse oximetry, should be employed to ensure the adequacy of ventilation and oxygen delivery. Normal $Pa\text{CO}_2$ levels range from 3.7 to 5.3 kPa (28 to 40 mmHg); a level over 6 kPa (45 mmHg) represents significant hypercapnia. A $Pa\text{O}_2$ of 7.3–10.7 kPa (55-80 mmHg) is optimal for most infants. Levels over 10.7 kPa (80 mmHg) persisting beyond 12 hours may predispose the preterm infant to the retinopathy of prematurity (retrolental fibroplasia)[133] but is only one of many inciting factors.[133,134] Indeed, the predominant factor is neither oxygen nor disease, but the degree of retinal immaturity associated with the postconceptual age of the infant.[135]

Whenever tachypnoea, thoracic retraction and cyanosis develop in the neonate postoperatively, a chest x-ray should be obtained to exclude pneumothorax, atelectasis or aspiration pneumonitis. Arterial pH and gas tensions are necessary to exclude acute respiratory failure (see below). Persistent effects of anaesthetics, residual neuromuscular blockade and hypothermia can cause subtle airway obstruction, hypoventilation and apnoeic episodes. Quick detection, reintubation of the infant's trachea, and mechanical ventilation until the cause can be determined and corrected prevent such complications from resulting in major morbidity (see 'Complications of anaesthesia', below).

In neonates recovering from general anaesthesia and operation, hypothermia also causes intense peripheral vasoconstriction and metabolic acidosis from the metabolism of free fatty acids liberated during non-shivering thermogenesis.[44] The paediatric postanaesthetic care unit and infant intensive care unit, therefore, must be equipped to rewarm and maintain the body temperature of small infants. Overhead radiant warmers, incubators, circulating warm water or air blankets, and plastic wraps will effectively rewarm cold patients.

Acute respiratory failure, a condition in which ventilation and pulmonary gas exchange fail to meet basic respiratory requirements of the vital organs, causes death and severe morbidity in susceptible neonates following anaesthesia and operation. Infants at greatest risk include:

1. *preterm neonates*, especially those under 1500 g weight; and those with *respiratory distress syndrome* (hyaline membrane disease) or its occasional sequela, bronchopulmonary dysplasia (BPD);
2. infants following repair of *congenital diaphragmatic hernia* (with persistent pulmonary hypertension and lung hypoplasia), *oesophageal atresia with tracheo-oesophageal fistula* (with acid aspiration pneumonitis), *omphalocele or gastroschisis* (with raised diaphragm due to high intra-abdominal pressure), and *congenital*

cardiovascular disease (with excess or inadequate pulmonary blood flow, atelectasis, tracheomalacia);

3. infants with *anomalies of the airway*, the *brain and spinal cord*, and the *thoracic cage and diaphragm*;

4. congenital myopathies and central hypoventilation which, although rare, can precipitate respiratory failure in the postoperative period.

The diagnosis of acute respiratory failure can be made if an infant has two, or more, of the following:

- recurrent apnoeic episodes (> 15 seconds)
- tachypnoea (> 60 breaths/minute)
- severe thoracic retraction and thoracoabdominal discoordination during inspiration
- PaO_2 < 8 kPa (60 mmHg) in > 60 per cent oxygen
- $PaCO_2$ > 6.7 kPa (50 mmHg)

This diagnosis calls for immediate intervention with reintubation of the trachea, mechanical ventilation and other components of neonatal respiratory care. To delay until conditions deteriorate leads to an unacceptably high risk of eventual cardiopulmonary arrest. Details of neonatal respiratory care can be found in the works suggested under 'Further reading'.

An important component in the increasing survival of critically ill neonates is intravenous alimentation. Sufficient energy substrates should be provided intravenously, usually through a catheter in a major central vein, if an infant's gastrointestinal tract cannot absorb the nutrients for more than 2–3 days. Without adequate nutrition the infant will suffer progressive loss of tissue protein and fat, with impairment of thermal control, ventilation and other vital functions. After the first 4 postnatal days, the neonate requires 500–585 J/kg (120–140 cal/kg) per day to support vital functions, wound healing and growth.[136]

Fluid intake and output and metabolic care should be monitored continually in the immediate post-anaesthetic period as in pre-anaesthetic and intraoperative phases. A urine output of 1–2 ml/kg per hour, an arterial pH between 7.35 and 7.45, with normal blood glucose, electrolyte, ionized calcium and bilirubin levels signify metabolic stability

Complications of anaesthesia

A recent prospective survey of 40 000 anaesthetics administered to children in France found a rate of serious complications of 4.3/1000 anaesthetics among 2100 anaesthetics administered in infants under the age of 12 months; this was significantly higher than the rate of 0.5/1000

anaesthetics observed in older children.[137] In addition, cardiac arrests related to anaesthesia occurred in infants at a rate ten times that in children (2.0/1000 vs 0.2/1000 anaesthetics).[137] Clearly, the neonate has the highest rate of intraoperative anaesthetic and post-anaesthetic complications, including intraoperative cardiac arrest, compared to older infants and children.[138,139] Data on the incidence in neonates of specific adverse events and their outcomes remain limited to a few institutions.

The most common, potentially life-threatening, complications of anaesthesia in neonates relate to ventilation, and include misplacement of the tracheal tube, severe upper airway obstruction or ventilatory depression due to a relative anaesthetic overdose. As mentioned above, post-anaesthetic prolonged apnoea can occur in 37 per cent of infants born at less than 37 weeks' gestation who are less than 60 weeks post-conceptual age.[123] Circulatory problems are the next most frequent category, and often involve systemic hypovolaemia with cardiovascular depression and systemic arterial hypotension. These can result in cardiac arrest and cerebral hypoxic damage during and following anaesthesia.

Failure of a neonate to regain consciousness or normal neuromuscular function for many hours post-anaesthesia, in the absence of residual effects of narcotics or muscle relaxants, raises the question of severe hypoxic encephalopathy with possible intracranial haemorrhage or cerebral thrombosis. Although impaired ventilation or abnormal shunts causing hypoxic hypoxia can cause cerebral damage, the most severe neuronal injury results from ischaemic hypoxia which deprives the cells of energy substrates as well as oxygen.[140] Therefore, maintenance of cerebral blood flow within the normal range, as should occur with mean systemic arterial pressures of 40–60 mmHg and $PaCO_2$ levels of 3.3–6 kPa (25–45 mmHg), remains the major therapeutic priority.

Hypoxic encephalopathy associated with anaesthesia is unlikely without an identifiable event, provided the monitoring guidelines described above have been employed. However, preoperative or postoperative intracranial haemorrhage or cerebral ischaemia may occur without a clear inciting cause, most often in preterm neonates under 1500 g weight.[140] Intraventricular haemorrhage (IVH) may be precipitated by elevated intracranial vascular pressures due to severe hypoxia, by sudden arterial hypertension or by coagulopathy. Diagnosis can usually be made by ultrasound techniques. IVH does not lead to severe permanent handicap in all infants, although many incur some disability.

Protracted and severe hypoglycaemia (blood glucose < 30 mg/dl) or hyponatraemia (serum sodium < 120 mmol/l or mEq/l), or the rare inherited amino acid metabolic deficiencies, also cause stupor and coma. Unless treated promptly, these conditions can result in permanent brain damage. There is no discernible harm from mild hyperglycaemia, and recent preliminary animal data suggest that hyperglycaemia may even have a protective effect on the cerebrum during hypoxia.[141] The data and conclusions concerning the potential harmful effects of hyperglycaemia in adults with cerebral hypoxia should not be applied to the neonate until the issue has been thoroughly studied in infants.

Bacteraemia and sepsis, usually with Gram-negative organisms, remain among the most common causes of severe morbidity in newborns, including postoperative neonates. The incidence of sepsis rises dramatically in preterm infants under 1500 g in weight; in term infants, 1/1000 will have positive blood cultures compared to 4/1000 in preterm infants over 1500 g, and 300/1000 infants under 1500 g weight.[142] Sources of infection include intravascular catheters, tracheal airways and surgical wounds or implanted devices. Multiple organ failure, including pulmonary lesions and pathophysiology similar to adult respiratory distress syndrome, and a coagulopathy can develop, with a mortality well over 50 per cent. Management consists of specific multiple antibiotic therapy, controlled mechanical ventilation with positive end-expiratory pressure, maintenance of circulating blood volume with Ringer's lactate, plasma and 5 per cent albumin solution, support of cardiac output and systemic perfusion with inotropic agents (e.g. dopamine and dobutamine), and correction of the coagulopathy.

Conclusion

Research in cell biology, biochemistry, molecular pharmacology and applied physiology related to the clinical issues discussed in this chapter should create the foundation upon which the next important advances in neonatal anaesthesia will occur. Of particular relevance will be studies of the impact of growth and development on the transfer of oxygen from the environment to the mitochondria. Also of importance will be knowledge of those factors which affect the cell's utilization of oxygen and energy substrates, and the effects of anaesthetic and adjuvant drugs on these functions. The key organs to command our attention in these studies will be the developing brain, lung and heart.

Clinical investigation will be necessary to convert the new knowledge into practical benefit for the infant. Epidemiological studies of morbidity and mortality associated with anaesthesia in infants will help to point investigators toward the most important questions. The challenge for the subspecialty of paediatric anaesthesiology in the 1990s will be to meet the increasing clinical demands for excellent anaesthetic care under constrained funding, and yet pursue the research agenda. This must be accomplished, however, if we are to improve significantly the safety and efficacy of anaesthesia for sick infants by the end of the century.

REFERENCES

1. Weibel ER. Design and development of the mammalian lung. In: Weibel ER, ed. *The pathway for oxygen*. Cambridge, MA: Harvard University Press, 1984: 175–210.

2. Murray JF. Prenatal growth and development of the lung. In: Murray JF, ed. *The normal lung*. Philadelphia: WB Saunders, 1986: 1–24.

3. Murray JF. Postnatal growth and development of the lung. In: Murray JF, ed. *The normal lung*. Philadelphia: WB Saunders, 1986: 25–60.

4. Scarpelli EM. Pulmonary solute and liquid balance. In: Scarpelli EM, ed. *Pulmonary physiology: fetus, newborn, child, adolescent*. Philadelphia: Lea & Febiger, 1990: 215–32.

5. Hand IL, Krauss AN, Auld PAM. Pulmonary physiology of the newborn infant. In: Scarpelli EM, ed. *Pulmonary physiology: fetus, newborn, child, adolescent*. Philadelphia: Lea & Febiger, 1990: 405–20.

6. Fisher BJ, Carlo WA, Doershuk CF. Pulmonary function from infancy through adolescence. In: Scarpelli EM, ed. *Pulmonary physiology: fetus, newborn, child, adolescent*. Philadelphia: Lea & Febiger, 1990: 421–45.

7. Hoffman JIE, Heymann MA. Normal pulmonary circulation. In: Scarpelli EM, ed. *Pulmonary physiology: fetus, newborn, child, adolescent*. Philadelphia: Lea & Febiger, 1990: 233–56.

8. Rudolph AM, Yuan S. Response of the pulmonary vasculature to hypoxia and H^+ ion concentration changes. *Journal of Clinical Investigation* 1966; **45**; 399–411.

9. Anderson PAW. Physiology of the fetal, neonatal and adult heart. In: Polin RA, Fox WW, eds. *Fetal and neonatal physiology*. Philadelphia: WB Saunders 1992: 729–30; 736; 740–5.

10. Talner NS, Lister G, Fahey JT. Effects of asphyxia on the myocardium of the fetus and newborn. In: Polin RA, Fox WW, eds. *Fetal and neonatal physiology*. Philadelphia: WB Saunders, 1992: 759–69.

11. Serwer GA. Postnatal circulatory adjustments. In: Polin RA, Fox WW, eds. *Fetal and neonatal physiology*. Philadelphia: WB Saunders, 1992: 718–19.

12. Perlman JM, Volpe JJ. Episodes of apnea and bradycardia in the preterm newborn: impact on cerebral circulation. *Pediatrics* 1985; **76**: 333–8.

13. Koch G, Wendel H. Adjustment of arterial blood gases and acid–base balance in the normal newborn infant during the first week of life. *Biology of the Neonate* 1968; **12**: 136–61.

14. Oski F, Delivoria-Papadopoulos M. The red cell–2,3-diphosphoglycerate and tissue oxygen release. *Journal of Pediatrics* 1970; **77**: 941–56.

15. Albert MS, Winters R. Acid–base equilibrium of blood in normal infants. *Pediatrics* 1966; **37**: 728–32.

16. Dong SH, Liu HM, Song GW, Rong ZP, Wu YP. Arterialized capillary blood gases and acid–base status in normal individuals from 29 days to 24 years of age. *American Journal of Diseases of Children* 1985; **139**: 1019–22.

17. Yee WFH. Developmental physiology of respiratory control. In: Scarpelli EM, ed. *Pulmonary physiology: fetus, newborn, child and adolescent*. Philadelphia: Lea & Febiger, 1990: 370–1.

18. Altman DI. Cerebral blood flow in premature infants. In: Polin RA, Fox WW, eds. *Fetal and neonatal physiology*. Philadelphia: WB Saunders, 1992: 686–7.

19. Laterra JJ, Stewart PA, Goldstein GW. Development of the blood–brain barrier. In: Polin RA, Fox WW, eds. *Fetal and neonatal physiology*. Philadelphia: WB Saunders, 1992: 1528.

20. Nelson NM. Respiration and circulation after birth. In: Smith CA, Nelson NM, eds. *The physiology of the newborn infant*, 4th ed. Springfield, IL: CC Thomas, 1976: 177.

21. Anand KJS, Sippell WG, Aynsley-Green A. Randomized trial of fentanyl anaesthesia in preterm babies undergoing surgery: effects on the stress response. *Lancet* 1987; **1**: 243–8.

22. Anand KJS, Hansen DD, Hickey PR. Hormonal–metabolic stress responses in neonates undergoing cardiac surgery. *Anesthesiology* 1990; **73**: 661–70.

23. Costarino AT, Baumgart S. Neonatal water metabolism. In: Cowett RM, ed. *Principles of perinatal–neonatal metabolism*. New York: Springer-Verlag, 1991: 623–49.

24. Taeusch HW, Ballard RA, Avery ME, eds. *Diseases of the newborn*, 6th ed. Philadelphia: WB Saunders, 1991.

25. Friis-Hansen B. Body water compartments in children: changes during growth and related changes in body composition. *Pediatrics* 1961; **28**: 169–81.

26. Nelson NM. Respiration and circulation after birth. In: Smith CA, Nelson NM, eds. *The physiology of the newborn infant*, 4th ed. Springfield IL: CC Thomas, 1976: 159.

27. Mollison PL, Engelfriet CP, Contreras M. *Blood transfusion in clinical medicine*, 8th ed. London: Blackwell Scientific, 1987: 89–90.

28. Costarino AT, Gruskay JA, Corcoran L, Polin RA, Baumgart S. Sodium restriction versus daily maintenance replacement in very low birthweight premature neonates: a randomized, blind therapeutic trial. *Journal of Pediatrics* 1992; **120**: 99–106.

29. Vampee M, Blennow M, Linne T, Herin P, Aperia A. Renal function in very low birth weight infants – normal maturity reached during early childhood. *Journal of Pediatrics* 1992; **121**: 784–8.

30. Heimler R, Doumas BT, Jendrezjczak BM, Nemeth PB, Hoffman RG, Nelin LD. Relationship between nutrition, weight change and fluid compartments in preterm infants during the first week of life. *Journal of Pediatrics* 1993; **122**: 110–14.

31. Usher R, Lind J. Blood volume of the newborn premature infant. *Acta Paediatrica Scandinavica* 1965; **54**: 419–31.

32. Mollison PL, Veall N, Cutbush M. Red cell and plasma volume in newborn infants. *Archives of Disease in Childhood* 1950; **25**: 242–53.

33. Brans YW, Shannon DL, Ramamurthy RS. Neonatal polycythemia. II. Plasma, blood, and red cell volume in relation to hematocrit levels and quality of intrauterine growth. *Pediatrics* 1981; **68**: 175–82.

34. Taeusch HW, Ballard RA, Avery ME, eds. *Diseases of the newborn*, 6th ed. Philadelphia, WB Saunders, 1991: 1079

35. Behrman RE, Kliegman RM, Nelson WE, Vaughn VC, eds. *Nelson's textbook of pediatrics*, 14th ed. Philadelphia, WB Saunders, 1992: 1800–25.

36. Holliday MA, Seegar WE. The maintenance need for water in parenteral fluid therapy. *Pediatrics* 1957; **19**: 823–32.

37. Adamsons K, Grandy GM, James LS. The influence of thermal factors upon oxygen consumption of the newborn human infant. *Journal of Pediatrics* 1965; **66**: 495–508.

38. Fisher BJ, Carlo WA, Doershuk CF. Pulmonary function from infancy through adolescence. In: Scarpelli EM, ed. *Pulmonary physiology: fetus, newborn, child, adolescent*. Philadelphia: Lea & Febiger, 1990: 429.

39. Nelson NM. Respiration and circulation after birth. In: Smith CA, Nelson NM, eds. *The physiology of the newborn infant*, 4th ed. Springfield, IL: CC Thomas, 1976: 176.

40. Polk DH. Disorders of carbohydrate metabolism. In: Taeusch WH, Ballard RA, Avery ME, eds. *Diseases of the newborn* Philadelphia: WB Saunders, 1991: 965–71.

41. Ogata ES. Carbohydrate metabolism in the fetus and neonate and altered neonatal glucoregulation. *Pediatric Clinics of North America* 1986; **33**: 25–45.

42. Roberts JD, Todres ID, Coté CJ. Neonatal emergencies. In: Coté CJ, Ryan JF, Todres ID, Goudsouzian NG, eds. *A practice of anesthesia for infants and children*, 2nd ed. Philadelphia: WB Saunders, 1993: 230.

43. Adamsons K, Towell ME. Thermal homeostasis in the fetus and newborn. *Anesthesiology* 1965; **26**: 531–48.

44. Heiser MS, Downes JJ. Temperature regulation in the pediatric patient. *Seminars in Anesthesia* 1984; **3**: 37–42.

45. Reed MD, Besunder JB. Developmental pharmacology: ontogenic basis of drug distribution. *Pediatric Clinics of North America* 1989; **36**: 1053–74.

46. Nagourney BA, Aranda JV. Physiologic differences of clinical significance. In: Polin RA, Fox WW, eds. *Fetal and neonatal physiology*. Philadelphia: WB Saunders, 1992: 169–77.

47. Cook DR, Davis PJ. Pharmacology of pediatric anesthesia. In: Motoyama EK, Davis PJ, eds. *Smith's anesthesia for infants and children*. St Louis: CV Mosby, 1990: 157–97.

48. Salanitre E, Rackow H. The pulmonary exchange of nitrous oxide and halothane in infants and children. *Anesthesiology* 1969; **30**: 388–94.

49. Lerman J, Schmitt-Bantel BI, Gregory GA, Willis MM, Eger EI. Effect of age on the solubility of volatile anesthetics in human tissue. *Anesthesiology* 1986; **65**: 307–11.

50. Lerman J, Robinson S, Willis MM, Gregory GA. Anesthetic requirements for halothane in young children 0–1 month and 1–6 months of age. *Anesthesiology* 1983; **59**: 421–4.

51. Nicodemus HF, Nassiri-Rahimi C, Bachman L, Smith TC. Median effective doses (ED_{50}) of halothane in adults and children. *Anesthesiology* 1969; **31**: 344–8.

52. Friesen RH, Henry DB. Cardiovascular changes in preterm neonates receiving isoflurane, halothane, fentanyl, and ketamine. *Anesthesiology* 1986; **64**: 238–42.

53. Cameron CB, Robinson S, Gregory GA. The minimum

anesthetic concentration of isoflurane in children. *Anesthesia and Analgesia* 1984; **63**: 418–20.

54. Murray DJ, Forbes RB, Mahoney LT. Comparative hemodynamic depression of halothane versus isoflurane in neonates and infants: an echocardiographic study. *Anesthesia and Analgesia* 1992; **74**: 329–37.

55. Friesen RH, Lichtor JL. Cardiovascular effects of inhalation induction with isoflurane in infants. *Anesthesia and Analgesia* 1983; **62**: 411–14.

56. Glenski JA, Friesen RH, Berglund NL, Hassanein RS, Henry DB. Comparison of the hemodynamic and echocardiographic effects of sufentanil, fentanyl, isoflurane and halothane for pediatric cardiovascular surgery. *Journal of Cardiothoracic Anesthesia* 1988; **2**: 147–55.

57. Bikhazi GB, Davis PJ. Anesthesia for neonates and premature infants. In: Motoyama EK, Davis PJ, eds. *Smith's anesthesia for infants and children*, 4th ed. St Louis: CV Mosby, 1990: 436.

58. Coté CJ. Practical pharmacology of anesthetic agents, narcotics and sedatives. In: Coté CJ, Ryan JF, Todres ID, Goudsouzian NG, eds. *A practice of anesthesia for infants and children*, 2nd ed. Philadelphia: WB Saunders, 1993: 113–14.

59. Lockhart CH. Maintenance of general anesthesia. In: Gregory GA, ed. *Pediatric anesthesia*, 2nd ed. New York and London: Churchill Livingstone, 1989: 556.

60. Lindgren L, Saarnivaare L. Cardiovascular responses to enflurane induction followed by suxamethonium in children. *British Journal of Anaesthesia* 1983; **55**: 269–72.

61. Neigh JL, Garman JK, Harp JR. The electroencephalographic pattern during anesthesia with ethrane: effects of depth of anesthesia, $PaCO_2$, and nitrous oxide. *Anesthesiology* 1971; **35**: 482–7.

62. Taylor RH, Lerman J. Minimum alveolar concentration of desflurane and hemodynamic responses in neonates, infants and children. *Anesthesiology* 1991; **75**: 975–9.

63. Fisher DM, Zwass MS. MAC of desflurane in 60% nitrous oxide in infants and children. *Anesthesiology* 1992; **76**: 354–6.

64. Zwass MS, Fisher DM, Welborn LG, *et al.* Induction and maintenance characteristics of anesthesia with desflurane and nitrous oxide in infants and children. *Anesthesiology* 1992; **76**: 373–8.

65. Eger EI. Partition coefficient of I–653 in human blood, saline and olive oil. *Anesthesia and Analgesia* 1987; **66**: 971–3.

66. Yaster M. The dose response of fentanyl in neonatal anesthesia. *Anesthesiology* 1987; **66**: 433–5.

67. Murat I, Levron JC, Berg A, Saint-Maurice C. Effects of fentanyl on baroreceptor reflex control of heart rate in newborn infants. *Anesthesiology* 1988; **68**: 717–22.

68. Collins C, Koren G, Crean P. Fentanyl pharmacokinetics and hemodynamic effects in preterm infants during ligation of patent ductus arterious. *Anesthesia and Analgesia* 1985; **64**: 1078–80.

69. Koehntop DE, Rodman JH, Brundage DM. Pharmacokinetics of fentanyl in neonates. *Anesthesia and Analgesia* 1986; **65**: 227–32.

70. Lynn AM, Slattery JT. Morphine pharmacokinetics in early infancy. *Anesthesiology* 1987; **66**: 136–9.

71. Goudsouzian NG, Shorten G. Myoneural blocking agents in infants: a review. *Pediatric Anesthesia* 1992; **2**: 3–16.

72. Goudsouzian NG, Ryan JF, Savarese JJ. The neuromuscular

73. Fisher DM, Miller RD. Neuromuscular effects of vecuronium (ORG–NC45) in infants and children during N_2O, halothane anesthesia. *Anesthesiology* 1983; **58**: 519–23.

74. Fisher DM, O'Keefe C, Stanski DR, Cronnelly R, Miller RD, Gregory GA. Pharmacokinetics and pharmacodynamics of *d*-tubocurarine in infants, children, and adults. *Anesthesiology* 1982; **57**: 203–8.

75. Meretoja OA. Is vecuronium a long-acting neuromuscular blocking agent in neonates and infants? *British Journal of Anaesthesia* 1989; **62**: 184–7.

76. Cook DR, Fischer DG. Neuromuscular blocking effects of succinylcholine in infants and children. *Anesthesiology* 1975; **42**: 662–5.

77. Jonmarker C, Westrin P, Larsson S, Werner D. Thiopental requirements for induction of anesthesia in children. *Anesthesiology* 1987; **67**: 104–7.

78. Tiballs J, Malbegin S. Cardiovascular responses to induction of anesthesia with thiopentone and suxamethonium in infants and children. *Anesthesia and Intensive Care* 1988; **16**: 278–84.

79. Lockhart CH, Nelson WL. The relationship of ketamine requirement to age in pediatric patients. *Anesthesiology* 1974; **40**: 507–8.

80. Bain JA, Spoerel WE. A streamlined anaesthetic system. *Canadian Anaesthetists' Society Journal* 1972; **19**: 426–35.

81. Rose DK, Froese AB. The regulation of $PaCO_2$ during controlled ventilation of children with a T-piece. *Canadian Anaesthetists' Society Journal* 1979; **26**: 104–13.

82. Mizushima A, Wardall GJ, Simpson DL. The laryngeal mask airway in infants. *Anaesthesia* 1992; **47**: 849–51.

83. Markakis DA, Sayson SC, Schreiner MS. Insertion of the laryngeal mask airway in awake infants with the Robin sequence. *Anesthesia and Analgesia* 1992; **75**: 822–4.

84. Todres ID, Crone RK. Experience with a modified laryngoscope in sick infants. *Critical Care Medicine* 1981; **9**: 544–5.

85. Brandstater B. Dilatation of the larynx with Cole tubes. *Anesthesiology* 1969; **31**: 378–9.

86. Eckenhoff JE. Some anatomic considerations of the infant larynx influencing endotracheal anesthesia. *Anesthesiology* 1951; **12**: 401–10.

87. Goudsouzian NG, Morris RH, Ryan JF. The effects of a warming blanket on the maintenance of body temperatures in anesthetized infants and children. *Anesthesiology* 1973; **39**: 351–3.

88. Bissonnette B, Sessler DI, LaFlamme P. Passive and active inspired gas humidification in infants and children. *Anesthesiology* 1989; **71**: 350–4.

89. Greenwood RD, Rosenthal A, Nadas AS. Cardiovascular abnormalities associated with congenital diaphragmatic hernia. *Pediatrics* 1976; **57**: 92–7.

90. Greenwood RD, Rosenthal A, Nadas AS. Cardiovascular malformations associated with omphalocele. *Journal of Pediatrics* 1974; **85**: 818–21.

91. Greenwood RD, Rosenthal A. Cardiovascular malformations associated with tracheoesophageal fistula and esophageal atresia. *Pediatrics* 1976; **57**: 87–91.

92. Greenwood RD, Rosenthal A, Nadas AS. Cardiovascular

malformations associated with imperforate anus. *Journal of Pediatrics* 1975; **86**: 576–9.

93. Kitterman JA, Phibbs RH, Tooley WH. Catheterization of umbilical vessels in newborn infants. *Pediatric Clinics of North America* 1970; **17**: 895–912.

94. Horgan MJ, Bartoletti A, Polansky S, Peters JC, Manning TJ, Lamont BM. Effect of heparin infusates in umbilical arterial catheters on frequency of thrombotic complications. *Journal of Pediatrics* 1987; **111**: 774–8.

95. Anand KJS, Hickey PR. Pain and its effects in the human neonate and fetus. *New England Journal of Medicine* 1987; **317**: 1321–9.

96. Harries JT, Fraser AJ. The acidity of the gastric contents of premature babies during the first fourteen days of life. *Biology of the Neonate* 1968; **12**: 186–93.

97. Kimble KJ, Darnall RA, Yelderman M, Ariagno RL, Ream AK. An automated oscillometric technique for estimating mean arterial pressure in critically ill newborns. *Anesthesiology* 1981; **54**: 423–5.

98. Task Force on Blood Pressure Control in Children (Horan MJ, Chair). Report of the Second Task Force on Blood Pressure Control in Children – 1987. *Pediatrics* 1987; **79**: 1–25.

99. Edmonds JF, Barker GA, Conn AW. Current concepts in cardiovascular monitoring in children. *Critical Care Medicine* 1980; **8**: 548–53.

100. Baumgart S. Reduction of oxygen consumption, insensible water loss, and radiant heat demand with use of a plastic blanket for low-birth-weight infants under radiant warmers. *Pediatrics* 1984; **74**: 1022–8.

101. Coté CJ. Blood, colloid and crystalloid therapy. *Anesthesiology Clinics of North America* 1991; **9**: 865–84.

102. Gross JB. Estimating allowable blood loss: corrected for dilution. *Anesthesiology* 1983; **58**: 277–80.

103. Andrew M, Paes B, Milner R, Johnston M, Tollefsen DM, Powers P. Development of the human coagulation system in the full-term infant. *Blood* 1987; **70**: 165–72.

104. Lee KWT, Downes JJ. Pulmonary edema secondary to laryngospasm in children. *Anesthesiology* 1983; **59**: 347–9.

105. Warner LO, Martino JD, Davidson PJ, Beach TP. Negative pressure pulmonary edema: a potential hazard of muscle relaxants in awake infants. *Canadian Journal of Anaesthesia* 1990; **37**: 580–3.

106. Mason LJ, Betts EK. Leg lift and maximum inspiratory force, clinical signs and neuromuscular blockade reversal in neonates and infants. *Anesthesiology* 1990; **52**; 441–2.

107. Handler SD, Keon TP. Difficult laryngoscopy/intubation: the child with mandibular hypoplasia. *Annals of Otology, Rhinology and Laryngology* 1983; **92**: 401–4.

108. Stiles CM. A flexible fiberoptic bronchoscope for endotracheal intubation of infants. *Anesthesia and Analgesia* 1974; **53**: 1017–19.

109. Krucylak CP, Schreiner MS. Orotracheal intubation of an infant with hemifacial microsomia using a modified lighted stylet. *Anesthesiology* 1992; **77**: 826–7.

110. Hancock PJ, Peterson G. Finger intubation of the trachea in newborns. *Pediatrics* 1992; **89**: 325–7.

111. Bohn D, Tamura M, Perrin D, Barker G, Rabinovitch M. Ventilatory predictors of pulmonary hypoplasia in congeni-

tal diaphragmatic hernia, confirmed by morphologic assessment. *Journal of Pediatrics* 1987; **111**: 423–31.

112. Butt W, Taylor B, Shann F. Mortality prediction in infants with congenital diaphragmatic hernia: potential criteria for ECMO. *Anesthesia and Intensive Care* 1992; **20**: 439–42.

113. O'Rourke PP, Lillehei CW, Crone RK, Vacanti JP. The effect of extracorporeal membrane oxygenation on the survival of neonates with high-risk congenital diaphragmatic hernia: 45 cases from a single institute. *Journal of Pediatric Surgery* 1991; **26**: 147–52.

114. Roberts JD, Polaner DM, Lang P, Zapol WM. Inhaled nitric oxide in persistent pulmonary hypertension of newborn. *Lancet* 1992; **340**: 819–20.

115. Koop CE, Schnaufer L, Broennle AM. Esophageal atresia and tracheoesophageal fistula: supportive measures that affect survival. *Pediatrics* 1974; **5**: 558–64.

116. Templeton JM, Templeton JJ, Schnaufer L, Bishop HC, Ziegler MM, O'Neill JA. Management of esophageal atresia and tracheoesophageal fistula in the neonate with severe respiratory distress syndrome. *Journal of Pediatric Surgery* 1985; **20**: 394–7.

117. Karl HW. Control of life-threatening air leak after gastrostomy in an infant with respiratory distress syndrome and tracheoesophageal fistula. *Anesthesiology* 1985; **62**: 670–2.

118. Yaster M, Buck JR, Dudgeon DL, Manolio TA, Simmons RS, Zeller P. Hemodynamic effects of primary closure of omphalocele/gastroschisis in human newborns. *Anesthesiology* 1988; **69**: 84–8.

119. Gauntlett IS, Fisher DM, Hertzka RE, Kuhls E, Spellman MJ, Rudolph C. Pharmacokinetics of fentanyl in neonatal humans and lambs: effects of age. *Anesthesiology* 1988; **69**: 683–7.

120. Yaster M, Maxwell LG. Pediatric regional anesthesia. *Anesthesiology* 1989; **70**: 324–38.

121. Abajian JC, Mellish RWP, Browne AF, Perkins FM, Lambert DH, Mazuzan JE. Spinal anesthesia for surgery in the high-risk infant. *Anesthesia and Analgesia* 1984; **63**: 359–62.

122. Webster AC, McKishnie JD, Kenyon CF, Marshall DG. Spinal anesthesia for inguinal hernia repair in high-risk neonates. *Canadian Journal of Anesthesia* 1991; **38**: 281–6.

123. Kurth CD, Spitzer AR, Broennle AM, Downes JJ. Postoperative apnea in preterm infants. *Anesthesiology* 1987; **66**: 483–8.

124. Gleason CA, Martin RJ, Anderson JV, Carlo WA, Sanniti KJ, Fanaroff AA. Optimal position for a spinal tap in preterm infants. *Pediatrics* 1983; **71**: 31–5.

125. Wright TE, Orr RJ, Haberkern CM, Walbergh EJ. Complications during spinal anesthesia in infants: high spinal blockade. *Anesthesiology* 1990; **73**: 1290–2.

126. Welborn LG, Rice LJ, Hannallah RS, Broadman LM, Ruttimann UE, Fink R. Postoperative apnea in former preterm infants: prospective comparison of spinal and general anesthesia. *Anesthesiology* 1990; **72**: 838–42.

127. Maxwell LG, Yaster M, Wetzel RC, Niebyl JR. Penile nerve block for newborn circumcision. *Obstetrics and Gynecology* 1987; **70**: 415–19.

128. Gunter JB, Watcha MF, Forestner JE, *et al.* Caudal epidural anesthesia in conscious premature and high-risk infants. *Journal of Pediatric Surgery* 1991; **26**: 9–14.

129. Desparmet JF. Total spinal anesthesia after caudal anesthesia in an infant. *Anesthesia and Analgesia* 1990; **70**: 665–7.

130. Watcha MF, Thach BT, Gunter JB. Postoperative apnea after caudal anesthesia in an ex-premature infant. *Anesthesiology* 1989; **71**: 613–15.

131. Welborn LG. Post-operative apnea in the former preterm infant: a review. *Pediatric Anesthesia* 1992; **2**: 37–44.

132. Kurth CD, LeBard SE. Association of postoperative apnea, airway obstruction, and hypoxemia in former premature infants. *Anesthesiology* 1991; **75**: 22–6.

133. Flynn JT, Bancalari E, Snyder ES, Goldberg RN. A cohort study of transcutaneous oxygen tension and the incidence and severity of retinopathy of prematurity. *New England Journal of Medicine* 1992; **326**: 1050–4.

134. Flynn JT. Oxygen and retrolental fibroplasia: update and challenge. *Anesthesiology* 1984; **60**: 397–9.

135. Flynn JT. Retinopathy of prematurity. *Pediatric Clinics of North America* 1987; **34**: 1487–516.

136. D'Harlingue AE, Byrne WJ. Nutrition in the newborn. In: Taeusch HW, Ballard RA, Avery ME, eds. *Diseases of the newborn*, 6th ed. Philadelphia: WB Saunders, 1991: 71.

137. Tiret L, Nivoche Y, Hatton F, Desmonts JM, Vourc'h G. Complications related to anaesthesia in infants and children. *British Journal of Anaesthesia* 1988; **61**: 263–9.

138. Cohen MM, Cameron CB, Duncan PG. Pediatric anesthesia morbidity and mortality in the perioperative period. *Anesthesia and Analgesia* 1990; **70**: 160–7.

139. Keenan RL. Anesthetic mortality. *Seminars in Anesthesia* 1992; **11**: 89–95.

140. Allen WC, Riviello JJ. Perinatal cerebrovascular disease in the neonate. *Pediatric Clinics of North America* 1992; **39**: 621–50.

141. DiGiacomo J, Marro PJ, Lajevardi N, Mishra OP, Delivoria-Papadopoulos M. Effect of acute hyperglycemia (HG) on brain cell membrane function in newborn piglets. *Pediatric Research* 1991; **29**: 129A.

142. Cole FS. Bacterial infections of the newborn. In: Taeusch HW, Ballard RA, Avery ME, eds. *Diseases of the newborn*, 6th ed. Philadelphia: WB Saunders, 1991: 352–7.

FURTHER READING

Cook DR, Marcy JH, eds. *Neonatal anesthesia*. Pasedena: Appleton Davies, 1988.

Coté CJ, Ryan JF, Todres ID, Goudsouzian NG, eds. *A practice of anesthesia for infants and children*, 2nd ed. Philadelphia: WB Saunders, 1993.

Fanaroff AA, Martin RJ, eds. *Neonatal–perinatal medicine – diseases of the fetus and infant*, 5th ed. St Louis: Mosby–Year Book, 1992.

Hatch DJ, Sumner E. *Neonatal anaesthesia and perioperative care*, 2nd ed. London: Edward Arnold, 1986.

Motoyama EK, Davis PJ, eds. *Smith's anesthesia for infants and children*, 5th ed. St Louis: CV Mosby, 1990.

Chapter 31

Electrical, Fire and Compressed Gas Safety for the Patient and Anaesthetist

Jeffrey B. Gross and Harry A. Seifert

Electrical safety	Fire safety in the operating theatre
Electrocautery	Compressed gas safety
Magnetic resonance imaging	Summary

Electrical safety

The electromagnetic force, involving the attraction and repulsion of 'charged' particles, is one of the four primary forces of nature (the others being gravity, and the strong and weak nuclear forces). In the human body, many signal transmission and transduction processes are critically dependent upon electrical interactions. For example, the resting and action potentials of neuromuscular tissue depend upon the intra- and extracellular concentration of positively charged sodium and potassium ions, as well as the relative permeability of membranes to those ions. By interfering with the body's internal communication systems, externally applied electrical energy can have significant biological effects. When the site and timing of electrical stimulation are appropriate, the effects may be beneficial (e.g. cardiac pacemakers, transcutaneous electrical nerve stimulation). In contrast, when applied randomly, electrical stimulation is likely to be detrimental, causing burns, tetanic muscle contrations or ventricular fibrillation.

Until the 1960s, the most common uses of electrical power in the operating room were for illumination, electrocautery, and portable suction apparatus.[1] Many procedures were performed without ECG monitoring, and mechanical ventilators were pneumatically operated. Because of the widespread use of flammable anaesthetics such as ether and cyclopropane, the primary electrical hazard was generation of sparks which could cause an explosion.[2] Much effort was directed to reducing the buildup of static electricity on equipment and operating room personnel by the use of conductive flooring, shoes and hoses.[2] Explosion-proof electrical outlets (in which the electrical connections are broken before the plug is fully removed from its socket) reduced the risk that sparks

occurring at the time of disconnection would ignite flammable gases.

The risk of an anaesthetic explosion has been largely eliminated because flammable anaesthetics are only rarely used. However, in the UK the Department of Health and Social Security continues to recommend precautions.[3] The dangers have been reviewed and it has been suggested that the region within 25 cm of the breathing system is a zone of risk.[4] The use of electrical devices in the operating suite has become much more widespread. Most modern anaesthetic delivery systems and ventilators are electrically operated, and patients are connected to many electrically powered devices, including electrocardioscopes, pulse oximeters, automated blood pressure cuffs, warming mattresses and blood warmers. In addition, an increasing number of patients have intracardiac monitoring lines or pacemakers, which offer direct electrical access to the heart. As a result, the possibility of inadvertent electrical contact between equipment and patient or staff is always present; if equipment is faulty, the risk of electrocution is very real. In the section that follows, the mechanisms underlying electrical hazards in the operating room, as well as the ways in which engineering practice and equipment design can reduce these risks, are discussed.

Basic principles

Electric current involves the movement of charged particles; the most basic charged particles, electrons and protons, have charges of $-/+ 1.6/10^{19}$ coulomb (C), respectively. Charged particles exert forces upon each other. The magnitude of the force is proportional to the product of their charges, and inversely proportional to the square of distance between the particles. Two positively (or negatively) charged objects will repel each other, whilst

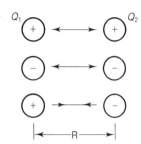

Fig. 31.1 Objects of like charge repel, while those of opposite charge attract. The force of repulsion/attraction is given by: $F = k \times Q_1 \times Q_2 \times R^{-2}$ where k is a constant, Q represents the amount of charge and R is the distance between the centres of the objects.

objects having opposite charges will attract (Fig. 31.1). This explains why electrons (negative charge) tend to remain in the vicinity of atomic nuclei (positive charge).[5]

For an electric current to flow, charged particles must be able to move. In some substances (metals, carbon) the outermost electrons are shared among neighbouring atoms; such substances are called *conductors* of electricity. The composition and dimensions of a conductor determine how readily electrons can move through it. The degree to which the flow of electrical charge is impeded is called the *resistance* (R), and is measured in units of ohms (Ω). In other substances (e.g. glass, myelin) electrons are bound tightly to individual atoms and cannot readily move between them; such materials are called *insulators*, and their resistance is essentially infinite. Solutions of ions (e.g. salt water, plasma) can also conduct electricity. In this case, rather than electrons moving from atom to atom, there is a bulk flow of ions through the solution in response to electric forces.

Electric current (denoted by the symbol I) is measured in terms of the quantity of electrical charge passing a point in a given time; by analogy, one might quantify the current of a river in terms of the quantity of water passing a point in a given time. One ampere (A) corresponds to a flow of one coulomb (6.25×10^{18} electrons or singly charged ions) past a given point in one second. The principle of conservation of mass (i.e. matter is neither created nor destroyed) requires that the total number of electrons in a physical system remains constant.* Therefore, electrical charges must flow in a closed loop or *circuit* (Fig. 31.2a).† Interrupting or 'opening' the circuit at any point will cause the flow of charged particles to cease. This is analogous to a decorative indoor fountain, where water is

---

* Strictly speaking, an electron may be annihilated by interacting with a positron, resulting in the production of electromagnetic energy according to the formula $E = mc^2$. However, this is not germane to the present discussion.
† Although electrical current generally involves the flow of negatively charged electrons, electrical diagrams conventionally denote current as the flow of *positive* charge (in the opposite direction). The diagrams in this chapter conform to this convention.

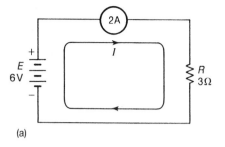

(a)

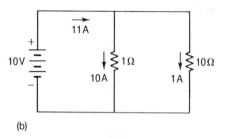

(b)

Fig. 31.2 (a) Electromotive force ($E = 6$ volts) from a battery causing a flow of electric current ($I = 2$ amps) to flow through a resistance ($R = 3$ ohms). In most conductors, current is carried by *negatively* charged electrons, which flow from the negative to the positive battery terminal. However, on schematic diagrams, current is conventionally denoted as the flow of *positive* charge in the opposite direction. (b) When there are two *parallel circuits*, most of the current follows the 'path of least resistance'.

pumped to the top, and after passing over the fountain, is collected by a drain and recycled; interruption of the flow of water at any point will cause the fountain to stop operating.

The force that causes electrons or ions to move through conducting media is called *electromotive force* (e.m.f. or E), and is measured in volts (V). Because the electromotive force is a measure of the amount of energy available to propel charged particles around a circuit, its units are those of energy (joules) per unit charge (coulomb): volts = joules/coulombs. The power required to move charged particles around a circuit is equal to the product of the electromotive force and the current. Thus, electrical power is given by:

$$P = I \times E \qquad (1)$$

The unit of power is the watt, where 1 watt = 1 joule/second. (Note that multiplying the current in amperes (coulombs/second) by the e.m.f. in volts (joules/coulomb) gives the power in watts (joules/second).)

Batteries produce electromotive force as a result of chemical reactions. The e.m.f. of a given battery depends upon the specific chemical reaction involved. Because the reaction is constant (at least until the battery is exhausted), the resulting e.m.f. is constant; when a circuit is attached to the battery, current always flows in the same direction. This *direct current* or *d.c.* is essential for the proper operation of most electronic devices, but is impractical for power distribution systems (see p. 655). In contrast, electrical generators convert mechanical energy

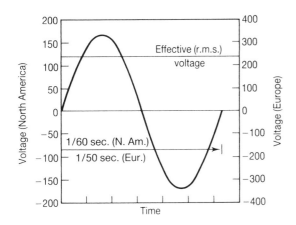

Fig. 31.3 Sinusoidal alternating current used for power distribution. The effective or root-mean-square (r.m.s.) voltage is 120 V in North America and 240 V elsewhere, whilst the frequency is 60 cycles/second (hertz, Hz) in North America and 50 Hz elsewhere. The peak voltage is 1.41 times the r.m.s. voltage.

to electrical energy by rotating coils of wire within intense magnetic fields; in the process, electrons are essentially forced through the coil, creating an electromotive force. Because of the continuously changing orientation of the coil relative to the magnetic field, the resulting electro-motive force varies sinusoidally with time (Fig. 31.3). The direction or *polarity* of the voltage alternates at a rate determined by the rotational speed of the generator – hence the term *alternating current* or *a.c.* In Europe, 50 of these 'cycles' are completed each second, corresponding to a frequency of 50 Hz (1 Hz = 1 cycle/second); in North America, power systems operate at 60 Hz. In a.c. systems, the mean voltage is zero, because the positive and negative parts of each cycle are equal in magnitude and duration but opposite in polarity. However, the square of the voltage is always positive; the effective voltage is determined by taking the square root of the mean of the squared instantaneous voltages (hence the term *root mean square* or *r.m.s.* voltage). The peak voltage is 1.41 times the r.m.s. voltage.* Therefore, in a 120-volt power line, the peak voltage is close to 170 volts; similarly, 240-volt power systems have peak voltages approaching 340 volts (Fig. 31.3). Because many of the detrimental effects of electricity depend upon peak rather than the effective voltages, contact with a given (r.m.s.) *a.c.* voltage is more likely to be damaging than contact with the same *d.c.* voltage.

The current (I) that flows in a conductor is proportional to the electromotive force (E), and inversely proportional to the resistance (R). This is known as Ohm's law, and is stated mathematically as:

---

* This results from the trigonometric identity $\sin^2 x = 0.5 - 0.5 \cdot \cos 2x$. Since the mean value of $\cos (2x)$ is 0, the mean value of $\sin^2 (x)$ is 0.5, and its square root (the root mean square) is 0.707. Thus, the ratio of the peak value of $\sin (x)$ [1] to its r.m.s. value [0.707] is 1.414.

$$I = E/R \text{ or } E = I \times R \tag{2}$$

A direct corollary of Ohm's law is that electricity follows the 'path of least resistance': If there are two alternative circuits for current to follow, the majority of the current flows through the circuit having the lower resistance (see Fig. 31.2b).

When current flows through a resistor, energy is dissipated in the form of heat. Substituting (2) into (1) gives:

$$P = I^2 \times R \tag{3}$$

Although an uninterrupted circuit is necessary for an electrical current to flow, there are two situations in which electrical energy can be transferred without a direct connection between conductors. Figure 31.4(a) denotes a circuit consisting of a battery (a source of e.m.f.) and two insulated but closely spaced parallel metal plates, forming a *capacitor*. When the switch is closed (Fig. 31.4b), electrons will be 'pushed' out of the negative battery terminal and will begin to collect on the attached plate of the capacitor Y, giving it a negative 'charge'. Because like charges repel, electrons will be repelled from the opposite plate of the capacitor X and return to the positive terminal of the battery. Thus, immediately after the switch is closed, it appears that current is flowing between the capacitor plates, despite the fact that they are electrically insulated from each other. However, as the plates become 'charged', a point is reached where the negative plate can accept no more electrons, the positive plate has no additional electrons to be repelled, and the current flow ceases (Fig. 31.4c).

Typically, the larger and closer the capacitor plates, the more charge they can accept (i.e. the greater their capacity). Before any additional current can flow 'through' the capacitor, some of the charge must be removed from the plates. This could be accomplished by reversing the battery; the flow of current would then remove excess electrons from plate Y and deposit them on plate X (Fig. 31.4d). Rather than manually changing the polarity of the battery, one might use an *a.c.* source, so that the flow of electrons reverses itself many times each second. In this case, the more rapidly the current alternates, the more readily current will *appear* to pass through the capacitor. The impedance (the a.c. equivalent of resistance, symbolized by Z) of a capacitor is inversely proportional to both the capacitance (C) and the frequency (F) of the alternating current:

$$Z = 1/(2\pi FC) \tag{4}$$

Passing a coil of wire through a magnetic field creates an electromotive force; a *changing* magnetic field passing through a coil of wire has the same effect. The device shown in Fig. 31.5, known as a transformer, takes advantage of this principle. A transformer consists of two coils of wire, wound around a common iron core. The coils

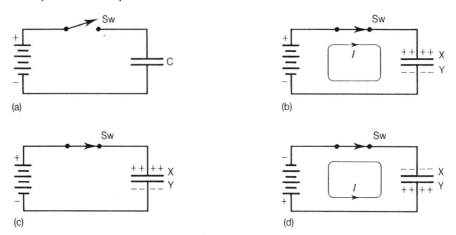

Fig. 31.4 (a) Electrical circuit consisting of a battery (source of unidirectional or *d.c.* voltage), two closely spaced but insulated conductors forming a capacitor (C), and a switch (Sw). (b) When the switch is closed, the battery causes electrons to accumulate on plate Y. These repel electrons from plate X, leaving it positively charged. The net effect is a transient flow of current (*I*). (c) Within a short time, plate Y can accept no more electrons, and plate X has no more electrons to be repelled. At this time, the flow of current ceases. (d) By reversing the battery, the electrons on plate X are replenished, while those on plate Y are repelled, leaving X negatively charged and Y positively charged. In the process, there is a transient flow of current (*I*) in the opposite direction.

are connected magnetically but *not* electrically. The coil on the left (known as the primary coil) is attached to a source of alternating current, resulting in a varying magnetic field, which *induces* an alternating e.m.f. in the coil on the right (known as the secondary). The relative voltage in the two coils depends upon the ratio of the number of turns in the respective coils: if the secondary has three times more turns than the primary, the output voltage will be three times the input voltage. Thus, transformers serve two functions: (1) changing voltages and (2) electrically isolating the primary and secondary circuits. The fact that transformers can only function with alternating current explains why essentially all electric power grids operate with *a.c.*

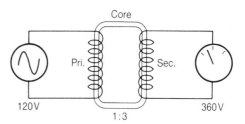

Fig. 31.5 Two electrically insulated coils of wire wound around a single iron 'core', to form a transformer. Alternating current flowing through the primary coil produces a varying magnetic field which generates or 'induces' an alternating current in the secondary coil. The primary/secondary voltage ratio is the same as the ratio of turns in the primary and secondary coils.

Biological effects of electrical current

Many factors contribute to the effect of electrical current on the body. The most important of these are the path of current through the body, the total current, the current density (i.e. the amount of current passing through a given tissue cross-sectional area) and, in the case of alternating currents, the frequency (frequencies between 5 and 5000 Hz being the most dangerous).[6]

Because of the large numbers of ions that are present, most body tissues readily conduct electricity. The body's primary protection from electrical injury is the resistance of the skin, which is determined by the size and location of electrical contact. Dry skin has a resistance of 10 000 ohms or more; this may drop to less than 1000 ohms in the case of moist skin or mucous membranes, or when there has been significant previous passage of electrical current (e.g. prior defibrillation).[1]

Table 31.1 Effects of electrical current on the body

0.01–0.05 mA	Microshock (arrhythmias if applied directly to heart via intracardiac catheters or electrodes)
0.05–0.15 mA	Threshold of perception
10–100 mA	Skeletal muscle contraction: 'let go' current
100–5000 mA	Ventricular fibrillation and/or burns
>5000 mA	Complete cardiac depolarization (defibrillation)

As shown in Table 31.1, the minimum perceptible electrical current is approximately 0.1 mA.[7] Currents of this magnitude flowing through the skin or mucous membranes locally depolarize nerve membranes, resulting in the familiar 'tingling' sensation, which can be reproduced by touching the terminals of a 9-volt battery to the tongue. The presence of this sensation when one touches the case of an electrical device suggests that power line voltage is 'leaking' to the chassis, and that the device should be checked for electrical leakage (see p. 656). Anaesthetized patients cannot report these sensations; this is one of the reasons for requiring special protective measures to be taken in the operating theatre.

Currents in the 10–100 mA range typically cause synchronized depolarization of motor neurons, resulting in contraction of the associated muscles. Clinically, this phenomenon is used by peripheral nerve stimulators, which have an output typically ranging from 10 to 60 mA. Flexor muscles tend to be stronger than the corresponding extensors, therefore electricians who are inadvertently exposed to currents of this magnitude may find themselves unable to release the energized conductors; as a result of this inability to 'let go', these are known as 'let go' currents.[8]

Skin resistance is 10 000 ohms; a.c. power lines operate at 120 and 240 r.m.s. volts which, using Ohm's law, will result in peak currents of 17 and 34 mA (see Fig. 31.3). These are within the 'let go' range. However, as current continues to flow, skin resistance tends to decrease as the epidermis breaks down; when the skin resistance has fallen to 1000 ohms, the current will increase to 170 and 340 mA respectively. These fall within the range of currents which may cause ventricular fibrillation. If the current flows between two fingers of the same hand, the risk of ventricular fibrillation is minimal, because the current density at the heart is low. In contrast, if current flows between two different extremities, significant current passes through the chest, producing a significant risk of arrhythmia.

Currents in excess of 5 A tend to synchronously depolarize much of the myocardium. Subsequently, a normal cardiac rhythm is often restored; this is the basis of a.c. and d.c. cardioversion and defibrillation. Ohm's law dictates that a defibrillator must deliver 5000 V to provide a 5 A shock to a patient whose skin resistance is 1000 Ω. This emphasizes the need to stand 'clear' when a defibrillator is discharged.

In addition to its effect on the cardiac conducting system, electricity passing through the body produces a significant amount of heat; as with any other resistor, the rate of heat production is given by the formula $P = I^2 \times R$ (3). Although the resulting burns superficially may appear to be confined to the sites of current entrance and exit, there is often extensive occult damage to blood vessels, muscle and visceral tissues.[9]

The cardiac conducting system is exquisitely sensitive to externally applied electrical fields. Currents as low as 20 µA applied to the endocardium may cause canine hearts to fibrillate[10]; the corresponding figure for human hearts is approximately 45 µA.[11] Such tiny currents, well below the threshold of perception, are said to cause 'microshock'. A precondition of microshock is a *direct* electrical connection to the endocardium; such connections potentially exist in patients with artificial pacemakers (especially temporary units with electrical connections outside the body) or saline-filled pulmonary artery catheters. The risk of arrhythmia is proportional to the current density (current per unit area). Thus, for a given total current, the risk of arrhythmia increases as the area of the contact decreaess.[10] Isolated power systems and ground fault circuit interrupters (p. 659) are designed to detect currents in excess of 1 mA. Thus, they cannot be relied upon to prevent microshock in vulnerable patients. All monitoring equipment used for patients with intracardiac catheters or wires should have electrically isolated patient connections in order to minimize the risk of microshock (see p. 661).

Electric power systems

Electric power systems are designed to deliver energy from the generating station to the consumer as efficiently as possible. Unfortunately, a certain amount of power is inevitably lost in the delivery process. For a power transmission line of resistance R, this is given by $P = I^2 \times R$. Because the power delivered to the customer is given by the product of the current and voltage ($P = I \times E$), it is apparent why power transmission is most efficient at high voltages: for a given amount of delivered power, doubling the transmission line voltage halves the current, resulting in a fourfold decrease in power loss. Transformers enable voltages to be 'stepped up' for long distance transmission, and then 'stepped down' to 120 or 240 volts for safe use in hospitals and homes.

Because of its large cross-sectional area and corresponding large capacity to absorb or release excess electrons, the earth acts as a reasonably good conductor of electricity (the exact conductivity depends on the terrain and soil moisture content). As a result, power distribution systems generally use the earth or 'ground' as one of their conductors (Fig. 31.6a). This is done for two reasons: (1) grounding provides a safe pathway for dissipation of lightning through the electrical wiring; and (2) if a transformer's insulation were to fail, the secondary coil could become electrically connected to the primary – in the absence of grounding, both conductors of the secondary would now carry several thousand volts directly into the home (Fig. 31.6b). Because the insulation of most domestic electrical appliances will withstand only a few hundred volts, there would be a very real electrical shock hazard. 'Grounding' or 'earthing' is accomplished by attaching one of the power conductors, known as the *neutral*, to a water main or other buried conductor at or near the point where the electrical lines enter a building. The other conductor, known as the *hot* (*live*) or *high* wire, is said to have a voltage of 120 or 240 volts 'above ground'.[12]

Whilst ground or earth referenced power systems minimize the risk that consumers will be exposed to transmission line voltages, they do create another hazard. Consider the situation shown in Fig. 31.7. Under normal

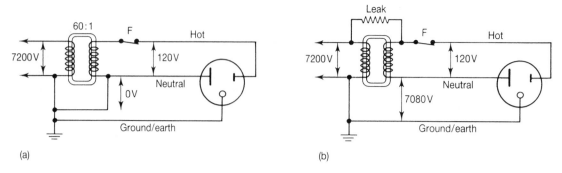

Fig. 31.6 (a) Power distribution system showing 7200 V from the power transmission line being 'stepped down' to 120 (or 240) V for domestic use. For safety, one terminal (the neutral) of the secondary coil is attached to earth or ground at the point where the power line enters a building. The fuse or circuit breaker (F) discontinues the flow of current if the current flowing in the 'hot' conductor exceeds the capacity of the wiring (typically 15–30 A). This protects the building from fire caused by overheated wiring, but offers no protection from macro- or microshock. (b) If the neutral conductor were not grounded, electrical leakage within the stepdown transformer (which is often subject to rain and snow outdoors) could result in lethal voltages on both the 'hot' and the 'neutral' conductors.

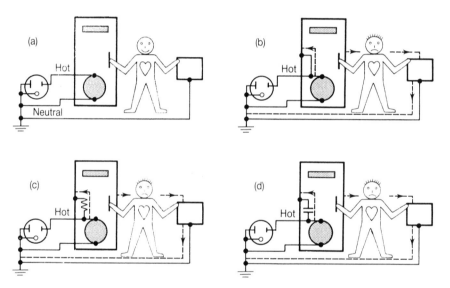

Fig. 31.7 (a) In this normally functioning refrigerator, the compressor receives 120 (or 240) volts from the 'hot' and the 'neutral' sides of the receptacle. The cabinet is not in contact with either of the power conductors, so an individual can contact the handle and a grounded object simultaneously without receiving an electrical shock. (b) Short circuit between the 'hot' side of the power line and the refrigerator cabinet. The compressor continues to operate normally, as it continues to be connected to the 'hot' and 'neutral' conductors. An individual contacting the handle and a grounded object simultaneously will receive a shock because current can flow from the 'hot' side of the power line, through his body, to the grounded neutral conductor. (c) Resistive 'leakage' from the 'hot' side of the power line to the refrigerator cabinet, caused by moisture permeating the wiring. Again, an individual can complete a circuit between the 'hot' and 'neutral' sides of the power line. The hazard current is inversely proportional to the leakage resistance. (d) Capacitative leakage between the 'hot' side of the power line and the refrigerator cabinet, resulting from the proximity of the wiring to the metal chassis. At power line frequencies (50–60 Hz), such leakage currents are relatively small; however, they may play a role in microshock.

circumstances (Fig. 31.7a), the motor of the refrigerator is electrically insulated from its enclosure, and the user cannot come into contact with live conductors. However, there are several ways in which an electrical connection could develop between the wiring and the chassis of the refrigerator. For example, the metal enclosure of the refrigerator could be dented, bringing it into direct contact with an electrically 'hot' (live) conductor[13] (Fig. 31.7b). Alternatively, if the wiring within the refrigerator were to

become wet, the moist insulation could act as resistor (Fig. 31.7c), allowing some current to flow from live conductors to the enclosure. This condition is known as 'leakage'. Another source of electrical leakage from the power line to the equipment enclosure is capacitive coupling (p. 654); depending upon the circuit geometry, alternating current may flow between adjacent conductors even though they are insulated from each other (Fig. 31.7d).[8] If any of these electrical faults were to develop, an individual could complete an electrical circuit by simultaneously touching the refrigerator and a grounded object such as a sink. If the leakage current were sufficient, electrocution could result. For this reason, all line powered electrical equipment used for patient care should be checked at 6-monthly intervals to ensure that the maximum current leakage from the equipment to ground is less than 100 μA.[14] Unfortunately, were a fault to develop between tests, there would be no warning of a potentially hazardous situation. Although the case would be electrically 'live', the refrigerator would continue to operate normally, because its electrical components would still be supplied with the proper voltage. Neither the fuse nor the circuit breaker provides protection in this situation. These are designed to reduce the risk of fire by breaking the circuit in the presence of a very high current flow (typically 15–30 A), and will be unaffected by potentially lethal leakage currents in the 0.1–5.0 A range, as illustrated in Fig. 31.7 (b–d).

Grounding or earthing

To minimize the risk associated with electrical leakage, the cases and/or chassis of many electrically operated appliances are attached directly to earth through a separate electrical conductor. This connection is commonly made via a third 'prong' on the electrical plug, which mates with a corresponding hole in the electrical outlet. The grounding prong is longer than the current-carrying conductors, ensuring that the case is attached to earth before electrical power is connected. This connection provides a direct, low-resistance path through which current can flow from the chassis of the refrigerator to ground (Fig. 31.8a). As a result, even in the presence of electrical leakage from the live (hot) side of the power line to the refrigerator chassis, the refrigerator does not become electrically live, and it does not pose an electrical hazard. Note that, in the case of a direct connection between the live (high) side of the power line and the grounded refrigerator case, a very high current would flow in the circuit (Fig. 31.8b). This would cause the fuse or circuit breaker protecting the wiring to the refrigerator to 'blow', disconnecting the power and providing a warning of the defect. However, in the case of 'leakage', the current flowing in the ground circuit would be relatively small, the fuse would not 'blow' and there would be no indication of the defect (Fig. 31.8c). If the ground connection were subsequently broken (because of a defect in the grounding wire, because of an improperly wired 'extension cord' or because of a 'cheater' used to adapt the three-prong plug to a two-prong ungrounded receptacle), the hazard would recur (Fig. 31.8d).[15] For this reason 'hospital grade' plugs and receptacles must be used in patient care settings, ground connections should be tested at 6-monthly intervals to ensure that they have a resistance of less than 0.50 Ω, and extension cords should not be used.[14–16]

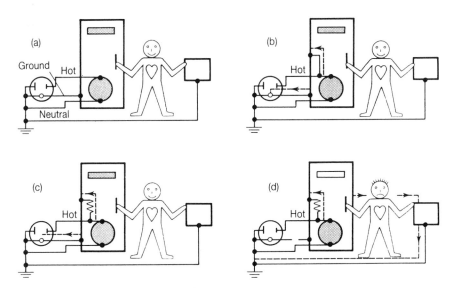

Fig. 31.8 (a) Under normal circumstances, grounding or earthing the refrigerator cabinet via a separate conductor within the power cord does not affect its operation. (b) In the case of a direct connection between the 'hot' side of the power line and the refrigerator cabinet, current will follow the path of least resistance via the grounding conductor to earth. This minimizes the risk to an individual who simultaneously contacts the cabinet and ground. Because the direct connection from the 'hot' side of the line via the cabinet and ground to the 'neutral' allows a high current flow, the fuse or circuit breaker protecting the outlet will open the circuit, providing an indication that the refrigerator is faulty. (c) If there is resistive or capacitive leakage from the 'hot' side of the line to the refrigerator cabinet, grounding the cabinet minimizes the risk to an individual contacting both the refrigerator and ground. However, the current flow is generally inadequate to open the fuse or circuit breaker. Under these circumstances there will be no warning of the defect. (d) If the ground connection is subsequently broken (defective plug, wiring or extension cord), the cabinet of the refrigerator becomes electrically 'live' without warning, and individuals are once again at risk.

Although grounding of equipment cases significantly reduces the risk of an electric shock, it does not, by itself, provide sufficient protection in the operating room, for the following reasons:

1. In the case of a direct connection between the live (hot) side of the power line and the chassis of the equipment, the fuse or circuit breaker will cut off power to all equipment attached to the same circuit as the defective device, potentially compromising patient safety.

2. In the absence of such a connection, there is no warning of electrical leakage from the power line to the equipment chassis.

3. Anaesthetized patients are unable to voluntarily pull away from any electrically charged objects that they might inadvertently contact.

4. Grounding alone provides no protection in the case of two simultaneous electrical faults, such as a defective ground connection *plus* a connection or leakage from the live (hot) side of the power line to the equipment chassis (Fig. 31.8d).

At present there are two mechanisms available to address these safety concerns, which are suitable for operating theatres that are restricted to non-flammable anaesthetic agents.

Ground fault circuit interrupters

A ground fault circuit interrupter (GFCI) can be installed to protect each electrical outlet in the operating theatre. This device constantly compares the current flow in live and neutral conductors of the power line; if these currents differ by more than 6 mA, the GFCI will disconnect the outlet from the mains within 25 ms. If electrical leakage occurs between the live side of the power line and grounded chassis, some current will return to ground via the grounding wire rather than via the neutral side of the power line (Fig. 31.9a). As a result, the current flow in the live and neutral conductors will be unequal, and the GFCI will interrupt the power. If the case were not connected to ground (e.g. defective ground wire), power would not be immediately disconnected, because there would be no alternative path for current to return to ground via the case (Fig. 31.9b). However, if a person were to touch the electrically live (hot) case and a grounded object simultaneously, current would flow through the body from the case to ground; because the currents in the live and neutral conductors would no longer be equal, the GFCI would interrupt power quickly enough to minimize the risk of electrocution (Fig. 31.9c). Thus, GFCIs provide protection in the event of two simultaneous failures: excessive current leakage *plus* a defective ground connection. As a result, most electrical codes require their use in places where

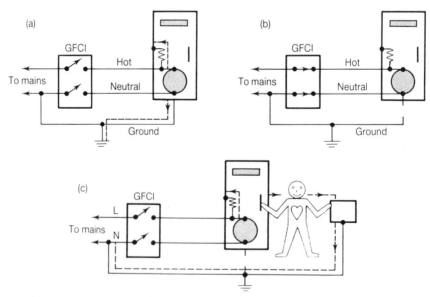

Fig. 31.9 (a) A ground fault circuit interrupter (GFCI) cuts off the power if the current in the live (hot) and neutral conductors are unequal. In this Figure, there is leakage from the live (hot) side of the line to the grounded refrigerator cabinet. Because the current flowing in the live (hot) conductor is equal to the sum of that returning via the neutral conductor and that returning via the leak and grounding conductor, the live (hot) and neutral currents are unequal, and the power has been cut off. (b) If the chassis of the appliance is not grounded, no current flows via the leakage resistance and ground to the neutral side of the power line. Because the currents in the live (hot) and neutral lines are equal, the GFCI does not cut off the power. (c) When an individual contacts the ungrounded cabinet and ground simultaneously, a path is completed from the live (hot) side (L) of the line via the leak, cabinet and person to ground and the neutral side (N) of the mains. If the current in the live (hot) conductor exceeds that in the neutral by more than 6 mA, the GFCI will open within 25 ms, minimizing the risk of electrical injury.

Table 31.2 Response sequence when ground fault circuit interrupter (GFCI) 'trips', cutting off power

1. Disconnect all equipment from the GFCI circuit in question.
2. Reset the GFCI. If it trips again, either the GFCI or associated wiring is defective.
3. Sequentially test each suspected device by plugging it in, turning it on and determining if the GFCI 'trips'. Turn off and unplug before testing the next device.
4. Any device which, by itself, trips the GFCI, is faulty and should *not* be used until the source of the current leakage is located and repaired.
5. Plug in the remaining equipment simultaneously. If the GFCI trips, this would suggest that although no individual device has significant leakage, the total leakage current is excessive. Temporarily, the problem can be alleviated by plugging the equipment into separate GFCI-protected circuits. However, the equipment should be checked for electrical leakage as soon as possible.
6. If the GFCI appears to trip randomly, when a device is activated or touched, the equipment may be improperly grounded, significant leakage current passing through the person contacting it. The device should therefore be removed from service immediately.

electrical leakage poses a significant risk. In addition to operating rooms, outlets in wet locations such as lavatories, kitchens, basements and outdoors should be protected by GFCIs.

GFCIs should be tested monthly for proper operation.

This may be accomplished by pressing the 'test' button or using a special test 'plug' with intentional electrical leakage from the live (hot) side to ground. GFCIs may occasionally 'trip' because of current surges when electric motors are turned on or off. If this occurs, power is easily restored by pressing the 'reset' button. If a GFCI trips repeatedly, the circuit must be tested as shown in Table 31.2.

Isolated power systems

Another method for minimizing electrical hazards in the operating room is the use of an isolated power system. As described earlier, the electrical distribution system used in hospitals and dwellings is 'ground' or 'earth' referenced; that is, for safety reasons, one side of the power line (the neutral) is attached to ground at the point where the electrical service enters the building. Thus, a significant electrical hazard exists when an individual completes a circuit between an electrically live device and any grounded object. In contrast, if neither of the power conductors were grounded, an individual could simultaneously contact either one of the live conductors and ground without completing a circuit (Fig. 31.10a). In such an *electrically isolated* power system, shock hazards exist only if one were to simultaneously contact both power conductors.

Electrical isolation can be accomplished by using a transformer having a 1:1 turns ratio. The input and output voltages are the same (120 or 240 volts in North

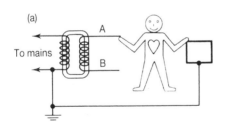
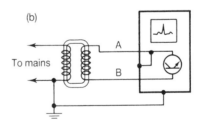

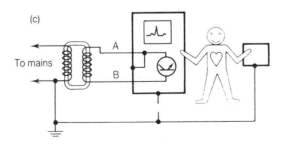

Fig. 31.10 (a) In an isolation transformer, the primary and secondary coils are coupled magnetically but *not* electrically; note that neither conductor of the secondary coil is attached to ground. Thus, an individual can simultaneously contact either one of the secondary lines and ground without completing a circuit. (b) A defect has caused a direct connection between one side of the isolated power system and the case of the electrocardioscope. The electrocardioscope continues to function normally (line voltage continues to be present between conductors A and B). However, the power system is no longer isolated from ground. A second defective unit could create a hazard by bringing a patient into simultaneous contact with conductor B and ground. (c) Even in the presence of a direct connection from one side of the line to the equipment case *and* a simultaneous faulty ground connection, an individual can safely contact the electrocardioscope and a grounded object without completing a circuit.

America and Europe, respectively). However, although the primary and secondary coils are coupled magnetically, they are electrically insulated from each other. Thus while the input to the isolation transformer is 'ground referenced' the output has no connection to ground (Fig. 31.10a). Note that, even with isolated power systems, the *cases* of line operated equipment are attached to ground via the separate grounding or earthing pin of the electrical plug and receptacle.

Consider a situation where, because of insulation failure, the case of an operating room monitor has become electrically connected to one side of the power line (Fig. 31.10b). The result of this single fault is to electrically ground one side of the previously isolated power system. One can imagine a circuit from side A of the power line through the equipment case to ground and then. . . ; as side B of the isolation transformer's output is not connected to ground, no current flows because a complete circuit has not been formed. However, the electrical system is no longer isolated from ground, and all the potential hazards associated with ground-referenced power systems now exist. Therefore, a monitoring system is necessary to determine whether the isolated status of the power system has been compromised (see below).

Let us now introduce a second, simultaneous, fault – a broken grounding connection on the defective monitor (Fig. 31.10c). If an individual were to contact simultaneously the defective equipment and ground, one could imagine a circuit from side A of the power line, through the equipment, operator and ground to . . .; since side B of the power line is *not* connected to ground, the circuit is incomplete and *no* current will flow. Thus, like ground-fault circuit interrupters, isolated power systems effectively minimize the risk of electrical shock even in the presence of two simultaneous faults.

Unlike GFCIs, however, isolated power systems *do not* interrupt electrical power in the presence of current leakage from the power line to ground. In the operating room, this offers the obvious advantage of allowing critical equipment to continue to operate safely in the presence of a single fault. However, unlike a GFCI, an isolation transformer alone will *not* indicate that a such a 'ground fault' exists. For this purpose, isolated power systems incorporate separate *line isolation monitors*, which continually determine the integrity of the isolated power system.

Line isolation monitors

A line isolation monitor (LIM) is essentially a milliammeter attached alternately between each side of the power line and ground. If the power system is, in fact, isolated from ground, no current will flow through the meter (Fig. 31.11a). However, if a direct or leakage connection exists from one side of the power line to ground (Fig. 31.11b), current will flow from one side of the isolated power line through the LIM to ground, returning to the other side of the isolation transformer output via the connection or leak. The amount of current flowing through the meter indicates the 'leakage hazard' – the maximum current that would flow through an individual if he were to touch side A of the power line and ground simultaneously. Because it is impossible to determine *a priori* which side of the power line will inadvertently be grounded, the LIM automatically cycles from side A to side B and back, either mechanically or electronically.[17]

A LIM is designed to sound an alarm when the leakage hazard exceeds 5 mA (2 mA in the presence of flammable gases). As in the case of GFCIs, this limit is sufficiently low to prevent macroshock but not microshock. It is important to recognize that the sounding of the LIM's alarm does not turn off electrical power or remove the potentially hazardous condition. Rather, it serves as a warning that a single fault exists, and that patients and personnel could be at risk if a second fault were to develop. Line isolation monitors should be tested monthly by activating the 'test' button, and semi-annually by attaching each side of the power line to ground through a resistance that will allow an r.m.s. leakage current of 5 mA at the nominal line voltage (24 kΩ for 120-volt systems, 48 kΩ for 240-volt systems).[18]

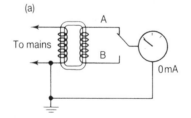

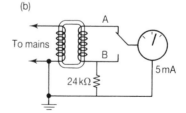

Fig. 31.11 (a) The line isolation monitor operates by alternately measuring the current flow from each side of the isolated power supply to ground. This represents the *maximum* current that could flow through an individual who simultaneously contacted one side of the isolated supply and ground. Note that the isolation monitor *per se* partially defeats the isolation. In this case, there is a high-resistance connection from A to ground via the meter; therefore, a small current could flow through an individual contacting B and ground simultaneously. (b) Leakage from side B of the power line to ground is detected by the line isolation monitor. Current flows from A through the meter to ground, returning to B via the 24 kΩ leakage resistance.

Table 31.3 Response sequence when line isolation monitor (LIM) alarm sounds, indicating leakage current hazard in excess of 5 mA

1. Each device connected to the isolated power system should be sequentially turned off and disconnected from the mains supply, starting with the equipment most recently plugged in. Leave each device unplugged until the fault is located.
2. When the defective equipment is unplugged, the LIM reading will return to the safe range. The faulty equipment should be repaired before being restored to service.
3. Reconnect the remaining equipment sequentially, to ensure that a second fault is not present.
4. If the defective equipment is essential to on-going patient care (e.g. bypass machine during open heart surgery), it may be used for the remainder of the surgical procedure before being taken out of service for repair. However, the power system is no longer isolated from ground, and a second faulty device could result in a shock hazard.
5. If the LIM indicates intermittent leakage, occurring only when a device is activated or touched, it is possible that the chassis of the equipment is inadequately grounded, and that leakage current is passing through the operator or patient. Such equipment should be unplugged and serviced immediately.

When the line isolation monitor alarm sounds, the sequence of steps shown in Table 31.3 should be taken.

Prevention of microshock

To minimize effectively the risk of microshock, any patient connections that could conceivably become connected to endocardial monitoring lines must be *electrically isolated*. Examples of such connections include electrocardioscope leads, pressure transducer inputs, thermodilution cardiac output monitors, temperature probes and external cardiac pacemakers.[19]

Electrical isolation requires that, under normal operating conditions, no more than 10 µA of current can pass through the 'isolated' connection into the patient. Under worst-case conditions, the standards allow a maximum of 20 µA (USA) and 50 µA (Europe).[14,20] Note that electrically grounding the vulnerable connection is *not* an acceptable substitute for electrical isolation, because such a connection could serve as a return path for leakage current from another device (Fig. 31.12a).

There are several methods by which electrical isolation can be obtained. Temporary pacemakers and some cardiac output monitors are electrically isolated by virtue of being battery powered, completely independent of the a.c. electrical power system. However, a hazard can develop if the circuitry or batteries of such devices inadvertently contact other operating room equipment.[21] Line-powered devices achieve electrical isolation through the use of internal low-leakage power transformers to supply electricity to the circuitry associated with critical patient connections. Signals coming from these connections are often transmitted to the remaining circuitry of the monitoring equipment using optical isolators. These devices convert electrical signals into a variable-intensity infrared beam, which is then detected by a phototransistor (Fig. 31.12b). Thus, there is no electrical connection from the patient connection to any component of the electrical power system, minimizing the risk of microshock.

Electrocautery

The basic principle of the electrocautery is that rapid, intense heating of tissues tends to 'coagulate' or obliterate the lumina of blood vessels, thus minimizing bleeding. One means of achieving this is to pass an electric current through a thin wire, causing it to heat to incandescence (600–1200°C). Such 'hot wire' cautery devices are occasionally used in ocular surgery, but the exposed hot wire may damage surrounding tissues and can be a source of ignition for operating room fires.[22] Another method of providing haemostasis is to rapidly heat tissues through the use of a high-voltage radiofrequency (500–3000 kHz) electrical spark.[23] This technique, used by modern electrosurgical units (ESUs), offers several distinct avantages. Because there is no filament to heat up or cool down, on–off control is instantaneous. By including a haemostat in the circuit, blood vessels held within the instrument can be obliterated, reducing the need to 'tie off' small bleeders. Finally, the ability to generate sparks under water makes it possible to obtain haemostasis during transurethral surgery (provided a non-ionic irrigating medium is used).

Two basic physical principles are critical to the operation of modern electrocautery units. First, as demonstrated graphically by Nikolai Tesla, high-frequency alternating currents (with frequencies in excess of 500 kHz) are carried on the surfaces of conductors rather than passing through them. As a result, high-power electrosurgical units can be used without risk of causing cardiac arrhythmias or thermal damage to internal organs. Second, although the total current is the same at all points in a circuit, the heat generated by an electrical current is proportional to the current *density*. The use of a small 'active' electrode at the site of surgery allows current density and heat production to be maximized. In contrast, by using a large return electrode, the current is dispersed over a wide area. The resulting low current density reduces the risk of unintentional thermal damage at the site of the return electrode (Fig. 31.13a). Many of the complications associated with the use of the cautery result from

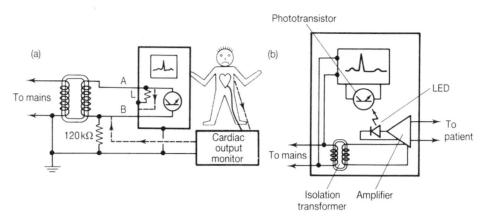

Fig. 31.12 (a) Grounding of an intracardiac monitoring lead does not provide protection from microshock. Here, there is leakage (L) from side A of the isolated power system to the case of the electrocardioscope, which also has a defective ground connection. Side B of the line has a 1 mA (120 kΩ) leak to ground, which is not enough to activate the line isolation alarm. Microshock currents can flow from A through leakage (L) to the electrocardioscope case, entering the patient's arm and exiting via the grounded intracardiac electrode before returning to B through the 120 kΩ ground leak. (b) Electrical isolation of patient connections is accomplished by supplying power to patient-connected amplifiers (ECG, pressure transducers, etc.) via a low-leakage isolation transformer. The electrical signals from these devices are coupled to the remainder of the monitoring circuitry via an optical isolator consisting of a light-emitting diode (LED) and a phototransistor.

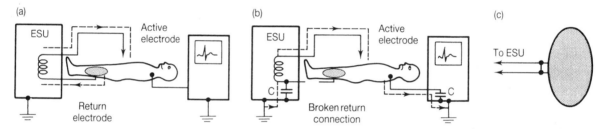

Fig. 31.13 (a) Normally functioning electrosurgical unit (ESU). Cutting/coagulating current is concentrated at the active electrode, generating the necessary heat. The large return electrode disperses the current to minimize the heating of underlying tissues. Little current returns via the higher resistance ECG pad. (b) In the event of a broken return connection, current flows from the active electrode via the patient to the ECG pad, and returns to the ESU through the ECG lead and ground. Although they are electrically isolated from ground at line current frequency (50–60 Hz), capacitive coupling (C) between the wires and cases of the equipment results in functional grounding of the ECG leads and the ESU return connection at radio frequencies (0.5–3.0 mHz). Because the ECG pad is relatively small, the return current is concentrated and a burn may develop. (c) Dual wire is used by some ESUs to ensure that the return pad has been *plugged into* the ESU. The ESU will not operate unless a complete circuit is detected between the two wires. With this scheme, the ESU can be activated whenever the grounding pad is plugged in, regardless of whether the pad has been affixed to the patient.

inadequate return connections or 'grounding' of the patient. (In the original electrosurgical units, the patient return electrode was literally connected to earth or 'ground'. In more recent units, a capacitor is used to isolate the return plate from ground at line current frequencies (50–60 Hz).[24] Although this reduces the risk of electrical shock associated with directly connecting the patient to ground,[25] it does not eliminate the risk of burns if the return path is inadequate.[26])

Return electrodes

The patient return electrode should have an area of at least 0.66 cm^2 per watt of ESU output;[23] assuming a maximum generator output of 500 W, a contact area of at least 330 cm^2 should be provided. As long as the return plate is placed over well-perfused muscular tissue, the resulting heat should be readily dissipated by tissue blood flow. However, if the return plate is placed over a bony prominence (e.g. the sacrum), heat dissipation may be inadequate, and a burn may ensue.[27] Similarly, if the contact between the return electrode and the skin is not uniform, 'hot spots' may develop. The likelihood of this can be minimized by applying conductive paste or by using 'pre-gelled' self-adhesive return electrodes. Additionally, the return electrode should not be positioned over an electrically conductive prosthetic implant (e.g. hip arthroplasty), because the current density may then be non-uniform and a burn may result.

In theory, if the return electrode were not attached to both the patient and the electrosurgical unit, one would expect that no current would flow, as the circuit would be incomplete. However, when two insulated conductors are in proximity to one another, they form a capacitor (p. 653), the impedance of which decreases as the frequency of the alternating current source increases. At the frequencies used by electrosurgical units, even patient connections that are 'electrically isolated' at power line frequencies (50–60 Hz), such as temperature probes and electrocardioscope leads, can serve as alternate return pathways (Fig. 31.13b).[28] Since the contact area of such connections is small, patients may be burned because of the resulting high current densities.[29]

Because of the hazards associated with failure to provide a proper return path for electrosurgical current, several schemes have been devised to ensure that the grounding pad is properly connected. For instance, the grounding pad may be attached to the ESU via two separate wires (Fig. 31.13c); the cautery cannot be activated unless special circuitry detects a complete circuit between them. This ensures that the grounding plate is attached to the ESU, but does not guarantee that it is affixed to the patient. In a more sophisticated design, the grounding pad, itself, is split into two conductive areas, each of which is attached to the ESU via a separate wire. The electronics within the ESU must detect a complete circuit via the patient's body before the cautery can be activated.[30] Even this system is not risk-free: the continuous sensing current between the two halves of the grounding electrode may cause electrolysis of the conductive paste, and the resulting caustic breakdown products may damage the underlying skin.

Despite these safeguards, anaesthetists must constantly be aware that electrosurgical return connections can become partially or completely disrupted. Signs that this may have occurred include excessive interference on the ECG or pulse oximeter, a tingling or 'static-like' sensation when the patient is touched with an ungloved hand, and repeated requests by the surgeon to 'turn up the ESU'. The need to increase the ESU output results from the increased total resistance of the circuit; by Ohm's law, a higher voltage is necessary to obtain the same coagulating current.

Monopolar vs bipolar

The usual ESU is a *monopolar* device; the surgeon uses the 'active' electrode, whilst the returning current follows the path of least resistance to the return or grounding plate. This can be detrimental under circumstances in which the return path is confined. For example, when a fallopian tube is cauterized using a monopolar ESU, all of the current necessarily flows via the tube to its uterine attachment, at which point it disperses through the body. This may cause thermal damage to tubal structures. Similarly, in neurosurgery, monopolar electrosurgery may cause undesirable heating of adjacent areas of the brain. To alleviate this problem, *bipolar* ESUs have been developed. The bipolar ESU does not depend on a large, dispersive return electrode. Rather, the surgeon uses a forceps-like instrument; each of the blades is attached to one of the output terminals of the ESU generator. As a result, the coagulating electrical field is confined to the space between the electrodes, and damage to adjacent tissues is minimized. A further benefit is that less power is required. If, as a result of an electrical defect, one of the output terminals of a bipolar ESU were to become electrically connected to its case, the patient and the surgeons would be subject to the same risks associated with a monopolar ESU having an inadequate return path.[31]

Cardiac pacemakers

Radiofrequency electrosurgical units pose special risks to patients with implanted pacemakers. Most modern pacemakers use sensing circuitry, which inhibits pacing when intrinsic cardiac activity is detected. For example, the output of a VVI pacemaker (ventricle pace, ventricle sense, inhibited pacemaker) will be inhibited if a ventricular depolarization is detected. Unfortunately, the wire leading from the pacemaker to the heart may act as an antenna, picking up the radiofrequency energy 'transmitted' by the cautery. In the presence of such interference, the pacemaker may act as if electrocardiographic 'R' waves are being detected when the ventricle is not, in fact, depolarizing. The risk of 'turning off' a pacemaker with the electrocautery may be minimized by placing the return electrode as close as possible to the site of surgery and using the minimum effective power output from the ESU. Because of their lower output power and direct return path, bipolar ESUs are less likely than monopolar units to interfere with pacemaker function.

In pacemaker-dependent patients, it is necessary to carefully observe the ECG (and listen to heart tones, especially if electrical interference is present) to determine whether the pacemaker is temporarily disabled when the ESU is activated. If this is the case, there are several possible options. One is to ask the surgeon to limit his use of the cautery to short, intermittent bursts, allowing the pacemaker to function between them. Another option is to place a magnet over the pacemaker generator unit. This will convert the pacemaker to VOO pacing (continuous, uninhibited ventricular pacing at a fixed rate of approximately 72 beats/min). The latter option has two potential risks. The first is that there is a small but finite possibility of a paced beat occurring during the vulnerable period following a previous native intrinsically conducted beat, resulting in ventricular tachycardia or fibrillation. The second potential risk is that many modern pacemakers are programmable: their rate, sensing thresholds and outputs can be remotely controlled by electrical and magnetic signals applied via the chest wall. The combination of a magnet and random electrical signals from the ESU may reprogramme the pacemaker in unpredictable ways.[32] However, the new programme will not take effect as long as the magnet is in place. Therefore, when removing the magnet one must be aware that the pacemaker programme may have been altered; if the ECG does not show normal pacemaker operation, the magnet should be replaced over the pacemaker until the appropriate programming device is available. (Note that placing the programming device over the pacemaker generator and pressing the red 'panic button' will return the unit to the VVI mode with a rate of 70 beats/min and an output of 5 volts; this should be adequate for most patients until the original programme can be restored.

Because of experience with pacemakers, there has been some concern that automatic implantable cardioverter defibrillators (AICDs) might be subject to ESU interference. In this case, the AICD might be expected to falsely interpret interference from the cautery as evidence of ventricular dysrhythmia, resulting in unnecessary and potentially hazardous automatic defibrillation.[33] Therefore, the manufacturers recommend that the AICD be turned off preoperatively by placing a magnet over the unit for 30 seconds. However, evidence obtained in both animals and patients suggests that this may not be necessary, since automatic gain control circuitry within the AICD seems to prevent sensing of electrocautery signals.[34]

Electrical interference

Other electronic devices for patient care may also be susceptible to interference from the electrocautery. The effect of the electrocautery on pulse oximeters and ECG monitors is well known to most anaesthetists. Good circuit design and proper shielding can help to reduce or eliminate such interference.[35] Similarly, infusion pumps may malfunction during ESU operation, resulting in unintended changes in infusion rate.[36] Power cords and patient connections may act as antennae, conducting the interference into the devices. This risk can be minimized by keeping the pump as far from the active ESU electrode as possible, by using the lowest possible ESU output, by using battery rather than a.c. mains power and by careful monitoring of pump operation during electrosurgery.[36]

Magnetic resonance imaging

Although not requiring the use of ionizing radiation, magnetic resonance imaging (MRI) equipment presents several potential hazards to patients. MRI scanning operates by applying an intense magnetic field (10^5 times as strong as the earth's) to the part of the body being examined. This aligns the nuclei of atoms which have an odd number of protons, such as hydrogen. This alignment is then perturbed by an intense radiofrequency (RF) field, with data collected as the nuclei 'relax', reassuming their initial alignment. The intense magnetic field will attract any ferromagnetic materials in its proximity; thus, monitoring and gas delivery systems used in the vicinity of MRI scanners must be constructed of non-magnetic materials. If they are ferromagnetic, implanted medical devices such as staples and haemostasis clips will be attracted by the magnetic field, with potentially disastrous consequences.[37] The magnetic field may also disable pacemaker generating units and dislodge their intracardiac connections. For this reason, patients with implanted pacemakers should not undergo MRI scanning.[38]

The powerful radiofrequency electrical field used by MRI scanners also introduces safety concerns. As with the electrocautery, there may be significant interference with monitoring equipment such as electrocardioscopes and pulse oximeters. ECG interference may be minimized by braiding the proximal ends of the lead wires together, by keeping the electrodes as close together as possible and by avoiding any loops in the wiring. Pulse oximeter interference can be reduced by placing the sensor on an extremity that is as far as possible from the scan site. Depending on their design, infusion pumps may be susceptible to electrical interference from MRI scanners, resulting in unpredictable changes in drug administration rates.[39]

Patients undergoing MRI scanning may be subject to a significant burn hazard. Any wires within the radiofrequency field will act as receiving antennae, much of the received energy being dissipated as heat. To minimize the risk of thermal injury, essential monitoring wires should be thermally insulated from the patient by draping them over a sheet or towel. Loops of wire should be avoided, because they are especially efficient as antennae.

Fire safety in the operating theatre

Defining the surgical fire hazard

In the past, fires in the operating theatre were most commonly associated with the use of flammable anaesthetic agents. As operating theatre personnel became more aware of the risks of these agents and of appropriate safety measures, the estimated incidence of operating room fires decreased from 1 in 80 000 anaesthetics in 1955 to 1 in 150 000 anaesthetics in 1963.[40] However, while the use of flammable anaesthetics has markedly decreased, the number of ignition sources, such as fibreoptic light sources and cables, lasers and electrosurgery units, has increased dramatically. Unfortunately, in the USA, even the National Fire Protection Association (NFPA) (Quincy, MA) is unable to provide reliable statistics on the present incidence of such events, partly because most surgical fires are small and immediately controlled by the operating room staff, and partly because the fear of litigation may have discouraged reporting. Whilst the incidence of fires in the operating room is undefined and may be small,[41] the potential for injury to patients and health care providers is undeniable. It is therefore essential for anaesthetists to understand the causes and prevention of such potential tragedies.

The fire triad

For a fire to occur, fuel, an oxidizer and a source of heat or ignition must be present simultaneously and in the correct proportions. It follows that fires can be prevented or extinguished by containing or removing any one of the elements of the so-called 'fire *triad*'. Because fuels (e.g. drapes and flammable liquids), sources of heat (e.g. electrocautery and lasers) and oxidizers (e.g. oxygen and nitrous oxide) are in close proximity in the operating room, an understanding of the fire triad and safe management of its elements will help the surgical team to reduce the risk of fire.

Products of combustion

Complete combustion converts the fuel into water, carbon dioxide and other oxides. However, most fires that occur in the operating theatre result in incomplete combustion, producing a variety of byproducts. The compounds produced by a given fire depend upon many factors, including the temperature of the fire and the nature and proportions of the fuel(s) and oxidizer(s). In general, the products of combustion include carbon dioxide, carbon monoxide, water, unoxidized carbon (soot) and a variety of small organic molecules.

Although few studies have examined the specific combustion products of materials commonly found in the operating room, it has been suggested that plastics, including polyvinylchloride (PVC) and Silastic, are especially hazardous.[42] Burning plastics produce acids such as hydrogen chloride and hydrogen sulphide (sulfide), toxins such as cyanide and phenol, as well as other organic compounds, carbon monoxide, carbon dioxide and water.

The pulmonary toxicity of combustion products, in addition to the consumption of oxygen by the fire and

reduced oxygen availability resulting from carbon monoxide, makes asphyxia a major risk during surgical fires. Therefore, when confronted with a fire in the operating theatre, members of the surgical team must be concerned with more than simply the management of heat and flames. It is also necessary to keep the patient adequately oxygenated and ventilated while preventing and managing injury from the products of combustion.

Preventing and managing operating room fires

Prevention

Clearly, the most effective method of fighting fires is to prevent them. This is accomplished by keeping the elements of the fire triad separated, either spatially or temporally. In practical terms, this means controlling heat sources by following basic laser and electrical safety guidelines; minimizing the presence of flammable materials in the surgical field; and minimizing the concentrations of oxygen and nitrous oxide in the vicinity of ignition sources. The responsibility for fire safety is shared among all members of the surgical team.

Sources of heat and ignition

Most institutions have rigid protocols for laser safety; this topic is covered in Chapter 70. Similar policies should be enacted for safe use of electrosurgical units (ESUs) and cardiac defibrillators. These can generate sufficient heat to ignite many fuel sources, especially when used in oxygen-enriched environments.

Many ESUs emit an audible warning tone when activated. These tones fulfil an important safety function and therefore should not be disabled or rendered inaudible. The responsibility for the operation of foot switches should be clearly assigned, and these switches should be disconnected when the ESU is not in use. Placing the ESU probe or forceps in an electrically insulated, non-flammable 'holster' rather than resting it on the patient or drapes reduces the risk of arcing from its tip if accidentally activated, in addition to isolating the device from the patient and flammable material. The tip of the ESU probe should be kept clean to minimize the risk of igniting adherent material. Finally, allowing the probe to cool before removing it from the operative site will minimize the risk of igniting drapes, tubing or other materials as it is placed in its holster.

Cardiac defibrillation requires that the heart receive an electrical current of sufficient magnitude to depolarize a critical mass of myocardium. The delivery of this current is often through large, externally applied electrodes, which are electrically coupled to the patient's chest with a combination of manual pressure and conducting gel or pads. It is important that the electrodes make a direct electrical connection with the patient's chest, without intervening wires, bedlinen or clothing. Otherwise, sparks may occur when the defibrillator is discharged; these may ignite adjacent fuel sources such as surgical drapes, especially in the oxygen-enriched atmosphere adjacent to the patient's head. Some conductive gels, sold for use during defibrillation, may liquefy and lose conductivity with repeated attempts at restoring cardiac rhythm;[43] in addition to the potential risk of spark generation, this will decrease the effectiveness of subsequent defibrillation attempts. Alcohol or alcohol-soaked pads should *never*, of course, be used as the conducting medium between defibrillator electrodes and the patient. Nitroglycerin patches and ointment have also been shown to ignite in this manner.[44] Thus, despite the urgency of treating life-threatening cardiac arrhythmias, it is essential that the cardiac resuscitation team pay careful attention to ensure the safe use of the defibrillator.

Small sparks may occur at the electrical outlet when equipment is unplugged before it is turned off. In addition to being a potential electrical hazard, these can act as ignition sources. In the UK the DHSS, while allowing standard wall-mounted switches and sockets, recommended that, when switches are situated close to anaesthetic apparatus, gas-tight sparkless mercury switches and sockets should be used.[45] Simply turning equipment off before it was unplugged could have prevented two recent fires in a critical care unit.[46]

Fuels

The presence of fuel sources at the intended site of laser or electrosurgery should be minimized. Flammable disinfecting or degreasing solutions, such as alcohol and acetone, must evaporate, and their fumes must be completely eliminated before surgery begins. Care must be taken to ensure that these solutions have not saturated surgical drapes or accumulted beneath occlusive drapes or dressings, forming reservoirs of flammable material. Tissue contents, such as bowel gas, may also be a fuel source; in anticipation of this, the use of electrocautery to enter the unprepared colon is discouraged.[47]

Electrosurgery in the airway, where high concentrations of oxygen are invariably present, introduces special hazards. Most tracheal tubes, oesophageal stethoscopes and nasogastric tubes, are made of polyvinylchloride, which is flammable when the oxygen concentration exceeds 19 per cent.[48–50] The ease of ignition and rapidity of combustion increase with increasing concentrations of oxygen; this may occur if the cuff of a tracheal tube is pierced by the ESU or laser during airway surgery. Nitrous oxide provides no

protection, as it, too, serves as an oxidant. Electrocautery should be used sparingly, if at all, under these circumstances. If cautery is necessary, the lowest power possible should be used, and use of a bipolar cautery (p. 663) to minimize the risk of unintended arcing should be considered.[51]

Drapes and towels should be removed from the field and/or saturated with water or saline before the ESU or laser is activated. It must be emphasized that virtually all surgical drapes, both disposable and reusable, are flammable, especially in oxygen-enriched environments.[52–54] The position and condition of these fuels must be assessed repeatedly during laser or ESU use, to prevent an unsafe situation from developing.

The introduction of forced-air patient warming blankets may represent a new potential fire hazard in the operating theatre. These devices generally consist of heaters that blow warmed air into disposable, paper and polyethylene blankets. The manufacturers claim that the blankets are flame-retardant, and that the rapid movement of air makes it unlikely that they will ignite. However, once ignited, rapid combustion is likely because these devices combine fuel sources (the drape material) with a continuing supply of oxidizer (forced air).

When choosing a supplier for surgical drapes or similar materials, it is important to ask the manufacturer about the flammability or fire resistance of the material in question. The manufacturer should specify the ignition source, the oxygen concentration(s) tested, the orientation of the material, and the duration of the exposure. Independent confirmation of these results, especially if flame resistance is claimed, is important. Finally, regardless of manufacturers' claims, no drape material should be regarded as completely flame-proof, and all surgical drapes and towels should be protected from direct contact with ignition sources.[54]

Oxidizers

Minimizing the concentration of oxidization sources by controlling the escape of oxygen and nitrous oxide will also reduce fire risk. High concentrations of oxidizers will enhance the flammability of most fuels. Generally, this is of greatest concern during surgery close to the patient's head, neck and airway. For this, oxygen concentrations should be as low as the patient's condition permits. *Selective* use of supplemental oxygen is essential. In some instances, the concomitant use of air or an inert gas, such as helium or nitrogen, may allow reduced oxygen concentrations and effective ventilation and oxygenation.[55] In other circumstances (e.g. during airway surgery) the use of electrosurgical devices may be contraindicated during dissection in and around the trachea and pharynx.[56]

Table 31.4 Fire safety education for operating room personnel

Activation of fire alarm and communication systems

Location and operation of electrical and medical gas controls

Location and use of fire-fighting tools

Management of small fires, to prevent their spread

Appropriate responses when fires spread beyond control

When and how to evacuate a room, even when it is crowded with equipment and people

Appropriate routes to take for evacuation, especially when smoke or flames block the normal entry and exit routes

The location of 'safety zones', both inside and outside the building, to which patients and personnel should be evacuated.

Preparation: fire drills

Preparation for a fire requires that all members of the perioperative care team, on all shifts and from all departments, know their responsibilities in the case of a fire. In the USA, the Joint Commission on Accreditation of Healthcare Organizations (JCAHO) requires that member organizations conduct regular fire safety education, as detailed in Table 31.4. However, although there is no JCAHO requirement for practice in fighting fires during surgery, this has been strongly recommended.[42,57]

Fire drills should account for the crowded rooms, blocked exits and medically unstable patients that will invariably complicate an actual fire. Practice sessions should include instruction in the use of fire-fighting equipment, the operation of the valves that control medical gases and the switches for electrical systems. Finally, fire drills serve to remind operating room staff of their shared commitment to the safety of the patient and each other.

Destinations, or safety zones, for evacuation from the operating room must be established. For limited fires, evacuation to a nearby, easily accessible hospital area, such as the post-anaesthetic care unit ('recovery room') or intensive care unit may be a better alternative. The destination site should have electrical and medical gas systems that are independent of those in the fire zone, so that shutting off these utilities to the area of the fire does not interrupt the supply to the safety zone. It is not expected that surgical procedures will be completed in the safety zone, but rather that patients will be stabilized there before transportation to another facility.[58,59]

More distant evacuation destinations must be assigned in anticipation of more extensive fires. Medical and surgical nursing units must be able to function at these alternative safety zones, though with the general understanding that surgical procedures will be completed after patients are transferred to another facility. Finally, a hospital-wide disaster plan should allow for evacuation of the entire facility, as well as transportation of its patients to nearby hospitals. These broader efforts are best coordinated by regional hospital consortia, fire departments and disaster planning agencies.

Management of operating room fires

Fire-fighting strategies rely on principles similar to those of fire prevention: fires are controlled by limiting or eliminating the availability of at least one component of the fire triad. For smaller fires, this can be accomplished by removing the fuel source, such as removing a burning gauze sponge from the operative field and placing it in a metal basin. In the case of airway fires, the flow of oxidizers (i.e. oxygen and nitrous oxide) should be interrupted, and the fuel source (e.g. tracheal tube) removed from the patient. For larger fires, controlling the components of the fire triad may require the use of one or more types of fire extinguishers.

Fire extinguishers

Simply stated, a fire extinguisher is any device that disrupts combustion. Different types of fire extinguishers are available for different fire circumstances. To understand the uses of the various types of extinguishers, one should understand the different types of fires. Fire extinguishers are classified according to their utility in fighting three types of fires:

Class A – wood, paper, cloth and most plastics
Class B – flammable liquids and grease
Class C – fires involving electricity

There are five types of fire extinguishers generally available: blankets, water, halon, carbon dioxide and dry chemical.

Fire blankets are mats of flame-retardant material. They are designed to control fires by limiting the flow of oxidizer – air – to the site of combustion. Blankets are best used as class A extinguishers, and should not be used for electrical or solvent fires. They are safe, inexpensive and non-toxic.

Water for fighting fires is dispensed from pressurized extinguishers or from large fire hoses. These devices are best for fighting class A fires, and may be effective for some class B fires. Water should never be used for class C fires because of the risk of electrical shock. Water disrupts combustion by interrupting the flow of air to the flames and by dispersing heat. Pressurized water extinguishers are heavy, and require that a spray rather than a stream is used, so that the flames are quenched rather than fanned. Water from fire hoses is generally delivered at rates of about 200 litres (40–50 gallons) per minute, and is best reserved for extreme situations.[42]

Halon can be one of several bromofluorohydrocarbons that act by cooling and smothering the fire. The specific type(s) of halon to be used are selected to minimize the adverse effects for a given application – in this case, to reduce the production of toxic products of combustion in an enclosed space. Halon is relatively inert chemically and disperses rapidly after application. It does not leave a residue after application and thus does not damage electrical equipment. Although halon extinguishers are most effective for class B and C fires, they are also effective for class A fires. These extinguishers are available in small, easily portable sizes. This combination of low toxicity, effectiveness and portability make halon extinguishers especially suited for use in the operating room.

Carbon dioxide fire extinguishers are, like halon, most effective for class B and C fires, although they are also useful for small class A fires as well. They emit a mixture of gaseous and liquid CO_2 that cools and smothers a fire. Unlike halon, most CO_2 extinguishers are heavy, bulky devices. For this reason, they are not the best choice for use in the operating room.

Dry chemical is a mixture of compounds, mainly ammonium phosphate. The chemical acts mainly by interrupting the flow of oxidizer, and to a lesser extent by cooling and disrupting the chemical reactions of combustion. Dry chemical leaves a residue that can irritate eyes and mucous membranes, contaminate wounds and damage equipment, and is of uncertain toxicity. Although effective for all classes of fires, the problems associated with its use makes the dry chemical extinguisher a poor choice for the operating room.

Fire fighting strategies

Most operating room fires are said to begin either *on* or *in* the patient.[42] Because of the speed with which a fire can spread, it is essential to act quickly and decisively to control the fire. Protection of the patient is the primary responsibility of the operating room staff; personal well-being is a secondary consideration.

Communication among the members of the operative care team – surgeons, anaesthesists, nurses and technicians – is essential for fire prevention and management. Initially, most operating room fires are small, and so the first signs of combustion may be observed by a single member of the team. It is his/her responsibility to communicate this to the other members of the care team. It then becomes their joint responsibility to respond appropriately and continue the communication.

For small fires external to the patient, the most effective strategy is usually to interrupt the flow of oxidizer by smothering the flames with a wet towel or fire blanket. For fires within the patient, or near the patient's airway, the flow of respiratory gases – oxygen, air and nitrous oxide – should be discontinued immediately. Whilst this may seem counter-intuitive, it should be remembered that almost all patients can tolerate momentary interruptions of oxygenation; furthermore, it is likely that the combustion process is actually consuming most of the available oxygen before it is delivered to the alveoli.

Isolation of fuel sources can be accomplished by removing the burning or burned materials from the patient. Small items can be placed in a metal basin; larger items, such as surgical drapes, should be removed from the surgical field. In either instance, the flames should be extinguished immediately after removal of the fuel from the patient, using the appropriate fire extinguisher. Care of the patient, including the resumption of ventilation and oxygenation, control of haemorrhage and support of circulation, can then be resumed.

Sometimes the effort to extinguish the fire is not immediately successful, and smoke or flames endanger the patient and other occupants of the operating theatre. The room should then be evacuated and the fire alarm activated. It should be remembered that the hot products of combustion are less dense than air, and form a layer near the ceiling, fresh air forming a layer near the floor. Operating room personnel should therefore keep as low as possible. The patient should be evacuated, preferably on the operating table, with ventilation continued by a manual resuscitator. As the staff leave the room, the flow of medical gases, the electrical system and the room ventilation system should be turned off. After it is verified that all personnel have evacuated the room, the doors should be closed.

If the fire has progressed to this point, personnel from the other operating theatres as well as the local fire department should be notified. Fire-fighting efforts should be left to these trained individuals, allowing the operating room staff to concentrate on patient care.

Reporting the fire

Many US states have regulations that require that all hospital fires be reported to the local fire department. The JCAHO also requires that member institutions document all fires, and that such documentation be available for inspection at the time of a site visit. Furthermore, the NFPA Standard for Health Care Facilities requires that activation of a fire alarm automatically notify the local fire department. When the facility housing an operating room meets these standards, activation of the alarm will automatically initiate documentation of the fire.

Guidelines from professional organizations, including the JCAHO, do not carry the force of law. However, these regulations can be introduced as evidence of the standard of care. Violation of a state or municipal regulation could be considered negligence *per se*.[42] Thus, it is especially important to notify immediately the hospital safety officer and risk manager in the case of a fire. These invidividuals can help operating room staff to complete the documentation that is necessary for compliance with local reporting requirements. In the USA, the circumstances of the fire should be analysed to determine if a report to the US Food and Drug Administration is required by the Safe Medical Devices Act of 1990.[42] Finally, the event should be analysed in an objective and dispassionate manner, so that the causes can be identified and further fires prevented.

Table 31.5 Identifying colours and filling pressures for medical gases[60,65]

Cylinder contents	Colour code		Filling pressure at 21℃	
	Europe, Canada	USA	kPa	lb/in^2
Oxygen	White*	Green	15 000	2 200
Nitrogen	Black	Black	15 000	2 200
Air	Black/white	Yellow	15 000	2 200
Nitrous oxide	Blue	Blue	5 000	745
Carbon dioxide	Grey	Grey	5 700	830
Helium	Brown	Brown	11 400	1 650

* UK, black with white collar.

Compressed gas safety

There are three potential hazards associated with the use of medical gases. First, the gases may be improperly identified, resulting in hypoxia or toxicity. Second, the gas supply may fail, with the resulting risk of hypoxia (loss of O_2) or inadequate anaesthesia (loss of N_2O). Finally, compressed or liquefied gas storage tanks may fail, resulting in a rapid release of energy. These risks can be minimized by following appropriate safety procedures.[60]

In addition to printed labels that identify their contents, tanks of medical grade gases are also identified by colour codes (Table 31.5). It is important to note that, in the USA, oxygen cylinders, valves and gauges are green but in the UK the cylinders are black with a white collar, whilst in some countries they are white. Cylinders also have non-interchangeable fittings, depending upon their contents, as a further safeguard against misconnection during filling or use. Large 'H' cylinders are fitted with outlets whose thread diameter, gender (inside vs outside threads) and direction (clockwise vs counterclockwise tightening) vary according to their contents. Portable or machine-mounted 'D' and 'E' cylinders are identified by the 'pin index safety system', which requires that pins in the cylinder yoke engage with mating holes in the cylinder valve. If an extra *o*-ring gasket is present on the yoke, it may be possible to obtain a gas-tight connection without engaging the pins, rendering pin indexing ineffective.

Central gas supplies are also subject to cross-connections. For this reason, during their installation, both sides of every junction point should be labelled with the appropriate colour code. Following pressure testing for continuity, leakage and cross-connections, the identity and purity of gases flowing from each gas outlet in the system should be verified by mass spectrometry or another suitable means before the central gas supply is certified for patient use.[61] Gas outlets should have check valves to prevent inadvertent supply contamination via equipment such as anaesthetic gas machines and intensive care ventilators that are simultaneously attached to two different gases.

In addition to prominent labelling and colour coding, medical gas outlets have unique connectors depending upon the gas being supplied. These fall into two categories: the Diameter Index Safety System is used for semi-permanent (screw-in) connections between medical gas outlets and flexible gas delivery tubing, whilst quick connect systems are commonly used for connections that are changed frequently (e.g. attachment of gas machines or flowmeters to wall outlets). The latter systems are not foolproof, however. There have been instances when the wrong connector has been forced into a gas outlet, allowing an incorrect gas to be administered to patients.[62] Therefore, before equipment is attached to central gas outlets, it is essential to verify both the label and the colour codes, and not to force an adapter into an outlet where it does not fit smoothly.[63]

Central gas supply systems are equipped with automatic backup systems. These are designed to switch to a reserve gas supply when the primary source is depleted. At the same time, an alarm is activated, indicating that the supply needs to be replenished; if this alarm is located in the operating theatre, anaesthetists should be familiar with the procedures for notifying the medical gas supplier. Anaesthetic machines are also equipped with backup supplies of compressed oxygen (and often N_2O and air). These should be checked at the beginning of each day; before connecting the machine to the central gas supply the cylinder (tank) valves should be closed and the system purged until the pressure gauge reads '0'. This sequence ensures that the reserve supply is not inadvertently depleted while the pipeline supply is in use.

Although its walls are less than 1 cm thick, a portable 'E' cylinder filled with a compressed (non-liquefied) gas such as air or oxygen has a potential energy of about 35 000 J, enough to illuminate a 100 W lamp for almost 6 minutes; the potential energy of the gas in an 'H' cylinder is ten times greater. If a cylinder ruptures, the instantaneous release of this energy can, of course, cause tremendous damage;

cylinders can spin out of control, or 'rocket' through masonry walls. To minimize the risk of cylinder rupture, several safeguards have been devised. Compressed gas cylinders are fitted with safety valves that are designed to release cylinder contents in the event that heating during transportation, fire or hot weather causes the tank pressure to exceed safe limits. Additionally, within the USA, tanks are tested every 10 years to ensure their integrity, and in the UK every 5 years. Because tanks are most vulnerable at their valve fittings, 'H' cylinders are equipped with valve covers to minimize the risk of valve damage during transportation. Regardless of their size, when compressed gas cylinders are in use they should never be allowed to stand free; were a cylinder to fall on its regulator, the valve stem could snap off, causing a catastrophe. Rather, they should always be attached to a wall, carrier or other similar support (note that cylinders should *never* be attached to a radiator or similar heat source, because the pressure within the cylinder may rise excessively).[64] If this is not practical (e.g. when portable cylinders are used for patient transportation) they should be kept horizontally on the bed or floor.

Rapid compression of a gas produces a significant amount of heat. In fact, this type of 'adiabatic' heating serves to ignite the fuel–air mixture used by diesel engines. If traces of oil or similar fuels are present when a compressed oxidizer (e.g. oxygen, air, N_2O) tank is opened, an explosion can result. Therefore, oil and grease should never be used in the vicinity of compressed gas fittings or gauges,[64] and tanks should be opened slightly to blow contaminants from their fittings before they are attached to the appropriate regulators. To minimize the risk of personal injury, high-pressure gauges should be of certified design, and regulators with defective gauges (e.g. broken or missing cover glass) should not be used. Gauges should face away from the operator when a cylinder is opened.

Summary

Electrical and compressed gas supplies as well as ignition sources such as electrosurgical units and lasers are essential to patient care in the modern operating theatre. However, their use presents potential risks to patients and operating personnel. Because of our responsibility for overall patient safety, as well as our role in the design and outfitting of operating theatres, anaesthetists can play an important role in minimizing these risks. To do this effectively, it is important that anaesthetists understand the rationale underlying the safety procedures that have been established to minimize the risks of fire, electrical shock and compressed gas hazards.

REFERENCES

1. Bruner JMR. Hazards of electrical apparatus. *Anesthesiology* 1967; **28**: 396–425.

2. Department of Health and Social Security. Health Technical Memorandum: *Anti-static precautions: flooring in anaesthetic areas*, 4th ed. London: HMSO, 1977.

3. Department of Health and Social Security. *Anaesthetic machines: use of flammable agents*. Safety Information Bulletin (87) 79, 1987.

4. Kermit E, Staewen WS. Isolated power systems (historical perspective and update on regulations). *Biomedical Technology Today* 1986; **1**: 86–90.

5. *The ARRL handbook for radio amateurs*, 71st ed. Newington CT: American Radio Relay League, 1994: 2.1–2.33.

6. Hopps JA. Shock hazards in operating rooms and patient-care areas. *Anesthesiology* 1969; **31**: 142–55.

7. Tan KS, Johnson DL. Threshold of sensation for 60-Hz leakage current: results of a survey. *Biomedical Instrumentation and Technology* 1990; **24**: 207–11.

8. Leonard PF. Characteristics of electrical hazards. *Anesthesia and Analgesia* 1972; **51**: 797–809.

9. Artz CP, Yarbrough III DR. Burns: including cold, chemical and electrical injuries. In: Sabiston DC, ed. *Davis-Christopher textbook of surgery*. Philadelphia: WB Saunders, 1977; 319–21.

10. Roy OZ. 60 Hz ventricular fibrillation and rhythm thresholds and the nonpacing intracardiac catheter. *Medical and Biological Engineering* 1975; **13**: 228–33.

11. Hull CJ. Electrocution hazards in the operating theatre. *British Journal of Anaesthesia* 1978; **50**: 647–57.

12. Leeming MN. Protection of the 'electrically susceptible patient': a discussion of systems and methods. *Anesthesiology* 1973; **38**: 370–83.

13. Chambers JJ, Saha AK. Electrocution during anaesthesia. *Anaesthesia* 1979; **34**: 173–5.

14. *National fire codes*. Quincy, MA: National Fire Protection Association, 1990: 99–76–7, §7–5.1.3.5.

15. Bruner JMR. Common abuses and failures of electrical equipment. *Anesthesia and Analgesia* 1972; **51**: 810–20.

16. Nieman M, Shaw A, Railton R. Electrical distribution boards in the operating theatre. *Anaesthesia* 1988; **43**: 584–6.

17. Bernstein MS. Isolated power and line isolation monitors. *Biomedical Instrumentation and Technology* 1990; **24**: 221–3.

18. *National fire codes*. Quincy, MA: National Fire Protection Association, 1990: 99–37, §3–4.3.3.6.

19. Titel JH, El Etr AA. Fibrillation resulting from pacemaker electrodes and electrocautery during surgery. *Anesthesiology* 1961; **29**: 845–6.

20. Lipschultz A. Should US leakage current limits conform to the international standard? *Biomedical Instrumentation and Technology* 1989; **23**: 68–70.

21. Cooper JB, DeCesare R, D'Ambra MN. An engineering critical incident: direct current burn from a neuromuscular stimulator. *Anesthesiology* 1990; **73**: 168–72.

22. Chestler RJ, Lemke BN. Intraoperative flash fires associated with disposable cautery. *Ophthalmic Plastic and Reconstructive Surgery* 1989; **5**: 194–5.

23. Battig CG. Electrosurgical burn injuries and their prevention *Journal of the American Medical Association* 1968; **204**: 1025–9.

24. Schneider AJL, Apple HP, Braun RT. Electrosurgical burns at skin temperature probes. *Anesthesiology* 1977; **47**: 72–4.

25. Finlay B, Couchie D, Boyce L, Spencer E. Electrosurgery burns resulting from use of miniature ECG electrodes. *Anesthesiology* 1974; **41**: 263–9.

26. Mitchell JP. The isolated circuit diathermy. *Annals of the Royal College of Surgeons of England* 1979; **61**: 287–90.

27. Neufeld GR. Principles and hazards of electrosurgery including laparoscopy. *Surgery, Gynecology and Obstetrics* 1978; **147**: 705–10.

28. Parker EO. Electrosurgical burn at the site of an esophageal temperature probe. *Anesthesiology* 1984; **61**: 93–5.

29. Rolly G. Two cases of burns caused by misuse of coagulation unit and monitoring. *Acta Anaesthesiologica Belgica* 1978; **29**: 313–16.

30. Becker CM, Malhotra IV, Hedley-Whyte J. The distribution of radiofrequency current and burns. *Anesthesiology* 1973; **38**: 106–21.

31. Gilbert TB, Shaffer M, Matthews M. Electrical shock by dislodged spark gap in bipolar electrosurgical device. *Anesthesia and Analgesia* 1991; **73**: 355–7.

32. Domino KB, Smith TC. Electrocautery-induced reprogramming of a pacemaker using a precordial magnet. *Anesthesia and Analgesia* 1983; **62**: 609–12.

33. Carr CME, Whiteley SM. The automatic implantable cardioverter–defibrillator. Implications for anaesthetists. *Anaesthesia* 1991; **46**: 737–40.

34. Wilson JH, Lattner S, Jacob R, Stewart R. Electrocautery does not interfere with the function of the automatic implantable cardioverter defibrillator. *Annals of Thoracic Surgery* 1991; **51**: 225–6.

35. Alexander CM, Teller LE, Gross JB. Principles of pulse oximetry: theoretical and practical considerations. *Anesthesia and Analgesia* 1989; **68**: 368–76.

36. Kawasaki H, Egawa H, Takasaki M, Yokoyama K. Malfunctioning of infusion pumps due to interference from an electrosurgical unit. *Japanese Journal of Anesthesiology* 1991; **40**: 997–1002.

37. Gangarosa RE, Minnis JE, Nobbe J, Praschan D, Genberg RW. Operational safety issues in MRI. *Magnetic Resonance Imaging* 1987; **5**: 287–92.

38. Patterson SK, Chesney JT. Anesthetic management for magnetic resonance imaging: problems and solutions. *Anesthesia and Analgesia* 1992; **74**: 121–8.

39. Engler MB, Engler MM. The effects of magnetic resonance imaging on intravenous infusion devices. *Western Journal of Medicine* 1985; **143**: 329–32.

40. Nicholson MJ, Crehan JP. Fire and explosion hazards in the operating room. *Anesthesia and Analgesia* 1967; **46**: 412–24.

41. Moxon MA, Ward ME. Fire in the operating theatre. Evacuation pre-planning may save lives. *Anaesthesia* 1986; **41**: 543–6.

42. Emergency Care Research Institute. 'The patient is on fire!' *Health Devices* 1992; **21**: 19–34.

43. Hummel RS, Ornato JP, Weinberg SM, Clark AM. Spark-generating properties of electrode gels used during defibrillation. *Journal of the American Medical Association* 1988; **260**: 3021–4.

44. Wrenn K. The hazards of defibrillating through nitroglycerin patches. *Annals of Emergency Medicine* 1990; **19**: 1327–8.

45. Department of Health and Social Security. Switches and socket outlets in anaesthetising areas. Letter G/H 39/6. London: DHSS, 1969.

46. Sankaran K, Roles A, Kasian G. Fire in an intensive care unit: causes and strategies for prevention. *Canadian Medical Association Journal* 1991; **145**: 313–15.

47. Joyce FS, Rasmussen TN. Gas explosion during diathermy gastrotomy. *Gastroenterology* 1989; **96**: 530–1.

48. Rita L, Seleny F. Endotracheal tube ignition during laryngeal surgery with resectoscope. *Anesthesiology* 1982; **56**: 60–1.

49. Wolf GL, Simpson JI. Flammability of endotracheal tubes in oxygen and nitrous oxide enriched atmosphere. *Anesthesiology* 1987; **67**: 236–9.

50. Simpson JL, Wolf GL. Flammability of esophageal stethoscopes, nasogastric tubes, feeding tubes, and nasopharyngeal airways in oxygen- and nitrous oxide-enriched atmospheres. *Anesthesia and Analgesia* 1988; **67**: 236–9.

51. Bailey MK, Bromley HR, Allison JG, Conroy JM, Krzyzaniak W. Electrocautery-induced airway fire during tracheostomy. *Anesthesia and Analgesia* 1990; **71**: 702–4.

52. Cameron BDG, Ingram GS. Flammability of drape materials in nitrous oxide and oxygen. *Anaesthesia* 1971; **26**: 281–8.

53. Ott AE. Disposable surgical drapes – a potential fire hazard. *Obstetrics and Gynecology* 1983; **61**: 667–8.

54. Emergency Care Research Institute. Laser ignition of surgical drapes. *Health Devices* 1992; **21**: 15–16.

55. Pashayan AG, Gravenstein JS. Helium retards endotracheal tube fires from carbon dioxide lasers. *Anesthesiology* 1985; **62**: 274–7.

56. Aly A, McIlwain M, Duncavage JA. Electrosurgery-induced endotracheal tube ignition during tracheotomy. *Annals of Otology, Rhinology and Laryngology* 1991; **100**: 31–3

57. Vidor KK, Puterbaugh S, Willis CJ. Fire safety training. *AORN Journal* 1989; **49**: 1045–9.

58. Nagel EL, Perdue M, Hayes JD, Kennedy W. Drill prepares OR for fire emergency. *Hospitals* 1973; **47**: 99–105.

59. Green WS. Safety of the surgical patient. *Proceedings of the Royal Society of Medicine* 1976; **69**: 603–4.

60. Eichhorn JH, Ehrenwerth J. Medical gases: storage and supply. In: Ehrenwerth J, Eisenkraft JB, eds. *Anesthesia equipment: principles and application*. St Louis: Mosby, 1993: 6–7.

61. Feeley TW, Hedley-Whyte J. Bulk oxygen and nitrous oxide delivery systems: Design and dangers. *Anesthesiology* 1976; **44**: 301–5.

62. Lane GA. Medical gas outlets – a hazard from interchangeable 'quick-connect' couplers. *Anesthesiology* 1980; **52**: 86–7.

63. Dorsch JA, Dorsch SE. Medical gas piping sytems. In:

Understanding anesthesia equipment, 2nd ed. Baltimore: Williams and Wilkins, 1984: 16–35.

64. *National Fire Codes*. Quincy, MA: National Fire Protection Association, 1990: **99**–58–9, §4–6.2.1.2.

65. *Minimum performance and safety requirements for components and systems of continuous-flow anesthesia machines for human use* (ANSI Z79.8–1979). New York: American National Standards Institute, 1979: 15.

Chapter 32

Anaesthesia for Infants and Children

George Meakin

Fundamental differences	Management of anaesthesia
Paediatric anaesthetic pharmacology	Postoperative care
Preoperative preparation	

The practice of paediatric anaesthesia demands an appreciation of the differences between paediatric patients and adults. The most obvious of these is size, which alone can present considerable practical problems. The paediatric anaesthetist must develop a high level of skill in the performance of practical procedures and the use of specialized equipment. The second major difference between paediatric patients and adults is the former's ability to grow and mature. Maturation occurring at different times in different organ systems can have complex effects on the responses to anaesthetic drugs. Finally, it is important to understand the nature of the child's emotional responses and the need for adequate preparation for hospitalization, anaesthesia and surgery. The special problems of neonatal anaesthesia are dealt with in Chapter 30. The present chapter reviews the important differences between paediatric patients and adults and outlines current methods of anaesthesia for infants aged 1–12 months and children aged 1–12 years.

Fundamental differences

Anatomical differences

These differences are of size, the venous system and the airway.

Size

The size of an individual may be expressed in terms of weight, surface area or length. Depending on which of these criteria is chosen, we get a different impression of the difference in size between patients of different ages. For example, whilst normal neonates are approximately one-twentieth the weight of an adult, they measure more than one-tenth of the adult's surface area and one-third of the length.[1] Basal metabolic rate, fluid and ventilation requirements are essentially constant when related to body surface area. However, as surface area is difficult to measure, these requirements are normally calculated on a body weight basis using formulae that give proportionately greater amounts to smaller patients.

Venous system

A knowledge of the venous system is necessary in order to select suitable sites for intravenous induction of anaesthesia, infusion of fluids and measurement of central venous pressure. The most useful sites for intravenous induction of anaesthesia are the small veins on the front of the wrist, although the scalp veins may also be used in infants. These veins can be entered easily using small gauge needles, and their use at induction of anaesthesia preserves larger veins for subsequent placement of intravenous cannulae. When these are required, the veins on the back of the hand or the long saphenous vein of the leg may be used. The latter is a large constant vessel which, although it may not be immediately visible, is readily palpated as a fluctuant non-pulsatile swelling just in front of the medial malleolus at the ankle. Central venous cannulation is most often performed via the internal jugular vein or the subclavian vein.

The airway

In infants and young children the head is relatively large and the neck shorter than in the adult. These factors together with the relatively large tongue predispose to upper airway obstruction, and probably account for the greater use of tracheal intubation in these patients.

The infant glottis is situated opposite the C3–C4 intervertebral disc. By the age of 3 years it has descended to the C4–C5 interspace, where it remains until puberty, when it descends further to lie opposite the body of C5.[2]

The epiglottis of the infant is longer and usually U-shaped, as opposed to the flat leaf shape of the adult; before puberty the narrowest part of the larynx is at the level of the cricoid ring. Whilst it is often stated that the larynx in infants occupies a more anterior position compared with the adult, the difference is not striking.

Respiratory system

Respiratory rate

Oxygen consumption in the neonate is 7 ml/kg per minute, or about twice the adult value. This is reflected in an increase in the resting minute ventilation, which is 200 ml/kg per minute in the newborn, compared with 100 ml/kg per minute at puberty. As tidal volume remains constant at 7 ml/kg throughout life, the increased ventilation in younger patients is brought about by an increase in respiratory rate; this is approximately 30 breaths/min at birth, 24 breaths/min at 1 year and 12 breaths/min in the adult.

Lung capacities

Normal lung capacities are the result of a balance between the elastic forces of the chest wall and the lung. In children the chest wall is more compliant than the lungs, with the result that functional residual capacity (FRC) is reduced and small airways have an increased tendency to close at end-expiration.[3] There is evidence that, in the awake state, infants use active mechanisms such as glottic closure (laryngeal breaking) and premature cessation of expiration to maintain FRC above its true resting value. However, these mechanisms are not available during anaesthesia and airways closure may result in absorption atelectasis with ventilation/perfusion imbalance and hypoxaemia. These effects can be prevented by the use of controlled ventilation with large tidal volumes (12 ml/kg) and applying up to 5 cmH$_2$O end-expiratory pressure.

Control of respiration

After the neonatal period, control of breathing is generally similar to that in adults. Prematurely born infants are a notable exception to this rule. In these infants, maturation of neuronal respiratory control is related to postconceptual age rather than postnatal age. Immaturity of respiratory control, together with an increased susceptibility to fatigue of the respiratory muscles, may be responsible for the increased risk of postoperative apnoea in preterm infants with a gestational age of less than 46 weeks.[4]

Cardiovascular system

Ventricular size

In the fetus the work of providing a circulation is shared equally between the right and left ventricles by virtue of the communications between the pulmonary and systemic circulations (see Chapter 30). As a result, the development of the two ventricles is similar, with approximately equal wall thicknesses. At birth, the establishment of the adult pattern of circulation, with a relatively high systemic pressure, places a greater burden on the left ventricle. Consequently, there is progressive hypertrophy of the myocardium of the left ventricle which attains relative adult proportions by the age of 3 months.

Cardiac output

Changes in cardiac output during infancy and childhood mirror those in pulmonary ventilation. At birth, resting cardiac output is 200 ml/kg per minute, declining gradually to 100 ml/kg per minute by adolescence (Fig. 32.1).[5] As resting stroke volume remains fairly constant at about 1 ml/kg, the increased cardiac output in younger patients is achieved mainly by an increase in the heart rate; this is approximately 120–160 beats/min at birth, falling to 70–80 beats/min by puberty.

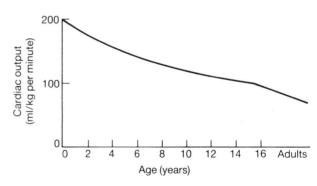

Fig. 32.1 Variation in cardiac output with age. (After Rudolph,[5] with permission.)

Blood pressure

Average systolic blood pressure at birth is 80 mmHg; it increases to 100 mmHg by about 1 year, where it remains until 6–7 years. Thereafter there is a gradual increase in systolic blood pressure to 120 mmHg at puberty. Thus, a systolic blood pressure of less than 70 mmHg is cause for concern at any age.

Autonomic control

Evidence from animal studies indicates that, whilst parasympathetic control of the cardiovascular system is well developed at birth, sympathetic control is immature. For example, it has been shown that the heart of the newborn rabbit is not fully innervated with sympathetic fibres, and the noradrenaline (norepinephrine) content of the cardiac sympathetic nerves is less than in the adult.[6] Decreased sympathetic neural output may explain the normally low blood pressure in human infants and their increased susceptibility to reflex bradycardia and hypotension. Furthermore, a low level of baroreceptor activity in infants may reduce their ability to adapt to hypotension by an increase in heart rate.[7]

Causes of bradycardia

Potent causes of reflex bradycardia and hypotension in infants during anaesthesia include laryngoscopy, tracheal intubation, tracheal suctioning and traction on eye muscles and viscera. Bradycardia may also be caused by a variety of anaesthetic drugs including suxamethonium (succinylcholine), halothane and neostigmine. These effects can be prevented or treated with intravenous atropine 20 µg/kg.

Blood

Blood volume

Average blood volume ranges from 80 to 90 ml/kg at birth depending on whether the umbilical cord is clamped early or late. By 1 month of age the blood volume is 70–80 ml/kg, as in the adult.

Haemoglobin concentration

At birth the haemoglobin concentration ranges from 16 to 19 g/dl and the red cell count is 3.74–6.54×10^{12}/l. These values are greater than the corresponding values in adults because the relatively hypoxic environment of the uterus stimulates the production of erythropoietin, which in turn triggers red cell production. After birth there is a sharp fall in erythropoietin activity due to the greater availability of oxygen. As a result, the haemoglobin concentration and red cell count decline steadily to reach 9.5–11.0 g/dl and 3.4×10^9/l respectively by 7–9 weeks, after which recovery begins. By 6 months, a mean haemoglobin of 12.5 g/dl is achieved, which is maintained until 2 years of age. Thereafter there is a gradual increase up to puberty.[8]

Oxygen delivery

At full term, 70–80 per cent of haemoglobin in the circulating red cells is in the form of fetal haemoglobin (HbF), the remainder being adult haemoglobin (HbA). The change-over from HbF to HbA synthesis is probably genetically determined, and occurs at around 32 weeks gestational age regardless of the time of birth. HbF has a higher affinity for oxygen, which is advantageous in the hypoxic intrauterine environment. However, after birth the persistence of HbF becomes a problem because oxygen cannot be unloaded so easily in the tissues. In term infants, HbF is virtually all replaced by HbA by the age of 6 months and there is an increase in levels of 2,3-diphosphoglycerate (2,3-DPG); these changes shift the oxygen dissociation curve to the right. The net result is that, although haemoglobin concentration in the normal infant decreases in the first 3 months of life, oxygen delivery to the tissues increases progressively from birth to 8 months of age when it achieves adult values.[8] These observations make the point that anaemia should be judged by its effect rather than by any arbitrary concentration of haemoglobin. Increases in cardiac output or elevations in blood lactate are rarely seen in normal individuals until the haemoglobin concentration falls below 6 g/dl.[9]

Renal function and fluid balance

Renal function

The kidneys are immature at birth and both glomerular filtration and tubular function are reduced. However, the age at which renal function can be considered to have matured depends on how it is measured. If related to surface area, glomerular filtration rate (GFR) does not attain adult values for 3–8 years; if related to weight, it reaches adult values in 1–2 weeks.[10] Distal tubular reabsorption of sodium is very active in early postnatal life, and proximal tubular reabsorption of sodium and water increases in line with the increase in GFR.[11] Judged by its ability to handle a normal load of fluid and solute, the infant kidney functions adequately 2–4 weeks after birth.

Body fluids

At birth, total body water constitutes 80 per cent of body weight, but this falls dramatically to around 60 per cent by the end of the first year (Fig. 32.2).[12] Most of this reduction is accounted for by a decrease in the extracellular fluid volume, which declines from 46 per cent of the body weight at birth to 26 per cent at 1 year of age. There is a further gradual reduction in extracellular fluid volume throughout childhood, such that in the adult it constitutes about 16 per

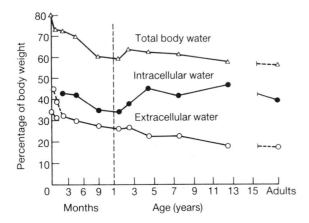

Fig. 32.2 Variation in body water compartments with age. (Reproduced, with permission of *Pediatrics*, from Friis-Hansen.[12])

cent of the body weight. The increased volume of extracellular fluid in infants compared with adults enables them to tolerate a somewhat greater degree of dehydration before developing clinical symptoms. It also provides a greater volume of distribution for highly ionized drugs such as muscle relaxants.

Fluid and electrolyte requirements

Maintenance fluid and electrolyte requirements are related to metabolic rate. In general, 1000 ml of fluid is required for every 1000 kcal expended. Requirements for sodium, potassium and chloride are usually quoted as 30, 20 and 20 mmol/1000 kcal, although quantities considerably in excess of these can readily be handled. Thus, maintenance requirements for both fluid and electrolytes will be satisfied by infusing a solution of 0.18 per cent saline with 20 mmol KCl/l at a rate equal to caloric expenditure. It is customary to make this solution isotonic by adding 4 per cent dextrose. The dextrose provides only 16 per cent of the

Table 32.1 Maintenance fluid requirements[13]

0–10 kg	4 ml/kg per h
10–20 kg	40 ml + 2 ml/kg per h for each kg over 10
20–70 kg	60 ml + 1 ml/kg per h for each kg over 20

total energy requirement but this is sufficient to prevent ketosis.

Holliday and Segar[13] related maintenance fluid and caloric requirements of hospitalized patients to body weight by a simple formula (Table 32.1). This formula is now the standard method of calculating maintenance fluid requirements for infants and children.

In infants and small children it is useful to employ an intravenous infusion pump to regulate the small hourly infusion volume. Earlier models designed to count the drops delivered from special paediatric giving sets have largely been superseded by volumetric pumps which can be set to deliver volumes as low as 1 ml per hour.

Treatment of dehydration

Dehydration develops rapidly in infants and small children because of the larger turnover of body fluid (Table 32.1). Depending on the relative amounts of water and electrolyte (Na$^+$) lost, it may be classified as iso-, hypo- or hypernatraemic.

In infants mild, moderate and severe dehydration correspond to a loss of 5, 10 and 15 per cent of the body weight respectively (Table 32.2).[14] In older children and adults these same degrees of dehydration correspond to a loss of 3, 6 and 9 per cent of body weight. The total fluid deficit in litres equates with the weight loss in kilograms. Treatment consists of replacing the deficit with a solution of appropriate ionic composition.

Severe dehydration, regardless of type, must be treated initially by rapid expansion of the extracellular fluid

Table 32.2 Correlation of the intensity of clinical signs with the degree of dehydration

	Degree of dehydration		
	Mild	Moderate	Severe
Weight loss (fluid deficit)	5%	10%	15%
Skin turgor	↓	↓↓	↓↓↓
Mucous membranes	Dry	Very dry	Parched
Skin colour	Pale	Grey	Mottled
Urine	Slight oliguria	Oliguria	Marked oliguria and azotaemia
Pulse	± ↑	↑	↑↑
Blood pressure	Normal	± Normal	Reduced

The correlation shown applies to isotonic dehydration in an infant. Modifications are required for hypertonic and hypotonic dehydration as well as for age. Reproduced, with permission of *Pediatrics*, from reference 14.

volume with 0.9 per cent saline. Up to 50 per cent of the calculated deficit may be required in the first hour with 20 mg/kg in the form of salt-poor albumin. Once the circulation has been restored, one can proceed more slowly to treat the specific type of dehydration on the basis of serum electrolyte values.

Maintenance of body temperature

Maintenance of body temperature in homoeothermic animals ensures that the enzyme systems responsible for body functions operate optimally over a wide range of environmental temperatures. This control is achieved by reflex mechanisms integrated in the hypothalamus which balance heat production and heat loss. Thus, a fall in body temperature is countered by an increase in metabolic rate, peripheral vasoconstriction and shivering.

Effects of anaesthesia

During anaesthesia, temperature control mechanisms are less effective or absent, and body temperature tends to fall towards that of the environment. Hypothermia is associated with reductions in oxygen consumption, metabolic rate, respiratory rate, heart rate, cardiac output and blood pressure. The potency of inhaled anaesthetics is increased and the duration of action of muscle relaxants and opioids is prolonged. Postoperatively, metabolic activity is increased to restore body temperature. However, if the depressed respiratory and cardiovascular systems are unable to meet the greatly increased oxygen demands of cold stress, hypoxaemia and lactic acidosis will appear.

Prevention of heat loss

Infants are particularly at risk of accidental hypothermia during anaesthesia owing to their small size, relatively large surface area and increased ventilation requirements. These factors increase heat losses by conduction, convection, radiation and evaporation. Heat loss from the body surface can be minimized by increasing the temperature of the operating room to 22–24°C and keeping the infant covered as much as possible. Heat loss by evaporation of water in the respiratory tract can be reduced by warming and humidifying the anaesthetic gases. Blood and other fluids for infusion should be warmed to 37°C.

Heating devices

The prevention of heat loss alone is rarely sufficient to maintain the body temperature of anaesthetized infants above 36°C. Overhead radiant heaters are useful after induction of anaesthesia and before the surgical drapes have been applied. Exposure of the body surface is often greatest during this period when monitoring equipment is being attached and intravenous infusions started. Intraoperatively, a heated water mattress should be used[15] and additional heat may be supplied by a fan heater circulating warm air under the surgical drapes.[16] The use of these devices often allows the temperature of the operating room to be reduced to a more comfortable level for operating room personnel. In order to prevent thermal injury, radiant heaters should not be placed too close to the patient's skin and the working temperature of other heating devices should not exceed 40°C.

Psychological differences

Development

Psychological development proceeds rapidly after birth. During the second month the social smile appears, to be followed at 3 and 4 months by cooing and laughing respectively. By 6–8 months infants recognize their mothers and become upset when separated from them. The second year is frequently a time of heightened frustration as the child submits to increasing social pressures to control his bodily functions and behaviour. Temper tantrums and breath-holding spells are common during this period and may reappear in times of stress in later preschool years. Emotional lability declines gradually throughout childhood, but this trend may reverse itself during adolescence when the child must cope with the problem of sexual maturation and the desire for independence from parental influences.

Stress of hospitalization

Almost all children hospitalized for anaesthesia and surgery experience stress. Most commonly this is based on fear of separation from loved ones, exposure to the strange hospital environment, fear of painful procedures, fear of the operation itself or fear of anaesthesia. Fear of separation is the main focus of anxiety in preschool-aged children, whilst older children may be equally worried about the prospect of painful procedures, surgery or anaesthesia.[17]

Many children are emotionally upset by their experiences in hospital and show behavioural changes when they return home. These changes include increased bed-wetting, nightmares, phobias, temper tantrums, hostility and rebellion. The incidence of these problems is greatest in preschool children. The stress experienced by children in hospital and the subsequent behavioural changes can be reduced by appropriate psychological preparation, effective sedative premedication and the supportive care of nursing staff and parents.[17,18]

Paediatric anaesthetic pharmacology

Paediatric drug dosage is determined by a large number of pharmacokinetic and pharmacodynamic factors, which are changing independently of one another during development.

Pharmacokinetics

The pharmacokinetic factors affecting paediatric responses to drugs include alterations in cardiac output, plasma protein binding, membrane permeability, fluid volumes, tissue volumes, and hepatic and renal function. The increased cardiac output (p. 674) leads to more rapid distribution of drugs to and from their sites of action. In young infants, reduced protein binding increases the effective plasma concentration of some drugs (e.g. thiopentone), and increased permeability of the blood–brain barrier allows increased passage of partially ionized drugs into the brain (e.g. morphine). The increased volume of extracellular fluid (pp. 675–676) provides an increased volume of distribution for highly ionized drugs such as muscle relaxants.

Liver

The liver is the principal organ of drug metabolism. The main processes involved are oxidation, reduction, hydrolysis (phase I reactions) and conjugation (phase II reactions). The aim of these processes is to transform lipid-soluble drugs into water-soluble metabolites which can be excreted readily by the kidney. In the neonate the microsomal enzyme systems responsible for many of these reactions are deficient or absent. Oxidation and reduction are weak but increase to adult levels within a few days; conjugation reactions take 1–3 months to develop. From the age of 3 months to 3 years, the rate of drug metabolism may actually be increased compared with that in adults, due either to increased microsomal enzyme activity or to the relatively large hepatic mass.

Kidney

Drugs and their metabolites are eliminated mainly by the kidney. The processes involved are glomerular filtration, active tubular secretion and passive tubular reabsorption. Many drugs are simply filtered by the kidney, in which case their rate of elimination is determined by the glomerular filtration rate (GFR). When normalized for body weight, GFR in newborns is less than one-half of that in adults; this difference disappears at 1–2 weeks.[10]

Pharmacodynamics

Pharmacodynamic differences between paediatric patients and adults are less well defined, although such differences are often inferred when altered dose requirements cannot be explained by altered pharmacokinetics. Possibly the clearest example of a pharmacodynamic difference is the threefold sensitivity of the neonatal neuromuscular junction to non-depolarizing muscle relaxants. The same sensitivity has recently been demonstrated in young rats[19] and shown to be due to a threefold reduction in acetylcholine release.[20]

Intravenous anaesthetics

Thiopentone

Thiopentone remains the standard intravenous induction agent for children. Intravenous injection is painless and produces smooth induction of anaesthesia in one arm–brain circulation time. Recovery occurs in 5–10 minutes by redistribution of the drug. There are no significant differences in distribution kinetics[21] or apparent recovery times between children and adults.

The dose of thiopentone varies with age (Fig. 32.3).[22,23] In neonates the ED_{50} sleep dose is only 3.5 mg/kg, but increases rapidly to around 7 mg/kg in infants aged 1–6 months, thereafter declining gradually throughout infancy and childhood. Thus, it appears that neonates require 4–5 mg/kg, infants 7–8 mg/kg and children 5–6 mg/kg of thiopentone for fast reliable induction of anaesthesia. These doses may be reduced by up to 50 per cent with sedative premedication. The reduced requirements in neonates

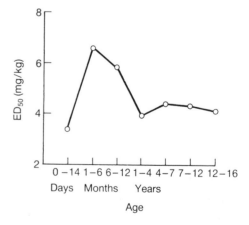

Fig. 32.3 Variation in the ED_{50} sleep dose of thiopentone with age. (After Westrin et al.,[22,23] with permission.)

compared with older infants can be explained by a decrease in plasma protein binding.[24] The increased requirements in infants and children compared with adults (usual adult dose 4 mg/kg) may be due to their increased cardiac output, as this would be expected to reduce the first-pass concentration of thiopentone arriving at the brain.

Methohexitone

Methohexitone is used for intravenous induction when rapid recovery is required (e.g. day-case procedures). The usual dose for unpremedicated children is 2.5 mg/kg. Recovery occurs in 5–10 minutes because of rapid redistribution of the drug.[25] Pain on injection can be reduced by adding 50 mg of lignocaine (lidocaine) to 50 ml of 1 per cent methohexitone solution. Muscular twitching and hiccough may occur owing to central nervous system excitation.

Propofol

Propofol is a newer intravenous induction agent with recovery characteristics similar to those of methohexitone. Induction of anaesthesia is smoother than with methohexitone, but not as smooth as with thiopentone. The usual dose for children is 3 mg/kg. Pain on injection can be overcome by adding 40 mg of lignocaine to every 20 ml of 1 per cent propofol.

Opioids

Morphine

Morphine is the standard opioid analgesic for children. It is also the only poorly lipid-soluble opioid (octanol/H_2O coefficient 1.4), hence it penetrates the intact blood–brain barrier with difficulty.

After a single intravenous dose of 150 µg/kg, plasma morphine concentration declines rapidly during distribution to 45 µg/l at 15 minutes, and then more slowly during elimination to reach 10 µg/l at 2 hours (Fig. 32.4).[26] Because clinical signs of overdose requiring pharmacological reversal are uncommon at morphine concentrations less than 80 µg/l,[27] and postoperative pain is relieved at concentrations of 12 µg/l,[28] the duration of significant respiratory depression after intravenous morphine is brief, whilst the duration of analgesia is long. In the author's experience a loading dose of 100 µg/kg with maintenance doses of 25 µg/kg per hour are convenient for supplementing balanced anaesthesia in children. This regimen produces few problems with respiratory depression at the conclusion of surgery in patients aged over 6 months, while producing excellent postoperative analgesia.[28]

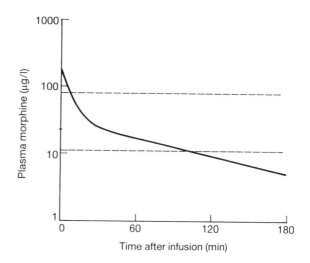

Fig. 32.4 Variation in plasma concentration of morphine with time in 4 children given 150 µg/kg i.v. (Curve redrawn from Olkola *et al.*[26] with permission.) Upper broken line = threshold for respiratory depression;[27] lower broken line = threshold for analgesia.[28]

Neonates and infants aged less than 6 months appear to be sensitive to the respiratory depressant effects of morphine owing to increased permeability of the blood–brain barrier.[29] A similar mechanism may be responsible for morphine sensitivity in neurosurgical patients. In neonates, a loading dose of 25 µg/kg followed by maintenance doses of 5–10 µg/kg per hour are suitable for supplementing balanced anaesthesia.

Fentanyl

Fentanyl is a synthetic opioid which is 50–100 times more potent than morphine owing to its greater lipid solubility (octanol/H_2O coefficient 813). After a single dose of 1–2 µg/kg, the duration of analgesia and respiratory depression is brief because the plasma concentration declines rapidly owing to redistribution. However, after multiple doses the plasma concentration falls more slowly as the drug is eliminated, and significant respiratory depession may be present at the conclusion of anaesthesia.[30]

Neuromuscular blocking drugs and antagonists

Tubocurarine

A dose of 0.5 mg/kg of tubocurarine produces about 45 minutes of neuromuscular blockade in older children, as in adults. Plasma clearance (Cl) of tubocurarine does not vary with age, but the volume of distribution (V_d) is greater in infants, reflecting their greater volume of extracellular fluid.

As a result, the half-time of elimination (0.693 V_d/Cl) and the clinical duration of action of tubocurarine are longer in infants than in older age groups.[31]

In neonates, the plasma concentration (C_p) corresponding to 50 per cent depression of the muscle twitch is only one-third of that found in older children and adults.[31] This suggests that neonates have a threefold sensitivity to tubocurarine at the level of the neuromuscular junction. However, because of the increased volume of distribution in younger patients the dosage of tubocurarine ($V_d \times C_p$) should not vary significantly with age. This is probably also true of other non-depolarizing muscle relaxants.

Atracurium

Atracurium 0.5 mg/kg produces about 30 minutes of neuromuscular blockade in children and adults. In contrast with other non-depolarizing muscle relaxants, the plasma clearance of atracurium is greater in infants compared with older children.[32] This unique feature relates probably to the fact that a significant proportion of the drug is eliminated by non-organ routes such as Hofmann hydrolysis. As with other relaxants, the volume of distribution is increased in infants compared with older children, but the net result of these changes is a reduction in elimination half-time. This reduction in elimination half-time correlates with a slightly reduced duration of action in infants, which makes atracurium a very attractive drug for use in this age group.[33]

Suxamethonium (succinylcholine)

There have been no conventional studies of the pharmacokinetics of suxamethonium in children owing to the lack of a suitable assay. However, Cook and colleagues[34] estimated elimination rate constants [k_{el}] for infants, children and adults from dose–duration data. Applying the formula $t_{\frac{1}{2}\beta} = 0.693/k_{el}$ the following elimination half-times can be calculated: infants, 1.7 minutes; children, 1.8 minutes; adults, 4.3 minutes. The reduction in half-times of suxamethonium correlates with a greatly reduced duration of action in infants and children.

The widely quoted doses of suxamethonium (2 mg/kg for infants and 1 mg/kg for children) are extrapolations from experiences with 1.0 mg/kg in infants and 0.5 mg/kg in children.[35] Recent work with dose–response curves suggests that infants require at least 3 mg/kg and children 2 mg/kg of suxamethonium to produce reliable conditions for intubation.[36] The duration of action of these increased doses is the about the same or somewhat less than that produced by 1 mg/kg in adults.[37]

Neostigmine

After a standard dose of neostigmine, antagonism of non-depolarizing neuromuscular blockade is faster in paediatric patients than in adults.[38] Increasing the dose beyond 35 µg/kg does not increase the rate of reversal significantly, and prompt reversal is unlikely in the absence of any response to train-of-four stimulation.

Inhaled anaesthetics

Uptake and elimination of inhaled anaesthetics are more rapid in children than in adults. The principal reason for this is that paediatric patients have a larger minute volume of ventilation and a lower functional residual capacity (p. 674), so more of the gas in the lungs is exchanged with each breath. Also, the increased cardiac output in children increases the rate of anaesthetic equilibration in the tissues. The latter may account for the more rapid appearance of cardiovascular side effects of halothane such as bradycardia and hypotension.[39]

Halothane

Halothane is the standard volatile anaesthetic for children. It has a sweet, non-irritant odour, allowing smooth induction and maintenance of anaesthesia.[40] Its relatively low solubility in blood and high potency permit rapid onset and recovery. Hepatotoxicity following halothane is extremely rare in paediatric patients.

The potency of halothane varies with age (Fig. 32.5). In neonates the MAC of halothane is about 0.9 per cent, but increases rapidly to a maximum of 1.2 per cent at 6 months, after which it declines gradually to about 0.8 per cent in the adult.[41,42] The similarity between the variation in the MAC of halothane and that of the ED_{50} of thiopentone with age is

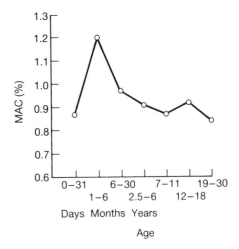

Fig. 32.5 Variation in MAC of halothane with age. Data from Lerman *et al.*[41] and Gregory *et al.*[42]

striking, but it may be coincidental. The reduced requirements of halothane in neonates compared with older infants may be related to immaturity of the central nervous system and attenuation of the pain response. The increased requirements in infants and children compared with adults may reflect an increase in brain water.[43]

Isoflurane

Isoflurane has an irritant, ethereal odour which may cause laryngospasm during induction and coughing during maintenance and recovery from anaesthesia. Thus, despite its lower blood solubility compared with halothane, induction of anaesthesia is significantly slower and total procedure time is longer.[40] The MAC of isoflurane varies with age in a manner similar to that of halothane, but it is approximately half as potent.

Preoperative preparation

Psychological preparation

This covers the preadmission period as well as the day of admission and the anaesthetist's preoperative visit.

Preadmission period

Psychological preparation for anaesthesia and surgery begins with the child at home. Parents should inform the child of the intended hospitalization and surgery with complete honesty, and emphasize that any separation will be temporary. Most children's hospitals and many district hospitals with large paediatric practices organize a weekly club to familiarize patients with the hospital and to provide an opportunity for parents and children to meet medical and nursing staff. These sessions usually include a tour of the hospital and refreshments. Information sheets and booklets are distributed, and a video presentation of a child's experiences in hospital may be shown.[44]

Day of admission

On admission to hospital, every effort must be made to assist the child to adjust to his or her new environment. Decorations should be appropriate and a selection of toys should be available in every ward. In addition, the child should be encouraged to bring a favourite toy or treasure from home. The child's nickname, as used at home, should be entered in the nursing notes and used by his carers.

Preoperative visit of anaesthetist

The preoperative visit of the anaesthetist affords an important opportunity to provide reassurance to both child and parents. The child will be reassured by an anaesthetist who has a sympathetic manner, explains the proposed procedure simply and avoids words that may cause alarm. For example, if intravenous induction of anaesthesia is planned, it is more tactful to describe this as a 'scratch on the hand' rather than a 'prick from a needle'. Parents will be reassured by an anaesthetist who appears to be genuinely interested in the child's well-being. One of the parents may be invited to accompany the child during induction of anaesthesia.

Anaesthetic assessment

History

Normally, a full history and examination will be recorded in the patient's case notes by a junior member of the surgical team; the anaesthetist will concentrate on finding specific risk factors for anaesthesia. The case notes should be reviewed at the bedside, with particular attention to previous illnesses and operations, current medications, allergy and any unusual syndrome with anaesthetic implications. Parents of infants should be asked whether the child was born at term and whether he breathed immediately after birth. Infants born prematurely, especially those with a history of apnoeic spells, are more likely to develop apnoea following anaesthesia and should not be accepted for day-case procedures until they are at least 46 weeks' post-gestational age and 4 months' postnatal age.[4] Delay in establishing sustained respiration may indicate birth asphyxia.

In older children it is important to note whether the weight given on the child's notes is appropriate for his or her age. This will give an indication of growth and may influence the selection of anaesthetic equipment (e.g. size of tracheal tube). The average weights of children aged from 1 to 12 years may be calculated using the formula:

$$\text{Weight (kg)} = \frac{(\text{Age (years)} + 3) \times 5}{2}$$

Respiratory infection

Several studies have shown that children with a history of upper respiratory tract infection (URTI) within 4 weeks of operation, or who have symptoms of URTI preoperatively, are at increased risk of respiratory complications such as laryngospasm, bronchospasm, atelectasis and hypoxaemia during or after anaesthesia.[45] In view of the seriousness of these complications, it is usual to postpone elective surgery for 4–6 weeks. However, bearing in mind that some small children have five to ten URTIs per year, this may not always be practical. Moreover, many of the children with recurrent symptoms will be suffering from allergy. In these cases it is appropriate to discuss the situation with the parents and the surgeon to obtain a balanced view of the risks and benefits of proceeding with the surgery. Where a decision is made to proceed, it would seem wise to administer a drying agent preoperatively, and plan to intubate the patient to minimize the risk of coughing or laryngospasm during anaesthesia. The patient will require supplemental oxygen and monitoring of oxygen saturation during recovery.

Examination

The head and neck should be examined to detect loose deciduous teeth and signs of potentially difficult tracheal intubation. If it appears that a deciduous tooth is so loose that it may become dislodged during anaesthesia or recovery, the parents' permission should be obtained to remove it during anaesthesia. Signs of potentially difficult intubation include limited mouth opening, micrognathia, a large tongue or noisy breathing.

The presence of high temperature (38°C), cough, malaise and audible chest signs suggests lower respiratory tract infection. Elective anaesthesia should be postponed for a least 4–6 weeks to allow hyperactive airways to return to normal.[46]

When intravenous induction is planned, the veins on the backs of the hands and fronts of the wrists should be inspected for possible venepuncture sites. One or two of these should be marked for application of a eutectic mixture of local anaesthetics (EMLA) at least 1 hour preoperatively. The use of EMLA is not recommended in infants because of the risk of possible ingestion of the cream or aspiration of the occlusive plastic dressing.

Investigations

Preoperative investigations should not be required in healthy children undergoing relatively minor surgery.[47] Groups of patients at risk of significant anaemia will require a haemoglobin estimation. These include infants, patients with chronic disease, children with signs and symptoms suggesting anaemia and children at risk of having a haemoglobinopathy. The last should also have a Sickledex test, which, if positive, should be followed by haemoglobin electrophoresis to determine the type of haemoglobin present. Haemoglobin estimation and cross-matching will be required prior to major surgery in which significant blood loss is anticipated. Other investigations, such as chest x-ray, ECG, serum urea and electrolytes,

coagulation tests, should be ordered when clinically indicated. Children with complex medical problems may need referral to an appropriate specialist for investigation or treatment.

Assessment of risk

Based on the medical history, examination and investigations, a formal assessment of the risk of anaesthesia should be made and entered into the patient's notes. This is most easily accomplished using the American Society of Anesthesiologists (ASA) classification (see Chapter 29). A note should also be made of any special precautions the anaesthetist intends to take.

Fasting and premedication

The duration of preoperative fasting required to ensure minimum volume of gastric contents at induction of anaesthesia is undergoing review. Several studies indicate that it is safe to give children clear fluids up to 3 hours before induction, and that patient comfort is increased.[48] Solids should continue to be withheld for at least 6 hours.[49]

Premedication will vary according to the needs of the patient. Infants do not require sedative premedication, but atropine 20 µg/kg may be given preoperatively or preferably intravenously at induction of anaesthesia to prevent reflex bradycardia (p. 675). In older children, sedative premedication reduces the incidence of pre-operative anxiety and post-hospitalization behavioural changes.[18] In view of the child's fear of needles the oral route is preferred. In the UK, trimeprazine syrup (2 mg/kg) is popular for inpatients weighing less than 30 kg. A derivative of phenothiazine, trimeprazine is an excellent antiemetic and drying agent but sedation is variable. In children weighing more than 30 kg, a combination of diazepam (0.25 mg/kg; max. 10 mg) and droperidol (0.25 mg/kg; max. 10 mg) tablets produces satisfactory sedation and antiemesis. In day-case patients, midazolam 0.5 mg/kg (max. 20 mg) given orally 30–45 minutes preoperatively produces effective sedation and does not delay discharge from hospital.

Management of anaesthesia

Almost all methods of anaesthesia can be adapted for use in paediatric patients. The choice of a particular technique will depend on the age of the child, the nature of the surgery to be performed and the particular skills and preference of the anaesthetist.

Induction

Induction of anaesthesia is usually accomplished by intravenous injection of a short-acting hypnotic or inhalation of a volatile anaesthetic agent. Intravenous induction of anaesthesia is quicker, creates less operating room pollution and may be less disturbing to the child than inhalational methods. The use of EMLA cream reduces the fear of needles and produces satisfactory local anaesthesia provided it is applied at least 1 hour preoperatively.

Intravenous induction

All drugs should be prepared before the patient is brought to the operating suite, and needles and syringes should be kept out of sight. If a parent is to be present, the anaesthetist must have a plan for 'stage managing' the events at induction of anaesthesia. Normally, the parent is asked to stand on the opposite side of the patient from the anaesthetist, and encouraged to cuddle the child or hold his hand while venepuncture is performed. As soon as the child loses consciousness the parent is escorted from the operating suite by the ward nurse.

A 25 s.w.g. winged infusion set with 30 cm of extension tubing and a removable cap is suitable for insertion into small veins on the dorsum of the hand or front of the wrist (Abbot Ltd). Initially, the cap should be removed and the extension tubing filled with sterile water or intravenous anaesthetic solution. An assistant is essential to immobilize the forearm and apply gentle pressure to distend the veins. The anaesthetist should take the hand and stretch the skin to stabilize the vein. Successful venepuncture is confirmed by blood syphoning out of the vein into the extension tubing. No attempt should be made to advance the needle or fix it with adhesive tape before the intravenous drugs are injected, as these manoeuvres may dislodge it from the vein.

Inhalation induction

Inhalation induction may be performed if the patient has no visible veins or specifically requests it. It is also mandatory when signs of respiratory obstruction are present or when difficult tracheal intubation is anticipated. Halothane is used most commonly, as it is non-explosive and less irritant than isoflurane or enflurane. Initially, the patient is allowed to breathe 66 per cent nitrous oxide in oxygen from the breathing system, which may be held in the anaesthetist's cupped hand or fitted with a facemask. After a few breaths, halothane is introduced and the concentration is gradually increased to 2 per cent.

Tracheal intubation

Tracheal intubation is indicated far more frequently in paediatric patients owing to their predisposition to airways obstruction (p. 674) and the large number of operations that are performed on the head and neck. It is also a prerequisite for controlled ventilation. It may be accomplished during deep anaesthesia with 4 per cent halothane or, preferably, after the administration of a muscle relaxant. For short procedures in children, when spontaneous ventilation is planned, intubation may be facilitated with suxamethonium 2 mg/kg. For longer procedures, when controlled ventilation is required, a non-depolarizing muscle relaxant such as atracurium 0.5 mg/kg may be given.

Positioning for intubation

Infants and children are intubated with their head in a neutral position. Raising the head on a low pillow does not improve the view of the larynx in these patients, because there are fewer intervertebral joints above the larynx that can be flexed.[2] The most effective manoeuvre is the application of external pressure at the level of the cricoid cartilage to push the larynx into view. In the first instance this pressure should be applied by the anaesthetist himself. Pressure applied blindly by an assistant is more likely to push the small larynx out of view. Having achieved the best view using his own hand, the anaesthetist should ask the assistant to press in the same place and adjust the assistant's hand if necessary before attempting intubation.

Laryngoscopes

For infants, a flat laryngoscope blade such as the infant Magill which passes posterior to the epiglottis may be more suitable than a curved one, because it flattens out the curvature of the epiglottis and can be used to lift it forwards to expose the larynx. In children aged over 1 year, laryngoscopy can usually be accomplished using a medium-sized Macintosh blade with the tip placed in the vallecula. The laryngoscope should be inserted gently into the mouth to avoid disturbing any loose deciduous teeth, and care should be taken to avoid trapping the lips between the teeth and the laryngoscope blade.

Tube size

The narrowest part of the larynx before puberty is the cricoid ring, so cuffed tracheal tubes are not usually required in infants and children. The use of an excessively large tube may result in postintubation croup. The correct size of tube for an infant or child is one that passes easily through the cricoid ring and allows a slight leak when

20–25 cmH_2O pressure is applied.[50] The following formula may be used as a guide in children aged 2 years and over:

$$\text{Tube size (mm i.d.)} = \frac{\text{Age (years)}}{4} + 4.5$$

Using this formula, any quarter sizes should be rounded down. Tube sizes for infants and children aged less than 2 years have to be memorized. A useful aide memoire is that a normal neonate weighing 3 kg requires a 3 mm tracheal tube; premature and low weight babies may require a 2.5 mm tube. Other tube sizes can be interpolated.

Tube length

Tracheal tubes should be cut to a length that allows the tip of the tube to be placed in the mid-trachea, while 2–3 cm protrudes from the mouth for fixation. The following formula may be used to estimate orotracheal tube length in children aged over 2 years:

$$\text{Orotracheal tube length (cm)} = \frac{\text{Age (years)}}{2} + 13$$

Orotracheal tube lengths for patients aged less than 2 years have to be memorized. The appropriate length for a neonate is 10 cm, and that for a 1 year old is 12 cm; other tube lengths can be interpolated. Some disposable tracheal tubes have depth markings which, when located at the vocal cords, indicate correct placement of the tip of the tube in the mid-trachea. The position of the tracheal tube should always be checked by auscultation of the lung fields.

Maintenance of anaesthesia

Infants

In general, infants are poor candidates for spontaneous ventilation during anaesthesia, because of their poor pulmonary mechanics (p. 674) and increased susceptibility to the cardiovascular depressant effects of volatile anaesthetic agents (p. 675). In these patients, the combination of tracheal intubation and balanced anaesthesia with full doses of muscle relaxants, controlled ventilation, minimum concentrations of volatile anaesthetics and reduced doses of opioids (p. 679) is recommended. This regimen provides ideal surgical conditions with minimal cardiovascular depression and rapid return of laryngeal reflexes at the conclusion of anaesthesia.

Children

Children aged over 1 year undergoing operations of more than 45 minutes' duration will also benefit from balanced anaesthesia with full doses of muscle relaxants, opioids and controlled ventilation. However, for many children undergoing operations lasting less than 30–40 minutes, simple inhalation anaesthesia with 66 per cent nitrous oxide in oxygen and halothane 1–2 per cent may be adequate. When appropriate, this may be combined with an opioid analgesic (e.g. 0.1 mg/kg morphine i.m.), local infiltration or a regional block to provide analgesia in the postoperative recovery period.

Caudal block

Caudal block is technically easy to perform in anaesthetized children and effective for a wide range of procedures. The level of the block is determined largely by the volume of local anaesthetic injected. A simplified dosage schedule based on plain bupivacaine 0.19–0.25 per cent is shown in Table 32.3.[51] In order to reduce the incidence of motor block when volumes exceed 20 ml, one part water or saline is added to three parts of bupivacaine 0.25 per cent, resulting in a concentration of 0.19 per cent.

Table 32.3 Paediatric caudal dosage*

Area	Example	Dose (ml/kg)
Lumbosacral	Circumcision	0.5
Thoracolumbar	Inguinal hernia	1.0
Mid-thoracic	Orchidopexy	1.25

* Concentrations of plain bupivacaine: 0.25% for volumes up to 20 ml; 0.19% for volumes over 20 ml.
Reproduced, with permission, from reference 51.

Anaesthetic breathing systems

Most modern breathing systems incorporating low resistance valves can be adapted for paediatric use.

Carden system

The standard anaesthetic breathing system in use at the Royal Manchester Children's Hospital is the Carden system (MIE Ltd).[52] This system is an example of an enclosed afferent reservoir (EAR) system (see Chapter 22), hence it is highly efficient in both spontaneous and controlled ventilation owing to conservation of dead space gas. Absence of rebreathing ensures that the inspired concentration of the anaesthetic is essentially the same as that set at the vaporizer. It is adapted for use in infants and children by fitting a 600 ml bellows supplied by the manufacturer, and employing lightweight, small-bore breathing hoses (Fig. 32.6). Using this type of system in children, a fresh gas flow equal to the normal alveolar ventilation ($0.6 \times \sqrt{wt}$ (kg) l/min) is sufficient to ensure normocapnia during controlled ventilation[53] and prevent rebreathing during spontaneous ventilation.[54] For infants, a minimum flow rate of 1.8 l/min is recommended.

T-piece systems

Two modifications of Ayre's T-piece, the Bain and the Jackson Rees systems, are also in use. The compact Jackson Rees system is especially convenient for induction of anaesthesia in infants (Fig. 32.7). During controlled ventilation in children, these systems require a fresh gas flow approximately equal to the normal minute volume ($0.8 \times \sqrt{wt}$ (kg) l/min) to produce normocapnia.[53,55] A fresh gas flow rate greater than twice the minute volume is required to eliminate rebreathing completely during spontaneous ventilation. However, in a study of 12 children anaesthetized with nitrous oxide and halothane, Meakin and Coates[56] found no physiological changes associated with rebreathing until fresh gas flow rate was reduced below the normal minute volume. In practice, the use of a fresh gas flow 25 per cent greater than the normal minute volume (i.e. $\sqrt{wt}$ (kg) l/min) simplifies calculations and allows a margin of safety.[57] For infants, a minimum fresh gas flow of 3 l/min is recommended.

Monitoring

The published guidelines for minimum standards of monitoring during anaesthesia (Chapter 23) apply equally to children as to adults.

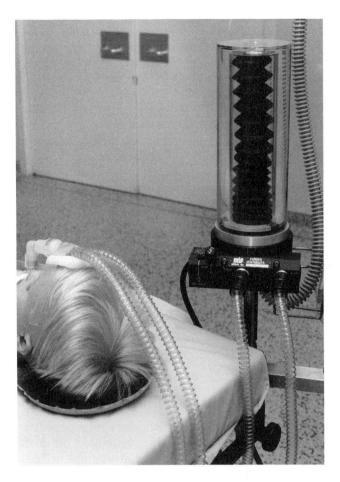

Fig. 32.6 The Carden breathing system (Engström MIE Ltd). The system has been adapted for use with infants and children by fitting a 600 ml bellows and small-bore breathing hose.

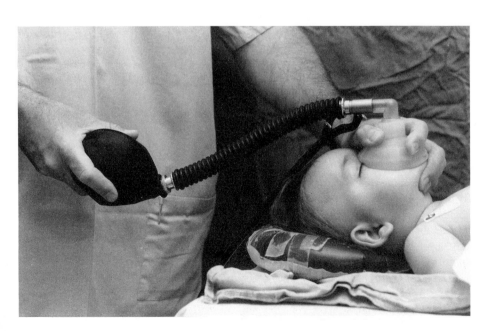

Fig. 32.7 Manual ventilation of an infant, using the Jackson Rees T-piece.

Clinical monitoring

Direct observation of the patient is the most important method of monitoring. This is more difficult in small children, who can easily be 'lost' under a mountain of drapes, hidden by an overhead table, or surrounded by surgical and nursing staff. The anaesthetist must frequently use a degree of ingenuity and tact to ensure that contact with the patient is maintained. In infants especially, disposable clear plastic drapes are helpful in obtaining an adequate view. Monitoring the colour, chest movements, breathing sounds and arterial pulse provides information about the patient's oxygenation, ventilation and circulation. Limb movement, lacrimation or sweating may indicate that anaesthesia is too light.

Instrumental monitoring

The ECG, heart rate and blood pressure should be monitored in all patients and recorded at intervals on the anaesthetic chart. The indirect methods of blood pressure measurement require a cuff that covers two-thirds of the upper arm. A pulse oximeter is particularly valuable in younger children, in whom desaturation can occur with alarming rapidity. The capnograph is useful for continuous fine tuning of ventilation and is useful in detecting disconnection. Rectal, oesophageal or axillary temperature should be monitored routinely to detect hypothermia (p. 677) or, rarely, hyperthermia. Axillary temperature is usually 0.5–1°C lower than core temperature. Neuromuscular transmission monitoring is useful in view of the marked variation in response to muscle relaxants.

Intraoperative fluid therapy

An intravenous infusion should be established for all but minor procedures. A 22 s.w.g. cannula can generally be sited in a peripheral vein in an anaesthetized infant, whilst a 20 s.w.g. cannula is suitable for children.

Crystalloids

As paediatric patients are usually kept 'nil-by-mouth' for 3–4 hours, maintenance fluids (p. 676) should be increased threefold in the first hour and twofold in the second and third hours to replace the fluid deficit. In addition, a balanced salt solution, such as Ringer lactate should be given to replace extracellular fluid translocated into the surgical site (third space loss). The magnitude of the third space loss will depend upon the degree of trauma associated with the surgery. Estimates vary from 1 to 10 ml/kg per hour, being highest for intra-abdominal procedures. Translocation is greatly reduced after 2–3 hours.[58]

Blood

Accurate measurement of blood loss is essential. The many methods employed include graduated suction bottles, swab weighing, colorimetry and visual assessment. In general, blood loss less than 10 per cent of the blood volume (estimated at 80 ml/kg) either requires no replacement or can be replaced by crystalloid solution. Losses of 10–20 per cent can be replaced by colloids or blood, but losses over 20 per cent must be replaced by blood. Many anaesthetists add 10 per cent to the measured blood loss to allow for unmeasured losses on surgical drapes, etc. The adequacy of blood replacement should be assessed with regard to the patient's blood pressure, pulse and the central venous pressure.

Emergence from anaesthesia

Balanced anaesthesia

When a balanced anaesthetic technique has been used, the volatile anaesthetic should be discontinued just before the conclusion of surgery. Provided that at least one twitch of the train-of-four is visible, residual neuromuscular blockade can be reversed with neostigmine 35 µg/kg together with glycopyrrolate 10 µg/kg or atropine 20 µg/kg.[38] if no response can be elicited to train-of-four stimulation, there is a danger of incomplete reversal with neostigmine. In this case, anaesthesia should be continued with 66 per cent nitrous oxide in oxygen until spontaneous recovery of the single twitch occurs, at which time the reversal drugs may be given. Subsequently, the patient should be ventilated with 100 per cent oxygen, the oropharynx should be suctioned and the patient should be extubated when fully awake, as indicated by spontaneous eye opening or grimacing.

Inhalation anaesthesia

Following anaethesia with spontaneous ventilation, extubation may be performed with the child deeply anaesthetized or fully awake. Extubation during light anaesthesia is likely to result in laryngeal spasm. Therefore, if extubation is to be performed 'deep', it is probably safer to continue the volatile anaesthetic until after the child has been extubated, as elimination of volatile anaesthetic occurs rapidly (p. 680). The tube is removed after oropharyngeal suctioning, an oropharyngeal airway is inserted and the child is given 100 per cent oxygen to breathe.

Laryngeal spasm

If, despite the precautions outlined above, laryngeal spasm occurs following extubation, it can usually be overcome by administering oxygen by continuous positive pressure via a firmly applied facemask. Laryngeal spasm that fails to respond to this treatment should be managed by giving suxamethonium 2 mg/kg, followed by oxygenation and reintubation.

Postoperative care

Once the anaesthetist is satisfied that adequate spontaneous ventilation is present following cessation of anaesthesia, the child should be turned into the lateral position and transported, breathing oxygen, to a fully equipped recovery room (see Chapter 40).

Recovery room protocol

On arrival in the recovery room, details of the operative procedure and any special instructions should be given to the recovery nurse assuming care of the child. Recovery room care should proceed according to an agreed protocol, which should include attention to the airway, provision of oxygen therapy, monitoring of pulse, respiration and blood pressure, and the completion of a postanaesthetic recovery chart. Recovery nurses should be trained to detect early changes in respiration or circulation and be able to initiate treatment while the anaesthetist is summoned. The duration of stay in the recovery room will vary with the condition of the patient, but is not usually less than 30 minutes. When the child is awake, one or both parents may be invited to join him in the recovery room prior to his eventual return to the surgical ward.

Problems during recovery

Hypoxaemia

Introduction of the pulse oximeter has shown that transient hypoxaemia is common in apparently healthy infants and children during recovery from anaesthesia.[59] It is therefore recommended that oxygen be administered routinely to all paediatric patients until they are awake and able to maintain an oxygen saturation greater than 95 per cent in air. In order to reduce the risk of retinopathy in newborn and premature infants with a gestational age less than 44 weeks, the inspired oxygen concentration should be adjusted to maintain an oxygen saturation between 90 and 95 per cent ($PaO_2 \approx 9.3$ kPa or 70 mmHg).

Postoperative apnoea

Postoperative apnoea (> 20 s) is a particular problem in newborn and preterm infants with a gestational age less than 46 weeks.[4] In addition to standard monitoring, these patients will require apnoea alarms for 12–24 hours postoperatively. The majority of apnoeic episodes in infants following anaesthesia respond readily to manual stimulation or brief manual ventilation with oxygen. Mechanical ventilation is required occasionally to treat repeated prolonged apnoea.

Nausea and vomiting

Nausea and vomiting may occur in the recovery room or later in the surgical ward. The incidence is particularly high after strabismus surgery, ear surgery and following opioid administration. The incidence of vomiting may be reduced by premedication with trimeprazine or droperidol (p. 682). Metoclopramide is a useful antiemetic for persistent cases. It may be given orally, intramuscularly or by intravenous injection in a dose of 0.15 mg/kg every 8 hours.

Postintubation croup

The incidence of postintubation croup has declined dramatically since the introduction of disposable, sterile, implant-tested tracheal tubes. The most common cause is the use of a tube that is too large for the trachea, and signs of obstruction are usually obvious within 1 hour of extubation.[50] Initial treatment consists of inhalation of cold mist and oxygen. If moderate symptoms persist, inhalation of nebulized adrenaline (epinephrine), 0.5 ml of 2.25 per cent racemic adrenaline diluted to 3 ml with 0.9 per cent saline, or 5 ml of 1/1000 adrenaline (BP), may give transient relief.[60,61] Very few patients will require reintubation, but all require observation in hospital for 12–24 hours.

Postoperative pain relief

The use of wound infiltration, regional blocks and morphine to supplement anaesthesia, as described above, will usually provide analgesia for several hours postoperatively. Subsequent analgesic requirements will depend upon the extent of the surgery. Following minor surgical procedures, paracetamol (acetaminophen) 15 mg/kg 8-hourly should suffice; after more major surgery an opioid analgesic such as morphine will be required. For children and infants aged over 6 months, morphine 0.1 mg/kg may be given 4-hourly i.m. or, preferably, by continuous intravenous infusion (25 µg/kg per hour). Younger infants are sensitive to the respiratory depressant effects of morphine and require careful monitoring. In neonates a continuous infusion of 5–10 µg/kg per hour should provide adequate plasma levels of analgesia without respiratory depression.[29]

Postoperative fluids

Normal maintenance fluid and electrolytes should be given as described earlier (p. 686). In addition, loss of extracellular fluid due to nasogastric aspiration or fistulous drainage should be replaced with a balanced salt solution. Blood loss should be monitored and replaced if necessary. Potassium replacement should not be necessary for the first 24 hours.

REFERENCES

1. Harris JS. Special pediatric problems in fluid and electrolyte therapy in surgery. *Annals of the New York Academy of Sciences* 1957; **66**: 966–75.
2. Westhorpe RN. The position of the larynx in children and its relationship to the ease of intubation. *Anaesthesia and Intensive Care* 1987; **15**: 384–8.
3. Mansell A, Bryan C, Levinson H. Airway closure in children. *Journal of Applied Physiology* 1972; **33**: 711–14.
4. Liu LM, Coté CJ, Goudsouzian NG, et al. Life–threatening apnea in infants recovering from anesthesia. *Anesthesiology* 1983; **59**: 506–10.
5. Rudolph AM. *Congenital diseases of the heart.* Chicago: Year Book Medical Publ, 1974: 28.
6. Friedman WF, Pool PE, Jacobowitz D, Seagren SC, Braunwald E. Sympathetic innervation of the developing rabbit heart. *Circulation Research* 1968; **23**: 25–32.
7. Gregory GA. The baroresponses of preterm infants during halothane anaesthesia. *Canadian Anaesthetists' Society Journal* 1982; **29**: 105–7.
8. Letski EA. Anaemia in the neonate. In: Hann IM, Gibson BES, Letsky EA, eds. *Fetal and neonatal haematology.* London: Baillière Tindall, 1991: 51–86.
9. Oski FA. Designation of anaemia on a functional basis. *Journal of Pediatrics* 1973; **83**: 353–4.
10. Morselli PL. Clinical pharmacokinetics in neonates. In: Gibaldi M, Prescott L, eds. *Handbook of clinical pharmacokinetics.* New York: Adis Health Science Press, 1983: 79–97.
11. Spitzer A. The role of the kidney in sodium homeostasis during maturation. *Kidney International* 1982; **21**: 539–45.
12. Friis-Hansen B. Body water compartments in children: changes during growth and related changes in body composition. *Pediatrics* 1961; **28**: 169–81.
13. Holliday MA, Segar WE. The maintenance need for water in parenteral fluid therapy. *Pediatrics* 1957; **19**: 823–32.
14. Dell RB. Pathophysiology of dehydration; normal acid base regulation. In: Winters RW, ed. *The body fluids in pediatrics.* Boston: Little, Brown, 1973: 142.
15. Goudsouzian NG, Morris RH, Ryan JF. The effects of a warming blanket on the maintenance of body temperature in anesthetized infants and children. *Anesthesiology* 1973; **39**: 351.
16. Nightingale P, Meakin G. A new method for maintaining body temperature in children. *Anesthesiology* 1986; **65**: 447–8.
17. Visintainer MA, Wolfer JA. Psychological preparation for surgical pediatric patients: the effect on children's and parent's stress responses and adjustment. *Pediatrics* 1975; **56**: 187–202.
18. Eckenhoff J. Relationship of anesthesia to postoperative personality changes in children. *American Journal of Diseases of Childhood* 1953; **86**: 587–91.
19. Meakin G, Morton RH, Wareham AC. Age-dependent variation in response to tubocurarine in the isolated rat diaphragm. *British Journal of Anaesthesia* 1992; **68**: 161–3.
20. Wareham AC, Morton RH, Meakin GH. Low quantal content of the endplate potential reduces safety factor for neuromuscular transmission in the diaphragm of the newborn rat. *British Journal of Anaesthesia* 1994; **72**: 205–9.
21. Sorbo S, Hudson RJ, Loomis JC. The pharmacokinetics of thiopental in pediatric surgical patients. *Anesthesiology* 1984; **61**: 666–70.
22. Westrin P, Jonmarker C, Werner O. Thiopental requirements for induction of anaesthesia in neonates and in infants 1–6 months of age. *Anesthesiology* 1989; **71**: 344–6.
23. Jonmarker C, Westrin P, Larsson S, Warner O. Thiopentone requirements for induction of anesthesia in children. *Anesthesiology* 1987; **67**: 104–7.
24. Kingston HG, Kendrick A, Sommer KM, Olsen GD, Downes H. Binding of thiopental in neonatal serum. *Anesthesiology* 1990; **72**: 428–31.
25. Björkman S, Gabrielsson J, Quanor H, Corby M. Pharmacokinetics of intravenous and rectal methohexitone in children. *British Journal of Anaesthesia* 1987; **59**: 1541–7.
26. Olkkola KT, Maunuksela E-L, Korpela R, Rosenberg PH. Kinetics and dynamics of postoperative intravenous morphine in children. *Clinical Pharmacology and Therapeutics* 1988; **44**: 128–36.
27. Chinyanga HM, Vandenburghe H, MacLeod S, Soldin S. Assessment of immediate postanaesthetic recovery in young children following intravenous morphine infusions, halothane and isoflurane. *Canadian Anaesthetists' Society Journal* 1984; **31**: 28–35.
28. Lynn AM, Opheim KE, Tyler DC. Morphine infusion after paediatric cardiac surgery. *Critical Care Medicine* 1984; **12**: 863–6.
29. Lynn AM, Slattery JT. Morphine pharmacokinetics in early infancy. *Anesthesiology* 1987; **66**: 136–9.
30. Andrews CJH, Prys-Roberts C. Fentanyl – a review. In: Bullingham RES, ed. *Opiate analgesia: Clinics in anaesthesiology,* vol 1, no 1. London: WB Saunders, 1983: 97–122.

31. Fisher DM, O'Keefe C, Stanski DR, Cronnelly R, Miller RD, Gregory GA. Pharmacokinetics and pharmacodynamics of *d*-tubocurarine in infants, children and adults. *Anesthesiology* 1982; **57**: 203–8.

32. Brandom BW, Stiller RL, Cook DR, Woelfel SK, Chakravorti S, Lai A. Pharmacokinetics of atracurium in anaesthetized infants and children. *British Journal of Anaesthesia* 1986; **58**: 1210–13.

33. Meakin G, Shaw EA, Baker RD, Morris P. Comparison of atracurium-induced neuromuscular blockade in neonates, infants and children. *British Journal of Anaesthesia* 1988; **60**: 171–5.

34. Cook DR, Wingard LB, Taylor FH. Clinical pharmacology of succinylcholine in infants, children and adults. *Clinical Pharmacology and Therapeutics* 1976; **20**: 493–8.

35. Cook DR, Fischer CG. Neuromuscular blocking effects of succinylcholine in infants and children. *Anesthesiology* 1975; **42**: 662–5.

36. Meakin G, McKiernan EP, Morris P, Baker RD. Dose–response curves for suxamethonium in neonates, infants and children. *British Journal of Anaesthesia* 1989; **62**: 655–8.

37. Meakin G, Walker RWM, Dearlove OR. Myotonic and neuromuscular blocking effects of increased doses of suxamethonium in infants and children. *British Journal of Anaesthesia* 1990; **65**: 816–18.

38. Meakin G, Sweet PT, Bevan JC, Bevan DR. Neostigmine and edrophonium as antagonists of pancuronium in infants and children. *Anesthesiology* 1983; **59**: 316–21.

39. Brandom BW, Brandom RB, Cook DR. Uptake and distribution of halothane in infants: *in vivo* measurements and computer simulations. *Anesthesia and Analgesia* 1983; **62**: 404–10.

40. Fisher DM, Robinson S, Brett CM, *et al*. Comparison of enflurane, halothane, and isoflurane for diagnostic and therapeutic procedures in children with malignancies. *Anesthesiology* 1985; **63**: 647–50.

41. Lerman J, Robinson S, Willis MM, Gregory GA. Anesthetic requirements for halothane in young children 0–1 month and 1–6 months of age. *Anesthesiolgy* 1983; **59**: 421–4.

42. Gregory GA, Eger EI, Munson ES. The relationship between age and halothane requirements in man. *Anesthesiology* 1969; **30**: 488–91.

43. Cook DR, Brandom BW, Shiu G, Wolfson BW. The inspired median effective dose, brain concentration at anesthesia, and cardiovascular index for halothane in young rats. *Anesthesia and Analgesia* 1981; **60**: 182–5.

44. Meakin G. Preparation and anaesthesia for day surgery in children. *Surgery* 1988; **60**: 1432–4.

45. Cohen MM., Cameron CB. Should you cancel the operation when a child has an upper respiratory tract infection? *Anesthesia and Analgesia* 1991; **72**: 282–8.

46. Empey W, Laitinen LA, Jacobs L, *et al*. Mechanisms of bronchial hyperreactivity in normal subjects after upper respiratory infections. *American Review of Respiratory Disease* 1976; **113**: 131–9.

47. Steward DJ. Screening tests before surgery in children. *Canadian Journal of Anaesthesia* 1991; **38**: 693–5.

48. Splinter WM, Stewart JA, Muir JG. Large volumes of apple juice preoperatively do not affect gastric pH and volume in children. *Canadian Journal of Anaesthesia* 1990; **37**: 36–9.

49. Meakin G, Dingwall AE, Addison GM. Effects of fasting and oral premedication on the pH and volume of gastric aspirate in children. *British Journal of Anaesthesia* 1987; **59**: 678–82.

50. Koka BV, Jeon IS, Andre JM, MacKay I, Smith RM. Postintubation croup in children. *Anesthesia and Analgesia* 1977; **56**: 501–5.

51. Armitage EN. Regional anaesthesia. In: Sumner E, Hatch DJ, eds. *Textbook of paediatric anaesthesia*. London: Baillière Tindall, 1989: 217.

52. Fletcher IR, Carden E, Healy TEJ, Poole TR. The MIE Carden ventilator. *Anaesthesia* 1983; **38**: 1082–9.

53. Meakin G, Jennings AD, Beatty PCW, Healy TEJ. Fresh gas requirements of an enclosed afferent reservoir breathing system during controlled ventilation of children. *British Journal of Anaesthesia* 1992; **68**: 43–7.

54. Meakin G, Jennings AD, Beatty PCW, Healy TEJ. Fresh gas requirements of an enclosed afferent reservoir breathing system in anaesthetized spontaneously ventilating children. *British Journal of Anaesthesia* 1992; **68**: 333–7.

55. Nightingale DA, Lambert TF. The behaviour of the Jackson Rees circuit with controlled ventilation. In: *Proceedings of the Association of Paediatric Anaesthetists Annual Scientific Meeting* 1978: 21–4.

56. Meakin G, Coates AL. Evaluation of rebreathing with the Bain circuit during anaesthesia with spontaneous ventilation. *British Journal of Anaesthesia* 1983; **55**: 487–96.

57. Meakin G. Fresh gas requirement of the T-piece systems. *British Journal of Anaesthesia* 1986; **58**: 935.

58. Bennet EJ. Fluid balance in the newborn. *Anesthesiology* 1975; **43**: 210–24.

59. Motoyama EK, Glazener CH. Hypoxaemia after general anaesthesia in children. *Anesthesia and Analgesia* 1986; **65**: 267–72.

60. Jordon WS, Graves CL, Elwyn RA. New therapy for post-intubation laryngeal edema and tracheitis in children. *Journal of the American Medical Association* 1970; **212**: 585–8.

61. Remington S, Meakin G. Nebulised adrenaline 1:1000 in the treatment of croup. *Anaesthesia* 1986; **41**: 272–5.

FURTHER READING

Berry FA. *Anesthetic management of difficult and routine pediatric patients*. New York: Churchill Livingstone, 1990.

Brown TCK, Fisk GC, eds. *Anaesthesia for children*, 2nd ed. Oxford: Blackwell Scientific, 1992.

Motoyama EK, Davis PJ, eds. *Smith's anaesthesia for infants and children*, 5th ed. St Louis: CV Mosby, 1990.

Sumner E, Hatch DJ, eds. *Textbook of paediatric anaesthetic practice*. London: Baillière Tindall, 1989.

Chapter 33

Total Intravenous Anaesthesia

John W. Sear

What are the ideal drug properties required for TIVA?	Effects of hypnotic and analgesic drugs on systemic responses to anaesthesia and surgery
What are the indications for TIVA?	
Pharmacokinetics and dynamics of TIVA	Concluding remarks

In 1994 we are still without any one intravenous drug that can *alone* provide all the requirements of anaesthesia (i.e. unconsciousness, analgesia and muscle relaxation). Hence there is need to administer several different agents to produce the desired end result. This may, in turn, lead to important and significant drug interactions.

The availability of rapid and short-acting sedative–hypnotics (e.g. propofol, etomidate), analgesics (alfentanil, sufentanil, remifentanil, A-3665) and muscle relaxants (atracurium, vecuronium, mivacurium) has refocused the anaesthetist's attention on the complete provision of anaesthesia by the intravenous route: total intravenous anaesthesia (TIVA).

The use of intravenous agents to achieve these goals began with the introduction of the rapidly acting barbiturates in 1934. However, the kinetics of these early barbiturates (see later) did not render the drugs as ideal for use for the *maintenance* of anaesthesia. For example, the provision of analgesia could not be achieved by the barbiturates alone, and the addition of either pethidine (meperidine) or morphine (both drugs having slow blood–brain equilibration) led to overdosing and hence poor clinical conditions, especially in the spontaneously breathing patient. Thus, the availability of modern *volatile* agents (beginning with halothane in 1956) with their easy titratability encouraged the anaesthetist to turn away from the intravenous agents for the maintenance of anaesthesia.

Probably one of the more important studies in the development of TIVA was that reported by Savege and colleagues in 1975 using the steroid agent Althesin (alphaxalone:alphadolone acetate) and pethidine to supplement oxygen-enriched air in the spontaneously breathing patient.[1] Subsequent developments included the use of the carboxylated imidazole, etomidate and fentanyl, and infusions of ketamine.

The use of morphine to provide anaesthesia as a *sole anaesthetic agent* in patients undergoing cardiac surgery was first described by Lowenstein and colleagues in 1969, where the hypnotic side effects of large doses of morphine were used to provide sedation.[2] However, this method of anaesthesia (more popular in the USA and Europe than in the UK) was not reliably effective. It was associated with a high incidence of episodes of awareness during anaesthesia, as well as intraoperative episodes of hypotension due to histamine release, resulting in increased intraoperative and postoperative fluid and blood requirements. The more selective and highly potent μ-receptor opioids (fentanyl, alfentanil and sufentanil) have the advantage of less cardiovascular depression, as well as the ability to obtund the haemodynamic responses to laryngoscopy and intubation. However, this apparent stability was less evident during surgery and, in particular, during the period of sternotomy and aortic root dissection. These phases were often accompanied by episodes of significant hypertension and tachycardia, which were not always remediable by increased doses of the opioids. Thus it became clear that use of high doses of opioids *alone* was inappropriate for the provision of anaesthesia. Attempts have been made to control the haemodynamic effects of surgical stimulation and to reduce the incidence of awareness, by the use of either low inspired concentrations of volatile agents or infusions of a sedative–hypnotic agent (mainly methohexitone, propofol, midazolam).

However, the TIVA techniques are also not without difficulties: combinations of fentanyl and diazepam can result in significant reductions in both cardiac output and systemic vascular resistance. The efficacy of the benzodiazepines in reducing the circulating levels of catecholamines may be important because greater stability has been found in cardiac surgical patients when the combination of sufentanil with ketamine or midazolam has been employed. This may be due to the ketamine interacting to counteract the peripheral effects of the other two drugs.

Opioids are also popular for non-cardiac surgery. They are used as the basis of neurosurgical anaesthetic techniques because they do not alter the carbon dioxide reactivity of the cerebral blood vessels, whilst the kinetics and dynamics of alfentanil have allowed the anaesthetist to titrate more closely the dose requirements for both abdominal and other major body surface surgical procedures.[3]

In this chapter we consider the properties of the sedative–hypnotics and analgesics used in TIVA.

What are the ideal drug properties required for TIVA?

The properties of the ideal drug for use during continuous infusion anaesthesia may be summarized as follows.

1. Soluble in water so that the use of a solvent is avoided.
2. Stable in solution, and on exposure to light for prolonged periods of time.
3. No adsorption onto plastic tubing such as giving sets.
4. No venous damage (pain on injection, venous phlebitis or thrombosis) or tissue damage when administered either extravascularly or intra-arterially.
5. Sleep produced in one arm–brain circulation time.
6. Short duration of action, and inactivation by metabolism in either the liver, blood or other organs of the vessel-rich group of tissues.
7. Inactive, non-toxic, water-soluble metabolites.
8. Minimal cardiovascular and respiratory side effects.

It will be instantly appreciated that *none* of the present hypnotic and analgesic agents fulfils all of these criteria, but that some are more suitable than others under different circumstances.

What are the indications for TIVA?

The indications for TIVA are:

1. By infusion as an alternative to volatile agents used to supplement nitrous oxide in oxygen anaesthesia.
2. To provide sedation during local or regional anaesthetic techniques.
3. For ambulatory surgery, when the speed and completeness of recovery are important.
4. For situations in which conventional anaesthetics may be difficult to administer: due to a lack of resources such as nitrous oxide; at sites of military or

non-military trauma; and anaesthesia at increased ambient pressure.
5. In circumstances in which nitrous oxide may either be undesirable (e.g. due to the need for high inspired oxygen concentrations) or relatively contraindicated (e.g. one-lung anaesthesia, middle ear surgery, some neuroanaesthesia, prolonged abdominal surgery, relief of cardiac tamponade, bronchoscopy, laryngoscopy and bronchotracheal surgery).

However, the use of TIVA techniques is not without some important disadvantages. These include the possibility of awareness, the likelihood of postoperative respiratory depression due to the persistent effects of concurrently administered narcotic agents, the requirement for a separate dedicated intravenous access site and appropriate infusion pumps, and the inability to control depth of anaesthesia as well as is possible with the volatile agents.

Pharmacokinetics and dynamics of TIVA

Hypnotic agents

In this section we consider barbiturates, etomidate, ketamine and propofol.

Barbiturates

Infusions of the barbiturates (especially thiopentone and methohexitone) have been used by several authors for the maintenance of anaesthesia.

Thiopentone
Early studies of the properties of thiopentone by Lundy in Rochester, Minnesota, and Waters in Madison, Wisconsin, indicated the safety of this first thiobarbiturate. However, the subsequent experience, especially in the context of anaesthesia for the traumatized patient as occurred at Pearl Harbor, led Halford to describe the use of thiopentone in the presence of shock and hypovolaemia as 'an ideal form of euthanasia'.[4] This disaster led to the realization that reduced doses of the barbiturate were indicated under such circumstances, and that oxygen was an important adjunct to thiopentone anaesthesia – particularly in patients with hypovolaemic shock.

Despite the passage of over 55 years since the introduction of thiopentone into clinical practice, there have been few studies critically evaluating the relation between concentration and effect, in either healthy or infirm patients, when thiopentone is administered either as a sole agent or to supplement an opioid given by an infusion or by incremental doses.

When thiopentone is used in conjunction with nitrous oxide (67 per cent) infusion rates of the order of 150–300 µg/kg per minute are required. These will produce plasma drug concentrations of between 15 and 25 µg/ml.[5] In another study, Becker determined the plasma concentration, associated with abolition of the response to squeezing the trapezius muscle (equatable with the initial surgical stimulus), to be 42.2 µg/ml in the absence of nitrous oxide.[6] EEG burst suppression occurs at thiopentone concentrations of around 40–50 µg/ml.

One argument against the use of thiopentone is that continuous infusion causes saturation of the peripheral tissue storage sites, resulting in slow recovery. This is due to the low hepatic clearance rate for thiopentone. The major factor in the decline of the plasma thiopentone concentration is redistribution to peripheral tissues rather than elimination. Drug elimination based solely on zero-order elimination is not seen below concentrations of the order of 50–60 µg/ml. If the concentrations are retained in the range 15–25 µg/ml (as described by Crankshaw and colleagues), a prompt decline occurs in the drug concentration over the first 15 or so minutes after cessation of the infusion, and the patient awakens. This can be explained on a kinetic basis. The clearance rate of thiopentone from the central compartment (volume: 5 litres) is about 200 ml/min; thus the initial fall in the plasma drug concentration (30–50 per cent) will bring the brain concentration below that associated with hypnosis.

At infusion rates greater than about 300 µg/kg per minute, thiopentone concentrations will increase exponentially as the peripheral stores become saturated. Under these circumstances, use of the drug is associated with prolonged recovery. Accumulation of the drug is the result of two separate processes: a change in the kinetics to zero-order elimination (as outlined above), and the biodegradation of thiopentone to pentobarbitone, an active metabolite. After a 2-hour infusion, about 20 per cent of the total drug dose will be present as pentobarbitone.[7] The elimination half-life of this barbiturate is longer than that of thiopentone and results in persistent drowsiness into the immediate postoperative period.

There are also other disadvantages of giving thiopentone by infusion: a low therapeutic index; porphyrinogenicity; and, in large doses, possible hepatotoxicity. However, there are some potential pluses: minimal cardiovascular depression and an effectiveness in offering brain protection during episodes of cerebral ischaemia (see later).

Methohexitone

Methohexitone as a sole agent, or as supplement to nitrous oxide or opioids, has been used for body surface, intra-abdominal and neurosurgical operations.

There are few data on the kinetics of methohexitone when given by continuous infusion. In a study in volunteers, Breimer infused the barbiturate at a dose of 3 mg/kg over 60 minutes, and calculated the following mean disposition parameters: elimination half-life 97 minutes, and clearance 826 ml/min.[8] However, this is an underestimate of the half-life, based on bolus dose kinetics described by Hudson and colleagues.[9] A more recent infusion kinetic study has been reported by le Normand et al. in which methohexitone was infused at 60 and 90 µg/kg per minute to patients for 14 hours, and the post-infusion decay followed for a further 12 hours.[10] The derived kinetic estimates were similar to those of Hudson and colleagues: half-life, 420–460 minutes; apparent volume of distribution at steady state, 4.5–4.7 l/kg; and clearance, 9.6–9.8 ml/kg per minute.

Few dynamic relations between concentration and effect have been published for methohexitone. In a series of dose–response studies in patients premedicated with either diazepam or morphine–atropine, the ED_{50} and ED_{95} infusion rates have been determined when methohexitone was used to supplement nitrous oxide for patients aged 20–60 years.[11,12] The values for those rates, when the applied stimulus was the initial surgical incision, were 66.0 and 80.7 µg/kg per minute for diazepam-premedicated patients, and 48.8 and 75.9 µg/kg per minute respectively after opioid premedication. There were no reported data relating drug concentration to effect in these patients. In a separate study, in which methohexitone was used as the sole anaesthetic for neurosurgical anaesthesia, Todd et al. infused this drug at higher infusion rates (400 µg/kg per minute).[13] The EEG showed burst suppression at concentrations between 8 and 10 µg/ml, and became isoelectric at between 9 and 13 µg/ml.

Unlike thiopentone, this methoxybarbiturate has no active metabolites. Clinical studies in which methohexitone is infused with an opioid infusion utilize rates between 50 and 120 µg/kg per minute. Recovery is prompt unless the total drug dose administered is in excess of 500–600 mg.[12] However, its use may be accompanied by some unwanted side effects, such as excitatory movements, pain on injection and a predisposition to convulsions.

Comparative studies of methohexitone with other intravenous and inhalational agents suggest that recovery is faster after etomidate or propofol when these are given by incremental doses or by infusion, although the quality of recovery after the initial wake-up period requires to be further evaluated to clarify whether real differences exist between the separate agents. There are no reported adverse sequelae following methohexitone with regard to liver function tests or effects on the adrenal cortex.[12,14]

Table 33.1 Incidence of adverse hypersensitivity reactions to intravenous hypnotic agents

	Incidence
Thiopentone	1/14 000–1/20 000
Methohexitone	1/1600–1/7000
Propofol (as Cremophor formulation)	5/1131
Propofol (as emulsion)	20+ cases (estimate 1/80 000–1/100 000)
Etomidate	10+ cases (estimate 1/100 000–1/450 000)
For comparison	
Althesin (in Cremophor)	1/400–1/11 000
Propanidid (in Micellophor)	1/500–1/17 000
Neuromuscular blocking drugs	1/5000
Penicillin	1/2500–1/10 000
Dextrans	1/3000
Gelatins	1/900
Hydroxyethyl starch	1/1200

Etomidate

The carboxylated imidazole etomidate was introduced into clinical practice in the UK in 1978. It was considered by many anaesthetists to be the ideal agent for the induction and maintenance of anaesthesia in the cardiac-compromised or hypovolaemic patient. It has been formulated in four separate solvents to date (and a fifth, Intralipid, has also recently been studied), but none of these has resulted in a significant reduction in the incidence of pain on injection or in the accompanying venous sequelae of thrombosis and phlebitis. Other minor side effects include involuntary muscle movements (especially in the unpremedicated patient), coughing and hiccoughs.

When it was given by continuous infusion, de Ruiter determined an elimination half-life of 170 minutes, volume of distribution at steady state of 310 litres, and clearance 1280 ml/min.[15] Similar data have been reported in volunteers by Schuttler and colleagues (terminal half-life, 87–126 minutes; clearance, 1175–2550 ml/min).[16] Anaesthesia was associated with a plasma concentration between 300 and 500 ng/ml, and burst suppression of the EEG occurred at concentrations greater than 1.0 µg/ml. Etomidate is biotransformed by enzymes present in both the plasma and the liver with the formation of inactive, water-soluble metabolites (the main one being the corresponding carboxylic acid). It is bound to plasma proteins (76 per cent; binding to both albumin and α_1-acid glycoprotein).

Schuttler and colleagues reported a significant decrease in etomidate clearance in the presence of a steady state concentration of fentanyl (10 ng/ml).[17] Although they hypothesized that this was due to an alteration in the

volume of drug distribution, there is little support for this from other studies. However, this drug–drug interaction is clearly important to the clinical anaesthetist (see later).

When given at high infusion rates, etomidate causes a decrease in both cardiac output and liver blood flow, resulting in greater plasma drug concentrations than might otherwise have been predicted.

Etomidate has many of the ideal properties required for a hypnotic agent for TIVA: sleep in one arm–brain circulation time; no detectable histamine release and low incidence of allergic reactions (Table 33.1); minimal cardiovascular and respiratory depressive effects with no inhibition of the hypoxic pulmonary vasoconstrictor reflex; reduction in cerebral metabolism, cerebral blood flow and intracranial pressure; and rapid recovery after single doses or continuous infusions. It also offers advantages for anaesthesia in patients with poor cardiac reserve, hypovolaemia or acute intermittent porphyria.

However, its use is associated with a high incidence of nausea and vomiting, excitatory movements and a dose-related inhibition of adrenal steroidogenesis by reversible interaction with mitochondrial cytochrome P_{450} (affecting 11β-, 17α- and 18-hydroxylases, and 20,22-lyase).

Watt and Ledingham reported that the use of this drug by infusion for the provision of sedation in the intensive care unit was associated with an increased mortality.[18] They also noted that many of the patients receiving etomidate exhibited low plasma cortisol levels (< 100 nmol/l; normal range, 260–550 nmol/l). Their hypothesis was that the increased mortality associated with etomidate by infusion was due, at least in part, to adrenocortical suppression.

In vitro data show etomidate at normal therapeutic concentrations to be a more potent inhibitor of steroid synthesis than metyrapone. Other hypnotic agents (e.g. thiopentone and propofol) can also inhibit steroidogenesis but only at supra-therapeutic concentrations. This effect of etomidate as an inhibitor of 11β-hydroxylation is related to both dose and concentration (Table 33.2). Crozier *et al.*

Table 33.2 Dose–response for the effects of etomidate on the suppression of cortisol and aldosterone synthesis in female patients undergoing lower abdominal surgery

Etomidate dose (mg/kg)	Cortisol (nmol/l)	Aldosterone (pmol/l)
0	+697 (42)	442 (83)
0.3	−33 (75)*	107 (41)*
1.15	−75 (22)*	3 (14)**

P < 0.01 vs control group.
Response recorded as difference in plasma hormone concentration between baseline value and value 4 hours after the start of surgery. Control group received thiopentone (3–5 mg/kg) for induction of anaesthesia; etomidate groups received either 0.3 mg/kg for induction, or 0.3 mg/kg for induction followed by an intraoperative infusion of 10 µg/kg per minute (total dose 1.15 mg/kg). Data shown as mean (SEM).

determined the ED_{50} plasma etomidate concentration for inhibition of steroidogenesis to be 110 nmol/l (20 ng/ml).[19] On the basis of these data and the known kinetics of the drug, the expected duration of the inhibition will be about 4–8 hours.

Etomidate is no longer licensed in many countries (including the UK) for administration by continuous infusion.

Ketamine

This phencyclidine derivative is the only presently available hypnotic agent that also possesses defined analgesic properties. It is a racemic mixture (the stoichiometry of $R(-) : S(+)$ being 1 : 1). However, the two stereoisomers $R(-)$ and $S(+)$ have different potencies and kinetics.[20,21] $S(+)$ ketamine is about twice as potent in terms of anaesthesia and is associated with faster recovery compared with the racemic mixture, and four times compared with the $R(-)$ isomer. However, cardiovascular stimulation and psychotomimetic effects are seen with both stereoisomers.

There are few kinetic data on the plasma concentrations of ketamine when given by continuous infusion. Idvall *et al.* and Clements and colleagues established the relation between concentration and effect for both analgesia and hypnosis.[22,23] The analgesic threshold is about 200 ng/ml, whilst the concentration required for hypnosis as a supplement to nitrous oxide is 1.5–2.5 µg/ml. Idvall *et al.* achieved clinical anaesthesia using a loading dose of 2 mg/kg followed by an infusion of 40 µg/kg per minute. The corresponding steady state drug concentration was between 1.7 and 2.4 µg/ml. Disposition kinetics indicate an elimination half-life of 80–180 minutes, clearance of 980–1500 ml/min, and volume of distribution (V_{ss}) 1.8–3.1 l/kg. Metabolite I appeared in the plasma by 5 minutes, and metabolite II by 20 minutes after the start of the infusion. The concentrations of these metabolites increased still further after cessation of the infusion. When used as the sole agent, infusion rates of 60–80 µg/kg per minute provide clinical anaesthesia.

The metabolic fate of ketamine is complex, but one metabolite (norketamine; metabolite I) is pharmacologically active, with a potency of around 30 per cent of the parent drug and a longer elimination half-life. Another metabolite (metabolite II: dehydronorketamine) has about 1 per cent of the activity of the parent compound. The main excretory metabolites are ketamine and metabolite I and II glucuronides; hence efficacy may be enhanced in patients with renal impairment.

The disadvantages of ketamine for TIVA include cardiovascular stimulation (due to peripheral vasoconstriction and increased heart rate), intra- and postoperative dreaming and hallucinations, and excessive salivation.

These effects may be attenuated by benzodiazepine premedication, although these drugs prolong the elimination half-life and increase the duration of effect. Ketamine is contraindicated in patients with a history of cerebrovascular disease, hypertension, ischaemic or valvular heart disease, and in the presence of increased intracranial or intraocular pressure.

There is some debate over the effects of ketamine on hepatic function. Dundee *et al.* found significant increases in alanine aminoaspartate transferase, alkaline phosphatase and γ-glutamyl transferase after infusions of ketamine.[24] Sear and McGivan examined the significance of these observations *in vitro* by studying the effects of ketamine on isolated rat hepatocytes. The ED_{50} doses for inhibition of gluconeogenesis was around 450 µmol/l.[25] This is well in excess of concentrations normally achieved during clinical anaesthesia.

Propofol (di-isopropylphenol)

This sterically hindered phenol was first administered to volunteers in 1977, but initial problems with severe pain on injection and then with complement-mediated adverse reaction to a Cremophor formulation have resulted in three solvents being investigated. The present one is a 1 per cent oil in water emulsion formulation containing 10 per cent soyabean oil, 1.2 per cent egg phosphatide and 2.25 per cent glycerol. Propofol has a therapeutic index similar to thiopentone (3.4 vs 3.91), but recovery in the mouse given twice the median hypnotic dose was faster after propofol than after thiopentone. The difference in recovery becomes more exaggerated when repeated doses are administered.

Initial studies evaluating the induction of anaesthesia with propofol probably overestimated the induction dose because of the delayed loss of the eyelash reflex (the commonly used endpoint for such studies). More recently Naguib *et al.* have reported the ED_{50} and ED_{95} values for both loss of response to verbal commands and loss of the eyelash reflex in a comparison between thiopentone and propofol.[26] Drugs have been administered over 15 seconds, the attainment of the defined endpoints made at 30, 60 and 90 seconds after the end of drug injection (Fig. 33.1). The shapes of the dose–response curves for loss of verbal command differ for the two drugs although they are parallel and similar shaped for loss of the eyelash reflex. Thus the potency of different hypnotic agents clearly depends on the chosen endpoint and on the time after the end of the injection at which the comparisons are made.

Propofol has three advantages when compared with thiopentone rapid clearance resulting in short duration of effect, and no interaction with either steroidogenesis or haem synthesis. Thus, as an induction agent, propofol's *only* real advantage over thiopentone is in day-case surgery,

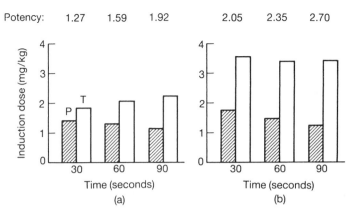

Fig. 33.1 ED_{50} induction doses of propofol (P) and thiopentone (T) (mg/kg) for abolition of reaction to verbal command (a) and to eyelash stimulation (b) at 30, 60 and 90 seconds after induction of anaesthesia. The differing potency ratios (expressed as $ED_{50}T/ED_{50}P$) are also calculated for the two endpoints at the three time points shown. (Adapted, with permission, from reference 26.)

with its faster recovery and low incidence of nausea and vomiting. The latter is thought to be due to an effect on the chemoreceptor zone. Thiopentone is probably the induction agent of choice for other surgery.

Propofol is a highly lipophilic drug, and is rapidly metabolized to inactive metabolites (the glucuronide and corresponding quinol glucuronides and sulphates). Clinical anaesthesia requires drug concentrations of around 4–6 μg/ml when propofol supplements either nitrous oxide or an opioid infusion.[27,28] Plasma concentrations at recovery are about 1.0 μg/ml.

Pharmacokinetics and dynamics

Kinetic data following bolus dose studies give values for elimination half-life of 226–674 minutes, clearance of 1.66–1.91 l/min, a central volume of distribution of 13–56 litres and a volume of distribution at steady state of 171–781 litres. The large volume of distribution is due to a high octanol:water partition coefficient (log P, 3.7). All kinetic studies show considerable interindividual variability.

More recent studies from Campbell et al. in Melbourne have suggested an even greater variability in the various kinetic parameters (terminal half-life, 3.6–63 hours; clearance, 870–2140 ml/min; and an apparent volume of distribution (V_β) 180–1730 litres).[29] This longer terminal half-life is probably representative of the rate of drug clearance from the peripheral lipid tissues. The high clearance rate for propofol has been commented on by many authors to be in excess of the liver blood flow, and hence begs the question of whether there are other sites of metabolism of the hindered phenol (e.g. lungs or kidney). Indeed, data from Lange et al. in Göttingen proposed that extrahepatic metabolism may be the primary method of drug elimination.[30]

However, in a more recent study, Veroli et al. have compared the clearance of a bolus dose of propofol before and during the anhepatic phase of orthotopic liver transplantation and confirmed that the liver is the major site of elimination.[31] This finding has been supported by data from Mather et al. at Flinders University, Adelaide, who found similar evidence of extrahepatic clearance using their chronically instrumented sheep model.[32]

The high clearance characteristics and rapid initial decline in drug concentration in the blood make propofol an ideal agent for administration as part of a TIVA technique.

Infusion of propofol

Although the kinetics of propofol given by infusion are well defined, most of these parameter estimates have been obtained in ASA 1 or 2 patients, and there is urgent need for more data on the influence of ageing, obesity, disease states and intercurrent therapy on the kinetics of propofol. Coupled with the high clearance rate (1000–1800 ml/min) is a short blood-effector site equilibration rate constant (k_{eo}) of 0.24 per minute, so allowing the drug to be titrated more readily to patient response.

The relation between concentration and effect for infusions of propofol Vuyk et al. have determined the EC_{50} values for propofol, in the absence of either premedication or anaesthetic agents, to be 2.02 μg/ml for the loss of eyelash reflex and 3.40 μg/ml for loss of consciousness.[33]

Other data provided by Prys-Roberts have determined the ED_{50} and ED_{95} infusion rates and their associated drug concentrations (EC_{50} and EC_{95}) for infusions of propofol when given to supplement 67 per cent nitrous oxide. The applied stimulus in all cases was the initial surgical incision. Their data are summarized in Fig. 33.2. Similar data have been obtained by Davidson et al. using a microcomputer-controlled infusion system to achieve predetermined target concentrations of propofol.[34] They found the EC_{50} for the initial incision during propofol and nitrous oxide anaesthesia to be 5.36 μg/ml, and 8.1 μg/ml during propofol and oxygen. These were significantly higher than the values defined in Prys-Roberts' studies. There is no obvious explanation for the differences.

Other aspects of the relation between concentration and effect have been defined by Shafer et al.[27] in ventilated patients receiving an infusion of propofol to supplement 70 per cent nitrous oxide and given incremental doses of pethidine (meperidine). The average blood propofol concentration associated with an obtunding of the autonomic responses during abdominal surgery was 4.05 μg/ml, and 2.97 μg/ml during body surface surgery – results closer to those of Prys-Roberts.

Data for TIVA regimens are fewer. Vuyk et al., at the University of Leiden, examined the dynamics of alfentanil

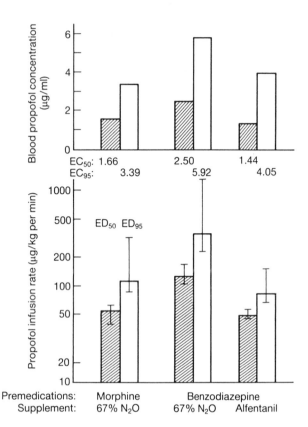

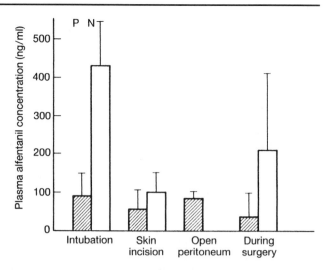

Fig. 33.3 EC_{50} values for alfentanil (ng/ml) concentration–effect relationships as supplement to propofol–oxygen (P: mean concentration, 4.01 µg/ml) or nitrous oxide (67 per cent) (N). Data shown as mean (SD). (Adapted, with permission, from reference 35.)

Fig. 33.2 ED_{50} and ED_{95} infusion rates (µg/kg per minute) (and their associated blood concentrations – µg/ml) for propofol when given to supplement 67 per cent oxide in oxygen, or an alfentanil infusion and oxygen-enriched air. Data shown as mean (95 per cent CI).

when used to supplement either an infusion of propofol (measured concentration 4 µg/ml) or nitrous oxide in oxygen in patients undergoing lower abdominal surgery.[35] The EC_{50} of alfentanil needed for intubation was 92 ng/ml in the TIVA group compared with 429 ng/ml in the nitrous oxide group. Other data are shown in Fig. 33.3. Thus alfentanil requirements appear to be less when used to supplement propofol (4 µg/ml) compared with nitrous oxide (67 per cent).

Different anaesthetists hold dissimilar views on whether the hypnotic (propofol) component of TIVA regimens or the analgesic component should be altered according to patient response. There are but few controlled studies comparing the two components. However, Monk and colleagues have used either alfentanil or propofol to control acute haemodynamic responses during radical prostatectomy under TIVA.[36] The aim was to return the arterial pressure back to within 10 per cent of the pre-incision value. Both techniques were effective but control was achieved faster with alfentanil. Plasma hormone levels in response to anaesthesia and surgery were similar in the two treatment groups.

TIVA in children

There are few reports about the use of TIVA in children. Browne and colleagues have recently described the use of a three-stage stepped infusion regimen of propofol (based on that reported in 1988 by Roberts *et al.*[37]) as a supplement to alfentanil in children aged 3–12 years.[38] The ED_{95} dose was about twice that required in adults (10.5 mg/kg per hour compared with 5 mg/kg per hour) – although the achieved blood propofol concentrations were similar. This supports the known kinetic profile of the drug in children, in whom there is a higher systemic clearance. The dynamic effects of the alfentanil–propofol regimen in children resulted in only minor decreases in blood pressure and heart rate (again less than those seen in adults with comparable blood propofol concentrations).

Analgesia for TIVA techniques

A number of different approaches have been adopted for the provision of analgesia during TIVA. The need to develop techniques for use in the spontaneously breathing patient has necessitated investigation of analgesic drugs with either limited respiratory depressive effects (e.g. the partial opioid agonists) or the careful titration of drug dose to effect. The common practice of administering a large loading dose of an opioid such as fentanyl at the beginning of surgery is inefficient – as the drug concentrations in both the plasma *and the biophase* (effect compartment) will initially exceed those of the therapeutic window for the opioid, and then rapidly decline to subtherapeutic concentrations (often at times of maximal surgical stress) by the combined processes of redistribution and metabolism.

Traditional agents include morphine and pethidine, but these are being displaced by newer drugs having higher potencies and fewer side effects (e.g. anilino-piperidine derivatives).

For TIVA, fentanyl, alfentanil and sufentanil have all been used – given either by multiple increments or by continuous infusion. Alfentanil is less potent than the parent drug, fentanyl, but its kinetic properties and more rapid blood-biophase equilibration allow the anaesthetist to alter blood (and brain) concentrations rapidly in response to different surgical or other noxious stimuli. Hence it has many of the ideal properties for the analgesic component of intravenous techniques (half-life, 90–120 minutes; clearance, 200–500 ml/minute; apparent volume of distribution at steady state, 30–40 litres; blood-biophase equilibration time 1–1.5 minutes). The major disadvantages of alfentanil include its emetic effects and the potential for producing laryngeal and chest wall rigidity.

In contrast to alfentanil, sufentanil is seven to ten times more potent than fentanyl, with a different kinetic profile (half-life, 140–200 minutes; clearance, 1000–1400 ml/min; blood-biophase equilibration time, 4–6 minutes). There are few data on the relations between dose and effect and between concentration and effect for this opioid when used as supplement (0.4–2.0 µg/kg) to nitrous oxide, volatile agents or an infusion of a hypnotic. Both alfentanil and sufentanil (as well as fentanyl) have been reported to cause postoperative respiratory depression.

Pharmacodynamics of opioids in TIVA

Concentration–effect data suggest a requirement for a concentration of 3–5 ng/ml for superficial body surface surgery when *fentanyl* is used to supplement either nitrous oxide or an infusion of a hypnotic, although concentrations of 4–8 ng/ml may be needed for intra-abdominal surgery.[39,40] These are achievable by regimens of 5–10 µg/kg loading dose, and 3 µg/kg per hour and 2 µg per minute maintenance rates respectively. Although these early studies were based on the administration of fixed dose infusions, there is now good evidence that titration of dose-to-effect provides better control of both blood pressure and heart rate, reduced drug requirements and improved recovery. After fentanyl infusion, the duration of analgesia and the extent of respiratory depression appear to be less than when the drug is given by incremental bolus dosing.

The relation between concentration and effect for alfentanil during nitrous oxide and propofol infusion anaesthesia are more easily defined as there is faster equilibration of blood, brain and effector site concentrations.[41] In a comparison of computer-assisted infusions of

alfentanil with intermittent bolus dosing (as a supplement to nitrous oxide anaesthesia),[42] the infusion group showed a lower incidence of responses to intubation and surgery, with greater haemodynamic stability and a reduced requirement for postoperative naloxone to initiate adequate spontaneous ventilation. There was also a tendency for the incidence of side effects to be reduced.

Dose requirements for alfentanil are affected by a number of different factors, such as protein binding or gender but not age or alcohol intake.

There are few data comparing fentanyl and alfentanil as part of a TIVA technique. Jenstrup *et al.* have assessed haemodynamic stability and recovery characteristics in 30 patients receiving alfentanil or fentanyl to supplement an infusion of propofol.[43] Cardiovascular changes were minimal and comparable in both groups, and both showed rapid emergence (1.8 minutes to eyes opening after propofol–alfentanil and 3.0 minutes after fentanyl–propofol). However, contrary to expectation, Jenstrup has demonstrated that the performance of post-anaesthetic psychomotor tests was *better* in the fentanyl compared with the alfentanil group! Clearly other double-blind comparative studies of a similar type need to be undertaken to examine this further, as there is no evidence of differences in the adequacy of anaesthesia in the two groups nor of differences in the duration of anaesthesia. If we examine the *effect-site* kinetics, this is not what Shafer and Varvel would predict for opioid infusions lasting about 120 minutes[44] (Fig. 33.4). Are there other explanations?

If a number of different intravenous drugs are employed in TIVA techniques, there is clearly a possibility for drug interaction to occur – these could be either kinetic or dynamic. Janicki and colleagues studied the effects of clinical concentrations of propofol on the microsomal metabolism of sufentanil and alfentanil (Fig. 33.5).[45] Propofol inhibits in a dose-dependent manner the oxidative degradation of both opioids. The IC_{50} values for man were

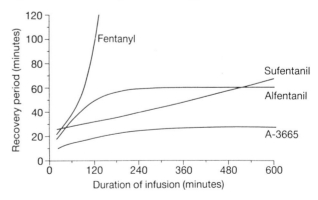

Fig. 33.4 Recovery curves for fentanyl, alfentanil, sufentanil and A-3665 showing the time required for a 50 per cent decrease in the *effect-site concentration* (recovery period) of the opioid after termination of the infusion. (Adapted, with permission, from reference 44.)

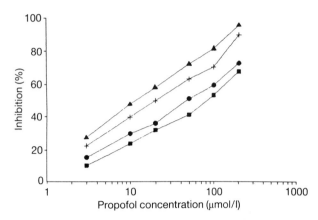

Fig. 33.5 Inhibition by propofol of oxidative metabolism of alfentanil and sufentanil by liver microsomes *in vitro*. Each point represents the mean (SD) of five separate experiments. IC_{50} values for propofol were determined by interpolation of semilog plots against percentage inhibition of alfentanil and sufentanil degradation. Pig liver: mean (SD) IC_{50} values of propofol were 32.6 (8.3) μmol/l for alfentanil (▲) and 22.1 (7.9) μmol/l for sufentanil (+). Human liver: mean (SD) IC_{50} values were 62.8 (16.4) μmol/l for alfentanil (●) and 52.9 (15.4) μmol/l for sufentanil (■). (Reproduced from reference 45, with the permission of the authors, Editor and publishers of the *British Journal of Anaesthesia*.)

52.9 μmol/l and 62.8 μmol/l respectively. (Put in perspective, peak clinical concentrations of propofol are in the range 40–60 μmol/l, and concentrations for maintenance of TIVA 10–30 μmol/l.) This study provides *in vitro* support for the observations of Gepts *et al.* of increased concentrations (comparing measured and predicted) of alfentanil during opioid–propofol anaesthesia.[28] Gepts *et al.* suggest that the cause might be a reduction in the apparent volume of drug distribution (similar to the etomidate–fentanyl interaction described by Schuttler *et al.*[17]

More recently, Baker and colleagues have provided data suggesting that propofol can inhibit phase I metabolism by phenobarbitone-inducible cytochrome P_{450} (P 450IIB1) at concentrations of 0.02–0.03 mmol.[46] Although its own clearance is altered by a *kinetic* interaction with fentanyl, etomidate has no effect on the disposition or analgesic potency of alfentanil.[17,47]

Dynamic interactions are also found between hypnotic and opioid drugs. Using the endpoint of failure to open eyes to command, Short and colleagues have shown synergism between midazolam, propofol and alfentanil when any two of these drugs are given together for induction of anaesthesia.[48] However, the response to all three agents together was *less* than that expected from the effects of the individual drugs and their two-drug interactions. Hence, in deriving new TIVA regimens, the clinician must clearly define the dose requirements for any given endpoint – and not assume that calculated requirements from single drug studies are additive! This is of greater importance when TIVA is used to provide anaesthesia in the paralysed and ventilated patient.

Effects of hypnotic and analgesic drugs on systemic responses to anaesthesia and surgery

Effects of intravenous induction agents on central haemodynamics in healthy patients

In this section we consider: barbiturates, etomidate and propofol; benzodiazepines; and ketamine.

Barbiturates, etomidate and propofol

Induction of anaesthesia with a sleep dose of a hypnotic drug decreases arterial blood pressure owing to both myocardial depression and peripheral vasodilation. The decreases in systolic (SAP) and diastolic pressures (DAP) are maximal within 1 minute of injection, and are sustained for at least 5 minutes. The magnitude of response varies between patients and with the different intravenous agents.

The decrease in SAP with either of the barbiturates is accompanied by a compensatory tachycardia, resulting in a maintenance of cardiac output. Compared with the barbiturates, etomidate causes less haemodynamic perturbation when used for induction of anaesthesia. At doses of 0.3 mg/kg, there is no significant change in systemic or coronary haemodynamics in both healthy patients and those with ischaemic heart disease. Maximum decreases in systolic and diastolic pressures are between 12 and 20 per cent, with a smaller reduction in cardiac output.

Most studies investigating the induction of anaesthesia with propofol in unpremedicated patients have shown only minimal increases in heart rate. As a result of this and the decrease in stroke volume (10–15 per cent), blood pressure and cardiac output fall in both normotensive and hypertensive patients. These decreases are more pronounced in patients concurrently receiving opioids, and in hypertensive patients receiving β-adrenoceptor blocking drugs. These haemodynamic responses may be attenuated by induction of anaesthesia with the combination of a subhypnotic bolus dose (1 mg/kg) and a low dose infusion (10 mg/ker per hour) instead of the usual induction dose of 2 mg/kg.[37] By reducing the rate of administration, the induction dose of most hypnotic drugs can be reduced by about 50 per cent. This is explained again by *effect–site kinetics* – with titration of dose so that the biophase *hypnotic* concentration is just achieved. The higher the biophase concentration, the greater is the resulting cerebral depression.

Laryngoscopy and intubation evoke strong sympathomimetic responses which are suppressed to a variable extent by different anaesthetics and their blood concentrations achieved at the time of laryngoscopy. Propofol appears to be effective in suppressing the haemodynamic consequences of laryngoscopy and intubation, in contrast to anaesthetic doses of induction agents such as thiopentone, methohexitone and etomidate.[13,49,50] The addition of fentanyl 2 μg/kg prior to induction of anaesthesia resulted in arterial pressures lower than those observed with the induction agent alone, and attenuation, although not abolition, of the response to laryngoscopy and intubation.

These changes in blood pressure are accompanied by comparable increases in intraocular pressure which, on occasion, can lead to disastrous consequences.[51] Compared with thiopentone and etomidate, propofol causes a greater decrease in intraocular pressure (IOP), and following a second smaller dose is more effective in preventing the rise in IOP secondary to suxamethonium (succinylcholine) and tracheal intubation.

Benzodiazepines

In healthy patients, there are no significant differences between the cardiovascular effects of midazolam and thiopentone for induction of anaesthesia.[52] The decrease in blood pressure following induction is due to a reduction in systemic vascular resistance. Although the intravenous benzodiazepines are considered by many anaesthetists to be the ideal drugs for induction of anaesthesia in the 'sick patient', significant reductions in both cardiac output and systemic vascular resistance occur with both diazepam and midazolam (just as with thiopentone), the effects of the latter being more marked. The cardiovascular effects of midazolam are not influenced by the addition of 50 per cent nitrous oxide in oxygen, but the addition of nitrous oxide to diazepam may cause significant increases in right atrial pressure (indicating some degree of myocardial depression).

In a recent comparison of propofol, etomidate and midazolam plus fentanyl 5 μg/kg as induction agents in patients with chronic coronary arterial disease, Lischke *et al.* found no influence of induction agent on the incidence of ST segment deviations during induction, laryngoscopy and intubation.[53] Thus, although propofol with fentanyl cause significant decreases in blood pressure and heart rate (and hence a possible reduction in coronary blood flow), there appears to be a compensatory reduction in myocardial oxygen consumption.

Ketamine

Induction of anaesthesia with ketamine (2–10 mg/kg i.v.) is almost always associated with hypertension, tachycardia, an increase in cardiac output and an increased left ventricular force of contraction. In the absence of premedicant drugs, the systolic pressure increases by 20–40 mmHg, with a smaller diastolic increase. These effects are probably mediated through sympathomimetic stimulation, and may be obtunded by a number of other intravenous drugs, including thiopentone, diazepam, flunitrazepam, and midazolam. The effects of ketamine during induction of anaesthesia are similar in patients with healthy and impaired myocardia.

Cardiovascular effects of infusions of hypnotic agents as part of TIVA

There are few studies comparing the cardiovascular effects of different TIVA techniques. In patients with coronary artery disease, Lepage *et al.* compared infusions of methohexitone or propofol infused at 100 μg/kg per minute during controlled ventilation with 100 per cent oxygen.[54] Both drugs decreased mean arterial pressure by 15–20 per cent owing to a reduction in cardiac output. Systemic vascular resistance was unaltered. There were differences in heart rate responses, with an increase of 10–25 per cent during methohexitone anaesthesia.

In those patients receiving propofol, the decrease in cardiac output was accompanied by a reduction in preload. The latter was unchanged with methohexitone infusions. However, in the group receiving methohexitone there was a decrease in ejection fraction and an increase in left ventricular end-diastolic volume, indicating a negative inotropic effect of the drug.

All studies with propofol by infusion confirm that it does *not* impair baroreflex sensitivity but rather resets the control gain, so allowing lower arterial pressures for a given heart rate. In comparison, the sensitivity of the baroreflex is depressed by methohexitone, *and* there is also a resetting to allow more rapid heart rates at lower arterial pressures during anaesthesia when compared with the pre-anaesthetic state.

When propofol is infused with fentanyl in the patient with ischaemic heart disease, there is a further reduction in mean arterial pressure (up to 35 per cent) due to reductions in both cardiac output and systemic vascular resistance.[55] Heart rate also decreases but stroke volume and left ventricular end-diastolic volume are left unchanged. Thus, again, propofol (this time in combination with fentanyl) does not alter left ventricular performance in patients with good LV function. Nevertheless, the significant reduction in coronary perfusion pressure could be prejudicial in patients with severe ischaemic heart disease. However, data from Stephan *et al.* show propofol alone (at infusion rates of 200 µg/kg per minute) to reduce perfusion by a reduction in preload.[56] One of the 12 patients in this study showed myocardial lactate production. This is of additional interest as it suggests that propofol-based TIVA techniques may, on occasion, lead to the development of myocardial ischaemia in patients with ischaemic heart disease.

In a comparison of propofol–fentanyl and etomidate–midazolam–fentanyl for coronary artery bypass surgery, Seitz *et al.* found that induction with rapid infusions of propofol (100 µg/kg per minute) was associated with significant decreases in both arterial pressure and heart rate, such that coronary perfusion might be jeopardized.[57] However, stable circulatory parameters were observed during the maintenance phase. With both groups, there were significant reductions in circulating basal and stimulated catecholamine concentrations, together with a peroperative decrease in cortisol secretion. However, none of the patients showed a peripheral lactic acidaemia (an indirect index of tissue perfusion), cardiac ischaemia or liver dysfunction. The comment is made again that the combination propofol–fentanyl may not be the most appropriate induction regimen in patients with limited coronary perfusion.

Schuttler *et al.* have studied the cardiovascular effects and recovery characteristics of the combination propofol–ketamine when infused to achieve a median EEG frequency of approximately 3 Hz.[58] This was associated with few changes in heart rate and blood pressure; the associated plasma propofol and ketamine concentrations were 2.8 µg/ml and 1.04 µg/ml respectively. The efficacy of this regimen needs to be studied further before any clear conclusion can be drawn.

Infusion of midazolam has also been used for maintenance of TIVA.[59,60] In both studies, regimens of about 0.125 mg/kg per hour achieved good clinical anaesthesia when supplementing a variable rate infusion of alfentanil. The main advantage of the technique appears to be the cardiovascular stability in terms of the responses to laryngoscopy and intubation and to noxious surgical stimuli. However, recovery was delayed when compared with propofol–alfentanil, and use of a single dose of the antagonist flumazenil to improve immediate recovery may be followed by later re-sedation.[61]

Effects of hypnotic agents on respiration and ventilatory performance

Induction of anaesthesia

Induction of anaesthesia by intravenous drugs is frequently followed by a period of apnoea (the incidence varying with

drug and the rate of administration by between 4 and 30 per cent). This rate is increased in patients premedicated with a benzodiazepine or an opioid. In addition, bronchospasm and laryngospasm are common after induction with barbiturates, probably owing to the attempted early instrumentation of the airway in a lightly anaesthetized patient. With propofol, the incidence and duration of apnoea is dose-related and may be exacerbated by ventilation with 100 per cent oxygen prior to induction. Most comparative studies looking at the acute ventilatory effects of thiopentone and propofol reveal similar depressions of tidal volume and ventilation rate. Apnoea and respiratory upsets are also seen in patients induced with midazolam, in whom they are more pronounced and of longer duration in the patient with chronic obstructive airways disease.

In contrast, ketamine has a minimal effect on central respiratory drive, with an unaltered ventilatory response to carbon dioxide. It also acts as a smooth muscle relaxant, resulting in bronchodilatation and improved pulmonary compliance. These effects are mediated both by a sympathomimetic action and by a direct antagonism of carbachol and histamine-induced constriction. One adverse effect on the airway, especially in the unpremedicated and in children, is increased salivation. This may result in upper airway obstruction or laryngospasm. Although the majority of upper airway reflexes are maintained intact during ketamine anaesthesia, silent aspiration has been reported.

Volatile agents all depress the hypoxic pulmonary vasoconstrictor (HPV) reflex with resultant $\dot{V}/\dot{Q}$ mismatching. In dogs, Naeije *et al.* found no effect of propofol on HPV during either hyperoxia or hypoxia, nor did it interfere with pulmonary vascular tone.[62] Clinical data similarly suggest that the intravenous agents used as part of TIVA have no effect on HPV.[63] There are few data on the influence of TIVA on lung function during one-lung anaesthesia. Spies *et al.* compared 1 MAC enflurane to supplement 100 per cent oxygen with an infusion of propofol in patients undergoing pulmonary resection.[64] During the one-lung phase of the study, there was a greater shunt fraction with enflurane–oxygen anaesthesia (38.5 per cent compared with 35.9 per cent) owing to the inhibition of HPV. There were no episodes of hypoxaemia Pao_2 <9.3 kPa or <70 mmHg) with propofol, whilst 4 of 14 patients receiving enflurane had low arterial oxygen tensions. Volatile anaesthetics and ketamine also cause bronchodilatation by both direct effects on airway smooth muscle and effects on the autonomic nervous system.

Effect of hypnotic infusions on respiratory activity

All agents cause dose-dependent respiratory centre depression, which may persist after recovery from anaesthesia. The latter is potentially dangerous when opioids have been administered as part of TIVA. There is also depression of the chemoreceptor responses. Both intravenous and volatile agents exert little effect *at clinical concentrations* on either the diaphragmatic or intercostal EMG. At clinical concentrations, both volatile and intravenous agents cause a decrease in tidal volume, an increased frequency of breathing and a reduction in the percentage time of the inspiratory phase;[65] moreover, it is likely that at least some of the changes seen during anaesthesia are due to the effects of sleep *per se* rather than being agent specific.

When equipotent infusion rates of the different hypnotic components of TIVA are studied as a supplement to 67 per cent nitrous oxide in spontaneously breathing patients in the absence of surgery, there are differences in the degree of ventilatory depression (as assessed by the arterial carbon dioxide tension). One limitation of most TIVA techniques is that doses of agents needed by the spontaneously ventilating patient to prevent responses to surgery may result in apnoea. Opioids influence the timing of inspiration to expiration by prolongation of the latter. Although the peripheral chemoreceptor responses to hypoxia are blunted during halothane anaesthesia,[66] they appear to retain their responsiveness during infusions of propofol.[67]

The usual response to isocapnic sustained hypoxia is an initial period of hyperventilation followed by decline. In the presence of midazolam (1 mg/70 kg per hour), there was *no* inhibition of these hyperventilatory responses.[68]

Effects of intravenous induction agents on cerebral haemodynamic function and cerebral pressures

Hypnotic induction agents all affect CSF pressure, which is the resultant of its rate of formation and its rate of reabsorption. Thiopentone, midazolam and etomidate all reduce CSF formation when used in high doses, but have little effect on its reabsorption.

In patients presenting with severe head injuries, these three agents increase cerebrovascular resistance and therefore reduce intracranial pressure – presumably due to a reduction in cerebral blood flow and a decrease in cerebral blood volume. They also decrease $CMRO_2$.

There are few data on the effects of intravenous induction agents in patients with raised intracranial pressure (ICP). Thiopentone and etomidate reduce ICP. The effects of propofol are ambiguous: some writers suggest that it acts to reduce ICP through a simple reduction in arterial pressure, whilst others propose maintenance of autoregulatory responsiveness to changes in arterial pressure.

Effects of infusion anaesthesia on cerebral dynamics

Nitrous oxide increases cerebral blood flow in healthy humans, and this may be undesirable in neurosurgical practice – thus limiting the usefulness of balanced anaesthetic techniques.

Barbiturates (either by bolus dosing or continuous infusion) have been shown to effectively reduce ICP, and have other potential advantages in terms of improving outcome after cardiopulmonary bypass and in affording protection after regional ischaemia (e.g. during aneurysm clipping, and in the control of post-hypoxic convulsive activity). However, their use is associated with prolonged recovery (especially after thiopentone) and liability to seizures (after large doses of methohexitone).

Propofol and etomidate both decrease CSF pressure while maintaining cerebral perfusion pressure. The reactivity of the cerebral vessels to changes in Pa_{CO_2} is unaltered during propofol anaesthesia.

Thiopentone is a satisfactory agent for use during somatosensory evoked potential monitoring, the evoked responses being elicited even after administration of large doses of thiopentone that result in a silent EEG. However, thiopentone does cause a dose-dependent change in the brain auditory evoked response. The action of propofol on brainstem auditory evoked potentials is controversial. Chassard *et al.* have reported increased latency *but* no change in amplitude,[69] whilst Thornton *et al.* have found that the majority of intravenous agents affect only early cortical responses, and appear to *spare* the brainstem responses[70] (Fig. 33.6).

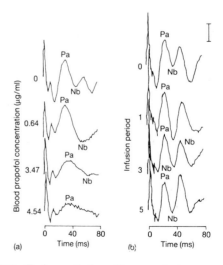

Fig. 33.6 Early cortical auditory evoked responses being depressed by increasing concentrations of propofol (a) compared with saline controls (b). Vertical bar = 0.5 µV. These data were obtained in the absence of surgery but in the presence of 70 per cent nitrous oxide. Nb, negative b; Pa, positive a. (Reproduced from reference 70, with the permission of the authors, Editor and publishers of the *British Journal of Anaesthesia*.)

High-dose infusions of most intravenous agents cause EEG burst suppression. The concentrations necessary are 40 μg/ml, 10 μg/ml and 10 μg/ml for thiopentone, methohexitone and propofol infusions respectively.

There are few outcome studies comparing the protective properties of hypnotic agents and the incidence of cerebral ischaemia. Nussmeier and colleagues investigated post-operative neurological sequelae in 182 patients undergoing cardiopulmonary bypass:[71] 89 received thiopentone (39.5 mg/kg) during bypass to maintain EEG silence; the other 93 patients received fentanyl alone. Enflurane was used in both groups to control intraoperative hypertension. Neurological sequelae occurred in both groups when assessments were made on the first postoperative day; however, by the tenth postoperative day there was evidence of persistent dysfunction in 7 of the 93 patients in the control group, compared with none in the thiopentone group. The use of thiopentone resulted in longer sleeping times, delayed extubation and increased postoperative sedation.

In a follow-up study, Metz and colleagues compared thiopentone by infusion during cardiopulmonary bypass with a smaller bolus (16 mg/kg) of the barbiturate given just prior to aortic declamping.[72] The latter regimen afforded the same protection, and reduced the awakening time. However, it did not alter the need for inotropic support during weaning from bypass.

It is still uncertain whether other presently available intravenous agents (such as propofol and midazolam) afford any cerebral protection in humans.

Awareness during TIVA has been a major concern of many anaesthetists and has led to reluctance by some to adopt TIVA. Bethune et al. assessed auditory perception and recall during TIVA used to supplement opioid-based anaesthesia for coronary arterial surgery.[73] After propofol anaesthesia there was a greater incidence of patients recalling auditory applied stimuli than was the case with methohexitone by infusion. Whether this is a drug-specific effect is uncertain, as the authors made no attempt to achieve equipotent anaesthetic input rates. However, the message is clearly stated – auditory perception and recall are possible during TIVA in patients who appear otherwise to be adequately clinically anaesthetized.

The major cause of true awareness during TIVA is undoubtedly the result of infusing drugs according to cookbook recipes rather than titration of dose according to effect!

Influence of drugs used during TIVA on liver, kidney and other tissues

The dose-dependent decreases in cardiac output with halothane and enflurane result in a reduction in liver blood flow; all volatile agents decrease renal blood flow, glomerular filtration rate and urinary volume. The greatest reduction is with isoflurane. There are decreases in renal blood flow, glomerular filtration rate and urinary volume during nitrous oxide–opioid anaesthesia despite any fall in blood pressure (probably attributable to an opioid-induced increase in ADH).

Intravenous hypnotic agents given by infusion (with the possible exception of ketamine) cause dose-related decreases in liver and renal blood flow. There are, however, few data on the rates of recovery of normal haemodynamic homoeostasis after TIVA. Observations in chronically cannulated merino sheep suggest that liver blood flow returns to pre-anaesthesia control values within 30 minutes of the end of anaesthesia. Propofol infusions were followed by a faster recovery of renal blood flow towards the awake baseline values than was seen with thiopentone infusions; with the latter, sheep took up to 5 hours to return to baseline pre-anaesthetic values.[74]

In a small study of 13 patients undergoing head and neck reconstructive surgery, Murray and Trinick found no effect of propofol–alfentanil anaesthesia or surgery on indocyanine green clearance, or plasma glutathionine S-transferase activity.[75] This was in contrast to an earlier study by the same group using halothane anaesthesia. The data from Murray and Trinick also confirmed the absence of hepatotoxic effects of long-term infusion of propofol, so confirming previous data.[76]

The problems of organ toxicity seen with the volatile agents (e.g. liver damage, renal failure) do not seem to occur with intravenous agents. The effects of intravenous agents on adrenal steroidogenesis and pituitary–adrenocortical control have been discussed earlier.

Of more recent concern is the effect of drugs (or their solvents) on the blood or blood components. Etomidate is currently formulated in propylene glycol, which has a high osmolality (4965 mmol/kg). This may contribute to the aetiology of the vascular inflammatory response after injection and to red cell haemolysis. Reformulation of etomidate as a lipid emulsion reduces the osmolality (to 400 mmol/kg) and, in doses of 0.3 mg/kg, leads to less reduction in the plasma haptoglobin concentration than when it is formulated to propylene glycol.

Anaphylactic and anaphylactoid reactions to intravenous hypnotic agents

All intravenous agents are liable to initiate allergic-type responses. The overall incidence of reactions to intravenous agents is between 1/5000 and 1/20 000; for barbiturates, the incidence of reactions is 1/23 000–1/30 000. There is the possibility of cross-sensitivity existing between methohexitone and the thiobarbiturate. In contrast, the incidence of reactions to both etomidate (about 1/450 000) and ketamine (2 cases) is low (see Table 33.1).

When formulated in Cremophor EL, at least 5 cases of hypersensitivity were reported among the 1131 patients who received propofol. In its present formulation of soyabean oil–egg phosphatide emulsion, there have been over 30 cases categorizable as 'allergic, anaphylactoid or anaphylactic', with one associated death. Other side effects that occur with a high incidence (and which may be related to histamine or other mediator release) include erythematous rashes, bronchospasm, flushing and hypotension. Although the frequency of reports of allergic-type responses to propofol seems to be increasing, the concurrent administration of neuromuscular blocking drugs may be a significant causative factor. When all of these hypersensitivity reactions are taken together, the overall incidence for propofol is probably between 1/80 000 and 1/100 000 administrations.

Use of the benzodiazepines has also been associated with reports of clinically overt adverse reactions. These followed the use of diazepam when formulated in propylene glycol (Valium; Roche) or in a Cremophor formulation (Stesolid).

More recently, there has been a report of a 20-year-old female who developed signs of flushing, urticaria and upper body oedema after receiving Diazemuls (an emulsion formulation) as sedation for gastroscopy. This 'histaminoid' picture was not associated with activation of either complement pathway. Thus, all three current formulations of this commonly prescribed benzodiazepine have been implicated in adverse reactions.

There are no reports of allergic reactions to any other of the benzodiazepines commonly used in anaesthetic practice (midazolam, lorazepam, temazepam).

New hypnotic and analgesic agents potentially useful in TIVA

Intravenous hypnotic agents

There are at present a number of new formulations of some intravenous hypnotic agents undergoing laboratory and/or clinical investigation. These include emulsion formulations of propanidid, methohexitone and etomidate, as well as etomidate formulated in 2-hydroxypropyl, β-cyclodextrin and a liposomal preparation of propanidid. Whether these offer advantages to current formulations remains to be seen.

Steroid anaesthetic agents

These have all had higher therapeutic indices than thiopentone but had important side effects resulting in their withdrawal from clinical use (hydroxydione: thrombophlebitis; Althesin: high incidence of allergic reactions to the solubilizing agent, Cremophor EL; minaxolone: problems relating to long-term toxicological testing).

5β-Pregnanolone – eltanolone

This 5β-pregnane steroid with anaesthetic properties has been evaluated in various animal species, and shown to have a high therapeutic index (TI >40). Currently the drug is being investigated in humans.

Eltanolone is water-insoluble and has been formulated as an emulsion. In rats, induction led to minimal cardiovascular depression but recovery was not as rapid as following propofol or Althesin. Preliminary kinetic data indicate an elimination half-life between 47 and 376 minutes, and clearance between 1.25 and 2.34 1/kg per hour.

There are now two studies defining the ED_{50} induction dose in either benzodiazepine or opioid premedicated patients (0.33 mg/kg and 0.44 mg/kg respectively). In both studies, induction caused only minimal haemodynamic depression (mean decrease in SAP and DAP −12 per cent, and increase in heart rate +9 per cent). Pain on injection was not observed, the main side effects being involuntary movements, mild apnoea and hypertonus.[77,78]

The relative potency of propofol (compared to eltanolone) is 0.313 in benzodiazepine-premedicated patients, but recovery appears to be slower after eltanolone. Cardiovascular responses to induction of anaesthesia, laryngoscopy and intubation are similar following either of these two agents.

Opioids

Potential new opioid drugs include remifentanil which, in volunteer and clinical studies, has a rapid onset of effect (similar to alfentanil) but an even shorter elimination half-life of 6–11 minutes. It is an ester, and undergoes rapid degradation by non-specific plasma and tissue esterases. The estimated clearance is between 40 and 70 ml/kg per minute and the blood–brain equilibration half-time is 1.3 minutes. Recent data in cats have shown a very good correlation between changes in blood concentration and EEG effect. There are few clinical data published to date.

Another opioid is A-3665, which has a potency similar to that of alfentanil. Its kinetic–dynamic profile shows an elimination half-life of 143 minutes (i.e. similar to alfentanil) and a clearance of 453 ml/min. The blood–brain equilibration time is 1.2 minutes. Effect compartment modelling suggests that recovery after A-3665 will be rapid even after an infusion of up to 8 hours' duration.[79]

α2-Adrenergic agonists

α2-Receptors are widely located on presynaptic terminals of adrenergic nerves in both the central and the peripheral nervous systems, and postjunctional in vascular smooth muscle, liver, platelets, kidneys and adipose tissue. The prejunctional receptors are not restricted to the adrenergic system, but also occur on nerve terminals releasing other transmitters (ACh, 5-HT). At least three α_2-isoreceptors have been identified, each coded by a different allele.

Several different agents have been used in veterinary practice as total anaesthetics (xylazine, detomidine, medetomidine, dexmedetomidine), but they show significant side effects at anaesthetic doses – bradycardia, hypotension and arrhythmias. They also inhibit renin and ADH release, and block the effects of ADH on the kidney tubules, causing a diuresis.

Premedication with clonidine or dexmedetomidine (DMD) reduces the anaesthetic requirements of both intravenous and inhalational agents but has no effect on neuromuscular blockade. They also diminish the haemodynamic and adrenergic responses to noxious stimuli and bring about a reduction in systemic vascular resistance in patients with impaired myocardial function (so increasing cardiac output), and attenuation of postoperative shivering. Peroperative infusion of DMD reduces isoflurane requirement in humans but, unlike in animals, these drugs cannot be used as the sole anaesthetic agent.

Concluding remarks

The clinical anaesthetist now has drugs and infusion equipment available such that the conduct of TIVA is a viable reality. Why should they be used?

Some of the possible advantages of TIVA include: rapidity of recovery (although studies comparing TIVA with desflurane anaesthesia are still lacking); a significant decrease in the incidence of nausea and vomiting when avoiding both volatile agents *and* nitrous oxide; a reduction in atmospheric pollution by exhaled or expelled fluorohydrocarbons; avoidance of the effects of nitrous oxide (during both anaesthesia and chronic occupational exposure) on vitamin B_{12} metabolism; as well as avoiding many of those known factors that can trigger malignant hyperthermia.

On the other hand, we know little about drug–drug interactions, and hence there are the risks of over- or underdosing; there is a need to titrate dosing to effect, *not* to depend on kinetic recipes as requirement changes with surgical stress and type of surgery; and *at present* it is not possible to depend on the feedback microcomputer control of drug administration until we know more about the way indices of adequacy of anaesthesia change with drug concentration and surgery.

REFERENCES

1. Savege TM, Ramsay MAE, Curran JPJ, Cotter J, Walling PT, Simpson BR. Intravenous anaesthesia by infusion. A technique using alphaxalone/alphadolone (Althesin). *Anaesthesia* 1975; **30**: 757–64.

2. Lowenstein E, Hallowell P, Levine FH, Daggett WM, Austen WG, Laver MB. Cardiovascular response to large doses of intravenous morphine in man. *New England Journal of Medicine* 1969; **281**: 1389–93.

3. Ausems ME, Hug Jr CC, Stanski DR, Burm AGL. Plasma concentrations of alfentanil required to supplement nitrous oxide anesthesia for general surgery. *Anesthesiology* 1986; **65**: 362–73.

4. Halford FJ. A critique of intravenous anesthesia in war surgery. *Anesthesiology* 1943; **4**: 67–9.

5. Crankshaw DP, Edwards NE, Blackman GL, Boyd MD, Chan HNJ, Morgan DJ. Evaluation of infusion regimens for thiopentone as a primary anaesthetic agent. *European Journal of Clinical Pharmacology* 1985; **28**: 543–52.

6. Becker KE. Plasma levels of thiopental necessary for anesthesia. *Anesthesiology* 1978; **49**: 192–6.

7. Chan HNJ, Morgan DJ, Crankshaw DP, Boyd MD. Pentobarbitone formation during thiopentone infusion. *Anaesthesia* 1985; **40**: 1155–9.

8. Breimer DD. Pharmacokinetics of methohexitone following intravenous infusion in humans. *British Journal of Anaesthesia* 1976; **48**: 643–9.

9. Hudson RJ, Stanski DR, Burch PG. Pharmacokinetics of methohexital and thiopental in surgical patients. *Anesthesiology* 1983; **59**: 215–19.

10. le Normand Y, de Villepoix C, Pinaud M, *et al.* Pharmacokinetics and haemodynamic effects of prolonged methohexitone infusion. *British Journal of Clinical Pharmacology* 1988; **26**: 589–94.

11. Sear JW, Phillips KC, Andrews CJH, Prys-Roberts C. Dose–response relationships for infusions of Althesin or methohexitone. *Anaesthesia* 1983; **38**: 931–6.

12. Prys-Roberts C, Sear JW, Low JM, Phillips KC, Dagnino J. Hemodynamic and hepatic effects of methohexital infusion during nitrous oxide anesthesia in humans. *Anesthesia and Analgesia* 1983; **62**: 317–23.

13. Todd MM, Drummond JC, U HS. The hemodynamic consequences of high-dose methohexital anesthesia in humans. *Anesthesiology* 1984; **61**: 495–501.

14. Crozier TA, Beck D, Wuttke W, Kettler D. Endocrinological changes following etomidate, midazolam or methohexital for minor surgery. *Anesthesiology* 1987; **66**: 628–35.

15. de Ruiter G, Popescu DT, de Boer AG, Smeekens JB, Breimer DD. Pharmacokinetics of etomidate in surgical patients.

Archives Internationales de Pharmacodynamie et de Therapie 1981; **249**: 180–8.

16. Schuttler J, Schwilden H, Stoeckel H. Infusion strategies to investigate the pharmacokinetics and pharmacodynamics of hypnotic drugs: etomidate as an example. *European Journal of Anaesthesiology* 1985; **2**: 133–42.

17. Schuttler J, Wilms M, Stoeckel H, Schwilden H, Lauven PM. Pharmacokinetic interaction of etomidate and fentanyl. *Anesthesiology* 1983; **59**: A247 [abstract].

18. Watt I, Ledingham IMcA. Mortality amongst multiple trauma patients admitted to an intensive therapy unit. *Anaesthesia* 1984; **39**: 973–81.

19. Crozier TA, Beck D, Wuttke W, Kettler D. *In vivo* suppression of steroid synthesis by etomidate is concentration-dependent. *Anaesthesist* 1988; **37**: 337–9.

20. Schuttler J, Kloss S, Ihmsen H, Pelzer E. Pharmacokinetic–dynamic properties of *S*(+) ketamine vs racemic ketamine: a randomized double-blind study in volunteers. *Anesthesiology* 1992; **77**: A330 [abstract].

21. Schuttler J, Stanski DR, White PF, *et al.* Pharmacodynamic modeling of the EEG effects of ketamine and its enantiomers in man. *Journal of Pharmacokinetics and Biopharmaceutics* 1987; **15**: 241–53.

22. Idvall J, Ahlgren I, Aronsen KF, Stenberg P. Ketamine infusions: pharmacokinetics and clinical effects. *British Journal of Anaesthesia* 1979; **51**: 1167–73.

23. Clements JA, Nimmo WS, Grant IS. Bioavailability, pharmacokinetics and analgesic activity of ketamine in humans. *Journal of Pharmaceutical Sciences* 1982; **71**: 539–42.

24. Dundee JW, Fee JPH, Moore J, McIlroy PDA, Wilson DB. Changes in serum enzyme levels following ketamine infusions. *Anaesthesia* 1980; **35**: 12–18.

25. Sear JW, McGivan JD. The cytotoxicity of intravenous anaesthetic agents on the isolated rat hepatocyte. *British Journal of Anaesthesia* 1979; **51**: 733–9.

26. Naguib M, Sari-Kouzel A, Seraj M, El-Gammal M, Gomma M. Induction dose-response studies with propofol and thiopentone. *British Journal of Anaesthesia* 1992; **68**: 308–10.

27. Shafer A, Doze VA, Shafer SL, White PF. Pharmacokinetics and pharmacodynamics of propofol infusions during general anesthesia. *Anesthesiology* 1988; **69**: 348–56.

28. Gepts E, Jonckheer K, Maes V, Sonck W, Camu F. Disposition kinetics of propofol during alfentanil anaesthesia. *Anaesthesia* 1988; **43**: suppl, 8–13.

29. Campbell GA, Morgan DJ, Kumar K, Crankshaw DP. Extended blood collection period required to define distribution and elimination kinetics of propofol. *British Journal of Clinical Pharmacology* 1988; **26**: 187–90.

30. Lange H, Stephan H, Rieke H, Kellerman M, Sonntag H, Bircher J. Hepatic and extrahepatic disposition of propofol in patients undergoing coronary bypass surgery. *British Journal of Anaesthesia* 1990; **64**: 563–70.

31. Veroli P, O'Kelly B, Bertrand F, Trouvin JH, Farinotti R, Ecoffey C. Extrahepatic metabolism of propofol in man during the anhepatic phase of orthotopic liver transplantation. *British Journal of Anaesthesia* 1992; **68**: 183–6.

32. Mather LE, Selby DG, Runciman WB, McLean CF. Propofol: assay and regional mass balance in the sheep. *Xenobiotica* 1989; **19**: 1337–47.

33. Vuyk J, Engbers FHM, Lemmens HJM, *et al.* Pharmaco-dynamics of propofol in female patients. *Anesthesiology* 1992; **77**: 3–9.

34. Davidson JAH, Macleod AD, Howie JC, White M, Kenny GNC. Effective concentration 50 for propofol with and without 67% nitrous oxide. *Acta Anaesthesiologica Scandinavica* 1993; **37**: 458–64.

35. Vuyk J, Lim T, Engbers FHM, Burm AGL, Vletter AA, Bovill JG. Pharmacodynamics of alfentanil as a supplement to propofol or nitrous oxide for lower abdominal surgery in female patients. *Anesthesiology* 1993; **78**: 1036–45.

36. Monk TG, Ding Y, White PF. Total intravenous anesthesia: effects of opioid versus hypnotic supplementation on autonomic responses and recovery. *Anesthesia and Analgesia* 1992; **75**: 798–804.

37. Roberts FL, Dixon J, Lewis GTR, Tackley RM, Prys-Roberts C. Induction and maintenance of propofol anaesthesia. A manual infusion scheme. *Anaesthesia* 1988; **43**: suppl, 14–17.

38. Browne BL, Prys-Roberts C, Wolf AR. Propofol and alfentanil in children: infusion technique and dose requirement for total i.v. anaesthesia. *British Journal of Anaesthesia* 1992; **69**: 570–6.

39. Andrews CJH, Sinclair M, Prys-Roberts C, Dye A. Ventilatory effects during and after continuous infusion of fentanyl or alfentanil. *British Journal of Anaesthesia* 1983; **55**: suppl 2, 211s–216s.

40. McQuay HJ, Moore RA, Paterson GMC, Adams AP. Plasma fentanyl concentrations and clinical observations during and after operation. *British Journal of Anaesthesia* 1979; **51**: 543–9.

41. Scott JC, Ponganis KV, Stanski DR. EEG quantitation of narcotic effect: the comparative pharmacodynamics of fentanyl and alfentanil. *Anesthesiology* 1985; **62**: 234–41.

42. Ausems ME, Vuyk J, Hug Jr CC, Stanski DR. Comparison of a computer-assisted infusion *versus* intermittent bolus adminis-tration of alfentanil as a supplement to nitrous oxide for lower abdominal surgery. *Anesthesiology* 1988; **68**: 851–61.

43. Jenstrup M, Nielsen K, Fruergard K, Moller A-M, Wiberg-Jorgensen F. Total i.v. anaesthesia with propofol–alfentanil or propofol–fentanyl. *British Journal of Anaesthesia* 1990; **64**: 717–22.

44. Shafer SL, Varvel JR. Pharmacokinetics, pharmacodynamics, and rational opioid selection. *Anesthesiology* 1991; **74**: 53–63.

45. Janicki PJK, James MFM, Erskine WAR. Propofol inhibits enzymatic degradation of alfentanil and sufentanil by isolated liver microsomes *in vitro*. *British Journal of Anaesthesia* 1992; **68**: 311–12.

46. Baker MT, Chadam MV, Ronnenberg WC. Inhibitory effects of propofol on cytochrome P450 activities in rat hepatic microsomes. *Anesthesia and Analgesia* 1993; **76**: 817–21.

47. Shuttler J, Schwilden H, Stoekel H. Pharmacokinetics as applied to total intravenous anaesthesia: practical implications. *Anaesthesia* 1983; **38**: suppl, 53–6.

48. Short TG, Plummer JL, Chui PT. Hypnotic and anaesthetic interactions between midazolam, propofol and alfentanil. *British Journal of Anaesthesia* 1992; **69**: 162–7.

49. Harris CE, Murray AM, Anderson JM, Grounds RM, Morgan M. Effects of thiopentone, etomidate and propofol on the haemodynamic response to tracheal intubation. *Anaesthesia* 1988; **43**: suppl, 32–6.

50. Sanders DJ, Jewkes CF, Sear JW, Verhoeff F, Foëx P. Thoracic

electrical bioimpedance measurement of cardiac output and cardiovascular responses to the induction of anaesthesia and to laryngoscopy and intubation. *Anaesthesia* 1992; **47**: 736–40.

51. Mirakhur RK, Shepherd WFI, Darrah WC. Propofol or thiopentone: effects on intraocular pressure associated with induction of anaesthesia and tracheal intubation (facilitated with suxamethonium). *British Journal of Anaesthesia* 1987; **59**: 431–6.

52. Nilsson A, Lee PFS, Revenas B. Midazolam as induction agent prior to inhalational anaesthesia: a comparison with thiopentone. *Acta Anaesthesiologica Scandinavica* 1984; **28**: 249–51.

53. Lischke V, Probst S, Behne M, Kessler P. Myocardial ischaemia during induction of anaesthesia in patients with chronic coronary artery disease: a comparison of propofol, etomidate and midazolam. *Anaesthesist* 1993; **42**: 435–40.

54. Lepage J-YM, Pinaud ML, Helias JH, Cozian AY, Le Normand Y, Souron RJ. Left ventricular performance during propofol or methohexital anesthesia: isotopic and invasive cardiac monitoring. *Anesthesia and Analgesia* 1991; **73**: 3–9.

55. Lepage J-YM, Pinaud ML, Helias JH, *et al.* Left ventricular function during propofol and fentanyl anesthesia in patients with coronary artery disease: assessment with a radionuclide approach. *Anesthesia and Analgesia* 1988; **67**: 949–55.

56. Stephan H, Sonntag H, Schenk HD, Kettler D, Khambatta HJ, Effects of propofol on cardiovascular dynamics, myocardial blood flow, and myocardial metabolism in patients with coronary artery disease. *British Journal of Anaesthesia* 1986; **58**: 969–75.

57. Seitz W, Lubbe N, Schaps D, Haverich A, Kirchner E. Propofol for induction and maintenance of anesthesia in cardiac surgery. Results of pharmacological studies in humans. *Anaesthesist* 1991; **40**: 145–52.

58. Schuttler J, Schuttler M, Kloos S, Nadstawek J, Schwilden H. Total intravenous anesthesia with ketamine and propofol with optimized dosing strategies. *Anaesthesist* 1991; **40**: 199–204.

59. Nilsson A, Persson MP, Hartvig P, Wide L. Effect of total intravenous anaesthesia with midazolam/alfentanil on the adrenocortical and hyperglycaemic response to abdominal surgery. *Acta Anaesthesiologica Scandinavica* 1988; **32**: 379–82.

60. Vuyk J, Hennis PJ, Burm AGL, de Voogt J-WH, Spierdijk J. Comparison of midazolam and propofol in combination with alfentanil for total intravenous anesthesia. *Anesthesia and Analgesia* 1990; **71**: 645–50.

61. Nilsson A, Persson P, Hartvig P. Effects of the benzodiazepine antagonist flumazenil on postoperative performance following total intravenous anaesthesia with midazolam and alfentanil. *Acta Anaesthesiologica Scandinavica* 1988; **32**: 379–82.

62. Naeije R, Lejeune P, Leeman M, Melot C, Dellof T. Effects of propofol on pulmonary and systemic arterial pressure–flow relationships in hyperoxic and hypoxic dogs. *British Journal of Anaesthesia* 1989; **62**: 532–9.

63. Van Keer L, Van Aken H, Vandermeersch E, Vermaut G, Lerut T. Propofol does not inhibit hypoxic pulmonary vasoconstriction in humans. *Journal of Clinical Anesthesia* 1989; **1**: 284–8.

64. Spies C, Zaune U, Pauli MHF, Boeden G, Martin E. Comparison of enflurane and propofol during thoracic surgery. *Anaesthesist* 1991; **40**: 14–18.

65. Goodman NW, Black AMS, Carter JA. Some ventilatory effects of propofol as sole anaesthetic agent. *British Journal of Anaesthesia* 1987; **59**: 1497–503.

66. Knill RL, Gelb AW. Ventilatory responses to hypoxia and hypercapnia during halothane sedation and anesthesia in man. *Anesthesiology* 1978; **49**: 244–51.

67. Dow AC, Goodman NW. Effect of hyperoxia on the breathing of patients anaesthetized with infusions of propofol. *British Journal of Anaesthesia* 1993; **70**: 532–5.

68. Dahan A, Ward DS. Effect of i.v. midazolam on the ventilatory response to sustained hypoxia in man. *British Journal of Anaesthesia* 1991; **66**: 454–7.

69. Chassard D, Joubard A, Colson A, Guiraud M, Dubreuil C, Banssillon V. Auditory evoked potentials during propofol anaesthesia in man. *British Journal of Anaesthesia* 1989; **62**: 522–6.

70. Thornton C, Konieczko KM, Knight AB, *et al.* Effect of propofol on the auditory evoked response and oesophageal contractility. *British Journal of Anaesthesia* 1989; **63**: 411–17.

71. Nussmeier NA, Arlund C, Slogoff S. Neuropsychiatric complications after cardiopulmonary bypass: cerebral protection by a barbiturate. *Anesthesiology* 1986; **64**: 164–70.

72. Metz S, Slogoff S, Keats AS. Single dose thiopental compared to infusion for central nervous system protection during cardiopulmonary bypass. *Anesthesia and Analgesia* 1990; **70**: S266 [abstract].

73. Bethune DW, Ghosh S, Gray B, *et al.* Learning during general anaesthesia: implicit recall after methohexitone or propofol infusion. *British Journal of Anaesthesia* 1992; **69**: 197–9.

74. Runciman WB, Mather LE, Selby DG. Cardiovascular effects of propofol and of thiopentone anaesthesia in the sheep. *British Journal of Anaesthesia* 1990; **65**: 353–9.

75. Murray JM, Trinick TR. Hepatic function and indocyanine green clearance during and after prolonged anaesthesia with propofol. *British Journal of Anaesthesia* 1992; **69**: 643–4.

76. Sear JW, Prys-Roberts C, Dye A. Hepatic function after major vascular reconstructive surgery. A comparison of four anaesthetic techniques. *British Journal of Anaesthesia* 1983; **55**: 603–9.

77. Powell H, Morgan M, Sear JW. Pregnanolone: a new steroid intravenous anaesthetic. Dose-finding study. *Anaesthesia* 1992; **47**: 287–90.

78. Hering W, Biburger G, Rugheimer E. Induction of anaesthesia with the new steroid intravenous anaesthetic eltanolone (pregnanolone). Dose finding and pharmacodynamics. *Anaesthesist* 1993; **42**: 74–80.

79. Lemmens HJM, Dyck JB, Shafer SL, Stanski DR. The application of pharmacokinetics, dynamics and computer simulations to drug development: A-3665 versus fentanyl and alfentanil. *Anesthesiology* 1992; **77**: A456 [abstract].

FURTHER READING

Dundee JW, Sear JW, eds. Intravenous anaesthesia – what is new? *Clinical Anaesthesiology: International Practice and Research* 1991; **5** (2).

Morgan M. Total intravenous anaesthesia. *Anaesthesia* 1983; **38**: suppl, 1–9.

Prys-Roberts C. Cardiovascular effects of continuous intravenous anaesthesia compared with those of inhalational anaesthesia. *Acta Anaesthesiologica Scandinavica* 1982; **26**: suppl 75, 10–17.

Schwilden H, Stoeckel H, Schuttler J, Lauven PM. Pharmacological models and their use in clinical anaesthesia. *European Journal of Anaesthesiology* 1986; **3**: 175–208.

Sear JW. General kinetic and dynamic principles and their application to continuous infusion anaesthesia. *Anaesthesia* 1983; **38**: suppl, 10–25.

Sear JW. Toxicity of i.v. anaesthetics. *British Journal of Anaesthesia* 1987; **59**: 24–45.

Sear JW. Intravenous anaesthetics. In: Foex P, ed. Anaesthesiology of the Compromised Heart: *Balliere's Clinical Anaesthesiology* 1989; **3** (1): 217–42.

Sear JW. Continuous infusion of hypnotics for anaesthesia. In: Kay B, ed. *Total intravenous anaesthesia*, Monographs in Anaesthesia. Amsterdam: Elsevier 1991: 15–55.

Sear JW. Drug interactions and adverse reactions. In: White PF, ed. Kinetics of Anaesthetic Drugs in Clinical Anaesthesiology: *Clinical Anaesthesiology: International Practice and Research* 1991; **5** (3): 703–33.

Chapter 34

Regional Anaesthetic Techniques

I. McConachie and J. McGeachie

Central neural blockade	Peripheral regional blocks

The most important prerequisite is a thorough knowledge of anatomy. The facilities required for successful and safe performance of regional anaesthesia depend on whether regional anaesthesia is to be provided in isolation or in combination with general anaesthesia. However, sole performance of a regional block is no excuse for poor standards and poor preparation. Full resuscitation facilities must be immediately available. The requirements for the safe performance of regional anaesthesia are shown in Table 34.1. Also shown are the 'optional extras' for which the need may be a matter of personal preference. Certain aspects are controversial; for example, routine monitoring during the injection of the block, although this can be useful – particularly for central neural blockade.[1] Many anaesthetists do not routinely wear a sterile gown or facemask for all regional techniques. However, the majority of anaesthetists wear a gown and mask for insertion of an extradural catheter which may be difficult to control while maintaining sterility.

Needles

Standard sharp-pointed, long-bevelled needles are suitable for many blocks. However, short-bevelled 'blunt' needles enable tissue planes to be identified more easily. It is logical that standard needles will cause less damage if the nerve is pierced. This is presumably because the fibres are separated rather than torn. However, there is histological evidence to the contrary.[2] Needles designed for central neural blocks are discussed in the relevant section. Unless otherwise stated, all blocks described here are performed with a 23 s.w.g. needle.

Peripheral nerve stimulators

Nerve stimulators can be used in anaesthetized patients to identify nerves with a motor component. Eliciting paraesthesiae may be unpleasant for an awake patient,

and may result in an increased incidence of neurological complications.[3] Nerve stimulators may not solve all the problems, and are not helpful if an anaesthetized patient has received muscle relaxants. Care must be taken that the response observed with stimulation is not a result of direct local muscle stimulation. The response should be sought in an area where direct local stimulation could not occur.

The value of electrical stimulators is confirmed by evidence that their use increases the success rate for some blocks.[4] Contrary to some reports,[5] special sheathed needles that are electrically insulated apart from the tip are not required;[6] the point of maximum current density lies just in front of the tip of the needle. Nerve stimulation should occur before the nerve is touched by the needle, as the point of the needle is behind the point of maximum stimulation.

Practical points in the use of nerve stimulators

1. A low current of approximately 0.5 mA should be selected, or stimulation may occur when the needle tip is still some distance away from the nerve.
2. The negative electrode or cathode must be attached to the stimulating needle.
3. The positive electrode or anode must be attached to a site distant to the block site.
4. The first 1–2 ml of anaesthetic should abolish the evoked contraction immediately if a successful block is to be ensured. This implies that the needle is immediately adjacent to the nerve fibres. This mechanism is not due to a pharmacological action of the local anaesthetic but is probably due to physical deformation of the nerve fibres by the small volume injected (saline will have the same effect).

Factors influencing the choice of local anaesthetic drug

A full discussion of the pharmacology of local anaesthetic drugs is given in Chapter 10. In general, a highly lipid-soluble drug will have a prolonged duration of action; for example, bupivacaine is more lipid soluble than lignocaine (lidocaine). The degree of ionization is inversely related to the lipid solubility: poorly ionized drugs will be present in a higher free form and will therefore exhibit a rapid onset of anaesthesia. Lowering the pH in the tissues (as with local infection) will increase the amount of ionized drug and reduce the amount of free lipophilic drug present, and retard the development of anaesthesia. Conversely, if the local pH of the drug is increased (e.g. by adding sodium bicarbonate), the free portion of drug should increase (all local anaesthetic drugs have a pK_a greater than 7) and hasten the onset of anaesthesia.[7] The results of studies are not all convincing[8] and the additional effect of doubtful practical significance. The degree of protein binding also influences the duration of action; for example, amethocaine (tetracaine) and bupivacaine are highly protein bound and produce prolonged blockade compared to procaine which is weakly protein bound.

From a practical point of view the greater the concentration of anaesthetic used, the shorter the duration of onset, the longer the duration of action and the more profound the block (i.e. motor as well as sensory). The duration of blockade is increased if the local anaesthetic remains at the required site, i.e. if vascular uptake is inhibited by the addition of adrenaline (epinephrine) to the anaesthetic solution. The reduced rate of absorption and consequent lowering of blood concentration by the addition of adrenaline raises the 'quoted' maximum safe dose of lignocaine from 3 to 7 mg/kg. However, there are factors involved in systemic toxicity other than the relation between the amount of drug and the patient's weight.[9]

Preparation and management of the patient

The patient should be prepared exactly as one would prepare a patient who is to have a general anaesthetic, especially with regard to fasting. All regional techniques have a small failure rate, and the patient who is to have a regional anaesthetic may require general anaesthesia. Indeed, all patients who are to have a regional anaesthetic should be forewarned of this possibility. However, despite this pessimistic counsel it is important to project an air of confidence to the patient. Many patients will prefer to 'go to sleep'. These patients should be reassured that sedation will be available if required. Other patients have fears of permanent neurological complications (particularly paralysis following extradural or intrathecal anaesthesia). These patients should be truthfully reassured of the safety of regional anaesthesia. Appropriate sedative premedication may be used. Lastly, many techniques especially extradural and brachial plexus blocks require up to 20 minutes after

Table 34.1 Facilities required

Gloves, cleaning fluid, towels, gowns, masks, etc.
Resuscitation drugs
Appropriate needles, syringes and catheters
Tipping trolley
Access to circulation
Thiopentone, suxamethonium and other drugs
Intravenous fluids
Trained assistance (e.g. for positioning)
Oxygen and means of administering under pressure
Means of assisted ventilation
Laryngoscopes and tracheal tubes
Suction
Monitoring
All facilities for general anaesthesia
Appropriate local anaesthetic drugs

Optional extras
Peripheral nerve stimulator
Music and headphones
Screen to prevent patient viewing operation
Sedation drugs if used
Oxygen facemask if sedating
Skin marker

their completion for surgical anaesthesia to be achieved. Thus, planning of the operating schedule may be required to prevent delays. Impatience and/or pressure from the surgeons may result in too early testing of the quality of the block, which may lead to loss of confidence by the patient.

There are three cardinal rules to follow when preparing the patient and performing the block:

1. The site of injection of the block must be cleaned with an appropriate agent, according to local policy. After skin cleansing, either the skin must not be touched again during the injection ('no touch technique') or sterile gloves should be worn. Gown and facemask may be advisable for catheter techniques (Table 34.1).

2. Prior to injection, aspiration *must* be performed to detect blood and/or cerebrospinal fluid depending on the site. This is essential to minimize the risk of inadvertent intravascular or intrathecal injection of local anaesthetic. The only exception is the 'fan-wise' infiltration of local anaesthetic into the tissues. On these occasions it is sufficient to aspirate as injection commences and then to continually move the needle point during injection to ensure that, even if a small vessel is entered, only a very small portion of anaesthetic will be injected intravascularly.

3. If the patient is awake, the procedure is made more tolerable by subcutaneous infiltration with lignocaine 1 per cent, and then by allowing several minutes for its action to take effect.

Table 34.2 Keys to success

Careful selection of patients, avoiding those with absolute contraindications and exercising great care with those with relative contraindications

Preoperative visit with explanation of procedure, and giving opportunity for questions and reassurance of techniques available if failure (complete or partial for any reason) should occur

Meticulous attention to detail in anaesthetic room, in particular intravenous access, positioning of patient and preparation of patient

Aseptic technique at all times

Maintain dialogue with patient or at least keep talking to patient and say what you are about to do

Careful and considered use of supplementation technique(s) when indicated

Some keys to successful regional anaesthesia are presented in Table 34.2.

Choice of suitable patients

Despite the use of sedation there are some patients who are unsuitable for a sole regional technique. These include the extremely anxious patient, the psychotic patient or the confused and uncooperative patient. However, they may still be suitable for a combined regional and general technique. Patients who cannot tolerate the positioning or injection of the block because of pain are also unsuitable for a regional technique without general anaesthesia. The systemic or local contraindications to regional anaesthesia described below should be noted.

Contraindications to regional anaesthesia

As for any technique or therapy, there are both relative and absolute contraindications to regional techniques. Relative contraindications may on occasion be outweighed if the technique proposed carries appropriate benefits. However, it must be emphasized that none of the contraindications should be taken lightly. There are also contraindications that are specific for the central neural blocks.

Absolute contraindications

1. Patient refusal or lack of cooperation or understanding.
2. Definite allergy to the local anaesthetic drugs.
3. Absence of resuscitation equipment.
4. No intravenous access for resuscitation.
5. Coagulopathy, either therapeutic or pathological (see below).
6. Local sepsis at the point of injection.

General sepsis is a contraindication for central neural blocks, as an infection may occur in an extradural haematoma. In addition, hypovolaemia or continuing rapid blood loss is an absolute contraindication for a central block. The induced sympathetic block may cause rapid and catastrophic cardiovascular collapse.

Relative contraindications

1. *Neurological disease.*
2. *Orthopaedic disorder,* especially backache if central block is proposed. There is no confirmed scientific evidence that performing a block will exacerbate a neurological or orthopaedic disorder but coincidental exacerbation of the condition may be attributed to the block. Certainly if a neurological impairment is present the nature and extent of the disease must be documented before performing any regional anaesthetic technique.
3. *Performing blocks on awake children.* These patients may not understand or be able to cooperate with the procedure. The occasional paediatric anaesthetist should not perform intrathecal or lumbar extradural anaesthesia on small children and infants. Specialized centres, however, report the value of intrathecal anaesthesia in neonates.[10]

Additional relative contraindications to central neural blocks are recognized:

4. *Severe cardiac disease* when cardiac output depends on either vascular tone or an adequate preload; a sympathetic block could lead to a marked fall in cardiac output. This may be particularly relevant in pregnant women with valvular heart disease. Careful titration of anaesthetic to achieve a controlled extradural block is safer than single-shot intrathecal anaesthesia, after which the extent of sympathetic block is often unpredictable.
5. *High abdominal surgery.* The difficulty here is to achieve an adequate height of block with adequate muscle relaxation but without extensive sympathetic block.
6. *Increased intracranial pressure.* Deliberate or inadvertent intrathecal puncture may result in a sudden fall in cerebrospinal fluid pressure, with herniation of the brainstem through the foramen magnum.
7. *Potential massive haemorrhage* during caesarean section for placenta praevia may quickly result in hypovolaemia with cardiac decompensation in the presence of a sympathetic block.

Regional anaesthesia and disorders of coagulation

It is generally accepted that some regional anaesthetic techniques would be hazardous in the presence of a coagulopathy such as thrombocytopenia. In particular, during central blockade (intrathecal, extradural and caudal anaesthesia), bleeding is contained in a space restricted by bone. This may be hazardous and lead to permanent neurological damage if not recognized and treated promptly. Warning signs include intense localized back pain and signs of cord compression. These signs may, of course, be masked by the blockade, especially if the technique is used to provide postoperative analgesia. It should be remembered that there is an incidence of spontaneous spinal haematoma formation[11] and a spinal haematoma has presented following general anaesthesia without the use of any regional technique.[12]

Further difficulties are posed by the now increasingly widespread use of pharmacological measures, such as low dose heparin, to reduce the incidence of deep venous thrombosis after surgery, sometimes combined with the use of low dose aspirin therapy both prophylactically and following incidences of cerebral and myocardial ischaemia and infarction. There is also debate surrounding the use of central blockade in patients who are fully anticoagulated.[13] Until large well-controlled studies are available, the advice offered in Wildsmith and McClure's editorial[14] should be followed. For a summary and practical guidelines see Table 34.3.

Intraoperative management of a regional technique alone

The anaesthetist's task is not finished once a successful block has been performed. Other aspects must be considered to ensure that the overall experience for the patient is as pleasant as possible. For example, attention must be paid to the patient's overall comfort while lying on a hard table, the patient's warmth and the provision of a relaxed environment. The patient should be protected from viewing the operation by the use of screens and towels.

Sedation and supplementation

Most patients, except those undergoing very minor surgery, expect to be rendered unconscious for their surgical procedures. A notable exception is the obstetric patient.

Sedative supplementation of a regional anaesthetic technique may be by one of several methods.

Table 34.3 Regional anaesthesia and disorders of coagulation (therapeutic and pathological)

Preoperative anticoagulant therapy/pathological coagulative disorders
Established therapeutic anticoagulation should be terminated before all but the most minor elective surgery, regardless of the anaesthetic technique. If a central regional anaesthetic technique is considered, there must be preoperative evidence of normal clotting.

Central blockade would be hazardous in patient with established (untreated) pathological coagulopathy. If reliable correction is possible, one may be able to proceed with caution.

Low dose heparin therapy (5000 units subcutaneously two or three times daily)
For central blockade avoid preoperative administration. First dose of subcutaneous heparin can be administered after establishment of block. In the postoperative period, catheter removal should take place 1 hour before the dose of heparin is administered.

Aspirin therapy (75 mg daily)
Considerable controversy in this area. Bedside platelet function test (Ivy bleeding time) may provide guidance as to safety (should be less than 10 minutes, according to some workers) but there are practical difficulties with standardization and reproducibility of test and wide normal range.

Perioperative anticoagulant therapy (heparin bolus or infusion)
Generally safe, as block performed well in advance of starting therapy. Therapy may be continued into postoperative period and thus caution during removal of epidural catheter.

Postoperative therapeutic anticoagulant therapy (heparin infusion/ warfarin tablets)
For single-shot blocks no problems.
Removal of epidural catheter may be hazardous. Consult with surgeons regarding cessation of heparin for several hours prior to removal. If warfarin is started or considered necessary, catheter must be removed before warfarin becomes effective.

Distraction therapy

The simple action of diverting the patients' attention from their worries may be of great benefit during surgery performed under regional anaesthesia. The use of personal stereos is a particularly good method,[15] as it should be possible to accommodate the musical taste of most patients provided there is a large enough choice. Alternatively, a member of the theatre staff may talk to the patient, offering reassurance and encouragement when required.

General anaesthesia

Whilst it may seem perverse to deliberately render a patient insensible when the regional anaesthetic is providing perfect operating conditions, any feelings of anxiety or doubt regarding the extent of the anaesthesia provided will be removed. Only a very light general anaesthetic may be needed.

Relative analgesia

Patients' anxiety can be reduced and cooperation increased and additional analgesia may be provided by the use of mixtures of nitrous oxide and oxygen, with the concentration titrated against the individual patient's response. Whilst this technique is particularly popular with dental surgeons, it has yet to become commonplace in the hospital environment.

Intravenous supplementation

There is no doubt that advances in pharmacology have provided anaesthetists with safer and more predictable sedative agents which can be used to supplement regional anaesthesia. However, the extensive use of sedation in conjunction with regional anaesthesia is illogical at best and dangerous at worst, and the anaesthetist must be able to justify the use of sedative drugs. There is no doubt, however, that the provision of amnesia for the events surrounding the time spent in the operating theatre can be an advantage.

The choice of sedative drugs may be wide in theory, but in practice only a few drugs are used. The water-soluble benzodiazepine midazolam offers considerable advantages over its predecessor diazepam, and may be administered by intravenous bolus (0.07–0.15 mg/kg) or by continuous infusion (0.25 mg/kg per hour). Care and caution are required with midazolam, as there is a wide variation in dose response and these doses are for guidance only. The newer anaesthetic agent propofol can also be used, although a continuous infusion is probably the best method of administration for sedation throughout the whole of the surgical procedure.[16] An infusion of propofol of 3–4 mg/kg per hour will provide sedation with a rapid recovery. Care with regard to the cardiorespiratory systems is required with this drug. The amnesic effect of propofol is not as powerful as that provided by midazolam.

Small doses of short-acting opioid drugs such as alfentanil (0.25–0.5 mg) and fentanyl (50–100 µg) can be used for analgesia in addition to or instead of sedative drugs. Combinations of sedative and analgesic drugs are not without their dangers,[17] and it is mandatory that supplemental oxygen is administered and arterial oxygen saturation is monitored continuously whenever sedative and/or analgesic drugs are administered to supplement any regional anaesthetic, especially extradural and intrathecal anaesthesia.

The inadequate block

If the resultant block is clearly inadequate for the proposed surgery, it can occasionally be supplemented more peripherally; for example, a wrist block to supplement a brachial plexus block or local infiltration by the surgeon. Care must be taken with the total dose of anaesthetic. Systemic opioids may on occasion be useful but their overenthusiastic use can be dangerous. It is probably wiser to admit defeat and convert to controlled general anaesthesia than to administer excessive amounts of 'sedation'. The 'failed' block may still provide residual postoperative analgesia. Persistance with surgery in the face of a clearly inadequate block may be grounds for later complaint.

Pros and cons of regional vs general anaesthesia

There is no consensus on whether certain types of patient undergoing certain types of surgery benefit more from a regional, a general or a mixture of the two types of anaesthesia. Obvious indicators such as mortality have not proved to be helpful. Different generations of anaesthetists may have different priorities and preferences; an example is the fall in popularity of intrathecal anaesthesia in the UK in the 1950s and 1960s following the Woolley and Roe court case.[18] For many anaesthetists the benefits of regional anaesthesia appear to be self-evident and worth attaining. Others perform virtually no regional techniques throughout their entire practice. However, it is the authors' belief that an ability to perform safe and effective regional anaesthesia is an essential part of an anaesthetist's armamentarium.

Despite the lack of consensus there are certain advantages and disadvantages of the two main forms of anaesthesia that are generally accepted universally, and these are shown in Table 34.4. Some of these are expanded upon in the next section, on the high-risk patient. Combinations of regional and general techniques or regional techniques with accompanying sedation will tend to blur the advantages and disadvantages of particular techniques.

Regional anaesthesia and the high-risk patient

Many believe that (if the proposed surgery is suitable) a form of regional anaesthesia is safer than general anaesthesia for the patient with severe respiratory disease. Peripheral regional blocks (e.g. brachial plexus block) are probably safer than general anaesthesia for patients with severe cardiac disease. However, central blocks where hypotension may not always be preventable are not necessarily a safer alternative to careful general anaesthesia, particularly in inexperienced hands.[19] In the special circumstance of fractured hips in the elderly, intrathecal anaesthesia is associated with a reduced early mortality compared with general anaesthesia but no

Table 34.4 Regional vs. general anaesthesia

Advantages of regional anaesthesia	*Advantages of general anaesthesia*
Decreased blood loss for certain surgery	Patent airway may be assured
Decreased DVT rate for certain surgery	Adequate oxygenation may be assured
Decrease in metabolic changes following surgery	Cardiovascular effects usually titratable
Better for severe respiratory impairment	Familiarity for most anaesthetists
Preserves airway (if no sedation)	Patient preference
Cardiovascular stability if no sympathetic blockade	
Residual postoperative analgesia	
Allows verbal communication: warning of, for example, hypoglycaemia	
Excellent muscle relaxation	
No atmospheric pollution	
Less expensive than many GA techniques	
Avoidance of rare complications of GA (e.g. malignant hyperthermia)	
Can drink and/or eat immediately after surgery (depending on type of operation)	
Disadvantages of regional anaesthesia	*Disadvantages of general anaesthesia*
Discomfort, especially for long operations	Cardiovascular depression
Discomfort eliciting paraesthesiae	Decreased protective reflexes
Contraindicated in confused patients	Prolonged psychomotor impairment
Potential for nerve damage (rare)	Nausea, vomiting, headache, etc.
Profound cardiovascular effect with central neural blockade	Possibility of inadvertent awareness
Leg weakness and urinary retention with central neural blockade	
Toxic doses for some combinations of blocks	
Toxicity with high serum levels of anaesthetic or accidental i.v. injection	
Poor communication of intentions if deaf	
Patient preference to be asleep	
Embarrassment for certain positions (e.g. lithotomy) if awake – especially in elderly	

significant difference in late mortality[20] (where other factors, including pre-existing medical disease, have the major influence). However, less well proven is the hypothesis that regional anaesthesia and particularly intrathecal and extradural anaesthesia are 'better' than general anaesthesia for high-risk, major general and vascular surgical procedures in medically compromised patients and will result in a lower mortality. One oft-quoted paper[21] demonstrated improved morbidity and mortality in high-risk surgical patients with a combined extradural and general anaesthetic technique. However, the study was small and the results have not been widely confirmed.[22]

Undoubtedly, the excellent postoperative analgesia with continuous extradural analgesia leads to a reduction in respiratory complications.[23] (Despite this reduction in respiratory complications the impairment in pulmonary function following laparotomy is not abolished with extradural analgesia.) However, it is possible that equivalent analgesia from other methods (e.g. patient-controlled analgesia) may also lead to a reduction in complications.[24] The provision of adequate analgesia may be more important than the method of analgesia employed.

In addition to the postoperative respiratory benefits from central neural blockade other potential benefits have been demonstrated:

1. The reduction in the metabolic and hormonal stress response to surgery, as described below.
2. Reduced blood loss, especially during hip surgery[25] and transurethral prostatectomy.[26] The mechanism may be a reduction in venous pressure leading to reduced venous oozing and a reduction in arteriolar bleeding from relaxation of capillary sphincters, with added contribution from a reduced mean blood pressure if present. The reduction in blood loss during caesarean section is discussed in Chapter 64.
3. A reduction in the rate of postoperative deep venous thrombosis (DVT) following surgery of the lower body has been demonstrated with intrathecal and extradural anaesthesia, particularly after hip surgery.[27] The incidence of pulmonary embolism is similarly reduced. Suggested mechanisms include increased leg blood flow, maintenance of calf muscle temperature and inhibition of coagulation and fibrinolysis.[28] There may also be a direct pharmacological effect.[29]
4. In major bowel surgery the incidence of anastomotic leaks is reduced with central neural blockade,[30]

possibly secondary to increased colonic blood flow.[31] Postoperative nitrogen balance after bowel surgery is improved by extradural anaesthesia.[32]

5. Vascular surgery is discussed in detail in Chapter 56, but a recent study demonstrated reduced cardiac and other complication rates when extradural anaesthesia was performed.[33] In addition, graft patency was improved in the extradural group owing to the effects of extradural anaesthesia on thrombotic complications and on maintenance of leg blood flow.[34]

The magnitude and duration of all these effects is probably related to the duration of the block achieved. The effect of extradural and intrathecal opioids on morbidity and mortality in high-risk surgical patients is less well studied but is likely to be less significant than following central neural block with local anaesthetics.

The evidence for a reduction in mortality with regional anaesthesia compared to general anaesthesia with good postoperative analgesia in high-risk surgical patients is not conclusive, but many anaesthetists are convinced by their clinical experience with such patients. Owing to the low mortality rate, even after major surgery, the beneficial effects on complication rates and morbidity as detailed above may ultimately be more persuasive than overall effects on mortality in deciding to choose a regional technique.

Influence of regional techniques on the metabolic response to surgical trauma

This subject is fully explored in Chapter 17. Briefly, the increase in circulating catabolic hormones and catecholamines, the negative nitrogen balance, and the sodium and water retention, all of which constitute the 'stress' response, may be influenced by neural blockade, especially central neural blockade.

Many anaesthetists believe that the increase in patient well-being following major surgery in those in whom a regional technique has been successfully employed is, in part, due to the suppression of this response. Certainly there is evidence that postoperative morbidity may be reduced by the use of regional anaesthetic techniques.[35] However, objective proof that *mortality* from surgery is improved because of suppression of the hormonal and metabolic stress response is not available. The influence of anaesthetic technique on immune function after surgery is less well studied than the effects on metabolic responses. In general, the modification of the immune response to surgery by regional compared to general anaesthesia is probably of less clinical importance than the different effects on the hormonal and metabolic changes.[36]

It must be remembered that the termination of the block means the end of the suppression of the stress response. Thus, with the possible exception of prolonged extradural blockade, the response is delayed rather than prevented.[37]

Table 34.5 Some differences between extradural and intrathecal anaesthesia

	Extradural		Intrathecal
	Caudal	*Lumbar/thoracic*	
Technique	Single injection	Single injection	Single injection
	Catheter possible (but very rare)	Catheter usual	Catheter possible (but problems)
Onset	5–20 minutes	5–20 minutes	5 minutes
Duration	Single shot: 120–240 minutes	Single shot: 120–240 minutes	150–180 minutes
		Catheter: Potentially days	
Volume (dose)	10–20 ml 25–100 mg (bupivacaine)	10–20 ml 25–100 mg (bupivacaine)	1.0–4.0 ml 5–20 mg (bupivacaine)
BP	Depends on area/spread of block Fall in BP slower than spinal but can be profound		Can be a problem Fall can be sudden and profound
Headache	Very unlikely unless accidental dural tap occurs; then likely and may need aggressive treatment		Incidence from 0 to 20%; affected by many factors, especially needle size

Central neural blockade

There are essentially two forms of central blockade: intrathecal (spinal or subarachnoid) anaesthesia and extradural (epidural) anaesthesia. Caudal anaesthesia is a form of extradural anaesthesia.

The spinal cord and surrounding structures are enclosed within the bones of the vertebral column. To provide central blockade the tip of the anaesthetist's needle needs to be in close proximity to these structures.

Successful regional anaesthesia depends largely on a sound knowledge of the anatomy and physiology of the central and peripheral nervous systems. Intrathecal anaesthesia depends on an appreciation of the physiology of cerebrospinal fluid (CSF) and the anatomy of the vertebral column. Because of the clearly defined endpoint (CSF dripping from the hub of the needle), there is a high degree of success in establishing an intrathecal anaesthetic block. Failure is, however, not entirely unknown. Intrathecal anaesthesia is probably the most widely employed regional anaesthetic technique. Extradural anaesthesia is widely utilized during labour. Some differences between intrathecal and extradural anaesthesia are shown in Table 34.5.

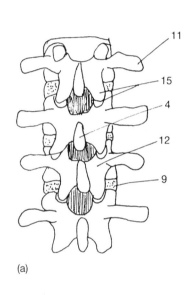

(a)

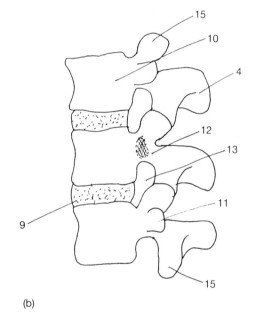

(b)

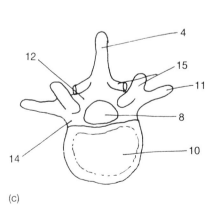

(c)

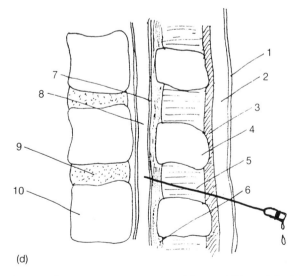

(d)

Fig. 34.1   Anatomy of the lumbar spine: (a) posterior aspect; (b) lateral aspect (transverse process removed from middle vertebra); (c) superior aspect; (d) midline section. 1, skin; 2, subcutaneous tissue; 3, supraspinous ligament; 4, spinous process; 5, interspinous ligament; 6, ligamentum flavum; 7, dura; 8, spinal canal, subarachnoid space; 9, intervertebral disc; 10, vertebral body; 11, transverse process; 12, lamina; 13, intervertebral foramen; 14, pedicle; 15, articular process.

Intrathecal anaesthesia

This block is achieved by the deposition of the local anaesthetic agent in the CSF. Figure 34.1 shows the anatomy of the spinal canal at the level of the lumbar vertebrae. Thecal puncture is performed below the level of the first lumbar vertebra because of the risk of damage to the spinal cord if puncture is attempted above this point. The final choice of which interspace to use for the lumbar puncture will often depend upon the presentation and anatomy of the patient once positioned.

Technique

Before positioning the patient, it is mandatory that reliable intravenous access be obtained with a large-bore cannula. The administration of 500–1000 ml of crystalloid solution to limit the hypotension that may result from the sympathetic block produced by the spinal anaesthetic has some merit, but great care may be required with this manoeuvre in patients with heart disease.

The choice of position will depend on a number of factors. The proposed surgery is probably the most important, although there is some evidence that the spread of the sensory block is little delayed if the patient is quickly returned to the supine position following the deposition of the local anaesthetic in the CSF. On the other hand, if the patient is kept in one position (sitting, for example), the spread of the block is limited once the local anaesthetic has become fixed to the nervous tissue. The advantage of this is that not only is the sensory block kept to the minimum required for surgery but also the sympathetic block (and therefore the hypotensive effect of the anaesthetic) can be reduced. With the patient in the sitting position the relationship of the vertebrae (and therefore the spinous processes) is more predictable, with the spinous processes more likely to remain in the midline. The practice of placing the patient in a position where the operation site is dependent may not be pleasant for the patient (fractured neck of femur, for example) and should be carefully considered. The lateral position may offer advantages over the sitting position when a unilateral procedure (e.g. total hip replacement) is proposed, or when it is inappropriate or impossible for the patient to sit for the lumbar puncture.

Sitting position

The patient should be asked to swing his or her legs over the edge of the trolley or operating table. Care may be required if premedication has been administered, and there should be assistance with this manoeuvre. The patient's feet may be rested on a stool or chair to help flex the lumbar spine. The patient should be encouraged to flex the spine and the assistant can provide support. The lower back should be examined at this time. A few moments spent on careful consideration of the most appropriate interspinous space to use for the lumbar puncture may save much time, increases the chances of successful lumbar puncture and will ensure that the procedure is as comfortable as possible for the patient.

Lateral position

The patient should be asked to turn or be assisted onto his or her side. If a unilateral procedure (e.g. total hip replacement) is proposed, the operative side should be dependent if the patient will tolerate this position. The patient should then be asked to bring the knees up towards the chest, to encourage opening of the interspinous spaces in the lumbar region. It may be helpful if a pillow or blanket is placed between the patient's legs to maintain the horizontal relationship of the vertebral bodies. Although the anatomy is the same as that for the lumbar puncture with the patient in the sitting position, the anaesthetist must be aware of the different approach and possible problems. As with the sitting position, careful consideration of the anatomy and choice of interspace will increase the chances of successful lumbar puncture.

The lumbar puncture must be conducted with a strict aseptic technique and all the equipment required should be placed on a sterile towel, preferably on a trolley. The patient's lower back should be cleaned with an antiseptic solution and allowed to dry. The anaesthetist should place a sterile towel at the base of the spine and may infiltrate the skin with local anaesthetic solution before inserting the spinal needle. There is a wide choice of needles available for the performance of the lumbar puncture from the traditional Quincke point to the more recently introduced Whitacre and Sprotte needles. The needles introduced more recently have been shown to produce a lower incidence of postdural headache and so are indicated for spinal anaesthesia in younger patients who can be mobilized early following surgery. The advantage of using these needles in older patients who may be in bed for a day or more following surgery (e.g. major joint replacement) is debatable. In the UK the choice of anaesthetic drugs available for spinal anaesthesia has been severely reduced over the last ten years. Bupivacaine 0.5 per cent in 8 per cent dextrose solution is the drug most frequently used. The plain, slightly hypobaric, solutions of bupivacaine 0.5 per cent can be used, but the spinal block that is produced can be less predictable and less dense than that produced by the hyperbaric solution. The volume of solution required will be determined by the surgery to be undertaken, the position that is adopted for the lumbar puncture and the patient's general condition (Table 34.6).

Some anaesthetists believe that the local anaesthetic solution that will be injected into the CSF should be filtered through a 20 μm filter to remove all particles of glass and other debris that may be present in the opened ampoule.

Table 34.6 Spinal anaesthesia

Site of surgery	Preferred position of patient	Volume of local anaesthetic*
Perineum	Sitting	1.0–1.5 ml
Lower limb	Lateral (then supine)	2.0–2.5 ml
Lower abdomen	Lateral (then supine)	2.5–3.0 ml
Above umbilicus	Lateral (then supine)	3.0–4.0 ml

* The volume required, bupivacaine (hyperbaric), will be affected by a number of factors apart from the position that the patient adopts during and immediately after injection. These include barbotage, height of the patient, baricity of the drug, speed of injection and level of injection.

Passage of the needle may be assisted (especially the Sprotte needle) by the use of an introducer or by a small incision in the skin with a no. 11 scalpel blade. The needle should be inserted in the midline in the middle of the interspace, with the tip pointed slightly in the cranial direction. If a needle with a bevelled end is used for the lumbar puncture, it is recommended that the bevel is parallel to the fibres of the ligamentum flavum and dura so as to part these fibres rather than cut them as the needle is advanced. The fibres of the ligamentum flavum run in the craniocaudal direction as do most of the fibres of the dura. Inserting the needle in this fashion reduces the chances of a post-dural puncture headache.[38] As it is pushed towards the intrathecal space the needle passes through the subcutaneous tissues, the supraspinous ligament, the interspinous ligament, the ligamentum flavum and, finally, the dura. Often there is a characteristic 'give' when the needle passes through the dura. The appearance of CSF at the hub of the needle (this may take some time if a very fine needle is used), when the needle stylet is removed, confirms that the tip of the needle is in the correct position and the filtered local anaesthetic solution may be injected. It is good practice to check that the tip of the needle has not migrated by aspirating CSF at some point during the injection. If there is doubt regarding the authenticity of the fluid, the use of a bedside glucose testing strip may be helpful.

Possible problems

Unfortunately, not every attempt at lumbar puncture is successful and there are several possible problems that may be encountered.

Incorrect direction of the needle is a common cause of failure. The needle can pass in the wrong direction and either miss the spinal canal completely or be stopped by one of several bony parts of the vertebral body. Withdrawal of the needle, reappraisal of the anatomy and a change of direction (sometimes surprisingly small) may be all that is required. Repositioning of the patient and change of site to

another interspace may be needed in other cases. Rarely, it can be impossible to locate the intrathecal space. The anaesthetist should be wary of many attempts at different interspaces, as the patient's comfort and dignity should be respected.

The various ligaments that the needle passes through can become calcified in older patients and it can be difficult to push a thin needle through ligaments so affected. The use of a spinal needle introducer or, if unavailable, a larger needle may help. Post-dural puncture headache is, of course, more likely if a larger needle is used for the lumbar puncture.

Complications of intrathecal anaesthesia

Hypotension

Approximately one-third of patients will suffer from hypotension as a result of the intrathecal anaesthetic;[39] this is more likely in the older patient, in patients with higher resulting blocks (T5 and above) and in higher lumbar puncture sites. The definition used by some – systolic less than 90 mmHg – is clearly challengeable, and it may be more appropriate to consider a change of 30 per cent from the preoperative blood pressure as a more practical level at which to consider treatment.

Whilst the treatment of hypotension will depend on whether a bradycardia is present at the same time, the initial management should be:

1. Increase the inspired oxygen concentration to at least 40 per cent.
2. Increase the rate of administration of the intravenous fluids. Often a bolus of 500–1000 ml of crystalloid will restore the blood pressure.
3. Look for a cause (other than the block) for the fall; for example, blood loss, packs and retractors in abdominal surgery.
4. Consider a change in patient posture (e.g. raise legs or head-down tilt), although this may not be possible if surgery has commenced.
5. Vasopressors.

Ephedrine is the traditional first choice of vasopressor. Administered intravenously in a dose of 3–6 mg, the effect of this drug is more a β-adrenergic agonist than an α agonist and produces an increase in heart rate rather than an increase in peripheral resistance. This is advantageous in pregnancy when uterine blood flow is important. Methoxamine has exclusive α-adrenergic agonist activity and increases peripheral resistance. It may restore pressure at the expense of flow but is preferable to ephedrine when a tachycardia is present. Caution should be exercised when the heart rate is normal, as a reflex bradycardia may follow.

The initial dose is 1–2 mg. Methoxamine has a longer duration of action than ephedrine.

Bradycardia

Bradycardia is more likely with a high block (T5 and above), in patients with a normal heart rate less than 60, ASA 1 patients and patients on β-adrenergic antagonist drugs. If the heart rate falls to less than 60 and certainly to 50 beats per minute, treatment should be considered – especially in the elderly and in patients with heart disease. Intravenous glycopyrrolate 0.3 mg or atropine 0.3 mg should be administered. A tachycardia is more likely with atropine. The circulation time may be much delayed by the bradycardia and patience may be required to avoid giving further, unnecessary, doses of the drugs. If the hypotension and bradycardia prove resistant to treatment, it has been recommended that intravenous adrenaline be considered sooner rather than later.[40]

Nausea and vomiting

Nausea and vomiting are commonly associated with hypotension, bradycardia and a high sensory block. Treatment of the cardiovascular problems may not relieve these symptoms in every case and the use of conventional antiemetics can sometimes be unrewarding.

Post-dural puncture headache

Headache has always been recognized as a side effect of dural puncture and therefore of intrathecal anaesthesia. The aetiology of the headache is believed to result from leakage of CSF both at the time of the dural puncture and, probably more importantly, continuing leak afterwards. Factors known to increase the likelihood of post-dural puncture headache include the size of needle used for the dural puncture (the larger the needle, the higher the incidence), the age of the patient (younger patients more likely to have headache than are older patients) and early ambulation.[41] Newer needle designs (Sprotte and Whitacre) have a lower incidence of headache, especially in the higher risk groups, and their use should be considered in the obstetric population. Needles of very small diameter (27 to 32 s.w.g.) have been produced but practical difficulties have limited their use. Management with periods of bed rest and high fluid intake to prevent the development of the headache is disappointing.

Characteristically the headache is throbbing in nature, eases quickly on lying down and returns on standing. It is unusual for the headache to present more than 48 hours after the lumbar puncture. The management of the post-dural puncture headache can be conservative (bed rest, adequate analgesia and a fluid intake of 1.5–3 litres daily)

but is not successful in all cases. A severe post-dural puncture headache may render the patient bedbound and merits more aggressive and invasive treatment, and conservative management should be abandoned if it is ineffective after 24 hours. Extradural infusion of Hartmann's solution is effective both for treatment and for prophylaxis (especially when associated with an inadvertent dural puncture by a Tuohy needle).

The most reliable and effective method of treatment for post-dural puncture headache is the autologous extradural blood patch. Twenty ml of blood is removed aseptically from the patient and this blood then injected into the extradural space. It is usual for two doctors to be involved in the manoeuvre, one to remove the blood and the other to perform the injection. The headache usually disappears within minutes of the injection, with an 89 per cent success record and a good long-term safety record. If the patient is pyrexial it is inadvisable to use the technique.[42,43]

Neurological sequelae

These are extremely rare[44] but if any neurological problem is identified it should be investigated and documented thoroughly and a neurological opinion sought. Urgent neurosurgical treatment is required if the cause is a space-occupying lesion in the spinal canal (haematoma, abscess or tumour).

High block (total spinal)

Prediction of the final height of an intrathecal anaesthetic depends on a number of factors,[45] including the position of the patient (both at the time of the injection and afterwards), the volume of local anaesthetic injected, the speed of the injection and the age, height and sex of the patient.

Despite consideration of these known variables and their influence over the final height of the block, problems can develop. These relate mainly to the cardiovascular system (hypotension and bradycardia), which may require very aggressive management with intravenous fluids and vasopressors. If the block progresses to the cervical area, there may be respiratory difficulty which may, as a result of phrenic nerve paralysis, require intubation and ventilation. The ventilation should continue until the intrathecal block regresses, which may take some hours. It should be remembered that, although patients are unable to breathe spontaneously, they may be conscious and so appropriate steps should be taken to ensure that distress is kept to a minimum.

Urinary retention

Urinary retention can be a problem, especially in the elderly male patient and in patients who have had a large volume of intravenous fluid as either prophylaxis or treatment for hypotension. The loss of bladder sensation is transient and recovers when the intrathecal anaesthetic recovers, but the sympathetic block produced may regress before the return of bladder sensation and some patients may require catheterization within this time.

Extradural anaesthesia

The techniques of extradural anaesthesia and analgesia have become widespread following their introduction and acceptance by patients and staff in the labour suite and obstetric theatre. The anatomy for the lumbar and (to some extent) the thoracic extradural techniques is the same as that for intrathecal anaesthesia. The anatomy for caudal anaesthesia is, however, different and is described later in the section dealing with this technique. Spinal nerve roots pass through the extradural space, and anaesthesia is obtained by injecting local anaesthetic into that space.

Indications for extradural anaesthesia

The indications for extradural anaesthesia are similar to those for intrathecal anaesthesia. It can also be used to provide good-quality postoperative analgesia by intermittent injection or by continuous infusion of the local anaesthetic solution through an extradural catheter. (For further information on extradural analgesia and postoperative analgesia, see Chapter 42.)

Extradural anaesthesia may have an advantage over intrathecal anaesthesia in that it can be added to during the course of the procedure by injections through a catheter, although some anaesthetists also employ the catheter technique for intrathecal anaesthesia.

Technique

For the technique to be a success the tip of the anaesthetist's needle must be placed accurately in the extradural (epidural) space. Whilst there is a definite endpoint to the technique (such as the loss of resistance) it is not always easy to be certain and failure does occur.

The applied anatomy is similar to that of intrathecal anaesthesia, except that the needle stops short of the dura in the extradural space. The local anaesthetic is given either by a single injection or by multiple injections through a catheter.

Care with the positioning of the patient is vital. The awake patient can be positioned sitting or in the lateral position for insertion of the needle. If anaesthetized, the lateral position is used. Reliable large-bore intravenous access should be established before extradural anaesthesia

is attempted. The use of intravenous preloading with 500–1000 ml of crystalloid solution is common practice.

Positioning the patient

The same considerations for positioning discussed for intrathecal anaesthesia pertain to extradural anaesthesia. Again, the lateral position may offer advantages over the sitting position. Unlike intrathecal anaesthesia, the choice of side to lie on is probably not important. If the patient is anaesthetized, an assistant will be required to flex the spinal column.

Extradural anaesthesia must be conducted with a strict aseptic technique. If the patient is awake, the skin and subcutaneous tissues should be infiltrated with local anaesthetic solution (such as lignocaine) before the needle is inserted, as the needle used (Tuohy) is large. Patients should be warned that they may be aware of insertion despite the generous use of local anaesthesia.

Extradural anaesthesia and analgesia can be commenced via an entry point above the level of the first lumbar vertebra. The choice of site for introduction of a catheter into the extradural space is influenced largely by the site proposed for surgery. Ideally, the tip of the extradural catheter should be located in the middle of the area to be rendered anaesthetic. There is a danger during thoracic extradural anaesthesia of direct damage to the spinal cord if the needle accidently punctures the dura. In general, the higher up the vertebral column, the larger the cord becomes and therefore the greater the risk. Some controversy exists as to whether or not the patient should remain awake while a thoracic extradural needle is inserted.[46]

Lumbar extradural technique

Once the sterile field has been established and local anaesthetic infiltrated, a small incision in the skin should be made to ease the passage of the Tuohy needle through the skin. The needle should be inserted in the midline for lumbar extradural anaesthesia, although the lateral approach (described in the next section, on thoracic extradural technique) can be considered if difficulty is encountered with the midline approach.

The needle passes through the subcutaneous tissues, the supraspinous ligament and the interspinous ligament and reaches the ligamentum flavum. Often there is a characteristic feel to the needle at this point as it is gripped by the tough fibres of the ligamentum flavum. As the average distance from the skin to the extradural space is 4–5 cm (although this may be as little as 3 cm in a thin subject), once the needle has been introduced to this depth, care should be taken because the ligamentum flavum can be soft in some individuals and the gripping of the needle may not be so obvious.

Once the ligamentum flavum has been identified the stylet should be removed and a 5 or 10 ml syringe containing air or sterile 0.9 per cent saline should be attached to the Tuohy needle. It is (almost) impossible to inject into the ligamentum flavum with a syringe attached to the needle at this point and this phenomenon is central to the 'loss of resistance technique' for identification of the extradural space. Some anaesthetists recommend a well-lubricated glass syringe, others an ordinary plastic syringe and yet others the use of specially designed plastic 'loss of resistance' syringes.

Regardless of whether air or saline is used, once the syringe is attached to the Tuohy needle, the plunger is advanced slowly and carefully until it is possible to inject down the needle. This sudden loss of resistance to injection is usually very obvious and the tip of the needle should now be in the extradural space.

The syringe can now be removed from the needle and either a single slow injection of local anaesthetic made through the needle or a plastic catheter can be fed into the extradural space.

The catheter should pass easily into the extradural space. If there is resistance and the catheter has not passed the tip of the epidural needle, the catheter can be withdrawn. However, if any length of catheter has passed the tip of the needle, the catheter should not be withdrawn as there is a small chance that this manoeuvre will lead to part of the catheter shearing off and being left behind in the extradural space. The catheter and needle must be withdrawn together.

If difficulty is encountered in passing the catheter, a number of manoeuvres can be tried. First, one may reattach the loss of resistance syringe and attempt to reinject air/saline. If the injection is easy, try again. If there is difficulty, it may be necessary to advance the needle a small distance to obtain easy injection. Try again. Having the patient inhale deeply during the attempt to advance the catheter may be helpful. If the catheter still cannot be passed, it may be necessary to advance the needle again, as only part of the bevel tip of the needle may be in the extradural space, thus preventing the catheter from advancing. If this does not result in successful catheter passage, it may be prudent to remove the needle and try another interspace.

There is no advantage in passing more than 2–4 cm of catheter into the extradural space.

Before securing the catheter to the skin with sterile gauze swab(s) and plastic waterproof adhesive tape, the anaesthetist must aspirate the catheter to ensure that no blood or CSF is obtained. If CSF is obtained, the catheter cannot be used and must be removed. If accidental dural puncture has taken place, the anaesthetist should try to secure passage of a catheter in an adjacent interspace to assist in the prevention of a post-dural puncture headache.

Saline vs air for loss of resistance

Air is compressible whereas saline is far less compressible. The identification of the extradural space is therefore more convincing when saline is used for the loss of resistance. However, saline may flow out of the needle when the syringe is removed and mimic a dural tap. The flow of saline that may take place is usually transient and will feel cold on the back of the ungloved hand whereas CSF will feel warm. There should be no difficulty in correctly identifying a dural tap, especially when this occurs in the sitting position; the flow of CSF that follows is usually obvious and continuous. If there is any doubt, a bedside glucose analysis strip will change colour if the liquid is CSF. Obviously, if air is used for the loss of resistance test, there is a very high probability that any fluid appearing is cerebrospinal and a dural tap has occurred.

After the catheter has been secured to the skin, the first dose of local anaesthetic may be injected into the extradural space. The advantage of a test dose is that, if a small volume of a mixture of local anaesthetic and adrenaline is injected (e.g. 4 ml of 2 per cent lignocaine and adrenaline 1/200 000), an intravascular injection will be indicated by the appearance of a tachycardia. A sudden drop in blood pressure (with sensory loss as in intrathecal anaesthesia, if the patient is awake) may indicate that the tip of the catheter lies intrathecally. However, it is important that a false sense of security does not develop because the test dose had no adverse effects. Although it is rare, it is not unknown for extradural catheters to migrate into blood vessels and intrathecally after unremarkable response to test doses. The anaesthetist must be vigilant at all times.

Selection of the therapeutic dose and choice of local anaesthetic will be influenced by a number of factors. If the primary function of the catheter is to provide postoperative analgesia, the anaesthetist may decide to use bupivacaine 0.25 per cent with or without an opioid analgesic drug. If the primary purpose is anaesthesia for a short procedure (e.g. repair of a hernia), 15–20 ml of lignocaine 1.5 per cent with adrenaline may be appropriate.

Thoracic extradural technique

For upper abdominal surgery and thoracic surgery the thoracic route can be considered for extradural anaesthesia and analgesia. The advantage of placing the catheter higher in the vertebral canal is that a lower volume (and lower dose) of local anaesthetic may be needed because the tip of the catheter will be nearer the central area of the sensory innervation of the incision. However, there are disadvantages. The incidence of cardiovascular side effects can be higher, and placement of the catheter can be more difficult, because of the more acute angulation of the spinous processes of the vertebral bodies higher up the spine.

Technique

Patient positioning and preparation are as for the lumbar approach. The needle is introduced through the skin at the level of the appropriate thoracic spinous process. Correct identification of the spinous processes at this level can be difficult and it may be easier to count down from the seventh cervical spinous process. With the midline approach to the extradural space the needle will have to be angled far more cranially than for the lumbar approach. The ligamentum flavum is also narrower in the thoracic region and great care must be taken to prevent dural puncture.

There is a case to be made for thoracic extradural catheterization to be performed with the patient awake to minimize the chances of direct spinal cord damage, as needle contact with the spinal cord will be instantly recognized.[46] However, having an anaesthetized patient, who will remain still, may be an advantage.

If the midline approach is unsuccessful, the lateral (or paramedian) approach may be successful. This avoids the difficulties presented by the angles of the thoracic spinous processes and calcified interspinous ligaments. Some anaesthetists consider the paramedian approach to be the preferred method for performing thoracic extradural blockade. The needle is introduced about 1 cm from the midline and deliberately directed into the lamina of a thoracic vertebra. The depth at which this occurs should be noted. The needle is then withdrawn about 1 cm and redirected in a cranial and medial direction until the ligamentum flavum and the extradural space are identified in the usual fashion. If the needle were to continue in the original direction, the extradural space might be missed altogether or approached in its lateral aspect, thus increasing the chances of a unilateral block.

Complications of extradural anaesthesia

Dural puncture

A dural puncture, if unrecognized, can place the patient at great risk. Large volumes of local anaesthetic injected intrathecally have rapid and dramatic effects on the cardiovascular and respiratory systems. The diagnosis is usually not in doubt (see above). The management of a dural tap involves reintroduction of an extradural catheter at another level and use of this catheter to provide anaesthesia and/or analgesia. Care is required with monitoring these patients as there is a possibility of catheter migration and spread of local anaesthetic into the CSF through the dural hole. If the local anaesthetic can be administered continuously for 24 or 48 hours, the incidence of severe headache will be low, especially if the patient is well hydrated.

If a post-dural puncture headache does develop despite the above management or if an extradural catheter cannot be placed at the time of the puncture, a blood patch should be considered (see earlier, 'Complications of intrathecal anaesthesia').

Blood vessel damage

It is unusual for the Tuohy needle to damage blood vessels in the extradural space, but damage is more likely when the catheter is passed. Damage is more likely when intra-abdominal pressure is raised (ascites, large tumours); the catheter should not be passed during uterine contractions when extradural anaesthesia is attempted in labour, as the extradural veins are dilated as blood is squeezed out of the uterus. If blood flows back through the catheter or is aspirated from the catheter, the catheter should be pulled back 1–2 cm and aspirated again. If there is still blood aspirated, the catheter should be removed and another interspace used. Provided the patient has normal coagulation the chances of any serious sequelae are very small.

Hypotension

The sympathetic blockade that is provided by the extradural blockade can result in hypotension in the same fashion as during intrathecal anaesthesia. The management is similar, with the careful use of intravenous crystalloid and colloid, and the use of vasopressors if needed.

Urinary retention

Urinary retention may be more likely than with intrathecal anaesthesia, as many extradural catheters are left *in situ* to provide postoperative analgesia. Incidence may be reduced by the use of minimum volumes to provide analgesia and lower concentrations of local anaesthetic drugs (e.g. 0.125 per cent bupivacaine). The use of lower concentrations with opioid drugs also is associated with problems, as extradural opioids are associated with urinary retention.

Neurological sequelae

Neurological complications are rare. If suspected or identified, they must be approached and managed aggressively.

Catheter problems

Very rarely, an extradural catheter is difficult to remove. Although the catheters available now are manufactured to high standards and are very durable, it is still possible for them to be damaged on removal. Cases have been reported in which several centimetres of catheter have been left in a patient's back. If difficulty is encountered despite spine flexion, spine extension may sometimes make removal easier. If there are problems and a piece of the catheter is left behind, there is little chance of any long-term sequelae. There is debate as to whether the patient should be informed.

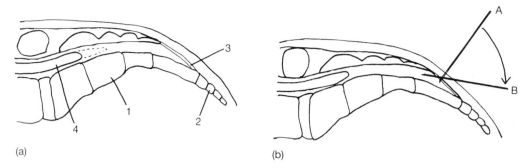

Fig. 34.2 Caudal extradural technique. (a) 1, sacrum; 2, coccyx; 3, sacrococcygeal membrane; 4, dural sac. (b) A, initial angle of needle for entry; B, position of needle for injection of anaesthetic.

Backache

Recent work[47] has raised the issue that there may be a significant incidence of backache following extradural anaesthesia and analgesia during labour. There has, however, been some criticism of the methods used in this study, and further work is awaited. There are no data to support the same conclusions in the non-obstetric population.

Caudal anaesthesia and analgesia

Any surgical procedure on the perineum, anus, male and female genitalia and lower urinary tract can be performed under caudal anaesthesia and can benefit from the postoperative analgesia that the caudal may provide.

The anatomy is outlined in Fig. 34.2. A needle inserted through the sacrococcygeal membrane will lie in the most distal part of the extradural space, and injection of local anaesthetic solution will provide anaesthesia and analgesia in the same way as an injection in the lumbar or thoracic region. It is unusual for catheter techniques to be adopted because the lumbar extradural site is easier to secure and nurse postoperatively. Unfortunately, there is more variety in the normal anatomy than in other parts of the spinal column and the technique is not always easy.

Technique

The patient can be placed in either the prone or, more commonly, the lateral position for the injection. An ordinary 21 s.w.g. needle can be used. For the lateral approach the patient should be placed on the side and positioned with the legs drawn up into the abdomen. In the obese patient the upper buttock can overlie the midline and make palpation of the coccyx difficult.

The sacral hiatus can usually be palpated by identifying the coccyx at the base of the spine and carefully moving cranially. The sacral cornua lying laterally will confirm the location of the hiatus. Once identified the skin should be throughly cleaned and sterile towel(s) placed to protect the area. If the patient is awake, local anaesthetic can be infiltrated into the skin and subcutaneous tissue. Large volumes of infiltration should be avoided, as identification of the landmarks may be made difficult. The needle should be inserted through the hiatus and the membrane at right angles. Once the ligament is penetrated, some anaesthetists advance the needle until bone is struck, then withdraw the needle slightly and angle the needle up the sacral canal; others recommend that as soon as the membrane has been punctured the needle angle should be changed and directed up the canal. Once in place, the local anaesthetic may be injected after careful aspiration to check for blood and CSF (rare). If blood is aspirated, a slight withdrawal of the needle and reaspiration will usually rectify the problem.

The needle may be misplaced subcutaneously, causing a swelling on injection. Periosteal injection should be easily recognizable because the injection will be very difficult to accomplish. Repositioning the needle should correct these problems.

In the adult 20 ml of local anaesthetic solution should produce a good block of the sacral and some lower lumbar nerves. On occasions the block may extend higher.

Complications

Dural puncture

Some workers recommend abandoning the caudal approach and considering alternative sites of injection. The possibility of post-dural puncture headache should be remembered.

Hypotension

Hypotension is uncommon unless the block extends to a high lumbar level. Standard management of hypotension caused by central neural blockade as outlined above should be adopted.

Urinary retention

As with intrathecal and lumbar extradural blockade, urinary retention may be troublesome and may require catheterization to relieve the problem. This plus the potential lower limb weakness that may be produced has led to individual anaesthetists searching for alternative methods of postoperative analgesia (e.g. penile block).

Peripheral regional blocks

Simple nerve blocks

To describe the following blocks as 'simple' is not to decry the significant effects that they can have for minor surgery and also in promoting good postoperative analgesia, especially for day-case surgery. Simple surgery on the digits can be performed under 'ring' block of the digital nerves or the block can be performed for postoperative analgesia. Each digit is supplied by two dorsal and two palmar or plantar nerves running along the digit. All four branches need to be blocked for successful analgesia. Small digits will probably be blocked by instilling 2 ml of anaesthetic containing *no* vasoconstrictor on each side of the base of the digit using a fine, short needle. Larger digits will benefit from 1–2 ml injected both superficially and deeply on both sides of the digits to ensure that all four branches are reached. Care should be taken and smaller volumes used if the digit is already traumatized, as the circulation may be more readily compromised.

Simple infiltration along the wound margins is remarkably effective in promoting postoperative analgesia after surgery[48–50] although it may not be as useful following major abdominal surgery.[51] The only limitation is from the maximum safe dose according to the patient's weight. A prolonged effect may be obtained from local anaesthetic irrigation via a catheter twice a day.[52] As to whether infiltration performed prior to incision reduces the amount of pain compared with infiltration after skin suture (the concept of pre-emptive analgesia[53]), the results are contradictory and futher study is needed.

Many anaesthetists routinely infiltrate lignocaine prior to venous cannulation with large cannula, an approach recently supported by a study showing significant reductions in pain scores.[54]

Intercostal blocks

The three sensory branches of each intercostal nerve are:

1. A posterior cutaneous branch supplying skin and muscle in the paravertebral region
2. A lateral cutaneous branch, arising anterior to the midaxillary line, supplying skin on most of the thorax and abdominal wall

3. An anterior cutaneous branch supplying the front of the thorax, the sternum and front of the abdominal wall.

The areas of innervation of each nerve undergo significant overlap both vertically and across the midline. One must remember that the neurovascular bundle lies in the costal groove behind the inferior margin of the rib. The nerve lies below the artery and vein respectively.

Intercostal blocks are generally performed at either the angle of the rib or in the anterior axillary line. For blockade at the angle of the rib, just lateral to the sacrospinalis muscle, the patient is turned either three-quarters or fully prone. At the anterior axillary line the block can be performed with the patient supine or on the opposite side. There is evidence that distal spread along the rib may occur whichever site of injection is chosen.[55] Adjacent ribs may also be blocked by one large injection,[56] implying perhaps subpleural spread. At the angle of the rib there may also be spread into the extradural space via the paravertebral gutter.[57]

For both approaches the cardinal rule when performing an intercostal block is that one must be able to palpate the rib; otherwise, the chance of a successful block is low but the chance of pneumothorax is high. However, if the rib is palpated and the technique is sound, the success rate is very high. When blocking multiple ribs, most practitioners start with the most inferior rib and work upwards. The safest technique which allows full control of the point of the needle at all times is as follows (see also Figure 34.3):

1. Standing alongside the patient, the uppermost hand (for most people the left hand while standing to the patient's right) firmly retracts the skin superiorly. The other (right) hand then places the point of the needle with attached syringe directly onto the rib.
2. As the skin retraction is slackened, the needle 'walks' off the rib. As the needle slides off the rib, the point is advanced 2–3 mm only and 3–5 ml of anaesthetic is injected. The aspiration test must be scrupulous in this very vascular area.
3. After injection, the needle is placed back on the rib just blocked while the upper hand palpates the next rib. This is necessary as otherwise it is surprisingly easy to block some ribs twice and others not at all!
4. The procedure is repeated as often as desired. For the upper ribs the point of injection should be more medial, with the shoulder abducted in order to avoid the scapula.
5. There is no need to change sides when performing bilateral blocks.

Intercostal blocks are indicated for pain relief following fractured ribs and for intra- and postoperative analgesia for operations on the abdominal wall (e.g. cholecystectomy) and chest wall (e.g. mastectomy). For prolonged analgesia

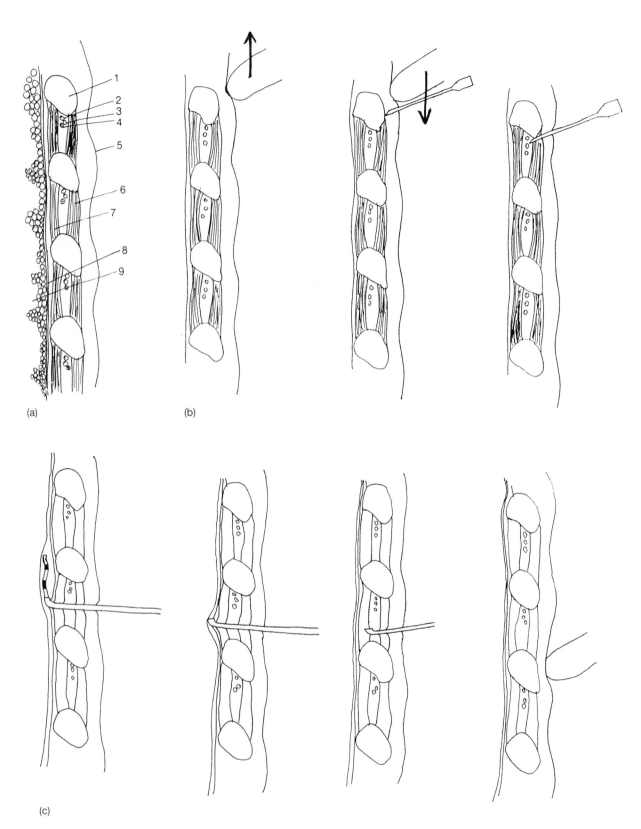

Fig. 34.3 (a) Intercostal space in the mid-axillary line. 1, rib; 2, intercostal vein; 3, intercostal artery; 4, intercostal nerve; 5, skin; 6, external intercostal muscle; 7, internal intercostal muscle; 8, parietal pleura; 9, lung. (b) Intercostal nerve block (see text for details). (c) Intrapleural catheter technique (see text for details).

after rib fractures, a catheter may be inserted.[58] For an indication of the extent of block required, a dermatome chart should be consulted. Bilateral blocks are required for abdominal surgery. Less commonly considered but very effective is the use of intercostal blocks for lower abdominal surgery (e.g. appendicectomy).[59] The role of T12 block for analgesia after hip surgery is discussed below. Some practitioners combine bilateral intercostal block with coeliac plexus block for complete analgesia for intra-abdominal surgery.

Pneumothorax is not as common a complication as one might expect (as long as the technique is sound) – a reported incidence of 0.073 per cent.[60] Blood levels of anaesthetic are high after intercostal blocks and adrenaline-containing solutions may be indicated in view of the vascularity of the area.

Intrapleural catheter block

This technique was first reported in 1984[61] for management of postoperative pain and has been used sucessfully for post-traumatic pain, especially in the management of fractured ribs.[62] The expected pain should be in the distribution of the intercostal nerves that will be bathed by the local anaesthetic agent once it has been injected into the pleural cavity. This effectively means that the area from the second to the tenth intercostal nerves can be reliably blocked. As bilateral block is not recommended, midline incisions will require alternative methods of pain relief.

Special contraindications include obesity, which renders palpation of the ribs difficult or impossible, and previous lung disease that increases the likelihood of pleural adhesions.

When the technique is to be used for postoperative pain management the catheter may be inserted once the patient is anaesthetized but generous local infiltration with lignocaine may be needed if the patient is awake while the catheter is inserted.

Technique
(see Fig. 34.3)

The patient should be positioned in the lateral position with the affected side uppermost. As with the intercostal block, it is mandatory that the ribs are palpable for this technique. Aseptic technique is also mandatory. In the anterior axillary line at the level of the third to the sixth ribs, depending upon which is best palpable, a Tuohy needle is inserted superior to the rib with the needle angle slightly upwards. A small incision with a scalpel blade will make insertion of the needle easier and offer better control of the needle as it passes through the chest wall. The negative pressure that exists inside the pleural space can be utilized to safely identify the endpoint in the insertion of the needle (the loss of resistance technique is not recommended).

Once the needle tip has passed into the subcutaneous tissues, a well-lubricated 10 ml glass syringe filled with sterile saline solution is attached to the needle. The two are then advanced slowly through the remaining chest wall. Artificial ventilation should be suspended at this point until the catheter has been inserted, to minimize the chances of damage to the visceral pleura. If the patient is awake and is able to do so, he or she should be asked to breathe out and remain in expiration until the needle is in the correct position. Once the posterior intercostal membrane is punctured (often with a 'click') the parietal pleural will be punctured shortly afterwards and the barrel of the syringe pulled down by the negative intrathoracic pressure.

The syringe should be removed and, as quickly as possible, an extradural catheter inserted into the pleural space. At least 10 cm of catheter should be inserted before removing the Tuohy needle over the catheter. The catheter should then be covered with a sterile swab and waterproof tape dressing. After negative aspiration of the catheter for air or blood, 15–20 ml of 0.5 per cent bupivacaine with adrenaline can be injected slowly over 5 minutes. Pain relief from a single injection can last for 6–9 hours. A continuous infusion of local anaesthetic, at a rate of 5–10 ml an hour, can provide longer-lasting pain relief which is particularly effective with unilateral fractured ribs. When used for postoperative analgesia the catheter can be left for 48 hours, but longer periods may be considered if required.

The catheter can also be placed in the thorax and passed out from the inside of the chest at thoracotomy by the surgeon.

Complications and side effects

Rapid injection of local anaesthetic may lead to high or toxic plasma levels of the local anaesthetic. Lack of attention to detail may permit air to enter the pleural cavity during insertion of the catheter. Certainly, small amounts may be introduced if air is used instead of saline in the glass syringe. These small pneumothoraces can be demonstrated radiologically but, provided there has been no damage to the lung or visceral pleura, they will resolve spontaneously.

The significance of the introduction of a small amount of air at the beginning of a general anaesthetic that features nitrous oxide is also a matter of debate and novel methods have been described to minimize its occurrence.[63] In theory, the visceral pleura can be damaged by the Tuohy needle,[64] and a pneumothorax requiring underwater seal drainage may be required.

Inguinal block

Traditionally employed for postoperative analgesia following surgery for an inguinal hernia, this block may be useful for other operations for which the incision lies along the L1 dermatome (e.g. lower segment caesarean section[65] and hysterectomy).

The inguinal region is supplied by the ilioinguinal and iliohypogastric nerves arising from T12, L1 and L2. The medial end of the lower part of the inguinal canal is supplied by the genital branch of the genitofemoral nerve arising from the same spinal levels. The ilioinguinal and iliohypogastric nerves may easily be blocked by generously instilling up to 20 ml of anaesthetic between the external and internal oblique abdominal muscles. The point of injection is approximately 2 cm medial and 2 cm inferior to the anterior superior iliac spine. When a short-bevelled needle is used, a definite 'click' will be felt as the appropriate tissue plane is identified. If no click is identified, the solution should be infiltrated in a 'fan-wise' manner through the abdominal layers in several directions from the point of entry.

Some specialists routinely block the genital branch of the genitofemoral nerve for hernia surgery by a similar 'fan-wise' infiltration of 10 ml of anaesthetic at the inferior end of the inguinal canal. However, others do not, as postoperative analgesia seems effective in its absence. Midline infiltration of local anaesthetic may be required if an overenthusiastic surgeon has extended the incision across the midline.

Penile blocks

In many centres penile blocks are increasingly employed for intraoperative and postoperative analgesia for surgery of the penis. The technique is simple and may be preferred to caudal analgesia owing to the lack of effect on motor power and avoidance of urinary retention. The dorsal nerve of the penis is a terminal branch of the pudendal nerve and passes under the symphysis pubis to pass along the penis on the dorsal surface of the corpora cavernosa. At the symphysis pubis a branch is given off to supply the ventral surface and frenulum. A commonly used technique is to 'walk off' the symphysis pubis with a fine, short needle and deposit 5–10 ml of 0.25 per cent bupivicaine (less in infants and children) approximately 5 mm below the bone.[66] Occasionally, a click will be felt as Buck's fascia is penetrated. Other practitioners inject half the total dose on each side of the midline at an angle of approximately 15 degrees. A third technique is to perform a superficial, circumferential block around the base of the penis.

Complications mainly arise if the dorsal vessels are punctured or the corpora cavernosum entered, leading to failed block, intravascular injection or haematoma formation. Haematoma or excessive volumes of anaesthetic can result in ischaemia from compression. A total dose of 10 ml in adults should not be exceeded, to minimize the risk of compression whichever technique is employed. The use of vasoconstrictor-containing anaesthetics at the base of the penis must be avoided because spasm of the vessels and profound ischaemia may ensue. A recent report recommended the performance of a caudal block if such accidents occur, to promote sympathetic block and vasodilation.[67]

Cervical plexus block

The anterior primary rami of C2, C3 and C4 unite to form the cervical plexus as they leave the intervertebral foramina and pass between the anterior and medial scalene muscles posterior to the sternocleidomastoid muscle approximately half way down its length. The superficial plexus is easily blocked by a fan-wise injection of 10 ml of anaesthetic at the point where the external jugular vein crosses the posterior edge of the sternocleidomastoid muscle. This will provide surface analgesia in the distribution of the C2–C4 nerve roots – that is, the neck above the clavicle, upper shoulder area and a lateral area of the cranium around the ear.

Both the superficial and deep plexuses may be blocked by injecting 2–3 ml of anaesthetic at each transverse process of the cervical vertebra or, less reliably, by injecting a larger volume of 6–10 ml at just one site. (Anaesthetics containing vasoconstrictors should not be used because of the proximity of the vertebral artery and other vessels.) This will provide deeper analgesia, for example, for thyroid surgery or tracheostomy. Deep cervical plexus block will also cover the posterior rami, providing analgesia around the back of the neck and occiput. However, extreme caution must be exercised because of the risk of both vascular and dural puncture. Convulsions will be produced by injection of even very low doses of local anaesthetic into the vertebral artery.

Bilateral blocks may be complicated by respiratory distress, as the stellate ganglion, recurrent laryngeal nerve and phrenic nerve can all be affected during deep cervical plexus block. Cervical plexus block is a recognised 'complication' of the interscalene approach to the brachial plexus.

Brachial plexus block

Brachial plexus block is a valuable addition to a general anaesthetic for surgery of the upper limb or a suitable alternative to general anaesthesia in unfit patients. The

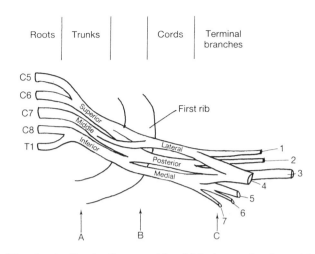

| Roots | Trunks | | Cords | Terminal branches |

Fig. 34.4 Simple diagram of brachial plexus, showing roots, trunks, divisions, cords and the important terminal branches. For simplicity, high branches have been omitted. 1, musculocutaneous nerve; 2, axillary nerve; 3, radial nerve; 4, median nerve; 5, ulnar nerve; 6, medial cutaneous nerve of forearm; 7, medial cutaneous nerve of arm. A, interscalene approach; B, supraclavicular approach; C, axillary approach.

anatomy of the brachial plexus (Fig. 34.4) is complex and should be referred to in a standard anatomy textbook for more detailed description. Successful blockade of the branches of the plexus relies on the fact that these branches are enveloped in a tubular sheath of fascia. Thus if one branch is identified by eliciting paraesthesiae or by using a nerve stimulator, and a reasonably large volume of anaesthetic is injected, blockade of the entire plexus may be predicted. Only the three main approaches to the plexus are discussed here: interscalene, supraclavicular and axillary. For a more complete discussion the reader should refer to the textbook by Winnie (see 'Further reading').

Note. Owing to the large volumes of anaesthetic required, many anaesthetists attach an extension set to the needle so that the needle point may be held still during injection, syringe removal and syringe reattachment.

Interscalene approach

The roots of the plexus exit from their intervertebral foramina and pass laterally and inferiorly between the anterior and middle scalene muscles, uniting to form three main trunks. The interscalene groove can usually be palpated posterior to the sternocleidomastoid muscle. With the patient's head turned to the opposite side, the needle tip should be inserted at the level of the transverse process of the sixth cervical vertebra (the level of the cricoid cartilage – often the level where sternocleidomastoid is crossed by the external jugular vein) in a slightly posterocaudal direction until paraesthesiae or muscle contractions with the nerve stimulator are elicited. The transverse process is often only about 2 cm below the

skin, and care should be taken to prevent damage to deeper structures. A large volume of up to 40 ml of anaesthetic is required to ensure blockade of the entire plexus.[68] Lower volumes may result in sparing of the lower trunk (i.e. chiefly the ulnar nerve). Cervical plexus block is also achieved, resulting in analgesia of the shoulder region.

Of all the approaches to the brachial plexus, the interscalene approach is most commonly associated with blockade of:

1. the phrenic nerve, resulting in unilateral diaphragmatic paresis reducing pulmonary function by up to 25 per cent
2. the sympathetic chain, resulting in Horner's syndrome
3. the recurrent laryngeal nerve, resulting in hoarseness

and, more rarely:

4. accidental intrathecal or epidural injection
5. direct injection into the vertebral artery, causing convulsions.

Supraclavicular block

Of the three common approaches to brachial plexus block, this is the most hazardous owing to the risk of pneumothorax. This approach may therefore be relatively contraindicated in the obese patient or any patient in whom the landmarks are difficult to palpate.

However, at the level of the supraclavicular approach the brachial plexus is arranged in a very compact space as the plexus crosses the first rib, and thus inclusion of all three trunks in the block may be predicted. If surgical conditions permit, one may ask the awake patient to try to stretch the arm on the side to be blocked down towards the patient's foot. With the patient's head turned to the opposite side, the subclavian artery may be palpated about 1 cm posterolateral to the midpoint of the clavicle. If this is so, the point of insertion of the needle is immediately lateral to this point at about 80 degrees to the skin. If paraesthesiae are not obtained, the needle is advanced until the first rib is contacted. The needle is then 'walked' in the longitudinal direction of the rib until paraesthesiae or evoked muscle contractions with the nerve stimulator are obtained, at which stage the needle point is taken to lie within the sheath. If the artery is not palpated, the first rib should be sought approximately 1 cm posterolateral to the midpoint of the clavicle.

When the needle point is thought to lie within the sheath, 30 ml of anaesthetic should be injected after aspiration. Up to 30 minutes may be required for complete anaesthesia to be obtained.

With careful technique the incidence of pneumothorax is probably as low as 0.5 per cent. Horner's syndrome and phrenic nerve palsy may also develop. Bilateral supraclavicular blocks are contraindicated by the risks of bilateral pneumothoraces and phrenic nerve paralysis (and by the toxic doses of local anaesthetic that would be required). Damage to the lung or pleura may provoke a cough at the time of injection. A chest x-ray should be performed if the patient complains of cough, pleuritic pain or dyspnoea at or after performance of the block. However, it should be remembered that a small leak may take some time to result in a pneumothorax large enough to cause symptoms or show on an x-ray. Thus, if the x-ray is performed immediately, a false sense of security may ensue. Many specialists do not routinely obtain screening chest x-rays in asymptomatic patients.

Axillary block

Owing to the lack of risk of pneumothorax and the ease of execution, this is the favoured approach to the brachial plexus for many anaesthetists. This block is performed most easily when the head is turned to the opposite direction and the arm abducted to 90 degrees. The axillary artery is palpated as proximal as possible in the axilla. A short-bevelled needle inserted over the pulse towards the axilla will result in a characteristic 'click' as it penetrates the sheath (often surprisingly superficial). Oscillation of the needle with the pulse beat should then occur.

Paraesthesiae or evoked muscle contractions with the nerve stimulator may be elicited but are not strictly necessary and, indeed, may give rise to an increased incidence of peripheral neuritis.[3] If blood is aspirated, this is good evidence that the needle tip is within the sheath and the needle should be withdrawn until the needle tip is no longer within the vessel. Deliberate puncture of the artery is a required component of this technique.

Up to 40 ml of anaesthetic is required, especially if spread to the more proximal parts of the plexus is desired. This may be encouraged by applying digital pressure over the sheath distal to the point of injection both during and after the injection. A tourniquet should not be used during needle placement as this may cause collapse of the vessels, a false negative aspiration test and inadvertent intravascular injection of anaesthetic. A catheter may be placed in the sheath for prolonged analgesia.

Note. The musculocutaneous nerve will not be blocked as it leaves the sheath proximal to the site of axillary block, so an area on the radial side of the forearm may not be covered by axillary block unless this nerve is blocked separately. As the nerve passes through the coracobrachialis muscle, it can be infiltrated with anaesthetic at the level of the needle insertion for the axillary block. With a large volume of local anaesthetic injected by the axillary approach and massaged proximally up the sheath, this separate injection may not be necessary.

Some advantages and disadvantages of the three main

Table 34.7   Comparisons of different approaches to brachial plexus blocks

	Advantages	*Disadvantages*	*Special complications*
Interscalene	Landmarks usually easy to identify, even in obese patients Does not require change in patient position Shoulder analgesia included Minimal pneumothorax Rapid onset (usually in 10 minutes)	Ulnar nerve may be missed Potential rare but serious complications	Intrathecal/epidural Vertebral artery injection Phrenic nerve paralysis Horner's syndrome Recurrent laryngeal nerve paralysis
Supraclavicular	Reliable, complete block of all three trunks No movement of arm during performance	Slow onset without paraesthesiae Risk of pneumothorax Relatively contraindicated in respiratory cripples due to risk of pneumothorax and phrenic nerve paralysis	Phrenic nerve paralysis Horner's syndrome
Axillary	No risk of pneumothorax Suitable for most hand and forearm surgery Catheters may be inserted for prolonged analgesia Paraesthesiae not always necessary	Arm abducted during performance Usually inadequate for upper arm surgery Circumflex nerve and musculocutaneous nerve often missed as left sheath more proximal	Intravascular injection Haematoma due to artery puncture

approaches to brachial plexus blockade are shown in Table 34.7.

Distal arm blocks

Blockade of the more distal nerves of the arm may be useful as a supplement for an inadequate brachial plexus block or for surgery when a brachial plexus block is inappropriate (e.g. difficult anatomy, bilateral surgery). Because the origins of the innervation of the forearm are proximal to the elbow, the functional result in the hand will be similar if the nerves are blocked at either the wrist or elbow. Paraesthesiae or evoked muscular contractions with the nerve stimulator may be used. Alternatively, fan-wise infiltration in the appropriate areas will usually suffice.

Ulnar nerve

Five ml of anaesthetic can be injected between the olecranon and medial epicondyle of the elbow along the ulnar groove. The nerve is very superficial at this point and care should be taken to avoid intraneural injection. At the wrist, 5 ml of anaesthetic can be infiltrated from the deep fascia and bone to the subcutaneous tissues between the ulnar side of the ulnar artery and the radial side of flexor carpi ulnaris.

Median nerve

Five to ten ml of anaesthetic can be infiltrated in fan-wise manner at several levels just medial to the brachial artery in the antecubital fossa. At the wrist, 5 ml of anaesthetic is infiltrated in a similar fashion to ulnar nerve block at the wrist, between palmaris longus and flexor carpi radialis.

Radial nerve

It is uncommon to block this nerve at the level of the elbow. At the wrist, 5–10 ml of anaesthetic solution can be infiltrated subcutaneously along the radial border of the wrist for a simple and effective block.

Note. For all wrist blocks care must be taken to prevent damage to or accidental injection into the arteries and numerous small veins in this area.

Intravenous regional anaesthesia (Bier's block)

Intravenous regional anaesthesia (IVRA) can be used for manipulations of fractures of the forearm, and minor surgery of the hand and forearm. A peripheral cannula is inserted as distal as possible in the forearm to be operated upon. A second cannula must be inserted in the opposite arm to permit access to the circulation if required. A double pneumatic tourniquet cuff is applied to the upper arm and then an exsanguinator is used to exsanguinate the arm below. If the arm is too painful to permit exsanguination, the arm may be elevated for 2 minutes. The proximal cuff is then inflated to 50 mmHg above the systolic blood pressure. Up to 40 ml of anaesthetic is then injected into the vein in the exsanguinated arm. A characteristic mottling of the skin will be seen. Prilocaine 0.5 per cent is the preferred choice of agent because of its high margin of safety. Once anaesthesia is obtained after 10–15 minutes, the distal cuff can be inflated on the anaesthetic upper arm and the proximal cuff then deflated. This will tend to ameliorate 'tourniquet pain'. Anaesthesia is limited by the allowable duration of occlusion of circulation to the arm, but in any case the distal cuff should not be deflated for at least 20 minutes. Hopefully, by this time most of the anaesthetic will have 'fixed' to the tissues and there should be no overt toxicity. Intermittent deflation followed by full reinflation of the cuff to limit wash-out of anaesthetic into the circulation has also been advocated. Once the circulation is restored, residual analgesia is short lived (which tends to contradict the assumption that the anaesthetic is fixed to the tissues). The use of anaesthetic solutions containing adrenaline in an attempt to prolong the block is dangerous. IVRA is contraindicated when tourniquets are contraindicated (e.g. sickle cell or Raynaud's disease). Although theoretically possible, lower limb IVRA is not often performed owing to the large volumes (60–80 ml) required.

Although it is a very simple technique, many surgeons find the relaxation obtained to be poor, with better conditions the more distal the procedure. IVRA has been superseded for most anaesthetists by new short-acting general anaesthetics and/or the use of brachial plexus blocks.

Many doctors, particularly in the emergency department, perform IVRA without the assistance of anaesthetists. However, there is always the potential for severe toxic reactions, especially if there is sudden, early or unexpected cuff deflation. Therefore, any doctor performing IVRA must be skilled in airway management, resuscitation and recognition of signs of local anaesthetic toxicity. Even if such skills are possessed, the use of IVRA by a single operator/anaesthetist is to be deprecated. The double tourniquet must be checked regularly and short cuts in technique avoided. IVRA is always a potentially dangerous technique owing to the large volumes of anaesthetic injected directly into the circulation, and avoidable deaths have occurred.[69]

There have been reports of the use of ketamine[70] and muscle relaxants[71] as part of an IVRA technique. Whilst undoubtedly of interest pharmacologically (in the case of ketamine) or improving muscle relaxation (muscle relaxants), these cannot be recommended because the, perhaps obvious, potential side effects when the cuff is deflated.

Lumbar plexus or psoas sheath block

The lumbar plexus forms within the psoas sheath and as such can be blocked if a large volume of anaesthetic (up to 40 ml) is deposited after loss of resistance or a 'pop' is felt on entering the compartment. The patient is positioned fully or three-quarters prone and the tip of a Tuohy needle is 'walked off' the transverse process of the L3 vertebra. The correct compartment is usually reached approximately 1–3 cm beneath the transverse process. A nerve stimulator may be helpful in producing evoked contractions of the quadriceps muscles.

In view of the large volume required, the variability found for the site of injection[72] and the resultant potential for complications, some workers do not advocate the block.[72] However, other workers report success in providing analgesia for hip surgery.[73] The requirements for turning the patient and the large volumes required have led to the development of an inguinal approach to the main components of the lumbar plexus – the so-called '3 in 1' block.

Femoral and '3 in 1' blocks

Femoral nerve block

The femoral nerve is the main product of the lumbar plexus and runs into the leg deep to the inguinal ligament in the groove between the psoas major and iliac muscles. In the inguinal region it lies approximately 1 cm lateral to the femoral artery, which is the main landmark sought for its effective blockade. The femoral nerve divides into a superficial branch supplying the anterior surface of the thigh and sartorius and pectineus muscles, and a deep branch supplying the quadriceps muscles, articular branches to the hip and knee joints and terminating as the saphenous nerve. This division is variable and may occur above the inguinal ligament. It is important to realize that the femoral nerve is deep to both fascia lata and fascia iliaca while the artery and vein are only deep to fascia lata – that is, they are not all in the same neurovascular compartment.

Blockade is achieved by injection of 10 ml of local anaesthetic with one of two techniques. Using a nerve stimulator, the nerve may be located lateral to the artery in anaesthetized or sedated patients. Alternatively, if two 'pops' are felt using a short-bevelled needle just lateral to the artery, blockade can be predicted with or without eliciting paraesthesiae.

Blockade of the femoral nerve provides analgesia for knee surgery[74] (but not complete analgesia unless the other nerves supplying the knee, especially the sciatic nerve, are blocked), relieves muscle spasm and pain from fractures of the femur,[75] and can be used as the sole anaesthetic for muscle biopsy as an investigation into malignant hyperthermia susceptibility[76] or for the donor site for skin grafts.[77]

Complications (e.g. intravascular injection, haematoma, nerve damage) are uncommon with careful technique. Femoral nerve block is contraindicated in the presence of artificial grafts to the femoral artery.

3 in 1 block

Winnie described the '3 in 1' block[78] whereby injection of up to 30 ml of local anaesthetic by one of the above methods enables blockade of the other two main components of the lumbar plexus, namely the lateral cutaneous nerve of thigh (supplying the skin on the lateral aspect of the thigh) and the obturator nerve (supplying the adductor muscles of the thigh and articular branches to both the hip and knee joints). This technique relies on the large volume of anaesthetic filling up the fascial canal and travelling upwards towards the lumbar plexus (cf. brachial plexus block). However, either lateral cutaneous nerve block or obturator nerve block may not be achieved, and this block has been described as 'femoral block with overdosage'.[72] In view of the importance of blocking both the femoral and the lateral cutaneous nerve for hip surgery, specific blockade of each with 10 ml of anaesthetic may be preferred, although single '3 in 1' block can be effective in producing analgesia following surgery for fractured neck of femur.[79]

Continuous low-dose '3 in 1' block has been shown to significantly improve postoperative pain relief following knee replacement.[80]

Lateral cutaneous nerve of thigh block

This nerve enters the thigh by passing under the lateral attachment of the inguinal ligament 1–2 cm medial to the anterior superior iliac spine. The anterior superior iliac spine thus acts as the chief landmark for individual blockade of this nerve. Using a short-bevelled needle, 10 ml of anaesthetic can be deposited under the fascia lata 1–2 cm both inferior and medial to the anterior superior iliac spine. A 'pop' and loss of resistance indicate puncture of the fascia. The success rate can be improved by employing a fan-wise injection of a larger volume of anaesthetic around this point. A nerve stimulator is of no help for performing this block in the anaesthetized patient because the nerve is purely sensory. In addition to surgery for fractures of the femur, this block is also useful for analgesia of skin graft donor sites.

T12 or the subcostal nerve supplies fibres to the ilioinguinal and iliohypogastric nerves and also supplies a small area of skin laterally at the extreme top of the thigh. It has been shown[81] that analgesia following hip surgery is improved by blocking T12 by subcutaneously infiltrating anaesthetic across the top of the iliac crest.

Sciatic nerve block

The sciatic nerve arises from the sacral plexus and is the largest peripheral nerve in the body. The nerve leaves the pelvis through the sciatic notch or foramen, and passes through the saddle area beneath gluteus maximus to enter the thigh approximately equidistant between the ischial tuberosity and the greater trochanter of the femur.

The *posterior approach* is most popular and reliable. A needle at least 10 cm in length will be needed even in thin patients, owing to the distance from skin to the nerve. Paraesthesiae or evoked muscular contractions with the nerve stimulator should be sought.

The patient is placed on the opposite side to be blocked, with the uppermost leg flexed at the hip and knee joints (i.e. the recovery position). A line may be drawn or imagined between the greater trochanter of the femur and the posterior superior ischial spine. The point of skin entry is approximately 4 cm caudally from the midpoint of this line and the nerve is usually found at 8–10 cm of depth. The needle point may have to be 'walked' until paraesthesiae or evoked muscular contractions are elicited (movements of the foot should be sought to avoid confusion with direct muscle stimulation).

The *anterior approach* has the advantage that it can be performed with the patient supine without painful movements. A line is visualized or drawn from the greater trochanter to the lesser trochanter, parallel to the inguinal ligament. A 10 cm needle should be inserted at this level such that it can be 'walked' off the medial border of the femur. Paraesthesiae or evoked muscular contractions should be sought approximately 5 cm beyond the femur. Unfortunately, owing to variations in anatomy, the success rate is lower with this approach.

For both approaches, 20 ml of anaesthetic should be injected. With such a volume the posterior cutaneous nerve of thigh is usually blocked as well. Owing to the size of the sciatic nerve, up to 30 minutes may be required to achieve full analgesia.

Popliteal fossa block

A simple approach to the sciatic nerve for surgery below the knee is to perform the block at the level of the popliteal fossa behind the knee.[82] The sciatic nerve usually bifurcates at the popliteal fossa into the tibial and common peroneal nerves, although it may bifurcate higher up the thigh. With the patient in the prone or lateral position, the popliteal fossa can be thought of as a diamond-shaped, fat-filled space with the popliteal vessels and the above-mentioned nerves contained within it. The tibial nerve will usually be found with the aid of the nerve stimulator at the apex of the fossa in the midline, lateral to the popliteal artery. The common peroneal nerve lies slightly more lateral, the nerve eventually winding round the head of the fibula. A 'pop' may be felt as the fascia is punctured. Results are better if both nerves are located, either by paraesthesiae or with the nerve stimulator, but even if only one is found injection of 30 ml of anaesthetic will usually produce a satisfactory block. The combination of a popliteal fossa block and a femoral/saphenous nerve block will provide complete analgesia of the lower leg.

Ankle block

Five main nerves supply the foot, and can all be easily blocked by appropriate injections around the ankle. Apart from the saphenous nerve which is the terminal branch of the femoral nerve, it should be noted that these nerves are all branches of the sciatic nerve.

Tibial nerve

The anaesthetic (5 ml) is injected just anterolateral to the posterior tibial artery, if this is palpable, at the level of the upper border of the medial malleolus. If the artery is not palpable, 10 ml of anaesthetic should be injected at the same level medial to the Achilles tendon. Branches of the tibial nerve supply the majority of the sole of the foot.

Saphenous nerve

The anaesthetic (5 ml) is injected subcutaneously around the saphenous vein anterior to the medial malleolus. The saphenous nerve supplies the medial malleolus and the medial edge and dorsum of the foot.

Deep peroneal nerve

The anaesthetic (5 ml) should be injected lateral and slightly beneath the dorsalis pedis artery. If the artery is not palpable, the injection should be given about 1 cm lateral to the extensor hallucis longus tendon. This block is necessary only if anaesthesia of the first two toes is required.

Superficial peroneal nerve

This nerve branches higher up the leg and must therefore be blocked by subcutaneously infiltrating up to 10 ml of anaesthetic along the area between the anterior border of the tibia and the lateral malleolus. The nerve supplies the majority of the dorsal aspect of the foot.

Sural nerve

The anaesthetic (5 ml) is injected between the lateral malleolus and the border of the Achilles tendon. The sural nerve supplies the heel, lateral malleolus and lateral edge of foot and the posterolateral part of the sole of the foot. Alternatively, this block can be usefully combined with a superficial peroneal nerve block by infiltrating subcutaneously down the entire lateral border of the ankle.

These blocks are easily performed after general anaesthesia has been induced (eliciting paraesthesiae is difficult and may be painful for the patient – but will hasten the onset of analgesia). It seems wise to avoid adrenaline-containing solutions for blocks around the two major arteries in the presence of peripheral vascular disease.

REFERENCES

1. McConachie I. Vasovagal asystole after spinal anaesthesia. *Anaesthesia* 1991; **46**: 281–2.
2. Selander D, Dhuner KG, Lundborg G. Peripheral nerve injury due to injection needles used for regional anaesthesia. *Acta Anaesthesiologica Scandinavica* 1977; **21**: 182–8.
3. Selander D, Edshage S, Wolff T. Paresthesiae or no paresthesiae? Nerve lesions after axillary block. *Acta Anaesthesiologica Scandinavica* 1979; **23**: 27–33.
4. Smith BE, Allison A. Use of a low-power nerve stimulator during sciatic block. *Anaesthesia* 1987; **42**: 296–8.
5. Ford DJ, Pither C, Raj PP. Comparison of insulated and uninsulated needles for locating peripheral nerves with a peripheral nerve stimulator. *Anesthesia and Analgesia* 1984; **63**: 925–8.
6. Montgomery SJ, Raj PP, Nettles D, Jenkins MT. The use of the nerve stimulator with standard unsheathed needles in nerve blockade. *Anesthesia and Analgesia* 1973; **52**: 827–31.
7. DiFazio CA, Carron H, Grosslight KR. Comparison of pH adjusted lidocaine solutions for epidural anesthesia. *Anesthesia and Analgesia* 1986; **65**: 760–4.
8. Bedder MD, Kozody R, Craig DB. Comparison of bupivacaine and alkalinized bupivacaine in brachial plexus anesthesia. *Anesthesia and Analgesia* 1988; **67**: 49–52.
9. Scott DB. 'Maximum recommended doses' of local anaesthetic drugs. *British Journal of Anaesthesia* 1989; **63**: 373–4.
10. Abajian JC, Mellish PWP, Browne AE, Perkins FM, Lambert DH, Mazuzan Jr JE. Spinal anesthesia for surgery in the high risk infant. *Anesthesia and Analgesia* 1984; **63**: 359–62.
11. Findlay GFG. Compressive and vascular disorders of the spinal cord. In: Miller JD, ed. *Northfield's surgery of the central nervous system*, 2nd ed. Edinburgh: Blackwell Scientific, 1987: 707–59.
12. Newbery JM. Paraplegia following general anaesthesia. *Anaesthesia* 1977; **32**: 78–9.
13. Odoom JA, Sih IL. Epidural analgesia and anticoagulant therapy: experience with one thousand cases of continuous epidurals. *Anaesthesia* 1983; **38**: 254–9.
14. Wildsmith JAW, McClure JH. Anticoagulant drugs and central nerve blockade. *Anaesthesia* 1991; **46**: 613–14.
15. Van Nest RL. Radio headset for use with regional anaesthesia. *Anesthesiology* 1979; **50**: 275.
16. Mackenzie N, Grant IS. Propofol for intravenous sedation. *Anaesthesia* 1987; **42**: 3–6.
17. Smith DC, Crul JF. Oxygen desaturation following sedation for regional anaesthesia. *British Journal of Anaesthesia* 1989; **62**: 206–9.
18. Cope RW. The Woolley and Roe case. Woolley and Roe versus Ministry of Health and Others. *Anaesthesia* 1954; **9**: 249–70.
19. Holland R. Trends recognised in cases reported to the New South Wales special committee investigating deaths under anaesthesia. *Anaesthesia and Intensive Care* 1987; **15**: 97–8.
20. McKenzie PJ, Wishart HY, Smith G. Long term outcome after repair of fractured neck of femur. *British Journal of Anaesthesia* 1984; **56**: 581–6.
21. Yeager MP, Glass DD, Neff RK, Brinck-Johnsen T. Epidural anesthesia and analgesia in high risk surgical patients. *Anesthesia and Analgesia* 1987; **66**: 729–36.
22. Seeling W, Bruckmooser KP, Hufner C, Kneitinger K, Rigg C, Rockemann M. Keine verminderung potoperativer komplikationen durch katheterepiduralanalesie nach grossen abdominellen eingriffen. *Anaesthesist* 1990; **39**: 33–40.
23. Spence AA, Logan DA. Respiratory effects of extradural nerve block in the postoperative period. *British Journal of Anaesthesia* 1975; **47**: 281–3.
24. Seeling W, Bothner U, Eifert B, *et al.* Patientenkontrollierte Analgesie versus Epiduralanalgesie mit Bupivacain oder Morphin nach grossen abdominellen Eingriffen. *Anaesthesist* 1991; **40**: 614–23.
25. Modig J. Beneficial effects on intraoperative and postoperative blood loss in total hip replacement when performed under lumbar epidural anaesthesia. *Acta Chirurgica Scandinavica Supplementum* 1988; **550**: 95–103.
26. McGowan SW, Smith GFN. Anaesthesia for transurethral prostatectomy. *Anaesthesia* 1980; **35**: 847–53.
27. Modig J, Borg T, Karlstrom G, Maripuu E, Sahlstedt B. Thromboembolism after total hip replacement. Role of epidural and general anesthesia. *Anesthesia and Analgesia* 1983; **62**: 174–80.
28. Donadoni R, Baele G, Devulder J, Rolly G. Coagulation and

fibrinolytic parameters in patients undergoing hip replacement: influence of the anaesthesia technique. *Acta Anesthesiologica Scandinavica* 1989; **33**: 588–92.

29. Cooke ED, Bowcock SA, Lloyd MJ, Pilcher MF. Intravenous lignocaine in prevention of deep venous thrombosis after elective hip surgery. *Lancet* 1977; **2**: 797–9.

30. Ryan P, Schweitzer S, Collopy B, Taylor D. Combined epidural and general anesthesia in patients having colon and rectal anastomoses. *Acta Chirurgica Scandinavica Supplementum* 1988; **550**: 146–51.

31. Aitkenhead AR, Gilmour DG, Hothersall AP, Ledingham IMcA. Effects of subarachnoid spinal nerve block and arterial PcO_2 on colon blood flow in the dog. *British Journal of Anaesthesia* 1980; **52**: 1071–6.

32. Vedrinne C, Vedrinne JM, Guirard M, Patricot MC, Bouletreau P. Nitrogen sparing effect of epidural administration of local anaesthetics in colon surgery. *Anesthesia and Analgesia* 1989; **69**: 354–9.

33. Tuman KJ, McCarthy RJ, March RJ, DeLaria GA, Patel RV, Ivankovich AD. Effects of epidural anesthesia and analgesia on coagulation and outcome after major vascular surgery. *Anesthesia and Analgesia* 1991; **73**: 696–704.

34. Haljamae H. Effects of anesthesia on leg blood flow in vascular surgical patients. *Acta Chirurgica Scandinavica Supplementum* 1988; **550**: 81–7.

35. Scott NB, Kehlet H. Regional anaesthesia and surgical morbidity. *British Journal of Surgery* 1988; **75**: 299–304.

36. Salo M. Effects of anaesthesia and surgery on the immune response. *Acta Anaesthesiologica Scandinavica* 1992; **36**: 201–20.

37. Brandt MR, Fernandes A, Mordhurst R, Kehlet H. Epidural analgesia improves postoperative nitrogen balance. *British Medical Journal* 1978; **1**: 1106–8.

38. Mihic DN. Postspinal headache and relationship of needle bevel to longitudinal dural fibres. *Regional Anesthesia* 1985; **10**: 76–81.

39. Moore DC. Spinal (subarachnoid) block. *Journal of the American Medical Association* 1966; **195**: 123–8.

40. Keats AS. Anaesthesia mortality – a new mechanism. *Anesthesiology* 1988; **68**: 2–4.

41. Reid JA, Thorburn J. Headache after spinal anaesthesia. *British Journal of Anaesthesia* 1991; **67**: 674–7.

42. Abouleish E, de la Vega S, Blendinger I, Trio T. Long term follow up of epidural blood patch. *Anesthesia and Analgesia* 1975; **57**: 459–63.

43. Crawford JS. Experiences with epidural blood patch. *Anaesthesia* 1980; **35**: 513–15.

44. Dripps RD, Vandam LD. The long-term follow-up of patients who received 10,098 spinal analgesics: failure to discover major neurological sequelae. *Journal of the American Medical Association* 1954; **156**: 1486–91.

45. Lambert DH. Factors influencing spinal anaesthesia. *International Anesthesiology Clinics* 1989; **27**, 1: 13–20.

46. Bromage P. The control of post thoracotomy pain. *Anaesthesia* 1989; **44**: 445.

47. MacArthur C, Lewis M, Knox EG, Crawford JS. Epidural anaesthesia and long-term backache after childbirth. *British Medical Journal* 1990; **301**: 9–12.

48. Hashenis K, Middleton MD. Subcutaneous bupivacaine for postoperative analgesia after herniorraphy. *Annals of the Royal College of Surgeons of England* 1983; **65**: 38–9.

49. Owen H, Galloway DJ, Mitchell KG. Analgesia by wound infiltration after surgical excision of benign breast lumps. *Annals of the Royal College of Surgeons of England* 1985; **67**: 114–15.

50. Chester JF, Stanford BJ, Gazet J–C. Analgesic benefit of locally injected bupivacaine after hemorrhoidectomy. *Diseases of the Colon and Rectum* 1990; **33**: 487–9.

51. Adams WJ, Avramovic J, Barraclough BH. Wound infiltration with 0.25% bupivacaine not effective for postoperative analgesia after cholecystectomy. *Australian and New Zealand Journal of Surgery* 1991; **61**: 626–30.

52. Thomas DFM, Lambert WG, Lloyd-Williams K. The direct perfusion of surgical wounds with local anaesthetic solutions. An approach to postoperative pain. *Annals of the Royal College of Surgeons of England* 1983; **65**: 226–9.

53. McQuay HJ. Pre-emptive analgesia. *British Journal of Anaesthesia* 1992; **69**: 1–3.

54. Harrison N, Langham BT, Bogod DG. The appropriate use of local anaesthesia for venous cannulation. *Anaesthesia* 1992; **47**: 210–12.

55. Moore DC. Intercostal nerve block: spread of India ink injected to the rib's costal groove. *British Journal of Anaesthesia* 1981; **53**: 325–9.

56. Kirno K, Lindell K. Intercostal nerve blockade. *British Journal of Anaesthesia* 1986; **58**: 246.

57. Nunn JF, Slavin C. Posterior intercostal nerve block for pain relief after cholecystectomy. *British Journal of Anaesthesia* 1980; **52**: 253–60.

58. O'Kelly E, Garry B. Continuous pain relief for multiple rib fractures. *British Journal of Anaesthesia* 1981; **53**: 989–91.

59. Bunting P, McGeachie JF. Intercostal nerve blockade producing analgesia after appendicectomy. *British Journal of Anaesthesia* 1988; **61**: 169–72.

60. Moore DC. Intercostal nerve block for postoperative somatic pain following surgery of thorax and upper abdomen. *British Journal of Anaesthesia* 1975; **47**: 284–6.

61. Kvalheim L, Reiestad F. Intrapleural catheter in the management of postoperative pain. *Anesthesiology* 1984; **61**: A231.

62. Rocco A, Reiestad F, Gudman J, McKay W. Intrapleural administration of local anaesthetics for pain relief in patients with multiple rib fractures. *Regional Anesthesia* 1986; **12**: 12–14.

63. Scott PV. Interpleural regional analgesia: detection of the interpleural space by saline infusion. *British Journal of Anaesthesia* 1991; **66**: 131–3.

64. Ananthanarayan C, Kashtan H. Pneumothorax after interpleural block in a spontaneously breathing patient. *Anaesthesia* 1990; **45**: 342.

65. Bunting P, McConachie I. Ilioinguinal block for analgesia after lower segment caesarean section. *British Journal of Anaesthesia* 1988; **61**: 773–5.

66. Bacon AK. An alternative block for post circumcision analgesia. *Anaesthesia and Intensive Care* 1977; **5**: 63–4.

67. Berens R, Pontus SP. A complication associated with dorsal penile nerve block. *Regional Anaesthesia* 1990; **15**: 309–10.

68. Winnie AP. Interscalene brachial plexus block. *Anesthesia and Analgesia* 1970; **49**: 455–66.

69. Heath M. Deaths after intravenous regional anaesthesia. *British Medical Journal* 1982; **288**: 913–14.

70. Durrani Z, Winnie AP, Zsigmond EK. Ketamine for intravenous regional anesthesia. *Anesthesia and Analgesia* 1989; **68**: 328–32.

71. McGlone R, Heyes F, Harris P. The use of a muscle relaxant to supplement local anesthetics for Bier's blocks. *Archives of Emergency Medicine* 1988; **5**: 79–85.

72. Smith BE, Haycock JC. Local anaesthesia for surgery of the lower limb. *Current Anaesthesia and Critical Care* 1992; **3**: 37–41.

73. David BB, Lee E, Croitow M. Psoas block for surgical repair of hip fracture: a case report and description of a catheter technique. *Anesthesia and Analgesia* 1990; **71**: 298–301.

74. Ringrase NH, Cross MJ. Femoral nerve block in knee joint surgery. *American Journal of Sports Medicine* 1984; **12**: 398–402.

75. Berry FR. Analgesia in patients with fractured shaft of femur. *Anaesthesia* 1977; **32**: 576–7.

76. Ellis R. Evaluation of '3 in 1' lumbar plexus block in patients having muscle biopsy. *British Journal of Anaesthesia* 1989; **62**: 515–17.

77. Thorburn J, Rogers KM. Orthopaedic surgery. In: Henderson JJ, Nimmo WS, eds. *Practical Regional Anaesthesia*. Oxford: Blackwell Scientific, 1983: 184–214.

78. Winnie AP, Ramamurthy S, Durrani Z. The inguinal paravascular technique of lumbar plexus anaesthesia. 'The 3-in-1 block'. *Anesthesia and Analgesia* 1973; **52**: 989–96.

79. Hood G, Eldbrooke DI, Genish SP. Postoperative analgesia after triple nerve block for fractured neck of femur. *Anaesthesia* 1991; **46**: 138–40.

80. Dahl JB, Christiansen CL, Dangaard JJ, Schultz P, Carlsson P. Continuous blockade of the lumbar plexus after knee surgery. Postoperative analgesia and bupivacaine plasma concentration: a controlled clinical trial. *Anaesthesia* 1988; **43**: 1015–18.

81. Smith BE, Allison A. Nerve block for hip surgery. *Anaesthesia* 1987; **42**: 1016–17.

82. Rorie DK, Byer DE, Nelson DO, Sittipong R, Johnson KA. Assessment of block of the sciatic nerve in the popliteal fossa. *Anesthesia and Analgesia* 1980; **59**: 371–6.

FURTHER READING

Bromage PR. *Epidural anaesthesia*. Philadelphia: WB Saunders, 1978.

Cousins MJ, Bridenbaugh PO, eds. *Neural blockade*, 2nd ed. Philadelphia: JB Lippincott, 1988.

Ellis H, Feldman S. *Anatomy for anaesthetists*, 6th ed. Oxford: Blackwell Scientific, 1992.

Lee JA, Atkinson RS, Watt MJ, eds. *Sir Robert Macintosh's lumbar puncture and spinal analgesia*, 5th ed. Edinburgh: Churchill Livingstone, 1985.

Moore DC. *Regional block*, 4th ed. Springfield: CC Thomas, 1976.

Raj P, Nolte H, Stanton-Hicks M. *Illustrated manual of regional anaesthesia*. Berlin: Springer-Verlag, 1988.

Wildsmith JAW, Armitage EN, eds. *Principles and practice of regional anaesthesia*. Edinburgh: Churchill Livingstone, 1987.

Winnie AP. *Plexus anesthesia*, vol 1, *Perivascular techniques of brachial plexus block*. Edinburgh: Churchill Livingstone, 1984.

The Difficult Airway

Martin L. Norton

In this era of high technology, we still face an ageless problem in anaesthesia – the difficult airway. Recent attention to this issue has been based on a risk management approach. Studies in the UK, Australia and the USA have shown that significant anaesthetic-related mortality and morbidity are related to airway management.[1-5] The report of the Confidential Enquiry into Peri-Operative Deaths (CEPOD) published for England and Wales, 1986[5], revealed that a difficult or failed intubation contributed to six of the 4034 deaths reported. Indeed, as many as 600 deaths may occur each year as a result of a difficult intubation.[6]

The complex tests described later in this chapter cannot be applied to all surgical patients. Simple tests which may be able to identify those who are apparently normal but who will prove difficult to intubate should, therefore, be available for all patients. A number of bed-side tests have been proposed.[7-9] Patil and colleagues[7] considered the thyromental distance. Mallampati reported on a visual examination of the posterior wall of the pharynx.[8] Frerk has combined these tests[10] and shown that when the posterior pharyngeal wall can, from the front, be seen below the soft palate with the patient's mouth open wide, intubation should be straightforward (grade I or II). On the other hand, if the posterior pharyngeal wall cannot be seen, the patient should be graded III or IV and may be expected to be difficult to intubate, particularly if the thyromental distance is short (i.e. less than 7 cm). Fortunately, when an unexpected difficulty arises, flexible fibreoptic endoscopy may often provide a solution.

No problem is more frustrating than the anaesthetic management of a patient of apparently normal appearance, in whom it is impossible to visualize the glottis and insert a tracheal tube.

Defining the difficult airway

Definition of the 'difficult airway' is usually related solely to tracheal intubation or problems with mask ventilation. Central to management must be an understanding of the anatomical, biomechanical or pathophysiological mechanisms responsible for airway complications. For example, an understanding of mucopolysaccharidosis or fibrofascial myositis ossificans can warn of potential airway disaster just as much as retrognathia or micrognathia (separate but overlapping entities).

In approaching the difficult airway, it is useful to consider the following broad categories:

1. Previous history of difficulty or failure in establishing control and access to the patient's airway.
2. Presence of anatomical or physiological pathology which might predictably be associated with difficulty in airway management. Although previous documented success in such a patient may put the practitioner at ease, it does not necessarily guarantee that the planned anaesthetic will be uneventful.
3. Limited experience in evaluating, planning and managing the difficult airway.

Criteria

Various reports to date have attempted to focus on single, and sometimes simplistic, factors such as macroglossia or micrognathia.[11] These studies have all been based on a search for prospective reproducible criteria of value to the clinical practitioner. Another suggestion (Table 35.1) is to score a number of criteria in attempting to predict whether there will be difficulty with airway management. This approach has yet to be fully validated, but it must not be forgotten that the best of predictive methods may fail to draw attention to the patient in whom difficulties become obvious only after the induction of anaesthesia. Thus, as in all aspects of practice, the anaesthetist must continually be

Table 35.1 Signs suggesting difficult endoscopy*

Sign	Score
Micrognathia with acute mandibular angles	5
Glossoptosis or basal macroglossia	4
Difficulty with prior laryngoscopy	3
Long, high-arched palate associated with long, narrow dental arch	2
Temporomandibular joint limitation	4
Short muscular neck with full dentition	2
Protruding maxillary dentition with premaxillary overgrowth	2
Increased alveolomental depth	1
Limited extension of the upper cervical vertebrae	2
Limited motion of lower cervical vertebrae	3
Pathological signs of airway obstruction	3
Decreased distance between hyoid and thyroid cartilages	2
Decreased distance of epiglottis from posterior wall of pharynx	3
Skill of the anaesthetist	1–5
Planning for endoscopy	1–5
Müller's sign (3–4)	3

*Scale ranges from 1 to 5, 1 being best and 5 being worst.
Reproduced, with permission, from Norton ML, Brown ACD, eds. *Atlas of the difficult airway*, St Louis: Mosby Year Book, 1991: 38.

prepared to manage unexpected and potentially life-threatening loss of control of an airway.

Whilst it is tempting to believe that the use of regional anaesthesia will obviate such problems, this is certainly not a universal solution. Indeed, what happens when the solution suddenly becomes a new problem? Inadequate regional anaethesia, too high a spinal level, a suddenly excited patient, toxic reactions – all potentially require emergency airway management. Failure to be able to guarantee an airway can risk a patient's life. It always has been – and remains – our obligation to be certain that we can control the patient's airway.

Philosophy

The skilled clinician may have justifiable pride in his or her abilities but a degree of objectivity is essential for optimum patient management. Thus, whenever a difficult intubation is anticipated, the presence of another skilled practitioner may be invaluable in providing intelligent and definitive assistance. Furthermore, there will be occasions when consultation with colleagues, referral of patients to others with more experience in a particular aspect of airway abnormality, or even deferral of surgery for additional planning will be in the patient's best interest. The anaesthetist may face situations involving various external pressures; these should be recognized and dealt with maturely. There is no shame in being unable to solve a problem at the specific moment; there must be no reluctance to step back to rethink and plan.

Psychology of the problem airway

Patients can be hurt by insensitive physicians. An individual who has a craniofacial dysostosis, gargoylism or other distortion is often sensitive to the slightest suggestion of revulsion. They can understand referral to 'experts' but not the feeling that they are being rejected. They have done nothing to deserve feeling that the problem is 'their fault'.

The doctor

The truism that an ounce of compassion is worth a pound of medication must never be forgotten. Similarly, exaggerated effusive expression, pretending, is easily recognized by patients. That one ministers to the whole patient is not the least part of being a physician.

Trust

It is sometimes necessary to intubate an awake patient. Thorough explanation of the reasons as well as the process is part of the anaesthetist's responsibility to establish rapport. Explanations must not be rushed. Just as topical anaesthesia requires meticulous time and effort, so too does development of a state of confidence require empathy and understanding. Once trust has been established, patient cooperation will be enhanced, leading to a smoother, successful procedure. Remember that a patient will probably come to surgery again. Present obligations to the patient therefore go far beyond the immediate anaesthetic.

Follow-up

Responsibility does not end with placement and ultimate removal of the tracheal tube. Consideration must be given to the fact that the patient may have a future need for tracheal intubation. The patient, as well as the referring practitioner, must be informed about the 'whys' and 'wherefores' of their problem airway. Because, with the passage of time, memory becomes inaccurate, a written report must be provided.[12]

Technology

Useful technology is now available to us. Most valuable is the facility afforded by the flexible fibreoptic endoscope (Fig. 35.1). This instrument is a complex of glass fibres that enables the target to be illuminated and the channel and its components to be visualized. The problems encountered with this new tool are:

1. Cost
2. Importance of visual anatomical knowledge
3. Lubrication
4. Surface anaesthesia
5. Clearing the field (suction)
6. Light intensity transmitted (source and lumen density)
7. Skill level

Some anaesthetists advocate early use of the flexible fibreoptic endoscope for intubation but others recommend other techniques such as the light wand, intubation guides, the Brain laryngeal mask or even 'tactile laryngoscopy'. An anaesthetist must be able to handle all approaches. If only 1–3 per cent of patients present with difficult airways, an anaesthetist may be confronted with only a few hundred such problems in a professional lifetime. Physicians in the Airway Clinic at the University of Michigan see more than 150 referred patients, and at least 60 additional unexpected patients, each year. This experience provides a major resource which gives the interested anaesthetist the opportunity to gain experience in thoroughly examining the nasal/oral passages, thereby providing skills that can be brought to daily practice.[13] Preoperative assessment of the airway should minimize the occurrence of this pathology presenting unexpectedly during a 'routine' general tracheal anaesthetic.

Management planning

Schemata for the expected and the unexpected problem airway range from the complex[14] to the simple.[15]

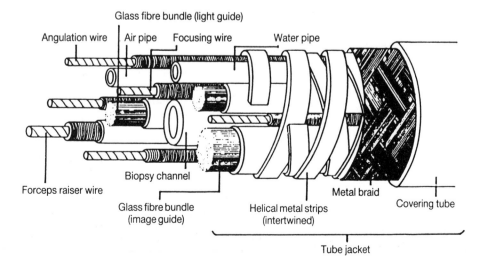

Glass fibre bundle (light guide)

Angulation wire   Air pipe   Focusing wire   Water pipe

Forceps raiser wire

Biopsy channel

Glass fibre bundle (image guide)

Helical metal strips (intertwined)

Metal braid

Covering tube

Tube jacket

Fig. 35.1 Internal construction of a flexible fibreoptic endoscope.

Algorithms represent a valid training aid but do not completely satisfy the needs of the situation in that they do not provide the rationale. It is necessary to constantly seek criteria – those factors that can be recognized are reproducible and can be solved.

Similarly, technology may provide a tool but an understanding of the essentials of medicine and knowledge of the anatomy, physiology, pharmacology, biomechanics and pathology are required. The flexible fibreoptic endoscope has made it possible for visual approaches to be the norm, reducing but not obviating the need for 'blind' techniques.

More esoteric approaches as described by Norton and Londy (below) using the C-arm fluoroscope for dynamic airway studies and management have demonstrated how, in different positions (e.g. sitting vs supine), the airway changes.

Financial cost of instrumentation is a significant factor but it must not be the only consideration. This will certainly be less than the physical cost accruing to a patient with a mismanaged airway or, from a pragmatic point of view, the financial cost to a defendant in a medical negligence suit.

Finally, the skilled clinician must be familiar with all approaches to the difficult airway and must, therefore, develop skills that do not only depend on the most advanced technology.

The difficult airway clinic

A new concept has been developed – the difficult airway clinic.[16] Its purpose is to concentrate experience in complex examinations and to provide clinical management advice for the solution of demonstrated or potential upper airway problems. Planning the best first approach, with back-up options, can eliminate the need for tracheostomy, or reduce the incidence of anaesthetic induction trauma as well as operating room delays or case cancellations. Some patients who may benefit from these services include those with severe arthritis, craniofacial dysostoses, Hallerman–Streif, macroglossia, amyloidosis, achondroplasia, acromegaly, fibromyositis ossificans dystrophica progressiva, mucopolysaccharidoses, temporomandibular dysfunction, obesity, post-radiation pharyngolaryngeal fibrosis and other conditions involving the airway.

Premedication

Ideally, the patient should be brought to the operating room without any premedication that can suppress ventilation. Even small doses of narcotic, tranquillizer or barbiturate

can turn a partial obstruction into a complete obstruction with alarming rapidity.

The use of antisialogogues (glycopyrollate, hyoscine (scopolamine)) is controversial. A reduction in the massive mucoid secretions, which may obstruct the view, is desirable. However, the resulting increase in viscosity may make it more difficult to suction away mucoid collections. Furthermore, those very secretions may serve as a lubricant and facilitate the passage of instruments.

Perhaps of greater importance is the decision to continue or discontinue the patient's usual medication. Unless it presents a significant risk in the presence of anaesthetic agents or procedures greater than the underlying pathophysiology for which these drugs were prescribed, the patient's usual medication should be continued.

Critique

Anaesthetic techniques designed solely for efficiency are inappropriate in the management of a difficult airway. In

Table 35.2 Complications and constraints

Technique	Complications and constraints
Jet (Venturi)	Pneumolarynx and trachea, pneumomediastinum, pneumothorax, also infection
Retrograde	Trauma to vocal folds, infection, other laryngeal and tracheal trauma, etc.
Light wand	Tissue burn from heat of bulb
Fibreoptic (nasal)	Haemorrhage, nasal diameter
Fibreoptic (oral)	[Access via intubation airway]
Intubation guide (Norton Teflon or Eschmann® woven or Sheridan TXX® tubular)	[Access via oral route only]
Bullard® laryngoscope with intubation guide	[Access via oral route only]
Tracheostomy	Tracheal infection, stenosis, loss of Valsalva mechanism, communication problems, cicatrix, stenosis, haemorrhage
Nasal airway	Haemorrhage
Laryngeal mask	Leakage of liquids (e.g. regurgitant HCl), and disruption of oesophagopharyngeal introitus

The above listing does not include reactions to topical anaesthetics and adjuvants required for 'shrinking' mucosae, reduction of sensation to passage of instruments, depression of gag reflex, and psychotropic effects. Entries in brackets denote a modifier to the title.

such patients, rapid induction of anaesthesia by administration of a precalculated dose of barbiturate followed by a rapidly acting muscle relaxant is likely to be devastating and require urgent or emergency action. When airway difficulties are anticipated, the selection of a technique should include consideration and anticipation of all potential complications (Tables 35.2 and 35.3). Muscle relaxants must be withheld until the anaesthetist is certain that the airway can be controlled.

Monitoring

All patients being evaluated for, or undergoing, control of the airway require systemic monitoring. The pulse oximeter provides the most useful information. The effects of the Valsalva manoeuvre increase in airway obstruction, and changes in cardiac rate are rapidly identified by this device.

The continuous electrocardiogram is also used, particularly for patients with a history of cardiac disease or arrhythmias. Lastly, the end-tidal $P\text{CO}_2$ monitor[17] assists in confirming correct intubation and is considered mandatory in patients with a history of bronchospastic disease.

Iatrogenic problems

Opening the mouth

The mechanics of opening the mouth involve both muscular and bony components. Trismus, a pathological muscle spasm secondary to pain, limits mouth opening. It usually responds to local anaesthetic infiltration, analgesia or general anaesthesia. However, what appears to be trismus may not necessarily result from muscular spasm. Thus, if difficulty in opening the mouth is caused by bony pathology, the above therapy will obviously not be effective.

The principal bony component to be considered is the temporomandibular joint. The contour of the zygomatic arch and temporal bone is that of an elongated 'S' lying on its side. The joint is located between the mandibular (glenoid) fossa of the inferior surface of the temporal bone and the condylar process of the mandibular bone. The condyloid process of the mandible seats into the glenoid sulcus with the intervention of the bursa. The mechanical pattern of mouth opening involves a hinge (ginglymus motion) accompanied by rotation and translocation ventrally. This joint can be temporarily dislocated,

Table 35.3 Examples of soft tissue conditions causing airway difficulty

Congenital
Glottic webs
Haemangiomas
Hypoplastic mandible
Laryngomalacia

Infection and progessive airway obstruction
Epiglottis
Croup
Ludwig's angina
Retropharyngeal abscess
Diphtheria

Sudden
Foreign body aspiration
Trauma
Retrophyaryngeal haematoma
Large neck haematoma

Tumours
Tongue
Epiglottis
Larynx

allowing the mouth to be opened further with the use of muscle relaxants.

Physician-induced difficulty occurs with the usual technique of opening the patient's mouth by pushing downward on the lower jaw with the patient supine. This manoeuvre impacts the coronoid process into the sulcus, limiting the translocation motion of the joint, and inhibits mouth opening. The required movement is a downward, then outward, then lifting of both the mandible and base of the tongue in a C-shaped motion. This can be accomplished by the following manoeuvres.

1. Thumb over index finger (scissors) – placing the index finger as far posteriorly on the occlusal surface of the upper dentition, and the thumb on the lower dentition. This is followed by a scissor motion with the thumb extending downward, outward and upward.
2. Inserting the thumb over the tongue base as far posteriorly as possible – and grasping the tongue base and mandible (with the remaining fingers). The motion, again, is a lifting downward, outward and upward – thus *lifting* the tongue base and mandible *on to* the rigid laryngoscope blade. This manoeuvre is best for an edentulous patient and mandatory for those with unstable cervical vertebrae.

Extending the head

Anaesthetists have been indoctrinated into placing the head in the 'sniffing position'.[18] Emphasis on this is based on alignment of the oral, pharyngeal and tracheal axes for insertion of rigid laryngoscope blades.[19] Unfortunately, this has been overemphasized to the point of resultant flexion (by too many blankets or pillows behind the head) which also stresses the restraining ligaments of the ventrally placed odontoid process (C1–C2 complex), most important in atlantoaxial instability. On the other hand, we frequently observe the anaesthetist placing one hand on the chin, with the other on the occipitoparietal area of the cranium and pushing the head into extreme extension. This not only distorts the airway axis, artificially displacing the tracheal cartilage ventrally, but also forces the odontoid process against the body of C1 vertebra. This is particularly hazardous in patients with osteoporosis, Paget's disease or achondroplasia in whom there is a risk of odontoid fracture and displacement against the spinal cord.

Blade placement

Visual obstruction can occur because of improper placement of a laryngoscope blade. The lateral flange of the blade was so designed to deflect the bulk of the tongue to the side. Placement of the blade directly in the midline induces the tongue on the opposite side to curl around into the channel of the blade, thus obstructing vision. Proper placement requires that the instrument is placed in the lateral part of the mouth, thus deflecting the tongue by the flange of the laryngoscope to the opposite paramedian position.

Sellick manoeuvre

The Sellick manoeuvre[20] has become almost a reflex activity during intubation. However, it is not an absolute assurance against the aspiration of liquids; furthermore, it prevents the hyoid bone from moving forward with the tongue during laryngoscopy. It may cause lingual nerve damage by stretching the nerve as it crosses the hyoglossus,[21] and lead to hyperaesthesia of the tongue.[22] Misplaced pressure can drive the thyroid up under the wings of the hyoid cartilage, reducing the visualization available for insertion of the tracheal tube.

Unexpected airway problems in the operating room

An unexpected problem is quite different from that faced by the anaesthetist with a patient whose airway pathology has been diagnosed prospectively. Most anaesthetists have, however, been confronted with an apparently normal patient in whom the situation unexpectedly 'gets out of hand'. Thus, such an eventuality should be kept in mind when planning the induction and intraoperative anaesthetic management of all patients.

Induction

First plan for all eventualities in airway management. This includes having readily available the equipment that might be needed if a problem arises. This need not be complex but certainly should include intubation guides (Norton, Eschmann, Sheridan), light wand and a prepared 'airway trolley' for specialized techniques (fibreoptic, Bullard, jet). One should look through the lumen of chosen tracheal tubes for obstructive blebs or membranes produced during manufacture or a constricted lumen on inflation of the cuff. The use of the long intubation guide has not been given enough attention recently because of the introduction of instruments such as fibreoptic laryngoscopes, the Bullard laryngoscope and jet (Venturi) techniques. In fact, the simplest approaches are often the best.

Intraoperative

During operation, other factors can be important. The tubes, or tracheal tube cuffs, may be cut by the surgeon. Plugs of mucus or blood may be desiccated in the lumen of the nasal or orotracheal or endobronchial tube, or in the tracheobronchial tree – especially when using dry, cold gases. Kinking, torsion and compression of the airways also occur, as does improper placement or movement so that the distal tip becomes occluded by soft tissue. Merely passing a suction catheter through the lumen may be insufficient because of the ball-valve or trap-door action of some plugs, in which case it is necessary to pass a fibreoptic flexible endoscope to locate and determine the nature of the plug and remove it. It may require special forceps (for highly viscous or solid materials), changing the airway device or, in very rare instances, rigid bronchoscopy for removal.

Positioning for endoscopy

The standard position for rigid endoscopy is with the patient supine and the anaesthetist above the head of the patient. Flexible fibreoptic intubation can be performed in the same way but may not be the ideal position, particularly for the patient with a difficult airway.[23] The optimum position in this case is with the patient sitting, and the endoscopist standing or sitting beside and facing the patient. The advantages of this position are:

1. The patient's face and reaction can be observed.
2. The light reflection (nasal or cervical), which indicates the position of the fibrescope, can be seen.
3. The awake patient does not feel disorientated and remote vis-à-vis the endoscopist (face to face communication).
4. Blood or secretions draining from the pharynx can be swallowed.

When surgery is performed using suspension laryngoscopy, the patient must be asleep in order to tolerate the force of suspension. However, when the airway is difficult to visualize the problem is quite different. The configuration of the oropharynx and laryngopharynx as well as the forces affecting that configuration must be considered.

The normal larynx is protected from above by the base of the tongue, the epiglottis and the funnel action of the pharyngeal muscle. This chapter does not include a discussion of pharyngeal biomechanics; textbooks are available on the topic.[24,25] The larynx is suspended by the stylohyoid and omohyoid muscles along with the digastric muscles which assist in positioning the hyoid ventrally or dorsally. When the patient is supine, the suspensory structures tend to relax, decreasing the airway channel to one-half to one-third its erect diameter. This reduction has been demonstrated with radiological views using C-arm dynamic fluoroscopy, even in the awake subject. However, when the patient sits up, the larynx tends to uncoil (spring action discussed by Fink)[24,25] as the suspensory structure becomes functional. The hyoid separates further from the tracheal cartilage, permitting the normal adult tracheal cartilage to open up and become more heart shaped.

The chin and the base of the tongue should be brought forward. This motion, via the pulley of the hyoid (anterior and posterior digastric and omohyoid muscles), pulls the hyoid wings cephalad, away from the thyroid cartilage, further expanding the aditus.

The phrase 'anterior larynx' is often used. It would be hard to find anything more anteriorly placed than the thyroid prominence just under the skin. A better description would be 'superiorly shifted larynx' with a relative posterior displacement (glossoptosis) of the tongue base and epiglottis. These are the movements considered when the phrase 'anterior larynx' is used and imply that there will be difficulty in visualizing the lumen of the larynx at the level of the vocal folds. In fact, in the supine position there is a cephalad displacement of the tracheal cartilage

wings such that the larynx rises within the limits of the hyoid, narrowing the aperture.

The sitting position which reduces biomechanical obstruction is therefore more appropriate than the supine for opening up the airway for endoscopy.

Where the flexible fibreoptic endoscope is very useful for safe airway management, in the unexpected emergency it may only be helpful in the hands of the expert. Presence of blood, viscous mucus or massive tissue oedema will lead to failed airway visualization. Furthermore, the supine position, with its reduced airway support, will contribute to the difficulty.

Use of the fibrescope

The flexible fibreoptic endoscope can be used to guide the tracheal tube through either the oral or nasal passages. The keys are:

1. Knowledge of anatomy
2. The sitting position
3. Repeated practice to develop skill

A review of the anatomy of the airway, demonstrated by Fink,[24,25] reveals the position that opens the airway maximally.

The guide wire

A guide wire (e.g. Cook, type ST: diameter 0.38 in, length 145 cm, flexible tip, straight stock PWG 2588) may prove useful in difficult situations. There are times when advancing the endoscope is not possible in spite of the aditus laryngis being in view. The wire should have a flexible tip which can be guided, either by direct vision or by radiological assistance, through the larynx. The flexible endoscope tip can be used to direct the wire through the laryngeal aperture. The wire is then advanced into the lower trachea to reduce the chance of its being pulled out of the trachea by a 'whipping' action as the endoscope is advanced over the wire guide down to the carina under wire guide and direct vision control. Finally, the previously placed warm-water-softened tracheal tube can be advanced over the probe of the endoscope into the trachea. The fibrescope should then be withdrawn to allow a period of oxygenation through the tracheal tube. The fibrescope may then be reinserted, via a fibreoptic bronchoscope swivel adapter, down to the carina to enable the anaesthetist to count four to five rings up from the carina to ensure that the tracheal tube can be placed so that it will not inadvertently advance into the right main bronchus.

Airway anaesthesia

Topical anaesthesia may be combined with a regional nerve block for airway anaesthesia, especially of the larynx. Use of the Jackson (Krause) right-angle toothed forceps with cocaine-impregnated packing of the pyriform fossae is the usual approach. Superior laryngeal nerve block and transtracheal instillation of local anaesthetic agent through the cricothyroid membrane are also used.[26]

Another technique is to ask a sitting patient to inhale deeply and slowly while the 'spray' from a local anaesthetic nebulizer, with a variable direction tip and oxygen as the carrier, is activated.[27]

It is important to spray both transnasally (inferior ethmoid passage) and orally in order to anaesthetize the dorsal surface of the uvula. This is equally important for both nasal and oral intubation using fibreoptic endoscopy. In the case of fibreoptic oral intubation, it is necessary to spray the hard and soft palate to prevent stimulation by the oral intubation airway.

The agents suitable for topical anaesthesia are lignocaine (lidocaine) 4 per cent and cocaine 2 per cent in a 10 : 2 ratio. If there is a history of thyrotoxicosis, hypertension or other contraindication, cocaine should be omitted. For nasal intubation, the nasal route may be packed with anaesthetic-soaked nasal packs using Magill, Cohen or Hartman forceps. On occasion, it may be necessary to dilate the nasal passage using progressively increased sizes of nasal trumpet.

When there is excessive mucoid secretion, this may be reduced by intravenous glycopyrollate or hyoscine. However, these drugs may increase the viscosity of the secretions, making it harder to suction them effectively through the small fibreoptic endoscope suction channel. This problem can often be resolved by passing a suction catheter through the contralateral nasal passage and down to the larynx. Intermittent suction can then be used as required.

Transnasal intubation

Transnasal intubation may be accomplished 'blindly' or by using Magill forceps or with a flexible fibreoptic scope. In spite of frequency of use, the nasal route is the least understood by anaesthetists from the viewpoint of anatomy and pathology. Severe nasal bleeding (Kiesselbach's plexus in Little's area), amputation of the ethmoid or inferior turbinate (Fig. 35.2) and impaction of foreign material into the orifice of the eustachian (pharyngotympanic) tube have occurred.

The reason for the choice of the nasal route is the relatively gradual curve, through which the tracheal tube passes to the glottic opening, in comparison with the oral route.

(a)

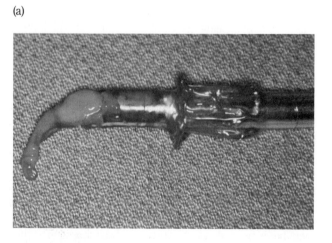

(b)

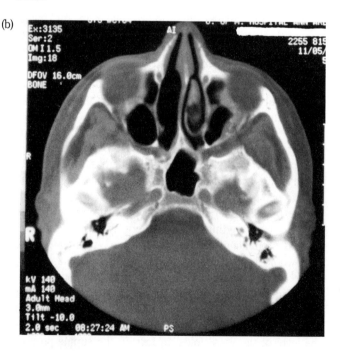

Fig. 35.2   (a) Extirpated ethmoid in tracheal tube tip. (b) Extirpated right ethmoid.

The nasal airway should be examined to determine whether there is an anterior obstruction due to septal deviation, anomaly of the inferior turbinates or floor of nasal passage, and presence of polyps. An otic speculum of appropriate size with otoscope can be used for this purpose. A deeper examination requires a nasopharyngoscope or the fibreoptic endoscope.

A long moistened cotton wool bud ('Q-tip') should then be inserted to obtain the tactile sense of a posterior obstruction as well as the direction and depth of the inferior ethmoid passage. If the depth (from maxillary entrance to posterior pharyngeal wall) is much less than 9–9.5 cm in the normal adult, suspect a posterior passage abnormality. As the Q-tip is withdrawn it should be swept in a circular motion to get a tactile sense of the canal diameter. Note that the size of the nares contributes nothing to this evaluation. One cannot rely on the patient's account of the ease of breathing through either side, unless there is an almost total obstruction.

Finally, the nasopharynx should be included in the examination as the flexible fibreoptic endoscope is inserted. Unsuspected polyps or other pathology may be revealed. Mirror nasopharyngoscopy (Fig. 35.3) can readily reveal posterior canal polyps and adenoidal masses, as well as lesions on the base of the tongue.

Following local anaesthetic preparation mentioned above, a Codman pack (2.5 × 10 cm) soaked in cocaine 2 per cent may be inserted for anaesthesia, vasoconstriction and canal dilatation, after which nasal trumpets, thoroughly lubricated with an anaesthetic gel, may be inserted to dilate and lubricate the canal. At this stage a previously prepared tracheal tube that has been softened in a warmed solution of saline or water may be inserted. If flexible fibreoptic intubation is to be used, it is imperative not to insert the tracheal tube too far because it may limit movement of the tip of the endoscope to be passed through it or, more frequently, it may impact against the posterior pharynx. Some anaesthetists feel that, although this is time consuming and requires meticulous attention, the result is a more comfortable patient, almost complete elimination of bleeding and improved intubation success.

Nasal intubation is not without other complications (i.e. maxillary sinusitis). Studies indicate that bacterial maxillary sinusitis is found in more than 60 per cent of recently intubated patients.[28–30]

Oral intubation guides

The oral route is usually used for fibreoptic intubation when the supine position is chosen. However, this approach does require the mouth to open sufficiently to insert an intubating oral airway. Thus, it is not possible when the mouth cannot be opened, as in temporomandibular joint ankylosis.

An oral intubation airway of choice is the Berman intubation airway which has a side slit so that it can be removed. Another guide is the Williams airway which has the disadvantage of not being similarly slotted and removable. A third frequently used approach is the Ovassapian airway which is particularly useful for a patient with a large broad tongue; its disadvantage is the open top (proximal to the hard and soft palate). When using oral intubation airways, the tracheal tube should be preplaced in the airway lumen but not beyond the distal opening analogous with the transnasal technique.

Lately, another device, the laryngeal mask, has made possible airway access for either the intubation guide or the flexible fibreoptic.[31] The main disadvantage of the oral approach in the supine position is that the diameter of the upper airway diminishes to approximately one-third of the sitting diameter when the patient is placed supine. In fact, this is why the Flexilume light wand approach may be equally efficacious, taking advantage of the funnel mechanics of the oral and laryngopharynx.

The technique for using this is similar to the nasal procedure, described above.

Sublaryngeal difficult airways

Other important uses of the fibreoptic endoscope include: (1) positioning of tracheal and endobronchial tubes; and (2) placement of tracheal tubes below areas of tracheal compression.

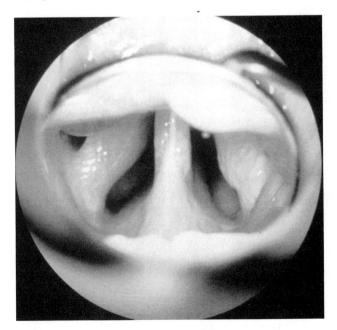

Fig 35.3 Mirror view of the nasopharynx.

Tracheal tube positioning

The value of the fibreoptic endoscope for the insertion of tracheal tubes through the airway has been discussed elsewhere. It is often forgotten that the usual techniques for checking tube placement are fraught with error. Auscultation does not guarantee proper placement. However, direct visualization via a tracheal fibreoptic swivel adapter can readily be accomplished, visualizing the carina and then counting the number of rings up to the tip of the tracheal tube, and is an absolute assurance of the tube position.

Similarly, the fibreoptic techniques taught by Ovassapian[32] for placement of endobronchial double lumen tubes and blockers provide absolute assurance of proper placement.

The compressed trachea

The usual cause for a compressed trachea is a massively enlarged thyroid. The most frequent effects may be lateral displacement, significant compression of the trachea and,

less often, of the main stem bronchi (substernal goitre). An obvious concern is tracheal torsion or further obstruction during surgical manipulation, and possible tracheal collapse after the tracheal tube is removed.

Tracheomalacia is a problem that must be considered following all long-term tracheal compression. It is relatively rare, being seen principally following surgery for Riedel's struma. This hard, enlarged thyroid can erode the tracheal cartilages. However, its effect is not obvious until the supportive tumour structure is removed, at which time the weathered tracheal cartilages collapse. These possibilities must be anticipated and the tracheal tube must be placed beyond the point of compression or potential collapse. This can best be achieved using fibreoptic evaluation.

The examples discussed in this section are the common sublaryngeal causes of the difficult airway. Other examples include mucopolysaccharidoses infiltrating the trachea and post-traumatic tracheal disruption. For these conditions, visual extubation of the trachea is of value in determining whether tracheal collapse will make extubation inadvisable at this time, while providing a visual guide for immediate reintubation.

(b)

(a)

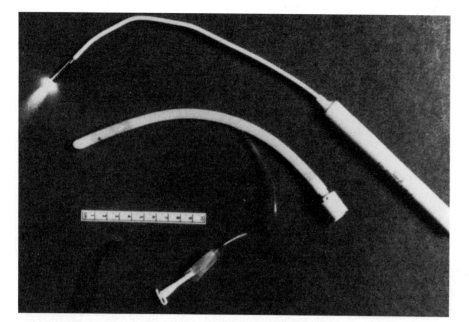

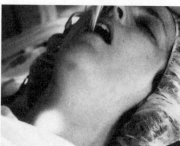

Fig. 35.4   (a) A light wand. (b) Light wand technique.

Instrumentation

Guides and light wands

The two simplest aids for problem intubations are intubation guides[33] and the light wand.[34]

Intubation guides

Intubation guides include the Sheridan TXX (hollow tubing), Eschmann woven and Norton teflon rod. Occasionally, these are called, erroneously, bougies. A bougie is a tapered instrument for introduction into tubular structures for dilatation and diagnosis of strictures. On the other hand, a guide is just that – an instrument to guide or direct another instrument along a particular pathway. Guides are long instruments that are passed through a tracheal tube and over which the tube is passed through the oropharynx and larynx into the trachea. They are often used at time of extubation in reverse manner. The guide is gently inserted through the tracheal tube and the tube is then carefully removed, leaving the guide in place. This ensures that reintubation can be accomplished if a patient with a difficult airway develops post-extubation oedema or other complication requiring reintubation.

Light wand

Light wand techniques (Fig. 35.4) involve inserting a light through the tracheal tube. The bulb of the 'wand' is illuminated. The room can be partly darkened and the reflection of light sought in the patient's neck. When the reflection is noted superficially in the midline, the tube is advanced into the trachea.[35]

Intubation with radiographic C-arm assistance

The following discussion is presented in some detail to describe a technique that is now available to the anaesthetist in some modern operating theatres.

Many techniques, ranging from simple to computer-assisted, are now available for radiological examination of the neck and airway. The traditional upper airway film, a lateral view of the soft tissue of the neck, is useful for a quick limited review of the upper airway. Xeroradiography,

laryngography and tomography are other plain film or 'cut film' imaging modalities of the upper airway.

Computer-driven modalities such as computed tomography (CT) can provide a valuable insight in their imaging of both the cervical spine and soft tissues of the neck. Digitized CT scans now make it possible to study tongue volume.[36] A more recent development, magnetic resonance imaging (MRI) offers detailed pictures. Similarly, the combination of cineradiographic observation with simultaneous polysomnography (somnofluoroscopy) provides a new technique for the diagnosis of sleep apnoea[37] and this also permits observation of the changes occurring during general anaesthesia.

C-head dynamic fluoroscopy, while not itself a new technique, offers a flexible, expeditious technique for assessment of the upper airway and for assisting tracheal intubation. The C-arm enables airway configuration, over a wide range of patient motion, to be quickly determined.

The C-arm is a fixed radiographic tube that is aimed at a phosphorus image receptor. The image receptor is then photographed using a small camera within the C-arm head. The photograph is displayed by the monitor on the C-arm monitor trolley. The final image can be adjusted by a change in radiation factors or standard brightness/contrast controls on the monitor itself.

The C-arm computer allows options for the operator such as fluorography with or without pixel averaging. (Pixels are the basic image units, the little dots that comprise the image on the monitor screen; when they are averaged, the 'graininess' of the image is reduced.) The patient is placed on the table in the sitting position for intubation. The radiologist and the anaesthetist determine the placement of the C-arm so as to insure that lateral and anterior/posterior radiographic positions are available and that the appropriate operating room table or patient position is used. Occasionally the neurosurgical Mayfield Swivel Horseshoe Head Rest may be very useful. The radiological monitor is placed so that it can be seen by both the radiologist and the anaesthetist.

During the intubation, the anaesthetist will view through the fibreoptic scope. The radiologist will provide a report of the image on the C-arm monitor screen. Using this method it has been possible to guide the fibreoptic scope when blood, mucus, redundant tissue, etc., have temporarily obscured direct fibreoptic visualization. The use of the C-arm for patients who have suffered trauma enables the airway to be visualized with the fibreoptic scope within it.

The use of the C-arm enables the anaesthetist or radiologist to know when the fibreoptic scope is being misdirected. Anterior/posterior (A/P) positioning of the fibrescope can be seen using the lateral view. The lateral view also identifies when the fibrescope is being directed to the left/right midline. Assuming that the image is a true lateral of the airway, the image of the scope, if in midline,

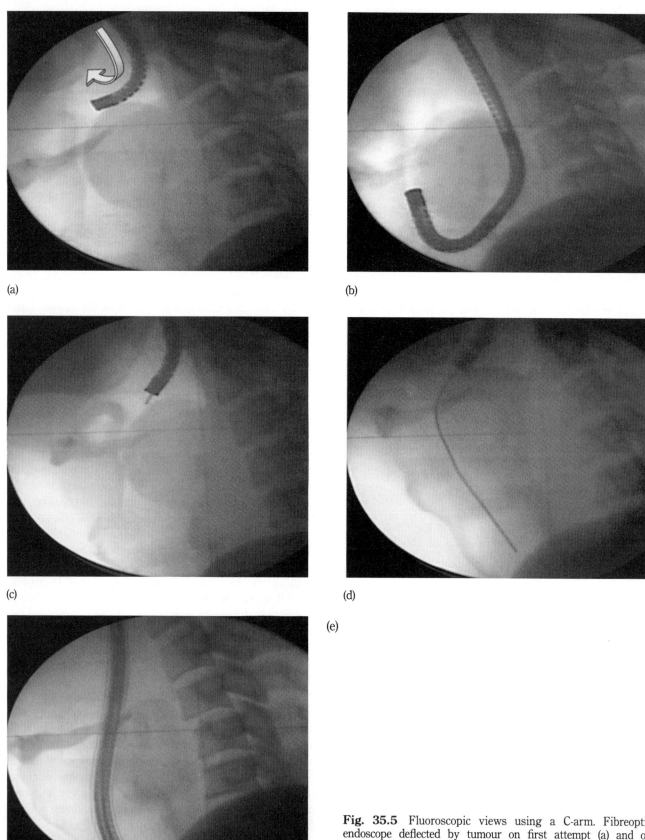

(a)

(b)

(c)

(d)

(e)

Fig. 35.5 Fluoroscopic views using a C-arm. Fibreoptic endoscope deflected by tumour on first attempt (a) and on second attempt (b). (c) The wire guide. (d) The wire guide in the trachea. (e) The fibreoptic endoscope in the tracheal tube, advancing.

should have a flat tip. Therefore, the first indication of a rounded appearance of the tip of the scope will alert the operator to the fact that the scope is moving away from the midline. Confirmation of left/right positioning can be made with the A/P view.

The most difficult intubations can often be performed with wire-guided intubation (Fig. 35.5). Under C-arm guidance, a tracheal tube is positioned in the nasal or oral passage, though not too deeply. The fibrescope is then passed through the tracheal tube and guided by direct vision and/or with the C-arm so as to be pointed at the tip of the epiglottis and in view of the glottic opening. A flexible-tip guide wire is passed through the fibrescope suction channel and, under real-time fluorographic vision, advanced along the posterior surface of the epiglottis, through the vocal folds and into the trachea to the carina. The fibrescope is then advanced (using the wire as a guide) into the trachea down to the point of the carina. The wire is then removed and the tracheal tube is advanced over the fibrescope. The fibrescope is used to confirm positioning of the tracheal tube tip at approximately four or five rings above the carina. The use of the C-arm need not be confined to those patients in whom the airway is known to be difficult; it may be used when no difficulty has been foreseen but when the intubation proves unexpectedly impossible.

Attention must be paid to radiation exposure dosage to both patients and staff. Patient exposure dosages average 300–450 mrad/min. This is very much in line with, or below, other ionizing imaging modalities. The average examination exposure time should be under 2 minutes' actual fluorographic time for diagnosis.

Jet ventilation

The term 'jet ventilation' has been used for different forms and means of introducing gas (oxygen) into the tracheobronchial tree. The technique may be used for laser surgery.[38] Another use has been advocated, i.e. for airway management of the difficult airway.[39] The latter involves the introduction of large-bore needles into the trachea.

It is essentially a blind technique and incorrect placement or excessive pressure can produce a pneumothorax, pneumomediastium, pneumotrachea, pneumolarynx or pneumopericardium. However, in extreme situations when time is crucial the technique can buy time.

In this era of technological advances, flexible fibreoptic endoscopes, the Bullard laryngoscope, the Bellhouse laryngoscope, the non-invasive light wand and intubation guides are recommended.

Jet ventilation via a 'needle' placed within the lumen of a

Dedo or Jako otolaryngological laryngoscope may permit laser surgery without a tracheal tube.[40] This is particularly suitable for patients with glottic, subglottic and tracheal stenosis or other obstruction.

Retrograde intubation

This approach, originally suggested by Waters, involves passing a needle via the cricothyroid membrane.[41] An extradural catheter is advanced cephalad through the needle, the larynx and out of the mouth. The tracheal tube is either passed over this guide or tied to it via the Murphy eye. The major difficulty is guiding the tracheal tube past the epiglottis and vestibular/vocal folds.

There are many variations on this theme. Unfortunately, there are also a number of complications to be considered, including infection, needle laceration of the posterior tracheal wall, needle trauma to the vocal folds, amputation of a vocal fold by a sawing motion when attempting to pull the tethered tracheal tube past the epiglottis, and haemorrhage. This technique, as in the case of the transtracheal jet needle, should be used only as a last resort.

The Bullard laryngoscope

The recent modification of the Bullard laryngoscope (Fig. 35.6) using a stylet/guide has increased its application. Its main advantage is that this device permits visualization of the upper airway, although limited to the orolaryngeal structures.

The Bullard is an excellent aid for diagnostic evaluation of glottic/supraglottic airway obstruction or other malfunction, post-thyroidectomy evaluation of the vocal folds and problems involving glossoptosis and basal macroglossia. However, it is inadequate for infraglottic diagnoses, placement of double lumen tubes or determining the position of the distal tip of a tracheal tube. In practised hands it can be used to introduce tracheal tubes through the glottic aperture.

Facial trauma

The patient with facial trauma may present problems of distorted anatomy, massive oedema, haemorrhage, urgency of management, shock and air hunger. These pathologies

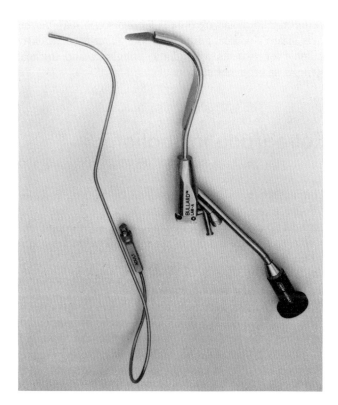

Fig. 35.6 The Bullard laryngoscope.

are seen primarily following automobile accidents and gunshot (especially shotgun) wounds. Facial trauma is often accompanied by patient restlessness and inability to cooperate. The surgical option of tracheostomy may be limited by massive trauma of soft tissue of the neck. In this circumstance the Bullard laryngoscope can be a useful aid in intubation.

The laryngeal mask

The laryngeal mask (Brain)[42–44] is now widely used in the UK and its use in the USA is increasing. The concept requires oral insertion of the device to cover the laryngeal aperture. It has proved most successful for short procedures such as uterine dilatation and curettage. However, concern has been expressed about the potential complications. There is confirmation (in at least one case based on legal action) of the following problems:

1. The need to open the mouth widely enough to insert the device, thus obviating its use in any situation with temporomandibular joint dysfunction or oropharyngeal obstruction.
2. Leakage of liquid gastric contents (regurgitation) around the rim of the mask and into the laryngotrachea.

3. Inadvertent placement into the oesophagous, at least in part, with disruption of the oesophageal introitus on inflation of the cuff (rim) of the mask.

Perhaps the most risky application is its use in obstetrics when the anaesthetist must be constantly alert to the 'full stomach'. Despite these concerns, this device appears to have valuable application in limited situations. After placement of the mask, an intubation guide or flexible fibreoptic endoscope may be placed via the laryngeal mask lumen into the laryngotrachea which will serve as a guide during introduction of the tracheal tube. However, the technique is limited by the mechanics of removing the laryngeal mask to allow advancement of the tracheal tube.

Endobronchial intubation

Endobronchial intubation may be required to control intrapulmonary or lower endobronchial bleeding, to permit one-lung anaesthesia, and for pulmonary lavage in cases of alveolar proteinosis. The technique, described by Ovassapian,[32] is ideal for the elective situation. Dependency on visualization through the endoscope will obviate its use in the bleeding patient. In this latter circumstance, the Fogarty embolectomy catheter is of value if properly placed via a rigid bronchoscope. Vigorous suction may be required.

The rigid bronchoscope

Foreign bodies in the airway as well as 'clothes line' and 'handle bar' (motorcycle, bicycle) injuries to the trachea represent a specific indication for use of the rigid bronchoscope. The terms 'handle bar' and 'clothes line' refer to trauma to the neck involving tracheal ring dislocation and, in rare cases, disruption. The injury is seen most frequently after motorcycle, bicycle or snowmobile accidents during which the rider is propelled toward the handle bar. Similar trauma is seen in touring car (convertible, cabriolet) accidents in which the top of the automobile is retracted, leaving a windshield extending upwards. The 'clothes line' cognomen refers to runners or cyclists who literally run into a clothes line, producing a transection-like injury. The important factor to keep in mind is that there may be little external sign of trauma, but the injury must be suspected from the history of the accident. The patient usually has discomfort on swallowing and breathing. There often is minimal stridor or fremitus. On looking into the larynx and trachea, one may see arytenoid displacement (high injury) with skewing of the horizontal axis of the arytenoid/interarytenoideus complex. More often, an appearance of a cricoid shelf is noted below the level of the cricoid, which is a manifestation of tracheal ring dislocation. If the mucosa is swollen and discoloured, disruption with haemorrhage and oedema is evident. In this latter circumstance, immediate intubation and internal stenting is required to prevent near-total airway occlusion. Note particularly that these patients rarely tolerate the supine position. This is explained by the biomechanics of airway support.[25]

This situation is encountered primarily in the emergency room or in the patient rushed to the operating theatre. The techniques for management are as follows:

1. Foreign bodies in the upper airway can usually be visualized using a nasal speculum or nasopharyngeal mirror (nasal foreign bodies), or by indirect and direct laryngoscopy. In the latter situations, the prime loci are tonsillar fauces, vallecula of the epiglottis, pyriform sinuses, vestibule of the larynx (between vestibular and vocal folds) and at the level of the cricoid ring.

2. Foreign bodies in the trachea (including cricoid ring and extending into the main stem and lower bronchi) are most frequently found on the right side because of the angulation of the bronchi. These require a rigid bronchoscope for removal. In the anaesthetic setting the problem commonly occurs in the obstetric delivery room following aspiration of solid matter.

   The bronchoscope should be inserted using direct laryngoscopy to rule out upper airway obstruction. The bevel of the bronchoscope should initially be facing posteriorly, visualizing any potential crico-pharyngeal lesion (e.g. Zenker's pouch or diverticulum). When entering the larynx, the bevel should be directed laterally to approximate the long axis of the rima glottidis and vocal folds. On passage, the level should again be directed posteriorly, watching for pulsations suggestive of aneurysmal encroachment on the soft wall of the trachea (especially in the elderly patient).

   When the foreign body is visualized, it should be impacted into the distal end of the bronchoscope (by grasping forceps or use of a Fogarty catheter vascular balloon). Then both the object and the bronchoscope are removed as one unit. No attempt should be made to pull the object through the bronchoscope.

3. 'Clothes line' or 'handle bar' injuries present a very different problem. There may or may not be a true 'fracture' of the tracheal cartilages. Usually, there is more of a disruption of continuity of the trachea with dislocations of the 'rings'. This then presents a figure of 8 configuration with some overlapping. The mucosae may not be disrupted, but may have an internal shelf-like configuration. The objective then is to restore the anatomical pattern of the rings and splint them for optimum channel integrity. A bronchoscopic telescope (0 or 180 degrees) is of great value in delineating the lumen. The bronchoscope itself can be used as a tracheal stent or passed through a tracheal tube. Once in place the bronchoscope can be removed, leaving the tracheal tube in the tracheal lumen as a stent. An alternative approach is rigid bronchoscopy, followed by insertion of an intubation guide to assist the passage of a tracheal tube. This technique must be used with extreme caution because the tip of the tracheal tube may impinge on the dislocated tracheal ring, causing further disruption of the channel.

The essential warning in these situations is never to advance an instrument without visualizing the distal pathway, and certainly never against resistance.

Transmission of infection

Some patients with upper airway problems have chronic infection, particularly in the nasal passages and paranasal sinuses. Patients with relatively flat facies, a depressed nasal bridge and a deviated nasal septum or post-traumatic obstruction fall into this category, as do those with herpetic lesions.

The problem that arises is the transmission of infection via the endoscopes or tracheal tubes during the process of 'controlling' the airway. Concern should be directed towards disinfection or sterilization of the instruments.[45] Instruments, including suction channels, should be flushed with cold water as soon as possible after use to wash away blood and other debris before it dries. Insertion of a fresh channel brush in fibreoptic suction channels to remove particulate matter is also advised. (There remains some question of residual contamination on channel brushes that are re-used.)

Postoperative management

Eventually, the patient must be extubated. In people who have had problems with airway management, very serious consideration should be given to using the intubation guide as an extubation/re-intubation guide. This can be done with ease in adults, and the option is now readily available for children using the Sheridan TXX paediatric guide.

When the decision is made to extubate the patient, the tracheal tube connector (proximal end) should be removed. Then the guide is passed intraluminally to a depth below the distal end of the tracheal tube. The tube is removed, leaving the guide in place for 1.5–2 hours. Experience has confirmed that situations requiring re-intubation occur well within this period. Simultaneously, the larynx and trachea should be moisturized with appropriate nebulization (e.g. racenephrine (Vaponefrin) in saline) and dexamethasone may be administered to counteract oedema development. (Patients may be given dexamethasone 4 mg before intubation, 4 mg more during surgery and 4 mg in the recovery room.)

An ice collar is often applied, although this may have more psychological effect than physical value.

Perhaps the most useful ancillary aid is the sitting position. This promotes venous drainage and biomechanically opens up the airway, as described above.

Paediatrics

Airway management for paediatric patients requires an understanding of developmental anatomy linked with an understanding of the mechanics of the airway function of the tongue, pharynx, epiglottis, larynx and subglottic tissues (cricoid ring). The bony configurations of the mandible, temporomandibular joints, palate, nose and dentition must also be considered.

Patients may present with morphological abnormalities or abnormal function due to a variety of disease processes. The most common morphological anomalies involve mandibular dysostoses such as the Pierre Robin anomaly

as well as with associated macroglossia (Bechwith–Wiedemann syndrome and cystic hygroma, among others). Careful assessment is crucial, as many of these syndromes have coexistent mid-facial hypoplasia (Stickler, Treacher Collins, Goldenhar) or velocardiofacial syndromes with pharyngeal hypotonia.

The mucopolysaccharidoses (Hurler and Hunter syndromes) represent aberrations of both structure and function. In these patients a combination of facial, upper airway and cervical spine (often with thoracic components) anomalies produces major difficulties in airway management.

Apart from the above, airway obstruction may occur for a variety of reasons even in morphologically normal patients. The anteroposterior diameters of head and thorax which influence neck extension, as well as relative macroglossia and glossoptosis, contribute to our concerns. Thus, variation in the patient's oxygenation and ventilation requirements will also affect flow patterns across the site of a relative obstruction.

Management of these patients is superficially analogous to that of the adult. However, the margin of safety and tendency to convert from urgent to emergency airway conditions is increased.

Reviews, references and criteria

Reviews discussing airway problems can be found in Latto and Rosen's excellent book *Difficulties in tracheal intubation* and Norton's *Atlas of the difficult airway* (see Further Reading).

REFERENCES

1. American Society of Anesthesiologists, Committee on Professional Liability. Preliminary study of closed claims. *ASA Newletter* 1988; **4**, 52: 8–10.
2. Caplan RA, Ward RJ, Posner K, *et al.* Unexpected cardiac arrest during spinal anesthesia: closed claims analysis of predisposing factors. *Anesthesiology* 1988; **68**: 5–11.
3. Caplan RA, Posner K, Ward RJ, *et al.* Adverse respiratory events in anesthesia: a closed claims analysis. *Anesthesiology* 1990; **72**: 828–33.
4. Holland R. Anesthesia related mortality in Australia. In: Pierce EC, Cooper JB, eds. *Instructional anesthesia clinics.* Boston: Little, Brown, 1984: 61–71.
5. Department of Health and Social Security. *Report on confidential enquiries into maternal deaths in England and Wales, 1979–1981*, Report on Health and Social Security Subjects, 29. London: HMSO, 1986.
6. King TA, Adams AP. Failed tracheal intubation. *British Journal of Anaesthesia* 1990; **65**: 400–14.
7. Patil VU, Stehling LC, Zaunder HL. *Fiberoptic endoscopy in anesthesia.* Chicago: Year Book Medical, 1983.
8. Mallampati SR, Gatt SP, Gugino LD, *et al.* A clinical sign to predict difficult tracheal intubation: a prospective study. *Canadian Anaesthetists' Society Journal* 1985; **32**: 429–34.
9. Wilson ME, Spiegelhalter D, Robertson JA, Lesser P. Predicting difficult intubation. *British Journal of Anaesthesia* 1988; **61**: 211–16.

10. Frerk CM. Predicting difficult intubation. *Anaesthesia* 1991; **46**: 1005–8.

11. White A, Kander PL. Anatomical factors in difficult direct laryngoscopy. *British Journal of Anaesthesia* 1975; **47**: 468.

12. Mark LJ, Beattie C, Ferrell CL, *et al.* The difficult airway: mechanisms for effective dissemination of critical information. *Journal of Clinical Anesthesia* 1992; **4**, 3: 247–51.

13. Benjamin BNP. *Diagnostic laryngology*. Sydney: Saunders, 1990.

14. Latto IP, Rosen M, eds. Management of difficult intubation. In: *Difficulties in tracheal intubation*. Eastbourne: Baillière Tindall, 1985: Ch. 7.

15. Norton ML, Brown ACD, eds. The difficult airway clinic. In: *Atlas of the difficult airway*. St Louis: Mosby/Year Book, 1991: Ch. 4.

16. Norton ML, Wilton N, Brown AC. The difficult airway clinic. *Anesthesiology Review* 1988; **15**: 25–8.

17. Smalhout BA. *Quick guide to capnography and its use in differential diagnosis*. Böblingen, Germany: Hewlett Packard Medical Products Group, 1983.

18. Elze CJ. Anatomie des Kehlkopfes. In: Denker A, ed. *Handbuch des Hals – Nasen Ohrenteilkunde*. Berlin: Springer, c. 1868.

19. Dripps RD, Eckenhoff JE, Vandam LD. *Introduction to Anaesthesia*, 6th ed. Philadelphia: WB Saunders, 1982.

20. Sellick BA. Cricoid pressure to control regurgitation of stomach contents during induction of anaesthesia. *Lancet* 1961; **2**: 404.

21. Jones BC. Lingual nerve injury: a complication of intubation. *British Journal of Anaesthesia* 1971; **43**: 730.

22. Terchner RL. Lingual nerve injury: a complication of orotracheal intubation. *British Journal of Anaesthesia* 1971; **43**: 413–44.

23. Bock-Utney JG, Jaffo RA. Tracheal intubation with the patient in a sitting position. *British Journal of Anaesthesia* 1991; **67**: 225–6.

24. Fink BR. *The human larynx – a functional study*. New York: Raven Press, 1975: 42.

25. Fink BR, Demarest RJ. *Laryngeal biomechanics*. Cambridge MA/London: Harvard University Press, 1978.

26. Norton ML, Brown ACD, eds. Fig. 15–4. In: *Atlas of the difficult airway*. St Louis: Mosby/Year Book, 1991: 184.

27. Norton ML, Brown ACD, eds. Fig. 15–3. In: *Atlas of the difficult airway*. St Louis: Mosby/Year Book, 1991: 184.

28. Gosgnach M, Ghedira S, Xiang J, *et al.* Incidence of maxillary sinusitis in recently intubated critically ill patients. *Anesthesiology* 1990; **73**: A1224.

29. Leguillou JL, Rouby JJ, Xiang J, *et al.* Endotracheal intubation induced sinusitis and nosocomial pneumonia in critically ill patients. *Anesthesiology* 1991; **75**: A1019; and Endotracheal intubation induced maxillary sinusitis: a prospective randomized study. 1020.

30. Rouby JJ, Rossignon MD, Nicolas MH, *et al.* A prospective study of protected bronchoalveolar lavage in the diagnosis of nosocomial pneumonia. *Anesthesiology* 1989; **71**: 679–85.

31. Allison A, McCrory J. Tracheal placement of a gum elastic bougie using the laryngeal mask. *Anaesthesia* 1990; **45**: 419–20.

32. Ovassapian A, Schrader SC. Fiber-optic-aided bronchial intubation. *Seminars in Anesthesia* June 1987; **7**: 2; 133–42.

33. Benger RC, Chang J. A jet-stylet catheter for difficult airway management. *Anesthesiology* 1987; **66**: 221–3.

34. Ducrow M. Throwing light on blind intubation. *Anesthesia* 1978; **33**: 827–9.

35. Norton ML, Brown ACD, eds. Fig. 15–11. In: *Atlas of the difficult airway*. St Louis: Mosby/Year Book, 1991: 193.

36. Lowe AA, Gionhaku N, Takeuchi K, *et al.* Three-dimensional CT reconstructions of tongue and airway in adult subjects with obstructive sleep apnea. *American Journal of Orthodontics and Dentofacial Orthopedics* 1986; **90**: 364–73.

37. Katsantonis GP, Walsh JK. Somnofluoroscopy: its role in the selection of candidates for uvulopalatopharyngoplasty. *Orolaryngology, Head and Neck Surgery* 1966; **94**: 1.

38. Norton ML, Strong S, Vaughan C. Endotracheal intubation and Venturi (jet) ventilation for laser microsurgery of the larynx. *Annals of Otology, Rhinology and Laryngology* 1976; **85**: 656–64.

39. Benumof JL. Management of the difficult adult airway. With special emphasis on awake tracheal intubation. *Anesthesiology* 1991; **75**: 1087–110.

40. Spink LK. *Principles and practice of flow meter engineering*, 8th ed. Foxboro MA: Foxboro Co, 1958.

41. Waters DJ. Guided blind endotracheal intubation. *Anaesthesia* 1963; **18**: 158–62.

42. Brain AI, McGhee TD, McAteer EJ, Thomas A, Abu-Saad MA, Bushman JA. The laryngeal mask airway. Development and preliminary trials of a new type of airway. *Anaesthesia* 1985; **40**: 356–61.

43. Brain AI. Three cases of difficult intubation overcome by laryngeal mask airway. *Anaesthesia* 1985; **40**: 353–5.

44. Brain AI. The laryngeal mask – a new concept in airway management. *British Journal of Anaesthesia* 1983; **55**: 801–5.

45. Centers for Disease Control. Nosocomial infections and pseudoinfection from contaminated endoscopes and bronchoscopes. *Wisconsin and Missouri MMWR* 1991; **40**: 675–8. Reprinted in *Journal of the American Medical Association* 1991; **266**: 2197–8.

FURTHER READING

Latto IP, Rosen M, eds. *Difficulties in tracheal intubation*. Eastbourne: Baillière Tindall, 1985.

Norton ML, ed. *Atlas of the difficult airway*, 2nd ed. St Louis: Mosby/Year Book, 1991.

Chapter 36

Anaesthesia for the Patient with Renal Disease

Jennifer M. Hunter

Anaesthesia for patients with renal dysfunction has become commonplace since the advent of renal transplantation, but such patients may also require anaesthesia for unrelated conditions. Their medical history may be complex, with many systems affected, particularly if they have had renal dysfunction for several years. It is therefore important to take a full preoperative history and examination and to evaluate the drug therapy, often complex, that the patient is receiving. This chapter includes a discussion about the 'chronic' (long-term) dialysis techniques available to renal patients and the medical complications that these patients may suffer. The required preoperative preparation is outlined and the most appropriate anaesthetic techniques are suggested. It deals mainly with the patient in end-stage renal failure undergoing supportive dialysis treatment, although the same basic principles are applicable to the patient with deteriorating renal function who has not yet reached this stage. The special anaesthetic problems of paediatric patients presenting for renal transplantation are dealt with in Chapter 59.

Chronic dialysis techniques

Patients in end-stage renal failure either are maintained on haemodialysis for periods of 4–6 hours two to three times per week or use continuous ambulatory peritoneal dialysis (CAPD). Haemodialysis removes fluid and metabolic products from the plasma, replacing them with isotonic fluid and, if necessary, blood or plasma. The patient uses either an arteriovenous fistula in the arm or a prosthetic vein graft to furnish venous access for the haemodialysis machine by an indwelling cannula. For short-term purposes, a two-channel central venous line may be inserted into a subclavian or jugular vein.

CAPD involves the insertion of a peritoneal dialysis catheter, which must be cared for daily using a strict aseptic technique. The patient slowly runs three or four 1-litre bags of dextrose solution (1.36 per cent) into the peritoneal cavity every 24 hours, draining the fluid out before a new bag is commenced. The regular introduction of fluid into the peritoneal cavity irritates the omentum, increasing tissue vascularity. This produces a large area for the passive diffusion of products of metabolism from the plasma into the peritoneal fluid. CAPD is easier for the patient to use than haemodialysis, allows the patient more freedom to travel and is a much cheaper maintenance therapy. It is therefore becoming an increasingly popular method of dialysis. Insulin can be added to the litre bags for diabetic patients, producing better control of blood sugar than intermittent injections. Excess fluid can be removed from the patient by the use of one high osmolality dextrose (3.86 per cent) bag every 24 hours.

The main disadvantage of CAPD is the risk of peritonitis; adhesions form around the omentum and decrease the efficacy of the technique. If infection cannot be cleared up, resort must be made to haemodialysis. CAPD has, because it is cheaper, allowed more patients access to maintenance therapy. It is particularly useful in the elderly with cardiovascular disease, in whom haemodialysis may cause large swings in blood pressure and possibly induce attacks of angina.[1] CAPD has been shown to be as efficient as haemodialysis.[2]

The medical problems of patients with renal disease

This section discusses the cardiovascular, biochemical and haematological problems experienced by patients with renal disease.

Cardiovascular

Hypertension

Over 50 per cent of adult patients with renal disease need treatment for hypertension, which is thought to be caused in part by excess renin production from the diseased k e) The increased blood pressure is often resistant to c 1 ipy, and the patient will often be taking two or more antihypertensives. Calcium antagonists such as nifedipine and diltiazem are frequently employed, as are β-blocking drugs, usually those with specific β1-blocking properties such as atenolol or acebutolol. Angiotensin-converting enzyme (ACE) inhibitors such as captopril and enalapril have become increasingly popular in the management of patients with renal dysfunction, although, as these drugs are excreted in part through the kidney, the dose should be decreased. Vasodilators such as hydralazine may also be used, although this particular drug is better reserved for intravenous use in the acute state because of the high risk of side effects when used long term (e.g. symptoms similar to systemic lupus erythematosus). Hypertension is thought to be less marked in patients undergoing dialysis who have had a bilateral nephrectomy, possibly because of the subsequent lack of renin production from the native kidneys.

Rarely, hypertension in the patient with renal dysfunction may be due to renal artery stenosis, especially if the patient has received a renal transplant. Hypertension from this cause is particularly difficult to treat with drug therapy; indeed, ACE inhibitors are known to precipitate a deterioration in renal function in these circumstances.[3] The possibility of renal artery stenosis being the cause of an intractable rise in blood pressure should be excluded using radiological techniques. It may also be possible, if a narrowing is found, to perform an angioplasty to reduce the stenosis. This is an invaluable radiological technique, as surgical intervention in such cases, although increasingly common, can be hazardous.[4] There is a significant risk of a cerebrovascular accident during the perioperative period, owing to difficulty in controlling the very high and labile blood pressure.

Peripheral vascular disease is also more common, and occurs at an earlier age, in patients with renal dysfunction.

Myocardial infarction

Myocardial infarction is ten times more common in the patient with a renal transplant than in the general population of the same age and sex.[5] Episodes of angina are frequent, and intercurrent anaemia is usually a contributory cause (see below). Myocardial arrhythmias, which are exacerbated by metabolic problems such as hyper- and hypokalaemia and hypocalcaemia, may also ensue.

Pulmonary oedema

Pulmonary oedema is a common problem in patients with renal dysfunction, for two reasons. The first is that patients may simply be overloaded with fluid which they are unable to clear through the diseased kidney. The fluid therefore enters the interstitial tissues from the overloaded intravascular compartment; the process is potentiated by hypoproteinaemia. Increased ventricular filling pressures lead to left heart failure and pulmonary oedema.

The second reason is that acute pulmonary oedema may also develop as a consequence of a myocardial infarction or myocardial ischaemia, which are frequent incidents during the perioperative period in the renal patient.[5] More than half the patients with renal failure who present for transplantation are also in congestive (right) heart failure.[6]

Pericarditis

Pericarditis is not uncommon in patients with renal dysfunction and is sometimes accompanied by a pericardial effusion. Such problems are more common in the poorly controlled patient, who is overloaded with fluid, and can occur in the absence of infection. Such clinical signs must be taken seriously, and anaesthesia should if possible be delayed until the causal condition has been treated, because of the risk of severe myocardial depression per- and postoperatively.

Biochemical

Potassium

Decreased excretion of this mainly intracellular electrolyte by the patient with compromised renal function frequently results in a chronic rise in total body potassium, which is largely resistant to treatment with drug therapy (e.g. glucose and insulin or calcium polystyrene sulphonate) which only induces a transient improvement. Patients who present for surgery with a high serum potassium should be dialysed before anaesthesia is induced. Surgical trauma, suxamethonium (succinylcholine) and hypercapnia all cause the serum potassium to rise, sometimes at very fast rates, with the risk of cardiac arrhythmias and even ventricular fibrillation during surgery. High concentrations of potassium in some of the older preservation fluid used for transport of cadaveric kidneys (e.g. Collin's solution) could also induce hyperkalaemia sufficient to cause cardiac arrest during renal transplantation.[7]

Higher plasma potassium levels are usually recorded in patients receiving peritoneal rather than haemodialysis, probably because the technique is slightly less efficient.

Hypokalaemia is unusual in the patient with renal dysfunction; it can be interpreted as being due to fluid overload in the anephric patient, or to loop diuretics in the patient who still produces urine or has recently had a renal transplant. It must be remembered that non-depolarizing muscle relaxants are potentiated by hypokalaemia.

Sodium

In contrast to potassium, sodium is mainly an extracellular cation. The serum level is therefore a more accurate estimate of the total body content. With proper dietary control of intake, appropriate diuretic therapy and dialysis if necessary, the serum sodium can be managed easily. A rise in serum sodium is usually an indication of dehydration rather than an excess of sodium ions; a fall is usually due to fluid retention, caused by excessive fluid intake and/or inadequate dialysis.

Calcium and phosphate

Patients with renal failure usually suffer from hypocalcaemia and hyperphosphataemia, sometimes accompanied by the deposition of calcium in body tissues, in particular the blood vessels. Heavily calcified arteries may be seen throughout the body on x-ray, and the coronary arteries may be involved. Affected vessels lose their power of autoregulation, and patients with widespread vessel calcification are often intractably hypertensive and at high risk of myocardial infarction.

In contrast, the patient with a functioning renal transplant may develop hypercalcaemia and hypophosphataemia in response to hypertrophy of the parathyroid glands and the release of excess parathyroid hormone – *tertiary hyperparathyroidism*.[8] Hypercalcaemia potentiates acetylcholine release from the presynaptic vesicles of the neuromuscular junction; thus it potentiates neuromuscular transmission and decreases the potency of non-depolarizing neuromuscular blocking drugs, making adequate relaxation difficult to achieve when such patients present for surgery such as parathyroidectomy. Hypocalcaemia, in

contrast, by impairing acetylcholine release, might potentiate neuromuscular block.

Aluminium and magnesium

Although *aluminium* is not present in the body in large amounts, it may accumulate in patients in chronic renal failure, especially if they are undergoing regular haemodialysis in an area with a high aluminium content in the tap water. This is more common in hard water, which contains a higher level of aluminium salts. As patients with renal disease are also more prone to increased gastric acidity, they often consume large amounts of antacids. If these contain aluminium (e.g. aluminium trisilicate), the plasma level may also rise. Aluminium may be deposited in the brain, causing confusion[9] and even epileptic fits. It decreases myocardial contractility and, in contrast to calcium, release of acetylcholine from the presynaptic membrane is reduced. Thus neuromuscular block might be expected to be potentiated.

An excess of *magnesium* ions, also consumed in antacid therapy, may produce similar effects.

Acid–base changes

Renal impairment is associated with a failure to excrete hydrogen ions adequately. Accumulation results in a fall in intra- and extracellular pH, thus decreasing myocardial contractility and renal blood flow. The entry of hydrogen ions into the cells is accompanied by the movement of potassium ions out of the cells into the plasma, causing a further rise in serum potassium. Plasma bicarbonate levels fall – useful indications of the severity of disease and of inadequate dialysis.

Glucose

The decreased excretion of glucose by the diseased kidney makes control of blood glucose difficult, especially in the insulin-dependent diabetic patient in renal failure, or in the patient receiving drugs that increase blood glucose (e.g. corticosteroid therapy). Hyperglycaemia increases the risk of infection in the postoperative patient and is in part the reason why, before the advent of the immunosuppressant drug cyclosporin, when steroid therapy was commonplace, renal transplantation was less successful in the diabetic patient. A rise in blood sugar may be accompanied by a rise in serum potassium, making cardiac arrhythmias more likely.[10]

The diabetic patient in renal failure will probably have significant cardiac disease. They are a high anaesthetic risk, and need detailed preoperative assessment (see below). They may also have impaired vision and, because of this, more effort is required to explain the details of anaesthesia and surgery.

Albumin

Plasma albumin levels are often reduced in patients with renal disease. This may be due to haemodilution, to impaired synthesis if the patient is catabolic or malnourished, to an increased loss of protein through the diseased nephron, or to combinations of these factors. Alterations in plasma albumin may alter the degree of binding of many drugs. This is particularly relevant with drugs that are highly protein bound (98 per cent), such as warfarin, diazepam and phenytoin. Here only 2 per cent or less of the drug in the plasma is unbound and thus free to have an effect. A small decrease in plasma albumin, and thus drug binding, can cause a significant increase in available free drug; were unbound drug to increase from 2 to 3 per cent, this would represent a 50 per cent increase, markedly potentiating drug effect. Most anaesthetic agents are not predominantly bound to plasma albumin: muscle relaxants are mainly bound to plasma globulins, although not to a significant degree.[11]

Haematological

Anaemia

Patients with renal disease are commonly anaemic. The anaemia is of the normochromic, normocytic type and unresponsive to oral iron therapy. It is more severe in patients on haemodialysis than on peritoneal dialysis, possibly because of the increased trauma to the red blood cells that occurs with the former technique and the inevitable blood loss that occurs, to a small degree, every time a patient is subjected to haemodialysis. The anaemia is thought to be due to decreased erythropoiesis resulting from decreased release of the hormone erythropoietin by the diseased kidney. Patients with polycystic renal disease, in which the kidneys are replaced by large cysts, may, however, continue to produce the hormone and are less anaemic.

Erythropoietin (EPO) is available commercially.[12] This substance is genetically engineered and expensive to administer on a long-term basis. It rapidly raises the haemoglobin concentration, depleting iron stores if iron and folic acid are not administered simultaneously. The haemoglobin concentration will return to normal values over a period of 2–3 weeks, increasing the patient's sense of well-being, exercise tolerance, appetite and libido. Such rapid rises in haemoglobin concentration are not without problems, however; in 30–35 per cent of patients, hypertension worsens, probably due to increased blood viscosity and peripheral resistance, and arteriovenous fistulae, used for haemodialysis, may thrombose. There is an increased risk of a cerebrovascular accident[13] and 5 per cent of patients develop epileptiform convulsions.[14] Liver function tests may become abnormal[15] and the serum potassium may rise,[16] possibly due to the improved appetite, although dialyser clearance may be reduced by the increased haematocrit (packed cell volume). A 'flu-like illness may develop on starting therapy, though this does not seem to be due to antibody formation. The haemoglobin concentration should be raised to only about 10 g/dl, to optimize oxygen delivery to the tissues without increasing blood viscosity excessively.

Determining a maintenance dose of erythropoietin, once an optimal haemoglobin has been achieved, is difficult. Although usually administered by the intravenous route, a recent report has described the successful use of smaller doses of erythropoietin administered subcutaneously, with a significant reduction in cost.[17]

Haemolysis

The red blood corpuscles of patients with renal disease have a shorter half-life than those of a healthy person who is not uraemic; thus if the cells of a renal failure patient were to be given to a healthy patient, their half-life would increase. The increased haemolysis potentiates the normochromic anaemia and depletes iron stores further.

Platelet dysfunction

Although the platelet count may be normal in the uraemic patient, platelet dysfunction may occur (thrombasthenia), prolonging bleeding times and thus increasing the risk of haemorrhage during surgery, especially in the patient who has not been adequately dialysed. The surgeon must be careful to control bleeding sites in the patient with renal failure.

Heparin

The use of this drug to control the bleeding time during haemodialysis may also potentiate blood loss if surgery is carried out soon afterwards. A heparin titre should be estimated before an antagonist is used, for there is always the risk of excessive coagulability if protamine is given in the absence of heparin.

Blood transfusion

It was once considered inappropriate to transfuse regularly the anaemic patient with renal disease. The benefit of the (transient) rise in haemoglobin was thought to be outweighed by the risks of transmission of blood-borne infection, such as serum hepatitis, and of the presence of unwanted antigens on the surface of the transfused blood cells which could stimulate antibody production and subsequently lead to rejection of an allogenic renal transplant. With improved techniques for screening donated blood, which reduce the risk of transmission of infection, and following studies reporting a better transplant survival in patients who had been regularly transfused with third party blood in the months preceding transplantation,[18] preoperative transfusion is used more frequently. However, in some transplant centres, where live related donor transplantation is carried out, only blood from the future donor is used for preoperative transfusion, and third party blood is avoided, except in life-threatening circumstances, to reduce the risk of organ rejection. The kidney survival rates in these circumstances are impressive – over 90 per cent at 1 year.[19]

Tissue oxygenation

It is thought that, in the anaemic patient with renal disease, there is a shift to the right of the oxygen dissociation curve, increasing oxygen delivery to the tissues, despite the low haemoglobin concentration. Increased delivery of oxygen may also be due to an increase in cardiac output which may occur in the renal failure patient; this is a further reason for not aiming to raise completely the haemoglobin concentration to normal values with erythropoietin because this will increase cardiac workload.

Preoperative assessment

As well as a routine preoperative assessment, the following points need to be given particular attention in the patient with renal disease.

Haemoglobin, platelet count and white blood cell count

These must be noted. If the platelet count is less than 50 000 preoperatively, this should be investigated. It may be that more adequate dialysis is necessary, or that covert infection is present, which needs treatment before surgery.

Blood cross-match

As patients with renal disease are frequently anaemic, it is often necessary to have blood cross-matched for surgery in which it would not normally be required for a healthy patient. A haemoglobin of over 8 g/dl may be considered

normal in renal patients; these patients can lose up to 500 ml of blood during surgery before blood is given; for a lesser loss, crystalloids are sufficient. If the haemoglobin is in the range 5–8 g/dl, intraoperative blood loss should be replaced with blood; if the level is below 5 g/dl preoperatively, transfusion is required before surgery. This is usually carried out during haemodialysis if the patient is receiving such treatment, to decrease the effect of the volume load, as fluid can be removed as necessary. Otherwise, care must be taken that the transfusion load does not precipitate left ventricular failure. A loop diuretic should accompany blood transfusion if the patient is still responsive to frusemide (furosemide); in the patient on peritoneal dialysis, a high-dextrose (3.86 per cent) bag may need to be used to remove excess water.

Plasma sodium, potassium, bicarbonate and urea

These should always be checked on the day of surgery – yesterday is not good enough. Rapid changes can occur in these variables, making an immediate preoperative level essential. If the serum sodium is low (<125 mmol/l) or the potassium level high (>5.5 mmol/l), the patient needs further dialysis before surgery. If the plasma bicarbonate is reduced, the drug or dialysis therapy is inadequate. Acidotic patients will have a prolonged effect from neuromuscular blocking drugs, increased myocardial depression from anaesthetic agents and a higher risk of postoperative complications. The acidosis should be treated prior to surgery.

Plasma albumin

Plasma albumin should be measured. If this is low, it may be due to the dilutional effect of fluid overload or the presence of infection. Anaesthesia carried out in such circumstances carries an increased risk: drugs may have a prolonged action, and pulmonary oedema, especially at the end of anaesthesia after a period of artificial ventilation, is more common.

Chest x-ray

A chest x-ray is mandatory before anaesthesia in a patient with severe renal dysfunction, whatever the age. It will give an indication of pulmonary congestion, especially in the upper lobes, well before frank pulmonary oedema is present (Fig. 36.1). Increased hilar markings may also be present before any clinical symptoms can be detected. Areas of pulmonary collapse may be present which may be asymptomatic, especially in the patient receiving immunosuppressive therapy.

Cardiac enlargement, common in patients with renal disease (Fig. 36.2), is usually associated with hypertension

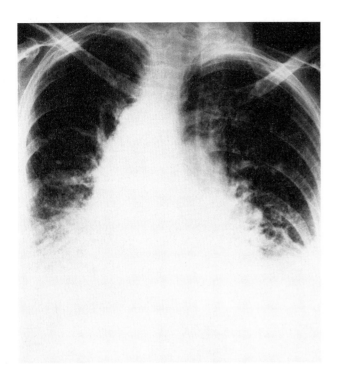

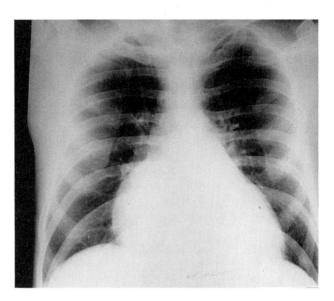

Fig. 36.2 A preoperative chest x-ray of a man presenting for renal transplantation. Note the right and left ventricular enlargement.

Fig. 36.1 A preoperative chest x-ray of a woman with a failing renal transplant, presenting for insertion of a CAPD cannula under general anaesthesia. Note the increased pulmonary markings and bilateral pleural effusions, together with cardiac enlargement.

but sometimes is due to ischaemic disease or to a pericardial effusion.

Electrocardiography

An ECG is necessary in these patients preoperatively. Cardiac enlargement, myocardial ischaemia, evidence of previous myocardial infarction and arrhythmias are common.

Drug therapy

This should be studied in detail. Most patients with significant renal disease are receiving oral iron, folic acid, sodium bicarbonate and calcium therapy. They may also be receiving diuretics (in which case, note the electrolyte status) and antihypertensive agents – if on β-blocking drugs, particularly note the pulse rate. Be aware that more than two antihypertensive agents suggests that the problem is refractory and that the blood pressure is likely to be labile peroperatively and into the immediate postoperative period, especially if blood loss is significant.

Note if the patient is receiving steroid therapy or other immunosuppressant drugs. Steroid cover will need to be increased peroperatively: it is useful to give the normal morning oral dose early on the day of surgery and then start hydrocortisone 100 mg 6-hourly, intravenously or intramuscularly, at induction of anaesthesia.

It should also be noted if the patient is receiving the immunosuppressant cyclosporin[20] prior to surgery, or if it is planned to introduce this drug peroperatively by the intravenous route. Cyclosporin potentiates non-depolarizing neuromuscular blocking agents[21] and therefore necessitates the use of neuromuscular monitoring peroperatively.

Cyclosporin, which is a more specific immunosuppressant than its predecessors, azathioprine and the steroids, acts mainly on the T-lymphocyte to suppress the immune response. Its use has improved graft survival significantly.[19,20] It is a fungal peptide which is metabolized in the liver by the cytochrome P_{450} enzyme system. It potentiates the action of other drugs metabolized by this enzyme system, such as nifedipine, propofol, phenytoin, phenobarbitone and cimetidine. Unlike cortisone, it does not affect the control of blood sugar and has improved the success of transplantation in diabetic patients. But cyclosporin does have side effects. It is nephrotoxic, and this can cause problems in achieving an appropriate maintenance dose and in differentiating toxicity from acute rejection; it also causes gum hypertrophy and hirsutism.

Postoperative bacterial infections were common when steroids and azathioprine were used for immunosuppression after renal transplantation, but with the advent of cyclosporin, atypical or viral infections are more frequently reported. Chest infections in renal transplant recipients taking cyclosporin, due to cytomegalovirus, *Mycoplasma pneumoniae*, legionella pneumonia, or influenzal pneumonia, may be sufficiently severe to require admission to an

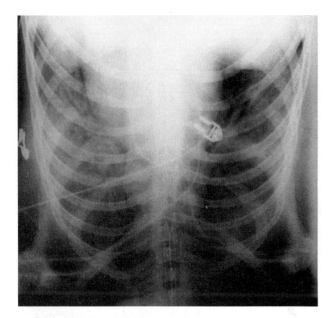

Fig. 36.3 A 17-year-old transplant patient receiving cyclosporin and prednisolone, who developed cytomegalovirus pneumonia in both lungs, requiring artificial ventilation. Air bronchograms, typical of adult respiratory distress syndrome, are visible in the left lung.

intensive therapy unit and the use of artificial ventilation (Fig. 36.3).

Cyclosporin is a much more expensive maintenance immunosuppressant than are steroids or azathioprine, and, together with erythropoietin, has significantly increased the drug expenditure by renal transplant units.

Long-term immunosuppression in transplant patients carries a higher risk of tumour formation than in the general population, with uncommon malignancies, such as lymphomas, presenting at an earlier age than would be expected (Fig. 36.4).

Diabetes mellitus

Patients with insulin-dependent diabetes mellitus (IDDM) and significant renal impairment need careful perioperative supervision. If these patients are unable to respond to a fluid load with a diuresis, the Alberti regimen is unsuitable as, with this, the patient receives 100 ml per hour of fluid. The additional potassium is also inappropriate. It is suggested that, for major surgery, the patient is established on a constant infusion of human insulin (1 unit/ml) via a separate pump, although possibly through the same intravenous cannula, from an infusion of dextrose, given at a slow rate. If dextrose 20 per cent is tolerated without uncontrollable blood sugar levels, 40 ml per hour should be given, but it may well be necessary to use the same volume of dextrose but only 10 or 5 per cent during the stressful perioperative period to achieve good blood sugar control. The blood sugar should be measured at induction of

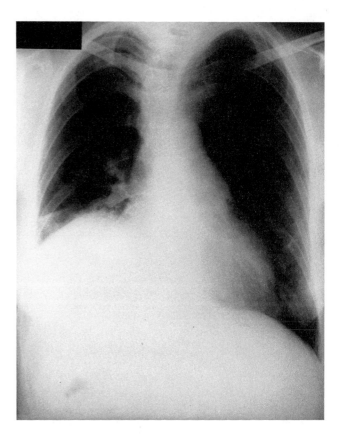

Fig. 36.4 Persistent right lower lobe collapse due to carcinomatosis in a 38-year-old renal transplant patient, with good renal function, who had been receiving steroid immunosuppression for 5 years. A vulval carcinoma with widespread metastases was present.

anaesthesia, then at least every hour, and on recovery. It should be rechecked every 2–4 hours in the postoperative period and a sliding scale introduced for subcutaneous or intravenous administration.

If the patient is undergoing only minor surgery, the morning dose of insulin may be omitted and an infusion of dextrose 5 per cent started at 40 ml per hour, aiming to give the usual evening dose of insulin. If the patient has IDDM and is on continuous ambulatory peritoneal dialysis, and using injection of insulin into the dialysis bags for control, the insulin should be omitted from the overnight bag prior to surgery the next day.

If the blood sugar level goes out of control in a patient with renal failure, the serum potassium level may also change significantly. A rise in blood sugar will be accompanied by a rise in extracellular potassium, a fall in blood sugar by a fall in serum potassium.[10]

Patient's weight

It is essential that this be noted prior to surgery. These patients will know their 'dry' weight and that, if their weight is raised, there is a need for further dialysis and fluid removal prior to surgery. Patients with chronic renal disease are often catabolic, especially if treatment is inadequate, and they have a reduced muscle mass. The dose of neuromuscular blocking drug required may therefore need to be reduced. All other anaesthetic agents that are excreted through the kidney or which depress the myocardium should be used with caution (see below).

When the patient was last dialysed

This is a useful fact to note – if patients finished haemodialysis only a few hours before surgery, they may well be relatively hypovolaemic, with the increased risk of hypotension on induction of anaesthesia. If haemodialysis is due in the immediate future, the patient may be relatively overhydrated and dialysis may be preferable prior to surgery, because of the increased risk of pulmonary oedema in the perioperative period.

Previous general anaesthesia

Previous general anaesthetic details should be noted, in particular if any problems of residual curarization occurred (see below). The previous anaesthetic record should be obtained if possible.

Allergic reactions

Allergic reactions must also be detailed, although these are less common in the immunosuppressed, uraemic patient than in the general population.

Perioperative fluid therapy

Perioperative fluid balance in the patient with renal disease must be guided by the degree of remaining renal function. It should not be forgotten that in all patients with renal disease it is still essential to replace insensible fluid loss. At least 40 ml per hour of dextrose 5 per cent is advisable if the patient is apyrexial. This should be increased by 5 ml per hour for each degree Celsius increase in core temperature. We can identify two groups of patients: (1) the patient with a raised serum creatinine and continuing urine output – usually less than 2 litres per day; and (2) the anephric patient.

The patient with raised serum creatinine

These patients have a limited ability to concentrate urine. An adequate fluid volume must therefore be given postoperatively and probably accompanied, if an added solute load was administered during surgery, by a dose of a loop diuretic such as frusemide. It is essential to maintain a good urine output throughout the perioperative period. The use of dextrose 5 per cent, at a rate of at least 80 ml per hour postoperatively, will provide this water load without giving excess solute, in particular sodium ions, which will inhibit free water excretion. Potassium supplementation may be necessary.

It is preferable to pass a urinary catheter and to insert a central venous line in such patients prior to major surgery. This facilitates management of perioperative fluid and electrolyte balance.

The anephric patient

These patients are being maintained on dialysis and either pass no urine or empty their bladder only once every 2–3 days. They only require insensible loss to be replaced, together with any fluid loss suffered intra- or postoperatively. It is preferable to replace blood loss with blood or colloid, rather than with crystalloid solutions, to avoid the possibility of excess fluid load and pulmonary oedema. A central venous catheter is essential if major surgery is being undertaken.

During renal transplantation it is important to maintain a good intravascular volume, allowing adequate perfusion of the transplanted kidney once the arterial clamps have been removed. A central venous pressure of at least +10 cmH_2O from the anterior axillary line is recommended during intermittent positive pressure ventilation; left ventricular end-diastolic pressures and arterial blood flow should then be increased. It is usual to administer an osmotic diuretic such as mannitol (0.5 g/kg) intravenously before the transplanted kidney is perfused, or a dose of a loop diuretic such as frusemide, to promote urine output. A 'low-dose' intravenous infusion of dopamine may also be used to increase renal blood flow (2 µg/kg per minute).

Specific anaesthetic problems in renal patients

Gastric emptying

Uraemia delays gastric emptying. Patients with renal disease, especially if they have not been well managed preoperatively, should be considered to have a full stomach. The use of a rapid sequence induction technique should be considered before the start of anaesthesia, although the use of suxamethonium must be balanced against the rise in serum potassium that this drug will produce (see below).

Gastric acidity

Uraemia also produces an increase in gastric acidity. This, combined with the delayed gastric emptying, increases the risk and severity of aspiration pneumonia in these immunocompromised patients. Patients with renal dysfunction are often receiving regular antacid therapy. It must be remembered, if H_2-antagonists are being administered, that these drugs are excreted in part through the kidney and have a greater intensity of effect and duration of action in renal failure. *Cimetidine* may cause confusion in these patients and may also be contraindicated because it is a rare cause of bone marrow depression. This drug inhibits cytochrome P_{450}, thereby potentiating many other drugs metabolized by this enzyme system, including cyclosporin, β-blocking agents, propofol and anti-epileptic drugs. *Ranitidine* is preferable, in reduced dosage. Antacids containing aluminium or magnesium should also be avoided in the renal failure patient (see above).

Immunosuppression

Uraemic patients with renal disease must be considered to be immunosuppressed, even if they are not receiving immunosuppressive agents such as steroids, azathioprine or cyclosporin; this relates directly to their disease. This makes them much more liable than the healthy patient to infection, possibly of an atypical variety, in the perioperative period. It is usual to administer a broad-spectrum antibiotic on induction of anaesthesia, prior to major surgery. As Gram-positive cocci are a common commensal on the skin of renal failure patients, it is important that such an agent as amoxycillin and clavulanic acid (Augmentin), which 'covers' this group of bacteria as well as Gram-negative organisms, is used.[22] Scrupulous attention to an aseptic technique is essential throughout surgery, and during the insertion of monitoring lines by the anaesthetist.

Tuberculosis was more common in these patients than in the general population, especially when they were receiving steroid therapy. Although this infection can now be adequately treated, patient compliance is still necessary. This disease must be considered in all renal patients with a persistent chest infection, even if the chest x-ray findings are inconclusive. A suitable anaesthetic technique, which allows adequate postoperative cleaning of all the equipment used, must be employed in a patient with tubercle in the sputum.

Serum hepatitis has also long been a threat to both the patient receiving haemodialysis and attendant staff, because of the repeated handling of blood-stained tubing. The regular screening of these patients, at 6-monthly intervals, together with the use of disposable tubing on haemodialysis machines, has much reduced the risk of this potentially fatal disease.

Care of arteriovenous fistulae

The patient on regular haemodialysis usually has an arteriovenous fistula established, if possible on the non-dominant arm. This is the patient's 'life line', which must be diligently cared for by the anaesthetist. It should not be used for the injection of anaesthetic drugs, which may cause venous irritation or thrombosis, nor for infusion of intravenous fluids. It is recommended that the fistula be well covered with gauze padding during surgery and that the arm be allowed to rest alongside the patient without being flexed, as this could also promote congestion and thrombosis. Significant decreases in blood pressure may also decrease perfusion of the fistula and promote thrombosis.

Venous access

As patients with renal disease experience frequent blood sampling and may well have had repeated vascular access surgery, venepuncture may be difficult. It is essential to have good venous access throughout anaesthesia. The long-term use of central venous cannulae for haemodialysis in these patients, a recent development, is causing an increased incidence of stenosis in central vessels, such as the subclavian or internal jugular veins.

Anaesthesia for patients with renal disease

Premedication

These patients have often had repeated anaesthetics, and they may well be anxious prior to surgery, so premedication is indicated. They know the perioperative routines, and therefore cooperation may be difficult, especially, for instance, if preoxygenation is desired. As with all patients, a preoperative visit with a full explanation of the planned procedure is more effective than any drug that may be administered to calm the patient and allay anxiety. It is recommended that all the patient's routine antihypertensive therapy, and other important drugs, be taken as usual on the day of surgery, even if the patient is being starved.

Sedation

Sedatives are a suitable premedication for these patients because they can be given orally, which is less unpleasant than when parental routes are used. Benzodiazepines are particularly useful, and may be accompanied by night sedation the evening before surgery. Antihistamine drugs are partially excreted by the kidney and will be potentiated in renal failure, and may cause excessive sedation.

Narcotic analgesics

Narcotic analgesics usually require the use of an intramuscular injection, which is more likely to produce a large haematoma, because of attendant clotting defects in these patients. These drugs are also more potent in the patient with renal failure (see below). They are not, therefore, the premedication of choice in this condition.

Vagal blockade

Vagal blockade is not appropriate in these hypertensive patients, even though they may be receiving a β-blocking agent. It is preferable to have atropine drawn up, ready for use intravenously in the anaesthetic room, than to administer such drugs prophylactically. Only if these drugs are specifically indicated, as with fibreoptic intubation, to dry up secretions, should they be used. The resulting tachycardia promotes the pre-existing hypertension and may decrease coronary artery perfusion. In addition, the drying effect is unpleasant for the patient.

Steroid therapy

Steroid therapy may need to be increased immediately prior to induction of anaesthesia, to cover the stress response. It is preferable to avoid an intramuscular injection and to give these drugs intravenously on arrival in the anaesthetic room.

Induction of anaesthesia

A rapid sequence induction must be considered (see above), as should preoxygenation, if this can be accomplished without distressing the patient. Preoxygenation is particularly indicated if the patient is very anaemic or has severe coronary artery disease.

Intravenous induction agents

Barbiturates are more potent in renal disease, due in part to the pre-existing myocardial problems.[23] Nevertheless, *thiopentone*, in reduced dosage and administered slowly, is the agent of choice. *Propofol* is likely to cause a greater fall in blood pressure than thiopentone in all patients, but this decrease does not appear to be greater in patients with renal failure.[24] The plasma levels of propofol are similar after a bolus dose in renal compared with healthy patients (Fig. 36.5), as are the pharmacokinetic parameters.[24] *Etomidate* may be considered if hypotension is a particular risk. This drug is not thought to have any significant immunosuppressive effect when used as a bolus dose, and has less effect on the blood pressure than the barbiturates. It is more expensive than thiopentone. *Ketamine* is not recommended in renal patients because

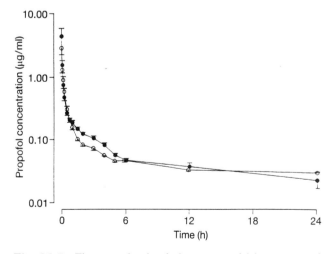

Fig. 36.5 The mean levels of plasma propofol in a group of renal failure (o) compared with healthy (•) patients. No significant difference is evident.[24] (Reproduced with kind permission of the Editor of the *British Journal of Anaesthesia*.)

its sympathomimetic properties may worsen the pre-existing hypertension and tachycardia.

An *inhalational induction* of anaesthesia should be considered in those patients who have significant myocardial disease, possibly accompanied by an intravenous sedative or opioid analgesic. The more gradual onset of anaesthesia allows more control of the blood pressure than when an intravenous induction is used, although it has the disadvantage that the patient may not cooperate with the close application of a facemask. The use of a clear Perspex mask may help.

Inhalational agents

Cyclopropane, ether and *trichloroethylene* are known to cause renal vasoconstriction, even in the denervated organ,[25,26] but fortunately are not often used today. *Methoxyflurane*, which is metabolized significantly in the liver, with the production of nephrotoxic inorganic fluoride ions, is not therefore recommended.[27]

Although *halothane* potentiates the bradycardia produced by β-blocking drugs, it will not alter renal perfusion unless the patient is hypotensive, hypovolaemic or hypercapnic, when renal blood flow will decrease further.[28] Halothane may also potentiate arrhythmias in the patient with myocardial ischaemia and is a potent myocardial depressant. Despite this, it was used successfully as an adjuvant to artificial ventilation for many years in patients with renal disease.

Enflurane and *isoflurane* are less likely to cause myocardial depression than is halothane. Both agents undergo some hepatic metabolism to produce fluoride ions, which are nephrotoxic, but this is significantly less with isoflurane than with enflurane.[29] There have been cases of enflurane producing renal failure in patients with renal dysfunction,[30] and higher levels of plasma inorganic fluoride have been reported in transplant patients receiving enflurane compared with halothane anaesthesia.[31] Isoflurane is probably the inhalational agent of choice in these patients, despite the risk of hypotension and coronary artery steal in the presence of coronary artery disease.

Neuroleptanalgesia has the benefits of maintaining cardiovascular stability and a well maintained renal blood flow.[32] *Fentanyl* only decreases renal blood flow in very high doses, of the order of 10–15 μg/kg.[32] The α-blocking activity of *droperidol* may be thought to be advantageous in maintaining renal blood flow, although in the doses usually used in clinical practice, this effect is probably not significant. Droperidol is excreted through the kidney and this long-acting drug may have a greater effect in patients with renal failure.

Table 36.1 Morphine kinetics in renal transplant compared with healthy patients[33]

	Renal	Healthy
Elimination half-life (min)	290	286
Clearance (ml/min)	533	741
Vd_{ss} (l)	141	241*

* $P<0.002$
Vd_{ss} = volume of distribution at steady state.

Narcotic analgesics

Although these drugs are primarily metabolized in the liver, the metabolites, some of which have potent analgesic properties, are excreted through the kidney and may cumulate in the presence of renal impairment, be it transient, as in hypovolaemia, or permanent, as in end-stage renal failure. *Morphine sulphate* is metabolized in the liver to morphine-3-, morphine-3,6- and morphine-6-glucuronide. The pharmacokinetics of morphine itself are, therefore, little changed in the presence of renal impairment (Table 36.1), although the volume of distribution may be reduced, possibly because of hypoproteinaemia.[33] Morphine-6-glucuronide is, however, a very potent analgesic and respiratory depressant, and is thought to contribute significantly to the enhanced effect of morphine in the renal patient.[33] The dose of this drug should be reduced in these circumstances and care is required, particularly if repeated doses are used.

The main metabolites of *pethidine* (meperidine), including norpethidine and norpethidine acid, are also excreted through the kidney. Norpethidine possesses analgesic and respiratory depressant properties,[34] although not to as marked a degree as morphine-6-glucuronide. Pethidine is probably the analgesic of choice in these patients; morphine can be used if pethidine proves inadequate. Both drugs should be used in reduced dosage.

The clearance of *fentanyl* is little changed in renal failure,[35] and, as this drug has a shorter half-life than morphine or pethidine and its metabolites are inactive, it is the peroperative analgesic of choice. *Alfentanil* would appear to have a similar disposition in renal failure to that in healthy patients[36] but, because of a short half-life, needs to be given by constant infusion for major surgery. *Buprenorphine*, which is also mainly metabolized in the liver, does not appear to cumulate in renal failure, although the inactive metabolites do so.[37]

Neuromuscular blocking agents

Suxamethonium (succinylcholine)

Although it has long been recognized that a single dose of suxamethonium will increase the serum potassium by about 0.5 mmol/l during halothane anaesthesia,[38] this increase is thought not to be exaggerated in renal disease. The problem is that such patients may have a higher plasma potassium level before the suxamethonium is given and a rise may therefore be sufficient to cause cardiac arrhythmias and even, rarely, cardiac arrest. The serum potassium must always be measured before suxamethonium is given to the patient with renal dysfunction, both in the operating theatre and in the intensive care unit. Suxamethonium remains the best muscle relaxant for a rapid sequence intubation – it is the only such agent that allows tracheal intubation within 1 minute,[39] and this is particularly relevant in the uraemic patient with decreased gastric emptying.

Non-depolarizing muscle relaxants

The percentage excretion of these drugs through the kidney is given in Table 36.2. The earlier non-depolarizing agents were excreted to a significant degree through the kidney, and cumulation was a recognized problem, especially if repeated doses had been given or an increment was used to allow the surgeon to close the abdomen. These drugs are more potent in an acidotic state. If reversal was inadequate after surgery, respiratory distress leading to carbon dioxide retention might cause respiratory acidosis in addition to the pre-existing metabolic acidosis, and thus apparently worsen any residual neuromuscular block. This recurarization has been reported in renal patients after the use of tubocurarine[40]

Table 36.2 Percentage excretion of various muscle relaxants* in the urine over 24 hours

	Excretion (%)
Suxamethonium (succinylcholine)	<10
Tubocurarine	66
Gallamine	>95
Pancuronium	60
Metocurine	52
Alcuronium	80
Atracurium	10
Vecuronium	20–30
Pipecuronium	64
Doxacurium	60
Mivacurium	<10
Rocuronium	33

* Listed in chronological order of their first clinical usage.

and *pancuronium*.[41] As *gallamine*[42] is excreted almost entirely through the kidney, it is not surprising that there were reports of persistent curarization lasting for several days after the use of this drug in patients with renal failure. The use of gallamine in this condition is contraindicated. *Alcuronium*[43] and *metocurine* are also excreted through the kidney to a significant degree (Table 36.2).

With the introduction of *atracurium* and *vecuronium*, the management of neuromuscular block in the patient with renal disease became more simple. Only 10 per cent of a dose of atracurium is excreted by the kidney[44] and, although the relative contributions of Hofmann degradation and ester hydrolysis to the breakdown of atracurium in humans are uncertain, it is thought that Hofmann degradation acts as a form of safety net, in the absence of any organ function. Pharmacodynamic[45] and pharmacokinetic[46] studies have demonstrated no cumulation of atracurium in patients with renal failure, even if the drug is given by constant infusion for several days.[44] Atracurium is the neuromuscular blocking agent of choice in the patient with renal disease.

Concern has been expressed that the metabolite of Hofmann degradation, laudanosine, which is known to produce cerebral irritation in animals and to be excreted in part through the kidney, may accumulate in the presence of renal dysfunction. Fahey and colleagues did show a significant increase in plasma laudanosine levels in patients with renal failure, given a bolus dose of atracurium, compared with a control healthy group.[47] The plasma laudanosine level was, however, only of the order of 0.2 μg/ml in the renal group, and that required to produce cerebral irritation in dogs is of the order of 14 μg/ml.[48] There is no risk of important cumulation of laudanosine during general anaesthesia. If atracurium is given by constant infusion for many days to the critically ill patient with multiple organ failure in an intensive care unit, plasma levels of laudanosine will inevitably be higher, at up to 5 μg/ml.[44] No clinical reports have as yet appeared of laudanosine causing convulsions in such circumstances, but it is not impossible to imagine that these may occur if large doses of atracurium are given to such patients for many days, perhaps in the presence of a low epileptic threshold, hypoxaemia and metabolic disturbances.

Although vecuronium is excreted through the kidney to a greater extent than is atracurium, it is still at a rate significantly less than in the case of the earlier non-depolarizing drugs.[49] Cumulation is unlikely unless many repeated doses are used. Potentiation of block may occur, however, if the patient is very acidotic. There have been several reports of persistent block when vecuronium has been given repeatedly to patients in renal failure[50] or when it has been used by constant infusion in the intensive therapy unit for patients with impaired renal function.[51] This may be due in part to cumulation of the active

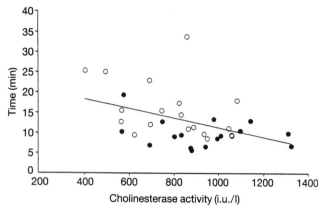

Fig. 36.6 The time to 5 per cent recovery of the first twitch of the train of four (t_1) is plotted against plasma cholinesterase activity in a group of renal (○) and healthy (●) patients. There was a significant correlation between these variables ($r = -0.42$; $P<0.02$).[55]

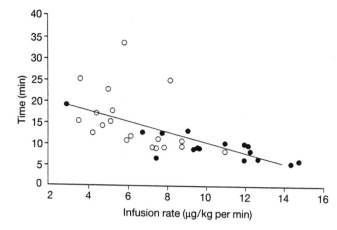

Fig. 36.7 The time to 5 per cent recovery of t_1 is plotted against the infusion rate of mivacurium required to maintain t_1 at 5 per cent in renal (○) and healthy (●) patients ($r = -0.447$; $P<0.01$).[55]

metabolite of vecuronium, 3-desacetylvecuronium, which is excreted through the kidney, has a long half-life and has about 40 per cent of the neuromuscular blocking potency of the parent drug.[52] It is preferable to give vecuronium in bolus doses to the critically ill.

Pipecuronium, a recently introduced steroidal muscle relaxant, and *doxacurium*, a benzylisoquinolinium deriva-tive, although both free of cardiovascular effects, are excreted to a significant degree by the kidney (Table 36.2) and do not represent advances in the management of the patient in renal failure.[53,54]

Mivacurium is thought to be metabolized in the plasma, by plasma cholinesterase, at 88 per cent of the rate of suxamethonium. As cholinesterase levels are reduced in patients with chronic renal disease, prolongation of block may be anticipated when these patients are given mivacurium. A direct relationship has been demonstrated in both healthy and renal failure patients, for the time to 5 per cent recovery of the first twitch of the train of four (t_1) after mivacurium (0.15 mg/kg) and both plasma cholin-esterase activity (Fig. 36.6) and the infusion dose of mivacurium required to keep the height of t_1 at 5 per cent (Fig. 36.7).[55]

Rocuronium (Org 9426), another steroidal non-depolariz-ing muscle relaxant with a rapid onset of action, has also been shown in early studies to have an increased duration

of action in renal disease,[56] although other workers have not confirmed these findings.[57]

Whichever muscle relaxant is used during anaesthesia for patients with renal dysfunction, it is essential to monitor neuromuscular block throughout surgery.

Reversal of residual neuromuscular blockade

As anticholinergic agents, such as *atropine* and *glycopyr-rolate*, and anticholinesterases, such as *neostigmine* and *edrophonium*, are all highly ionized, water-soluble sub-stances, they are excreted to a significant extent through the kidney.[58,59] Their length of action is therefore prolonged in patients with renal disease (Table 36.3). As the anticholinergics have shorter half-lives than the anticholinesterases, but the half-lives of both types of agents are approximately doubled in the presence of renal disease, the period of bradycardia produced by the anticholinesterases will be exaggerated. It will be further potentiated in those patients receiving β-blocking drugs.

Glycopyrrolate has less cardiac effect than atropine, so it may be preferable to use this drug in the patient with

Table 36.3 The effect of renal failure on the elimination half-life and clearance of neostigmine and edrophonium[58,59]

	Neostigmine		*Edrophonium*	
	Elimination half-life (min)	*Clearance* (ml/kg per min)	*Elimination half-life* (min)	*Clearance* (ml/kg per min)
Healthy	77.8	16.7	110	9.6
Renal failure	181	7.8	304	3.9

Table 36.4 A suggested general anaesthetic technique for the patient with renal disease

Premedication
Nitrazepam 5–10 mg orally the evening before surgery
Diazepam 5–10 mg orally, 2–3 hours preoperatively

Induction of anaesthesia
Fentanyl 1.0 µg/kg
Thiopentone 4–5 mg/kg
Atracurium 0.5 mg/kg
(or suxamethonium 1 mg/kg if stomach full)

Maintenance of anaesthesia
N_2O/40% O_2/isoflurane
Increments of fentanyl and atracurium

Reversal of residual block
Atropine 1.2 mg or glycopyrrolate 0.5 mg
Neostigmine 2.5 mg

cardiac pathology, as well as renal disease. It is three times more expensive than atropine.

A suggested general anaesthetic technique for the patient with renal disease presenting for major surgery is given in Table 36.4.

Intraoperative monitoring

It is essential that electrocardiographic monitoring be commenced before induction of anaesthesia and continued until the patient leaves the recovery room. Arterial oxygen saturation should also be monitored using a pulse oximeter throughout this period, together with heart rate and blood pressure. As these patients are anaemic, higher inspired oxygen concentrations may be required (e.g. 40 per cent) to provide adequate oxygen supplies to the tissues. End-tidal carbon dioxide concentration should be kept in the normal range to optimize renal blood flow.[28] Hypovolaemia should also be corrected, for the same reason. Accurate measurement of blood loss during surgery will aid the maintenance of correct fluid balance. For major surgery, central venous pressure monitoring is also essential. Rarely, insertion of a pulmonary artery catheter at induction of anaesthesia may be useful in optimizing left ventricular output, and hence renal blood flow, in a patient whose central venous filling pressure is not a good reflection of left atrial pressure, because of severe lung disease. If the patient is expected to pass any urine during the procedure, a urinary catheter should be inserted and hourly measurements recorded throughout the perioperative period. Neuromuscular monitoring should be instigated after induction of anaesthesia and continued until the patient is awake.

If major blood loss occurs, it is useful to check the patient's haemoglobin, serum potassium and sodium peroperatively as well as performing blood clotting studies. The fall in haemoglobin may not be as great as anticipated, because of pre-existing haemodilution. If the patient is an insulin-dependent diabetic, blood sugar should also be checked regularly (see above).

Postoperative monitoring in the recovery room

Because renal patients are usually anaemic, oxygen therapy should continue into the postoperative period for as long as is feasible. It is now clear that periods of arterial desaturation are not infrequent during the first two postoperative nights,[60] and as these patients are not only anaemic, with decreased oxygen carrying capacity, but also frequently have coronary artery disease, they must be at considerable risk of developing myocardial ischaemia during this time. Electrocardiographic monitoring should be continued in the recovery room, together with pulse oximetry.

Fluid balance should be charted at least every hour, taking into account the central venous pressure, urine output (if any) and fluid drainage. As many patients with end-stage renal failure live constantly on the brink of pulmonary oedema, they rarely lie flat; they sleep propped up on several pillows. It is preferable to place them in their usual sitting position on recovery from anaesthesia. If there is considerable risk of pulmonary oedema, a chest x-ray is indicated.

The recovery room nurse should be informed of the site of an arteriovenous fistula, to prevent the blood pressure inadvertently being taken on the same arm. Every attempt should be made to keep the patient warm because shivering, which is common in all patients postoperatively, frequently potentiates hypertension in the anephric patient during this unstable period.

Local anaesthesia in the patient with renal disease

Local analgesic drugs have a shorter duration of action in the acidic tissues of a patient with renal disease and therefore appear to be less efficacious, particularly when used for local infiltration.[61] The increased risk of haemorrhage makes the benefits of using local blocks in restricted areas such as the epidural space or spinal canal questionable. There is undoubtedly a higher risk of a haematoma than in the healthy patient, with possible permanent neurological damage a consequence. In addition, these anxious patients often wish to be given a general anaesthetic.

Vascular surgery is often required in the arm of renal patients, at a time, perhaps, when dialysis is proving inefficient because of difficulty with pre-existing vascular access; hence brachial plexus block may be useful for intraoperative management. The supraclavicular approach has the advantage of producing analgesia over the whole arm, which is particularly important if the surgeon works in the antecubital fossa or deltoid region. Infraclavicular blocks may fail to establish adequate analgesia over these areas. The disadvantage of the supraclavicular approach is the risk of puncturing the pleura, producing a pneumothorax in a high-risk patient.

Bupivacaine is more useful than lignocaine if a brachial block is to be instituted, because of its longer duration of action; *lignocaine* may be of insufficient duration to cover the period of what may become prolonged surgery. Bupivacaine does, however, have the disadvantage of a prolonged onset of action: 20 minutes or more may elapse before an adequate block is established, especially if the drug is not injected very close to the brachial plexus. Lignocaine may be mixed with the bupivacaine to reduce the delay in onset. For the supraclavicular approach, bupivacaine 0.25 per cent, 30 ml, in three 10 ml aliquots, injected along the first rib, as the three branches of the brachial plexus traverse it, is recommended.

REFERENCES

1. Williams AJ, Antao AJO. Referral of elderly patients with end-stage renal failure for renal replacement therapy. *Quarterly Journal of Medicine* 1989; **268**: 749–56.
2. Burton PR, Walls J. Selection-adjusted comparison of life expectancy of patients on continuous ambulatory peritoneal dialysis, haemodialysis and renal transplantation. *Lancet* 1987; **1**: 1115–19.
3. Legendre C, Saltiel C, Kreis H, Grunfeld J-P. Hypertension in kidney transplantation. *Klinische Wochenschrift* 1989; **67**: 919–22.
4. Weaver RA. Restoration of renal function by renal revascularisation. *Western Journal of Medicine* 1989; **151**: 189.
5. Gunnarsson R, Lofmark R, Norlander R, Nyquist O, Groth C-G. Acute myocardial infarction in renal transplant recipients: incidence and prognosis. *European Heart Journal* 1984; **5**: 218–21.
6. Heino A, Orko R, Rosenberg PH. Anaesthesiological complications in renal transplantation: a retrospective study of 500 transplantations. *Acta Anaesthesiologica Scandinavica* 1986; **30**: 574–80.
7. Soulilou JP, Fillaudeau F, Keribin JP, Guenel J. Acute hyperkalaemia risks in recipients of kidney graft cooled with Collins solution. *Nephron* 1977; **19**: 301–4.
8. Horl WH, Riegel W, Wanner C, *et al.* Endocrine and metabolic abnormalities following kidney transplantation. *Klinische Wochenschrift* 1989; **67**: 907–18.
9. Smith I, Hyland K. Disturbance of cerebral function by aluminium in haemodialysis patients. *Lancet* 1989; **2**: 501–2.
10. Goldfarb S, Cox M, Suiger I, Goldbert M. Acute hyperkalaemia induced by hyperglycaemia; hormonal mechanisms. *Annals of Internal Medicine* 1976; **84**: 426–32.
11. Hunter JM. Resistance to non-depolarising muscle relaxants. *British Journal of Anaesthesia* 1991; **67**: 511–14.
12. Hambley H, Mufti GH. Erythropoietin: an old friend revisited. *British Medical Journal* 1990; **300**: 621–2.
13. Raine AEG. Hypertension, blood viscosity and cardiovascular morbidity in renal failure: implications of erythropoietin therapy. *Lancet* 1988; **1**: 97–9.
14. Eschbach JW, Egrie JC, Downing MR, Browne JK, Adamson JW. Correction of the anemia of end-stage renal disease with recombinant human erythropoietin. *New England Journal of Medicine* 1987; **316**: 73–8.
15. Bommer J, Alexiou C, Muller-Butil U, Eifert J, Ritz E. Recombinant human erythropoietin therapy in haemodialysis patients – dose determination and clinical experience. *Nephrology, Dialysis and Transplantation* 1987; **2**: 238–42.
16. Casati S, Passerini P, Campse MR, *et al.* Benefits and risks of protracted treatment with human recombinant erythropoietin in patients having haemodialysis. *British Medical Journal* 1987; **295**: 1017–20.
17. Stevens ME, Summerfield GP, Hall AA, *et al.* Cost benefits of low dose subcutaneous erythropoietin in patients with anaemia and end-stage renal disease. *British Medical Journal* 1992; **304**: 474–7.
18. Opelz G. The role of HLA matching and blood transfusions in the cyclosporine era. *Transplantation Proceedings* 1989; **21**: 609–12.
19. Sells RA, Hillis A, Bone MJ, Evans P, Evans CM, Scott MM. Donor specific transfusion with and without cyclosporin A – a controlled clinical trial. *Transplantation Proceedings* 1988; **10**, suppl 3: 270–3.
20. Calne RYC, White DJG, Thiru S, *et al.* Cyclosporin A in patients receiving renal allografts from cadaveric donors. *Lancet* 1978; **2**: 1323–7.
21. Gramstadt L, Gjerlow A, Hysing ES, Rugstad HE. Interaction of cyclosporin and its solvent cremophor with atracurium and vecuronium. Studies in the cat. *British Journal of Anaesthesia* 1986; **58**: 1149–55.
22. Evans CM, Purohit S, Colbert JW, *et al.* Amoxycillin–clavulanic (Augmentin) antibiotic prophylaxis against wound infections in renal failure patients. *Journal of Antimicrobial Chemotherapy* 1988; **22**: 363–9.

23. Dundee JW, Annis D. Barbiturate narcosis in uraemia. *British Journal of Anaesthesia* 1955; **27**: 114–23.

24. Kirvela M, Olkkola KT, Rosenberg PH, Yli-Hankala A, Salmela K, Lindgren L. Pharmacokinetics of propofol and haemodynamic changes during induction of anaesthesia in uraemic patients. *British Journal of Anaesthesia* 1992; **68**: 178–82.

25. Deutsch S, Pierce EC, Vandam LD. Cyclopropane effects on renal function in normal man. *Anesthesiology* 1967; **28**: 547–58.

26. Hamelburg W, Sprouse JH, Mahaffey JE, Richardson JE. Catecholamine levels during light and deep anesthesia. *Anesthesiology* 1960; **21**: 297–302.

27. Cousins MJ, Mazze RI, Kosek JC, Hitt BA, Love FV. The etiology of methoxyflurane toxicity. *Journal of Pharmacology and Experimental Therapeutics* 1974; **190**: 530–41.

28. Hunter JM, Jones RS, Utting JE. Effect of acute hypocapnia on renal function in the dog artificially ventilated with nitrous oxide, oxygen and halothane. *British Journal of Anaesthesia* 1980; **52**: 197–8.

29. Chase RE, Holaday DA, Fiserova-Bergesova V, Saidman L, Macks FE. The biotransformation of ethrane in man. *Anesthesiology* 1971; **35**: 262–7.

30. Eichhorn JH, Hedley-White J, Steinman TL, Kaufmann JM, Laasberg LH. Renal failure following enflurane anesthesia. *Anesthesiology* 1976; **45**: 557–60.

31. Wickstrom I. Enflurane anaesthesia in living renal transplantation. *Acta Anaesthesiologica Scandinavica* 1981; **25**: 263–9.

32. Hunter JM, Jones RS, Utting JE. Effects of acute hypocapnia on some aspects of renal function in anaesthetised dogs. *British Journal of Anaesthesia* 1979; **51**: 725–31.

33. Sear JW, Hand CW, Moore RA, McQuay HJ. Studies on morphine disposition: influence of renal failure on the kinetics of morphine and its metabolites. *British Journal of Anaesthesia* 1989; **62**: 28–32.

34. Armstrong PJ, Bersten A. Normeperidine toxicity. *Anesthesia and Analgesia* 1986; **65**: 536–8.

35. Corall IM, Moore AR, Strunin L. Plasma concentrations of fentanyl in normal surgical patients and those with severe renal and hepatic disease. *British Journal of Anaesthesia* 1980; **52**: 101P.

36. Sear JW, Bower S, Potter D. Disposition of alfentanil in patients with chronic renal failure. *British Journal of Anaesthesia* 1986; **58**: 812P.

37. Hand CW, Sear JW, Uppington J, Ball MJ, McQuay HJ, Moore RA. Buprenorphine disposition in patients with renal impairment: single and continuous dosing with special reference to metabolites. *British Journal of Anaesthesia* 1990; **64**: 276–82.

38. Paton WDM. The effects of muscle relaxants other than muscle relaxation. *Anesthesiology* 1959; **20**: 453–4.

39. Hunter JM, Jones RS, Utting JE. Use of atracurium during general surgery monitored by the train-of-four stimuli. *British Journal of Anaesthesia* 1982; **54**: 1243–51.

40. Riordan DD, Gilbertson AA. Prolonged curarisation in a patient with renal failure. *British Journal of Anaesthesia* 1971; **43**: 506–8.

41. Rouse JM, Galley RLA, Bevan DR. Prolonged curarisation following renal transplantation. *Anaesthesia* 1977; **32**: 247–51.

42. Jenkins IR. Three cases of apparent curarisation. *British Journal of Anaesthesia* 1961; **33**: 314–18.

43. Buzello W, Agoston S. Comparative clinical pharmacokinetics of tubocurarine, gallamine, alcuronium, pancuronium. *Anaesthesist* 1978; **27**: 313–18.

44. Shearer ES, O'Sullivan EP, Hunter JM. Clearance of atracurium and laudanosine in the urine and by continuous venovenous haemofiltration. *British Journal of Anaesthesia* 1991; **67**: 569–73.

45. Hunter JM, Jones RS, Utting JE. Use of atracurium in patients with no renal function. *British Journal of Anaesthesia* 1982; **54**: 1251–4.

46. Fahey MR, Fisher DM, Miller RD, et al. The pharmacokinetics and pharmacodynamics of atracurium in patients with and without renal failure. *Anesthesiology* 1984; **61**: 699–702.

47. Fahey MR, Rupp SM, Canfell C, et al. Effects of renal failure on laudanosine excretion in man. *British Journal of Anaesthesia* 1985; **57**: 1049–51.

48. Hennis PJ, Fahey MR, Miller RD, Canfell C, Shi W-Z. Pharmacology of laudanosine in dogs. *Anesthesiology* 1984; **61**: A305.

49. Lynam DP, Cronnelly R, Castagnoli KP, et al. The pharmacodynamics and pharmacokinetics of vecuronium in patients anesthetized with isoflurane with normal renal function or with renal failure. *Anesthesiology* 1988; **69**: 227–31.

50. Cody MW, Dormon FM. Recurarisation after vecuronium in a patient with renal failure. *Anaesthesia* 1987; **42**: 993–5.

51. Smith CL, Hunter JM, Jones RS. Vecuronium infusions in patients with renal failure in an ITU. *Anaesthesia* 1987; **42**: 387–93.

52. Segredo V, Matthay MA, Sharma ML, Gruenke LD, Caldwell JE, Miller RD. Prolonged neuromuscular blockade after long term administration of vecuronium in two critically ill patients. *Anesthesiology* 1990; **72**: 566–70.

53. Caldwell JE, Canfell PC, Castagnoli KP, et al. The influence of renal failure on the pharmacokinetics and duration of action of pipecuronium bromide. *Anesthesiology* 1987; **67**: A612.

54. Cashman JN, Luke JJ, Jones RM. Neuromuscular block with doxacurium (BW A938U) in patients with normal or absent renal function. *British Journal of Anaesthesia* 1990; **64**: 184–92.

55. Phillips BJ, Hunter JM. The use of mivacurium chloride by constant infusion in the anephric patient. *British Journal of Anaesthesia* 1992; **69**: 492–8.

56. Cooper RA, Maddineni VR, Mirakhur RK, Wierda JMKH, Brady MM, Fitzpatrick KJT. Time course of neuromuscular effects and pharmacokinetics of rocuronium bromide (Org 9426) during isoflurane anaesthesia in patients with and without renal failure. *British Journal of Anaesthesia* 1993; **71**: 222–6.

57. Szenohradszky J, Segredo V, Caldwell JE, Sharma M, Gruenke LD, Miller RD. Pharmacokinetics, onset and duration of action of Org 9426 in humans: normal vs absent renal function. *Anesthesia and Analgesia* 1991; **72**: S290.

58. Morris R, Cronnelly R, Miller RD, Stanski DR, Fahey MR. Pharmacokinetics of edrophonium and neostigmine when antagonizing *d*-tubocurarine neuromuscular blockade in man. *Anesthesiology* 1981; **54**: 399–402.

59. Cronnelly R, Stanski DR, Miller RD, Sheiner LB, Sohn YJ.

Renal function and the pharmacokinetics of neostigmine in anesthetized man. *Anesthesiology* 1979; **51**: 222–6.

60. Reeder MK, Goldman MD, Loh L, *et al.* Postoperative hypoxaemia after major abdominal vascular surgery. *British Journal of Anaesthesia* 1992; **68**: 23–6.

61. Orko R, Pitkanen M, Rosenberg PH. Subarachnoid anaesthesia with 0.75% bupivacaine in patients with chronic renal failure. *British Journal of Anaesthesia* 1986; **58**: 605–9.

FURTHER READING

Bastron RD, Deutsch S. *Anaesthesia and the kidney*. New York, San Francisco, London: Grune & Stratton, 1976.

Bevan DR, ed. *Renal function in anaesthesia and surgery*. London: Academic Press; New York: Grune & Stratton, 1979.

Hunter JM. Recent advances in the management of renal disease: their relevance to the anaesthetist. *Current Opinion in Anaesthesiology* 1990; **3**: 452–6.

Priebe H-J, ed. The Kidney in Anaesthesia. *International Anesthesiology Clinics*, vol 22, no 1. Boston: Little, Brown, Spring 1984.

Chapter 37

The Liver and Anaesthesia

P. Hutton and M. H. Faroqui

Hepatic anatomy and physiology
Drug metabolism and the liver
Anaesthesia in patients with coexisting liver disease
Anaesthesia for liver transplantation and liver surgery

Acute liver failure and its intensive care management
Viral infections of the liver: implications for anaesthesia and anaesthetists
The investigation and management of postoperative jaundice
The liver in pregnancy and its implications for anaesthesia

Hepatic anatomy and physiology

Macroscopic anatomy

The liver is the largest single organ in the body, weighing 1.2–2.0 kg. It is divided into right and left lobes, the left being the smaller and constituting approximately one-sixth of the whole. It is related by its domed upper surface to the diaphragm, which separates it from the pleura, lungs, pericardium and heart. Its posteroinferior (or visceral) surface abuts against the abdominal oesophagus, the stomach, duodenum, hepatic flexure of colon, the right kidney and suprarenal gland. Anteriorly it is covered by the lower ribs and anterior abdominal wall. These relations are shown in Fig 37.1.[1] There are therefore many adjoining structures at risk during liver surgery and biopsy.

The biliary system arises from a condensation of fine bile capillaries which originate in the liver lobules. The right and left hepatic ducts fuse in the porta hepatis to form the common hepatic duct. This in turn joins with the cystic duct which drains the gall bladder to form the common bile duct (CBD) (Fig. 37.2).[1] The CBD begins 2.5 cm above the duodenum and then passes behind it to open at a papilla on the medial aspect of the duodenum itself. For part of its course the CBD lies behind or within the head of the pancreas. The commonest arrangement is shown in Fig. 37.2 but variations are not rare. The functioning gall bladder typically holds about 50 ml of bile and acts as a bile concentrator and reservoir. More detailed anatomy can be found in standard reference textbooks.[2]

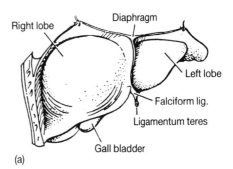

(a)

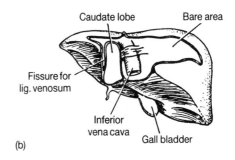

(b)

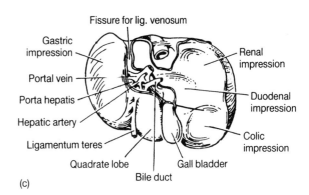

(c)

Fig. 37.1 The topographical anatomy of the liver. (Reproduced, with permission, from Ellis H. *Clinical anatomy*, 5th ed. Oxford: Blackwell Scientific, 1974.)

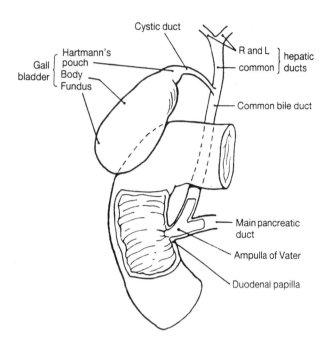

Fig. 37.2 The gall bladder and its duct system. (Reproduced, with permission, from Ellis H. *Clinical anatomy*, 5th ed. Oxford: Blackwell Scientific, 1974.)

Microstructure and histology

The liver architecture is arranged as a very large number of polyhedral *hepatic* lobules. In histological section these appear hexagonal and measure approximately 1 mm across. Each has a small central vein (which is a tributary of the hepatic vein) running down its central axis, with groups of three tubes called portal triads at its edges. Each portal triad contains a branch of the portal vein, a branch of the hepatic artery (perilobular arteriole) and an interlobular bile ductule (Fig. 37.3).[3] All these three structures are surrounded by a connective tissue sheath called the perivascular fibrous capsule. A *portal* lobule, on the other hand, consists of the adjoining parts of three hepatic lobules which centre on and surround a portal triad. Under normal anatomical conditions the hepatic lobular structure predominates, but this can be changed to a portal lobular structure by alteration in the relative pressure in the portal and hepatic venous systems.[2]

Hepatocytes have a spheroidal nucleus and the cytoplasm has numerous ribosomal clusters many of which are attached to endoplasmic reticulum. There are high concentrations of mitochrondia, lysosymes and a well developed Golgi complex; in short, the cells are well prepared for metabolic activity. Hepatocyte function is not, however, uniform. Periportal hepatocytes receive blood with the highest oxygen saturation and show the highest metabolic activity. These cells have highly active transaminase enzymes and are thought to be mainly involved in

protein anabolism and catabolism. Centrilobular hepatocytes receive blood with a lower oxygen saturation. These cells are the site of drug biotransformation and have a high content of cytochrome P_{450}. Cells between these two sites are intermediate in oxygen supply and function. Between the hepatocytes that are arranged radially around a central vein are vascular sinusoids. These sinusoids transfer blood from the perilobular arterioles and portal vessels to the centre of the lobule where it drains via the central vein to the hepatic vein. The role of hepatocytes in intermediary metabolism has been reviewed in detail elsewhere.[4]

Some of the cells lining the venous sinusoids are hepatic macrophages (or Kupffer cells) of the reticuloendothelial system. These phagocytose bacteria passing to the liver from the gastrointestinal (GI) tract via the portal vessels; phagocytose endotoxins; denature foreign proteins and remove ferritin and haemosiderin.[4] During fetal life, the liver acts as one of the main haemopoietic organs, producing both red and white blood cells in the mesenchyme covering the endothelium of the sinusoids.[2]

Blood supply and innervation

The liver receives blood from two sources and at rest its total blood flow is equal to approximately 25 per cent of the cardiac output (1100–1800 ml/min). Although there are considerable variations, the hepatic artery normally arises from the coeliac trunk.[2] In the healthy adult the hepatic artery delivers about 30 per cent of the total hepatic blood flow and 40–50 per cent of the total hepatic oxygen supply and the flow rate is controlled by sphincter mechanisms[4] (see later).

The portal vein is formed from a number of venous tributaries (splenic vein, superior mesenteric vein, left and right gastric veins, cystic vein and paraumbilical vein), and enters the liver via the porta hepatis.[2] The portal vein supplies 70 per cent of the total liver blood flow and 50–60 per cent of the oxygen supply. The oxygen saturation of the portal venous blood ranges from 60 to 75 per cent, depending on the activity of the GI tract. The portal vein is a valveless system and resistance to flow is about 6–10 per cent of that in the hepatic artery. The pressure within the portal system depends on the pressure within the liver and the tone of the sphincter-like sections of the hepatic vein.[4] The upper and lower hepatic veins drain into the inferior vena cava.

Flow within the liver is very dependent on the pressure gradients present across the sinusoids. Typical figures in health are a pressure of 35 mmHg in the hepatic arterioles, 5–13 mmHg in the portal vein and 6 mmHg in the hepatic vein.[4] In the cat model, substances cleared from the liver are extracted equally well from both hepatic artery and portal blood flows.[5]

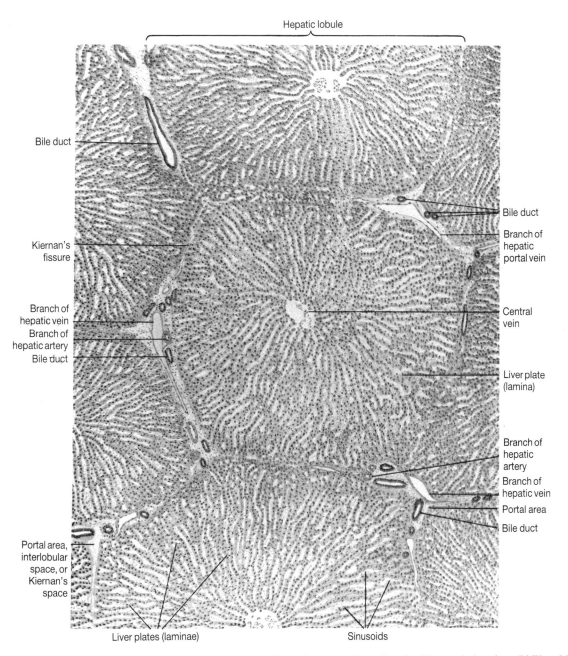

Fig. 37.3 The microscopic architecture of the liver. Haematoxylin–eosin, × 45. (Reproduced, with permission, from Di Fiore MSH. *Atlas of human histology.* Philadelphia: Lea & Febiger, 1973.)

The nerve supply to the liver comes from the hepatic plexus which enters via the porta hepatis. The hepatic plexus is derived from branches of the coeliac plexus, the right and left vagus nerves and the right phrenic nerve. It has afferent and efferent sympathetic and parasympathetic functions. The termination of many fibres is not known with certainty but some control the gall bladder and sphincter of Oddi.[2]

Control of liver blood flow

Total liver blood flow is influenced by many factors, as summarized in Table 37.1.[6] These can be broadly grouped as intrinsic and extrinsic.

Intrinsic

There are two main intrinsic regulatory mechanisms: autoregulation and arteriovenous reciprocity.

Table 37.1 Factors influencing hepatic blood flow

Increase liver blood flow	Decrease liver blood flow
Hypercapnia	IPPV + PEEP
Acute hepatitis – viral, alcoholic	Surgery
	Hypocapnia; hypoxia
Supine posture	Upright posture
Food	Cirrhosis
Drugs:	α-Adrenostimulation
β-adrenostimulation	β-Adrenoceptor blockade
phenobarbitone	Ganglion blockade
other enzyme inducers	Cimetidine (ranitidine)
drugs	Pitressin
	Anaesthetics – volatile and intravenous

Reproduced, with permission, from Sear JW. Hepatic physiology. *Current Anaesthesia and Critical Care* 1990; **1**: 196–203.

Autoregulation of arterial flow occurs down to a pressure of about 80 mmHg; below that, flow becomes pressure dependent.[6,7] Portal pressure is determined by flow from the spleen and intestines and the resistance of the vascular bed in the liver. In congestive heart failure there is a decrease in hepatic arterial flow as well as in portal venous flow.[8,9]

Arteriovenous reciprocity is a phenomenon by which decreases in portal blood flow reduce hepatic arterial vascular resistance and hence increase arterial flow and vice versa.[10] Changes in arterial flow do not, however, result in similar reciprocal responses in the portal venous system. Although the exact mechanism is unknown, it is thought to depend on adenosine.[11] Total occlusion of one inflow circuit reduces the resistance to flow in the other by approximately 20 per cent:[12] these changes are therefore insufficient to provide total compensation of oxygen delivery. With portal hypertension the liver becomes very dependent on hepatic arterial flow.[4]

Extrinsic

Surgery

Surgery can have a variety of effects, depending upon its exact nature. Clamps and packs can all directly affect the blood flow but, in addition, mesenteric traction[13] and the standard 'surgical stress' of laparotomy reduce splanchnic blood flow in a variable manner.[14] Surgical intervention rather than anaesthesia is the main determinant of splanchnic blood flow: the type of anaesthesia then modulates these surgical effects.[15–18]

Ventilation

Because the liver is predominantly a 'low pressure' tissue bed, its total flow is significantly dependent on the difference between intra-abdominal and intrathoracic pressures. During spontaneous ventilation there is a fairly steady venous return from the inferior vena cava,[19] but with IPPV the rise in intrathoracic pressure during inspiration produces an increased resistance. IPPV, particularly with large tidal volumes or high expiratory pressures, decreases splanchnic blood flow.[20,21] The reductions in mesenteric arterial and portal blood flow have been found to parallel reductions in cardiac output.[22,23] Although high-frequency positive-pressure ventilation did not alter liver blood flow under normal conditions, as soon as airway pressure was raised, hepatic blood flow fell.[24]

Hypoxia

The response of the circulation to hypoxia is inconsistent in that the true effect of hypoxia is to depress cardiac output; indirectly, however, the hypoxia-mediated release of catecholamines does the opposite. The literature is inconclusive on the overall effect, which would, of course, be modified anyway by the depth and type of anaesthesia.

Hyper- and hypocapnia

Hypercapnia causes vasodilation of vascular beds as a direct effect of carbon dioxide together with systemic sympathoadrenal stimulation. The net effect of hypercapnia is to increase portal flow while producing little change in hepatic artery flow.[25,26] Hypocapnia can reduce total hepatic flow by over 30 per cent.[27] Hypocapnia is, of course, produced by overventilation so the effects of IPPV are often present as well.

Haemorrhage

Haemorrhage produces a greater fall in portal venous than in hepatic artery flow. Initially oxygen supply is maintained by increased extraction; during blood loss the liver functions as a vascular reservoir, as it contains up to 25–30 ml of blood per 100 g of tissue. If hypotension occurs, sympathetic stimulation can result in the mobilization of about 50 per cent of the reservoir blood into the systemic circulation.[6]

Hormones

Richardson and Withrington[28] have reviewed the effects of hormones on hepatic blood flow. The hepatic arterial vessels contain both α- and β-receptors but the portal vein has only α-receptors. At physiological concentrations, dopamine has no effect on liver haemodynamics but physiological concentrations of adrenaline (epinephrine) do. The arterial supply shows an initial α vasoconstriction followed by vasodilation but there is only vasoconstriction in the portal vasculature.

Effect of drugs used during anaesthesia

Anaesthetics can alter splanchnic circulation by direct effects on the splanchnic vasculature, by neural and humeral mechanisms, and by changes in systemic haemodynamics. Their actions have been reviewed by Gelman and Frenette,[29] to which the reader is referred for a detailed review. Over all, the majority of anaesthetics decrease total splanchnic blood flow and in so doing can alter the ratio of hepatic artery to portal venous flow.[30]

The direct effect of halothane on intestinal vasculature is vasodilatation[31] but its more general effect of lowering the blood pressure and cardiac output are the overriding influences on hepatic blood flow. If, however, the cardiac output or blood pressure does not fall by more than 30 per cent, hepatic artery blood flow is well maintained.[32–34] Halothane anaesthesia also has an effect on other drugs administered simultaneously, causing a reduced clearance of lignocaine (lidocaine), verapamil, propranolol and fentanyl.[35–38] The effects of enflurane on liver blood flow are dominated, as are those of halothane, by changes in blood pressure and cardiac output.[39,40]

Isoflurane, like halothane and enflurane, dilates the intestinal vasculature and decreases its oxygen uptake.[31,33] Hepatic artery flow is substantially preserved during isoflurane anaesthesia, possibly owing to the preservation of autoregulation or possibly from the profound vasodilatory effect of isoflurane on the hepatic vasculature.[33,34,39,41,42] Experiments on miniature pigs demonstrated that isoflurane was the optimal inhalational anaesthetic for providing a balance between hepatic oxygen supply and demand.[43] Compared with other anaesthetics, isoflurane is the only agent which preserves hepatic arterial blood flow in cirrhotic rats.[44]

When halothane, isoflurane and sevoflurane were compared under similar conditions, the hepatic oxygen supply/uptake ratio was worst with sevoflurane, which suggests that it is not the drug of choice if the liver is likely to be subject to physiological insult.[29,45] Nitrous oxide alone causes little change in cardiac output or blood pressure, which possibly explains why the hepatic circulation is not disturbed significantly when nitrous oxide is used in trained animals without any baseline anaesthesia or surgical intervention.[29,42]

Work on barbiturates has shown conflicting results in both animal and human studies.[29,46,47] In chronically cannulated sheep, Runciman and colleagues[47] found that hepatic blood flow was reduced by 17 per cent during propofol anaesthesia; Diedericks and colleagues (who found similar results) attributed this to the fall in cardiac output.[48]

Small doses of morphine are accompanied by dilatation of the intestinal vasculature whilst higher doses cause vasoconstriction.[49] In a pig model, fentanyl alone increased both hepatic oxygen supply and extraction with little subsequent change in hepatic venous blood oxygen content.[18,29] Alfentanil, on the other hand, decreased hepatic artery blood flow to a greater extent than it

reduced mean arterial pressure or cardiac output.[50] Unfortunately, no simultaneous measurements were made of portal blood flow so no conclusions can be drawn concerning total liver blood flow.

Pancuronium and vecuronium have little effect on the splanchnic circulation, but the 'less clean' drugs with autonomic ganglion blocking properties, histamine release, myocardial depression etc. all exert an influence indirectly via their effect on cardiac output and blood pressure.[29]

Very few studies have looked at the effect of regional anaesthesia on hepatic blood flow, but spinal and epidural anaesthesia in humans has produced falls in total hepatic blood flow that are in line with the reductions in arterial blood pressure.[51–54]

Drug metabolism and the liver

An outline of hepatic drug metabolism

The liver transforms lipophilic pharmacological compounds into hydrophilic excretion products by a combination of non-synthetic and conjugative reactions; the relevance of these processes to anaesthesia has been the subject of review articles.[55,56] The effect of anaesthetic drugs *per se* in liver disease is only part of what concerns the anaesthetist because in the perioperative period these patients will receive a wide variety of other medications. There are recent detailed reviews of liver disease and drug disposition.[57–59] Drug disposition is influenced by many factors in patients with liver disease, the most important of which are abnormal (usually reduced) liver blood flow, portasystemic shunting, impaired metabolic capacity and abnormal binding.[59]

A primary function of the liver is the deactivation of exogenous and endogenous compounds. Although this may be regarded as a detoxification process, in certain cases hepatically generated metabolites may be active drugs in their own right, or may themselves be a generated toxin. The liver, because of this and because it receives unprocessed molecules direct from the digestive tract, is at substantial risk of toxic damage.

The liver microstructure (see above) with the hepatic sinusoids is ideally designed to allow the free diffusion of blood proteins. This enables protein-bound drugs to come into direct contact with hepatocyte membranes. Usually, metabolic processes are divided into two groups: phase I and phase II reactions.

Table 37.2   Some examples of phase I and II reactions

Reaction	Example	Product
Phase I		
Hydroxylation	Phenobarbitone to *p*-hydroxyphenobarbitone	Inactive
N-Demethylation	Imipramine to desipramine	Active
Hydrolysis	Acetylsalicylic acid to salicylic acid	Active
Desulphuration	Thiopentone to pentobarbitone	Active
Phase II		
Glucuronidation	Paracetamol (acetaminophen)	Inactive
	Morphine	Active
Sulphation	Minoxidil	Active
Acetylation	Procainamide	Active
	Sulphanilamide	Inactive

Reproduced, with permission, from Jorm CM, Stamford JA. Hepatic metabolism of xenobiotics with reference to anaesthesia. *Ballière's Clinical Anaesthesiology* 1992; **6**: 751–79.

Phase I reactions

Phase I reactions alter the existing functional groups of a drug molecule to increase water solubility or hydrophilicity. Reductase and hydrolase enzymes are located predominantly in the cytoplasm, and most oxidations (hydroxylations) are carried out by the cytochrome P_{450} system located mainly in the smooth endoplasmic reticulum of the centrilobular cells. Each person may be considered to have his or her own P_{450} 'fingerprint' derived from a combination of genetic and environmental factors, which leads to considerable individual variation in drug handling.[60,61] Some examples of phase I reactions are given in Table 37.2.

Phase II reactions

Phase II metabolism consists of conjugation reactions occurring primarily in the cytosolic fraction of hepatocytes. The conjugation of polar compounds occurs by the attachment of endogenous hydrophilic groups to pre-existing or newly formed $-OH$, $-COOH$, $-NH_2$, or $-SH$ groups. Although phase II reactions usually follow phase I reactions, phase II reactions can, if the substrates are already polar, act directly on drugs. The most common phase II reaction is glucuronidation.[62] This is particularly important in the metabolism of morphine, whose primary metabolite (morphine-6-glucuronide) may be responsible for the majority of its action.[62,63] The vast majority of phase II reactions produce an inactive water-soluble compound although there are a few exceptions. Some examples of phase II reactions are given in Table 37.2.

Clearance

The clearance of a drug (Cl), is defined as the volume of blood from which the drug is removed in unit time. For simplicity, hepatic clearance is often represented as:[55]

$$Cl = \dot{Q}E$$

where Cl is the clearance, $\dot{Q}$ is liver blood flow and E is the extraction ratio (i.e. the proportion of drug extracted in one pass through the liver). Drug metabolism may therefore be controlled either by blood flow (the control of which has been described above) or the metabolic capacity of the liver. Classically, drugs may be classified into three main groups with reference to hepatic metabolism: 'high risk', 'limited risk' and 'low risk' drugs[59] The degree of risk depends largely upon the dependence of drug metabolism on liver blood flow.

Three factors determine the extraction of drugs from the liver: the activity of enzyme systems, the fraction of free drug in the plasma and the rate at which drug is presented to the liver. When the extraction ratio is very high, clearance approaches liver blood flow. When the extraction is small, clearance depends on the rate of metabolism and is independent of liver blood flow.[56] Drugs can therefore be classified into two main types: flow-limited drugs ($E > 0.7$) for which clearance depends on liver blood flow; and capacity-limited drugs ($E < 0.3$) for which clearance depends on the rate of liver metabolism. Examples of these effects are given in Table 37.3.

Table 37.3   Effect of extraction ratio (*E*) and protein binding on elimination of drugs by the liver

	Approximate extraction (E)	Protein binding (%)	Comments on effects of liver disease
Flow-limited drugs			Changes in liver blood flow and
Labetalol	0.85	40	intrinsic clearance associated with liver
Lignocaine (lidocaine)	0.60	65	disease affect these drugs. The
Morphine	0.75	35	shunting of blood around the liver has
Propranolol	0.65	95	important effects on the bioavailability
Verapamil	0.80	92	of these drugs
Flow/enzyme-sensitive drugs			Changes in liver blood flow, free
Paracetamol (acetaminophen)	0.30	20	intrinsic clearance, and free fraction of
Chlorpromazine	0.30	95	drug in blood may be important for this
Pethidine (meperidine)	0.50	70	class of drugs
Methohexitone	0.53	–	
Metoprolol	0.56	10	
Quinidine	0.27	85	
Ranitidine	0.28	15	
Capacity-limited, binding-insensitive			This class of drugs is most sensitive to
Antipyrine (phenazone)	0.05	10	changes occurring in the free intrinsic
Caffeine	0.04	31	drug clearance with liver disease
Hexobarbitone	0.15	47	
Theophylline	0.05	62	
Capacity-limited, binding-sensitive			This class of drugs will be influenced
Chlordiazepoxide	0.02	96	by changes in free fraction of drug in
Diazepam	0.02	97	blood and free intrinsic drug clearance.
Diphenylhydantoin (phenytoin)	0.03	92	The overall change in drug clearance
Warfarin	0.005	99	will be governed by which one of these factors changes the most as a result of the disease process.

Reproduced, with permission, from Eagle CJ, Strunin L. Drug metabolism in liver disease. *Current Anaesthesia and Critical Care* 1990; **1**: 204–12.

Protein binding

The degree of protein binding is important because it is only the unbound drug fraction that diffuses freely, reaches receptor sites and takes part in reactions. A reduction in binding from 95 to 90 per cent, if a drug is highly bound, results in a doubling of the free fraction of the drug, whereas for a poorly bound drug a reduction from 10 to 5 per cent has much less effect. Consequently, those drugs with extensive binding to proteins will have a clearance that is affected by changes in binding but those with less extensive binding will not be significantly affected. Examples of the protein binding of drugs relevant to anaesthesia are shown in Table 37.4;[55,64] its effects are included in Table 37.3.

Effects of liver disease on drug metabolism

In a patient with end-stage liver disease and gross hepatocellular failure the action of drugs is very unpredictable and can be dramatic. The majority of patients seen in anaesthetic practice are, however, well compensated with good preservation of hepatocellular function. Despite this, their response to drugs is not that of a normal subject and, from clinical observations, authors have made suggestions for modifications of drug dosage in patients with cirrhosis. Information on drugs used in anaesthetic practice is summarized in Table 37.5.[55,56]

For reasons that are not clear but are probably related to redistribution, patients with alcohol-induced cirrhosis may require an increased dose of thiopentone.[56] It is now over 40 years since 'resistance' to curare in cirrhotics was described

Table 37.4 Plasma protein binding of some drugs used in anaesthesia

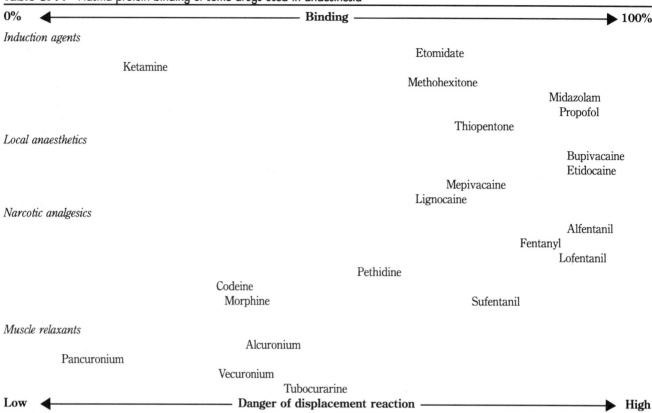

0% ◄─────────────────────────── **Binding** ───────────────────────► 100%

Induction agents

Ketamine ... Etomidate ... Methohexitone ... Midazolam ... Propofol ... Thiopentone

Local anaesthetics

Bupivacaine ... Etidocaine ... Mepivacaine ... Lignocaine

Narcotic analgesics

Alfentanil ... Fentanyl ... Lofentanil ... Pethidine ... Codeine ... Morphine ... Sufentanil

Muscle relaxants

Alcuronium ... Pancuronium ... Vecuronium ... Tubocurarine

Low ◄─────────────────── **Danger of displacement reaction** ───────────► High

Lignocaine = lidocaine *USP*; pethidine = meperidine *USP*.
Data from references 55 and 64.

by Dundee and Gray. This may well be due to sequestration of the drug in the liver and spleen, giving it an increased volume of distribution; more recently, pancuronium and vecuronium have shown an increased half-life in cirrhotic patients.[65,66] Atracurium, which breaks down spontaneously in plasma, does not appear to be affected by cirrhosis.[67]

The pharmacokinetics of opioids has been studied extensively in cirrhosis. After single-dose administration the elimination half-lives of fentanyl and sufentanil are unaffected,[68,69] but both the elimination half-life and the plasma-free fraction of alfentanil are increased.[70] Morphine appears to be safe in well compensated cirrhosis if renal function is normal, perhaps because the glucuronide pathway is spared until late in liver disease.[71] Pethidine, in contrast, has a markedly prolonged half-life in cirrhotics.[72] It should, however, be remembered that when opioids and other CNS depressants are used in cirrhotics (especially in those with end-stage disease), receptor density changes and altered permeability of the blood–brain barrier may result in an unpredictable and enhanced effect. Some patients with liver disease are especially sensitive to benzodiazepines in a way that appears to be a result of altered pharmacological responsiveness rather than derangement of pharmacokinetics.[73]

Volatile anaesthetics undergo metabolism to both volatile and non-volatile metabolites by both oxidative and reductive pathways.[56] Although there are good animal models demonstrating metabolic pathways, their relevance to damage in humans is unclear.[74] All available evidence suggests that when volatile agents are administered to patients with liver disease the outcome after operation is related to the extent of pre-existing damage and surgical trauma and that no one anaesthetic agent or technique has been shown to be safer than another.[56] The specific problem of postoperative jaundice and the effects of volatile anaesthetics is dealt with later in the section 'The investigation and management of postoperative jaundice'.

Table 37.5 The characteristics of drug disposition in normal subjects and recommendations for dose adjustment in patients with liver disease for a variety of drugs to anaesthetists

Drug	Protein binding (%)	Volume of distribution (V_d) (l/kg)	Half-life ($t_{1/2}$) (h)	Hepatic/renal elimination	Effect of liver disease on drug disposition	Adjustment of dose
Drugs commonly used in anaesthetic practice						
Thiopentone	85	2.3	9.0	>99% hepatic <1% renal	$t_{1/2}$ (u); V_d (u); Cl (u); f_p (i)	Uncertain; may need to decrease dose
Methohexitone	–	61	2.0	>90% hepatic <10% renal	No data; assume Cl (d), $t_{1/2}$ (i)	Probably decrease dose
Suxamethonium (succinylcholine)	–	–	0.1	plasma pseudo-cholinesterase	$t_{1/2}$ (u); Cl (u)	Single dose unchanged
Atracurium	–	0.16	0.33	Hofmann elimination; autometabolism	$t_{1/2}$ (u); V_d (i); Cl (u); long $t_{1/2}$ of metabolite	Decrease dose if long-term use
Pancuronium	0.416	3.5	1.5		$t_{1/2}$ (i), V_2 (i), Cl (d)	Increase loading dose
Cimetidine	20	1.1	2.3	40% hepatic 60% renal	$t_{1/2}$ (u); V_d (i) or (d) or (u); Cl (u) or (d); f_p changes assumed unimportant. Drug associated with increased incidence of mental confusion in cirrhotics	Decrease dose in severe liver disease
Ranitidine	–	1.2	2.7	50% renal 30% hepatic	$t_{1/2}$ (i); Cl (d)	Changes seen only with advanced disease
Narcotics						
Pethidine (meperidine)	65	4.5	4.5	>95% hepatic <5% renal	$t_{1/2}$ (i); V_d (u); Cl (d) 50%; f_p (u)	Decrease oral dose by 50% in cirrhosis or acute viral hepatitis
Morphine	35	3.7	2.0	90% GI tract and liver, 10% renal; glucuronidation	$t_{1/2}$(u); V_d (u); Cl (u); f_p (u), by some reports f_p (i)	None, but avoid in severe liver disease
Alfentanil	90	0.28	1.5	99% hepatic 1% renal	$t_{1/2}$ (i); V_d (u); Cl (d); f_p (i) (dose-dependent)	Decrease dose
Fentanyl	80	3.5	4.0	92% hepatic 8% renal	$t_{1/2}$ (u); V_d (u); Cl (u)	None
Sufentanil	90	4.0	4.1		$t_{1/2}$ (u) V_d (u) Cl (u)	None
Cardiovascular						
Atenolol	<5	0.55	6.5	10% hepatic 90% renal	$t_{1/2}$ (u); V_d (u); Cl (u)	None
Digitoxin	95	0.60	180	70% hepatic 30% renal	$t_{1/2}$ (u) or (d); Cl (i) or (u); f_p (i)	None
Digoxin	30	6.0	35	30% hepatic 70% renal	Appears negligible	None
Disopyramide	80 non-linear	1.0	8	45% hepatic 55% renal	No data; would not expect a tremendous change in liver disease	Probably slight decrease

(continues)

Table 37.5 Continued

Drug	Protein binding (%)	Volume of distribution (V_d) (l/kg)	Half-life ($t_{1/2}$) (h)	Hepatic/renal elimination	Effect of liver disease on drug disposition	Adjustment of dose
Cardiovascular continued						
Labetalol	50	11.5	3.0	>95% hepatic <5% renal	$t_{1/2}$ (u); V_d (d); Cl (u) or (d); f_p (?)	Decrease oral dose; decrease i.v. dose to much smaller extent
Lignocaine (lidocaine)	65 non-linear	1.1	2.0	97% hepatic 3% renal	$t_{1/2}$ (i); V_d (i) or (u); Cl (d) ~ 50%; f_p? Low therapeutic ratio. Decrease in Cl depends on severity of disease	Decrease dose by 50% in severe liver disease
Lorcainide	70	12.9	8.0	98% hepatic 2% renal	$t_{1/2}$ (i); V_d (u); Cl (d) 29%; f_p (i) slightly. Cl_{int} exhibits a very large decrease	Decrease dose
Metoprolol	10	3.2	4.0	95% hepatic 5% renal	$t_{1/2}$ (i); V_d (i) slightly; Cl (d) 23%; f_p?	Decrease dose slightly
N-Acetyl-procainamide	10	1.4	8.0	20% hepatic 80% renal	No data; except little change unless renal function altered	None
Nifedipine	98	1.0	3.0	100% hepatic	$t_{1/2}$ (i); V_d (u); Cl (d)	Decrease dose
Pindolol	57	6.2	3.5	70% hepatic 30% renal	Not affected by AVH. Cirrhosis Cl (d) slightly and renal excretion of drug is increased	Some decrease in severe liver disease
Prazosin	97	1.3	3.0	95% hepatic 5% renal	No data – would expect $t_{1/2}$ (i); Cl (d); f_p (i)	Decrease dose
Procainamide	15	2.2	3.0	45% hepatic 55% renal Drug acetylated	$t_{1/2}$ (i); V_d?; Cl? probably decreased slightly	Some minor decrease in dose
Propranolol	95	4.0	4.0	>95% hepatic <5% renal	$t_{1/2}$ (i); V_d (i); Cl (d) ~ 60%; f_p (i). Tremendous decrease in Cl_{int}. Flow/enzyme-limited in cirrhosis	Decrease dose depending on extent of damage
Quinidine	85	3.0	6.0	80% hepatic 20% renal	$t_{1/2}$ (i); V_d (i); Cl (u); f_p (i); Cl_{int} decreased significantly	Decrease dose
Frusemide (furosemide)	95	0.15	1.0	35% hepatic 65% renal	$t_{1/2}$ (i) or (u); V_d (i) or (u); Cl (u); f_p (i); the change in f_p decrease in Cl_{int} of liver	None or slight decrease in severe cases
Sedative/hypnotic						
Diazepam	99	1.2	45	>97% hepatic <3% renal	$t_{1/2}$ (i); V_d (i); Cl (d) 50%; f_p (i). AVH and cirrhosis increase $t_{1/2}$. Large therapeutic index-safe.	Single dose, no change; chronic, decrease dose

Table 37.5 Continued

Drug	Protein binding (%)	Volume of distribution (V_d) (l/kg)	Half-life ($t_{1/2}$) (h)	Hepatic/renal elimination	Effect of liver disease on drug disposition	Adjustment of dose
Sedative/hypnotic continued						
Lorazepam	90	1.3	12.0	>98% hepatic <2% renal Extensive glucuronidation	$t_{1/2}$ (i); V_d (i); Cl (u); f_p (i). Neither AVH nor cirrhosis affects drug dosing	None
Midazolam	–	1.3	1.6	>95% hepatic <5% renal	$t_{1/2}$ (i); V_d slightly (i) Cl (d)	Decrease dose
Oxazepam	90	1.6	6.0	>95% hepatic <1% renal Extensive glucuronidation	$t_{1/2}$ (i); V_d (u); Cl (u); f_p (u). Neither AVH nor cirrhosis alters disposition significantly	None

AVH, acute viral hepatitis; *Cl*, clearance; d, decreased; f_p, plasma fraction; i, increased; u, unchanged.
Reproduced, with permission, from Eagle CJ, Strunin L. Drug metabolism in liver disease. *Current Anaesthesia and Critical Care* 1990; **1**: 204–12.

Anaesthesia in patients with coexisting liver disease

Preoperative assessment of hepatic function

Biochemically, the liver has a large functional reserve and it needs to be significantly compromised before any impairment manifests itself clinically. The exception to this is obstructive jaundice which is usually produced by extrahepatic pathology. Both anaesthesia and surgery may affect liver function adversely, and preoperative occult disease may only become manifest postoperatively. In patients known to have compromised hepatic function, thorough preoperative assessment and the forward planning of postoperative care are the key elements of success.

The presentation of liver disease is similar for several different aetiologies; because of this, clinical examination usually has to be augmented by liver function tests and other special procedures. In addition, it is of the greatest importance to remember that liver disease often has effects on other organs such as the kidney, heart, lungs and brain and that preoperative assessment should concentrate as much on these extrahepatic manifestations as on the liver itself. Diagnostic laparotomy in patients with liver disease carries a high complication rate,[75] so it is imperative to establish a diagnosis by other less invasive means.

Liver function tests and diagnostic procedures

When liver function tests are requested, hospital laboratories usually supply a battery of results similar to those shown in Table 37.6.[76] It is normal practice to screen for hepatitis at the same time. The results are useful as a form of baseline screen rather than as a diagnostic tool. Minor asymptomatic changes in liver function tests are not uncommon,[77,78] and can be related to alcohol intake, obesity or diabetes. If, however, serial tests are abnormal over an extended period, it is likely that hepatocellular damage is occurring. The combination of an elevated bilirubin, AST, APT and ALT has a predictive accuracy of over 90 per cent for the presence of liver disease.[79]

The enzymes present in liver cells (Table 37.6) are released into the plasma when the cells become diseased. AST and ALT are raised with hepatocellular damage and necrosis but the clinical picture must be taken into account because they are also released after damage to other cells and after trauma and myocardial infarction. There has been a long-standing attempt to find a single marker to determine the degree of liver damage in specific pathologies. Glutathione-*S*-transferase has been used to assess damage from anaesthetic agents,[80,81] γ-glutamyl transpeptidase rises after alcohol- and drug-induced liver damage[82] and ALP increases early in biliary obstruction. It must not be forgotten that, although enzyme assays indicate the loss of integrity of liver cells, they indicate nothing about the remaining metabolic or synthetic capacity of the liver. When the liver is failing acutely, initially high enzyme concentrations may fall, indicating total cellular failure rather than recovery.

When the serum concentration of bilirubin rises above

Table 37.6 Liver function tests

Test	Normal range	If abnormal, may indicate:
Aspartate aminotransferase	10–40 u/l	Hepatocellular damage/necrosis
Alanine aminotransferase	10–37 u/l	Hepatocellular damage/necrosis
Alkaline phosphatase	35–100 u/l	Cholestasis
γ-Glutamyl transpeptidase	11–64 u/l	Alcohol- or drug-related damage
Serum albumin	30–44 g/l	Decreased synthetic function
Prothrombin time	1–1.3 INR	Decreased synthetic function
Serum bilirubin:		
total	4–17 µmmol/l	Assessment of jaundice/primary biliary cirrhosis
unconjugated	<0.3 µmmol/l	Gilbert's disease
5'-Nucleotidase	1–18 u/l	Cholestasis
Total serum bile acids	0–6 µmol/l	Cholestasis/portasystemic shunt
Serum iron	8–31 µmol/l	Increased ferritin and transferritin suggests haemochromatosis
Transferritin saturation	<50%	–
Ferritin	10–250 µg/l	–
Ceruloplasmin	1.1–2.9 µmol/l	Low with low copper suggests Wilson's disease
Serum copper	11–24 µmol/l	–
α_1-Antitrypsin	25–64 µmol/l	Low in deficiency disease
α-Fetoprotein	<0.29 nmol/l	Increased in hepatoma

INR, international normalized ratio
Reproduced, with permission, from Strunin L. Preoperative assessment of hepatic function. *Baillière's Clinical Anaesthesiology* 1992; **6**: 781–93.

20 µmol/l there is usually sufficient bilirubin sequestered in the tissues to cause jaundice. The serum bilirubin can be divided into conjugated and unconjugated fractions, but these are not normally assayed separately by the laboratory. An increase in the unconjugated fraction (which cannot enter the urine and produces acholuric jaundice) results from an abnormal bilirubin load (haemolysis) or from a biochemical defect at, or proximal to, the site of conjugation on the smooth endoplasmic reticulum of the hepatocyte. The latter is typified in Gilbert's disease, a benign condition of intermittent jaundice affecting 5 per cent of the population.

An elevation of the serum conjugated bilirubin is always evidence of liver pathology and indicates defective excretion via the normal route into the biliary tree. Once conjugated, bilirubin is water soluble and can enter the urine. Traditional teaching suggests that disordered liver function tests, interpreted with the conjugated/unconjugated bilirubin ratio, can specify whether the problem is pre-, intra- or post-hepatic. Although neat in theory, this can rarely be done in practice because parenchymal disease ultimately possesses an obstructive component and obstructive disease produces cellular dysfunction.

Defective excretion may occur either within the bile canaliculi and small ducts within the liver (intrahepatic cholestasis) or in the main bile ducts between the liver and the duodenum (extrahepatic cholestasis). Extrahepatic obstruction results from gall stones, strictures, and carcinomas of the bile duct and head of pancreas. Intrahepatic cholestasis may, for instance, occur as a result of widespread hepatocellular damage (viral, alcohol, cirrhosis), cellular reactions around the ductules from drug reactions (e.g. phenothiazines), and inflammatory reactions around the intralobular and septal bile ducts (primary biliary cirrhosis, ascending cholangitis). Causes that have a significant inflammatory or autoimmune component may have the bilirubin level dramatically reduced by steroids.

In the presence of biliary obstruction, intravenous cholangiography will not produce adequate images; in recent years, ultrasound has become the most important imaging technique for characterizing the cause of cholestasis and for determining the pathway for further specialized tests. These include CT scanning, MRI scanning, and invasive procedures such as angiography, ERCP (with or without papillotomy) and percutaneous transabdominal cholangiography.

The synthetic ability of the liver can be indexed by the levels of albumin and the prothrombin ratio. A serum albumin level below 25 g/l indicates substantial liver damage provided that the nephrotic syndrome and excessive protein losses from gastrointestinal disease have been excluded. Most of the clotting factors are synthesized in the liver, and the synthesis of factors II, VII, IX and X depends on vitamin K. Deficiency of these prolongs the prothrombin time. This used to be expressed in seconds compared with a control but is now usually expressed as the international normalized ratio (INR), which is unity when there is no abnormality. In obstructive jaundice, vitamin K is not absorbed from the intestine because bile salts are absent. Parenteral vitamin K begins to act within a few hours and usually will return the prothrombin time to normal within 24 hours. If it remains elevated (vitamin K resistant), this indicates cellular damage, the hepatocytes being unable to synthesize the clotting factors in the presence of active substrate. In these circumstances, fresh-frozen plasma will correct the clotting defect temporarily to allow diagnostic procedures (e.g. liver biopsy) to be undertaken.

Liver biopsy will demonstrate definitive evidence of cirrhosis. Cirrhosis is the final result of many conditions and represents the endpoint in the sequence cellular necrosis, fibrosis and nodular regeneration with a distortion of the architecture. Histologically there are further subdivisions based on the nodule size and the location of the fibrosis but the functional result is the same.

All patients with hepatic disease need to be screened for hepatitis (see later). α-Fetoprotein is raised in 70–90 per cent of patients with primary hepatocellular carcinoma (which can also be used postoperatively as a marker for recurrence), and over 90 per cent of patients with primary biliary cirrhosis will have antinuclear antibodies in the serum. Patients with Wilson's disease have low levels of copper and ceruloplasmin, and those with haemochromatosis have increased concentrations of ferritin and transferritin.

Clinical assessment of the patient

The anaesthetist may be called upon to care for patients with liver disease who are having surgical treatment for an unrelated condition or for those undergoing specific procedures concerned with the abnormal liver itself. Apart from the relief of simple extrahepatic biliary obstruction, the latter is dealt with separately in this chapter.

The preoperative assessment needs to achieve three objectives:

1. To assess the type and degree of liver dysfunction.
2. To assess the general condition of the patient and to look for the possible effects liver disease has had on other organ systems.
3. To ensure that in high-risk cases there are appropriate postoperative facilities available to care for the patient.

Clinical assessment of liver function

Jaundice This is best seen in the sclera under white light and can easily be missed in the artificially illuminated ward. Jaundice is never a trivial finding and an accurate preoperative diagnosis of the cause is very important. Even if the diagnosis has been definitely made it is prudent to ask about a family history of jaundice, contact with jaundiced individuals, blood transfusions, tattooing, acupuncture, drug addiction, sexual orientation and contacts,

foreign travel, recent immigration, alcohol intake, drug ingestion, general health and occupational hazards.

In a mildly jaundiced patient with simple biliary obstruction there are few additional physical signs. Unexplained jaundice of 4 weeks' duration or longer will prove to be caused by obstruction in approximately 75 per cent of patients. Gall stones in the common bile duct are usually accompanied by pain and pyrexia and may present as pancreatitis, with or without jaundice. Weight loss suggests an underlying neoplasm of the bile ducts or the pancreas. Weight gain or abdominal distension should raise the possibility of ascites. Malaise, lethargy, nausea and pruritus are common and related to obstructive jaundice; personality and mental changes are rare unless there is associated hepatocellular failure.

Jaundice always carries a risk of renal impairment, the management of which is described under 'Renal function'. Long-standing biliary obstruction is often accompanied by infection of the biliary tree, and endotoxin (which may be responsible for renal failure) has been found in the systemic circulation.[83] If the serum bilirubin is very high (over 140 µmol/l), it has been recommended that percutaneous drainage of the biliary tree be undertaken together with appropriate antibiotic therapy preoperatively to try to prevent postoperative renal problems.[84] Preoperative oral bile salts have also been given in an effort to reduce the incidence of endotoxaemia in obstructive jaundice, and this may help to prevent the associated renal failure.[85]

Cholecystectomy for gall stones is a common operation which in uncomplicated cases carries a low morbidity and a mortality well below 1 per cent. Morbidity and mortality do, however, increase substantially if there is to be exploration of the common duct, or if there is pre-existing sepsis, malignancy, diabetes, anaemia or jaundice.[84,86,87]

Cirrhosis Cirrhosis can range from the unsuspected to the life threatening: the anaesthetist is primarily interested in the degree of hepatocellular failure, the biochemical hepatic reserve and the presence of portal hypertension. The causes of cirrhosis are summarized in Table 37.7. The activity of the cirrhotic process is assessed by clinical, biochemical and histological observations and subsequently classified as progressive, stationary or regressive. In many cases cirrhosis is compatible with good health and may be clinically and biochemically undetectable. On the other hand, when severe, cirrhosis can run an aggressive course culminating in liver failure and death. In cirrhosis the portal vascular bed is distorted and the portal blood flow is mechanically obstructed. The normal portal venous pressure of 7 mmHg can be raised up to 50 mmHg with the creation of alternative collateral venous channels. The clinical features of portal hypertension are oesophageal varices, prominent collateral veins radiating from the umbilicus, dilated rectal veins and splenomegaly.

Table 37.7 The causes of cirrhosis

Unknown or cryptogenic (30% in UK)
Alcoholic (30% in UK and rising)
Viral hepatitis
Prolonged cholestasis, intra- or extrahepatic obstruction
Hepatic venous outflow obstruction
Metabolic disease (e.g. haemochromatosis, Wilson's disease,
  α_1-antitrypsin deficiency, storage diseases)
Autoimmune:
  primary biliary cirrhosis
  chronic active hepatitis
  'lupoid hepatitis'
Drugs (e.g. methotrexate, isoniazid, methyldopa)
Intestinal bypass surgery for obesity

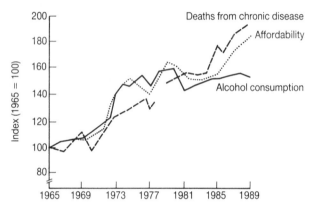

Fig. 37.4 Alcohol consumption, price and deaths from chronic disease. (Reproduced, with permission, from Anderson P. The health of the nation: responses; alcohol as a key area. *British Medical Journal* 1991; **303**: 766–9.)

It should also be noted that excess alcohol intake is now the commonest cause of cirrhosis and that, when this is the case, the associated problems of alcoholism and withdrawal will need attention. The relation between alcoholism, affordability and chronic disease is shown in Fig. 37.4.[88]

The natural history of chronic liver disease is not necessarily one of unremitting and progressive deterioration, and it may only be the stress of intercurrent surgery or other illness that renders it apparent.[76] When symptoms do occur they tend to be vague, such as malaise, dyspepsia, weight loss, loss of libido etc. Physical signs can be completely absent or there may be skin signs (white spots, paper money skin, white nails, palmar erythema). Finger clubbing and pigmentation are non-specific. Perhaps the most useful sign of cirrhosis is spider naevi on the face, arms and upper body. They may occur transiently in pregnancy but are particularly noticeable in alcoholic and chronic active hepatitis.

The cirrhotic liver is firm to touch, and in macronodular disease it may be possible to feel the surface irregularities. The liver tends to be large in alcoholic cirrhosis and haemochromatosis but shrunken in cryptogenic and

primary biliary cirrhosis. Hepatocellular failure is characterized by jaundice (see above), ascites and encephalopathy (see later). The two most important factors in the development of ascites are a lowered plasma oncotic pressure (because of the failure of albumin synthesis) and portal venous hypertension (see above). When more fluid enters the peritoneal cavity than leaves it, ascites develops. This results in depletion of the effective intravascular volume which causes the renal tubules to retain sodium and water. This in turn encourages the further formation of ascites. Ascites may develop suddenly or insidiously over a period of months. Long-standing ascites can also lead to the hepatorenal syndrome (see below).

General condition and other organ systems
The ascites and tissue oedema in cirrhosis often mask an appalling nutritional state with considerable loss of muscle mass and power. Special attention needs to be given to the effects of hepatic disease on the cardiovascular, renal, respiratory and central nervous systems.

Cardiovascular system Patients with advanced cirrhosis may present in a hyperkinetic circulatory state with decreased peripheral vascular resistance, increased cardiac output and increased shunting both peripherally and in the lungs. There can be a tachycardia, flushed extremities, bounding pulse, capillary pulsation and an ejection systolic murmur. Patients with alcoholic liver disease may have poor left ventricular function.[89] If the blood pressure is low, it is important to differentiate between low peripheral resistance, hypovolaemia from occult blood loss and ventricular impairment.

Respiratory system Intrapulmonary shunting may be associated with hypoxaemia, impaired hypoxic pulmonary vasoconstriction and hyperventilation.[90] Approximately one-third of patients with decompensated cirrhosis have a reduced Pa_{O_2} and are cyanosed. Chest infections are common. If the patient has hepatomegaly or ascites, there will be an increase in the closing volume of the lung with basal atelectasis. Pleural effusions sometimes coexist with ascites. Bronchospasm may be an additional feature, and this can be an important problem if the patient requires treatment with a β-blocker for portal hypertension.[90]

Renal function The effect that jaundice can have on the kidney has already been commented upon, and the link between preoperative jaundice and postoperative renal dysfunction has been recognized for many years. In a series of 114 patients undergoing surgery on the liver, biliary tract and pancreas, 36 patients (an incidence of 17 per cent) developed acute renal failure (with a mortality rate of 100 per cent), compared with an incidence of renal failure of 1 per cent in 78 patients who were not jaundiced

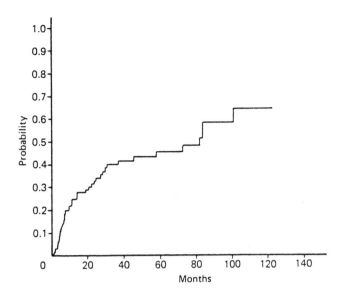

Fig. 37.5 Cumulative probability of developing hepatorenal syndrome from the onset of ascites in a series of 136 cirrhotic patients. (Reproduced, with permission, from Ginès P, Arroyo V, Rodes J. Treatment of ascites and renal failure in cirrhosis. *Baillière's Clinical Gastroenterology* 1989; **3**: 165–86.)

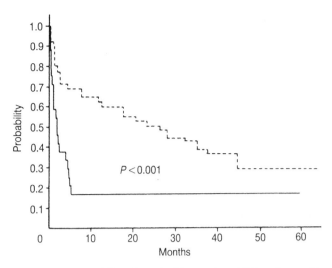

Fig. 36.6 Probability of survival in a series of 136 patients with cirrhosis and ascites, classified according to the presence (—) (*n* = 24) or absence (- - -) (*n* = 112) of hepatorenal syndrome (HRS). (Reproduced, with permission, from Ginès P, Arroyo V, Rodes J. Treatment of ascites and renal failure in cirrhosis. *Baillière's Clinical Gastroenterology* 1989; **3**: 165–86.)

preoperatively.[91] In addition to jaundice, the preoperative serum creatinine is a factor of major prognostic importance in identifying patients at risk of developing renal failure who are to undergo biliary tract surgery.[92]

Separate from the problems of jaundice is the presence of the hepatorenal syndrome. This can be defined as 'a decrease in renal function which develops in a patient with serious liver disease in whom all other causes of renal dysfunction are excluded'.[93] As liver disease progresses, the chance of developing the hepatorenal syndrome increases (Fig. 37.5);[94] the longer the syndrome has been present, the greater the chance of dying (Fig. 37.6).[94] It is present in 85 per cent of patients seriously ill with cirrhosis. The hepatorenal syndrome is a condition of complex aetiology with progressive water retention, hyponatraemia, azotaemia, and oliguria due to reduced renal plasma flow.[76] Its management is difficult and needs to be undertaken in specialist units

Intraoperative measures for protecting the kidney are described later.

Haemostatic function Patients with liver disease may have compromised haemostatic function (because of failure to produce clotting factors) and have a prolonged bleeding time.[95] Factors II, VII, IX and X can all be reduced but may respond to vitamin K (see above). Acutely, the bleeding tendency may be reduced by the administration of fresh-frozen plasma and platelets but there is no lasting effect if liver function does not improve. There are also other factors involved in this haemorrhagic tendency, including

dilutional coagulopathy, an exaggerated fibrinolytic response and metabolic problems such as hypothermia, acidosis and hypocalcaemia.[90] The increased tendency towards fibrinolysis in cirrhotic patients may be the result of an exaggerated response to vasoactive substances resulting from delayed hepatic clearance of plasminogen, decreased synthesis of naturally occurring inhibitors and activation of protein C.[90]

The preoperative coagulation status of the patient and the pathology has been related to the amount of intraoperative blood loss.[96] Patients with post-necrotic cirrhosis, fulminant hepatitis and sclerosing biliary cirrhosis required a significantly greater number of blood products than patients suffering from primary biliary cirrhosis or carcinoma of the liver.

Central nervous system (CNS) Cerebellar ataxia and peripheral neuropathy may occur in alcoholics without encephalopathy. CNS signs which suggest that chronic liver disease has become decompensated include a flapping tremor and disturbances of conscious level.[97] The stages of hepatic encephalopathy are described later in the section 'Acute liver failure and its intensive care management'.

Gastrointestinal system Melaena may indicate bleeding varices. Dilated veins around the umbilicus and on the abdominal wall demonstrate portal hypertension. The specific problems of the gastrointestinal system in acute liver failure are covered later.

Table 37.8 Clinical and laboratory classification of patients with cirrhosis in terms of hepatic functional reserve

	Group A	Group B	Group C
Serum bilirubin (μmol/l)	<40	40–50	>50
Serum albumin (g/l)	>35	30–35	<30
Ascites	None	Easily controlled	Poorly controlled
Neurological disorder	None	Minimal	Advanced coma
Nutrition	Excellent	Good	Poor, wasting
Risk of operation	Good (5%)	Moderate (10%)	Poor (50%)

Reproduced, with permission, from Child III CG, Turcotte JG. Surgery and portal hypertension. In: Child III CG, ed. *The liver and portal hypertension*. Philadelphia: WB Saunders, 1964: 50.

Table 37.9 Grading of severity of liver disease

Clinical and biochemical measurement	Points scored for increasing abnormalities		
	1	2	3
Encephalopathy (grade)	None	1 and 2	3 and 4
Bilirubin (μmol/l)	<25	25–40	>40
Albumin	35	28–35	<28
Prothrombin time (seconds prolonged)	1–4	4–6	>4

4–6 points = Child group A; 7–9 points = Child group B; 10–12 points = Child group C
Reproduced, with permission, from Pugh RNH, Murray-Lyon IM, Dawson JL. Transection of the oesophagus for bleeding oesophageal varices. *British Journal of Surgery* 1973; **60**: 646–9.

Perioperative management

The principles of management of any patient are essentially the same whether the patient is well compensated or decompensated, but the latter has a much reduced margin of safety. The most useful information for the anaesthetist is an estimation of the hepatic reserve and thus, by inference, the risk associated with operation. Tables 37.8, 37.9 and 37.10 show the scoring systems developed by Child and Turcotte,[98] Pugh and colleagues[99] and Garrison *et al.*[100] These schemes may be helpful in making decisions about elective surgery.[101] The Child and Pugh classifications are easy to apply and it has been recommended[76] that as a working rule only patients in Child's group A should be recommended for elective

Table 37.10 Preoperative risk factors associated with increased postoperative mortality

Serum albumin <3 g/dl
Presence of infection
White blood cell count >10 000
Treatment with more than two antibiotics
Prothrombin time >1.5 s over control
Serum bilirubin >50 mmol/l
Presence of ascites
Malnutrition
Emergency surgery

Data from references 90 and 100.

surgery. Those in group B may be acceptable but every effort must be made to correct abnormalities of nutrition, to raise the plasma albumin and to correct anaemia and clotting abnormalities. Those in group C should only be considered for emergency operations.

The preoperative visit

The exact nature of the preoperative visit will depend on the severity of the liver dysfunction present and the extent of the planned operation. It is important at this stage to check that the appropriate preoperative investigations have been done and that the results are available in the notes. A checklist prior to major surgery is given in Table 37.11.[90] It is essential to have this data as baseline information so that postoperative results and complications can be interpreted and treated in relation to the preoperative state. The current state of the veins, arteries and hydration can be assessed and any drug therapy such as diuretics, steroids, antibiotics and antacids noted. In addition to this the anaesthetist must make sure that the appropriate blood and blood products have been ordered and that there are suitable postoperative facilities available.

At the preoperative visit any expected postoperative intensive or high-dependency care procedures such as ventilatory support (continuous positive airway pressure etc.), monitoring and analgesia can be explained to the patient and relatives. It is particularly important to prepare

Table 37.11 Preoperative investigations

Cardiorespiratory	Chest x-ray
	Exercise ECG
	Pulmonary function tests
	Blood gases
Metabolic	Serum glucose
	Serum electrolytes
	Serum urea
	Serum creatinine
	Urinary electrolytes
	Urinary urea
	Urinary creatinine
Haematological	Haemoglobin
	White blood cell count
	Platelet count
	Clotting factors
	TT, APTT, INR
Liver function	Serum albumin/globulin
	Serum bilirubin
	Liver enzymes
Hepatic imaging	
Microbiological	Cultures
	Hepatitis B markers
	Viral antibodies

Reproduced, with permission, from Browne DRG. Anaesthesia in impaired liver function. *Current Anaesthesia and Critical Care* 1990; **1**: 220–7.

the relatives for the possible postoperative complications if the patient has suboptimal cerebration or is confused.

Premedication and preoperative preparation

If the patient has normal neurological status and desires it, an anxiolytic premedication may be prescribed. Short-acting benzodiazepines (e.g. temazepam) have been used successfully by the oral route (this avoids the bleeding and haematoma associated with intramuscular injections), although only small doses should be prescribed. Oral H$_2$-receptor antagonists or cytoprotective agents should also be prescribed to reduce the risk of stress ulceration. The use of vitamin K has been described earlier.

Patients who are jaundiced and those with the hepatorenal syndrome are at risk of renal damage if they are allowed to become dehydrated or hypotensive as a result of the preoperative fast. They should therefore have an intravenous infusion established at the commencement of the preoperative fast to give crystalloid at the rate of 1–2 ml/kg per hour. In some centres it is the policy to recommend that diuretics are also prescribed to enhance urine flow. The problem with this approach is that if the diuretics are given but the intravenous infusion stops then drug action dehydrates the patient, thereby compounding the problem it was intended to prevent. It is useful to give fluids but not to commence diuretics until the patient reaches the operating theatre.

Anaesthetic technique

The anaesthetic technique used and the degree of monitoring obviously depend on the combination of patient and procedure, and considerable clinical judgement is required. Different anaesthetists have used a variety of techniques and drugs to produce a successful outcome. It is therefore the way drugs are used rather than what they are that is most important. Whatever operation is undertaken and whatever anaesthetic technique is employed, it is imperative to minimize the physiological insult on the liver and kidney.

As already described, the cirrhotic liver derives a greater proportion of its blood supply from the hepatic artery and the regeneration nodules are supplied mainly with arterial blood. Systemic hypoxaemia and hypotension may result in severe liver necrosis and must be prevented at all costs. The kidney, which is always at high risk in liver disease, similarly needs to be protected from hypotension and hypoxia, and also from a low filtration rate secondary to hypovolaemia (this concentrates toxins in the filtrate). Fluid balance therefore must be meticulous. It is imperative to avoid hypotensive anaesthesia in cirrhotic patients for elective surgery (e.g. middle ear surgery, plastic procedures).

Minor surgical procedures (e.g. drainage of abscess, inguinal hernia repair) can often be successfully undertaken with patients breathing spontaneously using a face or laryngeal mask but there should be a low threshold for intubation and ventilation. What is described below is the anaesthetic technique for a patient with hepatic disease undergoing a major procedure with a substantial anticipated blood loss. It can be adapted by the reader using clinical judgement to meet the needs of any particular individual patient.

The regimen recommended by Browne[90] for major surgery is given in Table 37.12. Temazepam may be substituted for lorazepam, a combination of etomidate and fentanyl for thiopentone, and alfentanil for fentanyl. For elective starved cases, suxamethonium (succinylcholine) may be omitted.

Monitoring

The degree of monitoring above the minimal requirements will depend on the extent of the surgery; that recommended by Browne[90] is given in Table 37.13. It should be remembered that repeated measurements of blood pressure with an upper arm cuff can produce considerable bruising in patients with altered haemostatic function, and there may be considerably less morbidity produced from the insertion of an arterial line. If an arterial line is to be used, it

Table 37.12 Anaesthetic technique

Ensure that BLOOD PRODUCTS are available in the operating room before surgery starts

Premedication	Lorazepam ± ranitidine
Induction	Thiopentone
Intubation	Suxamethonium (succinylcholine) Low-pressure cuffed oral tracheal tube
Maintenance	Fentanyl 1 μg/kg p.r.n. Isoflurane, O₂, N₂O/air IPPV ± PEEP (aim for $P\text{aco}_2$ 3.5–4.6 pKa
Drug infusions	Atracurium 0.5 mg/kg per h Dopamine 2 μg/kg per min (Mannitol 10% 1 g/kg per h)
Venous access	Peripheral cannulae (two) (13 G or 9 G SG introducers into antecubital fossae) Central venous lines (two): one for monitoring one for drug infusions (triple lumen)

Arterial line

Urinary catheter

Reproduced, with permission, from Browne DRG. Anaesthesia in impaired liver function. *Current Anaesthesia and Critical Care* 1990; **1**; 220–7.

is best inserted under local anaesthesia at the outset so that the haemodynamic effects of induction can be monitored. If more than one puncture is required, do not forget to apply pressure to the puncture site for several minutes to prevent haematoma formation.

Induction

After preoxygenation the induction agent should be given slowly to minimize reductions in cardiac output and to allow time for the effects to become apparent. In the sick patient (especially with encephalopathy), only very small doses will be required to produce hypnosis, but in the compensated cirrhotic increased binding to globulins, or enzyme induction, may lead to an increase in requirements. Although the intrinsic clearance of thiopentone may be delayed in liver disease, in practice recovery is not a problem.[102] Similarly, although cholinesterase levels may be deficient in patients with liver disease, the prolonged action of suxamethonium is not usually a problem.[103] With very sick encephalopathic patients the arterial blood pressure needs to be monitored carefully during induction and treated to prevent rises which would produce increases in intracranial pressure.

Table 37.13 Intraoperative monitoring/equipment

ECG

Blood pressure:
  arterial line
  Dinamap

Central venous pressure (Swan–Ganz catheter)

Oximeter
Capnograph

Nerve stimulator

Core temperature:
  warming blanket
  heated humidifier

Urinary output

Blood loss:
  swabs
  suction
  floor

Input/output charts

Rapid infusion system

Reproduced, with permission, from Browne DRG. Anaesthesia in impaired liver function. *Current Anaesthesia and Critical Care* 1990; **1**; 220–7.

After induction but before surgery is the usual time to insert central venous and (if required and appropriate) pulmonary artery catheters. If cannulation is problematical, the anaesthetist should not forget to press on puncture sites. The jugular route is preferred to the subclavian because haemorrhage is easier both to see and to control.

Maintenance

A convenient plan for the arrangement of the equipment, transfusion, monitoring and infusion lines for a major abdominal procedure is shown in Fig. 37.7.[90]

For the reasons given at the beginning of this chapter, ventilation should maintain normocapnia. In an abdominal operation, the surgeon should be reminded to avoid compressing the major hepatic vessels whenever possible.

Atracurium is the muscle relaxant of choice for maintenance of relaxation during anaesthesia because it is broken down spontaneously by Hofmann degradation. It is, however, important to monitor the degree of relaxation carefully because patients with liver disease show increased resistance to the drug and a much shorter duration of action.[103] This may be due to the increased binding of atracurium to the raised levels of globulin present in these patients. It is important to note that vecuronium, which depends on the liver for its elimination, even in mild disease has a recovery time that is twice as long as normal.[104] It has been shown that in cirrhotic patients fentanyl pharmacokinetics are not dissimilar from

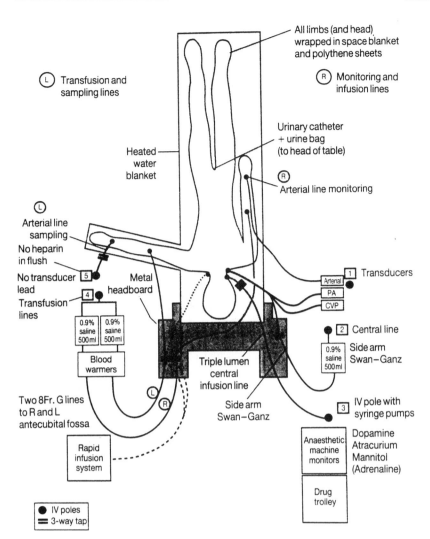

Fig. 37.7 The layout for major surgery on a patient with severe liver disease. CVP, central venous pressure; PA, pulmonary artery. (Reproduced, with permission, from Browne DRG. Anaesthesia in impaired liver function. *Current Anaesthesia and Critical Care* 1990; **1**: 220–7.)

normal,[105] but in end-stage liver disease the action becomes very variable and cannot be predicted with accuracy.[106] Isoflurane is the inhalational agent of choice because hepatic oxygen supply is thought to be better preserved when this agent is used, compared with other agents.[34]

Temperature should be monitored throughout and appropriate measures taken to prevent heat loss. These will obviously depend upon the surgery, but in long cases with exposure of organs should include warming blankets, humidified gases, space blanket, polythene limb bags and warmed intravenous fluids.

Fluid balance must be managed meticulously and the patient kept normovolaemic with a haemoglobin of 10 g/dl. Our technique is to optimize filling pressures using CVP and PA (pulmonary artery) catheters and, if the BP and cardiac output are insufficient, to move earlier rather than later to inotropic support. It is vital to prevent the patient becoming hypovolaemic and to maintain a steady urine output. If in the presence of mannitol and dopamine (see Table 37.12) and no hypovolaemia or hypotension the urine

flow drops below approximately 2 ml/kg per hour, a frusemide infusion should be started.

Anaesthesia for patients with liver dysfunction is unusual in that considerable attention has to be paid to the intraoperative monitoring of haematological and biochemical parameters (Table 37.14).[90] When there is considerable blood loss and this has been replaced by crystalloids, colloids and packed red cells, it is important to replace the clotting factors. A practical rule of thumb is to give one unit of fresh-frozen plasma for every unit of packed cells, or 250 ml of 0.9 per cent saline or colloid transfused. Red blood cell replacement is unnecessary if the haemoglobin level is 10 g/dl or greater.

Routine laboratory tests for monitoring coagulation include measurement of the platelet count, fibrin degradation products, the prothrombin time, the partial thromboplastin time (PTT) and the activated clotting time (ACT) all of which measure hypocoagulability. All these measurements do, however, end with the formation of the first fibrin strands in plasma.[107,108] Furthermore, the physiological

Table 37.14 Biochemical/haematological intraoperative monitoring

Electrolytes Na$^+$, K$^+$, Ca^{2+}
Blood sugar
Arterial blood gases
Haemoglobin
Platelet count
Bleeding/clotting studies (TT, APTT, INR, FDPs, ACT)
Thromboelastograph recordings

ACT, activated clotting time; APTT, activated partial thromboplastin time; FDPs, fibrin degradation products; INR, international normalized ratio, TT, thromboplastin time
Reproduced, with permission, from Browne DRG. Anaesthesia in impaired liver function. *Current Anaesthesia and Critical Care* 1990; **1**; 220–7.

status of the patient (e.g. hypothermia, acidosis, hypocalcaemia) is not taken into account. In addition, they do not measure the quality of the coagulation factors and there is a lag time of 30–60 minutes before the results become available.[109] More recently the thromboelastogram (TEG) has re-emerged as a tool for use during surgery on patients with poor hepatic function.[108–110] The TEG measures the haemostatic process in the whole blood from the start of clotting to the final stages of clot lysis, thereby monitoring both hypocoagulable states and fibrinolysis. The envelope of oscillation of the sensor in the clotting blood produces characteristic shapes. Although sometimes ambiguous, it is possible to tell whether the patient is suffering from lack of platelets, other clotting factors or fibrinolysis (Fig. 37.8).[90] Unfortunately, the time taken for the whole of the envelope to appear can be well over 30 minutes and this too limits its usefulness. It is also easy to produce artefactual results if there is heparin contamination of the test sample. Thromboelastography has recently been reviewed in detail.[110]

The problem of fibrinolysis in patients with cirrhosis was first described by Goodpasture in 1914.[111] He reported that in normal people a blood clot took 3 days to dissolve, whereas in those with cirrhosis it dissolved within 3 hours. To counter fibrinolytic activity, ε-aminocaproic acid[112] and antithrombin III[113] have been shown to have some beneficial effect. Schipper and Ten Cate[113] confirmed the presence of an increased turnover of ^{125}I-labelled fibrinogen in patients with clinically stable cirrhosis, and correction of the antithrombin III deficiency decreased the rapid fibrinogen turnover. They concluded that their treatment reduced the haemorrhagic diathesis by controlling disseminated intravascular coagulation.

Thrombocytopenia and platelet dysfunction occur in liver disease,[114] and it is sometimes necessary to administer a platelet transfusion if the count falls below 70 000. Desmopressin has been given to promote platelet adhesiveness and it has been shown to have some success in

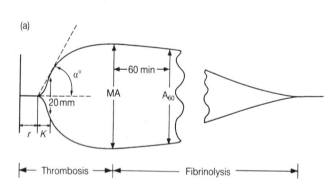

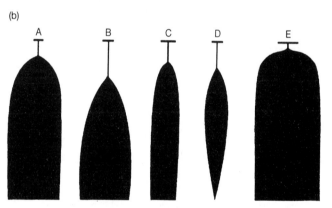

Fig. 37.8 Thromboelastography (TEG). (a) Quantification of TEG variables. Analysis of the thromboelastograph. r = Reaction time (time from sample placement in the cuvette until TEG tracing amplitude reaches 2 mm (normal range 6–8 min)). This represents the rate of initial fibrin formation and is related functionally to plasma clotting factor and circulating inhibitor activity (intrinsic coagulation). Prolongation of the r time may be a result of coagulation factor deficiencies, anticoagulation (heparin) or severe hypofibrinogenaemia. A small r value may be present in hypercoagulability syndromes. K = clot formation time (normal range 3–6 min); measured from r time to the point where the amplitude of the tracing reaches 20 mm. The coagulation time represents the time taken for a fixed degree of viscoelasticity to be achieved by the forming clot, as a result of fibrin build-up and cross-linking. It is effected by the activity of the intrinsic clotting factors, fibrinogen and platelets. Alpha angle ($\alpha°$) (normal range 50–60°) = angle formed by the slope of the TEG tracing from the r to the K value. It denotes speed at which solid clot forms. Decreased values may occur with hypofibrinogenaemia and thrombocytopenia. Maximum amplitude (MA) (normal range 50–60 mm) = greatest amplitude on the TEG trace and is a reflection of the absolute strength of the fibrin clot. It is a direct function of the maximum dynamic properties of fibrin and platelets. Platelet abnormalities, whether *qualitative* or quantitative, substantially disturb the MA. A$_{60}$ (normal range = MA - 5 mm) = amplitude of the tracing 60 min after MA is achieved. It is a measure of clot lysis or retraction. The clot lysis index (CLI) (normal range >85%) is derived as A$_{60}$/MA × 100 (%). It measures the amplitude as a function of time and reflects loss of clot integrity as a result of lysis. (b) Specific haemostatic defects produce characteristic TEG traces. A, Normal trace. B, Haemophilia: marked prolongation of r and K times. Decreased alpha angle. C, Thrombocytopenia: normal r and rK times, decreased MA (<40 mm). D, Fibrinolysis: CLI <85%. E, Hypercoagulability: short r time, increased MA and steep clot formation rate. (Reproduced, with permission, from Mallett SV, Cox DJA. Thromboelastography. *British Journal of Anaesthesia* 1992; **69**: 621–30.)

reducing blood loss in cardiac surgery[115] but the role of the drug in patients with liver disease undergoing surgery has yet to be fully evaluated.[90]

The most important electrolyte to monitor intraoperatively is ionized calcium. This requires calcium-balanced heparin to be used or the results will be incorrect. With normal liver function, the risk of lowering the serum ionized calcium is small unless blood or fresh-frozen plasma is transfused at a rate greater than one unit per 5 minutes or more than four units are given in total. These limits are eroded in hepatocellular failure, and calcium supplements may need to be given earlier. ECG changes of hypocalcaemia are seen as a change in the Q–T segment. The problems of blood transfusion and citrate toxicity have recently been reviewed.[116] The administration of calcium supplements is described in the section on transplantation. Patients with cirrhosis may have reduced glycogen stores, and serum glucose needs to be monitored and corrected as necessary.

Postoperative management

If the procedure has been relatively minor the neuromuscular blockade can be reversed (if required) and the patient allowed to regain consciousness and extubated (if required). There must be an adequate recovery area with well trained staff and adequate monitoring facilities. The patient needs oxygen-enriched air at least for the first postoperative night after all but the most trivial cases.

Following major surgery or if there is severe liver dysfunction, the patient may need IPPV to be continued postoperatively. During this period the fluid and electrolyte imbalance can be corrected, cardiovascular stability achieved and hypothermia corrected. As the patient rewarms peripherally, appropriate intravenous fluids need to be given to fill the expanding vascular volume and to ensure an adequate urine output (over 1 ml/kg per hour). Blood loss should be measured at frequent intervals and replaced with blood, fresh-frozen plasma and platelets as required. The maintenance of oxygenation, blood pressure and cardiac output are just as important in the postoperative period as in the intraoperative period. Hypovolaemia can also result from invisible protein losses into the abdominal cavity and may need correction with salt-poor albumin or fresh-frozen plasma.

The period of recovery to consciousness may be very prolonged in the presence of cirrhosis, and postoperative confusion occurs on occasion. There may need to be a period of elective ventilation until the patient is stable and gas exchange is adequate. As soon as the patient has returned to the intensive care or high dependency unit, a chest x-ray should be taken to check the position of the central venous and pulmonary arterial catheters, nasogastric tube (if present), tracheal tube (if present), and the lung fields, looking especially for atelectasis, pulmonary oedema and effusions. The patient should be extubated when fully conscious and able to maintain good gas exchange on 40 per cent or less inspired oxygen. Throughout this time adequate analgesia must be given, preferably in small doses to assess its effect. Physiotherapy is vital to ensure proper ventilation of the lungs, and continuous positive airway pressure (CPAP) may be necessary. Antibiotics must be continued together with H_2-receptor antagonists. Drains should be removed as soon as clinically possible to reduce the risk of infection.

Common problems in the postoperative period are chest infection and a deterioration in hepatic and renal function. When these occur, the patient will need to be treated as if he or she has acute liver failure (see section 'Acute liver failure and its intensive care management', later).

Anaesthesia for liver transplantation and liver surgery

Anaesthesia for liver transplantation

The first liver transplant in humans was carried out by Thomas Starzl in Denver, Colorado, in 1963,[117] and in May 1967 Roy Calne[118] performed the first liver transplant in the UK at Addenbrooke's Hospital in Cambridge. Over the subsequent 17 years the number of centres performing these procedures increased several-fold both in the USA and in Europe. It was, however, not until 1980 when cyclosporin A was used experimentally in humans as an immunosuppressant in renal transplantation that a significant improvement in graft survival was noticed. Cyclosporin increased the liver graft survival rate from 35 to 78.6 per cent at 1 year.[119,120]

If the donor organ is transplanted to the normal anatomical position the procedure is called orthotopic, and if transplanted to an abnormal position it is called heterotopic. A variety of heterotopic techniques have been described by Starzl[121] but none is ideal. It is difficult to accommodate an extra organ in the abdomen in a way that it receives arterial and portal blood and drains hepatic venous blood and bile. There is also the added danger of multiple anastomoses getting kinked or otherwise compromised. Since the 1983 NIH Congressional Conference,[122] orthotopic liver transplantation has ceased to be experimental and is now recognized as a therapeutic procedure for the treatment of end-stage liver disease both in adults and in children.

As described earlier, the liver is an unpaired but complex

organ. Whilst there is dialysis for renal replacement, cardiopulmonary bypass for the heart and ventilation for the lungs, the complexities of hepatic function have, as yet, rendered this organ irreplaceable. It is therefore imperative for survival that a liver diseased beyond its capacity to recover should be replaced with a healthy one.

Indications for liver transplantation are:

Adults
1. Primary biliary cirrhosis
2. Chronic active hepatitis
3. Primary hepatoma
4. Carcinoma of the hepatic ducts
5. Primary sclerosing cholangitis
6. End-stage cirrhosis

Children
1. Congenital biliary atresia
2. Familial intrahepatic cholestasis
3. Arteriohepatic dysplasia
4. Neonatal hepatitis
5. α_1-Antitrypsin deficiency
6. Wilson's disease
7. Tyrosinaemia
8. Glycogen storage disease types I and IV

With major advances in both anaesthetic and surgical management, liver transplantation has become increasingly successful, the indications have broadened and the contraindications have narrowed. Whereas extremes of age were a contraindication in the past, recent developments have seen successful graft survival in the old and the very young.

The contraindications are now limited to extrahepatic malignancy, active substance abuse and active extrahepatic infections. However, published reports demonstrating successful transplants even in these groups continue to reduce the contraindications. Transplantation for alcoholic liver disease, viral hepatitis and hepatic neoplasm remains controversial because there is a high incidence of recurrence of the original disease.[123]

Surgical technique

Liver transplantation is accomplished through a wide bilateral subcostal ('Mercedes Benz') incision in the recipient. The xiphoid process is removed to improve exposure and access to the suprahepatic abdominal vena cava and the main hepatic veins. A self-retaining retractor (Rochard) is often used to further enhance surgical access. The diseased organ is excised by dividing the hepatic artery, the portal vein, and the suprahepatic and the infrahepatic vena cava. The donor liver is then placed in the same anatomical position as the removed liver, the vascular anastomoses made and liver reperfused. The operative

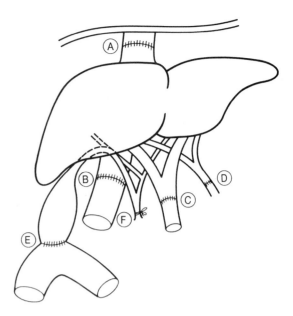

Fig. 37.9 Schematic representation of an orthotopic liver transplant. Local differences in anatomy (e.g. absent gall bladder, bifid artery) are accommodated by minor variations of surgical technique. A–E: Anastomosis of (A) suprahepatic inferior vena cava; (B) infrahepatic inferior vena cava; (C) portal vein; (D) hepatic artery; and (E) gall bladder to duodenum. F, Ligation of common bile duct.

procedure is completed by biliary anastomosis either duct to duct or Roux-en-Y choledochojejunostomy (Fig. 37.9).

The donor organ may be split into two grafts for smaller recipients or paediatric patients. Sometimes the left lobe of the liver is removed from a healthy living related donor and used as a graft.

The surgical process can be divided into three stages: (1) initial dissection phase; (2) anhepatic phase while the new organ is anastomosed; and (3) reperfusion phase. Even though the preoperative condition of the patient undergoing liver transplantation poses difficult problems that are common throughout these three phases, some specific anaesthetic problems pertain to each of these, and will be dealt with in detail later.

Anaesthetic management

The clinical features of liver disease have been described above, and are present in patients for liver transplantation. In particular, the involvement of other organ systems is important and is often seen as the presence of a hyperkinetic circulation, pulmonary hypertension, intrapulmonary shunts, coagulation disorders including thrombocytopenia, renal impairment, portal hypertension and its associated sequelae. Not infrequently, if there has been recent deterioration, signs suggestive of acute liver failure such as hypoglycaemia and deteriorating conscious level may be present.

Cardiomyopathy can be associated with alcoholic liver disease or haemosiderosis but it may be difficult to detect without an endomyocardial biopsy. Serum albumin level is often low, resulting in fewer binding sites for the drugs, an increase in the volume of distribution and impaired drug metabolism and excretion (see earlier).

Anaesthetic variants specific to liver transplantation

Technique The basic principles of anaesthetizing a patient with liver disease have been described above. For a patient undergoing transplantation, adequate venous access with large-bore cannulae should be established in the upper part of the body. Fluids administered through lower body cannulae are lost in the surgical field. When possible an indwelling arterial catheter should be placed in the right radial or right brachial artery before induction. The mechanical compression of the left axillary artery with the 'return' limb of the venovenous bypass (see below) may lead to distortion of the arterial waveform and hence the arterial pressures measured from the left arm.

Special attention should be paid to positioning the patient correctly with adequate padding because surgery can last a long time. Arm abduction not exceeding 90 degrees with forearm supination and anterior flexion should minimize injury to the brachial plexus. Liver transplant patients have a high incidence of brachial plexus neuropathy. The aetiology, though multifactorial, includes cyclosporin toxicity and nerve compression from the axillary limb of a venovenous bypass circuit.[124]

The anaesthetic drugs used differ from centre to centre, and a typical regimen is that described above from Browne.[90] For induction, an agent such as etomidate with minimal cardiovascular effects is appropriate. 'Rapid sequence' technique may be required to reduce the risk of aspiration of gastric contents in patients with a full stomach. The muscle relaxant of choice is atracurium. Ventilation with added positive end-expiratory pressure of up to 5 cmH$_2$O is used to prevent atelectasis and to reduce the risk of air embolism from the hepatic bed. Anaesthesia is maintained with a mixture of oxygen-enriched air and isoflurane supplemented with fentanyl or alfentanil. Isoflurane is considered to be the ideal inhalational agent in patients with liver disease, as there is evidence that hepatic artery blood flow is substantially maintained, possibly by the preservation of autoregulation or possibly as a result of the profound vasodilatation of the hepatic vasculature.[33,39] Nitrous oxide is avoided because it can cause distension of the bowel and increase the size of air emboli if they occur.

Benzodiazepines should be used with caution in cirrhosis owing to an increased risk of excessive sedation resulting from reduced drug elimination and increased sensitivity.[125] They are relatively contraindicated in patients with incipient encephalopathy in whom GABA-ergic neurotrans-

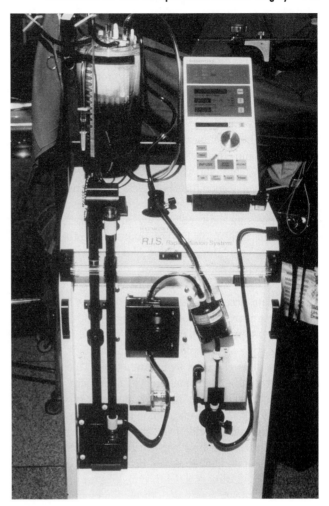

Fig. 37.10 A Hemonetics Rapid Infusion System.

mission may be increased by elevated levels of benzodiazepines.[126]

Body temperature is difficult to maintain owing to the wide and prolonged exposure of a large surface area. The use of efficient fluid warmers, heating blankets, heated humidifiers and forced air warming devices is highly recommended. Large volumes of warmed blood and blood products may be required at high flow rates during the dissection phases of both liver transplantation and liver resection. Devices such as Rapid Infusion System (RIS) (Hemonetics Corporation, Braintree, MA) or Level I warming device (Level I Technologies Inc, Rockland, MA) meet these requirements. RIS incorporates a 3-litre fluid reservoir, a 170 m filter, a heat exchanger, a roller pump, a 40 m filter, an infusion rate controller and air detectors (Fig. 37.10). When large blood loss is anticipated, a blood salvage system (Cell-Saver, Hemonetics, Braintree, MA) will help reduce the demand on the blood bank for red cells. This can be connected directly to the reservoir of the RIS in a life-threatening emergency, and salvaged blood transfused directly into the patient.

Monitoring   The principles of monitoring a patient with liver disease have been described above. During liver transplantation, not only may hepatic blood flow be adversely affected by the surgical procedure but also hepatic surgery usually entails significant blood loss, large fluid shifts, cardiovascular instability, coagulation disorders, electrolyte and acid–base imbalance, and difficulty in maintaining body temperature. It is therefore essential that monitoring and therapy are based on a knowledge of hepatic physiology and pathophysiology.

Cardiovascular monitoring needs to include multichannel electrocardiogram, invasive arterial blood pressure, central venous pressure and pulmonary artery pressure, with facilities for measuring cardiac output and other derived parameters. Ventilatory function should be monitored by pulse oximetry, end-tidal carbon dioxide analysis and frequent arterial blood gas measurements.

Immediate access to laboratory services is essential to monitor the rapid changes in electrolytes and coagulation profile. It is preferable to have a 'mini-lab' within the operating suite for a rapid analysis and a quick 'turnover' of results. It is essential to follow ionized calcium levels closely (every 15–20 minutes), as citrate toxicity can occur rapidly with transfusion of large amounts of blood and blood products. It is not uncommon to have to give up to 60 mmol of calcium chloride over an hour as an infusion to keep the ionized calcium levels within the normal range. Coagulation state should ideally be monitored by measuring prothrombin time, partial thromboplastin time, fibrinogen level, platelet count and thrombin clotting time. A relatively rapid (30 minutes) and qualitative assessment of the state of coagulation can be obtained within the operating suite by using a thromboelastograph (TEG)[90,110] (see above). The TEG provides a graphic display of the clot formation and lysis. Calculated variables – reaction time, maximum amplitude, and clot formation and lysis rate – are all dislayed along with the normal ranges on a video monitor. Diagnosis of altered coagulation and its effective treatment with blood components should be based on a balanced judgement of the results of standard tests and the TEG.

Neuromuscular monitoring with a nerve stimulator is recommended, particularly if muscle relaxants are administered by continuous intravenous infusion.

Blood sugar level should be monitored regularly. Hypoglycaemia, though rarely seen, is an indication of severe liver failure.

Kidney impairment (which has been described earlier) may be intensified by physiological insults during the surgical procedure, by massive haemorrhage, by profound vasoconstriction, and by clamping of the major abdominal vessels such as the aorta and the inferior vena cava. Efforts should be made to preserve renal function by meticulous surgical technique, adequate and appropriate fluid administration and maintenance of adequate cardiac output to ensure end-organ perfusion. Measures should include the use of frusemide (furosemide), mannitol and dopamine or dopexamine. Dopamine is commonly believed to possess renal preservatory effects in dosages of 2–3 µg/kg per minute, though this effect has not been confirmed.[127–129] Dopexamine, on the other hand, has been shown to improve renal function, due possibly to the agonistic action of this drug at the renal vascular and tubular DA_1 receptors.[130]

Specific factors during the three phases of transplantation

Dissection phase   This includes the dissection and isolation of the hepatic vasculature, the hilar structures and the liver. Venovenous bypass is established by cannulation of the portal and the femoral veins. The draining blood passes through a centrifugal pump and is returned to the patient via the axillary vein (Fig. 37.11). The use of heparin and heparin-bonded tubing differs from unit to unit. Blood loss, both insidious and obvious, can occur especially in patients with portal hypertension. Adhesions can be very extensive, tedious and the cause of substantial blood loss. Large volumes of ascitic fluid may be lost, with resultant fluid shifts which need to be anticipated and corrected with appropriate fluids. In patients with clinically significant coagulopathy and abnormal TEG and clotting results, attempts should be made to correct the defect as soon as possible. However, it should be borne in mind that a platelet count as low as $20\,000/mm^3$ has been tolerated in some patients. Surgical manipulation, especially around the major vessels, can lead to vascular compression and a fall in cardiac output due to a reduction in the venous return. The patient is considered anhepatic once the hepatic vasculature is ligated.

Anhepatic phase   Interruption in the venous return at the time of clamping of the inferior vena cava can cause a major fall in the cardiac output in patients without a venovenous bypass. A reduction in renal perfusion with oliguria is not uncommon at this stage. This fall is usually less with long-standing portal hypertension because of a well developed portasystemic collateral circulation. The decrease in cardiac output can be compensated by fluid loading with or without vasopressors. However, this exposes patients to the additional risk of dilutional coagulopathy and, in the reperfusion phase, to the development of pulmonary oedema.

The physiological changes induced by venovenous bypass which runs throughout the anhepatic phase include a fall in body temperature, a decrease in both heart rate and arterial pressure, and an increase in central venous pressure.[131,132] A prolonged bypass period leads to an increase in haematocrit (packed cell volume), colloid oncotic pressure and serum osmolality. A severe and

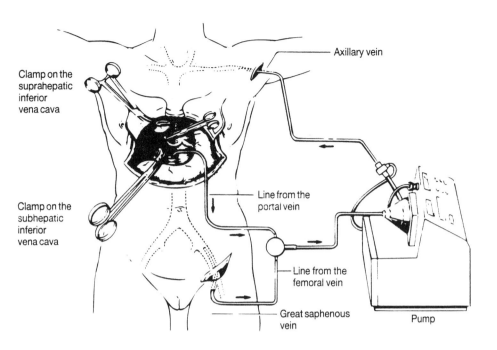

Fig. 37.11 Diagrammatic representation of venovenous bypass to reduce back-pressure on the gastrointestinal tract during inferior vena caval cross-clamping.

Table 37.15 Composition of preservative solutions used to preserve donor livers

	UW solution	*Eurocollins solution*
Impermeants	Raffinose (30 mmol/l) Lactobionate (100 mmol/l)	Glucose (194 mmol/l)
Hydrogen ion buffers	KH_2PO_4 (25 mmol/l)	KH_2PO_4 (15 mmol/l) K_2HPO_4 (42 mmol/l) $NaHCO_3$ (10 mmol/l)
Colloids	Hydroxyethyl starch (5 g/dl)	
Metabolites and others	$MgSO_4$ (5 mmol/l) Adenosine (5 mmol/l) Glutathione (3 mmol/l) Allopurinol (1 mmol/l) Insulin (100 u/l) Dexamethasone (8 mg/l)	KCl (15 mmol/l)
Osmolarity	320 mmol/l	335 mmol/l

Reproduced, with permission, from Cofer JB, Klinkmalm GB, Howard TK, *et al.* A comparison of UW with Eurocollins preservation solution in liver transplantation. *Transplantation* 1990; **49**: 1088–93.

progressive metabolic acidosis suggests less than optimal perfusion of the lower half of the body.[133] Markers of renal perfusion have been shown to be preserved with the use of bypass.[134,135] Air entrainment is likely if the draining cannulae are not secured tightly, and air embolism has been reported with venovenous bypass.[136] Following caval anastomosis, portal bypass is discontinued to allow the portal anastomosis to be completed. Venovenous bypass is then discontinued and the new graft reperfused. The anhepatic phase ends with washing out of the preservative solution prior to unclamping of vessels.

It is worthwhile considering the organ preservation and the mechanism of reperfusion injury before describing the reperfusion phase in detail.

Preservation Eurocollins solution containing a high concentration of potassium was used until 1988 to preserve the donor organs. The donor livers could be allowed a maximum period of ischaemia of 8 hours. In September 1987, Jamieson *et al.*[137] reported that cold ischaemia of canine and human liver homografts could be extended safely to 1 day with an infusate first described by Wahlberg et al.[138] in experiments on pancreas preservation. This University of Wisconsin preservation fluid (UW solution) has allowed liver transplantation to become a semi-elective procedure. Cofer et al.[139] have demonstrated improved enzyme function, decreased blood usage, and shorter hospital stays with UW solution. Table 37.15 compares the composition of these preservative solutions.

Reperfusion injury and syndrome   Meticulous care at the time of harvesting the donor liver, improved preservation during cold ischaemia and therapeutic measures aimed specifically at reducing reperfusion injury all help to improve graft function and survival. Free radical scavengers such as superoxide dismutase, catalase[140] and mannitol have been used to reduce reperfusion injury. Greig et al.[141] have reported the successful use of prostaglandin E$_1$ (PGE$_1$) in livers that appeared to be compromised at reperfusion.

Reperfusion phase   Relatively minor physiological changes are noticed when the inferior vena cava flow is re-established through the new graft. However, reperfusion through the portal vein marks the beginning of critical time with the potential to develop the reperfusion syndrome. This is the name given to haemodynamic instability characterized by bradycardia, hypotension, dysrhythmias and, ultimately, circulatory arrest. Right ventricular dysfunction with an increase in pulmonary artery pressures has been reported.[142] This is in response to the sudden release into the circulation of cold acidic hyperkalaemic preservative solution. We have observed 'first flush' serum potassium levels as high as 35 mmol/l despite careful flushing of the donor liver prior to reperfusion. Using transoesophageal echocardiography, Roberts et al.[143] have recorded a stream of air bubbles entering the heart. Although minor changes require no treatment, major instability demands vigorous resuscitative measures. Prophylactic atropine decreases the incidence of bradycardia and sinus arrest. Calcium chloride is the drug of choice for acute management of acute profound hyperkalaemia. These changes can be clearly seen on the ECG. In very severe cases of reperfusion syndrome, bolus administration of adrenaline may be required for ventricular assistance. Noradrenaline (norepinephrine) has been shown to be effective in some cases of severe cardiovascular collapse.

Hepatic artery and biliary anastomoses complete the surgical procedure during this stage. Rewarming and haemostasis are the two main concerns for the anaesthetist. In some patients, reperfusion coagulopathy with fibrinolysis may occur. Fresh-frozen plasma, platelets and cryoprecipitate are then needed to correct the acute coagulation disorder. Once the new graft starts to function effectively there is an increase in body temperature,

acidosis is corrected, coagulation state returns to normal, hyperkalaemia reduces and an increase in carbon dioxide production is noticed.[144] We do not recommend vigorous correction of intraoperative acidosis if the graft is working well, as this leads to postoperative metabolic alkalosis. The optimal haematocrit level to aim for without a risk of postoperative hepatic artery thrombosis is between 30 and 35 per cent.

Postoperative management

Patients are electively ventilated in an intensive care unit during the immediate postoperative period. With a well functioning graft and adequate analgesia, routine monitoring of vital signs and fluid balance is all that is required. Ventilation is continued until the patient recovers fully from the effects of anaesthesia. Routine postoperative monitoring includes blood sugar estimations, coagulation studies and liver function tests. The patency of the hepatic artery is determined by Doppler ultrasonography. Early intervention in the event of hepatic artery thrombosis can salvage the graft and prevent the need for regrafting.[145] Graft malfunction due to injury or acute rejection is associated with a rise in liver enzymes (particularly AST), failure to correct the coagulation disorder and a subsequent rise in serum bilirubin levels.

A major cause of death after transplantation continues to be infection,[146] usually associated with graft and/or multiple organ failure. It is therefore essential to continue an aggressive prophylactic regimen that is effective against both bacterial and fungal organisms.

Other postoperative problems include continuing haemorrhage, and development of hypertension in about 50 per cent of the patients. The latter requires treatment with adrenergic receptor blockers, ACE inhibitors and diuretics.

Although the survival of the grafts has improved with the introduction of cyclosporin, so also has the risk of renal failure because of its renal toxicity. Hence a careful balance of cyclosporin dosage against renal function must be undertaken.[147] Various other types of immunosuppressants are under investigation, including OKT3 monoclonal antibody and FK-506.

Major non-transplant liver surgery

Apart from liver transplantation, an anaesthetist may be called upon to provide anaesthesia for procedures such as segmental resection of the liver or repair following hepatic trauma. In both these situations, the basic anaesthetic considerations are similar to the ones outlined above for patients with coexisting liver disease and liver transplantation, but there are some significant differences

Hepatic resection

Based on the distribution of the portal pedicles and the location of the hepatic veins, five anatomical segmental resections can be undertaken. These are right and left trisegmentectomy, right and left lobectomy, and left lateral segmentectomy. For smaller lesions it is possible to undertake subsegmental or local wedge resections.

Intraoperative ultrasonography is used to further define the blood supply and accurately outline the position of the intrahepatic lesions.[148]

The indications for hepatic resection are:

1. Benign hepatic tumours
2. Metastatic lesion limited to resectable areas
3. Primary malignant tumours in resectable areas
4. Split liver grafts in living related transplants

Haemorrhage, coagulation disorders and metabolic changes are the primary problems in these patients.

In patients with advanced tumours, two other approaches for liver resection have been described. These are *ex-situ* and the *in-situ* hypothermic technique. In the *ex-situ* procedure the patient is rendered anhepatic, as in the liver transplantation procedure, and the resection is completed on the bench. Despite venovenous bypass, significant metabolic and coagulation disorders are associated with this prolonged anhepatic phase.[149] In the *in-situ* technique the liver is isolated and perfused with hypothermic preservative solution, the hepatic artery is clamped and rewarming is prevented by wrapping the liver in silver foil. Venovenous bypass is used to minimize the haemodynamic disturbances.

The major complications of hepatic resections are hepatic failure, haemorrhage, biliary leak and infection, with postoperative death largely attributable to hepatic failure.[150]

Hepatic trauma

Even though the mortality from hepatic trauma has decreased considerably from 66 per cent in World War I to 9 per cent in Vietnam,[151] uncontrolled haemorrhage continues to be the primary cause of death.[152] Ruptured liver is still the commonest missed diagnosis in multiple trauma.[153]

Clinical presentation depends upon severity of the hepatic trauma, and other associated injuries. Some patients may require detailed diagnostic work-up, including peritoneal lavage, computed tomography and arteriography, whilst those with severe injury may present with profound hypotension and a distended abdomen requiring immediate surgical intervention. It is imperative that adequate resuscitation is undertaken as soon as possible. Facilities for rapid fluid administration with large-bore

intravenous catheters in the upper half of the body must be obtained before surgery commences.

Oxygenation and control of the airway (especially in an unconscious patient) are of paramount importance. Anaesthetic agents that cause minimal cardiovascular depression (e.g. etomidate or ketamine) combined with an opioid–relaxant technique are the drugs of choice.

Massive blood transfusion may be necessary to replace the blood loss once the abdomen is opened. Transfusion alone may not be enough to maintain adequate perfusion of vital organs, and vasopressors may be required.

Other areas of concern during surgery include coagulation disorders, hypothermia, acidosis and problems associated with any other injuries the patient may have. It is therefore essential that arterial blood gas analyses are carried out frequently and that electrolytes, glucose, ionized calcium, haematocrit and clotting factor estimations are measured on a regular basis.

Depending on the severity of the hepatic injury, the treatment options range from perihepatic packing to total hepatectomy and liver transplantation.

Postoperative complications include persistent coagulopathy, renal failure, local sepsis, systemic sepsis with or without adult respiratory distress syndrome and features of multiple organ failure.

Acute liver failure and its intensive care management

Acute liver failure (ALF) is defined as severe hepatic dysfunction leading to encephalopathy or coagulopathy within 6 months of the first symptom. Rapid progression of the disease with encephalopathy developing within 8 weeks is referred to as fulminant hepatic failure (FHF).[154]

Causes of acute liver failure (ALF)

The commonest cause of ALF world wide is viral hepatitis, although in the UK the commonest cause is paracetamol (acetaminophen) overdose taken with suicidal intent. Hepatotoxicity has, however, been reported in patients with therapeutic doses of paracetamol who are concurrently taking enzyme-inducing drugs such as phenytoin or who are chronic users of alcohol.[155]

Halothane hepatitis, although rare, recurs as an easy target for the cause of postoperative jaundice. This typically develops 5–15 days after second exposure to halothane. Because of its importance in anaesthetic practice, it is described in detail later (in the section 'The investigation and management of postoperative jaundice') along with a description of other drug reactions.

Hepatitic viruses cause ALF as outlined in the next main section. ALF can also be part of generalized infection with cytomegalovirus, coxsackie B, echovirus, Epstein–Barr and herpes simplex.

Rare causes of ALF include mushroom poisoning with *Amanita phalloides*, acute fatty liver of pregnancy, Wilson's disease and Budd–Chiari syndrome.

Clinical features and management of ALF

The basic requirement for the management of acute liver failure is to minimize any secondary damage to the other organs, which might lead to multiple organ failure, and to allow time for the regeneration and recovery of liver function or to procure a suitable graft for transplantation.

Encephalopathy and cerebral oedema

Encephalopathy is graded according to its severity into four grades (Table 37.16).[156] Encephalopathy grades I–II, which do not progress, carry the best prognosis. It takes 3–4 days for the signs of encephalopathy to appear following paracetamol overdose. During the subsequent 24–48 hours there is a rapid progression through the various stages. However, the onset is much more variable in patients with acute liver failure of viral origin. Cerebral oedema, a major cause of death, is seen in 80 per cent of the patients with FHF.[157] The signs and symptoms of raised intracranial pressure include systemic hypertension with episodes of bradycardia, hyperventilation, decerebrate posture, abnormal pupillary reflexes and impaired brainstem reflexes.

The precise pathogenic mechanisms responsible for encephalopathy in FHF are still not well understood, though various factors have been implicated. These include benzodiazepine agonists, increased aromatic amines,

Table 37.16   Clinical features and grade of encephalopathy

I. Mild or episodic drowsiness, impaired intellect, concentration and psychomotor function, but rousable and coherent

II. Increased drowsiness with confusion and disorientation; rousable and conversable

III. Very drowsy, disorientated, responds to simple verbal commands, often agitated and aggressive

IV. Responds to painful stimuli at best, but may be unresponsive; may be complicated by evidence of cerebral oedema

Reproduced, with permission, from O'Grady JG, Williams R. Management of acute hepatic failure. *Current Anaesthesia and Critical Care* 1990; **1**: 213–19.

ammonia, phenols, mercaptans and altered γ-aminobutyric acid levels. Basile *et al.* have demonstrated raised brain concentrations of 1,4-benzodiazepines in patients with FHF.[158] Mullen *et al.* have identified an endogenous benzodiazepine ligand with a molecular weight less than 10 000 in the cerebrospinal fluid of FHF patients.[159] Flumazenil, a benzodiazepine antagonist, might therefore be considered to be of use in reversing the encephalopathy. However, it is recommended that, until a beneficial role for such agents is established, they should not be used in the presence of cerebral oedema as they are likely to aggravate the condition.[156]

Patients with grade III or IV encephalopathy should be electively intubated and ventilated, and nursed in a head-up position with minimal stimuli. The main aim of therapy is to maintain a cerebral perfusion pressure greater than 50 mmHg. Raised intracranial pressure can be controlled by hyperventilation in the acute stages but sustained hyperventilation should be avoided because ultimately this may lower the perfusion pressure. Bolus administration of mannitol 0.5 g/kg over 10 minutes is also effective in reducing ICP, and can be repeated in patients with normal renal function if the plasma osmolarity is below 320 mmol/l. In patients with renal impairment, ultrafiltration or intermittent haemofiltration may be required to remove two or three times the volume of mannitol infused. Sodium thiopentone in bolus doses of 185–500 mg followed by an infusion of 50–250 mg/h has been shown to be effective for patients with mannitol-resistant cerebral oedema.[160] The use of sodium thiopentone does, however, make the interpretation of cerebral signs difficult and can preclude the diagnosis of brainstem death.

Coagulopathy

Most coagulation factors apart from factor VIII are produced by the liver (see above). In acute liver failure, circulating levels of these factors are reduced, as is antithrombin III. The latter is an inhibitor of coagulation. In addition, there is an increase in the peripheral consumption of clotting factors leading to disseminated intravascular coagulation.[161] Thrombocytopenia with increased platelet adhesiveness but impaired aggregation is also a feature of ALF.

Prolongation of prothrombin time is commonly used as an index of liver damage, though a more sensitive marker might be a specific factor V assay because this has the shortest half-life of all the clotting factors synthesized by the liver.

Coagulopathy can be corrected in the short term by appropriate use of fresh frozen plasma and platelet concentrates. The incidence of gastrointestinal bleeding can be significantly reduced by the use of H_2 antagonists or cytoprotective agents.

Cardiovascular effects

Patients in ALF have high cardiac output with a low peripheral vascular resistance. They always need intra-arterial cannulation and many benefit from the information obtained from pulmonary artery catheterization. Persistent hypotension despite adequate volume replacement requir-ing ionotropic support is usually a poor prognostic indicator. Ionotropes may, however, be needed to allow therapeutic measures such as haemodialysis or transplantation. Adrenaline, noradrenaline and vasopressin infusions have all been used to increase systemic blood pressure but their use should be guided by derived haemodynamic parameters.

Cardiac dysrhythmias, though rare, are usually due to hypo- or hyperkalaemia, hypoxia, acidosis or mechanical irritation owing to the presence of central vascular catheters.

Metabolic effects

It is essential to monitor blood glucose level because a hypoglycaemia sufficient to cause brain damage is common in ALF. Infusion of 10–20 per cent glucose may be required to maintain normoglycaemia. Metabolic acidosis is seen more commonly in patients with paracetamol-induced ALF. Progressive metabolic acidosis carries a high mortality. Impaired tissue oxygen extraction leads to raised serum lactate levels.[162] Poor tissue oxygen extraction is an indication of a poor prognosis.

Alkalosis of mixed origin is often associated with hypokalaemia. Other metabolic changes that have been reported in patients with normal renal function include dilutional hyponatraemia and hypophosphataemia.

Renal failure

O'Grady and Williams reported an incidence of oliguric renal failure of around 75 per cent in patients with grade IV encephalopathy due to paracetamol overdose, compared with 30 per cent of other cases of AHF.[163] Renal failure after paracetamol overdose may well be due to direct nephrotoxicity and often precedes the encephalopathy. However, in cases of ALF due to other causes, renal impairment develops when the encephalopathy is advanced. If oliguria (urine output < 300 ml/24 h) persists despite adequate intravascular volume, dopamine infusion (2–4 µg/kg per min) may be useful to slow the deterioration in renal function. Intermittent haemodialysis may be needed to correct hyperkalaemia and acidosis, and to remove excess fluid. The haemodynamic changes associated with haemodialysis may lead to surges in intracranial pressure with consequent fall in cerebral perfusion pressure.[164] Continuous haemodiafiltration (continuous arteriovenous haemodialysis, CAVHD) is an effective alternative and causes minimal haemodynamic disturbances. Despite coexisting coagulopathy, low dose heparin with prostacyclin infusion is needed for CAVHD.

Respiratory complications

Alkalosis commonly seen in these patients is due to hyperventilation. Hypoxaemia is also a frequent complication of ALF. The aetiology of arterial hypoxaemia is multifactorial and includes aspiration of gastric contents (especially in patients with grade II–III encephalopathy), atelectasis and intrapulmonary bleeding. Non-cardiogenic pulmonary oedema has also been seen in patients who have metabolic acidosis following paracetamol overdose. Mechanical ventilation usually corrects arterial hypoxaemia.

Sepsis

Rolando et al. reported an 80 per cent incidence of bacterial infection, the predominant organisms being Staphylococcus aureus, streptococci and coliform.[165] Fungal infection, mainly due to Candida species, has also been reported.[166] This increased susceptibility in ALF is due to a compromised immune system with impaired neutrophil and Kupffer cell function and to a deficiency in complement components and fibronectin.

Tight surveillance with daily cultures and appropriate antibiotics for positive cultures is indicated though the role of prophylactic broad-spectrum cover and selective decontamination of the digestive tract has yet to be established.

Other treatments

Various specific measures aimed at improving hepatocyte regeneration have included charcoal haemoperfusion, therapy with insulin and glucagon, interferon and prostaglandin E_1. Despite theoretical benefits, conclusive evidence has yet to be demonstrated. Measures such as plasmaphaeresis and exchange transfusion with transient improvement in encephalopathy have convinced some authors about their efficacy. N-Acetylcysteine, a specific antidote for paracetamol, has been shown to be effective when given within 8 hours of an overdose.[167] There is evidence that it is effective in improving the outcome even when administered late.[168,169]

There is growing evidence to recommend orthotopic liver transplantation as a therapeutic option in ALF, and survival rates of up to 80 per cent have been reported.[170–172] Criteria for patient selection need to be carefully defined for good long-term prognosis.

The altered physiological changes of ALF (raised intracranial pressure, coagulopathy etc.) persist for some time into the post-transplant period; it is therefore essential to continue the aims of the preoperative therapy for some time after surgery.

Viral infections of the liver: implications for anaesthesia and anaesthetists

The past decade has seen significant advances in the understanding and management of viral hepatitis. These have been well summarized in review articles.[173,174] Many viruses such as herpes simplex, Epstein–Barr, yellow fever, cytomegalovirus and HIV all cause hepatitis or liver failure as part of their clinical syndrome, but this section focuses on viruses that specifically affect the liver.

Viral hepatitis is a worldwide public health problem infecting hundreds of millions of people. It affects patients as an acute infection and subsequently from chronic sequelae which include chronic active hepatitis, cirrhosis and primary liver cancer. World wide, hepatocellular carcinoma is one of the ten most common cancers.

Types of viral hepatitis

The various types of viral hepatitis are referred to by alphabetical letters.

Hepatitis A

Hepatitis A virus (HAV) is a small RNA virus. It is the cause of infectious or epidemic hepatitis transmitted by the faecal–oral route and is usually associated with poor standards of sanitation and hygiene. People at risk include staff and residents of mental homes, children in day care centres, active male homosexuals, narcotic drug abusers, sewerage workers, health care workers, military personnel and low socioeconomic groups. The incubation period for HAV is approximately 4 weeks, and the virus replicates in the liver. Passive protection may be obtained by the use of human immunoglobulin, which may still be effective post-exposure if given early.

The severity of the illness ranges from the asymptomatic to acute liver failure (ALF). ALF due to HAV is, however, uncommon, occurring in less than 0.5 per cent of hospitalized cases.[156] Exposure to the virus later in life has been associated with a higher mortality both in England[175] and in the USA.[173]

Hepatitis B

Hepatitis B virus (HBV) is a double-stranded DNA virus that causes the most common form of parenterally transmitted viral hepatitis. At-risk groups are listed later. HBV is an important cause of acute and chronic liver disease. The acute disease has an incubation period of 1–6 months and clinically ranges from an asymptomatic, anicteric infection to acute liver failure. The latter develops in up to 4 per cent of hospitalized patients.[156]

In up to 10 per cent of immunocompetent adults and 90 per cent of infants infected perinatally the virus persists and the patient continues to be an infection risk. Persistent carriage of hepatitis B is defined as the presence of the hepatitis B surface antigen in the serum for more than 6 months. Long-term continuing viral replication may progress to cirrhosis and hepatocellular carcinoma. A serum marker of continuing replication which is secreted by infected hepatocytes is hepatitis Be antigen (HBeAg). Carriers of HBV who are also HBeAg positive have high concentrations of virus in their blood and are therefore likely to be sources of occupational transmission.

There is now an effective vaccine against HBV. It is imperative that when this is administered it is delivered intramuscularly because vaccine injected into fat does not produce an adequate antibody response. The vaccine is therefore best given into the deltoid muscle rather than into the gluteal region. Immunization against hepatitis B in high-risk situations and patient groups is now recognized as a high priority in preventive medicine throughout the world. Vaccination policies are constantly being reviewed and over 20 countries (including the USA) now offer hepatitis B vaccine to all newborn,[173] both for their own safety and as a preventive measure to control the transmission of the infection.

The current indications for giving hepatitis B vaccine to those working and living in low prevalence areas are summarized below.[173]

Medical staff and patients

1. All health care personnel in frequent contact with blood or needles; groups at the highest risk in this category include:
   (a) personnel, including teaching and training staff, directly involved in patient care over a period of time in residential institutions for the mentally handicapped where there is a known risk of hepatitis;
   (b) personnel directly involved in patient care over a period of time in units giving treatment to those known to have a high risk of hepatitis B infection;
   (c) personnel directly involved in patient care work-ing in haemodialysis or haemophilia, and in other

centres regularly performing maintenance treatment of patients with blood or blood products;

(d) laboratory workers regularly exposed to increased risk from infected material;

(e) health care personnel on secondment to work in areas of the world where there is a high prevalence of hepatitis B infection, if they are to be directly involved in patient care;

(f) dentists and ancillary dental personnel with direct patient contact.

2. Patients:

(a) on first entry into those residential institutions for the mentally handicapped where there is a known high incidence of hepatitis B;

(b) treated by maintenance haemodialysis;

(c) before major surgery who are likely to require repeated blood transfusions and/or treatment with blood products.

3. Contacts of patients with hepatitis B – the partners and other sexual contacts of patients with acute hepatitis B or carriers of HBV, and other family members in close contact.

Other indications for immunization

1. Infants born to mothers who are persistent carriers of HBsAg or who are HBsAg positive as a result of recent infection, particularly if HBeAg is detectable, or born to HBV-positive mothers without antibody to e antigen (anti-e). The optimum timing for hepatitis B immunoglobulin (HBIg) to be given is immediately at birth or within 12 hours.

2. Health care workers who are accidentally pricked with a needle used for patients with hepatitis B. The vaccine may be used alone or in combination with HBIg as an alternative to passive immunization with HBIg only. Studies on the efficacy of these different schedules of immunization are nearing completion.

3. Individuals who frequently change sexual partners, particularly promiscuous male homosexuals and prostitutes.

4. Narcotic and intravenous drug abusers.

5. Staff at reception centres for refugees and immigrants from areas of the world where hepatitis B is very common, such as south-east Asia.

6. Although they are at 'lower risk', consideration should also be given to long-term prisoners and staff of custodial institutions, ambulance and rescue services, and selected police personnel.

7. Military personnel are included in some countries.

The recommendations for immunization in intermediate and high prevalence regions include the universal immunization of infants.

If immediate protection is required, for example after accidental inoculation, active immunization with the vaccine should be combined with simultaneous administration of HBIg at a different site. Passive immunization with up to 600 iu of anti-HBs does not interfere with an immune response. If infection has already occurred at the time of vaccination, viral multiplication is unlikely to be inhibited completely, but severe illness and the development of the carrier state may be prevented.

Hepatitis C

Hepatitis C is a single-stranded RNA virus responsible for the majority of what used to be, and still is, called non-A, non-B (NANB) hepatitis. The diagnosis of non-A, non-B hepatitis is presumptive on the basis of exclusion of other causes, but relatively recently an assay for circulating antibodies to what is called hepatitis C virus (HCV) has been described.[176] Although this test is positive in 80–90 per cent of chronic NANB hepatitis patients and in 58 per cent of sporadic cases, its diagnostic value is limited by the low (15 per cent) positivity rate during the acute phase. False positive results have also been recorded in patients with elevated serum globulins.[177]

Fulminant hepatic failure occurs in up to 5 per cent of hospitalized cases of NANB hepatitis, and NANB hepatitis is the predominant cause of late-onset hepatic failure.[156,178] Current data suggest that approximately 50 per cent of infections with HCV progress to chronicity: histological examination of liver biopsies from asymptomatic blood donors who were HCV carriers revealed that none had normal histology and up to 70 per cent had chronic active hepatitis and/or cirrhosis.[173] Almost all cases of NANB hepatitis are sporadic except the water-borne variety occurring on the Indian subcontinent.[156]

Hepatitis D

Hepatitis δ (D) virus is a single-stranded RNA virus diagnosed by the presence of δ antigen or IgM antibody to δ in the serum. It requires the presence of HBV for it to become effective. An individual can be infected simultaneously with HBV and HDV, resulting in a more severe form of hepatitis B. Vaccination against hepatitis B also prevents co-infection with HDV. Alternatively, an individual chronically infected with HBV may become superinfected with HDV: this can accelerate the course of chronic disease or render previously asymptomatic disease apparent.[173] Evidence of HDV infection was found in up to 43 per cent of patients with HBV fulminant hepatic failure, as compared with up to 19 per cent of less severe cases.[179,180]

Hepatitis E

Hepatitis E virus (HEV), is a single-stranded RNA virus which is thought to be the cause of enterically transmitted NANB hepatitis. HEV is an important cause of large epidemics of acute hepatitis in India, central and south-east Asia, the Middle East, parts of Africa and the former USSR.[173] The highest incidence is found in young adults, and mortality rates of up to 25 per cent have been recorded in the third trimester of pregnancy.[1]

Occupational hazards from viral hepatitis

About 1 in 500 of the adult population of the UK is a carrier of HBV; some of them will be asymptomatic and their carrier state will therefore not be realized.[181] It is therefore important that normal, everyday anaesthetic practice anticipates this possibility and that sensible precautions are taken. Whilst this is usually the case in the USA and parts of Europe, compliance is probably poorer in the UK.[182] Anaesthetists are at risk of occupational infection by blood-borne viruses through contact with infected blood and blood-stained body fluids. This may occur through needle-stick and other 'sharps' injury or through cuts and abrasions of the anaesthetist's skin and mucous membranes. The risk of transmission of HBV following occupational inoculation is reported to be between 5 and 30 per cent.[181]

Guidelines for the protection against HIV and hepatitis viruses were issued in 1990.[183] Precautions of particular relevance to anaesthetists are as follows.[181]

1. Gloves must be worn during the induction of anaesthesia, inserting intravenous cannulae, setting up intravenous infusions, and inserting and removing airways and tracheal tubes.

   Where substantial spillage of blood may occur, as, for example, in setting up an intra-arterial line, a plastic apron, mask and eye protection should be worn.

   Gloves should normally be discarded on taking the patient into the operating theatre and a fresh pair donned when the above procedures are carried out during or at the end of the anaesthetic. Equipment, notes and other articles must not be handled with contaminated gloves.

2. Needles that have been in contact with the patient should not be resheathed. Needles, syringes with needles attached and other 'sharps' *must not* be handed directly from one person to another. They should always be placed in a tray and if necessary picked up from the tray by the other person. All needles and 'sharps' should be disposed of in an appropriately tough disposal bin.

3. Whilst intact skin is impermeable to blood-borne viruses, cuts and abrasions on the anaesthetist's skin that might become contaminated by a patient's blood or body fluids must be covered with a waterproof dressing. Care should be taken to prevent contamination of such skin with blood or blood-stained fluids. An anaesthetist with considerable skin lesions such as eczema, chapping or several scratches is particularly at risk of being infected.

4. As these infections are not airborne, those parts of the breathing system outside the patient do not constitute a risk either to anaesthetists and their assistants or to other patients unless they become contaminated with blood. Oropharyngeal airways, laryngeal mask airways, nasopharyngeal airways, tracheal tubes, other instruments used in the airway and contaminated breathing systems must either be disposed of or sterilized between patient use.

5. Where possible, non-disposable contaminated equipment should be autoclaved. Where this is not possible the equipment should be thoroughly washed with detergent and water. It should then be left for the recommended period of time in 2 per cent freshly prepared glutaraldehyde or other agent recommended by local infection control policies.

6. Floors and surfaces contaminated with blood or blood-stained fluids should be washed with a solution of hypochlorite containing 10 000 p.p.m. available chlorine. The floors and surfaces should then be washed with detergent and water. Gloves must be worn.

7. Anaesthetists have a particular responsibility towards those who assist them in their clinical work, to ensure that they also observe these precautions routinely.

The hepatitis-B-infected anaesthetist

The transmission of HBV from infected health care workers to patients is well documented.[184] Apart from renal dialysis units, outbreaks have been mainly associated with surgical operations, transmission occurring during invasive procedures defined as follows by the UK Health Departments:[181,185]

surgical entry into tissues, cavities or organs or repair of major traumatic injuries, cardiac catheterisation and angiography, vaginal or Caesarean deliveries and other obstetric procedures during which bleeding may occur; the manipulation, cutting or removal of any oral or perioral tissues, including tooth structure during which bleeding may occur.

The document goes on to point out that:

the risk of injury to a health care worker depends on a variety of factors which include the type of procedure, skill of the operator, circumstances of the operation and physical condition of the patient. Examples of procedures where infection might be transmitted are those in which hands may be in contact with sharp instruments or sharp tissues (spicules of bone or teeth), inside a patient's body cavity or open wound particularly when the hands are not completely visible.

It is obvious that such procedures must not be performed by an anaesthetist who is HBeAg and/or HBsAg positive. Apart from this, on current evidence a hepatitis-infected anaesthetist may continue in clinical practice.[181]

The investigation and management of postoperative jaundice

Incidence

Jaundice or raised liver enzymes occurring after surgery and anaesthesia will be encountered by all anaesthetists sometime during their clinical practice. The investigation and management of postoperative liver dysfunction have been reviewed in monographs.[186,187] Minor and transient abnormalities are common in the early postoperative period and are not accompanied by any histological abnormality.[188] Mild jaundice follows approximately 17 per cent and marked jaundice 4 per cent of major operations,[189] and occurs in 25 per cent of patients after open heart surgery.[190] There are a number of causes of postoperative elevation of bilirubin with or without normal aminotransferase concentrations but, despite this, it is often alleged that it is due to

the use of an inhaled anaesthetic agent. It is therefore vital that a careful evaluation is made of all the possible causative factors so that an incorrect and misleading label is not applied to the condition: this is particularly so when the diagnosis of drug-related injury is essentially one based on exclusion of other causes, temporal association and past drug history. As discussed above, those patients most at risk of developing postoperative hepatic problems are those with compromised preoperative liver function,[186,187] especially if subjected to hypotension, hypoxia or sepsis.

Pathogenesis

The causes of postoperative jaundice can be divided into two main categories – patient factors and perioperative factors – and they are listed in Table 37.17.[187] The part of the pathway at which they exert their effects is shown in Fig. 37.12.

When hyperbilirubinaemia presents without hepatic aminotransferase elevation, it is due to a defect in bilirubin metabolism or to overproduction of bilirubin. The commonest defect in metabolism is that due to Gilbert's disease, which affects 3–7 per cent of the population.[191] Rare causes are Dubin–Johnson and Rotor's syndromes and Crigler–Najjar disease. Normal adults produce approximately 250–350 mg of bilirubin per day and can handle up to three times this load.[186] However, with excessive extravascular and/or intravascular haemodestruction, bilirubin loads may exceed this level. Extravascular breakdown of red blood cells from a 1 litre haemotoma produces 5000 mg of bilirubin. Intravascular destruction of red blood cells can occasionally result from haemolytic conditions (G6P-dehydrogenase deficiency, cardiopulmonary bypass, artificial valves, sickle cell disease etc.) but is more commonly due to (usually multiple) blood transfusions. Approximately 10 per cent of 2-week-old red blood cells in a unit of banked blood will undergo haemolysis.[192] Mismatched transfusion reactions are usually apparent from other accompanying signs but haemolysis may be delayed

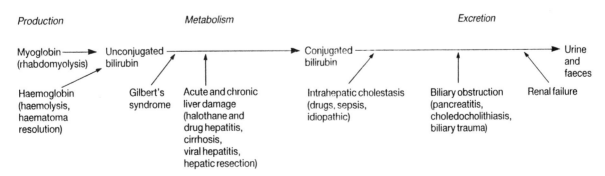

Fig. 37.12 Mechanisms of hyperbilirubinaemia. (Reproduced, with permission, from Hayes PC, Bouchiér IAD. Postoperative hepatic dysfunction. *Baillière's Clinical Gastroenterology* 1989; **3**: 485–505.)

Table 37.17 Causes of postoperative jaundice

Patient factors
Congenital haemolytic disorders
Acquired haemolytic disorders
Pre-existing liver disease with or without portal hypertension
Coagulopathy
Gilbert's syndrome
Sepsis

Perioperative factors
Anaesthetic-induced reduction in liver blood flow
Bleeding
Hypotension
Blood transfusion
Biliary tree trauma
Viral hepatitis
Drugs:
   halothane
   antibiotics
   non-steroidal agents
   polypharmacy and drug interactions
   antihypertensives
   potential haemolytic drugs

Reproduced, with permission, from Hayes PC, Bouchiér IAD. Postoperative jaundice. *Baillière's Clinical Gastroenterology* 1989: **3**: 485–505.

for from 2 to 21 days.[193] Delayed transfusion reactions occur following a second stimulus of transfused blood and may produce jaundice in the postoperative period several days after anaesthesia.[186]

Hyperbilirubinaemia with mild to moderate elevations of transaminase occurs in relation to cholestasis, which can be intra- or extrahepatic. Early recognition of extrahepatic biliary obstruction and/or an associated acute pancreatitis is important so that treatment can be initiated as soon as possible. If the obstruction is because of surgery of the biliary tree, hyperbilirubinaemia and elevated levels of alkaline phosphatase occur within 2–3 days of surgery; the aminotransferase levels follow later if obstruction persists.[186] Serum amylase levels of over 1000 u confirm the presence of pancreatitis. Postoperative cholecystitis and/or pancreatitis may, however, occur following non-biliary surgery between 3 and 30 days postoperatively. This condition is poorly understood, the cholecystitis is often acalculous, it may be complicated by gangrene of the gall bladder and there may be a poor outcome.[194]

Postoperative intrahepatic cholestasis is often called benign postoperative cholestasis. It can follow major surgery, is often associated with multiple blood transfusions, is not due to drugs or viruses[186] and there may be contributory factors such as hypotension and hypoxia. Abnormalities in liver-associated enzymes are characteristically cholestatic with elevation of conjugated bilirubin between 2 and 7 days postoperatively, moderate increases in alkaline phosphatase (two or four times normal), and little disturbance of transaminase levels, at least in the early stages.[187] Postoperative liver biopsies taken in such patients typically show dilated biliary canaliculi and bile casts with minimal hepatocellular damage and inflammation.[188,195] Resolution of the hyperbilirubinaemia and intrahepatic cholestasis occurs within 3 weeks. The main danger from this condition is probably that of misdiagnosis and the failure to relieve extrahepatic biliary obstruction.

Hyperbilirubinaemia with significant elevation of aminotransaminases is the most sinister form of postoperative liver dysfunction because it implies release of enzymes from damaged hepatocytes. Although it can be caused by a variety of insults in any part of the perioperative period (not just during surgery and anaesthesia), the histological picture of centrilobular necrosis is constant. This is because the hepatocytes in this region are at a lower oxygen tension than in the periportal area and are thus most prone to ischaemia and toxaemia (see earlier).

Sepsis, hypotension and hypoxia, both separately and together, produce centrilobular damage. Sepsis works both by toxin damage and by the systemic effects of septic shock. The severity of the injury may often be correlated with the degree of hypoxaemia and the duration of hypotension.[188,196] Vasopressors used to correct the systemic blood pressure may themselves further impair liver perfusion.[197] If ischaemia is corrected, hepatocellular function will be rapidly restored; if it persists, permanent injury and hepatic necrosis may follow.

Passive hepatic venous congestion in right-sided and congestive heart failure and in association with pneumonia can also produce a less acute and florid but mechanistically similar form of liver damage.[198]

Viral hepatitis has been described above and can be acquired in the perioperative period. The most important cause of post-transfusion hepatitis in western countries is NANB; some studies in the past, before there was effective screening for hepatitis C, have put it as high as 10 per cent of those receiving blood transfusions.[199] Traditionally it has been taught that a high mortality occurs in patients undergoing surgery in the presence of acute viral hepatitis[200] and that only life-saving surgery should be undertaken. This is probably still wise counsel but there have, to the authors' knowledge, been no more recent reports that take into account the changes in clinical practice that have occurred in recent years.

Drug-induced liver injury does not depend upon pre-existing liver dysfunction, and when it occurs has a spectrum from the asymptomatic to the life threatening. A very large number of drugs (or their metabolites), self or therapeutically administered in the perioperative period, may produce hepatotoxicity. Up to 7 per cent of reported adverse drug reactions have been due to presumed drug-induced liver disease.[201] A list of some drugs that have been implicated in the production of liver disease, together with the mechanism, is given in Table 37.18.[186] Under

Table 37.18 Drugs known to produce or suspected of producing liver damage and the resulting histological injury

Drug	Injury pattern
Isoniazid, phenytoin, α-methyldopa	Viral hepatitis-like injury
Oxacillin, aspirin	Non-specific hepatitis
Quinidine, allopurinol, phenylbutazone	Granulomatous hepatitis
α-Methyldopa, methotrexate, amiodarone, dantrolene	Chronic hepatitis
Chlorpromazine, erythromycin, oestrogen	Cholestasis
Tetracycline, valproic acid, ethanol	Fatty liver changes
Oral contraceptives	Hepatic vein thrombosis
Anabolic steroids	Peliosis hepatitis

Reproduced, with permission, from Frink EJ, Brown BR. Postoperative hepatic dysfunction. *Baillière's Clinical Anaesthesiology* 1992; **6**: 931–52.

certain circumstances, hepatotoxicity may occur with therapeutic drug doses if there is underlying liver damage.[155,202] Acute or chronic alcohol ingestion may produce alcoholic hepatitis which can easily go unrecognized unless suspected. Caution is required in anaesthetizing these patients because, even without anaesthesia and surgery, mortality for acute-on-chronic alcoholic hepatitis can approach 40 per cent.[203] Because of their importance, adverse drug reactions are dealt with separately below.

Adverse drug reactions

Description and mechanisms

Adverse drug reactions affecting the liver are classically divided into two types: predictable and idiosyncratic.[204] Predictable drug reactions are usually dose dependent and can be illustrated by paracetamol: above a certain level of ingestion all patients will be affected in a similar way. These reactions are normally reproducible in animal models and hence can be minimized by therapeutic rationale. In

contrast to this, idiosyncratic reactions do not have a simple and predictable dose-dependent relationship with toxicity. The interval between taking the drug and the manifestation of liver damage is both wider and more variable. Idiosyncratic reactions may be due to metabolic idiosyncracy or to activation of the immune system.

Idiosyncratic reactions tend to be seen more commonly in clinical practice, as those drugs with a predictable hepatotoxicity are usually given in a defined dose and liver damage is seen only after overdose, either iatrogenic or self-administered.[204] Characteristics of predictable, idiosyncratic and immune-mediated damage are given in Table 37.19.[204] From the point of view of anaesthetists, the most important group of drugs in relation to liver damage are the volatile anaesthetic agents (see below).

Inhalational agents and liver damage

Halothane
Hepatitis secondary to halothane has been the subject of several review articles.[186,187,204,205] By 1963, 5 years after the introduction of halothane, at least 350 cases of so-called 'halothane hepatitis' had been recorded.[206] General concern produced a large number of individual case reports and anecdotal retrospective surveys which added to the literature rather than to the knowledge base of the subject. Because of this, the National Halothane Study was set up to review cases of hepatic necrosis occurring within 6 weeks of anaesthesia in 34 hospitals in the USA.[207] Out of 850 000 administrations of general anaesthesia, 9 cases of inexplicable hepatic necrosis were identified, of whom 7 had received halothane; of these 7, 4 had received halothane on more than one occasion within the last 6 weeks. Other retrospective studies confirmed that halothane was associated with severe liver dysfunction, with an incidence ranging from 1 in 6000 to 1 in 20 000; in addition, it became apparent that there was a greater problem with repeated exposure to halothane, especially at short intervals.[208–211]

Prospective studies of a condition with such a low incidence are bound to be difficult to arrange. They have,

Table 37.19 Classification of drug hepatotoxicity

	Dose-dependent	Onset	Response to challenge	Animal model	Immune features
Predictable	+	Rapid (days)	Rapid	+	−
Idiosyncratic metabolic	+/−	Variable (up to 1 year)	Delayed	+/−	−
Immune-mediated	−	Variable (1 day to 12 weeks)	Rapid	−	+

Reproduced, with permission, from Neuberger JM. Halothane and the implications of hepatitis. In: Kaufman L, ed. *Anaesthetic review, 8.* London: Churchill Livingstone, 1991: 179–94.

however, yielded important information and demonstrated that there are in fact two patterns of liver dysfunction: one is common, minor and predictable; the other is rare, severe, idiosyncratic and may be immune mediated.[212,213] Although not all reports have published similar findings,[214,215] there now seems to be agreement that there is a rise in transaminase levels in up to 25 per cent of patients following anaesthesia and that this rise is much more common following halothane than other anaesthetics.[216–219] The rises rarely exceed three times the upper limit of normal, and where liver histology was available it showed features of a focal hepatitis;[217] it was also shown that abnormal liver function tests may not be detectable until well into the second postoperative week.[218] More recently,[220,221] increased plasma concentrations of glutathione-S-transferase (a sensitive and specific index of acute, drug-induced, hepatocellular dysfunction) have been demonstrated after halothane but not after isoflurane anaesthesia. These studies also showed that halothane is associated with less liver damage when given under high (100 per cent) oxygen tensions than at lower (30 per cent) oxygen tensions. There now seems to be an acceptance that this mild, subclinical form of liver dysfunction could be caused by the toxic products of halothane metabolism, possibly influenced by genetic and pre-existing enzyme factors[222] or by hepatic hypoxia in relation to oxygen demand.[205]

Although the mild reactions described above fit into the definition of halothane hepatitis as 'otherwise unexplained liver damage occurring within 28 days of halothane exposure in a patient with a previously normal liver',[204] the more usual interpretation of 'halothane hepatitis' relates to a severe, unpredictable and idiosyncratic reaction leading to liver failure. Halothane hepatitis represents the commonest iatrogenic cause of fulminant hepatic failure. The risk factors for halothane hepatitis which have been commonly found in published series are listed in Table 37.20.[204] Of particular concern is that a previous episode of jaundice goes unnoticed. In one study of 40 patients,[223] 11 had previously documented adverse reactions to halothane. It should also be noted that, even if halothane is not used for such a patient, the agent can be present as a contaminant of anaesthetic equipment and significant levels of halothane metabolites can be detected in the urine.[224] There have been only very rare reports of halothane hepatitis occurring in children and there is no doubt, given its frequent use in paediatrics, that it is much rarer in children than in adults.[205] It has been concluded by some authors that 'it would seem unreasonable to restrict the use of this drug in paediatric patients' but 'should fever, eosinophilia, and liver enzyme elevation occur following a halothane anaesthetic in a child, it would be prudent to restrict further halothane anaesthetics'.[186]

The clinical features of halothane hepatitis have been

Table 37.20 Risk factors for halothane hepatitis

High	Recent previous exposure
	Previous adverse reaction
Uncertain	Obesity
	Female
	Drug allergy
	Lymphocyte sensitivity to phenytoin
	Family history of halothane hepatotoxicity

Reproduced, with permission, from Neuberger JM. Halothane and the implications of hepatitis. In: Kaufman L, ed. *Anaesthetic review*, 8. London: Churchill Livingstone, 1991: 179–94.

well described in review articles.[204,205] Many patients initially complain of malaise, anorexia and non-specific gastrointestinal symptoms such as nausea and upper abdominal discomfort. Although not invariable, the onset of fever is characteristically the first sign of the illness. The latent period before the onset of pyrexia is 7 days for a first exposure, reducing to 4 for a re-exposure. Jaundice almost always occurs, the onset being a few days to up to 4 weeks; as with the fever, multiple exposures are associated with a shorter delay.

The presence of eosinophilia and autoantibodies has been proposed as a diagnostic marker. Recent work has supported this hypothesis.[225–229] The majority of patients with a clinical diagnosis of halothane hepatitis have serum antibodies that react with one or more specific liver microsomal proteins that have been covalently altered by the trifluoroacetyl chloride metabolite of halothane. These serum antibodies are not seen in patients with other types of liver failure[229] but neither are they present in all patients for whom the clinical diagnosis of halothane hepatitis has been made.[228] Halothane hepatitis is therefore still essentially a clinical diagnosis for which, on occasion, there will be antibody data in support.

Biochemical tests of liver function generally reflect changes typical of hepatocellular damage with gross rises in serum transaminases (500–2000 u/l), and alkaline phosphatase activity generally less than twice the upper limit of normal. The incidence of some of the features of halothane hepatitis is shown in Table 37.21.[204]

The most practical way of reducing the incidence of halothane hepatitis is almost certainly to take a good history of previous anaesthetic exposure; in one author's experience, one-third of patients with halothane hepatitis have had a previous episode of unexplained postoperative jaundice following halothane exposure.[204] Preoperative screening for halothane antibodies is likely to be very expensive and, given the very low incidence of the disease, is unlikely to be economical. Furthermore, there is no information on how long these antibodies persist and thus the value of preoperative screening is unknown.[204] The

Table 37.21 Features of halothane hepatitis

Female/male	1.6 : 1
Previous exposure	78%
Drug allergy	15%
Eosinophilia	21%
Autoantibodies	29%

Reproduced, with permission, from Neuberger JM. Halothane and the implications of hepatitis. In: Kaufman L, ed. *Anaesthetic review, 8*. London: Churchill Livingstone, 1991: 179–94.

Table 37.22 Halothane exposure guidelines (from the Committee on the Safety of Medicines)

Avoid halothane exposure if:
Previous exposure within 3 months
Previous adverse reaction to halothane
Family history of adverse reaction to halothane
Adverse reaction to other halogenated hydrocarbon anaesthetic
Pre-existing liver disease

current guidelines issued in the UK by the Committee on the Safety of Medicines (CSM) are shown in Table 37.22. It should, however, be noted that although the shorter the interval between exposures, the more severe is the reaction and the shorter the latency, there have been cases when the penultimate exposure was years previously. This must bring into question the 3-month recommendation by the CSM.[205]

Other inhalational agents
There is not the same concern regarding the hepatotoxicity of other inhalational agents compared with halothane. A number of cases of enflurane hepatitis have been published but in some of these viral hepatitis and other forms of hepatic injury could not be ruled out. It has been concluded that in rare patients it cannot be denied that unexplained liver damage follows enflurane anaesthesia but the known incidence is of the order of 1 in 800 000, which is less than the spontaneous 'attack rate' of viral hepatitis.[205]

Isoflurane has been studied in cases of hepatic dysfunction reported to the Food and Drug Administration in the USA but, after analysis of this data, current evidence suggests that it is highly unlikely that isoflurane is even rarely responsible for postoperative hepatotoxicity.[230,231] Direct damage from nitrous oxide is also considered unlikely[205] and it is at present too soon to comment on the hepatotoxic potential of sevoflurane or desflurane.[205] Despite these apparently reassuring conclusions, good practice nevertheless dictates that a good drug history should always be taken and any treatment followed by ill effects should not normally be repeated.

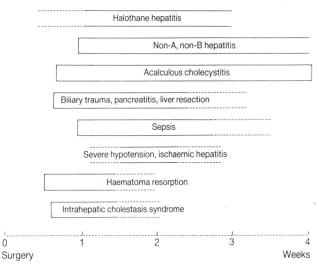

Fig. 37.13 Timecourse of onset of postoperative jaundice. (Reproduced, with permission, from Hayes PC, Bouchiér IAD. Postoperative hepatic dysfunction. *Baillière's Clinical Gastroenterology* 1989; **3**: 485–505.)

Clinical management of postoperative jaundice

The management of postoperative jaundice requires cooperation between the anaesthetist, surgeon and hepatologist. In the majority of patients with postoperative jaundice the cause can be identified easily from the history, the clinical examination, the pattern of hepatic injury as assessed by liver function tests, the nature of the surgical operation, the time of onset of the jaundice (Fig. 37.13) and a knowledge of the drugs used.[187] The algorithm in Fig. 37.14 outlines a logical process of diagnosis but it should be emphasized that the single most important investigation is the liver function tests from which the dysfunction can be characterized into a hepatitic, cholestatic, mixed or isolated hyperbilirubinaemia pattern.[187]

Although it has been emphasized before, important causes to identify are sepsis and extrahepatic biliary obstruction, as early treatment has a major effect on outcome. Undoubtedly, in patients at risk because of pre-existing liver disease or who have a history of previous drug exposure, prevention is the best treatment. It should also not be forgotton that when the liver fails there is often associated failure of the kidneys, heart and lungs – all of which need treatment to optimize the best conditions in which the liver can recover.

In summary, a balanced conclusion is that which adheres to the philosophy of Hayes and Bouchier:

Postoperative jaundice is common, particularly after operations such as open-heart surgery. The mechanism in many patients is multifactorial but a predominant

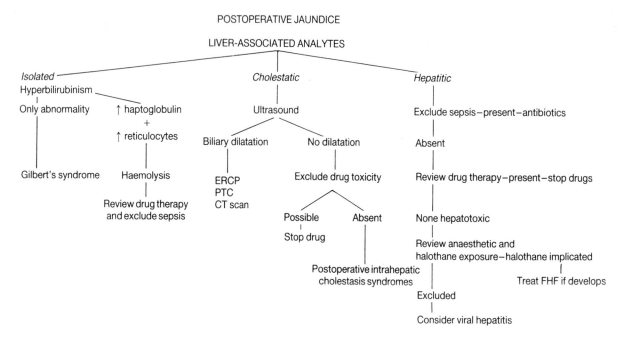

Fig. 37.14 Algorithm for investigation of postoperative jaundice. ERCP, endoscopic retrograde cholangiopancreatography; FHF, fulminant hepatic failure; PTC, percutaneous transhepatic cholangiogram. (Reproduced, with permission, from Hayes PC, Bouchiér IAD. Postoperative hepatic dysfunction. *Baillière's Clinical Gastroenterology* 1989; **3**: 485–505.)

cause should be sought and treated. They [causes] can be divided into patient and perioperative factors, and can usually be determined from a knowledge of the liver-associated analysis abnormalities, time of onset of jaundice, operative procedure, existence of pre-existing liver disease and drug therapy. Every effort to arrive at a specific diagnosis should be made as treatment varies widely with the underlying cause, ranging from reassurance to treatment of fulminant hepatic failure. The diagnosis of benign postoperative intrahepatic cholestasis is one of exclusion in a likely clinical setting (i.e. with multiple blood transfusion, hypotension, etc.) and should not be made without certain exclusion of biliary obstruction or sepsis. Many cases of post-operative jaundice can be prevented by refraining from operating on patients with liver disease, by identifying previous halothane sensitivity, and avoiding wherever possible potentially hepatotoxic drugs.[187]

The liver in pregnancy and its implications for anaesthesia

This subject has been reviewed in detail elsewhere[232] so in this section only those features relevant to the practice of anaesthesia in pregnancy will be described. The major implications for all these hepatic conditions during pregnancy relate to coagulation problems (and hence to the contraindications for regional blockade), blood loss and premature delivery.

Normal pregnancy

The liver does not alter significantly in size, weight or blood flow during normal pregnancy. The finding of an enlarged liver during pregnancy is therefore presumptive evidence of liver disease. The proportion of cardiac output delivered to the liver falls from 35 per cent in non-pregnant females to 29 per cent during late pregnancy. Serum alkaline phosphatase levels rise slowly during the early months, with a sharp rise in the last trimester; concentrations of twice normal are frequently found at term. Care must therefore be taken with alkaline phosphatase as a marker of liver disease in pregnancy. In contrast, serum transaminases (AST, ALT) are not affected by pregnancy, and elevation of these is a reliable indicator of hepatocellular damage. Serum cholinesterase activity is reduced by 30 per cent at term but prolonged neuromuscular block with appropriate doses of suxamethonium is rare;[233] if it does occur in those with a normal genotype, it rarely lasts more than 20 minutes.[234]

Hyperemesis gravidarum

In severe cases (usually those requiring hospitalization), serum transaminases may rise by up to twice normal, but jaundice is rare. The dysfunction is never severe and death from hepatic failure has never been reported. Liver function tests improve as vomiting subsides.

Intrahepatic cholestasis of pregnancy

Intrahepatic cholestasis of pregnancy (IChP) is second only to acute viral hepatitis as a cause of jaundice in pregnancy but less than 1 per cent of pregnant women develop symptoms or signs of this disorder. The distinct possibility exists, however, that IChP has been poorly recognized and therefore under-reported.[235] A diagnosis of this condition of unknown aetiology should be considered in any pregnant patient with pruritus or jaundice.[232] The diagnosis is confirmed with the elevation of serum bile acids by 10- to 100-fold.[235] AST and ALT may be normal or elevated up to four times normal. The prothrombin time may be prolonged and can be treated by vitamin K. Maternal prognosis is excellent but several studies have reported an increased incidence of premature labour, fetal distress and neonatal death.[232] Some obstetricians advocate induction of labour before the 38th week to prevent fetal distress and its sequelae.

Acute fatty liver of pregnancy

Acute fatty liver of pregnancy (AFLP) is a rare and serious disorder of unknown aetiology which manifests itself in late pregnancy. It has an estimated incidence of one case per million pregnancies, only 140 cases having been reported up to 1985.[236] Symptoms appear suddenly in the third trimester and include abdominal pain, anorexia, nausea, vomiting, fatigue, headaches and fever. Progression is rapid with bleeding, jaundice, renal failure and encephalopathy and right upper quadrant tenderness. There is hyperbilirubinaemia and elevation of the transaminases. Coagulation disturbances may range from mild extension of the prothrombin time and/or thrombocytopenia to disseminated intravascular coagulation. Hypoglycaemia can be life threatening and there may be raised serum creatinine if renal failure is present.

AFLP poses a severe threat to both the mother and the child, and the disease runs a relentless downhill course until after delivery. Prompt delivery is therefore advisable as soon as the diagnosis has been made. Patients with AFLP may develop hepatic failure within a few days, with a marked coagulopathy and a high risk of severe postpartum haemorrhage.[232] Treatment of coagulation disorders, as described in earlier sections, is essential.

Toxaemia and the HELLP syndrome

Patients with pre-eclampsia do not usually have clinical or biochemical evidence of liver disease but mild liver abnormalities have been reported in up to 10 per cent. The term HELLP, introduced in 1982,[237] is used to describe the combination of **h**aemolysis, **e**levated **l**iver enzymes and **l**ow **p**latelet count. It has, however, been argued that this syndrome is not a separate entity but 'merely a cluster of signs seen in some patients with hypertensive disorders of pregnancy'.[238] Symptoms include epigastric and right upper quadrant pain, nausea and vomiting. Fibrin is deposited in hepatic sinusoids which may occlude them and lead to necrosis and periportal haemorrhage. Macroscopically, in severe cases one may see focal or diffuse necrosis together with subcapsular and/or intrahepatic haemorrhage. This intrahepatic haemorrhage may be accompanied by severe right upper quadrant/epigastric pain, and signs of hypovolaemia or shock should suggest intraperitoneal rupture.[232] Diagnosis can be made by abdominal paracentesis, ultrasound or abdominal CT scan. Immediate treatment of hypovolaemia by volume expansion and correction of coagulopathy and management of other aspects of pre-eclamptic toxaemia is essential. Intrahepatic haemorrhage may be managed conservatively, but capsular rupture requires laparotomy in association with caesarean section. Prepare for difficulties in both anaesthetizing a pregnant patient and one with liver rupture (which see). Despite improved diagnosis and management, maternal mortality remains high at 50–60 per cent and fetal mortality at 60 per cent.[232]

REFERENCES

1. Ellis H. *Clinical anatomy*, 5th ed, Oxford: Blackwell Scientific, 1974: 92, 94.
2. Warwick R, Williams PL. *Gray's anatomy*, 35th ed. London: Longman, 1973: 1302–14.
3. Di Fiore MSH. *Atlas of human histology*. Philadelphia: Lea & Febiger, 1973: 155, plate 71.
4. Sear JW. Anatomy and physiology of the liver. *Baillière's Clinical Anaesthesiology* 1992: **6**, no 4; 697–727.

5. Lautt WW, Legare DJ, Daniels TR. The comparative effect of administration of substances via the hepatic artery or portal vein on the hepatic arterial resistance, liver blood volume and hepatic extraction in cats. *Hepatology* 1984; **4**: 927–32.

6. Sear JW. Hepatic physiology. *Current Anaesthesia and Critical Care* 1990; **1**: 196–203.

7. Norris CP, Barnes GE, Smith EE, Granger HJ. Autoregulation of superior mesenteric blood flow in fasted and fed dogs. *American Journal of Physiology* 1979; **237**: H174–7.

8. Lutz J, Pieper U, Bauereisen E. Appearance and size of veno-vasomotor reactions in the liver circulation. *Pflügers Archiv; European Journal of Physiology* 1968; **299**: 311–25.

9. Hanson KM, Johnson PC. Local control of hepatic arterial and portal venous flow in the dog. *American Journal of Physiology* 1966; **211**: 712–20.

10. Lautt WW. Mechanism and role of intrinsic regulation of hepatic arterial blood flow: the hepatic arterial buffer response. *American Journal of Physiology* 1985; **249**: G549–56.

11. Lautt WW, Legare DJ, d'Almeide MS. Adenosine as putative regulator of hepatic arterial flow (the buffer response). *American Journal of Physiology* 1985; **248**: H331–8.

12. Richardson PDI, Withrington PG. Pressure flow relationships and the effects of noradrenaline and isoprenaline on the simultaneously perfused hepatic arterial and portal venous vascular beds of dogs. *Journal of Physiology* 1978; **282**: 451–70.

13. Seltzer JL, Ritter DE, Starsnic MA, Marr AT. The hemodynamic response to traction on the abdominal mesentery. *Anesthesiology* 1985; **63**: 96–9.

14. Gelman S. Disturbances in hepatic blood flow during anesthesia and surgery. *Archives of Surgery* 1976; **111**: 881–3.

15. Clarke RSJ, Doggart JR, Lavery T. Changes in liver function after different types of surgery. *British Journal of Anaesthesia* 1976; **48**: 119–28.

16. Bohrer SL, Rogers EL, Koehler RC, Traystman RJ. Effect of hypovolemic hypotension and laparotomy on splanchnic and hepatic arterial blood flow in dogs. *Current Surgery* 1981; **38**: 325–8.

17. Harper MH, Collins P, Johnson BH, Eger EI, Biava CG. Postanesthetic hepatic injury in rats: influence of alterations in hepatic blood flow, surgery, and anesthesia time. *Anesthesia and Analgesia* 1982; **61**: 79–82.

18. Gelman S, Dillard E, Bradley EL. Hepatic circulation during surgical stress and anesthesia with halothane, isoflurane or fentanyl. *Anesthesia and Analgesia* 1987; **66**: 936–43.

19. Moreno AH, Burchell AR. Respiratory regulation of splanchnic and systemic venous return in normal subjects and in patients with hepatic cirrhosis. *Surgery, Gynecology and Obstetrics* 1982; **154**: 257–67.

20. Manney J, Justice R, Hechtman HB. Abnormalities in organ blood flow and its distribution during end-expiratory pressure. *Surgery* 1979; **85**: 425–32.

21. Halden E, Jakobson S, Janeras L, Norlen K. Effects of positive and end-expiratory pressure on cardiac output distribution in the pig. *Acta Anaesthesiologica Scandinavica* 1982; **26**: 403–8.

22. Bonnet F, Richard C, Glaser P, Lafay M, Guesde R. Changes in hepatic flow induced by continuous positive pressure ventilation in critically ill patients. *Critical Care Medicine* 1982; **10**: 703–5.

23. Brendenberg CE, Paskanik AM. Relation of portal hemodynamics to cardiac output during mechanical ventilation with PEEP. *Annals of Surgery* 1983; **198**: 218–22.

24. Gioia FR, Harris AP, Traystman RJ, Rogers MC. Organ blood flow during high frequency ventilation at low and high airway pressure in dogs. *Anesthesiology* 1986; **65**: 50–5.

25. Thompson IA, Fitch W, Hughes RL, Campbell D. Effect of increased concentrations of carbon dioxide during halothane anaesthesia on liver blood flow and hepatic oxygen consumption. *British Journal of Anaesthesia* 1983; **55**: 1231–7.

26. Fujita Y, Takayuki S, Akiyuki O. Effects of hypocapnia and hypocarbia on splanchnic circulation and hepatic function in the beagle. *Anesthesia and Analgesia* 1989; **69**: 152–7.

27. Cooperman LH, Warden JC, Price HL. Splanchnic circulation during nitrous oxide anesthesia and hypocarbia in normal man. *Anesthesiology* 1968; **29**: 254–8.

28. Richardson PDI, Withrington PG. Liver blood flow. 2. Effects of drugs and hormones on liver blood flow. *Gastroenterology* 1981; **81**: 356–75.

29. Gelman S, Frenette L. Effects of anaesthetics on liver blood flow. *Baillière's Clinical Anaesthesiology* 1992; **6**, no 4: 729–50.

30. Strunin L. The splanchnic, hepatic and portal circulation. In: Prys-Roberts C, ed. *The circulation in anaesthesia*. London: Blackwell Scientific, 1980: 241–51.

31. Tverskoy M, Gelman S, Fowler KC, Bradley EL. Intestinal circulation during inhalation anesthesia. *Anesthesiology* 1985; **62**: 462–9.

32. Tranquilli WJ, Manohar M, Parks CM, Thurmon JC, Theodorakis MC, Benson GJ. Systemic and regional blood flow distribution in unanesthetized swine and swine anesthetized with halothane + nitrous oxide, halothane, or enflurane. *Anesthesiology* 1982; **56**: 369–79.

33. Gelman S, Fowler KC, Smith LR. Regional blood flow during isoflurane and halothane anesthesia. *Anesthesia and Analgesia* 1984; **63**: 557–65.

34. Gelman S, Fowler KC, Smith LR. Liver circulation and function during isoflurane and halothane anesthesia. *Anesthesiology* 1984; **61**: 726–30.

35. Burney RG, DiFazio CA. Hepatic clearance of lidocaine during N_2O anesthesia in dogs. *Anesthesia and Analgesia* 1976; **55**: 322–5.

36. Borel JD, Bently JB, Nenad RE, Gillespie TJ. The influence of halothane on fentanyl pharmacokinetics. *Anesthesiology* 1982; **57**: A239.

37. Chelly JE, Rogers K, Hysing ES, Taylor A. Cardiovascular effects of and interaction between calcium blocking drugs and anesthetics in chronically instrumented dogs. I. Verapamil and halothane. *Anesthesiology* 1986; **64**: 560–7.

38. Reilly C, Merrell J, Wood A, Koshakij RB, Wood M. Comparison of the effects of isoflurane or fentanyl–nitrous oxide anaesthesia on propranolol in dogs. *British Journal of Anaesthesia* 1988; **60**: 791–6.

39. Frink E, Morgan S, Coetzee A, Conzen P, Brown B. The effects of sevoflurane, halothane, enflurane and isoflurane on hepatic blood flow and oxygenation in chronically instrumented greyhound dogs. *Anesthesiology* 1992; **76**: 86–90.

40. Hughes RL, Campbell D, Fitch W. Effects of enflurane and halothane on liver blood flow and oxygen consumption in the greyhound. *British Journal of Anaesthesia* 1980; **52**: 1079–86.

41. Miller C, Fitch W, Thompson A. Effect of isoflurane on the canine hepatic circulation and hepatic oxygen balance. *British Journal of Anaesthesia* 1990; **65**: 698–703.

42. Lundeen G, Manohar M, Parks C. Systemic distribution of blood flow in swine while awake and during 1.0 and 1.5 MAC isoflurane anesthesia with and without 50% nitrous oxide. *Anesthesia and Analgesia* 1983; **62**: 499–512.

43. Nagano K, Gelman S, Parks D, Bradley L. Hepatic oxygen supply uptake relationships and metabolism during anesthesia in miniature pigs. *Anesthesiology* 1990; **72**: 902–10.

44. Debaene B, Goldfard G, Braillon A, Jolis P, Lebrec D. Effects of ketamine, halothane, enflurane and isoflurane on systemic and splanchnic hemodynamics in normovolemic and hypovolemic cirrhotic rats. *Anesthesiology* 1990; **76**: 86–90.

45. Fujita Y, Kumura K, Hamada H, Takaori M. Comparative effects of halothane, isoflurane and sevoflurane on the liver with hepatic artery ligation in the beagle. *Anesthesiology* 1991; **76**: 86–90.

46. Thompson IA, Fitch W, Hughes RL, Campbell D, Watson R. Effects of certain i.v. anaesthetics on liver blood flow and hepatic oxygen consumption in the greyhound. *British Journal of Anaesthesia* 1986; **58**: 69–80.

47. Runciman WB, Mather LE, Selby DG. Cardiovascular effects of propofol and of thiopentone anaesthesia in the sheep. *British Journal of Anaesthesia* 1990; **65**: 353–9.

48. Diedericks J, Sear JW, Foex P. The effect of propofol on hepatic blood flow. *Anesthesiology* 1990; **73**: A624.

49. Tverskoy M, Gelman S, Fowler KC, Bradley EL. Influence of fentanyl and morphine on intestinal circulation. *Anesthesia and Analgesia* 1985; **64**: 577–84.

50. Kien ND, Reitan JA, White DA, Wu CH, Eisele JH. Hemodynamic responses to alfentanil in halothane-anesthetized dogs. *Anesthesia and Analgesia* 1986; **65**: 765–70.

51. Kennedy W, Everett G, Cogg L, Allen G. Simultaneous systemic and hepatic hemodynamic measurements during high epidural anesthesia in normal patients. *Anesthesia and Analgesia* 1971; **50**: 1069–77.

52. Hendolin H, Penttilla M. Liver enzymes after retropubic prostatectomy in patients receiving continuous lumbar epidural or general anesthesia. *Annals of Clinical Research* 1982; **14**: 1–6.

53. Grieg T, Andreen M, Irestedt L. Hemodynamics and oxygen consumption in the dog during high spinal epidural block with special reference to the splanchnic region. *Acta Anaesthesiologica Scandinavica* 1983; **27**: 211–17.

54. Loft S, Boel J, Kyst A, Rasmussen B, Hansen S, Dossing M. Increased hepatic microsomal enzyme activity after surgery under halothane or spinal anesthesia. *Anesthesiology* 1985; **62**: 11–16.

55. Jorm CM, Stamford JA. Hepatic metabolism of xenobiotics with reference to anaesthesia. *Ballière's Clinical Anaesthesiology* 1992; **6**, no 4: 751–79.

56. Eagle CJ, Strunin L. Drug metabolism in liver disease. *Current Anaesthesia and Critical Care* 1990; **1**: 204–12.

57. Arns PA, Wedlund PJ, Branch RA. Adjustment of medications in liver failure. In: Chernow B, ed. *The pharmacologic approach to the critically ill patient.* Baltimore and London: Williams & Wilkins, 1988: 85–111.

58. Howden CW, Birnie GG, Brodie MJ. Drug metabolism in liver disease. *Pharmacology and Therapeutics* 1989; **40**: 439–74.

59. Hayes PC. Liver disease and drug disposition. *British Journal of Anaesthesia* 1992; **68**: 459–61.

60. Watkins PB. Role of cytochromes P450 in drug metabolism and hepatoxicity. *Seminars in Liver Disease* 1990; **10**: 235–50.

61. Andersson T. Omeprazole drug interaction studies. *Clinical Pharmacokinetics* 1991; **21**: 195–212.

62. Mulder GJ. Glucuronidation and its role in regulation of biological activity of drugs. *Annual Review of Pharmacology and Toxicology* 1992; **32**: 25–49.

63. Portenoy RK. Thaler HT, Inturrisi CE, Friedlander-Klar H, Foley KM. The metabolite morphine-6-glucuronide contributes to the analgesia produced by morphine infusion in patients with pain and normal renal function. *Clinical Pharmacology and Therapeutics* 1992; **51**: 422–31.

64. Wood M. Plasma drug binding: implications for anesthesiologists. *Anesthesia and Analgesia* 1986; **65**: 786–804.

65. Lebrault C, Berger JL, D'Hollander AA, Gomeni R, Henzel D, Duvaldestin P. Pharmacokinetics and pharmacodynamics of vecuronium (ORG NC45) in patients with cirrhosis. *Anesthesiology* 1985; **62**: 601–5.

66. Duvaldestin P, Saada J, Berger JL, D'Hollander AA, Desmonts JM. Pharmacokinetics, pharmacodynamics, and dose–response relationship of pancuronium in control and elderly subjects. *Anesthesiology* 1982; **56**: 36–40.

67. Ward S, Neill EA. Pharmacokinetics of atracurium in acute hepatic failure (with acute renal failure). *British Journal of Anaesthesia* 1983; **55**: 1169–72.

68. Hammerer JP, Schaeffler P, Couture E, Duvaldestin P. Fentanyl pharmacokinetics in anaesthetized patients with cirrhosis. *British Journal of Anaesthesia* 1982; **54**: 1267–70.

69. Chauvin M, Ferrier C, Haberer JP, *et al.* Sufentanil pharmacokinetics in patients with cirrhosis. *Anesthesia and Analgesia* 1989; **68**: 1–4.

70. Ferrier C, Marty J, Buffered Y, Hammerer JP, Levron JC, Duvaldestin P. Alfentanil pharmacokinetics in patients with cirrhosis. *Anesthesiology* 1985; **62**: 480–4.

71. Patwardhan RV, Johnson RF, Hoympa A, *et al.* Morphine metabolism in cirrhosis. *Gastroenterology* 1981; **81**: 1006–11.

72. McHorse TS, Wilkinson GR, Johnson RF, Shenker S. Effect of acute viral hepatitis in man on the disposition and elimination of meperidine. *Gastroenterology* 1975; **68**: 775–80.

73. Branch RA, Morgan MH, James J, Reid AE. Intravenous administration of diazepam in patients with chronic liver disease. *Gut* 1976; **17**: 975–83.

74. Stock JGL, Strunin L. Unexplained hepatitis following halothane. *Anesthesiology* 1985; **63**: 424–39.

75. Powell-Jackson P, Greenway B, Williams R. Adverse effects of exploratory laparotomy in patients with suspected liver disease. *British Journal of Surgery* 1982; **68**: 449–51.

76. Strunin L. Preoperative assessment of hepatic function. *Baillière's Clinical Anaesthesiology* 1992; **6**, no 4: 781–93.

77. Schemel WH. Unexpected hepatic dysfunction found by multiple laboratory screening. *Anesthesia and Analgesia* 1976; **55**: 810–14.

78. Hulcrantz R, Glauman H, Lindberg G, Nilsson LH. Liver investigation in 149 asymptomatic patients with moderately

elevated activities of serum aminotransferases. *Scandinavian Journal of Gastroenterology* 1986; **21**: 109–13.

79. Henry DA, Kitchingman G, Langman MJS. ^{14}C-aminopyrine breath analysis and conventional biochemical tests as predictors of survival in cirrhosis. *Digestive Diseases and Sciences* 1985; **30**: 813–18.

80. Hussey AJ, Aldridge LM, Paul D, *et al*. Plasma glutathione-*S*-transferase concentrations: a measure of hepatocellular integrity following a single anaesthetic with halothane, enflurane or isoflurane. *British Journal of Anaesthesia* 1988; **60**: 130–5.

81. Murray JM, Trinick TR. Hepatic function and indocyanine green clearance during and after prolonged anaesthesia with propofol. *British Journal of Anaesthesia* 1992; **69**: 643–4.

82. Gluud C, Anderson I, Dietrichson O, *et al*. Gamma-glutamyl-transferase, aspartate aminotransferase and alkaline phosphatase as markers of alcoholic consumption in outpatient alcoholics. *European Journal of Clinical Investigation* 1981; **11**: 171–6.

83. Bailey ME. Endotoxin, bile salts, and renal failure in obstructive jaundice. *British Journal of Surgery* 1976; **63**: 774–8.

84. Pain JA, Cahill CJ, Bailey ME. Perioperative complications in obstructive jaundice. Therapeutic considerations. *British Journal of Surgery* 1985; **72**: 942–7.

85. Cahill CJ. Prevention of postoperative renal failure in patients with obstructive jaundice – the role of bile salts. *British Journal of Surgery* 1983; **70**: 590–5.

86. Pitt HA, Cameron JL, Postier RG, *et al*. Factors affecting mortality in biliary tract surgery. *American Journal of Surgery* 1981; **141**: 66–9.

87. Dixon JM, Armstrong CP, Duffy SW, *et al*. Factors affecting morbidity and mortality after surgery for obstructive jaundice: a review of 373 patients. *Gut* 1983; **24**: 845–52.

88. Anderson P. The health of the nation: responses; alcohol as a key area. *British Medical Journal* 1991; **303**: 766–9.

89. Limas CJ, Guiha NH, Lekagul O, Cohn JN. Impaired left ventricular function in alcoholic cirrhosis with ascites. *Circulation* 1974; **49**: 755–60.

90. Browne DRG. Anaesthesia in impaired liver function. *Current Anaesthesia and Critical Care* 1990; **1**: 220–7.

91. Allison MEM. The kidney and the liver. Pre- and post-operative factors. In: Blumgart LH, ed. *Surgery of the liver and biliary tract*. London: Churchill Livingstone, 1988: 405–22.

92. Blamey SL, Fearon KCH, Gilmour WH, Osborne DH, Carter DC. Prediction of risk in biliary surgery. *British Journal of Surgery* 1983; **70**: 535–8.

93. Boyer TD. Major sequelae of cirrhosis. In: Wyngaaden JB, Smith LH, eds. Cecil, *Textbook of medicine*, 16th ed. Philadelphia: WB Saunders, 1982; 804–8.

94. Ginès P, Arroyo V, Rodes J. Treatment of ascites and renal failure in cirrhosis. *Baillière's Clinical Gastroenterology* 1989; **3**: 165–86.

95. Blake JC, Sprengers D, Grech P, McCormick PA, McIntyre N, Burroughs AK. Bleeding time in patients with hepatic cirrhosis. *British Medical Journal* 1990; **301**: 12–15.

96. Bontempo FA, Lewis JH, Ragni MV, Starzl TE. The preoperative coagulation pattern in liver transplant patients. In: Winter PM, Kang YG, eds. *Hepatic transplantation.*

Anesthetic and perioperative managment. Westport CT: Praeger Publishers, 1986: 135–41.

97. O'Grady JG, Williams R. Acute liver failure. *Baillière's Clinical Gastroenterology* 1989; **3**: 75–89.

98. Child III CG, Turcotte JG. Surgery and portal hypertension. In: Child III CG, ed. *The liver and portal hypertension*. Philadelphia: WB Saunders, 1964: 50.

99. Pugh RNH, Murray-Lyon IM, Dawson JL. Transection of the oesophagus for bleeding oesophageal varices. *British Journal of Surgery* 1973; **60**: 646–9.

100. Garrison RN, Cryer HM, Howard DA, Polk HC. Clarification of risk factors for abdominal operations in patients with hepatic cirrhosis. *Annals of Surgery* 1984; **199**: 648–55.

101. Alber I, Hartmann H, Bircher J, Creutzfeldt W. Superiority of the Child–Pugh classification to quantitative liver function tests for assessing prognosis of liver cirrhosis. *Scandinavian Journal of Gastroenterology* 1989; **234**: 269–76.

102. Pandele G, Chaux F, Salvadori C, Farinotti M, Duvaldestin P. Thiopental pharmacokinetics in patients with cirrhosis. *Anesthesiology* 1983; **59**: 123–6.

103. Gyasi HK, Naguib M. Atracurium and severe hepatic disease: a case report. *Canadian Anaesthetists' Society Journal* 1985; **32**: 161–4.

104. Hunter JM, Parker CJR, Bell CF, Jones RS, Utting JE. The use of different doses of vecuronium in patients with liver dysfunction. *British Journal of Anaesthesia* 1985; **57**: 758–64.

105. Haberer JP, Schoeffler P, Couderc E, Duvaldestin P. Fentanyl pharmacokinetics in anaesthetised patients with cirrhosis. *British Journal of Anaesthesia* 1982; **54**: 1267–9.

106. Kang YG, Uram M, Shiu GK, *et al*. Pharmacokinetics of fentanyl in end-stage liver disease. *Anesthesiology* 1984; **61**: 3A, A380.

107. Howland WS, Schweizer O, Gould P. A comparison of intraoperative measurements of coagulation. *Anesthesia and Analgesia* 1974; **53**: 657–63.

108. Zuckerman L, Cohen E, Vagher JP, Woodward E, Caprini JA. Comparison of thrombelastography with common coagulation tests. *Thrombosis and Haemostasis* 1981; **46**: 752–6.

109. Kang YG. Monitoring and treatment of coagulation. In: Winter PM, Kang YG, eds. *Hepatic transplantation. Anesthetic and perioperative management*. Westport CT: Praeger Publishers, 1986: 151–73.

110. Mallett SV, Cox DJA. Thromboelastography. *British Journal of Anaesthesia* 1992; **69**: 621–30.

111. Goodpasture EW. Fibrinolysis in chronic hepatic insufficiency. *Bulletin of the Johns Hopkins Hospital* 1914; **25**: 330–6.

112. Kang YG, Lewis JH, Navalgund A, *et al*. Epsilon-amino-caproic acid for treatment of fibrinolysis during liver transplantation. *Anesthesiology* 1987; **66**: 766–73.

113. Schipper HG, Ten Cate JW. Antithrombin III transfusion in patients with hepatic cirrhosis. *British Journal of Haematology* 1982; **52**: 25–33.

114. Thompson JN, Rotoli B. Endotoxins. The liver and haemostasis. In: Blumgart LH, ed. *Surgery of the liver and biliary tract*. London: Churchill Livingstone, 1988: 133–44.

115. Salzman EW, Weinstein MJ, Weintraub RM, *et al*. Treatment with desmopressin acetate to reduce blood loss after cardiac surgery. *New England Journal of Medicine* 1986; **314**: 1402–6.

116. Donaldson MDJ, Seaman MJ, Park GR. Massive blood transfusion. *British Journal of Anaesthesia* 1992; **69**: 621–30.

117. Starzl TE, Marchioro TL, Von Kaulla KN, *et al.* Homotransplantation of the liver in humans. *Surgery, Gynecology and Obstetrics* 1963; **117**: 659–76.

118. Calne RY, Williams R. Liver transplantation in man. 1. observations on technique and organisation in five cases. *British Medical Journal* 1968; **4**: 535–40.

119. Starzl TE, Klintmalm GBG, Porter KA, *et al.* Liver transplantation with use of cyclosporin A and prednisolone. *New England Journal of Medicine* 1981; **305**: 266–9.

120. Calne RY, Williams R, Lindop M, *et al.* Improved survival after orthotopic liver grafting. *British Medical Journal* 1981; **283**: 115–18.

121. Starzl TE, Marchioro TL, Porter KA. Progress in homotransplantation of the liver. [Review] *Advances in Surgery (Chicago)* 1966; **2**: 295–370.

122. National Institute of Health. Consensus Development Conference Statement: Liver transplantation *Hepatology* 1984; **4**, suppl: 107–10.

123. Van Thiel DH, Carr BI, Watsuki S, *et al.* Liver transplantation for alcoholic liver disease, viral hepatitis and hepatic neoplasms. *Transplantation Proceedings* 1991; **23**: 1917–21.

124. Whitten CW, Ramsay MAE, Paulsen AW, *et al.* Upper extremity neuropathy after orthotopic hepatic transplantation: a retrospective analysis. *Transplantation Proceedings* 1988; **20**: 628–9.

125. Bakti G, Fisch HU, Karlaganis G, *et al.* Mechanics of the excessive sedative response of cirrhotics to benzodiazepines: model experiments with triazolam. *Hepatology* 1987; **7**: 629–38.

126. Basile AS, Hughes RD, Harrison PM, *et al.* Elevated brain concentrations of 1,4-benzodiazepines in fulminant hepatic failure. *New England Journal of Medicine* 1991; **325**: 473–8.

127. Polson RJ, Park GR, Lindop MJ. The prevention of renal impairment in patients undergoing orthotopic liver grafting by infusion of low dose dopamine. *Anaesthesia* 1987; **42**: 15–19.

128. Arcas M, Pensado A, Gomez-Arnan J, *et al.* Renal failure in immediate post-operative phase after liver transplantation. In: *Proceedings of first symposium of the international society for perioperative care in liver transplantation*, 7th September, Pittsburgh, Pennsylvania. 1990.

129. Swygert TH, Roberts LC, Valek TR, *et al.* Effect of intra-operative low-dose dopamine on renal function in liver transplant recipients. *Anesthesiology* 1991; **75**: 571–6.

130. Lokhandwala MF. Renal actions of dopexamine hydrochloride. *Clinical Intensive Care* 1990; **1**: 163–74.

131. Paulsen AW, Whitten CW, Ramsay MAE, *et al.* Consideration for anesthetic management during veno-venous bypass in adult hepatic transplantation. *Anesthesia and Analgesia* 1989; **68**: 489–96.

132. Khoury GF, Kaufman RD, Musich JA. Hypothermia related to the use of veno-venous bypass during liver transplantation in man. *European Journal of Anaesthesiology* 1990; **7**: 501–3.

133. Ramsay MAE, Swygert TH. Anaesthesia for hepatic trauma, hepatic resection and liver transplantation. *Baillière's Clinical Anaesthesiology* 1992; **6** (4): 863–94.

134. Brown M, Gunning T, Roberts C, *et al.* Biochemical markers of renal perfusion are preserved during liver transplantation with veno-venous bypass. *Transplantation Proceedings* 1991; **23**: 1980–1.

135. Gunning TC, Brown MC, Swygert TH, *et al.* Peri-operative renal function in patients undergoing orthotopic liver transplantation. *Transplantation* 1991; **51**: 422–7.

136. Khoury GF, Mann ME, Porot MJ, Abdul-Rasool IH, Busuttil RW. Air embolism associated with veno-venous bypass during orthotopic liver transplantation. *Anesthesiology* 1987; **67**: 848–51.

137. Jamieson NV, Sundberg R, Lindell S, *et al.* Successful 24 to 30 hour preservation of the canine liver: a preliminary report. *Transplantation Proceedings* 1988; **20**: 945–7.

138. Wahlberg JA, Love R, Landegaard L, *et al.* Seventy-two hour preservation of the canine pancreas. *Transplantation* 1987; **43**: 5–8.

139. Cofer JB, Klinkmalm GB, Howard TK, *et al.* A comparison of UW with Eurocollins preservation solution in liver transplantation. *Transplantation* 1990; **49**: 1088–93.

140. Adkinson D, Hollworth ME, Benoit JN, *et al.* Role of free radicals in ischaemic reperfusion injury to the liver. *Acta Physiologica Scandinavica* 1986; **548**: suppl, 101–7.

141. Greig PD, Woolf GM, Abecassis M, *et al.* Prostaglandin E for primary non-function following liver transplantation. *Transplantation Proceedings* 1989; **21**: 3360–1.

142. Lichtor JL. Ventricular dysfunction does occur during liver transplantation. *Transplantation Proceedings* 1991; **23**: 1924–6.

143. Roberts LC, Duke PK, Gottlich CM, *et al.* Transesophageal echocardiography during orthotopic liver transplantation. *Anesthesiology* 1991; **75**: A388.

144. Paulsen AW, Brajtbord D, Klintmalm GB, *et al.* Intra-operative measurements related to subsequent hepatic graft failure. *Transplantation Proceedings* 1989; **21**: 2337–8.

145. Klintmalm GB, Olson LM, Nery JR, *et al.* Treatment of hepatic artery thrombosis after liver transplantation with immediate vascular reconstruction: a report of three cases. *Transplantation Proceedings* 1988; **20**: 610.

146. Park GR, Gomez-Arnan J, Lindop MJ, *et al.* Mortality during intensive care after orthotopic liver transplantation. *Anaesthesia* 1989; **44**: 959–63.

147. de-Groen PC. Cyclosporin: a review and its specific use in liver transplantation. *Mayo Clinic Proceedings* 1989; **64**: 680–9.

148. Maknuchi M. Ultrasonically guided liver surgery. *Japanese Journal of Ultrasonic Medicine* 1980; **7**: 45–9.

149. Grosse H, Pichlmayr R, Hansen B, *et al.* Specific anaesthetic problems in ex-situ resections of the liver. *Anaesthesia* 1990; **45**: 726–31.

150. Thompson HH, Thompkins RK, Longmire Jr WP. Major hepatic resections. A 25 year experience. *Annals of Surgery* 1983; **197**: 375–88.

151. Hardaway RM. Vietnam wound analysis. *Journal of Trauma* 1978; **18**: 635.

152. Beal SL. Fatal hepatic haemorrhage: an unresolved problem in the mangement of complex liver injuries. *Journal of Trauma* 1990; **30**: 163–9.

153. Royal College of Surgeons of England. *Report of the working party on the management of patients with major injuries*, London: RCSE, 1988.

154. Trey C, Davidson C. *The management of fulminant hepatic failure*. New York: Grune & Stratton, 1970: 282–98.

155. Bray G, Mowat C, Muir D, Tredger J, Williams R. The effect of chronic alcohol intake on prognosis and outcome in paracetamol overdose. *Human and Experimental Toxicology* 1991; **10**: 435–8.

156. O'Grady JG, Williams R. Management of acute hepatic failure. *Current Anaesthesia and Critical Care* 1990; **1**: 213–19.

157. O'Grady JG, Gimson AES, O'Brien CJ, *et al.* Controlled trials of charcoal haemoperfusion and prognostic factors in fulminant hepatic failure. *Gastroenterology* 1988; **94**: 1186–92.

158. Basile A, Hughes R, Harrison P, *et al.* Elevated brain concentrations of 1,4-benzodiazepines in fulminant hepatic failure. *New England Journal of Medicine* 1991; **325**: 473–8.

159. Mullen KD, Martin JV, Mendelson WB, *et al.* Could an endogenous benzodiazepine ligand contribute to hepatic encephalopathy? *Lancet* 1988; **1**: 457–9.

160. Forbes A, Alexander GJM, O'Grady JG, *et al.* Thiopental infusion in the treatment of intracranial hypertension complicating fulminant hepatic failure. *Hepatology* 1989; **10**: 306–10.

161. O'Grady JG, Langley PG, Isola LM, Aledort LM, Williams R. Coagulopathy of fulminant hepatic failure. *Seminars in Liver Disease* 1986; **6**: 159–63.

162. Bihari DJ, Gimson AES, Williams R. Cardiovascular, pulmonary and renal complications of fulminant hepatic failure. *Seminars in Liver Disease* 1986; **6**: 119–28.

163. O'Grady JG, Williams R. Management of acute liver failure. *Schweizerische Medizinische Wochenschrift* 1986; **116**: 541–4.

164. Davenport A, Will E, Davidson A, *et al.* Changes in intracranial pressure during machine and continuous haemofiltration. *Critical Care Medicine* 1989; **12**: 439–44.

165. Rolando N, Harvey F, Braham J, Fagan E, Williams R. Prospective study of bacterial infection in acute liver failure: an analysis of fifty patients. *Hepatology* 1990; **11**: 49–53.

166. Rolando N, Harvey F, Braham J, Fagan E, Williams R. Fungal infections: a common unrecognised complication of acute liver failure. *Hepatology* 1991; **12**: 1–9.

167. Prescott L, Illingworth R, Critchley J, Stewart M, Adam R, Proudfoot A. Intravenous *N*-acetyl cysteine: the treatment of choice for paracetamol poisoning. *British Medical Journal* 1979; **2**: 1097–1100.

168. Harrison PM, Keays R, Bray GP, Alexander GJM, Williams R. Improved outcome of paracetamol induced fulminant hepatic failure by late administration of acetyl cysteine. *Lancet* 1990; **1**: 1572–3.

169. Keays R, Harrison PM, Wendon JA, Gimson AES, Alexander GJM, Williams R. Intravenous acetyl cysteine and paracetamol induced fulminant hepatic failure: a prospective controlled trial. *British Medical Journal* 1991; **303**: 1026–9.

170. Vickers C, Neuberger J, Buckels J, McMaster P, Elias E. Transplantation of the liver in adults and children with fulminant hepatic failure. *Journal of Hepatology* 1988; **7**: 143–50.

171. Emond JC, Aran PP, Whitington PF, Broelsch CE, Baker AL. Liver transplantation in the management of fulminant hepatic failure. *Gastroenterology* 1989; **96**: 1583–8.

172. Schafer DF, Shaw BW. Fulminant hepatic failure and orthotopic liver transplantation. *Seminars in Liver Disease* 1989; **9**: 189–94.

173. Harrison TJ, Zuckerman JN, Zuckerman AJ. Viral hepatitis. *Baillère's Clinical Anaesthesiology* 1992; **6**, no 4: 795–817.

174. Alexander G. Treatment of acute and chronic viral hepatitis. *Baillère's Clinical Gastroenterology* 1989; **3**: 1–20.

175. Forbes A, Williams R. Increasing age – an important adverse prognostic factor in hepatitis A virus infection. *Journal of the Royal College of Physicians of London* 1988; **22**: 237–9.

176. Kuo G, Choo Q-L, Alter HJ, *et al.* An assay for circulating antibodies to a major etiologic virus of human non-A, non-B hepatitis. *Science* 1989; **244**: 362–4.

177. McFarlane IG, Smith HM, Johnson PJ, Bray GP, Vergani D, Williams R. Significance of possible anti-HCV antibodies in autoimmune chronic active hepatitis. *Lancet* 1994 (*in press*).

178. Gimson AES, O'Grady J, Ede RJ, *et al.* Late-onset hepatic failure: clinical, serological and histological features. *Hepatology* 1986; **6**: 288–94.

179. Smedile A, Farci P, Verme G, *et al.* Influence of delta infection on severity of hepatitis B. *Lancet* 1982; **2**: 945–7.

180. Govindarajan S, Chin KP, Redecker AG, Peters RL. Fulminant B viral hepatitis: role of delta agent. *Gastroenterology* 1984; **86**: 1417–20.

181. Association of Anaesthetists. *HIV and other blood-borne viruses*. London: Assoc. of Anaesthetists, 1992.

182. O'Donnell NG, Asbury AJ. The occupational hazard of immunodeficiency virus and hepatitis B infection. 1. Perceived risks and preventative measures adopted by anaesthetists: a postal survey. *Anaesthesia* 1992; **47**: 923–8.

183. UK Health Departments. *Guidelines for clinical health care workers: protection against HIV and hepatitis viruses*. London: HMSO, 1990.

184. Breuer J, Jeffries DJ. HIV and hepatitis B virus infection in health care workers: a risk to patients? *Reviews in Medical Microbiology* 1992; **3**: 1–8.

185. UK Health Departments. *AIDS – HIV-infected health care workers. Occupational guidance for health care workers, their physicians and employers*. Recommendations of the Expert Advisory Group on AIDS. London: Dept of Health, 1991.

186. Frink EJ, Brown BR. Postoperative hepatic dysfunction. *Ballière's Clinical Anaesthesiology* 1992; **6**, no 4: 931–52.

187. Hayes PC, Bouchiér IAD. Postoperative jaundice. *Ballière's Clinical Gasteroenterology* 1989; **3**: 485–505.

188. LaMont JT, Isselbacher K. Postoperative jaundice. In: Wright R, Alberti KGMM, Karran S, Millward-Sadler GH, eds. *Liver and biliary disease*, 2nd edn. London: Ballière Tindall/WB Saunders, 1984; 1367–77.

189. Evans C, Evans M, Pollock AV. The incidence and causes of postoperative jaundice. *British Journal of Anaesthesia* 1974; **46**: 520–5.

190. Chu C-M, Chang C-H, Liaw Y-F, Hsieh M-J. Jaundice after open heart surgery: a prospective study. *Thorax* 1984; **39**: 52–6.

191. Owens D, Evans J. Population studies on Gilbert's syndrome. *Journal of Medical Genetics* 1975; **12**: 152–6.

192. Valeri CR. Viability and function of preserved red cells. *New England Journal of Medicine* 1971; **284**: 81–6.

193. Solanski D, McCurdy PR. Delayed hemolytic transfusion

reactions: an often missed clinical entity. *Journal of the American Medical Association* 1978; **239**: 729–31.

194. Johnson LB. The importance of early diagnosis of acute acalculous cholecystitis. *Surgery, Gynecology and Obstetrics* 1987; **164**: 197–203.

195. LaMont JT, Isselbacher KJ. Postoperative jaundice. *New England Journal of Medicine* 1973; **288**: 305–8.

196. Banks JG, Foulis AK, Ledingham IMcA, McSween RNM. Liver function in septic shock. *Journal of Clinical Pathology* 1982; **35**: 1249–52.

197. Richardson PDI, Withrington PG. Alpha- and beta-adreno-ceptors in the hepatic portal vascular bed of the dog. *British Journal of Pharmacology* 1978; **62**: 376–7.

198. Dunn GD, Hayes P, Breenk S. The liver in congestive heart failure: a review. *American Journal of the Medical Sciences* 1973; **265**: 174–89.

199. Aach RD, Szmuness W, Mosley JW, et al. Serum alanine amino-transferase of donors in relation to the risk on non-A, non-B hepatitis in recipients. The transfusion-transmitted viruses study. *New England Journal of Medicine* 1981; **304**: 989–94.

200. Harville DP, Summerskill WH. Surgery in acute hepatitis, causes and effects. *Journal of the American Medical Association* 1963; **184**: 258–61.

201. Bass HM, Ockner RK. Drug-induced liver disease. In: Zakim D, Boyer TD, eds. *Hepatology – a textbook of liver disease*. Philadelphia: WB Saunders, 1990: 754–80.

202. Seeff LB, Cuccherini BA, Zimmerman JH, et al. Acetamino-phen hepatotoxicity in alcoholics. A therapeutic misadven-ture. *Annals of Internal Medicine* 1986; **104**: 399–404.

203. Brown Jr BR. Hepatitis. In: Brown Jr BR, ed. *Anesthesia in hepatic and biliary tract disease*, Philadelphia: FA Davis, 1988: 189, 197–8.

204. Neuberger JM. Halothane and the implications of hepatitis. In: Kaufman L, ed. *Anaesthetic review, 8*. London: Churchill Livingstone, 1991: 179–94.

205. Ray DC, Drummond GB. Halothane hepatitis. *British Journal of Anaesthesia* 1991; **67**: 84–99.

206. Brown Jr BR, Gandolfi AJ. Adverse effects of volatile anaesthetics. *British Journal of Anaesthesia* 1987; **59**: 14–23.

207. Bunker JP, Forrest WH, Mosteller F, Vandam LD, eds. *National Halothane Study: a study of the possible association between halothane anesthesia and postoperative hepatic necrosis*. Washington DC: US Government Printing Office, 1969.

208. Böttiger LE, Dalén E, Hallén B. Halothane-induced liver damage: an analysis of the material reported to the Swedish Adverse Drug Reaction Committee 1966–73. *Acta Anaesthe-siologica Scandinavica* 1976; **20**: 40–6.

209. Inman WHW, Mushin WW. Jaundice after repeated exposure to halothane: an analysis of reports to Committee on Safety of Medicines. *British Medical Journal* 1974; **1**: 5–10.

210. Inman WHW, Mushin WW. Jaundice after repeated exposure to halothane: a further analysis of reports to Committee on Safety of Medicines. *British Medical Journal* 1978; **2**: 1455–6.

211. Walton B, Simpson BR, Strunin L, Doniach D, Perrin J, Appleyard AJ. Unexplained hepatitis following halothane. *British Medical Journal* 1976; **1**: 1171–6.

212. Neuberger J, Williams R. Halothane anaesthesia and liver damage. *British Medical Journal* 1984; **289**: 1136–9.

213. Farrell GC. Mechanism of halothane induced liver injury. *Journal of Gastroenterology and Hepatology* 1988; **3**: 465–82.

214. Allen PJ, Downing JW. A prospective study of hepatocellular function after repeated exposures to halothane or enflurane in women undergoing radium therapy for cervical cancer. *British Journal of Anaesthesia* 1977; **49**: 1035–9.

215. McEwan J. Liver function tests following anaesthesia. *British Journal of Anaesthesia* 1976; **48**: 1065–70.

216. Fee JPH, Black GW, Dundee JW, et al. A prospective study of liver enzyme and other changes following repeat adminis-tration of halothane and enflurane. *British Journal of Anaesthesia* 1979; **51**: 1133–41.

217. Trowell J, Peto R, Crampton-Smith A. Controlled trial of repeated halothane anaesthetics in patients with carcinoma of the cervix treated with radium. *Lancet* 1975; **1**: 821–4.

218. Wright R, Chisholm M, Lloyd B, et al. A controlled prospective study of the effect on liver function of multiple exposures to halothane. *Lancet* 1975; **1**: 817–20.

219. Stock JGL, Strunin L. Unexplained hepatitis following halothane. *Anesthesiology* 1985; **63**: 424–39.

220. Allan LG, Hussey AJ, Howie J, et al. Hepatic glutathione S-transferase release after halothane anaesthesia: open randomised comparison with isoflurane. *Lancet* 1987; **1**: 771–4.

221. Hussey AJ, Aldridge LM, Paul D, Ray DC, Beckett GJ, Allan LG. Plasma glutathione S-transferase concentration as a measure of hepatocellular integrity following a single general anaesthetic with halothane, enflurane or isoflurane. *British Journal of Anaesthesia* 1988; **60**: 130–5.

222. Nomura F, Hatano H, Ohnishi K, Akikusa B, Okuda K. Effects of anticonvulsant agents on halothane-induced liver injury in human subjects and experimental animals. *Hepatology* 1986 **6**: 952–6.

223. Kenna JG, Neuberger J, Williams R. Specific antibodies to halothane-induced liver antigens in halothane-associated hepatitis. *British Journal of Anaesthesia* 1987; **59**: 1286–90.

224. Varma RR, Whitesell RC, Iskandarani MM. Halothane hepatitis without halothane: role of inapparent circuit contamination and its prevention. *Hepatology* 1985; **5**: 1159–62.

225. Kenna JG, Neuberger J, Williams R. Identification by immunoblotting of three halothane-induced liver microsomal polypeptide antigens recognised by antibodies in sera from patients with halothane-associated hepatitis. *Journal of Pharmacology and Experimental Therapeutics* 1987; **242**: 733–40.

226. Satoh H, Gillette JR, Davies HW, Schulick RD, Pohl LR. Immunochemical evidence of trifluoroacetylated cytochrome P-450 in the liver of halothane-treated rats. *Molecular Pharmacology* 1985; **28**: 468–74.

227. Satoh H, Martin BM, Schulick AH, Christ DD, Kenna JG, Pohl LR. Human anti-endoplasmic reticulum antibodies in sera of patients with halothane-induced hepatitis are directed against a trifluoroacetylated carboxylesterase. *Proceedings of the National Academy of Sciences of the USA, Medical Sciences* 1989; **86**: 322–6.

228. Bird GLA, Williams R. Detection of antibodies to a halothane

metabolite hapten in sera from patients with halothane associated hepatitis. *Journal of Hepatology* 1989; **9**: 366–73.

229. Martin JL, Kenna JG, Martin BM, Thomassen D, Reed GF, Pohl LR. Halothane hepatitis patients have serum antibodies that react with protein disulphide isomerase. *Hepatology* 1993; **18**: 858–63.

230. Stoelting RK, Blitt CD, Cohen PJ, Merin RG. Hepatic dysfunction after isoflurane anaesthesia. *Anesthesia and Analgesia* 1987; **66**: 147–53.

231. Stoelting RK. Isoflurane and postoperative hepatic dysfunction. *Canadian Journal of Anaesthesia* 1987; **34**: 223–6.

232. Lunzer MR. Jaundice in pregnancy. *Baillière's Clinical Gastroenterology* 1989; **3**: 467–83.

233. Schnider SM. Serum cholinesterase activity during pregnancy, labour and puerperium. *Anesthesiology* 1965; **26**: 335–9.

234. Blitt CD, Petty WC, Albertenst EE, Wright BJ. Correlation of plasma cholinesterase activity and duration of action of succinylcholine during pregnancy. *Anesthesia and Analgesia* 1977; **56**: 78–83.

235. Lunzer M, Barnes P, Blyth J, O'Halloran M. Serum bile acid concentrations during pregnancy and their relationships to obstetric cholestasis. *Gastroenterology* 1986; **91**: 825–9.

236. Kaplan MM. Acute fatty liver of pregnancy. *New England Journal of Medicine* 1985; **313**: 367–70.

237. Weinstein L. Syndrome of hemolysis, elevated liver enzymes and low platelet count: a severe consequence of hypertension in pregnancy. *American Journal of Obstetrics and Gynecology* 1982; **142**: 159–67.

238. MacKenna J, Dover NL, Brame RG. Pre-eclampsia associated with haemolysis, elevated liver enzymes and low platelets – an obstetric emergency? *Gynaecology* 1983; **62**: 751–4.

Chapter 38

The Geriatric Patient

Stanley Muravchick

Concepts of ageing and geriatrics
Ageing and organ function
Cardiopulmonary function
Hepatorenal function
Metabolism and body composition

Central nervous system
Peripheral nervous system
Autonomic nervous system
Analgesic and anaesthetic requirement
Anaesthetic management and outcome

In the twentieth century, advances in nutrition, public health, education and social services have produced major changes in human longevity. Life expectancy for adult males has increased by more than 30 years in just a few generations. Therefore, unless intentionally limited to paediatrics or obstetrical patients, every anaesthetist in contemporary practice eventually becomes a subspecialist in geriatric medicine. Current demographic analyses of industrialized societies confirm that one-quarter or more of surgical patient populations are 65 years of age or older, with even greater representation anticipated in the next two decades. The sections that follow define the current concepts of ageing that are relevant to anaesthetic practice, discuss the distinction between ageing and age-related disease, and strategies for perioperative assessment of the elderly patient, and review some practical aspects of anaesthetic management and outcome in geriatric surgical patients.

Concepts of ageing and geriatrics

Clinical experience amply demonstrates that the elderly are neither a medically nor a physically homogenous patient subpopulation. In fact, increased interpatient *variability* has long been recognized as an important characteristic of geriatric medicine, perhaps more so than for any other medical specialty. As they age, adult patients exhibit an increasingly complex array of physical responses to environmental and socioeconomic conditions and to concurrent disease states, as well as the effects of traumatic injury. Longevity also enables the complete expression of intrinsic genetic qualities in all their subtle manifestations.

Therefore, establishing a rigid and finite chronological definition of the term 'geriatric' has little value other than for administrative, actuarial or epidemiological applications.

Discrete and quantifiable 'biological markers of ageing' such as visual acuity or the racemization of structrual amino acids in connective tissue have been useful, but, thus far, only for highly specialized non-clinical applications such as evaluation of 'anti-ageing' therapies administered to laboratory animals. Consequently, there is no consensus as to whether the 'geriatric' era has a discrete beginning, or whether any one physiological marker can identify a patient as 'elderly'. For clarity and consistency within this chapter, the term 'elderly' and 'geriatric' are used synonymously to describe human subjects who, by arbitrary convention, are simply 65 or more years of age. Because clinical experience suggests that there is some predictive value in subdividing elderly patients into chronological subgroups, the author reserves the term 'aged' for individuals over 80 years of age.

Life span is an idealized, species-specific biological parameter that quantifies maximum attainable age under optimal environmental conditions. Historical anecdote suggests that human life span has remained constant at 110–115 years for at least the past 20 centuries.[1] In contrast, the term *life expectancy* describes an empirical estimate of typical longevity under prevailing societal circumstances; it represents a biological version of the 'mean time before failure' (MTBF) used to describe the reliability of computer hardware. Advances in medical science and health care have improved life expectancy dramatically in industrialized societies and increased their relative 'agedness' to the point that the economics and the politics of ageing have, for the first time in history, assumed roles of major importance.

Studies of human ageing are often difficult to interpret. *Cross-sectional* studies measure physiological parameters simultaneously in young and in elderly subjects. Although straightforward in design and in execution, the cross-sectional approach will not exclude patients with subtle or preclinical manifestations of disease. Changes due to *age-related disease* may be erroneously attributed to age itself. Similarly, this experimental design cannot be controlled for cohort-specific factors such as nutritional and environmental history, for individual physical activity level, genetic background or prior exposure to infectious agents. In effect, the two patient groups being compared may differ not only in age but also in terms of their overall physiological, anatomical and biochemical characteristics. Therefore, data from cross-sectional studies rarely permit unambiguous conclusions regarding the effect of age itself on the measured parameter.

In contrast, *longitudinal* studies of ageing require the investigators to obtain repeated measurements over several decades in an ageing population. In effect, each subject becomes his own control for comparison with subsequent measurements made when he is older. If any of the subjects in a longitudinal study eventually manifest signs of age-related disease, they can be excluded from the study, thereby leaving behind a smaller study group which can be considered free of defined disease. Although difficult and expensive to organize, longitudinal studies have already produced enough data to make it clear that much of what was once assumed to be age-related impairment of organ function actually reflects *disease*, and not ageing itself.

The details of experimental studies attempting to demonstrate the effects of ageing are now also known to be of great importance. The published conclusions of many 'classic' cross-sectional studies of ageing are simply no longer valid. The gerontological literature must be reviewed and reinterpreted to make clear distinctions between ageing *per se* and age-related disease. Consequently, some of the statements and conclusions presented in this chapter represent substantial revision of the 'conventional wisdom' taught only a few years ago.

Despite considerable efforts, the mechanisms that control the ageing process remain unknown. The impressive consistency of observed life span may not, as was once believed, necessarily imply a genetic or centrally coordinated, hypothalamic 'biological clock' for each species.[2] Age-related decline in organ and tissue function may simply be the inevitable accumulation of non-specific, degenerative phenomena such as ionizing radiation.[3] These changes may just have become more obvious and more prevalent as a greater proportion of patients survive diseases and injuries that, only a few decades ago, were lethal. From this perspective, the unique life span demonstrated by each species could be explained solely as a complex interaction between environment and genetically determined biochemical and anatomical attributes. Is ageing a carefully programmed physiological process, or simply the reflection of the wear and tear we

experience during a lifetime? This controversy has yet to be resolved.

Ageing and organ function

For the reasons given above, the only currently valid definitions of ageing are conceptual rather than quantitative. Ageing appears to be a universal and progressive physiological phenomenon characterized by degenerative changes in both the structure and the function of organs and tissues. At one time, the functional consequences of ageing were represented graphically as linear declines of pulmonary vital capacity, or oxygen consumption, the decrement from maximum function beginning in young adulthood and continuing inexorably downward thereafter. However, more current concepts describe a complex, non-linear process which is first apparent following the peak of somatic maturation in the fourth decade of human life. Decrements of function in most organs are now believed to be relatively subtle during the middle adult years, but become progressively more dramatic during the traditional years of geriatric senescence, in the seventh decade of life and beyond.

The competence of integrated organ system function varies considerably from one elderly patient to the next, even in the absence of disease. Elderly patients who maintain greater than average functional capacities are said to be 'physiologically young'; when function declines at an earlier age, or at a more rapid rate, elderly patients can be considered to be 'physiologically old' (Fig. 38.1). In healthy geriatric patients, maximum organ system function at all ages is greater than the basal level of activity, at rest. The difference between this maximum capacity and basal levels of function is organ system *functional reserve* (Fig. 38.2), a

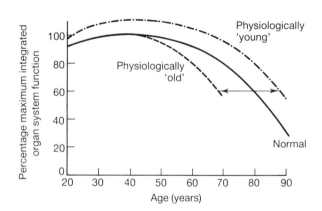

Fig. 38.1 Variability in the rate at which organ system function changes with increasing age explains the presentation of patients as physiologically 'young' or 'old'.

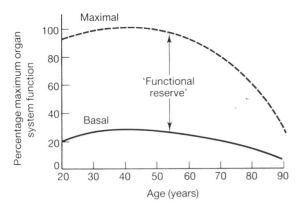

Fig. 38.2 Organ system 'functional reserve' represents the difference between resting basal and maximal organ system function. Organ system functional reserve declines markedly within increasing age.

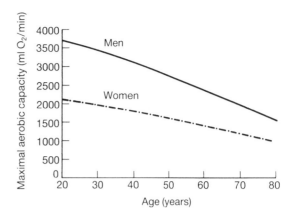

Fig. 38.3 Maximal aerobic capacity, expressed as ml O_2 per minute to reflect changes in total body weight, declines with increasing age in both men and women.[110]

'safety margin' of organ function which is available to meet the additional demands for cardiac output, carbon dioxide excretion or protein synthesis imposed upon the patient by trauma or disease or by surgery and convalescence. Cardiopulmonary functional reserve can be quantified and assessed clinically using various exercise or stress tests. However, there is, at present, no comparable approach to assessment of renal, hepatic, immune or nervous system functional reserve. Consequently, it has simply been assumed that the functional reserve of these organ systems is progressively and significantly reduced in elderly patients. Presumably, this is the mechanism that increases the susceptibility of elderly patients to stress- and disease-induced organ system decompensation.

Cardiopulmonary function

Classic data suggesting that ageing produces an irreversible depression of cardiac output are contradicted by more recent studies of fit and active elderly subjects in whom demand for cardiopulmonary function is maintained through daily exercise.[4] In young and elderly alike, therefore, integrated cardiopulmonary function and functional reserve increase in response to vigorous physical activity and to variations in overall metabolic demand. Consequently, the modest decrease in resting cardiac index observed in healthy elderly subjects is not evidence of degenerative change, but rather represents an appropriate response to the lessened perfusion and reduced metabolic requirements associated with age-related atrophy of skeletal muscle and the loss of tissue mass in many major organs with high intrinsic metabolic rates.[5] By reducing

lean body mass, ageing simply produces a progressively smaller aerobic machine (Fig. 38.3).

Under conditions of submaximal demand, rates of myocardial shortening and ventricular pressure generation, two indices of myocardial contractility, appear to remain uncompromised by increasing age, at least until the eighth decade. Ageing does, however, impose significant limitations upon maximal heart rate and maximal cardiac output achieved during physical exercise. It also reduces the cardiac end-organ response to intrinsic adrenergic stimulation and to inotropic drugs, particularly β-agonists.[6]

Short-term increases in cardiac output in the elderly patient are accomplished largely by increased left ventricular diastolic volume and augmented ventricular preload.[4] These mechanical compensatory mechanisms predominate over changes in heart rate mediated by baroreflex activity, and there is little enhancement of ejection fraction.[7] However, a stiffer, less compliant ventricular and atrial myocardium also implies a narrow margin between the extremes of inadequate filling pressure and volume overload. Timely and aggressive management of the rapid variations in central blood volume which occur commonly during anaesthesia and surgery therefore becomes even more critical in elderly patients. Relatively small decreases in venous return such as those produced by positive pressure ventilation, surgical haemorrhage or venodilator drugs may significantly compromise stroke volume; with limited reflex-mediated ability to increase heart rate,[8] arterial hypotension is predictably frequent and severe.[9] Conversely, rates of intravenous fluid administration that would be modest for young adults may, in the geriatric patient, produce increases in atrial and pulmonary artery pressure sufficiently large to disrupt the balance of forces that control lung water, precipitating congestive heart failure and pulmonary oedema.

Virtually all longitudinal studies confirm the universal nature of progressive, gender-independent arterial hypertension, a phenomenon due to fibrotic replacement of elastic tissues within the walls of arteries. The ability of the aorta and large arteries to store hydraulic energy is compromised, increasing the vascular impedance to cardiac ejection and raising cardiac workload, ultimately producing symmetrical hypertrophy of the left ventricular wall. For any given stroke volume, loss of compliance increases arterial pulse pressure, particularly the systolic pressure component. In fact, the large stroke volumes characteristic of ageing hearts further amplifies this phenomenon to produce the familiar 'overshoot' characteristics of radial artery waveform tracings in geriatric patients, with large discrepancies between blood pressure values obtained by invasive techniques and those measured by more traditional occlusive cuff techniques. Nevertheless, in the absence of disease, neither intrinsic myocardial contractility nor integrated cardiac function limits cardiac output at rest or during moderate exercise. Although elderly patients clearly have some compromise of maximal cardiac output and aerobic capacity, there remain many aged individuals who can still compete successfully in a variety of strenuous athletic events. Ageing itself, therefore, should not be considered to be a major determinant of cardiac function in the clinical setting.

Age-related *loss of tissue elasticity* appears to be rather ubiquitous, involving the lung as well as the cardiovascular system. Loss of lung elastic recoil is a primary anatomical mechanism by which ageing exerts deleterious effects on pulmonary gas exchange.[10] With increasing age, lung elastin content declines, and there is a proportionate increase in fibrous connective tissue within the lung parenchyma. As a result, elderly individuals experience virtually inevitable emphysema-like increases in lung compliance. Calcification and stiffening of the thorax reduces chest wall compliance, however, so total pulmonary compliance is little changed, and the loss of lung recoil does not cause a dramatic increase in functional residual capacity (FRC), the volume of the lung at rest.[11] However, vital capacity is significantly and progressively compromised because residual volume is increased at the expense of inspiratory and expiratory reserve volumes. In elderly subjects, there may actually be small airway closure at lung volume close to FRC as the patency of small airways is compromised by loss of normal tethering effects.[12]

Age-related breakdown of alveolar septa reduces total alveolar surface area, limiting gas exchange and progressively increasing both anatomical and alveolar dead space. These changes in the physical properties of the lungs are non-uniform, so they also severely impair the normal matching of ventilation and perfusion, increasing physiological shunting and thereby reducing the efficiency of oxygenation. Decline in arterial oxygen saturation and partial pressure is significant and progressive, pre- and postoperatively (Fig. 38.4), and may be further exacerbated during general anaesthesia by increases in ventilation/

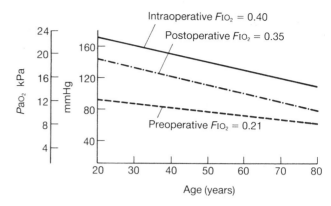

Fig. 38.4 Preoperative, intraoperative and postoperative arterial oxygenation decrease progressively with increasing age despite supplementation of inspired oxygen. Ageing reduces the safety margin between ambient and minimally acceptable levels of arterial oxygen saturation in geriatric surgical patients.[111]

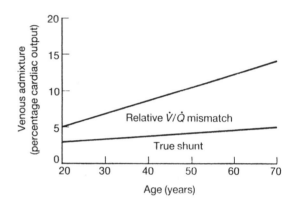

Fig. 38.5 During general anaesthesia, both true intrapulmonary shunting and relative $\dot{V}/\dot{Q}$ mismatch increase in patients in direct proportion to age. (Data courtesy of Dr G. Hedenstierna.)

perfusion mismatch (Fig. 38.5), perhaps because many anaesthetics depress active hypoxic pulmonary vasoconstriction. Consequently, virtually all surgical patients 70 or more years of age require perioperative supplemental oxygen if gross hypoxaemia following sedation, diagnostic procedures or surgical intervention is to be avoided.

Ageing significantly compromises ventilatory mechanics. The costochondral calcification which makes the thorax more rigid and decreases the compliance of the chest wall also increases the work of breathing. Total pulmonary compliance is reduced only insignificantly, but limited thoracic flexibility and increased airway resistance associated with partial obstruction of small airways produce a modest increase in the total work of breathing of elderly subjects.

The moment-to-moment control of ventilation and the response to changes in pH and respiratory gases appear to be unchanged in healthy and alert elderly subjects. However, the cardiovascular and the ventilatory stimulation normally mediated by reflex mechanisms in response to imposed hypoxia or hypercapnia are delayed in onset, and of considerably smaller magnitude, in geriatric patients.[13] Elderly subjects also experience a higher incidence of transient apnoea and episodic respiration when given narcotics, for reasons that are not completely understood, and may be more sensitive to the respiratory depression produced by non-narcotic drugs such as the benzodiazepines. Narcotic-induced rigidity of the chest wall occurs more frequently in older than in younger adults.[14] Protective laryngeal reflexes require a greater threshold stimulus magnitude,[15] increasing the risk of aspiration pulmonary injury in older patients, especially if level of consciousness is depressed. Consequently, geriatric surgical patients are clearly at greater risk of unrecognized respiratory failure in the typical postoperative setting of residual anaesthetics and pain therapy. Supplemental inspired oxygen, enhanced monitoring and continued observation of the adequacy of oxygenation and ventilation are essential in this surgical subpopulation.

Hepatorenal function

Microsomal and non-microsomal enzymatic activities, expressed per gram of hepatic tissue, are virtually the same in biopsy specimens taken from elderly and from young adults.[16] Therefore, at least on a biochemical level, ageing produces little qualitative change in hepatocellular function. However, in clinical studies, elderly women appear to metabolize benzodiazepines at rates close to that of their younger counterparts, yet elderly men do not,[17] suggesting that this organ system may undergo subtle physiological changes that are both age- and gender-specific. Elderly men, but not women, also are frequently found to have significant reductions in the plasma activity of the enzyme cholinesterase,[18] and therefore may require a smaller dose of suxamethonium (succinylcholine) than their younger counterparts in order to achieve an equal effect or a comparable duration of effect.

Ageing depresses smoking-induced enhancement of hepatic microsomal enzyme activity.[19] Hepatic metabolism and drug biotransformation may also be significantly altered in this patient subpopulation by their sustained exposure to the polypharmacy used to treat age-related disease. Cimetidine, widely used in geriatric patients to relieve gastritis that occurs because of age-impaired secretion of protective gastric mucus, further depresses the hepatic biotransformation of benzodiazepines; the metabolism of terfenadine and astemizole is impaired with reduced hepatic function, increasing the risk of polymorphous ventricular tachycardia in the elderly

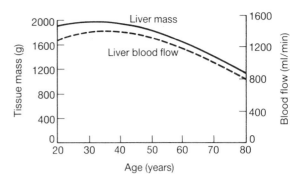

Fig. 38.6 Both liver tissue mass and blood flow fall at an accelerating rate with increasing age.

patient exposed to erythromycin, and perhaps to other drugs as well.

Although plasma concentrations of transaminase and other hepatocyte-derived enzymes are comparable to those of young adults,[20] the bromsulphothalein (BSP; sulphobromophthalein) retention test used as a general index of hepatic function is prolonged with increasing age, approaching the upper limit of normal in the seventh decade of life in individuals who demonstrate no other evidence of hepatic dysfunction.[21] Although adequate to meet the requirements for normal coagulation, hepatic capacity for protein synthesis is nevertheless significantly reduced by the eighth decade of life.

Age-related changes in hepatic and splanchnic anatomy are far less subtle than the biochemical alterations described above. Liver tissue mass declines about 40 per cent by the age of 80 years,[22] and splanchnic and hepatic blood flow are reduced proportionally (Fig. 38.6). Liver blood flow is compromised both in absolute terms and also as a percentage of cardiac output; therefore, age, in effect, produces a redistribution of blood flow away from the splanchnic vascular bed. In fact, it is these *quantitative losses of hepatic mass and blood flow*, and not the qualitative impairment of liver function, that are primarily responsible for altered perioperative hepatic function of clinical significance to the anaesthetist.[23] Hepatic biotransformation and protein synthesis, adequate for modest basal demand, may easily be overwhelmed by the circumstances of maximal stress, disease and surgical intervention, especially if associated with arterial hypotension, low cardiac output, hypothermia or any form of direct hepatic injury. Similarly, age-related loss of well-perfused hepatic tissue mass is generally thought to be responsible for the significantly delayed elimination and subsequent prolonged clinical effects of the narcotics, anaesthetic agents and adjuvants for which hepatic biotransformation is needed for ultimate pharmacokinetic disposition (Table 38.1).

Age-related organ atrophy is also clearly evident in anatomical and postmortem observations of the kidneys. About 30 per cent of the maximum young adult value for bilateral renal tissue mass is lost by the eighth decade (Fig.

Table 38.1 Adjustments of anaesthetic and adjuvant drugs in elderly patients

Drug group	Adjustment needed
Potent inhalational agents	Decrease inspired concentration; allow more time for emergence
Barbiturates, etomidate, propofol, Althesin	Small to moderate decrease in initial dose; smaller maintenance doses; reduced infusion rate
Narcotics	Marked decrease in initial dose; anticipate increased duration of systemic and epidural effects, greater incidence of rigidity
Local anaesthetics for spinal/epidural anaesthesia	Small to moderate decrease in segmental dose requirement; anticipate prolonged effects
Benzodiazepines	Modest decrease in initial dose; anticipate marked increase in duration (except midazolam)
Suxamethonium (succinylcholine)	Slightly reduced dose in elderly men
Non-depolarizing relaxants	Same or slight increase in initial dose; anticipate increased duration (except atracurium or mivacurium)
Neostigmine, edrophonium	No change in dose or efficacy; slightly prolonged effect
Atropine	Increased dose for equal heart rate response; anticipate central anticholinergic syndrome
Adrenaline (epinephrine), isoprenaline (isoproterenol), adrenergic agonists; β-antagonists	Increased doses for equal cardiovascular responses

38.7), declining from 270 to 185 g of tissue. Cellular attrition is especially marked in the renal cortex, although the extent of parenchymal loss is masked to some extent by a reciprocal increase in renal fat and by a diffuse and generalized process of interstitial fibrosis.[24] The effects of ageing on the renal microarchitecture is even more dramatic: more than one-third of glomeruli and their associated nephron tubular structures will disappear by the age of 80 years.[25] In 10–20 per cent of the glomeruli that remain, a process of sclerosis impairs effective filtration by producing dysfunctional continuity between afferent and efferent glomerular arterioles. Tubular diverticula are common, and permeability to water may be increased, compromising urinary concentrating ability.[26]

Equally dramatic is the effect of age on renal perfusion. Total renal blood flow falls almost 50 per cent, a decline of about 10 per cent per decade beginning in early adulthood. The renal cortex appears to be particularly compromised by a progressive reduction of tissue vascularity, but there is relative sparing of the renal medulla. Both renal plasma flow (RPF) and, therefore, glomerular filtration rate (GFR) decline more rapidly than would be expected from the loss of renal tissue mass alone. In elderly subjects, GFR is reduced less than RPF because the shifting of renal perfusion from cortex to medulla appears to produce a slight compensatory increase in filtration fraction.[27]

Despite these marked reductions of absolute rates of perfusion and filtration, serum creatinine concentrations, a common marker used to assess renal function, remain

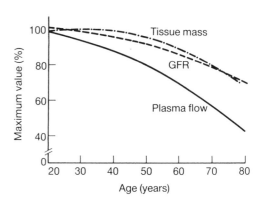

Fig. 38.7 Renal plasma flow declines more rapidly than does renal tissue mass with increasing age. Glomerular filtration rate (GFR) falls somewhat more slowly than plasma flow in those individuals in whom filtration fraction is increased.

within the normal range in elderly patients because of a decline in skeletal muscle mass, a change in body morphology that is particularly evident in men. The loss of skeletal muscle generates a progressively reduced creatinine load which requires less renal function to maintain metabolic equilibrium. Basal renal function is sufficient to avoid uraemia and to maintain normal plasma osmolarity and electrolyte concentrations. However, renal functional reserve may be inadequate to withstand gross disruptions of water and electrolyte balance or to respond promptly or appropriately to major contraction or expansion of intravascular volume.

Elderly patients have an exaggerated release of

antidiuretic hormone (ADH) following a hypertonic saline load,[28] but demonstrate a reduced end-organ response to this hormone, as well as a decreased maximal reabsorption rate for glucose and significant impairment of the ability to conserve sodium or to concentrate the urine when dehydrated. Excretion of a free water load is also markedly delayed. Diminished thirst, poor diet and the use of diuretic agents to decrease age-related hypertension also make intravascular and intracellular dehydration a more common finding in the preoperative examination of elderly patients. Geriatric surgical patients do not appear to require a unique fluid replacement protocol, but their limited renal functional reserve requires meticulous calculation and monitoring of fluid and electrolyte balance.

Reduced absolute renal blood flow and loss of the total available mass of nephron units in the elderly delay drug clearance and prolong the clinical effects of the many injectable drugs used perioperatively which require final elimination by renal pathways (see Table 38.1). The renal effects of ageing therefore require a suitable adjustment of drug choice, dose and timing in geriatric surgical patients. In addition, the reduced intrinsic tissue vascularity of this surgical subpopulation may expose them to an increased risk of renal ischaemia if cardiac output is depressed, especially geriatric patients taking cyclo-oxygenase inhibitors such as non-steroidal anti-inflammatory agents, drugs that reduce the vasodilator activity of renal prostacyclin. Acute renal failure is responsible for at least one-fifth of perioperative mortality in elderly surgical patients.[29]

Metabolism and body composition

Ageing produces a progressive and generalized loss of skeletal muscle mass (especially in men) and selective atrophy within the most metabolically active areas in brain, liver and kidney, as well as significant, but extremely variable, increases in the lipid fraction of total body mass. These changes in body composition reduce the basal metabolic requirements for ageing patients by 10–15 per cent compared to their young adult counterparts.[30] The reduction in rate of body heat production and simultaneous age-related impairment of the thermosensitivity and efficiency of autonomic thermoregulation increase the risk of inadvertent intraoperative hypothermia: decreases in core body temperature average almost 1°C per hour, about twice as great as observed in young adults under comparable circumstances.[31] In addition, the time needed for spontaneous rewarming postoperatively appears to increase in direct proportion to the patient's age.[32]

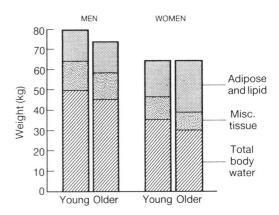

Fig. 38.8 Age-related changes in body composition are gender specific. Increases in body lipid offset bone loss and intracellular dehydration in women, whilst accelerated loss of skeletal muscle mass produces a modest drop in total body weight and intracellular water in men.[34]

Loss of skeletal muscle mass is also associated with progressive impairment of the ability to handle an intravenous glucose challenge, in large part because this tissue and liver normally provide storage for carbohydrates. However, age-related glucose intolerance[33] may also reflect, at least in part, a progressive impairment of insulin function or subtle antagonism of its effect on these target tissues, since both the timing and the magnitude of insulin release itself appear to remain normal. In any case, intravenous fluid replacement using glucose-containing solutions should be limited to environments that permit frequent measurement of blood sugar levels in elderly patients.

The change in total body mass seen during and following the young adult years is biphasic, with the well-known increases in body mass commonly described as 'middle age spread'. However, in aged men, accelerated *loss of muscle and central organ atrophy* produce a significant fall in body weight. From young adulthood on, most men initially gain about 12 kg of adipose tissue while they lose about 8 kg of skeletal muscle mass, a change that may result in the loss of as much as 15 per cent of their total body water (Fig. 38.8). In women, dehydration and bone loss due to osteoporosis are largely offset by increasing body fat. Contrary to classic assumptions, however, virtually all of the age-related changes in body water remain limited to intracellular compartments. Plasma volume, red cell mass and extracellular fluid volumes are well maintained in non-hypertensive elderly individuals with adequate levels of daily physical activity.[34]

These age-related changes in body composition are universal, progressive and relatively irreversible, although the rate at which skeletal muscle mass is lost may be minimized by vigorous exercise, and the progressive increase in adiposity tempered by changes in diet or by maintenance of high caloric (energy) expenditure. Nevertheless, many of these physiological and anatomical changes reflect the unavoidable consequences of reduced testosterone, thyroid hormone and other endogenous modulators of tissue metabolism at least as much as they reflect individual life style and the social environment. Therefore, as ageing increases the lipid fraction of body mass relative to the proportion of aqueous, well-perfused body tissues, it also enlarges the distribution volumes, expressed per kilogram of body mass, which function as reservoirs for anaesthetic and for other drug molecules that are preferentially soluble in lipid.[35–37] Consequently, age-related changes in body composition have a significant impact on the pharmacokinetic processes that occur during anaesthesia, and may delay emergence from anaesthesia to a degree greater than that which would be expected solely from reduced rates of hepatic and renal drug elimination.[38]

Central nervous system

Many age-related changes in both the structure and the function of the human brain and nervous system are well described, yet their clinical significance and their inter-relationship remain unclear. Some of the ambiguity simply reflects inability to distinguish clearly between ageing and age-related disease. Neurofibrillary tangles, for example, were once considered to be non-specific stigmata of the ageing process within the nervous system. However, they are now known to be the microscopic sequelae of a pathological cerebrocortical acetylcholine deficiency, which is clinically apparent as Alzheimer's disease. Other 'senile' forms of neurological dysfunction, in particular those with a neurohumoral basis, may actually turn out to be age-related diseases rather than the expression of inevitable generic characteristics of an ageing nervous system.

There are some unequivocal hallmarks of ageing within this organ system, however. Ageing reduces brain size: average adult brain weight is about 20 per cent less by the age of 80 years than respective values measured *post mortem* in young adults. The fraction of intracranial volume occupied by brain tissue falls from 92 to 82 per cent over the same time period, the most rapid reduction in brain mass and the greatest rate of compensatory increase in cerebrospinal fluid occurring after the sixth decade (Fig. 38.9). Ageing, in effect, produces a low pressure hydrocephalus.

Most of this loss of nervous system tissue reflects attrition of neurons, particularly in the grey matter, and not atrophy of the supportive, non-neuronal glial cells which normally constitute almost half of total brain mass.[39] The oft-cited average rate of neuronal cell death, 50 000 per day from an initial pool of about 10 billion,[40] is probably a misleading generalization because neuronal loss is highly

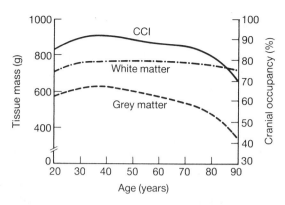

Fig. 38.9 Ageing is associated with marked reductions in grey matter tissue mass within the central nervous system, producing a progressive decline in cranial occupancy as quantified by the craniocerebral index (CCI).[39]

selective, and the actual rate of loss probably varies greatly at different ages and under different circumstances. Unfortunately, the most metabolically active, highly specialized neuronal subpopulations, particularly those that synthesize neurotransmitters, appear to suffer the most severe degree of loss. Fully 30–50 per cent of the neuronal population of the cerebral and cerebellar cortices, thalamus, locus ceruleus and basal ganglia disappear in the aged subject, and those neurons that remain demonstrate a markedly simplified pattern of synaptic interconnection.[41] There are generalized depletions of dopamine, noradrenaline (norepinephrine), tyrosine and serotonin, and a simultaneous increase in the activity of enzymes such as monoamine oxidase and catechol-*o*-methyltransferase which are essential to the destruction of the neurotransmitter substances.[42] The 'up-regulation' or increase in number of neurotransmitter receptor sites that appear in young neural tissues in response to a reduction of neurotransmitter activity may be slow and incomplete in the ageing brain.[43]

Nevertheless, the intrinsic mechanisms coupling cerebral electrical activity, tissue metabolism and cerebral blood flow (CBF) appear to persist intact in healthy elderly subjects.[44] Absolute CBF falls in proportion to reduced brain tissue mass, while specific CBF in the grey matter, expressed per 100 g of brain tissue, declines 20–30 per cent from the maximal values seen in the young adult,[45] a rate that parallels age-related declines in the neuronal density of cortical tissues.[46] Regional cortical and subcortical adjustments to variation of local metabolic demands occur in the same manner in elderly as in young adults. Similarly, autoregulation of cerebrovascular resistance (CVR) in response to changes in arterial blood pressure is well maintained, and the cerebral vasoconstrictor response to hyperventilation remains intact in the healthy geriatric patient.[47] Therefore, hyperventilation to produce intentional cerebral vasoconstriction and the guidelines for arterial blood pressure generally recommended for use in neuro-surgical anaesthesia are also appropriate for use in elderly patients, at least for those individuals who are free of established cerebrovascular disease. In fact, the relatively low incidence of intraoperative stroke in the geriatric surgical patient population[48] supports the conclusion that ageing is not inevitably associated with 'hardening of the arteries' and inadequate cerebral perfusion. In contrast, in the healthy elderly individual, decreased cerebral blood flow is a consequence, not a cause, of brain tissue atrophy.

The effect of ageing on complex intellectual function has also been studied. Higher, more complex aspects of 'crystallized intelligence' such as language skills, aesthetics and personality do not appear to decline with increasing age.[49] Despite the long-established bias that ageing produces inevitable deterioration of mental function, most recent studies suggest that information store, comprehension and long-term memory are well maintained, even in aged individuals.[50] There may, however, be an inevitable decline in short-term memory, visual and auditory reaction time, and other aspects of 'fluid intelligence' that require immediate processing or rapid retrieval of information, although much of this may actually reflect age-related compromise of attention span.[51] In general, however, the level of complex nervous system function across the entire range of adult life span remains close to that seen in young adults. Anatomical and functional redundancy appear to compensate adequately for the attrition of cellular elements, reduction of neurotransmitter concentrations, and the simplification of the complex system of neuronal interconnections within the neuropil that are characteristic of the ageing brain.

Peripheral nervous system

The threshold stimuli needed to initiate all forms of perception, including vision, hearing, touch, joint position sense, smell, and peripheral pain and temperature, are progressively elevated in the geriatric era (Fig. 38.10). This generalized and progressive process of *deafferentation* may be accelerated by degenerative changes at specialized sense organs; for example, there is an almost exponential fall in the density of pain-sensing Meissner's corpuscles in the skin of elderly individuals. However, there are also anatomical changes at more central sites: loss of afferent conduction pathways in the peripheral nervous system and spinal cord, and reduced velocity and amplitude of evoked electrical potentials in remaining pathways.[52]

Ageing also produces deterioration of electrical conduction along efferent motor pathways: peripheral motor nerve conduction velocity falls by approximately 0.15 m/s per year,[53] and impairment of corticospinal transmission increases the time needed between intention and onset of

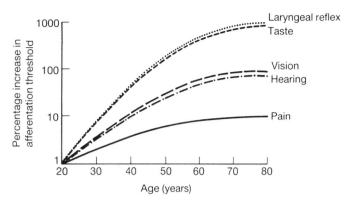

Fig. 38.10 Age-related processes of deafferentation produce a logarithmic increase in the thresholds for virtually all sensory modalities.

voluntary motor activity. There are no changes in skeletal muscle metabolism, although mitochondrial volume is reduced. Neurogenic skeletal muscle atrophy causes the dynamic strength, control and ability to maintain steadiness of the skeletal muscles in the extremities to decline 20–50 per cent by the age of 80 years, although isometric muscle strength appears to be well maintained.[54]

Motor neuron degeneration also produces the *disseminated neurogenic atrophy* responsible for age-related changes at the neuromuscular junction. Failing proximodistal protoplasmic transport in ageing motor neurons reduces the myotrophic support normally provided to skeletal muscle.[55] Thickening of the post-junctional membrane and its spread beyond the usual end-plate areas because of age-related deficiency of cyclic GMP generate atypical 'extrajunctional' cholinergic receptors.[56] The increase in number and variety of cholinergic receptors at the end plate and surrounding areas offsets the age-related decline in the number and the density of motor neuron/end-plate units. Therefore, despite loss of skeletal muscle mass, dose requirements for competitive neuromuscular blocking drugs are not reduced, and, in fact, are frequently slightly increased.

Autonomic nervous system

Viewed from a neuroendocrine perspective, ageing can be thought of as a *progressive hyperadrenergic state*. Plasma concentrations of adrenaline and noradrenaline are elevated two- to fourfold both at rest and in response to exercise-induced physical stress in elderly subjects.[57] Nevertheless, neurons in sympathoadrenal pathways, like the rest of the peripheral nervous system, are subject to cellular attrition, and adrenal tissue mass and cortisol secretion decline at least 15 per cent by the age of 80.[58] In elderly patients, high levels of catecholamines are rarely apparent clinically, however, because ageing also produces a marked depression of autonomic end-organ responsiveness. There is a significant impairment of the ability of β-agonists to enhance the velocity and force of cardiovascular contraction, and the general impairment of maximal chronotropic responses to isoprenaline (isoproterenol) is well known in elderly human subjects.[59] This *endogenous β-blockade of ageing* may reflect simple attrition of adrenoceptors, their reduced affinity for agonist molecules or compromise of adenylate cyclase activation because of decreased cell membrane fluidity.[60] The demonstration of decreased affinity of the ageing adrenoceptor for both agonist and antagonist molecules, however, suggest that qualitative, not quantitative, changes are predominant.[59]

The more complex, integrated autonomic reflex responses which maintain cardiovascular and metabolic homoeostasis so closely in young adults are also progressively impaired in elderly individuals. Baroreflex responsiveness, the vasoconstrictor response to cold stress and beat-to-beat heart rate responses following postural change in elderly subjects grow progressively less rapid in onset, smaller in magnitude and less effective in stabilizing blood pressure under a variety of circumstances.[61] The autonomic nervous system in the elderly patient is, in effect, 'underdamped' and less effectively self-regulated, permitting wider variation from homoeostatic set-points and delayed restabilization during stress. Therefore, anaesthetic agents that disrupt end-organ function or reduce plasma catecholamines, or techniques such as spinal or epidural anaesthesia that produce rapid pharmacological sympathectomy, are more likely to cause arterial hypotension in elderly than in young patients.[62,63] If exaggerated endogenous autonomic activity compensates for significant end-organ disease as, for example, in elderly patients with congestive heart failure, an anaesthetic-related 'disintegration' of autonomic homoeostasis can have abrupt and catastrophic consequences.

Analgesic and anaesthetic requirement

The simple age-related structural and functional changes within the nervous system described above are measurable and consistent, but their net effect on consciousness and on generalized, pain-related neurological function in healthy elderly subjects remains controversial. Some recent data[64,65] do not support the traditional assumption that the *deafferentation* associated with ageing is necessarily responsible for increased pain threshold or a decreased

need for analgesic or local anaesthetic agents. Decreased segmental dose requirements for local anaesthetics during epidural analgesia[66] probably reflect arteriosclerotic changes in the anatomy of the epidural space,[67,68] although slightly higher sensory levels do occur in the elderly patient undergoing spinal anaesthesia, even when a fixed drug dose and volume protocol is followed.[69,70] In any case, the role of amplification, modulation and selectivity of afferent input within the spinal cord, thalamus and perhaps at other locations within the ageing nervous system has yet to be studied adequately to permit broad generalizations regarding ageing and susceptibility to pain. Clinical evidence actually suggests that elderly patients have elevated thresholds for discrete and superficial discomfort[71] but experience increased vulnerability and sensitivity to severe or visceral pain, especially if associated with illness, injury or the prospect of long-term debility.

The effect of nervous system ageing on requirements for general anaesthesia is less equivocal. With increasing age, relative minimum alveolar concentration (MAC) declines progressively by as much as 30 per cent from young adult values.[72] The reason for this increase in sensitivity to anaesthetic agents remains unknown, but the observed consistency of this phenomenon for a variety of anaesthetic agents with markedly different chemical characteristics (Fig. 38.11) suggests that it is the result of fundamental neurophysiological, rather than purely pharmacological, processes.

It is also of interest that the decline in anaesthetic requirements parallels the rate of attrition of cortical neurons, the reduction in neuronal density within the cortex, the decline of absolute cerebral metabolic rate and blood flow[73] and, in particular, the reduction in brain neurotransmitter activity that occurs with increasing age. The direct relationship between levels of brain catecholamines and anaesthetic requirement as quantified by MAC has long been established for patients who have been

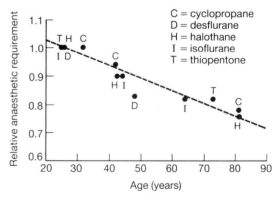

Fig. 38.11 Linear age-related declines in relative anaesthetic requirements (MAC or ED_{50}) for a variety of inhaled and injected anaesthetic agents in unsedated human subjects suggest that they reflect a common, fundamental physiological phenomenon rather than pharmacologically unique explanations.[104,112]

treated with drugs known to deplete or to enhance these substances. Whatever the mechanism, changes in anaesthetic requirement may be a useful measure of the ability of this organ system to resist the disruption of its normal function, consciousness. In this regard, anaesthetic requirement may be a clinically relevant indicator of nervous system functional reserve; as in other organ systems, ageing therefore would appear to produce a progressive, age-related decline in nervous system functional reserve.

The data for the effect of ageing on the pharmacodynamics, or dose requirement, for narcotics, barbiturates and benzodiazepines is less consistent than for inhalational anaesthetics. In fact, there is considerable controversy as to whether the apparent age-related increase in the potency of these drugs is truly a pharmacodynamic phenomenon, or whether it simply reflects a pharmacokinetic aberration, with unusually high 'α-phase' plasma drug concentrations appearing immediately after drug injection.[74,75] Poor mixing due to delayed intercompartmental transfer of drug,[76] rather than to decreased initial volume of distribution, may explain these observations.[77,78] Interpretation of these data is further complicated because of the difficulty in defining the anaesthetic end point in a process that, in contrast to the conditions under which MAC is determined, does not represent an equilibrium or a pseudo-equilibrium pharmacokinetic state. For example, complex analysis of the electroencephalogram (EEG) suggests that ageing increases brain sensitivity to narcotics,[79] yet the same methodology demonstrated no change in the median effective plasma concentrations of barbiturate or etomidate in elderly as compared to young adults.[80] Nevertheless, clinical experience and many measurements of dose–response relationships suggest that age is associated with a reduction in the dose requirements for thiopentone[81] and virtually all other agents that depress the central nervous system.

The dosages of competitive non-depolarizing neuromuscular blocking drugs such as pancuronium, vecuronium, tubocurarine or atracurium have been shown to be either unchanged or increased only slightly in the elderly patient.[82] These dosage requirements obviously reflect the quality and quantity of cholinergic receptors at the neuromuscular junction, and do not decline in parallel with the loss of skeletal muscle mass itself. The duration of their clinical effect is significantly prolonged if their elimination is dependent upon organ function, and drug clearance for almost all neuromuscular blocking drugs declines dramatically with increasing age.[83] However, the dynamics and the efficacy of antagonism or 'reversal' by neostigmine or by edrophonium of non-depolarizing neuromuscular blockade are unchanged.[84] It is the choice of reversal agent, and not patient age, that actually determines the speed and the completeness of return of neuromuscular transmission for any given level of neuromuscular blockade.

Anaesthetic management and outcome

It is obvious that overall perioperative mortality and major morbidity increase with advancing age,[85–87] beginning as early as the third decade of life (Fig. 38.12). However, assuming competent standards of anaesthetic practice, the same classic epidemiological data can be reshaped to demonstrate that the risk of major complications reflects the physical status of the patient, and not age itself (Fig. 38.13). *Age-related disease*, and not ageing, is primarily responsible for the progressive increase in morbidity and mortality characteristics of any elderly surgical population.[88,89] Morbidity and mortality rates are higher in elderly patients

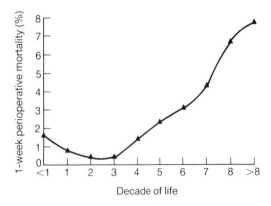

Fig. 38.12 Data for 1-week overall perioperative mortality, originally described by Marx in 1973 for a large series ($n = 34\,145$) of unselected patients, shows a dramatic increase at the extremes of age.[113]

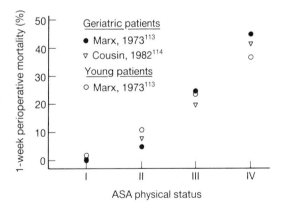

Fig. 38.13 When classic data for overall perioperative mortality presented in Fig. 38.12 is replotted according to American Society of Anesthesiologists (ASA) physical status rather than age itself, both younger and geriatric patients show essentially the same relationship between physical status and perioperative mortality.[113,114]

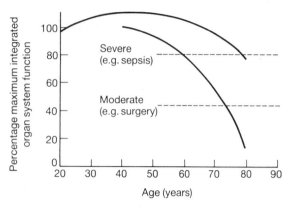

Fig. 38.14 Patients with a dysfunctional organ system curve (lower solid line) may be unable to meet even moderate demands for maximal organ system function, whilst geriatric patients with good physical status and high levels or organ system function can survive even the severe demands imposed by perioperative complications.

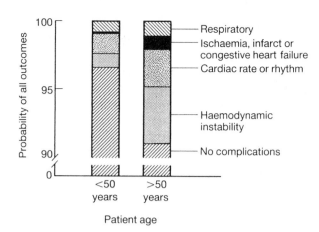

Fig. 38.16 Adverse outcomes observed in a recent large-scale prospective study ($n = 17\ 201$)[115] suggest that older surgical patients are primarily subject to cardiovascular complications, which make up the majority of all perioperative complications observed.

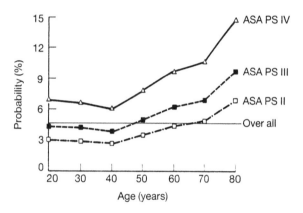

Fig. 38.15 The probability of any severe cardiovascular or pulmonary adverse outcome, estimated prospectively from a recent analysis of a large series ($n = 639/13\ 696$) of randomized patients,[115] suggests that physical status, rather than age, is the primary factor.

simply because this surgical patient subpopulation has a far higher average incidence and severity of concurrent disease than is found in younger adults; recent prospective studies confirm that age itself is only a minor predictor of postoperative mortality.[90] Although there probably are no patients 'too old' for anaesthesia, the accelerated loss of organ system functional reserve seen in aged patients may further increase intrinsic risk, especially for those with multiple organ system diseases[91,92] in whom integrated organ system function has fallen below those levels needed to survive moderate to severe stress (Fig. 38.14). Perioperative outcome in the elderly, therefore, as in the young adult, is largely determined by the site of operation, by the urgency of the procedure, and by the patient's physical status as determined by the number and complexity of underlying disease processes (Fig. 38.15). In a geriatric patient population, the nature of adverse outcomes shows a

relative predominance of disorders of cardiac rate or rhythm, myocardial ischaemia, or general haemodynamic instability due to *loss of intrinsic homoeostatic competence* (Fig. 38.16).

Adequate time for diagnosis, treatment and preparation of the anaesthetic plan is essential if the rate and severity of complications experienced by geriatric surgical patients are to be reduced.[93,94] In fact, two-thirds of a geriatric surgical population summarily 'cleared' for major surgery by medical consultants, if they are evaluated preoperatively with invasive monitoring, are nevertheless found to have mild to severe cardiopulmonary functional deficits. Less than 15 per cent of these patients will be completely normal physiologically, and almost one-third of the dysfunction discovered represents such profound abnormalities as congestive heart failure and hypoxaemia.[88]

In the absence of ventricular dysfunction or severe lung disease, however, an elderly patient, even an aged patient, can be an excellent candidate for any kind of surgery. There is no single 'best' anaesthetic for elderly patients. However, some complications may be more frequently associated with one form of anaesthesia than another. For example, deep vein thromboembolism is more common with general anaesthesia than with regional anaesthesia in an elderly population undergoing repair of hip fracture,[95] and emboli and other pulmonary complications represent a significant fraction of perianaesthetic morbidity and mortality.[96,97] Nevertheless, when overall outcome for any given type of surgery is determined over a period that encompasses both immediate and longer term complications, a period of at least 30 days, it is clear that the choice of anaesthetic agent or technique itself is not a major determinant of overall outcome.[95,98–100] Numerous retrospective and prospective clinical studies have arrived at the same conclusion: there are no significant differences in perioperative survival or major morbidity that are attributable directly to the choice of anaesthetic agents or to the anaesthetic technique selected.

A few principles unique to the design of the anaesthetic plan for elderly patients have been well established, however. Prompt and complete postoperative recovery of mental function is particularly important in elderly patients whose mentation is already compromised by age-related disease, a goal that becomes progressively more difficult in the face of a pharmacologically complex anaesthetic.[101,102] In fact, the most common cause of failure to emerge promptly from anaesthesia is simply the use of too much anaesthesia or too many anaesthetic agents. Even the relatively sparing use of intravenous sedation has been shown to retard the rapid recovery of mental function that otherwise occurs in elderly patients following spinal or epidural anaesthesia.[103] When anaesthetic management is appropriate and surgical convalescence is uncomplicated, the recovery of mental status to preoperative levels after a prolonged general anaesthetic may nevertheless require 5–10 days.[104] The neurophysiological and pharmacological mechanisms responsible for this subtle but prolonged disruption of organ system function remain unknown, but may simply reflect impaired nervous system functional reserve.

The physical management of elderly patients also requires specific precautions. Aged skin and bones are fragile, joints are stiff and the range of motion is limited. The elderly surgical patient requires gentle and expert routine care if traumatic injuries from improper positioning, bandaging or enforced bed rest are to be prevented. In the elderly patient, bleeding diatheses, hypercoagulable states[105] and a predisposition to bacterial infection appear to be more likely than in younger adults,[106] especially if sustained sympathoadrenal stress responses occur during a protracted and difficult recovery. For this reason, an anaesthetic plan that includes postoperative epidural sympathectomy[107] and analgesia[108] may be of special value, accelerating rewarming,[109] promoting healing, reducing cardiovascular and pulmonary demands, and eliminating the long periods of inadequate analgesia sometimes imposed upon geriatric patients because of fears of narcotic side effects.

REFERENCES

1. Schneider EL, Reed Jr JD. Life extension. *New England Journal of Medicine* 1985; **312**: 1159–68.
2. Hayflick L. The biology of human aging. *American Journal of Medical Science* 1973; **265**: 433–45.
3. Hayflick L. Biologic and theoretical perspectives of human aging. In: Katlic MR, ed. *Geriatric surgery: comprehensive care of the elderly patient.* Baltimore: Urban and Schwarzenberg, 1990: 3–21.
4. Rodeheffer RJ, Gerstenblith G, Becker LC, Fleg JL, Weisfeldt ML, Lakatta EG. Exercise cardiac output is maintained with advancing age in healthy human subjects: cardiac dilatation and increased stroke volume compensate for diminished heart rate. *Circulation* 1984; **69**: 203–13.
5. Tzankoff SP, Norris AH. Effect of muscle mass decrease on age-related BMR changes. *Journal of Applied Physiology* 1977; **43**: 1001–6.
6. Guarnieri T, Filburn CR, Zitnik G, Lakatta EG. Contractile and biochemical correlates of beta-adrenergic stimulation of the aged heart. *American Journal of Physiology* 1980; **239**: H501–8.
7. Aronow WS, Stein PD, Sabbah HN, Koenigsberg M. Resting left ventricular ejection fraction in elderly patients without evidence of heart disease. *American Journal of Cardiology* 1989; **63**: 368–9.
8. Duke PC, Wade JG, Hickey RF, Larson CP. The effects of age on baroreceptor reflex function in man. *Canadian Anaesthetists' Society Journal* 1976; **23**: 111–24.
9. Cahalan MK, Hashimoto Y, Aizawa K, *et al.* Elderly, conscious patients have an accentuated hypotensive response to nitroglycerin. *Anesthesiology* 1992; **77**: 646–55.
10. Cohn JE, Donoso HD. Mechanical properties of lung in normal men over 60 years old. *Journal of Clinical Investigation* 1963; **42**: 1406–10.
11. Mittman C, Edelman NH, Norris AH, Shock NW. Relationship between chest wall and pulmonary compliance and age. *Journal of Applied Physiology* 1965; **20**: 1211–16.
12. Wahba WM. Influence of aging on lung function – clinical significance of changes from age twenty. *Anesthesia and Analgesia* 1983; **62**: 764–76.
13. Peterson DD, Pack AI, Silage DA, Fishman AP. Effects of aging on ventilatory and occlusion pressure responses to hypoxia and hypercapnia. *American Review of Respiratory Diseases* 1981; **124**: 387–91.
14. Bailey PL, Wilbrink J, Zwanikken P, Pace NL, Stanley TH. Anesthetic induction with fentanyl. *Anesthesia and Analgesia* 1985; **64**: 48–53.
15. Pontoppidan H, Beecher HK. Progressive loss of protective reflexes in the airway with the advance of age. *Journal of the American Medical Association* 1960; **174**: 2209–13.
16. Woodhouse KW, Mutch E, Williams FM, Rawlins MD, James OFW. The effect of age on pathways of drug metabolism in human liver. *Age and Ageing* 1984; **13**: 328–34.
17. Divoll M, Greenblatt DJ, Ochs HR, Shader RI. Absolute bioavailability of oral and intramuscular diazepam: effects of age and sex. *Anesthesia and Analgesia* 1983; **62**: 1–8.
18. Shanor SP, Van Hees GR, Baart N, Erdos EG, Foldes FF. The influence of age and sex on human plasma and red cell cholinesterase. *American Journal of Medical Science* 1961; **242**: 357–61.
19. Wood AJJ, Vestal RE, Wilkinson GR, Branch RA, Shand DG.

Effect of aging and cigarette smoking on antipyrine and indocyanine green elimination. *Clinical Pharmacology and Therapeutics* 1979; **26**: 16–20.

20. Kampmann JP, Sinding J, Moller-Jorgensen I. Effect of age on liver function. *Geriatrics* 1975; **30**: 91–5.

21. Thompson EN, Williams R. Effect of age on liver function with particular reference to Bromsulphalein excretion. *Gut* 1965; **6**: 266–69.

22. Geokas MC, Haverback BJ. The aging gastrointestinal tract. *American Journal of Surgery* 1969; **117**: 881–92.

23. Wilkinson GR, Shand DG. A physiological approach to hepatic drug clearance. *Clinical Pharmacology and Therapeutics* 1975; **18**: 377–90.

24. McLachlan MSF. The ageing kidney. *Lancet* 1978; **2**: 143–6.

25. Boss GR, Seegmiller JE. Age-related physiological changes and their clinical significance. *Western Journal of Medicine* 1981; **135**: 434–40.

26. Rowe JW, Shock NW, DeFronzo R. The influence of age on the renal response to water deprivation in man. *Nephron* 1976; **17**: 270–8.

27. Epstein M. Effects of aging on the kidney. *Federation Proceedings* 1979; **38**: 168–72.

28. Helderman JH, Vestal RE, Rowe JW, Tobin JD, Andres R, Robertson GL. The response of arginine vasopressin to intravenous ethanol and hypertonic saline in man: the impact of aging. *Journal of Gerontology* 1978; **33**: 39–47.

29. Kumar R, Hill CM, McGeown MG. Acute renal failure in the elderly. *Lancet* 1973; **1**: 90–1.

30. Morrison RC. Hypothermia in the elderly. *International Anesthesiology Clinics* 1988; **26**: 124–33.

31. Goldberg MJ, Roe CF. Temperature changes during anesthesia and operations. *Archives of Surgery* 1966; **93**: 365–9.

32. Frank SM, Beattie C, Christopherson R, Norris EJ, Rock P, Parker S. Epidural versus general anesthesia, ambient operating room temperature, and patient age as predictors of inadvertent hypothermia. *Anesthesiology* 1992; **77**: 252–7.

33. Davidson MB. The effect of aging on carbohydrate metabolism: a review of the English literature and a practical approach to the diagnosis of diabetes in the elderly. *Metabolism* 1979; **28**: 688–705.

34. Fulop Jr T, Worum I, Csongor J, Foris G, Leovey A. Body composition in elderly people. *Gerontology* 1985; **31**: 6–14.

35. Kaiko RF, Wallenstein SL, Rogers AG. Narcotics in the elderly. *Medical Clinics of North America* 1982; **66**: 1079–89.

36. Christensen JH, Andreasen F, Jansen JA. Thiopentone sensitivity in young and elderly women. *British Journal of Anaesthesia* 1983; **55**: 33–9.

37. Klotz U, Avant GR, Hoyumpa A. The effects of age and liver disease on the disposition and elimination of diazepam in adult man. *Journal of Clinical Investigation* 1975; **55**: 347–59.

38. Richey DP, Bender AD. Pharmacokinetic consequences of aging. *Annual Review of Pharmacology and Toxicology* 1977; **17**: 49–65.

39. Creasey H, Rapoport SI. The aging human brain. *Annals of Neurology* 1985; **17**: 2–10.

40. Schjeide OA. Relation of development and aging; pre- and postnatal differentiation of the brain as related to aging. *Advances in Behavioral Biology* 1975; **16**: 37–83.

41. Feldman ML. Aging changes in the morphology of cortical dendrites. In: Terry RD, Gershon S, eds. *Neurobiology of aging.* New York: Raven Press, 1976: 211–27.

42. McGeer EG, McGeer PL. Age changes in the human for some enzymes associated with metabolism of catecholamines, GABA, and acetylcholine. *Advances in Behavioral Biology* 1975; **16**: 287–305.

43. Greenberg LH. Regulation of brain adrenergic receptors during aging. *Federation Proceedings* 1986; **45**: 55–9.

44. Lassen NA, Ingvar DH, Skinhoj E. Brain function and blood flow. *Scientific American* 1978; **239**: 62–71.

45. Kety SS. Human cerebral blood flow and oxygen consumption as related to aging. *Journal of Chronic Diseases* 1956; **3**: 478–86.

46. Devaney KO, Johnson HA. Neuron loss in the aging visual cortex of man. *Journal of Gerontology* 1980; **35**: 836–41.

47. Yamamoto M, Meyer JS, Sakai F, Yamaguchi F. Aging and cerebral vasodilator response to hypercarbia. *Archives of Neurology* 1980; **37**: 489–96.

48. Cogbill CL. Operation in the aged. Mortality related to concurrent disease, duration of anesthesia, and elective or emergency operation. *Archives of Surgery* 1967; **94**: 202–5.

49. Katzman R, Terry RD. Normal aging of the central nervous system. In: Katzman R, Terry RD, eds. *The neurology of aging.* Philadelphia: FA Davis, 1983: 15–50.

50. Macht ML, Buschke H. Speed of recall in aging. *Journal of Gerontology* 1984; **39**: 439–43.

51. Knoefel JE, Albert ML. Neurologic changes in the elderly. In: Katlic MR, ed. *Geriatric surgery: comprehensive care of the elderly patient.* Baltimore and Munich: Urban and Schwarzenberg, 1990; 153–64.

52. Beck EC, Dustman RE, Schenkenberg T. Life span changes in the electrical activity of the human brain as reflected in the cerebral evoked response. *Advances in Behavioral Biology* 1975; **16**: 175–92.

53. Dorfman LJ, Bosley TM. Age-related changes in peripheral and central nerve conduction in man. *Neurology* 1979; **29**: 38–44.

54. Larsson L. Morphological and functional characteristics of the aging skeletal muscle in man. *Acta Physiologica Scandinavica, Supplement* 1978; **457**: 1–36.

55. Gutmann E, Hanzlikova V, Jaboubek B. Changes in the neuromuscular system during old age. *Experimental Gerontology* 1968; **3**: 141–6.

56. Betz H, Changeux JP. Regulation of muscle acetylcholine receptor synthesis *in vitro* by cyclic nucleotide derivatives. *Nature* 1979; **278**: 749–52.

57. Ziegler MG, Lake CR, Kopin IJ. Plasma noradrenaline increases with age. *Nature* 1976; **261**: 333–4.

58. Kaack B, Ordy JM, Trapp B. Changes in limbic, neuroendocrine and autonomic systems, adaptation, and homeostasis during aging. *Advances in Behavioral Biology* 1975; **16**: 209–31.

59. Vestal RE, Wood AJJ, Shand DG. Reduced adrenoceptor sensitivity in the elderly. *Clinical Pharmacology and Therapeutics* 1979; **26**: 181–6.

60. Dax EM. Receptors and associated membrane events in aging. *Review of Biological Research on Aging* 1985; **2**: 315–36.

61. Collins KJ, Exton-Smith AN, James MH. Functional changes in autonomic nervous responses with ageing. *Age and Ageing* 1980; **9**: 17–24.

62. Coe AJ, Revanas B. Is crystalloid preloading useful in spinal anaesthesia in the elderly? *Anaesthesia* 1990; **45**: 241–3.

63. Carpenter RL, Caplan RA, Brown DL, Stephenson C, Wu R. Incidence and risk factors for side effects of spinal anesthesia. *Anesthesiology* 1992; **76**: 906–16.

64. Hapidou EG, DeCatanzaro D. Responsiveness to laboratory pain in women as a function of age and childbirth pain experience. *Pain* 1992; **48**: 177–81.

65. Harkins SW, Chapman CR. Detection and decision factors in pain perception in young and elderly men. *Pain* 1976; **2**: 253–8.

66. Bromage PR. Ageing and epidural dose requirements. Segmental spread and predictability of epidural analgesia in youth and extreme age. *British Journal of Anaesthesia* 1969; **41**: 1016–22.

67. Sharrock NE. Epidural dose responses in patients 20 to 80 years old (clinical report). *Anesthesiology* 1978; **49**: 425–8.

68. Park WY, Hagins FM, Rivat EL, MacNamara TE. Age and epidural dose response in adult men. *Anesthesiology* 1982; **56**: 318–20.

69. Veering BT, Burm AGL, Vletter AA, van den Hoeven RAM, Spierdijk J. The effect of age on systemic absorption and systemic disposition of bupivacaine after subarachnoid administration. *Anesthesiology* 1991; **74**: 250–7.

70. Cameron AE, Arnold RW, Ghoris MW, Jamieson V. Spinal analgesia using bupivacaine 0.5 per cent plain: variation in the extent of block with patient age. *Anaesthesia* 1981; **36**: 318–22.

71. Procacci P, Bozza G, Buzzelli G, Della Corte M. The cutaneous pricking pain threshold in old age. *Gerontology Clinics* 1970; **12**: 213–18.

72. Quasha AL, Eger II EI, Tinker JH. Determination and application of MAC. *Anesthesiology* 1980; **53**: 315–34.

73. Gregory GA, Eger II EI, Munson ES. The relationship between age and halothane requirement in man. *Anesthesiology* 1969; **30**: 488–91.

74. Homer TD, Stanski DR. The effect of increasing age on thiopental disposition and anesthetic requirement. *Anesthesiology* 1985; **62**: 714–24.

75. Singleton MA, Rosen JI, Fisher DM. Pharmacokinetics of fentanyl in the elderly. *British Journal of Anaesthesia* 1988; **60**: 619–22.

76. Avram MJ, Krejcie TC, Henthorn TK. The relationship of age to the pharmacokinetics of early drug distribution: the concurrent disposition of thiopental and indocyanine green. *Anesthesiology* 1990; **72**: 403–11.

77. Hull CJ. Editorial. How far can we go with compartmental models? *Anesthesiology* 1990; **72**: 399–402.

78. Stanski DR, Maitre PO. Population pharmacokinetics and pharmacodynamics of thiopental: the effect of age revisited. *Anesthesiology* 1990; **72**: 412–22.

79. Scott JC, Stanski DR. Decreased fentanyl and alfentanil dose requirements with increasing age. A simultaneous pharmacokinetic and pharmacodynamic evaluation. *Journal of Pharmacology and Experimental Therapy* 1987; **240**: 159–65.

80. Arden JR, Holley FO, Stanski DR. Increased sensitivity to etomidate in the elderly: initial distribution versus altered brain response. *Anesthesiology* 1986; **65**: 19–27.

81. Muravchick S. Effect of age and premedication on thiopental sleep dose. *Anesthesiology* 1984; **61**: 333–6.

82. Shanks CA. Pharmacokinetics of the nondepolarizing neuro-

muscular relaxants applied to calculation of bolus and infusion dosage regimens. *Anesthesiology* 1986; **64**: 72–86.

83. McLeod K, Hull CJ, Watson MJ. Effects of ageing on the pharmacokinetics of pancuronium. *British Journal of Anaesthesia* 1979; **51**: 435–8.

84. Muravchick S. Age does not change the dynamics of atracurium reversal. *Anesthesiology* 1990; **73** suppl: A384.

85. Greenburg AG, Saik RP, Farris JM, Peskin GW. Operative mortality in general surgery. *American Journal of Surgery* 1982; **144**: 22–8.

86. Pedersen T, Eliasen K, Henriksen E. A prospective study of mortality associated with anaesthesia and surgery: risk indicators of mortality in hospital. *Acta Anaesthesiologica Scandinavica* 1990; **34**: 176–82.

87. McSherry CK, Glenn F. Incidence and causes of death following surgery for nonmalignant biliary tract disease. *Annals of Surgery* 1980; **191**: 271–5.

88. Lewin I, Lerner AG, Green SH, Del Guercio LRM, Siegel JH. Physical class and physiologic status in the prediction of operative mortality in the aged sick. *Annals of Surgery* 1971; **174**: 217–31.

89. Gerson MC, Hurst JM, Hertzberg VS, *et al.* Cardiac prognosis in noncardiac geriatric surgery. *Annals of Internal Medicine* 1985; **103**: 832–7.

90. Davis FM, Woolner DF, Frampton C, *et al.* Prospective, multi-centre trial of mortality following general or spinal anaesthesia for hip fracture surgery in the elderly. *British Journal of Anaesthesia* 1987; **59**: 1080–8.

91. Tiret L, Hatton F. Complications associated with anaesthesia – a prospective survey in France. *Canadian Anaesthetists' Society Journal* 1986; **33**: 336–44.

92. Filzwieser G, List WF. Morbidity and mortality in elective geriatric surgery. In: Lunn JN, Mushin WW, eds. *Mortality associated with anaesthesia*. London: Nuffield Provincial Hospital Trusts, 1982: 75–82.

93. Gracey DR, Divertie MB, Didier EP. Preoperative pulmonary preparation in patients with chronic obstructive pulmonary disease. *Chest* 1979; **76**: 123–9.

94. Cogbill CL. Operation in the aged. Mortality related to concurrent disease, duration of anesthesia, and elective or emergency operation. *Archives of Surgery* 1967; **94**: 202–5.

95. Davis FM, Laurenson VG. Spinal anaesthesia or general anaesthesia for emergency hip surgery in elderly patients. *Anaesthesia and Intensive Care* 1981; **9**: 352–8.

96. Van Demark GE, Van Demark RE. Hip nailing in patients of eighty years or older; experiences in 104 consecutive personal cases. *American Journal of Surgery* 1953; **85**: 664–8.

97. Davie IT, MacRae WR, Malcolm-Smith NA. Anesthesia for the fractured hip: a survey of 200 cases. *Anesthesia and Analgesia* 1970; **49**: 165–70.

98. Gauthier JL, Hamelberg W. Hip fractures: influence of anesthesia on patient's hospital course. *Anesthesia and Analgesia* 1963; **42**: 609–15.

99. Valentin N, Lomholt B, Jensen JS, Hejgaard N, Kreiner. Spinal or general anesthesia for surgery of the fractured hip? *British Journal of Anaesthesia* 1986; **58**: 284–91.

100. Wickstrom I, Holmberg I, Stefansson T. Survival of female geriatric patients after hip fracture surgery: a comparison of

5 anesthetic methods. *Acta Anaesthesiologica Scandinavica* 1982; **26**: 607–14.

101. Hole A, Terjesen T, Breivik H. Epidural versus general anaesthesia for total hip arthroplasty in elderly patients. *Acta Anaesthesiologica Scandinavica* 1980; **24**: 279–87.

102. Chung F, Meier R, Lautenschlager E, Carmichael FJ, Chung A. General or spinal anesthesia: which is better in the elderly? *Anesthesiology* 1987; **67**: 422–7.

103. Chung FF, Chung A, Meier RH, Lautenschlager E, Seyone C. Comparison of perioperative mental function after general anaesthesia and spinal anaesthesia with intravenous sedation. *Canadian Journal of Anaesthesia* 1989; **36**: 382–7.

104. Muravchick S. Immediate and long-term nervous system effects of anesthesia in elderly patients. *Clinics in Anaesthesiology* 1986; **4**: 1035–45.

105. Yardumian A, Machin SJ. Hypercoagulable states. *International Anesthesiology Clinics* 1985; **23**: 141–55.

106. Harbrecht PJ, Garrison RN, Fry DE. Role of infection in increased mortality associated with age in laparotomy. *American Surgeon* 1983; **49**: 173–8.

107. Stefansson T, Wickstrom I, Haljamae H. Effect of neurolept and epidural analgesia on cardiovascular function and metabolism in the geriatric patient. *Acta Anaesthesiologica Scandinavica* 1982; **26**: 386–92.

108. Ready LB, Loper KA, Nessly M, Wild L. Postoperative epidural morphine is safe on surgical wards. *Anesthesiology* 1991; **75**: 452–6.

109. Carli F, Gabrielczyk M, Clark MM, Aber VR. An investigation of factors affecting postoperative rewarming of adult patients. *Anaesthesia* 1986; **41**: 363–9.

110. Smith D, Bierman E, Robinson N, eds. *The biological ages of man*, 2nd edn. Philadelphia: WB Saunders, 1978.

111. Kitamura H, Sawa T, Ikezono E. Postoperative hypoxemia: the contribution of age to the maldistribution of ventilation. *Anesthesiology* 1972; **36**: 244–52.

112. Rampil IJ, Lockhart SH, Zwass MS, *et al.* Clinical characteristics of desflurane in surgical patients: minimum alveolar concentrations. *Anesthesiology* 1991; **74**: 429–33.

113. Marx GF, Mateo CV, Orkin LR. Computer analysis of postanaesthetic deaths. *Anesthesiology* 1973; **39**: 54–8.

114. Cousin MT. Evaluation of operative risk in the aged. In: Conseille C, Cousin MT, eds. *Complications of anaesthesia: operative risk*. New York: Elsevier, 1982: 257–71.

115. Forrest JB, Rehder K, Cahalan MK, Goldsmith CH. Multicenter study of general anesthesia. III. Predictors of severe perioperative adverse outcomes. *Anesthesiology* 1992; **76**: 3–15.

Chapter 39

Induced Hypotension

Alan J. McLintic and J. Gordon Todd

Intraoperative bleeding may be reduced by using locally applied techniques (tourniquets, vasoconstrictors, sympathetic blockade, posture) or by inducing systemic hypotension. A reduction in bleeding leads to two major benefits: large volume losses are reduced and the surgical field is improved. The advantages of the former, in terms of haemodynamic and metabolic stability, transfusion requirements and infection transmission are clear. The 'dry' operative field promotes accurate identification and dissection of structures, thus reducing trauma to surrounding tissues and improving the precision of surgery. Operative time may also be reduced[1,2] and, because there is less cauterized tissue and haematoma formation, wound healing may be improved.[3] Induced hypotension is employed in procedures which are made difficult by even minor bleeding as in many ENT, ophthalmic and neurosurgical operations or when blood loss is likely to be large, as in much orthopaedic, maxillofacial and pelvic surgery.[4-9] In addition, while the main application for induced hypotension is to control bleeding, it is also used to improve safety in intracranial vascular surgery. Applications for induced hypotension are listed in Table 39.1.

Table 39.1 Applications for induced hypotension

ENT surgery	*Ophthalmic*
Mastoid surgery	Intraocular tumours
Middle ear surgery	Vitrectomy, lens extraction
Septorhinoplasty	Lacrimal, oculoplastic, orbital surgery
Head and neck	
Faciomaxillary tumours	*Orthopaedic*
Block dissection of the neck	Shoulder and hip arthroplasty
Laryngectomy	Spinal surgery
Neurosurgery	*Pelvic*
Intracranial aneurysm surgery	Major vulvar, vaginal, uterine surgery
Arteriovenous malformations	Cystectomy
Vascular tumours	Rectal resection
Trans-sphenoidal hypophysectomy	*Jehovah's Witness*

Similarly, whilst all efforts must be made to prevent or treat coagulopathy, even a normal haemostatic response is incapable of controlling bleeding from a surgical incision. Bleeding must therefore be controlled by reducing the flow from cut vessels.

Bleeding

The volume of blood lost from a wound depends on the number of vessels cut, the rate of flow from the transected vessels and the speed and effectiveness of haemostasis. The former depends on the size of the incision and the vascularity of the tissue and may not be modifiable.

Arterial and arteriolar bleeding

The rate of blood flow from a damaged vessel is related to the size of the vessel and its intraluminal pressure. The pressure can be lowered by reducing either vascular resistance or cardiac output, although most believe that the latter should be avoided because organ perfusion may be compromised.[4,10-13] The principle of controlled hypotension is therefore to induce a low-pressure, vasodilated circulation.

With most hypotensive techniques, as blood pressure falls, the decline in baroreceptor activity causes a reflex rise in heart rate, myocardial contractility and vasomotor tone, resulting in increasing vasodilator doses to maintain the same level of hypotension. The result is a low-pressure but hyperdynamic circulation in which some regional blood flows may be increased. Bleeding is still attenuated in this situation[14] and allows, for example, choroidal dissection to be carried out in the face of a greatly increased cardiac output (own unpublished data). Similarly, lumbar sympathetic blockade reduces blood loss in lower limb surgery despite increasing arterial flow in the leg.[15] There is little consensus as to why bleeding should still be reduced in the presence of increased flow but there are several possible contributing factors. First, vasodilatation opens previously dormant vascular beds so that increases in flow become evenly distributed and intraluminal pressures minimally changed. When, in addition, the operative site is raised above heart level, the inflow to vascular beds is reduced, the drainage improved and local intraluminal pressure lowered. This has been termed 'postural ischaemia'.[16] Lastly, as intraluminal pressure falls, there is a tendency for small vessels to gradually close under the collapsing forces of vessel wall tension. The relative importance of the above factors is not yet established.

Bleeding is influenced by two additional factors. Arterial bleeding is worsened by tachycardia, possibly because there is less time for effective arteriolar run-off,[17] and incisions through inflammatory tissue cause particularly troublesome bleeding because of the high vascularity of the tissue.

Venous bleeding

Bleeding from capacitance vessels is a major component of surgical haemorrhage. Venodilatation alone does little to significantly reduce venous bleeding and must be combined with wound elevation to ensure optimal venous drainage and pooling of blood away from the operation site. In addition, attention must be paid to preventing any obstruction to venous flow from surgical retraction, poor patient positioning or excessive rises in intrathoracic pressure. The last includes the prevention of coughing or straining and the use of appropriate settings when intermittent positive pressure ventilation (IPPV) is used.

Safe hypotension

As blood pressure is reduced, blood flow within vital organs is maintained by autoregulatory vasodilatation.

When this is maximal, further reductions in pressure are directly related to falls in organ blood flow. This point is reached in the coronary and cerebral circulations when mean arterial pressure (MAP) falls below 60 mmHg, and in the kidney when MAP falls below 80 mmHg.[18–20] How is it then that, during induced hypotension, MAP is frequently reduced below this limit without adverse effect?

Standard estimates of the limits of autoregulation are based on experimental hypotension induced by haemorrhage – a situation accompanied by intense sympathetic stimulation. The latter prevents maximal cerebral vasodilatation and may also worsen outcome if ischaemic damage occurs.[21,22] During induced hypotension, cerebral blood flow is preserved at lower pressures than in haemorrhagic hypotension because cerebral vasculature is relaxed by vasodilating hypotensive agents.[23,24] A point is still reached at which flow becomes pressure dependent but flow can fall some way below this level before ischaemic thresholds are reached and cell function is compromised.[12,23] Anaesthetic agents reduce cerebral metabolism and are often considered to confer cerebral 'protection' in hypoperfused states. This must not lead to a false sense of security, however, as the clinical significance of the effect is now thought to be, at best, marginal.[25,26] Thus, safety is most dependent on maintaining the oxygen supply to the brain rather than reducing demand. This strengthens the argument against techniques that primarily reduce cardiac output, as they may also impair cerebral blood flow.[13]

The margins of safety for myocardial perfusion are also greater during well-managed controlled hypotension than during haemorrhagic hypotension because myocardial oxygen demand is reduced and coronary blood flow is preserved to lower pressures.[27]

It is important to remember that safety is very dependent on the way in which hypotensive agents are used. Blood pressure must be reduced slowly so that compensatory mechanisms in the heart and brain have time to develop,[28,29] tachycardia must be prevented as this increases oxygen demand and reduces coronary filling, and sudden swings in pressure must be avoided because autoregulation is impaired by vasodilators.[24]

Techniques of induced hypotension

Patient selection and preparation

Surgical requirements must be discussed in advance of admission to theatre. If a surgeon prefers to operate under normotensive conditions, this must be respected or haemostasis may be inadequate. Each patient must be carefully assessed before the appropriate technique is selected. Because of the (understandable) lack of controlled studies, there is some debate as to what constitutes an absolute contraindication to deliberate hypotension. In certain circumstances (e.g. intracranial aneurysm clipping) it may be considered that the increased risk associated with inducing hypotension in a debilitated patient is outweighed by the benefits of a low-pressure circulation. For the most part, however, a conservative policy should be adopted, as hypotension is seldom an absolute requirement for surgery. Under most circumstances the following should be regarded as contraindications to the deliberate reduction of blood pressure:

1. 'Fixed' cardiac output (e.g. aortic stenosis, obstructive and restrictive cardiomyopathy, constrictive pericarditis and left ventricular failure)
2. Hypertension that is inadequately treated or associated with left ventricular hypertrophy
3. Angina; myocardial infarction within previous 6 months
4. Carotid artery stenosis; previous cerebrovascular accident
5. Hypovolaemia
6. Pregnancy

Induced hypotension is frequently carried out in the following situations but the attendant risks are higher than with normotensive techniques.

Treated hypertension   Although some outcome analyses of hypotensive anaesthesia have shown minimal morbidity in hypertensive patients, caution must be exercised even when blood pressure is adequately controlled.[30] Arteriolar hypertrophy, β-blockade and diuretic therapy all reduce effective blood volume and lead to an exaggerated response to hypotensive agents. Atheromatous disease is common and the non-compliant vascular tree means that autoregulation fails at a higher MAP.

Quiescent ischaemic heart disease   Coronary artery disease (CAD) is not regarded by all as an absolute contraindication to induced hypotension and, to a certain extent, this view is supported by studies during normotensive anaesthesia in patients with CAD which show no increase in myocardial ischaemia during inadvertent hypotensive episodes.[31] Safety margins are, however, clearly reduced. Autoregulation is impaired by CAD and coronary 'steal' can be precipitated by the pure arterial vasodilators often used for induced hypotension.[32] In addition, whilst the controlled reduction of blood pressure may reduce myocardial oxygen demand, emergence from hypotension and anaesthesia has the opposite effect, particularly if there is rebound hypertension. Careful risk–benefit evaluation is required.

Insulin-dependent diabetes mellitus (IDDM)   As with other forms of anaesthesia, meticulous monitoring and adjustment of blood glucose is essential. Ganglion blockers and β-adrenergic antagonists should be avoided because they may potentiate hypoglycaemia. There is some evidence that cerebrovascular reactivity is impaired in IDDM,[33] and it has been suggested that pressure reduction should be limited because of this.[34]

Respiratory disease   Provided that no other system is significantly diseased, if a patient with a respiratory disorder is regarded as fit for elective general anaesthesia, hypotension may also be induced. Close intraoperative monitoring of arterial gases is, however, mandatory and hypotension must be ceased and reversed if oxygenation cannot be maintained. Isoflurane may be the most suitable agent in view of its minimal effect on pulmonary shunt (see later). The adverse effect of induced hypotension on gas exchange increases with age.[35]

Premedication

Anxiolytic and analgesic premedication usefully obtunds the early catecholamine response to preoperative anxiety, intubation and pain. Atropine should be avoided because it induces tachycardia, and the routine use of hypotensive premedication is not recommended (see below).

Anaesthesia

Deep anaesthesia combined with 'postural ischaemia' and a sound analgesic technique will produce operating conditions that are satisfactory for most procedures. The effect on bleeding is, however, inconsistent and limited (particularly in the young) and there is little room for manoeuvre if bleeding is brisk. These limitations must be anticipated and accepted, and attempts should not be made to reduce pressure with excessively high concentrations of anaesthetic, as cardiac output and safety will be compromised. However, a meticulous anaesthetic technique must always be the bedrock on which any hypotensive technique is based. No single anaesthetic technique has been shown to be superior to any other in this context, and it is the authors' choice to induce anaesthesia with thiopentone, to provide muscle relaxation with a non-depolarizing agent such as vecuronium and to intubate the trachea after topical anaesthesia of the larynx. Anaesthesia is maintained, usually by IPPV, using nitrous oxide and isoflurane, and analgesia is provided by systemic opiates and, where possible, local or regional nerve blockade. Historically, tubocurarine has been popular because of the 15–20 per cent fall in blood pressure it may induce as a consequence of ganglion blockade and histamine release.[36] The latter may, however, cause undesirable side effects, and the long duration of action (which may be further extended during induced hypotension by deep volatile anaesthesia and trimetaphan) reduces flexibility in blood pressure and relaxant control.[37,38]

IPPV vs. spontaneous respiration

Opinions differ strongly over the relative merits of controlled or spontaneous ventilation for induced hypotension. This generally suggests that there is little to choose between them and, indeed, there is no firm statistical evidence to show that one is associated with less bleeding than the other. Although there are certainly specific situations, such as intraocular and neurosurgery, in which IPPV is strongly indicated, the choice is often made on personal preference.

The lower intrathoracic pressure associated with spontaneous respiration confers advantages in terms of lower venous pressure and better maintained cardiac output, although the differences may be slight.[39] Paradoxically, some supporters of IPPV cite the reduction in cardiac output as being a useful adjunct to pharmacologically induced hypotension, particularly if sympathetic reflexes have been obtunded by ganglion blockade or adrenoreceptor antagonists and positive end-expiratory pressure applied. This practice can, however, lead to venous engorgement and may compromise organ perfusion. Ventilator settings are therefore best adjusted to produce minimal rises in intrathoracic pressure.

The pattern of spontaneous breathing has been advocated as a useful monitor of medullary perfusion but recognition and interpretation of abnormality requires considerable experience.[40] Furthermore, although there is inherent safety in avoiding muscle paralysis, there are also advantages in employing it to eliminate the risk of coughing or straining during delicate surgery. Probably the strongest argument for using IPPV during induced hypotension is to prevent hypercapnia because this increases circulating catecholamines and promotes cardiovascular stimulation and arrhythmogenicity.[41] In addition, hypercapnia itself may increase pulmonary dead space.[35] Overventilation must also be prevented, however, as cerebral vasoconstriction can reduce blood flow to critical levels.[23]

General considerations

During hypotensive anaesthesia, ventilation/perfusion (V/Q) mismatch impairs arterial oxygenation by a degree that is related to the age of the patient, preoperative lung function, degree of hypotension and the technique used (see below). Inspired oxygen concentrations of 40–50 per cent (or higher) are, therefore, commonly required.

The importance of careful positioning cannot be overemphasized. The production of 'postural ischaemia' and optimal venous drainage improves operating conditions and thereby reduces the degree of systemic hypotension required. For operations on the head and neck this may mean a head-up tilt. Caution is always required when using this position, however, as cerebral arterial pressure is reduced, $\dot{V}/\dot{Q}$ mismatch increased[42] and stroke volume decreased because of reduced venous return. For pelvic operations, the benefit to venous drainage of a small degree of head-down tilt outweighs any impairment to the induction of hypotension caused by the increased venous return to the heart. It must be remembered that raising the operation site above heart level also increases the risk of air embolism, particularly when the patient is breathing spontaneously and when occipital, intracranial or pelvic veins are exposed. Constant vigilance is required by both anaesthetist and surgeon.

During hypotension, maintenance fluid therapy helps to preserve right heart filling pressures and cardiac output and minimizes the rise in pulmonary dead space.[43,44] Care must be taken to prevent overhydration, as this may precipitate hypertension or left heart failure in susceptible patients when vasodilators are withdrawn. Deliberate haemodilution has been successfully used to reduce homologous blood transfusion requirements.[45] Haemodilution is associated with improved tissue perfusion and should not impair haemostasis unless the haematocrit (packed cell volume) is reduced below 20 per cent.[46] When haemodilution is used in conjunction with hypotension induced by sodium nitroprusside, transfusion requirements are reduced still further[47] but cardiac performance may also be significantly impaired.[48]

Monitoring

Meticulous monitoring of patient, apparatus and breathing system is essential. Association of Anaesthetists of Great Britain and Ireland have issued guidelines for minimal standards of monitoring. Four areas of monitoring are of particular importance, as follows.

Electrocardiography

Continuous monitoring of the electrocardiogram (ECG) is mandatory. β-Adrenoceptor antagonist premedication increases the risk of bradyarrhythmia,[7] reflex tachycardia may require directed therapy and ST-segment changes or arrhythmias may indicate myocardial ischaemia. For dependable recording, time must be spent preoperatively preparing the skin, critically placing the electrodes and firmly securing leads. A unipolar exploring electrode in the V_5 position is the most sensitive single lead for the detection of myocardial ischaemia[49] and is most closely represented in a bipolar system by a CC_5 arrangement (positive electrode–V_5; negative electrode–V_5R).[50] The greatest sensitivity with a bipolar system is, however, achieved with a CM_5 arrangement where the positive electrode is placed at V_5 and the negative at the manubrium. Multi-lead systems increase sensitivity further, and displays of the ST-segment trend make it easier to assess the significance of subtle changes.

Blood pressure measurement

Blood pressure measurements made by manual oscillometry bear very close comparison to those of an intra-arterial device, even during significant hypotension.[51] In addition, when the cuff is set at a target pressure, the oscillometer can be used as a continuous monitor to detect 'beat to beat' deviation from the chosen pressure. However, the technique requires considerable expertise and can be very time consuming.

Automated oscillometric devices provide regular and, for the most part, reliable digital displays of blood pressure without necessitating continuous involvement of the anaesthetist. However, they are easily upset by movement artefact, may be inaccurate at low pressure and cannot function as continuous monitors.[52] They are therefore unsuitable for anything other than mild hypotensive procedures.

Two non-invasive devices do have potential, however. Arterial tonometry using the Colin CBM-3000 instrument (Colin Electronics, Komaki) can provide an accurate, reliable, real-time monitor of pressure during moderate hypotension,[53] and the Finapres monitor (Finapres, Ohmeda), which is based on the volume-clamp method of pressure measurement, can provide a near-continuous pressure display with systolic accuracy even during profound hypotension. With the latter device, mean and diastolic pressures tend to be overestimated during hypotension and, in some patients, the inaccuracies are large.[54,55] For this reason the Finapres cannot be recommended for profound hypotension.

The 'gold standard' for blood pressure monitoring is still direct intra-arterial measurement and must be employed whenever evanescent agents are used to titrate pressures or when large blood losses or pressure swings are anticipated. The indwelling cannula also allows repeated arterial sampling for blood gas analysis. Placing the transducer at brain level provides an indication of cerebral perfusion pressure.

Monitoring of central venous pressure is not necessary for routine purposes but should be used to ensure that right heart filling pressure is adequately maintained when large blood losses are anticipated.

Gas exchange

Capnography and pulse oximetry should be regarded as mandatory during induced hypotension. Pulse oximetry allows the inspired oxygen concentration to be adjusted appropriately in order to compensate for $\dot{V}/\dot{Q}$ mismatch induced by hypotension and hypotensive agents. Capnography is essential both as a disconnection alarm and as an indicator of air embolism. Its value as an index of arterial carbon dioxide tension is limited because of the rise in pulmonary dead space during hypotension.

Cerebral function

When cerebral blood flow (CBF) falls below a critical level, cerebral electrical activity first becomes abnormal (at a CBF of approximately 20 ml/100 g per minute) and then extinct (approximately 15 ml/100 g per minute).[56] Irreversible cell damage will follow unless neuronal perfusion is quickly improved. During induced hypotension, monitoring the electroencephalogram (EEG) and evoked potential may therefore be used to identify cortical and deep sensory pathway hypoperfusion and to provide warning signals that pressure must be immediately increased. Such monitors are not, however, a practical consideration for routine hypotensive anaesthesia and are usually reserved for neurosurgical procedures, cardiopulmonary bypass or severe hypotension.

Regional anaesthesia

Regional sympathetic blockade reduces surgical bleeding even if blood pressure is maintained at normal levels.[57, 58] The exact mechanism is not clear but intrinsic to success is correct positioning of the operation site above heart level to encourage venous drainage and reduce local arterial pressure. It has been suggested that the surgical field may be further improved during regional anaesthesia by reducing cardiac output through the use of a steep anti-Trendelenburg tilt, IPPV with positive end-expiratory pressure (PEEP) or regional cardiac blockade (T1 to T4). This is, however, unnecessary and undesirable. The advantages of a regional technique over a purely general anaesthetic technique, in terms of blood loss and transfusion requirements, have been confirmed for many procedures, including total hip replacement[59] and prostatic,[57] gynaecological,[60] lower limb[61] and abdominal surgery.[62] With a general anaesthetic technique, hypotension usually has to be profound to make similar reductions in blood loss. Other advantages of regional anaesthesia include a reduction in the incidence of deep venous thrombosis and pulmonary embolism,[63,64] improved viability of large bowel anastomosis[65] and the ability to continue complete analgesia into the postoperative period. Hypotension may precipitate vasovagal episodes in the conscious patient; therefore, unless blood pressure is to be kept at normal levels, it is a kindness to combine sympathetic blockade with sedation or a light general anaesthetic.

Specific hypotensive agents

Hypotensive agents must be infused through a dedicated line to prevent an inadvertent bolus being given during the administration of another drug. Infusion rates must be precisely controlled using constant infusion devices. Those that incorporate cyclical chamber refilling do not, strictly speaking, produce a constant infusion and can lead to small fluctuations in pressure control. Infusion should be commenced after the initial response to the surgical stimulus has settled and the infusion rate increased slowly until satisfactory operative conditions or a target pressure is attained. Current literature reveals the frequent use of MAP targets between 40 and 60 mmHg, with the lower pressure usually reserved for clipping an intracranial aneurysm. Through vast clinical experience, Enderby has found a systolic pressure of 60 mmHg to be safe in most patients but it is vital to remember that, in the overwhelming majority of cases, optimal operating conditions are the goal, not a specific pressure.[34] To this end, there must be good communication between anaesthetist and surgeon so that the pressure can be titrated against operating conditions. Consideration should always be given to, and provision made for, the use of adjuvant intravenous therapy (usually a β-adrenoceptor blocker for a vasodilator technique) so that troublesome reflex responses can be obtunded. The advantages and disadvantages of individual agents are dealt with later.

Sodium nitroprusside

Pharmacology

Sodium nitroprusside (SNP) is a direct acting vasodilator whose action stems from release of nitric oxide (NO) during metabolism. SNP is rapidly degraded by non-enzymatic processes in the red blood cell, and to a much lesser extent

the plasma, to release free cyanide (CN), NO and methaemoglobin. The last has a high affinity for CN and mops up one of the CN molecules to form cyanmethaemoglobin. Most CN is converted to thiocyanate by a rhodanase/thiosulphate reaction and excreted in the urine. A little is also combined with vitamin B_{12} or excreted unchanged in the kidneys and lungs. The drug is extremely evanescent, with a speed of onset and duration of action of only 60–120 seconds.

Administration

SNP is usually infused in 0.01–0.04 per cent solution. Once prepared, the solution must be protected from light, as photodegradation of SNP will liberate CN and increase the risk of toxicity.[66] The initial dose is usually 0.2–0.5 μg/kg per minute.

Toxicity

SNP toxicity results in histotoxic hypoxia with systemic acidosis and is caused by inhibition of the cytochrome oxidase system by the CN liberated during metabolism. The safe maximum dose of SNP is thought to be 1.5 mg/kg but recommendations differ as to the safe maximum infusion rate.[67] Roche recommend 1.5 μg/kg per minute (duration unspecified) for hypotensive anaesthesia and up to 8 μg/kg per minute (for several hours) for hypertensive crises.[68] Others have cited a limit of 10 μg/kg per minute for 1-hour infusions.[38] Whilst the lower dose rate will often prove inadequate in the young patient, if the higher dose rates are being approached, adjuvants are clearly required. Plasma CN concentration of 3 μmol/l should not be exceeded,[68] although as little as 1.53 μmol/l can cause lactic acidosis.[38] Toxicity should be suspected if there is resistance to high infusion rates or if tachycardia persists after the infusion is discontinued. The finding of metabolic acidosis, raised lactate level or elevated venous oxygen saturation strengthens the diagnosis.

Pharmacodynamics

Cardiovascular

SNP reduces blood pressure primarily by arteriolar vasodilatation and has a profound effect on both diastolic and systolic pressure. Venous pressure is also lowered and may contribute to the hypotensive effect by reducing right heart filling.[69] Although SNP has no direct effect on myocardial contractility, when filling pressures are adequately maintained cardiac output may rise as a consequence of reduced left ventricular afterload and baroreceptor-mediated reflex sympathetic stimulation.[43, 70] The sympathetic response is rapid, and cardiac output and heart rate may begin to rise within seconds of starting an SNP infusion. The resulting tachyphylaxis is a serious disadvantage of the drug and is compounded by both vasopressin release and, after 15–30 minutes, by an activated renin–angiotensin axis.[71] Angiotensin II opposes the action of SNP by direct vasoconstriction and by stimulating release of catecholamines.[72] As the latter also enhance the release of renin, a positive feedback mechanism develops and can result in sustained high levels of plasma renin for some time after SNP is discontinued.[73–75] For this reason, an SNP infusion should be tailed off gradually over 10–30 minutes to prevent rebound hypertension.[68] Although SNP reduces myocardial oxygen consumption (by its effect on afterload) and has been shown in animal models to preserve myocardial blood flow during induced hypotension,[27,76,77] human studies have shown that 'ischaemic type' ECG changes are more likely with SNP than with GTN hypotension.[78] This is probably due to the adverse effect that diastolic hypotension and tachycardia have on coronary filling, as well as a 'steal' effect when there is coronary disease.

Respiratory

SNP inhibits hypoxic pulmonary vasoconstriction (HPV)[79–81] and increases shunt fraction[82,83] and $\dot{V}/\dot{Q}$ scatter.[35] Alveolar dead space will also rise if pulmonary perfusion is compromised. This can be minimized by ensuring that the patient is adequately hydrated.[43]

Central nervous system

SNP is a cerebral vasodilator and increases cerebral blood flow and intracranial pressure at normotension.[84] Because of this, if control of intracranial pressure is critical to neurosurgical outcome, the SNP infusion should not be commenced until the dura is open. Cerebral blood flow is better maintained during SNP hypotension than during glyceryl trinitrate- (GTN), trimetaphan- (TMP) or adenosine-induced hypotension and aerobic metabolism is preserved to lower pressures.[23,85,86]

Renal and hepatic function

SNP-induced hypotension may reduce renal blood flow,[87] urine output[11] and creatinine clearance[88] but a long-standing deleterious effect has not been described. Liver blood flow and function are generally well preserved.[1,89]

Trimetaphan

Pharmacology

Trimetaphan camsylate is a presynaptic ganglion blocker with a rapid onset of action (1–3 minutes). Its potency is increased by additional direct vasodilator action and α-adrenergic blockade.[90] TMP-induced histamine release appears to be related to the rate of administration but does not play an important part in its haemodynamic effect.[91] TMP is partly metabolized by enzymatic hydrolysis and partly excreted unchanged in the urine. This occurs rapidly and accounts for its short duration of action (5–15 minutes). Delay of 30 minutes or more in recovery of blood pressure is, however, well recognized.

Administration

TMP is usually infused in a 0.1 per cent solution starting at 20–50 µg/kg per minute.

Adverse effects

Trimetaphan potentiates the action of depolarizing and non-depolarizing agents and, in overdose, can cause respiratory muscle paralysis.[92,93] Potential side effects from the ganglion blockade include paralytic ileus, retention of urine and increased sensitivity to insulin. Because of histamine release, caution is required in subjects with allergy.

Pharmacodynamics

Cardiovascular

Hypotension is produced primarily through arteriolar vasodilatation but venodilatation with blood pooling may also contribute. Ganglion blockade obtunds the sympathetic[94] and renin–angiotensin[95] responses to hypotension although tachycardia and resistance still occur because of vagal blockade and vasopressin release.[96] The longer duration of action of TMP in comparison with SNP explains why blood pressure is less easy to control and recovery is slower. Rebound hypertension is, however, less common.

Respiratory

As with SNP, TMP-induced hypotension impairs gas exchange by causing $\dot{V}/\dot{Q}$ mismatch.[97] This may be compounded by an increase in bronchial smooth muscle tone induced by histamine.[38]

Central nervous system

Because TMP produces little cerebral vasodilatation, intracranial pressure is more stable and cerebral 'steal' is less likely than with SNP.[84] TMP may therefore be preferred in the presence of focal cerebral ischaemia.[23] In most circumstances, however, hypotension induced by TMP results in autoregulation failure, cortical hypoxia, cerebral lactic acidosis and EEG abnormality at a higher MAP than with SNP.[23,56,85,94] A further disadvantage of TMP is that the mydriasis caused by ciliary ganglion blockade prevents the pupil being used as an indicator of cerebral perfusion and anaesthetic depth.

Renal and hepatic function

Renal and splanchnic blood flow may fall during TMP-induced hypotension.[98]

The trimetaphan–sodium nitroprusside mixture

The potency and evanescent nature of SNP coupled with its ability to maintain organ blood flow at low pressures would make it the ideal hypotensive agent if it were not for the problems of reflex activity and cyanide release. MacRae has pioneered the use of the TMP–SNP mixture, which enables the dose of SNP to be reduced while still retaining a powerful and evanescent vasodilator effect.[39,99] Cyanide levels are reduced and reflex sympathetic responses are blunted.[100,101] Tachyphylaxis and rebound hypertension still occur but are obtunded in comparison to those of SNP alone.

Administration

The most effective mixtures have a TMP:SNP ratio of 2.5–5:1[71,101] and it is our practice to use 250 mg trimetaphan and 50 mg SNP in 500 ml of 5 per cent dextrose. This is commenced at 1–2 ml per hour and increased as required.

Glyceryl trinitrate

Pharmacology

GTN is an organic nitrate and causes direct relaxation of vascular smooth muscle by reacting with tissue sulph-hydryls to produce, eventually, nitric oxide (NO). During prolonged infusion, tolerance to GTN may occur due to saturation of the tissue sulphhydryl groups.[102] GTN is rapidly metabolized in the liver, red blood cells and blood vessels, and its active metabolites are excreted in the urine. The onset of action and half-life are both approximately 2 minutes, and effects are usually reversed within 10 minutes of discontinuing the infusion.[78]

Administration

It is recommended that GTN be used in a concentration of 400 µg/ml or less.[103] Infusion is commenced at 10–20 µg per minute and increased at 2-minute intervals; 400–500 µg per minute is sometimes required to produce a significant pressure drop.[78]

Toxicity

Prolonged use of high dose intravenous GTN may lead to acute methaemoglobinaemia and impaired oxygen transport despite normal arterial oxygen tensions. If this occurs, the infusion must be stopped and methylene blue therapy commenced.

Pharmacodynamics

Cardiovascular

Venous dilatation predominates at low doses but reflex sympathetic stimulation often prevents a significant reduction in pressure unless arteriolar dilatation is established at higher doses.[104] Target pressures are achieved more slowly and are more difficult to maintain than with SNP but titrations are smoother[78] and there is a reduced risk of overshoot and rebound hypertension.[105] GTN is also superior to SNP in the preservation of myocardial perfusion and is the agent of choice in ischaemic heart disease.[78,106,107]

Respiratory

GTN is a pulmonary vasodilator, inhibits the HPV response and increases shunt fraction.[80,82] Unless pulmonary blood flow is maintained during hypotension, dead space will also increase. An increased inspired oxygen concentration is required.[78]

Central nervous system

GTN causes a greater rise in intracranial pressure than SNP at normotension[108] and is less effective in extending the lower limit of cerebral blood flow preservation during hypotension.[23]

Isoflurane

Pharmacology

Isoflurane is a methyl ethyl ether with relatively low blood solubility. Emergence from anaesthesia is therefore faster than with halothane or enflurane. It is mainly excreted unchanged in the lungs, so the potential for renal or hepatic toxicity from metabolites is small.[109]

Pharmacodynamics

Cardiovascular

In concentrations of up to 1.9 MAC, isoflurane produces little change in cardiac output but induces a dose-related reduction in blood pressure by lowering systemic vascular resistance.[110,111] This makes it suitable as a sole agent for induced hypotension, and safer, in this context, than enflurane and halothane which decrease pressure primarily by reducing cardiac output.[111] When profound hypotension is required, the use of isoflurane is less valid because the high concentrations that are usually required can significantly impair cardiac output. Isoflurane requirements can be minimized in this situation by using adjuvant therapy.[112] Isoflurane (and other volatile anaesthetics) inhibits baroreceptor pathways, so reflex cardiovascular stimulation from its own or adjuvant's action is muted.[113–115] There have been concerns over the propensity of isoflurane to cause coronary 'steal'.[116,117] This now appears to be no more common than with enflurane or halothane,[31] and much less common than with the non-anaesthetic arterial vasodilators.[32]

Respiratory

Gas exchange is better maintained during isoflurane-induced hypotension than during SNP hypotension. Shunt fraction and arterial oxygenation are not significantly altered, dead space is usually unchanged because of the preservation of pulmonary artery pressure, and the reduction in functional residual capacity is no more than during normotensive anaesthesia.[44,83,118,119] The HPV reflex is probably preserved at concentrations of up to 1.5 per cent but may be inhibited at higher concentrations.[120,121]

Central nervous system

Although isoflurane may increase cerebral blood flow at concentrations above 1 MAC at normotension,[122] during isoflurane-induced hypotension CBF is usually maintained at normal levels.[123–125] Isoflurane markedly reduces cerebral oxygen consumption and, with respect to global cerebral oxygenation, has advantages over halothane, TMP, SNP, GTN and adenosine for induced hypotension in neurosurgery.[10,86,123,126] Isoflurane may, however, cause 'steal' from underperfused regions and does not improve outcome once ischaemic cell damage has occurred.[25,127,128]

Adenosine and adenosine triphosphate (ATP)

Although, at present, adenosine and ATP are not licenced for use in induced hypotension there is much interest in their potential for this purpose. Adenosine is a potent vasodilator whose actions are mediated through specific receptors.[129] It has a faster onset of action than SNP and, with a plasma half-life of less than 20 seconds, an extremely short duration of action.[130] The actions of ATP stem from its rapid degradation to adenosine in the circulation.[131]

The features of adenosine-induced hypotension that are particularly advantageous are the ease of pressure control, an enhanced cardiac output without troublesome tachycardia,[132] well-maintained organ perfusion[89] with a reduction in metabolic demand,[133,134] possible analgesic properties[135] and, because of inhibition of catecholamine and renin release,[136] minimal tachyphylaxis and rebound hypertension.[133] On the negative side, there are concerns over the effect of adenosine on renal function[137] and cerebral oxygenation at low pressure[86,138] and its potential to cause coronary 'steal'.[139,140] In addition, ATP administration is associated with lactic acidosis and the accumulation of phosphate which can potentiate arrhythmias.[141,142]

Adrenergic antagonists

Labetalol

Labetalol is an adrenoceptor antagonist with actions at α- and β-receptors. The action at β-receptors is up to seven times greater than that at α-receptors and lasts for 90 as opposed to 30 minutes.[143] The overall effect is rapid, persistent reduction in blood pressure without concomitant tachycardia. If several increments are required, β-blockade can become unnecessarily profound and may lead to bradyarrhythmias or a delay in pressure recovery for several hours. Despite this, labetalol is very useful in induced hypotension, either on its own[144] or as an adjuvant to reduce SNP requirements and combat tachycardia, tachyphylaxis and rebound hypertension. Similar benefits have been shown using bolus doses of propranolol.[145]

Inhalational anaesthetics potentiate labetalol and, with halothane anaesthesia, the recommended starting dose in adults is 10–20 mg by intravenous bolus.[146,147] The maximal effect usually occurs by 5 minutes, after which, if the goal is not achieved, further increments may be given.[147]

Esmolol

There has been much interest in the use of esmolol, a β1-adrenoceptor antagonist, in controlled hypotension. It has the advantage over other β-antagonists in having a short elimination half-life (9 minutes) and produces excellent operating conditions without tachycardia or rebound.[148] This is achieved, however, at the expense of myocardial performance and possibly tissue oxygen balance, and is recommended as adjuvant therapy only.[149,150]

Phentolamine and phenoxybenzamine

Phentolamine is a short acting, competitive α-adrenoceptor antagonist which has additional direct vasodilator action. The predominant effect is vasodilatation but blockade of α2-receptors also causes noradrenaline release and cardiac stimulation. Although phentolamine is occasionally used to achieve rapid control of pressure, tachycardia reduces its usefulness in controlled hypotension.

Phenoxybenzamine is a relatively selective α1-adrenergic antagonist but binds irreversibly with the receptor and has an effective half-life of up to 24 hours. Its main use is in anaesthesia for phaeochromocytoma surgery and in preventing sympathetic overdrive after paediatric cardiac surgery.

Others

Hydralazine

Hydralazine is a direct-acting vasodilator whose action develops relatively slowly over 15–30 minutes. It may be used by bolus (5–10 mg at 20- to 30-minute intervals) as an adjuvant to anaesthesia-induced hypotension. Reflex sympathetic stimulation leads to a hyperdynamic circulation.

Clonidine

Clonidine is an α2-adrenoceptor agonist that decreases central and peripheral sympathetic outflow and inhibits the release of renin and vasopressin.[151,152] Clonidine reduces SNP requirements for induced hypotension in animals and may also have a cerebral protective effect.[22,153] In the clinical setting it has been used to treat rebound hypertension after isoflurane withdrawal.[112]

Calcium channel blockers

A single bolus of verapamil (0.07 mg/kg) has been used to produce hypotension without tachycardia,[154] and of diltiazem to reduce dose requirements in SNP hypotension.[155] Nicardipine infusions induce comparable hypotensive conditions to SNP but, apart from the absence of toxicity, do not appear to confer any significant advantages and are associated with slower blood pressure recovery.[115]

Drugs used in premedication

β-Adrenoceptor antagonists

Pretreatment with oral propranolol modifies renin and catecholamine release in response to SNP-induced hypotension, reduces nitroprusside dosage and suppresses reflex tachycardia and rebound hypertension.[75,156] Atenolol (own unpublished data), metoprolol and oxprenolol premedications have similar effects.[7] β-Blocker premedication is, however, associated with an increased risk of brady-arrhythmias after induction of anaesthesia.[7] Whilst this might be prevented by including atropine in the premedication, the risks of inducing β-blockade in a largely unmonitored situation usually outweigh the potential benefits.[7] β-Blocker premedication should be used only when profound hypotension is required and only in conjunction with an anticholinergic agent.

Angiotensin-converting enzyme (ACE) inhibitors

Pretreatment with captopril reduces SNP requirements during induced hypotension and may prevent rebound hypertension.[157,158] Tachycardia still occurs, and the potential adverse effects of decreased CBF and delayed pressure recovery confirm that their routine use in premedication cannot be recommended.[158–160]

Problems and complications

Mortality

Reviews of hypotensive anaesthesia in the 1950s and early 1960s often revealed high mortality rates and led to much concern over safety.[161] At that time, however, experience was limited, monitoring basic and patient selection often poor. Since then, there have been vast improvements in these areas and in the understanding and provision of low-pressure circulations. This has been reflected in the wider application and improved outcome associated with modern techniques. Enderby has reviewed 20 558 hypotensive anaesthetics between 1950 and 1979.[34] A sample analysis estimated that 11 per cent of patients had been taken to systolic pressure of 50–60 mmHg and 78 per cent to 60–80 mmHg. Of the ten deaths within the whole series, only two were attributable to poor blood pressure control, both occurring in the first decade. There was no death among 700 cases reported by Kerr in 1977.[162] There is, unfortunately, a lack of large controlled studies of hypotensive techniques and, although smaller ones have failed to show that there is any increase in mortality (or morbidity) associated with induced hypotension, conclusions have limited value because of the small patient numbers.[163,164]

Central nervous system

Short-term impairment of higher cerebral function is a relatively common complication of normotensive general anaesthesia, particularly in the elderly, but there is, as yet, no strong evidence to suggest that it is exacerbated by properly applied hypotensive techniques.[1,165–167] Much of the early morbidity and mortality was due to severe cerebral ischaemia but this is seldom reported today.[161] In Enderby's series there were only two cases of cerebral thrombosis in over 20 000 hypotensive procedures, one being due to postoperative airway obstruction.[34]

Anterior spinal artery thrombosis is a very rare but catastrophic complication. Factors that have been implicated in its development include aortic surgery, arterial disease, regional anaesthesia with adrenaline-containing local anaesthetics and hypotension.[168] Concerns over the effect of hypotension on spinal cord perfusion during scoliosis surgery have led to the evaluation of hypotensive agents and spinal cord distraction on animal spinal cord blood flow. Flows were not compromised by spinal distraction[169,170] and were well maintained during hypotension induced by GTN,[89,170] SNP[77,89] and adenosine.[77,89] Trimetaphan-induced hypotension has compromised spinal cord blood flow in dogs and cannot be recommended for spinal surgery.[169,171]

Myocardial ischaemia

Although, in Enderby's series, myocardial infarction was seen only once during the course of an anaesthetic, this statistic should be treated with caution because ECG diagnosis was not available in the majority of cases. Nevertheless, serious cardiac morbidity is clearly uncommon during hypotensive anaesthesia because of careful patient selection, the pharmacodynamic reasons mentioned earlier and because ECG monitoring allows early detection of myocardial ischaemia. It is increasingly recognized that perioperative cardiac morbidity is associated more frequently with the postoperative period than with the intraoperative period.[172,173] This emphasizes the importance of very careful postoperative care.

Resistance and rebound

Resistance to the acute reduction in blood pressure is due to the opposing actions of activated homoeostatic mechanisms. These are most dramatic in the young patient and are mainly comprised of the neurohumeral sympathetic response, vasopressin release and activation of the renin–angiotensin axis. Resistance is more than just a frustration to the anaesthetist. An activated stress response increases tissue oxygen consumption, tachycardia reduces coronary filling time and promotes bleeding, and repeated increases in the dose of the hypotensive agent may lead to toxicity.

Rebound hypertension occurs when, at the end of a procedure, the abrupt withdrawal of an evanescent hypotensive agent leads to the above mechanisms (particularly angiotensin II) being unopposed. Because the return of normal cerebral blood flow autoregulation is delayed for some time after a vasodilator (including isoflurane) is stopped, the brain is especially vulnerable to this pressure surge, and cerebral hyperaemia and oedema may result.[23,24,174–176] In addition to this, there may be renewed bleeding from the wound, and increased myocardial oxygen demand may lead to ischaemia in marginally perfused tissue. Hypotensive agents must, therefore, be tailed off gradually, guided by meticulous pressure monitoring. Both resistance and rebound are attenuated by ensuring adequate anaesthesia, analgesia and warmth and by using adjuvant therapy preoperatively (e.g. β-adrenergic agents, ACE inhibitors) or intraoperatively by either infusion (the TMP–SNP mixture) or bolus (e.g. labetalol, esmolol, clonidine). Isoflurane, TMP and adenosine hypotensive techniques have advantages over SNP in this context (see above).

Other complications

Postoperative, reactionary haemorrhage is promoted by inadequate surgical haemostasis, rebound hypertension and abnormalities of the haemostatic mechanism. These can all be favourably influenced in a well-managed technique. Although the induction of hypotension *per se* does not cause a clinically significant haemostatic abnormality, fibrinolytic activity may be raised by prolonged use of tourniquets,[177] thrombotic mechanisms are altered by regional anaesthesia[178] and platelet function may be adversely affected by SNP.[179] The clinical significance of the last is largely untested.

Tissue trauma from pressure may be more severe during hypotensive states and may be seen, for example, in the occurrence of retraction injury during neurosurgery or the mixed-tissue damage in the contralateral leg during hip surgery.[180] Meticulous surgical technique and careful positioning will minimize this problem.

Postoperative care

It is clear from the above and from the early reviews that much of the morbidity and mortality associated with induced hypotension occurs in the postoperative period.[34,166] The patient must therefore be closely observed during this period and meticulous attention must be paid to the maintenance of a clear airway and the provision of supplemental oxygen, analgesia and a warm environment. The ECG, blood pressure, respiratory rate, oxygen saturation and wound drainage must be monitored and recorded until stability is achieved. Recovery staff must be aware that patients may be slower to awaken and regain airway control if a prolonged deep isoflurane technique has been used (although this is not a universal finding[114]). Muscle relaxation may also be prolonged by induced hypotension, and neuromuscular transmission should therefore be objectively assessed using a nerve stimulator before the patient leaves the operating theatre. If there is rebound hypertension, a degree of head-up tilt may reduce the risk of cerebral oedema. On the other hand, orthostatic hypotension and cerebral hypoxia may occur if the patient is propped up too soon, especially if there is residual sympathetic blockade from a regional or ganglion blocker technique.[23] In these circumstances, therefore, the patient must be moved from the supine position as little as possible and the blood pressure response observed if a change in position is unavoidable. The complications of hypotensive anaesthesia that may present during the recovery period are listed in Table 39.2.

Table 39.2 Problems arising in the recovery period and mechanisms specifically related to hypotensive anaesthesia

Cardiovascular	*Respiratory*
Delayed pressure recovery (labetalol, TMP; inadequate fluid replacement; myocardial ischaemia)	Airway obstruction
	Hypoxia (persistent hypotension, vasodilatation, $\dot{V}/\dot{Q}$ scatter; increasing oxygen consumption)
Postural hypotension (TMP, regional block)	
Rebound hypertension	
Myocardial ischaemia	*Wound*
Arrhythmias (β-blockade; myocardial ischaemia)	Reactionary bleeding
	Metabolic
Nervous system	Hypothermia (prolonged vasodilatation)
Delayed awakening (cerebral ischaemia; cerebral oedema; deep anaesthesia; metabolic; [prolongation neuromuscular blockade])	Lactic acidosis (SNP, ATP toxicity; excessive hypotension; inadequate fluid replacement)
Impaired higher cerebral function (general anaesthesia; cerebral ischaemia)	
Neurological deficit (cerebral, spinal ischaemia; peripheral nerve injury)	

Specific applications

Detailed descriptions of neurosurgical or other specialist anaesthetic techniques are beyond the scope of this chapter and are dealt with elsewhere in this volume. This section outlines the role of hypotensive anaesthesia in selected situations.

Neurosurgery

For many years, induced hypotension has played a major role in the provision of safe operating conditions for arteriovenous malformation excision, trans-sphenoidal hypophysectomy and, in particular, cerebral aneurysm clipping. With the improvement in surgical techniques and the advent of the operating microscope, hypotension is now used more selectively and with less frequent employment of profound levels. The technique is, however, still widely practised and regarded as an invaluable tool for high-risk cases.

In cerebral aneurysm surgery, a low-pressure circulation not only reduces blood loss and improves visualization in the operative field but also reduces aneurysm transmural pressure. This favours safe application of a clip and probably reduces the risk of intraoperative rupture. In addition, if an aneurysm ruptures during surgery, profound hypotension reduces bleeding and facilitates rapid identification and control of the bleeding point.[9]

Vasodilator infusions are commenced after the dura has been opened, to avoid producing a surge in intracranial pressure, and the dose is increased slowly to produce moderate hypotension. When the dissection nears the aneurysm, the MAP is reduced to between 50 and 60 mmHg and maintained there until the aneurysm is clipped. The pressure is then allowed to recover slowly. It is recommended that in chronic hypertension the MAP be reduced by only 30 per cent of the preoperative level.[9,181]

Certain qualifications must be made. Hypotension must be avoided if there is preoperative evidence of cerebral vasospasm of if a feeding artery is to be temporarily clipped: the former because autoregulation of blood flow is lost during spasm and the latter in order to preserve collateral perfusion of the ischaemic area.[182,183] Cerebrovascular sensitivity to carbon dioxide is lost during induced hypotension, as it is in the presence of intracranial pathology.[23,184] Therefore, although hyperventilation is still recommended to try to reverse pathological hyperperfusion, cerebral 'steal' and brain swelling, the actual benefit is uncertain. It has been suggested that esmolol may have a role to play when hypotension is induced in patients with cerebral hyperperfusion but the premise is unproven.[13]

Ophthalmic surgery

The primary role of the anaesthetist in ophthalmic procedures is to provide safe anaesthesia, a stationary eye and cardiovascular stability. For intraocular procedures the additional aims must be to prevent the extrusion of intraocular contents by lowering intraocular pressure (IOP) and to prevent bleeding when uveal dissection is undertaken.

The major influences on IOP during surgery are aqueous volume dynamics, choroidal blood volume, vitreous volume and intra- and extraocular muscle tone. Deep, balanced general anaesthesia reduces IOP, probably through a combination of central IOP control centre inhibition, facilitation of aqueous outflow and both intra- and extraocular muscle relaxation.[185] Choroidal blood volume can be readily reduced by using an anti-Trendelenburg tilt to improve venous drainage and hyperventilation to produce choroidal vasoconstriction. An impeccable anaesthetic technique can therefore produce reductions in IOP that are satisfactory for most procedures, without the use of significant hypotension. Choroidal blood flow undergoes autoregulation and the point at which flow becomes pressure dependent is influenced by IOP itself.[186–188] The relation between pressure, flow and IOP is therefore complex, and recent work using a prone pig model has questioned any direct relation between IOP and systemic arterial pressure.[189] However, if uveal dissection is to be carried out, profound hypotension must be induced to prevent bleeding into the vitreous. With tumours of the ciliary body it is usually sufficient to reduce MAP to between 50 and 60 mmHg but resection of a choroidal melanoma requires a period of choroidal ischaemia, and the MAP must be transiently reduced to approximately 35 mmHg.[8] This is, clearly, highly specialized surgery and anaesthesia, although the principle of a slow, controlled induction of a low-pressure vasodilated circulation remains the same. The TMP–SNP mixture is the mainstay of the technique and a single oral dose of atenolol with the premedication obtunds the heart rate response. Cerebral function monitoring is essential when patients are subjected to this degree of hypotension, and the compressed spectral array has proved very satisfactory in this context.[8] Over 350 choroidal resections have now been carried out in this centre without significant morbidity (and no mortality) attributable to the hypotensive technique. Occasionally, because of intraoperative ECG or EEG abnormalities, the pressure reduction is halted and reversed and the eye, inevitably, sacrificed. This possibility must be clearly understood and accepted by all parties before the operation is undertaken.

Scoliosis surgery

Anaesthetic considerations for spinal fusion include the provision of general anaesthesia for patients in the prone position who frequently have respiratory dysfunction, the monitoring of spinal cord function using either a 'wake-up' test or somatosensory evoked potentials (SSEPs) and the control of bleeding. Blood loss can be massive and, whilst this will be somewhat attenuated by anaesthesia, local infiltration of vasoconstrictors and careful positioning, it can be markedly reduced by lowering the blood pressure.[190-192] There are occasional reports in which induced hypotension has been implicated as the cause of spinal cord ischaemia;[193,194] for this reason, it is emphasized that pressure must be reduced slowly (preferably under SSEP monitoring) and only to the point at which operating conditions become satisfactory or a MAP of 60–70 mmHg is reached.[195] The use of TMP on its own cannot be recommended (see above), and hyperventilation must be prevented as this may induce vasoconstriction of spinal vessels.

Pregnancy

Unless it is considered to be potentially life saving, hypotension must not be induced during pregnancy because fetal asphyxia may result from decreased uterine blood flow.[196,197] The one circumstance in which it is very occasionally used is for intracranial aneurysm clipping.[198] The fetoplacental unit is freely permeable to all the commonly used hypotensive agents and can lead to high levels of fetal cyanide during SNP infusions.[38] Clearly, this puts the fetus at considerable risk, although there are reports of normal fetal outcome and development after high maternal SNP dosage.[199] TMP can cause fetal ileus and may also inhibit compensatory mechanisms if there is inadvertent maternal aortocaval compression.[200] Although commonly used in pre-eclampsia, labetalol is occasionally associated with perinatal and neonatal distress and should not be given in prolonged high dose.[147] Deep isoflurane anaesthesia has resulted in fetal acidosis in an animal model.[201] GTN may be the least toxic of the agents mentioned and its safe use in pregnancy has been reported.[202,203] Close fetal heart rate monitoring is mandatory and serves as a guide to the safe level for hypotension and hyperventilation.[198]

REFERENCES

1. Thompson GE, Miller RD, Stevens WC, Murray WR. Hypotensive anesthesia for total hip arthroplasty: a study of blood loss and organ function (brain, heart, liver and kidney). *Anesthesiology* 1978; **48**: 91–6.

2. Brodsky JW, Dickson JH, Erwin WD, Rossi CD. Hypotensive anesthesia for scoliosis surgery in Jehovah's Witness. *Spine* 1991; **16**: 304–6.

3. Beare R. Indications for hypotensive anaesthesia. In: Enderby GEH, ed. *Hypotensive anaesthesia*. Edinburgh: Churchill Livingstone, 1985: 99–108.

4. Lawson NW, Thompson DS, Nelson CL, Flacke JW, North ER. Sodium nitroprusside-induced hypotension for supine total hip replacement. *Anesthesia and Analgesia* 1976; **55**: 654–62.

5. Ward CF, Alfery DD, Saidman LJ, Waldman J. Deliberate hypotension in head and neck surgery. *Head and Neck Surgery* 1980; **2**: 185–95.

6. Powell JL, Mogelnicki SR, Franklin III EW, Chambers DA, Burrell MO. A deliberate hypotensive technique for decreasing blood loss during radical hysterectomy and pelvic lymphadenectomy. *American Journal of Obstetrics and Gynecology* 1983; **147**: 196–202.

7. Simpson DL, MacRae WR, Wildsmith JAW, Dale BAB. Acute beta-adrenoreceptor blockade and induced hypotension. *Anaesthesia* 1987; **42**: 243–8.

8. Todd JG, Colvin JR. Ophthalmic surgery. In: MacRae WR, Wildsmith JAW, eds. *Induced hypotension*. Amsterdam: Elsevier Science Publ, 1991: 257–69.

9. Black S. Cerebral aneurysm and arteriovenous malformation. In: Cucchiara RF, Michenfelder JD, eds. *Clinical neuroanesthesia*. New York: Churchill Livingstone, 1990: 223–54.

10. Newberg LA, Milde JH, Michenfelder JD. Systemic and cerebral effects of isoflurane-induced hypotension in dogs. *Anesthesiology* 1984; **60**: 541–6.

11. MacRae WR. Induced hypotension. *British Journal of Hospital Medicine* 1985; **33**: 341–3.

12. Lam AM. Monitoring methods during induced hypotension. *Current Opinion in Anaesthesiology* 1988; **1**: 101–4.

13. Ornstein E, Young WL, Prohovnik I, Ostapkovich N, Ree GT, Stein BM. Effect of cardiac output on CBF during deliberate hypotension. *Anesthesiology* 1990; **73**: A169.

14. Sivarajan M, Amory DW, Everett GB, Buffington C. Blood pressure, not cardiac output, determines blood loss during induced hypotension. *Anesthesia and Analgesia* 1980; **59**: 203–6.

15. Modig J, Malmberg P, Karlström G. Effect of epidural versus general anaesthesia on calf blood flow. *Acta Anaesthesiologica Scandinavica* 1980; **24**: 305–9.

16. Enderby GEH. Historical review of the practice of deliberate hypotension. In: Enderby GEH, ed. *Hypotensive anaesthesia*. Edinburgh: Churchill Livingstone, 1985; 75–91.

17. Wright MO. The cardiovascular system. In: MacRae WR, Wildsmith JAW, eds. *Induced hypotension*. Amsterdam: Elsevier Science Publ, 1991: 11–86.

18. Shipley RE, Stady RS. Changes in the renal blood flow, extraction of inulin, GFR, tissue pressure and urine flow with acute alterations in renal artery pressure. *American Journal of Physiology* 1951; **167**: 676–88.

19. Mosher P, Ross J, McFate PA, Shaw RF. Control of coronary blood flow by an autoregulatory mechanism. *Circulation Research* 1964; **14**: 250–9.

20. Lassen NA, Christensen MS. Physiology of cerebral blood flow. *British Journal of Anaesthesia* 1976; **48**: 719–34.

21. Harper AM, Deshmukh VD, Rowan JD, Jennett WB. The influence of sympathetic nervous activity on cerebral blood flow. *Archives of Neurology* 1972; **27**: 1–6.

22. Hoffman WE, Cheng MA, Thomas C, Baughman VL, Albrecht RF. Clonidine decreases plasma catecholamines and improves outcome from incomplete ischemia in the rat. *Anesthesia and Analgesia* 1991; **73**: 460–4.

23. McDowall DG. Induced hypotension and brain ischaemia. *British Journal of Anaesthesia* 1985; **57**: 110–19.

24. Stånge K, Lagerkranser M, Sollevi A. Nitroprusside-induced hypotension and cerebrovascular autoregulation in the anesthetized pig. *Anesthesiology* 1991; **73**: 745–52.

25. Todd MM, Warner DS. A comfortable hypothesis reevaluated. Cerebral metabolic depression and brain protection during ischemia. *Anesthesiology* 1992; **76**: 161–4.

26. Sano T, Drummond JC, Patel PM, Grafe MR, Watson JC, Cole DJ. A comparison of cerebral protective effects of isoflurane and mild hypothermia in a model of incomplete forebrain ischemia in the rat. *Anesthesiology* 1992; **76**: 221–8.

27. Jupa M, Ducardus R. Autoregulation of myocardial blood flow under controlled hypotension in the dog. *Acta Anaesthesiologica Scandinavica* 1982; **26**: 199–204.

28. Rollason WN, Hough JM. A re-examination of some electro-cardiographic studies during hypotensive anaesthesia. The effect of rate of fall of blood pressure. *British Journal of Anaesthesia* 1969; **41**: 985–93.

29. Patel H. Monitoring hypotensive anaesthesia. Cerebral electrical activity. In: Enderby GEH, ed. *Hypotensive anaesthesia*. Edinburgh: Churchill Livingstone, 1985: 214–25.

30. Sharrock NE, Urquhart BS. Outcome analysis of hypotensive anesthesia in hypertensive patients. *Anesthesiology* 1990; **73**: A68.

31. Slogoff S, Keats AS, Wayne ED, *et al*. Steal-prone coronary anatomy and myocardial ischemia associated with four primary anesthetic agents in humans. *Anesthesia and Analgesia* 1991; **72**: 22–7.

32. Lillehug SL, Tinker JH. Why do 'pure' vasodilators cause coronary steal when anesthetics don't (or seldom do)? *Anesthesia and Analgesia* 1991; **73**: 681–2.

33. Dandona P, James AM, Newbury ML, Beckett AG. Cerebral blood flow in diabetes mellitus: evidence of abnormal cerebrovascular reactivity. *British Medical Journal* 1978; **2**: 325–6.

34. Enderby GEH. Safe hypotensive anaesthesia. In: Enderby GEH, ed. *Hypotensive anaesthesia*. Edinburgh: Churchill Livingstone, 1985: 262–75.

35. Wildsmith JAW, Drummond GB, MacRae WR. Blood-gas changes during induced hypotension with sodium nitroprusside. *British Journal of Anaesthesia* 1975; **47**: 1205–11.

36. Vickers MD, Schneiden H, Wood-Smith FG, eds. *Drugs in anaesthetic practice*, 6th ed. London: Butterworths, 1984: 280–3.

37. Miller RD, Way WL, Dolan WM, Stevens WC, Eger II EI. The dependence of pancuronium- and *d*-tubocurarine-induced neuromuscular blockades on alveolar concentrations of halothane and Forane. *Anesthesiology* 1972; **37**: 573–81.

38. Adams AP, Hewitt PB. Clinical pharmacology of hypotensive agents. *International Anesthesiology Clinics* 1981; **20**: 95–109.

39. Wildsmith JAW, Sinclair CJ, Thorn J, MacRae WR, Fagan D,

Scott DB. Haemodynamic effects of induced hypotension with a nitroprusside–trimetaphan mixture. *British Journal of Anaesthesia* 1983; **55**: 381–9.

40. Holmes F. Induced hypotension in orthopaedic surgery. *Journal of Bone and Joint Surgery [Br]* 1956; **38**: 846–54.

41. Atkinson RS, Rushman GB, Lee JA. *A synopsis of anaesthesia*. Bristol: John Wright, 1987: 762.

42. Simpson P. Perioperative blood loss and its reduction: the role of the anaesthetist. *British Journal of Anaesthesia* 1992; **69**: 498–507.

43. Khambatta HJ, Stone JG, Matteo RS. Effect of sodium nitroprusside-induced hypotension on pulmonary deadspace. *British Journal of Anaesthesia* 1982; **54**: 1197–9.

44. Nicholas JF, Lam AM. Isoflurane-induced hypotension does not cause impairment in pulmonary gas exchange. *Canadian Anaesthetists' Society Journal* 1984; **31**: 352–8.

45. Fahmy NR, Chandler HP, Patel DG, Lappas DG. Hemodynamics and oxygen availability during acute hemodilution in conscious man. *Anesthesiology* 1980; **53**: S84.

46. Messmer K, Sunder-Plassmann L. Hemodilution. *Progress in Surgery* 1974; **13**: 208–45.

47. Fahmy NR. Techniques for deliberate hypotension. Haemodilution and hypotension. In: Enderby GEH, ed. *Hypotensive anaesthesia*. Edinburgh: Churchill Livingstone, 1985: 164–83.

48. Crystal GJ, Salem MR. Myocardial and systemic hemodynamics during isovolemic hemodilution alone and combined with nitroprusside-induced controlled hypotension. *Anesthesia and Analgesia* 1991; **72**: 227–37.

49. Blackburn H, Katigbak R. What electrocardiographic leads to take after exercise? *American Heart Journal* 1964; **67**: 184–5.

50. MacFarlane PW. Lead systems. In: MacFarlane PW, Lawire TDV, eds. *Comprehensive electrocardiology. Theory and practice in health and disease*. New York: Pergamon Press, 1989: 315–52.

51. Enderby DH. Monitoring hypotensive anaesthesia. Blood pressure and other systems. In: Enderby GEH, ed. *Hypotensive anaesthesia*. Edinburgh: Churchill Livingstone, 1985; 193–213.

52. Gourdeau M, Martin R, Lamarche Y, Tetreault L. Oscillometry and direct blood pressure: a comparative clinical study during deliberate hypotension. *Canadian Anaesthetists' Society Journal* 1986; **33**: 300–7.

53. Kemmotsu O, Ueda M, Otsuka H, *et al*. Blood pressure measurement by arterial tonometry in controlled hypotension. *Anesthesia and Analgesia* 1991; **73**: 54–8.

54. Aitken HA, Todd JG, Kenny GNC. Comparison of the Finapres and direct arterial pressure monitoring during profound hypotensive anaesthesia. *British Journal of Anaesthesia* 1991; **67**: 36–40.

55. Epstein RH, Bartkowski RR, Huffnagle S. Continuous noninvasive finger blood pressure during controlled hypotension. A comparison with intraarterial pressure. *Anesthesiology* 1991; **75**: 796–803.

56. Prior PF. EEG monitoring and evoked potentials in brain ischaemia. *British Journal of Anaesthesia* 1985; **57**: 63–81.

57. Abrams PH, Shah PJR, Bryning K, Gache CGC, Ashken MH, Green NA. Blood loss during transurethral resection of the prostate. *Anaesthesia* 1982; **37**: 71–3.

58. Thorburn J. Subarachnoid blockade and total hip replacement.

The effect of ephedrine on intraoperative blood loss. *British Journal of Anaesthesia* 1985; **57**: 290–3.

59. Modig J. Beneficial effects on intraoperative and postoperative blood loss in total hip replacement when performed under lumbar epidural anesthesia. An explanatory study. *Acta Chirurgica Scandinavica Supplement* 1989; **550**: 95–103.

60. Loudon JD, Scott DB. Blood loss in gynaecological operations. *Journal of Obstetrics and Gynaecology of the British Empire* 1960; **67**: 561–5.

61. Cook PT, Davies MJ, Cronin KD, Moran P. A prospective randomised trial comparing spinal anaesthesia using hyperbaric cinchocaine with general anaesthesia for lower limb vascular surgery. *Anaesthesia and Intensive Care* 1986; **14**: 373–80.

62. Worsley MH, Wishart HY, Peebles Brown DA, Aitkenhead AR. High spinal nerve block for large bowel anastomosis. A prospective study. *British Journal of Anaesthesia* 1988; **60**: 836–40.

63. Thorburn J, Louden JR, Vallance R. Spinal and general anaesthesia in total hip replacement: frequency of deep vein thrombosis. *British Journal of Anaesthesia* 1980; **52**: 1117–20.

64. Modig J, Hjelmstedt A, Sahlstedt B, Maripuu E. Comparative influences of epidural and general anaesthesia on deep venous thrombosis and pulmonary embolism after total hip replacement. *Acta Chirurgica Scandinavica* 1981; **147**: 125–30.

65. Aikenhead AR, Wishart HY, Peebles Brown DA. High spinal nerve block for large bowel anastomosis. *British Journal of Anaesthesia* 1978; **50**: 177–85.

66. Ikeda S, Schweiss JF, Frank PA, Homan SM. *In vitro* cyanide release from sodium nitroprusside. *Anesthesiology* 1987; **66**: 381–5.

67. Vesey CJ, Cole PV, Simpson PJ. Sodium nitroprusside in anaesthesia. *British Medical Journal* 1975; **3**: 229.

68. Association of the British Pharmaceutical Industry. *ABPI data sheet compendium 1991–1992*, London: Datapharm, 1991: 1268–70.

69. Beierholm EA, Sorensen MB, Sroczynski Z, Spotoft H, Gothgen I, Thorshauge C. Haemodynamic changes during sodium nitroprusside-induced hypotension and halothane/nitrous oxide anaesthesia. *Acta Anaesthesiologica Scandinavica* 1983; **27**: 99–103.

70. Wildsmith JAW, Marshall RL, Jenkinson JL, MacRae WR, Scott DB. Haemodynamic effects of sodium nitroprusside during nitrous oxide/halothane anaesthesia. *British Journal of Anaesthesia* 1973; **45**: 71–4.

71. MacRae WR. Principles of management. In: MacRae WR, Wildsmith JAW, eds. *Induced hypotension*. Amsterdam: Elsevier Science Publ, 1991: 211–40.

72. Peach MJ. Renin–angiotensin system: biochemistry and mechanisms of action. *Physiological Reviews* 1977; **57**: 313–70.

73. Pettinger WA. Anesthetics and the renin–angiotensin–aldosterone axis. *Anesthesiology* 1978; **48**: 393–6.

74. Khambatta HJ, Stone JG, Kahn E. Hypertension during anesthesia on discontinuation of sodium nitroprusside-induced hypotension. *Anesthesiology* 1979; **51**: 127–30.

75. Khambatta JH, Stone JG, Khan E. Propranolol alters renin release during nitroprusside-induced hypotension and prevents hypertension on discontinuation of nitroprusside. *Anesthesia and Analgesia* 1981; **60**: 569–73.

76. Bloor BC, Fukunaga AF, Ma C, *et al.* Myocardial hemodynamics during induced hypotension: a comparison between sodium nitroprusside and adenosine triphosphate. *Anesthesiology* 1985; **63**: 517–25.

77. Kien ND, White DA, Reitan JA, Eisele JH. Cardiovascular function during controlled hypotension induced by adenosine triphosphate or sodium nitroprusside in the anesthetized dog. *Anesthesia and Analgesia* 1987; **66**: 103–10.

78. Fahmy NR. Nitroglycerin as a hypotensive drug during general anesthesia. *Anesthesiology* 1978; **49**: 17–20.

79. Benumof JL. Hypoxic pulmonary vasoconstriction and infusion of sodium nitroprusside. *Anesthesiology* 1979; **50**: 481–3.

80. D'Olivera M, Sykes MK, Chakrabarti MK, Orchard C, Keslin J. Depression of hypoxic pulmonary vasoconstriction by sodium nitroprusside and nitroglycerine. *British Journal of Anaesthesia* 1981; **53**: 11–17.

81. Naeije R, Lejeune P, Leeman M, Mélot C, Deloof T. Pulmonary arterial pressure-flow plots in dogs: effects of isoflurane and nitroprusside. *Journal of Applied Physiology* 1987; **63**: 969–77.

82. Casthely PA, Lear S, Cottrell JE, Lear E. Intrapulmonary shunting during induced hypotension. *Anesthesia and Analgesia* 1982; **61**: 231–5.

83. Bernard JM, Pinaud M, Ganansia MF, Chatelier H, Souron R, Letenneur J. Systemic haemodynamic and metabolic effects of deliberate hypotension with isoflurane anaesthesia or sodium nitroprusside during total hip arthroplasty. *Canadian Journal of Anaesthesia* 1987; **34**: 135–40.

84. Turner JM, Powell D, Gibson RM, McDowall DG. Intracranial pressure changes in neurosurgical patients during hypotension induced with sodium nitroprusside or trimetaphan. *British Journal of Anaesthesia* 1977; **49**: 419–24.

85. Ishikawa T, McDowall DG. Electrical activity of the cerebral cortex during induced hypotension with sodium nitroprusside and trimetaphan in the cat. *British Journal of Anaesthesia* 1981; **53**: 605–11.

86. Seyde WC, Longnecker DE. Cerebral oxygen tension in rats during deliberate hypotension with sodium nitroprusside, 2-chloroadenosine, or deep isoflurane anesthesia. *Anesthesiology* 1986; **64**: 480–5.

87. Lagerkranser M, Andreen M, Irestedt L. Central and splanchnic haemodynamics in the dog during controlled hypotension with sodium nitroprusside. *Acta Anaesthesiologica Scandinavia* 1984; **28**: 81–6.

88. Behnia R, Siqueira EB, Brunner EA. Sodium nitroprusside-induced hypotension: effect on renal function. *Anesthesia and Analgesia* 1978; **57**: 521–6.

89. Norlen K. Central and regional haemodynamics during controlled hypotension produced by adenosine, sodium nitroprusside and nitroglycerin. Studies in the pig. *British Journal of Anaesthesia* 1988; **61**: 186–93.

90. Harioka T, Hatano Y, Mori K, Toda N. Trimethaphan is a direct arterial vasodilator and an α-adrenoceptor antagonist. *Anesthesia and Analgesia* 1984; **63**: 290–6.

91. Fahmy NR, Soter NA. Effects of trimethaphan on arterial blood histamine and systemic hemodynamics in humans. *Anesthesiology* 1985; **62**: 562–6.

92. Dale RC, Schroeder ET. Respiratory paralysis during treatment of hypertension with trimethaphan camsylate. *Archives of Internal Medicine* 1976; **136**: 816–18.

93. Nakamura K, Koide M, Imanaga T, Ogasawara H, Takahashi M, Yoshikawa M. Prolonged neuromuscular blockade following trimetaphan infusion. *Anaesthesia* 1980; **35**: 1202–7.

94. Michenfelder JD, Theye RA. Canine systemic and cerebral effects of hypotension induced by hemorrhage, trimethaphan, halothane or nitroprusside. *Anesthesiology* 1977; **46**: 188–95.

95. Jones RM, Hantler CB, Knight PR. Use of pentolinium in postoperative hypertension resistant to sodium nitroprusside. *British Journal of Anaesthesia* 1981; **53**: 1151–4.

96. Hiwatari M, Nolan PL, Johnston CI. The contribution of vasopressin and angiotensin to the maintenance of blood pressure after autonomic blockade. *Hypertension* 1985; **7**: 547–53.

97. Skene DS, Sullivan SF, Patterson RW. Pulmonary shunting and lung volumes during hypotension induced with trimetaphan. *British Journal of Anaesthesia* 1978; **50**: 339–43.

98. Rowbothom DJ, Nimmo WS. Pharmacology of vasodilator drugs. In: MacRae WR, Wildsmith JAW, eds. *Induced hypotension*. Amsterdam: Elsevier Science Publ, 1991: 87–147.

99. MacRae WR, Wildsmith JAW, Dale BAB. Induced hypotension with a mixture of sodium nitroprusside and trimetaphan camsylate. *Anaesthesia* 1981; **36**: 312–15.

100. Fahmy NR. Nitroprusside versus nitroprusside–trimethaphan mixture for induced hypotension. A comparison of hemodynamic effects and cyanide release. *Anesthesiology* 1984; **61**: A40.

101. Nakazawa K, Taneyama C, Benson KT, Unruh GK, Goto H. Mixtures of sodium nitroprusside and trimethaphan for the induction of hypotension. *Anesthesia and Analgesia* 1991; **73**: 59–63.

102. Needleman P, Johnson EM. Mechanism of tolerance development to organic nitrates. *Journal of Pharmacology and Experimental Therapeutics* 1973; **184**: 709–15.

103. Association of the British Pharmaceutical Industry. *ABPI data sheet compendium 1991–1992*. London: Datapharm, 1991: 438–40.

104. Abrams J. Hemodynamic effects of nitroglycerin and long-acting nitrates. *American Heart Journal* 1985; **110**: 216–23.

105. Todd MM, Morris PJ, Moss J, Philbin DM. Hemodynamic consequences of abrupt withdrawal of nitroprusside or nitroglycerin following induced hypotension. *Anesthesia and Analgesia* 1982; **61**: 261–6.

106. Goldstein RE, Michaelis LL, Morrow AG, Epstein SE. Coronary collateral function in patients without occlusive coronary artery disease. *Circulation* 1975; **51**: 118–25.

107. Chiariello M, Gold HK, Leinbach RC, Davis MA, Maroko PR. Comparison between the effects of nitroprusside and nitroglycerin on ischemic injury during acute myocardial infarction. *Circulation* 1976; **54**: 766–73.

108. Rogers MC, Traystman RJ. Cerebral haemodynamic effects of nitroglycerine and nitroprusside. *Acta Neurologica Scandinavica* 1979; **60**, suppl 72: 600–1.

109. Stoelting RK, Blitt CD, Cohen PJ, Merin RG. Hepatic dysfunction after isoflurane anesthesia. *Anesthesia and Analgesia* 1987; **66**: 147–53.

110. Stevens WC, Cromwell TH, Halsey MJ, Eger II EI, Shakespeare TF, Bahlman SH. The cardiovascular effects of a new inhalation anesthetic, Forane, in human volunteers at constant arterial carbon dioxide tension. *Anesthesiology* 1971; **35**: 8–16.

111. Eger II EI. Isoflurane: A review. *Anesthesiology* 1981; **55**: 559–76.

112. Van Aken J, Leusen I, Lacroix E, De Somer A, Rolly G, Calliauw L. Influence of converting enzyme inhibition on isoflurane-induced hypotension for cerebral aneurysm surgery. *Anaesthesia* 1992; **47**: 261–4.

113. Seagard JL, Elegbe EO, Hopp FA, *et al*. Effects of isoflurane on the baroreceptor reflex. *Anesthesiology* 1983; **59**: 511–20.

114. Lam AM, Gelb AW. Cardiovascular effects of isoflurane-induced hypotension for cerebral aneurysm surgery. *Anesthesia and Analgesia* 1983; **62**: 742–8.

115. Bernard JM, Passuti N, Pinaud M. Long-term hypotensive technique with nicardipine and nitroprusside during isoflurane anesthesia for spinal surgery. *Anesthesia and Analgesia* 1992; **75**: 179–85.

116. Becker LC. Conditions for vasodilator-induced coronary steal in experimental myocardial ischemia. *Circulation* 1978; **57**: 1103–10.

117. Reiz S, Bålfors E, Sørenson MB, Ariola S, Friedman A, Truedsson H. Isoflurane – a powerful coronary vasodilator in patients with coronary artery disease. *Anesthesiology* 1983; **59**: 91–7.

118. Rehder K, Mallow JE, Fibuch EE, Krabill DR, Sessler AD. Effect of isoflurane anesthesia and muscle paralysis on respiratory mechanics in normal man. *Anesthesiology* 1974; **41**: 477–85.

119. Fusciardi J, Guggiari M, Retamal O, Pertuiset B, Philippon J, Viars P. Clinical use of isoflurane as a hypotensive agent for controlled hypotension during cerebral aneurysm surgery. *Anesthesiology* 1985; **63**: A106.

120. Domino KB, Borowec L, Alexander KM, *et al*. Influence of isoflurane on hypoxic pulmonary vasoconstriction in dogs. *Anesthesiology* 1986; **64**: 423–9.

121. Carlsson AJ, Bindslev L, Hedenstierna G. Hypoxia-induced pulmonary vasoconstriction in the human lung. The effect of isoflurane anesthesia. *Anesthesiology* 1987; **66**: 312–16.

122. Murphy Jr FL, Kennell EM, Johnstone RE, *et al*. The effects of enflurane, isoflurane and halothane on cerebral blood flow and metabolism in man. *Abstracts of Scientific Papers of the Annual Meeting of the American Society of Anesthesiologists* 1974; 61–2.

123. Newman B, Gelb AW, Lam AM. The effect of isoflurane-induced hypotension on cerebral blood flow and cerebral metabolic rate for oxygen in humans. *Anesthesiology* 1986; **64**: 307–10.

124. Madsen JB, Cold GE, Hansen ES, Bardrum B. The effect of isoflurane on cerebral blood flow and metabolism in humans during craniotomy for small supratentorial cerebral tumors. *Anesthesiology* 1987; **66**: 332–6.

125. Madsen JB, Cold GE, Hansen ES, Bardrum B, Kruse-Larsen C. Cerebral blood flow and metabolism during isoflurane-induced hypotension in patients subjected to surgery for cerebral aneurysms. *British Journal of Anaesthesia* 1987; **59**: 1204–7.

126. Newberg LA, Michenfelder JD. Cerebral protection by

isoflurane during hypoxemia or ischemia. *Anesthesiology* 1983; **59**: 29–35.

127. Nehls DG, Todd MM, Spetzler RF, Drummond JC, Thompson RA, Johnson PC. A comparison of the cerebral protective effects of isoflurane and barbiturates during temporary focal ischemia in primates. *Anesthesiology* 1987; **66**: 453–64.

128. Gelb AW, Boisvert DP, Tang C, *et al.* Primate brain tolerance to temporary focal cerebral ischemia during isoflurane- or sodium nitroprusside-induced hypotension. *Anesthesiology* 1989; **70**: 678–83.

129. Laing BT. Adenosine receptors and cardiovascular function. *Trends in Cardiovascular Medicine* 1992; **2**(3): 100–8.

130. Fredholm BB, Sollevi A. The release of adenosine and inosine from canine subcutaneous adipose tissue by nerve stimulation and noradrenaline. *Journal of Physiology* 1981; **313**: 351–67.

131. Sollevi A, Lagerkranser M, Andreen M, Irestedt L. Relationship between arterial and venous adenosine levels and vasodilatation during ATP- and adenosine-infusion in dogs. *Acta Physiologica Scandinavica* 1984; **120**: 171–6.

132. Öwall A, Gordon E, Lagerkranser M, Lindquist C, Rudehill A, Sollevi A. Clinical experience with adenosine for controlled hypotension during cerebral aneursym surgery. *Anesthesia and Analgesia* 1987; **66**: 229–34.

133. Sollevi A, Lagerkranser M, Irestedt L, Gordon E, Lindquist C. Controlled hypotension with adenosine in cerebral aneurysm surgery. *Anesthesiology* 1984; **61**: 400–5.

134. Waaben J, Husum B, Hansen AJ, Gjedde A. Regional cerebral blood flow and glucose utilization during hypocapnia and adenosine-induced hypotension in the rat. *Anesthesiology* 1989; **70**: 299–304.

135. Fukunaga AF, Fukunaga BM, Kiruta Y. Assessment and characterization of the anesthetic effects of intravenous adenosine in the rabbit. *Anesthesia and Analgesia* 1992; **74**: S103.

136. Lam AM, Winn HR, Grady MS. Catecholamines and plasma renin activity during adenosine and sodium nitroprusside induced hypotension. *Anesthesiology* 1990; **73**: A75.

137. Zäll S, Eden E, Winso I, Volkman R, Sollevi A, Ricksten S-E. Controlled hypotension with adenosine or sodium nitroprusside during cerebral aneurysm surgery: effects on renal hemodynamics, excretory function and renin release. *Anesthesia and Analgesia* 1990; **71**: 631–6.

138. Newberg LA, Milde JH, Michenfelder JD. Cerebral and systemic effects of hypotension induced by adenosine or ATP in dogs. *Anesthesiology* 1985; **62**: 429–36.

139. Zäll S, Milocco I, Ricksten S-E. Effects of adenosine on myocardial blood flow and metabolism after coronary artery bypass surgery. *Anesthesia and Analgesia* 1991; **73**: 689–95.

140. Pelleg A, Porter RS. The pharmacology of adenosine. *Pharmacotherapy* 1990; **10**: 157–74.

141. Dedrick DF, Mans AM, Campbell PA, Hawkins RA, Biebuyck JF. Does ATP-induced hypotension cause potentially serious metabolic complications? *Anesthesiology* 1982; **57**: 3A.

142. Saito T, Ishihara K, Yamada K, Sakamoto Y, Inoue T, Ogawa R. Blood biochemistry during acidosis associated with ATP-induced hypotensive anesthesia. *Masui* 1991; **40**: 936–41.

143. Vickers MD, Schneiden H, Wood-Smith FG, eds. *Drugs in Anaesthetic Practice*, 6th ed. London: Butterworths, 1984: 382.

144. Goldberg ME, McNulty SE, Azad SS, *et al.* A comparison of labetalol and nitroprusside for inducing hypotension during major surgery. *Anesthesia and Analgesia* 1990; **70**: 537–42.

145. Bedford RF, Berry Jr FA, Longnecker DE. Impact of propranolol on hemodynamic response and blood cyanide levels during nitroprusside infusion: a prospective study in anesthetized man. *Anesthesia and Analgesia* 1979; **58**: 466–9.

146. Scott DB. The use of labetalol in anaesthesia. *British Journal of Clinical Pharmacology* 1982; **13**, suppl 1: 133S–135S.

147. Association of the British Pharmaceutical Industry. *ABPI data sheet compendium 1991–1992*. London: Datapharm, 1991: 413–15.

148. Blau WS, Kafer ER, Anderson JA. Esmolol is more effective than sodium nitroprusside in reducing blood loss during orthognathic surgery. *Anesthesia and Analgesia* 1992; **75**: 172–8.

149. Quinn TJ, Zayas VM, LeDonne K, McDonald M, Pickering T. Esmolol and controlled hypotension: adverse effects on oxygen extraction. *Anesthesiology* 1990; **73**: A94.

150. Ornstein E, Young WL, Ostapkovich N, Matteo RS, Diaz J. Deliberate hypotension in patients with intracranial arteriovenous malformations: esmolol compared with isoflurane and sodium nitroprusside. *Anesthesia and Analgesia* 1991; **72**: 639–44.

151. Pettinger WA, Keaton TK, Campbell WB, Harper DC. Evidence for a renal alpha-adrenergic receptor inhibiting renin release. *Circulation Research* 1976; **38**: 338–46.

152. Reid IA, Ahn JN, Trinh T, Shackelford R, Weintraub M, Keil LC. Mechanism of suppression of vasopressin and adrenocorticotropic hormone secretion by clonidine in anesthetized dogs. *Journal of Pharmacology and Experimental Therapeutics* 1984; **229**: 1–8.

153. Ghignone M, Calvillo O, Caple S, Quintin L. Clonidine reduces the dose requirement for nitroprusside induced hypotension. *Anesthesiology* 1986; **65**: A51.

154. Zimpfer M, Fitzal S, Tonczar L. Verapamil as a hypotensive agent during neuroleptanaesthesia. *British Journal of Anaesthesia* 1981; **53**: 885–9.

155. Bernard J-M, Moren J, Demeure D, Hommeril J-L, Pinaud M. Diltiazem reduces the nitroprusside doses for deliberate hypotension. *Anesthesiology* 1992; **77**: A427.

156. Khambatta HJ, Stone JG, Matteo RS, Khan E. Propranolol premedication blunts stress response to nitroprusside hypotension. *Anesthesia and Analgesia* 1984; **63**: 125–8.

157. Fahmy NR. Impact of oral captopril or propranolol on nitroprusside-induced hypotension. *Anesthesiology* 1984; **61**: A41.

158. Jacob L, Bonnet F, Sabathier C, Chiron B, Lhoste F, Viars P. Hormonal response to captopril pre-treatment during sodium nitroprusside-induced hypotension in man. *European Journal of Anaesthesiology* 1987; **4**: 101–12.

159. Jensen K, Bunemann L, Riisager S, Thomsen LJ. Cerebral blood flow during anaesthesia: influence of pretreatment with metoprolol or captopril. *British Journal of Anaesthesia* 1989; **62**: 321–3.

160. Mirenda JV, Grissom TE. Anesthetic implications of the renin–angiotensin system and angiotensin-converting enzyme inhibitors. *Anesthesia and Analgesia* 1991; **72**: 667–83.

161. Larson AG. Deliberate hypotension. *Anesthesiology* 1984; **25**: 682–706.

162. Kerr A. Anaesthesia with profound hypotension for middle ear surgery. *British Journal of Anaesthesia* 1977; **49**: 447–52.

163. Warner WA, Shumrick DA, Caffrey JA. Clinical investigation of prolonged induced hypotension in head and neck surgery. *British Journal of Anaesthesia* 1970; **42**: 39–64.

164. Hugosson R, Hogstrom S. Factors disposing to morbidity in surgery of intracranial aneurysms with special regard to deep controlled hypotension. *Journal of Neurosurgery* 1973; **38**: 561–7.

165. Eckenhoff JE. Observations during hypotensive anaesthesia. *Proceedings of the Royal Society of Medicine* 1962; **55**: 942–4.

166. Rollason WN, Robertson GS, Cordiner CM, Hall DJ. A comparison of mental function in relation to hypotensive and normotensive anaesthesia in the elderly. *British Journal of Anaesthesia* 1971; **43**: 561–6.

167. Townes BD, Dikmen SS, Bledsoe SW, Hornbein TF, Martin DC, Janesheski JA. Neuropsychological changes in a young, healthy population after controlled hypotensive anesthesia. *Anesthesia and Analgesia* 1986; **65**: 955–9.

168. Reynolds F. Adverse effects of local anaesthetics. *British Journal of Anaesthesia* 1987; **59**: 78–95.

169. Kling TF, Wilton N, Hensinger RN, Knight PR. The influence of trimethaphan (arfonad)-induced hypotension with and without spine distraction on canine spinal cord blood flow. *Spine* 1986; **11**: 219–24.

170. Spargo PM, Tait AR, Knight PR, Kling TF. Effect of nitroglycerine-induced hypotension on canine spinal cord blood flow. *British Journal of Anaesthesia* 1987; **59**: 640–7.

171. Sperry RJ, Longnecker DE. Regional blood flow changes during induced hypotension. *Current Opinion in Anesthesiology* 1988; **1**: 94–100.

172. Mangano DT, Browner WS, Hollenberg M, London MJ, Tubau JF, Tateo IM, SPI Research Group. Association of perioperative myocardial ischemia with cardiac morbidity and mortality in men undergoing noncardiac surgery. *New England Journal of Medicine* 1990; **323**: 1781–8.

173. Marsch SCU, Schaefer H-G, Skarvan K, Castelli I, Scheidegger D. Perioperative myocardial ischemia in patients undergoing elective hip arthroplasty during lumbar regional anesthesia. *Anesthesiology* 1992; **76**: 518–27.

174. Keany NP, McDowell DG, Turner JM, Lane JR, Okuda Y. The effects of profound hypotension induced with sodium nitroprusside on cerebral blood flow and metabolism in the baboon. *British Journal of Anaesthesia* 1973; **45**: 639P.

175. Klatzo I. Brain oedema following brain ischaemia and the influence of therapy. *British Journal of Anaesthesia* 1985; **57**: 18–22.

176. Van Aken H, Fitch W, Graham DI, Brüssel T, Themann H. Cardiovascular and cerebrovascular effects of isoflurane-induced hypotension in the baboon. *Anesthesia and Analgesia* 1986; **65**: 565–74.

177. Enderby GEH. Haemostatic mechanisms. In: Enderby GEH, ed. *Hypotensive anaesthesia*. Edinburgh: Churchill Livingstone, 1985; 66–72.

178. Odeom JA, Bovill JG, Hardeman MR, Oosting J, Zuurmond WWA. Effects of epidural and spinal anaesthesia on blood rheology. *Anesthesia and Analgesia* 1992; **74**: 835–40.

179. Hines R, Barash PG. Infusion of sodium nitroprusside induces platelet dysfunction *in vitro*. *Anesthesiology* 1989; **70**: 611–15.

180. Smith JW, Pellicci PM, Sharrock N, Mineo R, Wilson Jr PD. Complications after total hip replacement. The contralateral limb. *Journal of Bone and Joint Surgery [Am]* 1989; **71**: 528–35.

181. Colley PS. Intracranial aneurysms. Anesthesia. In: Newfield P, Cottrell JE, eds. *Neuroanesthesia: handbook of clinical and physiologic essentials*. Boston: Little, Brown, 1991: 183–214.

182. Farrar JK, Gamache FW, Ferguson GG, Barker J, Varkey GP, Drake CG. Effect of profound hypotension on cerebral blood flow during surgery for intracranial aneurysms. *Journal of Neurosurgery* 1981; **55**: 857–64.

183. Batjer H, Samson D. Intraoperative aneurysmal rupture: incidence, outcome, and suggestions for surgical management. *Neurosurgery* 1986; **18**: 701–7.

184. Michenfelder JD. Cerebral blood flow and metabolism. In: Cucchiara RF, Michenfelder JD, eds. *Clinical neuroanesthesia*. New York: Churchill Livingstone, 1990: 1–40.

185. Murphy DF. Anesthesia and intraocular pressure. *Anesthesia and Analgesia* 1985; **64**: 520–30.

186. Adams AD, Barnett KC. Anaesthesia and intraocular pressure. *Anaesthesia* 1966; **21**: 202–10.

187. Friedman E. Choroidal blood flow. 1. Pressure–flow relationships. *Archives of Ophthalmology* 1970; **83**: 95–9.

188. Tsamparlakis J, Casey TA, Howell W, Edridge A. Dependence of intraocular pressure on induced hypotension and posture during surgical anaesthesia. *Transactions of the Ophthalmological Societies of the United Kingdom* 1980; **100**: 521–6.

189. Jantzen J-P, Hennes HJ, Rochels R, Wallenfang T. Deliberate arterial hypotension does not reduce intraocular pressure in pigs. *Anesthesiology* 1992; **77**: 536–40.

190. Mandel RJ, Brown MD, McCollough NC, Pallares V, Varlotta R. Hypotensive anesthesia and autotransfusion in spinal surgery. *Clinical Orthopaedics and Related Research* 1981; **154**: 27–33.

191. Malcolm-Smith NA, MacMaster MJ. The use of induced hypotension to control bleeding during posterior fusion for scoliosis. *Journal of Bone and Joint Surgery [Br]* 1983; **65**: 255–8.

192. Patel NJ, Patel BS, Paskine S, Laufer S. Induced moderate hypotensive anesthesia for spinal fusion and Harrington-rod instrumentation. *Journal of Bone and Joint Surgery [Am]* 1985; **67**: 1384–7.

193. Grundy BL, Nash CL, Brown RH. Arterial pressure manipulation alters spinal cord function during correction of scoliosis. *Anesthesiology* 1981; **54**: 249–53.

194. Grundy BL, Nash CL, Brown RH. Deliberate hypotension for spinal fusion: prospective randomized study with evoked potential monitoring. *Canadian Anaesthetists' Society Journal* 1982; **29**: 452–61.

195. Horlocker TT, Cucchiara RF, Ebersold MJ. Vertebral column and spinal cord surgery. In: Cucchiara RF, Michenfelder JD, eds. *Clinical neuroanesthesia*. New York: Churchill Livingstone, 1990; 325–50.

196. Pevehouse BC, Boldrey E. Hypothermia and hypotension for

intracranial surgery during pregnancy. *American Journal of Surgery* 1960; **100**: 633–4.

197. Aitken RR, Drake CG. A technique of anesthesia with induced hypotension for surgical correction of intracranial hemorrhages. *Clinical Neurosurgery* 1974; **21**: 107–14.

198. Rosen MA. Cerebrovascular lesions and tumors in the pregnant patient. In: Newfield P, Cottrell JE, eds. *Neuroanesthesia: handbook of clinical and physiologic essentials*. Boston: Little, Brown, 1991: 230–48.

199. Donchin Y, Amirav B, Sahar A, Harkoni S. Sodium nitroprusside for aneurysm surgery in pregnancy. *British Journal of Anaesthesia* 1978; **50**: 849–51.

200. Association of the British Pharmaceutical Industry. *ABPI data sheet compendium 1991–1992*. London: Datapharm, 1991: 1236.

201. Palahniuk RJ, Shnider SM. Maternal and fetal cardiovascular and acid–base changes during halothane and isoflurane anesthesia in the pregnant ewe. *Anesthesiology* 1974; **41**: 462–72.

202. Snyder SW, Wheeler AS, James III FM. The use of nitroglycerin to control severe hypertension of pregnancy during cesarean section. *Anesthesiology* 1979; **51**: 563–4.

203. Hood DD, Dewan DM, James III FM, Bogard TD, Floyd HM. The use of nitroglycerin in preventing the hypertensive response to tracheal intubation in severe preeclamptics. *Anesthesiology* 1983; **59**: A423.

Chapter 40

Recovery from Anaesthesia: Assessment and Management

M. Herbert

Measures of immediate recovery	Measures of longer-term recovery
Measures of intermediate recovery	Selection and management considerations

Interest in the speed with which patients recover post-operatively, the type of deficits that anaesthesia may induce in them and the overall profile of their physiological and psychological states has grown rapidly over the last two decades. Of course, the prevention and alleviation of negative postoperative sequelae have always been of concern to practising health care providers, who naturally wish to prevent unnecessary discomfort, pain and distress in their patients. However, the increasing concern with the patient's post-anaesthetic state has probably been fuelled by the rapid development of day-case, outpatient, 'ambulatory' surgery.

From relatively modest use in the 1970s, day-case procedures, in which patients are admitted into hospital for surgery and discharged on the same day, have increased remarkably quickly. In the USA, hospital-based ambulatory surgery expanded by 86 per cent between 1979 and 1983.[1] Korttila has observed that the number of specialist surgical centres in that country grew from 39 in 1982 to almost 1000 over the course of 5 years and that it is likely that 50 per cent of all surgery will be done as day cases in the near future.[2] A similar trend is observable in the UK.[3]

The reasons for this growth in ambulatory surgery are numerous but can be seen to be dependent mainly on efficiency and economic grounds. Ogg,[3] for example, listing the positive aspects of one-day surgery, cited the advantages of being able to treat a large number of patients, with a corresponding reduction in the number of patients on waiting lists. Fewer nurses are required to attend patients; moreover, one-day surgery provides good nursing recruitment, probably because shifts and working hours are relatively predictable. In addition to direct economic benefits such as the decreased demand for inpatient beds and 'hotel' facilities, Ogg also saw other advantages such as reducing the time that children spend away from their home and caretakers, and reduced rates of cross-infection.

However, in the present age of medical audit, one has to evaluate the negative as well as the positive outcomes of any procedural change. Again, Ogg[3] considered that the disadvantages of one-day surgery included the fact that it was regarded as a second-class service by some surgeons and that good preoperative selection of appropriate patients was essential. In his eyes, there are also problems in dealing with minor postoperative sequelae and with the need to consider the issue of transferring the economic savings from hospitals on to the community services. He also drew attention to the problems of what patients do once they are discharged. In an earlier study in the UK, the same author reported that 73 per cent of patients who owned cars had driven them within 24 hours following minor outpatient anaesthesia and that a surprising 9 per cent had actually driven themselves home after surgery.[4] As anaesthetic agents are used to render patients unconscious and insensate, the problem of the extent to which these substances and gases continue to exert their effects, even when patients have regained consciousness, is clearly a matter of some importance. There would seem to be a strong case, based on legal and ethical concerns, for anaesthetists and their colleagues to provide clear and firm guidelines about what activities patients can be expected to carry out during the immediate postoperative period and what activities – particularly ones that are hazardous, such as car driving or operating other dangerous machinery – they should either avoid or require special precautions.

It is for reasons such as these that increasing attention has been paid to monitoring not only the physical and kinaesthetic abilities of postsurgical patients but also their 'higher mental functions' such as concentration, memory and other cognitive abilities. This chapter will examine the work that has attempted to assess patients' postoperative states. It will also draw attention to some of the difficulties facing those who wish to interpret that evidence in an effort to provide the best advice for their patients.

Measures of immediate recovery

It seems reasonable to expect that the physical and mental state of patients will be compromised maximally at the point at which they have just regained consciousness after a general anaesthetic. Measures that have been used to assess the patient's state during this immediate recovery period have tended to focus predominantly on physiological or vestibular–motor functioning.[5] Clearly, patients experiencing nausea and vomiting or those who have difficulties in focusing visually or those with an unsteady gait or other clinical problems will not be fit for discharge from the recovery area until the symptoms have subsided. A variety of techniques have therefore been recommended to assess patients' functional capabilities immediately after surgery has ended.

Binocular coordination

Divergence of the eyes is relatively easy to measure using a device such as the Maddox wing test which requires people to align pointers on an ocular measuring instrument. Because anaesthetic agents generally result in a decrease in muscle tone, it has been proposed that this test could be used as an effective means of measuring early recovery from general anaesthesia.[6] Following an inhalational induction with halothane, oxygen and nitrous oxide, it was observed that patients had regained preoperative ocular coordination about 20 minutes following surgery.[6]

The Maddox wing test has continued to attract interest as a measure of early post-anaesthetic recovery. Zuurmond and van Leeuwen[7] used it to compare the after effects of sufentanil and isoflurane, but found no difference between the two agents although full recovery of ocular function took up to 142 minutes. Effects have also been found lasting up to 120 minutes after anaesthesia using buprenorphine,[8] and a similar duration has been reported following halothane or alfentanil.[9] Zuurmond and colleagues reported that midazolam resulted in impairment lasting 3 hours.[10]

Postural stability

The ability to maintain a steady upright posture would also seem to be an important aspect of functioning to be regained before discharge can be contemplated. Experimentally, body sway can be measured in a variety of ways.[11] One of the first researchers to use it to study anaesthesia was Vickers,[12] who attached a thread to the back of a subject's collar and then to pointers which moved with the amount of sway and which therefore gave objective readings. Although there was some evidence for a dose-related effect of thiopentone and methohexitone on this measure, the lower doses having a short-lasting effect of 15–30 minutes and the larger doses extending the recovery time to 105–120 minutes, Vickers concluded that other variables such as suggestion and the problems of achieving a reliable baseline made it unreasonable to place too great a reliance on the quantitative aspects of this test. Nevertheless, other authors have gone on to incorporate this measure in their assessments.

Eriksen et al[13] asked patients to stand on a force-plate which measured body sway in all directions. They found that, whereas propanidid had little effect, thiopentone continued to affect balance up to 3 hours later. However, others have failed to find an effect of thiopentone in healthy volunteers after 1 hour had elapsed, although diazepam was still exerting adverse effects up to 7 hours later.[14] A simpler measure of postural stability has been advanced in which patients are asked to walk back and forth along a straight line of 4 metres, their gait being rated as 'normal', 'staggering', 'side steps' or 'unable'.[15] Compared with patients administered isoflurane, those receiving propofol were significantly better able to walk at 30 minutes after the infusion, although there was still some impairment compared with normal walking ability.

Clinical observations

A range of measures aimed at assessing the patients' clinical state rather than their functional abilities has also been suggested by several authors. Aldrete and Kroulik, for example, devised a scale in which the patient's activity (defined as movement of extremities on command), respiration, circulation, alterations in systolic blood pressure from preoperative to postoperative levels, levels of consciousness (measured by the ability to answer questions) and normal skin colour were assigned a score of 0, 1 or 2.[16] Measurements were taken on arrival in the recovery room postoperatively and at 1, 2 and 3 hours thereafter. The sensitivity of this measure may be open to question, as 78.4 per cent of a large group of patients obtained a score of 8 or more on arrival in the recovery room 10 minutes after surgery had ceased. One hour later, 80 per cent of patients were receiving the maximum score.

A similar scale has been devised using assessments of breathing, movement and wakefulness.[17] Sanders et al reported that this measure discriminated between patients given either propofol or thiopentone up to 30 minutes after arrival in the recovery room.[18] Propofol permitted a quicker recovery, a finding that has been endorsed by others.[19,20]

Yet another version of a rating scale assessing immediate postoperative recovery has been put forward.[21] Scores ranging from either zero to 5 or, on some measures, zero or 1, were assigned on the basis of how vigilant patients were, whether they could understand simple orders, their postoperative orientation, short-term memory and a subjective evaluation of his/her condition. The authors were able to show little difference between patients premedicated with a benzodiazepine or clonidine, and reported that this measure appeared to be adversely affected for up to 2 hours postoperatively when compared with preoperative scores.

Subjective estimates of the patient's clinical state such as these appear to have a useful role to play in assessing early recovery from anaesthesia. However, researchers must bear in mind the problems of inter-rater reliability. It is not always easy to ensure that the criteria used to assign scores to various categories will remain the same if different individuals do the rating. Furthermore, the obvious pitfalls of non-independent ratings being carried out by investigators must be borne in mind when trying to evaluate the results of evidence relying on such data.

In general terms, many of the ways of assessing immediate recovery from anaesthesia appear to show that patients generally make a speedy return to normal functioning although such a conclusion will, of course, depend upon the type of anaesthetic agent used, the duration of surgery and other intraoperative procedures. But even if measures do show a relatively rapid recovery rate, this may not necessarily mean that residual impairments cease to exist. It may be that, although patients appear to be clinically normal, there is a potential danger that impairments that are not overtly obvious will continue to exert adverse effects upon their functioning.[6] It is therefore unwise to assume that, once initial measures have returned to normal, the patient is fit to be discharged from the hospital. An important distinction has been made between the concepts of 'home readiness' and 'street fitness'.[2,22] The former term implies some partial degree of recovery and that the patient can be discharged into the care of a responsible person at home. The criteria that might be used to define 'home readiness' are that the patient can walk on a straight line, can correctly carry out pencil and paper tests and can perform simple measures of coordination and reaction time. 'Street fitness', on the other hand, refers to the patient's ability to carry out safely everyday activities such as driving a car. Batteries of psychomotor tests have been recommended for assessing this level of postoperative recovery.[22] That issue is addressed in the next section.

Measures of intermediate recovery

Attempts to measure the speed with which patients have returned to unimpaired levels of functioning and which would therefore indicate their relative safety in normal activities are numerous. Most of the studies carried out in this area have used psychomotor tests either in isolation or as part of a test battery. The range of tests that has been used to measure post-anaesthetic functioning is daunting and a comprehensive survey of the often contradictory and conflicting findings is beyond the scope of this chapter; however, brief attention can be drawn to the use of a measure that may be familiar to many anaesthetists. This is the critical flicker fusion (CFF) test and its variant, the auditory flutter fusion (AFF) test. In essence, these measures require patients to determine when a series of separate visual or auditory signals, presented at an increasingly faster rate, can no longer be discriminated as separate stimuli but fuse into a subjectively continuous light or tone. It has been suggested that these measures may be a reasonable indicator of sedation levels induced, for example, by diazepam.[23] The CFF and AFF tests do have the advantages of being easy to administer and cheap to construct, but they also have disadvantages. The visual version may be influenced by the degree of light adaptation, and the auditory version, which can be performed with the eyes closed, may be subject to inattention rather than to an inability to discriminate stimuli. Furthermore, the practical implications of a changed threshold are not clear. The interested reader is referred to a review of these and other techniques by Hindmarch and Bhatti.[24] Typical examples of work that has been undertaken and recommendations that have emerged from a number of studies based predominantly on indices reflecting psychomotor coordination or manual dexterity are now examined.

The pegboard test

This test has gained some popularity in the literature as a measure of post-anaesthetic impairment of function. Essentially, it involves placing or replacing pegs into predetermined holes or slots. The degree of manual dexterity that the test requires can be assessed either by the number of transfers made in a given time period or by the time needed to complete the task. Its acceptance as a useful instrument probably stems from the fact that it does not need expensive or elaborate equipment, that it is relatively easy to administer and that it can be completed in a reasonably short time.

One of the earliest reports of the pegboard test being considered as a potential measure of performance in an anaesthetic setting is that of Vickers.[12] He examined the effect of an injection of either 2.5 mg/kg or 5 mg/kg of thiopentone and either 1 mg/kg or 2 mg/kg of methohexitone on a 45-second test. His conclusion that the highest dose of thiopentone delayed recovery to pre-injection baseline levels by 105–120 minutes must be considered tentative, as the few volunteers used precluded statistical analysis. However, the author was among the first investigators to note that, without prior adequate training on the test, the results could be obscured by practice effects.

Vickers' report was followed closely by a study that gave a battery of tests, including a pegboard test, before and after surgery to groups of elderly male and female patients.[25] Only the male patients showed a statistically significant postoperative increase in the time taken to place 16 pegs into their slots. The performance of the female patients, and that of a control group before and after moving to special accommodation, was not affected. Interpretation of this study is somewhat hampered by the lack of information about the nature of the surgery or anaesthesia and by the uncertainty surrounding when and how the tests were given.

Subsequent work improved to some extent on methodological deficits in research design. Carson and colleagues[26] gave varying doses of Althesin, thiopentone or methohexitone to 150 female patients premedicated with 0.6 mg of atropine and who were undergoing short and minor gynaecological surgery. Patients performed a preoperative baseline measure on the pegboard test and repeated measures were taken at several postoperative times. On the basis of the time taken to regain preoperative levels of performance, the authors concluded that thiopentone delayed full recovery more than did Althesin or methohexitone. It is worth noting that, even with quite large doses of thiopentone (12 mg/kg), patients had returned to baseline levels within 90 minutes of induction of anaesthesia.

A spate of reports using the pegboard test or variants of it emerged during the early to mid 1980s. A detailed study by Bahar et al.[27] gave patients scheduled for gynaecological surgery 2 or 3 mg/kg of propofol, or 4 or 6 mg/kg of thiopentone, or 1.5 mg/kg of methohexitone. A six-peg version of the pegboard test was given before surgery and was repeated during the recovery period until patients had returned to preoperative levels of performance. The highest doses of propofol and thiopentone resulted in the longest recovery times but, nevertheless, were still relatively short (23 minutes after induction).

A different version of the test was used in a study of midazolam (0.2 mg/kg) and thiopentone (3.5 mg/kg).[28] Here, the use of the Purdue pegboard required people to remove and invert as many pegs as possible in a 30-second period. The measure of recovery was the mean time taken by 75 per cent of their sample of minor, elective surgery patients to regain pre-anaesthetic baseline values. Using that criterion, midazolam patients took 320 minutes, which was significantly longer than the 250 minutes for patients given thiopentone. The differential adverse response between these two compounds was supported by a study using a manual dexterity test requiring patients to place dots in small squares.[29] In a sample of cystoscopy patients, midazolam (0.125 mg/kg) resulted in more impairment as compared with baseline 1 hour after induction than did 5 mg/kg of thiopentone.

A number of reports have used yet another variation on the basic pegboard task described by Vickers.[12] That test is commonly referred to as the 'postbox (or 'mailbox') test' and is based essentially on the children's toy that requires people to 'post' differing shapes of objects through corresponding slots in a container. Craig and co-workers gave this test to a group of gynaecological patients undergoing surgery who were given methohexitone, or Althesin or etomidate.[30] The times taken to reach preoperative levels of functioning on the test were, once again, relatively brief, varying from 43.8 to 76.2 minutes across the various groups. The authors also reported that they formed a control group of 20 healthy females who repeated the test 1 or 2 hours later. Unfortunately, the data provided by that control group were not compared with the anaesthetic subgroups, so the extent of the drug-induced impairment cannot be assessed.

The same research group, using the same test, have also investigated the differential effects of Althesin and thiopentone in patients aged above and below 50 years.[31] After induction, anaesthesia was maintained in all patients with a mixture of 67 per cent nitrous oxide in oxygen, supplemented by 1.5 per cent halothane. Althesin prevented a return to preoperative values on the postbox test for about 30 minutes in both age groups, whereas thiopentone resulted in a somewhat longer impairment (45 minutes) in the older patients as compared with 28 minutes in the younger group.

Postoperative recovery has also been investigated following the supplementation of intravenous induction agents with short-acting opioids.[32,33] No differences were found between varying doses of alfentanil and fentanyl supplementing methohexitone induction in the time taken to recover preoperative values on the postbox text. Most patients had recovered their manual dexterity according to this criterion by 36–45 minutes after induction. Similarly, no differences in performance were observed after alfentanil was given as a supplement to either methohexitone or Althesin. This report also noted a speedy recovery in functioning, patients regaining preoperative values approximately 25 minutes later. Steib et al. reported that patients were within control values on the postbox test after 45 minutes following total intravenous anaesthesia with propofol but that midazolam was still impairing performance up to 3 hours later.[34]

On the other hand, Korttila et al. demonstrated that, at least in some situations, impairment could last for a relatively long time after anaesthetic procedures.[35] Their battery of psychomotor tests included a pegboard test in which pegs had to be turned. This test was given before anaesthesia and was required to be done twice with the right hand, twice with the left hand and once with both hands simultaneously. Volunteer student subjects were premedicated with atropine sulphate 5 minutes before anaesthesia was induced with either halothane or enflurane. Anaesthesia lasted slightly over 5.5 minutes and was followed by re-administration of the various performance tests 1 and 5 hours later. Compared with unanaesthetized controls, both experimental groups subjected to anaesthesia displayed significantly slower performance 5 hours later, particularly when the test was being done with both hands.

Apart from that study, the bulk of the evidence considered so far suggests that for a wide variety of anaesthetic regimens recovery, as judged by the pegboard test and associated measures of manual dexterity, is relatively quick. However, the pegboard test has been considered unreliable as a measure of recovery on the grounds that the most effective measures of post-anaesthetic performance are provided by those tests that do not reveal changes over time when given to unanaesthetized control subjects – i.e. those which do not display changes due to practice.[36] Because practice effects have been observed on the pegboard test,[12,37] the value of this measure was considered to be suspect.

The implicit assumption lying behind the criticism that performance that does remain stable over time is consequently free from training effects has been questioned,[38] as there is evidence for the existence of a 'masked training effect' which can confound post-drug performance levels. The need to employ tests that are sensitive enough to detect residual impairment of functioning which may not be clinically obvious has long been recognized[6] but it is precisely those tests sensitive enough to detect subtle changes in performance that are most susceptible to practice effects.[39] The problem therefore becomes one of taking into account the improvement in performance due to repeated encounters with the test. Ways of overcoming the potentially obscuring consequences of the practice effect in clinical research is discussed later.

Reaction time tests

Measurement of the speed with which individuals can react to simple or complex stimuli has a long history in psychopharmacological research. As with the pegboard test, reaction time tests are relatively easy to administer and usually require inexpensive equipment. It is not surprising, therefore, that several investigators have used reaction time as an index of psychomotor functioning after anaesthesia.

Early interest focused on levels of carbon dioxide achieved during anaesthesia, as a result of the report showing that low levels of the gas could exert adverse long-term effects on reaction time.[40] All patients in this study underwent routine surgery and were given a standard anaesthetic procedure involving premedication with pethidine (meperidine) and hyoscine (scopolamine), induction with thiopentone, muscle relaxation using tubocurarine and anaesthetic maintenance with 70 per cent nitrous oxide and 30 per cent oxygen. Carbon dioxide levels were varied by adding 1 or 2 per cent of the gas to randomly selected patients. A test of simple reaction time showed that patients in whom $PaCO_2$ was below 3.2 kPa (24 mmHg) were significantly slower up to 6 days postoperatively than patients with a $PaCO_2$ of 3.2 kPa (24 mmHg) or greater.

The study was repeated on a small group of healthy volunteers who were exposed to normocapnia ($PaCO_2$ 5–5.5 kPa; 38–41 mmHg) or hypocapnia ($PaCO_2$ 2.3–2.5 kPa; 17–19 mmHg).[41] A battery of tests given before and after anaesthesia included simple reaction time and a measure of disjunctive reaction time in which one of two lights acted as the stimulus requiring a differential response. Both normo- and hypocapnia resulted in a slowing of disjunctive reaction times 90 minutes after anaesthesia as compared with pre-induction measures, but there was no difference between the carbon dioxide conditions in the extent of that slowing. No further effects were evident on tests given up to 4 days later, and simple reaction time remained unaffected at all post-anaesthetic points. A further complicated experiment compared performance on a battery of tests, including reaction time, in two groups of young and older female patients exposed to hyper-, normo- or hypocapnia.[42] Differences between preoperative baseline and postoperative points showed the hypocapnic condition to result in poorer speed of reactions up to 24 hours after induction.

Many other reports have concentrated on the assessment of typical anaesthetic regimens and have taken a return of performance levels to preoperative baselines as the criterion of recovery. Azar and co-workers,[43] for example, tested simple reaction time 1 and 2 hours after anaesthetic supplementation with fentanyl, enflurane or isoflurane and found that only enflurane exerted any deleterious effect 2 hours after anaesthesia. On the other hand, evidence exists for postoperative slowing of reaction time being still present 3 hours after surgery in patients receiving fentanyl, whereas after enflurane, they had regained preoperative reaction time values at that time.[44] The discrepancy between these two reports may be ascribed to the fact that, in the first, thiopentone was used to induce anaesthesia in the fentanyl group, whereas the enflurane group had an inhalation induction with halothane.

Since the introduction of propofol as a new anaesthetic agent, a number of studies have been mounted to examine its post-anaesthetic consequences. Sanders *et al.*, for example, reported that reaction times were back to baseline levels within 2 hours following TIVA using this agent,[18] a finding that has received support by others.[45]

A number of studies have incorporated control groups into their research designs in an attempt to measure the extent of post-anaesthetic performance impairment. Unfortunately, having included such groups, very often no direct comparison is made between the control groups and groups undergoing anaesthesia. In a study of the postoperative sequelae of anaesthesia induced by either propofol or methohexitone, patients and controls were tested on a choice reaction time test at times corresponding to preoperative baseline and 30, 60, 120 and 240 minutes after awakening.[46] Declines from baseline values were significant at 30 minutes for both anaesthetic groups and at 60 minutes postoperatively for the methohexitone group. Thereafter, performance was statistically indistinguishable from initial levels. Little change was observed within the control group's performance across similar times, but no direct comparisons were made between that group and the anaesthetized ones. A further study examining thiopentone as an induction agent found that it impaired choice reaction time up to 90 minutes into the postoperative period.[47] After methohexitone, performance returned to normal within 60–90 minutes. The control group once more showed no evidence of change over time but, again, their performance was not contrasted directly with the two other groups. Other research using similar methodological procedures has concluded that reaction times have recovered to normal values within 50–75 minutes after methohexitone and 100 minutes after Althesin.[48,49]

Studies in which the reaction times of anaesthetized patients have been directly compared with the performance of a control group have tended to suggest that, at least for longer-established anaesthetic agents, reaction time may be impaired for longer than the studies cited above would suggest. Korttila *et al.*[35] included in their study of halothane and enflurane anaesthesia, postoperative measures of a two-choice reaction time test and a similar test requiring a response to either lights or sounds. Compared with the control group, both anaesthetic agents produced significantly impaired performance 5 hours after anaesthesia on the more complex of the two tests. Using a relatively demanding test of continuous reaction time lasting minutes, Scott and colleagues[50] tested performance following either thiopentone, nitrous oxide and halothane anaesthesia or methohexitone and fentanyl/nitrous oxide anaesthesia. Again, compared with unanaesthetized controls, both groups still displayed impaired performance between 6

and 8 hours after surgery. Changes in a choice reaction time test have also been detected 4 hours after anaesthesia preceded by hyoscine premedication as compared with a saline placebo injection.[51] However, in that study, no such impairment in performance was evident 1 hour after regaining consciousness.

However, not all studies using control groups have been able to show a longer duration of adverse effects. Propofol, with or without isoflurane, may have no effects lasting longer than an hour after surgery.[52,53] This finding might be attributable to the characteristics of the agent.

Simulated driving tests

The tests described so far have been used as easily administered analogues of some of the skills needed in everyday activities such as car driving, but many investigators have been attracted to the notion of measuring such skills more closely by using driving simulators.

In what appears to be the first report on the use of such a simulator in anaesthetic research, Egbert and colleagues recorded the times taken to turn a wheel or apply the brake in response to appropriate signals.[54] Fourteen healthy volunteer subjects were studied on each of four occasions, before which they received either 1 or 2 per cent of methohexitone and either 2.5 or 5 per cent of thiopentone to induce very brief periods of unconsciousness ranging from 2.9 to 20 minutes. The times taken by the volunteers to return to their pre-anaesthetic control reaction times were relatively quick, varying from 14.2 minutes after a low dose of methohexitone to 65 minutes after the high dose of thiopentone. Comparisons between the two compounds showed that thiopentone was consistently worse in terms of affecting performance. The conclusion that thiopentone exerts more deleterious effects than methohexitone was upheld by showing that student volunteers 'drove' significantly slower 75 minutes after being given 2.64 mg/kg of thiopentone than they did after receiving 0.88 mg/kg of methohexitone.[55,56] Nevertheless, it was concluded that the two drugs investigated 'impair driving performance only very slightly'.

An extensive series of studies examining the effects of anaesthetic, analgesic and sedative compounds on simulated driving skills has been carried out by a Finnish research group led by Korttila.[2] Their intricate system, which involved projecting a moving roadway representing both urban and rural conditions, allowed a wide variety of variables to be monitored, including brake reaction times and driving errors such as collisions, neglected instructions and driving off the road. This system was used to monitor the performance of volunteer students 2, 4, 6 and 8 hours after intravenous anaesthesia had been induced with 6 mg/

kg of thiopentone, or 2 mg/kg of methohexitone, or 6.6 mg/kg of propanidid or 85 µl/kg of Althesin. Compared with an unanaesthetized control group, driving performance was significantly poorer 6 hours after receiving thiopentone and 8 hours after being given methohexitone. The separate measure of brake reaction time was still poorer than controls at the final 8-hour post-anaesthetic test session in the thiopentone group. Propanidid did not result in any impairment at any time during the experiment.

The same research group has also investigated brief, inhalational anaesthesia lasting 3.5 minutes, using either halothane or enflurane. Although both anaesthetic agents adversely affected manual dexterity and reaction time 5 hours later, as outlined earlier, no adverse effects of enflurane on driving were found during the postoperative period. Nevertheless, halothane did result in more driving errors being made 4.5 hours after anaesthesia.

Research limitations

The bulk of the evidence from studies reviewed so far has suggested that post-anaesthetic impairment of functioning as measured by tests of manual dexterity, speed of reaction or simulated driving is relatively short-lived. Frequently, test performance has recovered to baseline values within a few hours of induction of anaesthesia, a pattern of results implying that everyday tasks based on those skills may not be grossly affected for very long. Even where relatively prolonged effects have been noted, it has been suggested that patients should be advised not to drive for 24 hours after surgery.[2]

That suggestion, in combination with the general findings of relatively brief deterioration on performance tests, appears to have exerted a strong influence on anaesthetic opinion. Early recommendations[57] that patients should be warned against driving for 48 hours post-operatively were rejected on the grounds that it was against current anaesthetic practice and opinion.[58] Recent assertions have also been made that mental, sensory and motor functions return at some point between the patient being able to maintain an airway without assistance and the time when all the anaesthetic is eliminated.[5] The important question is whether these functions return to *normal* levels. In the interests of patients and of the medical profession, it is therefore necessary to examine in more detail the reliability of the experimental evidence on which such contentions are based.

On the face of it, giving people tests of mental efficiency after they have experienced general anaesthesia seems to be a relatively simple procedure. However, when it comes to placing firm reliance on experimentally derived data, a number of problems arise, which include the following.[59]

Patient selection and anaesthetic procedures

Women undergoing minor gynaecological procedures have often been used to study psychomotor consequences of anaesthesia. However, to draw comparisons between the performance of such a group with performance of convenient samples of patients drawn from a wide variety of surgical procedures or with healthy student volunteers not experiencing surgery may be misleading.

A large number of other variables must also be considered when researching in the area of general anaesthesia and mental functioning.[60] These include the characteristics of the surgery, intraoperative events such as hypoxia, variations in premedication, the effects of the hospital environment in which testing takes place and the level of motivation of the patients taking part in the study. To this list might be added the generally unrecognized importance of the 'experimenter effect' by which levels of performance on psychomotor tests given by male or female patients might differ depending on the sex of the data collector.

Probably for reasons such as these, there are many confusions and contradictions in the findings of studies looking at post-anaesthetic recovery.[61] In clinical as opposed to laboratory research it may be difficult to control all or many of these variables, but an awareness of the potentially confounding nature of these factors should be present in all investigators.

Criterion of recovery

It has already been observed above that tests which are sufficiently sensitive to reflect subtle changes in performance are also, unfortunately, subject to marked practice effects. The more often people encounter the tests, the better they do. Such improvement has already been evident in both the pegboard test and the test of serial reaction time,[12,50] and is important in deciding the yardstick by which full recovery is deemed to have taken place. If performance on a given measure does improve with repeated exposure to it, then to adopt a return to preoperative baseline levels as the criterion of recovery will artificially bias the data towards showing relatively fast recovery.

One way to control for the confounding effects of practice is to expose patients preoperatively to a sufficient number of trials in order for them to reach asymptote on the performance test. In clinical practice such an approach is, for the most part, effectively impossible. Some tests require considerable practice beforehand in order to prevent further improvement, and it is highly unlikely that patients would volunteer to attend for extensive preoperative sessions before surgery. Other methods exist to control for practice effects,[62] among which is a recommendation to use a measure that is not affected by practice. However, to reiterate the point made earlier,[39] tests that are subtle enough to reflect changes in performance that are not clinically obvious are precisely those tests that are likely to be susceptible to practice. An alternative approach to the problem is therefore to use a control group design in which volunteers or patients are tested at times equivalent to the anaesthetized patients but who have not undergone any form of anaesthesia. The value of adopting this approach has been emphasized by Marshall and his team.[53]

Control groups

Matching of unanaesthetized controls with the experimental groups should be done for as many variables as possible. Age and sex in particular would seem to be very important. Matching on those two variables will help reduce the problems of the experimenter effect outlined above, and ensuring that groups are similar in age profile will help reduce the confounding of data due to the strong effect of increasing age on the slowing of psychomotor functioning. As there can be up to a 10 per cent variation in performance efficiency across the day,[63] it is also important to ensure that controls are tested at the same clock time as the anaesthetized group. Controls should also be tested in the same physical conditions as the anaesthetic group. If patients are to be tested in bed on a ward several times postoperatively, there is a strong case for selecting as controls a different patient group who are already hospitalized in similar circumstances, because bed rest alone can have a detrimental effect on performance.

Selecting the performance test

There are obvious attractions to choosing tests that are easy to construct, easy to administer and which take up little time. As outlined earlier, it may be for such reasons that the pegboard test and reaction time tests have been employed widely. It is also tempting to incorporate such tests into a large test battery in order to monitor a wide range of potential postoperative mental impairments. However, the choice of an appropriate test should not be a haphazard one.

Several considerations should be taken into account when selecting tests to evaluate the effects of subtle environmental stressors on performance.[64] The longer the test lasts, the more sensitive it becomes to various kinds of stressors. If performance is restricted to only a few seconds or minutes, the chances of both missing any effects and not assessing maximum deterioration are increased. There are also pragmatic reasons for extending the test period. The major concern of post-anaesthetic performance testing is to present analogues of real life. But it is highly probable that most patients drive their cars for longer than a few seconds or even minutes. Prolonging the performance measures for a relatively long time probably reflects real life situations more closely.

The problems posed by motivation levels are also overcome to some extent by extended periods of testing. Many people given a test of psychological ability will regard that test as a challenge and will be tempted to mobilize their resources as much as possible. As a result, brief tests may demonstrate little or no impairment. For reasons such as these, the administration of a battery of tests may also be counterproductive. Going from one novel test to another may also serve to enhance interest and increase motivation, and consequently to obscure any minor, residual impairment that may be present.

Experimental design

The problems of using a within-group experimental design and adopting a return to baseline as a criterion of recovery have been discussed earlier. However, a further problem faces experimenters who adopt a within-group repeated-measures approach to data collection.

The problem of asymmetrical transfer in such experimental designs has received considerable attention over the years.[65] In order to study the effects of different combinations or amounts of different anaesthetic agents, it might seem an attractive strategy to use the same volunteers for testing under each experimental condition after suitable periods for recovery and washout of the drug have been given. A frequent strategy is to counterbalance the order in which volunteers receive the drug or experimental conditions. A recent example of this methodology in anaesthetic research is provided by Korttila et al.[66]

The difficulty with this approach is that the performance of individuals on subsequent occasions can be influenced by their performance following compounds received earlier. If, for example, a particular anaesthetic agent results in remarkably poor performance and is given first, the volunteers will approach the second and perhaps subsequent occasions starting from a different baseline level. Such interactions between drugs and conditions exert a confounding and conservative influence on the data so that potentially significant differences may be obscured. Between-group experimental designs may be necessary in order to prevent this particular problem.

Assumption of permanent recovery

Few investigators have prolonged postoperative performance testing for longer than a few hours after anaesthesia has ceased. The assumption appears to be that, once performance levels have returned to unimpaired levels, no subsequent worsening will take place. That assumption ignores the fact that, on a wide variety of performance tests, people function better as the day goes on. The degree of psychomotor impairment following anaesthesia may simply be masked by this diurnal improvement. It is possible that, at times of day when performance is naturally 'poorer', residual impairment may re-emerge.

Statistical analysis

Many practitioners reading research reports and many investigators carrying it out will be aware of the potential dangers of accepting conventional levels of statistical significance when a wide variety of tests, or sub-measures within a single test, are analysed. Giving patients a battery of tests to carry out on a number of occasions after anaesthesia and then to perform multiple statistical tests on the results is to run the risk of finding some of the differences to be significant by chance alone. For every 20 tests carried out there is a 5 per cent chance that one of them will reach conventional levels of significance. An alternative approach to serial data analysis argues that data collected from repeated measures should be examined by a strategy that looks at the area under the curve.[67] It has also been argued that a particular difficulty facing clinicians in general and anaesthetists in particular is posed by the need to know whether an individual patient remains affected by medical interventions for longer than the average value.[68] It may be that placing reliance on group averages serves only to obscure individual behaviour. In other words, 'when assessing street fitness after day case anaesthesia, the individual's worst performance is more important than that of a theoretical group'.[68]

There is much merit in these arguments, but they should not *preclude* analysis of group performance because it is precisely that which probably forms the basis for general advice about risks of postoperative impairments. An awareness that within general trends there will be individual variation probably adds to the quality of advice about exercising caution postoperatively.

Measures of longer-term recovery

An attempt to account for at least some of the variables discussed above has been made by Herbert *et al.* in studies lasting across two postoperative days.[69,70] In the first study, the effects of routine clinical anaesthetic procedures were investigated in groups of male patients hospitalized for elective surgical repair of hernia. Patients were premedicated with a standard dose of 10 mg of diazepam 1 hour before induction of anaesthesia. Anaesthesia was induced in one group with thiopentone (250 mg) and was maintained with 0.5–1.5 per cent of halothane with nitrous oxide and oxygen. Controlled ventilation was performed on 22 patients who were given suxamethonium and alcuroniun as muscle relaxants. A group of 21 patients breathed spontaneously and a third group was formed from patients in whom anaesthesia was induced with halothane and who breathed spontaneously during surgery.

Performance was measured using a serial, four-choice reaction time test lasting 5 minutes. Patients were required to press one of four buttons corresponding geometrically to one of four lights that could be illuminated. Pressing the button extinguished the light and brought on another in random order. Pressing an incorrect button resulted in the same effect, but was recorded on magnetic tape as an error response. After familiarization with the test, patients carried out a preoperative assessment; they repeated the test 90 minutes after returning to the ward from the recovery room and at 08:30, 11:00, 13:30 and 16:30 on the 2 days following surgery. Patients were tested by the same investigator on each occasion and carried out the test in bed with the screens drawn to prevent distraction. A control group was formed from similar aged, hospitalized, male orthopaedic patients who had not undergone an operation for at least 2 weeks. They were tested at times corresponding to those of the anaesthetized patients.

All groups were comparable in their levels of preoperative reaction time, but when compared with controls, the three groups given anaesthesia showed slowed reaction times 90 minutes after regaining consciousness. That impairment of reaction speed persisted into the first postoperative day in patients in whom anaesthesia was induced with halothane. Their performance was significantly poorer during the two morning tests and the one given early in the afternoon. Performance during the second postoperative day was also impaired, compared with control groups, in patients who had breathed spontaneously during anaesthesia and in patients given an inhalational induction. On the basis of this pattern of results the authors concurred with recommendations[57] that patients be advised not to drive cars for at least 48 hours postoperatively.

In the second of Herbert's studies, hernia patients were induced into anaesthesia with either 2.5 mg/kg of propofol or 5 mg/kg of thiopentone. Anaesthesia was maintained by a mixture of halothane, nitrous oxide and oxygen. Performance on the four-choice reaction time test was given at times similar to those in the first study. When compared with controls, thiopentone resulted in impaired performance 90 minutes after regaining consciousness, throughout the first postoperative day and into the morning of the second day. Propofol exerted no significant adverse effect on reaction times at any point.

It appears then that, with the use of sufficiently sensitive tests, administered in appropriate conditions and compared with control groups, the deleterious effects of some anaesthetic procedures in current clinical use can be shown to persist for longer than is supposed. Further inspection of the results reveals that, whereas the control group displayed a consistent improvement in performance over time, patients receiving thiopentone seemed unable to benefit from practice. Whether this is a reflection of the impairment in memory produced by the action of barbiturates is only one of the intriguing questions posed by the investigation of post-anaesthetic mental functioning and clearly is one that is open to further research.

Selection and management considerations

Clearly, the answer to what appears to be a very simple question about the extent to which patients are likely to be impaired after anaesthesia turns out to be quite complex. It depends upon the agents used during anaesthesia as well as on a multitude of factors governing experiment design and analysis. In the light of these difficulties, what can be done to help staff in day-case units provide appropriate management of and advice to patients undergoing such procedures?

An important way of overcoming potential postoperative problems obviously lies in the application preoperatively of appropriate selection criteria to determine the type of patient likely to have difficulties afterwards. Guidelines issued by the Royal College of Surgeons of England[71] include both physical and social criteria, such as the exclusion of the elderly or infirm, although it is recognized that 'age or fitness are hard to define, and clinical judgement is all-important'. In a later section the Guidelines observe that 'the upper age limit of about 65–70 years should be judged on biological grounds rather than chronological age'. Ogg is more definite in his recommendation that people aged over 70 years should be excluded.[3]

Physical status is also important. Recommended inclusion criteria suggest that patients with ASA grade 1 or 2 are suitable for day-case procedures but those who are obese, diabetic or who have chronic respiratory or cardiovascular disease are not.[3] The Royal College suggests that, in most circumstances, a routine measurement of pulse, respiratory rate and blood pressure will be adequate.[71] Further additions to these physical criteria have been suggested,[3] including recommendations that anticipated postoperative pain or haemorrhage or a planned duration of surgery longer than 30 minutes should serve to exclude patients from this type of procedure.

The patient's social circumstances must also be considered in the selection process. Adequate housing conditions including toilet facilities, a responsible adult being available to care for the patient day and night, access to a telephone in case of emergencies and living within easy reach of the hospital by official or private transport have been seen to be necessary.[71]

Having attempted in this way to pre-empt many potential postoperative difficulties, the next issue concerns the state of patients who have undergone day-case procedures. Korttila's distinction between 'home readiness' and 'street fitness' is a potentially valuable one.[2,22] His recommendations for considering someone ready for safe discharge include the stability of vital signs for at least 1 hour and no evidence of respiratory depression. Patients must be orientated to person, place and time, and able to dress themselves and walk unaided. He also suggests that, after spinal or epidural anaesthesia or for patients who have had surgery near the pelvic area, the ability to maintain oral fluids and to void may form important criteria. Other measures of the patient's physical condition include the absence of more than minimal nausea or vomiting, bleeding and excessive pain. It is, however, not clear what constitutes minimal nausea or excessive pain, and one probably has to rely on the patient's own evaluation of these.

A substantial part of the present chapter has focused on the problems inherent in attempts to define Korttila's concept of 'street fitness' because it is this issue that will shape the advice which patients are offered about their potential readiness to return to everyday activities. Korttila's personal opinion[2] is that patients should be advised not to drive for 24 hours if the duration of anaesthesia is less than 2 hours. For anaesthesia lasting longer than that, he recommends a 48-hour abstinence period.

Whilst it is true that many of the studies cited earlier have generally included relatively brief durations of anaesthesia and have usually demonstrated a fairly rapid return to unimpaired levels of functioning, at least two studies[69,70] have shown adverse effects on psychomotor function persisting into the second postoperative day following general anaesthesia lasting for only 20 minutes or so. The re-emergence of deleterious consequences occurring after apparent full recovery is also an issue of some importance. It is the present author's opinion, therefore, that one should adopt a cautious approach and advise patients of potential impairments lasting up to at least 48 hours postoperatively. That suggestion, however, may have to be tempered in the light of the kind of anaesthetic agents used. Some of the newer compounds may have fewer postoperative adverse sequelae.

Whichever yardstick the practising clinician chooses to adopt, it is an ethical and legal concern to ensure that patients, and their carers, are given information about pre-anaesthetic requirements such as avoiding foods or liquids, potential postoperative side effects as well as emergency telephone contact numbers. Because human memory is fallible, simple reliance on verbal explanations may be inadequate. The literature contains examples of forms currently in use, giving not only standard information but also containing sections for completion by patients and their carers signifying that instructions have been explained to them and have been understood.[3,72] To adopt precautions such as these may serve to protect the patient from misadventure and ensure that ambulatory anaesthesia is maximally efficient.

REFERENCES

1. Lagoe RJ, Bice SE, Abulencia PB. Ambulatory surgery utilization by age level. *American Journal of Public Health* 1987; **77**: 33–7.

2. Korttila K. Recovery from day case anaesthesia. *Ballière's Clinical Anaesthesiology* 1990; **4**: 713–32.

3. Ogg TW. An anaesthetist's view of day case surgery. In: Hindmarch I, Jones JG, Moss E, eds. *Aspects of recovery from anaesthesia*. Chichester: John Wiley, 1987: 9–15.

4. Ogg TW. An assessment of post-operative outpatient cases. *British Medical Journal* 1972; **4**: 573–6.

5. Curtis D, Stevens LOC. Recovery from general anesthesia. *International Anesthesiology Clinics* 1991; **29**: 7–11.

6. Hannington-Kiff JG. Measurement of recovery from outpatient general anaesthesia with a simple ocular test. *British Medical Journal* 1970; **3**: 132–5.

7. Zuurmond WWA, van Leeuwen L. Recovery from sufentanil anaesthesia for outpatient arthroscopy: a comparison with isoflurane. *Acta Anaesthesiologica Scandinavica* 1987; **31**: 154–6.

8. Manner T, Kanto J, Salonen M. Use of simple tests to determine the residual effects of the analgesic component of balanced anaesthesia. *British Journal of Anaesthesia* 1987; **59**: 978–82.

9. Moss E, Hindmarch I, Pain AJ, Edmondson RS. Comparison of recovery after halothane or alfentanil anaesthesia for minor surgery. *British Journal of Anaesthesia* 1987; **59**: 970–7.

10. Zuurmond WWA, van Leeuwen L, Helmers JHJH. Recovery from fixed-dose midazolam-induced anaesthesia and antagonism with flumenazil for outpatient arthroscopy. *Acta Anaesthesiologica Scandinavica* 1989; **33**: 160–3.

11. McClelland GR. Body sway. In: Klepper ID, Sanders LD, Rosen M, eds. *Ambulatory anaesthesia and sedation*. Oxford: Blackwell Scientific, 1991: 106–16.

12. Vickers MD. The measurement of recovery from anaesthesia. *British Journal of Anaesthesia* 1965; **37**: 296–302.

13. Eriksen J, Jansen E, Larsen RE, Olesen MB. Postanaesthetic postural stability following thiopental or propanidid anaesthesia. *Acta Anaesthesiologica Scandinavica* 1978; **22**: 323–6.

14. Korttila K, Ghonheim MM, Jacobs L, Lakes RS. Evaluation of instrumental force platform as a test to measure residual effects of anesthetics. *Anesthesiology* 1981; **55**: 625–30.

15. Valanne J. Recovery and discharge of patients after long propofol infusion vs isoflurane anaesthesia for long ambulatory surgery. *Acta Anaesthesiologica Scandinavica* 1992; **36**: 530–3.

16. Aldrete JA, Kroulik D. A postanesthetic recovery score. *Anesthesia and Analgesia* 1970; **49**: 924–33.

17. Steward DJ. A simplified scoring system for the post-operative recovery room. *Canadian Anaesthetists' Society Journal* 1975; **22**: 111–13.

18. Sanders LD, Clyburn PA, Rosen M, Robinson JO. Propofol in short gynaecological procedures. *Anaesthesia* 1991; **46**: 451–5.

19. Paut O, Guidon-Attali C, Viviand X, Lacarelle B, Bouffier C, Francois G. Pharmacodynamic properties of propofol during recovery from anaesthesia. *Acta Anaesthesiologica Scandinavica* 1992; **36**: 62–6.

20. Ravussin P, Tempelhoff R, Modica PA, Bayer-Berger MM. Propofol vs thiopental – isoflurane for neurosurgical anesthesia: comparison of hemodynamics, CSF pressure and recovery. *Journal of Neurosurgical Anesthesiology* 1991; **3**: 85–95.

21. Bellaiche S, Bonnet F, Sperandio M, Lerouge P, Cannet G, Roujas F. Clonidine does not delay recovery from anaesthesia. *British Journal of Anaesthesia* 1991; **66**: 353–7.

22. Korttila K. Recovery and home readiness after anesthesia for ambulatory surgery. *Seminars in Anesthesia* 1990; **9**: 182–9.

23. Healy TEJ, Lautch H, Hall N, Tomlin PJ, Vickers MD. Interdisciplinary study of diazepam sedation for outpatient dentistry. *British Medical Journal* 1970; **3**: 13–17.

24. Hindmarch I, Bhatti JZ. Recovery of cognitive and psychomotor function following anaesthesia. A review. In: Hindmarch I, Jones JG, Moss E, eds. *Aspects of recovery from anaesthesia*. Chichester: John Wiley, 1987: 113–65.

25. Blundell E. A psychological study of the effects of surgery on eighty-six elderly patients. *British Journal of Social Clinical Psychology* 1967; **6**: 297–303.

26. Carson IW, Graham J, Dundee JW. Clinical studies of induction agents. XLIII: recovery from Althesin – a comparative study with thiopentone and methohexitone. *British Journal of Anaesthesia* 1975; **47**: 358–64.

27. Bahar M, Dundee JW, O'Neil MP, Briggs LP, Moore J, Merrett JD. Recovery from intravenous anaesthesia: comparison of disoprofol with thiopentone and methohexitone. *Anaesthesia* 1982; **37**: 1171–5.

28. Reitan JA, Porter W, Braunstein M. Comparison of psychomotor skills and amnesia after induction of anesthesia with midazolam or thiopental. *Anesthesia and Analgesia* 1986; **65**: 933–7.

29. Boas RA, Newson AJ, Taylor KM. Comparison of midazolam with thiopentone for outpatient anaesthesia. *New Zealand Medical Journal* 1982; **96**: 210–12.

30. Craig J, Cooper GM, Sear JW. Recovery from day-case anaesthesia: comparison between methohexitone, Althesin and etomidate. *British Journal of Anaesthesia* 1982; **54**: 447–51.

31. Sear JW, Cooper GM, Kumar V. The effect of age on recovery: a comparison of the kinetics of thiopentone and Althesin. *Anaesthesia* 1983; **38**: 1138–61.

32. Cooper GM, O'Connor M, Mark J, Harvey J. Effect of alfentanil and fentanyl on recovery from brief anaesthesia. *British Journal of Anaesthesia* 1983; **55**: 179–81S.

33. Sinclair ME, Cooper GM. Alfentanil and recovery. *Anaesthesia* 1983; **38**: 435–7.

34. Steib A, Freys G, Jochum D, Ravenello J, Schaal JS, Otteni JC. Recovery from total intravenous anaesthesia. Propofol versus midazolam–flumazenil. *Acta Anaesthesiologica Scandinavica* 1990; **34**: 632–5.

35. Korttila K, Tamisto T, Ertama P, Pfaffli I, Blomgren E, Hakkinen S. Recovery, psychomotor skills and simulated driving after brief inhalational anaesthesia with halothane or enflurane combined with nitrous oxide and oxygen. *Anesthesiology* 1977; **46**: 20–7.

36. Denis R, Letourneau JE, Londorf D. Reliability and validity of psychomotor tests as measures of recovery from isoflurane or enflurane anesthesia in a day-care unit. *Anesthesia and Analgesia* 1984; **63**: 653–6.

37. Letourneau JE, Denis R. The modified GATB (M) as a measure of recovery from general anesthesia. *Perceptual and Motor Skills* 1983; **56**: 451–8.

38. Millar K. Psychomotor tasks and recovery from anesthesia. *Anesthesia and Analgesia* 1986; **65**: 543–4.

39. Herbert M. Assessment of performance in studies of anaesthetic agents. *British Journal of Anaesthesia* 1978; **50**: 33–8.

40. Wollman SB, Orkin LR. Postoperative human reaction time and hypocarbia during anaesthesia. *British Journal of Anaesthesia* 1968; **40**: 920–5.

41. Blenkarn GD, Briggs G, Bell J, Sugioka K. Cognitive function after hypocapnic hyperventilation. *Anesthesiology* 1972; **37**: 381–6.

42. Horvorka J. Carbon dioxide homeostasis and recovery after general anaesthesia. *Acta Anaesthesiologica Scandinavica* 1982; **26**: 498–504.

43. Azar I, Karambelkar DJ, Lear E. Neurologic state and psychomotor function following anesthesia for ambulatory surgery. *Anesthesiology* 1983; **60**: 47–9.

44. Horvorka J, Lehtinen AM, Kalli I. Recovery from general anaesthesia for laparoscopy. *Acta Anaesthesiologica Scandinavica* 1983; **27**: 369–9.

45. Nightingale JJ, Lewis IH. Recovery from day-case anaesthesia: comparison of total i.v. anaesthesia using propofol with an inhalational technique. *British Journal of Anaesthesia* 1992; **68**: 356–9.

46. Mackenzie M, Grant IS. Comparison of propofol with methohexitone in the provision of anaesthesia for surgery under regional blockade. *British Journal of Anaesthesia* 1985; **57**: 1167–72.

47. Mackenzie M, Grant IS. Comparison of the new emulsion formulation of propofol with methohexitone and thiopentone for induction of anaesthesia in day cases. *British Journal of Anaesthesia* 1985; **57**: 725–31.

48. Gale GD. Recovery from methohexitone, halothane and diazepam. *British Journal of Anaesthesia* 1976; **48**: 691–7.

49. Grant IS, Smith G, Shirley AW. The audio visual reaction time test: use in assessment in recovery from Althesin anaesthesia. *Anaesthesia* 1980; **35**: 869–72.

50. Scott WAC, Whitwam JG, Wilkinson RT. Choice reaction time: a method of measuring postoperative psychomotor performance decrements. *Anaesthesia* 1983; **38**: 1162–8.

51. Anderson S, McGuire R, McKeown D. Comparison of the cognitive effects of premedication with hyoscine and atropine. *British Journal of Anaesthesia* 1985; **57**: 169–73.

52. Milligan KR, O'Toole DP, Howe JP, Cooper JC, Dundee JW. Recovery from outpatient anaesthesia: a comparison of incremental propofol and propofol–isoflurane. *British Journal of Anaesthesia* 1987; **59**: 1111–14.

53. Marshall CA, Jones RM, Bayorek PK, Cashman JN. Recovery characteristics using isoflurane or propofol for maintenance of anaesthesia: a double-blind controlled trial. *Anaesthesia* 1992; **47**: 461–6.

54. Egbert LD, Oech SR, Eckenhoff JE. Comparison of the recovery from methohexital and thiopental anesthesia in man. *Surgery, Obstetrics and Gynecology* 1959; **109**: 427–30.

55. Elliott CJR, Green R, Howells TH, Long HA. Recovery after intravenous barbiturate anaesthesia: comparative study of recovery from methohexitone and thiopentone. *Lancet* 1962; 68–71.

56. Green HR, Long HA, Elliott CJR, Howells TH. A method of studying recovery after anaesthesia. *Anaesthesia* 1963; **18**: 190–200.

57. Havard JA. *Medical aspects of fitness to drive.* London: Medical Commission on Accident Prevention, 1976.

58. Baskett BJF, Vickers MD. General anaesthesia and driving. *Lancet* 1979; **1**: 490.

59. Herbert M. Measuring post-anaesthetic impairment of mental functioning. In: Klepper ID, Sanders LD, Rosen M, eds. *Ambulatory anaesthesia and sedation.* Oxford: Blackwell Scientific, 1991: 47–59.

60. Flatt JR, Birrell PC, Hobbes SA. Effects of anesthesia on some aspects of mental functioning of surgical patients. *Anesthesia and Intensive Care* 1984; **12**: 315–24.

61. Epstein BS. Recovery from anesthesia. *Anesthesiology* 1975; **43**: 285–8.

62. Brooks DN. Measuring neuropsychological and functional recovery. In: Levin HS, Grafman J, Eisenberg HM, eds. *Neurobehavioural recovery from head injury.* New York: Oxford University Press, 1987.

63. Folkard S. Diurnal variation in logical reasoning. *British Journal of Psychology* 1975; **66**: 1–8.

64. Wilkinson RT. Some factors influencing the effect of environmental stressors upon performance. *Psychological Bulletin* 1969; **72**: 260–72.

65. Millar K. Clinical trial design: the neglected problem of asymmetrical transfer in cross-over trials. *Psychological Medicine* 1983; **13**: 867–73.

66. Korttila K, Nuotto EJ, Lichtor JL, Ostman PL, Apfelbaum J, Rupani G. Clinical recovery and psychomotor function after brief anesthesia with propofol or thiopental. *Anesthesiology* 1992; **76**: 676–81.

67. Matthews JNS, Altman DG, Campbell MJ, Royston P. Analysis of serial measurements in medical research. *British Medical Journal* 1990; **300**: 230–5.

68. Hickey S, Asbury AJ, Millar K. Psychomotor recovery after

outpatient anaesthesia: individual impairment may be masked by group analysis. *British Journal of Anaesthesia* 1991; **66:** 345–52.

69. Herbert M, Healy TEJ, Bourke JB, Fletcher IR, Rose JM. Profile of recovery after general anaesthesia. *British Medical Journal* 1983; **286:** 1539–42.

70. Herbert M, Makin SW, Bourke JB, Hart EA. Recovery of mental abilities following general anaesthesia induced by propofol ('Diprivan') or thiopentone. *Postgraduate Medical Journal* 1985; **61:** suppl 3, 132.

71. Royal College of Surgeons of England. *Commission on the provision of surgical services.* London: RCSE, 1985.

72. Korttila K. Practical discharge criteria. *Problems in Anaesthesia* 1988; **2:** 144–51.

FURTHER READING

Hindmarch I, Jones JG, Moss E, eds. *Aspects of recovery from anaesthesia.* Chichester: John Wiley, 1987.

Klepper ID, Sanders LD, Rosen M. *Ambulatory anaesthesia and sedation.* Oxford: Blackwell Scientific, 1991.

Chapter 41

Awareness and the Depth of Anaesthesia

C. J. D. Pomfrett and T. E. J. Healy

Awareness is not an inevitable consequence of light anaesthesia, but light anaesthesia is the main reason for awareness; deeply anaesthetized patients do not report awareness during surgery. Awareness will still have occurred even if the patient has been given amnesic agents and cannot explicitly recall the event, and so methods of disrupting the storage of long-term memory are discussed only briefly here.

Incidence of awareness

In order to quantify the problem of awareness under anaesthesia, one must first use some standard against which to describe the incidence. Four levels of awareness have been described.[1]

I. Conscious awareness without amnesia
II. Conscious awareness with amnesia
III. Subconscious awareness with amnesia
IV. No awareness

Stage I is normal wakefulness; the patient volunteers explicit memories of events during surgery. Implicit expression of memory is encountered during stage III, when the patient has no voluntary recall of events. However, events such as morbid conversations by the surgical team are memorized subliminally and recalled as subsequent adverse behavioural reactions.[2] The Medical Defence Union in the UK reported that, of their total claims during 1990, amounting to £50 million, 12.2 per cent were related to awareness during general anaesthesia.[3] This was a greater proportion of claims than for perioperative death (11.6 per cent). Seventy per cent of awareness was attributed to a faulty technique, 20 per cent to a failure to check equipment, and only 2.5 per cent of claims for awareness were attributed to apparatus failure, spurious claims, justified risks or unknown causes.

Determination of the incidence of stage I, explicit awareness, requires careful experimental design. Public awareness of the possibility of conscious awareness has been heightened in recent years and it is important to discriminate awareness from dreams, hallucinations or false claims. Properly conducted trials with structured interviews of subjects have been conducted and revealed that the incidence of awareness ranged from around 0.2 per cent to 1.6 per cent.[4] The occurrence of dreaming ranged from 0.9 to 26 per cent in the same studies. Other trials without structured interviews place the incidence of awareness at around 8 per cent for caesarean section.[5] The incidence extends up to 45 per cent during accident and emergency surgery, when there had been gaps in maintenance of adequate levels of anaesthetic depth.[6] It has been estimated that approximately 10 per cent of patients who suffer from awareness experience pain.[3]

There have been attempts to convert stage I, conscious awareness with recall, into stage II, awareness with amnesia, by the use of benzodiazepines. However, it should be remembered that removing the memory of awareness is not the same as preventing awareness. It is quite possible that implicit recall of an event may still be present, even after the retrograde disruption of explicit memory with an amnesic agent. The target of all anaesthetists should be the prevention of awareness. The use of morphine and diazepam significantly reduced the incidence of awareness and unpleasant dreams in one study of 68 patients undergoing caesarean section.[7] The morphine and diazepam combination was found to cause both anterograde and retrograde amnesia. The authors also reported that neither the size of the pupils nor the blood pressure changes were indicative of awareness. The loss of pupillary reactivity is not surprising, given that the administration of morphine has been shown to lead to a rapid decrease in pupil diameter.[8]

Explicit awareness is a problem associated with the use of muscle relaxation; spontaneously breathing patients move if too lightly anaesthetized. There has been only one case report of a spontaneously breathing patient who experienced awareness during anaesthesia.[9,10]

Subconscious awareness with amnesia can be demonstrated by testing for a behavioural response to an intraoperative suggestion. One such study was performed on a group of 33 patients in whom anaesthesia was induced with thiopentone and then maintained with nitrous oxide–oxygen and halothane or enflurane.[11] A significant number (9 of 11) of subjects exhibited postoperative ear touching behaviour in response to intraoperative verbal suggestions. None of the patients was able to recall the suggestions during subsequent hypnosis, and none of the patients complained of awareness during anaesthesia. Word sequences presented during anaesthesia may be recalled postoperatively.[12] Detection of ten key words out of a forty-word sequence was significantly better in patients who had listened to a tape of the key words intraoperatively, rather than a tape of radio static.[12] The authors recommended the use of earplugs to prevent recall of words heard during anaesthesia. The use of earplugs obviously should not be taken as a way of preventing awareness under anaesthesia, but it does remove the possibility that a patient who suffers from awareness may recognize key phrases from the operating theatre.

It has been suggested that positive verbal suggestions presented intraoperatively, and subject to implicit recall, may reduce the duration of postoperative recovery. In a study involving 91 patients, exposure to intraoperative positive suggestions was reported to protect elderly patients against prolonged postoperative stays.[13] None of the patients demonstrated explicit recall after anaesthesia of the verbal stimuli. However, a subsequent study failed to demonstrate any evidence of an improved postoperative outcome or a reduced duration of stay by patients who had received verbal therapeutic suggestions during a total abdominal hysterectomy.[4]

The consequences of any statistically measurable level of awareness, even one as low as 0.2 per cent, are striking when the number of uses of general anaesthesia every day are considered; i.e. 1 in 500 patients will suffer from conscious awareness during their surgery. It is therefore desirable for the anaesthetist to be able to identify impending light anaesthesia in order to prevent conscious awareness.

Postoperative visits by the anaesthetist are recommended as part of good clinical practice. One recent study has highlighted the use of the postoperative visit to calm and reassure patients who had suffered from awareness.[14] One of the most disturbing recollections by such patients is that the anaesthetist did not care about them. It is obvious that such an impression harms not only the hospital team but also the practice of anaesthesia. Postoperative review of well-kept notes often does not reveal an obvious cause for awareness, as was the case in this study, where blinded anaesthetic notes from patients who had suffered from awareness were deliberately concealed among notes from those who had not. A panel of senior anaesthetists scrutinized the notes, and subsequent statistics showed that they were unable to distinguish those patients who had suffered from awareness.

Relatively few incidences of awareness have been reported in the literature as a result of using total intravenous anaesthesia (TIVA) without nitrous oxide. This is surprising, because one of the major attractions of TIVA is the smooth, rapid recovery from anaesthesia associated with modern intravenous agents, such as propofol. Such a rapid recovery in the event of failure of an infusion pump due to either operator error or equipment failure would supposedly facilitate awareness. Five cases of awareness out of approximately 2500 patients ($\sim$ 0.2%) have been reported in one study in which propofol was used with alfentanil.[15] Two of the patients recalled some pain. However, the authors simply asked the patients if they had slept well, and this question did not form part of a properly structured interview. There is clearly a need for a detailed study, including structured postoperative interviews, of the incidence of awareness with different anaesthetic agents. This is the only way in which potential problems of awareness associated with particular agents or techniques may be identified objectively.

The incidence and implications of implicit awareness of subliminal suggestions are more difficult to assess. It is of concern to the anaesthetist that a negative suggestion regarding the health of the patient may be subconsciously learnt and implicitly recalled, affecting the eventual prognosis. Conversely, positive suggestions that help recovery should be used if they can be proved to be beneficial. There are obviously serious ethical reasons for not allowing negative suggestions to be presented deliberately to the subconscious patient. One study, on ten patients who were anaesthetized with thiopentone, nitrous oxide and ether, required the anaesthetist to deliberately make statements to the effect of the following, 'Stop the operation. I don't like the patient's colour. His/her lips are too blue. I'm going to give a little oxygen.' These statements were followed with reassuring phrases: 'There, that's better now. You can carry on with the operation.'[16] None of the patients exhibited explicit recall of the phrases. However, on interviewing them under hypnosis 1 month later, four patients were able to recall the phrases exactly, four had partial recall and two could not recall the operation at all. This study is of particular relevance to contemporary research, since all patients exhibited irregular, slow EEG waveforms (monitored using six-channel scalp recordings); the presence of 'adequate' anaesthesia as defined by the EEG did not prevent implicit recall.

A further study, described by the same author, gives rise to further concerns. Levinson was present when a patient was anaesthetized with thiopentone, nitrous oxide, oxygen and halothane during removal of a cyst from the mouth. A surgical colleague stated 'Good gracious! This may not be a simple cyst; it may be cancer.' The heart rate varied in response to the statement, to which Levinson scribbled the words 'Reassure her please!' The surgeon removed the cyst and stated 'On second thoughts, it was only a simple cyst.' After recovery the patient was shown the laboratory report confirming that the cyst was non-malignant. Under subsequent hypnosis the patient was able to recall the words spoken by the surgeon, and stated that she had not believed the surgeon's reassurances. Two years later, the patient died of a uterine cancer. It is, of course, coincidental that there was such an outcome to this particular experiment, but the very fact that the patient could recall a negative suggestion, and disbelieve a positive reassurance, suggests the need for operating theatre staff to be very careful about what they say. The potentially terrible consequences of a poorly guarded statement and the subsequent feelings of guilt in the perpetrator are worth avoiding at all costs.

Depth of anaesthesia – history of concept

Anaesthesia is a tendency towards death. Progressive loss of consciousness was staged by Snow, who described 'five stages of narcotism'. These were subsequently refined by Guedel.[17] Guedel's sleep stages from the basis of assessment for most modern, volatile anaesthesia, even though developed with ether and before the advent of muscle relaxants. Most importantly, staging the depth of anaesthesia is not the same as staging potential levels of awareness. The first stage in Guedel's classification was analgesia, followed by a second stage of excitement. Stage three was defined as surgical anaesthesia, originally comprising four sequential planes: I, from automatic respiration or regular breathing to cessation of eye movement; II, to commencement of intercostal paralysis; III, to complete intercostal paralysis; IV, to diaphragmatic paralysis. Guedel's fourth stage was overdosage, ranging from the onset of diaphragmatic paralysis to apnoea and death. Ether anaesthesia conveniently results in a progressive dilation of the pupil from stage 3, plane I, through to death, and a loss of pupillary light reflex during stage 3, plane III. White[18] adopted the concept of a continuum, increasing anaesthesia providing a progressive loss of physiological control. Sedation can be considered as an intermediate stage between unconsciousness and wakefulness.

Muscle relaxants removed the need for deep anaesthesia to reduce muscle tone. Unfortunately, muscle relaxation is not anaesthesia. Concern about inadequate levels of anaesthesia, and the incorrect use of muscle relaxants as anaesthetics, has resulted in the maintenance of adequate anaesthetic depth becoming a legal requirement during animal experimentation in the UK (Scientific Procedures Act 1986).[19] Similar concern has not led to minimum legal standards for monitoring anaesthetic depth in humans.

Measurement of anaesthetic depth – windows on the brain

Dose-related estimates of anaesthetic adequacy are empirical and assume the correct operation of the equipment used for the delivery of the agent. Vaporizers, even when regularly calibrated, do not measure the amount of anaesthetic actually present in the patient. Mechanical failure, airway obstruction or abnormal metabolism may all conspire to reduce the level of anaesthesia to a potentially inadequate level. End-tidal gas analysers are costly, and do not measure the levels of intravenous agents. No monitor based on anaesthetic levels will monitor the physiological response of a patient during anaesthesia, nor will such a monitor detect the effects of surgical stimulation on lightening of anaesthesia.

There are both subjective and objective methods of assessing anaesthetic depth. Subjective estimates rely on the opinion and experience of an anaesthetist to state whether the patient is adequately anaesthetized. Some sense of objectivity will be introduced when an operating theatre monitor can be used to give an independent estimate of the level of anaesthetic depth. Until a simple, reliable monitor of anaesthetic depth is commercially available, the anaesthetist must rely on easily detectable, physiological presentations of brain function such as the pupillary light reflex and changes in haemodynamic variables in order to estimate the level of anaesthesia.

Subjective measurement

Autonomic changes

Sudden changes in blood pressure or heart rate may indicate lightening of anaesthesia to levels that result in awareness.[15] Unfortunately, such haemodynamic changes may also be associated with a wide range of other events, and it may be impossible eliminate all of the other causes in the time before light anaesthesia turns into conscious awareness. Attention to changes in the heart rate and blood pressure may vary with the psychological state of the anaesthetist.[20] Poor design of monitoring equipment may lead to incorrect usage, especially when the operating room is unfamiliar to the anaesthetist. There is the tendency to switch off alarms that are unduly distracting, or to set the ranges of alarms to wider limits than physiologically appropriate. As yet there are no dynamic alarms with a sensitivity that changes depending on the progress through the anaesthetic procedure. Such an alarm would be part of an expert system, trained originally by the subject matter experts (anaesthetists!).

Pupil diameter (Fig. 41.1)

Classic staging of ether anaesthesia shows how the pupil dilates, after initial contraction, with progressively deeper anaesthesia.[17] This dilation is due to changing levels of circulating catecholamines as well as a direct effect on the central nervous system. Contemporary anaesthetics do not give the same change in catecholamines, and so the pupil diameter does not change reliably with the depth of anaesthesia. Opioid premedication induces miosis, and atropine and hyoscine (scopolamine) induce mydriasis, which further disturb changes in the pupil diameter due to anaesthesia. Similarly, the pupillary light reflex may also be affected by intraoperative opioid analgesics[8] and anoxia, and hence rendered unreliable; the presence of an

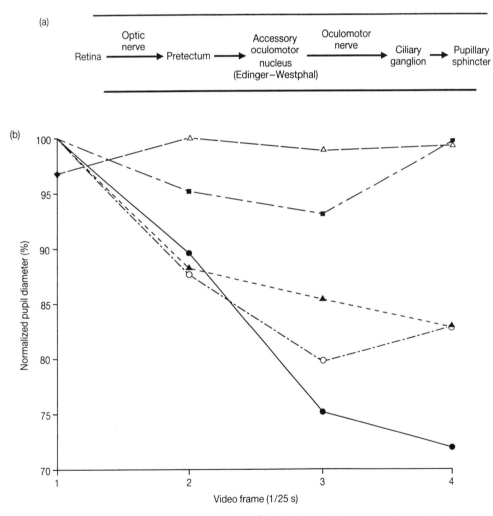

Fig. 41.1 (a) Neural pathway of the pupillary light reflex. (b) The reduction in pupillary contraction velocity for a single patient during anaesthesia, and the effect of fentanyl. Data were obtained before, during and after total intravenous anaesthesia with propofol. A high resolution video camera was directed at the eye during exposure to a brief period of strong illumination from a photographic flashgun. A computer was then used to measure the reduction in pupil area after the flash for four 1/25th-second consecutive video frames. After induction, the pupil contraction velocity (gradient of the curve) declined, and then started to recover after the bolus induction dose (2.5 mg/kg) of propofol had worn off, and a steady propofol infusion commenced (10 mg/kg per hour). However, the light reflex was disrupted when fentanyl was given, and had only slightly recovered 5 minutes after anaesthesia ceased. ●, preinduction; ▲, propofol induction; ○, propofol infusion; △, fentanyl; ■, post-fentanyl recovery.

unresponsive pupil does not necessarily mean that the patient is deeply anaesthetized.

Isolated forearm technique

The isolated forearm technique (IFT) is a simple method of preventing the suppression of movement by competitive neuromuscular blockade.[21] A tourniquet is placed on an arm (not the arm with the intravenous infusion) of a patient before paralysis. The arm is therefore free to move during the surgical procedure, and purposeful movements in response to instruction indicate excessively light anaesthesia. Ischaemia has to be prevented by periodically releasing the tourniquet, usually before topping up the level of neuromuscular blockade, when the levels of circulating neuromuscular blocking agent are at a minimum.

Despite the ease of this technique, relatively few anaesthetists employ the IFT. Surgeons express concern about patients moving during the procedure, and disturbing the surgical field; the levels of anaesthesia needed to prevent movements in all patients during the IFT are significantly higher than those routinely used since the advent of muscle relaxants. It has been reported that the IFT does not act as an indicator of subsequent recall; patients have reported that they have heard commands to move the isolated arm, but have been unable to do so, even though electrical nerve stimulation suggested that the arm was not paralysed.[22]

Objective measurement

An ideal objective monitor of anaesthetic depth should meet the following requirements in order to be of value to the ordinary anaesthetist:[23] it should indicate the stage during light anaesthesia preceding conscious awareness; it should closely reflect changing concentrations of anaesthetic agents and be sensitive to stimuli of different modalities, especially surgical stimulation; it should have a high temporal resolution; it should be able to stage the depth of anaesthesia. In addition, the monitor should meet the criteria that anaesthetists have come to expect of contemporary medical equipment, including ease of use and short set-up time. One additional requirement set by most anaesthetists is that any putative monitor of anaesthetic depth should work for all modern anaesthetic agents, and the characteristics of the response should be the same with different anaesthetics at equivalent levels of anaesthesia.

Frontalis electromyogram

The use of competitive neuromuscular blockade to induce paralysis has meant that it is impossible to gauge the level

of anaesthesia simply by looking at the level of voluntary movement. However, one easily studied muscle that is less affected by neuromuscular blockade is the frontalis muscle of the forehead. The frontalis is innervated by a branch of the facial nerve. A stick-on electrode positioned over the frontalis muscle can record the frontalis electromyogram (FEMG). It has been noted that the level of FEMG activity declines during anaesthesia, and rises to pre-induction levels at recovery. This measurement has formed the basis of one commercially available monitoring system (Datex, ABM®) which also records EEG indices via the same electrodes.[24] An example of a typical ABM record is shown in Fig. 41.2. The ABM uses three electrodes, two placed unilaterally on the mid-forehead and one behind the ear, at the mastoid process. One study examined the routine use of the ABM and noted that clinical doses of neuromuscular blocking agents did not suppress the FEMG. The level of the FEMG was observed to fall during anaesthesia, and to rise to pre-anaesthetic levels just before awakening, in 72 patients undergoing propofol anaesthesia.[25] However, the authors concluded that no isolated ABM value was of interest and that the scales were not absolute; relative changes in combinations of EEG zero crossing frequency, EEG mean integrated voltage and FEMG together allowed staging of anaesthesia against the Guedel scale.

Lower oesophageal contractility (LOC)

The oesophagus consists of striated muscle in the upper portion, and changes to smooth muscle in the lower region. During wakefulness, the oesophagus exhibits regular, spontaneous contractions (SLOC) of several types, under the control of the autonomic nervous system. During anaesthesia, these contractions reduce in both frequency and amplitude. It is possible to measure SLOC using a specialized probe inserted into the oesophagus. The probe may also have an inflatable cuff in order to provoke contractions of the lower oesophagus. These provoked contractions (PLOC) may continue as anaesthesia is deepened, even when the spontaneous contractions have ceased. Despite much study, neither SLOC nor PLOC has been widely adopted as a method of measuring the depth of anaesthesia. One study in particular has reported that SLOC and PLOC do not accurately reflect anaesthetic depth with certain combinations of anaesthetic.[26]

Electroencephalogram (EEG) and derived indices

EEG
The raw EEG is a small (1–500 μV) voltage deflection recorded from silver/silver chloride electrodes affixed to the scalp. An active electrode is positioned over a site of neuronal activity, and an indifferent electrode is positioned at a site on the scalp some distance away. It is important to

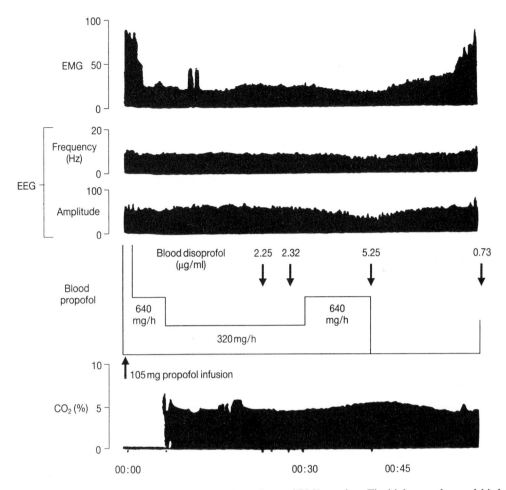

Fig. 41.2   Graph of frontalis EMG, EEG and CO_2 obtained using a Datex ABM® monitor. The high rate of propofol infusion between 30 and 40 minutes resulted in a fall in EEG frequency and amplitude and increased end-tidal CO_2 levels, which were reversed when the infusion was discontinued. The level of frontalis EMG activity was reduced after induction and rose at recovery. (Redrawn, with permission, from reference 24.)

understand the mechanisms behind the formation of the EEG, so that observations made regarding changes in the EEG due to anaesthesia may be placed into the correct context. The EEG is the sum of ionic currents from thousands of neurons around the site of the active electrode. The spatial organization of pyramidal neurons, with dendrites perpendicular to the surface of the cerebral cortex, means that they contribute the most towards the EEG. In addition, slow wave synaptic potentials rather than action potentials contribute more towards the EEG, since, being slower potentials, they more readily summate at the scalp. The nature of the deflection in the EEG waveform may differ according to the site of the neuronal event in the cerebral cortex. Excitatory postsynaptic potentials cause an upward deflection in the scalp-recorded EEG if the synapse is in superficial layers of the cerebral cortex, but cause a downward deflection if originating in deeper layers. Conversely, inhibitory postsynaptic potentials cause a downward deflection if originating in superficial layers, and an upward deflection if present in deeper layers.[27]

EEG recording is non-invasive, but it is essential that the recording electrodes are attached carefully to the scalp in order to ensure minimal impedance and the largest possible amplitude of EEG waves. This is usually achieved with sequential application of an abrasive paste and an adhesive gel along with the use of an electrode impedance meter to ensure good connections. This increases the amount of time spent in the anaesthetic room attaching the electrodes. Poorly screened electrical equipment and diathermy contribute to interference in the EEG, which at best make the waveform difficult to interpret and at worst may destroy the sensitive recording amplifiers. Care should be taken before anaesthetists embark on *ad hoc* projects, because many amplifiers designed for research purposes were never intended for use in the operating theatre. Advice should always be sought from the local medical physics department to ensure patient safety.

Visual inspection of the EEG clearly shows it to be a rapidly fluctuating waveform, varying from periods of intense, high-frequency fluctuations to an almost flat line.

Classic identification of components in the EEG requires classification into α (8–13 Hz), β (13–30 Hz), δ (0.5–4.5 Hz) and θ (4–7 Hz) frequencies.[27] Wakefulness with eyes closed is associated with higher levels of α frequencies, in contrast to lower levels of δ. Sleep is associated with more low-frequency activity, including δ and θ waves.

Spectral analysis

The raw EEG waveform is complex, and changes during anaesthesia are due not only to the level of anaesthesia but also to the type of anaesthetic used and other factors, such as the haemodynamic state of the patient. It takes an expert to interpret these changes. Considerable efforts have been directed toward some automatic interpretation of the EEG state. Early efforts employed electronic filtering of the EEG, with the integrated amplitude of the EEG waveform indicating the level of brain activity. The cerebral function monitor (CFM) and cerebral frequency analysing monitor[28] (CFAM) are two examples of this approach. The CFM gives a single trace of integrated EEG amplitude, increasing levels of cerebral activity appearing as a broadening of the trace, which ranges from 5 to 18 μV peak to peak amplitude. As with all EEG-based recordings, the CFM is affected by diathermy and other artefacts, frequently emphasized by poor electrode contact with the scalp. As a result, the CFM is almost unique in incorporating a simultaneous trace of electrode impedance which clearly demonstrates periods of poor electrode contact and diathermy, so that the anaesthetist may reject these artefacts. The CFAM filters the EEG into five frequency bands and adds one extra trace demonstrating periods of burst suppression, as encountered during deep anaesthesia with certain combinations of anaesthetics.

The advent of affordable microcomputers during the 1980s saw an increase in research using algorithms, rather than electronic components, to digitally filter the EEG waveform into its spectral components. The most common algorithm is the fast Fourier transform (FFT). By squaring the results of the FFT, the power spectrum of the EEG may be obtained, which, when plotted as power against frequency, gives a frequency distribution of the EEG, as shown in Fig. 41.3. The individual distributions can be considered as 'time-slices' and joined together into a three-dimensional plot, most frequently termed the compressed spectral array (CSA) (Fig. 41.4). Several proprietary versions of the CSA have appeared in monitors for determining the level of brain function during procedures such as carotid clamp, when the object is to detect ischaemia and prevent resulting damage. As with all EEG-based techniques, a number of limitations have to be placed on the nature of the signal. The FFT algorithm assumes that the data sample, interval, or epoch, of EEG data is

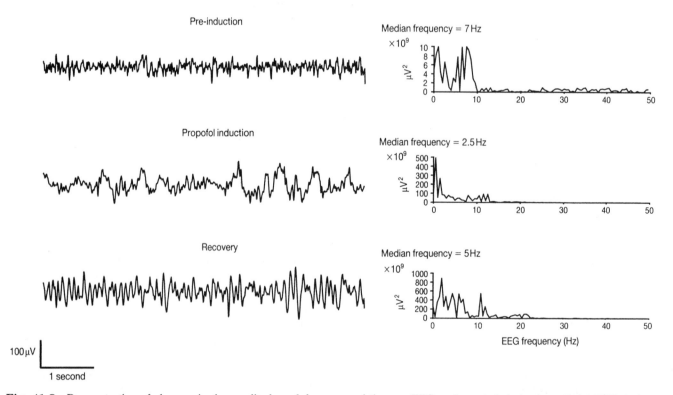

Fig. 41.3 Demonstration of changes in the amplitude and frequency of the raw EEG and spectral derivatives of the EEG during anaesthesia. Propofol induction (2.5 mg/kg) caused an increase in low frequency activity, which was reflected in a reduction of the median frequency from 7 Hz to 2.5 Hz. Five minutes after recovery, some high frequency activity had returned, but the EEG was not the same as at preinduction. This was seen in the median frequency rising from 2.5 Hz to only 5 Hz.

stationary – i.e. there is no change in the characteristics of the waveform during the epoch. This means that, in order to give a reliable indicator during changing levels of anaesthesia, epochs should be kept as short as possible. However, artefacts within the EEG may affect the FFT analysis and appear as erroneous, and potentially misleading, peaks in the power spectrum, especially within short epochs. Ideally, the raw EEG should be visually inspected for artefacts before spectral analysis. However, this cannot be achieved by a real-time system. Appropriate filtering must also be performed on any waveform to be analysed using FFT-based techniques. Unless low pass filtering is applied to the data at half the original sampling frequency of the waveform (the Nyquist frequency), false aliasing and harmonics may appear in the data. Therefore, if the highest frequency of interest is 40 Hz, the sampling frequency must be at least 80 Hz and a 40 Hz low pass filter must be applied to the data. Substantially higher sampling frequencies allow the detection and rejection of artefacts such as electrical spikes in the data. Other artefacts, such as ECG complexes appearing on the EEG, may also adversely affect any automated spectral frequency analysis of the EEG, and must be excluded from the analysis.

Characteristic changes in the CSA during anaesthesia have been documented.[29] There is considerable variability between the characteristics of the CSA depending on the type of anaesthesia employed. However, the general trend is for deeper anaesthesia to result in low frequency activity, shifting the peaks of the CSA from higher frequencies. At recovery, there is a progressive increase in the amount of high frequency activity, with a corresponding decrease in low frequency activity. However, the spectral components of the EEG do not quickly return to exactly the same characteristics seen before exposure to the anaesthetics. For example, N_2O has been shown to change characteristics of the EEG at recovery, and to give a different EEG response on subsequent exposure within 30 minutes of the initial exposure, leading the authors to suggest that patients could develop a measurable tolerance to N_2O.[30] In addition, it has been suggested that atropine may affect the frequency distribution of the EEG.[31] This effect appears

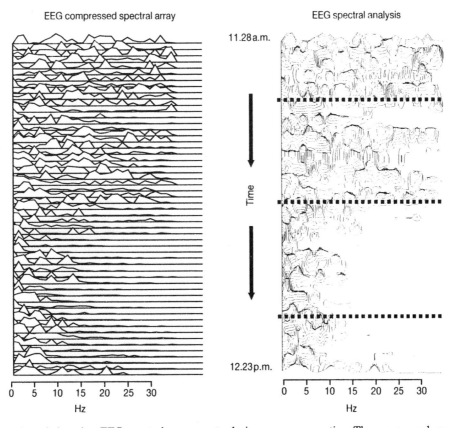

Fig. 41.4 Demonstration of changing EEG spectral components during caesarean section. The compressed spectral array (CSA) and more clear contour plot of iso power lines demonstrate that induction of anaesthesia with thiopentone was associated with a reduction in high frequency components of the EEG. Light, 0.7 MAC isoflurane anaesthesia during surgery resulted in an increase in high frequency EEG components. Deeper, 1.2 MAC isoflurane anaesthesia resulted in more low frequency EEG components.

to be independent of the changes in EEG frequency observed due to varying the depth of anaesthesia.[32]

Although the CSA is considerably more compact than the raw EEG, it is still a complex display that takes time to comprehend, and changes within it are difficult to quantify. Several single-figure numeric indices have been derived from power spectral analysis of the EEG. These include mathematical expressions of the spectral distribution (mean,[33] median[33] and spectral edge[34] frequencies) and analyses of coincident spectral frequencies (e.g. bispectral index[35]). Original studies using the spectral edge frequency[34] described the use of a pattern-matching technique to detect the highest frequency at which significant levels of power were present in the EEG. Most workers have simplified the methodology, and now describe the spectral edge frequency as that below which 95 per cent of the EEG power is contained. The mean frequency describes the arithmetic mean of the power distribution. However, the mean frequency tends not to accurately represent high frequency (β) outliers in the power distribution. The median frequency, above and below which 50 per cent of the power spectrum is distributed, is generally accepted as a good indicator of changing distributions of EEG power during anaesthesia.

The median frequency has been examined during a number of types of anaesthesia. With propofol total intravenous anaesthesia, 5 Hz has been suggested as an empirical guideline for adequate surgical anaesthesia, above which anaesthesia is too light.[33]

A derivation of conventional power spectral analysis is bispectral analysis.[35–37] In bispectral analysis, rather than considering each component sine wave of a signal to be independent as in conventional power spectral analysis, each potential interaction between components is analysed in order to detect the presence of component harmonics. It has been suggested that such analysis adds a predictive element to EEG analysis, and further research may reveal bispectral analysis to be a useful technique for measuring the depth of anaesthesia.

Evoked potentials

Whereas analysis of the EEG shows the responses of many thousands of neurons in the cerebral cortex, evoked potentials (EPs) show the response of more localized areas of the brainstem, midbrain and cerebral cortex to specific stimuli. Most research has concentrated on the use of EPs to elucidate the integrity of functional pathways, but it is also possible to look at changes in activity in these pathways due to the effects of anaesthesia. EP research relies on time-locking the stimulus relative to small changes in the raw EEG in response to the stimulus. These time-locked signals are then averaged in order to minimize the effects of other EEG components, such as random noise,

which cancel out because they are not time-locked to the stimulus. The duration of the sampling epoch is a major drawback of any EP-based technique for use as a monitor of depth of anaesthesia. Stimulus presentation rates of up to 10 Hz, and averaging over 1024 samples, require a sampling epoch of at least 102 seconds. The adoption of advanced signal processing techniques will undoubtedly produce displays of evoked potentials that can be used to assess the depth of anaesthesia. One promising approach is monitoring the coherent frequency of the EEG to a high frequency auditory stimulus (10–46 Hz).[38]

Standard silver chloride EEG scalp electrodes are used to obtain the EP, selection of positioning depending on the type of EP to be evoked. The EP signals are small, and are easily swamped by electrical interference in the operating room, especially the diathermy employed by surgeons. The need for filtering, isolation, low noise, and high gain of EEG amplifiers has resulted in high cost and complicated operation. Recent EEG amplifiers employ digitization of the EEG signal in a computer-controlled amplifier close to the head of the patient, and it is hoped that such technology will make EP recordings cheaper and more accessible in electronically hostile environments such as the operating room. Examples of typical EP traces are shown in Fig. 41.5.

Auditory evoked potentials

Considerable research has been conducted to determine the changes in the characteristics of auditory evoked potentials (AEPs) during anaesthesia. Newton and others have demonstrated the changes in specific components of the AEP during anaesthesia and recovery from anaesthesia.[39] Halothane, isoflurane and enflurane tended to increase the latencies of the brainstem AEP waves III and V as anaesthesia deepened. Furthermore, increases in the depth of anaesthesia also increased the latency, and decreased the amplitude, of the early cortical AEP components Pa and Nb. The changes were similar for the volatile agents, and were present across a range of inspired concentrations appropriate for general anaesthesia. Although intravenous barbiturates also increased the latency of brainstem components III and V, other intravenous anaesthetics (i.e. etomidate, Althesin and propofol) did not affect the brainstem response, but did change cortical Pa and Nb latency and amplitudes in a manner similar to that observed for the volatile anaesthetic agents. The brainstem response did not change with some intravenous agents and, therefore, it has been dismissed as a putative method for determining the depth of anaesthesia. The early cortical AEP remains as a potentially useful method for determining the depth of anaesthesia; this is especially so since surgical stimulation has been shown to affect the amplitude of the early cortical wave Nb during light halothane anaesthesia (0.3 per cent end-tidal, in 67 per cent N_2O).[39] However, there are still a number of potential problems that

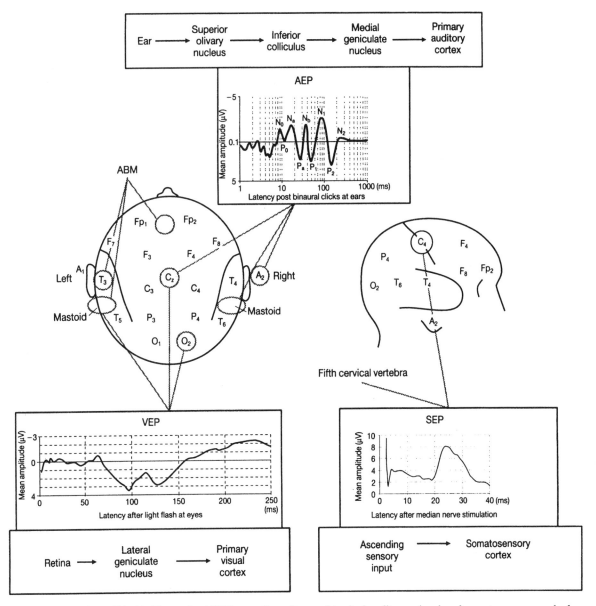

Fig. 41.5 Examples of possible 10–20 standard EEG recording sites, and typical auditory, visual and somatosensory evoked responses for unanaesthetized patients. The sensory pathways effecting the evoked responses are shown. Anaesthesia generally results in a progressive increase in latency, and reduction in amplitude, of the components of each evoked potential. Typical electrode positions for the ABM, from which EEG and frontalis EMG recordings are made, are also shown.

need to be addressed before the AEP can be used routinely. Two of these problems are the significant change in evoked potentials observed during mild hypoglycaemia[40] and its applicability in patients with a hearing defect.

Visual evoked potentials

Visual evoked potentials (VEPs) have been studied during anaesthesia with isoflurane, enflurane and thiopentone.[41] The VEP is commonly generated by the presentation of bright flashes through the closed eyelid using light-emitting diodes mounted within goggles. As the eyelids are commonly taped shut during anaesthesia, the VEP is an attractive putative method for measuring the depth of

anaesthesia. Three components of the VEP waveform have been studied: P60, N70 and P100. The origin of P100 has been described as the primary visual cortex, but the origin of the earlier waves has not been characterized. As the earlier components have a shorter latency than P100, it is logical to assume that they originate from earlier stages in the visual pathway, such as the lateral geniculate nucleus. Single neuron recordings in the visual cortex of cats have shown that a range of velocity and spatial frequency coding properties, including the identification of textured visual stimuli, still occur in the primary visual cortex during light halothane anaesthesia.[42] The primary visual cortex is therefore well able to process simple flash stimuli during anaesthesia. However, the P60 and N70 components of the human VEP are not reliably reproduced in all subjects under anaesthesia, and the P100 component does not change in latency or amplitude through the range of volatile anaesthetic concentrations used for deep surgical anaesthesia. Despite this, the VEP shows promise as a technique for assessing the level of sedation in patients in the intensive care unit,[43] and could still be used to warn of excessively light anaesthesia.

Somatosensory evoked potentials

Relatively few studies have considered the somatosensory evoked potential (SEP) as a putative method for determining anaesthetic depth. The SEP takes some 80 seconds to calculate, and requires time-locked averaging of responses to stimuli presented to sites such as the median nerve. Typical recording sites are the fifth cervical vertebra and the scalp above the contralateral somatosensory cortex. One carefully designed study described a small increase in latency (21.1 ms $\pm$ 1.8 SD to 21.9 ms $\pm$ 1.8 SD) and reduction in amplitude (4.6 μV $\pm$ 2.5 SD to 2.9 μV $\pm$ 1.5 SD) as a result of induction of anaesthesia with thiopentone.[44] However, this trend was not observed in all patients from the group ($n = 22$). In two patients there was an increase in SEP latency, and in one patient there was no change in latency with anaesthesia. No correlation could be determined between changes in SEP and either heart rate or blood pressure, although these autonomic indices are by no means 'gold standards' for determining anaesthetic depth. Unfortunately, SEPs were not recorded during recovery of the patient because of contamination of the recordings by muscle artefacts. The results do not discount SEPs as a putative index of anaesthetic depth, but, as with all evoked potentials, considerable progress needs to be made in the field before a commercial depth-of-anaesthesia monitor employing the techniques could be made available to the anaesthetic community.

Heart rate variability

Changes in autonomic tone are the basis of most subjective estimates of anaesthetic depth (e.g. changes in heart rate and blood pressure). Ideally, one needs a measurement of autonomic tone that is not affected by factors other than anaesthetic depth, and which is easy to measure.

Recent research using animal models has cast doubt on the value of the measurements of cerebral cortical activity, such as the EEG, to estimate the efficacy of anaesthesia. Specifically in rats, minimum alveolar concentration (MAC) has been shown to depend on neural circuitry within the brainstem and not higher cortical influences.[45] Analysis of specific components of the rat electromyogram has revealed that MAC equivalent responses to noxious stimuli can be explained by the effect of anaesthetic agents either directly

or indirectly on motor neurons, and not on higher regions of the central nervous system.[46] If anaesthesia does originate in the brainstem, and then subsequently affects higher cortical centres, it seems logical to propose that the first signs of lightening or wakening due to light anaesthesia or inadvertent recovery will be determined by opening a window on not just the cerebral cortex but also onto the brainstem. Such a window should include areas of the brainstem responsible for mediating both afferent and efferent information, rather than primarily afferent sensory pathways via regions such as the auditory nuclei.

A possible explanation for the anaesthetic effects on the brainstem is provided by activation of inhibitory glycine synapses, which play a major role in the brainstem, in contrast to GABA, the main inhibitory neurotransmitter in the brain.[47] Anaesthetic drugs appear to act on the brainstem by enhancing inhibitory glycine-mediated activity. Anaesthesia that originates in the brainstem probably inhibits the cerebral cortex via ascending, efferent projections from the midbrain.[45]

Objective measurement of brainstem-mediated autonomic tone has been an area of particular interest for workers screening for autonomic neuropathy, especially in the study of diabetes. The heart rate is not constant from beat to beat in subjects with a healthy autonomic nervous system; there is a balance between sympathetic excitation and parasympathetic inhibition of the sinoatrial and atrioventricular node of the heart, leading to heart rate variability (HRV). Subjects exhibiting autonomic neuropathy show a reduction, and eventual loss, of HRV. Spectral analysis of HRV has revealed three components: a low frequency fluctuation, generally believed to be circadian; a medium frequency variation, attributed to the baroreceptor reflex; and a high frequency component, which coincides with the frequency of ventilation. The high frequency variation is respiratory sinus arrhythmia (RSA), and is typically characterized as a greater than 10 per cent variation in the ECG P-wave interval over 5 minutes. RSA is easily visible on an ECG monitor that is time-locked to an ECG R-wave peak (Fig. 41.6), but is difficult to distinguish using modern monitors with a rolling display. During RSA, heart rate increases during inspiration and decreases during expiration under the control of a predominately parasympathetic reflex in the supine subject.

Since 1985 there have been a number of reports that the level of RSA changes during anaesthesia. One study, using off-line analysis of 10-minute epochs of ECG, suggested that RSA diminished during isoflurane anaesthesia, and then increased again at recovery.[48] The authors suggested that a real-time system could form the basis of a monitor for measuring the level of anaesthesia. A later study also suggested that the technique could be used for anaesthetic agents other than isoflurane.[49] A dose–response relation has been demonstrated between RSA and the level of

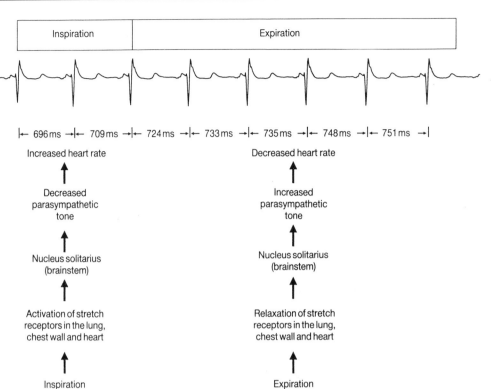

Figure 41.6 The mechanism responsible for respiratory sinus arrhythmia (RSA), a cyclic variation in heart rate coincident with respiration. The variations in ECG R–R wave intervals may be small and difficult to visualize with a conventional ECG monitor using a rolling time axis.

isoflurane anaesthesia measured at different MAC levels[32,50]

Methods of measuring HRV

A number of techniques have been used to screen for autonomic neuropathy based on studies of heart rate variability (HRV). The simplest estimate is to determine the mean heart rate and the standard deviation from the mean, the standard deviation giving an estimate of the amount of heart rate variability. Unfortunately, ectopic beats and the level of the mean heart rate both have a direct effect on the standard deviation to the extent that it cannot be used in isolation as a measurement of HRV. Spectral analysis of HRV has been used in several studies. However, in order to resolve the low frequency components, epochs of at least 5 minutes have to be used. FFT-based spectral analysis requires that data are stationary during an epoch, and it is difficult to guarantee that the ECG will not change during the course of an epoch during surgery, when surgical stimulation or haemodynamic changes may impose a change in heart rate.

A circular statistical technique has been used to quantify the level of uncontaminated RSA using ECG R-wave times correlated against respiratory cycles. This has the advantage, when compared with other techniques such as determining the standard deviation or power spectrum of ECG R–R intervals, that ectopic beats and other irregularities not related to breathing are separated from those variations caused by breathing. The results obtained using this technique have been reported following studies

performed during anaesthesia with total intravenous propofol[51] and isoflurane maintenance,[32] as well as during propofol sedation in the intensive care unit,[52] caesarean section[53] and as a result of surgical stimulation during light enflurane anaesthesia.[54] Some results are summarized in Fig. 41.7. It appears likely that the level of RSA reflects the level of anaesthetic depth, deeper anaesthesia resulting in lower levels of RSA. In addition, surgical stimulation during light anaesthesia elicits a greater increase in RSA than seen during lightening anaesthesia alone. RSA is a convenient window on the brainstem, which needs only standard ECG and respiratory signals.

The future

It is unlikely that any single method will be found to measure the depth of anaesthesia reliably for all patients and all anaesthetic agents. The different molecular structure of anaesthetic agents suggests the possibility that they will elicit responses in different populations of neuronal membrane receptors and must, therefore, act in subtly different ways on different regions of the brain.[47] In addition, all of the methods for determining the depth of anaesthesia proposed to date have some potential exclusion criteria. For example, evoked potentials depend on an intact neural pathway from which to obtain the response to

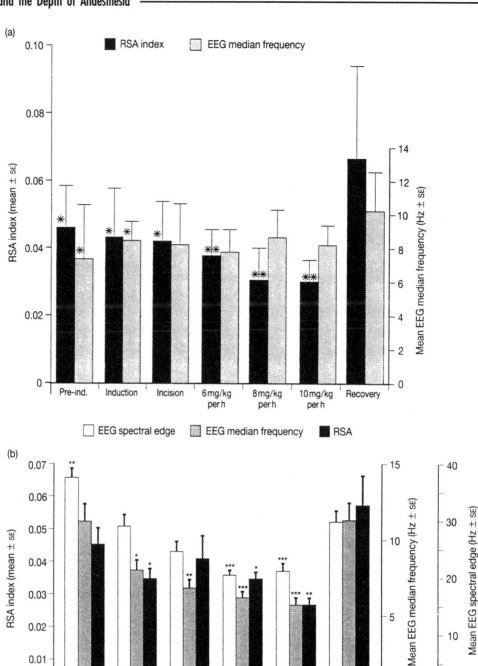

Fig. 41.7 Graphs comparing reductions in respiratory sinus arrhythmia (RSA) and derived indices of the EEG during anaesthesia. Significant differences between values obtained during anaesthesia and at recovery are shown (* = $P < 0.05$, ** = $P < 0.01$, † = $P < 0.001$). (a) RSA and the median frequency of the EEG were reduced during total intravenous anaesthesia with propofol in ten patients. (b) RSA, EEG median and spectral edge frequencies reduced with increasing levels of isoflurane anaesthesia in ten patients. Redrawn from data in references 32 and 51.

stimulation. Any break in the pathway, such as visual impairment with VEPs or loss of hearing in AEPs, will change the response characteristics before an anaesthetic is given. Similarly, techniques based on monitoring the autonomic nervous system, such as monitoring HRV, will be impaired in the event of autonomic neuropathy. The only reliable way of determining the depth of anaesthesia will require a non-invasive measure of cerebral activity and a localization of the activity to specific cortical regions, and areas of the brainstem, in real-time.

Positron emission tomography (PET) has been used in a limited number of studies to monitor brain function during anaesthesia. However, the technique is invasive to the extent that isotopes of glucose with a short half-life are used and these have to be prepared close to the recording site with a cyclotron, and cannot therefore form the basis of a cheap operating theatre monitor for use by the anaesthetist. However, PET scanning has revealed that propofol anaesthesia has a widespread suppressive effect on cerebral metabolism, and it is probable that future research will allow the localization of specific regions in the brain that are affected by anaesthesia.[55] Ultimately, this approach may lead to the adoption of a standard, absolute scale of anaesthetic depth, against which future researchers may calibrate their putative tests of anaesthetic depth.

Ultrasensitive superconducting quantum interference devices (SQUIDs) have been used to provide a new insight into cerebral function by non-invasively measuring not only structures, as resolved with more conventional magnetic resonance imaging, but also functional activity in the brain. Although prohibitively expensive at the moment, new technology may lead to the provision of relatively cheap, direct-imaging brain function monitors based on SQUID technology. This will provide the anaesthetist with the ultimate monitor, being capable of determining not just anaesthetic depth but also awareness, anoxia, ischaemia and unusual pathology. Although the routine use of such devices seems unlikely in the near future, one has only to consider the rapid adoption of microcomputer technology over the last ten years to realize that new monitoring technologies may be available in the relatively near future.

REFERENCES

1. Jones JG. Depth of anaesthesia. *Ballière's Clinical Anaesthesiology* 1989; **3** (3): ix–xii.

2. Dixon NF. Unconscious perception and general anaesthesia. *Ballière's Clinical Anaesthesiology* 1989; **3** (3): 473–85.

3. Aitkenhead AR. Awareness during anaesthesia: what should the patient be told? *Anaesthesia* 1990; **45**: 351–2.

4. Liu WHD, Thorp TAS, Graham SG, Aitkenhead AR. Incidence of awareness with recall during general anaesthesia. *Anaesthesia* 1991; **46**: 435–7.

5. Schultetus RR, Hill CR, Dharamraj CM, Banner TE, Berman LS. Wakefulness during caesarean section after anesthetic induction with ketamine, thiopental, or ketamine and thiopental combined. *Anesthesia and Analgesia* 1986; **65**: 723–8.

6. Bogetz MS, Katz JA. Recall of surgery for major trauma. *Anesthesiology* 1984; **61**: 6–9.

7. Abouleish E, Taylor FH. Effect of morphine–diazepam on signs of anesthesia, and dreams of patients under N_2O for caesarean section. *Anesthesia and Analgesia* 1976; **55**: 702–5.

8. Miller CD, Asbury AJ, Brown JH. Pupillary effects of alfentanil and morphine. *British Journal of Anaesthesia* 1990; **65**: 415–17.

9. Saucier N, Walts LF, Moreland JR. Patient awareness during nitrous oxide, oxygen, and halothane anesthesia. *Anesthesia and Analgesia* 1983; **362**: 239–40.

10. Aitkenhead A. Awareness: incidence and medico-legal implications. In: *Awareness in anaesthesia and awareness of current issues*, Report of a Symposium held at Queen's College, Cambridge, 28–29 June, 1993. Oxford: Oxford Clinical Communications: 1993.

11. Bennett HL, Davis HS, Giannini JA. Non-verbal response to intraoperative conversation. *British Journal of Anaesthesia* 1985; **57**: 174–9.

12. Millar K, Watkinson N. Recognition of words presented during general anaesthesia. *Ergonomics* 1983; **26**: 585–94.

13. Bonke B, Schmitz PIM, Verhage F, Zwaveling A. Clinical study of so-called unconscious perception during general anaesthesia. *British Journal of Anaesthesia* 1986; **58**: 957–64.

14. Moerman N, Bonke B, Oosting J. Awareness and recall during general anaesthesia. Facts and feelings. *Anesthesiology* 1993; **79**: 454–64.

15. Sandin R, Norström O. Awareness during total i.v. anaesthesia. *British Journal of Anaesthesia* 1993; **71**: 782–7.

16. Levinson B. The states of awareness in anaesthesia in 1965. In: Bonke B, Fitch W, Millar K, eds. *Memory and awareness in anaesthesia*. Amsterdam: Swetz & Zeitlinger, 1990: 11–18.

17. Guedel A. Third-stage ether anesthesia: a subclassification regarding the significance of the position and movements of the eyeball. *American Journal of Surgery, Quarterly Supplement in Anaesthesia* 1920; **34**: 53–7.

18. White DC. Anaesthesia: a privation of the senses. An historical introduction and some definitions. In: Rosen M, Lunn JN, eds. *Conscious awareness and pain in general anaesthesia*. London: Butterworths, 1987: 1–9.

19. Flecknell PA. Anaesthesia of animals for biomedical research. *British Journal of Anaesthesia* 1993; **71**: 885–94.

20. Stollery BT, Broadbent DE, Lee WR, Keen RI, Healy TEJ, Beatty P. Mood and cognitive functions in anaesthetists working in actively scavenged operating theatres. *British Journal of Anaesthesia* 1988; **61**: 446–55.

21. Tunstall ME. Detecting wakefulness during general anaesthesia for caesarean section. *British Medical Journal* 1977; **1**: 1321.

22. Russell IF. Auditory perception under anaesthesia. *Anaesthesia* 1979; **34**: 211.

23. Simons AJR, Boesman EHJF, Pronk RAF. Automatic EEG monitoring of anaesthesia. *Ballière's Clinical Anaesthesiology* 1989; **3** (3): 623–46.

24. Kay B. The anaesthesia and brain monitor (ABM). *Acta Anaesthesia Belgica* 1984; **S35**: 167–74.

25. Herregods L, Rolly G, Mortier E, Bogaert M, Mergaert C. EEG and SEMG monitoring during induction and maintenance of

anesthesia with propofol. *International Journal of Clinical Monitoring and Computing* 1989; **6** (2): 67–73.

26. Issac PA, Rosen M. Lower oesophageal contractility and detection of awareness during anaesthesia. *British Journal of Anaesthesia* 1990; **65**: 319–24.

27. Kandel ER, Schwartz JH, Jessell TM. *Principles of neural science*, 3rd ed. New York: Elsevier, 1991.

28. Lloyd-Thomas AR, Cole PV, Prior PF. Isoflurane prevents EEG depression during trimetaphan-induced hypotension in man. *British Journal of Anaesthesia* 1990; **65**: 313–18.

29. Pichmayer I. *EEG atlas for anaesthesiologists*. Berlin: Springer-Verlag, 1987.

30. Avramov MN, Shingu K, Mori K. Progressive changes in electroencephalographic responses to nitrous oxide in humans: a possible drug tolerance. *Anesthesia and Analgesia* 1990; **70**: 369–74.

31. Pickworth WB, Herning RI, Koeppl B, Henningfield JE. Dose-dependent atropine-induced changes in spontaneous electro-encephalogram in human volunteers. *Military Medicine* 1990; **155**: 166–70.

32. Pomfrett CJD, Sneyd JR, Barrie JR, Healy TEJ. Respiratory sinus arrhythmia: comparison with EEG indices during isoflurane anaesthesia at 0.65 and 1.2 MAC. *British Journal of Anaesthesia* 1994; **72**: 397–402.

33. Schwilden H. Use of the median EEG frequency and pharmacokinetics in determining depth of anaesthesia. *Ballière's Clinical Anaesthesiology* 1989; **3** (3): 603–21.

34. Rampil IJ, Sasse FJ, Smith NT, Hoff BH, Flemming DC. Spectral edge frequency – a new correlate of anesthetic depth. *Anesthesiology* 1980; **53**: S12.

35. Vernon J, Bowles S, Sebel PS, Chamoun N. EEG bispectrum predicts movement at incision during isoflurane or propofol anesthesia. *Anesthesiology* 1992; **77**: A502.

36. Kearse L, Saini V, deBros F, Chamoun N. Bispectral analysis of EEG may predict anesthetic depth during narcotic induction. *Anesthesiology* 1991; **75**: A175.

37. Sebel PS, Bowles S, Saini V, Chamoun N. Accuracy of EEG in predicting movement at incision during isoflurane anesthesia. *Anesthesiology* 1991; **75**: A446.

38. Munglani R, Andrade J, Sapsford DJ, Baddeley A, Jones JG. A measure of consciousness and memory during isoflurane administration: the coherent frequency. *British Journal of Anaesthesia* 1993; **71**: 633–41.

39. Thornton C, Newton DEF. The auditory evoked response: a measure of depth of anaesthesia. *Ballière's Clinical Anaesthesiology* 1989; **3** (3): 559–85.

40. Jones TW, McCarthy G, Tamborlane WV, *et al.* Mild hypoglycemia and impairment of brain stem and cortical evoked potentials in healthy subjects. *Diabetes* 1990; **39**: 1550–5.

41. Chi OZ, Field C. Effects of enflurane on visual evoked potentials in humans. *British Journal of Anaesthesia* 1990; **64**: 163–6.

42. Hammond P, Pomfrett CJD. Direction and orientational tuning of feline striate cortical neurones: correlation with neuronal class. *Vision Research* 1989; **29**: 653–62.

43. Wang DY. Indices of Sedation in the Intensive Care Unit. PhD Thesis, University of Manchester, 1994.

44. Sebel PS. Somatosensory, visual and motor evoked potentials in anaesthetized patients. *Ballière's Clinical Anaesthesiology* 1989; **3** (3): 587–602.

45. Rampil IJ, Mason P, Singh H. Anesthestic potency is independent of forebrain structures in the rat. *Anesthesiology* 1993; **78**: 707–12.

46. Rampil IJ. 'F-Waves' – a nonsynaptic, but sensitive indicator of anesthetic effect in rats. *Anesthesia and Analgesia* 1994; **78**: S350.

47. Franks NP, Lieb WR. Molecular and cellular mechanisms of general anaesthesia. *Nature* 1994; **367**: 607–14.

48. Donchin Y, Feld JM, Porges SW. Respiratory sinus arrhythmia during recovery from isoflurane–nitrous oxide anaesthesia. *Anesthesia and Analgesia* 1985; **64**: 811–15.

49. Porges SW. Vagal mediation of respiratory sinus arrhythmia. Implications for drug delivery. *Annals of the New York Academy of Sciences* 1991; **618**: 57–66.

50. Kato M, Komatsu T, Kimura T, Sugiyama F, Nakashima K, Shimada Y. Spectral analysis of heart rate variability during isoflurane anesthesia. *Anesthesiology* 1992; **77**: 669–74.

51. Pomfrett CJD, Barrie JA, Healy TEJ. Respiratory sinus arrhythmia: an index of light anaesthesia. *British Journal of Anaesthesia* 1993; **71**: 212–17.

52. Wang DY, Pomfrett CJD, Healy TEJ. Respiratory sinus arrhythmia: a new, objective sedation score. *British Journal of Anaesthesia* 1993; **71**: 354–8.

53. Healy TEJ, Bellman MH, Pomfrett CJD. Respiratory sinus arrhythmia indicates light anaesthesia during caesarean section. *Anesthesia and Analgesia* 1994; **78**: S156.

54. Pomfrett CJD, Barrie JR, Healy TEJ. Respiratory sinus arrhythmia reflects surgical stimulation during light enflurane anaesthesia. *Anesthesia and Analgesia* 1994; **78**: S344.

55. Alkire MT, Barker SJ, Haier RJ, *et al.* A positron emission tomography study of cerebral metabolism in a volunteer during propofol anesthesia. *Anesthesia and Analgesia* 1994; **78**: S5.

Chapter 42

Postoperative Analgesia: the Pain Service

Wilhelmina C. Korevaar

Since the beginning of this century, surgeons have been aware of the importance of acute pain relief, particularly with regard to the effect of the patient's responses to injury on postoperative morbidity and mortality.[1] Recently, anaesthetists have become increasingly involved in the provision of postoperative analgesia and the development of pain management services.[2] With further social progress, anaesthetists will be able to fill the gap between in-hospital surgical services and primary care by providing analgesic care, enhancing awareness of the need and justification for pain management through evaluation and research, and by enabling communication between surgeons and primary care physicians.[3]

The emphasis in this chapter is on the provision of postoperative analgesia services for both in-hospital and day surgery settings.

Acute pain: nociception and response

Pain receptors are not anatomically specialized. Instead the bare or free nerve endings associated with small, unmyelinated C and small myelinated Aδ fibres may respond to a variety of mechanical, chemical and thermal stimuli. The extent of the response (frequency and spatial distribution) in the absence of modifying factors determines the perception of acute pain, or nociception. C and Aδ fibres transmit nociceptive stimuli from somatosensory structures (skin, muscle, bone). The frequency of firing in visceral fibres determines whether visceral function is associated with discomfort.[4]

C fibres are considered to be polymodal because they may respond to a variety of stimuli. Aδ fibres may respond to mechanical or thermal stimuli, or both. C fibres follow the vasculature and Aδ fibres follow a more dermatomal distribution. The mechanical distortion and altered temperature first associated with tissue injury will initiate firing in the Aδ system. Aδ afferents are distal projections of bipolar neurons located in the dorsal root ganglion which project onto Rexed laminae I, II and V of the dorsal horn.

Localized responses to injury increase the frequency of C-fibre activity. Local chemical environmental changes following tissue injury such as altered pH, the release or synthesis of endogenous algesic substances (prostaglandin, serotonin, histamine, bradykinin, substance P, etc.) and altered local temperature and microcirculation decrease the threshold of C fibres. Any additional mechanical distortion may result in high-frequency bursts of impulses from C fibres which have their cell bodies located in the dorsal root ganglion and project onto the Rexed laminae I and II (substantia gelatinosa).

Peripheral nociceptive activity results in segmental and suprasegmental responses. Segmental responses result in localized muscle spasm mediated via a simple sympathetic reflex. Muscle spasm is the most important active response to tissue injury because immobilization of the injured area promotes haemostasis and tissue rebuilding.[5]

Suprasegmental pain transmission depends on the interaction between spinal cord inhibitory and facilitatory pathways. Pain transmission is mediated by peptides, including substance P, neurotensin, the gut peptides vasoactive intestinal peptide (VIP) and cholecystokinin (CCK). Inhibitory pathways include the opioid-dependent and opioid-independent (γ-aminobutyric acid, GABA) systems. Peak activity within the opioid-dependent pathways occurs approximately 3 days after injury, and coincides with inflammatory responses. Activation of opioid-independent pathways (GABA system) attenuates central anxiety responses to injury.[6]

Responses to acute tissue injury are processed in the

thalamus. Pain behaviour is predetermined by upbringing, environmental factors and expectations that the duration of injury will be finite.[7]

Stress responses to injury

Tissue injury results in well-characterized human responses to stress, which are central, peripheral and immunological. The pituitary/hypothalamic axis is activated following tissue injury. Large prohormones are synthesized and released, resulting in elevated levels of ACTH and β-endorphin. There are also increased levels of growth hormone, antidiuretic hormone, and other mediators of energy consumption. Peripherally, the increased release of adrenaline (epinephrine) and noradrenaline (norepinephrine) from the adrenal glands, as well as the decreased pancreatic release of insulin, causes a transient hyperglycaemic response and a metabolic tendency to catabolism (the consumption of proteins and fats).[8]

The increased tissue oxygen consumption results in increases in respiratory demand. The central stress responses are mediated by primary afferent input to the central nervous system, while peripheral stress responses are modulated by the central hormonal release and by the sympathetic nervous system. Immunocompetence following injury is indirectly proportional to the degree of catabolism.

Analgesia in the postoperative period rarely results in measurable changes in stress responses to injury unless there are predisposing factors, such as cardiac or respiratory compromise or immune disease, that make an individual less likely to survive an injury.[9] The segmental muscle spasm associated with nociceptive activity promotes immobility and, depending on the site of pain, impedes volitional increases in ventilatory effort. Tissue oxygen delivery suffers, with a consequent increase in morbidity.[10] Beyond the first 2–3 postoperative days, tissue oxygenation is less likely to be a determining factor in postoperative morbidity. However, if ambulation cannot be resumed because of overwhelming pain responses, thrombophlebitis is more likely. Pain relief during recovery from acute injury improves survival and speeds rehabilitation by promoting volitional activity.[11]

Modification of segmental responses to nociception

At a segmental level, nociceptive activity promotes stiffness and a decreased range of joint motion. There is a loss of graded muscle contraction associated with decreased activity of the large fibres. A practical concern with stiffness is its effect on voluntary pulmonary function. Upper abdominal and thoracic incisions result in an altered relation between closing volume (CV) and functional residual capacity (FRC). As the FRC declines in relation to the CV, there is an increase in alveolar-to-arterial oxygen difference. The FRC falls more than the CV in the first 48 postoperative hours due to alterations in diaphragmatic activity. Patients have fewer respiratory complications when the FRC remains above 60 per cent of the preoperative value. The decrease in FRC may follow that of vital capacity by up to 24 hours. Any form of analgesia following upper abdominal surgery helps to restore vital capacity and results in measurable improvements in effort-dependent pulmonary function tests. There are no data that confirm a restorative effect of analgesia on the relation of FRC and CV. Because effort-dependent pulmonary function reflects the segmental response to nociception, pulmonary function testing can be used to compare the efficacy of various pain-relieving modalities in the postoperative period.

Segmental responses to nociception can be attenuated in a variety of ways. The threshold and responsiveness of C fibres can be affected by decreasing the amount of peptide byproduct of the inflammatory response. Primary afferent input to dorsal horn neurons associated with nociception can be altered with local anaesthetic axonal blockade from the periphery to the spinal cord. Central neuronal activity is affected by administration of narcotic peripherally or at the neuroaxis.

Anti-inflammatory drugs

Peripheral nociceptor activity can be modified by systemic drugs which alter the synthesis and release of algesic substance synthesized or released in response to tissue injury. Cyclo-oxygenase inhibition by acetylsalicylic acid slows or stops the synthesis of prostaglandin and other byproducts of arachidonic acid. The non-steroidal anti-inflammatory drugs have a similar but less profound action on cyclo-oxygenase. Corticosteroids inhibit the metabolism of a precursor of prostaglandin. Capsaicin depletes fine cutaneous nerve endings of substance P, an algesic substance that is antidromically transported following injury.

Acetylsalicylic acid is associated with a decrease in platelet adhesiveness for the life of the platelet (10–14 days). Non-steroidal anti-inflammatory drugs are associated with decreased platelet activity for 2–3 days. Non-steroidal anti-inflammatory drugs have been used to decrease the postoperative pain associated with upper abdominal surgery,[12] dental surgery[13] and laparoscopic surgery.[14]

Local anaesthetics

Local anaesthetics impede axonal conduction and have been used in conjunction with other modalities to decrease intraoperative stress responses. Blockade of primary afferent input to the central nervous system is effective in the management of decreased voluntary activity due to postoperative pain. Local anaesthetic infiltration or flooding of the operative field is used to diminish early postoperative pain, and can be effective for patients undergoing a variety of procedures in which a large superficial surface area has been violated (face lift, cystoscopy, laparoscopy).[15,16] Interpleural bupivacaine instillation can effectively preserve voluntary respiratory efforts in patients with flail chest in whom intercostal nerve blocks or a thoracic epidural cannot be accomplished safely.[17] Peripheral nerve blocks can be used to advantage when there has been a lateral incision. Improvement in effort-dependent pulmonary function is demonstrated following local anaesthetic intercostal nerve blocks after thoracotomy or upper abdominal (cholecystectomy) incisions.[18] Ilioinguinal and iliohypogastric nerve blocks improve ambulation and decrease time to discharge following inguinal procedures in paediatric and adult populations.[19] Penile nerve blocks have been used to minimize pain for the first few hours following circumcision. Brachial plexus nerve block has been used to provide analgesia and improved blood flow following upper extremity procedures. The improved ambulation, voluntary respiratory function and decreased hospital stay associated with local anaesthetic infiltration, instillation or peripheral nerve block are associated with improved outcome and patient satisfaction.

Local anaesthetics injected or infused into the epidural space cause segmental blockade of afferent and efferent fibres. The extent of the block is directly proportional to the amount of drug instilled. The time to onset and the relative degree of small vs large fibre blockade depends on lipophilicity of the drug chosen. From the epidural space,

the local anaesthetic must cross the dura and arachnoid membranes. Individual anatomy, including pregnancy, obesity and spinal stenosis, helps determine the concentration gradient of local anaesthetic and influences the intensity and extent of neural blockade. Direct intrathecal injection of local anaesthetic results in a steeper gradient between CSF and the spinal cord. Nociceptive input to the spinal cord arrives via dorsal root ganglion neurons which project onto neurons in the dorsal horns. Smaller C fibre input is projected onto the more superficial Rexed laminae I and II from Lissauer's tract. Regardless of epidural or intrathecal injection site, the conduction along C fibre and B (autonomic) pathways in the dorsolateral superficial tracts will be affected first. Penetration to deeper axonal structures and conduction blockade depend on the available concentration gradient and drug lipophilicity. Improved survival following blunt, hindlimb trauma in dogs was reported in 1914 when induction of spinal analgesia preceded the trauma. There have been numerous subsequent reports of diminished stress responses to surgery in the presence of local anaesthetic neuraxial blockade.[20] Attenuation of stress responses is demonstrated only following the establishment of a sensory level to T10 for lower extremity surgery and to T8 for lower abdominal (pelvic/visceral) surgery at least 30 minutes before surgical incision. The somatosensory level and conditions requisite for decreased stress responses following upper abdominal and thoracic surgery are less well described. Preincisional establishment of sensory blockade extending from S5 through T4 resulted in prolonged nitrogen sparing following colonic surgery if neural blockade was maintained through the first 48 hours postoperatively and patients were continuously provided basal caloric intake.[21]

Improved pulmonary function and ambulation are repeatedly demonstrated following upper abdominal and thoracic incisions after local anaesthetic neuraxial blockade in the postoperative period.[11] Important procedural risks include: epidural haematoma or subarachnoid bleeding and permanent neurological sequelae when performed in the presence of coagulation abnormalities;[22] the evolution of pressure sores in patients rendered immobile and insensible; the possibility of incomplete analgesia; and the chance that epidural infection might complicate recovery.

Narcotics

Since the late 1970s descending opioid inhibitory pathways from the dorsal brainstem to the dorsal horn neurons have been biochemically characterized. There are at least two central nervous endogenous opioid systems: one is localized in the pituitary hypothalamic limbic axis; the other involves midbrain structures and the spinal cord dorsal horn neurons. Both are activated in response to stress and produce a vegetative response associated with analgesia. Opioid receptors are also found peripherally, with activity paralleling inflammation.[23]

Narcotics are delivered using a variety of administration modes.[24] The least invasive delivery of narcotics is accomplished using iontophoresis. The creation of a subcutaneous steady state of narcotic depends on drug lipophilicity and half-life. The fentanyl patch for post-operative patients who are not regular users of narcotics has failed to achieve popularity because of unpredictable dose delivery and occasional respiratory depression requiring reversal of the analgesic.[25]

Oral narcotics are increasingly popular for the treatment of chronic intractable cancer pain, and require an intact and normally functioning gastrointestinal tract. Gut function is erratic during the first 2 postoperative days, unless there has been bowel trauma. In the absence of normal gastrointestinal function a sudden bolus delivery of drug via the rectal circulation can occur after the first bowel movement on the second or third postoperative day. The rectum itself has a unique circulation, only the proximal one-third draining into the portal system such that narcotics administered rectally will bypass hepatic degradation on the first pass providing higher initial blood levels. Rectal administration of narcotics has not gained widespread acceptance in the perioperative setting, but is popular among patients with cancer in whom the gastrointestinal tract does not function normally.

Narcotics may be delivered subcutaneously by infusion using a very fine needle and infusion pump. A subcutaneous reservoir of narcotic assures sustained blood levels. To achieve adequate postoperative analgesia using narcotics, the extracellular fluid space must be filled, therapeutic blood levels be maintained and bolus doses be available to allow ambulation, breathing and coughing with minimal discomfort.

The most widely prescribed mode of narcotic administration is intramuscular. Patients suffer when analgesic blood levels peak and trough, a common occurrence when a standard 'as required' (PRN) regimen is followed.[26] Equal analgesia can be achieved with intramuscular narcotics prescribed in a 'reverse PRN' schedule. The patient is offered a stated dose of narcotic at prescribed intervals rather than allowing the patient to initiate a cycle of request, wait and pain. Education of both physicians and nurses is still important in erasing the perception that adequate narcotic analgesia will lead to drug-seeking behaviour in postoperative patients.

Patients can be educated to give themselves anticipatory bolus infusions of narcotic to facilitate mobility during recovery. Patient-controlled analgesia devices have become a popular addition to most postoperative analgesia services.[27] These computerized devices can be programmed to provide a basal infusion of narcotic after the administration of a loading dose to achieve a steady state level of drug. Bolus doses and lock-out intervals can be programmed in addition to or instead of the infusion to provide added analgesia when necessary.

The very specialized 'intraspinal' administration of narcotics has been developed since the early 1980s, following the demonstration of decreased heat sensitivity in the rat after intrathecal administration of morphine.[28] Anaesthetists desiring to produce segmental analgesia without sympathetic or sensory blockade have described dose ranges and side effects for epidural and intrathecal narcotics in humans.[29] Exogenously administered opioids effect anti-nociception at neuronal receptors. At least five types of opioid receptors exist in human nervous tissue; not all of them are associated with analgesia.[30] Stressors, such as repeated electrical shocks, heat, cold or inflammation, are associated with an increased opioid production in rats. Human labour and inflammation have been associated with enhanced opioid receptor activity.

Hydrophilic narcotics such as morphine, administered into the epidural space (using approximately 10 per cent of the parenteral dose) or intrathecally (using approximately 10 per cent of the epidural dose) do not result in an incidence of respiratory depression greater than is produced by intramuscular narcotics. Morphine is still the ideal substance for a single intraspinal dose, with an average duration of 18 hours for a single epidural dose and 22 hours for a single intrathecal dose. Using a maximum epidural dose under 4 mg and an intrathecal dose less than 0.25 mg, the incidence of delayed respiratory depression is 0.09 per cent (epidural) and 0.36 per cent (intrathecal).[31]

Most patients can be nursed safely on the wards following single dose intraspinal morphine administration, particularly when staff have been educated to observe for somnolence and a drop in respiratory rate below 11 during the first 24 hours after administration. This is rapidly reversed by intravenous naloxone. Despite the lack of sympathetic blockade following epidural or intrathecal narcotic administration, urinary retention is a common side effect. Patients often require catheterization during the period of intraspinal opioid administration. Urinary retention does not reliably reverse with the administration of low dose naloxone, unlike the intense pruritus which may accompany intraspinal opioid activity.

Narcotics, regardless of mode of administration, do not appear to attenuate perioperative hormonal responses to stress.[32,33] A central attenuation of sympathetic activity has been proposed in patients undergoing aortic surgery when arterial blood pressure is evaluated in the immediate postoperative period.[34]

Epidural infusion using a percutaneously placed

cathether has gained popularity in the management of postoperative pain for up to 3 days. Mixtures containing low concentrations of a local anaesthetic (bupivacaine) and a narcotic (morphine or fentanyl) are titrated to allow respiratory effort and movement without pain.[35] The relatively sophisticated equipment and intensive surveillance for potential side effects make epidural analgesic infusions more suitable for relatively sicker patients who also require other special unit care. Aggressive attention to postoperative analgesia is associated with a reduction in morbidity and hospital stay in this population, giving an improved short-term outcome.[36]

Indwelling epidural or intrathecal catheters are associated with local skin contamination after 3 days. Epidural space infection or abscess is a small risk when cathethers are left in place for longer than 3 days at a time.

Freezing

Cryoanalgesia is useful to provide analgesia following thoracotomy in patients with pre-existing respiratory impairment.[37] The nerve is cooled to −70°C for seconds and then thawed. A number of freeze–thaw cycles ensures complete neuronal destruction. The integrity of the epineurium is maintained during cryoneurolysis, so nerve regeneration may occur without neuroma formation. The procedure may be accomplished under direct vision during the thoracotomy. The intercostal nerves are frozen at the level of incision as well as two segments above and below it. Intercostal nerves are somatosensory, and patients may still experience discomfort related to pleural irritation, particularly when there is a chest tube in place during the postoperative period. Loss of intercostal muscle function is not clinically a problem for maintenance of adequate postoperative ventilation. Incomplete neurolysis can result in neuritis-like pain, similar to that seen with alcohol and phenol. Nerve regeneration will occur over 6–12 months. Patients may experience some sensory loss in the interim, corresponding to the area of freezing, but there is a reportedly decreased incidence of chronic post-thoracotomy pain syndrome.[38]

Electrical stimulation

Transcutaneous electrical nerve stimulators (TENS) deliver frequency- and intensity-controlled electrical impulses to the skin. When the stimulator is placed over superficial peripheral nerves, large fibre input to the spinal cord is enhanced. Restoration of balanced small and large fibre activity is believed to close the segmental gating mechanism. Application of TENS units in the postoperative period was initially limited by an increased incidence of wound infection following the use of non-sterile pads within the surgical dressing. Sterile strip conducting pads are now available. Patients should receive preoperative instruction in the proper use of TENS. Sterile pads are placed along the length of the incision in the operating room underneath the sterile dressing. Postoperatively, high frequency stimulation along lateral incisions results in decreased pain perception and improved mobility. Effective postoperative analgesia is reported in more than 30 per cent of the patients.[39]

Positive imagery and therapeutic relations

Appreciation of the therapeutic power of placebo responses,[40] which occur in 30–35 per cent of interventions, and the efficacy of hypnosis suggests that these may be an untapped resource in the management of pain and stress associated with surgery. Patients desire a heightened sense of control over their pain and analgesia. Patient control is restored only by education, and the importance of preoperative teaching cannot be overemphasized.

In the UK and the USA, as health care expenditures are increasingly determined by government agencies, evaluation of outcome will be an important factor in determining payment. Duration of stay in hospital is likely to play a significant role in these considerations. The resting state of the sympathetic nervous system (preoperative anxiety) has tremendous impact on surgical morbidity. Reassurance that pain will be controlled postoperatively can be therapeutic and may lead to improved induction, intraoperative course and emergence from anaesthesia.

Adequate postoperative analgesia begins during the preoperative interview, which encompasses the psychological and behavioural identity of the patient. Preoperative medication can be used to prepare the patient for surgery and can add to postoperative analgesia. With individualized planning and education, the analgesia provided during the first 48 postoperative hours can facilitate mobilization and recovery. Anaesthetists may serve as a liaison between the members of the surgical staff and members of the primary care staff in ensuring education and continuity of pain management. In recognition of the multidisciplinary nature of acute pain management, postoperative pain services have been organized to include members from several departments in addition to anaesthesiology.[41]

REFERENCES

1. Crile GW, Lower WE. *Anoci-association.* Philadelphia: WB Saunders, 1914: 22–225.

2. Ready LB, Oden R, Chadwik HS, *et al.* Development of an anesthesiology-based postoperative pain service. *Anesthesiology* 1988; **68**: 100–6.

3. Cousins MJ. Acute pain and the injury response: immediate and prolonged effects. *Regional Anesthesia* 1989; **14**: 162–79.

4. Fields HL. *Pain.* New York: McGraw-Hill, 1987: 13–40.

5. Bonica JJ. Anatomic and physiologic basis of nociception and pain. In: Bonica JJ, Loeser JD, Chapman CR, Fordyce WE, eds. *The management of pain.* Philadelphia: Lea & Febiger, 1990: 73–94.

6. Bonica JJ, Yaksh T, Liebeskind JC, Pechick RN, DePaulis A. Biochemistry and modulation of nociception and pain. In: Bonica JJ, Loeser JD, Chapman CR, Fordyce WE, eds. *The management of pain.* Philadelphia: Lea & Febiger, 1990: 95–121.

7. Melzack R, Wall PD, Ty TC. Acute pain in an emergency clinic: latency of onset and descriptor patterns related to different injuries. *Pain* 1982; **1**: 33–43.

8. Michie HR, Wilmore DW. Sepsis, signals, and surgical sequelae (a hypothesis). *Archives of Surgery* 1990; **125**: 531–5.

9. Reinhart K, Foehring U, Kersting T, *et al.* Effects of thoracic epidural anesthesia on systemic hemodynamic function and systemic oxygen supply–demand relationship. *Anesthesia and Analgesia* 1989; **69**: 360–9.

10. Yeager MP, Glass DD, Neff RK, Brink-Johnsen T. Epidural anesthesia and analgesia in high risk surgical patients. *Anesthesiology* 1987; **66**: 728–36.

11. Tuman KJ, McCarthy RJ, March RJ, DeLaria GA, Patel RV, Ivankovich AD. Effects of epidural anesthesia and analgesia on coagulation and outcome after major vascular surgery. *Anesthesia and Analgesia* 1991; **73**: 696–704.

12. Power I, Noble DW, Douglas E, Spence AA. Comparison of i.m. ketorlac trometamol and morphine sulfate for pain relief after cholecystectomy. *British Journal of Anaesthesia* 1990; **65**: 448–55.

13. Frame JW, Evans CRH, Flaum GR, Langford R, Rout PGL. A comparison of ibuprofen and dihydrocodeine in relieving pain following wisdom teeth removal. *British Dental Journal* 1989; **166**: 121–4.

14. Rosenblum M, Weller RS, Conrad PL, Falvey EA, Gross JB. Ibuprofen provides longer lasting analgesia than fentanyl after laparoscopic surgery. *Anesthesia and Analgesia* 1991; **73**: 255–9.

15. Ejlersen E, Andersen HB, Eliasen K, Mogensen T. A comparison between preincisional and postincisional lidocaine infiltration and postoperative pain. *Anesthesia and Analgesia* 1992; **74**: 495–8.

16. Tverskoy M, Cozacov C, Ayache M, Bradley EL, Kissin I. Postoperative pain after inguinal herniorrhaphy with different types of anesthesia. *Anesthesia and Analgesia* 1990; **70**: 29–35.

17. Carli P, Duranteau J, Mazoit X, Gaudin P, Ecoffey C. Pharmacokinetics of interpleural lidocaine administration in trauma patients. *Anesthesia and Analgesia* 1990; **70**: 448–53.

18. Kaplan JA, Miller ED, Gallagher EG. Postoperative analgesia for thoracotomy patients. *Anesthesia and Analgesia* 1975; **54**: 773–7.

19. Hinkle AJ. Percutaneous inguinal block for the outpatient management of postherniorrhaphy pain in children. *Anesthesiology* 1987; **67**: 411–13.

20. Kehlet H. Influence of regional anesthesia on postoperative morbidity. *Annales Chirurgiae et Gynaecologiae* 1984; **73**: 171–6.

21. Vedrinne C, Vedrinne JM, Guiraud M, Patricot MC, Bouletreau P. Nitrogen-sparing effect of epidural administration of local anesthetics in colon surgery. *Anesthesia and Analgesia* 1989; **69**: 354–9.

22. Horlocker TT, Wedel DJ, Offord KP. Does preoperative antiplatelet therapy increase the risk of hemorrhagic complications associated with regional anesthesia? *Anesthesia and Analgesia* 1990; **70**: 631–4.

23. Stein C, Comisel K, Haimerl E, *et al.* Analgesic effect of intraarticular morphine after arthroscopic knee surgery. *New England Journal of Medicine* 1991; **325**: 1123–6.

24. McLeskey CH, Shafter SL. Transdermal opioids. Literature scan. *Anesthesiology* 1991; **5**: 2 and 27.

25. Hare BD. The opioid analgesics: rational selection of agents for acute and chronic pain. *Hospital Formulary* 1987; **22**: 64–84.

26. Austin KL, Stapleton JV, Mather LE. Multiple intramuscular injections: a major source of variability in analgesic response to meperidine. *Pain* 1980; **8**: 47.

27. White PF. Patient-controlled analgesia: a new approach to the management of postoperative pain. *Seminars in Anesthesia* 1985; **4**: 255–66.

28. Yaksh TL, Rudy TA. Analgesia mediated by a direct spinal action of narcotics. *Science* 1976; **192**: 1357–8.

29. Cousins MJ. Comparative pharmacokinetics of spinal opioids in humans: a step toward determination of relative safety. *Anesthesiology* 1987; **67**: 875–6.

30. Pasternak GW. Multiple morphine and enkephalin receptors and the relief of pain. *Journal of the American Medical Association* 1988; **259**: 1362–7.

31. Rawal N, Arner S, Gustafsson LL, Allvin R. Present state of extradural and intrathecal opioid analgesia in Sweden. *British Journal of Anaesthesia* 1987; **59**: 791–9.

32. Rutberg H, Hakanson E, Anderberg B, Jorfeldt L, Martensson J, Schildt B. Effects of the extradural administration of morphine, or bupivacaine, on the endocrine response to upper abdominal surgery. *British Journal of Anaesthesia* 1984; **56**: 233–7.

33. Hakanson E, Rutberg H, Jorfeldt L, Martensson J. Effects of the extradural administration of morphine or bupivacaine, on the metabolic response to upper abdominal surgery. *British Journal of Anaesthesia* 1985; **57**: 394–9.

34. Breslow MJ, Jordan DA, Christopherson R, *et al.* Epidural morphine decreases postoperative hypertension by attenuating sympathetic nervous system hyperactivity. *Journal of the American Medical Association* 1989; **261**: 3577–81.

35. Dahl JB, Rosenberg J, Hansen BL, Hjortso NC, Kehlet H. Differential analgesic effects of low-dose epidural morphine and morphine–bupivacaine at rest and during mobilization

after major abdominal surgery. *Anesthesia and Analgesia* 1992; **74**: 362–5.

36. Baron JF, Bertrand M, Barre E, *et al.* Combined epidural and general anesthesia versus general anesthesia for abdominal aortic surgery. *Anesthesiology* 1991; **75**: 611–18.

37. Nelson KM, Vincent RG, Bourke RS, *et al.* Intraoperative intercostal nerve freezing to prevent post-thoractomy pain. *Annals of Thoracic Surgery* 1974; **18**: 280–4.

38. Maiwand MO, Makey AR, Rees A. Cryoanalgesia after thoracotomy. Improvement of technique and review of 600 cases. *Journal of Thoracic and Cardiovascular Surgery* 1986; **92**: 291–5.

39. Tyler E, Caldwell C, Ghia JN. Transcutaneous electrical nerve stimulation: an alternative approach to the management of postoperative pain. *Anesthesia and Analgesia* 1982; **61**: 449–56.

40. Lasagna L. The placebo effect. *Allergy and Clinical Immunology* 1986; **78**: 161–5.

41. Acute Pain Management Guideline Panel. *Acute pain management: operative or medical procedures and trauma. Clinical practice guideline.* AHCPR Pub. No. 92-0032. Rockville, MD: Agency for Health Care Policy and Research, Public Health Service, US Department of Health and Human Services. Feb. 1992; 145 pages.

Nutritional Support

S. P. Allison

The anaesthetist may be faced with providing nutritional support in two situations: first, in the perioperative period, for patients who come to surgery in a malnourished state or who become malnourished as a result of postoperative complications; second, in the intensive care unit. Although the average length of stay of patients in intensive care units is only a few days, there are a few whose illness is so severe that they require prolonged ventilatory support and are consequently unable to eat, for long periods.

Malnutrition

Form and function

Obese patients are often required by the surgeon to lose weight in order to facilitate surgery. The loss of 10 per cent of the body weight with a properly supervised reducing diet may have a beneficial effect in such patients and even improve surgical outcome. On the other hand, the initially slim or mesomorphic patient who is unable to eat properly because of disease is in an entirely different category. Total starvation for 3 weeks results in 15 per cent weight loss.[1] Starvation for 9 weeks causes 38 per cent weight loss and a 30 per cent mortality from malnutrition alone.[2,3] Periods of total or partial starvation for 2–3 weeks are not uncommon in hospital practice and may be superimposed upon weight loss prior to admission. The accelerated catabolism associated with acute illness or injury (Chapter 17) may further exacerbate tissue loss. Just as important, however, weight loss beyond 8 per cent results in increasing impairment of function in ways that handicap recovery from disease and multiply its complications.[4] The classic Minnesota study of Keys and his colleagues in the 1940s[5] showed that young men who underwent a 24-week period of voluntary partial starvation lost 24 per cent of their body weight and suffered profound physical and psychological changes, which had not returned to normal after a 16-week period of refeeding. They became depressed and irritable, and suffered muscle weakness with a severe reduction in their fitness score. Muscle wasting and weakness from starvation affects particularly the fast twitch fibres and hence the respiratory muscles, including the diaphragm, impairing respiratory function and the ability to cough and clear secretions.[6–14] Malnutrition also reduces respiratory drive.[15] In contrast, nutritional support has facilitated weaning from ventilators[16,17] as well as improving muscle strength[6,18,19] and allowing more rapid recovery from orthopaedic injuries.[20] Increased susceptibility to infection has always been a prominent feature of famines, highlighted in the studies carried out by the Jewish physicians in the Warsaw ghetto[21] and by more recent studies showing impaired immune function and increased infection rates with undernutrition.[22–24] The Warsaw studies described reduced cardiovascular and gastrointestinal function, with lowered blood pressure, decreased cardiac output and achlorhydria among other features. Acute illness and malnutrition may impair not only the digestive functions of the gut but also the barrier it imposes to infection.[25–28] Both functions may be protected by enteral but not by parenteral feeding unless supplemented by glutamine.[29–33] Cardiovascular reflexes have been studied following injury and shown to be impaired,[34] thereby inhibiting compensatory responses to blood loss. More than 48 hours' starvation may also impair the vasoconstrictor response to cold, reducing heat conservation.[35] Loss of body mass also impairs the thermogenic response to cold, whereas feeding and recovery in weight restore these thermoregulatory responses to normal.[36,37] Malnutrition also contributes to increased surgical risk, poorer wound healing and slower recovery from surgery.[10,24,38–42]

As weight loss proceeds, the different compartments of the body change at different rates.[10] Keys and colleagues[5] showed that, by 24 weeks of partial starvation, fat mass was reduced by over 60 per cent and fat free mass by 16 per cent. Because of accelerated protein catabolism, the loss of lean body mass in the injured and acutely ill may be proportionately greater than this. The Minnesota study[5] also showed that the absolute extracellular fluid volume remains unchanged, though, as body mass shrinks, it increases as a proportion of body weight. Both prolonged starvation and the response to injury confer an inability to excrete an excess salt and water load,[5,21,43–47] and put both depleted and sick patients at risk of salt and water overload unless their intake is carefully controlled. The kidneys of such patients are incapable of compensating for the errors of the doctor prescribing intravenous fluids.

As well as starvation and the response to injury, immobility also contributes to muscle wasting. Studies have shown how, in healthy normal subjects, immobilization in bed can produce negative nitrogen balance and muscle wasting.[48] Conversely, mobilization and exercise enhance protein synthesis and restore wasted muscles. This consideration applies particularly to patients who spend prolonged periods paralysed on ventilators in the intensive care unit. Muscle anabolism in such patients cannot be achieved by feeding alone and the sooner the patient can be weaned from the ventilator and mobilized, the sooner will muscle mass be restored.

Measurement

The simple practical measurements that may be carried out in any hospital are discussed here. For the more sophisticated measurements used for research purposes, the reader is referred to appropriate texts.

Weight

The serial measurement and recording of weight is of the greatest value.[10,38,49,50] If this can be related to the patient's remembered or previously recorded weight, the percentage weight loss can be estimated. Weight can also be related to skeletal length using measurement of height or demi-span[50] (measured from the suprasternal notch to the web between the third and fourth fingers with the arm outstretched). Dividing weight in kilograms by height^2 in metres, a body mass index can be derived with a normal range of 18–25. Alternatively, the patient's weight and height can be related to tables from relevant normal populations of the same age and sex. The wide variation in normal values makes them valuable for epidemiological purposes but not always for the assessment of the individual. Weight is also affected in the short term by the state of hydration. In the postoperative period, fluid retention and oedema may conceal true tissue weight loss unless such factors are allowed for. During treatment, long-term weight gain in the absence of oedema may reflect true tissue gain, whereas day-to-day changes in weight may be the best measure of fluid balance. Knowing the weight has additional value, not only for calculating the dose of drugs but also for estimating the patient's nutritional requirements.

Voluntary oral intake

The simple estimation of voluntary oral intake postoperatively has been validated by Hessov[51] and shown to be a useful clinical tool.[40,52,53] Meguid[53] has described the 'inadequate oral nutrient intake period', or IONIP, and shown that, in the postoperative period, those who fail to achieve an oral intake greater than 60 per cent of their estimated requirements by the 10th day postoperatively tend to do badly and should be considered for nutritional support. The help of an experienced dietitian is invaluable in this respect.

Anthropometrics and body composition

The simple measurement with a tape of mid-arm circumference, half way between the acromion and the olecranon, combined with the measurement of triceps skinfold thickness, gives an estimate of body composition and the respective amounts of fat and muscle.[10,54–56] Triceps skinfold thickness, particularly, requires some skill in its measurement and there can be a 20 per cent interobserver variation in unskilled hands. Mid-arm circumference is a fairly robust measurement, however, and can be useful in clinical practice. Both may be affected by the patient's state of hydration. A bedside estimate of body composition may be obtained by the bioimpedance method, although the errors that occur in unstable or very sick patients can affect its accuracy.[57] In clinical practice, it probably adds little to the measurements already described.

Plasma proteins

The concentration of the serum albumin and of the shorter half-life protein thyroid-binding prealbumin (TBPA) have been cited as nutritional parameters, despite the fact that with starvation alone they remain unchanged. They fall postoperatively or in the presence of acute illness or inflammation due to changes in distribution rather than metabolism.[3,20,58] Recovery to normal levels may, however, be delayed by malnutrition, and assisted by nutritional support. Such measurements are therefore useful in monitoring the response to treatment, though of minimal value in deciding whether nutritional support is needed in the first place.

Table 43.1 Calculation of energy expenditure using Schofield equations[61] (weight in kg; BMR in kcal/24 hours*)

Age (years)	Male	Female
15–18	$BMR = 17.6 \times wt + 656$	$BMR = 13.3 \times wt + 690$
18–30	$BMR = 15.0 \times wt + 690$	$BMR = 14.8 \times wt + 485$
30–60	$BMR = 11.4 \times wt + 870$	$BMR = 8.1 \times wt + 842$
> 60	$BMR = 11.7 \times wt + 585$	$BMR = 9.0 \times wt + 656$

To BMR:

1. Add stress factor:	Postoperative	+ 10%
	Multiple injury	+ 25–30%
	or with sepsis, for each 1°C rise	+ 10%
2. Add activity factor:	Bedbound awake	+ 10%
	Sitting out	+ 20%
	Mobile in ward	+ 30%
3. Add for thermogenic action of food:		+ 10%
4. Reduce if on ventilator:		− 15%

Example

40-year-old ventilated male with multiple fractures (weight 60 kg, temperature 39°C)

BMR = $11.4 \times 60 + 870$	=	1554 kcal
Add stress factor +30%	= +	466
Reduce ventilator factor −15%	= −	233
Add food factor + 10%	= +	155
Total	=	1942 kcal

* Multiply kilocalories by 4.184 to obtain kilojoule equivalents.

Nutritional or risk indices

The parameters of weight, haemoglobin, plasma proteins and skin immune response have been variously combined in the form of so-called nutritional indices which have been shown to correlate with postoperative complications.[59] Such derived values should more properly be designated as 'risk' indices, as they reflect the severity of disease and improve with the treatment of disease and only sometimes with nutritional support.

Energy requirements

In health, the basal metabolic rate (BMR) or the less stringent resting energy expenditure (REE) may be predicted from equations relating metabolic rate to body size, age and sex. The commonly used Harris–Benedict equations[60] or those suggested by Schofield[61] (Table 43.1) are still of value in the clinical setting, although they underestimate nutritional requirements in disease. Kinney's group[62] measured REE in 237 patients and 37 normal subjects, comparing measured and predicted (Harris–Benedict) values in different groups. Measurement was carried out using the canopy method of indirect calorimetry. These authors found an increase of up to 28.6 per cent between normal subjects and those with multiple injuries, and a 30 per cent increase of measured over predicted.

They also found a variation of 15 per cent between individuals. The thermogenic effect of feeding was also apparent, with increases of up to 12 per cent with parenteral nutrition. They conclude that, in patients, accurate figures for energy expenditure can be obtained only by competent indirect calorimetry but, because the equipment is expensive and demands expert technical skill for its use, most centres have to rely on the predictive formulae and the use of correction factors to correct for disease. We have found the Schofield equations used in this way to be adequate for clinical purposes and for the calculation of the appropriate energy requirements for nutritional support, despite the fact that this may give a slight overestimate in some patients. In the past, attempts were made to bring about positive energy and nitrogen balance with the use of high calorie and carbohydrate 'hyperalimentation' regimens. These proved toxic[63] and were abandoned in favour of a more defensive approach, i.e. just meeting expenditure during the flow phase of injury, reserving higher intakes for the anabolic or convalescent phase. It is rare for postoperative patients to require more than 8400 kJ (2000 kcal) per day and catabolic patients can be managed using 8400–12 500 kJ (2000–3000) kcal) per day.

Nitrogen balance

The accurate measurement of nitrogen balance is difficult and essentially a research tool. It is not necessary in order to conduct satisfactory nutritional support. Nitrogen losses correlate positively with increased energy expenditure and reflect the severity of illness. An intake of nitrogen 0.15 g/kg body weight per day may suffice for a basal feed, increasing to 0.3 g/kg in the most catabolic patients.

Creatinine/height index

The 24-hour urinary creatinine excretion correlates with muscle mass and is reduced with muscle wasting. This index, which is defined as the 24-hour creatinine excretion of the patient divided by the expected 24-hour excretion of a normal adult of the same sex and height, can be a useful measure of nutritional depletion.

Functional parameters

Measurements such as hand dynamometry and respiratory function and immunological tests may be of value in monitoring the effect of treatment, as it is improvement in function rather than form that is the prime aim.[16] They are not essential for the initial decision to feed.

Summary

Measurements of weight, anthropometrics and oral intake form the basic parameters upon which the decision to give nutritional support is based. Plasma proteins, functional measurements and changes in haematological and biochemical parameters are important in monitoring the response to treatment. A knowledge of the likely natural history of the underlying disease process is also vital in sensible decision making.

Benefits of treatment

It is one thing to show that starvation and weight loss are associated with measurable changes in structure and function. It is another to show that nutritional support not only reverses these changes but also results in significantly improved clinical outcome at reasonable cost in resources and without excessive complications. It is also necessary to establish what degree of malnutrition is significant, and therefore which patients are likely to benefit from treatment. Feeding may produce improvements in nitrogen balance, in anthropometric measurements and in weight (chiefly due to fluid gain) without producing any measurable improvement in outcome. In the severely malnourished, however, improvements in function due to feeding may be more clinically significant and occur before any increase in body mass. Reviews and meta-analyses published between 1984 and 1987[64-66] concluded that there was no evidence of benefit from nutritional support in patients with mild to moderate nutritional depletion and that the evidence for benefit in severely malnourished patients was suggestive but required confirmation. Many trials were criticized for their design, methodology, patient selection and inadequate numbers.

One of the earliest trials to show benefit was in elderly women after surgery for fractured neck of femur.[20] Patients were divided by anthropometric means into normal, thin (1–2 SD below the reference range) and very thin (>2 SD below the reference range). The thin and very thin groups were randomized to receive ward diet or ward diet plus 1000 kcal in the form of a supplementary overnight nasogastric tube feed. Rehabilitation to mobility was 10 days in the normal group, not significantly affected by feeding in the thin group, but reduced from 23 to 16 days in the very thin group. These results have recently been confirmed in fractured femur patients using oral supplements.[67]

Other more recent trials have also shown benefit from oral supplements among elderly patients in a geriatric unit,[68,69] with reductions in mortality, in morbidity from pressure sores and in hospital stay. Oral supplements given as early as possible postoperatively have been shown not only to preserve weight and muscle strength but also to reduce infectious complications.[70] Two major trials of perioperative feeding[71,72] have also shown clear benefit from feeding by the intravenous or enteral route, but only in severely malnourished patients with at least 10 per cent weight loss preoperatively. The Veterans trial[71] showed actual harm in the slightly malnourished group, with an increase in non-catheter-related infections. This is a salutary reminder that nutritional support must follow the same rules as any other form of treatment, with a careful appraisal in each patient of the likely benefit or harm to be expected from it. If the provision of perioperative nutritional support in selected patients is now based on firmer evidence, there is a paucity of such evidence among intensive care patients in whom there is an even greater possibility of doing harm.[63] The short stay of most patients and the very heterogeneity of the patient population makes adequate trials difficult. However, as the laws of thermodynamics have yet to be repealed, similar criteria for treatment should be applied to patients in the intensive care unit as elsewhere. What evidence there is suggests that improved muscle strength and weaning from ventilators may be expected with judicious nutritional support, whilst the increased oxygen consumption and carbon dioxide production resulting from the administration of excess carbohydrate or total calories may exacerbate respiratory failure.

The clinical decision to treat

From a consideration of the evidence outlined thus far, the following indications for nutritional support are suggested:

1. Weight loss greater than 10 per cent and continuing
2. Continuing inadequate oral intake
3. The presence of disease whose known natural history is associated with accelerated catabolism and poor food intake for 10 days or more. There is no substitute for clinical experience in this respect. In a burned patient with ileus, for example, no time should be wasted in starting nutritional support after the shock phase is over. It is not necessary to wait and see.

Treatment

Methodology

Having made the clinical decision to provide nutritional support, how should it be given and what should be the composition of the feed? Before answering such questions, it is first necessary to consider how nutritional support should be organized, as the sporadic and occasional use of the more technically demanding forms of artificial nutrition can be fraught with complications when conducted by the inexpert. Any anaesthetist wishing to take an interest in this field should be associated with a hospital nutrition team, whose members should include doctors, nurses, dietitians and pharmacists. With parenteral nutrition, for example, it has been shown that the formation of a nutrition team working to agreed standards and protocols results in a reduction in catheter sepsis rates from 28 per cent to under 3 per cent, of mechanical problems from 24 to 6 per cent and of metabolic problems from 30 to 12 per cent.[3,73,74] Costs per day and per patient are also reduced. A standard recording system should be employed to allow regular audit of the equipment and feed used, and of complications and costs. By careful attention to such detail, septic complications may be reduced to zero and satisfactory cost-effectiveness and cost-benefit demonstrated.[3,73,74]

In deciding upon the route of administration of feed, the rules are simple. If the gut works, try to use it. If the patient can swallow, try oral supplements or, failing this, some form of enteral tube feeding. Parenteral nutrition is the treatment of prolonged gastrointestinal failure or

dysfunction, in the same sense that dialysis is the treatment of renal failure.

Oral supplements

The use of the traditional postoperative drip and suck regimens in patients undergoing abdominal surgery has been challenged by Hessov and colleagues from Denmark.[75,76] They showed that patients who had undergone cholecystectomy or large bowel surgery could start to feed orally very early in the postoperative period, resulting in earlier return of bowel function and no increase in vomiting or complications such as anastomotic breakdown. Controlled trials demonstrating the clinical benefit of modest amounts of oral supplement, given postoperatively, have already been mentioned.[70] Further studies are required but a more aggressive and effective approach by nurses and dietitians towards oral feeding may be the most cost-effective way of providing nutritional support in most postoperative patients, though this method is clearly impractical in the intensive care unit.

Enteral tube feeding

Many elderly patients, in particular, are slow to recover their appetite postoperatively, even though bowel function has returned. Whilst many will respond to oral supplements, a few benefit from a period of fine-bore nasogastric tube feeding. Patients with swallowing difficulties may be managed similarly, although, if tube feeding needs to be continued for longer than 1–2 weeks, a feeding gastrostomy or jejunostomy should be considered.[77,82] With obstructive lesions of the upper gastrointestinal tract, causing severe weight loss, there may be a case for a period of preoperative tube feeding, as described in the Maastricht study.[72] The insertion of a jejunostomy or gastrostomy allows feeding to be carried out at home, which is particularly important in patients with inoperable obstructive lesions.

The technique of using nasogastric or nasoenteral fine-bore tubes is well known[83] but a few points are worth emphasizing. It is important to select a good quality tube, preferably of polyurethane with side holes and a flexible guide wire which is removed easily. The internal diameter of the tube need not be greater than 2–3 mm for adequate feeding. After insertion of the tube, an x-ray to ascertain its position is necessary only in those with impaired consciousness or gag reflex. Auscultation for bubbles in the epigastrium when air and water are syringed down the tube is usually sufficient in the conscious patient. Reversal of the Luer fittings with the male fitting on the tube and the female fitting on the giving set prevents accidental connection of the feeding line with an intravenous catheter. We favour continuous pumped feeding, although this may be carried out intermittently both for convenience and to allow the gastric pH to fall to bactericidal levels.[84] Overnight nasogastric tube feeding is a valuable technique in some patients, allowing freedom to exercise and take oral feed during the day. We have found that this method may result in a disinhibition of appetite in anorectic patients, accelerating the return to a full oral diet.[85] This is quite contrary to the inhibition that might be expected and its mechanism is uncertain, although gastrointestinal hormones may have some part to play, since intravenous feeding has the opposite effect. A sterile feed should be used and the giving sets and reservoirs changed daily, particularly for intensive care patients, because enteral feeding has been incriminated in cross-infections and even fatalities.

Gastrostomy and jejunostomy

Gastrostomies and jejunostomies[77–82] may be inserted during surgery. Many a patient undergoing upper gastrointestinal surgery could have been saved from having to undergo intravenous feeding by the prophylactic insertion of one of these tubes at operation. With prolonged gastric outlet obstruction, for example, it is invaluable to have a jejunostomy in place to help with postoperative management. The Delaney type of fine-bore jejunostomy[82] is to be preferred, inserted via a wide-bore needle and tunnelled in the bowel wall, with the entry site sewn to the anterior abdominal wall. Such tubes can be used for long periods of time and are effective and unobtrusive. Infection is the major complication, but this is uncommon in skilled hands and usually responds to antibiotics.

Gastrostomies may also be inserted by the percutaneous endoscopic gastrostomy technique described by Ponsky and colleagues.[77,78,82] A gastroscope is passed and a second operator observes the gastroscope light in the epigastrium, directing the local anaesthetic to this area, which is then punctured with a large cannula until the endoscopist observes the point of the cannula within the stomach. A nylon thread is passed through this, grasped by the endoscopy forceps and brought up through the mouth. To the thread is attached a narrow-bore gastrostomy tube with a flange. The second operator then pulls the nylon thread back through the epigastric cannula until the gastrostomy tube appears on the outside and the flange fits snugly against the gastric mucosa. An external flange is then put in place to anchor the tube and a hub attached for connection with the giving set. The internal part of the gastrostomy tube may be short and end in the stomach itself. Some tubes have a long tail, however, which can then be threaded by the endoscopist through the pylorus and round into the jejunum, overcoming problems of gastric outlet obstruction and hopefully diminishing the amount of reflux.

Type of feed

For most enteral tube feeding, a standard defined formula whole protein feed is all that is required. Such feeds contain a balanced proportion of protein, carbohydrate and fat, with adequate supplies of vitamins, minerals and trace elements. Where constipation is a problem, fibre-containing feeds may be advantageous, and in some cases diarrhoea is improved by the addition of fibre. We have not used an elemental diet (amino acids rather than whole protein) for many years, although there are some patients, particularly those with a jejunostomy, in whom a peptide feed appears to be tolerated better. There is some evidence that high fat, low carbohydrate feeds give a lower respiratory quotient (RQ), make less demands for gas exchange and are therefore advantageous in weaning patients off ventilators. Others maintain that a reduction in caloric intake is just as effective. As with parenteral nutrition, it is customary to supply half the energy as fat and half as carbohydrate. The use of disease-specific feeds and of special substrates such as glutamine and ornithine α-ketoglutarate and $\omega3$ fatty acid, has created a whole new discipline of pharmacological nutrition (see below). Although early clinical studies are promising, the role of these substrates in practice has not yet been fully defined.

Complications[83–92]

Whichever technique is used, the slow continuous or intermittent administration by pump diminishes many of the complications of reflux, nausea and diarrhoea associated with the bolus method. If diarrhoea still occurs, it is often in association with broad-spectrum antibiotics, which should be stopped if possible. Otherwise the addition of loperamide or codeine phosphate syrup is effective. In cases both of constipation and of diarrhoea, the use of a fibre-containing feed may also be helpful. In the intensive care unit, there has been a shift of emphasis from parenteral to enteral feeding on the grounds of cost effectiveness and the reported complications of the former method.[87]

Enteral feeding has been advocated to maintain bowel function and the integrity of the mucosal barrier to infection.[33] On the other hand, it is not without its problems, apart from those which have already been mentioned. The most potentially serious problem is that of aspiration pneumonia. Aspiration has been variously reported as occurring in 4–74 per cent of patients.[84–88] The different rates reported are due to the differences in the sensitivity of the methods employed, and the type of patients studied. Ibanez et al.,[88] using a highly sensitive isotope method that picks up even minor degrees of aspiration, studied the difference in aspiration rate between the supine and the semi-recumbent (30°–45°) position. Without a nasogastric tube, 50 per cent of supine patients had some aspiration. This was reduced to 12 per cent by semi-recumbency. With a 5.5 mm diameter nasogastric tube in place, the aspiration rate in the supine position rose to 81 per cent, compared with 35 per cent in the semi-recumbent position. It is not clear whether rates of aspiration would have been reduced by narrower bore tubes, used in many units. Another important study, by Strong et al.[89] using physical signs and chest x-ray, showed an aspiration rate of 30–40 per cent of patients, and also that there was no improvement when nasoduodenal rather than nasogastric tubes were used. Another recent study, by Mullen and colleagues,[90] again using physical signs supported by chest x-ray, found only a 4.4 per cent aspiration rate and remarked that not only was the condition uncommon but it was also usually benign in its effect. Our own practice is to use a wide-bore tube initially and to infuse 30–60 ml per hour, aspirating 4-hourly to ensure that gastric emptying is taking place. Once this is established, the tube is replaced by a fine-bore tube and the feed administered at gradually increasing rates. Much is made of the osmolality of feeds, and it has been suggested that a dilute feed should be used initially. It is, however, moles delivered per unit time rather than per unit volume that are important, and a slowly delivered feed is swiftly diluted by the large volume of gastrointestinal juice. Contaminated feeds have been reported as a cause of significant infection in enterally fed patients.[91,92] It is important, therefore, particularly for intensive care patients, to use sterile feeds straight from the container and to change the reservoir and giving set each day. Metabolic complications are unusual, provided that an appropriate feed is given and fluid balance is maintained. Hyperosmolar states have been reported, due to inadequate water intake, excess nitrogen administration or excess carbohydrate in glucose-intolerant patients.

Parenteral feeding

When Wretlind introduced his amino acid solution, Aminosol, and his fat emulsion, Intralipid, in the 1950s, he designed the feeds to be isotonic and to be given via a peripheral vein. Because fat emulsions were unavailable in the USA, concentrated glucose was the only energy source available for intravenous feeding. It was Dudrick's contribution[93] to introduce Aubaniac's method of central venous catheterization to parenteral feeding, and this is the standard method now adopted throughout the world, although, with improved technique, there has been a swing back to peripheral vein feeding in some centres.[94]

This method of course requires that the patient has adequate and accessible peripheral veins. New techniques using fine paediatric Silastic cannulae, nitrate patches to dilate the vein and heparin with hydrocortisone to prevent thrombophlebitis have allowed the use of the same catheter site for many days whereas, previously, the cannula had to be re-sited every 24–48 hours. It is not yet clear whether, with the new cannulae, the use of hydrocortisone and heparin is strictly necessary. Central venous catheterization via the jugular, or via the subclavian, is a technique that is too well known to warrant repeating here. For details of the technique, the reader is referred to appropriate texts.[95] Some points need emphasis, however. A low level of mechanical complications at the time of insertion depends very much on the skill and experience of the operator. It has been the custom in our nutrition team to confine the insertion of lines to two or three doctors with particular experience of the technique. With the use of strict aseptic technique and protocols, central venous catheters may be safely inserted on the ward. It is not where it is done that matters but who does it and how it is done.

The development of new plastic materials has greatly improved the complication rate of central lines. The modern polyurethane catheters and the infusion of less hypertonic material has reduced trauma and thrombosis in the large veins, to the point where these are now an almost negligible problem. For long-term feeding (i.e. more than 3 months), we use the slightly wider bore cuffed Hickman or Broviac lines. The lines are always tunnelled, not to prevent infection but to anchor them firmly and prevent displacement. The addition of a small extension tubing allows the changing of giving sets to take place well away from the insertion site. If possible, lines should be dedicated to feeding and not used for other purposes, although this is an ideal that is sometimes difficult to achieve in unstable patients in the intensive care unit. Trials have shown that the use of triple lumen lines does not really reduce the infection rate when they are used for multiple purposes.[96] Such devices may be convenient, but it should be recognized that they are likely to become infected and withdrawn within 7–14 days. The alternative is to have one line dedicated to feeding and a second line on the opposite side for other purposes. Apart from the skill and meticulous care of the operator who inserts the line, the success of intravenous feeding depends crucially upon the subsequent nursing management. No one, not even medical staff, should be allowed to touch the line apart from the nurses who have been specially trained in the protocols required. By adopting this policy, we and other centres have achieved a catheter sepsis rate of zero on the nutrition ward, while in the intensive care unit it is running at about 5 per cent. Careful audit over the years has allowed us to modify our protocols so that, with prolonged postoperative feeding, the catheter entry sites are cleaned and dressed and the giving set changed only once a week, without any increase in complications and with a considerable reduction in costs.[3] In the intensive care unit, especially if lines are used for more than one purpose, changes of giving set and site dressing may need to be more frequent. New techniques in the pharmacy have also contributed to improved efficiency.

Following the pioneering work of Solassol and colleagues in France,[97] it has become possible to mix glucose, amino acids, fat emulsions and micronutrients in a single large plastic bag and for the resulting mixture to be stable over several days. This is not only much more convenient than the multiple bottle technique used previously but also greatly reduces both the demands on nursing time and the opportunities for introducing infection.

In contrast to enteral feeding, amino acid rather than protein solutions are much more effective. Some of the earlier hydrolysates contained peptides, and recent work has shown that these are not only effective but also allow the delivery of important but relatively insoluble or unstable amino acids such as cystine and glutamine. Glucose solutions and fat emulsions form the mainstay of energy supply, each supplying approximately 50 per cent of the energy required to avoid carbohydrate overload and in recognition of the fact that, with sepsis, fat may be oxidized preferentially.[98,99] The administration of glucose at a rate in excess of 6 mg/kg per minute results in excess CO_2 production and may worsen respiratory failure.[100] Both the response to injury and diabetes are associated with glucose intolerance; the administration of insulin may be necessary. Insulin may also help to reduce protein catabolism[101,102] and enhance the excretion of excess salt and water[103] in very sick patients. The algorithm described by Woolfson[104] is a useful guide, relating changes in blood glucose to the appropriate insulin infusion rate. The majority of stable patients can be fed with a standard feed, an example of which is shown in Table 43.2. Requirements for micronutrients and minerals are better understood,[105] and excellent preparations are available commercially. The provision of nutrients or the amounts of water, electrolytes and minerals may have to be varied according to the clinical circumstances, for example inpatients with gastro-intestinal fistulae or in hypercatabolic subjects (e.g. burns).

Complications[95,106]

The subject of preventing catheter infections has already been addressed. If the patient develops a fever and catheter sepsis is suspected, it is important to take blood cultures through the line as well as peripherally. The line is then heparin locked and systemic antibiotic therapy with vancomycin and netilmicin is commenced, pending the catheter culture results. If *Staphylococcus epidermidis* (the usual contaminant) is identified, the line can be sterilized using antibiotic. If *Staphylococcus aureus* or fungi are identified, the line should always be removed. Fibrinous obstructions to the line can usually be cleared with a urokinase lock, but, if in doubt, remove the catheter and

Table 43.2 A typical standard parenteral feed for maintenance feeding to give 2400 kcal and 14 g nitrogen in 2.5 litres

1000 ml aminoacid solution to give 14 g nitrogen and 400 calories

500 ml 20% lipid emulsion to give 1000 kcal and 7.5 mmol phosphate

1000 ml of 26% glucose/electrolyte/mineral solution, to give:

Energy	1000 kcal
Na^+	120 mmol
K^+	80 mmol
Mg^{++}	7.5 mmol
Phosphate	10 mmol
Calcium	7.5 mmol
Chloride	240 mmol
Fluoride	0.05 mmol
Iodide	0.001 mmol
Zinc	0.12 mmol
Copper	0.02 mmol
Manganese	0.005 mmol
Molybdenum	0.002 mmol
Selenium	0.0005 mmol

Additional fat and water-soluble vitamins

Multiply kilocalories by 4.184 to obtain kilojoule equivalents.

replace it. Venous thrombosis is now uncommon in short-term feeding.

Lack of stimulation of the gallbladder results in the formation of sludge and gallstones, which may present problems with prolonged intravenous feeding. This can be prevented by the intermittent use of cholecystokinin or by the infusion of fatty food into the duodenum. Excess carbohydrate or total calorie loads may also be associated with abnormalities of liver function tests, fatty liver and cholestasis. Amino acids, a balance of fat and carbohydrate and the avoidance of overfeeding are protective. Intra-abdominal sepsis is frequently associated with abnormal liver function tests. The finding of such abnormalities does not constitute a reason for stopping the feeding, but should stimulate a review of the feeding prescription. With appropriate prescription of feeds and adequate monitoring, other metabolic complications are infrequent. In stable patients, regular weighing, fluid balance charts and once or twice weekly haematology and biochemical screening are usually sufficient, although more frequent measurements may be required in the intensive care unit.

The pharmacology of nutrition

The problems of trying to reverse the catabolic changes of acute illness have already been discussed and the hazards of hyperalimentation regimens described. Attention has switched to try to reduce the stimuli to catabolism, and this has proved a very fruitful approach. In a series of studies,

Kehlet[107] compared the effect of spinal versus general anaesthesia and showed that spinal blockade could block the neuroendocrine and metabolic responses to injury in patients undergoing lower abdominal surgery. Improved anaesthetic and surgical techniques have done much to lessen the impact of surgery. Improved fluid management and blood volume replacement have also been important, as has been the more progressive control of infections and of pain and anxiety. Nursing patients in a thermoneutral environment also lessens the cold stimulus to hypermetabolism.[108] For these reasons, the reported metabolic rates of seriously ill patients appear to be lower now than those reported many years ago.

Attempts have also been made to reverse catabolism by hormonal means. Insulin[101,102,109] and growth hormones[109,110] have been shown to diminish protein catabolism, but improved clinical outcome has not been demonstrated. Trials are currently in progress assessing the effect of cytokine blockers, but here again improved clinical outcome has yet to be reported.

One of the most interesting developments in recent years has been the introduction of new substrates. The long chain fatty acid emulsions have been criticized for their effects on the lung and the immune system, although the importance of these has probably been exaggerated.[111] Unlike long chain triglycerides, medium chain triglycerides can enter the mitochondria without the mediation of carnitine. The ω3 fatty acids have been advocated for their effects on prostaglandin metabolism. Whether the new tailor-made fat emulsions containing these new components will prove clinically superior at acceptable cost remains to be seen.

The importance of amino acids such as glutamine and arginine has been increasingly recognized. Trials of glutamine, and of its ketoanalogue α-ketoglutarate combined with ornithine, are beginning to show not only important biochemical effects but also clinical advantage.[112–114] Work in these areas will lead to a whole new generation of intravenous amino acid and lipid preparations, and the reader should be encouraged to keep abreast of this field, which may have important implications for the critically ill.

Examples

A few examples may help to illustrate some of the practical points discussed in this chapter.

Example 1

A 55-year-old obese woman, weighing 80 kg, was admitted to hospital complaining of a week's history of increasing muscle weakness. A diagnosis of Guillain–Barré syndrome was made and, because of increasing difficulty in breathing due to muscle weakness, the patient was transferred to the intensive care unit for artificial ventilation, which from our knowledge of the disease, we expected to last for several weeks. Swallowing would therefore be difficult, although the gastrointestinal tract would continue to function. The patient was initially overweight but, although this might be some protection against starvation, it would not prevent the catabolic effects of the disease upon her lean body mass. Muscle wasting would also be exacerbated by the neuropathy and by immobilization.

From the equations in Table 43.1, her basal metabolic rate was calculated to be 6200 kJ (1490 kcal) per day. An increase of 10 per cent was allowed for the catabolic effect of the disease and a further 10 per cent for the thermogenic effect of the feed. A reduction of 15 per cent for artificial ventilation gave a net increase of 5 per cent. Her maintenance food intake should therefore have been 6500 kJ (1565 kcal) per day.

A nasogastric tube was passed, and first 30 and then 60 ml of water were infused; 4-hourly aspirations produced only 50 ml in 4 hours, and adequate gastric emptying was therefore likely. The tube was then replaced by a fine-bore feeding tube and the feed was pumped continuously, initially at a rate of 30 ml per hour, increasing until 1600 ml per day was achieved. The patient was nursed in a semi-recumbent position to minimize reflux, and the feed given for 8-hour periods with a gap of up to 4 hours in between.

Example 2

A young man was admitted with acute asthma of such severity as to require ventilation. It seemed likely from previous experience that this would be necessary for only 24–48 hours before his bronchospasm responded to medical treatment. It was likely, therefore, that he would be back on a normal oral intake within a few days, and no nutritional support needed to be considered.

Example 3

A woman of 23 with a 2-year history of ulcerative colitis developed a severe exacerbation, during which she lost 15 per cent of her body weight and failed to respond to medical treatment. Abdominal distension and discomfort was observed, with continuing blood-stained diarrhoea. It was considered that she was in danger of perforation, and, rather than giving preoperative nutritional support, she was taken to the operating theatre for emergency colectomy, with the intention of giving intravenous feeding postoperatively. However, on the second postoperative day she began to take fluids by mouth and appeared well and apyrexial. It was therefore decided to monitor her oral intake over the next few days and to ensure that it reached adequate level, not only for maintenance but also to restore lost body tissue. Full restoration of lean body mass and full restoration of physical fitness were likely to take between 2 and 3 months.

Example 4

A man of 28, weighing 65 kg and previously fit, suffered 40 per cent full-thickness burns and inhaled toxic fumes, resulting in respiratory failure shortly after admission. Ventilatory support was necessary for a period of 3 weeks before recovery, and for the first week an ileus was present. In view of his catabolic state and the likelihood that he would be unable to take food by mouth for more than 10 days, no time was wasted following the shock phase in starting nutritional support intravenously. From the formulae in Table 43.1, his basal metabolic rate was calculated as 7000 kJ (1665 kcal) per day. A stress factor of 50 per cent was added for his burns and 10 per cent for the specific dynamic action of the feed; 15 per cent was subtracted for being on a ventilator, giving a metabolic expenditure of 10 130 kJ (2424 kcal) per day. The nitrogen intake would need to be at the upper end of the range, at 0.3 g/kg per day – which equalled a total of 19.5 g of nitrogen. Because he had an ileus, intravenous feeding was necessary, and a central line was passed. Using the prescription in Table 43.2 as a basis, 4180 kJ (1000 kcal) could be found from 500 ml of 20 per cent fat emulsion, the nitrogen intake from 1.5 litres of the amino acid solution gave 21 g of nitrogen and an additional 2110 J (504 cal), while the 1 litre of 26 per cent glucose would make up the total to 10 500 kJ (2504 kcal), which was near enough to the estimated target.

Despite an apparently adequate intake being maintained for the 3 weeks in the intensive care unit, the patient was returned to the burns unit suffering from muscle wasting and weakness. This was mistakenly assumed to be metabolic in origin, whereas it was almost certainly mechanical and due to being paralysed on a ventilator for 3 weeks. An attempt was made to reverse this by a high nitrogen nasogastric feed – apparently given with insufficient water, for the patient developed a rising blood urea and a hyperosmolar syndrome. This resolved rapidly when the nitrogen intake was reduced and the water intake increased. Adequate oral intake was resumed soon after this, and the patient made an uninterrupted recovery.

Example 5

A 70-year-old woman of normal weight (45 kg) had lost 10 kg in weight following a laparotomy, complicated by intra-abdominal sepsis and prolonged inability to eat. For convenience, she was given the standard intravenous feed shown in Table 43.2, but in the middle of the night developed shortness of breath which was mistakenly diagnosed as pulmonary embolus. The diagnosis became clear when the appropriate calculation showed that this patient was receiving 290 kJ/kg (69 kcal/kg), more than twice her estimated requirement. Apart from the effect of this caloric load on increasing her metabolic rate, she was receiving glucose 5.2 mg/kg per minute. The combined effect of these was to so increase the demands for gas exchange that, with her weakened respiratory muscles, the sensation of shortness of breath became intolerable. The moral of this story is that the calculation should be made at the beginning.

Example 6

A 60-year-old woman of normal weight (50 kg) developed bowel symptoms, losing 3 kg in weight. A sigmoid carcinoma was resected but anastomotic breakdown led to postoperative intra-abdominal sepsis and colonic fistula. Three weeks postoperatively she had received no food by mouth and only dextrose and salt solutions intravenously. Her weight had fallen to 46 kg, but she was noted to be extremely oedematous with a serum albumin of 18 g/l. From this picture and clinical evidence it was concluded that she had a reduced plasma volume but an expanded interstitial fluid volume. It was also estimated that, with the oedema, her true weight was near 40 kg, giving a total weight loss of 20 per cent. Accordingly, salt-free albumin and frusemide (furosemide) were given intravenously, a central venous catheter was inserted and a low salt feed begun. Her basal metabolic rate was calculated, on an estimated real weight of 40 kg, as 4250 kJ (1016 kcal) per day. She was bed-bound and awake with a raised temperature of 38°C. Another 20 per cent was therefore added for these factors and a further 10 per cent for the specific dynamic action of her feed. Although the feed needed to be introduced cautiously, it was necessary to add a further 20 per cent to allow for positive balance and weight gain. In view of the very obvious protein deficit, a nitrogen intake of just over 0.2 g/kg was given. The initial prescription consisted of 2 litres of salt-free solution containing 12 g of nitrogen, 4180 J (1000 cal) from fat emulsion and 2500 J (600 cal) from glucose. During the first 10 days there was a steady fall in weight and rise in serum albumin, as the excess salt and water load was excreted and fluid was redistributed. Subsequently, after the sepsis had been treated and the patient mobilized, the feed was continued and resulted in a continuous and steady real tissue weight gain of 1 kg per week. After 4 weeks, the fistula had resolved and the patient was able to take sufficient by mouth, allowing the central venous catheter to be withdrawn.

Summary and conclusions

Evidence has been put forward that progressive weight loss is associated with increased morbidity and mortality from disease. This becomes highly significant beyond 10 per cent weight loss. Trials of efficacy show that benefits from nutritional support can be demonstrated, resulting in improved clinical outcome. A practical approach to nutritional assessment is the use of changes in weight, simple anthropometric measures and voluntary food intake. With such measurements and a knowledge of the natural history of the disease, a rational decision can be made when to start nutritional support, which is then carried out in the simplest and most effective way possible. With adequate training, protocols and care, enteral and parenteral feeding can be carried out, in the perioperative period or in the intensive care unit, effectively and without undue complications or excessive cost. When things go wrong, it is not usually because of failure to understand the complexities but because the simple rules have not been observed.

REFERENCES

1. Benedict FG. *A study of prolonged fasting,* Carnegie Inst Publ 203. Washington DC: Carnegie Institute, 1915.

2. Love G. Weight loss during hunger strike. 1992 (Personal communication).

3. Allison SP. The uses and limitations of nutritional support. *Clinical Nutrition* 1992; **11**: 319–30.

4. Kinney JM. The influence of calorie and nitrogen balance on weight loss. *British Journal of Clinical Practice* 1988; **42**: suppl 63, 114–20.

5. Keys A, Brozek J, Henschel A. *The biology of human starvation.* Minneapolis: University of Minnesota Press, 1988.

6. McRussel DR, Pendergast PJ, Darby PL, *et al.* A comparison between muscle function and body composition in anorexia nervosa: the effect of refeeding. *American Journal of Clinical Nutrition* 1983; **38**: 229–37.

7. Russell DM, Walker PM, Leiter LA, *et al.* Metabolic and structural changes in muscle during hypocaloric dieting. *American Journal of Clinical Nutrition* 1984; **39**: 503–13.

8. Pichard C, Jeejeebhoy KN. Muscle dysfunction in malnourished patients. *Quarterly Journal of Medicine* 1988; **260**: 1021–45.

9. Church JM, Choong SY, Hill GL. Abnormalities of muscle metabolism and histology in malnourished patients awaiting surgery: effect of a course of intravenous nutrition. *British Journal of Surgery* 1984; **71**: 563–9.

10. Hill GL. *Disorders of nutrition and metabolism in clinical surgery.* Edinburgh: Churchill Livingstone, 1992.

11. Arora NS, Rochester DF. Effect of body weight and muscularity on human diaphragm muscle mass, thickness and area. *Journal of Applied Physiology* 1982; **52**: 64–70.

12. Kelsen SG, Ference M, Kapoor S. Effects of prolonged undernutrition on structure and function of the diaphragm. *Journal of Applied Physiology* 1985; **58**: 1354–9.

13. Lewis MI, Sieck GC, Fournier M, Belman MJ. Effect of nutritional deprivation on diaphragm contractility and muscle fibre size. *Journal of Applied Physiology* 1986; **60**: 596–603.

14. Lewis MI, Sieck GC. Effect of acute nutritional deprivation on diaphragm structure and function. *Journal of Applied Physiology* 1990; **68**: 1938–44.

15. Doekel Jr RC, Zwillich CW, Scroggin CH. Clinical semi-starvation: depression of hypoxic ventilatory response. *New England Journal of Medicine* 1976; **295**: 358–61.

16. Kelly SM, Rosa A, Field S, *et al.* Inspiratory muscle strength and body composition in patients receiving total parenteral nutrition therapy. *American Review of Respiratory Disease* 1984; **130**: 33–7.

17. Bassili HR, Deitel M. Effect of nutritional support on weaning patients off mechanical ventilation. *Journal of Parenteral and Enteral Nutrition* 1981; **5**: 161–3.

18. Jeejeebhoy KN. Bulk or bounce – the object of nutritional support. *Journal of Parenteral and Gut Nutrition* 1988; **12**: 539–49.

19. Murciano D, Armengauk MH, Rigand D, *et al.* Effect of renutrition on respiratory and diaphragmatic function in patients with severe mental anorexia. *American Review of Respiratory Disease* 1990; **141**: A547.

20. Bastow MD, Rawlings J, Allison SP. Benefits of supplementary tube feeding after fractured neck of femur: a randomised controlled trial. *British Medical Journal* 1983; **287**: 1589–92.

21. Winick M, ed. *Hunger disease: studies by the Jewish physicians in the Warsaw ghetto.* New York: Wiley, 1979.

22. Shizgal HM. Nutrition and immune function. *Surgery Annual* 1981; **12**: 15–29.

23. Chandra RK. Immunity and infection. In: Kinney JM, Jeejeebhoy KN, Hill GL, Owen OE, eds. *Nutrition and metabolism in patient care.* Philadelphia: WB Saunders, 1988: 598–604.

24. Windsor JA, Hill GL. Risk factors for postoperative pneumonia: the importance of protein depletion. *Annals of Surgery* 1988; **208**: 209–14.

25. Deitch EA, Winterton J, Li M, Berg R. The gut as a portal of entry for bacteremia: role of protein malnutrition. *Annals of Surgery* 1987; **205**: 681–90.

26. Berg RD. Translocation of indigenous bacteria from the intestinal tract. In: Hentges DJ, ed. *Human intestinal microflora in health and disease.* New York: Academic Press, 1983.

27. Fiddian-Green RG. Studies in splanchnic ischaemia and multiple organ failure. In: Marston A, Bulkley G, Fiddian-Green RG, Haglund U, eds. *Splanchnic ischaemia and multiple organ failure.* London, Melbourne, Auckland: Edward Arnold, 1989: 339–48.

28. Alexander JW. Nutrition and translocation. *Journal of Parenteral and Enteral Nutrition* 1990; **14**: suppl, 170–4S.

29. Levine GM, Deren JJ, Steiger E, Zinno R. Role of oral intake in maintenance of gut mass and disaccharidase activity. *Gastroenterology* 1974; **67**: 975–82.

30. Burke DJ, Alverdy JC, Aoys E, Moss GS. Glutamine supplemented TPN improves gut immune function. *Archives of Surgery* 1989; **124**: 1396–9.

31. Souba WW, Henskowitz K, Sallourn RM, Chen MK, Austgen TR. Gut glutamine metabolism. *Journal of Parenteral and Enteral Nutrition* 1990; **14**: suppl, 45–50S.

32. Alverdy JC, Aoys E, Moss GS. TPN promotes bacterial translocation from the gut. *Surgery* 1988; **104**: 185–90.

33. Moore F, Feliciano D, Andrassy R, *et al.* Enteral feeding reduces postoperative septic complications. *Journal of Parenteral and Enteral Nutrition* 1991; **15**: suppl, 22S.

34. Little RA, Stoner HB. Effect of injury on the reflex control of pulse rate in man. *Circulatory Shock* 1983; **10**: 161–71.

35. Macdonald IA, Bennett T, Sainsbury R. The effect of a 48 hour fast on the thermoregulatory response to graded cooling in man. *Clinical Science* 1984; **67**: 445–52.

36. Fellows IW, Macdonald IA, Bennett T, Allison SP. The effect of undernutrition on thermoregulation in the elderly. *Clinical Science* 1985; **69**: 215–22.

37. Mansell PI, Fellows IW, Macdonald IA, Allison SP. Restoration of normal thermoregulation following weight gain in undernourished patients. *Quarterly Journal of Medicine* 1990; **76**: 817–29.

38. Studley HO. Percentage of weight loss, a basic indication of surgical risk in patients with chronic peptic ulcers. *Journal of the American Medical Association* 1936; **106**: 458–60.

39. Windsor JA, Hill GL. Weight loss with physiologic impairment – a basic indication of surgical risk. *Annals of Surgery* 1988; **207**: 290–6.

40. Windsor JA, Knight GS, Hill GL. Wound healing response in surgical patients: recent food intake is more important than nutritional status. *British Journal of Surgery* 1988; **75**: 135–7.

41. Christiansen T, Kehlet H. Postoperative fatigue and changes in nutritional status. *British Journal of Surgery* 1984; **71**: 473.

42. Schroeder DS, Hill GL. Postoperative fatigue; a prospective physiological study of patients undergoing major abdominal surgery. *Australian and New Zealand Journal of Surgery* 1991; **61**: 774–9.

43. Wilkinson AW, Billing BH, Nagy C, *et al.* Excretion of chloride and sodium after surgical operations. *Lancet* 1948; **1**: 640.

44. Moore FD. *Metabolic care of the surgical patient.* Philadelphia: WB Saunders, 1959.

45. Starker PM, Lasala PA, Askanazi J, *et al.* The influence of preoperative total parenteral nutrition upon morbidity and mortality. *Surgery, Gynecology and Obstetrics* 1986; **162**: 569–74.

46. Clark RG. Postoperative water and sodium metabolism. In: Little RA, Frayn KN, eds. *The scientific basis for the care of the critically ill.* Manchester: Manchester University Press, 1986.

47. Sitges-Serra A, Arcas G, Guirao X, Garcia-Domingo M, Gil MH. Extracellular fluid expansion during parenteral refeeding. *Clinical Nutrition* 1992; **11**: 63–9.

48. Schønheyder F, Heilskov NSC, Olesen K. Isotopic studies on mechanism of negative nitrogen balance. *Scandinavian Journal of Clinical and Laboratory Investigation* 1954; **6**: 178–88.

49. Hill GL, Beddore AH. Dimensions of the human body and its compartments. In: Kinney JM, Jeejeebhoy KN, Hill GL, Owen OE, eds. *Nutrition and metabolism in patient care.* Philadelphia: WB Saunders, 1988: 89–118.

50. Lehmann AB, Bassey EJ, Morgan K, Dalloso HM. Normal values for weight, skeletal size and body mass indices in 890 men and women aged over 65 years. *Clinical Nutrition* 1991; **10**: 18–22.

51. Hessov I. Detecting deficient energy and protein intake in hospital patients: a simple recording method. *British Medical Journal* 1988; **1**: 1667–8.

52. Hackett AF, Yeung CK, Hill GL. Eating patterns in patients recovering from major surgery: a study in voluntary food intake and energy balance. *British Journal of Surgery* 1979; **66**: 415–18.

53. Meguid MM, Campos ACL, Meguid V, Debonis D, Terz JJ. IONIP, a criterion of surgical outcome and patient selection for perioperative nutritional support. *British Journal of Clinical Practice* 1988; suppl 63: 8–14.

54. Durnin JVGA, Womersley J. Body fat assessed from body density and its estimation from skinfold thickness: measurements on 481 men and women aged from 16 to 72 years. *British Journal of Nutrition* 1974; **32**: 77–97.

55. Collins JP, McCarthy ID, Hill GL. Assessment of protein nutrition in surgical patients – the value of anthropometrics. *American Journal of Clinical Nutrition* 1979; **32**: 1527–30.

56. Bastow MD. Anthropometrics revisited. *Proceedings of the Nutrition Society* 1982; **41**: 381.

57. Elia M. Body composition analysis: an evaluation of 2 component models, multicomponent models and bedside techniques. *Clinical Nutrition* 1992; **11**: 114–27.

58. Fleck A. Plasma proteins as nutritional indicators in the perioperative period. *British Journal of Clinical Practice* 1988; **42**: suppl 63, 20–4.

59. Clark RG, Karatzas T. Preoperative nutritional status. *British Journal of Clinical Practice* 1988; **42**: suppl 63, 2–7.

60. Harris JA, Benedict FG. *A biometric study of basal metabolism in man.* Carnegie Inst Publ 279. Washington DC: Carnegie Institute, 1919.

61. Schofield WN. Predicting basal metabolic rate, new standards and review of previous work. *Human Nutrition: Clinical Nutrition* 1985; **39C**: suppl 1.5, 41.

62. Shaw-Delanty SN, Elwyn DH, Askanazi J, Iles M, Schwarz Y, Kinney JM. Resting metabolic expenditure in injured septic and malnourished adult patients on intravenous diets. *Clinical Nutrition* 1990; **9**: 305–12.

63. Askanazi J, Carpentier YA, Elwyn DH, *et al.* Influence of total parenteral nutrition on fuel utilisation in injury and sepsis. *Annals of Surgery* 1980; **191**: 40–6.

64. Koretz RL. Is perioperative support of demonstrated value? Negative. In: Gitnick G, ed. *Controversies in gastroenterology.* Edinburgh: Churchill Livingstone, 1984: 253–61.

65. Koretz RL. Nutritional support: how much for how much? *Gut* 1986; **27**: suppl 1, 85–95.

66. Detzky AS, Baker JP, O'Rourke K, Goel V. Perioperative parenteral nutrition; a meta analysis. *Annals of Internal Medicine* 1987; **107**: 195–203.

67. Delmi M, Rapin C-H, Bengoa JM, Delmas PD, Vasey H, Bonjour JP. Dietary supplementation in elderly patients with fractured neck of femur. *Lancet* 1990; **335**: 1013–16.

68. Larsson J, Unosson M, Ek A-C, Nilsson L, Thorslund S, Bjurulf P. Effect of dietary supplement on nutritional status and clinical outcome in 501 geriatric patients – a randomised study. *Clinical Nutrition* 1990; **9**: 179–84.

69. Ek A-C, Unosson M, Larsson J, von Schenck H, Bjurulf P. The development and healing of pressure sores related to the nutritional state. *Clinical Nutrition* 1991; **10**: 245–50.

70. Rana SK, Bray J, Menzies Gow N, *et al.* Short term benefits of postoperative oral dietary supplements in surgical patients. *Clinical Nutrition* 1992; **11**: 337–44.

71. Veterans Affairs Total Parenteral Nutrition Study Group. Perioperative total parenteral nutrition in surgical patients. *New England Journal of Medicine* 1991; **325**: 525–32.

72. Von Meyenfeldt MF, Meijerink WJHJ, Rouflart MMJ, Buil-Maassen MTHJ, Soeters PB. Perioperative nutritional support: a randomised clinical trial. *Clinical Nutrition* 1992; **11**: 180–6.

73. Elia M. Artificial nutritional support in clinical practice. *Journal of the Royal College of Physicians, London* 1993; **27**: 8–15.

74. Lennard-Jones JE, ed. *A positive approach to nutrition as treatment.* King's Fund Report. London: King's Fund Centre, 1992.

75. Wara P, Hessov I. Nutritional intake after colorectal surgery: a comparison of a traditional and a new postoperative regimen. *Clinical Nutrition* 1985; **4**: 225–8.

76. Hessov I. Oral feeding after uncomplicated abdominal surgery. *British Journal of Clinical Practice* 1988; **42**: suppl 63, 75–9.

77. Gauderer MWL, Ponsky JL, Izant RJ. Gastrostomy without laparotomy: a percutaneous endoscopic technique. *Journal of Pediatric Surgery* 1980; **15**: 872–5.

78. Ponsky JL, Gauderer MWL. Percutaneous endoscopic gastrostomy: a non-operative technique for feeding gastrostomy. *Gastrointestinal Endoscopy* 1981; **27**: 9–11.

79. Park RHR, Allison MC, Lang J, *et al.* Randomised comparison of percutaneous endoscopic gastrostomy and naso-gastric tube feeding in patients with persisting neurological dysphagia. *British Medical Journal* 1992; **304**: 1406–9.

80. Wicks C, Gimson A, Vlavianos P, *et al.* Assessment of the percutaneous endoscopic gastrostomy feeding tube as part of an integrated approach to enteral feeding. *Gut* 1992; **33**: 613–16.

81. Hull MA, Rawlings J, Murray FE, *et al.* An audit of outcome of long-term enteral nutrition using percutaneous endoscopic gastrostomy. *Lancet* 1993; **341**: 869–72.

82. Rombeau JL. Gastrostomy. In: Rombeau JL, Caldwell MD, Forlaw L, Guenter PA, eds. *Atlas of nutritional support techniques.* Boston and Toronto: Little, Brown, 1989: 107–82.

83. Silk DBA. Towards the optimisation of enteral nutrition. *Clinical Nutrition* 1987; **6**: 61–74.

84. Jacobs S, Chang RWS, Lee B, *et al.* Continuous enteral feeding: a major cause of pneumonia among ventilated intensive care patients. *Journal of Parenteral and Enteral Nutrition* 1990; **14**: 353–6.

85. Bastow MD, Rawlings J, Allison SP. Overnight nasogastric tube feeding. *Clinical Nutrition* 1985; **4**: 7–11.

86. Bastow MD. Complications of enteral nutrition. *Gut* 1986; **27**: suppl 1, 51–5.

87. Maynard ND, Bikari DJ. Postoperative feeding. *British Medical Journal* 1991; **303**: 1007–8.

88. Ibanez J, Penafiel A, Raurich JM, Marse P, Jorda R, Mata F. Gastroesophageal reflux in intubated patients receiving enteral nutrition: effect of supine and semi-recumbent positions. *Journal of Parenteral and Enteral Nutrition* 1992; **16**: 419–28.

89. Strong RM, Condon SC, Solinger MR, Namihas N, Ito-Wong LA, Leuty JE. Equal aspiration rates from post-pylorus and intragastric placed small-bore nasoenteric feeding tubes: a randomized prospective study. *Journal of Parenteral and Enteral Nutrition* 1992; **16**: 59–63.

90. Mullan H, Roubenoff RA, Roubenoff R. Risk of pulmonary aspiration among patients receiving enteral nutrition support. *Journal of Parenteral and Enteral Nutrition* 1992; **16**: 160–4.

91. Bastow MD, Greaves P, Allison SP. Microbial contamination of enteral feeds. *Human Nutrition: Applied Nutrition* 1982; **36A**: 213–17.

92. Casewell MW. Nasogastric feeds as a source of Klebsiella infection for intensive care patients. *Research and Clinical Forums* 1979; **1**: 101–5.

93. Dudrick SJ, Wilmore DW, Vars HM, *et al.* Long term total parenteral nutrition with growth, development and positive nitrogen balance. *Surgery* 1968; **64**: 134–41.

94. Macfie J, Nordenstrom J. Full circle in parenteral nutrition. *Clinical Nutrition* 1992; **11**: 237–9.

95. Caldwell MD, Pomp A. Percutaneous central venous catheterization. In: *Atlas of nutritional support techniques.* Rombeau JL, Caldwell MD, Forlaw L, Guenter PA, eds. Boston and Toronto: Little, Brown, 1989: 193–299.

96. Clark-Christoff N, Watters VA, Sparks W, Snyder P, Grant JP. Use of triple-lumen subclavian catheters for administration of total parenteral nutrition. *Journal of Parenteral and Enteral Nutrition* 16: 403–7.

97. Solassol C, Joyeux H. Ambulatory parenteral nutrition. In: Fischer JE, ed. *Total parenteral nutrition.* Boston and Toronto: Little, Brown, 1976.

98. Greenberg GR, Marliss EB, Anderson GH, *et al.* Protein sparing therapy in postoperative patients. *New England Journal of Medicine* 1986; **294**: 1411–16.

99. Nordenstrom J, Askanazi J, Elwyn DH, *et al.* Nitrogen balance during total parenteral nutrition. *Annals of Surgery* 1983; **197**: 27–33.

100. Burke JF, Wolfe RR, Mullaney CJ, Mathews DE, Bier DM. Glucose requirements following burn injury. *Annals of Surgery* 1979; **190**: 274–85.

101. Hinton P, Allison SP, Littlejohn S, Lloyd J. Insulin and glucose to reduce catabolic response to injury in burned patients. *Lancet* 1971; **1**: 767.

102. Woolfson AMJ, Heatley RV, Allison SP. Insulin to inhibit protein catabolism after injury. *New England Journal of Medicine* 1979; **300**: 14.

103. Hinton P, Allison SP, Littlejohn S, Lloyd J. Electrolyte changes after burn injury and the effect of treatment. *Lancet* 1973; **1**: 218.

104. Woolfson AMJ. An improved method for blood glucose

control during nutritional support. *Journal of Parenteral and Enteral Nutrition* 1981; **5**: 436–40.

105. Shenkin A, Fell GS. Micronutrients. In: Woolfson AMJ, ed. *Biochemistry in hospital nutrition*. Edinburgh: Churchill Livingstone, 1986: 83–122.

106. Pennington CR. Parenteral nutrition: the management of complications. *Clinical Nutrition* 1991; **10**: 133–7.

107. Kehlet H. Modification of responses to surgery and anesthesia by neural blockade. In: Cousins MJ, Bridenhagh PO, eds. *Clinical anesthesia and management of pain*. Philadelphia: JB Lippincott, 1987.

108. Bessey PQ, Wilmore DW. The burned patient. In: Kinney JM, Jeejeebhoy KN, Hill GL, Own OE, eds. *Nutrition and metabolism in patient care*. Philadelphia: WB Saunders, 1988; 672–700.

109. Liljedahl SO, Gemzell CA, Plantin LO, *et al.* Effect of human growth hormone in patients with severe burns. *Chirurgica Scandinavica Acta* 1961; **122**: 1.

110. Manson, JM, Wilmore DW. Positive nitrogen balance with human growth hormone and hypocaloric intravenous feeding. *Surgery* 1986; **100**: 188–7.

111. Palmblad J. Intravenous lipid emulsion and host defense – a critical review. *Clinical Nutrition* 1991; **10**: 303–7.

112. Ziegler TR. L-Glutamine enriched parenteral nutrition in catabolic patients. *Clinical Nutrition* 1993; **12**: 65–6.

113. Moukarzel A, Goulet O, Cynober L, Ricour C. Positive effects of ornithine alpha-ketoglutarate in paediatric patients on parenteral nutrition and failure to thrive. *Clinical Nutrition* 1993; **12**: 59–60.

114. Donati L, Signorini M, Grappolini S. Ornithine alpha-ketoglutarate administration in burn injury. *Clinical Nutrition* 1993; **12**: 70–1.

Chapter 44

Pharmacology of Analgesia

T. Andrew Bowdle

Opioid receptor pharmacology
Relations between structure and activity
Therapeutic drug effects
Relations between dose and effect
Non-therapeutic drug effects
Drug interactions

Pharmacokinetics and pharmacodynamics
Partial agonists and agonist–antagonists
Antagonists
Peripheral mechanisms of analgesia: opioids and non-steroidal anti-inflammatory drugs

Among the remedies which it has pleased Almighty God to give to man to relieve his sufferings, none is so universal and so efficacious as opium. Thomas Sydenham, English physician (1680)

The neurophysiology of nociception is complicated. Many neuronal pathways and different neurotransmitters are involved (see Chapter 5). Despite this, a single large class of drugs, opioid analgesics, have been the mainstay of pain therapy since ancient physicians discovered the useful effects of the alkaloids derived from the opium poppy, *Papaver somniferum*. A specific opium alkaloid was first isolated from crude opium in 1806 and named morphine, after Morpheus, the Greek god of dreams. A major breakthrough in the understanding of opioid pharmacology occurred in 1973 with the simultaneous announcement from several laboratories of the discovery of stereospecific binding sites for opioids. The first endogenous opioid receptor ligands (neurotransmitters), the pentapeptides, met- and leuenkephalin, were identified in 1975. Subsequently, three families of endogenous opioid peptides have been identified (enkephalins, endorphins, dynorphins), and three distinct types of opioid receptors, designated μ, κ and δ; there is also evidence for subtypes of these receptors.

While our understanding of opioid drugs and the manner in which they affect the nervous system has grown tremendously in recent years, the molecular structure and function of the opioid receptors is not well understood. The molecular biology of opioid receptors is likely to emerge in the near future, as the receptor genes are cloned. This will undoubtedly result in modification of some of the current concepts about opioid receptors and their ligands.

The main purpose of this chapter is to present the clinical pharmacology of opioid drugs. While opioids are by far the most important analgesic drugs used in surgical patients, other classes of drugs are used as well, particularly the non-steroidal and anti-inflammatory drugs (NSAIDs). The NSAIDs are becoming more popular for treatment of postoperative pain, often in combination with opioids. Therefore the pharmacology of NSAIDs will also be included in this chapter.

There is some confusion about the terminology of drugs with morphine-like actions. The older term, opiate, continues to be used occasionally. However, the predominant practice is to use the term, opioid, to refer to all drugs with morphine-like effects. The term, narcotic, has no specific pharmacological meaning and should be avoided; derived from the Greek word for stupor, it has been applied to any drug that produces sleep. The term opioid will be used in this chapter.

Opioid receptor pharmacology

In order to understand opioid receptor pharmacology and particularly to understand partial agonist and agonist–antagonist opioids it is necessary first to define certain terminology and review relevant aspects of drug–receptor theory.

'Potency' refers to the quantity of drug required to produce a particular degree of receptor occupancy and is related to the affinity of the drug for the receptor. In clinical pharmacology potency is usually expressed in terms of the dose administered, often in mg or mg/kg; in molecular pharmacology potency is considered in terms of concentration. A reduction in potency shifts the dose–effect curve (to the right) but does not change its shape.

'Efficacy', or 'intrinsic activity' refers to the shape of the dose–effect curve and is related to the consequences of the molecular interaction between the drug and the receptor. An opioid has a spectrum of intrinsic activity ranging from a maximum (full agonist) to zero (antagonist). Agonists are capable of producing the maximum effect possible by binding to the receptor. Antagonists bind to the receptor but cause no direct effects. Between these two extremes are partial agonist drugs with intermediate activity that produce some effect but are incapable of producing the maximum effect of the full agonist.

Partial agonists produce a dose–effect curve that is less steep and has a lower maximum effect. A lower maximum effect does not necessarily imply a lack of potency; hypothetically, partial agonists may be quite potent, such that very small quantities of drug may produce a significant (albeit submaximal) effect.

An interesting and somewhat complicated phenomenon occurs when a full agonist and a partial agonist are applied together. If the concentration of the full agonist is low, introduction of the partial agonist results in an additive effect. However, if the concentration of the full agonist is relatively high, introduction of the partial agonist results in a net reduction in the effect. The opioid, nalorphine (*N*-allylnormorphine), an *N*-allyl substituted derivative of morphine, is an example of a partial agonist drug. It is a potent analgesic; yet, it is capable of antagonizing morphine.

When nalorphine was evaluated as a possible therapeutic agent, investigators realized that its pharmacological properties could not be explained entirely on the premise that it acted as a partial agonist at the morphine receptor. The subjective effects of nalorphine in human subjects differed from morphine, and the abstinence syndrome associated with nalorphine dependence was qualitatively different from the morphine abstinence syndrome. W.R. Martin proposed in 1967 that there must be multiple opioid receptors.[1] Subsequently, Martin postulated a family of opioid receptors consisting of three members: μ, standing for the prototype agonist, morphine; κ, from the prototype agonist ketocyclazocine; and σ, for the prototype agonist SKF 10,047 (*N*-allylnormetazocine). All opioid drugs were thought to have a variable degree of activity at one or more of these receptors.

Since 1967 an enormous amount of research has resulted in many changes to Martin's original proposal. However, the postulated existence of multiple opioid receptors has been confirmed. The opioid receptor family is now recognized to be extremely complex. Research in this area is very active and opioid receptor pharmacology undergoes frequent revision. Thus, the presentation below (also see Partial agonists and agonist–antagonists) is somewhat simplified and very much subject to modification.

Opioid receptors are currently classified into three major types based upon bioassays and binding studies, μ, κ, and δ; each receptor type also has subtypes. These receptors appear to exert inhibitory actions primarily by presynaptic modulation of the release of excitatory neurotransmitters. Their molecular mechanism of action is by coupling to G-proteins (guanine nucleotide-binding proteins) that ultimately affect the activity of various ion channels. All three types of receptors are involved in regulating nociception through extremely complex interrelations that are not well understood. There are probably additional types and subtypes of opioid receptors that are not currently identified.

The fentanyl analogues (e.g. fentanyl, sufentanil, alfentanil), morphine, and pethidine (meperidine) probably exert their clinically relevant effects mainly through μ-receptors. The agonist–antagonist opioids (e.g. pentazocine, nalbuphine, butorphanol) exert their effects through both μ- and κ-receptors. Additional discussion of opioid receptor pharmacology is found below in the section, Partial agonists and agonist–antagonists.

Relations between structure and activity

The structures of the opioid drugs discussed in this chapter are shown in Fig. 44.1. Morphine, a naturally occurring opium alkaloid, is the prototype compound. Many clinically useful opioid drugs are semisynthetic derivatives of morphine or thebaine, another natural opium alkaloid. Other opioid drugs, including the fentanyl analogues, are synthetic. Opioids can be classified based on their derivation from a basic parent structure. The major structural classes include: opium alkaloids (and their semisynthetic derivatives), morphinans, benzomorphans, methadone, and phenylpiperidines. Semisynthetic derivatives of morphine or thebaine of interest in anaesthesia include nalorphine (obsolete, but of great historical significance), nalbuphine, buprenorphine, naloxone and naltrexone. Butorphanol is a morphinan and pentazocine is a benzomorphan. Methadone is a unique structure, thus it constitutes a class of its own. Pethidine and the fentanyl analogues are phenylpiperidines.

Opioid receptors are stereoselective. The laevorotatory (−) isomer is the active enantiomer at opioid receptors. Dextrorotatory (+) enantiomers may have pharmacological effects (see 'σ-Receptor'), but they are not reversible by naloxone (not mediated by opioid receptors).

Modifications corresponding to the 3, 6, and 17 positions of morphine (see Fig. 44.1) are particularly important for the relations between structure and activity. Particular substituents on the tertiary amine (position 17) result in antagonist activity. The first opioid with significant antagonist activity was nalorphine (*N*-allylnormorphine),

MORPHINE OR THEBAINE DERIVATIVES

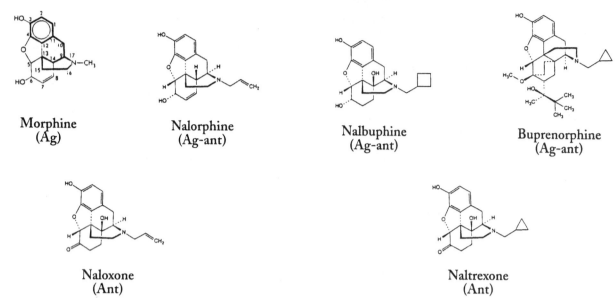

Morphine
(Ag)

Nalorphine
(Ag-ant)

Nalbuphine
(Ag-ant)

Buprenorphine
(Ag-ant)

Naloxone
(Ant)

Naltrexone
(Ant)

PHENYLPIPERIDINES

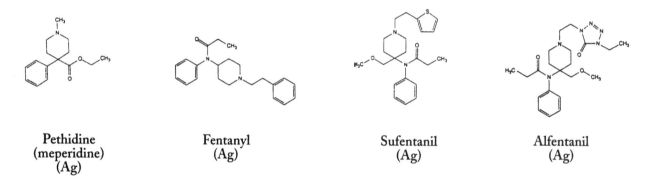

Pethidine
(meperidine)
(Ag)

Fentanyl
(Ag)

Sufentanil
(Ag)

Alfentanil
(Ag)

MORPHINAN

BENZOMORPHAN

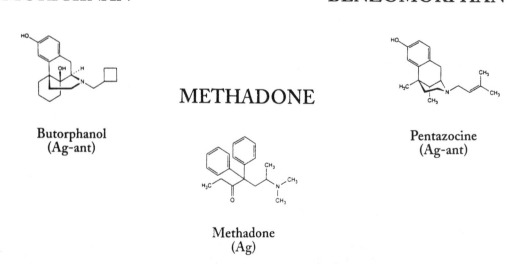

Butorphanol
(Ag-ant)

METHADONE

Pentazocine
(Ag-ant)

Methadone
(Ag)

Fig. 44.1 Chemical structures of opioids. Ag, agonist; Ag-ant, agonist–antagonist; Ant, antagonist.

the N-allyl analogue of morphine. Incorporation of N-allyl substituents in almost all series of synthetic opioids that contain three of the five rings present in morphine result in antagonist activity. N-Cyclopropylmethyl (naltrexone, buprenorphine) and N-cyclobutylmethyl (butorphanol, nalbuphine) substituents have also resulted in clinically useful compounds with agonist–antagonist or pure antagonist activity.

In general, the opioids in clinical use do not have an extremely high degree of opioid receptor specificity. However, morphine, pethidine, methadone and the fentanyl analogues probably interact primarily with μ-receptors. Sufentanil is a particularly potent μ-receptor agonist. Nalorphine, pentazocine, nalbuphine, and butorphanol are probably partial μ-agonists and partial κ-agonists (see Partial agonists and agonist–antagonists). Buprenorphine is a partial μ-agonist; it has activity at κ-receptors as well. Naloxone is a pure antagonist at μ-, κ- and δ-receptors, although it is most potent at μ-receptors.

Therapeutic drug effects

Endogenous opioid peptide neurotransmitters appear to exert subtle modulatory effects on the central nervous system under normal conditions. The administration of naloxone to normal subjects produces remarkably few noticeable effects. This is in dramatic contrast to the effects of administering therapeutic doses of opioid agonists. Opioid agonists produce analgesia, sedation, and in large doses, unconsciousness and anaesthesia.

The capability of opioids to produce general anaesthesia has been debated. It is useful to consider general anaesthesia in terms of its component parts: amnesia, analgesia, unconsciousness, immobility (absence of movement in response to a surgical stimulus), muscle relaxation (surgical exposure), and control of autonomic and endocrine responses to surgery. No currently existing intravenous anaesthetic agent provides all these features in every situation. Therefore, it would be inappropriate to recommend the routine use of any particular drug as a sole anaesthetic agent. However, opioids are capable of producing many of the features of general anaesthesia, especially analgesia, unconsciousness and control of autonomic and endocrine responses to surgery; from this perspective it is reasonable to regard opioids as anaesthetic agents.

There is clear evidence that opioids alone are capable of producing unconsciousness in humans (see relations between dose and effect). In rats, sufentanil virtually replaced halothane for preventing movement in response to a tail clamp.

The capability of opioids to produce amnesia in conscious subjects has not been well studied; generally the assumption has been that smaller doses of opioids do not effectively produce amnesia. It is interesting that butorphanol has been reported to have a small but significant anterograde amnesic effect.[2]

High doses of opioids blunt the neuroendocrine stress response to surgery; plasma concentrations of catecholamines, cortisol, ADH, glucose, insulin and growth hormone can be maintained at normal levels. However, the endocrine stress response to cardiopulmonary bypass is not completely controlled by opioids. Opioids are useful for reducing the hyperdynamic cardiovascular response to a surgical stimulus (see Cardiovascular effects).

Many anaesthetists use opioids in large doses as primary anaesthetic agents, in combination with other intravenous agents or inhaled agents. The term 'balanced anaesthesia' (as currently used) refers to the combination of substantial doses of opioids with inhaled anaesthetics.

Electroencephalographic effects

Opioids produce high voltage slow (δ) waves. Higher doses of opioids do not produce a progression to burst suppression and a flat EEG as with volatile anaesthetics. The EEG effects of opioids provide a convenient, continuous measurement of drug effect that has been exploited for numerous pharmacodynamic studies. When monitoring cerebral function with EEG (e.g. during carotid endarterectomy) it is important to realize that the appearance of slow waves produced by opioids may be indistinguishable from slowing produced by cerebral ischaemia; therefore it is advisable to avoid bolus doses of opioids at critical times (such as carotid cross-clamping) when confusion with cerebral ischaemia could arise.

Shivering

Postanaesthetic shaking or shivering that is not related to hypothermia can be effectively abolished by certain opioids. Pethidine is effective for this purpose. Morphine is not effective. There is evidence that butorphanol and tramadol are also effective. The reason that certain opioids are effective while others are ineffective is unknown.

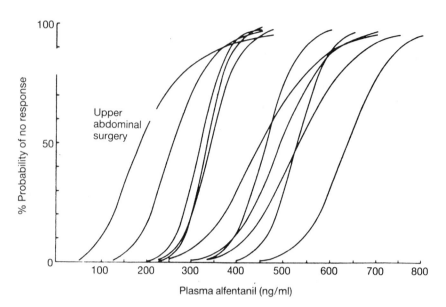

Fig. 44.2 Alfentanil plasma concentration vs effect curves for probability of not responding to a surgical stimulus during upper abdominal surgery. Anaesthesia consisted of alfentanil and 66 per cent nitrous oxide. Each curve represents one patient. The tremendous variation in plasma concentration required to prevent response to a surgical stimulus is evident. (From Ausems M, Hug C, Stanski D, Burm A. Plasma concentrations of alfentanil required to supplement nitrous oxide anesthesia for general surgery. *Anesthesiology* 1986; **65**: 362–73, with permission.)

Relations between dose and effect

The relations between dose and effect are extremely valuable for answering an important practical question in therapeutics: 'How much drug should be given in order to obtain a particular, desired effect?' Unfortunately high quality dose–effect data for humans are relatively scarce, probably because of the difficulty in performing clinical studies that typically require large numbers of patients and many data 'points'.

There is tremendous variability in opioid effects among individuals. Thus the dose of opioid must always be adjusted for each patient. Information from the literature can be a useful guide to dosing; however, there is no substitute for a thoughtful and observant anaesthetist. Figure 44.2 provides an example of the large variation in individual responses.[3]

McDonnell *et al.*[4] and Vinik *et al.*[5] have reported dose–effect studies of alfentanil for induction of anaesthesia. McDonnell *et al.* determined the ED_{50} and ED_{90} for unconsciousness in unpremedicated young adults using both loss of response to verbal commands and loss of response to a nasopharyngeal airway as endpoints to identify unconsciousness (Fig. 44.3). The ED_{50} and ED_{90} were 92 and 111 µg/kg respectively for loss of response to a verbal stimulus and 111 and 169 µg/kg respectively for loss of response to a nasopharyngeal airway. Several practical aspects of opioid clinical pharmacology are made evident by this study. An opioid in sufficient dose can produce unconsciousness when used as the sole agent for induction of anaesthesia. However, the clinical endpoint to identify unconsciousness during induction is not entirely clear. This

author's experience is that the loss of the 'eyelash reflex', commonly used to confirm unconsciousness during induction with barbiturates, is not a useful endpoint during induction with opioids. McDonnell *et al.* demonstrated that responsiveness to a verbal stimulus may not be an adequate endpoint during opioid induction and some more intense test stimulus may be necessary. This chapter author has occasionally seen patients who did not respond to verbal or gentle tactile stimuli during opioid induction, but opened their eyes when an oral airway was inserted; the practice of inserting an airway (or providing some other

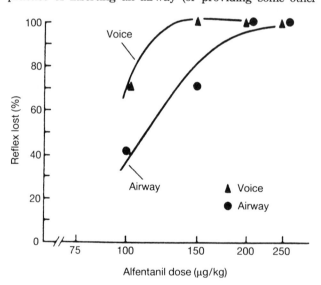

Fig. 44.3 Dose–effect curves for induction of anaesthesia with alfentanil in unpremedicated young adults. Alfentanil alone, in sufficient dose, produced unconsciousness. Absence of response to placement of a nasopharyngeal airway was a more reliable sign of unconsciousness than a verbal stimulus. (From McDonnell T, Bartowski R, Williams J. ED_{50} of alfentanil for induction of anesthesia in unpremedicated young adults. *Anesthesiology* 1984; **60**: 136–40, with permission.)

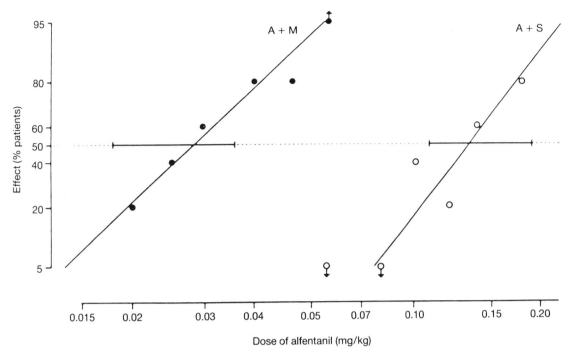

Fig. 44.4 Dose–effect curves for induction of anaesthesia with alfentanil (right side of figure) or alfentanil plus midazolam (left side of figure) in unpremedicated patients. Data are plotted on a log-log scale. Midazolam, 0.07 mg/kg, reduced the alfentanil ED_{50} for unconsciousness by about 80 per cent. (From Vinik H, Bradley E, Kissin I. Midazolam–alfentanil synergism for anesthetic induction in patients. *Anesthesia and Analgesia* 1989; **69**: 213–17, with permission.)

comparable stimulus) prior to the administration of an intubating dose of neuromuscular blocker may be advisable when opioids are the primary induction agent. Vinik *et al.* reported an ED_{50} for induction of anaesthesia with alfentanil of 130 µg/kg (Fig. 44.4), remarkably similar to that found by McDonnell *et al.*

Ausems *et al.*[3] have determined alfentanil plasma concentration–effect curves for patients anaesthetized with alfentanil/N_2O undergoing lower abdominal gynaeco-logical, upper abdominal or breast surgery (see Fig. 44.2).

Non-therapeutic drug effects

Depression of breathing

µ-Agonist opioids produce dose-related depression of breathing. Resting minute ventilation and tidal volume are decreased. Respiratory rate was found to be decreased in some studies, but not in others; slow breathing does occur, but a normal rate of breathing does not ensure that the central control of breathing is unaffected. The ventilatory responses to hypercapnia and hypoxia are blunted. Opioids may also alter the normal rhythm of breathing, resulting in irregular breathing and apnoeic

intervals. Sufficient doses of opioids produce complete apnoea, although apnoeic patients that remain conscious may continue to breathe on command.

Analgesic effects and depression of breathing caused by µ-agonists are generally assumed to be proportional. However, Bailey *et al.*[6] have suggested that sufentanil produces less depression of breathing than fentanyl, relative to analgesia. The existence of µ-receptor subtypes provides a hypothetical explanation (see µ-Receptor); µ-agonists having less activity at the µ-receptor subtype(s) mediating effects on breathing could produce less depres-sion of breathing relative to analgesia. This hypothesis remains to be tested.

The agonist–antagonist opioids are probably partial µ-agonists; therefore the maximum effects on breathing are expected to be less than full µ-agonists such as morphine. Nalorphine, nalbuphine and dezocine were found to have a 'ceiling' for depression of breathing equivalent to about 30 mg/70 kg of morphine. At lower doses, the depression of breathing was dose related. Higher doses produced no additional depression in the ventilatory response to hypercapnia. The pentazocine dose–effect curve was flatter than for morphine, but a plateau could not be demonstrated because intense dysphoria in the volunteers limited the maximum dose. The dose–effect curve for butorphanol was flatter than for morphine but doses large enough to establish a true ceiling were not studied. The effects of buprenorphine on breathing have not been fully characterized. An important clinical observation is that the depression of breathing caused by buprenorphine cannot always be antagonized by naloxone (see Partial agonists and agonist–antagonists); doxapram, an analeptic drug that stimulates ventilation, has been suggested as an alternative antidote for buprenorphine-induced depression of breathing. Very large doses of naloxone (5–10 mg) have also been tried and may be effective.

There are a number of case reports of severe, delayed depression in breathing following balanced anaesthesia with fentanyl analogues.[7] Neuromuscular blockers were adequately reversed in most cases and patients were awake and breathing spontaneously on arrival in the recovery room. After an interval of apparently adequate ventilation, ranging from minutes to about an hour, severe respiratory depression supervened. Prompt response to naloxone has confirmed the suspected relation to opioids in many cases. Because these events followed an interval of apparently adequate ventilation, the depression of breathing has often been described as delayed or recurrent. The mechanism is not known with certainty. Enterohepatic circulation of opioids sequestered in the stomach by acid trapping has been postulated to cause fluctuating plasma concentrations

in the postoperative setting. However, studies of fentanyl and alfentanil distribution in rats and other pharmacokinetic considerations have suggested that this is an unlikely mechanism of delayed depression of breathing. An appealing alternative is the possibility that opioids stored in 'peripheral compartments', such as muscle, are mobilized as patients begin to move about, resulting in transient increases in plasma concentrations. Delayed depression of breathing may not be the result of a single pharmacokinetic phenomenon, but rather the common endpoint of several possible mechanisms, including excessive opioid plasma concentrations at the time of extubation, the relatively short duration of action of naloxone (given to some patients at the time of extubation), drug interactions (e.g. benzodiazepines) or the onset of sleep (adding to the depression of breathing caused by opioids). Whatever the aetiology may be, recovery room personnel should be well aware of the possibility for severe depression of breathing in patients who initially appear to be breathing adequately.

Patients receiving opioids for postoperative analgesia on the surgical wards may also undergo episodes of depressed breathing and hypoxaemia. This problem has not been thoroughly studied, but available data suggest that postoperative hypoxaemia related to opioids is remarkably common.[8]

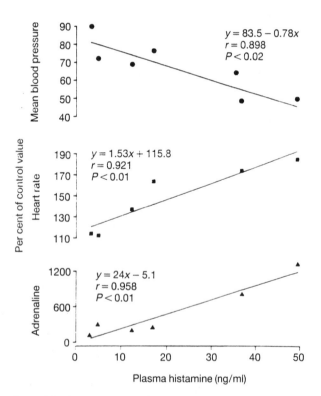

Fig. 44.5 Histamine release was stimulated by administration of morphine or pethidine, resulting in a reduction in blood pressure and elevation of heart rate and plasma adrenaline. The changes in vital signs and adrenaline levels were significantly correlated to plasma histamine concentration. (From Flacke J, Flacke W, Bloor B, Van Etten A, Kripke B. Histamine release by four narcotics: a double-blind study in humans. *Anesthesia and Analgesia* 1987; **66**: 723–30, with permission.)

Cardiovascular effects

The cardiovascular system is extraordinarily complex. While it is relatively simple to measure drug effects on vital signs, such as blood pressure and heart rate, determination of the mechanism of drug action can be very difficult. In general the effects of opioids on the cardiovascular system are not completely understood. Nevertheless, certain basic aspects of opioid cardiovascular pharmacology are understood well enough to be clinically useful.

Stimulation of histamine release by morphine or pethidine, but not by fentanyl analogues, can provoke significant hypotension and tachycardia.[9] Flacke *et al.* administered incremental, equipotent doses of pethidine (4.3 ± 0.1 mg/kg), morphine (0.6 ± 0.02 mg/kg), fentanyl (7.0 ± 0.4 μg/kg) or sufentanil (1.3 ± 0.1 μg/kg) for induction of anaesthesia. Subsequent plasma histamine levels were elevated in 31 per cent of patients receiving pethidine, 10 per cent of patients receiving morphine, and in no patients receiving fentanyl or sufentanil. Elevations of plasma histamine were significantly correlated with reduction in mean blood pressure, elevation of heart rate and elevation of circulating adrenaline (epinephrine) (Fig. 44.5). The relation between dose and effect for histamine release by pethidine or morphine are unknown, but slow administration of morphine, $\leq$ 5 mg/min, appears to minimize histamine release.

Opioids depress contractility of isolated heart muscle. However, the concentrations of morphine or fentanyl analogues necessary to produce myocardial depression are greatly in excess of concentrations achieved in patients. Thus, morphine and fentanyl are believed not to produce significant myocardial depression in the clinical setting. On the other hand, pethidine produces substantial myocardial depression at clinically relevant concentrations; pethidine was estimated to be 200 times more potent as a myocardial depressant than morphine or fentanyl. Severe hypotension limits the usefulness of pethidine as a primary anaesthetic agent. Pethidine, 2.5 mg/kg, produced severe cardiovascular depression in dogs; higher doses (e.g. 35 mg/kg) produced rapid cardiac arrest.

Morphine and the fentanyl analogues typically reduce heart rate by a vagomimetic action that is antagonized by atropine. The reduction in heart rate is variable, but severe bradycardia or even asystole is possible, especially with the fentanyl analogues (see Drug interactions). Reduction in heart rate induced by opioids may have a beneficial effect on the balance between coronary oxygen supply and demand in patients with coronary disease, provided that bradycardia does not result in severe hypotension; this author's opinion is that the heart rate lowering effects of the fentanyl analogues are among their most useful properties. Pethidine, paradoxically, tends to increase heart rate.

Fentanyl analogues often produce minor reductions in blood pressure and occasionally severe hypotension, due primarily to a reduction in systemic vascular resistance. Flacke et al.[10] have suggested that the mechanism may involve a centrally mediated reduction in sympathetic tone. They showed that fentanyl (100 µg/kg) produced no cardiovascular depression in dogs previously deprived of autonomic tone by a combination of vagal transection and subarachnoid block.[10] Flacke et al. also showed that the cardiovascular depressant effects of fentanyl in dogs were antagonized by naloxone and that clonidine prevented naloxone from reversing fentanyl's cardiovascular effects, presumably by decreasing sympathetic outflow from the central nervous system.[11] Severe hypotension following administration of fentanyl analogues can probably be avoided by paying close attention to the patient's volume status (opioids may be hazardous in hypovolaemic patients) and by avoiding certain drug interactions that exacerbate haemodynamic effects (see Drug interactions). Opioid-induced hypotension can usually be treated effectively by intravenous fluid and/or an α-adrenergic agonist drug (e.g. phenylephrine) to restore vascular tone.

While studies of the cardiovascular effects of opioids usually focus on cardiovascular depression, there is evidence of cardiovascular stimulation in some circumstances. Thomson et al.[12] reported that induction of anaesthesia with fentanyl, 50 µg/kg, resulted in substantial increases in heart rate, blood pressure, pulmonary artery pressure and cardiac index in some patients. It is interesting that certain agonist–antagonist opioids, notably pentazocine, produce cardiovascular stimulation more often than not (see below). The incidence of cardiovascular stimulation and the mechanism are unknown; this author has certainly observed such responses by patients in his own practice.

From the foregoing discussion it is evident that morphine and pethidine are more likely to produce cardiovascular problems than the fentanyl analogues. Pethidine in particular has potentially hazardous cardiovascular effects. It is interesting that the proclivity for cardiovascular depression is inversely related both to the potency and the therapeutic index of opioids. The therapeutic indices (LD_{50}/ED_{50}; ratio of lethal dose to effective dose) in rats are 4.8 for pethidine, 70 for morphine, 277 for fentanyl, and 26 700 for sufentanil; potency is in the same rank order (sufentanil is most potent).

Cardiovascular effects of agonist–antagonist opioids

The cardiovascular effects of the agonist–antagonist opioids have not been studied nearly as extensively as the full µ-agonists discussed above. The older agonist–antagonist drugs, nalbuphine and pentazocine, have considerable cardiovascular stimulating effects; pentazocine produces increases in blood pressure, heart rate, systemic vascular resistance, pulmonary artery pressure and left ventricular end-diastolic pressure. The newer agents generally appear to have relatively minor cardiovascular effects (except meptazinol and dezocine; see below). Butorphanol has been shown in some studies to cause a small increase in pulmonary artery pressure (reminiscent of pentazocine); however, other studies found no change in pulmonary pressures.

Meptazinol appeared to produce only minor haemodynamic effects when given in analgesic doses. Larger doses of meptazinol, 2, 3, or 4 mg/kg, given to a small number of patients anaesthetized with etomidate, produced substantial reductions in heart rate and blood pressure. Pulmonary artery pressures were unaffected. These results suggest that the haemodynamic effects of meptazinol may be of clinical importance. Additional studies are warranted.

Dezocine produced serious cardiovascular depression in dogs. Marked cardiovascular effects are generally not found in humans given analgesic doses of dezocine. Additional studies are needed to elucidate the mechanism of cardiovascular depression in dogs and to determine if similar effects are possible in humans.

Rigidity and other neuroexcitatory effects

Muscle rigidity occurs frequently during induction of anaesthesia with larger doses of opioids. A simple experiment convincingly demonstrates that opioid rigidity is central in origin. After a tourniquet was applied to the arm, preventing the entry of drugs, administration of alfentanil produced rigidity in the entire body, including the isolated extremity. Studies in rodents suggest that opioid rigidity is mediated by µ-receptors in brainstem midline nuclei; a naloxone analogue applied by microinjection to these brain areas antagonized alfentanil induced rigidity.[13] The basal ganglia have also been implicated in opioid rigidity.

Because opioid rigidity often prevents ventilation of the lungs with a bag and a mask, opioid rigidity was traditionally associated with the chest wall muscles. However, Benthuysen et al.[14] have shown that opioid rigidity affects not only the chest muscles, but virtually all the major muscle groups in the body. They quantified rigidity during induction of anaesthesia with alfentanil, 175 µg/kg, by recording electromyograms (EMG) from sterno-cleidomastoid, deltoid, biceps, forearm flexors, intercostal, rectus abdominus, vastus medialis/lateralis and gastrocnemius muscles. Rigidity occurred in all the muscles, beginning first in the upper body (sternocleidomastoid, deltoid, biceps, forearm). Rigidity was often sudden in onset and could be provoked by stimulation, such as passive movement of an extremity, manipulation of the anaesthesia mask, or a loud sound. Stereotyped postures were noted: flexion of the upper extremity, extension of the lower extremity, rigid immobility of the head, flexion of the neck with chin on chest or severe rigidity of the abdomen and chest wall.

Opioid-induced rigidity is usually observed during induction of anaesthesia; however, there are case reports of rigidity occurring either immediately upon emergence from anaesthesia, or delayed until 3–5 hours postoperatively.[15] Naloxone administration dramatically relieved the rigidity in those cases where it was tried, implicating an opioid mechanism. The explanation for delayed rigidity is unknown, much as the explanation for delayed depression of breathing from opioids is unknown.

Currently, the only reliable treatment for opioid-induced rigidity is the administration of neuromuscular blocking drugs, in doses large enough to facilitate intubation. A very small dose of pancuronium, 1.5 mg/70 kg (e.g. for defasciculation), was reported to attenuate rigidity; however, it is difficult to explain how a dose of neuromuscular blocker that is generally too small to prevent voluntary movement could significantly affect the considerable motor activity necessary to produce rigidity. Another study using electromyography to quantify rigidity found that small doses of neuromuscular blockers did not significantly attenuate rigidity. Considering the neuroanatomical basis of opioid rigidity, drugs that act on GABAergic, serotonergic or adrenergic pathways might be expected to affect rigidity. The serotonin receptor antagonist, ketanserin, and the α_2-adrenergic agonist, dexmedetomidine, prevented opioid-induced rigidity in rats. Benzodiazepines and thiopentone (thiopental) (GABAergic mechanisms of action) have also been reported to attenuate rigidity in rats and humans, although this finding has not been consistent.

Opioids have been associated with tonic–clonic movements, or myoclonus, in addition to tonic rigidity. Opioid-induced myclonus may be quite dramatic, resembling generalized epileptic convulsions. Opioids in very large doses have produced seizure activity in laboratory animals. However, there appears to be no EEG evidence that opioids produce generalized, cortical seizure activity in humans, despite many studies of opioids in which EEG recordings have been made (with the exception of norpethidine (normeperidine), see below). To the contrary, myoclonus has been observed during simultaneous EEG recording, in the absence of EEG evidence of seizure activity. Fentanyl (mean dose 25 µg/kg) has been reported to produce localized temporal lobe electrical seizure activity in patients with complex partial epilepsy, but this was not accompanied by motor activity.[16] Thus it is tempting to speculate that the mechanism of opioid-induced myoclonus may be closely related to opioid-induced tonic rigidity.

The N-demethylated metabolite of pethidine (norpethidine) is capable of producing CNS excitation and generalized seizures, not antagonized by naloxone. Norpethidine is eliminated by the kidney and by hydrolysis to norpethidinic acid. The half-life of norpethdine is 15–40 hours, thus it may accumulate with prolonged pethidine administration. Patients with impaired renal function are probably at increased risk for norpethidine toxicity.

Cerebral blood flow and intracranial pressure

Until recently it was generally believed that opioids produced no change or modest reductions in cerebral blood flow and cerebral metabolic oxygen consumption. However, Milde et al.[17] reported that sufentanil caused substantial increases in cerebral blood flow in normocapnic dogs anaesthetized with small doses of halothane; for the most part the increase in cerebral blood flow was not explained by concomitant changes in cerebral oxygen consumption, implying that sufentanil was dilating cerebral vessels directly.

Sperry et al.[18] studied intracranial pressure in patients with severe head trauma. The patients were paralysed with vecuronium and their intracranial pressure was controlled using standard clinical procedures, including hyperventilation, elevation of the head, sedation with midazolam and osmotic agents. Intracranial pressure was monitored by a subarachnoid bolt device. Administration of fentanyl (3.0 μg/kg) or sufentanil (0.6 μg/kg) resulted in a mean increase in intracranial pressure of about 1.3 kPa (10 mmHg) (Fig. 44.6). Mean blood pressure declined by about 1.3 kPa (10 mmHg). The maximum increase in intracranial pressure occurred about 5 minutes after opioid administration and persisted for about 20 minutes.

The explanation for the apparent increase in cerebral blood flow or intracranial pressure in these studies is unknown. It is interesting that feline cerebrovascular smooth muscle was reported to dilate in response to direct microapplication of μ- or δ-agonists.[19] There is also some evidence for neuroexcitatory effects of opioids, including localized electrical (but not motor) seizure activity in neurosurgical patients (see Rigidity and other neuroexcitatory effects)[16] that could be associated with increases in regional cerebral blood flow.

The subject of opioid effects on the cerebral circulation has become very complex and controversial because of unexplained discrepancies between studies. Some recent studies have not found increased cerebral blood flow or increased intracranial pressure following opioid administration. The effects of opioids on the cerebral circulation may be highly variable depending on the circumstances. Until this issue is resolved, opioids should be used cautiously in patients with critically reduced intracranial compliance.

Gastrointestinal effects

Potency varies between particular drugs, but virtually all commonly used opioids produce gastrointestinal side effects, by a combination of central and peripheral actions. Intestinal motility is generally reduced and constipation can be a problem. Increased tone of the sphincter of Oddi can produce increased pressure in the biliary ducts, occasionally producing pain. Opioid-induced spasm of the sphincter may prevent radiographic dye from entering the duodenum during intraoperative cholangiography, simulating the presence of a gallstone in the common bile duct. Some anaesthetists prefer to avoid opioid balanced anaesthesia for cholecystectomy because of the possibility of interfering with the cholangiogram. However, this is a relatively uncommon problem[20] and may be resolved by administering glucagon (alternatively, naloxone or nalbuphine) which relaxes the sphincter of Oddi. Obstruction of the common bile duct that persists after glucagon administration is unlikely to be related to opioid effects.

Nausea and vomiting

Opioids commonly produce nausea and vomiting. The neurophysiology and pharmacology of nausea and vomiting are very complex. The vomiting centre in the medulla receives input from the cerebral cortex, the gut, the vestibular system and the chemoreceptor trigger zone. The chemoreceptor trigger zone is located in the floor of the fourth ventricle, outside the blood–brain barrier. The chemoreceptor trigger zone contains many types of receptors, including opioid receptors, that promote vomiting. Experiments in cats have suggested that opioids actually have an antiemetic effect on the vomiting centre that is reversible by naloxone. Thus opioids appear to have opposite effects in the vomiting centre (antiemetic) and the chemoreceptor trigger zone (emetic). Intracerebroventricular injection of naloxone prevented opioid-induced emesis, but intravenous administration of naloxone did not, suggesting that the vomiting centre was much more sensitive to intravenous naloxone than the chemoreceptor trigger zone. Intravenous naloxone appeared to reverse the antiemetic effects of opioids on the vomiting centre, but not the emetic effect of opioids on the chemoreceptor trigger zone, thereby promoting emesis; for example, fentanyl did not produce emesis in cats until they were pretreated with

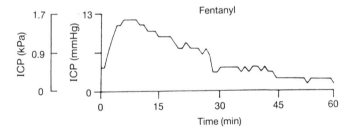

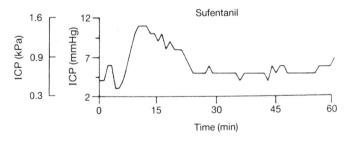

Fig. 44.6 Intracranial pressure was elevated significantly within minutes of administration of fentanyl (3 μg/kg) or sufentanil (0.6 μg/kg) to patients with severe head trauma. Each plot represents a single patient. (From Sperry R, Bailey P, Reichman M, Peterson J, Petersen P, Pace N. Fentanyl and sufentanil increases intracranial pressure in head trauma patients. *Anesthesiology* 1992; **77**: 416–20; with permission.)

intravenous naloxone.[21] This is consistent with clinical observations by Longnecker *et al.*[22] that naloxone frequently resulted in vomiting when administered to antagonize morphine.

Many remedies for postoperative, anaesthetic-related nausea and vomiting have been tried. Effective antiemetic drugs include antagonists of dopaminergic, histaminic, cholinergic or serotinergic receptors. Droperidol, a neuro-leptic drug with dopamine antagonist activity, is effective for reducing postoperative nausea when given prior to emergence from anaesthesia. More recently, transdermal hyoscine (scopolamine) (cholinergic antagonist), metoclo-pramide (significant effects at several receptor types) and ondansetron (serotonin antagonist) have also been shown to be effective. Whether a particular antiemetic drug is superior to another for prophylaxis or treatment of opioid-induced nausea and vomiting is not clear.

Some patients may experience significant differences between particular opioid drugs in the severity of nausea and vomiting; it may be reasonable to switch to an alternative opioid if nausea and vomiting are particularly troublesome.

Pruritis

Pruritis is a common opioid side effect, especially with neuraxial opioid administration. Pruritis is generally quite sensitive to very small doses of µ-antagonists (either naloxone or agonist–antagonist drugs such as nalbuphine); it may be possible to antagonize the pruritis selectively while preserving analgesia in many patients.

Drug interactions

Despite massive volumes of information documenting various drug interactions, most are clinically insignificant. However, there are several drug interactions between opioids and other anaesthetic agents of vital significance in clinical practice. The interactions of greatest impor-tance include those with intravenous induction agents, benzodiazepines, and neuromuscular blocking drugs.

Although most of the clinically significant opioid drug interactions apply broadly to all opioids, there appears to be a unique interaction between pethidine and monoamine oxidase inhibiter antidepressants (MAOIs). There are a number of case reports describing a syndrome of hyperpyrexia, hypotension and coma following administra-tion of pethidine to patients taking MAOIs; some fatalities have been recorded. The mechanism of this interaction is

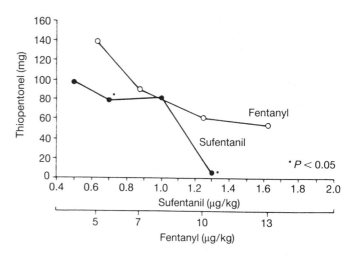

Fig. 44.7 Sufentanil or fentanyl boluses were administered over 1 minute. Three minutes later, thiopentone was administered in 25 mg increments every 30 seconds until the patient was unconscious. Sufentanil and fentanyl doses were plotted on an equipotent scale (potency ratio of 1:8, based on the EEG effects). There was an inverse relation between opioid dose and thiopentone requirement. (From Bowdle T, Ward R. Induction of anesthesia with small doses of sufentanil or fentanyl: dose versus EEG response, speed of onset and thiopental requirement. *Anesthesiology* 1989; **70**: 26–30, with permission.)

unknown; a synergistic effect on neuronal serotonin uptake has been postulated. The apparent rarity of this syndrome, in the face of widespread use of both pethidine and MAOIs, suggests an idiosyncratic mechanism (i.e. affecting only uniquely susceptible patients). Apparently no opioid drugs other than pethidine have been implicated.

Concomitant use of opioids generally reduces the dose of intravenous induction agent (e.g. thiopentone, etomidate, propofol) or benzodiazepine required to induce anaesthesia (produce unconsciousness). Bowdle and Ward[23] studied induction of anaesthesia with a moderate dose of sufentanil or fentanyl followed by incremental doses of thiopentone. They found an inverse relation between the dose of opioid and the dose of thiopentone required to produce uncon-sciousness (Fig. 44.7). The mean thiopentone requirement following sufentanil, 0.5 µg/kg, was 100 mg, but following sufentanil, 1.3 µg/kg, the mean thiopentone requirement was only 5 mg (9/10 subjects required no thiopentone; 1 subject required 50 mg).

Fentanyl appears to attenuate the usual sympathomi-metic effects of ketamine. Ketamine, 1.5 µg/kg, following fentanyl, 50 µg/kg, administered to patients undergoing coronary artery bypass surgery, produced no significant haemodynamic changes.

Vinik *et al.*[5] determined the effect of midazolam on the dose of alfentanil required for induction of anaesthesia and found a synergistic interaction (see Fig. 44.4). The ED_{50} (for unconsciousness) for alfentanil was 0.13 mg/kg without midazolam and 0.028 mg/kg after midazolam, 0.07 mg/kg, an 80 per cent reduction in dose requirement for alfentanil.

The interaction of opioids and benzodiazepines may also produce significant haemodynamic effects. Tomicheck *et al.*[24] studied the haemodynamic consequences of administration of diazepam, 0.125–0.5 mg/kg, followed by fentanyl, 50 µg/kg to patients undergoing coronary artery surgery. They found a dramatic reduction in mean blood pressure due entirely to reduction in systemic vascular resistance in patients receiving diazepam and fentanyl (Fig. 44.8). Haemodynamic changes were insignificant in control patients receiving only fentanyl. Diazepam alone, prior to fentanyl, also produced insignificant changes. Flacke *et al.*[10] have suggested that this interaction occurs as a consequence of a reduction in sympathetic tone; they found that the combination of fentanyl and diazepam did not produce cardiovascular depression in dogs previously deprived of autonomic tone. Some investigators have suggested that the interaction of benzodiazepines and opioids may produce myocardial depression; however, this has not been found consistently.

Interaction between opioids and benzodiazepines may also affect ventilation. Bailey *et al.*[25] administered midazolam (0.05 mg/kg) and fentanyl (2.0 µg/kg) to healthy, young volunteers, measured CO_2 responsiveness and oxygen saturation by pulse oximetry, and observed the pattern of breathing. They found that fentanyl without midazolam produced hypoxaemia in 6/12 of the subjects (breathing room air) and apnoea in no subjects. The addition of midazolam to fentanyl produced hypoxaemia in 11/12 subjects and apnoea in 6/12. Midazolam alone did not produce hypoxaemia or apnoea. The authors concluded that the interaction between an opioid and a benzodiazepine may produce an exceptionally high risk of hypoxaemia and apnoea and that patients receiving these drugs should be monitored appropriately.

The interaction of opioids and neuromuscular blockers has received considerable attention because of the effects on heart rate. The fentanyl analogues often produce reductions in heart rate, occasionally resulting in severe bradycardia or even asystole. Some anaesthetists favour the use of pancuronium to facilitate intubation, based on the premise that pancuronium tends to elevate heart rate, thereby opposing the heart rate lowering effect of the opioid. However, the combination of fentanyl, 100 µg/kg, and pancuronium, 0.1 mg/kg, produced ST segment depression in some patients undergoing coronary artery surgery, associated with significant increases in heart rate; no ST segment changes occurred in control groups receiving metocurine or a metocurine–pancuronium combination. The opinion of this author is that choice of a combination of opioid and muscle relaxant depends upon the clinical priority. If the priority is to avoid bradycardia, pancuronium is a reasonable choice. On the other hand, if the priority is to avoid increases in heart rate, an alternative

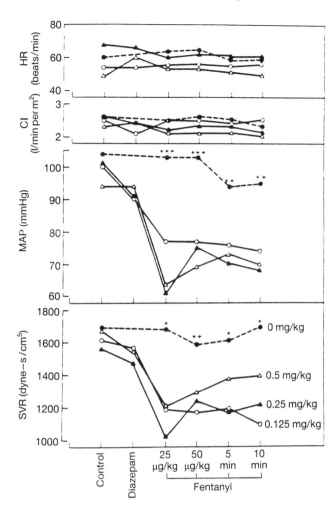

Fig. 44.8 Fentanyl or diazepam separately produced no significant changes in (from top of figure) heart rate (HR), cardiac index (CI), mean arterial pressure (MAP) or systemic vascular resistance (SVR) when administered to cardiac surgery patients for induction of anaesthesia. The combination of fentanyl and diazepam (0.5, 0.25 or 0.125 mg/kg) resulted in substantial reductions in mean arterial blood pressure and sytemic vascular resistance. (From Tomicheck R, Rosow C, Philbin D, Moss J, Teplick R, Schneider R. Diazepam–fentanyl interaction – hemodynamic and hormonal effects in coronary artery surgery. *Anesthesia and Analgesia* 1983; **62**: 881–4, with permission.)

muscle relaxant devoid of cardiac stimulating properties may be preferable.

An interesting interaction has been reported between epidural morphine and the local anaesthetic, 2-chloroprocaine. A test dose of either lignocaine (lidocaine) or 2-chloroprocaine was administered to confirm the proper placement of epidural catheters, prior to bupivacaine epidural anaesthesia for caesarean section. The duration of action of epidural morphine, given for postoperative pain control, was significantly shorter in patients receiving the 2-chloroprocaine test dose (16 hours in the 2-chloroprocaine groups vs 24 hours in the lignocaine group; $P < 0.05$).[26] Based upon these results, the authors recommended avoiding 2-chloroprocaine when administration of epidural opioids is contemplated. The mechanism of this interaction is unknown.

Epidural analgesia for labour with combinations of dilute local anaesthetics and opioids produced similar analgesia with less motor block, compared to more concentrated local anaesthetic solutions without opioids.

Pharmacokinetics and pharmacodynamics

Pharmacokinetics and pharmacodynamics constitute the study of the time course of drug disposition and the time course of drug effects, respectively. Clinicians are ultimately most interested in pharmacodynamics; however, pharmacokinetic events often have significant impact on the time course of drug effects. Therefore, opioid pharmacokinetics and pharmacodynamics must be considered together.

The two pharmacokinetic/pharmacodynamic questions of greatest importance for clinicians are: What is the time interval between drug administration and onset of effects? What is the duration of the effects?

Speed of onset

The precise explanation for the speed of onset of an intravenously administered drug is unknown. There are several factors that could be important, including molecular size, lipid solubility, plasma protein binding, tissue binding, and distributional phenomena such as first-pass tissue binding by the lung. Lipid solubility is probably quite

important for drugs that traverse lipid membranes to gain entry to the site of action; the blood–brain barrier contains the lipid membranes of significance for opioids. The degree of ionization is an important determinant of lipid solubility, ionized molecules being much less lipid soluble than unionized molecules. The opioids are weak bases with pK_a ranging from 6.5 (alfentanil) to 9.3 (methadone) (Table 44.1). Based on the pK_a, and assuming a blood pH of 7.4, the percentage of uncharged opioid molecules (per cent 'free base') can be calculated, ranging from 1.4 (methadone) to 90 per cent (alfentanil). Overall lipid solubility has often been determined from the distribution of a drug between a mixture of octanol (an organic solvent) and water. The ratio of the drug concentration in octanol to the drug concentration in water at equilibrium is the octanol:water partition coefficient; this coefficient is usually adjusted to reflect the degree of ionization at pH 7.4. The apparent octanol:water partition coefficient at pH 7.4 ranges from 1778 (sufentanil; most lipid soluble) to 1.4 (morphine; least lipid soluble). The propensity to cross lipid barriers and enter the CNS has been estimated by multiplying the octanol:water partition coefficient by the free fraction in plasma; this hybrid parameter ranges from 0.91 for morphine (less likely to penetrate CNS) to 130 and 123 for fentanyl and sufentanil, respectively (more likely to penetrate CNS). Based solely upon pH-adjusted lipid solubility and plasma protein binding, morphine should have the slowest onset of action and fentanyl and sufentanil the fastest. Indeed morphine has a very slow onset (see below); however, alfentanil, with intermediate lipid solubility, is considerably faster in onset compared to fentanyl or sufentanil. The explanation for this is not known with certainty, but several factors may be involved.

The lungs temporarily take up significant amounts of lipophilic, basic drugs, on the 'first pass' through the pulmonary venous circulation, and then release them back into the arterial circulation. The first-pass retention of

Table 44.1 Physical properties and pharmacokinetic data of opioids

	Morphine	Pethidine	Fentanyl	Sufentanil	Alfentanil	Methadone
pK_a	8.0	8.5	8.4	8.0	6.5	9.3
% unionized, ph 7.4	23	7	8.5	20	90	1.4
Octanol:water partition coefficient (pH = 7.4)	1.4	39	813	1778	145	116
% bound to plasma protein	35	70	84	93	92	89
V_{dss} (l/kg)	3–5	3–5	3–5	2.5–3.0	0.4–1.0	6.1
Clearance (ml/kg per min)	15–30	8–18	10–20	10–15	4–9	2.7
$t_{1/2\beta}$ (h)	2–4	3–5	2–4	2–3	1–2	35
Hepatic extraction ratio (E)	0.8–1.0	0.7–0.9	0.8–1.0	0.7–0.9	0.3–0.5	<0.2
Potency (in mg, relative to morphine, 10 mg)	10	75	0.10	0.010	1.0	10

V_{dss}, volume of distribution at steady state; $t_{1/2\beta}$, elimination half-life.
Numbers in this table are rough estimates, intended to facilitate comparisons between drugs. Actual values may vary considerably in specific situations.
Relative potency is for parenteral administration.

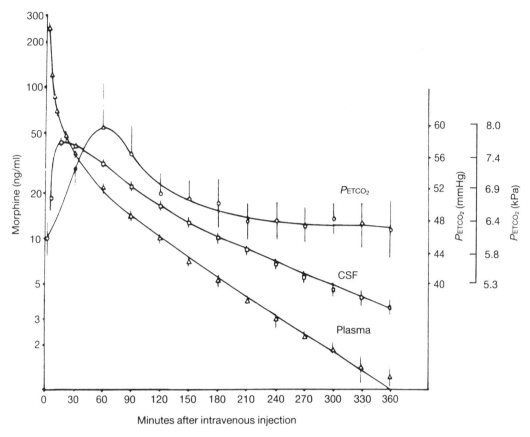

Fig. 44.9 An intravenous bolus of morphine was administered to dogs. Plasma and CSF morphine concentrations and end tidal carbon dioxide were measured. Peak morphine concentrations did not occur in CSF until 15–30 minutes after morphine administration. End tidal carbon dioxide did not reach a peak until about 60 minutes. The delay in the peak effect is probably related to the low lipid solubility of morphine and possibly to the active metabolite, morphine-6-glucuronide. (From Hug C, Murphy M, Rigel E, Olson W. Pharmacokinetics of morphine injected intravenously into anesthetized dogs. *Anesthesiology* 1981; **54**: 38–47, with permission.)

fentanyl and sufentanil are in the order of 60–75 per cent of the administered dose, while the first-pass retention of alfentanil is much smaller, about 10 per cent. Alfentanil was also shown to be released back into the circulation more quickly than sufentanil. Thus, Boer *et al.*[27] suggested that the lung may act as a 'capacitor', tending to reduce the peak arterial concentration of fentanyl and sufentanil, but not alfentanil. Possibly this contributes to the faster speed of onset of alfentanil; further study is needed to understand the pharmacodynamic significance of first-pass lung uptake.

After penetration of the blood–brain barrier, drugs must diffuse through a partially aqueous extracellular environment to gain access to receptor sites. Lipid solubility may be relatively disadvantageous at this point. Bernards and Hill[28] studied diffusion of drugs through spinal meninges, containing both lipophilic and lipophobic regions, and found that very lipophilic or very lipophobic drugs penetrated less well than drugs with intermediate lipid solubility. Interestingly, alfentanil penetrated much better than either fentanyl or sufentanil.

Morphine has a very slow onset of action, probably

because of poor lipid solubility. Hug *et al.*[29] administered an intravenous bolus of morphine to dogs and showed that the peak concentration in CSF did not occur for approximately 15–30 minutes (Fig. 44.9). By comparison, more lipid-soluble drugs, such as lignocaine and fentanyl, attain peak CSF concentrations within a few minutes of an intravenous bolus. They also measured end tidal CO_2 as a measure of morphine-induced ventilatory depression. End tidal CO_2 did not peak until about 60 minutes following drug administration, a remarkably delayed onset of action. However, the slow development of ventilatory depression may have been due in part to the action of the pharmacologically active metabolite of morphine, morphine-6-glucuronide (see below).

The fentanyl analogues are much faster in onset than morphine. The EEG effects of the fentanyl analogues have been used to quantify their action in the brain. Scott *et al.*[30] fitted EEG data to a pharmacodynamic model and reported a half-time for onset of 6.4 minutes for fentanyl and 1.1 minutes for alfentanil; as mentioned, alfentanil is remarkably faster in onset than fentanyl or sufentanil, perhaps because of its intermediate lipid solubility. Similar data for sufentanil suggest that the speed of onset of sufentanil and fentanyl are comparable. Bowdle and Ward[23] reported that peak EEG effects of fentanyl or sufentanil occurred less than 2 minutes after an intravenous bolus, the 2 minutes including 1 minute taken to administer the drug. They also observed that speed of onset was dose related; larger doses had earlier peak effects.

Duration of action

Plasma concentrations of morphine may not correspond closely to morphine effects. The lipid insolubility that retards penetration of morphine into the brain also tends to prevent it from leaving the brain once it has arrived. In addition, the morphine metabolite, morphine-6-glucuronide, is pharmacologically active and probably extends the effective duration of action of morphine. Within 6 minutes of intravenous administration of morphine to human volunteers, the concentration of morphine-6-glucuronide in plasma exceeded that of morphine.[31] Morphine-6-glucuronide had a potency 45 times greater than morphine when administered intracerebrally to mice, and its ventilatory depressant effects in dogs were 5–10 times more potent than morphine itself. It can cross the blood–brain barrier (albeit slowly because of its polarity), and it is eliminated more slowly than morphine from the CSF. Considering these facts, it appears likely that morphine-6-glucuronide contributes substantially to the pharmacological effects of morphine. Because morphine-6-glucuronide is cleared by the kidney, renal failure is expected to promote accumulation and prolong its action.

There is a close correspondence between plasma concentrations of the lipid-soluble opioids, such as the fentanyl analogues, and pharmacological effects. Therefore, pharmacokinetic data are useful for predicting the duration of action. These drugs all have relatively large volumes of distribution. Because of this, plasma concentrations fall rapidly following an intravenous bolus as drug diffuses from the circulation into tissues. After distribution has reached pseudoequilibrium, plasma concentrations continue to decline because of metabolic clearance, but more slowly than during the initial distribution phase.

The duration of action of drugs that undergo extensive tissue distribution is dose related in the following manner. If plasma concentrations fall below the threshold for significant pharmacological effects during the distribution phase, as with a relatively small dose, the duration of action is short, because distribution takes place rapidly. This is the reason that thiopentone has a short duration of action when given as a bolus for induction of anaesthesia, despite having an elimination half-life of many hours; this principle applies to most intravenous anaesthetic agents, as well as to thiopentone. On the other hand, if the plasma

concentrations remain above the effect threshold after distribution is completed, as with a relatively large dose, the duration of action is longer, because clearance takes place less rapidly than distribution (Fig. 44.10).

Following repeated doses or continuous infusions, predicting duration of action becomes considerably more complicated because the tissue compartments begin to 'fill up', altering the kinetic profile. Hughes et al.[32] have simulated by computer the relation between duration of a continuous infusion and the time required for drug concentrations to fall by 50 per cent after the infusion is turned off. They have referred to this as the 'context-sensitive half-time', not to be confused with elimination half-life; the context refers to the duration of the infusion (Fig. 44.11). One of the interesting results of these simulations was that sufentanil had the shortest context-sensitive half-time following infusions lasting less than

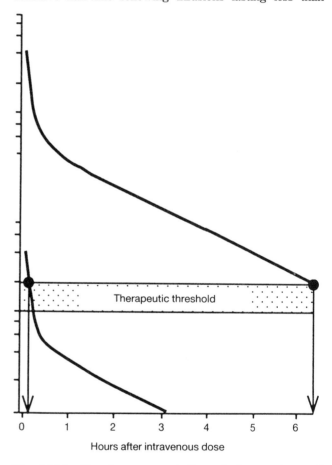

Hours after intravenous dose

Fig. 44.10 The plasma concentrations of a large and a small intravenous bolus dose of a hypothetical anaesthetic drug are shown. After the smaller dose, distribution causes plasma concentrations to fall below the therapeutic threshold within minutes. After the larger dose, plasma concentrations are well above the therapeutic threshold after distribution pseudoequilibrium is reached. Eventually (after many hours), clearance lowers plasma concentrations below the therapeutic threshold. Thus, the duration of action of this drug is dose dependent. Most intravenous anaesthetics agents are subject to this effect.

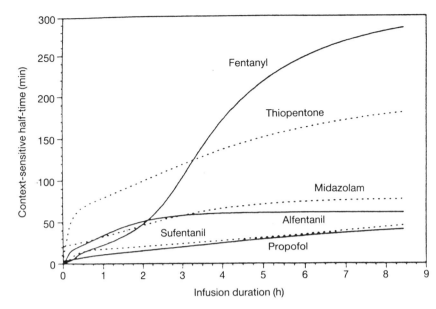

Fig. 44.11 The time required for plasma concentration to fall by 50 per cent after stopping an infusion (context-sensitive half-time) is plotted against the duration of the infusion by computer simulation. The context-sensitive half-times are heavily influenced by the distribution kinetics of the drugs. See text for details. (From Hughes A, Glass P, Jacobs J. Context-sensitive half-time in multicompartment pharmacokinetic models for intravenous anesthetic drugs. *Anesthesiology* 1992; **76**: 334–41, with permission.)

about 10 hours, despite having a substantially longer elimination half-life than alfentanil (the fentanyl analogue with the shortest elimination half-life). This occurs because sufentanil has a much larger volume of distribution than alfentanil. Even after a prolonged infusion, this large volume of distribution acts as a 'sink' for sufentanil, causing plasma concentrations to fall relatively rapidly, due to continued distribution. By contrast, the smaller volume of distribution for alfentanil is relatively 'filled up', and reduction in alfentanil plasma concentration depends more heavily on metabolic clearance.

Thus the duration of action of certain drugs may be dramatically affected by distribution kinetics. Predicting the duration of action of such drugs under conditions of continuous infusions or repeated bolus doses is complex, and not necessarily intuitive.

Pharmacokinetic properties of particular agents

A summary of pharmacokinetic parameters is provided in Table 44.1. The values in Table 44.1 are averages from representative studies. There is enormous variation in pharmacokinetic parameters among patients due to normal biological diversity and disease processes. Further comments on pharmacokinetic properties of particular agents follow.

Unique pharmacokinetics of morphine spinal analgesia

The pharmacokinetics and pharmacodynamics of morphine have been described extensively above. Distinguishing characteristics of morphine include low lipid solubility and

biotransformation to a pharmacologically active metabolite, morphine-6-glucuronide. Slow onset and prolonged duration of action are normally disadvantageous for an anaesthetic agent. However, these same properties confer on morphine a unique role in spinal analgesia (the term spinal analgesia, as used here, includes intrathecal and epidural administration). The systemic absorption of morphine after intrathecal or epidural administration is very slow, resulting in long duration of analgesia and low plasma concentrations. Intrathecal morphine, 0.3 mg, produced a peak plasma concentration of 4.5 ng/ml (minimum analgesic concentration in plasma, 20–40 ng/ml), while concentration in CSF was 6410 ng/ml. Cephalad spread of morphine in CSF may result in significant effects in the brain, including depression of breathing. Thus, the effects of spinal morphine are mediated by direct penetration into the CNS, not by absorption into the circulation and subsequent redistribution to the CNS.

Highly lipid-soluble opioids, such as the fentanyl analogues, are much more rapidly absorbed into the circulation after neuraxial administration, resulting in relatively short duration of action and clinically significant plasma concentrations. When systemic absorption of an opioid from the neuraxis becomes rapid enough, the plasma concentrations and concomitant effects resemble intravenous administration; under these circumstances, the effective spinal dose is similar in size to the intravenous or intramuscular dose, and neuraxial opioid administration may not be advantageous. For example, epidural infusion and intravenous infusion of fentanyl following knee surgery produced similar plasma fentanyl concentrations and comparable analgesia. Camman *et al.*[33] have reviewed the literature and concluded that the effects of lipophilic opioids in general are substantially the same whether given intravenously or spinally. On the other hand, single bolus doses (as opposed to continuous infusions) of lipid-soluble opioids probably produce selective spinal analgesia for brief periods of time and may be advantageous under particular circumstances.

Many drugs, including some opioids, have been given by spinal routes without adequate animal toxicology to determine whether or not the drugs or their vehicles produce direct spinal cord irritation or injury. Such practices may be unwise.

Fentanyl and sufentanil are high extraction drugs

Fentanyl and sufentanil have hepatic extraction ratios approaching unity. The intrinsic capacity of the liver to metabolize these drugs (intrinsic clearance) is so great that the blood passing through the liver is virtually completely cleared of drug. Plasma protein binding does not affect clearance in this situation because the liver 'strips' the drug from plasma protein. Clearance of fentanyl or sufentanil will be similar to liver blood flow, and consequently clearance will be diminished if liver blood flow is reduced. Clearance of these drugs is relatively insensitive to changes in activity of liver enzymes (because there is a substantial excess), unless they are extreme. Consequently, the clearance of fentanyl or sufentanil was not altered by hepatic cirrhosis (hepatic blood flow was apparently maintained in the patients studied).

Alfentanil is an intermediate extraction drug

The hepatic extraction ratio of alfentanil was estimated to be 0.3–0.5. This intermediate extraction ratio corresponds to a relatively low intrinsic capacity of the liver to metabolize alfentanil and also reflects the high degree of alfentanil plasma protein binding; under these circumstances clearance of bound drug tends to be 'restricted'. Thus clearance of alfentanil is determined by a complex relation between intrinsic clearance, plasma protein binding, and liver blood flow. Reductions in plasma protein binding will increase clearance, while reductions in liver enzyme activity or liver blood flow will decrease clearance. The clearance of alfentanil was reduced substantially in hepatic cirrhosis, and even mild hepatic dysfunction may have a significant effect. Shafer et al.[34] found that patients without a clinical history of liver disease, but with mild abnormalities of liver function tests, had alfentanil clearance 50 per cent lower than patients with normal liver function tests. Patients with liver function test abnormalities frequently required naloxone to restore spontaneous ventilation at the conclusion of balanced anaesthesia with alfentanil.[34]

Methadone

Methadone has an unusual pharmacokinetic profile due to relatively low clearance and a large volume of distribution, resulting in a very long elimination half-life, averaging 35 hours in surgical patients.[35] Gourlay et al. reported that methadone, 20 mg intravenously, produced a median duration of analgesia of 27 hours in surgical patients.[35]

Effect of ageing on the pharmacokinetics of opioids

Numerous studies have documented alteration in pharmacokinetics or pharmacodynamics of opioids in elderly patients. Taken as a whole these studies suggest that the elderly will be more sensitive to opioids. However, it is difficult to be specific because the results of several studies are in conflict. An interesting study by Lemmens et al. found a negative correlation between alfentanil clearance and age in female surgical patients but not in males,[36] a reminder that gender can make a significant difference in the disposition of particular drugs.

Partial agonists and agonist–antagonists

Understanding the pharmacology of partial agonist and agonist–antagonist opioids requires a more detailed consideration of basic opioid receptor pharmacology than was presented earlier in this chapter. Therefore, the pharmacology of the opioid receptors will be elaborated, followed by a presentation of the partial agonist and agonist–antagonist opioids.

μ-Receptor

μ-Receptors appear to mediate many of the clinical effects of morphine-like drugs. Two subtypes of the μ-receptor have been identified in rodents, μ_1 and μ_2. μ_1-Receptors appear to mediate supraspinal analgesic effects of morphine-like opioids; the endogenous ligands (i.e. neurotransmitters) of μ_1-receptors appear to be opioid peptides. μ_2-Receptors appear to mediate respiratory depression and spinal analgesia.[37] Significant species differences exist, but assuming that different μ-receptor subtypes mediate analgesia and respiratory depression in humans as well as in rodents, there would be a hypothetical possibility that highly specific μ-receptor subtype agonists would produce analgesia without respiratory depression.

κ-Receptor

κ-Receptors were named for the agonist—antagonist drug, ketocyclazocine. The endogenous ligands of this receptor are probably the opioid peptides known as dynorphins. The traditional concept was that κ-receptors mediated analgesia in the spinal cord but not in the brain. Currently it is believed that both spinal and supraspinal κ-receptors are involved in antinociception. There appear to be three major κ-receptor subtypes, designated κ_1, κ_2 and κ_3. κ_1-Receptors appear to mediate spinal analgesia in rodents. κ_3-Receptors are found in the rodent brain with a density twice that of μ- or δ-receptors. The agonist—antagonist drug, nalorphine, has been shown to produce analgesia mediated by κ_3-receptors, and this receptor has been regarded as the 'original' κ-receptor proposed by Martin.[37]

Although μ- and κ-agonists both produce analgesia, the subjective effects of κ-agonists are distinct from μ-agonists. While μ-agonists tend to produce euphoria, κ-agonists tend to produce sedation and dysphoria (the opposite of euphoria). Perhaps because of these differences, experimental animals and human subjects tend to prefer self-administration of μ-agonists to κ-agonists. This implies that κ-agonists should produce a lower potential for abuse and dependence.

κ-Receptors appear to mediate psychotomimetic (hallucinosis, delirium) effects[38] that are prominent with certain opioids, especially benzomorphan derivatives (see Relations between structure and activity); highly selective κ-agonists produce naloxone-sensitive dysphoria and psychotomimetic effects in humans.

An interesting neuroendocrine role of κ-receptors involves the regulation of vasopressin release from the posterior pituitary. The κ-receptor peptide ligand, dynorphin, is contained with vasopressin in the neurosecretory vesicles of the neuron terminals in the posterior pituitary. Dynorphin appears to mediate an inhibitory feedback loop in which dynorphin released with vasopressin activates κ-receptors, inhibiting further release. Thus κ-agonist drugs tend to produce a diuresis by inhibiting release of vasopressin. This is an example of a bioassay that can be used to detect the κ-agonist properties of an opioid drug.

Highly selective κ-agonists have not produced respiratory depression in animals.

δ-Receptor

Delta refers to the opioid receptors with greatest selectivity for the endogenous opioid peptides known as enkephalins.

There is evidence for two δ-receptor subtypes, designated δ_1 and δ_2.[37] Because of their relative selectivity for enkephalins, δ-receptors may be somewhat less important for the effects of exogenously administered opioids. On the other hand, many opioid drugs bind δ-receptors to some degree, thus δ-receptors may have some part in mediating drug effects. δ-Receptors may be involved in opioid-induced depression of breathing.

σ-Receptor

The σ-receptor was one of the original triad of opioid receptors proposed by W.R. Martin in 1976.[1] Martin postulated such a receptor to explain the psychotomimetic and cardiovascular effects of N-allylnormetazocine (SKF 10,047), and related agonist—antagonist opioids. Opioids with such properties were regarded as 'sigma opioids'. Subsequently an enormous effect was made to identify and characterize the σ-receptor, an effort that is still continuing. While the σ-receptor is not entirely understood, the results of many studies now suggest that it does not actually mediate the psychotomimetic effects of agonist—antagonist opioids, as originally proposed. Natural opioids (such as morphine) are laevorotatory (−) isomers, and dextrorotatory (+) enantiomers do not have opioid activity. Many synthetic opioids are racemic mixtures (others are non-chiral). (+/−)SKF 10,047 (as originally used by Martin) is a racemic mixture that is now recognized to bind (at least) three types of receptors: (−)SKF 10,047 binds mainly to μ- and κ-opioid receptors; (+)SKF 10,047 binds to phencyclidine (PCP) receptors (ketamine also acts at this receptor) and to another receptor currently designated as σ. The σ-receptor shows a preference for dextrorotatory compounds and is not sensitive to naloxone (which is laevorotatory). Because the psychotomimetic effects of agonist—antagonist opioids are mediated by laevorotatory enantiomers and can be antagonized by naloxone, it appears that neither PCP nor sigma receptors are involved, contrary to Martin's original scheme.[39] Presumably, opioid psychotomimetic effects are mediated by κ-receptors or by some unknown receptors. The physiological significance, if any, of actions by dextrorotatory opioids on σ- or PCP receptors, is unknown.

Nalorphine-like drugs: pentazocine, nalbuphine and butorphanol

The so-called nalorphine-like drugs, including nalorphine, pentazocine, nalbuphine and butorphanol were originally believed to be μ-antagonists and κ-agonists. These drugs are probably more accurately described as partial agonists at both μ- and κ-receptors, with the intensity of agonist activity varying somewhat from one drug to another. Some of the reasons for classifying nalorphine-like drugs as partial μ- and partial κ-agonists are interesting to examine. While nalorphine-like drugs uniformly depress breathing (see Depression of breathing), selective κ-agonists do not; the selective κ-agonist, bremazocine, did not produce depression of breathing in dogs or rats and the highly selective prototype κ-agonist drug, U50,488, did not depress breathing when given by intrathecal or intracerebroventricular routes to rats. These results imply that κ-receptors do not mediate the respiratory effects of opioids and suggest that nalorphine-like drugs mediate depression of breathing at μ-receptors. β-Funaltrexamine (β-FNA) is a highly selective μ-receptor antagonist. β-FNA markedly shifts to the right, in a non-parallel fashion, the dose–response curves of nalorphine, nalbuphine and butorphanol in the mouse abdominal contraction assay, a bioassay that reflects opioid-mediated analgesia. This further supports the notion that nalorphine-like drugs are partial μ-agonists. Thus, the ability of nalorphine-like drugs to antagonize morphine-like μ-agonists is probably because of μ-partial agonist activity.

Suppression of vasopressin levels and resulting diuresis has been a useful bioassay for classifying κ-agonist activity. Results of such studies suggest that nalorphine-like drugs are partial κ-agonists; nalorphine and butorphanol suppress vasopressin levels but do not have the full efficacy of prototype κ-agonists such as U50,488.

Although nalorphine-like drugs share many similarities, there are significant differences between the particular drugs. Nalorphine and pentazocine produce more intense dysphoria and psychotomimetic effects than nalbuphine or butorphanol, and have cardiovascular stimulating properties. While butorphanol and nalbuphine do not produce as much dysphoria or cardiovascular stimulation, they do produce sedation. Butorphanol in particular appears to have a marked sedative effect.[2] All the nalorphine-like drugs have antagonist activity towards full μ-agonists, but nalbuphine appears to have the most prominent antagonist properties (see below).

Buprenorphine

Buprenorphine appears to be a partial agonist at μ- and κ-receptors; to this extent it resembles the nalorphine-like drugs. However, slow dissociation from μ-receptors and an unusual 'bell'-shaped dose–response curve give buprenorphine a unique pharmacological profile (Fig. 44.12).

Several studies have demonstrated that buprenorphine dissociates very slowly from opioid receptors in rat brain and guinea-pig ileum. Clinical reports suggest an analogous phenomenon in humans; naloxone does not always completely reverse buprenorphine effects, such as depressed breathing.

Dum and Herz examined the antinociceptive effects of buprenorphine in rats and found that doses of buprenorphine up to about 0.5 mg/kg produced the expected dose-related analgesia but doses greater than 0.5 mg/kg resulted in a paradoxical reduction in antinociception (Fig. 44.12).[40] The opioid antagonist naltrexone shifted the buprenorphine dose–effect curve symmetrically to the right, suggesting that the bell shape is related to opioid receptor activity. The explanation for the unusual buprenorphine dose–response curve is unknown.

Comparable dose–effect studies in humans are not available; however, it is interesting to consider a case report of two patients with severe postoperative pain after balanced anaesthesia with buprenorphine and nitrous oxide.[41] The anaesthetists were aware of the bell-shaped dose–effect curve and reasoned that the patients had received too much buprenorphine, placing them on the downsloping side of the curve (dose-related reduction in

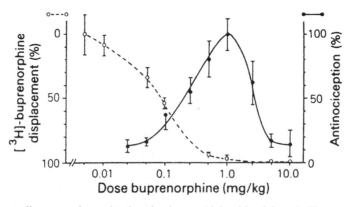

Fig. 44.12　Buprenorphine dose–effect curve for antinociception in rats (right side of figure). The curve is 'bell shaped'. The extent of buprenorphine binding to receptors in rat brain is also shown for the same range of buprenorphine doses (left side of figure). (From Dum JE, Herz A. In vivo receptor binding of the opiate partial agonist, buprenorphine, correlated with its agonist and antagonistic actions. *British Journal of Pharmacology* 1981; **74**: 627–33, with permission.)

analgesia). Administration of naloxone to these patients resulted in 'total pain relief', consistent with a shift to the right of a bell-shaped dose–effect curve.

Buprenorphine binds κ-receptors in rat brain and it antagonizes the diuretic effects of the κ-agonist, bremazocine, in morphine-tolerant rats, suggesting antagonist or partial agonist activity at κ-receptors.

Buprenorphine pharmacology is complicated and not completely understood. Under some circumstances naloxone may not completely antagonize buprenorphine agonist effects or naloxone may actually cause a paradoxical increase in buprenorphine agonist effects.

Dezocine

Dezocine is a partial μ-agonist. Subjective effects (sedation or dysphoria) in humans suggest κ-activity and dezocine binds κ-receptors *in vitro*. Dezocine antagonized morphine effects and produced an abstinence syndrome in morphine-dependent animals. Dezocine appeared to be a more efficacious μ-agonist than other agonist–antagonist drugs, judging from its analgesic and respiratory effects in animals and humans. A plateau of analgesic activity and depression of breathing could be demonstrated in humans, but the level of the plateau appeared higher than for butorphanol or nalbuphine. The anaesthetic sparing effect (reduction of enflurane MAC; see below) of dezocine in dogs was similar to morphine, and much greater than nalbuphine or butorphanol.

Meptazinol

Meptazinol binds μ-receptors in rat brain with moderate affinity and behaves clinically as a partial μ-agonist; it is moderately selective for μ_1-receptors. It has very low affinity for rat brain κ-receptors, and therefore may not have appreciable κ-effects. The antinociceptive effects of meptazinol in animals appeared to be mediated in part by a cholinergic mechanism.

Agonist–antagonists as anaesthetic agents

There is an interesting literature describing the use of agonist–antagonist opioids in various regimens of balanced anaesthesia. Murphy and Hug studied the enflurane sparing effect (MAC reduction) of morphine, fentanyl, butorphanol and nalbuphine in dogs.[42] Morphine or fentanyl reduced enflurane MAC by about 65 per cent, while butorphanol or nalbuphine reduced enflurane MAC

by only 11 or 8 per cent respectively. These authors suggested that the smaller MAC reduction with the agonist–antagonists was the result of a 'ceiling' for analgesia. Hall *et al.*[43] studied dezocine in the same dog model and found a 58 per cent reduction in enflurane MAC; the maximum dose of dezocine was limited by cardiovascular depression. These studies could be interpreted to suggest that butorphanol and nalbuphine would not be efficacious as analgesic supplements to general anaesthesia. However, this conclusion may not be warranted. There are substantial differences in opioid pharmacology between various species; experiments in dogs may not apply directly to humans. The contribution of antinociception to general anaesthesia is unclear and many drugs that produce general anaesthesia (e.g. most intravenous induction agents [except ketamine] and potent volatile anaesthetics) do not produce analgesia in conscious subjects. Even with a 'ceiling' for antinociceptive effects, other opioid effects, such as sedation mediated by κ-receptors, might contribute to the anaesthetic state.

MAC reduction studies with agonist–antagonists have not been performed in humans. However, a substantial number of clinical studies of agonist–antagonist opioids, including blinded, randomized comparisons to full μ-agonists, suggest that agonist–antagonists can be efficacious as the opioid component of balanced anaesthesia. For example, Aldrete *et al.*[44] compared butorphanol to morphine; either opioid was combined with diazepam, pancuronium and nitrous oxide in a randomized, blinded study of anaesthesia for coronary artery bypass surgery. The average dose of butorphanol was 25 mg compared to 127 mg of morphine. The average dose of diazepam was 22 mg in both groups. The two opioids were equally satisfactory and there were no significant differences in haemodynamic events during surgery. The 'blinded' anaesthetists were unable to discern which drug they were using. In three out of four cases of patient movement, intraoperative awareness or haemodynamic instability (the blinded anaesthetists were convinced the opioid was butorphanol), the opioid was morphine.

The use of agonist–antagonists to antagonize full agonists

The agonist–antagonist opioids can antagonize full agonists under certain conditions. The antagonist properties of nalbuphine have been studied most often. Nalbuphine (15 mg/70 kg) actually worsened depressed breathing produced by a small dose of morphine (15 mg/70 kg), as predicted by theory (see Opioid receptor pharmacology). However, several studies have shown that nalbuphine effectively antagonized depressed breathing from large doses of μ-agonists. The dose–response for nalbuphine

Table 44.2 Pharmacokinetic data for partial agonists and agonist–antagonists

	Pentazocine	Nalbuphine	Butorphanol	Buprenorphine	Dezocine	Meptazinol
Clearance (l/min or *l/kg per h)	1.38	1.5	2.7	1.3	3.0*	1.3
V_{dss} (l) or $V_{d\beta}$* (l/kg)	396	315	350	188	11*	5.0*
$t_{1/2\beta}$ (min)	203	222	160	184	156	120
Potency in mg relative to morphine, 10 mg	30	10	2	0.2	10	100

V_{dss}, = volume of distribution at steady state; $V_{d\beta}$, volume of distribution during terminal elimination ($V_{d\beta} > V_{dss}$); $t_{1/2\beta}$, = elimination half-life.
Numbers in this table are rough estimates, intended to facilitate comparisons between drugs. Actual values may vary considerably in specific situations. Relative potency is for parenteral administration.

antagonist effects has not been rigorously determined and a wide range of doses has been used, from about 1–20 mg/70 kg, intravenously; 2.5–5.0 mg is probably a reasonable starting dose.

There is considerable interest in selective opioid antagonists that would preserve analgesia and not elevate sympathetic tone, because of the problems associated with naloxone (see Antagonists). Nalbuphine did not elevate circulating catecholamines in dogs anaesthetized with fentanyl and enflurane, in contrast to naloxone. Bailey et al. suggested that nalbuphine preserved analgesia better than naloxone.[45] On the other hand, the use of nalbuphine as an antagonist has been associated with a spectrum of naloxone-like adverse effects, including pulmonary oedema.[46] Nalbuphine is an interesting alternative to naloxone in the clinical setting; whether it is superior to naloxone is not clear.

Butorphanol[47] and buprenorphine[48] have also been used experimentally to antagonize depressed breathing produced by fentanyl; however, insufficient information is available to recommend their use in clinical practice.

Pharmacokinetics of partial agonist and agonist–antagonist opioids

The pharmacokinetic properties of partial agonist and agonist–antagonist opioids have not been studied nearly as extensively as for full agonist drugs. Pharmacokinetic data for agonist–antagonist opioids are summarized in Table 44.2.

Antagonists

Naloxone is a pure antagonist of all three types of opioid receptors, although higher doses are required to antagonize κ- and δ-receptors than μ-receptors. Naloxone has a relatively short duration of action compared to many of

the opioid agonists; the terminal half-life is about 1–1.5 hours. Thus there may be a risk of recurrence of opioid agonist effects when naloxone is given in small bolus doses. Longnecker et al. were able to prolong the action of naloxone by giving it intramuscularly, from 80 minutes (0.35 mg/70 kg, intravenously) to 6 hours (0.7 mg/70 kg, intramuscularly).[22]

Naloxone has been given by intravenous infusion in attempts to antagonize opioid side effects such as pruritis and depressed breathing while preserving analgesia. Some authors have reported successful results. However, interference with analgesia is always a possibility when naloxone is used.

The use of naloxone to antagonize opioid agonists can be associated with serious cardiovascular problems. Severe hypertension, cardiac arrhythmias (including ventricular fibrillation), and pulmonary oedema have been reported; a few deaths have been attributed to naloxone administration. These adverse effects may not be dose related; pulmonary oedema has been reported following a small dose (0.1 mg). The mechanism of cardiovascular side effects is believed to be a sudden increase in sympathetic tone, as demonstrated in dogs by Flacke et al.[11] The mechanism of pulmonary oedema is presumed to be similar to neurogenic pulmonary oedema.

Naloxone has been used to treat cardiovascular shock states, stroke, spinal cord injury and a variety of miscellaneous conditions. There is also evidence that naloxone may antagonize some anaesthetic drugs that are not opioids, such as nitrous oxide, benzodiazepines and barbiturates. However, these miscellaneous uses of naloxone are not firmly established.

There are remarkably few pure opioid antagonists, compared to the number of opioid agonists that are available. Naltrexone and nalmefene are pure opioid antagonists that are structurally related to naloxone. Naltrexone is the N-cyclopropylmethyl analogue of oxymorphone, whereas naloxone is the N-allyl analogue. Several antagonists have resulted from modification of the C-6 keto group of naltrexone, including nalmefene, naltrindole and norbinaltorphimine (nor-BNI). Naltrindole and norbinaltorphimine are selective antagonists for δ- and κ-receptors, respectively (not available for clinical use).

Naltrexone and nalmefene, like naloxone, are most potent at µ-receptors but have significant activity at all three types of opioid receptor; they are both much longer acting than naloxone (terminal half-life in excess of 8 hours) and have significant oral bioavailability. The clinical role of the newer pure antagonists is unclear.

Peripheral mechanisms of analgesia: opioids and non-steroidal anti-inflammatory drugs

Traditional pharmacological strategies for producing analgesia are largely concerned with blunting nociceptive afferent stimuli inside the CNS. An alternative strategy is to prevent the afferent stimulus from ever reaching the CNS. Sensory nerve conduction block by local anaesthetics is an example of this approach. Recently there has been considerable interest in methods to block nociception at the most peripheral possible site, the afferent nerve ending. Two classes of drugs have been tried with some success, the opioids and the non-steroidal anti-inflammatory agents; these drugs have significant peripheral, as well as central effects on nociception.

Non-steroidal anti-inflammatory drugs

Non-steroidal anti-inflammatory drugs (NSAIDs) are a heterogenous group of agents that mediate anti-inflammatory, analgesic, antipyretic, and platelet inhibitory effects. There are over 100 NSAIDs marketed around the world. NSAIDs are also referred to as 'aspirin-like', after the prototype drug of the class, acetylsalicylic acid (aspirin); aspirin was introduced into clinical use in 1899. The term, non-steroidal anti-inflammatory, is probably preferable to 'aspirin-like' because many of the drugs are chemically unrelated to aspirin. The anti-inflammatory, analgesic, antipyretic and platelet inhibitory effects vary among specific NSAIDs. For example, paracetamol (acetaminophen) is antipyretic and analgesic but lacks significant anti-inflammatory activity.

NSAIDs are particularly effective for producing analgesia in settings where inflammatory mediators, such as prostaglandins, have sensitized afferent sensory nerve endings. NSAIDs interfere with prostaglandin biosynthesis by inhibition of the enzyme prostaglandin synthase (cyclo-oxygenase). NSAID inhibition of prostaglandin synthase is highly stereospecific; the (+) isomers are more potent.

The mechanism of inhibition of prostaglandin synthase varies with different agents. Some are competitive inhibitors, while others irreversibly block the enzyme; for example, aspirin inhibits prostaglandin synthase by irreversible acetylation of a serine associated with the active site. Platelets are particularly vulnerable to irreversible inhibition of prostaglandin synthase because they have no cellular apparatus for producing a new enzyme. Thus, a single dose of aspirin will inhibit platelet aggregation and prolong bleeding time until sufficient numbers of normal platelets enter the circulation to restore normal platelet function; this process typically requires 3 or 4 days.

Prostaglandin synthase inhibition explains many of the anti-inflammatory effects of NSAIDs. Adverse effects of NSAIDs, including platelet dysfunction, gastrointestinal ulceration, and decreased renal function, are also related to inhibition of prostaglandin synthesis. However, other mechanisms of action are probably important. The potency of NSAIDs to inhibit prostaglandin synthesis is not well correlated to anti-inflammatory activity, and a number of other NSAID effects on the inflammatory process have been identified. Also there is evidence that NSAID actions in the CNS may affect peripheral inflammation. Catania et al.[49] found that lysine acetylsalicylate and sodium salicylate injected into the lateral cerebral ventricle of mice inhibited inflammatory oedema in the mouse ear. Indomethacin did not have this effect; the lack of anti-inflammatory activity with intracerebroventricular indomethacin implied that prostaglandins were not involved in the central anti-inflammatory activity of salicylates, because both indomethacin and salicylates inhibit prostaglandin synthesis.

NSAIDs can produce serious side effects. NSAIDs decrease renal blood flow, a reversible, harmless effect in most patients. However, renal failure can be precipitated in patients with impaired renal circulation from low cardiac output, chronic renal disease or hypovolaemia. Gastric and intestinal ulceration is a common side effect mediated both by local irritation (with oral administration) and by systemic action. Platelet inhibition and prolongation of bleeding time has been mentioned; whether this results in clinically significant haemorrhage depends upon specific circumstances, including the type of surgery and the status of other aspects of the patient's haemostatic system.

An idiosyncratic reaction to NSAIDs occurs in some patients with asthma, nasal polyps or chronic urticaria. These patients may respond to NSAIDs with a potentially life-threatening syndrome resembling anaphylaxis. Although this syndrome is often associated with aspirin, it may occur with any NSAID. The mechanism is unknown.

Treatment is the same as for anaphylaxis; adrenaline is the drug of choice.

NSAIDs appear to be moderately effective for the treatment of postoperative pain. Interest in these drugs has grown enormously in the United States with the advent of ketorolac, an NSAID that has been promoted specifically for postoperative pain. However, ketorolac is not the only NSAID available for parenteral use and whether it has any advantages over other NSAIDs (even including aspirin) is unknown. Ketorolac or other NSAIDs are often used in combination with opioids. The required dose of opioids may be reduced in this circumstance.[50] Such 'opioid sparing' may be beneficial by virtue of minimizing opioid side effects, such as depressed breathing; on the other hand, NSAIDs also have their own side effects. Additional clinical research is needed to clearly establish the role of NSAIDs in the treatment of postoperative pain.

Peripheral opioids

Until recently, opioid-mediated analgesia was thought to occur entirely within the central nervous system. Now there is evidence that opioid receptors on peripheral afferent neurons may mediate analgesia, especially in the setting of inflammation.[51] Opioid peptides produced by immune cells in the vicinity of the neurons may be the source of endogenous ligands for these receptors. Exogenous opioid agonist drugs can also mediate peripheral antinociceptive effects; there is evidence for peripheral analgesic activity of μ-, δ- and κ-agonists, suggesting that all three types of opioid receptors may be involved. The few available clinical studies of peripheral administration of opioid drugs are inconclusive but positive results from some studies are encouraging. For example, Khoury et al.[52] instilled morphine, 1 mg, into the knee joint at the conclusion of arthroscopic surgical procedures and compared the resulting analgesia to intra-articular bupivacaine. The onset of analgesia after morphine was delayed compared to bupivacaine, but the duration of analgesia was substantially longer; from the fourth postoperative hour until 2 days later (the end of the study period), analgesia was superior with morphine. A combination of bupivacaine and morphine produced rapid onset and prolonged analgesia.

REFERENCES

1. Martin WR. History and development of mixed opioid agonists, partial agonists, and antagonists. *British Journal of Clinical Pharmacology* 1979; **7**: 273S–9S.

2. Dershwitz M, Rosow CE, DiBiase PM, Zaslavsky A. Comparison of the sedative effects of butorphanol and midazolam. *Anesthesiology* 1991; **74**: 717–24.

3. Ausems M, Hug C, Stanski D, Burm A. Plasma concentrations of alfentanil required to supplement nitrous oxide anesthesia for general surgery. *Anesthesiology* 1986; **65**: 362–73.

4. McDonnell T, Bartowski R, Williams J. ED$_{50}$ of alfentanil for induction of anesthesia in unpremedicated young adults. *Anesthesiology* 1984; **60**: 136–40.

5. Vinik H, Bradley E, Kissin I. Midazolam–alfentanil synergism for anesthetic induction in patients. *Anesthesia and Analgesia* 1989; **69**: 213–17.

6. Bailey P, Streisand J, East K, *et al.* Differences in magnitude and duration of opioid-induced respiratory depression and analgesia with fentanyl and sufentanil. *Anesthesia and Analgesia* 1990; **70**: 8–15.

7. Krane B, Kreutz J, Johnson D, Mazuzan J. Alfentanil and delayed respiratory depression: case studies and review. *Anesthesia and Analgesia* 1990; **70**: 557–61.

8. Catley D, Thornton C, Jordan C, Lehane J, Royston D, Jones J. Pronounced, episodic oxygen desaturation in the postoperative period: its association with ventilatory pattern and analgesic regimen. *Anesthesiology* 1985; **63**: 20–8.

9. Flacke J, Flacke W, Bloor B, Van Etten A, Kripke B. Histamine release by four narcotics: a double-blind study in humans. *Anesthesia and Analgesia* 1987; **66**: 723–30.

10. Flacke J, Davis L, Flacke W, Bloor B, Van Etten A. Effects of fentanyl and diazepam in dogs deprived of autonomic tone. *Anesthesia and Analgesia* 1985; **64**: 1053–9.

11. Flacke J, Flacke W, Bloor B, Olewine S. Effects of fentanyl, naloxone and clonidine on hemodynamics and plasma catecholamine levels in dogs. *Anesthesia and Analgesia* 1983; **62**: 305–13.

12. Thomson I, Putnins C, Friesen R. Hyperdynamic cardiovascular responses to anesthetic induction with high-dose fentanyl. *Anesthesia and Analgesia* 1986; **65**: 91–5.

13. Weinger M, Smith N, Blasco T, Koob G. Brain sites mediating opiate-induced muscle rigidity in the rat: methylnaloxonium mapping study. *Brain Research* 1991; **544**: 181–90.

14. Benthuysen J, Smith N, Sanford T, Head N, Dec-Silver H. Physiology of alfentanil-induced rigidity. *Anesthesiology* 1986; **64**: 440–6.

15. Bowdle T, Rooke G. Postoperative myoclonus and rigidity after anesthesia with opioids. *Anesthesia and Analgesia* 1994; **78**: 783–6.

16. Tempelhoff R, Modica P, Bernardo K, Edwards I. Fentanyl-induced electrocorticographic seizures in patients with complex partial epilepsy. *Journal of Neurosurgery* 1992; **77**: 201–8.

17. Milde L, Milde J, Gallagher W. Effects of sufentanil on cerebral circulation and metabolism in dogs. *Anesthesia and Analgesia* 1990; **70**: 138–46.

18. Sperry R, Bailey P, Reichman M, Peterson J, Petersen P, Pace N. Fentanyl and sufentanil increase intracranial pressure in head trauma patients. *Anesthesiology* 1992; **77**: 416–20.

19. Wahl M. Effects of enkephalins, morphine and naloxone on pial arteries during perivascular microapplication. *Journal of Cerebral Blood Flow and Metabolism* 1985; **5**: 451–7.

20. Jones R, Detmer M, Hill A, Bjoraker D, Pandit U. Incidence of choledochoduodenal sphincter spasm during fentanyl-supplemented anesthesia. *Anesthesia and Analgesia* 1981; **60**: 638–40.

21. Costello D, Borison H. Naloxone antagonizes narcotic self-blockade of emesis in the cat. *Journal of Pharmacology and Experimental Therapeutics* 1977; **203**: 222–30.

22. Longnecker D, Grazis P, Eggers G. Naloxone antagonism of morphine induced respiratory depression. *Anesthesia and Analgesia* 1973; **52**: 447–52.

23. Bowdle T, Ward R. Induction of anesthesia with small doses of sufentanil or fentanyl: dose versus EEG response, speed of onset and thiopental requirement. *Anesthesiology* 1989; **70**: 26–30.

24. Tomicheck R, Rosow C, Philbin D, Moss J, Teplick R, Schneider R. Diazepam–fentanyl interaction – hemodynamic and hormonal effects in coronary artery surgery. *Anesthesia and Analgesia* 1983; **62**: 881–4.

25. Bailey P, Pace N, Ashburn M, Moll J, East K, Stanley T. Frequent hypoxemia and apnea after sedation with midazolam and fentanyl. *Anesthesiology* 1990; **73**: 826–30.

26. Eisenach J, Schlairet T, Dobson C, Hood D. Effect of prior anesthetic solution on epidural morphine analgesia. *Anesthesia and Analgesia* 1991; **73**: 112–18.

27. Boer F, Bovill J, Burm A, Mooren R. Uptake of sufentanil, alfentanil and morphine in the lungs of patients about to undergo coronary artery surgery. *British Journal of Anaesthesia* 1992; **68**: 370–5.

28. Bernards C, Hill H. Physical and chemical properties of drug molecules governing their diffusion through the spinal meninges. *Anesthesiology* 1992; **77**: 750–6.

29. Hug C, Murphy M, Rigel E, Olson W. Pharmacokinetics of morphine injected intravenously into anesthetized dogs. *Anesthesiology* 1981; **54**: 38–47.

30. Scott J, Ponganis K, Stanski D. EEG quantitation of narcotic effect: the comparative pharmacodynamics of fentanyl and alfentanil. *Anesthesiology* 1985; **62**: 234–41.

31. Osborne R, Joel S, Trew D, Slevin M. Morphine and metabolite behavior after different routes of morphine administration: demonstration of the importance of the active metabolite morphine-6-glucuronide. *Clinical Pharmacology and Therapeutics* 1990; **47**: 12–19.

32. Hughes A, Glass P, Jacobs J. Context-sensitive half-time in multicompartment pharmacokinetic models for intravenous anesthetic drugs. *Anesthesiology* 1992; **76**: 334–41.

33. Camann WR, Loferski BL, Fanciullo GJ, Stone ML, Datta S. Does epidural administration of butorphanol offer any clinical advantage over the intravenous route? *Anesthesiology* 1992; **76**: 216–20.

34. Shafer A, Sung M-L, White P. Pharmacokinetics and pharmacodynamics of alfentanil infusions during general anesthesia. *Anesthesia and Analgesia* 1986; **65**: 1021–8.

35. Gourlay G, Wilson P, Glynn C. Pharmacodynamics and pharmacokinetics of methadone during the perioperative period. *Anesthesiology* 1982; **57**: 458–67.

36. Lemmens H, Burm G, Hennis P, Gladines M, Bovill J. Influence of age on the pharmacokinetics of alfentanil. *Clinical Pharmacokinetics* 1990; **19**: 416–22.

37. Pasternak G. Pharmacological mechanisms of opioid analgesics. *Clinical Neuropharmacology* 1993; **16**: 1–18.

38. Millan MJ. Kappa-opioid receptors and analgesia. *Trends in Pharmacological Sciences* 1990; **11**: 70–6.

39. Musacchio J. The psychotomimetic effects of opiates and the sigma receptor. *Neuropsychopharmacology* 1990; **3**: 191–200.

40. Dum JE, Herz A. In vivo receptor binding of the opiate partial agonist, buprenorphine, correlated with its agonistic and antagonistic actions. *British Journal of Pharmacology* 1981; **74**: 627–33.

41. Pedersen JE, Chraemmer-Jorgensen B, Schmidt JF, Risbo A. Naloxone – a strong analgesic in combination with high-dose buprenorphine? *British Journal of Anaesthesia* 1985; **57**: 1045–6.

42. Murphy MR, Hug CC. The enflurane sparing effect of morphine, butorphanol and nalbuphine. *Anesthesiology* 1982; **57**: 489–92.

43. Hall RI, Murphy MR, Szlam F, Hug Jr CC. Dezocine – MAC reduction and evidence for myocardial depression in the presence of enflurane. *Anesthesia and Analgesia* 1987; **66**: 1169–71.

44. Aldrete JA, de Campo T, Usubiaga LE, Renck R, Suzuki D, Witt WO. Comparison of butorphanol and morphine as analgesics for coronary bypass surgery: a double-blind, randomized study. *Anesthesia and Analgesia* 1983; **62**: 78–83.

45. Bailey PL, Clark NJ, Pace NL, *et al.* Antagonism of postoperative opioid-induced respiratory depression: nalbuphine versus naloxone. *Anesthesia and Analgesia* 1987; **66**: 1109–14.

46. DesMarteau JK, Cassot AL. Acute pulmonary edema resulting from nalbuphine reversal of fentanyl-induced respiratory depression. *Anesthesiology* 1986; **65**: 237.

47. Bowdle TA, Greichen SL, Bjurstrom RL, Schoene TB. Butorphanol improves CO_2 response and ventilation after fentanyl anesthesia. *Anesthesia and Analgesia* 1987; **66**: 517–22.

48. Boysen K, Hertel S, Chraemmer-Jorgensen B, Risbo A, Poulsen NJ. Buprenorphine antagonism of ventilatory depression following fentanyl anaesthesia. *Acta Anaesthesiologica Scandinavica* 1988; **32**: 490–2.

49. Catania A, Arnold J, Macaluso A, Hiltz M, Lipton J. Inhibition of acute inflammation in the periphery by central action of salicylates. *Proceedings of the National Academy of Sciences, USA* 1991; **88**: 8544–7.

50. Gillies G, Kenny G, Bullingham R, McArdle C. The morphine sparing effects of ketorolac tromethamine. *Anaesthesia* 1987; **42**: 727–31.

51. Stein C. Peripheral mechanisms of opioid analgesia. *Anesthesia and Analgesia* 1993; **76**: 182–91.

52. Khoury G, Chen A, Garland D, Stein C. Intraarticular morphine, bupivacaine, and morphine/bupivacaine for pain control after knee videoarthroscopy. *Anesthesiology* 1992; **77**: 263–6.

Chapter 45

Critical Incidents during Anaesthesia

J. M. Davies

The anaesthetic system
Concepts about the genesis of critical incidents
Review of the literature of critical incident studies in and relevant to anaesthesia

How critical incidents fit into 'audit' or 'quality assurance'
Prevention and management
Summary

The anaesthetic system

To understand critical incidents during anaesthesia, certain concepts about anaesthetic systems must be defined.

Definition and components

A system is a set of interrelated components that act together in an environment to achieve a particular outcome – in this case, the provision of safe, efficient, cost-effective anaesthetic services. Systems with both human and technical components are known as 'man–machine systems',[1] and anaesthesia is a typical example. Components of the anaesthetic system include the anaesthetist, patient, anaesthetic machine/monitors, operating room personnel (surgeon, nurses, technicians, porters), operating room equipment and facilities, and the hospital/health care authority. All possible components must be considered, to ensure inclusion of all factors that might influence or be influenced by the system. This concept must be remembered, not only when investigating anaesthetic accidents but also, and perhaps more importantly, when evaluating optimal performance of the total system; for example, to identify potential faults within the system as part of the prevention of problems.

Purpose

Systems have a function or purpose; for example, provision of anaesthetic services to a day care surgical unit. Achieving this function starts with assembly of the system components, proceeds through the activity, and ends with the product or results. (Using quality assurance terms, this corresponds to Structure, Process and Outcome.[2]) Thus, Structure/components consist of where (environment), with what tools (equipment) and by whom (personnel) the function is to be carried out. Process/activities consist of what will be done (tasks) and how it will be done (methods) – that is, the *course* of the endeavour. Outcome/results consist of knowing what was done (audit) and how well it was done (evaluation) – that is, the *consequence* of the endeavour.[3]

System characteristics

Systems are characterized by the number of their components and how these components interact (the degree of complexity) and by how the system functions (the degree of coupling).

'Complexity' is a term often used in describing types of systems; for example, 'complex man–machine system'. Strictly speaking, the term applies to the nature, number and familiarity (or visibility) of possible interactions of system components. If interactions are *complex*, they are of 'unfamiliar sequences, or unplanned and unexpected sequences, and either not visible or not immediately comprehensible'. In contrast, if interactions are *simple*, they are those 'in expected and familiar production or maintenance sequence, and those that are quite visible even if unplanned'. Most systems contain more linear than complex interactions,[4] although to most anaesthetists the majority of interactions in the operating room would appear to be complex.

How a system functions is described by the *degree of coupling*, which describes the nature of the interaction between system components: whether or not what happens in one part of the system has a direct effect on another part. 'What happens in one directly affects what happens in the other.' The degree of coupling ranges from 'loose' to 'tight'.[4] Tightly coupled systems are less likely to tolerate

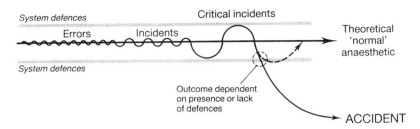

Fig. 45.1 Progression of a theoretical normal anaesthetic, showing the relation between critical incidents and system defences.

perturbations than are loosely coupled systems. In general, tightly coupled systems are recognized to have problems with Structure, Process and Outcome. Structural problems include limitations of supplies of personnel and equipment, and components with consequent high potential for failure. Procedural problems include limited options for action, inability to substitute personnel or materials, poorly executed tasks because of suboptimal ergonomics, and lack of variability of process. Outcome problems include a low tolerance for delays in process, for this failure to be reflected in outcome, and difficulty in obtaining information directly about the functioning of the system.[5] (Each of these characteristics applies to the anaesthetic system.)

'Normal' anaesthetics vs. incidents, critical incidents and accidents

Although there is a theoretical 'normal' course for an anaesthetic (see Fig. 45.1), in fact this never actually occurs. The performance of the humans in the anaesthetic system is never perfect, and, therefore, during most anaesthetics there will be many errors (of omission and commission). Take an example of a patient undergoing elective arthroscopic examination of the knee under general anaesthesia. After injection of the induction agent, the anaesthetist injects a neuromuscular blocking agent that lacks sympathomimetic properties. The anaesthetist plans then to inject slowly 1 ml of an opioid solution (which happens to have vagotonic properties). Because of distraction, the anaesthetist forgets to slow the rate of infusion before the planned (slow) injection of the opioid (an error of omission) and, because of the same distraction, injects 2 ml of the opioid (error of commission). Meanwhile, the surgeon starts the operation without asking the anaesthetist. The two (common) anaesthetic errors, in combination with the intrinsic condition of the patient (a nervous but fit, large male), and the premature surgical stimulus result in a decrease in heart rate. This is an 'incident', the result of several conditions and events, which in this case combine to provoke bradycardia. An *incident* simply implies that the course of the anaesthetic was not as planned or predicted. Incidents should be considered the normal result of human behaviour, and their occurrence should not imply any malicious intent or moral failing on

the part of an anaesthetist. Indeed, many of these 'unsafe acts'[5] will not be recognized or reported, neither will their significance be realized.

If an incident occurs, it may then evolve into, or contribute to the evolution of, a 'critical incident'[6,7] (Fig. 45.1). The transition to *critical incident* is characterized by the development of the *potential* for harm to a patient. (However, neither 'incident' nor 'critical incident' implies that the outcome for a patient will be less than optimal.) In the example, the patient's heart rate declines rapidly from 65 to 40 to 10 beats per minute. A blank line is then seen on the electrocardiograph.

A critical incident may then evolve to an 'accident' or 'complication'.[6,7] Only when the outcome of an anaesthetic is not as planned or predicted is a *complication* said to have occurred. For example, if the period of asystole results in prolonged lack of coronary or cerebral perfusion, end-organ damage will occur.

What allows the evolution of incident to critical incident or accident? Certain systems tend to evolve toward *dysfunction* (failure or catastrophe) because of the behaviour of the humans and the particular system characteristics, which are poorly tolerant of errors. In contrast, other systems seem to function with few problems, because of protective *defences*. In all of these systems there are three levels at which dysfunction may occur. As shown in Fig. 45.2, these three levels are Structure (behaviour-shaping factors), Process (event-shaping factors) and Outcome (severity-shaping factors). The associated defences prevent or decrease the probability of errors occurring (Structure), allow the detection, absorption or recovery from errors (Process), or use risk-reduction methods to decrease the consequences of errors (Outcome).

For example, consider a patient undergoing a diagnostic laparoscopic examination. The anaesthetist wishes to avoid hurting the patient (a behaviour-shaping factor). He therefore chooses to insert a small intravenous cannula. He knows that the gynaecologist is well experienced. (The provision of well-qualified personnel is a system defence intended to decrease the probability of error.) Unfortunately, the surgeon places the laparoscopic trochar into the abdominal aorta, with resultant major blood loss. (This is an event-shaping factor.) The anaesthetist responds by inserting a large-bore intravenous cannula, infusing large volumes of crystalloid, and calling for help. (These are

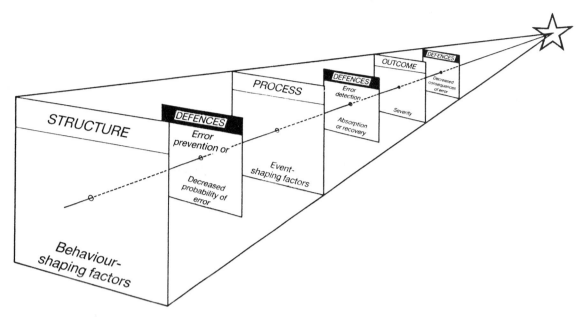

Fig. 45.2   An integrated model of system performance showing the relation between Structure, Process and Outcome. (Davies JM & Lee RB, unpublished 1991, revised 1993)

defences to decrease the consequences of the error.) The patient is a healthy 20-year-old and she tolerates the period of hypotension much better than if she had been 50 years old and with coronary artery disease (severity-shaping factors).

In general, the potential for incidents to occur is most influenced by the Structural elements of the system, or the behaviour-shaping factors. These will be countered by defences that decrease the probability of errors occurring. The potential for critical incidents to occur will be determined by the Procedural elements of the system, or event-shaping factors. These will be balanced by defences to decrease the consequences of errors. The adequacy of these defences will determine the probability of a complication occurring, the degree of adversity of which will be determined by the severity-shaping factors. Other defences modify the effects of problems on Outcome. In general, the more complex a system, the greater the need for defences.

In summary, *errors* occur so frequently that they are often not recognized – often being thought to be an intrinsic part of human behaviour. The concurrence of errors leads to *incidents*, which occur commonly (but less often than errors), although their incidence is also not known. They have low potential for severity of outcome and their occurrence is most influenced by behaviour-shaping factors. *Critical incidents* (CI) occur less frequently than incidents, but, in one study of 451 adult patients, CI were noted in 46.6 per cent of patients in the operating room (OR) and in 38.6 per cent of patients in the recovery room (RR).[8] CI have a higher potential for severity of outcome and are most influenced by event-shaping factors. *Accidents* or

complications are rare events, with an incidence ranging, for example, from about 1 in 10 for all types of non-lethal complications[9] to 1 in 185 000 for anaesthetic-related death occurring within 30 days of operation.[10]

Concepts about the genesis of critical incidents

Over the past four decades, three major concepts have been developed about why critical incidents occur.

Human error

This concept was based on the proposition that incidents and critical incidents result from human behaviour in complex systems. Humans make errors and these 'human errors' can be categorized as mistakes, slips and lapses. Mistakes occur at the level of planning an action and have therefore been defined as *planning failures*, where the intention is not appropriate.[5] They may be *rule-based*, where familiar problems are solved with *if . . . then* rules, and the mistake results from misapplication of a good rule or application of a bad rule.[5] For example, Eagle and colleagues[11] have described a case of fatal aspiration of gastric contents. The surgeon had booked the patient to undergo elective cystoscopy under local anaesthesia, using the 'rule' that 'elderly = local', although the patient was demented and unable to cooperate. Mistakes may also be *knowledge-based*, where they result from an unsuccessful attempt to solve novel problems when there are no appropriate 'rules'. The attempt is made using conscious analysis of stored knowledge of the related subject matter.[5] For example, in the case above describing bradycardia after injection of the muscle relaxant and opioid, a trainee anaesthetist might not know of the interaction between the two drugs and the consequent propensity for bradycardia. In contrast, slips and lapses are characterized as *execution failures* where the action was unintended. Slips are due to attentional failures and may be verbal ('Give me the pentothal' instead of 'pancuronium' or active (injecting pancuronium instead of reversal agents). Lapses are due to memory failures, such as forgetting planned intentions[5] (e.g. forgetting to turn on the ventilator).

Much has been written over the years about human error, and this knowledge was used by anaesthetists in an attempt to describe and explain poor outcome from anaesthesia. Indeed, some held the view that the most effective way to improve anaesthetic safety was to eliminate human error,[12] a task that is, unfortunately, not possible because errors are a normal characteristic of human behaviour. Furthermore, although the phrase 'human error' well describes what happened in a critical incident or accident, it does not explain it. This inadequacy therefore led to further analysis of what factors might facilitate human error.

Human factors

The second concept about the genesis of critical incidents was based on the study of human factors. For example, in the 1978[13] and 1984[14] studies by Cooper and colleagues, certain critical incidents occurred 'because there was a clearly described failure to recognize a hazardous condition or a failure to act on available information'.[14] The authors described some of these failures as being due to human error and then listed associated factors which 'conceivably could have contributed to the occurrence of an error or a failure to promptly detect an error'. Examples of these associated factors included 'first experience with situation' and 'conflicting equipment designs'.

In 1972, Edwards developed a conceptual model of human factors, with its four components, Software, Hardware, Environment and Liveware, giving rise to the acronym, SHEL. The fourth component, Liveware, represents man or personnel, which is now considered the central factor in the model.[15] Research into human factors recognized the importance of the variable interaction of man with each of the components. The major task of man

was seen as a processor of information, and the aim of research was to develop ways to match the capabilities and demands of the human with the task and to design systems to minimize consequences of errors.

Great strides in the understanding of human factors have been made, much arising from studies of human performance in civil and military aviation. The interested reader is referred to the comprehensive review of the subject in anaesthesia by Weinger and Englund.[16] One of the most well-studied factors is fatigue.[7,16] Although the physiological consequences of lack of sleep affect all workers, they are magnified in those whose tasks require sustained vigilance together with intensive cognitive and physical skills. Certain workers, such as airline pilots, operate under strict guidelines as to hours of duty and rest, because of recognition of the possible effects of fatigue on their performance. Similarly in anaesthesia, more than three decades ago, Kok and Mullan described the role of the human factor – lack of experience, judgement and sleep – in contributing to deaths. They stated that 'no anaesthetist should be obliged to do a long routine list the next morning when he had been up all night'.[17] This statement also acknowledges the difficulty in maintaining vigilance under conditions of low workload (stimulus), that is 'routine lists'. (Intraoperatively, anaesthetists rely more heavily on cognitive skills than do surgeons, who apply more psychomotor skills.) Recently, Wilson and Weston applied the rules for hours of work for airline pilots to junior doctors. The work of two groups of three junior anaesthetists was recorded over 3 months and ranged between 48 and 112 hours each week, with an average time worked of 75 hours. After applying the aviation work rules, 'as though scheduling air crew', calculations showed that 26 doctors would be required to cover the work of the six, were the guidelines to be followed.[18]

Other important factors particularly applicable to anaesthesia include hurry, inattentiveness and careless-ness,[19] as well as ageing. With the last, cognitive failure may lead to a decrement in knowledge, skills or behaviour. Unfortunately, physicians may carry on practising because they or their communities believe their services to be a necessity, and they may be subjectively unaware that their level of performance has declined. However, although the discipline of human factors was able to explain error on the part of human operators at the individual or small group level, the explanations did not go far enough. Further understanding requires asking (and answering) questions such as, 'why does the system allow a trainee to work in a situation where "first experience" leads to a critical incident (or to an accident)?', 'why do anaesthetists tolerate working with machines and monitors with "conflicting equipment designs"?', and 'why does the system allow such a situation to continue when non-standard and poorly designed equipment is known to contribute to operator error?'.

Contributory factors

The study of human factors has expanded knowledge and resulted in effective measures to decrease the risk of progression of incident through critical incident to accident. But to say that human factors alone are responsible for incidents is to ignore all other potential components of the event. Like small streams, each of which feeds into an increasingly large and turbulent river, so multiple factors contribute to the evolution of critical incidents, resulting in system failure, or accidents. For example, in Cooper and colleagues' 1978 study of critical incidents, those which did not contribute to actual complications had 2.5 associated factors, whereas those which were associated with a complication had 3.4 associated factors.[13]

This appreciation of the multiplicity of factors contributing to the occurrence of an incident, critical incident or complication has led in some quarters to rejection of the term 'cause'. Investigators have moved away from the position of seeking 'cause' to that of identifying those factors which 'play a significant part in bringing about an end or result'.[20] For example, the Australian Bureau of Air Safety Investigation (BASI) seeks 'significant factors', where a 'factor' is an 'element in the circumstances of an incident/accident without which the incident/accident would not have occurred'. BASI discarded 'cause' because it is often difficult (or impossible) to derive cause–effect relation in incidents and accidents. For example, bad weather may have been a significant factor in an accident, but it could not have been the cause. Although some countries still require safety investigations to nominate a single 'most probable cause' of accidents, BASI does not because it is not productive in achieving prevention of similar events. Nominating a single 'cause' tends to focus corrective action on one factor only, although many others typically are identified in an investigation. (However, this does not imply rejection of the term 'causal relations' which are intrinsic to science and medicine.) In addition, the term 'cause' has legal implications, including the concept of blame, which is not helpful. This is particularly so in the investigation of an incident or accident (in aviation or in anaesthesia), as the purpose of investigation should not include apportioning blame or liability.

Actually, the idea of contributory factors is not a new one in anaesthesia. One of the earliest to consider interrelated factors was Henry Ruth, who encouraged the establishment of 'anaesthesia study commissions'. Their purpose included review of complications and fatalities, and determination of 'factors which may have initiated the death'.[21] In 1961, Dripps and colleagues attempted to determine if the anaesthetic 'contributed to the death of a patient'.[22] More recently, the difficulty in selecting a single cause of anaesthetic-related mortality was described. Harrison argued that most 'anaesthetic deaths' have little to do with anaesthesia and that at least 90 per cent of deaths are actually the result of the operation and the patient's preoperative condition.[23] This larger contribution of patient and procedure to postoperative mortality, in comparison with anaesthesia, has been shown by others.[24] Thus, in anaesthesia, the most commonly identified contributory factors would be the patient, the operation and the anaesthetic, although many studies, of both critical incidents and complications, continue to identify a sole cause, for example 'human error' or 'machine failure'. As a result, the effectiveness of such studies in improving safety in anaesthesia is limited.

The most recent delineation of contributory factors to the genesis of critical incidents and accidents comes from research into large-scale technological disasters; for example, the Bhopal pesticide plant, the Challenger space shuttle, the Chernobyl and Three Mile Island nuclear power plants, and the Herald of Free Enterprise car ferry. At the same time, research in human engineering and industrial psychology has provided better understanding of the role of humans in these accidents.[11] This understanding has allowed researchers to move away from the 'traditional' view that most of the imperfection in innately flawed systems lies with the individuals.

In the model developed by Reason,[5,25] complex socio-technological systems (such as anaesthesia) are afflicted with two different but complementary types of flaws or failures: active and latent. The first type, *active failures*, represents the actions of the operator, in particular those that occur when humans interact with the other components of the complex system in which they work. These 'front-line operators' often appear to be the principal instigators of system breakdown. (This is where the term 'human error' was often applied.) However, rather than just instigators, their role is catalytic, providing many of the local triggering conditions necessary to expose the second type of system flaws. These crucial weaknesses, or *latent failures*, represent flaws already present or built-in to the system. Structural flaws may be found within the administration (decision-makers, line management), and staff and equipment (preconditions). Procedural flaws may be found within operational procedures. Outcome-influencing flaws may be found within intrinsic and extrinsic recovery mechanisms or system defences.

Thus, although the role of human factors is well recognized in the genesis of system failure, this will not originate solely from human error. Rather, systems will fail when active failures combine with or produce a particular set of circumstances necessary to reveal the latent failures (also described as 'resident pathogens'). When active failures combine with latent failures, the potential for system dysfunction will depend on the degree of malignancy of the latent failures – defined by their number, complexity and position in the system. All systems contain some pathogens and the more complex a system, the greater the number.[5,25]

As an example, consider the problem of difficult intubation. The anaesthetist performs laryngoscopy but is unable to view the cords. This is an *incident*, unplanned but common in anaesthesia. But the incident may become a critical incident or an accident, depending on the subsequent actions of the anaesthetist and other latent failures in the system. The anaesthetist is fatigued, having had to work the previous night and the following day because of staff shortages (a latent failure). The anaesthetist repositions the patient's head, repeats laryngoscopy and inserts the tube into the oesophagus (an active failure). This is now a *critical incident* – although the presence of the tube in the oesophagus is not inherently harmful, there is potential for harm. If the anaesthetist, although fatigued, is able to recognize oesophageal intubation, immediately removes the tube and oxygenates and ventilates the patient, no harm has come to the patient at this point. (The action of the anaesthetist is an example of an event-shaping factor.) Similarly, the anaesthetist may not recognize oesophageal intubation until a flat trace appears on the capnograph (another event-shaping factor). However, if a capnograph is not available or not working (latent failure), and the anaesthetist does not recognize oesophageal intubation (active failure), the patient will suffer a *complication*, hypoxic brain damage. The severity of outcome will depend in part on the patient's ability to withstand a period of hypoxia (a severity-shaping factor).

Review of the literature of critical incident studies in and relevant to anaesthesia

Like many other aspects of anaesthetic safety, the term 'critical incident' comes from the aviation industry. During World War II, the US Army Air Forces established the Aviation Psychology Program to select and classify air crews. This was carried out using the critical incident technique, which was a method of studying human performance. The technique consisted of a 'set of procedures for collecting direct observations of human behaviour in such a way as to facilitate their potential usefulness in solving practical problems and developing broad psychological principles'. Additional emphasis was placed on the 'importance of factual reports on performance made by competent observers'. Distinct criteria were used for the observations, with an 'incident' being 'any observable human activity that is sufficiently complete in itself to permit inferences and predictions to be made about the person performing the act'. 'Critical' required the incident to occur 'in a situation where the purpose or intent of the act seems fairly clear to the observer and where its consequences are sufficiently definite to leave little doubt concerning its effects'.[26]

These definitions are taken from Flanagan's 1954 review of the critical incident technique.[26] Although the method dates back to Sir Francis Galton in the previous century, most of the work cited by Flanagan came from US Aviation and Civil Aeronautics publications, as well as research theses. (The review thus made this early work accessible.) These first analyses of critical incidents led to criteria and recommendations for improvements in the selection and training of pilots, requirements for combat leadership, designs for cockpits and instrument panels, and procedures to prevent and overcome in-flight vertigo. When the paper was published, Flanagan (and his colleagues) had had about eight years' experience with the technique, which was considered applicable to a wide variety of situations. These included equipment design, measures of proficiency (standard samples), selection and classification, job design and purification, operating procedures, training, measures of typical performance (criteria), motivation and leadership (attitudes), and counselling and psychotherapy.

Flanagan also described the procedure for 'gathering certain important facts concerning behaviour in defined situations'. However, he emphasized that the critical incident technique was not a 'single rigid set of rules governing such data collection', but rather a 'set of principles which must be modified and adapted to meet the specific situation at hand'. Central to the technique was the use of qualified observers, making agreed-upon observations and only simple types of judgements. Stress was placed on objectivity, or the 'tendency for a number of independent observers to make the same report'. This requirement reflects back onto the initial 'precision' with which the observation was defined, as well as the 'competence of the observer'. Thus, the technique followed general scientific principles, requiring definition of the 'general aims' (or hypothesis), instructions for observers (or methods), data collection techniques, data analysis, and conclusions based on observations. (An outline of the CI technique is shown in Table 45.1.)

Since then, the United Nations' aviation safety regulatory agency, the International Civil Aviation Organization (ICAO), has defined an 'incident' as 'an occurrence, other than an accident, associated with the operation of an aircraft, which affects or could affect the safety of operation'.[27] Note that this definition applies to Process only, with the term 'safety of operation' implying only the possibility of change in Outcome. An 'accident' is defined by ICAO as 'an occurrence associated with the operation of an aircraft ... in which a person is fatally or seriously

Table 45.1 Outline of the critical incident technique*

General aims
Specify activity as a brief statement
Recognize points of view of workers vs supervisors

Plans and specifications
Observers: knowledge of activity, relation to those observed, training requirements
Observed: general description, personnel, location, time
Activity: general type, specific type, relevance, importance to general aim

Data collection
Interviews: individual and group
Questionnaires: narrative and checklist

Data analysis (incident classification)
Selection of frame of reference: uses of the data, pre-existing formats
Formulation of categories: inductive
Relation to general behaviours: logical, meaningful

Interpretation and reporting
Limitations of observations
Value of judgements
Formulation of procedures for improving performance

---

* Derived from reference 26.

injured . . . the aircraft sustains damage or structural failure . . . or the aircraft is missing or completely inaccessible'. Thus, an accident requires a change in Outcome from that predicted.

Application of the critical incident technique from aviation to anaesthesia was first made by Blum.[28] He reported difficulties with oxygen supply resulting from ambiguity of the pressure gauge design. He described how the gauge could be misread as 'full' (when it was empty) because of a double-ended indicator needle, the 'full' reading being sited almost opposite 'empty', and the small size of the dial. Noting that misreading was more likely to occur if an operator was under stress, Blum argued for the 'simplest design to supply the necessary information'. He cited the early work of the US Air Force in ergonomics and the importance of analysis of 'patterns of man–machine interface in order to adjust mechanical devices to the limitations of human nature'. Blum's appreciation of the importance of equipment ergonomics included a new design not only for the gauge but also for the layout of the anaesthetist's workplace (with swivel chair). He recognized the concept of 'vigilance' with respect to human perception and 'negative transfer', where carrying out one manoeuvre could be influenced or even negated by previously established patterns of manoeuvres. He even criticized the equipment manufacturers for continuing to produce their own (different) designs while being in need of 're-evaluation of responsibilities and recognition of the importance of ergonomic principles'. His statements about the influence of human perception and reaction on the

effectiveness of equipment and the use of appropriate guidelines to ensure safety remain as appropriate now as then.

Wide-scale application of a version of the critical incident technique to anaesthetic practice was made by Cooper and co-workers.[13] In anaesthesia, 'critical incident' has been used to refer to both Process and Outcome. For example, Cooper and colleagues reviewed critical incidents that 'could have led (if not discovered or corrected in time) or did lead to an undesirable outcome, ranging from increased length of stay to death or permanent disability'. Thus, their definition of 'critical incident' was a mixture of Process (could have led) and Outcome (did lead). They reviewed critical incidents which both did and did not lead to a change in Process of care (increased length of stay) or in Outcome (permanent disability or death). (However, the authors also stated that outcome was not considered a criterion for inclusion of an incident.) In addition, only 'clearly preventable' incidents involving 'human error or equipment malfunction' were included – thus fulfilling one of the requirements of the CI technique, that of specific standards for evaluation. Information about preventable mishaps was collected during a tape-recorded interview from both staff and resident anaesthetists who were selected at random. Although experienced with interview techniques, the observer had training only in the rudiments of anaesthetic practice. During the 60- to 90-minute interview, prompting questions (from a prepared list) were offered, but no suggestions were made as to a specific type of incident. Between September 1975 and April 1977, 552 incidents were recorded, of which 359 met all criteria. Incidents, as well as circumstances contributing to their occurrence, were then classified into 23 major categories. The most commonly occurring incidents involved problems with connection of the breathing circuit, gas supply, tracheal intubation, syringe swap and hypovolaemia (and, despite the authors' definition of critical incident, the listed problems included only incidents and not accidents). Readers should note that all but the last category of problems reflect the interface between the anaesthetist and equipment whereas hypovolaemia reflects the interface between the anaesthetist and the patient. In addition, the participants described associated factors that they felt contributed to each incident.

Certain findings were made. First, more than 80 per cent of critical incidents were thought to be due to 'human error' and only 14 per cent were related to 'equipment failure'. This ratio is remarkably similar to the finding in other systems. Second, more of the reported incidents occurred during the day (79 per cent) than at night (21 per cent), although this finding did not quite parallel the distribution of work (85 per cent day vs 15 per cent night). The authors did not discuss this finding but one interpretation is that a greater proportion of daytime incidents was reported in contrast to night-time incidents – a potential problem with voluntary reporting schemes. Third, despite a preconception by the authors that most problems would occur at the beginning or end of cases, only 26 per cent of incidents were noted during induction and less than 10 per cent at emergence, with 42 per cent during maintenance. The authors correctly suggested that the large numbers of incidents reported during maintenance could be due to the fact that many errors proceed to a point at which they are finally recognized as incidents. They also postulated that anaesthetists who expected most problems to occur during induction or emergence might anticipate, and therefore detect, fewer problems during maintenance than those not trapped by this preconception. However, the authors did not consider the weighting of the number of incidents to duration of each phase of the anaesthetic. In a case lasting 60 minutes, induction should take about 6 minutes. Thus, the occurrence of 26 per cent of incidents in only 10 per cent of the time would suggest an 'incident-rich' period. Fourth, a relief anaesthetist often led to the discovery of an incident. This highlights the importance of protocols for the taking over of cases, as is done in the cockpit. Fifth the associated factors cited most frequently included inadequacy of total experience and lack of familiarity with device/equipment. However, the study was carried out in a large American teaching hospital with incidents reported by staff and resident anaesthetists of widely varying experience, and where the staff anaesthetists acted mainly in a supervisory capacity. The authors noted that many of the incidents were those which occur as part of training. Thus, the results of this study may not be applicable to an institution with only staff anaesthetists.

The next application to anaesthesia was in 1981 – in Craig and Wilson's 'survey of anaesthetic misadventures'.[29] Over 6 months, consultants and trainees contributed anonymous reports of anaesthetic misadventures. This term was equated to critical incident and defined as 'any anaesthetic mishap, major or minor, which was harmful or potentially harmful to the patient, and associated with human error or equipment failure'. As with our studies, the distinction between critical incident and accident was not clear, as the list of 'incidents' included 'postoperative respiratory depression', 'crown dislodged at intubation', 'crown broken on Guedel airway' and 'damage to eyes under adhesive tape'. During the study period, 8312 anaesthetics were given, of which 6422 were elective cases, 1765 were emergency cases and 125 were epidurals for obstetric patients. A total of 81 'misadventures' were reported, of which 36 represented 'problems with transmission of anaesthetic gases and vapours to the patient', 21 represented problems with 'tracheal tubes', 12 represented problems with 'drug administration', and 12 were labelled as 'miscellaneous'. Three of these last incidents involved the

airway, emphasizing the magnitude of problems with this area.

The authors also attempted to assign 'associated factors', either human error (67 per cent) or equipment failure (16/81), or both (10/81). The report described one case where neither factor could be assigned, thus reflecting, like other papers of its time, the prevalence of the then-current model of human error in the genesis of a critical incident. For example, disconnections were thought to be related to a 'combination of faulty or ill-fitting equipment plus inadequate care in assembly'. Today, these problems would be recognized as 'latent failures' in the system and reflect administrative problems.

In 1984, adaptations of the CI technique were again used by Cooper and colleagues, in a two-phase study.[14] In phase one, a trained (non-anaesthetist) interviewer elicited details from volunteer anaesthetists (staff, residents, nurses) about preventable incidents involving human error or equipment failure. Participants worked in one of four hospitals, two large institutions with training schemes and two smaller ones with private practice anaesthetists supervising nurses. In phase two, volunteers were questioned (from a list) about specific types of incidents, including those involving a relieving anaesthetist. These volunteers were then considered 'trained observers' and asked to report future incidents as soon as possible after their occurrence. The definition of critical incident was similar to that used in the previous study, and again represented a mixture of Process and Outcome: 'a human error or equipment failure that could have led (if not discovered or corrected in time) or did lead to an undesirable outcome, ranging from increased length of stay to death'. Classification of causal patterns was related to subsets of incidents, one of which was described as 'substantive negative outcome' (SNO). An SNO was defined as 'mortality, cardiac arrest, cancelled operative procedure, or extended stay in the recovery room, in an intensive care unit or in the hospital'.

Of a total of 1887 reports, 1089 qualified for analysis, criteria for which again included preventability. Separate analysis was carried out for phase one (retrospective reporting) and phase two (instant reporting). In the 1978 study, incidents were analysed for human error or equipment failure. In this study, a third category, 'disconnection', was tabulated. Thus, depending on how the data were analysed (with disconnection included or counted separately), the attribution of human error may be compared with the 1978 results and between phase one and phase two of the 1984 study. With disconnection included, human error represented 83 per cent in phase one and 77 per cent in phase two (unchanged from 1978). Equipment failure was cited at 11 per cent in phase one and 19 per cent in phase two (slight change from 1978). The authors tabulated both the most frequently occurring critical incidents and those which led to an SNO. The authors

Table 45.2 Areas for system improvement for prevention or detection of critical incidents, classified according to the SHEL model*

Software
Specific protocol development
More complete preoperative assessment
Equipment or apparatus inspection

Hardware
Equipment/human factors
Additional monitoring instrumentation

Environment
Specific organizational factors
Communication

Liveware
Personnel selection
Additional training
Supervision/second opinion

* Derived from reference 14.

reported an increased incidence of adverse outcomes of errors in less healthy patients, reflecting both slower discovery of incidents and a smaller margin for safety in the seriously ill. A list of 'potential strategies for prevention or detection' of incidents was developed and is presented in Table 45.2, but with each strategy classified according to the SHEL model. Perhaps most important in the paper was the conclusion that 'perhaps the most insidious hazard of anesthesia is its relative safety'. This statement thus gives recognition to the problem of lack of sustained vigilance (a behaviour-shaping factor) in the genesis of critical incidents.

Next to report a critical incident study were Williamson and colleagues who carried out two trials, first for 6 months in 1980 in hospitals in Townsville, Queensland, and second for 4 months in 1984 in selected hospitals in Brisbane, Sydney, Townsville and Wollongong.[30] The criteria for participation (voluntary) and definition of critical incidents were virtually the same as in the Boston studies. The Australian authors also did not differentiate between Process and Outcome of care. They described the CI technique as a method of sampling, not only 'mistakes' but also morbidity and mortality. (However, the critical incidents reported were only events, not outcomes.) The authors stated also that the CI technique offered simplicity and was easily applied in different centres. An incidence of 80 per cent for human error as a factor was noted, although a large number of equipment problems was noted in one centre. (This was followed by improvements in equipment.) The authors concluded that, although the CI technique could not detect an incidence for any problem, it offered the additional advantage of 'perceived freedom from punitive threat for the great majority of incidents'.

The critical incident studies described above are examples of large studies that have provided important information. However, studies on a smaller scale, such as might be carried out by a single anaesthetic department, have proved to be equally useful. For example, Kumar and associates undertook a prospective survey of critical incidents between April 1984 and January 1985 and between April 1985 and January 1986.[31] They defined a critical incident as an 'incident or mistake which could be harmful or potentially harmful to the patient during management of anaesthesia'. (Note that this definition conforms to the concept of potential for harm only.) Data were collected using a simple form with space for narrative comment about the nature of the event and outcome and a checklist for associated factors. As a result of reports from the first phase, an equipment checklist was placed on each anaesthetic machine before each scheduled procedure and the anaesthetist asked to complete it. These forms were collected at the end of the case. The anaesthetists included staff, residents, interns from other specialties and senior medical students. Although the attribution of incidents to human error (80 per cent) and to equipment problems (20 per cent) remained unchanged between the two periods, the number of reports was halved. In particular, there was a marked increase in certain associated factors – 'failure to perform a normal check' and 'lack of experience or familiarity with equipment'. The authors suggested that the decline in reported events might have resulted from replacement of old anaesthetic machines, introduction of an equipment checklist, and increased awareness of problems from regular discussion at departmental meetings. They recognized that under-reporting of incidents was inevitable. However, they emphasized the importance of the CI technique to identify specific problems and of the need for a national or regional database.

Another example of a single study was that of McKay and Noble.[32] They used the CI technique to define indications for monitoring with a pulse oximeter during anaesthesia, specifically to study when and where the pulse oximeter shortened the time to detection of critical events. (This study was carried out before use of oximetry was routine.) Between October 1986 and February 1987, anaesthetists were asked if the oximeter was the first warning of an unexpected problem and if false alarms occurred. If the answers were positive, the chart was examined and the anaesthetist interviewed, within 2 days of the event. A critical incident was recorded if an unexpected physiological deterioration severe enough to require intervention by an anaesthetist to prevent a likely bad outcome was signalled first by the oximeter. Reports of events requiring intervention were made in two-thirds of the 4797 anaesthetics given at the regional specialist care (tertiary-care) centre. A critical incident occurred in 191 (6 per cent) of these, the most common event being desaturation ($n = 151$), followed by a change in heart rhythm ($n = 45$). Although this study did not determine if the rate of complications was decreased by use of oximetry, it did show that the time to discovery of an increased risk of complication was shortened (in 4 per cent). The authors calculated the cost of pulse oximetry to be about $C2.40 per case, in contrast to the cost of more than $C1 million (1988) for a patient with hypoxic brain damage. The CI technique was considered to be advantageous in permitting inexpensive study of a rare event (hypoxia) in a large database.

In 1989, a very large critical incident monitoring survey was instituted in Australia. At a conference 2 years previously, the need for a national study was determined and the Australasian Incident Monitoring Study (AIMS) was established. AIMS continues to be coordinated by the Australian Patient Safety Foundation (APSF). This is a non-profit-making organization founded to promote, organize, fund, conduct research into and establish mechanisms for advancing patient safety. The APSF has links with various bodies, including bioengineering, consumer, medical and paramedical groups, representatives of the health care industries, standards associations, morbidity and mortality committees, the Bureau of Air Safety Investigation and the American Patient Safety Foundation.[33]

The purpose of AIMS is to provide a database to allow the identification of commonly occurring errors and other potentially hazardous incidents in daily anaesthetic practice. Both critical and non-critical incidents are reported, a critical incident being 'any untoward event, or mishap, not necessarily harmful' but clearly preventable. The database should allow development, evaluation and refinement of strategies to reduce or eliminate errors or hazards. Facilitation of voluntary reporting of critical incidents occurs with a coordinator or 'person on the spot' (POS), who ensures that each incident is identified only by a form number. Thus, confidentiality and anonymity of reporter, patient and anaesthetist are maintained.[34] AIMS emphasizes that the reporting of incidents is voluntary and that a non-threatening atmosphere in the department is essential to reporting of events. Questionnaires are used to gather data. The reporting anaesthetist writes a narrative summary of the event and then makes multiple-choice-style responses for other details while the POS supplies key words for coding (Table 45.3).[35] Data are collated at a central office and the degree of confidentiality is such that not even the identity of more than 90 participating hospitals and private group practices can be determined. This is of particular importance when reporting incidents from geographically distinct areas. AIMS coordinators provide regular feedback reports, as well as more detailed descriptions of particular incidents from which all identifying details have been removed. AIMS has received more than 2500 reports and the first 2000 incidents have now been analysed. A wealth of data are now available, with two of the 20 papers cited here.

Table 45.3   AIMS anaesthetic incident report*

Item 1: Description of incident

Item 2: Keywords

Item 3: What happened?
  circuitry incident
  circuitry involved
  equipment involved
  multiple incident
  pharmacological incident
  airway incident

Item 4: Why it happened
  factors contributing to incident
  factors minimizing incident
  suggested corrective strategies

Item 5: Anaesthesia and procedure
  procedure category
  elective admission
  anaesthesia at time
  monitors in use

Item 6: When and where it happened
  phase when alerted
  duration of incident
  location

Item 7: Patient outcome
  immediate effects
  final outcome

Item 8: To whom it happened
  patient classification (ASA)
  emergency
  patient age group

* Derived in part from reference 35.

In the first 2000 reports analysed, 144 (7 per cent) involved 'wrong drug' incidents, where 'wrong drug' included 'wrong ampoule', 'syringe swap' and 'other'. The first two of these were assessed as to 'wrong action by the anaesthetist' and 'correct action but a wrong result in circumstances which misled the anaesthetist'. If the wrong ampoule was chosen, the risk of giving the drug was 58 per cent; if the drug was drawn up in a syringe, this risk rose to 93 per cent. Factors influencing choice of a wrong ampoule included similarity of ampoule (18 per cent), ampoule in wrong location (8 per cent) and 'getter not giver' (5 per cent). Factors influencing a syringe swap included similar labelling (24 per cent), same size syringe (20 per cent) and 'getter not giver' (11 per cent). 'Other' incidents included mistakes with vaporizers, wrong line injection, label mix-up with intravenous sets, and syringe driver mistakes. Factors that contributed to these critical incidents included similarity of labelling, shape or colour of syringes or ampoules, inattention, haste, fatigue and communication problems.[36]

Another serious problem detected in the reports was that of interruption of delivered gas, in 317 incidents.

Disconnections were cited in 148 incidents, leaks in 129 incidents and misconnections in 36 incidents, with complete failure to ventilate in 143 cases. Disconnections were most often detected first by a monitor (61 per cent) – the low pressure alarm in 55 incidents (37 per cent). However, this alarm 'failed to warn of non-ventilation in 12 incidents (in 6 because it was not switched "on" and in 6 because of a failure to detect the disconnection)'. This failure was usually associated with ventilator bellows descending in exhalation (i.e. 'hanging bellows'). In one-third of cases, disconnection was associated with 'interference to the anaesthetic circuit by a third party' and, in nearly one-half of the cases, with head and neck surgery. Leaks were commonly associated with 'seal failure of the absorber system' whilst misconnections most commonly involved the scavenging system. Over all, 'failure to check' was the main contributing factor in 108 of 317 incidents.[37]

As well as evaluation of anaesthetic care, critical incident studies have been used to evaluate the clinical performance of trainee anaesthetists (anaesthetic residents).[38] The Clinical Anesthesia System of Evaluation (CASE) involves quantifying daily comments describing the performance of trainee anaesthetists. Advantages of CASE include providing specific real-time evaluations, improving sampling techniques of evaluation, enhancing the discriminatory power of the evaluation process, and improving documentation.

How critical incidents fit into 'audit' or 'quality assurance'

Since the specialty was established almost 150 years ago, anaesthetists have been concerned about the safety of their practice. From John Snow's audit of deaths after ether[39] to the CEPOD study with its anaesthetic and surgical cooperation,[10] the specialty has led the way in what is now termed medical quality assurance (QA). With its components of Structure, Process and Outcome, QA is a method of determining whether a system is organized, working and producing to its optimum level.[3] This requires giving consideration to the contribution to system function by other operating room personnel, the operating room environment, the organization of the hospital, and local, regional and national health policies.[11] Traditionally, however, anaesthetists have focused on the endpoint of their care – Outcome – rather than Process or Structure. Thus, anaesthetists have tended to study complications, the spectrum of which ranges from death (mortality) to dysfunction (morbidity), with some studies of mortality including cases of hypoxic cerebral damage.[22,40] Non-fatal

complications have been described as 'unplanned, unwanted, undesirable consequence(s) of anaesthesia',[41] and may be subdivided into disease, disability, discomfort and dissatisfaction. A fifth 'D', dollars, represents the cost of complication and its treatment.[42] Major complications, i.e. the first two 'D's, are most often detected and diagnosed by physicians, nurses and/or health record personnel. Minor complications, or discomfort, include such problems as nausea, sore throat and dental damage. These complaints, although not life-threatening, may become the reason for (and constitute the majority of) law suits.[43] Lack of satisfaction with care is not only a valid measure of outcome but may also provide useful information for review of the anaesthetic process. Such patient-described otucome[44] has the advantage of providing data that lack the bias of those who actually provided the care.

Studies of complications provide results as a ratio of the number of anaesthetics, thus giving some indication of the number of individuals (in a similar population and under similar conditions) at risk of that complication. For example, mortality requires tabulation of the number of deaths (numerator) as a proportion of the number of patients who could have died (denominator). Morbidity is generally expressed per 1000 individuals at risk for frequently occurring problems, and per 10 000 anaesthetics for events of low frequency.[9] In contrast, critical incident studies provide a numerator, but rarely a denominator, usually because of the voluntary reporting systems used with most studies. Lack of a denominator remains one of the major problems with most critical incident studies, with a resultant inability to predict the possibility (or risk) of that incident occurring in future. Another problem, also related to the intrinsic nature of critical incident studies is the lack of randomness of the sample being studied. The purpose of using a random sample is to ensure equal representation of all components of the population being studied. Thus, when analysing the results of critical incident studies, extrapolation to larger populations must be made cautiously.

The majority of data about complications come from retrospective studies which are relatively easy and inexpensive to carry out and do not require large amounts of time to complete. There are certain potential disadvantages: bias at the stage of study design; need for complete and legible records; inadequate recording of information at the time of documentation; a change in pattern of medical practice over the period of study; and insufficient numbers of cases, requiring pooling of data with a difference in assigned values. Application of the technique of meta-analysis, for example, as done by Pace[45] in reviewing prevention of suxamethonium (succinylcholine) myalgias, may solve some of these problems but carrying out well-defined prospective studies is the preferred technique. These require careful planning with formulation of a hypothesis, are complicated and expensive to carry out, and the actual study may influence its own outcome through the Hawthorne effect. In addition, prospective studies are not immune from bias that may occur during the process of observation: not observing, not recognizing or not recording. A variation on prospective studies is that of 'occurrence screening', in which patients are enrolled and observed. This type of study allows consideration and inclusion of any adverse event and is less prone to bias in observing and reporting. The reports of the Canadian four-centre study of anaesthetic outcomes are recently published examples.[46–48]

Other sources of Outcome data include isolated (or anecdotal) case reports and reports from medical protection societies. The former are of usually rare events (but with a resultant lack of denominator and therefore of calculated incidence). However, they are useful in providing case details and the wide variation possible in pathophysiology, and certain reports have led to the recognition of specific syndromes, such as malignant hyperthermia.[49] Reports from medical protection societies or malpractice insurance companies usually detail outcomes that have led to litigation, the denominator being the number of anaesthetists insured rather than number of patients at risk. However, trends in problems leading to litigation may reflect changes in anaesthetic practice. For example, recent data from the Canadian Medical Protective Association, which represents more than 95 per cent of Canadian physicians, suggest a decline in problems associated with anaesthetic equipment but continuing problems with perioperative ulnar neuropathy (associated with positioning) and dental damage.[43]

These studies of anaesthetic outcome have been the mainstay of the epidemiology of anaesthesia, detailing the end result of what can go wrong. However, restricting study to mortality is to look only at the end of a complex and intricate process. Indeed, postoperative deaths have been described as the tip of an iceberg of clinical mismanagement.[50] More recently there has been a need to understand why things go wrong, which is partly filled by the study of critical incidents. As well stated by Bunker,[51] 'if the ultimate goal is to be able to predict postoperative outcome when patients are counseled for an operation, ... the quality of care offered by the surgeon, anesthesiologist and hospital must also be included in the analysis'. In this sense, critical incident studies show the 'normal' pattern of what can occur in the course of an anaesthetic, without attempting to predict outcome.

Prevention and management

Probably the most important step in preventing critical incidents is accepting that incidents or unsafe acts occur

and may progress to critical incidents or accidents, unless recognized and corrected. In addition, everyone concerned with anaesthesia must accept that a certain amount of variation in practice is normal. Unfortunately, the degree of variation is not known. What is needed in anaesthesia is a 'black box' to monitor both patient and anaesthetic machine, thus demonstrating what actually occurs during anaesthesia. In the aviation industry, monitoring devices now detect variation from recognized 'exceedance limits'. Similar monitors are required in anaesthesia. These would not be used to identify 'bad apple'[11] anaesthetists but to develop a database demonstrating the true degree of variability of anaesthetic practice and of patient response.

What should an anaesthetist do if a critical incident develops, either in the operating theatre or the recovery room? The first step is to accept that critical incidents may occur – to even the brightest and the best – because of the influence of other factors such as the patient, the operation and the system.[52] Refusal to recognize the occurrence of a critical incident is to contribute to the 'chain of accident evolution'.[6] Second, a rapid assessment should be made to delineate exactly what the problem is. Most often, rechecking airway/breathing/circulation (ABC) will provide a diagnosis, and, in doing so, may facilitate treatment and devolution of the critical incident. Third, the anaesthetist should ask for help – for assessment, management and record keeping. Depending on where and when the critical incident occurs, this may also mean having the surgeon stop operating (if possible). Fourth, assessment and treatment should be continued, following ABC and then following an appropriate protocol.[53] For example, protocols are available (and should be practised) for many problems, such as failed intubation, malignant hyperthermia and cardiac arrythmias/arrest. (This is where simulators, such as those described by Gaba and colleagues, may have their best practical application.[6,52,53]) Fifth, at various stages of treatment/resuscitation, a pause should be made to reassess. Who actually does this will depend on the availability of help – sometimes it is better to have a 'fresh eye', that is an individual who was not involved (physically or psychologically) with the case from the start.[54] Sixth, when the critical incident (or accident) has ended, and the patient's condition stabilized, attention should turn to the completion of records. The anaesthetic record should be filled in with as much detail as possible. The law recognizes the problems of such record keeping during times of difficulty – for example, cardiac arrest, massive haemorrhage. Annotation after the fact is perfectly acceptable, as long as the timing of the annotation is also documented and a statement made to this effect. Use of a different colour of ink is helpful. A narrative summary of the facts of the events should also be written or dictated,[54] as other anaesthetists may be the only physicians able to interpret a complex anaesthetic record. Copies should then be made of all records.

If complications occur, it is important that investigation of contributory factors be carried out immediately, while the memory of the case is still fresh.[6] Investigation should be carried out systematically, with attention paid to personnel as well as to equipment and the environment.[54–56] The importance of developing a database about anaesthetic practice cannot be overemphasized,[57] because of the benefits of 'allowing hazards to be pinpointed and warnings issued to prevent repetition'.[29] However, the information collected must be accurate, as the passage of time obliterates not only physical evidence but also memory of circumstances and events.

Summary

Anaesthesia is a complex system centring on the interaction between man (anaesthetist, patient), machine (anaesthetic machine and monitors) and the environment (surgeon, nurses, the operating room and the hospital/ health care authority). Anaesthesia is also a dynamic system, seen as a continuum from the Structural components (equipment, personnel), the Procedural components (the actual anaesthetic and operation), and the Outcome components (what is predicted). How the continuum proceeds will depend, as in other systems, on the interaction of each of these components (active factors) and the intrinsic flaws in the system (latent factors). During conditions of normal function of the system, many incidents will occur, with most of these related to human performance. The outcome of the patient should be optimal if the defences of the system are appropriate. However, if the system contains design and administrative flaws (latent factors) which overcome system defences (particularly at the Procedural level), incidents may develop into critical incidents – with potential for less than optimal outcome of the patient. Furthermore, a decrease in or negation of certain defences (particularly at the Outcome level) will facilitate progression of the critical incident to a complication or accident. Knowledge of the importance of each of the system components, and especially of human factors, should lead to understanding and prevention of critical incidents and accidents and to improved safety of anaesthesia.

REFERENCES

1. Leplat J. Occupational accident research and systems approach. In: Rasmussen J, Duncan K, Leplat J, eds. *New technology and human error*. Chichester: John Wiley, 1987.
2. Donabedian A. Evaluating the quality of medical care. Part 2. *Milbank Memorial Fund Quarterly* 1966; **11**: 166–206.
3. Davies JM. Quality assurance: learning from the American experience. *Hospimedica*. The International Journal of Hospital Medicine 1991; Jan/Feb: 14–16.
4. Perrow C. *Normal accidents: living with high-risk technologies*. New York: Basic Books, 1984.
5. Reason J. *Human error*. Cambridge: Cambridge University Press, 1990.
6. Gaba DM, Maxwell M, DeAnda A. Anesthetic mishaps: breaking the chain of accident evolution. *Anesthesiology* 1987; **66**: 670–6.
7. Gaba DM. Human error in anesthetic mishaps. *International Anesthesiology Clinics* 1989; **27**: 137–47.
8. Vaughan RW, Vaughan MS, Hagaman RM, Cork RC. Predicting adverse outcomes during anesthesia and surgery by prospective risk assessment. *Anesthesiology* 1983; **59**: A132.
9. Cohen MM, Duncan PG, Pope WDB, Wolkenstein C. A survey of 112,000 anaesthetics at one teaching hospital (1975–83). *Canadian Anaesthetists' Society Journal* 1986; **33**: 22–31.
10. Buck N, Devlin HB, Lunn JN. *The Confidential Enquiry into Perioperative Deaths*. London: Nuffield Provincial Hospital Trust/Kings Fund, 1987.
11. Eagle CJ, Davies JM, Reason J. Accident analysis of large-scale technological disasters applied to an anaesthetic complication. *Canadian Journal of Anaesthesia* 1992; **39**: 118–22.
12. Cooper JB, Gaba DM. A strategy for preventing anesthesia accidents. *International Anesthesiology Clinics* 1989; **27**: 148–52.
13. Cooper JB, Newbower RS, Long CD, McPeek B. Preventable anesthesia mishaps: a study of human factors. *Anesthesiology* 1978; **49**: 399–406.
14. Cooper JB, Newbower RS, Kitz RJ. An analysis of major errors and equipment failures in anesthesia management: considerations for prevention and detection. *Anesthesiology* 1984; **60**: 34–42.
15. Hawkins FH. *Human factors in flight*. London: Gower Technical Press, 1987.
16. Weinger MB, Englund CE. Ergonomic and human factors affecting anesthetic vigilance and monitoring performance in the operating room environment. *Anesthesiology* 1990; **73**: 995–1021.
17. Kok OVS, Mullan BS. Deaths associated with anaesthesia and surgery. A review of 1,573 cases. *Medical Proceedings* 1962; **15**: 31–5, 55–61, 76–82, 91–8.
18. Wilson AM, Weston G. Application of airline pilots' hours to junior doctors. *British Medical Journal* 1989; **299**: 799–81.
19. Harrison GG. Anaesthesia accidents. *Clinics in Anaesthesiology* 1983; **1**: 415–29.
20. *Webster's ninth new collegiate dictionary*. Markham, Ontario: Thomas Allen, 1989.

21. Ruth HS. Anesthesia Study Commissions. *Journal of the American Medical Association* 1945; **127**: 514–17.

22. Dripps RD, Lamont A, Eckenhoff JE. The role of anesthesia in surgical mortality. *Journal of the American Medical Association* 1961; **178**: 261–6.

23. Harrison GG. Anaesthetic contributory death – its incidence and causes. I. Incidence. *South African Medical Journal* 1968; **42**: 514–18.

24. Cohen MM, Duncan PG, Tate RB. Does anesthesia contribute to operative mortality. *Journal of the American Medical Association* 1988; **260**: 2859–63.

25. Reason J. The contribution of latent failures to the breakdown of complex systems. *Philosophical Transactions of the Royal Society of London* 1990; **B327**: 475–84.

26. Flanagan JC. The critical incident technique. *Psychological Bulletin* 1954; **51**: 327–58.

27. International Civil Aviation Organization. Annex 13, Chapter 1, Convention on International Civil Aviation. International Standards and Recommended Practices. Aircraft Accident Investigation Montreal, May 1988.

28. Blum LL. Equipment design and 'human' limitations. *Anesthesiology* 1971; **35**: 101–2.

29. Craig J, Wilson ME. A survey of anaesthetic misadventures. *Anaesthesia* 1981; **36**: 933–6.

30. Williamson JA, Webb RK, Pryor GL. Anaesthesia safety and the 'critical incident' technique. *Australian Clinical Review* 1985; **5**: 57–61.

31. Kumar V, Barcellos WA, Mehta MP, Carter JG. An analysis of critical incidents in a teaching department for quality assurance. *Anaesthesia* 1988; **43**: 879–83.

32. McKay WPS, Noble WH. Critical incidents by pulse oximetry during anaesthesia *Canadian Journal of Anaesthesia* 1988; **35**: 265–9.

33. Runciman WB. The Australian Patient Safety Foundation. *Anaesthesia and Intensive Care* 1988; **16**: 114–16.

34. Runciman WB. Report from the Australian Patient Safety Foundation: Australian Incident Monitoring Study. Anaesthesia and Intensive Care 1989; **17**: 107–8.

35. Webb RK, Currie L, Morgan CA, *et al.* The Australian Incident Monitoring Study – an analysis of 2000 incident reports. *Anaesthesia and Intensive Care* 1993; **21**: 520–8.

36. Currie L, Mackay P, Morgan C, *et al.* The 'wrong drug' problem in anaesthesia – an analysis of 2000 incident reports. *Anaesthesia and Intensive Care* 1993; **21**: 596–601.

37. Russell WJ, Webb RK, Van der Walt J, Runciman WB. Problems with ventilation – an analysis of 2000 incident reports. *Anaesthesia and Intensive Care* 1993; **21**: 617–20.

38. Rhoton F. A new method to evaluate clinical performance and critical incidents in anesthesia: quantification of daily comments by teachers. *Medical Education* 1990; **24**: 280–9.

39. Snow J. *On chloroform and other anaesthetics: their action and administration.* London: John Churchill, 1858.

40. Tiret L, Desmonts JM, Hatton F, Vourc'h G. Complications associated with anaesthesia – a prospective survey in France. *Canadian Anaesthetists' Society Journal* 1986; **33**: 336–44.

41. Lunn JN. Preventable anaesthetic mortality and morbidity. *Anaesthesia* 1985; **40**: 79.

42. Cohen MM. Using epidemiology to study adverse outcomes in anaesthesia. *Canadian Journal of Anaesthesia* 1990; **37**: Sxlv–xlviii.

43. McIntyre RW. Medico-legal consequences. *Canadian Journal of Anaesthesia* 1991; **38**: 1035–6.

44. Pagenkopf D, Davies JM, Bahan M, Cuppage A. A complementary approach to outcome analysis in the parturient. *Quality Assurance in Health Care* 1991; **3**: 241–5.

45. Pace NL. Prevention of succinylcholine myalgias: a meta-analysis. *Anesthesia and Analgesia* 1990; **70**: 477–83.

46. Cohen MM, Duncan PG, Tweed WA, *et al.* The Canadian four-centre study of anaesthetic outcomes. I. Description of methods and populations. *Canadian Journal of Anaesthesia* 1992; **39**: 420–9.

47. Cohen MM, Duncan PG, Pope WDB, *et al.* The Canadian four-centre study of anaesthetic outcomes. II. Can outcomes be used to assess the quality of anaesthesia care? *Canadian Journal of Anaesthesia* 1992; **39**: 430–9.

48. Duncan PG, Cohen MM, Tweed WA, *et al.* The Canadian four-centre study of anaesthetic outcomes. III. Are anaesthetic complications predictable in day surgical practice? *Canadian Journal of Anaesthesia* 1992; **39**: 440–8.

49. Denborough MA, Lovell RRH. Anaesthetic deaths in a family. *Lancet* 1960; **2**: 45.

50. Harrison GG. Death attributable to anaesthesia. A 10-year survey. *British Journal of Anaesthesia* 1978; **50**: 1041–6.

51. Bunker JP. Predicting outcome. *Anesthesiology* 1979; **51**: 1–2.

52. Runciman WB. Crisis management. *Anaesthesia and Intensive Care* 1988; **16**: 86–8.

53. Gaba DM, DeAnda A. The response of anesthesia trainees to simulated critical incidents. *Anesthesia and Analgesia* 1989; **68**: 444–51.

54. Howard SK, Gaba DM, Fish KJ, Yang G, Sarnquist FH. Anesthesia crisis resource management training: teaching anesthesiologists to handle critical incidents. *Aviation, Space and Environmental Medicine* 1992; **63**: 763–70.

55. Davies JM. On-site risk management. *Canadian Journal of Anaesthesia* 1991; **38**: 1029–30.

56. Armstrong JN, Davies JM. A systematic method for the investigation of anaesthetic incidents. *Canadian Journal of Anaesthesia* 1991; **38**: 1033–5.

57. Lee RB. Why accidents happen. *Canadian Journal of Anaesthesia* 1991; **38**: 1030–1.

Chapter 46

Inherited Disease Affecting Anaesthesia

P. M. Hopkins and F. R. Ellis

Malignant hyperthermia	Inherited disorders of metabolism
Other inherited muscle diseases	Inherited abnormalities of plasma cholinesterase

In order to appreciate fully the significance of an inherited condition, having elicited a positive family history, the clinician must have a basic understanding of the genetic basis of the disease as well as its implications for anaesthesia. Some important genetic concepts will therefore be briefly described.

The genetic code that determines the various characteristics of the individual is carried in the double helical structures of deoxyribose nucleic acid (DNA) that make up the chromosomes of the nuclei of cells. The normal human chromosomal complement is 23 pairs, 22 of which are autosomes with one pair of sex chromosomes (XX in females, XY in males). Genes are regions of chromosomal DNA which, by nature of the order of the constituent nitrogenous bases, are responsible for the arrangement of the 20 different types of amino acids which form the various peptide chains that are the building blocks of proteins. The site of a gene on its chromosome is termed the locus for that gene. Through the process of mutation over many millions of years, each gene exists in one or more forms, or alleles. In many instances different alleles of the same gene will produce a protein with identical function, but in others some alleles will produce proteins with altered function which may give rise to pathology – an inherited disease. Whether the presence of a disease allele leads to the clinical expression of the condition (disease phenotype) is determined by the pattern of inheritance of the condition.

A dominant condition is one that occurs when only one of the pair of alleles is the disease allele, that is the disease phenotype occurs with heterozygous genotype. A recessive condition, however, requires two disease alleles or homozygous genotype for the disease phenotype to be expressed. Inherited diseases are further classified according to whether the abnormal gene is present on the sex chromosomes (X-linked, as very little functioning genetic material is present on the Y chromosome) or absent (autosomal). Family pedigrees illustrating the three commonest patterns of inheritance found with single gene disorders (autosomal dominant, autosomal recessive and X-linked recessive) are shown in Fig. 46.1. Table 46.1 lists the more common inherited disorders according to their mode of inheritance.

As can be seen from Table 46.1, there are inherited diseases affecting each of the systems of the body, with many being multisystemic. The relevance of some of these to anaesthesia is described in other chapters (20, 36, 37). Many common disorders described in other sections of the book, such as diabetes mellitus and coronary heart disease, have a genetic component to a multifactorial aetiology. Here we describe other inherited conditions that have a direct bearing on perioperative management. Two of these, malignant hyperthermia and plasma cholinesterase disorders, are discussed in greatest detail because their adverse effects are virtually confined to an abnormal

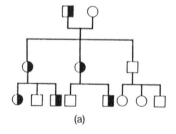
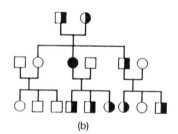
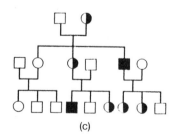

(a)   (b)   (c)

Fig. 46.1 The three most common patterns of inheritance. (a) Autosomal dominant: ◖, affected female; ◯, non-affected female; ◖, affected male; ☐, non-affected male. (b) Autosomal recessive: ●, affected female; ◯, non-affected female; ◖, carrier female (non-affected); ■, affected male; ☐, non-affected male; ◧, carrier male (non-affected). (c) X-linked recessive: ◯, non-affected female; ◖, carrier female (non-affected); ■, affected male; ☐, non-affected male.

Table 46.1 Mode of inheritance of some of the more common inherited disorders

Autosomal dominant	Autosomal recessive	X-linked recessive
Huntington's chorea	Cystic fibrosis	Duchenne muscular dystrophy
Neurofibromatosis	Phenylketonuria	Haemophilia
Myotonic dystrophy	Sickle cell anaemia	Fragile X syndrome
Malignant hyperthermia	Mucopolysaccharidoses	Childhood blindness
Polycystic kidneys	Friedreich's ataxia	
Polyposis coli	Cystinuria	
Hypercholesterolaemia		
Acute intermittent porphyria		

response to certain drugs used in anaesthesia. Also included in this chapter are the inherited myopathies and metabolic disorders.

Malignant hyperthermia

There can be no doubt that malignant hyperthermia (malignant hyperpyrexia, MH) has been responsible for a proportion of anaesthetic deaths almost since the introduction of ether and chloroform in the middle of the nineteenth century. Yet it was not until 1960 that MH, as a distinct familial condition associated with unexpected near death under anaesthesia, was first described in an Australian family by Denborough and Lovell.[1] It was presumably the fact that, until quite recently, death due to anaesthesia was not uncommon that a relatively rare cause (estimated incidence of MH in the UK population is 1/8000) was not recognized earlier. As anaesthesia has become progressively safer, so morbidity and mortality due to MH have become relatively more significant. From 1987 to 1992 there were 0.5 severe reactions per million of the UK population each year.

Pathophysiology of MH

From the response of muscle from MH-susceptible (MHS) patients to halothane and caffeine *in vitro* it was postulated that the clinical syndrome resulted from an acute rise in skeletal muscle intracellular calcium ion concentration. Yet it was not until quite recently that this hypothesis was verified.[2,3] It is the mechanism of this acutely elevated $[Ca^{2+}]_i$ that remains to be elucidated.

Porcine MH   Much of the physiological and biochemical evidence for the site of the primary defect in MH has been obtained using the pig model of MH. Porcine MH is also known as porcine stress syndrome, a description that

reflects triggering of the condition in susceptible pigs not only by anaesthetic drugs but also by high environmental temperatures and physical activity. Great interest was caused by publications reporting altered ryanodine binding to sarcoplasmic reticulum (SR) fractions in MHS pigs compared to controls,[4,5] implicating a biochemical abnormality of the SR calcium release channel as the cause of MH. Such an abnormality is supported by evidence for a mutation in the gene coding for this channel in MHS pigs.

Human MH   Fill and colleagues have looked for an abnormality of human MH SR calcium channel function.[6] A defect was identified in some of the channels isolated from only some of the patients. Rather than study SR fractions, the preparation of which could produce artefactual differences between samples from MHS and normal individuals, Hopkins and colleagues investigated the sensitivity of muscle bundles to contracture development in the presence of ryanodine.[7] Muscle from MHS patients developed contractures earlier than normal muscle. However, further studies using muscle bundles from patients with other neuromuscular disorders indicate that this effect of ryanodine is not necessarily the result of its action on an abnormally sensitive SR calcium release channel.[8]

Previous reports that suggested a crucial role for extracellular calcium in the genesis of the MH response[9,10] have been confirmed by Adnet *et al.*[11] The same group has also demonstrated the influence of dihydropyridine agonists and antagonists on the development of *in vitro* contracture responses of muscle from MHS individuals.[12,13] Hopkins has demonstrated a caffeine-mediated influx of extracellular calcium in normal rat muscle.[14] This provides a physiological basis for proposing that differences in response to caffeine between muscle from normal and MHS individuals may be due to a sarcolemmal, rather than or in addition to, an SR abnormality.

Genetics of MH

Early family studies indicated an autosomal dominant inheritance for MH.[1] This mode of inheritance has been confirmed by the results of family screening using *in vitro* contracture tests. With the revolution in molecular biological techniques over the past ten years it has been possible to study the association of MH susceptibility with DNA markers of known chromosomal location in individual MH families. Malignant hyperthermia susceptibility was first reported to map to chromosome 19q in Irish[15] and North American[16] families. Similar results have been obtained by groups in Germany, the Netherlands and Britain.[17,18]

The ryanodine receptor locus (RYR), which lies on chromosome 19 in man and codes for the protein constituting the calcium-release channel in skeletal muscle, has recently been implicated in porcine stress syndrome[19] (see above). However, despite intensive investigation, involvement of this gene in the human condition has not yet been proved.

Indisputable evidence for genetic heterogeneity has recently been published. Reports from North America[20] indicate that in some families malignant hyperthermia susceptibility did not show linkage to chromosome 19q. Similar results have been obtained with European pedigrees.[21–23] Current genetic investigations are directed towards searches for further candidate genes. The genetic loci of the subunits of the skeletal muscle dihydropyridine receptor are being sought in this regard.

The fact that MH is a genetically heterogeneous condition means that the development of a DNA-based blood test for screening purposes will prove complex. The genetic abnormality responsible in each family may be identifiable but it will need to be identified to a high degree of certainty, which, for the near future at least, means that large numbers in each family will have to be studied both genetically and by muscle biopsy and *in vitro* contracture tests before a genetic test alone will become statistically reliable.

Clinical presentation

At the time MH was first described, monitoring of the anaesthetized patient was invariably purely clinical. Under these circumstances the most obvious abnormality of an MH reaction was a progressive rise in body temperature followed by death, hence the name malignant hyperthermia. With the description of further cases, however, it became apparent that hyperthermia is a relatively late manifestation of the reaction.

In the absence of a family history of MH, the first indication that an individual may be susceptible to the condition is the development of an exaggerated initial response to the depolarizing neuromuscular blocking drug suxamethonium (succinylcholine). It has been well demonstrated[24,25] that in most people, along with the more easily recognized fasciculations, suxamethonium causes an increase in tension of the jaw muscles. In some cases this is clinically apparent as resistance to opening of the mouth for laryngoscopy. When this jaw stiffness is severe and especially when it is prolonged, the condition is called masseter spasm. Masseter spasm may be accompanied by generalized muscular rigidity. Having made a diagnosis of masseter spasm the anaesthetist is bound to refer the patient for investigation of MH status (whether or not other signs of MH are detected) because masseter spasm is associated with MH susceptibility in up to 50 per cent of cases.

The 'true' MH response, as opposed to exaggerated suxamethonium-induced muscle rigidity, is characterized by metabolic stimulation, the consequences of the loss of sarcolemmal integrity (hyperkalaemia and myoglobinaemia) and a later-onset muscle rigidity. Such a response may develop within 15 minutes of the introduction of one of the volatile anaesthetic drugs, or it may be delayed for several hours in the course of a long procedure. If the patient is carefully monitored, one or more abnormalities will always be detected during the course of the anaesthetic itself; that is, MH reactions do not start after the discontinuation of the triggering drugs, although if they have progressed beyond the earliest stages they will not necessarily be stopped by the discontinuation of triggering drugs in the absence of other therapeutic measures.

The early signs of hypermetabolism are a rising end-tidal CO_2 concentration (in the presence of appropriate ventilation and fresh gas flow for the particular circuit in use) and inappropriate tachycardia. These may be accompanied, or followed, by hypertension, hypoxaemia, tachypnoea (in the spontaneously breathing patient), acidosis, early exhaustion of soda-lime in a circle system, and subsequently a rising body temperature. Hyperkalaemia may lead to cardiac arrhythmias, including asystole, whilst myoglobinaemia can cause acute renal failure. If the acute reaction is not terminated the hypermetabolism and cellular destruction may lead to disseminated intravascular coagulation.

From this description it can be seen that the most essential monitoring devices for the early detection of an MH reaction are a capnograph, a pulse oximeter and an ECG monitor. A prerequisite to early diagnosis is, of course, an appropriate level of suspicion for the condition by the vigilant anaesthetist. It is important that an early diagnosis is made because, as mucle rigidity develops, the blood supply to the muscle cells is restricted, thus impeding the supply of therapeutic agents to the muscle. It is also likely that a stage of the cellular pathological process is reached when the process becomes irreversible.

Management of the acute MH reaction

The management of an acute MH reaction, like other medical emergencies, requires several therapeutic interventions to be employed as rapidly as possible: written descriptions of such actions cannot convey the urgency required.

Measures to halt the MH process

Stop the administration of trigger drugs   As well as actually turning off the vaporizer, the elimination of volatile drugs from the body can be enhanced by hyperventilation, and from the anaesthetic breathing system by increasing fresh gas flow.

Active body cooling   This not only prevents the potentially harmful effects of hyperthermia on heat-labile cellular constituents but, if sufficiently effective, may also cool the body sufficiently to inhibit the MH response at the cellular level. Cooling can be achieved by converting warming blankets to cooling blankets, intravenous infusion of refrigerated crystalloid solutions, peritoneal lavage with cold saline, applying icepacks to the axillae and groins, and in extreme cases (where facilities exist) a blood heat exchanger can be used. It is important to remember to try to prevent peripheral vasoconstriction as this will prevent heat loss. An adequate fluid load may be all that is required to do this, but, if not, methylprednisolone (10 mg/kg) is an effective vasodilator and has been shown to reduce halothane-induced muscle contracture *in vitro*.[26]

Dantrolene sodium   This drug is effective in reducing muscle activity resulting from an increase in intracellular calcium ion concentration and is therefore efficacious in the treatment of MH (see 'Pathophysiology of MH', above). It acts at the site of excitation–contraction coupling and is relatively sparing on cardiac and diaphragmatic muscle. Dantrolene for intravenous injection is prepared as an alkaline mixture with mannitol to increase its solubility; even so, its preparation for use is slow. The initial dose is 1–2 mg/kg, repeated until there is evidence that hyper-metabolism is reversing, and further doses given if trends reverse. Up to 10 mg/kg has been necessary.

Treatment of the effects of MH

Acidosis   The hyperventilation used to increase elimination of volatile drugs will partially compensate for the metabolic acidosis. Further correction can be achieved by titrating sodium bicarbonate against arterial pH and base excess.

Hypoxaemia   Hyperventilation should initially be with 100 per cent oxygen. This should be maintained until the reaction is brought under control unless lower inspired oxygen fractions can be demonstrated to provide for adequate oxygen delivery in the face of the greatly increased oxygen consumption.

Hyperkalaemia   Serum potassium concentrations should be measured frequently from the onset of an MH reaction. Intravenous calcium should be used only when asystole has occurred or is considered to be imminent, as it is possible that influx of extracellular calcium is responsible for the generation of the MH response. Otherwise, hyperkalaemia should be corrected with infusion of dextrose and insulin, ion exchange resins or, if especially refractory, by haemodialysis.

Cardiac arrhythmias   These should be treated symptomatically along the usual lines. Antiarrhythmic drugs that may have an additional effect of reducing the MH response are verapamil and procainamide. Lignocaine (lidocaine) in very high concentrations causes muscle damage *in vitro* (PM Hopkins, 1990, unpublished observations). It could therefore exacerbate the MH response (although it is known not to trigger the reaction) and so is perhaps best avoided with the availability of alternatives such as amiodarone.

Disseminated intravascular coagulation (DIC)   Blood samples should be sent for clotting studies and analysis of fibrin degradation products. DIC should be managed along standard lines, with transfusion of platelets, fresh frozen plasma and cryoprecipitate. The use of tranexamic acid has not been reported in MH.

Myoglobinaemia   Even before myoglobinuria is detected, significant myoglobinaemia must be assumed to have occurred during an MH response. The bladder should be catheterized and a diuresis established using mannitol and frusemide (furosemide). A high renal blood flow should be encouraged during the acute phase of the illness with dopamine.

Other measures

The surgery should be completed as rapidly as possible to allow optimal resuscitation. Anaesthesia should be maintained with non-triggering drugs. An accurate record of the reaction, with special regard to the timing of events, is essential information for the MH referral centre in their decision as to whether the patient warrants further investigation. Also in this regard, it is most helpful if the results of blood gases, serum potassium, creatine kinase (daily estimation for 3 days) and urinary myoglobin assays are made available.

Laboratory diagnosis of MH susceptibility

In 1970 and 1971 two *in vitro* tests for the diagnosis of MH susceptibility (MHS) were described by Kalow *et al.* (using caffeine)[27] and Ellis *et al.* (using halothane).[28] These tests form the basis of current diagnostic methods which involve open muscle biopsy and the exposure of live strips of muscle to halothane and to caffeine.[29] MHS muscle develops an abnormal contracture response to these agents. Responses of muscle bundles from MHS and normal individuals to halothane are shown in Fig. 46.2.

All patients suspected of developing an MH response should be referred to a recognized MH centre (accuracy of results are dependent on strict quality control) for confirmation of diagnosis and follow-up of family members. The exception to this rule is prepubertal children in whom the amount of muscle tissue required might cause a functional deficit as well as being cosmetically unacceptable. There are also concerns that the contracture tests may yield false negative results in young children; such a diagnosis has potentially fatal consequences.

Anaesthesia for the patient susceptible to MH

The key to safe anaesthesia for MHS individuals is the avoidance of triggering drugs. Table 46.2 lists triggering drugs and those drugs known to be safe to use in MH; also listed are drugs that are thought to be safe but which should be used with caution. This last group mainly consist of drugs that have been implicated as MH triggers in early case reports; doubts must exist as to whether many of these cases were MH reactions. Also, it can be seen that these

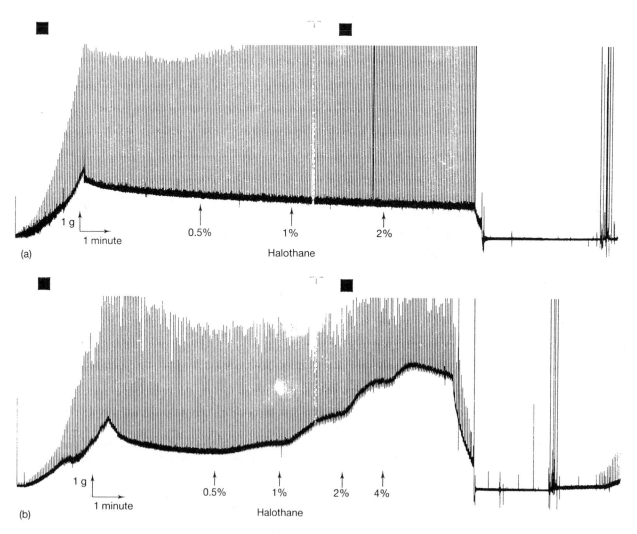

Fig. 46.2 *In vitro* responses to halothane of skeletal muscle bundles from (a) a normal individual and (b) a patient susceptible to malignant hyperthermia. Freshly excised samples of vastus medialis muscle are mounted in a muscle bath and a baseline tension of 2 g is applied. After a short stabilization period, incremental concentrations of halothane are vaporized into the gas mixture aerating the muscle bath. There is a dose-dependent increase in baseline tension (a contracture) in the specimen from the MH patient in response to halothane which is not observed in normal muscle.

Table 46.2 Classification of drugs according to their potential for triggering malignant hyperthermia

Triggering drugs	Drugs that are probably safe	Safe drugs
Halothane	Phenothiazines	Benzodiazepines
Enflurane	Haloperidol	Thiopentone
Isoflurane	Tricyclic antidepressants	Propofol
Desflurane	Monoamine oxidase inhibitors	Ketamine
Sevoflurane	Cyclopropane	Metoclopramide
Ether		Droperidol
Chloroform		Opioids
Suxamethonium (succinylcholine)		Non-depolarizing neuromuscular blocking drugs
		Atropine
		Neostigmine
		Nitrous oxide
		All local anaesthetics

drugs are in common general use and it would be expected that if they were MH triggers many 'awake' MH reactions would have been reported with their use. Appropriate techniques are local anaesthesia and intravenous general anaesthesia (Chapters 34 and 33).

Many hospitals have a dedicated 'vapour-free' anaesthetic machine for use with MHS patients. If such a machine is not available, a machine can be prepared for use by removing the vaporizers and running maximum flows of oxygen through it for 2 hours. As volatile drugs adsorb onto plastics and rubber, new circuits and facemasks should be used.

If triggering drugs are meticulously avoided the use of prophylactic dantrolene is not required. Although hepatitis, pleural effusions, pericarditis and convulsions are associated only with chronic use of dantrolene, its short-term use does provoke nausea and vomiting, and muscle weakness in a high proportion of recipients as well as unpleasant CNS effects including euphoria and disorientation. Its use is therefore not warranted as part of a 'belt and braces' approach. The relatively weak effects of dantrolene on cardiac and diaphragmatic muscle may also become significant in patients with pre-existing cardiac or respiratory disease, respectively.

Other inherited muscle diseases

The muscular dystrophies

The two main types of muscular dystrophy are Duchenne's and Becker's.

Duchenne muscular dystrophy (DMD)

DMD, or severe X-linked dystrophy, is the commonest of the muscular dystrophies, with an incidence of up to 35 per 100 000 live male births. The disease is characterized by rapidly progressive generalized muscle weakness with pseudohypertrophy of the calf muscles and occasionally of other groups. Symptoms usually become apparent in the third year of life, progressing until around the age of 10 when the child requires a wheelchair. A wheelchair existence is often associated with kyphoscoliosis, fixed joint contractures, obesity, impaired circulation to the legs and feet with dependent oedema, along with psychological problems. The second decade of these boys' lives is punctuated by frequent chest infections due to impaired ventilatory effort. Death, due to pneumonia or cardiac disease, is usual by the age of 20. Symptoms of heart failure are deceivingly uncommon because of the enforced sedentary existence but myocardial dysfunction is common, as is evident from echocardiographical studies.[30]

Anaesthesia is required in DMD patients for muscle biopsy, which is essential to provide a prognosis and to exclude rarer but more treatable myopathies. Orthopaedic procedures for releasing fixed joint contractures to improve the duration of mobility are becoming more common. Anaesthetic problems are related to myocardial involvement, respiratory function and the effects of drugs on the diseased skeletal muscles. Sethna et al.[31] reported their 5-year experience of anaesthesia for 25 DMD patients, 4 of whom had complications which, as it happens, typify most adverse events in other reports. In the first case, induction was followed within 10 minutes by cardiac arrest associated with acute hyperkalaemia (8.9 mmol/l). The anaesthetic in this case involved only halothane with nitrous oxide and oxygen. Even more dramatic hyperkalaemic cardiac arrests have occurred when gaseous induction with halothane has been followed by administration of suxamethonium.[32]

Their second case illustrates the need for awareness

regarding cardiac involvement even in asymptomatic individuals. In this patient the echocardiogram was unremarkable but the boy died of cardiac failure during a long procedure. The postmortem revealed biventricular dilatation and hypertrophy with numerous fibrotic bands. Increased incidence of arrhythmias is another potential problem which can be exacerbated by hypercarbia secondary to depressant drugs at any stage of the perioperative period.[33] The last two cases of interest in Sethna's report both relate to a rise in body temperature and tachycardia which resolved when halothane was withdrawn. Tachycardia and hyperthermia are signs found during a malignant hyperthermia (MH) reaction. These and other signs found in MH, such as hyperkalaemia, acidosis, myoglobinuria and raised creatine kinase (CK) levels, have been reported in DMD patients anaesthetized with halothane and/or suxamethonium.[32] These authors wrongly conclude, as do Sethna *et al.*, that their patients have MH. The possible responses of skeletal muscle to disease are limited in their expression, and the response of DMD muscle to halothane and suxamethonium in its similarity to MH is an example of this. The use of suxamethonium has also been implicated as the common factor in six DMD patients who required postoperative ventilation because of ventilatory insufficiency.[34]

An approach to anaesthetizing the patient with DMD might be as follows:

- Use local/regional techniques when possible and acceptable to the patient (spinal anaesthetic techniques which can cause sudden changes in afterload may not be appropriate in the patient with significant cardiac involvement).
- Avoid heavily depressant premedicants.
- Avoid suxamethonium.
- Consider carefully the choice of maintenance drugs for general anaesthesia in individual patients. Despite the reports of hypermetabolic reactions to volatile drugs, there is a great experience of the safe use of halothane in DMD patients: such reactions, if detected early in their course with the aid of appropriate monitoring, appear to be readily reversed by discontinuing the volatile drug. Of greater concern is the possibility of sudden massive rhabdomyolysis. These disadvantages of volatile drugs must be weighed against the detrimental cardiovascular effects of intravenous agents in patients with a cardiomyopathy.

Becker muscular dystrophy (BMD)

BMD, in its clinical characteristics, closely resembles DMD but with a later onset (mean, 11 years of age) and slower progression. Like DMD, it is an X-linked recessive condition, and lately DMD and BMD have both been shown to be due to defects of the gene coding for a sarcolemmal protein, dystrophin. Within a BMD family the clinical course of affected individuals is similar, but there is a wide spectrum of symptom onset and rate of disease progression between different BMD families.

There have been no recent reports of anaesthetic problems in BMD. This probably reflects its rarity compared to DMD, the rarity of significant cardiac involvement despite ECG changes and the milder nature of the skeletal muscle abnormality. However, the nature of this disease is so similar to DMD that the potential for abnormal responses to suxamethonium, halothane and related drugs is likely to be present. Indeed, muscle from a BMD patient has been shown to have an abnormal *in vitro* response to halothane.[35]

Other muscular dystrophies

These are generally less severe than the X-linked dystrophies discussed above and are given descriptive titles indicating the distribution of the muscle weakness and wasting, although other muscle groups may be involved as the disease progresses. Those diseases with special anaesthetic implications are listed below.

Oculopharyngeal dystrophy   Bulbar muscle involvement leads to a high risk of pulmonary aspiration.

Facioscapulohumeral dystrophy   There is facial weakness but with *sparing* of bulbar muscles. Dresner and Ali[36] described a patient who was successfully anaesthetized using alfentanil, nitrous oxide and atracurium. They observed a normal sensitivity to atracurium with a more rapid than usual recovery from neuromuscular blockade. This is difficult to attribute to the muscle pathology but perhaps reflects the contribution to neuromuscular blockade usually provided by a volatile anaesthetic agent.

Humeroperoneal (Emery–Dreifuss) dystrophy   Early contractures involve the posterior cervical muscles. A consistent feature is atrioventricular block, which leads to recurrent episodes of syncope and embolic phenomena.

Myotonic disorders

Myotonia may be defined as the delayed relaxation of muscles. Myotonia following voluntary contraction usually improves with repeated activity, is worsened by cold and improved by warmth. Myotonia occurring during surgery may, in the least, hinder the surgeon but, if the diaphragm and other respiratory muscles are involved, may prove fatal. This is such a potentially dangerous event because myotonia can be induced by mechanical stimulation of the muscles and yet, because the causative biochemical defect lies in the muscle cell, it cannot be prevented by neuromuscular blocking drugs or by nerve conduction block with local anaesthetics. Simple measures, such as keeping the patient warm and careful handling of the muscles, are the best prevention. If myotonia does occur the following drugs may be tried:

- procainamide,
- phenytoin,
- dantrolene,
- an ester local anaesthetic, e.g. procaine, by direct injection into the affected muscle.

Some of the myotonic disorders pose other anaesthetic problems, which need separate consideration.

Myotonic dystrophy (dystrophia atrophica)

This is a multisystem disease affecting the eyes (cataracts), hair (frontal baldness), bones (skull thickening), endocrine and metabolic function (hypogonadism, impaired glucose tolerance), immune system and intelligence, as well as skeletal, cardiac and smooth muscle. The skeletal muscle involvement is predominantly of the facial, oropharyngeal and distal limb muscles. Muscle weakness is invariably the major symptom; myotonia is relatively mild in the early stages of the disease and becomes even less apparent as the dystrophic process progresses.

Aspiration pneumonia is a great hazard as the bulbar muscle weakness is compounded by oesophageal dilatation caused by smooth muscle involvement and by prolonged somnolence following the use of thiopentone.[37,38] In one of these reports, etomidate was used in combination with thiopentone.[38] This same patient had previously been anaesthetized with propofol, which resulted in prompt postoperative awakening. A normal recovery has also been reported following a 4-hour infusion of propofol.[39] Postoperative respiratory problems may also be caused by respiratory muscle weakness.

Because of the muscle weakness, neuromuscular blocking drugs must be used with care. Vecuronium and atracurium have both been successfully used when the dose has been carefully titrated against response:[39,40] the doses used were about half the dose one would predict for a normal patient. Suxamethonium and other drugs that stimulate muscle (e.g. neostigmine) should be avoided in all myotonic disorders. Heart muscle involvement leads to dysrhythmias and/or conduction defects in about 60 per cent of cases, and is progressive.

Local and regional anaesthetic techniques overcome many of the problems of general anaesthesia in myotonic dystrophy, with the notable exception of myotonia. Such a technique has been used for anaesthesia for caesarean section.[41]

The provision of safe anaesthesia can usually be achieved for patients with myotonic dystrophy, but relies on the diagnosis having been made preoperatively. This is not always the case, as the number of typical features varies considerably, even between members of the same family, because of differences in gene expressivity. Therefore, to avoid unforeseen problems, the anaesthetist should remember the features of this disease and, if any are present, a detailed family history should be taken.

Myotonia congenita (Thomsen's disease)

This rare condition is characterized by the unique combination of diffuse muscle hypertrophy and generalized myotonia. The myotonia, described as muscle cramps or stiffness, is exacerbated by cold and inactivity. Muscle hypertrophy occurring within a limited fascial compartment may cause ischaemic damage and intermittent claudication. Mild weakness may occur in the rarer autosomal recessive variety (Becker) but is usually confined to the limbs. Other body systems are not involved in the disease.

Paramyotonia congenita

Generalized myotonia develops in response to cold, vigorous exercise and, in some cases, hyperkalaemia. A flaccid paresis, most prominent in the feet and cranial musculature, may follow the myotonia and last longer than 15 minutes. As with myotonia congenita, there is no multisystem involvement. Paramyotonia congenita has many similarities to hyperkalaemic periodic paralysis (see below) and it may well be that the two conditions lie at opposite ends of a spectrum of disorders with clinical features of myotonia and flaccid paralysis with exposure to cold, exercise and hyperkalaemia as exacerbating conditions. Streib[42] has reported two very interesting cases from the same paramyotonia family with known *hyper*kalaemic-induced weakness, both of whom developed *hypo*kalaemic paralysis following general anaesthesia which required postoperative mechanical ventilation and large doses of potassium chloride by intravenous infusion. Unfortunately, knowledge of the details of the anaesthetic agents used in each case is limited but our inference is that suxamethonium and a volatile agent were used in both.

The periodic paralyses

Hypokalaemic periodic paralysis

Attacks of flaccid paralysis of varying frequency, severity and duration occur which are associated with a fall in the serum potassium concentration. A severe attack may render the patient quadriplegic but able to speak, swallow and move the eyes with minimal effect on respiratory function. The occurrence of most attacks is unpredictable but they are commonly precipitated by heavy exercise, excessive carbohydrate intake, stress, trauma and cold. These last three factors are obviously associated with anaesthesia and surgery, and cannot always be avoided. Perioperatively, an acute episode may be prevented by maintaining normokalaemia using an infusion of potassium chloride with monitoring of the serum potassium level.

It should be noted that hypokalaemic periodic paralysis can, rarely, be an acquired condition. It can occur in association with thyrotoxicosis, or secondary to renal or gastrointestinal potassium wasting.

Hyperkalaemic periodic paralysis

As noted above, this condition appears to be related to paramyotonia congenita. However, the predominant symptom is episodic weakness associated with a rise in the serum potassium level (although not necessarily above the upper limit of normal). Myotonic symptoms may be present at all times, only preceding or during an attack, or not at all.

It is of considerable significance that, in addition to precipitating factors in common with paramyotonia congenita, an attack of hyperkalaemic periodic paralysis can be induced by fasting, sleep or anaesthesia. Anaesthetic considerations include the administration of intravenous glucose (with no potassium!) during fasting, general measures for myotonic disorders (avoidance of suxamethonium, neostigmine and cold; careful handling of muscles) and monitoring of the serum potassium concentration. Aarons and colleagues have reported the successful use of vecuronium as part of a balanced anaesthetic technique with fentanyl and isoflurane in a patient with hyperkalaemic periodic paralysis.[43] An attack can be treated with intravenous glucose and insulin.

Normokalaemic periodic paralysis

Episodic attacks varying from mild extensor weakness to quadriplegic flaccid paralysis lasting from minutes to days occur every few months. They usually occur during a period of rest following exertion but may also be precipitated by anxiety, ingestion of alcohol and excessive sleep. General anaesthesia must therefore be considered as a potential precipitating factor. Attacks are not related to changes in the serum potassium concentration and myotonia is not a feature. Treatment is with intravenous sodium chloride.

Morphologically distinct myopathies

Rod (nemaline) myopathy

Rod myopathy exists as a familial congenital non-progressive proximal weakness and an adult-onset progressive weakness starting with a scapuloperoneal distribution. The more common congenital form is associated with secondary bony deformities such as pes cavus, kyphoscoliosis and a high arched palate.

Specific anaesthetic problems are related to the restrictive respiratory defect associated with scoliosis and the effects of drugs on postoperative muscle function. Cunliffe and Burrows found the use of neuromuscular blocking agents unnecessary in establishing controlled ventilation,[44] whilst Head and Kaplan used suxamethonium followed by pancuronium in their patient without untoward effect.[45]

Type 1 fibre hypotrophy with central nuclei

This is also known as centronuclear or myotubular myopathy. It covers a wide range of clinical presentations from a fatal congenital form to a slowly progressive late-onset form. The predominant clinical feature is weakness.

Central core disease

Central core disease may present in infancy with hypotonia but in general the weakness is not progressive. Sometimes it does not become clinically apparent until late adult life, and some patients who are known to have the morphological abnormality never develop symptoms.

There is a strong association between central core disease and malignant hyperthermia although the nature of the relation is not clear.

Metabolic muscle diseases

The glycogenoses

The glycogenoses are a group of disorders characterized by defective glycogen and/or glucose metabolism. They can be classified as in Table 46.3.

Of the five glycogenoses that involve skeletal muscle, muscle symptoms predominate in three (types II, V and VII), whereas in types III and IV the dominant features are liver dysfunction and hypoglycaemia.

McArdle's disease (type IV)

Myophosphorylase deficiency leads to an inability to utilize skeletal muscle glycogen stores. The muscle then becomes

Table 46.3 Classification of the glycogenoses

Type	Name	Deficient enzyme	Skeletal muscle involvement	Liver involvement
I	von Gierke's	Glucose-6-phosphatase	No	Yes
II	Pompe's	Acid maltase	Yes	Yes
III	Forbes', Cori's	Amylo-1,6-glucosidase (debranching enzyme)	Yes	Yes
IV	Andersen's	Amylo-1,4-1,6-transglucosidase (branching enzyme)	Yes	Yes
V	McArdle's	Myophosphorylase	Yes	No
VI	Hers'	Liver phosphorylase	No	Yes
VII	Tarui's	Phosphofructokinase	Yes	No

dependent on the entry of substrates such as glucose and fatty acids. Under conditions of increased energy requirements (exercise) or decreased substrate supply (fasting, ischaemia) the muscle is liable to break down, releasing myoglobin which can precipitate renal failure.

Perioperatively, an intravenous infusion of glucose should be maintained throughout, and the use of tourniquets avoided. If myoglobinuria occurs, a diuresis must be established.

Phosphofructokinase deficiency (type VII)

This presents a clinical picture similar to McArdle's disease. The principles of anaesthetic management are also the same with the exception that a fatty acid solution is required instead of the glucose infusion as the enzyme defect is distal to the entry of glucose in the Embden–Meyerhof pathway.

Acid maltase deficiency (type II)

The infantile variety (Pompe's disease) is associated with massive accumulation of glycogen in all tissues, death from cardiac failure usually occurring within the first year of life.

The adult-onset form is restricted to skeletal muscle and eventually involves the respiratory muscles in most cases.

Disorders of muscle lipid metabolism

The energy for prolonged muscular effort is derived mainly from the oxidation of fatty acids in the mitochondria. The transport of long chain fatty acids into the mitochondria is dependent upon a carrier, carnitine, which is synthesized in the liver and actively taken up by muscle cells, and an enzyme, carnitine palmityl transferase 1.

Carnitine deficiency

Systemic This is due to defective hepatic synthesis. Slowly progressing hepatorenal failure precedes a myopathy which produces a variable degree of weakness. Fasting may produce a fatal acidosis and therefore should be accompanied by an infusion of glucose.

Muscle This type is due to a defective muscle uptake. It usually presents in childhood with proximal muscle weakness. A fatal cardiomyopathy may occur.

Carnitine palmityl transferase deficiency

Muscle cramping, pain, necrosis and myoglobinuria are associated with prolonged exercise. Anaesthetic management includes constant provision of usable substrate (intravenous glucose).

Mitochondrial myopathies

These are a group of rare, clinically heterogeneous conditions associated with many large and abnormal mitochondria in skeletal muscle cells. In some patients a deficiency of one or more of the mitochondrial respiratory chain components has been demonstrated. Myopathic symptoms are usually weakness and excessive fatigue; other abnormalities may include central and peripheral nervous system problems, and cardiac conduction defects (these may be severe enough to warrant permanent cardiac pacing).

Experimental mitochondrial myopathies have been produced by giving animals 2,4-dinitrophenol (2,4-DNP) which uncouples mitochondrial oxidative phosphorylation. 2,4-DNP has also been proposed as a tool to produce an animal model of malignant hyperthermia, as when it is administered with halothane animals become pyrexial, stiff and then die. This implies that patients with defects of mitochondrial oxidative phosphorylation might exhibit a potentially dangerous response to the volatile anaesthetic agents.

Inherited disorders of metabolism

The porphyrias

The porphyrias are a group of diseases resulting from defects in the synthesis of haem. They are characterized by overproduction and excretion of haem precursors, including

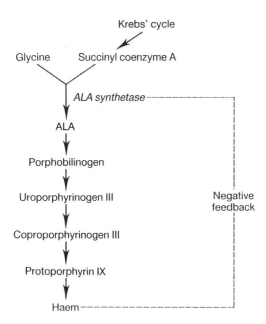

Fig. 46.3 The synthesis of haem.

Table 46.4 Drugs that are potentially porphyrinogenic

Drugs that have precipitated acute porphyric attacks	Other drugs that have been implicated
barbiturates	etomidate
sulphonamides	propofol
griseofulvin	pentazocine
chlordiazepoxide	lignocaine
imipramine	enflurane
phenytoin	
chlorpropamide	
steroids	
oral contraceptives	
ergot preparations	
alcohol	

porphyrins which are oxidized by ultraviolet light to form the purple-coloured uroporphyrin and coproporphyrin. These can easily be detected in urine and faeces. The rate-limiting step in the synthesis of haem is the formation of δ-aminolaevulic acid (ALA) from succinyl coenzyme A and glycine – a reaction catalysed by ALA synthetase (Fig. 46.3).

The significance of the porphyrias to anaesthetists is that some of them (the hepatic porphyrias: acute intermittent, variegate and hereditary coproporphyria) are exacerbated by certain lipophilic drugs. Acute exacerbations produce acute neurological disturbances involving central, peripheral sensory and motor, and autonomic nervous systems. The symptoms are dramatic and are as follows:

- *Central.* Agitation is a common early sign, but succeeding symptoms can mimic any psychotic illness.
- *Peripheral.* A progressive peripheral neuropathy may develop rapidly. Muscle weakness may be severe enough to necessitate artificial ventilation.
- *Autonomic.* Abdominal pain, vomiting, constipation, tachycardia, hypertension, postural hypotension, urinary incontinence or retention may feature. The abdominal pain can mimic almost any acute surgical intra-abdominal emergency, and such a patient may therefore be presented to the anaesthetist for laparotomy. Biochemical changes can result from vomiting and polyuria.

The key to uneventful anaesthesia for porphyric patients is the avoidance of precipitating factors. It is important to

remember those factors other than drugs that are imposed routinely perioperatively such as fasting. Infection is also a strong stimulus to increased porphyrin excretion: this may be mediated by steroids or other mediators of 'the stress response'. It would therefore seem logical to attempt to suppress the stress response associated with major surgery.

It is the lack of significant stress response with most elective surgical procedures that is perhaps the reason why drugs that are known to increase activity of hepatic ALA synthetase, such as etomidate, have been used safely in prophyric patients undergoing such elective surgery.[46] The absence of an acute reaction in these circumstances means that safe use of potentially porphyrinogenic drugs may be reported. The reported safe use of propofol[47,48] may be an example of this, especially as others have demonstrated increased excretion of porphyrins with its use.[49,50] As barbiturates are strongly porphyrinogenic, the choice of intravenous induction agent is limited. Ketamine has no effect on hepatic ALA synthetase activity and has been used safely in emergency anaesthesia.[51] In fact, there are many therapeutic groups of drugs, of which one or more members have been implicated (Table 46.4). Anaesthesia can be safely given with halothane, nitrous oxide, opioids and neuromuscular blocking drugs. Local anaesthesia should be with bupivacaine or prilocaine rather than with lignocaine.

The mucopolysaccharidoses

These are a group of disorders resulting from defects in the breakdown of proteomucopolysaccharides, which are structural components of cells in many tissues. All the disorders are inherited as autosomal recessive traits, other than Hunter's syndrome which is X-linked recessive. The pathological findings are due to deposition of proteomucopolysaccharides in the tissues involved. They have in common a progressive course, skeletal and joint involvement, and coarse facial features. Other tissues involved in some are the brain (causing mental retardation), the heart and blood vessels, the liver, the spleen and the cornea.

Anaesthesia requires careful planning. Difficulties with airway control should be anticipated because of the large head, short neck, abnormal mandible, large tongue, hyperplastic lymphoid tissue, inelastic tissues, and abnormal laryngeal and tracheal cartilages. In addition, the cervical spine may be unstable in the Morquio and Maroteaux–Lamy syndromes. To add to these structural factors is an excessive production of viscid respiratory tract secretions. Other problems relate to potential failure of other organs involved in the disease.

The glycogenoses

These have been classified earlier in the section on muscle diseases (see Table 46.3). Types I, II, III, IV and VI are characterized by liver failure and hypoglycaemia due to their inability to utilize glycogen. Type I (von Gierke's) patients also have a tendency to develop acidaemia.

The anaesthetic implications are those of the liver failure, avoidance of excessive starvation and administration of sufficient exogenous glucose to prevent hypoglycaemia.

Hereditary fructose intolerance

Symptoms result acutely from hypoglycaemia caused by administration of fructose. Longer term problems result from deposition of fructose-1-phosphate accumulation in the liver, kidneys and intestinal mucosa. It is the failure of these organs that presents the only problem to the patient in whom the diagnosis has been made, as acute symptoms are completely controlled by diet.

Fructose-1,6-diphosphatase deficency

Fructose-1,6-diphospatase is essential for gluconeogenesis. Any stimulus to gluconeogenesis, such as starvation or surgery, will therefore result in hypoglycaemia and lactic acidosis. Perioperatively, acid–base status and blood sugar should be monitored carefully and exogenous glucose supplied intravenously. A diet free of fructose for 48 hours prior to surgery also seems to limit the occurrence of hypoglycaemia.

Disorders of amino acid metabolism

The commonest of these is *cystinuria*, which is an epithelial amino acid transport defect leading eventually to chronic renal failure. While some renal function remains, it is important to maintain a high urine flow to prevent the formation of stones.

Perioperative problems common to disorders of amino acid degradation, such as homocystinuria, maple syrup urine disease and phenylketonuria, include avoidance of excessive breakdown of endogenous protein which may increase the progression of the disease. The increased levels of amino acids so produced may also cause hypoglycaemia through several possible mechanisms. Periods of starvation should therefore be kept to a minimum; intravenous dextrose should be given; measures to limit the catabolic response to surgery may be beneficial.

Homocystinuria is also associated with an increased tendency to form intravascular thromboses. This may be due to a platelet abnormality, in which case aspirin may be beneficial prophylactically. Circulatory stagnation should be prevented in both arterial and venous systems, and a degree of haemodilution is warranted. Low dose subcutaneous heparin is advisable.

Familial dysautonomia

Otherwise known as Riley–Day syndrome, familial dysautonomia is an autosomal recessive condition of highest incidence in ethnic Jews. It is a rare condition but provides major anaesthetic problems. The disorder is thought to be due to a deficiency in dopamine-β-hydroxylase, which is responsible for the conversion of dopamine to noradrenaline (norepinephrine) in sympathetic (and probably central) nerve endings. This would explain why these patients display a 'denervation hypersensitivity' to circulating catecholamines and only a limited sympathetic nervous system response to hypotensive stimuli such as inhalational anaesthetic drugs.

The other features of the disease with implications for the anaesthetist are not necessarily predictable from the term 'dysautonomia'. Respiratory secretions are excessive and thick, leading to recurrent chest infections. Respiratory drive is also abnormal and is reflected in reduced hypoxaemic and hypercapnic responses, especially following opioid administration. This increased depressant effect by opioids may be exacerbated by the lower analgesic requirements of these patients, who are relatively insensitive to pain. Cyclical vomiting and fever are typical features of the disease and may be triggered by anaesthesia and surgery.

Inherited abnormalities of plasma cholinesterase

Plasma cholinesterase (pseudocholinesterase) is an enzyme for which no physiological role has been identified: people without detectable enzyme activity have no disease or physiological defect. Indeed, it was only following the introduction of suxamethonium, which depends on rapid destruction by plasma cholinesterase for its short duration of action, that genetically inherited functional variants of the enzyme became apparent. This realization followed assessment of enzyme activity in the plasma of individuals (and their blood relatives) in whom suxamethonium had a prolonged action.

Characterization of allelic variants

Rather than rely on plasma cholinesterase activity, which is a measure of the amount of enzyme present in a sample as well as the nature of the enzyme, cholinesterase investigators have used percentage inhibition of enzyme activity by inhibitors to characterize variants. Kalow and Genest[52] first described such a method using the local anaesthetic dibucaine (cinchocaine, nupercaine) and termed the percentage inhibition the dibucaine number. They were thus able to identify patients homozygous for the usual allele (E_1^u, E_1^u), patients homozygous for the atypical allele (E_1^a, E_1^a) and heterozygous patients. Using analogous techniques with sodium fluoride, Harris and Whittaker[53] used the fluoride and dibucaine numbers to distinguish a fluoride-resistant allele, E_1^f. Family studies subsequently implicated a fourth allelic variant, which was termed the silent allele, E_1^s, because dibucaine and fluoride numbers in the heterozygous state were identical to those in the homozygote for the other allele.

There are families whose cholinesterase activities and inhibitor numbers cannot be explained by the alleles already described. Two of these have been called the 'j' and 'k' variants. If they are a real entity, their occurrence is exceedingly rare.

Suxamethonium apnoea

This is the name of the condition exhibited by individuals who have a genetically determined reduction in plasma cholinesterase activity sufficient to produce a clinically apparent prolongation of the action of suxamethonium. Depending on the enzyme activity, the period of apnoea

Table 46.5 Non-inherited causes of reduced plasma cholinesterase activity

Physiological	Acquired	Iatrogenic
Peripartum women	Liver disease	Ecothiopate
Infants < 6 months	Collagen diseases	Chlorpromazine
of age	Cachexia	Pancuronium
	Chronic anaemias	Procaine
	Uraemia	Neostigmine
	Malnutrition	Oral contraceptives
	Hypothyroidism	Cancer chemotherapy
	Burned patients	Radiotherapy
		Organophosphorus insecticides

may be anything from 10 minutes to 2 hours. The condition is life-threatening only if not recognized. Treatment is artificial ventilation, with appropriate sedation until the return of spontaneous ventilation. Active enzyme is present in fresh frozen plasma (unless the donor had abnormal cholinesterase activity), but spontaneous activity will often return before the plasma becomes available.

Investigation of suxamethonium apnoea

Non-inherited causes of reduced cholinesterase activity should be excluded (Table 46.5) and plasma samples sent to a reference laboratory. If an inherited abnormality is confirmed, other family members should be similarly investigated.

Other drugs metabolized by plasma cholinesterase

Ester local anaesthetics These are now used infrequently, but repeated doses may cause unexpected toxicity in patients with low plasma cholinesterase levels.

Mivacurium This new non-depolarizing neuromuscular blocking drug is metabolized primarily by plasma cholinesterase. It is therefore to be anticipated that patients susceptible to suxamethonium apnoea may have prolonged paralysis following administration of mivacurium. The authors are aware of one such case (R. Mirakhur, 1993, personal communication).

REFERENCES

1. Denborough MA, Lovell RRH. Anaesthetic deaths in a family. *Lancet* 1960; **2**: 15.
2. Lopez JR, Alamo L, Jones D, *et al.* Dantrolene reverses the syndrome of malignant hyperthermia by reducing the level of intracellular Ca^{2+}. *Biophysical Journal* 1985; **47**: 313a.
3. Iaizzo PA, Klein W, Lehmann-Horn F. Fura-2 detected myoplasmic calcium and its correlation with contracture force in skeletal muscle from normal and malignant hyperthermia susceptible pigs. *Pflügers Archives* 1988; **411**: 648–53.

4. Mickelson JR, Gallant EM, Litterer LA, Johnson KM, Rempel WE, Louis CF. Abnormal sarcoplasmic reticulum ryanodine receptor in malignant hyperthermia. *Journal of Biological Chemistry* 1988; **263**: 9310–15.

5. Ervasti JM, Strand MA, Hanson TP, Mickelson JR, Louis CF. Ryanodine receptor in different malignant hyperthermia susceptible porcine muscles. *American Journal of Physiology* 1991; **260**: C58–66.

6. Fill M, Stefani E, Nelson TE. Abnormal human sarcoplasmic reticulum Ca^{2+} release channels in malignant hyperthermic skeletal muscle. *Biophysical Journal* 1991; **59**: 1085–90.

7. Hopkins PM, Ellis FR, Halsall PJ. Ryanodine contracture: a potentially specific *in vitro* diagnostic test for malignant hyperthermia. *British Journal of Anaesthesia* 1991; **66**: 611–13.

8. Hopkins PM, Ellis FR, Halsall PJ. Comparison of *in vitro* contracture testing with ryanodine, halothane and caffeine in malignant hyperthermia and other neuromuscular disorders. *British Journal of Anaesthesia* 1993; **70**: 397–401.

9. Nelson TE, Bedell DM, Jones EW. Porcine malignant hyperthermia: effects of temperature and extracellular calcium concentration on halothane-induced contracture of susceptible skeletal muscle. *Anesthesiology* 1975; **42**: 301–6.

10. Hopkins PM, Ellis FR, Halsall PJ. Extracellular calcium and *in vitro* contracture of human MH muscle. *Journal of the Neurological Sciences* 1990; **98S**: 510.

11. Adnet PJ, Krivosic-Horber RM, Adamantidis M, Reyford H, Cordonnier C, Haudecoeur G. Effects of calcium-free solution, calcium antagonists, and the calcium agonist BAY K8644 on mechanical responses of skeletal muscle from patients susceptible to malignant hyperthermia. *Anesthesiology* 1991; **75**: 413–19.

12. Adnet PJ, Krivosic-Horber RM, Haudecoeur G, *et al.* Use of the calcium agonist Bay K 8644 for *in vitro* diagnosis of suceptibility to malignant hyperthermia. *British Journal of Anaesthesia* 1990; **65**: 791–5.

13. Adnet PJ, Krivosic-Horber RM, Haudecoeur G, Reyford GH, Adamantidis MM, Dupuis BA. Diltiazem and nifedipine reduce the *in vitro* contracture response to halothane in malignant hyperthermia susceptible muscle. *Canadian Journal of Anaesthesia* 1990; **37**: 556–9.

14. Hopkins PM. The effects of caffeine and halothane on isolated rat skeletal muscle $[Ca^{2+}]_i$ in the absence and presence of extracellular Ca^{2+}. *Journal of Physiology* 1993; **459**: 14P.

15. McCarthy TV, Healy JMS, Heffron JJA, *et al.* Localization of the malignant hyperthermia susceptibility locus to human chromosome 19q12–13.2. *Nature* 1990; **343**: 562–4.

16. MacLennan DH, Duff C, Zorzato F, *et al.* Ryanodine receptor gene is a candidate for predisposition to malignant hyperthermia. *Nature* 1990; **343**: 559–61.

17. Ropers HH, Pericak-Vance MA. Report of the committee on the genetic constitution of chromosome 19. *Human Gene Mapping 11: Cytogenetics and Cell Genetics* 1991; **58**: 751–84.

18. Ball SP, Dorkins HR, Ellis FR, *et al.* Genetic linkage analysis of chromosome 19 markers in malignant hyperthermia. *British Journal of Anaesthesia* 1993; **70**: 70–5.

19. Fuji J, Otsu K, Zorzato F, *et al.* Identification of a mutation in the porcine ryanodine receptor that is associated with malignant hyperthermia. *Science* 1991; **253**: 448–51.

20. Levitt RC, Nouri N, Jedlicka AE, *et al.* Evidence for genetic heterogeneity in malignant hyperthermia susceptibility. *Genomics* 1991; **11**: 543–7.

21. Deufel T, Golla A, Iles D, *et al.* Evidence for genetic heterogeneity of malignant hyperthermia susceptibility. *American Journal of Human Genetics* 1992; **50**: 1151–61.

22. Fagerlund T, Islander G, Ranklev E, *et al.* Genetic recombination between malignant hyperthermia and calcium release channel in skeletal muscle. *Clinical Genetics* 1992; **41**: 270–2.

23. Ball SP, Dorkins HR, Ellis FR, *et al.* Is muscle biopsy for malignant hyperthermia diagnosis still necessary? *British Journal of Anaesthesia* 1992; **69**: 222P–3P.

24. van der Spek AFL, Fang WB, Ashton-Miller JA, Stohler CS, Carlson DS, Schork MA. The effects of succinylcholine on mouth opening. *Anesthesiology* 1987; **67**: 459–65.

25. Leary NP, Ellis FR. Masseteric muscle spasm as a normal response to suxamethonium. *British Journal of Anaesthesia* 1990; **64**: 488–92.

26. Cain PA, Ellis FR. Anaesthesia for patients susceptible to malignant hyperpyrexia: a study of pancuronium and methylprednisolone. *British Journal of Anaesthesia* 1977; **49**: 941–4.

27. Kalow W, Britt BA, Terreau ME, Haist C. Metabolic error of muscle metabolism after recovery from malignant hyperthermia. *Lancet* 1970; **2**: 895–8.

28. Ellis FR, Harriman DGF, Keaney NP, Kyei-Mensah K, Tyrrell JH. Halothane induced muscle contracture as a cause of hyperpyrexia. *British Journal of Anaesthesia* 1971; **43**: 721–2.

29. European Malignant Hyperpyrexia Group. A protocol for the investigation of malignant hyperpyrexia (MH) susceptibility. *British Journal of Anaesthesia* 1984; **56**: 1267–9.

30. Goldberg S, Stern L, Feldman L. Serial two-dimensional echocardiography in Duchenne muscular dystrophy. Neurology 1982; **10**: 1101–5.

31. Sethna SF, Rockoff MA, Worthen HM, Rosnow JM. Anesthesia-related complications in children with Duchenne muscular dystrophy. *Anesthesiology* 1988; **68**: 462–5.

32. Wang JM, Stanley TH. Duchenne muscular dystrophy and malignant hyperthermia – two case reports. *Canadian Anaesthetists' Society Journal* 1986; **33**: 492–7.

33. Boba A. Fatal postanesthetic complications in two muscular dystrophic patients. *Journal of Pediatric Surgery* 1970; **5**: 71–5.

34. Smith CL, Bush GH. Anaesthesia and progressive muscular dystrophy. *British Journal of Anaesthesia* 1985; **57**: 1113–18.

35. Heiman-Patterson TD, Natter HM, Rosenberg H, Fletcher JE, Tahmoush AJ. Malignant hyperthermia susceptibility in X-linked muscle dystrophies. *Pediatric Neurology* 1986; **2**: 356–8.

36. Dresner DL, Ali HH. Anaesthetic management of a patient with facioscapulohumeral muscular dystrophy. *British Journal of Anaesthesia* 1989; **62**: 331–4.

37. Aldridge LM. Anaesthetic problems in myotonic dystrophy. *British Journal of Anaesthesia* 1985; **57**: 1119–30.

38. Pollard BJ, Young TM. Anaesthesia in myotonia dystrophica. *Anaesthesia* 1989; **44**: 699.

39. White DA, Smyth DG. Continuous infusion of propofol in dystrophia myotonica. *Canadian Journal of Anaesthesiology* 1989; **36**: 200–3.

40. Wruck G, Tryba M. Vecuronium bei Dystrophia myotonica (Curscmann–Steinert). *Anaesthesist* 1989; **38**: 255–8.

41. Cope DK, Miller JN. Local and spinal anesthesia for cesarian section in a patient with myotonic dystrophy. *Anesthesia and Analgesia* 1986; **65**: 687–90.

42. Streib EW. Hypokalaemic paralysis in two patients with paramyotonia congenita and known hyperkalaemic exercise-induced weakness. *Muscle and Nerve* 1989; **12**: 936–7.

43. Aarons JJ, Moon RE, Camporesi MD. General anesthesia and hyperkalemic periodic paralysis. *Anesthesiology* 1989; **71**: 303–4.

44. Cunliffe M, Burrows FA. Anaesthetic implications of nemaline rod myopathy. *Canadian Anaesthetists' Society Journal* 1985; **32**: 543–7.

45. Heard SO, Kaplan RF. Neuromuscular blockade in a patient with nemaline myopathy. *Anesthesiology* 1983; **59**: 588–90.

46. Famewo CE. Induction of anaesthesia with etomidate in a patient with acute intermittent porphyria. *Canadian Anaesthetists' Society Journal* 1985; **32**: 171–3.

47. Meissner PN, Harrison GG, Hift RJ. Propofol as an i.v. anaesthetic induction agent in variegate porphyria. *British Journal of Anaesthesia* 1991; **66**: 60–5.

48. Mitterschiffthaler G, Theiner A, Hetzel H, Fuith LC. Safe use of propofol in a patient with acute intermittent porphyria. *British Journal of Anaesthesia* 1988; **60**: 109–11.

49. Kantor G, Rolbin SH. Acute intermittent porphyria and caesarean delivery. *Canadian Journal of Anaesthesia* 1992; **39**: 282–5.

50. Weir PM, Hodkinson BP. Is propofol a safe agent in porphyria? *Anaesthesia* 1988; **43**: 1022–3.

51. Capouet V, Dernovoi B, Azagra JS. Induction of anaesthesia with ketamine during an acute crisis of hereditary coproporphyria. *Canadian Journal of Anaesthesia* 1987; **34**: 388–90.

52. Kalow W, Genest K. A method for the detection of atypical forms of human serum cholinesterase. Determination of dibucaine numbers. *Canadian Journal of Biochemistry* 1957; **35**: 339.

53. Harris H, Whittaker M. Differential inhibition of human serum cholinesterase with fluoride: recognition of two new phenotypes. *Nature* 1961; **191**: 496.

FURTHER READING

Stanbury JB. *The metabolic basis of inherited disease*, 5th ed. New York: McGraw-Hill, 1983.

Weatherall DJ. *The new genetics and clinical practice*, 3rd ed. Oxford: Oxford University Press, 1990.

Chapter 47

Hyperbaric Physiology and Medicine

Roderic G. Eckenhoff

Physiological implications of hyperbaric exposure	Hyperbaric oxygen therapy

The number of people being exposed to hyperbaric environments has increased. This is a result of an increase in the use of hyperbaric oxygen therapy, an increase in the popularity of underwater diving, and the continued necessity of pressurized (caisson) construction techniques. An understanding of human tolerance of hyperbaric environments and an appreciation of the unique aspects of medical care under pressurized conditions is therefore desirable. Because of the anaesthetist's expertise in gas physics and physiology, oxygen delivery and acute care monitoring, it is both natural and necessary that the anaesthetist have a basic understanding of the physiological and practical considerations of man's exposure to hyperbaria. Indeed, anaesthetists have been intimately involved in the operation and management of hyperbaric medicine centres world wide.

This introductory chapter is directed at the anaesthetist with no background in hyperbaric medicine. Its first section covers the physiology of exposure to hyperbaric environments; the second discusses the indications, rationale and risks of hyperbaric oxygen therapy, including the practicalities of patient care in a hyperbaric chamber.

Physiological implications of hyperbaric exposure

Exposure to hyperbaric conditions necessarily involves three phases – compression, stable pressure and decompression – and the physiological implications follow this organization reasonably well. This section will divide the important physiological effects into those occurring as a result of compression and habitation at increased pressure, and those occurring as a result of decompression.

Compression and stable pressure phase

The physical alterations present in hyperbaric conditions include (1) an increase in hydrostatic pressure, (2) an increase in the partial pressure of the surrounding and inspired gases and (3) an increase in gas density. Although this division is useful for purposes of discussion, it is important to note that these physical alterations are not easily separable, and the physiological consequences represent a summation of their influences. For example, the neurological consequences of exposure to very high pressure represent the effects of both increased partial pressures of the inspired gases and increased hydrostatic pressure itself.

Increased hydrostatic pressure

Hydrostatic or hydraulic pressure is mechanical pressure by itself, to be distinguished from gas partial pressure (below). The effect of hydrostatic pressure depends on magnitude and the site affected. Table 47.1 lists the commonly used units for pressure expression.

One of the more common problems experienced by those undergoing hyperbaric exposure, and which is primarily due to hydrostatic pressure, is *barotrauma*.[1] This is a consequence of the omnidirectional conduction of pressure in liquids that contain gas spaces (such as the body). If this gas space communicates freely with the external pressurized environment, molecules of gas or liquid can be admitted to equalize the pressure gradient. However, if this communication does not exist, the unequalized pressure gradient will attempt to compress the gas space, the significance of which depends on the anatomy. For example, because intestinal walls are highly deformable, compression of any contained gas has minimal consequences. On the other hand, because gas spaces in the head or chest have limits on compression (or expansion; see below), important consequences exist. A pressure gradient across the bony sinus walls can be transmitted only through the blood or lymphatics. Therefore a painful, blood-filled sinus is the frequent result of pressurization

Table 47.1 Pressure units and conversions

1 atm	=	1.013247 bar
1 atm	=	101.3247 kPa
1 atm	=	14.6959 p.s.i.
1 atm	=	760.00 torr[a]
1 atm	=	33.08 f.s.w.
1 atm	=	10.13 m.s.w.
1 bar	=	100.000 kPa
1 bar	=	100,000 Pa[a]
1 bar	=	14.50377 p.s.i.
1 bar	=	750.064 torr
1 bar	=	32.646 f.s.w.[b]
1 bar	=	10.00 m.s.w.
1 f.s.w.	=	3.063 kPa
1 f.s.w.	=	22.98 torr
1 m.s.w.	=	10.000 kPa[c]
1 m.s.w.	=	1.450 p.s.i.
1 m.s.w.	=	75.01 torr
1 p.s.i.	=	6894.76 Pa[a]
1 p.s.i.	=	51.7151 torr
1 p.s.i.	=	2.251 f.s.w.
1 MPa	=	10.000 bar
1 torr	=	133.322 Pa[a]

[a] Signifies a primary definition from which the other equalities were derived

[b] Primary definition for f.s.w.; assumes a density for seawater of 1.02480 at 4°C (the value often used for depth gauge calibration)

[c] Primary definition for m.s.w.; assumes a density for seawater of 1.01972 at 4°C

Reprinted with permission from *Undersea Biomedical Research* ©.

without osteal patency. A similar injury may occur in the middle ear but, because only the thin tympanic membrane separates pressurized and non-pressurized space, rupture may occur when eustachian (pharyngotympanic) tube dysfunction does not allow pressure equalization from the nasopharynx. Patients with marginal eustachian tube function may develop serous otitis media with repetitive exposure to pressure.[2] Systemic or topical decongestants may be useful to maintain eustachian tube and osteal patency, but occasionally surgery (myringotomy, for example) is required in circumstances where pressure exposure is necessary.

Compression barotrauma of the chest is infrequent, owing to the absolute requirement for airway patency and to the compressible nature of the chest and lung. In the absence of ventilatory activity (during a breath-hold dive, for example), the limits of chest wall compression require a cephalad movement of the diaphragm, and translocation of blood into the chest in an attempt to equalize the transthoracic pressure gradient. However, as the duration of such gradients is usually limited to only a few minutes, the physiological consequences are small. Pulmonary barotrauma is most dangerous during decompression, discussed below.

A more subtle form of compression barotrauma is commonly referred to as *compression arthralgia*. As one is compressed to greater than about 6 atmospheres absolute (ATA), increased joint crepitance and arthralgias may occur. The mechanism for this effect is obscure, but the rapidity of development implies that a compressible space (gas?) normally exists in or around the joint, the collapse of which can sufficiently distort the joint anatomy to produce pain and crepitance.

The final effect of hydrostatic pressure to be discussed here is the neurological consequence of exposure to very high pressure. Both animals and humans experience a generalized excitation of the central nervous system (CNS) on exposure to pressures greater than about 10 ATA. This syndrome, called the high pressure neurological syndrome (HPNS), is characterized by nervousness, tremor and ultimately convulsions, and has been a significant impediment to people's tolerance of high pressure environments.[3] The mechanism of this pressure-induced disorder is thought to be the result of specific pressure-sensitive proteins or receptor complexes in the CNS. HPNS can, in part, be controlled or masked by CNS depressant agents, and the anaesthetic 'inert' gases (see below). HPNS may be closely related to, or possibly the mechanism of, pressure reversal of anaesthesia,[4] a functional competition between excitatory and depressant sites in the CNS. HPNS (and pressure reversal of anaesthesia) is rarely encountered in clinical practice, as it is a feature of pressure exposures in excess of those normally experienced by patients undergoing hyperbaric oxygen therapy (<6 ATA), by underwater divers (<7 ATA) or by pressurized construction workers (<3 ATA).

Increased partial pressure of inspired gases

Henry's law predicts that exposure to increased atmospheric pressure will increase the concentration (tension) of gas dissolved in the liquid phase. Because the biological activity of any gas is related to its tension in the blood or tissues, exposure to increased pressure should have significant physiological effects. The gases to be considered in this section are oxygen, carbon dioxide, the inert gases and contaminant gases.

Oxygen

One of the primary reasons for exposing people to hyperbaric conditions is to increase the concentration of dissolved oxygen in the blood and tissue for the treatment of certain disease states (see 'Hyperbaric oxygen therapy', later). In this case, arterial oxygen tensions in excess of 200 kPa (1500 mmHg) may be produced by the inhalation of 100 per cent oxygen at 2–3 ATA. Tissue oxygen tension is not elevated as much, however, because of diffusion barriers, oxygen utilization, oxygen-induced vasoconstriction and the poor solubility of oxygen in plasma. Table 47.2 lists *in*

Table 47.2 Tissue oxygen tension, in kPa and (mmHg), with increasing inspired oxygen pressure

		Rat tibia[6]		Human[7,8]		
ATA O_2	Rat brain[5]	Normal	Osteomyelitis	Muscle	Subcut. tissue	Myonecrosis phlegmon
0.2	4.5 (34)	6.0 (45)	2.8 (21)	3.9 (29)	4.9 (37)	6.7 (50)
1.0	12.0 (90)	–	–	7.9 (59)	7.1 (53)	14.7 (110)
2.0	32.5 (244)	42.8 (321)	13.9 (104)	29.5 (221)	29.5 (221)	33.3 (250)
3.0	60.3 (452)	–	–	–	–	44.0 (330)

Reproduced, with permission, from Thom SR. Hyperbaric oxygen therapy. *Journal of Intensive Care Medicine* 1989; **4**: 58–74. (values in kPa added)

vivo oxygen tension measurements in a variety of tissues with increasing inspired oxygen partial pressure, demonstrating the modest increase in oxygen tension, especially in diseased tissues. Even though oxygen is poorly soluble in plasma, sufficient quantities can be dissolved at 3 ATA to meet metabolic needs in the complete absence of haemoglobin. In addition to producing a therapeutic benefit, these increased oxygen tensions can be toxic to several organ systems with prolonged exposure.

Pulmonary toxicity

The lung is continuously exposed to the highest PO_2 of any organ, and is the first to demonstrate toxicity with normobaric (1 ATA) exposures. Normobaric oxygen inhalation for as little as 6–8 hours decreases tracheal mucous velocity, and non-specific respiratory symptoms begin in as little as 12 hours in normal humans.[9–11] Pulmonary function changes, nausea, vomiting, anorexia and occasionally orthostasis are prominent symptoms with exposures of greater than 24 hours. Survival time of normal primates in 100 per cent oxygen is in excess of a week, with death the result of pulmonary oedema and, ironically, hypoxia. The development and rate of progression of pulmonary oxygen toxicity are directly related to the inspired oxygen partial pressure (PIO_2), such that substantial signs and symptoms are apparent in normal human subjects after 3–5 hours of exposure to 2.0 ATA oxygen.[12] Substantial individual differences in oxygen sensitivity exist.

The maximum non-toxic PIO_2 has not been established, but prolonged exposures to 0.5 ATA produce few symptoms in normal subjects, and even allow recovery from acute lung injury in animals and humans.[13,14] Between 0.21 and 0.50 ATA, physiological, biochemical and anatomical changes occur, but are considered to be adaptive rather than pathological. Absorption atelectasis is not an important contributor to pulmonary oxygen toxicity, demonstrated by the usual progression of toxicity in humans breathing air at 5.0 ATA.[14] Therefore, prolonged hyperbaric exposure at pressures greater than about 3 ATA should have an FIO_2 below that of air, so that a PIO_2 of less than 0.5 ATA is produced.

The treatment of pulmonary oxygen toxicity relies on decreasing the PIO_2 and providing supportive measures. There are currently no specific pharmacological approaches to treatment. Dramatic improvements in oxygen tolerance have been produced with prior exposure to sublethal hyperoxia in some species;[15] paradoxically, hypoxic pre-exposure can also result in increased resistance to the toxic effects of oxygen.[16] Human oxygen tolerance has been shown to be significantly increased by brief interruptions of oxygen inhalation with an approximately normoxic gas,[17] a technique commonly employed in hyperbaric oxygen therapy. In animals, adaptation has been associated with increased cellular antioxidant enzyme levels, proliferation of alveolar type II cells and increased alveolar surfactant levels.[18–20]

Central nervous system

Central nervous system oxygen toxicity is not usually observed with inhaled oxygen partial pressures of less than 2.0 ATA.[21] It may occur before pulmonary toxicity at pressures above 3.0 ATA, and, like pulmonary toxicity, there is wide variation in individual sensitivity. Oxygen toxicity of the CNS is characterized by convulsions, which may be preceded by visual symptoms or muscular twitching. Exercise and hypercapnia accelerate the onset of symptoms, probably due to cerebrovasodilatation and increased delivery of oxygen to the brain. CNS oxygen toxicity is rapidly reversible with decreases in the PIO_2, and permanent or residual sequelae have not been reported.

In addition to toxicity, increased oxygen partial pressures may have subtle physiological effects. Oxygen inhalation at 1 ATA or above causes a small and immediate respiratory depression in normal subjects, presumably due to loss of tonic chemoreceptor activity.[22] However, this is soon followed by an increase in ventilation, due to an increase in tissue carbon dioxide tensions.[23] The increased tissue CO_2 tension is a result of decreased elimination of carbon dioxide; the increased concentration of oxyhaemoglobin in the venous blood during oxygen inhalation implies a decreased concentration of deoxyhaemoglobin, an important component of venous CO_2 buffering. Oxygen inhalation may also cause absorption atelectasis in lung regions with poor ventilation because of the absence of nitrogen.

The cardiovascular effects of oxygen inhalation are

usually of little consequence. Heart rate and cardiac output are slightly reduced with little change in blood pressure, presumably because of oxygen-induced vasoconstriction. Pulmonary artery pressures are decreased somewhat by the removal of that component of pulmonary artery tone maintained by regional alveolar hypoxia.

A final and more practical result of increased oxygen concentration and partial pressure relates to fire risk, and is discussed in more detail later, under 'Risks of hyperbaric oxygen therapy'.

Carbon dioxide

At constant atmospheric CO_2 fractional concentration, the P_ICO_2 and related physiological effects increase with elevated pressure. For example, a normally well-tolerated atmosphere containing a CO_2 concentration of about 1 per cent (~1.1 kPa or ~8 mmHg), if compressed and inhaled at 5 ATA, would produce approximately the same physiological effect as breathing 5 per cent CO_2 at 1 ATA. However, this does not mean that expired CO_2 is more dangerous at 5 ATA than at 1 ATA; end-tidal CO_2 concentration would be about 1 per cent at 5 ATA, because the *partial pressure* of about 5.3 kPa (40 mmHg) is maintained.

Increased CO_2 partial pressures (over about 7.3 kPa or 55 mmHg) produce subjective hyperpnoea and dyspnoea, as well as headache due to increased cerebral blood flow. Higher partial pressures can produce CNS depression, and are often accompanied by significant acidosis. Inhalation of atmospheres with increased P_ICO_2 may have opposing direct and indirect effects on the cardiovascular system. Activation of the sympathetic nervous system with increases in blood catecholamine concentration occurs at moderate P_ICO_2, while opposing direct depressant effects on tissues, such as cardiac and vascular muscle, may occur simultaneously. The cerebral circulation, which does not have functionally important sympathetic innervation, undergoes dilation with increased P_ICO_2 This cerebral vasodilation may explain the increased incidence of CNS oxygen toxicity in high CO_2 atmospheres, and during exercise, a matter of importance during hyperbaric oxygen therapy. In general, however, the threshold for convulsions from other causes is elevated by increased P_ICO_2. Because of these important physiological effects of breathing increased P_ICO_2 atmospheres, CO_2 concentration is usually monitored in chamber atmospheres, and most hyperbaric chambers are ventilated frequently.

Nitrogen, helium and other inert gases

Most of the 'inert' gases produce CNS depression at high partial pressure. For example, nitrogen is essentially anaesthetic at pressures greater than about 30 ATA, and other gases have an anaesthetic potency which approximately follows their solubility in lipid (Table 47.3).

Table 47.3 Narcotic potency of gases

Gas	Mol wt	Lipid solubility*	Estimated MAC†
He	4	0.017	>180
Ne	20	0.022	>150
H₂	2	0.036	81
N₂	28	0.076	44
Ar	40	0.140	19
Kr	84	0.430	6.2
Xe	131	1.700	0.8

* Ostwald coefficient, 37°C.
† Minimum alveolar concentratiuon for lack of movement with noxious stimulation. For Ne and He, this has not been observed, and the values given are estimates based on lipid solubility. For H₂, N₂, Ar and Kr, the values determined in animals;[3] Xe MAC determined in humans.[24]

Although these anaesthetic partial pressures are produced by hyperbaric exposures of much greater magnitude than those used for hyperbaric oxygen therapy or for underwater diving, more subtle CNS effects may be observed at lesser pressures. For example, nitrogen narcosis, a well-known malady among SCUBA divers breathing air at depths greater than about 100 feet of sea water (f.s.w.) (4 ATA), includes euphoria and an impairment of judgement.[3] To minimize these CNS effects at high pressure, other inert gases have been substituted for nitrogen. Helium, in part because of its very low solubility, has such a low anaesthetic potency that the pressures required to produce CNS depression exceed those required to produce HPNS in animals; depression has not been observed. Low concentrations of nitrogen (5 per cent) have been added to the helium/oxygen atmospheres used for very high pressure exposures to control or mask the CNS excitation of HPNS. Such 'trimixes' have been used successfully for manned pressure exposures of about 70 ATA.[3] The use of hydrogen as a diluent gas for exposure to high pressure has received recent attention, because of an intermediate narcotic potency between nitrogen and helium, and a lower density than nitrogen. Unfortunately, it is dangerous because mixtures containing more than about 5 per cent oxygen are explosive.

A final and important implication of an increased partial pressure of the inert gases is that the blood and tissue content of dissolved gas is also increased. The magnitude of the increase depends on the pressure and duration of exposure and on the solubility of the inert gas. The kinetics and route of elimination of this additional inert gas are of fundamental importance to tolerance of the decompression (see later, under 'Decompression sickness').

Contaminants

Carbon monoxide and various hydrocarbons sometimes contaminate compressed gas mixtures, and their toxicity will be magnified when compressed and inhaled. Rigorous standards for scrubbing and filtering compressed gases destined for human inhalation have greatly reduced the probability of significant intoxication. Compressor air intakes need to be well away from sources of contaminants, such as motor and fume hood exhausts; to reduce hydrocarbons, compressor lubrication is often accomplished with water or soaps.

Increased gas density

An increase in pressure leads to an increase in inspired gas density. The physiological considerations of increased gas density simplify to two: respiration and thermal regulation. More practical sequelae, such as speech distortion, may also occur, but are not discussed further here.

The respiratory effects of increased gas density are primarily flow-related. Gas flow in the proximal airways (and of course in any external gas circuits such as tracheal tubes) is largely turbulent, and turbulent gas flow is inversely related to gas density. The relationship between gas flow and density is exponential:

$$Y = BX^{-0.45}$$

where Y is the flow rate at density X (g/l), B is the flow rate at 1 ATA (1 ATA density: 1.1 g/l) and the exponent is an experimentally derived value for humans over a wide range of inhaled gas densities.[14,25] As an example, peak flow rates are decreased by about 50 per cent at 5 ATA as compared to 1 ATA, reducing maximal ventilatory rates, and thereby the maximal sustainable workload. Because of helium's low density, its substitution for nitrogen in high pressure atmospheres allows a preservation of peak flow rates. This application is similar to the rationale for using helium/oxygen mixtures at 1 ATA in the therapy of large airways obstruction. However, it is important to note that the viscosity of helium is somewhat higher than nitrogen, and therefore flow under viscosity-dependent conditions (laminar – small distal airways) might be compromised. For example, helium inhalation may actually worsen gas flow in patients with severe bronchoconstriction.

Gas exchange in distal airways and alveoli occurs primarily by diffusion, and an increase in gas density may have important effects on diffusion. This would manifest as an increase in the A–a gradient for oxygen, due primarily to an imprecise calculation of P_{AO_2} and increased heterogeneity (layering) in the respiratory regions of the airways. Human studies have been unable to demonstrate this effect, although it is likely to exist at very high gas density.

Heat capacity and conductivity of a gas are proportional to the density. Thus, pressurized environments will cause an increase in heat loss from patients. The use of helium in such environments exacerbates this situation because of its uniquely high thermal conductivity. However, a common misconception is that the high thermal conductivity of helium increases *respiratory* heat loss in addition to that lost from the skin. Because respiratory gas is generally heated to body temperature, regardless of composition, the heat *capacity*, and not *conductivity*, is the relevant index. The heat capacity of helium is less than that of air (taking density into account), so inhalation of helium/oxygen mixtures may actually reduce the respiratory component of overall heat loss as compared to nitrogen/oxygen.

Decompression

Returning from pressure to 1 ATA (or from 1 ATA to altitude) can result in at least two problems: barotrauma and decompression sickness. Whilst the former is more common, the two may be confused in their more serious forms.

Barotrauma

After equalization of gas spaces at pressure (see above), an increased number of gas molecules exist in the gas space. These molecules must then leave on decompression, or the space will be at increased pressure relative to the environment. Gas egress normally occurs more easily than entry, but occasionally the sinus ostia, eustachian tube or bronchiole becomes obstructed and gas becomes trapped. The increased pressure resulting from trapped gas produces effects that depend on the anatomy. Elevated pressure in a sinus causes pain, mucosal ischaemia and occasionally hypaesthesia in areas supplied by nerves traversing the spaces. In a space bounded by more flexible walls, such as an alveolus, elevated pressure may cause distension and possibly rupture. The resultant pulmonary barotrauma may then cause pneumothorax, or the gas may dissect through tissue planes to cause pneumomediastinum, subcutaneous emphysema or gas embolism.[1,26] The last is most dangerous because the gas may enter a pulmonary vein and become directly arterialized. Depending on gas volume, position and haemodynamics, the emboli may become widely distributed; of most concern is gas that enters the cerebral or coronary circulations (see below, under 'Hyperbaric oxygen therapy'). Such gas emboli may cause ischaemia directly, as well as trigger a variety of mediator and humoral effects through bubble–blood and bubble–endothelium interactions. Cerebral gas embolism is usually characterized by a sudden onset of unconsciousness or hemiplegic symptoms during or rapidly following a decompression. Immediate hyperbaric oxygen therapy is usually curative, but delay may reduce the chance of success. It is difficult to identify a predisposition to pulmonary barotrauma, but any condition that can produce 'air-trapping' in the lung, such as the presence of blebs, bullae or very poorly ventilated regions, should be considered to increase the risk of pulmonary barotrauma during decompression. Inadequate training and panic may also predispose to pulmonary barotrauma because of breath holding during decompression.

Decompression sickness

As stated above, the exposure to increased pressure leads to an increased tissue content of inert gas. The quantity depends on the magnitude and duration of the pressure exposure, and the concentration and species of inert gas. Brief (1–2 hour) exposures usually allow sufficient time for only the well-perfused compartments to equilibrate with the new partial pressure of inert gas, whilst exposures of greater than about 6–12 hours are required for all of the tissues to equilibrate (saturate) with the more soluble inert gases such as nitrogen. Complete saturation of all tissues may take much longer, but it is not certain whether gas in these very slowly equilibrating compartments contributes to the pathophysiology of decompression sickness.

On decompression, the inert gas gradient is reversed, and the tissues eliminate gas. This normally occurs via release as dissolved gas, in the same manner as uptake, but it is possible that the reduction in barometric pressure is large and rapid enough to produce tissue or blood inert gas tensions that exceed their solubility at that pressure and temperature. This supersaturation may result in the formation of a gas phase, either stationary 'autochthonous' gas spaces in tissue or mobile blood or lymph bubbles. Initially, mobile bubbles are probably small (~100 µm), but grow and aggregate in areas of sluggish blood flow. There is now general agreement that both the stationary and the mobile gas bubbles are the aetiology of the decompression sickness syndromes,[27] but the mechanisms linking bubbles and symptoms are poorly appreciated.

Before a discussion of the pathogenesis, symptomatology and treatment of decompression sickness, a brief review of its prevention is appropriate. Considerable effort has gone into the development of decompression routines that limit the degree of supersaturation to some arbitrary point that correlates with a low incidence of symptoms. This is generally accomplished by controlling the rate of decompression so that a given inert gas tension (maximum or 'M' value) is never exceeded in some mathematically defined compartment – often the 360 minute half-time tissue. This concept allows the construction of decompression schedules or tables that specify the maximum allowable time at any pressure for which normal decompression rates can be used (no-decompression limits), and which also specify how decompression must be modified when this time limit is exceeded. The success of such routines is shown by the low incidence of decompression

sickness in sport diving; estimated at about 1 in 10 000 exposures.[28] A more tightly regulated (but possibly more stressed) diving community, the military, is estimated to have a higher overall incidence of about 1 in 100; underreporting in the sport diving community may be a significant problem. Nevertheless, military scientists are establishing new decompression schedules and tables that are based on a statistical treatment of accumulated world experience and that are flexible depending on the acceptable degree of risk.[29] Thus, fairly simple models for decompression have worked well in the prevention of the disorder, not only for divers but also for workers in hyperbaric chambers.

Aside from the use of controlled decompression rates to prevent decompression sickness, enhanced removal of inert gas from tissues may also be accomplished by the use of elevated inspired concentrations of oxygen. An example of this are the denitrogenation procedures used by astronauts prior to hypobaric exposures (analogous to decompression of those exposed to hyperbaric exposures). In this case, the gradient for inert gas elimination is optimized without a change in barometric pressure, and thus inert gas is eliminated while still in the dissolved form. This concept is also routinely used in hyperbaric exposures. For example, chamber attendants who have been exposed to pressure (breathing air) for prolonged periods may breathe oxygen for an hour prior to decompression to reduce the tissue content of dissolved inert gas. Oxygen decompression procedures are also used in underwater diving, but usually limited to highly regulated situations, such as with military or industrial operations.

The introduction of Doppler ultrasound techniques to monitor for mobile gas phase has led to the discovery that many, and perhaps most, decompressions are accompanied by easily detected venous bubbles in humans.[30,31] Arterial bubbles have been detected as well, although rarely. Because most of these bubbles are unaccompanied by symptoms, the term 'silent bubbles' has been applied – although it has not been established whether subclinical gas phase formation is actually benign. The dose–response relation for decompression magnitude and bubble formation has been recently described for humans, and it is remarkable how small an inert gas gradient is necessary.[31] The decompression ED_{50} for detection of bubbles is equivalent to only about 12 f.s.w. Importantly, bubbles are produced in about 10 per cent of people decompressing from exposures where supersaturation does not exist. This may have important implications to the mechanism of bubble formation, suggesting that at least some of the bubbles produced during decompression arise from pre-existing gas spaces, perhaps micronuclei stabilized in hydrophobic crevices.[32] Regardless, it is important to recognize that the initial aetiology of decompression sickness, the gas bubble, is present in most decompressions, and that the appearance of symptoms probably depends on the rate and volume of evolved gas, and also on location. Finally, it should be emphasized that, despite the apparent invariant physical forces responsible for bubble formation, there is pronounced individual variation in the ability to form gas phase after a decompression. The source of this differential tendency is poorly understood, but clearly moves the mechanism of bubble formation out of the realm of simple physical models that incorporate only exposure parameters.

The pathophysiological link between bubbles and symptoms is not at all clear, but is likely to be multifactorial. Gas may expand *in situ*, causing tissue distortion, dysfunction and pain, or may become lodged in the vasculature, causing ischaemia. Such obstruction may be venous, as in the case of low flow systems such as the spinal venous plexus, or may be arterial through the (uncommon) passage of gas emboli through the pulmonary circulation or through right to left shunts. Aside from a direct mechanical disruption of blood flow, the gas bubbles may cause a variety of indirect effects on the vasculature and tissue through interactions at the blood–bubble interface (complement and/or coagulation activation, for example) or between the bubble and endothelium. The importance of these indirect factors is suggested by the positive correlation between decompression sickness symptoms and the predetermined reactivity of the complement system in animals.[33] Thus, another layer of physiological heterogeneity with respect to *reaction to a gas phase* is superimposed on the heterogeneity of *gas phase formation*, resulting in such large variation in response to a decompression that the response appears to be essentially random. The apparent randomness of decompression outcome is the motivation for the current statistical approach to designing decompression procedures, until such time when the pathophysiology of decompression sickness is more completely understood.

The clinical presentation of decompression sickness, as expected from the above, is highly variable. The symptoms may range from vague, non-specific complaints such as unusual fatigue, to more specific and catastrophic complaints such as paraplegia. In between these extremes is cutaneous decompression sickness, characterized by a mottled rash and pruritus; pain-only decompression sickness, characterized by deep-seated pain in or around the major joints (particularly knees and shoulders); and inner ear decompression sickness, characterized by tinnitus, vertigo and hearing loss. Uncommon is pulmonary decompression sickness, which is a fulminant and life-threatening pulmonary disorder characterized by dyspnoea, substernal pain and pulmonary oedema. A wide variety of other symptoms may occur, the result of arterial embolization of bubbles to heart, brain or other vital organs. In addition, a chronic form of decompression sickness has been described, and is more common with repetitive or prolonged low pressure (2–3 ATA) exposures, such as experienced by pressurized caisson workers. This form presents as an aseptic osteonecrosis, usually of the long bones (humerus, femur), years after the pressure exposures. The patient's history may contain repeated diagnoses of decompression sickness. The mechanism is not completely clear, but probably involves bubble formation in the marrow cavity, elevating medullary pressure and causing ischaemia.[34] It has been suggested that pain-only decompression sickness is the acute manifestation of such a process, and, if untreated (see below), may result in future osteonecrosis; evidence for this suggestion is lacking.

Because the therapy of decompression sickness is specific and unique, and because the symptoms are often non-specific, accurate and rapid diagnosis is desirable but difficult. Of most importance is a recent history of hyperbaric exposure. Decompression sickness symptoms do not start during the compression or stable pressure phase of an exposure – they start during the decompression (usually late) or soon afterwards. In general, if symptoms have not occurred before 12 hours after decompression, they will not. Also, the later symptoms develop after a decompression, the better the prognosis.[35] Although the parameters of the hyperbaric exposure represent important historical information by which the chance of having decompression sickness and its expected severity can be gauged, they cannot establish a diagnosis. Likewise, the presence of Dopper-detected gas phase cannot establish the diagnosis, because of the high incidence of false positives. False negatives, on the other hand, should be rare. The diagnosis of decompression sickness rests primarily on the history of symptom onset during or shortly after a decompression.

Hyperbaric oxygen therapy is the primary mode of treatment of the decompression sickness syndromes, and this indication is discussed further under 'Hyperbaric oxygen therapy'. Rapidity of administration remains an important determinant of a successful outcome, but symptom resolution coincident with hyperbaric oxygen therapy as long as a week after the pressure exposure, has been reported.[36]

Hyperbaric oxygen therapy

The inhalation of oxygen at increased pressure is a therapy with several well-established indications. Its use has been somewhat limited owing to the expense of constructing and maintaining the complex facility, the lack of recognized training pathways for personnel, and incomplete acceptance by the general medical community. This last problem

Table 47.4 Current indications for hyperbaric oxygen therapy

Regional hypoxia
Compromised surgical graft or flap
Osteoradionecrosis
Problem wounds and ulcers
Crush injuries
Thermal burns

Global hypoxia
CO, CN intoxication
Severe anaemia

Infections
Clostridial myonecrosis
Necrotizing fasciitis
Refractory osteomyelitis
Rhinocerebral mucormycosis

Gas lesion conditions
Gas embolism
Decompression sickness

stems from a lack of well-controlled studies to support some of the indications for hyperbaric oxygen therapy, and to inappropriate application of hyperbaric therapy in the past. The lack of supporting evidence is, in part, due to the difficulty in conducting valid prospective randomized trials, to the small numbers of hyperbaric medicine centres, and to the small patient populations with the indications in question. Thus, theory and animal studies are used as the rationale for prescribing hyperbaric oxygen in some of its current indications. Nevertheless, several of the accepted indications for hyperbaric oxygen therapy are now well supported by appropriate studies, and the list of indications is growing. This section provides a brief summary of the indications and rationale, and a description of equipment, procedures, risks and some practical considerations of care in a pressurized environment. Detailed protocols are available in the 'Further reading'.

Indications

The principal goal of hyperbaric oxygen therapy is to increase tissue oxygen tension, so it is reasonable that the primary indications are conditions that include either regional or global hypoxia (Table 47.4). Another group of indications take advantage of the fact that specific micro-organisms are oxygen intolerant. Lastly, the increase in hydrostatic pressure inherent in hyperbaric oxygen therapy provides an important part of the rationale for use in gas lesion diseases. Each of these groups of indications is discussed below.

Regional hypoxia

The rationale for use of hyperbaric oxygen in cases of regional hypoxia is twofold. First, the large oxygen gradient possible with an inspired oxygen partial pressure of 2.0–2.5 ATA allows some degree of oxygen delivery and oxygen tension elevation in the hypoxic zone (see Table 47.2), unless there is a complete absence of blood flow, as occurs in occlusion of proximal large vessels. This acute rise in tissue oxygen tension, even if only a few millimetres of mercury, permits survival of the tissue and improved function of immune and repair processes. Second, a more long-term benefit of repeated daily hyperbaric oxygen therapies is a stimulation of angiogenesis in previously hypoxic areas, allowing a more permanent improvement in oxygenation and survival in normobaric air.

An example of such regional hypoxia is the *compromised surgical graft or flap*. Although not indicated for the routine surgical graft or flap, hyperbaric oxygen has been shown to improve the survival of those which are clearly compromised after surgery. Numerous well-controlled animal and human studies have demonstrated improved graft/flap survival with intermittent hyperbaric oxygen treatment for 1–3 days.[37,38]

Hyperbaric oxygen has also been shown to be of benefit in more chronic forms of regional hypoxia, such as *osteoradionecrosis*, particularly of the head and neck. Osteoradionecrosis occurs as a result of the radiation-induced obliterative endarteritis, and has been estimated to occur in up to 10 per cent of patients receiving radiation therapy for carcinoma.[39] It is characterized by aseptic necrosis, loss of tissue integrity and occasionally pathological fractures. Infection is generally secondary to the loss of tissue integrity. Repeated daily hyperbaric oxygen has been shown to promote repair and neovascularization of necrotic bone, and to significantly improve covering of exposed bone and the efficacy of any subsequent surgery.[40] The mechanism for enhanced angiogenesis is not yet clear, but is likely to involve improved macrophage and fibroblast function.[41]

Other examples of regional hypoxic conditions for which hyperbaric oxygen has been used are *problem wounds*, particularly microvascular ulcers, where hyperbaric oxygen reduces infection, improves healing and promotes a granulation tissue base for subsequent skin grafting.[42] *Crush injuries* and traumatic vascular compromises of the extremities have also been successfully treated with hyperbaric oxygen using the rationale that, in addition to improved oxygenation, the high arterial oxygen tension has the added benefit of causing a modest vasoconstriction (without compromising oxygen delivery) which may reduce the past-traumatic oedema and the possibility of compartment syndrome.[43] A reduction in oedema, combined with increased resistance to infection, has also served as the rationale for using hyperbaric oxygen in patients with *thermal burns.* A decrease in hospital stay and in cost has been reported for this application of hyperbaric oxygen.[44,45]

Global hypoxia

There are two major categories of global hypoxia for which hyperbaric oxygen has been shown to be a useful therapeutic tool. The first, and most important, are intoxications with carbon monoxide (CO) and cyanide (CN). CO is an extremely common poison. It interferes with oxygen delivery and utilization because of its high affinity for haemoglobin, the leftward shift of the oxyhaemoglobin dissociation curve, and the interaction with specific cellular proteins (e.g. cytochrome oxidase), especially under hypoxic conditions. The result is tissue hypoxia, manifest usually as neurological depression and occasionally haemodynamic instability. Patients who have had symptomatic CO exposures, and who recover, are at risk for delayed (2–20 days) neurological symptoms, presumably due to an ischaemia and reperfusion injury in the central nervous system.[46] The increased concentration of oxygen molecules achieved with hyperbaric oxygen competes with the CO molecules for haemoglobin (and other protein) binding sites and speeds the elimination of CO. For patients with a history of unconsciousness, haemodynamic instability or at the extremes of age, current evidence suggests that the early use of hyperbaric oxygen significantly reduces the incidence of delayed neurological sequelae.[47,48] The fetus is considered to be very sensitive to CO-mediated injury. Whilst hyperbaric oxygen therapy has been suspected to adversely influence fetal physiology (premature closure of ductus arteriosus, for example), the demonstrated benefits to the mother and possibly the fetus in symptomatic CO poisonings are considered to outweigh the theoretical risk of hyperbaric oxygen to the fetus.

The benefits of hyperbaric oxygen in CO intoxication include mechanisms other than increasing the rate of CO elimination, as it is now well appreciated that a poor correlation exists between CO_{Hb} and presenting symptoms or ultimate prognosis. This lack of correlation suggests that treatment with hyperbaric oxygen should be based on history and clinical findings instead of the CO_{Hb} value. A similar rationale exists for the use of hyperbaric oxygen in cyanide poisoning but, because pure cyanide exposure is rare and effectively treated by conventional antidote therapy, hyperbaric oxygen should be reserved for refractory cases, or cases of mixed CO/CN intoxication, not uncommon following smoke inhalation.[49]

Severe, acute anaemia is a clear example of global hypoxia, but one in which hyperbaric oxygen has had only a limited therapeutic role.[50] At 2.5–3.0 ATA inspired oxygen, sufficient arterial oxygen content is dissolved in plasma to meet metabolic needs, making hyperbaric oxygen useful to support life until red cells become available for transfusion, or until sufficient red cell mass is generated endogenously. This therapy might be useful for the occasional patient who is a Jehovah's Witness and who develops life-threatening anaemia, or the patient in whom cross-matching difficulties exist. However, because the toxicity of oxygen (see below) depends on its partial pressure and not content, this indication for hyperbaric oxygen is limited to only several hours of continuous therapy, or to cases where intermittent treatment is sufficient to bolster a marginal oxygen delivery.

Infections

There is substantial overlap between this group of indications and that labelled 'regional hypoxia'. Anaerobic infections often develop in hypoxic areas because of a lack of adequate host response, and ischaemic, hypoxic areas often develop because of an infection. Thus, part of the rationale for hyperbaric oxygen therapy remains the relief of ischaemia and improvement of host response to the infection. The leucocyte oxidative killing mechanisms operate only when the oxygen tension is above 4 kPa (30 mmHg).[51] However, some micro-organisms, or their effects on the host, are sensitive to oxygen tensions attainable with hyperbaric oxygen therapy, and this provides the second part of the rationale for its use.

Clostridial myonecrosis (gas gangrene) is an example of infection with an oxygen-sensitive micro-organism, and for which hyperbaric oxygen has been used successfully. It is a rapidly progressive and life-threatening infection, in large part because of the series of toxins produced by the clostridial micro-organism. While the toxins are oxygen stable, their production by the bacteria is inhibited by elevation of the tissue oxygen tension to 40 kPa (300 mmHg) or above.[52] This slows the progression and permits a host response to the normally overwhelming infection. Rapid diagnosis is essential and hyperbaric oxygen should be used early and frequently in the treatment regimen, which also includes parenteral antibiotics and surgery.

Other necrotizing infections, which are more slowly progressive, and for which the role of hyperbaric oxygen is less well established, are *necrotizing soft-tissue infections*, such as necrotizing fasciitis.[53] The infection is due to a variety of organisms, and is treated adequately in most cases with antibiotics and surgical debridement. In extensive or more rapidly progressive cases, typified by Fournier's disease,[54] hyperbaric oxygen may produce a sufficient increase in local oxygen tension to allow a host response, but, as in the case of gas gangrene, is still to be considered adjunctive to conventional care.

Osteomyelitis that has not responded to conventional therapy is labelled *refractory osteomyelitis*, and has been successfully treated with hyperbaric oxygen.[55] The rationale for treatment is similar to the above in that oxygen tension in the area of infection is increased, leucocyte and osteoclast function improved and bacterial susceptibility to antibiotics improved. In general, oxygen is not directly bacteriostatic to most micro-organisms associated with osteomyelitis at less than 10 ATA, although anaerobes are occasionally involved. A classification system based on anatomical location and host factors has been developed[56] and is useful to determine which types of osteomyelitis that hyperbaric oxygen therapy may benefit.

Hyperbaric oxygen has also been used to treat life-threatening fungal infections, such as rhinocerebral mucormycosis, based on a similar rationale as above. Although the published reports of this application are few, the results have been encouraging in a disease with high mortality.

Gas lesion disease

There are two principal types of gas lesion disease for which hyperbaric oxygen is a primary mode of therapy. The first, *decompression sickness*, has been discussed extensively above, and will be combined here with the second, *gas embolism*, because of the similarity of treatment rationale, despite the different initiation and pathophysiology.

Gas embolism is an acquired condition whereby gas is admitted to the vasculature and circulation. This may occur as a result of pulmonary barotrauma or a wide variety of iatrogenic or traumatic causes (surgery, catheters, trauma, abortions, orogenital sex).[57] Decompression sickness is a unique form of gas (micro)embolization due to *endogenous* production of the gas bubbles.[27] The effects of gas embolism are related to the gas volume, rate of administration and its ultimate location. For example, small volumes of air admitted through an intravenous line (<50 ml) will most likely be trapped efficiently by the pulmonary circulation, and cause little or no effect. On the other hand, much smaller volumes of gas admitted to the arterial side can cause catastrophic effects if distributed to the coronary or cerebral circulations. Venous emboli may be arterialized either through functional right-to-left shunts or in the presence of pulmonary hypertension, which may be precipitated by the venous embolism itself. Vascular obstruction and ischaemia are the primary results of gas embolization, but humoral influences arising from interactions at the blood–bubble interface or at the endothelial surface may play an important vasoactive role.

The rationale for hyperbaric oxygen therapy is founded in the two principal components of the treatment: pressure and hyperoxia. Hydrostatic pressure will cause a decrease in volume of the emboli, and the hyperoxia may improve oxygen delivery to tissue downstream of the obstructing emboli. The hyperoxia also maximizes the gradient for elimination of the gas (generally nitrogen) in the emboli. Of course, for hyperbaric therapy to be effective, rapid institution is essential – most treatment failures are probably due to the late use of hyperbaric therapy. This raises a dilemma in the post-surgical patient with suspected gas embolism in that the lack of emergence from anaesthesia may be multifactorial; gas embolism may be relegated to a late diagnosis of exclusion.

For understandable ethical reasons, randomized clinical trials of hyperbaric oxygen for this indication (either gas embolism or decompression sickness) have not been conducted. Nevertheless, the rapidity of symptom resolution coincident with the application of pressure, combined with numerous animal studies of both decompression sickness and gas embolism, is compelling evidence for the continued use of hyperbaric oxygen in the gas lesion diseases.

Hyperbaric chambers

Hyperbaric oxygen is administered in a hyperbaric chamber, of which there are two basic types – monoplace and multiplace. The modern monoplace chamber is transparent acrylic, can accommodate a single patient and is usually pressurized with oxygen; the patient does not require a mask. The primary advantages of the monoplace chamber are cost and space requirements. The multiplace chamber is usually steel (some are aluminium), can accommodate more than two people, and is pressurized with air while the patient breathes oxygen from a tight-fitting mask or circuit. The multiplace chamber is suitable for critically ill patients requiring ventilation, monitoring and constant attendance. However, even the monoplace chamber can be equipped with a full range of monitoring and critical care capability. The pressure and duration of hyperbaric oxygen therapy depends on the indication, and ranges from 2 to 6 ATA for 2–6 hours. Serious decompression sickness or gas embolism may require prolonged, continuous 'saturation' protocols when symptoms worsen during the decompression phase of conventional regimens. The P_{IO_2} in most hyperbaric oxygen protocols rarely exceeds 2.8 ATA. Emergency indications for hyperbaric oxygen therapy, such as CO poisoning or decompression sickness, generally require only two or three separate chamber treatments, whereas problem wounds often require 40 or more daily sessions in the hyperbaric chamber. Detailed treatment protocols may be found in the entries listed in the Further reading.

Practical aspects of care in a chamber

Monitoring of patients undergoing hyperbaric therapy is usually accomplished by maintaining the electrical components outside the chamber (for fire safety) and passing the electrode or transducer cables through electrically insulated passthroughs in the chamber wall. Necessary electrical equipment in the chamber should never be connected to or disconnected from line voltage during pressurization, to prevent sparks. Such equipment may also be purged continuously with low flows of nitrogen or helium to surround the electrical components with a low oxygen atmosphere. Cathode ray tube (CRT) monitors may not tolerate the additional pressure, although they may be placed outside a chamber viewport for viewing by inside personnel. The small battery-operated, liquid crystal display monitors appear to tolerate hyperbaric conditions well and have a minimal fire risk, and are now being used inside chambers.

Electrical defibrillation in a hyperbaric chamber is controversial, because of the possibility of poor skin contact, arcing and risk of fire. In addition, because of the largely metal environment, attendants are at risk of shock. Many hyperbaric centres require that the chamber be decompressed prior to the use of a defibrillator. This causes some unique decompression problems for the attendants if the therapy has been prolonged. However, the latency of bubble formation and onset of decompression sickness symptoms is sufficient to allow a brief (5–10 minute) excursion to 1 ATA for defibrillation, with a subsequent return to pressure.[58]

The use of intravenous infusions and invasive monitoring in monoplace chambers is difficult, but appropriate equipment and experience are accumulating. This is much simplified in the multiplace chamber, where normal equipment may be used with minor modification. For example, the flexible bags are preferred over glass bottles for intravenous transfusion so that a pressure gradient between the chamber atmosphere and the fluid reservoir does not occur. If bottles must be used (a nitroglycerin infusion, for example), the attendant must ensure that the gas space above the liquid is in constant communication with the chamber atmosphere. Battery-driven syringe pumps are probably the best alternative for intravenous drug infusion in the hyperbaric chamber. Gas-filled pressure bags for arterial catheters must be checked during and after compression (refilled) and vented during decompression. Other closed, gas-filled devices which must be carefully monitored are the tracheal tube cuff and facemask seal, both of which may be filled with an incompressible medium such as water or saline instead of air so that over- or under-expansion does not occur and result in injury or leakage. An exception is the gas-filled balloon on flow-directed pulmonary artery catheters; this balloon is best left deflated with the filling port open to atmosphere during hyperbaric therapy.

Positive pressure ventilation may be performed with self-inflating bags, although volume-cycled ventilators are preferred.[59] A means of monitoring expired minute volume is required, as changes may occur with the alterations in ambient pressure. Peak ventilatory pressures should also be monitored, because the increased gas density at pressure may require large increases in the ventilatory pressure to achieve reasonable flow rates and minute volumes. Expired gas from all such patient breathing circuits should be collected and exhausted directly to the outside of the chamber to maintain a chamber oxygen concentration of less than about 23 per cent (multiplace chamber).

Blood sampling at pressure is conducted as at sea level, and the sample slowly decompressed to 1 ATA in an airlock. Blood gas analysis is optimally conducted with the analyser in the chamber and calibrated at pressure. When in-chamber equipment is not available, the sample can be analysed after decompression. However, even small bubbles in the sample may lead to large errors in the Pa_{O_2} after decompression, because of the insolubility of oxygen and the huge gradient between the dissolved and normobaric air P_{O_2}. The Pa_{CO_2} in such samples is normally well preserved, because of the normal tension and large buffering and solubility of this gas.

Although the effects of intravenous sedative/hypnotics, analgesics and muscle relaxants are probably more predictable and their use more intuitive than with inhalational agents in these unusual environments, it is useful to understand how the delivery of an anaesthetic vapour at pressure may differ from that at sea level. The vapour pressure of a liquid is independent of barometric pressure, so the saturated vapour pressure of isoflurane, for example, is about 260 mmHg at room temperature whether at 10 000 feet altitude or at 2.8 ATA in a hyperbaric chamber. Therefore, depending on the density and viscosity dependence of the proportioning valve in the agent-specific vaporizers, approximately the equivalent *partial pressure* of the anaesthetic gas as that indicated on the calibrated dial for sea level will be delivered at increased pressure.[60,61] A similar situation exists for vaporizers of the copper-kettle variety; normal anaesthetic effects will be obtained by using standard flow ratios to the vaporizer. The delivered gas *concentration* (volume per cent) will depend directly on atmospheric pressure but, as the pharmacologically relevant parameter is the partial pressure, the change in concentration is unimportant. Whilst calculations predict that a slight deviation from indicated partial pressures may exist, it is important to remember that individual variation in response to the volatile anaesthetic agents is substantially larger than the variation in vaporizer output in altered pressure environments, making their use possible with normal vigilance. Rotameter flowmeters, because of their density dependence, will overestimate gas flows slightly during hyperbaric therapy. Nitrous oxide is rarely used in chambers because the primary indication for hyperbaric oxygen therapy is the delivery of 100 per cent oxygen. In general, the indications for general anaesthesia in a hyperbaric chamber are few and injectable agents are preferred. The pressure range used for clinical hyperbaric oxygen therapy has not been associated with any demonstrable change in the pharmacokinetic or pharmacodynamic properties of the drugs likely to be used in such settings.

Risks of hyperbaric oxygen therapy

Barotrauma

Aural barotrauma is the most common risk or complications of hyperbaric therapy, and is almost always due to the compression phase. It can be minimized by the use of topical or systemic vasoconstrictors, or, in specific cases, by myringotomy. Pulmonary barotrauma during hyperbaric therapy is rare, but should be suspected when any significant chest or haemodynamic symptoms occur during or shortly after the decompression; if suspected, decompression should be stopped, and pneumothorax treated with a chest tube. If evidence of gas embolism exists, the appropriate hyperbaric oxygen protocol should be initiated. Pre-existing pneumothorax should be treated with chest tube drainage prior to starting hyperbaric therapy, to prevent the development of tension pneumothorax during decompression due to continued air leak.

Decompression sickness

Decompression sickness is unlikely to occur to the patient who breathes 100 per cent oxygen during most of the therapy; little nitrogen (or other inert gas) uptake occurs. The chamber attendants, on the other hand, generally breathe air during the therapy, so substantial nitrogen uptake occurs and they are at risk of decompression sickness. However, because hyperbaric oxygen therapy protocols have been designed to limit nitrogen uptake by the attendants, this risk is small. If an attendant becomes unavoidably involved in a prolonged therapy, the decompression rate can be decreased to meet the decompression obligation, or oxygen inhalation can be used prior to and during decompression.

Oxygen toxicity

Central nervous system oxygen toxicity is possible with the exposures normally used for hyperbaric oxygen therapy. Generally, an inspired oxygen pressure of less than 3 ATA is used, and the incidence of CNS symptoms in resting patients is reported to be less than 1 in 10 000 patient therapies.[21] The risk may be higher with exertion or in hypercapnic patients, as discussed above. The primary manifestation of CNS oxygen toxicity is seizure activity which may be preceded by muscular twitching or visual symptoms (narrowing of visual fields). Termination of oxygen inhalation and protection from injury are usually sufficient therapy. Because of the possibility of airway obstruction and breath holding during seizure activity, concurrent decompression may increase the risk of pulmonary barotrauma and gas embolism; changes in chamber pressure should await cessation of seizures. In susceptible patients, anticonvulsant therapy may reduce the probability of oxygen-induced convulsions during hyperbaric oxygen therapy.

Pulmonary symptoms of oxygen toxicity are more predictable than those of the CNS, and are more slowly progressive. Significant symptoms in healthy subjects may develop in 8–10 hours of continuous oxygen inhalation at 2 ATA[9] but this is shortened to 3–4 hours at 3 ATA. Intermittent air breathing during hyperbaric oxygen therapy has been shown to substantially extend pulmonary oxygen tolerance, and it is now standard practice to include 5–15 minute air 'breaks' at regular intervals during oxygen breathing. Oxygen inhalation limits have been established so that the risk of toxicity in prolonged or unusual hyperbaric therapy protocols can be estimated, and modified if necessary. Pulmonary symptoms attributable to oxygen toxicity are readily reversible when the P_{IO_2} is decreased below 0.5 ATA,[14] and long-term sequelae have not been reported. Similarly, pulmonary sequelae of chronic daily therapy have not been detected.[62]

Visual function

Progressive myopia has been observed in some patients during prolonged daily therapy, presumably due to effects of either oxygen or pressure on lens shape or refractive index. It is usually reversed within days to weeks after completion of therapy.[63] A baseline ophthalmological examination is generally obtained prior to the initiation of a prolonged course of therapy. Cataract development has been associated with unusually long courses of therapy, such as 150–200 daily exposures to 2–2.5 ATA, and does not reverse after cessation of therapy. Finally, the neonatal retina is sensitive to oxygen; prolonged exposure to even 1 ATA may induce retrolental fibroplasia.[64]

Other risks and side effects

Claustrophobia may be a significant problem in isolated patients, especially in monoplace chambers, and may require sedation for successful therapy. Sedation must be used cautiously in the spontaneously breathing patient receiving hyperbaric oxygen therapy, however, because hypoventilation and hypercapnia may lower the threshold for oxygen-induced seizures.

The increased concentration and partial pressure of oxygen increases the risk of *fire*. In monoplace chambers, which are compressed with oxygen, scrupulous attention to elimination of sources of ignition is required. Because the multiplace chamber is compressed with air, and as the rate of combustion is related primarily to *fractional concentra-tion* (vol %) and not *partial pressure* of oxygen, the risk is less. In any case, limitation of flammable items and sources of ignition is necessary. Although most chambers have automatic fire-control systems, these should not be relied upon to stop a fire in an oxygen-enriched atmosphere.

REFERENCES

1. Wolf HK, Moon RE, Mitchell PR, Burger PC. Barotrauma and air embolism in hyperbaric oxygen therapy. *American Journal of Forensic Medicine and Pathology* 1990; **11**: 149–53.

2. Fernau JL, Hirsch BE, Derkay C, Ramasastry S, Schaefer SE. Hyperbaric oxygen therapy: effect on middle ear and eustachian tube function. *Laryngoscope* 1992; **102**: 48–52.

3. Bennett PB. Inert gas narcosis and the high pressure nervous syndrome in man. In: Bennett PB, Elliott DH, eds. *The physiology and medicine of diving*, 3rd ed. San Pedro: Best, 1982: 239–96.

4. Wann KT, MacDonald AG. Actions and interactions of high pressure and general anesthetics. *Progress in Neurobiology* 1988; **30**: 271–307.

5. Jamieson D, VanDen Brenk HAS. Measurement of oxygen tensions in cerebral tissues of rats exposed to high pressures of oxygen. *Journal of Applied Physiology* 1963; **18**: 869–76.

6. Mader JT, Brown GL, Guckian JC, *et al.* A mechanism for the amelioration by hyperbaric oxygen of experimental staphylococcal osteomyelitis in rabbits. *Journal of Infectious Diseases* 1980; **142**: 915–22.

7. Brummelkamp WH. Considerations on hyperbaric oxygen therapy at three atmospheres absolute for clostridial infections, type welchii. *Annals of the New York Academy of Sciences* 1965; **117**: 688–99.

8. Wells CH, Goodpasture JE, Harrigan DJ, Hart GB. Tissue gas measurements during hyperbaric oxygen exposure. In: Smith G, ed. *Proceedings of the sixth international conference on hyperbaric medicine*. Aberdeen: University Press, 1977: 118–24.

9. Clark JM. Pulmonary limits of oxygen tolerance in man. *Experimental Lung Research* 1988; **14**: 897–910.

10. Deneke SM, Fanburg BL. Normobaric oxygen toxicity of the lung. *New England Journal of Medicine* 1980; **303**: 76–86.

11. Sackner MA, Landa J, Hirsch J, Zapata A. Pulmonary effects of oxygen breathing. *Annals of Internal Medicine* 1975; **82**: 40–3.

12. Clark JM, Lambertsen CJ. Rate of development of pulmonary oxygen toxicity in man during oxygen breathing at 2.0 ATA. *Journal of Applied Physiology* 1971; **30**: 739–52.

13. Cheney FW, Huang TW, Gronka R. The effects of 50% oxygen on the resolution of pulmonary injury. *American Review of Respiratory Disease* 1980; **122**: 373–9.

14. Eckenhoff RG, Dougherty JH, Messier A, Osborne SF, Parker JW. Progression of and recovery from pulmonary oxygen toxicity in humans exposed to 5 ATA air. *Aviation, Space and Environmental Medicine* 1987; **58**: 658–67.

15. Kravetz G, Fisher AB, Forman HJ. The oxygen-adapted rat model: tolerance to oxygen at 1.5 and 2 ATA. *Aviation, Space and Environmental Medicine* 1980; **51**: 775–7.

16. Frank L. Protection from oxygen toxicity by pre-exposure to hypoxia: lung antioxidant enzyme role. *Journal of Applied Physiology* 1982; **53**: 475–82.

17. Hendricks PL, Hall DA, Hunter WL, Haley PJ. Extension of pulmonary oxygen tolerance in men at 2 ATA by intermittent oxygen exposure. *Journal of Applied Physiology* 1977; **42**: 593–9.

18. Crapo JD, Barry BE, Foscue HA, Shelburne J. Structural and biochemical changes in rat lungs occurring during exposures to lethal and adaptive doses of oxygen. *American Review of Respiratory Disease* 1980; **122**: 123–43.

19. Coursin DB, Cihla HP, Will JA, McCreary JL. Adaptation to chronic hyperoxia. Biochemical effects and the response to subsequent lethal hyperoxia. *American Review of Respiratory Disease* 1987; **135**:1002–6.

20. Holm BA, Matalon S, Finkelstein JN, Notter RH. Type II pneumocyte changes during hyperoxic lung injury and recovery. *Journal of Applied Physiology* 1988; **65**: 2672–8.

21. Davis JC. Hyperbaric medicine: patient selection, treatment procedures, and side effects. In: Davis JC, Hunt TK, eds. *Problem wounds: the role of oxygen*. New York: Elsevier, 1988: 225–35.

22. Robertson WG, Hargreaves JJ, Herlocher JE, Welch BE. Physiologic response to increased oxygen partial pressure. II. Respiratory studies. *Aerospace Medicine* 1964; **35**: 618–22.

23. Lambertsen CJ. Chemical control of respiration at rest. In: Mountcastle VB, ed. *Medical physiology*, 14th ed, vol 2. St Louis: CV Mosby, 1980: 1774.

24. Cullen SC, Eger EI, Cullen BF. Observations on the anaesthetic effect of the combination of xenon and halothane. *Anesthesiology* 1969; **31**: 305–9.

25. Wood LDH, Bryan AC. Effect of increased ambient pressure on flow volume curve of the lung. *Journal of Applied Physiology* 1969; **27**: 4–8.

26. Dutka AJ. Air or gas embolism. In: Camporesi EM, Barker AC, eds. *Hyperbaric oxygen therapy: a critical review*. Bethesda: Undersea and Hyperbaric Medical Society, 1991: 1–10.

27. Hallenbeck JM, Anderson JC. Pathogenesis of the decompression disorders. In: Bennett PB, Elliott DH, eds. *The physiology and medicine of diving*. London, New York: Baillière Tindall, 1982: 435–60.

28. Gilliam BC. Evaluation of decompression sickness incidence in multi-day repetitive diving for 77 680 sport dives. In: Lang MA, Vann RD, eds. *Repetitive diving workshop*, Proceedings of the American Academy of Underwater Sciences. Costa Mesa: American Academy of Underwater Sciences, 1992: 219–26.

29. Kelleher PC, Thalmann ED, Survanshi SS, Weathersby PK.

Verification trial of a probabilistic decompression model. *Undersea Biomedical Research* 1992; **19**: A123.

30. Spencer MP. Decompression limits for compressed air determined by ultrasonically detected blood bubbles. *Journal of Applied Physiology* 1976; **40**: 229–35.

31. Eckenhoff RG, Olstad CE, Carrod GE. Human dose–response relationship for decompression and endogenous bubble formation. *Journal of Applied Physiology* 1990; **69**: 914–18.

32. Tikuisis P. Modeling the observations of *in vivo* bubble formation with hydrophobic crevices. *Undersea Biomedical Research* 1986; **13**: 165–80.

33. Ward CA, McCullough D, Fraser WD. Relation between complement activation and susceptibility to decompression sickness. *Journal of Applied Physiology* 1987; **62**: 1160–6.

34. Lehner CE. Dive profiles and adaptation: pressure profiles target specific tissues for decompression injury. In: Lang MA, Vann RD, eds. *Repetitive diving workshop,* Proceedings of the American Academy of Underwater Sciences. Costa Mesa: American Academy of Underwater Sciences, 1992: 203–17.

35. Francis TJR, Dutka AJ, Flynn ET. Experimental determination of latency, severity and outcome in CNS decompression sickness. *Undersea Biomedical Research* 1988; **15**: 419–28.

36. Myers RAM, Bray P. Delayed treatment of serious decompression sickness. *Annals of Emergency Medicine* 1985; **14**: 254–7.

37. Bowersox JC, Strauss MB, Hart GB. Clinical experience with hyperbaric oxygen therapy in the salvage of ischemic skin flaps and grafts. *Journal of Hyperbaric Medicine* 1986; **1**: 141–9.

38. Perrins DJ. The effect of hyperbaric oxygen on ischemic skin flaps. In: Grabb WC, Myers MB, eds. *Skin flaps.* Boston: Little, Brown, 1975: 53–63.

39. Epstein JB, Wong FW, Stevenson-Moore P. Osteoradionecrosis: clinical experience and proposal for classification. *Journal of Oral and Maxillofacial Surgery* 1987; **45**: 104–10.

40. Marx RE, Johnson RP. Problem wounds in oral and maxillofacial surgery: the role of hyperbaric oxygen. In: Davis JC, Hunt TK, eds. *Problem wounds: the role of oxygen.* New York: Elsevier, 1988: 65–123.

41. Knighton DR, Hunt TK, Scheuestuhl H, *et al.* Oxygen tension regulates the expression of angiogenesis factor of macrophages. *Science* 1983; **221**: 1283–5.

42. Baroni G, Porro T, Faglia E, *et al.* Hyperbaric oxygen in diabetic gangrene treatment. *Diabetes Care* 1987; **10**: 81–6.

43. Nylander G, Lewis D, Nordstrom H, Larsson J. Reduction of the post-ischemic edema with hyperbaric oxygen. *Plastic and Reconstructive Surgery* 1985; **76**: 596–603.

44. Waisbren BA, Schultz D, Collentine G, Banaszak E. Hyperbaric oxygen in severe burns. *Burns* 1987; **8**: 176–9.

45. Cianci P, Lueders H, Lee H, *et al.* Adjunctive hyperbaric oxygen reduces the need for surgery in 40–80% burns. *Journal of Hyperbaric Medicine* 1988; **3**: 97–101.

46. Choi IS. Delayed neurologic sequelae in carbon monoxide intoxication. *Archives of Neurology* 1983; **40**: 433–5.

47. Norkool DM, Kirkpatrick JN. Treatment of acute carbon monoxide poisoning with hyperbaric oxygen: a review of 115 cases. *Annals of Emergency Medicine* 1985; **14**: 1168–71.

48. Mathieu D, Nolf M, Durocher A, Saulnier F. Acute carbon monoxide poisoning: risk of late sequelae and treatment by hyperbaric oxygen. *Clinical Toxicology* 1985; **23**: 315–24.

49. Thom SR. Smoke inhalation. *Emergency Medicine Clinics of North America* 1989; **7**: 371–87.

50. Hart GB, Lennon PA, Strauss MB. Hyperbaric oxygen in exceptional acute blood-loss anemia. *Journal of Hyperbaric Medicine* 1987; 2: 205–10.

51. Mandell G. Bacteriocidal activity of aerobic and anaerobic polymorphonuclear neutrophils. *Infection and Immunity* 1974; **9**: 337–41.

52. Bakker DJ, Clostridial myonecrosis. In: Davis JC, Hunt TK, eds. *Problem wounds: the role of oxygen.* New York: Elsevier, 1988: 153–72.

53. Bakker DJ. Pure and mixed aerobic and anaerobic soft tissue infections. *Hyperbaric Oxygen Review* 1985; **6**: 65–96.

54. Riegels-Nielsen P, Hesselfeldt-Nielsen J, Ganz-Jensen E, Jacobsen E. Fournier's gangrene: 5 patients treated with hyperbaric oxygen. *Journal of Urology* 1984; **132**: 918–20.

55. Strauss MB. Refractory osteomyelitis. *Journal of Hyperbaric Medicine* 1987; **2**: 147–59.

56. Cierny G, Mader JT, Pennick JJ. A clinical staging system for adult osteomyelitis. *Contemporary Orthopedics* 1985; **10**: 17–37.

57. Pierce EC. Cerebral gas embolism with special reference to iatrogenic accidents. *Hyperbaric Oxygen Review* 1980; **1**: 161–84.

58. Eckenhoff RG, Parker JW. Latency in the onset of decompression sickness on direct ascent from air saturation. *Journal of Applied Physiology* 1984; **56**: 1070–5.

59. Moon RE, Bergquist LV, Conklin B, Miller JN. Monaghan 225 ventilator use under hyperbaric conditions. *Chest* 1986; **89**: 846.

60. Satterfield JM, Russell GB, Graybeal JM, Richard RB. Anesthetic vaporizer performance under hyperbaric conditions. *Undersea Biomedical Research* 1989; **16**: A43.

61. Severinghaus J (Committee Report). Hyperbaric oxygenation: anesthesia and drug effects. *Anesthesiology* 1965; **26**: 812–24.

62. Rusca F, Garetto G, Ambrosio F, Schiavon M. Giron GP. HBO therapy and pulmonary function tests. *Undersea Biomedical Research* 1991; **18**: A196.

63. Lyne AJ. Ocular effects of hyperbaric oxygen. *Transactions of the Ophthalmological Society of New Zealand* 1978; **98**: 66–8.

64. Patz A. Effect of oxygen on immature retinal vessels. *Investigative Ophthalmology* 1965; **4**: 988–99.

FURTHER READING

Bennett PB, Elliott DH, eds. *The physiology and medicine of diving* 3rd ed. San Pedro: Best, 1982: 239–96.

Camporesi EM, Barker AC. *Hyperbaric oxygen therapy: a critical review.* Bethesda: Undersea and Hyperbaric Medical Society, 1991.

Thom SR. Hyperbaric oxygen therapy. *Journal of Intensive Care Medicine* 1989; 4: 58–74.

Thom SR (Chairman). *Hyperbaric oxygen therapy: a committee report.* Bethesda: Undersea and Hyperbaric Medical Society, 1992.

Chapter 48

Trauma Associated with Patient Transfer and the Positioning Process

J. M. Anderton

Accidents in the operating theatre and recovery rooms can never be totally prevented. Inadequate staffing quotas, fatigue and relative inexperience are undoubtedly contributing factors. It is possible at any stage of the positioning process for a patient to fall from the operating table or to be dropped by attendants involved in either positioning or transfer duties. Anecdotal reports of such mishaps undoubtedly occur from time to time. The potential for serious harm is considerable. It is not unreasonable to anticipate that ribs, long bones or hips could be fractured or that spinal or even intracranial damage could ensue. Such an accident is likely to happen in the following circumstances.

Inadequate patient supervision

This can occur during both the immediate pre-induction period and the early recovery phase. Premedication and confusion with unfamiliar surroundings may make the presence of more than one attendant essential prior to induction of anaesthesia. Narrow operating tables or transfer trolleys without guard rails erected further increase the hazard. If the recovery phase is complicated by restlessness it must be the primary duty of all available theatre staff to assist in the management of the patient until safety can be ensured. The first and most important step in this process is undoubtedly transfer off the operating table either to the patient's bed or to a safe transfer trolley.

Use of the prone position

For maximum safety employing the prone position requires skill, experience and a team with an adequate number of assistants. The larger the patient, the greater the importance of this last requirement. The use of a 'frame' or other more sophisticated propping system will further increase the risks. Methods of turning the patient vary. The change from supine to prone can be achieved after transfer to the operating table; alternatively, the patient can be turned in the process of transfer from the trolley to the operating table. The former is probably the safer method. Having achieved the prone position, stability must be ensured. This is not a problem with the simple horizontal position supported by pillows but, where more complex methods are required, the choice should certainly be made with this in mind. The Tarlov knee–chest[1] or seated prone position is both reliable and applicable to a wide range of body shapes and sizes and has much to recommend it[2] (Fig. 48.1) (see later).

At the completion of surgery, returning the patient to the supine position again requires the help and coordination of trained attendants. It is a wise precaution to maintain a good depth of anaesthesia until the patient is safely in the supine position. The situation can become rapidly out of control if a lightly anaesthetized patient, still intubated and in the prone position, moves spontaneously on a narrow operating table.

Patient transfer procedures

Although patient transfer systems have been devised that reduce direct handling to the minimum, there is almost always one stage at which the patient must be moved manually. Commonly, a stretcher canvas and poles are used or, alternatively, a roller or 'slide' may be employed. The first step is to ensure that the brakes are applied to any potentially mobile part of the system. Secondly, the task must be easily within the capabilities of the staff involved. Inadequacy in either of these prerequisites can result in problems; inadequacy in both is courting disaster.

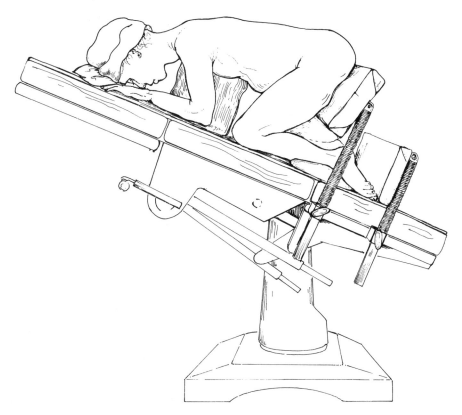

Fig. 48.1 Tarlov[1] described this as the 'knee–chest' position. However, the use of the term 'seated prone' distinguishes it from similar but less satisfactory versions.[2]

There are two major hazards associated with the use of stretcher canvases. Repeated use can leave a weakness in the sleeves, which may cause one pole to tear-out under the weight of the patient. With luck, the patient may fall only a few centimetres to the surface of the trolley or table. If less fortunate, a fall between the two could occur. A second hazard is the accidental placement of two canvases beneath the patient. The inefficiency of placing a pole in one sleeve of each canvas is revealed only on attempting to lift the patient. Unfortunately, some movement can be generated but ultimately the patient will fall between the poles; again the patient may finish up on the theatre floor if there is a gap, or potential gap, between table and trolley.

Badly designed operating tables

Operating tables designed on the principles of a separate theatre-sited pedastal with mobile 'transfer tops' are hazardous if the locking mechanism for the two parts fails. The author has seen such a table top fall off the pedestal after the transfer vehicle had been removed. Fortunately for the patient the foot end fell first and the table top remained propped against the pedestal at an angle of approximately 60 degrees. The patient was able to be retained in place and was unharmed. The foot end of transfer top made a 0.5 cm groove in the antistatic tiled flooring and would certainly have amputated part of an attendant's foot that might have intervened.

Trauma from the positioning process

The joints

A danger of positioning patients under anaesthesia is that joints may be subjected to a range of passive movements that would not be tolerated actively. Patients with any history of congenital or degenerative joint disease or who have had a prosthetic joint replacement should have their *normal range of movement elicited prior to anaesthesia*. The range of movement should not be exceeded in the positioning process. The lithotomy position is particularly hazardous, and associated sacroiliac strain is a well-recognized sequela of failure to coordinate movement of the legs. Patients with prosthetic hip replacements may suffer such severe postoperative pain from stretching of the capsular ligaments that x-ray investigations may be necessary to rule out dislocation.

In other positions, the knees, shoulders and elbows

Fig. 48.2 The 'Mohammedan prayer' position described by Lipton.[4] There is obvious potential for serious congestion and muscle damage in the lower limbs.

should be protected. Problems encountered with the spinal axis are dealt with later in this chapter.

Muscular trauma

Serious damage to muscle groups can occur either because of direct external pressure or from interference with venous drainage. The simplest example of the former is a case of deltoid muscle necrosis encountered in the underlying arm of a laterally positioned patient.[3] Cases of 'crush' syndrome following thigh and calf muscle compression in the Mohammedan prayer position, at one time advocated for use in lumbar intervertebral disc surgery, are of much greater importance[4] (Fig. 48.2). A death from acute renal failure was reported in 1952,[5] and a similar but non-fatal case in 1969.[6]

Direct external pressure on muscles of the anterolateral aspect of the leg can lead to oedema and swelling, and is one of the causes of 'compartment syndrome'.[7] As these muscles are enclosed within a well-defined fascial sheath, capillary blood perfusion is reduced below a level necessary for tissue viability. This in turn will lead to further ischaemia and oedema formation. Experimental evidence shows that, as the pressure increases, different components of the compartment are affected. Muscle damage with a rise in creatine phosphokinase occurs at an early stage, as does damage to the nerve; damage to the latter becomes irreversible after 12 hours. Postoperatively the legs will be very painful (out of proportion to the physical signs), tender and somewhat swollen; there may also be numbness and tingling in the distribution of the nerve of the compartment. The only reliable method of terminating the ischaemic cycle is for surgical decompression to be performed (i.e. fasciotomy). This condition has been reported following surgery in the lithotomy position.[8,9] It may be of significance that in both reported cases the leg support system involved pressure below the popliteal fossae and the surgery was prolonged. It has also occurred in a patient in the lithotomy position using Lloyd Davies supports in a procedure taking only 4 hours.[10] Compression by the leg

supports and elevation above heart level would cause reduced perfusion pressure, which was exacerbated in this case by head-down tilt and a systolic blood pressure of 80 mmHg for the first half of the operation.

The dangers of failing to diagnose and treat the condition are postoperative renal failure and permanent loss of function in the muscles affected.

Trauma to internal organs

Patients who have been admitted to hospital for treatment of multiple injuries may be at further risk of trauma to internal organs at any time from the commencement of their transfer from the scene of the accident. Great care must be taken to ensure the integrity of the spinal cord and to minimize the risks of unstable rib fractures perforating the internal organs of the chest and abdomen when patients are transferred to, and positioned in, the operating theatre.

Skin trauma

In the *supine* position, pressure necrosis over the sacrum, heels and occiput may easily occur unless effective support is provided. With advances in the safety of anaesthesia this has become more of a problem as complex surgery extends operating times beyond previous expectations. The use of warming blankets to maintain body temperature, though necessary, adds a further risk to the sacral area. The use of a 'fleece' or silicone-gel padding between skin and warming blanket is strongly recommended. Diabetic patients, especially those with neuropathy or absent foot pulses, are particularly prone to develop ulcers on the heels if there has been inadequte protection during their time on the operating table. Connor states quite categorically that these ulcers are not due to misfortune or misadventure,[11] but are preventable. Foam leg-troughs or ample padding with cotton wool are recommended but sheepskin heel-muffs do not appear to be of value. External pneumatic compression apparatus on the legs will prevent calf muscle compression and stasis, and certainly in non-diabetic patients protects the skin of the heels. Postoperative pressure alopecia has been reported following prolonged operations in which deliberate hypotensive techniques were employed.[12] Cardiopulmonary bypass operations in particular have been implicated in this problem by Lawson and colleagues.[13]

Skin damage caused by pneumatic tourniquets applied to the limbs is said to be uncommon. Pressure necrosis or shearing due to inadequate padding or poor application in patients with loose or thin skin is an obvious cause. Perhaps less well known is the danger of chemical burns occurring when spirit solutions commonly used in skin preparations are allowed to seep beneath the tourniquet and are held tightly against the skin.[14] Both full- and partial-thickness skin loss in the areas concerned have been described. It would seem a sensible precaution to use a self-adhesive surgical drape to seal off the tourniquet from the operative field before skin preparation commences.

With the patient in the *prone* position, damage to the skin of the malar region is a risk if the horseshoe headrest is used. Blistering and even full-thickness skin loss may result from excessive pressure, abrasive forces or both. Craniotomy involving the use of the Gigli saw is a particular risk, as the reciprocating movement inevitably rubs the malar skin against the headrest. In cases not involving cranial surgery the skin of the tip of the nose can be very sore postoperatively if the head was placed face down on a pillow. Special props or frames occasionally cause superficial skin blistering at the areas of bony contact, particularly the iliac crests. In contrast, the seated knee–chest prone position very rarely causes anything more than a slight erythematous reaction to the skin of the knees.

Trauma to appendages

The fingers can be crushed or even amputated in accidents involving the moving sections of the operating table. Courington and Little[15] graphically described such an incident where the arms of a patient in the lithotomy position were at her sides. Fingers were amputated as the foot section of the table was raised. Similar accidents occurring at the side of the table are possible with operating tables designed on the 'pedestal' and 'transfer-top' principle. Probably the safest place for the patient's hands is across the chest; they should most certainly not be 'tucked under the buttocks' to prevent the arm from falling off the sides of the operating table! (Unfortunately, elbow flexion may put the ulnar nerve at risk; see p. 978.) The longer and more complex an operation is, the more important it is that the hands are within the view of the anaesthetist. In this way, problems associated with arterial cannulation or failure of intravenous lines will be immediately apparent. A considerable volume of blood can be pumped subcutaneously if an intravenous cannula becomes dislodged. Splinting the hand and wrist using 2 cm Velcro tapes in order to maintain an extended position following radial artery cannulation has caused intensely painful wounds where the tapes cut into the flesh below the index and little fingers. It was thought that the prolonged period of hypotension which was necessary under cardiopulmonary bypass and postoperatively for 12 hours may have exacerbated the ischaemic effect of external pressure.[16]

Courington and Little drew attention to the possibility of an ear becoming necrosed if caught up and folded either by a facemask harness or between the head and a supporting headring.[15] The dangers to the genitalia and pudendal nerves in contact with the vertical post of the fracture table are dealt with fully elsewhere in this chapter.

Potential damage to breast tissue may be a problem in prone positioning. The first priority must always be the prevention of any traumatic damage due to pressure. If the breasts are so large as to cause any instability of the patient as a whole on the operating table or supporting frame, it would be wise to opt for the lateral position. Patients having had a mastectomy present a different problem, and prolonged pressure on a tender scar may result in postoperative pain or even ischaemic damage.

Smith and colleagues,[17] after discussion with conscious volunteer subjects, advised attempting where possible to displace the breasts laterally during the positioning process. Martin,[18] following a similar but unpublished review, disagrees and states that 'medial placement seems more satisfactory than lateral placement for most breast sizes during surgical procedures done in the prone position'.

Backache following anaesthesia and surgery

Attention has sporadically been drawn to this for at least half a century.[15,19–21] The commonly perceived explanation is that anaesthesia relaxes paraspinal muscles, allowing the supine position to flatten the convexity of the lumbar spine. This stresses the interlumbar and lumbosacral ligaments. In the lithotomy position, backward rotation of the pelvis flattens the convexity of the lumbar spine to an even greater extent than does the supine position. Unfortunately, it is still very common for anaesthetists to ignore recommendations that have been made to reduce the incidence of these problems. Courington and Little[15] first described modifying the operating table so that the patient 'no longer lies to attention supine' and Martin[18] has both renamed it the 'lawn-chair' position and quite rightly

Fig. 48.3 The lawn-chair position. (After Martin,[18] with permission of the publishers.)

attempted to popularize it (Fig. 48.3). It may be of benefit to a great many patients.

Alternatively, simply supporting the lumbar spine has been suggested by other authors.[22–24] An inflatable 3 litre urological irrigation bag linked via a three-way tap connector to a sphygmomanometer bulb and aneroid pressure gauge can easily be assembled for such use. O'Donovan et al.[22] assessed comfort level on conscious volunteers and found that the most satisfactory pressures were 25 mmHg for supine and 30 mmHg for lithotomy patients. One hundred and fifty-five patients were divided into study and control groups; the incidence of postoperative backache was 8.5 per cent and 38 per cent respectively (P < 0.001). A further more complex analysis[23] from the same department showed a reduction of first-day back pain from 46 per cent in the unsupported group to 21 per cent in the supported group. Patients were also shown to have benefited if they previously had a history of backache or if they were anaesthetized for more than 40 minutes. The incidence and severity of the backache was, not surprisingly, also shown to be increased if the surgery was performed on a very firm operating table surface than on a softer one.

In practice, a fixed contour support would probably be more convenient to use and would not be prone to accidental deflation. However, the normal variation in degree of lumbar lordosis might require that a range of differently shaped supports would need to be available. More widespread use of the lawn-chair position or lumbar supports is to be recommended. Postoperative backache is a largely unrecognized form of trauma resulting when the spine is unsupported in the supine or the lithotomy position, and is due to a serious error of omission.

Head and neck: some specific problems

Care of the patient's eyes

Probably every textbook of practical anaesthesia contains some advice on this subject. Whatever the position of the head during surgery, the eyelids should be closed and taped down so that they stay closed. Additional cover of the orbit with an occlusive dressing is a sensible additional safeguard in most cases. Failure to tape the eyelids together beneath such an occlusive dressing can result in corneal trauma. Any temptation to omit these precautions in minor surgical cases should be resisted. The risks to the corneal epithelium are ever present, irrespective of the length of surgery. Contamination with acid or bile-containing vomit can cause corneal ulceration and easily occurs with the lateral decubitus position for tracheal extubation. Any head-down tilt applied to prevent aspiration into the trachea virtually guarantees contamination of one or even both eyes. Aerosol-sprayed wound dressings and skin antiseptics may also gain access to an unprotected cornea and conjunctiva. A further danger, often unappreciated by anaesthetists, is that approximately 10 minutes' corneal exposure without eyelid movement is sufficient to produce surface dehydration. Permanent corneal scarring can then result from epithelial breakdown and infection. A secondary iridocyclitis may also develop. Immunosuppressed patients from special care units in which virulent organisms are frequently encountered are at particular risk of these complications. Oedema of the periorbital tissues is occasionally seen following prone positioning. It does not appear to be of any importance.

External pressure on the eye can result from a variety of causes. A badly fitting facemask that extends too high up the bridge of the nose and presses into the medial aspect of the orbits is an obvious hazard. Four cases of central retinal artery occlusion resulting from this have been reported.[25] Hypovolaemic hypotension was thought to have been a contributory factor.

The effect of the head-down tilt in the supine position was reported on by Tarkannen and Leikola in 1967.[26] Significant elevations in intraocular pressure were shown in tilt of up to 70 degrees from the horizontal. Fortunately, this is rarely necessary in modern surgery. Friberg and Sanborn[27] investigated optic nerve function in subjects volunteering to be completely inverted, and concluded that the direct effects on intraocular pressure were of primary importance. It can be concluded that steep head-down tilting for any length of time is undesirable and may cause optic nerve dysfunction.

The incidence of acute glaucoma following non-ophthalmic surgery is probably only in the order of 1 in 12 000, and appears to have decreased markedly in the last 20 years. Two cases recently described certainly had no predisposing factors caused by positioning or direct pressure.[28] However, as this condition has catastrophic ocular effects if not diagnosed and treated promptly, all postoperative cases of acute painful, red-eye require ophthalmological assessment.

The avoidance of external pressure on the eyeball is of particular importance in patients with an intraocular lens implant of the older type that clips onto the iris. Displacement of the lens can cause intraocular haemorrhage, or damage to the generative epithelium on the posterior corneal surface. The newer posterior chamber intraocular lenses are much more stable and are unlikely to be dislodged, though interactions with the coats of the eye remain a possibility.

Significant external pressure on the eye is most likely to occur when the patient is placed in the prone position. It is less common, but has been recorded, with lateral positioning. Walkup and colleagues[29] first described two cases of retinal ischaemia with *resulting unilateral blindness* occurring during thoracotomy carried out in the prone position. Undetected rotation of the head on a 'horseshoe' support resulted in protracted direct pressure on the globe. Eight similar cases were later reported from the Mayo Clinic.[30] Animal studies by the same authors suggested that induced hypotension and hypovolaemia were necessary, in addition to the external pressure, in order to reproduce findings similar to those observed in humans. A modification in the design of the headrest was also suggested in which the crossbar supporting the forehead was widened, the lateral bars were further separated and the padding in the orbital region was removed. Initial experience of its use was encouraging but further reports have not been found. It has, of course, been widely superseded by the use of a skull clamp. Implication of the horseshoe headrest was again a factor in a case of postoperative blindness described by Jampol and colleagues[31] and in another case described by Cooper and Ingham (personal communication) at a National Neurosurgical Anaesthetists Meeting in 1989.

Damage to the cervical spine and the consequences of head rotation

An unconscious patient loses the protection of tone in the musculature supporting the cervical vertebrae. Damage is therefore easily produced by passive movements, especially if they are poorly controlled. It can range from a simple musculotendinous 'strain' to intervertebral disc herniation or even facet-joint fracture and dislocation. Within the spinal canal itself, the delicate structure of the spinal cord, nerve roots and, most importantly, its blood supply, may also be compromised. Even the young patient with a capacious spinal canal (Fig. 48.4) may suffer from the effects of muscle or ligamentous damage, or even traction

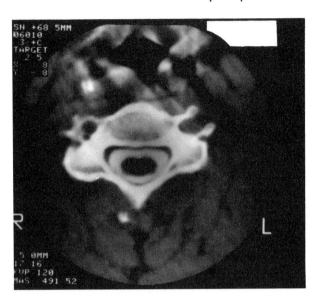

Fig. 48.4 A normal cervical spinal canal with no compression of the CSF and spinal cord.

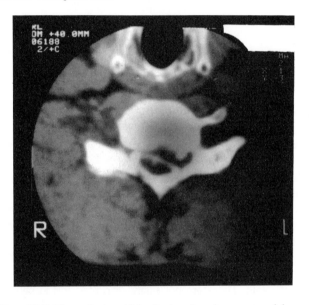

Fig. 48.5 The spinal cord is displaced and compressed by a bulging cervical intervertebral disc and large osteophyte. The rim of CSF is considerably narrowed.

injury to the nerve roots. The more elderly patient, who may have a spinal canal narrowed by osteophytes and degenerated intervertebral disc herniation, is at much greater risk from mishandling of the neck (Fig. 48.5).

In positioning the patient, neck injury can be produced in a variety of ways. Probably the best-known one is 'whiplash' injury which occurs when support for the head is suddenly withdrawn (Fig. 48.6). The classic example of this may be produced by lifting a patient with poles and stretcher canvas where the latter has been incorrectly placed and does not support the patient's pillow. Lifting

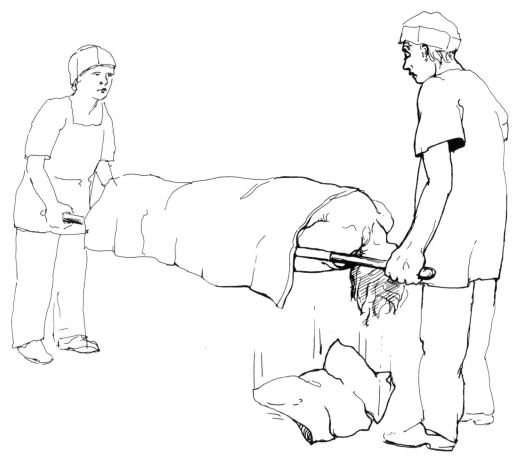

Fig. 48.6 It is essential that the stretcher canvas supports the head and neck.

the patient allows the pillow to drop swiftly away and the head to swing violently downwards. Turning the patient into the prone or lateral positions also carries similar risks unless the person supporting the head makes a positive effort to avoid hyperextension of the neck.

Head and neck rotation with the neck in forward flexion is particularly dangerous with respect to the integrity of the facet joints. These are not easily disrupted in the neutral or extended position but can readily become dislocated, with fracture of the delicate articulating surfaces by such a manoeuvre. The adjacent vertebral arteries running through their bony canals can easily become damaged or occluded by such injury. Any method of lifting or turning the unconscious patient can result in damage to the structures of the neck, and errors of omission in handling the head are largely responsible. Excessive head rotation can compromise the circulation to the spinal cord and brainstem even in the *unanaesthetized* patient. Cerebrovascular accidents secondary to obstruction of the vertebral artery have been reported following swimming,[32] ceiling painting[33] and yoga exercises.[34] Chiropractic manipulation of the neck has a particularly bad record in this respect.[35,36] Radiological studies in healthy men have shown that complete occlusion of the contralateral vertebral artery can

occur when the head is turned to one side.[37] Abrupt changes in the head position, whether active or passive, may injure the intima of the vertebral artery at the level of the atlanto-occipital joint, giving rise to thrombosis, embolism and progressive brainstem infarction.[38] Kim[39] monitored EEG changes in anaesthetized patients having an internal jugular vein catheter inserted with approximately 70 degrees of head rotation. This demonstrated adverse effects in 15 out of 28 patients studied, and these were bilateral in 5 of the affected patients. McPherson[40] noted complete disappearance of cortical somatosensory evoked potential recordings in a patient positioned with the head flexed and laterally rotated in the 'park-bench' position. The tracings returned on restoration of the head to the neutral position. Cases of mid-cervical tetraplegia have been reported following neurosurgical procedures carried out with the head and neck in this position.[41] The probable aetiology is stretching of the spinal cord and compromise of its vascular supply. It is very important always to limit the degree of flexion so that two fingers can be placed between the chin and the sternum or clavicle.

Evidence of this nature can only lead to the conclusion that positioning of the patient's head and neck in as 'neutral' or sagittal position whenever possible will always be the safest. Short periods of head and neck rotation may have an adverse effect on cerebral blood flow; as monitoring of the latter is not yet a commonplace practice in the operating theatre, these positions should be avoided whenever possible.

Serious postoperative *airway complications* have also been reported and attributed to extremes of head flexion with or without cervical spine rotation.[42,43] They have occurred in patients operated on in the sitting position, and it is thought that impaired venous and lymphatic drainage occurs and is responsible for the gross oedema of the face, soft palate and tongue. Some cases have required tracheostomy.

Dangers of some special head supports

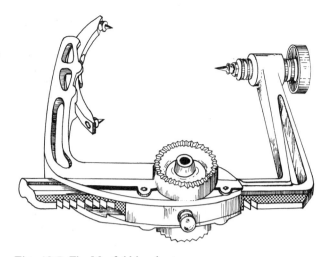

Fig. 48.7 The Mayfield headrest.

The horseshoe headrest

Surgical procedures on the head and neck may require the head to be supported so that pressure can be applied without the danger of any movement occurring. This is particularly important in some of the intracranial microsurgical procedures which have become more common in recent years. Probably one of the earliest such devices was the 'horseshoe headrest'. It must be stressed that there is no longer any justification for using this device for head support in patients in either the lateral or the prone position when the head or neck is *not* the site of the surgical operation. The risks of the head becoming displaced and blindness ensuing are too great. Safer methods such as silicone-filled 'rings', foam padding or pillows are available. Extreme care must be exercised in their use when the face is in the prone position for operative procedures on the cervical spine or posterior cranial fossae. The eyes in particular and the skin of the malar areas are vulnerable (see pp. 972, 971).

Skull clamps

Alternatively, a three-pin neurosurgical skull clamp may be used for accurate and very stable fixation of the head (Fig. 48.7). Although it is an instrument of barbaric appearance, it virtually eliminates the risk of pressure-related complications involving the face. It is essential for operations in the sitting position and highly desirable for cervical or intracranial procedures in the prone or lateral position. Despite its disadvantages, it appears to have had a very good overall safety record until recently (see below).

Disadvantages
Its application is associated with a quite severe surge of systemic and intracranial hypertension. This is undesirable in any patient, but particularly so in those presenting for

intracranial aneurysm surgery or with raised intracranial pressure. The use of local anaesthetic at the pin sites, or intravenous lignocaine (lidocaine) or thiopentone, has been advocated to attenuate the response.[44,45]

The second physiological risk is that of venous air embolism. Because the pins are designed to pierce the outer table of the skull, diploic veins are likely to be damaged by their insertion. Air can enter these veins when the pins are removed if the head is at a higher level than the heart. It is important, when applying the clamp, to do so with the patient horizontal and to obtain the correct siting of the headrest at the first attempt. Multiple unfavourable attempts to do so increase the risk of venous air embolism. At the termination of the case, especially if the sitting position has been used, the clamp should not be removed until the patient has been placed supine. The use of an antibacterial ointment around the entry points of the pins may help to seal them. Haemorrhage from the site of skin puncture can occasionally be a problem, and for control may require suturing.

Skin necrosis can occur if a small area of skin is trapped between the bone and the shoulder of the pin. This is likely if paediatric size pins are used in an adult, or if the correct size pins are applied at an angle rather than at 90 degrees to the skin surface. If, on application, this occurs and there is 'blanching' of an area of skin around the puncture site, the clamp should be reapplied correctly.

Scalp laceration can occur as a result of inadequate bony penetration and subsequent head movement.

The most serious danger associated with the use of this skull clamp relates to the use of too great a pressure in its application, with subsequent puncture of both the outer and inner tables of the skull. This can result in loosening of the remaining pins, which may allow the head to slip out of the clamp. Other potential complications relate to the depth of intracranial structures penetrated. Epidural haematoma, meningitis, cerebrospinal fluid fistula and cortical brain injury are possible. A fatality from serious intracranial haemorrhage occurred recently, and has resulted in a UK Department of Health 'Hazard Warning Notice'[46] being issued. The Gardner models 19–1020 and 19–1026 of this headrest have temporarily been withdrawn from clinical use pending further investigations.

The Boyle–Davis gag

The Boyle–Davis combined mouth gag and tongue depressor is widely used for intraoral access, especially by ENT surgeons (Fig. 48.8). There are some obvious dangers inherent in its use, such as damage to the teeth and bruising of the lips, dislocation of the mandible, and tracheal tube occlusion or disconnection. Hyperextension of the cervical vertebrae must be prevented. Failure to pay attention to the latter during the course of curetting adenoid tissue may result in damage to the ligaments overlying the anterior surface of the vertebral bodies. For this reason, some surgeons prefer to hold the gag by hand rather than use the supporting rods for this procedure.

Postoperative peripheral nerve lesions

The brachial plexus

Probably the best known and most feared nerve injury associated with surgery or anaesthesia is damage to the brachial plexus. It is highly susceptible to damage from poor positioning during anaesthesia. Its anatomical fixation in the neck and axilla and its close proximity to the bony structures of the clavicle, first rib and head of the humerus are well-established predisposing factors to traction injury. In recent years this has been almost entirely associated with the practice of having one arm abducted on an armboard for ease of access to intravenous infusion sites and blood pressure measurements. Limiting the abduction to approximately 80 degrees, pronating the hand and turning the head towards the abducted arm will considerably reduce the risks (Fig. 48.9). Failing to pronate the forearm has been associated with brachial plexus damage when the arm was abducted to only 60 degrees. Changing practices associated with the availability of more sophisticated and reliable monitoring techniques appear to be

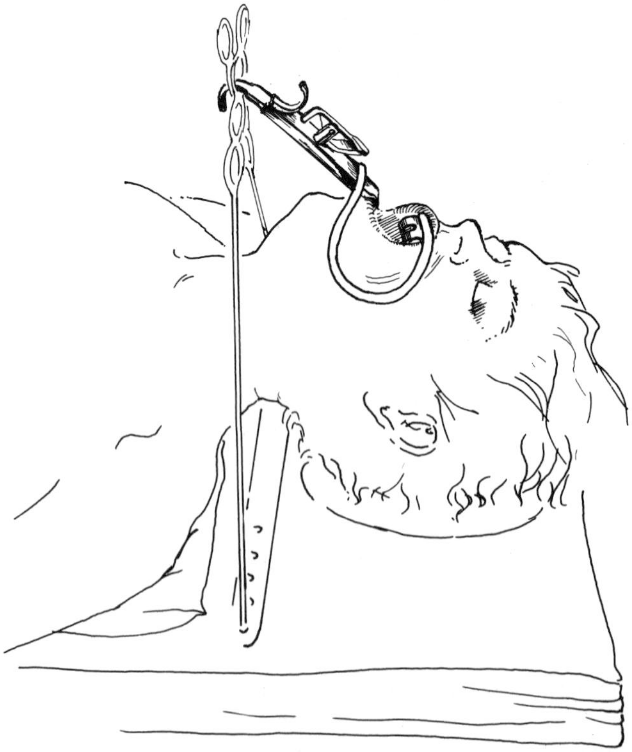

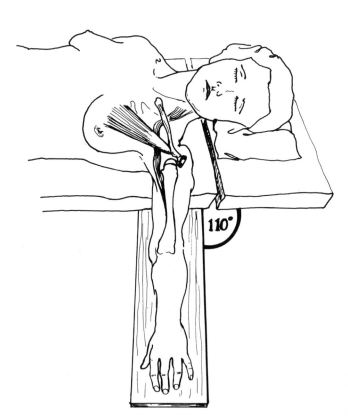

Fig. 48.8 The Boyle–Davis gag, commonly used for tonsillectomy.

Fig. 48.9 The correct position for the head, arm and hand when an arm-board is used.

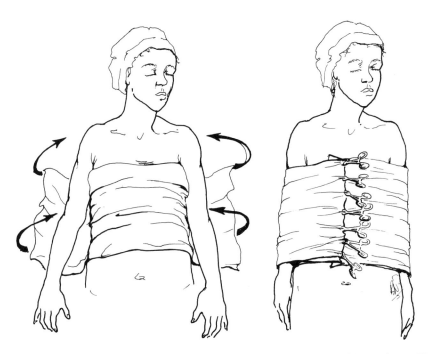

Fig. 48.10 A useful method of supporting the arms using a length of towelling. The same principle can be used in either the supine or the prone patient. Damage to main nerve trunks from external pressure by arm retainers or the surface of the table is prevented.

leading to arm positioning either fully adducted (Fig. 48.10) or adducted with the forearm across the chest. It is to be anticipated that this will result in a reduced incidence of brachial plexus injury.

The clinical presentation of a traction injury is of a motor deficit, usually painless, affecting the C5–C7 roots. This physiological deficit is suggestive of a focal conduction block, and a case described by Trojaborg in which nerve conduction tests were performed confirmed this hypothesis.[47] Most patients recover within 3 months, although some prolonged recoveries requiring up to 1 year after injury have been recorded. Surgical exploration of the plexus is not indicated. It is to be hoped that such possible causes of brachial plexus injury as restraining 'Trendelenburg' patients by the wrists or preventing them slipping off the operating table head first by the use of shoulder retainers have been long abandoned.

Direct compression of the plexus is probably a very rare cause of brachial plexus injury. It may be the culprit when patients are placed in the full lateral position without a protective air cushion or 'foam cylinder' supporting the rib cage adjacent to the axilla. Failure to provide this support can result in compression of the axillary neurovascular bundle between the head of the humerus and the lateral aspect of adjacent ribs. (This support also has an important function in preventing deltoid muscle necrosis.)

Some surgical procedures carried out in the *prone position*, particularly those on the dorsal or lumbar spine, allow for either the arms to be fully adducted or the shoulders to be extended so that the arms lie above the head resting on the operating table (Fig. 48.11). The former position is clearly quite safe for the brachial plexus provided that the security of the retaining system is reliable. The extended position was thought to be safe, having been advocated by Smith following investigations on conscious volunteers.[17] The illustrations in that article show the upper arms to be in close proximity to the sides of the face. Recent experience has shown that this can be associated with isolated axillary nerve damage and brachial plexus neuropraxia.[48,49] The position should be avoided if preoperative questioning reveals any symptoms of cervical spondylosis, or paraesthesiae when the arms are held above the head. Even when no such symptoms are present, it would be wise to avoid excessive arm abduction.

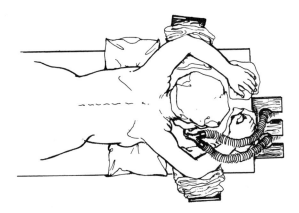

Fig. 48.11 Avoid having the hands any further forward than this. It is unwise to have the upper arms abducted against the sides of the head.[48,49]

There are reports of brachial plexus lesions in patients operated on in the sitting position through failure to support adequately the weight of the arm.[50,51]

Brachial plexus injury due to open heart surgery

Although this is not a positioning problem for which, at first sight, the anaesthetist can be held responsible, awareness of its occurrence and possible aetiology is important. The brachial plexus injury is related to the median sternotomy approach and occurs without arm abduction or any of the other potential causes of stretch injury. It differs from the latter in that the lower (C8–T1) roots of the plexus are most affected and it may be bilateral, although not necessarily symmetrical. Pain is a prominent feature of some cases and recovery may be protracted and incomplete.[52] A possible cause of damage by attempts at internal jugular vein cannulation has been suggested in a large prospective study from the Cleveland Clinic.[53] In 19 (73 per cent) of the 26 patients with brachial plexopathies the plexus lesion occurred on the same side as the cannulation of the internal jugular vein. In all instances arms were positioned at the patient's sides throughout the surgical procedure. Direct needle trauma is the obvious inference but compression by anticoagulant-induced haematomas is another possible causative factor. Left internal mammary dissection requiring greater degrees of retraction might have been a cause of excessive traction on the plexus or may have compressed it between the clavicle and thoracic rib cage. In this study, patients had less pain and the majority resolved within 2 months.

The ulnar nerve

The ulnar nerve is vulnerable to both external and internal compressive forces. The former is well recognized and occurs where the nerve lies superficially in the groove behind the medial epicondyle of the humerus. The edge of the operating table mattress is a frequently quoted culprit. Pronation of the extended forearm leaves the nerve more vulnerable at this site than does supination.[54] Internal compression can occur as the nerve passes more distally between the two heads of the flexor carpi ulnaris muscle. Full flexion of the elbow alone can cause entrapment here as it enters this cubital tunnel. As the elbow is flexed from zero to 45 degrees the arcuate ligament tightens because the distance between its points of attachment increase by 5 mm.[55] A combination of both external and internal compressive forces is, of course, possible.

Three criteria for the diagnosis have now been postulated: onset after general anaesthesia, persistent neural deficit, and clinical and electrophysiological evidence of a pure ulnar neuropathy. The symptoms may consist of numbness, tingling or pain. It is unusual for all three to be present. Onset ranges from immediate to 3 days, but may initially be unreported owing to preoccupation with the primary illness. Careful follow-up of the condition is important. Only half the patients can be expected to show improvement or complete resolution at 6 months, and further exacerbation with development of motor nerve signs affecting the intrinsic muscles of the hand will occur in some. Surgical decompression of the nerve may be required.

In recent years there has been a striking increase in the incidence of litigation associated with its occurrence. In a situation where the patient appears to have been symptom-free beforehand, it is easy to suggest that negligent practice in positioning the limb is a clear and obvious cause. Alvine and Schurrer, in an extensive prospective study, demonstrated that this is most certainly not the case.[56] A total of 6538 patients were followed through convalescence, and 17 (0.26 per cent) developed an ulnar nerve palsy during the postoperative period. When investigated at 6 months the most important finding was that nerve conduction was abnormal not only on the affected side *but also on the opposite side*. The authors concluded 'that many patients may have a subclinical ulnar neuropathy that may become symptomatic as a result of the many maneuvers and manipulations that are associated with surgical procedures'. This is supported by evidence from examination of ulnar nerves obtained during routine autopsy of 12 subjects without known disease of the peripheral nervous system.[57] Localized histological changes were found in 5, indicating subclinical entrapment.

Further study by Alvine and Schurrer showed that preventive measures have only a limited degree of success. Preoperative screening for symptoms of neuropathy, the use of elbow pads during the procedure and careful positioning of the arms were all strictly adhered to during the subsequent four years. Nevertheless, a further 11 patients with postoperative ulnar nerve lesions were identified. The authors stress the importance of eliciting a preoperative history of nocturnal paraesthesiae or dysaesthiae, or any evidence of ulnar neuropathy in surgical patients. When found, the patients should be warned of the risks that the surgical procedure may aggravate this condition.

The axillary nerve

The axillary nerve is derived from the fifth and sixth cervical nerve roots and is a branch of the posterior cord of the brachial plexus. Anatomically it has a close relation to the lowest part to the articular capsule of the shoulder joint and then runs below the surgical neck of the humerus. Anterior and posterior branches supply the deltoid and teres minor muscles and an area of skin on the lateral surface of the arm.

Isolated lesions are usually the result of either inferior dislocation of the shoulder joint or adjacent fracture of the humerus. Pollock and Davis first observed that stretching during prolonged sleep or anaesthesia could also be a cause.[58] Gwinnutt[48] has recently described a case resulting from prone positioning of the patient for revision spinal surgery in which the shoulders were extended and the arms placed above the head, resting on the operating table. Postoperatively the patient complained of weakness and numbness in her left arm. Examination revealed an inability to abduct the arm more than 50 degrees against gravity, and an area of decreased skin sensation affecting the lateral aspect of the upper arm. Despite the fact that both arms were similarly placed the lesion was unilateral and had occurred similarly some 15 years previously at her first operation. Both incidents resolved uneventfully within 2 months. Although this is a rare nerve injury, the increased incidence of surgical treatment for spinal column problems may result in further cases being recognized.

The radial nerve

Arising from cervical segments 6, 7 and 8 and leaving the brachial plexus as its posterior cord, the radial nerve runs through the axilla before winding round the lateral aspect of the mid third of the humerus in the spiral groove. Proximally it is at risk from external pressure in the axilla if the arm is permitted to 'overhang' the side of the operating table. The posterior cord of the plexus then rubs against the table edge.[59] The classic operating room injury is that of a misplaced 'ether screen' pressing against the lateral mid third of the arm. Wrist drop due to paralysis of the extensor muscles of the forearm and loss of sensation in the posterior surface of the forearm and posterolateral aspect of the hand result.

Damage to the nerve from misplaced intramuscular injections into the lower part of the deltoid muscle has resulted in this practice being discontinued.

The suprascapular nerve

This large motor nerve arises from the posterior border of the upper trunk of the brachial plexus and is derived from the fifth and sixth cervical nerves. In its course to supply the posterior muscles arising from the scapula it passes through the suprascapular notch and is tethered to adjacent structures at this position. Forced passive adduction of an extended arm across the midline may result in traction injury (in the proximal segment between its origin from the plexus and the suprascapular notch). Entrapment neuropathies of this nerve can also occur, and may be unrecognized as a source of upper extremity pain. The

pain is of a deep, poorly localized nature and is accentuated by shoulder movement involving the scapula. There may be secondary pain radiation in the distribution of the radial nerve.[60] Surgical division of the transverse scapular ligament may be necessary to resolve the entrapment neuropathy. Medicolegally it may be important to differentiate between the naturally occurring neuropathy and claims of traumatic damage due to faulty positioning technique.

The median nerve

Postoperative median nerve neuropathy resulting from faulty positioning of the arm is extremely uncommon. Parks, reviewing 50 000 operative procedures over a 13-year period, discovered only two cases of paresis.[61] Both patients were in the supine position; one had the arms abducted and the other had both arms at the sides. Traumatic damage from venepuncture attempts in the antecubital fossa or brachial artery catheterization is well recognized, as is severance of the palmar cutaneous branch during carpal tunnel decompression surgery.

The lower spinal cord and cauda equina

Unexpected paraplegia or symptoms of cauda equina compression that follow unrelated surgery are rare but are an extremely distressing occurrence. In 1969 Ehni published four cases of paraplegia secondary to operating table position (lordosis) in patients with spinal stenosis.[62] He believed that this was a poorly understood phenomenon and that it was frequently blamed on spinal anaesthesia or vascular complications. A further case report also implicated the operating table position as a causative factor.[63] A 61-year-old patient with a pre-existing but asymptomatic spinal canal stenosis was placed in the lateral position with some spinal extension and lateral bend for hip surgery. This position combined with severe spinal stenosis (proven later at operation) caused enough compression on the neural elements to produce permanent irreversible paraplegia. It is well known in patients with symptomatic spinal stenosis that flexion gives relief, and that extension exacerbates the symptoms. The implications for surgical positioning of elderly patients are therefore clear. The supine position with the spine at least straight, if not moderately flexed, would be the safest. This is a group of patients in whom it might be unwise to provide lumbar support in order to prevent postoperative backache.

In 1968 Moiel and Ehni described a case of postoperative cauda equina syndrome which was due to compression of that structure between intact fifth lumbar vertebral laminae and the upper sacral centrum where spondylolisthesis had occurred.[64] The aetiology of the neurological sequelae was almost certainly associated with the lithotomy position. Decompressive laminectomy produced considerable improvement in the neurological deficit.

Acute lumbar intervertebral disc disruption and even sequestration of a fragment within the spinal canal do not appear to have been reported as a cause of postoperative pain or neurological deficit. It is, however, easy to see that this might well occur as a result of positioning, turning or recovery room transfer procedures. Indeed, the ease with which this seems to occur in the conscious subject makes the absence of such reports all the more remarkable. In any case of unexpected postoperative spinal neurological deficit, urgent surgical exploration and decompression of the spinal canal may offer the only hope of alleviating the condition. The appropriate advice on this should be speedily obtained.

The sciatic nerve

This main branch of the lumbosacral plexus exits from the pelvis via the greater sciatic notch. From this point onwards it is vulnerable to damage by pressure, trauma or stretching. Very rarely, placing an emaciated patient in the lateral position with insufficient soft padding beneath the buttock can result in damage to the underlying sciatic nerve.[61] The prone position has also been similarly implicated, although in this case there was a predisposing diabetic peripheral neuropathy and the surgical operation extended to 8 hours.[65] The main source of direct trauma is from needlestick injury and disruption by misplaced intramuscular injections. In this context it is worth mentioning that flexion of the thigh to 90 degrees or more alters the course of the nerve through the gluteal region to a more lateral and superficial position. This renders the conventional 'upper outer quadrant' advice less reliable and it would be safer to avoid this area altogether unless the hip is in the neutral position. The pneumatic tourniquet is a well-recognized potential cause of damage to the sciatic nerve in its course through the thigh. Unnecessarily high pressure should not be used and the time of application should be carefully monitored.

Stretching of the main length of the nerve can easily occur with the thigh flexed as in the lithotomy position. Anatomical studies have drawn attention to the fact that it is relatively fixed to underlying structures both at the sciatic notch and where its common perineal branch passes round the neck of the fibula[66] (Fig. 48.12). It is therefore important to minimize flexion of the thighs, to reduce tension on the nerve between these two points. Variations in the lithotomy position in which the legs were held fully extended and almost vertical to the body have been associated with (reversible) sciatic nerve damage.[67,68] External

rotation of the hip with the legs in the more conventional position again puts the sciatic nerve 'on the stretch'. Surgical assistants should be discouraged from leaning against the knees or lower legs with the patient in lithotomy as this might exacerbate the hip rotation and thus stretch the nerve. With the patient in the full lateral position it is occasionally necessary to increase stability by using a broad leather or canvas strap running beneath the table top and over the patient's gluteal region. Sciatic nerve damage can ensue if insufficient padding is used or the strap is too tight.

The sciatic nerve divides into its two main terminal branches at the level of the knee joint. The tibial branch takes a deep course through the popliteal fossa and lower leg and there are no recorded dangers to it from positioning. The common perineal branch extends laterally around the head of the fibula where it is extremely superficial and vulnerable to external pressure. Foot drop and loss of cutaneous sensation on the lateral aspect of the leg and dorsum of the foot are the consequences of damage. This may be seen in the underlying limb when the patient has been in the lateral decubitus position, and possibly after use of some of the less common methods of lithotomy positioning. The method of leg positioning for surgical exposure of the saphenous veins in patients who undergo coronary artery bypass graft surgery has resulted in two case reports of the complication.[69] Incorrect application, or malfunction of external pneumatic compression appa-

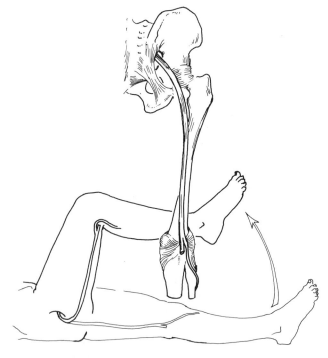

Fig. 48.12 Anatomical fixation of the sciatic nerve at the sciatic notch and the head of the fibula may predispose to damage from tension.[66]

ratus, has recently been suspected of causing similar damage.

The femoral nerve

This nerve is susceptible to damage from positioning only as it passes beneath the tough inguinal ligament. It can become kinked around the latter when the patient is placed in some forms of lithotomy position (Fig. 48.13). It becomes a problem either when there is excessive flexion of the thighs or when extremes of abduction and external rotation of the hip are permitted. The use of straight leg sling systems have been implicated as one cause of the latter.[70]

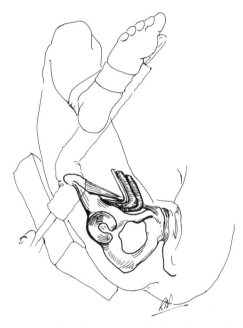

Fig. 48.13 The femoral neurovascular bundle may be damaged by kinking around the inguinal ligament when there is extreme flexion of the thigh.[70]

Branches of the femoral nerve trunk arising above the pelvic bones supply the psoas muscle and are intact if the thigh can be flexed towards the trunk. Damage occurring at or below the inguinal ligament affects the ability to extend the knee joint and the integrity of the knee jerk reflex. It is worth noting that the self-retaining abdominal retractors commonly used in pelvic surgery have been incriminated as a cause of femoral nerve damage.[71]

The saphenous nerve

With the patient in the lithotomy position the saphenous nerve may be compressed against the medial tibial condyle if the legs are placed outside the vertical stirrup support.[59] Usually, however, the stirrup is positioned more distally. Numbness or paraesthesiae of the medial side of the leg may occur.

The pudendal nerves

The pudendal nerves emerge from the pelvis into the ischiorectal fossae passing round the spine of the ischium (the classic site for infiltration of local anaesthetic nerve blockade). Each divides into branches supplying the skin of the external genitalia. A smaller posterior branch becomes the inferior haemorrhoidal nerve, with the very important function of external anal sphincter control. Sexual impotence and dysfunction following fractures of the pelvis have been well documented but it is probably not widely appreciated that pudendal nerve neurapraxia can occur as a complication of traction against the perineal post on the fracture table.[72] Hoffman and colleagues[72] described two patients operated on in the *prone* position for pin fixation of subcapital femoral fractures and muscle pedicle grafting. These patients developed numbness of the shaft and bulb of the penis and inability to obtain an erection for 3–4 months following surgery. In the prone position the genitalia hang dependently across the symphysis pubis, and both dorsal nerves of the penis are vulnerable to compression between the bone and the perineal post.

They also described two cases operated on in the *supine* position for simple pinning of fractured neck of femur, who developed unilateral anaesthesia of the penis and scrotum.[72] The anaesthetic area in each case was on the same side as the fracture. In one of these cases the perineal post had been positioned against the ipsilateral pubic ramus. Cadaver dissections have shown that the nerve is skeletally protected from the fracture table post by the ischial tuberosity and the inferior pubic ramus but it becomes vulnerable distally after trifurcation.

Pudendal nerve neurapraxia may be more common than the literature indicates. Hoffman cited five other cases in the orthopaedic literature. A case of inferior haemorrhoidal nerve damage resulting in incontinence of faeces has also been reported. The sensitive nature of the symptoms may inhibit patients from disclosing the occurrence of these complications. Prevention is clearly very important. The perineal post must be wide and well padded to distribute the traction pressure evenly. It should be placed between the genitalia and the uninjured limb when the patient is in the supine position, and should not be in the exact midline of the symphysis or in contact with the genitalia when the patient is in the prone position. Patients can be reassured that sensation will return but complete return of sexual function seems to be more prolonged.

The lateral femoral cutaneous nerve of thigh

Damage to this nerve can result in the condition of meralgia paraesthetica; there is numbness and an annoying hyperalgesia of the upper lateral thigh. When patients are positioned prone using any of the frames that support the anterior superior iliac spines there is theoretically the risk of compression or stretching of this nerve. Other surgical causes of this condition include iliac bone procurement for grafting and groin-flap procedures.[52]

Summary of perioperative nerve lesions

In analysing the effects of positioning, compression, traction or ischaemia are the most likely causes of these lesions. Accidental surgical damage, from either direct trauma or compression by a haematoma, occasionally occurs. Aetiology is important because the prognosis for recovery depends on the nature of the neurological deficit. It is necessary to know whether there is functional disturbance of nerve fibres that remain in continuity, or complete axonal loss. In the early postoperative period this may be difficult to assess clinically, but an expert in electrophysiological testing should be able to elucidate this as well as the exact localization of the nerve lesion. The functional integrity of both *motor* and *sensory* fibres must be assessed and the test repeated at appropriate intervals to monitor progress. The diagnosis is more difficult if evaluation by conduction studies and electromyography is delayed until concern is expressed over the lack of spontaneous recovery. After many weeks the electrophysiological changes of chronic denervation differ from those seen in the acute stages. The opportunity for early surgical exploration and possible reconstruction will have been missed if complete denervation has occurred.

Usually the precise mechanism of nerve injury is unclear. In the case of the brachial plexus, stretching seems the most probable explanation. Mild stretching effects have been studied experimentally in the cat and showed that epineural haemorrhage occurred. The resulting ischaemia was thought to be the most likely explanation of the ensuing transient muscle weakness. Other workers found signs of myelin disruption at moderate degrees of nerve stretch. An increase of 15 per cent in the resting length of the nerve stops blood flow completely. Conduction failure and even nerve degeneration may result if it is maintained for long periods. With very severe traction injuries of the brachial plexus, wallerian degeneration occurs with the inevitable pain, long-standing paralysis and sensory changes.

With lesions of the ulnar and peroneal nerves, compression between bone and a hard surface is usually the cause.

In patients with diabetes mellitus, ischaemic changes already present in the nerves make them more susceptible to the effects of injury. In the acute situation, there is displacement of the nodal myelin followed by paranodal demyelination and conduction block. Recovery can take 2 months following remyelination. A much longer period for recovery was described in a case in which oedema between axon and the myelin was a prominent feature. In the most serious compression injuries, wallerian degeneration is found.[52]

Only by constant attention to detail in the positioning of surgical patients will the incidence of these distressing iatrogenic injuries be reduced.

Some physiological dangers of positioning the surgical patient

These dangers can be divided into those affecting the respiratory system and those affecting the cardiovascular system.

The respiratory system

Diagnosis and management impeded by difficult access to the patient

Any position that seriously limits the anaesthetist's access to the patient's head and thorax is undoubtedly a source of potential danger. Neurosurgical procedures in which the anaesthetist is confined to the foot of the operating table are a classic example. Simple complications such as the migration of a tracheal tube into the right main bronchus can be difficult to assess and correct. Serious situations such as acute tension pneumothorax or the life-threatening onset of severe bronchospasm are challenging enough crises without the extra complication of impeded access. Positioning the patient should always be carried out so that such dangers, though admittedly rare, can be dealt with expeditiously; delay can be fatal.

Position-related changes in respiratory function

Altered lung mechanics and changes in blood flow in both the systemic and the pulmonary circulations have been studied in patients under general and regional anaesthesia and have been well reviewed.[73] These may be either improved or made worse by positioning the patient for surgery. This is particularly the case if extremes of positioning (e.g. Trendelenburg or sitting position) are employed.[74,75] In terms of patient management, it may sometimes be difficult to separate the dangers arising from changes to the respiratory or the cardiovascular systems. A recent case anaesthetized by this author illustrates the point. General anaesthesia was induced and tracheal intubation performed uneventfully. The patient was turned on to the Tarlov prone position and within 3 minutes the end-tidal CO_2 alarm was activated, indicating 'apnoea'. This was perplexing because the tracheal tube was clearly still in place and both sides of the chest were unmistakably expanding in time with the ventilator. However, the next automatic indirect blood pressure recording revealed a catastrophic fall in arterial blood pressure. Both the hypotension and the low CO_2 output responded to immediate treatment with a small dose of vasopressor and intravenous fluids.

Taken in isolation, changes to respiratory function are almost always in practice corrected by anaesthesia with tracheal intubation and intermittent positive pressure ventilation. Indeed, blood gas analysis figures that improve on the patient's preoperative status are often produced.

Tipping the patient into the Trendelenburg position has been shown radiologically to cause upward displacement of the carina in 40 per cent of patients. In half of these this will be sufficient to cause a tracheal tube previously positioned with the tip just above the carina to enter the right main bronchus. It is wise therefore to re-check for adequate aeration of both sides of the chest after the patient has been positioned in this way.[76]

The recovery period

This is probably the most dangerous in respect of risk of pulmonary complications. Two investigations have shown that about 20 per cent of all cases of death and serious neurological damage occur at this time.[77,78] The danger of aspiration can be minimized by correct lateral positioning, which would almost certainly have prevented some of these tragedies. Nurses in the recovery room should accept patients into their care in the supine position only if they are already fully conscious or if there is some overriding surgical or medical reason for avoiding the lateral position. Even after recovery of consciousness, there is evidence that the glottic reflexes can be obtunded for up to 2 hours.[79]

Once consciousness has been regained and vital functions are stable a change in posture may be thought to be appropriate, particularly as ventilatory excursion is impaired by narcotics and pain. However, placing the patient supine with a backrest raised to approximately 45 degrees has been shown by Russell to have no advantage,[80] despite previous work by Hsu Ho and Hickey[81] that functional residual capacity of the lungs is increased. Nineteen patients were studied by Russell at approxi-

mately 30 and 60 minutes following the completion of surgery. There was a statistically significant deterioration in arterial oxygenation in most patients.

The cardiovascular system

The patient's position may result in occlusion of major arteries

Occlusion of a major artery is quite a considerable danger with many positions used for surgery. With the patient placed supine or prone, cervical spine rotation can result in occlusion of the contralateral vertebral artery (dealt with elsewhere in this chapter). It is quite remarkable that many long ENT procedures carried out with the head at almost 90 degrees to the spinal axis appear to cause no neurological sequelae.

The arterial supply to the lower limbs can be compromised in several different positions. The femoral neurovascular bundle can become kinked around the inguinal ligament in the lithotomy position if too great a degree of flexion, abduction and lateral rotation of the thighs is allowed. In the prone position when props or 'frames' for the iliac crests are used they must be carefully adjusted to ensure that they are directly beneath the bone and that no lateral movement is possible. Stability throughout the whole surgical procedure must be guaranteed. Any lateral movement of either the support or the patient's body will result in direct pressure on the patient's groin and occlusion of the femoral artery. Even with the degree of adjustment afforded by the Relton and Hall 'four-poster frame',[82] it is still possible for the patient's pelvis to tip sideways on the pelvic supports. Pressure can then come to bear on the femoral vessels. In the prone 'Mohammedan prayer' position,[4] occlusion of the femoral and popliteal arteries is a near certainty and only one of the reasons for declaring the position to be totally unacceptable.

It is most unusual for the blood supply to the upper limbs to be compromised by the position of the arms. Perhaps it is the well-known concern to protect the brachial plexus which inadvertently subserves the same function. The position of the arms when the patient is placed in the prone position was studied by Smith who concluded that the arms should be resting on the operating table above the head[17] (see Fig. 48.11). However, the author has heard of one anecdotal report where bilateral loss of the peripheral pulses occurred from such positioning of the arms and was followed by severe postoperative upper limb weakness for 48 hours. As the symptoms resolved quite rapidly, it is unlikely that this was due to brachial plexus damage. This position for the arms can occasionally result in neurological sequelae and has already been discussed. It is not therefore as safe for the neurovasculature as was previously thought. The degree of abduction should not be extreme and peripheral pulses should be checked when it is used. It may be safer, if surgically acceptable, to position the arms at the sides of the body. This can be achieved as illustrated in Fig. 48.10. Unfortunately, this method has the disadvantage of making access to monitoring probes and intravenous infusion more difficult. A case can be made for the use of adjustable arm supports that attach to the side rails of the operating table.

Patients who have previously had coronary artery bypass surgery may be at some risk if positioned prone. A case has been reported where pressure under the sternum resulted in intraoperative ECG evidence of myocardial ischaemia.[83] In the recovery room massive anterolateral ischaemia was diagnosed and emergency coronary revascularization was undertaken. At operation, both saphenous vein bypass grafts were found to be patent and pulsatile. A further graft was inserted and a good recovery was obtained. It would appear that the thin-walled veins used as coronary grafts easily became compressed by pressure on the anterior chest wall. This mechanical cause of graft obstruction is reversible. Full 12-lead intraoperative ECG monitoring is now technically possible and should be considered in such patients. It would be unwise to allow surgery to begin until the adverse effects of positioning can be confidently excluded.

The abdominal aorta and its branches are compressed by the pregnant uterus in the supine position.[84] During the last trimester both its course and its size are disrupted, being pushed posteriorly, narrowed and applied closely to the vertebral column. It is usually displaced to the left below the origin of the renal arteries. Compression takes place at the first lumbar vertebra, being most marked at the level of the third and extending to the bifurcation. Systemic hypotension exaggerates these phenomena. The dangerous effect on placental blood flow is discussed further in the next section.

Obstruction of venous return to the heart

Obstruction of the inferior vena cava from any cause will impede filling of the right ventricle and may seriously reduce cardiac output. Credit for first recognizing the association between arterial hypotension and the supine position in late pregnancy is due to Hansen.[85] Recognition for investigation and further publicizing the problem must go to McRoberts[86] and Howard, Goodson and Mengert,[87] who described 'postural shock in pregnancy' and 'supine hypotensive syndrome' respectively. An excellent review of the 50-year history of aortocaval compression syndrome has recently been published by Marx.[88]

The important physiological sequelae of inferior venocaval compression depend on whether blood can bypass the obstruction via enlarged paravertebral venous path-

ways, and on the capacity of the systemic vascular resistance to maintain arterial blood pressure. From the anaesthetist's point of view all near-term pregnant patients are at risk from the dangers of this problem. However, they fall into two distinct groups. First, there are the majority who are symptom free and appear to have no obvious discomfort or adverse effects when they lie in the supine position. This can lead to a false sense of security and a very serious fall in arterial blood pressure following induction of general or regional anaesthesia, the sympathetic blockade induced by either method abolishing the ability to compensate by vasoconstriction. Routine use of the 'obstetric wedge' beneath the right buttock has become standard practice and is of undoubted value. However, it has recently been shown that the standard 15 degrees of tilt usually used in the UK is insufficient to relieve aortocaval compression completely. Both left and right pelvic tilt failed to reverse the decreased blood flow in the leg associated with the supine position.[89] Other workers have demonstrated that at caesarean section cardiac output increased towards the value found in the left lateral position when left-sided table tilt was supplemented with uterine displacement.[90] In the author's experience, simple 'wedging' can delay, by a few minutes, the onset of serious hypotension. When such a dramatic fall in blood pressure occurs, it is vital that there are an adequate number of operating theatre assistants available to reposition the patient fully lateral. If the obstetrician has already opened the peritoneum the crisis can be resolved, either by the very rapid delivery of the infant or by the surgeon holding the gravid uterus forward off the underlying great vessels until adequate cardiac output has been restored.

A second group of patients complain of feeling faint when supine, and indeed cannot usually be persuaded to stay in this position. Acute hypotension, pallor and sweating rapidly ensue and tachycardia gives way to a marked bradycardia. Complete relief occurs after these women turn fully lateral. At first sight it appears that a vasovagal or emotional fainting attack is superimposed in some patients who have aortocaval compression syndrome. The rational explanation for this seemed obscure for many years. More recent studies into circulatory collapse and acute blood loss offer an explanation. In any situation in which blood is lost to the circulation, whether externally, by obstruction or by extreme peripheral vasodilation (as in 'emotional' fainting), there is an initial increase in vasoconstrictor drive. This can be measured directly by both microneurography and estimation of plasma noradrenaline (norepinephrine) concentration. When the volume of blood loss, either true or simulated, exceeds approximately 25 per cent of the circulating volume, peripheral resistance, arterial pressure and heart rate fall profoundly. The light-headedness or loss of consciousness which follows is secondary to cerebral hypoxia. Sympathetic vasoconstrictor drive has stopped abruptly and the cardiovascular reflexes that maintained constant blood pressure suddenly seem to have been 'switched off'. At the same time, another vasoconstrictor mechanism is 'switched on'. There is a massive outpouring of antidiuretic hormone into the bloodstream; this is responsible for the nausea and skin pallor that persist for some 30–40 minutes after cardiovascular haemodynamics have been restored. This dramatic change in physiological response probably acts in a protective role. The initial high sympathetic drive and falling venous return will induce powerful, rapid, myocardial contractions around small volumes of blood. The supervening bradycardia may prevent myocardial damage and allow improved ventricular filling during a falling preload.[91,92]

The small number of antenatal patients who exhibit these serious symptoms of 'supine hypotensive syndrome' should not present too great a management problem to an experienced obstetric anaesthetist, especially when diagnosed beforehand. A regional anaesthetic technique for caesarean section, performed with the patient tipped as far lateral as practicable, is probably the management of choice. At the other end of the scale, general anaesthesia with the patient in the lithotomy position, even with 'wedged' pelvis, is likely to be extremely hazardous. Without a 'wedge' it could rapidly result in a maternal fatality.

If, for any reason, cardiopulmonary resuscitation is required in a pregnant patient near to term, it will be obvious from the foregoing discussion that a successful outcome will be very unlikely if the patient is left in the standard supine position. In such circumstances, it has been shown that an assistant kneeling on the floor and sitting on his or her heels can provide support for the patient's back. This 'human wedge' then uses one arm to stabilize the patient's shoulder and the other to stabilize the pelvis. Performance assessed using resuscitation mannikins has been favourable.[93] Marx has advocated that, if manual displacement of the uterus to the left and slightly cephalad does not achieve haemodynamic stability within a few minutes, immediate abdominal delivery of the infant must be performed to give the mother any chance of surviving without brain damage.[94]

Position for lumbar regional anaesthetic techniques
The ideal haemodynamic position for the insertion of a lumbar epidural catheter in the obstetric patient has, until recently, received little consideration. Individual preference usually determines the choice of lateral or sitting position.

The danger of an emotional vasovagal syncopy is present in *both* positions.[95] In the former, the patient is at least already horizontal and will require only head-down tilt and possibly intravenous vasoconstrictors for resuscitation (provided that she has not been so tightly 'curled up'

with 'knees to chest' that aortocaval compression is not itself the cause of the syncope). In the sitting position the danger depends on the speed with which the syncopal reaction supervenes and whether there is adequate support to prevent her falling forwards. Fortunately, there is usually time to take action and allow the patient to adjust into the lateral decubitus position.

Non-invasive assessment of cardiac output using the principle of thoracic bioimpedance (BOMED NCCOM3 Monitor) has provided a method for comparison of the different positions. Chadwick and co-workers[96] studied 20 healthy patients at term and 20 controls who were placed sequentially in four positions: supine; 15 degree wedge position; left lateral with hips and spine flexed; and sitting with the spine and neck flexed. The 'wedge' position was taken as baseline. The most striking change in the pregnant population occurred in the left lateral position, in which stroke index and consequently cardiac index were reduced by 21.8 and 23.4 per cent respectively. In comparison, in the sitting position, stroke index and cardiac index were reduced by 13.1 and 8.5 per cent respectively (Figs. 48.14 and 48.15). Systolic blood pressure was significantly reduced in the lateral position, by 8.6 per cent (Fig. 48.16). The study demonstrated that, of the two positions chosen by anaesthetists for the insertion of epidurals, the flexed left lateral position has significnatly more adverse effects on maternal cardiac output. Therefore, in the already compromised patient it may be preferable to perform the regional block in the sitting position. Elstein and colleagues have reported the use of similar non-invasive methods of cardiac output measurement to determine the optimal posture during caesarean section for a gravida

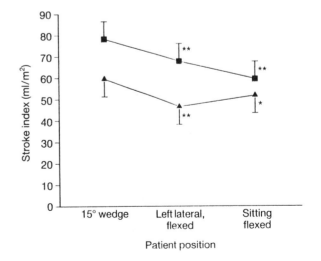

Fig. 48.14 Effect of posture on the stroke index. Mean with 95% confidence limits (ml/m²). ■, Control group; ▲, pregnant group; * $P < 0.05$ compared to 15° wedge position; ** $P < 0.01$ compared to 15° wedge position. (Reproduced, with permission, from reference 96).

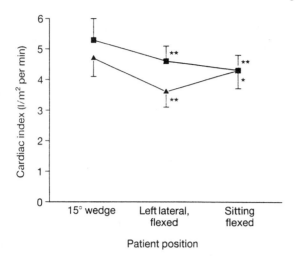

Fig. 48.15 Effect of posture on cardiac index. Mean with 95% confidence limits (1/m² per min). ■ Control group; ▲ pregnant group; * $P < 0.05$ compared to 15° wedge position; ** $P < 0.01$ compared to 15° wedge position. (Reproduced, with permission, from reference 96).

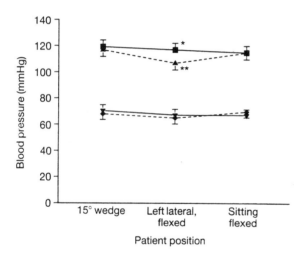

Fig. 48.16 Effect of posture on systolic (■, ▲) and diastolic (▼, ◆) blood pressure. Mean with 95% confidence limits (mmHg). ■, ▼, Control group; ▲, ◆, pregnant group; * $P < 0.05$ control group (left lateral flexed position compared to pregnant group; ** $P < 0.01$ pregnant group (left lateral position) compared to baseline 15° wedge position. (Reproduced, with permission, from reference 96).

with triplet gestation complicated by severe 'supine hypotensive syndrome'.[97] With more widespread availability of such apparatus it may, in the future, be possible to alleviate scientifically the dangers in positioning the obstetric patient.

Dangers to the fetus of the maternal supine position
The fact that the majority of mothers appear superficially to show no adverse effects from the unsupported supine

position, and indeed often produce apparently healthy babies, has made acceptance of 'lateral tilt' in the labour room an uphill struggle over the last 25 years. Evidence now accumulated leaves no room for doubt that the fetus is at risk, and the only correct management is to treat all mothers with this in mind. One of the earliest investigations reviewing placentograms showed diminution of placental perfusion with the mother in the unsupported supine position. Direct catheter studies showed reduced arterial pressure measurements in the different branches of the aorta distal to its obstruction by the gravid uterus.[98] Simultaneous brachial and femoral blood pressure measurements have shown that 60 per cent of women had significant femoral hypotension whilst only 18 per cent had similar brachial hypotension.[99] Kauppila's studies suggest that the autoregulation system for uteroplacental circulation is operative only in the non-placental component of the pregnant uterus.[100] Evidence from the examination of the fetus during labour and both its physiological and physical status at birth further condemn failure to ensure prevention of aortocaval compression. It has been demonstrated that deceleration of the fetal heart rate occurring while the mother was lying on her back could be reversed by turning her into the lateral position regardless of whether or not there was maternal hypotension. Umbilical blood gas and acid–base studies in patients having vaginal deliveries in either the supine or the supine/tilted position demonstrated that in the offspring of the supine mothers there was a time-related decrease in pH, whereas acid–base status was maintained within the normal range in the tilted group.[101] Evidence seems to be accumulating that, wherever possible, the full lateral position is to be preferred.[102]

In contrast, the use of lateral tilt at caesarean section became well established much earlier. Ansari and co-workers, measuring umbilical vein and artery oxygen saturation, demonstrated statistically significant advantages.[103] Crawford's work on Agpar scores at 1 and 5 minutes showed the cardiorespiratory status in infants of tilted mothers to be less depressed.[104] Although it is more inconvenient to the obstetrician, the superiority of left lateral tilt as opposed to right tilt has been demonstrated by Buley *et al.* on the basis of pH and blood gas values.[105] This is to be expected because the inferior vena cava, anatomically, lies marginally to the right of the vertebral column. Elevation of the hip with a foam rubber wedge or similar device seems to have become the method of choice over mechanical devices designed to attach to the operating table and displace the uterus to the left.

Peripheral venous obstruction or stasis and the danger of postoperative deep vein thrombosis

Deep vein thrombosis (DVT) remains a common event in hospital patients and carries the serious risk of pulmonary embolism. Although its aetiology is multifactorial, it is not difficult to argue that positioning for surgery and the lack of muscular activity while under anaesthesia will make a significant contribution. Intermittent pneumatic compression of the legs in general surgical patients has been shown to reduce the incidence of DVT from 27 per cent with no prophylaxis, to 18 per cent. When combined with graduated compression stockings, there is further reduction to 4.5 per cent.[106]

The position of the patient and the risk of venous air embolism

In 1983 *Anesthesiology* carried an editorial by Maurice Albin in which he wrote 'It is to be remembered that venous air embolism (VAE) was a much feared complication during the 19th century in the spontaneously ventilating (anesthetized or non-anesthetized) patient, with many procedures carried out in the sitting or semi-sitting position. In fact more than 150 articles, reviews and books were published on air embolism in the 1800s.' Furthermore, by this time a great deal of the pathophysiology of the condition had been elucidated. More recent advances in the knowledge of the subject include bubble detection by the use of ultrasonics and end-tidal CO_2 monitoring, an understanding of the enhancement of bubble size by nitrous oxide and management of VAE via the use of right-sided catheter. More recently the transoesophageal Doppler ultrasound and echocardiography have presented further potential for monitoring and investigating this condition.[107]

Whilst no one could maintain that the risks are as prevalent as they were, VAE can still cause a crisis in the operating theatre, severe morbidity and the occasional fatality. Although most anaesthetists are familiar with its classic association with posterior cranial fossa surgery in the sitting position, technology is revealing the presence of air emboli in less familiar situations. To some extent it is unfortunate that VAE has been associated with such a specific surgical operation, as this may have hindered the formulation of a correct diagnosis in less likely situations. Quite simply, VAE can occur whenever the patient is positioned so that open veins or a traumatized venous sinus are situated at a site above that of the right atrium of the heart. Other factors are given below.

The first factor is the inability of a venous channel to collapse under the influence of gravity or pressure of adjacent tissues. Venous sinuses and diploic or emissary veins whose adventitious coat is fused to adjacent periosteum are the classic examples. Self-retaining surgical retractors may have a similar effect, and a less widely appreciated example is the risk that is incurred with both removal and insertion of central venous catheters. Although predictable with their insertion, it has only latterly been appreciated that fibrin tracks form consistently around catheters, *sometimes within 24 hours of insertion*, creating a potential portal for VAE after catheter removal.[108] It is therefore important to remove the catheter with the patient slightly head down and to ensure immediate, complete cover of the wound with an air-occlusive dressing. Over a 20-year period, Peters has collected data in which VAE associated with central venous catheterization has been responsible for 50 fatalities and 151 cases of cardiorespiratory distress or considerable neurological morbidity.[109]

The second most important factor is the level of venous pressure present in the veins concerned. For air to enter, this must be low or even negative. The effects of *posture* on the pressure within the venous sinuses of the skull were measured by Iwabuchi *et al.*[110] They found that, when the upper half of the body was raised, the confluens sinuum pressure decreased to reach zero at +25 degrees. When the angle was +90 degrees, a marked negative pressure of 12.7 ± 3.0 cmH_2O was observed in adults. To obtain a confluens sinuum pressure of zero or a little higher the most favourable position is with the sinus 15 cm higher than the right atrium. In children under 6 years of age, however, such negative pressure was not observed even at an angle of +90 degrees. A low venous pressure at the operation site may be potentiated by negative intrathoracic pressures generated by spontaneous ventilation. Albin *et al.* drew attention to the fact that in the prone knee–chest position a pendulous free-moving abdomen could also exert a suction effect on veins opened during the course of spinal axis surgery. Gradients as low as 5 cmH_2O have been shown to allow large amounts of air, up to 200 ml, to become entrained.[111]

The widespread use of muscle relaxants and IPPV during anaesthesia may have reduced the incidence of VAE, and this may have made anaesthetists less likely to suspect its occurrence. It is a risk that may not have been appreciated in the UK where there has been a resurgence of regional anaesthesia for the conscious patient. Manual removal of a morbidly adherent placenta under spinal anaesthesia recently resulted in a rapidly fatal air embolism. This could have been due to a combination of deep inspiratory efforts, a low inferior vena cava pressure due to haemorrhage and repeated introduction of air into the uterus by the hand of the obstetrician.

Apart from the well-known VAE problems of neurosurgical operations performed in the sitting position, some recent reports have drawn attention to similar dangers in the following procedures.

Caesarean section

During caesarean section carried out under regional anaesthesia, patients not uncommonly complain of quite severe chest pain after delivery of the baby. On occasion, this sounds indistinguishable from angina. The ECG may appear quite unaltered. SpO_2 monitoring may show coincident falls in saturation to around 90 per cent, but equally may not be associated with any symptoms. Vartikar and colleagues prospectively investigated 78 patients with precordial Doppler ultrasound, ECG and peripheral pulse oximetry.[112] Both elective and emergency cases were studied, and anaesthesia was either regional or general. Sixty-five per cent had Doppler changes consistent with a diagnosis of VAE, and three-quarters of these had a decrease in SpO_2 averaging 5.2 per cent (range 3–20 per cent). Chest pain or ECG changes were not seen unless both abnormal Doppler recordings and SpO_2 depression were present simultaneously. Both the SpO_2 and ECG changes resolved within the following 3 hours. Rather unexpectedly, most of the Doppler changes were seen during the repair of the hysterotomy wound. Exteriorization of the uterus and traction on the wound edges increase the risks. The Trendelenburg position should now most certainly be avoided for caesarean section or any form of uterine manipulation, and patient positioning should be adopted that ensures that abdominal and uterine incision is always below heart level. Slight breaking of the table with head-up tilt to increase inferior vena cava pressure is recommended. Younker has reported a case of massive air embolism during caesarean section performed under general anaesthesia; resuscitation was successful.[113]

The dangers of air embolism in posterior spinal surgery

Almost all these procedures are carried out in the prone position and the risks of VAE occurring are extremely small when routine, uncomplicated lumbar disc surgery is being performed. This type of operation is currently the most commonly encountered reason for spinal-axis surgery. Reconstructive orthopaedic procedures on the thoracic spine and posterior fossa explorations, also carried out in the prone position, obviously carry a great potential for all manner of operative complications.

The necessity for a high level of complex patient monitoring is self-evident in these latter procedures, and may well include facility for detecting and managing VAE. There are lessons to be learnt from looking at the development of techniques for prone positioning for lumbar disc surgery.[2] Some methods advocated in the period between 1945 and 1965 took no account of the already well-documented anatomical connection between the vena cava and the extradural veins via the lumbar plexus. Inevitably, back-pressure from compression by pillows in the epigastrium or misplaced props in the groins could cause so much haemorrhage at the site of operation that VAE was clearly no risk whatsoever. In fact, excessive bleeding could result in premature termination of the surgery.[114] With greater recognition of the aetiology of this problem, positioning with well-designed frames or props to support the iliac crests was advocated. Probably the most significant advance was in the method of propping the 'knee–chest' positioned patient with a 'seat' under the ischial tuberosities[1] (see Fig. 48.1). With the back horizontal in this position it has been shown that venous pressure is usually small – approximately 0–2 cmH_2O[115] (Table 48.1). Inferior vena cava pressures measured in different positions by DiStefano confirmed these findings[116] (Table 48.2). It seems, therefore, that the search for more satisfactory surgical operating conditions and greater patient safety may theoretically set the stage for the development of VAE. In 1969 the first report of a fatality from air embolism in the prone position described a case in which surgery was being performed for a large arteriovenous malformation in the posterior cranial fossa.[117] In 1977 the 'questions and answers' section of the *Journal of the American Medical Association* posed the question 'is right atrial catheterization a prerequisite for lumbar disc surgery in the knee–

Table 48.1 Effects of position on central venous pressure in patients undergoing laminectomy[115]

Patient weight			Central venous pressure (cmH_2O)		
kg (lb)		Operation	Supine under general anaesthesia	Prone on laminectomy frame	Sitting prone position
91.7	(200)	Thoracic	1	14	1
49.9	(110)	Lumbar	3	10	2
68.0	(150)	Thoracic	10	16	2
91.7	(200)	Lumbar	5–6	8–10	1
77.1	(170)	Thoracic	3–4	12–14	0–1
113.4	(250)	Lumbar	1	16	1
79.4	(175)	Lumbar	2	13	0
68.0	(150)	Lumbar	6	7.5	0
91.7	(200)	Lumbar	4	6	0
81.6	(180)	Lumbar	6–7	10–12	0–1
77.1	(170)	Lumbar	4	8	0
81.6	(180)	Lumbar	3–4	8	0

Table 48.2 Central venous pressure values in a variety of positions[116]

Arrangement	Average CVP (cmH₂O)
Georgia prone	20.7
Prone chest rolls	15.9
Lateral decubitus	11.1
Wilson frame	10.1
Tuck position	9.4
Hastings frame	1.4

chest position?[118] Tarlov, replying, stated that in his experience of several hundred patients it was not necessary for lumbar surgery.[119] He made the point that spinal veins, unlike the large intracranial veins and dural sinuses, are not kept open by bony and fascial attachments. In further correspondence, Albin *et al.* disagreed and advocated both the use of right atrial catheters and Doppler ultrasonic transducers even for lumbar surgery.[111] They quoted detection of VAE in the lateral, supine and prone positions with gradients between the wound and the heart as small as 5 cm. At that time, therefore, it appeared that, on the American scene at least, there were two dramatically opposed schools of thought. In the UK there was no evidence of the routine use of atrial catheters for uncomplicated lumbar disc surgery, and it can be assumed that the morbidity from jugular or supraclavicular vein puncture for monitoring central venous pressure (CVP) on a routine basis would almost certainly have been expected to outweigh the advantages.

Very recently, Albin and colleagues[120] have described two fatalities in three case reports of VAE occurring during lumbar laminectomy in the prone position. Both fatalities occurred in patients positioned on the 'four poster' Relton and Hall frame. The surviving patient had been positioned in the seated knee–chest position (Tarlov). The cases described were, however, at an extreme end of the spectrum of surgical and anaesthetic difficulty. All were revision procedures requiring fusion and all three weighed over 100 kg. Blood loss was excessive in all cases, and the operation times when the incidents occurred were 5 hours in two cases and 3.5 hours in the third. On the basis of these three cases, it would be unwise to make general recommendations regarding invasive CVP monitoring, with multi-orificed catheters sited in the right atrium for all routine lumbar laminectomies. The great majority of these cases are primary procedures, blood loss should not exceed 150–200 ml and an operating time of 1.5–2 hours suffices.

As always, more complex cases will justify more complex monitoring, which could well include a thermodilution catheter as advocated by Backofen and Schauble.[121] These workers point out that the value of a right atrial catheter lies more in its propensity as a route of aspiration than for the information it may provide. They recommend that poor-risk patients who would be unlikely to tolerate a reduction in cardiac index of around 20 per cent should have arterial and thermodilution catheters inserted if they are to be positioned prone.

Conclusion

The dangers of failing to position the surgical patient safely may result in greater morbidity than that of the condition for which the patient is receiving surgery. A fatality is not unknown as the ultimate complication of this process. An appreciation of the potential problems, vigilance and a high standard of patient monitoring are all necessary to minimize the risks.

REFERENCES

1. Tarlov IM. The knee–chest position for lower spinal operations. *Journal of Bone and Joint Surgery [Am]* 1967; **49**: 1193–4.

2. Anderton JM. The prone position for the surgical patient: a historical review of the principles and hazards. *British Journal of Anaesthesia* 1991; **67**: 452–63.

3. Betlach DW. Wisconsin Anesthesia Study Commission of the Wisconsin Society of Anesthesiologists. *Wisconsin Medical Journal* 1958; **57**: 185–6.

4. Lipton S. Anaesthesia in the surgery of retropulsed vertebral discs. *Anaesthesia* 1950; **5**: 208–12.

5. Gordon BS, Newman W. Lower nephron syndrome following prolonged knee–chest position. *Journal of Bone and Joint Surgery [Am]* 1952; **35**: 764–8.

6. Keim HA, Weinstein JD. Acute renal failure. A complication of spine fusion in the tuck position. *Journal of Bone and Joint Surgery [Am]* 1970; **52**: 1248–50.

7. Julian D, Scott A, Allen MJ. Acute and chronic compartment syndromes. *Surgery* 1987; **46**: 1088–93.

8. Leff RG, Shapiro SR. Lower extremity complications of the lithotomy position: prevention and management. *Journal of Urology* 1979; **122**: 138–9.

9. Lydon JC, Spielman FJ. Bilateral compartment syndrome following prolonged surgery in the lithotomy position. *Anesthesiology* 1984; **60**: 236–8.

10. Stoodley NG, Thomson WHF. Compartment syndrome: a cautionary tale. *British Journal of Surgery* 1989; **76**: 1297.

11. Connor H. Iatrogenic injuries in theatre. *British Medical Journal* 1992; **305**: 956.

12. Abel RR, Lewis GM. Postoperative (pressure) alopecia. *Archives of Dermatology* 1960; **81**: 34–42.

13. Lawson NW, Mills NL, Ochsner JL. Occipital alopecia following cardiopulmonary bypass. *Journal of Thoracic and Cardiovascular Surgery* 1976; **71**: 342.

14. Dickenson JC, Bailey BN. Chemical burns beneath tourniquets. *British Medical Journal* 1988; **297**: 1513.

15. Courington FW, Little DM. Clinical anaesthesia. 3. The role of posture in anaesthesia. In: Jenlans MT, ed. *Common and uncommon problems in anaesthesiology.* Philadelphia: FA Davis, 1968: 24–54.

16. Luscombe F, Teasdale A, Powell J. Ischaemic lesions related to pressure from the bindings of a wrist splint and cardiopulmonary bypass. *Anaesthesia* 1989; **44**: 610–11.

17. Smith RH, Gramling ZW, Volpitto PP. Problems related to the prone position for surgical operations. *Anesthesiology* 1961; **22**: 189–93.

18. Martin JT, ed. *Positioning in anaesthesia and surgery,* 2nd ed. Philadelphia: WB Saunders, 1987.

19. Magill IW. Postoperative care: anaesthetic prospects. *Practitioner* 1937; **138**: 247.

20. Schleyer-Sanders E. Prevention of backache in women. *British Medical Journal* 1954; **1**: 28.

21. Brown EM, Elman DS. Postoperative backache. *Anesthesia and Analgesia – Current Research* 1961; **40**: 683–5.

22. O'Donovan N, Healy TEJ, Faragher EB, Wilkins RG, Hamilton AA. Postoperative backache: the use of an inflatable wedge. *British Journal of Anaesthesia* 1986; **58**: 280–3.

23. Hickmott KC, Healy TEJ, Roberts P, Faragher EB. Back pain following general anaesthesia and surgery. *British Journal of Surgery* 1990; **77**: 571–5.

24. Carrie LES. An inflatable obstetric anaesthetic wedge. *Anaesthesia* 1982; **37**: 745–7.

25. Givner I, Jaffe N. Occlusion of the central retinal artery following anaesthesia. *Archives of Ophthalmology* 1950; **43**: 197–201.

26. Tarkannen A, Leikola J. Postural variations of the intraocular pressure as measured with the Mackay–Marg Tonometer. *Acta Ophthalmologica* 1967; **45**: 569–75.

27. Friberg TR, Sanborn G. Optic nerve dysfunction during gravity inversion; pattern reversal evoked potentials. *Archives of Ophthalmology* 1980; **103**: 1687–9.

28. Eldor J, Admoni M. Acute glaucoma following non-ophthalmic surgery. *Israel Journal of Medical Sciences* 1989; **25**: 652–4.

29. Walkup HE, Murphy JD, Oteen NC. Retinal ischaemia with unilateral blindness – complications during pulmonary resection in prone position: two cases. *Journal of Thoracic Surgery* 1952; **23**: 174–5.

30. Hollenhorst RW, Svien HJ, Benoit CF. Unilateral blindness occurring during anaesthesia for neurosurgical operations. *Archives of Ophthalmology* 1954; **52**: 819.

31. Jampol LM, Goldbaum M, Rosenberg M, Bahr R. Ischaemia of ciliary arterial circulation from ocular compression. *Archives of Ophthalmology* 1975: **93**: 1311–17.

32. Tramo MJ, Hainline B, Petito F, *et al.* Vertebral artery injury and cerebellar stroke while swimming. Case report. *Stroke* 1985; **16**: 1039.

33. Okawar S, Nibbelink D. Vertebral artery occlusion following hypertension and rotation of the head. *Stroke* 1974; **5**: 640.

34. Hanus SH, Homer TD, Harter DH. Vertebral artery occlusion following yoga exercises. *Archives of Neurology* 1977; **34**: 574.

35. Krueger BR, Okazaki H. Vertebral–basilar distribution infarction following chiropractic cervical manipulation. *Mayo Clinic Proceedings* 1980; **55**: 322.

36. Schmidley JW, Koch T. The noncerebrovascular complications of chiropractic manipulation. *Neurology* 1984; **34**: 684.

37. Faris AA, Poser CM, Wikmore DW, Agnew CH. Radiologic visualization of neck vessels in healthy men. *Neurology* 1963; **13**: 386.

38. Sherman DD, Hart RG, Easton JD. Abrupt change in head position and cerebral infarction. *Stroke* 1981; **12**: 2.

39. Kim BY, Ngeow JYF, Kitahata LM, Swift CA. EEG changes with lateral rotation of the head. *Anesthesiology Review* 1985; **12**: 36–7.

40. McPherson RW, Szymanski J, Rogers MC. Somatosensory evoked potential changes in position-related brain stem ischemia. *Anesthesiology* 1984; **61**: 88–90.

41. Hitselberger WE, House WF. A warning regarding the sitting position for acoustic tumour surgery. *Archives of Otolaryngology* 1980; **106**: 69.

42. Tattersall MP. Massive swelling of the face and tongue. A complication of posterior fossa surgery in the sitting position. *Anaesthesia* 1984; **39**: 1015–17.

43. Munshi CA, Dhamee MS, Ghandi SK. Postoperative unilateral facial oedema: a complication of acute flexion of the neck. *Canadian Anaesthetists' Society Journal* 1984; **31**: 197–9.

44. Colley PS, Dunn R. Prevention of blood pressure response to skull-pin holder by local anesthesia. *Anesthesia and Analgesia* 1979 **58**: 241.

45. Bedford RF, Marshall WK, Persing JS. Rapid reduction of intracranial pressure: thiopentone versus lidocaine. *Anesthesia and Analgesia* 1980; **59**: 528.

46. *Hazard warning notice.* HC (Hazard) (91) 15. London: Department of Health, 16 July 1991.

47. Trojaborg W. Electrophysiologic findings in pressure palsy of the brachial plexus. *Journal of Neurology, Neurosurgery and Psychiatry* 1977; **40**: 1160–7.

48. Gwinnutt CL. Injury to the axillary nerve. *Anaesthesia* 1988; **43**: 205–6.

49. Anderton JM, Schady W, Markham DE. An unusual cause of postoperative brachial plexus palsy. *British Journal of Anaesthesia* 1994; **72**: 605–7.

50. Patel RI, Thein RMH, Epstein BS. Costoclavicular syndrome and the sitting position during anesthesia. *Anesthesiology* 1980; **53**: 341–2.

51. Saady A. Brachial plexus palsy after anesthesia in the sitting position. *Anesthesiology* 1981; **36**: 194.

52. Dawson DM, Krarup C. Perioperative nerve lesions. *Archives of Neurology* 1989; **46**: 1355–60.

53. Hanson MR, Breuer AC, Furlan AJ, *et al.* Mechanism and frequency of brachial plexus injury in open heart surgery: a prospective analysis. *Annals of Thoracic Surgery* 1983; **36**: 675–9.

54. Wadsworth TG, Wiliams JR. Cubital tunnel external compression syndrome. *British Medical Journal* 1973; **1**: 662–6.

55. Wadsworth TG. The external compression syndrome of the ulnar nerve at the cubital tunnel. *Clinical Orthopaedics* 1977; **124**: 189–204.

56. Alvine GF, Schurrer ME. Postoperative ulnar nerve palsy. Are

there predisposing factors? *Journal of Bone and Joint Surgery [Am]* 1987; **69**: 255–9.

57. Neary D, Ochoa J, Giliatt RW. Sub-clinical entrapment neuropathy in man. *Journal of the Neurological Sciences* 1975; **24**: 283–98.

58. Pollock LJ, Davis L. Peripheral nerve injuries. *American Journal of Surgery* 1932; **17**: 461–71.

59. Britt BA, Gordon RA. Peripheral nerve injuries associated with anaesthesia. *Canadian Anaesthetists' Society Journal* 1964; **11**: 514–36.

60. Kopell HP, Thompson WAL. *Peripheral entrapment neuropathies*. New York: Robert E Krieger, 1976.

61. Parks BJ. Postoperative peripheral neuropathies. *Surgery* 1973; **74**: 348–57.

62. Ehni G. Significance of the small lumbar spinal canal. Cauda equina compression syndromes due to spondylosis. 4. Acute compression artificially induced during operation. *Journal of Neurosurgery* 1969; **31**: 507–12.

63. Wilkes LL. Paraplegia from operating position and spinal stenosis in non-spinal surgery. *Clinical Orthopaedics* 1980; **146**: 148–9.

64. Moiel R, Ehni G. Cauda equinal compression due to spondylolisthesis with intact neural arch. Report of two cases. *Journal of Neurosurgery* 1968; **28**: 262–5.

65. Massey EW, Pleet AB. Compression injury of the sciatic nerve during prolonged surgical procedure in a diabetic patient. *Journal of the American Geriatric Society* 1980; **28**: 188–9.

66. Sunderland S. Relative susceptibility to injury of the medial and lateral division of the sciatic nerve. *British Journal of Surgery* 1953; **41**: 300–2.

67. Burkhart FL, Daly JW. Sciatic and peroneal nerve injury. A complication of vaginal operations. *Obstetrics and Gynecology* 1966; **28**: 99–102.

68. McQarrie HG, Harris SW, Elsworth HS, Stone RD, Anderson AE. Sciatic neuropathy complicating vaginal hysterectomy. *American Journal of Obstetrics and Gynecology* 1972; **113**: 223–32.

69. Hatono Y, Arai T, Iida H, Soneda J. Common peroneal nerve palsy. A complication of coronary artery bypass grafting surgery. *Anaesthesia* 1988; **43**: 568–9.

70. Roblee MA. Femoral neuropathy from the lithotomy position. Case report and a new leg holder for prevention. *American Journal of Obstetrics and Gynecology* 1967; **97**: 871–2.

71. Kinges KG, Wilbanks GD, Cole GR. Injury to the femoral nerve during pelvic operations. *Obstetrics and Gynecology* 1965; **25**: 619–23.

72. Hoffman A, Jones RE, Schoenvogel R. Pudendal nerve neuropraxia as a result of traction on the fracture table. A report of four cases. *Journal of Bone and Joint Surgery [Am]* 1982; **64**: 136–8.

73. Corvino BG, Fossard HA, Relider K, Strichers G, Bethseda C, eds. *Effects of anaesthesia*. Bethesda MD: American Physiological Society, 1985.

74. Coonan TJ, Hope CE. Cardiorespiratory effects of change in body position. *Canadian Anaesthetists' Society Journal* 1983; **30**: 424–37.

75. Marshall WK, Bedford RF, Miller ED. Cardiovascular responses in the seated position – impact of four anesthetic techniques. *Anesthesia and Analgesia* 1983; **62**: 648–53.

76. Heinonen J, Takki S, Tammisto T. The effect of Trendelenburg tilt and other procedures on the position of endotracheal tubes. *Lancet* 1968; **1**: 850–3.

77. Green RA. A matter of vigilance. *Anaesthesia* 1986; **41**: 129–30.

78. Lunn JN, Mushin WW. *Mortality associated with anaesthesia*. London: Nuffield Provincial Hospitals Trust, 1982: 21–2.

79. Tomlin PJ, Howarth FH, Robinson JS. Postoperative atelectasis and laryngeal incompetence. *Lancet* 1968; **1**: 1402–5.

80. Russell WJ. Position of the patient and respiratory function in the immediate postoperative period. *British Medical Journal* 1981; **283**: 1079–80.

81. Hsu Ho, Hickey RF. Effect of posture on the functional residual capacity postoperatively. *Anesthesiology* 1976; **44**: 520–1.

82. Relton JES, Hall JE. An operation frame for spinal fusion – a new apparatus designed to reduce haemorrhage during operation. *Journal of Bone and Joint Surgery [Br]* 1967; **49**: 327–32.

83. Weinlander CM, Coombs DW, Plume SK. Myocardial ischemia due to obstruction of an aortocoronary bypass graft by intraoperative positioning. *Anesthesia and Analgesia* 1985; **64**: 933–6.

84. Bieniarz J, Crottogini JJ, Curuchet E, *et al*. Aortocaval compression by the uterus in late human pregnancy. *American Journal of Obstetrics and Gynecology* 1968; **100**: 203–17.

85. Hansen R. Ohnmacht und schwangerschaft. *Klinische Wochenschrift* 1942; **21**: 241–5.

86. McRoberts WA. Postural shock in pregnancy. *American Journal of Obstetrics and Gynecology* 1951; **62**: 627–32.

87. Howard RK, Goodson JH, Mengert WF. Supine hypotensive syndrome in late pregnancy. *Obstetrics and Gynecology* 1953; **1**: 371–7.

88. Marx GF. Aortocaval compression syndrome: its 50 year history. *International Journal of Obstetric Anaesthesia* 1992; **1**: 60–4.

89. Kinsella SM, Lee A, Spencer JAD. Maternal and foetal effects of the supine and pelvic tilt positions in late pregnancy. *European Journal of Obstetrics, Gynecology and Reproductive Biology* 1990; **36**: 11–17.

90. Secher NJ, Arnsbo P, Heslet Anderson L, Thomson A. Measurements of cardiac stroke volume in various body positions in pregnancy and during caesarean section: a comparison between thermodilution and impedance cardiography. *Scandinavian Journal of Clinical and Laboratory Investigation* 1979; **39**: 569–76.

91. Ludbrook J. Faint heart. Opioids have a role in circulatory collapse due to acute blood loss. *British Medical Journal* 1989; **298**: 1053–4.

92. Anderson ID. Faint heart. [letter] *British Medical Journal* 1989; **298**: 1449–50.

93. Goodwin APL, Pearce AJ. The human wedge: a manoeuvre to relieve aortocaval compression during resuscitation in late pregnancy. *Anaesthesia* 1992; **47**: 433–4.

94. Marx GF. Cardiopulmonary resuscitation of late-pregnant women. *Anesthesiology* 1982; **56**: 156.

95. Verrill PJ, Aellig WH. Vasovagal faint in the supine position. *British Medical Journal* 1970; **4**: 348.

96. Chadwick IS, Eddleston JM, Candelier CK, Pollard BJ. Haemodynamic effects of the position chosen for the insertion of

an extradural catheter. *International Journal of Obstetric Anaesthesia* 1993; **2**: 197–201.

97. Elstein ID, Schwalbe SS, Marx GF. Cardiac output measurements during and after triplet gestation. *Obstetrics and Gynecology* 1989; **74**: 452–3.

98. Abitol MM. Aortic compression by pregnant uterus. *New York Journal of Medicine* 1976; **76**: 1470–5.

99. Eckstein KL, Marx GF. Aortocaval compression and uterine displacement. *Anesthesiology* 1974; **40**: 92–6.

100. Kauppila A, Koskinin M, Puolokka J, *et al*. Decreased intervillous and unchanged myometrial blood flow in supine recumbency. *Obstetrics and Gynecology* 1980; **55**: 203–5.

101. Humphrey MD, Chang A, Wood EC, *et al*. A decrease in foetal pH during the second stage of labour when conducted in the dorsal position. *Journal of Obstetrics and Gynaecology of the British Commonwealth* 1974; **81**: 600–2.

102. Kinsella SM, Whitwam JG, Spencer JAD. Reducing aortocaval compression: how much tilt is enough? Do as much as possible in the lateral position. *British Medical Journal* 1992; **305**: 539–40.

103. Ansari I, Wallace G, Clemetson CAB, *et al*. Tilt caesarean section. *Journal of Obstetrics and Gynaecology of the British Commonwealth* 1970; **77**: 713–21.

104. Crawford JS, Burton M, Davies P. Time and lateral tilt at caesarean section. *British Journal of Anaesthesia* 1972; **44**: 477–84.

105. Buley RJR, Downing JW, Brock-Utne JG, *et al*. Right versus left lateral tilt for caesarean section. *British Journal of Anaesthesia* 1977; **49**: 1009–15.

106. Colditz GA, Tuden RL, Oster G. Rates of venous thrombosis after general surgery; combined results of clinical trials. *Lancet* 1986; **2**: 143–6.

107. Albin MS. The sights and sounds of air. *Anesthesiology* 1983; **58**: 113–14.

108. Meninim P, Coyle CF, Taylor JD. Venous air embolism associated with removal of central venous catheter. *British Medical Journal* 1992; **305**: 171–2.

109. Peters JL. Removal of central venous catheter and venous air embolism. *British Medical Journal* 1992; **305**: 524–5.

110. Iwabuchi T, Sobata E, Susuki M, Susuki S, Yamashita M. Dural sinus pressure as related to neurosurgical positions. *Neurosurgery* 1983; **12**: 203–7.

111. Albin MS, Newfield P, Paulter S, *et al*. Atrial catheter and lumbar disc surgery. [Correspondence] *Journal of the American Medical Association* 1978; **239**: 496.

112. Vartikar JV, Johnson MD, Datta S. Precordial doppler monitoring and pulse oximetry during caesarean section delivery; detection of venous air embolism. *Regional Anaesthesia* 1989; **14**: 145–8.

113. Younker D, Rodriguez V, Kavanagh J. Massive air embolism during cesarean section. *Anesthesiology* 1986; **65**: 77–9.

114. Hunter AR. Anesthesia for operations in the vertebral canal. *Anesthesiology* 1950; **12**: 367–73.

115. Cook AW, Siddiqui TS, Nidzgorski R, *et al*. Sitting prone position for the posterior surgical approach to the spine and posterior fossa. *Neurosurgery* 1982; **10**: 232–5.

116. DiStefano VJ, Klein KS, Nixon JE, Andrews TE. Intraoperative analysis of position and body habitus on surgery of the low back. *Clinical Orthopaedics* 1974; **99**: 51–6.

117. Shenkin HN, Goldfedder P. Air embolism from exposure of posterior cranial fossa in prone position. *Journal of the American Medical Association* 1969; **210**: 726.

118. Goldberg D. Lumbar disc surgery in the knee–chest position: preanaesthetic atrial catheter unnecessary. Questions and Answers. *Journal of the American Medical Association* 1977; **238**: 253.

119. Tarlov I. Lumbar disc surgery in knee–chest position. Preanaesthetic atrial catheter unnecessary. Questions and Answers. *Journal of American Medical Association* 1977; **238**: 253.

120. Albin MS, Ritter RR, Pruett CE, Kalff K. Venous air embolism during lumbar laminectomy in the prone position: a report of three cases. *Anesthesia and Analgesia* 1992; **73**: 346–9.

121. Backofen JE, Schauble JF. Hemodynamic changes with prone position during general anesthesia. [Abstract] *Anesthesia and Analgesia* 1985; **64**: 194.

Section Three

Anaesthetic Management

Chapter 49

Anaesthetic Management of Patients with Endocrine Disease

Burnell R. Brown Jr and Edward J. Frink Jr

Diabetes mellitus	Hyperthyroidism
Phaeochromocytoma	Hypothyroidism
Hyperparathyroidism	Summary
Carcinoid syndrome	

Surgical patients with either hyperfunctional or hypofunctional pathology of endocrine glands provide the anaesthetist with complex problems. Surgery may be required for extirpation of a functioning tumour, as in the case of phaeochromocytoma, or a patient with a hypofunctional endocrine disorder may require non-endocrine surgery such as limb amputation in a diabetic. This chapter presents current concepts of pathophysiology for a variety of endocrinopathies, discusses risk factors and formulates a concise plan of perioperative care. Obviously, it cannot be comprehensive. The variety of endocrine diseases presented to the anaesthetist can range from the banal to the bizarre.

Diabetes mellitus

This is the most common endocrine problem. Diabetes represents a group of disorders with the unifying link of a relative or absolute lack of insulin. As many as 5 per cent of adults in Western societies may be diabetic using this criterion but, because of the complications of the disease, they require more surgery than other individuals.

Two distinct types of diabetes are recognized: type I, also known as insulin-dependent diabetes mellitus (IDDM) and juvenile onset diabetes; and type II, or non-insulin-dependent diabetes mellitus (NIDDM) and maturity onset diabetes. Type I patients are insulin deficient and can progress to ketoacidosis. Type II patients are often obese, may have normal or even elevated insulin levels with insulin resistance, and generally do not progress to ketoacidoses although they may develop a hyperglycaemic, hyperosmolar state. Type I diabetes is treated with insulin therapy. Type II diabetes is initially treated with diet only but may require oral hypoglycaemic medications which act by stimulating insulin release from pancreatic β

cells and extrapancreatic actions potentiating insulin-mediated activity. Such drugs include tolazamide and tolbutamide or newer drugs glipizide and glibenclamide (glyburide) which are second generation sulphonylureas with durations of 18–24 hours. However, stress of surgery may require type II patients to be treated temporarily with insulin. Most of the discussion in this chapter focuses on type I diabetes.

Risk factors for the diabetic surgical patient

An early study demonstrated that diabetics have greater postoperative mortality and serious morbidity than others.[1] Problems in the diabetics were infection, sepsis and complications of the accelerated atherosclerosis seen in this population. More recent studies[2] demonstrated that 11 per cent of diabetics had non-cardiac complications in the postoperative period, primarily infection and pneumonia. The major preoperative factors likely to be associated with problems were the pre-existence of end-organ complications (neuropathy, retinopathy or nephropathy), heart failure or peripheral vascular disease accompanied by infection. Cardiac complications occurred in 7 per cent of the diabetics. Over all, postoperative mortality was 4 per cent, primarily in the group with pre-existing cardiac disease. Other studies have demonstrated that surgery in diabetics may be associated with a tenfold increase in mortality.[3,4] This mortality is due to:

1. Sepsis
2. Autonomic neuropathy and its coincident problems
3. Complications of atherosclerosis (coronary artery disease, stroke, peripheral vascular disease)
4. Ketoacidosis and hyperglycaemic hyperosmolar coma.

Diabetes (type I) has features of autoimmune disease process that can cause destruction of the autonomic nervous system and give rise to autonomic neuropathy. Manifestations of autonomic neuropathy include hypohidrosis, lack of cardiac beat-to-beat variability (R–R variability), decreased heart rate response to a Valsalva manoeuvre (<5 beats/min) and orthostatic hypotension (>30 mmHg blood pressure fall with change to an upright posture). Patients with defined autonomic neuropathic features can suffer profound hypotension following administration of anaesthetic drugs.[5] In addition, the presence of neuropathy confirms the increased risk of gastroparesis and aspiration, of postoperative hypoxic episodes and of urinary retention. Preoperative treatment with metoclopramide may be beneficial in the patient with delayed gastric emptying. Hypertension may occur in as many as 50 per cent of diabetic patients with autonomic neuropathy. The incidence of diabetic neuropathy probably varies directly with the number of years the disease has been present. Pirart[6] noted a rate of 7 per cent within 1 year of diagnosis to as many as 50 per cent for those diagnosed more than 25 years previously. Burke et al.[7] found that 1.4 per cent of their clinic patients experienced abnormal heart rate variation. In general, autonomic dysfunction increases with both age and duration of the diabetes, and there appears to be a relation between deteriorating cardiovascular reflex tests and glycaemic control. Some diabetics with autonomic neuropathy have died suddenly. Possible aetiological explanations include an abnormal response to hypoxia, sleep apnoea or cardiac arrhythmias, but none is as yet convincingly documented. Patients with autonomic neuropathy are at particular risk from anaesthesia.

Renal disease is a frequent finding with advanced diabetes, particularly with type I. Microalbuminuria and pathology of glomerular filtration as manifested by altered creatinine clearances are the hallmarks of the condition. One of the benefits of strict control of blood sugar in diabetes is apparently improved preservation of kidney function.[8] The finding that antihypertensive therapy may delay development of renal complications is of further note.[9] Infection and sepsis play a major role in postoperative mortality and morbidity in diabetics. One of the cogent findings related to this problem is the fact that leucocyte function is markedly impaired in diabetes.[10] When blood sugars are tightly controlled and glucose levels are kept below 14 mmol/l (250 mg/dl) the phagocytic function of leucocytes is restored.[11] The implications of such studies for consideration for anaesthesia would seem clear, but unfortunately no outcome studies are available to validate the point that 'tight' control of blood glucose during surgery reduces morbidity.

Hogan et al.[12] reported an increased incidence of difficult intubations in diabetic patients. The reason is that some diabetics may have the stiff joint syndrome. The stiff joint syndrome is believed to be due to abnormal collagen tissue that is formed in periarticular tissues and can be suspected

in association with progressive microangiopathy and thin, tight and waxy skin. The abnormality in collagen may be related to non-enzymatic glycosylation of protein. Many patients with this association may have the 'prayer sign', which is an inability to approximate the palmar surfaces and joints of the fingers. The overall incidence in type I diabetes may be as high as 30 per cent.

Preoperative evaluation

Obviously the anaesthetist should identify the type of diabetes and recognize that type I will be associated with the need for insulin therapy whereas type II may or may not require additional insulin for the period of perioperative stress. Ketosis can develop in patients with type II diabetes during severe stress.[13] Ideally, the preoperative level of glycaemic control should be identified. An analysis of plasma glycosylated haemoglobin levels is of help in this regard. Proteins combine with glucose, at a rate proportional to plasma glucose concentration, to form glycosylation products eventually bonding covalently to proteins. The glycosylated haemoglobin (HbA_1C) provides information correlating with blood sugar control in the preceding 6–8 weeks. Cardiovascular complications (coronary artery disease, congestive heart failure, hypertension) are associated with increased mortality in the diabetic and should be excluded in patients requiring surgery. Autonomic neuropathy and evidence of renal disease should be assessed using appropriate clinical and laboratory testing.

Perioperative metabolic control

The basic objectives are to:

1. Correct acid–base, fluid and electrolyte abnormalities prior to surgery.
2. Provide sufficient carbohydrates to inhibit catabolic metabolism and ketoacidosis.
3. Determine insulin requirements to prevent hyperglycaemia.

Adult onset type II diabetes probably holds little in the way of increased risk per se and thus requires less alteration in pre-existing treatment. Certainly preoperative slimming is ideal. There is question as to whether insulin therapy should be instituted in these patients during the perioperative period. For relatively minor surgery, short-acting oral hypoglycaemics should be omitted on the day of surgery and long-acting drugs 2 days prior to surgery. For major surgery, small doses of human insulin (10 units or so twice a day) may be needed to control blood glucose and glycosuria. This clinical decision should be undertaken on an individual basis. Following are conditions in type II diabetic patients that may act as indications for insulin as recommended by Gavin.[13]

1. Current fasting blood sugar >10 mmol/l (180 mg/dl)
2. Glycosylated haemoglobin 0.08–0.1
3. Duration of surgery >2 hours

Minor surgery in the type I diabetic has been handled in two fashions. In one, the well-controlled fasting patient can be given one-half the usual dose of subcutaneous intermediate-acting insulin early in the morning prior to surgery. A dextrose 5 per cent intravenous solution is started at this time. Regular insulin can be given as a supplement, as indicated by measurement of glucose levels.[14] A problem noted with this regimen is the fact that intermediate insulin has a peak hypoglycaemic response at a time the patient is not well observed[15] (e.g. 6–10 hours following subcutaneous injection). To avoid this problem, such a regimen should be used when there is high confidence the patient will be able to resume feeding and self-care shortly after the surgery. Another approach for brief surgical procedures is to withhold all insulin preoperatively. The patient is kept fasting and no glucose given. After the surgery, oral food intake is resumed and the patient may be given 50 per cent of the usual insulin dose. Although increased losses of nitrogen were observed in patients admitted to this regimen no adverse clinical reactions were noted.[16]

Major surgery in these patients requires careful planning for optimal control. A technique recommended by Hirsh et al.[17] follows. Glucose 5–10 g per hour equivalent to 100–200 ml dextrose 5 per cent per hour is administered intravenously. Potassium may be added but care should be employed in patients with compromised renal function. Another infusion is added through the same cannula, as follows:

1. Mix 50 units of regular insulin into 500 ml 0.9 per cent saline
2. Infuse at a rate of 0.5–1.0 u/h (5–10 ml/h) using a standard infusion pump
3. Measure glucose levels in plasma hourly and make the following adjustments to the insulin drip rate based on the findings:

Glucose level in mmol (mg/dl)	Insulin requirements
4.4 (80)	Turn off pump; give glucose i.v.
4.4–6.6 (80–120)	Decrease insulin to 0.2–0.7 u/h
6.6–9.9 (120–180)	Continue insulin at 0.5–1.0 u/h
9.9–13.2 (180–240)	Increase infusion rate to 0.8–1.5 u/h
13.75 (>250)	Insulin at 1.5 u/h or greater

Another technique for insulin–glucose infusion is the intermittent bolus technique. This technique compares favourably with a continuous infusion technique reported by Raucoules et al.[18] An intravenous infusion of 1.25 u/h regular insulin was compared to an intravenous bolus of 10 u/2 h with the administration of 125 ml of 5 per cent dextrose per hour. The only problem observed was in the intravenous bolus group, in which one patient became hypoglycaemic. Obesity and severe infection will increase the requirements for insulin 1.5–2.0 times.

Effects of surgery and anaesthesia on metabolism in general

Surgical stress causes a profound catabolic response. Not only are there increased secretions of catecholamines, glucagon and cortisol but there is also decreased secretion of insulin. Thus surgery fosters hyperglycaemia, decreased utilization of glucose and increased gluconeogenesis. There is protein catabolism with decreased would healing and decreased resistance to infection. Hyperglycaemia is far more common than hypoglycaemia with stress. Many of these responses are triggered not only by pain (afferent neuronal stimulation) but also by secretion of peptides such as interleukin-1 and various hormones, including β-endorphins, growth hormone and prolactin. The effects of anaesthesia on these responses have been varied. Profound analgesia by itself seems to have only a limited action for modifying the overall catabolic response to surgery.[19] High epidural anaesthesia may block catabolic responses to surgery by means of dual afferent and autonomic nervous blockade.[20] High dose narcotic techniques (fentanyl 50 μg/kg i.v.) partially prevents the stress response[21] whereas general anaesthesia has less inhibitory action even when administered in high concentrations (2.1 MAC halothane).[22] It is unfortunate that at this time studies correlating decreased stress-response and anaesthetics with outcome in diabetic surgical patients have not been reported.

Selection and timing of anaesthesia in diabetic patients

It is probably worthwhile mentioning that often it is impossible to control completely the metabolic aberrations of diabetes prior to surgery. For example, in the presence of infection it is very difficult to produce complete metabolic homoeostasis because the infection amplifies the diabetes. In fact, the most important factor precipitating diabetic coma is infection, with inadequate insulin therapy only second.[23] Thus the clinician must decide when the endocrinopathy is as well controlled as possible and thus allow curative surgery to proceed without impediment predicated upon generation of 'perfect numbers'.

Some diabetics presenting for emergency surgery with infection may have a significant aberration in glucose levels and may also have ketoacidosis. Delaying surgery in order to achieve total correction of ketoacidosis is not warranted. However, correction of fluid deficit and significant electrolyte imbalance may be achieved within several hours. Isotonic saline is used to replace fluid deficits. Lactated Ringer's solution may adversely affect hyperglycaemic status because lactate is a gluconeogenic substrate. Volume deficits are generally several litres and infusion rates of 15–20 ml/kg per hour may be required. Often, serum sodium is reduced (fictitiously) by elevated serum glucose (1.6 mEq/decrease per 5.5 mmol/increase in glucose levels). Approximately one-third to one-half of the estimated volume deficit should be replaced within the first 8 hours.

Hyperglycaemia should be corrected by insulin infusion with rates determined by the serum level of glucose. Therapy may be initiated using 10 units of regular insulin intravenously followed by an infusion determined most readily by dividing the glucose level (mmol) by 8. During the first several hours, glucose levels may fall significantly as a result of volume replacement. Plasma glucose levels should be reduced by 4.1–5.5 mmol per hour. Once the plasma levels have fallen to 13.75–16.5 mmol, insulin infusion rates should be decreased and infusion of dextrose-containing solutions may be initiated.

With ketoacidosis an increased anion gap acidosis is present along with hyperkalaemia and a reduction in total body potassium stores. The use of bicarbonate therapy in ketoacidosis is somewhat controversial. Although a pH of less than 7.1 may impair myocardial function and response to catechols, rapid correction of acidosis with bicarbonate may produce excessive carbon dioxide, resulting in CNS acidosis. Over-zealous correction of acidosis would therefore seem unwarranted. Although serum potassium levels may initially be elevated, correction of hyperglycaemia with insulin therapy will result in a decline in serum potassium within 1–3 hours. Potassium replacement should commence once urine output is established. Phosphorus deficiency due primarily to renal losses may also be present and replacement required if levels fall below 0.3 mmol.

Anaesthetic drugs produce rather insignificant metabolic effects compared to the stress of surgery. There is evidence that potent halogenated inhalation anaesthetics can impair the uptake of glucose into both skeletal muscle and hepatic tissue. This has been termed the 'anti-insulin' action of volatile anaesthetics but the clinical significance, if any, is totally unexplored. Spinal and epidural anaesthesia probably play a significant role in patient safety. In older literature there were admonitions against the use of regional anaesthesia because of the theoretical possibility that peripheral diabetic neuropathy could be exacerbated.

There is no evidence to support this contention. In fact, regional analgesia with its potential for stress reduction can be favoured over general anaesthesia for many circumstances.

Whatever glucose–insulin infusion regimen is selected, the anaesthetist must consistently monitor hourly glucose plasma levels. When there is concern, ketone bodies must be sought. In prolonged surgery and in major cases, frequent determinations of electrolytes ('renal battery') are in order.

During the postoperative period the intravenous insulin–glucose infusion is continued until the patient can tolerate oral intake. When dietary intake is achieved the patient can be managed with his or her usual insulin doses coupled with a sliding scale to cover catabolic demands. Type II diabetics may require such a regimen for 1–3 days postoperatively before reverting to their usual oral hypoglycaemic drugs.

Phaeochromocytoma

Phaeochromocytoma is a tumour of chromaffin cells which secretes catecholamines. Autopsies indicate that the incidence of these tumours may be as high as 1/2000. This endocrine problem is frequently unrecognized, and if surgery and anaesthesia are performed in an individual unsuspected of harbouring a phaeochromocytoma, mortality can be quite high.[24] For diagnosed phaeochromocytoma, overall mortality rate is no higher than 2–3 per cent, in contrast to as high as 50 per cent several decades ago. This is due to better understanding of the haemodynamic changes that occur with phaeochromocytoma and a better knowledge of the pharmacology of the adrenergic nervous system, which have led to better preparation of the patient prior to surgery.

Pathophysiology of phaeochromocytoma

The symptomatic triad of phaeochromocytoma consists of hypertension, paroxysmal tachycardia, and paroxysmal sweating and headache. This symptom complex occurs with great frequency in patients with this tumour and may have a very high diagnostic accuracy if placed in context with other signs and symptoms. Constipation and neuropsychiatric disorders can also alert the clinician's suspicion. Diagnosis is confirmed by urinary and/or plasma catecholamines, and localization is determined by radiographic procedures such as computed axial tomography.

Death from phaeochromocytoma results from stroke due to uncontrolled hypertension, cardiac failure due to excessive peripheral resistance, or cardiac failure due to catecholamine myocardiopathy.

Although grey areas exist, patients with a phaeochromocytoma can be divided into two groups, each with its own distinctive characteristics. The first group usually harbour small tumours (<25 g) which are frequently multiple. These persons often have a strong family history of phaeochromocytoma, and may have a family history of multiple endocrine neoplasia syndrome (MEN syndrome). Approximately one in every ten cases of phaeochromocytoma occurs as part of a familial syndrome.[25] Such patients will have the classic paroxysmal hypertensive attacks with normotension between episodes. It must be remembered that paroxysms are brief because of the relatively short half-life of catecholamines in the circulation. This is due to the rapid uptake of catecholamines into postganglionic adrenergic neurons. Thus a single quantal release of catecholamines into the bloodstream from an autonomous tumour (phaeochromocytoma) is terminated within 8–12 minutes owing to rapid uptake. This group of patients must be investigated for associated endocrine abnormalities. Perhaps the most worrisome concomitant endocrinopathy for the anaesthetist is parathyroid adenoma because of its metabolic effects. If the patient is found to have combined phaeochromocytoma and parathyroid adenoma the operation is divided into a two-stage procedure. The phaeochromocytoma is the more life-threatening tumour and hence is operated upon first, followed 4–6 weeks later by excision of the parathyroid adenoma. Plasma calcium determinations should be a routine part of the work-up of patients with suspected phaeochromocytoma. A second grey area of patients with phaeochromocytoma are older individuals (age >35 years) with large tumours (weight >50 g). These tumours may have a high degree of malignancy (10–15 per cent). Patients in this category may have sustained hypertension in the periods between paroxysms as if their baroreceptors have been 'reset'.

Regardless of type of phaeochromocytoma, the disease has several common features. First, there is often diminution of red cell mass and plasma volume. Thus chronic hypovolaemia may exist and is the major reason for preoperative therapy with adrenergic blocking drugs which encourage volume re-expansion. This constricted volume may be due to chronic vasoconstriction leading to diminished red cell and plasma volume or to reduced red cell production by the elevated catecholamine levels.

Several other phenomena are noted in patients with phaeochromocytoma. They often exhibit mild hyperglycaemia (usually less than 8.5 mmol or 150 mg/dl) secondary to catecholamine-induced glycogenolysis. Psychiatric problems are also observed in many cases.

Pretreatment regimens

A multiplicity of therapeutic manoeuvres have been suggested to prepare the patient for surgery. Perhaps the most commonly utilized is treatment with the α-adrenergic blocking drug phenoxybenzamine. It should be noted that this drug has a long half-life (48–72 hours) and provides blockade of both α_1- and α_2-receptors in a non-competitive fashion. The drug is given to adults as 10 mg tablets three or four times a day initially. The following observations are made to determine final dosage:

1. Blood pressure is taken daily during therapy until orthostatic hypotension develops. The drug dosage is then maintained or decreased 10 mg/day depending on severity of the hypotension.
2. Serial haematocrits (packed cell volume). As re-expansion of plasma volume occurs, the haematocrit will decrease with a reduction in the threat of hypotension following tumour excision.
3. Fasting blood glucose falls.

A fall in haematocrit, an increase in general well-being and a reduction of symptoms and ablation of paroxysms all signal that the patient is prepared for surgery. β-Adrenergic blockers should be employed cautiously for persistent resting tachycardia (<100/min) and persistent ventricular arrhythmias present after adequate phenoxybenzamine treatment. β-Blocker drugs are actually dangerous in patients not receiving prior α-blockade because of amplified hypertension secondary to unopposed α-adrenergic agonism. Labetalol is not recommended for this very reason because its α-blocking activity is much weaker than its β-blocking activity and can precipitate a hypertensive crisis.[26] α-Methyl-*m*-tyrosine, prazosin and phentolamine have all been suggested for pretreatment but none has demonstrated marked superiority over phenoxybenzamine. Prazosin is a selective α_1-blocking agent and has been demonstrated to be effective. There are some concerns that its more selective blockade may not be adequate in patients with significant hypertension. α-Methyltyrosine inhibits the enzyme tyrosine hydroxylase in catecholamine synthesis. It is currently used for non-surgical cases and metastatic therapy.

Suggested preoperative laboratory investigations are listed in Table 49.1.

Diagnosis

Diagnosis of phaeochromocytoma is based on the clinical findings and urinary and plasma catecholamine values. Localization of the tumour is usually by CT scan. Table 49.2 gives the normal urinary excretion of catecholamines.

Table 49.1 Suggested preoperative laboratory examinations for phaeochromocytoma

1. Serial haematocrits
2. Urinalysis
3. Creatinine clearance
4. Urinary catecholamines
5. Electrolytes including calcium
6. ECG
7. Chest x-ray
8. CT scan
9. Hepatic function studies

Table 49.3 Suggested monitoring during surgery for phaeochromocytoma

1. Invasive arterial blood pressure, generally a radial arterial cathether
2. Back-up sphygmomanometer
3. Urinary catheter
4. Electrocardiogram
5. Central venous pressure catheter
6. Swan–Ganz (pulmonary artery) catheter in selected cases
7. Neuromuscular blockade monitor
8. Temperature monitor

Table 49.2 Normal 24-hour urinary excretion of catecholamines and catecholamine metabolites

Noradrenaline (norepinephrine)	10–70 µg
Adrenaline (epinephrine)	0–20 µg
Metanephrine	1.3 mg
Vanillylmandelic acid	1.8–9.0 mg

Vanillylmandelic acid (VMA) is the final product of catecholamine metabolism. Urinary metanephrines, including the o-methylated derivatives of both noradrenaline (norepinephrine) and adrenaline (epinephrine), are very accurate for diagnosis. Metanephrine elevation is more sensitive than VMA, which can give 15 per cent false positives. Urinary dopamine and the dopamine metabolite homovanillic acid may be elevated in the presence of malignant phaeochromocytoma.[27] Plasma catecholamines are also useful for diagnosis but can be falsely elevated by discomfort, diet, body position and radiographic contrast media. Localization is effected by CT scan or the newer technique of magnetic resonance imaging (MRI). An ^{131}I–m-iodobenzylguanidine (IMIBG) nuclear medicine scan has also been employed in some centres for diagnosis or localization of extra-adrenal tumours. It uses a radioactive guanidine analogue concentrated in catecholamine storage sites.

Anaesthetic management of phaeochromocytoma

Ideally, the patient should be well controlled prior to surgery, with significant amelioration of symptoms. The authors prefer heavy premedication primarily with morphine and benzodiazepines. Invasive monitoring is useful in these patients and is outlined in Table 49.3. An intra-arterial catheter for following instantaneously blood pressure changes is generally required. Use of the pulmonary artery catheter depends a great deal on the preoperative assessment of the patient; if there is evidence

of catecholamine cardiomyopathy, the reasons for use are obvious. Young individuals and others with no evidence of heart disease can be safely and successfully anaesthetized without employing this catheter. A central venous line is advocated for all cases because fluid and blood administration and fluid shifts may be great. Two large-bore peripheral intravenous lines are essential.

The selection of general anaesthesia techniques may be relatively liberal. The authors' preference is to use thiopentone for induction followed by an isoflurane-nitrous oxide sequence with vecuronium for neuromuscular blockade. The newer inhalation anaesthetics desflurane and sevoflurane should be satisfactory because neither causes significant 'sensitization of the myocardium' to circulating catecholamines. Most phaeochromocytomas are found in the adrenal glands, and rather profound myoneural blockade is required for this difficult retroperitoneal surgical exposure. Although a narcotic such as fentanyl is often quite useful to slow the cardiac rate produced by isoflurane, primary narcotic techniques do not produce sufficient degrees of cardiovascular depression to ensure a smooth anaesthetic. Hypertensive responses produced by surgical manipulations of the tumour are treated either by increases in the inspired concentrations of the potent volatile anaesthetic or by use of an intravenous vasodilator drug. Choice of vasodilator is at the preference of the anaesthetist. The authors employ phentolamine mixed as 10 mg phentolamine/100 ml dextrose 5 per cent. The action of this specific α_1-antagonist is brisk and of short duration and is thus quite useful. Sodium nitroprusside is an excellent alternative and may be a drug with which many anaesthetists have greater familiarity.

There have been advocates for the use of regional anaesthesia (spinal or epidural analgesia) for phaeochromocytoma removal. The practical difficulties of these techniques frequently outweigh any utility. It must be remembered the pathophysiology of phaeochromocytoma is such that catecholamines are liberated at a distance from the α- and β-receptors which are eventually activated. Spinal and epidural blockade produces sympathetic inhibition at the preganglionic level of the autonomic nervous system, not at the receptor sites. This implies that no receptor alteration occurs with regional analgesia. Another problem is that surgical exploration for phaeochromocytoma requires extensive subdiaphragmatic dissection. If regional analgesia techniques are used, a very high somatic level of block must be maintained for hours. The insertion of an epidural catheter for postoperative pain control, on the other hand, has much to recommend it.

When the offending tumour is removed, profound hypotension may occur. Prior to this event the anaesthetist has been concerned with treating hypertension and now must totally 'reverse gears'. The problem of hypotension is twofold. First, the patient's α-receptors are still blocked owing to the prolonged action of drugs such as phenoxybenzamine. In addition, during the hypertensive phase of the surgical procedure, hypovolaemia due to blood loss and/or extensive third space fluid shifts may have developed which were completely disguised by the high circulating levels of catecholamine produced by manipulation of the tumour. Treatment is usually with a combination of saline-containing fluids and blood and blood products. During this sequence the central venous pressure measurements can be very informative. Urinary output should be maintained at 50 ml per hour. If necessary, small amounts of a sympathomimetic amine such as phenylephrine can be employed *pro tempore* until such time as volume resuscitation is complete.

Hyperparathyroidism

In general, primary hyperparathyroidism due to parathyroid adenoma or hypertrophy is a relatively benign disease. Primary hyperparathyroidism may occur as a component of the syndrome of multiple endocrine neoplasia. Hypercalcaemia (>10.5 mg/dl) found in routine laboratory analysis is the usual diagnostic clue. Long-standing hypercalcaemia can be quite destructive, with involvement of the cardiac, renal, skeletal and neuromuscular systems. Routine preparation for the relatively asymptomatic patient scheduled for parathyroid surgery should include a cardiac assessment, including an electrocardiogram. QT and PR intervals are often shortened with hypercalcaemia. Renal function should also be investigated. Calcium, phosphorus and other blood electrolytes should be analysed. Generally, anaesthetic management of these cases is not particularly dramatic. At times, a rather prolonged exploration of the neck and mediastinum is required. The potential for postoperative pneumothorax may be quite high.

Hypercalcaemia (calcium level >14 mg/dl) is not benign, and constitutes a medical and surgical emergency. Hypercalcaemia occurs when renal excretory capacity is overwhelmed by increased calcium release into the circulation.[28] Symptoms of hypercalcaemia are quite nonspecific and include polydipsia, nausea, muscle weakness and malaise. Primary hyperparathyroidism, a condition that may require surgery, accounts for 50 per cent of cases of hypercalcaemia.[29] Severe hypercalcaemia is a life-theatening circumstance and requires rapid intervention. Generally speaking, primary parathyroidism is eventually a surgical disease and must be treated by surgery. However, hypercalcaemia and associated dehydration must be corrected as much as possible before the patient arrives in the operating theatre. Therapy consists of administration of copious quantities of isotonic saline solution (not Ringer's lactate because it contains calcium). Diuretics such as frusemide are also in order (to increase urinary calcium excretion); however, care must be taken that dehydration does not occur. Hypokalaemia can occur[30] but potassium therapy must be based on adequate knowledge of renal function. The basic features of treatment needed to prepare the patient for surgery are replacement of extracellular volume with feedback from central venous pressure measurements to maintain sodium and calcium diuresis without provoking hypokalaemia. As most conditions of severe hypercalcaemia are associated with enhanced bone resorption, the anaesthetist may observe agents being employed in these patients that inhibit bone resorption. Such drugs include the bisphosphonates such as clondronate and pamidrasate, which are potent hypocalcaemic drugs.[31] Calcitonin is also effective in some cases of hypercalcaemia but is rather slow in action. It may be particularly useful in patients with impaired renal function. Calcitonin is given in doses of 4–8 iu/kg subcutaneously. Mithramycin (plicamycin) is now considered unsuitable because of its cellular toxicity, and has been replaced by bisphosphonates. During the hypercalcaemic period cardiac arrhythmias may occur and so the ECG should be monitored.

Emergency parathyroid surgery for severe life-threatening hypercalcaemia is rare but dramatic. Successful therapy begins by reduction of plasma calcium levels to below 12 mg/dl prior to surgery by the methods described above. There are no outcome or other studies indicating that one anaesthetic is better than another. The authors found that a nitrous oxide–relaxant sequence, enabling moderate degrees of hyperventilation to be used, can dramatically lower serum ionized calcium.

Carcinoid syndrome

The anaesthetist may, on rare occasions, be called upon to anaesthetize a patient with carcinoid syndrome. Not all carcinoid tumours are functional and not all functional carcinoid tumours produce the carcinoid syndrome. The carcinoid syndrome consists of flushing, diarrhoea, bronchospasm and endocardial fibroelastosis. The syndrome occurs in 6–18 per cent of patients with carcinoid tumours.[32] Patients with the carcinoid syndrome invariably have hepatic metastasis from primarily midgut origins and may require debulking procedures.

Heart disease in these cases is caused by fibrosis of the endocardium; it involves the right side of the heart and the tricuspid valve, with eventual symptoms of right-sided cardiac failure. The bronchospasm may be severe and can be triggered by anaesthesia. The mediators involved have not been clearly identified although serotonin (5-HT), kallikrein, histamine, substance P, prostaglandin, dopamine and neurotensin have all been indicated.[33] Diagnosis of carcinoid syndrome can be established by measuring an increased urinary excretion of 5-hydroxyindole acetic acid (5-HIAA), a serotonin metabolite. Patients generally present for anaesthesia and surgery for resection of the primary tumour or isolated hepatic metastasis. They may also present for tricuspid or pulmonary valve repair secondary to endocardial fibrosis.

Intraoperative problems primarily revolve around hypotension, hypertension, bronchospasm, hypoglycaemia (the 'carcinoid crisis') and manifestations of cardiac disease. A variety of drugs have been employed in the preoperative phase to blunt the carcinoid crisis. These drugs include steroids, aprotinin (an inhibitor of the kallikrein cascade), H_1- and H_2-blocking antihistamines, and the serotonin receptor blocking drugs methysergide, cyproheptadine and ketanserin.[34] Somatostatin is a cyclic tetrapeptide that impairs cellular release of various vasoactive substances, and has been advocated as an excellent drug for intravenous use to combat side effects during surgery of patients with carcinoid syndromes. Somatostatin has a short half-life and must be given via continuous infusion. Patients undergoing surgery may be treated with octreotide (a somatostatin analogue) at a dose of 200–250 µg subcutaneously every 6–8 hours for 48 hours before the procedure. In addition, for the treatment of intraoperative carcinoid crisis octreotide can be given intravenously 100–500 µg as a bolus followed by continuous infusion.[35] It should be noted that catecholamines can cause serotonin release and are generally contraindicated in these patients. Hypotension has been treated with angiotensin infusion. Carcinoid syndrome victims may actually have hypovolaemia, and adequacy of preoperative hydration should be evaluated. No single anaesthetic sequence is judged better than any other; however, avoidance of histamine-releasing agents is advisable.

Hyperthyroidism

Hyperthyroidism or Graves' disease is usually caused by a multinodular diffuse enlargement of the thyroid gland and presents with weight loss, muscle weakness, nervousness, heat intolerance, tachycardia, heart failure and cardiac arrhythmias. In previous times the only effective therapy was surgical excision of the thyroid gland. At present, therapy is generally medical by the use of oral ^{131}I. There has been some recent resurgence in the popularity of surgical treatment. Prime candidates for surgery are uncooperative individuals, pregnant patients and younger persons. Hyperthyroid patients scheduled for surgery can be rendered medically euthyroid, signalled by the absence of tremor and a resting pulse rate of <80 beats per minute. Propylthiouracil or methimazole is given preoperatively to decrease synthesis of thyroid hormones. A β-adrenergic blocker such as propranol is added to mask symptoms and enhance well-being. It should be remembered that thyroid storm can develop despite adequate β-adrenergic blockade.[35] Iodine as oral sodium iodide is added to therapy just prior to surgery, primarily to decrease vascularity of the thyroid gland. The anaesthetist should note that the use of this halogen produces a 'window' which is 'open' about 3–5 days after initiation of iodine therapy and 'closes' after 7–8 days. At this time the gland escapes and actually increases vascularity. Surgery should not, therefore, be cancelled for a minor issue during the 'open' period.

Problems for the anaesthetist relate to the size of the gland and difficulty with the airway. Tracheomalacia may occur with large glands associated with tracheal deviation. Thus fibreoptic or other manoeuvres may be required for tracheal tube placement. Repeat endoscopy may be necessary after surgery to identify tracheal collapse, which may require a tracheal tube to be inserted or tracheostomy performed.[36] Postoperative problems also result from laryngeal nerve injury, particularly if a total thyroidectomy has been performed. Unilateral injury to the recurrent

Table 49.4 Treatment of thyroid storm

1. Ensure oxygenation ($FIO_2 = 1.0$)
2. Cool the patient with iced intravenous solutions, cooling blankets, icepacks
3. Establish an indwelling arterial line to follow metabolic derangements and aid in diagnosis
4. Steroids: 100–300 mg hydrocortisone i.v.
5. Propranolol i.v.: as needed during careful monitoring (calcium channel blockers are also acceptable)
6. Propylthiouracil: orally as soon as possible
7. Correct dehydration and electrolyte abnormalities

laryngeal nerves is generally well tolerated; however, bilateral injury may produce significant tracheal obstruction, with the requirement for reintubuation. Pneumothorax is rare but can occur during thyroid surgery. Hypocalcaemia due to parathyroid gland damage or removal may occur as early as 3–4 hours postoperatively.

Thyroid storm is a life-threatening emergency which can occur in known hyperthyroid patients or in undiagnosed hyperthyroid individuals who present for incidental surgery. Thyroid storm results from decompensation of a hyperthyroid state and has four features:[37]

1. Increase in body temperature to as high as 41°C.
2. Hallucinations (if awake).
3. Sinus tachycardia to 140 per minute, with arrhythmias such as atrial fibrillation.
4. Caused by a precipitating factor such as surgery.

Although thyroid storm can occur during anaesthesia, it also occurs in the postoperative period. Clinically, thyroid storm is more common in patients with Graves' disease than in those with toxic multinodular goitre or toxic adenoma. The anaesthetist may be confronted with an important differential diagnosis between thyroid storm and the better publicized hyperpyrexic state of malignant hyperpyrexia (MH). The pathophysiology and treatment of the two processes are totally different, but both are immediately life threatening and require urgent and appropriate therapy. Although serum T_4 assay and T_3 resin uptake tests are highly accurate for the diagnosis of thyroid storm, they are very slow analyses and therapy cannot wait for the results. Although hyperpyrexia is a common feature, diagnosis can usually be distinguished by:

1. Appearance: thyroid storm is common in young adults, particularly those with Graves' disease (exophthalmos) and an enlarged thyroid.
2. Metabolic acidosis and carbon dioxide levels are higher in MH.
3. Muscle destruction is greater in MH, as noted by greater rises in creatinine phosphokinase (CPK).

When the diagnosis of thyroid storm is made, therapy as noted in Table 49.4 should begin.[38]

Congestive heart failure can result and must be treated in conventional fashion. In the past, the mortality rate for thyroid storm approached 50 per cent[39] but is lower today.

Hyperthyroid patients are often quite resistant to various drugs.

Hypothyroidism

Patients with severe myxoedema who require surgical procedures are rare. Myxoedema coma is characterized by hypothermia, slow mentation progressing to coma and high serum creatinine phosphokinase (CPK) levels (>500 units). The patient is invariably myxoedematous and presents with macroglossia. Low voltage ECG and bradycardia are seen. Thyroid-stimulating hormone (TSH) levels are elevated and T_4 levels depressed.

If these patients in the untreated state require urgent or emergency surgery, mortality can be high.[40] Treatment before surgery includes large doses of T_4 (500 mg i.v.) or T_3 (40 µg i.v.), which are required to saturate thyroid-binding globulin (TBG) before free hormone is available for response. If T_3 is unavailable in an intravenous form, it may be obtained direct from the manufacturer for use in the treatment of myxoedema coma. Maintenance therapy for T_4 is 50 µg intravenously per 24 hours and for T_3 10–20 µg per 24 hours. Cortisone therapy and rewarming are used. Constant pulse oximetry and ECG monitoring must be used during this therapy; cardiac arrhythmias and myocardial infarction can and do occur. When such patients are brought to surgery it is well recognized that they show increased sensitivity to narcotics, anaesthetics and tranquillizers.[41]

Summary

This chapter has described features of several of the more important endocrinopathies of concern to the anaesthetist; it is not meant to be comprehensive. Although modes of therapy are changing, knowledge of basic pathophysiology of the endocrinopathies is important for safe anaesthetic management.

REFERENCES

1. Galloway JA, Shuman CR. Diabetes and surgery: a study of 667 cases. *American Journal of Medicine* 1963; **34**: 177–91.
2. MacKenzie CR, Charlson ME. Assessment of perioperative risk in the patient with diabetes mellitus. *Surgery, Gynecology and Obstetrics* 1988; **167**: 293–9.
3. Hjortrup A, Rasmussen BF, Kehlet H. Morbidity in diabetic

and non-diabetic patients after major vascular surgery. *British Medical Journal* 1983; **287**: 1107–14.

4. Walsh DB, Eckhauser FE, Ramsburgh SR, *et al.* Risk associated with diabetes mellitus in patients undergoing gallbladder surgery. *Surgery* 1982; **91**: 254–9.

5. Burgos LG, Ebert TJ, Asiddao L, *et al.* Increased intraoperative cardiovascular morbidity in diabetics with autonomic neuropathy. *Anesthesiology* 1989; **70**: 591–7.

6. Pirart J. Diabetes mellitus and its degenerative complications: a prospective study of 4,400 patients observed between 1947 and 1973. *Diabetes Care* 1978; **1**: 168–88.

7. Burke CM, O'Doherty A, Flanagan A, *et al.* Autonomic neuropathy in a diabetic clinic. *Irish Medical Journal* 1981; **77**: 202–5.

8. Milaskieuricz RM, Hall GM. Diabetes and anaesthesia: the past decade. *British Journal of Anaesthesia* 1992; **68**: 198–206.

9. Walker JD. Implication of microalbuminuria in diabetes. *British Journal of Hospital Medicine* 1991; **45**: 41–4.

10. Nolan CM, Beatty HN, Bagdade JD. Further characterization of the impaired bactericidal function of granulocytes in patients with poorly controlled diabetes. *Diabetes* 1978; **27**: 889–94.

11. McMurray JR. Wound healing with diabetes mellitus. Better glucose control for better wound healing in diabetes. *Surgical Clinics of North America* 1984; **64**: 769–84.

12. Hogan K, Rusy D, Springman SR. Difficult laryngoscopy and diabetes mellitus. *Anesthesia and Analgesia* 1988; **67**: 1162–5.

13. Gavin LA. Management of diabetes during surgery. *Western Journal of Medicine* 1989; **151**: 525–9.

14. Campbell PR, Hoar CS, Wheelock FC. Carotid artery surgery in diabetic patients. *Archives of Surgery* 1984; **119**: 1404–10.

15. Walts LF, Miller J, Davidson MD, *et al.* Perioperative management of diabetes mellitus. *Anesthesiology* 1981; **55**: 104–9.

16. Albert KG, Thomas DJ. The management of diabetes during surgery. *British Journal of Anaesthesia* 1979; **51**: 693–9.

17. Hirsh IB, McGill JB, Cryer PE, White PF. Perioperative management of surgical patients with diabetes mellitus. *Anesthesiology* 1991; **74**: 359–64.

18. Raucoules M, Ichai C, Lugrine D, Jambore P, Grimand P. Comparison of two methods of intravenous insulin administration in the diabetic patient during the perioperative period. *Anesthesiology* 1989; **71**: suppl, A328.

19. Hall GM. The anaesthetic modification of the endocrine and metabolic response to surgery. *Annals of the Royal College of Surgeons of England* 1985; **67**: 25–9.

20. Kehlet H. The modifying effect of general and regional anesthesia on the endocrine–metabolic responses to surgery. *Regional Anesthesia* 1982; **7**: 538–48.

21. Hall GM, Young C, Holdcraft A, Alaghband-Zadeh J. Substrate mobilization during surgery: a comparison between halothane and fentanyl anesthesia. *Anesthesia* 1978; **33**: 924–30.

22. Lacoumanta S, Paterson JL, Burrin J, Causon RC, Brown MJ, Hall GM. Effects of two differing halothane concentrations on the metabolic and endocrine responses to surgery. *British Journal of Anaesthesia* 1986; **58**: 844–50.

23. Bergen W, Keller U. Treatment of diabetic ketoacidosis and non-ketotic hyperosmolar diabetic coma. *Baillieres Clinical Endocrinology and Metabolism* 1992; **6**: 1–22.

24. Krane NK. Clinically unsuspected phaeochromocytomas. *Archives of Internal Medicine* 1986; **146**: 54–7.

25. Samaan NA, Hicky RC. Phaeochromocytoma. *Seminars in Oncology* 1987; **14**: 297–305.

26. Sever PS, Roberts JC, Small ME. Phaeochromocytoma. *Clinics in Endocrinology and Metabolism* 1980; **9**: 543–68.

27. Tippett PA, McEwan AJ, Achery DM. A re-evaluation of dopamine excretion in phaeochromocytoma. *Clinical Endocrinology* 1986; **25**: 401–10.

28. Bonjour JP, Rizzoli R. Pathophysiological aspects and therapeutic approaches of tumoral osteolysis and hypercalcaemia. *Recent Results in Cancer Research* 1989; **116**: 30–9.

29. Fisken RA, Heath DA, Somers S, *et al.* Hypercalcaemia in hospital patients. *Lancet* 1981; **1**: 202–7.

30. Aldinger KA, Samaan NA. Hypokalaemia with hypercalcaemia, prevalence and significance in treatment. *Internal Medicine* 1977; **87**: 571–3.

31. Sleebom HP, Bijvoet OLM, Van Oosterom KB, *et al.* Comparison of intravenous (3-amino-1-hydroxypropylidine)-1,1-bisphosphonate and volume repletion in tumour induced hypercalcaemia. *Lancet* 1983; **2**: 239–43.

32. Feldman JM. Carcinoid tumors and syndrome. *Seminars in Oncology* 1987; **14**: 237–46.

33. Conlon JM, Deacon CF, Richter G, *et al.* Circulating tachykinins (substance P, neurochronin A, neuropeptide A) and the carcinoid flush. *Scandinavian Journal of Gastroenterology* 1987; **22**: 95–105.

34. Longnecker DM, Roizen MF. Patients with carcinoid syndrome. *Anesthesiology Clinics of North America* 1987; **5**: 313–37.

35. Roy RC, Carter RF, Wright PD. Somatostatin, anesthesia and the carcinoid syndrome. *Anesthesiology* 1987; **42**: 627–32.

35. Strube H. Thyroid storm during beta-blockade. *Anaesthesia* 1984; **39**: 343–6.

36. Stehling LC, Roizen MF. Endocrine surgery. In: Nunn JF, Utting JE, Brown Jr BR, eds. *General anaesthesia,* 5th ed. London: Butterworths, 1989: 880–6.

37. Burger AG, Philippe J. Thyroid emergencies. *Baillieres Clinical Endocrinology and Metabolism* 1992; **6**: 77–93.

38. Stehling L. Anesthetic management of the patient with hyperthyroidism. *Anesthesiology* 1974; **41**: 585–95.

39. Mackin JF, Canary JJ, Pittman CS. Thyroid storm and its management. *New England Journal of Medicine* 1974; **291**: 1396–8.

40. Lindberger K. Myxoedema coma. *Acta Medica Scandinavica* 1975; **198**: 87–90.

41. Levelle JP, Jopling MW, Sklar H. Perioperative hypothyroidism: an unusual post anesthetic diagnosis. *Anesthesiology* 1985; **63**: 195–7.

Anaesthesia for Trauma and Emergencies

Cedric Bainton

Trauma is the largest killer of people between the ages of 1 and 36 years in the USA. It is the third leading cause of death overall and accounts for more years lost than heart disease and cancer combined. It is expensive – it is estimated to cost US$150 billion each year.[1] It is not surprising that the issue has attracted attention – particularly with regard to reducing the number of accidents, and devising ways to minimize mortality and morbidity.

Military experience has shown that the mortality and mobidity of soldiers decreased from World War I to World War II to the Korean war and finally to the Vietnam conflict as support staff were able to shorten the time for transport of the injured from the site of injury to the field hospital.[2] Thus speed of delivery from the street to the operating theatre has become the cornerstone in the management of traumatic injuries.

The problem is essentially one of efficient time management in which a large number of essential tasks must be completed quickly. Diagnostic and therapeutic priorities must be assigned and updated frequently. Trauma systems organized to compress the time for transport, evaluation and resuscitation, with operating rooms standing by, have demonstated the wisdom of this concept. These centres save a high proportion of patients with traumatic injuries who would otherwise die in conventional hospitals.[3]

There are several unique features to this concept.

1. There must be an organized system to collect these patients with well defined triage criteria to select those at risk from serious injury.

2. All resource groups must work in harmony to achieve this goal. Thus the concept of a team which involves paramedics or emergency medical technicians (EMTs), hospital administration, nursing, surgery, anaesthesia, emergency medicine, radiology, laboratories, blood banking, etc., becomes clear.

3. A set pattern must be in place to control examination of the patient and to facilitate obtaining historical facts. There will be little time for deriving first principles in the heat of the moment, so protocols and patterns of thinking must be established for evaluation of the patient and consideration of all possible injuries. Patterns of behaviour in response to recurring problems must be thought out thoroughly ahead of time. However, such protocols for nursing, surgery and anaesthesia must be the same where functions overlap. They must not be derived in isolation but rather with full participation, such that groups will not work at cross-purposes.

4. The work will be divided along traditional lines among nursing, surgery, emergency room (ER) medicine, and anaesthesia, but ideally there will be an overlap of capabilities to fill essential functions when and if the need should arise (e.g. surgeons and ER physicians capable of intubating the trachea; anaesthestist, ER physicians and perhaps even nurses capable of performing a cricothyrotomy).

5. Obtaining the history and physical examination for the anaesthetist will be fragmented by time constraints. The anaesthestist will not have the luxury of obtaining these data independently as he/she would for elective cases. Rather, the physical examination may be a composite of the anaesthetist's own observation, watching others perform the examination, perusing the chart, and ultimately personally filling in details which appear missing. Clearly, in addition to chart review, the history will be drawn from conversations with surgeons, observers and family to determine essential details where they are missing. What is absolutely essential is that the anaesthetist comes to an independent

conclusion based on available information. This will then generate a differential of obvious as well as occult problems to be discussed at length with responsible surgeons such that plans for therapy and after care do not omit important detail or work at cross-purposes.

These discussions not only add to the consistency of care but offer an opportunity to incorporate anaesthetic concerns into the management plan – consistent with the concept of a 'well oiled' team approach. Blunt trauma problems, in particular, tend to evolve with time. Any early thoughts about the epidemiology of the accident as well as essential physical and diagnostic findings, which enable those organs at risk to be identified, can reduce the element of surprise in the subsequent anaesthetic and postoperative course. One must then keep in mind the possible injuries which may declare themselves and not relax one's guard until all the injuries have been identified.

6. Emergency surgery becomes a 'no frills' approach to diagnostic detail and therapy for underlying problems. There is no time to: optimize drug therapy, obtain complete cardiac and pulmonary evaluations, wait for alcohol and recreational drugs to dissipate, let stomachs empty of food content, let an abused cirrhotic liver 'cool off', etc. Rather, risks are acknowledged and then dealt with by surgeon and anaesthetist.

'Old-fashioned' diagnosis based on physical examination is very much a part of the 'no frills' approach. The precision of CT scans cannot be denied but nothing matches the quick accumulation of information derived from what one can see, hear and feel.

7. The operating suite will need to be organized such that a room is always available for the immediate or STAT emergency. This may appear to be a waste of resources when the room is idle but there is no other way to accept a STAT patient unless this room is always ready. A corollary to this is that nursing and anaesthetic staff must have the resources to staff this room at all times.

8. All staff, including surgeons, must adjust to a fluid schedule. In elective surgery the schedule can be well ordered with reasonable predictions as to when cases can be expected to start and be completed. In emergency surgery the schedule is fluid. Surgeons must defer other responsibilities to take care of emergencies or delay elective cases while they wait for other surgeons to complete cases of greater priority. This takes a willingness to accommodate and a change in attitude when switching from the elective to the emergency mode.

9. Equipment must be simple and supply lines streamlined. It will not be possible to have an infinite variety of complex tools and supplies available. Uniformity will ensure familiarity and give greater assurance that items are always working and understood by those in charge of their maintenance. Well organized supply rooms and carts with a constant location for all items eliminates the need to search for these when the situation is moving fast. Maintenance must be well organized so that equipment is working at all times.

10. Plans for disasters must be designed with mechanisms for extra staffing to accommodate the increased case load.

These concepts are a challenge because they force all parties to cooperate to plan, educate, and discuss so that time is compressed and the patient is delivered to the operating room safely, intelligently, and efficiently to improve the chance of a successful outcome. When the system works it also produces a unique camaraderie.

It also means that each group must take an interest in administrative decisions required to make the system work more smoothly, and participate actively in the committee process for medical staff, operating room, trauma and hospital quality assurance.

A service of this type is expensive. It is analogous to a fire brigade standing by, ready to go into action at a moment's notice. It must have financial support such that it is available 24 hours a day. This is a community resource, it is necessary therefore to convince the public that it is both sensible and cost-effective to invest in this concept for the benefit of individuals and the community at large. Thus, no member of the *trauma team* can be excluded from the political process which must be appealed to for financial support.

To review, the unique feature of care for the injured is compression of time with speedy delivery of the patient from the street to the operating theatre. This begins with triage in the field and transport to a hospital with trauma capabilities. Once the patient is received in the emergency room, time will be compressed for evaluation and resuscitation without sacrificing completeness. This phase will be managed by a team, with its principal players being anaesthetists, surgeons, nurses and emergency room physicians. Each team member will have specific responsibilities. More than one activity may be carried out simultaneously. All team members will have a clear understanding of the role of others so that therapeutic options can be anticipated and quickly implemented. Surgeons may spearhead the evaluation process although this will be determined by local practice. Interactions of team members assure review of patient status, priorities, plans and timing.

Pre-hospital trauma care

Pre-hospital trauma care focuses on the safe extraction of injured patients from the accident site, careful stabilization of spine and extremity fractures, securing an adequate airway and management of obvious bleeding sites with local pressure. Exsanguinating haemorrhage is best managed by rapid transport to hospital facilities. If the distance is short, some advise the 'scoop and run' approach with an attempt made to start an intravenous infusion in the moving ambulance. If the distance is greater, or transportation time prolonged, an intravenous infusion will be essential. MAST suits can effectively stabilize pelvic fractures and tamponade retroperitoneal bleeding.[4]

EMTs may be the principal providers at the accident scene. It is essential that they be skilled in resuscitation, airway management and intubation. Anaesthetists may have varying degrees of involvement in this pre-hospital phase. They may participate at the scene, help design programmes for EMT services, or teach the skills described. It is important that anaesthetists remain interested and involved in pre-hospital care since it is the crucial first step in the successful resuscitation of the injured patient.

Predictions of injury based on mechanisms

Any trauma care system with multiple demands placed upon it must be able to gather its resources and direct them where they are needed most. All traumatized patients have life-threatening injury until proved otherwise. Thus a prediction about the severity of injury is an important first step in this mobilization. Traumatic injuries can be broadly considered as penetrating or blunt.

The severity of *penetrating injury* is more quickly defined because explorative surgery can follow the path of the specific trajectory. This type of injury can, of course, be lethal and in the case of shotgun blasts, particularly devastating. However, the injuries are often trivial, particularly in the case of sharp weapons, providing vascular injury has not occurred.

With *blunt injuries* the manifestations can be abrupt or delayed. Multiple organs are usually involved. Diagnosis is more subtle because injury is hidden in the thorax, abdomen and pelvis. Attempts to locate the injury are based on epidemiology, physical examination and x-rays. The patient with blunt trauma can appear well compensated

Table 50.1 Triage criteria

- Ejection from automobile
- Death in same passenger compartment
- Rollover
- Crash speed 20 miles/h (32 km/h) or more
- 30 inch (76 cm) deformity of automobile
- Rearward displacement of front axle
- Passenger compartment intrusion: 18 inches (46 cm) patient side, 24 inches (61 cm) on opposite side of car
- Falls 20 feet (6 m) or more
- Auto vs pedestrian injury >5 miles/h (8 km/h) impact
- Motorcycle crash >20 miles/h (32 km/h) or with separation of rider and bike
- Small children and older people can have injuries with lesser impact or simply have less reserve to compensate for injuries

for a very long time and then deteriorate quite insidiously. Physicians cannot lower their guard but must make frequent observations for deterioration – pulmonary, CNS, or continued bleeding which may be due to coagulation defects.

Triage criteria are useful because in each case they define a magnitude of force, capable of inflicting serious injury. The list in Table 50.1 is an example of such criteria. Such lists tend to vary with local preference.

Examination of the motor vehicle can reveal some clues. Indentations in the dashboard are caused by impact of the knees coming 'down and under' which may lead to dislocation of the knees, fractures of ankles and femur and posterior dislocation of the femur from the acetabulum. 'Spider web windshield' implies head injury until proved otherwise and results from an 'up and over' trajectory of the body. Cervical neck (C-spine) fractures are common and are caused by compression, hyperflexion and/or hyperextension. Obese victims are more likely to have rib, pelvic and extremity fractures as well as pulmonary contusion, but curiously less head and liver injury.[5] It is as if the obese abdomen acts as a built-in airbag.

Broken or bent steering wheels raise suspicion of rib fractures, contusion of thoracic and abdominal organs, and rupture of any hollow viscus. Seatbelts can cause intestinal and lumbar spine injury.[6] Those which leave abdominal or blank wall ecchymosis are particularly suspect.

Rear impact subjects the victim to hyperextension and hyperflexion of the cervical cord. Side impact can fracture clavicle, ribs, head of the femur, which in turn can drive through the acetabulum, resulting in pelvic fracture and retroperitoneal bleeding in the psoas and iliac muscles.

Free-fall injuries will be affected by the height; whether the patient hits head, feet, side or buttocks first; the nature of the surface, i.e. cement, mud or water; and whether the victim is an adult or a child.[7]

All tethered organs are at risk of being torn from vascular attachments in free fall: the lung, heart, aorta, liver, spleen, kidney and sometimes bowel. Spinal fractures are very common. Long bones and pelvis are at risk. Joint articulations will be injured in feet first falls but in turn can absorb considerable energy and thereby lessen the injury to the spine. Children in free fall may be more resilient because of greater cartilage and unossified bone.

Patients ejected from vehicles are propelled horizontally and subjected to similar injuries.

Direct blows to the abdomen are like hitting beef steak with a hammer, fracturing liver, spleen, kidney and pancreas. Saccular organs can burst: gallbladder, urinary bladder and bowel. The diaphragm can rupture at its weak posterior attachment.

It takes considerable force to imprint the pattern of clothing on the body. When this is seen, major injuries must be suspected.

Bleeding is a major issue in all traumatic injuries. External bleeding can be seen and dealt with expeditiously. Occult bleeding can only be suspected; it occurs with the fracture of solid organs and avulsion of the vascular pedicle of tethered organs or the aorta. Fractures of the transverse process of lumber vertebrae can bleed retroperitoneally into the iliac and psoas muscles. Three or more units of blood can surround long bone fractures. Pelvic fractures can easily sequester six or more units. Facial fractures, LeFort II and III, can bleed 6–8 units but the swallowed blood goes unnoticed. The scalp is very vascular and unsuspected scalp lacerations can bleed profusely while hidden under the sheets or drapes. In a child, this can cause significant hypotension. The thorax can hide several units of blood.

It takes considerable force to fracture the first and second ribs and scapula. Thus, vascular and aortic injuries must be suspected when these fractures occur. Contusion of heart and great vessels is associated with fractures of the sternum.

Emergency room

Assignment of tasks

Patients with life-threatening injuries will need the following tasks completed as quickly as possible:

1. airway control
2. adequate ventilation, monitored with arterial blood gas
3. general monitoring
4. i.v. access
5. fluid resuscitation and cardiac stabilization
6. blood samples for: type and screen, haematocrit

(Hct), platelets, prothrombin time (PT), partial thromboplastin time (PTT), electrolytes, osmolality
7. neurological assessment
8. bladder catheterized, Foley catheter
9. chest x-ray
10. exclude cervical spine (C-spine) fracture/dislocation
11. patient completely undressed in preparation for physical examination
12. nasogastric (NG) tube insertion

These tasks will be followed quickly by:

13. complete physical examination and history
14. decisions in regard to such items as: additional imaging (x-rays, CT scan, angiography) and pericardial tap
15. summary of the problem as to status, timing for surgery, re-evaluation

These tasks can be assigned to surgery, anaesthesia, nursing, or ER physicians depending on local arrangements. It is extremely important that one individual facilitates the process such that time is not wasted and valuable information is not lost in transfer.

In some countries anaesthetists take the major responsibility for these initial evaluations. In others, the surgeons may 'take charge', sharing responsibility with physicians and anaesthetists for special aspects of care. In my institution, anaesthetists take full responsibility for the airway and ventilatory management and the logistics and timing of delivery of patients to the operating room. Anaesthetists oversee the monitoring of patients who need sedation and paralysis for CT and angiography.

As the patient progresses to various locations these assignments will change depending upon available team members (Table 50.2). It is equally important that good communication is maintained along the way, between anaesthesia, nursing, surgery and surgical subspecialities. Note that the role of anaesthesia changes dramatically in this process. In the ER the primary responsibility will be airway and breathing, with physical examination and history left to others. In the operating theatre anaesthesia takes responsibility for all support functions. This frees the surgeons to focus on technical considerations – later responsibility is shared as the patient reaches the post-anaesthesia recovery (PAR).

This table emphasizes once again why the anaesthetist must formulate a clear and independent picture of the problem in order to focus attention on the plan and direction of therapy and formulate a differential of possible problems which might need attention.

Although the anaesthetist may have absolute responsibility for only airway management and ventilation in the emergency room, it is clear that his work will be severely compromised if any element of the essential list of tasks is omitted. Thus he/she must run his own check of these

Table 50.2 Tasks for patient evaluation and management

	In ER	In OT	In RR
Airway	A	A	A
Ventilation	A	A	A
Monitoring	N	A	N,A
Intravenous access	ER,N,S	A	N,A
Blood samples	N	A	N,A
Fluid and blood administration	ER,S,N	A	A,S
Cardiovascular support	ER,S,N,A	A	A,S
Neurological evaluation	S,A	A,S	A,S
Foley catheter	S,N	A,N,S	S,N
Chest film	S	A,S	A,S
Nasogastric tube	S	A	S,N
Cervical-spine clearance	S,A	A,S	S,A
Physical examination and history	S,A	A,S	S,A
Diagnostic work-up	S	S	S
Coagulation management		A	A,S
Arterial line placement		A	A,S
CVP placement		A	A,S
Pain management	S	A	A,S
Transport unstable patient	A	A	A
Formulate plans: define plans, timing for surgery, and post-op care	S,A	A,S	S,A

A, anaesthesia; S, surgery; N, nursing; ER, accident and emergency physicians; OT, operating theatre (room); RR, recovery room (post-anaesthetic room).

essentials with a reminder to responsible parties if omissions are found. If not, he/she will be struggling in the operating theatre to catch up. Although speed is the 'watch word' in the management of traumatic injuries, it is not speed at all cost. Not all patients will be saved and those on the brink will certainly be lost unless the essential tools for intelligent management are provided, namely the task list.

Patient evaluation

The Committee on Trauma of the American College of Surgeons in their course on advanced trauma life support (ATLS)[8] provides protocols for the completion of essential tasks both in regard to questions to be asked and evidence to be gathered. Lists must be committed to memory and applied in a reflex fashion to remove the temptation to derive solutions repetitively from first principles.

The Committee recommends that evaluation be divided into a primary survey and resuscitation phase, in which the severity of the problem is defined and resuscitation begun concurrently. When and only when these two are complete should a secondary survey begin which consists of a head to toe physical examination. A history will then be obtained and diagnostic studies undertaken as felt necessary: x-rays, peritoneal lavage, CT and contrast studies. Finally, a disposition will be determined defining the problem, the status of the patient and the timing of surgery or further evaluation.

The authors of the manual are quick to point out that prioritized assessment, listed as sequential steps, does not describe the integrated approach which actually happens. With specific tasks assigned to team members, primary survey and resuscitation will go on simultaneously. History will be obtained as the patient is able to communicate or as family members become available to provide details.

Anaesthetists may be called to the ER in anticipation of the patient's arrival, after medical care has started, or later when a supposedly stable patient turns 'sour'. Thus, the anaesthetist must thoroughly understand the ATLS protocols used and under way such that they can quickly include themselves in the evaluation process.

In this organized plan, the physician with primary responsibility, the trauma surgeon, will usually be the only one to perform a complete and independent physical examination and history review. This does not relieve the anaesthetist from formulating an independent summary of the problem, real and potential. The anaesthetist, however, standing at the head of the patient, is in a unique position to assist the surgeon with the complete physical examination, to listen for important historical clues from patient or on-site observers and to consider a preliminary survey of the problem. As the condition of the patient is clarified, the discussions with the surgeon will influence the planning of subsequent diagnostic steps to identify organs at risk and clarify the need for, or timing of, surgery.

The history and physical examination obtained by the anaesthetist will very much rely on this patchwork approach. For example, when suddenly summoned to the ER to evaluate a patient, I approach the patient from the feet. On circling the patient, making my way to the head, I observe the colour of the patient's skin, compress a toe to observe capillary refill, feel the temperature of the legs, running my hand proximally from the feet to above the knee. As I assemble the mask and oxygen delivery system, I check the neck for venous distension, palpate the carotid pulse and correlate both with reported blood pressure. Observing the head, neck and chest, I note traumatic deformities, tracheal deviation and the character of respiratory movements if present. I palpate for subcutaneous emphysema in the neck. The patient may or may not respond to my question: 'How do you feel? Does your neck hurt while lying there or if you move your head? I am going to place a mask over your face. Please take a big breath of oxygen.' If the patient does not respond, I can press on the supraorbital nerve to elicit a painful response and note any movement of extremities. I quickly check pupils for equality and response to light. If the patient does not respond, and is not breathing, I attempt mask ventilation. All this, while the EMTs, family and friends may be contributing details of medical history and a description of the accident scene.

In this 30-second inteval, I have:

1. established whether the airway is patent;
2. established whether the patient needs ventilation;
3. looked for the presence of subcutaneous emphysema indicating that the larynx or trachea may be injured;
4. considered whether the patient will be easy or difficult to intubate based on anatomical features;
5. estimated the adequacy of circulation and blood volume status based on patient skin colour, capillary refill, temperature of extremities (cold knees, mean low blood volume), quality of the pulse, and measured blood pressure;
6. considered the possibility of spinal shock indicated by warm legs and dilated leg veins in the presence of a low blood pressure and weak pulse;
7. considered the implications of distended neck veins and low blood pressure in regard to cardiac tamponade, tension pneumothorax, myocardial failure or pulmonary outflow obstruction;
8. completed a basic neurological examination; AVPU – A, alert; V, responds to voice; P, responds to pain; U, unresponsive; as well as examining for localizing eye signs, localizing extremity movement, and the suspicion of C-spine injury based on neck pain;
9. considered a list of associated problems based on medical history; and
10. begun to formulate a summary of the injuries with regard to severity, status, and timing for surgery or additional evaluation.

This is a list-oriented method of gathering information but not in an ordered sequence as described in the ATLS protocol. Rather, the list is completed by including essential information at random as it becomes available.

Rapid overview

With the assignment of specific tasks to team members, the ATLS evaluation and resuscitation protocol can be started promptly, but not without a certain flurry of activity. Markison and Trunkey[9] make the point that before this process is set in motion it is important that a single individual establish a rapid overview to answer the simple questions. Is the patient stable, unstable, dead or dying? The overview should quickly synthesize visual observations of spontaneous breathing or ventilatory support, the patient's colour, gross asymmetries, and the presence or absence of body movements. This brief glance can answer the questions posed and then set the tone, tempo, and direction of subsequent events and the team can go about its work. This is an important point because a team with duties compartmentalized for speed and efficiency may be blind to the overall direction it is taking or should take. Valuable time and resources may be wasted on the obviously moribund. Although it is important to give patients every chance for survival, it is also important to be reasonable when faced with overwhelming odds and a clinical course which is clearly going in the wrong direction.

Primary survey and resuscitation plan

The primary survey is designed to identify life-threatening conditions and begin management simultaneously. It has five components, ABCDE, listed in relative priority. These stand for: **A**irway, **B**reathing, **C**irculation, **D**isability (neurological), and **E**xposure (undress the patient in preparation for a complete physical examination). The resuscitation phase goes on simultaneously with monitoring, i.v. placement, fluid resuscitation and addresses each disability as identified in the primary survey.

Airway

The airway takes precedence over all other problems. The airway must be assessed as to patency. In the alert cooperative patient, this will be easy to establish. In the obtunded patient it will be necessary to clear the airway of tissue, blood and vomitus using a tonsil sucker to support the tongue and to attempt mask ventilation. If mask ventilation is not possible the airway must be secured by tracheal intubation, cricothyrotomy or tracheostomy. Any patient with multiple trauma has a C-spine injury until proved otherwise. The spine must be immobilized until the C-spine is cleared. If tracheal intubation is mandatory before C-spine clearance can be obtained, an assistant can stabilize the head while intubation is performed.

The anaesthetist must take full responsibility for airway assessment and intubation, seeking assistance from surgical colleagues for cricothyrotomy or tracheostomy. If surgeons are detained, the anaesthetists must be fully prepared to institute cricothyroid catheter ventilation until surgeons are free or proceed directly to cricothyrotomy if indicated. If the patient is already intubated when the anaesthetist arrives, he/she assumes full responsibility for the accurate placement of the tracheal tube.

The trachea is intubated to relieve obstruction, protect the airway from aspiration, to treat ventilatory and pulmonary inadequacy, and to protect the injured brain by lowering $P\text{CO}_2$ and elevating $P\text{O}_2$. In addition, the hypovolaemic patient with systolic blood pressure 80 mmHg will tend to improve with intubation, control of ventilation and better oxygenation.

If there is any hint that the hypovolaemic patient is deteriorating, it is best to intubate at that time rather than during transport to the operating room. The confused or agitated patient may need intubation for safe sedation in order to complete CT scans or angiography.

However, all patients cannot and need not be approached with the same intensity of effort. Priorities constantly change and should be based on the immediacy of the situation. Judging immediacy is greatly helped by experience but it can be learned by taking time to make simple observations first before plunging in. I suggest to residents, at the start of their training, as they approach the patient that they ask themselves: 'What would happen if I did nothing?' The intent is, of course, not to 'do nothing' but rather to do what is appropriate. Not every patient needs a tracheal tube immediately. Rather they need an anaesthetist who takes time to observe carefully, consider the options deliberately and then move expeditiously once the course is chosen. One possible course is to do nothing but observe, and on occasions this is entirely appropriate.

If the situation is deteriorating rapidly then intubation is the first priority regardless of C-spine, open globe or other concerns. If the C-spine damage has not been excluded then the head should be stabilized by an assistant and the anaesthetist should proceed with direct laryngoscopy. A rapid sequence approach should be used: oxygen by mask, sedation, suxamethonium (succinylcholine) for paralysis, cricoid pressure and intubation. Sedatives will not be

necessary in the comatose patient and should be used sparingly if at all when the blood pressure is marginal. Muscle relaxants will not be necessary if the patient's muscles are flaccid. The patient who cannot be intubated after three quick attempts at laryngoscopy and cannot be ventilated by mask will need an emergency cricothyrotomy. This may be preceded by placement of a cricothyroid catheter for oxygen insufflation to reduce concern for hypoxia during the cricothyrotomy.

For cricothyrotomy, most surgeons suggest a transverse incision over the cricothyroid membrane. However, if the neck is swollen and oedematous, a rostrocaudal skin incision is more likely to be successful. The incision should be made over what is judged to be the cricothyroid membrane extending the length and depth until the cricoid cartilage and lower margin of the thyroid cartilage are visualized. The cricothyroid membrane should then be identified and incised perpendicular to the skin incision (transverse) and the trachea entered. The incision is dilated with the back of the scalpel blade until a number 5 tracheal tube can be introduced. The cuff is inflated and ventilation instituted. The surgeon is the obvious person to perform this procedure but if not available, the anaesthetist should be prepared to perform it.

Cricothyroid catheter placement for oxygen insufflation or ventilation has its own hazards. The worst of these is subcutaneous emphysema, mediastinal emphysema and pneumothorax which may occur if the catheter is misplaced or dislodged from the trachea. To minimize this, a catheter 20 cm (8 inches) or longer should be used which is securely taped or sutured to the skin at the point it passes through the skin. Shorter catheters (9 cm; 3.5 inches) are often recommended but these, even if sutured to the skin, are easily dislodged from the trachea by simply pulling on the skin. The skin under tension is sufficiently elastic to displace 7.5–10 cm (3–4 inches) and dislodge the catheter tip from the tracheal lumen. Catheters can be placed either through a large needle or alternatively over a wire. In all cases it is extremely important to assure oneself that the catheter is intratracheal by aspirating tracheal air through the catheter before applying oxygen through the catheter hub. One must remember that this form of ventilation depends upon egress of thoracic gas through the larynx. If the larynx is totally obstructed oxygen can only be given to meet metabolic needs (200–250 ml/min). A cricothyrotomy is urgently indicated. Otherwise overinflation of the lungs may occur, causing barotrauma.

Trauma creates unique airway problems. Trauma to the face can be very disfiguring but can often be managed quite easily with direct laryngoscopy and continuous suctioning with a rigid tonsil sucker. Tubular blades[10] help displace distorted and oedematous tissue and create space such that glottic structures can be visualized. Pushing on the thorax or a constant oxygen flow from a cricothyroid catheter can produce a path of air or oxygen bubbles which can be followed to the glottis. A wire placed through the cricothyroid membrane in a retrograde direction into the pharynx can be a similar visual guide to the glottis.

Several lesions encroach on pharyngeal space. The tongue distended by a traumatic haematoma can swell to fill the posterior pharyngeal space. Bilateral mandibular fractures create an analogous situation. In this case, mandibular support for the tongue is pushed to the posterior pharynx. Lacerations of laryngeal support muscles may lead to malposition of the glottis and obstruction of the upper airway.

The mucous membranes of the pharynx can be distended with blood, air or fluid, as a result of overenthusiastic hydration. Bleeding in the neck and an expanding haematoma can also encroach on the pharyngeal space by direct pressure. Intubation for these issues can usually be solved by use of a tubular laryngoscopy blade[10] but surgical cricothyrotomy may be necessary.

The mouth may open poorly. Fractures of the zygoma and temperomandibular (TM) joint due to masseter spasm may limit mouth opening and if the mandibular fracture is impacted into the TM joint the mouth may not open at all. Maxillary fractures may result in the teeth floating free and obstructing vision.

Fractures of the nose, cribriform plates and basilar skull are contraindications for nasal intubation since the tracheal tube may enter nasal sinuses or cranium.[11]

Rents in the pharyngeal mucosa permit air to enter the submucosal space with positive pressure ventilation and entend into the cranium if fractures communicate.

Traumatic asphyxia[12] is caused by heavy compressive weights on the thorax and results in venous congestion and petechiae of the neck and face. Although striking in appearance, ventilation is generally reinstituted as soon as chest compression is relieved. Patients may be subject to complications of fat embolization at a later time.

Subcutaneous and mediastinal emphysema can occur with open mandibular fractures;[13] however, until proved otherwise, emphysema implies injury to the larynx or trachea. Lesions must be identified with fibreoptic visualization such that a tracheal tube can be placed without entering or extending the injury and the lesion is isolated from positive pressure airway ventilation. Evaluation of the lesions will necessitate preplacement of chest tubes to relieve gas which escapes through the rents and into the thorax. If intubation must proceed rapidly, i.e. there is insufficent time for fibreoptic evaluation, then intubation should be attempted with a small tracheal tube passed gently through the glottis. If any resistance is met, the tube should be removed and a surgical airway established. Once intubated, high frequency ventilation may be used to reduce intratracheal airway pressure and minimize a gas leak until the lesions can be bypassed.

If laryngotracheal avulsion is discovered, surgeons must be prepared to enter the chest quickly to secure the free end of the trachea.

Submandibular compliance can be dramatically changed by an expanding haematoma which may make direct laryngoscopy impossible. If time permits a good alternative is to use fibreoptic visualization.

Atlanto-occipital extension may have been limited before the trauma or by concern related to cervical spine fractures and instability. The patient with ankylosing spondylitis with acute trauma will need fibreoptic visualization, if time permits, or a surgical airway if time has run out. Blind techniques can be attempted but one should not persist unsuccessfully until tissues are masserated, oedematous and obstructed. Exclusion of cervical neck fractures is a major concern in trauma. If the patient is awake and denies any neck pain, spontaneous or with movement, the patient does not have a cervical neck injury. If the patient cannot communicate or complains of pain then anteroposterior, lateral (swimmers' view) and open mouth (odontoid) x-ray views must be taken. If these x-rays cannot resolve the question, CT or finally flexion-extension views may be necessary (only on the awake patient). It is extremely important to maintain cervical neck stabilization until a conclusive decision is reached. Cervical neck injury, real or potential, complicates laryngoscopy and intubation. If the airway is obstructed and asphyxia is imminent, the first priority must be to intubate the trachea after direct laryngoscopy as the quickest, most reliable means. An assistant, stabilizing the head without retraction, minimizes atlanto-occipital extension. This makes laryngoscopy more difficult but in most cases intubation can be achieved successfully with minimum neck movement.

Massive vomiting puts the patient at risk from aspiration. If tracheal intubation is mandatory, a tracheal tube can be first placed deliberately in the oesophagus and the cuff inflated. Vomitus can then be diverted from the pharynx, permitting a clear view of the glottis.

When direct laryngoscopy is found to be extremely difficult, yet mask ventilation is possible, it is reasonable to place a laryngeal mask[14] to first secure good ventilation and follow this by placing a tracheal tube through the mask. Others feel the pharyngolaryngeal airway[15] can adequately solve this problem.

Facial burns lead one to suspect the likelihood of an airway burn and are of great concern as pharyngeal mucous membranes may rapidly take on fluid and obstruct the airway. The trachea should be intubated sooner rather than later to protect the patient.

Penetrating neck injuries may lead to bleeding from laceration of major vessels. Surgeons may wish to explore these lesions surgically and often precede exploration with angiography or even embolization, particularly if an artery is cut in an inaccessible place. Intubation will be necessary. It is best to leave protruding foreign bodies in place (knives, sticks, etc.) before intubating. However, if the airway is jeopardized and there is no other course, the foreign body should be removed before proceeding. If the trachea has clearly been lacerated and the wound is sufficiently large it is quite reasonable to intubate the trachea directly through the open wound.

Breathing

This is the second priority of the primary evaluation. Those patients with pneumothorax, several fractured ribs, haemothorax and a pneumomediastinum will routinely need a tube thoracostomy.

Open pneumothorax

This should be quite obvious and is easily treated by covering the hole with Vaseline gauze and dressing, and placing a tube thoracostomy to evacuate the lung under vacuum.

Tension pneumothorax

Examination of the neck will reveal distended neck veins and deviated trachea away from the affected lung. Breath sounds may be absent or diminished on the affected side, although these may be difficult to hear in a noisy emergency room. Blood pressure may be reduced due to venous outflow obstruction caused by the mediastinal shift. A chest film, preferably upright, should confirm the diagnosis, but this should be bypassed if the patient is in distress and rapidly deteriorating. Rather, a plastic needle should be inserted in the second intercostal space at the midclavicular line followed quickly by a tube thoracostomy. The tube should be placed in the midaxillary line at the level of the nipple or higher and directed posterior and superior in the chest.

Flail chest

Broken ribs can be identified in the conscious patient by pushing on the sternum and laterally on each chest. Chest x-ray should confirm the fractures but will not identify costochondral fractures. Initial treatment to relieve the pain can be to roll the patient affected side down until the rib blocks can be performed. If ventilation is compromised then intubation will be necessary. Epidural analgesia can be very useful for pain relief but only instituted when time and the patient's condition permits.

Haemothorax

The patient may be in respiratory distress with poor arterial oxygen saturation. Dullness to percussion of the chest may be present. The chest x-ray will be diagnostic. Tube thoracostomy may be necessary. Initial evacuation of 1000–1500 ml blood or persistent bleeding of 200–300 ml/h may necessitate thoracotomy.

Injury to the laryngotracheal tree and oesophagus

The chest must be palpated for subcutaneous emphysema. Subcutaneous and/or mediastinal emphysema represents injury to the laryngotracheal tree until proved otherwise. If there is a compression injury of the trachea oesophageal injury must be excluded. Laryngeal injury should be suspected if the patient has a hoarse voice and palpation for fractures should be performed. Most tracheobronchial injuries occur within 2.5 cm (1 inch) of the carina and must be suspected following haemoptysis and/or a persistent air leak from a large chest tube. Oesophageal injuries should be considered when pain and shock are out of proportion to the apparent injury, particulate matter is coming from the chest tube, or there is a left chest pneumo/haemothorax without rib fractures.

Ruptured diaphragm

Blunt trauma can rupture the hemidiaphragm, generally the left hemidiaphragm. Either hemidiaphragm can be affected following penetrating injury. The chest x-ray should be examined for this. Many herniations may have minimal symptoms and may not be identified until days later. It may be a source of poor pulmonary compliance and result in a large (A a) O_2 difference during subsequent surgery. It can be confirmed by the elevated left diaphragm of acute gastric dilatation and a loculated pneumo/haemothorax. A nasogastric tube curled up in the chest may be the first intimation of this diagnosis.

Lung contusion

Lung contusion may be identified on x-ray, and if the (A a) O_2 difference is large, can be reason to intubate the trachea. Early lesions may progress during the subsequent operation, necessitating the use of PEEP or a ventilator that can deliver an adequate tidal volume as pulmonary compliance falls.

Inadequate ventilation

Estimating the adequacy of breathing can be difficult. Standing at the patient's feet, one can more easily observe the assymetry of a flail chest, paradoxical movement of

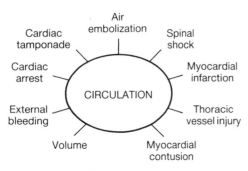

Fig. 50.1   Factors affecting circulation.

chest and abdomen in spinal injury, and the over-expanded chest of emphysema. Blood gas determinations define adequacy of ventilation but the final judgement must be based on the clinical prediction of a deteriorating condition and an estimate of the level of exhaustion of the patient.

Circulation

This is the third priority in the primary survey and must quickly address the restoration of circulating blood volume, external bleeding and adequate pump function. In doing so, the possibilities depicted in Fig. 50.1 must be considered.

Resuscitation and diagnostic procedures will be going on simultaneously as seen in Fig. 50.2.

Volume

Circulating blood volume deficit can be categorized as mild (up to 20 per cent), moderate (20–40 per cent), or severe (greater than 40 per cent).

Patients with mild deficit complain of thirst and being cold. Their pulse may increase and blood pressure fall, if they are tilted into the head-up position, but both measurements remain quite stable if the patient remains horizontal. Capillary refill is diminished and the skin feels cold. This coolness starts in the periphery, progressing centrally as shock becomes more profound. Patients with cold knees have significant blood loss.

In moderate volume deficit, a pulse increase is sustained and blood pressure is reduced. Urine output falls to less

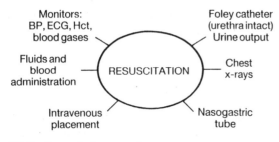

Fig. 50.2   Resuscitation procedures.

than 0.5 ml/kg per hour. Thus insertion of a urinary catheter is an extremely useful guide to fluid therapy. A urinary catheter should be inserted unless there is the possibility of urethral transection. This should be suspected with: (1) trauma to the perineum; (2) blood at the urethral meatus or in the scrotum; (3) a prostate that is high riding or cannot be palpated on rectal examination. If these signs are present, a radiopaque dye should be injected into the first few centimetres of the urethra and a urethrogram obtained to establish urethral continuity.

Patients with a severe volume deficit appear grey. Blood pressure is not sustained and pulse is consistently greater than 120 beats/min. As blood pressure reaches 30–50 mmHg, patients become confused, and agitated, progressing to unconsciousness. They may manifest a cardiac arrhythmia and ischaemic myocardial changes as coronary perfusion falls.

There are exceptions to these statements. Older patients and those with spinal shock will not demonstrate sympathetic compensation. Thus blood pressure falls progressively with volume loss and without a tachycardia. This is also true for the patient with a fixed rate pacemaker. Children, conversely, disguise blood loss with intense vasoconstriction. The clue to their depleted volume lies in a fast heart rate. Athletes have greater reserve with a volume expansion up to 20 per cent. It is as well to remember that the hypotensive patient with cerebral trauma is also hypovolaemic. Head trauma does *not* cause hypotension *per se*, rather hypertension.

Patients in shock need rapid blood volume replacement. Large bore intravenous infusions using 14-gauge catheters or larger must be placed in the upper and lower extremities with the only proviso that there must be vascular continuity between the point of insertion of the cannula and the right heart. In this regard, one should remember that injuries to the IVC or liver will make lower extremity i.v. infusions less effective if the IVC is clamped or if infused blood simply pours out through a gaping hole. Upper extremity access is preferable for the anaesthetist, in all cases, because the site is close to the working area and in full view if problems arise, and the length of i.v. lines can be kept short. A central venous line, via the internal jugular or subclavian approach, should not be attempted until venous pressure is sufficient to distend the vein. Attempts on collapsed veins run the significant risk of haemo- and/or pneumothorax. When i.v. infusions are placed, blood should be obtained for type, and cross-match, CBC, PT, PTT, platelet count, electrolytes, BUN, creatinine, amylase and serum osmolarity.

Vigorous fluid resuscitation should start with a balanced salt solution. Advantages are that it is effective, inexpensive, available and can be eliminated in the urine if overhydration occurs and the kidneys are working. If the patient cannot sustain a reasonable blood pressure after a

2000–3000 ml infusion, then universal O-positive packed cells should be given. O-negative blood should be reserved for women of childbearing age. This should be followed up with type-specific and finally cross-matched blood as it becomes available.

The ideal haematocrit (packed cell volume) for the traumatized patient is uncertain. Reduced blood use because of possible HIV and hepatitis transmission is a worthy goal. The suggestion that a Hct of 21–25 per cent in patients for elective surgery without ischaemic heart disease is reasonable. Less certain is the suggestion that impaired hearts may suffer from hypoperfusion, contusion and the release of cytokines after trauma. Thus, a haematocrit of 30 per cent seems justified when myocardial function is in question, and is analogous to the reasoning underlying an increased haematocrit for patients with ischaemic heart disease.

Certain patients are at risk of developing a coagulopathy from the start and should be given whole blood as soon as possible: massive head trauma (release of thromboplastin), prolonged by hypotension, liver trauma or pre-existing liver disease (decreased factor production), massive transfusion (factor and platelet dilution), and amniotic fluid embolization (consumptive coagulopathy). These patients need measurement of haematocrit, platelets, PT and PTT on admission to the emergency room, and to be checked frequently thereafter.

External bleeding

Rolled gauze pressed directly over the external site will control most bleeding. Tourniquets should be reserved for the stump of traumatic amputations. Blind application of clamps into rapidly bleeding wounds may preclude repair of vessels and nerves because of the tissue damage which may result.

Cardiac arrest

The patient without pulse or blood pressure must be assumed to have cardiac arrest. An ECG may support the need for treatment of fibrillation or other arrhythmia, but if hypovolaemia is the cause the need is for rapid volume replacement. It is at this point that the anaesthetist may be asked to assist surgeons in performing an emergency room thoracotomy. More often the incision will involve the left thorax with extension along the left sternum, but sometimes it will include both chests with an incision across the sternum.

The purpose of this procedure is: (1) to relieve pericardial tamponade; (2) to compress vascular or bronchial disruption manually and make an initial repair; (3) to support blood flow to head and heart by obstructing the descending aorta; and (4) to provide internal cardiac massage if warranted.

The emergency room thoracotomy demands that a capable surgeon is provided with good lights, sterile instruments, and skilled nurses – all the elements of a well staffed operating room. The demands, perhaps overstated, are to create a first class operating room on the street corner. A few hospitals approach the problem by maintaining a complete operating room in the emergency room. Others, without such facilities, sharpen the criteria for ER thoracotomy and concentrate on reducing the time taken to deliver the patient to the operating room once primary survey and initial resuscitation are complete.

At one time, ER thoracotomy was used widely for any patient in hypovolaemic cardiac arrest. Dismal results after blunt trauma have essentially reserved the procedure for penetrating trauma of right and left chest[16] where some impressive successes have been possible.

Pericardial tamponade

Pericardial tamponade must be suspected in patients who are 'in shock' and have distended neck veins and when pneumothorax has been treated or excluded. Pericardiocentesis may be beneficial as a first step but the problem may need to be resolved by surgical pericardiotomy. The anaesthetist must remember the precautions required to minimize positive airway pressure during IPPV and to sustain a fast heart rate until the problem is resolved.

Air embolization

Air embolization generally has a dismal outcome because signs which identify the problem often indicate that a lethal dose of air has entered the arterial circulation: (1) froth in an arterial blood sample; (2) air *in fundi* or focal neurological signs; (3) sudden cardiovascular collapse on instituting positive pressure ventilation. Treatment consists of immediate thoracotomy and cross-clamping the hilum of the injured lung. In such cases, the anaesthetist should consider the use of high frequency ventilation to minimize intratracheal positive pressure. If available, hyperbaric therapy may be of use (see Chapter 47).

Spinal shock

Spinal shock must be considered in any patient at risk from cervical neck trauma who is not responding to the usual fluid resuscitation. Slow pulse and warm legs with dilated leg veins are suggestive. Other signs to look for are flaccid areflexia, flaccid rectal sphincter, diaphragmatic breathing, flexion but not extension of the elbow, grimace to pain above but not below the clavicle, and priapism.

Myocardial infarction

Myocardial infarction may have been caused by the traumatic event or may be secondary to shock and coronary hypoperfusion. Treatment must be designed to concentrate on cardiac arrhythmias and to improve myocardial performance.

Thoracic vessel injury

Great vessel injury is suspected in any form of rapid deceleration and should be considered when impact forces have fractured the first and second rib and/or scapula. A left haemothorax without rib fracture is also suspicious. A chest x-ray showing wide mediastinum and wide aortic knob suggests the need for arteriography. Deviation of the trachea or nasogastric tube to the right are also suggestive. Only 20 per cent of patients with aortic rupture survive for 1 hour and almost half of these will be dead in 24 hours. A successful outcome may result when the rupture is contained, thereby permitting time for a definitive diagnosis and the care of concurrent problems of higher priority. It is important to check all extremity pulses to detect right-to-left arm discrepancies and possible acute coarctation syndrome.

Myocardial contusion

Patients with myocardial contusion are at risk of cardiac arrhythmias. In general, they do well and can be properly monitored in a hospital bed with ECG telemetry.

Disability: neurological status

A rapid neurological evaluation should be performed to establish the patient's level of consciousness, pupil size and reaction to light. The AVPU method can be remembered as analogous to the eye component of the Glasgow coma scale: A, alert; V, responds to vocal stimuli; P, responds to painful stimuli; U, unresponsive.

The neurosurgeons will be particularly interested in the triad of (1) unilateral dilated pupil with sluggish or no response to light; (2) unilateral extremity weakness, either ipsi- or contralateral in a patient; with (3) deteriorating consciousness. This triad may indicate an urgent need for temporal burr hole surgery on the side of the dilated pupil, to relieve transtentorial herniation and brainstem dysfunction.

Head injury takes a very high priority for immediate care. A unilateral dilated pupil unresponsive to light may demand an immediate need for burr hole surgery combined with surgical exploration of chest and abdomen if there are concurrent injuries. If the patient is haemodynamically stable, the surgeon may consider it more desirable to do a head CT scan to delineate the cranial lesions. The anaesthetist may be required to provide a motionless patient during these diagnostic procedures.

Exposure

Exposure refers to the removal of all the patient's clothing such that complete head to toe examination can begin.

Secondary survey

The purpose of the secondary survey is to identify all problems in a systematic head to toe examination which will evolve into an orderly diagnostic plan. I list those findings which will have a special implication for the anaesthetist's management of the patient.

The head examination may reveal depressed skull fractures which will need elevation and lacerations which will need to be sutured. Basilar skull fractures should be suspected if there are blood and CSF in the auditory canal, ecchymoses over the mastoid bone and periorbital soft tissue. Gastric tubes must not be placed through the nose in such patients.

Neck inspection will be performed once more for pain, crepitus, distended neck veins, and tracheal deviation. Penetrating neck wounds may result in vascular injury, AV fistula, and expanding haematomas which can compress the pharynx. Cervical spine injuries have not only the obvious implications for movement during laryngoscopy but also bleeding from fractures can find access to the retropharyngeal space and pharyngeal mucosa, swelling these membranes and totally obstructing the pharynx.

On the chest examination one should examine systematically for fractures of the clavicles and ribs which can ultimately lead to pneumothorax, intrathoracic bleeding or be a warning of a contused lung. Anteroposterior compression of the sternum can identify rib fractures and flail segments. The patient should be carefully 'log rolled' to examine the back for ecchymoses and entrance/exit wounds, and signs of occult problems.

The abdomen should be examined for distension, changes in girth, and evidence of peritoneal irritation. The examination often appears benign, and occult bleeding may necessitate peritoneal lavage or CT scan for definitive diagnosis. Penetrating wounds will demand a minimum of local exploration. A nasogastric tube should be placed not only to remove stomach contents but to examine for blood.

Pelvic fractures are a source of troublesome bleeding.

Pressure on the symphysis pelvis can often reveal fractures as can the rectal examination. Injury to the pelvis may also involve urethral injury which may preclude a Foley catheter and necessitate a cystostomy.

Long bone fractures must be identified to allow for stabilization to reduce occult bleeding. Vascular injury can occur around fracture sites and arterial pulses must be carefully checked for this possibility.

All extremity arterial pulses must be compared right to left, upper to lower body to evaluate for discrepancies which may suggest injury, or compartment syndromes. Bruits over penetration sites may indicate AV fistulae. Penetration into inaccessible areas such as the vertebral artery will necessitate angiography for definitive diagnosis.

Spinal vertebrae should be examined for possible fractures which can lead to occult bleeding and spinal instability. Until these fractures are ruled out the patient must be carefully moved and positioned to minimize risk of spinal cord injury. Palpation may reveal pain or local 'step off' deformities. These may be pain with motion or radiation into extremities, chest or abdomen. Muscles may be in spasm.

A complete neurological examination should now be completed, focusing on hemispheric, brainstem and spinal cord function. This examination should be completed by a neurosurgeon.

Diagnostic studies

History

The history will be obtained from ambulance crew, police, family or friends and should reveal details of the epidemiology of the injury. Some past medical history can hopefully be pieced together from these sources as well as history of alcohol and drug use which may have an impact on the patient's physiological status.

Lavage

Peritoneal lavage is simple, quick, inexpensive and 97 per cent accurate. It cannot be interpreted, however, if there is the possibility of contamination from retroperitoneal blood or pelvic bleeding. It has largely been replaced by abdominal CT but retains an important role where CT is not available or in the unstable head-injured patient who must proceed rapidly to the operating theatre. The test requires that a Foley catheter be in place to avoid injury to the bladder. A positive test is one in which free blood, bile or intestinal contents return to the lavage fluid or 75 000 red blood cells or 500 white blood cells per millimetre are present.

x-Rays

Chest x-rays and cervical spine films may be the only films obtained before transporting a critical patient to the operating theatre. If spinal cord injury is suspected, plain films of the rest of the spine are worth the time. Routine films of pelvis and long bones are important in unconscious or obtunded patients who have suffered blunt trauma.

Contrast studies

Angiography is indicated for:

1. chest injuries with first rib fracture, mediastinal widening, and deviation of the trachea to the right;
2. penetrating neck injury above the jaw line;
3. pelvic fractures and massive haemorrhage;
4. IVP for patients with gross haematoma or major blunt or penetrating abdominal trauma;
5. non-visualization of the kidney by intravenous pyelogram;
6. penetrating injuries of the extremities when close to major vessels;
7. dislocation of the knee;
8. all fractures associated with abnormal arterial pulses;
9. oesophagograms for potential oesophageal wounds as a result of blunt chest injuries or penetrating mediasinal wounds;
10. cystograms and urethrograms for bladder and urethral injuries with particular concern if associated with pelvic injuries.

Computed tomography

CT scanning is crucial for the evaluation of many trauma problems. The extent of serious head injury can be delineated best by CT. Scan of the neck is useful for the evaluation of soft tissue problems and lesions extending from the pharynx to the tracheal tree. Spinal trauma scans can determine cord encroachment and nerve root disruption. Abdominal CT is essentially 100 per cent accurate in defining the need for surgical intervention. But CT scanning takes time, is not always available, and unstable patients need to be observed carefully. Thus, its use must be carefully weighed against the immediate necessity for surgery.

Assigning priorities

The goal is to manage the injury which presents the greatest threat to life and function. Airway management takes the highest priority.

In the difficult choice between coincident airway obstruction and C-spine injury, the airway must take precedence. Severe external bleeding, chest injuries causing cardiorespiratory failure, and head injuries with an increase in intracranial pressure follow close behind.

Stable cranial injuries, burns, bleeding from retroperitoneal or solid intra-abdominal organs, and extensive soft tissue injuries can wait while issues which are unstable are attended to.

Loss of function may occur if the following are not attended to within hours: vascular, tendon, nerve and open eye injuries, partial amputation, and open fractures. Treatment of closed fractures and dislocations can wait longer.

The operating theatre (room)

With lesions identified and priorities established, patients are brought to the operating theatre. Personnel must be prepared to accept all patients regardless of priority, skilfully juggling available resources to meet the needs. One operating theatre ('crunch room') should be set aside and in constant readiness for the most acute serious cases. However, if this theatre is being used another should be set up to accept the next urgent case. Thus several rooms need to have identical operating theatre capability with regard to basic anaesthetic equipment and monitoring. The only unique feature of the 'crunch room' is that an arterial strain gauge is set up and calibrated and a rapid infusion device is similarly primed and standing by. To meet this need, more than one rapid infusion device must be functioning at all times.

Case management

It is quite clear that a single person cannot manage a complex trauma case in the early phase. It will take two people, preferably each with assistants, making up two teams. Team members clearly understand all essential tasks. Responsibility is divided, but ultimately the Team 1 leader who is 'in charge' must be provided with all available information. Once the situation has settled down and surgeons have gained control of the clinical issues, a single anaesthetist can manage quite easily and resources can be diverted to other critical areas.

This approach is depicted in Fig. 50.3. Team 1, directed by the anaesthetist 'in charge' must manage the anaesthetic care with respect to airway, drugs, monitoring including transoesophageal echo if available, maintain a dialogue with the surgeons as to patient status, and formulate immediate intraoperative and postoperative care plans. Team 2 members, directed by the 'facilitating' anaesthetist, have responsibility for establishing high rate perfusion lines, the rapid transfusion apparatus, administration of fluids and blood products, interaction with technicians who manage blood salvage, and nurses who help with procurement. In addition, the team will monitor haematocrit, blood gases, and coagulation status. Teams 1 and 2 may share responsibility for placement and maintenance of arterial and additional i.v. lines and for central venous line placement. Cardiac resuscitation is also shared since this is both volume and drug dependent. The 'facilitator' will inform the Team 1 leader of any change in status of haematocrit, blood gas or coagulation values and his/her ability to meet demands for blood products. In this way, a complex case can be managed expeditiously and with the best chance to gain control of the situation quickly.

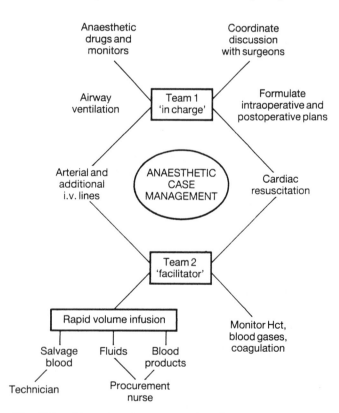

Fig. 50.3 Diagrammatic representation of case management.

Recovery area and critical care unit

As surgery is completed in the operating theatre the patients should be moved to the critical care units including the anaesthesia recovery area. Surgeons will gradually assume the responsible role for continued care but it is advisable that the anaesthetist retains responsibility for airway and ventilator management and keeps a close eye on continued volume resuscitation until the acute phase is resolved. Frequently these patients will need a rapid return to the operating theatre and the anaesthetist must help with organizing the case into the theatre schedule.

Anaesthesia in trauma

Concerns for anaesthesia in trauma are in large part concerns for anaesthesia in the hypovolaemic patient who has compensated with a massive stimulation of the sympathetic system. Heart rate is increased through reflex mechanisms. Adrenal secretion is stimulated with activation of the renin–angiotensin system. Vasopressin and histamine are also released. Arterioles are constricted and venous capacitance is abolished. Blood flow is sustained to the head and heart at the expense of all other organs. If we attempt to anaesthetize such a patient we disrupt this delicate physiological balance and thereby court disaster.

We would hope to avoid such disasters but the medical literature of World Wars I and II describes just this. Spinal anaesthesia was popular in World War I because it reportedly preserved laryngeal reflexes, made pulmonary disease less feared, provided excellent muscular relaxation, and decreased hospitalization time. Admiral Gordon-Taylor of The Royal Navy disagreed.[17] He considered spinal anaesthesia the ideal form of euthanasia in war surgery and believed general anaesthesia safer in the traumatized patient.

F.J. Halford[17] reported on the use of thiopentone and hexabarbitone at Pearl Harbour in 1941. Repeated deaths, preceded by 'cyanosis decolletage', followed administration of what he believed to have been a small dose (500 mg) of thiopentone. This led him to state that intravenous anaesthesia is also an ideal method of euthanasia in trauma and he returned to open drop ether as the drug of choice. Adams and Gray,[18] however, reported that they could give thiopentone safely in 25 mg doses slowly over several minutes to successfully anaesthetize a man in haemorrhagic shock following a gunshot wound. An unsigned editorial that followed these two articles places the issue in perspective: '. . . the drug was not dangerous but the method of administration was . . . [small] doses . . . administered slowly, with intervals between injections of sufficient length to allow the full effect to take place, is the only rational scheme of dosage.'[19]

In the ensuing years, we have learned a great deal about:

1. the contracted volume of distribution of drugs in hypovolaemia which explains why small doses have such profound effects;
2. the disastrous effects of shock on organ viability and survival and thus the need to reverse this process as soon as possible;
3. the release of cytokines, leukotrienes, myocardial depressant factors, and free radical damage to ischaemic tissue in shock which may begin to explain why the myocardium may lose effective function.

However, the basic understanding of what anaesthetics do in shock, i.e. a disruption of sympathetic compensation, and the pragmatic approach to the administration of anaesthetics, is presented in these war experiences and has not changed over the years, namely:

1. when you give anaesthetic drugs to the hypovolaemic patient, think small;
2. only give what you need;
3. give drugs slowly and wait to observe the effect.

However, with the requirement to use drugs sparingly or not at all we may end up paralysing a patient ventilating with oxygen without amnesic drugs and a real risk of patient recall. Although patients may recall surgical events and sometimes discomfort I have never had one object except that they survived the event, given the alternatives.

Thus, the priority of anaesthetic objectives in the shocked hypovolaemic patient are:

1. provide surgical conditions without killing the patient;
2. gain control of circulation such that amnesic analgesic drugs can be given safely;
3. gradually increase the level of anaesthesia such that adequate blood flow is established to vasconstricted organs;
4. wake the patient up at the end.

Choice of anaesthetic agent

The use of drugs will vary with local availability. The following are recommended.

Muscle relaxants

Suxamethonium is preferred for its rapid onset and quick dissipation. Effects on intravascular and intracranial pressure are not significant. It should be avoided with significant muscle denervation for more than 2 days, and for large areas of burns more than 7 days old. Vecuronium is useful because it has few side effects. It can have a rapid onset if used in doses greater than 0.25 mg/kg but unfortunately action persists.

Anaesthetics and amnesics

These must be used sparingly until some control of blood volume status has been achieved. Midazolam provides high quality amnesia. Diazepam does as well but its slower in onset and long in duration. Hyoscine (scopolamine) runs less risk of sympatholysis but responses to it are not consistent.

Thiopentone works well but must be given in small doses because of its effect on peripheral vasodilitation and myocardial depression.

Ketamine gives intense analgesia and produces a dissociative state. It is popular, because it maintains blood pressure and pulse rate. However, when the sympathetic system is impaired, myocardial depression may occur. It raises intracranial and intraocular pressure and is therefore contraindicated for intracranial and open globe surgery.

Etomidate produces less sympathetic ablation but produces troublesome movements.

N_2O is a myocardial depressant and reduces available oxygen delivery. It should not be used when there is concern for pneumothorax, pneumocephalus, or pulmonary contusion.

Propofol has no significant advantage over thiopentone and may act as a bacterial culture medium when held for some hours in open syringes.

Narcotics

Fentanyl is well tolerated but can produce some venodilatation. Sufentanil and alfentanil offer no special advantage. Morphine can cause hypotension.

Vasodilating anaesthetics

Halothane and isoflurane have excellent properties in this regard, but cannot be given until blood volume is restored.

Concurrent 'street' drugs can complicate the anaesthesia. Alcohol will decrease anaesthetic requirements. Cocaine may induce intense tachycardia and hypertension. Phenylcyclidine (PCP) can cause hypertension, tachycardia and severe rage reactions. Droperidol works well to control this situation.

Conclusion: the role of the anaesthetist in trauma cases

More than any other medical speciality, anaesthetists are involved in the efficient management of time and resources in the case of the traumatized patient. From the training of EMT technicians in intubation skills to the triage of critical case flow in the operating room, to the management of the patient in the operating room and finally the intensive care unit, anaesthesia is the glue that sticks the whole process together. There is risk in delay. Care must be titrated with responses made in seconds. Thus, thinking must be preorganized and rehearsed. But thinking must also be collegiate between anaesthetists, surgeons and nurses. Each group must clearly know, understand, and appreciate the important role each plays. More than in any other area of their work, anaesthetists must think like surgeons in order to anticipate timing to implement diagnostic and therapeutic options. In no other area is there quite the imperative to understand collaboration, have mutual respect, and resist fragmentation of purpose. Anaesthesia becomes broadly redefined by trauma care. It is the science and clinical practice of resuscitation, amnesia, anaesthesia, and reanimation practised in that order of priority. The anaesthetist must have expert skill in all these areas and a pragmatic approach which pays close attention to what he/she can see, hear and feel. It requires vigilance for the occult problem hidden and yet to be declared. Anaesthetists are thrust to the front lines and must be there and involved at all times.

REFERENCES

1. *Accident facts*. Chicago: National Safety Council, 1990: 2.
2. Trunkey DD. Trauma. *Scientific American* 1983; **249**: 28–35.
3. West JG, Williams MJ, Trunkey DD, Wolferth CC. Trauma systems: current status, future challenges. *Journal of the American Medical Association* 1988; **259**: 3597–600.
4. Flint LM, Brown A, Richardson JD, Polic HC. Definitive control of bleeding from severe pelvic fractures. *Annals of Surgery* 1979; **189**: 709–14.
5. Boulanger BR, Milzman D, Mitchell K, Rodriguez A. Body habitus as a predictor of injury pattern after blunt trauma. *Journal of Trauma* 1992; **33**: 228–32.
6. Newman KD, Bowman LM, Eichelberger MR, *et al*. The lap belt complex: intestinal and lumbar spine injury in children. *Journal of Trauma* 1990; **30**: 1133–8.
7. Warner KG, Demling RH. The pathophysiology of free-fall injury. *Annals of Emergency Medicine* 1986; **15**: 1088–93.
8. Committee on Trauma. *Advanced trauma life support course for physicians*. Chicago, IL: American College of Surgeons, 1989.
9. Markison RE, Trunkey DD. Establishment of care priorities. In: Capan LM, Miller SM, Turndorf H, eds. *Trauma: anesthesia and intensive care*. Philadelphia: JB Lippincott Company, 1991; **29**: 29–42.
10. Bainton CR. A new laryngoscope blade to overcome pharyngeal obstruction. *Anesthesiology* 1987; **67**: 767–70.
11. Muzzi DA, Losasso TJ, Cucchiara RF. Complications from a nasopharyngeal airway in a patient with a basilar skull fracture. *Anesthesiology* 1991; **74**: 366–8.
12. Jongewaard WR, Cogbill TH, Landercasper J. Neurologic consequences of traumatic asphyxia. *Journal of Trauma* 1992; **32**: 28–31.
13. Minton G, Tu HK. Pneumomediastinum, pneumothorax, and cervical emphysema following mandibular fractures. *Oral Surgery* 1984; 490–2.
14. Heath ML, Allagain J. Intubation through the laryngeal mask. A technique for unexpected difficult intubation. *Anaesthesia* 1991; **46**: 545–8.
15. Bartlett RL, Martin SD, McMahon JM, Schafermeyer RW, Vakich DJ, Hornung CA. A field comparison of the pharyngeotracheal lumen airway and the endotracheal tube. *Journal of Trauma* 1992; **32**: 280–4.
16. Lorenz HP, Steinmetz B, Liberman J, Schechter WP, Macho JR. Emergency thoracotomy: survival correlates with physiologic status. *Journal of Trauma* 1992; **32**: 780–8.
17. Halford FJ. A critique of intravenous anesthesia in war surgery. *Anesthesiology* 1943; **4**: 67–9.
18. Adams RC, Gray HK. Intravenous anesthesia with pentothal sodium in the case of gunshot wounds associated with accompanying severe traumatic shock and blood loss: report of a case. *Anesthesiology* 1943; **4**: 70–3.
19. Editorial. The question of intravenous anesthesia in war surgery. *Anesthesiology* 1943; **4**: 74–7.

Chapter 51

Management of Head Injury

Laurence Loh

The nature of head injury
Aspects of head injury management
Anaesthesia and the acute head injury

The practical control of ICP
Intensive care management of head injury
Brainstem death

Head injury, which in the UK accounts for 9 deaths per 100 000 population per year, is one of the commonest causes of death (15 per cent) in young adults.[1] Most are the result of road traffic accidents, and the wastage of human potential is enormous. Not only is there a tragic loss of life but there is also an even greater morbidity due to permanent mental and physical disablement. The eventual outcome of a head injury depends on several factors:

1. The condition of the patient before impact; in particular, age and general state of health.
2. The extent of the primary injury over which the physician has little control.
3. The extent of the secondary injury which might be significantly influenced by medical management.

This chapter therefore concentrates on the aspects of management aimed at reducing the severity of secondary injury.

Head injuries are the result of road traffic accidents, assault, falls, domestic and work-related incidents and sport or leisure activity. A high proportion are alcohol related. The frequency of occurrence varies according to age group and the local environment. The incidence of head injury is different in the two sexes, being twice as common in males than in females. The true incidence of head injury in the general population is difficult to define. Many fatal accident victims may not reach hospital and therefore not be recorded as having head injuries, and the majority of minor head injuries are probably not reported at all. Of the head injuries seen in accident and emergency departments, only about 20 per cent ($270/10^5$ population per year) are admitted to hospital.

The nature of head injury

It is important to appreciate the concept of primary and secondary brain damage although, in practice, it may not be possible to make a clear distinction between them. *Primary brain injury* is the damage sustained as a direct result of the impact on the skull and intracranial contents. This ranges from a neuronal injury with microscopic disruption of neuronal fibres and blood vessels to gross macroscopic lacerations and contusion of the brain tissue with or without skull fracture and scalp lacerations. The injury can be both diffuse and focal.

Diffuse neuronal injury, which is the result of shear forces disrupting mainly white matter at the time of impact, may be difficult to recognize immediately following injury. It frequently occurs without skull fracture and affects the white matter of the cerebral hemispheres, corpus callosum and the superior cerebellar peduncles. There is rarely a rapid recovery of consciousness, and initially the computed tomography (CT) scan may appear relatively normal. However, the neurological damage and associated vascular damage may trigger a sequence of events that lead to cerebral oedema, cerebral ischaemia and death or a persistent vegetative state. This cascade of events can be described as a *secondary injury.*

More focal lesions can occur as a result of the impaction of brain tissue on the rigid inner surfaces of the cranial cavity following external percussion to the skull (particularly at the site of a fracture) and from a contre-coup effect. In this instance gross contusion, laceration and haemorrhage are likely to occur. Because of the uneven nature of the floor of the temporal fossa, the undersurfaces of the temporal lobes are particularly prone to this kind of damage. As there is a close proximity of the temporal lobes to the brainstem, such damage and subsequent swelling and haemorrhage can lead directly to brainstem compression. In a similar manner to that following diffuse neuronal

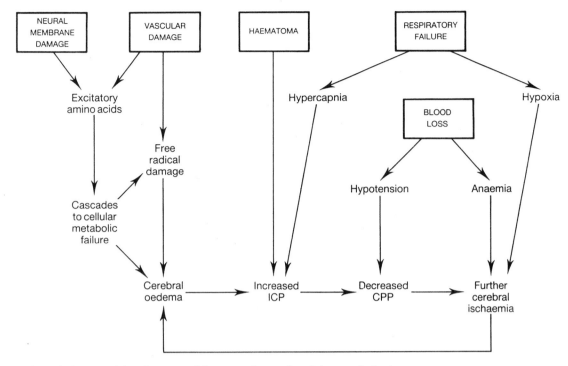

Fig. 51.1 The main intracranial and extracranial causes of secondary injury to the brain.

injury, focal neural tissue damage, with free blood in the tissues and ischaemia, triggers a series of changes that lead to secondary brain injury.

Whilst the primary injury in the brain can be fatal or give rise to severe disability due to neuronal destruction, in many head injuries the primary lesion can be relatively trivial – as demonstrated by many patients who regain consciousness and are lucid and talking shortly after injury. But a number will subsequently deteriorate owing to cerebral ischaemia from brain swelling, haematoma formation, hypoxia and hypotension. These are the main causes of secondary brain injury. Figure 51.1 shows a simplified systematic diagram of secondary injury, which can be divided into intra- and extracranial causes.

Intracranial secondary injury

Neurotoxic cascades

The importance of maintaining cerebral perfusion and oxygenation to prevent ischaemia in severe head injury should be emphasized. Unfortunately, this is not the complete picture, as much of the danger of secondary injury is the result of the liberation of chemicals that cause disruption of the blood–brain barrier, oedema and neural death either from direct trauma and ischaemia or following the restoration of blood flow, the *reperfusion injury*. Much animal experimentation employing various ischaemic stroke and head injury models has resulted in the most

likely mechanisms and mediators of cellular damage being identified. These are not unique to the central nervous system but are common to most inflammatory processes. There is much data on the possible benefits of specific antagonists to these processes in experimentally induced lesions but relatively scarce data in the human clinical situation. Nevertheless, one cannot help but be optimistic that some therapeutic benefit will come from this work. The three most likely mechanisms of damage are the liberation of excitatory amino acids, platelet-activating factor and oxygen free radicals with the ubiquitous nitric oxide radical just coming into the picture.

Calcium channel disturbance

It is believed that local tissue damage releases excitatory amino acids (EAAs), which in turn stimulate, for instance, N-methyl-D-aspartate (NMDA) glutamate receptors of calcium channels in the surrounding cells. There then follows a massive influx of calcium ions into the cells which leads to metabolic failure of the cells and cellular oedema. Antagonists to NMDA glutamate or blockers of the NMDA glutamate receptor such as dizocilpine (MK-801) have been successful in preventing brain injury in animals. Similarly, voltage-operated calcium channel blockers such as nimodipine have also been shown to have some brain protective effects.[2] Ketamine is an NMDA receptor antagonist that has been shown to improve neurological outcome in a rat brain injury model, but whether this will have practical value in a clinical setting remains to be seen.[3] Cellular metabolic failure is associated with the generation of free radicals of oxygen. It is also associated with the release of platelet-activating factor from damaged cells and blood vessels.

Oxygen free radical production

Oxygen free radicals can be generated from several sources.[4] The metabolic failure of cells generates free radicals. Ischaemia and vascular damage can stimulate the arachidonic acid cascade, leading to prostaglandin and prostacyclin release and also leukotrienes with free radical generation. Platelet-activating factor may play an important role in free radical generation and the destruction of superoxide dismutase. It was suggested in one clinical study that the outcome in severely head-injured patients was improved using polyethylene glycol conjugated superoxide dismutase (PEG-SOD). Presumably PEG-SOD mopped up oxygen radicals of the superoxide anion.[5]

Nitric oxide may also be important in the production of free radicals and, by blocking nitric oxide synthase, outcome is improved.[6] Experimental evidence suggests that antagonists to platelet-activating factor[7] and leucocyte antibody treatment[8] may also limit secondary brain injury.

The release and diffusion of oxygen free radicals to surrounding normal cells extends the secondary injury, causing further damage and swelling. The free radicals also cause further vascular damage, again leading to an increase in vascular permeability and vasogenic oedema.

All these processes feed on each other, and the spiral of vascular damage, ischaemia, cellular metabolic failure and more ischaemia eventually leads to further brain swelling, raised intracranial pressure, a decrease in cerebral perfusion pressure and more global cerebral ischaemia.

Haematoma formation

A skull fracture is frequently accompanied by rupture of a meningeal vessel. Quite often this is a meningeal vein; less frequently, a meningeal artery. Such a bleed gives rise to an *extradural haematoma* which forms between the inner, bony table of the skull and the dura. This haematoma will compress the brain and cause local ischaemia, a shift of midline structures and possible fatal brainstem damage.

Subdural or *subarachnoid haemorrhage* is the result of traumatic rupture of cerebral vessels, causing a haematoma within the brain substance or in the space between the brain surface and the dura. The problem with subdural and subarachnoid haematomata is that not only can they cause local compression and swelling of the brain substance and an increase in intracranial pressure, but also blood in the subarachnoid space can cause vasospasm and further cerebral ischaemia.

Extracranial secondary injury

Respiratory failure

Loss of consciousness following head injury may be accompanied by a period of central apnoea. If the state of unconsciousness persists, the airway is in jeopardy and obstruction of the airway, because of loss of pharyngeal muscle tone, can lead to severe hypoxia. Aspiration of vomit can cause further problems, as can injury to the chest wall and lungs impairing ventilation. Any hypoxia will aggravate cerebral ischaemia and increase secondary injury. Hypercapnia increases cerebral blood flow and cerebral blood volume, thus increasing intracranial pressure. Thus any degree of respiratory failure is particularly hazardous for the patient with head injury.

Blood loss

Cerebral perfusion is partly determined by the cerebral perfusion pressure (CPP), which is the difference between mean arterial pressure (MAP) and intracranial pressure (ICP) when ICP is greater than cerebral venous pressure. In a situation in which the ICP is raised, a fall in MAP may produce cerebral ischaemia. Not all areas of the brain will necessarily be equally affected but those areas that have been damaged and have lost autoregulation may be the most vulnerable. Hypotension from blood loss is not uncommon in multiple injury, and should be strenuously avoided and corrected. Blood loss can lead to anaemia and make cerebral ischaemia more likely.

Other causes of secondary injury

Infection

A major source of concern in open fractures of the skull is infection. This can arise from the scalp wound but often meningitis is the result of infection through a basal skull fracture into the nasopharynx, ear or nasal sinuses. Any patient with a cerebrospinal fluid (CSF) leak or who has air in the intracranial cavity has an open fracture of the skull and should be given an appropriate prophylactic antibiotic regimen. Often a CSF leak can be difficult to detect and all basal skull fractures should be viewed with suspicion.

Epileptic seizures

Some 5 per cent of adult cases of head injuries admitted to hospital develop epilepsy during the first week after injury. This early epilepsy is most likely to be associated with intracranial haematomata and depressed skull fracture.

Some 60 per cent of epileptic fits occur in the first 24 hours and about 10 per cent lead to status epilepticus. If the seizures are not controlled, they can cause cerebral hypoxia and increase the secondary injury already present. Any patient who has had a fit should be treated with anticonvulsants (200–300 mg phenytoin by slow intravenous injection or nasogastrically or orally). Status epilepticus may have to be treated with other anticonvulsants in addition. The incidence of epilepsy in children is double that in adults. There is a debate as to the virtues of prophylactic anticonvulsant treatment in all head injuries.

Aspects of head injury management

The sooner that medical and paramedical treatment are available at the scene of an accident the greater the chance of survival from head injury. Patients with moderate and severe head injury should be transferred to an accident and emergency (A&E) department for assessment as soon as possible.

As already mentioned, the unconscious patient's airway is not secure. Hypoxia and hypercapnia can occur through a central apnoea or from obstruction of the upper airway. Thus one must first assess the airway and establish a clear airway with adequate ventilation. Oxygen should be administered to any patient who is unconscious. Then attention should be turned to cerebral perfusion and the support of the cardiovascular system. Hypotension should be corrected by whatever means appropriate. (It is not appropriate to cover the spectrum of first aid and resuscitation methods relevant to the management of multiple trauma in this chapter.)

When a patient with head injury is admitted to an A&E department there is then the opportunity to make a thorough assessment, including history, examination and special investigations. Thereafter a process of triage can take place.

History

It is important to have as clear a picture of the nature of the incident as possible. Eye witness accounts and information from police and ambulance personnel are very valuable.

Loss of consciousness

A history of even a momentary loss of consciousness indicates a significant head injury. The duration of loss of consciousness probably relates to the severity of the primary injury. Continued loss of consciousness is most likely to be the result of diffuse brain injury. Focal signs of weakness or paralysis indicate focal brain damage but must be distinguished from immobility due to injury and pain. Post-traumatic amnesia is also a good index of the severity of head injury, but often it is possible to assess this only after recovery from injury. It is important to enquire into a history of drug and alcohol abuse and to judge if any alteration in consciousness is due to this cause.

Impaired consciousness

The Glasgow Coma Scale (GCS) in fact assesses impaired consciousness,[9] coma being a severe degree of such impairment. The GCS is particularly valuable because it can be applied by personnel without a lot of training. It has the advantage of being simple, objective and consistent between individual assessors. The GCS scores the best eye opening, verbal and motor responses, and can be particularly useful in assessing whether a patient is improving or deteriorating neurologically. Whilst the GCS scores the best response, clinicians are also interested in the worst responses. The motor responses in each of the four limbs should be recorded as well as the size and reaction of the pupils.

Other injuries

There should be a careful examination of the body for evidence of other injury in any patient with a head injury. The detection of an injury to the spine is particularly important and the suspect spine should be immobilized until radiological evidence has cleared the spine of instability. Equally important is the need to exclude major chest and abdominal trauma. Blood loss from wounds should be arrested; careful attention should be paid to bleeding from the nose and ears, and a search made for a CSF leak. Any scalp lacerations should be explored for evidence of a fracture and foreign body inclusion.

Pulse rate and blood pressure should be monitored, and blood and fluid replaced to maintain cerebral perfusion pressure and oxygenation.

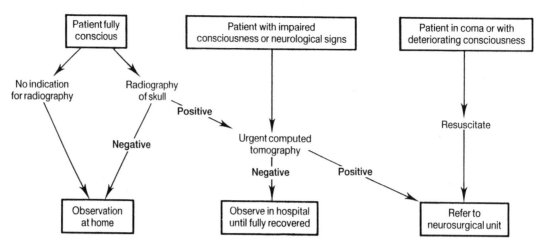

Fig 51.2 The management of patients with head injury in a general hospital with facilities for computed tomography. (Reproduced, with permission, from reference 11.)

Triage

Once the initial assessment has been made and the necessary resuscitation and stabilization have been achieved, certain decisions have to be made on where best to send the patient. Much depends on the facilities of the hospital to which the patient has been admitted and whether CT scanning and neurosurgical management are readily available.

The advent of CT scanning has made a major difference to the diagnosis and early treatment of intracranial haematomata and a reduction in morbidity and mortality. In 1984 a group of neurosurgeons produced a set of guidelines for the initial management after head injury in adults[10] which rationalized the use of CT scanning and the workload of regional neurosurgical centres in the UK. Since then, CT scanners have proliferated and are now more accessible; the guidelines should be revised to ensure early scanning of more patients with head injury (Fig. 51.2; Tables 51.1 and 51.2).[11] In particular, it is the low-risk patient rather than the severely injured who is likely to benefit most from early diagnosis and treatment of an intracranial complication.[12,13]

Head-injured patients may be divided into three groups:

1. Those who are fully alert and orientated on arrival at the A&E department with or without a history of altered consciousness (GCS = 15).
2. Those with impaired consciousness (GCS = 8–14).
3. Those in coma or with deteriorating consciousness (GCS = 3–7).

Patients in group 1 in whom there is no history of loss of consciousness are at minimal risk from haematoma formation and may be allowed to go home with specific instructions about what to look for and what to do in the event of clinical deterioration.

Table 51.1 Indications for computed tomography in general hospitals

Full consciousness but with a skull fracture
Confusion persisting after initial assessment and resuscitation
Unstable systemic state precluding transfer to neurosurgery
Diagnosis uncertain

Reproduced, with permission, from reference 11.

Patients with evidence of loss of consciousness and amnesia and those who have some bruising or swelling but who are perfectly lucid on admission should have a skull x-ray. If there is no fracture, they too will be at minimal risk and should be observed at home.

Because the risk of haematoma is considerably increased in the case of a skull fracture, patients with a fracture on skull x-ray should be referred for an urgent CT scan. Any patient with a depressed skull fracture or who shows a haematoma on CT scan is referred to the neurosurgical unit.

Table 51.2 Indications for referral to neurosurgical unit

Without preliminary computed tomography
Coma persisting after resuscitation
Deteriorating consciousness or progressive focal neurological deficits
Open injury: depressed fracture of vault or basal skull fracture
Patient fulfils criteria for computed tomography in a general hospital when this cannot be performed within a reasonable time – e.g. 3–4 hours

After computed tomography in general hospital
Abnormal tomogram (after neurosurgical opinion on images transferred electronically)
Tomogram considered to be normal but patient's progress is unsatisfactory

Reproduced, with permission, from reference 11.

Those with a fracture but without haematoma should be observed in the admitting hospital.

Patients in group 2 with a history of impaired consciousness or with some neurological signs should have an urgent CT scan, bypassing skull x-ray. If the CT scan reveals haematoma, the patient should be referred to the neurosurgical unit. A negative CT scan will allow the patient to be observed in the hospital until recovered.

Group 3 patients who are in coma or whose level of consciousness is deteriorating and those with depressed or open fractures of the skull, including basal fractures, should be referred directly, without a CT scan, to the neurosurgical unit. With the proliferation of CT scanners and other forms of imaging and the spread of computer networks so that images can be easily transmitted between hospitals, neurosurgeons will be able to review scans of patients in other hospitals and advise on transfer and management. The above guidelines will then have to be modified.

It is well established that the outcome of head injuries with haematomata is a function of the time between the injury and the evacuation of the haematoma; therefore, transfer should not be delayed unnecessarily.

Inter- and intra-hospital transfer

Inter-hospital transfer can be a hazardous procedure. Several studies have shown that secondary insults to the traumatized brain not infrequently occur during transfer of the head-injured patient from hospital to hospital or even from one department to another within the same hospital.[14,15] The most common insults are hypoxia due to airway problems and systemic hypotension. Although initial management at the site of the accident is important, suboptimal care within the hospital system accounts for most avoidable deaths after head injury.

Although transferring a patient to a neurosurgical unit may be of some urgency, it is more important to ensure that the patient does not come to grief during the transfer. Table 51.3 is a useful checklist which should be applied before transfer. Any patient who is comatose or who may have a problem with the airway should be intubated and if necessary sedated, paralysed and ventilated. Any hypotension should be corrected with fluid replacement. Extracranial injury should be stabilized, and nursing and medical staff should be sufficiently experienced to recognize and deal with any problem that may arise during transfer. Sufficient monitoring equipment should be available. A recent paper[16] outlines the lamentable state of hospital transfer in the south-east of England, and in particular the standard of monitoring and experience of the medical personnel (often the least experienced doctors are sent with these critical patients).

Table 51.3 Checklist before transfer of severely head-injured patient to a neurosurgical unit

Respiration	Is the P_{O_2} at least 13.3 kPa (100 mmHg)? Is the P_{CO_2} no more than 5.3 kPa (40 mmHg)? Is the airway clear? Is it adequately protected for the journey? Does the patient need intubation or ventilation?
Circulation	Is the systolic BP at least 120 mmHg? Is the pulse rate no more than 100 per minute? Is there peripheral perfusion? Is there a reliable intravenous line? Has enough blood been given to replace losses?
Head injury	What is the conscious level on the Glasgow Coma Scale? What has its trend been – impaired/deteriorating? Are there any focal neurological signs? Is there a skull fracture?
Other injuries	Has a cervical spine injury, chest injury, pneumothorax, broken rib been excluded? Could there be intrathoracic or abdominal bleeding? Are there any pelvic or long bone fractures? Have extracranial injuries been splinted? (e.g. cervical collar, limb splints)
Escort	Are the escorting doctor and nurse adequately experienced? Have they been instructed about what to do for this particular case? What equipment and drugs will be sent? Have the case notes and x-rays been sent?

Reproduced, with permission, from reference 14.

Anaesthesia and the acute head injury

The approach to anaesthesia in head injury, although similar to that of elective neurosurgical procedures, can differ in many respects:

1. The presence of a full stomach.
2. Associated neck injury and facial injury with a potentially difficult intubation.
3. Extracranial injury, shock, anaemia, aspiration, chest injury, and the need for resuscitation.
4. Evidence of raised ICP.
5. Likely postoperative intensive care management.

Induction and intubation

In the acute head injury, there is seldom an indication for premedication. Anticholinergic drugs can be given intravenously at induction if necessary. There is often doubt about the last oral intake and it might be best to regard all accident victims as having a potential full stomach. A rapid sequence induction would therefore be appropriate in the majority of cases. Although neurosurgical anaesthetists usually prefer a reinforced tube, this is not strictly necessary. The aim at induction is to rapidly gain control of the airway and to prevent coughing and straining. As these patients frequently end up in an intensive care unit (ICU) being intubated and ventilated for a day or so, the normal tracheal tubes used in that unit should be adequate.

Maintenance

Once the airway and ventilation are secured, a longer acting muscle relaxant can be given without necessarily waiting for the depolarizing muscle relaxant to wear off. It is more important to prevent coughing and straining than to ensure that the patient does not have a suxamethonium (succinylcholine) apnoea. If needed, analgesics may then be provided by potent short-acting intravenous narcotics such as fentanyl and alfentanil. This is reasonable provided ventilation is controlled. Anaesthesia can then be maintained with intravenous thiopentone or propofol or with a volatile agent. Isoflurane is probably the volatile agent of choice, as it appears to have least effect on the cerebral vasculature. If there is concern about raised ICP, a total intravenous anaesthetic technique would be most appropriate. Ventilation should be controlled to keep the $Pa\text{co}_2$ slightly below normal (4.5–4.0 kPa or 34–30 mmHg). Excessive hyperventilation is best avoided if ICP is raised, as the cerebral vasoconstriction produced may jeopardize cerebral perfusion. Hypotension should be corrected and fluids should be given to maintain a reasonable perfusion pressure.

The majority of head-injured patients who require anaesthesia do so because of the need to evacuate a haematoma or for the management of other major injuries. Postoperatively these patients will be cared for in an intensive care unit and sometimes electively ventilated and sedated for a period to stabilize their condition, to allow any swelling of the brain to subside and to reduce the chances of further bleeding.

The practical control of ICP

The fundamental cause of brain damage is ischaemia of brain tissue.[17] Local tissue damage increases vascular permeability and vasogenic oedema, and liberates excitatotoxins and various cascades already mentioned, with free radical liberation and cellular oedema. Much of the swelling is reversible given time. The major danger is that such swelling will increase ICP and produce compression and ischaemia of the brainstem through a tentorial or medullary cone, the outcome being brainstem death with no hope of recovery. The whole aim of management is to limit ischaemic damage and prevent irretrievable brainstem compression. Thus, it is pertinent to consider the methods available to modify intracranial hypertension.

The brain is enclosed in the rigid capsule of the skull and cushioned by the cerebral spinal fluid and perfused with blood in the cerebral circulation. Although the skull itself is inelastic, it is perforated at several points so that some of the intracranial contents (blood and CSF) can move in and out of the skull when the volume of the brain alters. Thus, small changes in the volume of the brain and blood compartments are easily compensated for by shifts in CSF volume. This accounts for the relatively flat part of the intracranial compliance curve. It is when this compensation mechanism is exhausted that the major problems of raised ICP arise and the compliance curve then rises sharply. The control of ICP is based on methods of altering the volume of the three major components of the intracranial contents, namely brain bulk, CSF and blood.

Alterations in brain bulk

Surgical decompression

A significant haematoma can be aspirated through a limited craniotomy. Severely contused brain tissue can be surgically removed as it is probable that the tissue would not survive to serve any useful function. In certain circumstances the neurosurgeon may decide to amputate a frontal or temporal lobe to allow room for the inevitable cerebral oedema, and not infrequently the bone flap raised at craniotomy is not replaced to allow for expansion of the brain bulk and the prevention of a disastrous rise in ICP.

Reduction of cerebral oedema

Diuretics
Mannitol is probably the main osmotic diuretic used to reduce ICP. It is a hypertonic solution (20 per cent mannitol)

and is given as a bolus infusion of 0.25–1.0 g/kg.[18] Mannitol shifts water from the interstitial and intracellular compartments into the intravascular compartment, increasing the circulating blood volume and lowering blood viscosity. Although this may transiently increase cerebral blood flow (CBF) and cerebral blood volume (CBV), over all the effect is a reduction in ICP within minutes. The extraction of water is from normal as well as from oedematous tissues. In areas where the blood–brain barrier is damaged, mannitol can diffuse into cells. As plasma osmolality falls with the metabolism and excretion of mannitol, water can be drawn back into the cellular space. This rebound phenomenon is least likely if mannitol is given as a single bolus to treat an acute episode of intracranial hypertension.

Several other hypertonic solutions have been used to promote a diuresis. None so far has been shown to be superior to mannitol. However, hypertonic saline has been beneficial in two cases in which mannitol has failed,[19] and hypertonic sodium lactate and hypertonic saline/dextran 70 have been proposed as possibly useful agents for reducing secondary brain injury.[20,21]

Frusemide (furosemide) will also cause some dehydration of brain tissue, and benefit ICP by mildly reducing CSF formation and encouraging CSF absorption. In combination with mannitol, it produces a marked diuresis which can lead to sodium depletion, unwelcome dehydration and pre-renal failure.[22]

Fluid restriction

Many neurosurgeons believe that fluid restriction is important and will dehydrate patients to reduce oedema. In the case of acute head injury, when fluid loss due to injury is a problem and hypovolaemic hypotension is a major cause of cerebral ischaemia, fluid restriction is not a good strategy and care must be taken to maintain intravascular fluid volume.

Steroids

Although steroids can be useful in reducing oedema that occurs around various intracranial tumours, they have not been found to be effective in reducing the oedema of acute trauma and therefore are not used for this purpose after head injuries.

Methods of altering CSF volume

CSF is produced by an active secretory mechanism of the choroid plexus situated in the ventricles. This active secretory process is not easily arrested in the clinical setting and thus there is not a great deal that can be done to prevent the production of CSF. Hypertonic solutions, frusemide and acetazolamide reduce CSF production but probably not enough to be clinically important.

CSF is drained from the intracranial compartment by two mechanisms. First, CSF passes into the venous sinuses in the skull by a filtration-like process through the arachnoid granulations that line the venous sinuses. The flow of CSF into the venous system therefore depends on the pressure difference between the ICP and the pressure in the venous sinuses. CSF drainage is encouraged if the sinus venous pressure is low. This can be achieved by nursing the patient in the head-up posture and decreasing the hydrostatic pressure in the venous sinuses. Care must be taken, however, to prevent systemic hypotension. Fluid restriction or venodilatation to reduce the venous pressure in the venous sinuses is inappropriate. It is also important to prevent a rise in venous pressure in the head due to obstruction of the jugular vein by tight clothing or excessive flexion or rotation of the neck.

Second, CSF can also pass out of the intracranial compartment into the spinal compartment through the foramen magnum. Gravity will assist the drainage of CSF into the spinal canal if the patient is positioned head up. The volume of CSF held in the spinal compartment largely depends on the compressibility of the venous plexus in the spinal canal.

If the volume of the brain continues to increase and more and more CSF is expelled from the intracranial compartment, there comes a stage when the surface of the brain occludes the arachnoid granulations and cuts off the escape of CSF from this route. Then, as the brain swelling progresses, the cerebellar tonsils are forced down through the foramen magnum, cutting off the movement of CSF out of the intracranial compartment into the spinal canal. CSF then has no means of escape but continues to be secreted. The ICP rises inexorably. This reduces cerebral perfusion pressure, increases cerebral ischaemia and further forces the cerebellar tonsils into the foramen magnum. Local pressure on the brainstem causes brainstem ischaemia and brainstem death.

This horrific scenario may be unavoidable. But not infrequently CSF can be removed from the intracranial compartment through a drain. The commonest method is through a ventricular drain. A soft, fine-bore tube is directed into one of the lateral ventricles and CSF is allowed to drain from the ventricles into an external drainage receptable. A ventricular drain has the advantage of also enabling ICP and compliance to be monitored. However, in many instances of head injury there is so much swelling and oedema of the brain that the ventricles are obliterated and a ventricular drain cannot be safely inserted.

Methods of altering intracranial blood volume

Intracranial blood volume is quite small (60–80 ml), but in the condition of raised ICP, when the patient is on the steep part of the compliance curve, even small changes in the intracranial blood volume can have dramatic effects on ICP.

The relations between cerebral blood flow, cerebral blood volume, cerebral vascular resistance, intracranial compliance and intracranial pressure are complex and not fully understood. They are influenced by the presence or absence of cerebral autoregulation. Because autoregulation is not necessarily uniform throughout the brain substance, the correct management in any particular circumstance is difficult to define, as one strategy that may benefit one part of the brain may not be appropriate for another.

Systemic blood pressure and autoregulation

Cerebral blood flow is a function of the cerebral perfusion pressure and cerebral vascular resistance. In normal brain there is an intrinsic autoregulatory mechanism which maintains cerebral blood flow relatively constant and sufficient for cerebral metabolic requirements. When arterial pressure rises the autoregulatory mechanism constricts the cerebral arterioles and increases vascular resistance, maintaining blood flow unchanged. Arteriolar dilatation occurs when systemic pressure falls. Autoregulation has a normal lower limit of 50–80 mmHg and upper limit of 130–170 mmHg mean arterial pressure, and the range is shifted upwards in hypertensive patients.

Therefore, if autoregulation is present, a rise in systemic pressure causes constriction of the cerebral arterioles, a decrease in cerebral blood volume and a fall in ICP. On the other hand, if autoregulation is absent, the rise in blood pressure increases cerebral blood volume, which increases ICP.[23] If the rise in blood pressure is excessive, the capillaries, no longer protected by arteriolar constriction, are exposed to high pressures and an increase in capillary permeability and breakthrough oedema can occur.

Hypotension is always detrimental because vasodilatation and a rise in ICP occur if autoregulation is intact, whereas decreased CBF with the danger of cerebral ischaemia occur if autoregulation is defective. The general consensus is that one should steer a middle course and maintain CPP at around 85 mmHg[24] and avoid marked hypertension.

Cerebral venous blood volume

About two-thirds of the cerebral blood volume is held in the venous capillary bed. Venous pressure significantly affects cerebral blood volume. Posture, coughing and straining, jugular venous and airway obstruction, positive pressure ventilation and many other factors that affect central venous pressure will alter cerebral blood volume. In a similar manner to CSF drainage, head-up tilt and careful positioning of the head will reduce ICP and will be beneficial provided CPP is maintained.

Dihydroergotamine increases peripheral vascular resistance largely through venoconstriction. Grande[25] gave 0.25 mg of dihydroergotamine to head-injured patients and noted a decrease in ICP for over an hour, presumably the result of a decrease in cerebral venous blood volume.

Indomethacin, a potent inhibitor of prostaglandin synthesis, is also a vasoconstrictor of cerebral vessels and has been used to decrease cerebral blood volume and ICP.[26] It remains to be seen whether indomethacin is of value in the treatment of head injury but it offers another interesting and novel therapy.

Volatile anaesthetic agents

The majority of volatile anaesthetic agents as well as nitrous oxide cause cerebral vasodilatation through a decrease in the cerebrovascular resistance and thus increase intracranial blood flow. Therefore, in acute head injury it might be best to avoid such agents, although isoflurane is the least problematical provided that ventilation is controlled.

Hyperventilation

It has been established that over the clinical range of arterial carbon dioxide there is a direct linear relation with cerebral blood flow.[27] $Paco_2$ and also pH directly affect cerebrovascular resistance. Therefore, it is essential to avoid hypercapnia in patients with head injury as this will raise ICP. This is one reason why opioids and other respiratory depressant drugs should not be administered to spontaneously breathing head-injury patients. One might also conclude that hyperventilation and the fall in ICP it produces would be beneficial, and for many years deliberate hyperventilation was one of the mainstays of the control of ICP.[28] Recent studies have failed to show any consistent benefit,[29] and the evidence suggests that hyperventilation is associated with a poorer outcome. In one study[30] comparing normocapnia, hypocapnia and hypocapnia with THAM buffer, hypocapnia alone had a poorer result. It was suggested that if hypocapnia had to be used to control ICP, THAM might help to reduce the adverse effects.

Whilst hyperventilation is an effective way of managing acute intracranial hypertension, it should be remembered that it can produce a dangerous decrease in cerebral blood flow, especially if CPP is low. Prolonged hypocapnia is associated with adaptation of the CSF pH and shift in CSF bicarbonate, such that the benefits of hyperventilation are lost. This, however, may take several hours.

Hypoxia

Hypoxia increases cerebral blood flow and cerebral blood volume and increases intracranial pressure. Hypoxia should be strenuously avoided.

Cerebral metabolic rate

There is also a direct relation between cerebral metabolic rate, cerebral oxygen consumption and cerebral blood flow and volume. Any decrease in cerebral metabolic rate will decrease ICP.

Induced hypothermia was used as a technique for cerebral protection in neurosurgical anaesthesia many years ago but, because the technique is complicated and troublesome, it is now seldom used. There is, nevertheless, a body of animal work that continues to demonstrate the effectiveness of hypothermia in brain protection.[31] How effective induced hypothermia is in human head injury in the post-ischaemic phase is as yet unknown. In animal experiments, hypothermia strongly suppresses the production of excitatory amino acids such as glutamate, aspartate and glycine.[32] Hypothermia also preserves protein kinases which are otherwise destroyed in normothermic ischaemia. These protein kinases limit intracellular calcium accumulation and protein phosphorylation.[33] The time may be approaching when the role of induced hypothermia should be seriously reviewed.[34,35]

Intravenous sedatives and analgesics

Intravenous opioids and barbiturates decrease cerebral metabolic rate and lower ICP. At the same time they may cause a fall in blood pressure. They should be used with controlled ventilation to prevent hypercapnia from respiratory depression. Because barbiturates are known to be free radical scavengers, there was a vogue for deep barbiturate coma in the treatment of head injury. Rather disappointingly, this form of therapy has not proved to be clearly beneficial.[36]

Intensive care management of head injury

The majority of severely head-injured patients will be referred to a neurosurgical centre and nursed on an intensive care unit which will provide facilities for neurological observations and monitoring, artificial ventilation and care of the unconscious patient.

A brief summary of the practical management of a severe head injury is now outlined.

Posture

The patient should be nursed for the majority of the time in the head-up posture, care being taken not to obstruct the jugular veins. This is to reduce cerebral venous pressure and intracranial venous blood volume and to encourage CSF drainage into the venous system and spinal compartment.

Diuretics

Mannitol should be given to produce an osmotic diuresis and reduce cerebral oedema. Frusemide may enhance this and also decrease CSF production and increase CSF absorption through the ventricular wall. Bladder catherization is advisable as it prevents urinary retention and allows urine output to be monitored.

Cerebral perfusion

Cerebral perfusion pressure should be maintained at about 85 mmHg by fluid replacement with blood or colloid as necessary, and inotropic agents may be needed. Intra-arterial and central venous pressure measurements are useful guides in the control of the fluid replacement.

Security of the airway

The unconscious patient is at risk from hypoxia and hypoventilation and also from aspiration of secretions. It is safest to maintain a clear airway with tracheal intubation. The oral route is the easiest method of intubation in an emergency but is more likely to stimulate coughing and straining and to increase the requirement for sedation, analgesia and muscle relaxation. Nasal intubation, although more comfortable, requires more nursing expertise to manage well and may be precluded because of a basal skull fracture. There is an increased incidence of nasal sinus infection with nasal intubation but this can be minimized by using smaller sized tubes and frequent nursing care of the nasal passages. Tracheostomy may be advisable if assisted ventilation is likely to be needed for more than a few days and also if there is major facial injury.

Tracheostomy has many advantages. It is easier to nurse a patient with a tracheostomy. It is more comfortable for the patient and requires minimal sedation. There is no pressure to extubate the patient at the earliest possible moment, and weaning can be performed in an unhurried, safe manner. With the increasing popularity of percutaneous dilatational techniques, tracheostomy can be performed safely with the minimum of fuss at the bedside.

Sedative and analgesic drugs

Sedation and analgesia may be necessary for the patient to tolerate intubation and artificial ventilation. In addition, analgesics and sedatives may help in reducing cerebral oxygen consumption and lowering ICP and may also have some brain-protective effects. Currently the agents of choice are propofol and midazolam as sedatives and fentanyl or alfentanil as analgesics. These are usually given as continuous intravenous infusions. The major disadvantage of sedation and analgesia is the difficulty in making an accurate neurological assessment. In general, the shorter acting the agent the better, so that neurological assessment can be made more frequently.

Muscle relaxants

The use of muscle relaxants in intensive care is declining. However, one indication for their use is to prevent the chance of coughing and straining in intubated and ventilated patients with critically high ICP. The shorter acting muscle relaxants such as atracurium and vecuronium, administered as continuous intravenous infusions, are the drugs of choice because rapid, reliable reversal allows easier neurological assessment.

Ventilation

It is important to ensure adequate oxygenation. As discussed above, hypocapnia has not been shown to be beneficial in recent studies and the general consensus is that there is a risk of inducing cerebral ischaemia with excessive hyperventilation. $Pa\text{CO}_2$ should be maintained above 4.0 kPa (30 mmHg).

Weaning

Having stabilized the patient's condition and allowed time for the surgical decompression of the brain and the resolution of swelling, there should then come a stage when cerebral perfusion is more normal and autoregulation has returned. This stage may be indicated by a stable low ICP and a reduction in brain swelling on the CT scan. Careful clinical observation may indicate the return of more normal neurological responses. A decision has then to be made as to whether sedation, analgesia and, if necessary, muscle relaxants should be reduced and a process of weaning from artificial ventilation initiated.

It is probably unwise to allow the $Pa\text{CO}_2$ to rise to normal levels too fast, and a gradual reduction in mandatory ventilation is required. This is to allow compensatory shifts in CSF pH and bicarbonate to take place over a period of a few hours. The more modern and versatile range of ventilators have particular advantages in allowing synchronized intermittent mandatory ventilation and inspiratory pressure support to produce a gradual transition from controlled to spontaneous ventilation. Some degree of light sedation may still be required to prevent the patient from objecting to the presence of a tracheal tube. Once spontaneous ventilation is established and secure, and provided there is an adequate gag reflex to protect the airway, extubation can take place. Throughout the period of intensive care, general management of the unconscious patient continues, special attention being paid to nutrition, skin, muscle and joint care as well as physiotherapy to the chest.

Neurological observations and monitoring

The patient with a head injury who is being sedated and who has also been given analgesics and muscle relaxants cannot be accurately assessed neurologically. The Glasgow Coma Scale is meaningless and most reflex responses will be altered in some way. Probably the only reliable response is that of the pupil to light, which will continue to signify the integrity of the brainstem. Flexion and extension movements of the limbs in response to pain may still be obtained and can be used to identify hemiplegia and other focal damage. But clinical signs alone may not indicate the true intracranial state. Other forms of investigation are required. The two main methods are repeated CT scanning and ICP monitoring.

CT scanning

CT scanning should be readily available in all neurosurgical centres, and these days scanning only takes a few minutes. It will reveal the position and size of a haematoma, the extent of cerebral oedema and infarction, the size of the ventricles, any shift of midline structures, compression of the brainstem and much other useful information. For these reasons, CT scanning is often regarded as being more important than ICP monitoring.

ICP measurements

The 'gold standard' for measuring ICP is the use of an intraventricular cannula introduced through a burr hole in the skull. The cannula is then connected to an external transducer. This allows accurate zeroing and calibration of the transducer. It also allows small volumes of fluid to be withdrawn or introduced into the ventricles to determine intracranial compliance. The technique does, however, have two major drawbacks. First, the ventricles may be so compressed that an intraventricular cannula cannot be introduced to start with and, second, there is the hazard of introducing intracranial infection.

Alternatives to intraventricular cannulation include ICP measurement in the subdural or subarachnoid space. The drawback of this technique is that with cerebral swelling the surface of the brain may occlude the opening of the transducer and give a false pressure reading. Several devices have been designed which screw or bolt in place through a burr hole in the skull. None has been entirely satisfactory and the risk of infection is still present.

Extradural transducers, which lie between the dura and the inner table of the skull, have also been used to reduce the risk of intracranial infection. But these too have not been satisfactory because the tension of the dura itself provides a variable barrier to the transmission of pressure from the intracranial to the extradural compartments.

Newer technology, for example the use of fibreoptics, has provided other types of transducers which can be calibrated easily but whether they will replace the ventricular cannula remains to be seen.

ICP monitoring can offer the following:

1. Provide a continuous display of ICP so that one can detect at an early stage any undesirable increases in ICP and initiate treatment to decrease the pressure. It will also demonstrate the effectiveness of the treatment.
2. It allows the accurate calculation of cerebral perfusion pressure and can therefore guide the use of fluids and inotropic agents to maintain this.
3. The ICP waveform may show certain pressure waves described by Lundberg which are of prognostic significance.
4. Enable the assessment of intracranial compliance by either an alteration in intracranial fluid volume or permit the processing of the waveform amplitude with each pulsatile change in intracranial blood volume.

Other forms of monitoring

A-mode doppler ultrasound is a useful and sensitive method for detecting a shift in midline structures. The technique can be easily applied at the bedside but has not been popular since the advent of CT scanning, as the information derived is limited. Also, the equipment is costly and requires some expertise.

Jugular venous Po_2 can be monitored using a catheter introduced retrograde up the jugular vein to the jugular bulb; recently, fibreoptic probes that enable the oxygen saturation to be monitored continuously have become available.[37] These can demonstrate cerebral hypoxia. However, the interpretation of the changes can be difficult, and areas of totally unperfused brain are not detected by the technique. The combination of jugular venous saturation and lactate level may be more meaningful. The oxygen saturation probes are rather prone to artefact, for instance when positioned against a vessel wall; furthermore, when cerebral perfusion is low, the jugular venous blood may be contaminated with extracerebral venous blood which could give high oxygen saturation values.

Visual evoked responses or *somatosensory evoked responses* have been used for the assessment of brainstem damage following head injury but are not commonly used to monitor the head injury itself.

The EEG is of value in the detection of seizures in the head-injured patient. The cerebral function monitor or cerebral function analysing monitor is an easier system to use in the intensive care unit than the raw EEG signal, as the system does not require expert interpretation.

Outcome

The outcome following head injury can be defined crudely in terms of death and survival. The mortality in severe head injury (GCS <9) has fallen from about 50 per cent to 30–40 per cent in recent years. However, often it is the quality of survival that is more relevant. Jennett and Bond[38] devised a scheme, the Glasgow Outcome Scale, which attempts to describe the overall social capability or dependence of the individual, using five main categories:

1. Death.
2. Vegetative state (absence of awareness).
3. Severely disabled (conscious but disabled and dependent).
4. Moderately disabled (disabled but independent).
5. Good recovery (resumption of normal life).

Categories 3–5 can be subdivided into further degrees of disability.

Vegetative state
This can be described as non-sentient survival, in which the cortex is not functional. There is an absence of awareness with no evidence of psychologically meaningful activity. However, brainstem function is often well preserved with spontaneous breathing, swallowing and other brainstem reflexes present. Although the majority of vegetative patients will die within the first year, many such patients are capable of long-term survival if adequate nutrition and nursing care are provided. The condition is usually the result of extensive diffuse damage to the white matter or the consequence of severe cerebral hypoxia or cardiac arrest.

Severe disability
This term is used to describe a patient who is conscious but dependent on others for daily living because of mental or physical handicap or both. The patient may be cared for in an institution or at home with major family support.

Moderate disability

This applies to patients who are able to lead an independent existence, travel on public transport and perhaps work in a sheltered environment. Their disabilities would include hemiplegia, ataxia, dysphasia and some intellectual, memory or personality disorder.

Good recovery

This implies that any neurological or psychological deficits do not prevent the resumption of a normal life with normal family relationships and leisure activities.

The Glasgow Outcome Scale is a practical scale to assess disability, which can be applied with consistency by a wide range of clinicians. With time, the degree of disability may change and a patient may move from one outcome grade to another. One value of such a scale is the objective assessment of ill-health. This can be used as an index for compensation for injuries. When making an estimate of the cost of injury, one should also take into consideration the age of the patient, the potential duration of the disability and the humanitarian costs as well as the material costs. Used in conjunction with the Glasgow Coma Scale, the Glasgow Outcome Scale can form the basis for the comparison and cost–benefit analysis of different regimens of treatment.

Ethical issues

The management of head injury often presents medical and nursing staff with difficult moral and ethical decisions. The majority of the difficult decisions are focused on the withholding or withdrawal of life-sustaining therapy.

A patient who fulfils the criteria for the diagnosis of brainstem death does not pose an ethical dilemma. The clinical situation is clear and the decisions that have to be made are concerned with the wishes of the relatives regarding the time of withdrawal of life support and their permission for the use of the patient as an organ donor.

The decision 'not to resuscitate' if a cardiorespiratory arrest occurs is also a relatively easy decision which should be based on the likely outcome of the illness and the predicted quality of life. It is made easier if the patient has expressed a wish not to be resuscitated under specific circumstances. In each case it is important to discuss the decision with a close relative or friend of the patient and inform those caring for the patient and also to document it in the patient's notes so that there is no ambiguity.

The more difficult decisions are whether to embark on a course of investigation and treatment that is unlikely to benefit the patient or to withdraw treatment when it is considered 'futile'.[39] In recent years there has been an increase in interest in these matters, for several reasons. A

number of such dilemmas have been highlighted in the popular press, and public and medical attitudes are changing. The data available concerning the outcome of specific clinical situations are quite extensive and the early prediction of outcome is now improved. The current cost of intensive care is so high that doctors are obliged to be more selective in their criteria for admission to and discharge from intensive care units.

It is generally accepted that the use of critical care resources should be governed by four basic considerations: patient autonomy, beneficence, non-maleficence and justice.[40]

It is a right of patients, who are fully sentient and well informed of the facts of their situation, to refuse treatment offered by their doctors. In the case of the severely head injured, this *patient autonomy* is unlikely and it is necessary to appoint a surrogate decision-maker. This should be a close relative or friend who is aware of the likely wishes and attitudes of the patient. The surrogate decision-maker should be fully informed of the clinical situation, the likely outcome and consequences of any course of action and should participate in the decision-making process.

It is the doctors' responsibility to *maximize beneficence*. Everything should be done in the early stages to save life and preserve the brain function and stabilize the patient's condition. However, it may be apparent within a few hours that the injuries are irrecoverable. At this point it is helpful to write a 'not for resuscitation' order and thus avoid 'futile' life-sustaining intervention. The term 'futile' may be used when any intervention is highly unlikely to result in a meaningful survival for the patient. It is important to resist beginning what Jennett calls 'the vicious cycle of commitment'[41] when, having made one decision to intervene, it becomes more difficult to avoid further interventions. It is not beneficent to prolong a patient's biological life if that life has no value to the patient. In fact, it could be interpreted as being *maleficent* to expose the patient to an extended, undignified process of dying. Certainly it is unkind to subject the patient's relatives to a prolonged period of distress. Whenever possible, a surrogate decision-maker should be identified and the position discussed in an objective manner. The withholding or withdrawal of life-sustaining therapy does not mean that all therapy should cease. Much can be done to relieve any distress and allow a peaceful and dignified death and to comfort the relatives.

If the situation is futile, the physician has no responsibility to prolong survival. Careful consideration should be given to the justification for admitting a futile case to the intensive care unit. In the current climate, when resources are limited, expensive treatment administered to a patient who cannot benefit from it is at the same time denying teatment to one who might benefit. Similarly, if a decision to withdraw treatment is made when the patient is in the intensive care unit, on the basis of *justice* alone, the patient should be transferred out of the unit. This can be a hard decision to make because of the anxiety of the relatives who should be given reassurance that all care necessary will continue to be provided.

The decision-making can be greatly helped by having guidelines for the admission to and discharge from the intensive care unit. These guidelines should be agreed by the clinicians and nursing staff and approved by the hospital management and in particular their legal department. A set of written guidelines would have the advantage of forming the basis for decision-making on sound ethical grounds, representing a group judgement outside the emotional context of an individual case. Such a document would ideally be part of a wider educational programme to support doctors, nurses and other professionals in making ethical decisions on the withholding and withdrawal of life-sustaining therapy in other spheres apart from head injury.[42]

Brainstem death

The concept of brainstem death is based on the belief that the ascending reticular activating system (ARAS), which connects the cerebral cortex to the rest of the body and the external environment, is essential for sustaining awareness or consciousness. If it can be demonstrated that there is permanent and irreversible disruption of the ARAS, awareness can never again occur and this can thus be regarded as a state of death. The ARAS is a diffuse core of cells which passes through the brainstem and is intimately connected with the cranial nerve nuclei. The complete absence of cranial nerve activity is believed to be congruent with complete inactivity of the ARAS. In order to be sure that the situation is irreversible, certain preconditions have to be fulfilled; these preconditions form a crucial base for the procedure to diagnose brainstem death outlined by the Medical Royal Colleges in the UK in 1976.[43–45] All patients in whom the diagnosis is to be contemplated will have been in unresponsive coma for several hours, requiring artificial ventilation. Frequently the autonomic and temperature-regulating systems are impaired, and active warming and inotropic drugs and fluid are required to maintain adequate tissue perfusion.

Pre-conditions

1. An intracerebral lesion that will account for the comatose state has to be identified. If no anatomical cause can be shown, the diagnosis should not be made. This means that a CT scan is usually required.
2. There should be no drug acting at the time of testing which might depress the central nervous system. If in doubt, screening for commonly used drugs may be necessary and blood levels of drugs known to have been administered may need to be measured. Sufficient time should be allowed for drug elimination to take place before testing for brainstem death.
3. There should be no neuromuscular blocking agent acting at the time of testing, which might mask the motor signs to be elicited in the test. This can be confirmed using peripheral nerve stimulation.
4. The body temperature should be above 33°C but better above 35°C. Active warming is frequently necessary, as hypothermia alone can suppress brainstem reflexes.
5. There should be no endocrine or metabolic cause for a comatose state. Renal and hepatic failure, myxoedema and diabetic coma and similar conditions should be excluded, and blood electrolytes should be within the normal range.

If the preconditions are fulfilled, one can be satisfied that any reflex activity in the cranial nerve distribution should be detectable. That being so, it is a relatively simple matter to choose a set of reflex responses that will involve the cranial nerve nuclei from the upper to the lower brainstem. These tests should be simple to perform without elaborate equipment and have distinct and easily identifiable endpoints.

Table 51.4 shows the commonly used tests of cranial nerve function. Each reflex has an afferent and efferent part involving the cranial nerve nuclei. Olfactory function (cranial nerve I) cannot be tested in a comatose patient.

For a diagnosis of brainstem death, no response at all should be elicited with the tests. Any reflex motor response detected in the cranial nerve distribution would indicate some activity in the brainstem and the diagnosis of brainstem death cannot be made.

If no activity is detected, then, provided the preconditions have been satisfied, there should be absolute confidence that the brainstem is completely and irreversibly disrupted and the diagnosis can be made. There is no medical reason to perform a second set of tests. However, conventionally, largely for social reasons, a second test is usually required. An advantage of having a second set of tests performed is that it gives relatives time to appreciate the significance of the results of the first test and to come to some conclusion regarding how to proceed once the result of the second test has been declared. The options are to (1) withdraw artificial ventilation immediately, (2) continue life support until cardiac function inevitably ceases (usually within 72 hours of brainstem death) and (3) consent to organ donation.

Table 51.4 Commonly used tests in the diagnosis of brainstem death

Test	Pupillary reflex	Corneal reflex	Caloric test	Gag reflex	Tracheal stimulation	Pain response	Respiratory movement
Stimulus	Shine bright light into one eye	Wipe cotton wool whisp over cornea	Syringe auditory canal with cold water	Stimulate pharynx with probe or move tracheal tube	Stimulate carina with suction catheter down tracheal tube	Pinch or pressure to head, neck or upper trunk	Allow Pa_{CO_2} to rise to >6.7 kPa (50 mmHg) (Decrease ventilation or add 5% CO_2)
Afferent pathway	Optic nerve II	Ophthalmic branch of trigeminal Va	Vestibular nerve VIII	Glossopharyngeal nerve IX	Vagus nerve sensory X	Afferent nerves to upper body	Respiratory chemoreceptor drive
Normal response	Bilateral pupillary constriction	Blink or grimace	In unconscious subject: tonic deviation of eyes towards side of syringing	Gag or swallow	Cough response	Grimace or movement in cranial nerve distribution	Spontaneous respiratory movement
Efferent pathway	Edinger–Westphal nucleus Parasympathetic III	Facial nerve VII	Nerves to extraocular muscles III, IV, VI	Vagus nerve motor X	Respiratory neurons in medulla	Via brainstem to motor cranial nerves	Respiratory neurons in medulla
Comment	Test bilaterally	Test bilaterally	Test bilaterally Check canal not blocked	—	—	Look only for movement in head and neck Ignore possible spinal reflex	Preoxygenate and apnoeic oxygenation during test. Check Pa_{CO_2} with arterial blood sample

The opportunity to obtain organs for transplantation from a brainstem-dead individual should not be missed, as from one donor a large number of individuals will be benefited. Two kidneys will help two patients directly and allow a further two onto a dialysis programme. The heart and lungs potentially will benefit two more patients, and liver and pancreas and bowel may help others. Also two corneas will benefit a further two patients. It is not just the recipients of organs who benefit but also their families and dependants. Society as a whole benefits because, for instance, a successful renal transplant is much more cost-effective than dialysis, and restoring health and enabling someone to return to work reduces the load on social services.

Brainstem death should be clearly distinguished from cortical brain death which, at the moment, can be satisfactorily diagnosed only if no cortical perfusion can be demonstrated during cerebral angiography.

REFERENCES

1. Jennett B, MacMillan R. Epidemiology of head injury. *British Medical Journal* 1981; **282**: 101–4.
2. Uematsu D, Araki N, Greenberg JH, Sladky J, Reivich M. Combined therapy with MK-801 and nimodipine for protection of ischemic brain damage. *Neurology* 1991; **41**: 88–94.
3. Shapira Y, Artra AA, Lam AM. Ketamine decreases cerebral infarct volume and improves neurological outcome following experimental head trauma in rats. *Journal of Neurosurgical Anesthesiology* 1992; **4**: 231–40.
4. Ikede Y, Long DM. The molecular basis of brain injury and brain edema: the role of oxygen free radicals. *Neurosurgery* 1990; **27**: 1–11.
5. Muizelaar JP, Marmarou A, Young HF, *et al*. Improving the outcome of severe head injury with the oxygen radical scavenger polyethylene glycol-conjugated superoxide dismutase: a phase 2 trial. *Journal of Neurosurgery* 1993; **78**: 375–82.
6. Nowicki JP, Duval D, Poignet H, Scatton B. Nitric oxide mediates neuronal death after focal cerebral ischemia in the mouse. *European Journal of Pharmacology* 1991; **204**: 339–40.
7. Bielenberg GW, Wagener G, Beck T. Infarct reduction by the platelet activating factor antagonist apafant in rats. *Stroke* 1992; **23**: 98–103.
8. Clark WM, Madden KP, Rothlien R, Zivin JA. Reduction in central nervous system ischemic injury in rabbits using leukocyte adhesion antibody treatment. *Stroke* 1991; **22**: 877–83.
9. Teasdale G, Jennett B. Assesment of coma and impaired consciousness. A practical scale. *Lancet* 1974; **2**: 81–5.
10. Briggs M, Clarke P, Crockard A, *et al*. Guidelines for initial management after head injury in adults. Suggestions from a group of neurosurgeons. *British Medical Journal* 1984; **188**: 983–5.
11. Teasdale GM, Murray G, Anderson E, *et al*. Risks of acute traumatic intracranial haematomata in children and adults: implications for managing head injuries. *British Medical Journal* 1990; **300**: 363–7.
12. Klauber MR, Marshall LF, Luerssen TG, Frankowski R, Tabaddor K, Eisenberg HM. Determinants of head injury mortality: importance of the low risk patient. *Neurosurgery* 1989; **24**: 31–6.
13. Stein SC, Ross SE. Moderate head injury: a guide to initial management. *Journal of Neurosurgery* 1992; **77**: 562–4.
14. Gentleman D, Jennett B. Audit of transfer of unconscious head-injured patients to a neurosurgical unit. *Lancet* 1990; **335**: 330–4.
15. Andrews PJD, Piper IR, Dearden NM, Miller JD. Secondary insults during intrahospital transport of head-injured patients. *Lancet* 1990; **335**: 327–30.
16. Vyvyan HAL, Kee S, Bristow A. A survey of secondary transfers of head injured patients in the south of England. *Anaesthesia* 1991; **46**: 728–31.
17. Graham DI, Adams JH, Doyle D. Ischemic brain damage in fatal non-missile head injuries. *Journals of the Neurological Sciences* 1978; **39**: 213–34.
18. Marshall LF, Smith RW, Rauscher LA, Shapiro HM. Mannitol dose requirements in brain-injured patients. *Journal of Neurosurgery* 1978; **48**: 169–72.
19. Worthley LIG, Cooper DJ, Jones N. Treatment of resistant intracranial hypertension. *Journal of Neurosurgery* 1988; **68**: 478–81.
20. Vassar MJ, Perry CA, Gannaway WC, Holcroft JW. 7.5% Sodium chloride dextran for resuscitation of trauma patients undergoing helicopter transport. *Archives of Surgery* 1991; **126**: 1065–72.
21. Shackford SR, Zhuang J, Schmoker J. Intravenous fluid toxicity; effect on intracranial pressure, cerebral blood flow, and cerebral oxygen delivery in focal brain injury. *Journal of Neurosurgery* 1992; **76**: 91–8.
22. Schettini A, Stahurski B, Young HF. Osmotic and osmotic-loop diuresis in brain surgery. Effects on plasma and CSF electrolytes and ion excretion. *Journal of Neurosurgery* 1982; **56**: 679–84.
23. Bouma GJ, Muizelaar JP, Bandoh K, Marmarou A. Blood pressure and intracranial pressure–volume dynamics in severe head injury: relationship with cerebral blood flow. *Journal of Neurosurgery* 1989; **77**: 15–19.
24. Schmidt JF, Waldemar G, Vorstrup S, Andersen AR, Gjerris F, Paulson OB. Computerized analysis of cerebral blood flow autoregulation in humans: validation of a method of pharmacologic studies. *Journal of Cardiovascular Pharmacology* 1990; **15**: 983–8.
25. Grande P-O. The effects of dihydroergotamine in patients with head injury and raised intracranial pressure. *Intensive Care Medicine* 1989; **15**: 523–7.
26. Jensen K, Ohrstrom J, Cold GE, Astrup J. The effects of indomethacin on intracranial pressure, cerebral blood flow and cerebral metabolism in patients with severe head injury

and intracranial hypertension. *Acta Neurochirurgica (Wien)* 1991; **108**: 116–21.

27. Harp JR, Wollman H. Cerebral metabolic effects of hyperventilation and deliberate hypotension. *British Journal of Anaesthesia* 1973; **45**: 256–62.

28. Gordon E. Non-operative treatment of acute head injuries (the Karolinska experience). *International Anesthesiology Clinics* 1979; **17**: 181–99.

29. Jennett B, Teasdale G, Fry J, *et al.* Treatment of severe head injury. *Journal of Neurology, Neurosurgery and Psychiatry* 1980; **43**: 289–95.

30. Muizelaar JP, Marmarou A, Ward JD, *et al.* Adverse effects of prolonged hyperventilation in patients with severe head injury: a randomised clinical trial. *Journal of Neurosurgery* 1991; **75**: 731–9.

31. Sano T, Drummond JC, Patel PM, Grafe MR, Watson JC, Cole DJ. A comparison of the cerebral protective effects of isoflurane and mild hypothermia in a model of incomplete forebrain ischemia in the rat. *Anesthesiology* 1992; **76**: 221–8.

32. Baker AJ, Zornow MH, Grafe MR, *et al.* Hypothermia prevents ischemia-induced increases in hippocampal glycine concentrations in rabbits. *Stroke* 1991; **22**: 666–73.

33. Yamamoto H, Fukunaga K, Lee K, Soderling TR. Ischemia-induced loss of brain calcium/calmodulin-dependent protein kinase II. *Journal of Neurochemistry* 1992; **58**: 1110–17.

34. Milde LN. Clinical use of mild hypothermia for brain protection: a dream revisited. *Journal of Neurosurgical Anesthesiology* 1992; **4**: 211–15.

35. Schubert A. Should mild hypothermia be routinely used for human cerebral protection? The flip side. *Journal of Neurosurgical Anesthesiology* 1992; **4**: 216–20.

36. Ward JD, Becker DP, Miller JD, *et al.* Failure of prophylactic barbiturate coma in the treatment of severe head injury. *Journal of Neurosurgery* 1985; **62**: 383–8.

37. Sheinberg M, Kanter MJ, Robertson CS, Contant CF, Narayan RK, Grossman RG. Continuous monitoring of jugular venous oxygen saturation in head-injured patients. *Journal of Neurosurgery* 1992; **76**: 212–17.

38. Jennett B, Bond M. Assessment of outcome after severe brain damage. A practical scale. *Lancet* 1975; **1**: 480–4.

39. Jennett B. Severe head injuries: Ethical aspects of management. *British Journal of Hospital Medicine* 1992; **47**: 354–7.

40. Luce JM. Conflicts over ethical principles in the intensive care unit. *Critical Care Medicine* 1992; **20**: 313–15.

41. Jennett B. Inappropriate use of intensive care. *British Medical Journal* 1984; **289**: 1709–11.

42. American Thoracic Society. Withholding and withdrawing life-sustaining therapy. *Annals of Internal Medicine* 1991; **115**: 478–85.

43. Conference of Medical Royal Colleges and their Faculties (UK). Diagnosis of brain death. *Lancet* 1976; **2**: 1069–70.

44. Conference of Medical Royal Colleges and their Faculties (UK). Diagnosis of death. *British Medical Journal* 1979; **1**: 322.

45. Pallis C. *ABC of brainstem death* London: British Medical Journal, 1983.

Anaesthesia for Neurosurgery

J. M. Turner

Patients requiring neurosurgery present specific problems for the anaesthetist. Some of these relate to the disordered physiology produced by neurosurgical disease and some to the cerebral effects of anaesthetic drugs, which produce major alterations in cerebral and cerebrovascular function in ways that may be beneficial or deleterious. Other problems are presented by the special requirements of neurosurgery.

Detailed knowledge of the effects of drug action on the brain has allowed anaesthetists to control brain bulk and the cerebral circulation in ways that can aid surgery. Special positioning may be required, as may the use of techniques such as hypotension. The anaesthetist must provide safety for the patient and also good intracranial operating conditions. A badly administered anaesthetic may make neurosurgical procedures difficult or impossible, and may be associated with brain damage.

The patient

Intracranial space occupation

Intracranial tumours may be benign or malignant, primary or secondary. The changes brought about by the occupation of intracranial space, of whatever sort, are well recognized. If the tumour is growing in an 'eloquent' area of the brain, such as the motor cortex, it will produce symptoms or signs at an early stage of development, whereas, if it is affecting a relatively 'silent' area of the brain, it may grow to considerable size and produce symptoms and signs related to the general effects of space occupation rather than to its specific position. An example of the impact of tumour size is seen in Fig. 52.1, which shows the major distortion produced by a large, 'benign' calcified bifrontal meningioma.

The effects of occupation of intracranial space

The skull contains the brain (about 1400 g in the adult), cerebrospinal fluid (CSF) (140 ml in total, of which half is in

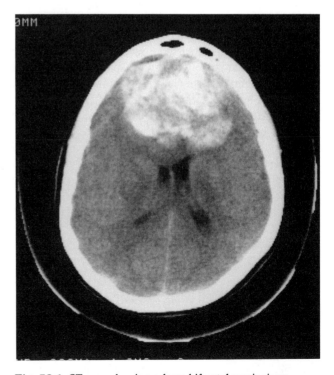

Fig. 52.1 CT scan showing a large bifrontal meningioma.

the cranial space and half in the spinal space), arterial blood (about 50 ml/100 g brain per minute) and venous blood. As blood is forced into the closed box of the skull, a distinct pressure is generated; this is called the intracranial pressure (ICP) and has a normal value of 15 mmHg or less. ICP is not a static pressure but varies with arterial pulsation, with breathing and with events such as coughing. Initially, when space occupation develops, compensation takes place as CSF is moved from the cerebral space to the spinal space and venous blood is lost from the thin-walled cerebral veins to the great veins in the chest, so that the pressure in the skull, the ICP, does not change. Clearly, as the volume of the space occupation continues to increase, more CSF and/or venous blood is lost from the intracranial space and eventually the compensation mechanisms will become exhausted. At this point the ICP may rise considerably. The relation between the volume of intracranial space occupation and the intracranial pressure is displayed graphically, as in Fig. 52.2. It should be noted that, even in the early phase of space occupation, compensation is not perfect, some rise in ICP being noted from the start. Sullivan et al.[1] also pointed out that, as the individual intracranial constituents have slightly different responses to volume loading, the ICP curve shown represents a simplified situation and will change as other factors change and between patients. Addition of a small amount of fluid to the ventricular system will produce a pressure rise, which will be small when the mechanisms for intracranial volume compensation are intact, but large when the capacity to compensate is exhausted. This response is termed the volume–pressure response (VPR).

The intracranial contents, being essentially liquid, are incompressible, but the brain is able to deform in a plastic fashion when space occupation develops. This allows for some accommodation for intracranial space occupation, and it is not uncommon to see shift of the midline structures on CT scan. A supratentorial tumour may force part of the cerebral hemisphere, usually the temporal lobe, to become impacted beneath the falx cerebri.

When the ICP rises to high values, the medulla and cerebellar tonsils may be forced from the posterior fossa into the narrow confines of the foramen magnum. The development of such a process, the medullary pressure cone, was described by Cushing.[2] Its significance is that the vital centres in the medulla are so compressed that cardiovascular function and ventilation are affected, producing the classic picture of hypertension, bradycardia and ventilatory irregularity.

Cerebrospinal fluid

Cerebrospinal fluid is formed from the choroid plexuses in the cerebral ventricles at a constant rate, about 0.4 ml/min, so that an amount of CSF equal to the total CSF volume is formed in approximately 4 hours.[3] The formation of CSF is an active process, requiring sodium–potassium-activated ATPase and carbonic anhydrase. The rate of production stays constant over a wide range of ICP.[4] Some CSF is also formed by the passage of brain tissue water across the ependymal lining of the ventricles and along perivascular channels into the subarachnoid space. CSF circulates through the third ventricle, along the aqueduct and into the fourth ventricle, to reach the subarachnoid space through the exit foramina.

Resorption of CSF takes place through the arachnoid villi, as a passive process, requiring a pressure gradient between the CSF and the blood in the venous sinus. If the sinus pressure is raised, CSF reabsorption is slowed.[5]

Tumours may also cause a rise in ICP by obstructing the flow of CSF through the ventricular system and by allowing the formation of oedema. Oedema formation may be extensive and related to the abnormal vascularity of the tumour. A small tumour may produce a serious degree of intracranial space occupation if oedema production is extensive (Fig. 52.3).

Tumours of the posterior fossa are particularly dangerous. The posterior fossa has a small volume, so the tumour mass will produce significant pressure effects at an earlier stage than a supratentorial tumour (Fig. 52.4). The patient in this instance presented with symptoms of high ICP and required an urgent ventriculoperitoneal shunt because of the CSF obstruction, demonstrated in Fig. 52.5. Local pressure effects produced by the tumour may interfere with the function of the lower cranial nerves, resulting, for example, in laryngeal incompetence. Figure 52.6 shows an acoustic neuroma distorting the midbrain. When this patient presented, he showed signs of laryngeal incompetence as well as the deafness related to the acoustic neuroma.

Symptoms and signs of raised ICP

The symptoms and signs of raised ICP need to be distinguished from those due to the lesion producing the raised ICP. Miller[6] suggested that the symptoms and signs due to raised ICP alone are headache, vomiting,

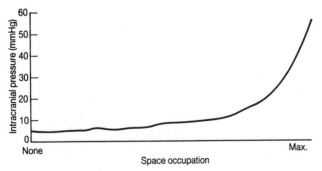

Fig. 52.2 The relation between intracranial space occupation and intracranial pressure.

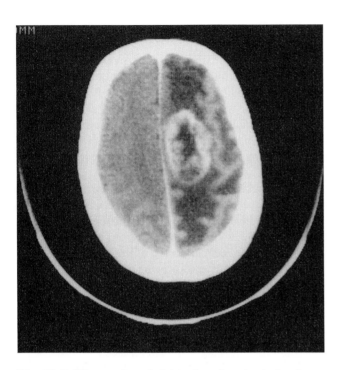

Fig. 52.3 CT scan after administration of contrast, showing an enhancing tumour of moderate size with an extensive degree of oedema production.

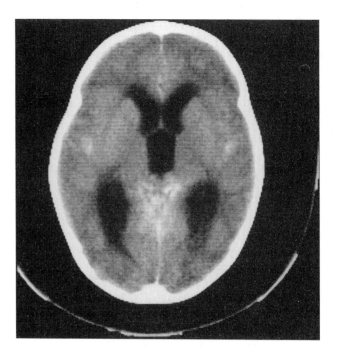

Fig. 52.5 CT scan from the same patient as Fig. 52.4, showing the hydrocephalus caused by the posterior fossa tumour.

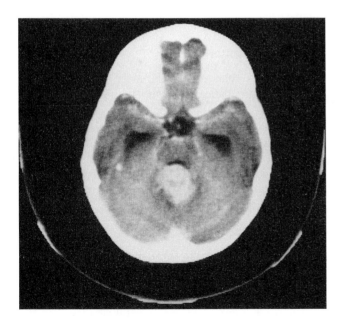

Fig. 52.4 CT scan showing a midline tumour in the posterior fossa. The fourth ventricle is visible.

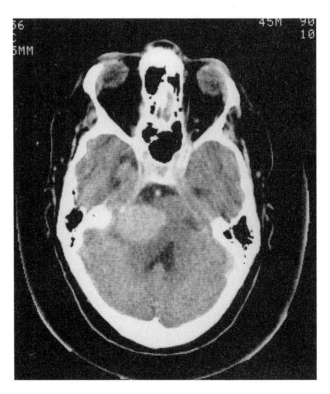

Fig. 52.6 CT scan showing an acoustic neuroma causing midbrain distortion.

papilloedema and drowsiness, whereas other signs, including bradycardia, arterial hypertension and pupillary changes, although often associated with raised ICP, may arise from brainstem distortion or ischaemia.

Bilateral papilloedema is the one sign that appears to be directly related to raised ICP, but takes 12–24 hours to develop.

Cerebral blood flow

The blood supply to the brain, about 15 per cent of the cardiac output, is provided mainly by the internal carotid (which provides two-thirds of the flow) and the vertebral arteries (which provide one-third). The basilar artery, formed from the two vertebral arteries and the internal carotid arteries, join to form the circle of Willis at the base of the skull. The cerebral arteries arise from the circle of Willis.

Cerebral veins are thin walled and contain no valves. They drain into the cerebral venous sinuses, which, being contained within two layers of dura, are somewhat protected from the pressure changes in the skull. CSF drainage takes place into the venous sinuses through the arachnoid villi. The venous drainage is through the internal jugular veins or the vertebral venous plexus.

Normal control of cerebral blood flow

Metabolism

The dominant factor controlling cerebral blood flow (CBF) is cerebral metabolism. Cerebral metabolism as revealed by oxygen utilization ($CMRO_2$) is normally about 3.0–3.5 ml/100 g per minute, though regional differences exist. If metabolism increases, a cerebral vasodilatation produces a rise in CBF to meet the increased need for oxygen; if metabolism falls, as for example with thiopentone, CBF also falls. Another measure of metabolism is the cerebral metabolic rate with respect to glucose (CMR_{gluc}).

The normal cerebral circulation responds to changes in CO_2, but some situations, such as injury or subarachnoid haemorrhage and some drugs or techniques, modify the relation. There is normally a sigmoid relation between $PaCO_2$ and CBF. Between 2.7 kPa (20 mmHg) and 10.7 kPa (80 mmHg) $PaCO_2$ the CBF increases fourfold. At lower levels of $PaCO_2$, cerebral vasoconstriction does not increase and CBF does not fall to any great extent. Similarly, at high values of $PaCO_2$ maximum cerebral vasodilatation appears to occur. The reactivity of the cerebral circulation to CO_2 may be reduced by intracranial disease and injury, especially when local cerebral ischaemia produces a local acidosis.

Low levels of PaO_2 (below 6.7 kPa, 50 mmHg) produce a cerebral vasodilatation whilst hyperbaric oxygen produces a vasoconstriction.

Autoregulation

Autoregulation maintains CBF at a remarkably constant value between 50 and 150 mmHg mean arterial pressure (MAP), by vasodilatation when the MAP falls and by vasoconstriction when the MAP rises. Outside the limits, CBF passively follows MAP. Intracranial disease or injury may abolish or impair autoregulation, so through such areas CBF may be pressure passive even at normal blood pressures. When autoregulation is impaired, there is a danger of cerebral ischaemia at relatively small reductions of blood pressure, and at higher blood pressures the danger of damage to the blood–brain barrier and therefore cerebral oedema formation. Cerebral perfusion pressure (CPP), which is normally defined as the difference between MAP and mean ICP, appears to be the source of the autoregulatory process rather than simple MAP. It seems that, when CPP is reduced by a rise in ICP, CBF is maintained at a lower CPP than when CPP is reduced by lowering the MAP.[7, 8] As CPP falls with increasing ICP, the brain is less compliant and the ICP pulse pressure increases, such that the volume of blood forced into the skull with each cardiac stroke provokes a greater pressure response.

It should be noted that the autoregulatory process is relatively slow to act in response to a change in CPP and that it may take up to 2 minutes.

The effectiveness of autoregulation also appears to be easily damaged, for example by injury[9] or periods of hypotension.[10]

Neurogenic control

The maximum change of CBF under the influence of sympathetic stimulation is about 10 per cent.[11,12] The change is not great in magnitude, but is of some clinical significance; if blood loss is allowed to occur, such that the blood pressure is reduced, it is likely that the degree of sympathetic stimulation that will exist as a result of haemorrhagic shock will add to the effects of blood pressure reduction on CBF, producing a significant degree of cerebral ischaemia.

Aneurysm and arteriovenous malformations

Aneuryms may develop on the cerebral circulation in the young adult, but most patients are in the 40–60 years age group. An aneurysm usually develops where an artery branches. Cerebral aneuryms present by causing a subarachnoid haemorrhage (SAH), which may be confined to a minimal leakage into the subarachnoid space or may involve massive bleeding with extensive cerebral disruption and death. The initial bleed carries a distinct mortality, and subsequent bleeds demonstrate an increasing mortality. Operation to clip the neck of the aneurysm may be required relatively urgently.

The SAH sets in train various complications which affect the neurological status of surviving patients. Botterell[13] suggested the use of a grading system; this was subsequently modified by Hunt and Hess.[14]

Grade 1. Patients are asymptomatic or have mild headache or neck stiffness.

Grade 2. Patients have headache and neck stiffness but no neurological abnormalities other than cranial nerve palsies.

Grade 3. Patients are drowsy, confused or have mild focal deficit.

Grade 4. Patients are stuporous and have moderate or severe hemiparesis.

Grade 5. Patients are in deep coma and moribund.

A major cause of morbidity after SAH is cerebral vasospasm. Neurological signs developing after an initially good recovery from SAH suggest the development of vasospasm, which occurs in 30 per cent of SAH patients. The commonest time at which signs referable to vasospasm develop is the 8th day following the initial bleed. Once cerebral vasospasm has developed, blood supply to an area of the brain may be inadequate and a neurological deficit may be noticed, which may develop into a completed stroke. The British Aneurysm Nimodipine Trial[15] reported that the administration of nimodipine soon after SAH, continuing through the operative period, produced a 34 per cent reduction in completed stroke and a 40 per cent reduction in poor outcome.

The mechanism for the improvement is not certain. Calcium entry blockers are associated with the occurrence of cerebral vasodilatation, but nimodipine appears not to reverse the angiographically demonstrated vasospasm. The alternative explanation is that nimodipine has cerebral protective effects.

SAH produces other systemic effects, which may affect the patient's fitness for anaesthesia, resulting in the need for particular care in assessment. Bed rest is a central part of the management of the patient after SAH, who may therefore be exposed to dehydration, to decubitus pneumonia and to the development of deep vein thrombosis.

Hypertension may be noted and is difficult to evaluate because it may have been present before the SAH and may have been a contributory factor. In other patients the sympathetic overactivity which not infrequently follows SAH may be responsible. ECG abnormalities may be present in 60 per cent of patients preoperatively and may be suggestive of ischaemia, including ST segment depression or elevation, T wave inversion or Q–T prolongation.[16] These changes, unless supported by other evidence of myocardial disease, do not constitute a contraindication to surgery.

Drug effects

Anaesthetic drugs have major effects on cerebral function, and much work has been done to identify these effects and to evaluate their clinical significance. Use of inappropriate drugs may worsen the operative conditions, making the operation difficult or impossible, but other effects may be beneficial for the patient undergoing neurosurgery.

Induction agents

Thiopentone
Thiopentone has played a valuable part in neuroanaesthesia. Barbiturates lower the $CMRO_2$ and the CBF falls in parallel. The effects are dose dependent; when enough thiopentone has been given to produce an iso-electric electroencephalogram (EEG), the $CMRO_2$ is about 50 per cent of awake values. No further fall in $CMRO_2$ occurs if more thiopentone is given.[17] The decrease in CBF is associated with a lowering of ICP, and the use of thiopentone during induction of anaesthesia in patients with intracranial space occupation is particularly valuable as it may prevent increases in ICP resulting from laryngoscopy and tracheal intubation.

Propofol
The effects of propofol on $CMRO_2$ and CBF are similar to those produced by thiopentone: propofol 1.5 mg/kg produced a 32 per cent fall in CSF pressure 2 minutes after induction of anaesthesia.[18] The reduction in MAP may be greater with propofol, and there is evidence that propofol protects more effectively against the pressor response to intubation.[19]

Etomidate
Etomidate has a minimal effect on the cardiovascular system. Renou[20] gave etomidate and found that it had a depressant effect on $CMRO_2$ and CBF similar to the barbiturates, producing a reduction in $CMRO_2$ of 45 per cent and in CBF of 34 per cent. Milde and colleagues[21] also showed the reduction of $CMRO_2$ and CBF, with etomidate being proportional to dose, but they added that with an

infusion of etomidate the CBF fell faster than the $CMRO_2$. The lack of any major cardiovascular depression means that CPP is well preserved with etomidate.

Maintenance of anaesthesia

Anaesthetic agents

Many inhalational agents have been used in neuro-anaesthesia.

Nitrous oxide

Much interest has centred on the cerebral effects of nitrous oxide, which, although it is the least potent of the inhalational agents, has weak cerebral vasodilating actions. Some workers[22] reported that in dogs a slight increase in $CMRO_2$ occurred whereas others[23] showed a 25 per cent reduction in $CMRO_2$ when 70 per cent nitrous oxide was used. In humans Greenbaum et al.[24] found that significant further increases in ICP could occur on induction of anaesthesia using nitrous oxide, as did Henriksen and Jorgensen.[25] More recently, Hansen and colleagues[26] have shown that adding 0.5 MAC nitrous oxide to a background of anaesthesia with 0.5 MAC halothane or enflurane produced a greater increase in CBF in rats than would have been expected by doubling the concentration of the background agent.

Halothane

Halothane uncouples the relation between $CMRO_2$ and CBF: 1 per cent halothane causes a fall in $CMRO_2$ by 26 per cent but an increase in CBF of 27 per cent and therefore a rise in ICP.[27] Low concentrations of halothane (0.6 MAC) have relatively little effect on CBF, but 1.1 MAC may triple CBF.

Adams et al.[28] studied the relation between halothane, hyperventilation and ICP. They found that if halothane was introduced to the anaesthetic simultaneously with hyperventilation, then, even if the $PaCO_2$ was reduced to 3.3 kPa (25 mmHg), several patients suffered significant increases in ICP. In a second group of patients hyperventilation was induced first and 1 per cent halothane added subsequently. In this second study no meaningful increases in ICP occurred. The general application of this work is difficult. The degree of space occupation varies between patients; because of the characteristics of the volume–pressure curve, the patients who are worst affected by intracranial space occupation will produce the greatest rise in ICP.

Any possible protection afforded by hyperventilation depends on the integrity of the reactivity of the cerebral circulation to CO_2, which may be lost, for example, in trauma.

Enflurane

Enflurane causes less depression of the $CMRO_2$ than halothane, and, used in clinical concentrations in humans, no change in ICP when ICP is normal nor an increase when ICP is raised. Some of the effects on ICP seem to be modified by the tendency of enflurane to lower the MAP. Enflurane appears also to cause seizure-like activity on the EEG, but the incidence of postoperative convulsions seems to be no higher than following isoflurane.

Isoflurane

Isoflurane produces a reduction in $CMRO_2$ greater than that produced by equivalent MAC values of halothane; administration of 2.5 MAC isoflurane is accompanied by an isoelectric EEG. The CBF increase with isoflurane is minimal below 1.1 MAC.[29] Todd and Drummond[30] found that, up to 1.5 MAC, isoflurane did not increase CBF but did produce a small increase in ICP; there was also a reduction in cerebrovascular resistance.

Some studies have shown that ICP can be increased with isoflurane, especially if the ICP is already raised, or if the patient is normocapnic, or if a preoperative CT scan shows that a mass lesion has produced marked shift of the midline structures.[31,32]

Desflurane

Desflurane also increases ICP at normocapnia,[33] and CO_2 responsiveness appears to be maintained even when 1.5 MAC is associated with hypotension to 60 mmHg MAP in dogs.

Muscle relaxants

The muscle relaxants, being highly charged molecules, do not pass the blood–brain barrier. They ought therefore to have no direct effect on the brain, and their use in neurosurgery depends more on their systemic effects than on any cerebral effects.

Tubocurarine

Tubocurarine has been used for many years, and its limitations, which include hypotension and histamine release, are well known. It is these effects that limit its usefulness in neurosurgery. Some studies in animals[34] and humans[35] showed various cerebral and systemic effects due to histamine release, including an increase in ICP and an increase in CSF production.

Pancuronium

Pancuronium causes an increase in heart rate and blood pressure.[36] This should have no discernible cerebral effect in the patients with normal intracranial dynamics. There is, however, a possibility that the pressor response to laryngoscopy and intubation will be greater with pancuronium than with other relaxants, so greater care must be taken to avoid it when pancuronium is used.

Atracurium and vecuronium

Atracurium has some potential for histamine release but, as with curare, this is unlikely to have significant cerebral effects.[37] Its metabolite, laudanosine, is well known as a convulsant, but plasma levels do not seem to reach dangerous levels. *Vecuronium* would seem to have many of the properties of the ideal relaxant, as it has no cerebral effects and few systemic side effects.[38]

Suxamethonium (succinylcholine)

The use of suxamethonium has been said to raise ICP; one explanation is that the fasciculations cause an increase in intra-abdominal pressure and therefore an increase in CVP. This may not be the entire story. Minton *et al.*[39] gave suxamethonium to patients with an intracranial tumour, both before and after muscle paralysis had been established with vecuronium. When given before vecuronium, suxamethonium produced a consistent increase in ICP from a mean of 15 mmHg to 20 mmHg. After vecuronium had been given, the increase generated by suxamethonium was still present, but of much smaller extent, the maximum increase being only 3 mmHg. The evidence suggests that increased muscle spindle activity increases the afferent neuronal traffic and produces an increase in cerebral activity and therefore rCBF. The observations were confirmed by Lanier and colleagues.[40]

Narcotic analgesics

The major danger of the analgesic drugs is that, if ventilation is not controlled when they are used, respiratory depression will lead to a rise in $PaCO_2$ and therefore CBF and ICP. If normocapnia is maintained the narcotic analgesics have little effect on $CMRO_2$ and CBF. There is some species difference: in dogs, morphine and pethidine caused reductions in $CMRO_2$ and CBF,[41,42] whereas in humans Jobes *et al.*[43] showed no change in CBF or vascular CO_2 reactivity when morphine was given during nitrous oxide anaesthesia.

The effects of fentanyl, alone or in combination with droperidol, have been studied by several groups. In humans, Sari and colleagues[44] showed that, in combination with droperidol, no significant changes were seen in CBF and $CMRO_2$. In dogs, however, Michenfelder and Theye[45] had noted increases in both variables for 30 minutes. High doses of fentanyl (100 µg/kg) in rats produced falls in both $CMRO_2$ and CBF, of 35 per cent and 50 per cent respectively.[46]

ICP has also been studied in humans.[47] In this study by Fitch *et al.* in patients with space-occupying lesions during controlled ventilation, ICP either did not alter or decreased only slightly when fentanyl and droperidol were given.

There are reports of significant increases in ICP when phenoperidine is used, and some authors[48] have suggested a difference between fentanyl and phenoperidine as far as their cerebral effects are concerned.

Benzodiazepines

Diazepam finds a major use in neurosurgery as a sedative and anticonvulsant. Information about its cerebral effects is conflicting and may reflect species differences. Carlsson *et al.*[49] gave diazepam to rats during nitrous oxide anaesthesia; they found that CBF and $CMRO_2$ were reduced by 40 per cent, but that this change did not occur when the nitrous oxide was withdrawn. Cotev and Shalit[50] did demonstrate a 25 per cent fall in $CMRO_2$ in head-injured patients after diazepam, but a smaller fall in CBF.

Midazolam (0.15 mg/kg) has been shown to reduce $CMRO_2$ by 21–30 per cent and CBF by 26 per cent.[51] In patients with space occupation, one study showed that induction of anaesthesia with midazolam (0.32 mg/kg) was associated with no change in ICP.[52]

Anaesthesia

Anaesthesia may be required in the imaging departments, x-ray, CT scan or magnetic resonance imaging as well as in the operating theatre.

Premedication

Heavy premedication, especially with narcotic analgesics, is inappropriate for the neurosurgical patient, because the conscious level may be further depressed. Any such depression of conscious level may lead to an increase in $PaCO_2$ from respiratory depression and therefore cerebral vasodilation and increased ICP. The use of oral benzodiazepines (temazepam 10–20 mg or diazepam 10–20 mg) has proved valuable.

It is worth remembering that some patients presenting for spinal surgery with an acute disc prolapse may well be in pain preoperatively and an analgesic premedication is valuable for such patients.

Induction

At all stages of the anaesthetic, cerebral vasodilatation produced by hypoxia or hypercapnia, or the injudicious use of vasodilating anaesthetic agents, should be avoided. It is also important to produce a low central venous pressure. This means ensuring profound muscle relaxation, with no tendency to cough or strain. Sudden rises in blood pressure are potentially dangerous and should be avoided.

Intravenous induction of anaesthesia is preferred; both thiopentone and propofol may be used. The depression of $CMRO_2$ produced by both drugs and the resultant reduction of CBF and ICP are of particular value when a space-occupying lesion is present. Etomidate can be used and the depression of $CMRO_2$ is about the same, but the greater cardiostability characteristic of the drug is associated with a greater rise in blood pressure on laryngoscopy than with thiopentone or propofol. Whilst use of bolus injection is commonplace, propofol may also be used by infusion for maintenance of anaesthesia, and the infusion may usefully be started at induction.

Muscle relaxation is required and can be provided with any of the common muscle relaxants. Some anaesthetists recommend the use of suxamethonium for tracheal intubation. They point out that it has the advantage of providing rapidly good conditions for tracheal intubation, and that its effects on ICP can be modified by pretreatment with a small dose of non-depolarizing relaxant to abolish fasciculations. The use of one of the intravenous induction agents immediately before suxamethonium is given will also modify the ICP effect.

It is important to ensure that profound muscle relaxation has developed before laryngoscopy and tracheal intubation are attempted. The coughing and straining that result from premature attempts at intubation may produce a rise in intrathoracic pressure and, therefore, in intracranial pressure. The introduction of the laryngoscope and the traction necessary to view the vocal folds may be associated with jugular venous distortion, such that venous blood flow out of the skull is impaired, producing a rise in ICP.

Laryngoscopy and intubation may also be complicated by a hypertensive response. This is most likely to be a problem for patients who present with an arterial weakness, such as an aneurysm or arteriovenous malformation. Clearly these patients, who may well have had a previous subarachnoid haemorrhage, may suffer a further haemorrhage if the blood pressure increases, especially if the rise is rapid. Many manoeuvres have been recommended to prevent or obtund the pressor response. Fentanyl 200–300 µg at induction is a partial protection. Premedication with β-blockers has been recommended, as has the use of lignocaine (lidocaine) intravenously 90 seconds before intubation.[53] The use of a second bolus of the intravenous induction agent before intubation is commonplace, especially if the intubation is being performed with non-depolarizing muscle relaxants. In the last case, the depth of anaesthesia has been reducing as the

degree of muscle relaxation develops and therefore a second bolus of the induction agent is appropriate, though the magnitude of the second bolus has not been determined. An infusion of an induction agent protects very efficiently against the pressor response to laryngoscopy. Clearly, although an agent must protect against the pressor response, it should not produce a worrying degree of hypotension in doing so. It is worth pointing out that propofol (3 mg/kg), with its greater effect on the cardiovascular system than thiopentone, protects more efficiently against the pressor response.[19]

In patients with vascular tumours, the avoidance of a pressor response may also be important, because sudden hypertension and increased blood flow through the tumour may produce an increase in its size. Clearly an intubation that involves an inadequately paralysed patient and an incompletely controlled pressor response is very likely to cause an increase in ICP, because two mechanisms are at work simultaneously.

Maintenance of anaesthesia

Neurosurgical operations may last several hours. The airway, venous access and monitoring must be secured in such a way as to be completely reliable for the whole of the operating period. In particular, the airway needs to be secured, because access to the head may be impossible for several hours. Normally an armoured tracheal tube should be used, and fixed securely in such a way as to leave cerebral venous drainage free. Some anaesthetists also secure the tracheal tube by an oropharyngeal pack, especially if the patient is prone or sitting.

During the surgery the anaesthetist should choose drugs and techniques so as to:

1. Maintain adequate cerebral perfusion
2. Maintain stable MAP
3. Avoid any factors that lead to increased ICP
4. Reduce high ICP
5. Provide induced hypotension if appropriate
6. Be ready for any situation which may suddenly produce cerebral ischaemia

At the end of surgery the anaesthetist must ensure rapid recovery of consciousness. Other techniques, such as hypothermia, may occasionally be required.

Maintenance of a stable MAP

A sudden rise in MAP arising from a painful stimulus needs to be prevented. If an aneurysm is present, the sudden hypertension may burst the aneurysm; if a vascular tumour is present, it may increase in bulk as the blood pressure rises, the tumour vessels not being capable of autoregulation. A sustained hypertension may lead to the acceleration of oedema formation.

The head is often held in a headrest, in which pins mounted in a frame are fixed into the outer table of the skull; clearly this is a painful procedure and should be blunted by analgesia.

The initial part of a craniotomy is also painful, particularly cutting the skin and reflecting the galea. As the burr holes are drilled, there may be pressure on the skull, raising the intracranial pressure. Similarly, if the dura has to be stripped from the inner table of the skull (most likely if the surgeon is using a saw guide and a Gigli saw), an intensely painful stimulus is produced as the periosteum is removed from the bone. At the same time there may be a marked sympathetic stimulation. Both factors (pain and sympathetic stimulation) tend to raise blood pressure. Surgery inside the skull is not markedly painful, unless the surgeon stretches the dura, which is most likely to happen when near the points of dural attachment.

Once the bone flap is cut, the surgeon may need to retract the brain to obtain access to deeper structures. During this period it is crucial to prevent any increase in brain bulk, which might increase the pressure on the retractors, leading to a greater degree of neuronal damage.

Blood loss in a craniotomy is usually less than 500 ml but may be marked in certain circumstances. Some tumours, such as meningiomas, are vascular, and surgery can produce fast and extensive blood loss. Not only may such tumours be vascular, but, having a low vascular resistance, they may also generate relatively high pressure in the veins draining them. Rapid blood loss may therefore occur if the draining veins are damaged, as can happen within a meningioma as the bone flap is removed. The ability to replace blood loss and control blood pressure is essential and usually means that both a large venous cannula and a central venous catheter are in use.

Urethral catheterization is essential, particularly if osmotic diuretics are to be used. Urine flow should be noted regularly to check that a diuresis has developed after use of osmotic diuretics.

The agents used during maintenance of anaesthesia have been subject to controversy for some time. All the volatile agents increase cerebral blood flow, to differing extents, and this may mean that the bulk of the brain will increase. This makes retraction of the brain more difficult, and so surgical difficulties increase and perhaps greater pressure is needed on the retractors. Aneurysm surgery, in particular, requires that the brain may be retracted with ease.

Some authors have pointed out that volatile agents may be used quite successfully with hyperventilation, when the increase in CBF generated by the volatile agent is to a certain extent overcome by the reduction in CBF produced by the hyperventilation. Other authors have suggested that this may not be so, especially when the mechanisms compensating for intracranial space occupation are near exhaustion, when a further, even small, addition of space occupation to the intracranial space may be critical. Studies have suggested, particularly in relation to isoflurane, that ICP can be increased, especially if the patient is normocapnic or if a preoperative CT scan shows that a mass lesion has produced shift of the midline structures.[31,32]

This topic has received further study recently from the work of Ravussin et al.[54] and Todd et al.[55] The latter group examined prospectively patients undergoing supratentorial craniotomy for tumour. Their patients were carefully assessed. ICP was measured and the extent of brain swelling was noted. The speed of emergence from anaesthesia was also recorded. The group also measured total stay and hospital costs. Patients were assigned to one of three groups. The first group received propofol induction and maintenance of anaesthesia with fentanyl 10 µg/kg load followed by 2–3 µg/kg per hour. The propofol infusion was set between 50 and 300 µg/kg per minute. In the second group anaesthesia was induced with thiopentone and maintained with N_2O, oxygen and isoflurane; fentanyl up to 2 µg/kg was given after bone flap replacement. The third group also had thiopentone induction of anaesthesia, but with fentanyl as in group one and 'low-dose' isoflurane. It was reported that ICP before craniotomy was 12 ± 7 mmHg in group one, 15 ± 12 mmHg in group two and 11 ± 8 mmHg in group three. Two patients in groups one and three but 9 patients in group two had ICP greater than 24 mmHg. The authors suggest that all three anaesthetic regimens were acceptable, pointing out that only small differences in ICP were observed between the three groups.

It would appear that isoflurane is indeed usable in neuroanaesthesia, but the incidence of high ICP and brain swelling reminds the anaesthetist of the need to avoid deleterious drugs or techniques in the most severely affected patients.

Propofol infusion

Propofol infusion avoids the need for vasodilating anaesthetic agents for maintenance of anaesthesia and is in use in a number of centres. Ravussin et al.[54] and Todd et al.[55] both reported on the successful use of propofol infusion for maintenance anaesthesia. Ravussin et al. considered in their study (in patients without space occupation) that propofol gave better control of responses to painful stimulus and faster recovery than thiopentone–isoflurane, whereas Todd et al. thought the differences in their patients were minimal. Another group[56] used 2.0–2.5 mg/kg of propofol for induction of anaesthesia and 12 mg/

kg per hour for 10 minutes followed by 9 mg/kg per hour for another 10 minutes, then 3–6 mg/kg per hour for the rest of the study. They showed that the responsiveness of the cerebral circulation was well maintained, the slope of CBF against $PaCO_2$ being 1.56 ml/100 mg per minute per mmHg. In this study the patients were ventilated with nitrous oxide and oxygen and had no neurosurgical disease.

Hyperventilation

Hyperventilation has been an important part of neuro-anaesthesia for many years, because of the fall in ICP produced by the cerebral vasoconstriction. The vasoconstriction reduces the intracapillary hydrostatic pressure and therefore shifts the Starling balance at the capillary in favour of fluid reabsorption from the extracellular space. Hyperventilation is therefore associated with a reduction in brain extracellular fluid, this process taking about 3 hours to develop.

The level of hyperventilation is of interest. Harp and Wollman,[57] who found no objective evidence of ischaemia with $PaCO_2$ values as low as 1.3 kPa (9.8 mmHg), recommended a conservative level of 3.3 kPa (25 mmHg). When induced hypotension is employed, it would seem prudent to avoid low levels of $PaCO_2$; McDowall[58] has suggested that, in patients with significant vasospasm or atheroma, the addition of hypocapnia might reduce flow to low values when hypotension is employed.

Analgesia

Regular doses of analgesics such as fentanyl are required throughout the anaesthetic, and the aim should be to produce cardiovascular stability, particularly with the heart rate below 90 beats per minute. Notably painful parts of the operation, such as fixing the pins of a head clamp or stripping the periosteum, may usefully be preceded by analgesic increments.

Muscle relaxants

Use of adequate doses of muscle relaxants will aid in the provision of good intracranial operating conditions. It is important to emphasize that, as recovery from relaxants begins, abdominal and thoracic muscle tone will recover faster than the ability to perform coordinated movements and also faster than neuromuscular function tested with a nerve stimulator on the arm or leg. The recovery of muscle tone produces a rise in intra-abdominal and intrathoracic pressure and therefore in increased central venous pressure

which may be transmitted to the cerebral veins, thus raising ICP.

All the commonly used muscle relaxants may be employed, none having any particular advantage for neurosurgery. In view of the problems associated with recovery of muscle tone, it is important to use atracurium and vecuronium by infusion, guided by nerve stimulation. Other relaxants, such as pancuronium, which show a slower recovery profile, may be given intermittently. At the end of surgery, recovery of muscle function must be complete so as to avoid in the neurosurgical patient any chance of hypoxia or hypercapnia due to persistent neuromuscular block.

Mannitol may usefully be given during the early part of the operation; 0.5–1 g/kg, used as a 20 per cent solution and given over 30 minutes, will begin to lower ICP and also the volume/pressure ratio. When mannitol is used, the urine output should be noted so that appropriate replacement fluids may be prescribed. This is especially important when cerebral vasospasm is present. Mannitol is considered in more detail later.

The tracheal tube should be removed with as little disturbance as possible. After the head dressing has been applied, which often involves skull movement, the relaxant should be reversed with an appropriate dose of anti-cholinesterase. Ventilation should be continued until the nerve stimulator shows neuromuscular function returning to normal, at which point lignocaine 1.5 mg/kg may be given, followed 90 seconds later by oropharyngeal suction and extubation if ventilation is adequate. Use of lignocaine[53] in this manner appears to reduce the hypertensive response to extubation (Cooper A, Turner J. 1986, unpublished observations).

The length of neurosurgical operations may be associated with some cooling, so active measures to maintain the patient's body temperature are required.

Special problems

Aneurysm surgery

Great care should be taken to prevent a sudden increase in blood pressure, but if there is evidence (persistent hypertension with arrhythymia) suggesting that the aneurysm may have burst, the high ICP and hypertension may be controlled acutely with thiopentone or propofol. If the aneurysm bursts during the surgical dissection, the surgeon may be able to control the bleeding by placing a clip on the feeding vessel. Again, covering this period with thiopentone or propofol may be of value.[59]

If surgical control of a bleeding aneurysm is not immediately possible, the anaesthetist may lower the MAP to reduce the rate of bleeding; this may be done in many ways. If hypotension is in use, the infusion may be increased; if not, then labetalol, thiopentone, propofol or isoflurane may be used. The MAP may be lowered to 40–50 mmHg for short periods. One report of a retrospective study[60] has suggested that lowering the MAP in this way may produce a poorer outcome than the application of a temporary clip.

Vasospasm

Fluid management should avoid the risk of dehydration. The urine volume following the diuresis caused by mannitol should be replaced and colloid used to maintain the haematocrit (packed cell volume) between 0.3 and 0.35.

Pituitary surgery

Patients presenting for pituitary surgery may show evidence of endocrine deficiency and require replacement therapy before operation. If bleeding has taken place into a pituitary tumour and the sudden increase in size of the tumour causes blindness, urgent operation is required. Such patients are likely to be acutely endocrine deficient.

Patients may also suffer from hypersecreting tumours. Patients with Cushing's disease, who have been accustomed to high plasma cortisol levels, may require higher postoperative doses of steroids. Acromegalic patients may be difficult to intubate, though with earlier diagnosis the problem appears to be less threatening. The arterial line should preferably not be placed in the radial artery.[61]

Patients with pituitary disease may also develop hypertension, diabetes mellitus, thyroid and adrenal insufficiency.

The surgical approach may be transcranial or trans-sphenoidal. The advantages of the trans-sphenoidal route are that, because it minimizes direct neuronal damage, the postoperative morbidity and the incidence of seizures are low. The disadvantages of the trans-sphenoidal route mainly centre around surgical difficulties if the tumour extends outside the sella. Operations by either approach should be treated as major intracranial procedures. Production of mucosal vasoconstriction with cocaine and adrenaline (epinephrine) is common. There may, of course, be bleeding into the airway and therefore a throat pack is of value.

The trans-sphenoidal operation is performed using the operating microscope and with x-ray control, so access to the head is difficult for the anaesthetist.

Hypophysectomy requires steroid replacement therapy and possibly thyroid replacement therapy. Diabetes insipidus may also follow, so hourly urine output needs to be charted. Fluid balance during the operation should include maintenance fluids operative losses. Diabetes insipidus may persist for 7–10 days and should be treated with desmopressin (DDAVP), up to 2 µg initially.

Postoperatively, there may be bleeding into the airway and the blood may be swallowed. The surgeon may use nasal packs to control bleeding.

Posterior fossa surgery

Most operations in the posterior fossa are for space-occupying lesions. The relatively small capacity of the posterior fossa, and the ease with which these tumours may cause hydrocephalus, means that surgery should not be unduly delayed once space occupation has been diagnosed. CSF obstruction may occur early and ventriculoperitoneal shunting may be required urgently to reduce ICP.

As well as looking for signs of generally raised ICP, the lower cranial nerves need to be examined particularly for signs of bulbar palsy. Any patient who presents with largyngeal incompetence or difficulty in swallowing will need specific care in the postoperative period.

It is also important to assess the proximity of the tumour to the vital centres in the floor of the fourth ventricle and to look for signs of cardiovascular disturbance during surgery. The older approach of maintaining spontaneous ventilation with a midline fourth ventricular tumour has been replaced by the near-universal use of controlled ventilation of the lungs in such cases. If midbrain disturbance occurs during surgery, the patient should be considered for positive pressure ventilation of the lungs in the postoperative period.

The choice of anaesthetic is influenced by the position required for surgery, which is described later.

Special techniques

Induced hypotension

Induced hypotension is the subject of Chapter 39.

The use of hypotension in neurosurgery is changing. It has been used extensively for surgery on vascular tumours, such as meningiomas, and for aneurysm surgery. As in any form of anaesthesia, the technique should be used only after a careful assessment of the risk to the individual patient, taking into account the state of the patient's cardiovascular system.

The use of deliberate hypotension for aneurysm surgery

has been recommended because of the danger of intraoperative rupture of the aneurysm. If the MAP is lowered, the intraluminal pressure in the aneurysm is lower and the wall less tense. It is suggested that this makes surgical dissection in the vicinity of the aneurysm easier, not only because the aneurysm is less likely to burst but also because it is more mobile.

Much work has been done to elucidate the relation between CBF and pharmacologically induced hypotension, in an attempt to define the level of MAP at which cerebral perfusion pressure is critically lowered. The initial studies of haemorrhagic hypotension produced a figure of 60 mmHg MAP,[62] but Fitch[63] showed that during hypotension produced by halothane, sodium nitroprusside or trimetaphan, CBF was maintained to a lower MAP than with haemorrhagic hypotension.

The response to reduced CBF has been well studied. Cerebral electrical activity begins to reduce until at a CBF of 15–20 ml/100 g per minute both EEG and evoked responses become impaired.[64] At a lower CBF (around 10 ml/100 g per minute) ion changes become apparent, there being an influx of Ca^{2+} into the cell and an efflux of K^+.[65,66] Finally, at lower flows, histological evidence of cerebral damage appears. The damage appears to be concentrated in zones on the margins of supply of the main cerebral arteries, often called the 'boundary zones'.[67]

The concern about using hypotension to a MAP of 60–65 mmHg relates to the fact that where cerebral vasospasm exists, or where autoregulation is impaired, as for example after SAH, even such moderate hypotension may produce focal cerebral ischaemia. Patients who are hypertensive or who have marked atheroma may also be sensitive to moderate hypotension. However, Campkin et al.,[68] using sodium nitroprusside or trimetaphan for moderate hypotension, were unable to show biochemical evidence of ischaemia.

Safety features

If induced hypotension is employed, the patient must be examined for significant ischaemic disease in all vascular beds, and the existence of hypertension or ECG signs of myocardial ischaemia should be noted as contraindications to induced hypotension. The level of blood pressure should be chosen so as to maintain an adequate cerebral blood flow and the duration of the hypotensive period must be as short as possible. Hypotension should not be started until the dura is opened and preferably left until dissection is nearing the aneurysm.

Oximetry must be in use to avoid the hypoxaemia that may result from the increased oxygen extraction in the periphery as the peripheral circulation slows. Intra-arterial measurement of blood pressure is essential and electrical monitoring of cerebral function is valuable.

Methods of induced hypotension

Induced hypotension is made easier if an appropriate anaesthetic technique is used, particularly an anaesthetic that avoids the occurrence of a tachycardia, whether arising as a result of drug action or of light anaesthesia.

Some anaesthetic agents, notably isoflurane and propofol, lower the MAP. Campkin and Flinn[31] described the use of isoflurane in such a way, as did Meyer and Muzzi.[69] Halothane and enflurane when used to produce hypotension gain their effect by peripheral vasodilatation together with cardiovascular depression. The cardiac output is often low. Isoflurane, on the other hand, tends to spare the cardiac output.

Labetalol, a combined α- and β-adrenoceptor antagonist, is a valuable adjunct to other agents for producing hypotension and appears to have little intracranial effect.

Sodium nitroprusside

Sodium nitroprusside (SNP) was for a while the most widely used agent for induced hypotension. It is a direct acting vasodilator with a rapid onset of action and very short half-life. SNP causes cerebral vasodilatation[70] and therefore a rise in ICP. Once the MAP has been reduced to 70 per cent of the control value, the ICP is reduced. Other groups have also shown this effect.[71]

Metabolism of SNP produces cyanide which combines with sulphhydryl groups from erythrocytes to form thiocyanate and cyanmethaemoglobin. Either a large total dose of SNP or a rapid infusion rate may result in the formation of excessive amounts of cyanide.[72] Clearly the agent must be given by carefully controlled infusion into a central vein and excessive infusion rates avoided. If the patient shows resistance to the SNP, another agent (e.g. labetalol) should be added to produce hypotension.

Trimetaphan

Trimetaphan is a short-acting ganglion blocking drug, which also has the potential for causing histamine release. Normally it has no effect on the cerebral circulation, but if the superior cervical ganglion is blocked, some increase in CBF will take place owing to the removal of sympathetic vasoconstriction.[70] Trimetaphan may cause pupillary dilatation.

Nitroglycerin

Nitroglycerin is similar to SNP in many respects, but has no toxic metabolites. In animals it causes elevation of the ICP. In one series using dogs with artificial intracranial space occupation, ICP increased further when nitroglycerin was started; pupillary dilatation developed, suggesting the existence of transtentorial pressure gradients and 'coning'.[71,73]

The potential dangers of induced arterial hypotension have led to an increase in the use of an alternative technique for reducing the pressure in the aneurysm. The surgeon can apply a temporary clip to the main supply vessel to the aneurysm. Although this technique needs to be used with care, short-term protection with thiopentone or propofol is widely used while the temporary clip is in place.

Hypothermia

The introduction of induced hypothermia followed from the knowledge that, as the body temperature falls, the metabolism of all tissues, including the brain, falls, and the tolerance to ischaemia is increased. Moderate hypothermia has had a long history in neurosurgery, often combined with circulatory arrest in one form or another. The reduction of core temperature to 31°C decreases $CMRO_2$ by 25 per cent. Hypothermia is used less often today, but the special nature of cerebral metabolic depression produced by hypothermia rather than by drugs may make it a valuable technique again in aneurysm surgery. Todd and Warner[74] have discussed cerebral protection against ischaemia and have described the differences between drug and hypothermia induced cerebral protection.

Deep hypothermia (core temperature 15°C) was used in the 1960s.[75] It requires cardiopulmonary bypass and therefore anticoagulation, but has recently been reported to be of value in the repair of complex aneurysms.[76]

Positioning

Supine

Much neurosurgery is performed with the patient supine. All patients undergoing intracranial surgery should be positioned slightly head up (about 10 degrees usually suffices), so that cerebral venous drainage is free. Frequently the head will be turned to one or other side and it is important to ensure that the degree of neck rotation is not excessive, usually by placing a support under one shoulder. Pressure points need particular care, because operations may be prolonged. The eyes should be closed and covered.

Prone

Spinal surgery and posterior fossa surgery may be carried out with the patient prone. The airway must be carefully secured, using an armoured tracheal tube and appropriate fixing. Positive pressure lung ventilation is required and great care needs to be taken to ensure that no increase in central venous pressure results from inappropriate setting of the ventilator or compression of the abdomen by faulty positioning. There are many ways of supporting the patient in the prone position, so the method chosen needs to be related to the needs of the operation planned. Sometimes firm supports under chest and pelvis, but leaving the abdomen free, are appropriate. On other occasions, such as in the case of lumbar spine surgery, the surgeon needs the vertebral column flexed and so supports such as the Wilson frame are required. The eyes need to be closed and covered and the arms padded, either by the patient's side or on arm boards as appropriate for the surgery. Care should be taken to avoid pressure on the femoral neurovascular bundle.

Sitting

The sitting position is still used in some centres for surgery of the posterior fossa or the posterior cervical spine and has distinct advantages and disadvantages. The disadvantages include difficulties in maintaining the blood pressure and the danger of venous air embolism. The advantages include a low venous pressure, unrestricted chest movement and, facilitating surgery, the fact that blood flows away from the surgical field rather than collecting in the base of the incision.

Maintenance of blood pressure

As an anaesthetized patient is placed in the sitting position, blood tends to pool in the lower extremities, such that venous return and cardiac output are reduced. The peripheral vascular resistance rises to maintain blood pressure.[77,78] Anaesthesia therefore needs to be chosen so as to maintain the integrity of the baroreceptor reflex arc. A fluid load may be helpful and hyperventilation should be avoided. Light anaesthesia, with muscle relaxants, using analgesics such as fentanyl and low doses of volatile agents allows the blood pressure to be maintained at adequate levels. The use of pancuronium, atracurium or vecuronium is acceptable, though the cardiovascular effects of pancuronium may be of value.

Venous air embolism

The reported incidence, from separate centres, of venous air embolism is variable, but in the case of sitting patients figures of up to 40 per cent have been reported.[79] The use of specific methods to detect early embolism is essential. When the patient is sitting, the hydrostatic level of the central venous pressure is likely to be considerably lower than the operation wound, so the pressure in the veins in the area of the operation will be subatmospheric. If, in that situation, a vein is opened by the surgeon, air will be drawn into the vascular system and passed on to the heart and lungs.

A method that is effective in detecting air embolism involves using a Doppler flow probe with the crystal positioned at the right sternal edge between the third and fourth interspace. The signal is audio modulated and air bubbles as small as 0.5 ml may be heard passing the probe. The major limitation of the Doppler method is that the radiofrequency signal of the surgical diathermy interferes with the Doppler signal and, as the greatest danger from air embolism is during the muscle dissection and bone work, the diathermy is in use for quite long periods of time. The precordial probe must be positioned with care, and checked when the patient is sat up.

End-tidal CO_2 measurement with a recorder may be used. If the air enters the circulation gradually and passes on to embolize in the pulmonary circulation, the peak value of expired CO_2 gradually reduces. If a large volume of air enters the circulation quickly, the fall in observed end-tidal CO_2 may be rapid. The measurement is not as sensitive as is the Doppler technique, but is quite adequate for clinical purposes. The only major disadvantage is that if the cardiac output falls for any other reason, the CO_2 trace will also fall.

Both methods of air detection require that, as soon as a positive signal is obtained, the surgeon is informed and neck compression started so as to occlude the neck veins. This prevents further embolism and causes bleeding in the wound, so that the surgeon can visualize and seal the bleeding point where air may be entering. There is some value in pressing on the neck at agreed times during the operation to detect any veins opened inadvertently, preferably during the periods of greatest risk, while the muscle is being dissected and during bone work.

Transoesophageal echocardiography[80] is also used for air bubble detection.

The risk from air embolism can be reduced to acceptable proportions by using such techniques carefully. A central venous catheter should always be placed in sitting patients so that air can be aspirated easily if the simple methods for preventing large amounts of air entering the circulation fail. Bunegin et al.[81] showed that more air may be aspirated from a catheter placed in the superior vena cava than from one placed in the right atrium. The use of a flow-directed pulmonary artery catheter is advocated by some because it helps to identify patients at risk of hypotension as they are moved into the sitting position. Using these catheters may also identify patients at risk of paradoxical air embolism from probe-patent foramen ovale; in addition, they can be employed to aspirate air.

Auscultation of the heart sounds will demonstrate the classic millwheel murmur only after a large embolism has taken place.

Positioning problems

With many operations in the sitting position, including operations on the neck, the spinal cord may be compromised by excessive flexion as the head is fixed in the support. This is perhaps most common if a central prolapsed intervertebral disc is present in the cervical region.

The arms need careful support, especially to prevent brachial plexus or ulnar nerve injury.

The back should be carefully supported in the lumbar area and the legs need to be positioned with both hips and knees flexed, so that no undue strain is placed on the back or sciatic nerve in a patient who is paralysed.

Intraoperative treatment of raised ICP

Intracranial pressure is the resultant of many factors. When the skull is open, as at craniotomy, the ICP on the side of the operation falls to atmospheric pressure. If processes that tend to increase the bulk of the brain continue, many problems may result. For example, part of the brain may be pushed out through the dural incision, and the tight edges of dura may interfere with the cerebral blood flow, thus producing infarction. If the brain is bulky and difficult to retract, the operation may become impossible, or at least increased pressure on the brain retractors may be required, thus producing nerve cell damage.

It is therefore important to avoid any factors that may increase brain bulk. Such factors may include faulty technique (Table 52.1), the choice of drugs and positioning. Though there are methods that may be used to lower intracranial pressure, the factors that raise ICP need to be corrected first when intraoperative ICP is high.

Table 52.1   Faults in technique

Inadequate muscle relaxation
Poor cerebral venous drainage
Jugular venous obstruction
Raised mean intrathoracic pressure
Positive end-expiratory pressure (PEEP)
Inadequate head-up position
Hypercapnia
Hypoxia
Overtransfusion

Treatment of raised ICP

Hyperosmolar diuretics

In the UK the standard agent used to treat raised ICP is mannitol, which should be given as a 20 per cent solution. Mannitol has many systemic and cerebral effects. When given in doses of 0.5–1.0 g/kg, it raises serum osmotic pressure so that water is drawn from the tissues into the vascular system. Therefore, initially, the circulating blood volume rises and the haematocrit falls.[82] The blood volume remains elevated for about 15–30 minutes and, during this time, blood pressure and CVP may also be elevated. The renal diuresis that occurs limits the extent of the rise in blood volume. The increased plasma oncotic pressure draws water from the brain and therefore reduces brain bulk. Mannitol, in addition to reducing ICP, also reduces the volume–pressure response.[83] The decreased haematocrit, as a result of a reduced blood viscosity, also allows a greater CBF; in patients with intact autoregulation, this leads to cerebral vasoconstriction (oxygen supply being kept in balance with demand), which also lowers ICP. This effect is not seen in patients in whom autoregulation is impaired and the increased CBF persists.[84]

Loop diuretics

Some diuretics have actions on the brain that may be useful when combined with other drugs. Acetazolamide, frusemide (furosemide) and spironolactone decrease the rate of CSF production.[3]

Frusemide, given in doses of 1.0 mg/kg, produces a fall in ICP similar to that produced by 1 g/kg of mannitol, but without the electrolyte changes that result from mannitol.[85] It acts by the inhibition of sodium and chloride reabsorption in the ascending limb of the loop of Henle and has a separate action in reducing CSF production by suppressing sodium transport. It is able to lower ICP by mobilizing normal brain extracelluar fluid and cerebral oedema.

The reduction in blood volume that occurs as a result of the diuresis also contributes to the reduction in ICP because the generation of a low CVP promotes, through the related low sinus venous pressure, resorption of CSF. Frusemide appears not to reduce the VPR.

Spironolactone, the carbonic anhydrase inhibitor, reduces the rate of CSF production and therefore may be of value when combined with other drugs. Ethacrynic acid, alone or in combination with mannitol, decreased ICP in experimental studies.[86]

Corticosteroids

Steroids have a marked effect in reducing cerebral oedema. This is seen most dramatically in patients who have focal lesions around which there is extensive oedema. Steroids are much less effective when there is widespread brain injury. Following the administration of steroids to patients with oedema surrounding a tumour, there is often a rapid improvement in conscious level and neurological state. The mechanism of action is not clear, but stabilization of the cell membranes so that intracellular–extracellular gradients for water and electrolytes are preserved is one possible explanation. Steroids also reduce extrachoroidal production of CSF.[87]

Steroids have little place in acute control of intraoperative raised ICP, but may be given intraoperatively to reduce oedema postoperatively. Dexamethasone, 8 mg i.v. as a single dose, followed by 4 mg 6-hourly, is an appropriate regimen.

Hyperventilation

Hyperventilation has been used for many years to lower ICP. The reduction in systemic $Pa\text{CO}_2$ is associated with a reduction in CBF and therefore a fall in ICP. If brain swelling exists during neurosurgery, it is wise to confirm the end-tidal CO_2 values with an arterial sample and to reduce $Pa\text{CO}_2$ to 3.3 kPa (25 mmHg).

CSF removal

CSF removal may be achieved in a number of ways. The surgeon may choose to cannulate the cerebral ventricles, either through the incision or through a separate burr hole. Placement of a supratentorial burr hole used to be routine before starting posterior fossa surgery, so that if pressure rose in the posterior fossa during the operation, the general ICP could be lowered without delay by tapping the ventricles.

In selected patients a lumbar subarachnoid catheter may be placed in the CSF at the start of the anaesthetic. Drainage of CSF should not start until after the dura is opened, and then CSF should be removed at about 5 ml/min. CSF removal in this way gives good surgical exposure.

Metabolic suppression

Cerebral metabolic suppression has been used during neurosurgery. This was initially produced by hypothermia used as part of an operative technique during circulatory arrest to enable cerebral aneurysms to be clipped. Hypothermia finds an occasional use, during surgery, for aneurysms such as those on the basilar artery.

Shapiro[88] described the use of barbiturates which further

depressed the cerebral metabolic activity and, with the associated cerebral vasoconstriction, lowered ICP intra-operatively. Other anaesthetic agents that have been used to lower ICP include propofol and etomidate. These agents lower ICP, but may also have major depressant effects on cardiovascular function. Lignocaine (1.5 mg/kg i.v.) may also be used to lower ICP, especially in the patient with cardiovascular instability. It has been suggested that 1.5 mg/kg of lignocaine is as effective in reducing ICP as 3 mg/kg of thiopentone.[89]

Monitoring

Oximetry

Oxygen saturation measurements have been accepted as routine in all forms of anaesthesia. The neurosurgical patient is especially sensitive to arterial hypoxaemia, both because of the brain swelling that will result from the hyperaemia and the oedema it is likely to cause and because many of the conditions for which neurosurgery is required imply some degree of regional cerebral ischaemia, to which the addition of hypoxaemia is obviously dangerous.

Arterial pressure, CVP, large venous cannula, temperature

Neurosurgery is likely at times to produce considerable and rapid blood loss. Some operations, especially those in the posterior fossa may involve cardiovascular disturbance. Hypotension may be required, as may other methods to control cardiovascular function. It seems obvious, therefore, that patients undergoing neurosurgery require arterial cannulation for all operations inside the dura. They also require a large venous cannula capable of rapid blood transfusion, and benefit from central venous cannulation and pressure measurements.

Core temperature, preferably nasopharyngeal or oeso-phageal, needs to be measured. The length of neurosurgical operations may be associated with quite large falls in temperature, and the dangers of postoperative shivering producing hypoxaemia, ventilatory impairment and increased intrathoracic pressure are obvious.

End-tidal CO$_2$ measurements

End-tidal CO$_2$ measurements are essential to ensure accurate levels of ventilation, and should be compared, at the start of surgery, with arterial blood samples.

EEG, cerebral function and analysing monitor, evoked potential

The electrical monitoring of brain function is a rapidly developing science, of value for estimating drug dosage, for the recognition of cerebral ischaemia and for the detection of impending surgical damage (as in surgery for acoustic neuroma when the facial nerve is close to the operative field) (see Chapter 24).

Jugular venous oxygen saturation

Measurements of jugular venous oxygen saturation and cerebral arteriovenous oxygen content difference have been used in a number of situations, both in the operating theatre and in the intensive care unit.[90] Although indwelling catheters that enable jugular bulb oxygen saturation to be recorded continuously are available, they only reflect global changes in cerebral function. Normal values for cerebral arteriovenous oxygen content difference are 5–7 ml/dl; values above 9 ml indicate global cerebral ischaemia and below 4 ml indicate hyperaemia. As well as being of use for clinical management, this technique has also been employed to investigate hypotension.[91]

Transcranial near-infrared spectroscopy is of value before the skull is opened and when arterial insufficiency is a problem, as in carotid endarterectomy.

Transcranial Doppler flow analysis

The ability of Doppler flow probes to pick up flow in arteries of the circle of Willis, usually the middle cerebral artery, has permitted the non-invasive measurement of an index of CBF. As with many methods of Doppler flow analysis, the readings are subject to problems, such as change in diameter of the artery under study and of the geometry of the measurement system, so readings must be interpreted with care.[92]

Postoperative care

Good postoperative care is essential for the neurosurgical patient. Many of the disease processes, such as cerebral oedema or vasospasm, which were present before the operation or which start during the operation persist into the postoperative period. Additional complications may arise after surgery, and the detection and appropriate treatment of all these problems is essential for the well-being of the patient. Barker reviewed postoperative care.[93]

Leech and colleagues[94] investigated the early postoperative state of the patient. They measured ICP at the end of surgery, before and after the reversal of the muscle relaxants and after removal of the tracheal tube and found the ICP to be surprisingly high. After surgery, but before reversal of the muscle relaxants, the mean ICP was 11 mmHg, but after reversal, with the patient breathing spontaneously, the mean ICP was 21 mmHg. Just before the patient's return to the ward, the ICP was still high at 19 mmHg. These are surprisingly high values, considering that craniotomy had been performed and CSF drainage had taken place. The study emphasizes that special care is required for the neurosurgical patient in the early postoperative period.

Postoperative complications

Haematoma formation

The incidence of postoperative haematoma formation should be low, because of the care taken to achieve haemostasis before closure of the dura. The surgeon should always be aware when hypotension is being used, so that he or she may be meticulous in securing haemostasis. Following hypotension the systolic blood pressure should be at least 100 mmHg during closure.

Recovery from muscle relaxants and extubation should be managed carefully so that there is no coughing, straining or airway obstruction, as these may promote venous bleeding.

Cerebral oedema

Oedema is often present before operation. A certain amount of oedema formation always follows intracranial surgery and may increase for up to 24–36 hours postoperatively, even affecting conscious level.

Oedema in the vicinity of the vital centres in the brainstem may lead to respiratory irregularity or apnoea in the postoperative period.[95] The impairment of breathing may be recognized immediately on recovery from anaesthesia, or it may be delayed by several hours. Such patients require elective positive pressure ventilation of the lungs postoperatively.

Respiratory failure

Postoperative ventilation should be considered when the operation has been unduly long, if there has been a raised ICP in the operating theatre, if surgery has been close to the vital centres (especially if vasomotor instability was noted during surgery) and when oedema may spread through the midbrain, affecting respiratory function. Furthermore, when major blood loss or periods of ischaemia have occurred during the operation, the patient may benefit from positive pressure lung ventilation.

Vasospasm

If vasospasm exists after an operation for clipping an aneurysm, fluid balance must be controlled carefully so that hypotension does not occur. Nifedipine infusion should be continued and therapy to raise the blood pressure, such as dopamine infusion, may be considered if there are neurological signs suggesting the occurrence of vasospasm.

Seizures

Seizures constitute a major emergency because they cause increased cerebral activity and metabolism at a time when the oxygen supply may be reduced by airway obstruction or respiratory insufficiency. Use of appropriate anticonvulsants such as phenytoin is usual.

Nerve dysfunction

Lower cranial nerve paresis may occur after the removal of a tumour in the cerebellopontine angle, though much surgical effort is applied to minimizing the damage. If lower cranial nerve damage is suspected, the movement of vocal folds should be inspected by gentle laryngoscopy after extubation and, if necessary, a nasogastric tube inserted.

Postoperative monitoring

The level of response, as indicated in an objective, repeatable way by such scales as the Glasgow Coma Scale,[96] is central to the detection of postoperative complications. It is essential, therefore, that the neurosurgical patient recovers rapidly and completely from anaesthesia and that observations made in the recovery room are comparable with those made later, when the patient has returned to the ward.

There must be regular assessment of localizing signs. Many neurological signs can be attributable to supratentorial lesions, but the detection of muscle weakness on one side is clinically the most valuable. The pupils must also be examined because pupillary changes such as unilateral dilatation are good indicators of dysfunction.

Accurate measurement of heart rate, blood pressure, central venous pressure, and intracranial pressure when indicated, are important, so that the disease processes that are still active may be modified. Hypotension should be avoided in patients with vasospasm. Blood pressure may be raised if hypotension has been employed,[97] and this may need to be controlled with drugs such as propranolol or labetalol.

Analgesia will be required, though most patients after cranial surgery can be managed with codeine phosphate. After spinal surgery, patients may require narcotic analgesics and may benefit from patient-controlled analgesia.

Cerebral protection

The limitation of the damage caused by cerebral ischaemia has provoked much research work of interest to anaesthetists.[98,99] Some research has related to drug therapy, some to hypothermia.[100] Hypothermia has been used during neurosurgery for many years.[75,76] The effects of drugs and techniques depend on whether focal or global ischaemia is present, the duration and severity of the ischaemia and blood glucose values.

Many techniques have been used to limit cerebral damage, including hyperventilation, which, it was suggested, by causing an increase in cerebrovascular resistance in normal brain, would shift CBF towards the ischaemic area where the vasculature would remain dilated. Other therapies attempt to increase oxygen supply, such as inducing hypertension when vasospasm is a problem. Mannitol[101] reduced the infarct size after middle cerebral artery occlusion, presumably by increasing rCBF to the area.

Michenfelder[98,99] indicated that there is a difference between the effects of hypothermia and barbiturates on the brain. He suggested that hypothermia acted by decreasing the rates of all biochemical reactions, a function of temperature rather than of the electrical activity of the brain. Barbiturates, on the other hand, stabilize electrically active membranes, thus decreasing the work of the brain and, therefore, its energy requirements. Todd and Warner[74] have reviewed the mechanisms of cerebral protection, suggesting that the simple idea that metabolic depression will protect the brain needs modification. They point out that the ability of agents to protect the brain does not parallel their ability to suppress the EEG or the $CMRO_2$, and that the protective effects of hypothermia may not be related to the degree of cerebral metabolic depression.[100]

The barbiturates may have a place in cerebral protection when focal ischaemia is likely. This is specially so in such situations as the clipping of a supply vessel to an aneurysm to gain control of that aneurysm, when it is possible to establish the effective plasma level of barbiturate before the ischaemia begins. Propofol, it has been suggested, may have little protective effect.[74] Drugs that depress metabo-

lism may certainly slow the rate at which the high energy phosphates are depleted and therefore slightly delay membrane depolarization.

Other drugs, particularly the *N*-methyl-D-aspartate antagonists such as phencyclidine and MK-801 (dizocilpine), are being extensively investigated. It is thought that they block the actions of the excitatory amino acids (including glutamate) produced when membrane depolarization occurs in ischaemia and which produce further neuronal damage..Calcium entry blockers may also be of value in preventing the influx of calcium that occurs. Many other factors are under investigation and our ability to treat effectively cerebral ischaemia may improve in the next few years.

Spinal surgery

Surgery may be required on the vertebral column for disorders of the intervertebral disc, for spinal canal stenosis, for the removal of tumours, or for the effects of trauma. Occasionally, surgery is required for the relief of intractable pain. Children may be operated on for correction of spinal deformity. Injury or compression of the spinal cord may, depending on the level, be associated with circulatory or respiratory disturbance.

In the cervical area, degenerative conditions can cause progressive cord compression leading to symptoms and signs in the long tract. Rheumatoid arthritis may be complicated by instability of the vertebrae, which may require fixation.

The patient may be positioned prone, lateral or sitting; the movement necessary to position the patient must not be allowed to cause further damage to the cord. Similarly, for patients with cervical spine disease, tracheal intubation needs to be managed in such a way as to eliminate dangerous movements of the neck. Fibreoptic bronchoscopy for intubation is a valuable technique, used after the neck is fixed in a position acceptable to the patient (e.g. with a collar, sandbags or traction). Intubation may also be performed under local anaesthesia.

Patients who have suffered spinal trauma exhibit a distinct set of problems. Spinal shock,[102] which may last from a few days to a few weeks, may be present and the patient may be hypotensive, develop pulmonary oedema and, if the lesion is high enough (particularly above C6), show ventilatory insufficiency. In such patients a pulmonary artery catheter is most valuable. If urgent surgery is required, great care needs to be taken during movement and tracheal intubation. Suxamethonium should be avoided because of the danger of a major potassium flux.[103]

All the factors relevant to intracranial surgery apply to

anaesthesia for spinal surgery. The spinal cord blood flow is under the same controlling mechanisms as the CBF and responds to drugs in the same way. Achieving a low venous pressure is as important as for intracranial surgery, but is rendered difficult because of the susceptibility of the epidural venous plexus to the pressure changes in the thorax. High mean intrathoracic pressure may cause engorgement of the epidural veins, with consequent surgical difficulty. Positioning the patient so that there is pressure on the abdomen will also increase the pressure in the epidural veins.

Anaesthesia may be given in the same way as for intracranial surgery, but the freedom allowed by the fact that there will be no intracranial space-occupying lesion permits greater use of the volatile anaesthetic agents. It is not usually necessary to induce hypotension. Pain may be marked after spinal surgery, and intraoperative analgesia followed postoperatively by patient-controlled analgesia is most valuable.

Summary

A careful assessment of the patient and an appropriate anaesthetic technique with the correct use of drugs are central to successful neurosurgery. New measurement techniques and new treatment possibilities are developing and offer improved hope for patients with neurosurgical disease in the next few years.

REFERENCES

1. Sullivan HG, Miller JD, Griffith RL, Becker, DP. CSF pressure transients in response to epidural and ventricular volume loading. *American Journal of Physiology* 1978; **234**: R167.
2. Cushing H. The blood pressure reaction of acute cerebral compression, illustrated by cases of intracranial hemorrhage. *American Journal of the Medical Sciences* 1903; **125**: 1017.
3. Plum F, Siesjo BK. Recent advances in CSF physiology. *Anesthesiology* 1975; **42**: 708.
4. Cutler RWP, Pale L, Galicich J, Watters GV. Formation and absorption of the cerebrospinal fluid in man. *Brain* 1968; **91**: 707.
5. Potts DG, Gomez DG. Arachnoid villi and granulations. In: Lundberg N, Ponten U, Brock M, eds. *Intracranial pressure II*. Berlin: Springer Verlag, 1975: 42.
6. Miller JD. Intracranial pressure monitoring. *British Journal of Hospital Medicine* 1978; **19**: 497.
7. Miller JD, Stanek AE, Langfitt TW. Concepts of cerebral perfusion pressure and vascular compression during intracranial hypertension. *Progress in Brain Research* 1972; **35**: 411.
8. Grubb RL, Raichle ME, Eichling JD. The effects of changes in $PaCO_2$ on cerebral blood volume, blood flow and vascular mean transit time. *Stroke* 1974; **5**: 630.
9. Lewelt W, Jenkins LW, Miller JD. Autoregulation of cerebral blood flow after experimental fluid percussion injury of the brain. *Journal of Neurosurgery* 1980; **53**: 500.
10. Keaney NP, Pickerodt VWA, McDowall DG, Coroneos NJ, Turner JM, Shah ZP. The cerebral effects of hypotension produced by deep halothane anaesthesia. *British Journal of Anaesthesia* 1973; **44**: 623.
11. Meyer MW, Klassen AC. Regional brain blood flow during sympathetic stimulation. *Stroke* 1973; **4**: 370.
12. Salanga VD, Waltz AG. Regional cerebral blood flow during the stimulation of the seventh cranial nerve. *Stroke* 1973; **4**: 213.
13. Botterell EH, Lougheed WM, Scott J, Vandewater SL. Hypothermia and interruption of carotid or carotid and vertebral circulation in the surgical management of intracranial aneurysms. *Journal of Neurosurgery* 1956; **13**: 1.
14. Hunt WE, Hess RM. Surgical risk as related to time of intervention in the repair of intracranial aneurysms. *Journal of Neurosurgery* 1968; **28**: 14.
15. Pickard JD, Murray GD, Illingworth R, *et al.* Effect of oral nimodipine on cerebral infarction and outcome after subarachnoid haemorrhage: British Aneurysm Nimodipine Trial. *British Medical Journal* 1989; **298**: 636.
16. Galloon S, Rees GAD, Briscoe CE, Davis S, Kilpatrick GS. Prospective study of electrocardiographic changes associated with subarachnoid haemorrhage. *British Journal of Anaesthesia* 1972; **44**: 511.
17. Michenfelder JD. The interdependency of cerebral function and metabolic effects following maximum doses of thiopentone in the dog. *Anesthesiology* 1974; **41**: 231.
18. Ravussin P, Guinard JP, Ralley F, Thorin D. Effect of propofol on cerebrospinal fluid pressure and cerebral perfusion pressure in patients undergoing craniotomy. *Anaesthesia* 1988; **43**: suppl, 37.
19. Harris CE, Murray AM, Anderson JM, Grounds RM, Morgan M. Effects of thiopentone, etomidate and propofol on the haemodynamic response to tracheal intubation. *Anaesthesia* 1988; **43**: suppl, 32.
20. Renou AM, Vernhiet J, Macraz P, *et al.* Cerebral blood flow and metabolism during etomidate anaesthesia in man. *British Journal of Anaesthesia* 1978; **50**: 1047.
21. Milde LN, Milde JH, Michenfelder JD. Cerebral functional, metabolic and hemodynamic effects of etomidate in dogs. *Anesthesiology* 1985; **63**: 371.
22. Theye RA, Michenfelder JD. The effect of nitrous oxide on canine cerebral metabolism. *Anesthesiology* 1968; **29**: 1119.
23. Wollman H, Alexander SC, Cohen PJ, Smith TC, Chase PE, Molen RA. Cerebral circulation during general anesthesia

and hyperventilation in man: thiopental induction to nitrous oxide and *d*-tubocurarine. *Anesthesiology* 1965; **26**: 329.

24. Greenbaum R, Cooper R, Hulme A, Macintosh IP. The effects of the induction of anaesthesia on intracranial pressure. In: Arias A, ed. *Recent progress in anaesthesiology and resuscitation*. Amsterdam and Oxford: Excerpta Medica, 1975: 794.

25. Henriksen HT, Jorgensen PB. The effect of nitrous oxide on intracranial pressure in patients with intracranial disorders. *British Journal of Anaesthesia* 1973; **45**: 486.

26. Hansen TD, Warner DS, Todd MM. Nitrous oxide is a more potent vasodilator than either halothane or isoflurane. *Anesthesiology* 1988; **69**: A537.

27. Christensen MS, Hoedt-Rasmussen K, Lassen NA. Cerebral vasodilatation by halothane anaesthesia in man and its potentiation by hypotension and hypocapnia. *British Journal of Anaesthesia* 1967; **39**: 927.

28. Adams RW, Gronert GA, Sundt Jr TM, Michenfelder JD. Halothane hypocapnia and cerebrospinal fluid pressure in neurosurgery. *Anesthesiology* 1972; **37**: 510.

29. Eger EI. *Isoflurane (Forane). A compendium and reference.* Madison WI: Anaquest, a Division of BOC Inc, 1986: 1.

30. Todd MM, Drummond JC. A comparison of the cerebrovascular and metabolic effects of halothane and isoflurane in the cat. *Anesthesiology* 1984; **60**: 276.

31. Campkin TV, Flinn RM. Isoflurane and cerebrospinal fluid pressure – a study in neurosurgical patients undergoing intracranial shunt procedures. *Anaesthesia* 1989; **44**: 50.

32. Grosslight K, Foster R, Colohan AR, Bedford RF. Isoflurane for neuroanesthesia: risk factors for increases in intracranial pressure. *Anesthesiology* 1985; **63**: 533.

33. Weiss MH, Kurze T, Apuzzo ML, Heiden JS. Effect of curare in intracranial dynamics. *Surgical Forum* 1974; **25**: 458.

34. Lutz LJ, Milde JH, Milde LN. The response of the canine cerebral circulation to hyperventilation during anesthesia with desflurane. *Anesthesiology* 1991; **74**: 504.

35. Tarkkanen L, Laitinen L, Johansen G. Effects of *d*-tubocurarine on intracranial pressure and thalamic electrical impedance. *Anesthesiology* 1974; **40**: 247.

36. Coleman AJ, Downing JW, Leary WP, Moys DG, Styles M. The immediate cardiovascular effects of pancuronium, alcuronium and tubocurarine in man. *Anaesthesia* 1972; **27**: 415.

37. Minton MD, Stirt JA, Bedford RF, Haworth C. Intracranial pressure after atracurium in neurosurgical patients. *Anesthesia and Analgesia* 1985; **64**: 1113.

38. Rosa G, Sanfilippo M, Vilardi V, Orfei P, Gasparetto A. Effects of vecurinium bromide on intracranial pressure and cerebral perfusion pressure. A preliminary report. *British Journal of Anaesthesia* 1986; **58**: 437.

39. Minton MD, Grosslight K, Stirt JA, Bedford RF. Increases in intracranial pressure from succinylcholine. Prevention by prior non-depolarizing blockade. *Anesthesiology* 1986; **65**: 165.

40. Lanier WW, Milde JH, Michenfelder JD. Cerebral stimulation following succinylcholine in dogs. *Anesthesiology* 1986; **65**: 165.

41. Messick JM, Theye RA. Effects of pentobarbital and meperidine on canine cerebral and total oxygen consump-

tion rates. *Canadian Anaesthetists' Society Journal* 1969; **16**: 321.

42. Takeshita H, Michenfelder JD, Theye RA. The effects of morphine and *N*-allylnormorphine on canine cerebral metabolism and circulation. *Anesthesiology* 1972; **37**: 605.

43. Jobes DR, Kennell E, Bitner R, Swenson E, Wollman H. Effects of morphine–nitrous oxide anesthesia on cerebral autoregulation. *Anesthesiology* 1975; **42**: 30.

44. Sari A, Okuda Y, Takeshita H. The effects of thalamonal on cerebral circulation and oxygen consumption in man. *British Journal of Anaesthesia* 1972; **44**: 330.

45. Michenfelder JD, Theye RA. Effects of fentanyl, droperidol and innovar on canine cerebral metabolism and blood flow. *British Journal of Anaesthesia* 1971; **43**: 630.

46. Carlsson C, Smith DS, Keykah MM, Englebach I, Harp JR. The effects of high dose fentanyl on cerebral circulation and metabolism in rats. *Anesthesiology* 1982; **57**: 375.

47. Fitch W, Barker J, Jennett WB, McDowall DG. The influence of neuroleptanalgesic drugs on cerebrospinal fluid pressure. *British Journal of Anaesthesia* 1969; **41**: 800.

48. Bingham RM, Hinds CJ. Influence of bolus doses of phenoperidine on intracranial pressure and systemic arterial pressure in traumatic coma. *British Journal of Anaesthesia* 1987; **39**: 927.

49. Carlsson C, Hagerdal M, Kaasik AE, Siesjo BK. The effect of diazepam on cerebral blood flow and oxygen consumption in rats and its synergistic reaction with nitrous oxide. *Anesthesiology* 1976; **45**: 319.

50. Cotev S, Shalit MN. Effects of diazepam on cerebral blood flow and oxygen uptake after head injury. *Anesthesiology* 1975; **43**: 117.

51. Forster A, Juge O, Morel D. Effects of midazolam on cerebral blood flow in human volunteers. *Anesthesiology* 1982; **56**: 453.

52. Giffin JP, Cottrell JE, Shwiry B, Hartung J, Epstein J, Lim K. Intracranial pressure, mean arterial pressure and heart rate following midazolam or thiopental in humans with brain tumors. *Anesthesiology* 1984; **60**: 491.

53. Bedford RF, Persing JA, Pobereskin L, Butler A. Lidocaine or thiopental for rapid control of intracranial hypertension? *Anesthesia and Analgesia* 1980; **59**: 435.

54. Ravussin P, Tempelhoff R, Modica PA, Bayer-Merger M-M. Propofol vs. thiopental–isoflurane for neurosurgical anesthesia: comparison of hemodynamics, CSF pressure, and recovery. *Journal of Neurosurgical Anesthesiology* 1991; **3**: 85.

55. Todd MM, Warner DS, Sokoll MD, *et al.* A prospective comparative trial of three anesthetics for elective supratentorial craniotomy. Propofol/fentanyl, isoflurane/nitrous oxide and fentanyl/nitrous oxide. *Anesthesiology* 1993; **78**: 1005.

56. Fox J, Gelb AW, Enns J, Murkin JM, Farrar JK, Manninen PH. The responsiveness of cerebral blood flow to changes in arterial carbon dioxide is maintained during propofol–nitrous oxide anesthesia in humans. *Anesthesiology* 1992; **77**: 453.

57. Harp JR, Wollman H. Cerebral metabolic effects of hyperventilation and deliberate hypotension. *British Journal of Anaesthesia* 1973; **45**: 256.

58. McDowall DG. Induced hypotension and brain ischaemia. *British Journal of Anaesthesia* 1985; **57**: 110.

59. McDermott MW, Durity FA, Borozny M, Mountain MA. Temporary vessel occlusion and barbiturate protection in cerebral aneurysm surgery. *Neurosurgery* 1989; **25**: 54.

60. Gianotta SL, Oppenheimer JH, Levy MI, Zelman V. Management of intraoperative rupture of aneurysm without hypotension. *Neurosurgery* 1991; **28**: 531.

61. Campkin TV. Radial artery cannulation: potential hazard in patients with acromegaly. *Anaesthesia* 1980; **35**: 1008.

62. Harper AM. Autoregulation of cerebral blood flow: influence of the arterial blood pressure on the blood flow through the cerebral cortex. *Journal of Neurology, Neurosurgery and Psychiatry* 1966; **29**: 398.

63. Fitch W, Ferguson GG, Sengupta D, Garibi J, Harper AM. Autoregulation of cerebral blood flow during controlled hypotension in baboons. *Journal of Neurology, Neurosurgery and Psychiatry* 1976; **39**: 1014.

64. Symon L. Flow thresholds in brain ischaemia and the effects of drugs. *British Journal of Anaesthesia* 1985; **57**: 34.

65. Astrup J, Symon L, Branston HM, Lassen NA. Cortical evoked potential and extracellular K^+ and H^+ at critical levels of brain ischemia. *Stroke* 1977; **8**: 51.

66. Harris RJ, Symon L, Branston HM, Bayhan M. Changes in extracellular calcium activity in cerebral ischemia. *Journal of Cerebral Blood Flow and Metabolism* 1981; **1**: 203.

67. Brierley JB, Brown AW, Excell BJ, Meldrum BS. Brain damage in the rhesus monkey resulting from profound arterial hypotension. 1. Its nature, distribution and general physiological correlates. *Brain Research* 1969; **13**: 68.

68. Campkin TV, Baker RG, Pabari M, Grove L. Acid–base changes in arterial blood and cerebrospinal fluid during craniotomy and hypotension. *British Journal of Anaesthesia* 1974; **46**: 263.

69. Meyer FB, Muzzi DA. Cerebral protection during aneurysm surgery with isoflurane anesthesia. *Journal of Neurosurgery* 1992; **76**: 541.

70. Turner JM, Powell DG, Gibson R, McDowall DG. Intracranial pressure changes in neurosurgical patients during hypotension induced with sodium nitroprusside or trimetaphan. *British Journal of Anaesthesia* 1977; **49**: 419.

71. Morris PJ, Todd M, Philbin D. Changes in canine intracranial pressure in response to infusion of sodium nitroprusside and trinitroglycerin. *British Journal of Anaesthesia* 1982; **54**: 991.

72. McDowall DG, Keaney NP, Turner JM, Lane JR, Okuda Y. Toxicity of sodium nitroprusside. *British Journal of Anaesthesia* 1974; **46**: 327.

73. Burt DER, Verniquet AJW, Homi J. The response of canine intracranial pressure to hypotension induced with nitroglycerin. *British Journal of Anaesthesia* 1982; **54**: 665.

74. Todd MM, Warner DS. A comfortable hypothesis reevaluated. Cerebral metabolic depression and brain protection during ischemia. *Anesthesiology* 1992; **76**: 161.

75. Campkin TV, McNeil WT. Hypothermia for neurosurgery. *British Journal of Anaesthesia* 1964; **36**: 77.

76. Solomon RA, Smith CR, Raps EC, Young WI, Sone JG, Fink ME. Deep hypothermic circulatory arrest for the management of complex anterior and posterior circulation aneurysms. *Neurosurgery* 1991; **29**: 732.

77. Albin MS, Babinski M, Maroon JC, Janetta PJ. Anaesthetic management of posterior fossa surgery in the sitting position. *Acta Anaesthesiologica Scandinavica* 1976; **20**: 117.

78. Dalrymple DG, MacGowan SW, Macleod GF. Cardiorespiratory effects of the sitting position in neurosurgery. *British Journal of Anaesthesia* 1979; **51**: 1079.

79. Michenfelder JD, Miller RH, Gronert GD. Evaluation of an ultrasonic device (Doppler) for the diagnosis of venous air embolism. *Anesthesiology* 1972; **36**: 164.

80. Cucchiara RF, Nugent M, Seward JB, Messick JM. Air embolism in upright neurosurgical patients: detection and localization by two-dimensional transesophageal echocardiography. *Anesthesiology* 1984; **60**: 353.

81. Bunegin L, Albin MS, Helsel PE, Hoffman A, Hung K. Positioning the right atrial catheter. *Anesthesiology* 1981; **55**: 343.

82. Muizelaar JP, Wei EP, Kontos HA, Becker DP. Mannitol causes compensatory cerebral vasoconstriction and vasodilatation in response to blood viscosity changes. *Journal of Neurosurgery* 1983; **59**: 822.

83. Miller JD, Leech PJ. Effects of mannitol and steroid therapy on intracranial volume–pressure relationships in patients. *Journal of Neurosurgery* 1975; **42**: 274.

84. Muizelaar JP, Lutz HA, Becker DP. Effect of mannitol on ICP and CBF and correlation with pressure autoregulation in severely head injured patients. *Journal of Neurosurgery* 1984; **61**: 700.

85. Cottrell JE, Robustelli A, Post K, Turndorf H. Furosemide and mannitol induced changes in ICP and serum osmolality and electrolytes. *Anesthesiology* 1977; **47**: 28.

86. Wilkinson HA, Wepsie JG, Austin G. Diuretic synergy in the treatment of acute experimental cerebral edema. *Journal of Neurosurgery* 1971; **34**: 203.

87. Martins AM, Ramirez A, Soloman LS, Weise GM. The effect of dexamethasone on the rate of formation of cerebrospinal fluid in the monkey. *Journal of Neurosurgery* 1974; **41**: 550.

88. Shapiro HM. Intracranial hypertension: therapeutic and anesthetic considerations. *Anesthesiology* 1975; **43**: 445.

89. Bedford RF, Winn HR, Park TS, Jane JA. Lidocaine prevents increased ICP after endotracheal intubation. In: Schulman K, Marmarou A, Miller JD, Becker DP, Hochwald GM, Brock M, eds. *Intracranial pressure 4*. New York: Springer Verlag, 1980: 28.

90. Andrews PJD, Dearden NM, Miller JD. Jugular bulb cannulation: description of a cannulation technique and a validation of a new continuous monitor. *British Journal of Anaesthesia* 1991; **67**: 553.

91. Turner JM, Baker RG, Patel A. Cerebral arteriovenous oxygen differences in hypotension induced by trimetaphan or sodium nitroprusside. *British Journal of Anaesthesia* 1981; **53**: 308.

92. Aaslid R, Markwalder T-M, Nornes H. Non-invasive transcranial Doppler ultrasound recording of flow velocity in basal cerebral arteries. *Journal of Neurosurgery* 1982; **57**: 769.

93. Barker J. Postoperative care of the neurosurgical patient. *British Journal of Anaesthesia* 1976; **48**: 797.

94. Leech PJ, Barker J, Fitch W. Changes in intracranial pressure during the termination of anaesthesia. *British Journal of Anaesthesia* 1974; **46**: 315.

95. Artru AA, Cucchiara RF, Messick JM. Cardiorespiratory and cranial nerve sequelae of surgical procedures involving the posterior fossa. *Anesthesiology* 1980; **52**: 83.

96. Teasdale G, Jennett WB. Assessment of coma and impaired consciousness. A practical scale. *Lancet* 1974; **2**: 81.

97. Khambatta HJ, Stone JG, Khan E. Hypertension during anesthesia on discontinuation of sodium nitroprusside induced hypotension. *Anesthesiology* 1979; **51**: 127.

98. Michenfelder JD. Cerebral protection with barbiturates; relation to anesthetic effect. *Stroke* 1978; **9**: 140.

99. Michenfelder JD. Hypothermia plus barbiturates: apples plus oranges? *Anesthesiology* 1978; **49**: 157.

100. Sano T, Drummon J, Patel P, Grafe M, Watson J, Cole D. A comparison of the cerebral protective effects of isoflurane and mild hypothermia in a model of incomplete forebrain ischemia in the rat. *Anesthesiology* 1992; **76**: 221.

101. Little JR. Modification of acute focal ischemia by treatment with mannitol and high dose dexamethasone. *Journal of Neurosurgery* 1978; **49**: 517.

102. Desmond J. Paraplegia: problems confronting the anaesthesiologist. *Canadian Anaesthetists' Society Journal* 1970; **17**: 435.

103. Gronert GA, Theÿe RA. Pathophysiology of hyperkalaemia induced by succinylcholine. *Anesthesiology* 1975; **43**: 89.

Chapter 53

The Severely Burnt Child

William David Lord

Thermal injury to the skin is a major source of morbidity and mortality in children. The resulting physical and cosmetic disability leaves a permanent scar not only on the child but also on the whole family.

Thermal injury in children falls into two main groups. Scalds predominate in the toddler age group, whereas flame burns with or without detonation occur in the young adolescent.

Morbidity and mortality are further increased by the presence of smoke inhalation, intercurrent sepsis or additional trauma or illness; for example, fractures, metabolic or endocrine disease.

In hospitals remote from the regional burns centre, the anaesthetist may be the person most experienced in resuscitating children and should be involved at an early stage along with the casualty specialist and general surgeon.

Adequate comfortable accommodation for parents should be available close to the unit and, given the stressful environment in which burns are treated, provision should be made for staff rest and recuperation near to but separate from the clinical area.

Individually plenum ventilated rooms with the facility to change temperature independently permit better temperature homoeostasis and isolation of infection.

The anaesthetist's role must be considered in the context of the pathophysiology, therapeutic and surgical management, intensive or invasive intervention and discussion of the recognized complications such as sepsis and the rare yet devastating toxic shock syndrome. Smoke inhalation with or without cutaneous injury is of particular concern to the anaesthetist.

All thermally injured children require intensive care to a greater or lesser extent.

The burns unit

The severe burn is more easily and successfully managed in a unit specifically designed for the purpose. In addition to reception, treatment and semi-convalescent facilities, there should be an intensive care provision for at least two patients since paediatric smoke inhalation victims rarely occur singly. An integral operating theatre facilitates transfer without temperature loss and also isolates potentially infected patients from non-burns patients and vice versa. Because anaesthesia or sedation may be required for wound toilet in the bath, specialized equipment and monitoring equipment may be required.

The cutaneous thermal injury

The extent of the injury will dictate the magnitude of the pathophysiological change and the timing and degree of surgical intervention.[1]

A superficial burn involves only the epidermis but may be accompanied by pain, hyperpyrexia, dehydration and confusion. The outcome is influenced by its extent and the adequacy of intravascular volume replacement. Surgical intervention is not usually required and healing progresses without scarring. Prolonged exposure to the sun can produce extensive superficial burns in the young child, producing heat stroke and dehydration.

A dermal burn destroys the epidermis and a variable amount of dermis. The depth of dermal destruction dictates the degree of metabolic disturbance and the rate of re-epithelialization from residual skin structures (e.g. hair follicles and sweat glands). Early surgery is recommended for deep dermal burns, to reduce the duration of the danger period and ensure minimal scarring.

A full-thickness burn destroys all skin structures and possibly the underlying fat down to deep fascia and beyond. Muscle injury is associated with damage to major vessels and usually results in the loss of a limb or major underlying organ dysfunction. A similar injury follows electrical injury in which the electrical flux follows the neurovascular bundle, thus devitalizing the limb. Because all skin structures are destroyed, such injuries will not heal spontaneously and will require excision and grafting at the earliest opportunity; electrical injuries are usually accompanied by major destruction of muscle and loss of the limb. Because all nerve endings are destroyed, the child may require less analgesia than the extent and degree of injury would suggest. This does not apply when, later in the course of management of severe full-thickness burns, dressings are changed: significant analgesia may then be required.[2]

The intravascular volume

This is depleted in all burns but to the greatest degree when the burn involves the dermis. The loss of plasma from the circulation takes place not only in the thermally damaged area but also from the rest of the circulatory system.

Burn oedema

Endothelial destruction in the area of thermal damage permits exudation of plasma from the intravascular space in the region of the burn. The microcirculation in the burned area is likely to be affected by the release of locally released mediators. Histamine and vasodilator prostaglandins may contribute to increased blood flow. Thromboxane A_2 will reduce blood flow to the damaged yet potentially viable area, resulting in deepening of the burn. Oxygen radicals may also be implicated in extending the damage.

The net effect is an increase in extracellular fluid pressure, resulting in reduced oxygen transport to damaged cells. Hypotension will augment any ischaemia caused by locally mediated shunting or vasoconstriction.

Oedema under circumferential burns will cause venous and then arterial compression, producing acute ischaemia

and possibly limb loss. This obstruction to perfusion will be relieved only when the inelastic full-thickness burn is incised down to deep fascia by escharotomy.

The vessels under the burn are equally permeable to crystalloids and colloids; therefore the timescale of oedema formation depends on cardiac output rather than the type of resuscitation fluid.

Non-burn oedema

This can be demonstrated in most serious burns in children. Cricoid oedema can be demonstrated in most children, so the tracheal tube size appropriate for anaesthesia and intensive care may be reduced by about 0.5–1 mm.

Hypoproteinaemia has been implicated as the main cause of oedema distant from the burns site. Plasma is therefore the colloid of choice for resuscitation, in order to minimize the formation of oedema remote from the burn. The oedema will be augmented by inappropriate crystalloid infusion or inconsistent resuscitation.

The combination of burn and non-burn oedema can be particularly dangerous around the head and neck, where massive and progressive oedema will cause swelling of the face, tongue and supraglottic structures. The oedema reaches its peak at 24–36 hours; early intervention is required to secure the airway while it is still possible.

When the burn is associated with serious tissue destruction, there is a systemic decrease in cell transmembrane potential, resulting in an intracellular shift of sodium and water. This has been demonstrated especially in muscle, but if it occurs in other organs may be responsible for significant dysfunction.

Metabolism

There is a massive and sustained release of catecholamines, glucagon and corticosteroids from the time of the injury. Insulin production may be overwhelmed and result in hyperglycaemia.

Unlike the adult, in whom increases in metabolism are marked from the onset of the 'flow' phase at 5 days, the child demonstrates an increase in metabolism much earlier. Infants and children with burns to 10 per cent or more of the body surface show a characteristic pattern of body temperature which appears to be unrelated to the size of the burn. After a short period of normothermia, there is a rapid rise 5–8 hours after the accident, which reaches a peak (usually in excess of 40°C) by 10–12 hours. The temperature remains at this elevated level over the next 12–24 hours.

This pyrexia is not associated with cutaneous infection and does not appear to be directed by a change in the hypothalamic temperature setting. High levels of the endogenous pyrogen interleukin-6 have been found at the height of the fever, and it is likely that cytokines are involved in the development of this early fever.

The stimulus to the development of this fever may be from gut-derived organisms, bacterial fragments or toxins. Alterations in gut permeability have been demonstrated in burned children as well as in adults.[3]

The rise in temperature appears to be benign in itself, the only apparent consequences being increased oxygen requirements and carbon dioxide production, the latter being on average 26 per cent above control values. This will aggravate any respiratory or neurological injury. Antipyretics have a variable effect and in general fail to return the temperature to within normal limits.[4]

Respiratory function

In the child the narrowest part of the airway is at the cricoid ring. Any oedema of the glottic or supraglottic structures secondary to oedema of the neck, pre-existing croup or direct thermal injury from swallowed fluids is poorly tolerated in the child. In the presence of a greatly increased metabolic rate the airway may become inadequate.

On admission, the patient demonstrates the classic hypoxia associated with reduced pulmonary perfusion; thereafter a marginal reduction in saturation may result from increased pulmonary shunting secondary to the dilated circulation. As temperature rises and metabolic rate increases, the child responds to the increasing CO_2 load by an increase in respiratory rate rather than depth, thus increasing the amount of work expended in dead space ventilation.

If resuscitation is erratic or the crystalloid load excessive, the lungs will absorb their share of the excess water. In the conscious child it is likely that symptoms of cerebral oedema will be noticed before the effects of excess lung water.

Cardiovascular function

The large losses from the intravascular compartment are compensated for by progressive peripheral and splanchnic vasoconstriction, and cardiac output is initially maintained. However, without rapid resuscitation, a fall in cardiac output with the development of anaerobic metabolism ensues and collapse may occur rapidly in the child. On the other hand, with effective resuscitation, peripheral vascular beds open up, peripheral resistance decreases and cardiac output increases dramatically to meet the needs of the massive rise in metabolism. During this time, particularly in the ventilated child with a severe burn, peripheral

resistance may suddenly increase and blood pressure rise dramatically. Such hypertensive events may be isolated or intermittent in presentation. In the presence of adequate analgesia and sedation their aetiology may be secondary to catecholamine output, or, if more sustained, due to renin or ADH secretion. The use of α-blockade with phenoxybenzamine allows a fall in peripheral resistance and an increase in renal blood flow and urine output.

Children who have a healthy myocardium and coronary arteries free from occlusive disease are capable of sustaining a high cardiac output and rapid pulse rates in excess of 180 beats per minute, for weeks if necessary. In developed septicaemic shock and toxic shock states, cardiac output can deteriorate catastrophically when the circulation responds poorly to administered inotropic catecholamines. Prognosis depends on early diagnosis and treatment before this shock state is developed.

Renal function

The kidneys will be damaged primarily by sustained hypovolaemia. This will be compounded by any systemic hypoxia. After an electrical burn or a burn with extensive muscle damage, myoglobinuria or haemoglobinuria will reduce renal function by tubular blockage.

The quantity and quality of urine output must be measured at all stages of treatment because the kidney not only provides homoeostasis but is also a valuable monitor of resuscitation. Renal blood flow should be maintained with low dose dopamine, and urine flow stimulated by appropriate colloid and crystalloid infusion with the administration of a diuretic such as frusemide if necessary.

At a variable time from 3 days after injury, with adequate renal function a 'normal' and huge diuresis will ensue as the 'flow' phase of the injury becomes established.

Hepatic function

The liver is also affected by hypoxia and hypovolaemia. Although its function must be markedly altered by metabolic and nutritional effects, altered protein synthesis and toxic or pharmacological stress, its basic functions are remarkably well preserved. Marked liver dysfunction accompanies the multisystem failure that occurs terminally when the therapeutic process has 'run out of steam'.

Gastrointestinal function

A period of hypovolaemia will be associated with gastric atony and a degree of paralytic ileus. Narcotics, relaxants and the ventilated state may perpetuate this. A nasogastric tube should be passed in all patients with burns in excess of 10 per cent, to reduce pooling of gastric secretions. Severe gastric ulceration has been reported to be more common in children than in adults. As a result of better resuscitation, clinical evidence of extensive ulceration of the gastric mucosa, as described at postmortem by Curling, is rarely seen. Major bleeding rarely occurs, but occult blood in nasogastric aspirate may be found. Its incidence may be reduced by the administration of antacids and by nasogastric feeding which should be commenced as soon as possible following restoration of intravascular volume. This provides a degree of protection to the gastric mucosa even in the absence of complete absorption. H_2-receptor antagonists are routinely prescribed to ventilated children but, because of the increased renal clearance, the dosage has to be increased to in excess of 60 mg/kg per day to maintain gastric pH greater than 4.0.[5,6]

Haemopoiesis

Haemoglobin will fall from the time of the burn and will be proportionate to the area of the burn. In burns up to 50 per cent of body surface area, the fall does not usually exceed 10 per cent of circulating red cells. The reasons for loss include:

1. A small proportion lost by heat destruction at the time of the burn; this may be of a magnitude sufficient to produce haemoglobinuria.
2. Loss by consumption and clotting in a full-thickness burn.
3. Damage at the time of the burn renders red corpuscles abnormally fragile, leading to removal by the reticuloendothelial system shortly following the injury.
4. Continued removal by the reticuloendothelial system for an unknown reason unrelated to fragility.

The destruction of red blood cells may continue following a massive burn, to be associated with consumption of clotting factors and hypocoagulability, and continued infusion with replacement of clotting factors and platelets may be required.

This red cell loss is compounded by suppression of haemopoiesis until the burn is healing.

There are definite theoretical and practical advantages to replacing red cell loss with red cells suspended in optimal additive (sodium chloride, adenine, glucose, mannitol; SAGM) solution. SAGM blood releases its oxygen to the tissues more readily and has a lower level of microaggregates. The 100 ml of water used for suspension can be accounted for in the fluid regimen and, if necessary, removed by diuresis and then replaced by fresh-frozen plasma.

Pharmacology

Drug behaviour and metabolism are markedly altered. This is not surprising, given the massively expanded redistribution volume, altered cardiac output, altered glomerular and renal function, altered protein binding and the possibility of enzyme induction. The effectiveness of any drugs should be closely monitored clinically and by laboratory investigation and estimation wherever possible.[7]

At different stages, depending on the drug, its mode of action and method of excretion, accumulation, tachyphylaxis or resistance may be found.

Adequate analgesia and sedation can be judged only by the clinical response and not by recommended dosage. When therapeutic or toxic levels can be affected by renal function (e.g. gentamicin), blood level estimation will ensure appropriate therapy.

Anaesthetic drugs

Intravenous induction agents

All currently available induction agents, including propofol, have been used in burned children. The dose may have to be significantly increased because of the altered cardiovascular responses and increased redistribution volume. Provided that hypovolaemia is adequately treated before induction, there are no additional contraindications to their use in the burned child. Propofol would not appear to offer any advantage over thiopentone for induction except for repeated dressings.

Ketamine has been widely recommended for intravenous induction and is used in many centres as the main agent for changing of burns dressings. This unique agent provides amnesia and analgesia when given in low doses (1.5–2.0 mg/kg i.m.), with rapid return of consciousness. If given in higher doses, prolonged anaesthesia can be produced. Experience with two children who demonstrated extremely distressing emergence phenomena followed by prolonged psychiatric disturbance resulted in its abandonment in my unit, in which inhalational and intravenous techniques are used.

Neuromuscular blocking agents

The behaviour of neuromuscular blocking agents is affected by the changes in cardiac output, redistribution volume and plasma protein disturbances that occur in the resuscitation phase. Abnormalities in the response to depolarizing and non-depolarizing relaxants in thermally injured children have been recognized for many years.[8] In addition, an increase in the number of skeletal muscle acetylcholine receptors occurs, which is clinically demonstrable from day 3 after injury and lasts for up to 3 years following the injury. This increase in receptors is thought to be responsible for the sensitivity to suxamethonium (succinylcholine) with concomitant excessive potassium release and for the development of resistance to non-depolarizing relaxants, as succinctly reviewed by Martyn *et al.*[9]

Muscle relaxation is required to facilitate control of ventilation in intensive care and during anaesthesia for surgical excision. There are several specific concerns affecting their behaviour and use in the child with a thermal injury.

In the casualty situation, intubation is required to facilitate airway control when there is a danger of obstruction. Inhalational induction with halothane is indicated rather than rapid sequence induction, and suxamethonium is usually avoided. From day 3 after the burn, there occurs extreme sensitivity to the effects of suxamethonium,[10] with the risk of massive hyperkalaemic response which precludes the use of suxamethonium for a prolonged period following injury.

In the first week following injury, changes in cardiac output, plasma proteins and redistribution volume may increase requirements for all non-depolarizing relaxants. Following this acute period, progressive resistance is experienced with all relaxants proportionate to the size of cutaneous burn. Before the advent of vecuronium and atracurium, dimethyl tubocurarine (metocurine) in a dose of 1 mg/kg was the relaxant of choice because of its low intrinsic anticholinergic (muscarinic) activity. The absence of anticholinergic (muscarinic) activity and completeness of reversal have established atracurium and vecuronium as the currently most frequently used agents. Atracurium will be required in a dose of 0.3 mg/kg in the first week, rising to twice this dose in burns of 20–60 per cent of surface area. Burns greater than 70 per cent may require 3 mg/kg.[11]

The timecourse of effect of atracurium is similar to that in non-burned patients, as are the plasma clearance and half-life (Fig. 53.1). The maximum twitch depression is smaller, not being abolished in many cases although intubating conditions are unaffected,[12] with recovery to 50 per cent of twitch height in a significantly shorter period.[13] Similar resistance has been demonstrated to vecuronium.[14]

Management of the fluid loss

There are many successful regimens for the management of the huge fluid losses in the thermally injured patient. Some are more appropriate to the management of children than adults, and all form a sound basis for management within their particular units.[15] Isotonic salt solutions, hypertonic salt solution and a combination of either with plasma have been described. This would suggest that the quality of fluid monitoring and patient response is the most important feature.[16,17]

In the UK, plasma is the most frequently employed colloid in the resuscitation of burnt children. The current management and steps in decision-making as practised in this unit have been expressed in algorithmic form.[18] The prime requirement is for immediate correction of hypovolaemia, with an assessment of the continued loss and continuous or repeated adjustment of the assessment in response to therapy. Attention should be paid to the amount of administered fluid that becomes sequestered outside the circulation and its effects on tissue oxygenation. In the case of cerebral insult or airway injury, this sequestration is of critical importance.

Isotonic saline or Ringer lactate solution will support the circulation only if given in quantities greatly in excess of the plasma deficit. Fluid will not only move into the extracellular space but will also tend to move intracellularly. Hypertonic saline solutions are required in reduced amounts and will also move extravascularly but, because of their increased osmotic nature, will reduce the intracellular drift of water. Human albumin or any colloid such as modified gelatine solution will stay within the circulation, impose an osmotic gradient and reduce intracellular drift of water. However, they have no clotting factors and their total volume should be limited if early surgery is to be contemplated; furthermore, if they move extravascularly,

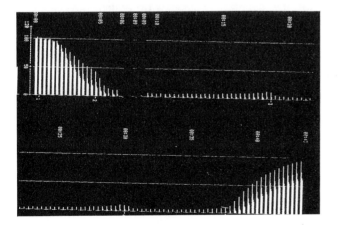

Fig. 53.1 An electromyographic (Relaxograph) tracing showing the typical response with atracurium 3 weeks following injury. Intubating dose of 1 mg/kg with a bolus of 0.5 mg/kg at 5 minutes and at 18 minutes. The neuromuscular response is abolished only briefly.

they impose an equal and opposite osmotic force, prolonging the oedema.

Fresh-frozen plasma contains albumin and clotting factors with the exception of platelets. It is the most complete colloid for volume replacement when early intervention is proposed after major cutaneous loss.

It would thus appear to be most reasonable to correct immediate losses with human albumin solution or modified gelatine solution, depending on availability, and replace continued losses with human plasma protein fraction or fresh-frozen plasma when the patient has reached the burns centre. Crystalloid solutions should not be used to support the intravascular volume in children in the casualty situation, especially when smoke inhalation is suspected.

Assessment of the damaged area

In adult practice, the extent of the damaged area can be assessed with reference to the 'rule of nine' – i.e. by comparing the area of the body divided up into multiples of 9 per cent. This does not accurately reflect the changing proportions of head, trunk and limbs in the child. Wherever possible, Lund and Browder charts should be used to accurately assess area (Fig. 53.2). In the casualty situation, the area covered by the patient's palm and fingers represents roughly 1 per cent of body surface area.

Mount Vernon regimen

This regimen was tested extensively at Mount Vernon Hospital and forms the basis of the treatment widely practised throughout the UK.[19] The regimen takes into account:

1. The proportional relation between the intravascular support required and the area burned.
2. The high initial requirement for intravascular volume support, which falls gradually over the following 2 days.
3. The need to regularly assess the response to therapy, any alterations to the rate or quality of infusion being continued forward throughout the resuscitation period.
4. The requirement for support of the red cell volume in a significant proportion of severe burns.

From the time of the injury, the resuscitation period is divided into six periods. Equal volumes of plasma are given in each of the successive periods of 4, 4, 4, 6, 6 and 12 hours. The amount of plasma required is assessed as follows. The area of burn excluding superficial erythema is estimated. The volume of plasma required per period is calculated from the formula:

$$\frac{\text{Total percentage of burn} \times \text{weight in kg}}{2} = \frac{\text{ml of plasma}}{\text{per period}}$$

Because there will inevitably have been some delay between the injury and commencement of infusion, the first aliquot of plasma will be administered in a much shorter period than 4 hours. It is likely that, even if the assessment of area is initially correct, the volume calculated will prove to be an underestimation rather than an overestimation. If the child initially demonstrates any signs of hypovolaemia, this should be corrected immediately by infusing plasma in boluses of 10 per cent of a blood volume each over 15 minutes until the pulse rate has settled and central and peripheral perfusion have improved.

It is suggested that an allowance for metabolic water should be prescribed as 5 per cent dextrose. As the patient will be maximally concentrating urine and displaying glucose intolerance, it would seem more appropriate to administer one-half to two-thirds of normal requirements as NaCl 0.45 per cent + dextrose 5 per cent. After 36 hours, metabolic water requirements can be increased to the normal:

100 ml/kg per 24 hours   for the first 10 kg
50 ml/kg per 24 hours   for the next 10 kg
20 ml/kg per 24 hours   for the rest up to adult

This fluid can be used as a carrier for the measured potassium requirements.

Losses for the gastrointestinal tract should be replaced with 0.9 per cent saline with added potassium or Hartmann's solution.

Monitoring of intravascular volume

The effectiveness of treatment depends on the assessment of adequate perfusion to vital organs.

The skin

Whether pink with capillary flush, white with vasoconstriction or blue from stagnation, the skin is a valuable indicator

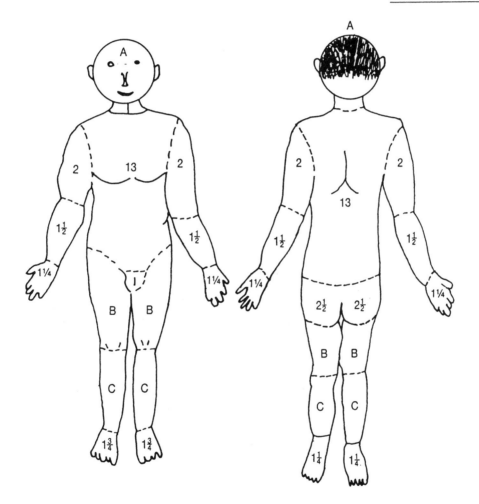

BURN RECORD (Birth – Adult)

Name _____

Age _____

Unit no. _____

Date of observation _____ _____

Total % burn _____

% deep dermal _____

% full thickness _____

RELATIVE PERCENTAGE OF AREAS AFFECTED BY GROWTH

Area	Age					
	0	1	5	10	15	Adult
A = ½ of head	9½	8½	6½	5½	4½	3½
B = ½ of one thigh	2¾	3¼	4	4½	4½	4¾
C = ½ of one leg	2½	2½	2¾	3	3¼	3½

Fig. 53.2   The Lund and Browder chart is used to assess the change in area in proportion to body size with increasing age.

of the fullness of the circulation, provided myocardial function is adequate. General changes in the skin are more difficult to interpret in non-caucasian children, but the oral mucosa, palms and nail beds can provide similar information. The skin-to-core temperature difference, measured dermally from a digit and centrally from rectum, nasopharynx or tympanic membrane, provides a quantitative measurement of perfusion. The difference should not exceed 4°C and preferably be within 2°C.

The kidney

In the human, the effects of hypovolaemia will be manifested by deteriorating renal function at an early stage. Unless action is taken to reverse these effects, acute tubular necrosis will ensue. In the management of the burnt patient the quantity and quality of urine reflect the adequacy of resuscitation.

An hourly urine volume of 0.5 ml/kg is the minimum acceptable volume. Concentrating ability is assessed by serial specific gravity readings supported by regular estimations of urine osmolarity. Refractometry at the bedside can be performed on a single drop of urine, but its accuracy must be qualified by osmolarity estimation. A high specific gravity in the range of 1030 indicates maximal concentration, whereas 1005 indicates a dilute urine. A small volume of high specific gravity indicates under-replacement, whereas a larger volume at lower gravity indicates excess water load. In the presence of seemingly adequate resuscitation and excessive oedema formation, dopamine 2 µg/kg per minute usually improves and stabilizes urine output. Dopamine in renal dose is administered to all children with burns in excess of 20 per cent.

Occasionally a low urine output occurs in the presence of high peripheral resistance, as judged by a wide toe–core temperature difference and sustained blood pressure; this condition has been attributed to prolonged catecholamine release. In these cases, phenoxybenzamine 1 mg/kg has resulted in increased urine output. Care should be taken to support the intravascular volume with plasma as it expands.

Routine measurement of serum and urine electrolytes, urea and creatinine will indicate any progression to renal failure, which requires the involvement of a paediatric nephrologist at an early stage.

In established renal failure associated with burns in children, peritoneal or, preferably, arteriovenous haemofiltration or dialysis has improved outcome considerably.

The cerebral state

Having excluded hypoxia, irritability may be an indication of hypovolaemia or of crystalloid imbalance. Excessive water load in a child will present as irritability leading to convulsions before respiratory signs are obvious. The susceptibility to febrile convulsions is the same as in the child without thermal injury, the rate of rise in temperature rather than the degree being the triggering factor.

Respiratory function

Tachypnoea will be a sign of hypovolaemia or water overload if other causes such pain, anxiety, pre-existing respiratory disease and increased CO_2 load from pyrexia have been excluded.

Haemoglobin concentration

Blood transfusion may be required to maintain the haematocrit (packed cell volume) above 30 per cent. However, after 48 hours, when the diuretic phase is established, the haematocrit will climb. This suggests that it may not be necessary to attempt to maintain higher haematocrit levels during resuscitation.

Invasive monitoring

The increased risks of infection associated with central venous lines precludes their *routine* use for the management of burns. Central venous access may be indicated:

- When monitoring of intravascular volume by the above methods proves inadequate, particularly in renal shutdown.
- When the patient is demonstrating the cardiovascular effects of septicaemia.
- When there is an associated traumatic or inhalational pulmonary injury, increased pulmonary vascular resistance dictates that right atrial filling pressures are no longer an acceptable monitor of left ventricular filling pressures; in this situation a pulmonary artery flotation catheter will be required.
- When peripheral venous access is impossible.

Pulmonary arterial flotation catheters are available with or without thermodilution facilities to suit children of all ages.

Scrupulous aseptic techniques and management must be maintained and the central line withdrawn at the earliest opportunity.

The surgery of cutaneous burns

The aim is to remove devitalized and potentially infected tissue, replacing it with split autologous skin, thus converting the burn into a surgical wound. Full-thickness injury is excised down to deep fascia. This is best performed as soon as any hypovolaemia has been corrected and cardiac, respiratory and renal function have been stabilized, usually between 6 and 20 hours after injury.

Partial-thickness injury is treated by tangential excision of devitalized tissue down to an actively bleeding surface. The shaved area is then covered with autologous split skin from any available site – including scalp but excluding face and the region of joints. When available donor sites have been exhausted, homograft from relations or cadavers or lyophilized animal skin has been used as a biological dressing awaiting healing of donor sites prior to recropping.

Early tangential excision and grafting is associated with lower morbidity and mortality from infection, better management of the metabolic response, shorter periods of hospitalization and a better cosmetic result with less contracture formation. This technique was first described in 1970 by the Yugloslav surgeon Zora Janzekovic. Although the beneficial effects in paediatric management are becoming accepted, controversy remains in adult practice.

From the anaesthetic point of view, surgery for partial-thickness injury should also be carried out as soon as possible following restoration of intravascular volume. Unfortunately, at this early stage it is difficult to assess an adequate excision, as further extension of the injury due to thrombosis in previously damaged yet still patent blood vessels may occur.

From the end of the initial resuscitation period, tangential excision is accompanied by increasing blood loss. After 7 days this becomes prohibitive, necessitating delay in surgery until the third week, when organization of the burn has occurred.

Analgesia and sedation

The provision of adequate analgesia and sedation is of paramount importance at the time of admission, following surgery and to facilitate dressing changes. The variability of injury and the patient response, both pharmacologically and emotionally, necessitate close monitoring and adjustments and modification of technique, and include pharmacological and non-pharmacological methods.[20]

On admission the child is not only in pain but frightened as well. The most suitable management is narcotic administration and support from parents and staff trained in the management of acutely distressed children and their parents.

Morphine sulphate should be given intravenously initially in a dose of up to 0.2 mg/kg. Intramuscular administration should be avoided because it will be associated with variable initial absorption and effect to be followed by a cumulative effect on return of perfusion of the muscles. Repeated intramuscular injection will reinforce the child's fear of needles. If the first dose proves to be insufficient, a further increment will be required within 20 minutes, when the initial dose has had sufficient time to take effect. Once adequate sedation and analgesia have been attained in the acute phase, further analgesia can be maintained by intravenous bolus injection, continuous background infusion at 0.05 mg/kg per hour or background infusion with nurse-administered bolus administration via a patient-controlled analgesia pump. This latter device has the benefit of providing further analgesia instantly the nurse deems it necessary while providing an exact record for monitoring the adequacy of pain relief.

A similar regimen may be used in the postoperative period. Provided the child is sufficiently conscious and of an age to comprehend the principles of patient-controlled analgesia, there appears to be no reason to deny the benefits that it provides. Children as young as 4 years have successfully used this technique, which has been shown to have a high index of safety.[21,22]

In addition to analgesia, the uncomfortable and fractious child may require background sedation. This can be provided by trimeprazine 2–3 mg/kg three times daily if required. This drug is effective in reducing the itching experienced in the recovery phase. Benzodiazepines are very variable in their effects and if trimeprazine proves inadequate, chloral hydrate may be more suitable. The aim is not to render the child unconscious but to provide a comfortable calm child who will respond to and benefit from the presence of parents who are also in a tense and emotional state.

Paracetamol (acetaminophen) has been administered in response to rises in temperature. Studies have shown that paracetamol administered orally or rectally is ineffective in reducing the pyrexia when associated with a burn injury.

It is difficult to assess the requirements for analgesia and sedation in the child requiring prolonged ventilation. Our current practice is to administer a continuous infusion of morphine and midazolam. The morphine infusion is discontinued after 24 hours, when it is assumed that the acutely painful state has abated. Further morphine by bolus or adjustment of midazolam is given after assessment of the haemodynamic response to physiotherapy and tracheal suction or motor response during recovery from neuromuscular blockade.

Indications for ventilation

Prolonged tracheal intubation by the nasal route and intermittent positive pressure ventilation (IPPV) may be associated with its own morbidity. Tracheostomy is associated with greatly increased mortality secondary to infection. The following are specific indications:

- Suspected inhalational injury.
- Scalds and burns of head and neck, in which rapidly developing oedema suggests that the airway should be secured before obstruction occurs.
- Electrical burns with an entry or exit point on the head or neck.
- Incipient septicaemia with pulmonary shunting.
- Following massive intravascular fluid replacement associated with excision and grafting to stabilize ventilation while the child organizes the fluid shifts between compartments.

IPPV will be associated with an increase in peripheral oedema due to increased venous pressure secondary to reduced venous return. Spontaneous ventilation should be encouraged as soon as possible to enable mobilization of the oedema. In addition, spontaneous coughing will promote more efficient bronchial toilet and a reduced incidence of lobar collapse.

Premedication

Preoperative sedation is desirable to reduce the stress of fasting, transport to theatre and induction of anaesthesia. Sedation, amnesia and sufficient drying of secretions will be provided by a premedicant dose of trimeprazine, supported if necessary by a further dose of intravenous morphine.

Atropine is not required to dry secretions or for its vagolytic effect. The routine administration of atropine to pyrexial and tachycardic children may be deleterious and will confuse the response to changing intravascular volume and temperature control.

Eutectic mixture of lignocaine (lidocaine) and prilocaine (EMLA) applied to a suitable area of skin will reduce pain if venepuncture is necessary.

Parental presence is encouraged whenever possible during induction. We have found the presence of a parent nothing but supportive provided there is adequate time for preparation. A greater proportion of parents become upset once the child has lost consciousness as compared with non-burns surgery. We therefore suggest the presence of a second parent or supporter during induction. Active involvement of the parents throughout the management

of the child with burns is regarded as essential, and contributes to the reduction of morbidity and mortality.[23]

On occasion, a child will exhibit extreme distress at the prospect of induction of anaesthesia. This is commonest in a child around adolescence and is associated with a degree of poor graft survival and extended hospitalization. Such a child will react noisily and aggressively at any movement or therapy. Following discussion with the child, confidence may be improved by the administration of a light general anaesthetic, with full airway control, in the bed before transport to theatre.

Anaesthesia

Anaesthesia will be required to establish airway control, grafting of the burned area or for major dressing changes.

Escharotomy of limbs or digits to prevent ischaemia, or of neck or thorax to prevent constriction, does not require anaesthesia except to establish control of ventilation.

Preoperative assessment

The aim is to take a patient, stabilized in intensive care, remove the source of metabolic stress by surgical excision and return the patient with minimal physiological disturbance. This can usually be achieved.

Given the maximum 7-day 'window of opportunity' for early surgical management without excessive blood loss, a good reason is required for postponement. Absolute contraindications for surgery are:

1. Inadequate resuscitation; abnormal plasma electrolytes, respiratory or metabolic acidosis; abnormal clotting.
2. Incipient septicaemia, unless surgery is necessary to remove the focus of infection; unexplained rapid rise in temperature, change in level of consciousness, alteration in respiratory or cardiovascular function.
3. A culture of β-haemolytic streptococcus or *Staphylococcus aureus*, both of which are known to lyse grafted skin.
4. Any condition that would benefit from a further 24 hours of active treatment.

There is a high incidence of upper respiratory tract infection among the toddler age group; this may be an aetiological factor in the accidental scald. Surgery should not be postponed because of upper respiratory tract or chest infections. Pyrexia of 39°C is not uncommon and fleeting or maculopapular rashes are also frequently associated with the thermally injured child.

Preoperative anaemia will in part be dilutional but levels below 10 g/dl should be corrected before surgery.

If the patient has suffered inhalational damage and is being ventilated or if the patient is in the state of renal shutdown, there remains the same urgency to remove non-viable skin so that normal physiological processes can be restored.

Adequate supplies of blood and blood products should be available in the hospital. Modern cross-matching techniques provide fully compatible products in 30 minutes. Provided there is an adequate specimen for further cross-match, a conservative estimate of blood requirements can be made which should be in the order of one unit for every 10 per cent of cutaneous burn.

Anaesthesia for surgery

The aim of the anaesthetist is to provide optimal conditions for surgery with minimal physiological disturbance.

The maintenance of a full hyperdynamic circulation, avoiding peripheral vasoconstriction, is essential. It will allow accurate discrimination between superficial and deep dermal injury and thus excessively deep excision will be avoided.

This state will exist only in the presence of adequate fluid replacement and good anaesthesia. Attention to suppression of response to surgical stimulus, temperature maintenance and avoidance of hypocapnia is essential.

Following surgical excision and grafting, it is advantageous to nurse the child with as much of the grafted area exposed as possible, to allow for toileting of any collected haematoma which will prevent graft adhesion and survival. This requires adequate postoperative sedation and analgesia which can be managed only by staff familiar with techniques of high dependency monitoring and care.

Venous access should be secured before any anaesthetic intervention. Adequate peripheral veins may be difficult to find. Primary sites for intravenous induction, in order of ease of access, are dorsum of hand, ventral wrist, dorsum of foot, antecubital fossa and, in the baby, finally the scalp. A larger vein will be required for volume replacement and, again in order of ease, the long saphenous, the antecubital fossa, the external jugular and then the central sites, internal jugular, subclavian and finally femoral. A cutdown should be performed following failure to gain an adequate percutaneous venous access capable of sustaining rapid infusion rates.

Inhalational induction is indicated when there is actual or potential airway obstruction. Pre-oxygenation and partial denitrogenation are advisable. Halothane, because of its potency and low incidence of laryngeal irritation, is the agent of choice.[24] A note should be made of any oedema, erythema, sloughs or soot in the supraglottic or glottic area.

Monitoring that may be attached to the child without causing distress should be commenced prior to induction. A minimum level of monitoring would include pulse oximetry, ECG and non-invasive blood pressure. A urinary catheter will already be in place. Unless there is a specific indication for inhalational induction or the child displays an educated preference for inhalational induction, intravenous induction is the most rapid and convenient method and, in the absence of hypovolaemia, provides rapid and safe induction. Thiopentone 6–8 mg/kg is followed by neuromuscular blockade with either atracurium (0.7–1 mg/kg) or vecuronium (0.15–2 mg/kg).

Tracheal intubation is performed orally using a disposable tracheal tube, avoiding trauma to the cricoid mucosa as previously noted. A cuffed tube will only be required in the child after puberty. Because of the likelihood of facial burns or greasy skin from topical creams, great care should be taken to ensure adequate stabilization of the tracheal tube. A mouthpack will provide further security.

Anaesthesia is maintained by intermittent positive pressure with nitrous oxide with at least 33 per cent oxygen, ensuring an oxygen saturation of 99 per cent. End tidal carbon dioxide should be maintained at between 35 and 40 mmHg.

Supplements

No single agent fulfils all the requirements for adequate anaesthesia. Volatile agents are potent and convenient but their vasodilator effects are terminated with the end of the anaesthetic, to be replaced by vasoconstriction.

Halothane has been used for repeated anaesthesia in the management of burns in children, without adverse effects. The reduced incidence of laryngeal irritation compared with other volatile agents makes halothane the only agent suitable for inhalational induction. On theoretical grounds, halothane produces peripheral vasodilatation and depresses cardiac output. In the presence of the slightest hypovolaemia, peripheral perfusion will be seriously compromised. Similarly, in adults, isoflurane and enflurane are associated with a decrease in oxygen requirement, cardiac index, mean arterial pressure and pulmonary artery pressure without a change in systemic and pulmonary vascular resistance.[25]

Isoflurane protects cardiac output and peripheral blood flow in children.[26] Its ability to cause laryngeal irritation renders it unacceptable as an agent for inhalational induction in an emergency when the airway must be secured in the shortest possible time.

Narcotics suppress the haemodynamic response to a surgical stimulus, and reduce the requirements for volatile agents and neuromuscular blocking agents. Fentanyl administered in a dose of 10 µg/kg before the onset of surgery ensures suppression of tachycardia during the harvesting of donor skin. Adequate spontaneous breathing will return after 1 hour, with no evidence of postoperative respiratory depression. Patients must be routinely and closely monitored in a high care situation. If postoperative controlled ventilation is to be continued, repeated fentanyl should be administered by bolus to prevent a tachycardia.

When a high inspired oxygen is required, unconsciousness should be ensured with a combination of volatile agent and narcotic or narcotic and propofol infusion to maintain suppression of haemodynamic responses.[26,27] Initial studies appear to indicate that the dose requirements of propofol for induction and maintenance of adequate anaesthesia are greater than those for non-burned children.

Heat conservation

The greatest care must be taken to prevent the patient cooling below the preoperative temperature because metabolic activity with concomitant calorie loss and increased oxygen consumption will be necessary in the postoperative period. The patient should be covered with drapes, Gamgee or foil and protected from draughts. Radiant heaters provide heat during the insertion of intravascular lines or debridement prior to surgery.

Depending on the size of the burn and the degree of preoperative pyrexia, the ambient theatre temperature should be maintained between 29 and 34°C. Local warming is provided by a warming blanket maintained at no greater than 1°C above skin temperature. Further heat and moisture can be conserved by using a condenser humidifier for short procedures, or a hot water humidifier for more prolonged procedures. All fluids should be warmed prior to infusion.

Postoperatively the patient is returned to a room with an ambient temperature of between 30 and 34°C to maintain a thermoneutral environment.

Control of intravascular volume

The blood loss during tangential excision can reach at least 30 ml per minute regardless of the size of the child, necessitating frequent pauses in the surgery to allow adequate volume replacement.

Having excluded preoperative hypovolaemia and ensured suppression of the response to surgical stimulus, pulse rate should be directly proportional to intravascular volume. With continuing blood loss, baseline pulse rate will rise by 10–15 beats per minute with 10 per cent blood loss, and 25–30 approaching 20 per cent loss before any significant fall in blood pressure. Further loss will be accompanied by hypotension and cardiovascular collapse. Colloid infusion in excess of normovolaemia will be associated with fall in pulse rate as stroke volume increases to approximately 20 per cent above normovolaemia, when signs of circulatory overload will be evident. The pulse rate rises, peripheral perfusion decreases and there is central desaturation, to be followed by cardiac arrhythmias as the circulation progresses into failure. Thus there is a band of 40 per cent of a blood volume where intravascular volume can be assessed by relatively simple means. In the adult and often elderly patient the response to changing intravascular volume is not as predictable.

In the management of major burns, accurate replacement should, if possible, be aided by direct measurement of arterial pressure and central venous pressure. During periods of rapid blood loss it is unlikely that there will be sufficient time or stability to perform meaningful estimations of pulmonary wedge pressure and cardiac output. Arterial blood gas estimation will detect when correction of any metabolic acidosis is required.

Because there is likely to be a significant water load from drugs and blood products, care should be taken not to produce water overload. Maintenance crystalloid infusion can usually be restricted for the duration of the excision.

In the presence of adequate haemoglobin concentration, losses of up to 20 per cent of blood volume may be treated with 0.9 per cent saline, lactated Ringer's solution or, preferably, human albumin solution or a colloid of choice. If continued blood loss is expected, it should be replaced with fresh-frozen plasma balanced with SAGM blood to maintain the haematocrit at around 30 per cent. It should be remembered that, for each unit of blood and FFP, 100 ml and 65 ml of water are imposed respectively. A corresponding water diuresis is required to maintain balance. Fresh-frozen plasma contains the greater amount of citrate per unit volume. At low rates of infusion, ionized calcium is not lowered because calcium is rapidly mobilized and the citrate metabolized. Cardiovascular collapse associated with rapid reductions in ionized calcium has been described following infusions of fresh-frozen plasma in excess of 1 ml/kg per minute in the presence of halothane anaesthesia. Faster rates of infusion should be accompanied by calcium chloride 2.5 mg/kg or calcium gluconate 7.5 mg/kg to increase the ionized calcium concentration.[29]

Indication of satisfactory coagulation can be gained by observation of clot formation from blood lost, supported after each blood volume replacement by coagulation studies and platelet count. Coagulation can be expected to be normal when blood replacement does not exceed two blood volumes. Hypocoagulation and thrombocytopenia of less than 50 000/mm^3 should be treated by platelet infusion, and surgical excision halted until adequate coagulation can be restored.[30] Up to seven blood volumes have been replaced in this unit without continued bleeding in the postoperative period.

Postoperatively, particular attention should be paid to

continued oozing from the graft site. Persistent coagulation abnormalities should be detected and treated.

Monitoring

There can be no substitute for the fullest possible monitoring as would be available to any intensive care patient. Emphasis is placed on the change in pulse rate, changes of blood pressure being secondary and an indication of delayed reaction to changes in pulse rate. This may be limited by available monitoring sites, but because areas of intact skin may be required as donor sites, the monitoring sites and emphasis may change throughout the procedure from ECG to arterial pressure to pulse oximetry to CVP as sites become unavailable. As normal adhesives for probes and ECG pads may prove inadequate, superglue can be used to maintain atraumatic monitoring contact for skin probes and ECG electrodes, even on mucous membranes or debrided areas.

Monitoring blood loss

Colorimetric analysis of swab washing gives an accurate estimation of blood loss but provides information much too slowly and involves staff in handling biological products. Individual swab weighing using inexpensive spring or electronic balances with estimation of blood on saline pre-wetted swabs provides an acceptable and immediate estimate of blood loss. An allowance of 30 per cent of measured loss is made for blood lost on the drapes. This has been shown to be accurate to within 1 g of predicted haemoglobin at 24 hours after the burn.

Anaesthesia for change of dressings

The requirement for general anaesthesia for the changing of dressings or wound toilet has declined in recent years. Techniques of analgesia using sedatives, narcotics or inhalational or intravenous agents prove adequate in most situations. Ketamine has been widely recommended for anaesthesia during a change of dressings but we find the duration of sedation and a number of disturbing psychological sequelae to be unacceptable in practice. Pre-mixed nitrous oxide 50 per cent/oxygen 50 per cent (Entonox) by continuous flow in the young child or demand flow in the older child, supplemented if necessary

by narcotics, provides adequate conditions accepted by patient and nursing staff in most cases. Children familiar with this technique will more readily accept gaseous induction of anaesthesia. This can be used to advantage if further general anaesthesia is required.

General anaesthesia is reserved for major burns, in the very young child or for intricate dressings (e.g. to the hand) taking a considerable time. In this last situation anaesthesia may be required on several occasions, separated by 2–3 days. Excessive sedation following anaesthesia is neither necessary nor desirable, as oral nutrition can be interfered with.

A small number of patients who are healing in part by granulation may require sequential soaking and dressing in the bath (Fig. 53.3). These patients exhibit a great degree of physical and mental debility. The anaesthetist plays a vital supportive role, ensuring adequate analgesia and stress relief. This re-establishes patient morale, which seems to directly aid the rate of healing. Continuous propofol sedation has proved not only satisfactory but also beneficial in this respect.[31]

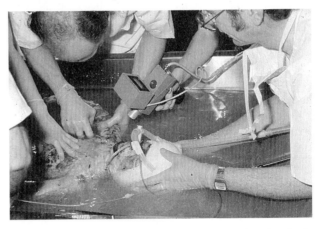

Fig. 53.3 Toileting grafted areas of a ventilated patient in the bath. Note the rigid tracheal tube support incorporated in the head dressing. An infrared temperature monitor is used to check skin temperature.

The laryngeal mask

This recent innovation facilitates airway control, particularly for minor repeated procedures and especially when facial burns with minimal oedema are present. It has been used as a guide for fibreoptic laryngoscopy and intubation under anaesthesia; when used in this manner it will prove invaluable in the process of intubation when contractures around the mouth and neck make formal laryngoscopy impossible.[32]

Smoke inhalation

The presence of an inhalational injury will greatly complicate the management of any cutaneous burn and is a significant cause of mortality in itself. The pathophysiology, diagnosis and management have been reviewed.[33,34]

In practice there are two types of inhalation injury. One is in the victim of a house fire, an incident occurring in a domicile not connected with industrial chemicals and usually not involving a detonation or conflagration. In this situation the child behaves completely differently from the adult, tending to hide from the trauma rather than trying to escape. This generally results in a more severe inhalational injury because the victim is exposed to a noxious environment for a greater length of time.

The second presentation of inhalational injury usually occurs in the adolescent experimenting with volatile flammable agents. Detonation and flash burns can occur such that the injury approximates to that of the adult industrial injury with extensive superficial burns associated with thermal damage to the airway. Because the injury usually takes place in a semi-closed space, the pulmonary damage is usually less and of a different nature than in the case of house fire.

In the UK, the majority of children involved in house fires sustaining inhalational injury die before reaching hospital. It therefore follows that, of the children reaching hospital, many will be suffering from the secondary effects of smoke inhalation. If there is a cutaneous burn, the effects of the injury are additive and, as will be shown, therapeutically competitive.

Noxious agents

Heat

The temperatures reached in the fire will determine the quality, quantity and timing of release of the toxic products produced. Thermal injury from dry heat is restricted to the skin and oral mucosa. Injury below the larynx is unusual, owing to the low specific heat capacity of dry gas and the ability of the well-vascularized tracheal mucosa to transfer heat. Laryngeal injury can occur but is less frequent than in oral scalds.

Hypoxic environment

As the fire becomes established, oxygen is consumed and can fall to as low as 10 per cent. The hypoxia so produced is additive to the chemically produced hypoxia.

Gaseous toxins

Over 200 organic and inorganic degradation products have been demonstrated following combustion of household materials at the temperatures found in house fires. Many, particularly the hydrocarbons, will have effects on cerebral function, myocardial function and vascular stability.

Carbon monoxide is present in all fires. Carbon monoxide produces hypoxia by forming carboxyhaemoglobin (COHb), shifting the haemoglobin dissociation curve to the left, and by inhibiting mitochondrial cytochrome oxidase. At sublethal levels it causes disorientation and incapacity. In greater concentration it is rapidly fatal through tissue hypoxia. The levels of COHb depend on the fractional concentration in the atmosphere, the duration of exposure and the physical activity of the victim in the smoke-filled environment. Carboxyhaemoglobin levels of 30–40 per cent are rapidly incapacitating but are more commonly seen following malfunction of heating appliances. Raising the level of inspired oxygen increases the dissociation of COHb in proportion to oxygen concentration, from a half-life of 4–5 hours in room air down to 40–120 minutes in 100 per cent oxygen. This reduces the usefulness of COHb estimation as a measure of severity of inhalation because a low level does not exclude damage from other inhaled agents. Hyperbaric oxygen therapy is beneficial in the management of poisoning with pure carbon monoxide but the reduced access to the patient in hyperbaric chambers makes its use impractical in the management of the victim of smoke inhalation with cutaneous burns.

Hydrogen cyanide is produced as a degradation product of polyurethane foam. Its importance lies in its ability to produce sudden and unexpected unconsciousness quickly followed by death. High temperatures are required to release cyanide, which may account for its high levels in fatalities at the scene. Much lower levels are found in surviving casualties. Cyanide binds to mitochondrial cytochrome oxidase, therefore inhibiting oxidative phosphorylation. Dissociation is speeded up by inhaled oxygen. The half-life of hydrogen cyanide is such that the specific antidotes sodium nitrite and sodium thiosulphate are inappropriate in children reaching hospital suffering from smoke inhalation. Successful management of severe cyanide poisoning may be achieved only by early pre-hospital treatment.[35]

Irritants

Irritants may be organic or inorganic compounds. Organic compounds such as the aldehyde acrolein, styrene, methanol and benzene are responsible for initial laryngospasm and bronchospasm.

Inorganic radicals such as chloride, nitrogen dioxide and sulphur dioxide become hydrated, to produce acids in the lung, and are responsible for mucosal oedema and later necrosis.

Soot

The carbon particles in soot cause physical problems of small airway blockage and alveolar collapse. Whilst the particles are not in themselves toxic, organic and inorganic radicals are adsorbed, carried throughout the lung and ensure continued contact with the mucosa. Blind suction down a tracheal tube produces little soot during the first 12 hours, raising false hopes of minor inhalation. Aspirated soot increases subsequently with increasing bronchorrhoea.

Because of the small airways, the child is likely to be more seriously affected than the adult.

Diagnosis

Diagnosis can sometimes be difficult because there are few constant findings that guarantee the presence or absence of inhalational injury.

Inhalation should always be presumed unless excluded by the obvious well-being of the child.

History

Information from fire, ambulance and police officers is of vital importance. The situation and atmosphere in which the child was found provide an accurate guide to the possibility of severe inhalation. Unconsciousness, history of convulsions, respiratory or cardiac arrest with successful resuscitation suggest severe injury.

Presenting signs

Bronchospasm, smoke around the nares, reduced level of consciousness or irritability suggest inhalation. Cutaneous burns around the face suggest severe inhalation.

Some hours into resuscitation, petechiae may develop over the head, arms and upper trunk. This suggests extreme straining against a closed glottis, which may be expected as an early response to intensely irritant vapours. Although of little diagnostic value, this common finding suggests that a severe hypoxic state has been present.

Laboratory evidence

This is usually disappointing. Carboxyhaemoglobin levels may be high or low in the presence of severe injury, as carbon monoxide is produced late in the history of a fire. Metabolic acidosis with or without respiratory alkalosis and a degree of hypoxia may accompany an isolated cutaneous lesion. Cyanide levels are usually not available as an emergency investigation and are therapeutically unhelpful; furthermore, there is a lack of correlation between carboxyhaemoglobin and cyanide levels in smoke inhalation injury.[36]

A chest x-ray is unhelpful in the acute phase.

The pulmonary injury

The initial hypoxia and acidosis are usually corrected with adequate resuscitation. Bronchospasm resolves without pharmacological measures in the majority of cases. Following this short period of initial stability a progressive deterioration of pulmonary function often occurs within hours of the insult. Mucosal cilial activity ceases from the time of inhalation, possibly only returning well into the recovery phase. Increasing airway resistance secondary to bronchiolar oedema and patchy atelectasis require ventilation with increased inflation pressures and high levels of positive end-expiratory pressure. Pulmonary vascular resistance gradually rises, increasing the differential between right and left atrial filling pressures. Substantially greater infusions of colloid may be required to raise left atrial filling pressure sufficiently to sustain adequate cardiac output and mean arterial pressure.

A period of stability should be attained by 24 hours and continued for 2–3 days, when improvement can be expected. Overoptimism should be avoided at this stage because the acute onset of mucosal sloughing can typically occur, resulting in catastrophic deterioration, segmental collapse and the development of pneumothoraces. If the necrosis is limited, bronchoscopy and aspiration of necrotic sloughs may improve the situation.

The extensive pulmonary damage renders the child susceptible to sudden and overwhelming bronchopneumonia at any stage until full ambulation has been achieved.

Management

The patient should be given 100 per cent oxygen from the time of retrieval from the fire. This speeds up the dissociation of carboxyhaemoglobin and the breakdown of cyanide. Restlessness must be assumed to be of cerebral hypoxic origin and on no account should be treated with sedatives. If there is a high index of suspicion of serious inhalation, the patient should be intubated .and ventilated and managed as for a hypoxic brain injury.

Laryngoscopy and bronchoscopy

Bronchoscopy will provide definite confirmation of inhalation. This investigation should be delayed until the child reaches the specialist centre and hypoxia and metabolic acidosis have been corrected. Fibreoptic bronchoscopy is, as yet, possible only through tracheal tubes of larger size. Following a period of denitrogenation by ventilation with 100 per cent oxygen in excess of 30 minutes, bronchoscopy can be performed using 100 per cent oxygen with the Storz system. For those unfamiliar with this equipment, it resembles a cystoscope allowing IPPV through the outer sleeve using any combination of gas or vapour mixture. Intermittent propofol has provided excellent anaesthesia in these circumstances.

The presence of bronchial soot, erythema or unusual pallor and bronchial oedema provides conclusive proof of inhalation. Unlike the adult inhalational injury involving moist gases, it is unusual to find any evidence of thermal damage below the larynx in cases of house fire.

When there is heavy soiling of the bronchial tree it is the practice in my unit to perform bronchial lavage in an attempt to reduce the amount of soot and neutralize its acid load. Small aliquots of isotonic 1.4 per cent sodium bicarbonate are instilled directly into each segmental bronchus in turn, starting with the basal segments. Lavage is continued until the bronchial tree looks at least 50 per cent cleaner. Following bronchoscopy, IPPV with a positive end-expiratory pressure in excess of 4 cmH$_2$O is instituted. Instillations of 2–5 ml of isotonic sodium bicarbonate or 0.9 per cent saline are continued at hourly intervals with physiotherapy until clear aspirates are returned and the pH of bronchial secretions remains at 7.3. This management has been successful in preventing pulmonary deterioration in a number of severely affected children.

IPPV must be continued until gas exchange and compliance have returned to within normal limits. A prolonged period of 2–3 days' spontaneous breathing against continuous positive airway pressure is usually required prior to extubation because of persistent bronchiolar oedema and tendency to small airway collapse.

Serial chest x-rays will usually demonstrate the clinical situation occurring some hours previously. Patchy atelectasis develops some hours after the inhalation and will still be present radiologically when respiratory function has clinically improved.

Management of the cerebral injury

All children rescued from a house fire will have suffered a degree of hypoxic insult. This may have been compounded by respiratory or cardiac arrest. These children should be ventilated to maintain the PaCO$_2$ between 3.7 and 4.7 kPa (28–35 mmHg); crystalloid fluids should be restricted and the patient managed as for a hypoxic brain injury.

When there is an accompanying large cutaneous burn, a conflict occurs between management of the cerebral oedema and the balance of the large fluid losses and their replacement. We have found the insertion of an intracranial pressure monitor helpful in this situation.[31] Careful management of pulmonary wedge pressure and intracranial pressure allows support of the mean arterial pressure in the early phase of resuscitation and control of intracranial pressure at the peak of oedema formation at around 24 hours. We have not experienced any incidence of infection of the ICP line when it is removed at 72 hours.

All survivors from severe inhalation demonstrate a variable period of disinhibition and antisocial behaviour, which may require continued psychological counselling.

Pharmacological support

Antibiotics

Because bacterial colonization will undoubtably occur, antibiotics should be delayed as long as possible. When clinically indicated, the appropriate combination of antibiotics should then be given in full bactericidal doses. The routine use of antibiotics will only increase the degree of resistance of the infecting organisms.

Anti-inflammatory agents

There is no place for the routine administration of steroids. Prospective trials when steroids have been used in the management of the pulmonary injury have shown increased mortality from overwhelming infection.

Toxic shock syndrome

This relatively rare condition is important because it is a cause of mortality in superficial cutaneous burns often of quite small area. It is more commonly associated with menstruation and tampon use, but it should be considered when any child suffers a deterioration in general condition out of proportion to the cutaneous injury.

Diagnosis

Diagnosis depends on fulfilling five out of the six following criteria:[37]

1. Temperature of 38.9°C or above.
2. Erythematous macular rash (occasionally petechial).
3. Systolic blood pressure less than 90 mmHg or less than 5th percentile for age in children.
4. Involvement of three or more organ systems:
   Gastrointestinal: vomiting and/or watery diarrhoea at onset of illness
   Mucous membranes: hyperaemic, vaginal, oropharyngeal or conjunctival hyperaemia
   Muscular: severe myalgia and or creatine phosphokinase level twice normal
   Renal: blood urea or creatinine level twice normal and/or urinary sediment with sterile pyuria (more than 5 white blood cells per high-power field)
   Hepatic: total bilirubin, SGOT or SGPT at least twice the upper limit of normal
   Haematological: thrombocytopenia of less than 100 000/mm^3
   CNS: disorientation or alterations in consciousness without focal findings in the absence of hypotension and hyperpyrexia
   Cardiopulmonary: hypoxia, adult respiratory distress syndrome or myocardial depression (cardiac index less than 3.3 l/m^2 per min)
   Metabolic: serum calcium less than 7.0 mg/dl, serum phosphate less than 2.5 mg/dl and total serum protein less than 5 mg/dl.
5. Reasonable evidence of absence of systemic infection.
6. The presence of desquamation of palms and soles during recovery phase.

The initially superficial burn will be found to have increased in depth.

Diagnosis may initially be difficult because of a high primary temperature, but cerebral and gastrointestinal signs present earlier than cardiovascular signs.

The illness is secondary to colonization of the superficial perfused burn with *Staphylococcus aureus* of phage type 59/29 in patients who have no natural immunity. Blood cultures for the organism are negative, the condition being produced by absorption of toxin.

Management

Successful treatment depends on recognition before cardiovascular deterioration is established. β-Lactamase-resistant antistaphylococcal antibiotics should be prescribed but may alter the course of the illness only when given early.

Prophylaxis with antistaphylococcal ointment or parenteral antistaphylococcal antibiotics has been used in some centres, with intensive monitoring or cardiovascular parameters and support of the circulation initially with fresh-frozen plasma. In the presence of adequate colloid infusion, dopamine, dobutamine, noradrenaline (norepinephrine) or adrenaline (epinephrine) may be required, depending on cardiovascular responses. Unfortunately, in severe cases, tachyphylaxis to these catecholamines may occur with fulminating metabolic acidosis.

The future

Modern surgical techniques and intensive care have provided a future for children who would otherwise not have survived. Severely burned and cosmetically disfigured children adapt and develop their abilities and adapt to their disabilities.

In children suffering from severe thermal injury, early, aggressive and persistent management will be rewarded with far greater survival and better cosmetic results than in the case of a similarly injured adult. With expert aftercare and support for the child and family, the only lasting 'scars' are the parents' misplaced feelings of guilt.

REFERENCES

1. Demling RH. Management of the burn patient. In: Shoemaker WC, Ayres S, Grenvik A, Holbrook PR, Thompson WL, eds. *Textbook of critical care*, 2nd ed. Philadelphia: WB Saunders, 1989: 1301–16.
2. Atchison NE, Osgood PF, Carr DB, Szyfelbein SK. Pain during burn dressing change in children. *Pain* 1991; **47**: 41–5.
3. Childs C, Watson SB, Fisher MI, Ward ID, Davenport PJ,

Little RA. Gut permeability in the acutely burned child. In: *Proceedings of the Association of Clinical Biochemists.* 1992: 52.
4. Childs C, Little RA. Paracetamol (acetaminophen) in the management of burned children with fever. *Burns* 1989; **14**: 343–8.
5. Martyn JAJ. Cimetidine and/or antacid for the control of gastric

acidity in pediatric burn patients. *Critical Care Medicine* 1985; **13**: 1–3.

6. Martyn JA, Greenblatt DJ, Hagen J, Hoaglin DC. Alteration by burn injury of the pharmacokinetics and pharmacodynamics of cimetidine in children. *European Journal of Clinical Pharmacology* 1989; **36**: 361–7.

7. Martyn J. Clinical pharmacology and drug therapy in the burned patient. *Anesthesiology* 1986; **65**: 67–75.

8. Bush GH. The use of muscle relaxants in burnt children. *Anaesthesia* 1964; **19**: 231–8.

9. Martyn JAJ, White DA, Gronert GA, Jaffe RS, Ward JM. Up-and-down regulation of skeletal muscle acetylcholine receptors. *Anesthesiology* 1992; **76**: 822–43.

10. Brown TC, Bell B. Electromyographic responses to small doses of suxamethonium in children after burns. *British Journal of Anaesthesia* 1987; **59**: 1017–21.

11. Mills AK, Martyn JA. Evaluation of atracurium neuromuscular blockade in paediatric patients with burn injury. *British Journal of Anaesthesia* 1988; **60**: 450–5.

12. Cunliffe M, Lord WD. Atracurium in the management of early desloughing of burns in the paediatric patient. At: Annual meeting of the Association of Burns and Plastic Surgical Anaesthetists, Manchester, 1987.

13. Marathe PH, Dwersteg JF, Pavlin EG, Haschke RH, Heimbach DM, Slattery JT. Effect of thermal injury on the pharmacokinetics and pharmacodynamics of atracurium in humans. *Anesthesiology* 1989; **70**: 752–5.

14. Mills AK, Martyn JAJ. Neuromuscular blockade with vecuronium in paediatric patients with burn injury. *British Journal of Clinical Pharmacology* 1989; **28**: 155–9.

15. Rubin WD, Mani MM, Hiebert JM. Fluid resuscitation of the thermally injured patient. *Clinics in Plastic Surgery* 1986; **13**: 9–20.

16. Demling RH. Fluid replacement in burned patients. *Surgical Clinics of North America* 1987; **67**: 15–30.

17. Graves TA, Cioffi WG, McManus WF, Mason Jr AD, Pruitt Jr BA. Fluid resuscitation in infants and children with massive thermal injury. *Journal of Trauma* 1989; **29**: 1261–7.

18. Miller JG, Carruthers HR, Burd DA. An algorithmic approach to the management of cutaneous burns. *Burns* 1992; **18**: 200–11.

19. Muir IFK, Barclay TL, Settle JAD. Treatment of burns shock. In: *Burns and their treatment*, 3rd ed. London: Butterworth Scientific, 1987: 14–54.

20. Osgood PF, Szyfelbein SK. Management of burn pain in children. *Pediatric Clinics of North America* 1989; **36**: 1001–13.

21. Gaukroger PB, Chapman MJ, Davey RB. Pain control in paediatric burns – the use of patient-controlled analgesia. *Burns* 1991; **17**: 396–9.

22. Wilder RT, Berde CB, Troshynski TJ, Cahill CA, Sethna NF. PCA in children and adolescents – outcome among 1589 patients. *Anesthesiology* 1992 **77**: A1187.

23. Benians RC. The influence of parental visiting on survival and recovery of extensively burned children. *Burns Including Thermal Injury* 1988; **14**: 31–4.

24. Black GW. Which volatile anaesthetic? *Paediatric Anaesthesia* 1992; **2**: 175–7.

25. Gregoretti S, Gelman S, Dimick A, Bradley Jr EL. Hemodynamic changes and oxygen consumption in burned patients during enflurane or isoflurane anesthesia. *Anesthesia and Analgesia* 1989; **69**: 431–6.

26. Wolf WJ, Neal MB, Peterson MD. The hemodynamic and cardiovascular effects of isoflurane and halothane anesthesia in children. *Anesthesiology* 1986; **64**: 328–33.

27. Morton NS, Johnston G, White M, Marsh BJ. Propofol in paediatric anaesthesia. *Paediatric Anaesthesia* 1992; **2**: 89–97.

28. Demling RH, Lalonde C. Oxygen consumption is increased in the postanesthesia period after burn excision. *Surgery* 1989; **10**: 381–7.

29. Coté CJ, Drop LJ, Hoaglin DC, Daniels AL, Young ET. Ionized hypocalcemia after fresh frozen plasma administration to thermally injured children: effects of infusion rate, duration, and treatment with calcium chloride. *Anesthesia and Analgesia* 1988; **67**: 152–60.

30. Coté CJ, Liu LMP, Szyfelbein SK. Changes in serial platelet counts following massive blood transfusion in pediatric patients. *Anesthesiology* 1985; **62**: 197.

31. Mills DC, Lord WD. Propofol for repeated burns dressings in a child. *Burns* 1992; **18**: 58–9.

32. Kay S, Samba Siva Rao G, Lord D. Intracranial pressure monitoring in the burnt and asphyxiated child. *Burns* 1986; **12**: 212–13.

33. Traber DL, Linares HA, Herndon DN, Prien T. The pathophysiology of inhalation injury – a review. *Journal of Burn Care and Rehabilitation* 1988; **14**: 357–64.

34. Herndon DN, Barrow RE, Linares HA, Rutan RL, Prien T, Traber DL. Inhalation injury in burned patients: effects and treatment. *Burns Including Thermal Injury* 1988; **14**: 349–56.

35. Jones J, McMullen MJ, Dougherty J. Toxic smoke inhalation: cyanide poisoning in fire victims. *American Journal of Emergency Medicine* 1987; **5**: 317–21.

36. Barrillo DJ, Goode R, Rush Jr BF, Lin RL, Freda A, Anderson Jr EJ. Lack of correlation between carboxyhemoglobin and cyanide in smoke inhalation injury. *Current Surgery* 1986; **43**: 421–3.

37. Fisher Jr CJ, Panacek EA. Toxic shock syndrome. In *Textbook of critical care*, 2nd ed. Philadelphia: WB Saunders, 1989: 1003–6.

Chapter 54

Management of the Adult with Severe Burns

Joan M. Brown

An adult who sustains a skin burn of 15 per cent or more body surface area (BSA) will require treatment with intravenous fluids and is classified as having a major burn. Some other young and elderly patients and those with smoke inhalation injury, electrical burns, or burns of the face, hands or perineum will also require treatment in a special unit.

Most special units now have dedicated anaesthetists involved in resuscitation and intensive care as well as anaesthesia and pain relief. Other anaesthetists, however, may meet patients with thermal injury during the primary treatment in an accident and emergency department. Optimal treatment at this time will undoubtedly improve the outcome for the patient.

The burn injury

The depth of the thermal injury determines the manner of healing and the nature of any intervention required, and how ill the patient may become (Fig. 54.1).

Very superficial injury of the outer epidermis only (USA first degree) will heal within days and requires little treatment. A superficial partial-thickness burn destroys the epidermis and the outer part of the dermis, may produce blisters filled with extravasated fluid, and is very painful. Many epidermal remnants remain, such as lining hair follicles and other skin organs. These will proliferate and

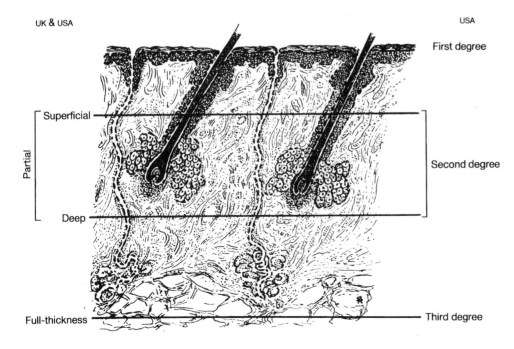

UK & USA

USA

First degree

Second degree

Superficial

Partial

Deep

Third degree

Full-thickness

Fig. 54.1 Diagrammatic representation of depths of burn, with terminology used in the UK and the USA. (Reproduced, with permission, from Muir IFK, Barclay TL. *Burns and their treatment*. London: Lloyd-Luke, 1974.)

achieve healing in 14–21 days. A deeper partial-thickness burn will have few such remnants and, without intervention, will heal badly and slowly. A full-thickness burn destroys all the dermis and epidermis; without intervention, virtually no healing can occur.

Following a full-thickness burn, a coagulum is formed. Beneath this, cells may be viable but injured. Some small vessels may thrombose. This vulnerable zone can easily be rendered more ischaemic by reduced perfusion, thus increasing the severity of injury. The microvessels near the burn become porous. It is presumed that vasoactive substances released by the injured tissue, such as histamine, prostaglandins, oxygen-free radicals or leukotriene, cause this change in the endothelium. A fluid containing plasma solutes and albumin, but not the larger protein molecules, is rapidly lost from the circulating volume into the extracellular space. This loss of 'plasma' amounts to many litres following an extensive injury, but may not be immediately obvious. Hypovolaemic shock ensues within hours if the patient is not treated. 'Plasma' is sometimes exuded from the burn surface.

Cell membrane malfunction allows sodium and water to enter cells in the general vicinity of the injury. Circulating kinins and toxic radicals adversely affect tissue remote from the burn; this probably explains the generalized body oedema that occurs in patients with large burns. This oedema is greater if fluid resuscitation is with fluids of low oncotic pressure.

Hypovolaemia, hypoxaemia and suboptimal oxygen delivery as well as later sepsis will of course increase the likelihood of organ failure developing later.

Some pulmonary changes are associated with large skin burns even in the absence of smoke inhalation injury. Arterial desaturation is commonly present. Lung perfusion will be abnormal if the circulating blood volume is significantly reduced. Increased pulmonary artery pressure and increased airways resistance have been reported.[1] It is presumed that these changes are due to the liberation, from the injured tissue, of bronchoconstrictor and vasoactive substances. There is no increase in interstitial lung water in the absence of smoke inhalation, but chest movement can be seriously restricted if there is a circumferential chest burn.

Some thermal injury is sustained by red cells, particularly in deep burns, and this shortens their life. Bone marrow function and the immune system are also depressed.

Burns of muscle and deeper tissue cause the liberation of haemoglobin and myoglobin which are excreted by the kidney and are manifested as port-wine-coloured urine. The products of these molecules are toxic to the renal tubules.

Destruction of skin results in a loss of water by evaporation and heat as latent heat of evaporation. Heat conservation by vasoconstriction is also compromised. The barrier to bacterial invasion is removed.

Electrical injury

Accidents involving high voltage current (from the national electric grid system) may cause extensive but invisible tissue damage depending on the resistance and diameter of the structures encountered by the current. Ventricular standstill can occur. Myocardial injury should be suspected.[2]

Fluid and circulatory changes

Immediately after injury, cardiac output starts to fall, seemingly preceding the fall in circulating volume. Existence of a myocardial depressant factor has been postulated but its clinical significance is disputed. After a few hours cardiac output rises and is sustained at a high level as the hypermetabolic phase is initiated and remains until healing is complete.

The outflow of 'plasma' is greatest during the first 4 hours or so, decelerating over the course of the subsequent 24 hours as capillary integrity returns towards normal. On some occasions the increase in hydrostatic pressure, in tissue beneath an inelastic eschar, may to some extent constrain the leak.

In normal patients the pulse rate rises to maintain the systolic blood pressure. Peripheral perfusion and peripheral temperature fall when the blood volume is not adequately maintained by fluid therapy. Most patients receiving standard therapy tend nevertheless to have low central venous and pulmonary artery wedge pressures.

Urine output, although not directly reflecting glomerular filtration rate, is universally used as a measure of the adequacy of organ perfusion and, therefore, fluid infusion rate.

Fluid therapy

Adults with more than a 15 per cent BSA burn, and children with more than 10 per cent, need intravenous fluid replacement. Patients with lesser burns can be managed with balanced salt solutions such as Dioralyte by mouth. (Powder reconstituted as directed provides sodium 60 mmol, potassium 20 mmol, chloride 60 mmol, citrate 10 mmol and glucose 90 mmol per litre.) In disaster situations when supplies of fluid are short, quite large burns can be treated in this way. Fluid given at this stage is usually called 'resuscitation' although it is given to correct the vascular fluid losses more or less as they occur before clinical shock becomes apparent.

It has long been known that the amount of fluid that is effective is related both to the size of the patient and to the size of the burn injury, and that a generous volume must be given rapidly at first, tailing off later.

Type of solution

Because one cannot directly measure the rate of loss from the circulation, many rather didactic 'formulae' have evolved in different centres. These formulae have been developed as a result of a combination of both theory and experience with the outcomes. Most people agree that the fluid should have electrolyte concentrations, in particular sodium, similar to that of plasma.[3–7] Those who support the 'crystalloid only' theory suggest that if colloid is given during the first 8 hours, a loss of albumin molecules will occur through the permeable capillary walls which will perpetuate the burn oedema. In the case of crystalloid therapy, faster infusion rates and larger volumes are given than is necessary when a colloid regimen is used. Crystalloid therapy tends to be associated with more generalized oedema remote from the burn, and a higher than normal haematocrit (packed cell volume), as the solution does not remain long within the vascular compartment. It is a less effective therapy than colloid for very large area burns.

In order to replace the obligatory sodium loss without waterlogging the patient a balanced solution with a very high sodium concentration has been advocated (Table 54.1).[8] Smaller volumes may then be used. Plasma electrolyte levels must be carefully monitored, as hypernatraemia may sometimes occur and require treatment. Very high concentrations of sodium are not usually suitable when treatment has been delayed, but may be advantageous following massive burn injury.

Many North American centres use a mixture of crystalloid and colloid, the colloid often being added after the first hours. In Britain there is a long experience with the use of normal colloid solution. Initially this was freeze-dried plasma but now 4.5 per cent human albumin solution is used. Good quality supplies, free of cost in the UK, have been available, which may partly explain the historical differences in practice. The infusion can certainly be manipulated to maintain a normal haematocrit, and the blood volume is usually better maintained than when crystalloid therapy is used. Colloid is considered by some to be preferable following smoke inhalation injury as this may minimize interstitial lung water.[9]

Gelatin solutions, widely used as first-line resuscitation fluids for hypovolaemia following trauma, have been used in war situations but they leave the circulation fairly quickly, causing a confusing osmotic diuresis; furthermore, there is a limit to the amount that can be given. The use of hetastarch and other starch solutions awaits evaluation, but whether such solutions can safely be given in large quantities is not yet known. (A man with a 50 per cent burn will need 11 or more litres over 36 hours.) Long-term side effects have not been defined.

Whilst the detailed requirements for protein and water vary with the formula used, there is general agreement with respect to the sodium needed – i.e. about 0.5 mmol/kg for each per cent BSA. Clearly none of these regimens is likely to be consistently and markedly better than the others. They are merely guidelines, the important message being that the patient's status must be frequently reassessed and treatment modified accordingly. The volume and content necessary to be effective can vary quite considerably from that estimated using the formulae.

Examples

The 'Parkland' regimen

This formula recommends that Ringer's lactate 4 ml/kg for each per cent BSA should be given over 24 hours. Half the volume should be given over the first 8 hours. Towards the end of the 24 hours fresh-frozen plasma, 0.5 ml/kg for each per cent BSA, should also be infused.[10]

The 'Mount Vernon' regimen (Muir and Barclay)

Human albumin solution 4.5 per cent (3 ml/kg for each per cent BSA) should be given over 36 hours. The infusion rate should initially be rapid but then tailed off, i.e. designed to mimic the rate of loss of 'plasma'.[5,6] The 36-hour period is

Table 54.1 Composition of resuscitation fluids

	Na^+ (mmol/l)	K^+ (mmol/l)	Ca^{2+} (mmol/l)	Cl^- (mmol/l)	HCO_3^- (mmol/l)	Lactate (mmol/l)	Albumin (g/l)
Ringer's or Hartmann's	130	4	3	110	28	–	–
Hypertonic lactated saline	254	–	–	154	–	100	–
Human albumin	130	< 2	–	< 160	–	–	46

divided into time periods of 4, 4, 4, 6, 6 and 12 hours. A volume (ml) calculated using the formula

$$\frac{\text{kg body weight} \times \% \text{ BSA burned}}{2}$$

is the guideline for each period. A larger volume is often needed, probably because the formula was developed for use with freeze-dried plasma.

Some intravenous electrolyte-free fluid is also recommended, as this would normally be ingested for 'metabolic needs'. However, unlimited water drinking is not recommended because some patients, especially children, are at risk of becoming hyponatraemic, with a form of inappropriate ADH secretion. The amount of water lost may vary as a result of exposure, whether nursed on an air loss bed or whether pulmonary positive pressure ventilation is used.

Blood transfusion

Some red cells are injured at the time of a deep burn, but administration of blood during the 'resuscitation' phase is no longer used as this may confuse interpretation of haematocrit trends. However, if escharotomies are performed, a large volume of blood can be lost and this must be replaced.

Table 54.2 First aid

Cool burn with water
Remove hot clothes
Cover with clean material or cling-film
Keep warm
Chemical burn: prolonged water rinse
Electrical burn: disconnect patient safely
Smoke inhalation: oxygen

Practical first line management

First aid measures are summarized in Table 54.2.

The priorities for treatment of a newly burned patient are listed in Table 54.3. Establish an intravenous infusion and then treat urgent problems of the airway, respiration or other injury.

The provisional infusion rate based on the guidelines described above should be calculated from the time of the accident. There will thus be some catching up to do. The patient's weight and the percentage of affected body surface must be estimated. It can be quite difficult to assess the latter accurately. The 'rule of nines' (Fig. 54.2) is traditional and easy to remember. There are other charts that are more detailed and assist calculation for patients of

Table 54.3 Priorities in treating a major burn

	Action	Need to know
Circulation	Set-up fast i.v. infusion of crystalloid or colloid	
Airway	Treat appropriately	Arterial gases
Breathing	Treat appropriately	COHb
Unconscious?	Treat appropriately	History
Other serious injury?	Treat appropriately	
Alcohol/drugs?	Treat appropriately	
Medical problems?	Treat appropriately	Biochemistry
Pain	Give i.v. narcotic/anxiolytic	
Distress	Reassure	
	Keep patient warm	
'Shock'	Calculate infusion rate	Burn time
		Burn size
		Weight
	Insert urine catheter	Urine output
		Urine colour
	Modify infusion rate	Haematocrit
Circumferential burns:		
Chest	Consider escharotomies	Is breathing restricted?
Neck		Is airway compromised?
Limbs	Seek advice	Is peripheral circulation impeded?
	Dress with plain tulle gras	

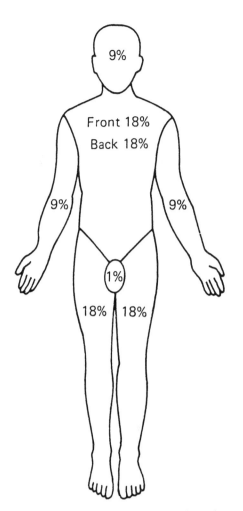

Fig. 54.2 'Rule of nines' body proportions in an adult. (Reproduced, with permission, from Wallace AB. The exposure treatment of burns. *Lancet* 1951; **1**: 502.)

different ages (Fig. 54.3). These charts must always be available in accident centres.

An immediate haemoglobin and haematocrit estimation will enable the fluid loss to be estimated and hence the rate of fluid replacement calculated. Urine output is used as a reasonable practical guide to the adequacy of the circulating volume and organ perfusion. An output of 0.5–1 ml/kg per hour is satisfactory. If there is no urine flow the infusion rate should be increased. This is nearly always effective in increasing urine output. A diuretic is usually not needed and its use eliminates the value of the urine output as a monitor of the adequacy of blood volume and organ perfusion.

Initial monitoring should include pulse rate, peripheral temperature and oxygen saturation. Arteries and central veins should not be cannulated unless this is essential; cannulation may constitute an infection risk and the vessels may be needed later.

Other first line treatment (see Table 54.3) follows

standard principles. Smoke inhalation injury is described later.

Myoglobinuria/haemoglobinuria

The appearance of port-wine-coloured urine has grave significance and immediate treatment is essential. Haemoglobin and myoglobin will be excreted as a result of the thermal disruption of red cells and muscle following deep burns. If the urine is acid (below pH 5.6) both haemoglobin and myoglobin dissociate into ferrihaemate and globin. Ferrihaemate is directly toxic to the renal tubule, and may result in acute renal failure which is made worse if the patient is hypovolaemic.[11] Large volumes of fluid and sodium bicarbonate 1 mmol/kg per hour must be given intravenously to dilute and raise pH of the urine. Frusemide (furosemide) and mannitol may restore tubular function at cellular level and one or other should be given.

Continuing management

The patient will usually require transfer to a burns or intensive care unit.

Fluids and monitoring

The administration of resuscitation fluids must be carefully reassessed at defined time intervals. The rate of infusion must be guided by haematocrit, urine output and general evidence of good perfusion. One reason why fluid requirements may vary significantly from the predicted need can be an error in the assessment of burn size. Haemoglobin, oxygen saturation and respiratory function require monitoring in order to provide early warning of septicaemia, fluid overload, chest infection or acute respiratory distress syndrome (ARDS). All indwelling cannulae carry the risk of infection. Their management should be aseptic and they should be changed every few days. In a patient with an extensive injury, invasive arterial cannulation is necessary for pressure measurement and for blood sampling. Routine cases (i.e. those with less serious injury) can usually be managed without measurement of central venous pressure, and only in a patient with myocardial disease or other complicating factor is it necessary to consider the use of a pulmonary artery flotation catheter.

The aim must be as far as possible to restore and maintain homoeostasis and to promote optimal oxygen

Name_____ Ward_____ Number _____ Date_____
Age _____   Admission weight _____

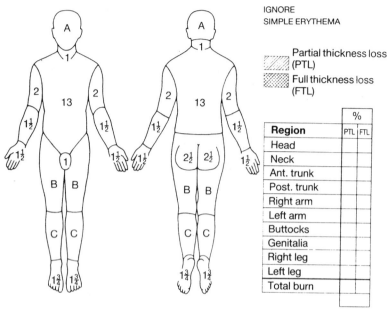

LUND AND BROWDER CHARTS

IGNORE
SIMPLE ERYTHEMA

Partial thickness loss (PTL)
Full thickness loss (FTL)

Region	% PTL	FTL
Head		
Neck		
Ant. trunk		
Post. trunk		
Right arm		
Left arm		
Buttocks		
Genitalia		
Right leg		
Left leg		
Total burn		

Relative percentage of body surface area affected by growth

Area	Age 0	1	5	10	15	Adult
A = ½ of head	9½	8½	6½	5½	4½	3½
B = ½ of one thigh	2¾	3¼	4	4½	4½	4¾
C = ½ of one leg	2½	2½	2¾	3	3¼	3½

Fig. 54.3 Chart for estimating surface area burned in patients of all ages. (Reproduced, with permission, from Lund CC, Browder NC. The estimation of the area of burns. *Surgery, Gynecology and Obstetrics* 1944; **79**: 356–7.)

delivery to the tissues to pre-empt later organ failure. It has not been established in these burned patients whether a pharmacologically induced increase in cardiac output would improve outcome. Cardiac output is in any case usually raised and there is some evidence that administered catecholamines are not effective in burned patients.[12]

Loss of water, salt and protein from the burn wound continues and is largely unmeasurable, so the management of the 'fluid balance' becomes progressively more difficult than following other injuries.

Renal failure

Urine osmolality and urine volume should be recorded. The presence of pigment in the urine can be recognized and the treatment has been described, but non-anuric renal failure is not uncommon, particularly in the elderly or when there is a large burned surface. A reduced urine output with a low osmolality, usually around 350 mmol/l, will be observed before plasma urea and creatinine rise. This development can sometimes be reversed by careful treatment, including dopamine. In view of the catabolic nitrogen load,

haemodialysis or haemofiltration must be started early if there is no improvement.

Prophylactic renal dose dopamine should be considered for the patient with extensive burn injury.

Hypermetabolism and feeding

A serious burn triggers by far the most intense stress response of any injury. From about the third day after the burn there is an outpouring of the catabolic hormones: catecholamines, cortisol, glucagon and growth hormone. Resistance to insulin develops and therefore there is a preferential breakdown of protein and fat to produce glucose. A raised blood glucose and glycosuria are common. This hormone response is associated with a raised metabolic rate augmented by increased calorie consumption linked to heat loss, especially from evaporation, and to a mild pyrexia (38°C) associated with hypothalamic changes. The metabolic rate may be twice normal; therefore generous calories and protein must be supplied. Temporary insulin therapy may be needed. Most patients should commence enteral feeding during the first or second post-burn day. This helps to prevent villous atrophy from disuse. Parenteral nutrition is used only if the gut has ceased to function.

In some units energy expenditure is measured by indirect calorimetry at the bedside.

Stress ulcer

Curling's acute gastric ulcer is now a rare occurrence, since patients are currently treated with antacids or H_2-receptor blockers such as ranitidine.

Anaemia

A tendency to anaemia is invariable until healing is complete. Significant numbers of red cells are destroyed by heat and the life of others may be shortened.[13] Bone marrow function may be presumed to be depressed. Repeated transfusions must be given to maintain the haemoglobin level and optimize oxygen delivery.

Low plasma albumin

A tendency to a low plasma albumin remains until healing has occurred. Values in the 22 g/l range are common; they are not improved by protein feeding and only very temporarily improved by the infusion of albumin.

Infection

The wound that contains dead tissue soon becomes colonized with bacteria. Topical application of antiseptics and bactericidal agents are usually used but practice varies from unit to unit. Routine prophylactic antibiotics must not be used. Septicaemia with or without shock, and often without evidence of bacteria, is a common and sometimes terminal event. The infection may stem from the wound or, more often, from the gastrointestinal tract. Preliminary reports of improved outcomes using selective decontamination are promising.[14] Instant intravenous administration of a combination of antibiotics spanning the bacterial spectrum is indicated if septicaemia is suspected. Antibiotics to cover the bacteraemia of surgery are usually given. Early surgery to remove dead tissue is the best preventive measure.

Smoke inhalation injury

Smoke is the product of aerobic or anaerobic combustion of synthetic or natural materials, sometimes at high temperatures. In a closed space visibility is rapidly reduced and therefore the hazard is increased. Hot air, flames and particles may be inhaled, but actual thermal injury is largely limited to the oropharynx and larynx. A variety of chemicals injurious to mucous membranes and alveoli may also be present (Table 54.4). Carbon monoxide will be generated when there is incomplete combustion. Another systemic poison, hydrogen cyanide, is formed from nitrogen-containing substances (wool, polyurethane) at high temperatures. The burning tends to produce an environment low in oxygen and high in carbon dioxide. Therefore carbon monoxide and possibly cyanide poisoning may be present and cause morbidity or death after house fires. Upper airway obstruction may also occur. Chemical injury to the bronchial tree or alveoli may also be present. The pathology presumably depends on what was burning and under what conditions.

Diagnosis

Patients who have inhaled smoke often need intensive care and respiratory support. The combination of smoke inhalation injury and a skin burn creates a particularly complicated clinical problem and the prognosis must be guarded.

A history of being trapped in a smoke-filled space (house, car, plane) is usual. Burns of the face are common following other types of accident, but should arouse suspicion that airway injury has occurred. Smoke staining of the face, nose and oropharynx are diagnostic, but do not indicate the severity of the trauma. Injury is suggested by hoarseness or

Table 54.4 Possible constituents of smoke

Heat
Particles
Toxic chemicals:
    Hydrogen cyanide
    Hydrogen chloride
    Acetonitrile
    Acrolein
    Di-isocyanates
    Oxides of nitrogen
    Acetaldehydes
Gases:
    Carbon monoxide
    Reduced oxygen
    Increased carbon dioxide

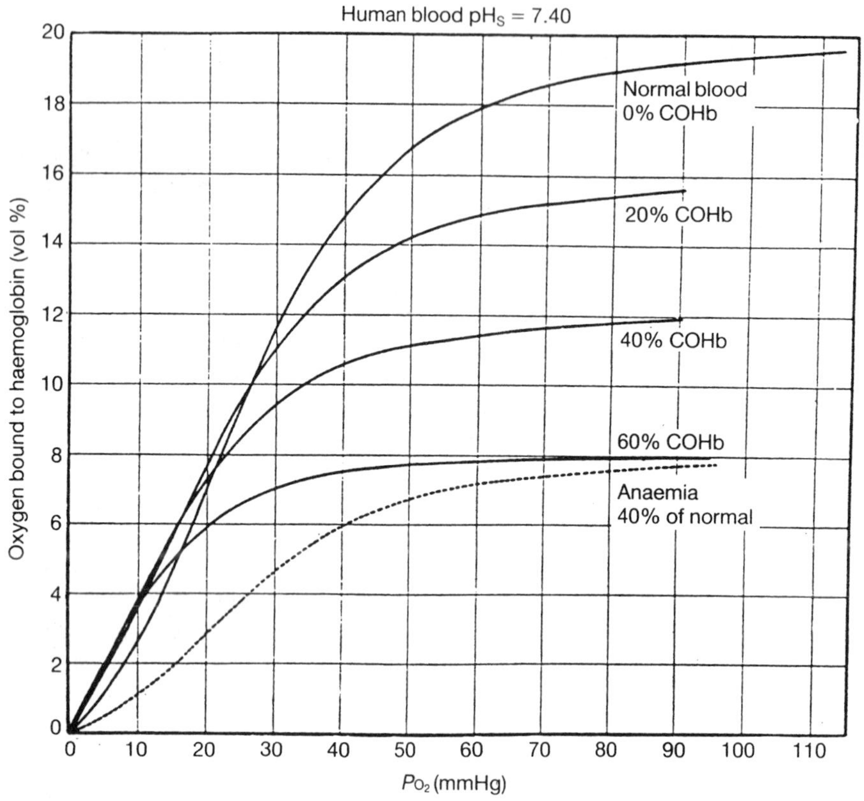

Fig. 54.4 Calculated oxygen-dissociation curves of human blood containing varying amounts of carboxyhaemoglobin, plotting the absolute amounts of bound O_2 rather than the percentage of available haemoglobin bound to O_2. (Reproduced, with permission, from Roughton FJW, Darling RC. The effect of carbon monoxide on the oxyhemoglobin dissociation curve. *American Journal of Physiology* 1944; **141**: 28.)

voice change, upper airway obstruction or drooling. Wheezing due to chemical irritation and narrowing of airways by injured mucosal cells may predominate. The sputum may often contain soot particles. Early respiratory failure including florid pulmonary oedema may be present, especially if there has been a severe skin burn.[15] A higher than normal carboxyhaemoglobin (COHb) level confirms that smoke has been inhaled, but does not indicate the severity of the lung injury.

Carbon monoxide poisoning

The traditional cherry red hue of the skin is always obscured in burned smoky shocked patients.

Carbon monoxide has a great affinity for the haemoglobin molecule, forming COHb and preventing the formation of oxyhaemoglobin. The haemoglobin molecule is also influenced such that the oxygen-dissociation curve is moved to the left, reducing the release of oxygen in the tissues (Fig. 54.4). About 60 per cent COHb is lethal. Hypoxic cerebral oedema and unconsciousness can occur. Carboxyhaemoglobin dissociates fairly rapidly and is known to have a natural half-life of about 4 hours. This period can be reduced by increasing the arterial oxygen tension. Increased inspired oxygen concentration or hyperbaric pressure are the principal forms of treatment.

When carbon monoxide occupies the haemoglobin

molecules the plasma P_{O_2} is unaffected. Calculation of oxygen saturation and content by a blood gas analyser makes erroneous assumptions from the measured P_{O_2}, and will be wrong. The amount of carboxy- and oxyhaemoglobin must be measured using spectrophotometry. Pulse oximeters may also be unable to distinguish between oxy- and carboxyhaemoglobin.

A casualty who has suffered smoke inhalation must receive a high concentration of oxygen to breathe at the scene of the accident as soon as is safely possible. A more efficient method of delivery can be substituted later if carbon monoxide poisoning is confirmed. The value of hyperbaric oxygen therapy is disputed. The factors to consider are the half-life of COHb, and the special triage and transport that may be needed unless a hyperbaric unit is available nearby.[16]

The maximum level of COHb at the time of exposure can be estimated using a standard nomogram based on the half-life of COHb.[17]

Cyanide poisoning

High levels of plasma cyanide have been found in some victims of house fires. These levels are difficult to measure and too time consuming to be of clinical value. Symptoms caused by a degree of generalized cellular hypoxia are vague. In theory, a high mixed venous oxygen and lactic acidosis are likely to be present. Such poisoning might contribute to the multi-organ failure that so often ensues. The antidotes used to treat poisoning from ingested cyanide, such as cobalt edetate, are too toxic for use in this unclear situation.[18]

Upper airway obstruction

The tissues of the glottis are prone to become oedematous if the oropharynx or larynx is thermally damaged and also if the face and neck are burned or scalded. Oedema develops over the first few hours and increases greatly for 12 hours or so, remaining significant for 2–3 days (Fig. 54.5). It is crucial that the airway is secured early rather than late. There is no more potentially disastrous scenario than attempting tracheal intubation when the neck is swollen and rigid, the glottic opening cannot be identified, and tracheostomy or cricothyroid puncture more or less impossible to perform. The anaesthetist should make a plan of campaign for a potentially difficult intubation of an obstructed airway, according to his or her skills, with full preparation and monitoring. An awake fibreoptic intubation can be performed by an experienced operator, or an inhalation induction followed by deep anaesthesia is appropriate. No muscle relaxant should be used. Whether intubation is through the nose or the mouth is immaterial in the first instance. In the long term a nasal tube is tolerated better unless the nose is itself badly burned. The airway mucosa should be traumatized as little as possible and low pressure cuffs should be used. A tracheostomy should only very rarely be performed through burned tissue, as lethal pulmonary infection would almost certainly ensue. It might, however, occasionally be indicated if the neck were intact and the face or larynx injured.

Lower airway injury

Some of the smoke chemicals are irritant to the small airways, causing bronchospasm, and some are toxic to the

(a)

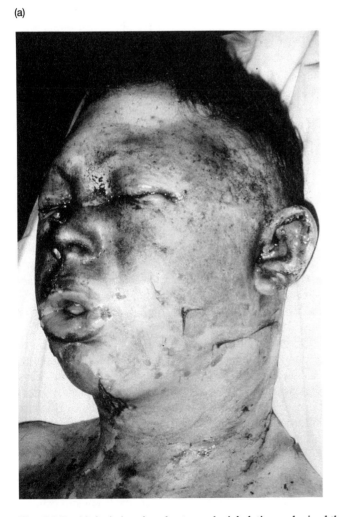

(b)

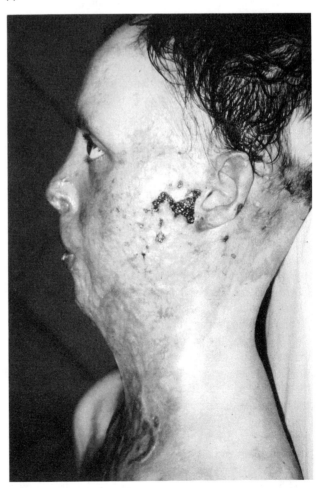

Fig. 54.5 (a) A victim of moderate smoke inhalation and mixed-thickness skin burn 6 hours after injury. (b) The same patient: potential problems for intubation some weeks later.

cells, causing airway plugging as aggregates of dead cells form. Ciliary function is impaired. Diagnosis may be helped by fibreoptic bronchoscopy to identify soot and mucosal sloughs in the trachea, and by ^{133}Xe scan (if available) to confirm the closure of small airways.

Treatment is directed towards the removal of sputum and casts, the treatment of the 'bronchospasm' and the detection of patchy lung collapse. Overenthusiastic repetitive tracheal suction which traumatizes the mucosa of the airways should, however, be avoided.

Lung parenchymal injury

Patients who have inhaled smoke without sustaining a skin burn rarely develop failure of gas exchange and pulmonary oedema. When a skin burn is also present, respiratory failure may develop progressively, and the severity may be related to the size of the skin burn. A patient with a severe scald may also succumb to respiratory failure after some days. Clearly, many factors that might contribute to ARDS are present (Table 54.5). The acute smoke injury is compounded by at least some facets of the syndrome.

Attempts are being made to explain the significance of the vasoactive and destructive mediators produced by the initial burn, the interaction of leucocytes and dead tissue, and the altered behaviour of macrophages in the injured lung tissue, and the triggering of the alternative complement pathway. The occurrence of an increased permeability of the pulmonary capillaries after smoke inhalation is disputed, but increased lung water may sometimes be found.[19]

The clinical picture is of progressive respiratory failure with some decrease in lung compliance and patchy x-ray changes or generalized pulmonary oedema. Metabolic rate rises at this time such that there is a need for increased oxygen and carbon dioxide exchange.

Treatment follows standard intensive care practice. There is a consensus that intravenous infusion of colloid rather than crystalloid should be used for resuscitation. On no account should the patient be allowed to be dehydrated. The fluid shifts in the lung are apparently obligatory and perfusion of other organs must not be jeopardized. Pulmonary artery pressure is usually normal, unaffected by the increased lung water. Continuous positive airway pressure with spontaneous ventilation may be effective in maintaining oxygen and carbon dioxide exchange or controlled ventilation may be required. Meticulous intensive care and also burns nursing is demanded.

It is to be hoped that in the future an effective therapy for ARDS will become available, such as a free radical scavenger or prostacyclin. Nitric oxide, which has local vasodilatory effects, may prove to be effective.[20]

Table 54.5 Risk factors for ARDS in burns, and possible consequences in the lung

Risk factors	Possible consequences
Skin burn	Many mediators Leucocyte aggregation
Low plasma oncotic pressure	Increased lung water
Low blood volume Anaemia Hypoxaemia	Reduced oxygen delivery
Septic shock Bacteraemia	Reduced oxygen delivery Added infection
Disseminated intravascular coagulation Blood transfusions	Sludging of microvessels
Smoke inhalation	Direct alveolar injury Increased capillary permeability Increased lung water

Other complications

Pneumonia and pulmonary atelectasis are common. Pulmonary emboli, however, seem to occur only rarely. Late morbidity in the form of bronchiolitis obliterans, bronchial strictures and laryngeal stenosis occurs.

Drugs for burned patients

Critically ill patients usually require smaller amounts of therapeutic drugs but this is not so in the case of those who have suffered burns. Pharmacokinetics and pharmacodynamics are influenced by fluid shifts, changes in metabolic rate and cardiac output and in other organ functions[21] (Table 54.6).

Drugs bound to albumin will have an increased free fraction of drugs due to low albumin levels throughout the healing phase, while those bound to the acute phase proteins especially α_1-globulin, which is raised early after a burn, will have a low available free fraction. This may influence the volume of distribution.

Cardiac output and glomerular filtration rate are usually raised, so drugs excreted by the kidney, especially aminoglycosides and cimetidine, are excreted rapidly.[22–24] It is important to use very large doses of aminoglycosides, and to monitor blood levels frequently. It is more common to find blood levels too low than too high. Gentamicin has been found in significant quantity in the wound exudate.

Table 54.6 Changes in the pharmacology of drugs in burn victims

	Pathology	Effects
Plasma proteins	↓ Albumin ↑ α_1-Acid glycoprotein	↑ Free fraction (e.g. diazepam, midazolam, phenytoin, salicylic acid) ↓ Free fraction (e.g. lignocaine (lidocaine), pethidine (meperidine), imipramine
Cardiovascular system	Early low output state Later hyperdynamic circulation	Poor redistribution of intramuscular or oral drugs ↑ Organ blood flow and drug clearance
Kidney	↑ Renal blood flow and GFR	↑ Excretion some drugs (e.g. aminoglycosides, cimetidine)
Liver	Enzyme induction	Accelerated metabolism some drugs
Wound exudate	–	Significant loss of drugs, especially in children
Neuromuscular junction	↑ Acetylcholine receptor sites	Hyperkalaemia after suxamethonium (succinylcholine) Diminished effect of non-depolarizers
High endogenous catecholamines	–	Cause of decreased effect of dopamine and topical adrenaline

The most important difference is the change in effectiveness of the muscle relaxant drugs. This is discussed below under 'Anaesthesia for the burned patient'.

Surgery for burns

It is not always possible to diagnose the thickness of a new burn, especially when the patches of burn have different depths. Experience with the outcome after burns caused by different agents (flame, scald, flash, radiation) may help. At some stage a full-thickness burn must be excised and covered with with a split-thickness skin graft taken from the patient. Obvious small full-thickness areas may have surgery immediately. Patients with large areas of full-thickness injury should ideally have at least some of these excised and grafted as soon as possible; however, early surgery of areas of full-thickness burn causes considerable bleeding and physiological disturbance. The management of these patients needs an experienced multidisciplinary team, appropriate equipment and good timing. The aim is to get the patient healed as fast as possible, to reduce the risks of infection and catabolism. If surgery cannot be initiated early, because the patient's general condition is unstable or because a 'wait and see' diagnostic policy has been adopted, or the facilities are unsuitable, it should be undertaken at least 3 weeks later when granulations have formed and bleeding will be less.

Anaesthesia and pain relief may therefore be needed for a wide range of interventions, and the timing of operations and their extent should be planned jointly by the surgeon and anaesthetist.

Anaesthesia for the burned patient

The most appropriate anaesthetic technique for a specific case must be decided by the anaesthetist taking into consideration the following problems and, in the case of the severely injured patient, the presence of the various physiological derangements already described.

Practical problems

Some ingenuity is required to solve the problems posed to the anaesthetist by surgery for extensive burns. Surgeons will be debriding or taking grafts from one or sometimes all four limbs and/or any other body part, possibly entailing changes in position of the patient and of the limbs.

There may be difficulty in finding an unburned unused vein for an intravenous infusion that will be undisturbed by the surgery. Access for monitoring devices will be restricted. It may be necessary to place the oximeter probe and ECG electrodes at unconventional sites. The blood pressure cuff may have to be placed over burned tissue (negotiate with the surgeons!) or be omitted altogether; a rectal temperature probe often falls out. A tympanic membrane or oesophageal temperature probe may be used. Arterial cannulation may also be difficult to site away from a burned area but is a valuable option for pressure measurement.

Clinical judgement is sometimes all that is available and this will always be so during war and catastrophe. In particular, the adequacy of peripheral perfusion, the pulse rate (in the presence of adequate analgesia) and the colour or pallor of the patient are important clinical signs of the circulatory state.

Compassion must be used before and during the induction of anaesthesia. Perhaps induction of anaesthesia can be on the patient's own bed, or without the patient changing position (e.g. prone). If the face is burned, thought must be given to managing the facemask and in securing the tracheal tube or laryngeal mask and protecting the burned areas.

The airway

If the face or neck is burned, obviously one must prepare for a difficult intubation. Beware also burns of the anterior chest wall which can tether the skin of the neck sufficiently to cause rigidity and limited neck extension. The laryngeal mask now often allows this kind of airway problem (and also later neck contractures) to be managed smoothly without a complicated difficult intubation scenario and without traumatizing the larynx. However, if the patient is going to be prone or lateral, or there is the slightest doubt about the position of the laryngeal mask, conventional tracheal intubation will be prudent.

Blood loss and replacement

In the techniques used for early surgery and tangential excision, a lot of blood is lost into drapes and gowns, onto the floor and into saline soaked packs. The total loss can be estimated but not measured. Even when tourniquets are used, some bleeding occurs after the tourniquet pressure is released. The use of adrenaline (epinephrine)-soaked packs on the bleeding surface, now widely employed, appears to be quite an effective manoeuvre. Dysrhythmias have not proved to be a problem, but the wounds continue to ooze quite substantially postoperatively.

For major surgery 6–8 units of blood may be needed. It is wise to start the transfusion promptly, although some haemodilution with Hartmann's solution until most of the bleeding has stopped can prevent waste of blood without hazard. When two teams of surgeons are operating, the loss of blood may be too fast for simultaneous replacement. This is one time when the extent of surgery and the rate at which it is performed must be adjusted to the patient's condition.

Heat loss

Considerable heat is lost if most of the patient is both exposed and wet. Heat conservation mechanisms are already compromised and a vasodilating anaesthetic militates against heat conservation. It is important that the patient should not shiver postoperatively, as this will cause hypoxia, bleeding and haematoma formation and the grafts may be dislodged.

Dedicated burn unit theatres have warm air temperature but all heat-conserving measures, in particular the warming of infusion fluids, must also be used. Anaesthetic circle circuits with low fresh gas flows deliver warm and humidified inspired gas. If core temperature does fall, surgery may be curtailed, but in practice this is rarely a problem. Reflective metallic blankets wrapped round the patient should be used to conserve heat in the vulnerable postoperative period.

Hypermetabolism

It must be remembered that if the patient has a raised metabolic rate, increased alveolar ventilation will be mandatory.

Repeated anaesthesia

Halothane was used for burned patients for years without any reports of related hepatitis. The newer more evanescent vapours enflurane and isoflurane, which are not metabolized, are now used when possible.

Depolarizing muscle relaxants

It is well known that it is inadvisable to administer suxamethonium (succinylcholine) to burned patients. There were reports of cardiac arrest during intubation of such patients in the 1960s. These occurred between the 20th and 50th days post-injury. The cause was found to be a high plasma potassium level immediately following the administration of the relaxant.[25,26] This phenomenon may occur a few days after a moderately severe burn and may last until after healing has occurred. A similar phenomenon has been reported in patients with denervation injury and following muscle trauma.[27] The experimental administration of very small doses (0.1 or 0.2 mg/kg) of suxamethonium to burned children has shown enhanced paralysis related to burn size, most exaggerated between the fourth and twelfth post-burn days.[28]

Generations of burned patients have received suxa-

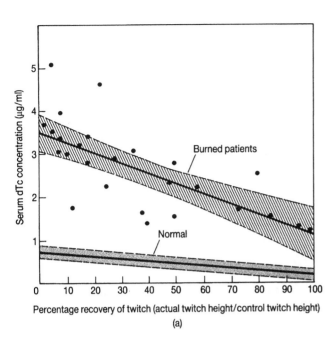

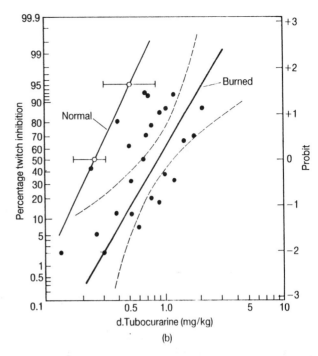

(a) (b)

Fig. 54.6 (a) A highly significant linear correlation ($r = 0.72$, $P < 0.001$) was found between serum tubocurarine concentration (y) and percentage recovery of twitch (x). The equation for the regression line is given by $y = 3.44 - 0.022x$. The regression for normal patients previously described by Matteo and colleagues[30] is at the bottom of the graph. The 95% confidence bands for the mean serum concentrations for normal subjects and burned patients are shown by the shaded areas. There is no overlap of the bands. (b) Comparative dose–response curves for normal[30] and burned patients. The regression equation is given by $y = 2.99 + 1.23$ probits ($r = 0.52$, $P = < 0.01$). The regression line for normal patients with the 95% confidence limits for ED_{50} and ED_{95} is indicated by the thin line on the left. The thick line and the dashed curved line are the regression line and the 95% confidence bands, respectively, for burned patients. Each dot represents the percentage twitch inhibition for each of the intravenous doses of tubocurarine administered. (Reproduced, with permission, from Martyn JAJ, Szyfelbein SK, Hassan HA, Matteo RS, Savarese JJ. Increased d-tubocurarine requirement following major thermal injury. *Anesthesiology* 1980; **52**: 352–5.)

methonium uneventfully, but there are virtually no indications for its use now. Rapid intubation can be achieved with drugs such as atracurium and vecuronium. Emergency intubation for burned patients is usually required for impending laryngeal obstruction when any muscle relaxant would be contraindicated.

Non-depolarizing relaxants

Curare-like drugs have to be given in large and frequent doses during surgery for burns. In 1980 it was found that in burned patients the serum levels of tubocurarine were about five times higher than in controls for the same level of twitch depression, and that the dose–response curve is shifted to the right (Fig. 54.6).[29,30] In burned patients, tubocurarine is somewhat more bound to protein than normal,[31] and specifically to globulins which may well be raised. However, this is not the main cause of the increased resistance to the drug, which is also found for pancuronium, which is highly albumin bound.[32] Similar results have been identified for metocurine (children),[33] atracurium (children and adults)[34–36] (Fig. 54.7) and vecuronium

(children)[37] (Fig. 54.8). The reduced sensitivity to atracurium reported in children[35] has been confirmed in adults (Table 54.7).[36]

The pharmacokinetics of tubocurarine[31] and of atracurium[36] have been found to differ a little between burned and unburned patients, but not sufficiently to explain this phenomenon. It has therefore been postulated that there is a profound change at the neuromuscular junction and that a proliferation of nicotinic acetylcholine receptors on the surface of the muscle, outside the end plate, take up the muscle relaxant. There are some conflicting data on this following rat studies but no confirmation yet of this phenomenon in humans.

The clinically significant facts are that higher than usual doses of non-depolarizing relaxants are required; the larger the burn, but greater the dose and the faster the recovery time. The phenomenon is not obvious in the first week post-burn, but continues thereafter at least until healing has occurred. It is not important in burns of less than 10 per cent. Normal doses of reversal agents can be used.

Both atracurium and vecuronium cause minimal cardio-vascular disturbance. Either can be used in a large dose

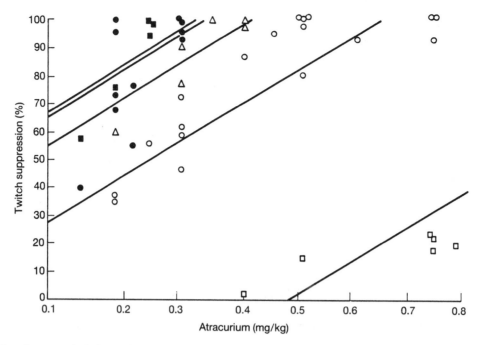

Fig. 54.7 Atracurium dose vs twitch depression in controls and acutely burned children. In the first week (●), the dose–response curves were similar to controls (■). After this time, there was a relative resistance to the effects of atracurium which was related to the size of the burn. The slopes of the curves are similar but the intercepts are significantly different between minor (△), moderate (○) and major (□) burns. These groups were also significantly different ($P < 0.001$) when compared with controls and burned patients studied within 1 week of injury. (Reproduced, with permission, from Mills AK, Martyn JAJ. Evaluation of atracurium neuromuscular blockade in paediatric patients with burn injury. *British Journal of Anaesthesia* 1988; **60**: 452.)

for intubation. In practice, atracurium 1.0 mg/kg or vecuronium 150 µg/kg are effective.

Regional anaesthesia

Both the burned area and the donor site have to be anaesthetized, so a peripheral nerve block is rarely useful. There is a potential infection risk for the patient with a sizeable burn, from skin contamination or blood-borne bacteria, when either extradural or intradural injections are used. Regional anaesthesia does have its proponents, and is useful for postoperative analgesia.[38]

Postoperative care

The patient should be warm and pain-free. Shivering or undue activity will cause grafts to sheer and bleeding to occur. The new graft site is a source of pain, and application of bupivacaine or lignocaine (lidocaine) gel to the area before dressing seems to be quite effective and without unwanted sequelae.

Patient-controlled analgesia devices may help some patients but the manipulation of the request button is often impractical. Much ingenuity is used in designing gadgets to allow the patient to be in control,[39] but a

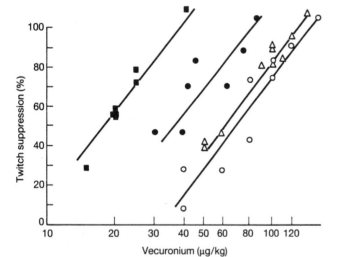

Fig. 54.8 Logarithm of dose vs twitch suppression for vecuronium in controls and burned children. In the presence of acute burn injury the vecuronium ED values increased with increasing burn size. The slopes of the curve were not different but the intercepts were significantly different ($P < 0.01$). ■, control; ●, minor; △, moderate; ○, major burns. (Reproduced, with permission, from Mills AK, Martyn JAJ. Neuromuscular blockade with vecuronium in paediatric patients with burn injury. *British Journal of Clinical Pharmacology* 1989; **28**: 157.)

Table 54.7 Summary of response in adult patients to 0.5 mg/kg atracurium

Patient	Percentage maximal twitch depression	Time to 50% recovery of pre-atracurium twitch force (minutes)
Control		
1	100	48
2	100	64
3	100	43
4	100	53
Mean	100	52
SD	0	9.0
Burned		
5	87.5	23
6	65.4	15
7	35.7	0
8	87.1	24
9	54.6	9
Mean	66.1*	14.2*
SD	22.1	10.0
P value*	0.027	0.0013

* Compared with control group by *t* test.
Reproduced, with permission, from Marathe PH, Dwersteg JF, Pavlin EG, Haschke RH, Heimbach DM, Slattery JT. Effect of thermal injury on pharmacokinetics and pharmacodynamics of atracurium in humans. *Anesthesiology* 1989; **70**: 752–5.

nurse-controlled infusion or other familiar method of delivering opioids will often be needed. Feeding should be restarted promptly.

An evening pyrexia frequently follows the bacteraemia that occurs during surgery.

Anaesthesia for dressing changes

The protocols for care of the burn surface vary between centres. In some, dressings are changed twice daily without general anaesthesia; in some, lengthy toilet and dressing changes are carried out under anaesthesia every few days. The procedure may include showering the (horizontal) patient, passive physiotherapy, splint making, and resiting venous and other lines. The dressing changes become less painful as healing proceeds.

Venous access may be very difficult. Monitoring is often limited because of access and because the patient is being moved a lot. A pulse oximeter should, however, be used. The main problems are the patient's apprehension and starvation time. It is vital that as far as possible the patient be assured of pain-free dressings and a comfortable wake-up. Apart from humanitarian reasons, apprehension adds to the stress response.

Sufficient food intake is difficult to maintain in the catabolic patient. Unduly lengthy fasting times before and after dressings must be avoided. Calorific liquid feeds can be continued until 2–3 hours preoperatively. It is important to predict correctly the timing of the treatment.

The anaesthetist can therefore design the anaesthetic or analgesic techniques to suit the circumstances. The patient may well develop strong views and of course should be consulted. As time passes, tachyphylaxis to narcotic and hypnotic drugs is often obvious and generous pain relief should be the rule.

Entonox (50 per cent oxygen/nitrous oxide) alone is sometimes effective, but often insufficient.

For many patients a totally intravenous technique is most satisfactory. It gives a quick induction and emergence; pain relief and depth of anaesthesia can be titrated against the activity in progress. Every combination of intravenous hypnotic and analgesic has probably been used. At present an infusion or bolus administration of alfentanil and propofol or midazolam is often used. The doses required are usually large and the problem is more likely to be that the anaesthesia will become too light rather than too deep.

Ketamine is a useful drug, providing profound analgesia while muscle tone and upper airway patency are relatively well maintained, usually with little assistance from the anaesthetist. An overwhelming advantage of intramuscular ketamine is that it acts rapidly and effectively, and can be used when there are no visible veins. The dose is 6–9 mg/kg i.m. A vein can usually be cannulated later. The intravenous dose is 1–2 mg/kg, repeated as required. Atropine must always be given to children, otherwise salivation may cause serious laryngospasm. A benzodiazepine such as midazolam is also given as an amnesic/anxiolytic agent and to eliminate postoperative hallucinations.

Later reconstructive surgery

Burned patients often need repeated operations to release contractures and to improve function and appearance. Empathy and reassurance are important, as a phobic element often develops. Venepuncture through or near scar tissue seems to be abnormally painful. Extreme contractures of the neck presenting 'impossible' intubation problems are the result of treatment delay and are usually seen only in countries with insufficient health resources. The tethered skin must be severed under local anaesthesia or possibly ketamine, though now in many cases the laryngeal mask will solve the problem of airway control for these patients.[40]

REFERENCES

1. Demling RH, Wong C, Jin L, Lalond C. Early lung dysfunction after major burns (role of edema and vasoactive mediators). *Journal of Trauma* 1985; **25**: 959–66.
2. Remensnyder JP. Acute electrical injuries. In: Martin JAJ, ed. *Acute management of the burned patient.* Philadelphia: WB Saunders, 1990: 66–86.
3. Evans EI, Purnell OJ, Bobinett PW, Batchelor ADR, Martin M. Fluid and electrolyte requirements in severe burns. *Annals of Surgery* 1952; **135**: 804–17.
4. Baxter CR. Early resuscitation of patients with burns. In: Welch CE, ed. *Advances in surgery, 4.* Chicago: Year Book Medical, 1970.
5. Muir I. The use of the Mount Vernon formula in the treatment of burn shock. *Intensive Care Medicine* 1981; **7**: 49–53.
6. Settle JAD. Resuscitation and fluid balance. In: Judkins KC, ed. *Clinical anaesthesiology: burns and plastic surgery.* London: Ballière Tindall, 1987: 575–95.
7. Pruit BA. Fluid resuscitation of extensively burned patients. *Journal of Trauma* 1981; **21**: suppl 8, 690–2.
8. Monafo WW, Halverson JD, Schechtman K. The role of concentrated sodium solutions in the resuscitation of patients with severe burns. *Surgery* 1984; **95**: 129–34.
9. Demling RH. Medical progress – burns. *New England Journal of Medicine* 1985; **313**: 1389–98.
10. Hunt JL, Purdue GF. Resuscitation of major burns. In: Rylah LTA, ed. *Critical care of the burned patient.* Cambridge: Cambridge University Press, 1992: 44–58.
11. Knochel JP. Rhabdomyolysis and myoglobinuria. *Seminars in Nephrology* 1981; **1**: 75–86.
12. Martyn JAJ, Farago LS, Burke JF. Hemodynamics of low dose dopamine in non-septic burned patients. *Chest* 1986; **89**: 357–60.
13. Baar S. Anaemia of burns. *Burns* 1979; **6**: 1–8.
14. Mackie DP, Van Hertum WAJ, Schumberg T, Kuyper EC, Knape P. Prevention of infection in burns: preliminary experience with selective decontamination of the digestive tract in patients with extensive injuries. *Journal of Trauma* 1992; **32**: 570–5.
15. Brown JM. Inhalation injury and progressive pulmonary insufficiency in a British burns unit. *Burns* 1977; **4**: 32–43.
16. Grube BJ, Marvin JA, Heimbach DM. Therapeutic hyperbaric oxygen: help or hindrance in burn patients with carbon monoxide poisoning. *Journal of Burn Care and Rehabilitation* 1988; **9**: 249–52.
17. Clarke CJ, Campbell D, Reid WH. Blood carboxyhaemoglobin and cyanide levels in fire survivors. *Lancet* 1981; **1**: 1332–5.
18. Langford RM, Armstrong RF. Algorithm for managing injury from smoke inhalation. *British Medical Journal* 1989; **299**: 902–5.
19. Head JM. Inhalation injury in burns. *American Journal of Surgery* 1980; **139**: 508–12.
20. Roissant R, Falke KJ, Lopex F, Slama K, Pison U, Zapol WM. Inhaled nitric oxide for the adult respiratory distress syndrome. *New England Journal of Medicine* 1993; **328**: 399–405.
21. Martyn JAJ. Clinical pharmacology in burns. In: Martyn JAJ, ed. *Acute management of the burned patient.* Philadelphia: WB Saunders, 1990: 180–200.
22. Loirat P, Rohan J, Baillet A, Beaufils F, David R, Chapman A. Increased glomerular filtration rate in patients with major burns and its effect on the pharmacokinetics of tobramycin. *New England Journal of Medicine* 1978; **299**: 915–19.
23. Zaske DE, Sawchuk RJ, Gerding DN, Strate RG. Increased dosage requirements of gentamicin in burn patients. *Journal of Trauma* 1976; **19**: 824–8.
24. Martyn JAJ, Abernethy DR, Greenblatt DJ. Increased cimetidine clearance in burned patients. *Journal of the American Medical Association* 1985; **258**: 1288–91.
25. Tolmie JD, Joyce TH, Mitchell GD. Succinylcholine danger in the burned patient. *Anesthesiology* 1967; **28**: 467–70.
26. Schaner PJ, Brown RL, Kirksey TD, Gunther RC, Ritchie CR, Gronert GA. Succinylcholine-induced hyperkalemia in burned patients. 1. *Anesthesia and Analgesia* 1969; **48**: 764–70.
27. Gronert GA, Theye RA. Pathophysiology of hyperkalemia induced by succinylcholine. *Anesthesiology* 1975; **43**: 89–99.
28. Brown TCK, Bell B. Electromyographic responses to small doses of suxamethonium in children after burns. *British Journal of Anaesthesia* 1987; **59**: 1017–21.
29. Martyn JAJ, Szyfelbein SK, Hassan HA, Matteo RS, Savarese JJ. Increased *d*-tubocurarine requirement following major thermal injury. *Anesthesiology* 1980; **52**: 352–5.
30. Matteo RS, Spector S, Horowitz PE. Relation of serum *d*-tubocurarine concentration to neuromuscular blockade in man. *Anesthesiology* 1975; **41**: 440–3.
31. Martyn JAJ, Matteo RS, Greenblatt DJ, Lebowitz PW, Savarese JJ. Pharmacokinetics of *d*-tubocurarine in patients with thermal injury. *Anesthesia and Analgesia* 1982; **61**: 241–6.
32. Martyn JAJ, Liu LMP, Szyfelbein SK, Ambalavanar ES, Goudsouzian NG. The neuromuscular effect of pancuronium in burned children. *Anesthesiology* 1983; **59**: 561–4.
33. Martyn JAJ, Goudsouzian NG, Matteo RS, Liu LMP, Szyfelbein SK, Kaplan RF. Metocurine requirements and plasma concentrations in burned paediatric patients. *British Journal of Anaesthesia* 1983; **55**: 263–8.
34. Dwersteg JF, Pavlin EG, Heimbach DM. Patients with burns are resistant to atracurium. *Anesthesiology* 1986; **65**: 517–20.
35. Mills AK, Martyn JAJ. Evaluation of atracurium neuromuscular blockade in paediatric patients with burn injury. *British Journal of Anaesthesia* 1988; **60**: 450–5.
36. Marathe PH, Dwersteg JF, Pavlin EG, Haschke RH, Heimbach DM, Slattery JT. Effect of thermal injury on pharmacokinetics and pharmacodynamics of atracurium in humans. *Anesthesiology* 1989; **70**: 752–5.
37. Mills AK, Martyn JAJ. Neuromuscular blockade with vecuronium in paediatric patients with burn injury. *Journal of Clinical Pharmacology* 1989; **28**: 155–9.
38. Wilson GR, Tomlinson P. Pain relief in burns – how we do it. *Burns* 1988; **14**: 331–2.
39. Kinsella J, Glavin R, Reid WH. Patient controlled analgesia in burns patients. *Burns* 1988; **14**: 500–3.
40. Gaukroger PB, Chapman MJ, Davey RB. Pain control in paediatric burns – the use of patient controlled analgesia. *Burns* 1991; **17**: 396–9.

FURTHER READING

Textbooks

Martyn JAJ, ed. *Acute management of the burned patient*. WB Saunders, 1990: 340 pp.

Rylah LTA, ed. *Critical care of the burned patient*. Cambridge: Cambridge University Press, 1992: 206 pp.

Ward handbook

Settle J. *Burns – The first five days*. Romford: Smith & Nephew Pharmaceuticals, 1986: 28 pp.

Review articles

Demling RH. Medical progress – burns. *New England Journal of Medicine* 1985; **313**: 1389–98.

Kinsella J. Smoke inhalation. *Burns* 1988; **14**: 269–79.

Kinsella J, Booth MG. Pain relief in burns. *Burns* 1991; **17**: 391–5.

Martyn JAJ. Clinical pharmacology and drug therapy in the burned patient. *Anesthesiology* 1986; **65**: 67–75.

Chapter 55

Anaesthesia for Cardiac Surgery

James A. DiNardo

Preoperative assessment	Anaesthetic technique
Monitoring	Cardiopulmonary bypass

Anaesthesia for cardiac surgery requires that the anaesthetist possess much of the knowledge conventionally within the domain of the cardiologist and cardiac surgeon. The cardiac anaesthetist must make clinical decisions based on a sound understanding of the cardiovascular pathophysiology present in each patient. This requires familiarity with invasive and non-invasive diagnostic techniques as well as with the proposed surgical procedure.

The scope of the subspeciality of cardiac anaesthesia has become so broad as to prohibit its complete coverage in one chapter. The goal of this chapter is to present a basic introduction to the anaesthetic considerations and techniques which are unique to the cardiac anaesthetist.

Preoperative assessment

It goes without saying that all patients undergoing cardiac surgery require a comprehensive evaluation of their cardiovascular system. However, these patients may have coexisting diseases which have contributed to the development of their cardiac dysfunction. Furthermore, organ dysfunction may develop as the result of cardiac dysfunction. Therefore it is important that cardiac surgical patients undergo a comprehensive, multisystem preoperative evaluation. Many patients will have undergone a comprehensive screening prior to being seen by the anaesthetist but others will require additional tests following preoperative evaluation. The cardiac anaesthetist must be capable of conducting a comprehensive preoperative evaluation and should be familiar with the methodology, limitations, and accuracy of the tests most commonly used in evaluation of cardiac surgical patients.

Endocrine evaluation

A careful evaluation for endocrine abnormalities should be sought in the history and physical examination. The following conditions deserve special consideration.

Diabetes mellitus

Diabetes mellitus (DM) is a risk factor for the development of coronary artery disease and thus perioperative management of DM is a common problem facing those who anaesthetize patients for cardiac surgery.

Cardiopulmonary bypass (CPB) is associated with changes in glucose and insulin homoeostasis in both diabetic and non-diabetic patients. During normothermic CPB, elevations in glucagon, cortisol, growth hormone and catecholamine levels produce hyperglycaemia via increased hepatic glucose production, reduced peripheral utilization of glucose, and reduced insulin production.[1,2] During hypothermic CPB hepatic glucose production is reduced and insulin production remains low such that blood glucose levels remain relatively constant.[1] Rewarming on CPB is also associated with increases in glucagon, cortisol, growth hormone, and catecholamine levels and is accompanied by enhanced hepatic production of glucose, enhanced insulin production, and insulin resistance.[1,3] The transfusion of blood preserved with acid–citrate–dextrose, the use of glucose solutions in the CPB prime, and the use of β-adrenergic agents for inotropic support further increase exogenous insulin requirements.[1,3,4] In non-diabetic patients these hormonally mediated changes usually result in mild hyperglycaemia.[2,4] In diabetic patients these changes may may produce significant hyperglycaemia and ketoacidosis.[2]

Because of the varying insulin requirements which occur during cardiac surgery and the unreliable absorption of subcutaneously administered insulin in patients undergoing large changes in body temperature and peripheral perfusion, insulin is best delivered intravenously in insulin-dependent diabetics undergoing cardiac surgery. Patients taking oral hypoglycaemic agents should have them discontinued at least 12 hours prior to surgery. They rarely require insulin infusion to maintain glucose homoeostasis.

Hypothyroidism

Hypothyroidism is characterized by a reduction in the basal metabolic rate. Patients with hypothyroidism have reductions in cardiac output which are appropriate for this reduction in metabolic rate. Cardiac output may be reduced by up to 40 per cent due to reductions in both heart rate and stroke volume.[5] In addition, both hypoxic and hypercapnic ventilatory drive are blunted by hypothyroidism.[6] Furthermore, hypothyroidism may be associated with blunting of baroreceptor reflexes, reduced drug metabolism and renal excretion, reduced bowel motility, hypothermia, hyponatraemia from syndrome of inappropriate secretion of ADH (SIADH) and adrenal insufficiency.[6,7]

To complicate matters further, antianginal drugs such as nitrates and propranolol may be poorly tolerated by the depressed cardiovascular system.[8] Hypothyroid cardiac surgical patients have been observed to exhibit delayed emergence from anaesthesia, persistent hypotension, tissue friability and bleeding, and adrenal insufficiency requiring exogenous steroids. Despite these problems thyroid replacement in cardiac surgical patients, particularly those with ischaemic heart disease, is not always desirable. In hypothyroid patients requiring coronary revascularization thyroid hormone replacement may precipitate myocardial ischaemia, myocardial infarction, or adrenal insufficiency.[7,9,10] Coronary revascularization has been successfully managed, in hypothyroid patients, with thyroid replacement withheld until the postoperative period.[9–12] Some degree of careful thyroid replacement is probably warranted in severely hypothyroid patients scheduled for cardiac surgery.[8]

Respiratory evaluation

A history of pulmonary disease should be sought. Emphasis should be placed on determining the extent and length of cigarette use, any history of asthma, the existence of recurrent pulmonary infection, and the presence of dyspnoea. Physical examination should focus on the detection of wheezes, a flattened diaphragm, air trapping, consolidation, and clubbing of the nails. The following conditions deserve special consideration.

Congenital heart disease

Lesions which produce excessive pulmonary blood flow (large ventricular septal defect, truncus arteriosus, *d*-transposition of the great arteries, patent ductus arteriosus) are associated with pulmonary dysfunction.

Rarely, compression of large airways may occur secondary to enlargement of the pulmonary arteries. More commonly these lesions produce pulmonary vascular changes which affect pulmonary function. The pulmonary vascular smooth muscle hypertrophy which accompanies increased pulmonary blood flow[13,14] produces peripheral airway obstruction and reduced expiratory flow rates characteristic of obstructive lung disease.[15,16] In addition, smooth muscle hypertrophy in respiratory bronchioles and alveolar ducts in patients with increased pulmonary blood flow may also contribute to this obstructive pathology.[16] These changes predispose the patient to atelectasis and pneumonia. Children with Down's syndrome appear to have a more extensive degree of pulmonary vascular and parenchymal lung disease than non-Down's children with similar heart lesions.[17] This predisposes such patients to greater postoperative respiratory morbidity and mortality.[18,19] Patients with lesions which reduce pulmonary blood flow (pulmonary atresia or stenosis, tetralogy of Fallot) also have characteristic pulmonary function changes. These patients have normal lung compliance as compared to the decreased compliance seen in patients with increased pulmonary blood flow. However, the large dead space/tidal volume ratio in these patients greatly reduces ventilatory efficiency and large tidal volumes are required to maintain a normal alveolar ventilation.[20]

Acquired heart disease

The presence of clinically and laboratory diagnosed pulmonary dysfunction has important implications in assessing operative morbidity in adults undergoing cardiac surgical procedures. In patients undergoing valvular surgery the presence of pulmonary dysfunction was associated with a 2.5-fold increase in perioperative mortality and a 2.5-fold increase in postoperative respiratory complications.[21] In patients undergoing coronary revascularization the presence of pulmonary dysfunction was not associated with an increase in morbidity or mortality.[21]

Cigarette smoking

Chronic cigarette use has several physiological effects which complicate a patient's anaesthetic management. Cigarette smoking may accelerate the development of atherosclerosis[22] and in combination with pre-existing stenoses may be responsible for acute reductions in coronary blood flow by inducing increases in blood viscosity, platelet aggregation and coronary vascular resistance.[23,24] Nicotine, through activation of the sympathetic nervous system, increases myocardial oxygen consumption by increasing heart rate and blood pressure.[22,23,25] Furthermore, the increased carboxyhaemoglobin levels which accompany chronic cigarette use reduce systemic and myocardial oxygen delivery.[25] It has been shown that the threshold for exercise-induced angina is reduced by carboxyhaemoglobin levels as low as 4.5 per cent.[26] Short-term abstinence (12–48 hours) is sufficient to reduce carboxyhaemoglobin and nicotine levels and improve the work capacity of the myocardium.[26]

Numerous studies have demonstrated that an increased incidence of postoperative respiratory morbidity exists in patients who smoke. In patients undergoing coronary revascularization, abstinence from smoking for 2 months or more reduced the incidence of postoperative respiratory complications to nearly that seen in non-smokers (11 per cent),[27]. On the other hand, abstinence for less than 2 months did not reduce the incidence of postoperative respiratory complications.[27]

Bronchospasm

Patients with asthma may present for cardiac surgery. In addition, patients with chronic obstructive pulmonary disease (COPD) and patients with pulmonary parenchymal disease secondary to valvular heart disease may have a bronchospastic component to their pulmonary dysfunction. Treatment of bronchospasm in the cardiac surgical patient is complicated by the fact that many bronchodilator drugs have cardiovascular side effects and that many cardiac drugs may worsen bronchospasm.

In those patients where bronchospasm is well controlled preoperatively it is essential to continue therapy in the perioperative period. β_2-Agonist metered dose inhaler or nebulizer therapy can be continued until arrival in the operating room and can be reinitiated again as soon as the patient is awake. Intraoperatively and postoperatively metered dose inhalation therapy can be delivered via the tracheal tube with the aid of a T-piece designed for this purpose. Theophylline therapy can be converted to aminophylline therapy until the patient's gastrointestinal tract is functioning normally. Likewise, intravenous hydrocortisone therapy may be used in place of parenteral prednisone.

Haematological evaluation

All patients scheduled for cardiac surgical procedures should have a careful history of bleeding taken with the emphasis placed on abnormal bleeding occurring after surgical procedures, dental extractions and trauma. Signs of easy bruising should be sought on physical examination. All patients should undergo laboratory screening for the presence of abnormalities in haemostasis. A platelet count, bleeding time (BT), partial thromboplastin time (PTT), and prothrombin time (PT) should be obtained. Any

abnormalities should be investigated prior to surgery so that post-CPB haemostasis is complicated by neither unknown nor unsuspected abnormalities.

Elevations in PTT and PT should be investigated with regard to factor deficiencies, factor inhibitors, and the presence of anticoagulants such as warfarin and heparin. It is important that documentation of a normal PTT and PT existing prior to warfarin or heparin administration is made so that other causes of an elevated PTT and PT are not overlooked. Deficiencies in factors VIII, IX and XI are those most commonly encountered.[28]

Coagulopathies in children with congenital heart disease have been well described.[29–32] The aetiology of these coagulopathies appears to be multifactorial.[33] Quantitative and qualitative platelet dysfunction plays a role in patients with both cyanotic and acyanotic lesions.[30,32] An acquired deficiency in the large von Willebrand factor multimers has been demonstrated in acyanotic patients.[34] In cyanotic patients hypofibrinoginaemia, low grade disseminated intravascular coagulation (DIC), deficiencies in factors V and VIII, and deficiencies in the vitamin K dependent factors (II, VII, IX, X) have all been implicated.[29–32,34] Furthermore, in patients who are cyanotic and polycythaemic the plasma volume and quantity of coagulation factors is reduced and this may contribute to the development of a coagulopathy.[29] Finally, factors synthesized in the liver may be reduced in both cyanotic and acyanotic patients in whom severe right heart failure results in passive hepatic congestion and secondary parenchymal disease.[29].

Renal/metabolic evaluation

Renal function and electrolyte balance should be evaluated prior to cardiac surgical procedures.

Renal dysfunction

Patients presenting for cardiac surgery may possess varying degrees of renal dysfunction ranging from a mild elevation in creatinine to dependence on dialysis. In addition, patients with a renal transplant may also present for surgery. In all these patients optimization of renal function prior to the operative procedure is imperative. The dialysis-dependent patient will require dialysis preoperatively. If the patient is too haemodynamically unstable to tolerate preoperative dialysis and surgery is urgent, dialysis can be managed intraoperatively. Dialysis will correct or improve the abnormalities in potassium, phosphate, sodium, chloride and magnesium. In addition the platelet dysfunction which accompanies uraemia will be improved. Desmopressin acetate (DDAVP) administration has also been shown to improve uraemia-induced platelet dysfunction[35] and should be considered if clinically significant post-dialysis platelet dysfunction exists. Dialysis will not favourably affect the anaemia, renovascular hypertension, or immune compromise associated with chronic renal failure.

In non-dialysis-dependent patients preoperative hydration is necessary to prevent pre-renal azotaemia from complicating the underlying renal dysfunction. This is particularly important following procedures such as cardiac catheterization and arteriography which utilize intravascular contrast agents. All patients appear to suffer a fall in creatinine clearance after contrast arteriography;[36] however, in patients with pre-existing azotaemia this reduction is much more likely to result in the clinical manifestations of an increased deterioration in renal function.[37]

Hypokalaemia

Hypokalaemia, defined as a serum potassium less than 3.5 mmol/l, is not an uncommon finding in the preoperative evaluation of the cardiac surgical patient. The incidence of atrial and ventricular dysrhythmias during anaesthesia for cardiac surgery was similar in normokalaemic, hypokalaemic (3.1–3.5 mmol/l), and severely hypokalaemic (< 3.0 mmol/l) patients.[38] Thus on the basis of the currently available information routine potassium administration does not appear to be warranted in all cardiac surgical patients with chronic hypokalaemia. It may be prudent to maintain a serum potassium above 3.5 mmol/l in cardiac surgical patients taking digitalis, those at high risk for myocardial ischaemia and in those who have suffered an acute reduction in serum potassium.

Cardiovascular evaluation

A myriad of non-invasive and evasive methods of evaluating cardiovascular function are currently available to cardiologists and there is great variation between institutions in the composition of a routine preoperative cardiovascular evaluation. Nevertheless, the cardiac anaesthetist must assemble a database which allows comprehensive evaluation of the following.

Diastolic function

Normal diastolic function depends on normal ventricular diastolic distensibility and compliance. Reduced distensibility occurs as the result of myocardial ischaemia in patients with coronary artery disease (CAD)[39–43] and in patients with aortic stenosis without CAD.[44] This diminished diastolic distensibility often precedes systolic dysfunction.[45] Reduced left ventricular distensibility may be caused by extrinsic limitations to ventricular expansion

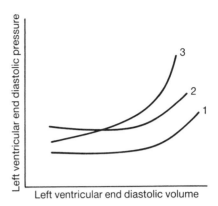

Fig. 55.1 A series of diastolic pressure–volume curves. Curve 1 represents the relation between the normal diastolic pressure and volume with a flat pressure–volume relation (high compliance) at low volumes and a steeper pressure–volume relation (reduced compliance) at higher volumes. Curve 2 has the same compliance characteristics as curve 1; however, the entire pressure–volume relation of curve 2 is shifted upward relative to curve 1 (reduced distensibility). Curve 3 represents reduced compliance. The relation between pressure and volume (curve 3) exhibits lower compliance at all volumes relative to curves 1 and 2.

in diastole as occurs with distension of the right ventricle and a leftward septal shift. In addition reduced distensibility of both ventricles may occur due to the presence of a diseased or fluid-filled pericardium. Reduced distensibility will produce an increased diastolic pressure in any given diastolic volume (Fig. 55.1).

Diastolic compliance may be reduced as the result of increased chamber stiffness as occurs in aortic stenosis or systemic hypertension. In these cases there is an increase in the quantity of myocardial tissue due to concentric hypertrophy. Diastolic compliance of the ventricle as a whole is diminished despite the fact that the compliance of the individual muscle units is normal. Patients with reduced compliance have diastolic pressure volume plots which exhibit steeper slopes than normal and which manifest this increase in slope at lower volumes than normal (Fig. 55.1).

Systolic function

Systolic function is not synonymous with contractility. Systolic function measures the ability of the ventricle to perform external work (generate a stroke volume) under varying conditions of preload, afterload and contractility. Ejection fraction (EF) and assessment of wall motion are the most commonly obtained assessments of systolic function. Because EF is an ejection phase index it depends on loading conditions. For this reason the presence of a low impedance outflow tract such as that which exists via the mitral valve in mitral insufficiency will cause the EF to overestimate systolic function. Ejection fraction is an assessment of global systolic function and will not be

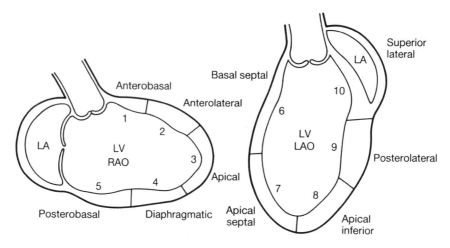

Fig. 55.2 Schematic delineation of the 5 wall segments seen in the RAO projection and the 5 wall segments seen in the LAO projection during left ventriculography. The following is a summary of the coronary arterial supply to these regions (LMCA, left main coronary artery; LAD, left anterior descending artery; CIRC, circumflex artery; RCA, right coronary artery; PDA, posterior descending artery): 1, anterobasal – LMCA; proximal LAD; 1st diagonal; 2, anterolateral – LMCA; proximal or mid LAD; 1st diagonal; 3, apical – LMCA; proximal, mid, or distal LAD; 2nd diagonal; 4, diaphragmatic (inferior) – proximal, mid, or distal RCA; PDA; 5, posterobasal – proximal, mid, or distal RCA; PDA; 6, basal septal – LMCA; proximal or mid LAD, 1st septal; 7, apical septal – LMCA; proximal, mid, or distal LAD; 8, apical inferior – proximal, mid, or distal RCA; 9, posterolateral – LMCA; proximal or distal CIRC marginals; 10, superior lateral – LMCA; proximal CIRC marginals.

depressed until a relatively large portion of ventricle exhibits compromised systolic function. An EF < 40 per cent is felt to represent poor systolic function. Wall motion abnormalities are analysed during ventriculography. When large areas of akinesis and dyskinesis exist, systolic function will be severely compromised. In patients with very low ejection fractions, such as those with end-stage cardiomyopathies presenting for heart transplantation, acute elevations in afterload are very poorly tolerated.

Coronary anatomy

The extent of coronary disease and the presence or absence of collateral flow to jeopardized areas should be assessed. This will allow intelligent assessment of the areas most at risk for developing ischaemia. Knowledge of these areas will allow ECG and transoesophageal echocardiography (TOE) monitoring to be directed appropriately (Fig. 55.2).

The coronary angiograms will also provide insight into the likelihood that a given area can be successfully revascularized. Distal disease in small vessels makes the technical success of the operative procedure less likely. Furthermore, bypassing a vessel supplying a dyskinetic or aneurysmal area of ventricle may yield little in terms of improved systolic function postoperatively because such areas have little salvageable myocardium.

In patients undergoing reoperation, evaluation of the extra cardiac conduits (internal mammary arteries and reverse saphenous veins) should also be made. A determination of how much myocardium is supplied by the conduits which are patent should be made. Nicking or cutting of these conduits during the difficult dissection

which often accompanies repeat surgery may severely compromise myocardial perfusion. Likewise, surgical manipulation of these conduits may result in embolization of atherosclerotic plaques into the distal coronary vascular bed with subsequent transmural myocardial ischaemia.[46,47]

Valvular function

The following factors must be considered when evaluating the patient with valvular heart disease.

Pathophysiology and adaptation

The nature of left ventricular loading Mitral regurgitation and aortic regurgitation present the left ventricle with volume overload while aortic stenosis results in left ventricular pressure overload. Mitral stenosis is unique because it results in left ventricular volume underload secondary to compromised diastolic filling.

Acute vs chronic lesions The haemodynamic consequences of acute valvular lesion are generally more severe than after gradual adaptation has occurred.

Mechanical vs functional lesions Mechanical valvular lesions are those in which the valve leaflets and/or supporting structures are damaged. These lesions may be chronic as in calcification of aortic and mitral valves or acute as in papillary muscle rupture or bacterial endocarditis. Functional lesions are those in which the valve apparatus is not damaged but where valve function is compromised acutely and reversibly. Acute ventricular dilation leading to mitral regurgitation, papillary muscle dysfunction leading to mitral regurgitation, and proximal aortic dissection leading to aortic regurgitation are examples.

Ventricular diastolic function and the atrial transport mechanism

Atrial transport via sinus rhythm is important in lesions where diastolic function is impaired.

Myocardial oxygen balance

It is necessary to be familiar with the unique relation between myocardial oxygen supply and demand in each valve lesion.

Ventricular systolic function

Careful assessment of ventricular systolic function is necessary to identify high risk patients and to screen for those patients unlikely to have improved ventricular function following valve replacement. The anaesthetic management of patients with valvular heart disease and poor ventricular function is a demanding challenge.

Right ventricular function and the pulmonary vasculature

Pulmonary artery hypertension may exist secondary to the primary valve lesion. It is important to know the extent of the pulmonary artery hypertension as this will affect right ventricular function. It is also necessary to know whether the pulmonary artery hypertension is likely to be reversible following surgery.

The effect of heart rate alterations on haemodynamics

Small alterations in heart rate may profoundly affect the haemodynamics of a particular lesion. For example, in aortic regurgitation, bradycardia may drastically increase the regurgitant fraction.

The effect of afterload alterations on haemodynamics

Alterations in afterload may worsen or improve the haemodynamics of a particular lesion. For example, in volume overload lesions such as mitral regurgitation, an increase in the impedance to ejection via the aorta may dramatically increase the regurgitant fraction.

Cardiac anatomy and shunt location

While some congenital heart lesions involve purely regurgitant or obstructive valve lesions, the presence of intra- and extracardiac shunt lesions is the hallmark of congenital heart disease. Anaesthetic management of the patient with congenital heart disease depends on a clear understanding of cardiac anatomy, the types of shunts and the dynamics of the shunting process. In addition the anaesthetist must understand the concepts of blending and intercirculatory mixing.

In a shunt, a communication (orifice) exists between pulmonary and arterial vessels or heart chambers. In a simple shunt there is no obstruction to outflow from the vessels or chambers involved in the shunt. The dynamics of simple shunting are summarized in Table 55.1. In a complex shunt, obstruction to outflow is present in addition to a shunt orifice. The obstruction may be at the valvular, supravalvular, or subvalvular level. In addition the obstruction may be fixed (as with valvular or infundibular stenosis) or variable (as with dynamic infundibular obstruction). The dynamics of complex shunting are summarized in Table 55.2.

Blending refers to the situation in which various proportions of systemic venous and pulmonary venous blood mix in a cardiac chamber or great vessel. Complete blending will occur in complex shunts with complete outflow obstruction (tricuspid atresia) or in simple shunts where a common chamber exists secondary to a very large defect (truncus arteriosus with a large VSD). When complete blending occurs the arterial saturation will

Table 55.1 Simple shunts (no obstructive lesions)

Restrictive shunts (small communications)	Non-restrictive shunts (large communications)	Common chambers (complete mixing)
Large pressure gradient	Small pressure gradient	No pressure gradient
Direction & magnitude more *independent* of PVR/SVR	Direction & magnitude more *dependent* on PVR/SVR	Bidirectional shunting
Less subject to control	More subject to control	Net $\dot{Q}p/\dot{Q}s$ totally depends on PVR/SVR
Examples: small VSD, PDA, Blalock shunts, small ASD	Examples: large VSD, large PDA, large Waterston shunts	Examples: single ventricle, truncus arteriosus, single atrium

Abbreviations: PVR, pulmonary vascular resistance; SVR, systemic vascular resistance; $\dot{Q}p$, pulmonary blood flow; $\dot{Q}s$, systemic blood flow; PDA, patent ductus arteriosus.

Table 55.2 Complex shunts (shunt and obstructive lesion)

Partial outflow obstruction	Total outflow obstruction
Shunt magnitude and direction largely fixed by obstruction	Shunt magnitude and direction totally fixed
Shunt depends less on PVR:SVR	All flow goes through shunt
Orifice and obstruction determine pressure gradient	Pressure gradient depends on orifice
Examples: tetralogy of Fallot, VSD and pulmonic stenosis, VSD with coarctation	Examples: tricuspid atresia, mitral atresia, pulmonary atresia, aortic atresia

Abbreviations: PVR, pulmonary vascular resistance; SVR, systemic vascular resistance.

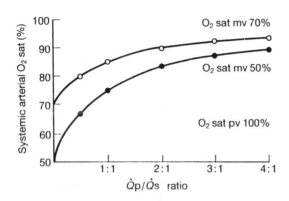

Fig. 55.3 Influence of the $\dot{Q}p/\dot{Q}s$ ratio, mixed venous (mv) oxygen saturation, and pulmonary venous (pv) oxygen saturation on arterial oxygen saturation in complete mixing lesions. Note the $\dot{Q}p/\dot{Q}s$ ratios above 3:1 do not substantially further increase arterial oxygen saturation. Sat, saturation.

depend on the following factors which are summarized in Fig. 55.3:

1. the ratio of pulmonary to systemic blood flow ($\dot{Q}p/\dot{Q}s$);
2. systemic venous saturation;
3. pulmonary venous saturation.

Mixing refers specifically to the exchange of systemic venous and pulmonary venous blood between the parallel pulmonary and systemic circulations which occurs in transposition of the great vessels (TGV). The magnitude of this intercirculatory mixing in large part determines the arterial saturation in TGV.

Pulmonary vascular pathophysiology

Pulmonary vascular occlusive disease (PVOD) may occur as the result of some congenital and acquired cardiac lesions. PVOD produces morphological changes in the pulmonary vasculature which results in progressive and irreversible elevations in pulmonary vascular resistance and pulmonary artery pressure. In particular three categories of abnormalities predispose to the development of PVOD:

1. exposure of the pulmonary vascular circuit to systemic arterial pressures and high pulmonary blood flows as is seen with a large VSD.

2. exposure of the pulmonary vascular circuit to high pulmonary blood flows in the absence of high pulmonary perfusion pressures as is seen with a large ASD.
3. obstruction of pulmonary venous drainage as is seen with stenotic pulmonary veins, in totally anomalous pulmonary venous return or as the result of chronically elevated left atrial pressures.

Monitoring

Electrocardiogram (ECG)

The gold standard for diagnosis of myocardial ischaemia is the presence of ECG changes. ECG changes occur relatively late in the temporal sequence of myocardial ischaemia following deterioration of ventricular diastolic and systolic function. In order for ECG monitoring to be effective in detecting ischaemia the appropriate leads must be monitored. Simultaneous monitoring of leads II and V_5 is commonly used because of the high sensitivity of this combination in detecting myocardial ischaemia. Monitoring with leads II and V_5 will not detect true posterior ischaemia and for this reason the use of an oesophageal ECG lead has been suggested to monitor right ventricular and posterior left ventricular ischaemia.[48,49]

Pulmonary capillary wedge (PCW) and central venous (CV) pressure traces

The CVP and PCWP traces via the pulmonary artery catheter have been described as a tool to make early diagnosis of right and left ventricular myocardial ischaemia.[49,50] Reduction in ventricular distensibility occurs in the early stages of ischaemia due to diastolic dysfunction. This reduced diastolic distensibility will have to be accompanied by a higher atrial pressure if diastolic filling of the ventricle is to be maintained. This elevation of atrial pressure will result in magnification of the normal CVP waveforms during right ventricular ischaemia and of the PCWP waveforms during left ventricular ischaemia. In addition, dilatation of the atrium may result in a more forceful atrial contraction and an enlarged A wave. Likewise ventricular dilatation or papillary muscle dysfunction may produce tricuspid or mitral regurgitation and a prominent V wave.

Unfortunately changes in PCWP and the PCWP waveform have poor sensitivity and specificity in detecting episodes of myocardial ischaemia.[51–54] It appears that in patients where only a small region of ventricular wall develops diminished distensibility with an ischaemic episode, the distensibility of the ventricle as a whole changes only slightly and subsequent changes in ventricular end-diastolic pressure (PCWP and CVP) are difficult to detect. The quantitative change in CVP or PCWP and the qualitative change in their waveforms necessary to define an ischaemic event have not been systematically defined.[53] Furthermore, it must be remembered that acute elevations in afterload in the absence of ischaemia can produce elevations in the CVP and PCWP.[53] Because these pressure trace changes are not seen in all episodes of myocardial ischaemia they can not be relied upon as the sole indicator of ischaemia. For this reason it is important to emphasize that suspicious ECG changes cannot be ignored simply because there is no change in the pressure tracings.

Transoesophageal echocardiography (TOE)

Segmental wall motion abnormalities during systole (diminished inward motion and thickening) are known to occur as a result of ischaemia and these wall motion and thickening abnormalities are known to precede ECG changes. In the instance of complete coronary occlusion wall motion abnormalities and ECG changes occur within 60 seconds of each other.[55] In cases where ischaemia is less severe, wall motion abnormalities may precede ECG changes by several minutes.[56] In fact, numerous studies have shown intraoperative TOE qualitative analysis of segmental wall motion and thickening to be a more sensitive detector of myocardial ischaemia than ECG changes and to be capable of detecting ischaemia prior to ECG changes.[56]

Unfortunately, recent data demonstrate a lack of concordance between ischaemia detected by ECG and that detected by TOE.[54,57] This may be due to several factors. Normally the TOE probe is placed at the mid-left ventricular level (level of the papillary muscles) where wall segments in the distribution of all three coronary arteries can be monitored.[56] Because a short axis view of the left ventricle can only be obtained at one level at a time, ischaemic changes occurring in the basal or apical

ventricular levels will be missed. Likewise, the mid-left ventricular short axis view does not provide a good short axis view of the right ventricle.[58] Ischaemic episodes may be missed because qualitative wall motion analysis is difficult in patients with pre-existing wall motion abnormalities.[57] Finally, the ECG may be positive with small areas of subendocardial ischaemia undetectable by TOE and all wall motion abnormalities (particularly severe hypokinesis) may not be ischaemic in origin.[57] These observations, combined with the fact that no evidence exists to support the contention that early detection and treatment of ischaemia alters outcome, temper enthusiasm for TOE as the ideal ischaemia monitor.

TOE is gaining popularity for intraoperative, online assessment of surgical repairs in both children[59] and adults.[60] Single plane (horizontal) TOE with Doppler colour-flow mapping provides excellent delineation of the important anatomical and flow features of most congenital heart lesions. Single plane TOE has limited utility in imaging the right ventricular outflow tract. Biplane (horizontal and sagittal) TOE probes allow better delineation of the right ventricular outflow tract and are now becoming available for use in infants and children.

Intraoperative TOE is useful following valve surgery. In adults both single and biplane TOE have proved useful for intraoperative assessment of mitral valve repair. Post-CPB assessment of mitral valve repair and ventricular function has proved to be a good predictor of postoperative outcome.[60] Only limited intraoperative assessment of aortic valve procedures is possible with single plane TOE. Single plane TOE and Doppler colour-flow mapping allows detection and quantification of aortic regurgitation. However, detection of aortic stenosis is unreliable and quantification is impossible. Biplane TOE is needed to obtain the long axis views of the aortic valve and aortic outflow tract necessary for accurate pulse wave or continuous wave Doppler investigation of aortic stenosis and regurgitation.

Anaesthetic technique

Premedication

Ample time should be taken in the preoperative period to reassure patients and address their concerns. Adults with preserved ventricular function are premedicated with morphine sulphate 0.1 mg/kg i.m., hyoscine (scopolamine) 0.005 mg/kg i.m. and diazepam 0.15 mg/kg p.o. approximately 1.5 hours before scheduled incision time. An alternative scheme is to substitute lorazepam 0.04 mg/kg p.o. for the hyoscine and diazepam. For elderly patients or patients with poor ventricular function the doses can be reduced and supplemental premedication given, under direct observation, in the operating theatre holding area. Supplemental O_2 with a facemask is started at the time of premedication. Ranitidine 150 mg p.o. is given at the time of premedication and the evening prior to surgery.

All cardiac medications are continued on schedule until the time of surgery and are taken with sips of water. Preoperative β-blocker therapy has been shown to blunt the haemodynamic responses to surgical stimulation during coronary revascularization[61] and to reduce the incidence of heart rate related intraoperative ischaemic events.[62,63] Continuation of preoperative β-blockade therapy does not compromise myocardial performance in the post-CPB period.[64,65] Furthermore, abrupt withdrawal of β-blockade therapy has been associated with myocardial ischaemia, hypertension, and tachydysrhythmias secondary to a β-blockade-induced increase in β-receptor density.[66]

Continuation of preoperative nifedipine therapy does not exacerbate the decreases in blood pressure seen with fentanyl–pancuronium and fentanyl–diazepam–pancuronium inductions in patients undergoing coronary revascularization.[67] Likewise, continuation of preoperative diltiazem therapy in combination with fentanyl–pancuronium anaesthesia is not associated with a reduced systemic vascular resistance (SVR) or a requirement for additional vasopressor or inotropic support in the pre- or post-CPB period.[68] Withdrawal of calcium channel blockers may be associated, on the contrary, with an increased need for vasodilator therapy in the post-CPB period.[69] Patients treated with diltiazem and nifedipine preoperatively do demonstrate a diminished but intact response to phenylephrine.[70] Thus treatment of intraoperative hypotension may require higher than usual doses of phenylephrine in patients continued on preoperative calcium channel blockers.[70] The incidence of perioperative ischaemia during coronary revascularization is greater in patients receiving just calcium channel blockers (nifedipine, diltiazem or verapamil) preoperatively than in patients receiving β-blockers or a combination of calcium channel and β-blockers.[62,63] This is probably due to better attenuation of tachycardia-induced ischaemia in patients receiving β-blockers.

A variety of premedication combinations have proved to be effective in paediatric patients.[71] No one technique is superior to another but it must be kept in mind that the goal is a haemodynamically stable patient who is easily separated from anxious parents. The neonate and young infant may need nothing more than an antisialagogue.

Anaesthetic goals

Heart transplantation (end-stage cardiomyopathy)

1. Avoid increased afterload; it results in ventricular dilatation and sustained reduction in stroke volume.
2. Maintain inotropic support; systolic dysfunction is usually severe.
3. Maintain preload; stroke volume depends on a large ventricular end-diastolic volume.
4. Mitral regurgitation may result from ventricular dilatation; cautious afterload reduction may be necessary.
5. Sinus rhythm when present should be maintained; it substantially augments end-diastolic volume.

Coronary artery disease

1. Avoid increases in myocardial oxygen consumption produced by hypertension and increased contractility.
2. Avoid tachycardia; it compromises oxygen delivery at any mean arterial pressure.
3. Avoid hypotension particularly in combination with tachycardia as this seriously compromises oxygen delivery.
4. Appreciate that a large proportion of ischaemic episodes are not related in time to any haemodynamic abnormality. These episodes should be treated when recognized.

Aortic stenosis (AS)

1. Sinus rhythm is essential; it substantially augments end-diastolic volume.
2. Maintain heart rate 70–90 beats/min.
3. Maintain a PCWP at 15–20 mmHg to guarantee adequate LVEDV.
4. Maintain afterload, decreases in diastolic blood pressure are to be avoided.
5. Maintain contractility.

Aortic regurgitation (AR)

1. Maintain sinus rhythm; this is not as important as in AS.
2. Maintain heart rate at 75–85 beats/min.
3. Avoid increase in afterload; pursue afterload reduction in acute AR and in chronic AR when LVEDP and arterial blood pressure are elevated and cardiac output and ejection fraction are depressed.
4. Maintain a PCWP high enough to guarantee adequate LVEDP.
5. Maintain contractility.

Mitral stenosis (MS)

1. Maintain sinus rhythm where possible.
2. Maintain heart rate at 60–80 beats/min.
3. Maintain PCWP high enough to guarantee as large an LVEDV as possible without pulmonary oedema (15–25 mmHg).
4. Maintain left ventricular (LV) afterload, increases are poorly tolerated. Decreasing afterload is warranted where LV systolic function is poor and afterload is high. In this instance preload *must* be maintained.
5. Maintain contractility.
6. When pulmonary vascular resistance is high and right ventricular (RV) systolic performance is compromised RV afterload reduction will improve RV and subsequently LV output. Caution must be exercised so that excessive systemic vasodilatation does not result.
7. Avoid hypercapnia, hypoxaemia and acidaemia which tend to cause pulmonary hypertension and may result in acute RV decompensation.

Mitral regurgitation (MR)

1. Maintain sinus rhythm where possible.
2. Maintain heart rate at 80–100 beats/min.
3. Maintain end-diastolic PCWP high enough to guarantee a large LVEDV without ventricular distension and increased valve area.
4. Decrease LV afterload, increases are poorly tolerated. Decreasing afterload will improve cardiac output by decreasing regurgitation but preload *must* be maintained.
5. Maintain contractility.
6. When pulmonary vascular resistance is high or when RV systolic dysfunction exists, RV afterload reduction will improve RV and subsequently LV output. The concurrent systemic vasodilatation will also directly reduce regurgitation.
7. Avoid hypercapnia, hypoxaemia and acidaemia which tend to cause pulmonary hypertension and may result in acute RV decompensation.

Induction and maintenance

Acquired cardiac lesions

A variety of anaesthetic techniques utilizing both intravenous and inhalation agents have been shown to result in comparable outcomes after coronary revascularization surgery.[72–74] The use of high doses of potent narcotics remains a popular method of inducing and maintaining anaesthesia for cardiac surgical procedures and will be discussed in detail here. Fentanyl and sufentanil generally provide stable haemodynamic performance and a favourable afterload which are essential in preventing imbalance in myocardial oxygen supply and demand. These agents have no effects on myocardial contractility and cause a reduction in peripheral vascular resistance only through a diminution in central sympathetic tone.[75–78] Generally 50–75 µg/kg or fentanyl of 10–15 µg/kg of sufentanil are used as the primary anaesthetic for induction and maintenance prior to cardiopulmonary bypass. A benodiazepine should be included as a part of the premedication or be administered intraoperatively to avoid awareness.

The interaction of muscle relaxants, premedicants and antianginal drugs influences the haemodynamic stability of a high dose narcotic anaesthetic.[79] In patients who are clinically well β-blocked, sufentanil–pancuronium or fentanyl–pancuronium provides stable induction haemodynamic performance while in poorly β-blocked patients sufentanil–vecuronium or fentanyl–vecuronium provides similar haemodynamic stability.[80,81] Severe bradycardia has been reported in patients who are clinically well β-blocked and who subsequently receive vecuronium in combination with high dose fentanyl[82] or sufentanil.[83] Patients taking β-blockers are protected against pancuronium-induced tachycardia while those taking a calcium channel blocker are not.[84] Blood pressure following high dose fentanyl is lower with lorazepam premedication while heart rate is higher with hyoscine premedication.[85] Hyoscine as part of the premedication augments the vagolytic actions of pancuronium when administered with high dose fentanyl.[82]

When surgical stimulation (skin incision, sternotomy, sternal spreading and aortic manipulation) produces hypertension additional doses of sufentanil (1–5 µg/kg) or fentanyl (5–25 µg/kg) have traditionally been used. Recent data do not support this practice. Once a patient has received high doses of sufentanil (10–30 µg/kg) or fentanyl (50–100 µg/kg), administration of an additional agent as an infusion or a bolus will not reliably blunt the haemodynamic response to noxious stimuli.[87] Control of hypertension will require the use of vasodilator agents or adjuvant anaesthetic agents. Sodium nitroprusside is a readily titratable, potent arteriolar dilator and can be used effectively to treat hypertension. An infusion can be started at 0.25 µg/kg per minute and titrated upward. However, because sodium nitroprusside is a potent arteriolar dilator it has potential to induce a coronary steal in the presence of the appropriate anatomy.[88]

Glyceryl trinitrate (nitroglycerin) dilates large coronary vessels but does not dilate arterioles and so is not implicated in the steal phenomena. Glyceryl trinitrate has it greatest systemic dilating capacity on the venous beds and arterial dilatation occurs only at higher doses. Despite this, when used in appropriate doses glyceryl trinitrate and nitroprusside have been shown to be equally effective in treatment of hypertension associated with coronary artery bypass surgery.[89,90] For treatment of hypertension glyceryl trinitrate is started at 0.5 µg/kg per minute and may have to be increased up to 20 µg/kg per minute.

Treatment with vasodilators may produce a reflex tachycardia which requires treatment. Propranolol in incremental doses of 0.5–1.0 mg to a total of 0.1 mg/kg may be used in patients without severe ventricular systolic dysfunction. In patients with a history of bronchospasm or reactive airway disease a β_1-selective blocking agent such as metoprolol is useful. Incremental doses of 2.5–5.0 mg to a total of 0.5 mg/kg can be used.

In some instances the ultrashort-acting β_1-selective blocker esmolol may be useful. Esmolol has an elimination half-life of 9 minutes due to metabolism by red cell esterases and is relatively β_1-selective.[91] Esmolol is started with a bolus of 0.5 mg/kg given over several minutes followed by an infusion of 50 µg/kg and titrated up to 300 µg/kg as necessary. Esmolol is useful in patients with poor ventricular function because if it is not tolerated therapy can be quickly terminated. Esmolol is well tolerated in patients with bronchospastic disease.[92] Furthermore, unlike longer acting β-blockers, esmolol can be used aggressively in the pre-CPB period without fear that it will compromise termination of CPB.[93]

The following adjuvant anaesthetic agents are also useful.

Benzodiazepines

Diazepam, midazolam and lorazepam administered in conjunction with fentanyl or sufentanil have, in causing decreases in peripheral vascular resistance, been implicated in subsequent hypotension.[94–96] Therefore they must be titrated with caution. Diazepam in increments of 2.2–5 mg or lorazepam in increments of 0.5–1 mg are reasonable.

Inhalation anaesthetics

Nitrous oxide in association with high dose opiates may decrease contractility in patients with coronary artery disease but this reduction does not appear to be ischaemia induced.[97,98] Nitrous oxide combined with high dose opiates does elevate pulmonary vascular resistance particularly in the presence of pre-existing pulmonary hypertension.[99] Isoflurane has been safely used as an adjuvant to high dose narcotic anaesthesia in concentrations as high as 2.0 per cent without evidence of steal-induced ischaemia.[100–102] In addition, isoflurane, enflurane and halothane have all been used as primary anaesthetics for cardiac surgery with outcomes comparable to a high dose narcotic technique when haemodynamics are well controlled.[72–74,103]

Congenital cardiac lesions

A variety of induction techniques have been shown to provide haemodynamic stability and improved arterial oxygen saturation in patients with cyanotic heart disease.[104–106]

The haemodynamic stability that has been well documented in adult cardiac surgical patients with high dose narcotic techniques is also seen in paediatric patients. Fentanyl (25–100 µg/kg) and sufentanil (5–20 µg/kg) with pancuronium and oxygen provide stable induction and maintenance for infants and children undergoing correction of all types of congenital heart lesions.[107–110] These agents generally prevent or attenuate increases in systemic and pulmonary vascular resistances in response to intubation, incision and sternotomy.[104,106,107] This attenuation of stress-induced increases in pulmonary vascular resistance (PVR) is of particular importance given the highly reactive nature of the pulmonary vasculature in patients with congenital heart disease. As with adults, supplementation with an inhalation agent or benzodiazepine may be necessary to attenuate hyperdynamic circulatory responses during maximal stimulation. Finally, in both adults with acquired heart disease[111] and children with congenital heart disease[112] there is evidence that post-operative intensive analgesia may reduce morbidity and mortality.

Management of specific congenital heart lesions is beyond the scope of this chapter but is covered in detail elsewhere.[113] However, it is important to point out that manipulation of the ratio of PVR to SVR is an important determinant of shunt magnitude and direction in many congenital heart lesions. Intra- and postoperative control of PVR independent of SVR is therefore extremely important in the management of these patients. The following interventions will alter PVR.

Po_2

Both alveolar hypoxia and arterial hypoxia induce pulmonary vasoconstriction.[114,115]

Pco_2

Hypercapnia increases PVR independent of arterial pH.[116] Hypocapnia, on the other hand, only reduces PVR through production of an alkalosis.[117] Maximal reduction in PVR occurs at an arterial Pco_2 of 2.7 kPa (20 mmHg) and a pH of near 7.60.[118]

pH

Both respiratory and metabolic alkalosis reduce PVR[114,117,118] while both respiratory and metabolic acidosis increase PVR.[114,115]

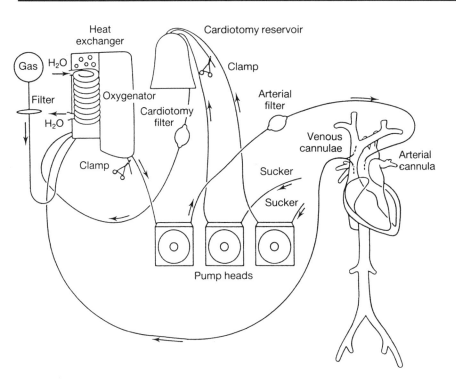

Fig. 55.4 Typical CPB circuit.

Vasodilator agents

At present no oral or intravenously delivered drug is available which acts as a selective pulmonary vasodilator. Drugs such as PGE_1, glyceryl trinitrate, sodium nitroprusside, and tolazoline all induce some degree of systemic vasodilatation as well. There is increasing enthusiasm for nitric oxide which, delivered as an inhalant, is proving to be a potent selective pulmonary vasodilator.[119,120]

Cardiopulmonary bypass

The CPB circuit is intended to isolate the cardiopulmonary system so that optimal surgical exposure can be obtained for operations on the heart and great vessels. In order for this isolation to be effective the CPB circuit must be able to perform the functions of the intact cardiopulmonary system for a finite period. The circuit must at least be capable of adding O_2 and removing CO_2 from blood and of providing adequate perfusion of all organs with this blood. In addition the circuit must be able to fulfil these requirements without doing permanent damage to the cardiopulmonary system, the blood, or any of the patient's end organs. The components of a typical CPB circuit are illustrated in Fig. 55.4. The following is a summary of the most important management issues associated with CPB.

Mean arterial blood pressure

Generally a mean arterial blood pressure less than 70 mmHg is sought to remain within the limits of cerebral autoregulation and to reduce non-coronary collateral blood flow once the aortic cross-clamp is applied. Because the non-coronary collaterals originate from the mediastinum and pericardium the blood they deliver to the heart will be at the same temperature as the patient (25–30°C). While the aortic cross-clamp is in place and the heart arrested with cardioplegia, optimal myocardial temperature is usually between 8 and 15°C. Excessive non-coronary collateral flow will cause the heart to rewarm and compromise myocardial preservation. Excessive non-coronary collateral flow will also compromise surgical exposure by causing bleeding from the incised coronary artery. There are several strategies available to reduce mean arterial pressure and each has its own advantages and disadvantages:

1. Reduction of pump flow is useful as a temporizing manoeuvre but a prolonged reduction may result in systemic and cerebral hypoperfusion.
2. Sodium nitroprusside is a very effective arteriolar dilator and thus very useful in reducing SVR. It is easily titratable, has no negative inotropic properties and has a short half-life. It has, however, been implicated in causing platelet dysfunction during CPB.[121]
3. Glyceryl trinitrate (nitroglycerin) does not cause platelet dysfunction, is easily titratable, has a short half-life and no negative inotropic properties but is very much less effective than nitroprusside. This

decreased effectiveness is due to the fact that glyceryl trinitrate is primarily a venodilator and actively adheres to the plastic components of the CPB circuit.[122]

4. Volatile anaesthetic agents such as halothane, enflurane and isoflurane are very effective, easily titratable arteriolar dilators. To be used the vaporizer must be spliced into a fresh gas flow line of the oxygenator. This gives the anaesthestetist one more vaporizer to worry about maintaining and filling on a piece of equipment which is not his primary responsibility (the pump). Damage has occurred when isoflurane has been accidentally spilled on both bubble[123] and membrane[124] oxygenators. In addition the volatile agents have negative inotropic effects which can compromise separation from CPB. During CPB with a bubble oxygenator isoflurane is completely eliminated from the circulation 10 minutes after it is discontinued.[125] Nevertheless it is prudent to discontinue these agents prior to removal of the aortic cross-clamp to allow adequate time for their elimination before coronary blood flow resumes.

The safe lower limits of mean arterial pressure (MAP) on CPB must be determined with all organ systems in mind. If the aortic cross-clamp is applied within 1–2 minutes following the start of CPB a mean arterial pressure greater than 30 mmHg with a pump flow of 2.0–2.4 l/m² per minute is acceptable for patients without cerebrovascular disease. Cerebral blood flow is autoregulated between mean arterial pressures of 30 and 110 mmHg during CPB in patients without cerebrovascular diseae or pre-existing hypertension.[126] In contrast, in the intact circulation of normal, non-hypertensive patients cerebral blood flow is autoregulated between mean arterial pressures of 50 and 150 mmHg. Unfortunately there are no data on the lower limit of cerebral autoregulation in CPB in patients with coexisting cerebrovascular disease. It may be prudent to maintain slightly higher mean arterial pressures (> 50 mmHg) on CPB in patients with known cerebrovascular disease although there is no evidence to support or refute this recommendation.

If there is a delay in placing the aortic cross-clamp then attention should be directed towards optimal perfusion of the myocardium. A mean arterial pressure of at least 50 mmHg should be maintained for perfusion of the beating, empty heart. If ventricular fibrillation develops and is likely to persist for more than a few minutes than mean arterial pressure must be elevated to at least 60–80 mmHg to assure subendocardial perfusion. Likewise, if ventricular fibrillation is likely to persist, myocardial temperature should be decreased to reduce oxygen consumption and improve perfusion by lowering wall tension. A vent must be incorporated if left ventricular distension occurs. If coronary stenoses or ventricular hypertrophy also exist

MAP must be elevated to 80–100 mmHg and the cross-clamp applied as soon as possible.

If high pump flows are needed to maintain the desired mean arterial pressure, the maximal flow capacity of the oxygenator may be exceeded. If this is likely then SVR can be increased and pump flow decreased to maintain mean arterial pressure. SVR can be increased by transfusion of blood into the venous reservoir to increase viscosity. Generally this strategy is reserved for those instances where the haematocrit is less than 20 per cent. SVR can also be increased by increasing arteriolar tone. The simplest way to accomplish this is by infusing an α-agonist such as phenylephrine into the pump. As a rule as CPB progresses arteriolar tone will increase and the phenylephrine can be stopped.

Acid–base management

Alpha stat pH regulation maintains electrochemical neutrality although pH appears alkalotic after being corrected for temperature. For practical purposes it is easiest to measure blood gases at 37°C, rather than correct for temperature, and keep pH and P_{CO_2} in the range considered normal at 37°C. It has been demonstrated that cerebral blood flow and oxygen consumption are appropriately coupled when alpha stat regulation is utilized.[127–129] pH stat regulation refers to maintaining pH and P_{CO_2} at normal values after correcting for temperature. These values will appear acidotic when not corrected for temperature. For practical purposes pH stat is maintained by adding CO_2 to the ventilating gas during hypothermic CPB to elevate the P_{CO_2} and decrease the pH. In contrast to alpha stat regulation where total CO_2 content is kept constant, pH stat regulation results in an increase in total CO_2 content. It has been demonstrated that hyperperfusion relative to cerebral oxygen demand (uncoupling of cerebral blood flow and metabolism) occurs when pH stat regulation is used with hypothermic CPB.[127,130,131] Despite the theoretical advantages of alpha stat regulation clinical studies to date have not demonstrated any difference in neurological outcome in patients managed with alpha stat versus pH stat.[132]

Low flow and circulatory arrest

Low pump flow rates may be required to reduce bleeding and improve exposure in the repair of complex lesions in neonates and small children. Cerebral blood flow and oxygen delivery are maintained with flow rates as low as 1.0 l/m² per minute[133] and perhaps as low as 0.5 l/m² per minute[134] when moderately hypothermic CPB and alpha stat regulation are used. There appears to be a preferential distribution of blood flow to the cerebral hemispheres when low pump flows are utilized. Despite this, cerebral oxygen at these flows is maintained at the expense of increased cerebral extraction of oxygen and a decreased jugular venous oxygen saturation.[135] Thus low flow rates may be tolerated for limited periods of time in patients without cerebrovascular disease.

Likewise, deep hypothermic circulatory arrest (DHCA) may be used to allow cannula removal and cessation of flow for improved surgical exposure. The neurological outcome following these techniques is in large part determined by the extent of cerebral metabolic reduction induced with hypothermia and anaesthetic agents and the duration of the reduced or arrested flow state. This subject is extensively reviewed elsewhere.[136]

In general, DHCA is limited to 60–70 minutes and should be undertaken only in centres well versed in its use. Systemic hypothermia with temperatures between 15 and 20°C is used. Specific recommendations for the safe conduct of DHCA have been made:[136]

1. surface cooling as well as core cooling with CPB;
2. haemodilution;
3. alpha stat pH regulation;
4. reduction of whole body oxygen consumption with muscle relaxants during arrest;
5. use of EEG to monitor suppression of brain activity with hypothermia and anaesthetic agents;
6. attenuation of stess responses;
7. avoidance of severe hyperglycaemia before and after arrest.

Surface cooling is usually accomplished in the pre-CPB period by placing the patient on a cooling blanket, using unwarmed intravenous fluids and unhumidified gases, reducing room temperature, and packing the head in ice. These techniques allow the patient to be cooled to 30–32°C prior to commencing CPB and core cooling. Ideally, profound systemic hypothermia dramatically reduces cerebral metabolic rate and results in an isoelectric EEG. However, residual cerebral electrical activity may persist despite the use of profound systemic hypothermia. In these instances further suppression of brain activity and cerebral metabolic rate can be accomplished with anaesthetic agents such as thiopentone or isoflurane. The efficacy of such interventions remains to be demonstrated in clinical trials.

Renal blood flow

Normally renal blood flow is autoregulated between renal artery pressures of 80 and 180 mmHg. Despite this, it has been demonstrated that systemic flows as low as 1.6–1.8 l/m^2 per minute and sustained mean arterial pressures as low as 50 mmHg during hypothermic, haemodiluted, non-pulsatile CPB do not result in a deterioration of post-CPB renal function.[137,138] In addition, no correlation between urine output on CPB and post-CPB renal function has been demonstrated.[137,138] In fact, the presence of post-CPB ventricular dysfunction and low cardiac output is a major risk factor in the development of post-CPB renal dysfunction and failure.[137,138]

Oliguria must be considered a harbinger of renal hypoperfusion and ischaemia. Urine output should be assessed every 15 minutes while on CPB. If urine output falls below 0.1 ml/kg per 15 minutes (0.5 ml/kg per hour) action is necessary. Aortic perfusion pressure should be increased to at least 50 mmHg. If this fails to increase urine output or if aortic perfusion pressure is already optimized a diuresis with frusemide (furosemide) 0.5–1.0 mg/kg or mannitol 0.5–1.0 g/kg should be instituted. The volume expansion which occurs following mannitol infusion does not present a problem while on CPB. Induction of a brisk diuresis will result in obligate potassium losses in urine. This can result in hypokalaemia prior to termination of CPB which may require potassium supplementation.

Termination of CPB

Termination of CPB requires very close cooperation between anaesthetist, surgeon, and perfusionist. CPB should not be terminated until each of the three individuals involved is satisfied with the status of the patient. The following basic principles should be remembered:

1. The range of filling pressures expected to provide optimal preload for a particular patient should be discussed prior to termination of CPB. It is best to err on the side of underfilling the heart to avoid ventricular distension. Distension will compromise subendocardial perfusion and elevate wall tension. Visual inspection of the heart will provide information on distension and wall motion but it must be remembered that it is primarily the free wall of the right ventricle which is visible through a median sternotomy. The functional status of the right ventricle is not necessarily the same as that of the left ventricle.

2. If inotropic support is anticipated to be necessary for termination of CPB, it is best not to start infusing the agent into the heart any earlier than 5 minutes prior to termination of CPB. Early infusion of intropic agents increases myocardial oxygen consumption while the heart is being reperfused on CPB. This compromises replenishment of energy stores and serves no useful purpose.

3. CPB should be terminated at a pump flow and SVR that are reasonable for the patient. If the perfusionist is pumping 5 l/min with a mean arterial pressure (MAP) of 50 mmHg then the patient will have to have a cardiac output of geater than 5 l/min to maintain a mean arterial pressure of 70 mmHg once CPB is terminated. If such an output is an unreasonable expectation then efforts must be made to increase SVR to a normal range prior to termination of CPB. This will allow termination of CPB at the desired MAP with a cardiac output that the patient can realistically produce. Recall that on CPB:

$$SVR = (MAP \times 80)/\text{pump flow}$$

with the normal range 1000–1500 dynes.s.cm^{-5}

4. CPB is terminated by gradually clamping the venous line so that venous return to the reservoir is impeded. Simultaneously the perfusionist slows the rate of the arterial pump head. This results in gradual transfusion of the patient via the arterial cannula. The volume of blood in the venous reservoir will slowly drop as the process continues. When adequate preload is obtained as assessed by both filling pressures (PCWP and CVP) and visual inspection of the heart, the transfusion is terminated.

5. As the heart fills with blood ejection should begin. The simplest way to assess this is by observing the arterial waveform. Elevation of filling pressures with evidence of poor ejection is indicative of afterload mismatch. There is either poor systolic function or a level of afterload which is excessive for the inotropic state of the heart. If SVR has been normalized prior to termination of CPB efforts can be directed toward increasing the contractile state of the heart with inotropes.

In some institutions calcium chloride 5–10 mg/kg is used as the first intervention in this setting. Calcium chloride increases the intropic state of the myocardium and induces an increase in SVR which outlasts the inotropic effects. Often this intervention is all that is required to complete termination of CPB. Others believe that calcium chloride administration is contraindicated at this time because of the compromised calcium homoeostasis which accompanies the insult of aortic cross-clamping. Administration of calcium may exacerbate ischaemic damage by causing an accumulation of intracellular calcium. β–Adrenergic agents, on the other hand, increase intracellular calcium but also promote its re-uptake into the sarcoplasmic reticulum and may be more appropriate in this setting.

6. Before therapy for a low arterial pressure is initiated the possibility that a discrepancy between the intra-arterial radial arterial pressure and the central aortic or femoral artery pressure exists should be investigated. This can be determined by simultaneous observation of the central aortic and radial artery pressures and will prevent unnecessary therapeutic interventions. It has been demonstrated that following CPB intra-arterial radial arterial systolic pressure may underestimate central aortic or femoral artery systolic pressure by 10–30 mmHg in a significant number of both children[139] and adults.[140–143] This discrepancy develops shortly after starting CPB[143] and may persist for up to 90 minutes after the termination of CPB.[140,142] Should such a discrepancy exist while the chest is open central aortic blood pressure may be monitored. Once the chest is closed it may be necessary to place a femoral arterial line for continued monitoring.

7. Should the patient separate from CPB with high filling pressures, ventricular distension, low cardiac index (< 2 l/m^2 per minute) and low MAP (< 40 mmHg) prolonged efforts to correct the haemodynamics off CPB should not be made. Reinstitution of CPB will prevent the downward spiral induced by subendocardial ischaemia and elevated wall tension. In addition it will allow reperfusion of the heart during a low energy consumption state. Once CPB is reinstituted several questions must be addressed:

- Was the ventricular failure biventricular or was only one ventricle involved?
- Was there pulmonary artery hypertension and RV failure?
- Was there RV or LV ischaemia?
- Is the prosthetic valve working properly?
- Is all the monitoring equipment functioning properly and giving accurate information?
- Are all the inotropes and vasoactive agents actually being infused?
- Are there unsuspected lesions, such as previously unsuspected mitral regurgitation?
- Is the repair of the congenital lesion complete and successful? Is there a residual defect which requires repair?
- Is the heart rate optimized?
- Is the appropriate type of pacing being done and is the pacemaker capturing the desired chambers?

Having addressed these questions, therapy directed toward specific problems can be instituted and another attempt to terminate CPB made.

Post-CPB management

Following termination of CPB it is necessary to stabilize the patient and to reverse heparinization to allow decannulation and chest closure.

Protamine administration

Protamine is a polyvalent cation derived from salmon sperm which is currently used to neutralize systemic heparinization. Protamine is normally given after stable cardiovascular function is maintained after termination of CPB. It should not be administered until the likelihood that having to reinstitute CPB is small. Once protamine neutralization of heparin begins, the cardiotomy suction should not be used and removal of the arterial and venous cannulae should proceed. This prevents contamination of the heparinized CPB circuit with protamine, should prompt reinstitution of CPB be necessary, and prevents thrombus formation on the cannulae.

There are several approaches to the neutralization of heparin with protamine, all with reportedly good clinical results.[144–147] The activated clotting time (ACT) should be checked following administration of the selected protamine dose. The goal is to return the ACT to its preheparin baseline. There are several adverse reactions associated with protamine administration. These protamine reactions are well documented in adults but it must be emphasized that they occur in paediatric patients as well.[148] The reactions associated with protamine administration are as follows.

Pulmonary hypertension

Pulmonary hypertension following protamine reversal of heparin may be so profound as to result in RV failure and circulatory collapse.[149,150] This reaction appears to be idiosyncratic and is associated with high levels of thromboxanes and C5a anaphylatoxins which result in bronchoconstriction and pulmonary vasoconstriction in susceptible patients following protamine administration.[151] To date there is no good evidence to suggest that patients with pre-existing pulmonary hypertension are at greater risk for this reaction.[152] Fortunately this reaction is rare, occurring in less than 3 per cent of cases.[151,152]

Systemic hypotension

This is a much more common reaction to protamine administration.[153,154] The decrease in systemic blood pressure is due to a decrease in SVR. Pulmonary histamine release has been implicated in causing this SVR reduction.[154,155] Systemic hypotension is more common in patients with poor ventricular function secondary to their inability to compensate for the decrease in SRV.[154] The incidence of this reaction is related to the speed of infusion.

Regardless of the infusion site (right atrium, left atrium, aorta, peripheral vein) this reaction can be avoided or at least decreased in severity when a slow infusion (> 3 minutes) of protamine is used.[153,156]

Protamine-induced platelet aggregation and release of vasoactive substances may also produce systemic hypotension, particularly when platelet concentrates are infused in close temporal proximity to protamine.[157]

True allergic (anaphylactic and anaphylactoid) reactions

An increased incidence of protamine reactions has been reported to occur in patients taking NPH insulin. These patients are presumed to be presensitized to protamine from the insulin preparation. An incidence of reactions as high as 27 per cent has been reported[158] but recent studies suggest an incidence of less than 2 per cent.[159,160] An increased incidence of reactions has also been suggested in patients with fish allergies[161] as opposed to those with shellfish allergies. Vasectomized men in theory have an increased risk of adverse reactions due to the presence of antisperm antibodies.[162] When allergic reactions occur they may present with any or all of the following: hypotension, flushing, bronchospasm, or pulmonary oedema secondary to capillary leak.

Rarely heparin rebound may occur following protamine administration. This phenomenon has received much attention and is defined as a reprolongation of the ACT which occurs 1–8 hours following adequate heparin neutralization and normalization of the ACT with protamine. Several aetiologies have been suggested. Heparin may be taken up by cells of the reticuloendothelial system where it is inaccessible to protamine. As heparin levels fall due to binding with protamine, this sequestered heparin leaves the reticuloendothelial cells to exert an anticoagulant effect. Concurrent protamine metabolism may also be a factor.[163] Heparin rebound may also be due to enhancement of residual heparin by infusion of blood components such as fresh frozen plasma which elevate antithrombin 3 levels[164] or to an anticoagulant activity associated with large doses of protamine.[165] Finally, the rate of protamine administration seems to be important in heparin reversal. A comparison of slow administration of protamine (30-minute infusion) and rapid administration (5-minute bolus) showed no difference in postoperative blood loss between the two groups but longer coagulation times 2 hours after protamine administration in the bolus group.[166]

Assessment of coagulation status

Bleeding or oozing which continues after appropriate protamine reversal of heparin must be investigated. The most likely causes follow.

Surgical bleeding

The surgical field must be carefully assessed for bleeding sites. This inspection must include the cut sternal edges, the harvest site of the mammary artery, and the back of the heart.

Platelet function defects

This is the most common cause of a bleeding problem following CPB after heparin is reversed and surgical bleeding is controlled.[167–169] Transient defects in platelet plug formation and aggregation are seen in all patients put on CPB.[167–169] Generally platelet function returns to near normal status 2–5 hours following CPB.[169] These defects are exacerbated and prolonged by drugs which inhibit platelet function. Aspirin ingestion inhibits platelet aggregation and thromboxane A_2 production for 5 days after ingestion while non-steroidal anti-inflammatory drugs (NSAIDs) have similar effects only as long as significant blood levels are maintained. Preoperative aspirin ingestion has been shown to increase blood loss after CPB by exacerbating CPB-induced platelet dysfunction.[170]

Platelet functional defects tend to compound haemostasis problems from all other causes as well. Whether CPB-induced platelet dysfunction results in clinical problems with haemostasis depends on the extent of the defects, the patient's pre-existing platelet function, and the presence or absence of additional coagulation defects.

Desmopressin acetate (DDAVP) has not been consistently effective in reducing intraoperative and total blood loss in cardiac surgical patients following CPB.[171–173] Recent data demonstrate that thromboelastography can be used to identify patients who will benefit from DDAVP following CPB.[174] Von Willebrand's factor is known to mediate platelet aggregation and adherence on thrombogenic surfaces. The increased levels of von Willebrand's factor which result from DDAVP infusion may be responsible for improving platelet function post-CPB but this has not been proved.[171]

Thrombocytopenia

Platelet counts do decrease on and after CPB but they generally do not fall below 100 000.[167–169] In general the fall in platelet counts on CPB is slightly greater than that expected from haemodilution alone. Because platelet function defects are so prevalent, thrombocytopenia, when it does exist, is an important cause of altered haemostasis.

Dilution of clotting factors

There is a dilutional reduction in the levels of factor II, factor V, factor VII, factor X, and factor XI for up to 48 hours following CPB.[167,169] Levels of factor VIII:C have been shown to decrease[167] and increase[169] during and following CPB. In general, factor levels rarely fall below the 30 per cent level which compromises haemostasis.[167,169] Likewise, the fibrinogen level is reduced by dilution during and following CPB but rarely falls below the critical 100 ml/dl level.[167,169] It is not surprising therefore that the prophylactic administration of fresh frozen plasma has not been shown to reduce bleeding after routine CPB procedures in adult patients without pre-existing coagulopathies.[175] Obviously if pre-existing deficiencies in factor levels or activities are present, CPB-induced reductions in these may become clinically significant and factor replacement may be required.

Infants and small children are at risk for significant dilution of coagulation factors when the prime volume is large relative to their blood volume. Dilution of coagulation factors is particularly likely to occur when the patient is polycythaemic (reduced plasma volume) and when large volumes of asanguinous or packed red cell primes are used in place of whole blood primes. Children whose transfusion requirements in the immediate post-CPB period are met with whole blood less than 48 hours old have total transfusion requirements 85 per cent less than those treated with component therapy.[176]

Primary fibrinolysis

It is generally agreed that enhanced fibrinolytic activity occurs during CPB[167,168,174] and that primary fibrinolysis is rarely the sole cause of post-CPB deficits in haemostasis.[167,168,174] Theoretically antifibrinolytic agents may have some utility in reducing bleeding following CPB. Recently the use of ε-aminocaproic acid (EACA) has been shown to result in a small but significant reduction in post-CPB bleeding in adults.[177] More impressive, however, are recent results obtained with aprotinin used in high doses.[178,179]

REFERENCES

1. Kuntschen FR, Galletti PM, Hahn C. Glucose–insulin interactions during cardiopulmonary bypass. Hypothermia versus normothermia. *Journal of Thoracic and Cardiovascular Surgery* 1985; **91**: 451–9.

2. Kuntschen FR, Galletti PM, Hahn C, Arnulf JJ, Isetta C, Dor V. Alterations of insulin and glucose metabolism during cardiopulmonary bypass under normothermia. *Journal of Thoracic and Cardiovascular Surgery* 1985; **89**: 97–106.

3. Elliott MJ, Gill GV, Home PH, Noy GA, Holden MP, Alberti GMM. A comparison of two regimens for the management of diabetes during open heart surgery. *Anesthesiology* 1984; **60**: 364–8.

4. McKnight CK, Elliott M, Pearson DT, Alberti KGMM, Holden

MP. Continuous monitoring of blood glucose concentration during open-heart surgery. *British Journal of Anaesthesia* 1985; **57**: 595–601.

5. Murkin JM. Anesthesia and hypothyroidism: a review of thyroxin physiology, pharmacology, and anesthetic implications. *Anesthesia and Analgesia* 1982; **61**: 371–83.

6. Becker C. Hypothyroidism and atherosclerotic heart disease: pathogenesis, medical management, and the role of coronary artery bypass surgery. *Endocrinology Review* 1985; **6**: 432–40.

7. Ladenson PW, Levin AA, Ridgway EC, Daniels GH. Complications of surgery in hypothyroid patients. *American Journal of Medicine* 1984; **77**: 261–6.

8. Gyermek L, Henderson G. Low ventilatory and anesthetic drug requirements during myocardial revascularization in a hypothyroid patient. *Journal of Cardiothoracic and Vascular Anesthesia* 1988; **2**: 70–3.

9. Hay ID, Duick DS, Vlietstra RE, Maloney JD, Pluth JR. Thyroxin therapy in hypothyroid patients undergoing coronary revascularization. A retrospective analysis. *Annals of Internal Medicine* 1981; **95**: 456–7.

10. Paine TD, Rogers WJ, Baxley WA, Russell RO. Coronary arterial surgery in patients with incapacitating angina pectoris and myxedema. *American Journal of Cardiology* 1977; **40**: 226–31.

11. Myerowitz PD, Kamienski RW, Swanson DK, *et al.* Diagnosis and management of the hypothyroid patient with chest pain. *Journal of Thoracic and Cardiovascular Surgery* 1983; **86**: 57–60.

12. Finlayson DC, Kaplan JA. Myxoedema and open heart surgery: anaesthesia and intensive care unit experience. *Canadian Anaesthetists' Society Journal* 1982; **29**: 543–9.

13. Rabinovitch M, Haworth SG, Castaneda AR, Nadas AS, Reid LM. Lung biopsy in congenital heart disease: a morphometric approach to pulmonary vascular disease. *Circulation* 1978; **58**: 1107–22.

14. Hoffman JIE, Rudolph AM, Heymann MA. Pulmonary vascular disease with congenital heart lesions: pathologic features and causes. *Circulation* 1981; **64**: 873–7.

15. Hordof AJ, Mellins RB, Gersony WM, Steeg CN. Reversibility of chronic obstructive lung disease in infants following repair of ventricular septal defect. *Journal of Pediatrics* 1977; **90**: 187–91.

16. Motoyama EK, Tanaka T, Fricker EJ. Peripheral airway obstruction in children with congenital heart disease and pulmonary hypertension. *American Review of Respiratory Disease* 1986; **133**: A10.

17. Yamaki S, Horiuchi T, Sekino Y. Quantitative analysis of pulmonary vascular disease in simple cardiac anomalies with the Down syndrome. *American Journal of Cardiology* 1983; **51**: 1502–6.

18. Morray JP, MacGillivray R, Duker G. Increased perioperative risk following repair of congenital heart disease in Down's syndrome. *Anesthesiology* 1986; **65**: 221–4.

19. Yamaki S, Horiuchi T, Takahashi T. Pulmonary changes in congenital heart disease with Down's syndrome: their significance as a cause of postoperative respiratory failure. *Thorax* 1985; **40**: 380–6.

20. Lindahl SGE, Olsson A-K. Congenital heart malformations and ventilatory efficiency in children. Effects of lung perfusion during halothane anaesthesia and spontaneous breathing. *British Journal of Anaesthesia* 1987; **59**: 410–18.

21. Warner MA, Tinker JH, Frye RL, *et al.* Risk of cardiac operations in patients with concomitant pulmonary dysfunction. *Anesthesiology* 1982; **57**: A57.

22. Klein LW. Cigarette smoking, atherosclerosis and the coronary hemodynamic responses: a unifying hypothesis. *Journal of the American College of Cardiology* 1984; **4**: 972–4.

23. Nicod P, Rehr P, Winniford MD. Acute systemic and coronary hemodynamic and serologic responses to cigarette smoking in long term smokers with atherosclerotic coronary artery disease. *Journal of the American College of Cardiology* 1984; **4**: 964–71.

24. Conti CR, Mehta JL. Acute myocardial ischemia: role of atherosclerosis, thrombosis, platelet activation, coronary vasospasm, and altered arachidonic acid metabolism. *Circulation* 1987; **75** (suppl 5): 84–95.

25. Pearce AC, Jones RM. Smoking and anesthesia: preoperative abstinence and perioperative morbidity. *Anesthesiology* 1984; **61**: 576–84.

26. Anderson EW, Andelman RJ, Strauch JM, Fortuin NJ, Knelson JH. Effect of low carbon monoxide exposure on onset and duration of angina pectoris. *Annals of Internal Medicine* 1973; **79**: 46–50.

27. Warner MA, Divertie MB, Tinker JH. Preoperative cessation smoking and pulmonary complications in coronary artery bypass patients. *Anesthesiology* 1984; **60**: 380–3.

28. Vander Woude JC, Milam JD, Walker WE, *et al.* Cardiovascular surgery in patients with congenital plasma coagulopathies. *Annals of Thoracic Surgery* 1988; **46**: 283–8.

29. Milam JD, Austin SF, Nihill MR, *et al.* Use of sufficient hemodilution to prevent coagulopathies following surgical correction of cyanotic heart disease. *Journal of Thoracic and Cardiovascular Surgery* 1985; **89**: 623–9.

30. Mauer HM. Hematologic effects of cardiac disease. *Pediatric Clinics of North America* 1972; **19**: 1083–93.

31. Ramsey G, Arvan DA, Stewart S, Blumberg N. Do preoperative laboratory tests predict blood transfusion needs in cardiac surgery? *Journal of Thoracic and Cardiovascular Surgery* 1983; **85**: 564–9.

32. Komp DM, Sparrow AW. Polycythemia in cyanotic heart disease – a study of altered coagulation. *Journal of Pediatrics* 1970; **76**: 231–6.

33. Colon-Otero G, Gilchrist GS, Holcomb GR, *et al.* Preoperative evaluation of hemostasis in patients with congenital heart disease. *Mayo Clinic Proceedings* 1987; **62**: 379–85.

34. Gill JC, Wilson AD, Endres-Brooks J, Montgomery RR. Loss of the largest von Willebrand factor multimers from the plasma of patients with congenital cardiac defects. *Blood* 1986; **67**: 758–61.

35. Mannucci PM, Remuzzi G, Pusineri F. Deamino-8-arginine vasopressin shortens the bleeding time in uremia. *New England Journal of Medicine* 1983; **308**: 8–12.

36. Mason RA, Arbeit LA, Giron F. Renal dysfunction after arteriography. *Journal of the American Medical Association* 1985; **253**: 1001–4.

37. D'Elia JA, Gleason RE, Alday M, *et al.* Nephrotoxicity from angiographic contrast material. A prospective study. *American Journal of Medicine* 1982; **72**: 719–25.

38. Hirsch IA, Tomlinson DL, Slogoff S, Keats AS. The overstated risk of preoperative hypokalemia. *Anesthesia and Analgesia* 1988; **67**: 131–6.

39. McLaurin LP, Rolet EL, Grossman W. Imperial left ventricular relaxation during pacing induced ischemia. *American Journal of Cardiology* 1973; **32**: 751–7.

40. Barry WH, Brooker JZ, Alderman EL, Harrison DC. Changes in diastolic stiffness and tone of the left ventricle during angina pectoris. *Circulation* 1974; **49**: 255–63.

41. Mann JT, Brodie RR, Grossman W, McLaurin LP. Effect of angina on left ventricular pressure-volume relationships. *Circulation* 1977; **55**: 761–6.

42. Grossman W, Mann JT. Evidence for impaired left ventricular relaxation during acute ischemia in man. *European Journal of Cardiology* 1978; **7**: 239–49.

43. Bourdillion PD, Lorell BH, Mirsky I, *et al.* Increased regional myocardial stiffness of the left ventricle during pacing-induced angina in man. *Circulation* 1983; **67**: 316–23.

44. Fifer MA, Bourdillon PD, Lorell BH. Altered left ventricular diastolic properties during pacing-induced angina in patients with aortic stenosis. *Circulation* 1986; **74**: 675–83.

45. Aroesty JM, McKay RG, Heller GV, *et al.* Simultaneous assessment of left ventricular systolic and diastolic dysfunction during pacing-induced ischemia. *Circulation* 1983; **67**: 889–900.

46. Keon WJ, Heggtveit HA, Leduc J. Perioperative myocardial infarction caused by atheroembolism. *Journal of Thoracic and Cardiovascular Surgery* 1982; **84**: 849–55.

47. Grondin CM, Pomar JL, Herbert Y. Re-operation in patients with patent atherosclerotic coronary vein grafts. *Journal of Thoracic and Cardiovascular Surgery* 1984; **87**: 379–85.

48. Kates RA, Zaidan JR, Kaplan JA. Esophageal lead for intraoperative electrocardiographic monitoring. *Anesthesia and Analgesia* 1982; **61**: 781–5.

49. Trager MA, Feinberg BI, Kaplan JA. Right ventricular ischemia diagnosed by an esophageal electrocardiogram and right atrial pressure trace. *Journal of Cardiothoracic and Vascular Anesthesia* 1987; **1**: 123–5.

50. Kaplan JA, Wells PH. Early diagnosis of myocardial ischemia using the pulmonary artery catheter. *Anesthesia and Analgesia* 1981; **60**: 789–93.

51. Haggmark S, Hohner P, Ostman M, *et al.* Comparison of hemodynamic, electrocardiographic, mechanical, and metabolic indicators of intraoperative myocardial ischemia in vascular surgical patients with coronary artery disease. *Anesthesiology* 1989; **70**: 19–25.

52. Kleinman B, Henkin RE, Glisson SN, *et al.* Qualitative evaluation of coronary flow during anesthetic induction using thallium-201 perfusion scans. *Anesthesiology* 1986; **64**: 157–64.

53. van Daele MERM, Sutherland GR, Mitchell MM, *et al.* Do changes in pulmonary capillary wedge pressure adequately reflect myocardial ischemia during anesthesia? A correlative preoperative hemodynamic, electrocardiographic, and transesophageal echocardiographic study. *Circulation* 1990; **81**: 865–71.

54. Leung JM, O'Kelly BF, Mangano DT, *et al.* Relationship of regional wall motion abnormalities to hemodynamic indices of myocardial oxygen supply and demand in patients undergoing CABG surgery. *Anesthesiology* 1990; **73**: 802–14.

55. Wohlgelernter D, Cleman M, Highman HA, *et al.* Regional myocardial dysfunction during coronary angioplasty: evaluation by two-dimensional echocardiography and 12 lead electrocardiography. *Journal of the American College of Cardiology* 1986; **7**: 1245–54.

56. Clements F, de Bruijn NP. Perioperative evaluation of regional wall motion by transesophageal two-dimensional echocardiography. *Anesthesia and Analgesia* 1987; **66**: 249–61.

57. London MJ, Tubau JF, Wong MG, *et al.*, The 'natural history' of segmental wall motion abnormalities in patients undergoing non cardiac surgery. *Anesthesiology* 1990; **73**: 644–55.

58. Schluter M, Hinrichs A, Their W, *et al.* Transesophageal two-dimensional echocardiography: comparison of ultrasonic and anatomic sections. *American Journal of Cardiology* 1984; **53**: 1173–8.

59. Muhiudeen IA, Roberson DA, Silverman NH, *et al.* Intraoperative echocardiography for evaluation of congenital heart defects in infants and children. *Anesthesiology* 1992; **76**: 165–72.

60. Sheikh KH, de Bruijn NP, Rankin JS, *et al.* The utility of transesophageal echocardiography and Doppler color flow imaging in patients undergoing cardiac valve operations. *Journal of the American College of Cardiology* 1990; **15**: 363–72.

61. Sill JC, Nugent M, Moyer TP, *et al.* Influence of propranolol plasma levels on hemodynamics during coronary artery bypass surgery. *Anesthesiology* 1984; **60**: 455–63.

62. Slogoff S, Keats AS. Does chronic treatment with calcium channel entry blocking drugs reduce perioperative myocardial ischemia? *Anesthesiology* 1988; **68**: 676–80.

63. Chung F, Houston PL, Cheng DCH, *et al.* Calcium channel blockade does not offer adequate protection from perioperative myocardial ischemia. *Anesthesiology* 1988; **69**: 343–7.

64. Heikkila H, Jalonen J, Laaksonen V. Metoprolol medication and coronary artery bypass grafting operation. *Acta Anaesthesiologica Scandinavica* 1984; **28**: 677–82.

65. Stanley TH, deLange S, Boxcoe MJ. The influence of chronic preoperative propranolol therapy on cardiovascular dynamics and narcotic requirements during operation in patients with coronary artery disease. *Canadian Anaesthestists' Society Journal* 1982; **29**: 319–24.

66. Miller RR, Olson HG, Amsterdam EA. Propranolol-withdrawal-rebound phenomenon. Exacerbation of coronary events after abrupt cessation of anti-anginal therapy. *New England Journal of Medicine* 1975; **293**: 416–18.

67. Roach GW, Moldenhauser CC, Hug CC, Schultz JD, Curling PE. Hemodynamic responses to fentanyl or diazepam-fentanyl anesthesia in patients on chronic nifedipine therapy. *Anesthesiology* 1984; **61**: A374.

68. Larach DD, Hensley FA, Pae LR, *et al.* A randomized study of diltiazem withdrawal prior to coronary artery bypass surgery. *Anesthesiology* 1985; **63**: A23.

69. Casson WR, Jones RM, Parsons RS. Nifedipine and cardiopulmonary bypass. *Anaesthesia* 1984; **39**: 1197–201.

70. Massagee JT, McIntyre RW, Kates RA, *et al.* Effects of preoperative calcium entry blocker therapy on alpha-

adrenergic responsiveness in patients undergoing coronary revascularization. *Anesthesiology* 1987; **67**: 485–91.

71. Nicolson SC, Betts EK, Jobes DR, *et al*. Comparison of oral and intramuscular preanesthetic medication for pediatric surgery. *Anesthesiology* 1989; **71**: 8–10.

72. Slogoff S, Keats AS. Randomized trial of primary anesthetic agents on outcome of coronary artery bypass operations. *Anesthesiology* 1989; **70**: 179–88.

73. Tuman KJ, McCarthy RJ, Spiess BD, *et al*. Does choice of anesthetic agent significantly affect outcome after coronary artery surgery? *Anesthesiology* 1989; **70**: 189–98.

74. Leung JM, Goehner P, O'Kelly BF, *et al*. Isoflurane anesthesia and myocardial ischemia: comparative risk versus sufentanil anesthesia in patients undergoing coronary artery bypass graft surgery. *Anesthesiology* 1991; **74**: 838–47.

75. Rosow CE, Philbin DM, Keegan CR, Moss J. Hemodynamics and histamine release during induction with sufentanil or fentanyl. *Anesthesiology* 1984; **60**: 489–91.

76. Stanley TH, Webster LR. Anesthesia requirements and cardiovascular effects of fentanyl–oxygen and fentanyl–diazepam–oxygen anesthesia in man. *Anesthesia and Analgesia* 1978; **57**: 411–26.

77. Sebel PS, Bovil JG. Cardiovascular effects of sufentanil anesthesia. *Anesthesia and Analgesia* 1982; **61**: 115–19.

78. Howie MB, McSweeney TD, Lingam Maschke SP. A comparison of fentanyl–O$_2$ and sufentanil–O$_2$ for cardiac anesthesia. *Anesthesia and Analgesia* 1985; **64**: 877–87.

79. Thomson IR, MacAdams CL, Hudson RJ, Rosenbloom M. Drug interactions with sufentanil. Hemodynamic effects of premedication and muscle relaxants *Anesthesiology* 1992; **76**: 922–9.

80. McDonald DH, Zaidan JR. Hemodynamic effect of pancuronium and pancuronium plus metocurine in patients taking propranolol. *Anesthesiology* 1984; **60**: 359–61.

81. Zahl K, Ellison N. Influence of beta-blockers on vecuronium/sufentanil or pancuronium/sufentanil combinations for rapid induction and intubation of cardiac surgical patients. *Journal of Cardiothoracic and Vascular Anesthesia* 1988; **2**: 607–14.

82. Paulissan R, Mahdi M, Joseph N, *et al*. Hemodynamic responses to pancuronium and vecuronium during high dose fentanyl anesthesia for coronary artery bypass grafting. *Anesthesiology* 1986; **65**: A523.

83. Starr NJ, Sethna DH, Estafanous FG. Bradycardia and asystole following rapid administration of sufentanil with vecuronium. *Anesthesiology* 1986; **64**: 521–3.

84. Estafanous FG, Williams G, Sethna D, Starr N. Effects of preoperative Ca channel blockers, beta blockers, and pancuronium or vecuronium on hemodynamics of induction of anesthesia in patients with coronary artery disease receiving sufentanil anesthesia. *Anesthesiology* 1986; **65**: A524.

85. Thomson IR, Bergstrom RG, Rosenbloom M, Meatherall RC. Premedication and high dose fentanyl anesthesia for myocardial revascularization: a comparison of lorazepam versus morphine–scopolamine *Anesthesiology* 1988; **68**: 194–200.

86. Gravlee GP, Ramsey FM, Roy R, *et al*. Rapid administration of a narcotic and neuromuscular blocker: a hemodynamic comparison of fentanyl, sufentanil, pancuronium, and vecuronium. *Anesthesia and Analgesia* 1988; **67**: 39–47.

87. Philbin DM, Roscow CE, Schneider RC, *et al*. Fentanyl and sufentanil anesthesia revisited: how much is enough? *Anesthesiology* 1990; **73**: 5–11.

88. Mann T, Cohn PF, Holman BL, *et al*. Effect of nitroprusside on regional myocardial blood flow in coronary artery disease. Results in 25 patients and comparison with nitroglycerin. *Circulation* 1978; **57**: 732–7.

89. Flaherty JT, Magee PA, Gardner TL, *et al*. Comparison of intravenous nitroglycerin and sodium nitroprusside for treatment of acute hypertension developing after coronary artery bypass surgery. *Circulation* 1982; **65**: 1072–7.

90. Kaplan JA, Jones EL. Vasodilator therapy during coronary artery surgery. Comparison of nitroglycerin and nitroprusside. *Journal of Thoracic and Cardiovascular Surgery* 1977; **77**: 301–9.

91. Newsome LR, Roth JV, Hug CC, Nagle D. Esmolol attenuates hemodynamic responses during fentanyl–pancuronium anesthesia for aortocoronary bypass surgery. *Anesthesia and Analgesia* 1986; **65**: 451–6.

92. Gold MR, Dec GW, Cocca-Spofford D, Thompson BT. Esmolol and ventilatory function in cardiac patients with COPD. *Chest* 1991; **100**: 1215–18.

93. Girard D, Shulman BJ, Thys DM, *et al*. The safety and efficacy of esmolol during myocardial revascularization. *Anesthesiology* 1986; **65**: 157–64.

94. Tomicheck RC, Rosow CE, Philbin DA, *et al*. Diazepam–fentanyl interaction – hemodynamic and hormonal effects in coronary artery surgery. *Anesthesia and Analgesia* 1983; **62**: 881–4.

95. Heikkila H, Jalonen J, Arola M. Midazolam as adjunct to high-dose fentanyl anesthesia for coronary artery bypass grafting operation. *Acta Anaesthesiologica Scandinavica* 1984; **28**: 683–9.

96. Heikkila H, Jalonen J, Laksonen V. Lorazepam and high-dose fentanyl anaesthesia: effects on hemodynamics and oxygen transportation in patients undergoing coronary revascularization. *Acta Anaesthesiologica Scandinavica* 1984; **28**: 357–61.

97. Mitchell MM, Prakash O, Rulf EN, *et al*. Nitrous oxide does not induce myocardial ischemia in patients with ischemic heart disease and poor ventricular function. *Anesthesiology* 1989; **71**: 526–34.

98. Cahalan MK, Praksah O, Rulf ENR, *et al*. Addition of nitrous oxide to fentanyl anesthesia does not induce myocardial ischemia in patients with ischemic heart disease. *Anesthesiology* 1987; **67**: 925–9.

99. Schulte-Sasse U, Hess W, Tarnow J. Pulmonary vascular response to nitrous oxide in patients with normal and high pulmonary vascular resistance. *Anesthesiology* 1982; **57**: 9–13.

100. O'Young J, Mastrocostopoulos G, Hilgenberg A, *et al*. Myocardial circulatory and metabolic effects of isoflurane and sufentanil during coronary artery bypass surgery. *Anesthesiology* 1987; **66**: 653–8.

101. Smith JS, Cahalan MK, Benefeil DJ, *et al*. Intraoperative detection of myocardial ischemia in high risk patients: electrocardiography versus two-dimensional transesophageal echocardiography. *Circulation* 1985; **72**: 1015–21.

102. Smith JS, Cahalan MK, Benefiel DJ, *et al*. Fentanyl versus

fentanyl and isoflurane in patients with impaired left ventricular function. *Anesthesiology* 1985; **63**: A18.

103. Samuelson PN, Reves JG, Kirklin JK, *et al.* Comparison of sufentanil and enflurane–nitrous oxide anesthesia for myocardial revascularization. *Anesthesia and Analgesia* 1986; **65**: 217–26.

104. Laishley RS, Burrows FA, Lerman J, Roy WL. Effect of anesthetic induction regimes on oxygen saturation in cyanotic congenital heart disease. *Anesthesiology* 1986: **65**: 673–7.

105. Greeley WJ, Bushman GA, Davis DP, Reves JG. Comparative effects of halothane, and ketamine on systemic arterial saturation in children with cyanotic heart disease. *Anesthesiology* 1986; **65**: 666–8.

106. Hensley FA, Larach DR, Martin DE, *et al.* The effect of halothane/nitrous oxide/oxygen mask induction on arterial hemoglobin saturation in cyanotic heart disease. *Journal of Cardiothoracic and Vascular Anesthesia* 1987; **1**: 289–96.

107. Hickey PR, Hansen DD, Wessel DL. Pulmonary and systemic responses to fentanyl in infants. *Anesthesia and Analgesia* 1985; **64**: 483–6.

108. Moore RA, Yang SS, McNicholas KW, *et al.* Hemodynamic and anesthetic effects of sufentanil as the sole anesthetic for pediatric cardiac surgery. *Anesthesiology* 1985; **62**: 725–31.

109. Davis PJ, Cook DR, Stiller RL, Davin-Robinson KA. Pharmacodynamics and pharmacokinetics of high-dose sufentanil in infants and children undergoing cardiac surgery. *Anesthesia and Analgesia* 1987; **66**: 203–8.

110. Hickey PR, Hansen DD. Fentanyl- and sufentanil-oxygen-pancuronium anesthesia for cardiac surgery in infants. *Anesthesia and Analgesia* 1984; **63**: 117–24.

111. Mangano DT, Siliciano D, Hollenberg M, *et al.* Postoperative myocardial ischemia. Therapeutic trials using intensive analgesia following surgery. *Anesthesiology* 1992; **76**: 342–53.

112. Anard KJS, Hickey PR. Halothane–morphine compared with high-dose sufentanil for anesthesia and postoperative analgesia in neonatal cardiac surgery. *New England Journal of Medicine* 1992; **326**: 1–9.

113. Strafford MA, DiNardo JA. Anesthesia for congential heart disease. In: DiNardo JA, Schwartz MJ, eds. *Anesthesia for cardiac anesthesia*. Norwalk, CT: Appleton & Lange, 1990: 117–72.

114. Fishman A. Hypoxia on the pulmonary circulation: how and where it acts. *Circulation Research* 1976; **38**: 221–31.

115. Abman SH, Wolfe RR, Accurso FJ. Pulmonary vascular response to oxygen in infants with severe bronchopulmonary dysplasia. *Pediatrics* 1985; **75**: 80–4.

116. Malik AB, Kidd BSL. Independent effects of changes in H$^+$ and CO$_2$ concentration on hypoxic pulmonary vasoconstriction. *Journal of Applied Physiology* 1973; **34**: 318–23.

117. Schreiber MD, Heyman MA, Soifer SJ. Increased arterial pH, not decreased P_{CO_2}, attenuates hypoxia-induced pulmonary vasoconstriction in newborn lambs. *Pediatric Research* 1986; **20**: 113–17.

118. Drummond WH, Gregory GA, Heyman MA. The independent effects of hyperventilation, tolazoline and dopamine on infants with persistent pulmonary hypertension. *Journal of Pediatrics* 1981; **98**: 603–11.

119. Frostell C, Fratacci MD, Wain JC, *et al.* Inhaled nitric oxide: a selective pulmonary vasodilator reversing hypoxic pulmonary vasoconstriction. *Circulation* 1991; **83**: 2038–47.

120. Girard C, Lehot JJ, Pannetier JC, *et al.* Inhaled nitric oxide after mitral valve replacement in patients with chronic pulmonary hypertension. *Anesthesiology* 1992; **77**: 880–3.

121. Hines R, Hannon C, Barach PG, *et al.* Sodium nitroprusside: does it cause platelet dysfunction in humans? *Anesthesiology* 1985; **63**: A4.

122. Dasta JF, Jacobi J, Wu LS. Loss of nitroglycerin to cardiopulmonary bypass circuit. *Circuit Care Medicine* 198; **11**: 50–2.

123. Maltry DE, Eggers GWN. Isoflurane-induced failure of the Bentley-10 oxygenator. *Anesthesiology* 1987; **66**: 100–1.

124. Cooper S, Levin R. Near catastrophic oxygenator failure. *Anesthesiology* 1987; **66**: 101–2.

125. Price SL, Brown DL, Carpenter RL, *et al.* Isoflurane elimination via a bubble oxygenator during extracorporeal circulation. *Journal of Cardiothoracic and Vascular Anesthesia* 1988; **2**: 41–4.

126. Govier AV, Reves JG, McKay RD, *et al.* Factors and their influence on regional cerebral blood flow during nonpulsatile cardiopulmonary bypass. *Annals of Thoracic Surgery* 1984; **38**: 592–600.

127. Murkin JM, Farrar JK, Tweed WA, *et al.* Cerebral autoregulation and flow/metabolism coupling during cardiopulmonary bypass: the influence of $P_{a CO_2}$. *Anesthesia and Analgesia* 1987; **66**: 825–32.

128. Prough DS, Stump DA, Roy RC, *et al.* Response of cerebral blood flow to changes in carbon dioxide tension during hypothermic cardiopulmonary bypass. *Anethesiology* 1986; **64**: 576–81.

129. Johnsson P, Messeter K, Ryding E, *et al.* Cerebral blood flow and autoregulation during hypothermic cardiopulmonary bypass. *Annals of Thoracic Surgery* 1987; **43**: 386–90.

130. Murkin JM, Farrar JK, Tweed WA, *et al.* Relationship between cerebral blood flow and O$_2$ consumption during high-dose narcotic anesthesia for cardiac surgery. *Anesthesiology* 1985; **63**: A44.

131. Lundar T, Lindegaard KF, Froysaker T, *et al.* Dissociation between cerebral autoregulation and carbon dioxide reactivity during nonpulsatile cardiopulmonary bypass. *Annals of Thoracic Surgery* 1985; **40**: 582–7.

132. Basheain G, Townes BD, Nessly ML, *et al.* A randomized study of carbon dioxide management during hypothermic cardiopulmonary bypass. *Anesthesiology* 1990; **72**: 7–15.

133. Murkin JM, Farrar JK, Cleland A, *et al.* The influence of perfusion flow rates on cerebral blood flow and oxygen consumption during hypothermic cardiopulmonary bypass. *Anesthesiology* 1987; **67**: A9.

134. Fox LS, Blackstone EH, Kirklin JW, *et al.* Relationship of brain blood flow and oxygen consumption to perfusion flow rate during profoundly hypothermic cardiopulmonary bypass. *Journal of Thoracic and Cardiovascular Surgery* 1984; **87**: 658–64.

135. Rebeyka IM, Coles JG, Wilson GJ, *et al.* The effect of low-flow cardiopulmonary bypass on cerebral function: an experimental and clinical study. *Annals of Thoracic Surgery* 1987; **43**: 391–6.

136. Hickey PR, Andersen NP. Deep hypothermic circulatory

arrest: a review of pathophysiology and clinical experience as a basis for anesthetic management. *Journal of Cardiothoracic and Vascular Anesthesia* 1987; **1**: 137–55.

137. Hiberman M, Derby GC, Spencer RJ. Sequential pathophysiological changes characterizing the progression from renal dysfunction to acute renal failure following cardiac operations. *Journal of Thoracic and Cardiovascular Surgery* 1980; **79**: 838–44.

138. Hiberman M, Myers BD, Carrie BJ. Acute renal failure following cardiac surgery. *Journal of Thoracic and Cardiovascular Surgery* 1979; **77**: 880–8.

139. Gallagher JD, Moore RA, McNicholas KW, Jose AB. Comparison of radial and femoral arterial blood pressures in children after cardiopulmonary bypass. *Journal of Clinical Monitoring* 1985; **1**: 168–71.

140. Stern DH, Gerson JI, Allen FB, Parker FB. Can we trust the direct radial artery pressure immediately following cardiopulmonary bypass? *Anesthesiology* 1985; **62**: 557–61.

141. Mohr R, Lavee J, Goor DA. Inaccuracy of radial arterial pressure measurement after cardiac operations. *Journal of Thoracic and Cardiovascular Surgery* 1987; **94**: 286–90.

142. Pauca AL, Meredith JW. Possibility of A-V shunting upon cardiopulmonary bypass discontinuation. *Anesthesiology* 1987; **67**: 91–4.

143. Rich GF, Lubanski RE, McLoughlin TM. Differences between aortic and radial artery pressure associated with cardiopulmonary bypass. *Anesthesiology* 1992; **77**: 63–6.

144. Culliford AT, Gitel SN, Starr N, et al. Lack of correlation between activated clotting time and plasma heparin during cardiopulmonary bypass. *Annals of Surgery* 1981; **193**: 105–11.

145. Esposito RA, Culliford AT, Colvin SB, et al. The role of the activated clotting time in heparin administration and neutralization for cardiopulmonary bypass. *Journal of Thoracic and Cardiovascular Surgery* 1983; **85**: 174–85.

146. Umlas J, Taft RH, Gauvin G, Swierk P. Anticoagulation monitoring and neutralization during open heart surgery – a rapid method for measuring heparin and calculating safe reduced protamine doses. *Anesthesia and Analgesia* 1983; **62**: 1095–9.

147. LaDuca F, Mills D, Thompson S, Larson K. Neutralization of heparin using a protamine titration assay and the activated clotting time. *Journal of Extra-Corporeal Technology* 1987; **19**: 358–64.

148. Ullman DA, Bloom BS, Danker PR, et al. Protamine-induced hypotension in a two-year-old child. *Journal of Cardiothoracic and Vascular Anesthesia* 1988; **2**: 497–9.

149. Lowenstein E, Johnson WE, Lappas DG, et al. Catastrophic pulmonary vasoconstriction associated with protamine reversal of heparin. *Anesthesiology* 1983; **59**: 470–3.

150. Kronenfeld MA, Garguilo R, Weinberg P, et al. Left atrial injection of protamine does not reliably prevent pulmonary hypertension. *Anesthesiology* 1987; **67**: 578–80.

151. Morel DR, Zapol WM, Thomas SJ, et al. C5a and thromboxane generation associated with pulmonary vaso- and broncho-constriction during protamine reversal of heparin. *Anesthesiology* 1987; **66**: 597–604.

152. Konstadt SN, Thys DM, Kong D, et al. Absence of prostaglandin changes associated with protamine administration in patients with pulmonary hypertension. *Journal of Cardiothoracic and Vascular Anesthesia* 1987; **1**: 388–91.

153. Horrow JC. Protamine: a review of its toxicity. *Anesthesia and Analgesia* 1985; **64**: 348–61.

154. Michaels IALM, Barash PG. Hemodynamic changes associated with protamine administration. *Anesthesia and Analgesia* 1983; **62**: 831–5.

155. Casthely PA, Goodman K, Fyman PN, et al. Hemodynamic changes after the administration of protamine. *Anesthesia and Analgesia* 1986; **65**: 78–80.

156. Horrow JC. Protamine allergy. *Journal of Cardiothoracic and Vascular Anesthesia* 1988; **2**: 225–42.

157. Bjoraker DG, Ketcham TR. *In vivo* platelet response to clinical protamine sulfate infusion. *Anesthesiology* 1982; **57**: A7.

158. Stewart WJ, McSweeney SM, Kellett MA. Increased risk of severe protamine reactions in NPH insulin-dependent diabetics undergoing cardiac catheterization. *Circulation* 1984; **70**: 788–92.

159. Levy JH, Zaidan JR, Faraj B. Prospective evaluation of risk of protamine reactions in patients with NPH insulin-dependent diabetes. *Anesthesia and Analgesia* 1986; **65**: 739–49.

160. Levy JH, Schwieger IA, Zaidan JR, et al. Evaluation of patients at risk for protamine reactions. *Journal of Thoracic and Cardiovascular Surgery* 1988; **98**: 200–4.

161. Knape JTA, Schuller JT, DeHaan P. An anaphylactic reaction to protamine in a patient allergic to fish. *Anesthesiology* 1981; **55**: 324–5.

162. Watson RA, Ansbacher R, Barry M, et al. Allergic reaction to protamine: a late complication of elective vasectomy? *Urology* 1983; **22**: 493–6.

163. Fabian I, Aronson M. Mechanism of heparin rebound. In vitro study. *Thrombosis Research* 1980; **18**: 535–42.

164. Soloway HB, Christansen TW. Heparin anticoagulation during cardiopulmonary bypass in an antithrombin 3 deficient patient: implications relevant to the etiology of heparin rebound. *American Journal of Clinical Pathology* 1980; **73**: 723–5.

165. Guffin AV, Dunbar RW, Kaplan JA, Bland JW. Successful use of a reduced dose of protamine following cardiopulmonary bypass surgery. *Anesthesia and Analgesia* 1976; **55**: 110–15.

166. Zaidan JR, Johnson S, Brynes R, et al. Rate of protamine administration: its effect on heparin reversal and thrombin recovery after coronary artery surgery. *Anesthesia and Analgesia* 1986; **65**: 377–80.

167. Mammen EF, Koets MH, Washington BC, et al. Hemostasis changes during cardiopulmonary bypass surgery. *Seminars in Thrombosis and Hemostasis* 1985; **11**: 281–92.

168. Bick RL. Hemostasis defects associated with cardiac surgery, prosthetic devices, and other extracorporeal devices. *Seminars in Thrombosis and Hemostasis* 1985; **11**: 249–80.

169. Harker LA, Malpass TW, Branson HE, et al. Mechanism of abnormal bleeding in patients undergoing cardiopulmonary bypass: acquired transient platelet dysfunction associated with selective alpha-granule release. *Blood* 1980; **56**: 824–34.

170. Ferraris VA, Ferraris SP, Lough FC, Berry WR. Preoperative aspirin ingestion increases operative blood loss after

coronary artery bypass grafting. *Annals of Thoracic Surgery* 1988; **45**: 71–4.

171. Salzman EW, Weinstein MJ, Weintraub RM, *et al.* Treatment with desmopressin acetate to reduce blood loss after cardiac surgery. A double-blind randomized trial. *New England Journal of Medicine* 1986; **314**: 1402–6.

172. Rocha E, Llorens R, Paramo JA, *et al.* Does desmopressin acetate reduce blood loss after surgery in patients on cardiopulmonary bypass? *Circulation* 1988; **77**: 1319–23.

173. Andersson TLG, Solem JO, Tengborn L, Vinge E. Effects of desmopressin acetate on platelet aggregation, von Willebrand factor, and blood loss after cardiac surgery with extracorporeal circulation. *Circulation* 1990; **81**: 872–8.

174. Mongan PD, Hosking MP. The role of desmopressin acetate in patients undergoing coronary artery bypass surgery. A controlled clinical trial with thromboelastographic risk stratification. *Anesthesiology* 1992; **77**: 38–46.

175. Roy RC, Stafford MA, Hudspeth AS, Meredith JW. Failure of fresh frozen plasma to reduce blood loss and blood

replacement after cardiopulmonary bypass. *Anesthesia and Analgesia* 1987; **67**: S190.

176. Manno CS, Hedberg KW, Kim HC, *et al.* Comparison of the hemostatic effects of fresh whole blood, stored whole blood and components after open heart surgery in children. *Blood* 1991; **77**: 930–6.

177. Vander Salm TJ, Ansell JE, Okike ON, *et al.* The role of epsilon-aminocaproic acid in reducing bleeding after cardiac operations: a double-blind study. *Journal of Thoracic and Cardiovascular Surgery* 1988; **95**: 538–40.

178. Dietrich W, Spannagl M, Jochum M, *et al.* Influence of high-dose aprotinin treatment on blood loss and coagulation patterns in patients undergoing myocardial revascularization. *Anesthesiology* 1990; **73**: 1119–26.

179. Bidstrup BP, Royston D, Sapsford RN, Taylor KM. Reduction in blood loss after cardiopulmonary bypass using high dose aprotinin (Trasylol): studies in patients undergoing aorto-coronary bypass surgery, reoperations and valve replacement for endocarditis. *Journal of Thoracic and Cardiovascular Surgery* 1989; **97**: 364–72.

Chapter 56

Anaesthesia for Vascular Surgery

Andrew J. Mortimer

Elective vascular surgical procedures are mostly performed for the relief of ischaemic symptoms resulting from atherosclerotic disease of the internal carotid, aortoiliac and femoropopliteal arteries. Emergency surgery is concerned with attempted treatment of the ruptured aorta and thromboembolic disease of the mesenteric and femoral arteries.

Both elective and emergency procedures carry a high morbidity and mortality which are attributable to two main factors. First, all the patients are classified in the American Society of Anesthesiologists (ASA) preoperative physical status groups III to V. These are the highest risk groups[1] – overall mortality approximately 5–50 per cent (Fig. 56.1) – the risk being greater when the patient is an emergency.[2] This results from their advanced chronological or biological age plus the presence of severe to incapacitating coexisting cardiac, pulmonary and renal disease. Second, repair of the arterial lesion requiring surgical treatment may cause ischaemia reperfusion injury.[3] This can result in varying degrees of local cellular hypoxia and tissue necrosis plus systemic organ failure involving the lungs, liver and kidneys.

Anaesthesia for vascular surgical procedures is therefore a high-risk speciality. This is recognized in the UK by the high level of premiums required from anaesthetists, vascular surgeons and intensive care physicians for the provision of medical indemnity insurance (Medical Defence Union Limited, London).

Before proceeding to discuss the special problems of vascular surgical patients in more detail, an outline of the pathogenesis and pathophysiology of the atherosclerotic plaque is useful because it is the consequences of this arterial lesion that result in the need for surgery to be performed.

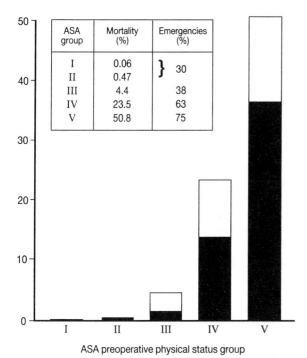

ASA group	Mortality (%)	Emergencies (%)
I	0.06	} 30
II	0.47	
III	4.4	38
IV	23.5	63
V	50.8	75

ASA preoperative physical status group

Fig. 56.1 Chart showing mortality rate (per cent) with increasing ASA preoperative physical status group. Mortality (full column) increases with ASA group number for both elective (white column) and emergency (black column) operations. Data from analysis of 34 145 consecutive surgical patients.[2]

Pathogenesis of atherosclerosis

The term atherosclerosis is formed from the Greek words *athere*, meaning porridge, and *scler*, meaning hard. This describes well the fatty deposits seen in the arterial wall at surgery (Fig. 56.2). The fatty deposits originate from the plasma lipids. The incidence of atheromatous plaque development appears to be related to the blood cholesterol level.[4]

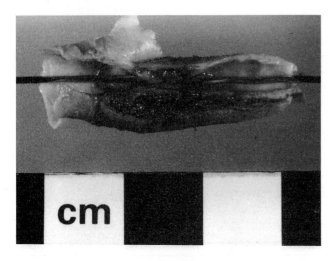

Fig. 56.2 Atheromatous plaque dissected from the internal carotid artery and the origin of the external carotid artery. Note that the major stenosis is located at the beginning of the internal carotid artery and is limited to between 1 and 2 cm in length.

Plasma lipids

The plasma lipids consist of cholesterol and triglycerides, both of which are insoluble in water. They are both rendered water soluble by combining with lipoproteins in the blood. Five classes of lipoproteins are recognized. The largest particles are the chylomicrons, which are triglycerides bound to lipoproteins. In health, these are detectable in blood only after a fatty meal. The remainder of the lipoproteins are divisible into two main groups: the low density lipoproteins (LDL), which account for 60–70 per cent of cholesterol in plasma, and the high density lipoproteins (HDL), which account for 20–30 per cent. The LDL-bound cholesterol originates from intestinal absorption and cellular synthesis, the plasma concentration correlating positively with the development of atherosclerosis.[5] On the other hand, the HDL-bound cholesterol appears to be cholesterol removed from peripheral tissues such as muscle and arterial wall, in transit to the liver for breakdown and elimination.

Risk factors

Factors that increase the plasma LDL concentration are positive risk factors (harmful) for atherosclerotic disease whereas an increased HDL concentration is a negative risk factor (beneficial).

The principal major and minor risk factors associated with atherosclerotic plaque formation are listed in Table 56.1.

Apart from age, sex and genetic hyperlipidaemia, these risk factors may be modified by a change in behaviour or controlled by drug treatment. Clinical trials confirm that

Table 56.1 Risk factors for atherosclerotic disease

Major	Minor
Cigarette smoking	Advanced age
Hypertension	Male sex
Hyperlipidaemia (LDL)	Obesity
Diabetes mellitus	Lack of regular exercise
Hypothyroidism	Persistent stress

the symptoms caused by atherosclerosis can be reduced by adopting these measures.[6, 7] Because of the success of risk factor intervention, the UK government is currently seeking reductions of up to 40 per cent in the incidence of arterial disease requiring treatment by the year 2000.[8]

Plaque formation

Results from numerous biochemical, animal and human epidemiological studies during the last decade indicate that atherosclerosis is caused by peroxidation of LDL. High plasma levels of LDL lead to increased LDL in the intima of the arterial wall. This LDL is oxidized by oxygen-derived free radicals, the oxidized LDL attracting circulating monocytes into the intimal space by chemotaxis. The monocytes ingest the oxidized LDL to form lipid-laden foam cells. These ultimately die, generating subendothelial fatty streaks.[9]

Plaque morphology

Fatty streaks

These represent the earliest form of atherosclerosis and consist of slightly raised yellow lesions found in the arteries of infants and children. The streaks enlarge and mature with age. This occurs by invasion with monocytes from the circulation and smooth muscle cells from the media of the arterial wall. Eventually, the smooth muscle cells undergo phenotypic change and synthesize collagen. An atherosclerotic plaque thus consists of four basic constituents: cholesterol, collagen, monocytes and smooth muscle cells.[7]

Fibrous plaques

Fibrous plaques represent more advanced lesions with a preponderance of connective tissue and muscle cells which overly the lipid laden core. Although the majority of fibrous plaques originate from fatty streaks, they can also arise from arterial trauma and from conversion of a mural thrombus. With increased age the plaques enlarge, become calcified and progressively occlude the arterial lumen. An

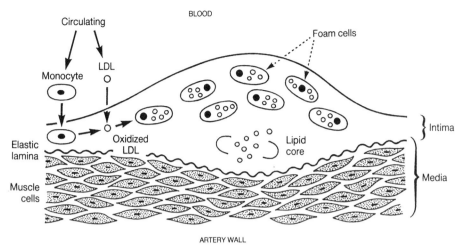

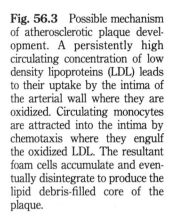

Fig. 56.3 Possible mechanism of atherosclerotic plaque development. A persistently high circulating concentration of low density lipoproteins (LDL) leads to their uptake by the intima of the arterial wall where they are oxidized. Circulating monocytes are attracted into the intima by chemotaxis where they engulf the oxidized LDL. The resultant foam cells accumulate and eventually disintegrate to produce the lipid debris-filled core of the plaque.

outline of the development of an atheromatous plaque is shown in Fig. 56.3.

Distribution of plaque

Haemodynamic factors such as shearing stress, turbulence and flow separation are implicated in the deposition of plaque. It is deposited in a non-uniform segmental fashion and is especially localized to the coronary arteries, carotid bifurcation, infrarenal aorta and iliofemoral vessels. This is why the majority of operations are performed in these locations. A digital subtraction aortogram illustrating the segmental nature of plaque deposition is shown in Fig. 56.4.

Complications of plaque

In addition to progressive arterial occlusion, fibrous plaque may develop acute complications.

Ulceration

The endothelial surface may ulcerate, allowing the lipid contents to escape and resulting in cholesterol embolization. Platelets are also attracted to the ulcerated area. These too may embolize following aggregation. Even more sinister is the development of a mural thrombus which may occlude the arterial lumen or embolize distally. These events account for the clinical syndromes of transient ischaemic attacks (TIAs) and stroke when occurring in the carotid artery and the blue toe syndrome when occurring in the legs.

Haemorrhage

Spontaneous haemorrhage may occur within the plaque, causing acute distension and plaque separation and embolization. As with ulceration, thrombus formation may follow.

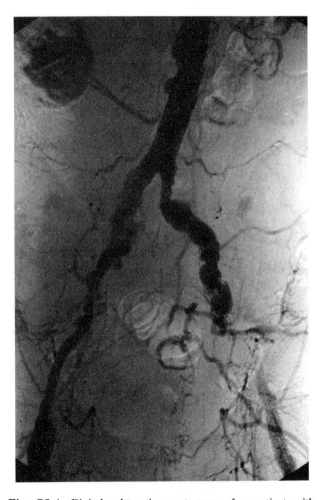

Fig. 56.4 Digital subtraction aortogram of a patient with aortoiliac occlusive disease. Atherosclerotic plaque is present throughout the aorta and iliac arteries, shown by the multiple small stenoses and variable arterial calibre. Note the major stenosis of the left common iliac artery, the complete occlusion of the left external iliac artery and the associated extensive collateral circulation that has developed.

Dissection

The plaque in the intima may separate from the media during dissection with partial or complete arterial occlusion. This is seen most often in the aorta.

Aneurysm formation

Why some patients develop thinning of the media and dilatation of the arterial wall is unknown. Aneurysms were previously considered to be a variant of atherosclerotic disease. Currently, however, it seems that aneurysmal disease may be a separate entity from atherosclerotic occlusive disease,[10] since it develops a decade later, is almost exclusively confined to males and may be determined genetically.[11] Furthermore, although the aneurysmal sac is filled with fatty deposits and haematoma, the cross-sectional area of the lumen of the artery is not usually reduced. The major hazards of aneurysms are thrombosis and embolization of the contents, and rapid enlargement and rupture. As with dissection, this occurs most often in the infrarenal aorta with embolization of debris into the femoral and tibial vessels.

Pathophysiology of atherosclerosis

The acute complications of plaque are superimposed on the chronic symptoms arising from the arterial stenosis. There are usually no symptoms arising from chronic arterial occlusion until the diameter is reduced by 50 per cent. This corresponds to a reduction in cross-sectional area of 75 per cent. Symptoms of angina and claudication develop as the diameter is reduced further to 70 per cent (area reduction of 90 per cent). These occur only during exertion when oxygen consumption ($\dot{V}O_2$) in the tissues distal to the obstruction is temporarily increased but oxygen delivery (DO_2) is limited by the stenosis. Above 70 per cent reduction in diameter (90 per cent area reduction), DO_2 is inadequate to match $\dot{V}O_2$, thereby resulting in rest pain (ischaemic legs). The principles of blood flow reduction with decrease in arterial lumen are shown in Fig. 56.5.

Ischaemia reperfusion injury

Ischaemia (hypoxia) reperfusion (reoxygenation) injury is the paradox whereby tissue damage caused by hypoxia is aggravated and amplified when the post-ischaemic tissue is reperfused with oxygenated blood.[12] Whilst improvement in or restoration of arterial blood flow is the therapeutic

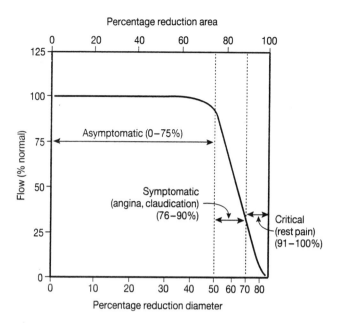

Fig. 56.5 The effect on arterial blood flow (percentage of normal) of reductions in vessel diameter and cross-sectional area. Flow reduction occurs only when the diameter is less than 50 per cent (75 per cent decrease in area) and critical flow (rest pain) when the diameter is less than 70 per cent (90 per cent reduction in area). These principles apply to all blood vessels.

goal of most vascular operations, the procedure may result in superimposed local and systemic tissue damage. This is especially so when the ischaemic tissue is large in mass, such as the intestines or lower limbs. This is an important consideration during aortic surgery.

The contribution of ischaemia to the total injury increases with the duration of ischaemia but appears to be less than the reperfusion component in most clinical situations. Even when a limb is non-viable owing to ischaemia, it may be safer for it to be amputated rather than be reperfused with oxygenated blood in the hope that recovery is possible.

The biochemical basis of ischaemia reperfusion injury is progressive cellular energy depletion during the hypoxic phase followed by the generation of oxygen-derived free radicals (ODFR) during the reoxygenation phase.

Biochemistry of ischaemia

Ischaemia resulting in cellular hypoxia occurs when blood flow is insufficient to supply aerobic oxygen consumption. Oxygen is utilized within the cell as the final electron acceptor in the mitochondrial cytochrome chain. High energy phosphate compounds such as adenosine triphosphate (ATP) are produced by the process of oxidative phosphorylation. Under hypoxic conditions *efficient aerobic* metabolism, which generates 38 moles of ATP per mole of glucose substrate,[13] is replaced with *inefficient anaerobic* metabolism, which generates only 2 moles of ATP from the

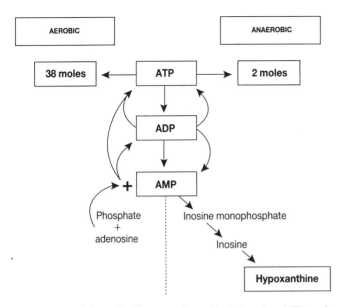

Fig. 56.6 Schematic diagram of aerobic (38 moles ATP) and anaerobic (2 moles ATP) metabolism of intracellular high energy tri- (ATP), di- (ADP) and mono- (AMP) phosphate compounds. Under persistent anaerobic conditions, hypoxanthine accumulates.

same amount of substrate. Maintenance of cellular activity is achieved by recruitment of the glycolytic, creatinine kinase and adenylate kinase pathways although at a much reduced local level of oxygen consumption. Oxygen consumption under these circumstances is not equivalent to $\dot{V}O_2$ under aerobic conditions.

Anaerobic glycolysis has the disadvantage of increasing the intracellular concentrations of phosphate, lactate and hydrogen ion. The pH is lowered as a result, and enzyme activity and hence metabolism are decreased. When the local buffering capacity is saturated, these ions diffuse freely into the circulation. They contribute greatly to the systemic hypotension that follows cross-clamp release during aortic surgery.

When cellular hypoxia persists, the purine metabolite hypoxanthine and the enzyme xanthine oxidase accumulate intracellularly. The hypoxanthine is derived from the catabolism of residual ATP (Fig. 56.6) whilst xanthine oxidase is produced from xanthine dehydrogenase under hypoxic conditions. Xanthine dehydrogenase is a constituent of most cells and highly concentrated in intestinal mucosal epithelial cells and vascular endothelium. Consequently, the intestines and lower limbs are locations where very high tissue concentrations of xanthine oxidase accumulate during ischaemia.

Biochemistry of reperfusion

When blood flow is restored to a post-ischaemic tissue, the accumulated xanthine oxidase and hypoxanthine react with the freshly supplied oxygen to form purine metabolites (xanthine and uric acid) plus free radicals.

Oxygen-derived free radicals

A free radical is a molecule or fragment thereof that possesses an unpaired electron. The most important free radicals in cells are all derived from oxygen. They are very reactive and have an extremely short half-life (microseconds).

During aerobic metabolism, oxygen is normally reduced to water by the addition of four electrons via the cytochrome chain. However, the oxygen molecule can also be reduced in univalent steps to yield three different reactive oxygen species (ROS), the superoxide anion radical ($O_2^{-\cdot}$), hydrogen peroxide (H_2O_2) and the hydroxyl free radical (OH$\cdot$). Hydrogen peroxide is the most important biologically because it is the precursor of the hydroxyl radical, which is regarded as the most toxic of the group.

The major toxic effects of ODFR are the destruction of cell and organelle membranes and breakage of DNA strands with resultant enzyme inactivation.[16] The membrane damage leads to increased intracellular calcium ion concentrations with activation of the membrane-bound enzyme phospholipase A_2 and initiation of the arachidonic acid cascades (prostaglandins, thromboxane and leukotrienes).[17]

All cells have a basal production of ODFR due to the spontaneous escape of electrons from the respiratory chain to molecular oxygen. This accidental production of ODFR is minimized by enzymic and non-enzymic defence mechanisms. The enzymic mechanisms involve inactivation of ODFR by naturally occurring intracellular enzymes such as superoxide dismutase (SOD), catalase and various peroxidases. The non-enzymic mechanisms involve the use of aqueous and membrane-bound antioxidants. Ascorbic acid (vitamin C) is the most important aqueous antioxidant present both in plasma and in intracellular fluid whereas β-carotene (pro-vitamin A) and α-tocopherol (vitamin E) are the most important membrane-bound antioxidants.

Role of neutrophils

The magnitude of reperfusion injury appears to be dependent on circulating neutrophils. The generation of ODFR and subsequent activation of the arachidonic acid cascades in the early phase of reperfusion results in chemotaxis of neutrophils to the reperfused tissue. More importantly, they then adhere to the vascular endothelium and are activated to produce proteolytic enzymes and more ODFR. In the later phase of reperfusion, further neutrophils are attracted to the previously ischaemic tissue with an increase in microvascular permeability and resultant tissue oedema. Coincident with the formation of peripheral oedema, some of the circulating neutrophils are sequestered in the lungs, causing non-hydrostatic pulmonary oedema and the clinical syndrome of acute lung injury.[18] The neutrophil thus aggravates and amplifies the local effects of ischaemia reperfusion injury and accounts for the systemic effects in the pulmonary circulation. A schematic

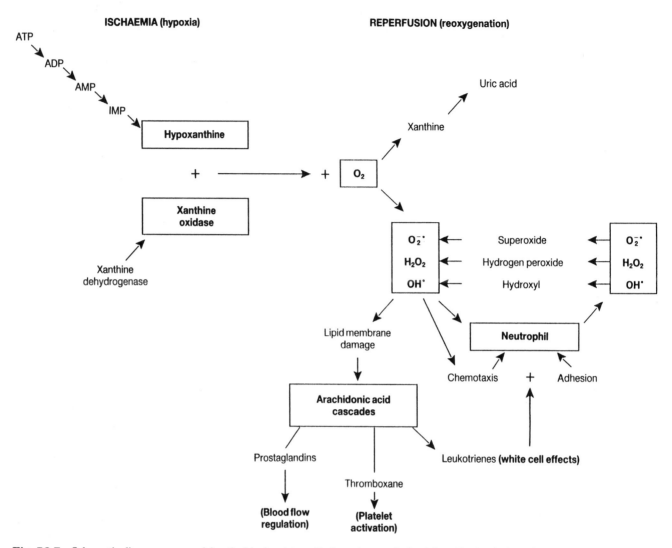

Fig. 56.7   Schematic diagram summarizing the biochemistry of ischaemia reperfusion injury. During ischaemic conditions, hypoxanthine and xanthine oxidase accumulate intracellularly. When oxygen supply is restored during tissue reperfusion, not only are the purine metabolites xanthine and uric acid produced but also toxic oxygen-derived free radicals (ODFR). These have two major deleterious effects. First, circulating neutrophils are attracted by chemotaxis to the ischaemic endothelium, resulting in the formation of additional ODFR. Second, the lipid membrane damage caused by ODFR results in activation of the arachidonic acid cascades with the synthesis of prostaglandins, thromboxane and leukotrienes with their local and systemic effects.

diagram of the basis of ischaemia reperfusion injury is shown in Fig. 56.7.

The pathogenesis and pathophysiology of atherosclerotic disease have been discussed in moderate detail to enable the reader to understand the effects of ischaemia and consequences of (surgical) reperfusion.

Patient characteristics

Although the majority of surgical procedures for the relief of ischaemic symptoms are performed at one of three

different locations in the arterial tree (carotid, aorta and femoropopliteal), most patients undergoing each procedure have broadly similar characteristics and coexisting medical conditions.

Age

Approximately two-thirds of all patients are between 61 and 70 years old and one-third between 71 and 80 years old. A small percentage exceed 81 years. This last group is likely to increase in the future as life expectancy increases in economically developed countries.

Patients younger than 60 years usually have advanced

atherosclerotic disease due to genetic predisposition, familial hyperlipidaemia or diabetes mellitus.

Sex

More males than females undergo vascular surgical procedures, despite the higher prevalence of elderly females; they account for around 80 per cent of aortic operations.[19] However, the number of females presenting for peripheral vascular surgery and carotid endarterectomy appears to be increasing. This may be related to the increase in cigarette smoking in females which occurred in the USA and Europe between 1950 and 1970.

Smoking

All patients, excluding those with genetic or endocrine conditions, can be regarded as previous or present cigarette smokers. This is not surprising, since cigarette smoking for several years' duration is a major risk factor in the pathogenesis of atherosclerotic plaque. Cigarette smoking also has deleterious acute effects. Tachycardia and hypertension result from the absorption of nicotine into the circulation, and this may further compromise an already ischaemic myocardium. The partial replacement of oxyhaemoglobin with carboxyhaemoglobin reduces tissue oxygen delivery, which is likely to impair tissue viability where the blood flow is critical – as may occur in the lower limbs.

Ideally, smoking should stop 6–8 weeks prior to elective surgery because this enables pulmonary secretions to be reduced and for complete clearance of nicotine and carboxyhaemoglobin from the circulation to occur. When this cannot be achieved, the next best advice is to insist that patients refrain from smoking overnight before their operation, as this will provide 12 hours or longer for the blood levels of nicotine and carboxyhaemoglobin to subside.[20]

Coexisting disease

The morbidity and mortality of vascular operations are determined not only by factors related to the site of the operation but also by the performance of the heart, lungs and kidneys. Coronary artery disease and hypertension are very common (Table 56.2). The incidence of fatal and non-fatal cardiovascular complications appears to be similar in patients undergoing carotid or aortic surgery. However, in patients undergoing revascularization procedures for lower limb ischaemia, the cardiac morbidity and mortality are

Table 56.2 Incidence of coexisting disease in vascular surgical patients

Condition	Incidence (%)
Coronary artery disease:	65
Angina	15
Previous myocardial infarction	50
Hypertension	50
Chronic obstructive airways disease	35
Congestive cardiac failure	15
Renal disease	10
Diabetes mellitus	10

Table compiled from data in references 19 and 29.

higher than for the other two groups.[21] As this fact does not appear to be widely appreciated, it does need to be taken into consideration when plans are made for the perioperative care of patients in this higher risk group.

Coronary artery disease

Radiologically significant atherosclerotic disease of the coronary arteries is present in almost two-thirds of all patients.[22] In the majority of these it is clinically significant, having resulted in myocardial infarction (50 per cent) or angina (15 per cent). Many of these patients are asymptomatic with silent myocardial ischaemia which is associated with a poor prognosis. Thus around 50 per cent of all patients have a major myocardial complication before surgery is undertaken! This is especially important because myocardial infarction is the principal cause of death in the perioperative period, accounting for 50 per cent of all early postoperative deaths. Consequently, determination of the functional severity of coronary artery disease is an essential part of the preoperative assessment.

The significance of previous myocardial infarction is emphasized by the high incidence of perioperative reinfarction in such patients which ranges from 6 to 13 per cent.[23, 24] The peak incidence is on the third postoperative day, possibly associated with nocturnal hypoxia and intravascular hypovolaemia. Its occurrence may be reduced by intensive perioperative cardiovascular monitoring and by avoiding surgery when possible within 6 weeks of a myocardial infarction, since by that time the myocardium will have developed adequate scar tissue.[25]

Hypertension

Fifty per cent of all patients have hypertension as defined by the World Health Organization and the International Hypertension Society.[26] The ranges of systolic and diastolic pressure in hypertension are given in Table 56.3. The high incidence of hypertension is to be expected in an elderly

Table 56.3 Values of systolic and diastolic blood pressure in hypertension (mmHg)

Classification	Systolic	Diastolic
Normal	< 140	< 90
Mild	140–180	90–105
Moderate to severe	> 180	> 105

Values taken from reference 25.

patient population and is also a major factor in the pathogenesis of atherosclerotic disease.

Hypertension is important because of its pathophysiological effects. The increase in peripheral (arteriolar) resistance increases the afterload on the left ventricle for a given cardiac output. Consequently, myocardial work and therefore oxygen demand are increased, necessitating greater oxygen delivery via the coronary arteries. Since these are partially occluded in the majority of patients, it is not surprising that the incidence of perioperative myocardial infarction is high. Untreated hypertension results in left ventricular hypertrophy and, eventually, dilatation as the heart fails. The hypertrophied left ventricle requires higher filling pressures because of the decrease in compliance. The practical consequences of these compensatory changes is labile blood pressure, especially during induction of anaesthesia or hypovolaemia (exaggerated hypotension) and surgical stimulation (exaggerated hypertension). Frequent non-invasive (every 3 minutes) or invasive (continuous) blood pressure monitoring, provision of adequate intravascular volume and slow administration of intravenous anaesthetic drugs are essential during anaesthesia for any vascular operation, or indeed for any patient with coronary artery disease or hypertension.

Another major effect of chronic hypertension is the shift to the right of blood flow autoregulation in the cerebral and renal circulations (Fig. 56.8). The significance of this rightward shift is that ischaemia of the brain or kidneys may occur at apparently normal blood pressures, hence the importance of frequent or continuous blood pressure monitoring.

Because the majority of patients with hypertensive disease are taking medication, it is essential that this is continued right up to the operation and afterwards if possible. In a patient with untreated hypertension in whom surgery is urgent, a single dose of β-adrenoceptor blocker given with the premedication has been shown to reduce the incidence of intraoperative myocardial ischaemia from 28 per cent to 2 per cent.[27]

Congestive cardiac failure

Cardiac failure is present in 10–15 per cent of patients preoperatively and up to 30 per cent of patients post-operatively.[28] Left ventricular failure resulting from

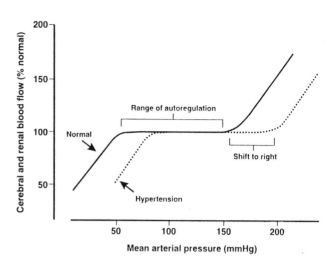

Fig. 56.8 The relation between cerebral and renal blood flow (% normal) and mean arterial pressure (MAP). In health (solid line), normal blood flow is maintained when the MAP is between approximately 50 and 150 mmHg. In hypertension (dotted line), the curve is shifted to the right. Consequently, an apparently normal MAP may result in reduced organ blood flow.

hypertension and coronary artery disease usually precedes right ventricular failure, which occurs in response to persistently elevated left ventricular end-diastolic pressures. Ultimately, biventricular or congestive cardiac failure is present with signs of pulmonary and peripheral oedema.

The presence of decompensated heart failure in a patient implies very severe cardiovascular disease which is associated with a perioperative mortality of 15–20 per cent. Consequently, *all* patients with heart failure must be treated vigorously with diuretics, inotropes and vasodilators prior to surgery.

Dysrhythmias

Atrial fibrillation is the dysrhythmia most frequently encountered, usually in association with coronary artery disease and hypertension. It may also arise *de novo* in the postoperative period in association with hypokalaemia, hypovolaemia or an intra-abdominal procedure. The goal of treatment is to reduce the ventricular rate to between 70 and 90 beats per minute by increasing atrioventricular conduction blockade. The significance of atrial fibrillation is that, if left uncontrolled, cardiac output is invariably reduced in the absence of the 30 per cent atrial contribution to ventricular filling. The resultant hypotension impairs coronary, cerebral and renal perfusion, with all the associated complications.

Pulmonary disease

Chronic obstructive airways disease (COAD) is present in around one-third of patients. The high incidence reflects the high frequency of previous and present cigarette smokers among this patient population. The problems resulting from COAD are twofold. First there is the potential for the development of right heart failure due to the increased afterload caused by chronic elevation of pulmonary artery pressures. Second is the increased production of mucus, which leads to small airway closure especially prevalent in the early postoperative period. This may be aggravated by immobility of the chest wall owing to pain and splinting of the diaphragm due to swelling after intra-abdominal surgery.

Basal atelectasis leading to bronchopneumonia and hypoxaemic respiratory failure are frequent complications of aortic surgery. When the airways are also filled with fluid in addition to mucus, as may occur in pulmonary reperfusion injury following aortic rupture, the hypoxaemic respiratory failure is especially severe.

Renal disease

Renal function is vulnerable in patients with hypertension or heart failure, and is manifest in 10–15 per cent of patients as elevated blood levels of urea and creatinine. In hypertensive patients, there is a marked increase in total body sodium and water retention, possibly mediated via aldosterone and atrial natriuretic peptide together with a rightward shift of blood flow autoregulation. In patients with cardiac failure, the low perfusion pressure decreases glomerular filtration and may lead to renal ischaemia because of the impaired autoregulation of blood flow, as previously discussed. For these reasons, together with the presence of atherosclerotic plaques that are found in the renal arteries of two-thirds of patients, the kidneys are especially vulnerable to hypotension that occurs during myocardial ischaemia, dysrhythmias or the rapid blood loss following aortic rupture.

Diabetes mellitus

Patients with diabetes mellitus comprise about 10 per cent of all patients with vascular disease[29] and a larger proportion of those with peripheral vascular disease. They have not only premature atherosclerotic occlusive disease of the large conducting arteries but also small vessel (arteriolar) occlusive disease. Consequently, the feet and toes are underperfused and often develop infection and gangrene. Diabetic patients are therefore frequently scheduled for femoropopliteal reconstructive surgery and, when that fails, progressive proximal amputation of the toes, feet and lower leg.

Points to be considered during anaesthesia are the likelihood of coronary artery disease,[30] awkward venous access due to obesity, maintenance of steady blood glucose concentrations by appropriate intravenous insulin

treatment (between 6 and 10 mmol/l) and adequate intravascular hydration to optimize tissue perfusion.

Long-term medication

Almost every patient scheduled for vascular surgery is receiving regular oral medication from one or more of the drug groups discussed below. The significance of these drugs, apart from the anti-thromboembolic and hypoglycaemic agents, is that they may interact with intravenous and volatile anaesthetics, causing undue hypotension. All cardiovascular drugs should be continued up to the day of surgery. Otherwise, the blood pressure may be more labile than expected.

Digitalis glycosides

Digitalis glycosides increase myocardial contractility and conduction blockade between the atria and ventricles. Digoxin is used in patients with atrial fibrillation to reduce the ventricular rate and as an inotrope in heart failure. The previous practice of routine administration of digoxin to elderly patients who may be at risk of developing heart failure or atrial fibrillation is no longer indicated.

α_2-Adrenoceptor antagonists

Clonidine is used for the treatment of hypertension and as a premedicant drug. It has two sites of action. In the central nervous system sympathetic outflow is reduced, whilst in the peripheral nervous system the release of noradrenaline is reduced. These effects combine to reduce the effects of stimuli which would cause tachycardia and hypertension.

Sudden withdrawal of clonidine treatment may cause severe rebound hypertension and this must be avoided.

β-Adrenoceptor antagonists

β-Blockers such as propranolol and atenolol are used in the management of ischaemic heart disease and hypertension. β-Adrenoceptor blockade attenuates the effects of exogenous and endogenous sympathetic nervous system stimulation. Consequently, tachycardia and hypertension are obtunded and myocardial ischaemic episodes reduced.[27]

Nitrates

Nitroglycerin (glyceryl trinitrate, GTN) and isosorbide mononitrate are vasodilators used in the treatment of myocardial ischaemia. They act directly on the coronary vasculature, improving subendocardial perfusion as well as reducing ventricular preload through dilatation of venous capacitance vessels. Glyceryl trinitrate given intravenously is very useful in the intraoperative treatment of acute myocardial ischaemia. It may also be administered as cutaneous patches to minimize the likelihood of myocardial ischaemia during induction of anaesthesia and tracheal intubation. The major drawback to this group of drugs is tachyphylaxis. Consequently, they should be used only intermittently in order to maintain their pharmacological effect.

Calcium channel antagonists

This diverse group of drugs antagonize the post-depolarization calcium influx into sinoatrial node cells (verapamil), cardiac muscle (verapamil, diltiazem, nifedipine) and vascular smooth muscle (nifedipine, nicardipine, nimodipine). They are used to treat a wide range of conditions but especially angina, hypertension and supraventricular dysrhythmias.[31] The major practical problem encountered is their hypotensive synergism with halogenated volatile anaesthetics, which must therefore be used with care.

Angiotensin-converting enzyme (ACE) inhibitors

ACE inhibitors are the most recent addition to the therapeutic regimens available for the control of hypertension and heart failure (captopril, enalapril, lisinopril). They act by preventing the conversion of angiotension I to angiotension II and also decease the synthesis of aldosterone. Consequently, ventricular afterload is reduced with considerable improvement in heart failure. Like calcium channel antagonists, marked hypotension may occur when high concentrations of volatile anaesthetics are used in patients receiving one of these drugs and so frequent blood pressure monitoring should be performed.

Anti-thromboembolic drugs

Anticoagulants

Patients on full anticoagulation with oral vitamin K antagonists such as warfarin should have their treatment changed to a continuous intravenous infusion of 1000 units of heparin per hour 48–72 hours prior to surgery. The dose is then adjusted to maintain either the activated partial thromboplastin time (APTT) or activated clothing time (ACT) about twice normal. In an emergency, coagulation factors will need to be administered in the form of fresh-frozen plasma and possibly factor concentrates, the quantity required being monitored by measurement of the prothrombin time (PT). Full anticoagulation is an *absolute* contraindication to regional anaesthetic techniques.

Anti-platelet drugs

The principal anti-platelet drug is aspirin, which reduces the aggregation of platelets (and therefore the formation of platelet emboli) by irreversible inhibition of platelet cyclo-oxygenase, the enzyme responsible for thromboxane A_2 (TXA_2) synthesis. Thromboxane A_2 is the substance that normally initiates and promotes platelet aggregation. A single dose of aspirin is effective as a cyclo-oxygenase inhibitor for the 7- to 10-day life span of a platelet.[32]

Aspirin is now given to patients with symptoms of atherosclerotic occlusive disease such as angina, previous myocardial infarction, transient ischaemic attack (TIA) and stroke. Indeed, the majority of patients scheduled for surgery in the author's practice are now taking this drug. The recently reported results of a meta-analysis of 300 controlled trials prepared by the Anti-Platelet Triallists Collaboration[33] revealed a 25 per cent reduction in vascular events in these high-risk patients. Significant improvements were also observed in postoperative coronary and peripheral arterial vein grafts. As a result, all or most vascular surgical patients will probably be taking aspirin for the duration of their lives.

The anaesthetic implications of aspirin are the possibility of increased intraoperative blood loss and the advisability of performing spinal and epidural anaesthesia with the attendant risk of spinal or epidural haematoma. Control of the blood loss is achieved by meticulous attention to haemostasis given by the surgical team. However, there appears to be no consensus of opinion regarding regional neural blockade in aspirin-treated patients. The safest answer is to regard aspirin as a *relative* contraindication and to consider the risks and benefits for each individual patient, having ensured that the bleeding time is less than 10 minutes.[34]

Other significant medication

Steroids

Patients with an inflammatory condition such as Takayasu's arteritis or severe COAD may be taking oral prednisolone. This should be changed to intravenous hydrocortisone in increased dosage during the perioperative period.

Hypoglycaemic drugs

The oral hypoglycaemic drugs must be stopped preoperatively. Long-acting insulin preparations should be changed to short-acting soluble insulin. In all diabetic patients the blood glucose concentration should be measured hourly and appropriate doses of insulin given with the goal of avoiding hypo- and hyperglycaemia.

Anticonvulsants

Long-acting barbiturates (phenobarbitone) or carbamazepine are occasionally encountered for the treatment of convulsions following TIAs or stroke. They must be continued perioperatively.

Preoperative assessment

The cornerstones of preoperative assessment are the medical history, clinical examination, routine laboratory tests and special investigations. The latter are focused on providing a detailed assessment of cardiovascular function. Only a brief survey of preoperative assessment will be given here because the topic is considered fully in Chapter 29.

History and examination

In addition to documenting the symptoms caused by the specific surgical lesion, questions must be asked about cerebral, cardiac, pulmonary and renal function.

Loss of cerebral function may have occurred following stroke or TIA. Details of any residual neurological deficit must be recorded prior to carotid endarterectomy. Otherwise, it may be difficult to determine whether an apparently new deficit was present preoperatively or occurred intraoperatively. Similarly, it is important to note crude spinal cord function prior to abdominal aortic surgery and absolutely essential with thoracic aortic procedures.

Symptoms of coronary artery disease and heart failure should be sought. Coronary artery disease manifests (after 75–90 per cent occlusion!) as angina or myocardial infarction. Anginal pain is typically crushing in nature and lasts 5–15 minutes in association with increased effort, emotion, sexual activity or eating. The pain is felt over the anterior chest wall (96 per cent) and may radiate into the left arm (30 per cent), lower jaw (22 per cent), back (17 per cent) or right arm (12 per cent), and eases when the initiating activity is stopped.[30] Myocardial infarction may be transmural (main coronary artery) or subendocardial, and is characterized by persistent pain usually necessitating emergency medical treatment and admission to hospital.

Assessment of ventricular function will be straightforward in a patient with frank congestive cardiac failure with ankle or sacral oedema and shortness of breath due to increased pulmonary extravascular fluid. Early left ventricular failure may be indicated by shortness of breath in the supine position which is eased by sitting up (orthopnoea). This results from the chronically raised left ventricular end-diastolic pressures.

A smoking history should be taken from every patient, and signs and symptoms of COAD sought. Shortness of breath, wheeziness and recurrent episodes of excessive production of mucus that readily becomes infected will be evident in one-third of all patients.

Mild to moderate renal impairment is usually asymptomatic but should be considered in any patient with diabetes mellitus, atrial fibrillation or heart failure. Severe renal impairment manifest as oliguria or anuria is usually associated with shock resulting from a ruptured aortic aneurysm.

The patient's weight should be recorded for use as a guide in planning intravenous anaesthetic drug dose requirements and in estimating circulating blood volume for operations where blood loss is anticipated (weight in kg × 70 ml). Additionally, the patient's height should be measured. This together with the weight can be used to estimate the body surface area in patients in whom a pulmonary artery flotation catheter (PAFC) may be used and cardiac index derived from measurements of cardiac output.

Laboratory investigations

Blood

A full blood count (haemoglobin, white blood cells and platelets) plus biochemical profile (calcium, potassium and creatinine in particular) should be performed in every patient. Mild to moderate anaemia is frequently discovered whereas polycythaemia usually indicates severe COAD.

Hypokalaemia (< 3.5 mmol/l) is occasionally observed in patients taking diuretics. Some believe that it is best corrected by intensive oral potassium replacement over 3–4 days. In emergencies, surgery should proceed because it is the ratio of extracellular to intracellular potassium that is physiologically important and not the absolute concentration.

As an index of liver congestion, hepatocellular enzymes and plasma proteins should be measured in patients with heart failure. When coagulation disorders may arise either by dilution due to anticipated blood loss or factor XII activation during tissue ischaemia, a coagulation screen to establish baseline values should be carried out.

In patients with dyspnoea at rest or abnormal spirometry, a single measurement of the arterial blood gases with the patient breathing air is a useful guide to the management of the patient's oxygenation and pulmonary ventilation in the intra- and postoperative periods. For example, a lowered PaO_2 (< 9 kPa; < 67.5 mmHg) would indicate the optimum that patients may achieve 1 week postoperatively without supplemental oxygen, whereas a raised $PaCO_2$ (> 7 kPa; > 52.5 mmHg) would indicate the likelihood of the need for postoperative ventilation.

Finally, the patient's blood group must be determined. For operations in which blood loss is inevitable, such as arterial trauma or aortic reconstruction, 4–6 units of fresh whole blood should be cross-matched in accordance with hospital transfusion policy. Blood transfusion is rarely required for other operations.

Lung function tests

Lung function tests confirm clinically obvious lung disease. In these patients simple spirometry such as measurement of forced vital capacity (FVC) and forced expiratory volume in 1 second (FEV_1) are useful as guides to what the patient can be expected to achieve in the postoperative period, especially after intra-abdominal surgery.

Chest radiography

All patients should have a preoperative chest radiograph taken, as the majority have significant cardiac and pulmonary disease. Whilst new conditions are rarely discovered,[35] the heart size and appearance of the lung fields provide a baseline with which to compare post-operative films.

Coronary artery disease and hypertension do not *per se* have any specific radiological signs until complications develop. An enlarged cardiac shadow will indicate an increase in chamber size due to valvular dysfunction (e.g. mitral stenosis) or ventricular aneurysm, hypertrophy or cardiomyopathy. When considered in conjunction with the clinical history and examination, the cause is usually clear.

The presence of basal pulmonary vascular engorgement, shadowing in the costophrenic angles and transverse fissure, together with pulmonary interstitial shadowing, suggests heart failure. A chest film showing a normal-sized heart and clear lung fields does not necessarily indicate that the heart is normal but it does suggest that it is functioning normally.

Electrocardiogram

A resting 12-lead electrocardiogram (ECG) provides information on heart rate, conduction disorders, hypertrophy of the atria and ventricles and previous transmural (Q wave) or subendocardial (persistent T-wave changes) infarction. Right ventricular hypertrophy indicates increased right ventricular afterload due to COAD or left heart failure, whereas left ventricular hypertrophy develops in response to aortic valve disease or hypertension.

The practical significance of the resting ECG is that it may be normal in 25–50 per cent of patients with coronary artery disease.[30] Consequently, special tests of cardiac function may be required.

Special investigations

Ambulatory ECG monitoring

Continuous ECG monitoring of the patient for 24 hours or so detects myocardial ischaemia from depression of the ST segments. Preoperative ischaemia appears to be a powerful predictor of postoperative ischaemia and subsequent cardiac events.[36] Because most of the ischaemic episodes are asymptomatic (silent myocardial ischaemia) and associated with a poor outcome, this investigation has the potential for much wider application, especially as it is non-invasive, safe and simple to perform.

Exercise ECG monitoring

The principle of this investigation is the detection of ST-segment depression with increased myocardial work as occurs during graded exercise on a treadmill. The heart rate and blood pressure responses are also measured. It is mostly used in the investigation of chest pain caused by coronary artery disease and is inappropriate and impractical in patients with severe carotid, aortic or peripheral vascular disease.

Echocardiography

This non-invasive investigation utilizes the principle of reflection of sound waves from the surfaces of the heart to produce a dynamic picture of the heart contracting. Precordial echocardiography uses sound waves at frequencies of 2–3 MHz to permit penetration of the anterior chest wall with resolutions of 2 mm. Transoesophageal echocardiography uses higher frequencies (3.5–5.0 MHz) which achieve higher resolutions.

Regional and global ventricular wall motion plus thickness can be displayed in addition to function of each valve and associated papillary muscles. Consequently, this technique provides very useful information about the effects of coronary artery disease or hypertension on the structure and function of the heart, unlike the static information obtained from the chest x-ray and resting ECG. Additionally, this technique enables estimation of left ventricular volume (normal 100 ml) and calculation of the ejection fraction (normal 65 ± 5 per cent). When this is less than 50 per cent, the heart is in significant failure. The use of a pulmonary artery flotation catheter should then be considered for intraoperative and postoperative haemo-dynamic monitoring.

Other investigations

Nuclear imaging and coronary angiography can be used to determine the extent and location of coronary artery plaques and the magnitude of infarction. The high cost makes them inappropriate investigations for most vascular surgical patients. Routine coronary angiography was carried out in all patients scheduled for reconstructive surgery at the Cleveland Clinic from 1978. However, no clinical benefits ensued so the practice has been abandoned.[37]

Table 56.4 Preoperative cardiac risk index in non-cardiac surgery

Risk factor	Points
History	
Age > 70 years	5
Myocardial infarction < 6 months	10
Examination	
Third heart sound or distended jugular vein	11
Significant aortic stenosis	3
ECG	
Rhythm other than sinus or supraventricular extrasystole	7
> 5 ventricular extrasystole per minute at any time	7
Poor general condition (any one of the following)	3
PaO_2 < 8 kPa (60 mmHg) or $PaCO_2$ > 6.5 kPa (49 mmHg)	
K^+ < 3.0 mmol/l or HCO_3^- < 20 mmol/l	
Creatinine > 3 mmol/l	
Increased hepatocellular enzymes	
Operation	
Abdominal, thoracic or aortic	3
Emergency	4
Maximum	53

Risk of death
　　< 6 = 0.2%
　6–25 = 2%
　> 25 = 56%

Data taken from reference 39.

Multifactorial risk assessment

Because the mortality of elective vascular surgery is high (4–8 per cent for elective aortic surgery) compared to similar patients undergoing non-vascular surgery (0.1–0.4 per cent),[38] attempts have been made to predict the risk of adverse postoperative outcome. If this were known with certainty, surgery could either be postponed or carried out after extensive myocardial investigation and further treatment of ischaemia with appropriate perioperative cardiovascular monitoring.

Goldman's multifactorial cardiac risk index[39] (Table 56.4) was the first major attempt to categorize risk by allocation of a point score to each of nine clinical and laboratory variables, the higher the total score the greater the risk of adverse cardiac outcome. The original cardiac risk index has since been modified and alternative risk indices published.[40] Although these risk indices have been validated in cohorts of patients, like all scoring systems they can never accurately predict outcome in an individual patient. The best use of risk indices appears to be as a research tool for comparison of different anaesthetic and surgical techniques in comparable patients. They are not used routinely in the UK.

Anaesthesia for specific procedures

It will by now be confirmed for the reader that anaesthesia for vascular operations is high risk because of the significant cardiac mortality even for elective surgery. Local anaesthetic methods with the patient awake and general anaesthetic techniques with spontaneous or controlled ventilation have all been employed for the various operations. Furthermore, all the different intrave-nous induction agents, analgesics, muscle relaxants and volatile anaesthetics have been used. As the advantages and disadvantages of the different techniques and drugs have been discussed in preceding chapters, what follows is a synopsis of the current anaesthetic techniques that are most widely used in the UK and the USA for carotid artery and aortic surgery. What matters most is the way the various drugs are used rather than which drugs are chosen.

Carotid endarterectomy

Carotid endarterectomy (CEA) was first described in 1954 as a procedure for the prevention of ischaemic stroke in a patient with atherosclerotic stenosis at the origin of the internal carotid artery.[41] The operation has since become performed 19- to 27-fold more frequently in the USA than in the UK.[42] This huge difference is likely to decrease now that the benefits of the operation have been proven in the recently published North American[43] and European[44] carotid artery surgery trials. The operation is superior to medical treatment in patients with TIA or stroke and greater than 70 per cent internal carotid artery stenosis on the relevant side, these being the current indications for surgery.

Cerebral circulation

Anatomy
The brain is supplied with blood from the two internal carotid and two vertebral arteries, all of which anastomose at the base of the brain to form the circle of Willis. Minor extracranial-to-intracranial anastomoses via the ophthalmic, trigeminal, intrapetrosal and intracavernous arteries also contribute to the cerebral blood supply. This collateral circulation of blood via the communicating arteries (Fig. 56.9) ensures that the cerebral hemispheres remain perfused despite reduction of blood flow due to plaque in one or both carotid arteries.

Pathophysiology
Cerebral blood flow (CBF) amounts to 50 ml/100 g brain per minute and accounts for 15 per cent of the resting cardiac output. Cerebral oxygen consumption (CMRO$_2$) is 3–5 ml/ 100 g of brain per minute, being higher in grey matter than white matter. Cerebral perfusion pressure (CPP) is determined by the mean arterial pressure (MAP) minus the intracranial pressure (ICP).

The autoregulation of CBF has already been discussed (cf. Fig. 56.8). However, it is important to emphasize that up to two-thirds of patients undergoing CEA are hypertensive and therefore subject to loss of autoregulation during anaesthesia, thereby aggravating blood flow that is already reduced by plaque. Consequently, hypotensive episodes *must* be avoided at all times.

Cerebral blood flow is affected by two other factors which may change during anaesthesia. The arterial Po$_2$ and Pco$_2$ have independent profound effects (Fig. 56.10). A reduction in Pao$_2$ below 7 kPa (52.5 mmHg) causes a massive increase in CBF by vasodilation whereas the CBF varies directly with the Paco$_2$. Clinically, whilst hypoxia must be avoided for obvious reasons, so too must hypocapnia resulting from hyperventilation of the lungs because this will impair cerebral perfusion.

Surgical principles

The cutaneous incision is made along the anterior border of the sternomastoid muscle. Subcutaneous fat, the platysma muscle and facial vein are divided, the sternomastoid muscle is retracted laterally, and the internal jugular vein and vagus nerve are separated from the common carotid artery. The fascia around the carotid bifurcation is then dissected, avoiding damage to the hypoglossal nerve. Clamps are then placed on the external, internal and common carotid arteries, thereby isolating the stenotic plaque from the circulation. An incision extending above and below the plaque is then made in the internal and common carotid arteries. The intimal plaque is then dissected, including removal of the plaque in the origin of the external carotid artery. The arteriotomy is then closed and the arteries unclamped following saline flushing to remove air and debris.[45]

Although the operation is intended to prevent future TIA and stroke, it may be complicated by stroke. This currently occurs in around 2 per cent of operations. Of these strokes, 80 per cent are mechanical and are due to plaque emboli arising from clamping and unclamping procedures whilst the remaining 20 per cent are haemodynamic and due to inadequate cerebral blood flow while the carotid arteries are clamped.[46]

Carotid shunting
To minimize the likelihood of cerebral ischaemia intra-operatively, some surgeons insert an intraluminal carotid shunt or bypass to maintain cerebral perfusion during the cross-clamping period. The decision to insert a shunt is surgical and appears to be based on personal preference as

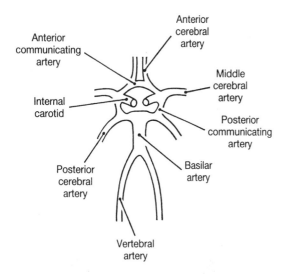

Fig. 56.9 Diagram of the circle of Willis. The brain is supplied by the paired anterior, middle and posterior cerebral arteries, which are themselves supplied via the paired internal carotid (60 per cent flow) and vertebral (40 per cent flow) arteries. The circle is completed by the anterior and posterior communicating arteries. Blood flow to the brain is thus maintained when an internal carotid or vertebral artery is partially or totally occluded.

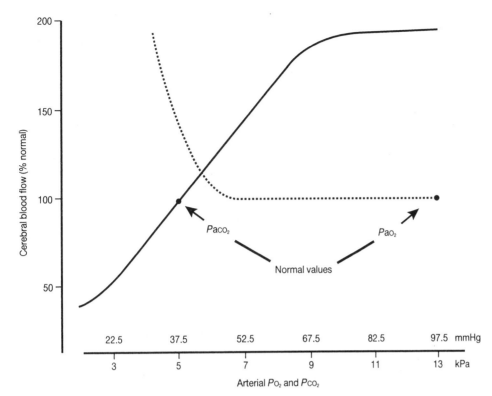

Fig. 56.10 The relation between cerebral blood flow (CBF) and arterial blood gas tensions. CBF increases and decreases linearly from normal at arterial P_{CO_2} tensions between 3 and 8 kPa (22.5 and 60 mmHg). The arterial P_{O_2} only increases CBF at tensions below 6.6 kPa (50 mmHg).

there are no precisely defined criteria. Measurement of the stump pressure[47] (internal carotid artery back-pressure when the external and common carotids are clamped) and the reduction in blood velocity in the middle cerebral artery (MCA) during cross-clamping measured by transcranial Doppler ultrasonography[48] have both been recommended but the safe lower limits of pressure and blood velocity are not known.

Techniques of anaesthesia

Because of the superficial location of the carotid bifurcation, CEA is a suitable operation for regional as well as general anaesthetic techniques.

Regional anaesthesia

Regional anaestheia by local infiltration or by deep cervical plexus block are methods favoured by a few surgeons. The major advantage claimed is a lower incidence of cerebral and myocardial complications (approximately 2 per cent for each), but these do not appear to be significantly different from those occurring when general anaesthesia is used.[49]

Disadvantages of regional techniques are, first, failure of the anaesthetic block and, second, the need for prolonged immobility required of an elderly patient while the operation is performed. Because of this and the likely need for intravenous sedation, general anaesthesia is the most preferred method owing to the greater degree of control over physiological variables. Regional techniques are not popular in the UK.

General anaesthesia

General anaesthesia ensures that the patient is unaware of the procedure, is in the ideal position to facilitate surgery and enables some degree of control over cardiovascular and respiratory variables. The major anaesthetic considerations are summarized in Table 56.5.

Premedication is desirable to decrease the potential development of tachycardia and aggravated hypertension. Drying agents are not required with modern anaesthetic agents. Mild sedation is the ideal and this can be achieved with a short to medium duration benzodiazepine (e.g. temazepam). Prolonged, deep central nervous system (CNS) depression should be avoided because this delays recovery and may make postoperative interpretation of CNS function difficult.

Induction With the patient awake, five-lead ECG and fingertip or earlobe oxygen saturation (SaO_2) monitoring are established. A five-lead ECG permits ST-segment monitoring (acute myocardial ischaemia) in leads II and V_5. When this is not possible, a three-lead recording should be made using the CM_5 configuration.

A crystalloid intravenous infusion is set up and a radial arterial cannula inserted following cutaneous anaesthesia with lignocaine (lidocaine). If this proves technically very difficult, it can be deferred until after induction of anaesthesia in order to reduce the patient's stress. In this situation, blood pressure should be measured temporarily with a non-invasive cuff method at frequent intervals.

Table 56.5 Anaesthetic considerations for carotid endarterectomy

Preoperative	
Coexisting conditions	Hypertension
	Coronary artery disease
	TIAs and stroke
Cardiovascular medication	Continue
	Light premedication
Peroperative	
Goals	Normotension
	Normocapnia
	Normovolaemia
	Optimize cerebral perfusion
	Maintain high arterial P_{O_2}
Monitoring	ECG (5-lead)
	Sa_{O_2}
	BP (direct radial)
	End-tidal P_{CO_2}
	Transcranial Doppler
	ultrasonography
Complications	Hypotension (carotid sinus
	dissection)
	Hypertension (cross-clamping)
Postoperative	
Assessment of neurological function	Stroke
Blood pressure instability	Monitor and treat
Haemorrhage with haematoma	Airway obstruction

Following preoxygenation, anaesthesia is slowly induced with intravenous thiopentone, etomidate or propofol. All these agents reduce $CMRO_2$ and therefore the demand for cerebral perfusion; thiopentone and propofol also reduce systemic blood pressure and must be administered with care, hence the merit of arterial cannulation before induction of anaesthesia, while the radial pulse is easily palpable.

Intubation of the trachea is essential to safeguard the airway and ensure adequate pulmonary gas exchange. Either depolarizing (suxamethonium (succinylcholine)) or non-depolarizing muscle relaxants may be used. Of the latter, the agents of choice are pancuronium, atracurium and vecuronium. Pancuronium has the advantage of causing mild tachycardia, which antagonizes vagal stimulation which may occur during dissection of the carotid bifurcation. The other two drugs have no significant cardiovascular effects but need to be given in repeated doses because of their short duration of action. Consequently, venous engorgement during coughing and straining is a possibility when these drugs wear off unexpectedly.

The sympathetic stress response to laryngoscopy and intubation may be obtunded in four ways. First, intubation should proceed rapidly and smoothly only after full muscle relaxation is achieved. Second, the vocal folds and trachea may be sprayed with lignocaine. Third, intravenous opioids such as fentanyl or alfentanil may be administered. Fourth, a short-acting β-adrenoceptor blocker such as esmolol may be given intravenously.[50]

The tracheal tube selected (e.g. cuffed oral RAE) should be positioned to leave the mouth at the side opposite that of the operation and fixed in place using adhesive tape. Tying the tube in place would interfere with the surgical incision. The eyes should be protected in the usual way. The patient should then be positioned for surgery. This usually entails a small pad being placed between the shoulder blades to produce moderate cervical extension with the head facing away from the operation site. The operating table is adjusted to head-up tilt to optimize venous drainage and minimize venous pooling at the operation site.

Maintenance   Anaesthesia is maintained with volatile agents using spontaneous or controlled ventilation. Spontaneous ventilation has a theoretical advantage of providing both a continuous monitor of brainstem perfusion intraoperatively and maintaining moderate to high arterial P_{CO_2} tensions. Its disadvantage is the need for moderate to high concentrations of volatile agent, and therefore myocardial depression, for the duration of the operation. It is a technique best reserved for procedures of short duration.

Volatile agents   Controlled ventilation is the most widely used method with isoflurane in an air–oxygen or oxygen–nitrous oxide mixture. The volatile anaesthetics, like the intravenous induction agents, all depress cerebral blood flow and $CMRO_2$ but isoflurane is the agent of choice because of its cerebral protective effects.[51,52] With isoflurane, cerebral ischaemia develops only when cerebral blood flow reaches 10 ml/100 g brain tissue per minute compared to 15–18 ml per minute with enflurane and 20 ml per minute with halothane. Against this must be considered the coronary vasodilator effects of isoflurane which are more potent than enflurane and halothane and may cause the coronary artery steal syndrome. However, this appears to be a theoretical problem and not clinically relevant.[53]

Ventilation is adjusted to achieve an end-tidal P_{CO_2} of 4.5–5.5 kPa (34–41 mmHg). This ensures an arterial P_{CO_2} of 5–6 kPa (37–45 mmHg) when the alveolar–arterial P_{CO_2} difference is allowed for.

Intravenous analgesia is usually achieved with incremental fentanyl or infusion of alfentanil. By using an analgesic, the concentration of volatile agent may be reduced so that the cardiovascular hypotensive effects may be minimized.

A practical goal for the intraoperative management of blood pressure is to maintain the systolic pressure within ± 40 mmHg of the preoperative values. This usually means keeping it between 120 and 200 mmHg.

Fluids Crystalloid intravenous fluids, avoiding dextrose-containing solutions except in diabetic patients, are administered at a moderate rate (100–200 ml hourly) following an initial 500–1000 ml bolus to counteract preoperative dehydration. It is essential to be wary of fluid overload in these elderly patients, many of whom have decreased left ventricular compliance owing to hypertension or scarring from previous myocardial infarction. Dextrose-containing solutions should be avoided because hyperglycaemia has been shown to aggravate ischaemic brain damage.[54]

Heparin Intravenous heparin is usually requested by the surgeon prior to cross-clamping the internal carotid artery. Up to 5000 units may be given. It is usual to allow the anticoagulant effect to decay spontaneously and not to use protamine for reversal.

Cerebral monitoring Although the adequacy of cerebral perfusion is uncertain intraoperatively and safe lower limits of blood flow are not yet defined, it is usual for some form of cerebral function monitoring to be used. This is most valuable during carotid cross-clamping which is when plaque embolization and resultant cerebral ischaemia are most likely to occur. Cerebral blood flow has been assessed indirectly from global hemisphere function in the awake patient, and stump pressure in the unconscious patient. Other forms of monitoring that have been used are the electroencephalograph (EEG), somatosensory evoked potentials (SSEP), transcranial Doppler ultrasonography (TCD) and, more recently, near infrared cerebral spectroscopy (NIRS).[55] The last provides a continuous display of cerebral oxygen saturation (CsO_2) and, when combined with TCD, gives simultaneous measurements of ipsilateral global blood flow (MCA blood velocity) and cerebral perfusion. Both of these methods are non-invasive, requiring only an ultrasound probe in a headband (TCD) and a flexible plate electrode (NIRS) to be attached to the skull. These techniques are used in the author's hospital.

The principle of CsO_2 measurement is to beam light of wavelengths between 650 and 1100 nm from an emitting electrode through the skull into the brain. The light may be transmitted, absorbed or reflected. The reflected light is detected by two detectors, one near and one more distant from the emitter. Light detected by the near detector represents light transmitted through the scalp and skull, whilst that detected by the far detector represents light transmitted through the scalp, skull and brain (Fig. 56.11).

Subtraction of the near signal from the far signal gives a value for the brain which is normally between 65 and 75 per cent. This value is lower than non-cerebral tissue because about 70 per cent of blood within the skull is venous. Typical signals for CsO_2 observed peroperatively are shown in Fig. 56.12. Carotid cross-clamping usually results in a fall in TCD peak blood flow velocity and CsO_2 because the

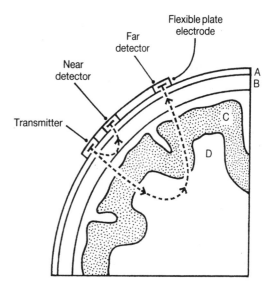

Fig. 56.11 The principle of near infrared cerebral spectroscopy. A flexible plate electrode is fixed to the scalp over the hemisphere to be monitored. Light between 650 and 1100 nm in wavelength is directed through the scalp (A), skull (B), grey matter (C) and white matter (D) from the transmitting electrode. Reflected light is detected by both near (A and B) and far (A, B, C and D) electrodes, subtraction of the former from the latter providing a value for cerebral oxygen saturation (65–75 per cent).

ipsilateral hemisphere is entirely dependent on collateral blood flow. When the cross-clamp is removed following closure of the arteriotomy, the TCD peak velocity and CsO_2 signals increase owing to the large rise in blood flow.

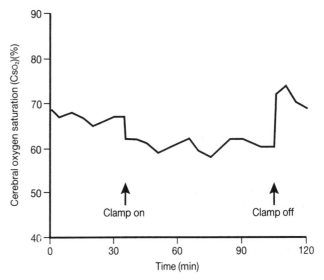

Fig. 56.12 Typical changes in cerebral oxygen saturation (CsO_2) during carotid endarterectomy in a patient with a poor collateral circulation. When the internal carotid artery is cross-clamped, CsO_2 falls by approximately 5 per cent and remains unchanged during the endarterectomy procedure. Following unclamping, CsO_2 increases and exceeds pre-cross-clamp values initially, indicating hyperaemic reperfusion of the ischaemic brain. In patients with a good collateral circulation, no significant change in CsO_2 is observed following application of the cross-clamp.

Reversal and recovery Following skin closure, the muscle relaxants are reversed as usual and the patient is given 100 per cent oxygen. Once spontaneous ventilation is established, the patient is transferred to the recovery room for postoperative ECG, SaO_2 and radial arterial pressure monitoring for up to 2 hours. When appropriate, the tracheal tube is removed after tracheal and oropharyngeal suction have been performed. A simple clinical assessment is made of the patient's neurological state by asking him to move each arm and leg in turn. Occasionally (1 patient in 50) an intraoperative stroke will have occurred, causing hemiplegia. This should be recorded in the clinical notes.

Postoperative pain is not a major concern because the surgery is superficial. Infiltration of the wound with long-acting local anaesthetic (bupivacaine) followed by simple oral analgesics are usually sufficient.

Complications

Intraoperative

Intraoperative complications are either hypotension or hypertension.

Hypotension

Intraoperative hypotension may develop gradually or suddenly. Gradual hypotension usually results from the combined effects of hypovolaemia and the vasodilator and cardiac depressant effects of volatile anaesthetic agents. Sudden hypotension usually follows baroreceptor stimulation.

The baroreceptors are located in the carotid sinus in the wall of the internal carotid arteries just above the bifurcation. During manipulation and dissection of this area, the carotid sinus nerve (a branch of the glossopharyngeal nerve) transmits impulses to the vasomotor centre in the brainstem with resultant vagal efferent activity which causes vasodilation plus bradycardia. When this occurs, the surgeon should be told to stop operating momentarily, the concentration of volatile anaesthetic should be reduced and ephedrine (α- and β-adrenoceptor activity) given intravenously in 3 mg doses until the previous blood pressure reading is restored. Other α-adrenoceptor agonists such as methoxamine or metaraminol may also be given. If bradycardia and hypotension persist, the carotid sinus nerve should be infiltrated with up to 3 ml of lignocaine. Surgery may proceed when the blood pressure has recovered.

Hypertension

Apart from light anaesthesia (empty vaporizer, etc.) hypertension may follow cross-clamping of the internal carotid artery.[56] Just as with hypotension, this is also presumed to be a baroreceptor-mediated reflex, but on this occasion the absence of blood flow in the ipsilateral carotid artery causes vasoconstriction and inhibition of vagal activity via the brainstem. The systolic pressure may rise rapidly and reach 300 mmHg within 5 minutes.

Such a hypertensive response may be beneficial in terms of cerebral perfusion but is dangerous and life threatening because of the cardiac complications. Increased afterload on the left ventricle and the high intraventricular pressures result in the conflicting conditions of increased myocardial demand for oxygen in the face of reduced myocardial perfusion. Profound myocardial ischaemia is likely to occur (ST-segment depression), leading to subendocardial infarction. Prompt treatment is therefore essential. Immediate intravenous glyceryl trinitrate to augment subendocardial perfusion and offload the left ventricle is the initial treatment, followed by intravenous labetalol (combined α- and β-adrenoceptor blockade) given in 25 mg boluses until the pressure is controlled. The response to treatment is rapid and usually permanent.

Postoperative

Postoperative complications are either surgical or medical.

Surgical

The principal surgical complications are haemorrhage with haematoma formation and stroke. The haemorrhage may be acute, arising directly from a flaw in closure of the arteriotomy in the carotid artery, or subacute, arising from all the wound surface as may occur in patients taking aspirin preoperatively. If a substantial haematoma develops within the first hour postoperatively the wound should be re-explored. Deviation and compression of the trachea with partial airway obstruction should be taken into account in planning induction of anaesthesia.

Stroke should be considered in any patient who is hemiplegic or remains sleepy and uncoordinated 30 minutes after reversal of anaesthesia (using the anaesthetic technique described). Assessment of cerebral blood flow with TCD or carotid angiography should be considered. Re-exploration of the carotid artery may be necessary to exclude an intimal flap, stenosis or kinking as a cause of the stroke.

Medical

Derangement of blood pressure is common in the early postoperative period, affecting between 19 and 68 per cent of all patients.[49] Hypertension (> 200 mmHg) is more common than hypotension (< 120 mmHg). These disorders are attributed to resetting of the baroreceptors, which may take up to 6 months to return to normal.[57]

In the recovery room, specific treatment may need to be given once adequacy of analgesia and circulating volume are confirmed. Systolic pressures greater than 200 mmHg usually settle spontaneously but may need to be reduced with hydralazine.

Low blood pressure, although less common, may need pharmacological treatment *having first excluded a myocardial cause*. Modest intravenous doses of an α-adrenoceptor agonist such as phenylephrine, metaraminol or methoxamine may be used. Methoxamine is probably the drug of choice since an increased incidence of myocardial infarction has been observed with metaraminol,[58] and phenylephrine is very short acting. The author's practice is to avoid intravenous hypertensive drugs if possible in the post-operative period by ensuring minimal cerebral depression by use of light premedication and anaesthetic agents that permit rapid recovery.

Abdominal aortic resection

Resection and grafting of the abdominal aorta is performed for the treatment of aneurysms and stenosing occlusive disease. These two conditions occur together in about 25 per cent of patients.[59] The goals of surgery are avoidance of rupture and relief of symptoms with restoration and maintenance of blood flow to the viscera and legs.

The operation requires temporary cross-clamping of the aorta which may cause indirect impairment of cardiac function owing to the potential massive increase in left ventricular afterload. Direct impairment of blood flow to the intestines and lower limbs causes ischaemia reperfusion injury, the magnitude of which varies with the collateral circulation. Renal and spinal cord perfusion may also be impaired. These pathophysiological effects are severely exacerbated by the acute blood loss associated with aortic rupture.

In addition to enabling surgery to be performed, the goals of anaesthesia are maintenance of cardiopulmonary function, preservation of renal and spinal cord perfusion and minimizing the effects of ischaemia reperfusion injury during and following aortic cross-clamping.

Regional, general and combined anaesthetic techniques have all been employed successfully. A balanced general anaesthetic with tracheal intubation, muscle relaxation and controlled ventilation combined with postoperative epidural infusion analgesia appears to be widely used and enables the anaesthetic objectives to be met.

The goals are similar in emergency anaesthesia for ruptured aorta but organ function is usually severely impaired owing to acute profound hypovolaemia. Only a light general anaesthetic technique is appropriate, with special emphasis on rapid correction of the intravascular volume deficit to minimize the tissue ischaemia.

Indications for surgery

Aneurysms

The majority of abdominal aortic aneurysms are asymptomatic, the pulsatile swelling being detected only incidentally during abdominal examination, radiography (Fig. 56.13) or ultrasonography. Symptomatic aneurysms manifest either as thromboembolic disease of the leg or severe pain and shock associated with rupture.

The natural history of an aneurysm is expansion at an average rate of 0.4 cm per annum with ultimate rupture; the mortality from rupture is 100 per cent unless surgically treated.[60] In the UK, only 40–50 per cent of all patients with a ruptured aorta reach hospital alive, and of those who undergo surgery the overall mortality may reach 75 per cent.[61] Thus the total survival rate of 10–15 per cent for surgical repair of the ruptured aorta compares very unfavourably with the 90–95 per cent survival following elective aortic repair. Surgery is therefore indicated to prevent inevitable rupture occurring.

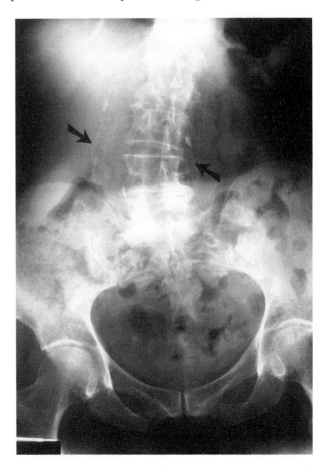

Fig. 56.13 Radiograph showing the calcified sac (arrows) of a large abdominal aortic aneurysm discovered by chance in a 68-year-old male patient. The anteroposterior diameter of the aneurysm was 9 cm.

Aneurysms are mostly fusiform or occasionally saccular in shape and infrarenal in origin in about 90 per cent of patients, and juxtarenal or suprarenal in the remainder.[62] Preoperative definition of the upper and lower limits of the aneurysm is performed by computed tomographic scanning. The upper limit determines the location of aortic cross-clamp application – i.e. whether it is above or below the origin of the renal arteries – whilst the lower limit, which may involve the iliac arteries, determines whether a tube or bifurcation graft is inserted.

The indications for operating on patients with small asymptomatic aneurysms await clarification following large-scale ultrasound screening programmes.[63] Currently, most surgeons would offer surgery to patients with a 4.5 cm diameter aneurysm because, when the ultrasound measurement error of ± 0.5 cm is taken into account, the aneurysm could be larger. Furthermore, since surgery will inevitably be required, the sooner it is performed, the lower the morbidity and mortality are likely to be because the coexisting cardiac, pulmonary and renal disease will be less advanced.

Occlusive disease

Atherosclerotic occlusive disease of the aorta and its branches tends to be progressive and results from chronic exposure to the risk factors previously discussed. Unlike the majority of patients with aneurysms who are asymptomatic, all patients with occlusive disease have symptoms. Approximately two-thirds have intermittent claudication, and the other third have either ischaemic rest pain or gangrene.[64] The other major differences from aneurysmal disease are, first, the peak incidence of surgery which is performed in patients 10 years younger, second, a decreased incidence of coexisting disease (because the patients are younger) and, third, a decreased requirement for blood transfusion.[65] Otherwise, the two groups are similar, with myocardial infarction the major perioperative complication.

Abdominal and lower limb circulation

The intestines and lower limbs are supplied with blood from the abdominal aorta which is a tapering, smooth-walled tube extending from the aortic hiatus in the diaphragm to the bifurcation into the left and right common iliac arteries (segmental level L4). Each common iliac artery divides into internal and external iliac arteries anterior to the upper end of the sacroiliac joint. The internal iliac supplies the pelvic organs, perineum and buttocks, whilst the external iliac runs round the pelvic brim and, after emerging under the inguinal ligament, becomes the common femoral artery which supplies the leg.

The blood supply of the intestines follows the embryo-

logical origins of the fore-, mid- and hindgut. The foregut, comprising the stomach and duodenum as far as the entry of the bile duct, is supplied from the coeliac artery (T12). The midgut, comprising the duodenum, ileum, and ascending and transverse colon as far as the splenic flexure, is supplied from the superior mesenteric artery (L1) whilst the hindgut, comprising the descending and sigmoid colon and first part of the rectum, are supplied by the inferior mesenteric artery (L3). There are anastomoses between the branches of these arteries which maintain continuity of blood supply to the intestinal mucosa when blood flow is compromised. This is especially significant for the hindgut because the inferior mesenteric artery is invariably totally or partially occluded in both aneurysmal and occlusive disease.

The renal arteries (L2) supplying the kidneys are the largest branches of the abdominal aorta (25 per cent of cardiac output). Other branches of significance in vascular disease are the four pairs of lumbar arteries that supply the back and anterior abdominal wall plus the anterior spinal artery (artery of Adamkiewicz) which is a small branch of an intercostal or lumbar artery which may originate anywhere between T8 and L2.[66]

Awareness of the origin of these various arteries from the aorta enables the anaesthetist to consider the ischaemic consequences of aortic cross-clamping and plan to minimize the effects. For example, suprarenal cross-clamp application renders not only the kidneys ischaemic but also most of the intestines in addition to the legs. This is a far greater pathophysiological insult than infrarenal aortic cross-clamping where only the legs may be ischaemic.

Surgical principles

Successful repair of an abdominal aortic aneurysm using preserved human artery (homograft) was first described in 1952[67] and subsequently in the same year using a textile prosthesis.[68] Currently available prostheses are made either from knitted or woven Dacron or polytetrafluorethylene (PTFE).

The abdominal aorta is retroperitoneal and is most often approached through the peritoneal cavity via either a vertical midline or transverse supraumbilical incision. The transperitoneal approach is favoured by most surgeons because it facilitates access to the aorta and its branches in addition to the abdominal organs. The disadvantages of this technique are: first, the severe postoperative pain which may limit chest movement and lung expansion; second, the loss of heat and fluid from the intestines; and third, the paralytic ileus which may persist for several days. To minimize these complications, which contribute significantly to the postoperative morbidity of the operation, a retroperitoneal approach is used by some surgeons.[69]

The abdominal contents are thoroughly explored,

following which the transverse colon, greater omentum and small intestine are lifted superiorly and placed outside the abdomen between moist packs.[70] The posterior peritoneum is incised and dissected to expose the aorta from the left renal vein superiorly to the common iliac arteries inferiorly. Dissection near the aneurysm is kept to a minimum to prevent bleeding.

Systemic heparin (5000 units) is given to prevent distal thrombus formation during the cross-clamp period. Vascular clamps are then applied, first to both common iliac arteries to prevent embolization of thrombus and debris into the legs and then to the aorta which is incised longitudinally. All atherosclerotic debris and thrombus are removed from the aortic lumen and the paired ostia of the lumbar arteries oversewn. The inferior mesenteric artery is dealt with similarly if it is still patent.

The chosen graft is sutured into the infrarenal aorta at its proximal end and the anastomosis tested by brief unclamping of the aorta. If necessary, repairs are made to the suture line. The distal anastomosis is then sutured (aorta–tube graft, or iliac arteries–bifurcation graft) and, before completion, the distal vascular clamps are released to permit back-bleeding and the vessel is flushed to remove any residual intraluminal debris. Restoration of blood flow to the lower limbs should be permitted to occur in a controlled fashion one leg at a time, to minimize unclamping hypotension.

The original aortic wall is then wrapped round the graft and sutured, the peritoneum is closed, the intestines are returned to the abdominal cavity and the abdominal wall is repaired.[71]

Anaesthetic technique

The anaesthetic technique is similar for all types of abdominal aortic disease. A controlled airway with tracheal intubation and ventilation is essential because of the huge abdominal incision. This may be supplemented with epidural analgesia, which is especially advantageous postoperatively. A regional anaesthetic technique alone is inappropriate for such major body cavity surgery. The goals of anaesthesia are summarized in Table 56.6.

Surgical anaesthesia

Premedication Anxiolytic premedication is desirable to minimize the release of endogenous catecholamines and decrease sympathetic nervous system stimulation that result from stress in the preoperative period. Oral benzodiazepines (temazepam, diazepam or lorazepam) given 60–90 minutes preoperatively are effective in maintaining stable cardiovascular conditions. Another approach is to include either α- or β-adrenoceptor blockade with the premedication using drugs such as clonidine[72] or

Table 56.6 Anaesthetic goals for abdominal aortic surgery

Provision of surgical anaesthesia	Controlled airway Analgesia and muscle relaxation Ventilate to normocapnia
Maintain body temperature	Heated blanket Heated i.v. fluids Humidified gases Intestinal packs
Maintain circulating volume	Crystalloids i.v. at 10–15 ml/kg per hour Monitor CVP (aim for + 10 mmHg) PAOP if indicated (aim for + 10–15 mmHg) Autologous blood Minimize homologous blood
Minimize cardiovascular instability	Cross-clamping – vasodilators Unclamping Volume loading Monitor coronary perfusion (ST segment)
Maintain renal and spinal cord perfusion	Optimize cardiac output Monitor BP (radial) Aim for 50 ml urine per hour
Minimize local and systemic reperfusion injury	Mannitol i.v. 0.2–0.5 g/kg before cross-clamp Short cross-clamp time

PAOP, pulmonary artery occlusion pressure

atenolol.[27] These drugs blunt the heart rate and hypertensive responses to the emotional stress of waiting for surgery and the physical stress of induction of anaesthesia.

Monitoring Cardiorespiratory monitoring is established prior to induction of anaesthesia. This includes ECG (five-lead for ST-segment monitoring), peripheral SaO_2 and direct measurement of blood pressure through a radial arterial cannula. A peripheral intravenous infusion of crystalloid is started. It is convenient if both arterial and venous cannulae are placed in the same arm to facilitate connections with the monitoring equipment.

Induction and maintenance Preoxygenation is performed for 3–5 minutes, after which anaesthesia is induced intravenously. The choice of induction agent is less critical than the way in which the drug is administered. The goal is to prevent hypotension followed by tachycardia and hypertension during laryngoscopy and intubation. Etomidate, thiopentone or propofol are equally suitable provided they are given slowly to minimize cardiovascular changes. Fentanyl is the most widely used analgesic, given as an initial bolus to minimize the stress response to intubation, with further increments to minimize the responses to skin incision and aortic cross-clamping.

As with carotid artery surgery, either depolarizing (suxamethonium) or non-depolarizing (pancuronium, atracurium or vecuronium) muscle relaxants may be used to facilitate tracheal intubation. Pancuronium has the advantage that a single dose may provide muscle relaxation for the duration of the operation whereas atracurium and vecuronium, although short acting, are suitable for a total intravenous anaesthetic technique (TIVA) This is the author's preferred method using a combination of propofol, atracurium and alfentanil, thereby avoiding the need to use a volatile anaesthetic agent to maintain anaesthesia. When a volatile anaesthetic is required, isoflurane appears to be the agent of choice. Although it is a more potent coronary artery vasodilator than either halothane or enflurane, causing a higher incidence of coronary artery steal syndrome in laboratory animals,[53] the incidence of myocardial ischaemia in patients is no higher with isoflurane.[73] Furthermore, this volatile agent provides the greatest cerebral protection (as previously discussed) and causes far less impairment of renal function than halothane during aortic cross-clamping.[74]

Once a nasogastric tube is passed, the trachea is intubated and the lungs are ventilated with an air–oxygen or oxygen–nitrous oxide mixture. The former mixture is preferable to the latter as there is no intestinal distension, which facilitates wound closure for the surgeon. Ventilation is adjusted to achieve normal values of arterial $P\text{CO}_2$ for each patient (usually 5.0–5.5 kPa; 37.5–41.3 mmHg), allowing for the alveolar–arterial $P\text{CO}_2$ difference. Maintenance of normocapnia throughout the operation contributes to the global cardiovascular stability. End-tidal $P\text{CO}_2$ monitoring is continued throughout the operation with this objective in mind.

After the patient is anaesthetized, central venous pressure (CVP) monitoring is established. A useful technique is to insert a large-bore (8.5 French gauge) percutaneous sheath in the right internal jugular vein. Identification of the vein is facilitated by the rapid infusion of crystalloid and tilting the patient 20–30 degrees head down. Using a wide-bore sheath permits use of this alone for CVP monitoring and the rapid infusion of fluids, or it enables a triple-lumen catheter or PAFC to be inserted at any time. A PAFC is indicated when the ejection fraction is less than 50 per cent or there is any disorder of myocardial function detected on preoperative echocardiography.

Finally, a lumbar epidural catheter is inserted and taped in position. Placement of the catheter at this stage ensures that any bleeding will have stopped before heparinization. There is no advantage in thoracic epidural catheter placement over lumbar when epidural opioids are used.[75] A member of the surgical team catheterizes the bladder at this stage. The hourly urine production can then be monitored.

Surgery proceeds as outlined previously. The interval up to aortic cross-clamping provides the anaesthetic team plenty of time to establish and control the normal values of cardiorespiratory variables for the patient. When a PAFC is used, measurements of pulmonary artery occlusion pressure (PAOP) and cardiac output (CO) are made and the derived variables of cardiac index (CI) and systemic vascular resistance (SVR) calculated. The relation of CVP to PAOP is observed and recorded since there is a poor correlation between the two when one ventricle is diseased and the other is not.[76] Arterial blood gases are also measured to enable pulmonary gas exchange to be optimized.

Body temperature   Core temperature is measured either by the thermistor on a PAFC or by an oesophageal probe. Because the operation involves a large abdominal incision with evisceration of the intestines, the core temperature may decline by up to 3–4°C.

Heat loss by conduction and radiation is reduced by ensuring that the operating room temperature is adequate (20°C minimum) and the patient is lying on a heated mattress. All intravenous fluids are also heated. Heat loss by evaporation is minimized by thorough wrapping of the intestines in moist gauze packs and use of a condenser humidifier in the patient's airway. If the patient is cold at the end of the operation, this may be an indication for postoperative ventilation to permit normal body temperature to recover gradually. This ensures that there are no sudden massive increases in oxygen consumption due to shivering, and minimal risk of hypoxia aggravating myocardial ischaemia.

Circulating volume   Adequacy of circulating volume is judged from measurements of the CVP and PAOP. The values of these measurements are similar in a healthy heart but often different in patients undergoing aortic surgery because two-thirds have significant coronary artery disease. Previous myocardial infarction, which affects 50 per cent of patients, results in some degree of ventricular wall scarring and higher values of PAOP than CVP, hence the value of monitoring both pressures. By using PAOP as the guide to filling pressures, hydrostatic pulmonary oedema is likely to be prevented if PAOP is kept below 18 mmHg. Suitable intraoperative goals for PAOP and CVP are 15 mmHg and 10 mmHg respectively. In the absence of the facility to measure PAOP, maintaining the CVP at around 10 mmHg throughout the operation ensures near-optimal ventricular function with no significant risk of pulmonary oedema developing.

Fluid and blood losses are large compared to other abdominal operations. Extracellular fluid is lost by evaporation, sequestration into the lumen of the intestines and the development of interstitial oedema (third space effect) due to handling and mobilization of a large mass of tissue. These losses, which are superimposed on preoperative dehydration and relative hypovolaemia due to the vasodilation caused by anaesthetic agents (propofol or isoflurane), are best replaced with a continuous infusion of a balanced electrolyte solution (Hartmann's solution, UK; Ringer's lactate, USA). The infusion rate ranges from 6 to 16 ml/kg per hour (average about 10 ml/kg per hour) to maintain a stable circulation and prevent oliguria.[19] Some 3–6 litres of crystalloid may thus be required in the perioperative period, and occasionally more.

Blood loss averages 2.15 ± 1.15 litres during aneurysm surgery, with an average transfusion requirement of 4.5 ± 2.4 units of blood.[19] However, blood transfusion with homologous blood is potentially dangerous, up to 10 per cent of patients experiencing either immunological (red and white cell incompatibility) or infectious (hepatitis B, C and HIV) complications. Consequently, in many institutions, homologous blood transfusions are now avoided unless necessary.

Until recently it has been customary to transfuse blood when the loss exceeds 20 per cent of the predicted circulating blood volume but there is no evidence to support a single criterion for transfusion.[77] Although there appear to be no major complications arising when the haemoglobin concentration is allowed to fall to 7 g/dl, it is recommended to aim for 10 g/dl in patients with CAD.[78]

The hazards of blood transfusion can be avoided by meticulous surgical techniques and use of autologous blood, and minimized by using as little homologous blood as necessary. Autologous blood can be removed from the patient preoperatively (predonation) or intraoperatively (normovolaemic haemodilution); when combined with intraoperative red cell autotransfusion, homologous transfusion is avoided in around two-thirds of patients.[79] Blood transfusion is discussed in detail in Chapter 18.

Cardiovascular stability

Cardiovascular stability is achieved by minimizing heat loss, maintaining normocapnia and adequate volume loading (CVP of 10 mmHg) and ensuring appropriate depth of anaesthesia in relation to surgical stimulation (e.g. deepening anaesthesia prior to skin incision). Major disturbances of the circulation may arise at three points in the operation: in association with placement of intraabdominal retraction and clamping and unclamping of the aorta.

Abdominal retraction

Removal of the intestines from the abdominal cavity followed by placement of large retractors to enable the aorta to be exposed may result in mechanical obstruction of venous return via the inferior vena cava. If hypotension develops rapidly in association with a low CVP, the retractors should be repositioned to correct the disturbance.

Another cause of hypotension in association with sudden tachycardia and facial flushing is the mesenteric traction syndrome described by Selzer et al. in 1985.[80] This syndrome occurs following mesenteric traction, lasts less than 30 minutes and is associated with increased plasma prostacyclin concentrations. Prostacyclin is a prostaglandin derived from arachidonic acid in vascular endothelium (cf. Fig. 56.7). It may be synthesized in response to intestinal ischaemia resulting from displacement of the intestines from the abdominal cavity. When cyclo-oxygenase, an enzyme involved in the synthesis of prostacyclin, is inhibited by preoperative treatment with either aspirin[81] or ibuprofen[82] the syndrome does not occur.

Aortic cross-clamping

The consequences of cross-clamping are similar regardless of the level at which the aorta is occluded, but the magnitude of the changes is greater the higher the level. Infrarenal cross-clamping is the norm (90 per cent of patients), with mechanical effects on the circulation and ischaemic consequences for the pelvis and lower limbs.

The circulatory effects result from the abrupt increase in left ventricular afterload which leads to increased PAOP, SVR and mean arterial pressure with decreases in cardiac output and preload. When coronary artery disease is present, these changes may result in the occurrence of subendocardial ischaemia (ST-segment depression) and arrhythmias.[83] These complications may need rapid intravenous treatment with coronary vasodilators such as GTN.

The ischaemic consequences for the pelvis and lower limb result from interruption of the arterial blood supply, which renders the tissues ischaemic during the crossclamping period. Anaerobic metabolism occurs with accumulation of lactic acid initially and subsequent production of reactive oxygen species, as outlined previously. In patients with a well-developed collateral circulation via the abdominal wall and back, ischaemia may be minimal. A good collateral circulation is more likely in patients with occlusive disease.[84]

Suprarenal and supracoeliac cross-clamping impose a much greater afterload on the left ventricle, resulting in cardiovascular changes of greater magnitude. A continuous intravenous infusion of vasodilator such as GTN or nitroprusside may be required to control blood pressure. The kidneys are rendered ischaemic and so too are the intestines, which appear to be especially sensitive to deprivation of their blood supply. For example, in laboratory animals superficial mucosal acidosis from anaerobic metabolism develops within minutes of occlusion of the mesenteric artery.[85] Disruption of the mucosal barrier has been proposed as the mechanism for development of septicaemia with intestinal organisms and death from multiple organ failure in patients following abdominal aortic surgery.[86]

Aortic unclamping

Aortic unclamping results in restoration of blood flow to the ischaemic tissues and vasodilated blood vessels of the lower torso. The cardiovascular changes are broadly the opposite of those occurring during cross-clamping. Thus the blood pressure and SVR decrease (unclamping shock). The cardiac output, however, may decrease further owing to decreased preload from pooling of blood in the hyperaemic lower extremities before returning to previous levels. There may also be acute blood loss from the graft anastomosis site.

Unclamping shock is usually of short duration (< 10 minutes) and its magnitude can be reduced in several ways. First, the surgeon can release the clamp slowly (over 2–3 minutes), thereby minimizing the rate at which the stagnant blood in the lower torso is washed into the trunk. Second, when a bifurcation graft is used, blood flow can be restored to one leg at a time, again reducing the rate at which acidic blood mixes with the circulation. Third, the anaesthetist can prepare for unclamping by stopping any infusion of vasodilator and reducing the concentration of volatile anaesthetic agent 5–10 minutes before the anticipated release of the clamp. Simultaneously with these manoeuvres, if the ventricular preload (CVP or PAOP) has not been rigorously maintained at the values suggested, it is raised rapidly by 5 mmHg or so by rapid infusion of colloid. These measures, when applied together, will usually minimize the complications of unclamping.

Renal and spinal cord perfusion

Renal   Renal function is impaired preoperatively in about 10 per cent of patients and this may be aggravated by hypovolaemia and hypotension intraoperatively. Despite infrarenal application of the aortic cross-clamp, renal blood flow may be reduced by about 40 per cent with an increase of up to 75 per cent in renal vascular resistance which is not affected by sympathetic blockade.[87] The reduction in blood flow reduces the glomerular filtration rate and urine formation. The mechanism for the reduction in blood flow is uncertain but may be related in part to the reduction in cardiac output during cross-clamping and possibly also a humoral mechanism, since concentrations of renin are increased following cross-clamping.

Provided that cardiovascular function is optimized with maintenance of cardiac filling pressure, renal function is usually adequately maintained with the production of 40–60 ml per hour of urine. Some authorities consider that pharmacological support of the kidney is essential and routinely administer intravenous mannitol (25–50 g), frusemide or dopamine by infusion (1–3 µg/kg per hour).

Mannitol acts in two ways. First, as an osmotic diuretic it is a renal vasodilator and promotes renal tubular urine flow, and, second, it is a hydroxyl radical scavenger and thereby lessens the effects of ischaemia reperfusion injury. To be most effective the mannitol must be administered before the ischaemic episode.[88] Mannitol is essential whenever one or both kidneys are likely to be ischaemic, as in suprarenal cross-clamping. Frusemide is also a renal vasodilator that may be used to augment urine flow.

Dopamine is very effective in increasing sodium, potassium and urine output and creatinine clearance during cross-clamping,[89] and should be used when renal impairment already exists.

Spinal cord   Complete flaccid paraplegia may follow aortic cross-clamping. It is attributed to spinal cord ischaemia due to interruption of flow in the anterior spinal artery, the segmental origin of which is not known to the surgeon before the clamp is applied.

The incidence of paraplegia is about 0.25 per cent in patients undergoing elective abdominal aneurysm surgery and about 2.5 per cent when the aorta has ruptured. The incidence is also higher during surgery on thoracic or thoracoabdominal aneurysms, because of the higher segmental level of cross-clamp application.[90]

Ischaemia reperfusion injury

Primary ischaemic injury occurs in the muscles of the pelvis and lower limbs during infrarenal aortic cross-clamping, with the addition of the kidneys and intestines during suprarenal cross-clamping. The same sites are also exposed to reperfusion injury, which follows restoration of blood flow together with systemic reperfusion injury in the lungs in particular.

Advances in the management of patients undergoing surgery for advanced atherosclerotic disease are likely to occur through attempts to minimize by pharmacological blockade the toxic effects of oxygen-derived free radicals. The role of mannitol in the prevention of ischaemia reperfusion injury of the kidney has already been discussed. The same drug in doses of 0.2 g/kg given before cross-clamp application also limits the acute lung injury and pulmonary oedema which follow aortic unclamping.[91, 92] The local and systemic effects of reperfusion injury are much more severe following aortic rupture. Therefore, mannitol should be administered to every patient undergoing aortic surgery and should always be given before the onset of ischaemia to be most effective.

Allopurinol, which inhibits the action of xanthine oxidase, has been studied in animals. Treatment of rabbits prior to hindlimb ischaemia reduced the skeletal muscle reperfusion injury.[93] The results of clinical trials are awaited with interest.

Postoperative management

The goals of postoperative management are similar to those pertaining intraoperatively. Good analgesia is essential to enable adequate pulmonary ventilation to occur, and this can be better achieved with epidural infusion analgesia using a combination of dilute local anaesthetic and opioid than intramuscular or intravenous opioids alone.[75]

Maintenance of adequate circulating volume must be continued to optimize cardiac output and mean arterial pressure, and this requires continued monitoring of CVP ± PAOP for 24 hours or so. A postoperative ileus will be present in most patients, which prevents oral fluid intake usually until the second postoperative day.

Cardiovascular stability is achieved by attention to pulmonary gas exchange, body temperature, pain relief and adequate circulating volume. Renal function is monitored by observation of the hourly urine output and creatinine clearance.

All these objectives can be met only if the patient is transferred postoperatively to a high dependency or intensive care unit where cardiorespiratory monitoring and intensive nursing care can be given. Many patients are suitable for reversal of anaesthesia and extubation at the end of surgery. However, when the operation is complicated or prolonged, postoperative ventilation should be considered and carried out until complete physiological stability has been achieved. Only then is extubation performed. By following these guidelines the potential cardiac,[94] pulmonary[95] and renal[96] complications can be prevented or minimized.

Ruptured aortic aneurysm

Leakage or rupture of an abdominal aortic aneurysm is fatal unless surgical treatment is attempted. Rupture is retroperitoneal in 75 per cent of patients and intraperitoneal in the remainder.[97] The latter is associated with more severe hypotension from blood loss because of the lack of tamponade effect which contains the haemorrhage when it is retroperitoneal. The outcome of surgery is determined by the interval between the onset of symptoms (abdominal or back pain plus collapse due to shock) and control of the haemorrhage by cross-clamping, plus the severity of the patient's coexisting cardiac, pulmonary and renal disease.

The goal of hospital management is resuscitation, rapid diagnosis, immediate laparotomy and control of the haemorrhage. All patients need massive volume replacement using crystalloids and colloids initially while a minimum of 10 blood group specific units of blood are

Table 56.7 Anaesthetic considerations in ruptured aortic aneurysm

Preoperative
Coexisting cardiac, pulmonary and renal disease
Pain and shock
Fluid resuscitation – 2 large-bore intravenous cannulae
Mannitol

Intraoperative
Preoxygenation – monitor ECG, SaO_2, non-invasive BP
Rapid-sequence induction
Hypotension – aggravated by laparotomy
CVP, PAOP, radial BP after induction
Light anaesthesia
Cross-clamping hypertension – vasodilators
Unclamping hypotension – vasopressors
Massive blood transfusion – coagulopathy

Postoperative
Ventilate until physiologically stable
Death:
    early – cardiac failure
    late – multiple organ failure (ischaemia reperfusion injury)

cross-matched. Time should not be wasted on elaborate laboratory investigations in any patient with shock, back pain and a pulsatile abdominal mass. The principles of anaesthetic management are dealing with the exaggerated cardiovascular responses to induction of anaesthesia, skin incision, cross-clamping and unclamping plus the major blood loss. The major anaesthetic considerations are summarized in Table 56.7.

Anaesthesia is induced in the operating room with the surgeon ready to make the incision. Two wide-bore intravenous infusions are established, the ECG and peripheral SaO_2 are monitored and the blood pressure is measured by a non-invasive cuff. Two or more anaesthetists are essential. The patient is given 100 per cent oxygen while basic monitoring is established, following which anaesthesia is induced with a rapid-sequence induction technique using drugs such as etomidate and suxamethonium. Once the trachea is intubated, surgery may begin. Anaesthesia is continued with ventilation of the lungs using 100 per cent oxygen, intravenous opioid and a muscle relaxant such as atracurium until the patient's response to stress indicates the need for deeper anaesthesia, when a volatile agent such as isoflurane may be added. Mannitol is given (50 g) intravenously in an attempt to minimize the systemic ischaemia reperfusion injury and augment urine production. A dopamine infusion is set up in an attempt to preserve renal function.

The main complication following induction is undue hypotension, which may be resistant to fluid and blood administration. Muscle relaxation and opening of the abdomen may also contribute to the hypotension by removal of the tamponade which has previously contained the haemorrhage. Once the aorta is cross-clamped, undue hypertension may be the next complication necessitating deepening of anaesthesia and a GTN infusion. When the primary haemorrhage is surgically controlled, the anaesthetic team can concentrate on establishing arterial, CVP and PAOP monitoring to attempt to optimize cardiac filling pressures, cardiac output and renal perfusion.

Anaesthesia and surgery then proceed in accordance with the descriptions given for an elective operation. During the operation a massive blood transfusion may be required and be associated with a dilutional coagulopathy. This is treated with fresh-frozen plasma and platelets based on the results of laboratory tests. The final intraoperative problem is persistent unclamping hypotension, which may occur despite adequate cardiac filling pressures. In this situation, intravenous vasopressors, either as a single dose (methoxamine) or by infusion (noradrenaline (norepinephrine), phenylephrine), are administered.

Postoperatively the patient is ventilated in intensive care until physiological stabilization is achieved and then extubation attempted. Epidural analgesia is inappropriate because of the coagulation disorder. Analgesia is therefore administered using traditional techniques. Of the patients who survive surgery, there are very few who do not have major complications affecting the heart, lungs, or kidneys, resulting from pre-cross-clamp ischaemia (shock) and post-cross-clamp reperfusion injury. Cardiac ischaemia is the usual cause of perioperative death whilst multiple organ dysfunction syndrome is the usual cause of late death.

REFERENCES

1. Saklad M. Grading of patients for surgical procedures. *Anesthesiology* 1941; **2**: 281–4.

2. Marx GF, Mateo CV, Orkin LR. Computer analysis of post anaesthetic death. *Anesthesiology* 1973; **39**: 54–8.

3. Welbourn R, Goldman G, Paterson IS, *et al.* Ischaemia–reperfusion injury: pathophysiology and treatment. *British Journal of Surgery* 1991; **78**: 651–5.

4. O'Brien BJ. *Cholesterol and coronary heart disease: consensus or controversy?* London: Office of Health Economics, 1991: 17.

5. Solberg LA, Strong JP. Risk factors and atherosclerotic lesions: a review of autopsy studies. *Arteriosclerosis* 1983; **3**: 187–98.

6. Blankenhorn DH, Nessim SA, Johnson RL, *et al.* Beneficial effects of combined colestipol–niacin therapy on coronary atherosclerosis and coronary venous bypass grafts. *Journal of the American Medical Association* 1987; **257**: 3233–40.

7. DePalma RG. The pathology of atheromas: theories of aetiology and evolution of atheromatous plaques. In: Bell PRF, Jamieson CCV, Ruckley CV, eds. *Surgical management of vascular disease.* London, New York, Toronto, Sydney, Tokyo: WB Saunders, 1992: 21–34.

8. *The health of the nation: a summary of the strategy for health in England.* London: HMSO, 1992.

9. Esterbauer H, Wag G, Puhl H. Lipid peroxidation and its role in atherosclerosis. *British Medical Bulletin* 1993; **49**: 566–76.

10. Tilson MD, Stansel HC. Differences in results for aneurysm vs occlusive disease after bifurcation grafts. *Archives of Surgery* 1980; **115**: 1173–5.

11. Dobrin PB. Pathophysiology and pathogenesis of aortic aneurysms. *Surgical Clinics of North America* 1989; **69**: 687–703.

12. Carden DL, Smith JK, Zimmerman BJ, *et al.* Reperfusion injury following circulatory collapse: the role of reactive oxygen metabolites. *Journal of Critical Care* 1989; **4**: 294–307.

13. Guyton AC. Metabolism of carbohydrates and formation of adenosine triphosphate. *Textbook of medical physiology*, 8th ed. Toronto, Montreal, Sydney, Tokyo, Philadelphia, London: WB Saunders, 1991: 744–53.

14. Gutierrez G, Marini C. Cellular response to hypoxia. In: Vincent JL, ed. *Update in intensive care and emergency medicine, 8.* Berlin, Heidelberg, New York, London, Paris, Tokyo: Springer-Verlag, 1989: 182–94.

15. Cheeseman KH, Slater TF. An introduction to free radical biochemistry. *British Medical Bulletin* 1993; **49**: 481–93.

16. Ernster L. Biochemistry of reoxygenation injury. *Critical Care Medicine* 1988; **16**: 947–53.

17. Morel DR. Role of arachidonic acid metabolism in ARDS. In: Vincent JL, ed. *Update in intensive care and emergency medicine, 8.* Berlin, Heidelberg, New York, London, Paris, Tokyo: Springer-Verlag, 1989: 115–24.

18. Beale R, Grover ER, Smithies M, Bihari D. Acute respiratory distress syndrome (ARDS): no more than a severe acute lung injury? *British Medical Journal* 1993; **307**: 1335–9.

19. Hessel EA. Intraoperative management of abdominal aortic aneurysms: the anesthesiologist's viewpoint. *Surgical Clinics of North America* 1989; **69**: 775–93.

20. Pearce AC, Jones RM. Smoking and anesthesia: preoperative abstinence and perioperative morbidity. *Anesthesiology* 1984; **61**: 576–84.

21. Weitz HH. Cardiac risk stratification prior to vascular surgery. *Medical Clinics of North America* 1993; **77**: 377–96.

22. Hertzer NR. Basic data concerning associated coronary disease in peripheral vascular patients. *Annals of Vascular Surgery* 1987; **1**: 616–20.

23. Rao TLK, Jacobs KH, El-Etr AA. Reinfarction following anesthesia in patients with myocardial infarction. *Anesthesiology* 1983; **59**: 499–505.

24. Ennix Jr CL, Lawrie GM, Morris Jr GC, *et al.* Improved results of carotid endarterectomy in patients with symptomatic coronary disease: an analysis of 1,546 consecutive carotid operations. *Stroke* 1979; **10**: 122–5.

25. Goldman L. Assessment of the patient with known or suspected ischaemic heart disease for noncardiac surgery. *British Journal of Anaesthesia* 1988; **61**: 38–43.

26. Subcommittee of WHO/ISH Mild Hypertension Liaison

Committee. Summary of 1993 World Health Organization–International Society of Hypertension guidelines for the management of hypertension. *British Medical Journal* 1993; **307**: 1541–6.

27. Stone JG, Föex P, Sear JW, *et al.* Myocardial ischemia in untreated hypertensive patients: effects of a single oral dose of beta-adrenergic agents. *Anesthesiology* 1988; **68**: 495–500.

28. Young AE, Sandberg GW, Couch NP. The reduction of mortality of abdominal aortic aneurysm resection. *American Journal of Surgery* 1977; **134**: 585–90.

29. Cunningham AJ. Anaesthesia for repair of abdominal aortic aneurysms. In: Atkinson RS, Adams AP, eds. *Recent advances in anaesthesia and analgesia.* Edinburgh, London, New York: Churchill Livingstone, 1992: 49–69.

30. Mangano DT. Pre-operative assessment of the patient with cardiac disease. *Baillière's Clinical Anaesthesiology* 1989; **3**: 47–102.

31. Föex P, Reeder MK. Anaesthesia for vascular surgery. *Baillière's Clinical Anaesthesiology* 1993; **7**: 97–126.

32. Vandermeulen EPE, Vermylen J, Van Aken H. Epidural and spinal anaesthesia in patients receiving anticoagulant therapy. *Baillière's Clinical Anaesthesiology* 1993; **7**: 663–89.

33. Underwood MJ, More RS. The aspirin papers. *British Medical Journal* 1994; **308**: 71–2.

34. Macdonald R. Aspirin and extradural blocks. *British Journal of Anaesthesia* 1991; **66**: 1–3.

35. Archer C, Levy AR, McGregor M. Value of routine preoperative chest x-rays: a meta-analysis. *Canadian Journal of Anaesthesia* 1993; **40**: 1022–7.

36. Raby KE, Barry J, Creager MA, *et al.* Detection and significance of intraoperative and postoperative myocardial ischemia in peripheral vascular surgery. *Journal of the American Medical Association* 1992; **268**: 222–7.

37. Beven EG. Routine coronary angiography in patients undergoing surgery for abdominal aortic aneurysm and lower extremity occlusive disease. *Journal of Vascular Surgery* 1986; **31**: 682–4.

38. Clark NJ, Stanley TH. Anesthesia for vascular surgery. In: Miller RD, ed. *Anaesthesia,* vol. 2, 3rd ed. New York, Edinburgh, London, Melbourne, Tokyo: Churchill Livingstone, 1990: 1693–736.

39. Goldman L, Caldera DI, Nussbaum SR, *et al.* Multifactorial index of cardiac risk in non-cardiac surgery procedures. *New England Journal of Medicine* 1977; **297**: 845–50.

40. Goldman L. Cardiac risk assessment in patients with arteriosclerotic disease. In: Kaplan JA, ed. *Vascular anaesthesia.* New York, Edinburgh, London, Melbourne, Tokyo: Churchill Livingstone, 1991: 1–20.

41. Eastcott HHG, Pickering GW, Robb CG. Reconstruction of internal carotid artery in a patient with intermittent attacks of hemiplegia. *Lancet* 1954: **2**: 994–6.

42. Garrioch MA, Fitch W. Anaesthesia for carotid artery surgery. *British Journal of Anaesthesia* 1993; **71**: 569–79.

43. North American Symptomatic Carotid Endarterectomy Trial Collaborators. Beneficial effect of carotid endarterectomy in symptomatic patients with high-grade carotid stenosis. *New England Journal of Medicine* 1991; **325**: 445–53.

44. European Carotid Surgery Trialists' Collaborative Group. MRC European Carotid Surgery Trials: interim results for symptomatic patients with severe (70–99%) or with mild (0–29%) carotid stenosis. *Lancet* 1991; **337**: 1235–43.

45. Sandmann W. Carotid endarterectomy. In: Bell PRF, Jamieson CCV, Ruckley CV, eds. *Surgical management of vascular disease.* London, New York, Toronto, Sydney, Tokyo: WB Saunders, 1992: 671–81.

46. Naylor AR, Bell PRF, Ruckley CV. Monitoring and cerebral protection during carotid endarterectomy. *British Journal of Surgery* 1992; **79**: 735–41.

47. Archie JP. Technique and clinical results of carotid stump back pressure to determine selective shunting during carotid endarterectomy. *Journal of Vascular Surgery* 1991; **13**: 319–27.

48. Bishop CCR, Powell S, Rutt D, Browse NL. Transcranial Doppler measurement of middle cerebral artery velocity: a validation study. *Stroke* 1986; **17**: 913–15.

49. Youngberg JA, Gold MD. Carotid artery surgery: perioperative anaesthetic considerations. In: Kaplan JA, ed. *Vascular anaesthesia.* New York, Edinburgh, London, Melbourne, Tokyo: Churchill Livingstone, 1991: 333–61.

50. Cucchiara RF, Benefiel DJ, Matteo RS, *et al.* Evaluation of esmolol in controlling increases in heart rate and blood pressure during endotracheal intubation in patients undergoing carotid endarterectomy. *Anesthesiology* 1986; **65**: 528–31.

51. Michenfelder JD, Sundt TM, Fode N, *et al.* Isoflurane when compared to enflurane and halothane decreases the frequency of cerebral ischemia during carotid endarterectomy. *Anesthesiology* 1987; **67**: 336–40.

52. Messick JM, Casement B, Sharbrough FW, *et al.* Correlation of regional cerebral blood flow (rCBF) with EEG changes during isoflurane anesthesia for carotid endarterectomy: critical rCBF. *Anesthesiology* 1987; **66**: 344–9.

53. Hobbhahn J, Hansen E, Keyl C. Volatile anaesthetics in patients with coronary artery disease and with heart failure. *Baillière's Clinical Anaesthesiology* 1993; **7** (4): 1057–77.

54. Pulsinelli WA, Levy DE, Sigsbee B, *et al.* Increased damage after ischaemic stroke in patients with hyperglycemia with or without established diabetes mellitus. *American Journal of Medicine* 1983; **74**: 540–4.

55. Harris DNF, Bailey SM. Near infrared spectroscopy in adults. Does the Invos 3100 really measure intracerebral oxygenation? *Anaesthesia* 1993; **48**: 694–6.

56. Kane CM, Le Cheminant D, Horan BF. Acute cardiovascular responses to internal carotid artery occlusion during carotid endarterectomy and to the restoration of internal carotid flow. *Anaesthesia and Intensive Care* 1987; **15**: 289–95.

57. Dehn TCB, Angell-James JE. Long term effect of carotid endarterectomy on carotid sinus baroreceptor function and blood pressure control. *British Journal of Surgery* 1987; **74**: 997–1000.

58. Riles TS, Kopelman I, Imparato AM. Myocardial infarction following carotid endarterectomy: a review of 686 operations. *Surgery* 1979; **85**: 249–52.

59. Johnston KW, Scobie TK. Multicenter prospective study of nonruptured abdominal aortic aneurysms. 1. Population and operative management. *Journal of Vascular Surgery* 1988; **7**: 69–81.

60. Reid DB, Welch GH, Pollock JG. Abdominal aortic aneurysm: a preventable cause of death? *Journal of the Royal College of Surgeons of Edinburgh* 1990; **35**: 284–8.

61. Lambert ME, Baguley P, Charlesworth D. Ruptured abdominal aortic aneurysms. *Journal of Cardiovascular Surgery* 1986; **27**: 256–61.

62. Kwitka G, Kidney SA, Nugent M. Thoracic and abdominal aortic aneurysm resections. In: Kaplan JA, ed. *Vascular anaesthesia*. London, New York, Edinburgh, Melbourne, Tokyo: Churchill Livingstone, 1991: 363–94.

63. Greenhalgh RM. Prognosis of abdominal aortic aneurysm. *British Medical Journal* 1990; **301**: 136.

64. Szilagyi DE, Elliott Jr JP, Smith RF, *et al.* A thirty year survey of the reconstructive surgical treatment of aortoiliac occlusive disease. *Journal of Vascular Surgery* 1986; **3**: 421–36.

65. Sumio BE, Traquina DN, Gusberg RJ. Results of aortic grafting in occlusive vs aneurysmal disease. *Archives of Surgery* 1985; **120**: 817–19.

66. Picone AL, Green RM, Ricotta JR, *et al.* Spinal cord ischemia following operations on the abdominal aorta. *Journal of Vascular Surgery* 1986; **3**: 94–103.

67. Dubost C, Allary M, Oeconomos N. Resection of an aneurysm of the abdominal aorta. Reestablishment of continuity by a preserved human arterial graft, with result after five months. *Archives of Surgery* 1952; **64**: 405–8.

68. Vorhees AB, Joretzki A, Blasmore AH. The use of tubes constructed from Vinyon 'N' cloth in bridging arterial defects. *Annals of Surgery* 1952; **135**: 332–6.

69. Sicard GA, Allen BT, Munn JS, Anderson CB. Retroperitoneal versus transperitoneal approach for repair of abdominal aortic aneurysms. *Surgical Clinics of North America* 1989; **69** (4): 795–806.

70. Imparato AM. Surgical management of abdominal aortic aneurysms. In: Bell PRF, Jamieson CCV, Ruckley CV, eds. *Surgical management of vascular disease*. London, New York, Toronto, Sydney, Tokyo: WB Saunders, 1992: 843–57.

71. Mannick JA, Whittemore AD, Couch NP. Abdominal aortic aneurysms. In: Greenhalgh RM, ed. *Vascular surgical techniques*. London, Boston, Toronto: Butterworths, 1984: 107–16.

72. Ghignone M, Calvillo O, Quintin L. Anesthesia and hypertension: the effect of clonidine on perioperative hemodynamics and isoflurane requirements. *Anesthesiology* 1987; **67**: 3–10.

73. Slogoff S, Keats A, Dear WE, *et al.* Steal-prone coronary anatomy and myocardial ischemia associated with four primary anesthetic agents in humans. *Anesthesia and Analgesia* 1991; **72**: 22–7.

74. Colson P, Capdevilla X, Cuchet D, *et al.* Does choice of the anesthetic influence renal function during infrarenal aortic surgery? *Anesthesia and Analgesia* 1992; **74**: 481–5.

75. Vercauteren MP. The role of the perispinal route for postsurgical pain relief. *Baillière's Clinical Anaesthesiology* 1993; **7**: 769–92.

76. Ansley DM, Whalley DG, Bent JM, Derbekyan V. The relationship between central venous pressure and pulmonary capillary wedge pressure during aortic surgery. *Canadian Journal of Anaesthesia* 1987; **34**: 594–600.

77. National Institutes of Health Consensus Development Conference. Perioperative red cell transfusion. *Transfusion Medicine Reviews* 1989; **3**: 63–8.

78. Carson JL, Willett LR. Is a hemoglobin of 10 g/dl required for surgery? *Medical Clinics of North America* 1993; **77**: 335–47.

79. Tulloh BR, Brakespear CP, Bates SC, *et al.* Autologous predonation, haemodilution and intraoperative blood salvage in elective abdominal aortic aneurysm repair. *British Journal of Surgery* 1993; **80**: 313–15.

80. Selzer JL, Ritter DE, Starsnic MA, Marr AT. The hemodynamic response to traction on the abdominal mesentery. *Anesthesiology* 1985; **63**: 96–9.

81. Gottlieb A, Skrinska VA, O'Hara P, *et al.* The role of prostacyclin in the mesenteric traction syndrome during anesthesia for abdominal aortic reconstructive surgery. *Annals of Surgery* 1989; **209**: 363–7.

82. Hudson JC, Wurm WH, O'Donnell Jr TF, *et al.* Ibuprofen pretreatment inhibits prostacyclin release during abdominal exploration in aortic surgery. *Anesthesiology* 1990; **72**: 443–9.

83. Cunningham AJ. Anaesthesia for repair of abdominal aortic aneurysms. In: Atkinson RS, Adams AP, eds. *Recent advances in anaesthesia and analgesia*. Edinburgh, London, New York: Churchill Livingstone, 1992: 49–69.

84. Cunningham AJ. Anaesthesia for abdominal aortic surgery – a review. Part 1. *Canadian Journal of Anaesthesia* 1989; **36**: 426–44.

85. Fiddian-Green RG. Splanchnic ischaemia and multiple organ failure in the critically ill. *Annals of the Royal College of Surgeons of England* 1988; **70**: 128–34.

86. Fiddian-Green RG, Gantz NM. Transient episodes of sigmoid ischaemia and their relation to infection from intestinal organisms after abdominal aortic operations. *Critical Care Medicine* 1987; **15**: 835–9.

87. Gamulin Z, Forster A, Simonet F, *et al.* Effects of renal sympathetic blockade on renal hemodynamics in patients undergoing major aortic abdominal surgery. *Anesthesiology* 1986; **65**: 688–92.

88. Ouriel K, Smedina NG, Ricott JJ. Protection of the kidney after temporary ischemia: free radical scavengers. *Journal of Vascular Surgery* 1985; **2**: 49–53.

89. Salem MG, Crooke JW, McLoughlin GA, *et al.* The effect of dopamine on renal function during aortic cross clamping. *Annals of the Royal College of Surgeons of England* 1988; **70**: 9–12.

90. Szilagyi DE, Hageman JH, Smith RF, *et al.* Spinal cord damage in surgery of the abdominal aorta. *Surgery* 1978; **83**: 38–56.

91. Paterson IS, Klausner JM, Pugatch R, *et al.* Noncardiogenic pulmonary edema after abdominal aortic aneurysm repair. *Annals of Surgery* 1989; **209**: 231–6.

92. Paterson IS, Klausner JM, Goldman G, *et al.* Pulmonary edema after aneurysm surgery is modified by mannitol. *Annals of Surgery* 1989; **210**: 796–801.

93. Oredsson S, Plate G, Qvarfordt P. Allopurinol – a free radical scavenger – reduces reperfusion injury in skeletal muscle. *European Journal of Vascular Surgery* 1991; **5**: 47–52.

94. Heagerty AM, Barnett DB. Postoperative cardiac complications following vascular surgery. In: Bell PRF, Jamieson CCV, Ruckley CV, eds. *Surgical management of vascular disease*. London, New York, Toronto, Sydney, Tokyo: WB Saunders 1992: 999–1011.

95. Coley S, Smith G. Respiratory problems in the postoperative period. In: Bell PRF, Jamieson CCV, Ruckley CV, eds. *Surgical management of vascular disease*. London, New York, Toronto, Sydney, Tokyo: WB Saunders, 1992: 1013–25.

96. Smithies M, Cameron JS. Renal failure following reconstructive arterial surgery. In: Bell PRF, Jamieson CCV, Ruckley CV, eds. *Surgical management of vascular disease.* London, New York, Toronto, Sydney, Tokyo: WB Saunders, 1992: 1027–48.

97. Roizen MF. Anesthesia for emergency surgery for abdominal aortic reconstruction. In: Roizen MF, ed. *Anesthesia for vascular surgery.* New York: Churchill Livingstone, 1990: 311–16.

Chapter 57

Anaesthesia for Thoracic Surgery

Jay B. Brodsky

Preoperative assessment and management

Prior to any major thoracic operation a complete medical history and physical examination, an electrocardiogram, a chest radiograph and an arterial blood gas measurement should be obtained from all patients.

The physical examination is directed at the pulmonary system. The breathing pattern should be observed. The presence of wheezing, râles, rhonchi or other abnormal breath sounds should be noted. Their presence may suggest the need for further medical intervention. Coughing may indicate increased bronchial irritability or secretions. Central cyanosis suggests severe arterial hypoxaemia. Clubbing may be seen in patients with chronic pulmonary disease or malignancy.

The trachea should be midline, and any deviation should alert the anaesthetist to a potentially difficult intubation or the possibility of airway obstruction during induction of anaesthesia.

Blood studies may reveal secondary polycythaemia, a reflection of inadequate oxygenation. Leucocytosis may indicate active pulmonary infection.

The ECG of the patient with chronic obstructive pulmonary disease may demonstrate evidence of right atrial and ventricular hypertrophy. An enlarged P wave in lead II ('P' pulmonale) indicates right atrial hypertrophy. There may be a low-voltage QRS complex due to hyperinflation of the lungs and poor precordial R-wave progression.

Because many patients have pre-existing pulmonary disease and because major alterations in pulmonary function are expected in all patients following thoracotomy, preoperative pulmonary function tests are useful to establish a baseline and quantify available reserves. This information will help to predict the patient's ability to maintain adequate spontaneous ventilation, deep breathing and coughing following surgery.

Flow–volume loop measurements are one means of assessing airway status. The shape and peak air flow rates during expiration at high lung volumes are effort-dependent and indicate the patency of the larger airways. Expiration at low lung volumes is effort-independent and reflects resistance in smaller airways.

The most readily available spirometric tests for routine preoperative evaluation are the forced vital capacity (FVC) and the forced expiratory volume (FEV). These simple measurements can be made at the patient's bedside with a spirometer.

Forced vital capacity can be compared with normal values based on sex, height and age, and can be expressed as a percentage of the predicted vital capacity (%FVC). A %FVC greater than 80 per cent is considered normal, 70–80 per cent borderline normal, 60–70 per cent suggests pulmonary disease and less than 60 per cent indicates significant restriction of pulmonary function.

The FEV in 1 second (FEV_1) is the volume of air forcefully expired in the first second of a FVC manoeuvre. The FEV_1 can be compared with the actual measured FVC (ratio FEV_1/FVC = %FEV_1). The %FEV_1 is useful in differentiating between restrictive and obstructive pulmonary disease. Normally, the %FEV_1 is greater than 70 per cent. In restrictive disease both FEV_1 and FVC decrease so %FEV_1 is normal. In obstructive disease only FEV_1 is reduced. A %FEV_1 less than 70 per cent suggests significant airway resistance which increases the work of breathing during stress. A 15 per cent improvement in %FEV_1 after bronchodilator therapy indicates reversibility of airway obstruction, and in these patients bronchodilators should be administered preoperatively.

The %FVC and %FEV_1 can be used to predict the

degree to which the pre-existing obstructive and restrictive components of pulmonary function may compromise the ability to ventilate adequately and to maintain clear lungs after thoracic surgery. A vital capacity at least three times greater than tidal volume is necessary for an effective cough. A preoperative FVC of less than 20 ml/kg, FEV_1 of less than 1.2 litres and a %FEV_1 of less than 35 per cent are each highly predictive of postoperative acute ventilatory failure.

More sophisticated split-lung ventilation/perfusion studies predict the amount of functional lung tissue remaining after pulmonary resection and are especially useful before pneumonectomy. Radioactive xenon is injected intravenously, and the radioactivity subsequently measured in each lung is proportional to regional perfusion. Ventilation is measured by the inhalation of radioactive gas. The radioactivity measured in each lung area is proportional to the degree of regional ventilation.

Although preoperative spirometric and ventilation/perfusion studies are the most reliable means of identifying patients with a high risk for pulmonary complications following thoracotomy, other indices have been used. These include diffusion capacity, pulmonary artery pressure, oxygen tension during pulmonary artery occlusion and presurgical exercise testing.[1] Results have been variable and usually are not as accurate as spirometry for predicting postoperative pulmonary function.[2]

Besides preoperative pulmonary disease, important predictive factors of postoperative pulmonary complications include the site and extent of the operative procedure, pre-existing cardiovascular or neurological dysfunction, and the interval between surgery and ambulation. Detection and correction of pre-existing reversible pulmonary disease, smoking, sepsis and obesity are particularly useful in reducing postoperative mortality and morbidity. Cessation of cigarette smoking, even for as little as 48 hours before surgery, decreases carboxyhaemoglobin levels and improves postoperative recovery.

A system to predict the risk of pulmonary complications following abdominal and thoracic operations has been suggested (Table 57.1). For the thoracic surgical patient with a preoperative score of zero, there is minimal expectation of postoperative pulmonary complications. Patients with a score of 1–2 will probably require supplemental oxygen and incentive spirometry. The high-risk patient (score 3 or greater) should remain in the intensive care unit for at least 1 day following surgery, to be monitored and assisted with pulmonary toilet.

Table 57.1 Predicting the risk of pulmonary complications after thoracic surgery*

Category	Point
I. Expiratory spirogram	
a. Normal (%FVC + %FEV_1/FVC > 150)	0
b. %FVC + %FEV_1/FVC = 100–150	1
c. %FVC + %FEV_1/FVC < 100	2
d. Preoperative FVC < 20 ml/kg	3
e. Postbronchodilator FEV_1/FVC < 50%	3
II. Cardiovascular system	
a. Normal	0
b. Controlled hypertension, myocardial infarction without sequelae for more than 2 years	0
c. Dyspnoea on exertion, orthopnoea, paroxysmal nocturnal dyspnoea, dependent oedema, congestive heart failure, angina	1
III. Arterial blood gases	
a. Acceptable	0
b. $Paco_2$ > 6.7 kPa (50 mmHg) or Pao_2 < 8 kPa (60 mmHg) on room air	1
c. Metabolic pH abnormality > 7.50 or < 7.30	1
IV. Nervous system	
a. Normal	0
b. Confusion, obtundation, agitation, spasticity, discoordination, bulbar malfunction	1
c. Significant muscular weakness	1
V. Postoperative ambulation	
a. Expected ambulation (minimum, sitting at bedside) within 36 hours	0
b. Expected complete bed confinement for at least 36 hours	1

* Reproduced, with permission, from reference 51. (kPa figures added in IIIb.)
Points are assigned at the preoperative assessment of the patient. The predictive risk of pulmonary complications following thoracic procedures is based on the total point score. Low risk, 0 points; moderate risk, 1–2 points; high risk, 3 points.

No preoperative management plan will completely eliminate postoperative pulmonary complications but, for patients with significant chronic pulmonary disease or retained secretions, a significant improvement in vital capacity or FEV_1 can be obtained within 48 hours with appropriate bronchial hygiene therapy (aerosol therapy with or without bronchodilator, chest physiotherapy, intermittent positive pressure breathing).[3] Every patient must be encouraged to cough, breathe deeply and get up and about as soon as possible after surgery. Preoperative instructions in breathing and coughing exercises can diminish the incidence and severity of postoperative pulmonary complications.

Choice of anaesthetic agent

The anaesthetic management of the patient undergoing thoracic surgery is challenging because haemodynamic stability and oxygenation must be maintained during selective one-lung ventilation (OLV). The choice of anaesthetic agent requires consideration of multiple and sometimes conflicting effects (Table 57.2).

Airway resistance

General anaesthesia normally increases airway resistance by reducing FRC. Patients undergoing thoracotomy are at further risk of increased airway resistance due to airway obstruction from excess secretions or tumour. Many patients have chronic obstructive pulmonary disease, asthma, cystic fibrosis or other lung diseases which also affect airway calibre and reactivity. Surgical trauma to the lung can cause haemorrhage and bronchospasm. Furthermore, endobronchial intubation can produce bronchospasm from direct mucosal stimulation, whilst ventilation through a single lumen of the double-lumen tube (DLT) produces an increase in airway resistance.

These effects can be partially alleviated by the use of inhalational anaesthetics. In experimental models halothane, enflurane and isoflurane are potent, direct bronchodilators. They also obtund bronchoconstrictive airway reflexes in patients with reactive airways. Clinically, isoflurane is preferred because it produces less myocardial depression and fewer ventricular arrhythmias than either halothane or enflurane.

Drugs that release histamine can produce bronchospasm. Thiopentone, thiamylal and propanidid cause dose-related histamine release, whereas methohexitone, etomidate and propofol do not.[4] Ketamine has bronchodilating properties and can be used for induction of anaesthesia in the patient

Table 57.2 The properties of anaesthetic agents used for thoracic procedures*

Anaesthetic	Desirable	Undesirable
Volatile	Permits use of high FIO_2 Bronchodilation Diminishes airway reflexes Readily eliminated	Inhibits HPV Myocardial depression
Narcotics	Do not inhibit HPV No myocardial depression when used alone Provide postoperative analgesia	Not general anaesthetics May depress ventilation in immediate postoperative period
Nitrous oxide	Readily eliminated Probably no effect on HPV	Reduces FIO_2
Ketamine	Diminishes airway irritability Does not inhibit HPV Cardiovascular stability during hypovolaemia	Myocardial ischaemia Emergence delirium
Thiopentone	Does not inhibit HPV	Minor potential for histamine release and bronchospasm
Muscle relaxants	Facilitate mechanical ventilation Enhance surgical exposure Minimize doses of general anaesthetics	Potential for postoperative weakness Possible histamine release and bronchospasm Need for use of reversal agent
Cholinesterase inhibitors	Reverse neuromuscular blockade	May produce acetylcholine-mediated bronchospasm

HPV, hypoxic pulmonary vasoconstriction.
* Reproduced, with permission, from Seigel LC, Brodsky JB. Choice of anesthetic agents. In: Kaplan JA, ed. *Thoracic anesthesia*, 2nd ed. New York: Churchill Livingstone, 1991.

with increased airway reactivity. For muscle relaxation, vecuronium and pancuronium are preferred. Curare releases histamine and should be avoided.

Hypoxic pulmonary vasoconstriction

Many factors influence the degree of arterial hypoxaemia during OLV, including inadequate inspired oxygen concentration (FIO_2), alveolar hypoventilation, and the large alveolar-to-arterial oxygen tension gradient that occurs from continued perfusion to the selectively deflated lung.[5] Normally, regional hypoxia in the lung causes arteriolar constriction with diversion of blood flow away from the hypoxic segment (the 'hypoxic pulmonary vasoconstrictive' (HPV) response). Under experimental conditions, HPV is an important regulator of blood flow to the atelectatic lung. The primary stimuli for the HPV reflex are alveolar oxygen tension (PAO_2) and mixed venous oxygen tension ($P\bar{v}O_2$).[6]

The HPV response can be modified by many factors. For example, vasoconstrictive drugs (dopamine, adrenaline (epinephrine), phenylephrine) will preferentially constrict the pulmonary vessels perfusing normoxic or hyperoxic lung segments. This increases pulmonary vascular resistance in the ventilated lung, causing redistribution of blood flow to atelectactic areas and therefore the lowering arterial oxygen content. Vascular smooth muscle relaxation by the direct action of sodium nitroprusside or nitroglycerin will blunt HPV in the atelectactic lung, thereby increasing blood flow to that lung and lowering PaO_2.

Similarly, surgical manipulation of the lung can transiently blunt HPV owing to thromboxane- and prostacyclin-mediated local vasodilation.[7] The effects of prostaglandins on HPV are complex. Because some prostaglandins inhibit HPV, it has been suggested that prostaglandin inhibitors might potentiate the HPV response. In an *in vitro* study of hypoxic rat lung, ibuprofen, a cyclo-oxygenase inhibitor, potentiated HPV and reversed the depression of HPV by halothane.[8] Other prostaglandins have an opposite effect. Prostaglandin $F_{2\alpha}$ is a potent pulmonary vasoconstrictor. In another animal study, direct infusion of this drug into the non-ventilated lung resulted in a significant increase in PaO_2 and a decrease in shunt.[9] There have been no clinical applications of these direct pharmacological manoeuvres to improve oxygenation during OLV.

The overall effects of anaesthetics on HPV are complicated.[10] *In vitro* animal studies show intravenous drugs (pentobarbitone, thiopentone, hexobarbitone, diazepam, droperidol, ketamine, opioids and pentazocine) do not alter the HPV response, whereas the inhaled anaesthetics (halothane, methoxyflurane, enflurane and isoflurane) all inhibit HPV in a dose-dependent manner.[11] In intact animals and patients there is no inhibition of HPV by intravenous anaesthetics (ketamine, thiopentone, pentobarbitone, fentanyl, pethidine (meperidine), lignocaine (lidocaine) and chlorpromazine), but a wide range of effects are seen with the inhalational anaesthetics. The conflicting results between *in vitro* and *in vivo* studies may be due to the complex effects of anaesthetics on cardiac output, oxygen consumption, shunt, $P\bar{v}O_2$, and to other mechanical effects – surgical manipulation of the lung, use of positive end-expiratory pressure (PEEP) – in the patient undergoing an operation.

Because they directly inhibit HPV and indirectly augment the HPV response, the overall clinical effect on HPV by inhalational agents is small.[12] Inhalational anaesthetics directly increase shunt through partial inhibition of HPV, producing modest reductions in PaO_2, but any inhalational anaesthetic that lowers cardiac output more than it decreases oxygen consumption will also lower $P\bar{v}O_2$, thereby producing a potent stimulus for HPV. The effectiveness of the HPV reflex varies inversely with cardiac output. For instance, an agent such as halothane, a myocardial depressant, will decrease cardiac output, thereby reducing blood flow to the collapsed lung and further neutralizing any direct depression of HPV. The level of oxygenation remains relatively unchanged.

Ventilation with 100 per cent oxygen, the application of continuous positive airway pressure (CPAP) to the non-dependent lung and intermittent positive pressure to the dependent (ventilated) lung counter any drug-mediated effects on HPV and maintain oxygenation during one-lung ventilation[13] (see 'Optimizing oxygenation during one-lung ventilation', later).

Respiratory drive

In addition to producing loss of awareness and analgesia, inhalational anaesthetics, opioids and sedatives act on the central nervous system to depress breathing. In the awake, spontaneously breathing patient ventilation increases linearly as arterial ($PaCO_2$) or alveolar carbon dioxide ($PACO_2$) concentration rises. Increasing the depth of anaesthesia results in an elevation of $PaCO_2$ and a diminished ventilatory response to carbon dioxide reflected as a decrease in the slope and a shift to the right of the carbon dioxide response curve. These effects on ventilatory control are of little concern during most intrathoracic procedures because ventilation is mechanically controlled.

Unimpaired control of ventilation is of paramount importance in the postoperative period. Following thoracotomy volatile anaesthetics are rapidly eliminated and do not contribute significantly to ventilatory depression in the immediate postoperative period.

At clinical concentrations both the hypercapnic and the hypoxic ventilatory drives are depressed by inhalational anaesthetics. Subanaesthetic concentrations of halothane and enflurane have *no* effect on the ventilatory response to hypercapnia, but depression of the ventilatory response to hypoxaemia is seen at levels as low as 0.05 MAC with all halogenated anaesthetics. This effect persists into the early postoperative period after the patient has regained consciousness and while he appears to be recovering. For patients with advanced pulmonary disease who are normally dependent on their hypoxic drive to breathe, this effect can be important.[14] The intraoperative use of parenteral opioids increases the risk of respiratory depression in the postoperative period to a greater extent than when an inhalational anaesthetic is used alone.

Ketamine has a rapid onset of action, maintains cardiovascular stability and is useful for induction of anaesthesia in unstable patients undergoing emergency thoracotomy. Ketamine has direct bronchodilatory effects and antagonizes bronchoconstriction from histamine release without depressing respiration. Continuous infusion of ketamine (combined with nitrous oxide and a muscle relaxant) has been used successfully during thoracic surgery without unpleasant emergence phenomena.[15]

Regional anaesthesia during thoracotomy usually has no clinical effect on ventilation unless the level of anaesthesia is high enough such that the intercostal and phrenic nerves become paralysed. High thoracic epidural anaesthesia causing mechanical impairment of rib cage movement may decrease the ventilatory response to carbon dioxide.

Incomplete reversal of muscle relaxants allowing residual motor block must always be considered as a potential cause of postoperative ventilatory depression following thoracotomy.

Monitoring

Routine monitoring for all thoracotomy patients should include a chest or oesophageal stethoscope, continuous non-invasive blood pressure, electrocardiography, temperature, end-tidal capnography and pulse oximetry. A urinary catheter should also be placed before long procedures and/or when epidural opioids are used.

Continuous monitoring of oxygenation is mandatory during one-lung ventilation, and for many patients pulse oximetry is all that is needed.[16] However, pulse oximetry is inaccurate during periods of hypothermia or marked hypotension. Oxygenation, ventilation and acid–base status are best measured by sampling arterial blood; therefore if blood gas analysis is desirable, an indwelling arterial line should be placed. Beat-to-beat haemodynamic monitoring with an arterial line is particularly useful because surgical retraction on the pericardium and heart

or great vessels can cause sudden hypotension and arrhythmias. Intra-arterial fibreoptic probes (optodes) which continuously measure PaO_2, display arterial pressure and allow arterial blood sampling are now becoming available.

As fluid volume shifts are not large during thoracotomy, central venous pressure (CVP) or pulmonary artery (PA) monitors are not needed routinely. For procedures with large volume shifts (oesophageal resection and pneumonectomy) they are helpful after surgery. With advanced pulmonary disease and/or left ventricular dysfunction, the CVP may not reflect left-sided filling pressures. A PA catheter permits monitoring of haemodynamic function (preload, afterload and cardiac output) and oxygen consumption (mixed venous oxygen saturation), so it is particularly useful in the patient with pre-existing myocardial dysfunction, ischaemia and valvular heart disease.

Open thoracotomy in the lateral decubitus position limits the usefulness of information derived from CVP or PA catheters. Normally the CVP reflects blood volume, venous tone and right ventricular performance. During thoracotomy the CVP may be altered by surgical retraction, mediastinal and diaphragmatic shifts, and the application of PEEP which changes intrathoracic pressure.

When a PA catheter is used, its position should be documented radiographically *before* surgery, as these catheters enter the right pulmonary artery 85 per cent of the time. Therefore, during right thoracotomy the catheter tip will be in the non-dependent pulmonary artery and during left thoracotomy the catheter tip will be in the vessel of the non-operated, dependent lung. In both situations the data obtained may be inaccurate. Haemodynamic measurements and mixed venous oxygen saturation measurements are lower during right thoracotomy than during left thoracotomy, as both are affected by reduced blood flow to the non-ventilated right lung.

Isolation of the lungs

Intentionally isolating and collapsing the operated lung provides optimal conditions for the surgeon because vigorous retraction is unnecessary. Enhanced surgical exposure is probably the most common indication for selective lung collapse. Isolating the lungs protects the non-operated, dependent lung from contamination by blood, mucus or tumour material. Independent distribution of ventilation to each lung is essential for certain procedures (see 'Special procedures', later). With the patient in the lateral decubitus position, selective large tidal volume ventilation to the dependent lung minimizes atelectasis of that lung during OLV.

Endobronchial blockade

Endobronchial blockade has been performed with gauze tampons, special cuffed rubber blockers and the balloon of embolectomy, pulmonary artery or urinary catheters. Lung tissue distal to the obstruction collapses.

Because the smallest double-lumen tube (size 28 Fr) is too large for small children, blockade with an embolectomy or pulmonary artery catheter is currently the only practical method for lung separation in paediatric patients. Larger embolectomy catheters (8–12 Fr) can be used in adults.[17] In adults the catheter is passed through a tracheal tube, whilst in children it is passed alongside the tracheal tube. Fibreoptic bronchoscopy is required to position the balloon at the appropriate site in the bronchus.

There are several disadvantages to the use of bronchial blockers. Bronchoscopy, either rigid or fibreoptic, is required for accurate placement. Blockers are easily displaced during change in patient position or from surgical manipulation. If the blocker slips into the trachea, it will obstruct ventilation to the non-operated lung. Lung tissue distal to the obstruction cannot be suctioned or re-expanded during the procedure.

The Univent tube (Fuji Systems Corp., Tokyo, Japan) overcomes some of these disadvantages. The tube consists of a conventional tracheal tube with an additional small lumen. This second lumen contains a thin tube which can be advanced as far as 8 cm past the larger lumen into the bronchus. A balloon located at the tip of this tube serves as a blocker. Accurate placement requires fibreoptic bronchoscopy.[18] Suctioning, pulmonary lavage and oxygen insufflation can be performed through the smaller lumen.[19] Ventilation to both lungs can be reinstituted at any time by deflating the balloon and withdrawing the blocker tube back into the body of the main tube, so the patient need not be reintubated if postoperative ventilation is planned.

Because the bronchial tube is so thin, its balloon must be inflated with two to ten times as much air as the bronchial cuff of a double-lumen tube in order to seal the airway. This generates high, potentially dangerous wall pressures and predisposing the balloon to carinal herniation[20] (Fig. 57.1).

Double-lumen tubes (DLTs)

The major advantage of DLTs compared to bronchial blockers is that each lung can be independently ventilated, collapsed and re-expanded at will.

Properties

As a positioning aid, the original Carlens and White DLTs had hooks to engage the carina. However, these hooks made

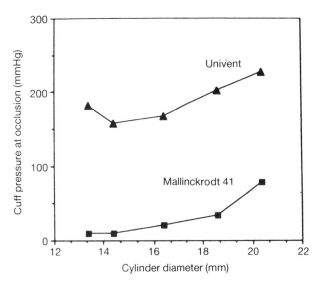

Fig. 57.1 Bronchial cuff pressures of a plastic double-lumen tube (size 41 Fr BronchoCath tube; Mallinckrodt, St Louis, MO) were compared with a bronchial blocker (size 9 mm, Univent tube, Fuji Systems, Tokyo, Japan). The estimated transmural wall pressures (mmHg) exerted at the minimum volume necessary to occlude the airway in several different size bronchial models are shown. The transmural wall pressure equals the total pressure generated at the minimal occlusion volumes minus the pressure needed to maintain inflation of the cuff itself in free air at the minimal occlusion volume. In all bronchial models the inflated cuff of the double-lumen tube exerted less transmural wall pressure than the inflated balloon of the bronchial blocker. (Reproduced, with permission, from reference 20.)

passage through the glottis difficult. Robertshaw introduced a DLT without a hook and with larger internal lumens. Modern plastic DLTs resemble the Robertshaw tube.

All DLTs have a cuff placed above the opening of the short tracheal lumen to prevent gas leaks during ventilation. Inspired gas can be diverted into either or both lumens by a second cuff on the long bronchial lumen. The bronchial cuff also isolates and protects each lung from contralateral contamination. In tubes designed for the right main bronchus, the lateral aspect of the bronchial cuff or bronchial lumen is fenestrated to allow gas exchange with the right upper lobe bronchus. The proximal end of each lumen is fitted to a special connector that distributes ventilation to either or both lungs. Each lumen can be independently opened to the atmosphere, collapsing the lung on that side while ventilation to the other lung continues. A suction catheter or fibreoptic bronchoscope can be passed down either lumen while ventilating the other lung.

Disposable plastic tubes have generally replaced rubber tubes. Reusable rubber DLTs are expensive and have a relatively short shelf life. They are easily damaged during cleaning and resterilization, which increases the risk of airway injury.[21]

Plastic DLTs are available in five sizes (28, 35, 37, 39 and

41 Fr). Their relatively large lumens allow independent suctioning or passage of a paediatric fibreoptic bronchoscope to either lung. During OLV the large D-shaped internal lumens present less resistance to air flow than the thicker walled, smaller lumens of red rubber tubes. The transparent plastic material allows continuous observation of moisture during ventilation. Secretions or blood coming from either lung will be obvious. The bronchial cuff of the plastic tubes is dyed blue so that it may be easily visualized during fibreoptic bronchoscopy. In contrast to the low volume/high pressure cuffs of rubber tubes, the bronchial cuffs of plastic tubes have high volume/low pressure characteristics, which reduces the danger of ischaemic pressure damage to the respiratory mucosa.[22]

Large DLTs are preferred (men, 41 Fr; women, 39 Fr). The bronchial cuffs of larger tubes require less air to seal the bronchus, which in turn decreases the risk of airway trauma or bronchial cuff herniation. Thin smaller tubes are easily advanced too far into the bronchus, increasing the chance of obstructing the upper lobe bronchus.[23] Small tubes are indicated if there is intrinsic or extrinsic obstruction of glottis, trachea, carina or main stem bronchus.

Choice of right or left tube

The human tracheobronchial tree is not symmetrical. The average length of the adult right main bronchus is 2.3 cm compared to the left main bronchus which is over 5.0 cm long. The 'margin of safety' is greater with a left DLT because there is less risk of the bronchial cuff obstructing the upper-lobe bronchus than when a right tube is used.[24] The choice between right versus left DLT is controversial: some anaesthetists prefer to intubate the bronchus of the

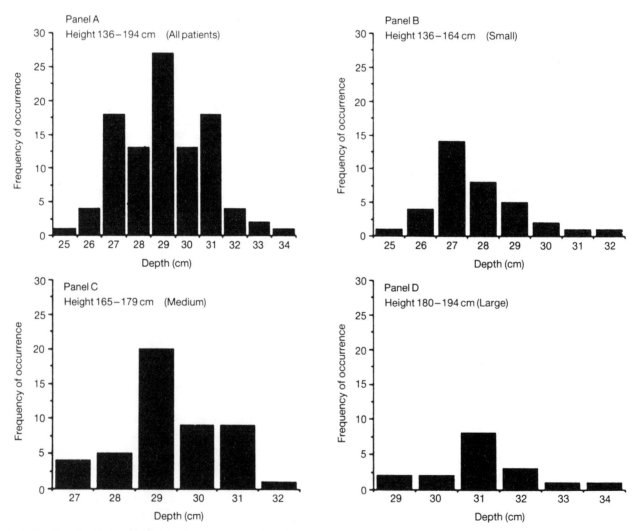

Fig. 57.2 The depth of insertion (cm) for left double-lumen tubes for all adult patients, and for three grouped intervals based on patient height (cm). The average depth of insertion for both males and females is 29 cm. (Reproduced, with permission from the International Anesthesia Research Society, from Brodsky JB, Benumof JL, Ehrenwerth J, Ozaki GT. Depth of placement of left double-lumen endobronchial tubes. *Anesthesia and Analgesia* 1991; **73**: 570–2.)

operated lung, others prefer to intubate the non-operated lung, while others use a left tube for either right or left procedures.

The rationale for intubating the operative lung is that a malpositioned DLT will usually obstruct the upper lobe and this will be obvious either when the chest is opened or when lung deflation is attempted.[23] Because the intubated lung is visible during surgery, bronchoscopy is not needed to confirm placement or to aid in repositioning. The surgeon can guide the tube manually if repositioning becomes necessary.

Opponents argue that intubation of the operative lung increases the risk of tube displacement from surgical retraction and manipulation, and complicates airway management during pneumonectomy.[25] They feel that, with the bronchial lumen in the dependent lung, the airway will be stented with less chance that the sagging mediastinum will obstruct the dependent bronchus. Unfortunately, the weight of the lung and mediastinum can compress the bronchus distal to the tip of the bronchial lumen and obstruct ventilation.

A left tube can be safely used to isolate and separate either lung. The margin of safety is greater with a left DLT because the wide bronchial cuff on a right tube can obstruct the right upper-lobe bronchus or herniate above the carina even if its ventilation slot is accurately positioned at the upper lobe opening.

Placement

A left-sided tube is usually used for either right or left thoracotomy, so only the steps for positioning a left tube are described.[26]

Care must be taken in passing the tubes because the cuffs are fragile and easily torn by the patient's teeth. A Macintosh laryngoscope blade is preferred because it provides the largest area in which to pass a DLT.

The tip of the tube is advanced just past the vocal cords, and the stylet in the bronchial lumen is removed. The tube is rotated 90 degrees counterclockwise (towards the left bronchus) and advanced until moderate resistance is encountered. The average depth of insertion for both males and females is 29 cm (Fig. 57.2)[27].

Once the tube is in the bronchus, *both* cuffs are inflated. When an appropriate size tube is used, only 1–2 ml of air in the bronchial cuff should seal the airway. The patient is ventilated through both lumens. Moisture should appear in each lumen, both sides of the chest should move and breath sounds should be present bilaterally. The tracheal lumen is then clamped. Breath sounds should be heard only over the intubated (left) lung. If breath sounds are present bilaterally, the tube is not deep enough and should be advanced further into the bronchus. If breath sounds are heard only over the right lung, the tube is in the right main bronchus (Fig. 57.3). In this situation, both cuffs should be

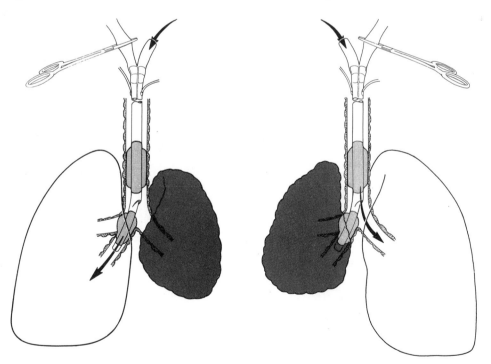

Fig. 57.3 After the double-lumen tube is advanced into the bronchus, both the tracheal (right) and bronchial (left) cuffs are inflated. First the tracheal lumen is occluded. Breath sounds should be heard only over the left lung. Breath sounds present over only the right lung indicate that the right bronchus has been intubated. This is confirmed by clamping the left lumen and ventilating through the right bronchial lumen. Breath sounds will be heard only on the left. (Reprinted, with permission, from Brodsky JB, Mark JBD. A simple technique for accurate placement of double-lumen endobronchial tubes. *Anesthesiology Review* 1983; **10** (8): 26–30.)

deflated and the tube withdrawn until its tip is above the carina. The tube is again rotated to the left and readvanced. Turning the patient's head and neck to the right while bending the head down will help direct the tube into the left bronchus.

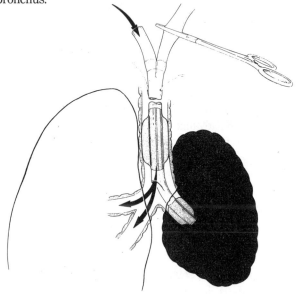

Fig. 57.4 Once the left double-lumen tube is in the left bronchus, the left lumen is occluded. Breath sounds should be present only over the right lung. (Reproduced, with permission, from Brodsky JB, Mark JBD. A simple technique for accurate placement of double-lumen endobronchial tubes. *Anesthesiology Review* 1983; **10** (8): 26–30.)

Once the tube is in the left bronchus the left lumen is clamped, and the patient is ventilated through the tracheal lumen. Breath sounds should now be heard over the right lung (Fig. 57.4). If there is difficulty ventilating the patient, deflate *only* the bronchial cuff while continuing to ventilate through the right lumen. If the tube is still not deep enough, breath sounds will now be present bilaterally (Fig. 57.5). If the tube is too deep, breath sounds will now be present only over the left lung (Fig. 57.6).

Tube position must be rechecked before the start of surgery because it may become displaced during the move to the lateral decubitus position. After the tube is first positioned it can be withdrawn a few millimetres. If the bronchial seal is maintained without requiring additional air, the tube is readvanced the initial distance back down the bronchus. This manoeuvre ensures some leeway against accidental movement of the bronchial cuff into the carina.

Verification of tube position is accomplished by physical examination of the chest, including auscultation and observation of chest wall movement, and measurement of peak inspiratory pressures during independent lung ventilation. The tension of the pilot balloon to the bronchial cuff should be noted after initial inflation. Subsequent softening of the pilot balloon usually means that the tube has been partially displaced out of the bronchus, and the bronchial cuff is herniating into the carina.

Following intubation and positioning, a fibreoptic

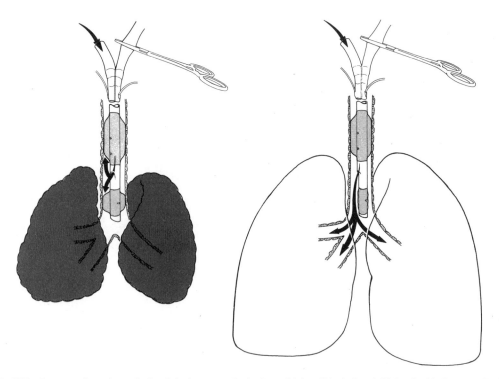

Fig. 57.5 If it is difficult to ventilate through the right lumen, *only* the bronchial cuff is deflated. If the double-lumen tube is still not deep enough, breath sounds will now be present over both lungs. (Reproduced, with permission, from Brodsky JB, Mark JBD. A simple technique for accurate placement of double-lumen endobronchial tubes. *Anesthesiology Review* 1983; **10** (8): 26–30.)

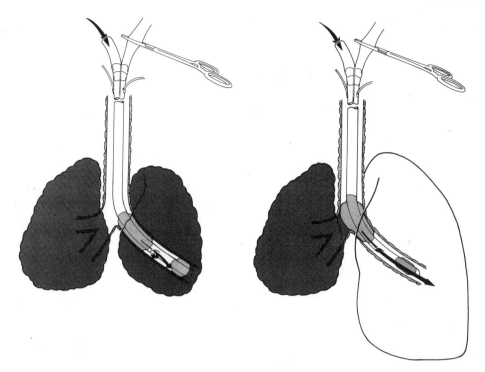

Fig. 57.6 If it is difficult to ventilate through the right lumen, *only* the bronchial cuff is deflated. If the double-lumen tube is too deep, breath sounds will now be present only over the left lung. (Reproduced, with permission, from Brodsky JB, Mark JBD. A simple technique for accurate placement of double-lumen endobronchial tubes. *Anesthesiology Review* 1983; **10** (8): 26–30.)

bronchoscope can be used to confirm tube position visually. Looking down the tracheal lumen, there should be an unobstructed view of the blue bronchial cuff immediately below the carina in the appropriate bronchus.[28] The bronchoscope should also be passed down the bronchial lumen to ensure patency of that lumen and to confirm that the upper-lobe orifice is not obstructed. A 4.9 mm diameter fibreoptic bronchoscope will pass down larger DLTs, but a 3.6 mm diameter bronchoscope is needed if smaller (35 Fr, 37 Fr) tubes are used. Bronchoscopy is not a substitute for, but is an adjunct to, careful auscultation. Although helpful, it can be time consuming, and is itself a potential source of airway trauma. Considerable expertise is needed for anything beyond identifying the carina through the tracheal lumen.[29]

When visualization of the glottis by direct laryngoscopy is not possible, a DLT can be placed over a fibreoptic bronchoscope for tracheal (and bronchial) intubation.

Complications

Airway damage and hypoxaemia can result from a misplaced DLT, whilst failure to collapse the operated lung can compromise the operation. The most common positioning problems are not advancing the tube far enough into the bronchus, intubating the wrong bronchus and passing the tube too deep into the correct bronchus.

Tubes may become malpositioned during surgery, so constant vigilance is important. Ventilation of the depen-

dent lung, through either the tracheal or the bronchial lumen, is never directly visualized. Ventilation must be monitored by observing changes in the peak inspiratory pressure to the dependent lung, end-tidal carbon dioxide, oxygen saturation and, most importantly, mediastinal movement during inspiration.

The airway can be injured during intubation and extubation with a DLT. Trauma ranges from ecchymosis of the mucous membranes to arytenoid dislocation and torn vocal cords. Airway trauma from red-rubber and plastic DLTs is relatively uncommon. Factors that increase the risk of injury include direct trauma during intubation, over-distension of the cuffs and pre-existing airway pathology. A damaged airway can present with an air leak, subcutaneous emphysema, haemorrhage or cardiovascular instability due to tension pneumothorax. With an incomplete laceration, air may dissect into the adventitia, producing an aneurysmal dilatation of the membranous wall. Nitrous oxide will further distend this air collection. The signs of injury may not be evident until many hours later when rupture into the mediastinum or pleural space occurs. Immediate surgical intervention is essential.

Airway rupture has been reported with plastic DLTs.[30] Their high volume/low pressure bronchial cuff mimics the dangerous high-pressure characteristics of red-rubber tubes when large volumes of air are used[22] (Fig. 57.7). The bronchial cuff will usually require only 1–2 ml of air. If a larger volume is needed a cuff leak may be present, but herniation of the cuff into the carina is more likely. When

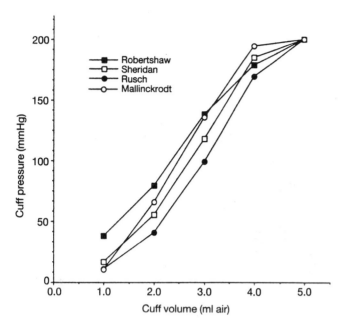

Fig. 57.7 The pressure/volume relationships of the bronchial cuffs of four left double-lumen tubes. The Mallinckrodt, Rusch and Sheridan plastic tubes were size 41 Fr and the red-rubber Robertshaw tube was size 'large'. The high volume/low pressure bronchial cuffs of plastic double-lumen tubes mimic the dangerous high-pressure characteristics of rubber double-lumen tubes when inflated with large volumes of air. (Reproduced, with permission from the International Anesthesia Research Society, from Brodsky JB, Adkins MO, Gaba DM. Bronchial cuff pressures of double-lumen tubes. *Anesthesia and Analgesia* 1989; **69**: 608–10.)

nitrous oxide is used, both cuffs should be deflated periodically to prevent excessive pressure on the mucosa. After initially confirming tube position, the bronchial cuff should be deflated and should be reinflated only when separation or collapse of the lungs is required.

Bronchial cuff underinflation can result in a cross-leak with failure to collapse the operated lung and contamination of the dependent lung. Overinflation can damage the airway or obstruct the trachea. Devices are available that can determine the exact endpoint for cuff inflation, but with periodic attention by digital examination of the bronchial cuff's pilot balloon these devices are seldom needed.

Optimizing oxygenation during one-lung ventilation

During two-lung ventilation in the lateral decubitus position approximately 40 per cent of the cardiac output flows to the non-dependent lung while the remaining 60 per cent goes to the dependent lung. There is normally perfusion of non-gas-exchanging areas in each lung, so

during two-lung ventilation with the patient in the lateral position approximately 35 per cent of the cardiac output participates in gas exchange in the non-dependent lung.

Following selective lung collapse the non-dependent lung continues to be perfused but is not ventilated. Several factors influence the degree of this wasted perfusion or 'shunt' (see 'Choice of anaesthetic agent', earlier). Blood flow to the dependent, ventilated lung is increased by the effects of gravity and increased vascular resistance in the non-dependent lung, owing to surgical retraction and compression, and by the mechanical effects of total atelectasis. The protective effect of the HPV reflex may play a small role, but HPV is less important during surgery than under experimental conditions because surgical manipulation of the lung releases vasoactive substances that blunt HPV. However, in the chronically diseased non-dependent lung, increased pulmonary vascular resistance may be present preoperatively, so blood flow may already be preferentially diverted to the dependent healthy lung. Under the best of conditions the shunt to the non-dependent lung during OLV is between 20 and 25 per cent of cardiac output.

At the commencement of OLV, minute ventilation should be continued unchanged. The ventilator rate should be adjusted to keep $PaCO_2$ between 4.8 and 5.3 kPa (36–40 mmHg). End-tidal carbon dioxide measurements can be used, but one must realize that end-tidal carbon dioxide during OLV in the lateral decubitus position may be as much as 1.3 kPa (10 mmHg) lower than actual $PaCO_2$. Hypocapnia should be avoided because it will increase pulmonary vascular resistance in the ventilated lung, redirecting blood flow to the non-ventilated lung. Hypercapnia is usually not a problem if minute ventilation is maintained.

In order to minimize hypoxaemia, efforts are directed toward optimizing ventilation to the dependent lung and increasing the oxygen content of the blood returning from the non-dependent, non-ventilated lung.

If hypoxaemia occurs, the position of the DLT or endobronchial blocker should immediately be reconfirmed. Other mechanical problems (tube obstruction, bronchospasm) should be considered. Clinically significant hypoxaemia usually does not occur immediately after initiation of OLV but after 10–15 minutes because it takes that long for the non-ventilated lung to completely collapse and for any remaining oxygen in that lung to be absorbed.

Because a large proportion of the cardiac output will be directed to the ventilated lung, the matching of ventilation to perfusion is important. The dependent lung should be ventilated with 100 per cent oxygen and a large tidal volume to maximize PaO_2.[5]

In the lateral decubitus position, the intra-abdominal contents shift the diaphragm cephalad, reducing FRC of the dependent lung. General anaesthesia further decreases FRC. Therefore, during lateral thoracotomy the dependent lung may have areas of low ventilation/perfusion ratios and areas that are atelectatic. The dependent lung must be ventilated with large tidal volumes (10–14 ml/kg) to recruit dependent lung alveoli. Tidal volumes of less than 8 ml/kg result in a further decrease in FRC with increased areas of dependent lung atelectasis. Larger tidal volumes (greater than 15 ml/kg) overdistend the alveoli and increase pulmonary vascular resistance, causing redistribution of blood flow to the non-dependent lung.

Even with a shunt of 25 per cent to the non-ventilated lung, an $F\text{IO}_2$ of 1.0 and large tidal ventilation to the dependent lung will usually produce a $Pa\text{O}_2$ of more than 20 kPa (150 mmHg).[13] At this oxygen tension, arterial haemoglobin is 100 per cent saturated. A high $F\text{IO}_2$ causes vasodilation of the vessels in the dependent lung,

increasing perfusion of that lung and further decreasing shunt.

A theoretical concern is that an $F\text{IO}_2$ of 1.0 can lead to absorption atelectasis in the dependent lung. Although the addition of 10–20 per cent nitrogen to the inspired gas mixture will decrease the propensity for the dependent lung to collapse, clinically this manoeuvre is unnecessary. Likewise, any concern about pulmonary oxygen toxicity has no clinical relevance, except perhaps in the patient with a history of bleomycin therapy in whom high oxygen concentrations should be avoided.

During OLV, peak inspiratory pressure will be high because the dependent lung will have decreased compliance while being inflated with a relatively large tidal volume. In

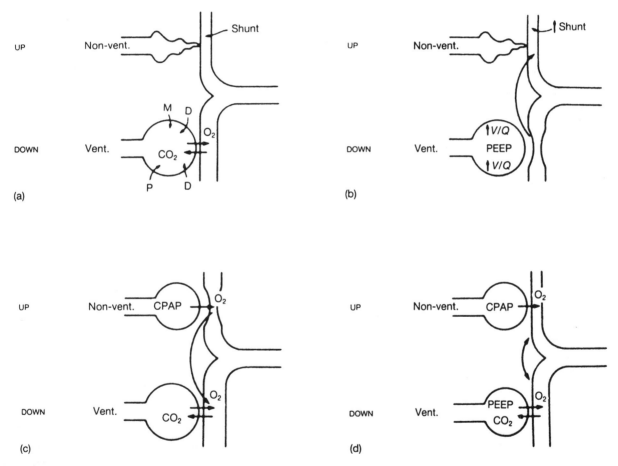

Fig. 57.8 The effects of various differential lung management approaches. (a) The one-lung ventilation (OLV) situation is depicted. The 'DOWN' (dependent) lung is ventilated (Vent.), but is compressed by the weight of the mediastinum (M) from above, the pressure of the abdominal contents against the diaphragm (D), and by positioning effects of rolls, packs and shoulder supports (P). The 'UP' (non-dependent) lung is not ventilated (Non-vent.), and blood flow (Shunt) to this lung does not participate in gas exchange. (b) The dependent lung is selectively treated with PEEP. This improves the ventilation to perfusion (V/Q) relationship in the dependent lung but also increases dependent lung vascular resistance, diverting blood flow and increasing shunt to the non-ventilated lung. (c) Selective application of CPAP to the non-dependent lung permits oxygen uptake from this lung and increases vascular resistance which reduces shunt and increases perfusion of the dependent ventilated lung. (d) With differential lung CPAP (non-dependent lung) and PEEP (dependent lung) blood perfusing either lung participates in gas exchange and oxygenation is maximized. (Reproduced, with permission from the author and the International Anesthesia Research Society, from Benumof JL. One-lung ventilation and hypoxic pulmonary vasoconstriction: implications for anesthetic management. *Anesthesia and Analgesia* 1985; **64**: 821–33.)

addition, airway resistance will increase through the single lumen of the DLT.

If hypoxaemia occurs, positive end-expiratory pressure (PEEP) must be used cautiously during OLV with large tidal volumes. Many factors cause a reduction in FRC during OLV, including the effects of general anaesthesia, the pressure of the abdominal contents on the diaphragm, the weight of the sagging mediastinum and perhaps improper positioning of the patient on the operating room table. In the presence of decreased FRC, PEEP (10 cmH$_2$O) will recruit collapsed and underinflated alveoli and improve oxygenation of blood perfusing the dependent lung. However, with normal or increased FRC, PEEP will increase alveolar airway pressure and dependent lung pulmonary vascular resistance, which in turn will divert blood flow to the non-ventilated lung and worsening the hypoxaemia (Fig. 57.8). Large tidal volume ventilation maintains FRC and precludes the need for PEEP.

The collapsed lung can be fully re-expanded during severe hypoxaemia. A single breath to the operated lung will temporarily correct the ventilation/perfusion imbalance. The lung will collapse from absorption atelectasis and must be re-expanded every 5 minutes. This manoeuvre will interfere with operative field exposure.

Insufflation with 100 per cent oxygen to the non-ventilated lung is inadequate treatment of hypoxaemia because the oxygen fails to reach and recruit collapsed alveoli. However, insufflation by continuous positive airway pressure (CPAP) with 100 per cent oxygen to the non-ventilated lung is an effective means of correcting hypoxaemia (Fig. 57.8c). CPAP maintains the patency of the non-dependent alveoli with oxygen, so unsaturated mixed venous blood perfusing that lung becomes oxygenated. Any increased airway pressure in the non-dependent lung from the CPAP may further increase pulmonary vascular resistance, which will divert blood flow to the dependent, ventilated lung. The non-ventilated lung remains partially distended but 'quiet' with CPAP.

Relatively high pressures are required when applying CPAP to an atelectatic lung initially in order to open the collapsed alveoli. CPAP should be applied during the deflation phase of a full breath to the non-dependent lung.[31] Once distended, relatively small levels of CPAP (5–10 cmH$_2$O) are all that is required to maintain satisfactory oxygenation. Higher levels of CPAP overdistend the lung and interfere with surgical exposure.

CPAP is effective only when there are no major disruptions in the operative bronchial tree, because a non-intact bronchus will not allow distending pressures to be maintained. CPAP is therefore not helpful with a bronchopleural fistula, during sleeve resection of the airway, or during massive pulmonary haemorrhage or bronchopleural lavage. Any airway obstruction by mucus,

blood or tumour may not permit adequate airway pressures to reach the lung.

The combination of PEEP (5–10 cmH$_2$O) applied to the ventilated lung and CPAP (5–10 cmH$_2$O) to the non-ventilated lung has been used to treat severe hypoxaemia but is seldom necessary when upper-lung CPAP is used alone[32] (Fig. 57.8d).

During pneumonectomy, ligation of the pulmonary artery completely eliminates shunt, maximizing the ventilation/perfusion relationship. Clamping the lobar pulmonary artery during lobectomy should increase vascular resistance in that lung segment, directing more blood to the ventilated lung. In animal models, inflation of a non-occlusive pulmonary artery catheter balloon located in the main pulmonary artery of the operated lung has been used as a means of mechanically redistributing blood flow to the non-operated lung. However, if the pulmonary artery catheter were to shift to the dependent lung, inflation of the balloon could potentially increase hypoxaemia by redirecting blood to the non-dependent lung. Occlusion of the main pulmonary artery during limited pulmonary resection is not advised because this manoeuvre will alter the activity of the alveolar lining material (surfactant) and will deleteriously affect lung re-expansion following surgery.

Ventilation at the completion of thoracotomy

Both lungs must be fully expanded and the mediastinum must be midline at the completion of OLV. Following total atelectasis the non-dependent lung will re-expand unevenly during inflation. Lung volume increases as collapsed alveoli are recruited. Alterations in pulmonary surfactant occur during OLV, necessitating the application of high sustained pressures in order to reopen the atelectatic lung. Once the lung has been fully reinflated and deflates to residual volume, subsequent inflation will require lower pressures consistent with normal surfactant activity.

Following lobectomy, the integrity of the bronchial repair must be tested before the chest is closed. The surgeon pours warm saline into the pleural cavity while the anaesthetist applies high (30–40 cmH$_2$O) inflation pressures by manually compressing the reservoir bag. This manoeuvre demonstrates air leaks and, when repeated several times, allows the previously collapsed lung to re-expand fully.

If re-expanded too rapidly, a lung that has been collapsed for several days can experience unilateral ('re-expansion') pulmonary oedema. This is extremely uncommon following the relatively short duration of selective lung collapse that occurs during surgery.

After pneumonectomy, the integrity of the bronchial suture line is tested in the same manner as for lobectomy. Saline is poured into the hemithorax and pressure is applied to the bronchial stump.

Prior to extubation, both lumens of the DLT should be suctioned to remove any mucus, blood or debris from each lung. Immediately after most thoracic procedures (including pneumonectomy) patients can be extubated and allowed to breathe spontaneously. After any pulmonary resection, positive pressure ventilation places stress on the bronchial suture line and exacerbates air leakage and formation of bronchopleural fistulae, so early extubation of the airway is preferred.

Postoperative complications

Atelectasis

The most common pulmonary complication following thoracotomy is atelectasis. It may develop as a result of pulmonary injury during surgery, from incomplete re-expansion following OLV or from bronchial obstruction by mucus due to inadequate clearance of secretions. The resulting atelectasis causes a ventilation/perfusion mismatch and hypoxaemia. Deep breathing, coughing and clearance of secretions with incentive spirometry, aerosolized bronchodilators, and early postoperative ambulation are important, but effective pain relief is essential (see 'Managing post-thoracotomy pain', later).

For lobar or whole-lung atelectasis due to mucus plugging, the patient should be placed in the lateral position with the non-operated lung dependent to improve ventilation/perfusion and promote mucus drainage from the operated lung. If oxygenation does not improve with lateral positioning and chest physiotherapy, fibreoptic bronchoscopy and lavage may be required.

Unresolved atelectasis can proceed to pneumonia, a potentially fatal condition following pneumonectomy or in patients with borderline pulmonary reserve. Associated respiratory failure may require reintubation and mechanical ventilation, which then increases the risk of stump disruption and formation of a bronchopleural fistula. Empyema (pus in the pleural space) may result from pneumonia.

Pneumothorax

In the presence of a bronchopleural communication the volume of air in the pleural space will increase. A tension pneumothorax will develop if air continues to enter the chest and is not effectively decompressed. The risk of pneumothorax following pulmonary resection is reduced by placement of a pleural drainage tube. However, a pneumothorax can still develop if the chest tube or suction apparatus malfunctions, or if the tube is occluded. The tube must have access to the air leak. Positive pressure ventilation, especially with PEEP, increases the risk of developing a pneumothorax. If the pneumothorax continues to enlarge it will displace mediastinal structures and compress the great vessels, leading to cardiovascular collapse.

Signs of a tension pneumothorax include decreased chest movement and diminished breath sounds, a unilateral wheeze, and hyper-resonance to percussion on the affected side. The peak inspiratory pressure during mechanical ventilation will be inappropriately high. A chest radiograph is diagnostic. The patient may experience increasing respiratory distress, tachypnoea, tachycardia and hypotension. A decrease in arterial oxygen saturation may occur before other signs become obvious. Immediate decompression with a large intravenous catheter or new chest tube may be life saving.

Cardiac herniation

Cardiac herniation can occur after pneumonectomy if the adjacent pericardium has been disrupted. If the pericardial defect is repaired intraoperatively the risk of herniation is low. Herniation is more common after right pneumonectomy because the great vessels and mediastinal structures provide more of a barrier to herniation on the left. The signs of cardiac herniation may include atrial and ventricular dysrhythmias, sudden hypotension and a superior vena caval syndrome. Cardiovascular collapse occurs due to acute angulation of the heart and great vessels. Haemodynamic stability can return only following surgical replacement of the heart to its normal position. If not immediately diagnosed and promptly treated, cardiac herniation is almost uniformly fatal, and even with aggressive treatment the mortality rate is 50 per cent. Until the chest is reopened, the patient should be positioned with the remaining lung dependent.

Haemorrhage

The clinical signs of major haemorrhage (tachycardia, hypotension, oliguria) are usually obvious if the patient develops hypovolaemia. Excessive chest tube drainage and a falling haematocrit (packed cell volume) are indications for surgical re-exploration.

Structural injuries

Structural injuries following thoracotomy include neurological injuries from direct surgical trauma or from positioning, and damage to the thoracic duct from surgery or central venous cannulation on the left side.

Dysrhythmias

Dysrhythmias (atrial tachycardia, atrial flutter and atrial fibrillation) are especially common post-pneumonectomy, but occur following lobectomy and wedge resection as well. Any resulting impairment of cardiac output may require fluid loading to enhance preload, which in turn may adversely affect pulmonary function. Prophylactic digitalization has been recommended, especially after pneumonectomy. Arrhythmias associated with profound hypotension require emergency cardioversion. Verapamil is effective for atrial flutter or multifocal atrial tachycardia, but may cause hypotension because of vasodilatory and negative inotropic effects. The simultaneous administration of phenylephrine will support cardiac output. β-Blockade may be necessary for rapid supraventricular dysrhythmias associated with hypertension and myocardial ischaemia, but β-blockers may cause bronchospasm. Esmolol is effective and short acting. Following a loading dose, a maintenance infusion is started.

Right heart failure

Extensive lung resection decreases pulmonary vascular cross-sectional area, resulting in increased pulmonary vascular resistance which may lead to acute right heart failure with or without pulmonary oedema. Prior to pneumonectomy, if temporary unilateral occlusion of the main pulmonary artery to the operated lung by a PA catheter increases mean PA pressure more than 30–40 mmHg, the risk of postoperative right heart failure is high. Even when predicted right heart function is adequate, postoperative hypoxaemia can precipitate right heart failure.

Clinical signs of right heart failure include supraventricular dysrhythmias, distended neck veins, hepatomegaly and peripheral oedema. The ECG may demonstrate left and right ventricular strain, and a chest radiograph may reveal right atrial and ventricular enlargement. Echocardiography will confirm the diagnosis.

The therapeutic goal is to support right ventricular preload and decrease pulmonary vascular resistance without lowering systemic blood pressure. Ventilatory support may be needed to correct reversible causes of increased pulmonary vascular resistance (hypoxaemia, hypercapnia and respiratory acidosis). Volume loading supports right ventricular preload. Pharmacological treatment includes dobutamine, an inotrope whose vasodilatory action decreases pulmonary vascular resistance. Amrinone, which has less chronotropic effect than dobutamine, is an alternative, especially in the presence of myocardial ischaemia, but systemic hypotension may limit its usefulness. For pulmonary vasodilatation and right ventricular afterload reduction, nitroglycerin is superior to sodium nitroprusside. Both agents cause systemic hypotension and blunt HPV.

Managing post-thoracotomy pain

The decrease in FRC and the ability to cough and to breathe deeply following thoracotomy are influenced by the degree of chest wall pain. If the patient experiences pain and is unable to clear secretions, airway closure and atelectasis will result. Analgesic regimens associated with the best post-thoracotomy pulmonary function are felt to be superior.

Systemic opioids

Although still a mainstay of post-thoracotomy pain management, systemic opioids have a very narrow therapeutic window. With inadequate pain relief the patient will be unwilling to cough and clear secretions; with too much the patient will be oversedated and be unable to cough and breathe deeply.

With intermittent administration of opioids there may be a significant time lag between the request for medication, the actual administration of the medication and its onset of action. If systemic opioids are used, patient-controlled analgesia (PCA) or continuous intravenous infusion are preferred.

Cryoanalgesia

Cryoanalgesia, the freezing of intercostal nerves, can reduce pain after thoracotomy. The nerve at the incision space and the two intercostal nerves above and below are frozen. The loss of sensory and motor function that follows cryoanalgesia usually lasts 1–6 months, and many patients experience painful neuralgias at the treatment sites.[33]

Cryoanalgesia does not significantly improve pulmonary function when compared to other pain treatment modalities. Cryoanalgesia may reduce, but does not eliminate, post-thoracotomy pain and is useful only as a supplement to other forms of pain control.

TENS

Electrodes can be placed on either side of the thoracic incision and electrical stimulation applied (transcutaneous electrical nerve stimulation, TENS). The effectiveness of this therapy in reducing opioid requirements and improving pulmonary function following thoracotomy is controversial. TENS does not provide complete analgesia and must be used as an adjunct to other therapy.

Intercostal nerve blocks

Intercostal nerve blocks can be performed intraoperatively with the lung collapsed or percutaneously following surgery. Most studies have demonstrated improved analgesia, reduced opioid requirements and improved pulmonary function following thoracotomy. Because of the short action of the local anaesthetics, continuous blocks are more practical than repeated individual blocks.[34] Continuous intercostal blockade can be accomplished by placing several catheters in the intercostal grooves during surgery. Systemic toxicity from intravascular absorption or direct injection and unrecognized dural puncture resulting in marked hypotension are two major complications of intercostal nerve blocks.

Interpleural blockade

Interpleural blockade is performed by injecting local anaesthetics into the thoracic cavity between the visceral and parietal pleura.[35] An epidural catheter is placed through the chest wall while the thorax is still open. Catheter placement must be completed before the chest is closed because, once the lung is re-expanded, the risk of pneumothorax and the inability to confirm catheter position limit the safety of this technique. Most studies report inadequate relief of post-thoracotomy pain when the interpleural technique is used alone. Clamping the chest tube following anaesthetic administration increases the success rate because there is a significant loss of local anaesthetic with an open chest drainage tube.

Epidural block (local anaesthesia)

Epidural analgesia with local anaesthetics can reduce pain and improve pulmonary function after thoracotomy. Either the thoracic or the lumbar approach is used. Epidural local anaesthetics can be used during surgery to supplement general anaesthesia, and then continued into the post-operative period for analgesia. Any concurrent hypotension and/or motor blockade limits the patient's ability to be up and about. The combination of epidural local anaesthetic and systemic or epidural opioids has been recommended but does not significantly improve the analgesia achieved when epidural opioids are used alone.[36]

Spinal (intrathecal and epidural) opioids

Post-thoracotomy regional analgesia with opioids minimizes impairment of pulmonary function, aids in recovery and prevents postoperative pulmonary complications.[37] When compared to parenterally administered opioids, spinal opioids improve post-thoracotomy pulmonary function and provide superior analgesia. Most opioid drugs have been given either intrathecally or epidurally following thoracic surgery.[38–42] Unlike local anaesthetics, spinal opioids are 'selective' as, other than potentiation of analgesia, they have no known haemodynamic or motor effects. Spinal opioids work by binding to opiate receptors in the spinal cord.

Single-shot intrathecal morphine produces analgesia for up to 24 hours, but the intrathecal route is limited by the need for repeated injections (unless a catheter is placed).[43] The epidural route is generally preferred because the incidence of respiratory depression is less, and the presence of a catheter allows for continuous, prolonged opioid administration following surgery.

Lipophilic agents rapidly diffuse across the dura into the cerebral spinal fluid, bind to the spinal opiate receptors and produce a rapid onset of action. Fentanyl and sufentanil provide satisfactory analgesia following thoracotomy but, because of their relatively short duration of action, both must be administered by constant infusion.

Hydrophilic agents (morphine, hydromorphone) diffuse more slowly into the cerebral spinal fluid.[44] Their onset of action is delayed, but the length of action is longer than the lipophilic agents, so they are preferred if a continuous infusion is not possible. These agents can be safely administered by constant infusion. Their low lipid solubility allows these drugs to be given at either the thoracic or the lumbar level for post-thoracotomy pain relief without any difference in analgesic efficacy.

The complications of intraoperatively administered epidural opioids occur in the postoperative period. All epidural opioids decrease the ventilatory response to carbon dioxide, but significant respiratory depression and apnoea are rare. The lipophilic drugs cause *acute* respiratory depression due to central effects from systemic absorption. The hydrophilic opioids are associated with *delayed* respiratory depression occurring hours after administration, presumably due to the rostral spread of the drug in the cerebral spinal fluid to the brain. The factors predisposing to delayed respiratory depression include concomitant administration of opioids or sedatives by other routes prior to or during surgery, high doses, the hydrophilicity of the agent, advanced age and the site of administration (intrathecal > thoracic epidural > lumbar epidural). Serious respiratory depression is uncommon, but as thoracotomy patients are at increased risk for pulmonary complications they should be monitored in a special care unit for the first 18–24 hours following surgery.

The frequency and occurrence of minor complications (pruritus, urinary retention, nausea and vomiting) vary with the specific opioid used and should not be a contraindication to their use following thoracotomy.

Non-steroidal anti-inflammatory drugs

Non-steroidal anti-inflammatory drugs (NSAIDs) may have a role in treating post-thoracotomy pain. When used in combination with other analgesics, they improve pain relief and postoperative pulmonary function.[45] NSAIDs have a peripheral site of action, have anti-inflammatory and antipyretic activity, and can be conveniently administered intravenously or by rectal suppository. They do not cause respiratory failure but other side effects, including increased bleeding time, gastrointestinal and central nervous system symptoms and worsening of renal function, may limit their use.

Special procedures

Bronchoscopy

Fibreoptic bronchoscopy can be performed under local or general anaesthesia. If local anaesthesia is chosen, sedatives should be used with caution, as they may not be well tolerated by the patient with pulmonary disease. An anticholinergic should be given to dry the airway and to blunt irritative reflexes. Glycopyrrolate is preferred to atropine or hyoscine (scopolamine). The upper airway is first anaesthetized either with a nebulizer which sprays local anaesthetic solution or with viscous lignocaine gargle. Tracheal anaesthesia is achieved by transtracheal injection of local anaesthetic. Superior laryngeal and glossopharyngeal nerve blocks can be performed. Because these blocks depress airway reflexes, the patient should take nothing by mouth for several hours *following* bronchoscopy.

Any general anaesthesic technique is acceptable, but nitrous oxide is usually avoided to allow a higher FIO_2.

During awake fibreoptic bronchoscopy the patient should receive supplemental oxygen. This is achieved using mouth-held nasal prongs or with a special facemask (Patil–Syracuse mask) that has a diaphragm through which the bronchoscope is passed.

A fibreoptic bronchoscope passed through a tracheal tube produces a significant PEEP effect. Barotrauma can result if ventilation is attempted through a small obstructed tube, so a large tracheal tube should be used. Suctioning the airway through the bronchoscope decreases alveolar oxygen, reduces FRC and can cause significant hypoxaemia.

For rigid bronchoscopy the patient should be paralysed and ventilation controlled in order to minimize the risk of trauma from movement. Patients with suspected malignancy may have an increased sensitivity to non-depolarizing muscle relaxants (myasthenic (Eaton–Lambert) syndrome) so relaxants should be given in small incremental doses and titrated using a blockade monitor.

A variety of ventilatory techniques can be used during rigid bronchoscopy. Following hyperventilation with 100 per cent oxygen to denitrogenate the lungs and to lower $PaCO_2$, oxygen can be delivered by insufflation at a high flow (10–15 l/min) ('apnoeic oxygenation') without actually ventilating the patient. Although satisfactory oxygenation can be achieved for long periods, apnoea should not extend beyond 5 minutes because of carbon dioxide accumulation. Oxygen and anaesthetic gases can be delivered through the side arm of the bronchoscope by intermittent ventilation. Ventilation is possible as long as the eyepiece is in place, but must be interrupted whenever biopsy or suctioning is performed. During long procedures, carbon dioxide accumulates and predisposes the patient to dysrhythmias, particularly in the presence of light anaesthesia. Intermittent hyperventilation lowers $PaCO_2$ and deepens the anaesthetic. High flows of fresh gas are needed to compensate for the leak around the bronchoscope. The use of the oxygen flush should be kept to a minimum because the flush bypasses the anaesthetic vaporizer and dilutes the inhalational agent. Oxygen can also be delivered by a Sanders system which uses the Venturi principle to deliver oxygen by jet ventilation. The presence of an eyepiece is not necessary with this technique, so continuous uninterrupted ventilation is possible.

The complications of bronchoscopy include damage to the teeth, haemorrhage, bronchospasm, subglottic oedema and airway perforation. A chest radiograph should be obtained following bronchoscopy to exclude the presence of pneumothorax or mediastinal emphysema from barotrauma.

Mediastinoscopy

Transcervical mediastinoscopy is usually performed on the right side of the chest with the patient in the reverse Trendelenburg position. The mediastinoscope is advanced following the anterior aspect of the trachea, passing near the innominate vessels and the aortic arch.

Mediastinoscopy can be performed under local anaesthesia. However, with the patient breathing spontaneously the risks of venous air embolism and mediastinal injury from patient movement are greatly increased. General anaesthesia with muscle paralysis and controlled positive pressure ventilation is preferred. An anticholinergic agent should be given to blunt autonomic reflexes (bradycardia) that result from compression or stretching of the trachea, vagus nerve or great vessels.

The major complications of mediastinoscopy include pneumothorax, recurrent laryngeal nerve damage, air embolism and haemorrhage. If a large mediastinal blood vessel is torn, fluids given through an intravenous line in the arm may enter the mediastinum through the tear in the vein, so in this situation a large-bore catheter should be placed in a lower extremity vein. An emergency lateral thoracotomy or sternotomy may be required to control bleeding. Although blood loss is usually quite small, blood should always be available in the operating room. Other complications of mediastinoscopy include acute tracheal collapse, tension pneumomediastinum, haemothorax and chylothorax. All patients should have a chest radiograph in the immediate postoperative period.

The mediastinoscope can compress the innominate artery, causing a loss of the right radial pulse in the presence of a normal ECG. This apparent 'hypotension' may lead to inappropriate aggressive treatment. During mediastinoscopy, blood pressure measurements should be obtained from the left arm, and the right radial pulse should be continuously monitored by plethysmography or by pulse oximetry. A decrease in the right radial pulse with normal blood pressure in the left arm indicate that the mediastinoscope needs to be repositioned. This is especially important if the patient has a history of impaired cerebral vascular circulation, as carotid artery perfusion can be compromised by the vascular compression.

Mediastinal mass

Anterior mediastinal masses may compress the superior vena cava, the major airways and even the heart. Complete or partial airway obstruction from an anterior mediastinal mass can occur following patient positioning and/or with relaxation of the muscles maintaining airway patency.[46]

Preoperatively, it is important to determine if the patient has experienced dyspnoea in the supine position. A chest CT scan and other radiological studies should be obtained to determine the extent of the tumour mass and any involvement with surrounding structures. Flow–volume loops are needed in both the upright and the supine positions. A marked decrease in FEV_1 and peak expiratory flow rate in the supine position suggests the potential for airway obstruction with anaesthesia. In this situation radiotherapy of the mass should be considered prior to surgery.

A rigid bronchoscope may be needed to bypass the obstruction and to ventilate the patient if the trachea becomes obstructed during induction of anaesthesia. The patient's position may need to be changed from supine to lateral or even prone to relieve the compression.

If biopsy of the mediastinal mass cannot be performed under local anaesthesia, an awake intubation with a fibreoptic bronchoscope followed by induction of general anaesthesia has been recommended. Spontaneous ventilation helps maintain the airway patency. The Trendelenburg position and positive pressure ventilation should be avoided if superior vena caval obstruction is present.

Thoracoscopy

Thoracoscopy, the insertion of an endoscope into the thoracic cavity, has been used to obtain pulmonary and pleural biopsies. The technique is now being extended for therapeutic procedures such as laser treatment of tumours, spontaneous pneumothoraces and bullous emphysema, and for limited lung and chest wall resections.[47]

Thoracoscopy can be performed under local, regional or general anaesthesia, the choice depending on the duration and extent of the operative procedure. When the incision is made, air enters the pleural cavity, causing a partial pneumothorax. If the procedure is performed using local anaesthetics with the patient awake, the pneumothorax is usually well tolerated because the skin and chest wall form a seal around the thoracoscope, limiting the degree of atelectasis. When general anaesthesia is indicated, a DLT is used to isolate and collapse the operative lung while ventilation to the other lung continues. If pleurodesis is performed for recurrent pneumothorax, general anaesthesia allows complete re-expansion of the lung without the severe pain associated with instillation of talc or tetracycline.

Pulmonary haemorrhage

Protecting the non-involved airway is the first priority in the management of massive pulmonary haemorrhage. Death usually results from drowning in blood, rather than from hypovolaemia or exsanguination. The methods used for isolating the lung during pulmonary haemorrhage include a DLT, placement of a single-lumen tracheal tube into the bronchus of the non-bleeding lung or bronchial blockade with the Univent tube. If bleeding is massive and visualization of the larynx is difficult, awake laryngoscopy and intubation is recommended. Advancing an uncut tracheal tube to its full length will usually intubate the right main bronchus. If bleeding is coming from the right lung, the tube may have to be passed over a fibreoptic bronchoscope into the left lung. If the left lung is bleeding, the single-lumen tube inserted into the right main bronchus will probably obstruct the right upper lobe as well as isolating the entire left lung from ventilation. In this situation, hypoxaemia will be greater than during conventional OLV.

Whichever method is used, confirmation of proper lung isolation using fibreoptic bronchoscopy is necessary before the patient is turned to the lateral decubitus position.

Once the airway is established and the contralateral airway is protected, attention is directed towards detecting and controlling the site of bleeding while maintaining adequate gas exchange. The site of haemorrhage is determined by bronchoscopy. The larger lumen of a rigid bronchoscope is better for suctioning greater volumes of blood, but fibreoptic bronchoscopy with an airway is safer.

Bronchopleural fistula

A bronchopleural fistula is a pathological communication between the airway and the pleural cavity. The fistula may extend to the skin (bronchopleural–cutaneous fistula). Bronchopleural fistulae occur following pulmonary resections (most common after pneumonectomy), from rupture of a bulla, bleb or cyst into the pleural space, from erosion of a carcinoma and from trauma.

If an empyema is present it should be drained prior to surgery on the fistula. The safest means is to drain the empyema under local anaesthesia with the patient sitting up and leaning toward the affected side. A drain, connected to an underwater seal, should be left in the cavity. The empyema may be loculated, so complete drainage of the empyema is not always possible.

The most important anaesthetic concerns in the management of a bronchopleural fistula are separation and isolation of the lungs to prevent contamination and to allow ventilation of the healthy side. These goals are best achieved with a DLT.

Prior to intubation the chest drainage tube should be left unclamped for continued drainage of pus and to prevent a tension pneumothorax. Intubation can be performed in an awake patient or under general anaesthesia as long as the patient breathes spontaneously. Controlled ventilation is

avoided until the lungs are separated in order to prevent a tension pneumothorax.

The endobronchial end of the DLT should be placed in the bronchus of the unaffected lung. A 'blindly' advanced tube can pass through the fistula. To prevent this complication a fibreoptic bronchoscope can first be passed down the endobronchial lumen into the non-involved side, and the tube can then be advanced over the bronchoscope into that bronchus.

Once in position the bronchial cuff should be immediately inflated to separate the lungs. Only the dependent, healthy lung is ventilated. If an empyema is present, immediately after intubation there may be a massive outpouring of pus through the tracheal lumen.

Bronchopleural fistulae can be treated without surgery using a DLT to ventilate each lung independently. The healthy lung is ventilated with normal tidal volumes while the affected lung is ventilated with a smaller volume or allowed to remain unventilated. Continuous positive airway pressure (CPAP) can also be applied at pressures below the opening pressure of the fistula. The critical opening pressure is found by adding small increments of CPAP to the affected bronchus until continuous bubbling appears in the underwater-seal chest drain. High-frequency jet ventilation is ineffective and may actually worsen the bronchopleural fistula.[48]

Lung cysts and bullae

Thin-walled, air-filled cavities in the lungs (cysts and bullae) may represent end-stage emphysematous lung changes or may be congenital and isolated findings. Repeated pneumothoraces from spontaneous rupture, infection and/or dyspnoea are the usual indications for surgery. The degree of functional impairment depends on the condition of the remaining lung, the size of the air space and the amount of compression of the surrounding healthy lung tissue by the cyst.

If the bulla is very compliant, during controlled ventilation a large portion of the applied tidal volume will be wasted in this additional dead space. Positive pressure ventilation should be used with caution because the bulla may rupture, leading to a tension pneumothorax. Nitrous oxide is avoided. Once the chest is opened the risk of tension pneumothorax disappears.

Intubation with a DLT in a spontaneously breathing awake or anaesthetized patient is recommended. Positive pressure ventilation with rapid small tidal volumes at pressures less than 15 cmH$_2$0 can be used during induction and maintenance of anaesthesia. Once the DLT is in place, the non-affected lung is ventilated with normal tidal volumes. After resection of each bulla, the operated lung should be re-expanded and checked for air leaks and for the presence of additional bullae.

Most patients can be extubated at the completion of surgery, but if postoperative controlled ventilation is needed, low positive pressures should be used to reduce the chance of developing a tension lung cyst or pneumothorax.

Patients with spontaneous pneumothoraces can be divided into two clinical groups. The first have apical blebs and usually are young with excellent pulmonary reserve. They tolerate open thoracotomy very well. The second group have emphysematous blebs, tend to be older and have advanced chronic obstructive pulmonary disease and carbon dioxide retention. These patients benefit by avoiding thoracotomy with thoracoscopic resection of their bullae and blebs.

Bronchopleural lavage

Irrigation of a whole lung is done in the treatment of alveolar proteinosis, bronchiectasis, refractory asthma, inhalation of radioactive dust and cystic fibrosis.[49] In situations where bilateral lung disease is present, ventilation/perfusion scans should be obtained preoperatively so that the lavage can be first performed on the more severely affected lung.

A DLT is placed with the patient under general anaesthesia in order to isolate the lungs. The endobronchial lumen should enter the bronchus on the side to be lavaged. One hundred per cent oxygen should be inspired throughout the procedure. The cuff of the bronchial lumen should maintain separation of the lungs at pressures as high as 50 cmH$_2$0 to prevent lavage fluid from leaking into the healthy, ventilated lung.

Following intubation the patient should be turned to the lateral decubitus position with the head slightly elevated. The DLT's position and its bronchial cuff seal are rechecked. Some prefer the lavaged lung to be dependent so that the risk of leakage to the good (non-dependent) lung is reduced. Others prefer the lavaged lung to be non-dependent because, in this position, perfusion will match more closely ventilation in the dependent lung. As a compromise, lavage may be performed in the supine position.

Warm heparinized isotonic saline is infused by gravity down the endobronchial lumen from a height of 30–40 cm above the chest. When the lavage fluid ceases to flow the patient is placed in the head-down position, and chest percussion and vibration are applied for 1 minute to loosen material. The infused fluid is then allowed to drain passively to a collecting system. This may have to be repeated as many as 20 times before the drained lavage fluid becomes clear. Accurate measurement of inflow and outflow volumes is important. After the last passive drainage, the lung is suctioned and then re-expanded using large tidal volumes and high pressures. Most patients can be extubated in the operating room.

The most serious complication of this procedure is aspiration of the lavage fluid into the ventilated lung. This may result in a marked decrease in oxygenation, requiring immediate reinstitution of two-lung ventilation with PEEP.

Bronchial obstruction

Secretions, blood, anatomical distortion, tumour or foreign bodies can each cause bronchial obstruction. Retention of secretions can occur distal to the obstruction. A preoperative chest radiograph or CT scan may reveal consolidation or an abscess cavity beyond the obstruction.

Isolation of the lungs is critically important because, once the obstruction is relieved, pus can flood the airway. Manipulation of the lung or simply turning the patient to the lateral decubitus position can dislodge a foreign body or secretions into the trachea or dependent lung. Therefore, the lungs should be isolated before turning the patient. A DLT is preferred to a bronchial blocker because it provides protection to the dependent lung while allowing the operated lung to be suctioned. Only the dependent lung is ventilated while the lumen to the non-dependent lung is left open for drainage or suction.

In some instances, the obstruction can behave as a ball valve. Positive pressure ventilation distends the bronchus and allows gas to pass the obstruction during inflation, but then the gas is trapped during deflation. Progressive rise in peak inspiratory pressure and a decrease in expired volume compared to inspired volume may indicate gas trapping. Ventilation of only the dependent lung is extremely important in this situation.

Nd-YAG laser

The neodymium–yttrium–aluminum–garnet (Nd-YAG) laser is used to debulk tumours and to tunnel through or widen obstructed airways.[50] The Nd-YAG laser beam can be transmitted through a flexible quartz monofilament, so it can be used with either rigid or flexible fibreoptic bronchoscopes.

Although the procedure can be performed under local anaesthesia, general anaesthesia with muscle relaxants is preferred because any movement could result in a misfired laser, causing damage to healthy tissue. With general anaesthesia there is improved airway control, and debris and blood are more easily removed to prevent aspiration. An inhalational anaesthetic agent, an FIO$_2$ of 0.4 or less, and 60 per cent or more nitrogen (from inspired air) reduce the chance of airway fire. Nitrous oxide, which is combustible, should be avoided. No special tracheal tube is needed because the laser is fired only beyond the tip of the tube. However, if misfired, tube ignition is possible. The laser should only be fired deep in the airway, but damage to the anaesthetist's eyes is possible from a misfired laser, so goggles or glasses are recommended.

Only the surface of the treated tissue is visible. Any underlying damage and oedema will not be apparent, and airway obstruction and haemorrhage can develop as late as 48 hours after the procedure. Hypoxaemia, perforation of the tracheobronchial tree and airway haemorrhage are major complications of Nd-YAG procedures.

REFERENCES

1. Boysen PG, Clark CA, Block AJ. Graded exercise testing and postthoracotomy complications. *Journal of Cardiothoracic Anaesthesia* 1990; **4**: 68–72.

2. Gracey DR, Divertie MB, Didier EP. Preoperative pulmonary preparation of patients with chronic obstructive pulmonary disease. A prospective study. *Chest* 1979; **76**: 123–9.

3. Gass GD, Olsen GN. Clinical significance of pulmonary function tests. Preoperative pulmonary function testing to predict postoperative morbidity and mortality. *Chest* 1986; **89**: 127–35.

4. Eisenkraft JB. Effects of anaesthetics on the pulmonary circulation. *British Journal of Anaesthesia* 1990; **65**: 63–78.

5. Kerr JH, Smith AC, Prys-Roberts C, Meloche R, Foëx P. Observations during endobronchial anaesthesia. II. Oxygenation. *British Journal of Anaesthesia* 1974; **46**: 84–92.

6. Domino KB, Wetstein L, Glasser SA, *et al.* Influence of mixed venous oxygen tension ($P\bar{v}O_2$) on blood flow to atelectatic lung. *Anesthesiology* 1983; **59**: 428–34.

7. Arima T, Matsuura M, Shiramatsu T, Gennda T, Matsumoto I, Hori T. Synthesis of prostaglandins TXA_2 and PGI_2, during one lung anesthesia. *Prostaglandins* 1987; **34**: 668–78.

8. Marshall C, Kim SD, Marshall BE. The actions of halothane, ibuprofen and BW755C on hypoxic pulmonary vasoconstriction. *Anesthesiology* 1987; **66**: 537–42.

9. Scherer RW, Vigfusson G, Hultsch E, Van Aken H, Lawin P. Prostaglandin $F_{2\alpha}$ improves oxygen tension and reduces venous admixture during one-lung ventilation in anesthetized paralyzed dogs. *Anesthesiology* 1985; **62**: 23–8.

10. Benumof JL. One-lung ventilation and hypoxic pulmonary vasoconstriction. Implications for anesthetic management. *Anesthesia and Analgesia* 1985; **64;** 821–33.

11. Marshall C, Lindgren L, Marshall BE. Effects of halothane, enflurane, and isoflurane on hypoxic pulmonary vasoconstriction in rat lungs *in vitro. Anesthesiology* 1984; **60**: 304–8.

12. Benumof JL, Augustine SD, Gibbons JA. Halothane and isoflurane only slightly impair arterial oxygenation during one-lung ventilation in patients undergoing thoracotomy. *Anesthesiology* 1987; **67**: 910–15.

13. Capan LM, Turndorf H, Chandrakant P, Ramanathan S, Acinapura A, Chalon J. Optimization of arterial oxygenation during one-lung anesthesia. *Anesthesia and Analgesia* 1980; **59**: 847–51.

14. Hirshman CA, McCullough RE, Cohen PJ, Weil JV. Depression of hypoxic ventilatory response by halothane, enflurane and isoflurane in dogs. *British Journal of Anaesthesia* 1977; **49**: 957–63.

15. Weinreich AI, Silvay G, Lumb PD. Continuous ketamine infusion for one-lung anesthesia. *Canadian Anaesthesia Society Journal* 1980; **27**: 485–90.

16. Brodsky JB, Shulman MS, Swan M, Mark JBD. Pulse oximetry during one-lung ventilation. *Anesthesiology* 1985; **63**: 212–14.

17. Ginsberg RJ. New technique for one-lung anesthesia using an endobronchial blocker. *Journal of Cardiovascular Surgery* 1981; **82**: 542–6.

18. MacGillivray RG. Evaluation of a new tracheal tube with a movable bronchus blocker. *Anaesthesia* 1988; **43**: 687–9.

19. Benumof JL, Gaughan S, Ozaki GT. Operative lung constant positive airway pressure with the Univent bronchial blocker tube. *Anesthesia and Analgesia* 1992; **74**: 406–10.

20. Kelley JG, Gaba DM, Brodsky JB. Bronchial cuff pressures of two tubes used in thoracic surgery. *Journal of Cardiothoracic and Vascular Anesthesia* 1992; **6**: 190–2.

21. Linter SPK. Disposable double-lumen tubes. A cost-effectiveness survey. *Anaesthesia* 1985; **40**: 191–2.

22. Brodsky JB, Adkins MO, Gaba D. Bronchial cuff pressures of double-lumen tubes. *Anesthesia and Analgesia* 1989; **69**: 608–10.

23. Brodsky JB, Shulman MS, Mark JBD. Malposition of left-sided double-lumen endobronchial tubes. *Anesthesiology* 1985; **62**: 667–9.

24. Benumof JL, Partridge BL, Salvatierra C, Keating J. Margin of safety in positioning modern double-lumen endotracheal tubes. *Anesthesiology* 1987; **67**: 729–38.

25. McKenna MJ, Wilson RS, Botelho RJ. Right upper lobe obstruction with right-sided double-lumen endobronchial tubes: a comparison of two tube types. *Journal of Cardiothoracic Anesthesia* 1988; **6**: 734–40.

26. Brodsky JB, Mark JBD. A simple technique for accurate placement of double-lumen endobronchial tubes. *Anesthesiology Review* 1983; **10** (8): 26–30.

27. Brodsky JB, Benumof JL, Ehrenwerth J, Ozaki GT. Depth of placement of left double-lumen endobronchial tubes. *Anesthesia and Analgesia* 1991; **73**: 570–2.

28. Smith GB, Hirsch NP, Ehrenwerth J. Placement of double-lumen endobronchial tubes. Correlation between clinical impressions and bronchoscopic findings. *British Journal of Anaesthesia* 1986; **58**: 1317–20.

29. Slinger PD. Fiberoptic bronchoscopic positioning of double-lumen tubes. *Journal of Cardiothoracic Anesthesia* 1989; **3**: 486–96.

30. Burton NA, Fall SM, Lyons T, Graeber GM. Rupture of the left main-stem bronchus with a polyvinylchloride double-lumen tube. *Chest* 1983; **83**: 928–9.

31. Slinger P, Triolet W, Wilson J. Improving arterial oxygenation during one-lung ventilation. *Anesthesiology* 1988; **68**: 291–5.

32. Cohen E, Eisenkraft JB, Thys DM, Kirschner PA, Kaplan JA. Oxygenation and hemodynamic changes during one-lung ventilation: effects of $CPAP_{10}$, $PEEP_{10}$, and $CPAP_{10}/PEEP_{10}$. *Journal of Cardiothoracic Anesthesia* 1988; **2**: 34–40.

33. Roxburgh JC, Markland CG, Ross BA, Kerr WF. Role of cryoanalgesia in the control of pain after thoracotomy. *Thorax* 1987; **42**: 292–5.

34. Sabanathan S, Smith PJB, Pradhan GN, Hashimi H, Eng J, Mearns AJ. Continuous intercostal nerve block for pain relief after thoracotomy. *Annals of Thoracic Surgery* 1988; **46**: 425–6.

35. Mann LJ, Young GR, Williams JK, Dent OF, McCaughan BC. Intrapleural bupivacaine in the control of postthoracotomy pain. *Annals of Thoracic Surgery* 1992; **53**: 449–54.

36. Badner NH, Komar WE. Bupivacaine 0.1% does not improve postoperative epidural fentanyl analgesia after abdominal or thoracic surgery. *Canadian Journal of Anaesthesia* 1992; **39**: 330–6.

37. Shulman M, Sandler AN, Bradley JW, Young PS, Brebner J. Postthoracotomy pain and pulmonary function following epidural and systemic morphine. *Anesthesiology* 1984; **61**: 569–75.

38. Patrick JA, Meyer-Witting M, Reynolds F. Lumbar epidural diamorphine following thoracic surgery. A comparison of infusion and bolus administration. *Anaesthesia* 1991; **46**: 85–9.

39. El-Baz NMI, Faber LP, Jensik RJ. Continuous epidural infusion of morphine for treatment of pain after thoracic surgery. A new technique. *Anesthesia and Analgesia* 1984; **63**: 757–64.

40. Gough JD, Williams AB, Vaughan RS, Khalil JF. The control of post-thoracotomy pain. A comparative evaluation of thoracic epidural fentanyl infusions and cryo-analgesia. *Anaesthesia* 1988; **43**: 780–3.

41. Whiting WC, Sandler AN, Lau LC, *et al.* Analgesic and respiratory effects of epidural sufentanil in patients following thoracotomy. *Anesthesiology* 1988; **69**: 36–43.

42. Brodsky JB, Chaplan SR, Brose WG, Mark JBD. Continuous epidural hydromorphone for postthoracotomy pain relief. *Annals of Thoracic Surgery* 1990; **50**: 888–93.

43. Gray JR, Fromme GA, Nauss LA, Wang JK, Ilstrup DM. Intrathecal morphine for post-thoracotomy pain. *Anesthesia and Analgesia* 1986; **65**: 873–6.

44. Brose WG, Tanelian DL, Brodsky JB, Mark JBD. CSF and blood pharmacokinetics of hydromorphone and morphine following lumbar epidural administration. *Pain* 1991; **45**: 11–15.

45. Perttunen K, Kalso E, Heinonen J, Salo J. IV diclofenac in post-thoracotomy pain. *British Journal of Anaesthesia* 1992; **68**: 474–80.

46. Neuman GG, Weingarten AE, Abramowitz RM, Kushins LG, Abramson AL, Ladner W. The anesthetic management of the patient with an anterior mediastinal mass. *Anesthesiology* 1984; **60**: 144–7.

47. Wakabayashi A. Expanded applications of diagnostic and therapeutic thoracoscopy. *Journal of Thoracic and Cardiovascular Surgery* 1991; **102**: 721–3.

48. Bishop MJ, Benson MS, Sato P, Pierson DJ. Comparison of high-frequency jet ventilation with conventional mechanical ventilation for bronchopleural fistula. *Anesthesia and Analgesia* 1987; **66**: 833–8.

49. Cohen E, Eisenkraft JB. Bronchopulmonary lavage: effects on oxygenation and hemodynamics. *Journal of Cardiothoracic Anesthesia* 1990; **4**: 609–15.

50. Hanowell LH, Martin WR, Savelle J, Foppiano LE. Complications of general anesthesia for Nd:YAG resection of endobronchial tumors. *Chest* 1991; **99**: 72–6.

51. Shapiro BA, Harrison RA, Kacxmarek RM, Cane RD. *Clinical application of respiratory care*, 4th ed. Chicago: Mosby-Year Book, 1989.

52. Seigel LC, Brodsky JB. Choice of anesthetic agents for intrathoracic surgery. In: Kaplan JA, ed. *Thoracic anesthesia*, 2nd ed. New York: Churchill Livingstone, 1991: chap 13.

FURTHER READING

Benumof JL. *Anesthesia for thoracic surgery.* Philadelphia, PA: WB Saunders, 1987.

Brodsky JB, ed. *Thoracic anesthesia. Problems in Anesthesia* 1990; **4**: No 2.

Kaplan JA, ed. *Thoracic anesthesia,* 2nd ed. New York: Churchill Livingstone, 1991.

Marshall BE, Longnecker DE, Fairley HB, eds. *Anesthesia for thoracic procedures.* Boston, MA: Blackwell Scientific, 1988.

Wolfe WG, ed. *Complications in thoracic surgery. Recognition and management.* St Louis, MO: Mosby Year Book, 1992.

Chapter 58

Anaesthesia for Transplant Surgery and Care of the Organ Donor

M. J. Lindop

History	Heart–lung transplant
Common features of transplant surgery	Lung transplant
Kidney transplant	Bone marrow transplant
Liver transplant	Corneal transplant
Pancreas transplant	Bowel transplant
Heart transplant	Management of the multi-organ donor

Transplants can offer spectacular transformation of life style to the patient suffering from major organ failure. Problems still remain when complications of grafting – rejection, infection and infarction – lead to graft failure, either complete or partial. Transplant surgery and anaesthesia are a brief but vital window of therapy in a long period of medical care – renal, hepatic, cardiac, diabetic, etc. The patient has usually been maintained through a period of declining organ function until the point of transplantation is reached. After the procedure, the patient returns to medical care for the supervision of graft function. Good liaison with medical teams, particularly for preoperative assessment protocols, is an essential part of anaesthetic management.

History

Following experimental work in animals the techniques for transplantation of major organs have been available since the beginning of the century. The problems in applying them to man were the availability of organs and the phenomenon of rejection.

Kidney transplants were attempted in animals in 1902. Carrel, whose name still attaches to the aortic cuff taken at the donor operation, was awarded the Nobel Prize in 1912 for work with kidney allografts in cats and dogs. He also reported a heart transplant in a dog. The first success in man occurred in 1946 before dialysis had been developed when a cadaver kidney was grafted onto the brachial vessels and functioned for a critical 48 hours until development of the diuretic phase of acute renal failure. Kidney grafts were reported from living related donors,

mother to son and identical twins, and from cadavers. The failure to control rejection delayed real progress until the 1960s when immune suppressive drugs were developed. Liver transplantation began in 1963, but hearts and lungs were not transplanted until the early 1980s.

Common features of transplant surgery

Urgency

Organs become available unpredictably. Recipients are usually on a waiting list, living at home, but those on an urgent list may well be inpatients, often in an intensive care unit. Transplant coordinators set in motion a complex organization to match the offer of a donor to a suitable recipient, and then set a timetable for surgery, keeping the numerous members of the team informed of the not infrequent changes that occur along the way to the start of surgery. It is important that preoperative assessment is started before the patient is called in on the day of surgery for by then there will be no time for subtle tests of cardiac and respiratory function.

Immediately before surgery the priority is to improve fitness within time constraints. The time is set by the rate of deterioration of the stored organs despite preservation solutions. The acceptable ischaemia time varies from organ to organ, short for lungs, long for kidneys. Baseline measurements are made. A period of dialysis, elective ventilation or inotrope support in the intensive care unit may be worthwhile, and in the more ill patient, careful discussion is required between all members of the team to determine the best timetable for surgery.

A period of starvation is not always possible and a rapid sequence induction, assuming that there is a full stomach, is wise. This may be difficult to manage safely in the patient with severe cardiovascular depression and careful choice of the minimum effective drug dose with as slow administration as possible will be critical.

Organ preservation

At removal organs are cooled and perfused with preservative solution which is designed to minimize storage damage. Solutions contain free radical scavengers such as glutathione and are isotonic, with electrolyte contents resembling intracellular fluid (e.g. K^+ 100 mmol/l) to minimize membrane ion and fluid shifts. If such a solution is flushed into the recipient there will be a sudden dangerous potassium load. Therefore, before connecting the venous drainage of the graft it is essential to flush it out carefully. The development of improved solutions (e.g. University of Wisconsin) has increased confidence to allow longer ischaemia times.[1] More time is available to prepare the recipient, and the operation may be delayed until daylight hours. A common transplant timetable is completion of donor brainstem death criteria during the day, with donor operation in the evening, and recipient operation either later that night or first thing next morning.

The immune response

The immune system of the recipient recognizes the antigens of the transplanted tissue as foreign. Some tissues stimulate a greater response than others – gut, heart and kidney are worse than liver, and all worse than cornea.

Control of the immune response

HLA matching

The immune response is reduced when the match of the donor to the recipient is good. This can be assessed preoperatively by human leucocyte antigens (HLA). This test takes several hours and because organ preservation (ischaemic time) is limited for many organs, especially heart and lungs, there is no time to use the test to choose the best match for a given recipient. HLA matching is used for kidney transplants but for other organs only ABO blood group compatibility can be achieved.

Drugs

The mainstay for control of the immune response is the development of powerful suppressive drugs. The principal drugs used are cytotoxic agents, steroids, cyclosporin, FK506, and monoclonal antibodies.

Azathioprine 1.5 mg/kg is a cytotoxic drug which acts by preventing the clonal proliferation of sensitized cells. It may cause bone marrow suppression, intestinal disturbances, fever, liver toxicity or pancreatitis.

Steroids may be given in very high dose to control acute rejection. These act in various ways – inhibiting macrophage function, lymphokine production and the generation of cytotoxic T-cells. They are not without many adverse effects such as poor healing, peptic ulceration, hypertension, osteoporosis, avascular necrosis of bone, cataract and diabetes.

Cyclosporin is a fungal peptide which inhibits the induction phase of the immune response, interfering with T-cell activation and function. It is nephrotoxic and cannot be used until renal function is well established. Cyclosporin 2–4 mg/kg is given in divided doses and blood level monitoring is used to establish the therapeutic level. Cyclosporin may cause convulsions. Bone marrow is depressed less than with the cytotoxic drugs. There is a 28-fold increase in the incidence of lymphomas, though this becomes a 59-fold increase if azathioprine and steroids are used.[2]

FK506 is derived, like cyclosporin, from a soil fungus. Its mode of action is similar to cyclosporin, and it is a promising potent new drug. It shares the toxic effects of cyclosporin.

Antibodies

Early polyclonal antibodies such as anti-lymphocyte globulin (ALG) and anti-thymocyte globulin (ATG) were produced in animals, and cause hypotension, fever and even major anaphylaxis which are extremely hazardous in the ill transplant recipient. These must be given with care – in a heart transplant, after the patient is on bypass.

Monoclonal antibodies are less variable in content and cause less cross-reactivity with other blood components. OKT3 is a monoclonal antibody raised against the CD3 surface antigen on T-lymphocytes. Campath also binds to adhesion molecules on lymphoid cells and limits the potential for an immune response.

These drugs are used in various combinations and continue to be evaluated.

The complications of immune suppressive therapy of most significance to the anaesthetist are bone marrow depression, nephrotoxicity and steroid-induced infection, tissue friability, peptic ulceration and hypertension.

Multiple organ grafts

Although the operative procedure will be more complex when several organs are grafted simultaneously the immune response is reduced.[3] Heart, lungs and liver have been simultaneously grafted for cystic fibrosis with cirrhosis.

Tolerance

Patients can develop insidious chronic rejection for which no treatment has yet been found. Often luckier patients develop tolerance which allows the dose of immune suppressant drugs to be steadily reduced so that, in a few cases, further treatment can be stopped. Much current research is directed to identify methods of inducing tolerance and to reduce the need for immune suppressive drugs with their attendant toxicity.

Effect of anaesthesia and surgery

General anaesthesia depresses immune response.[4] Various white cell responses have been described in association with anaesthesia and surgery. Decreased lymphocytic responses occur up to 2 weeks after cholecystectomy. White cell count falls after many anaesthetic agents – barbiturates, etomidate, fentanyl and nitrous oxide, while halothane and enflurane induce leucocytosis. The trauma of surgery itself also stimulates the mobilization of white cells, although this is less after regional anaesthesia. The *in vitro* cell-mediated immune capacity of lymphocytes is reduced by halothane, thiopentone, ketamine but not by nitrous oxide.[5]

Transfusion of stored blood improves the survival of grafts. The effect is small.[6]

Since powerful immune suppressant drugs have been introduced all these effects can be considered clinically insignificant.

Complications of immune suppressive therapy

Infection

Infection is the major threat to survival of the transplant patient. The principal factor is the depression of host resistance by immune suppressant drugs, but the risk is also increased during rejection.

Endogenous flora and latent infection must be reduced by identification and treatment of septic foci. The threat is greater for transplants involving the gut (liver, pancreas and gut). Selective bowel decontamination is described with wide spectrum antibiotics and antifungals.[7] Most units will use oral antifungals such as nystatin to control *Candida*. Donors and recipients are screened by swabbing for bacteria and fungi and by serology for viral hepatitis, cytomegalovirus, toxoplasma and human immune deficiency virus (HIV).

Infection control is essential. Disposable clean circuits with bacterial filters should be used and all invasive procedures must be carried out with full aseptic technique.

A prophylactic antibiotic protocol is always used. Postoperative pyrexia is common but usually no organism is isolated. Patients will often have had long and complex antibiotic regimens and a microbiologist's help in planning strategy is important to limit the establishment of multiresistant gastrointestinal flora.

Systemic candidiasis may develop and require amphotericin B which is also nephrotoxic, like cyclosporin. An alternative antifungal, ketoconazole, inhibits microsomal enzymes which degrade cyclosporin, risking a rise to toxic levels.

Cytomegalovirus (CMV) is the most important of the viral infections. The diagnosis is established by serology and will be either a primary infection or a reactivation of previous CMV infection. The route of transmission is either by the transplanted organ or via blood or blood products. Seronegative patients should receive seronegative blood where possible. Treatment is with ganciclovir. Diagnosis may be delayed until a rise in specific IgM antibody titres occurs. Herpes simplex, varicella-zoster and Epstein–Barr virus may also cause infections and acyclovir is the drug of choice. Patients who were hepatitis B (HBV) positive before liver grafting have developed further HBV infection of their grafts.[8] They may receive hepatitis B immunoglobulin for the first week post-transplant, with further treatment if required to maintain the anti-HBs greater than 100 iu/l for the next year.

Toxoplasma gondii can cause fatal infections and pyrimethamine prophylaxis is given to seronegative recipients from seropositive donors.

All immune suppressed patients must take co-trimoxazole (Septrin) prophylaxis against *Pneumocystis* for 6 months after their transplant.

Rejection

The main mechanisms of acute rejection consist of a humoral response with antibody being directed to antigens present on the surface of transplanted cells, and by cell-mediated immune mechanisms of delayed hypersensitivity or direct cytotoxicity. The latter respond well to immune suppression, but the humoral response, particularly that leading to fibrinoid necrosis of blood vessels, is more resistant.

Anti-rejection therapy uses the drugs already described

to achieve the optimum balance of risk between rejection and the complications of immune suppression.

Correct timing of therapy depends on graft assessment. The ease and accuracy of the assessment varies with the organ. Urine output is a guide to kidney graft function, alanine transferase level and prothrombin time to liver function. Biopsy may be helpful and can usually be obtained by Trucut needle or, for the heart, by endocardial transvenous catheter. Open biopsy may require general anaesthesia.

Further surgery

Patients undergoing transplantation commonly require further surgery. The transplant itself is a major vascular operation and postoperative bleeding may necessitate surgical exploration. If the graft fails completely and acutely it will need removal and, for heart, lung or liver, immediate replacement.

Surgery unrelated to the transplant may also be required. Problems may be anticipated principally with heart and with lung transplant recipients as described in those sections. If graft function is good problems are not normally encountered. Main elements in management are careful preoperative assessment of graft function and of drug therapy. Patients still on high doses of immune suppressive drugs will be susceptible to infection and particular care must be taken with asepsis.

Main elements of anaesthetic management

These are summarized in Table 58.1.

Kidney transplant

End-stage renal disease is the indication for surgery. Contraindications are few (sepsis, malignancy and systemic collagen vascular disease). More than 3500 patients are potential recipients in the UK, though less than half these will be grafted because of lack of donor organs. The 1-year graft survival rate is better than 80 per cent, and the perioperative mortality is less than 1 per cent.

Table 58.1 Main elements of anaesthetic management

Anaesthetic assessment prior to the day of transplant
Intensive preoperative therapy, if required
Scrupulous aseptic technique
Anaesthetic technique appropriate to identified organ failures, if possible, not relying on new graft function
Careful monitoring, particularly during graft revascularization
Avoidance of postoperative hypotension, to maximize graft perfusion

The operation

The new kidney is put heterotopically onto the iliac vessels extraperitoneally in the opposite iliac fossa of the recipient. It is a major vascular operation lasting approximately 2 hours. The potential for blood loss depends on the surgeon's skill in managing vascular anastomoses. Rapid blood replacement may occasionally be required when the vessels are unclamped. When arterial and venous anastomoses are completed and the graft is reperfused, the ureter must be tunnelled into the bladder.

Common problems

Cardiovascular

Hypertension is common and many patients will be on antihypertensive therapy which should be continued until surgery. Most patients will have severe atherosclerosis leading to peripheral vascular disease and ischaemic heart disease. Uraemia and dialysis can lead to cardiomyopathy and pericardial effusion. Normally a raised cardiac output compensates for chronic anaemia and haemodialysis arteriovenous shunts. Impaired myocardial function may lead to left ventricular failure. Careful assessment of dyspnoea and clinical examination will identify the cause and need for further treatment. Autonomic neuropathy may lead to cardiovascular instability in diabetic patients with renal failure.

Fluid and electrolyte balance

There is usually limited ability to excrete urine and overload may readily precipitate pulmonary oedema. On the other hand, preoperative dialysis may leave the patient hypovolaemic and hypokalaemic. Most renal patients know their 'dry' weight and any change from this can suggest the extent and direction of significant fluid changes. In the oliguric patient inadequate dialysis may lead to hyperkalaemia, metabolic acidosis, hyponatraemia and hypermagnesaemia. Obligatory urine loss in some forms of renal failure can lead to hypernatraemia or hypokalaemia. It is important to delay surgery to allow a brief period of dialysis to bring electrolyte values into the normal range with a normal blood volume. Hypovolaemia leads to cardiovascular instability. Control of serum potassium is particularly important as hypokalaemia causes cardiac arrhythmias and increased susceptibility to muscle relaxants. Hyperkalaemia may be increased during surgery by rapid transfusion of stored blood or by any preservation solution entering the circulation at reperfusion of the new kidney.

Haematological

A normocytic normochromic anaemia (4–9 g/dl) is common, the prime cause being reduced renal production of erythropoietin. Other contributing factors include uraemic marrow depression, increased red cell fragility and haemolysis, and increased blood loss from chronic gastric ulceration. The anaemia is managed by provision of human recombinant erythropoietin, by control of uraemia and gastric ulceration and by correction of any deficiencies in haematinics. Transfusion is not specifically indicated except as replacement for blood loss during surgery. Patients will normally have adapted to the chronically anaemic state by increasing cardiac output. Oxygen delivery to the periphery will be enhanced by increased 2,3-diphosphoglycerate and by any acidaemia. It is not essential to transfuse these chronically anaemic patients before administering general anaesthesia.

Haemostatic function is impaired. Although clotting times may be normal, the bleeding time is prolonged.[9] Significant thrombocytopenia is uncommon in end-stage renal failure but platelet function is impaired in uraemia. Drugs, such as aspirin, which further impair platelet function should be avoided. Desmopressin (DDAVP) 20 µg can improve platelet stickiness in the face of bleeding by increasing the release of von Willebrand factor from the vascular endothelium.

Paradoxically such changes as there are in the coagulation cascade are in the direction of promoting coagulation. Factor VIII and fibrinogen may be increased and antithrombin III is decreased.

Alimentary

Gastritis and duodenitis with chronic blood loss are common with high acid secretion. Aspiration risk at induction is significant since uraemia prolongs gastric emptying. Nausea and vomiting are common.

These patients have a high risk of developing transfusion-related hepatitis and should be screened for this prior to surgery. Liver function tests should be checked.

Nervous system

A peripheral neuropathy typified by burning feet is caused by uraemia. Diabetic patients will also be likely to have a peripheral and an autonomic neuropathy. If regional anaesthesia is planned any neuropathy must be identified. Too rapid dialysis can induce dialysis dysequilibration syndrome with headache, nausea and vomiting, hypertension, muscle twitching and seizures. These would need to be resolved before anaesthesia and surgery.

Pulmonary

These patients are prone to pulmonary oedema. There may also be splinting of the diaphragm associated with chronic ambulatory peritoneal dialysis. A preoperative chest x-ray is required.

Endocrine

A quarter of patients with end-stage renal failure will have diabetes. They are suitable for transplant but are a considerable challenge, often with severe peripheral vascular disease in addition to severe insulin-dependent diabetes. Uraemia increases insulin requirements. An insulin infusion is established at a rate to control the blood sugar perioperatively.

Phosphate retention causes hypocalcaemia, and gut absorption of calcium is reduced. Parathormone levels are high and may be part of the stimulus to a catabolic state.

Drugs

Renal excretion of drugs and their metabolites will be completely or almost absent. Renal metabolism will be reduced. There may be accompanying changes in hepatic intermediary metabolism, elimination, and an altered distribution volume, protein binding and lipid solubility of drugs. The choice of drugs should avoid any that might damage the graft, e.g. by metabolism to fluoride, and any that rely on graft function for their clearance.

Anaesthetic management

A preoperative visit allows proper assessment and allays anxiety which is present even in this very experienced group of hospital patients. There is always time to correct any fluid and electrolyte disorder prior to transplant. If the organ has already waited too long it certainly should not be put into a patient in suboptimal condition and it would be better discarded or a fitter recipient found.

Any arteriovenous fistula must be identified and checked. It will have to be carefully protected during surgery. Any sustained hypotension will cause clotting off of the fistula.

Ranitidine and metoclopramide and 0.3 per cent sodium citrate should be used as prophylaxis to reduce risk from aspiration.

Premedication can be prescribed. Vagolytics and opiates will slow gastric emptying further, and a modest oral dose of a short-acting benzodiazepine will suffice.

Choice of anaesthesia

The option of regional anaesthesia, epidural or subarachnoid block, is well established.[10] However, arguments against its use include the risk of epidural haematoma formation associated with the increase in bleeding time resulting from platelet dysfunction, the difficulty in handling major blood replacement, in the awake patient, in the event of a vascular anastomotic problem, the maintenance of the awake patient's well-being during a long procedure, the medicolegal complexity of a peripheral neuropathy, the instability when a regional autonomic block is added to an autonomic neuropathy with the possibility of significant blood loss, and the unpredictable response of a hypertensive patient on treatment to vasopressors. Although there is a safe record with the use of regional anaesthesia, the availability of new drugs makes general anaesthesia increasingly attractive.

Induction

Most cases will be managed by routine monitoring (ECG, non-invasive blood pressure, pulse oximetry, end tidal carbon dioxide) with the addition of central venous pressure measurement. The central line will prove useful in postoperative management, too. A pulmonary artery catheter should be used if cardiac function is severely impaired.

A rapid sequence induction is required. These patients are particularly prone to swings in blood pressure in response to induction, laryngoscopy and intubation. These can best be reduced by giving time for a preinduction dose of narcotic analgesic such as fentanyl to work and by choosing induction agents with which the anaesthetist is familiar. Thiopentone or etomidate in careful dosage is appropriate. β-Blockers such as esmolol will attenuate the peak[11] but may deepen the trough of blood pressure. Suxamethonium is used because serum potassium will be normal and any small rise will have no clinical significance. Arrhythmias may be precipitated. The patient is intubated and ventilated using an inhalation anaesthetic.

Maintenance

Nitrous oxide can be used but it inhibits methionine synthetase in patients whose later immune suppression will include marrow depressant drugs. It can also increase any bowel distension, and provoke nausea. If air is used as an alternative carrier gas, the higher concentration of volatile anaesthetic required may well have other disadvantages, such as greater cardiovascular depression.

All volatile agents have been reported to be safe. The use of halothane has decreased because of its potential for liver damage. Enflurane is metabolized to free fluoride, and significant, but subnephrotoxic, levels have been reported.[12] Isoflurane is not metabolized to fluoride significantly, and is presently the vapour of choice because it offers less myocardial depression, good rhythm stability and rapid uptake and clearance.

Narcotic analgesics should be titrated with care. The half-lives of fentanyl and alfentanil are prolonged but they remain the preferred drugs. There is a problem of cumulation when drugs with a short duration of action are used for long operations. However, the metabolites of agents with a longer duration of action, pethidine and morphine, accumulate in renal failure. Norpethidine is a convulsant, and morphine-6-glucuronide is an active metabolite.

Atracurium is the muscle relaxant of choice since its hydrolysis and Hofmann degradation are independent of renal function and its pattern of action remains unchanged.[13] All other agents have a prolonged action with the risk of cumulation. Unfortunately atracurium is more likely than vecuronium to cause a histamine-like reaction with a fall in blood pressure. All muscle relaxants (except gallamine) have been used in the past and so long as great care is taken to use minimal effective doses most patients recover adequate motor power postoperatively. Use of the peripheral nerve stimulator will allow more accurate dosage of relaxant drug and with the extra-peritoneal surgical approach only modest relaxation is required.

Significant events

It is essential to have an adequate circulating blood volume prior to release of the clamps to ensure reperfusion of the kidney. The anaesthetist must be alert to the problem, and, using central venous pressure measurements, must provide good preparation. If the patient is at all hypovolaemic, profound hypotension can follow. The new kidney is denervated and maximally vasodilated so perfusion will be determined by arterial pressure.

Early renal function is important. Most units will have a protocol such as giving diuretics at the time of unclamping – 0.5 g/kg of mannitol with 250 mg frusemide. A dopamine infusion (2 μg/kg per minute) may help renal perfusion.

Prophylactic antibiotics are given. Immune suppressive drugs will be started.

Postoperative care

Care is taken to ensure the complete reversal of muscle relaxation. Pain relief is provided by, for example, a patient-controlled analgesia system. Fluid management must be careful. A useful regimen is to infuse 0.45 per cent saline 20 ml hourly plus the previous hour's urine volume. Electrolytes must be monitored closely for when early function is good there may be a major diuresis in response to the mannitol and frusemide. Occult intra-abdominal bleeding can occur. Hourly urine flow is an excellent monitor. Most early oliguria is due to hypovolaemia and not to graft dysfunction. Central venous pressure measurement will distinguish between these possibilities.

Liver transplant

End-stage chronic parenchymal liver disease, liver tumours and fulminant hepatic failure form the main groups of recipients. The recurrence rate from malignant tumours has been high despite careful screening to identify any signs of metastasis before surgery. Most of the tumour patients presenting for transplant are not cirrhotic and are otherwise well and are good operative risks in marked contrast to those with end-stage parenchymal disease.

Fulminant hepatic failure provides some of the greatest challenges in medical management leading to transplantation. Common causes are toxins and drugs, e.g. paracetamol, and various types of viral hepatitis. Recovery from severe fulminant failure occurs spontaneously in 60 per cent of patients when the cause is paracetamol or hepatitis A or B, but in less than 20 per cent where the cause is halothane or non-A non-B hepatitis. Case selection to identify those with no hope of recovery who require transplant is critical.[14] Such selected patients will be in grade IV hepatic coma with cerebral oedema, hypotension, renal impairment and major coagulation defects.

End-stage parenchymal disease has many causes including chronic active hepatitis, primary biliary cirrhosis, primary sclerosing cholangitis, Budd–Chiari syndrome, haemochromatosis and Wilson's disease.

In England alone over 400 liver transplants are performed each year with a 1-year survival rate of about 75 per cent.

The operation

The recipient will be ABO compatible and about the right size for the new liver. Although liver segments have been put in heterotopically to support liver function, normally the entire liver is put in orthotopically so that the patient's diseased liver must first be removed. The procedure can be divided into four phases: (1) dissection and hepatectomy; (2) anhepatic; (3) reperfusion; and (4) post-reperfusion.

The dissection phase can be extremely difficult especially in the presence of portal hypertension and where there has been previous upper abdominal surgery. Some techniques (piggy-back) dissect the liver free from the vena cava, leaving the vessel intact; however, the liver is normally removed complete with the inferior vena cava which runs immediately posterior to it. This stops the venous return from the lower half of the body and can cause major cardiovascular instability.

Once the liver is removed and the patient anhepatic, haemostasis is attained and the new liver put into position. Suprahepatic and infrahepatic caval, and portal vein anastomoses are made. This takes about 40 minutes and is a time of stable reduced venous return and moderate hypotension. The graft is flushed clear of the preservative solution and reperfused.

At the time of reperfusion upper and lower caval anastomoses are opened up to restore venous return and improve cardiac output. The clamps are then removed from the hepatic and portal veins. There is often cardiovascular instability at this time.

During the post-reperfusion phase the hepatic artery anastomosis is made and then biliary drainage is established, either by direct anastomosis or by creating a Roux loop of bowel. The operation usually takes about 5 hours but may take considerably longer.

Common problems

Cardiovascular and respiratory

Despite impaired myocardial function many cirrhotic patients will have a high cardiac output with hypotension and a low systemic vascular resistance. Portal venous flow falls in patients with cirrhosis, reducing overall hepatic blood flow. Ascites interferes with circulation by raising intra-abdominal pressure. The ventilation/perfusion abnormality, commonly seen, is due in part to an impaired hypoxic pulmonary vasoconstrictive reflex and also to the reduction in functional residual capacity as the diaphragms are pushed up. This, together with intrapulmonary shunting and reduced diffusing capacity from increased intrapulmonary water, commonly leads to hypoxaemia in these patients. There is often an acidosis which will reduce the affinity of haemoglobin for oxygen.

Renal

Hepatorenal syndrome describes renal failure associated with severe liver disease and is a functional vasomotor nephropathy. Kidneys from such patients show no tubular damage, and can be transplanted with immediate return of normal function. When liver function improves, improvement of renal function follows rapidly. Increase in aldosterone secretion and impaired hepatic clearance with increased antidiuretic hormone (ADH) leads to water and sodium retention. Much of this is redistributed into interstitial fluid and ascites. Blood urea is not a good index of renal function as urea synthesis may be impaired. However, the levels may rise acutely after blood is lost into the intestine following a variceal haemorrhage. Creatinine clearance is a better guide to function, but high serum bilirubin can interfere with the accuracy of the serum estimation.

Severely ill patients may develop acute tubular necrosis and this may be prolonged. Such patients will have a higher urinary sodium secretion (>10 mmol/day) and an isosmotic urine.

Coagulation

There are two issues for consideration – thrombocytopenia and lack of clotting factors. Thrombocytopenia may be due to bone marrow depression, splenomegaly and hypersplenism, or disseminated intravascular coagulation. The liver is the main source of clotting factors – fibrinogen, prothrombin, and factors V, VII, IX and X (but not VIII). The rise in prothrombin time reflects any deficiency (and proves a sensitive monitor of liver graft function in the early postoperative period).

Fluid and electrolyte balance

Medical therapy with control of fluid intake and administration of diuretics can lead to marked derangement of fluid and electrolyte balance. Hyper- or hyponatraemia may occur. Hypokalaemia is a common result of secondary aldosteronism, treatment with loop diuretics and low intake. Acid–base disturbance is a mixture of respiratory alkalosis due to mild hyperventilation, and metabolic acidosis due to failure of the normally important role of the liver in metabolizing a wide range of acids (lactate, citrate, free fatty acids and others).

Blood glucose may be low enough to cause coma in patients with fulminant hepatic failure. Glycogen stores are depleted and hepatic gluconeogenesis is reduced.

Circulating levels of insulin may be high due to reduced hepatic clearance.

Central nervous system

Commonly there has been a history of episodes of encephalopathy in the end-stage liver failure patient. Such patients will be on lactulose and/or oral neomycin. In patients with fulminant hepatic failure cerebral oedema is common and intubation and controlled ventilation will be needed.

Anaesthetic management

The patient's condition must be improved as much as time allows following admission to hospital. Baseline measurements (Table 58.2) are taken and cultures made of any sputum and other secretions. A brief period of intensive care is valuable when the patient's condition is unstable, and especially if consciousness is impaired. Monitoring and potent drug therapy can be established. Inotropes, dopamine infusion to improve renal blood flow, intubation and ventilation, or a period of haemofiltration may prove beneficial. However, care must be taken not to correct hyponatraemia too rapidly as this has been associated with pontine demyelination. Ideally serum potassium will be at the low end of the normal range (3.5 mmol/l).

β-Blockers, given to control portal hypertension, should be discontinued.

Coagulation derangements should be corrected immediately before surgery. Fresh frozen plasma, cryoprecipitate (where fibrinogen is very low) and platelet concentrates are used, and since vascular access is often a problem it is often convenient to give the blood products into the first large intravenous line established after induction of anaesthesia. It is wise to have achieved correction of severe clotting defects before arterial and central lines are inserted.

A short-acting benzodiazepine, e.g. temazepam, given orally in modest dosage is valuable if anxiety requires treatment. Ranitidine and metoclopramide, and 0.3 per cent sodium citrate, are given to reduce the risk of aspiration.

Induction

Intravascular monitoring can usually be set up once the patient is asleep, but induction should be accomplished under monitoring appropriate to the medical state of the patient. A rapid sequence induction is standard using a sleep dose of thiopentone 5 mg/kg with fentanyl 2 µg/kg and suxamethonium to facilitate intubation. Although plasma cholinesterase can be reduced in severe liver failure, the recovery from suxamethonium is rarely affected.

Table 58.2 Preoperative investigation before liver transplantation

Haematological	*Hb, WCC, platelet count *Clotting times Blood group Cross-match 20 units blood
Metabolic	*Serum electrolytes, urea, creatinine, glucose, Ca^{2+}, PO_4 24-hour urine electrolytes, urea, creatinine
Liver function	Serum bilirubin, albumin, globulins, liver enzymes α-Fetoprotein Abdominal ultrasound
Cardiorespiratory	*Chest x-ray, arterial blood gases, *ECG ? Exercise MUGA scan, ?echocardiogram, respiratory function tests
Infection	*Cultures (nose, stool, sputum, etc.) Hepatitis markers (A,B,C, delta) Viral antibodies (HIV, CMV, toxoplasma)
Immunological	Immunoglobulins, autoantibody screen, CD4 count Tissue typing

* Repeated immediately preoperatively.

Maintenance

The guiding principle is that drugs are chosen that have no toxic effects on the graft, do not require hepatic function to be metabolized, and are easily managed. The muscle relaxation should be provided by atracurium which does not depend on the liver for detoxification though, since early postoperative ventilation is routine, longer-acting cheaper agents such as pancuronium can be used without influencing the long-term results.

All the anaesthetic vapours have been employed but at present isoflurane offers the best attributes with little myocardial depression, stable cardiac rhythm and no hepatotoxic metabolites. Nitrous oxide can be used but it interferes with haematopoiesis and bone marrow function will also be threatened later by immune suppressant drugs. These patients often have a distended bowel and nitrous oxide will increase this and impede surgery. Any bubbles of air embolus will also expand. Air–oxygen is usually used as the ventilating gas mixture.

Cardiovascular changes

A prime feature of the operation is the variety of cardiovascular changes that are met. In the dissection/hepatectomy phase hypotension is most commonly due to hypovolaemia from blood loss, but may arise also from hypocalcaemia and myocardial depression following too rapid blood transfusion, from traction on the liver impeding vena caval flow, and from direct compression of the inferior surface of the heart by retractors held below the diaphragm. The vena cava is cross-clamped above and below the liver before the liver is removed. The resulting reduction in venous return often halves cardiac output and profound hypotension can follow. This is much worse if blood volume replacement has been inadequate, though this may be difficult to assess when cardiac filling pressures are low, not because of hypovolaemia but due to reduction in venous return as a result of caval obstruction during hepatectomy. Careful measurement of blood loss is vital together with careful timing of the measurements of filling pressures so as to avoid long periods of surgical traction. When the liver is removed, any open vein may allow an air embolus if the venous pressure is not kept positive to atmosphere.

Reperfusion of the liver is a highly unstable phase. The graft is flushed through with plasma expander to remove any preservation solution. The vena caval anastomoses are opened first to restore venous return from the lower half of the body. The portal vein inflow and the hepatic vein outflow are then opened. The portal venous blood drains from gut that has had outflow obstruction throughout the anhepatic phase unless bypass has been used (see below). This effluent has a low pH with a high potassium level and contains toxic vasoactive and myocardial depressant factors. Cardiac output and peripheral resistance fall and severe hypotension follows. The most important protection against this is an adequate circulating blood volume and a generous infusion of calcium chloride 13.4 per cent to maintain high normal ionized calcium levels. Redistribution tempers the depressant effect over 4 or 5 minutes but sometimes the hypotension is persistent with a low systemic vascular resistance – the post-reperfusion syndrome. The cause of this is not clear. The use of bypass does not seem to reduce the incidence of these dangerous changes at reperfusion.

Bypass

Venovenous bypass has been used to drain blood from both the caval system below the liver and the portal system and return it to the heart via the superior vena caval system. A Biomedicus pump with heparin bonded tubing achieves flows of 1–3 l/min without systemic heparinization. The technique is used for all cases in some centres and for none in others. The advantages are:

1. a reduction in blood loss by a reduction in abdominal systemic and portal venous pressures, thus relieving intestinal congestion
2. maintenance of renal blood flow and venous return.

The advantages must be set against the disadvantages of:

1. vascular trauma during access,
2. risk of platelet consumption,
3. heat loss from the lines,
4. thromboembolism,
5. air embolism, and
6. right heart overload.

Many centres limit the use of bypass to those patients in whom:

1. there has been previous right upper quadrant surgery,
2. there is severe portal hypertension,
3. renal function is marginal,
4. left ventricular function is impaired or when
5. a period of hepatic vascular exclusion during the dissection phase (trial clamping) causes sustained hypotension unresponsive to volume replacement.

Blood loss

Blood loss is less than 5 litres in over 75 per cent of operations, but patients have survived after losses of over 90 litres. The factors which limit blood loss are:

1. good surgical technique,
2. avoidance of abdominal venous engorgement, and
3. maintenance of normal coagulation status.

Aprotinin 280 mg followed by an infusion 70 mg/h has been shown to reduce blood loss significantly.[15]

Monitoring and venous access

Two large (12G) venous cannulae are the minimum access required for blood replacement. An arterial line is set up for cardiovascular monitoring and a second for blood sampling is useful. A triple lumen central venous catheter allows

drug infusion and simultaneous monitoring of right-sided pressures. A pulmonary artery line will allow measurement of occlusion pressure, cardiac output and mixed venous oxygen saturation. The occlusion pressure trends usually follow the right-side pressures and may prove to be more helpful in postoperative management. Systemic vascular resistance can be calculated from cardiac output measurements and thus enable proper use to be made of α- and β-adrenergic agonists for the treatment of persistent hypotension in the post-reperfusion syndrome. Mixed venous oxygen saturation adds little insight.

Oesophageal temperature is monitored and care is taken not to push the probe too far down where it will lie close to the cold graft.

A bladder catheter is used to measure hourly urine volumes.

Coagulation changes

A major disruption in coagulation takes place peroperatively, particularly at the time of reperfusion of the graft. Fibrinolysis is a major feature,[16] especially where there is a preoperative disturbance of coagulation and fibrinolysis. There is an imbalance between plasminogen activators and inhibitors, causing primary fibrinolysis. Generous use of fresh frozen plasma and platelets at the time of graft reperfusion is the mainstay of treatment. Aprotinin has been shown to reduce fibrinolysis during thromboelastography, and to reduce the rise in tissue type plasminogen activator (tPA), and to reduce the fall in plasminogen activator inhibitor (PAI). When coagulation remains a problem antifibrinolytics such as ε-aminocaproic acid may be helpful, although aggravation of fibrinolysis may occur if disseminated intravascular coagulation is a factor in promoting secondary fibrinolysis. It is not used routinely. Transfused blood should be filtered to 40 μm to reduce microaggregates.

Dilutional coagulopathy may occur when large volumes of stored blood are transfused. Coagulation times can be kept in the normal range by transfusing fresh frozen plasma in response to changed coagulation times. These can now be measured in the operating theatre (512 Coagulation Monitor, Ciba-Corning).

Platelets must be transfused to maintain the count in the normal range. Desmopressin 20 μg may help to improve function.

Bleeding time is also prolonged by hypothermia[17] so special attention must be paid to maintenance of temperature.

Biochemical changes

Serum potassium changes most dramatically with a surge up to 10 mmol/l at the unclamping and revascularization of the graft. Rapid redistribution makes the cardiac depression transient. This surge will not be tolerated if the patient was preoperatively already hyperkalaemic. The transfusion of even massive amounts of stored blood does not cause a troublesome rise in potassium, but it does have a major effect due to its citrate content. In major blood loss cases a cell saver (Haemonetics) can be used to salvage red cells for reinfusion. It can also be used to wash the cells in stored blood packs to reduce the potassium and acid load.

Ionized calcium levels can be low because citrated blood products are given in large volume. With little or no liver function citrate is poorly metabolized. Electrolyte monitors must be available in the operating theatre to allow frequent measurement and correction by infusion of 13.4 per cent calcium chloride. Large amounts may be required. Normal ionized calcium levels are a major factor in promoting cardiovascular stability.

Hydrogen ion accumulates especially in the previously acidotic patient. Mild hyperventilation will provide some compensation. Bicarbonate is not used to correct the acidosis unless this is severe (base excess >10 mmol/l); however, bicarbonate may help to improve the response to inotropes.

Glucose rises remorselessly throughout the operation, even in the anhepatic phase. Insulin is given to control any rise above 15 mmol/l.

Renal function

Renal blood flow is promoted by peroperative infusion of dopamine 2 µg/kg per minute. Use of venovenous bypass helps to maintain urine production peroperatively, but has no effect on the incidence of renal failure postoperatively.

Temperature

Hypothermia increases the bleeding time and must be avoided. Obsessional attention to avoid exposure, especially at induction, use of warming blankets and plastic drapes, effective warming of all fluids and humidification of inspired gases all help to limit the fall in temperature. The Rapid Infusion System (Haemonetics) allows the blood to be warmed to 38°C, reversing the temperature fall when there has been major blood loss.

When the cold 'new' liver is revascularized there is always a fall of about 1°C.

Postoperative care

The patient should be electively ventilated until gas exchange is good and the circulation is stable. The wound is transverse with midline upward extension. Intercostal blocks provide effective pain relief with minimal need for

central sedation. Coagulation should be monitored carefully and fresh frozen plasma and platelets given to maintain normal values. Where a coagulopathy has led to persistent bleeding, abdominal distension has been limited by use of a MAST shock suit. An important strategy for such patients is early and sometimes repeated laparotomy to look for bleeding points and to evacuate blood clot before renal failure is induced by compression of the renal circulation.

The 'new' liver takes up potassium and therefore large supplements may be required over the first 12 hours to avoid hypokalaemia. Metabolic alkalosis is quite common as the citrate from blood products is metabolized. Urine flow is promoted by continuing the infusion of dopamine at 2 µg/kg per minute. Prophylactic antibiotics are continued for 48 hours. Fitter patients will move to a high dependency area within 2 days, but prolonged intensive care may be required for those who start with more severe debilitation.

Pancreas transplant

Diabetes mellitus is a common disease treated effectively by the use of insulin. Of those diabetics who die before the age of 40 one-fifth succumb as a result of renal failure. When diabetics receive kidney transplants the graft may deteriorate due to the development of microangiopathic changes. A concurrent pancreas transplant could improve carbohydrate balance and delay the diabetes-induced vascular changes. The commonest indication for pancreas transplant is, in association with a kidney graft, for the diabetic with renal failure. Rare indications can be diabetics who have marked intolerance to subcutaneous insulin.

Over 2000 pancreas transplants have been performed. The 1-year graft survival is about 60 per cent,[18] and patient 1-year survival is 85 per cent. Graft survival is improved when a kidney has been grafted at the same time.

The operation

A variety of surgical procedures have been tried. A major question relates to the organ's exocrine secretions. Either the pancreatic duct is occluded with latex or the secretions are led into the jejunum, the stomach or the bladder. The pancreas must be taken from a beating heart donor. It is placed in the iliac fossa and connected to the external iliac vessels on the opposite side from any kidney graft. Alternatively it can be placed next to the stomach and linked onto the splenic vessels. The procedure takes 2–3 hours, which may be in addition to the time taken for other simultaneous organ tansplants such as kidney or liver.

An alternative strategy is to take just the islets of Langerhans and implant them in the liver. Various coatings of the islets have been used to improve survival but these have yet to prove effective. Anaesthesia may be required to allow access to the portal venous system to be achieved for the injection of the cell preparation.

Common problems

Renal

Most pancreatic grafts will be in patients who also require a kidney graft, and it is the kidney graft which will determine most of the management protocol.

Diabetes

Insulin dependence is the rule; some patients present for transplant because of the difficulty experienced in establishing an insulin regimen that adequately controls blood sugar. These patients will have advanced diabetes with not only nephropathy, but also peripheral vascular disease, retinal disease and peripheral neuropathy. Typically they are often young but are gross arteriopaths. In Cambridge the mean age of 43 transplant patients was 38.1 years.

Anaesthetic management

Preoperatively, particular attention should be paid to the diabetic state. Evidence of an autonomic neuropathy which heralds risk of cardiovascular instability peroperatively should be sought by finding a history of diarrhoea, impotence or faintness on standing and by checking blood pressure lying and standing.

An intravenous infusion of 5 per cent dextrose is started and hourly blood glucose estimations made until the patient is clearly stable. An insulin infusion is started at a rate of 1–6 units/h to maintain blood glucose levels between 5 and 10 mmol/l.

The guidelines described for the management of kidney transplant anaesthesia should be followed. In addition the blood sugar should be monitored and controlled at least hourly throughout surgery. The major risk to the patient is hypoglycaemia due to excess insulin, and this must be avoided. The normal trend during anaesthesia is for a steady rise in blood glucose.

Significant events

The initial part of the operation is the identification and dissection of the vessels to which the graft will be attached.

This is not usually associated with blood loss. As with all transplants it is when the graft is revascularized that major blood loss can occur and good vascular access for blood transfusion is essential. This access can be difficult to obtain in this group of hospitalized patients with arterial disease with, perhaps, the vessels of one arm unavailable because of a haemodialysis shunt fistula. The most important management priority is to have an adequate circulating blood volume before release of the vascular clamps.

Any insulin infusion should be stopped at the onset of reperfusion as the graft begins to function virtually immediately. There will be further surgery related to the drainage of the exocrine secretions of the pancreas. Further blood loss may occur during this extended surgery. Blood loss for a combined kidney and pancreas transplant averages 1500 ml, but may be trivial.

Postoperative care

No further insulin will be required, but blood glucose should be monitored hourly until clearly stable. Otherwise postoperative management should follow the lines described for kidney transplant. It is unusual for patients to require postoperative ventilation unless the preoperative medical state was very poor, the surgery long, and the blood loss high.

Heart transplant

The first clinical heart transplant was in 1967. It is now established treatment for end-stage cardiac failure due to cardiomyopathies and ischaemic heart disease with a life expectancy less than 1 year. One-year survival after grafting is better than 80 per cent. Contraindications include significant pulmonary hypertension, severe peripheral or cerebrovascular disease, drug addiction and chronic lung disease.

The operation

The heart is placed orthotopically (in place of the recipient's own heart). The donor heart must be taken from a beating heart donor and carefully preserved by topical hypothermia and cold hyperkalaemic cardioplegia. There is then an upper limit for ischaemic time of 4 hours before the heart must be reperfused in the new donor. There needs to be very tight timing of donor and recipient operations to achieve this.

Cardiopulmonary bypass is established in the routine fashion. The old heart is removed by transecting across the right atrium proximal to the appendage and across to the aortic root above the coronary ostia. The pulmonary artery is transected and the left atrium cut around, leaving the posterior wall with its pulmonary veins.

The new heart is sewn in and bypass continued to allow myocardial recovery and rewarming of the patient. The whole procedure lasts about 4 hours.

Common problems

Patients will have end-stage cardiac failure NYHA class 4. They will often be receiving a variety of drug therapies, including digoxin, diuretics, ACE inhibitors, calcium channel blockers, other vasodilators, and antithrombotic agents. These may cause electrolyte disturbance, aggravated by some renal dysfunction. Chronic hepatic congestion may lead to abnormal liver function with low albumin and abnormal clotting factors. Baseline values should be established. There will be little time to improve the state of the patient who will usually be called in urgently from home.

Anaesthetic management

Constant care must be directed to maintain an aseptic technique. Prior to induction arterial and central venous pressure monitoring must be established. Pulmonary artery catheters are not usually used as they add to the risk of infection and will have to be removed before cardiectomy. Induction can be difficult in an orthopnoeic patient who has no cardiac reserve. The stomach must be assumed to be full and after preoxygenation, cricoid pressure should be used to prevent regurgitation. Vecuronium may be used to give rapid muscle relaxation with less circulatory disturbance than suxamethonium. All intravenous induction agents cause myocardial depression, though this problem is less with etomidate or diazepam. The dose and the risk of depression are reduced if the patient has been fit enough to allow some sedative premedication.

Reducing the load on the heart by the use of vasodilators such as sodium nitroprusside 1–3 μg/kg per minute may be critical. Filling pressures are high, and inotropes may be needed throughout induction. Bolus administration of calcium chloride can be used to treat hypotension.

Maintenance of anaesthesia is balanced between the cardiovascularly benign narcotic analgesics which cause little cardiovascular disturbance and slow elimination and the more depressant but rapidly cleared inhaled vapours.

Nitrous oxide is normally avoided as it adds to myocardial depression, may aggravate any air embolus and will interfere with bone marrow function.

Once cardiopulmonary bypass is established, haemofiltration and dialysis can be added to correct preoperative fluid balance disturbance. When coming off bypass the heart rate is maintained at 100–120/min using isoprenaline infusion to prevent overdilatation. Arrhythmias will be less if the serum potassium is kept in the high normal range. Offloading by sodium nitroprusside infusion may be helpful.

Fresh frozen plasma may be required to restore normal clotting if coagulation was abnormal preoperatively due to either anticoagulant therapy or liver disease.

Postoperative care

Overnight ventilation should be planned with early extubation when respiratory function is adequate. Management follows routine principles of postbypass surgery, with particular attention given to keeping up heart rate by isoprenaline infusion. Immune suppressive treatment is adjusted after myocardial biopsy.

Anaesthesia for non-cardiac surgery

Heart transplant recipients, 14 per cent in one series,[19] may present for any kind of surgery, e.g. total hip replacement for avascular necrosis. Atherosclerosis is common and assessment must be careful.[20] Ischaemia may be concealed since the denervated heart graft does not cause angina pain. Poor exercise tolerance, proneness to arrhythmias and shrinking ECG voltages are signs of a deteriorating graft. The ECG can normally show two P waves, one from the recipient remnant, one from the new heart. Opportunistic lung infections may occur.

Denervation prevents the reflex (but not necessarily the hormonal) response to laryngoscopy or carotid sinus pressure. Intrinsic mechanisms remain intact. An increase in rate cannot reflexly follow arterial hypotension, but cardiac output can be preserved, provided the preload is adequate, by maintaining stroke volume. β-Adrenergic receptor sensitivity is unaltered by denervation and a normal response to catecholamines can be expected. In the presence of a bradycardia atropine may be ineffective and direct acting catecholamines, e.g. isoprenaline, are preferred.

Drug therapy may be complex. Normal preoperative assessment will reveal any significant issues. General anaesthesia is usually preferred as there is a possibility of impaired response to hypotension following spinal or epidural anaesthesia. Tracheal intubation and controlled ventilation provide good surgical conditions without major cardiovascular stress.

Careful asepsis throughout the procedure is essential.

Heart–lung transplant

Combined heart–lung transplants have developed since 1982 to treat severe cardiac disease with pulmonary hypertension as well as end-stage parenchymal lung disease. Common indications include Eisenmenger's syndrome, primary pulmonary hypertension and cystic fibrosis. One-year survival is almost as good as for heart transplant alone at better than 75 per cent. The better results of heart–lung transplant make it preferable to lung transplant alone in some cases of primary lung disease. If the heart of the recipient is suitable it can be used alone for a subsequent heart transplant (domino) as a method of overcoming the lack of donors.

The operation

Careful timing of the start of the recipient operation seeks to reduce the ischaemia time of the graft, but there may be need for extensive dissection in some cases. The heart, left lung and then right lung are removed with careful preservation of vagal and phrenic nerves. Haemostasis is important at this stage when there is good visibility. Postoperative bleeding problems are much commoner than with heart transplant alone. The new graft is received in a block and the trachea, aorta and right atrium are anastomosed to their partners. The operation takes about 4 hours but may be longer.

Anaesthetic management

Patients suffering with pulmonary hypertension and right heart failure are often more hypoxaemic than those for heart transplant alone. Any improvable factors are treated urgently.

The problems and management are those described already in the section on heart transplantation with several additional factors to consider.

In patients with Eisenmenger's syndrome systemic vascular resistance must not be reduced nor pulmonary vascular resistance raised lest the right-to-left shunt is increased. Inhalation agents are used minimally and tidal volumes are kept small. A pulmonary vasodilating inotrope such as aminophylline may be helpful.

The tracheal tube must be inserted with care, so that the cuff is only just below the cords, and gently inflated. It may have to be withdrawn slightly during the tracheal anastomosis.

Complex coagulation problems can occur at the completion of surgery and these must be corrected with platelets and fresh frozen plasma. Aprotinin and fibrinolytics may be useful.

Postoperative care

Patients are weaned from ventilation and extubated as soon as respiratory function allows. Training for cough and clearance of secretions is important as there is no cough reflex from the lungs.

Great care is taken with fluid replacement since these patients tolerate fluid loads poorly, possibly due to the surgical interruption of lymphatics. This impairs fluid drainage from the lungs and also the scavenging of particulate matter.

Immune suppression continues but pulmonary rejection does not always occur in conjunction with cardiac rejection. Diagnosis of lung rejection can be difficult when dyspnoea might also be due to left ventricular failure or to acute onset pneumonia. Early pulmonary oedema is usually assumed to be due to rejection and treated with high dose steroids.

Lung volumes fall immediately following transplantation but there is steady improvement, as with gas exchange, over the first year. Later problems are increasing dyspnoea with recurrent pneumonia. A decreased peak flow is a good guide to acute lung rejection. Bronchiectasis and bronchiolitis obliterans may be due to chronic rejection or a permanently impaired cough.

Rejection of the heart is less frequent than after heart transplants alone – the tolerance of multi-organ grafts.

Anaesthesia for non-cardiac surgery

A constant aseptic technique is important. All the problems described in the section on heart transplant apply but, in addition, special attention must be given to placement of the tracheal tube above the level of the anastomosis. It must be remembered that the patient will still have no cough reflex and postoperatively will need special respiratory therapy.

Fluid replacement may be difficult, steering between maintenance of high filling pressures to promote stroke volume and the risk of pulmonary oedema from fluid overload. Central venous pressure must be monitored and, for complex procedures in patients with indifferent graft function, a pulmonary artery catheter should be used to assess occlusion pressures.

Lung transplant

Single lung transplant has been less successful (65 per cent 1-year survival) and experience is more limited than with heart and heart–lung transplants. Severe pulmonary fibrosis is the main indication.

The operation

The left lung is normally replaced with one judged to be slightly larger than the existing contracted fibrotic lung. If an omental patch is used in an attempt to improve the vascularity and healing of the bronchial anastomosis, the first stage is a limited laparotomy to take omentum up behind the xiphisternum. The patient is then turned into the lateral position and the pneumonectomy performed with careful haemostasis. The pulmonary veins of the new lung are sewn in as an atrial cuff and then the arterial anastomosis is made. The bronchial circulation cannot be restored. The bronchial anastomosis is made before the lung is reperfused. The omental patch, if used, is brought up over the bronchial anastomosis.

Anaesthetic management

The patient will have long-standing hypoxic lung disease. Preoperative baseline measurements should be made to determine if the progress of the disease state has changed from the initial assessment. Cardiac function is usually not abnormal, but patients with severe pulmonary hypertension will usually have heart–lung transplants rather than a lung transplant alone. Right ventricular function must be adequate (ejection fraction >20 per cent) to cope with the rise in pulmonary artery pressure after pneumonectomy. The patient will be receiving long-term drug therapy including steroids.

Preoperative sedation is unnecessary. Monitoring should be established, and this should include a pulmonary artery catheter with its tip in the contralateral lung. Careful intravenous induction after preoxygenation is followed by bronchial intubation using either a double lumen tube or a tracheal tube and a Fogarty catheter used as a bronchial blocker put in place at bronchoscopy. The patient will always be on the verge of hypoxia.

It may not be possible when oxygenation is critical to move to one lung anaesthesia during the dissection. As time passes compensatory reflexes can reduce the initial level of hypoxaemia. Positive end-expired pressure (PEEP) applied to the single lung may reduce hypoxaemia but can also impair circulation and increase shunting. Temporary clamping of the proximal pulmonary artery can improve oxygenation by directing blood to the ventilated lung though the rise in pulmonary artery pressure can induce right ventricular failure. Vasodilators and inotropes, hydralazine or prostacyclin, aminophylline or dopexamine, may help. If these are unsuccessful partial femoral venoarterial bypass is instituted.

Ventilation should be with an air–oxygen mixture and anaesthesia maintained by intravenous agents either totally or with a supplementary vapour, such as isoflurane.

Postoperative care

The patient is ventilated, often for several days, until respiratory function is stable. Careful attention to pain relief and cough encouragement prevent accumulation of secretions. The other bronchus and lower trachea retain sensation so that cough is preserved better than after a heart–lung transplant. The new lung is extremely prone to accumulation of interstitial fluid and therefore a tight control of fluid balance must be maintained, erring on the 'dry' side. Haemofiltration may help if resistant overload develops. Rejection can be difficult to distinguish from infection. Serial chest x-rays are useful.

Bone marrow transplant

The role of the anaesthetist is during the harvesting of bone marrow. The donor may be the patient in remission from a haematological malignancy, or a very closely HLA matched subject. In such cases the donors are in good health and have only the normal risks from anaesthesia.

The procedure is to perform multiple bone marrow aspirations of about 5–10 ml at a time until the total volume is obtained. The target volume will vary with the cell count of the initial marrow aspirate, but typically 700 ml (10 ml/ kg body weight) will be required. Replacement of blood volume with a plasma expander is appropriate. Adequate venous access is important to allow the volume to be replaced over the 30–40 minutes of the procedure.

Prime sites for aspiration will be the sacrum and ilium posteriorly and the iliac crest and the sternum anteriorly. The procedure will therefore normally start with the patient prone and then, if a further volume is required, the patient is turned supine. The key factor in the anaesthetic management will be control of the airway and to ensure adequate ventilation in the prone position. Tracheal intubation and IPPV are therefore required. Particular care is needed as the patient is turned back to the supine position, to maintain airway control and to check for sudden hypotension if inadequate fluid replacement has been given.

At the end of the procedure the haematologists may take the opportunity of inserting a central line for the administration of chemotherapy.

Corneal transplant

Eyes with corneal damage from trauma, keratoconus, keratitis or corneal dystrophy may require a corneal graft. Corneas are stored in tissue culture which reduces allergenicity. In the UK sufficient corneas are available from the National Transplant Services to meet national demand. The operation is entirely elective and lasts about 2 hours. The disc of cornea is removed and replaced with a disc of identical size from the donor cornea. Partial thickness (lamellar) grafts may also be used. In the future a larger disc may be used but the increased vascularity at the edge promotes possible rejection.

Patients are usually otherwise well and the anaesthetic management is aimed to create a 'soft eye' as for other intraocular surgery using general anaesthesia with tracheal intubation and controlled ventilation. The immune response is limited unless there is a marked vascular response at the edge of the graft. Any rejection response is readily visible and therapy can be instituted quickly with standard immune suppressant drugs. Long-term transparency is achieved in over 90 per cent of patients.

Bowel transplant

A small bowel transplant can be performed to treat gastrointestinal failure requiring parenteral nutrition with all its problems of long-term venous access. Indications have been short bowel syndrome, multiple bowel strictures and abdominal desmoid tumour. Experience is limited but there have been some graft survivals longer than a year.

The operation

A section of small bowel can either be transplanted separately or the procedure combined with a liver transplant. The donor tissue must be carefully preserved since the mucosa is delicate and readily autolysed. The bowel of both donor and recipient will have been decontaminated with antibiotic mixture. The graft is taken on the superior mesenteric artery pedicle with aortic conduit and anastomosed to the recipient aorta. The end of the donor jejunum is sewn to the recipient duodenum and the other end of the graft is brought to the skin as an ileostomy. If an ileocolic anastomosis is made, two ileostomies may be formed for postoperative control and feeding. If combined with a liver transplant, the liver is transplanted first followed by the bowel.

Anaesthetic management

The particular points to assess preoperatively are the potential problems with vascular access. Often these patients have been receiving intravenous feeding for years and all the large veins will have had many cannulations. A large bore cannula must be sited to ensure that a rapid blood transfusion is possible in the not uncommon event of serious bleeding, especially if a liver transplant is planned in association with the bowel transplant. This may require a surgical approach because a sternal split may be necessary to achieve this.

The procedure is not particularly complex from the anaesthetist's viewpoint. The physiological disturbance is similar to that following any laparotomy and this determines anaesthetic technique. Meticulous asepsis is essential. The main anxiety is the possibility of sudden blood loss associated with difficulty with the vascular anastomoses. There may have been many previous abdominal operations with a long and difficult dissection being required before the graft can be sewn in. It is likely to take at least 3 hours which may be in addition to a liver transplant. Warming blankets and warmed fluids should be used to minimize temperature fall during surgery.

Postoperative care

Maintenance of adequate circulating blood volume is essential to ensure early good perfusion of the graft. If there has been no simultaneous liver graft, the early recovery resembles that from any long and difficult laparotomy. The level of support will be determined primarily by the patient's preoperative fitness. Careful attention to pain relief will be important after this major upper abdominal operation. Patient-controlled analgesia systems are helpful. Epidural analgesia may be used but care must be taken to avoid hypotension which may threaten the perfusion of the graft.

The principal problem has been the severity of the immune response which causes shedding of the graft mucosa, thus destroying the barrier to micro-organisms and fungi which can then pass readily into the portal blood. Cytomegalovirus infection is common. The gut is rich in lymphocytes and, in successful grafts, the recipient population is replaced by host cells (chimerism) which do not provoke an immune response. Rejection leads to complex abdominal sepsis and intensive care may be prolonged.

Management of the multi-organ donor

Management begins from the moment that the patient is declared to be dead. Until then therapy has been directed at achieving recovery of the patient; suddenly the aim is to establish the best possible function of the various potential donor organs.

Table 58.3 Clinical tests for brainstem death

Prerequisite criteria must have been satisfied (diagnosis, absence of depressant and muscle relaxant drugs, temperature >35°C)

Pupils dilated and fixed in response to light

Absent corneal reflex

Absent vestibulo-ocular reflex

    The external auditory meatus is checked that there is no blockage. No eye movements must be seen during or after slow injection of 20 ml of ice cold water into each external auditory meatus in turn

No motor responses in the cranial nerve distribution

No gag reflex or response to bronchial suction

No respiratory response on apnoea

    The patient should be ventilated with 100% oxygen for 15 minutes and then disconnected from the ventilator. 100% oxygen at 3 l/min should then be administered into the open tracheal tube via a fine catheter. The patient must then be observed for 10 minutes for respiratory effort. A blood gas must be taken to confirm a Pa_{CO_2} greater than 6.7 kPa (50 mmHg). If blood gas levels cannot be measured the final 5 minutes of ventilation before disconnection must be with 5% CO_2 in oxygen.

Diagnosis of death

The process of dying is a continuum which is terminated by death, but the onset of death has no clear moment in time. As time elapses the diagnosis of death becomes increasingly unmistakable. All countries have struggled to determine criteria for the diagnosis of death that are socially acceptable, and ethically and clinically sound. Patients who are clinically dead should not be subjected to continuing therapy which only prolongs the distress of their relatives and their carers.

In the United States and the United Kingdom, irreversible loss of brain function has been accepted as diagnostic of death.[22] However, in the US the concept has been of irreversible cessation of all functions of the *entire* brain.[23] It is recognized that the key distinction between brain death and a persistent vegetative state is the absence of brain*stem* function. In the UK the emphasis has focused on cessation of brainstem function since this inevitably leads to death of all the other organs in the body.[24,25] Both countries use the same signs to establish the diagnosis of brain death. Debate continues about the relevance of supporting evidence such as absence of carotid and vertebral blood flow and absence of electroencephalographic (EEG) activity. These criteria are not considered necessary in the UK.

Brainstem death

The essential prerequisite for a diagnosis of brainstem death is a known cause for catastrophic brain injury such as intracerebral haemorrhage, head trauma, cerebral anoxia due to drug overdose, cardiac arrest, smoke inhalation or drowning, or primary brain tumour. The patient must be deeply comatose, requiring lung ventilation with no depressant drugs present; the core temperature must be >35°C; and metabolic or endocrine causes for coma must have been excluded. The diagnosis is not normally considered until at least 6 hours after the onset of coma or, if cardiac arrest was the cause of the coma, until 24 hours after the circulation has been restored.

The signs are shown in Table 58.3. The tests should be performed by an experienced doctor who is not part of the transplant team, and will be confirmed by another experienced doctor after a short interval. On completion of the confirmatory test, the patient is declared dead.

When a patient likely to fulfil these criteria is identified, the local transplant coordinator is notified. The patient must have no systemic infection or malignant disease (other than primary brain tumour). The organs for donation must be undamaged and have good function. Blood electrolytes and urea, full blood count, liver function tests, serum amylase, 12-lead ECG, chest x-ray and blood gases

Table 58.4 Guidelines for suitability of specific organs

Kidney
  Age less than 70 years
  No hepatic or renal damage
  No history of hypertension
  Urine output >0.5 ml/kg per hour
  HLA compatibility

Liver
  Age less than 55 years
  No liver damage, no drug or alcohol abuse
  Normal liver function tests
  Gallbladder present
  Liver looks normal

Pancreas
  Age less than 50 years
  Normal serum amylase
  No diabetes mellitus

Heart
  Age less than 50 years
  No heart disease, hypertension, or myocardial trauma
  No prolonged cardiac arrest
  Virtually no inotropic support needed
  Normal ECG and chest x-ray

Heart–lung or lung
  As for heart, but also:
  No history of heavy smoking or chronic lung disease
  Good gas exchange and lung compliance
  No pulmonary infection
  Good match of size with recipient
  Brief period of IPPV

should be essentially normal. Specific guidelines for acceptability of various organs are shown in Table 58.4. There is pressure to relax these criteria so that the number of donors can increase.

Formal consent for organ donation must be obtained from the nearest relative, and in the UK the approval of the coroner's officer is sought when the cause of death, e.g. road traffic accident, requires investigation by the coroner. Blood is obtained for identification of ABO blood group, HLA antigens, serological tests for cytomegalovirus, toxoplasma, Epstein–Barr virus, hepatitis and HIV.

Availability of donors

No country has sufficient donors to satisfy the transplant requirements of its population. In the UK 48 kidneys per million population per year would meet the need, but the actual donation rate is only about 28 per million. A major obstacle is the fear that distress may be caused to a potential donor's relatives, when approached for permission to remove organs. As public awareness increases and medical staff gain experience in making the request this fear is decreasing. Indeed, relatives may resent *not* being asked, when they recognize the situation. Some states in the US have now placed on the medical staff a legal obligation to seek permission for organ donation when a patient's condition fulfils the necessary criteria. Permission is more readily granted when the request is made on an occasion isolated from the initial 'bad news' and trained transplant coordinators can be particularly expert and effective in this role.[26]

Audit of intensive care deaths in England has revealed that failure to seek consent from relatives occurred in only 6 per cent of patients with brainstem death.[27] Up to a third of the patients in an intensive care unit who fulfil brainstem death criteria may be unsuitable for organ donation on other medical grounds.[28] In an attempt to improve the availability or organs, it has been suggested that patients with fatal brain injury should be managed with lung ventilation and intensive care even when the prognosis is certain and such treatment would not normally be considered.[29] The aim is to protect the possibility of organ donation, but this protocol raises difficult ethical issues which have prevented its widespread adoption.

Management

General management

Once death is declared it is important that organ donation is managed expeditiously for the sake of the relatives and the hospital staff and to allow successful donation. Fifty per cent of patients who fulfil the criteria may suffer cardiac arrest in the first 24 hours despite intensive management.[30]. Therapy for management of the injured brain can now be abandoned and attention directed to improving the general perfusion of the donor organs. Lung ventilation is adjusted to bring Paco$_2$ up into the normal range. Monitoring must be maintained and, if anything, intensified.

Cardiovascular management

Hypotension with a low systemic resistance is common. Hypovolaemia following the previous diuretic and fluid restrictive therapy often needs correction. Colloid solutions are preferred. Inotropes and vasoactive drugs may be useful in the restoration of cardiac output and to increase perfusion. Care must be taken not to overinfuse and thereby overload donors, especially of lungs, and it is preferable to accept systolic blood pressures of about 80 mmHg in a vasodilated patient.

Fluid management

Diabetes insipidus should be suspected if urine flows >4 ml/kg per hour, osmolality <300 mosmol/kg, and <10 mmol Na/l with rising plasma sodium and osmolality are found. Diabetes insipidus can lead to major fluid imbalance. Arterial and central venous pressure monitoring will therefore be needed. Electrolyte derangement should be corrected. Desmopressin (DDAVP) 1–10 μg given intramuscularly may be repeated or vasopressin 2 units/h given intravenously to control diabetes insipidus. In patients without diabetes insipidus, there may be oliguria. This is treated by blood volume correction and diuretics with a target urine flow of 0.5 ml/kg per hour.

Hormone therapy

Cardiovascular stability may be improved by hormone supplementation. Hydrocortisone 100 mg hourly should be given for cardiovascular instability and if adrenal insufficiency is suspected. Insulin may be needed for hyperglycaemia. Tri-iodothyronine (5 μg, then 3 μg/h) may help[31] though instability has not been correlated with measurable endocrine deficits.

Temperature

Heat loss with poor thermoregulation will lead to hypothermia. Blankets and warmed fluids should be used to maintain body temperature.

Prevention of infection

A scrupulous aseptic technique should be observed to prevent transmission of infection to the recipient. Special care is needed in the case of lung donors. Microscopy of material and culture from possible sites of infection is routine.

The donor operation

Kidneys, corneas and bone can all be successfully transplanted when taken from cadavers. Other organs must be removed before the circulation ceases. Multi-organ donation is a major operation which will last 3 hours or so. Intensive management to maintain cardiovascular stability will be required until the circulation stops. The intravascular monitoring used in the ICU should continue in the operating theatre. Extensive dissection of the donor organs is carried out whilst the circulation is maintained. Blood loss averages 2 litres in the adult and is usually replaced with colloid, or even blood. Spinal reflexes may give rise to movement and hypertension.[32] Movement is controlled by the use of muscle relaxants, and hypertension by fentanyl and small amounts of isoflurane. Donor drug protocols will vary but may include prophylactic antibiotics, heparin, methylprednisolone, phenoxybenzamine, dopamine, dobutamine, vasopressin, tri-iodothyronine, insulin, and prostacyclin. The abdominal organs (liver, kidney, pancreas) are perfused with cold preservative solution and removed before proceeding to the heart and lungs. Bone and eyes may also be taken later.

The success of transplant surgery depends on the quality of the donor organs, and therefore the care of the donor must match that given later to the recipient.

REFERENCES

1. Jamieson NV. A new solution for liver preservation. *British Journal of Surgery* 1989; **76**: 107–8.
2. Bowman H, Lennard TWJ. Immunosuppressive drugs. *British Journal of Hospital Medicine* 1992; **48**: 570–7.
3. Calne RY, Sells RA, Pena JR, Davis DR, Millard PR, Herbertson BR. Induction of immunological tolerance by porcine liver transplantation. *Nature* 1969; **223**: 474–6.
4. Moudgil GC. Update on anaesthesia and the immune response. *Canadian Anaesthetists' Society Journal* 1986; **33**: S54–60.
5. Salo M. Effects of anaesthesia and surgery on the immune response. In: Watkins J, Salo M, eds. *Trauma, stress and immunity in anaesthesia and surgery.* London: Butterworth, 1982.
6. Opelz G, Terasaki PI. Improvement of kidney graft survival with increased numbers of blood transfusions. *New England Journal of Medicine* 1978; **299**: 799–803.
7. Stoutenbeek OP, van Saene HKF, Miranda DR, Zandstra DF, Langrehr D. Nosocomial gram-negative pneumonia in critically ill patients. *Intensive Care Medicine* 1986; **12**: 419–23.
8. Starzl TE, Koep LJ, Halgrimson CG, *et al.* Fifteen years of clinical liver transplantation. *Gastroenterology* 1979; **77**: 375–88.
9. Remuzzi G. Bleeding in renal failure. *Lancet* 1988; **1**: 1205–8.
10. Linke CL, Merin RG. A regional anaesthetic approach for renal transplantation. *Anaesthesia and Analgesia* 1976; **55**: 69–73.
11. Sheppard S, Eagle CJ, Strunin L. A bolus dose of esmolol attenuates tachycardia and hypertension after tracheal intubation. *Canadian Anaesthetists' Society Journal* 1990; **37**: 202–5.
12. Wickstrom I. Enflurane anaesthesia in living related donor renal transplantation. *Acta Anaesthesiologica Scandinavica* 1981; **25**: 263–9.

13. Hunter JM, Jones RS, Utting JE. Use of atracurium in patients with no renal function. *British Journal of Anaesthesia* 1982; **54**: 1251–8.

14. O'Grady J, Alexander GJM, Thick M, Potter D, Calne RY, Williams R. Outcome of orthotopic liver transplantation in the aetiological and clinical variants of acute liver failure. *Quarterly Journal of Medicine* 1988; **69**: 817–24.

15. Mallett SV, Rolles K, Cox D, Burroughs A, Hunt B. Intraoperative use of aprotinin in orthotopic liver transplant. *Transplantation Proceedings* 1991; **23**: 1931–2.

16. Dzik WH, Arkin CF, Jenkins RL, Stump DC. Fibrinolysis during liver transplantation in humans – role of tissue type plasminogen activator. *Blood* 1988; **71**: 1090–5.

17. Valeri CR, Cassidy G, Khuri S, Feingold H, Ragno G, Altschule MD. Hypothermia induced reversible platelet dysfunction. *Annals of Surgery* 1986; **265**: 175–81.

18. Moudry-Munns KC, Gillingham K, Sutherland DER. Pancreas registry report: clinical pancreas transplantation. *Journal of Transplant Coordination* 1991; **1**: 11–14.

19. Steed DL, Brown B, Reilly JJ, *et al.* General surgical complications in heart and heart–lung transplantation. *Surgery* 1985; **98**: 739–45.

20. Shaw IH, Kirk AJB, Conacher ID. Anaesthesia for patients with transplanted hearts and lungs undergoing non-cardiac surgery. *British Journal of Anaesthesia* 1991; **67**: 772–8.

21. Wood RFM, Ingham Clark CL. Small bowel transplantation. *British Medical Journal* 1992; **304**: 1453–4.

22. Truog RD, Fackler JC. Rethinking brain death. *Critical Care Medicine* 1992; **20**: 1705–13.

23. Guidelines on the determination of death. Report of the medical consultants on the diagnosis of death to the President's Commission for the Study of Ethical Problems in Medicine and Biomedical and Behavioral Research. *Journal of the American Medical Association* 1981; **246**: 2184–6.

24. Conference of Medical Royal Colleges and their Faculties in the United Kingdom. Diagnosis of brain death. *British Medical Journal* 1976; **2**: 1187–8.

25. Jennett B, Gleave J, Wilson P. Brain death in three neurosurgical units. *British Medical Journal* 1981; **282**: 533–9.

26. Fisher RA, Alexander JW. Management of the multiple organ donor. *Clinical Transplantation* 1992; **6**: 328–35.

27. Gore SM, Hinds CJ, Rutherford AJ. Organ donation from intensive care units in England. *British Medical Journal* 1989; **299**: 1193–7.

28. Bodenham A, Berridge JC, Park GR. Brain stem death and organ donation. *British Medical Journal* 1989; **299**: 1009–10.

29. Feest TG, Riad HN, Collins CH, Golby MGS, Nicholls AJ, Hamad SN. Protocol for increasing organ donation after cerebrovascular deaths in a district general hospital. *Lancet* 1990; **335**: 1133–5.

30. Mackersie RC, Bronsther OL, Shackford SR. Organ procurement in patients with fatal head injuries. The fate of the potential donor. *Annals of Surgery* 1991; **213** (2): 143–50.

31. Novitzky D, Cooper DKC, Reichart B. Hemodynamic and metabolic responses to hormonal therapy in brain-dead potential organ donors. *Transplantation* 1987; **43**: 852–4.

32. Wetzel RC, Setzer N, Stiff JL, Rodgers MC. Hemodynamic responses in brain dead organ donor patients. *Anesthesia and Analgesia* 1985; **64**: 125–8.

FURTHER READING

Conacher ID. Isolated lung transplantation: a review of problems and guide to anaesthesia. *British Journal of Anaesthesia* 1988; **61**: 488–74.

Farman JV, ed. *Transplant surgery: anaesthesia and perioperative care.* Amsterdam: Elsevier, 1988.

Gelman S, ed. *Anaesthesia and organ transplantation.* Philadelphia: WB Saunders, 1987.

Sale JP, Patel D, Duncan B, Waters JH. Anaesthesia for combined heart and lung transplantation. *Anaesthesia* 1987; **42**: 249–58.

Chapter 59

Paediatric Organ Transplantation

Susan Firestone and Leonard Firestone

Organ transplantation in children has always followed from the pioneering work performed in adults. First, kidney transplantation, which dates from the 1970s, and more recently liver and heart transplantation were firmly established as therapeutic options for children. Lung, small bowel and pancreatic transplantation are currently in a process of evolution from 'experimental' modalities to standard therapies for children. Thus, while the number of transplants in adults reached the donor-limited maximums by the late 1980s, the growth in most paediatric transplantation continues unabated.[1,2]

As experience with paediatric transplantation increased, remarkable improvements followed closely in both patient and allograft survival. Recent statistics for heart transplantation demonstrate that, at centres of excellence, 1-year survival approaches 90 per cent even in children less than 1 year of age.[3] These results are comparable with those achieved in adults, and are considerably better than outcomes following repair of certain complex cardiac anomalies (e.g. hypoplastic left heart syndrome).[4,5]

Such survival statistics are remarkable, yet the ultimate value of transplantation for children must be based on long-term outcomes, as well as on quality of life and economic considerations. At present, such data are available only for renal transplantation for which it has been shown unambiguously that for children, growth, development and quality of life are superior after transplantation.[6,7] Until longitudinal studies can provide analogous information for the other major organs, enthusiasm for paediatric transplantation must be tempered.

Although the basic pathophysiology of end-stage organ disease, organ preservation and procurement, and immunology are similar in children and adults, there are numerous aspects of transplantation unique to the paediatric population, including brain death criteria in infants, donor organ sizing, surgical technique and aetiologies of end-stage organ disease. These are the focus of this chapter, particularly in relation to the most common major organs transplanted: the kidney, liver, heart and lung.

General principles

Brain death and organ donation

The publication, in 1981, of the recommendations of a special Presidential Commission[8] provided the basis for most of the laws (USA) governing the determination of brain death. However, issues raised in relation to the cessation of meaningful brain function in children were not explicitly discussed in the Commission's report. To remedy this omission, a special task force was convened by the American Academy of Pediatrics, and their 1988 report specifically addressed this problem.[9] Controversial areas such as criteria for brain death in neonates and young infants have received additional attention in more recent publications.[10,11] Given the fundamental lack of knowledge about the developing central nervous system, and its well-recognized ability to recover from ischaemic injury, only the most conservative criteria regarding brain death were universally acceptable.

Even with a consensus regarding general criteria for brain death in infants and children,[12] there are still areas of controversy. The debate regarding organ donation by anencephalic newborns illustrates some of the ambiguous circumstances unique to children. Despite the fact that these children have no potential to develop higher cortical function, and usually die within a few weeks of birth,[13] they do not fulfil the accepted criteria for brain death. There are many advocates for allowing organ donation from anencephalic infants;[14,15] however, most states still do not recognize anencephalics as legally acceptable organ donors.

Voluntary organ donation from living relatives is another source of organs which raises several ethical issues. Although adult-to-adult living-related (LR) kidney donation is common, most other LR transplantation involves a paediatric recipient. Ethical concerns include the unavoidable, subtle coercion to donate; the questionable ability to obtain informed consent in these circumstances; and the balance between risk to the healthy donor versus potential benefits to the recipient child. In the case of LR renal donation, the controversy has been resolved, as the donor is left with one intact kidney, and the mortality and morbidity for nephrectomy in this healthy population are quite low.[16] However, as the potential for causing harm in the donor increases (e.g. with partial hepatic or pulmonary lobectomies), the ethical conflicts increase. Moreover, whilst partial organ transplantation may forestall death, the long-term outcome for the recipients of partial organ transplants has not yet been determined. As a result, these forms of LR donation remain controversial, and at the present time only a few centres have active programmes.[17,18]

Anatomical considerations

The matching of transplanted organs by size is usually not a major issue in adults. In contrast, for small children, size considerations are of primary importance, both in accommodating the organ in the appropriate body cavity and in the necessary vascular anastomoses. Numerous innovative solutions to the technical problems that arise with significant size discrepancy have been developed. For example, in renal transplantation, the adoption of an intra-abdominal implantation technique allows the use of adult-sized kidneys in infants.[19] For liver transplantation, surgical reduction in the size of the donor organ ('reduced-size' liver graft) has increased the donor pool available for small children.[20,21] This was a particularly important development in liver transplantation, as the majority of paediatric candidates suffer from biliary atresia,[22] resulting in end-stage liver disease while patients are still quite small. Similar principles (i.e. lobar grafts) have been employed with some success in lung transplantation,[23] Yet, in spite of all these innovations, organ shortages remain a critical problem in paediatric transplantation.[16,24]

In addition to absolute size constraints, there are other anatomical challenges in paediatric transplantation. Many of the disease entities that cause end-stage organ disease in childhood are the result of developmental anomalies, such as biliary atresia, or congenital heart disease. In these instances, there are deviations from the normal anatomical relation or absence of structures integral to the transplantation procedure. Flexibility and a considerable degree of improvisational skill on the part of the surgical team are necessary in order to accomplish successful allograft placement. In these instances, the transplantation procedure is often more difficult, and lengthy, than in a 'typical' case. All these factors are reflected in the higher mortality and lower organ survival in paediatric transplantation as compared with adults,[1,2] and provide paediatric anaesthetists with some of their greatest intraoperative challenges.

Preanaesthetic considerations

The physiological consequences of end-stage organ disease are the same in children and in adults, but in children they are often a result of congenital anomalies rather than acquired diseases. Furthermore, many congenital disorders that result in end-stage organ disease (ESOD) are part of a constellation of associated anomalies, involving several organ systems. Recognition of other manifestations involved in these 'syndromes' is essential to develop an appropriate anaesthetic management plan for these patients. For example, children with Alagilles syndrome often develop cirrhosis early in childhood[25] but also have a high incidence of congenital heart disease (ventricular septal defects, pulmonary stenosis) as well as severe hyperlipidaemia (predisposing them to early coronary artery disease). Clinically important ramifications include the real possibility of perioperative myocardial ischaemia (usually not a consideration in children), and paradoxical arterial air emboli resulting from entrained air (common in liver transplantation) gaining access to the systemic circulation through an intracardiac shunt. Further discussion of common syndromes causing ESOD in childhood appears in the specific organ transplant sections.

Anaesthetic management

In addition to the monitoring equipment typical for any major surgical procedure, there are a number of special needs for paediatric transplantation. In particular, adaptations in standard equipment must be made for children under 10 kg in body weight. For example, to minimize the amount of heparin administered, arterial and central venous catheters should be connected to pressure transducers specifically designed to infuse low volumes of 'flush' solution. If such systems are not available, standard infusion pumps set at very low volumes (less than 5 ml/hour) can be substituted. Children with end-stage heart, liver or renal disease often have compromised lung function; therefore a ventilator appropriate for the accurate delivery of small volumes at rapid respiratory rates and potentially high inflation pressures should be available. The ventilators integral to most anaesthetic machines seldom possess these capabilities; a free-standing paediatric ventilator may therefore be necessary, although this may preclude the use of volatile anaesthetics.

The other category of special equipment is related to heat conservation, and includes warming blankets, high efficiency fluid warmers for *all* intravenous lines, inspired gas heaters and humidifiers, and wraps for the extremities. In cases where the need to transfuse several blood volumes can be anticipated, a rapid infusion device with integral heat exchanger should also be available (see below).

With the exception of living-related procedures, most transplant operations are scheduled as emergencies, so gastric aspiration precautions are usually indicated. Some form of 'rapid-sequence' induction appropriate to the patient's medical condition should be chosen. Although most children will tolerate sodium thiopentone as an induction agent, those with questionable plasma volume status or myocardial compromise may require agents that preserve haemodynamic stability, such as etomidate or ketamine. After induction, the primary issues are haemodynamic, acid–base and fluid management, so the choice of drugs for maintenance of anaesthesia must be selected accordingly. The most common technique combines opioids and long-acting muscle relaxants supplemented with a potent inhalation agent, as tolerated. Nitrous oxide is contraindicated in liver and kidney transplantation, because it can produce abdominal distension with deleterious haemodynamic and respiratory effects. Air embolism is a real threat during all major organ transplantation; therefore it is safer to avoid nitrous oxide altogether.[26] Intraoperative use of regional anaesthesia is relatively contraindicated in these cases, because the development of a significant coagulopathy is common during liver, heart and lung transplant surgery. This does not preclude later placement of epidural or caudal catheters for pain management in the postoperative period, after coagulation is normalized.

Invasive monitoring is as fundamental to paediatric transplantation as it is in adults, but size limitations affect the practical choices. Reliable arterial pressure monitoring requires the use of at least a 22 gauge catheter, usually placed in the radial artery either percutaneously or by cutdown. An upper limb site is preferable, as it is accessible throughout the procedure, but if this is not possible, placement of a femoral arterial line (3–4 Fr) is indicated. Likewise, the ideal is to obtain central venous access via the internal jugular vein with a large-bore, multi-port catheter, to allow for both monitoring and infusion administration. However, in small children, usually only one of these goals can be accomplished, and the availability of multiple ports should take precedence. In the majority of paediatric patients, central venous pressure and urinary output are usually sufficient to assess volume status. Pulmonary artery (PA) catheters are reserved for children with seriously compromised myocardial function or pulmonary hypertension, because introducers large enough to accommodate PA catheters are often difficult to insert in small children. Transoesophageal echocardiography has also proved valuable in distinguishing between left ventricular dysfunction and hypovolaemia; however, insertion of the probe past oesophageal varices carries considerable risk.

Large calibre intravenous (IV) catheters may be difficult to place in small children, but most transplant cases require additional vascular access. Although renal transplantation can proceed safely with two peripheral IVs, liver or heart transplants require a minimum of three moderate-sized IVs. To fulfil these requirements, it may be necesssary to insert appropriate-sized catheters in non-traditional sites; for example, 8 Fr infusion ports or 14 G catheters can be inserted in the axillary or internal jugular veins of small infants after exposure by cutdown. Ideally, vascular sites in an upper extremity are preferred, particularly for liver transplantation, because lower extremity venous continuity is interrupted during the procedure by clamping of the inferior vena cava. But this may be unrealistic for small children, so femoral venous access is an acceptable alternative.

During rapid transfusion, conventional fluid-warming devices can cause complications in small infants by delivering partially warmed, acidaemic and hyperkalaemic blood, or citrate-rich platelets or plasma, directly into the central circulation. This can result in significant hypothermia, as well as in hyperkalaemic, hypocalcaemic cardiac arrest. In cases that require massive transfusion, the use of a 'rapid infusion' system (RIS) can avoid these problems. This device is designed to deliver warmed mixtures of fluid and blood products at rates of up to 1500 ml per minute. Although two 8 Fr infusion ports are necessary to realize the full potential of the RIS,[27] clinically valuable infusion rates can still be achieved through two standard 18 G IV catheters. The advantages of an RIS include rigorous control of the transfusate composition, particularly with respect to temperature and electrolyte levels. Use of the RIS allows for dilution of potentially lethal potassium concentrations in older, stored red blood cells and citrate in platelets and fresh-frozen plasma. When a rapid infusion system is not available, it is best to request for transplantation procedures in small children either fresh whole blood or packed red blood cells stored for less than a week. Platelets and fresh-frozen plasma should be administered via peripheral sites at moderate rates of infusion to prevent hypocalcaemia.[28] Although most children require administration of at least some blood products, major blood losses (> one blood volume) during transplantation are not universal. Patients in whom significant transfusion can be anticipated include those who have had previous surgery near to the site of transplantation; liver recipients with severe portal hypertension or coagulopathy;[29] cardiac recipients with cyanotic congenital heart disease; and lung recipients with cystic fibrosis or other reasons for pleural scarring.

Finally, because most transplantation procedures are lengthy, exceptional precautions must be taken in positioning. Chronically ill young children have thin skin and minimal subcutaneous tissue, making them particularly vulnerable to pressure-induced injuries. Areas such as the scalp, sacrum, elbows and heels must be provided with extra protection, such as silicone-filled head rings and cushions, or foam-rubber padding. Wires, cables and intravenous tubing must not be allowed to remain under the weight of the trunk or limbs, nor to come in contact with bare skin. The scalp is at particular risk because the head is relatively heavy. Frequent position changes during the procedure will minimize the chance of ischaemia-induced ulceration or alopecia. Even the tracheal tube can cause damage by pressing against the corner of the mouth; the same is true for the nasogastric tube pulling on the ala nasi. Tissue damage from ischaemia in these areas may be very disfiguring, often to the point of requiring reconstructive plastic surgery. With some forethought and care, such complications are completely preventable.

Renal transplantation

Since the late 1960s, when kidney transplantation became a realistic form of therapy for end-stage renal disease (ESRD), there have been more than 300 000 cases performed world wide.[30] In the USA, each year there are nearly 10 000 renal transplants, of which some 200–300 are in children.[30] Although medical management of ESRD is life-sustaining, the negative impact of uraemia on growth and cognitive development,[6,7,31–33] as well as the progressive nature of the cardiovascular complications (hypertension, congestive heart failure), makes transplantation a viable alternative. The relatively high mortality and lower quality of life associated with chronic dialysis in young children[34–36] is felt to justify the risks of surgery and long-term immunosuppression.[6,37]

The overall incidence of ESRD in the USA is currently 5 cases per million children,[6] although throughout childhood the incidence increases with age.[33] Congenital disorders, including developmental anomalies (urinary tract malformations, renal dysplasias) and hereditary nephropathies (oxalosis, cystinosis) are the cause in as many as 50 per cent of the cases.[38] Some of the more common syndromes that affect other organ systems, as well as causing ESRD, are listed in Table 59.1. This contrasts with adults for whom acquired diseases such as diabetes-associated glomerulosclerosis or chronic glomerulonephritis are more important aetiological factors.[39] The most common indications for renal transplantation in children are listed in Table 59.2.

The physiological consequences of ESRD are fundamentally the same in children and adults, but abnormalities in calcium and phosphate metabolism obviously have more serious consequences in growing children (e.g. renal osteodystrophy).[40] Hypertension, secondary to high levels

Table 59.1 Common multi-system congenital disorders causing renal failure

Disorder	Manifestations in other organs
Alport's syndrome	Diabetes, thrombocytopenia, neurosensory hearing loss
Fabry disease	Cardiac and cerebral vascular disease, cardiac conduction abnormalities
Henoch–Schoenlein purpura	Coagulopathy, CNS and intestinal bleeding
Polycystic disease ('infantile')	Hepatic cysts, portal hypertension, systemic hypertension

Data from reference 85.

Table 59.2 Indications for renal transplantation in children

Indication	%
Obstructive uropathy	17
Renal hypoplasia	15
Glomerulonephritis	15
Nephrotic syndrome	16
Medullary cystic disease	5
Pyelonephritis	4
Haemolytic uraemic syndrome	4
Alport's syndrome	3
Oxalosis	2
Miscellaneous, including unknown	19

Reproduced, with permission, from Chavers B, Matas AJ, Nevins TE, et al. Results of pediatric kidney transplantation at the University of Minnesota. In: Terasaki P, ed. Clinical transplants. Los Angeles: UCLA Tissue Typing Laboratory, 1989: 253–66.

of renin and angiotensin, is also quite difficult to manage in paediatric patients, occasionally necessitating surgical removal of the affected kidneys. Pharmacological treatment usually requires high doses and combinations of angiotensin-converting enzyme inhibitors and calcium channel and β-blockers. It is important for the anaesthetist to recognize that such combinations can lead to serious intraoperative haemodynamic instability, by blunting the compensatory responses to hypovolaemia, and anaesthetic-induced myocardial depression and vasodilatation.

As with all other major organs, the number of available donor kidneys does not nearly meet the demand. It was to address this problem that LR kidney donation was first devised. In the case of renal transplantation, the benefits of LR donation for the recipients and their families are felt to outweigh the potential problems. For example, timing of transplantation can be controlled, allowing the child's medical condition to be optimized. Restoration of near-normal renal function is rapid because LR donor kidneys suffer minimal ischaemia, thus simplifying the postoperative course. In addition, the donor with the closest HLA match can be chosen. As a result of these factors, overall patient and graft survival for a LR organ is superior to the results achieved by cadaveric transplantation. At present, the 1- and 5-year LR graft rates are 93 per cent and 65 per cent respectively, whilst those for cadaveric organs are 85 and 50 per cent.[38,39] However, with recent changes brought about by combination immunosuppression regimens, these disparities in survival are beginning to narrow.[38] The advantage of LR organ donation could be eliminated by even more specific immunotherapy, but the overall shortage of kidneys will in all likelihood guarantee the persistence of LR transplantation for the foreseeable future.

Whether the kidney is from a relative or cadaverically derived, the vast majority of transplanted kidneys are from adult donors. The conventional adult surgical procedure involves extraperitoneal pelvic placement of the kidney, with vascular reanastomosis to the iliac vessels. A technique involving intra-abdominal implantation (Fig. 59.1) was devised to allow adult-sized kidneys to be used in children.[19] The abdomen has sufficient room and allows access to appropriately sized blood vessels (usually the aorta and inferior vena cava). However, transplanting adult kidneys into small children creates several other important problems. Placing an organ with a vascular bed volume of 150–200 ml into a child (whose entire blood volume may be only 800 ml) results in an acute volume shortage during reperfusion. This relative hypovolaemia is exacerbated by a chronic volume contraction following diuretic therapy and dialysis. Also, the newly transplanted kidney requires adult levels of perfusion pressures to function properly, and produces adult quantities of urine for several days after implantation. The clinical implications of these physiological perturbations are discussed below.

Renal transplantation is no longer an emergency procedure, since cadaveric organs can be preserved for more than 36 hours[41] and LR organ donation can be planned electively. Although some children receive a transplant before their kidneys fail completely,[42] the majority of recipients are on chronic dialysis. These recipients should be dialysed in anticipation of surgery, and full serum biochemical analysis repeated preoperatively. All antihypertensive medications should be continued up to the time of surgery to prevent perioperative hypertension. Although these patients can be kept 'nil by mouth',

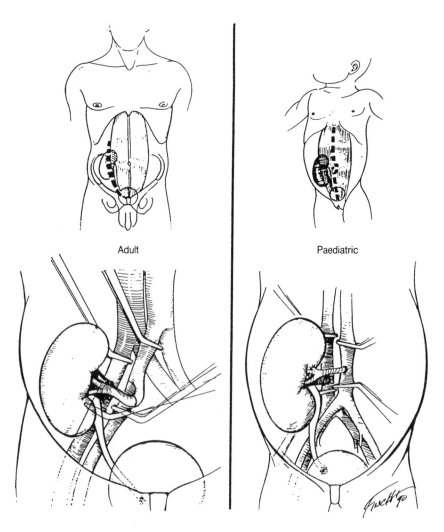

Adult   Paediatric

Fig. 59.1 Surgical procedure for renal transplantation in the adult (left) and child (right). (Reproduced, with permission, from Belani KG, Palahniuk RK. Kidney transplantation. *International Anesthesiology Clinics* 1991; **29**: 17–39.)

precautions against aspiration should be taken with all recipients, because uraemia can slow gastric emptying.[43]

Most of these procedures are from 3 to 4 hours in length, and rarely result in major blood loss or the development of coagulopathies. The use of regional anaesthesia (via epidural or caudal catheter) for both intra- and post-operative pain management, in combination with general tracheal anaesthesia, provides excellent operative conditions. Vecuronium and atracurium are the muscle relaxants of choice, because they do not depend primarily on the kidney for elimination.

In small children, the intra-abdominal procedure introduces additional risks of increased third-space fluid losses and a greater potential for blood loss. This justifies the routine use of invasive monitoring, including arterial and central venous pressure measurements, particularly when placing an adult kidney in a small child. Aggressive plasma volume expansion is necessary to prevent significant hypotension after unclamping the vascular supply to the new kidney. Colloid and crystalloid may be used for this purpose, but in small children, red blood cells may be needed to maintain an acceptable haematocrit (packed cell volume). The goal is to raise the central venous pressure by some 25 per cent prior to allograft reperfusion. In some cases, to provide appropriate perfusion pressure for the new kidney, it may be necessary to support the blood pressure by intentional hypervolaemia, or low-dose inotropes, such as dopamine. Transplantation of a LR kidney complicates fluid management, as significant volume losses from the immediate urinary output must be taken into account. So, in addition to an initial volume expansion, continuous fluid and electrolyte replacement may become necessary.

Extubation at the end of the renal transplant procedure is the rule, with the possible exception of procedures in very young infants. Patients are typically admitted to an intensive or high dependency unit for a brief period of time. Although patients who receive a cadaveric transplant may require dialysis for 7–10 days, recovery and discharge for LR organ recipients is frequently rapid.

Liver transplantation

After the introduction of cyclosporin to clinical practice in the mid-1980s, liver transplantation quickly made the transition from an experimental modality to standard therapy for end-stage liver disease (ESLD).[44] It is now the second most common viscus transplanted in the USA; 19 per cent of the nearly 3000 cases performed in 1991 were in paediatric recipients.[45] A further advance in hepatic transplantation was the development of superior organ

Table 59.3 Indications for liver transplantation in children

Indication	%
Biliary atresia	56
Inborn errors of metabolism	14
Fulminant hepatic failure	10
Cirrhosis (all causes)	10
Neonatal hepatitis	3
Malignant neoplasms	2
Cystic fibrosis	1.5
Miscellaneous	3.5

Reproduced, with permission, from UNOS Liver Transplant Registry. In: Cecka JM, Teraskai P, eds. *Clinical transplants*. Los Angeles: UCLA Tissue Typing Laboratory, 1992: 17–31.

Table 59.4 Common multi-system congenital disorders causing liver failure

Disorder	Manifestations in other organs
α_1-Antitrypsin deficiency	Pulmonary insufficiency
Wilson's disease	Fanconi's syndrome, aminoaciduria, cataracts, cardiac involvement
Tyrosinaemia	Renal tubular defects, hypoglycaemia, rickets, hypertension
Alagilles syndrome	CHD, hyperlipidaemia, coronary artery disease
Glycogen storage diseases	Cardiac, CNS, upper airway involvement

Data from reference 85.

preservation techniques (University of Wisconsin solution), which now safely permits at least 18 hours of ischaemic time.[46] This has transformed the emergent nature of hepatic transplantation while minimizing organ wastage.

There are numerous causes of ESLD in childhood (Table 59.3) leading to hepatic transplantation.[2,45] As is the case for renal disease, most causes of ESLD are congenital, not acquired disorders. The preponderance of these patients (56 per cent) have biliary atresia, a developmental abnormality causing progressive hepatic failure at a very young age. The category of inborn errors of metabolism, which includes α_1-antitrypsin deficiency, Wilson's disease and tyrosinaemia, accounts for a further 14 per cent. In contrast to adults, in whom hepatitis and alcoholic liver disease are common aetiologies,[45] infectious and toxic disorders are uncommon causes of liver failure in the paediatric population. Table 59.4 lists the more common congenital diseases that affect other organ systems as well as causing ESLD.

In children, the occurrence of ESLD is heavily weighted towards the younger age group, reflecting that manifestations of congenital disorders usually begin early in life. In the USA, more than 70 per cent of the paediatric liver transplant recipients are less than 5 years old.[2] The preponderance of younger, smaller recipients creates serious shortages of appropriately sized organs, because most paediatric donors are older children or adolescents.[47] Innovative surgical techniques devised to address these shortages include dividing an adult liver for use in two recipients[48,49] or simply reducing the graft size by performing a left hepatic lobectomy.[21,50] This leaves a right hepatic lobe allograft (Fig. 59.2) small enough to be transplanted into an infant. Another consequence of limited donor availability is the advent of living-related hepatic lobe donation.[17] However, because the potential for donor morbidity and mortality is significant, the number of institutions that offer this option is small.[16] Furthermore, long-term patient and graft survival with partial hepatic transplantation techniques have not been fully evaluated, although short-term graft survival rates with reduced-size grafts are comparable with those achieved with complete allografts.[47,51]

In general, outcome statistics for the smaller (< 10 kg) and younger (< 2 years) hepatic recipients are poorer than those for older children and adults. Current data document a 4-year survival of 67 per cent in the youngest age category, whilst that for older children approaches 85 per cent.[45] Reasons given for this increased mortality include a high incidence of hepatic artery thrombosis,[52] increased perioperative blood loss and reduced viability of 'split'

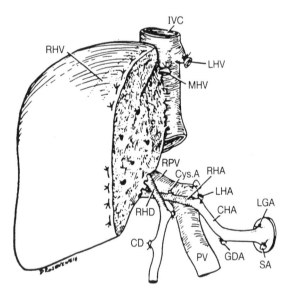

Fig. 59.2 Reduced-size hepatic graft (right lobe graft). CD, common duct; CHA, common hepatic artery; Cys.A, cystic artery; GDA, gastroduodenal artery; IVC, inferior vena cava; LGA, left gastric artery; LHA, left hepatic artery; LHV, left hepatic vein; MHV, middle hepatic vein; PV, portal vein; RHA, right hepatic artery; RHD, right hepatic duct; RHV, right hepatic vein; RPV, right portal vein. (Reproduced, with permission, from Broelsch CE, Emond JC, Thistlethwaite JR, *et al*. Liver transplantation, including the concept of reduced-sized liver transplants in children. *Annals of Surgery* 1988; **208**: 410–20.)

livers. Other technical problems related to transplantation in small children include the inability to use venovenous bypass to maintain venous return during the anhepatic phase, and a high rate of biliary tract complications.

A unique facet of transplantation for inborn errors of metabolism is that, once the allograft is functioning, the primary metabolic disorder is 'cured'. The classic example is Wilson's disease, which results from the absence of the copper-carrying protein, ceruloplasmin, normally synthesized in the liver.[53] Once the transplanted liver begins to function, the allograft hepatocytes produce ceruloplasmin, and the deposition of copper characteristic of the disease ceases.

In general, most anaesthetic considerations for liver transplantation in adults are equally applicable to children, including full invasive monitoring, routine use of precautions during induction to prevent aspiration and avoidance of nitrous oxide. An anaesthetic machine equipped with a versatile paediatric ventilator is necessary, because the surgical procedure severely compromises ventilation. One of the major challenges in paediatric liver transplantation is the maintenance of normal body temperature, so all available methods of heat conservation must be employed, including warming of intravenous fluid to 37–38°C. Intravenous access is usually easier in children with chronic liver disease and portal hypertension, because peripheral veins are also enlarged. Upper body sites for all IV catheters (minimum of four) are preferable, as venous return from the lower body is usually compromised during the anhepatic phase (see below).

The surgical procedure for liver transplantation has been divided into three sections: the pre-anhepatic (surgical dissection of the native liver), anhepatic (native liver removed, inferior vena cava and portal vein cross-clamping), and reperfusion (post-reperfusion) phases, each with particular clinical implications. During the pre-anhepatic phase there is significant potential for blood and fluid losses (ascites), so the primary tasks are to maintain intravascular volume and body temperature. This is often difficult because of the large surface area of an open peritoneal cavity and exteriorized intestines. Retraction on the diaphragm at this juncture can severely reduce lung volumes and compliance, thus compromising effective ventilation. Positive end-expiratory pressure can compensate to some extent. Haemodynamic changes during the anhepatic phase usually respond to volume loading. Progressive acid–base disturbances and oliguria are common, resulting from interruption of systemic and portal venous return, concomitant increases in splanchnic venous pressure and decreases in cardiac output. Most children with chronic liver disease have sufficient venous collaterals to withstand caval cross-clamping, although in children above 15 kg in weight, femoral to axillary venous bypass can ameliorate these problems.[54]

During the acute reperfusion period, it is usual to observe transient hypotension from volume redistribution and blood loss, as well as accelerated temperature losses of 1–2°C. Stagnant blood from the lower body, containing the metabolic products of ischaemia, can now return to the central circulation, so acidaemia, hyperkalaemia and hypocalcaemia may ensue. Hypothermia may result from reperfusion of the cold allograft. All of these factors in consort can result in significant myocardial depression and serious ventricular arrhythmias. Measurement of arterial blood gases and electrolytes every 5–10 minutes allows the detection and aggressive correction of serious abnormalities. Most liver transplant recipients have pre-existing coagulation disorders due to reduced levels of clotting factors (decreased hepatic synthesis) and thrombocytopenia (hypersplenism secondary to portal hypertension). Beyond these, additional obstacles to normal clotting during reperfusion include dilution and consumption of clotting factors and platelets, hypothermia and primary fibrinolysis.[55,56] Rapidly available serial clotting profiles (prothrombin time, partial prothrombin time, platelet count, fibrin split products) are essential for optimal management during this phase. In addition, many centres have found thromboelastography to be helpful in distinguishing between platelet or clotting factor deficiencies and fibrinolysis,[57] thus permitting specific, decisive therapeutic intervention.

The final phase of the procedure is characterized by the completion of the hepatic artery and biliary anastomoses, haemostasis, and then abdominal closure. As the newly reperfused graft reaches normal temperatures and begins to replenish hepatocellular energy stores, it will usually resume metabolic functions. Evidence for this includes the production of bile, and resolution of systemic metabolic acidosis with progressive decreases in the serum lactate concentration. Frequently, a demonstrably shorter dosing interval for muscle relaxants and other drugs is observed. Also during this phase, the various energy-requiring ion-pumping mechanisms that actively maintain cellular electrolyte balance will begin to restore normal extracellular fluid composition. In particular, after an acute rise during reperfusion, the serum potassium concentration can rapidly decline as hepatocytes recover their ability to transport potassium into cells. In many cases, replacement therapy is needed if hypokalaemia is to be prevented. Hypernatraemia is another common problem, which can result from infusing large quantities of sodium bicarbonate and 25 per cent albumin to correct acidaemia and hypovolaemia. Early substitution of THAM (trometamol, tromethamine) for sodium bicarbonate and allowing for spontaneous correction of moderate acid–base imbalances are effective strategies to minimize postoperative hypernatraemia. Overly rapid correction of hypernatraemia must be avoided, as it has been implicated as the causative factor in central pontine myelinolysis syndrome.[58] This syndrome is characterized by encephalopathy, seizures, permanent central neuronal demyelinization and rapid death from cerebral oedema.

Intestinal transplantation

Intestinal transplantation is still considered an experimental procedure in children, in whom it is most commonly employed for the treatment of 'short-gut' syndrome.[59,60] Most of these cases are caused by an intra-abdominal catastrophe in infancy, which results in ischaemic necrosis of the small intestine. Advanced widespread necrotizing enterocolitis, midgut volvulus or giant omphaloceles are among the antecedent surgical emergencies. In the past, these children usually succumbed shortly after the initial event; however, with the advent of total parenteral nutrition, the outlook has changed. Not only is it possible to sustain these children through the acute episode, but it has also become feasible to supply all nutritional requirements needed for growth and development for several years. Liver damage is associated with long-term hyperalimentation, so a high percentage of paediatric intestinal transplant candidates also require liver transplantation.

Most of the considerations for liver transplantation discussed above apply as well to combined small bowel and liver transplantation. However, because of the bowel's large surface area, fluid and temperature losses are usually much greater. The nearly universal history of previous intra-abdominal surgery and/or peritonitis in this group creates the potential for massive blood loss, and accounts for the prolonged duration (a minimum of 10 hours) of surgery.

Given such factors, the potential for complications relating to positioning and compromised pulmonary function is high. In addition, because the allowable ischaemic time for intestines is ideally under 8 hours, these procedures are scheduled as emergency cases. The long dissection time needed for preparation makes it imperative that surgery commence considerably before the donated organs arrive. This introduces the unavoidable possibility of anaesthetizing and operating on a patient, only to ultimately abort the procedure, if there is a problem with the donor organs.

Relatively low patient and graft survival[61] and chronic functional problems with the transplanted bowel have dampened enthusiasm for these procedures. Until solutions are found, intestinal transplantation will probably remain an experimental procedure.

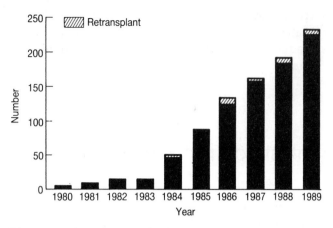

Fig. 59.3 Growth in paediatric heart transplantation, 1980–1989. (Reproduced, with permission, from Kriett JM, Kaye MP. Registry of the ISHLT: seventh official report, 1990. *Journal of Heart Transplantation* 1990; **9**: 323–30.)

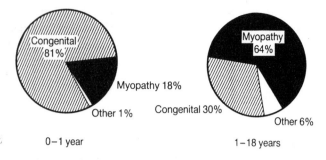

Fig. 59.4 Indications for heart transplantation in children (Reproduced, with permission, from Kaye MP. The registry of the ISHLT: tenth official report, 1993. *Journal of Heart and Lung Transplantation* 1993; **12**: 541–8.)

Heart transplantation

Paediatric recipients now comprise nearly 10 per cent of the 2700 heart transplants performed world wide annually.[1] Both the absolute number and the proportion of paediatric recipients steadily increased (Fig. 59.3), until 1991–92 when a plateau of 260–280 cases per year was reached.[1] The first children who received heart transplants were most often suffering from a viral or idiopathic cardiomyopathy, even though there was a far larger number of potential recipients with end-stage congenital heart disease (CHD). For many children with CHD, although all other surgical options had been exhausted, they were still not deemed good candidates for transplantation. The justification for this view was based on their histories of previous intrathoracic surgery, presence of major vascular anomalies and increased pulmonary vascular resistance. However, since 1989 this attitude against heart transplantation in children with end-stage CHD has changed significantly.[62] Technical modifications and further experience in intra- and postoperative care have resulted in marked reductions in morbidity and mortality. The most recent data available document an overall mortality rate of 16.6 per cent in such children,[1] as compared to 20–25 per cent in earlier publications.[63] Even better results have been reported from centres that specialize in paediatrics[3,64] Although not as good as those achieved in adults (8.5 per cent),[1] such statistics represent major progress in this inherently high-risk group of recipients.

At the present time, CHD has become the most common indication for heart transplantation in children under 1 year of age (Fig. 59.4), and this youngest age group comprises approximately 40 per cent of the total paediatric heart transplants now performed.[1] In several centres, transplantation has become the treatment of choice for neonates with selected congenital anomalies (e.g. hypoplastic left heart syndrome (HLHS), complex 'single ventricles').[62] As the management of patients with end-stage ischaemic or primary cardiomyopathy has already been discussed in relation to adults, this section focuses on aspects of heart transplantation unique to CHD.

Neonates with CHD are certainly the most haemodynamically challenging patients to support while awaiting transplantation. The vast majority of them have HLHS, which is uniformly lethal unless there is surgical intervention. There was no satisfactory palliative procedure to offer these children in the past, so they succumbed in the first month of life from low systemic cardiac output, when their ductus arteriosus closed.[64] In the past decade, with the advent of the Norwood procedure[65] and neonatal heart transplantation,[66] medical management strategies to support these children have been developed. The cornerstone is the infusion of prostaglandin E_1 (PGE_1) to maintain the patency of the ductus arteriosus, providing a route for systemic cardiac output. Some children also require enlargement of the foramen ovale by balloon septostomy (Rashkind procedure), to provide adequate mixing of the systemic and pulmonary venous return, at the atrial level.[67] Other techniques of interventional cardiac catheterization, such as placement of ductus arteriosus stents in children who are inadequately responsive to PGE_1, have increased the percentage of affected infants who can be resuscitated successfully and supported until surgery.[62,68]

These resuscitation techniques, used in combination with measures to maintain relatively high pulmonary vascular resistance – such as induced hypercapnia (hypoventilation, augmentation of $FICO_2$) and relative hypoxaemia (room air or lower FIO_2) – now allow the long-term maintenance of favourable systemic-to-pulmonary blood flow ratios. By aiming for systemic arterial oxygen saturations of 75–80 per cent and carbon dioxide concentrations of 6–8 kPa (45–60 mmHg), a satisfactory systemic cardiac output can be maintained without inotropic agents. These children can

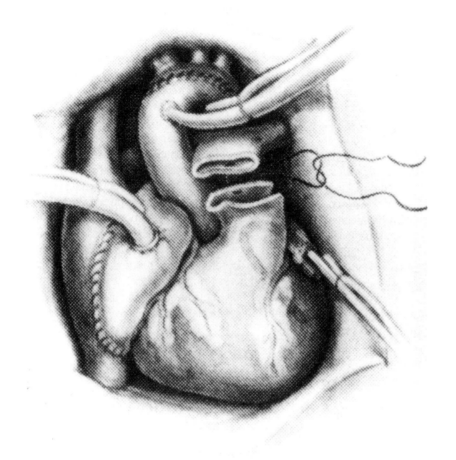

Fig. 59.5 Heart transplant procedure for hypoplastic left heart syndrome (note aortic arch reconstruction). (Reproduced, with permission, from Bailey L, Concepcion W, Shattuck H, *et al.* Method of heart transplantation for treatment of HLHS. *Journal of Thoracic and Cardiovascular Surgery* 1985; **92**; 1–5.)

now be supported for the prolonged intervals usually needed to procure an appropriately sized heart.[3]

Another unique aspect of neonatal transplantation relates to the procurement procedure for hearts destined for recipients with congenital heart disease.[62,69] Segments of donor aorta and great veins substantially longer than usual are often required to provide sufficient lengths for the necessary vascular reconstructions. For example, in the HLHS, the aorta is usually hypoplastic from the ascending portion to the insertion of the ductus arteriosus. During cardiac transplantation for this syndrome, the donor's aorta is used for enlargement (Fig. 59.5). In other lesions, it may be necessary to remodel the major venous connections or else perform intracardiac modifications, such as an interatrial baffle to redirect anomalous pulmonary venous return.[62,69,70]

The primary goal of anaesthetic management for neonatal transplantation is to maintain systemic cardiac output until cardiopulmonary bypass is initiated. Monitoring for these cases is the same as for other paediatric open heart surgery, with the addition of PA catheters, which must be placed after cardiac implantation. Transoesophageal echocardiography can provide continuous assessment of cardiac function, if necessary, prior to cardiopulmonary bypass. In order to achieve the haemodynamic goals, provision must be made to allow the safe delivery of gas mixtures that are mildly hypoxic (FIO_2 17–20 per cent) and/or hypercapnic ($FICO_2$ 3–5 per cent). Because modern anaesthetic machines are designed to prevent such mixture (via oxygen fail-safe devices, etc.), potentially dangerous modifications in the gas inlets are necessary, and reliable analysis of inspired gas composition (O_2, N_2, CO_2, etc.) at the tracheal tube becomes *absolutely essential*. In addition, avoidance of other measures that can lower pulmonary vascular resistance and exacerbate right-to-left shunting (drugs, hyperventilation, etc.) is also important. Paradoxically, an indication of impending decompensation in these patients is the presence of peripheral oxygen saturation of 100 per cent. If respiratory manoeuvres are unsuccessful in maintaining relatively high pulmonary vascular resistance (PVR), temporary surgical constriction of the branch pulmonary arteries is helpful. It takes only a very short period of increased pulmonary blood flow at the expense of decreased systemic cardiac output to induce pulmonary oedema and the metabolic effects of low systemic perfusion.

As a consequence of the additional reconstruction

needed in these infants, the length of cardiopulmonary bypass is usually prolonged, and in most instances necessitates the use of deep hypothermic circulatory arrest.[62] This introduces all the added risks of coagulopathy from platelet dysfunction, clotting factor consumption and, in some cases, primary fibrinolysis. Therefore, anaesthetists must be prepared for aggressive use of platelet and cryoprecipitate transfusion, and perhaps antifibrinolytics such as aminocaproic acid.

Provided that the allograft has been preserved properly, weaning from cardiopulmonary bypass can be accomplished without inotropic support, although low-dose dobutamine (5 µg/kg per min) is usually administered empirically. Occasionally, a chronotropic infusion (isoprenaline, isoproterenol) or a pacemaker will be needed transiently, to support heart rate to age-appropriate levels (e.g. 120–140 beats/min in infants). Strategies to control pulmonary vasospasm may also be needed, as most of these children have some degree of pre-existing pulmonary vascular disease or, in infants, there is always the potential for paroxysmal pulmonary hypertension. Newly transplanted hearts do not readily adapt to acute rises in PVR, and right heart failure can rapidly ensue.[69] The depth of anaesthesia can be an important factor in minimizing the episodes of pulmonary hypertension, so adequate doses of opioids should be given immediately prior to noxious stimuli (e.g. suctioning, skin closure). The traditional therapies for reducing PVR (administration of 100 per cent O_2; hyperventilation to induce alkalosis) are also effective. Nevertheless, if the pulmonary artery pressure remains high, vasodilator drug therapy may become necessary, which until recently consisted of either PGE_1 or prostacyclin infusion directly via the PA catheter. However, both of these drugs cause systemic as well as pulmonary vasodilation, thus limiting their usefulness. The introduction of inhaled nitric oxide (NO) as a 'selective' pulmonary vasodilator has provided the first rapidly acting, powerful tool for the treatment of pulmonary hypertension.[71,72] In concentrations as low as 5–10 parts per million, inhaled NO can effectively lower PVR, without concomitant decreases in arterial blood pressure or increases in pulmonary shunt fraction. The major drawback to NO use is the need for specialized equipment to deliver and, more importantly, monitor (via chemiluminescence) the concentration of NO, and the potentially toxic higher oxides of nitrogen, being delivered to the patient.

Size mismatch becomes most apparent after the heart is filled, and may presage serious haemodynamic compromise after sternal closure. Even when size matching is adequate, accumulation of parenchymal oedema fluid from preservation injury may result in cardiac swelling. As this effect reaches a maximum 12–24 hours after reperfusion, the sternum should not be approximated when it is apparent that there is insufficient room for expansion in the mediastinum. If the chest is closed in these cases, the resulting compromise to myocardial perfusion and performance that occurs may reach a critical point, causing cardiac arrest. However, to reduce the infection risk, the skin can be closed. It is usually possible to complete the sternal closure safely on the second or third post-transplant day.

One of the initial premises used to support transplantation in the neonatal period was that early transplantation confers an immunological advantage. This was based on evidence suggesting that neonates have blunted responses to foreign antigens, including those on the surface of transplanted organs. Recent data that some neonates can be maintained on lower doses of immunosuppressant agents than older children or adults[3] support this theory. Whether the reduced incidence of rejection will translate into longer organ survival or a lower incidence of accelerated coronary artery disease[73,74] or longer organ survival remains to be seen. In fact, although many of the technical problems of neonatal heart transplantation have been solved, the long-term outcome remains a question, because the first patients to undergo transplants as neonates are now only 7–8 years old.[68]

Heart–lung and lung transplantation

Although significant numbers of heart–lung transplantation procedures were performed in children throughout the 1980s, poor early results (< 50 per cent 1-year survival) discouraged widespread application of any form of lung transplantation.[75] A few centres persisted, and, with experience gained in donor selection, organ preservation, surgical techniques and post-transplant care, recent outcome statistics have improved substantially.[76] In one large series of lung transplantation (245 cases) reported from the University of Pittsburgh, overall survival increased from 53 per cent (pre-1991) to more than 70 per cent (1991–1993).[77]

The surgical procedures for pulmonary transplantation have undergone considerable evolution during recent years, leading directly to improved survival.[1] For example, *en bloc* double lung transplantation has virtually been supplanted by the bilateral sequential lung transplant (BSLT) procedure.[76,77] The former operation required cardiopulmonary bypass to accomplish tracheal, main PA and left atrial (LA) anastomoses, whilst the latter can avoid the need for cardiopulmonary bypass by employing bibronchial, branch PA and two LA cuff anastomoses. The advantages of the BSLT procedure have been confirmed by substantial survival benefits (85 per cent vs 59 per cent 1-year survival)[1] over those seen previously (Fig. 59.6). Likewise,

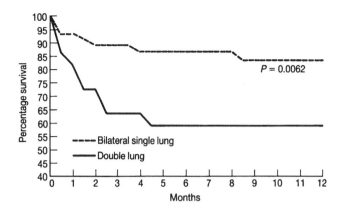

Fig. 59.6 One-year survival after paediatric lung transplantation; double lung vs bilateral sequential lung transplantation. (Reproduced, with permission, from Kaye MP. The registry of the ISHLT: tenth official report. 1993. *Journal of Heart and Lung Transplantation* 1993; **12**: 541–8.)

Table 59.5 Indications for paediatric lung transplantation

Indication	%
Cystic fibrosis	61
Primary pulmonary hypertension	12
Congenital heart disease	6
Other	12
Retransplantation	8

Reproduced, with permission, from Kaye MP. The registry of the ISHLT: tenth official report – 1993. *Journal of Heart and Lung Transplantation* 1993; **12**: 541.

when only single lung transplantation (SLT) is indicated (e.g. with chronic obstructive pulmonary disease), recent excellent outcome statistics are available to justify the use of this operation.[1,77]

In children, the most common cause of end-stage lung disease is cystic fibrosis.[78] Other inherited diseases, as well as acquired diseases that are indications for lung transplantation in children, are listed in Table 59.5. As in adults, the underlying aetiology of pulmonary disease determines which transplant operation is chosen. For example, one of the forms of double lung transplantation must be used for infectious or pulmonary vascular diseases, and because the majority of paediatric recipients carry diagnoses within one of these two categories, BSLT is the most common operation in children. In contrast to adults, SLT is rarely indicated in children.

Although the number of heart–lung transplants performed world wide has been decreasing since 1990, of the 49 cases performed in the USA in 1992, more than half were in children (Table 59.6).[1] The overall decline has been attributed to earlier referral for isolated lung transplantation, before cor pulmonale becomes irreversible. However, in children, there are still a number of uncorrectable types of CHD, which inevitably lead to Eisenmenger's syndrome,

Table 59.6 Indications for paediatric heart–lung transplantation

Indication	%
Congenital heart disease	46
Primary pulmonary hypertension	19
Cystic fibrosis	18
Miscellaneous	7
Retransplantation	9

Reproduced, with permission, from Kaye MP. The registry of the ISHLT: tenth official report – 1993. *Journal of Heart and Lung Transplantation* 1993; **12**: 541.

so the need for paediatric heart–lung transplantation will continue for the foreseeable future.[79,80]

In practical terms, since the maximum safe preservation time for lungs is currently 8 hours,[81] these procedures must be scheduled as emergencies. BSLT is performed in the supine position, through bilateral anterior thoracotomies/transverse sternotomy, known as the 'clam shell' incision.[77] *En bloc* double lung and heart–lung transplantation procedures are performed through a midline sternotomy.

Essential anaesthetic equipment includes a ventilator capable of high levels of positive end-expiratory pressure (PEEP), respiratory rates and inflation pressures; on a rare occasion, a second ventilator is needed for differential ventilation. In addition, the availability of appropriately sized fibreoptic bronchoscopes to facilitate therapeutic bronchoscopy, double-lumen tracheal tube placement, and assessment of anastomotic integrity are mandatory. Placement of an introducer that can accommodate a PA catheter (placed after unclamping of the pulmonary artery) is routine in paediatric lung transplantation. Availability of transoesophageal echocardiography to buttress other haemodynamic data has also proved valuable. Because lung transplantation can result in serious haemorrhage, provision for rapid transfusion (see the section, 'Liver transplantation', earlier) and ready availability of blood products and clotting factors should be made well in advance.

A unique aspect of lung transplantation is the airway management, which must include strategies to manage bronchial discontinuities and bleeding resulting from surgery. Maintenance of gas exchange during one-lung ventilation (OLV), when pulmonary function in the native lung is poor, is among the greatest challenges anaesthetists face. Although cardiopulmonary bypass provides a 'safety net', the increased risks of bleeding and other complications make this choice a last resort. Larger children (> 25–30 kg) are managed with small-sized double-lumen tracheal tubes, but the airways of smaller children are more safely secured by a cuffed single-lumen tracheal tube. In the BSLT procedure, during implantation of the first allograft (usually on the right lung), a single-lumen tracheal tube can be advanced to intubate selectively the appropriate bronchus. This rapid, simple technique is effective in children of all sizes, and provides a more secure airway than bronchial blockers. Furthermore, because of its size, a single-lumen tracheal tube allows more effective suctioning and bronchoscopy, which is particularly important with cystic fibrosis.

In spite of severe respiratory disease, it is usually possible to maintain adequate oxygenation during OLV in children; however, CO_2 elimination is often difficult. If the first allograft functions well after reperfusion, blood gases are usually better during implantation of the second lung. However, allografts that have suffered reperfusion or preservation injury will rapidly develop pulmonary oedema, pulmonary hypertension and deteriorating compliance. Increasing PEEP decreases oedema formation and improves gas exchange, and in most cases even these recipients are able to be weaned and extubated within several days. Thoracic epidural catheters are placed postoperatively, for pain relief, which in turn facilitates weaning and extubation. The postoperative course of lung recipients is often complicated by some degree of allograft dysfunction, infection and/or rejection. Aggressive surveillance for these problems, using fluoroscopic-guided trans-bronchial biopsies and bronchoalveolar lavage, is essential.

Recent published series of paediatric lung transplantation[82,83] have demonstrated that it is now possible to achieve 1-year survival of 85 per cent. Current experience with transplants for cystic fibrosis indicates that these recipients do not fare as well,[84] but protocols for infection prophylaxis may alter these results in the future.

REFERENCES

1. Kaye MP. The registry of the International Society for Heart and Lung Transplantation: Tenth Official Report – 1993. *Journal of Heart and Lung Transplantation* 12: 541–8.

2. Gordon RD, Bismuth H. Liver transplant registry report. *Transplantation Proceedings* 1991; **23**: 58–60.

3. Boucek MM, Kanakriyeh MS, Mathis CM, et al. Cardiac transplantation in infancy: donors and recipients. *Journal of Pediatrics* 1990; **116**: 171–6.

4. Franklin RCG, Spiegelhalter DJ, Anderson RH, et al. Double-inlet ventricle presenting in infancy. *Journal of Thoracic and Cardiovascular Surgery* 1991; **101**: 917–23.

5. Farrell PE, Cheng AC, Murdison KA, et al. Outcome assessment after the modified Fontan procedure for hypoplastic left heart syndrome. *Circulation* 1992; **85**: 116–22.

6. Turcotte JG, Campbell Jr DA, Dafoe DC, et al. Pediatric renal transplantation. In: Cerilli GJ, ed. *Organ transplantation and replacement*. Philadelphia: JB Lippincott, 1988; 349–60.

7. Davis ID, Chang P, Nevins TE. Successful renal transplantation accelerates development in young uremic children. *Pediatrics* 1990; **86**: 594–600.

8. President's Commission for the Study of Ethical Problems in Medicine and Biomedical and Behavioral Research. Guidelines for the determination of death. *Journal of the American Medical Association* 1981; **246**: 2184–6.

9. Task Force Report. Guidelines for the determination of brain death in children. *Pediatrics* 1987; **80**: 298–300.

10. Ashwal S. Brain death in early infancy. *Journal of Heart and Lung Transplantation* 1993; **12**: S176–8.

11. Ashwal S, Schneider S. Brain death in the newborn. *Pediatrics* 1989; **84**: 429–37.

12. Ashwal S, Schneider S. Pediatric brain death: current perspectives. In: Barness LA, ed. *Advances in pediatrics*. Chicago: Mosby–Year Book, 1991; 181–202.

13. Baird PA, Sadovnick AD. Survival in infants with anencephaly. *Clinical Pediatrics* 1984; **23**: 268–71.

14. Peabody JL, Emery JR, Ashwal S. Experience with anence-phalic infants as prospective organ donors. *New England Journal of Medicine* 1989; **321**: 344–50.

15. Girvin J. Brain death criteria must be revised so that society can readily benefit from families who offer anencephalic infants as organ donors. *Journal of Heart and Lung Transplantation* 12: S369–73.

16. Busuttil RW. Living-related liver donation: CON. *Transplantation Proceedings* 1991; **23**: 43–5.

17. Broelsch CE, Emond JC, Whitington PF, et al. Liver transplantation in children from living-related donors. *Annals of Surgery* 1991; **214**: 428.

18. Spray TL. Projections for pediatric heart and lung transplantation. *Journal of Heart and Lung Transplantation* 1993; **12**: 337–43.

19. Starzl TE, Marchioro TL, Morgan WW, Waddell WR. A technique for use of adult renal homografts in children. *Surgery, Gynecology and Obstetrics* 1964; **119**: 106–8.

20. Bismuth H, Houssin D. Reduced-size, orthotopic liver graft in hepatic transplantation in children. *Surgery* 1984; **94**: 367.

21. Broelsch CE, Emond JC, Thistlethwaite JK, Rouch DA, Whitington PF, Lichtor JL. Liver transplantation with reduced-size donor organs. *Transplantation* 1988; **45**: 519.

22. Terasaki PI, Cecka JM, Cho Y, et al. A report from the UNOS Scientific Renal Transplant Registry. *Transplantation Proceedings* 1991; **23**: 53–4.

23. Starnes VA, Oyer PE, Bernstein D, et al. Heart, heart–lung and lung transplantation in the first year of life. *Annals of Thoracic Surgery* 1992; **53**: 306.

24. Boucek MM, Mathis CM, Kanakriyeh M, et al. Donor shortage: use of the dysfunctional donor heart. *Journal of Heart and Lung Transplantation* 1993; **12**: S186–90.

25. Jones KL. *Smith's recognizable patterns of human malformation*. Philadelphia: WB Saunders, 1988: 526.

26. Ellis JE, Lichtor JL, Feinstein SB, et al. Right heart dysfunction, pulmonary embolism and paradoxic embolism during liver transplantation. *Anesthesia and Analgesia* 1989; **68**: 777–82.

27. Sassano JJ. The rapid infusion system. In: Winter PW, Kang Y, eds. *Hepatic transplantation*. Philadelphia: Praeger, 1986: 120–34.

28. Szyfelbein SK, Drop LJ, Martyn JA, *et al.* Persistent ionized hypocalcemia in patients during resuscitation and recovery phases of body burns. *Critical Care Medicine* 1981; **9**: 454–8.

29. Lichtor JL, Emond J, Chung MR, *et al.* Pediatric orthotopic liver transplantation: multifactorial predictions of blood loss. *Anesthesiology* 1988; **68**: 607–11.

30. Worldwide kidney transplant directory. In: Cecka MJ, Terasaki P, eds. *Clinical transplants 1992*. Los Angeles: UCLA Tissue Typing Laboratory, 1992: 590.

31. McGraw ME, Haka-Ikse K. Neurologic–developmental sequelae of chronic renal failure in infancy. *Journal of Pediatrics* 1985; **106**: 579–83.

32. Warady BA, Kriley M, Lovell H, Farrell SE, Hellerstein S. Growth and development of infants with end stage renal disease receiving long term peritoneal dialysis. *Journal of Pediatrics* 1988; **12**: 714–19.

33. Fine RN, Ettenger RB. Renal transplantation in children. In: Morris PJ, ed. *Kidney transplantation principles and practice*. Philadelphia: WB Saunders, 1988; 635–91.

34. Salusky IB, von Lilien T, Anchondo M, *et al.* Experience with continuous cycling peritoneal dialysis during the first year of life. *Pediatric Nephrology* 1987; **1**: 172–5.

35. Hull AR, Parker TF. Proceedings from the Morbidity, Mortality and Prescription of Dialysis Symposium, Dallas, September 15–17, 1989. *American Journal of Kidney Disease* 1990; **15**: 375–83.

36. ESRD Facility survey tables. *Contemp Dial Nephronol*, 11–12 August 1990.

37. Penn I. The changing pattern of posttransplant malignancies. *Transplantation Proceedings* 1991; **23**: 1101–3.

38. Chavers B, Matas AJ, Nevins TE, *et al.* Results of pediatric kidney transplantation at the University of Minnesota. In: Terasaki P, ed. *Clinical transplants*. Los Angeles: UCLA Tissue Typing Laboratory, 1989: 253–66.

39. UNOS Scientific Renal Transplant Registry. In: Cecka JM, Terasaki P, eds. *Clinical transplants 1992*. Los Angeles: UCLA Tissue Typing Laboratory, 1992: 1–16.

40. Avioli LV, Teitelbaum SL. Renal osteodystrophy. In: Edelmann Jr CM, ed. *Pediatric kidney disease*. Boston: Little, Brown, 1978: 366–401.

41. Baron P, Heil J, Condie R, *et al.* 96-hour renal preservation with silica gel precipitated plasma cold storage versus pulsatile perfusion. *Transplantation Proceedings* 1990; **22**: 464–5.

42. Migliori RJ, Simmons RL, Payne WD, *et al.* Renal transplantation done safely without prior chronic dialysis therapy. *Transplantation* 1987; **43**: 51–5.

43. Grodstein G, Harrison A, Robert C, *et al.* Impaired gastric emptying in hemodialysis patients. *Kidney International* 1979; **16**: 952A.

44. Starzl TE, Iwatsuki S, Malatack JJ, *et al.* Liver and kidney transplantation in children receiving cyclosporine and steroids. *Journal of Pediatrics* 1982; **100**: 681–6.

45. UNOS Liver Transplant Registry. In: Cecka JM, Terasaki P, eds. *Clinical transplants 1992*. Los Angeles: UCLA Tissue Typing Laboratory, 1992: 17–31.

46. Todo S, Nery J, Yanaga K, *et al.* Extend preservation of human grafts with UW solution. *Journal of the American Medical Association* 1989; **26**: 711–14.

47. Piper JB, Whitington PF, Woodle S, *et al.* Pediatric liver transplantation at the University of Chicago Hospitals. In: Cecka JM, Terasaki P, eds. *Clinical transplants 1992*. Los Angeles: UCLA Tissue Typing Laboratory, 1992: 179–80.

48. Thistlethwaite JR, Emond JC, Woodle ES, *et al.* Increased utilization of organ donors: transplantation of two recipients from single donor livers. *Transplantation Proceedings* 1990; **22**: 1485.

49. Emond JC, Whitington PF, Thistlethwaite JK, *et al.* Transplantation of two patients with one liver: analysis of a preliminary experience with 'split' livers. *Annals of Surgery* 1990; **212**: 14.

50. Broelsch CE, Emond JC, Thistlethwaite JK, *et al.* Liver transplantation, including the concept of reduced-sized liver transplants in children. *Annals of Surgery* 1988; **208**: 410–20.

51. Broelsch CE, Emond JC, Whitington PF, *et al.* Application of reduced-size liver transplants as split grafts, auxiliary orthotopic grafts, and living-related segmental transplants. *Annals of Surgery* 1990; **212**: 368–77.

52. Massaferro V, Esquivel CO, Makowka L, *et al.* Hepatic artery thrombosis after pediatric liver transplantation. *Transplantation* 1989; **47**: 971–7.

53. Pleskow RG, Grand RJ. Wilson's disease. In: Walker WA, Durie PR, Hamilton JR, Walker-Smith JA, Watkins JB, eds. *Pediatric gastrointestinal disease*. Philadelphia: Decker, 1991.

54. Shaw Jr BW, Martin DJ, Marquez JM, *et al.* Advantages of venous bypass during orthotopic transplantation of the liver. *Seminars in Liver Disease* 1985; **5**: 344–8.

55. Lewis JH, Bontempo FA, Kang Y, *et al.* Intraoperative coagulation in liver transplantation. In: Winter PW, Kang Y, eds. *Hepatic transplantation*. Philadelphia: Praeger, 1986: 142–50.

56. Porte RJ, Bontempo FA, Knott EAR, *et al.* Systemic effects of tissue plasminogen activator-associated fibrinolysis. *Transplantation* 1989; **47**: 978–84.

57. Kang YG, Martin DJ, Marquez JM, *et al.* Intraoperative changes in blood coagulation and thromboelastographic monitoring in liver transplantation. *Anesthesia and Analgesia* 1985; **64**: 888–96.

58. Estrol CJ, Faris AA, Martinez JJ, *et al.* Central pontine myelinolysis after liver transplantation. *Neurology* 1989; **39**: 493–8.

59. Williams JW, Sakary HN, Foster PF, *et al.* Splanchnic transplantation: an approach to the infant dependent on parenteral nutrition. *Journal of the American Medical Association* 1989; **261**: 1458–62.

60. Grant D, Wall W, Mimeault R, *et al.* Successful small bowel/liver transplantation. *Lancet* 1990; **335**: 181–4.

61. Starzl TE, Rowe MI, Todo S, *et al.* Transplantation of multiple abdominal viscera. *Journal of the American Medical Association* 1989; **261**: 1449–57.

62. Bailey LL. Heart transplantation techniques in complex congenital heart disease. *Journal of Heart and Lung Transplantation* 1993; **12**: S168.

63. Kaye MP, Kreitt JM. Pediatric heart transplantation: the world experience. *Journal of Heart and Lung Transplantation* 1991; **10**: 856–9.

64. Barber G. Hypoplastic left heart syndrome. In: Garson Jr A, Bricker JT, McNamara DG, eds. *The science and practice of pediatric cardiology.* Malvern, PA: Lea & Febiger, 1991: 1316–33.

65. Norwood WI, Lang P, Hansen D. Physiologic repair of aortic atresia and hypoplastic left heart syndrome. *New England Journal of Medicine* 1983; **308**: 23–6.

66. Bailey L, Concepcion W, Shattuck H, *et al.* Method of heart transplantation for treatment of HLHS. *Journal of Thoracic and Cardiovascular Surgery* 1985; **92**: 1–5.

67. Ruiz CE, Zhang HP, Larsen RL. The role of interventional cardiology in pediatric heart transplantation. *Journal of Heart and Lung Transplantation* 1993; **12**: S164–7.

68. Bailey LL, Grundy SR, Razzouk AJ, *et al.* Bless the babies: 115 late survivors of heart transplantation during the first year of life. *Journal of Thoracic and Cardiovascular Surgery* 1993; **105**: 805–15.

69. Menkis AH, McKenzie FN, Novick RJ, *et al.* Special considerations for heart transplantation in congenital heart disease. *Journal of Heart Transplantation* 1990; **9**: 602–7.

70. Chartrand C, Guerin R, Kangah M, Stanley P. Pediatric heart transplantation: surgical considerations for congenital heart diseases. *Journal of Heart Transplantation* 1990; **9**: 608–17.

71. Pepke-Zaba J, Higgenbottam TW, Tuan Ding-Xuan A, *et al.* Inhaled nitric oxide as a cause of selective pulmonary vasodilation in pulmonary hypertension. *Lancet* 1991; **338**: 1173–4.

72. Girard CJ, Neidecker MC, Laroux G, *et al.* Inhaled NO in pulmonary hypertension after total anomalous pulmonary venous return. [letter]. *Journal of Thoracic and Cardiovascular Surgery* 1993; **106**: 369.

73. Pahl EF, Fricker FJ, Armitage J, *et al.* Coronary arteriosclerosis in pediatric heart transplant survivors: limitation of long-term survival. *Journal of Pediatrics* 1990; **116**: 177–83.

74. Berry GJ, Rizeq MN, Weiss LM, Billingham ME. Graft coronary disease in pediatric heart and combined heart–lung transplant recipients: a study of fifteen cases. *Journal of Heart and Lung Transplantation* 1993; **12**: S309–19.

75. Kaye MP. Intrathoracic transplantation. *Transplantation Proceedings* 1991; **23**: 51–2.

76. Kaye MP. Pediatric thoracic transplantation: the world experience. *Journal of Heart and Lung Transplantation* 1993; **12**: S344–50.

77. Griffith BP, Hardesty RL, Armitage JM, *et al.* A decade of lung transplantation. *Annals of Surgery* 1993; **218**: 310–20.

78. Maclusky I, Levison H. Cystic fibrosis. In: Chernick V, ed. *Disorders of the respiratory tract in children.* Philadelphia: WB Saunders, 1990: 692–730.

79. Smyth RL, Scott JP, Whitehead B, *et al.* Heart–lung transplantation in children. *Transplantation Proceedings* 1990; **22**: 1470–1.

80. Starnes VA, Marshall SE, Lewiston NJ, Theodore J, Stenson EB, Shumway NE. Heart–lung transplantation in infants, children and adolescents. *Journal of Pediatric Surgery* 1991; **26**: 434–8.

81. Hardesty R, Aeba R, Armitage J, *et al.* A clinical trial of UW solution for pulmonary preservation. *Journal of Thoracic and Cardiovascular Surgery* 1993; **105**: 660–6.

82. Armitage JM, Fricker, FJ, Kurland G, *et al.* Pediatric lung transplantation: expanding indications, 1985 to 1993. *Journal of Heart and Lung Transplantation* 1993; **2**: S246.

83. Metras D, Kreitmann B, Shennib H, Niorclerc M. Lung transplantation in children. *Journal of Heart and Lung Transplantation* 1992; **11**: S282–5.

84. Armitage JM, Fricker FJ, Kurland G, *et al.* Pediatric lung transplantation: the years 1985 to 1992 and the clinical trial of FK 506. *Journal of Thoracic and Cardiovascular Surgery* 1993; **105**: 337–46.

85. In: Behrman RE, Kliegman RM, eds. *Nelson's textbook of pediatrics*, 14th ed. Philadelphia: WB Saunders, 1992.

Chapter 60

Anaesthesia for Dental and Faciomaxillary Surgery

Stuart A. Hargrave

General anaesthesia in dentistry
Outpatient dental anaesthesia
The future of dental anaesthesia

Sedation techniques for conservative dentistry
Anaesthesia for minor oral surgery and conservative dentistry
Anaesthesia for major oral and faciomaxillary surgery

In 1844 Horace Wells, a dentist in Hartford, Connecticut, after witnessing a public demonstration of the effects of nitrous oxide, had one of his own teeth removed while under the influence of the gas. He immediately realized the potential value of his discovery but his own public attempt to demonstrate the use of nitrous oxide to produce surgical anaesthesia ended in failure. In 1846 William T. G. Morton, who was also a dentist and had been in partnership with Wells, successfully anaesthetized a patient for the removal of a tumour of the mandible using ether and has therefore been generally credited with the discovery of anaesthesia. Some writers and historians feel that Wells deserves equal if not greater credit[1] and certainly since his discovery anaesthesia and the practice of dentistry have been inextricably linked.

In both the United Kingdom and the United States of America the use of general anaesthesia in dentistry increased rapidly so that by the 1960s some 1.5–2 million dental anaesthetics were being given every year in the UK and about 5 million in the USA. Over the last 25 years, however, the number of anaesthetics given in general dental practice in the UK has been steadily falling and by 1989 was down to 371 000. The reasons for this fall are several but more widespread use of fluoridation, improvements in local anaesthetics and a greater emphasis on conservative dental treatment have all played a major role.[2] If the number of general anaesthetics required continues to fall then it is possible that eventually general anaesthesia will no longer be practised in the general and community dental services in the UK.[3]

When judged by a very low mortality rate (1:226 000;[4] 1:300 000[5]) outpatient dental anaesthesia appears to have a good safety record. However, the fact that these deaths have occurred mostly in young, fit patients undergoing a brief, comparatively trivial procedure has been a major cause for concern. The majority of outpatient dental anaesthetics have been given by non-specialist anaesthetists, either medically or dentally qualified and in surgeries which are often poorly equipped for the purpose, particularly with regard to electrical monitoring devices.[6,7] Several reports[8–10] have made recommendations in attempts to improve training and equipment standards with limited success. The most recent report of an Expert Working Party[11] has been accepted by the UK Department of Health and funding has been made available to help implement its recommendations. There is no doubt that 'The Poswillo Report', as it has come to be known, will have a major impact on the provision of dental anaesthesia and sedation.

General anaesthesia in dentistry

When a dental practitioner formulates a treatment plan it should include not only the dental treatment required but also the way to carry it out that is most acceptable to the patient. An assessment of the patient's medical fitness should also be made and if general anaesthesia is required a decision taken as to where the treatment is to be carried out. This may be in the dental surgery, community dental clinic or in a hospital outpatient, day-stay or inpatient department. The decision will depend on the availability of anaesthetic skills and the organization of dental and oral surgical services in the dental practitioner's area. Most routine dental treatment can be performed using local anaesthesia alone and the excellent safety record makes this the method of choice. Some patients, however, find cooperation difficult or impossible because of fear and anxiety. A sympathetic approach coupled with behavioural management techniques, hypnosis or simple sedation can be very successful in helping some of these patients. General anaesthesia should only be used if there are very strong indications and it is a major recommendation of the Poswillo Report that 'the use of general anaesthesia should be avoided wherever possible'.

Indications for general anaesthesia

Children

In the UK about 75–80 per cent of outpatient dental general anaesthetics are given to children under the age of 16 years. The main indication is for extractions, either for caries or for orthodontic reasons. If a child requires multiple extractions general anaesthesia is probably kinder since it only involves one visit and no injections in the mouth. However, in expert hands the combination of inhalation sedation and local anaesthesia can be very successful.[12] Where children require extensive restorative treatment and cooperation is difficult then an initial comprehensive treatment under general anaesthesia on a day-stay basis with subsequent behavioural management can achieve a high degree of success as judged by acceptance of routine treatment under local anaesthesia at a later date.[13]

Mental handicap

Patients with mild degrees of mental handicap may be able to cooperate with dental treatment, especially if simple sedation is used. Patients with more severe handicap will require general anaesthesia and since it is likely that extensive treatment may be required this should be arranged on a day-stay or inpatient basis.

Acute infection

The presence of local infection will often render local anaesthetics ineffective due to altered tissue pH. There is also a risk of spreading infection with a local anaesthetic injection. If there is associated facial swelling and limited mouth opening then extreme caution must be exercised when administering a general anaesthetic.

Allergy to local anaesthetics

True allergy to local anaesthetics is very rare and is almost always to one of the amide type of agents. The preservative methyl paraben present in some local anaesthetic solutions can also provoke allergic reactions. Patients who claim to be allergic should be questioned carefully since it is much more common for patients to describe vasovagal fainting reactions, or flushing and palpitations as a result of absorption of adrenaline from the anaesthetic solution.

Failure of local anaesthetic

It is not uncommon for a general anaesthetic to be requested because an attempt to extract a tooth using local anaesthesia has been unsuccessful. It is unusual for modern local anaesthetics to fail completely unless they have been injected incorrectly but it is common for patients to retain some sensation of pressure around the tooth. Fear and anxiety may lead them to interpret this as pain and cooperation is lost. The use of a simple sedative technique may rectify the situation but if the services of an anaesthetist are readily available, as may be the case in hospital, and the patient is suitably prepared then a brief general anaesthetic may be preferable.

Persistent fainting reactions

Fainting at the sight or even thought of an intraoral needle happens occasionally in dental practice. Most often teenage and young adult patients are affected but it is most unusual for young children to faint. Careful handling plus the use of sedation where necessary will be successful in most patients and general anaesthesia only rarely required. Patients with a history of fainting should always be treated in the supine position but it must be remembered that some patients will still faint when lying down. In such a case the faint is often a severe one and requires active treatment in the form of oxygen and intravenous atropine. Full cardiovascular stability may take some time to be restored and if this happens immediately prior to induction of anaesthesia it is wise to abandon the anaesthetic on that day.

Failure of sedation

Most patients with a high degree of fear, anxiety and phobia related to dental treatment can be managed very successfully using the techniques of simple dental sedation combined with a kind and sympathetic approach. However, even with careful patient selection, a small number of patients will still prove to be extremely resistant to treatment. The temptation to give more sedative agent should be resisted because of the risks of oversedation, respiratory depression and loss of airway control. If the patient requires extensive dental treatment then, as with children (see above), it may be preferable to arrange this under general anaesthesia and then try to re-educate the patient once restoration has been achieved.

Contraindications to outpatient general anaesthesia

Only patients who are fit and well and satisfy the ASA categories I and II should be considered for outpatient anaesthesia. Patients with serious cardiopulmonary conditions, diabetes or other endocrinological disease, neuromuscular disorders or indeed any other potentially life-threatening illness must be referred to hospital. Patients

Table 60.1 An example of a questionnaire for outpatient anaesthesia which can be given to the patient for completion before the appointment

Have you had an anesthetic before?

Did you have any problems with an anaesthetic?

Are you in good general health?

Have you had any trouble with your heart?

Do you have any problems with your lungs?

Have you suffered from any chest pains, shortage of breath, do you have a cough? Do you cough and produce phlegm?

Have you had a cold recently or have you one now?

Do you find that you bleed or bruise easily?

Are you taking any medicines from your doctor?

Do you suffer from any allergies?

The questions should be kept simple but a YES to any question will prompt further enquiry.

taking significant amounts of drugs in order to control medical conditions should also be referred. Patients of African, Mediterranean or Middle Eastern origin should be screened for sickle cell disease or trait or other haemoglobinopathies. Any patient with a major degree of oral infection, particularly if there is marked orofacial swelling and trismus, must be referred to hospital because of the risk of airway problems. Since many patients referred for outpatient anaesthesia are children major congenital defects must be excluded. Children with congenital cardiac defects can be anaesthetized in a hospital outpatient or day-stay clinic provided the defect is a simple one and is not limiting the child in any way. If there is any doubt then a cardiological opinion must be obtained and the necessity for antibiotic cover confirmed.

Outpatient dental anaesthesia

Outpatient general anaesthesia is usually employed for dental extractions, mainly in children. Occasionally very minor oral surgery such as drainage of small abscesses or extraction or exposure of partially erupted or unerupted teeth can be carried out on this basis. The whole procedure is designed and expected to be very brief and ideally should not exceed 5 minutes in duration.[14] The problems presented to the anaesthetist are those common to all types of outpatient or day-stay anaesthesia with particular difficulties related to the nature of the surgery.

Medical assessment

This will usually have been performed by a dental practitioner or a dental member of the hospital staff and the anaesthetist is often the patient's first contact with a physician. A simple health questionnaire should be given to the patient at the time the appointment for anaesthesia is made (Table 60.1).

Anxiety and fear

Patients, particularly adults, presenting for outpatient dental treatment under anaesthesia are often very apprehensive and fearful and it is impossible to persuade them to accept treatment under local anaesthesia, even with sedation. Great care must be exercised with the patient who is obviously very frightened. High levels of circulating catecholamines can cause tachycardia and hypertension and vasoconstriction in the skin can make venous access very difficult. Fear can also increase the likelihood of fainting. In extreme cases it may be advisable to defer the appointment to another day and prescribe an anxiolytic to be taken by mouth beforehand. The patient must then be warned that recovery from anaesthesia may take longer and they should be prepared for this. Children are occasionally upset and frightened but a calm, soothing approach and the presence of a sensible parent or relative can be very helpful.

Adherence to instructions

Patients should be given written preoperative instructions at the time of making the appointment which include directions about eating and drinking and that the patient must be accompanied by a responsible adult. It is essential to check that all patients have been properly starved prior to anaesthesia. Occasionally a patient will forget to mention having had a drink, being under the impression that because the procedure is to be very brief a proper period of starvation is not necessary. Some parents may consider a sweet or mouthful of soft drink given to console a hungry child to be unimportant and will not mention it unless pressed. If the appointment for treatment is in the afternoon then a light breakfast and a drink can be allowed but a period of starvation of 6 hours thereafter is well tolerated. Children undergoing outpatient anaesthesia in the afternoon do not appear to develop hypoglycaemia even after long periods of fasting but care should be exercised with children below the age of 4 years.[15]

Rapid induction and recovery

Anaesthesia on an outpatient basis requires a rapid recovery and return to 'street fitness'. This can be aided by a rapid induction of anaesthesia and a very brief operating period but proper recovery facilities must be available.

Shared airway

Anaesthesia for simple exodontia is usually carried out without tracheal intubation. The dental practitioner or oral surgeon has to gain access to the patient's mouth and the oral and nasal airways may be compromised.

Patient preparation

Before induction of anaesthesia it is essential to ensure that preoperative instructions have been adhered to and that written consent to the procedure has been obtained. The medical fitness questionnaire should be checked and the patient or parent asked whether anything has changed since the appointment was made. In particular patients should be asked about recent or active upper respiratory tract infections and any drugs they may be taking. The nasal airway should be checked to ensure that it is patent. The bladder should have been emptied before entering the surgery and spectacles, earrings and dentures should be removed. The anaesthetist must check with the dentist or surgeon what procedure is proposed and estimate how long it will take. The patient's mouth should be examined for the presence of any loose teeth or food (e.g. chewing gum). The patient can then be settled in the chair, the head rest adjusted and tight clothing loosened. The electrodes for the electrocardiogram and the pulse oximeter probe should be applied. In the case of young children who are frightened and who will not sit in the chair, anaesthesia can be induced while the child is sitting or lying in the parent's lap and then the child can be transferred to the dental chair.

Equipment

All anaesthetic and emergency equipment must be thoroughly checked before the anaesthetic session commences. A syringe each of atropine and suxamethonium should be ready to use in the event of a sudden emergency. Oxygen, suction and a means of inflating the patient's lungs must be immediately available in the recovery area.

The dental chair

Whichever position is adopted for anaesthesia it must be possible to put the chair into a horizontal or even head-down position immediately in an emergency. If the chair is electrically powered there should be a manual release to achieve this in the event of a power failure. The chair should have an adjustable head rest to allow the head to be correctly positioned and the arm rests should be easily removable. In the hospital outpatient or day-stay unit, if the supine position is to be used, a tipping trolley or operating table can be used.

The anaesthetic machine

The traditional intermittent or demand flow machine originally developed for dental anaesthesia is now of historical interest only. A standard continuous flow anaesthetic machine is used in most situations although smaller machines which can be used for relative analgesia or general anaesthesia have proved to be very satisfactory.

Modern machines have an oxygen failure alarm and a shut-off mechanism for nitrous oxide plus a link between the oxygen and nitrous oxide rotameters so that hypoxic gas mixtures cannot be delivered. If the oxygen and nitrous oxide rotameters can be controlled independently then an oxygen monitor must be connected to the fresh gas outlet.[16] Machines such as the Quantiflex MDM are designed for providing relative analgesia and will not deliver less than 30 per cent oxygen (Fig. 60.1).

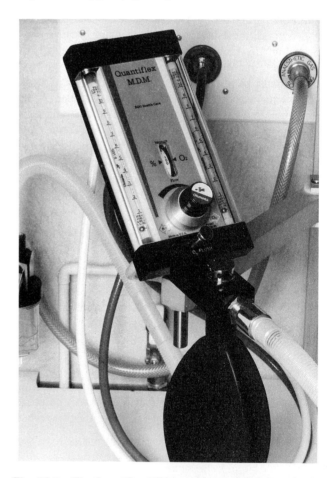

Fig. 60.1 The Quantiflex MDM machine for relative analgesia. A single rotary control sets the oxygen percentage of the gas mixture and the large black knob controls the total flow rate. The machine cannot deliver less than 30 per cent oxygen and has a nitrous oxide shut-off if the oxygen supply fails.

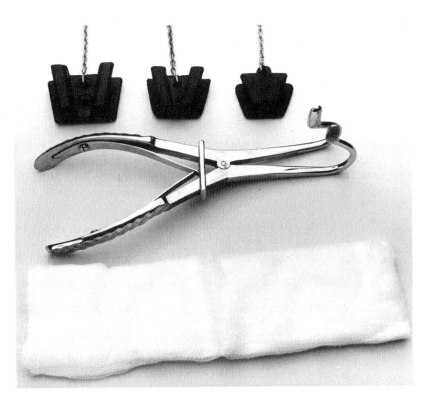

Fig. 60.2 Mouth props of different sizes made of black rubber, a Ferguson mouth gag and a gauze mouth pack.

Circuits and masks

Various breathing circuits can be used in dental anaesthesia in combination with nasal masks. Modifications of the standard twin tube Magill circuit to allow for scavenging have been described,[17] but hoods applied over the expiratory valve on the nasal mask can be cumbersome and require extra tubing close to the face. The Goldman nasal mask can be very satisfactorily combined with a coaxial circuit and has the benefits of low resistance to breathing with ease of connection to a scavenging system. Alternatively, the McKesson nasal mask can be used in combination with the Lack parallel or the Humphrey ADE breathing systems.[18,19]

Mouth props, packs and gags (Fig. 60.2)

The device used most commonly to keep the mouth open during anaesthesia is the mouth prop. The McKesson type comes in various sizes and is suitable for most patients. Mouth gags of the Ferguson type are used less commonly because of the possibility of causing trauma inside the mouth if incorrectly applied. It is useful to have a gag available for emergencies, however. An efficient mouth pack is absolutely essential for the safe practice of dental anaesthesia. The pack must be correctly positioned to isolate the quadrant of the mouth which is to be worked on so as to absorb blood and saliva and prevent aspiration, thus protecting the airway. Great care must be taken when inserting it to prevent the tongue being pushed back and causing airway obstruction (Fig. 60.3).

Most operators use a gauze pack although some prefer a butterfly sponge. If teeth are to be removed from both sides of the mouth then two packs should be used, the first pack being left *in situ* to absorb any blood from the exposed tooth sockets while a fresh one is inserted into the second side.

Nasopharyngeal airways

A selection of soft nasopharyngeal airways should be available and may be necessary to improve the nasal airway during anaesthesia.

As in any situation where general anaesthesia is to be carried out a reliable and powerful suction apparatus must be available.

Resuscitation equipment

A full range of tracheal tubes with all the accessories for tracheal intubation plus two working laryngoscopes must be immediately available. A full range of emergency drugs with needles and syringes must be available in the surgery along with all the equipment required for setting up an intravenous infusion. A defibrillator must be available in any situation where general anaesthesia is to be practised.[11] All dental anaesthetists must be familiar with the recommended resuscitation equipment and drugs and be able to perform tracheal intubation.

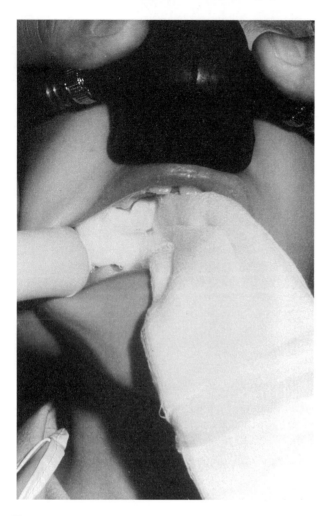

Fig. 60.3 Photograph to illustrate the position of the mouth pack. A mouth prop has already been placed in the other side of the mouth and is concealed by the pack. One end of the pack is carefully positioned in the lingual sulcus between the teeth and the tongue and behind the teeth to be extracted. Great care must be taken not to push the tongue backwards or to place the pack too far backwards and obstruct the nasopharynx.

Induction of anaesthesia

Anaesthesia can be induced using either an inhalation or intravenous technique depending largely upon the personal preference of the anaesthetist but the wishes of the patient must be taken into consideration. Inhalational induction is commonly used for younger children, especially if intravenous access is likely to be difficult. Older children and adults will usually accept an intravenous needle but occasionally request a gaseous induction. The prior application of a eutectic local anaesthetic cream mixture (Emla, Astra Pharmaceuticals) to the backs of the hands at least 1 hour beforehand makes venepuncture virtually painless and more acceptable.

Inhalational induction is usually commenced by holding the nasal mask close to the patient's face and cupping the hand below the chin to direct the gas mixture towards the mouth and nose. A high initial gas flow with nitrous oxide and not less than 30 per cent oxygen should be used. After a few breaths the volatile agent is commenced and the concentration increased by 0.5–1.0 per cent increments, again after every few breaths. The nasal mask should be applied as soon as the patient will accept it and the mouth should be closed to discourage mouth breathing. The child may struggle during induction and it is here that the presence of a parent can be very reassuring. The anaesthetist should talk to the child in a kind and soothing way but gentle restraint by assistants may occasionally be necessary. The most commonly used volatile agent is still halothane because of the smooth, rapid induction that can usually be achieved in children and the high level of patient acceptability. A concentration of 3–4 per cent may be necessary briefly during induction but once the jaw has relaxed sufficiently to insert a mouth prop the concentration can be reduced to 1–2 per cent for maintenance of anaesthesia. Enflurane is a useful alternative to halothane. It is readily accepted by children and a smooth induction can usually be achieved. Induction of anaesthesia can be almost as rapid as with halothane[20,21] but because of the lower potency of enflurane higher concentrations are required to produce adequate levels of anaesthesia. In some patients a vaporizer with a maximum output of 5 per cent is inadequate and maintenance of anaesthesia is less satisfactory than with halothane. Isoflurane has been shown to be inferior to halothane in outpatient paediatric dental anaesthesia. Isoflurane has a pungent smell and is irritant to the tracheobronchial tree; therefore gaseous induction may be complicated by a high incidence of coughing, salivation, laryngospasm and episodes of oxygen desaturation.[22–24] Induction of anaesthesia is therefore no quicker than with halothane and in spite of isoflurane having a lower blood solubility recovery may take slightly longer.[24] As soon as respiration becomes more regular and muscle tone diminishes, the jaw can be checked to assess whether there is sufficient relaxation to allow the insertion of a prop. If a very brief procedure is anticipated for the extraction of only one or two teeth then deeper levels of anaesthesia are not required and it is not necessary to wait until the pupils become central.

Intravenous induction of anaesthesia is the preferred method of many dental anaesthetists. It has the advantages of rapidity and avoidance of the facemask which some children and adults find frightening. Salivation is less and there is less atmospheric pollution with anaesthetic gases. The disadvantages are the higher incidence of cardiorespiratory depression with hypotension and hypoventilation or apnoea. Adult patients may faint at the sight of a needle. If an intravenous technique is used then an indwelling venous cannula should be inserted.[11]

Methohexitone is the agent most commonly used in a dose of about 1.5 mg/kg in adults and 2 mg/kg in children. It is a reliable agent with good recovery characteristics and a low incidence of nausea and vomiting. Following induction there is usually a fall in blood pressure of about 15 per cent as a result of reduction in peripheral vascular resistance. A compensatory tachycardia results and the blood pressure is restored once the prop is inserted and surgery commences. Disadvantages of methohexitone are pain on injection, involuntary muscle movements, hiccups and respiratory depression. Methohexitone should be avoided in patients with epilepsy. Thiopentone is not suitable for outpatient anaesthesia because of a slower recovery and a higher incidence of 'hangover' effects such as headache and dizziness. Propofol (di-isopropyl phenol) is the newest intravenous anaesthetic agent to come into clinical practice. It is insoluble in water and is presented in a 1 per cent emulsion with soya bean oil, egg phosphatide and glycerol. This formulation (Diprivan, ICI) produces pain on injection but, as with methohexitone this can be considerably reduced by the addition of a small amount of lignocaine to the solution immediately prior to injection. Induction of anaesthesia is very smooth with only occasional involuntary movements. Doses of 2–2.5 mg/kg are satisfactory in adults but children need more and 3–3.5 mg/kg may be required. Propofol causes greater degrees of cardiorespiratory depression than other intravenous agents but these are within acceptable limits.[25,26] In dental outpatients propofol produces operating conditions and recovery times which are very similar to those following methohexitone.[27,28] In adult patients the author found that propofol causes a greater fall in blood pressure and a lower maximum heart rate than methohexitone but there were no significant differences in these changes between the semi-supine and supine positions.[29] Older children and adults will usually tolerate the placing of a mouth prop just as induction starts. This helps to reduce the overall anaesthesia and surgery time since it is not necessary to wait for jaw relaxation.

Maintenance of anaesthesia

Once the mouth prop and pack have been inserted surgery can commence. If the dental extractions are likely to take only a few seconds or less than a minute then it may be possible to accomplish this during the effects of the i.v. agent and only oxygen should be given via the nasal mask. If an inhalation technique has been used then it may be possible to turn off the anaesthetic agents and perform the extractions during what is effectively the start of the recovery period. If the procedure is to take a little longer the anaesthesia is maintained using the nasal mask delivering oxygen (minimum 30 per cent), nitrous oxide and a low

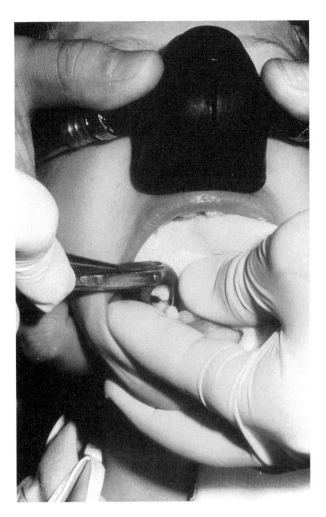

Fig. 60.4 When extracting lower teeth it is very easy to push the lower jaw downwards and cause airway obstruction. Both the anaesthetist and surgeon must support the lower jaw and keep it held forward to maintain a clear airway.

concentration of the volatile agent. The anaesthetist's prime responsibility is to maintain a clear airway by holding the mandible forward and upwards to prevent the tongue falling backwards. This can be difficult when the operator is pressing down on the lower teeth prior to extraction but the operator can help by holding the jaw correctly (Fig. 60.4).

The anaesthetist must also monitor the patient's condition by observing the colour of the lips and mucosal surfaces and watching the movements of the reservoir bag in the circuit. It may be possible to feel the carotid pulse in a child using the little finger or the facial artery underneath the mandible in an adult. Because of the potential for cardiorespiratory complications the ECG and oxygen saturation must also be monitored. Undoubtedly the best results and the safest conditions for the patient are achieved when the operator and the anaesthetist are used to working with each other and know what can be realistically undertaken.

Recovery

At the end of the surgery the prop is removed but the packs are left in place over the sockets. Whatever position has been adopted during the surgery the patient should be laid flat and in the lateral position during recovery. Even after very short procedures falls in oxygen saturation can occur in the recovery period as a result of residual respiratory depression, airway obstruction or the diffusion of nitrous oxide into the alveoli. The airway and breathing must be monitored continuously and suction used to keep the mouth clear of blood if necessary. Oxygen should be given until the patient has opened his or her eyes and expelled the mouth packs. Children are often very restless during the initial recovery period until fully awake. Gentle restraint may be necessary to prevent them from falling out of the dental chair or off the trolley but the temptation to stimulate them to try to speed up recovery should be resisted.

Patients should remain in the recovery area for at least 30 minutes before being discharged into the care of a responsible adult who must accompany them. They should be able to stand and walk unsupported and all bleeding must have stopped. Analgesics can be given as soon as the patient requests them and if there is obvious distress in a child. Soluble paracetomol is usually very effective. Written postoperative instructions must be given which emphasize that the patient must not drive or operate machinery for the rest of the day and that alcohol should be avoided.

Posture

For many years anaesthesia was administered to dental patients in the sitting position until Bourne in 1957[30] drew attention to the dangers of the unrecognized fainting attack occurring immediately prior to or during induction of anaesthesia. He postulated that many of the cases of cerebral damage or deaths related to dental anaesthesia could be explained by hypotension, bradycardia or cardiac arrest associated with a faint and recommended the routine adoption of the supine position. Studies of mortality related to dental anaesthesia have not fully confirmed this theory and have emphasized the dangers of hypoxia and sudden catastrophic cardiovascular collapse during anaesthesia or the recovery period.[4,5] Coplans and Curson[4] could find 'no evidence of a difference in safety between the supine and nonsupine position'.

In the supine position there is a greater risk of airway obstruction due to the tongue and soft palate falling backwards although this can be relieved by correct positioning of the head and mandible. There is a greater risk of pharyngeal soiling with blood and regurgitation of

stomach contents.[31] There is a reduction in expiratory reserve volume and in some patients a deterioration in gaseous exchange due to the relation between ERV and the closing volume of the lung.[32] Some dental anaesthetists maintain that it is more difficult to maintain the airway with the nasal mask when the patient is supine although Evans and Dawson[33] have shown no differences in oxygen saturation in children anaesthetized in the sitting or supine position. However, other studies have shown falls in oxygen saturation in both positions and have emphasized that the experience of the anaesthetist and operator are particularly important.[34,35]

Cardiovascular changes during dental anaesthesia do not appear to be influenced by posture. Falls in blood pressure and rises in heart rate occur with both intravenous and inhalational techniques and there are no significant differences between sitting and supine positions.[36,37] The incidence of arrhythmias is not influenced by posture.[37]

In order to achieve the best balance of respiratory and cardiovascular conditions the position commonly adopted for outpatient dental anaesthesia is the semi-supine. The head can be held erect which helps to keep blood and debris in the floor of the mouth and it is easier for the anaesthetist to hold the jaw and maintain the airway while standing behind the patient.

Complications

Respiratory

Airway obstruction can occur from a number of causes. The tongue or soft palate may fall backwards and the mandible will need repositioning. The nasal airway in children may be narrow especially if the child has been crying and the nasal mucosa is swollen. Adenoids can be a particular problem in small children. Insertion of a prop and pack must be very carefully performed. The pack must not push the tongue back or be placed so far back as to encroach upon the pharynx and occlude the airway. If it is difficult to obtain a clear nasal airway then insertion of a soft, well lubricated nasopharyngeal airway will usually solve the problem. Insertion should be gentle to avoid causing haemorrhage.

Mouth breathing will dilute the anaesthetic gases and the patient may become too lightly anaesthetized and start to struggle. The pack should be correctly positioned to seal off the mouth but again a nasopharyngeal airway may be required.

Apnoea may be due to breath-holding during induction and the patient will usually start to breathe again after a short period. It may be due to respiratory depression if an intravenous induction has been used in which case gentle inflation of the lungs may be required. Provided that an overdose of intravenous agent has not been given, the patient will start breathing as soon as the surgery commences.

Laryngospasm and coughing are always potentially very serious. If they occur during induction it may be necessary to reduce the inspired concentration of volatile agent and give more oxygen to prevent hypoxia. If they occur during the surgery it may be due to the combination of the surgical stimulus and too light a level of anaesthesia. More dangerously, laryngospasm may be the result of blood or saliva irritating the vocal cords. The surgery must stop and the pharynx must be sucked out. Oxygen must be administered, under pressure with a facemask if necessary. If the laryngospasm does not correct itself then intubation may be required.

Pulmonary aspiration is possible if blood or tooth fragments escape past the mouth pack and enter the pharynx. If the head is upright as in the semi-supine position then aspiration may be more likely. In the supine position regurgitation of stomach contents may occur more readily and pharyngeal soiling with blood from the mouth is more common than in the semi-supine position. If the supine position is used for any procedure other than a very brief one it is wise to protect the airway either by tracheal intubation or insertion of a laryngeal mask.

The laryngeal mask

Originally devised by Dr A. Brain, the laryngeal mask airway has had an enormous impact on the practice of anaesthesia. Use of the laryngeal mask in outpatient dental anaesthesia has been described and the advantages compared with the nasal mask emphasized.[38,39] The laryngeal mask provides a more secure airway and a much greater degree of protection of the larynx and trachea. Consequently episodes of airway obstruction with falls in arterial oxygen saturation occur much less frequently. The laryngeal mask is particularly useful for minor oral surgical procedures such as raising a small flap, removing bone, extraction of unerupted or very difficult teeth or indeed anything which is likely to take more than 2–3 minutes. The fact that tracheal intubation can be avoided may make the mask especially useful in outpatient anaesthesia. The standard version of the mask has a fairly rigid tube part but it can still be moved from side to side in the mouth to make access easier without disturbing the laryngeal cuff. A pack should be placed in front of the laryngeal mask to act as a throat pack. A new version of the mask with a reinforced flexible tube has very recently been introduced into clinical practice (Fig. 60.5).

This version has been specifically designed with the needs of oral and facial surgery and anaesthesia in mind and is now undergoing clinical evaluation.

Cardiovascular

Fear, anxiety and occasionally pain in patients presenting for outpatient dental anaesthesia can cause significant autonomic nervous system overactivity. High circulating catecholamine levels and an increased sympathetic component give rise to a labile blood pressure and tachycardia. Parasympathetic activity can cause bradycardia and sweating and will predispose to fainting.

Fainting is common in association with dental treatment under local anaesthesia. It is triggered by emotional factors and, as a result of stimulation of the limbic cortex and hypothalamus, produces reflex vasodilatation mainly in muscle. Increased parasympathetic activity causes bradycardia which also occurs due to the sudden fall in venous return produced by the vasodilatation and the ventricles having to beat on almost empty chambers. The fall in cardiac output and severe hypotension cause a rapid loss of consciousness which can usually be corrected by putting the patient in a head-down position or elevating the legs to restore venous return.[40] If fainting occurs immediately prior to or during induction of anaesthesia the patient is in danger of brain damage or cardiac arrest if the faint is not immediately recognized, especially if the patient is in a

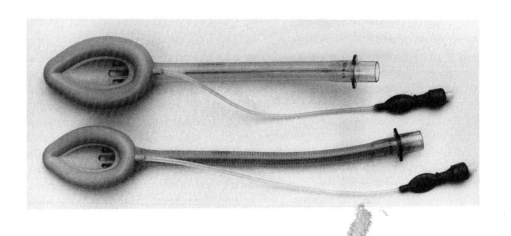

Fig. 60.5 Laryngeal airways. Above is the standard model. Below is the new version with flexible tube for use in anaesthesia for orofacial and head and neck surgery.

sitting or semi-supine position. The anaesthesia must be abandoned and the patient laid flat and given oxygen. If induction and maintenance of anaesthesia have proceeded normally then emotional fainting cannot occur but hypotension and bradycardia can result from other mechanisms.

Hypotension

Induction of anaesthesia in the dental chair whether by the inhalational or intravenous route usually produces a fall in blood pressure. The extent of the fall does not appear to be influenced by posture and the blood pressure and heart rate rise as soon as surgery commences. In fit, healthy outpatients undergoing a brief procedure these blood pressure changes are well tolerated. Occasionally brady-cardia and hypotension may occur during tooth extraction, possibly as a result of a reflex mediated through trigeminal and vagal nerve pathways. Bradycardia and nodal rhythm may also result from the use of halothane.

Dysrhythmias

Many studies have demonstrated a high incidence of cardiac dysrhythmias during dental and oral surgery. Nodal rhythm and ectopic beats, both unifocal and multifocal, are common but bradycardia much less so. Tachycardia is very common with heart rates over 120 beats/min during tooth extraction frequently seen, especially in children. Important factors in the aetiology of dysrhythmias include high preoperative catecholamine levels, surgical stimulation in a lightly anaesthetized patient and the presence of airway obstruction with hypoxia. Halothane sensitizes the myocardium to the action of circulating catecholamines and its use is associated with an incidence of dysrhythmias of about 30 per cent. Enflurane and isoflurane produce dysrhythmias in about 10 per cent of patients.

The significance of these dysrhythmias is controversial. In young, fit patients undergoing dental extractions or minor oral surgery the presence of ectopic rhythms does not lead to significant falls in cardiac output or blood pressure and the ectopic beats usually resolve spontaneously when surgical stimulation stops. In patients with unsuspected cardiac disease such as viral myocarditis or cardiomyo-pathy, however, the development of frequent, multifocal ectopic beats may be a precursor of ventricular fibrillation. Myocarditis may be present in the young patient suffering or recovering from an upper respiratory tract infection and such patients should be deferred until fully recovered. Bourne[41] reported ventricular fibrillation in four patients who died during dental anaesthesia and Coplans and Curson,[4] in a 10-year survey, identified 22 patients who died following a sudden cardiovascular collapse, often in the recovery period. Lack of ECG monitoring prevented an accurate diagnosis of the cause for the cardiac arrests but ventricular fibrillation is a strong possibility in some of these cases. The subject of cardiac dysrhythmias associated with dental anaesthesia has been extensively reviewed by Rodrigo[42] and cardiovascular changes by Thurlow.[37]

The future of dental anaesthesia

The Report of the Expert Working Party makes a number of very important recommendations with regard to the personnel, equipment and facilities which must be present wherever dental anaesthesia is practised. Essentially, there must be no difference in standards between hospital, community clinics and dental surgeries or offices. Full electronic monitoring including capnography, if intubation is employed, must be available, as well as a defibrillator. At present in the UK both non-specialist medical and dental practitioners can give dental anaesthetics but a scheme requiring refresher training is recommended and will probably be introduced following further discussions between the Royal College of Anaesthetists, the Faculty of Dentistry, the General Dental Council and other interested parties. It is likely that the result of these recommendations plus the financial cost of implementing them may reduce the number of dental practices offering general anaesthesia and ultimately all anaesthetics will be administered in hospital. The adoption of all the recom-mendations should lead to improved safety in dental anaesthesia but it may take some time to demonstrate that this is the case.

Sedation techniques for conservative dentistry

Many people, because of fear and anxiety, are not able to cooperate with dental treatment in the usual way using local anaesthesia alone. In the UK about 50 per cent of the population do not attend regularly for dental treatment and for many patients fear of pain or a previous unpleasant experience are the main reasons. Since the early work of Langa with nitrous oxide and Jorgensen using intravenous agents a variety of sedative techniques for conservative dentistry and minor oral surgery have been described. Patients have varied from being fully conscious when simple relaxation and behavioural management techniques are practised, to being almost anaesthetized when techniques of 'deep sedation' or 'ultra-light anaesthesia' are used. Unfortunately significant degrees of cardio-respiratory depression and occasional fatalities have been reported as a result of incorrectly administered and monitored dental sedation. Since dental treatment is rarely, if ever, life-saving, it is essential that sedation only be administered by skilled practitioners in a properly equipped environment and with skilled assistants in order to reduce the risk to the patient to the absolute minimum.

Definition

There have been many definitions of dental sedation. The most recent is that contained in the Report of an Expert Working Party[11] which has been accepted by the UK Department of Health. The Working Party defines simple dental sedation as:

> A carefully controlled technique in which a single intravenous drug, or a combination of oxygen and nitrous oxide, is used to reinforce hypnotic suggestion and reassurance in a way which allows dental treatment to be performed with minimal physiological and psychological stress, but which allows verbal contact with the patient to be maintained at all times. The technique must carry a margin of safety wide enough to render unintended loss of consciousness unlikely. Any technique of sedation other than as defined above be regarded as coming within the meaning of dental general anaesthesia.

The general principles of dental sedation have been well reviewed by Ryder and Wright.[43] It is not enough to rely purely on a pharmacological agent and successful sedation requires proper patient selection and management. Patients must want the dental treatment, be willing to cooperate and must understand that they will not be unconscious. The dental practitioner must provide psychological support with a strong element of hypnotic suggestion. The dental surgery should be arranged in a non-threatening way with equipment kept out of sight until the patient is sedated.

Indications

Most patients requiring sedation are those with a simple genuine fear or phobia of dental treatment. Children can present particular problems and often require very careful handling. Patients with mild systemic disorders such as controlled hypertension, angina and asthma which may be exacerbated by the stress of dental treatment represent medical indications. Patients with neuromuscular disorders such as spasticity, parkinsonism and involuntary movement conditions often wish to cooperate but physically cannot. These patients can frequently be helped by using the relaxant and anticonvulsant proper-

ties of the benzodiazepines. Dentally related problems such as gagging and trismus, persistent fainting and moderately difficult or prolonged surgery are good indications for sedation.

Contraindications

As for general anaesthesia only patients who satisfy ASA I and II criteria should be considered for sedation in the dental surgery. Patients with significant cardiorespiratory disease or neuromuscular weakness or wasting conditions should be referred to hospital. Severe psychiatric disorders or mental subnormality make cooperation difficult or impossible and general anaesthesia is more appropriate for these patients. Pregnant patients and lactating mothers should avoid sedation since the drugs may affect the developing fetus or be excreted in breast milk. Uncooperative, unwilling or unaccompanied patients are absolute contraindications and sedation should not be used to attempt very difficult or prolonged surgery. Sedation should not be attempted if the dental practitioner or his assistants have insufficient training or experience and lack the appropriate resources as recommended in the Poswillo Report.[11]

The main requirements of an ideal sedative agent or technique are shown in Table 60.2. The plethora of techniques which have been described confirms that the ideal technique does not exist and the main ones currently practised use drugs given by the oral, inhalational or intravenous routes.

Oral sedation

The powerful anxiolytic effects of the benzodiazepines make them the drugs of choice but there are problems with both the size of dose and the timing. Patients who are extremely nervous may be helped by taking a dose of diazepam before attending the surgery but must be accompanied. Alternatively faster acting benzodiazepines such as temazepam or midazolam can be given once the

Table 60.2 Desirable properties of the ideal sedative technique

Should be uncomplicated
Should be well tolerated by the patients
Should have a swift onset and recovery
Should have a predictable duration
Should produce effective anxiolysis
Should moderate the vomiting and gag reflexes
Should not obtund the protective reflexes
Should have no side effects
Should not be toxic
Should be inexpensive

patient arrives at the surgery and time allowed for them to work but the wide variability of dose–response often means that the sedation is not as effective as can be achieved by intravenous techniques.

Inhalational sedation

The technique of relative analgesia described by Langa[44] uses varying concentrations of nitrous oxide in oxygen via a nose mask and is very effective in many patients, particularly children. Usually a concentration of between 25 and 45 per cent nitrous oxide will produce a feeling of relaxation, sedation and mild euphoria. Constant reassurance by the dental practitioner is essential and local anaesthesia is required for potentially painful procedures.

Relative analgesia machines deliver a minimum of 30 per cent oxygen. It is essential that the circuitry employed allows for scavenging of waste and expired nitrous oxide since this may have effects on the dental team over a period of time. The main risk is of bone marrow depression due to inactivation of vitamin B_{12} and inhibition of methionine synthetase.[45] Nitrous oxide concentrations should be below 400 p.p.m.[46]

Subanaesthetic concentrations of isoflurane have also been shown to produce effective sedation for dentistry.[47]

Intravenous sedation

Intravenous sedation is often the most effective sedative technique, especially for extremely nervous patients. It allows careful titration of a drug against the effect on the patient under the direct observation of the sedationist. The original technique described by Jorgensen used a combination of pentobarbitone, pethidine and hyoscine which, while effective, often produced quite deep levels of sedation with prolonged recovery. Many other techniques have been described which often involve administering two or more drugs in combination. Without a doubt the intravenous route is potentially the most dangerous method of sedation, especially in inexperienced hands, since significant degrees of cardiorespiratory depression can easily be produced and fatalities have occurred.

For many years diazepam was the intravenous agent of choice but concerns about prolonged recovery, active metabolites and recirculation have led to it being superseded by midazolam which is more potent and shorter acting. Midazolam is an excellent anxiolytic with powerful anterograde amnesic effects. It must be titrated very carefully and slowly until the patient reaches a sedation 'endpoint' characterized by a delayed response to questions and commands and some slurring of speech. Verbal contact must be maintained. For most patients this will require

0.07–0.14 mg/kg which will provide useful sedation for about 45 minutes. Midazolam has no analgesic properties and local anaesthetic injections will be required for painful procedures. In sedative doses midazolam causes minimal cardiovascular depression but respiratory depression can be marked. Patients must be monitored very carefully by an assistant who is trained to do so and it is recommended that pulse oximetry be used. It has been suggested that all patients receiving intravenous sedation be given oxygen via nasal cannulae at 1–2 l/min to prevent hypoxia.

Flumazenil is a specific benzodiazepine antagonist and must be available in any situation where benzodiazepines are being administered intravenously in order to treat any inadvertent overdose. Flumazenil has also been used to reverse the residual effects of midazolam sedation following a variety of outpatient procedures including dental surgery. Provided recommended doses of midazolam have not been exceeded and a period of at least 15–30 minutes has elapsed since the administration of midazolam, significant resedation is most unlikely to occur since both agents have short plasma half-lives.[48] Patients must not be discharged too quickly, however, and must be accompanied home.

Anaesthesia for minor oral surgery and conservative dentistry

Procedures such as extraction of impacted third molar teeth and other complicated extractions, evacuation of simple cysts, exposure of unerupted teeth and other small operations make up the practice of minor oral surgery. There are also some patients who require general anaesthesia for conservative dental treatment because of mental handicap or for other reasons. Many of these operations are of short duration, taking only 20–40 minutes, and are suitable to be carried out on a day-stay basis.

Day-stay surgery

The main advantage of day-stay surgery for the patient is a rapid return home, avoiding admission to hospital while the advantage to the hospital or health authority is a considerable saving in cost. For the patient requiring minor oral surgery or dental treatment it may be possible to arrange this in the dental surgery providing all the recommended facilities including a proper recovery area are available. Increasingly, however, such surgery is being carried out in purpose-built day-stay centres. In selecting patients for treatment on a day-stay basis certain strict criteria must be met:

1. As for all outpatient general anaesthesia all patients must be fit and healthy and free from cardiorespiratory disease, satisfying ASA categories I and II.
2. The surgery should not be expected to last more than 1 hour and there should be no significant risk of complications which would require admission to hospital.
3. The patient must be able to return home after an appropriate period of recovery. The patient must be accompanied by a responsible adult and the home circumstances should be suitable for continuing postoperative care.

Screening by means of a health questionnaire and simple investigations such as full blood count and urinalysis should be performed at the time of the initial visit once a decision for day-stay surgery is made. Surgery should be carried out in the morning to allow a longer recovery period although shorter operations (15–30 minutes) can be undertaken in the early afternoon depending on when the day-stay unit closes. It is acceptable to allow clear fluids by mouth up to 4 hours prior to surgery and, except in the very anxious patient, it is probably better to avoid premedication so as not to delay recovery.

For minor oral surgery and conservative dentistry the operator requires easy access to the surgical field. Frequently several instruments, packs and suction apparatus are in the mouth at the same time and it is easy for the tongue to be displaced backwards. The anaesthetist requires a clear airway at all times with protection of the larynx and tracheobronchial tree against aspiration of blood, saliva, tooth fragments or water from the drill or irrigating solutions. The anaesthetic technique should be one which produces a rapid recovery and minimal postoperative morbidity.

The majority of patients requiring minor oral surgery are adults or older children and induction by the intravenous route is preferred. In younger children an inhalational induction may be used but an indwelling intravenous cannula must be inserted once the child is asleep. Propofol is often the intravenous agent of choice because it possesses excellent recovery characteristics[26,49] and has a lower incidence of postoperative nausea and vomiting when compared with other intravenous agents.[50] Maintenance of the airway is usually by tracheal intubation although it is probable that the laryngeal mask will be used with increasing frequency for this type of minor surgery.[39] Techniques using nasopharyngeal airways or the nasal mask have been described and in the hands of experts can be very successful.[51] With these techniques very careful

packing and suction must be employed and meticulous attention paid to the airway as well as the use of full monitoring.

The nasal route is usually preferred for tracheal intubation although for shorter operations involving only one side of the mouth an oral tube may be acceptable. Muscle relaxation for intubation can be achieved using suxamethonium but the well known problem of post-operative muscle pains in ambulant patients makes it a less than ideal agent. The muscle pains can be reduced in severity but not abolished by the prior administration of a small dose of non-depolarizing muscle relaxant. Alternatively, intubation can be performed using deep anaesthesia with a volatile agent in the spontaneously breathing patient or by using a shorter-acting non-depolarizing relaxant such as atracurium or vecuronium. Another interesting possibility is intubation using propofol, either by itself or in combination with a short-acting opioid such as alfentanil. Propofol produces good relaxation of the upper airway and sufficient suppression of reflexes to allow intubation.[52] For nasal tracheal intubation preformed plastic tubes should be used (Fig. 60.6).

Intubation can be performed 'blind' or using a laryngoscope to help guide the nasal tube into the larynx. Following intubation the pharynx should be packed with either a moistened gauze or special preformed throat pack. Maintenance of anaesthesia is usually achieved by means of a spontaneously breathing technique using nitrous oxide and a volatile agent. Halothane has the disadvantage of a slower recovery when compared with more modern agents and is associated with an incidence of cardiac dysrhythmias of about 30 per cent in several studies. Enflurane and isoflurane have both proved to be satisfactory in minor oral surgery and cardiac dysrhythmias occur in less than 5 per cent with both agents. They have similar recovery characteristics[53] although recovery from isoflurane is slightly faster.[54] Techniques using controlled ventilation with muscle relaxation and a volatile anaesthetic agent or intravenous opioid have been described for day-stay oral surgery. When compared with patients breathing halothane spontaneously the incidence of cardiac dysrhythmias is much less and recovery is faster. However, there is still a high incidence of postoperative morbidity in the form of nausea, vomiting, drowsiness and dizziness with both techniques.[55,56] Atracurium and vecuronium are the relaxants of choice for day-stay patients if suxamethonium is to be avoided because they are relatively short acting with rapid offset of neuromuscular blockade. Reversal of effect with neostigmine is easily achieved, provided an adequate period of time has been allowed.[57,58]

Once the surgery has been completed the pack should be removed and the pharynx carefully sucked out. Extubation is best performed with the patient turned on to one side. A period of 2–4 hours should be allowed for full recovery and

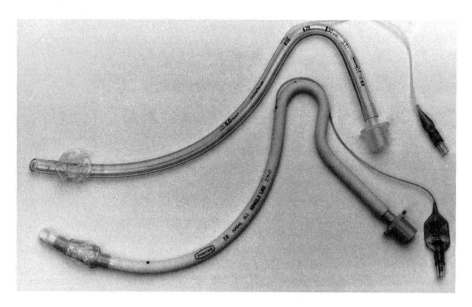

Fig. 60.6 Preformed cuffed nasotracheal tubes.

the patient must be carefully assessed before discharge. Morbidity in the form of sore throat, headache, nausea and vomiting seem to be fairly common in the initial recovery period and may require treatment. It should be remembered that minor oral surgery using local anaesthesia alone is associated with a surprisingly high incidence of these symptoms which may not, therefore, be caused entirely by a general anaesthetic.[59] Paracetamol is usually effective for moderate pain due to dental extractions but stronger analgesics such as the non-steroidal anti-inflammatory agents suprofen, diclofenac or ketorolac may also be used. Opiate analgesics including codeine are often not very effective in alleviating pain following oral surgery and are better avoided in day-stay patients. On discharge the patient must be accompanied and be given written postoperative instructions which must include information about what to do in the event of any bleeding or abnormal swelling.

Anaesthesia for major oral and faciomaxillary surgery

Patients requiring major oral or faciomaxillary surgery present a wide variety of conditions and diseases ranging from relatively simple operations to major resections with reconstruction. The main problem for both the anaesthetist and surgeon is the close proximity of the airway to the operative site and in some cases the airway itself may be affected by the disease process. An additional consideration for the anaesthetist is that the airway may still present a problem at the end of the operation, especially if the mandible and maxilla have been wired together.

Operations can be classified as elective or emergency procedures. Elective operations are either:

1. corrections for a variety of congenital orofacial deformities or abnormalities and will usually involve maxillary or mandibular osteotomies; or
2. resections for tumours of the mandible or maxilla or soft tissue structures which may require extensive surgery with reconstruction.

Emergency operations follow orofacial trauma and may be simple, as in nasal bone fractures, or complex, if extensive damage to the facial skeleton and soft tissues has occurred. Orofacial infections causing swelling and trismus may require surgical drainage.

Elective surgery

Patient assessment

Most operations for correction of congenital deformities are delayed until late teenage or early adulthood when the facial bones have stopped growing. In infancy and childhood, corrections required for cleft lip and palate form the largest single group of indications for operation. Occasionally neonates may have such severe degrees of mandibular hypoplasia, as in the Pierre Robin syndrome, that a tracheostomy may be necessary to maintain the airway. In older children mandibular osteotomies and bone grafting for hemifacial microsomia may be required. Adult patients should be assessed in the same way as for any major operation. A full medical history must be taken and physical examination carried out. Any significant medical problems must be treated and corrected as far as possible before operation. When assessing patients with tumours of the orofacial structures attention must be paid to any effects the tumour or associated swelling may have on the upper airway. There may be pain and trismus which limits mouth opening and the patient may be in a poor nutritional state if he or she has not managed to eat properly for some time. Patients with such tumours are usually in the older age groups and there is a link between squamous carcinoma of the mouth with both heavy cigarette smoking and heavy alcohol intake.

Operations for major orthognathic corrections or resections for tumours are often very extensive procedures and are associated with significant blood loss. However, major advances in faciomaxillary surgical techniques have done much to reduce operation time. The adoption of internal fixation of fractures or osteotomies using miniplates now means that it is seldom necessary to wire the jaws together or use external fixation. Nevertheless, it is essential that the anaesthetist confers with the surgeon so that the extent and nature of the operation is fully understood. In particular the anaesthetist must know if it is likely that the jaws will be wired together.

The airway must be carefully assessed preoperatively and a plan for securing the airway formulated.[60] A difficult intubation should always be planned for and never come as a surprise in this area of anaesthesia (see Chapter 35). Limited mouth opening, prominent upper teeth, a small receding lower jaw and a short neck with limited movements should raise suspicions and radiographs of the upper airway may be helpful. If it is suspected that intubation will be difficult then it is important to bear in mind certain points:

1. Never assume that it will be possible to inflate the lungs with a bag and facemask if intubation is impossible.
2. Never paralyse the patient until it has been fully established that it is possible to inflate the lungs with a bag and facemask.
3. Never assume that because the patient has been successfully intubated in the past, even recently, that intubation will be possible on the present occasion.
4. Always have a full range of equipment for dealing with a difficult intubation available including facilities for emergency tracheostomy or cricothyrotomy.

Airway management

The choice of technique for intubation will depend on several factors:

1. whether the patient has a normal or known difficult airway and in particular whether there is limited or fixed mouth opening;
2. what the requirements of the operation are and whether an oral or nasal tube is appropriate;
3. whether operative fixation with jaw wiring will be required for a significant postoperative period.

The skills and experience of the anaesthetist are also vitally important factors and the anaesthetist must always choose a course of action which is prudent and has patient safety as a prerequisite.

The normal airway

If it can be safely predicted that the airway is normal and there are no intubation problems then anaesthesia can proceed in a routine way as for any major surgery. Intubation is performed orally or nasally using muscle relaxation and laryngoscopy as required. If an oral tube is used then it should be of the non-kinkable armoured type. It is usual to place a throat pack to prevent blood and other material entering the stomach or settling above the tracheal cuff in the trachea.

The difficult airway

The management of a known or suspected difficult intubation is a major topic and is discussed in detail elsewhere (see Chapter 35). The main techniques which may be used can be divided into those used in the awake or sedated patient and those under general anaesthesia.

Patient awake or sedated

If the mouth can be opened normally or to a reasonable degree then laryngoscopy and intubation may be attempted using local anaesthesia. The mouth and larynx can be anaesthetized using a combination of oral sprays, gels or lozenges, superior laryngeal nerve blocks and cricothyroid puncture and injection of local anaesthetic. If mouth opening is limited the technique of retrograde catheterization described by Waters[61] may be tried. A cricothyroid puncture is made and a catheter passed up through the vocal cords and out through the mouth or nose. The catheter is then used as a guide for a tracheal tube. If mouth opening is extremely limited or fixed or there is gross distortion of the anatomy, tracheostomy under local anaesthetic must be considered. The method of choice, however, in most situations where awake intubation is required is to use a fibreoptic bronchoscope or laryngoscope, especially if a tracheostomy is not required postoperatively. Skill and constant practice in the use of the fibreoptic instruments is essential. It is usually easier to use the nasal route since the approach to the larynx is less angulated than via the oral route but the latter route can be used if the surgery requires an oral tube. Modern fibreoptic instruments come in various sizes but the thinnest have a diameter of less than 3 mm and will pass through tubes of 4 mm internal diameter.

Intubation under general anaesthesia

Children and uncooperative adults will often require general anaesthesia for intubation or tracheostomy. The presence of sepsis in the mouth may render local anaesthetics ineffective and hence prevent awake intubation. If premedication is indicated then anxiolytics may be prescribed but respiratory depressant drugs should be avoided. An anticholinergic is useful to dry secretions and if the nasal route is going to be used for intubation it is helpful to anaesthetize the nasal cavity with cocaine paste or solutions beforehand. The decision to use general anaesthesia must never be taken lightly and a technique which will 'fail-safe' should be chosen. The actual technique will depend on whether the patient can open his or her mouth.

After preoxygenation and if the mouth opens normally or at least wide enough to insert an oral airway, the induction can proceed using either an inhalation agent only or with the addition of small doses of an intravenous agent. Once anaesthesia is deep enough and inflation of the lungs by bag and mask confirmed a laryngoscope can be inserted and the airway assessed. If a sufficient view of the larynx or posterior arytenoids can be obtained then intubation can proceed under deep spontaneous ventilation or after a short-acting muscle relaxant has been given. If the larynx cannot be viewed then it may be possible to intubate using a stilette or bougie to guide the tracheal tube. Alternatively the fibreoptic scope can be used in combination with a facemask or modified oral airway.[62,63] Another possibility is to use the laryngeal mask airway if the mouth will open wide enough. The use of the laryngeal mask in difficult intubation has been described by Brain[64] and intubation through the mask by Heath and Allagain.[65] The fibreoptic scope can also be passed through the laryngeal mask and used to place a guide tube for tracheal intubation.[66]

If the mouth cannot be opened or if opening is very limited then general anaesthesia should only be considered if other techniques are definitely contraindicated. Following preoxygenation induction should be by the inhalational route which will allow the airway to be assessed constantly as anaesthesia is deepened. If there is a tendency for obstruction to occur it may be possible to relieve this by careful positioning of the head and the insertion of a nasopharyngeal airway. If the obstruction cannot be relieved or gets worse as anaesthesia is deepened then intubation under deep anaesthesia may have to be abandoned and an alternative technique employed. This may have to be tracheostomy under light anaesthesia with local anaesthesia or fibreoptic intubation using sedation with local anaesthetic.

The nasal or oral route can be used depending on whether the mouth can be opened sufficiently if a deep enough level of anaesthesia for intubation can be reached. If the nasal route is chosen then a blind technique may be tried or alternatively the nasal tube can be advanced to the pharynx and a fibreoptic instrument passed down the tube and on into the trachea. Anaesthesia can be maintained during this manoeuvre by connecting the anaesthetic circuit to the nasal tube by means of a special angle connector through which the fibrescope can be passed.

Maintenance of anaesthesia

Once the airway has been secured by intubation anaesthesia can be maintained as for any major operation. It is usual to employ artificial ventilation with muscle relaxation supplemented by a neuroleptanalgesic technique or a volatile anaesthetic agent. Careful positioning of the patient with a head-up tilt will encourage venous drainage from the operative site but the legs should be elevated to enhance venous return. If significant blood loss is expected then induced hypotension will help to reduce it but full monitoring including continuous direct arterial pressure, central venous pressure, end tidal carbon dioxide, inspired and expired oxygen concentrations and pulse oximetry should be used.

For long operations a urinary catheter should be inserted and every effort made to prevent a fall in body temperature. Inspired gases should be warmed and humidified, all intravenous fluids warmed and a heating mattress placed under the patient. Hypotension, hypocapnia, hypovolaemia and hypothermia must all be avoided if microvascular surgery with the use of free flaps is part of the operation.[67] If the surgery involves the floor of the mouth, tongue or pharynx with the possibility of oedema or haematoma formation and a postoperative risk to the airway then a tracheostomy should be performed. With modern surgical techniques it is often possible to retain a rim of mandibular bone during a resection for tumour, thus maintaining mandibular continuity, but if there is any doubt about the airway, especially if an emergency reintubation in the postoperative period is likely to be difficult, then a tracheostomy is indicated.

Extubation

The use of internal fixation means that it is now seldom necessary to wire the jaws together and if no problem with the airway is anticipated extubation can take place in the usual way. If there is any possibility of airway obstruction post-extubation then it is wise to let the patient wake sufficiently to remove the tube spontaneously. If the jaws are wired then wire cutters must be kept immediately to hand and must stay with the patient throughout the postoperative period.

Emergency surgery

Maxillofacial injuries range from relatively minor fractures of the nasal bones or zygomatic arch to major fractures of the maxilla and mandible with extensive soft tissue damage. Severe mandibular and midface injuries are often associated with significant haemorrhage and oedema which may cause acute airway obstruction. If the patient is conscious then it may be sufficient to nurse the patient in the prone position to keep the mouth clear of blood, saliva or bits of teeth or dentures. If the patient is unconscious then securing the airway may be an acute emergency. Insertion of an oral airway and administration of oxygen must be an immediate procedure and may be necessary in order to transport the patient to hospital. The patient may even be already intubated on arrival at hospital if paramedic ambulancemen have been called to the scene of the accident. If not then emergency intubation or tracheostomy may be required.

If the mouth can be opened then intubation following preoxygenation, an intravenous agent and suxamethonium should be tried. Cricoid pressure must be applied since the stomach will usually contain blood and food. If the mouth cannot be opened then a tracheostomy under local anaesthesia may be necessary. Alternatively it may be possible to intubate using the fibreoptic instrument. Extensive maxillofacial trauma is often associated with multiple injuries to the head, chest and abdomen. Life-threatening injuries must always be treated first and maxillofacial surgery delayed until the patient is stable unless urgent surgery to control bleeding is required. The more widespread use of internal fixation means that patients with only mild or moderate injuries may not require a tracheostomy but this should always be considered in major injuries. Indeed if it is thought that the patient might need a tracheostomy then he or she probably does.

REFERENCES

1. Laird WRE. 'The Yankee Dodge': some new observations on the discovery of anaesthesia. *British Dental Journal* 1990; **169**: 217–19.

2. Padfield A. The future of general anaesthesia in the dental surgery: discussion paper. *Journal of the Royal Society of Medicine* 1989; **2**: 30–2.

3. Parbrook GD. Death for anaesthesia in the general and community dental services? *British Journal of Anaesthesia* 1986; **58**: 369–70.

4. Coplans MP, Curson I. Deaths associated with dentistry. *British Dental Journal* 1982; **153**: 357–62.

5. Tomlin P. Death in outpatient dental anaesthetic practice. *Anaesthesia* 1974; **29**: 551–70.

6. Allen NA, Dinsdale RCW, Reilly CS. A survey of general anaesthesia and sedation in dental practice in two cities. *British Dental Journal* 1990; **169**: 168–72.

7. Shirlaw PJ, Scully C, Griffiths MJ, Levers BGH, Woodwards RTM. General anaesthesia, parenteral sedation and emergency drugs and equipment in general dental practice. *Journal of Dentistry* 1986; **14**: 247–50.

8. The Wylie Report. *British Dental Journal* 1981; **150**: 385–8.

9. The Seward Report. *British Dental Journal* 181; **150**: 389–91.

10. The Spence Report. *British Dental Journal* 1981; **150**: 392–5.

11. *General anaesthesia, sedation and resuscitation in dentistry: report of an Expert Working Party.* London: Department of Health, 1990.

12. Crawford AN. The use of nitrous-oxide inhalation sedation with local anaesthesia as an alternative to general anaesthesia for dental extractions in children. *British Dental Journal* 1990; **168**: 395–8.

13. O'Sullivan EA, Curzon MEJ. The efficacy of comprehensive dental care for children under general anaesthesia. *British Dental Journal* 1991; **171**: 56–8.

14. Goldman V. Symposium: anaesthesia for the ambulant patient. *British Dental Journal* 1964; **116**: 15–25.

15. Padfield A. Blood glucose concentrations in children undergoing outpatient dental anaesthesia. *British Journal of Anaesthesia* 1984; **56**: 1225–8.

16. *Standards of monitoring during anaesthesia and recovery.* London: Association of Anaesthetists of Great Britain and Ireland, 1988.

17. Young ER, DelCastilho R, Patell M, Kestenburg SH. A scavenging system developed for the Magill anaesthetic circuit for use in the dental office. *Anesthesia Progress* 1990; **37**: 252–7.

18. Blumgart C, Hargrave SA. Modified parallel 'Lack' breathing system for use in dental anaesthesia. *Anaesthesia* 1991; **47**: 993–5.

19. Humphrey D. Eliminating pollution in paediatric and dental anaesthesia. (Letter) *Anaesthesia* 1992; **47**: 640.

20. Strunin L, Strunin JM, Phipps JA, Corall IM. A comparison of halothane and enflurane (ethrane) for out-patient dental anaesthesia. *British Dental Journal* 1979; **147**: 299–301.

21. Simmons M, Miller CD, Cummings GC, Todd JG. Outpatient paediatric dental anaesthesia. A comparison of halothane, enflurance and isoflurane. *Anaesthesia* 1989; **44**: 735–8.

22. Cattermole RW, Verghese C, Blair IJ, Jones CJH, Flynn PJ, Sebel PS. Isoflurane and halothane for outpatient dental anaesthesia in children. *British Journal of Anaesthesia* 1986; **58**: 385–9.

23. Sampaio MM, Crean PM, Keilty SR, Black GW. Changes in oxygen saturation during inhalation anaesthesia in children. *British Journal of Anaesthesia* 1989; **62**: 199–201.

24. McAteer PM, Carter JA, Cooper GM, Prys-Roberts C.

Comparison of isoflurane and halothane in outpatient dental anaesthesia. *British Journal of Anaesthesia* 1986; **58**: 390–3.

25. MacKenzie N, Grant IS. Comparison of the new emulsion formulation of propofol with methohexitone and thiopentone for inducing anaesthesia in day cases. *British Journal of Anaesthesia* 1985; **57**: 725–31.

26. Wells JKG. Comparison of ICI 35868, etomidate and methohexitone for day-case anaesthesia. *British Journal of Anaesthesia* 1985; **57**: 732–5.

27. Logan MR, Duggan JE, Levack ID, Spence AA. Single-shot anaesthesia for outpatient dental surgery. Comparison of 2,6-di-isopropyl phenol and methohexitone. *British Journal of Anaesthesia* 1987; **59**: 179–83.

28. Pollock JSS, MacKenzie NM. General anaesthesia in the dental surgery: a comparison of propofol and methohexitone. *British Dental Journal* 1992; **173**: 207–9.

29. Hargrave SA, Stafford MA, Shetty GR. A comparison of the cardiovascular effects of methohexitone and propofol in outpatient dental anaesthesia. *Proceedings of the Association of Dental Anaesthetists* 1988; **6**: 22–3.

30. Bourne JG. Fainting and cerebral damage. *Lancet* 1957; **2**: 499–505.

31. Thurlow AC. The postural evolution. *Anaesthesia* 1981; **36**: 565 (Abstract).

32. Jones JG. Pulmonary function. In: Coplans MP, Green RA, eds. *Monographs in anaesthesiology. Anaesthesia and sedation in dentistry.* Amsterdam: Elsevier Science Publishers, 1983: 19–49.

33. Evans CS, Dawson ADG. Oxygen saturation during general anaesthesia in the dental chair. A comparison of the effect of position on saturation. *British Dental Journal* 1990; **168**: 157–60.

34. Bone ME, Galler D, Flynn PJ. Arterial oxygen saturation during general anaesthesia for paediatric dental extractions. *Anaesthesia* 1987; **42**: 879–82.

35. Walsh JF. Training for day-case dental anaesthesia. Oxygen saturation during general anaesthesia administered by dental undergraduates. *Anaesthesia* 1984; **39**: 1124–7.

36. Forsyth WD, Allen GD, Everett GB. An evaluation of the cardiorespiratory effects of posture in the dental outpatient. *Oral Surgery, Oral Medicine and Oral Pathology* 1972; **34**: 562–80.

37. Thurlow AC. Cardiovascular effects. In: Coplans MP, Green RA, eds. *Monographs in anaesthesiology. Anaesthesia and sedation in dentistry.* Amsterdam: Elsevier Science Publishers, 1983: 1–18.

38. Bailie R, Barnett MB, Fraser JF. The Brain laryngeal mask. A comparative study with the nasal mask in paediatric dental outpatient anaesthesia. *Anaesthesia* 1991; **46**: 358–60.

39. Young TM. The laryngeal mask in dental anaesthesia. *European Journal of Anaesthesiology* 1991; **4**: 53–9.

40. Bourne JG. The common fainting attack. Its danger in dentistry. *British Dental Journal* 1980; **149**: 101–4.

41. Bourne JG. Deaths with dental anaesthetics. *Anaesthesia* 1970; **25**: 473–81.

42. Rodrigo CR. Cardiac dysrhythmias with general anesthesia during dental surgery. *Anesthesia Progress* 1988; **35**: 102–15.

43. Ryder W, Wright PA. Dental sedation. A review. *British Dental Journal* 1988; **165**: 207–16.

44. Langa H. *Relative analgesia in dental practice.* London: WB Saunders, 1976.

45. Sweeney, B, Bingham B, Amos, R, Amess J, Petty AC, Cole PV. Toxicity of bone marrow in dentists exposed to nitrous oxide. *British Medical Journal* 1985; **291**: 567–9.

46. Yagiela JA. Health hazards and nitrous oxide: a time for reappraisal. *Anesthesia Progress* 1991; **38**: 1–11.

47. Parbrook GD, James J, Braid DP. Inhalational sedation with isoflurane: an alternative to nitrous oxide sedation in dentistry. *British Dental Journal* 1987; **163**: 88–92.

48. Whitwam JG, Hooper PA. Flumazenil – the first benzodiazepine antagonist and some implications in the dental surgery. *SAAD Digest* 1988; **7**: 97–104.

49. O'Toole DP, Milligan KR, Howe JP, McCollum JSC, Dundee JW. A comparison of propofol and methohexitone as induction agents for day case isoflurane anaesthesia. *Anaesthesia* 1987; **42**: 373–6.

50. Gunawardene RD, White DC. Propofol and emesis. *Anaesthesia* 1988; **43** (supplement): 65–7.

51. Buxton JD. Day-stay clinic. In: Coplans MP, Green RA, eds. *Monographs in anaesthesiology. Anaesthesia and sedation in dentistry.* Amsterdam: Elsevier Science Publishers, 1983: 323–41.

52. Keaveny JP, Knell PJ. Intubation under induction doses of propofol. *Anaesthesia* 1988; **43** (supplement): 80–1.

53. Shultz RE, Richardson DD, Nespeca JA. Comparison of recovery times of isoflurane and enflurane as a sole anesthetic agent for outpatient oral surgery. *Anesthesia Progress* 1989; **36**: 13–14.

54. Valanne JV, Kortilla K. Recovery following general anesthesia with isoflurane or enflurane for outpatient dentistry and oral surgery. *Anesthesia Progress* 1988; **35**: 48–52.

55. Heneghen C, MacAuliffe R, Thomas D, Radford P. Morbidity after outpatient anaesthesia. A comparison of two techniques of endotracheal anaesthesia for dental surgery. *Anaesthesia* 1981; **36**: 4–9.

56. Sale JP, Poobalasingam N, Dalal SD. Controlled ventilation in dental outpatients. Controlled ventilation with atracurium and alfentanil analgesia compared with halothane. *Anaesthesia* 1985; **40**: 3–7.

57. Pearce AC, Williams JP, Jones RM. Atracurium for short surgical procedures in day patients. *British Journal of Anaesthesia* 1984; **56**: 973–6.

58. Pearce AC, Hodge M, Jones RM. Vecuronium in day-stay surgery. (ARS) *British Journal of Anaesthesia* 1984; **56**: 794P.

59. Muir VMJ, Leonard M, Haddaway E. Morbidity following dental extraction: a comparative survey. *Anaesthesia* 1976; **31**: 171–80.

60. Berwick EP. Anaesthesia for faciomaxillary surgery. In: Coplans MP, Green RA, eds. *Monographs in anaesthesiology. Anaesthesia and sedation in dentistry.* Amsterdam: Elsevier Science Publishers, 1983: 363–93.

61. Waters DJ. Guided blind endotracheal intubation for patients with deformities of the upper airway. *Anaesthesia* 1963; **18**: 159–62.

62. Rogers SN, Benumof JL. New and easy techniques for

fiberoptic endoscopy-aided tracheal intubation. *Anesthesiology* 1983; **59**: 569–72.

63. Frankel R, Mizrahi S, Simon K. A modified airway to assist fibreoptic orotracheal intubation. (Letter) *Anaesthesia* 1990; **45**: 249–50.

64. Brain A. Three cases of difficult intubation overcome by the laryngeal mask airway. *Anaesthesia* 1985; **40**: 353–5.

65. Heath ML, Allagain J. Intubation through the laryngeal mask. A technique for unexpected difficult intubation. *Anaesthesia* 1991; **46**: 545–8.

66. Hasham F, Kumar CM, Lawler PGP. The use of the laryngeal mask airway to assist fibreoptic orotracheal intubation. (Letter) *Anaesthesia* 1991; **46**: 891.

67. MacDonald DJF. Anaesthesia for microvascular surgery. A physiological approach. *British Journal of Anaesthesia* 1985; **57**: 904–12.

Chapter 61

Anaesthesia for Ear, Nose and Throat Surgery

Allan C. D. Brown

Assessment of the patient for anaesthesia	Special anaesthetic techniques
Preparation of the patient for anaesthesia	The post-anaesthesia recovery unit
Conduct of anaesthesia	Conclusion

With the introduction of general anaesthesia on 16 October 1846, the first main obstacle to modern surgical practice was overcome. It is of interest that the first patient to officially submit to this new 'Yankee dodge' was a young man of 19 years, Gilbert Abbott, who had a tumour removed from under his jaw, probably a tuberculoma.[1] On the strength of this somewhat doubtful justification, our ENT colleagues have claimed to have been there at the beginning of the specialty of anaesthesia!

Prior to general anaesthesia, operations were few and many patients could expect to die on the operating table, of 'surgical shock' – not the hypovolaemic variety but rather as the result of an autonomic storm in the absence of effective intraoperative pain relief. Even with anaesthesia, the death rates from postoperative wound infection had to await the introduction of, first, antiseptic and then aseptic surgical technique before improvement was evident. Then, as the number and scope of operations increased, intraoperative haemorrhage became a significant contributor to surgical mortality. This problem was resolved only with Landsteiner's description of human blood groups and the subsequent introduction of organized blood transfusion services following the Spanish Civil War. The final advance that allowed modern surgery to develop its full potential was the introduction of the antimicrobial and then the true antibiotic agents from the 1930s onwards, which reduced surgical infections to a manageable level.

Deaths resulting from the anaesthetic process itself had been considered a small price to pay for the boon of patient insensibility and the expanded practice of surgery. However, by the mid 1930s, anaesthetic deaths had assumed increasing significance as other perioperative causes of death were progressively overcome. This prompted the organization of formal training in anaesthesia and the identification of that fundamental corpus of knowledge and expertise required to establish a specialty within the practice of medicine. Even so, in the 1950s Dornette and Orth reported a mortality rate in the USA of 1/2500 resulting from anaesthesia, which they considered amounted to a public health problem.[2]

Further studies suggested that anaesthetic causes of death could be attributed to three broad categories: human, patient and technological. Of the three, human factors were thought to play a decisive part in approximately 70 per cent of the incidents studied. With continuing improvement in anaesthesia training, pharmacology and equipment, Davies and Strunin reported 30 years later an improvement in anaesthetic death rates to about 1/10 000 which they felt was the experience in most Western countries.[3] However, a recent study of closed medical liability claims by the American Society of Anesthesiologists showed that human factors were still the major contributing factor to anaesthetic deaths; problems with the airway accounted for more than 30 per cent of the adult cases studied and 43 per cent in children.[4]

ENT patients are particularly at risk for airway problems because the specialty encompasses the treatment of the pathology and dysfunctions of the aerodigestive tract. Tumours, trauma, scarring and infection can all lead to varying degrees of airway obstruction requiring surgical intervention. These conditions can also interfere with the anaesthetist's efforts to secure a safe airway using standard laryngoscopy techniques. For some operations the ENT surgeon is actively competing with the anaesthetist for space within the airway, thereby increasing the risk of airway compromise and accidental disconnection of the circuit. The positioning of the patient and the use of closely applied surgical head drapes may compound such problems by delaying the detection and correction of any problems that occur. Therefore the major concerns with all anaesthetics for ENT operations may be summarized:

1. A reliably secure airway with minimal facial or tissue distortion.
2. A circuit set-up that permits the best possible surgical access.
3. Reliable physiological monitoring.

4. A contingency plan to deal with intraoperative failures with the original plans for 1, 2 and 3 above.

To achieve a safe routine practice of anaesthesia for the specialty of ENT surgery requires an unusually high level of cooperation between surgeon and anaesthetist. Fortunately for both patient and anaesthetist, such cooperation is a well established tradition within the specialty.

Assessment of the patient for anaesthesia

The detailed evaluation of a patient for anaesthesia is dealt with in Chapters 29 and 32. Here, only a general summary is offered, together with reference to some particular points related to patients undergoing ENT surgical procedures.

The American Society of Anesthesiologists (ASA) has sponsored several attempts to develop a classification of anaesthetic risk according to pre-existing diseases and conditions of the patient as well as the operation to be undertaken. Its efforts have met with only partial success, as a positive correlation between these factors is not evident.[5] What has developed from this work is the ASA physical status classification, which may or may not directly reflect the anaesthetic risk:

ASA 1 A normal healthy patient.
ASA 2 A patient with a mild systemic disease.
ASA 3 A patient with a severe systemic disease that limits activity but is not incapacitating.
ASA 4 A patient with an incapacitating systemic disease that is a constant threat to life.
ASA 5 A moribund patient not expected to survive 24 hours with or without operation.

Each class may be modified with the letter 'E' denoting an emergency presentation with all that that implies for the preparation of the patient prior to operation.

The problems associated with the preoperative assessment of risk are evident when considering the case of a young healthy adult with a minor deformity of the second branchial arch structures which, while not being apparent in everyday life, may pose significant anatomical problems for safe tracheal intubation. Similarly, the young healthy patient with a family history of malignant hyperthermia, although at possibly considerable risk from an imprudent anaesthetic, could still be classed as an ASA class 1 patient. Therefore the preoperative anaesthetic assessment is slanted not only towards the assessment of the general physical status of the patient, which at best is an indicator of the general state of the patient's integrated physiological systems, but also towards detecting specific problems

posed by coexisting conditions and disease, the requirements of the operation, any significant medication being taken, as well as problems and points in the history suggesting a true anaesthetic risk. Of particular significance is a history of previous problems with anaesthesia or a difficult intubation.

Assessment of physical status requires answers to two major questions: (1) what is the patient's general physiological ability to respond to stress (both anaesthetic and surgical) and (2) what disease conditions exist concurrently which might modify or limit that response? The history of the patient's response to the level of physical activity undertaken in everyday living is usually the best guide to the patient's probable response to the iatrogenic stresses of the operating theatre. However, some individuals appear to lead perfectly satisfying lives without exposing themselves to a stress sufficient to gauge the appropriateness of the response from the history. Asking the patient to walk with the interviewer or climb a flight of stairs is usually sufficient to form a valid clinical impression.[6] The heart rate and rhythm, blood pressure, respiratory pattern and general state of distress may suggest the need for further investigation. Pre-existing systemic disease would modify one's enthusiasm and the number of flights of stairs, but in general the approach to physical assessment is the same. It is the patient's integrated response to stress that is being assessed even if modified by disease, rather than any specific problems of management posed by the disease itself.

The presence of systemic disease requires answers to different questions:

1. Is the disease process under the best medical control possible or, in the case of an emergency situation, has everything reasonable been attempted to stabilize the patient's condition?
2. What drugs are being used to control the disease and do any of them interact with anaesthetic and adjuvant drugs?
3. Does the disease process pose any specific perioperative problems for surgical or anaesthetic management or does it impose any limits upon what may be undertaken in terms of positioning, technique and duration of the procedure?
4. Who will manage the disease during the perioperative period?
5. Have the expected gains of the surgical intervention been appropriately weighed against the estimated risks of anaesthesia associated with the systemic disease under this degree of control?
6. Does the anaesthetic technique pose any problems for the perioperative management of the disease?

Patients with a poor physical status due to disease generally have impaired homoeostatic control and are particularly vulnerable to the depressant effects of inhalational anaesthetic drugs given as the sole anaesthetic agent. This tends to dictate the use of balanced anaesthetic techniques with muscle relaxant drugs. In some ENT operations where the surgeon relies on the use of a nerve stimulator this might persuade the surgeon to delay elective surgery until the patient's physical status may be improved. If the anaesthetist, in consultation with the patient and the surgeon, elects to proceed with conduction anaesthesia or infiltration anaesthesia with the anaesthetist standing by, a new set of problems and limitations emerge. Will the patient's condition permit correct positioning for surgery? Can the patient breathe while lying flat? Can the patient tolerate lying on the operating table for sufficient time to allow completion of the proposed procedure?

These alternatives are not particularly promising in the face of significant pulmonary, cardiac or arthritic conditions. It is worth emphasizing that, when a patient for elective surgery with a significant systemic disease is not considered to be in the best possible medical condition, local–standby anaesthesia is not an appropriate choice to 'get away with surgery' on an improperly prepared patient. Long operations under local infiltration anaesthesia frequently require local anaesthetic drug dosages approaching or exceeding toxic levels. Even then, general anaesthesia may still be required if the patient cannot tolerate the duration of the procedure, thereby subjecting the patient to two sets of risks that could have been avoided. No anaesthetist should allow this situation to occur because of a misplaced concern for either the patient or surgical convenience.

The anaesthetist should be concerned not only about the effects of systemic disease on the conduct of anaesthesia but also about the impact of the anaesthetic on the subsequent course and management of the disease. The respiratory cripple imprudently paralysed and ventilated for surgery may not be able to be weaned from the ventilator postoperatively. Anaesthetic management may create problems in re-establishing the myasthenic and the diabetic on stable medication postoperatively. Is the withdrawal of chronic medication to permit the safe use of one anaesthetic technique really in the patient's interest when another, perhaps less ideal, technique would permit its continuation? These are all necessarily matters of judgement, to be decided only after consultation between the anaestheist, the surgeon, the appropriate medical specialist and the patient.

Systemic medication should always be reviewed at the preoperative visit, with particular attention to possible interactions with anaesthetic drugs. It is also important to remember that some patients will not volunteer information concerning 'over-the-counter' drugs because, as they were not prescribed by a physician, they are not considered to be 'real' drugs. Similarly, the patient admitted for surgery with a serious systemic disease may neglect to mention drugs being taken for an unrelated older chronic condition.

Examples of this encountered in practice include elderly patients who have had their mild hypertension managed successfully for many years with reserpine, and those patients who may be taking long-acting anticholinesterase drugs such as ecothiopate for glaucoma. Both drugs have significant interactions with drugs used in anaesthetic practice. Reserpine blocks peripheral adrenergic neurons by depleting synaptic noradrenaline (norepinephrine) stores; this may be associated with supersensitivity to direct-acting and insensitivity to indirect-acting sympathomimetic amines which may be used to counteract hypotension under anaesthesia.[7,8] Ecothiopate iodide, a long-acting anticholinesterase, interferes with the hydrolysis of the depolarizing muscle relaxant suxamethonium (succinylcholine) by pseudocholinesterase, thereby prolonging its clinical action significantly.[9]

As a general rule, if the patient is sufficiently ill to require systemic medication while awake, the same medication is required under anaesthesia. The anaesthetist's preoperative assessment should allow modifications to a proposed anaesthetic technique to take account of the actions and interactions expected from these systemic drugs. This is particularly true for most antihypertensive drugs and those used to control cardiac and pulmonary disease. However, a few drugs have potential interactions so devastating that their withdrawal prior to anaesthesia should be considered.[10] A factor also to be considered is the possible effect of systemic medication on the usual biological clearance mechanisms of anaesthetic drugs. Saturation of the normal enzyme detoxification pathways' capacity in the liver is thought to lead to detoxification of anaesthetic drug metabolites by alternative pathways. Alternative degradation pathways are thought to be one of the possible mechanisms leading to 'halothane hepatitis'.[11] Similarly, many drugs, such as the aminoglycoside antibiotics, are nephrotoxic and will impair renal clearance sufficiently to prolong the action of anaesthetic drugs relying on that route for excretion.

Airway and general anatomy require particularly careful assessment at the preoperative visit, as these two factors are likely to impose limitations on surgical positioning and on the choice of technique for anaesthetic induction and maintenance. An extremely obese patient may pose not only problems of access for the surgeon but also difficulty for the anaesthetist in intubation and choice of technique to minimize well recognized postoperative sequelae.[12] The elderly patient with limitation of movement or joint deformity requires careful attention to positioning and protection of pressure points; when the temporomandibular joints are affected, access to the airway may be a problem. Of concern in the elderly patient with systemic vascular disease is the presence of carotid plaques, with or without bruit. If the operation requires exaggerated rotation of the head, as for posterior auricular approaches, it is worth checking preoperatively that the patient can do it without fainting, to avoid a potentially fatal manoeuvre under anaesthesia.

Airway compromise is usually evident from the general history and examination. However, to detect occult impending airway compromise requires its detailed examination. Congenital anomalies of the airway may involve any part of the respiratory tract, but the most commonly underestimated ones occur in structures arising from the first and second branchial arch structures. The well recognized syndromes (Treacher Collins, Hallermann–Streif, etc.) are obvious but similar deformities of a lesser degree are not. These may occur in otherwise normal patients who compensate for their deformity with muscular effort. Premedication or induction of anaesthesia modifies or abolishes this compensatory effort, leading to airway obstruction.

Anatomical variations, scarring and tumour also contribute to technical difficulties with laryngoscopy and intubation. A larynx noted by the surgeon to be 'easily visualized with indirect laryngoscopy' *may not be able to be visualized as easily with direct laryngoscopy*. Of particular interest in anticipating such problems is the presence of micrognathia, microstomia, macroglossia or relative macroglossia, unusal angulation of the larynx, short thick neck or combinations of these. In the clinical examination of the upper airway, attention should be paid to the patency and size of the nasal passages for the potential use of nasal tracheal tubes. The full face should be examined for symmetry and a full lateral view examined for the size and position of the mandible relative to the maxilla. The state and presentation of dentition should be noted, as should the presence of a narrow high-arched hard palate. A high-arched palate in association with closely spaced, parallel, upper alveolar ridges can make the simultaneous manipulation of laryngoscope and tracheal tube impossible. The position of the larynx relative to the mandible should be identified. If the distance between the prominence of the thyroid cartilage and the lower border of the mandible admits one and a half finger-breadths (3 cm) or more, it is unlikely that a high larynx tucked under the base of the tongue will be encountered. The mobility of the larynx should be observed by asking the patient to swallow, and the presence of induration or scarring in the base of the tongue and hypopharynx should be sought by palpation of the soft tissues beneath the mandible. Palpation of the neck will confirm the midline position of the trachea and identify the possibility of an awkward laryngeal presentation due to oedema or scarring, together with the presence of any tumours impinging on the trachea, which might make the passage of a tracheal tube difficult. Finally, the full and free movement of the cervical spine and the atlanto-occipital joints should be confirmed as well as the patient's ability to open the mouth without limitation. If the inability to intubate or ventilate the patient is encountered unexpectedly at anaesthetic induction the presence of a short thick neck makes a safe rapid tracheostomy more difficult, thereby placing the patient's life in immediate jeopardy. The need for a thorough preoperative examination of the patient's airway cannot be overemphasized.

If the patient has previously received general anaesthesia, a well documented accurate record will be invaluable because it will make available information relating to the actual interaction of anaesthetic conditions with a possibly compromised airway. However, the airway may change for the worse in the meantime.

Laboratory studies

The use of the laboratory to 'document' the obvious and the irrelevant in the preoperative patient has increased, is still increasing, but should be diminished. There are very few tests that can be justified as a 'routine' and any test ordered to pander to the doctor's insecurities rather than further the patient's interest is to be avoided. A review of the cost-effectiveness of 'routine' preoperative testing has been presented elsewhere.[13] The question to be asked before ordering a laboratory test is: will the result, anticipated or unanticipated, be likely to lead the anaesthetist to change the anaesthetic plan suggested by his clinical assessment? If not, the test will serve no useful preoperative purpose and is therefore superfluous. This is not to suggest that the degree of already recognized impairment to physiological functions resulting from a patient's pre-existing medical conditions should not be quantified if, in the clinical judgement of the anaesthetist, the anaesthetic or the operation might be expected to cause a deterioration in the patient's condition, leading to management problems. However, it does preclude the routine preoperative chest x-ray in the absence of clinical indications as well as an ECG for most younger patients coming to operation. The prime purpose of anaesthetic preoperative laboratory tests is to define the limits of anaesthetic manoeuvres intraoperatively. Because the anaesthetist has two principle duties – to render the patient insensitive to pain and to maintain an adequate supply of oxygen to the tissues – preoperative testing should be limited to defining the basis for these two functions in those patients where these abilities are in question.

It suffices here to draw attention to a few problems associated with ENT patients. The ability to render the patient insensitive to pain depends on the choice of anaesthetic technique. Techniques using local anaesthetics are common in ENT practice but they do not require preoperative testing to confirm patient suitability in the absence of a history of agent sensitivity. If a history of sensitivity is obtained, the specific drug should be avoided using one from the other group (amide vs ester); if the history is not sufficiently precise, a conjunctival or patch test using the drug selected should be considered. Conduction blockade can pose a problem in the presence of a bleeding diathesis, particularly if the injection could result in a haematoma adjacent to the airway or bleeding into the airway. Similarly, where nasal intubation is required, any bleeding diathesis must be taken seriously. In such cases preoperative coagulation studies and therapy are justified.

The thorny question about the use of volatile halogenated agents for general anaesthesia is of continuing interest. 'Halothane hepatitis' is not unique to halothane and has now been described following the use of both enflurane and isoflurane.[14] It is a diagnosis made by exclusion of other possible causes of hepatitis, such as hypoxia, haemorrhage with transfusion and hypotension. No direct causal relation between halogenated agent exposure and 'halothane hepatitis' has been proved, other than in the rare case of a true immune response, neither is there a unique pathological picture that can be judged to be pathognomonic of the condition. However, there is sufficient circumstantial evidence to make the prudent anaesthetist hesitant to expose the same patient repeatedly to halogenated agents, particularly halothane, over a short period of time, and a history of jaundice following any exposure in the past suggests avoidance altogether in the future. The real problem is whether or not to give a halogenated agent in the face of potentially active liver disease. There does not appear to be an increased risk if such agents are given to a patient who has made a complete recovery from infectious hepatitis in the past, but common prudence would suggest its avoidance during active hepatitis.[15] Since a surprising number of patients presenting for elective surgery have been shown to be incubating viral hepatitis,[16] it is tempting to wonder whether it is exposure to halogenated agents under such circumstances that might explain the occurrence of the majority of cases of 'halothane hepatitis'. However, with the continued development of viral serological tests, it now appears that undetected viral hepatitis can account for only a small number of halothane hepatitis cases.[17] Whatever the final answer may be, it would appear reasonable to screen patients' liver function perhaps with nothing more than an estimation of serum glutamic pyruvate transaminase (SGPT) and avoid halogenated agents when the level is significantly raised. ENT operations rarely require the use of non-depolarizing muscle relaxants and many are a contraindication to their use; therefore, halogenated agents still have a major role in the specialty and still offer considerable advantages in many situations.

A recurring problem in ENT patients is the presenta-tion of respiratory distress with a partially obstructed airway. The condition of such patients can deteriorate rapidly while waiting for laboratory results and, similarly, the radiology department is not the place to manage an airway emergency if it can be avoided. It is more prudent to assemble the operating team in the theatre with the patient and gather whatever additional studies are required while the means of effective intervention are immediately to hand. Blood gas samples can be drawn on the operating table and any x-ray studies required for anaesthetic purposes can be obtained using a C-arm image intensifier which usually provides more useful information than a plain film.[18]

Informed consent

A California court has stated:

> A physician violates his duty to his patient and subjects himself to liability if he withholds any facts which are necessary to form the basis of an intelligent consent by the patient to the proposed treatment. Likewise the physician may not minimize the known dangers of a procedure or operation in order to induce his patient's consent.[19]

The opinion then goes on to recognize the dilemma that sometimes full disclosure may not be to the benefit of the patient's welfare by further alarming a patient who is already apprehensive. This dilemma has led to the existence of two different rules: the 'professional standard of consent', which holds that the amount of information disclosed is a matter of reasonable medical judgement; and the 'lay standard of reasonableness', where it is not what the physician thinks the patient should know but what the patient should know to make an informed decision. It is the lay standard of reasonableness that now largely governs informed consent in the USA although professional judgement is still recognized in other jurisdictions.

Anaesthesia has not had a high public profile in the past, and today patients still have only a vague idea of what anaesthesia entails. The last thing that most patients consider is the possibility of being maimed or killed by the anaesthetic process when they come to hospital for an operation. Since anaesthesia cures little or nothing, it should not injure. Unfortunately, it may. Therefore the anaesthetist has a particular duty to inform the patient of any possible serious adverse effects, even if the incidence is low. Since most of the serious adverse sequelae of anaesthesia are rare, the physician seeking consent must be sufficiently conversant with the anaesthetic literature to be aware of their existence and significance in the light of the condition of the individual patient.

Preparation of the patient for anaesthesia

Preparations on the ward

The anaesthetic preparation of the ENT patient begins with the anaesthetist's preoperative visit and evaluation. The anaesthetist has to try to establish rapport quickly with the previously unknown patient. Informed consent has to be obtained. Frequently any lingering questions and doubts concerning the proposed operation have to be addressed without contradicting or undermining what the surgeon has already told the patient. This is usually accomplished the evening before operation but, with the developing practice of 'day-of-surgery admission', care must be taken not to detract from the thoroughness of the preoperative visit.

Apart from familiarizing himself with the patient's medical problems and devising the appropriate anaesthetic plan, the anaesthetist may have a number of other objectives to be achieved through preoperative medication:

1. The prime objective is always the relief of the patient's anxiety when it is judged to be significant. Methods used may range from reassurance and 'light' anxiolytic medication, through time-consuming hypnotic suggestion, to 'heavy' premedication with sedatives or narcotics. Some studies have suggested that the relief of anxiety before surgery depends more on the rapport the anaesthetist establishes with the patient than on the choice of drugs for premedication.
2. When pain is present preoperatively, it is important to prescribe analgesics in sufficient dosage to minimize exacerbation from the movements involved in transporting the patient from the bed to the operating table.
3. When a light balanced anaesthetic technique is contemplated, a sedative/amnesic drug should form part of the premedication to reduce the chance of awareness.
4. The inclusion of an antisialagogue is frequently of importance for head and neck surgery and endoscopy, as the patient's airway will not be accessible for manual suctioning by the anaesthetist. The reduction in volume of secretions will also facilitate endoscopy.
5. When the suppression of cardiovascular reflexes is considered of importance for induction of anaesthesia or the surgical procedure, sedation and vagolytic agents are indicated.
6. Depressant premedicant drugs can also facilitate the anaesthetic technique itself by smoothing inhalation inductions and reducing the requirement for intra-operative agents.
7. Premedicant drugs can also be used to reduce the incidence of postoperative nausea and vomiting, although to achieve effective anti-emesis for operations on the inner ear, premedicant effects must usually be reinforced with drugs such as droperidol just before emergence from anaesthesia.

The patient should fast for 4–6 hours before anaesthesia, certainly with respect to solid food, although in any individual patient this may not guarantee an empty stomach. Even when 'empty' the stomach contains a sufficient volume of strong hydrochloric acid to pose an ever-present risk of acid aspiration pneumonitis at induction. This has led some anaesthetists to prescribe an H_2 histamine blocker (e.g. cimetidine) as a routine. When acid reflux, with or without hiatus hernia, is suspected the addition of an antacid by mouth on the morning of operation has been recommended. However, the question of whether a patient should be subjected to an absolute fast (nil by mouth after midnight; NPO), thereby depriving the patient of a significant fluid intake during the fast period, is now being questioned for adults as well as small children undergoing elective surgery.[20] Any relaxation of fasting rules should be approached with care as, despite claims to the contrary, some patients do become confused if offered even reasonable choices, and compliance with instructions diminishes.

The surgical outpatient

People for outpatient surgery require special mention, not because their preoperative evaluation differs significantly from that of the surgical inpatient but because certain additional requirements must be considered. Because there is insufficient time for the full evaluation of significant systemic disease on the day of operation, it falls to the surgeon to ensure that this is done thoroughly during the patient's outpatient clinic visit and any appropriate medical consultations are obtained at that time. As a large proportion of ENT anaesthetic practice involves outpatients, it is sometimes more efficient to see these patients routinely in a separate anaesthesia preoperative clinic.

As outpatient surgery has developed, the emphasis has been on ASA 1 patients to minimize the risks of postoperative adverse events. This approach has been overly conservative. Modern anaesthetic techniques, appropriately selected, permit rapid recovery and pose few problems for the patient discharged home. It is now reasonable to include ASA 2 patients together with carefully selected ASA 3 patients, where admission to hospital following operation is available if anything untoward develops. As a general rule, the following factors would militate against outpatient surgery:

1. The need for extensive medical preparation, rather than just 'tuning up' the medical control of chronic disease.
2. The need for intravenous therapy of any sort.
3. The need for invasive monitoring during operation.
4. The patient's home situation; for instance, the elderly patient living alone and the patient living at great distance from the hospital with difficult access to emergency care.
5. Procedures in which significant postoperative pain may be expected and which it is anticipated may not be amenable to management with oral analgesics.
6. Conditions in which extensive postoperative medical management is required.
7. The patient's lack of ability to comprehend and follow postoperative instructions.

It devolves on the surgeon to ensure that the selected patients arrive properly prepared on the morning of surgery, accompanied by a competent adult into whose care and protection the patient will be discharged. Premedication, if any, is usually given intravenously by the anaesthetist after the patient has been evaluated and informed consent obtained. All patients must be given both written and oral instructions before discharge, cautioning them against undertaking any activities that may pose a hazard to themselves or others due to the 'hangover' effect of anaesthetic drugs still being cleared from the body over the following 24–48 hours, which can affect their judgement and reflexes. Such warning should encompass driving, operating machinery, nursing young infants, the hazards of the domestic kitchen and signing any legal documents.

As the cost advantages of outpatient surgery become more widely appreciated, it is becoming more common. However, it must be understood that the usual operating theatre suite staffing levels and ratio of patient preparation, recovery and discharge space to operating theatres are not adequate for a safe, efficient outpatient surgery service. Outpatient surgery is space and labour intensive, particularly in what is required for patient preparation, recovery and discharge.

Preparations in the operating theatre

After equipment preparation and preinduction checks are complete and the patient arrives in the operating theatre, his identity is confirmed together with the site of operation. Then he is made comfortable on the operating table. An intravenous cannula is inserted under intradermal local anaesthesia to provide a drug injection route. An intravenous infusion may be started if intraoperative

hydration or blood replacement is anticipated. The appropriate clinical measuring instruments are applied according to the patient's condition, but a minimum standard of care now requires the use with every patient of at least:[21]

1. Indirect blood pressure measurement.
2. Electrocardiogram.
3. Temperature.
4. Oxygen analysis, at least on the circuit supply line.
5. Expired CO_2 analysis (capnometry or capnography).
6. Pulse oximetry.

While the final preparations are made for induction, the patient is 'preoxygenated' by mask with 100 per cent inspired oxygen to create an oxygen reserve in his functional residual capacity to increase the time available for manoeuvre in the event of a difficult induction. When all is prepared, if not already present, the surgeon should be summoned before induction begins.

Conduct of anaesthesia

Securing the airway

The single most important piece of information required before anaesthesia for an ENT patient is to know exactly what the surgeon proposes to do. Only when the objective of the operation and the area of the patient to be prepared for surgery is known can a rational plan for airway management be made. First it must be established that there is an indication for tracheal intubation. Even a straightforward intubation carries a risk for the patient.[22] The general indications for intubation are no different from any other branch of anaesthesia but certain indications are more commonly found with ENT patients:

1. To secure the airway when the anaesthetist is remote from the patient's head.
2. To protect the airway against contamination with blood, pus or regurgitated material.
3. To secure a marginal airway.
4. To support or control ventilation.
5. To bypass upper airway obstruction.
6. To permit surgical access to the aerodigestive tract.

There are three basic methods of securing the airway for an ENT operation: nasotracheal, orotracheal and sublaryngeal tracheal intubation via tracheostomy. Tracheotomy is the least desirable owing to associated complications, particularly in children. Orotracheal intubation is always the most desirable route, provided it is feasible in that

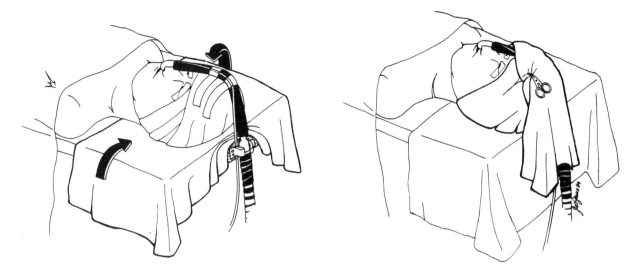

Fig. 61.1 A nasal tracheal tube offers security as an airway for many ENT operations, as the upper part of the tube is surrounded by bone. The stability of circuit connections is further enhanced by the use of a wrap-around head drape. (Reproduced, with permission, from reference 48.)

patient and compatible with the operation's objectives. However, an atraumatically placed nasotracheal tube offers a particularly secure airway when used with a wrap-around head drape (Fig. 61.1).

As a general guide, all operations on or above the upper lip require an oral tracheal tube; those involving the jaw or tongue require a nasal tracheal tube. If the operative site is confined to one side of the head or neck, an oral tracheal tube may be used if it is placed in the contralateral sulcus beside the tongue and properly secured. The anaesthetic indications for a tracheotomy are rare and are usually limited to those patients in whom it is judged that pathology or abnormal anatomy may make other intubation attempts dangerous or it is anticipated that their airway may be at risk from obstruction in the postoperative period (usually due to oedema). If the patient already has a tracheotomy when presenting for surgery, this is the route of choice for securing the airway. These guidelines are not sacrosanct, as compromises have to be made, for plastic procedures, to prevent facial distortion. Operations inside the mouth or the pharyngeal airway may require any of the three options, depending on individual circumstances.

Having decided to secure the patient's airway with a tracheal tube, a brief review of the types available and suited to ENT work is appropriate. The standard plastic disposable tracheal tube with a high-volume low-pressure cuff and a side port is the most suitable for tracheotomy and endoscopy work when the anaesthetist is by the patient's head. When a nasal tracheal tube is required, one designed for the purpose is more appropriate. These are available in both disposable and reusable forms, and are characterized by a longer bevel giving a softer tip, a streamlined cuff and no side port to act as a curette as it passes through the nose. The disposable variety satisfies

most requirements but some anaesthetists have found that the older Magill red rubber tubes work better with blind intubation techniques. For the majority of ENT operations requiring oral intubation when the anaesthetist is not in immediate proximity to the patient's head, the precurved disposable oral tubes (e.g. Rae™ oral tracheal tube) have been found particularly useful. This type permits the circuit to be led away towards the patient's side or feet while it presents a low profile as it emerges from the mouth, minimizing the potential for kinking and obstruction under the surgical drapes. It is secured much more easily and simply to the patient's jaw or teeth, depending on surgical requirements. However, when a mouth gag is to be used (e.g. tonsillectomy), these tubes are particularly vulnerable to displacement and kinking with placement and opening of the gag (Fig. 61.2). In the case of a tracheotomy, a standard tracheotomy tube is best avoided if it is to be covered by the surgical drapes, to avoid obstruction by rotation within the trachea. The patient's airway may be secured more reliably by inserting an armoured tracheal tube and suturing it to the skin. Recently, precurved plastic armoured tubes have become available; like the disposable oral variety, they give a low profile but also minimize the tendency of the old armoured tubes to spring out of the tracheotomy or to migrate into the right main bronchus.

Other specialized methods of controlling and ventilating the airway are discussed later.

Operating theatre set-up

The anaesthetic set-up for ENT surgery is dictated by access to, and competition for, the airway between the anaesthetist and the surgeon. In ophthalmological and

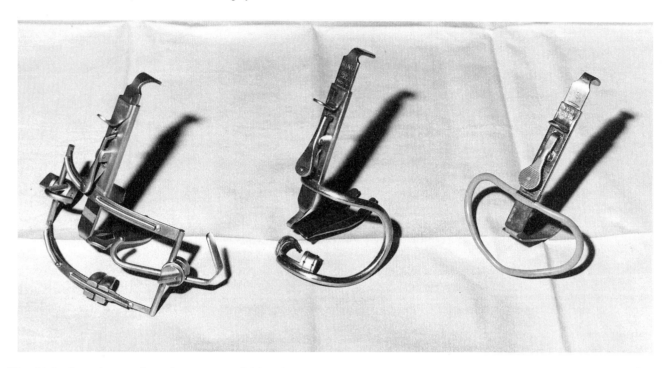

Fig. 61.2   Several types of mouth gags are available to facilitate surgical access. Three examples are shown here: from left to right, the Dingman, Davis and McIvor gags. Tongue blades with a central channel to hold the tracheal tube are preferred as they provide better stability and protection for the airway. They are shown here on the Dingman and Davis gags but are available for most types. (Reproduced, with permission, from reference 49.)

neurosurgery the anaesthetist usually has some restricted access to an oral tracheal tube. In otorhinolaryngology, even if the surgeon is not working in the aerodigestive tract itself, the patient's head is usually completely draped.

The clustering of the surgeons, the scrub nurse and the instrument tables closely about the patient's head usually means that the anaesthetist and his equipment are forced towards the foot of the table. This dictates a meticulous arrangement of equipment following induction with a well secured airway and close monitoring of the anaesthetic circuit's integrity throughout the case (requiring the setting of circuit pressure alarm limits depending on the equipment in use). The situation demands close cooperation from the surgeon. Any changes in the surgical field, such as increased oozing, dark blood or surgical manipulation around the carotid bodies, must be communicated to the anaesthetist immediately. If the anaesthetist in turn has any concern about the patient's condition, the surgeon should be informed and the matter resolved with good grace, by undraping the patient if necessary.

Anaesthetic induction is always undertaken with the anaesthetist at the head of the patient. If a tracheostomy is to be performed, the anaesthetist remains at the patient's head where he has control of the tracheal tube and can assist the surgeon. When the trachea is exposed the tube cuff is deflated and the tube is advanced towards the carina. This prevents the cuff being incised with the trachea, and enables the anaesthetist to seal the airway if

the surgeon encounters bleeding problems. Once the window in the trachea has been made without incident, the anaesthetist withdraws the tube until the surgeon states that the tip is level with the upper border of the window. The tube is never withdrawn from the larynx completely until after the surgeon has inserted and tested the correct position of the tracheostomy tube.

If the surgical intention is to undertake endoscopy or microsurgery of the larynx, the operating table is rotated through 90 degrees and the anaesthetist remains near, but below, the patient's head on the left side of the patient. This is to allow the anaesthetist to place and secure various ventilatory attachments to whatever endoscopes may be used, or to permit him to ensure manually that the tracheal tube is not inadvertently dragged from the larynx by the surgeon during endoscopic manipulation. The oral tracheal tube is secured in the left side of the mouth, as endoscopes are introduced on the right or in the midline.

For most other procedures the anaesthetist is to the side of the operating table near the foot or at the end of the table. The advantage of being off to one side is that this position permits the use of a standard circle circuit without modification, and some anaesthetists feel more comfortable if the airway is at least within reach, if not immediately accessible (Fig. 61.3). The disadvantage of the position is the crowding inherent if more than one surgeon is involved in the procedure. If the anaesthetist sits at the foot of the table, this latter problem is overcome but now special

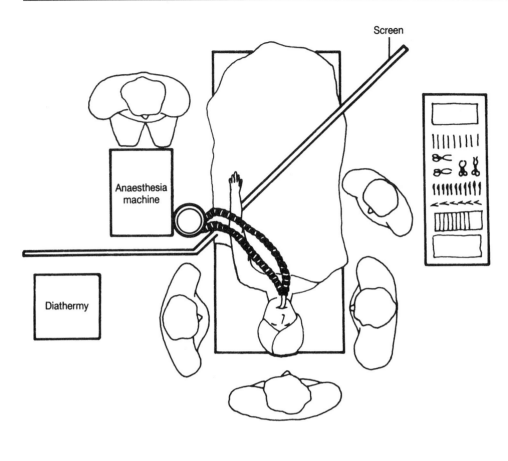

Fig. 61.3 The anaesthetist at the side of the operating table, using a standard circle system. (Reproduced, with permission, from reference 48.)

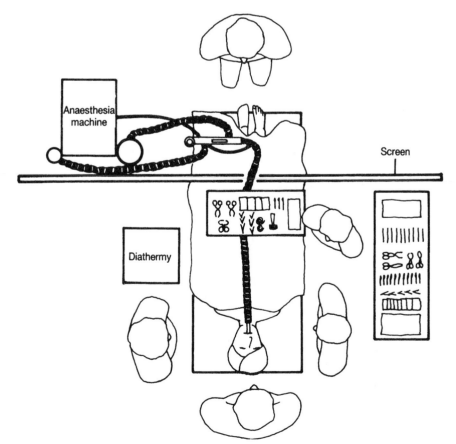

Fig. 61.4 The anaesthetist at the foot of the table, using a coaxial circuit. (Reproduced, with permission, from reference 48.)

circuits have to be used. A circle circuit may be used but, as the standard circuit is too short, either extension tubes have to be added or a long disposable circle circuit must be employed. The former puts additional connections in the circuit, all of which are potential sites for disconnection and gas leaks, and both increase the volume of the circuit – introducing additional compression and distension errors in ventilator settings. The latter may be compensated for in normal patients, but ventilatory management in those with reduced chest compliance may become difficult. These problems are largely overcome by the use of coaxial

circuits, such as the Bain (Fig. 61.4). The single tube is light and easy to secure but because rebreathing expired CO_2 is prevented by using high gas flows,[23] the use of the coaxial circuit requires more of the anaesthetic agents than the circle. Loss of both heat and water is also increased unless an artificial 'nose' is interposed between the tracheal tube and the circuit. An advantage of the coaxial circuit is that the metal 'head' containing the expiratory valve and various hose attachments can be clamped to the operating table itself, further minimizing the chances of circuit disconnection due to any changes in table position

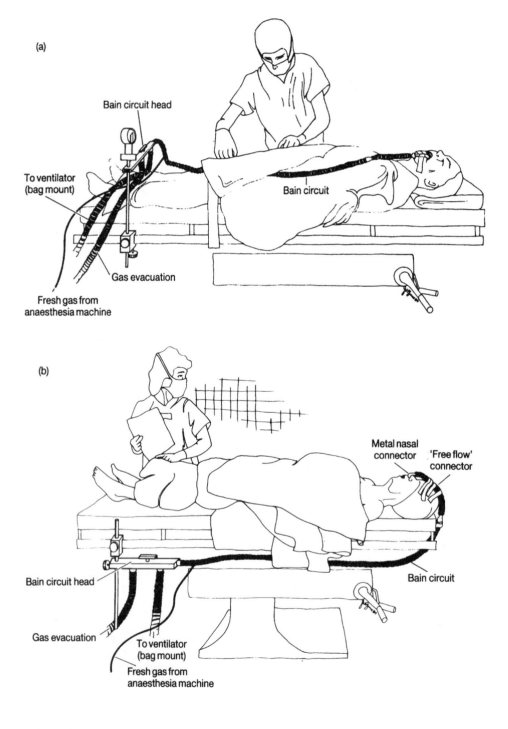

(a)

Bain circuit head

To ventilator (bag mount)

Gas evacuation

Fresh gas from anaesthesia machine

Bain circuit

(b)

Metal nasal connector

'Free flow' connector

Bain circuit head

Gas evacuation

To ventilator (bag mount)

Fresh gas from anaesthesia machine

Bain circuit

Fig. 61.5   The use of a coaxial anaesthetic circuit (Bain). (a) Set-up for an oral tracheal tube. (b) Set-up for nasal tracheal tube. Note the table mount for the circuit head to prevent disconnections with table position changes. (Reproduced, with permission, from reference 48.)

requested by the surgeon during the procedure (Fig. 61.5). The foot position affords a better, symmetrical view of chest wall movement during anaesthesia and gives access to both sides of the table for the anaesthetist.

Anaesthetic techniques

The four main groups of anaesthetic techniques are available to the ENT patient.

Topical anaesthesia

Topical anaesthesia is used mainly for nasal operations and for diagnostic or therapeutic endoscopy, particularly where patient cooperation is required as with Teflon injection of the vocal cords. Attention to total drug dosage is important and the careful selection of those patients who will tolerate the experience. The secret of consistent success is to take sufficient time at each step to allow full analgesia to develop. The lips and gums are 'painted' first and then the tongue, epiglottis and larynx in turn under direct or indirect vision. Rather than spraying the larynx, some anaesthetists may prefer to achieve laryngeal anaesthesia by applying cotton pledgets soaked in local anaesthetic to the pyriform fossae using Krause (Jackson) forceps. At the conclusion of paint-up a small amount of viscous lignocaine (lidocaine) is given to the patient to gargle and swallow. If conditions are still not satisfactory, resort has to be made to various conduction blocks.

Cocaine, an alkaloid of *Erythroxyum coca*, is still a popular agent – particularly for topical preparation of the nose as it is the only local anaesthetic with vasoconstrictor properties. Its analgesia precedes the vasoconstrictor effects by approximately 5 minutes and may last as long as 90 minutes without reapplication. It is absorbed from mucosal surfaces and, if applied to the pyriform fossae, blood levels equivalent to an intravenous injection are achieved quickly. Peak levels are usually reached within the first hour, and first order kinetics are followed with a serum half-life at 60–80 minutes. Most is metabolized by hydrolysis by plasma pseudocholinesterase. Cocaine inhibits the reuptake of endogenous noradrenaline (norepinephrine) and blocks the uptake of exogenous adrenaline (epinephrine), suggesting the reason for cocaine's ability to sensitize various end-organs to catecholamine effects. It is recommended that no greater than a 4 per cent solution should be used, and the maximum safe dosage in an adult to be used at one time should not exceed 3 mg/kg; although severe toxic reactions have been reported with a total dosage as low as 20 mg in adults, the usual fatal dose is approximately 1 g. Cocaine toxicity is manifested by a biphasic reaction in the central nervous system: first excitement, then depression with convulsions and unconsciousness. Respiratory and cardio-vascular signs will be evident and death can occur rapidly without treatment. The best protection is to avoid toxic doses but if treatment is required, intravenous injections of thiobarbiturates or benzodiazepines, together with labetalol to block both α- and β-adrenergic responses[24] are usually rapidly effective.

Iontophoresis is an unusual variation of topical anaesthesia, used for achieving analgesia of the tympanic membrane. It is usually reserved for the insertion of ventilation tubes in the membrane in the surgical clinic. Local anaesthetic is instilled into the ear and then a low voltage electrode is suspended in the solution in the ear canal. An electrical current is passed into solution, ionizing the drug and 'driving' it into the tympanum. It may take as long as 20 minutes to achieve useful analgesia and only the tympanum, not the lining of the external canal, is rendered analgesic. The technique is not particularly suited for use in a busy operating theatre list.

Infiltration anaesthesia

Infiltration anaesthesia, with or without standby for monitoring, sedation and/or resuscitation by the anaesthetist, is most commonly used for minor superficial operations or operations in the middle ear. The surgeon is solely responsible for the use and dosage of whatever local anaesthetic agent is selected and should ensure that the safe total dose with or without adrenaline is not exceeded,[25] but the anaesthetist should nevertheless keep a running check on the cumulative doses that have been given. Care should be exercised when selecting patients, to ensure that the procedure as planned can be completed in the time made available by the technique and that the patient can tolerate the experience. The safety of the technique depends on not exceeding safe drug dosages and frequent aspiration during injection with a constantly moving needle to minimize the risk of inadvertent intravenous injection. This is not a 'second best' technique to 'get away' with surgery on inadequately prepared patients already rejected for general anaesthesia, so the anaesthetist should be fully aware of what is to be attempted.

Conduction anaesthesia

Conduction anaesthesia is the technique whereby a local anaesthetic agent is introduced via a needle to the immediate proximity of a specific nerve, series of nerves or a nerve trunk in order to produce analgesia over the sensory distribution. The technique is used only infrequently for head and neck surgery. Two useful blocks for endoscopy of the upper airway are the percutaneous superior laryngeal nerve block and the glossopharyngeal nerve block as it passes through the pillars of the fauces. Spinal anaesthesia can theoretically be used for analgesia of the head and neck, and, indeed, has been used in the past.[26] However, to achieve a sensory block high enough to be of use to the surgeon usually results in a simultaneous motor block of the muscles of respiration. This requires control of the patient's respiration, necessitating general anaesthesia, which somewhat defeats the object of conduction blockade in the first place! Blockade of the cervical nerves is a more suitable technique which creates a very favourable operative field that is particularly useful for thyroid surgery, cervical lymph node dissections and surgery of the carotid vessels.

Apart from the specific complications of cervical blockade such as subarachnoid spread and involvement of other cranial nerves in the neck, two of the fundamental disadvantages of conduction blockade are encountered. The duration of surgical analgesia depends on the accuracy of initial injection and the properties of the drug and the concentration injected. Therefore, if the surgical procedure takes longer than anticipated and the block wears off, general anaesthesia with its concomitant risks has to be superimposed. The other major drawback is the time required for the block to be executed and for analgesia to develop. It is because of the induction time and the inflexibility of a 'one shot' technique that these blocks are not used more frequently. In spite of these considerations, conduction blockade still has two well established roles in head and neck surgery: blockade for surgery on the teeth and supporting tissues;[27] and the control of the chronic pain syndromes.[28]

General anaesthesia

In spite of the availability of a rapidly increasing variety and number of anaesthetic drugs, general anaesthetic techniques can still be divided into two broad categories: single agent techniques and balanced techniques. The fundamental difference between the two categories lies in the use of muscle relaxants to diminish or abolish skeletal muscle tone.

A single inhalation agent can achieve all three properties of the anaesthetic triad, but profound degrees of skeletal muscle relaxation are achieved only by a relative overdose of the agent compared with that required for narcosis and reflex suppression. This means that the maintenance of the circulation becomes more difficult with greater degrees of muscle relaxation. The problem is overcome with muscle relaxant drugs. However, these drugs mandate the full control of the patient's ventilation with a tracheal tube and, by definition, converts the normal negative intrathoracic pressure to positive, with the associated implication for venous pressure and surgical field oozing. With single agent anaesthesia, spontaneous breathing through a mask or tracheal tube can be maintained, albeit with some intermittent support of ventilation during the longer

surgical procedures to minimize peripheral lung atelectasis. However, even with spontaneous breathing, the problem of oozing in the surgical wound is not completely overcome as the anaesthetized respiratory centre requires a higher driving tension of carbon dioxide than normal (5.9–7.2 kPa as opposed to 5.0–5.5 kPa; 44–54 mmHg vs 38–42 mmHg), and hypercapnia leads to increased oozing. Because otolaryngology head and neck surgery always requires a secure airway, a tracheal tube is indicated for most procedures, no matter which type of technique is being used. Thus the choice of preferred technique for surgery, in the absence of specific patient medical problems, is reduced to a consideration of a few general advantages and disadvantages in different surgical situations (Table 61.1).

The list in Table 61.1 is not exhaustive, as a specific medical condition may favour the choice of a technique different from that which might have been chosen for a healthy patient undergoing the same surgical procedure. The use of muscle relaxants is further complicated by the possible use of a nerve stimulator or evoked potentials during the operation. If dissection of the facial nerve is required as in parotid surgery, the surgeon will expect nondepolarizing muscle relaxants to be avoided to enable a full visible response to his probing for the nerve and its branches. If a cochlear implant is to be attempted, the anaesthetist may be requested to paralyse the patient to eliminate extraneous electrical muscle background noise while the implant electrode is optimally sited.

One of the standard problems faced by the ENT anaesthetist is the use of vasoconstrictors by the surgeon even when the patient is anaesthetized with a volatile anaesthetic. It has long been recognized that the halogenated volatile anaesthetics 'sensitize' the myocardium to catecholamine-induced cardiac arrhythmias, with the possible risk of ventricular fibrillation and circulatory arrest. Sensitization of the myocardium is defined as a state in which the dose of adrenaline required to produce an arrhythmia is less than the dose in the awake state when the myocardium is free of drug effect.[29] Halothane has a far greater sensitizing effect than the other two commonly used agents, enflurane and isoflurane. Johnston and colleagues measured the dose of adrenaline that produced ventricular arrhythmia in 50 per cent of patients (ED_{50}) at 1.25 MAC with $PaCO_2$ controlled at 3.9–5.3 kPa (29–40 mmHg) finding the ED_{50} for halothane at 2.1 µg/kg, the ED_{50} for isoflurane at 6.7 µg/kg and the ED_{50} for enflurane at 10.9 µg/kg.[30] Children, however, appear to be much less susceptible to sensitization, for reasons that are still unclear.[31] The clinical implications of this work are not as obvious as they might seem, as many factors can modify the degree of sensitization observed. Joas and Stevens demonstrated that deepening anaesthesia in dogs raised the arrhythmogenic threshold, as did the use of lignocaine/adrenaline

Table 61.1 Comparison of two basic approaches to general anaesthesia

Single agent technique	Balanced technique
1. Simple: reduces number of potential adverse reactions to drugs	Complex: more drugs, more potential adverse reactions
2. Control of circulation more difficult: patient usually requires fluid 'loading'	Can be tailored to maintain circulation more easily
3. Cardiac dysrhythmia with exogenous catecholamines more common	Can be tailored to minimize catecholamine dysrhythmias
4. Nerve stimulator use unaffected	Response to nerve stimulator abolished or diminished
5. With stable state, patient immobility can be nearly guaranteed	Even with blockade monitoring, patient immobility cannot be guaranteed without excessive paralysis
6. Muscle relaxant reversal not required	Muscle relaxant reversal required; may be incomplete and 'recurarization' can occur postoperatively
7. Anaesthetic effects completely reversible as long as patient is breathing with a clear airway	Excretion of intravenous drugs depends primarily on liver and renal excretion, which may be impaired
8. Awareness during anaesthesia extremely rare	Awareness may occur
9. Spontaneous respiration an option with or without a mask	Controlled respiration; use of mask ill-advised
10. Particularly suited to initial management of difficult airway	Contraindicated for initial management of difficult airway
11. Involves use of halogenated hydrocarbon agents	Volatile agents can be excluded with hepatitis risk
12. Increases cerebral blood flow and intracranial pressure	Can be used to decrease intracranial pressure
13. Associated with risk of malignant hyperthermia	Can be tailored to avoid known triggers of malignant hyperthermia

mixtures,[32] and thiobarbiturate at induction of anaesthesia.[33] The role of $PaCO_2$ is not clear. Joas and Stevens observed an increased threshold with hypercapnia whereas Katz and Katz described the opposite.[34] However, the latter authors made three suggestions for clinical precautions that have stood the test of time:

1. Only solutions of adrenaline of 1/100 000 to 1/200 000 should be used (greater concentrations offer little additional vasoconstriction[35]).
2. The dose in adults should not exceed 10 ml of 1/100 000 adrenaline in any 10-minute period.
3. The dose in adults should not exceed 30 ml of 1/100 000 adrenaline in any given 60-minute period.

A single agent may be an inhaled drug or an intravenous one, depending on circumstances. The use of a single agent without muscle relaxants does not preclude the full control of ventilation if required.

Special anaesthetic techniques

Management for microsurgery of the ear

When using an operating microscope, the otologist requires a patient who remains still, a relatively bloodless field and consideration of the possible effects of nitrous oxide in the anaesthetic carrier gas mixture. The operating microscope magnifies everything that is seen. The slightest patient movement becomes an 'earthquake'. Blood oozing into the field looks like a flood. Therefore the anaesthetist must pay particular attention to maintaining an adequate depth of anaesthesia in the case of a single-agent technique, and an appropriate level of muscle blockade when using a balanced technique to ensure that the patient does not move. It is also best to avoid inadvertently leaning on or knocking the operating table during the procedure.

To prevent significant oozing into the wound, meticulous attention to the airway is required to ensure the absence of an inadvertent positive end-expiratory pressure. Airway pressure should equal atmospheric during the end-expiratory pause, whether the patient is breathing spontaneously or is being ventilated, in order to prevent a raised venous pressure. Venous drainage from the area can be facilitated by a slight head-up tilt of the patient, but this does increase the risk of occasional air embolism from the wound. If this position is adopted, the need for the precautions of a chest ultrasound monitor and a central venous line for aspiration should be considered and sterile water should be kept on the instrument table to flood the wound if the situation arises. Analysis of expired CO_2 and nitrogen will also give warning of the condition. It is useful with any technique to allow the blood pressure to fall somewhat from normal values when this is deemed safe for the individual patient. Some surgeons and anaesthetists consider that formal controlled hypotensive techniques have much to offer in improving the operative field further. It must be remembered that hypotensive techniques are not risk free and should not be used for the convenience of the surgeon but rather to increase the chances of a successful surgical outcome for a necessary operation. In the shorter operations on the middle ear, such as tympanoplasty where bleeding may lift a technically successful graft, the shorter acting agents such as sodium nitroprusside or trimetaphan by intravenous infusion are indicated. A similar approach may be employed for surgery on the inner ear and neural canals. Some anaesthetists use longer-acting drugs such as labetalol or pentolinium. However, the action of longer-acting drugs may not be reversed easily.

Air under pressure in the closed middle ear cavity is usually vented passively through the normal eustachian tube. Nitrous oxide in the anaesthetic carrier gas mixture, although relatively insoluble in blood, is still much more soluble than nitrogen, the major constituent of air. Therefore more nitrous oxide molecules will present themselves in the walls of the cavity and slide down the diffusion gradient into the middle ear gas than can be compensated for by the removal of nitrogen molecules down their own diffusion gradient into the blood. In fact, for every molecule of nitrogen leaving the cavity, ten molecules of nitrous oxide are available. This net increase in molecules leads, in time, to increasing intracavity pressure. If for any reason normal passive venting is impaired, pressure can rise rapidly to levels that can rupture the normal tympanum and disrupt tympanic grafts (Fig. 61.6). Because nitrous oxide is a relatively insoluble gas its uptake and washout from the blood is rapid, so it is usually sufficient to turn the gas off approximately 30 minutes before a tympanic graft is sited, in order to prevent the problem.

Similarly, the rapid diffusion of nitrous oxide out of the middle ear cavity may cause subatmospheric pressures and retraction of the tympanum or graft (Fig. 61.7). Occasional cases of irreversible sensorineural hearing loss in patients who have undergone a previous stapedectomy with prosthetic replacement have been attributed to the pressure effects of nitrous oxide driving the prosthesis into the inner ear.[36] Of more recent interest are the possible beneficial effects of the agent in forcing fluid from the middle ear and elevating retracted tympani preparatory to ventilation tube insertion.

Management for microsurgery of the larynx

When the surgeon has declined to operate on the conscious patient, several methods are available for the management of general anaesthesia. The major problem to be faced is the difficulty in securing the airway against contamination due to the close proximity of surgical activity. If a small

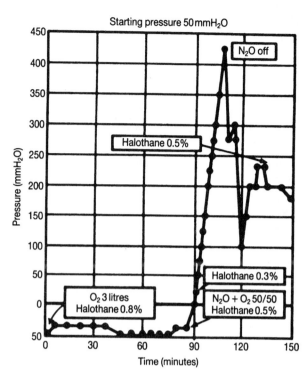

Fig. 61.6 Pressure measurements in a single patient show an immediate increase in middle ear pressure when nitrous oxide is added to the carrier gas. (Reproduced, with permission, from reference 36.)

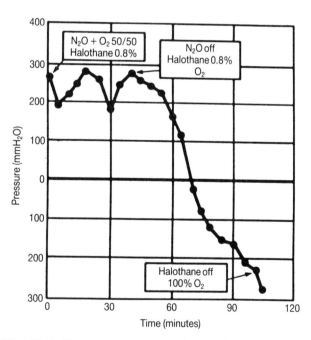

Fig. 61.7 Pressure measurements in a single patient show a progressive fall in middle ear pressure to subatmospheric levels when nitrous oxide is withdrawn. (Reproduced, with permission, from reference 36.)

standard cuffed tracheal tube is used, normal inhalation techniques are feasible. However, the surgeon's access to the structures of the larynx is restricted, which may be unacceptable if work is to be done on the posterior vocal folds where the tube comes to lie naturally. If laryngeal polyps are the reason for surgery, difficulty in working around the tube may be insurmountable; some anaesthetists would consider the risk of 'seeding' polyps down the trachea by passing the tube through such a glottis in the first place to be unacceptable.

An alternative is the use of Venturi jet ventilation as part of a balanced technique. Jet ventilation may be achieved by using a rigid injector attached to the laryngoscope and protruding below the vocal folds into the trachea or by using the cuffed Carden tube (Fig. 61.8) placed below the larynx independent of the laryngoscope. Either is attached to a high pressure gas source. Oxygen or a nitrous oxide/oxygen mixture from a high pressure blender and intermittent pulses of gas are employed to inflate the patient's lungs, using a manual or automatic pneumatic switch. The high pressure gas stream in theory entrains room air from the pharynx. In practice, entrainment is slight and when a nitrous oxide mixture is used, anaesthetic tensions can be achieved. The patient must be fully paralysed to facilitate ventilation and the surgeon must ensure a clear expiratory pathway at all times to prevent pressure injury to the lungs. Intravenous adjuvants are used to ensure an adequate depth of anaesthesia. The rigid jet used in modifications of the Toronto ventilating laryngoscope improves access relative to a standard tube but is still inflexible. The advantage of the Carden tube is that only the small cuff inflation tube and the jet tube

protrude above the larynx, and, as both are made of soft flexible material, the surgeon can lift them out of the way if required. With Venturi 'jet' techniques where it is elected to place the injector above the larynx, the surgeon must remember the risks of blowing blood and particulate matter into the unprotected tracheobronchial tree.

It is important to ensure with all jet ventilation techniques that the driving pressure is started at a low level and increased in small steps until adequate ventilation is achieved, all the while checking for full chest deflation during expiration. The injector should be aligned as parallel to the axis of the trachea as possible to avoid the jet acting as a pneumatic knife on the wall. The jet should be shut off immediately with any transient impairment of expiration. A close watch must be kept for any change in vital signs or progressive diminution or asymmetry of chest movement in the presence of constant jet settings together with any apparent increase in neck circumference. If there is any suggestion of barotrauma to the airway the procedure should be abandoned, the patient intubated immediately and given 100 per cent oxygen. As pneumothorax associated with jet ventilation is quite common, it is prudent to examine all patients undergoing the technique after the procedure and obtain a chest x-ray if judged appropriate.

Management for laser surgery

Light Amplification by Stimulated Emission of Radiation is finding an increasing number of applications as a tool for surgery. The lasers of particular interest in ENT surgery

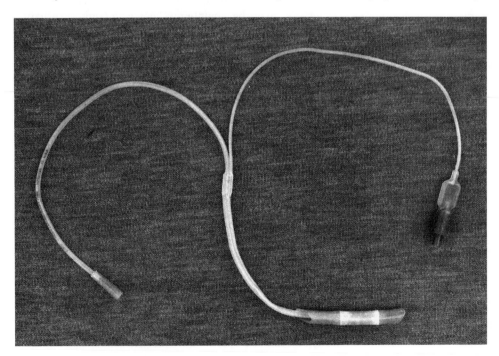

Fig. 61.8 The Carden jet-ventilation tube. High-pressure jet tube is to the left, the cuff inflation tube to the right. The body of the tube is inserted completely below the vocal folds. (Reproduced, with permission, from reference 48.)

are the CO_2 laser, the argon laser and the neodymium : YAG laser. Although each differs from the others in its physical properties and applications, the requirements for anaesthetic management are similar:

1. Complete patient immobility to ensure that normal tissue is not accidentally burned.
2. Unencumbered surgical access to and visibility of the target area.
3. Protection of the patient from stray laser radiation or reflection.

Where the use of a laser in the airway is contemplated, further precautions are needed:

4. Protection of the airway against blood and the products of vaporization.
5. The removal of smoke and debris.
6. The prevention of airway fires.

Although the use of lasers for stapedectomy, extirpation of base of skull tumours and operations on the nose has been described, these applications must be considered to be under development if not still experimental. However, applications in the airway have already established the laser as an important treatment modality.

The need for absolute immobility demands the use of muscle relaxants with blockade monitoring and hence general anaesthesia. The choice of method of ventilation and for maintaining the airway depends upon the site of the target lesion. The CO_2 laser is used with the operating microscope for susceptible lesions in the larynx. Here access for the surgeon is difficult and two methods for ventilation are available: the use of a small cuffed tracheal tube or, if that restricts access to the lesion, Venturi ventilation with the orifice of the injector attached to the laryngoscope sited below the cords. Positioning the injector below the target minimizes the risk of blowing blood into the lungs, and the expired gases are very effective in removing smoke. However, the relaxed vocal cords balloon with ventilation, requiring synchronization of surgical activity on the cords with the ventilatory cycle so as not to have to hit a moving target.

The use of a tracheal tube carries the risk of intratracheal fires. Initial efforts to minimize this risk involved wrapping standard tubes with dampened muslin or metallic tape to disperse or reflect the aberrant laser strike. This approach was not particularly successful as tubes still ignited. Energy would be transferred from the laser beam to the tube and, once the 'flash-point' for the material comprising the tube was reached, ignition occurred. The subsequent fire was fed by the anaesthetic carrier gases creating a reasonable facsimile of a blowtorch within the trachea. Different tube materials have been tried to minimize this risk, such as metal tape-wrapped red rubber tubes which do afford increased protection. However, it now seems clear that the all-metal tubes (Fig. 61.9) are the instrument of choice for ventilation.[37] Even these tubes are not without problems. They are designed to leak gas to facilitate cooling, which can lead to pollution of the operating theatre. With rough handling, metal burrs are created that can lead to direct trauma with placement, and the tube can still heat up with laser impacts and cause contact tissue burns. A latex cuff is necessary for full control of the airway, though this can ignite. This risk is minimized by cuff inflation with 1 per cent aqueous lignocaine rather than air, so that, if the cuff ignites and ruptures, the escaping liquid will tend to quench the fire and minimize any local tissue burn. However, with current concerns about patients with latex sensitivity, these cuffs are becoming more difficult to obtain; the disposable double-cuffed armoured silicone tubes that have recently become available are a reasonable alternative.

For palliative procedures to relieve neoplastic obstruction of the trachea, where it is not possible to get below the laser target, Venturi ventilation through an uncuffed metal tube is indicated. For endobronchial procedures using an Nd : YAG laser through a fibreoptic bronchoscope, the airway may be maintained using a rigid ventilating bronchoscope (Fig. 61.10) with a Sanders Venturi ventilation attachment.[38] If the patient already has a tracheostomy, the tube should be changed for a metal tracheostomy tube suitably protected.

Additional general precautions are required for the conduct of laser surgery. Care should be exercised to ensure that the patient is electrically grounded. The patient's eyes should be protected with wet eye patches secured with canvas tape that does not melt. The immediate

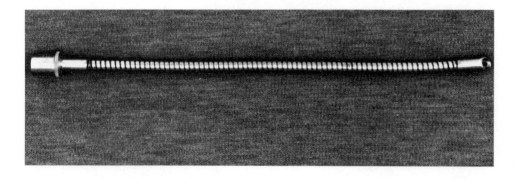

Fig. 61.9 The all-metal Norton tracheal tube for laser surgery.[37] (Reproduced, with permission, from reference 48.)

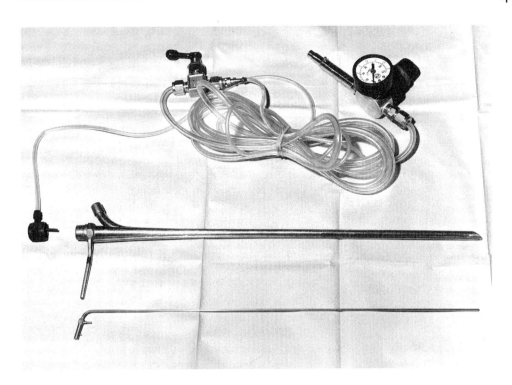

Fig. 61.10 Jet ventilation via a rigid bronchoscope. Note the adjustable pressure-reducing valve leading through a manual interrupter switch to the Sanders Venturi attachment for the head of the bronchoscope. (Reproduced, with permission, from reference 49.)

area surrounding the operative site should be draped with wet linen towels, as should all immediately adjacent reflective surfaces. All personnel in the operating theatre should wear protective eyeglasses with side guards appropriate to the laser type (Fig. 61.11).

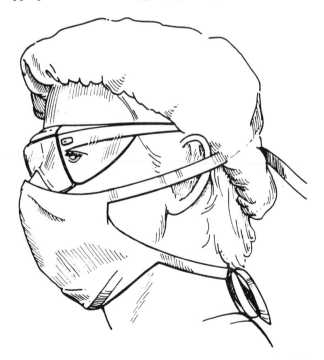

Fig. 61.11 Laser goggles with protective side-arms should be worn by all personnel in the operating theatre. Each laser type requires protective goggles made of appropriate materials. They are not interchangeable. (Reproduced, with permission, from reference 50.)

If, in spite of precautions taken, a laser fire still occurs, the flow of anaesthetic gases should be cut off and the laser discontinued immediately. The fire is quenched with water held in readiness on the instrument table for such an eventuality. The tracheal tube is withdrawn under direct vision and the patient is immediately reintubated with a new tube. The standard therapeutic and respiratory support measures for severe airway burns must then be instituted.[39] It is then profitable to check, with a fibreoptic bronchoscope through the tracheal tube, for any charred material in the airway and to remove it if the patient's condition permits (see also Chapter 70).

Management for general endoscopy

The 'paint-up' technique discussed earlier for intubating the larynx in the awake patient is equally suited to laryngoscopy and bronchoscopy, less so for oesophagoscopy owing to patient discomfort. Therefore it only remains to add here a few points that can facilitate the technique. When the procedure is to be undertaken electively, it is wise to starve the patient as his airway protective reflexes will be obtunded; a light premedication with a benzodiazepine or barbiturate will facilitate the 'paint-up' if not contra-indicated by an airway already compromised by disease. When the surgeon's practice requires large planned endoscopy clinics, the services of an anaesthetist in preparing the second and subsequent patients while the surgeon is working not only speeds up the operating list but also contributes to safety.

When the use of a topical technique is inappropriate the following methods of management under general anaesthesia are available.

Laryngoscopy

As far as the anaesthetist is concerned, laryngoscopy involves two types of surgical technique: suspension and manual. When the laryngoscope is suspended and manipulation kept to a minimum, Venturi ventilation with or without a Carden tube is indicated as giving the best possible access to the surgeon. The patient must be fully paralysed and maximum inflation pressures monitored with care. The surgeon must ensure a clear expiratory pathway at all times to prevent barotrauma to the lungs. Anaesthesia should be induced with an intravenous agent–depolarizing relaxant sequence in the usual way but maintained with intravenous agents because standard vaporizers cannot be used in high-pressure systems.

If a 'manual' laryngoscopic technique is used for diagnostic work, particularly when the surgeon is investigating the hypopharynx, Venturi ventilation should *not* be used because intermittent interruption of the expiratory pathway occurs too frequently. If biopsies are being taken when the larynx is not in continuous view, the additional risk of blood contaminating the tracheobronchial tree has to be recognized. In this situation a small cuffed tracheal tube is the better choice for securing the airway with anaesthesia maintained using inhaled vapours.

Bronchoscopy

Bronchoscopy allows four main variations in technique.

1. The rigid ventilating bronchoscope allows the maintenance of anaesthesia with a standard anaesthetic vapour through a side arm with a 15 mm connector which permits attachment to standard anaesthetic circuits. The surgeon derives some protection from anaesthetic gases by use of a glass eyepiece, but must ensure that it is firmly attached during positive pressure ventilation, otherwise it can blow off. An alternative is to use a Sanders Venturi attachment and proceed with a balanced intravenous technique.

2. Fibreoptic bronchoscopy permits the passage of the instrument through the lumen of a standard tracheal tube. This allows the airway to be secured against aspiration at all times while permitting standard anaesthetic maintenance techniques. The connection between the circuit and tube is made with a right-angled swivel connector with a perforated rubber diaphragm which permits access for the broncho-scope while maintaining a gas-tight seal. The limitation of the approach is the resistance to

expiration, which is determined by the cross-sectional area of the instrument relative to the area of the lumen of the tube. The size of the tube lumen is dictated by anatomy, but the size of the bronchoscope is a matter of technology. This latter has been advancing rapidly and now the technique may be considered even in small children.

3. Apnoeic oxygenation utilizes the principle of oxygenation by diffusion in a fully paralysed patient. Intravenous drugs are used to maintain anaesthesia, and paralysis is maintained throughout. Once anaesthesia is induced, profound hypocapnia is achieved with hyperventilation by mask. A small-gauge catheter is passed through the larynx to the carina under direct vision after the larynx has been sprayed with lignocaine. A 1–2 l/min flow of oxygen is attached to the catheter and the patient is presented to the surgeon. The duration of uninterrupted access is limited only by the rate of rise of endogenous CO_2 to significant levels, as long as the patient remains fully paralysed. The technique works well in patients without significant pulmonary disease, but the surgeon must be sparing and careful in his use of suction through the bronchoscope.

4. Deep inhalation anaesthesia with spontaneous respiration is another established technique that is particularly helpful in small children when the surgeon has to use a rigid bronchoscope. Anaesthesia is induced and the patient is allowed to breathe spontaneously an anaesthetic vapour in oxygen. Any of the standard agents is suitable but the most prolonged uninterrupted access to the airway can be achieved with the highly blood-soluble potent analgesic vapour of methoxyflurane. Anaesthesia is taken to the deepest level that prudence permits. When the mask is removed, the surgeon proceeds while the anaesthesia lightens slowly, the patient still breathing spontaneously. It is helpful to spray the larynx with lignocaine before the bronchoscope is passed. Both techniques 3 and 4 are equally suited to flexible or rigid bronchoscopy.

Oesophagoscopy

Oesophagoscopy under general anaesthesia requires full control of the airway. With a large rigid oesophagoscope, some difficulty may be experienced in passing the instrument past the posterior tracheal bulge of the tube cuff (not to be biopsied by inexperienced surgeons). Therefore the anaesthetist should be ready to deflate the cuff transiently to assist the surgeon. The occasional patient will develop profound reflex bradycardia, requiring intervention as the instrument passes the arch of the aorta. However, the major risk is perforation of the oesophagus, so the patient must remain immobile. Some surgeons maintain that, if the patient is breathing spontaneously, the residual muscle tone, particularly in the inferior pharyngeal constrictor, permits them to judge more accurately the amount of force they are using to pass the instrument.

Management of the mentally retarded patient

Otorhinolaryngology is one the specialties in which one has frequent contact with mentally retarded patients. Mental retardation is often associated with congenital abnormalities resulting in multiple otorhinolaryngological problems, particularly those associated with hearing defects.

Anaesthetic evaluation of these patients is difficult. The history obtained is often sketchy and unreliable, and a disproportionate reliance must be placed on the opinions of lay observers. Rapport may be impossible to establish and the physical examination can bear more resemblance to a wrestling match than a dignified medical procedure. Nevertheless, associated congenital deformities of the cardiovascular system, airway and skeletal system must be identified, if present, in order to formulate an anaesthetic plan. It is particularly important in institutionalized patients to remember their high risk of exposure to infective hepatitis.

The anaesthetist should develop a healthy scepticism concerning the adequacy of the standard preoperative safety precautions in these patients. The patient's stomach is never assumed to be empty even when fasted. The cunning of some patients will thwart the efforts of even the most dedicated ward nurse, and a very interesting variety of objects, other than food, has been retrieved from their stomachs during the course of anaesthesia.

Premedication also poses problems. Some form of sedation is usually indicated, but persuading the patient to take it is another matter. The severely retarded, like children, do not take kindly to needles, which should be reserved for use in the operating theatre. Many of them are already on behaviour-modifying medication, and enzyme induction is to be expected with unpredictable implications for the effects from standard premedicant drug dosages. The standard drug mixtures may or may not work. An orally administered mixture of a butyrophenone (haloperidol) and a benzodiazepine (diazepam) may prove effective when mixed with a small amount of fruit preserve if necessary, on the basis of trial and error only.

If the patient is tranquil on arrival in the operating theatre the choice of anaesthetic technique may be made in the usual way. However, if the patient is not cooperative the problem of how to prepare the patient for anaesthesia has to be faced. Reasoning with the patient is rarely an effective option. The surreptitious intramuscular use of ketamine (4

mg/kg) injected into whichever limb muscle may conveniently present itself for a few seconds is suggested. The niceties of removing clothing and skin preparation prior to injection may have to be dispensed with in the face of a patient counterattack. The assistance of able-bodied operating theatre staff is invaluable. Once the drug has taken effect, monitors and an intravenous infusion can be set up and a modified anaesthetic induction begun. Owing to the possibility of a full stomach, cricoid pressure should be used with all such induction.

Frequent visits to the hospital can result from the habit of some retarded patients of swallowing anything to hand. This means brief glory for the anaesthetist if he is able to retrieve the lost object from the hypopharynx at induction without surgical assistance. He should certainly be looking for it for fear of pushing it into a less favourable position. Multiple visits of this nature should not lead the patient's attendants to relax the normal standards of care. These patients are rarely legally competent, so time should be taken to ensure that the person signing the consent is in fact the legal guardian and not merely the patient's custodian.

Management for cancer surgery

The only factors that justify the separate consideration of anaesthesia for cancer surgery of the head and neck are the propensity for large blood loss, heat loss and the requirement for fine dissection around nerves and vessels. Not only do these patients require close attention to heat conservation, to intravenous infusion and the more direct measurement of blood pressure, but also the employment of controlled hypotension may be considered to limit blood loss and provide a drier surgical field. The use of controlled hypotension is still a matter of controversy even after some 70 years of experience.[40]

Many techniques have been described for the deliberate lowering of blood pressure under anaesthesia[41] but only a few are suitable for use in head and neck surgery. The advantages claimed for the technique may be summarized as follows:

1. Reduced bleeding leads to better visualization of the field for fine dissection, especially with the shallow depth of focus of an operating microscope. This statement has two implications:
   a. better visualization diminishes the risk of accidental damage to important structures (e.g. facial nerve in parotidectomy);
   b. better visualization encourages more definitive dissection, thereby improving the chances of success in cancer surgery.
2. Reduced bleeding reduces the need for blood transfusion with its associated risks. This claim has been clearly substantiated for selected operations.[42]

3. For some operations it makes the 'impossible' possible (e.g. resection of the juvenile angiofibroma).

4. It is also claimed that reduced bleeding results in less of the surgeon's time being spent in securing haemostasis. This has a number of advantages. There is less necrotic tissue in the wound from diathermy use, less foreign material in the form of haemostatic ligatures and less tissue handling in the first place. These factors are said to result in less wound infection and breakdown, and lead to better wound scars. It is further claimed that the reduced time spent on haemostatic manoeuvres results in shorter surgical procedures, but on this point the literature is equivocal.

5. The continuation of blood pressure control into the postoperative period is said to reduce the incidence of rebound hypertension and wound haematoma.

Before deciding to use controlled hypotension these advantages must be set against the risks for the individual patient. Anaesthetists who have experience in the technique have varied contraindications but one fact has been clear from its inception. Institutions that use controlled hypotension frequently have a far lower complication rate attributable to the technique itself than those that use it only occasionally. The complications associated with the improper use of the technique are serious and it should not be undertaken lightly.[43] For the details of the different drugs and techniques available the reader is referred to Chapter 39. Remarks here will be confined to the general standard of care applicable to its use in head and neck surgery.

All patients require a large volume line for fluid infusion and transfusion if necessary. A second low dead-space line is required for the administration of the potent vasoactive agents. The recent advances and general availability of sophisticated monitoring equipment now require the direct measurement of arterial pressure and electrocardiography including lead V_5. Arterial blood samples should be taken at frequent intervals for blood gas analysis, haematocrit (packed cell volume) and haemoglobin-oxygen saturation determinations. Core temperature should be monitored in all patients although no measures are taken to actively warm the patient with contact methods because of the risk of thermal burns over pressure points as skin circulation falls. It is better to raise the operating theatre temperature and rely on the meticulous warming of all intravenous fluids together with the active warming and humidification of anaesthetic gases. Particular care must be taken in older patients with atrophic skin as mere body weight on pressure points will produce damage more quickly with controlled hypotension; a synthetic sheepskin under the patient is a suitable precaution.

With improvement in our understanding of the changes that occur in ventilation/perfusion ratios in the lungs at low pulmonary perfusion pressures, the full control of ventilation with a cuffed tracheal tube and intermittent positive pressure ventilation by machine should be used. However, to avoid excessive cerebral vasoconstriction, care must be exercised to maintain $PaCO_2$ close to normal levels. This frequently requires an artificial dead space between the tube and the anaesthesia circuit. Patients at the extremes of age are 'vagal dominant' but those in between tend to respond to attempts to lower their blood pressure with a compensatory tachycardia. This may prove troublesome if it is allowed to become established so β-adrenergic blocking drugs are indicated early in the case. The availability of a fully staffed recovery room with nurses trained to care for these patients is a *sine qua non* of safe practice. The remaining points of technique are still a matter of opinion based on the anaesthetist's personal experience.

For the surgeon's purpose, it should be suggested that the procedures most likely to benefit from controlled hypotension are the major cancer dissections of the head and neck, the fine dissection entailed with operations on the parotid gland and some ear procedures such as tympanoplasty. The absolute indication is the intranasal juvenile angiofibroma. These operations require a reduced blood pressure held stable for an extended period of time. This requirement is best met with the longer acting ganglionic blockers with a slight degree of head-up tilt of the operating table.

The surgeon should be cautioned, however, that arteries still bleed profusely when cut and tissues are more susceptible to injury from careless handling. In particular, the surgeon must guard against 'retractor ischaemia' when nerves or vessels have been retracted. The technique has much to offer the competent surgeon but will not improve pedestrian efforts.

Although the specialist anaesthetist still receives surgical requests for formal hypotensive anaesthetic techniques, they are much less common in modern practice. One reason is the increased use of free flap grafting for primary closure of large defects immediately following cancer resection. Hypotension in any form tends to embarrass the blood supply to these flaps so is contraindicated, certainly during the second part of these dual procedures. The problems of blood loss and heat conservation remain, and are in fact magnified by the increase in operating time associated with complex primary repairs. The value of conservation of heat cannot be overemphasized and its continuation into the postoperative recovery period may prove of importance to those patients undergoing these extended operations.

Management of the difficult airway

Luckily, the truly difficult airway is rare but, since death or serious morbidity may result from mismanagement, the anaesthetist must master the techniques that may be used to avert disaster. Anaesthetists working in otorhinolaryngology gain much experience with difficult airways but at least they work with surgeons who are skilled in tracheotomy when the need arises.

It is useful to divide patients with airway problems into three distinct groups in order to discuss their management.

1. Patients *in extremis* with severe airway obstruction and hypoxia will have accompanying hypercapnia, delirium or unconsciousness. Cardiac arrest will occur if the airway is not cleared immediately.
2. Patients with respiratory distress represent the largest group that presents for anaesthesia. They may exhibit stridor, tracheal tug, intercostal retraction, laboured breathing and agitation. Although increasingly fatigued, they are able to compensate sufficiently to maintain adequate oxygenation and remain alert and cooperative.
3. Patients with occult impending obstruction volunteer little information in their history to suggest airway difficulties, and a cursory physical examination reveals little to warn of the management problems that are to follow the induction of anaesthesia.

The management of the first group is self-evident. Oxygenation must be re-established immediately. If the patient is unconscious and relaxed it is worth a few seconds to attempt rapid laryngoscopy and intubation, but no time should be wasted. If intubation is obviously not going to be easy, the surgeon or anaesthetist should perform an immediate cricothyroidotomy to re-establish the airway before moving the patient to the operating theatre for formal tracheotomy. There is probably no longer any place for the so-called 'emergency' tracheotomy done under unfavourable conditions by an inexperienced surgeon in the hospital environment. If the patient is still moving some air into his lungs laryngoscopy should be omitted, as this could lead to total obstruction. The patient should have his airway supported, and be transferred to the operating theatre breathing 100 per cent oxygen by mask and accompanied by the means of doing a cricothyroidotomy[44] if his condition deteriorates in transit.

The remaining two groups of patients permit the anaesthetist and surgeon the opportunity to plan airway management in advance. The purpose of the management plan is to achieve full control of the airway, to guarantee oxygenation, while sealing it against contamination. The ideal method is to achieve control from above the larynx by passing an oral tracheal tube, under direct vision, in a relaxed anaesthetized patient. This is assumed to cause the least trauma. However, when pathology intervenes the patient's best interests may be served by deviations from the ideal; the choices to be made are represented by the 'decision tree' in Fig. 61.12.

If the problems involved are very severe, the surgeon may elect to do a tracheotomy under local anaesthesia without an oral tracheal tube. But this is less desirable than doing a tracheotomy after the protection afforded by a tracheal tube cuff is in place, even in an awake patient. Similarly, visualized manipulations are always preferable to blind approaches, the latter with their increased risk of damage to the structures of the nasal cavity and the posterior pharyngeal wall with associated bleeding. The height of folly is to undertake surgery under general anaesthesia with a totally uncontrolled airway. In the second group of patients with respiratory distress it is the technical skill of the anaesthetist that determines both the airway control plan and its success. In patients with occult impending obstruction it is the anaesthetist's diagnostic ability in detecting that a problem is present in the first place which is paramount.

Most of the occult problems result from subtle congenital anatomical abnormalities for which the patient compensates with muscular effort. The only clue to the condition may be a history of obstructive sleep apnoea. However, the patient may only admit to early waking under direct questioning, and often the true sequence of events is obtained only from the patient's 'bedmate', who is awakened by heavy snoring and then hears the onset of obstruction or 'choking' followed by the patient's awakening. The normal anaesthetic induction abolishes all muscle tone and promptly precipitates complete obstruction, which

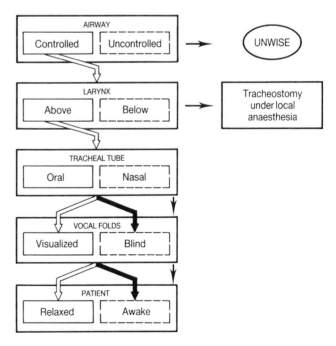

Fig. 61.12 A decision tree for the control of the patient's airway. The more desirable options are indicated by white arrows, the less desirable by black. Each decision (or compromise) is based upon the particular patient's problems, the surgical requirements and the skill of the anaesthetist. (Reproduced, with permission, from reference 48.)

is usually not amenable to the normal methods of re-establishing the airway due to the anatomical abnormalities involved. Premedication alone may be sufficient to precipitate the crisis. Besides congenital conditions, some carcinomas of the base of tongue behave in this manner, as may the patient with epiglottitis.

Having assessed the patient, the means of achieving airway control must be considered. If the desirable condition is oral intubation under general anaesthesia, alternatives have to be decided by a process of exclusion. Ludwig's angina with its large upwardly swelling tongue precludes the use of an oral tube, but passage of a nasal tube with fibreoptic visualization may be feasible. However, the leukaemic coagulopathy resulting in a large tongue would also render any nasal intubation attempt unwise. Similarly, although micrognathia might prevent oral visualization of the larynx and the passage of an oral tube, blind nasal intubation may be easily achieved. Any friable haemorrhagic tumour above the larynx or a large pointing abscess in the pharynx is an absolute contra-indication to any blind technique and could dissuade most from visualized techniques as well. This leaves tracheotomy under local anaesthesia as the only alternative.

Certain precautions should be observed for all. Patients with a compromised airway should receive no depressant premedicant drugs. The neck should be prepared for tracheotomy before any intubation procedures are started. The patient should be fully preoxygenated. The surgeon, fully scrubbed, must remain available in the operating theatre until the airway is secured. If general anaesthesia is to be induced, an inhalation technique is suggested, so that in the event of difficulties developing in maintaining airway patency the induction process may be reversed and the patient awakened. Intravenous drugs once given cannot be retrieved. Sudden obstruction can still occur and if a pharyngeal airway or rapid laryngoscopy does not resolve the problem, immediate tracheotomy is indicated. If the induction is successful, numerous methods are available to ensure that tubes pass into the trachea; these are described elsewhere[45] (see also Chapter 35), but the prerequisites of consistent success in the management of the difficult airway are thorough preoperative assessment, good planning and full cooperation between the anaesthetist and the surgeon.

The post-anaesthesia recovery unit

Patients recovering from general anaesthesia pass through the stages of anaesthesia in reverse and must be watched closely throughout the process. If still in third stage, without a tracheal tube or tracheostomy, continued airway support is needed. As stage 2 is reached, muscle tone progressively returns and the patient becomes irritable and salivates, which can lead to coughing and swallowing. External stimulation should be kept to a minimum to prevent the risk of laryngospasm or vomiting (which are not mutually exclusive). In stage 1 the patient is rousable but pain-free. However, some time may elapse before normal mental facilities with spatial–temporal orientation are regained. It is during this period that the perception of postoperative pain and nausea first intrudes.

All forms of anasthesia potentially reduce cardiac output and peripheral vascular resistance. This leads to increased ventilation/perfusion mismatch in the lungs and a reduced Pa_{O_2}. Therefore all postoperative patients must be supported with supplemental oxygen administered by mask giving an inspired oxygen concentration of at least 30–35 per cent. This is equally true for patients who are still intubated and ventilated. The results of this therapy should be monitored with pulse oximetry. All patients benefit from humidification of the inspired gas, particularly those with an irritable airway.

The effects of anaesthetics on the circulation at the end of an operation may be masked by the release of endogenous catecholamines in response to the final surgical manoeuvres. By the time the patient reaches the recovery room the catecholamine response has usually subsided, exposing the true degree of anaesthetic depression or inadequate fluid replacement. Both causes can usually be treated with a fluid load rather than vasopressors, always remembering that blood pressure, even with adequate circulating volume, will not return to normal preoperative values until the initial phases of anaesthetic recovery are complete.

Depression of respiration also continues into the recovery phase. Those patients who have received a single volatile agent may be relied upon to return progressively to normal as long as spontaneous respiration with a clear airway is maintained (Fig. 61.13). This may not be true for patients who have undergone a balanced technique where the timing of the last dose of muscle relaxant or narcotic drug is of importance. All non-depolarizing relaxants should be reversed but changes in body temperature or pH may lead to recurarization. The surgeon should also be aware of the risk associated with the use of some of the antibiotics prescribed postoperatively. Most of the 'mycin' antibiotics are weak neuromuscular blockers in their own right and can synergize with residual muscle relaxants to the extent that may pose a serious risk to the patient. If this situation develops, rather than attempt reversal of the block with additional anticholinesterases the patient should be reintubated and ventilated until normal excretion takes care of the problem.

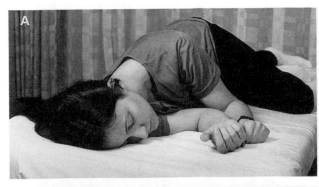

Fig. 61.13 The tonsil position can aid in maintaining a clear airway during the recovery period, particularly for patients with blood and secretions draining from the mouth. It is sometimes useful during inhalation induction of anaesthesia when similar problems are encountered. (a) Note the body's 45-degree inclination forward stabilized by arm and leg positions. (b) Note that the absence of a pillow under the head permits the width of the shoulder to incline the pharynx downwards, facilitating drainage away from the larynx.

The narcotics pose parallel difficulties in terms of circulatory and central respiratory depression, and injudicious use of additional postoperative analgesics can lead to profound depression. Therefore the need for additional narcotics should be carefully evaluated with small intravenous doses until the patient's overall clinical condition is better defined. If significant narcotic depression is suspected, narcotic antagonists such as naloxone may be used, but it must be remembered that the duration of action of the narcotic usually exceeds the duration of action of the antagonist, so late renarcotization can occur.

If a difficult intubation has been encountered at induction of anaesthesia, it is usual to transfer the patient from the operating theatre with the tracheal tube in place. The patient should be allowed to awaken fully before any consideration is given to removing the tube. If the patient fights the tube during second stage emergence, it must not be removed but rather the patient sedated and vital signs controlled as appropriate. Once the patient is fully conscious and orientated, and the anaesthetist is satisfied with the adequacy of the patient's spontaneous ventilation, a standard protocol for extubation should be followed:

1. Prepare equipment for reintubation, including a tracheal tube one full size smaller.
2. If reintubation is required and fails, *what is the plan?*
3. Instill 4 ml of 2 per cent lignocaine down the tracheal tube into the patient's trachea.
4. Suction tracheal and pharyngeal secretions if necessary and preoxygenate the patient.
5. The anaesthetist then deflates the tube cuff and 'stops' the tube (obstructs the lumen with his thumb) to see if the patient can breathe around the tube. If he can:
6. Pass a tube guide (e.g. Teflon or gum-elastic bougie) through the tube into the trachea.
7. Withdraw the tube over the guide, leaving the guide in place.
8. Secure the guide through an oxygen mask and observe the patient for developing obstruction.

Patients will tolerate the presence of a guide for extended periods and can talk around it (Fig. 61.14). If there is no evidence of developing obstruction, it is usually safe to remove it after 30 minutes. However, if there is any suggestion of deterioration in the function of the patient's airway, the guide is used immediately to 'railroad' a new tube of similar size or smaller back into the trachea. If the patient cannot breathe around the original tube when stopped, obstruction on extubation is more likely, so the patient's surgeon should be consulted before proceeding further. The surgeon may wish to be present during extubation attempts or he may decide to return the patient to the operating theatre for an elective tracheotomy with the existing tracheal tube in place. These precautions are no protection against the occasional late airway obstruction that does occur. Therefore if the patient is a 'difficult intubation', this should be communicated to the postoperative ward and clearly noted.

Once these potential 'normal' emergence problems have been overcome, the patient's continued presence in the recovery unit is not required. The criteria listed in Table 61.2 (modified from Aldrete and Kronlik[46]) are of use in determining readiness for discharge from the recovery ward. The score for each section is summed and those patients having a total score in the range of 8–10 are considered ready for discharge to general care. It is unlikely that a patient can achieve a top score in four sections with a zero in the fifth; the one exception is the patient with chronic respiratory paralysis, in whom continued ventilatory support would require admission to intensive care rather than general care for initial convalescence.

Several additional factors must be considered in the discharge of outpatients. The patient must not only be stable but also 'street ready'. Activity assessment must include the return of the postural reflexes required for standing and walking safely. Postoperative pain must be under good control. The patient must demonstrate the

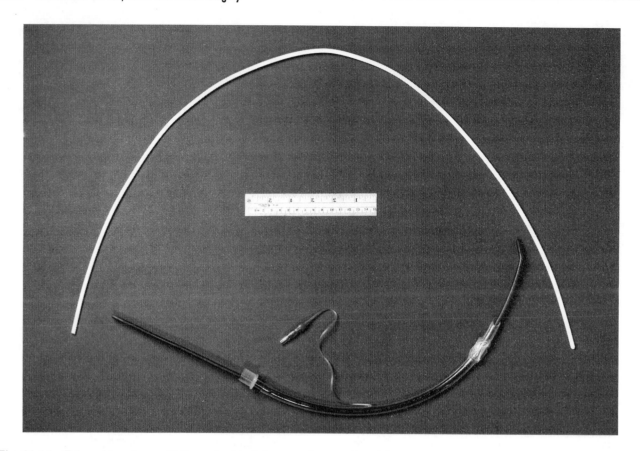

Fig. 61.14 Tube guides: above, a Teflon rod guide; below, a modern version of the gum-elastic bougie. These devices are well tolerated for awake extubation manoeuvres and are also useful as intubation guides when visualization of the larynx or space within the pharynx is restricted.

Table 61.2 Criteria to determine patient's readiness for discharge

System	Observation	Score
Respiration	Patient is able to breathe freely	2
	Respiratory effort limited (splinting or dyspnoea)	1
	Absence of spontaneous respiration	0
Blood pressure	Systolic pressure ± 20% of preoperative level	2
	Systolic pressure ± 20–50% of preoperative level	1
Consciousness	Fully alert and orientated	2
	Arousable only when called by name	1
	Auditory stimulus fails to produce response	0
Colour	Normal 'pink' mucous membranes and nailbeds	2
	Discoloration – anything but pink or blue	1
	Frank cyanosis	0
Activity on command	Able to move all four limbs	2
	Able to move only two limbs	1

ability to micturate before discharge. Immediate post-operative surgical problems must be excluded and significant nausea and vomiting controlled. Last but not least, the outpatient may be discharged only to the care of the competent adult for transport home and must remain under supervision at home for the first 24 hours following anaesthesia.

Conclusion

A survey[3] of the English-language literature on the safety of anaesthesia concluded that the risk of death attributable to anaesthesia has dropped from 1 in 2680 to about 1 in 10 000 over the last 30 years. Those who have studied the subject believe that all Western nations experience nearly the same mortality rate. If this rate is applied to the approximately 20 million anaesthetics administered annually in the USA alone, at least 2000 deaths may be anticipated. Various authorities have suggested, with the benefit of hindsight, that anywhere between 50 and 90 per cent of these deaths could be prevented.

The serious complications of anaesthesia following which the patient does not die are also rare, but when they do occur they tend to be devastating, because most involve variable degrees of hypoxic brain damage with the high costs of chronic dependency. The problem is not limited to short periods of anoxia associated with an acute airway or circuit accident but includes prolonged 'sub-oxygenation' which may occur in several ways when patients of poor physical status undergo major surgery. The line between adequate and inadequate tissue oxygenation may be very fine indeed. In spite of this depressing catechism of serious problems, the average patient still has less chance of sustaining a serious injury from anaesthesia than of being killed or seriously injured in a traffic accident.

What is of more interest to the patient are the chances of experiencing the minor complications of anaesthesia. Much work remains to be done in quantifying the true incidence of these complications. Existing studies show a wide range of variation in incidence.[47] This is not only due to differing patient populations undergoing different operations with varying anaesthetic techniques, but also appears to be related to the elapsed time between the end of anaesthesia and the patient interview. For example, the reported incidence of postoperative nausea possibly attributable to anaesthesia appears to vary from 27 per cent of patients interviewed within 24 hours of anaesthesia, up to 70 per cent of patients interviewed after 3 days. The most common irritating complications of general anaesthesia that the patient should be warned of are given in Table 61.3.

ENT patients undergoing middle and inner ear procedures are particularly prone to nausea from transient vertigo. All patients who have undergone surgery in the aerodigestive tract tend to swallow blood, and experience a high incidence of nausea and emesis. Therefore, such symptoms should not automatically be attributed to the anaesthetic.

In addition to the above, minor complications associated with specific techniques or surgical procedures must be remembered. These complications can result from poor

Table 61.3 Common irritating complications of general anaesthesia[47]

Complication	Reported incidence (%)
Headache	2–60
Sore throat	6–38
Muscle pains (following suxamethonium)	0–100
Nausea/vomiting	27–70
Venous complications (following i.v. drugs)	1–11

positioning and poor protection measures of the patient under anaesthesia. Nerve compressions (e.g. ulnar from poor elbow protection) and stretching (e.g. brachial plexus in the abducted arm) can result from poor positioning. Corneal abrasions from lack of adequate eye protection and careless mask technique, as well as jaw and neck pain from over-forceful laryngoscopy and head extension have all been reported. A recurring source of patient complaint is minor damage to the teeth and lips associated with laryngoscopy by either surgeon or anaesthetist. Most of these incidents can be prevented with care but the patient should still be warned of their possible occurrence in advance.

The anaesthetist should be particularly aware of the delayed complications that may develop. Manipulation of the airway, whether by surgeon or anaesthetist, is the commonest cause of laryngospasm and acute airway obstruction in the recovery room. Recurarization and renarcotization leading to hypoxia must always be considered in the restless patient. In fact, restlessness is usually due to pain, blood loss or hypoxia, in that order of frequency, but it is *assumed to be due to hypoxia until proved otherwise*. The persistence of nerve compression paraesthesia, hypotension, urinary retention and altered mental status into the postoperative period must always be taken seriously. A careful watch should be kept also for the development of postoperative pulmonary complications and the appearance of jaundice.

REFERENCES

1. Bigelow HJ. Insensibility during surgical operation produced by inhalation. *Boston Medical Surgical Journal* 1846; **35**: 309–16.

2. Dornette WHL, Orth OS. Deaths in the operating theatre. *Anesthesia and Analgesia* 1956; **35**: 545–69.

3. Davies JM, Strunin L. Anaesthesia in 1984: how safe is it? *Canadian Medical Association Journal* 1984; **131**: 437–41.

4. Murray JP, Geiduschek JM, Caplan RA, Posner KL, Gild WM, Cheney FW. A comparison of pediatric and adult anesthesia closed malpractice claims. *Anesthesiology* 1993; **78**: 461–7.

5. Keats AS. The ASA classification of physical status – a recapitulation. *Anesthesiology* 1978; **49**: 233–8.

6. Green NM. Evaluation of perioperative risk. *Anesthesia and Analgesia* 1981; **60**: 623–4.

7. Munson WM, Jenicek JA. Effects of anesthetic agents on patients receiving reserpine therapy. *Anesthesiology* 1962; **23**: 741–6.

8. Gelder MG, Vane JR. Interaction of the effects of tyramine, amphetamine and reserpine in man. *Psychopharmacologia (Berlin)* 1962; **3**: 231–41.

9. DeRoetth Jr A, Dettbarn WD, Rosenberg P. Effect of

phospholine iodide on blood cholinesterase levels. *American Journal of Ophthalmology* 1965; **59**: 586–92.

10. Smith RB. Drug interactions and drug reactions. *Otolaryngologic Clinics of North America* 1981; **14**: 615–29.

11. Ray DC, Drummond GB. Halothane hepatitis [review]. *British Journal of Anaesthesia* 1991; **67**: 84–99.

12. Putnam L, Jenicek JA, Allen CR, Wilson RD. Anesthesia in the morbidly obese patient. *Southern Medical Journal* 1974; **67**: 1411–17.

13. Roizen MF. Routine preoperative evaluation. In: Miller RD, ed. *Anesthesia.* New York: Churchill Livingstone, 1990: 743–72.

14. Gregoire S, Smiley RLC. Acute hepatitis in a patient with mild factor IX deficiency after anaesthesia with isoflurane. *Canadian Medical Association Journal* 1986; **135**: 645–6.

15. Powell-Jackson P, Greenway B, Williams R. Adverse effects of exploratory laparotomy in patients with suspected liver disease. *British Journal of Surgery* 1982; **68**: 449–51.

16. Zuckerman AJ. Hepatitis C virus: a giant leap forward. *British Medical Journal* 1989; **299**: 871–2.

17. Elliott RH, Strunin L. Hepatotoxicity of volatile anaesthetics. [review] *British Journal of Anaesthesia* 1993; **70**: 339–48.

18. Londy F, Norton, ML. Radiologic techniques for evaluation and management of the difficult airway. In: Norton ML, Brown ACD, eds. *Atlas of the difficult airway.* St Louis: Mosby Year Book, 1991: 55–66.

19. Informed consent: *Salgo v Leland Stanford Jr. Univ. Bd of Trustees,* 154 Cal App 2d560, 317 P.2d170 (1957).

20. Strunin L. How long should patients fast before surgery? Time for new guidelines. *British Journal of Anaesthesia* 1993; **70**: 1–3.

21. Standards for basic intra-operative monitoring. In *ASA directory.* Chicago: American Society of Anesthesiologists, 1993: 709–10.

22. Stauffer JL, Olson DE, Petty TL. Complications and consequences of endotracheal intubation and tracheostomy. A prospective study of 150 critically ill adult patients. *American Journal of Medicine* 1981; **70**: 65–76.

23. Bain JA, Spoerel MD. A streamlined anaesthetic system. *Canadian Anaesthetists' Society Journal* 1972; **19**: 426–35.

24. Gay GR, Loper KA. Control of cocaine-induced hypertension with labetalol. [letter] *Anesthesia and Analgesia, Current Research* 1988; **67**: 92.

25. Fink BR. Acute and chronic toxicity of local anaesthetics. *Canadian Anaesthetists' Society Journal* 1973; **20**: 5–16.

26. Koster H. Spinal anesthesia with special reference to its use in surgery of head and neck and thorax. *American Journal of Surgery* 1928; **5**: 554–70.

27. Martof AB. Anesthesia of the teeth, supporting structures and oral mucous membrane. *Otolaryngologic Clinics of North America* 1981; **14**: 653–68.

28. Carron H. Control of pain in the head and neck. *Otolaryngologic Clinics of North America* 1981; **14**: 631–52.

29. Gallo JA. Catecholamine anesthetic interactions in ENT surgery. In: Brown BR, ed. *Anesthesia and ENT surgery.* Philadelphia: FA Davis, 1987: 7–30.

30. Johnston RR, Eger II EI, Wilson CA. A comparative interaction of epinephrine with enflurane, isoflurane and halothane in man. *Anesthesia and Analgesia* 1976; **55**: 709–12.

31. Kahl HW, Swedlow DB, Lee KW, Downes JJ. Epinephrine–halothane interaction in children. *Anesthesiology* 1983; **58**: 142–5.

32. Joas TA, Stevens WC. Comparison of arrhythmic doses of epinephrine during forane, halothane and fluroxene anesthesia in dogs. *Anesthesiology* 1971; **35**: 48–53.

33. Atlee JL, Malkinson CD. Potentiation by thiopental of halothane–epinephrine-induced arrhythmias in dogs. *Anesthesiology* 1982; **57**: 285–8.

34. Katz RL, Katz GJ. Surgical infiltration of pressor drugs and their interaction with volatile anaesthetics. *British Journal of Anaesthesia* 1966; **38**: 712–18.

35. Scott DB, Jebson PJR, Braid DP, Örtengren B, Frisch P. Factors affecting plasma levels of lignocaine and prilocaine. *British Journal of Anaesthesia* 1972; **44**: 1040–9.

36. Patterson ME, Bartlett PC. Hearing impairment caused by intratympanic pressure changes during general anesthesia. *Laryngoscope* 1976; **86**: 399–404.

37. Norton ML, deVos P. A new endotracheal tube for laser surgery of the larynx. *Annals of Otorhinolaryngology* 1978; **87**: 554–7.

38. Sanders RD. Two ventilating attachments for bronchoscopy. *Delaware Medical Journal* 1967; **39**: 170–5.

39. Schramm VL, Mattox DE, Stool SE. Acute management of laser ignited intratracheal explosion. *Laryngoscope* 1981; **91**: 1417–26.

40. Enderby GEH. In: Special issue on controlled hypotension. *British Journal of Anaesthesia* 1975; **47**: 743–803.

41. Larson AG. Deliberate hypotension [review]. *Anesthesiology* 1964; **25**; 682–706.

42. Sataloff RT, Brown ACD, Sheets EE, Rubinstein MI. A controlled study of hypotensive anesthesia in head and neck surgery. *Ear, Nose and Throat Journal* 1987; **66**: 479–85.

43. Hampton LJ, Little PM. Complications associated with the use of controlled hypotension in anesthesia. *AMA Archives of Surgery* 1953; **67**: 549–56.

44. Weis S. A new emergency cricothyrotomy instrument. *Journal of Trauma* 1983; **23**: 155–8.

45. Brown ACD. Anesthetic management. In: Norton ML, Brown ACD, eds. *Atlas of the difficult airway.* St Louis: Mosby Year Book, 1991: 169–98.

46. Aldrete JA, Kronlik D. A post-anesthetic recovery score. *Anesthesia and Analgesia* 1970; **49**: 924–34.

47. Riding JE. Minor complications of general anaesthesia. *British Journal of Anaesthesia* 1975; **47**: 91–101.

48. Brown ACD. Anesthesia. In: Cummings X, Fredrickson X, Harker X, Kianse X, Schuller X, eds. *Otolaryngology – Head and Neck Surgery,* 2nd ed, vol 13. St. Louis: Mosby-Year Book 1993: 214–42.

49. Sataloff RT, Brown ACD. Special equipment in the operating room for otolaryngology – head and neck surgery. *Otolaryngologic Clinics of North America* 1981; **14**: 669–86.

50. Cork RC. Anesthesia for otolaryngologic surgery involving use of a laser. In: Brown BR, ed. *Anesthesia and ENT surgery.* Philadelphia: FA Davis, 1987: 127–40.

Chapter 62

Anaesthesia for Orthopaedic Surgery

Frank L. Murphy

Issues common to patients requiring orthopaedic procedures	Specific orthopaedic procedures

Orthopaedic operations rarely compromise the function of major organ systems in the way that invasive neurosurgical, abdominal, cardiac or thoracic procedures may. Therefore, relatively few patients die as an immediate result of orthopaedic surgery, provided that they receive minimally competent care from their surgeons and anaesthetists. Nevertheless, these patients gain much from skilful management of their anaesthetics. Enhanced patient comfort, fewer transfusions, a reduced rate of nerve damage, fewer and less severe episodes of hypotension and hypertension, lessened risk of thromboemboli and possibly even a reduced death rate may all result from sophisticated anaesthetic management directed at the special needs of orthopaedic patients.

Such management calls on knowledge and techniques found elsewhere in this book. Orthopaedic patients may be using steroids or suffer from endocrine diseases associated with osteoporosis (Chapter 49). With increasing frequency, they are geriatric patients (Chapter 38); children also require special techniques (Chapter 32). Orthopaedic procedures frequently require positions more hazardous than the supine (Chapter 48). Patients who undergo minor orthopaedic procedures often return home on the day of surgery (Chapter 68). This chapter addresses issues specific to patients requiring anaesthesia for orthopaedic procedures.

Issues common to patients requiring orthopaedic procedures

Preoperative evaluation

Preoperative evaluation of patients scheduled for orthopaedic procedures requires special attention to a number of matters. Difficulties in positioning on the operating table

may arise, owing to arthritis, fractures, joint deformities or unstable vertebrae; sometimes it is profitable to rehearse the surgical position while the patient is awake. Patients with aseptic necrosis of the femoral or humeral head may be alcoholics, use steroids or have suffered from barotrauma. If the patient has donated autologous blood in anticipation of a major operation, it is prudent to ensure the availability of these units before inducing anaesthesia.

Rheumatoid arthritis and ankylosing spondylitis

Patients suffering from rheumatoid arthritis and ankylosing spondylitis present for a wide variety of orthopaedic procedures, ranging from spinal surgery to prosthetic replacement of hip, knee and hand joints. Patients with rheumatoid arthritis are more likely to be women and to suffer peripheral joint disease, with sparing of the thoracic and lumbar spines, making spinal and epidural anaesthesia practical choices. The most common locus for instability of the cervical spine is at C1–C2; the cervical spine is usually stable in extension, but not in flexion. Ankylosing spondylitis occurs more often in men, is more likely to be confined to the spine, and affects the entire spine, making spinal and epidural anaesthesia often impractical, and leaves the spine unusually brittle. The cervical spine is especially susceptible to fracture and dislocation at the C5–C7 level.[1]

These diseases commonly produce some difficulties for the management of anaesthesia. Immobility and fragility of the cervical spine result from either, and spinal cord or cervical root damage may already be present. Arthritis affecting the temporomandibular joints may restrict opening of the mouth, whilst cricoarytenoid disease may leave the patient hoarse with limited vocal cord mobility and a restricted glottic opening. These problems require care in manipulating the head and neck for airway management; special techniques such as blind intubation, fibreoptic laryngoscopy or lighted stylet intubation may be required (see Chapter 35).

Both rheumatoid arthritis and ankylosing spondylitis may affect organ systems other than bones and joints, but remaining function is almost always sufficient to permit orthopaedic procedures. Restrictive pulmonary disease results from fibrosis and limitation of chest wall movement. Carditis, conduction defects, valvular disease and renal damage may all be found.

Corticosteroid therapy

In addition to patients with rheumatoid arthritis, many other patients coming for orthopaedic procedures use or have used corticosteroids in the past. It has become routine for many surgeons and anaesthetists to treat these patients with large doses of corticosteroid on the day of operation to avoid the risk of acute adrenal insufficiency (see Chapter 17).

This practice is open to question. The rare harmful effect of corticosteroid deficiency, unwanted hypotension, may be treated as it occurs. Also, large doses of steroids can impair wound healing. Thus it is possible that prophylactic corticosteroids do more harm than good; large-scale outcome studies are lacking. Furthermore, the dose or the duration of previous steroid therapy does not predict reliably the occurrence of adrenocortical deficiency. This makes it impossible to select accurately those patients who should receive prophylactic treatment without formal testing of adrenal responses.[2]

Sickle cell disease

The management of patients with sickle cell disease is presented in Chapter 20. These patients are prone to ischaemic necrosis of the femoral head and may require total hip arthroplasty as well as other orthopaedic procedures. It is logical to fear that the cooling, hypoxaemia and acidosis that occur distal to the occlusive tourniquets used to control blood loss in orthopaedics will provoke sickling. However, experience suggests that the use of a tourniquet does not increase the risk of sickle crises in these patients.[3]

Patients suffering multiple trauma

Some multiple trauma victims require immediate orthopaedic surgery for an open bone or joint injury (6 hours is a recommended maximum delay) or when an extremity suffers neurovascular compromise. The overall management of these patients follows the principles outlined in Chapter 50. There are a number of special concerns for the fracture patient.

Blood loss from pelvic and femoral fractures, or from multiple fractures of smaller bones, can amount to many litres, and is difficult to assess by examination. Those caring for severely traumatized patients must be prepared to deal with massive blood loss. In many cases, unreplaced blood loss and the difficulties in positioning the patient with multiple injuries make spinal or epidural anaesthesia impractical. Inducing general anaesthesia before transferring the patient from the stretcher or bed can provide welcome relief from the pain involved in movement. Respiratory distress unexplained by chest trauma may be due to fat embolism, which is prevalent in these patients. If hypocapnia is needed to control increased intracranial pressure, arterial blood gas determinations must be used to supplement measurement of end-tidal carbon dioxide when pulmonary injury widens the gap between the two values. Furthermore, augmented ventilation is needed to treat the hypercapnia that follows deflation of the pneumatic tourniquet.[4]

Other injuries do not require immediate operation; the visit to the operating room can be delayed for days or weeks, so that acute problems of cardiovascular instability or respiratory insufficiency may resolve. The patient's injuries or the use of skeletal traction may still limit positioning for spinal or epidural anaesthesia. If there is extensive muscle injury, suxamethonium (succinylcholine) may provoke dangerous hyperkalaemia. Thrombi that have accumulated in the injured leg may be mobilized during the skin washing, during wrapping with an elastic bandage to express blood, or even with simple tourniquet inflation.[5] In such cases, it is best to avoid using the elastic bandage and simply elevate the leg to allow blood to drain before inflating the tourniquet.

Blood loss

Expected blood loss

During orthopaedic procedures, most blood is lost from raw bone and muscle surfaces rather than from identifiable blood vessels. This limits the surgeon's ability to control bleeding directly, allows much of the shed blood to escape collection by suction catheters or gauze sponges, and ensures that bleeding continues after the wound is closed. Surgeons and anaesthetists almost always underestimate blood loss. Radioisotope study of blood lost in major orthopaedic procedures showed that estimates of loss were, on average, 50 per cent of the true measured losses.[6]

So many factors affect the volume of blood lost that tables of expected losses for various procedures are of little use. However, under some circumstances blood losses will be greater than usual during orthopaedic procedures:

1. Proximal surgery without a tourniquet
2. Large areas of raw bone
3. Previous surgery at operative site
4. Radiation therapy at operative site
5. Infection at operative site
6. Tumour at operative site
7. Surgical technique
8. Proliferative bone disease (e.g. Paget's disease)

Benefits of reduced blood loss

When blood loss is minimized, patients benefit because their exposure to blood donated by others is reduced. The two major viruses transmitted through blood transfusion are HIV and hepatitis viruses, the latter being by far the more common. Present screening methods reduce the risk of hepatitis transmission to 3 cases per 10 000 units transfused, making bank blood a relatively safe product if only the risk of serious or fatal viral infection is considered.[7]

The use of banked blood may have other harmful effects on patients undergoing orthopaedic operations. It is now clear that blood transfusion impairs the recipient's immunocompetence. This may explain a provocative study of 84 patients undergoing hip replacement surgery, among whom 16 of the 50 who received homologous blood suffered various postoperative infections, including wound infections, compared to 1 of the 34 who received only autologous blood.[8] Although these infections are not as dramatic as the far rarer HIV and hepatitis infections, they raise the fear of bacterial infection of implanted prostheses. Further studies of this issue are required before concluding that homologous transfusion increases the postoperative infection rate, but this complication may become a major contraindication to the use of bank blood.

Reduced bleeding in the operative field not only facilitates surgery but might also, in some cases, improve surgical results. Radiographic assessment (without long-term follow-up) of the cement–bone interface in the acetabular component of cemented total hip arthroplasties showed that patients managed with deliberate hypotension and who had operative blood losses less than 300 ml exhibited better penetration of the cement into the bone.[9]

Methods of limiting blood loss

A pneumatic tourniquet placed proximal to the site of a peripheral procedure and inflated to occlude arterial flow eliminates blood loss during the operation. However, significant losses still occur postoperatively, when the tourniquet is removed. Some surgeons routinely deflate the tourniquet before closing the wound to ligate or coagulate bleeding points. Whilst this may be of use in some procedures on the hand or other areas, when a haematoma may compromise the surgical result, it has been shown in one study not to reduce overall blood loss in total knee arthroplasty.[10]

Haemodilution, maintaining the patient's blood volume with cell-free fluids such as saline, pasteurized albumin or starch, is the most common method used for the management of blood loss without resorting to bank blood. Such haemodilution is limited by the levels of allowable anaemia. The 1988 NIH Consensus Conference suggested a lower limit for haemoglobin content of 8 g/dl, but much more severe anaemia has been tolerated without harm by Jehovah's Witnesses undergoing major orthopaedic procedures.

Autologous blood, collected from the wound or the surgical field or donated before surgery by the patient, is safer than banked blood. It is difficult to justify proceeding with an elective orthopaedic operation that entails significant blood loss until adequate supplies of the patient's own blood have been collected, unless the patient cannot provide blood for some reason. Bleeding during major orthopaedic procedures occurs by oozing from large surfaces; therefore, induced hypotension reduces blood losses.[11] Spinal or epidural surgery used alone reduces blood losses significantly during total hip arthroplasty or other forms of hip surgery. When high levels of epidural anaesthesia (T4) are combined with general anaesthesia, and mean arterial pressures are allowed to decrease to about 55 mmHg, total hip arthroplasty can be performed with blood losses of 300 ml or less.[12]

Many patients scheduled for such operations are old and may present relative contraindications to hypotensive anaesthesia. Nevertheless, Sharrock has reported a series of 1016 consecutive patients who underwent total hip arthroplasty, 987 of whom were managed with combined epidural and general anaesthesia, using an adrenaline (epinephrine) infusion to maintain mean arterial blood pressure of 52–55 mmHg. Among these patients, one-third of whom suffered preoperatively from hypertension, postoperatively there were no strokes, no renal failure, two myocardial infarctions (MI) and three deaths (one due to MI, one to aspiration and one to pulmonary embolism).[12] These results suggest that the combination of high levels of epidural anaesthesia with general anaesthesia may be safer than one would predict, and deserves wider application as a means of reducing blood loss during orthopaedic surgery.

Although early studies suggested that desmopressin (DDAVP) might reduce blood loss during surgery for the insertion of a Harrington rod, a recent controlled study revealed it did not reduce blood losses in healthy patients with no preoperative coagulopathy.[13]

Management of blood loss

The proper management of blood loss during major orthopaedic procedures begins with recommending regional anaesthesia when other factors permit, and the use of autologous blood whenever possible. The administration of cell-free fluids to prevent the signs of hypovolaemia, rather than basing fluid therapy on unreliable estimates of blood loss, will ensure that patients do not suffer from hypovolaemia and low cardiac output. Moderate degrees of hypotension and haemodilution are likely to be both safe and useful. Extremes of anaemia and reduced blood pressure, with or without epidural or spinal anaesthesia, appear to have been safe in the hands of some, but are not yet universally practised and certainly require careful selection of patients and the exercise of sophisticated judgement.

Tourniquets

Orthopaedic surgeons place occlusive pneumatic tourniquets round patients' arms or legs, usually after expressing the blood from the limb with a tightly wrapped elastic bandage. This may reduce total blood loss (see above), and it provides a dry field, making exposure and dissection more precise.

Effects of inflating tourniquets

The tourniquet must be placed over smoothly applied padding to prevent skin damage; antiseptic solutions used to prepare the skin must not be allowed under the tourniquet lest chemical burns occur. It is likely that overinflation increases tissue damage. Inflation pressure should be limited to that needed to block arterial flow, usually 50–100 mmHg greater than the patient's systolic pressure. The patient's blood pressure may increase after incision and therefore the blood pressure measured during the surgical preparation is not always a reliable guide to setting the pressure for the tourniquet. The time at which the cuff is inflated and deflated, as well as the pressure, should be recorded on the anaesthetic record. The pneumatic tourniquet, as with any medical device, requires periodic inspection and calibration.

Neurapraxias sometimes follow the use of a pneumatic tourniquet. Electrophysiological testing and a thorough neurological examination serve to distinguish these injuries from more proximal injuries that might be attributed to spinal or epidural anaesthesia. The immediate effect of applying the tourniquet is to increase the central blood volume; if both legs are exsanguinated at once, marked increases in central venous pressure (CVP) may occur, which may not be tolerated by patients with heart disease.

Within 10–20 minutes, conduction fails in nerves underlying the tourniquet, contributing to anaesthesia at the surgical field.

The isolation of the limb from the central circulation allows it to cool. Isolation also eliminates the exposed skin of the operated limb as a cooling site, conserving the patient's body heat. In children anaesthetized in a warm room, the use of a tourniquet can even permit the patient's temperature to increase during the anaesthetic.[14]

In unanaesthetized patients, dull aching pain at the site of the tourniquet rules out the use of this device if local anaesthesia alone is to be used. Unmedicated volunteers tolerated thigh tourniquets for 30 minutes, reporting pain not only under the tourniquet but elsewhere in the leg as well, suggesting that 'tourniquet pain' is a complex entity.[15]

In some but not all patients undergoing general anaesthesia, there appears within the first hour a progressive increase in blood pressure that may reflect the pain inflicted by the tourniquet. This hypertension is difficult to treat with the usual doses of narcotic or inhaled anaesthetics, especially because the operation is likely to end immediately after deflation of the tourniquet. Delayed emergence is likely if large doses of the drugs are used. If the patient is otherwise well anaesthetized, direct-acting vasodilators such as sodium nitroprusside or trimetaphan may be suitable choices for the treatment of hypertension induced by a pneumatic tourniquet.

In patients under intravenous regional, spinal or epidural anaesthesia (but usually not a formal brachial plexus block), tourniquet pain sometimes appears before the anaesthesia fails in respect of the surgery itself, or before the level assessed by pinprick has receded to the level of the tourniquet. This pain occurs inconsistently, coming on at various intervals after the inflation of the tourniquet.

Sympathetic fibres, once cited as the mediators of tourniquet pain, are not especially involved, as tourniquet pain is still experienced with blockade up to T4 spinal level. Similarly, stellate ganglion block does not ameliorate tourniquet pain experienced during intravenous regional anaesthesia of the arm.[16]

Administering greater doses of drug for spinal anaesthesia decreases the incidence of tourniquet pain.[17] Hypobaric and isobaric spinal techniques result in a lower incidence of pain than do hyperbaric techniques, even when the drugs, dose and dermatome levels are the same.[18] A given dose of near isobaric spinal bupivacaine provides better freedom from tourniquet pain than does the same dose of amethocaine (tetracaine), despite identical levels of anaesthesia.[19] This may be related to the fact that, in isolated rabbit C-fibre preparations, bupivacaine is the more potent drug in blocking higher-frequency repeated impulses.[20] Intrathecal narcotics reduce the severity of tourniquet pain.[21]

Taken together, these observations can be explained by

making the assumption that tourniquet pain represents a maximal pain stimulus, perhaps carried by C-fibres. Its appearance during spinal or epidural anaesthesia represents an inadequate or dissipating block. The fact that tourniquet pain can be relieved during continuous spinal or epidural anaesthesia by adding more drug is consistent with this hypothesis. This hypothesis suggests that, to reduce the incidence of tourniquet pain in patients undergoing spinal or epidural anaesthesia, one could use larger doses and higher concentrations of local anaesthetics, add opioids and supplement the anaesthetic through an indwelling catheter. Adding bicarbonate to local anaesthetic solutions speeds the onset of epidural anaesthesia, presumably by increasing the fraction of the drug present as the free base; it seems likely that this technique will also reduce the incidence of tourniquet pain.

Pneumatic tourniquets produce tissue ischaemia, cell damage and death inevitably following if ischaemia is prolonged. Recommendations as to the permissible inflation time have been based on electron microscopic evidence of cell disruption and depletion of high-energy phosphate stores. From 90 to 120 minutes seems to be an appropriate interval, with reperfusion periods of 5–10 minutes allowed between episodes of tourniquet inflation.[22]

Effects of deflating tourniquets

When the pneumatic tourniquet is deflated, the resulting effects depend on the volume of ischaemic tissue and the duration of the ischaemia. Sequelae are usually negligible when a single arm tourniquet is deflated after a brief procedure, and are unmistakable when double leg tourniquets are deflated after an operation exceeding 2 hours.

Products of anaerobic metabolism are released; the venous blood coming from the leg demonstrates mixed respiratory and metabolic acidosis. For about 5 minutes after deflation of a single leg tourniquet, the patient's excretion of carbon dioxide is increased by about 50 per cent.[23] Hypotension probably results from a combination of the abrupt decrease in peripheral resistance, the increase in venous capacitance, the release of vasoactive substances derived from ischaemic vascular endothelium and the onset of bleeding. Hypoxaemia may be caused by showers of platelet aggregates, thromboemboli, air emboli, or fat and other marrow emboli. The patient's core temperature decreases as the limb is rewarmed.

Although these transient effects usually are well tolerated, patients must be monitored closely in the minutes following deflation of a pneumatic tourniquet. Oxygen and vasopressors usually suffice to treat any severe responses, but staged or intermittent deflation can spread these effects out over time and moderate their severity.

Methacrylate cement

Methylmethacrylate cement is used as a space filler (not an adhesive) to improve the fit of implanted orthopaedic prostheses to bone. The vapours given off by the curing cement are pungent and have raised some concern about pollution of the air in the operating room. The polymerizing reaction is exothermic, suggesting that there may be some risk of thermal injury to patients' tissues. However, the major problem arising from the use of cement is the cardiovascular response.

When the cement is introduced, some patients develop hypotension and hypoxaemia of varying severity, and may sustain a cardiac arrest. These reactions are not as common or as severe with uncemented prostheses. These responses to cement are poorly understood and are probably mediated by several complex mechanisms. The direct effect of the liquid monomer produces vasodilation, but the doses required in animal models to produce these effects are greater than those typically measured after the clinical use of cement. A secondary effect of cement may be more important than its direct effects: by improving the seal between the prosthesis and the walls of the marrow cavity, the cement acts to increase intracavitary pressures as the prosthesis is inserted, thereby forcing marrow contents into the bloodstream. This may explain why inserting the stem of a femoral prosthesis, which can produce intraluminal pressures exceeding 1 atmosphere, produces much more severe effects than does the cementing of an acetabular prosthesis, where no such piston and cylinder effect is present. The typical increase in pulmonary vascular resistance seen at the time of inserting cemented femoral prostheses is consistent with the idea that emboli are an important part of the reaction to cement. Ibuprofen does not ameliorate the hypotension, hypoxaemia or pulmonary hypertension following cemented arthroplasty in a dog model, suggesting that thromboxane and prostacyclin are not the only mediators of this reaction.[24]

In patients undergoing total hip arthroplasty, the reaction occurs most often during insertion of the femoral prosthesis. In total knee replacement the reaction is delayed until deflation of the tourniquet. Reactions do not seem to be as severe when prostheses are cemented at other sites, as in the proximal humerus. Management includes ensuring that the patient's blood volume is well replaced before the cement is applied, as hypovolaemic patients seem to suffer the most severe hypotension. Close monitoring of the blood pressure and treatment with vasopressors and oxygen as needed usually suffice to treat the reaction.

Table 62.1 Criteria for the diagnosis of fat embolism syndrome

Major criteria	Minor criteria
Petechiae: conjunctiva, axilla	Tachycardia > 110
$PaO_2 < 8kPa$ (60 mmHg); $FIO_2 > 0.4$	Fever ($T > 38.5°C$)
	Emboli on fundoscopic examination
CNS depression	
Pulmonary oedema	Fat in urine
Thrombocytopenia	Unexpected anaemia
	Increased sedimentation rate
	Fat in sputum

Fat embolism

Fat embolism, the appearance of fat globules in the bloodstream, may occur in up to 90 per cent of the victims of major trauma, but fat embolism syndrome (FES) is much rarer, the diagnosis being made in 1–3 per cent of patients with a single major fracture. The clinical diagnosis of FES (Table 62.1) is based on criteria described by Gurd.[25]

Patient's reactions to fat embolism span a spectrum of severity, ranging from clinically imperceptible effects, to severe FES, to death. The gravity of the illness may depend on the dose of fat, concurrent medical problems and other aggravating factors. There is no known treatment for the dementia, which usually clears if the patient survives. The respiratory distress syndrome is treated in the usual way. If possible, it is best to delay all but emergency surgery in trauma patients who develop FES until the syndrome has progressed to its full severity and stabilized or improved. An exception is the patient who requires fixation of fractures to halt continuing fat embolization.

Fat embolization has been demonstrated to occur regularly during placement of cemented femoral prostheses in dogs unless the medullary cavity is vented.[26] Fulminant FES, including death, has occurred after total hip arthroplasty, surgery for hip fractures and total knee arthroplasty. During total hip arthroplasty and total knee arthroplasty, emboli of unknown nature are visible with transoesophageal echocardiography. In an autopsy study of patients dying with fractured hips, the incidence of fat embolism was 7 per cent in those treated with cemented femoral components and zero in those with uncemented prostheses. As suggested above, the cement may have improved the transmission of forces to the marrow and thereby worsened the fat embolism.[27] Thus, it appears that unrecognized minor fat embolism frequently accompanies major bone surgery, especially when the marrow cavities of long bones are invaded with long-stemmed instruments. The role that subclinical fat embolism plays in causing the postoperative hypoxaemia and deteriorated mental func-

tion so often seen in these patients has yet to be determined.

Venous thrombosis and thromboembolism

Deep venous thrombosis (DVT), with the associated risk of pulmonary embolism, occurs in 45–70 per cent of patients who undergo orthopaedic procedures on the lower extremities.[28] The factors responsible for this include postoperative immobilization, intraoperative reduction of limb blood flow (tourniquets or kinking of femoral vessels by dislocation and flexion of the hip during total hip arthroplasty), and the hypercoagulable state seen after surgery.

Many measures have been employed to prevent this complication; none is fully effective. Early mobilization, elevation of the foot of the bed and pneumatic compression stockings seem safe. After total hip arthroplasty, but not total knee arthroplasty, a device that compresses only the foot has markedly reduced the incidence of DVT.[29] Many anticoagulation regimens employing varying doses of coumarin, heparin, aspirin or dextran have been investigated. These therapies typically reduce the rate of DVT by 50 per cent, but haematomas are more frequent and the induced coagulopathy may discourage the use of spinal or epidural anaesthesia, for fear of producing an epidural haematoma.

Spinal or epidural anaesthesia has been shown in numerous trials to reduce the rate of DVT and thrombo-embolism in patients undergoing total hip prosthesis operations by about one-half.[30,31] For patients undergoing total knee arthroplasty, the beneficial effect of regional anaesthesia is less marked. In one study, the use of epidural anaesthesia served to decrease the incidence of DVT after total knee arthroplasty only from 64 per cent to 48 per cent as compared to general anaesthesia.[32]

Two mechanisms have been proposed for this effect. First, spinal and epidural anaesthesia increase blood flow to the lower extremities, which would be expected to decrease clot formation. In particular, during total hip arthroplasty, epidural anaesthesia is associated with better blood flow in the operative limb than is general anaesthesia.[33] Second, regional anaesthesia seems to alter the overall stress response to surgical trauma, including the generalized activation of clotting mechanisms and inhibition of fibrinolysis. This effect has been clearly demonstrated in patients undergoing lower limb revascularization procedures, in whom epidural anaesthesia has been shown to reduce the rate of thrombotic complications, and to moderate the hypercoagulable state seen in patients given general anaesthesia.[34] The reduced effectiveness of regional anaesthesia in preventing DVT during total knee arthroplasty may be due to the overwhelming effect of the mechanical obstruction to blood flow, produced by the tourniquet, unaltered by the effects of regional anaesthesia.

Although regional anaesthesia does reduce the rate of DVT and thromboemboli, especially in patients undergoing major hip surgery, it is as yet inappropriate to recommend the technique universally to everyone who is to undergo orthopaedic surgery below the waist. First, regional anaesthesia does not reduce the rate of DVT to zero; other methods of prophylaxis, listed above, are likely to be as effective as regional anaesthesia and can be recommended with as much confidence to patients who fear regional techniques. Second, the optimum combination of prophylactic techniques has yet to be defined; it seems likely that some combination of postoperative mechanical enhancement of lower limb flow, an anticoagulation regimen and regional anaesthesia will offer the greatest protection. Third, the effects, if any, of extending regional anaesthesia or analgesia into the postoperative period for orthopaedic patients remain undetermined. Last, it is not certain that reducing the rate of DVT by these various prophylactic measures will decrease the death rate from thromboemboli in these patients.

Air and gas embolism

Fatal air or gas embolism has been reported in patients undergoing a wide variety of orthopaedic operations, including surgery to the humerus, shoulder, spine, pelvis and femur. Minor air embolism is common during total hip arthroplasty, and may play a part in the cardiovascular reaction that follows insertion of cement into the femoral marrow cavity (see above). Air trapping beneath the cement and the extent of air embolism can be reduced by using a long nozzle to introduce cement into the bottom of the femoral canal.[35] Liquid nitrogen used to ablate bone tumours has produced gas emboli, as has nitrogen escaping from gas-powered tools.

Except in operations in which the operative field is elevated above the heart, so that local venous pressures are subatmospheric, air embolism is rare and therefore special precautions such as prophylactic insertion of a right heart catheter are not warranted. Nevertheless, gas embolism must be included among the possible diagnoses when patients undergoing orthopaedic operations suffer intra-operative cardiovascular distress.

Measures to prevent infection

Wound infections can be catastrophic, especially in patients with implanted prostheses, so orthopaedic surgeons omit no measure that might prevent infection. Some of these measures affect anaesthetic practice.

Laminar flow air filtration systems reduce the bacterial count in the air in the operating room, but increase surface cooling of the patient. Patients are therefore vulnerable to inadvertent hypothermia, especially when large areas of skin are soaked with antiseptic solutions. In addition to the usual risks of increased oxygen consumption, tachycardia and hypertension as the patient rewarms, hypothermia interferes with coagulation mechanisms, potentially increasing blood loss. To minimize this effect, the patient must be kept covered at all times, except for the operative field. Patients exposed on fracture tables are especially vulnerable and require careful draping to prevent cooling.

Prophylactic antibiotics are most effective when given shortly before the skin incision and are ineffective if given later.[36] In some cases, surgeons may withhold antibiotics until specimens can be obtained from the wound for culture. Antibiotics such as gentamicin may reduce the required dose of non-depolarizing neuromuscular blockers. Another popular drug, vancomycin, produces hypotension and flushing if administered too quickly.

The possibility of HIV infection has become a cause for concern for orthopaedic surgeons. This has led to adoption of Kevlar gloves, clear face-shields and even 'space-suits'. When blood and bone chips fly from the operative field throughout the room, anaesthetists must give thought to protecting themselves, as well.

Postoperative pain relief

The general principles of managing postoperative pain are given in Chapter 42. The physiological benefits of postoperative analgesia are not as striking in patients who have undergone orthopaedic surgery as in patients who have undergone operations on the chest or abdomen, because pain relief does not result in the same dramatic improvement in diaphragmatic mobility. Patient-controlled analgesia, epidural or subarachnoid opioids and other conventional measures have all been used successfully for orthopaedic patients. Some techniques have special application in orthopaedic patients.

Continuous (via an indwelling catheter) or intermittent blockade of the brachial plexus or the femoral nerve provides good postoperative pain relief without the drawbacks of epidural or spinal narcotics. Intra-articular bupivacaine, or morphine which has a less intense but longer lasting effect, or a combination of the two, is effective after arthroscopic surgery of the knee. Intra-articular local anaesthetics have also been used after arthroscopic surgery for other joints. The severe pain often experienced by patients at the donor site of iliac crest bone grafts can be relieved temporarily at the end of the procedure by infiltration of local anaesthetic solution, or by longer term infusion administered through a catheter placed in the wound.[37] Patients who are to be placed in passive motion devices to maintain joint mobility after surgery benefit especially from postoperative analgesia.

After some operations, patients require evaluation to detect neurovascular compromise. Inappropriately prolonged anaesthesia can hinder the prompt detection of nerve injury, compartment syndromes or tight casts. In such cases, confining analgesia to opioids, whether systemic or spinal, may be the most appropriate choice.

Specific orthopaedic procedures

Joint manipulation, reduction of dislocations

Patients with scars or adhesions limiting motion of a joint may require anaesthesia to enable the surgeon to put the joint through a range of motion that, otherwise, would be too painful to bear. A dislocation not managed successfully in the clinic or emergency room may require reduction under anaesthesia. The management of these two groups of patients is similar.

Because the procedure is brief, general anaesthesia is usually preferred to regional techniques. Profound muscle relaxation allows the surgeon to distinguish anatomical limitations of joint motion from guarding. Thus, intravenous anaesthetics with a short duration of action and suxamethonium (succinylcholine) are effective. If the surgical lesion permits, the injection of intra-articular local anaesthetic may provide some postoperative analgesia.

Closed treatment of fractures

Patients with closed fractures who require anaesthesia have often already undergone a trial of reduction with sedation or with the injection of a local anaesthetic into the haematoma. These procedures can be lengthy, with fluoroscopy, x-rays (radiographs), casts and sometimes percutaneous pins, making regional anaesthesia more appropriate. A Colles fracture is common in the elderly; intravenous regional anaesthesia or brachial plexus block can provide an elegant means of anaesthetizing an elderly patient with multiple medical problems and a full stomach. If general anaesthesia is chosen, it may be important to

avoid violent fasciculations caused by suxamethonium, which may disturb the fracture.

Fractures of the hip

In the USA alone, approximately 250 000 patients suffer hip fractures each year. A prospective study of the 6-month mortality of 531 patients (23 per cent) with subcapital hip fractures, found five significant predictors of death:[38]

- depressed mental function preoperatively
- postoperative chest infection
- neoplasia
- old age >85 years old
- deep wound infections

Operative repair of a proximal femoral fracture is undertaken in the hope of getting the elderly patient out of bed quickly in order to reduce the ill effects of prolonged bed rest. These include pneumonia, skin deterioration and pulmonary thromboemboli. Although immediate surgery is not required, the operation is usually performed within a day or two of the accident. This allows time for medical evaluation and treatment of intercurrent pathology. In a survey of 468 patients with hip fracture, delaying operation beyond 48 hours did not increase the mortality, but did increase the incidence of complications, most of which were pressure sores.[39]

Preoperative evaluation includes attention to several special issues. First, the use of psychotropic drugs is associated with falls and fractures in the elderly, and dementia is associated with delayed recovery and increased mortality in those with fractured hips; an examination of mental status is an important part of the preanaesthetic evaluation. Demented patients may not be as suitable for regional anaesthesia as are those who can understand and communicate. Second, it can be illuminating to ask why the patient fell. Was there a new stroke, a myocardial infarction, an arrhythmia? Third, because of diuretic use, impaired thirst recognition in the aged, the hours or days that sometimes pass before the patient is rescued and the blood loss into the fracture site (400 ml for intracapsular neck fractures, 800 ml for trochanteric fractures), these patients are often severely dehydrated. Correction may require vigorous intravenous therapy and even invasive cardiovascular monitoring in some cases. Fourth, these patients are often more hypoxaemic than one would expect for their ages; this may be due to fat embolism, although at autopsy only 1–3 per cent of those dying are found to have fat globules in the lungs.

Retrospective studies reported a decade or more ago suggested that spinal anaesthesia offered a marked reduction in acute surgical mortality when compared with general anaesthesia (7 per cent vs 24 per cent). Recent controlled prospective studies have demonstrated 1-month mortality rates of 6–8 per cent with both spinal and general anaesthesia.[40,41] There has been an impression that patients getting regional anaesthesia are less likely to be confused and demented after surgical repair of hip fractures. However, a randomized controlled study, in which epidural anaesthesia was compared with halothane anaesthesia for patients with fractured femoral necks, found that postoperative confusion correlated with pre-operative confusion, the use of drugs with anticholinergic effect and with early postoperative hypoxaemia, but not with the choice of anaesthesia.[42] Another prospective study showed that general and spinal anaesthesia produced the same minimal postoperative mental impairment in patients undergoing elective hip or knee replacement, whether assessed soon after surgery or several months later.[43]

The outcome of surgery to repair a fractured hip in an elderly patient depends not only on the anaesthetist and surgeon but also on nurses, other physicians and intensive care specialists. Regional anaesthesia can be expected to confer benefits of reduced blood loss and a reduction in DVT and thromboemboli (see above), and perhaps a more prompt return to the preoperative mental state. However, many of these benefits are lost if patients are oversedated, especially with long-acting agents, or allowed to become hypoxaemic.

If practical considerations such as difficulties in positioning or dementia make regional anaesthesia imprac-tical, general anaesthesia can be recommended with confidence, because bank blood is relatively safe and because DVTs can be prevented by means other than the use of regional anaesthesia. A well-conducted general anaesthetic for these patients resembles a well-conducted regional anaesthetic in that it is planned to prevent postoperative sedation and hypoxaemia. In the frailer patients, this goal is achieved with seemingly homoeo-pathic doses of the shortest-acting drugs and the least soluble inhaled anaesthetics.

Total hip arthroplasty

The considerations described above for managing patients with fractured hips apply to patients who undergo total hip arthroplasty, except that these elective patients are usually younger and healthier with greater muscle mass and less osteoporosis. Positioning is easier because there is no fracture. If a cemented femoral component is used, the cardiovascular sequelae described previously must be anticipated and treated. Fat, marrow and air embolism probably occur in all patients to a degree and produce recognizable clinical problems only in some patients; increased pulmonary arterial pressure at the time of cementing the femoral prosthesis is the usual finding.

The lateral position is almost always used. In addition to the usual precautions, the anaesthetist must collaborate with the surgeon to ensure proper placement of an anterior pelvic brace used to hold these patients in the lateral position. If it is placed too far caudad and exerts too much pressure on the femoral triangle of the dependent leg, venous or arterial obstruction can result. If it is placed too far cephalad and pressed into the patient's abdomen too forcefully, it can limit motion of the diaphragm and compress the vena cava.

Because DVT and thromboembolism are so prevalent after this procedure, the measures described above to prevent these complications have all been used in managing total hip arthroplasty. All offer some benefit, but none has emerged as the best prophylaxis.

Blood loss in patients managed with no special precautions is expected to be about 1500 ml for primary total hip arthroplasty and 2000 ml for revisions. These losses can be reduced substantially by deliberately inducing hypotension. Many anaesthetists, fearing that this group of patients is likely to suffer unrecognized peripheral arterial occlusive disease that increases the risks of profound deliberate hypotension, limit the extent of spinal or epidural anaesthesia or limit the doses of vasodilators so that only moderate degrees of hypotension are attained. At the extreme, as described above, Sharrock has reported using safely a combination of light general anaesthesia and continuous lumbar epidural anaesthesia, with anaesthetic levels to T4, produced by large doses of bupivacaine 0.75 per cent. Mean arterial pressures are maintained at 50–60 mmHg with the infusion of vasopres-sors so that blood losses average less than 300 ml. Risk/ benefit analysis based on large scale comparative studies not yet performed may show that this is a valuable technique, but it is probably not necessary to reduce blood loss to this extent. Autologous blood transfusion and moderate degrees of deliberate haemodilution are safe techniques that allow these patients to be managed without bank blood when total blood losses are limited to 1500–2000 ml. These losses are easily achieved without very high epidural anaesthesia, invasive cardiovascular monitoring or extremes of hypotension.

In the recovery room, continued bleeding must be expected and treated with appropriate fluids. Unexpected hypoxaemia, especially if accompanied by deteriorating mental status, must raise the question of fat embolism.

At present, in the absence of contraindications and with the patient's permission, the best anaesthesia plan seems to be a spinal or epidural anaesthetic (the latter allowing postoperative analgesia), with supplemental general anaes-thesia if the duration of the operation or the patient's comfort requires it. There is not enough proven benefit from regional anaesthesia to warrant demanding that patients agree to a form of anaesthesia that they find distasteful or frightening. Moderate depression of arterial blood pressure can be obtained with low thoracic levels of spinal or epidural anaesthesia, or with adequate depths of general anaesthesia supplemented with β-sympathetic blockade and vasodilators as required. In younger patients, more extreme degrees of hypotension may be appropriate. Invasive cardiovascular monitoring is employed as needed, depending on the patient's condition. Unless the patient is at special risk of myocardial ischaemia or failure, or of pulmonary hypertension, measuring blood pressure by an arm cuff and measuring urine output with a bladder catheter usually suffice for monitoring the cardiovascular system.

Total knee arthroplasty

The considerations involved in managing patients for total knee arthroplasty are similar to those for total hip arthroplasty. The supine position makes it easier to manage patients under regional anaesthesia alone. DVT and thromboemboli are prevalent, but prophylactic measures such as choosing regional anaesthesia or using anticoagulants do not seem to reduce the incidence of these problems as much as they do with total hip arthroplasty. This may be because of the outright obstruction of blood flow produced by the tourniquet usually employed to prevent bleeding. The management of tourniquet pain under regional or general anaesthesia is described above.

When cemented total knee arthroplasty is performed without a tourniquet, a progressive syndrome of systemic hypotension and pulmonary vascular obstruction begins immediately upon insertion of the prosthesis.[44] When a tourniquet is used, the onset of this syndrome is delayed until deflation of the tourniquet at the end of the operation. At this time, as patients emerge from anaesthesia, they suffer the combined effects, described separately above, of tourniquet deflation, blood loss and embolization of cement, thrombi, fat and other marrow contents. As with total hip arthroplasty, the use of cement aggravates the severity of the effect.[45] Emboli may be identified in the right heart with echocardiography, and arterial hypoxaemia sometimes occurs. During this period, vital signs must be measured frequently and the patient treated with fluids, pressors and oxygen as required. Patients with cardiovascular disease who might not be able to compensate adequately for these events may require invasive cardiovascular monitoring.

About 1500 ml of blood is lost into the knee after total knee arthroplasty, much of it in the first minutes after the tourniquet is deflated. Techniques of deliberate hypotension have not been applied in an effort to decrease these losses, perhaps because it would be impractical to do so during recovery from the anaesthetic.

The physiological insult that occurs at the end of a single total knee arthroplasty is usually tolerated well by patients without severe cardiovascular disease. Simultaneous bilateral total knee arthroplasty has been accomplished safely, but doubling the dose of bleeding and embolization has also sometimes resulted in fat embolism syndrome, respiratory failure and death. Selection of healthier patients, monitoring of pulmonary vascular resistance and staging the two procedures so that the second does not begin until it is seen that the patient has weathered the first successfully are all steps likely to improve the safety of bilateral total knee arthroplasty.

Because of the drastic reactions sometimes recorded at the conclusion of total knee replacement, patients benefit from a period of close monitoring (usually less than 5 minutes) in the operating room after deflation of the tourniquet. Treatment includes fluids, pressors and oxygen. Transfer to the recovery suite is delayed until heart rate, blood pressure and peripheral oxygen saturation have recovered. In the recovery room, any unexplained hypoxaemia or deterioration in mental status provides warning of possible fat embolism syndrome and demands evaluation and treatment.

Spinal or epidural anaesthesia offer several advantages to patients undergoing total knee arthroplasty. Hypertensive reactions to prolonged tourniquet inflation are controlled more easily and postoperative pain relief can be provided. There is some reduction in the rate of DVT and thromboemboli, but these complications are by no means eliminated. However, given careful attention to details, general anaesthesia can also provide safe satisfactory anaesthesia for these patients.

Arthroscopic surgery of the knee

In the past decade, the majority of knee operations once requiring arthrotomy have been possible through the arthroscope. These include ligament repairs, meniscectomy, removal of loose bodies and replacement of the anterior cruciate ligament. All forms of anaesthesia are suitable, but the less invasive and more brief of these procedures can be performed with the intra-articular injection of local anaesthetic solutions. Absorption of local anaesthetic solution is not rapid, although the addition of adrenaline (epinephrine) markedly decreases the maximum blood levels; 30 ml of bupivacaine 0.5 per cent, with or without adrenaline, is a safe dose.[46] Femoral nerve block also provides good analgesia.

Operations on the foot and the lower extremity

Operations on the lower leg can be accomplished using peripheral nerve blocks, either at the level of the femoral and sciatic nerves or by individual nerve blocks at the level of the knee. If a thigh tourniquet is not to be used, operations on the foot are amenable to ankle block. Instead of a thigh tourniquet, a tightly wrapped rubber bandage at the ankle can occlude arterial inflow.

Operations on the hand and distal arm

Operations on the hand and arm are well suited to regional anaesthesia. If the operation is limited in extent, a distal block of a specific nerve, such as the ulnar nerve at the elbow, may suffice. Intravenous regional anaesthesia is widely applicable, but does not allow for deflation of the tourniquet to obtain haemostasis and is limited in the duration of anaesthesia it can provide. Tourniquet pain does not usually occur with brachial plexus block if care is taken to anaesthetize the intercostobrachial nerve as it enters the upper arm superficial to the fascia overlying the axillary artery.

Procedures about the elbow can usually be accomplished under axillary block, but more proximal procedures require an approach to the brachial plexus above the level of the clavicle, such as an interscalene block. This is especially advantageous in elderly patients who suffer fractures of the proximal humerus.

Operations on the shoulder

The deep structures of the shoulder are innervated by C5–C6, so interscalene and other supraclavicular approaches to the brachial plexus provide good regional anaesthesia. Anaesthesia of the skin over the anterior shoulder may require infiltration of local anaesthetics because of innervation from thoracic dermatomes. As a rule, regional anaesthesia is more successful for anterior approaches to the shoulder than posterior.

Patients, surgeons and anaesthetists often prefer general anaesthesia for shoulder procedures because of the proximity of the surgical field to the patient's face. The combination of light general anaesthesia and a brachial plexus block may provide good postoperative pain relief along with intraoperative comfort for the patient. Intra-articular local anaesthetic solutions can be used alone for anaesthesia for arthroscopy, and for postoperative pain relief after more extensive procedures performed under general anaesthesia. Whether regional anaesthesia alone or in combination with general anaesthesia provides any special health benefits analogous to those seen with surgery on the lower extremity has not been investigated.

Regardless of the anaesthetic technique, three relatively infrequent complications require vigilance during these operations. Shoulder procedures are often carried out with

the patient in a semi-sitting position, increasing the possibility of venous air embolism. When dissection is carried to deep levels, pneumothorax is possible. Operations on the clavicle can result in injury to the underlying subclavian vessels, with abrupt, massive blood loss.

Operations on the spine

Operations on the spine carried out by orthopaedic surgeons are divided into several categories. Laminectomy and spinal fusions of limited extent, which are usually on the lumbar spine, present challenges – mostly because of the position required. More extensive operations, such as instrumented fusions such as the Harrington procedure, raise the possibility of significant blood loss or damage to the spinal cord. Such patients may also suffer from kyphoscoliosis, with attendant restrictive lung disease and even right heart failure from cor pulmonale. Sitting approaches to the cervical spine create the risk of air embolism (see Chapters 48 and 52). Anterior approaches to the spine require incisions in the anterior neck, thorax or abdomen.

Positioning

Patients in the prone position suffer pressure on the abdomen; in all but the most slender of individuals this elevates the diaphragm, interferes with diaphragmatic motion and obstructs the vena cava. Obstruction of the vena cava in turn increases flow and pressure in the veins of the epidural venous plexus, increasing operative blood loss and making many spinal operations technically more difficult. Devices such as the Andrews frame, Relton frame and Hastings frame all position the patient so that the abdomen is free, and thereby eliminate this problem. The efficacy of these positions in decreasing venous pressures may explain why air embolism has now been reported during lumbar spine surgery.[47] It is fortunate that serious air embolism due to this mechanism seems to be rare, because studies in a dog model with the abdomen hanging free suggest that it is difficult to retrieve significant quantities of air from the right heart under these circumstances.[48]

Further details of management of patients in the prone and sitting positions are described in Chapter 48.

Blood loss

Controlled hypotension reduces blood loss in patients undergoing extensive spinal surgery and has become a popular technique for these patients. Blood recovery, autologous blood transfusion and deliberate haemodilution reduce the need for bank blood in these patients.

Spinal cord injury

In patients in whom extensive surgery is carried out or in whom a marked kyphoscoliosis is to be corrected, intraoperative spinal cord ischaemia may produce permanent neurological damage. To prevent this, patients are monitored during the operation to detect spinal cord damage. Somatosensory evoked potentials (SSEP), produced by stimulating the lower extremities and monitoring the cerebral cortex or the spinal cord above the level of the operation, are ablated when ischaemia affects the sensory tracts in the spinal cord. When this technique is in use, the anaesthetist must try to maintain a fixed level of anaesthesia so that anaesthetic effects will not be confused with the effects of ischaemia.

An alternative method is the 'wake-up test', which assesses the integrity of motor tracts. As the spine is straightened or distracted, the patient is awakened partially and asked to move his toes or fingers. If the patient moves his hands but not his feet, it is presumed that there is spinal cord compromise. In order to accomplish this test, the anaesthetist must coach and reassure the patient carefully beforehand. An analgesic and an amnestic sedative, administered as part of the anaesthetic mixture, allow the patient to participate without discomfort or distressing memories. Spontaneous recovery of neuromuscular function to the level of one or two responses to train-of-four stimuli enables the patient to execute commands but not to move violently and disrupt the operation. Rapid lightening and deepening of the anaesthetic are easily accomplished by employing nitrous oxide, desflurane, isoflurane or propofol.

Choice of anaesthesia

The more extensive operations on the spine require general anaesthesia, but simple excision of a lumbar intervertebral disc or a limited laminectomy can also be performed under spinal or epidural anaesthesia. The usual advantages of regional anaesthesia obtain, but there are several unique dangers that have limited the popularity of the technique. First, the prone patient cannot be very heavily sedated with safety because apnoea and airway obstruction may be difficult to manage in this position. Second, obese patients, patients with respiratory disease or those at special risk of hypotension from the prone position, who would in other respects be natural candidates for regional anaesthesia, may require support of ventilation or circulation using intubation, controlled ventilation and high concentrations of inspired oxygen. Third, postoperative neural deficit due to surgical trauma may be thought to be due to the anaesthetic technique.

Epidural or spinal anaesthesia can be advantageous techniques for lumbar spine surgery, but their safe application requires care, skill and discretion by both surgeon and anaesthetist. If the duration of the surgery exceeds the duration of the block, the surgeon can easily give a supplemental subarachnoid injection in the surgical field.

Amputation

Amputation of a limb presents two special problems for the management of anaesthesia. First, the psychological trauma to the patient of losing a major part of his body may require that general anaesthesia or heavy sedation be employed. Second, many patients suffer phantom limb sensations or pain for months or years after the operation. The pain can be treated after it develops, with spinal narcotics.[49] More interesting, one controlled study demonstrated that, in elderly patients undergoing leg amputation for painful ischaemia, 3 days of preoperative epidural blockade with local anaesthetics and opioids markedly reduced the incidence of postoperative phantom limb pain.[50] It is likely that regional anaesthesia offers special benefits to patients undergoing limb amputations.

Reimplantation and free microvascular grafts

Reimplantation of severed parts and the *en bloc* grafting of bone and soft tissue with vascular reanastomoses are lengthy operations that present several special challenges to anaesthetic management. Careful attention to positioning, repeatedly verifying that the tracheal tube cuff is not overinflated, padding to prevent ischaemic necrosis of skin, maintenance of body temperature and close regulation of fluid balance are all required for operations that may take 24 or more hours to complete.

Patient comfort during a prolonged procedure virtually demands general anaesthesia. One case report described prolonged emergence from general anaesthesia after such a procedure, but this can be prevented by careful management of anaesthetic depth to prevent relative overdose. It may be best to avoid nitrous oxide during the majority of the procedure because of the bone marrow depression reported with prolonged administration of this agent.

Regional anaesthesia offers advantages of increased blood flow to the limb and can be employed along with general anaesthesia. Repeated blocks with long-acting local anaesthetics or indwelling catheters can extend these benefits to the postoperative period. Other measures to improve blood flow include maintaining an optimal haematocrit (packed cell volume) (probably 30 per cent), keeping the patient warm and administering heparin or dextran.

REFERENCES

1. Salathe M, Johr M. Unsuspected cervical fractures: a common problem in ankylosing spondylitis. *Anesthesiology* 1989; **70**: 869–70.

2. Schlaghecke R, Kornely E, Santen R, Ridderskamp P. The effect of long-term glucocorticoid therapy on pituitary–adrenal responses to exogenous corticotropin-releasing hormone. *New England Journal of Medicine* 1992; **326**: 226–30.

3. Stein RE, Urbaniak J. Use of the tourniquet during surgery in patients with sickle cell hemoglobinopathies. *Clinical Orthopaedics and Related Research* 1980; **151**: 231–3.

4. Conaty KR, Klemm MS. Severe increase of intracranial pressure after deflation of a pneumatic tourniquet. *Anesthesiology* 1989; **71**: 294–5.

5. Hofmann AA, Wyatt RWB. Fatal pulmonary embolism following tourniquet inflation. *Journal of Bone and Joint Surgery [Am]* 1985; **67**: 633–4.

6. Gardner RC. Blood loss in orthopedic operations: comparative studies in 19 major orthopedic procedures utilizing radioisotope labeling and an automatic blood volume computer. *Surgery* 1970; **68**: 489–91.

7. Donahue JG, Munoz A, Ness PM, et al. The declining risk of post-transfusion hepatitis C virus infection. *New England Journal of Medicine* 1992; **327**: 369–73.

8. Murphy P, Heal JM, Blumberg N. Infection or suspected infection after hip replacement surgery with autologous or homologous blood transfusions. *Transfusion* 1991; **31**: 212–17.

9. Ranawat C, Beaver WB, Sharrock NE, Maynard MJ, Urquhart B, Schneider R. Effect of hypotensive epidural anaesthesia on acetabular cement–bone fixation in total hip arthroplasty. *Journal of Bone and Joint Surgery [Br]* 1992; **73**: 779–82.

10. Lotke PA, Faralli VJ, Orenstein EM, Ecker ML. Blood loss after total knee replacement. *Journal of Bone and Joint Surgery [Am]* 1992; **73**: 1037–40.

11. Rosberg B, Fredin H, Gustafson C. Anesthetic techniques and surgical blood loss in total hip arthroplasty. *Acta Anaesthesiologica Scandinavica* 1982; **26**: 189–93.

12. Sharrock NE, Mineo R, Urquhart B. Haemodynamic effects and outcome analysis of hypotensive extradural anaesthesia in controlled hypertensive patients undergoing total hip arthroplasty. *British Journal of Anaesthesia* 1991; **67**: 17–25.

13. Guay J, Reinberg C, Poitras B, et al. A trial of desmopressin to reduce blood loss in patients undergoing spinal fusion for idiopathic scoliosis. *Anesthesia and Analgesia* 1992; **75**: 405–10.

14. Bloch EC, Ginsberg B, Binner RA, Sessler DI. Limb tourniquets and central temperature in anesthetized children. *Anesthesia and Analgesia* 1992; **74**: 486–9.

15. Hagenouw RRPM, Bridenbaugh PO, van Egmond J, Stuebing R. Tourniquet pain: a volunteer study. *Anesthesiology* 1986; **65**: 1175–80.

16. Farah RS, Thomas PS. Sympathetic blockade and tourniquet pain in surgery of the upper extremity. *Anesthesia and Analgesia* 1987; **66**: 1033–5.

17. Egbert LD, Deas TC. Cause of pain from a pneumatic tourniquet during spinal anesthesia. *Anesthesiology* 1962; **23**: 287–90.

18. Bridenbaugh PO, Hagenouw RPM, Gielen MJM, Edstrom HH. Addition of glucose to bupivacaine in spinal anesthesia increases incidence of tourniquet pain. *Anesthesia and Analgesia* 1986; **65**: 1181–5.

19. Concepcion MA, Lambert DH, Welch KA, Covino BG. Tourniquet pain during spinal anesthesia: a comparison of plain solutions of tetracaine and bupivacaine. *Anesthesia and Analgesia* 1988; **67**: 828–32.

20. Stewart A, Lambert DH, Concepcion MA, et al. Decreased incidence of tourniquet pain during spinal anesthesia with bupivacaine. *Anesthesia and Analgesia* 1988; **67**: 833–7.

21. Tuominen M, Valli H, Kalso E, Rosenberg PH. Efficacy of 0.3 mg morphine intrathecally in preventing tourniquet pain during spinal anaesthesia with hyperbaric bupivacaine. *Acta Anaesthesiologica Scandinavica* 1988; **32**: 113–16.

22. Sapega AA, Heppenstall RB, Chance B, Park YS, Sokolow D. Optimizing tourniquet application and release times in extremity surgery. *Journal of Bone and Joint Surgery [Am]* 1985; **67**: 303–14.

23. Bourke DL, Silberberg MS, Ortega R, Willcock MM. Respiratory responses associated with release of intraoperative tourniquets. *Anesthesia and Analgesia* 1989; **69**: 541–4.

24. Byrick RJ, Wong PY, Mullen JB, Wigglesworth DF. Ibuprofen pretreatment does not prevent hemodynamic instability after cemented arthroplasty in dogs. *Anesthesia and Analgesia* 1992; **75**: 515–22.

25. Gurd AR. Fat embolism: an aid to diagnosis. *Journal of Bone and Joint Surgery [Br]* 1970; **52**: 732–7.

26. Kallos T, Enis JE, Gollan F, Davis JH. Intramedullary pressure and pulmonary embolism of femoral medullary contents in dogs during insertion of bone cement and a prosthesis. *Journal of Bone and Joint Surgery [Am]* 1974; **56**: 1363–7.

27. Sevitt S. Fat embolism in patients with fractured hips. *British Medical Journal* 1972; **2**: 257–62.

28. NIH Consensus Development Conference on Prevention of Venous Thrombosis and Pulmonary Embolism. Prevention of venous thrombosis and pulmonary embolism. *Journal of the American Medical Association* 1986; **256**: 744–9.

29. Fordyce MJF, Ling RSM. A venous foot pump reduces thrombosis after total hip replacement. *Journal of Bone and Joint Surgery [Br]* 1992; **74**: 45–9.

30. Thorburn J, Louden JR, Vallance R. Spinal and general anaesthesia in total hip replacement: frequency of deep vein thrombosis. *British Journal of Anaesthesia* 1980; **52**: 1117–21.

31. Modig J, Borg T, Karlstrom G, Maripuu E, Sahlstedt B. Thromboembolism after total hip replacement: role of epidural and general anesthesia. *Anesthesia and Analgesia* 1983; **62**: 174–80.

32. Sharrock NE, Haas SB, Hargett MJ, Urquhart B, Insall JN, Scuderi G. Effects of epidural anesthesia on the incidence of deep-vein thrombosis after total knee arthroplasty. *Journal of Bone and Joint Surgery [Am]* 1992; **73**: 502–6.

33. Davis FM, Laurenson VG, Gillespie WJ, Foate J, Seagar AD. Leg blood flow during total hip replacement under spinal or general anaesthesia. *Anaesthesia and Intensive Care* 1989; **17**: 136–43.

34. Tuman KJ, McCarthy RJ, March RJ, DeLaria GA, Patel RV, Ivankovich AD. Effects of epidural anesthesia and analgesia on coagulation and outcome after major vascular surgery. *Anesthesia and Analgesia* 1991; **73**: 696–704.

35. Evans RD, Palazzo MGA, Ackers JWL. Air embolism during total hip replacement: comparison of two surgical techniques. *British Journal of Anaesthesia* 1989; **62**: 243–7.

36. Classen DC, Evans RS, Pestotnik SL, Horn SD, Menlove RL, Burke JP. The timing of prophylactic administration of antibiotics and the risk of surgical-wound infection. *New England Journal of Medicine* 1992; **326**: 281–6.

37. Brull SJ, Lieponis JV, Murphy MJ, Garcia R, Silverman DG. Acute and long-term benefits of iliac crest donor site perfusion with local anesthetics. *Anesthesia and Analgesia* 1992; **74**: 145–7.

38. Wood DJ, Ions GK, Quinby JM, Gale DW, Steven J. Factors which influence mortality after subcapital hip fracture. *Journal of Bone and Joint Surgery [Br]* 1992; **74**: 199–202.

39. Parker MJ, Pryor GA. The timing of surgery for proximal femoral fractures. *Journal of Bone and Joint Surgery [Br]* 1992; **74**: 203–5.

40. Valentin N, Lomholt B, Jensen JS, Hejgaard N, Kreiner S. Spinal or general anaesthesia for surgery of the fractured hip? *British Journal of Anaesthesia* 1986; **58**: 284–91.

41. Davis FM, Woolner DF, Frampton C, *et al.* Prospective, multi-centre trial of mortality following general or spinal anaesthesia for hip fracture surgery in the elderly. *British Journal of Anaesthesia* 1987; **59**: 1080–8.

42. Berggren D, Gustafson Y, Eriksson B, *et al.* Postoperative confusion after anesthesia in elderly patients with femoral neck fractures. *Anesthesia and Analgesia* 1987; **66**: 497–504.

43. Jones MJT, Piggott SE, Vaughan RS, *et al.* Cognitive and functional competence after anaesthesia in patients aged over 60: controlled trial of general and regional anaesthesia for elective hip or knee replacement. *British Medical Journal* 1990; **300**: 1683–7.

44. Samii K, Elmelik E, Goutalier D, Viars P. Hemodynamic effects of prosthesis insertion during knee replacement without tourniquet. *Anesthesiology* 1980; **52**: 271–3.

45. Samii K, Elmelik E, Mourtada MB, Debeyre J, Rapin M. Intraoperative hemodynamic changes during total knee replacement. *Anesthesiology* 1979; **50**: 239–42.

46. Carnes RS, Butterworth JF, Poehling GS, Samuels MP. Safety and efficacy of intra-articular bupivacaine and epinephrine anesthesia for knee arthroscopy. *Anesthesiology* 1989; **71**: A729.

47. Albin MS, Ritter RR, Pruett CE, Kalff K. Venous air embolism during lumbar laminectomy in the prone position: report of three cases. *Anesthesia and Analgesia* 1991; **73**: 346–9.

48. Artru AA. Venous air embolism in prone dogs positioned with the abdomen hanging freely: percentage of gas retrieved and success rate of resuscitation. *Anesthesia and Analgesia* 1992; **75**: 715–19.

49. Chabal C, Jacobson L. Prolonged relief of acute postamputation phantom limb pain with intrathecal fentanyl and epidural morphine. *Anesthesiology* 1989; **71**: 984–5.

50. Bach S, Noreng MF, Tjellden NU. Phantom limb pain in amputees during the first 12 months following limb amputation, after preoperative lumbar epidural blockade. *Pain* 1988; **33**: 297–301.

Chapter 63

Anaesthesia for Ophthalmic Surgery

Rajinder K. Mirakhur and Joseph H. Gaston

The past few decades have seen an increase in the scope of ophthalmic surgery with developments in the treatment of diabetic retinopathy, the use of lasers and the treatment of many conditions during infancy. The range of the patients who are subjected to this type of surgery encompasses the whole span of life from infants a few days old to the very elderly. Furthermore, not only is the number of elderly people increasing but they are living longer, too. This has implications for anaesthetists, who must become proficient in administering anaesthesia for this ageing population which has quite marked physiological differences from healthy adults. These patients frequently have a high incidence of intercurrent disease: coronary artery disease, hypertension, diabetes and lung disease are frequently present and need proper control. They are often on medication, and may be deaf, confused and anxious as well as being blind.

At the other end of the scale are the very young in whom temperature control and fluid balance are particularly important. The young may also have other associated congenital anomalies that have important implications for the well-being of the child, cause problems with conduct of anaesthesia and produce abnormal responses to drugs.

It is important that anaesthetists have a good knowledge of the factors concerning ophthalmic surgery and the considerations they might have to bear in mind as in anaesthesia for other types of surgery. Some of these are considered here, although anaesthesia in the elderly is discussed in Chapter 38. Before discussing anaesthesia for various types of ophthalmic surgery it is appropriate to consider briefly the anatomy of the eye and the factors that affect intraocular pressure.

Ophthalmic surgery can be divided broadly into intraocular and extraocular procedures. The control of intraocular pressure (IOP) is paramount during intraocular surgery, in which the globe is opened. Indeed the success of intraocular surgery depends on a stable IOP. Absence of vascular congestion is equally important. Although the control of IOP is not of such importance in the case of extraocular surgery, control of vascular congestion and bleeding are as important as in intraocular surgery. The oculocardiac reflex is a frequent occurrence during extraocular surgery but it is equally important to prevent or treat it even when it occurs during intraocular surgery. Some of these factors are discussed below.

Anatomy of the eye

The eyeball (Fig. 63.1), with its ligaments, fasciae and extraocular muscles, is placed in the orbit. The orbit is a pyramid-like space composed of the frontal, zygomatic, sphenoid, maxillary, palatine, lacrimal and ethmoid bones. The upper margin of the orbit has a notch or a canal near the medial end for the transmission of the supraorbital

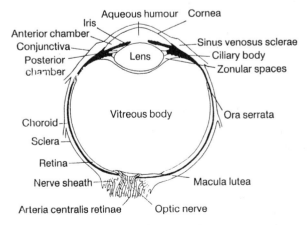

Fig. 63.1 Diagrammatic representation of the longitudinal section of the eye.

nerve and has a foramen below the lower margin for transmission of the infraorbital nerve. These landmarks are used as guides during injection of the local anaesthetic agent to produce a block of these nerves.

The globe itself is composed of three layers – the outer fibrous sclera, the middle vascular choroid and the inner nervous retina. The sclera is continuous in front with the cornea and both are covered by the conjunctiva which is reflected on to the inner surface of the lids. The function of the sclera is mainly protective, providing rigidity to keep the shape of the globe, and is pierced by the optic nerve and the central retinal artery and vein. The middle vascular layer is formed by the choroid posteriorly and the ciliary body and the iris anteriorly; the three structures together form the uveal tract. The choroid consists mostly of small arteries and veins which service a dense anastomotic network known as the choriocapillaris. The ciliary body is the site of the production of aqueous.

The iris with its central hole, the pupil, is essentially composed of stroma containing blood vessels and pigment cells that give colour to the eyes. The pupil includes the sphincter (circular) and dilator (radial) muscles in it. These receive the parasympathetic (oculomotor) and the sympathetic (cervical) innervation respectively and also the sensory nerves from the trigeminal. The iris divides the anterior segment of the eye into the anterior chamber (between it and the cornea) and the posterior chamber (between it and the lens with its supporting ligament). The anterior chamber in an average adult is 2.5–3.5 mm deep and its peripheral part contains the angle formed by the corneoscleral junction and the root of the iris. The canal of Schlemm lies here within the inner layers of the sclera and is the outlet for the aqueous humour through the meshwork at the periphery of the angle of the anterior chamber.

The retina, the innermost layer, is the delicate, highly cellular, photosensitive layer of the eye which is closely applied to the inner surface of the choroid. The internal surface of the retina is in contact with the posterior face of the vitreous body. The nerve fibre layer of the retina is continuous with the optic nerve and contains the macula lutea in which is the fovea centralis, the most sensitive part of the retina. The retina has ten layers and there is a small potential space between the outer pigment epithelial layer and the inner nine layers, which may fill with fluid when there is a retinal detachment. The inner layers of the retina are supplied by the central retinal artery and the pigment epithelial cell layer and other areas directly from the choroid.

The posterior segment of the eye from the lens to the retina is filled with the vitreous. The vitreous body is a colourless gel 4–5 ml in volume containing a few cells and salts but largely made up of water and enclosed in the hyaloid membrane. Normally there are no blood vessels in the vitreous but these may appear in disease states such as

diabetes and lead to vitreous haemorrhages. Any vitreo-retinal adhesions will transmit traction to the retina, producing holes, tears and the condition of retinal detachment.

The sclera is supplied by the anterior and posterior ciliary nerves, and the cornea, choroid and iris by fibres from the ophthalmic division of the trigeminal nerve. Blood supply to the eye and the orbital structures is via the internal and external carotid arteries, and the venous drainage by the multiple anastomoses or the ophthalmic veins and the central retinal vein.

The important structures that surround the eye and impart mobility to it are the extraocular muscles. These include the superior, inferior, medial and lateral recti and the superior and inferior oblique muscles which are responsible for the movement of the eye in various directions. The four recti arise from the fibrous ring surrounding the eye and the dural sheath of the optic nerve at the apex of the orbit, and are inserted into the sclera about 6–7 mm behind the corneoscleral junction. These form the muscle cone into which the local anaesthetic solution is usually deposited when instituting a retrobulbar (intraconal) block. The lateral rectus is supplied by the abducent, the superior oblique by the trochlear and the remaining extraocular muscles by the oculomotor nerve.

The eyelids protect the eye from injury and, by constant blinking, spread lacrimal and other secretions over the cornea. They are made up of skin, subcutaneous tissue, muscles, fascia, the tarsal plate and the tarsal conjunctiva. The lacrimal apparatus consists of the lacrimal gland in the upper and outer part of the orbit, its collecting ducts, the canaliculi, the lacrimal sac lying medial to the medial canthus and the nasolacrimal duct which opens into the nose. The lacrimal sac is located in a fairly confined space.

Intraocular pressure and factors affecting it

The normal IOP varies between 10 and 20 mmHg in adults although its range is now thought to be narrower, at 12 ± 2 mmHg.[1] It differs slightly in the two eyes and there is a diurnal variation of 5–7 mmHg, with higher values in the morning. The pressure is lower in infants and young children, which must be borne in mind when interpreting IOPs.[2]

The normal IOP is maintained by a balance between the production of aqueous and its drainage into the episcleral veins through the trabecular meshwork and the canal of Schlemm. The aqueous is produced from the ciliary plexus in the outer layer of epithelium covering the ciliary processes, at a rate of approximately 2 μl/min, and this is facilitated by the enzyme carbonic anhydrase. The volume of the vitreous, the choroidal blood volume and the scleral elasticity also exert important influences on the IOP. Reduced drainage of the aqueous is the main reason for increased IOP and glaucoma. Factors that contribute to the maintenance of normal IOP include the volume of the aqueous, the volume of the vitreous, the scleral elasticity and the choroidal blood volume. Although the IOP is effectively atmospheric (or zero) when the globe is opened, an increase in choroidal blood flow will have the same effect as an increased IOP, pushing the posterior segment forward and making the intraocular contents bulge out.

Intraocular pressure during health and during anaesthesia and surgery may be affected by physiological factors, by drugs (including anaesthetic agents) and by the technique of anaesthesia itself.[3–5] These effects may be mediated by their influence on the volumes of the aqueous, the vitreous or the choroidal blood flow, or occasionally by distorting the shape of the eye.

Arterial pressure

Variations in arterial pressure have, owing to a system of autoregulation, only a minimal effect on IOP. Sudden increases in arterial pressure, however, may cause a small and transient rise in IOP. Although a recent study in pigs showed no significant change in IOP with induced hypotension, studies in humans have shown that a decrease in arterial pressure produces a reduction in IOP once it is well below 90 mmHg systolic.[6–8] This is believed to be due to a reduction in the choroidal volume. The IOP is 3–4 mmHg at systolic arterial pressures of about 60 mmHg.[9]

Venous pressure and posture

IOP changes follow the changes in the central venous pressure in an almost parallel manner (Fig. 63.2).[10] In addition, the effects of increased venous pressure last longer than those of arterial pressure.[11] Therefore coughing, straining, breath holding or obstruction to respiration and venous drainage which raise the venous pressure raise the IOP because of reduced drainage from the episcleral veins. A bout of coughing may raise the IOP to 40 mmHg.[11] An increase in venous pressure is the mechanism underlying the increase in IOP following application of positive end-expiratory pressure.[12]

A head-up tilt produces a significant reduction in IOP, and vice versa. A 15 degree head-up tilt produces the same effect on IOP as hyperventilation to a Pa_{CO_2} of 3.5–4.0 kPa (26–30 mmHg).[10] The reason for this is a reduction in CVP

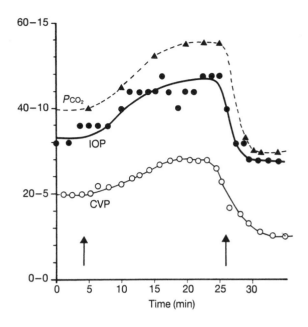

Fig. 63.2 Influence of Pa_{CO_2} on intraocular pressure. The arrows indicate the institution of hypo- (left) and hyperventilation (right). (Reproduced, with permission, from reference 10.)

with enhanced venous drainage and a consequent reduction in the choroidal blood volume.

Respiration and carbon dioxide tension

Changes in the arterial carbon dioxide tension exert significant effects on IOP. Hypoventilation and an increase in Pa_{CO_2} cause a rise in IOP, and the reverse occurs with hyperventilation (Fig. 63.2).[10] The relation between the Pa_{CO_2} and IOP is almost linear. In addition, IOP decreases quite rapidly with hyperventilation. These IOP changes are mediated by changes in the choroidal blood flow, which is reduced as the Pa_{CO_2} is lowered.[13] Hyperventilation is thus a very valuable method of keeping the IOP low when the eye is open during surgery.

Changes in Pa_{O_2} without alteration in Pa_{CO_2} have little effect on IOP.

Effect of anaesthetic agents and adjuvants

Although orally administered diazepam has no effect on IOP, both diazepam and midazolam when given intravenously reduce it.[14,15] Opioids given intramuscularly produce only a moderate reduction in IOP.[11] Anticholinergic drugs used for premedication have little effect on IOP.[16] However, as a precaution, a drop or two of pilocarpine may be put into the eye if there is concern about administering atropine to patients with severe glaucoma.[17] In any case, the use of anticholinergic premedication is now uncommon.[18]

Intravenous administration of potent opioids alfentanil and fentanyl results in a significant reduction in IOP, the effects of the former being somewhat greater.[19] A combination of fentanyl and droperidol is also effective in reducing the IOP.[20]

Administration of thiopentone and methohexitone results in a significant although short-lived reduction in IOP.[21,22] Non-barbiturate agents etomidate and propofol produce greater reductions in IOP than does thiopentone even though etomidate is associated with muscle movements.[23,24] The use of propofol is particularly advantageous for limiting the increases in IOP due to intubation.[25,26] The reports about the effects of ketamine on IOP are conflicting. Whereas Yoshikawa and Murai[27] reported a significant increase in IOP with the use of ketamine, Peuler and colleagues did not observe any such effect.[28] Ketamine is still a useful agent to use during measurement of IOP in small glaucomatous children, as the results are believed to reflect the IOP that is closer to the awake values.[29] Apart from this application, its use is limited in ophthalmology because of side effects such as nystagmus with contraction and squeezing of the eyelids.

Most of the inhalational agents, with the possible exception of nitrous oxide,[11] produce significant reduction in IOP. Older agents such as diethyl ether, cyclopropane, trichloroethylene and methoxyflurane have been associated with a reduction in IOP, although none is used currently in the developed world.[4] There is no doubt that halothane, enflurane and isoflurane reduce IOP, although there is conflicting evidence as to whether the effects are dose-related.[30–32] The reduction in IOP is greater under conditions of controlled ventilation. Although the exact mechanism for the reduction of IOP by both intravenous and inhalational agents is not known, this may be due to a reduction in the tone of the extraocular muscles, some reduction in the production of aqueous or an effect on the hypothalamic centres in the brain.

A low IOP is associated with the use of total intravenous anaesthesia with etomidate or propofol.[33,34]

The depolarizing muscle relaxant suxamethonium (succinylcholine) has significant effects on IOP. It increases IOP significantly both during a routine induction of anaesthesia following an induction agent and when administered during steady state anaesthesia.[35,36] The increase in IOP is maximal within 1 minute and is almost dissipated by 5–6 minutes, long before the surgeon is ready to open the eye (Fig. 63.3). The main reason for the rise in IOP following suxamethonium is believed to be the contracture of the extraocular muscles. These muscles contain both twitch and tonic fibres, the increased tension

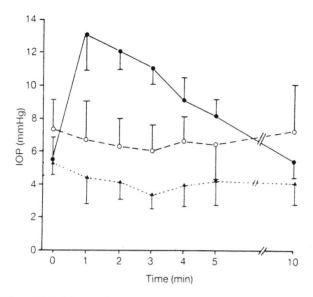

Fig. 63.3 Effect of suxamethonium (●——●), atracurium (○——○) and vecuronium (▲——▲) on intraocular pressure. (Reproduced, with permission, from reference 36.)

in the tonic fibres being the chief explanation for the increase in IOP.[3,5] Other possible mechanisms are an increase in the choroidal blood flow or distortion of the globe with axial shortening.[6] Several prophylactic measures have been suggested to reduce or abolish this increase in IOP.[3,5,37] These include pretreatment with intravenous acetazolamide, non-depolarizing relaxants, a small 'self-taming' dose of suxamethonium and the intravenous administration of diazepam, lignocaine (lidocaine), opioids such as alfentanil and the nitrates. An additional measure suggested recently is the use of propofol for the induction of anaesthesia followed by a second smaller dose immediately prior to intubation.[25] It must be stressed, however, that no method designed to reduce or abolish the rise in IOP has been shown to work consistently.

Non-depolarizing relaxants either reduce IOP or have no effect on it. Tubocurarine reduces IOP whilst pancuronium and alcuronium have little effect. Atracurium has no significant effect whilst vecuronium produces a small but significant reduction (Fig. 63.3).[36,38] It should be remembered that muscle relaxants are not used in isolation in anaesthesia, and other drugs used in anaesthesia may modify their effects. In addition, even with the use of non-depolarizing relaxants, the IOP may increase as a result of laryngoscopy and intubation.

Effect of other agents on IOP

Other agents that are often used to reduce IOP include the osmotic agents mannitol, urea, sucrose and glycerol. These reduce the IOP mainly by withdrawing fluid from the vitreous body, and their effects last for 4–6 hours. They are most often used in closed angle glaucoma prior to surgery, when the IOP is very high. Mannitol is pharmacologically inert and is administered as a 20 per cent solution in a dose of 1–1.5 g/kg over a period of 30–60 minutes. Sucrose is often used in a 50 per cent solution in a dose of 0.5–1 g/kg. It is believed to act more rapidly than other osmotic agents, and diuresis is not believed to be such a great problem.[11] Glycerol is usually administered orally as a 50 per cent solution in a dose of 1–1.5 g/kg. Urea, which is not commonly used now, was administered as a 30 per cent solution in a dose of 1–2 g/kg.

All the osmotic agents are hypertonic solutions and should be administered into larger veins. Glycerol given by mouth may induce nausea and vomiting, and should not be administered immediately prior to anaesthesia. In addition, all osmotic agents cause expansion of the intravascular volume, which may precipitate cardiac failure in susceptible subjects.

Another agent that is commonly used for reducing IOP is the carbonic anhydrase inhibitor acetazolamide. It acts by decreasing the production of aqueous by interfering with the secretory activity of the ciliary body. Its dosage is 125–250 mg by mouth 6-hourly, and in acute situations it can be administered intravenously in a dose of 500 mg. Its main untoward effect is metabolic acidosis.

The β-adrenergic receptor blocking agent timolol is used by topical administration for the treatment of glaucoma and is particularly effective when co-administered with adrenaline (epinephrine).[39] More recently, nitrates, calcium entry blockers, angiotensin-converting enzyme inhibitors and α_2-agonist drugs have also been shown to be effective in lowering IOP by topical or intravenous administration.[40–43] Intravenous administration of dexmedetomidine has been found to be useful in obtunding the IOP response to tracheal intubation.[43] Retrobulbar block, gentle massage of the eye and application of pressure using Honan's pressure balloon also help to reduce the IOP.

Other external factors that may have an influence on IOP include pressure from extraocular structures, forced closure of the eyelids or improper use of instruments such as lid retractors. IOP may also rise, to a certain extent, as a result of extubation.[44] Institution of extracorporeal circulation has been shown to cause an immediate increase in IOP which, however, decreases subsequently under the influence of other factors.[45]

The success of intraocular surgery depends on the maintenance of a low IOP and the presence of a soft eye. Anaesthetists can achieve this by judicious use of pharmacological agents and manipulation of many of the physiological factors, summarized in Table 63.1. Obviously, the suitability of a particular method depends on the circumstances that are prevalent at the time.

Table 63.1 Methods of maintaining good control of IOP

Avoidance of coughing and straining at induction of anaesthesia
Hyperventilation and lowering the Pa_{CO_2}
Reduction in arterial pressure
Head-up tilt
Unobstructed ventilation
Low venous pressure
Deep anaesthesia
Use of specific pharmacological agents

Side effects of drugs used by ophthalmologists

The conjunctiva is very permeable and the absorption of drugs through it can be both rapid and efficient but a proportion of the drug may go down the nasolacrimal duct into the nose and be swallowed. This may lead to systemic effects following absorption from the stomach. Drugs with implications for the anaesthetist include cycloplegics, mydriatics, miotics, local anaesthetics and β-receptor blocking agents.

Phenylephrine and adrenaline (epinephrine) are often used before surgery for mydriasis and capillary decongestion. Phenylephrine may cause hypertension, tachycardia, cardiac arrhythmias, fainting and headaches, and even death.[46,47] These are more common with the 10 per cent solution, and it is now considered advisable to use a 2.5 per cent solution – particularly in children and those with cardiovascular disease. Instillation of adrenaline during halothane anaesthesia may result in cardiac arrhythmias. However, toxicity following administration of phenylephrine is infrequent as long as the dose is less than 68 μg/kg.[48,49]

Cyclopentolate, which is popular as a mydriatic, is generally believed to be without side effects. However, central nervous toxicity even proceeding to convulsions may occur following its use in 1–2 per cent solution into the eye.[50] It is therefore advised, particularly in paediatric patients, to use a 0.5 or 1 per cent solution.

Cocaine was originally introduced into ophthalmology as a local anaesthetic. It is no longer employed for this purpose because of the risk of corneal damage, but it is quite often used for surgery of the nasolacrimal system. The drug is absorbed very rapidly from the mucosal surfaces, attains high levels and has a stimulating effect on the sympathetic nervous system.[51] It is therefore prudent to avoid the use of cocaine in hypertensive patients or in patients receiving tricyclic antidepressants or monoamine oxidase inhibitors. It may occasionally be necessary to use a β-adrenergic blocking agent to counteract some of the effects of ocular administration of cocaine.

The anticholinesterase ecothiopate used in the treatment of glaucoma may cause prolongation of the effect of suxamethonium, following systemic absorption over a prolonged period, owing to a reduction in plasma cholinesterase activity.[52] It may take several weeks for the cholinesterase activity to return to normal after stopping ecothiopate therapy. Additional side effects include abdominal pain, nausea, vomiting and hypotension.

More recently, timolol has become very popular in the treatment of glaucoma but it may precipitate an asthmatic attack in susceptible subjects and may also cause a persistent bradycardia.[53,54] The bradycardia may be resistant to atropine. Timolol eye drops should therefore be given with great caution in patients with cardiac conduction defects. A recently introduced, more oculospecific, anti-glaucoma drug, betaxolol, has fewer systemic side effects.

Atropine, although absorbed only to a limited extent, may be absorbed in a sufficient amount to give rise to a dry mouth, dry skin and tachycardia in susceptible people, particularly small children.

Acetylcholine or other powerful cholinergic drugs are used following lens extraction to produce constriction of the pupil. Systemic effects of such drugs include bradycardia and increased bronchial secretions, and occasionally bronchospasm and hypotension.[55] These effects can be readily reversed with intravenous atropine.

Anaesthetic technique

Traditionally, most ophthalmic surgery in the UK has been undertaken under general anaesthesia; the situation is now changing.[56] The reasons for selecting a local or a general anaesthetic technique are frequently obvious; in situations where this is not so the decision should be made only after a full discussion between the surgeon and anaesthetist and after taking into account the wishes of the patient as far as possible. Young people under 15 years of age, because of their poor cooperation, are very rarely suitable for local anaesthesia for either intraocular or extraocular surgery.

Local anaesthesia may be contraindicated in patients with bleeding disorders or for those on anticoagulant therapy. The confused and disorientated patient and those with a marked tremor are generally unsuitable for local anaesthesia, as are those with an uncontrollable cough. The latter should be treated prior to surgery. Many patients are very anxious about eye surgery under local anaesthesia and need reassurance and a proper explanation. A discussion with a postoperative patient who has had successful local anaesthesia is also helpful. General anaesthesia is probably the technique of choice for procedures lasting longer than 90 minutes, particularly for the elderly and the arthritic patient. General anaesthesia has often been considered the technique of choice for the patient who cannot lie flat because of cardiovascular or respiratory disease; nevertheless, with optimum treatment of the cardiac condition prior to surgery and adjustment of the operating table, many of these patients can have successful surgery under a local anaesthetic. It is worth remembering, however, that although there may be a higher incidence of perioperative and immediate post-operative complications with general anaesthesia, there is no significant difference in the postoperative psychomotor function or long-term morbidity or mortality.[57]

Following the introduction of cocaine, the first retro-bulbar block for ophthalmology was performed in 1884 but local anaesthesia for ophthalmic surgery remained relatively uncommon until the mid 1930s.[58] The development of safer and more predictable local anaesthetic agents has been associated with an increase in the use of local anaesthesia. The use of local anaesthesia has also been popularized by the establishment of special eye centres.[59] It is convenient, economical and particularly suitable for the elderly. It is also a very suitable technique for anaesthesia in day surgery units.[60]

Most local anaesthesia was traditionally provided by the surgeon, the patient rarely being monitored or having an intravenous access. The situation is now changing and anaesthetists not only monitor these patients but also administer the local anaesthetic block. This practice has gained widespread acceptance since the introduction of peribulbar anaesthesia in 1986.[56,61,62]

Full facilities for resuscitation must be immediately available before establishing any local anaesthetic block. The axial length of the eye should be known and care taken with eyes that are longer than 26 mm. During the surgical procedure the patient should have ECG, arterial pressure and oxygen saturation monitored. Oxygen 2–4 l/min should be given via the nasal cannulae or a modified facemask or by simple flushing under the drapes to prevent the patient feeling suffocated.

Local anaesthesia for intraocular surgery

The most common elective intraocular operations are for the removal of cataracts with or without the implantation of an intraocular lens, operations for the treatment of glaucoma, and corneal grafting (keratoplasty). Cataract removal may be intracapsular or, more frequently, extra-capsular and the surgical treatment of glaucoma includes iridectomy, iridotomy, flap sclerotomy, iridencleisis, trabe-culectomy or cyclodialysis. The first two may be carried out in both the elderly and the young whilst corneal grafting may be carried out over a wider age span. The choice for elective surgery in the adult is between the local and general anaesthetic techniques whilst general anaesthesia is necessary for surgery in children and adolescent patients.

Local anaesthesia for intraocular surgery, in particular cataract and glaucoma surgery, requires anaesthesia and akinesia of the conjunctiva and the globe, eyelids and the orbicularis muscle.

Topical anaesthesia of the conjunctiva and cornea can be obtained by instillation of amethocaine (tetracaine) 0.5–1 per cent or oxybuprocaine 0.4 per cent. Although cocaine 2–4 per cent produces more intense analgesia, it clouds the cornea and is therefore not used for this purpose any more. Conjunctival and corneal analgesia is required for most ophthalmic surgery and should precede the actual nerve blocks.

Anaesthesia and akinesia may be achieved by retrobulbar (intraconal) or peribulbar (periconal, periocular) blocks.

Retrobulbar (intraconal) block

Anaesthesia and akinesia of the globe is, with this method, obtained by depositing the local anaesthetic solution into the muscle cone surrounding the optic sheath. The sheath also contains the ciliary nerves through which pass the pain fibres from the eye. Retrobulbar block requires only 3–4 ml of local anaesthetic solution to produce anaesthesia and akinesia of the globe but it will not normally produce akinesia of the lids or the orbicularis. A facial nerve block is required to achieve this.

The traditional method for a retrobulbar block was to ask the patient to look upwards and inwards and to insert through the skin or transconjunctivally a 4 cm long needle at the outer and inferior angle of the orbit and direct it

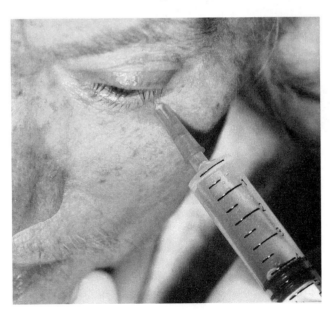

Fig. 63.4 Retrobulbar block. The needle is directed upwards, backwards and medially, the patient looking straight ahead.

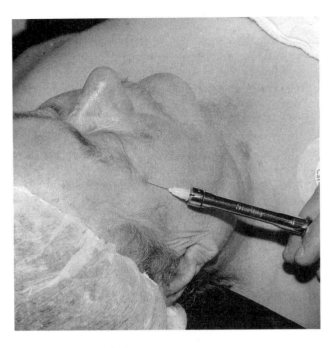

Fig. 63.5 Facial nerve block by the van Lint method. (Reproduced, with permission, from reference 4.)

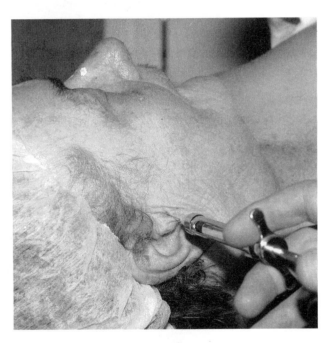

Fig. 63.6 Facial nerve block by the O'Brien method. (Reproduced, with permission, from reference 4.)

upwards, backwards and medially towards the apex of the orbit. This may, however, make the optic nerve more prone to damage by the needle. It is therefore better to make the patient adopt the neutral gaze position while carrying out the injection (Fig. 63.4). The direction of the needle is the same but the needle should not be inserted beyond 31 mm; 3–4 ml of the local anaesthetic solution is deposited within the muscle cone. The needle may be advanced a slightly shorter distance if a slightly larger volume (approximately) of the local anaesthetic solution is used.

Facial nerve block is obtained by the van Lint (Fig. 63.5) or the O'Brien methods (Fig. 63.6). In the van Lint method the terminal branches of the facial nerve are anaesthetized in the region of the lateral canthus by raising a skin weal and infiltrating the tissues down to the bone at the inferolateral angle of the orbit. A 4 cm needle is then advanced along the lateral and inferior margins of the orbit and about 2.0 ml of the local anaesthetic solution is deposited at each place. In the O'Brien method the condyle of the mandible is located with the patient's mouth open following infiltration in front of the tragus. The mouth is then closed, the needle advanced at right angles to the depth of the bone and 2.0 ml of the local anaesthetic deposited to block the temporofacial branches of the facial nerve. The superior rectus muscle may need to be blocked by injecting about 1.0 ml of the local anaesthetic solution into the belly of the muscle. The manoeuvres of facial nerve and the superior rectus blocks are quite uncomfortable.

Another method of facial nerve block is the Nadbath technique which is not employed very frequently in the UK but is used at some centres in the USA. With this method the main trunk of the facial nerve is blocked in the concavity between the anterior border of the mastoid process and the posterior border of the ramus of the mandible behind the ear. This may, however, also result in akinesia of the face.

The commonly used local anaesthetic agents include lignocaine (lidocaine), bupivacaine and mepivacaine, a combination of the first two being the most popular. Etidocaine is popular in some North American centres because of its rapid onset and long duration of action.[63,64] The addition of hyaluronidase improves the quality of anaesthesia, reduces the time taken to establish effective analgesia and enhances akinesia.[65,66]

Peribulbar (periconal) block

Peribulbar block differs from retrobulbar block in that the local anaesthetic solution is deposited around rather than behind the globe in the muscle cone. It requires a greater volume of local anaesthetic solution and the effect is slower in onset. This block is thought to work as a result of the

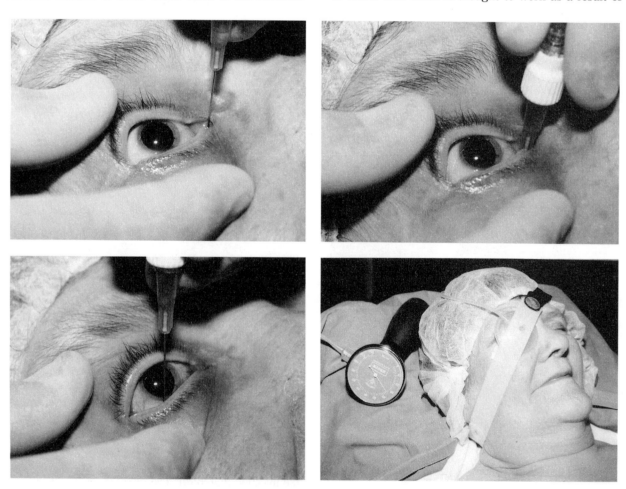

Fig. 63.7 Peribulbar block: (a, b) the needle going through the caruncle; (c, d) the inferolateral injection and the application of the pressure device.

local anaesthetic solution infiltrating through the interconnecting fatty tissue. Computed tomography following injection of local anaesthetic solution containing contrast medium has shown that the spread of the solution into the muscle cone occurs within 2 minutes.[67]

Several methods of performing the peribulbar block have been described.[59,61,62,68] The technique described here is a variation of that described by Mirakhur and Elliott.[68] Approximately 15–20 ml of local anaesthetic solution is required, which should be composed of equal volumes of 2 per cent lignocaine with 1/200 000 adrenaline and 0.5 per cent plain bupivacaine with 10 units/ml of hyaluronidase. A 25-gauge 2.5-cm needle is used for the injections; extension tubing is useful from a practical point of view. After application of topical anaesthetic and with the eye in the neutral gaze position the needle is introduced through the midpoint of the caruncle and advanced directly posteriorly for 2–2.5 cm (Fig. 63.7). Following aspiration, 5–8 ml of local anaesthetic solution is injected and the eye is gently massaged to spread the solution. After 1–2 minutes the needle is again introduced transconjunctivally at the inferolateral border of the orbit and advanced superomedially for the same distance and 4–5 ml of local anaesthetic solution is deposited. A pressure device such as Honan's Cuff inflated to 30–40 mmHg is applied over the eye for 10–15 minutes to reduce the IOP. The peribulbar technique causes relatively little pain, which can be reduced further if the solution is warmed. The sites of deposition of the local anaesthetic in the retrobulbar and the peribulbar injections are shown in Fig. 63.8.

The peribulbar injection produces akinesia of the superior oblique muscle, the eyelids and the orbicularis oculi muscle, thus obviating the need for the painful facial nerve block.

A combined retrobulbar and peribulbar technique has

also been described which produces good anaesthesia and total akinesia.[69,70]

Sedation for local anaesthesia

Considerable controversy exists as to the necessity or desirability for sedation during local anaesthesia. Most anaesthetists now accept that a calm, reassured patient without sedation is the better but some anaesthetists sedate the patient with a benzodiazepine such as midazolam. If sedation is employed, great care in dose selection is essential because the patient may become confused or uncooperative or wake up suddenly if sleep is induced. Some anaesthetists, particularly in North America, may administer methohexitone or propofol in a dose just sufficient to induce light sleep prior to institution of the local anaesthetic block.

Complications of local anaesthesia

Chemosis
Chemosis or the development of subconjunctival oedema may occur after either retrobulbar or peribulbar injections. This normally disappears completely in a few minutes, particularly with local anaesthetic solutions containing hyaluronidase and with the application of orbital compression.

Haemorrhage
Minor ecchymoses occur in association with all local anaesthetic injections and present no problems for surgery. Injections with fine sharp needles via the transconjunctival route help to reduce this complication. A periorbital haematoma may result after a periorbital block and may lead, rarely, to the need to postpone surgery.[67,71] Although retrobulbar haemorrhage occurs more frequently in association with the retrobulbar block and leads to the postponement of surgery, it is a rare complication, occurring in less than 1 per cent.[72] The bleeding will usually cease spontaneously but surgical intervention may occasionally be required in the case of arterial bleeding.

Perforation of the globe
All local anaesthetic injections around the globe, including infiltration of the eyelids, have been associated with perforation of the globe.[73] Perforation may go unnoticed but can cause a retinal tear or detachment requiring surgical correction. In extreme cases, visual acuity may be impaired although this tends to occur more commonly with inexperienced practitioners.

Neurological and muscular complications
Optic nerve damage, due either to direct trauma or to vascular occlusion, subarachnoid injection leading to

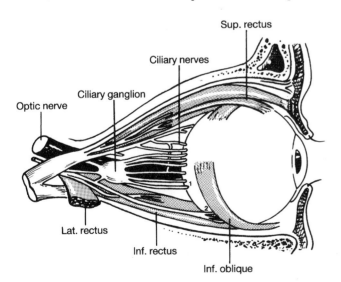

Fig. 63.8 Sites of deposition of the local anaesthetic solution during the retrobulbar (site 1) and the peribulbar (site 2) blocks.

respiratory or cardiac arrest, contralateral blindness and extraocular paralysis have all been associated with retrobulbar block.[74,75] Convulsions may occur as a result of intravascular injection of local anaesthetic solution.

Ptosis may result following both types of local blocks and may either clear quickly or persist for several days.[67,71] Paresis of the extraocular muscles may occur due to injection of local anaesthetic into the muscle.

Local anaesthesia in the form of infiltration of the eyelids is enough for minor surgical operations such as drainage of retention cysts, entropion and ectropion. The local anaesthetic solution is injected beneath the skin and superficial to the septum orbitale. However, full-thickness lid surgery requires local anaesthetic solution to be injected deep to the septum orbitale as well.

General anaesthesia for intraocular surgery

General anaesthesia provides greater control of IOP during surgery and allows greater surgical flexibility. On the other hand, IOP may increase during intubation and extubation, and postoperative vomiting is more common. The responses to drugs may be altered in the elderly; in particular, the benzodiazepines and opioids may have prolonged action, leading to postoperative sedation and postoperative hypoxaemia.

Anaesthesia is commonly induced with an intravenous agent such as thiopentone or propofol or by inhalational methods in smaller infants. Propofol has particular advantages in ophthalmic surgery, as it produces a greater reduction in IOP, is associated with a more rapid and clear-headed recovery and causes less nausea and vomiting.[23,25] A moderate dose of fentanyl (1–2 µg/kg) or alfentanil (10 µg/kg) at induction helps to attenuate the responses to intubation, to reduce the dose of the more cardiodepressant induction agents and provide some analgesia. Intubation may be facilitated with suxamethonium; although it raises the IOP, this is transient and of little consequence when the eye has not been opened. It does, however, suffer from other drawbacks (Chapter 9). Non-depolarizing relaxants such as atracurium and vecuronium are more acceptable. The increase in arterial and intraocular pressures associated with intubation may be attenuated by several methods. These include intravenous lignocaine, an additional dose of induction agent or deepening of anaesthesia prior to intubation.[37] The use of oral clonidine (5 µg/kg) has been shown to maintain a low IOP during laryngoscopy and tracheal intubation and to provide good intraoperative haemodynamic stability and to reduce the anaesthetic requirements.[76]

Anaesthesia is usually maintained with nitrous oxide in oxygen and one of the volatile agents. Controlled ventilation is employed, facilitated with a non-depolarizing relaxant. A moderate degree of hyperventilation to end-tidal CO_2 of about 4 kPa (30 mmHg) helps to control the IOP but this must not be overdone. Residual neuromuscular block is reversed at the end as appropriate.

An alternative to the volatile agents is a total intravenous technique using propofol for the maintenance of anaesthesia. Following an induction dose of 1.0–1.5 mg/kg of propofol the infusion is administered according to the scheme described by Roberts et al.[77] This is a three-step infusion scheme, initially administering 10 mg/kg per hour. The administration rate is then reduced at 10-minute intervals to 8 and 6 mg/kg per hour. The dosage is tailored to individual patient requirements and can sometimes be further reduced towards the end of surgery. Muscle relaxation and analgesia are as described previously, and ventilation may be maintained with oxygen-enriched air. This technique is followed by early recovery with a low incidence of nausea and vomiting.[78]

The laryngeal mask airway has been suggested as an alternative to tracheal intubation in patients undergoing elective ophthalmic surgery, even when controlled ventilation is employed.[79,80] Its advantages include a smaller rise in IOP and arterial pressure compared with those that follow tracheal intubation. There is also a lower incidence of coughing and breath holding at extubation. It is important to remember that the laryngeal mask does not protect against the hazards associated with regurgitation and aspiration, and may not be suitable for patients who are obese and who have a low chest compliance.

Anaesthesia for intraocular surgery in infants

Removal of congenital cataracts and treatment of glaucoma are common in infancy and childhood. Management of glaucoma in particular extends over many years, with regular assessment; surgery may be required more than once. Infants undergoing cataract extractions may have contact lenses fitted instead of intraocular implants, and these may need to be changed from time to time. General anaesthesia is managed as in the adult but particular attention must be paid in the postoperative period, especially in those who were born premature.

Children with glaucoma present for regular assessment and there has been considerable controversy about the ideal anaesthetic technique for this purpose. These examinations are normally carried out as day procedures. Adams has suggested the use of ketamine as, unlike the volatile agents, it does not reduce IOP and may give a true reflection of awake IOP.[29] Ketamine may be given either intramuscularly in a dose of 5–10 mg/kg or, more commonly, intravenously in a dose of 1–2 mg/kg.

Corneal grafting (keratoplasty)

Corneal grafting is frequently carried out as a semi-urgent procedure and is being increasingly performed on a day-stay basis. However, there is enough time for a proper pre-operative assessment and preparation. Corneal grafting may be partial or full thickness. General anaesthesia has been the technique of choice until recently but a local technique using a retrobulbar or a peribulbar injection can also be used. Nevertheless, general anaesthesia is preferable because the procedure may sometimes take up to 2 hours. Good control of IOP with a soft eye is essential, particularly for full-thickness grafts. This is best attained using muscle relaxants and controlled ventilation with deep anaesthesia. The techniques that employ volatile supplementation or total intravenous anaesthesia as described earlier are suitable.

Anaesthesia for perforated eye injuries

This is a challenge for the anaesthetist, particularly when the patient has a full stomach. The requirements in this situation are, on the one hand, the prevention of regurgitation and aspiration of gastric contents and, on the other hand, the avoidance of any marked increases in IOP which might further damage the eye. Surgery can often be delayed for some time in these patients, although this never guarantees an empty stomach. On the other hand, the presence of other injuries might require urgent intervention irrespective of the consequences for the eye. Obviously in these situations the more serious injuries will take precedence over a damaged eye. Occasionally the perforation is small and may even be plugged by a small portion of protruding iris.

The important consideration is whether to use suxamethonium. Suxamethonium is traditionally the muscle relaxant of choice for facilitating a rapid sequence induction and tracheal intubation. It has the advantage of a rapid onset of effect but it also increases the IOP. However, several measures have been suggested to limit the increase in IOP with its use. These include the administration of a small dose of non-depolarizing relaxant, diazepam, lignocaine, acetazolamide, and a small 'self-taming' dose of suxamethonium, although none is completely reliable. Tracheal intubation also increases the IOP but the increase is greater when suxamethonium is used for muscle relaxation in comparison with a non-depolarizing agent.[25,26]

Based on the findings of no change in the operator's report compared to the appearance of the eye prior to induction of anaesthesia, Libonati and colleagues concluded that suxamethonium was not contraindicated in patients with perforated eye injuries.[81] It was, however, a retrospective analysis, there was no control group and the criteria used were the lack of extrusion of the contents of the globe rather than the degree of postoperative visual acuity or the functional result. Non-depolarizing relaxants do not raise the IOP but have a relatively slow onset of action. Some anaesthetists, in an attempt to overcome this, have used a larger (0.2 mg/kg) dose of vecuronium and intubated patients at 60 seconds.[82] Nevertheless, some patients still exhibit an increase in IOP. Others have employed a modified rapid sequence technique using a non-depolarizing relaxant such as atracurium and vecuronium in moderately large doses given prior to thiopentone, and found it to be associated with a smaller increase in IOP during laryngoscopy and intubation.[36,38]

More recently, a technique using propofol and vecuronium has been described which maintains the stability of IOP throughout the process of induction of anaesthesia.[83] In this technique preoxygenation is carried out for 3–4 minutes with the patient holding the mask himself after a secure intravenous access has been established. Meanwhile, 2–3 µg/kg fentanyl is administered followed by vecuronium 0.15 mg/kg. This is followed by administration of 2–2.5 mg/kg propofol over 20–30 seconds as soon as any muscle weakness is perceived by the patient. Cricoid pressure is applied soon after administration of the induction agent, and tracheal intubation performed about 30 seconds later. Intubation is thus performed about 90 seconds after the administration of vecuronium and at the peak action of the induction agent. This particular technique has been associated with minimal or no increase in IOP following intubation and cuff inflation (Fig. 63.9). Propofol has the additional advantage that it reduces the reactivity of the airway.[84] The rest of the anaesthesia is maintained as appropriate.

It is a wise precaution to discuss the state of the injured eye with the surgeon, as on many occasions a slight increase in IOP is of no great consequence for the prognosis. In either case the anaesthetist has to make a judgement; if there is any doubt about the feasibility of intubation, it is a wise precaution to use suxamethonium and secure the airway rather than use any other technique. One can take additional precautions before induction of anaesthesia by administering one of the many agents that have been shown to reduce IOP.

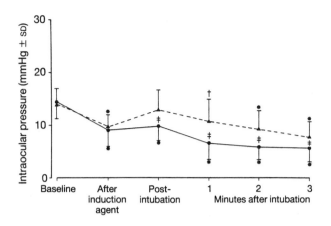

Fig. 63.9 Intraocular pressure changes during rapid sequence induction using vecuronium and thiopentone (▲- - - -▲) or propofol (●——●). * $P < 0.001$ in comparison to baseline; †$P <$ in comparison to baseline; ‡$P < 0.02$ between propofol and thiopentone groups. (Reproduced, with permission, from reference 37.)

Children with perforated eye injuries are particularly difficult to manage and may have extensive damage to the eye. An intravenous technique is preferable and pre-treatment with intravenous lignocaine has been advocated to control rises in IOP.[85]

Anaesthesia for detachment and vitreous surgery

Although vitrectomy is an essentially intraocular procedure and detachment repair extraocular, the two are dealt with together because they are longer procedures in comparison with other eye surgery and the two operations may be performed at the same time. This surgery may be carried out under either local or general anaesthesia, although a general anaesthetic technique is preferable owing to the longer duration of surgery.

The problems of this type of surgery relate to the longer duration, working in a darkened theatre, patients having repeated surgery and injection of air or sulphur hexa-fluoride into the eye.[86] Thus it is important to monitor fully the cardiovascular and respiratory responses during the procedure. Anaesthetic equipment should have appropriate alarm systems, which should be visible to the anaesthetist as there is a lot of surgical equipment in the area.

Gas exchange involving the instillation of sulphur hexafluoride and air into the eye requires that nitrous oxide is avoided. Nitrous oxide, because of its greater solubility, diffuses into the mixture and may treble the volume of the intravitreal bubble.[87] More importantly,

however, its withdrawal at the end may cause a reduction in the bubble volume and lead to a re-detachment. For this reason it is advisable to avoid the use of nitrous oxide or to withdraw it at least 20 minutes prior to the instillation of intravitreal gas.[88] Injection of silicone oil into the eye may occasionally, due to pupillary block, cause an increase in IOP in the postoperative period.

Anaesthesia is induced with an intravenous technique and maintained with an oxygen–air mixture supplemented with an analgesic and a volatile anaesthetic. Ventilation is facilitated by non-depolarizing muscle relaxants such as atracurium or vecuronium, which may be used initially as a bolus followed by repeated increments or by a continuous infusion.[89,90] In common with anaesthesia for cataract and glaucoma surgery, anaesthesia can be maintained using a total intravenous technique employing propofol, this method being associated with earlier recovery and minimal side effects.[91]

Anaesthesia for extraocular surgery

Strabismus (squint)

Strabismus or squint is a malalignment of the visual axis and may be either concomitant or paralytic. In concomitant strabismus, even though the eyes remain malaligned, they retain the same abnormal relation to each other. Con-comitant strabismus may be convergent or divergent, convergent strabismus being more common in children and divergent being more common in adult life. Divergent strabismus is usually secondary to poor vision in one eye or following previous strabismus surgery. Paralytic strabis-mus is normally the end result of a lower motor neuron lesion affecting the nerves of the extraocular muscles, and may lead to diplopia.

Surgery for strabismus involves shortening of one muscle (i.e. resection) and lengthening of the opposite muscle (i.e. recession). Occasionally an adjustable suture is inserted during surgery but not tied until the postoperative period when the patient is fully cooperative.

Most squint surgery is carried out under general anaesthesia, although some centres in North America use local anaesthesia for adults.[92] General anaesthesia is usually administered using tracheal intubation but laryn-geal mask airway has also been used more recently. An intravenous or inhalational induction is followed by suxamethonium and spontaneous ventilation or an inter-mediate-duration non-depolarizing muscle relaxant (e.g. vecuronium or atracurium) followed by controlled ventilation. Latent muscle disease must be considered a possibility in these patients and the use of suxamethonium may cause problems. Recent studies have, however, shown that children with squints do not behave in any unusual way regarding their response to suxamethonium.[93] Anaesthesia can be maintained as for other types of ocular surgery. Theoretically, suxamethonium can interfere with the interpretation of the forced duction test but this is not a problem after about 15 minutes following suxamethonium administration. Readjustment of adjustable sutures can usually be carried out using topical anaesthesia.

An accompaniment to all ocular surgery, but particularly strabismus surgery in children and adolescents, is the occurrence of the oculocardiac reflex. This is a marked slowing of the heart or the occurrence of dysrhythmias in response to traction on the extraocular muscles or pressure on the orbit.[94] The arrhythmias may be a nodal rhythm, premature ventricular contractions, bigeminy, atrioventricular block and, in extreme cases, cardiac arrest. The reflex is mediated via the trigeminal–vagal reflex arc, and sudden and sustained traction appears to produce more reflex activity than slow and gentle progressive traction. The reflex is terminated when the traction is stopped, and continued stimulation produces a diminished response. It is observed more commonly in anaesthetized subjects and may be seen in up to 90 per cent of children not receiving prophylaxis for its prevention.[95]

The prevention and treatment of oculocardiac reflex has aroused considerable interest and controversy. Apart from the anticholinergic drugs atropine and glycopyrrolate, which are the mainstay of prophylaxis and treatment, other methods suggested include retrobulbar block, use of gallamine and artificial ventilation. Retrobulbar block may sometimes itself evoke the reflex and the use of gallamine is not as effective as atropine.[4] A normal or low $PaCO_2$ has been suggested to be associated with a low incidence of oculocardiac reflex activation.[96] However, others have found artificial ventilation and a low $PaCO_2$, to be of no value.[97] Intramuscular administration of anticholinergic drugs provides only partial protection but intravenous atropine 0.02 mg/kg or glycopyrrolate 0.01 mg/kg at induction of anaesthesia or immediately prior to traction on the muscles provides the best protection.[95] Bradycardia may be prevented by application of gentle traction and is overcome rapidly if the traction ceases or, if necessary, by the administration of anticholinergic drugs. If the bradycardia is profound, atropine has the advantage of a more rapid effect than glycopyrrolate.

Pain following squint surgery is generally minimal and will respond to oral paracetamol although an opiate may occasionally be necessary. Nausea and vomiting appear to be more common after strabismus surgery and may justify the routine use of a prophylactic antiemetic.[98,99] Droperidol 15–25 µg/kg is considered to be very effective.[100] Newer agents such as ondansetron are expensive although their use may be justified in day-stay patients to prevent an unanticipated hospital admission.

Surgery on the lacrimal apparatus

Obstruction of the nasolacrimal system may result from a blockage of the canaliculi or obstruction in the lacrimal sac or the nasolacrimal duct resulting in excessive tears (epiphora). The duct may require probing through one of the canaliculi, followed by the injection of a small amount of saline often stained with fluorescein dye. Passage of the dye into the nose or the mouth will confirm the patency of the duct. Anaesthesia for this procedure is normally with either ketamine or an inhalational anaesthetic in infants and children. Tracheal intubation is rarely required.

If the nasolacrimal duct remains blocked and troublesome epiphora continues it will be necessary to perform a dacryocystorrhinostomy, in which silicone tubes are inserted into the nasolacrimal duct. This surgery is most common in adults but occasionally is necessary even in small children. Anaesthesia requires tracheal intubation, a pharyngeal throat pack and, occasionally, moderate hypotension (systolic arterial pressure of 80–90 mmHg) to control the venous bleeding in a small and confined space. This may be achieved with 0.25–0.5 mg/kg of labetalol. A head-up tilt of 10–15 degrees assists the venous drainage. Prepacking the nose with a cocaine-soaked swab and injection of local anaesthetic solution with 1/200 000 solution of adrenaline will reduce the necessity for a hypotensive technique in many cases. Anaesthesia is managed using an intravenous induction, administration of a muscle relaxant and controlled ventilation, and maintained with nitrous oxide in oxygen and isoflurane or enflurane.

Enucleation and evisceration

During enucleation the eye (globe) is completely removed, whereas in evisceration the scleral sheath is retained. These operations are generally carried out for malignancy (e.g. retinoblastoma or melanoma), for severe trauma with loss of intraocular contents or for a painful red eye which is generally secondary to advanced glaucoma or severe endophthalmitis. General anaesthesia is usually the method of choice unless there is a strong contraindication, when a retrobulbar or peribulbar block may be used. Any standard anaesthetic technique with intubation or the use of a laryngeal mask airway can be employed. It is important to remember that the oculocardiac reflex may be elicited from an empty orbit with pressure from a haematoma. There may be considerable postoperative pain and an opioid analgesic is often required.

Orbital surgery

Orbital surgery may be indicated for removal of orbital tumours or for repair of traumatic blow-out fractures of the orbital floor with herniation of the orbital contents into the maxillary sinus. The excision of tumours may be through an anterior or lateral orbitotomy or by approach through the floor of the cranial fossa. These conditions require general anaesthesia with controlled ventilation, and a controlled hypotensive technique may also be required. Considerable quantities of blood may be lost, and monitoring may have to include measurement of direct intra-arterial and central venous pressures.

Eyelid surgery

The repair of ectropions and entropions is often carried out using local anaesthesia unless extensive plastic surgery is required, when general anaesthesia may be employed.

Other procedures

These include curettage of chalazions and cysts, IOP measurements, examinations under anaesthesia, electro-physiological studies and the fitting of contact lenses. In such situations, consideration must be given to the presence of other congenital abnormalities. These procedures are usually brief and are often carried out on a day-stay basis. A suitable anaesthetic technique in these situations is either the use of ketamine or an inhalational technique using halothane.

Monitoring of patients undergoing ophthalmic surgery

Every patient under general anaesthesia must have basic cardiovascular and respiratory monitoring, including ECG, blood pressure and oxygen saturation and measurement of end-tidal CO_2. The ophthalmic patient receiving non-depolarizing muscle relaxants must have neuromuscular block monitored to ensure continued paralysis. Sudden patient movement during eye surgery may lead to permanent blindness and inevitable litigation.[101] The possibility of postoperative apnoea must be remembered, particularly in infants born prematurely. Detailed monitoring during anaesthesia is described elsewhere (Chapter 23).

REFERENCES

1. Martin XD. Normal intraocular pressure in man. *Ophthalmologica* 1992; **205**; 57–63.
2. Pensiero S, Da Pozzo S, Perissutti P, Caralleni GM, Guerra R. Normal intraocular pressure in children. *Journal of Pediatric Ophthalmology and Strabismus* 1992; **29**: 79–84.
3. Murphy DF. Anesthesia and intraocular pressure. *Anesthesia and Analgesia* 1985; **64**: 520–30.
4. Morrison JD, Mirakhur RK, Craig HJL. Intraocular pressure. In: *Anaesthesia for eye, ear, nose and throat surgery*, 2nd ed. London: Churchill Livingstone, 1985: 151–64.
5. Cunningham AJ, Barry P. Intraocular pressure – physiology and implications for anaesthetic management. *Canadian Anaesthetists' Society Journal* 1986; **33**: 195–208.
6. Adams AK, Barnett KC. Anaesthesia and intraocular pressure. *Anaesthesia* 1966; **21**: 202–10.
7. Tsamparlakis J, Casey TA, Howell W, Edridge A. Dependence of intraocular pressure on induced hypotension and posture during surgical anaesthesia. *Transactions of the Ophthalmological Society of the United Kingdom* 1980; **100**: 521–6.
8. Jantzen J-P, Hennes PJ, Rochels R, Wallenfang T. Deliberate arterial hypotension does not reduce intraocular pressure in pigs. *Anesthesiology* 1992; **77**: 536–40.
9. Dias PL, Andrew DS, Romanes GJ. Effect on the intraocular pressure of hypotensive anaesthesia with intravenous trimetaphan. *British Journal of Ophthalmology* 1982; **66**: 721–4.
10. Hvidberg A, Kessing SV, Fernandes A. Effect of changes in P_{CO_2} and body positions on intraocular pressure during general anaesthesia. *Acta Ophthalmologica* 1981; **59**: 465–75.
11. Holloway KB. Control of the eye during general anaesthesia for intraocular surgery. *British Journal of Anaesthesia* 1980; **52**: 671–9.
12. Nimmagudda UR, Joseph WJ, Salem MR, Villarreal JM, Lopez OI. Positive end-expiratory pressure increases intraocular pressure in cats. *Critical Care Medicine* 1991; **19**: 796–800.
13. Wilson TM, Strang R, MacKenzie ET. The response of the choroidal and cerebral circulations to changing arterial P_{CO_2} and acetazolamide in the baboon. *Investigative Ophthalmology and Visual Science* 1977; **16**: 576–80.
14. Al-Abrak MH. Diazepam and intraocular pressure. *British Journal of Anaesthesia* 1978; **50**: 866.
15. Artru AA. Intraocular pressure in anaesthetized dogs given flumazenil with and without prior administration of midazolam. *Canadian Journal of Anaesthesia* 1991; **38**: 408–14.

16. Cozanitis DA, Dundee JW, Buchanan TAS, Archer DB. Atropine versus glycopyrrolate. A study of intraocular pressure and pupil size in man. *Anaesthesia* 1979; **34**: 236–8.

17. Adams AK, Jones RM. Anaesthesia for eye surgery. General considerations. *British Journal of Anaesthesia* 1980; **52**: 663–9.

18. Mirakhur RK. Preanaesthetic medication: a survey of current usage. *Journal of the Royal Society of Medicine* 1991; **84**: 481–3.

19. Mostafa SM, Lockhart A, Kumar D, Bayoumi M. Comparison of effects of fentanyl and alfentanil on intra-ocular pressure. A double-blind controlled trial. *Anaesthesia* 1986; **41**: 493–8.

20. Presbitero JV, Ruiz RS, Rigor BM, Drouilhet JH, Reilly EL. Intraocular pressure during enflurane and neuroleptanesthesia in adult patients undergoing ophthalmic surgery. *Anesthesia and Analgesia* 1980; **59**: 50–4.

21. Joshi C, Bruce DL. Thiopental and succinylcholine: action on intraocular pressure. *Anesthesia and Analgesia* 1975; **54**: 471–5.

22. Verma RS. Self-taming of succinylcholine-induced fasciculations and intraocular pressure. *Anesthesiology* 1979; **50**: 245–7.

23. Mirakhur RK, Shepherd WFI. Intraocular pressure changes with propofol (Diprivan): comparison with thiopentone. *Postgraduate Medical Journal* 1985; **61**: suppl. 3, 41–4.

24. Calla S, Gupta A, Sen N, Garg IP. Comparison of the effects of etomidate and thiopentone on intraocular pressure. *British Journal of Anaesthesia* 1987; **59**: 437–9.

25. Mirakhur RK, Shepherd WFI, Darrah WC. Propofol and thiopentone: effects on intraocular pressure associated with induction of anaesthesia and tracheal intubation (facilitated with suxamethonium). *British Journal of Anaesthesia* 1987; **59**: 431–6.

26. Mirakhur RK, Elliott P, Shepherd WFI, Archer DB. Intraocular pressure changes during induction of anaesthesia and tracheal intubation; a comparison of thiopentone and propofol followed by vecuronium. *Anaesthesia* 1988; **43**: suppl, 54–7.

27. Yoshikawa J, Murai Y. The effect of ketamine on intraocular pressure in children. *Anesthesia and Analgesia* 1971; **50**: 199–202.

28. Peuler M, Glass DD, Arens JF. Ketamine and intraocular pressure. *Anesthesiology* 1975; **43**: 575–8.

29. Adams AK. Ketamine in paediatric ophthalmic practice. *Anaesthesia* 1973; **28**: 212–13.

30. Al-Abrak MH, Samuel JR. Further observations on the effects of general anaesthesia on intraocular pressure in man, halothane and nitrous oxide in oxygen. *British Journal of Anaesthesia* 1974; **46**: 756–9.

31. Zindel G, Meistelman C, Gaudi JH. Effects of increasing enflurane concentrations on intraocular pressure. *British Journal of Anaesthesia* 1987; **59**: 440–3.

32. Mirakhur RK, Elliott P, Shepherd WFI, McGalliard JN. Comparison of the effects of isoflurane and halothane on intraocular pressure. *Acta Anaesthesiologica Scandinavica* 1990; **34**: 282–5.

33. Thomson MF, Brock-Utne JG, Dean P, Walsh M, Downing JW. Anaesthesia and intraocular pressure; a comparison of total intravenous anaesthesia using etomidate with conventional inhalational anaesthesia. *Anaesthesia* 1982; **37**: 758–61.

34. Guedes Y, Rakotoseheno JC, Leveque M, Nimoumi F, Egreteau JP. Changes in intraocular pressure in the elderly during

35. Cook JH. The effect of suxamethonium on intraocular pressure. *Anaesthesia* 1981; **36**: 359–65.

36. Lavery GG, McGalliard JN, Mirakhur RK, Shepherd WFI. The effects of atracurium on intraocular pressure during steady state anaesthesia and rapid sequence induction. A comparison with succinylcholine. *Canadian Anaesthetists' Society Journal* 1986; **33**: 437–42.

37. Mirakhur RK. Use of relaxants in eye surgery. In: Rupp SM, ed. *Problems in anaesthesia: Neuromuscular relaxants*, vol. 3. Philadelphia: JB Lippincott, 1989: 510–21.

38. Mirakhur RK, Shepherd WFI, Lavery GG, Elliott P. The effects of vecuronium on intraocular pressure. *Anaesthesia* 1987; **42**: 944–9.

39. Drance SM, Douglas GR, Wijsman KJ, Schulzer M. Adrenergic and adrenolytic effects on intraocular pressure. *Graefes Archiv für klinische und experimentelle Ophthalmologie* 1991; **229**: 50–1.

40. Kelly SP, Walley TJ. Effect of the calcium antagonist nifedipine on intraocular pressure in normal subjects. *British Journal of Ophthalmology* 1988; **72**: 216–18.

41. Al-Sereiti MR, Turner P. Effect of captopril (an angiotensin converting enzyme inhibitor) on intraocular pressure in healthy human volunteers. *Journal of Ocular Pharmacology* 1989; **5**: 1–5.

42. Diestelhorst M, Hinzpeter B, Krieglstein GK. The effect of isosorbide mononitrate on the human intraocular pressure and aqueous humour dynamics. *International Ophthalmology* 1991; **15**: 259–62.

43. Jaakola ML, Ali-Melkkila D, Kanto J, Kallio A, Scheinin H, Scheinin M. Dexmedetomidine reduces intraocular pressure, intubation responses and anaesthetic requirements in patients undergoing ophthalmic surgery. *British Journal of Anaesthesia* 1992; **68**: 570–5.

44. Chari P, Sood J. Sustained intraocular tension during reversal of general anaesthesia. *Indian Journal of Ophthalmology* 1981; **29**: 5–8.

45. Bavbek T, Kazokoglu H, Temel A. Intraocular pressure variations during extracorporeal circulation and some influencing factors. *Thoracic and Cardiovascular Surgeon* 1991; **39**: 29–31.

46. Vaughan RW. Ventricular arrhythmias after topical vasoconstrictors. *Anesthesia and Analgesia* 1973; **52**: 161–5.

47. Fraunfelder FT, Scafidi AF. Possible adverse effects from topical ocular 10% phenylephrine. *American Journal of Ophthalmology* 1978; **85**: 447–53.

48. Lansche RK. Systemic reactions to topical epinephrine and phenylephrine. *American Journal of Ophthalmology* 1966; **61**: 95–8.

49. Smith RB, Douglas H, Petruscak J, Breslin P. Safety of intraocular adrenaline with halothane anaesthesia. *British Journal of Anaesthesia* 1972; **44**: 1314–17.

50. Kennerdell JS, Wucher FP. Cyclopentolate associated with two cases of grand mal seizure. *Archives of Ophthalmology* 1972; **87**: 634–5.

51. Meyers EF. Cocaine toxicity during dacryocystorhinostomy. *Archives of Ophthalmology* 1980; **98**: 842–3.

52. De Roeth A, Detbarn W, Rosenberg P, Wilensky JG, Wong A.

Effects of phospholine iodide on blood cholinesterase levels of normal and glaucomous subjects. *American Journal of Ophthalmology* 1965; **59**: 586–92.

53. Jones FL, Eckberg NL. Exacerbation of asthma by timolol. *New England Journal of Medicine* 1979; **301**: 270.

54. Kim JW, Smith TH. Timolol-induced bradycardia. *Anesthesia and Analgesia* 1980; **59**: 301–3.

55. Rongey KA, Weisman H. Hypotension following intraocular acetylcholine. *Anesthesiology* 1972; **36**: 412.

56. Rubin AP. Anaesthesia for cataract surgery – time for change. *Anaesthesia* 1990; **45**: 717–18.

57. Campbell DNC, Lim M, Muir MK, *et al.* A prospective randomised study of local versus general anaesthesia for cataract surgery. *Anaesthesia* 1993; **48**: 422–8.

58. Altman AJ, Albert DM, Fournier GA. Cocaine's use in ophthalmology: our 100 year heritage. *Survey of Ophthalmology* 1985; **29**: 300–7.

59. Hamilton RC, Gimbel HV, Strunin L. Regional anaesthesia for 12,000 cataract extraction and intraocular lens implantation procedures. *Canadian Journal of Anaesthesia* 1988; **35**: 615–23.

60. O'Sullivan G, Kerr-Muir M, Lim M, Davies W, Campbell N. Day care ophthalmic surgery: general or local anaesthesia. *Ophthalmology* 1990; **45**: 885–6.

61. Davis DB, Mandel MR. Posterior peribulbar anaesthesia; an alternative to retrobulbar anaesthesia. *Journal of Cataract and Refractive Surgery* 1986; **12**: 182–4.

62. Fry RA, Henderson J. Local anaesthesia for eye surgery. The peri-ocular technique. *Anaesthesia* 1989; **45**: 14–17.

63. Thornburn W, Thorn-Alquist A-M, Edstrom H. Etidocaine in retrobulbar anaesthesia. *Acta Ophthalmologica* 1976; **54**: 591–600.

64. Smith PH, Smith ER. A comparison of etidocaine and lidocaine for retrobulbar anesthesia. *Ophthalmic Surgery* 1983; **14**: 569–74.

65. Nicoll JMV, Treuren B, Acharya PA, Ahlen K, James M. Retrobulbar anesthesia: the role of hyaluronidase. *Anesthesia and Analgesia* 1986; **65**: 1324–8.

66. Abelson MB, Mandel E. The effect of hyaluronidase on akinesia during cataract surgery. *Ophthalmic Surgery* 1989; **20**: 325–7.

67. Ropo A, Nikki P, Ruusuvaara P, Kivisaari L. Comparison of retrobulbar and periocular injections of lignocaine by computerised tomography. *British Journal of Ophthalmology* 1991; **75**: 417–20.

68. Mirakhur RK, Elliott P. Anaesthesia for ophthalmological surgery. *Current Anaesthesia and Critical Care* 1992; **3**: 212–17.

69. Buttery B. Conal anaesthesia: a new approach to retrobulbar anaesthesia. *Australian and New Zealand Journal of Ophthalmology* 1989; **17**: 63–9.

70. Meyer D, Hamilton RC, Loken RG, Gimbel HV. Effect of combined peribulbar and retrobulbar injection of large volumes of anaesthetic agents on the intraocular pressure. *Canadian Journal of Ophthalmology* 1992; **27**: 230–2.

71. Ropo A, Ruusuvaara P, Paloheimo M, Maunuksela E-L, Nikki P. Periocular anaesthesia: technique, effectiveness and complications with special reference to postoperative ptosis. *Acta Ophthalmologica* 1990; **68**: 728–32.

72. Edge KR, Nicoll JMV. Retrobulbar haemorrhage after 12,500 retrobulbar blocks. *Anesthesia and Analgesia* 1993; **76**: 1019–22.

73. Duker JS, Belmont JB, Benson WE, *et al.* Inadvertent globe perforation during retrobulbar and peribulbar anesthesia. Patient characteristics, surgical management and visual outcome. *Ophthalmology* 1991; **98**: 519–26.

74. Nicoll JMV, Acharya PA, Ahlen K, Baguneid S, Edge KR. Central nervous system complications after 6,000 retrobulbar blocks. *Anesthesia and Analgesia* 1987; **66**: 1298–302.

75. Javitt JC, Addiego R, Friedberg HL, Libonati MM, Leahy JJ. Brain stem anaesthesia after retrobulbar block. *Ophthalmology* 1987; **94**: 718–24.

76. Ghignone M, Noe C, Calvillo O, Quintin L. Anesthesia for ophthalmic surgery in the elderly: the effects of clonidine on intraocular pressure, perioperative hemodynamics and anesthetic requirement. *Anesthesiology* 1988; **68:** 707–16.

77. Roberts FL, Dixon J, Lewis GTR, Tackley RM, Prys-Roberts C. Induction and maintenance of propofol anaesthesia: a manual infusion scheme. *Anaesthesia* 1988; **43**: suppl, 14–17.

78. Mirakhur RK, Elliott P, Stanley JC. Use of propofol in anaesthesia for ophthalmic surgery. In: Prys-Roberts C, ed. *Focus on infusion: intravenous anaesthesia.* London: Current Medical Literature, 1991: 134–9.

79. Holden R, Morsman CDG, Butler J, Clark GS, Hughes DS, Bacon PJ. Intraocular pressure changes during the laryngeal mask airway and tracheal tube. *Anaesthesia* 1991; **46**: 922–4.

80. Akhtar TM, McMurray P, Kerr WJ, Kenny GNC. A comparison of laryngeal mask airway with tracheal tube for intra-ocular ophthalmic surgery. *Anaesthesia* 1992; **47**: 668–71.

81. Libonati MM, Leahy JJ, Ellison N. The use of succinycholine in open eye surgery. *Anesthesiology* 1985; **62**: 637–40.

82. Abbott MA, Samuel JR. The control of intraocular pressure during the induction of anaesthesia for emergency eye surgery. A high-dose vecuronium technique. *Anaesthesia* 1987; **42**: 1008–12.

83. Mirakhur RK, Shepherd WFI, Elliott P. Intraocular pressure changes during rapid sequence induction of anaesthesia. Comparison of propofol and thiopentone in combination with vecuronium. *British Journal of Anaesthesia* 1988; **60**: 379–83.

84. McKeating K, Bali IM, Dundee JW. The effects of thiopentone and propofol on upper airway integrity. *Anaesthesia* 1988; **43**: 638–40.

85. Lerman J, Kiskis AA. Lidocaine attenuates the intraocular pressure response to rapid intubation in children. *Canadian Anaesthetists' Society Journal* 1985; **32**: 339–45.

86. Mirakhur RK. Anaesthetic management of vitrectomy. *Annals of the Royal College of Surgeons of England* 1985; **67**: 34–6.

87. Stinson TW, Donlon JV. Interaction of intraocular air and sulfur hexafluoride with nitrous oxide: a computer simulation. *Anesthesiology* 1982; **56**: 385–8.

88. Wolf GL, Capuano C, Hartung J. Nitrous oxide increases intraocular pressure after intravitreal sulfur hexafluoride injection. *Anesthesiology* 1983; **59**: 547–8.

89. Mirakhur RK, Ferres CJ. Muscle relaxation with an infusion of vecuronium. *European Journal of Anaesthesiology* 1984; **1**: 353–9.

90. Hunter JM, Kelly JM, Jones RS. Atracurium infusions in major ophthalmic surgery. *European Journal of Anaesthesiology* 1987; **4**: 9–15.

91. Mirakhur RK, Stanley JC. Propofol infusion for maintenance of anaesthesia for vitreous surgery: comparison with isoflurane. *Anesthesia and Analgesia* 1989; **68**: S197.

92. Diamond GR. Topical anesthesia for strabismus surgery. *Journal of Pediatric Ophthalmology and Strabismus* 1989; **26**: 86–90.

93. McLoughlin C, Mirakhur RK, Elliott P, Craig HJL, Trimble ER. Biochemical changes following administration of suxamethonium in children with and without strabismus. *Paediatric Anaesthesia* 1991; **1**: 101–6.

94. Alexander JP. Reflex disturbances of cardiac rhythm during ophthalmic surgery. *British Journal of Ophthalmology* 1973; **59**: 518–24.

95. Mirakhur RK, Jones CJ, Dundee JW, Archer DB. IM or IV atropine or glycopyrrolate for the prevention of oculocardiac reflex in children undergoing squint surgery. *British Journal of Anaesthesia* 1982; **54**: 1059–63.

96. Blanc VF, Hardy J, Milot J, Jacob J. The oculocardiac reflex: a graphic and statistical analysis in infants and children. *Canadian Anaesthetists' Society Journal* 1983; **30**: 360–9.

97. Mirakhur RK, Shepherd WFI, Jones CJ. Ventilation and the oculo-cardiac reflex. Prevention of oculocardia reflex during surgery for squints: role of controlled ventilation and anticholinergic drugs. *Anaesthesia* 1986; **41**: 825–8.

98. Nikki P, Pohjola S. Nausea and vomiting after ocular surgery. *Acta Ophthalmologica* 1972; **50**: 525–31.

99. Rowley MP, Brown TCK. Postoperative vomiting in children. *Anaesthesia and Intensive Care* 1982; **10**: 309–13.

100. Abramowitz MD, Elder PT, Friendly DS, Broughton WL, Epstein BS. Antiemetic effectiveness of intraoperatively administered droperidol in pediatric strabismus outpatient surgery. Preliminary report of a controlled study. *Journal of Pediatric Ophthalmology and Strabismus* 1981; **18**: 22–7.

101. Gild WM, Posner KL, Caplan RA, Cheney FW. Eye injuries associated with anaesthesia. A closed claims analysis. *Anesthesiology* 1992; **76**: 204–8.

Chapter 64

Anaesthesia and Analgesia for Obstetric Care

G. M. Cooper

Analgesia in labour	Anaesthesia for caesarean section
Epidural analgesia	Obstetric complications

The anaesthetic care of the obstetric patient differs from that of her non-pregnant counterpart because of the physiological changes in the parturient, the presence of a second individual (the fetus) who is also affected by the anaesthetic process and the fact that the majority of requests for anaesthesia are unplanned and urgent. Furthermore, the patients are often emotional and tired, and are also now better informed of the anaesthetic options available, of which they have higher expectations. These factors and the fact that the majority of patients are young and healthy with family responsibilities in whom a death would be an especial disaster mean that the obstetric anaesthetist needs to be skilled, well informed, sympathetic, tactful and diplomatic. Medicolegal consequences of dissatisfied parents are increasingly common, the defence of which requires accurate and complete record-keeping.

Placental transfer of drugs[1]

The placental membrane is covered with a continuous lipid layer which is freely permeable to lipid-soluble substances but less so to hydrophilic particles. Most drugs cross the placenta by simple diffusion. Lipophilic molecules of molecular weight up to 1000 diffuse readily whereas polar substances of molecular weight over 100 are able to diffuse only slowly across the placental membrane. Ionized substances diffuse even more slowly. The rate of placental transfer of polar molecules and ions is inversely dependent on their molecular size, whereas that of lipid-soluble drugs depends on the concentration gradient of its freely diffusible moiety and the blood flow on either side of the placental membrane. The concentration gradient depends on the rate of drug administration, its volume of distribution and its clearance, and on the ionization and protein binding on either side of the placental membrane. The pH of fetal plasma is lower than that of the mother; therefore weakly acidic drugs are more ionized in maternal plasma and weakly basic drugs are more ionized in fetal plasma. Because it is the non-ionized moiety that

equilibrates across the placenta, the effects of basic drugs such as pethidine (meperidine) or bupivacaine will be increased in the acidotic fetus with distress.

Clinical implications for the anaesthetist

A sedative drug given to a pregnant woman produces sedative effects in the baby unless given only shortly before delivery, and hence the use of barbiturates and phenothiazines in labour has now been abandoned. Diazepam, which is lipid soluble and bound to plasma albumin, crosses the placenta readily and, with its metabolite nordiazepam, produces prolonged effects including neonatal ventilatory depression, impaired thermoregulation, hypotonia and raised bilirubin serum concentration. Thus, although the use of diazepam as a routine premedicant cannot be justified, it may be indicated in the management of eclamptic fits.

Thiopentone, being highly lipid soluble, crosses the placenta readily, but frequently does not have the expected profound effects in the neonate that might be expected because of the short interval between its administration for induction of anaesthesia and delivery of the infant. Other induction agents do not appear to hold advantages over thiopentone.

Atropine, being lipid soluble, crosses the placenta and causes fetal tachycardia, whereas the quaternary ammonium compound glycopyrrolate does not. The placental transfer of inhalational anaesthetics is dose and time dependent, and thus, provided the induction to delivery time is not unduly prolonged, the effects on the neonate are unlikely to be severe or long lasting.

Non-depolarizing neuromuscular blocking drugs are fully ionized and water soluble and do not cross the placenta appreciably. Suxamethonium (succinylcholine), although not completely metabolized by placental pseudocholinesterase,[2] has no clinical effect on the neonate after maternal administration. Opioid analgesics have ventilatory depressant effects on the neonate and that of pethidine

Table 64.1 Physiological changes of pregnancy and the anaesthetic implications

System	Physiological change	Anaesthetic implication
Cardiovascular system	↑ cardiac output (4.5 → 6 l/min) Cardiac output up to 12 l/min in labour ↓ peripheral resistance ↓ blood pressure, especially diastolic ↑ organ blood flow, particularly uterus, kidney, skin ↑ coagulability of blood Aortocaval compression in supine position	Awareness more likely Important in presence of cardiac disease Less hypotensive effect of thiopentone and vasodilators Easier venous access ↑ risk of venous thrombosis Lateral tilt required to avoid fetal asphyxia
Respiratory system	↑ minute ventilation (7.5 → 10.5 l/min) ↓ Pa_{CO_2} 5.3 kPa (40 mmHg) → 4.0 kPa (30 mmHg) ↓ functional residual capacity ↑ oxygen consumption Congested nasal mucous membranes	Awareness more likely ↑ ventilation required ↓ oxygen stores, preoxygenation necessary Care with nasogastric or nasotracheal tubes
Gastrointestinal tract	↓ lower oesophageal sphincter tone ↑ extragastric pressure Delayed gastric emptying	↑ risk of gastric aspiration
Fluid status	↑ blood volume (by 1250 ml) Total body water ↑ by 6–8 litres ↑ red cell volume ↓ plasma concentration of sodium, potassium and urea ↓ concentration plasma proteins but ↑ total amount	↑ volume of distribution of drugs; some blood loss tolerated at delivery without requiring replacement Recognition of incipient renal failure with normal non-pregnant values ↑ drug binding
General	Breast enlargement Lumbar lordosis Ligamental softening	Difficulty introducing laryngoscope Difficulty opening up lumbar interspaces Difficulting recognizing ligamentum flavum and risk of dural tap

increases to have its peak effect about 3 hours after maternal administration.[3]

Pre-anaesthetic assessment

There is no reason why the obstetric patient should receive treatment inferior to that given to her non-pregnant counterpart requiring anaesthetic intervention. Assessment of any coexisting medical conditions and the influence of pregnancy on the patient and their bearing on anaesthetic procedures is vitally important. Many patients present as acute emergencies, and the middle of the night is not the optimum time for the anaesthetist to be confronted for the first time with an obscure condition complicated by the physiological and/or pathological changes of pregnancy. The majority of obstetric patients present initially for antenatal booking of their confinement, at the end of their first trimester. There are therefore nearly 6 months in which any coexisting medical conditions or potential anaesthetic difficulties can be fully assessed and the anaesthetic options researched and discussed before delivery is expected. This requires good communication from the obstetric staff.[4] Anaesthetists should record the problems and the outcome of discussions with the patient and obstetricians so that the anaesthetist faced with giving

an anaesthetic is aware of these deliberations. A distinctive label on the outside of the medical notes to warn anaesthetists of a serious problem has been suggested.

In addition to any coexisting medical condition, the physiological changes of pregnancy itself should be assessed. These, and their anaesthetic implications, are listed in Table 64.1. Assessment should include factors such as the patient's parity (i.e. whether the patient is primiparous or multiparous) and the gestation of pregnancy. A recent weight and haemoglobin concentration should be routinely available. Multiple births and poly-hydramnios will cause exaggerated physiological changes. Any pathological changes of pregnancy, such as those of pre-eclampsia (see below) should be fully assessed. The reason for anaesthetic intervention and the degree of urgency should be recorded. The duration of preoperative starvation must be asked because many patients may have had substantial recent intake, particularly in emergency cases. Factors in addition to those of pregnancy which may suggest difficult intubation need to be carefully assessed. These last two points may influence the choice between regional and general anaesthesia. It is also appropriate to make a psychological assessment of the patient in relation to her potential ability to remain calm if caesarean section is performed under regional anaesthesia.

Aspiration of gastric contents

The patients most at risk from gastric aspiration are those who undergo general anaesthesia, although it must be remembered that there is also a risk when protective airway reflexes are lost as a result of a convulsion or from a complication of regional anaesthesia. The risk of aspiration is reduced by performing fewer procedures under general anaesthesia and by the appropriate application of cricoid pressure before tracheal intubation is achieved when general anaesthesia is employed. Regurgitation is a particular hazard in the obstetric patient because of the lowered oesophageal sphincter tone and increased extragastric pressure. Although pregnancy itself probably does not delay gastric emptying, it is delayed in labour and this gastric stasis is markedly potentiated by the use of opioid analgesia.[5] Deaths have occurred from mechanical obstruction of the airway, so it is wise to have a labour ward policy of avoiding intake of solid food.

Deaths have also occurred from the chemical pneumonitis resulting from aspiration of gastric acid (Mendelson's syndrome).[6] The volume and the pH of gastric aspirate are both important in the causation of Mendelson's syndrome.[7] It is thus appropriate to take steps to reduce both the volume and the acidity of gastric contents in those women who may potentially require general anaesthesia. This is achieved by a period of starvation and by the administration of H_2 antagonists. The H_2 antagonist ranitidine is a popular choice and can be given orally if absorption is not expected to be adversely affected or parenterally if opioids have been given. It is obvious that women scheduled for elective caesarean section should receive this prophylaxis, even if regional block is the anticipated mode of anaesthesia because of the possibility of inadequate block or of complications requiring general anaesthesia. It is important to realize that the reduction in gastric acid secretion by H_2 antagonism will have effect only from the time it is given and will have no effect on acid already secreted into the stomach. Thus, in the majority of cases, it is still necessary to give an oral antacid such as sodium citrate, which is effective and relatively benign if inhaled.[8]

Those who wish for childbirth to be a natural process question why the majority of women who will require no anaesthetic intervention should receive drugs that they do not need. Unfortunately, midwives and obstetricians cannot predict accurately which patients will develop obstetric complications, fetal distress or a retained placenta. Although women can be categorized into high and low risk for requiring anaesthesia, it is a question of weighing up the risk/benefit/cost ratios of prophylactic treatment for all women. The benefit is reducing the number of maternal deaths nation wide, the risk is of interfering with a natural process but producing no ill-effects, and the cost is substantial. It is analogous to vaccination against infec-

tious diseases where only a minority will benefit, the risks are very low, and the cost is substantial.

Aortocaval compression

Up to 15 per cent of pregnant women near term develop symptoms of hypotension when lying supine. This was first described in 1953 and was termed the 'supine hypotensive syndrome';[9] in 1964 vena caval compression was demonstrated radiographically.[10] Four years later partial aortic compression was demonstrated[11] and the syndrome is now called aortocaval compression. Vena caval obstruction pools blood in the legs and reduces cardiac output because of the reduced venous return. Uterine arterial flow is decreased and placental perfusion is further reduced because of increased uterine venous pressure. Compression of the aorta which does not cause hypotension, as measured in the arm, nevertheless reduces uterine blood flow. Both the aortic and the vena caval effects result in fetal asphyxia and fetal distress. Aortocaval compression is particularly important to the anaesthetist because anaesthetic agents causing vasodilatation, such as thiopentone or volatile agents, or sympathetic blockade as in epidural or spinal anaesthesia, result in greater pooling of peripheral blood and hence greater hypotension.

Prevention of aortocaval compression is preferable to treatment and so the pregnant mother should never be allowed to lie supine. Most women choose not to lie on their back naturally in late pregnancy but it is especially important not to do so when there is peripheral vasodilatation as in regional blockade or general anaesthesia. Displacement of the uterus to the left by a 15 degree tilt, by use of a wedge or tilting the operating table, is effective prevention for most mothers. Right uterine displacement is more effective in about 10 per cent of mothers, and the efficacy can be checked by measuring maternal blood pressure and observing the fetal heart rate pattern. An unusually large uterus as seen in multiple pregnancy or polyhydramnios may necessitate a greater degree of lateral tilt.[12]

Analgesia in labour

Pain in labour can be the most intense pain known.[13] Nevertheless, it is not experienced as such by every woman and thus analgesia should be directed at the particular circumstances and the resources available. Apart from the humanitarian reasons for providing pain relief, a painful labour has detrimental effects on the mother and fetus. The maternal hyperventilation in response to pain shifts the maternal oxyhaemoglobin dissociation curve to the left, thus reducing fetal oxygenation. Hyperventilation is followed by hypoventilation between uterine contractions which, combined with decreased uterine blood flow (caused by catecholamine release in response to pain), worsens fetal hypoxaemia. Despite transient respiratory alkalosis, maternal acidosis supervenes as a result of lactic acid production from skeletal muscle activity and from free fatty acid production as a result of sympathetic activation. This produces incoordinate uterine action and a prolonged labour. The hypertensive response to sympathetic activation may be particularly detrimental in a mother who is already hypertensive.

In general, the contraction pains become more severe with the progress of labour.

Antenatal education

An important element in reducing fear is instruction in the elementary physiology of pregnancy and events of labour. This is appropriately carried out at childbirth preparation classes where the methods of analgesia available can be discussed. It should be made clear to the mother that she can change her chosen method of analgesia during labour without detriment to herself or her baby.

The importance of the psychological preparation for labour was introduced by Grantly Dick-Read,[14] who taught that a patient properly trained in relaxation techniques could have a natural delivery without the need for analgesia. It is important for the mother to appreciate that, whilst this may be helpful, she should not feel guilty or a failure if she finds that pain relief is required. The psychoprophylactic approach of Lamaze[15] involves the mother distracting herself from the pain by concentrating on specific objects or pictures and by breathing exercises. Conventional analgesic drugs are permitted with this method.

The French obstetrician, Leboyer,[16] advocated delivery into a dark quiet room and placing the newborn baby onto the mother's abdomen, with the umbilical cord being cut only after pulsations of the cord have stopped. With this method, there are the potential dangers of neonatal hypovolaemia with the baby being placed above the level of the placenta, unrecognized neonatal cyanosis and unrecognized maternal haemorrhage.

Transcutaneous electrical nerve stimulation (TENS)

TENS is the application of pulsed electrical current through surface electrodes placed on the skin at appropriate dermatomes (T11 to L1 for the first stage of labour; S2–4 for second stage), parallel with and close to the spine. It is thought to work by low frequency stimulation increasing endorphin production and by high frequency stimulation closing the 'gate' in the spinal cord to the transmission of pain information. Around 20 per cent of selected patients[17,18] gain substantial help from the use of TENS, although it can take up to 40 minutes to become effective. It also has the advantages of being non-invasive, not interfering with fetal heart monitoring and having no harmful effects on the mother or the fetus.

Entonox

This is the only inhalational method of analgesia in use now that methoxyflurane has been abandoned because of nephrotoxicity and approval for the use of trichloroethylene has been withdrawn.

Entonox is a 50 per cent mixture of nitrous oxide and oxygen and is delivered via a demand valve through a low resistance breathing system from cylinders after pressure reduction. During the first stage of labour, Entonox is breathed intermittently with the uterine contractions. Because it takes about 45 seconds for the analgesic effect to be attained it is important that the mother starts breathing Entonox with the onset of a contraction so as to obtain analgesia with the later painful phase of the contraction. Deep and relatively slow breathing should be encouraged and the patient should not use Entonox between contractions. In the second stage of labour, two or three deep breaths of Entonox can be inhaled before each expulsive effort, to attain some analgesia.

Mothers using Entonox should not be promised complete pain relief. It is not a suitable analgesic to be used alone for forceps delivery, perineal suturing or removal of retained placenta. Satisfactory analgesia is obtained in approximately 50 per cent of mothers who choose this form of analgesia for labour. Higher concentrations of nitrous oxide (70 per cent) improve the quality of analgesia but produce unconsciousness in a significant number.[19]

Small (500 litre) cylinders are suitable for home confinements. A recent recommendation[20] limiting the maximum permitted levels of nitrous oxide in a working environment to 50 parts per million may mean that the use of Entonox will have to be curtailed or methods of scavenging exhaled gas will have to be devised.

Opioid analgesia

The use of opioids in labour is a compromise between effective analgesia and unwanted side effects. Free placental transfer of opioids causes neonatal ventilatory depression and, whilst this effect can be reversed with antagonists, it is preferable that the need does not arise. Maternal effects also include nausea, vomiting, drowsiness, disorientation and disinhibition which led to dissatisfaction with the use of opioids, especially if pain relief is not particularly good. Nevertheless, some 50–60 per cent of mothers[21,22] receive useful if incomplete analgesia. Morphine has a more potent depressant effect on the neonate and hence pethidine is the more popular agent. The use of pethidine is authorized in the UK for intramuscular use by midwives. Continuous intravenous infusion or patient-controlled administration (PCA) may be used under medical supervision. PCA can provide useful analgesia, particularly when regional blockade is contraindicated, giving the mother some personal control.

Epidural analgesia

The most effective method of analgesia in labour is, undeniably, central neural blockade, most commonly provided by the epidural route. Obstetric epidural services have been developed in Great Britain since the late 1960s and most larger obstetric units are able to offer this service. Adequate trained staff, both midwifery and anaesthetic, are essential for the safety of such a service.

In the first stage of labour, pain is usually felt in the cutaneous distribution of the first lumbar and eleventh and twelfth thoracic dermatomes, in the lower abdomen, groins and lower back and is probably related to cervical dilatation. Backache may be severe, particularly when the baby is in the occipitoposterior position. The pain of the second stage of labour is conducted through the second, third and fourth sacral nerves. Epidural blockade is therefore directed towards these nerve roots.

The main indication for epidural analgesia is the relief of pain and as such the majority of requests come from the mother. Nevertheless, there are some mothers who will obtain particular benefit from epidural analgesia and who should have these benefits explained. Epidural analgesia is specifically indicated when tracheal intubation is anticipated to be difficult or impossible because the block can readily be extended should operative intervention be necessary. Mothers with respiratory disease benefit by avoiding the hyperventilation associated with painful contractions. Epidural analgesia has the advantage in hypertensive mothers of attenuating the hypertensive response to pain and facilitates blood pressure control in labour (and at caesarean section, if necessary). Thus epidural analgesia is indicated in mothers with pre-eclampsia provided there is no clotting disorder. Epidural blockade is also indicated in any condition in which an increase in intracranial pressure during labour or delivery would be dangerous. In this situation, the epidural catheter must be inserted with particular skill because an inadvertent dural puncture *must* be avoided. The reduction in the stress hormone response to labour may make the use of epidural analgesia beneficial in endocrine disorders, of which diabetes mellitus is a common example. Obstetric indications for epidural analgesia include prolonged labour, incoordinate uterine action, trial of labour and when there is a strong likelihood of operative or instrumental delivery. Fetal indications for epidural analgesia include prematurity, breech presentation and multiple pregnancy, because greater control of delivery is possible, and the depressant effects of opioids are avoided.

Epidural analgesia is contraindicated if the mother refuses. Other contraindications are uncorrected hypovolaemia, local or systemic sepsis, bleeding diathesis of any aetiology and fetal distress. The reasons are that a sympathetic block produces severe and unresponsive hypotension in the hypovolaemic mother, local sepsis risks the introduction of infective organisms to the central nervous system and systemic sepsis may produce an epidural abscess if an epidural vein is punctured. In the presence of a bleeding diathesis, epidural vein puncture would cause bleeding controlled only by pressure, which may cause permanent nerve damage. It is generally agreed that diseases causing bleeding diathesis (e.g. von Willebrand's, idiopathic thrombocytopenic purpura) or anticoagulant therapy (except low dose subcutaneous heparin) contraindicate epidural analgesia. An area of controversy concerns those patients who have been on aspirin therapy.[23] The anti-platelet effect of aspirin is irreversible and lasts 7–10 days. In view of the seriousness of the potential complications, many anaesthetists avoid extradural analgesia in those who have taken aspirin in the previous 10 days but might consider its use when the potential benefits outweigh the risks. This may be pertinent for those who have been on aspirin in an attempt to prevent the development or worsening of pre-eclampsia.

Epidural analgesia should be offered only if there are sufficient anaesthetists, midwives and obstetricians to care for the mother safely. This leads to difficulties in predicting staffing with shift changes and unpredictable admissions of women in labour. Resuscitation equipment should be available to every delivery room.

Insertion of an epidural catheter for labour has the added challenges of a woman who may be unable to keep still during contractions, of increased weight above the non-pregnant state and often of softening of the ligamentum flavum. These factors all make dural tap a more difficult complication to avoid. Identification of the epidural space is more likely to be successful if meticulous attention is paid to the correct positioning of the patient (see Chapter 34). With the sitting position the mother should sit squarely on the bed with her buttocks parallel to the edge of the bed and curl over a pillow. Alternatively, with the patient lying on her side, she should be on the edge of the bed with the back straight and vertical and her head on one pillow so that the shoulders are square, and she should be curled such that the upper leg is prevented from rolling forward.

Management of epidural analgesia

Before an epidural is sited the procedure must be explained to the mother and consent obtained. Although it might be ideal to obtain written consent, mothers often change their mind during labour about their choice of analgesia and hence written consent would frequently need to be obtained at a time of great pain and after the effects of nitrous oxide or opioids. Thus, consent should be obtained before labour. Otherwise it seems sensible to imply consent by the fact that the woman is willing to keep still enough to have the procedure performed.

An intravenous infusion must be sited before inserting the epidural so that potential complications can be managed and so that the circulation can be preloaded with approximately 1 litre of crystalloid solution to minimize hypotension. The technique of epidural space identification with the choices of methods is fully described in Chapter 34. There is no reason that a particular method is more suited to obstetrics, although there is a suggestion that loss of resistance to saline is associated with a lower incidence of dural puncture.[24] Anaesthetists should choose the technique with which they are most familiar. The need for asepsis is stressed.

Insertion at the level of L1/2, L2/3 or L3/4 is recommended, the choice of space being dependent on the ease of insertion.

Test dose

A test dose is intended to exclude intrathecal or intravenous placement following careful inspection of the catheter after aspiration has failed to reveal CSF or blood. Ideally, it should be done with a readily available drug and have few false positive or negative findings. A local anaesthetic should detect intrathecal placement provided 5 minutes are waited before testing for motor block and looking for hypotension; 3 ml heavy lignocaine (lidocaine) or bupivacaine would be most suitable agents but it is common for isobaric solutions to be used. Intravenous injection is more difficult to detect: a 3 ml injection of local anaesthetic containing 1/200 000 adrenaline (15 µg) i.v. produces an increase in heart rate of 20–30 beats per minute.[25] Detection of this requires continuous heart rate monitoring in the few minutes after injection. Tachycardia may also be caused by pain, and a false negative result may

be obtained in mothers who have a variable baseline heart rate. There are also possible adverse effects of intravenous adrenaline on uterine blood flow and fetal well-being.[26] Because of these difficulties even after a negative aspiration test, it is recommended that all doses of epidural local anaesthetic should be fractionated, no more than 5 ml being administered before observing the effects.

Top-ups

Bupivacaine is the local anaesthetic agent of choice because of its longer duration of action. The choice of concentration depends on how much the loss of motor power is considered important. The use of 0.5 per cent bupivacaine usually results in a fairly dense motor block, whereas the use of a weaker solution causes less immobility and hence may be preferred. Often the use of 0.25 per cent bupivacaine retains some motor power and allows the mother to move herself; 10 ml is usually sufficient volume. There is increasing popularity of 0.1 or 0.125 per cent solutions of bupivacaine. The blood pressure should be checked every 5 minutes for 20 minutes after each top-up. The effect of epidural analgesia on the course of labour and the mode of delivery has been hotly debated for many years. Many studies[27–29] have reported an increase in operative deliveries: it has to be remembered that mothers who choose epidural block are different from those who do not and that therefore any increase in instrumental delivery may not be a result of the analgesia. It seems now to be accepted that the second stage of labour may be prolonged during epidural analgesia as a result of the loss of the bearing down reflex and reduced efficacy of maternal expulsive effort. Now that it is no longer considered necessary to terminate the second stage of labour after a set time (1–2 hours), provided there is no evidence of fetal distress, the issue of malrotation seems to have disappeared from the literature.[30]

Continuous infusion

Continuous infusions with or without opioids have the advantage of providing a continuous stable anaesthetic level with less fluctuations in pain relief, but do not necessarily abolish the need for some top-ups. The use of dilute local anaesthetic solutions retains motor power and the mother is able to push more effectively in the second stage of labour. Continuous infusions have been compared with patient-controlled epidural analgesia (PCEA) and no differences were found in the duration of second stage or outcome of labour or other benefit of PCEA.[31]

Double catheter technique

Where selective blockade is thought to be an advantage, one catheter can be inserted in the upper lumbar region for use in the first stage of labour, and a second catheter inserted in the lower lumbar (or caudal) region for use in the second stage of labour. The advantages of this are the use of lower doses of local anaesthetic and the ability to feel contractions abdominally in the second stage, encouraging effective expulsive efforts.

Complications of epidural analgesia

The complications of epidural analgesia are outlined in Table 64.2. These are discussed more fully in Chapter 34, and the areas of particular relevance to obstetrics are pointed out here. The most common complications are inadequate analgesia and hypotension. Inadequate analgesia results from misplacement of the epidural catheter (it was never in the epidural space, has fallen out or has traversed through an intervertebral foramen). The specific diagnosis is made by assessment of the dermatomal sensory block and by inspection of the catheter in relation to the depth of the epidural space as recorded on the chart. If the problem is that the block simply does not extend as high as T10, a further top-up usually achieves analgesia. Other causes are usually most easily cured by reinsertion of the catheter, although pulling the catheter back can be effective if more than 3 cm was left in the epidural space. Persistent perineal pain (common when the fetus is in the occipitoposterior position) can be difficult to abolish but may be helped by adding an opioid (e.g. 50 µg fentanyl) to the top-up, by using a stronger local anaesthetic solution (e.g. 0.5 per cent bupivacaine) or by inserting a low lumbar or caudal epidural. A missed segment may be blocked by a further top-up with the unblocked side dependent and by ensuring that no more than 3 cm of catheter is in the space, but the catheter may need to be reinserted.

Preloading of the circulation with intravenous crystalloid helps prevent hypotension but this is nevertheless a common complication which is detected by routine frequent blood pressure measurement and by being alert to symptoms of light-headedness and nausea. Aortocaval

Table 64.2 The complications of epidural analgesia

Inadequate analgesia
Missed segment
Hypotension
Intravenous injection
Intrathecal injection
Urinary retention
Shivering
Dural puncture
Backache
Temporary neurological damage
Permanent neurological damage
Catheter breakage

compression must be prevented, and it is wise for the mother to adopt the full lateral position at all times. In the event of hypotension occurring, therapy is urgent: the other lateral position can be tried, the legs elevated, facial oxygen applied, the intravenous infusion increased and intravenous ephedrine administered. The effect of hypotension on the fetus needs to be monitored.

The administration of the test dose (see earlier) is designed to detect intravascular or intrathecal placement. A bloody tap occurs more frequently if the catheter is advanced into the epidural space during a contraction because of engorgement of the epidural veins. If blood is seen in the epidural catheter, it can be flushed and withdrawn until no further blood appears on aspiration. If sufficient catheter is left in the space, the test dose can be given cautiously, asking for symptoms of intravenous administration. Reinsertion of the epidural catheter is required if blood is seen in the catheter or if there are signs of intravenous placement. The reported incidence of convulsions during obstetric regional anaesthesia and resultant maternal morbidity and mortality is low, suggesting that, with proper treatment, local anaesthetic-induced toxic reactions should not cause permanent sequelae.[32]

Intrathecal placement implies dural puncture and management as below. The extent of block obtained by intrathecal administration depends on the volume of local anaesthetic used and, whilst the test dose recommended should not result in dangerously high blockade, larger volumes may result in a total spinal block. Respiratory arrest and profound sympathetic block occur, and if effective treatment is not instituted immediately cardiac arrest ensues. Artificial ventilation with oxygen, tracheal intubation, intravenous fluid administration and vasoconstrictors (ephedrine having least deleterious effect on uterine blood flow) are urgently indicated. Although the complications of intravenous or intrathecal placement occur most commonly on their insertion, it is possible for catheter migration to occur at a later stage. An exaggerated effect of a small volume of local anaesthetic is also seen after subdural injection.

When the epidural block affects the sacral segments the mother will not be aware of a full bladder; the midwife should encourage regular attempts at voiding and, if necessary, intermittent catheterization. If the bladder is left full the progress of labour may be impeded.

Shivering can be an annoying complication, which the mother can find distressing and which can be severe enough to interfere with blood pressure measurement. It can usually be abolished by the addition of an opioid to the epidural.[33]

After dural puncture the obstetric patient is more likely to develop a headache than her non-obstetric counterpart. Postdural puncture headache (PDPH) is frontal or occipital, is relieved by recumbency and exacerbated by the upright position. Relief of headache by abdominal compression is useful in confirming the diagnosis.[34] PDPH is particularly debilitating because the mother needs to care for her newborn baby. In the absence of therapy the incidence of PDPH is 75 per cent; it is usually self-limiting, lasting no more than 10 days.[24] The severity of PDPH can be reduced by the application of an epidural infusion of 1 litre 0.9 per cent saline after delivery.[24] The most effective cure remains an autologous blood patch (15–20 ml) sited in the epidural space near the original puncture; the relief is usually instantaneous[35,36] but a second patch may sometimes be needed and is usually effective.[24]

It is almost invariable for the mother to notice bruising at the site of epidural insertion. Backache of more than 6 weeks' duration has been found in 10.5 per cent of women after natural childbirth compared with an incidence of 18.9 per cent in women who had epidural analgesia for labour.[37] The severity of backache was not assessed in this survey which was part of the overall health of women after childbirth but the backache was ascribed to tolerance of abnormal positions and straining in the presence of segmental blockade. This is more likely to occur with more concentrated solutions of local anaesthetic which were in common use at this time. A more recent report also suggests that new long-term backache occurs more commonly after epidural analgesia in labour and that it tends to be postural but not severe.[38]

The incidence of neurological sequelae in obstetrics is difficult to estimate but a recent survey of non-fatal complications after epidural analgesia suggests that the figure lies somewhere between 1 in 7000 and 1 in 14 000, of which some are single nerve neuropathy likely to be due to obstetric causes.[39] Causes of chronic neurological sequelae include obstetric trauma, epidural haematoma, epidural or intrathecal infection and inadvertent injection of neurotoxic substances.

Removal of an epidural catheter is normally easy. If resistance is met, bending the patient's back may ease its removal. If the catheter is broken and part of it is left in the patient, she should be informed. Surgical removal is not indicated unless there is pain or neurological dysfunction.

Anaesthesia for caesarean section

The proportion of women delivered by caesarean section has been increasing steadily such that in the UK the overall incidence is 13 per cent [40] and in individual units is over 20 per cent, which compares with an overall rate of only 5 per cent some 20 years ago. In Brazil, one in three women is delivered by caesarean section.[41] The choice of anaesthetic technique rests between general and regional (epidural, spinal or combined epidural/spinal) and depends on patient preferences, coexisting medical conditions, the reason for surgery, the degree of urgency and the anaesthetist's judgement and experience. The aims of anaesthesia are to provide safety and comfort for the mother, minimal neonatal depression and optimal working conditions for the obstetrician.

General anaesthesia

It is desirable to have a routine general anaesthetic (which can be deviated from when indicated) so that no time is lost in genuine emergencies. General anaesthesia has the advantages of speed of onset, less hypotension, a lower failure rate, fewer contraindications, and being more acceptable to some patients, when compared with regional anaesthesia. Nevertheless, there are more deaths from general anaesthesia[42] and hence regional anaesthesia should be chosen where appropriate.

Preoperative assessment, antacid prophylaxis and lateral tilt need to be applied as discussed earlier. Preoxygenation lessens the likelihood of fetal or maternal hypoxia resultant from the reduced functional residual capacity and increased oxygen consumption (see Table 64.1). In addition, preoxygenation allows more time for tracheal intubation and lessens the likelihood of requiring ventilation from a mask, which might encourage gastric regurgitation. Cricoid pressure should be applied immediately before induction of anaesthesia. Thiopentone remains the induction agent of choice (see earlier) but in shocked states (e.g. antepartum haemorrhage) etomidate may be chosen although it suppresses cortisol production in the neonate.[43] Ketamine, in doses up to 1 mg/kg, is advantageous if there is severe hypovolaemia or asthma, but higher doses result in neonatal depression.[44] Tracheal intubation is required because of the likelihood of regurgitation. Caesarean section is normally completed within 20–40 minutes, and hence atracurium or vecuronium are suitable non-depolarizing neuromuscular blockers because of their ease of reversal.

Maintenance of anaesthesia with 50 per cent nitrous oxide in oxygen improves neonatal outcome when compared to lower inspired oxygen concentrations.[45] A higher inspired oxygen concentration may be beneficial to the compromised fetus.[46,47] Supplementation with a low dose of volatile agent (enflurane 1 per cent, isoflurane 0.75 per cent or halothane 0.5 per cent) helps prevent maternal awareness without causing neonatal depression or excessive uterine bleeding. Supplementation with intravenous opioids should be delayed until after delivery of the baby, at which time an oxytocic (e.g. oxytocin 5–10 units) is also given intravenously.

Monitoring should include maternal ECG, oxygen saturation, capnography, blood pressure, incision of the uterus to time of delivery and the time of delivery. The condition of the infant is directly related to the interval between incision of the uterus to delivery.[48] The lungs should be ventilated to maintain normocapnia of pregnancy (4.0 kPa or 30 mmHg) in order to maintain placental blood flow and oxygen transfer.[49]

Difficulties in tracheal intubation occur more frequently in obstetric anaesthesia, the incidence of failed intubation being approximately 1 in 280 compared with 1 in 2230 in surgical patients.[50] The reasons for this include difficulty in laryngoscope insertion because of enlarged breasts, abnormal positioning because of the lateral tilt and cricoid pressure application and, in some cases, laryngeal and pharyngeal oedema. Although difficulties may be predicted by Mallampati grading and thyromental distance (see Chapter 35), unexpected failure to intubate nevertheless occurs and the anaesthetist needs to be ready to deal with this emergency with a prepared plan.[51] Oxygenation *must* be maintained and a decision made as to whether to wake the patient and institute a regional block or awake fibreoptic intubation or whether to continue with general anaesthesia. This depends on the urgency of surgery and the ease of oxygenation. It is important to recognize difficult intubation early, accept possible failure and not continue futile attempts at intubation at the expense of oxygenation for which ventilation from a facemask will be required until neuromuscular blockade recovers after suxamethonium. Further assistance should immediately be summoned. Cricoid pressure should be maintained, spontaneous ventilation established and anaesthesia deepened using a volatile agent (halothane is frequently chosen because it is least irritant) if it is decided to continue with general anaesthesia. The role of the laryngeal mask has been contentious because it does not protect against aspiration; nevertheless, it has enabled safer oxygenation.[52] It is recommended that cricoid pressure be maintained with the laryngeal mask in place but this can sometimes obstruct the airway (MI Bowden, 1990, personal communication). In situations in which the airway cannot be maintained, oxygenation must be achieved with the choices of transtracheal ventilation through an intravenous cannula, cricothyroidotomy or tracheostomy being available.

Uterine relaxation may assist delivery in circumstances such as when the lower segment is unformed in preterm deliveries, when there is a transverse lie or when a breech presentation is in labour. This can be provided by increasing the inhaled volatile agent concentration (e.g. enflurane to 3 per cent) in the 2 minutes before uterine incision. Extra oxytocin may be needed to achieve adequate uterine contraction.

Regional anaesthesia

The fact that caesarean section can be performed very satisfactorily under regional anaesthesia,[53,54] giving both parents the opportunity of being present and aware of the birth, has resulted in a recent marked increase in its popularity. Regional anaesthesia avoids the problem of unintended awareness. Unless an excessively high block occurs there is no worry about difficult intubation, misplaced tracheal tube or aspiration of gastric contents. Advantages of regional anaesthesia are summarized in Table 64.3. Successful management is demanding of the anaesthetist, who must engender confidence to the parents. Thoughtful anticipation by the obstetricians can allow the anaesthetist to become involved early and to discuss the benefits of epidural insertion with potential candidates for emergency caesarean section.[55]

A block is required from S5 to T4,[53] and this should be established before surgery is begun. It is never a happy option to convert to general anaesthesia once surgery has commenced. Thus it is important that the extent and density of the block are assessed before surgery. With epidural blockade the volume of local anaesthetic required is variable (from 12 ml to 45 ml or more) and hence an incremental technique enables the upper extent of the block to be controlled. With a spinal anaesthetic the extent of the block depends on the volume of the local anaesthetic used, its baricity, the speed with which the mother lies down (if the spinal is performed sitting) and the posture adopted. A spinal block that does not extend sufficiently high can be extended by head-down tilt, which should be reversed before surgery starts in order to prevent collection of blood and amniotic fluid under the diaphragm and also further extension of the block. A much more dense block is usually obtained with a spinal block but the quality of anaesthesia with an epidural block can be greatly enhanced by the addition of 50–100 µg fentanyl to the local anaesthetic administered epidurally,[56,57] which does not appear to

Table 64.3 Advantages of regional anaesthesia in operative obstetrics

Parental participation and early bonding
Less depressant effect on neonate
Avoids tracheal intubation difficulties
Much less risk of gastric aspiration
Mother maintains normocapnia
Reduced blood loss
Early postoperative analgesia provided
Avoids hangover effect of general anaesthesia
Reduced risk of pulmonary embolism

cause any adverse effect on the neonate. Minor intraoperative discomfort can be alleviated by allowing the mother to breathe Entonox or by the administration of an intravenous opioid such as alfentanil, although this is usually viewed as a relative failure of the regional technique by the anaesthetist, if not the mother.

Meticulous detail should be paid to the maintenance of blood pressure and the circulation preloaded with at least 1 litre of fluid. There is no evidence that colloid is superior to crystalloid.[58] Avoidance of aortocaval compression is particularly important because the sympathetic block causes lower limb vasodilatation. Pharmacological vasoconstriction of the capacitance vessels is frequently required to prevent hypotension. Ephedrine, having a predominantly beta effect (and hence less uterine artery vasoconstriction), is the most popular agent to be used although there is recent work which suggests that phenylephrine, despite having α-adrenergic effects, may be an acceptable alternative.[59] The rapid onset of block after a spinal injection makes hypotension almost invariable, and it is common to give a vasoconstrictor by intravenous infusion as a routine. Even with frequent automated measurement of blood pressure, the earliest warning of hypotension is often the subjective sensation of nausea in the mother and this should always alert the anaesthetist to check and correct the blood pressure.

Supplementary oxygen should be given to the mother to breathe, either from nasal prongs or from a plastic face mask. If uterine relaxation is required, this can be achieved by using salbutamol (albuterol) inhaled about 2 minutes before uterine incision. Blood loss is generally about half that at caesarean section under general anaesthesia.[60] However, if severe bleeding does occur, the mother is unable to compensate because of profound sympathetic blockade and hence the anaesthetist has to be particularly meticulous about fluid replacement. Any blood given should be warmed because of the unpleasantness of a fast transfusion of cold blood.

The duration of block due to spinal anaesthesia may be insufficient if the surgery is unexpectedly prolonged, and this has led to the use of a combined epidural and spinal technique. A long spinal needle is passed through an epidural needle and, after withdrawal of the spinal needle, an epidural catheter is inserted. This combination of techniques offers the advantages of the rapidity and reliability of spinal block with the facility to modify and extend the block into the postoperative period.[61]

Spinal blockade for obstetric practice was previously unpopular because of the excessively high incidence of PDPH in this group of patients. The introduction of the ogival (Sprotte) and pencil (Whitacre) pointed needles (see Chapter 34) has reduced the incidence of headache to be on a par with the likelihood of inadvertent dural puncture and resultant headache.

Obstetric complications

Retained placenta

The placenta is normally delivered within 30 minutes. When the placenta needs to be removed manually, an epidural catheter in place for analgesia in labour can be topped up. In other instances the choice is between a spinal block (from S5 to T10) and a general anaesthetic, depending on how much blood has been lost and the mother's preference. Blood loss can be difficult to assess and the mother's heart rate and blood pressure should be taken into account. Uterine relaxation may be required if the cervix has clamped down; this can be provided with inhaled salbutamol during regional anaesthesia and with enflurane or halothane during general anaesthesia.

Inverted uterus

This is a rare (1 in 20 000 deliveries) but potentially fatal complication (up to 80 per cent mortality) of the third stage of labour. It presents as an obviously inverted uterus with severe pain and shock that is out of proportion to the blood loss. Prompt (within minutes) replacement of the uterus is required under general anaesthesia. A reduced dose of induction agent will be needed. The uterus may need to be relaxed to be replaced, and enflurane (5 per cent for 1 minute) or halothane (2 per cent for 1 minute) may be suitable.

Amniotic fluid embolus

This is a rare but serious complication of pregnancy, having a mortality of 80 per cent.[62] The presentation is usually sudden, during labour or delivery, or immediately postpartum, with acute dyspnoea, cyanosis, hypoxia, hypotension or cardiac arrest. There may be a coagulopathy and uterine haemorrhage. Convulsions occur in 10 per cent of cases. The pathophysiology is partly obstructive with mechanical obstruction of the pulmonary vascular bed by solid particles, and partly humoral with prostaglandin release causing pulmonary vasoconstriction. It needs to be distinguished from clot or air embolus, or aspiration of gastric contents and in the presence of convulsions, eclampsia or local anaesthetic toxicity. The diagnosis is confirmed by finding fetal debris in maternal central venous blood or in the sputum, or in lung tissue at autopsy. Investigations that should be performed are arterial blood gases, coagulation screen, full blood count and urea and electrolytes. A lung scan may be more helpful than a chest x-ray. Treatment is non-specific and supportive, ensuring oxygenation (by artificial ventilation if necessary), circulatory support and replacement of blood and clotting factors.

Pre-eclampsia

Pre-eclampsia is the commonest of the life-threatening complications of pregnancy, occurring in a mild form in 15 per cent of pregnancies.[63] It is characterized by hypertension (above 140/90 mmHg or if the systolic blood pressure increases more than 30 mmHg or the diastolic pressure more than 15 mmHg above the antenatal booking values) and proteinuria (more than 0.3 g/l). Oedema is frequently present, as are brisk tendon reflexes. Pre-eclampsia only rarely presents before the 24th week of pregnancy (unless associated with a hydatidiform mole) and is predominantly a disease of primigravidas. Uncontrolled severe disease progresses to eclampsia (convulsions).

The cause of pre-eclampsia is as yet unknown although it is thought that there may be a genetic predisposition which promotes an immune response within the placenta. Normal placental implantation fails, which leads to reduced uteroplacental perfusion and function. This results in a reduction in prostacyclin formation and a relative increase in the vasoconstrictor, thromboxane A_2, which gives rise to endothelial damage, reduced plasma volume, diminished organ perfusion and intravascular coagulation. The clinical presentation of severe disease is quite varied, some mothers presenting with aggressive hypertension, some with coagulopathy, some with predominantly cerebral signs or even eclampsia with few premonitory signs, some with incipient renal failure and some with gross peripheral oedema and, on occasion, laryngeal or pulmonary oedema. Headaches, vomiting, epigastric pain and visual disturbances are symptoms of severe disease which predate convulsions. The HELLP (haemolysis, elevated liver enzymes and low platelets) syndrome, a variant of the disease first described in 1982,[64] has a high maternal and fetal mortality. Once pre-eclampsia is established, abnormal maternal creatinine, platelet, bilirubin and aspartate transaminase concentrations detect the development of subclinical organ dysfunction. Elevated plasma urate concentration has been shown to correlate with the risk of fetal mortality and indicates early delivery.[65] Low-dose aspirin therapy is used to suppress thromboxane A_2 production from platelets in an attempt to prevent the condition from developing, but as yet the role of this treatment is not fully evaluated. Recent aspirin therapy has implications for the use of epidurals, as discussed earlier.

Cure of pre-eclampsia is effected by delivery of the placenta. In severe disease this may imply delivery of a preterm infant and often necessitates caesarean section.

The comparative rarity and complexity of severe disease and its prominence in the causes of maternal death have led to the suggestion[42] that each health region should establish an advisory team with special expertise. This team should include anaesthetists.

The severity and presentation of the disease requires careful evaluation before anaesthesia. Epidural anaesthesia has the advantages of avoiding the pressor response to intubation, of minimal neonatal depression in an already compromised fetus and of the ready provision of postoperative analgesia after caesarean section. It is also useful for vaginal delivery but is contraindicated in the presence of a coagulopathy and hence a coagulation screen should be routinely checked. Preloading of the circulation provides a particular problem in the presence of a low colloid osmotic pressure if pulmonary oedema is to be prevented. The use of albumin in conjunction with central venous pressure monitoring may be indicated, and has been found to increase cardiac output and decrease systemic vascular resistance with minimal effect on arterial blood pressure.[66] Spinal anaesthesia is contraindicated in all but the mildest forms of pre-eclampsia because of the potential for the rapid sympathetic block compromising placental perfusion in the presence of the concomitant reduced blood volume.

If general anaesthesia is chosen, the hypertensive response to intubation can be severe[67] and needs to be attenuated. The choices rest between pretreatment with opioids such as alfentanil or fentanyl (in which case the paediatrician needs to be informed that naloxone may be required), antihypertensives (e.g. hydralazine, labetalol, esmolol or sodium nitroprusside), lignocaine or magnesium. Whatever drug is used, close monitoring is required and the anaesthetist must be prepared to treat severe hypertension and, on occasion, hypotension. Tracheal intubation may be difficult because of laryngeal and pharyngeal oedema. Awake intubation or tracheostomy has been required to maintain airway patency.[68,69] Enflurane is preferably avoided in the pre-eclamptic patient because of its epileptogenic potential.

Although cure is effected by delivery, the patient frequently requires close monitoring and high-dependency care for a few days. The aims of therapy are directed at the particular presentation of each mother with close attention to the blood pressure (aiming at a diastolic pressure in the range 90–100 mmHg), maintaining adequate urine output, replacement of clotting factors as appropriate and adequate oxygenation. In severe disease, therapy is also directed at the prevention of convulsions. This is achieved by effective antihypertensive therapy and by specific anticonvulsant therapy. Numerous drugs have been used to prevent convulsions, including bromethol, paraldehyde, phenothiazines, lytic cocktails, chlordiazepoxide and clonazepam, but dissatisfaction with their results has led to their abandonment. The use of chlormethiazole has waned because of the need to administer a large volume of fluid and because oversedation with loss of airway reflexes is a potential problem.[70] Long-term therapy with infusion of diazepam has the disadvantage of loss of airway reflexes, oversedation and prolonged recovery time. There are currently three approaches to therapy. The first of these is to offer no prophylactic anticonvulsant but to monitor the mother closely and administer anticonvulsants only in the presence of frank eclampsia. The rationale of this is that it is difficult to predict which mothers will suffer grand mal seizures and other therapies have potential dangers. The second and third choices are prophylactic treatment with either magnesium or phenytoin.

Magnesium sulphate is the drug of choice for both the treatment and the prevention of convulsions in the USA and South Africa. Its mode of action is not anticonvulsant as such but it is thought to afford protection against cerebral damage by blocking calcium influx through the N-methyl-D-aspartate (NMDA) subtype of the glutamate channel.[71] Maternal blood concentration depends on the volume of distribution (usually increased in pregnancy and especially in pre-eclampsia) and renal excretion of the magnesium ion (decreased in severe disease). The therapeutic blood concentration of magnesium sulphate is 2–4 mmol/l. This is achieved by administering a loading dose of 40–80 mg/kg intravenously followed by a maintenance infusion of 2 g per hour. The early signs of magnesium toxicity are nausea, feeling of warmth, somnolence, double vision, slurred speech, weakness and loss of patellar tendon reflexes. These develop at plasma concentrations of 3.5–5 mmol/l. Muscular paralysis and respiratory arrest develop at plasma concentrations of 6.5–7.5 mmol/l. Cardiac arrest occurs at concentrations of 10–12 mmol/l.[72] Calcium gluconate (1 g) should be available to counteract any signs of magnesium toxicity. Despite these dangers, familiarity with the use of magnesium has a history of relative safety.

Over recent years, phenytoin has become the agent of choice in the UK. It is administered intravenously with a loading dose of 10 mg/kg, followed by a further dose of 5 mg/kg 2 hours later. A maintenance dose of 200 mg 8-hourly is initiated 12 hours later, and can be given intravenously or orally.[73] This should achieve the therapeutic concentration of 10–20 ng/ml, although it should be remembered that the plasma concentrations of phenytoin depend on maternal weight and serum albumin concentration. Intravenous administration may be associated with local venous irritation, flushing, tinnitus and nausea. At concentrations of 20 ng/ml, nystagmus is apparent. At 30 ng/ml, ataxia and incoordination are observed. The use of intravenous phenytoin is potentially cardiotoxic and hence electrocardiographic monitoring is essential during initial loading. Phenytoin has a saturable metabolism and hence small dosage adjustments lead to disproportionate increases in total and free plasma phenytoin concentrations.

Studies of the efficacy of anticonvulsant therapy have been in the prevention of further fits in those with established eclampsia. Three prospective controlled trials suggest that magnesium sulphate has possible advantages over diazepam[74] and phenytoin.[75,76] The numbers in these studies are small and further convulsions have also occurred after magnesium sulphate. In non-randomly allocated studies[77] further convulsions have occurred after administration of the recommended dose of phenytoin and also in mothers who had adequate serum phenytoin concentrations.[76]

Maternal deaths

Pregnancy-related deaths have been audited since 1952 in the UK (via triennial confidential reports) and since 1915 in the USA. Deaths are classified as direct (specifically pregnancy-related), indirect (e.g. related to concomitant medical disorders) or fortuitous (i.e. unrelated to pregnancy). Over the above periods the rate of deaths as a result of infection, pre-eclampsia and in-hospital haemorrhage has substantially reduced. The causes of direct deaths in England and Wales since 1970 and their estimated incidence are illustrated in Table 64.4. In the most recent triennial report (1985–87),[78] the two most prominent causes of death are thromboembolic disease and hypertensive diseases of pregnancy. The proportion of deaths caused by anaesthesia had been relatively constant at 10.8–13 per cent (mirroring the general decline in maternal deaths) until the two most recent reports (1988–90) in which anaesthesia was implicated in 2.8 and 4.4 per cent of direct maternal deaths. Although one is anxious for this improvement to be verified in subsequent enquiries, the encouraging findings are ascribed to increasing anaesthetic resources devoted to obstetric services, an increased awareness of the risks of anaesthesia and the increasing use of appropriately supervised regional anaesthesia.

There has been a decline in the numbers of women dying from acid aspiration, which could be ascribed to effective prophylactic regimens, to improved training of anaesthetists and their assistants, and to the increased use of regional anaesthesia. The most frequent cause of anaesthetic-related death in the latest triennium was failure to intubate the trachea. This itself should of course not be fatal but recognition of failure to site the tracheal tube is crucial. Expired carbon dioxide monitoring is the only certain method of ensuring that the tube is in the trachea.

During the periods 1982–84 and 1985–87, emergency caesarean section was approximately four times more likely to result in death than was elective surgery. Although the safety record of regional anaesthesia is laudable, there is no room for complacency.

Table 64.4 Direct deaths by cause, rates per million estimated pregnancies, England and Wales 1970–87*

	1970–72	1973–75	1976–78	1979–81†	1982–84	1985–87
Pulmonary embolism	17.6	12.8	18.5	9.0	10.0	9.1
Hypertensive diseases of pregnancy	14.9	13.2	12.5	14.2	10.0	9.4
Anaesthesia	12.8	10.5	11.6	8.7	7.2	1.9
Amniotic fluid embolism	4.8	5.4	4.7	7.1	5.6	3.4
Abortion	25.3	10.5	6.0	5.5	4.4	2.3
Ectopic pregnancy	11.5	7.4	9.0	7.9	4.0	4.1
Haemorrhage	10.4	8.1	10.3	5.5	3.6	3.8
Sepsis, excluding abortion	10.4	7.4	6.5	3.1	1.0	2.3
Ruptured uterus	3.8	4.3	6.0	1.6	1.2	1.9
Other direct causes	6.9	8.5	8.2	7.5	8.4	7.5
All deaths	118.7	88.0	93.4	70.0†	55.0	45.6

* Rates for the UK were not available as there was no information on pregnancies for Scotland and Northern Ireland.
† Includes two other direct deaths omitted in the 1976–78 report.
Reproduced, with permission of the Controller of Her Majesty's Stationery Office, from reference 78.

Cardiopulmonary resuscitation

The likelihood of cardiopulmonary resuscitation (CPR) being needed during pregnancy is considered to be 1 in 30 000 but this frequency may be an underestimate because of the improved medical management of young female patients with severe medical diseases, especially cardiac problems, who become pregnant.[79] The outcome of CPR is less successful in pregnancy than in the non-pregnant patient. A longer time is required to restore spontaneous circulation after successful resuscitation. Higher doses of both adrenaline and sodium bicarbonate are required and there is an increased mortality. The principal reason for this lower success rate of CPR in pregnancy is the hazard presented by aortocaval compression which parallels gestational age. Until the 25th gestational week the basic measures of CPR are identical with those applied to non-pregnant women (see Chapter 71) and can be directed almost exclusively towards maternal survival.

After the 25th gestational week the mother needs to be positioned so that aortocaval compression is avoided and yet effective thoracic compression is still achievable. This can be accomplished by placing a wedge (preferably 30 degrees) under the right flank. If these measures are not sufficient to generate adequate arterial blood pressures, emergency delivery by caesarean section, possibly including hysterectomy, is essential. The need for caesarean section is mainly determined by the size of the gravid uterus. Defibrillation is the primary therapy for ventricular fibrillation. The American Heart Association recommends that, for non-pregnant patients, sodium bicarbonate is not the drug of first choice during CPR.[80,81] However, in contrast to that in other patients, a maternal metabolic acidosis with a pH of 7.3 or less should be treated with bicarbonate, because the combination of acidosis and the administration of adrenaline causes uteroplacental vasoconstriction.[82,83]

Medicolegal implications

Patients in general, but obstetric patients in particular, now expect a high level of medical care. Failure to deliver this may result in litigation. In a survey of malpractice claims in the USA,[83] maternal death and newborn brain damage were the leading complications in obstetrics. Two essentials particularly pertinent for everyday obstetric practice are ensuring that awareness does not occur during caesarean section performed under general anaesthesia and ensuring that there is sufficient analgesia during caesarean section performed under regional blockade. The latter is now a more common cause for litigation. Patients also sue for more minor complications than they do in non-obstetric care[84] and therefore it is especially important that realistic expectations and knowledge of complications are conveyed by the anaesthetist. As with other areas of medical practice, clear documentation of activity helps to defend actions. It is important to document the height of a block before caesarean section is started.

REFERENCES

1. Reynolds F. Placental transfer of drugs. *Current Anaesthesia and Critical Care* 1991; **2**: 108–16.

2. Drabkova J, Crul JF, Van der Kleijn E. Placental transfer of ^{14}C-labelled succinylcholine in near-term *Macaca mulatta* monkeys. *British Journal of Anaesthesia* 1973; **45**: 1087–96.

3. Kuhnert BR, Kuhnert PM, Tu A-SL, Lin DCK. Meperidine and normeperidine levels following meperidine administration during labor. II. Fetus and neonate. *American Journal of Obstetrics and Gynecology* 1979; **133**: 909–14.

4. Cody MW, Johnston JA. Antenatal assessment for anaesthesia. *British Journal of Anaesthesia* 1992; **69**: 332.

5. Nimmo WS, Wilson J, Prescott LF. Narcotic analgesics and delayed gastric emptying in labour. *Lancet* 1975; **1**: 890–3.

6. Mendelson CL. The aspiration of stomach contents into the lung during obstetric anesthesia. *American Journal of Obstetrics and Gynecology* 1946; **52**: 191–204.

7. James CJ, Modell JH, Gibbs CP, Kuck JF, Ruiz BC. Aspiration: combined effects of pH and volume of the aspirate in the rat. *Anesthesia and Analgesia* 1983; **62**: 266–7.

8. Gibbs CP, Spohr L, Schmidt D. The effectiveness of sodium citrate as an antacid. *Anesthesiology* 1982; **57**: 44–6.

9. Howard BK, Goodson JH, Mengert WF. Supine hypotension syndrome in late pregnancy. *Obstetrics and Gynecology* 1953; **1**: 371–7.

10. Kerr MG, Scott DB, Samule E. Studies of the inferior vena cava in late pregnancy. *British Medical Journal* 1964; **1**: 532–3.

11. Bienarz I, Crottogini JJ, Curachet E. Aortocaval compression by the uterus in late human pregnancy. *American Journal of Obstetrics and Gynecology* 1968; **100**: 203–17.

12. Kim YI, Chandra P, Marx GF. Successful management of severe aortocaval compression in twin pregnancy. *Obstetrics and Gynecology* 1975; **46**: 362–4.

13. Melzak R. Labour is still painful after prepared childbirth training. *Canadian Medical Association Journal* 1981; **125**: 357.

14. Dick-Read G. In: Wessel H, Ellis H, eds. *Childbirth without fear*, 4th ed. New York: Harper & Row, 1984.

15. Lamaze F. *Painless childbirth. Psychoprophylactic method*. London: Burke, 1958.

16. Leboyer F. *Birth without violence*. Westminster MD: Alfred A Knopf, 1975.

17. Robson JE. Transcutaneous nerve stimulation for pain relief in labour. *Anaesthesia* 1979; **34**: 357–60.

18. Stewart P. Transcutaneous nerve stimulation as a method of analgesia in labour. *Anaesthesia* 1979; **34**: 361–4.

19. Committee on Nitrous Oxide Analgesia in Midwifery; Sir Dugald Baird, Chairman. Clinical trials of different concentrations of oxygen and nitrous oxide for obstetric analgesia. *British Medical Journal* 1970; **1**: 709–13.

20. Glass DC, Hall AJ, Harrington JM. *The control of substances hazardous to health. Guidance for the initial assessment in hospitals*. London: HM Stationery Office, 1992.

21. Grant AM, Holt EM, Noble AD. A comparison between pethidine and phenazocine (Narphen) for relief of pain in labour. *Journal of Obstetrics and Gynaecology of the British Commonwealth* 1970; **77**: 824.

22. Holdcroft A, Morgan M. An assessment of the analgesic effect in labour of pethidine and 50 per cent nitrous oxide in oxygen (Entonox). *Journal of Obstetrics and Gynaecology of the British Commonwealth* 1974; **81**: 603.

23. Macdonald R. Aspirin and extradural blocks. *British Journal of Anaesthesia* 1991; **66**: 1–3.

24. Stride PC, Cooper GM. Dural taps revisited. A 20-year survey from Birmingham Maternity Hospital. *Anaesthesia* 1993; **48**: 247–55.

25. Van Zundert AA, Vaes LE, DeWolf AM. ECG monitoring of mother and fetus during epidural anesthesia. *Anesthesiology* 1987; **66**: 584–5.

26. Hood DD, Dewan DM, James FM. Maternal and fetal effects of epinephrine in gravid ewes. *Anesthesiology* 1986; **64**: 610–13.

27. Johnson WL, Winter WW, Eng M, Bonica JJ, Hunter CA. Effect of pudendal, spinal and peridural block anesthesia on the second stage of labour. *American Journal of Gynecology* 1972; **113**: 166–73.

28. Robinson JO, Rosen M, Evans JM, Revill SI, David H, Rees JD. Maternal opinion about analgesia for labour: a controlled trial between epidural block and intramuscular pethidine combined with inhalation. *Anaesthesia* 1980; **35**: 1173–81.

29. Studd JWW, Crawford JS, Duignan NM, Rowbotham CJF, Hughes AO. The effect of lumbar epidural analgesia on the rate of cervical dilatation and the outcome of labour of spontaneous onset. *British Journal of Obstetrics and Gynaecology* 1980; **87**: 1015–21.

30. Miller AC, DeVore JS, Eisler EA. Effects of anesthesia on uterine activity and labour. In: Shnider SM, Levinson G, eds. *Anesthesia for obstetrics*, 3rd ed. Baltimore: Williams & Wilkins, 1993: 53–69.

31. Lysak SZ, Eisenach JC, Dobson CE. Patient-controlled epidural analgesia during labour: a comparison of three solutions with a continuous infusion control. *Anesthesiology* 1990; **72**: 44–9.

32. Shnider SM, Levinson G, Ralston DH. Regional anesthesia for labor and delivery. In: Shnider SM, Levinson G, eds. *Anesthesia for obstetrics*, 3rd ed. Baltimore: Williams & Wilkins, 1993: 135–53.

33. Sevorino FB, Johnson MD, Lema MJ, Datta S, Ostheimer GW, Naulty JS. The effect of epidural sufentanil on shivering and body temperature in the parturient. *Anesthesia and Analgesia* 1989; **68**: 530–3.

34. Gutsche BB. Lumbar epidural analgesia in obstetrics: taps and patches. In: Reynolds F, ed. *Epidural and spinal blockade in obstetrics*. London: Bailliere Tindall, 1990: 95–106.

35. Szeinfeld M, Ihmeidan TH, Moser MM, Machado R, Klose KJ, Serafini AN. Epidural blood patch: an evaluation of the volume and spread of blood injected into the epidural space. *Anesthesiology* 1986; **64**: 820–2.

36. Carrie LES. Epidural blood patch: why the rapid response? *Anesthesia and Analgesia* 1991; **72**: 129–30.

37. MacArthur C, Lewis M, Knox EG, Crawford JS. Epidural

anaesthesia and long-term backache after childbirth. *British Medical Journal* 1990; **301**: 9–12.

38. Russell R, Groves P, Taub N, O'Dowd J, Reynolds F. Assessing long-term backache after childbirth. *British Medical Journal* 1993; **306**: 1299–302.

39. Scott DB, Hibbard BM. Serious non-fatal complications associated with extradural block in obstetric practice. *British Journal of Anaesthesia* 1990; **64**: 537–41.

40. Savage W, Francome C. British caesarean section rates: have we reached a plateau? *British Journal of Obstetrics and Gynaecology* 1993; **100**: 493–6.

41. Notzon FC. International differences in the use of obstetric interventions. *Journal of the American Medical Association* 1990; **263**: 3286–91.

42. Department of Health. *Report on confidential enquiries into maternal deaths in England and Wales 1982–84.* London: HM Stationery Office, 1989.

43. Reddy BK, Pizer B, Bull PT. Neonatal serum cortisol suppression by etomidate compared with thiopentone for elective caesarean section. *European Journal of Anaesthesiology* 1988; **5**: 171–6.

44. Janeczko GF, El-Etr AA, Younes S. Low-dose ketamine anesthesia for obstetrical delivery. *Anesthesia and Analgesia* 1974; **53**: 828–31.

45. Marx GF, Mateo CV. Effects of different oxygen concentrations during general anaesthesia for elective caesarean section. *Canadian Anaesthetists' Society Journal* 1971; **18**: 587–93.

46. Bogod DG, Rosen M, Rees GAD. Maximum F_{IO_2} during caesarean section. *British Journal of Anaesthesia* 1988; **61**: 255–62.

47. Piggott SE, Bogod DG, Rosen M, Rees GAD, Harmer M. Isoflurane with either 100% oxygen or 50% nitrous oxide in oxygen for caesarean section. *British Journal of Anaesthesia* 1990; **65**: 325–9.

48. Crawford JS, Davies P. A return to trichloroethylene for obstetric anaesthesia. *British Journal of Anaesthesia* 1975; **47**: 482–9.

49. Levinson G, Shnider SM, de Lorimier AA, Steffenson JL. Effects of maternal hyperventilation on uterine blood flow and fetal oxygenation and acid–base status. *Anesthesiology* 1974; **40**: 340–7.

50. Samsoon GLT, Young JRB. Difficult tracheal intubation: a retrospective study. *Anaesthesia* 1987; **42**: 487–90.

51. Tunstall ME, Sheikh A. Failed intubation protocol: oxygenation without aspiration. In: Ostheimer GW, ed. *Clinics in anesthesiology, obstetric analgesia and anaesthesia, 4.* Philadelphia: WB Saunders, 1986: 171–87.

52. McClune S, Regan M, Moore J. Laryngeal mask airway for caesarean section. *Anaesthesia* 1990; **45**: 227–8.

53. Carrie LES. Epidural and spinal (subarachnoid) anaesthesia for caesarean section. *Current Anaesthesia and Critical Care* 1991; **2**: 78–84.

54. Russell IF. Subarachnoid (spinal) anaesthesia. *Current Opinion in Anaesthesiology* 1991; **4**: 340–4.

55. Morgan BM, Magni V, Goroszenuik T. Anaesthesia for emergency caesarean section. *British Journal of Obstetrics and Gynaecology* 1990; **97**: 420–4.

56. Preston PG, Rosen MA, Hughes SC, *et al.* Epidural anesthesia with fentanyl and lidocaine for caesarean section: maternal effects and neonatal outcome. *Anesthesiology* 1988; **68**: 938–43.

57. King MJ, Bowden MI, Cooper GM. Epidural fentanyl and 0.5% bupivacaine for caesarean section. *Anaesthesia* 1990; **45**: 285–8.

58. Ramanathan S, Masih A, Rock I, Chalon J, Turndof H. Maternal and fetal effects of prophylactic hydration with crystalloids or colloids before epidural anesthesia. *Anesthesia and Analgesia* 1983; **62**: 673–8.

59. Ramanathan S, Grant GJ. Vasopressor therapy for hypotension due to epidural anesthesia for caesarean section. *Acta Anaesthesiologica Scandinavica* 1988; **32**: 559–65.

60. Moir DD. Anaesthesia for caesarean section: an evaluation of a method using low concentrations of halothane and 50 per cent oxygen. *British Journal of Anaesthesia* 1970; **42**: 136–42.

61. Carrie LES. Extradural, spinal or combined block for obstetric surgical anaesthesia. *British Journal of Anaesthesia* 1990; **65**: 225–33.

62. Morgan M. Amniotic fluid embolus. *Anaesthesia* 1979; **34**: 20–32.

63. Mudie LL, Lewis M. Pre-eclampsia: its anaesthetic implications. *British Journal of Hospital Medicine* 1990; **43**: 297–300.

64. Weinstein L. Syndrome of hemolysis, elevated liver enzymes and low platelet count. A severe consequence of hypertension in pregnancy. *American Journal of Obstetrics and Gynecology* 1982; **142**: 159–67.

65. Redman CWG, Beilin LJ, Bonnar J, Wilkinson RH. Plasma urate measurements in predicting fetal death in hypertensive pregnancy. *Lancet* 1976; **1**: 1370–3.

66. Wasserstrum N, Kirshon B, Willis RS, Morse KJ, Cotton DB. Quantitative hemodynamic effects of acute volume expansion in severe pre-eclampsia. *Obstetrics and Gynecology* 1989; **73**: 545–50.

67. Connell H, Dalgleish JG, Downing JG. General anaesthesia in mothers with severe pre-eclampsia/eclampsia. *British Journal of Anaesthesia* 1987; **59**: 1375–80.

68. Keeri-Szanto M. Laryngeal oedema complicating obstetric anaesthesia. *Anaesthesia* 1978; **33**: 272.

69. Heller PJ, Scheider EP, Marx GF. Pharyngolaryngeal edema as a presenting symptom in pre-eclampsia. *Obstetrics and Gynecology* 1983; **62**: 523–4.

70. Fulton B, Park GR. Intravenous chlormethiazole. *British Journal of Hospital Medicine* 1992; **48**: 742–6.

71. Goldman RS, Finkbeiner SM. Therapeutic use of magnesium sulphate in selected cases of cerebral ischemia and seizure. *New England Journal of Medicine* 1988; **319**: 1224–5.

72. Sibai BM. Magnesium sulphate is the ideal anticonvulsant in pre-eclampsia–eclampsia. *American Journal of Obstetrics and Gynecology* 1990; **162**: 1141–5.

73. Appleton MP, Kuehl JJ, Raebel MA, Adams MR, Knight AB, Gold WR. Magnesium sulfate versus phenytoin for seizure prophylaxis in pregnancy-induced hypertension. *American Journal of Obstetrics and Gynecology* 1991; **165**: 907–13.

74. Crowther C. Magnesium sulphate versus diazepam in the management of eclampsia: a randomised controlled trial. *British Journal of Obstetrics and Gynaecology* 1990; **97**: 110–17.

75. Makeshwari JH, Desai SV, Hansotia MD, Walvekar VR.

Anticonvulsant therapy in eclampsia. *Journal of Postgraduate Medicine* 1989; **35**: 66–9.

76. Dommisse J. Phenytoin sodium and magnesium sulphate in the management of eclampsia. *British Journal of Obstetrics and Gynaecology* 1990; **97**: 104–9.

77. Tuffnell D, O'Donovan P, Lilford RJ, Prys-Davies A, Thornton JG. Phenytoin in pre-eclampsia. *Lancet* 1989; **2**: 273–4.

78. Department of Health. *Report on confidential enquiries into maternal deaths in the United Kingdom 1988–90*. London: HM Stationery Office, 1994.

79. Mauer DK, Gervais HW, Dick WF, Rees GAD and a Working Group on CPR of the European Academy of Anaesthesiology. Cardiopulmonary resuscitation (CPR) during pregnancy. *European Journal of Anaesthesiology* 1993; **10**: 437–40.

80. American Heart Association. Standards and guidelines for cardiopulmonary resuscitation and emergency cardiac care. *Journal of the American Medical Association* 1986; **255**: 2905–32.

81. American Heart Association: Emergency Cardiac Care Committee & Subcommittees. Guidelines for cardiopulmonary resuscitation and emergency cardiac care. *Journal of the American Medical Association* 1992; **268**: 2171–95.

82. Lee RV, Rodgers BD, White LM, Harvey RC. Cardiopulmonary resuscitation of pregnant women. *American Journal of Medicine* 1986; **81**: 311–18.

83. Rotmensch HH, Elkayam U, Frishmann W. Antiarrhythmic drug therapy during pregnancy. *Annals of Internal Medicine* 1983; **98**: 487–97.

84. Chadwick HS, Posner K, Caplan RA, Ward RJ, Cheney FW. A comparison of obstetric and non-obstetric anesthesia malpractice claims. *Anesthesiology* 1991; **74**: 242–9.

Chapter 65

Anaesthesia for Gastrointestinal Surgery

L. Strunin

Endoscopy	Laparoscopy
Preoperative fasting	Postoperative nausea and vomiting (PONV)
Anaesthetic considerations specific to the GIT	

The gastrointestinal tract is essentially a long tube that begins at the mouth and ends at the anus. The gastrointestinal tract traverses the thorax as the oesophagus; the rest of its tortuous path is within the abdomen. Here it receives 25 per cent of the cardiac output as the splanchnic circulation; thus any diminution in output as a result of anaesthesia or surgical manipulation[1] may have a dramatic effect on the gastrointestinal tract. Although the proper functioning of the gastrointestinal tract is essential for life, until relatively recently these functions could be assessed only in superficial and often inadequate ways. As a result, patients were subjected to unnecessary and occasionally fatal laparotomies carried out primarily for diagnostic purposes. The advent of flexible endoscopes, laparoscopy and the developments in imaging have made all parts of the gastrointestinal tract accessible. The day of the 'diagnostic laparotomy – just to have a look' are now gone and anaesthetists should resist any attempts to be inveigled into providing anaesthesia for such a venture.

Twenty years ago the commonest operations on the gastrointestinal tract were gastric and related to management of peptic ulceration. (For the purposes of this discussion, cholecystectomy – also a very common operation – has been excluded, although most consider the gall bladder as part of the gastrointestinal tract.) Emergency surgery for a perforated or bleeding peptic ulcer challenged the anaesthetists of the day with dehydrated, shocked and anaemic patients. Various forms of gastrectomy and later disruption of the vagus nerves in an effort to reduce gastric acid secretion were performed electively to try to prevent ulcer formation. The advent of H_2-receptor blocking drugs (e.g. cimetidine and ranitidine), and possibly changes in the prevalence of peptic ulcer disease, have dramatically reduced surgical management. The recent evidence that infection with *Helicobacter pylori* may be a major contributing factor, and its treatment with relatively simple antibiotic regimens, may further reduce the incidence of peptic ulcer problems. Today's surgery on the gastrointestinal tract is focused on endoscopy, laparo-scopy, colorectal surgery and oncology, the last including the oesophagus as well as the gastrointestinal tract within the abdomen.

Endoscopy

The gastrointestinal tract (GIT) is accessible by flexible endoscopy via the mouth – upper GIT – and the anus – sigmoidoscopy and colonoscopy. Endoscopy suites have been developed and are run by gastroenterologists interested in the upper GIT. Colonoscopy is within the remit of the colorectal surgeon.

Upper GIT

Most upper GIT endoscopy is carried out with sedation rather than a formal general anaesthetic. Benzodiazepines, most commonly midazolam, are used but some endosco-pists add 'short'-acting opioids such as alfentanil or fentanyl. These combinations can lead to cardiorespiratory depression in compromised patients, and most anaesthetic departments have had the experience of emergency calls to the endoscopy suite to resuscitate such patients. Endo-scopy suites, therefore, should have full resuscitation equipment, including facilities for artificial ventilation and defibrillation. Patients should be monitored by ECG, pulse oximetry and non-invasive blood pressure, and a qualified person, other than the endoscopist, should be responsible for observing the patient and the monitors. In addition, supplementary oxygen should be given as appropriate. Naloxone and flumazenil (specific reversal agents for opioids and benzodiazepines, respectively) must be available.

General anaesthesia may be required occasionally, most

commonly for children. From the anaesthetist's point of view, upper GIT endoscopy competes for the airway and operative procedures via the endoscope may lead to uncontrollable haemorrhage (e.g. injection of oesophageal varices in patients with portal hypertension[2]). Tracheal intubation with a non-kinking armoured tube is desirable, and there is some virtue in the anaesthetist also guiding the endoscope into the oesophagus to prevent dislodgement of the tracheal tube. Damage to the teeth and upper airway structures is more likely to occur in the unconscious patient. The general anaesthetic should therefore be sufficient to prevent straining and coughing while the endoscope is in the patient.

Colonoscopy

This may be performed for diagnostic purposes or for removal of polyps or biopsy in an endoscopy suite or as part of a laparoscopic or laparotomy operation. Fibreoptic colonoscopy has to a great extent replaced rigid sigmoidoscopy because the colonoscope may be passed more easily and further into the bowel. However, rigid instruments are still used for operative procedures on the rectum and anus. Many patients have malignant tumours and may be dehydrated and undernourished. Correction of these defects may require intravenous therapy and parenteral feeding prior to surgery. Diagnostic colonoscopy requires the bowel to be empty. Various purgative regimens are used which may compound pre-existing dehydration. Finally, colonoscopy can be a prolonged and painful procedure, and considerable quantities of sedative and analgesic drugs may be required. If a potentially difficult case is anticipated, general anaesthesia should be considered to allow better control of drug administration and maintenance of a clear airway. In either event, it should be noted that patients may be in the Trendelenburg or lateral positions for long periods, and care should be taken to prevent injury to peripheral nerves and other structures.

Preoperative fasting

Elective surgical patients have for many years been fasted from midnight on the night prior to surgery.[3] This has been required in the belief that the stomach would be empty when the patient arrived in the operating theatre some 8 hours later, and thus the patient would be protected against the risk of aspirating stomach contents during induction of anaesthesia. That such aspiration may be harmful was highlighted in the seminal paper by Mendelson in 1946.[4]

Mendelson reviewed retrospectively the records of 44 016 pregnant patients from his hospital between 1932 and 1945. He noted 66 cases of aspiration of stomach contents into the lungs. In 45 cases he was able to determine the aspirated material; in 40 it was liquid and the patients developed the characteristic chest radiographic changes of Mendelson's syndrome (acid aspiration syndrome). However, it is important to note that none of these patients died. Five patients inhaled solid material and died, in Mendelson's opinion from suffocation. In the same paper, Mendelson published his results of some rabbit studies designed to elucidate the cause of the radiographic changes. He found that injection of 0.9 per cent saline or distilled water at volumes of 5 ml/kg into the rabbit's trachea was harmless. However, when he took gastric aspirate from his patients, acidified it with hydrochloric acid and injected this into his rabbits he observed the same chest radiographic changes he had seen in his patients. Neutralized vomitus was not harmful unless it contained solid material which led to either partial obstruction with ensuing massive atelectasis or complete obstruction and suffocation. As a result of his observations, Mendelson offered the following advice for the management of obstetric patients:

1. No oral feeding during labour – intravenous fluids should be given
2. Wider use of local anaesthesia for operative obstetrics
3. Alkalinization and emptying of the stomach before general anaesthesia
4. Competent administration of general anaesthesia, with full appreciation of the dangers of aspiration during induction and recovery from anaesthesia
5. The use of transparent masks to see vomitus early, the facility to tip the operating table and consideration of tracheal intubation for all patients

These tenets form the basis of good practice today for both obstetric and emergency anaesthesia.

Pulmonary aspiration of 5 ml/kg (i.e. some 350 ml in a 70 kg human) is a large volume of fluid and substantially more than is likely to occur in clinical practice. Roberts and Shirley,[5] on the basis of a single experiment in an anaesthetized monkey, proposed that a volume greater than 25 ml (i.e. 0.4 ml/kg in a human) with a pH of less than 2.5 aspirated into the lungs was potentially harmful. More recent work in monkeys[6,7] suggests that 0.8 ml/kg and a pH approaching 3.5 is required before harm occurs. Clearly such studies cannot ethically be conducted in humans. Nevertheless, it is interesting to speculate why none of Mendelson's patients died, despite probably having aspirated considerable amounts, and yet subsequent studies[8] such as the triennial *Confidential Enquiries into Maternal Deaths in the United Kingdom* details such deaths over the past 20 years. Other factors such as repeated attempts at intubation despite obvious hypoxia and the use of positive pressure ventilation after inhalation of gastric contents may be responsible. The concept of reducing the residual gastric volume (RGV) to 25 ml or less and raising the pH above 2.5 has passed into the anaesthetic literature as a desirable feature of preoperative preparation to prevent acid aspiration. Is this achievable in elective patients by starvation? The simple answer is 'no'. The longer fluid is withheld, in general the larger the RGV and the lower the pH.

H_2-receptor antagonist drugs (such as cimetidine, ranitidine[9] and famotidine[10]) prevent the secretion of hydrogen ions into the stomach and also reduce gastric secretion. Although initially developed for the treatment of peptic ulceration, many studies have shown that H_2 antagonists are effective, if given as premedication, in reducing RGV and raising pH to ostensibly safe levels. However, the drugs require time to work (regardless of the route of administration; i.e. oral, i.m. or i.v.) – about 45 minutes – and are not effective in every patient. This probably reflects the inability of the drugs to affect acid already excreted into the stomach. Oral antacids are effective immediately, but the particulate varieties may be hazardous if inhaled and the non-particulate (e.g. sodium citrate) increase gastric secretion and are short acting. In summary, although drugs affecting gastric secretion and acidity are effective, they cannot be relied on in every patient.

An interesting observation in the studies of antacids was that administration of fluid (e.g. water, orange juice, tea or coffee with modest amounts of milk and sugar), far from increasing RGV, either had no effect or decreased the volume and, in addition, in some patients the gastric pH increased. The volumes usually ingested were in the range of 150–300 ml. Subsequently it has been shown that free access to fluid up to 2 hours before anaesthesia and surgery does not affect the RGV aspirated after induction, as compared with controls who fasted in the traditional manner. That these amounts were easily dealt with by the stomach is hardly surprising in the light of observations made in volunteers[11] who drank 500 ml of 0.9 per cent saline; the half-life in the stomach was only some 10–20 minutes. It should be noted that these observations have been made in elective surgical ASA 1 and 2 patients, both adults and children, unpremedicated (except for benzodiazepines) and with exclusions for known risk factors that affect the stomach. In the light of all these studies,[12–22] many hospitals have adopted new policies with regard to preoperative starvation of elective patients, allowing free access to fluid usually up to 3 hours before the scheduled time of operation.[17,18,23–29] This allows for the vagaries of operating lists and avoids unnecessary cancellations. Food should still be withheld prior to surgery, because the rate at which it leaves the stomach is too variable to predict; in this respect midnight seems an apposite time as it is rare in most hospitals to be able to acquire food after this time.

The patient with a full stomach

Patients with a full stomach may present in two circumstances. In one, the elective patient, sometimes a child, has misunderstood the instructions and has eaten prior to surgery. Anaesthesia should be postponed until the stomach has emptied. How long should this be? Normal physiology suggests that, in the absence of disease, some 6 hours will be adequate. It should be noted at this point that any opioid (e.g. as premedication or for pain relief) administered after food will add to the delay in emptying.

More commonly, the anaesthetist is faced with an emergency trauma case in which, from the history, it is clear that food has been ingested prior to injury and opioids may have been given for pain relief. In these cases it is impossible to predict how long it will be before the stomach is empty. (Anecdotally, patients have vomited solid material up to 24 hours after injury.) The decision when to operate lies with the surgeon. If it is felt that an operation is imperative, there is little point in delaying. Provided that the patient is adequately resuscitated with regard to fluid and/or blood products as required, the full stomach should be addressed as follows:

1. Consideration should be given to an awake tracheal intubation. The disposition of the patient, skill of the anaesthetist and availability of suitable equipment will dictate whether this is the best course.
2. H_2-receptor antagonists should be given intravenously if time permits, remembering that they take about 45 minutes to be effective. Oral non-particulate antacids are effective immediately but increase the gastric volume and may not mix well in the full stomach.
3. The airway should be assessed: if there is an anticipated major difficulty, consideration should be given to a local or regional technique if appropriate. However, if these fail, general anaesthesia may still be required.
4. If there are no airway problems, anaesthesia should be induced as follows:
   (a) suitable large-bore intravenous lines should be established under local anaesthesia;
   (b) monitoring consistent with the degree of patient injury should established, but must include pulse oximetry and the availability of immediate end-tidal carbon dioxide measurement;
   (c) a range of cuffed tracheal tubes, laryngoscopes, introducers and a powerful suction should be available;

(d) the patient should be preoxygenated;

(e) intravenous induction with a short-acting barbiturate followed by suxamethonium (succinylcholine) (rapid sequence induction[30]) will provide the best conditions for tracheal intubation; if there are contraindications to suxamethonium, a large dose of a short-acting non-depolarizing agent should be given;

(f) during induction the airway should be protected by cricoid pressure. This requires a dedicated assistant who understands the relevant anatomy. Compressing the oesophagus between the cricoid (from the Greek meaning a ring – the cricoid is the only complete tracheal cartilaginous ring and lies immediately below the thyroid cartilage) cartilage and the vertebral body behind will prevent regurgitation, but there is some concern that in the face of active vomiting continued pressure could result in oesophageal rupture. A further problem is that enthusiastically applied cricoid pressure may distort the larynx and make tracheal intubation impossible. The cricoid pressure should be maintained after tracheal intubation until end-tidal carbon dioxide monitoring demonstrates unequivocally that the tube is in the trachea and the tracheal cuff has been inflated. If the trachea cannot be intubated, cricoid pressure should be maintained. Consideration should be given to allowing the patient to awaken, or if surgery is truly imperative, the airway must be maintained with a simple oral device and the patient turned into the lateral position if appropriate.

Review of cases in which a full stomach has led to difficulties often reveals a lack of foresight and attempts at induction of anaesthesia and tracheal intubation by inexperienced staff. The patient with a full stomach is a high risk case and should be dealt with by senior staff. Trainees should not be inhibited from calling for assistance before commencing such a case, rather than waiting until matters are obviously out of control.

Anaesthetic considerations specific to the GIT

It is assumed for the purposes of this section that the normal considerations of patient assessment and preparation, and premedication if necessary, will have been carried out as appropriate for a patient scheduled for major surgery. In particular, the cardiovascular and respiratory systems should be carefully assessed along with electrolyte balance and nutritional status. Operations on the gastro-

intestinal tract may be lengthy and involve considerable fluid and blood loss. In addition to conventional monitoring, central venous pressure and temperature are important. Many patients may benefit from a period of intensive care postoperatively and there are advantages in arranging this electively.

Patient position and access

Laparoscopy (see below) both in the chest and in the abdomen is now an integral part of surgery of the gastrointestinal tract. Both closed and open operations may be performed simultaneously and the patient's position may alter radically many times during an operation: head-up perhaps for a laparoscopic approach, to head-down for an open assessment in the pelvis. In addition, the legs may be placed in stirrups (the Lloyd-Davis position) and care should be taken to protect the patient's peripheral nerves and hip joints. A Mayo table is commonly placed over the patient's head, and may restrict observation of the airway. Laparoscopic monitors and reduced lighting may all contribute to reduced patient access.

Oesophagectomy

Oesophagectomy is a major surgical procedure involving access to both the chest and the abdomen. A double-lumen tracheal tube (preferably, a left-sided endobronchial tube should be used) or a single-lumen tracheal tube with movable blocker (which can be left in place postoperatively after withdrawal of the blocker) will allow each lung to be controlled independently. The left lung will require deflation to allow access to the oesophagus. Once the oesophagus is removed, the stomach is pulled up into the chest to restore continuity of the gastrointestinal tract. Laparoscopic dissection of the oesophagus adds the dimension of a potential tension pneumothorax as well as prolonged lung deflation. In the author's experience these patients require a considerable period of artificial ventilation postoperatively to allow the lung to recover.

Neostigmine and intestinal anastomoses

In the 1970s there was great concern that neostigmine given to reverse the effects of non-depolarizing muscle relaxants caused anastomotic breakdown as a result of its effects on gut contractility.[31] There was considerable debate as to whether atropine was a contributory factor and about the modifying effects of other drugs such as pethidine (meperidine, Demerol) and halothane.[32] A study[33] of 68 cases of large bowel anastomosis compared the anastomotic breakdown rate in patients anaesthetized by a spinal or epidural block as well as a general anaesthetic and those given a general anaesthetic only. Most patients received halothane as part of their general anaesthetic. There was no significant difference between the two groups. In the intervening years, suture material has improved and the use of staples is now widespread. In addition, the relevance of infection and blood supply in the aetiology of anastomotic breakdown is acknowledged.[34] All patients having bowel surgery should receive prophylactic antibiotics (a combination of metronidazole and cefuroxime) intravenously at induction of anaesthesia and for 24–48 hours postoperatively.

The use of regional techniques is still debated.[33,35] Some surgeons do not like the bowel contraction that accompanies spinal or extradural block. However, epidural block with the passage of a catheter and infusion of a low concentration of bupivacaine with fentanyl allows general anaesthesia to be maintained at a light level. In addition, the catheter can remain in place for postoperative pain relief.

Morbid obesity

Patients who are grossly overweight (e.g. adults in excess of 136 kg or 300 lb) have been subjected to various operations on the gastrointestinal tract to try to reduce the gut's ability to absorb ingested food.[36,37] Gastroileal bypass is effective in producing spectacular weight loss. However, the procedure had a high complication rate and required further surgery to reverse the bypass after the patient's weight had stabilized. Many patients then put back all the weight they had lost! Partial gastrectomy or gastric stapling operations also lead to weight loss but are not successful in many patients who merely increase their intake.

Morbidly obese patients present an anaesthetic challenge. Intravenous access may be difficult, two operating tables may be required, large size blood pressure cuffs are needed and there may be airway difficulties. There may be cardiorespiratory problems and the patients are prone to venous thrombosis. Care should be taken when moving unconscious morbidly obese patients, to avoid injuring either the patients or those who may attempt to lift them without proper equipment.

Carcinoid tumour

Carcinoid tumour was first described as 'resembling carcinoma' and was thought to be benign. In 75 per cent of patients the carcinoid tumours are in the GIT (small bowel 36 per cent, appendix 13 per cent, and the rest in the lung). Tumours of the small bowel are more likely to have hepatic involvement. The tumour is not fast growing, but is capable of metastasizing. Histologically, carcinoid is a neoplasm of peptide- and amine-producing cells derived from the gut ectoderm. If the tumour has no hepatic metastases, there are often no symptoms and the tumour may be discovered accidentally at laparotomy. Carcinoid syndrome results when there is release of vasoactive peptide from hepatic metastases or a primary tumour which does not drain directly into the liver. The anaesthetic management[38] of this rare syndrome has been reviewed recently. Attention is drawn particularly to the necessity for expert handling of patients with marked carcinoid syndrome. Octreotide for controlling hypotension and labetalol or ketanserin for hypertension are recommended.

Closure of colostomy/ileostomy

Many patients undergoing major gastrointestinal surgery have a temporary colostomy or ileostomy performed to relieve pressure on their anastomosis. Many consider these to be minor operations and they are often left either to the most junior surgeon to finish off or, in the case of closure, are placed at the end of a busy surgical schedule. From the patient's perspective, the ostomy may be the most important and is clearly the most visible part of the operation. The patient should be carefully counselled in its care preoperatively by the ostomy nurse. This will prevent complications and make the closure more straightforward. The anaesthetist's contribution is to recognize that closure of ostomies is as important a part of the procedure as any other. Full muscle relaxation is required so that the gut is not damaged during closure. This will require tracheal intubation and artificial ventilation. Attempts to provide anaesthesia with spontaneous breathing and no muscle relaxation may make the closure impossible or lead to unnecessary damage to the gut.

Haemorrhoidectomy

Anal dilatation is extremely painful. Lightly anaesthetized patients may suffer bradycardia, and those who are unintubated, laryngospasm. Caudal block will prevent this, but produces anal sphincter relaxation which some surgeons dislike. However, a caudal block may be placed after surgery to provide postoperative pain relief.

Laparoscopy

Laparoscopy[39] is the endoscopic visualization of the peritoneal or thoracic cavity, achieved through either the chest or the abdominal wall after creation of a pneumothorax or pneumoperitoneum. Laparoscopic cholecystectomy is now one of the most commonly performed operations world wide. Sometimes it is called 'minimally invasive' or 'access' surgery on account of the apparent little disturbance as compared with an open operation. This may be misleading because, if things do not go well, the operation may be prolonged and extremely invasive in respect of complications.

History

Endoscopy for rectal examination was first described by the Kos School led by Hippocrates in 460–375 BC. Lens systems were developed towards the end of the nineteenth century. In 1902, Kelling performed abdominal endoscopy on a dog and suggested filling the abdominal cavity with air for a better view. In 1925 carbon dioxide was described to insufflate the peritoneum and Veress described his insufflating needle (still used today) in 1938. Transvaginal culdoscopy was introduced in the 1940s. In the 1970s Semm published an atlas of instruments and techniques developed for laparoscopic oöphorectomy, ovarian cystectomy and other gynaecological operations as well as appendicectomy and omental and bowel adhesiolysis. Despite this, the use of the laparoscope remained almost entirely in the hands of the gynaecologists until the late 1980s.

The change came with the development of improved optical systems, electronic insufflation systems and the microchip endocamera allowing video monitoring through the lens. The first laparoscopic cholecystectomy was performed by Mouret in Lyon in 1987. Since then the technique has spread around the world. Many other operations on the gastrointestinal tract are now routinely performed laparoscopically; for example, appendicectomy, hiatus hernia repair, vagotomy, colectomy, hernia repair and many others. The technique has been used in both adults and children.

Open cholecystectomy was first performed in Berlin in 1882. Since then the operation has become the 'gold standard' against which all other forms of therapy for symptomatic cholelithiasis should be compared. Removal of the gall bladder prevents disease recurrence and although open cholecystectomy is considered a safe operation the mortality is still of the order of 1 per cent. In addition, there is significant morbidity from the large incision, leading to a 4–6 week recuperation period. Laparoscopic removal of the gall bladder is carried out through small incisions rarely greater in length than 12 mm, thus preventing much of the tissue trauma associated with the open operation.

The relative safety and efficacy of laparoscopic surgery are still unknown. Although preliminary results suggest major benefits for the patient, valid comparisons with the open operation in similar patient populations have not been made and are probably now impractical. Minimal access surgery does not imply diminished risk. It is becoming clear that the complication rate is inversely related to the experience of the surgeon and that the learning curve is relatively steep. Some surgeons will either never develop or do not have the hand–eye coordination required for successful laparoscopic surgery. There is about a 5 per cent conversion rate to open operation for cholecystectomy. This does not imply failure, but means that laparoscopy should be undertaken only by surgeons capable of opening the abdomen if necessary.

The pneumoperitoneum

The first requirement for laparoscopy is the establishment of a pneumoperitoneum. This provides a clear view of the intraperitoneal structures by separating the anterior abdominal wall from the intra-abdominal contents, and allows the safe insertion of instruments for surgery. An ideal gas for insufflation should be colourless and physiologically inert, should not support combustion and have no explosive potential when used with electrocautery or a laser; the gas should be cheap, non-pollutant and non-irritant to the peritoneum. Air, oxygen, carbon dioxide, nitrous oxide and helium have been used. None is ideal. Although it is not physiologically inert, carbon dioxide is most commonly used. Nitrous oxide, air and oxygen all support combustion and are not suitable for electrocautery. Nitrous oxide is less irritant to the peritoneum than carbon dioxide but absorbs more slowly. Residual gas may give rise to shoulder pain postoperatively. In the event of gas embolism, carbon dioxide is rapidly dissolved and is safer than either nitrous oxide or air.

Physiological considerations

These have been studied extensively over the last 30 years in many clinical and laboratory models. The main effects are cardiorespiratory and depend on a number of factors, including the insufflating gas used, the increase of intra-abdominal pressure and the patient's position (degree of Trendelenburg or reverse Trendelenburg used). In addition, the anaesthetic technique (local, regional or general), whether the patient breathes spontaneously or is ventilated artificially, the depth of anaesthesia, the anaesthetic agents and the preoperative state of the patient all contribute to the effect of the pneumoperitoneum in a particular patient. Much of the data comes from studies during laparoscopy for gynaecological procedures. These patients tend to be younger and fitter than those presenting for cholecystectomy and other procedures on the gastroinstestinal tract. The latter may already have significant cardiorespiratory disease, which adds another variable to the pathophysiological response to the pneumoperitoneum.

Respiratory system changes

Hypercapnia occurs with carbon dioxide insufflation unless minute ventilation can be increased to maintain normocapnia. The hypercapnia is due to absorption from the peritoneal cavity as well as ventilation/perfusion mismatch due to reduction in functional residual capacity (FRC). The only practical method of increasing ventilation in anaesthetized patients is by intermittent positive pressure ventilation (IPPV). If nitrous oxide is used as the insufflating gas, hypercapnia does not occur.

If the patient is awake during laparoscopy under a regional or local block, carbon dioxide insufflation results in an increase in both respiratory rate and minute ventilation. In normal, fit patients this should be sufficient to prevent hypercapnia. However, the presence of cardiac or respiratory disease may be a contraindication to local or regional techniques, as they will prevent the patient from maintaining normocapnia. Therefore, in most circumstances, if carbon dioxide is to be the insufflating gas, general anaesthesia, tracheal intubation, muscle relaxation and IPPV comprise the safest combination. Unrestrained hypercapnia may result in cardiac arrhythmias, which in susceptible patients may prove fatal. An additional factor in these fatalities may have been concomitant hypoxia. Patient monitoring by pulse oximetry and end-tidal carbon dioxide analysis should alert the anaesthetist before potentially harmful hypercapnia and or hypoxia occurs.

Cardiovascular system changes

When carbon dioxide is the insufflating gas, the usual cardiovascular changes are tachycardia and hypertension. However, these are in direct relation to the degree of hypercapnia that occurs. An additional factor is the rate of gas insufflation and the intra-abdominal pressure. The modern insufflators control gas flow very carefully; most of the reported adverse cardiovascular changes associated with laparoscopy were associated with equipment that often had no pressure or flow controls. A high intra-abdominal pressure will reduce cardiac output by interfering with venous return.

Various cardiac arrhythmias have been reported. Most commonly, halothane was part of the anaesthetic and end-tidal carbon dioxide was not monitored. Adequate ventilation and avoidance of halothane should prevent any serious arrhythmias. Bradycardia may occur in patients with high vagal tone if a combination of fentanyl with atracurium or vecuronium coincides with peritoneal traction. Atropine or glycopyrrolate may be required in these patients.

Complications

Gastro-oesophageal reflux may occur. This may be due to the combination of atropine administration combined with Trendelenburg positioning. Tracheal intubation should be used to protect the lungs. The major complications include gas embolism, haemorrhage, cardiac arrhythmias and trauma to abdominal viscera. The last includes unobserved bowel perforation, which may occur as a result of direct trauma by the surgical instruments or from an electrocautery burn.

Haemorrhage from major vessels may cause morbidity and mortality. Vessels that have been traumatized include the aorta, inferior vena cava and the mesenteric vessels. Vessel trauma usually occurs during insertion of the Veress needle or other surgical instruments. Haemorrhage from the cystic duct is a common cause of conversion to open operation. Gastric perforation may be prevented by passing a nasogastric tube after induction of anaesthesia.

Gas embolus

Life-threatening gas embolus using modern laparoscopy equipment is rare, although acute cardiovascular collapse, thought to be due to gas embolus, is more frequent. The differential diagnosis includes ventricular arrhythmias and bradyarrhythmias, haemorrhage, compression of the inferior vena cava by raised intra-abdominal pressure and hypoventilation. Constant monitoring of end-tidal carbon dioxide concentration is the simplest way of detecting gas embolus early. An acute decrease may herald an embolus. Treatment includes termination of nitrous oxide and administration of 100 per cent oxygen, deflation of the pneumoperitoneum and release of any head-up tilt. If these measures are not successful, the right atrium should be aspirated via a cannula inserted into the right internal jugular vein; inotropic cardiac support should be initiated along with cardiac massage to push the gas embolus out of the right atrium. The introduction of lasers into the abdominal cavity as part of the surgical instrumentation should be viewed with some caution. Most lasers require cooling gas, and on occasion this is air. Enquiry should be made about the flow and pressure controls on such devices to prevent mishaps.

Postoperative pain

Pain is the commonest cause of postoperative morbidity after laparoscopic surgery. This may seem rather surprising as the surgical intervention is much less than in an open operation. An additional factor is that most surgeons believe that patients will have no pain after laparoscopic surgery and so conventional pain relief may be withheld. Shoulder pain, probably due to residual gas, is the commonest complaint and may persist for several days after surgery. Postoperative nausea and vomiting (PONV; see below) is also common after laparoscopy and reflects the higher incidence of women in the patient population as well as the direct effects of the pneumoperitoneum.

Pain relief should include attempting to release as much gas as possible from the pneumoperitoneum at the end of surgery, infiltrating all the surgical incisions with 0.5 per cent bupivacaine and prescribing adequate analgesics postoperatively.

Summary

Laparoscopic surgery with carbon dioxide insufflation is associated with a reduction in total respiratory compliance, a reduction in FRC, elevation of the diaphragm, increased peak inspiratory airway pressures, ventilation/perfusion inequality and hypercapnia. These changes may be more pronounced in patients with pre-existing cardiopulmonary disease and can lead postoperatively to impaired pulmonary function, especially in the elderly. In addition, the pneumoperitoneum is associated with tachycardia and hypertension which may adversely affect myocardial oxygen supply/demand, especially in those with ischaemic heart disease; a fall in cardiac output with a reduced preload and, in particular, an increase in afterload will be even less well tolerated in those with poor left ventricular function. Therefore, laparoscopic surgery should on no account be considered a suitable alternative treatment in a patient who is deemed unfit for an open operation. Indeed, open operation may be a more appropriate choice in patients with severe ischaemic heart disease or cardiac or respiratory failure.

Postoperative nausea and vomiting

The commonest causes of morbidity after anaesthesia and surgery are pain and postoperative nausea and vomiting (PONV).[40–43] Indeed the two are interrelated. Unrelieved pain is a common cause of PONV and opioids, widely used in pain relief, are also potent causes of PONV. As with pain, PONV is difficult to measure, is generally self-limiting, and until recently was accepted as 'due to the anaesthetic' and was something patients were expected to put up with

without complaining. However, both acute postoperative pain relief and prevention and treatment of PONV have recently assumed great importance in anaesthesia as the focus has shifted from mortality to morbidity. With regard to PONV, further driving influences have been (1) the shift to day surgery, in which the inability of patients to go home immediately because of nausea and/or vomiting may have major financial consequences, and (2) the recent development of highly specific 5-hydroxytryptamine-3 (5-HT$_3$) receptor antagonists (e.g. ondansetron and granisetron).

Historically, four phases may be identified with regard to the incidence of PONV. Prior to 1956, anaesthesia using agents such as diethyl ether combined with large quantities of opioid premedication (e.g. morphine) was associated with incidences of PONV of up to 75 per cent. When halothane was introduced in 1956, it seemed to be a universal anaesthetic, the use of opioids declined and the incidence of PONV was some 20 per cent. More recently, the addition of the potent opioids fentanyl and alfentanil to smooth out some of the vagaries of the newer volatile agents (e.g. isoflurane and enflurane) has increased PONV to an average of 30 per cent. Finally, the introduction of propofol and non-steroidal anti-inflammatory drugs (NSAIDs) for pain relief has reduced PONV in some day care settings to less than 10 per cent. The morphine-sparing effect of NSAIDs is well recognized, but there is not a concomitant reduction in PONV. This probably reflects the fact that the dose of morphine required to initiate PONV is considerably less than that required for analgesia. NSAIDs alone do not precipitate PONV.

PONV is multifactorial. A recent review of the physiology indicates that opioids and anaesthesia are only some of the factors.[44] The surgical procedure is also relevant, as can be demonstrated in procedures carried out under local or regional anaesthesia. In the latter, hypotension associated with sympathetic block is a potent cause of nausea and occasionally vomiting. Interestingly, ephedrine used to raise the blood pressure has also been shown to be effective in treating PONV. Table 65.1 shows patient and operative factors possibly associated with PONV.

Palazzo and Evans[45] used logistic regression analysis to determine the association of fixed patient factors with postoperative emesis. In a carefully controlled study they determined that the relation of logit postoperative sickness equalled 5.03 + 2.24 (use of postoperative opioids) + 3.97 (history of previous PONV) + 2.4 (gender) + 0.78 (history of motion sickness) − 3.2 (gender × previous PONV). A worked example can be expressed as follows:

A male patient with a previous history of PONV and motion sickness who also receives postoperative opioids: Logit postoperative sickness
$$= 5.03 + 2.24(1) + 3.97(1) + 2.4(0) + 0.78(1) - 3.2(0 \times 1)$$
$$= 5.03 = 2.24 + 3.97 + 0.78$$
$$= 1.97/e^{1.96} = 7.09$$

Table 65.1 Postoperative nausea and vomiting (PONV) –
possible risk factors

Patient factors
Gender: more common in women (may be associated with
hormonal differences)
Obesity: larger RGV, airway problems, need more anaesthetic, may
regurgitate
Age: more common in children; peak 11–14 years
History: previous history of motion sickness, PONV
Anxiety: sympathetic activity, air swallowing

Operative factors	*Miscellaneous*
Dental extractions	Antiemetic treatment
ENT/eye operations in children	Geographical factors
Head and neck surgery	Neuromuscular blocker
Knee arthroscopy	Volatile anaesthetic
Laparoscopy	
Neurosurgery	
Uterine dilatation and curettage	

Probability of sickness in 24 hours
= 7.09/8.09 = 0.876 or 87.6 per cent

This form of risk stratification should be used in studies of
antiemetics. Clearly there are some patients whose risk
factors are such that they will almost inevitably be sick,
whatever is done. Inclusion of a large number of such
patients in a control group (even in a randomized study)
will reflect badly on the active drug, or, if the situation is
reversed, may make a drug seem more effective than it
really is.

Antiemetics

PONV can be a very distressing complication of what is
often a seemingly straightforward and simple anaesthetic.
PONV is usually self-limiting and serious complications are
rare, therefore any treatment should have a low side-effect
profile. The multifactorial nature of PONV makes it
unlikely that any one therapy will be effective in all
cases. The introduction of ondansetron has caused some
confusion in this respect. Nausea and vomiting induced by
anti-cancer chemotherapy that stimulates 5-hydroxytrypta-
mine-3 (5-HT$_3$) receptors in the mucosa of the upper gut is
well controlled by 5-HT$_3$ antagonists such as ondansetron
or granisetron, with very high success rates. However, 5-
HT$_3$ antagonists have no effect on opioid-induced emesis or
motion sickness. As these last two factors are relevant in

PONV, it is hardly surprising that ondansetron is not as
effective in this regard as compared with its role in
chemotherapy-induced emesis.

Simple measures, such as avoiding gastric distension
with nitrous oxide and limiting patient movement until the
effects of the anaesthetic have passed, can markedly reduce
PONV. Adequate pain relief is important and alternatives to
opioids (NSAIDs and local analgesic blocks) should be
considered in high-risk cases. Acupuncture has some
efficacy in motion sickness and has been tried in
PONV;[47] unfortunately, the results have been equivocal.

The concept of a 'vomiting centre' in the brain with a
chemoreceptor trigger zone (CTZ) somewhere in the
medulla was described in the 1950s[48] and seemed a logical
arrangement for all the various stimuli that lead to PONV
to impinge on. However, it was not possible to demonstrate
the centre or the CTZ anatomically. More recently, attention
has focused on the area postrema[49] of the medulla
oblongata as the anatomical site of the 'CTZ'. Dopamine
is an important stimulant in this area and therefore drugs
with anti-dopaminergic activity (e.g. *gastrointestinal proki-
netics* – metoclopramide, domperidone; *phenothiazines* –
prochloperazine, perphenazine, and the *butyrophenones* –
droperidol) will have antiemetic activity. Central anticholinergic action is also associated with antiemetic
activity. Hyoscine and atropine have some activity both in
PONV and in motion sickness. In addition, antihistamine
receptor type 1 antagonists (e.g. cyclizine) also demonstrate
anticholinergic activity, and this may be responsible for
their antiemetic effects.[50]

The plethora of available antiemetics bears witness to
the fact that there is not one that is fully effective.[50] Studies
have not been stratified with regard to risk factors and are
often of small numbers. A recent review identifies cyclizine
and droperidol as most effective but both have significant
side effects in some patients. Prochlorperazine (Stemetil)
and metoclopramide (Maxolon), which are widely prescribed, have not been shown to be efficacious as compared
to a placebo in about 50 per cent of studies.[49] Ondansetron
has been studied and has been shown to be more effective
than a placebo but not markedly different from droperidol.
Ondansetron has few side effects but is considerably more
expensive than existing drugs. There are a number of new
5-HT$_3$ receptor antagonists under development, and
combination studies of these with existing drugs in risk
stratified groups should enable PONV to become a thing of
the past.

REFERENCES

1. Gelman S. Disturbances in hepatic blood flow during anesthesia and surgery. *Archives of Surgery* 1976; **111**: 881–3.

2. Ward ME, Davies TD, Strunin L. Anaesthesia for injection of oesophageal varices. *Annals of the Royal College of Surgeons of England* 1976; **58**: 315–17.

3. Strunin L. How long should patients fast before surgery? Time for new guidelines. *British Journal of Anaesthesia* 1993; **70**: 1–3.

4. Mendelson CL. The aspiration of stomach contents into the lungs during obstetric anesthesia. *American Journal of Obstetrics and Gynecology* 1946; **52**: 191–205.

5. Roberts RB, Shirley MA. Reducing the risk of acid aspiration during cesarian section. *Anesthesia and Analgesia* 1974; **53**: 859–68.

6. Raidoo DM, Rocke DA, Brock-Utne JG, Marszalek A, Engelbrecht HE. Critical volume for pulmonary acid aspiration: reappraisal in primate model. *British Journal of Anaesthesia* 1990; **65**: 248–50.

7. Rocke DA, Brock-Utne JG, Rout CC. At risk for aspiration: new critical values of volume and pH? *Anesthesia and Analgesia* 1993; **76**: 666.

8. *Report on confidential enquiries into maternal deaths in the United Kingdom, 1985–1987.* London: HM Stationery Office, 1991.

9. Durrant JM, Strunin L. Comparative trial of ranitidine and cimetidine on gastric secretion in fasting patients at induction of anaesthesia. *Canadian Anaesthetists' Society Journal* 1982; **29**: 446–51.

10. Vila P, Vallès J, Canet J, Melero A, Vidal F. Acid prophylaxis in morbidly obese patients: famotidine vs ranitidine. *Anaesthesia* 1991; **46**: 967–9.

11. Hunt JN. Some properties of an alimentary osmoreceptor mechanism. *Journal of Physiology* 1956; **132**: 267–88.

12. Cote CJ. NPO after midnight for children – a reappraisal. *Anesthesiology* 1990; **72**: 589–92.

13. Hutchinson A, Maltby JR, Reid CRG. Gastric fluid volume and pH in elective patients. 1. Coffee or orange juice *versus* overnight fast. *Canadian Journal of Anaesthesia* 1988; **35**: 12–15.

14. Phillips S, Hutchinson S, Davidson T. Preoperative drinking does not affect gastric contents: a prospective randomised trial. *British Journal of Anaesthesia* 1993; **70**: 6–9.

15. Maltby JR, Koehli N, Ewen A, Shaffer EA. Gastric fluid volume, pH and emptying in elective inpatients. Influences of narcotic–atropine premedication, oral fluid, and ranitidine. *Canadian Journal of Anaesthesia* 1988; **35**: 562–6.

16. The shortened fluid fast and the Canadian Anaesthetists Society new guide-lines for fasting in elective/surgical patients. *Canadian Journal of Anaesthesia* 1990; **37**: 905–6.

17. Scarr M, Maltby JR, Jani K, Sutherland LR. Volume and acidity of residual gastric fluid after oral fluid ingestion before elective ambulatory surgery. *Canadian Medical Association Journal* 1989; **14**: 1114–15.

18. Maltby JR, Lewis P, Martin A, Sutherland LR. Gastric fluid volume and pH in elective patients following unrestricted oral fluid until 3 hours before surgery. *Canadian Journal of Anaesthesia* 1991; **38**: 425–9.

19. Olsson GL, Hallen B, Hambraeus-Jonzon K. Aspiration during anesthesia: a computer aided study of 185,358 anaesthetics. *Acta Anaesthesiologica Scandinavica* 1986; **30**: 84–92.

20. Cote CJ, Goudsouzian NG, Liu LMP, Dedrick DF, Szyfelbein SK. Assessment of risk factors related to the acid aspiration syndrome in pediatric patients – gastric pH and residual volume. *Anesthesiology* 1982; **56**: 70–2.

21. Manchikanti L, Colliver JA, Marrero TC, Roush JR. Assessment of age-related acid aspiration risk factors in pediatric, adult, and geriatric patients. *Anesthesia and Analgesia* 1985; **64**: 11–17.

22. Sandhar BK, Goresky GV, Maltby JR, Shaffer EA. Effect of oral liquids and ranitidine on gastric fluid volume and pH in children undergoing outpatient surgery. *Anesthesiology* 1989; **71**: 327–30.

23. Sutherland AD, Stock JG, Davies JM. Effects of pre-operative fasting on morbidity and gastric contents in patients undergoing day stay surgery. *British Journal of Anaesthesia* 1986; **58**: 876–8.

24. Goodwin APL, Roe WL, Ogg TW. Oral fluids prior to day surgery. The effect of shortening the preoperative fluid fast on postoperative morbidity. *Anaesthesia* 1991; **46**: 1066–8.

25. Cobley M, Dunne JA, Sanders LD. Stressful pre-operative preparation procedures. *Anaesthesia* 1991; **46**: 1019–22.

26. Maltby JR, Sutherland AD, Sale JP, Shaffer EA. Preoperative oral fluids: is a five hour fast justified prior to elective surgery? *Anesthesia and Analgesia* 1986; **65**: 1112–16.

27. Miller M, Wishart HY, Nimmo WS. Gastric contents at induction of anaesthesia. Is a four hour fast necessary? *British Journal of Anaesthesia* 1983; **55**: 1185–7.

28. Schreiner MS, Triebwasser A, Keon TP. Oral fluids compared to preoperative fasting in pediatric outpatients. *Anesthesiology* 1990; **72**: 593–7.

29. Splinter WM, Stewart JA, Muir JG. The effect of preoperative apple juice on gastric contents, thirst and hunger in children. *Canadian Journal of Anaesthesia* 1989; **36**: 55–8.

30. Bogod DG. The postpartum stomach – when is it safe? *Anaesthesia* 1994; **49**: 1–2.

31. Bell CMA, Lewis CB. Effect of neostigmine on integrity of ileorectal anastomosis. *British Medical Journal* 1968; **3**: 587–8.

32. Wilkins JL, Hardcastle JD, Mann CV, Kaufman L. Effects of neostigmine and atropine on motor activity of ileum, colon, and rectum of anaesthetized subjects. *British Medical Journal* 1970; **1**: 793–4.

33. Aitkenhead AR, Wishart HY, Peebles-Brown DA. High spinal nerve block for large bowel anastomosis. *British Journal of Anaesthesia* 1978; **50**: 177–83.

34. Hawley PR. Infection – the cause of anastomotic breakdown. *Proceedings of the Royal Society of Medicine* 1970; **63**: 752.

35. Traynor C, Paterson JL, Ward ID, Morgan M, Hall GM. Effects of extradural analgesia and vagal blockade on the metabolic and endocrine response to upper abdominal surgery. *British Journal of Anaesthesia* 1982; **54**: 319–23.

36. Fox GS. Anaesthesia for intestinal short circuiting in the morbidly obese with reference to the pathophysiology of gross

obesity. *Canadian Anaesthetists' Society Journal* 1975; **22**: 307–15.

37. Vaughan RW, Vaughan MS. Morbid obesity: implications for anaesthetic care. *Seminars in Anesthesia* 1984; **3**: 218–27.

38. Veall GRQ, Peacock JE, Bax NDS, Reilly CS. Review of the anaesthetic management of 21 patients undergoing laparotomy for carcinoid syndrome. *British Journal of Anaesthesia* 1994; **72**: 335–41.

39. Healey M, Strunin L. Anaesthesia and laparoscopic cholecystectomy. In: Strunin L, Thomson S, eds. The Liver and Anaesthesia. *Baillière's Clinical Anaesthesiology, International Practice and Research*. London: Baillière Tindall, 1992: 819–45.

40. Palazzo MGA, Strunin L. Anaesthesia and emesis. I. Etiology. *Canadian Anaesthetists' Society Journal* 1984; **31**: 178–87.

41. Palazzo MGA, Strunin L. Anaesthesia and emesis. II. Presentation and management. *Canadian Anaesthetists' Society Journal* 1984; **31**: 407–15.

42. Watcha MF, White PF. Postoperative nausea and vomiting; its etiology, treatment and prevention. *Anesthesiology* 1992; **77**: 162–84.

43. Smith G, Rowbotham DJ. Supplement on postoperative nausea and vomiting. *British Journal of Anaesthesia* 1992; **69**: 1–68S.

44. Andrews PLR. Physiology of nausea and vomiting. *British Journal of Anaesthesia* 1992; **69**, suppl 1: 2S–19S.

45. Palazzo M, Evans R. Logistic regression analysis of fixed patient factors for postoperative emesis. *British Journal of Anaesthesia* 1993; **70**: 135–40.

46. Haigh CG, Kaplan LA, Durham LM, Dupeyron JP, Harmer M, Kenny GNC. Nausea and vomiting after gynaecological surgery: a meta-analysis of factors affecting their incidence. *British Journal of Anaesthesia* 1993; **71**: 517–22.

47. Dundee JW, McMillan CM. Positive evidence for P6 acupuncture antiemesis. *Postgraduate Medical Journal* 1991; **67**: 417–22.

48. Borison HL. Area postrema: chemoreceptor circumventricular organ of the medulla oblongata. *Progress in Neurobiology* 1989; **32**: 351–90.

49. Wang SC, Borison HL. The vomiting center, a critical experimental analysis. *Archives of Neurology and Psychiatry* 1950; **63**: 928–41.

50. Rowbotham DJ. Current management of postoperative nausea and vomiting. *British Journal of Anaesthesia* 1992; **69**, suppl 1: 46S–59S.

Chapter 66

Critical Care

Peter Nightingale and J. Denis Edwards

Intensive care has been defined as 'a service for patients with potentially recoverable diseases who can benefit from more detailed observation and treatment than is generally available in the standard wards and departments'.[1] The intensive care unit (ICU) is an area in the hospital designated for the treatment of actual or impending organ failure. It is supervised by nursing and medical staff with specialized knowledge, experience and skill not automatically acquired by training in other specialties.[2] These considerations necessitate a multidisciplinary approach. Standards have been set by the Intensive Care Society[3] (UK), the European Society of Intensive Care Medicine[4] and the Society of Critical Care Medicine[5] (USA). The technical facilities available include invasive monitoring, sophisticated mechanical ventilation, facilities for blood purification in renal or hepatic failure and, in some units, the facility to give mechanical assistance to a failing heart. Clearly in a single chapter such a vast field cannot be covered comprehensively; however, the major areas are addressed and some aspects of management reviewed.

It must be stressed that acute respiratory failure rarely occurs in the absence of circulatory derangements, and shock never occurs without a degree of respiratory dysfunction; thus, even in the acute phase of management, an integrated approach is necessary.

History

The philosophical concept of intensive care, that of grouping the most critically ill patients in a single area, was first suggested by Florence Nightingale during the Crimean War when she designed a system for keeping the most seriously wounded soldiers together adjacent to the nursing station. However, the major impetus to the modern concept of intensive care, especially the use of mechanical ventilation to treat acute respiratory failure, came from the epidemic of poliomyelitis that took place in Copenhagen in 1952–53. This was associated with a high incidence of bulbar paralysis and respiratory insufficiency, and occurred shortly after Astrup and Severinghaus had described their methods for measuring and calculating arterial blood gases. The sequence of events is summarized in a recent text.[6] An anaesthetist, Bjorn Ibsen, had noted that the afflicted children were not cyanosed and appeared to be well oxygenated immediately prior to death, but had a persistently elevated plasma bicarbonate. This had been interpreted by the paediatricians to be a metabolic alkalosis. At a case conference, Ibsen disagreed and proposed hypoventilation with hypercapnia as the primary mechanism, and the elevated plasma bicarbonate was a secondary phenomenon. Astrup was summoned back from his summer holidays and demonstrated arterial hypercapnia, thus proving Ibsen's case. The next day a young patient who was *in extremis* received a tracheostomy and was ventilated manually. The alkalosis disappeared and the patient improved. Ibsen went on to devise a bag system with a CO_2 absorber, and all further patients with alkalosis during the epidemic were treated in the same manner, with a dramatic improvement in outcome. All these patients were ventilated manually, eventually by teams of medical students working in shifts – and at one stage all academic activity in the University ceased. This encouraged the development of automatic mechanical ventilators for use in the ICU. Also stimulated was the application of measurements of pH, arterial oxygen and carbon dioxide tensions ($PaO_2, PaCO_2$) as a routine in clinical practice. This combination of clinical and laboratory cooperation typifies the modern approach to intensive care practice.

Availability, admission and outcome

The British Medical Association recommended in 1967 that 1–2 per cent of acute hospital beds should be designated for the intensive care of patients. In continental Europe 3–5 per cent of acute beds are so designated, and the figure in the USA may approach 15 per cent. In parts of Europe, and especially so in the UK, an ICU bed is not always readily available to admit a seriously ill patient. In a survey of European attitudes to intensive care medicine, 75 per cent of British intensive care units reported inability to admit because of bed unavailability as occurring 'generally or commonly'.[7]

Because of the potential human and monetary costs of admitting, refusing or being unable to admit a patient, it is appropriate constantly to seek answers to some very difficult questions about the practice of intensive care medicine.[8] The King's Fund[1] in 1989 bemoaned the lack of evidence regarding ICU costs and benefits, and recommended that admission was justified only for those expected to survive or with a good chance of recovery. Those whose prognosis was uncertain were also to be admitted but it was this class of patient for whom it was most urgent to conduct clinical trials to evaluate outcome from intensive care. Patients whose death was probable or imminent should not be admitted, except if organ donation were being considered. These stark judgemental decisions can, however, be extremely difficult to make, except perhaps with hindsight.

It is only recently that good quality data have become available for large cohorts of patients in the UK, on both outcome and monetary cost following ICU admission. The Intensive Care Society APACHE II study looked at ICU admission diagnosis and outcome in 11 612 consecutive ICU admissions. Hospital and long-term outcome, with an estimate of health status at 90 and 180 days following ICU admission, were also analysed[9] and are shown in Table 66.1. The majority (70 per cent) of survivors reported the same or a better level of functional status after intensive care than during the previous 3 months.

The cost of caring for ICU patients is substantial and is estimated to be three to five times more than conventional ward care.[10] Actual costs are reasonably easy to estimate but the costs to society are very difficult to quantify. These may be positive by returning a wage earner to employment, or negative by producing a dependent survivor. The human

Table 66.1   Mortality following ICU admission

	Mortality (%)
Deaths in ICU	17.9
Deaths in hospital	27.7
Deaths at 90 days	29.3
Deaths at 180 days	32.3

costs in terms of quality of life after surviving ICU care are equally important. Our attention should be directed to improving patient selection for ICU admission; those who are doomed to die should be allowed to do so with dignity.[8] But can these decisions be quantified?[11]

Scoring systems

Although the practice of intensive care is now well established, clinicians still have difficulty in describing how sick their patients are. Scoring systems are designed to enable a complex clinical situation to be expressed numerically by weighting certain variables, which are summed to give an overall score.[12] The use of such systems may include the following:

- to provide guidelines for admission
- to monitor the effects of transportation
- to allow comparisons between different cohorts of patients
- to compare outcome against expected mortality
- to stratify for severity of illness to reduce bias in research
- to demonstrate differences in mortality according to diagnosis
- to make outcome predictions that influence clinical decisions

Examples of the last use have included decisions to withdraw therapy, to identify the wasteful use of parenteral nutrition, to define the useless treatment of haematological malignancy, to avoid the expensive treatment of trauma patients who die, and to avoid the unnecessary admission to the ICU of low-risk patients.

Osler wrote that 'patients do not die from their disease, they die from the physiological consequences of the illness'. Many scoring systems, therefore, are based on deviations from normal of physiological variables. The variables to be recorded, and the degree of weighting applied to each, are usually defined by clinical consensus or multivariate analysis. Scoring systems must be validated prospectively on a separate cohort of patients because problems may arise if they are applied to patients on whom the system has not been validated. It is unlikely that every variable chosen will have a strong association with the eventual outcome, and likely that not all variables that influence outcome will have been recorded. Therefore, no scoring system can ever be wholly accurate at predicting outcome. Other interacting factors may influence outcome – for example, psychiatric illness and functional disability, although these are hard to quantify, especially in the elderly for whom outcome is related to premorbid status rather than biological age. Treatment factors such as expediency of therapy, its suitability and the patient's response imply that simply measuring physiological derangement will not be adequate to determine outcome. Scoring systems that take these factors into account may enable better outcome prediction in individual cases. However, because outcome is linked to underlying diagnosis, this too must be incorporated. As it is almost impossible to stratify for all variables, an experienced clinician may be just as accurate as any sophisticated mathematical prediction.

Most systems are known by their acronyms.

Severity of illness scoring

TISS (Therapeutic Intervention Scoring System)

TISS was introduced in 1974 as a method of quantifying nursing, medical and technological support activity, and for costing. Currently[13] it consists of 76 procedures assigned a point score of 1 to 4. It is used as a non-specific system for assessing ICU activity and expenditure, and may indicate nursing and medical dependency. TISS can therefore be used to indicate suitability for transfer from the ICU to a high dependency unit or general ward. A version of TISS intended for use in intermediate care units in the USA has been developed, and the Intensive Care National Audit and Research Centre is currently considering updating and modifying TISS for use in the UK.

APACHE (Acute Physiology And Chronic Health Evaluation)

This scoring system initially had 34 variables selected and weighted by clinical consensus. It proved unwieldy and was refined[14] in version II to 12 mandatory physiological variables giving the acute physiology score (APS). Further points are allocated for age and chronic health status. The score is computed after 24 hours of ICU care, using the worst values recorded of each of the 12 variables. APACHE II is the most commonly used system world wide, although originally validated only in the USA. The structure of the system is shown in Table 66.2.

The relation between score and hospital mortality is

Table 66.2   The components of the APACHE II scoring system

Variable	Maximum points
Temperature	4
Mean arterial pressure	4
Heart rate	4
Respiratory rate	4
Oxygenation	4
Arterial pH	4
Sodium	4
Potassium	4
Creatinine	8
Haematocrit	4
White cell count	4
Glasgow Coma Scale	12
Acute physiology score	60
Age	6
Chronic health evaluation	5
APACHE II score	71

impressive for large groups of patients, but outcome differs markedly between diagnostic groups with the same score. To predict outcome for individual cases we therefore need the primary admission category as determined by Knaus.[14] With a further weighting for whether emergency surgery has been performed, it is then possible to calculate the risk of death (ROD) for each patient. At 24 hours after ICU admission, outcome can be predicted correctly in 86 per cent of patients, although the sensitivity is only 47 per cent when using a ROD of 0.5 as the cut-off point.

The Intensive Care Society undertook a prospective validation of the APACHE II system in 35 hospitals in the UK and Ireland. Early reports[15,16] stress the variability of the case mix in the units studied, and this was important when comparing crude hospital mortality. The overall goodness of fit of the American APACHE II equation was good when applied to the British and Irish data, but did not adjust uniformly for specific subgroups of patients. Use of the American equation could advantage or disadvantage individual units in comparison with others, depending on their case mix. Although the APACHE II system has been accepted world wide as a general indicator of severity of illness for patients admitted to the ICU, it is recognized to have inherent problems.[17] These often reflect the lack of rules and definitions in data collection, such as lead time bias (therapy before admission) and assessment of the Glasgow Coma Scale. The diagnostic groups permitted by Knaus are extremely limited in scope, and variations in classification can have major effects on the calculated ROD.

APACHE III

The APACHE scoring system has been further refined and extended to overcome many of its shortcomings.[18] The number of variables has increased, statistical modelling has been used to derive and weight the variables of interest, functional physiological reserve is better delineated and resuscitation before admission is accounted for. APACHE III is used in a dynamic fashion with daily scoring and the use of TISS. The system is considerably more complex than its predecessor, but the improvement in correct classification rate is only modest. Unfortunately, the coefficients for the diagnostic categories are not in the public domain and the expense of purchasing the system from the authors will limit its widespread acceptance.

SAPS (Simplified Acute Physiology Score)

The original cumbersome APACHE system was simplified by French workers to only 14 variables. These are similar to those eventually used in APACHE II, but exclude any measure of oxygenation, chronic health data and diagnosis. However, SAPS does include urine output and the need for mechanical ventilation, factors missing from APACHE II.

SAPS II

The recently introduced SAPS II now has 17 variables derived and weighted by statistical modelling.[19] The type of surgery and presence of immunosuppressive illness are now included, and an arterial sample is required if the patient is on continuous positive airway pressure (CPAP) or mechanical ventilation. Definitions and instructions for data collection were provided, and centres from Europe, the USA and Canada took part in the data collection. Results from the initial evaluation have been promising and the simplicity of the system may well encourage its wider adoption.

Trauma scoring

This has been succinctly reviewed by Yates.[20] The Revised Trauma Score (RTS) utilizes the Glasgow Coma Scale, respiratory rate and systolic blood pressure. It is recorded on admission to hospital and gives an estimate of the physiological derangement of the patient which correlates with mortality.

The Abbreviated Injury Scale (AIS) scores each individual injury on a scale of 1 (minor) to 6 (fatal). The Injury Severity Score (ISS) is obtained by using the AIS to score each injury and then dividing the body into six regions:

- head and neck
- face
- chest
- abdomen and pelvic contents
- bony pelvis and limbs
- body surface

The maximum AIS score in the three worst-injured areas are squared and then summed. The maximum score is 75 ($5^2 + 5^2 + 5^2$); any injury with an AIS of 6 gives the maximum score automatically. ISS correlates closely with mortality, and an ISS $\geq$ 16 is deemed to represent major trauma. The ISS can only be recorded finally following death or discharge from the hospital.

The TRISS score is obtained by combining the Trauma Score and Injury Severity Score to produce a percentage estimate of survival. This can be used retrospectively for auditing outcome following trauma and is done nationally through the offices of the Major Trauma Outcome Study.[21]

Sepsis scoring

A problem in the ICU is that many sepsis scores have depended on the presence of a surgical incision. The more general score devised by Baumgartner[22] may be more useful. This uses admission values of physiological and laboratory data, haemodynamic data after fluid resuscitation, and weightings for underlying disease, infective focus and the organisms isolated. The score significantly improves the ability to predict hospital mortality compared to APACHE II.

Lung injury scoring

Murray et al.[23] have described a semi-quantitative acute lung injury severity score which is widely quoted, and certainly increases the number of patients who could be included in studies of adult respiratory distress syndrome (ARDS) but is not based on any formal data collection. The score is derived from the chest radiograph, PaO_2/FIO_2 ratio, and amount of PEEP and static respiratory compliance if ventilated. Each component scores from 1 to a maximum of 4 points, which are summed and divided by the number of components used. A score of > 2.5 indicates severe lung injury.

Although proposed as being a routine part of ICU practice, many questions remain to be answered regarding the use of scoring systems; these include:

1. Can biological age and physiological reserve be defined?
2. Can predictive systems cope with therapeutic advances?
3. How accurate should individual outcome predictions be?
4. Should nurses, doctors and the patient's family be told the predicted outcome?

Despite continued elaboration, ICU scoring systems should be used only for audit and research, not for outcome predictions in individual patients. As Hippocrates stated 'It is unwise to prophesy either death or recovery in acute disease'.

Invasive haemodynamic monitoring

Vascular pressures and waveforms are monitored from a systemic artery, and from the right side of the heart and pulmonary circulation using indwelling, intravascular catheters inserted percutaneously. There are technical problems and complications common to all these techniques[24] and some that are specific. Any percutaneous technique, whether arterial or central venous, involves the risk of haematoma formation which is especially troublesome for those patients with coagulation disorders. A major problem is infection, introduced either on insertion from the patient's own skin commensal organisms, often staphylococci which are highly resistant to antibiotics, or spread from other sites or other patients during procedures such as blood sampling.

Monitoring systemic intra-arterial waveforms

The commonest site for arterial cannulation is the radial artery but this may prove technically difficult to access in patients in shock. Initially the femoral artery is often used. Occasionally the brachial or axillary artery may be the only routes available; for instance in patients who have had aortofemoral grafts, or victims of major burns or acute exfoliative dermatitis. A meticulous, aseptic, Seldinger technique minimizes the risk of infection. Apart from infection of the cannula, the most serious complication is occlusion of the artery. This usually begins as vasospasm which proceeds to thrombus formation. Allen's test may be performed to limit the effects of this complication in the radial artery, though its value in confirming collateral circulation has been disputed. The closer the tip of the catheter is to the aortic arch, the closer is the accuracy of the detected mean pressure to the true central arterial pressure. Continuous display of the arterial waveform, with observation of the systolic and diastolic peaks and troughs and the dicrotic notch, is essential, both for quality control of the values obtained and for safety.

The arterial waveform is a complex composite of the transmitted pressure impulse in the blood vessel modified by the vessel wall, aortic valve closure (producing the dicrotic notch) and reflection from the periphery. The last effect is most pronounced in peripheral arteries such as the dorsalis pedis. The waveform is further modified during transmission via the catheter, connecting tubing and transducer to the bedside monitor.[25] Excessively compliant catheter or connecting material, thrombus formation or a large air bubble anywhere in the system can produce so-called 'damping' of the waveform. This is also the commonest and earliest manifestation of vasospasm. A damped waveform due to an air bubble in the three-way tap used for blood sampling is shown in the first part of Fig. 66.1; the undamped trace is shown after flushing the line.

Underdamping, with a pronounced systolic spike, overestimates systolic blood pressure. Mean arterial pressure is usually used in the ICU to guide treatment. The majority of pressure monitoring systems in clinical use are underdamped. Arterial lines are inserted primarily to monitor arterial pressure and guide therapy when non-invasive

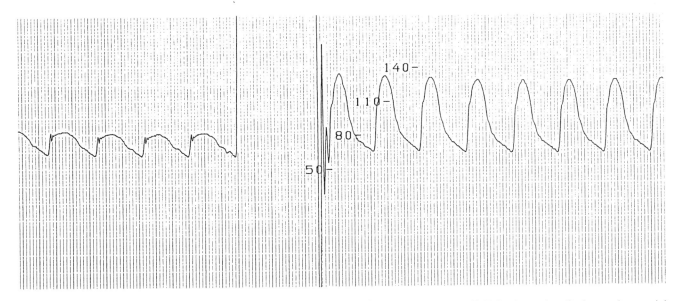

Fig 66.1 A femoral arterial waveform which initially shows overdamping with loss of detail. Following a fast flush test the arterial waveform becomes acceptable. The oscillations after the fast flush test show that the monitoring system is slightly underdamped.

measurements are difficult and inaccurate, especially in low cardiac output states or septic shock. They are also used for blood sampling.

Blood gas analysis

Accurate measurements of PaO_2 and $PaCO_2$ are needed to aid diagnosis and allow calculation of respiratory and oxygen transport variables.[24] Careful attention to sampling procedures is necessary; all dead space must be removed from the system, typically 4 ml for an arterial line and 5 ml for samples taken through the distal port of the pulmonary catheter. To avoid the problems of excess heparin, the use of preheparinized syringes is recommended. Otherwise, use heparin 1000 iu/ml to lubricate the syringe and expel as much as possible. After sampling, remove any froth and large air bubbles and then seal the syringe. A large bubble or froth will lead, within 3 minutes, to a significant change of PO_2 and PCO_2. If a delay of 10 minutes or more is expected, the samples should be stored on ice. In patients with extremely high white cell counts (> 50 000) this will not prevent a fall in PO_2 unless leucocyte metabolism is arrested totally. The advent of intra-arterial blood gas electrodes would of course obviate many of these difficulties but cost and problems with stability are problems yet to be overcome.[26] Interpretation of the PaO_2 requires knowledge of the fractional inspired oxygen concentration (FIO_2) but clinicians are notoriously optimistic about what FIO_2 patients are receiving. As calculation of alveolar PO_2 depends on the FIO_2, and oxygenation indices utilizing PAO_2 are in clinical use, this is not merely academic. With a variable performance device such as the Hudson mask and a fresh gas flow of 15 l/min the inspired oxygen will not exceed 50 per cent in a patient who is dyspnoeic.[27]

For calculation of oxygen content, saturation may be measured or estimated from the oxygen tension. Because of marked variability in the shape of the oxyhaemoglobin dissociation curve (OHDC) in ICU patients, saturations estimated from oxygen tensions, even when compensated for temperature, CO_2 and pH, can be inaccurate because there is no compensation for the 2,3-diphosphoglycerate level. This is especially true for mixed venous samples because these lie on the steep part of the OHDC. Only measured saturations using a co-oximeter are acceptable, and regular quality control procedures should be undertaken to ensure accuracy.[28] The co-oximeter is also invaluable for bedside estimation of haemoglobin, methaemoglobin and carboxyhaemoglobin levels. Direct measurement of oxygen content would avoid problems caused by the shape of the OHDC or by debates over which value should be used for the Hüfner factor (see equations). Combining the blood gas analyser with a co-oximeter allows convenient automatic calculation of arterial and mixed venous oxygen content difference and shunt fraction.

Right heart catheterization

This involves insertion of a catheter into the superior vena cava or right atrium, or placement of a pulmonary artery flotation catheter (PAFC). Of the available routes, direct central venous access is preferred but does impose additional complications. The commonest problem is accidental pneumothorax and/or haemothorax, with an incidence of 1–5 per cent related to the experience of the operator. It is highest when the subclavian route is employed and lowest with the internal jugular route using a relatively high approach. In patients with circulatory collapse or severe coagulopathy the femoral route may be the safest. All intravascular pressures are measured with the zero reference point at the junction of the mid-axillary line and the fifth intercostal space. The ideal position for the tip of a central venous pressure (CVP) catheter for pressure monitoring is the midpoint of the right atrium, but

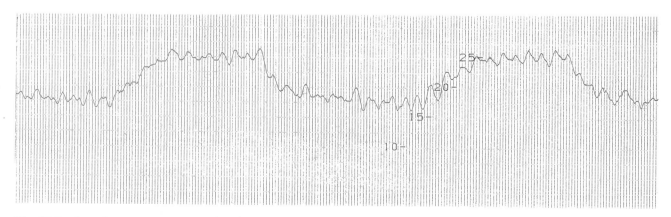

Fig. 66.2 Central venous pressure waveform in a mechanically ventilated patient. The atrial waves can be distinguished and there is an upward respiratory swing superimposed.

for long-term infusions the tip should be in the superior vena cava. The position must always be checked radiologically but a characteristic atrial waveform with 'a' and 'v' waves and appropriate respiratory movement should also be seen, as in Fig. 66.2. During insertion some diagnostic data can be obtained in specific clinical situations, such as the steep 'y' descent in the right atrial waveform seen in constrictive pericarditis or right ventricular infarction.

The accepted method of monitoring intravascular volume status by repeated fluid challenge and serial measurement of CVP can be misleading, except in the most straightforward cases of hypovolaemic shock. The CVP reflects venous return to the right side of the heart and its functional state. This may be affected by blood volume, sympathetic tone, exogenous vasoactive agents, tricuspid valve function and right ventricular loading conditions. The CVP bears little relation to blood volume in the critically ill patient[29] or to left ventricular filling pressures because of different right and left ventricular function curves.[30]

Pulmonary artery catheterization

Since the original description,[31] pulmonary artery catheterization has been used increasingly in critically ill patients. This is especially so for those with complex forms of circulatory shock, especially when combined with renal dysfunction or severe respiratory failure. The exact indications vary widely, and are disputed, but comprehensive guidelines are available.[32] If a patient is subjected to the additional risks of pulmonary artery catheterization, the maximum amount of information available from the procedure should be obtained and acted upon. Apart from those related to central venous access, the specific complications related to indwelling pulmonary artery catheters include pulmonary infarction and rupture of the pulmonary artery during balloon inflation. Arrhythmias and conduction disturbances may occur in unstable patients as the catheter traverses the heart. These risks have to be weighed against the advantages in each particular case.[33]

The technique was introduced to measure the so-called pulmonary artery occlusion pressure (PAOP), otherwise known as pulmonary artery wedge pressure (PAWP). This technique had been routinely performed in the cardiac catheter laboratory under fluoroscopic control for many years, but was introduced to the bedside in the early 1970s by incorporation of a balloon to float the catheter through the heart. The position of the tip of the catheter was indicated by display of the transduced waveform as it traversed the superior vena cava, right atrium, right ventricle and pulmonary artery. The principle of measurement of the PAWP is that, provided certain conditions are

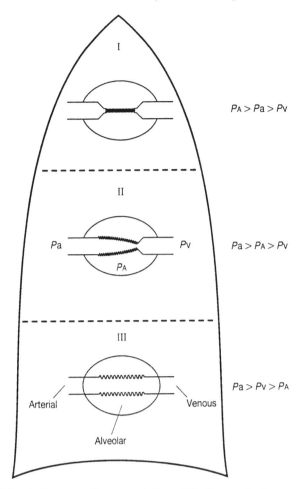

Fig. 66.3 A diagrammatic representation of West's zones in the erect position.

fulfilled, the pressure at the tip of the properly wedged catheter reflects that in the left atrium. There are numerous assumptions and possible confounding factors which need to be taken into account before this is possible.[34,35]

For accurate recording of pulmonary artery wedge pressure the catheter tip should be in West's zone III, where both pulmonary arterial and venous pressure exceed alveolar pressure, as shown in Fig. 66.3. In West's zones I and II the catheter tip will reflect alveolar pressure.

The wedged tracing should demonstrate 'a' and 'v' waves, and excessive respiratory artefact should not be present. It may be necessary to confirm that the tip of the catheter is below the level of the left atrium by a lateral chest radiograph. There must be no obstruction between catheter tip and left atrium because this will delay equilibration and PAOP will overestimate LAP. This may be caused by:

1. Mitral stenosis, the presence of artificial valves or left atrial myxoma
2. Catheter tip obstruction or clot
3. Severe pulmonary disease with fibrosis or capillary destruction

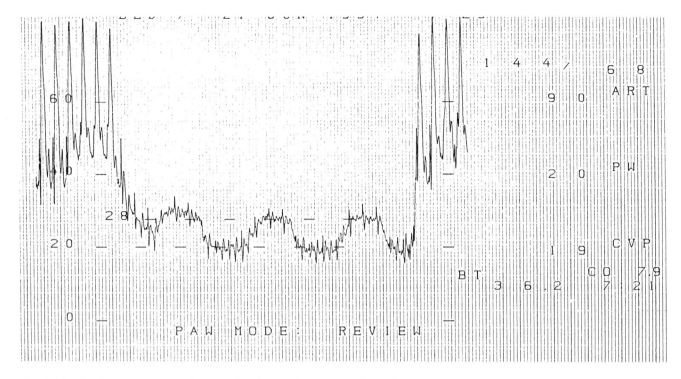

Fig. 66.4 A pulmonary artery wedge trace showing respiratory excursions. The end-expiratory PAWP is 20 mmHg; the bedside monitor erroneously placed the cursor at 28 mmHg, corresponding to end inspiration.

4. Mediastinal masses or emphysema
5. Pulmonary neoplastic processes

Because the pulmonary artery catheter is flow directed, it normally wedges in West's zone III, usually at the right base. However, West's zone III can be converted into zone II conditions by positional changes, hypovolaemia, and ventilator therapy, especially if PEEP is used or intrinsic PEEP develops.

Because of the wide variation in wedge pressure throughout the respiratory cycle, pressures should be displayed graphically and recorded at end expiration, for either mechanical or spontaneous breathing (Fig. 66.4). Large 'v' waves may occur in mitral regurgitation, and these may appear on the pulmonary artery trace as a second hump. During wedging, the pulmonary artery contour is lost but the large 'v' wave may remain visible. If this is not recognized as an atrial waveform, it may appear that wedging has not occurred and predispose to pulmonary artery rupture. Because large 'v' waves will cause the displayed numerical value of PAOP to be erroneously high, the displayed waveform should be used to estimate the amplitude of the 'a' wave as an index of left ventricular end-diastolic pressure.

Measurement of cardiac output (CO)

The other major benefit of pulmonary artery catheterization is the ability to measure cardiac output using thermodilution, a variation of the indicator dilution method. Temperature changes are sensed at the tip of the catheter in the pulmonary artery after injection of a bolus of 5 per cent dextrose into the right atrium. The higher the cardiac output, the smaller is the area beneath the thermodilution curve, and *vice versa*. Injectate at room temperature or cooled in iced water can be given satisfactorily if a 10 ml injection is used, but signal-to-noise ratio is improved if there is a large difference between the temperature of the injectate and that of the blood in the pulmonary artery. Use of in-line injectate temperature sensing ensures that the temperature is recorded accurately as close as possible to the site of injection. Injection should be rapid, less than 4 seconds. Modern cardiac output computers will monitor the temperature in the pulmonary artery for several seconds before signalling that an injection and calculation of cardiac output can be accepted. Errors in cardiac output estimation may be due to physiological disturbances such as the variations resulting from arrhythmias or mechanical ventilation. Other problems include injection of unequal amounts of dextrose, rapid infusion of intravenous fluids causing

Table 66.3 An example of the measured and derived haemodynamic variables produced by an ICU bedside monitoring system

	Patient's values	'Normal' values
CO (l/min)	12.0	4.0–7.0
HR (beats/min)	104	60–90
MAP (mmHg)	107	90
CVP (mmHg)	19	4
PAP (mmHg)	53	12
PAWP (mmHg)	17	8
BSA (m^2)	2.02	1.7
CI (l/min per m^2)	5.9	2.5–3.5
SV (ml)	115	70
SVI (ml/m^2)	57	40
SVR (dyn s/cm^5)	586	800–1500
PVR (dyn s/cm^5)	240	150
LVSWI (g m/m^2)	70	45
RVSWI (g m/m^2)	26	8

oscillations in baseline blood temperature, intracardiac shunts and tricuspid incompetence.

It is now possible to estimate right ventricular ejection fraction, and to monitor cardiac output continuously, using modifications of the original Swan–Ganz catheter.

Modern bedside haemodynamic monitoring systems incorporate the cardiac output into a large number of calculations to give a daunting list of derived variables. Table 66.3 is a typical example showing a patient with hyperdynamic sepsis (low systemic vascular resistance (SVR) and high cardiac index) requiring vasopressors, and who has severe acute lung injury (high mean pulmonary artery pressure). Extreme caution should be exercised when using the derived variables such as SVR to guide therapy, and the accuracy of the directly measured variables should be determined first.

Oxygen transport

Oxygen is the single most important substance carried by the circulation; it is the most flow dependent and has the lowest stores. The transfer of oxygen from the inhaled gas to the tissues is known as the oxygen transport system and involves an intimate relation between pulmonary gas exchange and the pulmonary and systemic circulations. Whole body oxygen consumption ($\dot{V}O_2$) was first utilized by Fick as a means of calculating cardiac output (see appendix to this chapter for abbreviations):

$$CO = \frac{\dot{V}O_2}{CaO_2 - C\bar{v}O_2} \text{ l/min} \tag{1}$$

As cardiac output is now easy to measure at the bedside using the thermodilution technique, the inverse Fick equation can be used to estimate $\dot{V}O_2$:

$$\dot{V}O_2 = CO \times (CaO_2 - C\bar{v}O_2) \text{ ml/min} \tag{2}$$

Nunn[36] describes the concept of oxygen flux or oxygen delivery (DO_2); that is, the total amount of oxygen delivered to the tissues:

$$DO_2 = CO \times 10 \times CaO_2 \text{ ml/min} \tag{3}$$

All the variables needed for the calculation of DO_2 and $\dot{V}O_2$ are available at the bedside of critically ill patients with a PAFC *in situ*. In clinical practice the values are indexed to body surface area, calculated from measurements of height and weight and a standard nomogram, and substituted in equations 2 and 3 above.

$$DO_2I = CI \times 10 \times CaO_2 \text{ ml/min per } m^2$$

$$\dot{V}O_2I = CI \times 10 \times C(a\text{-}\bar{v})O_2 \text{ ml/min per } m^2$$

Interest in the calculation and manipulation of these variables stems from a plethora of publications in the late 1980s,[37–39] which have resulted in a recognition of the relation between DO_2 and $\dot{V}O_2$ in different clinical scenarios. There is some evidence that manipulation of these variables may improve outcome.

When there is no defect in peripheral oxygen utilization, then, as DO_2 falls because of anaemia, hypoxia or reduction in cardiac output, there is an increase in the proportion of oxygen consumed to that delivered, the oxygen extraction ratio (OER). This is usually expressed as a percentage, the normal value being 25 per cent.

$$OER = \frac{CaO_2 - C\bar{v}O_2}{CaO_2} \times 100\%$$

In traumatic, haemorrhagic or cardiogenic shock, the OER may increase to 60 per cent, leading to critically low values of $S\bar{v}O_2$ which may limit cellular utilization of oxygen. Simplistically, there will be a switch from aerobic to anaerobic metabolism at the critical level, when supply cannot satisfy demand, and a rise in arterial blood lactate.[40] Oxygen consumption is independent of supply above this critical level but below it $\dot{V}O_2$ becomes dependent on DO_2. In some patients with ARDS and severe sepsis the ability to increase OER may be impaired (the exact mechanism being unclear), leading to dependence of consumption at much higher levels of delivery. These two situations have been termed physiological and pathological flow dependency respectively. These relations are usually depicted as in Fig. 66.5 with DO_2 on the 'x' axis and $\dot{V}O_2$ on the 'y' axis.

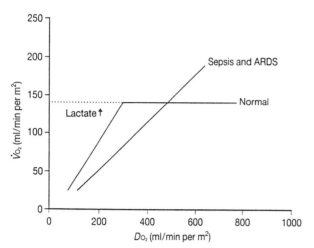

Fig. 66.5 A diagrammatic representation of the theoretical relation between oxygen delivery (DO_2) and oxygen consumption ($\dot{V}O_2$).

Another method of representing oxygen transport is by plotting OER on the 'x' axis and cardiac index on the 'y' axis.[41] This technique may be a useful guide to therapy and to detect pharmacological increases in cardiac index that are not necessary. Monitoring in this way is highly controversial and some of the debates have recently been summarized.[42] Oxygen consumption can also be measured from the inspired and expired gases by calorimetry. There are a variety of metabolic computers available for this purpose and the technology is constantly improving, but problems arise mainly from the use of a high FIO_2 (> 60 per cent) or air leaks. Direct calorimetry may fail to show an increase in $\dot{V}O_2$ when DO_2 is increased, even though calculated $\dot{V}O_2$ increases. It has been suggested that this is due to mathematical coupling of the equations for DO_2 and $\dot{V}O_2$. Direct measurement is to be preferred when investigating changes in $\dot{V}O_2$.[43] Measured and derived haemodynamic values, blood gas data, lactate and drug therapy can be combined to give a concise cardiorespiratory profile as shown in Table 66.4 for a patient with ARDS following sepsis. The patient has been resuscitated with fluids and vasoactive agents to maintain blood pressure and oxygen delivery during PEEP therapy.

Sepsis and multiple organ failure

The study of sepsis has been bedeviled by the lack of standard definitions. A consensus conference has now agreed on definitions for infection, bacteraemia, sepsis, severe sepsis, sepsis induced hypotension and septic shock.[44] The participants agreed that the terms 'septicae-

Table 66.4 An example of a concise cardiorespiratory profile

	Patient's values	'Normal' values
MAP (mmHg)	95	90
CI (l/min per m^2)	5.1	2.5–3.5
CVP (mmHg)	9	4
PAOP (mmHg)	12	8
SVR (dyn s/cm^5 per m^2)	747	800–1500
LVSWI (g m/m^2)	57	45
FIO_2	0.5	–
PEEP (cmH$_2$O)	15	–
DO_2 (ml/min per m^2)	785	560
$\dot{V}O_2$ (ml/min per m^2)	189	140
OER (%)	24	25
$\dot{Q}s/\dot{Q}t$ (%)	27	< 5
Lactate (mmol/l)	0.7	< 2
Dopamine (µg/kg per min)	3	–
Dobutamine (µg/kg per min)	12	–
Phenylephrine (µg/kg per min)	7	–

mia' and 'sepsis syndrome' be no longer used, and that two new terms were necessary.

Systemic inflammatory response syndrome (SIRS)

This is the systemic inflammatory response to a variety of severe clinical insults. The response is manifest by two or more of the following conditions:

Temperature > 38°C or < 36°C
Heart rate > 90 beats/min
Respiratory rate > 20 breaths/min or $PaCO_2$ < 4.3 kPa (32 mmHg)
WBC > 12 000 or < 4000 cells/mm^3 or > 10% immature forms

Sepsis is a category of SIRS in which patients have documented infection.

Multiple organ dysfunction syndrome (MODS)

This is the presence of altered organ function in an acutely ill patient such that homoeostasis cannot be maintained without intervention. MODS can develop as a primary event such as acid aspiration, or secondary to an insult that produces the above systemic response but which then leads to dysfunction of distant organs and eventually multiple organ failure (MOF). The syndrome of sequential failure of several organs was first described in 1971 by Tilney and

Table 66.5 Components of the multiple organ dysfunction syndrome

Acute lung injury
Circulatory failure
Renal impairment
Hepatobiliary dysfunction
Gastrointestinal failure
Nutritional and metabolic derangements
Neurological deterioration
Haematological failure
Musculoskeletal abnormalities

colleagues in patients following surgery for ruptured abdominal aortic aneurysms.[45] These systems are listed in Table 66.5.

The definitions for organ failure in different patient groups and from different centres have been in broad agreement, but rather frustratingly the exact definition has varied from one study to another, as highlighted by the different definitions for the degree of respiratory insufficiency seen in ARDS described later. All authors agree on two aspects of the multiple organ failure syndrome. First, it is increasing in frequency as a result of improved initial resuscitation combined with technical advances in artificial support for failing organ systems. Second, despite consuming an enormous amount of resources, the mortality remains high, although there are suggestions that it may be improving. Because MOF is usually described in patients with acute respiratory failure receiving mechanical ventilation, the first identifiable organ to fail is the lung. Following this there may be variations in the sequence of organ failures depending on the patient population studied, but a review of the available literature suggests the following to be the commonest recognized sequence:

1. Respiratory failure requiring ventilation with progressively higher levels of inspired oxygen and PEEP.
2. Hypotension requiring increasing amounts of replacement fluids.
3. Systemic vasodilatation or inadequate cardiac output unresponsive to fluid therapy and requiring vasoactive drug therapy.
4. Persisting oliguria despite attempts to induce a diuresis with frusemide (furosemide), mannitol and/or low dose dopamine, leading to established acute renal failure with the need for dialysis or haemofiltration. Renal failure may occasionally be non-oliguric.
5. Hepatobiliary failure with intrahepatic cholestasis, jaundice, elevated hepatic transaminases, acalculous cholecystitis, coagulopathy and ischaemic hepatic necrosis.
6. A hypermetabolic state with muscle wasting, hypoalbuminaemia and electrolyte and nutritional abnormalities, often despite apparently adequate intake.
7. Gastrointestinal failure with gastric stasis, stress ulceration, ileus, intolerance of enteral feeding, diarrhoea and pseudomembranous colitis.
8. Haematological failure with anaemia, thrombocytopenia and neutropenia in varying combinations.
9. Neurological complications with encephalopathy and peripheral neuropathy.
10. Musculoskeletal abnormalities with myopathy and eventually muscle calcification.

The majority of studies have been retrospective, although one prospective study in high-risk patients undergoing major surgery has defined the usual sequence of events in this patient population[46] in relation to estimated oxygen debt, as shown in Fig. 66.6.

The possible aetiologies of MOF have been extensively reviewed[47] and include prolonged or severe tissue hypoxia as a result of a catastrophic insult or delayed resuscitation, release of cytokines from infected or damaged tissues, and endotoxaemia or secondary bacteraemia due to loss of intestinal mucosal integrity. Apart from general mechanisms, there may be specific factors operating on individual organs. For instance, hepatic failure could be due to any combination of hypoxia, hypotension, low cardiac output, hepatotoxic drugs, the effects of PEEP therapy causing venous congestion or redistribution of blood flow, lipid deposition from excessive parenteral nutrition, cholestasis, multiple blood transfusion, systemic sepsis and direct hepatic trauma.

The management of MOF involves three aspects:

1. Amelioration or prevention by prompt initial resuscitation with rapid reversal of tissue hypoxia.
2. Removal of trigger factors by control of infection, prompt recognition and treatment of injuries with early fixation of fractures and removal of dead tissue, and early recognition and treatment of postoperative complications.
3. Supportive therapy; this is the most demanding aspect of management in terms of time and economic implications.

The typical patient will be on a mechanical ventilator, receiving two or three drugs for cardiovascular support and some form of renal replacement therapy. The patient will also be receiving parenteral nutrition, blood and blood factors and repeated courses of antibiotics. Multiple microbiological and radiological investigations will be undertaken to locate any source of sepsis. Abdominal ultrasound can be performed in the ICU, though specificity and sensitivity are poor, and transfer for CT scanning is more informative although both inconvenient and potentially dangerous.

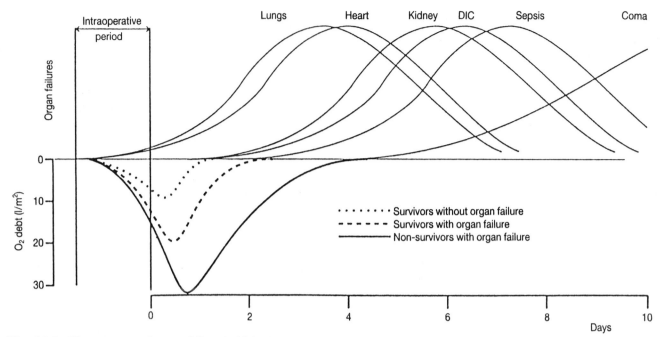

Fig. 66.6 The sequence of organ failure in high risk surgical patients in relation to cumulative oxygen debt. (Reproduced, with permission, from Shoemaker WC, Appel PL, Kram HB. Role of oxygen debt in the development of organ failure, sepsis and death in high risk surgical patients. *Chest* 1992; **102**: 208–15.)

The introduction of novel therapeutic agents has raised hopes that a reduction in mortality can be achieved. The value of the monoclonal antibody antagonists of endotoxin and tumour necrosis factor has yet to be confirmed. The use of an interleukin-1 receptor antagonist has not fulfilled its early promise. Other agents such as platelet-activating factor antagonists are also undergoing assessment. In view of the number of mediators involved in sepsis, it is not too surprising that therapy directed at only one has not proved clinically significant and cocktails of agents may need to be considered. Trials of these agents have been discussed thoroughly in the literature but at the present time they are not available for routine use.[48]

However, even failure of three or more organ systems does not preclude good quality survival. The prognosis in certain diagnostic groups is poor, and raises difficult ethical issues. These include irreversible and disabling pre-morbid pathology, the elderly patient, patients with neoplasia and persistent neutropenia following chemotherapy and clinical situations such as unwitnessed cardiorespiratory arrest, head trauma with a low admission Glasgow Coma Score and patients with extensive full-thickness cutaneous burns.

From this brief account it will be clear that, in the management of a patient with MOF, the multidisciplinary approach is very important. However, after the initial insult, both ARDS and sepsis remain intimately linked and medical intervention will be designed to maintain the circulation, to provide respiratory support and to identify and eliminate sources of infection.

Hospital-acquired infections

It is estimated that 30 000 patients per year die in the UK from severe sepsis, much of this acquired in hospital. In the European Prevalence of Infection in Intensive Care (EPIIC) Study, 21 per cent of patients had an infection directly related to their admission to the ICU. These infections prolong hospital stay and increase mortality and morbidity by approximately 300 per cent. The incidence continues to increase, and this is attributable to numerous factors:

1. More elderly patients are being treated
2. Patients are considered for care in the ICU despite greater severity of their illness
3. There is greater co-morbidity
4. The use of immunosuppressive therapy is more prevalent
5. More patients are undergoing repeat surgery
6. Patients are surviving hospitalization for longer periods
7. There is a greater use of prosthetic and invasive devices

Much of the nosocomial catheter-related infection can be traced to poor hand washing technique and poor technique during blood sampling and drug administration.

Use of any form of artificial airway may be associated with nosocomial pneumonia; although the literature suggests an average incidence of 30 per cent, this occurs in up to 60 per cent of high-risk patients. Infection is usually due to Gram-negative organisms but the diagnosis

Table 66.6 Likely organisms present in the ICU, by site of infection

Sepsis in the patient with burns
Colonization is common and need not be treated. Fatal sepsis is often associated with group A streptococci, *Pseudomonas*, *Staphylococcus aureus* and *Candida*.

Surgical wound infections
Bacteroides now rare. Often multi-resistant Enterobacteriaceae, especially *Klebsiella*, *Serratia* and *Enterobacter*. Consider *Acinetobacter*, *Pseudomonas* and enterococci as well.

Intravascular catheter-associated sepsis
Wide variability in diagnosis. Predominantly coagulase-negative staphylococci. Also Gram-negative aerobic bacilli and *Candida*.

Nosocomial pneumonia
May follow colonization of the oropharynx by enteric organisms which are multi-resistant Enterobacteriaceae, especially *Klebsiella*, *Enterobacter*, *Serratia* and *Acinetobacter*. Pseudomonas pneumonia and bacteraemia has a high mortality. Candida colonization is often seen.

Septicaemia in the surgical patient
Mostly Gram-negative, especially *Escherichia coli*, *Klebsiella*, *Enterobacter* and *Pseudomonas*. Consider *Candida* if there has been bowel perforation or pancreatitis, or if the patient is colonized, especially in the urine.

Sepsis due to sinusitis
Associated with nasal tubes. Normal respiratory pathogens plus *Pseudomonas*, Enterobacteriaceae, enterococci, staphylococci and *Candida*.

Sepsis in the neutropenic patient
No organism found in more than 60 per cent of febrile episodes. Usual organisms include Gram-negative aerobic bacilli, especially *E. coli* and *Pseudomonas*. Consider also Gram-positive organisms and invasive fungal infections.

of ventilator-associated pneumonia is fraught with difficulty. The presumed route of infection is by microaspiration of organisms from the upper gastrointestinal tract which have colonized the oropharynx.

Oropharyngeal colonization can be reduced by selective decontamination of the digestive tract (SDD). The technique combines the long-term application of an antimicrobial/antifungal paste plus a short course of systemic broad-spectrum antibiotics. This reduces the incidence of ventilator-associated pneumonia but has not been shown to definitively improve overall outcome. Many of the trials of SDD have been heavily criticized and the technique is not yet recommended for routine use.[49]

The incidence of respiratory infection can be minimized by regular attention to oropharyngeal toilet and pulmonary secretions, with humidification of inspired gases to reduce the incidence of retained secretions and preserve mucociliary function. The haemodynamic manifestations of sepsis are not related to the offending organism, but the site or presumed location of the infection can point to likely candidates (Table 66.6).

Infections such as acalculous cholecystitis and sinusitis will be treated by drainage procedures as well as appropriate antibiotics. Candida sepsis continues to be a cause of death in long-stay patients, and resistant species may thrive with the indiscriminate use of antifungal azoles such as fluconazole. The Gram-positive bacteria continue to be a major problem within the ICU, with methicillin-resistant *Staphylococcus aureus* now endemic in many

hospitals. Most worryingly, vancomycin-resistant enterococci are now also appearing in the UK, and although vancomycin-resistant coagulase-negative staphylococci have not yet been reported, it seems likely that this is only a matter of time.

Over the last few years the choice of antibiotics has widened considerably but rational prescribing is needed. The selection of an antibiotic regimen depends on a number of factors.

1. Site of infection
2. Known or presumptive organism
3. Underlying medical condition and hepatorenal function
4. Unit antibiotic policy
5. Cost

A distinction must be made between normal flora, colonization and infection. It is estimated that only 30–40 per cent of antibiotics are prescribed for a proper indication, and the introduction of a hospital antibiotics policy has been shown to reduce the emergence of multiply resistant organisms. An antibiotics policy should include guidelines for the blind treatment of serious infections and for infections caused by known pathogens. Clinicians should be aware of which organisms are prevalent in the hospital and in their ICU because blind empirical therapy may frequently be needed. The likelihood of fungal infection should never be underestimated.

Respiratory failure

Using values of PaO_2 and $PaCO_2$ in the arterial blood, Pontoppidan described two types of acute respiratory failure: type 1 with a low $PaCO_2$ and type 2 with a high $PaCO_2$.[50] This is an attractive theoretical concept but is not particularly useful in clinical practice. Many patients with so-called type 1 failure due to acute lung injury develop fatigue of the respiratory muscles, resulting in hypercapnia unless mechanically ventilated, thus becoming type 2 patients. Also, many patients who may be thought to be candidates for type 2 failure, for instance patients with neuromuscular diseases such as myasthenia gravis and Guillain–Barré syndrome, may develop a chest infection before complete respiratory muscle or bulbar paralysis and can present to the ICU with a type 1 arterial blood gas picture. However, this classification has some appeal, especially when the patient population has chronic lung disease, and is still widely quoted. Some authorities have distinguished between the terms 'respiratory distress' and 'respiratory failure'. For instance, a patient may feel dyspnoeic and have tachypnoea but normal blood gases. Given these caveats, there are two major classes of respiratory failure in Pontoppidan's classification. First, intrinsic lung disease causing primary hypoxaemia with reduced PaO_2, with at least initially a compensatory hyperventilation and reduction in $PaCO_2$ and, second, failure of the mechanisms of ventilation with impaired alveolar gas exchange and hence elimination of CO_2. Some of the more common causes of each of these types of respiratory failure are given in Table 66.7.

These lists are by no means exhaustive. Some disorders appear in both. In bronchial asthma there is alveolar hyperventilation initially, but as bronchospasm worsens and fatigue sets in the $PaCO_2$ will rise. Similarly, toxic gas or smoke inhalation can affect the upper or lower airways and the lung parenchyma, producing different blood gas patterns at various stages of evolution of the illness. Patients with a wide variety of acute and chronic respiratory problems may present to the ICU for intermittent positive pressure ventilation (IPPV). Available

Table 66.7 Some causes of type 1 and type 2 respiratory failure

Type 1	Type 2
Acute respiratory distress syndrome	Neuromuscular disease
Cardiogenic pulmonary oedema	Exacerbation of chronic pulmonary diseases
Severe acute asthma	Morbid obesity
Lobar pneumonia	Severe acute asthma
Smoke inhalation	Smoke inhalation

space here limits the description of the details of management of all forms of respiratory failure, but the principles discussed apply to the majority. The most important aspect for the clinician is the mechanism of the disturbance of pulmonary gas exchange because this has implications for the mode of mechanical ventilation and types of adjunctive therapy required.

Acute respiratory distress syndrome (ARDS)

This is the most severe form of acute lung injury. At a consensus conference, 'acute' rather than 'adult' was agreed as the preferred terminology because the same picture is also seen in children[51] and was the term used in the original description.[52] In view of its pre-eminence as a clinical problem, ARDS is discussed here in some detail. In clinical practice, ARDS may be relatively uncommon but assessment of incidence is made difficult by the wide number of different diagnostic criteria used, as shown in Table 66.8.

Probably the best study of prevalence comes from Villar and Slutsky[66] who, with a defined population in the Canary Islands, dealt with all patients who required mechanical ventilation in one ICU. This 3-year prospective study gave an overall incidence, depending on the definition used, of 1.5–3.5 cases per 100 000 population. The report from the National Heart and Lung Institute published in 1972 *estimated* an incidence of 75 cases per 100 000 population.[67]

An objective definition of severe acute lung injury would be:

1. The recognition of an underlying predisposing cause; however, in many series up to 10–15 per cent of cases are deemed to be idiopathic and attributed to viral pneumonitis although this is rarely confirmed.
2. Severe hypoxaemia unresponsive to supplemental inspired oxygen.
3. The need for mechanical ventilation with initially an $FIO_2 \geq 0.6$.
4. The need for positive end-expiratory pressure and/or inverse ratio ventilation.
5. The presence of bilateral alveolar infiltrates on chest x-ray.

In addition to the need for high FIO_2 and mechanical ventilation, it would seem sensible to employ a measure of the severity of disturbance of pulmonary gas exchange. These include:

$\dot{Q}s/\dot{Q}t$ the degree of pulmonary venous admixture
$P(A–a)O_2$ alveolar to arterial oxygen tension gradient
PaO_2/FIO_2 ratio of arterial oxygen tension to inspired oxygen concentration
PaO_2/PAO_2 arterial oxygen tension divided by estimated alveolar oxygen tension

Table 66.8 Some definitions of pulmonary insufficiency in ARDS

Definition	Authors	Year
Hypoxia, dyspnoea, pulmonary oedema	Ashbaugh[52]	1967
Arterial hypoxaemia after trauma and surgery	Powers[53]	1973
FIO_2 0.5, PaO_2 < 6.7 kPa (< 50 mmHg), IPPV	Danek[54]	1980
FIO_2 0.6, PaO_2 < 6.7 kPa (< 50 mmHg), IPPV or SV	Petty[55]	1982
FIO_2 0.5, PaO_2 < 6.7 kPa (< 50 mmHg), IPPV	Bell[56]	1983
FIO_2 1.0, 'severe hypoxaemia', IPPV	Mohsenifar[57]	1983
FIO_2 1.0, PaO_2 < 13.3 kPa (< 100 mmHg)	Kariman[58]	1984
FIO_2 0.4, PaO_2 < 6.7 kPa (< 50 mmHg), SV	Shoemaker[59]	1985
PaO_2 / FIO_2 < 150	Montgomery[60]	1985
FIO_2 0.4, PaO_2 8–13 kPa (60–100 mmHg), IPPV	Annat[61]	1986
No definition except IPPV	Smith[62]	1986
$A–aDO_2$ > 33.3 kPa (250 mmHg), IPPV	Jacobs[63]	1991
$\dot{Q}s/\dot{Q}t$ > 30%	Clarke[64]	1991
PaO_2 / PAO_2 ≤ 0.2	Suchyta[65]	1992

Table 66.9 Some causes of the acute respiratory distress syndrome

Direct	Indirect
Pulmonary aspiration	Multiple trauma
Chest trauma	Severe sepsis
Blast injury	Massive blood transfusion
Smoke inhalation	Fat embolism
Toxic gas inhalation	Amniotic fluid embolism
Near-drowning	Pancreatitis
Pneumonitis:	Prolonged or severe hypotension
Bacterial	Disseminated intravascular
Viral	coagulation
Protozoal	Malaria
	Poisoning:
	Paraquat
	Salicylate

None of these tension-based indices is reliable, however, and estimated or measured shunt fraction is preferable.[68]

It is usual to divide the causes of ARDS into those associated with a direct and those with an indirect pulmonary insult, as shown in Table 66.9. Naturally there will be a degree of overlap in many instances.

In the original series of Ashbaugh and Petty,[52] seven patients had multiple trauma, one had pancreatitis, and a diagnosis of presumed viral pneumonitis was given in four. Subsequent series have tended to include purely surgical or medical patients in either group.

Following a precipitating event such as shock, multiple trauma, burns or sepsis, there is often a latent interval of 12–72 hours. Respiratory failure then ensues with tachypnoea, intractable hypoxaemia and ventilatory muscle fatigue. Pathophysiologically there is a gross reduction in pulmonary functional residual capacity and a wide scatter of $\dot{V}/\dot{Q}$ ratios with an increase in both shunt fraction and alveolar dead space. There is a variable increase in extravascular lung water, but the defects in oxygenation are not simply due to non-cardiogenic pulmonary oedema from increased pulmonary capillary permeability. There are functional and anatomical abnormalities in the pulmonary vasculature, and complex histological abnormalities in the alveolar walls ranging from increased cellular infiltration, proliferation of type II alveolar cells, and hyaline membrane formation merging into a recovery phase or progressing to irreversible pulmonary fibrosis. In the acute phase, loss of surfactant activity leads to alveolar collapse. There is a reduction in elastic recoil forces, and grossly impaired pulmonary compliance, often < 30 ml/cmH_2O.

There are a wide number of peripheral circulatory defects associated with ARDS, and respiratory failure is just one aspect of the syndrome of MOF associated with generalized capillary endothelial damage. One manifestation of this defect in tissue oxygenation seen in ARDS is the phenomenon of pathological supply dependency of oxygen consumption on oxygen delivery. Therapy now centres on maintaining tissue oxygenation while preventing further lung damage.

Corticosteroids have no role in the prevention or initial treatment of ARDS. In the later fibroproliferative stage of ARDS, steroids can reduce inflammation in the lung, improve oxygenation and reduce mortality.[69] There is now a need for a large, controlled study to assess this mode of treatment.

Surfactant secretion by type II pneumocytes is reduced in ARDS. Loss of surfactant also occurs from inactivation by plasma enzymes and peroxidation by free radicals, and with ventilatory modes that allow alveolar collapse and overdistension. The role of exogenously delivered surfactant replacement therapy is being evaluated.[70,71] Although accepted practice in neonates, there are as yet no adult studies showing long-term improvement. One large-scale study was terminated early because of failure to influence mortality.[72]

The use of inhaled nitric oxide as a selective pulmonary vasodilator with no systemic effect has generated enormous interest. Inhalation of nitric oxide in acute respiratory failure, up to 2 parts per million, produces a dose-dependent reduction in pulmonary shunt and an increase in PaO_2.[73] Studies in patients with ARDS, many using larger doses, are encouraging.[74] Although this technique is promising, there are many technical factors militating against its routine use and prospective trials on long-term outcome are still awaited.

Indications for intermittent positive pressure ventilation (IPPV)

The need to institute IPPV is sometimes straightforward, as in the sudden onset of apnoea in a patient with head injury. However, more frequently the situation is less clear cut. A balance has to be struck between the clinical situation, the symptoms and signs, and the arterial blood gases. For instance, a patient who is about to undergo haemodialysis for acute renal failure, who has a normal PaO_2 on room air but a respiratory rate of 40 breaths per minute or an irregular respiratory pattern with paradoxical respiration, might be considered a candidate for intubation in view of the clinical signs of severe respiratory distress and the known deleterious effects of dialysis on cardiorespiratory function. The clinical diagnosis may lead to treatment that alters the initial need for ventilation. For instance, an accidental or intentional overdose of an opiate may respond to naloxone; relief of a haemopneumothorax in a case of chest trauma may dramatically improve both the clinical signs of respiratory distress and the arterial blood gases; a patient with acute left ventricular failure may be *in extremis* but respond dramatically to treatment with supplemental oxygen and vasodilators.

If there is any doubt, initial treatment is with high-flow oxygen therapy by facemask, preferably with a reservoir device to maximize FIO_2. Treatment is guided by repeated measurements of arterial blood gases, and, provided oxygen therapy is correctly prescribed, humidified and monitored, the dangers are minimal. The theoretical danger of inducing respiratory depression in patients with CO_2 retention has been grossly overemphasized. If the decision not to ventilate is made in an equivocal case, the patient must be observed closely in a high care area such as a resuscitation or anaesthetic recovery room or a high dependency unit. The use of peripheral pulse oximetry is indispensable, but may not warn of dangerous hypoventilation if supplemental inspired O_2 is given.[75]

Another reason to intubate is for airway protection. Even in the absence of respiratory failure, comatose or obtunded patients may aspirate gastric contents and some degree of artificial respiratory support may be needed until protective airway reflexes recover fully.

Patients with neuromuscular disorders, especially bulbar palsy or diaphragmatic weakness, may not be able to cough adequately to clear secretions, and tracheal intubation may be necessary for bronchial toilet.

Hypoxia refractory to high-flow inspired oxygen, with signs of respiratory distress, is a strong indication for mechanical ventilation. A rising $PaCO_2$ often indicates the need for IPPV, with the proviso that some patients may have irrecoverable chronic lung disease and mechanical ventilation may be deemed inappropriate. These considerations are summarized in Table 66.10.

Improved understanding of the mechanics of artificial ventilation and of the circulatory effects of positive intrathoracic pressure now often allow previous concerns about the side effects of IPPV to be outweighed by the potential benefits. There are no simple didactic rules to guide the decision to institute mechanical ventilation. However, any patient who previously had normal pulmonary function and who has hypoxia persisting despite high-flow supplemental oxygen (a PaO_2 of 8 kPa (60 mmHg) is close to the steep portion of the OHDC) should be considered a strong candidate. In any patient, a rise in $PaCO_2$ that causes a respiratory acidosis should alert the clinician, although for patients with exacerbations of chronic lung disease supportive therapy may avert IPPV.

Available modes of respiratory support

The techniques available for respiratory support have increased as a result of advances in technology, improved understanding of the pathophysiology of acute respiratory failure, and a realization of the possible benefits of various modes of ventilation. Unfortunately, many newer modes have been introduced merely on the basis of technical ability rather than as a result of a defined clinical need or demonstrable advantage to the patient. Even with the most popular modes of ventilation, currently believed by clinicians to improve outcome in severe ARDS, there are few controlled studies that demonstrate a reduction in morbidity and mortality.

Pressure-cycled ventilation

This was the first automatic mode to be employed: inflation ceased when a predetermined peak ventilatory pressure was reached, whatever the resulting tidal volume. Clinicians later considered this unsatisfactory as changes in airways resistance and pulmonary or thoracic compliance led to unpredictable and usually inadequate tidal volumes. The peak inspiratory flow rates were insufficient

Table 66.10 Some indications for intubation and mechanical ventilation in acute respiratory failure

Clinical	Biochemical
Impending apnoea	$PaO_2 < 8$ kPa (< 60 mmHg) on high FIO_2
Paradoxical respiration	Pulse oximeter < 90% on high FIO_2
Obvious fatigue	$PaCO_2 > 7$ kPa (> 53 mmHg) or rising from a lower level
Persistent tachypnoea	$PaCO_2$ not low with a metabolic acidosis
Variable respiratory rate	$PaCO_2$ persistently < 3.5 kPa (< 26 mmHg)
Absent protective reflexes	
Threat of airway obstruction	

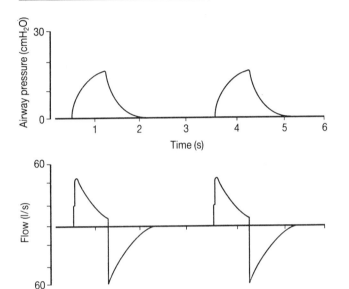

Fig. 66.7 Waveforms seen with pressure-cycled ventilation. Flow is initially high and then decelerates. Expiration is triggered when the preset inspiratory pressure is reached even though inspiratory flow is still occurring.

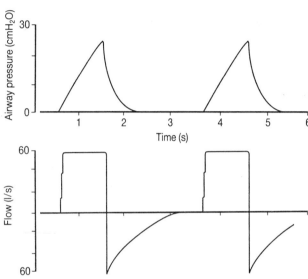

Fig. 66.8 Waveforms seen with volume-controlled ventilation. Flow is constant and expiration is time cycled.

to cope with the low levels of pulmonary compliance seen in ARDS. The pressure and flow waveform patterns in this mode are shown in Fig. 66.7.

Volume-controlled ventilation

In this mode a preset tidal volume is always delivered although very high levels of peak inspiratory pressure may be produced. Minute volume, in the absence of leaks in the respiratory circuit, is therefore guaranteed whatever the level of airways resistance or total thoracic compliance. The pressure and flow waveform patterns in this mode are shown in Fig. 66.8.

Recognition that very high levels of peak and mean inspiratory pressures were associated with barotrauma led to a degree of airway pressure regulation by setting an upper pressure alarm limit above which the inspiration was terminated.

Controlled mechanical ventilation (CMV)

This mode was, until recently, employed only with volume-controlled ventilation. All breaths are mandatory machine breaths and the estimated minute volume is delivered only by the ventilator. The disadvantage of this technique is that, if the patient tries to take a spontaneous breath or cough, commonly called 'fighting the ventilator', not only is it very uncomfortable but also surges in peak inspiratory pressure are produced. The term 'assist-control' is used when the patient can trigger a mandatory machine breath.

Synchronized intermittent mandatory ventilation (SIMV)

Intermittent mandatory ventilation (IMV) was introduced to allow the patient to breathe spontaneously between mandatory machine breaths. Mandatory breaths could be reduced in frequency, allowing the patient to add extra spontaneous breaths to maintain minute volume. However, the rate and timing of mandatory breaths were fixed and, once commenced, were continued until the preset tidal volume had been delivered, with potential hyperinflation if the mechanical breath came at the end of a patient breath. It was assumed that the patient would cooperate with the ventilator but this was rarely achieved in clinical practice. The technique was then refined to allow mandatory mechanical breaths to be initiated by and synchronized with patient effort (SIMV). This added considerably to patient comfort with less need for sedation and paralysis.[76] A problem remained, however, because patients with poorly compliant lungs tend to breathe rapidly. The increase in V_D/V_T commonly seen in these patients meant that dead space ventilation tended to increase even though minute volume was maintained. There is a strong possibility that premature attempts to wean patients using SIMV and which leads to this breathing pattern may contribute to sputum retention, atelectasis and pulmonary infection.[77]

Pressure support ventilation (PSV)

This mode is frequently used with SIMV as the basic ventilator strategy in the ICU.[78] With PSV, breaths are patient triggered and a variable flow inspiratory pressure support is then applied by the ventilator. The level of pressure support is preset to give an appropriate tidal volume. Usually, airway pressure is maintained at the preset level until inspiratory flow starts to fall, and at a certain level of flow the expiratory phase commences. As with all triggered modes, it is vital that the response time of the ventilator is rapid or patient work of breathing can be increased. Best response times are seen in those ventilators that are flow triggered.

Proportional assist ventilation (PAV)

This new mode of ventilation, in which the level of ventilatory support is proportional to the patient's inspiratory effort, has recently been described but there are few clinical studies.[79] Neither flow, nor pressure nor volume is set by the clinician! A gain control is adjusted to determine what proportion of the patient's inspiratory effort is supported by the ventilator.

Positive end-expiratory pressure (PEEP)

PEEP has been used and studied most extensively in patients with reduced pulmonary compliance due to diffuse lung disease. Airway pressure in the ventilator circuit is prevented from returning to atmospheric at the end of expiration. The usual therapeutic levels of PEEP range from 5 to 20 cmH$_2$O.[80] A representative pressure waveform is shown in Fig. 66.9.

The PEEP value set should be checked by direct observation of the pressure dial on the ventilator at end expiration. When beneficial, PEEP increases functional residual capacity by alveolar recruitment and improves the

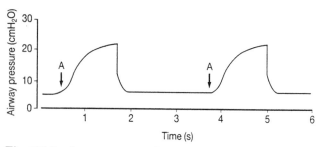

Fig. 66.9 A pressure waveform demonstrating PEEP. End-expiratory pressure is maintained at 5 cmH$_2$O until the next inspiration (point A).

distribution of pulmonary blood flow to ventilated lung units. This reduces pulmonary venous admixture and increases PaO_2 at any given FIO_2. There are several definitions for the optimum PEEP level, based on estimates of static compliance, efficiency of oxygenation and, more recently, values of tissue oxygen delivery, the tendency now being to use moderate values of 5–15 cmH$_2$O. The disadvantages of PEEP are an increase in peak inflation pressure and overdistension of lung units with short time constants, both of which contribute to pulmonary barotrauma. PEEP can reduce cardiac output, especially in hypovolaemic patients,[81] and the distribution of ventilation is often unpredictable in patients with asymmetric lung disease. For instance, if acute respiratory failure occurs as a result of consolidation in one lung, ventilation is preferentially distributed to the healthy lung. Depending on the exact level, PEEP therapy may reduce $\dot{V}/\dot{Q}$ abnormalities by opening collapsed alveoli in the consolidated lung and improve gas exchange. However, especially if high tidal volumes are used, there may be dramatic overdistension of the healthy lung with resultant barotrauma, and dead space will be increased. Pulmonary vascular resistance may be increased and compression of vessels around distended alveoli may divert blood to underventilated regions and worsen shunt fraction.

Inverse ratio ventilation (IRV)

In this mode the usual inspiratory to expiratory ratio (I : E ratio) of 1 : 2 or 1 : 3 is adjusted by prolonging inspiration so that the ratio becomes 1 : 1 and then 2 : 1 or greater. This theoretically allows for more even distribution of inspired gas to lung units with longer time constants. IRV can be produced in volume-controlled ventilation by an end-inspiratory pause, or by a slow or decelerating inspiratory flow rate. Because with IRV expiration is usually not complete before the next inspiration commences, air trapping occurs and intrinsic PEEP develops. Inverse ratio ventilation is also commonly employed with pressure-controlled ventilation (see below) by directly prolonging the inspiratory time.

Pressure-controlled ventilation (PCV)

This mode is currently becoming more popular because of its theoretical advantages compared to volume-controlled ventilation.[82] These include:

1. Peak airway pressure is lower for the same mean airway pressure
2. Regional overdistension may be prevented
3. Decelerating flow rate may be beneficial
4. Alveolar ventilation may be improved
5. Intrinsic PEEP develops regionally

The inspiratory pressure level is preset and expiration is time cycled. It is possible, but not usual, to let patients trigger pressure-controlled breaths. Flow rate is determined by algorithms within the ventilator and is initially high to pressurize the respiratory circuit, and then decelerates. It should be noted that this mode is completely different to, and must be distinguished from, pressure-cycled ventilation and is available only on the newest generation of mechanical ventilators. Representative waveforms are shown in Fig. 66.10.

In clinical practice, for those patients with the most severe forms of respiratory failure, PCV is frequently used to produce intrinsic PEEP by prolonging the inspiratory time. Modest amounts of extrinsic PEEP, usually 5–7.5 cmH_2O, are applied to stabilize the remaining relatively normal lung units. This method of ventilation is uncomfortable for the patient, and this has to be taken into account when prescribing sedation and muscle relaxants.

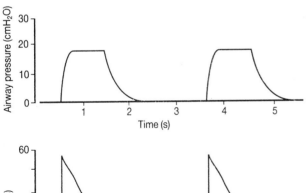

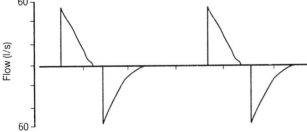

Fig. 66.10 Waveforms seen with pressure-controlled ventilation. As with pressure-cycled ventilation, and also with PSV, the flow rate is initially high and then decelerates. However, in PCV the preset pressure is held until expiration, which is time cycled.

High-frequency techniques

These vary from a rate of 60 to as high as 3000 ventilatory cycles per minute. They are arbitrarily subdivided into high-frequency positive pressure ventilation, high-frequency jet ventilation and high-frequency oscillation. Tidal volumes are often less than physiological dead space and there are various theories as to the mechanism of gas exchange, the most popular being brownian movement. It was originally thought that these modes would be superior to conventional mechanical ventilation in severe cases of ARDS but controlled trials have not confirmed this.[83] A new generation of machines capable of ultra-high-frequency jet ventilation at frequencies near the resonant frequency of the lung are, however, showing some promise.[84]

Continuous positive airway pressure (CPAP)

This is a means of providing the beneficial respiratory, and in some cases cardiovascular, effects of PEEP to a patient who is breathing spontaneously. CPAP by facemask may be used as a means of avoiding tracheal intubation and formal mechanical ventilation in patients whose respiratory failure may be expected to recover relatively rapidly, such as is seen in acute cardiogenic pulmonary oedema. The technique has two major requirements: a tight-fitting facemask to provide an air-tight breathing system, and a pressure generator that can supply a high inspiratory flow rate. In the correct circumstances CPAP can be valuable, but the patient should be closely supervised. Conscious patients find the facemask uncomfortable and prolonged use can cause oedema and necrosis of the soft tissues of the face. A recent technical advance is the ability to provide nasal CPAP, which patients may find more comfortable. In sick patients, airway protection cannot be guaranteed and gastric distension, which may occur with CPAP, can predispose to aspiration. In a patient recovering from a period of mechanical ventilation, CPAP with pressure support can be a useful adjunct to weaning.

Airway pressure release ventilation (APRV)

This mode of assisted ventilation uses two levels of CPAP, with the ventilator switching between them. This manoeuvre can also be synchronized with the patient's inspiratory efforts on some ventilators. The patient can breathe spontaneously at both CPAP levels, but the release of CPAP from a supra-ambient level to a lower level augments alveolar ventilation and CO_2 clearance. The advantage of this mode is the low peak airway pressure, but pulmonary overdistension must be prevented. The role of this technique in severely hypoxic patients is not yet clear, but early reports have been encouraging in mild acute respiratory failure.[85]

Extracorporeal techniques

Extracorporeal membrane oxygenation (ECMO) can produce survival in neonates and infants[86] who do not respond to conventional mechanical ventilation. It has also been successful in adults but generally the results are disappointing.[87] A prospective study[88] showed no improvement in mortality when patients receiving conventional ventilation were compared to patients who eventually underwent extracorporeal CO_2 removal. A major observation in that study was that the mortality in the control group was 58 per cent, in patients with a historical mortality of 12 per cent. It is suggested that perhaps the outcome from the most severe form of ARDS is improving.

Other techniques

Turning patients with severe lung injury into the prone position frequently improves their PaO_2 owing to reduction in shunt fraction. This may be due to a decrease in gravitational compressive forces and thus reduction in closing volume.[89] The manoeuvre perhaps needs to be more widely appreciated, for it is not as complex as one would imagine.

Perfluorocarbon-associated gas exchange (PAGE) has now reached the stage when clinical trials are likely to start.[90] In this technique, medical grade perfluorocarbons are instilled into the lung and volumes up to 30 ml/kg have been used to completely fill the functional residual capacity. This removes the air–fluid interface, and hence surface tension in the alveoli, allowing the recruitment and stabilization of alveolar units. Conventional ventilation is continued. The results of human studies are awaited with interest.

Establishing mechanical ventilation for acute respiratory failure

There are few if any published data on this aspect of intensive care practice and therefore a pragmatic approach is described. The basic principles apply to most forms of severe acute respiratory failure but, as described in detail below, the exact mode of ventilation and the ventilator settings may vary from one clinical circumstance to another. The most critical period is the introduction of controlled ventilation. This takes place in three phases: securement of the airway, initial ventilator settings and their subsequent adjustments.

Establishment of the airway

The most popular technique is oral tracheal intubation using a rapid sequence induction. Skilled assistance must always be available and pulse oximetry is indispensable. In the spontaneously breathing patient an alternative is awake intubation by the oral or nasal route. The latter may be useful if there is no suspicion of a fracture of the base of the skull[91] when a view of the normal landmarks is likely to be unobtainable by laryngoscopy; for instance, after upper airway burns with severe oropharyngeal oedema. Fibreoptic techniques can be useful, provided that the patient is not in extremis. If a potential upper airway obstruction is suspected, induction by inhalational anaesthesia may be necessary. Inability to intubate, or upper airway obstruction, may require a needle or surgical cricothyroidotomy. The final resort is tracheostomy, as both surgical and percutaneous techniques are difficult in a conscious, agitated and breathless patient. Hypoxic patients do not tolerate the supine position and may have to be intubated in a semi-erect position. Although pre-oxygenation using a tight-fitting anaesthetic facemask is essential, the reduced functional residual capacity often found in acute respiratory failure may limit its effectiveness. The choice of intravenous induction agent has to be made individually but etomidate is frequently chosen. Obviously, a completely obtunded patient may be intubated without anaesthesia but poor technique or case selection may increase the likelihood of awake intubation. In addition to the potentially unpleasant experience for the patient, operator and assistants, agitation, worsening hypoxia and coughing and gagging during laryngoscopy may stimulate vomiting. The risk of aspiration is still present, even in relatively awake patients, and effective cricoid pressure is necessary, especially in the presence of gastric dilatation. Cardiac arrhythmias, hypertension and cardiac ischaemia may also be precipitated. Suxamethonium (succinylcholine) is the drug of choice for muscle relaxation but may be contraindicated if serum K^+ is elevated or the patient has neuromuscular disease. In these circumstances, a non-depolarizing agent such as rocuronium or vecuronium may be considered, especially if it is known that the patient has been intubated previously. Hypotension due to induction agents may be minimized by fluid therapy, as the majority of patients with acute respiratory failure are hypovolaemic on presentation to the ICU. Not infrequently, judicious, small doses of a vasopressor are needed at this stage.

Initial ventilator settings

With modern ventilators, prolonged deep sedation is necessary only under specific circumstances (e.g. permissive hypercapnia, high levels of PEEP and IRV). However, it is common practice to sedate and if necessary administer muscle relaxants temporarily, until the ideal mode of ventilation and the individual ventilator settings are determined and circulatory stability is achieved. A reasonable starting point is a respiratory rate of 10 breaths/min, tidal volume of 10 ml/kg, inspiratory flow rate of 60 l/min and an I : E ratio of 1 : 2. Many clinicians also add a low level of PEEP at this stage, although this may not be indicated if the patient is hypovolaemic or if intrinsic PEEP is likely to develop. The FIO_2 should initially be high, at least 0.6 and possibly 1.0 depending on the clinical circumstances and pre-intubation pulse oximetry or PaO_2 when available. The position of the tracheal tube should be checked clinically and ultimately by chest x-ray, and one should be alert for the presence of a pneumothorax.

Definitive ventilator settings

Following arterial blood gas analysis, minute volume should be adjusted according to the $PaCO_2$ and FIO_2 adjusted according to the PaO_2. An acutely elevated $PaCO_2$ should not be normalized rapidly, but over at least 12 hours. Patients with acute or chronic CO_2 retention and respiratory acidosis should be ventilated to a normal pH, not to a normal $PaCO_2$. The danger of the latter is in inducing a severe metabolic alkalosis from chronic pre-existing bicarbonate retention, leading to cardiovascular and neurological sequelae, which can be profound.[92] The issue of oxygen toxicity is controversial but many clinicians reduce the FIO_2 after the period of resuscitation to keep PaO_2 at a minimum of 10–12 kPa (75–90 mmHg). When adjusting minute volume it must be remembered that most patients requiring mechanical ventilation will have a high dead space and it is not imperative to strive for a normal or low $PaCO_2$ except in specific circumstances such as raised intracranial pressure. An important practical point is that some ventilators such as the Siemens Servo 900 series are effectively minute volume dividers, minute volume is set directly and tidal volume selected by adjusting the respiratory rate. In practical terms, this means that, if minute volume is left unchanged and respiratory rate reduced, tidal volume will be increased. Alveolar ventilation may actually be increased depending on the VD/VT ratio.

Once the patient has been stabilized, usually SIMV with pressure support is instituted and decisions concerning the use of additional modes such as PEEP, PCV or IRV can be made. Before sedation has worn off, a nasogastric tube should be inserted, initially to empty the stomach and eventually to provide enteral nutrition or local antacid prophylaxis. Arterial and central venous catheters can also be conveniently inserted at this time if not already in place.

Two conditions have been chosen that require different approaches to mechanical ventilation, and are good examples of the versatility of the latest generation of ventilators.

Mechanical ventilation in severe acute asthma

The term 'severe acute asthma' is preferable to the traditional one of 'status asthmaticus' because two types of patients who may require IPPV are now recognized. The usual patient is a known asthmatic who has had repeated hospital admissions, intermittent courses of steroids and finally is admitted with severe airflow limitation as judged by clinical signs and a peak expiratory flow rate persistently less than 120 l/s. The PaO_2 can usually be maintained by increases in FIO_2. Despite optimal supportive therapy the patient shows signs of exhaustion with a rising $PaCO_2$ and falling pH, and eventually requires intubation. During intubation the tip of the tube should not irritate the carina as this may cause disastrous exacerbations of the bronchospasm.

There is, however, another type of asthmatic who may be referred for intensive care. These are patients who suffer sudden, severe, overwhelming asthma, often progressing to apnoea within hours or even minutes of the onset of symptoms, described by Perret and Feihl[93] as sudden asphyxia in asthma. Most are male under 30 years of age, often comatose on admission with no audible breath sounds, and bilateral pneumothoraces may be suspected. There are striking abnormalities of arterial blood gases on admission with a mean pH of 6.99 and $PaCO_2$ of 15 kPa (113 mmHg). In such patients there is no place for a trial of medical treatment; immediate intubation is indicated. Even in traditional status asthmaticus it can be difficult to interpret a normal value of $PaCO_2$ in isolation: it may be returning to normal as a sign of recovery or may indicate the onset of respiratory muscle fatigue. The blood gas patterns for these two distinct groups of patients are shown in Fig. 66.11.

A major problem with mechanical ventilation in severe asthma is that the high inspiratory resistance is not simply due to bronchospasm, but is also due to mechanical plugging of airway by secretions. Expiratory flow obstruction leads to overdistension with high levels of intrinsic PEEP which are not apparent from the displayed ventilator pressures unless specifically measured. With normal ventilatory tidal volumes into an already hyperinflated chest, high peak inspiratory and end-inspiratory pressures are generated and the patient is at high risk of barotrauma. A prolonged expiratory period is needed to allow for passive exhalation to occur in the presence of the severe limitation of expiratory airflow. Preferably, a ventilator should be used with display of pressure and flow waveforms to confirm that expiration is complete before the onset of the next inspiration. Failing this, close clinical observation at the bedside is essential. In order to allow a reduction in lung volume, relative hypoventilation

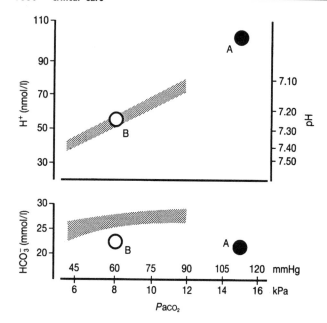

Fig. 66.11 Acid–base status of patients with sudden asphyxial asthma (point A) compared to those with progressive respiratory failure from muscle fatigue and exhaustion (point B).

has been described.[94] Low tidal volumes of 5–7 ml/kg are used with an I : E ratio of at least 1 : 4. This requires a respiratory rate of 6 breaths/min or lower. Full supportive measures, including volume expansion, corticosteroids, bronchodilators, physiotherapy and antibiotics, should be administered as indicated. In extreme cases, halothane or ether may be administered, but ether, in particular, poses technical problems in a general ICU. There is continuing debate about the role of controlled extrinsic PEEP to maintain airway patency in severe asthma.[95]

Mechanical ventilation in ARDS

The approach to mechanical ventilation in ARDS has undergone a complete revolution since the days when the recommended minimum tidal volume was 10–12 ml/kg or higher.[96,97] In patients with ARDS, the distribution of alveolar shadowing seen on the chest radiograph is not homogenous, and a number of thoracic CT scanning studies have confirmed that the consolidated areas tend to settle with gravity.[98] In severe ARDS, only about 30 per cent of the lung may actually be aerated, and the lung should be viewed as small and not merely as stiff. The delivery of normal tidal volumes into such a small alveolar volume will lead inevitably to alveolar rupture. It is now appreciated that conventional IPPV, with tidal volumes of 10 ml/kg or greater, produces damaging effects. Hickling[99] used lower tidal volumes and pressure-limited ventilation, demonstrating improved outcome in a retrospective study of 50 patients as predicted by APACHE II score (16 per cent vs 40 per cent). Many centres now reduce tidal volumes to keep

end-inspiratory or plateau pressure below 35 cmH$_2$O, and if necessary allow the PaCO$_2$ to rise. Arterial pH is normally maintained by renal mechanisms but occasionally bicarbonate has to be given. Although extrinsic PEEP may dramatically improve PaO$_2$ in severe ARDS, it frequently leads to a reduction in cardiac output.[100,101] This reduction can be modified but not prevented by volume loading, and additional inotropic support is often required. The use of inverse ratio ventilation[102] to produce intrinsic PEEP allows a reduction in extrinsic PEEP and hence in peak airway pressures. There is an increasing tendency to combine a modest amount of extrinsic PEEP with tidal volume limitation, often using PCV and some degree of inverse ratio ventilation. The level of intrinsic PEEP must be monitored closely in these circumstances but few ventilators are able to do this, which is perhaps one reason that PCV has yet to achieve more widespread acceptance.

Complications of mechanical ventilation

The major problems associated with positive pressure ventilation are circulatory depression and pulmonary barotrauma. The supine pneumothorax can be difficult to recognize and a high 'index of suspicion' is needed. The complications of prolonged tracheal intubation include damage to the larynx and vocal cords. Transient hoarseness of the voice is usual when a tracheal tube is *in situ* for more than 48 hours, and is exacerbated by movement of the tube irritating the vocal cords and laryngeal mucosa. These effects can be minimized by nasotracheal intubation, but this can be complicated initially by haemorrhage and in the long-term by purulent maxillary sinusitis.

Weaning from mechanical ventilation

Numerous indicators for successful weaning have been proposed although few are useful in clinical practice. If the patient develops rapid shallow breathing during weaning (frequency/tidal volume ratio > 100), weaning is unlikely to be successful in the long term.[103] It is more important to recognize those patients in whom weaning should not be attempted, particularly those with high cardiac filling pressures, as cardiac failure may be precipitated when positive intrathoracic pressure is removed.[104] Other factors to consider are given in Table 66.11.

Tracheostomy can be complicated by haemorrhage, which is rarely serious but can occasionally be catastrophic. In the long term, severe tracheal stenosis may occur in about 1 per cent of patients.[105] Tracheostomy is now frequently performed in the ICU, using the percutaneous technique.[106] Evidence is accumulating that early

Table 66.11 Important factors to consider during weaning

Control underlying illness, pain, infection and fever
Optimize cardiac function
Optimize lung function and ensure that $Pa\text{CO}_2$ is normal for the
    patient
Ensure normal serum potassium, phosphate
Recognize that patients may have muscle atrophy, myopathy or
    neuropathy
Avoid excess CO_2 production from parenteral feeding with glucose
Avoid exhausting patient and causing sleep deprivation
Consider tracheostomy

and late complications are less frequent and less severe than with conventional tracheostomy.

Many patients may be weaned rapidly, but for those who have undergone a long period of ventilation the newer generation of mechanical ventilators now allow more controlled and less stressful weaning than simply breathing on a T-piece. In the latter stages there is usually a transition from SIMV with pressure support, by reducing the mandatory breaths and the degree of pressure support, to CPAP. When the patient can clear secretions and swallow safely without the need for an artificial airway, the tracheal or tracheostomy tube may be removed.

Circulatory failure and shock

Shock has classically been recognized as a complex of symptoms and signs. There is agitation, apprehension, dyspnoea, thirst, tachycardia, weak peripheral pulses, low blood pressure, collapsed peripheral veins and cold peripheries.[107] Blood gas analysis may reveal an acidosis. However, shock is better defined as failure of the circulation either centrally or peripherally to meet the metabolic demands of the tissues. This may be due to an absolute reduction in cardiac output, but may also be due to a cardiac output of whatever level that is inadequate or maldistributed. This latter situation can be detected only by direct measurement or inferred from other measurements such as mixed venous oxyhaemoglobin saturation and blood lactate.

The arterial lactate concentration represents a balance between production and clearance. Although lactate rises in shock states, blood levels correlate poorly with $S\bar{v}\text{O}_2$, $D\text{O}_2$ and $\dot{V}\text{O}_2$ in critically ill patients. Arterial lactate may be normal in shock, even when an acidosis is present. An elevated lactate in a shocked patient indicates severe or prolonged circulatory failure, and failure of an elevated lactate to fall with treatment is a poor prognostic sign. The mildly elevated arterial lactate levels seen in hypermetabolic states in the ICU may not indicate tissue hypoxia.[108]

Shock is usually classified according to the underlying aetiology into hypovolaemic, cardiogenic, septic and obstructive, and treatment directed to the expected physiological disturbances. Simple monitoring can, however, misinform. When systemic vascular resistance is very high, as in haemorrhage or cardiogenic shock, blood pressure taken using a sphygmomanometer frequently underestimates that obtained from an indwelling arterial catheter.[109] The reverse may be true when systemic resistance is very low, as in septic or anaphylactic shock. The core–peripheral temperature difference has been shown to be an inadequate way of estimating the systemic vascular resistance, both in absolute terms and following therapeutic intervention.[110] Another method may be to analyse the results of invasive cardiovascular monitoring, allowing the specific cardiovascular abnormalities in the systemic and pulmonary circulations to be defined precisely. Treatment with volume replacement, vasodilators, inotropes or vasopressor can then be tailored accordingly. Both approaches to diagnosis have limitations, but immediate resuscitation based on measurements must be accompanied by a precise diagnosis; it would be irrelevant to try to reverse the inadequate tissue oxygenation seen in septic shock, for instance, without identifying the source and nature of the causative micro-organism(s) and providing appropriate surgical and/or medical treatment.

Frequently the pathophysiological findings are not in accord with the text-book descriptions. Generally, patients in septic shock have a high cardiac output, but only after hypovolaemia and hypoxia have been corrected. In some cases, for instance associated with long-standing bacterial peritonitis, biliary peritonitis or infarction of the large bowel, the systemic haemodynamic changes may simulate those of cardiogenic shock, especially in elderly patients with pre-existing cardiorespiratory disease.[111] It is sometimes very difficult to classify a patient as having cardiogenic or septic shock on the basis of clinical examination alone. Some patients with cardiogenic shock, possibly up to 50 per cent, may be suffering from hypovolaemia, in many cases due to diuretic therapy. Rigid diagnostic categories can also be unhelpful. A patient with multiple trauma may have chest injuries associated with myocardial contusion, a tension pneumothorax, plus

Table 66.12 Classification of shock by systemic haemodynamics

	CO	PAWP
Hypovolaemic	↓	↓
Septic	↓ ↔ ↑	↓ ↔ ↑
Cardiogenic	↓	↓ ↔ ↑
Obstructive	↓	↓ ↔ ↑

Table 66.13 A comparison of three artificial colloids

	Modified fluid gelatin (Gelofusine)	Hetastarch (Hespan)	Dextran 70 in saline
Concentration	4%	6%	6%
Weight average MW	30 000	450 000	70 000
Number average MW	22 600	70 000	35 200
Osmolarity (mmol/l)	279	310	320
pH	7.4	5.5	6.3
Sodium (mmol/l)	154	154	154
Potassium (mmol/l)	< 0.4	0	0
Chloride (mmol/l)	125	154	154
Calcium (mmol/l)	< 0.4	0	0
Daily maximum (ml)	No limit	1500	1500

MW = molecular weight

blood and fluid loss due to pelvic trauma and thus cannot be allocated to a single category. Despite all these limitations, a working guide to classification is shown in Table 66.12. These basic categories will be discussed individually whilst recognizing and emphasizing the overlaps in circulatory and pulmonary pathophysiology that occur between and within the groups.

Hypovolaemic shock

This is the simplest form of shock to consider. Venous return and, therefore, cardiac preload are reduced because of loss of circulating blood volume due to either internal or external haemorrhage or fluid losses associated with diarrhoea, vomiting, major burns or crush injury. A clinical diagnosis is often obvious, for example diabetic ketoacidosis or ruptured aortic aneurysm, and the amount and exact composition of fluid administered reasonably well estimated by clinical responses such as changes in conscious level, heart rate, non-invasive blood pressure and urine output.

Blood loss may well require blood transfusion, although blood is often supplied as plasma reduced and initial volume replacement will be necessary. There is a continual debate regarding the use of crystalloid or colloid for resuscitation.[112] In general terms, it must be accepted that crystalloids result in a smaller increase in circulatory plasma volume than equivalent volumes of colloid. Equally, there are differences in the chemical composition of individual fluids, for instance 0.9 per cent saline or Hartmann's solution, and in the physicochemical properties of the artificial colloids such as gelatins, starches and dextrans which may influence choice of fluid. Three popular types of colloid are compared in Table 66.13. Plasma volume replacement is inevitably accompanied by a degree of haemodilution and the fall in haemoglobin concentration has to be monitored and corrected.[113,114]

Even apparently simple hypovolaemic shock may be complicated by pre-existing chronic cardiorespiratory disease requiring treatment, and the underlying diagnosis will need to be confirmed because specific therapy may also be required. Examples include the need for oral vancomycin therapy for pseudomembranous colitis due to *Clostridium difficile* infection, physiological doses of corticosteroids in acute hypoadrenalism and surgery to control gastrointestinal haemorrhage.

Septic shock

The basic principles of airway, breathing and circulation apply to the management of septic as to all other forms of shock. True septic shock in patients referred to the ICU is invariably complicated by a degree of respiratory insufficiency ranging from mild to the severest forms of acute lung injury. High concentrations of oxygen should be administered and often the airway will need to be secured by tracheal intubation. Patients will require fluid replacement which may be up to 10 litres initially, and this rate of fluid therapy may continue until the patient is stabilized. Adequacy of resuscitation should be guided by appropriate monitoring, which in severe sepsis may include estimation of cardiac output. If the clinical manifestations of shock are adequately reversed, no further resuscitation is required and definitive management of the infection should proceed with appropriate diagnostic procedures such as swabs and blood cultures, serological tests for unusual infections such as leptospirosis and Legionnaire's disease, plus radiological investigation or surgical procedures as indicated. All necessary surgery should be performed promptly after resuscitation and administration of antibiotics. As the same clinical and haemodynamic picture is produced whatever the infecting organism, consultation with an experienced clinical microbiologist may be helpful when the causative organism cannot be immediately identified.

Table 66.14 Cardiorespiratory abnormalities in septic shock

Low systemic vascular resistance
High cardiac output, which may be inadequate
Impaired uptake of oxygen at the periphery
Hypoxaemia refractory to an increased FIO_2
Absolute or relative pulmonary artery hypertension
Right atrial pressure relatively higher than pulmonary artery
    wedge pressure

The cardiorespiratory abnormalities in septic shock are summarized in Table 66.14.

Patients who remain hypotensive after correction of hypoxaemia and/or hypovolaemia, or who have clinical and/or laboratory signs of impaired tissue perfusion such as oliguria, rising serum creatinine or persistently elevated blood lactate, will require vasoactive drugs to restore mean systemic arterial blood pressure and increase and/or maintain cardiac output and tissue oxygen delivery. Drugs that act as vasopressors in this situation include dopamine, adrenaline (epinephrine), noradrenaline (norepinephrine) and phenylephrine.[115] An unknown amount of infused dopamine is rapidly converted to noradrenaline and the effects of dopamine may be identical with those of noradrenaline.[116] When vasopressors are used, they are titrated to a mean systemic arterial pressure of 80 mmHg, or possibly higher in atherosclerotic patients. If calculated systemic vascular resistance is less than 800 dyn s/cm^5, excessive vasoconstriction and ischaemia of vital organs is avoided.[117,118] In fact, the increase in blood pressure with controlled vasopressor therapy improves renal function.[119] Directly or indirectly, all these agents (except phenylephrine, which is a pure α-agonist) may increase cardiac output by their β effects but the changes are unpredictable and often there are reflex falls in cardiac output as a result of vasoconstriction. A relatively pure β-agonist such as dobutamine may be a useful additional agent to maintain cardiac output. It is important to realize that the use of all inotropes is potentially dangerous, as they may cause sinus tachycardia, dysrhythmias and myocardial ischaemia. Because of its β_2 effects, dobutamine may cause peripheral vasodilatation and hypotension if used as the initial agent, especially in hypovolaemic subjects.

The myocardial depression seen in septic shock is multifactorial, with myocardial depressant substances probably playing the most important role. One aspect of therapy is the maintenance of an adequate coronary (artery) perfusion pressure (CPP) to ameliorate myocardial dysfunction.

CPP = aortic diastolic pressure − ventricular end-diastolic pressure

In septic shock there is a low aortic diastolic pressure, and ventricular dysfunction may be aggravated by poor coronary perfusion. This may be especially so for the right ventricle because patients with sepsis may have a relatively high central venous and right ventricular end-diastolic pressure.

If an adequate cardiac output cannot be obtained, other agents such as dopexamine or a phosphodiesterase inhibitor such as milrinone may be added.[115] The phosphodiesterase inhibitors may be synergistic with β-agonists, although they may exhibit profound chronotropic and vasodilating properties and have a relatively prolonged half-life. There are some data to suggest that dopexamine selectively improves splanchnic blood flow, but the issue of a selective increase in renal blood flow in response to dopamine remains controversial in critically ill patients.[120] The responses to vasoactive drugs are extremely variable and each agent has to be titrated to effect, carefully guided by clinical and haemodynamic monitoring. The dosages need to be continually adjusted to maintain the desired effect but should be withdrawn as the need for them diminishes. It is not uncommon for patients to depend on sympathomimetic agents for several days or even weeks. Increased recognition of the high metabolic requirements of these patients, and the need for a high cardiac output and thereby increased oxygen delivery, has led to the institution of therapeutic protocols aimed at achieving optimal values for these variables.[121–123] Corticosteroids are now no longer employed routinely but may be indicated in physiological doses if hypoadrenalism is suspected.

Cardiogenic shock

This is characterized by a low cardiac output state following, in the majority of cases, acute myocardial infarction.[124] The pulmonary artery wedge pressure is not always critically elevated,[125,126] and there are other causes of cardiac dysfunction that can produce a low cardiac output state; examples are shown in Table 66.15. It is important to exclude these cases because specific medical or surgical treatment may be needed; for instance, pericardial aspiration for pericardial effusion. The following discussion focuses on the intensive care aspects of shock following acute myocardial infarction.

Table 66.15 Causes of cardiogenic shock other than acute myocardial infarction

Acute myocarditis
Critical valvular stenosis, especially aortic
End-stage cardiomyopathy
Pericardial tamponade
Constrictive pericarditis

ECG shows features of inferior myocardial infarction

This is due to disease of the right coronary artery and is associated with predominant, but not isolated, infarction of the right ventricle. The jugular venous pressure is elevated and may move paradoxically with respiration (Kussmaul's sign), with an increase on inspiration. The right atrial pressure is equal to or greater than the pulmonary artery wedge pressure. Characteristic waveforms are seen on insertion of the pulmonary artery flotation catheter, the right atrial and ventricular waveforms being virtually indistinguishable in extreme cases. It is sometimes naively assumed that the wedge pressure is always low in this condition and that the low cardiac output will respond to volume expansion. This is not always the case unless critical hypovolaemia is iatrogenically induced by ill-advised use of diuretics or vasodilators. As the right ventricle contracts poorly, if at all, blood flow across the pulmonary circulation is often passive and depends on the pressure gradient between the right and left atria. There is frequently left ventricular involvement,[127] and the haemo-dynamic response to excessive volume loading may be disappointing or even undesirable, especially when central venous pressure exceeds 16–18 mmHg.[128] Inotropic support is usually required. Hypoventilation and hypoxaemia are common in all acute low cardiac output states even in the absence of pulmonary oedema, and mechanical ventilation may be necessary. Life-threatening arrhythmias and conduction disturbances are common and may require the use of antiarrhythmic drugs, cardioversion or temporary transvenous pacing.[124] Non-invasive confirmation of right ventricular infarction can be obtained within the first 2 hours of the episode by use of a right precordial electrode (V$_4$R) which may show S–T elevation or a Q wave but this sign is transient. With good basic management, survival can be as high as 50–70 per cent, and patients usually return to a good quality of life.

ECG shows features of anterior myocardial infarction

This is usually due to proximal or widespread occlusion of the left coronary arterial tree, often in association with widespread chronic coronary artery disease. At least two-thirds of the left ventricle is affected and there is usually severe cardiogenic pulmonary oedema. Initially the wedge pressure is ≥ 25 mmHg[125] but may be modified by diuretic or vasodilator therapy. CPAP by facemask can be useful both by ensuring a high F_{IO_2} and by providing a degree of positive intrathoracic pressure.[129] When the wedge pressure is excessively high with a failing heart, a positive intrathoracic pressure may actually increase cardiac output[130] and mechanical ventilation with PEEP should

be used if necessary. This will also reduce the oxygen demand of breathing and reduce cardiac work. Ventricular arrhythmias are particularly common and often resistant to pharmacotherapy or cardioversion. These measures are doomed to failure in the presence of persisting acidosis or hypoxaemia. Frequently the combination of high preload and excessive vasoconstriction further impairs cardiac output, and vasodilator therapy may be useful in reducing pulmonary oedema and increasing cardiac output, although attention must be paid to effective peripheral and coronary artery perfusion pressure.

Invasive measurement of blood pressure and cardiac output is necessary because clinical responses frequently lag behind haemodynamic changes. Most patients will have received thrombolytic therapy and this may make central venous access hazardous.

In either form of cardiogenic shock, echocardiography will exclude mechanical complications such as acute mitral regurgitation, ventricular septal defect and limited perforation of the free ventricular wall. Various attempts may be made to achieve reperfusion by angioplasty or acute coronary surgery but results in true cardiogenic shock have been poor. Mechanical circulatory assist may be useful as a 'bridge' to surgery. If inotropic support is used, then, in contradistinction to septic shock, only low normal values of cardiac output are sought, with a mixed venous saturation above 60 per cent.[127] Dobutamine is the initial agent of choice, possibly with the addition of a phosphodiesterase inhibitor. As in septic shock, the haemodynamic responses are unpredictable.

Obstructive shock

This is due to a mass or pressure effect obstructing venous return to the right or left side of the heart, or impeding outflow from the ventricles. Common causes are listed in Table 66.16.

Rapid diagnosis is essential, and definitive investigation is as important as resuscitation. Most of the common clinical conditions can be identified immediately by readily available non-invasive techniques such as electrocardiography, chest x-ray and echocardiography. Tension pneumothorax is the most important to exclude in patients receiving mechanical ventilation. It is often difficult to

Table 66.16 Common causes of obstructive shock

Tension pneumothorax
Massive pleural effusion/haemothorax
Massive pulmonary embolism
Pericardial tamponade
Constrictive pericarditis
Inferior vena caval compression
Acute aortic dissection

recognize on a supine portable chest film and the cardiorespiratory effects are not directly proportional to the visualized size of the pneumothorax.[131] Valuable clinical signs are diminished air entry and hyper-resonance to percussion on the affected side. The cardiorespiratory effects have, surprisingly, been documented in only a small number of cases, and consist of tachycardia and severely reduced cardiac output due to reduction in stroke volume despite apparently adequate filling pressures. Hypotension may not be as dramatic as is assumed because of a compensatory increase in systemic vascular resistance. Once a tension pneumothorax has been diagnosed, a thoracic drain should be inserted, using the fifth intercostal space in the mid-axillary line, and connected to an underwater seal drain.

Massive pulmonary embolism is a relatively uncommon cause of shock in ICU practice, but should be suspected in high-risk patients. The clinical picture is of circulatory collapse, tachypnoea, cyanosis, grossly elevated jugular venous pressure and dyspnoea that is relieved by lying the patient flat. Auscultation may reveal evidence of broncho-spasm in a small percentage of patients. There is a right ventricular third heart sound in up to 70 per cent of cases, though this may be difficult to hear. Most patients have an abnormal ECG but the changes are non-specific, with classic features of acute right heart strain seen in only 15 per cent of cases.[132] Blood gas abnormalities are always present, with hypoxia and hypocapnia the most consistent findings. Depending on the severity of the peripheral circulatory failure and reduction in oxygen delivery, there may be varying degrees of respiratory alkalosis or metabolic acidosis. The diagnosis can be confirmed by ventilation/perfusion lung scanning or pulmonary angio-graphy but in clinical practice there may be logistical problems in arranging these investigations urgently. If invasive monitoring is undertaken, the usual features are very high right atrial, right ventricular and pulmonary artery pressures, low cardiac output and high systemic vascular resistance. (Care must be exercised in interpreting the SVR in pulmonary embolus and right ventricular infarction, as the CVP is usually markedly elevated and this will tend to produce a spuriously low SVR.) Management consists of respiratory support to relieve hypoxia and reduce the work of breathing. Circulatory support, guided by haemodynamic measurements, usually involves fluid loading and inotropic support. Definitive therapy with thrombolytic agents is now given routinely and there is some evidence that these agents bring about a more rapid return of pulmonary artery pressures and cardiac output to normal levels, but evidence of improved survival is lacking. Surgical intervention is reserved for persistent shock despite all the measures described above. Occasional survival has been reported with immediate thoracotomy and pulmonary embolectomy when pulmonary embolus presents as cardiorespiratory arrest.

Cardiac tamponade and severe, rapidly progressing constrictive pericarditis are treated by drainage and pericardectomy respectively. In the interim, circulatory and respiratory support may be provided. In particular, many patients with tamponade may respond temporarily to careful volume loading. Arrhythmias, especially atrial fibrillation, are very common (up to 30 per cent of cases). Although echocardiography is the investigation of choice, the diagnosis can also be made by pulmonary artery catheterization when there is equalization of right atrial, right ventricular, mean pulmonary artery and wedge pressures.

The emergency treatment of aortic dissection includes recognition of haemorrhage into pleural or retroperitoneal cavities and blood transfusion. Usually control of hypertension is needed to limit further progression of the dissection.

In all the above cardiovascular emergencies expert advice should be sought urgently.

Renal failure

Acute renal failure remains common in the ICU: approximately 20 per cent in published reviews. Most cases seen in the ICU are related to surgery,[133] and the associated risk factors are shown in Table 66.17.

Oliguria may be defined as a urine output below 20 ml/h for 2 consecutive hours, and is most commonly due to renal hypoperfusion. Total anuria is usually mechanical in origin. Rarely should oliguria be attributed to lack of diuretic therapy. Renal hypoperfusion may not produce oliguria if the ability to concentrate is poor, as seen in the elderly, or in those with prior renal disease or on diuretic therapy. The factors listed in Table 66.18 should be considered, as usually more than one will be present.

A careful history and physical examination should be made for evidence of hypovolaemia such as thirst, tachycardia, hypotension or postural drop. Evidence of heart failure or circulatory overload should be sought, including tachypnoea with lung crepitations, a raised jugular venous pressure and peripheral oedema, and a third heart sound.

It is essential to exclude obstruction: check that the bladder is not palpable (!) and look for causes of obstruction such as an obstructed catheter. Catheterize the bladder if not done already. Ultrasound and x-rays can be used to check for kidney size and parenchymal pattern, hydro-nephrosis or renal vein thrombosis, and for the presence of calculi.

A tense abdomen, especially from intra-abdominal

Table 66.17 Factors related to the development of acute renal failure in the ICU

Patient factors	Perioperative factors
Advanced age	Hypotension
Aortic surgery	Hypovolaemia:
Atherosclerosis	Diuretic therapy
Biliary surgery/jaundice	Surgical oedema
Cardiac surgery	Preoperative starvation
Chronic renal disease	Gastric aspiration/vomiting
Cirrhosis	Peritonitis/ileus/obstruction
Diabetes	Diarrhoea/bowel preparation
Heart failure	Prolonged tissue exposure
Hypertension	Blood loss
Myeloma	Hypoxia
Nephrotoxic drugs	Tissue damage and inflammation:
Pre-eclampsia/eclampsia	Ischaemia and reperfusion
Sepsis	Major burns
	Multiple fractures
	Muscle breakdown
	Pancreatitis
	Transfusion reactions

bleeding following aortic or hepatic surgery, will cause a fall in urine output.[134] Rarely, one may see evidence of renal disease, or systemic disease associated with nephritis, with fever and a vasculitic rash. More commonly there will be evidence of poor perfusion with diminished femoral pulses and perhaps a vascular bruit over the abdomen. With brown or cloudy urine, suspect myoglobinuria.

Serial determination of creatinine clearance is the most sensitive test for detecting the onset of renal dysfunction and can be easily performed in the ICU.[135] Serum creatinine may not reflect glomerular filtration rate in patients with little muscle mass. Creatinine clearance can be reliably estimated on a 2-hour urine collection.[136] Isotope investigations, angiography and renal biopsy may occasionally be indicated but specialist advice should be sought.

Urinalysis should be undertaken for blood, protein and myoglobin. Urine microscopy, as shown in Table 66.19, may give clues to the underlying pathology. Urinary indices of function are relatively imprecise (Table 66.20) and become inaccurate after mannitol and loop diuretics. Urinary osmolarity and sodium concentration are insensitive discriminators between prerenal azotaemia and acute renal failure.

The fractional excretion of sodium (FE_{Na}) is a more accurate guide to renal integrity:

$$FE_{Na} = \frac{U_{Na}}{P_{Na}} \div \frac{U_{Cr}}{P_{Cr}}$$

(U, urine; P, plasma)

Invasive monitoring should be considered early in high-risk groups, especially if there is associated respiratory failure. Clinical estimates of haemodynamic and volume status in the seriously ill patient are often wrong, and pulmonary artery catheterization often leads to changes in therapy.[137] Invasive monitoring will aid assessment of any fluid deficit, which should be corrected with the appropriate fluid, but it must be remembered that both right and left sided cardiac filling pressures bear little relation to blood volume.[29] Postoperative patients have an impaired ability to excrete a water load, so avoid excessive dextrose administration.

Renal perfusion pressure should be maintained by keeping the mean arterial blood pressure above 80 mmHg, *or higher if previously hypertensive*. Efforts to

Table 66.18 Causes of oliguria in the ICU

Reduced renal blood flow	Intrinsic renal damage (in order of frequency)	Obstruction to flow
Hypovolaemia	Hypoxia:	Bladder neck obstruction
Hypotension	From prerenal causes	Blocked drainage system
Poor cardiac output	Renal vein thrombosis	Pelvic surgery
Pre-existing renal damage	Nephrotoxins:	Prostatic enlargement
Renal vascular disease (beware ACE-inhibitors)	Aminoglycosides	Raised intra-abdominal pressure
Renal vasoconstriction (beware NSAIDs)	Amphotericin	Renal or ureteric
Sepsis	Chemotherapeutic agents	calculi
	NSAIDs	clots
	Contrast media (beware diabetes/myeloma)	necrotic papillae
	Tissue injury:	
	Haemoglobinuria	
	Myoglobinuria	
	Uric acid (tumour lysis)	
	Inflammatory nephritides:	
	Glomerulonephritis	
	Interstitial nephritis	
	Polyarteritis	
	Myeloma	

Table 66.19 Diagnostic value of urine microscopy

Hyaline/granular casts	Underperfusion/chronic renal damage
Tubular casts	Acute intrinsic renal injury
Dysmorphic red cells/casts	Glomerulonephritis
White cell casts	Pyelonephritis/possible interstitial nephritis

Table 66.20 Urinary indices to distinguish prerenal from intrinsic renal dysfunction

	Reduced renal perfusion	Intrinsic renal failure
U/P osmolarity	> 1.5	< 1.1
U/P creatinine	> 40	< 20
U_{Na} (mmol/l)	< 20	> 40
FE_{Na} (%)	< 1	> 3

P, plasma; U, urine

induce a diuresis are worth while, as non-oliguric renal failure is easier to manage. Options include mannitol 20 g, frusemide up to 500 mg and dopamine 3 µg/kg per minute. There is no evidence in the literature that any of these therapies can reverse acute renal failure, and the use of dopamine may be associated with reduced gut perfusion.[138] If myoglobinuria is present, give sodium bicarbonate to alkalinize the urine, as this will increase the solubility of the filtered myoglobin and reduce tubular obstruction. Use mannitol and frusemide if necessary to maintain urine output above 100 ml per hour. If there is a urate nephropathy, add allopurinol.

During further treatment, continue to maintain intravascular volume but restrict fluids if oliguric acute renal failure becomes established. Because conventional haemodialysis is not available in every ICU, other methods of renal replacement therapy such as continuous haemofiltration and haemodiafiltration have evolved. With these techniques, blood flow may be driven by the patient's arterial blood pressure or by a pump. Specially trained dialysis nurses are not required, although pumped continuous venovenous haemodiafiltration approaches conventional dialysis in complexity. Continuous techniques are associated with less haemodynamic instability than intermittent haemodialysis, and this is probably their major advantage. The need for continuous anticoagulation is hazardous but the use of low molecular weight heparin or prostacyclin may be the safest approach. While lessening the risk of haemorrhage and prolonging the life of the filter, prostacyclin is a vasodilator and may have marked effects on oxygen transport variables. Bicarbonate haemodialysis, acetate-free biofiltration and sequential haemodialysis, with the separation of dialysis and ultrafiltration, may help to ameliorate the haemodynamic changes. Some cytokine mediators of sepsis and ARDS are removed by haemofil-

tration, and many clinicians choose to employ this as a method of treatment. There is no good evidence yet that this improves outcome.

It is hard to demonstrate that new techniques for renal support have reduced the mortality, which remains high. The commonest cause of death in these patients remains infection; however, most survivors recover renal function even after prolonged dialysis.[139]

Splanchnic failure

It is probable that the gut has a central role in the pathogenesis of the systemic inflammatory response syndrome and the subsequent development of multiple organ failure. It is frequently stated that failure of the gut barrier function leads to translocation of bacteria and endotoxin, which, if not cleared by the liver, lead to an ongoing systemic injury. Although this concept is appealing, most human data are indirect. Bacterial translocation to the mesenteric lymph nodes has been repeatedly demonstrated in humans during surgery under many circumstances, but has not been significantly related to the incidence of MOF. Conversely, in one study the incidence of translocation following trauma was low, even though the incidence of MOF was high.[140] It may be that gut ischaemia–reperfusion leads to MOF via release of inflammatory mediators and activated polymorphonuclear cells rather than via translocation of bacteria.

Techniques for assessing splanchnic perfusion have demonstrated an increased oxygen demand in sepsis, and ongoing splanchnic ischaemia despite apparently adequate global resuscitation.[141] Failure to match splanchnic oxygen delivery to demand leads to regional hypoxia and increased postoperative complications.[142] This led to the concept of gut-directed therapy. Splanchnic blood flow cannot be deduced from global measurements. However, indocyanine green clearance can now be measured repeatedly at the bedside using commercially available equipment, and such measurements of regional perfusion are likely to assume greater importance in the management of the critically ill patient.

Tonometry is a relatively non-invasive technique for the assessment of splanchnic perfusion.[143] This allows estimation of gastric or sigmoid intramucosal pH (pH_i) to be used as an indirect monitor of the adequacy of splanchnic oxygen delivery.[144] Saline introduced into a gas-permeable gastric balloon equilibrates with carbon dioxide in the gastric mucosa adjacent to the balloon. The bicarbonate concentration of an arterial blood sample and the PCO_2 of the aspirated saline are entered into the Henderson–Hasselbalch equation to derive gastric intramucosal pH.

$$pH_i = 6.1 + \log_{10} \frac{\text{arterial } [HCO_3^-]}{\text{tonometer } P_{CO_2} \times K}$$

Although there is some debate over the assumptions made in the calculations, there appears to be no doubt that pH_i can be a useful clinical monitoring tool.

Splanchnic perfusion and gut function can be optimized by ensuring an adequate mean arterial pressure and global oxygen transport. The use of vasodilators such as dopexamine may be beneficial.[145] Furthermore, the advantages of early enteral feeding and the positive benefits of luminal nutrients on the function of the enterocyte are now recognized.[146] Intestinal blood flow autoregulates to ensure maximum absorption. Enteral feeding enhances mesenteric blood flow and should ameliorate the tendency to splanchnic ischaemia in the critically ill. Enterocytes depend on an enteral source of glutamine for their energy requirements, and lack of glutamine in parenteral feeding solutions has been shown to worsen mucosal atrophy.[147] Enteral feeding preserves gut function and architecture in trauma patients, leading to fewer septic complications.

Enteral feeding is generally a more difficult technique in the critically ill patient. It has been neglected in the past because of the view that enteral feeding should be avoided when there is gastric stasis or ileus. However, early enteral feeding has been demonstrated to be well tolerated after trauma and surgery, and many of the problems have stemmed from poor gastric emptying. The use of jejunal feeding tubes circumvents this problem, and those with a gastric aspiration port ensure that gastric decompression is not compromised. Unfortunately, despite trials of prokinetic drugs such as metoclopramide and cisapride, it is still often necessary to use gastroscopy to guide jejunal tubes through the pylorus.

From historical studies, the incidence of stress ulceration in the ICU is high and the use of prophylactic measures is now widespread, with sucralfate probably the drug of choice. With modern intensive care, including early enteral feeding, the incidence of clinically important upper gastrointestinal bleeding from stress ulceration has been declining[148] and is now only about 1 per cent. Whether patients should have prophylaxis is now openly debated, as, even with no prophylaxis, the incidence of clinically important bleeding is low.[149] Although H_2-receptor antagonists, antacids and sucralfate, either alone or in combination, can reduce the incidence of bleeding from stress ulceration, this is not sufficient justification for their routine use. All these agents have potential hazards and definite costs, and a more targeted approach to prophylaxis is now warranted.

Previous inclusion criteria have included ventilator dependency at 48 hours, multiple trauma, hepatic failure, coagulopathy, peptic ulceration or upper gastrointestinal bleeding, head injury, organ transplants, sepsis and major burns. These indications may now need reviewing.

Sedation

Patients in the ICU will almost invariably require analgesic and sedative agents. Undersedation may cause pain, agitation and discomfort for the patient, with adverse psychological and physical effects. For example, hypertension and tachycardia may cause myocardial ischaemia, and ventilatory problems may lead to hypoxia and hypercapnia. However, oversedation because of excessive dosage or accumulation may also cause unwanted side effects, as seen in Table 66.21.

The problem of oversedation is frequently missed because formal assessment, with titration of the drugs used to the desired response, is rarely undertaken. Because patients after discharge have reported distressing experiences during their stay in the ICU, there has been a tendency by doctors and nurses to err on the safe side with sedation. Appreciation of the above adverse effects and with improvements in intensive care practice – notably the use of sophisticated ventilators which are easier to trigger and short acting, titratable agents – the preferred level of sedation has now changed. Patients are now kept comfortable but responsive unless neuromuscular blocking agents are in use, when deeper levels of sedation are necessary. The neuromuscular blocker should be allowed to wear off periodically, or be monitored by nerve stimulation, to prevent accumulation and also the danger of a paralysed but awake patient.

There have been major changes in the techniques used to sedate patients since a survey 1981 when diazepam, phenoperidine and pancuronium were the agents most commonly used.[150] Ten years later the use of newly introduced agents such as propofol was already widespread and this has continued.[151]

There is no ideal single drug for every patient, and we should consider sedation, analgesia and muscle relaxation

Table 66.21 Adverse effects of excessive sedation

Prolonged coma
Respiratory depression
Hypotension
Ileus
Immunosuppression
Hepatic and renal dysfunction
Hazards of immobility:
    Muscle wasting
    Venous thrombosis
    Pressure sores

separately. Sedation is usually provided by a benzodiazepine. Midazolam is the agent of choice because both it and its α-hydroxy metabolite have a relatively short half-life. However, it is not uncommon for the action of midazolam to be prolonged in the critically ill because of impaired metabolism and accumulation. Since the withdrawal of Althesin, propofol is now considered by many to be the agent of choice. It provides controllable sedation with rapid awakening, although hypotension may be a problem with its use. There can be a significant lipid load associated with its administration, and there may be elevations in levels of serum triglyceride and cholesterol. Chlormethiazole has occasionally been useful, especially for patients who are difficult to sedate. It may be of value during alcohol withdrawal because it has anticonvulsant properties. There is a significant fluid load during its use, and accumulation occurs. Ketamine is an analgesic sedative used occasionally in severe asthma because of its bronchodilator effects. It raises arterial and intracranial pressure, and reports of unpleasant hallucinations have limited its use. The use of isoflurane for sedation was described in 1987. There are practical and technological problems with administering isoflurane, the vaporizer needs to be accurate at very low concentrations and active gas scavenging is desirable. The serum fluoride is known to rise into the presumed nephrotoxic range with prolonged administration. Although popular in some centres, this technique has yet to achieve widespread acceptance.

Morphine remains the standard opioid for analgesia in the ICU. The morphine-6-glucuronide metabolite is 44 times more potent, and in patients with impaired renal function morphine and its metabolites accumulate, leading to prolonged narcotization. Alfentanil is now used widely because, unlike fentanyl, its action is terminated by clearance not initial redistribution. Its metabolites are inactive but accumulation may occur in patients with poor liver function.

Neuromuscular blocking agents may be indicated in certain circumstances, some of which are shown in Table 66.22. Because of the potential adverse effects of prolonged muscle relaxation, they should be used for as short a time as possible, and their effect monitored to prevent accumulation. Atracurium is currently the agent of choice, because its metabolism is independent of hepatic

and renal function. There are as yet no reports on the use of mivacurium or rocuronium in the ICU.

The use of sedation scoring has allowed the distinction between anxiolysis, hypnosis, analgesia and toleration of mechanical ventilation to be delineated. The assessment of sedation can be complex and there is a trade-off between simplicity of use and comprehensiveness. The number of systems described attests to the problem. The Addenbrooke's sedation score, shown in Table 66.23, is a relatively simple and commonly used system but does not distinguish between sedation and analgesia.[152] The system developed by Cook (Table 66.24) is more comprehensive but rather complex for routine use.[153] A useful algorithm combining assessment of sedation, anxiolysis, analgesia, confusion and neuromuscular blockade with management options has been published.[154]

During recovery from long periods of sedation, it is not uncommon for patients to experience episodes of confusion

Table 66.23 The Addenbrooke's sedation 'score'

Agitated
Awake
Roused by voice
Roused by tracheal suction
Unrousable
Paralysed
Asleep

Table 66.24 Cook's sedation score

Eyes open	
Spontaneously	4
To speech	3
To pain	2
None	1
Response to nursing procedures	
Obeys commands	5
Purposeful movement	4
Non-purposeful flexion	3
Non-purposeful extension	2
None	1
Cough	
Spontaneous strong	4
Spontaneous weak	3
On suction only	2
None	1
Respirations	
Extubated	5
Spontaneous intubated	4
Triggered respiration	3
Respiration against ventilator	2
No respiratory efforts	1

Table 66.22 Situations in which neuromuscular blocking agents are used

Initial intubation and placement on the ventilator
Stabilization period during insertion of catheters and monitoring lines
During some endoscopic procedures
For controlled ventilation in ARDS, asthma, head injury, etc.
To reduce oxygen demand

and they may exhibit frankly psychotic behaviour. This may be ameliorated by a reducing dose of benzodiazepines and/or methadone.

Conclusions

In this chapter we have tried to give some indication of the scope and application of intensive care practice. To provide a comprehensive support service the ICU should be consultant based with full multidisciplinary and diagnostic back-up. To ensure optimal patient care, there should be a full-time medical presence and a nurse/patient ratio of at least one-to-one.

Although modern intensive care can save lives, it may also prolong death at great cost both financially and in human terms. Scoring systems cannot be used to make individual decisions regarding admission, but can provide useful information for performance audit. Treatment must be instituted early to prevent or ameliorate organ failure, and perhaps the ICU should be used more frequently for preoperative assessment and monitoring rather than waiting for complications to ensue. This will require a profound change in attitudes from surgeons and managers. The current response to financial restrictions is to ration high technology services, especially to the elderly. This may be shortsighted because early monitoring and optimal treatment after major surgery are cost-effective.

REFERENCES

1. King's Fund. Intensive care in the United Kingdom. *Anaesthesia* 1989; **44**: 428–31.
2. Willatts SM. Development of intensive therapy. *Intensive Care Medicine* 1990; **6**: 474–6.
3. Intensive Care Society Standards Sub-Committee. *Standards for intensive care units*. London: Biomedica, 1984.
4. European Society of Intensive Care Medicine Task Force. Guidelines for the utilisation of intensive units. *Intensive Care Medicine* 1994; **20**: 163–4.
5. Task Force on Guidelines of the Society of Critical Care Medicine. Recommendations for intensive care unit admission and discharge criteria. *Critical Care Medicine* 1988; **16**: 807–8.
6. Astrup P, Severinghaus JW. *The history of blood gases, acids and bases*. Copenhagen: Munksgaard, 1986.
7. Vincent JL. European attitudes towards ethical problems in intensive care medicine: results of an ethical questionnaire. *Intensive Care Medicine* 1990; **16**: 256–64.
8. Jennett B. Inappropriate use of intensive care. *British Medical Journal* 1984; **289**: 1709–11.
9. Rowan KM. Outcome Comparisons of Intensive Care Units in Great Britain and Ireland using the APACHE II Method. PhD Thesis, University of Oxford, 1992.
10. Ridley S, Biggam M, Stone P. Cost of intensive therapy. *Anaesthesia* 1991; **46**: 523–30.
11. Suter P, Armaganidis A, Beaufils F, *et al*. Predicting outcome in ICU patients. *Intensive Care Medicine* 1994; **20**: 390–7.
12. Bion J. Severity scoring: principles, methods, and applications. In: *Recent Advances in Anaesthesia and Analgesia 17*. Edinburgh: Churchill Livingstone, 1992: 173–96.
13. Keene A, Cullen D. Therapeutic intervention scoring system: update 1983. *Critical Care Medicine* 1983; **11**: 13.
14. Knaus WA, Draper EA, Wagner DP, Zimmerman JE. APACHE II: a severity of disease classification system. *Critical Care Medicine* 1985; **13**: 818–29.
15. Rowan KM, Kerr JH, Major E, Short A, Vessey MP. Intensive Care Society's APACHE II study in Britain and Ireland – I: Variations in case mix of adult admissions to general intensive

care units and impact on outcome. *British Medical Journal* 1993; **307**: 972–7.
16. Rowan KM, Kerr JH, Major E, McPherson K, Short A, Vessey MP. Intensive Care Society's APACHE II study in Britain and Ireland – II: Outcome comparisons of intensive care units after adjustment for case mix by the American method. *British Medical Journal* 1993; **307**: 977–81.
17. Palazzo M, Patel M. The use and interpretation of scoring systems in the ICU: Part 2. *British Journal of Intensive Care* 1993; **3**: 286–9.
18. Knaus WA, Wagner DP, Draper EA, Zimmermann JE, *et al*. The APACHE III prognostic system. *Chest* 1991; **100**: 1619–36.
19. Le Gall JR, Lemeshow S, Saulnier F. A new simplified acute physiology score (SAPS II) based on a European–North American multicenter study. *Journal of the American Medical Association* 1993; **270**: 29057–68.
20. Yates DW. Scoring systems for trauma. *British Medical Journal* 1990; **301**: 1090–4.
21. Yates DW, Woodford M, Hollis S. Preliminary analysis of the care of injured patients in 33 British hospitals: first report of the United Kingdom Major Trauma Outcome Study. *British Medical Journal* 1992; **305**: 737–40.
22. Baumgartner J-D, Bula C, Vaney C, Wu M-M, Eggimann P, Perret C. A novel score for predicting the mortality of septic shock patients. *Critical Care Medicine* 1992; **20**: 953–60.
23. Murray JF, Matthay MA, Luce JM, Flick MR. An expanded definition of the adult respiratory distress syndrome. *American Review of Respiratory Disease* 1988; **138**: 720–3.
24. Nightingale P. Measurements, technical problems and inaccuracies. In: Edwards JD, Shoemaker WC, Vincent JL, eds. *Oxygen transport*. London: WB Saunders, 1993: 41–69.
25. Gardner RM. Direct blood pressure measurement – dynamic response requirements. *Anesthesiology* 1981; **54**: 227–36.
26. Shapiro BA. Evaluation of blood gas monitors: performance criteria, clinical impact, and cost/benefit. *Critical Care Medicine* 1994; **22**: 546–8.
27. Gibson RL, Comer PB, Beckham RW, McGraw CP. Actual

tracheal oxygen concentrations with commonly used oxygen equipment. *Anesthesiology* 1976; **44**: 71–3.

28. Beards SC, Edwards JD, Nightingale P. The need for quality control in measurement of mixed venous oxygen saturation. *Anaesthesia* 1994; **49**: 886–8.

29. Baek S-M, Makabali GG, Bryan-Brown CW, Kusek JM, Shoemaker WC. Plasma expansion in surgical patients with high central venous pressure (CVP); the relationship of blood volume to hematocrit, CVP, pulmonary wedge pressure, and cardiorespiratory changes. *Surgery* 1975; **78**: 304–15.

30. Raper R, Sibbald WJ. Misled by the wedge? The Swan–Ganz catheter and left ventricular preload. *Chest* 1986; **89**: 427–34.

31. Swan HJC, Ganz W, Forrester J, *et al.* Catheterization of the heart in man with use of a flow-directed balloon-tipped catheter. *New England Journal of Medicine* 1970; **283**: 447–51.

32. American Society of Anesthesiologists Task Force on Pulmonary Artery Catheterization. Practice guidelines for pulmonary artery catheterization. *Anesthesiology* 1993; **78**: 380–94.

33. Matthay MA, Chatterjee K. Bedside catheterization of the pulmonary artery: risks compared with benefits. *Annals of Internal Medicine* 1988; **109**: 826–34.

34. O'Quin R, Marini JJ. Pulmonary artery occlusion pressure: clinical physiology, measurement, and interpretation. *American Review of Respiratory Disease* 1983; **128**: 319–26.

35. Nadeau S, Noble WH. Misinterpretation of pressure measurements from the pulmonary artery catheter. *Canadian Anaesthetists' Society Journal* 1986; **33**: 352–63.

36. Nunn JF. Oxygen. In: *Nunn's applied respiratory physiology.* London, Boston: Butterworths, 1993: 247–305.

37. Shoemaker WC, Appel PL, Kram HB, *et al.* Prospective trial of supranormal values of survivors as therapeutic goals in high risk surgical patients. *Chest* 1988; **44**: 1176–86.

38. Cryer HM, Richardson JD, Longmire-Cook S, *et al.* Oxygen delivery in patients with adult respiratory distress syndrome who undergo surgery. *Archives of Surgery* 1989; **124**: 1378–85.

39. Russell JA, Ronco JJ, Lockhart D, *et al.* Oxygen delivery and consumption and ventricular preload are greater in survivors than in nonsurvivors of the adult respiratory distress syndrome. *American Review of Respiratory Disease* 1990; **141**: 659–65.

40. Schumacker PT, Cain SM. The concept of a critical oxygen delivery. *Intensive Care Medicine* 1987; **13**: 223–9.

41. Silance PG, Simon C, Vincent JL. The relation between cardiac index and oxygen extraction in acutely ill patients. *Chest* 1994; **105**: 1191–7.

42. Edwards JD. Clinical controversies concerning oxygen transport principles: more apparent than real? In: Vincent JL, ed. *Yearbook of intensive care and emergency medicine.* Berlin: Springer Verlag, 1993; 385–405.

43. Hanique G, Dugernier T, Laterre PF, *et al.* Significance of pathologic oxygen supply dependency in critically ill patients: comparison between measured and calculated methods. *Intensive Care Medicine* 1994; **20**: 12–18.

44. American College of Chest Physicians/Society of Critical Care Medicine Consensus Conference. Definitions of sepsis and organ failure and guidelines for the use of innovative therapies in sepsis. *Critical Care Medicine* 1992; **20**: 864–74.

45. Tilney NL, Bailey GL, Morgan AP. Sequential system failure after rupture of abdominal aortic aneurysms – an unsolved problem in post-operative care. *Annals of Surgery* 1973; **178**: 117–22.

46. Shoemaker WC, Appel PL, Kram HB. Role of oxygen debt in the development of organ failure, sepsis and death in high risk surgical patients. *Chest* 1992; **102**: 208–15.

47. Deitch EA. Overview of multiple organ failure. In: Prough DS, Traystman RJ, eds. *Critical care: state of the art.* Los Angeles: Society of Critical Care Medicine, 1993: 131–67.

48. Sprung CL, Eidelman LA. Monoclonal antibodies, new technologies and critical care medicine. *Critical Care Medicine* 1993; **21**: 1114–16.

49. Kollef MH. The role of selective digestive tract decontamination on mortality and respiratory tract infections. *Chest* 1994; **105**: 1101–8.

50. Pontoppidan H, Geffin B, Larenstein E. Acute respiratory failure in the adult. *New England Journal of Medicine* 1972; **287**: 690–8.

51. Bernard GR, Artigas A, Brigham KL, *et al.* Report on the American–European consensus conference on ARDS. *Intensive Care Medicine* 1994; **20**: 225–32.

52. Ashbaugh DG, Bigelow DB, Petty TL, *et al.* Acute respiratory distress in adults. *Lancet* 1967; **2**: 319–23.

53. Powers SR, Mannal R, Neclerio MS, *et al.* Physiologic consequences of positive end-expiratory pressure (PEEP) ventilation. *Annals of Surgery* 1973; **178**: 265–71.

54. Danek SJ, Lynch JP, Weg JG, *et al.* The dependence of oxygen uptake on oxygen delivery in the adult respiratory distress syndrome. *American Review of Respiratory Disease* 1980; **122**: 387–95.

55. Petty TL, Fowler AA. Another look at ARDS. *Chest* 1982; **82**: 98–104.

56. Bell RC, Coalson JJ, Smith LD, *et al.* Multiple organ system failure and infection in adult respiratory distress syndrome. *Annals of Internal Medicine* 1983; **99**: 293–8.

57. Mohsenifar Z, Goldbach P, Tashkin DP, *et al.* Relationship between O_2 delivery and O_2 consumption in the adult respiratory distress syndrome. *Chest* 1984; **84**: 267–70.

58. Kariman K, Burns SR. Regulation of tissue oxygen extraction is disturbed in adult respiratory distress syndrome. *American Review of Respiratory Disease* 1985; **132**: 109–11.

59. Shoemaker WC, Appel PI. Pathophysiology of adult respiratory distress syndrome after sepsis and surgical operations. *Critical Care Medicine* 1985; **13**: 166–72.

60. Montgomery AB, Stager MA, Carrico J, *et al.* Causes of mortality in patients with the adult respiratory distress syndrome. *American Review of Respiratory Disease* 1985; **132**: 485–9.

61. Annat G, Viale J-P, Percival C, *et al.* Oxygen delivery and uptake in the adult respiratory distress syndrome: lack of relationship when measured independently in patients with normal blood lactate concentrations. *American Review of Respiratory Disease* 1986; **133**: 999–1001.

62. Smith PEM, Gordon IJ. An index to predict outcome in adult respiratory distress syndrome. *Intensive Care Medicine* 1986; **12**: 86–9.

63. Jacobs S, Chang RWS, Lee B. Prognosis in the adult respiratory distress syndrome: comparison of a ventilatory

index with Riyadh Intensive Care Programme. *Clinical Intensive Care* 1991; **2**: 81–5.

64. Clarke C, Edwards JD, Nightingale P, *et al.* Persistence of supply dependency of oxygen uptake at high levels of delivery in adult respiratory distress syndrome. *Critical Care Medicine* 1991; **119**: 497–502.

65. Suchyta MR, Clemmer TP, Elliot CG, *et al.* The adult respiratory distress syndrome: a report of survival and modifying factors. *Chest* 1992; **101**: 1074–9.

66. Villar J, Slutsky AS. The incidence of the adult respiratory distress syndrome. *American Review of Respiratory Disease* 1989; **140**: 814–16.

67. National Heart and Lung Institute. *Respiratory diseases: taskforce report on problems*, DHEW Publication NIH 74-432. Washington DC: US Government Printing Office, 1972: 167–80.

68. Cane RD, Shapiro BA, Templin R, Walther K. Unreliability of oxygen tension based indices in reflecting intrapulmonary shunting in critically ill patients. *Critical Care Medicine* 1988; **16**: 1243–5.

69. Meduri GU, Chinn AJ, Leeper KV, *et al.* Corticosteroid rescue treatment of progressive fibroproliferation in late ARDS. *Chest* 1994; **105**: 1516–27.

70. Gommers D, Lachmann B. Surfactant therapy: does it have a role in adults. *Clinical Intensive Care* 1994; **4**: 284–95.

71. Wenstone R. Surfactant therapy in adult respiratory distress syndrome (acute lung injury). *British Journal of Intensive Care* 1994; **4**: 190–4.

72. Anzueto A, Baughman R, Guntupalli K, *et al.* An international, randomized, placebo-controlled trial evaluating the safety and efficacy of aerosolized surfactant in patients with sepsis-induced ARDS. *American Journal of Respiratory and Critical Care Medicine* 1995; (in preparation).

73. Puybasset L, Rouby JJ, Mourgeon E, *et al.* Inhaled nitric oxide in acute respiratory failure: dose response curves. *Intensive Care Medicine* 1994; **20**: 319–27.

74. Rossaint R, Falke KJ, Lopez F, Slama K, Pison U, Zapol WM. Inhaled nitric oxide for the adult respiratory distress syndrome. *New England Journal of Medicine* 1993; **328**: 399–405.

75. Hutton P, Clutton-Brock T. The benefits and pitfalls of pulse oximetry. *British Medical Journal* 1993; **307**: 457–8.

76. Downs JB, Klein EF, Desautels DA, *et al.* Intermittent mandatory ventilation: a new approach to weaning patients from mechanical ventilators. *Chest* 1973; **64**: 331–5.

77. Civetta JM. Nosocomial respiratory failure or iatrogenic ventilator dependency. *Critical Care Medicine* 1993; **21**: 171–3.

78. MacIntyre NR. Respiratory function during pressure support ventilation. *Chest* 1986; **89**: 677–81.

79. Younes M, Puddy A, Roberts D, *et al.* Proportional assist ventilation. *American Review of Respiratory Disease* 1992; **145**: 121–9.

80. Lachmann B. Open up the lung and keep the lung open. *Intensive Care Medicine* 1992; **18**: 319–21.

81. Qvist J, Pontoppidan H, Wilson RS, *et al.* Hemodynamic responses to mechanical ventilation with PEEP: the effect of hypervolemia. *Anesthesiology* 1975; **42**: 45–55.

82. Nightingale P. Pressure controlled ventilation – a true advance? *Clinical Intensive Care* 1994; **5**: 114–22.

83. Hurst JM, Brabson RD, Davies K, *et al.* Comparison of ventilation and high frequency ventilation. *Annals of Surgery* 1990; **211**: 486–91.

84. Keogh BF, Sim KM. High frequency ventilation in adult respiratory failure. *British Journal of Intensive Care* 1993; **3**: 263–71.

85. Rouby JJ, Ben Ameur M, Jawish D, *et al.* Continuous positive airway pressure (CPAP) *vs* intermittent mandatory pressure release ventilation (IMPRV) in patients with acute respiratory failure. *Intensive Care Medicine* 1992; **18**: 69–75.

86. Zapol WM, Snider MT, Hill JD, *et al.* Extracorporeal membrane oxygenation in severe adult respiratory failure. *Journal of the American Medical Association* 1979; **24**: 2193–6.

87. Brunet F, Dall'Ava-Santucci JD, Dhainaut JF. The place of extracorporeal CO_2 removal in the treatment of ARDS. In: Vincent J-L, ed. *Yearbook of intensive care medicine.* Berlin: Springer Verlag, 1992: 384–92.

88. Morris AH, Wallace CJ, Menlove RL, *et al.* Randomized clinical trial of pressure-controlled inverse ratio ventilation and extracorporeal CO_2 removal for adult respiratory distress syndrome. *American Journal of Respiratory and Critical Care Medicine* 1994; **149**: 295–305.

89. Albert RK. One good turn . . . *Intensive Care Medicine* 1994; **20**: 247–8.

90. Leach CL, Fuhrman BP, Morin FC, Rath MG. Perfluorocarbon-associated gas exchange (partial liquid ventilation) in respiratory distress syndrome: a prospective, randomized, controlled study. *Critical Care Medicine* 1993; **21**: 1270–8.

91. *Advanced Trauma Life Support*, 5th ed. Committee on Trauma, American College of Surgeons. 55 East Erie St, Chicago IL, 1993.

92. Rotheram EB, Safar P, Robin ED. CNS disorder during mechanical ventilation in chronic pulmonary disease. *Journal of the American Medical Association* 1964; **189**: 993–6.

93. Perret C, Feihl F. Respiratory failure in asthma: management of the mechanically ventilated patient. In: Vincent J-L, ed. *Yearbook of intensive care and emergency medicine.* Berlin: Springer Verlag, 1992: 364–71.

94. Darioli R, Perret C. Mechanical controlled hypoventilation in status asthmaticus. *American Review of Respiratory Disease* 1984; **129**: 385–7.

95. Marini JJ. Should PEEP be used in airflow obstruction? *American Review of Respiratory Disease* 1989; **140**: 1–3.

96. Bendixen HH, Hedley White J, Laver MB. Impaired oxygenation in surgical patients during general anesthesia with controlled ventilation. *New England Journal of Medicine* 1963; **269**: 991–6.

97. Suter PM, Fairley HB, Isenberg MD. Optimum end-expiratory airway pressure in patients with acute pulmonary failure. *New England Journal of Medicine* 1975; **292**: 284–9.

98. Gattinoni L, Pesenti A, Bombino M, *et al.* Relationships between lung computed tomographic density, gas exchange, and PEEP in acute respiratory failure. *Anesthesiology* 1988; **69**: 824–32.

99. Hickling KG, Henderson SJ, Jackson R. Low mortality associated with low volume pressure limited ventilation with permissive hypercapnia in severe adult respiratory distress syndrome. *Intensive Care Medicine* 1990; **16**: 372–7.

100. Lutch JS, Murray JF. Continuous positive-pressure ventilation: effects on systemic oxygen transport and tissue oxygenation. *Annals of Internal Medicine* 1972; **76**: 193–202.

101. Ashbaugh DG, Petty TL. Positive end-expiratory pressure: physiology, indications and contra-indications. *Journal of Thoracic and Cardiovascular Surgery* 1973; **65**: 165–70.

102. Poelaert JI, Visser CA, Everaert JA, *et al.* Acute hemodynamic changes of pressure-controlled inverse ratio ventilation in the adult respiratory distress syndrome. A transesophageal echocardiographic and Doppler study. *Chest* 1993; **104**: 214–19.

103. Yang KL, Tobin MJ. A prospective study predicting the outcome of trials of weaning from mechanical ventilation. *New England Journal of Medicine* 1991; **324**: 1445–50.

104. Lemaire F, Teboul JL, Cinotti L, *et al.* Acute left ventricular dysfunction during unsuccessful weaning from mechanical ventilation. *Anesthesiology* 1988; **69**: 171–9.

105. Heffner JE. Timing of tracheotomy in mechanically ventilated patients. *American Review of Respiratory Disease* 1993; **147**: 768–71.

106. Bennett MWR, Bodenham AR. Percutaneous tracheostomy. *Clinical Intensive Care* 1993; **4**: 270–5.

107. Cournand A, Riley RL, Bradley SE, *et al.* Studies of the circulation in clinical shock. *Surgery* 1943; **13**: 964–95.

108. Hotchkiss RS, Karl IE. Reevaluation of the role of cellular hypoxia and bioenergetic failure in sepsis. *Journal of the American Medical Association* 1992; **267**: 1503–17.

109. Cohn JN. Blood pressure measurement in shock. *Journal of the American Medical Association* 1967; **199**: 118–22.

110. Woods I, Wilkins RG, Edwards JD, *et al.* Danger of using core/peripheral temperature gradient as a guide to therapy in shock. *Critical Care Medicine* 1987; **15**: 850–2.

111. Vincent J-L, Weil MH, Puri V, *et al.* Circulatory shock associated with purulent peritonitis. *American Journal of Surgery* 1981; **142**: 262–70.

112. Shoemaker WC. Comparison of the relative effectiveness of whole blood transfusion and various types of fluid therapy in resuscitation. *Critical Care Medicine* 1976; **4**: 71–8.

113. Edwards JD, Nightingale P, Wilkins RG, *et al.* Hemodynamic and oxygen transport response to modified fluid gelatin in critically ill patients. *Critical Care Medicine* 1989; **17**: 996–8.

114. Haupt MT, Gilbert EM, Carlson RW. Fluid loading increases oxygen consumption in septic patients with lactic acidosis. *American Review of Respiratory Disease* 1985; **131**: 912–16.

115. Edwards JD. Management of septic shock. *British Medical Journal* 1993; **306**: 1661–4.

116. Martin C, Papazain L, Perrin G, *et al.* Norepinephrine or dopamine for the treatment of hyperdynamic septic shock? *Chest* 1993; **103**: 1826–31.

117. Desjars P, Pinaud M, Bubnon D, *et al.* Norepinephrine therapy has no deleterious renal effect in human septic shock. *Critical Care Medicine* 1989; **17**: 426–9.

118. Meadows D, Edwards JD, Wilkins RG, *et al.* Reversal of intractable septic shock with norepinephrine therapy. *Critical Care Medicine* 1988; **16**: 663–6.

119. Desjars P, Pinaud M, Potel M, *et al.* A reappraisal of norepinephrine therapy in human septic shock. *Critical Care Medicine* 1987; **15**: 134–7.

120. Thompson BT, Cockrill BA. Renal-dose dopamine: a siren song. *Lancet* 1994; **344**: 7–8.

121. Edwards JD, Brown GCS, Nightingale P, *et al.* Use of survivors' cardiorespiratory values as therapeutic goals in septic shock. *Critical Care Medicine* 1989; **17**: 1098–103.

122. Tuchschmidt J, Freid J, Astiz M, *et al.* Elevation of cardiac output and oxygen delivery improves outcome in septic shock. *Chest* 1992; **102**: 216–20.

123. Astiz ME, Rackrow EC, Falk JL, *et al.* Oxygen delivery and consumption in patients with hyperdynamic septic shock. *Critical Care Medicine* 1987; **15**: 26–8.

124. Creamer JE, Edwards JD, Nightingale P. Hemodynamic and oxygen transport variables in cardiogenic shock secondary to acute myocardial infarction and response to treatment. *American Journal of Cardiology* 1990; **65**: 1297–300.

125. Forrester JS, Diamond GA, Swan HJC. Correlative classification of clinical and hemodynamic function after acute myocardial infarction. *American Journal of Cardiology* 1977; **39**: 137.

126. Edwards JD, Whittaker S, Prior A. Cardiogenic shock without a critically elevated left ventricular end diastolic pressure: management and outcome in eighteen patients. *British Heart Journal* 1986; **55**: 549–53.

127. Creamer JE, Edwards JD, Nightingale P. Mechanism of shock associated with right ventricular infarction. *British Heart Journal* 1991; **65**: 63–7.

128. Berisha S, Kastrati A, Goda A, *et al.* Optimal value of filling pressure in the right side of the heart with acute right ventricular infarction. *British Heart Journal* 1990; **63**: 98–102.

129. Katz JA, Kraemer RW, Gjerde E. Inspiratory work and airway pressure with continuous positive airway pressure delivery systems. *Chest* 1985; **88**: 519.

130. Grace MP, Greenbaum MDM. Cardiac performance in response to PEEP in patients with cardiac dysfunction. *Critical Care Medicine* 1982; **10**: 358–60.

131. Bauman MH, Sahn SA. Tension pneumothorax: diagnostic and therapeutic pitfalls. *Critical Care Medicine* 1993; **21**: 177–8.

132. Ritz R, Marbet GA, von Planta M. Thrombolytic therapy in massive pulmonary embolism. In: Vincent J-L, ed. *Update in intensive care and emergency medicine*, vol 10. Berlin: Springer Verlag, 1990: 284–90.

133. Novis BK, Roizen MF, Aronson S, Thisted RA. Association of preoperative risk factors with postoperative acute renal failure. *Anesthesia and Analgesia* 1994; **78**: 143–9.

134. Cullen DJ, Coyle JP, Teplick R, Long MC. Cardiovascular, pulmonary, and renal effects of massively increased intra-abdominal pressure in critically ill patients. *Critical Care Medicine* 1989; **17**: 118–21.

135. Kellen M, Aronson S, Roizen MF, Barnard J, Thisted RA. Predictive and diagnostic tests of renal failure. *Anesthesia and Analgesia* 1994; **78**: 134–42.

136. Sladen RN, Endo E, Harrison T. Two-hour versus 22-hour creatinine clearance in critically ill patients. *Anesthesiology* 1987; **67**: 1013–16.

137. Connors AF, Dawson NV, Shaw PK, Montenegro HD, Nara AR, Martin L. Hemodynamic status in critically ill patients with and without acute heart disease. *Chest* 1990; **98**: 1200–6.

138. Baldwin L, Henderson A, Hickman P. Effect of postoperative

low-dose dopamine on renal function after elective major vascular surgery. *Annals of Internal Medicine* 1994; **120**: 744–7.

139. Spurney RF, Fulkerson WJ, Schwab SJ. Acute renal failure in critically ill patients: prognosis for recovery of kidney function after prolonged dialysis support. *Critical Care Medicine* 1991; **19**: 8–11.

140. Moore FA, Moore EE, Poggetti R, *et al*. Gut bacterial translocation via the portal vein: a clinical perspective with major torso trauma. *Journal of Trauma* 1991; **31**: 629–38.

141. Ruokonen E, Takala J, Kari A, Saxen H, Mertsola J, Hansen EJ. Regional blood flow and oxygen transport in septic shock. *Critical Care Medicine* 1993; **21**: 1296–303.

142. Mythen MG, Webb AR. Intra-operative gut mucosal hypoperfusion is associated with increased post-operative complications and cost. *Intensive Care Medicine* 1994; **20**: 99–104.

143. Mythen MG, Webb AR. The role of gut mucosal hypoperfusion in the pathogenesis of post-operative organ dysfunction. *Intensive Care Medicine* 1994; **20**: 203–9.

144. Björck M, Hedberg B. Early detection of major complications after abdominal aortic surgery: predictive value of sigmoid colon and gastric intramucosal pH monitoring. *British Journal of Surgery* 1994; **81**: 25–30.

145. Smithies M, Yee TH, Jackson L, Beale R, Bihari D. Protecting the gut and liver in the critically ill: effect of dopexamine. *Critical Care Medicine* 1994; **22**: 789–95.

146. Moore FA, Feliciano DV, Andrassy RV, *et al*. Early enteral feeding, compared with parenteral, reduces postoperative complications. *Annals of Surgery* 1992; **216**: 172–83.

147. Rennie MJ. Consensus conference on enteral feeding of ICU patients. *British Journal of Intensive Care* 1993; **3**(12): suppl.

148. Cook DJ, Fuller HD, Guyatt GH, *et al*. Risk factors for gastrointestinal bleeding in critically ill patients. *New England Journal of Medicine* 1994; **330**: 377–81.

149. Zandstra DF, Stoutenbeek CP. The virtual absence of stress-ulceration related bleeding in ICU patients receiving prolonged mechanical ventilation without any prophylaxis. *Intensive Care Medicine* 1994; **20**: 335–40.

150. Merriman HM. The techniques used to sedate ventilated patients. *Intensive Care Medicine* 1981; **7**: 217–24.

151. Reeve WG, Wallace PGM. A survey of sedation in intensive care. *Care of the Critically Ill* 1991; **7**: 238–41.

152. O'Sullivan G, Park GR. The assessment of sedation in critically ill patients. *Clinical Intensive Care* 1990; **1**: 116–22.

153. Cook S, Palma O. Diprivan as the sole sedative agent for prolonged infusion in intensive care. *Journal of Drug Development* 1989; **2**: 65–7.

154. Armstrong RF, Bullen C, Cohen SL, Singer M, Webb AR. Critical care algorithm. Sedation analgesia and paralysis. *Clinical Intensive Care* 1992; **3**: 284–7.

---APPENDIX---

Abbreviations

Cardiovascular

BSA	=	body surface area (m^2)
CI	=	cardiac index (l/min per m^2)
CO	=	cardiac output (l/min)
CVP	=	central venous pressure (mmHg)
HR	=	heart rate (beats/min)
LVSWI	=	left ventricular stroke work index (g m/m^2)
MAP	=	mean systemic arterial pressure (mmHg)
PAOP	=	pulmonary artery occlusion pressure (mmHg)
PAP	=	mean pulmonary artery pressure (mmHg)
PAWP	=	pulmonary artery wedge pressure (mmHg)
PVR	=	pulmonary vascular resistance (dyn s/cm^5)
PVRI	=	pulmonary vascular resistance index (dyn s/cm^5 per m^2)
RAP	=	right atrial pressure (mmHg)
RVSWI	=	right ventricular stroke work index (g m/m^2)
SV	=	stroke volume (ml)
SVI	=	stroke volume index (ml/m^2)
SVR	=	systemic vascular resistance (dyn s/cm^5)
SVRI	=	systemic vascular resistance index (dyn s/cm^5 per m^2)

Respiratory

CaO_2	=	arterial oxygen content (ml/dl)
$C(a-\bar{v})O_2$	=	arteriovenous oxygen content difference (ml)
$Cc'O_2$	=	end-capillary oxygen content (ml/dl)
$C\bar{v}O_2$	=	mixed venous oxygen content (ml/dl)
FIO_2	=	fractional inspired oxygen concentration
Hb	=	haemoglobin concentration (g/dl)
$PaCO_2$	=	arterial carbon dioxide tension (kPa or mmHg)
PaO_2	=	arterial oxygen tension (kPa or mmHg)
$P\bar{v}O_2$	=	mixed venous oxygen tension (kPa or mmHg)
$\dot{Q}s/\dot{Q}t$	=	pulmonary venous admixture (%)
SaO_2	=	arterial oxyhaemoglobin saturation (%)

$S\bar{v}O_2$ = mixed venous oxyhaemoglobin saturation (%)

Oxygen transport

DO_2 = oxygen delivery (ml/min)
DO_2I = oxygen delivery index (ml/min per m²)
OER = oxygen extraction ratio (%)
$\dot{V}O_2$ = oxygen consumption (ml/min)
$\dot{V}O_2I$ = oxygen consumption index (ml/min per m²)

Equations

Haemodynamic

$$CO = \frac{\dot{V}O_2}{CaO_2 - C\bar{v}O_2} \quad 1/min \qquad \text{the Fick equation}$$

$$CI = \frac{CO}{HR} \quad 1/min \ per \ m^2$$

$$LVSWI = SVI \times (MAP - PAWP) \times 0.0136 \ g \ m/m^2$$

$$PVR = \frac{(PAP - PAWP)}{CO} \times 80 \ dyn \ s/cm^5 \ per \ m^2$$

$$PVRI = \frac{(PAP - PAWP)}{CI} \times 80 \ dyn \ s/cm^5 \ per \ m^2$$

$$RVSWI = SVI \times (PAP - RAP) \times 0.0136 \ g \ m/m^2$$

$$SV = \frac{CO}{HR} \quad ml$$

$$SVI = \frac{CI}{HR} \quad ml/m^2$$

$$SVR = \frac{(MAP - RAP)}{CO} \times 80 \ dyn \ s/cm^5$$

$$SVRI = \frac{(MAP - RAP)}{CI} \times 80 \ dyn \ s/cm^5 \ per \ m^2$$

Oxygen transport

$$CaO_2 = Hb \times K \times SaO_2 + PaO_2 \times 0.0225 \ ml/dl$$

$$C\bar{v}O_2 = Hb \times K \times S\bar{v}O_2 + PaO_2 \times 0.0225 \ ml/dl$$

The Hufner factor, K, is usually taken to be 1.34 or 1.39 ml/g.

$$DO_2 = CO \times 10 \times CaO_2 \ ml/min$$

$$DO_2I = CI \times 10 \times CaO_2 \ ml/min \ per \ m^2$$

$$\dot{V}O_2 = CO \times 10 \times C(a\bar{v})O_2 \ ml/min$$

$$\dot{V}O_2I = CI \times 10 \times C(a\bar{v})O_2 \ ml/min \ per \ m^2$$

$$OER = \frac{CaO_2 - C\bar{v}O_2}{CaO_2} \times 100\%$$

$$\dot{Q}s/\dot{Q}t = \frac{Cc'O_2 - CaO_2}{Cc'O_2 - C\bar{v}O_2} \times 100\%$$

In clinical practice, DO_2 and $\dot{V}O_2$ are often used to denote indexed values. The ideal end-capillary oxygen content is calculated from the alveolar gas equation solved for oxygen. Pulmonary venous admixture is often referred to as shunt fraction or degree of intrapulmonary shunting.

Chapter 67

The Management of Chronic Pain

H. J. McQuay

Managing pain	Common non-cancer pain syndromes
Pain associated with cancer	

Managing pain

There are many ways to manage pain. These include drugs, injections, devices, procedures, complementary medicine and behavioural modification. In deciding what is best for the patient both the effectiveness and the appropriateness of the intervention should be considered. There is good evidence for the effectiveness of some, but not all, of the pain clinic interventions. Appropriateness may be yet more difficult to judge, depending as it does on the context of the patient and on the alternative strategies available. What is appropriate to relieve pain for a bedridden patient with a short life expectancy may be grossly inappropriate for an ambulant patient with normal life expectancy; even though the two patients have similar pains of similar intensity, what is appropriate management for one might be grossly inappropriate for the other.

The prevalence of chronic pain (pain resistant to 1 month of treatment) in the population at large is not known precisely. It may be surprisingly high.[1] Within the pain clinic population, one-quarter of patients have pain associated with cancer, three-quarters have non-cancer pain. Effective interventions are effective in both cancer pain and non-cancer pain. The appropriateness may be quite different.

There are those who argue that a good doctor, whatever his or her specialty, should be able to treat pain. Unfortunately this does not always happen, and the pain clinic provides a focus, a 'critical mass' of experience where, by combining pharmacology, nerve blocks and other measures, patients may be helped.

Pain associated with cancer

Patients with cancer do not necessarily have pain, but the prevalence increases with the progression of disease, so that with advanced disease 60–90 per cent of patients do have significant pain.[2,3]

Acute cancer-related pain is usually associated with diagnostic procedures or treatments. Management of this acute, and often self-limited, pain follows lines similar to management of any other acute pain.

Chronic cancer-related pain may be related to tumour progression (62 per cent) or tumour treatment (25 per cent), or may be a pre-existing chronic pain (10 per cent) unrelated to the tumour.[3] Pain due to cancer is often managed on a cooperative basis between general practitioner, oncology and radiotherapy departments, hospice and pain clinic. Whilst the specific role of the pain clinic is to provide pain-relieving procedures, most patients benefit from the combination of these procedures with skilful analgesic management, radiotherapy and appropriate control of symptoms other than pain. In our local experience, patients referred primarily because of pain tend to come to the pain clinic rather than the hospice; shared care with the hospice then makes available the full range of community and social support. Transient relief in cancer pain may be achieved by the same nerve blocks that are used in non-malignant pain, such as extradural local anaesthetic; extradural steroid may also be very effective. These 'simple' measures are often forgotten. Longer term relief can be achieved by using an extradural (or intrathecal) catheter and infusing a combination of local anaesthetic and opioid. Other procedures employed to obtain longer term relief include injections using neurolytic solutions, cryoanalgesia or radiofrequency and percutaneous cordotomy.

The pattern of referral of cancer pain problems to pain clinics falls into two main categories. The first relates to problems that are thought to be amenable to nerve-blocking procedures; the referral may come at an early stage in the patient's illness, or when he or she is terminally ill. The educative role of pain clinics is important here because others involved in the patient's care are sometimes unaware of what is or is not possible. The second category is the crisis intervention, often a despairing referral at a late stage

Table 67.1 Data from 1115 patients with chronic non-cancer pain

Pain condition*	Percentage of total	Sex ratio (M:F)	Age (years) mean and range	Duration of pain (years) mean and range
Low back pain	26	1:1.7	51.8 (23->80)	8.8 (0.5-32)
Post-herpetic neuralgia	11	1:1.6	73.0 (35->80)	2.6 (0.5-20)
Post-traumatic neuralgia	9	1:1.2	50.0 (19->80)	48 (0.5-36)
Atypical facial pain	6	1:2.0	48.2 (22-79)[†]	4.3 (0.5-17)
Intercostal neuralgia	5	1:1.5	55.0 (25->80)[†]	4.3 (0.5-30)
Trigeminal neuralgia	5	1:2.4	64.3 (32->80)	9.3 (0.5-52)
Perineal neuralgia	4	1:1.9	64.5 (32->80)	6.8 (0.5-35)
Abdominal neuralgia	4	1:1.2	56.2 (26->80)	5.6 (1-24)
Stump/phantom pain	3	1:0.4	62.6 (28->80)[‡]	12.5 (1-61)
Osteoarthritis of hip	3	1:1.2	74.4 (30->80)	4.3 (0.5-15)
Sympathetic dystrophy	2.4	1:1.1	59.1 (27->80)	4.2 (1-6)
Coccydynia	2.3	1:4.2	53.8 (26-73)	5.7 (0.5-31)[†]
Cervical spondylosis	2.1	1:1.5	52.6 (37->80)	5.9 (0.5-17)
Other conditions	18.5			

* All conditions with >2% incidence.
[†] Significant difference ($P<0.05$) in mean age or mean pain duration (female > male). [‡] Significant difference ($P<0.01$) in mean age or mean pain duration (female > male) (Student's t test). 'Other conditions' includes osteoarthritis (unspecified), osteoarthritis of the spine, causalgia, other nerve neuralgia, cord damage, thalamic syndrome, disseminated sclerosis, occipital neuralgia, claudication and neuroma.
Data from reference 4.

in the illness, when the pain is not responding well to management with drugs alone. It is important that the referring doctors and nurses be taught the factors that predict poor control with conventional analgesic management.

Non-cancer pain

The prevalence of the various pain syndromes presenting to the Oxford Unit is shown in Table 67.1.[4] The population served by the Oxford Region is 2.3 million. The Table summarizes the information (age, sex, duration of pain at first attendance) for pain syndromes with prevalence greater than 2 per cent. The pain syndrome with highest prevalence (26 per cent) was low back pain; only post-herpetic neuralgia also had prevalence greater than 10 per cent. For the purposes of this chapter, 'common' non-cancer pain conditions were those with greater than 3 per cent prevalence in Table 67.1.

The list of conditions in Table 67.1 is long; the doctor must therefore be able to manage a number of pain syndromes and no single clinic will see large numbers of patients with the low prevalence syndromes. Management of non-cancer pain management presents great resource problems, because for many syndromes there is no cure but treatment may provide short-term relief. Many of these patients have normal life expectancy and will continue to seek such short-term relief because it improves their quality of life. The ethos of the pain clinic, often the last medical resort, is very important to these patients and to their management.[1,5]

Seeing the patient

The most important principle is that both the patient and the physician are best served if the physician believes the patient's report.[3] Pain is necessarily subjective, and there are few objective signs that the doctor can use to judge the severity of reported pain. Many patients have no visible handicap and their problems may be ill-understood, even disbelieved, at work and at home. Much time and energy are wasted on procedures designed to 'catch the patient out'. Chronic pain changes people and affects their personal lives, their working lives and ultimately their personalities. Often such changes are reversible with successful treatment. Labelling patients as malingerers or the pain as psychogenic may be easier than admitting that there is no successful treatment.

Taking a pain history

A few questions added to the history-taking can provide clues both to diagnosis and to the likely effective treatment.

Site of pain:
Where do you feel this pain?
Does it go anywhere else?
Is it numb where you feel the pain?

Character of pain:
What sort of a pain is it – is it burning/shooting/stabbing/dull etc.?

History of pain:
How long have you had this pain?
How did it start?
Did it come on out of the blue or was it triggered by something?

Relieving factors:
Does anything make it better – position, drugs, distraction, alcohol etc.?

Accentuating factors:
Does anything make it worse – position/exercise/weather etc?

Pattern:
Is there any pattern to the pain? (frequency/severity)
Is it worse at any particular time of day?

Sleep disturbance:
Do you get off to sleep with no trouble?
Does the pain wake you up?

Activities:
What does the pain stop you doing which you would otherwise do?

Previous treatments:
What methods have been tried already?
Did they help the pain?

It is important to ask whether or not sensation is normal in the painful area and to ask whether the pain is shooting or stabbing in character. Answers to these questions may help to identify the pains that are known by a number of names – dysaesthetic, deafferentation and neuropathic. It is important to distinguish these pains from nociceptive pains because they may be less likely to respond to conventional analgesics. The pains often have little discernible pattern, but patients often report that the pain is less troublesome when they are distracted.

It is important to ask carefully about previous drugs that the patient has used. The doctor needs to know how effective the drugs have been, whether or not the drug caused adverse effects, and whether or not these adverse effects were manageable. This information should be sought for each particular drug class (major and minor analgesics, anticonvulsants etc.) (see Fig. 67.1) because it prevents the inept prescription of drugs that have failed previously, and may give important clues as to the kind of pain and its sensitivity to different classes of drug. Ideally the information will include the dose size, frequency and duration of prescription. Dose–response relations apply in analgesic management, and therapeutic failure should not be presumed if the dose was inadequate; a good example is the use of carbamazepine in trigeminal neuralgia. Is the patient taking drugs other than analgesics? Anticoagulation, for instance, is not only an (almost) absolute contra-indication to pain management by injection procedures, but also interacts with some anti-inflammatory drugs.

It is important to be sure whether nerve blocks used previously were technically effective (e.g. did the patient have any numbness after an extradural which included local anaesthetic?), before dismissing them as of no help to this patient. Equally, other measures, such as transcutaneous nerve stimulation, may not have been used correctly, and may succeed if the patient receives proper instruction in their use.

Some index of function is necessary as a baseline so that improvement or deterioration may be monitored. Useful clinical outcome measures are notoriously difficult; simple pain charts and indices of activity can work well.

Examination and investigations

Specific features of physical examination and investigations will be mentioned with the pain conditions. As patients with chronic pain may be seen over years, an accurate serial record of physical signs is important in deciding whether a new pathology has developed or whether new or repeat investigations are required.

What are the treatment options?

The major options (Fig. 67.1) are analgesics, procedures (including devices and surgery) and the alternatives such as behavioural management, hypnosis and acupuncture.

Analgesics

Drugs are the mainstay of chronic pain treatment. The simplest classification is conventional analgesics, from aspirin through to morphine, and unconventional, antidepressants and anticonvulsants (Fig. 67.1). Those used to treating acute pain rightly assume that most such acute pains can be managed with the conventional analgesics. One of the many differences between acute and chronic pain is that chronic pain may be far less straightforward. About one-third of our patients in both malignant and non-malignant categories have pains that respond poorly, if at all, to conventional analgesics, major or minor. There is nothing novel in this; carbamazepine is a classic 'unconventional' analgesic used successfully in the management of trigeminal neuralgia. A simple rule of thumb is that pains in numb areas, known variously as neuropathic and deafferentation pains, are unlikely to respond well to normal doses of opioids.[6] Good examples of deafferentation pain are phantom limb pain and the pain after brachial plexus avulsion; an example of a central neuropathic pain is post-stroke pain (thalamic pain). Opioids may make patients with these pains feel better, but analgesic effect will be poor. This distinction must be sought in the history, from the character of the pain and from previous response to conventional

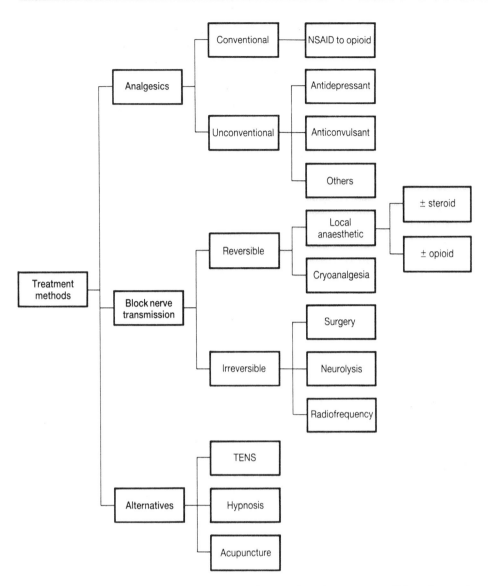

Fig. 67.1 Treatment options. TENS, transcutaneous electrical nerve stimulation.

analgesics. Is this a pain in a numb area? How has it responded to adequate doses of conventional analgesics?

Within the conventional analgesics, non-steroidal anti-inflammatory drugs (NSAIDs) at maximum recommended dose produce about 30 per cent more analgesia than aspirin or paracetamol 1000 mg. The combination of opioid and non-opioid produces additive analgesia; the two mechanisms may combine to give better pain control in contexts such as bony pain. Within the NSAID class it is necessary to find the best drug, best analgesia with minimal adverse effects, for the individual patient. Although there is little evidence from randomized controlled trials to suggest that one NSAID has any advantage over any other, either in terms of analgesic effect or in terms of adverse effect incidence at equianalgesic dosage, chronic pain patients certainly express preferences.

Using major opioids for non-malignant pain is controversial[7] because of fear that the patients may become addicts. This is extremely unlikely, but there is no point

prescribing drugs that the patient or others involved in their care do not appreciate. A simple rule is that where the pain is sensitive to opioids, and there is no other effective remedy, opioid prescription may be tried, but only with the agreement of all concerned. The opioid should be used by a non-injection route unless impracticable.

For pains insensitive to opioids the choice between antidepressants and anticonvulsants is empirical. For burning pains low-dose antidepressants are the first choice, preferably a non-selective tricyclic such as amitriptyline, at a dose of 25 mg nocte titrated as necessary to 150 mg nocte.[8] The analgesic effect occurs within a week, so dose increments can be made quickly. For shooting pains, anticonvulsants work best, and first choice is sodium valproate (200 mg b.d. or 500 mg nocte).

Nerve blocks

Nerve blocks can be thought of in two categories: reversible, as with local anaesthetic steroid or cryoanalgesia; and

Table 67.2 Common nerve blocks

Block	Common indications
Trigger point	Focal pain (e.g. in muscle)
Peripheral	Pain in dermatomal distribution Intercostal Sacral nerves Rectus sheath
Extradural	Uni- or bilateral pain (lumbosacral, cervical, thoracic etc.) Midline perineal pain
Intrathecal	Unilateral pain (neurolytic injection for pain due to malignancy, limbs, chest etc.) (Midline perineal pain)
Autonomic Intravenous sympathectomy	Reflex sympathetic dystrophy
Stellate ganglion	Reflex sympathetic dystrophy Arm pain Brachial plexus nerve compression
Lumbar sympathetic	Reflex sympathetic dystrophy Lumbosacral plexus nerve compression Vascular insufficiency lower limb Perineal pain
Coeliac plexus	Abdominal pain

irreversible, as with ablative nerve blocks and surgical or radiofrequency procedures (see Fig. 67.1). When the effect of reversible procedures wanes the pain is unchanged. If pain recurs after an irreversible procedure it may have altered character, as happens in anaesthesia dolorosa.

The common nerve-block procedures are summarized in Table 67.2. There is a trend away from procedures that destroy nerves, based on the idea that continued (peripheral) pain results in a 'memory' of the pain centrally in the nervous system. Disrupting the peripheral painful input may then make little difference. Continued painful stimuli alter the nervous system. Pain that had a peripheral origin may become 'central'.[9] An example is phantom limb pain; the peripheral neural basis for the pain no longer exists, yet the pain is felt. This concept of nervous system plasticity affects management in several ways.[10] With established chronic pain, conventional analgesics (see Fig. 67.1) may be relatively ineffective, because the neural basis of the pain has changed and no longer responds to those remedies. Plasticity also implies that attacking the cables that carry the (original) pain message may be ineffective in the short term if those cables are no longer the bearers of the message, and ineffective in the long term if the system can 'rewire'. These ideas are supported by both experimental work and clinical observation. Pre-emptive strategies, such as blocking pain before amputation to reduce phantom limb pain incidence, and the idea of modulating

the painful input, for instance by sustained local anaesthetic block as an alternative to irreversible attack on the cables, are being used increasingly.

The major indication for nerve blocking techniques is lack of response to pharmacological management and/or unacceptable adverse effects. Many of these blocks may be done 'diagnostically', with local anaesthetic, as a preliminary to making a more permanent block as with cryoanalgesia, radiofrequency lesions, phenol or surgery. Use of a steroid may convert the diagnostic block to a therapeutic block, as with extradurals in low back pain. Repeated blocks with local anaesthetic may in themselves be therapeutic.[11]

The technical aspects and the potential morbidity of these blocks are familiar to most anaesthetists. Image intensification should increase effectiveness and reduce the incidence of adverse effects for procedures such as lumbar sympathectomy[12] and coeliac plexus block. Many of these procedures are done on an outpatient basis, and the time between the block and when the patient goes home should be sufficient for any complications (most of which are 'early', viz. pneumothorax after intercostal or stellate blocks) to be apparent *before* the patient leaves. Two complications present particular problems. Hypotension occurs during the first few hours after coeliac plexus block, so patients may need to be admitted for the procedure. Lumbar sympathectomy with neurolytic agents may cause a troublesome neuralgic pain, often in the groin or on the anterior aspect of the thigh. Although this may be a severe pain, occurring in as many as 10 per cent of patients, it is self-limiting (6–8 weeks), and transcutaneous electrical nerve stimulation (TENS) may give very effective relief.

The place of many of these blocks in pain management is very ill-defined because, despite widespread use of the techniques, careful studies of quality and duration of relief and morbidity compared with other methods have not been done. Improvement in pain control generally, with better use of drugs, has reduced the numbers of blocks needed. The use of continuous spinal infusion of a combination of local anaesthetic and opioid is superseding neurolytic procedures.[13]

Using intercostal neuralgia as an example, diagnostic intercostal blocks with local anaesthetic may confirm that the pain could be controlled by blocking the relevant nerves. Extended duration of relief could then be achieved by cryoanalgesia to those intercostal nerves. If an intercostal block failed, extradural block with steroid should follow. In general, neurolytic blocks for non-malignant pain are not recommended because they are not permanent, and if the pain recurs it may be more difficult to manage, and because of the morbidity. In cancer pain with limited prognosis, an intrathecal neurolytic block may be used. For severe pains (particularly pelvic or perineal) these can be very rewarding; the limitation is the potential for motor and sphincter damage. This risk is higher with bilateral or repeat procedures, and also the lower the cord level of the block. Extradural neurolytics have limited efficacy. Whilst claims have been made that the paravertebral approach is preferable, patchy results may be attributed to unpredictable injectate spread. We have found the results of spinal infusion of a combination of local anaesthetic and opioid to be superior, providing good analgesia with minimal irreversible morbidity.

Similar distinction between cancer and non-cancer pain holds for coeliac plexus block for pancreatic pain. Pain associated with pancreatic cancer responds well to coeliac plexus block, and it may also help those with abdominal or perineal pain from tumour in the pelvis. In chronic pancreatitis, results are much less convincing.

Pain associated with cancer

The specific role of the pain clinic in pain associated with malignancy is the provision of nerve-blocking procedures. The use of neurolytic procedures has become much less common for patients who are still mobile and continent. This is because spinal infusion of local anaesthetic and opioid can provide better analgesia with minimal adverse effects. The place of the neurolytic blocks in pain management is becoming restricted to patients who are confined to bed and who have lost sphincter control.

Pharmacological management

Oral drug treatment

In the pain clinic many patients will not be terminal, and patients referred for a specific nerve block may also be helped by sensible use of analgesics (as well as other supportive measures), so the pain clinic doctor must be familiar with oral opioid analgesic use in the ambulant patient as well as in the terminal case.

Progressive increase in pain is managed by a progressive increase in the strength of analgesic. This progression is reflected in the World Health Organization ladder (Fig. 67.2). Within the category of conventional analgesics, NSAIDs at maximum recommended dose produce about 30 per cent more analgesia than aspirin or paracetamol 1000 mg. The combination of opioid and non-opioid produces additive analgesia; the two mechanisms may combine to give better pain control in conditions such as bony pain. Within the NSAID class it is necessary to find

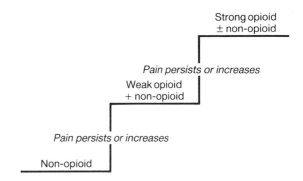

Fig. 67.2 The 'analgesic ladder' for cancer pain management.

the best drug – best analgesia with minimal adverse effects – for the individual patient.

The way in which oral opioids should be prescribed has been well established. Oral morphine sulphate solution or tablets are given every 4 hours with extra 'escape' doses of the same dose size, prescribed 2-hourly, to be taken if necessary. A 24-hour requirement is worked out by adding together the regular and the escape doses; this 24-hour requirement is then made into the regular prescription. This process of titration can take days; the 24-hour dose may then be administered as a slow release formulation if this is more convenient, giving half the 24-hour dose twice a day. If a 4-hourly morphine formulation is unavailable, the titration has to be with a sustained-release formulation. Initial drowsiness and nausea usually settle in 3 days. Laxatives should be started if required. The proportion of patients who require regular antiemetics varies with the clinical context; it is high in terminal hospice patients and much lower elsewhere. Doctors restrict the use of major opioids because of fears that the patients will become addicts. This is wrong. If the pain improves, the opioids can be reduced or stopped without problem.

For pains that are insensitive to opioids the choice between antidepressants and anticonvulsants is empirical. Classic teaching is that antidepressants are the first choice for burning pains, and anticonvulsants for shooting pains. This may be too simple. The non-selective tricyclics such as amitriptyline have an analgesic effect in pain due to diabetic neuropathy and in post-herpetic neuralgia, in both burning and shooting pain, at doses much lower than those used to treat depression, and the analgesic effect is apparent within a week. The non-selective drugs appear to be more effective than the selective. The use of these drugs for neuropathic cancer pain is an extrapolation from the chronic non-cancer pain experience, with a 25 mg single night-time dose, increasing by 25 mg increments at weekly intervals. The dose may need to be lower to prevent retention in the elderly. The adverse effects are dry mouth and drowsiness. Confusion or postural hypotension may occur in the sick patient at low doses (25–50 mg). The drowsiness is minimized by taking the drug at night. The analgesic effect is not simply due to an effect on mood.

The analgesic effect of anticonvulsants has been shown in trigeminal neuralgia, diabetic neuropathy and post-herpetic neuralgia. Traditionally, shooting pains respond better to anticonvulsants than to antidepressants. Our first choice in cancer pain is sodium valproate (200 mg b.d. or 500 mg nocte). Alternatives include carbamazepine, clonidine, mexiletine and flecainide.

When pain is due to nerve compression, oral steroids can be very effective; for example, oral dexamethasone 4 mg t.d.s. This should be effective within 3 days; if it is, trial of lower dosage may be appropriate. For a small number of patients, dexamethasone may be the only effective analgesic strategy. If each time the dose is reduced or the drug is stopped the pain returns and cannot be managed by other methods, the inevitable adverse effects of the steroids have to be balanced against the fact that the drug is the only effective remedy.

Alternative routes

Patients who cannot take opioids orally may manage with sublingual formulations; subcutaneous infusion[14] is the usual resort for patients with intestinal obstruction. High-solubility drugs are preferable for subcutaneous use because a high dose can be given in a small volume. Patients may be referred to the pain clinic for therapeutic trial of opioids given spinally. Selection of patients for extradural or intrathecal catheterization for opioids requires careful attention to detail for the procedures to be worth while.[15]

Problem pains

The two kinds of pain that respond badly to these simple management rules are movement-related pain and neuro-pathic pain. *Movement-related pain* can theoretically be controlled with oral opioid. In practice the dose of opioid required to control the patient's pain on movement is such that the patient is soundly sedated when not moving (not in pain). Conventional wisdom is that NSAIDs should be added if they have been omitted. Again, in practice this often has little impact. Some such pains, for instance due to vertebral metastases, can be helped by extradural steroid. The last resort is to use continuous extradural infusion of a combination of local anaesthetic and opioid. The synergism between the local anaesthetic and the opioid allows low doses to provide analgesia with little loss of mobility. We use percutaneous catheters and external syringe drivers. The usual starting dose is 10 ml of 0.5 per cent bupivacaine with 10 per cent of the previous 24 hour oral opioid dose, infused over 24 hours. This 'low-tech' approach is adequate for treatment lasting weeks to months; the staff are familiar with the devices because they are the same as those used for subcutaneous infusion, and the logistics of changing

syringes are less complicated than those required for bolus administration. More sophisticated devices may be implanted if longer-term use is contemplated, although few can deliver the necessary volume.

In general, we have found the use of spinal opioid alone to be merely a more complicated procedure than the use of subcutaneous opioid; we have not found any advantage in terms of a lower incidence of adverse effects at equianalgesic dosing or in terms of better analgesia for movement-related pain. The use of local anaesthetic combined with opioid, however, has altered our practice radically, because it can produce analgesia for pains poorly responsive to opioids alone. The logic is that pains poorly responsive to opioid given orally are unlikely to improve simply by changing the route by which the opioid is given. Spinal administration of local anaesthetic *and* opioid can produce the necessary analgesia.

The management of *neuropathic* cancer pain is often not straightforward. If such pain cannot be controlled by opioid, antidepressant or anticonvulsant, and steroids are inappropriate, we use the same technique – epidural infusion of a combination of local anaesthetic and opioid. An alternative is to add or to substitute clonidine to the epidural infusion.[16]

Nerve blocks

Lack of response to pharmacological management and/or unacceptable adverse effects are major indications for nerve blocking techniques for pain due to malignancy. The anticonvulsant and antidepressant drugs (see Fig. 67.1) may be useful alternatives if a block is impossible, or to manage pain at other sites.

Head and neck

Swallowing difficulties may limit oral opioid intake, and in this group subcutaneous, spinal and intracerebroventric-ular opioid infusions have been used. Pain in the trigeminal nerve distribution, accompanied at worst by trismus, dysphagia or halitosis, may be helped by blocking the relevant nerve branch; mandibular or maxillary blocks are the most useful.

Chest wall

Intercostal diagnostic blocks with local anaesthetic may be used. Cryoanalgesia to extend the duration of relief has superseded the use of neurolytics. If the pain does not respond to intercostal block, an intrathecal neurolytic block may be used. Extradural neurolytics have limited efficacy. Whilst claims have been made that the paravertebral approach is preferable, patchy results may be attributed to unpredictable spread of injectate.

Abdomen

Abdominal pain associated with pancreatic cancer responds well to coeliac plexus block,[17] and this block may also help those with abdominal or perineal pain from tumour in the pelvis. Postural hypotension, however, can be an unwelcome long-term adverse effect. The procedure may be technically more difficult, and less successful, if there is anatomical disruption. Abdominal pain from tumour in general is not an easy pain to manage with opioids alone, and opioids in conjunction with coeliac plexus block (if feasible) may provide better pain control. The splanchnic approach, placing the solutions dorsal rather than ventral to the diaphragmatic crura, has been claimed to enhance the technical precision of the block.[12] The good results with the classic method (technically improved by use of image intensification or CT scan imaging) are unlikely to be improved by the newer approaches but the low technical morbidity might be further reduced.

Sympathetic blocks (lumbar sympathectomy) may also be of value in rectal pain. Rectal 'phantom' pain after resection is not uncommon, and both this phantom pain and tenesmus may respond to sympathectomy.

Pain in the distribution of the sacral nerves may respond to sacral extradural block with local anaesthetic and steroid, and individual sacral nerves may also be blocked diagnostically at the foramina. Cryoanalgesia can be used to extend the duration of relief.

The 'last-resort' management for such pain remains the intrathecal neurolytic block, which for severe pelvic or perineal pain can be most rewarding. The limitation is the attendant higher risk of bladder dysfunction, particularly in the presence of pain requiring bilateral blocks.

Pain from bony metastases

Isolated metastases that occur in areas already maximally irradiated, or unresponsive to radiation, may respond to 'simple' blocks, such as extradural local anaesthetic and steroid for spinal metastases. Spinal continuous infusion may be necessary for pain in arm or leg long bones, particularly if the pain is movement-related. In general, movement-related pain is very difficult to manage;[18] oral analgesic regimens adequate to control the pain on movement tend to anaesthetize the patient when they are not moving. Cordotomy may be most useful for unilateral pelvic, hip or leg pain. Hypophysectomy used to be employed to manage the pain of multiple bony metastases; it is being superseded by hemi-body irradiation or strontium.

Pain from nerve plexus compression

Local spread of tumour, metastases or tumour treatment may result in compression of either the lumbosacral or the brachial plexus. Pelvic tumour (cancer of colon, rectum, cervix etc.) can produce direct nerve compression or compress the lumbosacral plexus at various levels. Carcinoma of the breast (and radiotherapy) and carcinoma of the lung (Pancoast's) are the commonest causes of brachial plexus involvement seen in the pain clinic.

The pain that results may be the most difficult pain of all to treat. In an advanced stage the pain is often in a numb (deafferented) area. One example is a breast cancer patient presenting with a swollen painful arm. Initial investigation should determine whether tumour spread has occurred (potentially treatable) or whether radiation fibrosis is the cause. In its early stages the swelling may be reduced by the simple techniques used to control lymphoedema; results in late cases are poor. The pain from a Pancoast's tumour may be more restricted to a particular dermatome (or dermatomes). In either condition, poor management with conventional analgesics may be improved by using an anticonvulsant and antidepressant regimen. If such regimens fail, our initial choice is spinal infusion of local anaesthetic and opioid, with intrathecal neurolytic block for the relevant dermatome if indicated clinically. Cordotomy is often not feasible if the pain is higher than the C6 level, and extradural neurolytic injection gives poor results. The response to intrathecal neurolytics is better for tumour damage than for radiation fibrosis.

Lumbosacral plexus involvement may present as root pain, pain from peripheral nerve involvement, or pain in widespread areas of numbness. Successful peripheral nerve diagnostic block with local anaesthetic may be extended as with cryoanalgesia. Cordotomy is practicable in this region for root pain or more widespread pain, and again spinal infusion of local anaesthetic and opioid or intrathecal neurolytic block are the major options.

Common non-cancer pain syndromes

Chronic back pain

Chronic back pain is by far the most common condition seen in the pain clinic (about 25 per cent of non-malignant cases; see Table 67.1). The patients fall into three main categories: back pain that has not responded to conservative measures, back pain recurring despite previous surgery and arachnoiditis.

The major pitfall is to miss a treatable cause of the pain in the rush to treat the symptoms. In the failed conservative management category the most difficult decision may be to identify which changes in signs and symptoms warrant further investigation. Pain despite surgery should become less common as awareness increases that pain alone may not justify laminectomy, because 40 per cent of these patients may have recurrent pain after 1 year.[19,20] With patients who still have pain despite having had a laminectomy it is important to remember that some may still have a surgically remediable lesion.

There are many different ways of classifying back pain; this section uses the common categories presenting at the pain clinic. For more general purposes the fundamental distinction of three main presenting problems (mechanical back pain, possible spinal pathology, nerve root pain) should be used.[21] Table 67.3 lists the common causes and rare pitfalls associated with back pain.

Back pain – causes

Prolapsed intervertebral disc

The site of the pain is in the back (lumbago), with pain down the leg (sciatica) if severe. The pain often occurs once or twice a year, with or without a history of provocation, and often settles on its own (conservative management). The pain is thought to come from swollen nerve roots, and then, after months, the disc retracts with scarring. A herniated disc compressing a nerve root may produce a limitation in one specific movement only, straight leg raising of the good leg may produce crossed pain in the bad leg and there may be a positive femoral stretch test.

Facet joint degeneration

The site of the pain is in the back with or without leg pain. The pain is said to be due to degeneration of the facet joints. It may be possible, from the history, to distinguish facet joint pain from disc disease (see Fig. 67.3). Facet joint degeneration may give pain on sitting, which may then be relieved by standing up and walking – the opposite of the usual story for disc problems. On examination, pain from the facet joints may be elicited by lying the patient prone and extending the facet joints by lifting the legs.

Pain recurring after previous surgery

Patients referred to the pain clinic because of back pain recurring after surgery present special diagnostic problems. The causes of such recurrent pain include failure to remove the disc that was causing the pain, pain persisting although the correct disc was removed and postoperative complications. If the disc that was causing the pain was not removed at surgery, this should be identifiable with straight x-ray to determine the level of the operation. Pain from a recurrent prolapsed intervertebral disc may be identifiable from a good history, a positive myelogram, a disc protrusion removed at operation and pain relief for several months or indeed years. History repeats itself,[19] and the pain (and any radiation) may be the same as before the operation. If a prolapsed intervertebral disc was found on myelography and then was removed at operation, and yet the pain persisted, that disc may not have been the cause of the pain. Original investigations (myelography) must include the conus in order to exclude other causes of the pain coexisting with a disc lesion. Patients in this category may respond to facet joint procedures, suggesting that this may have been the cause of the original pain. Postoperative complications causing pain include dural damage, arachnoid hernia, nerve root pressure, sciatica and sterile osteitis. Patients referred soon (6 weeks) after operation may have sterile osteitis, which presents as severe lumbar back pain coming on with the slightest movement. Translucency at the upper and lower margins of the adjacent vertebral bodies, raised ESR and white cell count confirm the diagnosis. The condition settles with rest and leads to bony fusion. Once fusion occurs the pain does not then recur.

Arachnoiditis

Intrathecal adhesions that occur after a variety of intrathecal insults may result in the clinical syndrome of arachnoiditis, which can cause grave long-term problems.[22] The causes include myelography, bleeding, rough or recurrent surgery, infection and idiopathic.

Half of the patients will develop symptoms and/or signs within a year of surgery; in a small number (15 per cent), 10 years may elapse before problems develop.[22] In that series of 80 patients, 43 had had myelography, and in many this had been a technically difficult procedure; 51 had had spinal surgery. A quarter of the patients had progressive disease. In a clinic with a high proportion of patients with back pain due to 'failed surgery', arachnoidits may be a common problem. The site of the pain is in the back, with root signs in 50 per cent. Classically the pain of

Table 67.3 Causes of back pain seen in the pain clinic

Common	Prolapsed intervertebral disc
	Facet joint degeneration
	Pain recurring after previous surgery
	Arachnoiditis
Rare pitfalls	Cauda equina claudication
	Cauda equina tumours
	Pelvic lesions
	Arteriovenous malformations of the cord
	Tumour:
	primary (e.g. myeloma)
	secondary (breast, prostate, melanoma etc.)
	compression of lumbosacral plexus
	Tuberculosis, osteomyelitis, anklyosing spondylitis

arachnoiditis is a burning constant pain with sciatic distribution to one or both legs, with signs of gradual and progressive loss of leg reflexes. In reality the pain may be difficult to distinguish from that caused by other back problems and the signs may be unconvincing. The diagnosis rests on the history, examination and myelographic evidence, where at least two of the signs of partial or complete block (i.e. narrowing of the subarachnoid space, obliteration of the nerve root sleeves and thickening of the nerve roots, irregular distribution and loculation of the contrast medium, fixity of previously inserted contrast medium or pseudo-cyst formation) should be sought to support the diagnosis.[22]

Rare pitfalls

Cauda equina claudication is caused by spondylosis producing narrowing of an already narrow lumbar canal, occurs mainly in those over 60 years old, and is often associated with a long history. The pain is sciatic in distribution, with pins and needles on standing and walking, relieved by sitting and lying down. The pain is often made worse by bending forward (flexion presumably accentuating the narrowing). It may be distinguished from claudication proper because stopping walking relieves claudication proper, but may not do so in the cauda equina condition. Some patients may be able to walk further by bending forward. Cycling often does not bring on the pain. On examination, one or both ankle jerks may be absent. On investigation, myelography shows complete block or marked stenosis.

Cauda equina tumours have the reputation for being classic diagnostic pitfalls.[23] The patients are often labelled hysterics. The site of pain is in the back, with a sciatic component in 50 per cent. There is characteristically a progressive history, the pain being worse at night. The patient has to get out of bed and then sit in a chair for the rest of the night. The pain may be made worse by jolting or jarring rather than by the twisting or bending that can bring on pain from disc problems. There may be micturition difficulties some time before motor or sensory signs are found. The clinical diagnosis rests largely on the history, because there may be few neurological signs.

Primary or secondary malignancy in the pelvis may produce back pain with or without leg pain. This may come from the disease mass or from lumbar plexus invasion. Arteriovenous malformation of the spinal cord can produce back pain, with or without leg pain. The pain may be worse on walking. Neurological signs appear progressively, and myelography may be reported as normal. Tumour and infection around the spine can produce back pain: causes include primary myeloma, bony secondaries, extradural tumour, tuberculosis, osteomyelitis and ankylosing spondylitis. The symptoms, signs and investigations distinguishing these diagnoses have been discussed by Waddell.[21]

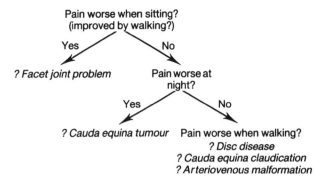

Fig. 67.3 Causes of back pain – clues from the history.

Back pain – clinical presentation

History

When taking the pain history, potentially the most useful questions for distinguishing the causes of the pain are those relating to potentiating and relieving factors (Fig. 67.3).

It is useful to have some idea about which activities are limited by the pain; their improvement may be the best guide as to the effectiveness of treatment.

The drug history may give a guide as to the type of pain. The patients will usually have tried non-narcotic analgesics, and often report that 'they just take the edge off the pain'. They may have distinct preferences between the drugs because of a differential incidence of adverse effects. This information is important when trying to choose the most effective drug, and particularly when injections fail to help and analgesics are all that is left. Some patients with arachnoiditis find little benefit with conventional analgesics, and this may point to a deafferentation type of pain.

Examination

The serial measurement and recording of neurological signs is very important in these patients because any change can be monitored objectively. The simple reminders of sensory innervation are that L1, L2 and L3 supply the anterior aspect of the thigh, L4 the medial malleolus and toes, and S1 the lateral malleolus. More weight should be placed on numbness with tingling than on the finding of numbness alone. L2 and L3 are involved in hip flexion, L3 and L4 in knee extension (knee jerk), other hip and knee movements being L5. Ankle movements involve S1 and S2 (ankle jerk, S1). Gauging the limitation of straight leg raising (SLR) may require common sense. Gross limitation of SLR is incompatible with the ability to sit comfortably at 90 degrees with legs outstretched.

Investigations

Straight x-ray of the spine, checking facet joints, should be carried out to exclude problems such as tumour, infection or spondylolisthesis. There is disagreement as to which of myelography, CT or MRI scans are 'best' in helping to decide if surgery is indicated. The lowest rate of false positive findings occurs with myelography with screening. If an abnormality at the conus is suspected, radiculograms will be inadequate. The combination of myelography and CT scanning may still be the final arbiter in problem cases.

Type and timing of treatment

Drug therapy of back pain is often unsatisfactory. The pain may be both sufficiently severe and sufficiently opioid-sensitive to justify the use of strong opioid analgesics, but this is rarely a socially acceptable solution. Non-narcotic analgesics may be used alone or in conjunction with injections (epidural or facet), or if injections have failed to help. In many patients such drugs provide inadequate relief even at maximum permitted dosage. Effervescent para-cetamol and ibuprofen (fewest NSAID adverse effects) may be the most useful. Where the pain is clearly a deaffer-entation pain, the unconventional analgesic regimens of antidepressants and anticonvulsants should be used (amitriptyline 25 mg nocte as first-line; then if no effect is achieved after an adequate trial, switch to valproate 200 mg b.d.).

Outpatient extradural (lumbar or sacral) steroid injec-tions or facet joint blocks are performed when appropriate, as first line treatment in many clinics. This is justified for the back pain sufferer because even short-term benefit after injections may be better than poor relief with drugs, and drugs usually cause more adverse problems than the injections. There are some careful studies of the effective-ness of extradural steroids (for reviews see references 20 and 24), which justify the technique. Better results may be achieved the earlier the patient is treated,[20] and when there is good evidence that the pain is due to nerve root irritation. When used as a first-line injection treatment for back pain in the Oxford Unit (mean duration of pain 9 years at first visit; see Table 67.1), we expect roughly half the patients to have short-term (4–8 weeks) benefit, and 10 per cent to have pain relief for 6 months or more. These figures improve in direct relation to the duration of the pain and are better in patients who have not had back surgery.

It may take as long as a week for benefit from the steroid to be felt,[20] so it is unwise to dismiss the injection as a failure after 1 hour. If the injection produces incomplete or short-lived relief, it is worth repeating, and a course of three injections is recommended. Usually no additional benefit accrues from more than three injections.

The steroid should be diluted in 5–10 ml diluent to prevent any toxicity from the glycol or phenol derivatives in the ampoule of steroid. The use of local anaesthetic with the steroid provides a technical 'marker' for correct extradural injection, so that a failure of the technique is not attributable to technical failure in placement.

Facet joint injection with local anaesthetic and steroid as

a diagnostic/therapeutic procedure may be indicated by a history of pain worse when sitting, and pain on lateral rotation and spine extension (Fig. 67.4). Short-lived success (less than 6 weeks) with local anaesthetic and steroid may be improved by use of cryoanalgesia or radiofrequency blocks to the nerves supplying the joints. Recent studies suggest that whether or not the injection is actually in the facet joint makes little difference,[25] and indeed cast some doubt on long-term utility.[26]

Some patients with long histories of back pain, and particularly where arachnoiditis is suspected, may have symptoms and signs suggestive of deafferentation. Diag-nostic lumbar sympathectomy with local anaesthetic may help such pains. Short-lived success may be prolonged with chemical sympathectomy.

Those who do not respond

Some patients with back pain, particularly those who have had back surgery, respond poorly, which is disheartening for both patient and doctor. It is important that they receive honest advice and treatment, and that patients who still have surgically remediable disease are identified (a tiny minority). Those with arachnoiditis are at risk of progressive disease.

Alternative methods such as transcutaneous electrical nerve stimulation, acupuncture or behavioural management are often tried at this stage, which is probably an unfair test of any treatment, and all can achieve short-term relief. Behavioural management techniques have better evidence of effectiveness than many of the other interventions. There is no simple way to manage these 'failed' back pain patients. It may be difficult to protect them from the worst excesses that may be offered to them as they go desperately from place to place seeking help. Their heterogeneity means that no single new treatment or drug is likely to be the complete answer.

Post-herpetic neuralgia

Post-herpetic neuralgia (PHN), like other neuralgias, is a pain in the distribution of a nerve. It follows an acute herpetic (shingles) attack, the pain of PHN being due to destruction of cells in the posterior nerve root. If enough are destroyed, there will be sensory loss and this is directly proportional to the degree of dysaesthesia. The natural history of PHN is intriguing and important for manage-ment. Shingles is an infection with the varicella zoster DNA virus. Some 90 per cent of the urban population have had chickenpox by early adulthood, and a prolonged carrier state results with intracellular latency in the dorsal root ganglion. Reactivation of the latent virus causes shingles. The overall incidence of 3–4 per 1000 masks the difference with increasing age; it is rare in children (fewer than 1 per

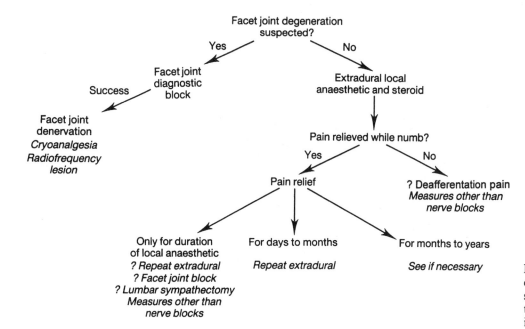

Fig. 67.4 The type and timing of treatment (if there is no surgically remediable lesion and there is no contraindication to injection).

1000 up to 9 years of age) and much more common in those over 80 years old (10 per 1000).[27] Whilst adults do not catch zoster by contact, children can acquire chickenpox by contact with zoster. In the general population, shingles is commonest in the thoracic region (more than 50 per cent), with 10–15 per cent of cases occurring in the trigeminal distribution (commonest in first division).

In a study by Hope-Simpson the incidence of PHN in general practice was 10–15 per cent of patients who had pain which lasted more than 1 month and 5–7 per cent of those who had pain for more than 3 months. A third of patients who were more than 80 years old when they contracted shingles then developed PHN.[27] PHN is thus more common in the elderly, but so is the incidence of shingles; it is unclear whether the elderly run an increased risk of PHN independent of their increased risk of shingles.

PHN in the pain clinic refllects the increased incidence with increasing age (see Table 67.1), such that 90 per cent of pain clinic PHN patients will be more than 60 years old. PHN in the trigeminal nerve distribution is more likely to be referred to the pain clinic than PHN affecting other sites (30 per cent trigeminal PHN in the pain clinic, 10–15 per cent in general practice).

The 'classic' accounts of the natural history of PHN claimed that it did not last more than 6 months.[28] However, median pain duration at first pain clinic visit was more than 2 years (see Table 67.1), and 15 per cent of patients

attending had had their PHN for more than 5 years, so those accounts were incorrect.

There is rarely diagnostic difficulty with PHN. The site of the pain is in the distribution of the shingles, and the scarring is often visible to confirm the diagnosis if this is in doubt. PHN patients are not usually woken by their pain.

Four qualities of pain may be distinguished: a burning pain initiated by light touch or the pressure of clothing and seen in about 25 per cent of pain clinic PHN patients; a constant deep aching pain with no dysaesthesia (50 per cent incidence); a crawling or scratching pain, just under the skin surface; and a stabbing or shooting pain (30 per cent incidence). Often, patients have several of these complaints.

There may be visceral involvement in the acute attack, the viscera involved being those supplied with afferent fibres by the posterior nerve roots corresponding to the affected skin areas.[29] Symptoms such as constipation, indigestion, frequency or dysuria, depending on the affected nerve, may persist, just like PHN.

Management

The depressing statement that 'the good results reported by one author are not confirmed subsequently by others, and no single method of therapy has produced more than temporary enthusiasm'[30] is unfortunately still true for PHN. There is a strong feeling that the success of any remedy is in inverse relation to the duration of the pain. Claims for putative remedies must be examined in this light.

It has long been held that effective management of the pain of the acute shingles attack reduces the incidence of post-herpetic neuralgia. Whilst sympathetic blocks (stellate, lumbar sympathetic or extradural) can undoubtedly be effective in relieving the pain of acute zoster,[31] there is as yet no controlled evidence to show whether such pain relief prevents the development of PHN. A control group in whom the intervention is not made is required.

The use of antiviral agents such as topical idoxuridine 35 or 40 per cent in dimethyl sulphoxide (DMSO) or acyclovir 800 mg 4-hourly orally can help 'abort' the acute zoster attack if given early.[32] It is not known yet if curtailing the acute attack in this way results in a lower incidence of PHN.

Conventional analgesics have a very limited role in PHN. The pain does not appear to be sensitive to opioids, although minor opioids may make the patient feel better. Of the unconventional analgesics, widespread use of the tricyclic antidepressants in PHN is supported by the positive results obtained in controlled studies.[33] Amitriptyline is the drug of choice, with an initial dose of 25 mg nocte reduced to 10 mg in the elderly or infirm. Its use may be inappropriate in elderly men with prostate problems. Persistent shooting or stabbing pain may respond to anticonvulsants, and valproate 200 mg b.d. or clonazepam 500 µg nocte (rising to b.d.) may be helpful. Local anaesthetic creams are of limited benefit, but may help the hyperaesthetic pain.

Transcutaneous electrical nerve stimulation may be helpful for the dysaesthetic pain, as indeed may a simple vibrator applied to the junction of the skin areas with normal and abnormal sensation.

There is little evidence from randomized controlled trials for the effectiveness of injections in providing effective long-term relief for PHN. Whilst sympathetic, regional or even local nerve blocks may provide short-term relief, there is little evidence that they alter the natural history of the disease process. If the affected area still has sensation, diagnostic nerve blocks with local anaesthetic may produce short-term relief, and this may be extended by measures such as cryoanalgesia. Neurolytic agents are not indicated because initial short-term relief may be followed by pain worse than that about which the patient had complained initially. The relevant blocks include stellate ganglion block for head and neck, lumbar sympathetic or extradural for trunk and lower limb and subcutaneous infiltration of local anaesthetic and steroid, repeated weekly for 3 weeks; this may be most effective in treating hyperaesthetic burning pain.

Facial pain

The more common causes of facial pain seen in the pain clinic are trigeminal neuralgia, atypical facial pain,

post-herpetic neuralgia, post-traumatic neuralgia, temporomandibular dysaesthesia and cancer. Post-herpetic neuralgia affecting the trigeminal division is not uncommon but rarely presents difficulties with diagnosis. Two other neuralgias affecting the face, trigeminal neuralgia and atypical facial pain, are relatively common in the pain clinic. Distinguishing them should not be difficult, but cases of atypical facial pain are often referred to the clinic as trigeminal neuralgia resistant to carbamazepine.

Trigeminal neuralgia

Trigeminal neuralgia is a primary neuralgia although many attempts have been made to explain its pathology on the basis of a somatic cause. Tumour, vascular malformations, dental disease, sinusitis or multiple sclerosis (3 per cent) can cause trigeminal neuralgia, but the aetiology of the bulk of the cases is unknown.[34] An abnormality in the pattern of the afferent transmission to the trigeminal nucleus is one explanation. Vascular elongation, local demyelination with age, local compression or cross-axonal discharges have all been proposed as causes of the abnormal transmission.

The condition occurs more often in the middle-aged and is twice as common in females as in males. This pattern of incidence is the same in the pain clinic (see Table 67.1); the striking feature is the long duration (mean 9 years) of pain prior to the first visit. Patients may be referred to a pain clinic with trigeminal neuralgia because the diagnosis is in doubt, because straightforward management with carbamazepine has failed, because of tolerance, adverse effects or allergy, or because previous invasive measures have failed.

The pain is, for practical purposes, strictly unilateral;[34] patients with multiple sclerosis constitute the bulk of the 2 per cent of patients with bilateral disease. Trigeminal neuralgia is twice as common on the right side. Combining published figures (8124 cases) the commonest (32 per cent) pain referral was to both the mandibular and the maxillary divisions. The figures for divisions 1, 2 and 3 separately were 4 per cent, 17 per cent and 15 per cent, respectively; 17 per cent had pain in all three divisions.[34] The pain may remain unchanged, but usually there is a spread to involve another division.

The pain in the face is characteristically sharp, severe (paroxysmal) and brief, lasting no more than a few seconds. Tic douloureux describes the facial contortions that may accompany the pain. A high frequency of these attacks may lead to a persistent pain that is duller in nature. The pain may be brought on by thermal, tactile or proprioceptive stimuli, but not by nociceptive stimuli. These intermittent attacks may last for 6–8 weeks. Long spontaneous remissions of months or even years may occur in the early stages, but tend to be shorter as the disease progresses. The severity and frequency of the attacks thus increase over the years.

The history and examination must exclude other potential pathology. There should be no abnormal neurological signs in trigeminal neuralgia, and the relief of pain with carbamazepine is taken to be diagnostic.

Drug treatment

Patients are often taking (or have already tried) the anticonvulsant carbamazepine when they come to the pain clinic. If the drug is well tolerated but ineffective, increasing the dose in divided doses to daily maximum of 1500 mg should be tried. If the drug is poorly tolerated (nausea and vomiting, ataxia, skin rash or blood dyscrasia) phenytoin 100 mg t.d.s. may be tried instead, or in addition to the lowest tolerated carbamazepine dose.

Nerve blocks

The role of nerve blocks is to provide pain relief when pharmacological management is unsuccessful, or as an adjunct to allow dose reduction. Individual divisions of the Vth nerve may be blocked with local anaesthetic where they leave the skull and enter the face and mouth (infraorbital, supraorbital, inferior dental and mental nerves). Besides their diagnostic value, it is not uncommon for these blocks to break a cycle of pain. Relief may then be obtained for much longer than the duration of local anaesthetic. Because the natural history of the disease is one of spontaneous remissions, it is difficult to quantify this relief in the absence of untreated controls; when such relief occurs the patient and the doctor accept it gratefully.

Short-lived success with local anaesthesia may be prolonged with cryoanalgesia.[35] In a study of 24 patients, median duration of relief was 186 days (range 0–1236), contrasting with median duration of sensory loss of 67 days (range 14–80) which indicates the 'reversible' nature of the cryoanalgesia lesion.[35] There were no complications in the series.

Cryoanalgesia, 'peripheral' radiofrequency lesions, neurolytic injections or avulsion and/or nerve section should not be regarded as curative; they are ineffective long-term. Relapse rates are higher the longer the follow-up. The advantage of cryoanalgesia is that the patient is no worse off if the pain returns, and the cryoanalgesia can be repeated successfully. This is not always the case with the other measures.

Surgical and other more central procedures

Over the years many surgical procedures have been used to treat trigeminal neuralgia. It takes years to establish the success of a novel procedure, because of the extended follow-up required to assess recurrence rate. Procedures need also to be judged on operative morbidity and mortality and long-term morbidity (sensory or motor loss, anaesthesia dolorosa, damage to other cranial nerves). It is a difficult balance. A patient being referred for such procedures may rightly expect to be cured. Unfortunately, the operations with the lowest recurrence rates are likely to be those with the highest complication rates. The most predictable manoeuvre is sensory root rhizotomy, usually performed in the middle cranial fossa, and although there may be some sparing of sensation, corneal ulceration, drooling and trophic lesions of the skin may follow. The recurrence rate with this procedure is low (4 per cent at 4 years). The most serious morbidity is anaesthesia dolorosa, a constant burning pain in the numb area. The incidence of this complication is probably 5–10 per cent, and the only treatment is symptomatic.

Vascular decompression procedures are currently fashionable, and time will tell if the low complication rate is balanced by a high recurrence rate (?50 per cent at 4 years). If that proves to be the case, such major surgery may not be justified. Radiofrequency lesions at the gasserian ganglion, where the electrode is inserted through the foramen ovale under x-ray control, use controlled temperatures to destroy smaller pain fibres selectively. If sensation is preserved, recurrence rates for the first division may be as high as 25 per cent, and there is also a risk of anaesthesia dolorosa (0.3–3 per cent). The injection of glycerol (glycerin) into Meckel's cave appears not to have been as successful as the early reports claimed.

Different clinics have different views on management, usually held very firmly, but trigeminal neuralgia is a condition in which the medical aphorism that the patient should not be made worse is particularly appropriate. The strategy proposed is that failed medical management is an indication for peripheral nerve diagnostic injection, with sustained relief if necessary provided by cryoanalgesia. If these peripheral measures fail, radiofrequency lesion of the gasserian ganglion or rhizotomy should be considered.

Atypical facial pain

The diagnosis of atypical facial pain is often made by exclusion of other causes of facial pain. Classically it affects middle-aged women.[36] Pain clinic patients in whom this is the ultimate (correct) diagnosis may have been referred because they were thought to have trigeminal neuralgia, which had failed to respond to carbamazepine, or because their facial pain was wrongly associated with dental problems, and had continued despite extraction(s). It should not be a difficult diagnosis to make, but the major confusion is with trigeminal neuralgia.

Although the site of pain is in the face, it may not link with the distribution of any particular nerve, and it may cross the midline. The features of the pain are that it is a dull pain (cf. the sharp pain of trigeminal neuralgia), which may be persistent or recurrent, uni- or bilateral, in the absence of any muscular or joint problems, and with no abnormal neurological signs, akin to trigeminal neuralgia.

Although the pain is episodic, it is unlike that of trigeminal neuralgia in that it builds up gradually to a climax and it may last for hours and even days. It is anatomically imprecise and covers a much larger territory than that supplied by an individual cranial nerve. The pain is never 'electric' and is usually described as tearing or crushing.

The pain may be associated with other symptom complexes, such as spastic colon, dysfunctional uterine bleeding, headaches and low back pain. These complexes may recur sequentially or simultaneously in response to stress.

It is wise to look for a hidden cause, and here teeth, sinuses and eyes may be incriminated. At least three 'parallel' conditions may be associated with atypical facial pain. Temporomandibular dysaesthesia (facial arthromyalgia) may be difficult to distinguish from atypical facial pain but there is usually a history of pain in muscles and/or joints, clicking, sticking or trismus, and a feeling of buzzing or fullness in the ear. On examination there may be ridged buccal mucosa, crenated tongue, or masseteric hypertrophy, and arthritic change may be seen on x-ray of the condylar surface. Atypical odontalgia may be distinguished from atypical facial pain because the continuous or throbbing pain is felt in the teeth (tooth), is hypersensitive to all stimuli and may move from tooth to tooth. Glossodynia or oral dysaesthesia is the sensation of a dry mouth, burning tongue and gums, with denture intolerance, disturbance of taste and salivation but no organic pathology.

Management and course

The pathogenesis remains mysterious, but the patients need to be reassured that it is a 'real pain', and is best managed medically rather than by surgery, which may make it more refractory. The association with adverse life events (80 per cent of patients) emphasizes that this condition may affect the emotionally fit as a stress response.

Tricyclic antidepressants are the treatment of choice. Treatment with dothiepin (75–150 mg nocte) proved effective and well tolerated.[37] It is not clear whether there is any difference between the various antidepressants. Treatment should be maintained for 3 months. A third of patients needed short courses (3 months or less) in response to stress, and 40 per cent needed 18–24 months' maintained therapy. Four years on from diagnosis, 70 per cent of the patients were pain-free, but half needed intermittent medication and 10 per cent still needed continuous treatment.

Post-traumatic neuralgia (including postoperative wound pain)

Pain after trauma and pain in or near the operative scar, occurring for months and even years after surgery, can be extremely difficult to manage. Because the principles of management are similar they are discussed together. Post-traumatic neuralgia may occur in a context where the patient is seeking compensation for the accident. Much has been made of the intransigence of the neuralgia until the compensation is settled; unfortunately for many of our patients the pain remains long after the compensation is paid. The aetiology of postoperative wound pain is unknown. Why are some patients affected and not others? As always when this is the case, doctors ascribe a psychogenic overlay.

The pain of post-traumatic neuralgia may be in any part of the body, and may be associated with damage to bone, nerve, muscle or other tissue. Specific information to be sought in the history and examination includes nerve damage to the affected area at the time of injury and the current neurological status of the painful area. Is it numb?, is there painful pins and needles? and is there any difference in temperature between affected and unaffected areas?

Three particular postoperative situations seem to be most likely to result in postoperative wound pain – thoracotomy (particularly thoracoabdominal incisions), nephrectomy and inguinal herniorrhaphy – but pain at or around scars from other operations is also seen. The quality of the pain is often described as burning, even if the area still has normal sensation (unusual). Commonly there is no hyperaesthesia. The aura of the pain (the extent of the area affected) may increase as time elapses. Conventional analgesics may have a very limited role in both post-traumatic and postoperative neuralgias referred to the pain clinic; indeed the patient was in all likelihood referred because such regimens had failed to help significantly. The use of antidepressants and anticonvulsants is more likely to help, particularly if the pain is burning and occurring in an area of altered sensation. A starting regimen might be amitriptyline 25 mg nocte. Alternatives are valproate 200 mg b.d. or clonazepam, starting at 500 µg nocte, and increasing the dose in the absence of side effects to 1.5 or 2 mg t.d.s. Diagnostic blocks with local anaesthetic should be tried when appropriate in these conditions. For the postoperative pains, intercostal blocks for thoracotomy or nephrectomy and ilioinguinal blocks for herniorrhaphy may produce relief. If the relief is short lived, the duration may be extended by repeating the block with cryoanalgesia. This may (especially for the ilioinguinal nerve) be best achieved with surgical exposure under direct vision rather than with a percutaneous approach. Stellate ganglion (for head, neck, upper limbs and upper chest wall) or lumbar sympathectomy may be the first line of management, particularly in post-traumatic neuralgia, and the chance of success is greatest if the pain is in an area of altered sensation. Extending the duration of pain relief with successful 'single-shot' stellate ganglion block may be achieved by repeating the block every 3–7 days on up to ten or more occasions. If there is undoubted success with local anaesthetic blocks but this cannot be sustained despite repeated injections, surgery may be appropriate, perhaps the best results being achieved with a transthoracic approach using the operating microscope. Short-lived relief from lumbar sympathectomy with local anaesthetic may be prolonged by using phenol.

Perineal neuralgia

Perineal neuralgia not associated with cancer occurs most commonly in women, often post-menopausal. Again it is a most difficult pain to treat successfully. The aetiology is unknown, but pelvic varicosities are thought to be a cause of pelvic pain;[38] however, the symptoms and signs (or lack of) in perineal pain would seem to be distinct from the pelvic pain syndrome.[38] The pain is often described as a 'tugging', 'pulling' or 'crawling' sensation, commonly in the clitoral area. Many women with this condition prefer to stand because the pain is worse when they sit. Conventional analgesics are rarely effective, and unfortunately antidepressants and anticonvulsants do not help most patients. Women are often prescribed local hormone therapy; few seem to be helped. Sacral extradural with local anaesthetic may be tried diagnostically. Many of these patients will still have their pain even when the relevant area is numb from the local anaesthetic. If the local anaesthetic does help the pain, the duration of relief may be prolonged with sacral extradural cryoanalgesia. Lumbar sympathectomy may be a logical procedure if the cause is pelvic varicosities, but has not proved successful.

Phantom and stump pain

The proportion of amputees suffering from phantom pain is high (78 per cent of the 55 per cent responding to a survey of 5000 amputees[39], although previous reports had given a much lower incidence of pain. There seem to be few predictors of disabling phantom pain; age at amputation, years since amputation, reason for amputation, site of amputation and pain before surgery were similar in those with and without problematic phantom pain.[39] The aetiology of the pain is unknown. It is thought that preventing pain before surgery may reduce the incidence of phantom pain.[40] Pain in the stump may occur early after surgery owing to surgical postoperative complications, but pain clinic patients with stump pain tend to be late referrals with well-healed stumps, and the origin of the pain is attributed to neuroma development.

At least 50 different methods of treating phantom limb pain are currently in use, but only 1 per cent of patients had long-term benefit from any of them.[39] Whichever method is used, best results come with early treatment. Neither minor (including NSAIDs) nor major conventional analgesics are of great analgesic benefit, but opioids may make the patient feel better. The use of anticonvulsants has been more successful, but with no controlled evidence to support their use. Severe shooting or lancinating pains are most likely to respond, and clonazepam 500 µg nocte rising to 1.5–2 mg t.d.s. if the drug is well tolerated may be the most effective of this group. Sympathetic blocks with local anaesthetic may be used diagnostically, and short-lived relief may be prolonged with repeat local anaesthetic blocks or chemical sympathectomy. Despite pain clinic claims for the efficacy of these procedures, there is little evidence for long-term benefit. Surgical procedures seem to be the least successful.

Stump pain often coexists with phantom pain,[39] and treatment of the phantom with anticonvulsants may also help the stump pain. A purely 'local' stump pain with neuroma development may benefit from local anaesthetic injection to the site, repeated if required, with either percutaneous or direct vision cryoanalgesia to extend the duration of relief. The dogma is that a neuroma should be excised once. If excision does not stop the pain, excising any recurrence is unlikely to help.

REFERENCES

1. Diamond A. The future development of chronic pain relief. *Anaesthesia* 1991; **46**: 83–4.
2. Twycross RG, Lack S. *Symptom control in far advanced cancer: pain relief.* London: Pitman Medical, 1983.
3. Foley KM. The treatment of cancer pain. *New England Journal of Medicine* 1985; **313**: 84–95.
4. McQuay HJ, Machin L, Moore RA. Chronic non-malignant pain: a population prevalence study. *Practitioner* 1985; **229**: 1109–11.
5. Editorial. Pain clinics. *Lancet* 1982; **1**: 486.
6. McQuay HJ. Pharmacological treatment of neuralgic and neuropathic pain. *Cancer Surveys* 1988; **7**: 141–59.
7. Jadad AR, Carroll D, Glynn CJ, Moore RA, McQuay HJ. Morphine responsiveness of chronic pain: double-blind randomised crossover study with patient-controlled analgesia method. *Lancet* 1992; **339**: 1367–71.
8. McQuay HJ, Carroll D, Glynn CJ. Dose–response for analgesic effect of amitriptyline in chronic pain. *Anaesthesia* 1993; **48**: 281–5.

9. Wall PD. Introduction. In: Wall PD, Melzack R, eds. *Textbook of pain.* London: Churchill Livingstone; 1989: 1–18.

10. McQuay HJ, Dickenson AH. Implications of nervous system plasticity for pain management. *Anaesthesia* 1990; **45**: 101–2.

11. Arner A, Lindblom U, Meyerson BA, Molander C. Prolonged relief of neuralgia after regional anesthetic blocks. A call for further experimental and systematic clinical studies. *Pain* 1990; **43**: 287–97.

12. Boas RA. The sympathetic nervous system and pain relief. In: Swerdlow M, ed. *Relief of intractable pain.* Amsterdam: Elsevier; 1983: 215–37.

13. Sjöberg M, Applegren L, Einarsson S, *et al.* Long-term intrathecal morphine and bupivacaine in 'refractory' cancer pain. I. Results from the first series of 52 patients. *Acta Anaesthesiologica Scandinavica* 1991; **35**: 30–43.

14. Dover SB. Syringe driver in terminal care. *British Medical Journal* 1987; **294**: 553–5.

15. Jadad AR, Popat MT, Glynn CJ, McQuay HJ. Double-blind testing fails to confirm analgesic response to extradural morphine. *Anaesthesia* 1991; **46**: 935–7.

16. McQuay HJ. Is there a place for alpha$_2$ adrenergic agonists in the control of pain? In: Besson JM, Guilbaud G, eds. *Toward the use of alpha$_2$ adrenergic agonists for the treatment of pain.* Amsterdam: Elsevier; 1992: 219–32.

17. Jones J, Gough D. Coeliac plexus block with alcohol for relief of upper abdominal pain due to cancer. *Annals of the Royal College of Surgeons of England* 1977; **59**: 46–9.

18. Banning A, Sjøgren P, Henriksen H. Treatment outcome in a multidisciplinary cancer pain clinic. *Pain* 1991; **47**: 129–34.

19. Martin G. The management of pain following laminectomy for lumbar disc lesions. *Annals of the Royal College of Surgeons of England* 1981; **63**: 244–52.

20. Benzon HT. Epidural steroid injections for low back pain and lumbosacral radiculopathy. *Pain* 1986; **24**: 277–95.

21. Waddell G. An approach to backache. *British Journal of Hospital Medicine* 1982; **28**: 187–233.

22. Shaw MDM, Russell JA, Grossart KW. The changing pattern of spinal arachnoiditis. *Journal of Neurology, Neurosurgery and Psychiatry* 1978; **41**: 97–107.

23. Fearnside MR, Adams CBT. Tumours of the cauda equina. *Journal of Neurology, Neurosurgery and Psychiatry* 1978; **41**: 24–31.

24. Kepes ER, Duncalf D. Treatment of backache with spinal injections of local anesthetics, spinal and systemic steroids. A review. *Pain* 1985; **22**: 33–47.

25. Lilius G, Laasonen EM, Myllynen P, Harilainen A, Grönlund G. Lumbar facet joint syndrome. *Journal of Bone and Joint Surgery [Br]* 1989; **71**: 681–4.

26. Carette S, Marcoux S, Truchon R, *et al.* A controlled trial of corticosteroid injections into facet joints for chronic low back pain. *New England Journal of Medicine* 1991; **325**: 1002–7.

27. Hope-Simpson RE. The nature of herpes zoster: a long-term study and a new hypothesis. *Journal of the Royal Society of Medicine* 1965; **58**: 9–20.

28. Burgoon CF, Burgoon JS, Baldridge GD. The natural history of herpes zoster. *Journal of the American Medical Association* 1957; **164**: 265–9.

29. Wyburn-Mason R. Visceral lesions in herpes zoster. *British Medical Journal* 1957; **1**: 678–81.

30. de Moragas JM, Kierland RR. The outcome of patients with herpes zoster. *Archives of Dermatology* 1957; **75**: 193–5.

31. Harding SP, Lipton JR, Wells JDC, Campbell JA. Relief of acute pain in herpes zoster ophthalmicus by stellate ganglion block. *British Medical Journal* 1986; **292**: 1428.

32. McKendrick MW, McGill JI, White JE, Wood MJ. Oral acyclovir in acute herpes zoster. *British Medical Journal* 1986; **2**: 1529–32.

33. Watson CP, Evans RJ, Reed K, Merskey H, Goldsmith L, Warsh J. Amitriptyline versus placebo in post-herpetic neuralgia. *Neurology* 1982; **54**: 37–43.

34. White JC, Sweet WH. *Pain and the neurosurgeon.* Springfield: Thomas; 1969: 123–78.

35. Barnard D, Lloyd JW, Evans J. Cryoanalgesia in the management of chronic facial pain. *Journal of Maxillofacial Surgery* 1981; **9**: 101–2.

36. Miller H. Pain in the face. *British Medical Journal* 1968; **2**: 577–80.

37. Feinmann C, Harris M, Cawley R. Psychogenic facial pain: presentation and treatment. *British Medical Journal* 1984; **288**: 436–8.

38. Beard RW, Reginald PW, Pearce S. Pelvic pain in women. *British Medical Journal* 1986; **293**: 1160–2.

39. Sherman RA, Sherman CJ, Parker L. Chronic phantom and stump pain among American veterans: results of a survey. *Pain* 1984; **18**: 83–95.

40. Bach S, Noreng MF, Tjellden NU. Phantom limb pain in amputees during the first 12 months following limb amputation, after preoperative lumbar epidural blockade. *Pain* 1988; **33**: 297–301.

Chapter 68

Anaesthesia for Day-stay (Come and Go) Surgery

Christopher D. Newson, Michael H. Nathanson and Paul F. White

In the earliest days of anaesthesia, both nitrous oxide[1] and ether[2] were used for day-case (outpatient) dental extractions. Paediatric outpatient anaesthesia was first reported by Nicoll in 1909[3] and in 1916 Ralph Waters opened the first outpatient clinic in Sioux City, Iowa. This freestanding clinic was so successful that in 1918 he was able to move into a prime city centre location.[4] However, it was not until the economic constraints of the 1970s and 1980s led to a large increase in outpatient surgery[5] that the subspecialty of outpatient (ambulatory) anaesthesia finally emerged. In 1984, the Society of Ambulatory Anesthesia (SAMBA) was formed in the USA and has grown to include over 2000 members. Between 1984 and 1988 day-stay surgery in New York doubled[5] and currently over 50 per cent of elective operations in the USA are performed on an ambulatory basis. This expansion has also occurred in the UK but not on the same scale. In the United States, many insurance companies are now reluctant to pay for inpatient treatment for certain operations such as cataract extractions.

Advantages of ambulatory anaesthesia

Hospital costs per patient may be reduced by the use of day-stay surgery for paediatric surgery,[6] laser mastectomy[7] and gynaecological procedures.[8,9] Following-day discharge (i.e. 23-hour stay) has produced similar cost reductions for radical mastectomy,[10] laparoscopic[11] and open cholecystectomy.[12] Patients are less likely to cancel immediately prior to admission for day-case than inpatient surgery.[13] However, the increased efficiency and reduction in waiting lists may actually increase total health service expenditure.[14]

Patients and their relatives experience less disruption to their personal lives and a more rapid return to daily activities. There is a reduced risk of wound infection, deep vein thrombosis and pulmonary embolism, and pneumonia.[15–18] However, some patients feel they have been discharged too soon. Following outpatient laparoscopy most patients require up to 5 days to recover completely.[19,20]

Patient selection criteria

Successful ambulatory surgery requires careful selection of patients, operative procedures and the facilities available. Proper planning will minimize unexpected admissions and allow easy access to inpatient services when necessary. Inappropriate candidates for day-stay surgery include patients undergoing operations associated with excessive fluid shifts or pain that requires parenteral analgesics.

Influence of physical status

Recent experience involving outpatients with chronic medical problems has led in the USA to the acceptance of American Society of Anesthesiologists (ASA) physical status III and IV patients for day surgery, providing that they have been medically stable for 3 months. In a prospective study of nearly 18 000 patients, Natof found those with pre-existing disease to have the same complication rate as healthy patients.[21] This surprising finding was attributed to good patient selection following thorough preoperative screening and close communication between the surgeon, anaesthetist and family doctor. In the Federated Ambulatory Surgery Association (FASA) multicentre study, pre-existing cardiovascular disease led to an increased risk of perioperative complications. However, the complication was directly related to the pre-existing disease

Table 68.1 Contraindications to day surgery

Medical contraindications to day surgery
  Infants at risk:
    Premature infant less than 50 weeks post-conceptual age
    Infant with apnoeic episodes, difficulty with feeding or
      failure to thrive
    History of respiratory distress syndrome
    Bronchopulmonary dysplasia
    Sibling who has died from sudden infant death syndrome
  History of susceptibility to malignant hyperthermia
  Uncontrolled epilepsy
  Unstable ASA physical status II, III or IV patients
  Morbidly obese with other systemic disease
  Monoamine oxidase inhibitor treatment
  Acute substance abuse
Social contraindications
  Patient refusal
  Patient unwilling to comply with instructions
  Lack of responsible person at home

Adapted, by permission, from Wetchler BV. Outpatient anesthesia. In: Barash PG, Cullen BF, Stoeltling RK, eds. *Clinical anesthesia*. 2nd ed. Philadelphia: JB Lippincott Company, 1992: 1389–416.

in less than 1 per cent of the patients.[18] Gold *et al.* studied 9616 patients and found that unanticipated admission was related to the type of anaesthesia and surgical procedure rather than the patient's clinical characteristics.[22]

Influence of age

Extremes of age should no longer be considered a deterrent to outpatient surgery. Few studies have shown any increase in morbidity with increasing age and by avoiding admission the confusion that occurs in up to half of elderly inpatients after surgery may be reduced.[23,24] However, recovery of fine motor function and cognitive skills following general anaesthesia or sedation is slower in the elderly and regional anaesthesia may be preferable.[25–27]

Children are usually well suited to day-case surgery and have less psychological disturbance than those admitted on the day before and discharged the day after surgery.[28] These children also require less additional attention in the first week after discharge home and have fewer sleep disturbances.[29] The very young and preterm infants are a special case. There is an increased risk of postoperative apnoea for up to 12 hours after general anaesthesia. Although this was initially reported in premature infants under 10 weeks postnatal age, it is present up to 46–55 weeks post-conceptual age.[30–32] Children in this age group should be admitted for postoperative apnoea monitoring. Children with bronchopulmonary dysplasia should be considered at risk as long as their symptoms persist.[33]

Influence of social factors

Patients should live within a reasonable distance of the unit and have a responsible person at home to care for them during the first postoperative night.

Preoperative assessment

Cancellations from inadequate preparation can be reduced if the patient is seen by an anaesthetist prior to the day of surgery. This may also reduce patient anxiety.[34] Those patients contacted by telephone during business or evening hours also have a lower cancellation rate and are less likely to need admission.[35] To save time, patients may be asked to complete a questionnaire on general health, past operations, drug history and family history.[36] This process may be automated using a computer.[37] These computer-based programs may facilitate the health screening process and provide recommendations regarding appropriate laboratory tests.[38] This approach has been shown to reduce the number of tests ordered and result in considerable cost savings.[39]

Preoperative testing should be based on the patient's age, history and examination.[40] Routine laboratory tests produce a large number of abnormal results, most of which are ignored by the anaesthetist; at most, 1 per cent of patients benefit.[41,42] Similarly, over 99 per cent of electrocardiograms (ECG) and chest radiographs currently requested have no influence on patient management.[43] The cancellation rates because of abnormal laboratory investigations are similar if testing is only performed when indicated or as a general screening procedure.

Healthy (ASA I) male outpatients under the age of 40 do not need routine laboratory testing. A haemoglobin or haematocrit (packed cell volume) is appropriate for females and possibly children under 5 years.[44,45] ECGs should only be performed on men over 40 years and women over 50 years. Chest radiographs should only be ordered in patients over 70 years, if at all.

Procedure-related factors

The number and type of procedures performed on outpatients continues to increase. In the UK the publication of a list of 83 suitable procedures has increased the scope of outpatient surgery (Table 68.2).[46] However, there are a number of areas of concern relating to the length of operation, requirements for blood transfusion, post-operative analgesia, physiotherapy and the risk of airway obstruction.

In the past, procedures expected to last longer than

Table 68.2 Examples of procedures suitable for day surgery

Ophthalmology
  Strabismus surgery
  Extracapsular cataract
    extraction

Ear nose and throat
  Myringotomy
  Septoplasty
  Nasal polypectomy
  Reduction nasal/zygoma/
    malar fractures
  Antral washout
  (Adeno) tonsillectomy
  Laryngoscopy

Oral surgery
  Odontectomy
  Dental clearance

Thoracic surgery
  Oesophagoscopy
  Bronchoscopy

Abdominal surgery
  Paracentesis
  Laparoscopy
  Herniorrhaphy
  Upper GI endoscopy
  Colonoscopy/sigmoidoscopy
  Pilonidal cystectomy
  Haemorrhoidectomy
  Anal dilatation
  Incision of fistulae

Genitourinary
  Urethroscopy and
    cystoscopy
  Orchidectomy
  Orchidopexy
  Hydrocele repair
  Vasectomy
  Circumcision

Gynaecology
  Laparoscopic sterilization
  Cervical polypectomy
  Cervical cone biopsy
  Dilatation and curettage
  Removal/insertion of IUCD
  Removal of vulval lesions

Orthopaedic surgery
  Removal of internal
    fixation
  Excision of exostosis/
    metatarsal heads
  Arthroscopy/meniscectomy
  Manipulation
  Arthrodesis of phalanges
  Arthroplasty (foot/hand)
  Excision of Baker's cyst
  Release trigger finger/
    carpal tunnel
  Ganglionectomy
  Tendon repair
  Fasciotomy

Other surgery
  Varicose vein ligation and
    stripping
  Lymph node biopsy
  Excision of lipoma/skin
    lesions
  Excision of breast mass
  Finger/toe nail removal
  Wart ligation
  Myelogram
  Examination under
    anaesthesia

Adapted, by permission, from Gabbay J, Francis L. How much day surgery? Delphic predictions. *British Medical Journal* 1988; **297**: 1249–52.

60–90 minutes were not scheduled on an outpatient basis because it was assumed that these patients would require a prolonged recovery period. However, oral, plastic and orthopaedic surgery procedures lasting 2–4 hours are currently being performed successfully in many outpatient facilities in the USA. Meridy, in a retrospective analysis, found no correlation between the duration of anaesthesia and the recovery time.[47] However, other studies have shown a direct relation between the length of anaesthesia and the recovery room stay.[48,49] The FASA Special Study found a

threefold increase in complications for operations lasting more than 2 hours compared with those lasting less than 1 hour.[18]

Procedures with excessive blood loss should be performed on an inpatient basis. Autologous blood transfusions have been used for more extensive outpatient plastic surgery (for example, reduction mammoplasty and liposuction).

Ambulatory patient-controlled analgesia (PCA) may become available and allow outpatient parental therapy in the future. Until then pain should be controllable with oral analgesics (for example, paracetamol with codeine) before discharge is considered.

Several series of outpatient tonsillectomy have reported very low rates of bleeding (0.28–0.49 per cent) or major complications (1.4–2.1 per cent).[50–53] Most of the bleeding occurred within the first 6–8 hours and very few patients (1/1082) required admission following discharge home.[54] However, the complication rate is much higher in children under 36 months for whom outpatient tonsillectomy is inappropriate.[55]

Facilities

Day surgery units may be either attached to a hospital (integrated unit) or freestanding. Some integrated units have a separate ward for pre- and postoperative care but use the main operating theatres. This may reduce initial capital costs but outpatients may be cancelled at short notice to make way for emergency operations. Integrated units may avoid this by having either a separate block of theatres or a separate building in the hospital grounds. Not all procedures need to be performed in the main outpatient facility; for example, vascular access surgery may be performed in renal units.[56] Some freestanding centres now have facilities for overnight observation following procedures such as laparoscopic cholecystectomy where a period of time close to trained assistance may be desirable.[11]

Admission rates vary from 1 to 6 per cent and 15–30 per cent of these are directly related to the anaesthetic.[18,57] Intractable nausea and vomiting, airway problems (for example stridor, bronchospasm), inability to void, dizziness, and delayed emergency are the most common problems in adults.* In paediatric outpatients, protracted vomiting accounted for 33 per cent of the unanticipated admissions after ambulatory surgery.[58] In a study of 2470 gynaecological patients, Meeks identified several factors associated with admission including previous abdominal surgery, significant medical illness, preoperative haemoglobin concentration and general anaesthesia. Postoperative emesis was the most common complication (23 per cent)

*White, PF. Unreported data from the Same Day Surgery Unit at Stanford University, 1986–1988.

Table 68.3 Multivariate logistic regression analysis of factors associated with an increased risk of unanticipated admission following day-case surgery

Factor	Odds ratio
General anaesthesia	5.2*
Postoperative emesis	3.0*
Abdominal surgery	2.9*
Operating theatre time > 1 h	2.7*
Age	2.6*
Laparoscopy	1.7
Journey time > 1 h	1.5

* $P < 0.05$.

Adapted, by permission, from: Gold BS, Kitz DS, Lecky JH, Neuhaus JM. Unanticipated to the hospital following ambulatory surgery. *Journal of the American Medical Association* 1989; **262**: 3008–10. Copyright 1989, American Medical Association.

leading to admission.[59] Gold's review of a 2.5-year period at one day-case facility identified seven factors associated with unexpected admissions (Table 68.3). Common surgical problems necessitating admission (30–50 per cent of cases) include intractable pain, excessive bleeding, surgical misadventure (for example bowel burn or perforation, uterine perforation), errors in diagnosis leading to more extensive surgery, and parenteral drug therapy (for example antibiotics).[47,60] Other causes of unanticipated admission after ambulatory surgery include the need for more intensive monitoring and social factors (for example the lack of appropriate transportation or a responsible escort).

Emergency admission for a life-threatening complication occurred in 1 in 12 500 patients in a recent multicentre study.[18] Procedures such as midtrimester abortions can be associated with serious life-threatening complications.[61] Close cooperation between surgeons and anaesthetists regarding patient selection, preoperative assessment and preparation, and the operations to be performed in the outpatient setting will minimize the number of unexpected admissions.

Anaesthetic management

Premedication

There is considerable debate on the use of premedication in outpatients. Premedication is not routinely used in most ambulatory surgery facilities in the United States. Many anaesthetists avoid using centrally active depressant premedication because they believe that recovery will be prolonged.[62,63] The indications for preoperative medica-

tions are similar to those for inpatients and include anxiolysis, sedation (especially for paediatric patients), analgesia, amnesia, vagolysis, and prophylaxis against postoperative emesis and aspiration pneumonitis. Anxiety is related to previous anaesthetic experience and the type of operation. Oral surgery patients may be particularly anxious and often desire anxiolytic premedication.[64] In healthy patients anxiety is not reduced by meeting the anaesthetist prior to the day of surgery or outside the operating room immediately prior to the procedure.[65,66]

The judicious use of premedication for outpatients can be beneficial and in most prospective studies premedication does not prolong recovery to 'street fitness'[47,67,68] although coordinative and reactive skills may be impaired for 5–12 hours.[69] Some studies have even found a decrease in early recovery times with analgesic or antiemetic premedication.[70,71] Premedication with sedative or analgesic drugs does not appear to increase the percentage of outpatients at risk of aspiration pneumonitis.[72]

Children

In children oral midazolam (0.5–0.75 mg/kg) provides good sedation and increases the number either asleep or awake and calm at induction of anaesthesia without increasing the time to discharge or overnight stay.[73,74] Intranasal and rectal midazolam are also highly effective routes of administration.[75,76] Oral transmucosal fentanyl citrate (OTFC) reliably induces preoperative sedation and facilitates inhalation induction of anaesthesia.[77] However, it may produce a significant decrease in respiratory rate and peripheral blood oxygen saturation, and a high incidence of postoperative nausea and vomiting not prevented by prophylactic droperidol.[78,79] Pruritus may also be a significant problem.[80] Chloral hydrate (40 mg/kg) increases the number of children who are calm or asleep at induction of anaesthesia compared with midazolam (0.05 mg/kg), alprazolam (0.005 mg/kg) or placebo. Postoperative behaviour and the incidence of vomiting were similar for all drugs.[81] Ketamine (2 mg/kg) intramuscularly in uncooperative children permits inhalation induction with halothane after 2–3 minutes. Although early recovery times were unaffected, home discharge was delayed by an average of 13 minutes.[82]

The expansion of outpatient surgery has made it difficult to provide an unhurried preoperative visit to the hospital. Children and their parents may visit when the facilities are otherwise unused. This may take the form of a 'Saturday club' and include a puppet show.[83] Other anaesthetists have made video tapes showing the operating theatres and explaining the risks of anaesthesia. Most parents reported that their concerns were helped and appeared to accept discussion of perioperative risks including death.[84]

Adults

Benzodiazepines

Midazolam's rapid onset of action and water solubility offer a number of advantages for outpatients. Midazolam may be given intramuscularly 30–60 minutes before surgery[85,86] or intravenously in the induction room.[87,88] Midazolam does not delay recovery after ambulatory surgery because of its relatively short elimination half-life. Midazolam (2 mg) given intravenously immediately prior to a propofol infusion reduces anxiety and increases amnesia without prolonging the recovery room stay.[89] However, it is associated with impaired postoperative psychomotor skills compared with placebo.[86] Although larger doses are required because of first-pass metabolism, midazolam given orally has been reported to be highly effective in adults as well as children.[90,91]

Temazepam is an effective oral premedication for outpatient surgery.[92,93] Temazepam 20 mg produced a similar degree of anxiolysis and sedation as midazolam 7.5–10 mg and did not affect recovery compared with placebo.[94] However, in a separate study temazepam produced less perioperative amnesia than did midazolam and although postoperative sedation was described as similar, the temazepam group were more sleepy and retired to bed earlier following surgery.[95] Lorazepam provides good amnesia and has been reported to have antiemetic effects but is generally considered too long acting for outpatient anaesthesia.[96] While oral triazolam can produce effective sedation and amnesia, it is less effective than diazepam or midazolam in decreasing anxiety and leads to more residual sedation at the time of discharge.[75,97]

Opioid analgesics

The routine use of opioid analgesics for premedication has been criticized unless the patient is experiencing acute or chronic pain. Use of traditional opioid premedicant combinations (for example, papaveretum–hyoscine, pethidine–atropine) may increase the incidence of postoperative nausea and vomiting.[98,99] The agonist–antagonist nalbuphine produces good sedation and smooth induction but has similar postoperative side effects.[100] The use of small doses of the potent opioid analgesics (for example, fentanyl 1–3 µg/kg, sufentanil 0.1–0.3 µg/kg) prior to induction of general anaesthesia reduces the intravenous anaesthetic requirement and may shorten early recovery times.[101–104] Sufentanil (0.15 µg/kg) reduced preoperative anxiety and provided more satisfactory induction, maintenance and recovery from anaesthesia than morphine, pethidine, or fentanyl.[105] However, the use of these potent, rapid acting opioids increases the incidence of postoperative nausea and vomiting.[102–104]

Preoperative fasting

Since prolonged fasting does not guarantee an empty stomach at the time of induction, several investigators have questioned the value of even a 4–5 hour fast prior to elective surgery.[106] About 50 per cent of outpatients complain of hunger or thirst following an overnight fast.[107] This may increase preoperative anxiety. More importantly one in seven young women coming to the operating theatre in the afternoon after an overnight fast had a serum glucose concentration of less than 2.5 mmol/l.[108] Children tend to go to bed earlier, producing a longer overnight fast; one in five fast for 16 hours or more and up to 9 per cent may be hypoglycaemic.[109] The ingestion by adults of 150 ml of either coffee or orange juice 2–3 hours before induction of anaesthesia has no significant effect on residual gastric volume or pH.[108] Also, the length of fasting has no effect; gastric fluid volume and pH were similar when intervals of less than 3 hours, 3–4.9 hours, 5–8 hours and nil by mouth after midnight were compared.[110] Similarly, in children, preoperative administration of apple juice (3 ml/kg) decreased gastric volume, thirst and hunger.[111] Furthermore, administration of ranitidine (2 mg/kg) with orange juice (5 mg/kg) 2–3 hours preoperatively resulted in a decrease in both volume and acidity of gastric contents.[112] Thus, the arbitrary restriction of fluids after midnight prior to an elective operation appears to be unwarranted. Only those patients suspected, or known, to be at risk of delayed gastric emptying require a prolonged fast which otherwise causes discomfort to outpatients without any apparent benefit.

Prevention of aspiration

The incidence of pulmonary aspiration in elective surgical patients without specific risk factors is very low (less than 1 in 35 000).[113] However, several studies have found that 40–60 per cent of outpatients would be defined as 'at risk' for aspiration pneumonitis by the traditional criteria (that is, gastric volume more than 25 ml with pH less than 2.5) despite an overnight fast.[106,107,114] Although outpatients have been reported to have significantly higher residual gastric volumes than inpatients,[115] this has not been confirmed in a more recent study.[116] These findings have led to the evaluation of a variety of premedication regimens to reduce the risk of aspiration pneumonitis. It has been suggested that all patients receiving anaesthesia by facemask should be protected against the acid component of pulmonary aspiration injury with an H_2-blocking drug.[117] However, less than 20 per cent of the outpatient centres surveyed in the United States routinely use prophylactic antacid or antisecretory medication. In elective gynaecological patients neither anxiety nor benzodiazepine premedication has a clinically important impact on gastric content.[118]

The H2-receptor antagonists cimetidine and ranitidine are both effective in decreasing the number of patients 'at risk' for pulmonary aspiration.[119] Ranitidine may be preferable because of its longer period of protection and fewer side effects than cimetidine.[120] Ranitidine may be given orally or parenterally, with peak effects occurring within 2 hours.[121] Oral ranitidine given with coffee or orange juice 2–3 hours prior to induction of anaesthesia produces lower residual gastric volumes, higher pH values, and less thirst compared with fasting alone.[122]

Metoclopramide has been shown to reduce gastric volume in outpatients without altering the pH.[114,123] Use of metoclopramide in combination with an H2-blocking drug has been advocated to decrease postoperative emesis and further reduce the risk of aspiration pneumonitis.[124] However, other studies have failed to demonstrate a significant advantage of this combination over an H2-receptor antagonist alone.[122,125] Metoclopramide may offer an additional protective effect as a result of its ability to increase lower oesophageal sphincter tone and thus have a role in diabetic and pregnant outpatients.[126,127]

Sodium citrate, a non-particulate oral antacid, is less effective in raising pH than the H2-antagonists and can increase gastric volume.[128] However, sodium citrate may be useful in combination with metoclopramide when little time is available prior to the operation.[129]

Prevention of nausea and vomiting

Postoperative nausea and vomiting remains a common problem after general anaesthesia, and can delay discharge or result in unexpected hospital admission.[23,108,130,131] Factors which increase the incidence of postoperative nausea and vomiting include the patient's body habitus and medical condition, the type of surgery performed (laparoscopy, orchidopexy, strabismus surgery), assisted ventilation using a facemask (especially by junior anaesthetists[132]), anaesthetic and analgesic medications (for example, fentanyl, etomidate, isoflurane and nitrous oxide), and postoperative hypotension.[133] Many different pharmacological and non-pharmacological regimens have been evaluated, including acupuncture and acupressure.[134–136]

Antiemetic drugs

The routine use of a prophylactic antiemetic drug may not be necessary as less than 30 per cent of patients will be nauseated or vomit postoperatively, and many of these will only experience one or two episodes which do not require treatment.[134] Most antiemetic drugs may cause sedation,

dysphoria or extrapyramidal side effects. However, prophylaxis should be considered for patients at high risk such as those with a history of motion sickness or previous postoperative vomiting, adults undergoing gynaecological laparoscopic procedures or extracorporeal shock wave lithotripsy, and children having strabismus surgery, otoplasty, adenotonsillectomy or orchidopexy.

Droperidol (5–15 μg/kg) intravenously is effective in decreasing the incidence of nausea and vomiting in children undergoing orthopaedic or dental procedures.[137,138] However, larger doses of droperidol (50–75 μg/kg) are required after strabismus surgery and although postoperative sedation is increased, recovery is not prolonged.[130,139,140] In adults, smaller doses of droperidol (7.5–15 μg/kg), given intravenously at induction or prior to extubation, may be effective in preventing postoperative nausea and vomiting.[137,141,142] However, in adults, these small doses of droperidol lead to sedative effects that may delay awakening and decrease psychomotor performance in the early postoperative period.[143,144] Droperidol (1.25 mg) intravenously may also produce anxiety or restlessness after discharge[145] and ultra-low doses of droperidol (0.25–0.5 mg) may be as effective as higher doses.[137,146]

Although Handley found metoclopramide (0.25 mg/kg) to be as effective as droperidol (0.075 mg/kg) in children following strabismus surgery,[147] other studies have failed to confirm metoclopramide's antiemetic action.[141,143,148] This variable response may be related in part to varying dosages, routes and timing of the administration. Metoclopramide may be most effective when given at the end of anaesthesia upon arrival in the recovery room[149] or in combination with other antiemetics[150] such as low dose droperidol (0.5–1.0 mg) with metoclopramide (10–20 mg).[70]

A transdermal hyoscine patch applied behind the ear the night prior to surgery can also decrease emetic sequelae and shorten discharge time.[151] However, its effectiveness is variable and side effects (for example, dry mouth, visual disturbances and dysphoria) can be troublesome.[152–154] The antihistaminic drug hydroxyzine also appears to be effective in decreasing emetic sequelae.[155] Other effective antiemetics include: prochlorperazine (5–10 mg); perphenazine (1–3 mg); benzquinamide (25–50 mg); promazine (5–15 mg); and trimethobenzamide (100–200 mg). Intravenous lignocaine has been claimed to decrease vomiting after strabismus correction surgery.[156]

Several newer drugs (for example, the serotonin antagonist ondansetron) are under active clinical investigation for the prevention and treatment of postoperative nausea and vomiting.[157,158] However, it is yet to be seen if their advantages will justify their cost in outpatient anaesthesia.[159]

Anaesthetic techniques

The ideal anaesthetic for outpatients would produce a rapid and smooth onset of action, intraoperative amnesia and analgesia, good surgical conditions, and a short recovery period free from side effects. Although many outpatient procedures are performed under general anaesthesia, regional techniques can be used for a wide variety of urological, gynaecological, and orthopaedic procedures. Sedative–analgesic drugs can be valuable supplements to local anaesthetic techniques during monitored anaesthesia care.

Monitoring

Outpatients require the same equipment as inpatients for delivery of anaesthetic drugs, monitoring and resuscitation.[160] Standard intraoperative monitoring equipment for outpatient operations includes an electrocardiogram, a non-invasive blood pressure machine, a pulse oximeter and a capnograph. The major risk factors for hypoxaemia during outpatient general anaesthesia include obesity, age greater than 35 years, lithotomy position, manual ventilation and light anaesthesia.[161] Temperature monitoring is useful for young adults, adolescents, and children undergoing general anaesthesia with known triggering agents of malignant hyperthermia.

In order to facilitate access to important patient data during the perioperative period, many outpatient centres have adopted anaesthesia record systems which combine preoperative, intraoperative, and postoperative information onto one chart.

Intravenous fluids

Children undergoing procedures lasting less than 15 minutes, which do not require the intravenous administration of drugs or fluid (for example myringotomies and eye examinations under anaesthesia) do not always require intravenous access. Hypoglycaemia does not occur in healthy children who fast for less than 15 hours.[162] However, during longer cases or if the patient has had a prolonged fast (over 15 hours), an intravenous line should be started and may be used for maintenance of fluid balance, glucose homoeostasis and administration of drugs during the perioperative period.[108]

The benefit of fluid administration during brief outpatient procedures is unproved. Whereas Cook et al. found that both compound sodium lactate (20 ml/kg) and glucose (1 g/kg) solutions produced significant improvement in variables that reflected hydration,[163] Ooi et al. were unable to demonstrate any significant benefit of 500 ml of intravenous fluid (although this may have been due to the size of the study).[164]

Local anaesthetic cream containing a eutectic mixture of lignocaine and prilocaine (Emla™) can decrease pain at the intravenous injection site allowing the use of intravenous induction techniques in children. It should be applied 60–90 minutes prior to induction of anaesthesia and covered with an occlusive dressing for maximum benefit.[165]

Airway management

Sore throat is a significant cause of morbidity following outpatient surgery. Sore throats are more common if the patient is allowed to breathe spontaneously after intubation but the incidence is not influenced by deep or light extubation.[166,167] Airway maintenance using a facemask or laryngeal mask airway reduces the incidence of postoperative sore throat.

Laparoscopy for sterilization or diagnosis of pelvic disease is a common outpatient procedure that may be performed under regional or general anaesthesia. In many centres all patients undergoing laparoscopy are intubated because of the risks of regurgitation and hypoventilation in the Trendelenburg position. However, in some units, patients are not intubated. By avoiding laryngoscopy and intubation, the amount of anaesthetic drugs administered is decreased and recovery is faster with fewer minor side effects.[168] Spontaneous ventilation via a facemask in the Trendelenburg position does not result in significant hypercapnia or acidosis in non-obese patients, nor is it associated with reflux of gastric contents except in patients with hiccoughs.[169]

The laryngeal mask airway (LMA) offers several advantages for outpatient anaesthesia. Insertion does not require muscle relaxation, may be easier than intubation with a tracheal tube and is associated with fewer sore throats. Use of the LMA is associated with less hypoxia than conventional mask anaesthesia.[170] The LMA is also suitable for children.[171] For dental surgery the LMA is preferable to conventional nasal mask as there are fewer episodes of hypoxaemia and surgical access is unaffected.[172] For examination under anaesthesia of the eye the LMA is associated with fewer interruptions to surgery than a conventional facemask anaesthetic.[173]

Regional anaesthetic techniques

Regional anaesthesia offers many advantages to the outpatient. The side effects of general anaesthesia and tracheal intubation are avoided, the risks of aspiration pneumonitis are minimized, postanaesthetic nursing care and patient recovery times may be reduced, and analgesia is provided into the early postoperative period.[23,174,175] Although some studies indicate that there may be fewer postoperative complications in patients receiving regional anaesthesia, there is continuing debate as to whether it is truly safer than general anaesthesia.

Following oral surgery there are relatively minor differences in postoperative morbidity following regional or general anaesthesia. Patients receiving local anaesthesia have greater difficulty eating on the day of the operation because of the residual numbness, while drowsiness and nausea are more common after general anaesthesia. However, 65 per cent of those who had local anaesthesia felt 'unfit' to return to work or school the following day compared with only 28 per cent in the group receiving general anaesthesia.[176]

In children, a regional block performed immediately after induction of general anaesthesia can reduce the anaesthetic requirement, provide postoperative analgesia and allow more rapid recovery.[177,178] Caudal epidural anaesthesia is an effective technique in children undergoing lower abdominal, perineal and lower extremity procedures. Combined ilioinguinal and iliohypogastric nerve block or a caudal reduce pain following herniotomy or herniorrhaphy, although rectal diclofenac or simple wound infiltration may be as effective.[179–182] Post-circumcision pain may be reduced by a caudal, a block of the dorsal nerve of the penis, subcutaneous ring block, or topical lignocaine ointment.[183–188]

Spinal (subarachnoid) blockade

Spinal anaesthesia is useful for lower extremity, urological, some gynaecological procedures (for example tubal ligation) and herniorrhaphy. However, the risk of postdural puncture headache (PDPH) has led many clinicians to limit the use of spinal anaesthesia to patients over the age of 60 years as the incidence is lower in this age group.[189] The use of smaller spinal needles can reduce the incidence of PDPH to 7.5 per cent with 26G or 1.8 per cent with 27G needles.[190,191] A period of enforced bedrest, prophylactic analgesics or a prophylactic epidural blood patch do not reduce the incidence of PDPH following spinal anaesthesia.[192] However, if a PDPH occurs and fails to respond to simple measures such as hydration and oral analgesics, an epidural blood patch is highly effective and may be performed on an outpatient basis.[193–196]

Epidural anaesthesia

Epidural anaesthesia has been advocated for lower extremity procedures, herniorrhaphy and extracorporeal shock wave lithotripsy.[197–200] Epidural blockade for diagnostic laparoscopy reduces the time to discharge and the incidence of postoperative nausea and vomiting compared with nitrous oxide–enflurane general anaesthesia.[201] However, combinations of either midazolam and alfentanil or propofol and fentanyl have shorter anaesthesia and recovery times than epidural anaesthesia for immersion lithotripsy.[202] Short-acting local anaesthetic agents, for example lignocaine (lidocaine), are sufficient for most outpatient procedures. For longer procedures the addition of adrenaline (epinephrine) or use of a longer acting agent may permit a 'single shot' technique. Continuous lumbar epidural anaesthesia with lignocaine has been used successfully for outpatient lithotripsy and arthroscopy.[203,204] Continuous thoracic epidural anaesthesia has been used for outpatient oncological and reconstructive breast surgery.[205]

Caudal epidural anaesthesia is a useful technique for anorectal surgery and some gynaecological procedures.[206] These procedures can also be performed with perianal infiltration or paracervical blocks, respectively.[207]

Brachial plexus blockade

Day-case procedures on the arm may be performed following a brachial plexus block. The axillary or interscalene approaches are preferable to the supraclavicular route because of the risk of pneumothorax. In a study of 543 brachial plexus blocks, 98 per cent were performed using the axillary approach and the success rate was similar using paraesthesia, transarterial fixation, nerve stimulation, or a combination for identification of the injection site; 7 per cent of the blocks were incomplete and intravenous sedation or general anaesthesia was required.[208]

Intravenous regional anaesthesia

For procedures on a single extremity, the intravenous regional block (Bier's block) is a simple and reliable technique. It may be used in children as well as adults.[209] Tourniquet pain may be reduced by using a double cuff. Alkalinization of the local anaesthetic solution may reduce pain on injection and improve the quality of anaesthesia.[210]

Miscellaneous nerve blocks

The 'three-in-one' block (femoral, obturator, and lateral femoral cutaneous nerves) provides adequate anaesthesia for outpatient knee arthroscopy and gives excellent postoperative analgesia.[211] Nerve block at the ankle is also simple and effective for surgery on the foot.[212]

Local anaesthetic infiltration

Infiltration of the operative site with a dilute solution of local anaesthetic is simple, safe, and provides satisfactory conditions for many operations including urological procedures such as vasectomy, orchidopexy, and hydrocele and spermatocele surgery.[213,214] Transurethral resection of the prostate may also be performed under local anaesthesia.[215] Inguinal herniorrhaphy under local anaesthesia has excellent patient acceptance and few postoperative complications.[216] Outpatient knee arthroscopy is commonly performed under local anaesthesia.[217] The use of a retrobulbar block instead of general anaesthesia for ophthalmological surgery has many advantages including a marked reduction in the incidence of postoperative emesis.[218] Combinations of local infiltration and intercostal nerve blocks have been used for lithotripsy and breast biopsy.[219,220] Although only superficial operations can be performed with local infiltration or a 'field block', wound infiltration with local anaesthesia will decrease incisional pain in the recovery room. The use of topical local anaesthetic as creams, aerosols and Emla™ also provides effective postoperative analgesia.[221,222]

Monitored anaesthesia care

In 1966, Shane described an 'intravenous amnesia' technique involving the use of small incremental doses of barbiturates, opioids, anticholinergics and ataractics.[223] The American Dental Association has distinguished between sedation and general anaesthesia (Table 68.4)[224]. Intravenous sedative and analgesic drugs are often used in conjunction with local anaesthesia, especially by oral and plastic surgeons, to reduce the discomfort of local anaesthetic injections and recall of intraoperative events.[225] Most oral surgery outpatients prefer local anaesthesia with sedation to local anaesthesia alone and patient satisfaction is directly related to the intraoperative level of sedation.[226,227]

Effective sedation–analgesia techniques can also be used as alternatives to general anaesthesia in the outpatient setting.[228] However, because of the risk of ventilatory depression, respiratory monitoring and supplemental oxygen are required. Clinically significant oxygen desaturation may occur in up to 40 per cent of patients who do not receive supplemental oxygen during sedation techniques although it has also been observed in patients who received local anaesthesia alone.[229] These risks must be taken into consideration when assessing the safety of 'local' anaesthesia. The FASA study found the incidence of perioperative complications to be lowest in those receiving local anaesthesia alone and significantly higher in those

Table 68.4 Definitions of conscious sedation and related terms proposed by the American Dental Association

Analgesia	Diminution or elimination of pain in the conscious patient
Local anaesthesia	Elimination of sensations, especially pain, in one part of the body by the topical application or regional injection of a drug
Conscious sedation	Minimally depressed level of consciousness that retains the patient's ability to independently and continuously maintain an airway and respond appropriately to physical stimulation and verbal command, produced by a pharmacological or non-pharmacological method, or a combination
General anaesthesia (including deep sedation)	Controlled state of depressed consciousness or unconsciousness, accompanied by partial or complete loss of protective reflexes, including the ability to independently maintain an airway and respond purposefully to physical stimulation or verbal command, produced by a pharmacological or non-pharmacological method, or a combination

Reproduced from McCarthy FM, Solomon AL, Jastak JT, *et al.* Conscious sedation: benefits and risks. *Journal of the American Dental Association* 1984; **109**: 546–57. Copyright © 1984, reprinted by permission of ADA Publishing Co., Inc.

receiving concurrent sedation.[18] Overall, morbidity following the use of intravenous sedation is higher in women than in men.[230] Achieving an optimal balance between patient comfort and safety requires careful titration of sedative and analgesic drugs, appropriate monitoring of the cardiovascular, respiratory and central nervous systems, and most importantly, good communication between the anaesthetist, patient and surgeon.

Benzodiazepines

Benzodiazepines are the most widely used drugs to produce sedation and amnesia in the operating room. Diazepam has a long elimination half-life (24–48 hours) which is increased in elderly patients. Recovery of cognitive and motor skills is slower after diazepam sedation than following either methohexitone or halothane anaesthesia.[231] Midazolam offers several advantages over other benzodiazepines. It is two to four times more potent than diazepam, has a faster onset of action and produces more effective amnesia.[232–234] Recovery from similar levels of intraoperative sedation is more rapid with midazolam than diazepam.[235] Oral midazolam has been used for conscious sedation in outpatients undergoing minor procedures and was associated with greater patient and physician acceptance than intravenous diazepam.[236] Sublingual lormetazepam is as effective as parenteral benzodiazepine for conscious sedation but is less acceptable to patients.[237] Lorazepam has a slow onset of action which precludes careful titration and its long duration of amnesia could delay discharge.

Flumazenil is a specific benzodiazepine antagonist that may be used to reverse the residual sedative and amnesic effects of benzodiazepines and permit earlier discharge.[238,239] However, the short half-life of flumazenil (1–2 hours) may allow recurrence of sedation after discharge.[240]

Opioid analgesics

The addition of an opioid analgesic to a benzodiazepine significantly improves patient comfort during local anaesthesia.[241] For example, midazolam and fentanyl may be combined by initially sedating the patient with midazolam (2–5 mg) intravenously followed by small aliquots of fentanyl (25–50 µg) given as needed to control pain. Because these drugs cause ventilatory depression by different mechanisms synergism may occur. Benzodiazepines reduce the hypoxic drive and opioid analgesics reduce the hypercapnic respiratory drive, resulting in a high incidence of apnoea.[242,243] Nalbuphine, an agonist–antagonist opioid, produces effective sedation with a reduced risk of severe ventilatory depression compared with fentanyl.[244] However the combination of nalbuphine and midazolam has a higher incidence of vomiting and hypoxia than midazolam alone.[245,246] When more profound sedation is required, a nalbuphine–methohexitone combination can be highly effective.[247,248] Carefully titrated, an alfentanil infusion can provide more profound analgesia than the agonist–antagonists and less ventilatory depression than the benzodiazepine–opioid combinations.[249]

Ketamine

Ketamine has been used for sedation[250] but may cause marked cardiovascular stimulation. Unpleasant psychomimetic emergence reactions may occur if the patient is not first sedated with a benzodiazepine.[251] Midazolam is preferable to diazepam as it produces more profound intraoperative sedation, anxiolysis and amnesia, and higher overall patient acceptance during low dose ketamine infusions.[232,252] However, amnesia after midazolam presedation may persist into the postoperative period.[232,253]

Propofol

Propofol infusions are highly effective in producing rapid, smooth and controllable sedation.[254,255] Following oral surgery propofol produced better amnesia and shorter recovery times than boluses of diazepam, and was preferred by 66 per cent of patients.[256] The overall quality

of intraoperative sedation is similar to a midazolam infusion but the use of propofol results in a more rapid recovery of cognitive function.[257] The propofol infusion should be given into a large vein (for example at the antecubital fossa) to minimize pain on injection.[258] Propofol, used alone in sedative–hypnotic doses does not cause respiratory depression.[259] However, care must be taken to avoid general anaesthesia with its associated complications.

Other sedation techniques

Barbiturates and etomidate have also been administered by infusion to provide sedation during local anaesthesia (for example methohexitone, 1–2 mg/min).[260,261] The dose of these drugs must be carefully titrated to avoid respiratory and cardiovascular depression.

Inhalational sedation with nitrous oxide (30–50 per cent), low dose enflurane (0.5 per cent), or isoflurane (0.5 per cent) can be used to supplement regional anaesthesia.[262–264] Frequent side effects (for example nausea, confusion), and concern over operating theatre pollution have limited the usefulness of these techniques and intravenous sedation is usually preferred.

General anaesthetic techniques

The delivery of safe and effective general anaesthesia with minimal side effects and a rapid recovery is essential in a busy outpatient surgery unit. General anaesthesia remains the most widely used technique for day-stay patients because of its popularity with patients, surgeons, and anaesthetists.

Intravenous induction

Induction of general anaesthesia is usually accomplished with a rapid-acting intravenous anaesthetic agent. Thiopentone (3–6 mg/kg), is usually associated with a rapid induction of anaesthesia without significant side effects.[265] However, poor psychomotor recovery and subjective feelings of tiredness and drowsiness limit its usefulness in day-case patients.[266] Methohexitone is associated with slightly shorter awakening and recovery times than thiopentone.[265,267,268] However, recovery of fine motor skills may not be complete until 6–8 hours after induction.[269] Methohexitone may also cause pain on injection, involuntary muscle movements and hiccoughing.[270] Use of a small dose of a rapid-acting opioid analgesic (for example, sufentanil 5–10 μg i.v.) can minimize these side effects without prolonging recovery.[104,271]

Induction of anaesthesia with propofol is associated with a greater decrease in blood pressure and heart rate than with either thiopentone or methohexitone but this is offset by a more rapid recovery (Fig. 68.1) and fewer post-operative side effects.[272–275] Most anaesthetists consider propofol to be the intravenous induction agent of choice for outpatient anaesthesia.[276] Propofol may be used in the elderly outpatient but the dose should be reduced by 25 per cent.[277] Pain on injection may be reduced by the addition of lignocaine or cooling the propofol to 4°C.[278] The direct antiemetic action of propofol may offer a further advantage.[279] However, propofol's improved recovery profile may be negated if anaesthesia is maintained with a volatile agent and nitrous oxide,[280,281] leading some

workers to question whether its advantages justify its expense.[282]

Etomidate may be used for induction and maintenance of general anaesthesia during short outpatient procedures.[265] Recovery is more predictable than after thiopentone and compares favourably with methohexitone.[102,103,283] Disadvantages of etomidate include pain on injection, post-operative nausea and vomiting, myoclonic movements, and transient suppression of adrenal steroidogenesis.[265,284,285] However, its haemodynamic stability offers an advantage over other available induction agents for outpatients with coronary artery or cerebrovascular disease.

Midazolam (0.2–0.4 mg/kg) alone is also an adequate intravenous induction agent.[286] However, its onset of action is slower and recovery is prolonged compared with the barbiturates.[287,288] Flumazenil given at the end of surgery speeds recovery following midazolam induction,[289] but its duration of reversal is limited to 60 minutes.[290] Recovery after midazolam is significantly slower and associated with worse psychomotor function than following propofol.[291,292]

Inhalational induction

In children, an inhalational induction is a useful alternative to the standard intravenous induction techniques. The combination of an intravenous agent for induction and an inhalational agent for maintenance of brief procedures is associated with a prolonged recovery compared with the use of an inhalational agent alone.[267] Unfortunately, inhalational inductions are frequently more time consuming and many children object to the facemask and the pungent smell of inhaled agents. In cooperative children these problems can be reduced by the use of the single breath induction technique.[293] Halothane remains the drug of choice as it is associated with shorter induction times and fewer respiratory problems than either isoflurane or enflurane.[294–297] Although dysrhythmias occurred more frequently during induction of anaesthesia with halothane, no serious rhythm disturbance was reported in any of these studies.

In unruly, frightened, or mentally retarded children, methohexitone (20–30 mg/kg) can be administered rectally prior to entering the operating theatre.[298] However, the recovery time may be prolonged. Rectal etomidate (6 mg/kg) or ketamine (50 mg/kg) can also produce a rapid onset of anaesthesia (less than 4 minutes) without cardio-respiratory depression and no delay in recovery.[299,300] Intramuscular ketamine (2–6 mg/kg) is useful for induction of an uncooperative child;[82,301] however, 'home-readiness' will be delayed when larger doses of ketamine (for example 6–10 mg/kg intramuscularly) are combined with a volatile anaesthetic. In addition, recurrent illusions (flashbacks) have been reported several weeks after

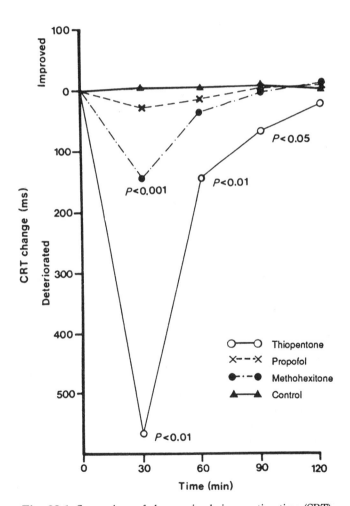

Fig. 68.1 Comparison of changes in choice reaction time (CRT) in patients following induction of anaesthesia with thiopentone, methohexitone, propofol and controls. (Reproduced, by permission, from Mackenzie N, Grant IS. Comparison of the new emulsion formulation of propofol with methohexitone and thiopentone for induction of anaesthesia in day cases. *British Journal of Anaesthesia* 1986; **57**: 725–31.)

ketamine administration in children who did not receive benzodiazepine premedication.[302]

Children's fears during the induction period may be minimized by play-oriented preoperative teaching. It is helpful to allow parents to comfort their child during induction to relieve the child's anxiety and produce a smoother induction of anaesthesia.[303]

Inhalational maintenance techniques

Volatile anaesthetic agents have generally been considered to be superior to intravenous agents for maintenance during outpatient surgery. The rapid uptake and elimination of volatile agents from the lungs allows rapid changes in the depth of anaesthesia and should theoretically provide for a faster recovery and discharge. The recovery times for adult outpatients undergoing brief procedures are similar with the three commonly used volatile agents (halothane, enflurane and isoflurane).[268,304,305] However, the two newer agents desflurane and sevoflurane should permit more rapid recovery times because of their lower blood–gas partition coefficients.

Halothane

Although children recover consciousness more rapidly after enflurane anaesthesia,[294,306] most studies have reported that halothane is associated with the lowest incidence of perioperative complications and does not prolong recovery times.[295–297] However, ventricular dysrhythmias are more likely during halothane anaesthesia, especially in women and children undergoing ENT and dental procedures.[296,307,308]

Enflurane

Enflurane is popular for outpatient anaesthesia because of its favourable recovery profile compared with halothane or isoflurane.[309,310] However, for procedures of over 90 minutes recovery times after isoflurane are shorter.[311] Most anaesthetists would avoid enflurane in outpatients with epilepsy as convulsions following discharge have been reported.[312]

Isoflurane

Despite its lower solubility, isoflurane may not offer any significant advantages over halothane or enflurane for short procedures. Outpatients receiving isoflurane rather than halothane or enflurane during minor gynaecological surgery have a higher incidence of postoperative complications.[304] Induction of anaesthesia in children with isoflurane

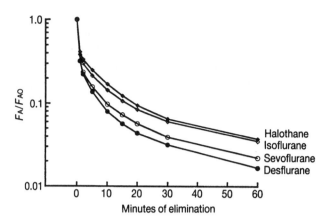

Fig. 68.2 Alveolar concentration (F_A) as a fraction of the last alveolar concentration during anaesthesia (F_{AO}) during the washout period. (Reproduced, by permission, from Eger II EI. Desflurane animal and human pharmacology: aspects of kinetics, safety, and MAC. *Anesthesia and Analgesia* 1992; **75:** S3–9.)

is associated with an increased incidence of coughing and laryngospasm compared to halothane or enflurane.[294–297]

Desflurane

Desflurane is the most recently introduced volatile anaesthetic agent. Its blood–gas partition coefficient (0.42) is similar to that of nitrous oxide and it has a rapid wash-in and wash-out (Fig. 68.2).[313] The time required for the end tidal concentration to decrease by 50 per cent is 2.5 minutes for desflurane compared to 9.5 minutes for isoflurane.[314]

Table 68.5 Recovery times after discontinuation of inhaled anaesthetic

	Isoflurane group	Desflurane group
End tidal agent concentration at end of surgery (%)	0.6 ± 0.2	2.9 ± 0.4*
Time to 50% decrease in end tidal concentration (min)	9.5 ± 3.4	2.5 ± 0.8*
Time to opening eyes (min)	10.2 ± 7.7	5.1 ± 2.4*
Time to following commands (min)	11.1 ± 7.9	6.5 ± 2.3*
Time to sitting up in a chair (min)	111 ± 27	95 ± 56
Time to phase 1 to 2 transfer (min)	118 ± 36	105 ± 49
Time to 'home ready' (min)	231 ± 40	207 ± 54

Values are mean ± SD.
* Significantly different from isoflurane group; $P < 0.05$.
Reproduced, by permission, from Ghouri AF, Bodner M, White PF. Recovery profile after desflurane–nitrous oxide versus isoflurane–nitrous oxide in outpatients. *Anesthesiology* 1991; **74:** 419–24.

Induction of anaesthesia with desflurane is rapid but associated with a high incidence of airway irritation, especially in children.[315–317] Compared with isoflurane, desflurane is associated with more rapid initial awakening and less impairment of cognitive function (Table 68.5).[313,318] Psychomotor recovery following desflurane for induction or maintenance of anaesthesia is better than following propofol for induction and maintenance.[315] However, desflurane is associated with a higher incidence of nausea than propofol.[316,319] Given its favourable early recovery profile, desflurane would appear to be a useful alternative to the established agents although its respiratory irritant properties make it unlikely to replace propofol as the induction agent of choice.[320]

Sevoflurane

Sevoflurane has not been introduced into the UK or USA although it has been licensed in Japan since 1990. Inhalational induction with sevoflurane and nitrous oxide in oxygen is rapid and not associated with coughing or laryngospasm. In children, induction with sevoflurane in oxygen produces more rapid emergence and a significantly shorter postoperative recovery time compared with halothane.[321] In adults, induction with sevoflurane and nitrous oxide is more rapid than with enflurane and nitrous oxide but recovery times are similar.[322] Emergence times following propofol or sevoflurane induction and sevoflurane maintenance are similar and more rapid than a propofol–isoflurane–nitrous oxide technique. However, late recovery times are similar.[323] Serum fluoride levels following sevoflurane anaesthesia correlate with exposure in MAC-hours although there does not appear to be any evidence of abnormal renal function following clinical use.[324]

Nitrous oxide

The use of a volatile agent in combination with nitrous oxide (60–70 per cent) in oxygen remains the most popular technique for maintenance of anaesthesia. The extremely low blood–gas partition coefficient of nitrous oxide contributes to the rapid onset and recovery of anaesthesia. Some investigators have found an association between nitrous oxide and postoperative nausea and vomiting,[325–328] but more recent studies have challenged these findings.[329–331] Certainly the combination of nitrous oxide and an opioid analgesic as part of a 'balanced' anaesthetic technique increases these side effects.[325]

Intravenous infusion techniques

An infusion of propofol is often used for maintenance of anaesthesia either in combination with an inhaled anaes-

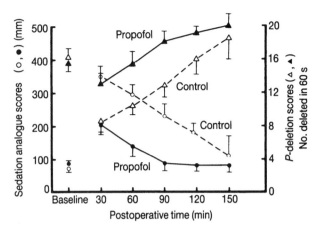

Fig. 68.3 Sedation analogue scores and p-deletion test scores following anaesthesia with propofol–nitrous oxide or thiopentone–isoflurane–nitrous oxide in outpatients. (Reproduced, by permission, from Doze VA, White PF. Comparison of propofol with thiopental-isoflurane for induction and maintenance of outpatient anesthesia. *Anesthesiology* 1986; 65: A544.)

thetic or as part of a total intravenous technique.[332–334] For very short cases infusion techniques may not offer any advantages over a conventional manual bolus method.[335] Total intravenous anaesthesia with propofol results in a more rapid recovery than with current inhalational techniques,[336] although the advent of desflurane and perhaps sevoflurane may alter this.

For outpatient procedures lasting less than 1 hour, early recovery times are shorter with propofol–nitrous oxide than with thiopentone–isoflurane–nitrous oxide anaesthesia[337] (Fig. 68.3) but no better than a methohexitone–nitrous oxide combination.[338] When used alone, a propofol infusion provides better surgical conditions, a faster recovery, earlier discharge and less postoperative morbidity than thiopentone–enflurane or thiopentone–isoflurane techniques.[339,340] Nevertheless, complete recovery of psychomotor function may require up to 3 hours after maintenance of anaesthesia with propofol.[341] Overall, propofol appears to improve the quality of outpatient care and reduces recovery room costs.[342]

Midazolam and alfentanil infusions have been used to provide general anaesthesia for outpatient surgery. However, even with flumazenil, recovery takes much longer and is less complete than following propofol[343] and resedation may occur.[344]

Opioid analgesics

Opioid analgesics given prior to or during general anaesthesia can reduce the requirement for sedative–hypnotic drugs and improve recovery times and psychomotor function after short procedures.[69,345,346] Although morphine and pethidine have been used in outpatient anaesthesia,[47] they have generally been replaced by the more potent, faster-onset, and shorter-acting opioid analgesics such as fentanyl, sufentanil and alfentanil. These drugs are usually used to supplement an inhaled anaesthetic agent during the maintenance period.[347] Opioids given at the time of induction decrease the incidence of pain on injection and involuntary motor activity associated with methohexitone, etomidate and propofol[346,348,349] and small doses (for example, fentanyl 2–4 µg/kg, sufentanil 0.25–0.5 µg/kg) can also effectively attenuate the cardiovascular response to laryngoscopy and intubation.[350,351]

The extremely rapid onset and short duration of action of alfentanil make it particularly useful in outpatient anaesthesia. Many studies have shown a more rapid emergence and recovery of psychomotor function after alfentanil compared to fentanyl.[352–356] Alfentanil's small volume of distribution and short terminal elimination half-life make it particularly suitable for administration by continuous infusion. The addition of alfentanil improves anaesthetic conditions and decreases postoperative morbidity after both propofol and thiopentone–enflurane anaesthesia.[339] Alfentanil may be combined with methohexitone or propofol to produce total intravenous anaesthesia.[332,357] However, combinations with etomidate or nitrous oxide produce a high incidence of postoperative nausea and vomiting.[356,358]

Fentanyl is commonly used as an adjunct for outpatient anaesthesia. However, for short gynaecological procedures its addition confers no advantage to a propofol–nitrous oxide technique.[359] Although a small dose of fentanyl (50 µg) does not increase postoperative nausea or vomiting, neither does it have any effect on analgesic requirements.[360] Use of a sufentanil infusion for maintenance of general anaesthesia with nitrous oxide has been associated with less nausea and reduced postoperative pain.[361] Replacing the inhalational agent with sufentanil may decrease the requirement for postoperative analgesics and shorten the recovery room stay.[362]

Opioid agonist–antagonist analgesics

It has been suggested that the semi-synthetic opioid agonist–antagonist compounds might offer advantages over the potent opioid compounds because of the 'ceiling effect' on respiratory depression. Unfortunately, there is a similar ceiling effect with respect to analgesic efficacy. Butorphanol (20–40 µg/kg) provides similar intraoperative conditions and better postoperative pain relief than fentanyl (1–2 µg/kg) during outpatient laparoscopy but the higher dose is associated with more postoperative nausea, sedation and a longer time to discharge.[363–365]

Buprenorphine (5 µg/kg) has been combined with propofol for total intravenous anaesthesia and provided haemodynamically stable operating conditions and rapid recovery without recall.[366] Nalbuphine (0.3–0.5 mg/kg) produces more unpleasant dreaming during surgery and more postoperative anxiety, drowsiness and nausea and vomiting compared with fentanyl (1.5 µg/kg).[367] However, lower doses of nalbuphine (0.25 mg/kg) can provide effective postoperative analgesia for 2 hours without increasing emetic sequelae.[368] Dezocine (6 mg) reduces postoperative analgesic requirements but increases postoperative nausea and recovery times compared with fentanyl.[369]

Non-opioid analgesic drugs

The high incidence of nausea and vomiting following day-case anaesthesia, which may in part be related to the use of opioids, has led to the search for alternative analgesic adjuvants (for example the non-steroidal anti-inflammatory drugs). Following arthroscopic procedures diclofenac (1 mg/kg) significantly reduces postoperative pain and reduces the need for further analgesic medication.[370,371] However, diclofenac (75 mg) had no significant effect on postoperative pain following day-case laparoscopy.[372] Both ketorolac (60 mg) and naproxen (550 mg orally) reduce postoperative pain, emetic side effects, and recovery times following laparoscopy and tubal ligation.[368,373]. Paracetamol–codeine combinations and ibuprofen have also been shown to reduce analgesic requirements following orthopaedic and dental surgery.[374] Infiltration of the mesosalpinx with bupivacaine (0.5 per cent) decreases the analgesic requirement after laparoscopic tubal ligation and likewise, intra-articular bupivacaine (0.5 per cent) reduces the opioid requirements and facilitates earlier mobilization following arthroscopy.[375,376]

Neuromuscular blocking agents

The use of muscle relaxants can decrease the total anaesthetic requirement and shorten recovery times.[377] Although it has been suggested that controlled ventilation with muscle relaxants might decrease postoperative emesis, Walsh et al. were unable to demonstrate any benefit in outpatients undergoing strabismus repair.[378]

Prior to the introduction of intermediate-acting non-depolarizing muscle relaxants, a suxamethonium infusion was frequently used during outpatient anaesthesia. Administration of suxamethonium may be associated with muscle pain lasting up to 4 days after the operation.[379,380] However, factors other than suxamethonium may also contribute to these myalgias. Although the use of non-depolarizing agents has been shown to reduce myalgias in several studies, this is not a universal finding.[381–383]

The neuromuscular blockade produced by the intermediate-acting non-depolarizing muscle relaxants atracurium and vecuronium can usually be reversed after brief surgical procedures.[384–388] Mivacurium has a shorter duration of action than either atracurium or vecuronium and, because of its high rate of spontaneous recovery, anticholinesterase agents may be avoided.[389] This may be an advantage as it has been suggested that the antagonism of neuromuscular blockade by neostigmine and atropine may lead to an increase in postoperative nausea and vomiting.[390] Mivacurium is metabolized by plasma cholinesterase and prolonged neuromuscular block may occur in individuals with reduced cholinesterase levels.[391] The onset of action of mivacurium is slower than suxamethonium.[392,393] Rocuronium produces rapid intubating conditions similar to suxamethonium[394,395] but its duration of action is the same as vecuronium.[396,397]

Post-anaesthetic care management

Experienced recovery room staff play an important role during the recovery process.[398] Episodes of oxygen desaturation may occur following short general anaesthetics despite the use of supplemental oxygen. However, these episodes may be reduced significantly when experienced staff trained in airway management are present.[399]

The incidence of postoperative symptoms after outpatient surgery is high irrespective of the anaesthetic technique used.[400] Tracheal intubation results in a higher incidence of airway-related complaints and greater morbidity.[401] Side effects after general anaesthesia may be reduced by simple manoeuvres, for example, the use of airway heating and passive heat and moisture exchangers to decrease postoperative shivering after longer outpatient procedures.[402] Central stimulants (for example, doxapram) have been used by some to hasten arousal after outpatient general anaesthesia.[403]

Discharge criteria

The accurate assessment of recovery of cognitive and psychomotor function is important in determining the appropriate time for discharge after ambulatory surgery.[404] The recovery room nurse, the surgeon, the anaesthetist, as well as the patient (and their responsible escort), all play

Table 68.6 Guidelines for discharge following day-surgery anaesthesia

1. Stable vital signs for at least 1 hour
2. No evidence of respiratory depression or airway obstruction
3. Orientated to time, place and person
4. Able to dress and walk at his/her preoperative level
5. Able to take fluid orally and void urine (optional)
6. Minimal nausea and vomiting
7. Pain controllable by simple analgesics
8. Responsible adult to escort the patient home and stay the first night
9. Patient has received instructions for postoperative care

Adapted, by permission, from Korttila K. The outpatient facility. In: Nunn JF, Utting JE, Brown BR, eds. *General anaesthesia*. 5th ed. London: Butterworth & Co. Ltd, 1989: 339–45.

important roles in determining when a 'home ready' state has been achieved. It is not always necessary for a physician to examine a patient immediately prior to discharge providing that the nursing staff apply specific discharge criteria which have been approved by the medical staff.

There are no standardized discharge criteria[405] although most units have developed their own (Table 68.6). A wide variety of psychomotor tests have been used to assess recovery following general anaesthesia or sedation.[406–410] Most of these tests are too complex and time consuming to use in a busy clinical setting but can provide information to

Table 68.7 Psychomotor tests used to assess cognitive recovery from anaesthetic and analgesia drugs

1. Simple reaction time	The time taken to press a button in response to a single light source flashed randomly at a constant place
2. Choice reaction time	The time taken to press the correct button corresponding to one of four coloured light sources flashed randomly at four constant places
3. Perceptive accuracy test	The percentage of correct responses in identifying a two digit number displayed on a computer screen for 0.5 seconds
4. Digit substitution test	The number and percentage of correct substitutions of symbols for numbers according to a key within 90 seconds
5. Bender Gestalt Track Tracer test	The percentage of time a stylus is in contact with the side whilst tracing narrow circular and square tracks
6. Trieger Dot test	The accuracy of joining dots on paper as judged by the number of dots missed

Adapted, by permission, from Gupta A, Larsen LE, Sjoberg F, Lindh ML, Lennmarken C. Thiopentone or propofol for induction of isoflurane-based anaesthesia for ambulatory surgery? *Acta Anaesthesiologica Scandinavica* 1992; **36**: 670–4. Copyright © 1992 Munksgaard International Publishers Ltd, Copenhagen, Denmark.

help develop practical discharge criteria.[411] Simple tests of memory and sensorimotor coordination appear to be the most useful indices of recovery (Table 68.7). The Bender Gestalt Track Tracer test is a reliable, valid, objective, non-invasive, inexpensive test which can be easily performed by outpatients in under a minute.[412]

Additional criteria must be applied to patients who undergo regional anaesthesia. Prior to discharge patients should have recovered full motor and sensory function. Once the sensory deficits have resolved, residual sympathetic blockade and orthostatic hypotension are rarely a problem on ambulation.[413] It has been suggested that patients may be safely discharged when they have two successive orthostatic mean arterial pressure decreases of 10 per cent or less after a spinal anaesthetic.[414] However, several studies have reported recovery of sympathetic activity before complete regression of the subarachnoid block.[413,415,416] Therefore, prior to ambulation, patients should have intact perianal sensation (S_{4-5}), the ability to plantar flex the foot and normal proprioception in the big toe.[413]

Whilst the use of a checklist is appropriate to determine when patients may be discharged from the day-surgery unit their content should be regularly reviewed. It is no longer considered necessary to drink prior to discharge: children forced to drink before discharge have a higher incidence of vomiting and a longer stay in the recovery unit.[417]

Prior to discharge pain should be controlled with oral medication. Following discharge the most important single factor in determining the delay in returning to normal activity is the total amount of pain experienced.[418]

Patients should be given written instructions on discharge including details of whom to contact if they experience any problems. In many day-surgery units in the USA patients are routinely contacted the following day by telephone to allow assessment of the post-discharge recovery.

Summary

The basic principles of ambulatory surgery were originally applied to healthy children and adults undergoing 'minor operations'.[3,419] There is increasing evidence that arbitrary limits placed on the type of surgery, age of patients, duration of operation, preoperative fasting period and selection of perioperative medication may be unwarranted. However, many controversies remain unresolved.[420] Over the last decade, an increasing number of 'high risk' patients have presented for outpatient surgery, making proper patient selection and evaluation increasingly important. Little is known about the optimal anaesthetic techniques for managing outpatients with pre-existing diseases or those at the extremes of age.

It is obvious that there is still much to learn about anaesthesia for ambulatory surgery.[421] In many situations, recovery times are still too long and the incidence of common side effects remain too high. A major limiting factor in outpatient surgery at present is inadequate postoperative pain management. The rational use of available combinations of anaesthetic drugs and equipment will provide for a rapid and smooth induction, excellent intraoperative conditions, and a rapid recovery with minimal side effects. The incidence of anaesthetic-related side effects (for example, drowsiness, headache, nausea, sore throat, myalgias, and dizziness) may be reduced depending on the premedication, the anaesthetic technique, and the skill of the anaesthetist. With the availability of more rapid and shorter-acting anaesthetic, analgesic, and muscle relaxant drugs, as well as improved techniques for administering these drugs (for example, the laryngeal mask airway and infusion pumps), the care which we provide to our expanding outpatient population should continue to improve in the future.

REFERENCES

1. Wells HA. *A history of the discovery of the application of nitrous oxide gas, ether and other vapors to surgical operations.* Hartford, Connecticut: J Gaylord Wells, 1847.

2. Biglow HJ. Insensibility during surgical operations produced by inhalation. *Boston Medical and Surgical Journal* 1846; **35**: 309–17.

3. Nicoll JH. The surgery of infancy. *British Medical Journal* 1909; **2**: 753–4.

4. Waters RM. The downtown anesthesia clinic. *American Journal of Surgery* 1919; **33**: 71–3.

5. Lagoe RJ, Milliren JH. Changes in ambulatory surgery utilization 1983–88: a community-based analysis. *American Journal of Public Health* 1990; **80**: 869–71.

6. Sadler GP, Richards H, Watkins G, Foster ME. Day-case paediatric surgery: the only choice. *Annals of the Royal College of Surgeons of England* 1992; **74**: 130–3.

7. Morris PB, Piper R, Reinke B, Young JR. Outpatient carbon dioxide laser mastectomy. Laser techniques, patient outcomes, cost containment. *AORN Journal* 1992; **55**: 984–92.

8. Wetchler BV, Brick J. Safety of outpatient cerclage. *Journal of Reproductive Medicine* 1990; **35**: 243–6.

9. Poindexter 3rd AN, Abdul-Malak M, Fast JE. Laparoscopic tubal sterilization under local anesthesia. *Obstetrics and Gynecology* 1990; **75**: 5–8.

10. Llorente J. Laparoscopic cholecystectomy in the ambulatory

surgery setting. *Journal of Laparoendoscopic Surgery* 1992; **2**: 23–6.

11. Clark JA, Kent 3rd RB. One-ay hospitalization following modified radical mastectomy. *American Surgeon* 1992; **58**: 239–42.

12. Ledet WP Jr. Ambulatory cholecystectomy without disability. *Archives of Surgery* 1990; **125**: 1434–5.

13. Strong NP, Wigmore W, Smithson S, Rhodes S, Woodruff G, Rosenthal AR. Daycase cataract surgery. *British Journal of Ophthalmology* 1991; **75**: 731–3.

14. Ashe RG, Corrigan DB, Myles TJ. Gynaecological day surgery at an area hospital 1983–1987. *Irish Journal of Medical Science* 1990; **159**: 280–2.

15. Zoutman D, Pearce P, McKenzie M, Taylor G. Surgical wound infections occurring in day surgery patients. *American Journal of Infection Control* 1990; **18**: 277–82.

16. Manian FA, Meyer L. Comprehensive surveillance of surgical wound infections in outpatient and inpatient surgery. *Infection Control and Hospital Epidemiology* 1990; **11**: 515–20.

17. Oesch A. Indications for and results of ambulatory varices therapy. *Therapeutische Umschau* 1991; **48**: 692–6.

18. Federated Ambulatory Surgery Association. *Special Study 1.* Alexandria, Virginia USA: Federated Ambulatory Surgery Association, 1986.

19. Michaels JA, Reece-Smith H, Faber RG. Case-control study of patient satisfaction with day-case and inpatient inguinal hernia repair. *Journal of the Royal College of Surgeons of Edinburgh* 1992; **37**: 99–100.

20. Collins KM, Docherty PW, Plantevin OM. Postoperative morbidity following gynaecological outpatient laparoscopy. A reappraisal of the service. *Anaesthesia* 1984; **39**: 819–22.

21. Natof HE. Pre-existing medical problems. Ambulatory surgery. *Illinois Medical Journal* 1984; **166**: 101–4.

22. Gold BS, Kitz DS, Lecky JH, Neuhaus JM. Unanticipated admission to the hospital following ambulatory surgery. *Journal of the American Medical Association* 1989; **262**: 3008–10.

23. Meridy HW. Criteria for selection of ambulatory surgical patients and guidelines for anesthetic management: a retrospective study of 1553 cases. *Anesthesia and Analgesia* 1982; **61**: 921–6.

24. Vandam LD. *To make the patient ready for anesthesia. Medical care of the surgical patient.* 2nd ed. Reading, Massachussets: Addison-Wesley, 1983.

25. Sear JW, Cooper GM, Kumar V. The effect of age on recovery. A comparison of the kinetics of thiopentone and althesin. *Anaesthesia* 1983; **38**: 1158–61.

26. Chung F, Lavelle PA, McDonald S. MMS: a screening test for elderly outpatients. *Anesthesiology* 1988; **69**: A900.

27. Chung F, Meier R, Lautenschlager E, Carmichael FJ, Chung A. General or spinal anesthesia: which is better in the elderly? *Anesthesiology* 1987; **67**: 422–7.

28. Campbell IR, Scaife JM, Johnstone JM. Psychological effects of day case surgery compared with inpatient surgery. *Archives of Disease In Childhood* 1988; **63**: 415–17.

29. Astfalk W, Warth H, Leriche C. Day case surgery in childhood from the parents' point of view. *European Journal of Pediatric Surgery* 1991; **1**: 323–7.

30. Steward DJ. Preterm infants are more prone to complications

following minor surgery than are term infants. *Anesthesiology* 1982; **56**: 304–6.

31. Liu LMP, Cote CJ, Goudsouzian NG, *et al.* Life-threatening apnea in infants recovering from anesthesia. *Anesthesiology* 1983; **59**: 506–10.

32. Kurth CD, Spitzer AR, Broennle AM, Downes JJ. Postoperative apnea in preterm infants. *Anesthesiology* 1987; **66**: 483–8.

33. Berry FA. Preexisting medical conditions of pediatric patients. *Seminars in Anesthesia* 1984; **3**: 24–31.

34. Egbert LD, Battit GE, Turndorf H, Beecher HG. The value of the preoperative visit by the anesthetist. *Journal of the American Medical Association* 1963; **185**: 553–7.

35. Patel RI, Hannallah RS. Preoperative screening for pediatric ambulatory surgery: evaluation of a telephone questionnaire method. *Anesthesia and Analgesia* 1992; **75**: 258–61.

36. Wilson ME, Williams NB, Baskett PJ, Bennett JA, Skene AM. Assessment of fitness for surgical procedures and the variability of anaesthetists' judgments. *British Medical Journal* 1980; **280**: 509–12.

37. Tompkins BM, Tompkins WJ, Loder E, Noonan AF. A computer-assisted preanesthesia interview: value of a computer-generated summary of patient's historical information in the preanesthesia visit. *Anesthesia and Analgesia* 1980; **59**: 3–10.

38. Lutner RE, Roizen MF, Stocking CB, *et al.* The automated interview versus the personal interview. Do patient responses to preoperative health questions differ? *Anesthesiology* 1991; **75**: 394–400.

39. Roizen MF, Coalson D, Hayward RS, *et al.* Can patients use an automated questionnaire to define their current health status? *Medical Care* 1992; **30**: MS74–84.

40. Kaplan EB, Sheiner LB, Boeckmann AJ, *et al.* The usefulness of preoperative laboratory screening. *Journal of the American Medical Association* 1985; **253**: 3576–81.

41. Golub R, Cantu R, Sorrento JJ, Stein HD. Efficacy of preadmission testing in ambulatory surgical patients. *American Journal of Surgery* 1992; **163**: 565–70.

42. Johnson Jr H, Knee-Ioli S, Butler TA, Munoz E, Wise L. Are routine preoperative laboratory screening tests necessary to evaluate ambulatory surgical patients? *Surgery* 1988; **104**: 639–45.

43. Wyatt WJ, Reed Jr DN, Apelgren KN. Pitfalls in the role of standardized preadmission laboratory screening for ambulatory surgery. *American Surgeon* 1989; **55**: 343–6.

44. Roy WL, Lerman J, McIntyre BG. Is preoperative haemoglobin testing justified in children undergoing minor elective surgery? *Canadian Journal of Anaesthesia* 1991; **38**: 700–3.

45. Hackmann T, Steward DJ, Sheps SB. Anemia in pediatric day-surgery patients: prevalence and detection. *Anesthesiology* 1991; **75**: 27–31.

46. Gabbay J, Francis L. How much day surgery? Delphic predictions. *British Medical Journal* 1988; **297**: 1249–52.

47. Meridy HW. Criteria for selection of ambulatory surgical patients and guidelines for anesthetic management: a retrospective study of 1553 cases. *Anesthesia and Analgesia* 1982; **61**: 921–6.

48. Kitz DS, Conahan TJ, Young ML, Lecky JH. Differences among 'short gyn procedures': selecting patient populations for

clinical research in ambulatory surgery. *Anesthesiology* 1988; **69**: A901.

49. Fahy A, Marshall M. Postanaesthetic morbidity in outpatients. *British Journal of Anaesthesia* 1969; **41**: 433–8.

50. Maniglia AJ, Kushner H, Cozzi L. Adenotonsillectomy. A safe outpatient procedure. *Archives of Otolaryngology – Head and Neck Surgery* 1989; **115**: 92–4.

51. Yardley MP. Tonsillectomy, adenoidectomy and adenotonsillectomy: are they safe day case procedures? *Journal of Laryngology and Otology* 1992; **106**: 299–300.

52. Colclasure JB, Graham SS. Complications of outpatient tonsillectomy and adenoidectomy: a review of 3,340 cases. *Ear, Nose, and Throat Journal* 1990; **69**: 155–60.

53. Guida RA, Mattucci KF. Tonsillectomy and adenoidectomy: an inpatient or outpatient procedure? *Laryngoscope* 1990; **100**: 491–3.

54. Helmus C, Grin M, Westfall R. Same-day-stay adenotonsillectomy. *Laryngoscope* 1990; **100**: 593–6.

55. Tom LW, DeDio RM, Cohen DE, Wetmore RF, Handler SD, Potsic WP. Is outpatient tonsillectomy appropriate for young children? *Laryngoscope* 1992; **102**: 277–80.

56. Didlake R, Curry E, Rigdon EE, Raju S, Bower J. Outpatient vascular access surgery: impact of a dialysis unit-based surgical facility. *American Journal of Kidney Diseases* 1992; **19**: 39–44.

57. Kinnard P, Lirette R. Outpatient orthopedic surgery: a retrospective study of 1996 patients. *Canadian Journal of Surgery* 1991; **34**: 363–6.

58. Patel RI, Hannallah RS. Anesthetic complications following pediatric ambulatory surgery: a 3-yr study. *Anesthesiology* 1988; **69**: 1009–12.

59. Meeks GR, Waller GA, Meydrech EF, Flautt Jr FH. Unscheduled hospital admission following ambulatory gynecologic surgery. *Obstetrics and Gynecology* 1992; **80**: 446–50.

60. Natof HE. Complications associated with ambulatory surgery. *Journal of the American Medical Association* 1980; **244**: 1116–18.

61. White PF, Coe V, Dworsky WA, Margolis A. Disseminated intravascular coagulation following midtrimester abortions. *Anesthesiology* 1983; **58**: 99–101.

62. Dawson B, Reed WA. Anaesthesia for day-care surgery: a symposium (III). Anaesthesia for adult surgical out-patients *Canadian Anaesthetists' Society Journal* 1980; **27**: 409–11.

63. Ogg TW. Use of anaesthesia. Implications of day-case surgery and anaesthesia. *British Medical Journal* 1980; **281**: 212–14.

64. Mackenzie JW. Daycase anaesthesia and anxiety. A study of anxiety profiles amongst patients attending a day bed unit. *Anaesthesia* 1989; **44**: 437–40.

65. Twersky RS, Lebovits AH, Lewis M, Frank D. Early anesthesia evaluation of the ambulatory surgical patient: does it really help? *Journal of Clinical Anesthesia* 1992; **4**: 204–7.

66. Arellano R, Cruise C, Chung F. Timing of the anesthetist's preoperative outpatient interview. *Anesthesia and Analgesia* 1989; **68**: 645–8.

67. Clark AJM, Hurtig JB. Premedication with meperidine and atropine does not prolong recovery to street fitness after out-

patient surgery. *Canadian Anaesthetists' Society Journal* 1981; **28**: 390–3.

68. Jakobsen H, Hertz JB, Johansen JR, Hansen A, Kolliker K. Premedication before day surgery. A double-blind comparison of diazepam and placebo. *British Journal of Anaesthesia* 1985; **57**: 300–5.

69. Korttila K, Linnoila M. Psychomotor skills related to driving after intramuscular administration of diazepam and meperidine. *Anesthesiology* 1975; **42**: 685–91.

70. White PF, Chang T. Effect of narcotic premedication on the intravenous anesthetic requirement. *Anesthesiology* 1984; **61**: A389.

71. Doze VA, Shafer A, White PF. Nausea and vomiting after outpatient anesthesia – effectiveness of droperidol alone and in combination with metoclopramide. *Anesthesia and Analgesia* 1987; **66**: S41.

72. Manchikanti L, Canella MG, Hohlbein LJ, Colliver JA. Assessment of effect of various modes of premedication on acid aspiration risk factors in outpatient surgery. *Anesthesia and Analgesia* 1987; **66**: 81–4.

73. Parnis SJ, Foate JA, van der Walt JH, Short T, Crowe CE. Oral midazolam is an effective premedication for children having day-stay anaesthesia. *Anaesthesia and Intensive Care* 1992; **20**: 9–14.

74. Feld LH, Negus JB, White PF. Oral midazolam preanesthetic medication in pediatric outpatients. *Anesthesiology* 1990; **73**: 831–4.

75. Forrest P, Galletly DC, Yee P. Placebo controlled comparison of midazolam, triazolam and diazepam as oral premedicants for outpatient anaesthesia. *Anaesthesia and Intensive Care* 1987; **15**: 296–304.

76. Wilson NCT, Leigh J, Rosen D, Pandit U. Intranasal midazolam premedication in pre-school children. *Anesthesia and Analgesia* 1988; **67**: S260.

77. Nelson PS, Streisand JB, Mulder SM, Pace NL, Stanley TH. Comparison of oral transmucosal fentanyl citrate and an oral solution of meperidine, diazepam and atropine for premedication in children. *Anesthesiology* 1989; **70**: 616–21.

78. Friesen RH, Lockhart CH. Oral transmucosal fentanyl citrate for preanesthetic medication of pediatric day surgery patients with and without droperidol as a prophylactic anti-emetic. *Anesthesiology* 1992; **76**: 46–51.

79. Ashburn MA, Streisand JB, Tarver SD, *et al.* Oral transmucosal fentanyl citrate for premedication in paediatric outpatients. *Canadian Journal of Anaesthesia* 1990; **37**: 857–66.

80. Feld LH, Champeau MW, van Steennis CA, Scott JC. Preanesthetic medication in children: a comparison of oral transmucosal fentanyl citrate versus placebo. *Anesthesiology* 1989; **71**: 374–7.

81. Anderson BJ, Exarchos H, Lee K, Brown TC. Oral premedication in children: a comparison of chloral hydrate, diazepam, alprazolam, midazolam and placebo for day surgery. *Anaesthesia and Intensive Care* 1990; **18**: 185–93.

82. Hannallah RS, Patel RI. Low-dose intramuscular ketamine for anesthesia pre-induction in young children undergoing brief outpatient procedures. *Anesthesiology* 1989; **70**: 598–600.

83. Rosen DA, Rosen KR, Hannallah RS. Preoperative characteristics which influence the child's response to induction of anesthesia. *Anesthesiology* 1985; **63**: A462.

84. Karl HW, Pauza KJ, Heyneman N, Tinker DE. Preanesthetic preparation of pediatric outpatients: the role of a videotape for parents. *Journal of Clinical Anesthesia* 1990; **2**: 172–7.

85. Raeder JC, Breivik H. Premedication with midazolam in out-patient general anaesthesia. A comparison with morphine–scopolamine and placebo. *Acta Anaesthesiologica Scandinavica* 1987; **31**: 509–14.

86. Shafer A, White PF, Urquhart ML, Doze VA. Outpatient premedication: use of midazolam and opioid analgesic. *Anesthesiology* 1989; **71**: 495–501.

87. White PF. The role of midazolam in outpatient anesthesia. *Anesthesiology Review* 1985; **12**: 55–60.

88. Lichtor JL, Korttila K, Lane BS, *et al.* The effect of preoperative anxiety and premedication with midazolam on recovery from ambulatory surgery. *Anesthesia and Analgesia* 1989; **68**: S163.

89. Taylor E, Ghouri AF, White PF. Midazolam in combination with propofol for sedation during local anesthesia. *Journal of Clinical Anesthesia* 1992; **4**: 213–16.

90. Nightingale JJ, Norman J. A comparison of midazolam and temazepam for premedication of day case patients. *Anaesthesia* 1988; **43**: 111–13.

91. McMillan CO, Spahr-Schopfer IA, Sikich N, Hartley E, Lerman J. Premedication of children with oral midazolam. *Canadian Journal of Anaesthesia* 1992; **39**: 545–50.

92. Beechey APG, Eltringham RJ, Studd C. Temazepam as premedication in day surgery. *Anaesthesia* 1981; **36**: 10–15.

93. Hargreaves J. Benzodiazepine premedication in minor day-case surgery: comparison of oral midazolam and temazepam with placebo. *British Journal of Anaesthesia* 1988; **61**: 611–16.

94. Turner GA, Paech M. A comparison of oral midazolam solution with temazepam as a day case premedicant. *Anaesthesia and Intensive Care* 1991; **19**: 365–8.

95. Short TG, Galletly DC. Double-blind comparison of midazolam and temazepam as oral premedicants for outpatient anaesthesia. *Anaesthesia and Intensive Care* 1989; **17**: 151–6.

96. Thomas D, Tipping T, Halifax R, Blogg CE, Hollands MA. Triazolam premedication. A comparison with lorazepam and placebo in gynaecological patients. *Anaesthesia* 1986; **41**: 692–7.

97. Pinnock CA, Fell D, Hunt PC, Miller R, Smith G. A comparison of triazolam and diazepam as premedication agents for minor gynaecological surgery. *Anaesthesia* 1985; **40**: 324–8.

98. Rita L, Seleny FL, Mazurek A, Rabbin SY. Intramuscular midazolam for pediatric preanesthetic sedation: a double-blind controlled study with morphine. *Anesthesiology* 1985; **63**: 528–31.

99. Wilton NCT, Burn JMB. Delayed vomiting after papaveretum in paediatric outpatient surgery. *Canadian Anaesthetists' Society Journal* 1986; **33**: 741–4.

100. Chestnutt WN, Clarke RSJ, Dundee JW. Comparison of nalbuphine, pethidine and placebo as premedication for minor gynecological surgery. *British Journal of Anaesthesia* 1987; **59**: 576–80.

101. White PF, Chang T. Effect of narcotic premedication on the intravenous anesthetic requirement. *Anesthesiology* 1984; **61**: A398.

102. Horrigan RW, Moyers JR, Johnson BH, Eger 2nd EI, Margolis A, Goldsmith S. Etomidate vs. thiopental with and without fentanyl – a comparative study of awakening in man. *Anesthesiology* 1980; **52**: 362–4.

103. Craig J, Cooper GM, Sear JW. Recovery from day-case anaesthesia. Comparison between methohexitone, althesin and etomidate. *British Journal of Anaesthesia* 1982; **54**: 447–51.

104. White PF, Sung M-L, Doze VA. Use of sufentanil in outpatient anesthesia – determining an optimal preinduction dose. *Anesthesiology* 1985; **63**: A202.

105. Pandit SK, Kothary SP. Intravenous narcotics for premedication in outpatient anaesthesia. *Acta Anaesthesiologica Scandinavica* 1989; **33**: 353–8.

106. Maltby JR, Sutherland AD, Sale JP, Shaffer EA. Preoperative oral fluids: is a five hour fast justified prior to elective surgery? *Anesthesia and Analgesia* 1986; **65**: 1112–22.

107. Sutherland AD, Stock JG, Davies JM. Effects of preoperative fasting on morbidity and gastric contents in patients undergoing day-stay surgery. *British Journal of Anaesthesia* 1986; **58**: 876–8.

108. Doze VA, White PF. Effects of fluid therapy on serum glucose levels in fasted outpatients. *Anesthesiology* 1987; **66**: 223–6.

109. O'Flynn PE, Milford CA. Fasting in children for day case surgery. *Annals of the Royal College of Surgeons of England* 1989; **71**: 218–19.

110. Scarr M, Maltby JR, Jani K, Sutherland LR. Volume and acidity of residual gastric fluid after oral fluid ingestion before elective ambulatory surgery. *Canadian Medical Association Journal* 1989; **141**: 1151–4.

111. Splinter WM, Stewart JA, Muir JG. The effect of preoperative apple juice on gastric contents, thirst and hunger in children. *Canadian Journal of Anaesthesia* 1989; **36**: 55–8.

112. Sandhar BK, Goresky GV, Maltby JR, Shaffer EA. Effect of oral liquids and ranitidine on gastric fluid volume and pH in children undergoing outpatient surgery. *Anesthesiology* 1989; **71**: 327–30.

113. Olsson GL, Hallen B, Hambraeus-Jonzon K. Aspiration during anaesthesia: a computer-aided study of 185 358 anaesthetics. *Acta Anaesthesiologica Scandinavica* 1986; **30**: 84–92.

114. Manchikanti L, Roush JR, Colliver JA. Effect of preanesthetic ranitidine and metoclopramide on gastric contents in morbidly obese patients. *Anesthesia and Analgesia* 1986; **65**: 195–9.

115. Ong BY, Palahnuik RJ, Cumming M. Gastric volume and pH in out-patients. *Canadian Anaesthestists' Society Journal* 1978; **25**: 36–9.

116. Manchikanti L, Roush JR. Effect of preanesthetic glycopyrrolate and cimetidine on gastric fluid pH and volume in outpatients. *Anesthesia and Analgesia* 1984; **63**: 40–6.

117. Coombs DW. Aspiration pneumonitis prophylaxis. *Anesthesia and Analgesia* 1983; **62**: 1055–8.

118. Haavik PE, Soreide E, Hofstad B, Steen PA. Does preoperative anxiety influence gastric fluid volume and acidity? *Anesthesia and Analgesia* 1992; **75**: 91–4.

119. Morrison DH, Dunn GL, Fargas-Babjak AM, Moudgil GC, Smedstat K, Woo J. A double-blind comparison of cimetidine and ranitidine as prophylaxis against gastric aspiration syndrome. *Anesthesia and Analgesia* 1982; **61**: 988–92.

120. Zeldis JB, Friedman LS, Isselbacher KJ. Ranitidine: a new H$_2$-receptor antagonist. *New England Journal of Medicine* 1983; **309**: 1368–73.

121. Manchikanti L, Collover JA, Roush JR, Canella MG. Evaluation of ranitidine as an oral antacid in outpatient anesthesia. *Southern Medical Journal* 1985; **78**: 818–22.

122. Maltby JR, Reid CRG, Hutchinson A. Gastric fluid volume and pH in elective inpatients. Part II: Coffee or orange juice with ranitidine. *Canadian Journal of Anaesthesia* 1988; **35**: 16–19.

123. Wyner J, Cohen SE. Gastric volume in early pregnancy: effect of metoclopramide. *Anesthesiology* 1982; **57**: 209–12.

124. Rao TL, Madhavareddy S, Chinthagada M, El-Etr AA. Metoclopramide and cimetidine to reduce gastric fluid pH and volume. *Anesthesia and Analgesia* 1984; **63**: 1014–16.

125. Pandit SK, Kothary SP, Pandit UA, Mirakhur RK. Premedication with cimetidine and metoclopramide. Effect on the risk factors of acid aspiration. *Anaesthesia* 1986; **41**: 486–92.

126. Hey VMF, Ostick DG, Mazumder JK, Lord WD. Pethidine, metoclopramide and the gastro-oesophageal sphincter. A study in healthy volunteers. *Anaesthesia* 1981; **36**: 173–6.

127. Stock JG, Sutherland AD. The role of H$_2$ receptor antagonist premedication in pregnant day care patients. *Canadian Anaesthetists' Society Journal* 1985; **32**: 463–7.

128. Foulkes E, Jenkins LC. A comparative evaluation of cimetidine and sodium citrate to decrease gastric acidity: effectiveness at the time of induction of anaesthesia. *Canadian Anaesthetists' Society Journal* 1981; **28**: 29–32.

129. Manchikanti L, Grow JB, Colliver JA, Hadley CH, Hohlbein LJ. Bicitra^R (sodium citrate) and metoclopramide in outpatient anesthesia for prophylaxis against aspiration pneumonitis. *Anesthesiology* 1985; **63**: 378–84.

130. Abramowitz MD, Oh TE, Epstein BS, Ruttimann UE, Friendly DS. The antiemetic effect of droperidol following outpatient strabismus surgery in children. *Anesthesiology* 1983; **59**: 579–83.

131. Metter SE, Kitz DS, Young ML, Baldeck AM, Apfelbaum JL. Nausea and vomiting after outpatient laparoscopy: incidence, impact on recovery room stay and cost. *Anesthesia and Analgesia* 1987; **66**: S116.

132. Hovorka J, Korttila K, Erkola O. The expertise of the person ventilating the lungs does influence post operative nausea and vomiting. *Acta Anaesthesiologica Scandinavica* 1990; **34**: 203–5.

133. White PF, Shafer A. Nausea and vomiting: causes and prophylaxis. *Seminars in Anesthesia* 1987; **6**: 300–8.

134. Watcha MF, White PF. Postoperative nausea and vomiting. Its etiology, treatment, and prevention. *Anesthesiology* 1992; **77**: 162–84.

135. Dundee JW, Chestnutt WN, Ghaly RG, Lynas AG. Traditional Chinese acupuncture: a potentially useful antiemetic? *British Medical Journal* 1986; **293**: 583–4.

136. Ghaly RG, Fitzpatrick KT, Dundee JW. Antiemetic studies with traditional Chinese acupuncture. A comparison of manual needling with electrical stimulation and commonly used antiemetics. *Anaesthesia* 1987; **42**: 1108–10.

137. O'Donovan N, Shaw J. Nausea and vomiting in day-case dental anaesthesia. The use of low-dose droperidol. *Anaesthesia* 1984; **39**: 1172–6.

138. Rita L, Goodarzi M, Seleny F. Effect of low dose droperidol on postoperative vomiting in children. *Canadian Anaesthetists' Society Journal* 1981; **28**: 259–62.

139. Lerman J, Eustis S, Smith DR. Effect of droperidol pretreatment on postanesthetic vomiting in children undergoing strabismus surgery. *Anesthesiology* 1986; **65**: 322–5.

140. Nicolson SC, Kaya KM, Bets EK. The effect of preoperative oral droperidol on the incidence of postoperative emesis after paediatric strabismus surgery. *Canadian Journal of Anaesthesia* 1988; **35**: 364–7.

141. Korttila K, Kauste A, Auvinen J. Comparison of domperidone, droperidol, and metoclopramide in the prevention and treatment of nausea and vomiting after balanced general anesthesia. *Anesthesia and Analgesia* 1979; **58**: 396–400.

142. Madej TH, Simpson KH. Comparison of the use of domperidone, droperidol and metoclopramide in the prevention of nausea and vomiting following gynaecological surgery in day cases. *British Journal of Anaesthesia* 1986; **58**: 884–7.

143. Cohen SE, Woods WA, Wyner J. Antiemetic efficacy of droperidol and metoclopramide. *Anesthesiology* 1984; **60**: 67–9.

144. Valanne J, Korttila K. Effect of a small dose of droperidol on nausea, vomiting and recovery after outpatient enflurane anaesthesia. *Acta Anaesthesiologica Scandinavica* 1985; **29**: 359–62.

145. Melnick B, Sawyer R, Karambelkar D, Phitayakorn P, Uy NT, Patel R. Delayed side effects of droperidol after ambulatory general anesthesia. *Anesthesia and Analgesia* 1989; **69**: 748–51.

146. Millar JM, Hall PJ. Nausea and vomiting after prostaglandins in day case termination of pregnancy. The efficacy of low dose droperidol. *Anaesthesia* 1987; **42**: 613–18.

147. Handley AJ. Metoclopramide in the prevention of postoperative nausea and vomiting. *British Journal of Clinical Practice* 1967; **21**: 460–2.

148. Spelina KR, Gerber HR, Pagels IL. Nausea and vomiting during spinal anaesthesia. Effect of metoclopramide and domperidone: a double blind trial. *Anaesthesia* 1984; **39**: 132–7.

149. Broadman LM, Ceruizzi W, Patane PS, Hannallah RS, Ruttiman U, Friendly D. Metoclopramide reduces the incidence of vomiting following strabismus surgery in children. *Anesthesiology* 1988; **69**: A747.

150. Palazzo MG, Strunin L. Anaesthesia and emesis II. Prevention and management. *Canadian Anaesthetists' Society Journal* 1984; **31**: 407–15.

151. Bailey PL, Streisand JB, Pace NL, *et al.* Transdermal scopolamine reduces nausea and vomiting after outpatient laparoscopy. *Anesthesiology* 1990; **72**: 977–80.

152. Gibbons PA, Nicolson SC, Betts EK, Rosenberry KR, Jobes DR. Scopolamine does not prevent post-operative emesis after pediatric eye surgery. *Anesthesiology* 1984; **61**: A435.

153. Uppington J, Dunnet J, Blogg CE. Transdermal hyoscine and postoperative nausea and vomiting. *Anaesthesia* 1986; **41**: 16–20.

154. Tigerstedt I, Salmela L, Aromaa U. Double blind comparison of transdermal scopolamine, droperidol and placebo against

postoperative nausea and vomiting. *Acta Anaesthesiologica Scandinavica* 1988; **32**: 454–7.

155. McKenzie R, Wadhwa RK, Uy NTL, *et al*. Antiemetic effectiveness of intramuscular hydroxyzine compared with intramuscular droperidol. *Anesthesia and Analgesia* 1981; **60**: 783–8.

156. Warner LO, Rogers GL, Martino JD, Bremer DL, Beach TP. Intravenous lidocaine reduces the incidence of vomiting in children after surgery to correct strabismus. *Anesthesiology* 1988; **68**: 618–21.

157. Bodner M, White PF. Antiemetic efficacy of ondansetron after outpatient laparoscopy. *Anesthesia and Analgesia* 1991; **73**: 250–4.

158. Monk TG, White PF, Lemon D. Ondansetron reduces nausea following outpatient lithotripsy. *Anesthesiology* 1992; **77**: A19.

159. White PF, Watcha MF. Are new drugs cost-effective for patients undergoing ambulatory surgery? *Anesthesiology* 1993; **78**: 2–5.

160. Blitt CD. Monitoring during outpatient anesthesia. *International Anesthesiology Clinics* 1982; **20**: 17–25.

161. Raemer DB, Warren DL, Morris R, Philip BK, Philip JH. Hypoxemia during ambulatory gynecologic surgery as evaluated by the pulse oximeter. *Journal of Clinical Monitoring* 1987; **3**: 224–8.

162. Stafford M, Jeon A, Pasucci R. Pre and post-induction blood glucose concentrations in healthy fasting children. *Anesthesiology* 1985; **63**: A350.

163. Cook R, Anderson S, Riseborough M, Blogg CE. Intravenous fluid load and recovery. A double-blind comparison in gynaecological patients who had day-case laparoscopy. *Anaesthesia* 1990; **45**: 826–30.

164. Ooi LG, Goldhill DR, Griffiths A, Smith C. IV fluids and minor gynaecological surgery: effect on recovery from anaesthesia. *British Journal of Anaesthesia* 1992; **68**: 576–9.

165. Manner T, Kanto J, Iisalo E, Lindberg R, Viinamaki O, Scheinin M. Reduction of pain at venous cannulation in children with a eutectic mixture of lidocaine and prilocaine (EMLA(R) cream): comparison with placebo cream and no local premedication. *Acta Anaesthesiologica Scandinavica* 1987; **31**: 735–9.

166. Fassoulaki A, Sarantopoulos C. Minor complications of general anesthesia in a series of 1220 patients: the influence of mode of ventilation. *Acta Anaesthesiologica Belgica* 1991; **42**: 157–63.

167. Patel RI, Hannallah RS, Norden J, Casey WF, Verghese ST. Emergence airway complications in children: a comparison of tracheal extubation in awake and deeply anesthetized patients. *Anesthesia and Analgesia* 1991; **73**: 266–70.

168. Kenefick JP, Leader A, Maltby JR, Taylor PJ. Laparoscopy: blood-gas values and minor sequelae associated with three techniques based on isoflurane. *British Journal of Anaesthesia* 1987; **59**: 189–94.

169. Roberts CJ, Goodman NW. Gastro-oesophageal reflux during elective laparoscopy. *Anaesthesia* 1990; **45**: 1009–11.

170. Smith I, White PF. Use of the laryngeal mask airway as an alternative to a face mask during outpatient arthroscopy. *Anesthesiology* 1992; **77**: 850–5.

171. Johnston DF, Wrigley SR, Robb PJ, Jones HE. The laryngeal mask airway in paediatric anaesthesia. *Anaesthesia* 1990; **45**: 924–7.

172. Bailie R, Barnett MB, Fraser JF. The Brain laryngeal mask. A comparative study with the nasal mask in paediatric dental outpatient anaesthesia. *Anaesthesia* 1991; **46**: 358–60.

173. Watcha MF, Garner FT, White PF. Perioperative conditions with face mask – oral airway or laryngeal mask airway during bilateral myringotomy in children. *Anesthesia and Analgesia* 1993; **76**: S456.

174. Bridenbaugh LD. Regional anaesthesia for outpatient surgery – a summary of 12 years experience. *Canadian Anaesthetists' Society Journal* 1983; **30**: 548–52.

175. Philip BV, Covino BG. Local and regional anesthesia. In: Wetchler BV, ed. *Anesthesia for ambulatory surgery.* 2nd ed. Philadelphia: JB Lippincott, 1985: 309–65.

176. Muir VMJ, Leonard M, Haddaway E. Morbidity following dental extraction. A comparative study of local analgesia and general anaesthesia. *Anaesthesia* 1976; **31**: 171–80.

177. Shandling B, Steward DJ. Regional anesthesia for post-operative pain in pediatric outpatient surgery. *Journal of Pediatric Surgery* 1980; **15**: 477–80.

178. Gunter JB, Dunn CM, Bennie JB, Pentecost DL, Bower RJ, Ternberg JL. Optimum concentration of bupivacaine for combined caudal–general anesthesia in children. *Anesthesiology* 1991; **75**: 57–61.

179. Moores MA, Wandless JG, Fell D. Paediatric postoperative analgesia. A comparison of rectal diclofenac with caudal bupivacaine after inguinal herniotomy. *Anaesthesia* 1990; **45**: 156–8.

180. Langer JC, Shandling B, Rosenberg M. Intraoperative bupivacaine during outpatient hernia repair in children: a randomized double blind trial. *Journal of Pediatric Surgery* 1987; **22**: 267–70.

181. Reid MF, Harris R, Phillips PD, Baker I, Pereira NH, Bennett NR. Day-case herniotomy in children. A comparison of ilio-inguinal nerve block and wound infiltration for postoperative analgesia. *Anaesthesia* 1987; **42**: 658–61.

182. Fell D, Derrington MC, Taylor E, Wandless JG. Paediatric postoperative analgesia. A comparison between caudal block and wound infiltration of local anaesthetic. *Anaesthesia* 1988; **43**: 107–10.

183. Lunn JN. Postoperative analgesia after circumcision. A randomized comparison between caudal analgesia and intramuscular morphine in boys. *Anaesthesia* 1979; **34**: 552–4.

184. May AE, Wandless J, James RH. Analgesia for circumcision in children. A comparison of caudal blockade and intramuscular buprenorphine. *Acta Anaesthesiologica Scandinavica* 1982; **26**: 331–3.

185. Yeoman PM, Cooke R, Hain WR. Penile block for circumcision? A comparison with caudal blockade. *Anaesthesia* 1983; **38**: 862–6.

186. Broadman LM, Hannallah RS, Belman AB, Elder PT, Ruttimann U, Epstein BS. Post-circumcision analgesia – a prospective evaluation of subcutaneous ring block of the penis. *Anesthesiology* 1987; **67**: 399–402.

187. Vater M, Wandless J. Caudal or dorsal nerve block? A comparison of two local anaesthetic techniques for post-

operative analgesia following day case circumcision. *Acta Anaesthesiologica Scandinavica* 1985; **29**: 175–9.

188. Tree-Trakarn T, Pirayavaraporn S. Postoperative pain relief for circumcision in children: comparison among morphine, nerve block and topical analgesia. *Anesthesiology* 1985; **62**: 519–22.

189. Flaatten H, Raeder J. Spinal anaesthesia for outpatient surgery. *Anaesthesia* 1985; **40**: 1108–11.

190. Quaynor H, Corbey M, Berg P. Spinal anaesthesia in day-care surgery with a 26-gauge needle. *British Journal of Anaesthesia* 1990; **65**: 766–9.

191. Kang SB, Goodnough DE, Lee YK, *et al.* Comparison of 26- and 27-G needles for spinal anesthesia for ambulatory surgery patients. *Anesthesiology* 1992; **76**: 734–8.

192. Carbaat PAT, van Crevel H. Lumbar puncture headache: controlled study on the preventive effect of 24 hours bed rest. *Lancet* 1981; **2**: 1133–5.

193. Flaatten H, Rodt S, Rosland J, Vammes J. Postoperative headache in young patients after spinal anaesthesia. *Anaesthesia* 1987; **42**: 202–5.

194. Paluhniak RJ, Cumming M. Prophylactic blood patch does not prevent post lumbar puncture headache. *Canadian Anaesthetists' Society Journal* 1979; **26**: 132–3.

195. DiGiovanni AJ, Galbert MW, Wahle WM. Epidural injection of blood for postlumbar-puncture headache. II. Additional clinical experiences and laboratory investigation. *Anesthesia and Analgesia* 1972; **51**: 226–32.

196. Ravindran RS. Epidural autologous blood patch on an outpatient basis. *Anesthesia and Analgesia* 1984; **63**: 962.

197. Aromaa U. Anaesthesia for short-stay varicose vein surgery. *Acta Anaesthesiologica Scandinavica* 1977; **21**: 368–73.

198. Abdu RA. Ambulatory herniorrhaphy under local anesthesia in a community hospital. *American Journal of Surgery* 1983; **145**: 353–6.

199. Ryan Jr JA, Adye BA, Jolly PC, Mulroy 2nd MF. Outpatient inguinal herniorrhaphy with both regional and local anesthesia. *American Journal of Surgery* 1984; **148**: 313–16.

200. Duvall JO, Griffith DP. Epidural anesthesia for extra-corporeal shock wave lithotripsy. *Anesthesia and Analgesia* 1985; **64**: 544–6.

201. Bridenbaugh LD, Soderstrom RM. Lumbar epidural block anesthesia for outpatient laparoscopy. *Journal of Reproductive Medicine* 1979; **23**: 85–6.

202. Monk TG, Boure B, White PF, Meretyk S, Clayman RV. Comparison of intravenous sedative-analgesic techniques for outpatient immersion lithotripsy. *Anesthesia and Analgesia* 1991; **72**: 616–21.

203. Kopacz DJ, Mulroy MF. Chloroprocaine and lidocaine decrease hospital stay and admission rate after outpatient epidural anesthesia. *Regional Anesthesia* 1990; **15**: 19–25.

204. Siler JN, Rosenberg H. Lidocaine hydrochloride versus lidocaine bicarbonate for epidural anesthesia in outpatients undergoing arthroscopic surgery. *Journal of Clinical Anesthesia* 1990; **2**: 296–300.

205. Jarosz J, Pihowicz A, Towpik E. The application of continuous thoracic epidural anaesthesia in outpatient oncological and reconstructive surgery of the breast. *European Journal of Surgical Oncology* 1991; **17**: 599–602.

206. Baker AB, Baker JE. Outpatient anaesthesia for dilatation and curettage. *Anaesthesia and Intensive Care* 1979; **7**: 362–6.

207. Landeen FH, Epstein L, Haas L. Special regional anesthetic techniques in ambulatory anesthesia. In: Brown Jr BR, ed. *Outpatient anesthesia.* Philadelphia: FA Davis, 1978: 71–7.

208. Davis WJ, Lennon RL, Wedel DJ. Brachial plexus anesthesia for outpatient surgical procedures on an upper extremity. *Mayo Clinic Proceedings* 1991; **66**: 470–3.

209. Olney BW, Lugg PC, Turner PL, Eyres RL, Cole WG. Outpatient treatment of upper extremity injuries in childhood using intravenous regional anesthesia. *Journal of Pediatric Orthopedics* 1988; **8**: 576–9.

210. Armstrong P, Watters J, Whitfield A. Alkalinization of prilocaine for intravenous regional anaesthesia. Suitability for clinical use. *Anaesthesia* 1990; **45**: 935–7.

211. Patel NJ, Flashburg MH, Paskin S, Grossman R. A regional anesthetic technique compared to general anesthesia for outpatient knee arthroscopy. *Anesthesia and Analgesia* 1986; **65**: 185–7.

212. Sarrafian SK, Ibrahim IN, Breihan JH. Ankle–foot peripheral nerve block for mid and forefoot surgery. *Foot and Ankle* 1983; **4**: 86–90.

213. Caldamone AA, Rabinowitz R. Outpatient orchiopexy. *Journal of Urology* 1982; **127**: 286–8.

214. Kaye KW, Clayman RV, Lange PH. Outpatient hydrocele and spermatocele repair under local anesthesia. *Journal of Urology* 1983; **130**: 269–71.

215. Orandi A. Urological endoscopic surgery under local anesthesia: a cost-reducing idea. *Journal of Urology* 1984; **132**: 1146–7.

216. Chang FC, Farha GJ. Inguinal herniorrhaphy under local anesthesia. A prospective study of 100 consecutive patients with emphasis of perioperative morbidity and patient acceptance. *Archives of Surgery* 1977; **112**: 1069–71.

217. Eriksson E, Haggmark T, Saartok T, Sebik A, Ortengren B. Knee arthroscopy with local anesthesia in ambulatory patients. Methods, results and patient compliance. *Orthopedics* 1986; **9**: 186–8.

218. Lawler RA, Larson C, Rudy T, Biglan A. The comparative incidence of postoperative vomiting in adult and teen unilateral strabismus surgeries performed under general anesthesia or retrobulbar blockade. *Anesthesiology* 1988; **69**: A370.

219. Malhotra V, Long CW, Meister MJ. Intercostal blocks with local infiltration anesthesia for extracorporeal shock wave lithotripsy. *Anesthesia and Analgesia* 1987; **66**: 85–8.

220. Huang TT, Parks DH, Lewis SR. Outpatient breast surgery under intercostal block anesthesia. *Plastic and Reconstructive Surgery* 1979; **63**: 299–303.

221. Tree-Trakarn T, Pirayavaraporn S, Lertakyamanee J. Topical analgesia for relief of post-circumcision pain. *Anesthesiology* 1987; **67**: 395–9.

222. Sinclair R, Cassuto J, Hogstrom S, *et al.* Topical anesthesia with lidocaine aerosol in the control of postoperative pain. *Anesthesiology* 1988; **68**: 895–901.

223. Shane SM. Intravenous amnesia for total dentistry in one sitting. *Journal of Oral Surgery* 1966; **241**: 27–32.

224. McCarthy FM, Solomon AL, Jatask JT, *et al.* Conscious

sedation: risks and benefits. *Journal of the American Dental Association* 1984; **109**: 546–57.

225. Vinnik CA. An intravenous dissociation technique for outpatient plastic surgery: tranquility in the office surgical facility. *Plastic and Reconstructive Surgery* 1981; **67**: 799–805.

226. Lundgren S, Rosenquist JB. Amnesia, pain experience, and patient satisfaction with intravenous diazepam. *Journal of Oral and Maxillofacial Surgery* 1983; **41**: 99–102.

227. Lundgren S, Rosenquist JB. Comparison of sedation, amnesia and patient comfort produced by intravenous and rectal diazepam. *Journal of Oral and Maxillofacial Surgery* 1984; **42**: 646–50.

228. Lundgren S. Sedation as an alternative to general anaesthesia. *Acta Anaesthesiologica Scandinavica* 1987; **32**(Suppl 88): 21–3.

229. White CS, Dolwick MF, Gravenstein N, Paulus DA. Incidence of oxygen desaturation during oral surgery outpatient procedures. *Journal of Oral and Maxillofacial Surgery* 1989; **47**: 147–9.

230. Campbell RL, Satterfield SD, Dionne RA, Kelley DE. Postanesthetic morbidity following fentanyl, diazepam and methohexital sedation. *Anesthesia Progress* 1980; **2**: 45–8.

231. Gale GD. Recovery from methoxitone, halothane and diazepam. *British Journal of Anaesthesia* 1976; **48**: 691–8.

232. White PF, Vasconez LO, Mathes SA, Way WL, Wender LA. Comparison of midazolam and diazepam for sedation during plastic surgery. *Plastic and Reconstructive Surgery* 1988; **81**: 703–12.

233. McGimpsey JG, Kawar P, Gamble JAS, Browne ES, Dundee JW. Midazolam in dentistry. *British Dental Journal* 1983; **155**: 47–50.

234. Dixon J, Power SJ, Grundy EM, Lumley J, Morgan M. Sedation for local anaesthesia. Comparison of intravenous midazolam and diazepam. *Anaesthesia* 1984; **39**: 372–6.

235. McClure JH, Brown DT, Wildsmith JAW. Comparison of the i.v. administration of midazolam and diazepam as sedation during spinal anaesthesia. *British Journal of Anaesthesia* 1983; **55**: 1089–93.

236. O'Boyle CA, Harris D, Barry H, McCreary C, Bewley A, Fox E. Comparison of midazolam by mouth and diazepam i.v. in outpatient surgery. *British Journal of Anaesthesia* 1987; **59**: 746–54.

237. O'Boyle CA, Barry H, Fox E, McCreary C, Bewley A. Controlled comparison of a new sublingual lormetazepam formulation and i.v. diazepam in outpatient minor oral surgery. *British Journal of Anaesthesia* 1988; **60**: 419–25.

238. White PF, Shafer A, Boyle 3rd WA, Doze VA, Duncan S. Benzodiazepine antagonism does not provoke a stress response. *Anesthesiology* 1989; **70**: 636–9.

239. Jensen S, Knudsen L, Kirkegaard L, Kruse A, Knusden EB. Flumazenil used for antagonizing the central effects of midazolam and diazepam in outpatients. *Acta Anaesthesiologica Scandinavica* 1989; **33**: 26–8.

240. Ghoneim MM, Dembo JB, Block RI. Time course of antagonism of sedative and amnesic effects of diazepam by flumazenil. *Anesthesiology* 1989; **70**: 899–904.

241. Boldy DAR, English JS, Lang GS, Hoare AM. Sedation for endoscopy: a comparison between diazepam, and diazepam plus pethidine with naloxone reversal. *British Journal of Anaesthesia* 1984; **56**: 1109–12.

242. Tucker MR, Ochs MW, White RP Jr. Arterial blood gas levels after midazolam or diazepam administered with or without fentanyl as an intravenous sedative for outpatient surgical procedures. *Journal of Oral and Maxillofacial Surgery* 1986; **44**: 688–92.

243. Bailey PL, Moll JWB, Pace NL, East KA, Stanley TH. Respiratory effects of midazolam and fentanyl: potent interaction producing hypoxemia and apnea. *Anesthesiology* 1988; **69**: A813.

244. Dolan EA, Murray WJ, Immediata AR, Gleason N. Comparison of nalbuphine and fentanyl in combination with diazepam for outpatient oral surgery. *Journal of Oral and Maxillofacial Surgery* 1988; **46**: 471–3.

245. Barclay JK, Hunter KM. A comparison of midazolam with and without nalbuphine for intravenous sedation. *Oral Surgery, Oral Medicine, Oral Pathology* 1990; **70**: 137–40.

246. Walton GM, Boyle CA, Thomson PJ. Changes in oxygen saturation using two different sedation techniques. *British Journal of Oral and Maxillofacial Surgery* 1991; **29**: 87–9.

247. Gilbert J, Holt JE, Johnston J, Sabo BA, Weaver JS. Intravenous sedation for cataract surgery. *Anaesthesia* 1987; **42**: 1063–9.

248. Hunter PL. Use of nalbuphine for analgesia in combination with methohexital sodium. *Anesthesia Progress* 1989; **36**: 150–68.

249. Schelling G, Weber W, Sackman M, Peter K. Pain control during extracorporeal shock wave lithotripsy of gallstones by titrated alfentanil infusion. *Anesthesiology* 1989; **70**: 1022–3.

250. Thompson GE, Moore DC. Ketamine, diazepam and Innovar^R. A computerized comparative study. *Anesthesia and Analgesia* 1971; **50**: 458–63.

251. White PF. Use of ketamine for sedation and analgesia during injection of local anesthetics. *Annals of Plastic Surgery* 1985; **15**: 53–6.

252. Scarborough DA, Bisaccia E, Swensen RD. Anesthesia for outpatient dermatologic cosmetic surgery: Midazolam-low-dosage ketamine anesthesia. *Journal of Dermatologic Surgery and Oncology* 1989; **15**: 658–63.

253. Philip BK. Hazards of amnesia after midazolam in ambulatory surgical patients. *Anesthesia and Analgesia* 1987; **66**: 97–8.

254. Mackenzie N, Grant IS. Comparison of propofol with methohexitone in the provision of anaesthesia for surgery under regional blockade. *British Journal of Anaesthesia* 1985; **57**: 1167–72.

255. Jessop E, Grounds RM, Morgan M, Lumley J. Comparisons of infusions of propofol and methohexitone to provide light general anaesthesia during surgery with regional blockade. *British Journal of Anaesthesia* 1985; **57**: 1173–7.

256. Valtonen M, Salonen M, Forssell H, Scheinin M, Viinamaki O. Propofol infusion for sedation in outpatient oral surgery. A comparison with diazepam. *Anaesthesia* 1989; **44**: 730–4.

257. White PF, Negus JB. Sedative infusions during local and regional anesthesia: a comparison of midazolam and propofol. *Journal of Clinical Anesthesia* 1991; **3**: 32–9.

258. McCulloch MJ, Lees NW. Assessment and modification of

pain on induction with propofol (Diprivan). *Anaesthesia* 1985; **40**: 1117–20.

259. Rosa G, Conti G, Orsi P, *et al.* Effects of low-dose propofol administration on central respiratory drive, gas exchanges and respiratory pattern. *Acta Anaesthesiologica Scandinavica* 1992; **36**: 128–31.

260. White PF, Dworsky WA, Horai Y, Trevor AJ. Comparison of continuous infusion of fentanyl or ketamine versus thiopental – determining the mean effective serum concentrations for outpatient surgery. *Anesthesiology* 1983; **59**: 564–9.

261. Urquhart ML, White PF. Comparison of sedative infusions during regional anesthesia: methohexital, etomidate and midazolam. *Anesthesia and Analgesia* 1989; **68**: 249–54.

262. Philip BK. Supplemental medication for ambulatory procedures under regional anesthesia. *Anesthesia and Analgesia* 1985; **64**: 1117–25.

263. Hallonsten AL. The use of oral sedatives in dental care. *Acta Anaesthesiologica Scandinavica* 1988; **32**(Suppl 88): 27–30.

264. Rodrigo MR, Rosenquist JB. Isoflurane for conscious sedation. *Anaesthesia* 1988; **43**: 369–75.

265. White PF. Continuous infusions of thiopental, methohexital or etomidate as adjuvants to nitrous oxide for outpatient anesthesia. *Anesthesia and Analgesia* 1984; **63**: 282.

266. Heath PJ, Ogg TW, Gilks WR. Recovery after day-case anaesthesia. A 24-hour comparison of recovery after thiopentone or propofol anaesthesia. *Anaesthesia* 1990; **45**: 911–15.

267. Hannington-Kiff JG. Measurement of recovery from outpatient general anaesthesia with a simple ocular test. *British Medical Journal* 1970; **3**: 132–5.

268. Cooper GM. Recovery from anaesthesia. *Clinics in Anaesthesiology* 1984; **2**: 145–62.

269. Korttila K, Linnoila M, Ertama P, Hakkinen S. Recovery and simulated driving after intravenous anesthesia with thiopental, methohexital, propanidid and alphadione. *Anesthesiology* 1975; **43**: 291–9.

270. Whitwam JG, Manners JM. Clinical comparison of thiopentone and methohexitone. *British Medical Journal* 1962; **1**: 1663–5.

271. Furness G, Dundee JW, Milligan KR. Low dose sufentanil pretreatment. Effect on the induction of anaesthesia with thiopentone, methohexitone or midazolam. *Anaesthesia* 1987; **42**: 1264–6.

272. MacKenzie N, Grant IS. Comparison of the new emulsion formulation of propofol with methohexitone and thiopentone for induction of anaesthesia in day cases. *British Journal of Anaesthesia* 1985; **57**: 725–31.

273. O'Toole DP, Milligan KR, Howe JP, McCollum JS, Dundee JW. A comparison of propofol and methohexitone as induction agents for day case isoflurane anaesthesia. *Anaesthesia* 1987; **42**: 373–6.

274. Johnston R, Noseworthy T, Anderson B, Konopad E, Grace M. Propofol versus thiopental for outpatient anesthesia. *Anesthesiology* 1987; **67**: 431–3.

275. Chittleborough MC, Osborne GA, Rudkin GE, Vickers D, Leppard PI, Barlow J. Double-blind comparison of patient recovery after induction with propofol or thiopentone for day-case relaxant general anaesthesia. *Anaesthesia and Intensive Care* 1992; **20**: 169–73.

276. Heath PJ, Kennedy DJ, Ogg TW, Dunling C, Gilks WR. Which intravenous agent for day surgery? A comparison of propofol, thiopentone, methohexitone and etomidate. *Anaesthesia* 1988; **43**: 365–8.

277. Dundee JW, McCollum JSC, Robinson FP, Halliday NJ. Elderly patients are unduly sensitive to propofol. *Anesthesia and Analgesia* 1986; **65**: S43.

278. Barker P, Langton JA, Murphy P, Rowbotham DJ. Effect of prior administration of cold saline on pain during propofol injection. A comparison with cold propofol and propofol with lignocaine. *Anaesthesia* 1991; **46**: 1069–70.

279. Borgeat A, Wilder-Smith OH, Saiah M, Rifat K. Subhypnotic doses of propofol possess direct antiemetic properties. *Anesthesia and Analgesia* 1992; **74**: 539–41.

280. Valanne J, Korttila K. Comparison of methohexitone and propofol ('Diprivan') for induction of enflurane anaesthesia in outpatients. *Postgraduate Medical Journal* 1985; **61**(Suppl): 138–43.

281. Sanders LD, Isaac PA, Yeomans WA, Clyburn PA, Rosen M, Robinson JO. Propofol-induced anaesthesia. Double-blind comparison of recovery after anaesthesia induced by propofol or thiopentone. *Anaesthesia* 1989; **44**: 200–4.

282. Cade L, Morley PT, Ross AW. Is propofol cost-effective for day-surgery patients? *Anaesthesia and Intensive Care* 1991; **19**: 201–4.

283. Miller BM, Hendry JGB, Less NW. Etomidate and methohexitone. A comparative clinical study in out-patient anaesthesia. *Anaesthesia* 1978; **33**: 450–3.

284. Boralessa H, Holdcroft A. Methohexitone or etomidate for induction of dental anesthesia. *Canadian Anaesthetists' Society Journal* 1980; **27**: 578–83.

285. Wagner RL, White PF. Etomidate inhibits adrenocortical function in surgical patients. *Anesthesiology* 1984; **61**: 647–51.

286. Crawford ME, Carl P, Andersen RS, Mikkelsen BO. Comparison between midazolam and thiopentone-based balanced anaesthesia for day-case surgery. *British Journal of Anaesthesia* 1984; **56**: 165–9.

287. Berggren L, Eriksson I. Midazolam for induction of anaesthesia in outpatients: a comparison with thiopentone. *Acta Anaesthesiologica Scandinavica* 1981; **25**: 492–6.

288. Verma R, Ramasubramanian R, Sachar RM. Anesthesia for termination of pregnancy: midazolam compared with methohexital. *Anesthesia and Analgesia* 1985; **64**: 792–4.

289. Zuurmond WWA, Van Leeuwen L, Helmers JH. Recovery from fixed-dose midazolam-induced anaesthesia and antagonism with flumazenil for outpatient arthroscopy. *Acta Anaesthesiologica Scandinavica* 1989; **33**: 160–3.

290. Philip BK, Simpson TH, Hauch MA, Mallampati SR. Flumazenil reverses sedation after midazolam-induced general anesthesia in ambulatory surgery patients. *Anesthesia and Analgesia* 1990; **71**: 371–6.

291. Forrest P, Galletly DC. Comparison of propofol and antagonized midazolam anaesthesia for day-case surgery. *Anaesthesia and Intensive Care* 1987; **15**: 394–401.

292. Norton AC, Dundas CR. Induction agents for day-case anaesthesia. A double-blind comparison of propofol and midazolam antagonized by flumazenil. *Anaesthesia* 1990; **45**: 198–203.

293. Wilton NCT, Thomas VL. Single breath induction of

anaesthesia using a vital capacity breath of halothane, nitrous oxide and oxygen. *Anaesthesia* 1986; **41**: 472–6.

294. Fisher DM, Robinson S, Brett CM, Perin G, Gregory GA. Comparison of enflurane, halothane, and isoflurane for diagnostic and therapeutic procedures in children with malignancies. *Anesthesiology* 1985; **63**: 647–50.

295. Pandit UA, Steude GM, Leach AB. Induction and recovery characteristics of isoflurane and halothane anaesthesia for short outpatient operations in children. *Anaesthesia* 1985; **40**: 1226–30.

296. Cattermole RW, Verghese C, Blair IJ, Jones CJ, Flynn PJ, Sebel PS. Isoflurane and halothane for outpatient dental anaesthesia in children. *British Journal of Anaesthesia* 1986; **58**: 385–9.

297. Kingston HGG. Halothane and isoflurane anesthesia in pediatric outpatients. *Anesthesia and Analgesia* 1986; **65**: 181–4.

298. Goresky GV, Steward DJ. Rectal methohexitone for induction of anaesthesia in children. *Canadian Anaesthetists' Society Journal* 1979; **26**: 213–15.

299. Linton DM, Thornington RE. Etomidate as a rectal induction agent. Part II. A clinical study in children. *South African Medical Journal* 1983; **64**: 309–10.

300. Saint-Maurice C, Laguenie G, Couturier C, Goutail-Flaud F. Rectal ketamine in paediatric anaesthesia. *British Journal of Anaesthesia* 1979; **51**: 573–4.

301. Carrel R. Ketamine: a general anesthetic for unmanageable ambulatory patients. *Journal of Dentistry for Children* 1973; **40**: 288–92.

302. Meyers EF, Charles P. Prolonged adverse reactions to ketamine in children. *Anesthesiology* 1978; **49**: 39–40.

303. Hannallah RS, Rosales JK. Experience with parents' presence during anaesthesia induction in children. *Canadian Anaesthetists' Society Journal* 1983; **30**: 286–9.

304. Tracey JA, Holland AJC, Unger L. Morbidity in minor gynaecological surgery: a comparison of halothane, enflurane and isoflurane. *British Journal of Anaesthesia* 1982; **54**: 1213–15.

305. Carter JA, Dye AM, Cooper GM. Recovery from day-case anaesthesia. The effect of different inhalational anaesthetic agents. *Anaesthesia* 1985; **40**: 545–8.

306. Davidson SH. A comparative study on halothane and enflurane in paediatric outpatient anaesthesia. *Acta Anaesthesiologica Scandinavica* 1978; **22**: 58–63.

307. Sigurdsson GH, Lindahl S. Cardiac arrhythmias in intubated children during adenoidectomy. A comparison between enflurane and halothane anaesthesia. *Acta Anaesthesiologica Scandinavica* 1983; **27**: 484–9.

308. Willatts DG, Harrison AR, Groom JF, Crowther A. Cardiac arrhythmias during outpatient dental anaesthesia: comparison of halothane with enflurane. *British Journal of Anaesthesia* 1983; **55**: 399–403.

309. Stanford BJ, Plantevin OM, Gilbert JR. Morbidity after day-case gynaecological surgery. Comparison of enflurane with halothane. *British Journal of Anaesthesia* 1979; **51**: 1143–5.

310. Simmons M, Miller CD, Cummings GC, Todd JG. Outpatient paediatric dental anaesthesia. A comparison of halothane, enflurane and isoflurane. *Anaesthesia* 1989; **44**: 735–8.

311. Valanne JV, Korttila K. Recovery following general anaesthe-

sia with isoflurane or enflurane for outpatient dentistry and oral surgery. *Anesthesia Progress* 1988; **35**: 48–52.

312. Fahy LT. Delayed convulsions after day case anaesthesia with enflurane. *Anaesthesia* 1987; **42**: 1327–8.

313. Jones RM, Cashman JN, Eger 2nd EI, Damask MC, Johnson BH. Kinetics and potency of desflurane (I-653) in volunteers. *Anesthesia and Analgesia* 1990; **70**: 3–7.

314. Ghouri AF, Bodner M, White PF. Recovery profile after desflurane–nitrous oxide versus isoflurane–nitrous oxide in outpatients. *Anesthesiology* 1991; **74**: 419–24.

315. Wrigley SR, Fairfield JE, Jones RM, Black AE. Induction and recovery characteristics of desflurane in day case patients: a comparison with propofol. *Anaesthesia* 1991; **46**: 615–22.

316. Van Hemelrijck J, Smith I, White PF. Use of desflurane for outpatient anesthesia. A comparison with propofol and nitrous oxide. *Anesthesiology* 1991; **75**: 197–203.

317. Taylor RH, Lerman J. Induction, maintenance and recovery characteristics of desflurane in infants and children. *Canadian Journal of Anaesthesia* 1992; **39**: 6–13.

318. Fletcher JE, Sebel PS, Murphy MR, Smith CA, Mick SA, Flister MP. Psychomotor performance after desflurane anesthesia: a comparison with isoflurane. *Anesthesia and Analgesia* 1991; **73**: 260–5.

319. Rapp SE, Conahan TJ, Pavlin DJ, *et al.* Comparison of desflurane with propofol in outpatients undergoing peripheral orthopedic surgery. *Anesthesia and Analgesia* 1992; **75**: 572–9.

320. White PF. Studies of desflurane in outpatient anesthesia. *Anesthesia and Analgesia* 1992; **75**: S47–53.

321. Naito Y, Tamai S, Shingu K, Fujimori R, Mori K. Comparison between sevoflurane and halothane for paediatric ambulatory anaesthesia. *British Journal of Anaesthesia* 1991; **67**: 387–9.

322. Saito S, Goto F, Kadoi Y, Takahashi T, Fujita T, Mogi K. Comparative clinical study of induction and emergence time in sevoflurane and enflurane anaesthesia. *Acta Anaesthesiologica Scandinavica* 1989; **33**: 389–90.

323. Frink Jr EJ, Malan TP, Atlas M, Dominguez LM, DiNardo JA, Brown Jr BR. Clinical comparison of sevoflurane and isoflurane in healthy patients. *Anesthesia and Analgesia* 1992; **74**: 241–5.

324. Smith I, Ding Y, White PF. Comparison of induction, maintenance, and recovery characteristics of sevoflurane–N_2O and propofol–sevoflurane–N_2O with propofol–isoflurane–N_2O anesthesia. *Anesthesia and Analgesia* 1992; **74**: 253–9.

325. Alexander GD, Skupski JN, Brown EM. The role of nitrous oxide in postoperative nausea and vomiting. *Anesthesia and Analgesia* 1984; **63**: 175.

326. Lonie DS, Harper NJN. Nitrous oxide anaesthesia and vomiting. The effect of nitrous oxide on the incidence of vomiting following gynaecological laparoscopy. *Anaesthesia* 1986; **41**: 703–7.

327. Melnick BM, Johnson LS. Effects of eliminating nitrous oxide in outpatient anesthesia. *Anesthesiology* 1987; **67**: 982–4.

328. Muir JJ, Warner MA, Offord KP, Buck CF, Harper JV, Kunkel SE. Role of nitrous oxide and other factors in producing postoperative nausea and vomiting: a randomized and blinded prospective study. *Anesthesiology* 1986; **66**: 513–18.

329. Korttila K, Hovorka J, Erkola O. Nitrous oxide does not

increase the incidence of nausea and vomiting after isoflurane anesthesia. *Anesthesia and Analgesia* 1987; **66**: 761–5.

330. Gibbons P, Davidson P, Adler E. Nitrous oxide does not affect post-op vomiting in pediatric eye surgery. *Anesthesiology* 1987; **67**: A530.

331. Sengupta P, Plantevin OM. Nitrous oxide and day-case laparoscopy: effects on nausea, vomiting and return to normal activity. *British Journal of Anaesthesia* 1988; **60**: 570–3.

332. Kay B, Hargreaves J, Sivalingam T, Healy TEJ. Intravenous anaesthesia for cystoscopy: a comparison of propofol or methohexitone with alfentanil. *European Journal of Anaesthesiology* 1986; **3**: 111–20.

333. de Grood PM, Harbers JB, van Egmond J, Crul JF. Anaesthesia for laparoscopy. A comparison of five techniques including propofol, etomidate, thiopentone and isoflurane. *Anaesthesia* 1987; **42**: 815–23.

334. Pace NA, Victory R, White PF. Anesthetic infusion techniques – how to do it. *Journal of Clinical Anesthesia* 1992; **4**: S45–52.

335. Brownlie G, Baker JA, Ogg TW. Propofol: bolus or continuous infusion. A day case technique for the vaginal termination of pregnancy. *Anaesthesia* 1991; **46**: 775–7.

336. Korttila K, Ostman P, Faure E, *et al.* Randomized comparison of recovery after propofol–nitrous oxide versus thiopentone–isoflurane–nitrous oxide anaesthesia in patients undergoing ambulatory surgery. *Acta Anaesthesiologica Scandinavica* 1990; **34**: 400–3.

337. Doze VA, Shafer A, White PF. Propofol–nitrous oxide versus thiopental–isoflurane–nitrous oxide for general anesthesia. *Anesthesiology* 1988; **69**: 63–71.

338. Doze VA, Westphal LM, White PF. Comparison of propofol with methohexital for outpatient anesthesia. *Anesthesia and Analgesia* 1986; **65**: 1189–95.

339. Millar JM, Jewkes CF. Recovery and morbidity after daycase anaesthesia. A comparison of propofol with thiopentone–enflurane with and without alfentanil. *Anaesthesia* 1988; **43**: 738–43.

340. Milligan KR, O'Toole DP, Howe JP, Cooper JC, Dundee JW. Recovery from outpatient anaesthesia: a comparison of incremental propofol and propofol–isoflurane. *British Journal of Anaesthesia* 1987; **59**: 1111–14.

341. Weightman WM, Zacharias M. Comparison of propofol and thiopentone anaesthesia (with special reference to recovery characteristics). *Anaesthesia and Intensive Care* 1987; **15**: 389–93.

342. Marais ML, Maher MW, Wetchler BV, Korttila K, Apfelbaum JL. Reduced demands on recovery room resources with propofol (Diprivan) compared to thiopental-isoflurane. *Anesthesiology Review* 1989; **16**: 29–40.

343. Steib A, Freys G, Jochum D, Ravanello J, Schaal JC, Otteni JC. Recovery from total intravenous anaesthesia. Propofol versus midazolam–flumazenil. *Acta Anaesthesiologica Scandinavica* 1990; **34**: 632–5.

344. Nilsson A, Persson MP, Hartvig P. Effects of flumazenil on post-operative recovery after total intravenous anesthesia with midazolam and alfentanil. *European Journal of Anaesthesiology* 1988; **2**: S251–6.

345. Cooper GM, O'Conner M, Mark J, Harvey J. Effect of alfentanil and fentanyl on recovery from brief anaesthesia. *British Journal of Anaesthesia* 1983; **55**: S179–82.

346. Wall RJ, Zacharias M. Effects of alfentanil on induction and recovery from propofol anaesthesia in day surgery. *Anaesthesia and Intensive Care* 1990; **18**: 214–18.

347. Sanders RS, Sinclair ME. Sear JW. Alfentanil in short procedures. *Anaesthesia* 1984; **39**: 1202–6.

348. Goroszeniuk T, Whitwam JG, Morgan M. Use of methohexitone, fentanyl and nitrous oxide for short surgical procedures. *Anaesthesia* 1977; **32**: 209–11.

349. Collin RIW, Drummond GB, Spence AA. Alfentanil supplemented anaesthesia for short procedures. A double-blind study of alfentanil used with etomidate and enflurane for day cases. *Anaesthesia* 1986; **41**: 477–81.

350. Martin DE, Rosenberg H, Aukburg SJ, *et al.* Low-dose fentanyl blunts circulatory responses to tracheal intubation. *Anesthesia and Analgesia* 1982; **61**: 680–4.

351. Cork RC, Weiss JL, Hameroff SR, Bentley J. Fentanyl preloading for rapid-sequence intubation of anesthesia. *Anesthesia and Analgesia* 1984; **63**: 60–4.

352. Kennedy DJ, Ogg TW. Alfentanil and memory function. A comparison with fentanyl for day case termination of pregnancy. *Anaesthesia* 1985; **40**: 537–40.

353. Kay B, Venkataraman P. Recovery after fentanyl and alfentanil in anaesthesia for minor surgery. *British Journal of Anaesthesia* 1983; **55**: S169–71.

354. Kallar SK, Keenan RL. Evaluation and comparison of recovery times from alfentanil and fentanyl for short surgical procedures. *Anesthesiology* 1984; **61**: A379.

355. Raeder JC, Hole A. Out-patient laparoscopy in general anaesthesia with alfentanil and atracurium. A comparison with fentanyl and pancuronium. *Acta Anaesthesiologica Scandinavica* 1986; **30**: 30–4.

356. White PF, Coe V, Shafer A, Sung ML. Comparison of alfentanil with fentanyl for outpatient anesthesia. *Anesthesiology* 1986; **64**: 99–106.

357. Dachowski MT, Kalayjian R, Angelillo JC, Dolan EA. Continuous infusion of methohexitone and alfentanil hydrochloride for general anesthesia in outpatient third molar surgery. *Journal of Oral and Maxillofacial Surgery* 1989; **47**: 233–7.

358. Kestin IG, Dorje P. Anaesthesia for evacuation of retained products of conception. Comparison between alfentanil plus etomidate and fentanyl plus thiopentone. *British Journal of Anaesthesia* 1987; **59**: 364–8.

359. Moffat AC, Murray AW, Fitch W. Opioid supplementation during propofol anaesthesia. The effects of fentanyl or alfentanil on propofol anaesthesia in daycase surgery. *Anaesthesia* 1989; **44**: 644–7.

360. Cade L, Ross AW. Is fentanyl effective for postoperative analgesia in day-surgery?. *Anaesthesia and Intensive Care* 1992; **20**: 38–40.

361. Phitayakorn P, Melnik BM, Vicinie 3rd AF. Comparison of continuous sufentanil and fentanyl infusions for outpatient anaesthesia. *Canadian Journal of Anaesthesia* 1987; **34**: 242–5.

362. Wasudev G, Kambam JR, Hazelhurst WM, Hill P, Adkins R, Coleman C. Comparative study of sufentanil and isoflurane

in outpatient surgery. *Anesthesia and Analgesia* 1987; **66**: S186.

363. Pandit SK, Kothary SP, Pandit UA, Mathai MK. Comparison of fentanyl and butorphanol for outpatient anaesthesia. *Canadian Journal of Anaesthesia* 1987; **34**: 130–4.

364. Philip BK, Scott DA, Freiberger D, Gibbs RR, Hunt C, Merray E. Butorphanol compared with fentanyl in general anaesthesia for outpatient laparoscopy. *Canadian Journal of Anaesthesia* 1991; **38**: 183–6.

365. Wetchler BV, Alexander CD, Shariff MS, Gaudzels GM. A comparison of recovery in outpatients receiving fentanyl versus those receiving butorphanol. *Journal of Clinical Anesthesia* 1989; **1**: 339–43.

366. Kamal RS, Khan FA, Khan FH. Total intravenous anaesthesia with propofol and buprenorphine. *Anaesthesia* 1990; **45**: 865–70.

367. Garfield JM, Garfield FB, Philip BK, Earls F, Roaf E. A comparison of clinical and psychological effects of fentanyl and nalbuphine in ambulatory gynecologic patients. *Anesthesia and Analgesia* 1987; **66**: 1303–7.

368. Bone ME, Dowson S, Smith G. A comparison of nalbuphine with fentanyl for postoperative pain relief following termination of pregnancy under day care anaesthesia. *Anaesthesia* 1988; **43**: 194–7.

369. Ding Y, White PF. Comparative effects of ketorolac, dezocine and fentanyl as adjuvants during outpatient anesthesia. *Anesthesia and Analgesia* 1992; **75**: 566–71.

370. McLoughlin C, McKinney MS, Fee JP, Boules Z. Diclofenac for day-care arthroscopy surgery: comparison with a standard opioid therapy. *British Journal of Anaesthesia* 1990; **65**: 620–3.

371. Laitinen J, Nuutinen L, Kiiskila EL, Freudenthal Y, Ranta P, Karvonen J. Comparison of intravenous diclofenac, indomethacin and oxycodone as postoperative analgesics in patients undergoing knee surgery. *European Journal of Anaesthesiology* 1992; **9**: 29–34.

372. Edwards ND, Barclay K, Catling SJ, Martin DG, Morgan RH. Day case laparoscopy: a survey of postoperative pain and an assessment of the value of diclofenac. *Anaesthesia* 1991; **46**: 1077–80.

373. Comfort VK, Code WE, Rooney ME, Yip RW. Naproxen premedication reduces postoperative tubal ligation pain. *Canadian Journal of Anaesthesia* 1992; **39**: 349–52.

374. Campbell WI. Analgesic side effects and minor surgery: which analgesic for minor and day-case surgery?. *British Journal of Anaesthesia* 1990; **64**: 617–20.

375. Alexander CD, Wetchler BV, Thompson RE. Bupivacaine infiltration of the mesosalpinx in ambulatory surgical laparoscopic tubal sterilization. *Canadian Journal of Anaesthesia* 1987; **34**: 362–5.

376. Smith I, Van Hemelrijck J, White PF, Shively R. Effects of local anesthesia on recovery after outpatient arthroscopy. *Anesthesia and Analgesia* 1991; **73**: 536–9.

377. Herbert M, Healy TEJ, Bourke JB, Fletcher IR, Rose JM. Profile of recovery after general anaesthesia. *British Medical Journal* 1983; **286**: 1539–42.

378. Walsh C, Smith CE, Ryan B, Polomeno RC, Bevan JC. Postoperative vomiting following strabismus surgery in paediatric outpatients: spontaneous versus controlled ventilation. *Canadian Journal of Anaesthesia* 1988; **35**: 31–5.

379. Brindle GF, Soliman MG. Anaesthetic complications in surgical out-patients. *Canadian Anaesthetists' Society Journal* 1975; **22**: 613–19.

380. Urbach GM, Edelist G. An evaluation of the anaesthetic techniques used in an outpatient unit. *Canadian Anaesthetists' Society Journal* 1977; **24**: 401–7.

381. Skacel M, Sengupta P, Plantevin OM. Morbidity after day case laparoscopy. A comparison of two techniques of tracheal anaesthesia. *Anaesthesia* 1986; **41**: 537–41.

382. Trepanier CA, Brousseau C, Lacerte L. Myalgia in outpatient surgery: comparison of atracurium and succinylcholine. *Canadian Journal of Anaesthesia* 1988; **35**: 255–8.

383. Zahl K, Apfelbaum JL. Muscle pain occurs after outpatient laparoscopy despite the substitution of vecuronium for succinylcholine. *Anesthesiology* 1989; **70**: 408–11.

384. Fragen RJ, Shanks CA. Neuromuscular recovery after laparoscopy. *Anesthesia and Analgesia* 1984; **63**: 51–4.

385. Pearce AC, Williams JP, Jones RM. Atracurium for short surgical procedures in day patients. *British Journal of Anaesthesia* 1984; **56**: 973–6.

386. Sengupta P, Skacel M, Plantevin OM. Post-operative morbidity associated with the use of atracurium and vecuronium in day-case laparoscopy. *European Journal of Anaesthesiology* 1987; **4**: 93–9.

387. Bailey DM, Nicholas AD. Comparison of atracurium and vecuronium during anaesthesia for laparoscopy. *British Journal of Anaesthesia* 1988; **61**: 557–9.

388. Zuurmond WWA, van Leeuwen L. Atracurium versus vecuronium: a comparison of recovery in outpatient arthroscopy. *Canadian Journal of Anaesthesia* 1988; **35**: 139–42.

389. Basta SJ. Clinical pharmacology of mivacurium chloride: a review. *Journal of Clinical Anesthesia* 1992; **4**: 153–63.

390. King MJ, Milazkiewicz R, Carli F, Deacock AR. Influence of neostigmine on postoperative vomiting. *British Journal of Anaesthesia* 1988; **62**: 403–6.

391. Petersen RS, Bailey PL, Kalameghan R, Ashwood ER. Prolonged neuromuscular blockade after mivacurium. *Anesthesia and Analgesia* 1993; **76**: 194–6.

392. Poler SM, Watcha MF, White PF. Mivacurium as an alternative to succinylcholine during outpatient laparoscopy. *Journal of Clinical Anesthesia* 1992; **4**: 127–33.

393. Goldberg ME, Larijani GE, Azad SS, *et al.* Comparison of tracheal intubating conditions and neuromuscular blocking profiles after intubating doses of mivacurium chloride or succinylcholine in surgical outpatients. *Anesthesia and Analgesia* 1989; **69**: 93–9.

394. Puhringer FK, Khuenl-Brady KS, Koller J, Mitterschiffthaler G. Evaluation of the endotracheal intubating conditions of rocuronium (ORG 9426) and succinylcholine in outpatient surgery. *Anesthesia and Analgesia* 1992; **75**: 37–40.

395. Cooper R, Mirakhur RK, Clarke RS, Boules Z. Comparison of the intubation conditions after administration of ORG 9246 (rocuronium) and suxamethonium. *British Journal of Anaesthesia* 1992; **69**: 269–73.

396. Mayer M, Doenicke A, Hofmann A, Peter K. Onset and recovery of rocuronium (ORG 9426) and vecuronium under

enflurane anaesthesia. *British Journal of Anaesthesia* 1992; **69**: 511–12.

397. Booth MG, Marsh B, Bryden FM, Robertson EN, Baird WL. A comparison of the pharmacodynamics of rocuronium and vecuronium during halothane anaesthesia. *Anaesthesia* 1992; **47**: 832–4.

398. Edilist G. Prophylaxis and management of post-operative problems. *Canadian Anaesthetists' Society Journal* 1983; **30**: 558–60.

399. Lanigan CJ. Oxygen desaturation after dental anaesthesia. *British Journal of Anaesthesia* 1992; **68**: 142–5.

400. Dhamee MS, Gandhi SK, Callen KM, Kalbfleish JH. Morbidity after outpatient anesthesia – a comparison of different endotracheal anesthetic techniques for laparoscopy. *Anesthesiology* 1982; **57**: A375.

401. Kurer FL, Welch DB. Gynaecological laparoscopy: clinical experience of two anaesthetic techniques. *British Journal of Anaesthesia* 1984; **56**: 1207–12.

402. Conahan 3rd TJ, Williams GD, Apfelbaum JL, Lecky JH. Airway heating reduces the recovery time (cost) in out-patients. *Anesthesiology* 1987; **67**: 127–30.

403. Freeman J. The effectiveness of doxapram administration in hastening arousal following general anesthesia in out-patients. *American Association of Nurse Anesthetists Journal* 1986; **54**: 16–20.

404. Korttila K. Psychomotor recovery after anesthesia and sedation in the dental office. In: Dionne RA, Laskin DM, eds. *Anesthesia and sedation in the dental office*. New York: Elsevier, 1986: 135–47.

405. Korttila K. Anaesthesia for ambulatory surgery: firm definitions of 'home readiness' needed. *Annals of Medicine* 1991; **23**: 635–6.

406. Korttila K. Postanesthetic cognitive and psychomotor impairment. *International Anesthesiology Clinics* 1986; **24**: 59–74.

407. Coachman JN, Power SJ, Jones RM, Adams AP. Assessment of recovery from anaesthesia: what tests should we use? *Anesthesiology* 1987; **67**: A434.

408. Fishburne Jr JI, Fulghum MS, Hulka JF, Mercer JP. General anesthesia for outpatient laparoscopy with an objective measure of recovery. *Anesthesia and Analgesia* 1974; **53**: 1–6.

409. Doenicke A, Kugler J, Laub M. Evaluation of recovery and 'street fitness' by EEG and psychodiagnostic tests after anaesthesia. *Canadian Anaesthetists' Society Journal* 1967; **14**: 567–83.

410. Steward DJ, Volgyesi G. Stabilometry: a new tool for the measurement of recovery following general anaesthesia for out-patients. *Canadian Anaesthetists' Society Journal* 1978; **25**: 4–6.

411. Korttila K. Recovery and driving after brief anaesthesia. *Anaesthesist* 1981; **30**: 377–82.

412. Denis R, Letourneau JE, Londorf D. Reliability and validity of psychomotor tests as measures of recovery from isoflurane or enflurane anesthesia in a day-care surgery unit. *Anesthesiology and Analgesia* 1984; **63**: 653–6.

413. Pflug AE, Aasheim GM, Foster C. Sequence of return of neurological function and criteria for safe ambulation following subarachnoid block (spinal blockade). *Canadian Anaesthetists' Society Journal* 1978; **25**: 133–9.

414. Alexander CM, Teller LE, Gross JB, Owen D, Cunningham C, Laurencio F. New discharge criteria decrease recovery room time after subarachnoid block. *Anesthesiology* 1989; **70**: 640–3.

415. Daos FG, Virtue RW. Sympathetic block persistence after spinal or epidural analgesia. *Journal of the American Medical Association* 1963; **183**: 285–7.

416. Roe CF, Cohn FL. Sympathetic blockade during spinal anesthesia. *Surgery, Gynecology and Obstetrics* 1973; **136**: 265–8.

417. Schreiner MS, Nicolson SC, Martin T, Whitney L. Should children drink before discharge from day surgery?. *Anesthesiology* 1992; **76**: 528–33.

418. Fraser RA, Hotz SB, Hurtig JB, Hodges SN, Moher D. The prevalence and impact of pain after day-care tubal ligation surgery. *Pain* 1989; **39**: 189–201.

419. Lahti PT. Early postoperative discharge of patients from the hospital. *Surgery* 1968; **63**: 410–15.

420. Paasuke RT, Davies JM. Anaesthesia for daycare patients – controversies and concerns. *Canadian Anaesthetists' Society Journal* 1986; **33**: 644–6.

421. White PF. Outpatient anesthesia – an overview. In: White PF, ed. *Outpatient anesthesia*. New York: Churchill Livingstone Inc., 1990: 1–48.

Anaesthesia for Laparoscopic-assisted Surgery

Timothy E. Black

Initiation of the operative procedure of laparoscopy	General anaesthesia
Insufflation	Regional anaesthesia
Gastric reflux	Local anaesthesia

Laparoscopy, once almost exclusively a technique practised by gynaecologists, has taken the surgical profession by storm. Laparoscopic tubal ligation, first reported in 1962, is commonplace today. Indeed, laparoscopy has evolved into an entirely new discipline of operative procedures, though it was not until 1989, when reports of laparoscopic-assisted cholecystectomy appeared, that the surgical subspecialties embraced its use enthusiastically.[1] Simultaneously, a major expansion in equipment technology and advancement in instrument development occurred.

Operations that were once extremely painful, debilitating and associated with large scars and prolonged recovery times now require only small limited incisions, very short hospital stays, and have faster recovery times. Likewise, health care costs are being reduced as the length of hospital stays is decreased with laparoscopic-assisted surgery. Even more importantly, society gains because patients have much reduced recovery times, thereby allowing them to return to useful employment and activity much sooner.

Consequently, both the medical profession and patients have recognized the tremendous benefits from laparoscopic-assisted surgery. Today, as techniques and equipment continue to evolve, a greater number as well as more involved surgical procedures are being performed through laparoscopy. As a result of the high frequency with which the anaesthetist will be presented with patients undergoing laparoscopic-assisted surgery, it is imperative that we obtain a clear understanding of the procedure, as well as its potential problems and the physiological changes that may occur.

Initiation of the operative procedure of laparoscopy

It is necessary to be familiar with the operative procedure by which pneumoperitoneum and laparoscopy are initiated.

One may then design an appropriate anaesthetic and in addition be familiar with the complications that may arise from the procedure itself.

Initiation of the pneumoperitoneum and placement of the cannula for access is the necessary first step. This has potential dangers. The urinary bladder and stomach should be decompressed using a Foley catheter and nasogastric tube respectively, if possible, to avoid injury during insertion of the Verees® needle and the trocar. A pneumoperitoneum is established with the patient in a slight Trendelenburg position. A small incision is made just above the umbilicus and a Verees® needle is introduced under control into the abdominal cavity in the midline at a 30-degree downward angle toward the pelvis. This aims the needle tip below the bifurcation of the aorta, thereby reducing the possibility of a major vascular puncture. Generally, a pop is felt as the needle penetrates through the fascia. The correct position of the needle tip in the abdominal cavity is ensured by several methods: the drop test, aspiration, and pressure. Several drops of sterile saline are placed in the hub of the needle and drawn into it by the negative intra-abdominal pressure – the tip of the needle is in the free abdominal cavity (positive drop test). The needle is aspirated to ensure that a vessel or bowel has not been entered.[2]

The Verees® is connected to the insufflator. A negative pressure of 0 to −5 mmHg as read on the insufflator pressure gauge reconfirms that the needle is in the peritoneal cavity. The insufflator upper pressure limit is set for 15 mmHg. Carbon dioxide is instilled at a low flow rate of 1 l/min into the peritoneal cavity through the Verees® needle. The quadrants of the abdomen are percussed gently to ensure an even distribution of gas. Once an adequate pneumoperitoneum has been established (about 1.5 litres), a higher flow rate of 2–3 l/min is used. When the upper pressure limit of 12–15 mmHg has been reached (usually 3–5 litres of CO_2), flow automatically ceases.[2]

The Verees® needle is removed and an 11 ml cannula is

inserted through the same umbilical incision using a trocar. Once in place, the trocar is removed and the insufflator is attached to the cannula and set to the maximum flow rate of 6–10 l/min. It is critical that the cannula delivering the pneumoperitoneum is maintained in the abdominal cavity and not in the subcutaneous tissue. Otherwise subcutaneous emphysema will occur rapidly.[2] The laparoscope is inserted and the abdominal contents inspected. Additional incisions may be made so that additional instruments can be placed in the abdominal cavity. Usually these are placed under the direct vision of the laparoscope. This basic operative procedure is the foundation for many operations which may now proceed through the laparoscope. Hopefully, a clear understanding of the initiation of this procedure will provide the anaesthetist with an insight into the potential risks and complications encountered during anaesthesia and surgery.

Insufflation

Laparoscopy of the upper and lower abdominal cavities makes use of peritoneal insufflation to obtain better views and access to surgical sites. Carbon dioxide became the standard for insufflation because of its valuable properties: it is non-flammable, permitting safe electrocautery, and highly soluble in blood and tissue, minimizing dangers of gas embolism.[3]

Pressure limiters are used during insufflation to decrease the likelihood of the deleterious effects of raised intra-abdominal pressure of 40 mmHg and higher. These are known to have serious side effects including circulatory impairment, respiratory embarrassment, as well as an increased risk of CO_2 embolism and peritoneal and subcutaneous emphysema.[4] Today, most equipment limits intra-abdominal pressure to 15 mmHg while maintaining appropriate gas volumes for adequate views.

It should be noted that during insufflation for lower abdominal laparoscopy, the Trendelenburg position is usually used, while upper abdominal laparoscopy uses reverse Trendelenburg. While this latter position may reduce some of the ventilatory abnormalities during insufflation, it may actually accentuate circulatory impairment.

Carbon dioxide homoeostasis

Much interest has surrounded the homoeostasis of carbon dioxide during CO_2 insufflation. Questions of serious derangement of respiratory parameters have arisen while many investigators have helped to clarify the situation.[4–8]

During controlled ventilation in the Trendelenburg position and with peritoneal insufflation giving intra-abdominal pressures of 15 mmHg, there are significant increases in $PaCO_2$ (10 mmHg), $PACO_2$ (4 mmHg) and with decreases in total lung compliance of 25 per cent.[5] Absorption of CO_2 occurs, as indicated by an increased end tidal CO_2, during stable controlled ventilation. In the clinical situation, the rate of CO_2 absorption appears to be relatively small and only a slight increase in minute volume is required to eliminate this extra CO_2.[6] Increased abdominal pressure leads to elevation of the diaphragm. This causes a decrease in FRC, total lung compliance and atelectasis, leading to less efficient minute ventilation contributing to a further increase in $PaCO_2$.[8] This situation is made worse by the Trendelenburg position and is somewhat diminished in the reverse Trendelenburg position. It should be noted that all respiratory parameters can be compensated for by the use of mechanical ventilation while monitoring the end tidal CO_2.

Large increases in $PaCO_2$ may occur during two other situations. First, $PaCO_2$ values as high as 70 mmHg have been reported for spontaneously breathing patients under general anaesthesia during insufflation.[9] Second, if CO_2 embolism occurs, abrupt increases in $PaCO_2$ and $PETCO_2$ are noted.[10,11] Careful attention to respiratory parameters and end tidal CO_2 will help identify the problems so that measures can be taken to avoid them.

Venous CO_2 embolism

Venous CO_2 embolism has been reported both intra- and postoperatively during CO_2 insufflation and must be considered in the event of a cardiovascular collapse.[10–12] One highly desirable property of CO_2 is that it is highly soluble. Small amounts of CO_2 that gain intravascular access are not harmful because they are quickly absorbed and excreted by the lungs. However, if a larger bolus is unrecognized, deleterious haemodynamic events may occur.

An important aetiological factor in the development of venous CO_2 is the use of high insufflating pressures (greater than 15 mmHg).[6] In the presence of open venous channels from surgical dissection and trauma, CO_2 under pressure may gain intravascular access. Obviously, this will also occur with unrecognized intravascular placement of the Verees® needle.

Continuous monitoring of heart sounds, and tidal CO_2 and systolic blood pressure will provide early clues to development of this situation. Early signs of significant CO_2 embolism include a falling systolic blood pressure, development of a mill wheel murmur, cyanosis, and desaturation evident by pulse oximetry. Initially with a small bolus of CO_2 the end tidal CO_2 may abruptly increase. However, if the bolus grows larger, causing a gas lock within the right heart or pulmonary outflow tract, the end tidal CO_2 will drop. Carbon dioxide embolism with cardiovascular collapse must be treated as any other gas embolism. Immediate deflation of the pneumoperitoneum may prevent further progression of the symptoms.[11,12]

Cardiovascular system

A relatively healthy patient with an induced pneumoperitoneum limited to 15 mmHg may have a cardiac output that is either unchanged or modestly increased.[4,5] Although most patients are very stable perioperatively during laparoscopy, hypotension and cardiovascular collapse have been reported.[10–12]

The changes in cardiac ouput during laparoscopy tend to parallel the changes in effective cardiac filling pressure which accompany abdominal inflation. As intra-abdominal pressure is increased to 20–30 mmHg there is a compensatory and a progressive increase in both cardiac output and effective cardiac filling pressure. When intra-abdominal pressure is increased further, filling pressure decreases accompanied by a fall in cardiac output, a response which is reversed at the end of endoscopy when intra-abdominal pressure is restored to normal.[4]

The biphasic change in cardiac output may be explained as follows: increased intra-abdominal pressure has two opposing effects on the cardiac vascular system. First, it forces blood out of the abdominal organs and IVC into the central venous reservoir. Second, it may prevent venous return from the legs and pelvic region, thereby decreasing the central venous blood volume.[4]

During pneumoperitoneum with the patient in the horizontal position, the CVP has been reported to increase by 8 mmHg when the intra-abdominal pressure is limited to 15 mmHg.[5] The Trendelenburg position itself will raise CVP. If we add a pneumoperitoneum, while in the Trendelenburg position, it has been reported that the CVP will increase by an additional 6 mmHg.[5] At low intra-abdominal pressures, the first effect dominates. However, as intra-abdominal pressures are increased further, the abdominal capacitance vessels empty. If the pressure is increased further, progressive blockade of venous return from the legs and pelvis effectively decreases central blood volume causing a fall in CVP and cardiac output.[4]

The Trendelenburg position is used during lower abdominal laparoscopy to keep small bowel and other contents from obstructing the view. For the same reason, upper abdominal laparoscopy uses the reverse Trendelenburg position. This may lead to less venous return due to gravity, thereby negating some of the increase in CVP associated with peritoneal insufflation. It may be necessary to volume load such patients (infuse plasma expander) to prevent this from occurring.[13]

It should be noted that most data obtained during laparoscopy are from relatively healthy individuals with stable cardiovascular systems.[4,5] As the use of laparoscopy evolves, one may be responsible for patients who are quite ill and who have a diminished cardiovascular reserve. The changes occurring during insufflation and position change may have dramatic and deleterious effects. Careful assessment including intensive monitoring may be required in certain circumstances.

Arrhythmias

Cardiac arrhythmias may occur during laparoscopy with pneumoperitoneum.[6,7,14] Respiratory acidosis leading to increased sympathetic outflow is thought to be the aetiology. The presence of halogenated anaesthetics may contribute to this problem. As the $PaCO_2$ exceeds the arrhythmic threshold for these agents, especially in the case of halothane, the onset of arrhythmias may result in hypotension.

Enflurane and isoflurane have a lower incidence of catecholamine-induced arrhythmias than does halothane.[14] Arrhythmias can be prevented in the presence of these anaesthetics by carefully controlling the $PaCO_2$.

Hypoxia can also contribute to the occurrence of arrhythmias. However, with supplemental oxygen and careful monitoring with pulse oximetry this can be completely avoided. The possibility of a severe bradyarrhythmia and cardiac asystole from a reflex increase in vagal tone associated with the peritoneal manipulation may be of importance.[14]

Gastric reflux

High intra-abdominal pressure due to the pneumoperitoneum during laparoscopy may increase the likelihood of an intraoperative reflux of gastric contents. With the use of cuffed tracheal intubation during general anaesthesia the risk is greatly reduced. In addition, the insertion of an oral or nasal gastric tube to decompress the stomach further helps to decrease the risk. However, if regional anaesthesia with sedation or general anaesthesia with mask ventilation are chosen, the problem of a reflux of gastric contents is more significant. Consideration of prophylactic therapy with agents such as cimetidine, ranitidine and metoclopramide, to increase gastric pH, decrease gastric volume, and increase lower oesophageal sphincter tone would be appropriate.[13,15,16]

Aspiration of gastric contents during mask anaesthesia may be catastrophic. However, in a series of 50 000 patients, of whom 5000 were not intubated, undergoing laparoscopy and studied prospectively in the UK, there was no case of inhalation of gastric contents.[17] However, to assess fully the true risk the population studied must be larger. Therefore, at present the risk remains real.

General anaesthesia

A well planned and executed general anaesthetic is by far the most common anaesthetic used for laparoscopy. It not only provides a complete anaesthetic, but also enables cardiovascular and respiratory stability to be maintained.[4,18]

Many laparoscopic operations are of short duration and patients are discharged soon thereafter. Anaesthetic techniques should be designed to accommodate the situation.

Anaesthetic monitoring of outpatients should follow standards set for inpatient anaesthetic care. The anaesthetist should keep in mind that in certain circumstances laparoscopic surgery may proceed to open laparotomy. One may need to alter the anaesthetic in progress at short notice. If a patient's overall physical condition or health is poor, invasive monitoring for closer perioperative care may be needed. Increasing numbers of surgical patients may present for laparoscopic surgery, compared with open laparotomy, because of the less invasive nature of the procedure, quick recovery times, and less debilitating postoperative course.

General anaesthesia with tracheal intubation and controlled mechanical ventilation while using end tidal CO_2 monitors and pulse oximetry offers the most stable physiological environment for the patient. Adjusting ventilation appropriately during the use of CO_2 insufflation will assure normocapnia and decrease the risk of cardiac arrhythmias. Use of controlled ventilation while adding supplemental FIO_2 will assure that patients do not become hypoxic.[4]

The use of tracheal intubation will also help to prevent intraoperative aspiration of gastric contents. While intubated, an oro- or nasal gastric tube can be easily inserted to decompress and empty the stomach. This also decreases the risk of stomach perforation by operative trocars.

Neuromuscular blockade with a non-depolarizing muscle relaxant can also be used during general tracheal anaesthesia. Relaxation of abdominal muscles will help to augment an adequate pneumoperitoneum at lower insufflation pressures and may reduce the untoward effects of raised intra-abdominal pressure and the complications resulting from high insufflation pressures. Short-acting agents such as vecuronium and atracurium are ideal for short procedures.[19]

The use of nitrous oxide as part of a general anaesthetic has been questioned due to its ability to produce bowel distension. However, its use has been shown to have no clinically significant effect on surgical conditions during laparoscopy.[18] Nitrous oxide also has been implicated in postoperative nausea and vomiting. However, this is by no means certain following laparoscopy under general anaesthesia and there is not always an increased incidence of this undesirable problem when nitrous oxide is used.

Of the halogenated general anaesthetics in use, isoflurane shows the least sensitivity to arrhythmias in the presence of raised catecholamines and therefore may be the most suitable agent to use during CO_2 insufflation.[14,20,21] Diprivan® (propofol) circumvents this problem completely and is an excellent choice for a general anaesthetic. Any number of combinations or choices of agents may be used when the physiological consequences of laparoscopy are considered.

Narcotic medication may be used to supplement analgesia when ventilation is controlled. This, however, may cause intraoperative spasm of the spinchtor of Oddi, making cholangiography difficult to interpret. If this occurs, intravenous glucagon or naloxone have been shown to be useful in reversing this effect.[22,23]

The use of spontaneous respiration (mask or tracheal tube) during laparoscopy is advocated by some.[9,20,24]

One criticism of spontaneous ventilation during laparoscopy with pneumoperitoneum is that hypercapnia, especially in combination with halothane, may lead to serious arrhythmias.[3,7] High levels of CO_2 tension have been confirmed in patients breathing spontaneously in this setting. Several factors contribute to this. First, there is depression of ventilation by some premedication drugs and anaesthetic drugs. Second, there is impairment of ventilation by mechanical factors such as abdominal distension and the use of a steep Trendelenburg position. Early reports of serious arrhythmias and cardiovascular collapse may indeed have been due to these problems.

Nevertheless, it has not been shown that spontaneous respiration during laparoscopy is unsafe. In healthy patients there is no difference in the surgical or the anaesthetic complications in groups breathing halothane spontaneously, by mask, and those in whom ventilation is controlled.[9] Spontaneous respiration with a mask using isoflurane can produce good operating conditions without hypercapnia, acidosis, and cardiac arrhythmias.[14]

Both techniques are used. The safety of the procedure probably depends less on the particular anaesthetic technique than on a short intraoperative time, close intraoperative monitoring, and the experience of the anaesthetist and surgeon.[20] General anaesthesia with spontaneous ventilation is best suited to lower abdominal laparoscopy of short duration. However, it must be remembered that spontaneous ventilation may be insufficient to deal with an increased CO_2 load and the respiratory consequences of a pneumoperitoneum, especially in the presence of inhalation or intravenous general anaesthetics. Spontaneous ventilation cannot be recommended for obese or less than healthy patients, nor when laparoscopy is to be prolonged.

Respiratory complications may occur after laparoscopy under general anaesthesia and especially following upper abdominal operative procedures such as cholecystectomy. Decreases in minute ventilation, FEV_1 and PaO_2 may persist for several days postoperatively, before returning to baseline.[25] Recent data concerning laparoscopic-assisted surgery, specifically cholecystectomy, also show a decreased respiratory reserve for a short time postoperatively. This is probably related to upper abdominal and diaphragmatic irritation and injury. However, it appears that this is 'self-correcting' after about 24 hours. The results overall are definitely an improvement over open laparotomy.

Regional anaesthesia

Epidural or spinal anaesthesia has been suggested to be a suitable alternative to general anaesthesia for laparoscopy.[24,25] During epidural anaesthesia for laparoscopy, respiratory factors may remain quite stable. Alveolar ventilation remains unaltered despite the increased respiratory workload and ventilation/perfusion inequality resulting from mechanical compression of lungs during pneumoperitoneum and Trendelenburg position. Patients increase their minute ventilation, by a greater increase in respiratory rate than tidal volume, in order to maintain an adequate alveolar ventilation.[26]

It has been shown that the mechanism which maintains adequate alveolar ventilation during CO_2 insufflation does not change during lignocaine epidural anaesthesia.[26,27] This, however, assumes that patients are not heavily sedated or narcotized.

One must be careful when sedative, narcotic, or other intravenous drugs are given in conjunction with an epidural anaesthetic, as this will decrease sensitivity to CO_2. These patients may then be unable to deal effectively with an increasing carbon dioxide load with serious consequences. The use of heavy sedation under epidural anaesthesia may also lead to significant hypoxia, especially in the absence of supplemental oxygen.

Laparoscopy under regional anaesthesia is probably best suited for lower abdominal procedures; however, upper abdominal procedures have been effectively managed using this technique. Patients, however, may have referred shoulder pain and difficulty breathing when laparoscopy and surgical sites are close to the diaphragm, as for example, laparoscopic cholecystectomy.

The effects of subarachnoid block can be assumed to be very similar to those of the epidural blockade used for laparoscopy. It is necessary to attain a block level up to T2 to be certain of adequate anaesthesia for abdominal laparoscopy.

Inhalation of gastric contents during anaesthesia may have catastrophic results. Pulmonary aspiration may occur when heavily sedated patients, during regional anaesthesia, are unable to protect their airways effectively.

Local anaesthesia

Diagnostic laparoscopy under local anaesthesia with i.v. sedation has been reported to be successful.[28] However, this technique should be limited to short diagnostic procedures. A more involved procedure would necessitate the use of a more complete anaesthetic.

REFERENCES

1. Dubois F, Icord P, Berthelot G. Coelioscopic cholecystectomy. Pre-liminary report of 36 cases. *Annals of Surgery* 1990; **211**: 60–2.

2. Talamani MA, Gadaz TR. Laparoscopic approach to cholecystectomy. *Advances in Surgery* 1991; **25**: 1–20.

3. Caverly RK, Jenkins LC. The anaesthetic management of pelvic laparoscopy. *Canadian Anaesthetists' Society Journal* 1973; **20**: 679–86.

4. Kelman GR, Swapp GH, Smith I, Benzie RJ, Gordon NLM. Cardiac output and arterial blood gas tension during laparoscopy. *British Journal of Anaesthesia* 1972; **44**: 1155–62.

5. Versichelen L, Serreyn R, Rolly G, Vanderkerckhove D. Physiologic changes during anesthesia administration for gynecologic laparoscopy. *Journal of Reproductive Medicine* 1984; **10**: 697–700.

6. Seed RF, Shakespear TF, Muldoon MJ. Carbon dioxide homeostasis during anaesthesia for laparoscopy. *Anaesthesia* 1970; **25**: 223–31.

7. Desmond J, Gordon RA. Ventilation in patients anaesthetized for laparoscopy. *Canadian Anaesthetists' Society Journal* 1970; **17**: 378–87.

8. Hodgson C, McClelland RMA, Newton JR. Some effects of the

peritoneal insufflation of carbon dioxide at laparoscopy. *Anaesthesia* 1970; **25**: 382–90.

9. Lewis DG, Ryder W, Burn N, Wheldon JT, Tacchi D. Laparoscopy – an investigation during spontaneous ventilation with halothane. *British Journal of Anaesthesia* 1972; **44**: 685–91.

10. Clark CC, Weeks DB, Gusdon JP. Venous carbon dioxide embolism during laparoscopy. *Anesthesia and Analgesia* 1977; **56**: 650–2.

11. Shulman D, Aronson HB. Capnography in the early diagnosis of carbon dioxide embolism during laparoscopy. *Canadian Anaesthetists' Society Journal* 1984; **31**: 455–9.

12. Root B, Levy MN, Pollack S, Lubert M, Pathek K. Gas embolism death after laparoscopy delayed by trapping in portal circulation. *Anesthesia and Analgesia* 1978; **57**: 232–7.

13. Hasnain JU, Matjasko MJ. Practical anesthesia for laparoscopic procedures. In: Zuker K, Bailey R, Reddick E, eds. *Surgical laparoscopy*. St Louis: Quality Medical Publishers, 1991; 77–86.

14. Harris MNE, Plantevin OM, Crowther A. Cardiac arrhythmia during anaesthesia. *British Journal of Anaesthesia* 1984; **56**: 1213–16.

15. Coombs DW, Hooper D, Colton T. Preanesthetic alteration of gastric fluid volume and pH. *Anesthesia and Analgesia* 1979; **58**: 183–8.

16. Schulze-Delrieu K. Metoclopramide. *Gastroenterology* 1979; **77**: 768–73.

17. Duffy BL. Regurgitation during pelvic laparoscopy. *British Journal of Anaesthesia* 1979; **51**: 1089–90.

18. Taylor E, Feinstein R, White P, Soper N. Anesthesia for laparoscopic cholecystectomy. *Anesthesiology* 1990; **73**: 1268–70.

19. Caldwell JE, Braidwood JM, Simpson DS. Vecuronium bromide in anaesthesia for laparoscopic sterilization. *British Journal of Anaesthesia* 1985; **57**: 765–9.

20. Kenefick JP, Leader JR, Maltby JP, Taylor JP. Laparoscopy: blood gas values and minor sequelae associated with three techniques based on isoflurane. *British Journal of Anaesthesia* 1987; **59**: 189–94.

21. Marco AP, Yeo CJ, Rock P. Anesthesia for the patient undergoing laparoscopic cholecystectomy. *Anesthesiology* 1990; **73**: 1268–70.

22. Jones RM, Fiddian-Green R, Knight P. Narcotic induced choledochoduodenal sphincter spasm reversed by glucagon. *Anesthesia and Analgesia* 1980; **59**: 946–7.

23. McCammon RL, Vieges OJ, Stoelting RK. Naloxone reversal of choledocholduodenal sphincter spasm associated with narcotic administration. *Anesthesiology* 1978; **48**: 432–9.

24. Kurer FL, Welch DB. Gynaecological laparoscopy: clinical experience of two anaesthetic techniques. *British Journal of Anaesthesia* 1984; **56**: 1207–12.

25. Lattimer RG, Dickman M, Day WC, Gunn ML, Schmidt CD. Ventilatory patterns and pulmonary complications after upper abdominal surgery determined by preoperative and postoperative computerized spirometry and blood gas analysis. *American Journal of Surgery* 1971; **122**: 622–31.

26. Ciofolo MJ, Clergue F, Seebacher J, Lefebuve G, Viars P. Ventilatory effects of laparoscopy under epidural anesthesia. *Anesthesia and Analgesia* 1990; **70**: 357–61.

27. Dohi S, Takeshima R, Hiroshi N. Ventilatory and circulatory responses to carbon dioxide and high level sympathectomy induced by epidural blockage in awake humans. *Anesthesia and Analgesia* 1986; **65**: 9–14.

28. Brown DR, Fishbourne JI, Robnerson VO, Hulka JF. Ventilatory and blood gas changes during laparoscopy with local anesthesia. *American Journal of Obstetrics and Gynecology* 1976; **124**: 741–5.

The Laser

Martin L. Norton

Use of the laser (light amplification by stimulated emission of radiation) is no longer a new concept.[1] The harnessing of light energy represents another tool in the armamentarium of surgeons. The radiation of the laser may be photoradiation (a beam of light) or ionizing radiation (x-rays). The anaesthetist is required to provide safe surgical conditions during the use of this device.

Laser principles

The laser requires the generation of an intense, controlled beam of light which does not generally lose its power over distance.

The properties of a laser depend on its place in the electromagnetic energy spectrum, i.e. its frequency and amplitude. Intrinsic properties include monochromaticity (one wavelength), coherence (all waves in phase), and collimation (waves in parallel). The energy produced can be conceptualized in terms of photons (particles) of light. When a suitable medium is bombarded with energy, an unstable potential energy level is induced. A spontaneously emitted photon then stimulates other photons to be released in a cascade in phase. The light emitted as the medium returns to the resting state may be narrowed to a single predominant wavelength.

In summary, the apparatus consists of a lasing medium which may be gas, solid or liquid. The medium is pumped with energy from an external source. The molecules under stress release a quantum of energy which is the laser beam. In order to produce a collimated parallel beam, the light is reflected in a long resonator and exits through a mirror as a series of parallel waves.

The tissues suffer thermal effects which depend directly on energy density (total energy absorbed per unit volume of tissue) and flux (rate of energy delivery). The resulting thermal effect on tissue varies with the temperature (see Table 70.1).

The primary anxiety during the use of a laser is combustion. Combustion results in fire or explosion depending on (1) the flammable substrate, (2) the oxidant and (3) the speed of combustion. This suggests the need to reduce the hydrocarbon base, to limit oxygen concentration and to slow or suppress the explosive force. Techniques which accomplish these objectives include replacing flammable substances with non-flammable substances, e.g. metal tracheal tubes (Norton) may be substituted for red rubber or plastic tubes. Another approach is to obviate the need for tracheal tubes by the use of Venturi* systems for jet ventilation.

Reducing the oxidant (oxygen) has its proponents but these do not recognize the clinical and the anaesthetic requirements. Many patients who benefit from laser

Table 70.1 The thermal effect on tissues

Temperature (°C)	Effect
Below 55°C	Slight destructive shrinking
Above 55°C	Welding of cell walls without damage to cell contents
60°C	Early irreversible denaturation of cell proteins
100°C	Cellular water begins to boil and vaporize; denaturation of proteins

* Giovanni Battista Venturi (1746–1822): Professor of Physics at Modena and later Pavia, Italy. His landmark paper *Recherches expérimentales sur le principe de communication latérale dans les fluides appliqué à l'explication de différents phénomènes hydrauliques*, (1797) Paris, was later translated and republished in English as a series in Nicholson's *Journal of Natural Philosophy*, 2–3 (1798–9), London.

surgery, e.g. to the respiratory tract, often have a great need for a high inspired O_2 concentration.

Helium has been used to reduce oxygen availability, thereby retarding the speed of combustion. The use of helium is, however, limited by the need to maintain the oxygen delivery to the patient.

Another consideration is the heat of combustion. The flash point can be avoided by maintaining the temperature below the critical temperature for combustion of the materials comprising the tracheal tube. This reduces the fire hazard. This, in part, is one of the approaches for which the Venturi system is useful. For excellent reviews see references 2, 3.

The reader will note that this discussion relates primarily to the use of lasers in the tracheolaryngopharyngeal tract.[4] In fact, the only application of the laser which directly affects anaesthetic management is its use for surgery on the structures of the airway. Other uses of the laser in fields such as neurosurgery, dermatology, ophthalmology, plastic surgery, gynaecology and vascular surgery require relatively little more than a quiet patient. Muscle relaxants or regional anaesthesia satisfy this requirement.

Categories of lasers

New laser substrates are being developed as well as variable (tunable) laser systems. The excellent article by Fisher[5] gives an overview of laser development from the first working laser (ruby) used by Maiman[6] in 1960.

Medical lasers may be classified into several categories, according to how they interact with tissue:

1. water absorbed (thermal effect);
2. pigment absorbed (photochemical);
3. absorbed by specific organic molecules within living tissues;
4. thermoacoustic;
5. ablative photodecomposition.

Choice of a laser is based in particular on the following advantages:

1. Tissues can be altered remotely (i.e. without direct contact).
2. Postoperative pain, oedema, bleeding is less than following scalpel or other modalities (electrosurgery, cryotherapy).
3. Microsurgery precision (use of microscope).
4. Easy transmission through endoscopes.
5. Selective destruction of tissue, using photosensitizers or irradiation or when using visible lasers.

Lasers used in medicine and surgery

Lasers used in medicine and surgery include:

1. Argon ion gas – for ophthalmic photocoagulation, skin lesions, angioplasty and stapedotomy.
2. Carbon dioxide gas – used for cutting and vaporizing surgery.
3. Liquid dyes – for laser angioplasty and colour absorption treatment (excitation of photosensitizers in cancer therapy).
4. Excimer gas – for laser angioplasty, radial keratotomy, bone.
5. Helium–neon gas – to serve as visual focus when using the CO_2 and other infrared (non-visual) lasers.
6. Krypton ion gas – ophthalmic and other photocoagulation.
7. KTP solid combined with neodymium-YAG – same applications as for neodymium-YAG alone, as well as argon applications.
8. Neodymium-YAG solid – coagulative and incisive heating of tissues for necrosis or haemostatis. Q-switched versions used for capsulotomies and other ophthalmic applications.

The anaesthetist must be aware of:

1. the specific laser medium being used,
2. its physical properties (absorption, spread below surface),
3. vital structures around the point of focus.

These considerations are important in terms of certain types of complication. For example, of concern to the surgeon is the use of the KTP for sinus surgery, especially when surgery is close to the cribriform plate and optic area. The heat shock wave ('iceberg effect') may extend distally, damaging vital structures. The same may be observed in laser tonsillectomy when the YAG is used for haemostasis. In this situation there is a risk of the 'iceberg effect' spreading distally towards the carotid sheath and its contents. Later, we will discuss similar relations occurring in the application of lasers to endobronchial surgery.

General safety considerations

When lasers are focused in the area of the head and neck, the patient's eyes, nose and moustache or beard (where applicable) must be protected. This can be achieved by the use of wet gauze and cloth adhesive tape further protected by wet towelling; plastic sheeting should be avoided. In addition, unguents and adhesives such as tincture of benzoin should not be used because of their combustibility.

Appropriate eye protection is necessary for patients and personnel. Each individual in the room should be protected by laser frequency-specific eye shields. While almost any lens will protect the eyes from CO_2 laser beams, wavelength-specific glasses are required for YAG and argon media. Similarly, windows and doors should have frequency-specific glass inserts since standard window glass will not protect the eyes of a curious individual peering into the operating room.

Reflective instrument surfaces can redirect laser beams to unintended targets. Appropriate signs such as those recommended by the American National Standards Institute, alarms and procedures should be used in meticulous detail when appropriate.

Finally, caution is required with respect to tissue cleansing solutions, most of which are extremely flammable.

Patient protection

Protection of the patient's eyes requires wet ophthalmic pads. Ophthalmic ointments should only be used as an adjunct to keep the eyelids closed. Even so, it must be recognized that ointments are combustible substrates. Metal eye cups should not be used since they readily absorb heat energy and can cause burns at the skin contact points. Similarly, adhesive tape used for fixation of eyepads or eyelids should be canvas and not plastic. Preference should be given to canvas tape with high porosity which absorbs moisture. The canvas will char, rather than burn, when the laser inadvertently strikes it. Plastic tape material as well as the adhesive contact substance is highly combustible.

Operating room drapes and towels should be made of absorbable material and thoroughly wet. Standard plastic drapes or towelling are extremely dangerous.

Perhaps the simplest precaution is to consider the physical properties of both the laser and any possible substrate. Anticipation is the key to patient protection.

Airway surgery

There are serious risks inherent in this type of laser surgery. Postsurgical oedema, respiratory movement during the 'burn', competition for airway space between the surgeon and the anaesthetist – all contribute to the problems.

The use of liberal doses of dexamethasone are said to decrease oedema. However, more important is postoperative positioning of the patient in the bolt-upright sitting position. This facilitates lymphatic drainage, reduces oedema and (because of the biomechanics of the airway system) opens the lumen to its maximum diameter.

With regard to competition for access to the airway, the use of a Venturi system provides a highly efficient solution, and a non-combustible approach.

Respiratory and cardiac motion affecting the airway requires synchronization of the laser beam impulse with the 'resting' phase of breathing. The routine use of muscle relaxants should make this relatively easy, enabling a quiet field to be provided when required.

Complications of laser bronchoscopic surgery

The most commonly used laser for endobronchial surgery is the Nd-YAG. The current CO_2 laser is attenuated during passage through the fibre used for transmission. Its use is therefore limited.

The YAG is a crystal composed of yttrium aluminium garnet doped with neodymium ions. This medium enables deeper penetration and scatter than with either the CO_2 or argon laser. Therefore, its ability to produce haemostasis is superior. Also it can be transmitted efficiently through flexible quartz fibres.

The crucial concerns are the need to:

1. maintain the patient in an oxygenated normocapnic anaesthetized state during endoscopic surgery;
2. be aware of the iceberg effect.

It is imperative to understand the differences in tissue effects of CO_2 and YAG media. The CO_2 produces a wedge-shaped intrusion into the tissue, with the apex at the base. The YAG produces a small entry point, but below the surface it spreads out (the iceberg effect). In addition the heat shock continues for a period of time *after* the laser treatment ceases. Thus the final diameter of the burn below the apparent surface is not always predictable, especially because different tissues have different constituents. This factor is of particular concern when recognition is given to the vital and relatively inaccessible structures surrounding the tracheobronchial tree. These include the aorta, branches of the pulmonary artery, the vagus nerve, the recurrent laryngeal nerve, the superior vena cava, the oesophagus, the branches of the carotid artery and the thyroid (Fig. 70.1). Since the power density of the Nd-YAG is concentrated *below* the surface, its spread may produce perforation of these structures. This may not be recognized at the time but leads to disastrous results.

The anaesthetist must be constantly alert for signs or symptoms of bleeding or air dissection in the mediastinum. Persistent bleeding into the tracheobronchial tree may lead to life-threatening hypoxia. Direct tamponade, with a rigid ventilating bronchoscope, distal to the bleeding site, should be immediately attempted. The patient should be placed in the lateral decubitus position with the healthy lung uppermost. Perforation of a major blood vessel (left pulmonary artery, brachiocephalic, etc.) usually has catastrophic consequences. Even with a trained thoracic surgeon present the prognosis is grave.

Anaesthetic management should be by topical application of local anaesthetics plus intravenous general anaesthetic agents. This will allow the respiratory tract to

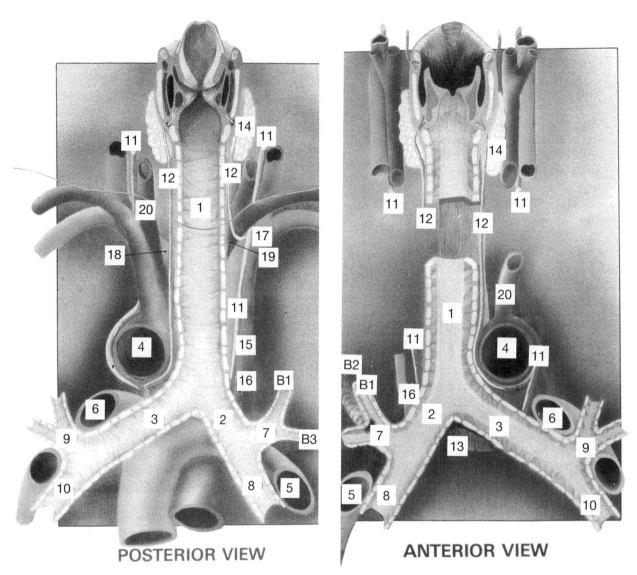

POSTERIOR VIEW   **ANTERIOR VIEW**

Fig. 70.1 The tracheobronchial tree showing related structures which may be damaged by the iceberg effect: 1, trachea; 2, right main stem bronchus; 3, left main stem bronchus; 4, aorta; 5, right pulmonary artery; 6, left pulmonary artery; 7, right upper lobe bronchus; 8, truncus intermedius; 9, left upper lobe bronchus; 10, left lower lobe bronchus; 11, vagus nerve; 12, recurrent nerve; 13, oesophagus; 14, thyroid gland; 15, superior vena cava; 16, azygos vein; 17, right innominate vein; 18, left innominate vein; 19, innominate artery; 20, left carotid artery.

be used solely to provide oxygen and remove CO_2 and tissue debris. Obviously pulse oximetry is mandatory.

One subject of major concern, to the anaesthetist, related to the use of the laser application in the tracheobronchial tree especially at and below the level of the carina, is bleeding. Perforation or delayed slough (e.g. 'iceberg effect') of the Nd-YAG and analogues can precipitate critical bleeding. In the absence of a surgeon competent to immediately diagnose and open the chest there is no hope for patient survival. This indicates the need for the routine use of large intravenous cannulae, the ready availability of blood and a rapid transfuser system, and a thorough knowledge of anatomy.

The postoperative period is the time when most fatal complications are manifest. Chest x-ray review should be routine. The availability of immediate rigid as well as flexible bronchoscopes is essential. The staff must be constantly alert to the cardiovascular sequelae. Hypoxia is usually related to retained secretions, respiratory depression, or secondary haemorrhage and perforation from delayed tissue necrosis. A close working relationship between all concerned with patient care (surgeon, anaesthetist and nurse) cannot be overemphasized. The delayed potential for these complications makes it necessary for these patients to be observed for an *extended* period after surgery.

Laser plumes

A laser plume is composed of solid particles and gases consisting of carbonized tissue, blood and virus. Its chemical composition may include water, carbon dioxide, formaldehyde, benzene, hydrogen cyanide, acrolein, light hydrocarbons, cyanates, polynuclear aromatic hydrocarbons, acetone, cyclohexane, toluene, fatty acid esters, ethanol, xylene, styrene, and even acetaldehyde.

These substances cannot be considered other than noxious (American Conference of Governmental Industrial Hygienists and the National Institute of Occupational Safety and Health). Particle sizes range from 0.1 to 0.8 μm. Note that the standard surgical masks do not filter out this range of particle sizes. Special masks and surgical gloves should be used to protect all personnel from a laser plume. Furthermore, efficient smoke evacuation must be maintained close to the operative field to remove the plume before it has been inhaled by operating room personnel.

Concern has been expressed about viable viral plumes when using lasers. They could transmit the virus further down the respiratory tract. Similarly, the plumes could be expelled (during the exhalation phase) to affect the operator. This possibility has been reported.[7,8] There is evidence that the plumes contain viruses that are associated with large (unspecified) particles. However, some particles travel only a few centimetres before settling. Concern associated with virus transmission has been most frequently expressed in relation to the use of Venturi (jet) ventilation during surgery of the tracheolaryngopharyngeal tract.

Vascular lasers and contact tip laser problems

Vascular applications are divided into two categories, laser welding and intravascular angioplasty.

Laser welding can be accomplished by the carbon dioxide gas used with nitrogen and/or helium. The concept is sutureless anastomosis. The tissue is heated to a gel-like state and then cooled. The technique, using the neodymium-YAG solid medium, has been applied to vessels, skin and vas deferens.

A more serious concern is the occurrence of air emboli, in some cases leading to death. This has occurred during cardiac laser angioplasty as well as in gynaecological procedures. The problem arises with Nd-YAG and KTP because of the tendency of contact tips to adhere to tissues. The tip must be cooled to release it from the tissue. Hyskon* or Rheomacrodex* is usually used for this purpose but occasionally nitrogen is used instead. The lower hydrostatic pressures of venous lakes increase the hazard.

Additional problems with contact tips include embolization of fragmented contact tips. Bowel explosion/fire has occurred secondary to ignition of methane.

Venturi (jet) ventilation

The various factors related to the safe use of lasers apply particularly to the use of tracheal tubes during laser surgery of the larynx. While there are many tubes purportedly with low susceptibility to combustion under varying situations, in reality there has been only one tube, commercially produced, that was absolutely combustion-proof;[9] however, even this tube is no longer available.

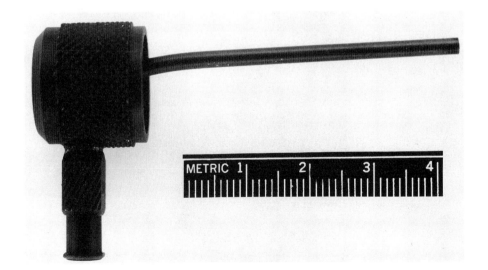

Fig. 70.2 The Woo-Pilling adapter for use with a tracheal tube.

* Hyskon, Dextran 70, Macrodex: average mol. wt 70 000, a polysaccharide composed of x-D-glucopyranoxyl units, are used primarily for plasma volume expansion. Dextran 40 or Rheomacrodex is used as blood flow adjuvant.

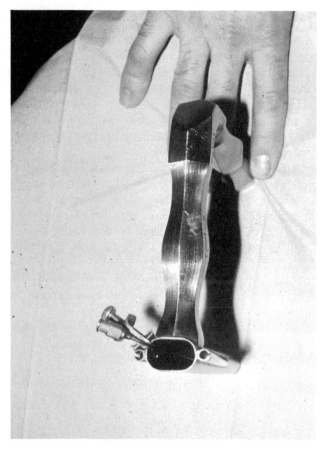

Fig. 70.3   An injector placed inside a suspension laryngoscope.

Another approach, first utilized at Toronto Children's Hospital, Canada, was later developed for general application.[10] The technique uses Venturi principles for ventilation without a tracheal tube. The application is most efficacious for:

1. suspension laryngoscopy,
2. bronchoscopy,
3. Norton or other uncuffed tracheal tube, using the Woo-Pilling adapter (Fig. 70.2).

The Venturi device involves the use of a high pressure oxygen source, entrainment of atmospheric (room) air, and rheological patterns of gas transmission. These principles are exemplified in the development of jet engines and garden hose nozzles. The closed tube Venturi relies on production of a vena contracta[11] with the vena contracta producing relative lateral wall negative pressures, i.e. significantly below atmospheric. The result is entrainment of air at atmospheric pressure, i.e. positive with regard to the lateral wall. An injector is then placed within the lumen of a suspension laryngoscope (Fig. 70.3). The jet system is controlled by the addition of a variable pressure reduction valve and gauge delivery system (Fig. 70.4). The tube within a tube acts as a safety feature. The outer tube is open at both ends to permit gas entrainment. It also allows the escape of excess gases during the exhalation phase of the respiratory cycle.

Recently, a practice has arisen of using a direct line from

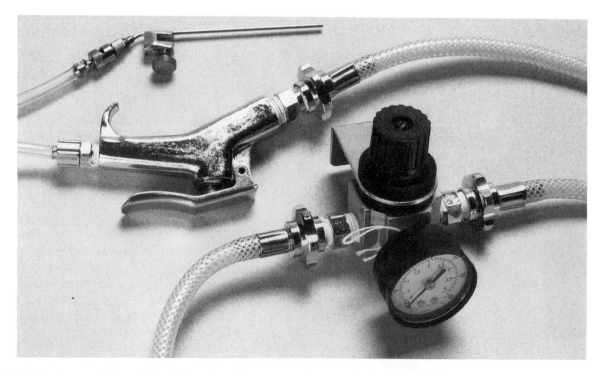

Fig. 70.4   Injector, gas flow control switch and variable pressure reduction valve and gauge delivery system.

the anaesthetic machine bypass oxygen flush mechanism for this purpose. This is extremely dangerous since neither a variable pressure reduction system nor regulatory gauge (manometer) is incorporated. Pressures applied in the case of adults (20–45 lb/in^2; 138–310 kPa) are absolutely contraindicated for usual paediatric usage which should be 4–10 lb/in^2; 27.6–69 kPa), and in the geriatric patient who may often have emphysematous lung bullae.

A further application is the use of the Woo-Pilling[12] adapter. A combustion-resistant tracheal tube is inserted as usual. No cuff is required. The adapter is attached to the proximal connector and the Venturi system attached.

Subglottic lesions pose a special problem. They are approached either with a rigid bronchoscope (Wolf–Dumon), a fibreoptic bronchoscope, or Storz–Venturi bronchoscope. In essence the Venturi bronchoscope is a standard bronchoscope to which a small bore tube (Venturi system) is attached. A gas (usually oxygen) then flows at a rapid rate through the small-bore tube. This creates a subatmospheric pressure within the bronchoscope itself. Since the ends of the bronchoscope are open to the atmosphere, room air is drawn in (entrained) and the patient is ventilated.

Occasionally a long jet ventilation cannula alone is placed below the vocal folds or lesion, but the absence of a patent channel for escape of excess gases at high ventilating pressures, especially when blocked by an obstructive lesion or the inferior aspects of the vocal folds which may fill out sail-like with gas from below, is a serious risk. The well known Venturi principle of a tube within a tube is required for safety, i.e. to avoid pneumothorax, pneumotrachea, pneumolarynx, and even pneumomediastinum.

The major advantage of the jet system is the absence of the tracheal tube, which allows the surgeon total access to laryngeal structures. However, the jet flow causes intermittent vocal fold distension as with sails. Further-more, the surgeon may wish to treat tissue in the anterior commissure. A tracheal tube may therefore be preferred especially in the latter situation because it tends to keep the vocal folds apart and stable. Whether there is a reduction in postsurgical anterior commissure webbing is still to be determined.

Monitoring and airway control

As with any surgical procedure, modern patient monitoring includes blood pressure, electrocardiography, end tidal $P\text{CO}_2$, pulse oximetry, pulse rate and, where indicated, blood gas analysis. Additionally urinary output measurement and the use of both cardiac and pulmonary auscultation may be included. The details relate more to the method of airway management than any other factor. If tracheal intubation is the choice, standard monitoring will be available. However, if the Venturi (jet) technique is used, cardiopulmonary auscultation and end tidal $P\text{CO}_2$ are difficult to monitor because of the mechanics of ventilation and the 'noise' produced. In this circumstance the pulse oximeter functions as the primary ventilation monitor (other than visual observation). An increased end tidal $P\text{CO}_2$, while not always amenable to measurement, is often part of a sympathetic reflex reaction to the potent stimulus of suspension laryngoscopy which may also cause hypertension. This should be controlled pharmacologically.

In another technique of airway management the standard tracheal tube is intermittently removed to allow the laser to be used. Obviously, controlled ventilation is not possible when the tube has been removed. The natural tendency of the surgeon is to want 'just a few more moments' to complete laser manipulations even when the pulse oximeter or ECG indicate a deteriorating situation (onset of hypoxaemia). There may, in addition, be mechanical trauma to the laryngeal tissues leading to increased postsurgical oedema. The suspension device may slip out of position and it may not be possible immediately to replace the tracheal tube in the case of patients who are difficult to intubate. This technique of airway management practice must, therefore, as it increases the risks for the patient, be condemned.

Anaesthetic agents and reduction of flammability

Intravenous anaesthetic agents should be used predominantly. All gases which support combustion (N_2O, O_2) should be used with caution, although there is an obvious conflict between hazards of fire and the patient's requirement for oxygen.

Helium has been used to reduce the possibilities of fires accompanying laser surgery. The ignition time was considerably prolonged when helium–oxygen mixtures replaced air–oxygen. Thus, with 30 per cent oxygen in air, some tracheal tubes ignited in 20–60 seconds. In contrast, none of these tubes ignited after as long as 60 seconds in the presence of 30 per cent oxygen in helium.[13] The clinical significance of these observations has yet to be settled, but intratracheal fires may lead to death. Therefore a better approach is to make all external tracheal tube surfaces highly laser resistant so that neither heat transfer nor combustion will occur.

Flammability is reduced as the oxygen concentration in the delivered gases is reduced. The use of online arterial

Table 70.2 Substrate change allows inspired oxygen concentration to be maintained

Substrate	Substitute
Tracheal tube	Venturi ventilation
Nitrous oxide anaesthetic and other flammable anaesthetic techniques	Intravenous techniques

oxygen concentration measurement and transcutaneous oxygen saturation measurement devices enables the anaesthetist to ensure adequate oxygenation while at the same time the oxygen in the inhaled gases is carefully reduced.

However, the oxygen needs of the patient must not take second place to the risk of airway fires. Fortunately, there are other ways by which this problem can be reduced.

First, the combustible substrate may be removed (Table 70.2).

Combustion may be suppressed by the addition of helium to the gas mixture in the respiratory tract. This, also, requires close monitoring by pulse oximetry (saturation) and arterial oxygen (concentration). The anaesthetist's concern must be to provide the patient with the necessary oxygen.

Management of intraluminal fires

There is an occasional need to manage tracheal tube fires resulting from laser surgery in the airway. Among the questions to consider are:

1. How may such fires be extinguished?
2. Should the tracheal tube be removed immediately; if so, should it be replaced?
3. How should tissue damage be treated?
4. If the initial intubation was difficult, how can control of the airway be re-established?

Fires may be quenched by removing the substrate (hydrocarbon), removing the agents that support combustion (O_2, N_2O), and reducing ignition–combustion temperature below the flash point.

Laser-initiated combustion may cause thermal injury by direct exposure to the flames. The severity of the burn will depend on the duration of exposure and the heat intensity. In addition, inhalation of smoke may produce a chemical burn with resulting bronchospasm, intra-alveolar haemorrhage, oedema and loss of surfactant leading to respiratory failure. The products of decomposition of rubber or other material used in tracheal tubes (e.g. hydrogen chloride, carbon monoxide, acrolein) may produce severe chemical pneumonitis and other toxic effects.

The usual advice is immediately to stop all anaesthetic agents and carriers (including nitrous oxide and oxygen) and to remove the combustible tracheal tube. This should be followed by flushing the area with water and the administration of steroids to reduce the oedema, followed by ventilation with 100 per cent oxygen (after the fire is quenched). Spontaneous ventilation should be the objective at the earliest possible moment. Fowler's position (sitting) offers effective lymphatic drainage as well as taking advantage of biomechanical suspension of the airway to provide an optimal opening of the airway channel.

However, additional problems may arise in the patient with a difficult airway. For the patient with a difficult airway, simple removal of the tracheal tube may prove fatal due to airway closure (oedema) and an inability to replace the original tube rapidly. Oedema may occur so rapidly and intensely that it may not be possible to perform a tracheostomy swiftly enough.

In such a situation a long intubation guide (Norton Teflon rod, Sheridan TXX, or Eschmann woven) may be helpful. The procedure should be modified as follows:

1. Discontinue the catalyst (oxygen) and gaseous substrate (nitrous oxide).
2. Flush the trachea with cold water.
3. Immediately insert the intubation guide through the lumen of the tracheal tube, if possible down to the carina.

Then, and only then, remove the tracheal tube. Immediately reinsert another tracheal tube using the intubation guide. The importance of the second tube is to maintain the integrity of the airway in the presence of inevitable severe mucosal oedema.

The laser and plastics

Occasionally the laser may be used for plastic trimming or removal (Teflon, methyl methacrylate). Consideration must be given to pyrolytic and other toxic breakdown products. Teflon trimming in the larynx is one example.

Teflon, polytetrafluoroethylene, is composed of at least 20 000 C_2F_4 monomer units linked into very long unbranched chains. While it is a highly stable thermoplastic homopolymer, it melts to an extremely viscous gel at 327 °C and reverts to the gaseous monomer at temperatures above 400 °C. The breakdown products are carboxyfluoride (COF_2), hydrogen fluoride (HF), polyfluoroisobutylene and carbon dioxide. Thus, polymer fume fever, with symptoms which resemble influenza, may occur. In addition, hydrofluoric acid gas is highly irritating and corrosive and soluble in water. Application of the laser to remove Teflon following vocal fold injection must therefore be accompanied by highly efficient targeted aspiration at the site of laser application to protect both the patient and the laser operator.

Similarly, laser removal of methyl methacrylate causes the cement to break down to acrolein and/or acrylic acid. In combination with solvents that contain nitrosyl chloride or chlorine, minute amounts of phosgene may also be produced under pyrolytic conditions. Methyl methacrylate is used in open orthopaedic procedures; the resulting fumes are, therefore, released into the room, affecting the surgeon and other operating room personnel as well as the patient.

Treatment depends on the severity of the airway reaction, and may include humidification of inspired gases, high dose steroids, and bronchial dilators. Frequent re-evaluation by fibreoptic endoscopy may suggest the need for other supportive measures such as intermittent positive pressure ventilation, positive end-expiratory pressure or even tracheostomy.

Photodynamic therapy

Photodynamic therapy involves introducing a cell or tissue-specific, light or heat-reactive dye into the body. This substance, once incorporated in the cell or tissue system, is then activated by particular light frequencies and/or thermal levels. Activation is manifest by physicochemical change in the tissue substrate causing a change in cell metabolism or even cell component destruction.

The use of this therapy has been limited by the occurrence of photosensitivity reactions and still limited knowledge about cell component dye specificity and the development of excimers. Excimers are diatomic molecules, which consist of a halogen atom and an atom of a noble gas (gases with low melting and boiling points which are very stable), existing in the excited state and which dissociate after emitting their characteristic ultraviolet radiation. Excimers cause rupture of the bonds in complex molecules, the components of which are ejected in a process resembling vaporization.

Photoplasmolysis

Another mode is 'optical breakdown'. This involves ionization of atoms by ultra-strong electric fields produced by very high intensity laser light. Thus electrons are pulled out of their atomic orbits and a plasma is produced. Plasmas are mixtures of free electrons and positive ions in about equal concentrations at very high temperatures.

Formation of a plasma can totally destroy biological architecture and thus vitality. The disintegrative effect of plasma formation combines with the microexplosive effect of plasma. The laser which is best suited for photoplasmolysis in living tissue is the Nd-YAG.

Newer applications

Newer laser substrates continue to be developed. Recently, erbium, holmium and thulium are of note, along with the more industrial copper and numerous dyes. Medical x-ray lasers are not yet commercially available and have as yet only very specialized research applications.

The holmium laser may be composed of Cr, Tm, Ho-YAG crystal combinations (which replaces the classical Nd-YAG). It is notable for its short wavelength, is analogous to Nd-YAG, but has water absorption properties similar to the CO_2 laser. Holmium combinations are either holmium-doped yttrium aluminium garnet (Ho-YAG) or holmium yttrium scandium gallium garnet (Ho-YSGG). The holmium laser has been used in arthroscopic soft tissue, transnasal sinus surgery and for dental applications. Experiments suggest that the Ho-YAG laser may prove useful for angioplasty procedures replacing bypass surgery and balloon angioplasty. This approach of removing the plaque by vaporization is significantly different from the use of a Nd-YAG laser to heat a metal cap that destroys the blockage in arteries.

Thulium combinations are thulium-doped YAG (Tm-YAG) and thulium scandium gallium garnet (Tm-YSGG). There is also the erbium-doped* YAG and YSGG (Er-YAG and Er-YSGG). The erbium laser will cut both soft tissue (skin, aorta, myocardium and cornea) and hard tissue (bone, dentine).

Tunable solid state lasers for medical application include alexandrite (chromium-doped chrysoberyl) in which laser activity has been noted in the Cr ion. Others are the titanium-doped sapphire laser and the cobalt-doped magnesium fluoride laser. The importance of tunable lasers is that they can be coupled to a tissue identification system. Thus, a smart laser can be tuned for the specific type of tissue being targeted.[14]

Dental lasers

Recently the dental profession has shown an interest in the holmium and erbium laser for 'curing' polymers used in

* Definition of dopant: a method of impregnation of ion which enhances or compensates the substrate. Doping of impurity, i.e. ion implantation added in small controlled amounts in order to produce the required wavelength.

dental fillings. It has also been used for root canal procedures. To date, however, the prime use of lasers remains in soft tissue procedures such as alveolectomy and gingivectomy as well as for removal of hypertrophic oral mucosa. It is important to note that laser training courses, oriented for the dental profession, are not generally available. This raises major laser safety concerns.

Future progress

No other modality of instrumentation is as well suited to endoscopic application as is the laser. Further development of endoscopic surgery will obviate the need, on many occasions, for major incisional surgery (arthroscopic, urological, cholecystoscopic, vascular). The limiting factors, to date, have been:

1. development of fibres that will efficiently transmit laser energy; and
2. development of a laser with a wavelength that can easily be varied over the entire spectrum of laser light similar to the theoretical potentials of the free-electron laser.

Another fruitful, yet less fully addressed, field of development is that of coherent, spectrally pure light effects of various wavelengths on biological systems. The work of Tina Karn[15] has shown the way to condense into an applicable science those effects which will extend the use of lasers beyond cutting, coagulation and ablation. The current pursuit is in the use of the photodynamic therapy (discussed above) and in the application of x-ray lasers.

Safety policy

Laser safety regulations are of concern to both the institution and to individuals such as the anaesthetist.

Personnel qualified to use and operate lasers (an inherently hazardous instrument) should be certificated trained in the use of lasers. The guidelines may include:

1. demonstrated understanding of basic laser physics, power density and laser tissue interaction;
2. prior experience, either in animal laboratory, or under the immediate and direct supervision of a laser surgeon certificated in human clinical applications;
3. demonstrated formal 'hands-on' course training.

Note that the training must include the specific class of instrument to be used. It is dangerous to assume that all

instruments are similar or that all media (CO_2, Nd-YAG, argon, ruby, krypton) are the same in effect or risk.

Laser safety programmes are as important for the anaesthetist as for the surgeon. The content should include:

1. fundamentals of laser operation (physical principles, construction);
2. bioeffects of laser radiation;
3. relation of specular and diffuse reflections;
4. non-radiation hazards (electrical, chemical, reaction byproducts);
5. control methods;
6. overall management and responsibilities;
7. medical surveillance practices;
8. ancillary problems in laser usage (airway management, avoidance of combustibles both solid and gaseous).

Since the anaesthetist is both a member of the team and the potential target of toxic effects (stray radiation) he or she has an ethical responsibility to become actively involved. Furthermore, it is the anaesthetist's professional duty to protect all who are in his/her care.

Practical application of these duties involve:

1. Avoidance of potentially hazardous materials (eye ointments, plastic tubes, drapes, adhesive, preparatory solutions, plastic eye patches).
2. Availability of water and a means of delivery (large syringe).
3. Protection of patient's eyes and neighbouring tissue or flammable substrate with wet pads, absorbent canvas tape kept wet throughout the procedure.
4. Special attention to keeping the areas around the laser, anaesthetic machine and operating table (floor, foot switch) dry, including sites of potential skin contact! The operator of the laser system, and the anaesthetist using electronic instruments, should take special precautions to avoid risk of electroshock or burns, both to himself or herself and the patient.
5. During surgery in the tracheolaryngopharyngeal tract, the tracheal tube and cuff, when used, should be protected by appropriate means. Cuffs should be filled with liquid* to quench any flame from inadvertent laser radiation. During laser treatment to the intestine, adequate suction should be used deep within the bowel to evacuate potentially lethal methane.
6. Wet cloth towels should be used to drape the immediate area instead of plastic or paper drapes.
7. The laser should not be used in the presence of flammable anaesthetics, surgical solutions, drying agents, ointments, methyl methacrylate or other plastics.

---

* Lignocaine is recommended to decrease tissue reactivity to the combustion.

Table 70.3 Some recurring complications of medical laser application

Complication	Cause
Mechanical	
Finger burn	Open shutter
Excessive burn depth	Timing malfunction
Intravascular embolization	Fibre tip fragmentation
Electrocution	Water leak from hose
Self-firing	Power surge defeating microprocessors
Operator error	
(physician, technician)	
Hypoxia	Reliance on visual cyanosis while wearing green glasses
Burn, undesired location	Failure to test focus of burn
Accidental burn	Failure to put machine on standby
Continuous fire	Use of nitrous oxide as cooling medium
Nitrogen emboli, convulsions, death	Excessive nitrogen gas pressures
Air embolism	Use of gas (nitrogen) to distend bladder instead of Dextran (YAG); also faulty hook-up of gas lines
Flashback burn, eyes	Failure to wear protective glasses
Superficial and eye burns of ancillary personnel (CO_2)	Failure to wear protective glasses including side protection; also failure to put machine on standby or off mode
Rupture of bowel	Ignition of methane in the bowel; failure to provide ventilation of the bowel
Fibre embolization	Faulty removal of fibre sheath; faulty fibre repairs
Perforation of bladder, bleeding	Failure to understand pattern of YAG burn delayed, death and delayed iceberg effect; use of continuous mode without interruption for cooling
Operating theatre (room) drape fire, burn	Failure to place device in standby or off mode

8. The laser machine must be locked in the OFF position when not in use. Use of standby modes should be severely limited.

9. Where available, instrumentation should be provided that will detract, scatter or defocus the laser beam rather than reflect or focus it inappropriately.

10. Above all, when the indicator (visible) laser beam is not in absolute conformity with the surgical beam (e.g. CO_2 burn site), the instrument must *not* be used and should be repaired before any further use.

11. Ventilation and suction systems must be adequate to remove any gases, vapours or particulate matter released during laser operation.

Lasers are medical devices subject to administrative regulation in the United States by agencies such as the Food and Drug Administration as well as the Customs Bureau.[16] Additionally, each State has a regulative authority. In the United Kingdom the Health and Safety Executive enforce most aspects of health and safety at work, and while there is no specific legislation governing the medical use of lasers, relevant national and international standards are taken into account. The most important of these are the European (CENELEC) and IEC standards embodied in the British Standards Institute documents BS EN 60825:1992[17] and BS EN 60601-2-22:1993,[18] together with guidance issued by the Department of Health.[19] Use, repair and maintenance of these instruments subjects the manufacturer, as well as technicians, physicians and health care institutions, to principles of tort law including negligence, product liability and workmen's compensation.

A review of US Food and Drug Administration medical device complication reporting files reveals a consistent pattern. Most laser complications resulted from human failure rather than mechanical failures, with the prime exception of shutter failures (including foot switch) (Table 70.3). Further, the medical specialities, gynaecology and urology, represented the highest number of complications with tracheal tube fires much further down the list. This suggests the need for continuing refresher education in these specialities. The range of complications included mostly careless or inept operative procedures.

A frequent error on the part of the practitioner is the assumption that 'all laser devices are the same', or that once the instrument has worked well for a preceding patient it does not need to be rechecked for focus, or calibration, for subsequent patients. Failure to recheck the instrument between cases is fraught with danger.

REFERENCES

1. Maiman TH. Stimulated optical radiation in ruby. *Nature* 1960; **187**: 493–4.
2. *Health Devices*, January 1992; **21**: 1.
3. Guidelines for laser safety and hazard assessment, Occupational Safety and Health Administration. *Instruction Pub 8–17*, Directorate of Technical Support, August 5, 1991, US Government Printing Office.
4. Jako GJ. Laser surgery of the vocal cords. *Laryngoscope* 1972; **82**: 2204.
5. Fisher JC. The current and future status of lasers in surgery. *Surgical Technology International* 1973; 29–35.
6. Maiman TH. *Physical Review Letters* 1960; **4**: 564.
7. *The Medical Devices Bulletin, 1988.* The Center for Devices and Radiologic Health. Food and Drug Administration, 1988.
8. *The Medical Devices Bulletin, 1990.* The Center for Devices and Radiologic Health. Food and Drug Administration, 1990.
9. Norton ML, DeVos P. New endotracheal tube for laser surgery of the larynx. *Annals of Otology, Rhinology and Laryngology* 1978; **87**: 554.
10. Norton ML, Vaughan S, Vaughan C. Endotracheal intubation and Venturi (jet) ventilation for laser microsurgery of the larynx. *Annals of Otology, Rhinology and Laryngology* 1976; **85**: 656–64.
11. Spink LK. *Principles and practice of flow meter engineering.* 8th ed. Foxboro, MA: Foxboro Co, 1958.
12. Woo P, Strong MS. Venturi jet ventilation through the metal endotracheal tube: a non-flammable system. *Transactions of the American Broncho-Esophagological Association 63rd Annual Meeting* 1983: 105–7.
13. Dumon JF, Fuentes P, Meric B. Recherche et enseignement sur l'endoscopic et la laser. Marseille: IMA Hôpital Salvator.
14. Acharekar MA. Solid state lasers for medical use. *Hospital Management International, London*, 1990: 347–50.
15. Karn TI. Photobiologic fundamentals of low-power laser therapy. *IEEJ Quantum Electronics* 1987; **QE23**: 1703.
16. Reyes-Akinbilege V. *Medical device regulation in the United States and the European Community.* Congressional Research Service, 20–22, 92–714 SPR, Sept. 17, 1992.
17. British Standards Institute. *Radiation safety of laser products, equipment classification, requirements and user's guide.* BS EN 60825:1992. London: BSI, 1992.
18. British Standards Institute. *Medical electrical equipment. Part 2. Particular requirements for safety Section 2.122. Specification for diagnostic and therapeutic laser equipment.* BS EN 60601-2-22:1993. London: BSI, 1993.
19. Department of Health. *Lasers. Guidance on the safe use in medical practice.* London: HMSO, 1984.

Chapter 71

Management of Cardiac and Respiratory Arrest

Charles W. Otto

The earliest developments in modern cardiopulmonary resuscitation (CPR) came as the result of attempts to treat cardiac arrest occurring in the operating theatre. The operating room remains the location where CPR has the highest rate of success. Cardiac arrest occurs approximately seven times for every 10 000 anaesthetics.[1] The cause for the arrest is anaesthesia related approximately 4.5 times for every 10 000 anaesthetics but mortality from these arrests is only 0.4 per 10 000 anaesthetics. Thus, resuscitation is successful approximately 90 per cent of the time in anaesthetic-related cardiac arrests. Application of CPR to the arrest victim has spread from the operating room to the rest of the hospital and to the community at large. Over the past 30 years, CPR has become widely practised, facilitated by the efforts of the American Heart Association, the British Heart Foundation and many other organizations around the world. Guidelines for the practice of CPR, encompassing the latest scientific evidence and expert opinions, are published periodically.[2] These organizations also develop courses at different levels of complexity for teaching CPR to lay people and to medical and paramedical personnel. The two levels of CPR care are referred to as basic life support (BLS) for ventilation and chest compressions without additional equipment and advanced cardiac life support (ACLS) for using all modalities available for resuscitation. As an expert in resuscitation, the anaesthetist is expected to be able to teach and lead these teams of CPR practitioners. To do so requires a thorough knowledge of the pathophysiology, pharmacology and techniques of CPR. It is not the purpose of this chapter just to review the protocols from BLS and ACLS courses, but to discuss the scientific background upon which current CPR practice is based. It will focus exclusively on the circumstances in which victims of cardiac and respiratory arrest are different from other cardiovascular crises.

Overview

In the United States, cardiovascular disease accounts for 50 per cent of the annual mortality in adults or approximately 1 million yearly deaths with one-half due to coronary artery disease.[2] The majority of deaths from coronary disease are sudden deaths with approximately two-thirds occurring outside the hospital. The severity of the underlying cardiac disease is the major determining factor in the success or failure of resuscitation of these victims. However, of the factors under control of the rescuers, poor outcomes are associated with: (1) long duration of arrest before CPR is begun; (2) prolonged ventricular fibrillation without definitive therapy; and (3) inadequate coronary and cerebral perfusion during CPR.

Brain adenosine triphosphate (ATP) is depleted after 4–6 minutes if there is no blood flow, although it returns to nearly normal within 6 minutes of starting CPR. Recent studies in animals suggest that good neurological outcome may be possible from 10 to 15 minute periods of normothermic cardiac arrest if good circulation is promptly restored. In clinical practice, survival depends on the rapid institution of resuscitation attempts, especially in out-of-hospital arrest. Outcome is improved when CPR is begun by bystanders. Optimum survival from ventricular fibrillation is obtained only if basic CPR is started within 4 minutes and defibrillation applied within 8 minutes.[3] The longer a heart fibrillates, the more difficult is defibrillation. Early defibrillation is so important that it should take precedence over all other resuscitative efforts if the diagnosis of ventricular fibrillation can be made and defibrillation equipment is available.[4]

In cities with an effective rapid response emergency medical system for out-of-hospital cardiac arrests, initial resuscitation rates of 40 per cent and survival to hospital discharge of 10–15 per cent may be possible.[4] A better outcome might be expected for in-hospital arrests because of rapid response times and expert personnel. However, overall rates for initial resuscitation and survival to discharge from in-hospital arrest are approximately 40 and 10 per cent respectively.[5] Intercurrent illnesses of hospitalized patients reduce the likelihood of survival and the arrest victim is more likely to be elderly, a factor that may reduce survival. Outside the operating suite, the best initial resuscitation rates are found in the ICU while the best survival rates are for patients arresting in the emergency department. The cause of arrest associated with the best outcome and the most common cause of out-of-hospital arrest is ventricular fibrillation secondary to myocardial ischaemia. This initiating event is less common in hospitalized patients. When applying CPR, attention to the details of effective resuscitation is important. However, it should always be remembered that CPR is only symptomatic therapy and therefore attention should not be paid to the mechanics of CPR while the search for a treatable cause of the arrest is forgotten.

Basic life support

Basic life support consists of those elements of resuscitation that can be performed without additional equipment: airway management, ventilation and chest compression. Common practice is to approach a victim with the ABC (airway–breathing–circulation) sequence although the CAB sequence has been used in some countries with comparable results. The full sequence for initiating CPR in an apparent arrest victim is given in Table 71.1. However, if complete resuscitation equipment and expertise in its use is available, the overall priorities of resuscitation may dictate altering this sequence (Table 71.2).

Table 71.1 Sequence of basic life support

Determine lack of response
Activate emergency medical services or code team
Position victim supine on firm surface
Open airway
Determine absence of breathing
Perform ventilation: two breaths
Determine absence of pulse
Initiate chest compression
Alternate 15 compressions with two breaths

Table 71.2 Procedure priorities during CPR*

Diagnose ventricular fibrillation and defibrillate
Open airway
Begin ventilation
Begin chest compressions
Administer supplemental oxygen
Obtain intravenous access
Administer adrenaline (epinephrine)
Assess adequacy of CPR efforts
Intubate trachea
Additional drug therapy

* Most important to least important interventions assuming all necessary equipment and expertise are available.

Airway management

The goal of airway management during cardiorespiratory arrest is the same as during general anaesthesia: to provide a clear path for respiratory gas exchange while minimizing gastric insufflation and the risk of pulmonary aspiration. Airway maintenance in an unconscious patient is a fundamental part of anaesthetic practice and all the techniques learned for use during anaesthesia are applicable to the cardiac arrest victim. Just as in the operating room, the most commonly used technique for opening the airway is the 'head tilt/chin lift' method. If this is ineffective, the 'jaw thrust' manoeuvre is frequently helpful. Oropharyngeal and nasopharyngeal airways are useful for helping maintain an open airway in patients that are not intubated. Care must be used to ensure that the airway is correctly inserted and that it does not increase airway obstruction. Insertion in the semiconscious patient can induce vomiting or laryngospasm.

Through the years it has become obvious that effective airway management during CPR is a major problem, even for medical professionals. Many individuals cannot manage effectively a self-inflating resuscitation bag and mask. Larger tidal volumes at lower pressures are delivered by mouth-to-mouth or mouth-to-mask ventilation.[6] The bag and mask apparatus is more effective if two individuals manage the airway, one to hold the mask and maintain the airway and one to squeeze the bag.[7] Tracheal intubation provides the best possible airway management. It should be carried out during every resuscitation lasting more than a few minutes if a skilled laryngoscopist is available. However, it should not be performed until adequate ventilation by other means (preferably with supplemental oxygen) and circulation by chest compressions have been established (Table 71.2). A number of alternative airways designed for blind placement have been described for use by individuals who are not skilled laryngoscopists. These include the oesophageal obturator airway, oesophageal gastric tube airway, pharyngotracheal lumen airway, combination oesophageal tracheal tube, and the laryngotracheal mask. None of these allows the degree of airway control obtained with a tracheal tube.

Foreign body airway obstruction

In 1989 in the United States, the National Safety Council reported that foreign body airway obstruction accounted for approximately 3900 deaths, less than 1 per cent of all sudden deaths.[2] However, few subjects in CPR have caused more controversy than the management of this emergency. Total airway occlusion by a foreign body must be considered in any victim who suddenly stops breathing, becomes cyanotic and unconscious. It occurs most commonly during eating and usually is due to food, especially meat, impacting in the laryngeal inlet, at the epiglottis or in the vallecula. Factors contributing to choking include poorly chewed pieces of food, poor dentition or dentures, and elevated blood alcohol levels. Sudden death in restaurants from this cause is frequently mistaken for a myocardial infarction, leading to the label 'cafe coronary'.

Total airway obstruction is characterized by the inability to speak, cough or breathe. There is complete lack of air movement in spite of respiratory efforts. Cyanosis and loss of consciousness occur quickly and cardiac arrest will follow within minutes. Partial airway obstruction is characterized by rasping or wheezing respirations accompanied by coughing. The voice may be weak or speaking may be impossible. If air exchange appears good and coughing is forceful, no intervention in the patient's attempts to clear the airway should be made. If the obstruction cannot be cleared, definitive care with extraction under controlled circumstances in a medical facility may be necessary. If the cough weakens or cyanosis intervenes, the patient should be treated as if there were complete obstruction.

Treatments for airway obstruction range from back blows to subdiaphragmatic abdominal thrusts (Heimlich manoeuvre) to sternal thrusts to finger sweeps to grasping with instruments. No one technique is always successful and each may be successful when another has failed.[8] Mothers and friends have been pounding on the backs of choking victims for centuries. In 1974, Heimlich proposed abdominal thrusts as a better method of relieving airway obstruction.[9] A subsequent report suggested that the sternal thrusts of cardiac compression were at least as effective as abdominal thrusts.[10] Experimental studies show that a normal cough produces better air movement and higher airway pressures than any artificial manoeuvre. Abdominal and sternal thrusts produce only modest elevations in airway pressure but can induce a high volume and flow of exhaled air that may move the obstructing object into the pharynx. Back blows produce little air flow or volume but result instantaneously in very high airway pressures, considered helpful in dislodging an impacted obstruction. There continues to be no convincing evidence that any single technique is superior to the others. However, because of confusion in teaching multiple techniques (especially to the lay public), the American Heart Association has chosen to emphasize the abdominal thrust manoeuvre (with chest thrusts as an alternative for pregnant and massively obese victims) and the finger sweep.[2] This recommendation is made on the basis that the abdominal thrust is at least as effective as other techniques and teaching one method simplifies education.

Abdominal thrusts in the awake victim are applied in the sitting or standing position. The rescuer reaches around the victim from behind, placing the fist of one hand in the epigastrium between the xiphoid and umbilicus. The fist is grasped with the other hand and a quick upward thrust is given in the epigastrium. In the unconscious, thrusts are applied by kneeling astride the victim, placing the heel of one hand in the epigastrium and the other on top of the first. The thrust should be in the midline and care should be taken that the xiphoid is not pushed into the abdominal contents.

With the massively obese or women in advanced pregnancy, sternal thrusts may be safer and more effective than abdominal thrusts. In the erect victim, the chest is encircled from behind as in the abdominal manoeuvre but the fist is placed in the midsternum. In the supine victim, thrusts are applied from the side with the hands in the same position as for external cardiac compressions.

If back blows are used, they should be applied directly over the thoracic spine between the scapulae. They must be delivered with force. The head-down position may help move the obstructing object into the pharynx with the manoeuvre.

Whatever technique is used, each individual manoeuvre must be delivered as if it will relieve the obstruction. If the first attempt is unsuccessful, repeated attempts should be made because hypoxia-induced muscle relaxation may eventually allow success. Complications of the thrust manoeuvres include lacerations of the liver and spleen, fractured ribs, gastric rupture, and regurgitation.

For the unconscious victim, if these manoeuvres are unsuccessful, an attempt should be made to dislodge the obstruction manually. The tongue and jaw should be grasped between the thumb and fingers and pulled forward. A finger of the other hand should then be inserted along the cheek and an attempt made to dislodge the object laterally. Care must be taken not to push the object deeper into the larynx. Direct visualization of the object can also increase the chance of success. In the absence of a laryngoscope and forceps, ordinary kitchen utensils (such as a tablespoon and ice tongs) may be used. However, blind grasping with instruments is rarely successful and may cause damage to tonsils or other tissue. Finally, if the object cannot be dislodged, a cricothyroidotomy may be life saving.

Ventilation

If the unconscious victim is not making respiratory efforts after an open airway is established, rescue breathing to assist ventilation must be started. Mouth-to-mouth or mouth-to-nose ventilation are the most efficient methods available for use without additional equipment. Mouth-to-mask ventilation using a simple anaesthetic facemask is also very effective. If the mask is fitted with a nipple adaptor, a flow of oxygen can also supplement inspired oxygen concentration. The self-inflating resuscitation bag, with mask, can also be used with or without supplemental oxygen.

When using the above technique (as well as those detailed in other sections of this chapter), the resuscitator should be aware of the possibility of the transmission of infection and take reasonable and appropriate precautions.

Physiology of ventilation during CPR

In the absence of a tracheal tube, the relative distribution of gas between the lungs and stomach during mouth-to-mouth or bag–valve–mask ventilation will be determined by the impedance to flow into each compartment. In the respiratory compartment, the lung–thorax compliance will be the major factor. Impedance to flow into the stomach will be determined primarily by oesophageal opening pressure. Although there are no specific data during human CPR, it is likely that the opening pressure of the oesophagus is no higher than it is under anaesthesia (approximately 2.0 kPa or 20 cmH$_2$O). In experimental animals, oesophageal pressure decreases from 3.7 kPa (28 mmHg) to 0.5 kPa (4 mmHg) during 15 minutes of cardiac arrest without CPR while the lung–thorax compliance decreases by 30 per cent over the same time.[11]

Insufflation of air into the stomach during resuscitation leads to gastric distension, impeding ventilation and increasing the danger of regurgitation and gastric rupture. If gastric insufflation is to be avoided, inspiratory airway pressures must be kept low. A major cause of increased airway pressures and gastric insufflation is partial airway obstruction by the tongue and pharyngeal tissues. Meticulous attention to maintaining a patent airway is necessary during rescue breathing. A tidal volume of 0.8–1.2 litres will be needed to cause an obvious rise in the chest of most adults. Even with an open airway, a relatively long inspiratory time is necessary to administer this volume at low pressure. Thus, rescue breaths should be given over 1.5–2.0 seconds during a pause between chest compressions. A useful aid to minimizing gastric insufflation is the use of cricoid pressure (Sellick manoeuvre). Pressure applied over the anterior arch of the cricoid cartilage can prevent air from entering the stomach at airway pressures up to 9.8 kPa (100 cmH$_2$O).

Technique of rescue breathing

Using the head tilt/chin lift method for maintaining an open airway, mouth-to-mouth ventilation is administered by using the hand on the forehead to pinch the nose. The rescuer takes a breath, seals the victim's mouth with his mouth and exhales, watching for the chest to rise. When both hands are being used with the jaw thrust, the cheek can be used to seal the nose. For mouth-to-nose ventilation, the rescuer's lips surround the nose and the lips are held closed. For exhalation, the rescuer removes his mouth from the victim, listening for escaping air and taking a breath. When initiating resuscitation, two breaths should be given and breathing should be continued at a rate of 10–12 breaths/min. During CPR with one rescuer, a pause for two breaths should be made after each 15 chest compressions. When there are two rescuers, a 1.5–2.0 second pause after every fifth chest compression will allow a breath to be given. Exhalation can occur during the subsequent compressions.

The best way to ensure adequate ventilation without gastric distension is tracheal intubation. It is indicated during any prolonged resuscitation. However, other aspects of the resuscitation that might lead to a restoration of spontaneous circulation should not be delayed for intubation (Table 71.2). Once a tracheal tube is in place, ventilation can proceed without concern for gastric distension or synchronizing ventilation with chest compression. Blood flow during CPR slows rapidly when chest compressions are stopped and recovers slowly when they are restarted. Consequently, following intubation, no pause should be made for ventilation and ventilation should be approximately 12 breaths/min without regard for the compression cycle.

Circulation

Physiology of circulation during closed chest compressions

Two theories have been proposed to explain the mechanism by which closed chest compression causes blood to flow through the circulatory system.[12,13] They are not mutually exclusive and which predominates in humans continues to be actively investigated.

Cardiac pump mechanism

Kouwenhoven, Jude and Knickerbocker,[12] in the original description of closed chest cardiac massage in 1960, suggested that the heart was compressed between the sternum and spine, resulting in increased intraventricular pressure, closing of the atrioventricular valves and ejection of blood into the lungs and aorta. During the relaxation phase, negative intrathoracic pressure caused by expansion of the thoracic cage facilitates blood return and aortic pressure results in aortic valve closure and coronary perfusion. This has come to be known as the cardiac pump theory of blood flow during CPR. Support for this theory comes from studies using echocardiography. These show a reduction in ventricular size and mitral valve closure with chest compression during the early stages of CPR. In addition, CPR techniques that incorporate direct sternal compressions, compared with techniques that raise intrathoracic pressure without sternal compressions, result in better tissue blood flow and survival in animals.[14,15]

Thoracic pump mechanism

A few early investigators questioned the cardiac pump mechanism and many practitioners have questioned the ability of rescuers to depress the sternum enough to compress the heart in very large victims. In 1976, Criley, Blaufuss and Kissel[16] reported a patient undergoing cardiac catheterization that simultaneously developed ventricular fibrillation and an episode of cough-hiccups. With every cough-hiccup, a significant arterial pressure was noted and the patient did not at any time lose consciousness even though CPR was not performed. This description of 'cough CPR' led to further investigations[17] and codification of the thoracic pump theory of blood flow during CPR.[13] According to this theory, all intrathoracic structures are compressed equally by the increase in intrathoracic pressure resulting from sternal compression. Backward flow through the venous system is prevented by valves in the subclavian and internal jugular veins and by dynamic compression of the veins at the thoracic outlet. Thicker, less compressible vessel walls prevent collapse on the arterial side. The heart acts as a passive conduit with the atrioventricular valves remaining open during chest compression.

A number of studies and observations support the thoracic pump theory.[18] Angiography during a cough has shown blood flowing through the left heart into the aorta without cardiac compression. Manoeuvres that raise intrathoracic pressure (such as simultaneous ventilation and chest compression or abdominal binding) increase arterial pressure and carotid blood flow compared to standard CPR. Artificial circulation adequate to maintain viability can be accomplished with simultaneous ventilation and inflation of vests surrounding the chest and abdomen in experimental animals.

It seems clear that fluctuations in intrathoracic pressure play a significant role in blood flow during CPR. It is also likely that the cardiac pump mechanism contributes under some circumstances. Which mechanism predominates probably varies from victim to victim and may vary even during the resuscitation of the same victim.

Distribution of blood flow during CPR

Whatever the actual mechanism of blood flow, cardiac output is severely reduced during closed chest compressions, ranging from 10 to 33 per cent of prearrest values in experimental animals. Total blood flow also tends to decrease with time during CPR although changes in technique and the use of adrenaline (epinephrine) may help sustain cardiac output. Nearly all of the cardiac output is directed to organs above the diaphragm. Brain blood flow is 50–90 per cent of normal and myocardial blood flow 20–50 per cent of normal while lower extremity and abdominal visceral flow is reduced to less than 5 per cent of normal. All flows tend to decrease with time but the relative distribution of flow does not change. Adrenaline improves flow to the brain and heart while flow to organs below the diaphragm is unchanged or further reduced.

Physiology of gas transport during CPR

During CPR, measurement of blood gases reveals an arterial respiratory alkalosis and a venous respiratory acidosis because the arterial $P\text{CO}_2$ is reduced and the venous $P\text{CO}_2$ is elevated. However, the cause is not respiratory in origin. The cause of these changes is the reduced cardiac output.

The no flow state of cardiac arrest and low flow state of CPR have severe effects on the transport of gases in the blood. To understand these changes, the relation between the volume content of gas in the blood and the partial pressure of the gas must be remembered. Molecules of carbon dioxide (CO_2) are produced at the tissue level, transported in the blood and excreted by the lung. These are measured in volumes of gas, e.g. ml of CO_2 produced or excreted and ml of $CO_2/100$ ml of blood. The partial pressure of CO_2 ($P\text{CO}_2$) in the blood is directly related to CO_2 volume content. The partial pressure is also related to the barometric pressure and influenced by buffers available to combine with CO_2, e.g. bicarbonate and haemoglobin. Thus, the same CO_2 content can result in different values for $P\text{CO}_2$.

The same considerations apply to oxygen (O_2) uptake, transport and consumption. Under usual conditions, tissue O_2 consumption and CO_2 production are equal to pulmonary O_2 uptake and CO_2 excretion because metabolism, circulation and ventilation are stable. In unstable conditions such as CPR, these relations may not hold.

When ventricular fibrillation occurs in a patient being mechanically ventilated, the cessation of blood flow causes O_2 uptake from the inspired gas to stop and CO_2 excretion in the exhaled gas rapidly approaches zero as the functional residual capacity is washed out by continued ventilation. Since end tidal CO_2 is a reflection of the CO_2 in the alveolar gas, it will also rapidly approach zero. At the tissue level, continued metabolism depletes meagre tissue O_2 stores resulting in a tissue O_2 deficit and CO_2 excess. Tissue pH will be reduced by the excess carbon dioxide. Anaerobic metabolism will contribute more CO_2 and, after many minutes, lactic acid to the acidosis. Since there is no blood flow, these changes in O_2 and CO_2 at the tissue level are not reflected in arterial or venous blood. Blood gases measured during cardiac arrest without CPR are virtually unchanged from prearrest values.

When normal circulation is restored by rapid defibrillation, cellular metabolism returns to normal but the tissue O_2 deficit and CO_2 accumulation remain. More O_2 will move into the tissues from the blood than usually occurs and more CO_2 will diffuse into venous blood. Mixed venous blood will have a lower O_2 content and higher CO_2 content than normal, resulting in increased pulmonary O_2 uptake and CO_2 excretion even though tissue O_2 consumption and CO_2 production may be normal. With constant ventilation, the increase in CO_2 being returned to the lungs causes a temporary rise in alveolar, end tidal and arterial CO_2. Within a short time, the excess CO_2 is removed and respiratory physiological variables return to normal.

During the low flow state of CPR, excretion of CO_2 (ml of CO_2/min in exhaled gas) is decreased from prearrest levels approximately to the same extent as cardiac output is reduced. This reduced CO_2 excretion is due primarily to shunting of blood flow away from the lower half of the body. The exhaled CO_2 reflects only the metabolism of the part of the body that is being perfused. In the non-perfused areas, CO_2 accumulates during CPR. When normal circulation is restored, the accumulated CO_2 is washed out and a temporary increase in CO_2 excretion is seen.

Although CO_2 excretion is reduced during CPR, the mixed venous partial pressure of CO_2 ($P\bar{v}CO_2$) usually is increased.[19] Two factors account for this elevation. Bicarbonate is consumed in buffering acid and results in a reduced serum bicarbonate, so that the same blood CO_2 content results in a higher $P\bar{v}CO_2$. In addition, the mixed venous CO_2 content frequently, but not inevitably, is elevated. When flow to a tissue is reduced, all the CO_2 produced fails to be removed and CO_2 accumulates, raising the tissue partial pressure of CO_2. This allows more CO_2 to be carried in each aliquot of blood and mixed venous CO_2 content increases. If flow remains constant, a new equilibrium is established in which all CO_2 produced in the tissue is removed but at a higher venous CO_2 content and partial pressure. In contrast to the venous blood, arterial CO_2 content and partial pressure ($PaCO_2$) are usually reduced during CPR. This reduction accounts for most of the observed increase in arterial–venous CO_2 content difference. Even though venous blood may have an increased CO_2, the marked reduction in cardiac output with maintained ventilation results in very efficient CO_2 removal.

Decreased pulmonary blood flow during CPR results in many non-dependent alveoli being unperfused. The alveolar gas in these lung units will therefore contain no CO_2. Consequently, mixed alveolar CO_2 (i.e. end tidal CO_2) will be very low and correlate poorly with arterial CO_2. However, end tidal CO_2 does correlate well with cardiac output during CPR. As flow increases, more alveoli become perfused, there is less alveolar dead space, and end tidal CO_2 measurements rise.

Technique of closed chest compression

Cardiac arrest must be assumed in the absence of a pulse in the major arteries (carotid, femoral, axillary) of an unconscious patient. After opening the airway and providing two ventilations, a search for a pulse should be made before starting chest compressions. If weak circulation exists in a patient with a primary respiratory arrest, pulses may return after adequate ventilation and chest compressions may not be necessary.

Standard chest compression technique consists of the rhythmic application of pressure over the lower half of the sternum. The patient must be on a firm surface with the head level with the heart for compressions to be effective in providing blood flow to the brain and heart. The rescuer should stand or kneel at the side of the patient so that the hips are on a level with the victim's chest. The heel of one hand is placed on the lower sternum and the other hand placed on top of the first. Care must be taken that the xiphoid is not pressed into the abdomen which can lacerate the liver. Pressure on the ribs or costal cartilages rather than the sternum increases the risk of rib fracture. The elbows should be locked in position with the arms straight and the shoulders over the hands. Using the weight of the entire upper body, the compression is delivered straight down with enough force to depress the sternum 3.5–5.0 cm. Following maximal compression, pressure is released completely from the chest but the hands stay in contact with the chest wall, maintaining proper hand position for the next thrust.

Chest compression should be performed at a rate of

80–100/min. These are most effective if the compression and relaxation phases of the cycle are equal in length. This 50 per cent compression time is easier to achieve at faster compression rates. If a single rescuer is providing CPR, it is recommended that 15 compressions be followed by a pause for two ventilations (1.5–2.0 seconds each) followed by 15 more compressions. If there are two rescuers, the person performing chest compressions should pause 1.5–2.0 seconds after every five compressions to allow the second rescuer to give a breath. When the airway is controlled by a tracheal tube, breaths at a rate of 12/min should be interposed between chest compressions without a pause for ventilation.

Alternative techniques of circulatory support

Better understanding of circulatory physiology during CPR, especially involving the thoracic pump mechanism, has generated several proposals for alternative techniques in recent years. Most are designed to provide better haemodynamics during CPR and, thus, extend the duration during which CPR can successfully support viability. Unfortunately, none has proved reliably superior to standard techniques and no improvement in survival from cardiac arrest has been consistently demonstrated.[20]

Closed chest techniques

According to the thoracic pump theory, manoeuvres that increase intrathoracic pressure during chest compression should improve blood flow and pressure.[17] Several methods for raising intrathoracic pressure during CPR have been studied, including simultaneous ventilation and compression, abdominal binding with compression and the pneumatic antishock garment. Early results indicated improved aortic pressures and carotid blood flows with these techniques.[17] However, subsequent studies failed to demonstrate consistently improved resuscitation success or survival in animals[14,15] or humans.[21] The increase in aortic pressure seen with these techniques was expected to improve myocardial and cerebral perfusion. Unfortunately, the elevation in right atrial, intraventricular and intra-cranial pressure is equal to or greater than the rise in aortic pressure. The net result is either no improvement or a diminution in myocardial and cerebral perfusion pressures and blood flows.

Another alternative technique, commonly called high-impulse CPR, is based more on the cardiac pump mechanism.[22] Using relatively short compression times with moderately high force results in cardiac output being directly related to the rate of compression. The optimal cardiac output and coronary blood flow are obtained with a compression rate of 120/min in this experimental model. This technique is quite similar to the standard technique and many of the results from studies with high-impulse CPR have been incorporated into the currently recommended chest compression technique. Emphasis is now placed on ensuring adequate sternal depression. Recommended compression rates have been increased to 80–100/min. The faster rate and focus on sternal depression result in a shorter compression time with the use of a relatively high force.

Three alternative techniques continue to be actively investigated in experimental animals and, to a limited extent, in humans. The pneumatic CPR vest relies entirely on the thoracic pump mechanism of blood flow.[18] With this technique, thoracic and abdominal vests containing pneumatic bladders are inflated simultaneously with positive pressure ventilation. Experimental animal studies have shown excellent haemodynamics and the ability to maintain viability for prolonged periods. Clinical data have focused primarily on haemodynamics. Improved outcome has not been demonstrated in animals or humans.[14]

The technique of interposed abdominal compression-CPR (IAC-CPR) uses an additional rescuer to apply manual abdominal compressions during the relaxation phase of chest compressions.[23] Abdominal pressure is released when chest compression begins. Although animal studies suggested better haemodynamics with this technique, there was no improvement in survival.[14] A large randomized trial of IAC-CPR in out-of-hospital human arrest found no improvement in survival compared to standard CPR[24] but a study of in-hospital cardiac arrest did demonstrate improved survival.[25] Further studies of the efficacy and safety of IAC-CPR are needed. The newest proposed alternative technique is called active compression–decompression CPR.[26] Very limited information is available about this technique. Further studies with outcome assessment are needed.

Invasive techniques

Much of the effort spent improving current CPR techniques and investigating new techniques was prompted by the hope that better blood flows would extend the time during which CPR can support viability. Unfortunately, the results have been disappointing. In spite of the occasional success achieved following prolonged attempts at resuscitation, it appears that closed chest compressions can sustain most patients only for 15–30 minutes. If successful restoration of spontaneous circulation has not occurred in that time, the results are dismal. In contrast to the closed chest techniques, two invasive manoeuvres have been shown to be able to maintain cardiac and cerebral viability during long periods of cardiac arrest. In animal models, open chest cardiac massage and cardiopulmonary bypass (through the femoral artery and vein using a membrane oxygenator) can provide better haemodynamic performance and myocardial and cerebral perfusion than closed chest techniques. Prompt restoration of blood flow and perfusion pressure with cardiopulmonary bypass can provide resuscitation with minimal neurological deficit after 20 minutes of fibrillatory cardiac arrest in canines. However, to be effective, these techniques must be instituted relatively early (probably within 20–30 minutes).[27] If open chest massage is begun after 30 minutes of ineffective closed chest compressions, there is no better survival even though circulatory function is improved.[28] The need to apply these manoeuvres early in an arrest obviously limits the application. In-hospital arrests occur in circumstances in which the necessary expertise may be available to apply these techniques. However, there is an appropriate reluctance to apply such invasive manoeuvres until it is clear that closed chest techniques are ineffective. Unfortunately, at that point it may be too late for invasive methods to be successful as well. Before invasive procedures can play a greater role in modern CPR, a method must be developed to predict, early in the resuscitation, which patients will and which will not respond to closed chest compression.

Assessing the adequacy of circulation during CPR

There is an obvious need to assess whether ongoing CPR is generating adequate myocardial and cerebral blood flow for viable resuscitation. The traditional method is to palpate the carotid or femoral pulse during chest compressions. However, a palpable pulse primarily reflects systolic blood pressure. Mean blood pressure which correlates better with cardiac output and diastolic pressure is the major determinant of coronary perfusion. Nevertheless, palpation of the pulse remains the only assessment tool available during basic life support.

Successful resuscitation in experimental models is associated with myocardial blood flows of 100–300 ml/min per 100 g.[29] Obtaining such flows relies on closed chest compressions generating adequate cardiac output and coronary perfusion pressure. During CPR, coronary perfusion occurs primarily during the relaxation phase (diastole) of chest compression. The critical myocardial blood flow is associated with aortic 'diastolic' pressure exceeding 40 mmHg and coronary perfusion pressure (aortic diastolic minus right atrial diastolic pressure) exceeding 25 mmHg in animal models.[29–32] One report has confirmed similar findings in humans, noting that all patients with successful return of spontaneous circulation had coronary perfusion pressures higher than 2.0 kPa (15 mmHg).[33] When invasive pressure monitoring is available

during CPR, it should be used to guide resuscitation efforts. If pressures are below these levels, adjustments should be made to improve chest compression and/or additional adrenaline should be administered. Obtaining pressures above the critical levels does not ensure success. Damage to the myocardium from underlying disease may preclude survival no matter how effective the CPR efforts. However, vascular pressures below these levels are associated with poor results even in patients who may be salvageable.

Although invasive pressure monitoring may be the ideal, exhaled end tidal CO_2 is an excellent non-invasive guide to the effectiveness of standard CPR. Carbon dioxide excretion during CPR with a tracheal tube in place depends primarily on flow rather than ventilation. Since alveolar dead space is large during low flow conditions, end tidal CO_2 is very low (frequently < 1.3 kPa; 10 mmHg). If cardiac output increases, more alveoli are perfused and end tidal CO_2 rises (usually to > 2.7 kPa; 20 mmHg during successful CPR). When spontaneous circulation resumes, the earliest sign is a sudden increase in end tidal CO_2 to greater than 5.3 kPa (40 mmHg). Within a wide range of cardiac output, end tidal CO_2 during CPR correlates with coronary perfusion pressure, cardiac output, initial resuscitation and survival.[34–36] End tidal CO_2 measured during human CPR has been used to predict outcome.[36] No patient with an end tidal CO_2 < 1.3 kPa (10 mmHg) could be successfully resuscitated. In the absence of invasive pressure monitoring, end tidal CO_2 monitoring can be used to judge the effectiveness of the chest compressions.[37] Attempts should be made to maximize the value by alterations in technique or drug therapy. Sodium bicarbonate administration results in the liberation of CO_2 in the venous blood and a temporary rise in end tidal CO_2. Therefore, end tidal CO_2 monitoring will not be useful for judging the effectivenes of chest compression for 3–5 minutes following bicarbonate administration.

Advanced cardiac life support

Advanced cardiac life support encompasses all those cognitive and technical skills that are necessary to restore spontaneous circulatory function when simple support does not result in resuscitation. In addition to BLS skills, it includes use of adjunctive equipment and techniques for assisting ventilation and circulation, electrocardiographic (ECG) monitoring with arrhythmia recognition and defibrillation, establishment of intravenous access, and drug therapy. A number of aspects of ACLS have been discussed in the preceding sections, especially regarding adjuncts for airway management, ventilation and alternative methods of chest compression.

Defibrillation

Duration and electrical pattern of fibrillation

Ventricular fibrillation is the most common ECG rhythm in adult cardiac arrest. The longer fibrillation continues, the more difficult it is to defibrillate and the less likely is successful resuscitation. The fibrillating heart has a high oxygen consumption, increasing myocardial ischaemia and decreasing the time to irreversible cell damage. The only effective treatment for this dysrhythmia is electrical defibrillation and the sooner it is applied the higher the rate of successful resuscitation.[4,38] Thus, conversion of ventricular fibrillation to a rhythm capable of restoring spontaneous circulation should be the first priority of any resuscitation attempt (see Table 71.2).

The amplitude (coarseness) of the fibrillatory waves on the ECG may reflect the severity and duration of the myocardial insult and, thus, have prognostic significance.[39] Low voltage fibrillation is associated with poor outcome. Increasing myocardial ischaemia results in less vigorous fibrillation, reduced amplitude electrical activity and more difficult defibrillation. Catecholamines with β-adrenergic activity, such as adrenaline, increase the amplitude of the electrical activity but have no influence on the ability to defibrillate.[38,40] Consequently, defibrillation should not be postponed for any other therapy but should be carried out as soon as the rhythm is diagnosed and the equipment available. The importance of early defibrillation has been demonstrated in numerous studies.[3,4] The application has been made much easier by the development of automatic external defibrillators (AEDs) that recognize ventricular fibrillation, charge automatically and give a defibrillatory shock. The AEDs allow minimally trained individuals to incorporate defibrillation into BLS skills.[4]

Defibrillators: energy, current and impedance

The defibrillator is a variable transformer that stores a direct current in a capacitor until discharged through the electrodes. Defibrillation is accomplished by the current passing through a critical mass of myocardium causing simultaneous depolarization of the myofibrils. However, the output of most defibrillators is indicated in energy units (joules or watt-seconds). The relations between energy, current and impedance (resistance) are given by the following equations (standard units are indicated):

$$\text{Energy(joules)} = \text{Power(watts)} \times \text{duration(seconds)} \quad (1)$$

$$\text{Power(watts)} = \text{Potential(volts)} \times \text{current(amperes)} \quad (2)$$

$$\text{Current(amperes)} = \frac{\text{Potential(volts)}}{\text{Resistance(ohms)}} \quad (3)$$

$$\text{Current(amperes)} = \sqrt{\frac{\text{Energy(joules)}}{\text{Resistance(ohms)} \times \text{duration(seconds)}}} \quad (4)$$

From these equations, it can be determined that as the impedance between the paddle electrodes increases, the delivered energy will be reduced. For consistency, the energy level indicated on most commercially available defibrillators is the output when discharged into a 50 ohm load. Even at a constant delivered energy, equation (4) indicates that delivered current will be reduced as impedance increases. A high impedance with a relatively low energy level could result in a current too low for defibrillation. Defibrillation is most successful when impedance is low.

Transthoracic impedance during human defibrillation has been measured between 15 and 143 ohms.[41] Many of the important factors which can minimize transthoracic impedance are under the control of the rescuers. Resistance decreases with electrode size, so large electrode paddles (> 8 cm diameter) should be used. The greatest impedance is between the metal electrode and skin. This can be reduced slightly by the use of saline-soaked gauze pads or ECG electrode cream, but the lowest resistance is obtained with the specially designed defibrillation gels or pastes. Self-adhesive defibrillation/monitor pads also work well when carefully applied. Firm paddle pressure of at least 11 kg reduces resistance by improving electrode–skin contact and by expelling air from the lung. Transthoracic impedance is reduced by successive shocks. This factor may partially explain why additional shocks of the same energy may succeed when previous shocks did not, although its clinical importance has been questioned.[41]

The average transthoracic impedance during human defibrillation is 70–80 ohms. If relatively high energy (> 300 J) shocks are used with reasonable attention to proper technique, impedance is probably of little clinical significance but when lower energy shocks are used great care should be taken to minimize resistance. Recently, defibrillators have been developed that measure transthoracic impedance, prior to the shock, by passing a low level current through the chest during the charge cycle. Although not yet widely available, this technology may allow current-based defibrillation by adjusting the delivered energy for the measured resistance.[42]

Energy requirements and adverse effects

The incidence and severity of myocardial damage due to defibrillation in humans is not clear. Repeated high level shocks in animals result in dysrhythmias, ECG changes and myocardial necrosis. Whether such injuries occur in humans is unknown although slight elevations in creatine kinase MB fractions have been reported after cardioversion using high energies. It would seem prudent to keep energy levels as low as possible during attempts at defibrillation.

There is a general relation between body size and the energy requirements for defibrillation. Children need lower energies than adults, perhaps as low as 0.5 J/kg although the recommended paediatric dose is 2 J/kg.[43] However, in the case of adults, body size does not seem to be a clinically important variable. Multiple studies in adults have now demonstrated that the use of relatively low level initial shocks is as successful as beginning with higher energy shocks.[44] Therefore, it is currently recommended that the initial shock be given at 200 J followed by a second shock at 200–300 J if the first is unsuccessful. If both fail to defibrillate the patient, additional shocks should be given at 300–360 J.

Pharmacological therapy

During cardiac arrest, drug therapy is secondary to other interventions (see Table 71.2). Chest compressions, airway management, ventilation and defibrillation, if appropriate, should take precedence over medications. Establishing intravenous access and giving drugs should come after other interventions are established. The reason for making drugs a secondary intervention is that they contribute little to resuscitation attempts in many circumstances. Effective chest compressions, ventilation and defibrillation are most important. Of the drugs used during CPR, adrenaline is universally acknowledged to be useful in the restoration of spontaneous circulation.

Routes of administration

The preferred route of administration of all drugs, during CPR, is intravenous and the doses listed in Table 71.3 are for intravenous use. The most rapid and highest drug levels occur with administration into a central vein. Therefore, when a central venous catheter is available during a cardiac arrest, it should be used for drug therapy. However, peripheral intravenous administration is also effective. The antecubital or external jugular vein should be the site of

Table 71.3 Advanced cardiac life support drug doses

	Adult	*Infant and child*
Adrenaline	1 mg	0.01 mg/kg
If dose fails, consider	3–7 mg	
2nd + subsq doses		0.1 mg/kg
Lignocaine	1.5 mg/kg	1 mg/kg
Atropine	1 mg	0.02 mg/kg
Sodium bicarbonate	1 mmol/kg	1 mmol/kg
Bretylium tosylate	5 mg/kg	5 mg/kg

first choice for initiating an infusion during resuscitation, because starting a central line usually necessitates stopping CPR. Sites in the upper extremity and neck are preferred because of the paucity of blood flow below the diaphragm during CPR. Following drug administration in the lower extremity, the response may be extremely delayed or the drug may not reach its site or sites of action. Even when injected in the upper extremity, a drug may require 1–2 minutes to reach the central circulation. The rate of onset of action may be speeded up if a drug bolus is followed by a 20–30 ml bolus of intravenous fluid.

Adrenaline, lignocaine and atropine do not injure the lungs and can be absorbed from the tracheal mucosa. Therefore, if intravenous access cannot be established, the tracheal route provides an alternative route for the administration of these drugs following intubation. Sodium bicarbonate should not be given by this route. The time to effect and drug levels achieved are very inconsistent using this route during CPR. Studies have demonstrated that volumes of 5–10 ml need to be delivered to have reasonable uptake. It is likely that higher doses of the drugs need to be used via this route; 2–2.5 times the intravenous dose is currently recommended. Studies conflict on whether deep injection is better than simple instillation in the tracheal tube.

Catecholamines and vasopressors

Mechanism of action

The only drugs universally accepted as being useful during CPR are the vasopressors.[45] Adrenaline has been used for resuscitation since the 1890s and has been the vasopressor of choice in modern CPR since the studies of Redding and Pearson in the 1960s.[46,47] The efficacy of adrenaline lies entirely in its α-adrenergic properties.[32] Peripheral vasoconstriction leads to an increase in aortic diastolic pressure causing an increase in coronary perfusion pressure and myocardial blood flow. It is tempting to invoke the β-adrenergic properties of cardiac stimulation to explain the success of adrenaline. However, animal studies have demonstrated that all strong α-adrenergic drugs (adrenaline, phenylephrine, methoxamine, dopamine, noradrenaline) are equally successful in aiding resuscitation regardless of the β-adrenergic potency. β-Adrenergic agonists without α-activity, such as isoprenaline (isoproterenol) and dobutamine are no better than placebo. α-Adrenergic blockade precludes resuscitation while β-adrenergic blockade has no effect on the ability to restore spontaneous circulation. It is generally believed that the ability of adrenaline to increase the amplitude of ventricular fibrillation (a β-adrenergic effect) makes defibrillation easier. In fact, animal studies have shown that adrenaline does not improve the success of or decrease the energy necessary for defibrillation.[38,40] Retrospective clinical studies have shown no effect of adrenaline on defibrillation success.[39]

The β-adrenergic effects of adrenaline are potentially deleterious during cardiac arrest. In the fibrillating heart, adrenaline increases oxygen consumption and decreases the endocardial/epicardial blood flow ratio, an effect not seen with methoxamine.[48] Myocardial lactate production in the fibrillating heart is unchanged after adrenaline administration during CPR, suggesting that the increased coronary blood flow does not improve the oxygen supply/demand ratio. Large doses of adrenaline increased deaths in swine early after resuscitation due to tachyarrhythmias and hypertension, an effect partially offset by metoprolol treatment.[49] In spite of these theoretical considerations, survival and neurological outcome studies have shown no difference when adrenaline is compared to a pure α-agonist (methoxamine or phenylephrine) during CPR in animals or humans.[50,51] Because of the long experience with adrenaline, it remains the vasopressor of choice in CPR. It should be administered whenever resuscitation has not occurred after adequate chest compressions and ventilation have been started and defibrillation attempted, if appropriate.

Adrenaline dose

The addition of adrenaline to chest compressions helps to develop the critical coronary perfusion pressure necessary to provide enough myocardial blood flow to restore spontaneous circulation. The adequacy of the myocardial blood flow can be judged by the vascular pressures if invasive monitoring is present during CPR. If the arterial diastolic pressure is less than 40 mmHg or the coronary perfusion pressure is less than 2.7 kPa (20 mmHg), a better chest compression technique and/or more adrenaline are needed. In the absence of such monitoring, the dose of adrenaline must be chosen empirically. The standard dose used in animals and humans has for many years been 0.5–1.0 mg i.v. On a weight basis, this dose is approximately 0.1 mg/kg in animals but only 0.015 mg/kg in humans. In swine, the human standard dose of adrenaline (0.02 mg/kg) is insufficient to improve coronary perfusion pressure and blood flow but a high dose (0.2 mg/kg) improves

haemodynamic performance to levels compatible with successful resuscitation.[52] Therefore, it has been suggested that higher doses of adrenaline in human CPR might improve myocardial and cerebral perfusion and improve success of resuscitation. There are several case reports and a series of children (with historical controls) that demonstrated return of spontaneous circulation when large doses (0.1–0.2 mg/kg) of adrenaline were given to patients that had failed resuscitation with standard doses.[53,54]

Outcome studies designed to compare prospectively standard and high dose adrenaline have not demonstrated conclusively that higher doses will improve survival. A single animal study has compared outcome with standard and high dose adrenaline.[49] There was no difference in 24-hour survival or neurological outcome but more of the high dose animals died in the early post-resuscitation period due to a hyperdynamic state. There are four published studies comparing standard doses (1–2 mg) with high doses (5–18 mg) of adrenaline in human CPR.[55–58] All are prospective randomized double-blind clinical trials in cardiac arrest victims, primarily out-of-hospital. Two of the studies[55,58] suggested that there may be an improvement in immediate resuscitation with high dose adrenaline while the other two found no difference. None of the studies found any improvement in survival to hospital discharge.

High doses of adrenaline are apparently not needed as initial therapy for most cardiac arrests and potentially could be deleterious under some circumstances. However, the successful case reports were in patients with prolonged CPR and the high doses were given as 'rescue' therapy when standard doses had failed. This may be the appropriate place for higher doses of adrenaline in CPR practice. Current recommendations are to give 1 mg intravenously every 3–5 minutes in the adult. If this dose seems ineffective, higher doses (3–8 mg) should be considered.

Lignocaine and bretylium

After adrenaline, the most effective drugs during CPR are those used to suppress ectopic ventricular rhythms. Lignocaine (lidocaine) and bretylium are used during a cardiac arrest to aid defibrillation when ventricular fibrillation is refractory to electrical countershock therapy or when fibrillation recurs following successful conversion. However, no antiarrhythmic agent has been shown superior to electrical defibrillation or more effective than placebo in the treatment of ventricular fibrillation. Consequently, electrical defibrillation should not be withheld or delayed for drug therapy but should be applied at the earliest possible opportunity when treating ventricular fibrillation.

Lignocaine is primarily an antiectopic agent with few haemodynamic effects. It tends to reverse the reduction in the threshold for ventricular fibrillation caused by ischaemia or infarction. It depresses automaticity by reducing the slope of phase 4 depolarization and reducing the heterogeneity of ventricular refractoriness. When ventricular tachycardia or ventricular fibrillation have not responded to or have recurred following adrenaline and defibrillation, lignocaine should be administered. Relatively large doses are necessary to rapidly achieve and maintain therapeutic blood levels during CPR. An initial bolus of 1.5 mg/kg should be given and additional bolus doses of 0.5–1.5 mg/kg can be given every 5–10 minutes during CPR up to a total dose of 3 mg/kg. Only bolus dosing should be used during CPR but an infusion of 2–4 mg/min can be started after successful resuscitation.

Bretylium has been called a primary antifibrillatory drug because it reduces the chances for re-entry to occur between ischaemic and normal areas of myocardium. Bretylium, as in the case of lignocaine, reverses the reduction in fibrillation threshold caused by ischaemia. Unlike lignocaine, bretylium has significant haemodynamic effects when administered intravenously. It causes the release of noradrenaline from adrenergic nerve endings. With a normal circulation, this results in tachycardia, hypertension and increased contractility. After approximately 20 minutes, blockade of the uptake and release of noradrenaline from the nerve terminal begins, an effect that peaks 45–60 minutes after drug administration. This blockade can lead to profound hypotension. Although bretylium has some theoretical advantages over lignocaine for use during cardiac arrest, direct comparisons of the drugs in clinical trials have found no differences in resuscitation success or survival.[59,60] Consequently, because of the tendency to cause hypotension, it is recommended for use in cardiac arrest only if ventricular tachycardia/fibrillation persists or recurs after administration of lignocaine. The initial dose of bretylium is 5 mg/kg by intravenous bolus. The dose can be increased to 10 mg/kg and repeated at 5-minute intervals for a total dose of 30–35 mg/kg.

Atropine

Atropine sulphate enhances sinus node automaticity and atrioventricular conduction by its vagolytic effects. It is used primarily during a cardiac arrest when the ECG shows a pattern of asystole or slow idioventricular rhythm. Animal and human studies provide little evidence that atropine actually improves outcome from asystolic or bradysystolic arrest.[61,62] The predominant cause of asystole and electromechanical dissociation (EMD) is severe myocardial ischaemia. Excessive parasympathetic tone probably contributes little to these 'rhythms' during cardiac arrest in adults. Even in children, the significance of

autonomic tone during an arrest is of doubtful importance. The most effective treatment for asystole or EMD is to improve the coronary perfusion and myocardial oxygenation by repeated chest compression, ventilation and adrenaline. However, cardiac arrest with these 'rhythms' has a very poor prognosis. Since atropine has few adverse effects, it can be tried during a cardiac arrest that is refractory to adrenaline and oxygenation. The recommended dose is 1.0 mg i.v., repeated every 3–5 minutes up to a total of 0.04 mg/kg. Full vagolytic doses may be associated with fixed mydriasis following a successful resuscitation and therefore confound neurological examination. Occasionally, a sinus tachycardia following resuscitation may result from the use of atropine during CPR.

Sodium bicarbonate

Although in the past sodium bicarbonate was frequently used during CPR, there is little evidence to support its efficacy. Current practice restricts its use primarily to cardiac arrests occurring in association with hyperkalaemia, severe pre-existing metabolic acidosis, and tricyclic or phenobarbitone overdose. It may be considered for use during protracted attempts at resuscitation and after other treatments have been used. The use of sodium bicarbonate during resuscitation has been based on the theoretical considerations that acidosis lowers fibrillation threshold and respiratory acidosis impairs the physiological response to catecholamines. One early animal study found improved success in resuscitation from ventricular fibrillation when bicarbonate therapy was given in addition to adrenaline.[47] Most subsequent studies have been unable to demonstrate an improvement in the success of defibrillation or resuscitation with the use of bicarbonate.[63–66] The observation that metabolic acidosis develops very slowly during CPR may explain the absence of effect of buffer therapy. Acidosis does not become severe for 15–20 minutes of cardiac arrest.[19,67,68]

Current recommendations that sodium bicarbonate should be restricted during CPR are based on the documented complications from excessive use. Metabolic alkalosis, hypernatraemia, and hyperosmolarity are well documented after administration of bicarbonate used during resuscitation.[68,69] These derangements are associated with low resuscitation rates and poor survival. However, if sodium bicarbonate is given judiciously in accordance with standard recommendations, no significant metabolic abnormalities should occur.[70] When bicarbonate is used during CPR, the usual dose is 1 mmol/kg (1 mEq/kg) initially with additional doses of 0.5 mEq/kg every 10 minutes. However, dosing with sodium bicarbonate should be guided by blood gas determination, whenever possible.

Studies describing the physiology of CO_2 transport during CPR have focused interest on the effects of administering bicarbonate during resuscitation.[19,67] Tissue acidosis during CPR is caused primarily by the low blood flow and accumulation of CO_2 in the tissues (see section on physiology of gas transport). Intravenous sodium bicarbonate combines with hydrogen ion to produce carbonic acid that dissociates into CO_2 and water. The PCO_2 in blood is temporarily elevated until the excess CO_2 is eliminated through the lungs. Carbon dioxide readily diffuses across cell membranes and the blood–brain barrier while bicarbonate diffuses much more slowly. Thus, it is possible that sodium bicarbonate administration could result in a paradoxical worsening of intracellular and cerebral acidosis by further raising intracellular and cerebral CO_2 without a balancing increase in bicarbonate. Direct evidence for this effect has not been found but one early study demonstrated an elevation in cerebral spinal fluid PCO_2 and reduction in pH when very large doses of bicarbonate were given.[71] A more recent study found no changes in spinal fluid acid–base status with clinically relevant doses.[72] Similarly, measurement of myocardial intracellular pH during bicarbonate administration did not detect an increasing acidosis.[73] Therefore, paradoxical acidosis following administration of sodium bicarbonate remains largely a theoretical concern.

Calcium salts

With normal cardiovascular responses, calcium increases myocardial contractility and enhances ventricular automaticity. Consequently, it has been advocated for years as a treatment for asystole and EMD. An early report of success in four children following open heart surgery[74] was bolstered by the animal studies of Redding[46] demonstrating moderate success with calcium chloride in asphyxial arrest. However, Redding's studies actually showed vasopressors to be more successful than calcium. More recent animal studies have not found calcium to improve resuscitation or electromechanical coupling. In 1981, Dembo questioned the efficacy of calcium used during a cardiac arrest and reported dangerously high serum calcium levels (up to 18.2 mg/dl) during CPR.[75] Subsequently, several retrospective studies and prospective clinical trials examined the efficacy of calcium during out-of-hospital human cardiac arrest.[76,77] Results showed that calcium was no better than placebo in promoting resuscitation and survival from asystole or EMD. Consequently, calcium should only be given during cardiac resuscitation if hyperkalaemia, hypocalcaemia or calcium channel blocker toxicity is present. There are no other indications for its use during CPR. If calcium is administered, the chloride salt (2–4 mg/kg) is recommended because it produces higher and more consistent levels of ionized calcium than other salts.

Calcium channel blockers

The reverse of administering calcium during CPR is the suggestion that calcium channel blockers should be used. Anoxia or ischaemia result in a rapid depletion of cellular high energy phosphate stores, allowing a large shift of calcium into cells. High intracellular calcium concentrations, especially during reperfusion, are associated with numerous adverse sequelae that might be counteracted by calcium channel blockers. In spite of the theoretical considerations, studies have shown the amelioration of ischaemic myocardial damage by calcium channel blockers to be disappointing. No studies have examined the effect of calcium channel blockers on the myocardium during or after cardiac arrest. Most studies have focused on the possibly improved neurological outcome when these drugs are given following resuscitation. Overall, there is no convincing evidence of a beneficial effect. There is currently no indication for the use of these drugs for the cardiac arrest victim.

Do not resuscitate orders

Over the past two decades, CPR has become the standard for medical care when an individual is found apparently dead. During the same time period, terminally ill patients have become increasingly concerned about the inappropriate application of life-sustaining procedures, including CPR. Consequently, do not attempt resuscitation (DNR) orders and other limitations of medical treatment have become more common. These instructions are generally accepted, even welcomed, by health care workers. However, the operating room is the one area of the hospital where DNR orders continue to cause ethical conflicts between medical personnel and patients.[78,79]

The patient's right to limit medical treatment including refusing CPR is firmly established in modern medical practice and is based on the ethical principle of respect for patient autonomy. A terminally ill patient can reject heroic measures such as resuscitation and still choose palliative therapy. There is no reason why a surgical intervention should be withheld if it will ameliorate symptoms or cure an additional problem that will improve the patient's quality of life. However, the same patient may reasonably desire to maintain the DNR status during surgery in order to avoid heroic measures that only prolong dying. An operative intervention may increase the risk of cardiac arrest. Thus, it is on these occasions that the DNR order provides that greatest protection against an unwanted intervention. The patient may not want the burden of surviving resuscitation in a worse condition than previously. In fact, he/she may view the possibility of death under anaesthesia as especially peaceful.

In spite of these rather strong arguments for treating a DNR status in the operating room in the same way it is treated elsewhere in the hospital, most operating room personnel are, at least, a little uneasy caring for these patients. Many assume or require that DNR orders be suspended during the perioperative period. There are multiple reasons for the reluctance to accept DNR status during surgery and anaesthesia. Many interventions used commonly in the operating room (mechanical ventilation, vasopressors, antiarrhythmics, blood products) may be considered forms of resuscitation in other situations. The only actions that may not be considered routine are cardiac massage and defibrillation. Therefore, the specific interventions included in a DNR status must be clarified with particular allowance made for the methods required to perform anaesthesia and surgery. Many, but not all, cardiac arrests in the operating room are related to a surgical or anaesthetic complication. In addition, resuscitation under these circumstances is highly successful.[1] Based on the ethical principle of non-maleficence (primum non nocere – first, do no harm), surgeons and anaesthetists have a tremendous sense of responsibility for what happens to patients in the operating room. Anaesthesia is an interventional procedure and therefore anaesthetists are highly diligent in monitoring and managing changes in the patient's status. Honouring a DNR under these circumstances can be viewed as failure to treat a reversible process and, therefore, is tantamount to killing. This can be an ethically sound view if the cause of arrest is readily identifiable, easily reversible and treatment is likely to allow the patient to fulfil the objectives of coming to surgery.[78]

There are ethically sound arguments on both sides of the issue in respect of whether or not DNR orders should be upheld in the operating room. For the individual patient, conflicts can usually be resolved by communication between the patient and the medical and nursing staff. A mutual decision can usually be reached to suspend or severely limit a DNR order in the perioperative period if the patient understands (1) the special circumstances of perioperative arrest; (2) that interventions are brief and usually successful; and (3) that the medical staff support the patient's goals in coming to surgery but respect his/her desire not to prolong dying.

Post-resuscitation care

The major factors contributing to mortality following a sucessful resuscitation are progression of the primary disease and cerebral damage suffered as a result of the arrest. Active management following resuscitation appears to mitigate post-ischaemic brain damage and improve neurological outcome without increasing the number of patients surviving in a vegetative state.[80] When flow is restored following a period of global brain ischaemia, three stages of cerebral reperfusion are seen in the ensuing 12 hours. Multifocal areas of the brain receive no reflow immediately following resuscitation. Within an hour there is global hyperaemia which is followed quickly by prolonged global hypoperfusion.

Post-resuscitation support is focused on the provision of oxygenation and stable circulatory responses in order to minimize any further cerebral insult. A comatose patient should be maintained on mechanical ventilation for several hours to ensure adequate oxygenation and ventilation. Restlessness, coughing or seizure activity should be treated aggressively with appropriate medications including neuromuscular blockers, if necessary. Arterial PaO_2 should be maintained above 13.3 kPa (100 mmHg) and moderate hypocapnia ($PaCO_2$ 3.3–4.6 kPa; 25–35 mmHg) may be helpful. Blood volume should be maintained normal and moderate haemodilution to a haematocrit (packed cell volume) of 30–35 per cent should be considered. A brief 5-minute period of hypertension to a mean arterial pressure of 120–140 mmHg may help overcome the initial cerebral no reflow. This frequently occurs and is secondary to the effects of adrenaline given during CPR. However, both prolonged hypertension (mean blood pressure > 110 mmHg) and hypotension are associated with a worse outcome. Hyperglycaemia during cerebral ischaemia is known to result in increased neurological damage. Thus, it seems prudent to control glucose in the 100–300 mg/dl range.

In contrast with general supportive care, specific pharmacological therapy, directed at brain preservation, has not been shown to have further benefit. The results of some trials using barbiturates (animal studies) were promising but a large multicentre trial using thiopentone reported no improvement in neurological status when this drug was given following cardiac arrest.[80] Similar results have been reported after the use of calcium channel blockers and no apparent improvement has been recorded in human trials. Currently, there is no evidence to suggest that any specific pharmacological agent will improve the neurological outcome following resuscitation from cardiac arrest.

REFERENCES

1. Olsson GI, Hallen B. Cardiac arrest during anaesthesia. A computer-aided study of 250 543 anaesthetics. *Acta Anaesthesiologica Scandinavica* 1988; **32**: 653–64.

2. Emergency Cardiac Care Committee and Subcommittees, American Heart Association. Guidelines for cardiopulmonary resuscitation and emergency cardiac care. *Journal of the American Medical Association* 1992; **268**: 2171–295.

3. Elsenberg MS, Bergner L, Hallstrom A. Cardiac resuscitation in the community. Importance of rapid provision and implications for program planning. *Journal of the American Medical Association* 1979; **241**: 1905–7.

4. Weaver WD, Hill D, Fahrenbruch CE, *et al.* Use of the automatic external defibrillator in the management of out-of-hospital cardiac arrest. *New England Journal of Medicine* 1988; **319**: 661–6.

5. Taffet BE, Teasdale TA, Luchi RJ. In-hospital cardiopulmonary resuscitation. *Journal of the American Medical Association* 1988; **260**: 2069–72.

6. Harrison RR, Maull KI, Keenan RL, *et al.* Mouth-to-mask ventilation: a superior method of rescue breathing. *Annals of Emergency Medicine* 1982; **11**: 74–6.

7. Hess D, Baran C. Ventilatory volumes using mouth-to-mouth, mouth-to-mask, and bag valve mask techniques. *American Journal of Emergency Medicine* 1985; **3**: 292–6.

8. Redding JS. The choking controversy: critique of evidence on the Heimlich maneuver. *Critical Care Medicine* 1979; **7**: 475–9.

9. Heimlich HJ. Pop goes the cafe coronary. *Emergency Medicine* 1974; **6**: 154–5.

10. Guildner CW, Williams D, Subtich T. Airway obstructed by foreign material: the Heimlich maneuver. *Journal of the American College of Emergency Physicians* 1976; **5**: 675–7.

11. Melker RJ. Recommendation for ventilation during cardiopulmonary resuscitation: time for change? *Critical Care Medicine* 1985; **13**: 882–3.

12. Kouwenhoven WB, Jude JR, Knickerbocker GG. Closed-chest cardiac massage. *Journal of the American Medical Association* 1960; **173**: 1064–7.

13. Babbs CF. New versus old theories of blood flow during CPR. *Critical Care Medicine* 1980; **8**: 191–5.

14. Kern KB, Carter AB, Showen RL, *et al.* Twenty-four-hour survival in a canine model of cardiac arrest comparing three methods of manual cardiopulmonary resuscitation. *Journal of the American College of Cardiology* 1986; **7**: 859–67.

15. Kern KB, Carter AB, Showen RL, *et al.* Comparison of mechanical techniques of cardiopulmonary resuscitation: survival and neurologic outcome in dogs. *American Journal of Emergency Medicine* 1987; **5**: 190–5.

16. Criley JM, Blaufuss AH, Kissel GL. Cough-induced cardiac compression. Self-administered form of cardiopulmonary resuscitation. *Journal of the American Medical Association* 1976; **236**: 1246–50.

17. Rudikoff MJ, Maughan WL, Effrom M, *et al.* Mechanisms of blood flow during cardiopulmonary resuscitation. *Circulation* 1980; **61**: 345–52.

18. Criley JM, Niemann JT, Rosborough JP, Hausknecht M. Modifications of cardiopulmonary resuscitation based on the cough. *Circulation* 1986; **74** (suppl IV): IV-42–IV-50.

19. Weil MH, Rackow EC, Trevino R, *et al.* Difference in acid–base state between venous and arterial blood during cardiopulmonary resusciation. *New England Journal of Medicine* 1986; **315**: 153–6.

20. Ewy GA. Alternative approaches to external chest compression. *Circulation* 1986; **74** (suppl IV): IV-98–IV-101.

21. Kirscher JP, Fine EG, Weisfeld ML, *et al.* Comparison of prehospital conventional and simultaneous compression–ventilation cardiopulmonary resuscitation. *Critical Care Mecicine* 1989; **17**: 1263–9.

22. Maier GW, Newton JR, Wolfe JA, *et al.* The influence of manual chest compression rate of hemodynamic support during cardiac massage: high-impulse cardiopulmonary resuscitation. *Circulation* 1986; **74** (suppl IV): IV-51–IV-59.

23. Babbs CF, Tacker WA. Cardiopulmonary resuscitation with interposed abdominal compression. *Circulation* 1986; **74** (suppl IV): IV-37–IV-41.

24. Mateer JF, Stueven HA, Thompson BM, *et al.* Pre-hospital IAC-CPR versus standard CPR: paramedic resuscitation of cardiac arrests. *American Journal of Emergency Medicine* 1985; **3**: 143–6.

25. Sack JB, Kesselbrenner MB, Bregman D. Survival from in-hospital cardiac arrest with interposed abdominal counterpulsation during cardiopulmonary resuscitation. *Journal of the American Medical Association* 1992; **267**: 379–85.

26. Cohen TJ, Tucker KJ, Lurle KG, *et al.* Active compression–decompression: a new method of cardiopulmonary resuscitation. *Journal of the American Medical Association* 1992; **267**: 2916–23.

27. Sanders AB, Kern KB, Atlas M, *et al.* Importance of the duration of inadequate coronary perfusioh pressure on resuscitation from cardiac arrest. *Journal of the American College of Cardiology* 1985; **6**: 113–18.

28. Kern KB, Sanders AB, Badylak SF, *et al.* Longterm survival with open-chest cardiac massage after ineffective closed-chest compression in a canine preparation. *Circulation* 1987; **75**: 498–503.

29. Ralston SH, Voorhees WD, Babbs CF. Intrapulmonary epinephrine during prolonged CPR. Improved regional blood flow and resuscitation in dogs. *Annals of Emergency Medicine* 1984; **13**: 79–86.

30. Redding JS. Abdominal compression in cardiopulmonary resuscitation. *Anesthesia and Analgesia* 1971; **50**: 668–75.

31. Sanders AB, Ewy GA, Taft TV. Prognostic and therapeutic importance of the aortic diastolic pressure in resuscitation from cardiac arrest. *Critical Care Medicine* 1984; **12**: 871–3.

32. Otto CW, Yakaitis RW. The role of epinephrine in CPR. A reappraisal. *Annals of Emergency Medicine* 1984; **13**: 840–3.

33. Paradis NA, Martin GB, Rivers EP, *et al.* Coronary perfusion pressure and the return of spontaneous circulation in human cardiopulmonary resuscitation. *Journal of the American Medical Association* 1990; **263**: 1106–13.

34. Sanders AB, Atlas M, Ewy GA, *et al.* Expired PCO_2 as an index

of coronary perfusion pressure. *American Journal of Emergency Medicine* 1985; **3**: 147–9.

35. Sanders AB, Ewy GA, Bragg S, *et al.* Expired Pco₂ as a prognostic indicator of successful resuscitation from cardiac arrest. *Annals of Emergency Medicine* 1985; **14**: 948–52.

36. Sanders AB, Kern KB, Otto CW, *et al.* End-tidal carbon dioxide monitoring during cardiopulmonary resuscitation. A prognostic indicator for survival. *Journal of the American Medical Association* 1989; **262**: 1347–51.

37. Kern KB, Sanders AB, Raife J, *et al.* A study of chest compression rates during cardiopulmonary resuscitation in humans. The importance of rate-directed compressions. *Archives of Internal Medicine* 1992; **152**: 145–9.

38. Yakaitis RW, Ewy GA, Otto CW, Taren DL, Moon TE. Influence of time and therapy on ventricular defibrillation in dogs. *Critical Care Medicine* 1980; **8**: 157–63.

39. Weaver SC, Cobb LA, Dennis D, *et al.* Amplitude of ventricular fibrillation waveform and outcome after cardiac arrest. *Annals of Internal Medicine* 1985; **102**: 53–5.

40. Otto CW, Yakaitis RW, Ewy GA. Effects of epinephrine on defibrillation in ischemic ventricular fibrillation. *American Journal of Emergency Medicine* 1985; **3**: 285–91.

41. Kerber RE, Grayzel J, Hoyt R, Marcus M, Kennedy J. Transthoracic resistance in human defibrillation. Influence of body weight, chest size, serial shocks, paddle size and paddle contact pressure. *Circulation* 1981; **63**: 676–82.

42. Lerman BB, DeMarco JP, Haines DE. Current-based versus energy-based ventricular defibrillation: a prospective study. *Journal of the American College of Cardiology* 1988; **12**: 1259–64.

43. Gutgesell HP, Tacker WA, Geddes LA, *et al.* Energy dose for defibrillation in children. *Pediatrics* 1976; **58**: 898–901.

44. Weaver WD, Cobb LA, Copass MK, Hallstrom AP. Ventricular defibrillation – a comparative trial using 175-J and 320-J shocks. *New England Journal of Medicine* 1982; **307**: 1101–6.

45. Otto CW. Cardiovascular pharmacology. II. The use of catecholamines, pressor agents, digitalis, and corticosteroids in CPR and emergency cardiac care. *Circulation* 1986; **74** (suppl IV): IV-80–IV-85.

46. Redding JS, Pearson JW. Evaluation of drugs for cardiac resuscitation. *Anesthesiology* 1963; **24**: 203–7.

47. Redding JS, Pearson JW. Resuscitation from ventricular fibrillation (drug therapy). *Journal of the American Medical Association* 1968; **203**: 255–60.

48. Livsay JJ, Follette DM, Fey KH, *et al.* Optimizing myocardial supply/demand balance with alpha-adrenergic drugs during cardiopulmonary resuscitation. *Journal of Thoracic and Cardiovascular Surgery* 1978; **76**: 244–51.

49. Berg RA, Otto CW, Kern KB, *et al.* High-dose epinephrine results in greater mortality after resuscitation from prolonged cardiac arrest in pigs. *Critical Care Medicine* 1994; **22**: 282–90.

50. Brillman JC, Sanders AB, Otto CW, *et al.* A comparison of epinephrine and phenylephrine for resuscitation and neurologic outcome of cardiac arrest in dogs. *Annals of Emergency Medicine* 1987; **16**: 11–17.

51. Silvast T, Saarnivaara L, Kinnunen A, *et al.* Comparison of adrenaline and phenylephrine in out-of-hospital CPR: a double-blind study. *Acta Anaesthesiologica Scandinavica* 1985; **29**: 610–13.

52. Brown CG, Werman HA, Davis EA. The effects of graded doses of epinephrine on regional myocardial blood flow during cardiopulmonary resuscitation in swine. *Circulation* 1987; **75**: 491–7.

53. Koscove EM, Paradis NA. Successful resuscitation from cardiac arrest using high-dose epinephrine therapy; report of two cases. *Journal of the American Medical Association* 1988; **259**: 3031–4.

54. Goetting MG, Paradis HA. High-dose epinephrine improves outcome from pediatric cardiac arrest. *Annals of Emergency Medicine* 1991; **20**: 22–6.

55. Linder KH, Ahnefeld FW, Prengel AW. Comparison of standard and high-dose adrenaline in the resuscitation of asystole and electromechanical dissociation. *Acta Anaesthesiologica Scandinavica* 1991; **35**: 253–6.

56. Stiell IB, Hebert PC, Weitzman BN, *et al.* High-dose epinephrine in adult cardiac arrest. *New England Journal of Medicine* 1992; **327**: 1045–50.

57. Brown CG, Martin DP, Pepe PE, *et al.* A comparison of standard-dose and high-dose epinephrine in cardiac arrest outside the hospital. *New England Journal of Medicine* 1992; **327**: 1051–5.

58. Callaham M, Madsen CD, Barton CW, *et al.* A randomized clinical trial of high-dose epinephrine and norepinephrine vs standard-dose epinephrine in prehospital cardiac arrest. *Journal of the American Medical Association* 1992; **268**: 2667–72.

59. Haynes RE, Chinn TL, Copas MK, Cobb LA. Comparison of bretylium tosylate and lidocaine in management of out of hospital ventricular fibrillation: a randomized clinical trial. *American Journal of Cardiology* 1981; **48**: 353–6.

60. Olson DW, Thompson BM, Darin JC, Milbrath MH. A randomized comparison study of bretylium tosylate and lidocaine in resuscitation of patients from out-of-hospital ventricular fibrillation in a paramedic system. *Annals of Emergency Medicine* 1984; **13**: 807–10.

61. Coon GA, Clinton JE, Ruiz E. Use of atropine for brady-asystolic prehospital cardiac arrest. *Annals of Emergency Medicine* 1981; **10**: 462–7.

62. Stueven HA, Tonsfeldt DJ, Thompson BM, *et al.* Atropine in asystole: human studies. *Annals of Emergency Medicine* 1984; **13**: 815–17.

63. Minuck M, Sharma GP. Comparison of THAM and sodium bicarbonate in resuscitation of the heart after ventricular fibrillation in dogs. *Anesthesia and Analgesia* 1977; **56**: 38–45.

64. Guerci AD, Chandra N, Johnson E, *et al.* Failure of sodium bicarbonate to improve resuscitation from ventricular fibrillation in dogs. *Circulation* 1986; **74**(suppl IV): IV-75–IV-79.

65. Federluk CS, Sanders AB, Kern KB, Nelson J, Ewy GA. The effect of bicarbonate on resuscitation from cardiac arrest. *Annals of Emergency Medicine* 1991; **20**: 1173–7.

66. Vukmir RB, Bircher NG, Radovsky A, *et al.* Sodium bicarbonate in cardiac arrest (Abstr). *Critical Care Medicine* 1992; **20**: S86.

67. Weil MH, Grundler W, Yamaguchi M, *et al.* Arterial blood gases fail to reflect acid–base status during cardiopulmonary resuscitation. A preliminary report. *Critical Care Medicine* 1985; **13**: 884–5.

68. Bishop RL, Weisfeldt ML. Sodium bicarbonate administration during cardiac arrest. Effect on arterial pH, pCO₂, and osmolality. *Journal of the American Medical Association* 1976; **235**: 506–9.

69. Mattar JA, Well MH, Shubin H, *et al.* Cardiac arrest in the critically ill. II. Hyperosmolal states following cardiac arrest. *American Journal of Medicine* 1974; **56**: 162–8.

70. White BC, Tintinalli JE. Effects of sodium bicarbonate administration during cardiopulmonary resuscitation. *Journal of the American College of Emergency Physicians* 1977; **6**: 187–90.

71. Berenyi KG, Wolk M, Killip T. Cerebrospinal fluid acidosis complicating therapy of experimental cardiopulmonary resuscitation. *Circulation* 1975; **52**: 319–24.

72. Sanders AB, Otto CW, Kern KB, *et al.* Acid–base balance in a canine model of cardiac arrest. *Annals of Emergency Medicine* 1988; **17**: 667–71.

73. Kette F, Weil MH, von Planta M, *et al.* Buffer agents do not reverse intramyocardial acidosis during cardiac resuscitation. *Circulation* 1990; **81**: 1660–6.

74. Kay JH, Blalock A. The use of calcium chloride in the treatment of cardiac arrest in patients. *Surgery, Gynecology and Obstetrics* 1951; **93**: 97–102.

75. Dembo DH. Calcium in advanced life support. *Critical Care Medicine* 1981; **9**: 358–9.

76. Stueven HA, Thompson BM, Aprahamian C, *et al.* Calcium chloride: reassessment of use in asystole. *Annals of Emergency Medicine* 1984; **13**: 820–2.

77. Stueven HA, Thompson BM, Aprahamian C, *et al.* The effectiveness of calcium chloride in refractory electromechanical dissociation. *Annals of Emergency Medicine* 1985; **14**: 626–9.

78. Cohen CB, Cohen PJ. Do-not-resuscitate orders in the operating room. *New England Journal of Medicine* 1991; **325**: 1879–82.

79. Walker RM. DNR in the OR. Resuscitation as an operative risk. *Journal of the American Medical Association* 1991; **266**: 2407–12.

80. Abramson NS, Safar P, Detre KM, *et al.* Randomized clinical study of cardiopulmonary-cerebral resuscitation: thiopental loading in comatose cardiac arrest survivors. *New England Journal of Medicine* 1986; **314**: 397–403.

Chapter 72

Management and Complications of Commonly Ingested and Inhaled Poisons

John A. Henry

General principles	Poisoning due to specific agents

General principles

The anaesthetist usually becomes involved in the management of the acutely poisoned patient in three sets of circumstances: first, when called on to assist in the assessment or management of a sick or unconscious patient in the accident and emergency department; second,

in resuscitation following cardiopulmonary arrest due to poisoning; and third, in the management of poisoned patients referred to the intensive therapy unit. Poison information or control centres are a useful resource found in many cities throughout the UK and the USA. They are available for emergency consultations regarding diagnosis and therapy. The centre should be contacted by the doctor who is not an expert or who is perplexed by a special case.

Table 72.1 Some major causes of blockade of the oxygen pathway due to poisoning, with mechanisms and typical values

Cause	Mechanism	Effect
Asphyxiant gases (e.g. butane, methane, carbon dioxide, nitrogen)	Hypoxic gas mixture	Reduced inspired oxygen fraction (F_{IO_2}) (normal 0.21)
Respiratory depression (e.g. opioids, barbiturates, other sedatives and hypnotics) Respiratory muscle disorders: paralysis (e.g. organophosphates, botulinus toxin) or spasm (e.g. strychnine, phencyclidine)	Failure of ventilation (type II respiratory failure)	Reduced alveolar oxygen tension (P_{AO_2}) (normal 13.3 kPa)
Aspiration pneumonitis Adult respiratory distress syndrome (e.g. paraquat)	Failure of oxygen transfer (type I respiratory failure)	Reduced arterial oxygen tension (P_{aO_2}) (normal 10–13.3 kPa)
Carboxyhaemoglobin (carbon monoxide) Methaemoglobin (e.g. nitrites) Haemolysis (e.g. arsine, stibine)	Loss of functioning haemoglobin	Reduced arterial oxygen content (C_{aO_2}) (normal 18–21 ml/dl)
Myocardial depressants (e.g. β-blockers, tricyclic antidepressants, dextropropoxyphene)	Reduced cardiac output	Reduced tissue oxygen delivery ($\dot{Q}_{O_2}$) (normal 12–16 ml/kg per min)
Chemical asphyxiants (cyanide, hydrogen sulphide)	Block of cytochrome enzyme chain	Reduced tissue oxygen consumption ($\dot{V}_{O_2}$) causing failure of oxidative metabolism (normal 3–4 ml/kg/min) Cell death

Poisons and the oxygen pathway

Most of the poisonings which the anaesthetist sees involve the oxygen pathway, from the inspired air to cellular respiration. A severe reduction in the oxygen content of the inspired air will cause acute hypoxaemia, producing collapse and coma. Interference with the mechanics of respiration by poisons may result in hypoxaemia and hypercapnia. Disturbance of oxygen transfer may produce hypoxaemia. Even if oxygen content of the air is not diminished, and respiration is functioning adequately, disturbances to either process may prevent oxygen reaching its intracellular site of action. The oxygen carrying capacity of the blood may be reduced by the presence of carboxyhaemoglobin or methaemoglobin, or more rarely by acute haemolysis. Cardiac output may be reduced by a number of poisons which cause cardiac arrhythmias or which depress the contractility of the heart or which cause extreme vasodilatation. The final step where poisons may interfere with the oxygen pathway is blockage of the cytochrome enzyme chain by toxins such as cyanide and hydrogen sulphide. These effects and the main poisons involved are outlined in Table 72.1.

Immediate assessment

The doctor should avoid the distraction of being so concerned about reversal of the poisoning as to overlook the immediate care of the patient. It is essential to check at once whether the patient's respiration and cardiac output are adequate. A rapid inspection will indicate whether the patient's life is in immediate danger.

Respiratory depression is a feature of poisoning by sedative, hypnotic and narcotic drugs, as well as many chemical toxins, and blood gas estimation and measurement of the minute and tidal volumes are important in the management of the comatose, collapsed or convulsing patient, as ventilation may be needed at any stage. Pupil size and reaction are also an important part of the diagnostic assessment. It should be noted that there are many causes of widely dilated or fixed pupils in poisoned patients (Table 72.2). The severely poisoned patient must be managed as any other critically ill patient. Cardiopulmonary stabilization is the immediate objective. Once the patient's immediate condition is stable, attention can be paid to assessing the type and severity of poisoning.

Diagnosis of poisoning

It is important to obtain a history from the patient before deterioration occurs or consciousness is lost. Relatives or

Table 72.2 Effects of poisons on pupil size

Widely dilated or fixed dilated pupils
  Cerebral hypoxia due to cardiorespiratory insufficiency or arrest
  Hypoxic cerebral damage
  Cerebral hypoxia due to cellular poisons: cyanide, hydrogen sulphide
  Severe hypothermia
  Poisons causing blindness: quinine, methanol
  Amphetamines, cocaine, monoamine oxidase inhibitors
  Anticholinergic agents

Small or pinpoint pupils
  Opioids
  Organophosphate
  Chloral hydrate
  Sodium valproate

ambulance crew may also be able to help, by describing the patient's behaviour or the circumstances where he or she was found, and providing hard evidence such as tablets, syringes or a suicide note. It is essential to search the patient's clothing for clues and to look for evidence of other metabolic, endocrine and neurological diseases (such as uraemia, myxoedema, diabetes and epilepsy).

The diagnostic features of poisoning have been reviewed.[1] Clinical examination may reveal a picture characteristic of poisoning by a certain type of agent. The main ones are summarized in Table 72.3. Evidence may be provided by the smell of the patient's breath. Solvents, cyanide, hydrogen sulphide, carbamates, acetone and many other agents produce characteristic odours which may help in diagnosis. The smell of alcohol is unreliable as it is mainly due to congeners and may be present at clinically insignificant levels and virtually absent at severely toxic levels. An alcohol meter may be helpful in diagnosis.

Further clues to poisoning include the evidence of self-injection: injection marks, 'tracks' and abscesses may be seen during the initial assessment. Inspection of the mouth and throat may show traces of tablet residues or signs of corrosion.

Laboratory diagnosis

In many cases, it will be unnecessary to carry out any specialized laboratory investigations to confirm or quantify the poison involved. The history and the clinical state of the patient may be sufficient, or the patient may be recovering when seen. In some cases, however, a definitive diagnosis of poisoning may have to be made by laboratory analysis of specimens taken from the patient or of material found on or beside the patient. The usual samples are blood (which should be separated and saved as plasma except when

Table 72.3 Analyses which may be required in the management of the poisoned patient

Toxin or measurement	Indications for measurement	Interpretation
Amphetamines (including MDA and MDMA)	Abuse	Tests can confirm exposure
Carbamazepine	1 Overdose 2 Therapeutic monitoring	Severe toxicity over 25 mg/l Therapeutic range 1.5–9 mg/l
Carboxyhaemoglobin	1 Carbon monoxide inhalation 2 Smoke exposed fire victims 3 Methylene chloride exposure	May confirm exposure and indicate severity (normal < 1%) Carboxyhaemoglobin over 12% indicates potential for pulmonary damage Carbon monoxide is a metabolite of methylene chloride (dichloromethane)
Chloroquine	Accidental or deliberate overdose	May confirm ingestion and indicate potential toxicity
Cholinesterase (RBC)	1 Acute poisoning with organophosphates 2 Chronic exposure to organophosphates	RBC cholinesterase < 20% of normal indicates significant exposure (usually by ingestion) RBC cholinesterase < 50% of normal helps to confirm exposure but is poor guide to severity
Digoxin	1 Digoxin or digitoxin overdose 2 Detecting digitalis immunoreactivity in cardiac glycoside plant ingestion 3 Detecting digitalis toxicity in therapeutic use	High levels shortly after ingestion indicate the need to obtain Fab antibody. The result from a sample taken over 6 hours after ingestion may be used to calculate the dose of Fab required May be used to confirm ingestion but is not a guide to severity To confirm toxicity or undertreatment. Hypokalaemia may cause signs of toxicity with digoxin levels in the therapeutic range
Drug abuse screen	1 Suspected acute toxicity 2 Medicolegal indications 3 Employment screening	May confirm diagnosis May be used to confirm abstinence (sample collection must be supervised) May exclude from certain types of employment
Ethanol	1 Severe intoxication 2 Intoxication (actual or possible) in head injured patients 3 Monitoring of treatment of methanol or ethylene glycol intoxication	Very high concentrations (over 4 g/l) in a severely obtunded patient may be an indication for haemodialysis Low concentrations (under 1.5 g/l) help to exclude ethanol as a cause of behavioural disturbance or altered conscious level Blood ethanol should be maintained between 1 and 2 g/l to inhibit alcohol dehydrogenase
Ethylene glycol	Suspected toxicity	High levels (over 500 mg/l) may be an indication for ethanol administration and possibly active elimination (haemodialysis)
Iron	Overdose	Serum iron over 55 µmol/l is potentially toxic. Desferrioxamine challenge test may be indicated
Lead	1 Clinical suspicion of poisoning 2 Monitoring industrial exposure	May confirm toxicity and indicate need for antidote therapy. Toxic concentration over 3.4 µmol/l. Normal values up to 1.0 µmol/l. Over 1.4 µmol/l, monitoring required. Over 3.4 µmol/l, toxic, unfit for work
Lithium	1 Acute overdose 2 Therapeutic monitoring	Plasma levels may confirm ingestion and indicate possible toxicity, but clinical manifestations are main guide to toxicity and need for haemodialysis (patient may be asymptomatic with plasma lithium over 5 mmol/l, depending on time since ingestion) Therapeutic range 0.7–1.2 mmol/l. Toxicity in therapeutic use can occur at over 1.5–2 mmol/l (often due to interactions with diuretic or NSAID therapy). Over 4 mmol/l severely toxic
Methanol	Suspected toxicity	High levels (over 200 mg/l) may be an indication for ethanol administration and possibly active elimination (haemodialysis) or folinic acid therapy

Table 72.3 Continued

Toxin or measurement	Indications for measurement	Interpretation
Opioids	1 Overdose 2 Abuse 3 Toxicity in renal failure patients	Tests confirm exposure Tests confirm exposure Requires measurement of morphine-6-glucuronide
Osmolality (plasma)	Suspected methanol or ethylene glycol ingestion	Depending on time since ingestion normal osmolality (preferably osmolar gap) can exclude potentially toxic ingestion. Osmolality of over 350 mosmol/l (methanol) or 370 mosmol/l (ethylene glycol) indicates need to commence ethanol regimen and measure implicated substance(s)
Paracetamol	1 Paracetamol overdose 2 Comatose patient with suspected drug overdose	High concentrations are an indication for antidote administration according to nomogram To exclude coingestion of paracetamol
Paraquat	Suspected ingestion (or other route of exposure)	Plasma paraquat levels can confirm ingestion and severity of exposure. Positive urine dithionite test confirms exposure. Dithionite test may be carried out on plasma – if positive confirms massive ingestion
Phenobarbitone	1 Overdose 2 Abuse 3 Therapeutic monitoring	Severe toxicity over 50 mg/l Levels may be in the range 30–60 mg/l with minimal impairment Normal range 5–30 mg/l
Phenytoin	1 Overdose 2 Following overdose to decide when therapy should be resumed 3 Therapeutic monitoring	Severe toxicity over 50 mg/l. Over 95 mg/l potentially fatal Normal range 7–20 mg/l, 50% of patients show side effects at over 30 mg/l
Quinine	1 Overdose 2 Monitoring of malaria therapy	Toxic levels > 10 mg/l Toxic levels > 15 mg/l
Salicylate	1 Salicylate overdose 2 Comatose patient with suspected drug overdose 3 Therapeutic monitoring in rheumatology	Provides an indication of severity and action to be taken. May indicate need for haemodialysis (over 700 mg/l) To exclude coingestion of salicylate Upper normal range 250 mg/l. Toxic levels > 500 mg/l
Solvents	1 Suspected poisoning in an unconscious patient 2 Behavioural disturbance or suspected abuse 3 Industrial exposure	May provide diagnosis May confirm exposure Tests can confirm excessive exposure
Theophylline	1 Acute overdose 2 Therapeutic monitoring	High concentrations (over 80 mg/l) in a severely symptomatic patient (convulsions, cardiac arrhythmias) may be an indication for haemoperfusion Severe toxicity may occur at 30 mg/l during regular therapy
Toxicology screen	1 Suspected acute poisoning in an unconscious patient 2 Behavioural disturbance due to suspected abuse or poisoning 3 Confirmation of brainstem death where drug administration is known or suspected	May provide diagnosis May provide diagnosis Presence of potentially toxic concentrations of drugs prevents diagnosis of brainstem death
Tricyclic antidepressants	1 Overdose 2 Therapeutic monitoring	Drug plus metabolite over 1 mg/l indicates potentially serious toxicity Drug plus metabolite between 50 and 300 μg/l is the usual therapeutic range. Some patients are fast metabolizers and have low levels despite high doses

Table 72.4 Commonly used antidotes in poisoning

Poison	Antidote	Mechanism of action	Dosage regimen
Anticholinergic agents	Physostigmine	Cholinesterase inhibitor	2 mg i.v. over 5 minutes, continue with an infusion of 4–6 mg hourly (adult dose)
Anticoagulants (warfarin type)	Vitamin K (phytomenadione)	Competitive antagonist at site of prothrombin manufacture in liver	2–5 mg i.v. adult, 0.4 mg/kg child
Benzodiazepines	Flumazenil	Competitive antagonist at benzodiazepine receptors	Initially 0.2 mg i.v. over 30 seconds. Further doses of 0.5 mg can be given over 30 seconds at 60-second intervals to a total dose of 3 mg
β-Blockers	Glucagon	Stimulates myocardial adenylate cyclase	5 mg i.v. over 1 minute followed by an infusion of 1–10 mg/h
	Isoprenaline	Competitive antagonist at β-receptors	10–50 mg/min i.v.
Carbon monoxide	Oxygen (normobaric or hyperbaric)	Competitive displacement of carbon monoxide from haemoglobin and cytochrome molecules	Administer as high an inspired oxygen as possible until carboxyhaemoglobin concentration falls below 5%. Consider hyperbaric oxygen in severe cases
Cyanide	Dicobalt edetate	Chelates cyanide ions	300 mg i.v. over 3 minutes
	Sodium nitrite	Forms methaemoglobin which combines with cyanide	10 ml of 30% solution i.v. over 10 minutes
	Sodium thiosulphate	Substrate for enzymatic detoxification of cyanide	50 ml of 25% solution i.v. over 10 minutes
	Hydroxocobalamin	Combines with cyanide to form cyanocobalamin	May be up to 4 g i.v.
	Oxygen	Competitive substrate binding	Administer a high inspired oxygen till clinical recovery occurs
Digoxin and digitoxin	Fab antibody fragments	Antidote forms an inert complex with poison	Dose should match the estimated dose of ingested digoxin
Ethylene glycol	Ethanol	Competitive substrate for alcohol dehydrogenase, slows toxic metabolite production	Dose given should be sufficient to maintain plasma ethanol level at 1–2 g/l
Heavy metals (lead, mercury, arsenic)	DMSA (2,3-dimercaptosuccinic acid)	Chelating agent	30 mg/kg 8-hourly for 5 days then 20 mg/kg 12-hourly for 14 days
	DMPS (sodium 2,3-dimercaptopropane-sulphonate)	Chelating agent	Chronic: 100 mg 3 times a day. Acute: 250 mg every 4 hours for 24 hours then 250 mg every 6 hours for the next 24 hours
	Sodium calcium edetate	Chelating agent	Up to 40 mg/kg twice daily by i.v. infusion repeated every 48 hours until lead level falls below toxic level
(NB – the latter three chelating agents are now seldom used. DMSA and DMPS are first choice	Dimercaprol	Chelating agent	Mercury: 2.5–3 mg/kg deep i.m. injection 4-hourly for 2 days, 2–4 times on third day. 1–2 times for up to 10 days
	Pencillamine	Chelating agent	Lead: 0.5–1.5 g per day orally for 1–2 months or until lead level falls below toxic level
Hydrofluoric acid	Calcium gluconate	Forms an inert complex (calcium fluoride)	For burns: calcium gluconate (10%) 0.25–0.5 mmol/kg, up to 25 mmol/kg
Iron salts	Desferrioxamine	Chelating agent	In severe iron poisoning (> 90 mmol/l) up to 15 mg/kg per hour reduced to keep the total i.v. dose under 80 mg/kg in each 24 hours

Table 72.4 Continued

Poison	Antidote	Mechanism of action	Dosage regimen
Methanol	Ethanol	Competitive substrate for alcohol dehydrogenase, slows toxic metabolite production	As for ethylene glycol
Methaemoglobin	Methylene blue	Cofactor for reduction of methaemoglobin by NADPH	0.2 ml/kg of 1% solution (i.e. 2 mg/kg) slowly i.v. over 5 minutes, repeated as necessary to a maximum of 6 mg/kg
	Ascorbic acid	Reducing agent	1 g/24 hours i.v. or orally
Narcotic analgesics (heroin, co-proxamol, etc.)	Naloxone	Competitive antagonist at opioid receptors	0.8–1.2 mg i.v. (children 0.2 mg). Repeat if respiratory depression (assessed by respiratory rate and minute volume) is not reversed within 1–2 minutes. Continue with half the amount required to produce a response as an infusion over 30 minutes
Organophosphates	Atropine	Competitive antagonist at acetylcholine receptors	2 mg i.v. (i.m. or s.c. in less severely poisoned patients) followed by further 2 mg doses at 5–10 minute intervals until clinical features of full atropinization become apparent (dry mouth is the most reliable sign)
	Pralidoxime	Cholinesterase reactivator	1 g i.v. (in 100 ml saline over 30 minutes) repeated every 4 hours for 24 hours in severe cases
Paracetamol	Acetylcysteine	Replenishes hepatic glutathione stores	300 mg/kg over 16 hours
	Methionine	Replenishes hepatic glutathione stores	2.5 g orally every 4 hours for 12 hours (total 10 g)
Thallium	Berlin blue	Chelating agent	250 mg/kg per day in divided doses. Ideally given until thallium level is < 10 mg/l in blood and urine

whole blood is required as for example in carbon monoxide, mercury and lead poisoning), gastric aspirate and urine. However, in all cases, patient samples should be obtained on admission and saved in case subsequent analyses are needed. Where there is adequate clinical evidence that a certain agent or group of agents is involved, and the knowledge acquired from chemical analysis is not going to alter the management of the patient, urgent analysis is not required. In cases where prolonged cardiopulmonary resuscitation is continuing, urgent toxicological analysis may confirm the substance involved and help in making a decision on how long to persevere with efforts at resuscitation. The usual indications for urgent analyses are in those cases in whom the management of the patient will be affected by the result, e.g. in the use of active procedures to remove the poison or in the use of antidotes (Table 72.4).

In every comatose patient suspected of self-poisoning by drugs, it is usual practice to obtain laboratory measurements of plasma salicylate and paracetamol (acetamino-phen), since these may have been taken together with another agent and may be overlooked. In very rare cases, massive overdose of paracetamol alone may be the cause of the coma.

Accident and emergency management

The anaesthetist may be called to the accident and emergency department for assessment of ventilatory status or to carry out intubation for protection of the airway while gastric lavage is carried out in the unconscious patient. Ventilatory status should be assessed as in Chapter 14, and the decision made as to whether the patient requires intubation and/or mechanical ventilation. The main problem to bear in mind is that the poisoned patient's condition may suddenly worsen at any time, with complications such as vomiting, convulsions or cardiac or respiratory deterioration; therefore, frequent monitoring is required.

Elimination procedures

Gastrointestinal decontamination

Emetics are of little use in preventing absorption of ingested poisons. Although emetics such as syrup of ipecacuanha induce vomiting very effectively, there is little evidence of their effectiveness in removing significant amounts of ingested poisons. Gastric lavage may be shown occasionally to remove large amounts of an ingested drug or poison, but there is little documentation of its effectiveness. However, gastric lavage is the only means of emptying the stomach in the unconscious patient, in which case the airway needs to be protected. Oral activated charcoal is widely used either alone or in combination with emesis or gastric lavage in order to reduce drug absorption. There is considerable evidence for its effectiveness, especially when given shortly after ingestion. It is accepted that the dose of charcoal should be at least ten times that of the dose of poison ingested; otherwise the dose is 5–10 g in an infant or child and 25–50 g for an adult. There is evidence that repeated doses of activated charcoal may be effective in removing poisons which have delayed absorption (sustained release preparations and salicylates) and those with a small volume of distribution, particularly theophylline, phenobarbitone, salicylates and carbamazepine. However, there has been no convincing demonstration that repeated doses of activated charcoal shorten the time spent in the intensive therapy unit.

Removal from the bloodstream

Haemodialysis is effective in removing many drugs and poisons from the bloodstream. The most important indications are salicylate poisoning, when haemodialysis additionally helps to correct metabolic disturbances, ethylene glycol, methanol, phenobarbitone, lithium and theophylline. Where possible, haemodialysis should be used in preference to peritoneal dialysis or continuous arteriovenous haemodialysis, which though effective at supporting renal function are less effective in removing toxins. Charcoal or resin haemoperfusion is effective in removing a wide range of drugs and poisons, but rarely indicated: the most important use is in severe theophylline overdose.

Cardiopulmonary resuscitation

In the patient who is apnoeic or pulseless, the most urgent measure is to re-establish ventilation and circulation by cardiopulmonary resuscitation. When this is done effectively, the opportunity may be taken to treat the cause of the problem. In some cases certain antidotes can be given at this stage which will be effective in reversing the effects of the poisoning during the course of cardiopulmonary resuscitation. Examples include naloxone, atropine, cyanide antidotes or digoxin-specific Fab antibodies (see Table 72.4). Other poisons are capable of producing cardiac arrest, intractable arrhythmias or electromechanical dissociation which cannot be reversed by the usual resuscitative measures. In these cases it may be worth persisting with resuscitation for several hours until spontaneous cardiac output occurs, as the drug or poison is metabolized by the body. There have been several case reports of the successful use of prolonged external chest compression in tricyclic antidepressant poisoning, when the patient had severe hypotension or asystole.[2] It has also been used successfully in poisoning by β-blockers, quinine, chloroquine and digoxin, local anaesthetic toxicity and also in hypothermia, which may complicate many types of poisoning. Prolonged resuscitative efforts are therefore indicated in many types of poisoning which are refractory to other treatments.

Management of hypotension and electromechanical dissociation (EMD)

Agents with a potent negative inotropic effect are sometimes taken as a means of self-poisoning. The usual clinical picture is one of profound hypotension, and the depression of myocardial function may be such that cardiac output is insufficient to maintain adequate cerebral perfusion. Many of these cases develop electromechanical dissociation (Chapter 71), and a variety of agents may be responsible. In these patients treatment should be aggressive, as electromechanical dissociation due to poisoning has a better prognosis than when due to other causes. While one study[3] showed total failure to resuscitate in 36 cases of EMD and 4 cases of asystole due to primary cardiac causes, successful reversal is not uncommon in acute poisoning. In many instances the drug or poison responsible has a membrane-stabilizing effect, and it has been suggested that this membrane-stabilizing property of drugs and chemicals may be a major cause of death.[4] Cardiogenic pulmonary oedema is not common, presumably because both left and right ventricles are equally depressed. Coma, hypoxaemia and a metabolic acidosis may persist despite adequate ventilation with a high inspired oxygen content. Convulsions may occur. Correction of hypovolaemia and inotropic support are required with mechanical support of the circulation by external chest compression in cases where cardiac output is judged to be inadequate and unresponsive to inotropic support.

In some cases of poisoning with cardiodepressant drugs, specific treatments are helpful. This is particularly true of poisoning with β-adrenergic antagonists, which respond readily to glucagon or isoprenaline. Isoprenaline has also been shown to be effective in reversing the profound depression of cardiac output seen in severe disopyramide poisoning.[5] Propoxyphene poisoning can cause severe cardiac depression unresponsive to naloxone; the patient presents with severe hypotension, seizures and electro-cardiographic abnormalities. Inotropic support is often required, and dopamine appears to be effective.[6] Many other drugs and poisons (Table 72.5) can cause severe myocardial depression, although the response to inotropoic agents is less well documented.

Hypothermia

Hypothermia presenting with hypotension, hypotonia, fixed dilated pupils and a measured acidosis must be included in the differential diagnosis of poisoning; profound hypo-thermia can be indistinguishable from death.[7] Hypothermia also commonly accompanies poisoning by narcotics, tricyclic antidepressants, and sedative and hypnotic drugs, especially where ethanol has also been ingested. The rectal temperature should be taken with a low-reading thermometer. Temperatures as low as 20–22°C are compatible with complete recovery by poisoned patients, and the prognosis is generally better than in hypothermia secondary to an underlying medical condition.

An important clinical feature is that the pupils are often fixed and dilated in severe hypothermia associated with poisoning even when opioids are responsible. This finding therefore does not necessarily indicate hypoxic cerebral damage and should not deter attempts at resuscitation. The management of hypothermia is controversial,[8,9] but relatively rapid rewarming should be the rule. Prolonged external chest compression has been used successfully when serious cardiac arrhythmias or asystole have occurred.[10,11] In injured patients hyperkalaemia indicates poor prognosis.[12]

Hyperthermia

The traditional method of treating hyperthermia is to attempt to accelerate heat loss by tepid sponging and fanning. Cooling using extracorporeal circulation and peritoneal dialysis have also been used and this is appropriate for severe anticholinergic poisoning. In many toxic hyperthermic states, the mechanisms involved are complex. Hyperthermia due to monoamine oxidase inhibi-tor poisoning is dealt with later in this chapter. Several other conditions deserve mention. Malignant hyperthermia, with hyperpyrexia, muscle rigidity and metabolic acidosis, can be triggered in susceptible people by a number of

Table 72.5 Factors contributing to hypotension and electro-mechanical dissociation in poisoning

Causes	Agents
Relative hypovolaemia	
Vasodilation	Central nervous depressants
	Vasodilators
	Nitrates
	Calcium antagonists
Anaphylaxis	Drug hypersensitivity
	Bee stings
Hypovolaemia	
Vomiting	Many agents
Haematemesis	Theophylline, iron
Sweating	Salicylates, MAOI, amphetamines
Hyperventilation	
Metabolic disorders	
Metabolic acidosis	Many causes
Hypoxaemia	
Hypothermia	
Negatively inotropic agents	
Cardiac drugs	Quinidine
	Disopyramide
	Flecainide
	β-Adrenergic antagonists
Antirheumatic/antimalarial	Chloroquine
	Quinine
Anaesthetics and analgesics	Dextropropoxyphene
	Local anaesthetic agents
Psychoactive agents	Tricyclic antidepressants
	Phenothiazines
Volatile substances	Toluene, petroleum vapour, etc.
Others	Digoxin
	Carbon monoxide
	Cyanide
	Terminally due to many causes

Hypovolaemia and relative hypovolaemia respond to volume repletion, while negatively inotropic agents can be managed with positively inotropic drugs, specific antidotes and, where necessary, external chest compressions.

agents including halothane and suxamethonium. In neuroleptic malignant syndrome, hyperthermia, muscle rigidity and mental disturbances occur in patients taking neuroleptic drugs, usually phenothiazines or butyrophe-nones. It has a mortality of 20–30 per cent[13] but treatment with dantrolene may be effective.[14] In each of these conditions, there appears to be a sudden loss of control of intracellular ionized calcium, which leads to the formation of short and rigid actomyosin. The reaction can become self-sustaining, since heat reduces the calcium requirement for excitation–contraction coupling. The excess calcium may also be absorbed by mitochondria where it can uncouple oxidative phosphorylation, leading to further heat production. The pharmacological approach is to decrease muscle contraction and heat production by giving dantrolene, 1 mg/kg repeated at 5–10 minute intervals, to a maximum of 10 mg/kg.[15] External cooling should be avoided if it will lead to cutaneous vasoconstriction which could be counterproductive in the patient with massive overproduction of heat.

Severe metabolic disturbances

Many poisons produce metabolic disturbances which can cause a deterioration in the patient's condition and which require immediate attention. An arterial blood gas estimation is therefore urgently indicated in every severely poisoned or deeply comatose patient. The most common abnormality is an anion-gap metabolic acidosis, which may be caused by a number of poisons. In addition to specific treatments, sodium bicarbonate should be given to correct the acidosis. Hyperkalaemia may result from a shift in potassium into the cells in chloroquine and theophylline poisoning. Hyperkalaemia may occur in digoxin poisoning, and the management is described later in this chapter. Hypocalcaemia is common in severe ethylene glycol poisoning due to the metabolic production of oxalic acid. Other metabolic and electrolyte disturbances should be dealt with as indicated.

Methaemoglobinaemia

While many drugs and chemicals can cause mild methaemoglobinaemia, the more important causes of life-threatening methaemoglobinaemia include aniline, sodium or potassium nitrite, sodium or potassium nitrate and sodium or potassium chlorate.

It should be noted that the cause may be iatrogenic, e.g. sodium nitrite given for cyanide poisoning can produce methaemoglobinaemia.[14] In most cases the proportion of methaemoglobin does not rise to 30 per cent and is not life-threatening. In severe acute methaemoglobinaemia, the patient becomes comatose and flaccid, may have cardiac arrhythmias and convulsions, and may progress to cardiorespiratory arrest. Methaemoglobinaemia should be suspected if the skin has a greyish cyanosed appearance; the blood may have a dark or chocolate-brown colour. This colour change occurs with methaemoglobin concentrations of 15–20 per cent, but clinical symptoms only appear at levels over 20–30 per cent, while consciousness is not likely to be lost at under 50 per cent. Death is common at methaemoglobin concentrations of over 70 per cent. In addition to supportive measures, which should include 100 per cent oxygen, methaemoglobinaemia can be reversed by the administration of methylene blue. It should only be

given in symptomatic cases, or those with methaemoglobinaemia over 40 per cent. When the cause is due to chlorate poisoning, methylene blue is unlikely to reverse the methaemoglobinaemia, though ascorbic acid (1 g per 24 hours) can be given. In severe cases, exchange transfusion can be used, and hyperbaric oxygen, if available, may also allow sufficient oxygen to be transported in the plasma to maintain life.

Poisoning due to specific agents

Sedative and hypnotic agents

Many drugs and poisons have a sedative or hypnotic effect, so that central nervous depression may occur in proportion to the severity of exposure. Symptoms may progress through drowsiness, hallucinations, nystagmus, coma, flaccidity, hypotension and respiratory depression. There may be further problems due to other properties of the poison, and convulsions, cardiac arrhythmias and cardiac depression may occur. Close observation and aggressive supportive care are the key features of management. Patients with no response to painful stimuli are at greatest risk of a fatal outcome, and obviously require airway protection and careful monitoring of respiratory function in case mechanical ventilation should be required. In contrast to the slowed respiratory rate associated with opioid drugs, respiratory depression due to sedative and hypnotic agents manifests itself not as a slowing of respiratory rate but as a reduction in tidal volume. A reduced minute volume and deteriorating blood gas values are an urgent indication for mechanical ventilation.

Benzodiazepines

The benzodiazepines are widely thought to be non-fatal in overdose, but deaths do occur mainly due to respiratory depression, aspiration of vomit or hypothermia. In most cases the coma amounts to a deep sleep with an adequate gag reflex, preserved tendon reflexes and response to painful stimuli. Concomitant toxicity from paracetamol and salicylate should be excluded. Patients rarely need management in the intensive therapy unit, but coma may be deep if the drug has been coingested with ethanol, opioids, barbiturates or tricyclic antidepressants. The question may arise as to whether flumazenil should be given in treatment or as a diagnostic test.[16] Although widely used for reversing benzodiazepine anaesthesia, one should be aware that reversal of benzodiazepine toxicity may provoke benzodiazepine withdrawal or convulsions due to a tricyclic antidepressant. In general it should not be given routinely, but only in carefully considered circumstances. A clinical response should be evident within 5–15 minutes, with recovery largely complete by 30 minutes; large doses (in excess of 5 mg) may sometimes need to be given.[17,18] Flumazenil is ineffective in ethanol poisoning.[19]

Chloral hydrate

Although regarded as a relatively mild hypnotic drug, chloral hydrate is rapidly reduced to trichloroethanol, which has an elimination half-life of 8–10 hours, and overdoses of 3 g or more may cause coma (often with meiosis), respiratory depression, convulsions, hypotension and cardiac arrhythmias, including multifocal ventricular tachycardia and ventricular fibrillation.[20] Repeated doses of oral activated charcoal may be effective in enhancing elimination of trichloroethanol, and charcoal haemoperfusion should be considered in severe cases.

Gas and smoke inhalation

The history provided by witnesses may indicate that the victim has been exposed to toxic gases, fumes, smoke or vapour. When this is unknown, 100 per cent oxygen should be administered while blood is assayed for carbon monoxide and other possible toxic agents, as well as arterial gases, haematology and blood chemistry.

It is important to establish the causative agent as soon as possible, as treatments differ. Poisoning by carbon monoxide and cyanide should be treated as described elsewhere in this chapter.

Exposure to asphyxiant gases demands full resuscitation and measures to prevent the sequelae of hypoxia. Simple asphyxiants act by reducing the proportion of oxygen in the inspired air. Many gases, such as hydrogen, helium, methane, propane, petroleum vapour, carbon dioxide and nitrogen, may be responsible. The duration and severity of hypoxia play a critical part in determining the symptoms and outcome. An atmosphere of less than 6–8 per cent oxygen will rapidly produce coma and cardiorespiratory collapse due to cerebral hypoxia. With gases heavier than air there may be several victims, as each tries to rescue the others without using breathing apparatus.

Exposure to other gases should be treated on their merits. Water-soluble irritant gases (such as CS or CN gas or ammonia) produce severe watering of the eyes and upper respiratory tract irritation and coughing, but resuscitation is not usually needed and symptomatic measures are sufficient. Water-insoluble gases such as acrolein or phosgene tend to reach the lungs and may cause bronchoconstriction and pulmonary oedema, often after a latent interval, so that the patient should be observed for 24 hours. Carboxyhaemoglobin levels and a careful clinical assessment are important in deciding the severity of smoke exposure. Cyanide toxicity may occur in fire victims; an elevated plasma lactate concentration may indicate toxicity in patients without severe burns.[21]

Gases such as arsine and stibine can cause minimal symptoms on inhalation but may produce severe haemolysis and haemoglobinuria, sometimes presenting as renal failure 24 hours after exposure. Accidental or intentional inhalation of volatile substances may result in deep coma, hyporeflexia and cardiac arrhythmias, which may progress to ventricular fibrillation. Treatment is symptomatic; a β-blocking drug such as esmolol or atenolol may help to control arrhythmias.

Carbon monoxide

Carbon monoxide exposure produces cerebral hypoxia secondary to its reversible combination with haemoglobin to form carboxyhaemoglobin. Symptoms are roughly related to the blood carboxyhaemoglobin content, but depend also on the duration of exposure, time since exposure and the treatment given. Throbbing headache, vomiting without diarrhoea, lethargy and hyperventilation are typical in the early stages and may progress to convulsions, coma, hypoventilation, bradycardia and cardiovascular collapse. When there is a history suggestive of exposure the diagnosis presents little difficulty, but in many cases the diagnosis may not be apparent and the possibility of carbon monoxide poisoning should be borne in mind especially when several people are affected with headache, vomiting and collapse. In the elderly, carbon monoxide toxicity may present as a cerebrovascular accident or myocardial infarction. The source of exposure is usually inhalation of products of combustion. In domestic incidents there may be a faulty appliance or a blocked flue. Motor exhaust gas inhalation is the cause of many suicidal deaths. The prognosis is grave in severe poisoning. Because of the high affinity of carbon monoxide for haemoglobin, a few breaths at high concentrations can be fatal; 1 per cent carbon monoxide in air can kill within a few minutes. There are no reports of recovery in patients who have required cardiopulmonary resuscitation following a cardiac or respiratory arrest.

Treatment consists of removal from the source of exposure and administration of 100 per cent oxygen immediately, after taking an anticoagulated whole blood sample for carboxyhaemoglobin estimation. The pregnant patient or any patient who has had symptoms suggestive of severe poisoning should be treated with hyperbaric oxygen if it is available. In addition to shortening the half-life of carboxyhaemoglobin, hyperbaric oxygen may prevent late neurological sequelae.

Cholinesterase inhibitors

Organophosphates inhibit cholinesterase, causing features of cholinergic toxicity. In peacetime the most serious poisoning is by ingestion of an organophosphate insecticide which may be attempted suicide. Cutaneous absorption and inhalation of sprays rarely cause serious toxicity. 'Nerve gases' such as Sarin and Tabun can cause severe toxicity after minimal exposure. Early symptoms are vague but the characteristic effects soon become apparent. Mild to moderate poisoning may cause headache, blurred vision, meiosis, excessive salivation, lacrimation, sweating, wheezing and lethargy. The patient should be kept under observation for at least 24 hours. Severe poisoning may cause coma, convulsions, respiratory muscle paralysis, bradycardia and hypotension.

Atropine should be given in very large doses: 2–4 mg intravenously (children 0.05 mg/kg) every 10 minutes till atropinization is achieved and it can be seen that secretions are inhibited or pupils have dilated. Atropine therapy should be continued for a minimum of 48 hours, and the patient should be observed for at least 72 hours, as relapse may occur due to delayed intestinal absorption.

Although the pulse rate is characteristically slow in this type of poisoning, it may be rapid due to sympathetic compensation, so that pulse rate is less useful as a guide than the other signs of atropinization. Pralidoxime (a specific cholinesterase reactivator) 1 g (children 20–40 mg/kg) in 100 ml saline over 30 minutes intravenously should ideally be started within 4 hours of exposure. After this, enzyme inactivation becomes less reversible, and pralidoxime is unlikely to have any effect after 24–36 hours. The dose should be reduced if there is renal impairment. Pralidoxime is ineffective in poisoning by carbamate-type cholinesterase inhibitors (e.g. neostigmine or pesticides such as carbaryl).

Treatment should obviously be given without waiting for laboratory confirmation of the diagnosis, but the results may indicate the severity of the poisoning and the likely duration of treatment. Plasma or red cell cholinesterase can be measured; the latter is a more reliable indicator. A reduction to 50 per cent of normal is diagnostic of organophosphate exposure but not of toxicity, and less than 10 per cent usually indicates severe poisoning. It has long been known that patients may relapse after apparent recovery. One reason for relapse is that treatment has not been given for long enough. The other main reason is because of muscle cell necrosis. This is now termed the 'intermediate syndrome'.[22] There may be widespread muscle weakness, and the muscles of respiration are often affected, leading to respiratory failure which requires mechanical ventilation. Recovery eventually occurs, but may take up to 2 weeks before the patient can be weaned off the ventilator.

Iron

Accidental ingestion of iron tablets is relatively common in young children, but serious toxicity and death are rare. Iron is highly corrosive, and in overdose destroys the mucosal barrier to iron absorption. This can lead to metabolic, cardiovascular and hepatic complications if the amount of free iron in plasma exceeds its iron binding capacity. The clinical course of iron poisoning can be divided into four phases.

The first phase, which occurs shortly after ingestion and usually resolves within 6–8 hours, consists of gastrointestinal symptoms, mainly acute vomiting and diarrhoea and sometimes haematemesis. This may be severe enough to lead to drowsiness, lethargy and shock with metabolic acidosis, but this phase usually resolves spontaneously.

The second phase is a 'silent interval', which may be followed by the third phase at around 24–48 hours after ingestion. This phase is characterized by metabolic acidosis, shock, coma, convulsions, hypoglycaemia, hepatic necrosis, cardiac failure and acute tubular necrosis. Several weeks after apparent recovery there may be stricture formation or pyloric stenosis resulting from corrosive injury (the fourth phase).

Treatment is aimed at preventing or reducing iron absorption, chelating absorbed iron with desferrioxamine, correcting metabolic disturbances and providing supportive care. The abdomen should be x-rayed and this may reveal the extent of ingestion, since undissolved iron salts are radiopaque. Activated charcoal does not adsorb iron, but orally administered desferrioxamine (5–10 g) may reduce toxicity. The serum iron concentration should be measured at around 4 hours post-ingestion so as to assess the severity of poisoning and the need for chelation therapy. A serum iron of 55–90 µmol/l is considered potentially toxic, between 90 and 180 µmol/l may be associated with severe toxicity and over 180 µmol/l is potentially fatal. A useful diagnostic test is to carry out a desferrioxamine challenge, which consists of giving desferrioxamine intramuscularly in a single dose of 50 mg/kg. The urine colour should be observed: if it develops a red or port-wine colour, this indicates that there is an excess of free iron in the body, and that desferrioxamine treatment should be started. The usual dose is 15 mg/kg per hour to a maximum of 80 mg in 24 hours. Treatment can be discontinued when the urine colour returns to normal. Desferrioxamine is not contraindicated after iron overdose in pregnant patients.

Cyanide

Cyanide poisoning can be rapidly fatal. It may be inhaled as hydrogen cyanide or enter the body by ingestion or cutaneous absorption of cyanide salts or intestinal hydrolysis of cyanogenic glycosides. Severe toxicity may develop within seconds if inhaled but may be delayed if salts are taken on a full stomach or following ingestion of cyanogenic glycosides or salts such as gold cyanide.

Cyanide acts as a chemical asphyxiant, combining with cytochromes and rapidly blocking cellular oxygen utilization so that cerebral function and circulation are soon impaired with the development of a metabolic acidosis. Venous blood tends to have a high oxygen saturation, because of reduced cellular oxygen uptake, and there may be little colour difference between retinal veins and arteries on ophthalmoscopy. Early signs include hyperventilation and tachycardia, but coma, cyanosis and convulsions soon follow. A high inspired oxygen (or hyperbaric oxygen if available) is an effective treatment, though its mode of action is unclear; it may be simple competition for binding sites. When the diagnosis is certain, dicobalt edetate 300 mg in 20 ml can be given intravenously over 3 minutes repeated as necessary, and is an effective antidote. However, if the patient is not poisoned, use of this antidote can produce severe anaphylactoid reactions with laryngeal oedema and convulsions. The safest course of action is to give the antidote only if the patient's level of consciousness is deteriorating. Another group of antidotes (nitrites such as sodium nitrite) act by producing methaemoglobin, each molecule of which can bind four cyanide ions in the bloodstream. Intravenous sodium thiosulphate (25 ml of a 50 per cent solution intravenously) is non-toxic and acts by providing a substrate for the enzyme rhodanese to neutralize cyanide, producing sodium thiocyanate. Subacute or chronic cyanide toxicity can also occur during prolonged nitroprusside therapy. The main clinical feature is a metabolic acidosis – if none is present then cyanide toxicity can be ruled out. A rapid improvement should occur after the intravenous administration of sodium thiosulphate. Other cyanide antidotes are not indicated in this case.

Methanol

Patients poisoned with methanol, though initially asymptomatic for up to 60 hours following ingestion, may present with profound coma, fixed dilated pupils with retinal haemorrhages or papilloedema, and profound metabolic acidosis. If the patient has reached this stage of toxicity, it is very probable that severe cerebral damage has already occurred, and recovery is unlikely. Recovery is virtually unknown in the patient who requires cardiopulmonary resuscitation.

In all cases of symptomatic methanol poisoning assessment and urgent treatment are needed. The important measures are to correct the metabolic acidosis with sodium bicarbonate and to protect the body from further enzymatic transformation of methanol to formic acid, which is the agent mainly responsible for its metabolic, cerebral and ocular toxicity. The usual treatment is to give ethanol by mouth or if necessary by intravenous infusion. Ethanol has a greater affinity for alcohol dehydrogenase than methanol and hence inhibits its metabolism to formaldehye which is converted, by aldehyde dehydrogenase, to formic acid. Haemodialysis can, if necessary, be used to remove methanol from the circulation.[23] Another approach is to speed metabolism of formic acid via the dihydrofolate reductase pathway by administration of folinic acid. Frusemide may be given to enhance formic acid excretion.

If the patient has taken a single dose of methylated spirits or surgical spirit, the main toxicity is that of ethanol and the small amount of methanol present will not cause problems as its metabolism will be slowed due to the ethanol present. On the other hand, prolonged or repeated drinking of alcoholic drinks containing even a small proportion of methanol can lead to serious poisoning due to the accumulation of methanol.[24] Epidemics of methanol poisoning have resulted from the use of illicit liquor containing excessive amounts of methanol.

Ethylene glycol

Ethylene glycol is a common constituent of automobile antifreeze. When ingested, it tends initially to cause signs of alcoholic intoxication followed later by tachycardia, pulmonary oedema, convulsions and acute renal failure. Metabolic acidosis, leucocytosis, hypocalcaemia and crystalluria are characteristic. The main principles of treatment are to correct the acidosis, delay the metabolism of ethylene glycol to toxic metabolites by administering ethyl alcohol (similar to the treatment of methanol poisoning above) and to hasten elimination by increasing the fluid output or by using haemodialysis in severe cases.

Chloroquine

In overdose this drug, which is used both as an antimalarial and an antirheumatic, can produce hypokalaemia (often <2 mmol/l), coma and cardiovascular collapse. While many drugs in overdose produce deep coma before cardiac toxicity occurs, chloroquine overdose may produce collapse from profound hypotension or cardiac arrhythmias with minimal or no depression of consciousness beforehand. Later severe hyperkalaemia may ensue, particularly if potassium salts have been given to correct the hypokalaemia. It was found by chance in Africa that patients who had taken an overdose of diazepam plus chloroquine fared better than those who had taken chloroquine alone, and although its mode of action is unclear, diazepam therapy is part of the treatment regimen. Resuscitation should be aggressive and the use of noradrenaline and diazepam may be life-saving.[25] Active removal methods such as haemoperfusion and haemodialysis are ineffective.

Cardioactive drugs

Cardiac glycosides

Acute poisoning with cardiac glycosides such as digoxin, digitoxin or plants such as foxglove, *Thevetia* or oleander can produce cardiac arrhythmias and hyperkalaemia, each of which can be fatal. If the patient presents within 4 hours of ingestion, gastric lavage should be performed and activated charcoal left in the stomach to adsorb the cardiac glycoside and thus reduce intestinal absorption. Plasma digoxin levels should be measured urgently, bearing in mind that the concentration may peak within 1–2 hours, and that levels cannot be readily interpreted until 6 hours after ingestion, when distribution is complete. Even in plant poisonings, a measured digoxin concentration by immunoassay, though of no help in quantifying the toxin, may be used to confirm ingestion. However, as soon as there is a real suspicion of a large overdose from the history, clinical features of toxicity (arrhythmias, heart block or a raised or increasing plasma potassium) or digoxin assay results, arrangements should be made to obtain digoxin-specific Fab antibodies urgently in case life-threatening toxicity should develop. These Fab fragments of sheep anti-digoxin immunoglobulin molecules bind with high affinity to digoxin and other cardiac glycosides, thus removing the poison from receptor sites in the tissues. The important investigations are the elecrocardiogram, digoxin assay and plasma potassium; a concentration of over 5 mmol/l is a strong indicator of potential toxicity.

β-Adrenoreceptor antagonists, lignocaine, amiodarone or phenytoin may be useful in treating cardiac arrhythmias, and insulin plus glucose can be used as a short-term measure for hyperkalaemia. A pacing wire should be inserted if second or third degree heart block develops. Direct current countershock should be used for ventricular fibrillation. If these measures fail to restore an adequate cardiac output, the circulation can be supported with external chest compression pending the arrival of Fab antibodies, which should reverse the clinical features of cardiac glycoside poisoning within 20–30 minutes.[26,27] The dose (in mg) of antibody can be calculated from the amount of digoxin ingested (dose in mg × 60) or the plasma digoxin concentration (concentration in g/l or ng/ml × 0.34 × weight of patient in kg). If the dose required cannot be calculated and treatment is indicated, a speculative dose of 200 mg (5 ampoules) can be given, and further doses of

200 mg added depending on the response. Overdosage with the antibody is at worst likely to cause hypokalaemia if treatment for hyperkalaemia has already been given or may precipitate fast atrial fibrillation or cardiac failure if the patient was taking digoxin for control of these disorders. Allergic reactions do not appear to be a problem. Once Fab antibodies have been given, plasma digoxin assay results are likely to be unreliable.[28]

β-Blocking drugs

In overdose, β-adrenergic antagonists can cause hypotension, bradycardia, cyanosis, coma and convulsions.[29] Deaths can occur with propranolol, oxprenolol, sotalol and acebutolol, but are extremely rare with other β-adrenergic blocking drugs. In severe cases hypotension may be marked and electromechanical dissociation may occur. Electrical conduction disturbances (first degree atrioventricular block and widened QRS complexes) are common. Sotalol can also cause ventricular arrhythmias, which respond to infusion of isoprenaline.

After correction of hypovolaemia, the most appropriate treatment is to give either an intravenous bolus of glucagon 5 mg, followed by an infusion of 1–5 mg/h, or an intravenous infusion of isoprenaline 10–100 g/min. The response should be assessed by the improvement in blood pressure, rather than by change in the pulse rate, and the isoprenaline infusion rate should be increased till the blood pressure improves. Where asystole occurs, external chest compression can be used. Once the patient comes under medical care full recovery should be the rule, provided hypoxic cerebral damage has not already occurred.

Theophylline

Theophylline is usually prescribed in sustained-release formulations, and symptoms of overdose may develop gradually over 6–12 hours. Vomiting, haematemesis, hyperventilation, tachycardia, cardiac arrhythmias, convulsions and coma may occur. Hypokalaemia and hyperglycaemia are the most common biochemical findings. Plasma theophylline levels do not correlate well with symptoms, and so are a poor guide to the severity of toxicity in acute overdose. Patients with levels of 150 mg/l may have minimal symptoms, whereas severe poisoning may be associated with levels of 60 mg/l. Accumulation from chronic overdosage may cause toxicity with theophylline levels of 30–60 mg/l.

When the patient presents early, severe toxicity should be prevented by giving oral activated charcoal and carrying out catharsis with magnesium sulphate, using metoclopramide or ondansetron in sufficient doses to inhibit vomiting. If the patient already has severe toxicity, the same measures should be used, plus full supportive care. In the non-asthmatic patient, intravenous propranolol (5–10 mg i.v. over 1 hour) will reverse the metabolic and cardiovascular changes, but in the asthmatic patient potassium replacement is more appropriate, as propranolol may precipitate bronchospasm. Charcoal haemoperfusion may be indicated in cases with intractable vomiting or severe metabolic or cardiovascular toxicity.

Antidepressants and lithium

Tricyclic antidepressants

Acute overdose with tricyclic antidepressants is the commonest cause of admissions following poisoning to intensive therapy units. Ironically, in most cases they have been prescribed to treat depressed and potentially suicidal patients.

The mechanisms of toxicity are complex and are due to at least three pharmacological effects: an anticholinergic effect, a 'quinidine-like' or 'membrane-stabilizing' effect, and a variety of other neurotransmitter effects, including blockade of noradrenaline uptake at adrenergic synapses, producing initial stimulation followed by α-adrenergic blockade. The most important of these is the quinidine-like effect which slows sodium flux into cells and is the mechanism underlying cardiac toxicity.[30]

Clinical features

There may be initial agitation and hallucinations, giving way to deepening coma. The neurological state may be confusing. The patient may be flaccid, but sometimes the reflexes are brisk, with an extensor plantar response and dysconjugate gaze.

Pupils are more often mid-size than widely dilated. Convulsions are common, and respiratory depression may occur. The most serious and potentially fatal complications are cardiac arrhythmias (most commonly atrioventricular block, ventricular tachycardia and ventricular fibrillation, which may be followed by asystole), and profound hypotension, which may amount to electromechanical dissociation. These can be further aggravated by hypoxaemia (Fig. 72.1). A sinus tachycardia and a widened QRS complex are common. The electrocardiogram (ECG) frequently shows a right bundle branch block pattern, and a QRS duration over 100 ms[31] is regarded as the best indicator of risk of cardiac toxicity. The electrocardiographic changes are so typical that a right bundle branch block pattern in a patient presenting with unconsciousness should raise the suspicion of a tricyclic antidepressant overdose. Plasma levels of tricyclic drugs do not correlate well with the severity of toxicity, though there is little

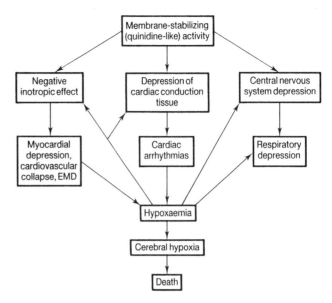

Fig. 72.1 Scheme for the mechanisms by which membrane stabilizing activity of drugs and poisons may lead to fatal toxicity in man.

danger of severe poisoning if concentrations of drug and metabolite combined are less than 1000 mg/l.

Management

Cardiopulmonary status must be assessed urgently and resuscitation commenced if necessary. The most important message about this type of poisoning is that it is potentially reversible, provided hypoxic cerebral damage has not occurred. Full recovery can be hoped for even after prolonged external chest compression, which may be necessary when profound hypotension or cardiac arrhythmias are unresponsive to treatment. Supportive care will be required for the comatose patient, and ventilatory support may be needed for respiratory depression.

Since the anticholinergic effects of these drugs cause delayed gastric emptying, it is worth performing gastric lavage up to 12 hours after ingestion in a seriously poisoned patient. Activated charcoal should then be given via the lavage tube, since as little as 10 g can absorb a potentially fatal amount of antidepressant. Further doses should also be given, since although they have a large volume of distribution and are unlikely to be removed by back-diffusion through the gut ('gastrointestinal dialysis'), there is considerable biliary excretion and reabsorption of tricyclic antidepressants, which can be interrupted by administration of repeated doses of activated charcoal.[32]

In the management of cardiac arrhythmias and hypotension, alkalinization to a pH of approximately 7.5 is effective and should constitute first line therapy.[33,34] A sodium bicarbonate infusion (0.5–2.0 mmol/kg given rapidly and repeated when necessary) may help to correct hypotension and control cardiac arrhythmias or convulsions. Its mechanism of action is controversial, but may involve decreased free drug concentrations due to pH-related change in protein binding or may be due to a direct cardiac effect. Hypotension not corrected by volume challenge may respond to sodium bicarbonate. An α-agonist such as noradrenaline may also be of value; mixed α- and β-agents such as dopamine may worsen hypotension. In EMD and intractable hypotension, physical support of the heart with external chest compression may be required. Cardiac arrhythmias may respond to phenytoin, atenolol or lignocaine, but negative inotropic drugs such as disopyramide, quinidine, and procainamide that prolong cardiac intervals are contraindicated. Direct current shock will be required for ventricular tachycardia or fibrillation, and cardiac pacing may be necessary for bradyarrhythmias causing hypotension. Patients without signs of cardiac toxicity, after 6 hours of observation, do not require extended cardiac monitoring. Monitoring may be discontinued when the patient has regained consciousness and has had a normal ECG for 24 hours. Metabolic disturbances (principally metabolic acidosis or hypokalaemia) should be corrected. Metabolic acidosis from convulsions or hypoperfusion may enhance cardiac toxicity; after ensuring adequate oxygenation, convulsions should be treated with diazepam in the first instance and phenytoin if they persist. Since the anticholinergic effects of tricyclic antidepressants do not cause serious toxicity, the use of physostigmine is no longer recommended. Administration of physostigmine in tricyclic poisoning has produced asystole in human beings[35] and has potentiated toxicity in an animal model.[36]

Monoamine oxidase inhibitors

Several drugs and foods are prohibited for patients taking monoamine oxidase inhibitors to prevent the well known 'cheese' reaction, which consists of a sudden and severe rise in blood pressure due to release of noradrenaline.[37] The hypertensive crisis may be accompanied by headache, vomiting and neurological signs, and is best managed by prompt antihypertensive therapy, ideally by α-adrenergic blockade with a drug such as phentolamine. Overdose of monoamine oxidase inhibitors does not usually produce a hypertensive crisis unless the patient deliberately provokes a reaction. Following an overdose symptoms usually build up over 12–24 hours, with muscle twitching progressing to widespread muscle spasms, trismus and opisthotonus. The blood pressure may vary between hypotension and moderate hypertension; there is usually a sinus tachycardia and the skin is hot to the touch, and sweating profuse. The pupils are fixed and dilated. The core temperature may rise steeply, leading to death from hyperthermia. The muscle spasms may lead to rhabdomyolysis, which can cause renal failure. Disseminated intravascular coagulation is a known complication.

Management

Rapid intravenous infusion of fluid usually restores the blood pressure and reduces the pulse rate. The continued muscle contraction causes excessive heat production, and the body temperature may rise to a point where hyperthermia is fatal, unless muscle relaxants are given. Sedation with diazepam is of little use. If the rectal temperature rises above 39°C the patient should be electively paralysed with a muscle relaxant such as alcuronium or pancuronium and mechanically ventilated for 12–24 hours to correct hyperthermia and prevent rhabdomyolysis. The temperature usually falls rapidly, without the need for any further measures, because the body's thermoregulatory mechanism is usually able to reduce body temperature once the overproduction of heat has been controlled.

External cooling is of little help as heat generation due to profound muscle contraction exceeds heat loss. Hypotension is usually due to hypovolaemia, but a low dose of dopamine can be tried if fluid replacement fails to restore blood pressure.

Amoxapine

Amoxapine (nor-loxapine) is a tricyclic agent which blocks dopamine receptors and has neuroleptic as well as antidepressant activity. Cardiovascular toxicity is usually mild, but this drug is associated with a high incidence of convulsions and CNS toxicity. Acute renal failure due to rhabdomyolysis may also occur.[38,39] Management includes aggressive anticonvulsant therapy and early intubation. Diazepam or thiopentone may be used to control seizures. Phenytoin given in animal models has been shown to prevent seizures when administered prophylactically.

Lithium

Lithium salts are widely used to treat bipolar or manic–depressive disorders. This metal is eliminated by the kidneys, and therefore reduced renal function can lead to accumulation. The therapeutic range is narrow (plasma concentrations between 1.0 and 1.4 mmol/l; blood should not be collected into a lithium heparin tube), and toxicity in therapeutic use can be caused by changes in fluid or electrolyte balance (particularly due to reduction in fluid intake or increased fluid loss from diarrhoea or vomiting). Diuretic therapy or non-steroidal anti-inflammatory drugs may also reduce lithium excretion and provoke toxicity. Symptoms of toxicity include confusion, agitation, drowsiness, tremor, hyperreflexia, hypertonia, ataxia, vomiting, convulsions and, rarely, electrocardiographic changes, diabetes insipidus and acute renal failure. In acute poisoning, plasma lithium levels of over 5 mmol/l may be associated with minimal symptoms, and the elimination half-life is relatively rapid.[40] The development of marked neurological symptoms such as tremor is an indication for active elimination following an acute overdose. Deaths from lithium poisoning are not common, but neurological impairment may be permanent.[41]

Management

Gastric lavage should be carried out up to 4 hours after ingestion to remove any lithium remaining in the stomach. Lithium is not adsorbed by activated charcoal. Lithium is excreted by the kidneys, and a good urine output should be ensured. A saline diuresis enhances lithium excretion. Urinary alkalinization does not increase excretion of lithium, and should therefore not be used.

Haemodialysis should be considered for a symptomatic patient, with a high plasma lithium concentration, following an acute overdose. Although well removed by haemodialysis, lithium has a large volume of distribution, a 'rebound' rise in plasma concentrations may occur, and repeated dialysis may be necessary. Peritoneal dialysis is not recommended.

Antipsychotic agents

These drugs include a number of phenothiazines and butyrophenones. Loxapine overdose can cause treatment-resistant convulsions and cardiac toxicity.[42] The main problems that occur with the phenothiazines and butyrophenones are: dystonic reactions, cardiac arrhythmias and sudden death, neuroleptic malignant syndrome and toxicity from overdose.

Dystonic reactions

Acute dystonia may occur following a single dose of a phenothiazine and often consists of spasm of the tongue and jaw muscles, torticollis and sometimes oculogyric crisis. It is commoner in young female patients. The patient is usually conscious but frightened. Reassurance is essential, and a small intravenous dose of diazepam may help. Procyclidine 5–10 mg intravenously usually causes resolution of the dystonic reaction.

Cardiac arrhythmias and sudden death

Sudden death can occur when high doses of neuroleptic drugs are used. It is most commonly recognized with chlorpromazine, but may occur with any of this group of drugs. It probably occurs because of ventricular fibrillation, but may also occur due to severe hypotension. The electrocardiogram in these patients may show a long Q–T interval, and torsade de pointes may occur. Increasing the heart rate by atrial overdrive pacing or infusion of

isoprenaline will decrease the risk of potentially fatal arrhythmias.

Neuroleptic malignant syndrome

This syndrome consists of relatively acute onset of confusion, muscle rigidity, hyperthermia, sweating, lactic acidosis and rhabdomyolysis, occurring hours to months after commencing therapy with a phenothiazine or butyrophenone drug. This syndrome may present acutely, but in some cases it may run a prolonged course with a differential diagnosis of febrile conditions. The mortality is 20 per cent. The drug treatment of choice is intravenous dantrolene.

Overdose

An acute overdose of antipsychotic agents may cause coma, convulsions, respiratory depression and hypotension. There may be hypothermia or hyperthermia. The electrocardiogram commonly shows a long Q–T interval, and arrhythmias may occur. Symptoms may only become severe 12–24 hours after ingestion. Management is mainly supportive. Intravenous sodium bicarbonate, as for tricyclic antidepressant poisoning, is often effective. Atrial overdrive pacing may correct torsades de pointes unresponsive to antiarrhythmic agents.

Anticonvulsants

Anticonvulsant overdose is most commonly encountered in epileptics. In this case, serial measurements of plasma concentrations of the drug should be made, so that dosage can be restarted, in order to prevent seizures, when the plasma concentration has fallen into the therapeutic range.

Phenytoin

This drug has dose-dependent elimination kinetics, and phenytoin toxicity can occur after regular therapy with inappropriately high doses, or from acute overdose. Clinical features include cerebellar ataxia, confusion and vomiting. Deep coma and respiratory depression are rare. Excessively rapid intravenous administration can cause cardiac arrest due to the propylene glycol vehicle. Gastric lavage, repeated doses of activated charcoal[43] and supportive care are the best means of treatment. Forced diuresis, haemodialysis and haemoperfusion are not necessary.

Phenobarbitone

Phenobarbitone overdose causes prolonged deep coma, and its elimination half-life in overdose may be from 50 to over

100 hours. Pupil size changes from time to time. There may be respiratory depression, vasodilatation and hypotension. Cardiac output is proportional to depth of coma. Ileus due to 'intestinal narcosis' may cause further absorption of drug and deepening of coma as signs of recovery become apparent. Treatment with repeated doses of activated charcoal may shorten the elimination half-life of phenobarbitone. Alkaline diuresis with dopamine has been used, but if poisoning is very severe, haemodialysis or haemoperfusion is the most effective way of removing the drug and shortening the duration of toxicity.

Sodium valproate

Overdose of this drug causes gastrointestinal symptoms, ataxia, irritability, drowsiness and coma. Pupils are constricted. A dose greater than 25 g (adult), or plasma concentrations exceeding 1000 mg/l, may result in convulsions and cerebral oedema. Management is supportive.[44]

Carbamazepine

In overdose there may be a delay in the onset of symptoms due to a cholinergic effect of the drug, enterohepatic circulation, bezoar formation or accumulation of an epoxide metabolite. Since late relapse may occur, close monitoring is necessary. Symptoms include drowsiness, ataxia, coma and myoclonus. The stomach should be emptied, and repeated doses of activated charcoal may increase elimination. Although many of the symptoms of carbamazepine overdose are similar to those of the tricyclic antidepressants, this drug has a small volume of distribution (1 l/kg), and charcoal haemoperfusion may be effective. Plasma levels of the drug should be obtained urgently, and repeated 4–6 hourly in case of delayed or prolonged absorption. Toxic effects may be noted at 12 mg/l, but patients with plasma concentrations of under 60 mg/l can be managed symptomatically. The epoxide metabolite should be measured if possible.

Analgesic drugs

The opioid drugs are considered below under drug abuse. Apart from these the most important cause of serious poisoning from analgesics occurs with paracetamol and the aspirin group of drugs. The other non-steroidal drugs have in general much lower toxicity. Ibubrofen has remarkably low toxicity in overdose. Mefanamic acid is unusual in that overdose may result in convulsions which are usually brief. The pyrazolones, phenylbutazone and oxypenylbutazone, though now rarely used, are toxic in overdose.

Paracetamol

Deaths are rare in comparison with its extremely wide use. The anaesthetist should be aware that although safe in therapeutic use paracetamol overdose may lead to severe hepatotoxicity and nephrotoxicity. Diagnosis of risk depends on the accepted nomogram for plasma paracetamol level against time since ingestion. However, paracetamol plasma concentrations may be misleading in patients with cytochrome enzyme induction by agents such as phenytoin, carbamazepine and phenobarbitone.

Drug and substance abuse

Narcotic analgesics

Abuse of any drug can cause a variety of serious medical complications but the opioids are a common problem. They are readily available as analgesics, cough medicines and antidiarrhoeals and are widely abused, mainly as heroin (diamorphine). Heroin abuse is common in the United Kingdom, and results in 200 deaths per year. Many of these are due to respiratory failure after the first use of the drug, but experienced users may die from acute respiratory depression or aspiration of vomit following a change in supplier (the purity of street heroin may vary from 10 to 80 per cent) or loss of tolerance after a period of abstinence. The dose requirement may fall by 10- to 100-fold over a period of 2–3 weeks of abstinence, so that a dose which previously produced euphoria may prove fatal.

The classic signs of opioid toxicity consist of depressed consciousness (which may amount to profound coma), small or pinpoint pupils (pethidine is an exception to this, and the atropine in Lomotil, a combination of diphenoxylate and atropine, may also cause dilated pupils), and respiratory depression. The respiratory depression is characteristic, and consists of a slowing of respiration with terminal apnoea. The initial management consists of immediate attention to the airway, ventilation and cerebral perfusion. Once the patient is not in immediate danger of death, naloxone should be given. A dose of 0.8–2.0 mg intravenously will usually produce a change in respiratory rate, pupil size and state of consciousness within a minute. A partial response may be related to the size of the overdose, in which case further intravenous bolus doses of naloxone to a maximum of 4–6 mg should be given. Failure to respond may be due to the presence of other central nervous depressant agents, hypothermia or hypoxic cerebral damage. Pethidine, buprenorphine and dextropropoxyphene may require large doses of naloxone. The fact that the patient may be a narcotic addict is not a contraindication to the use of naloxone; opioid withdrawal is not fatal, whereas opioid toxicity can easily kill. The presence of other sedative or hypnotic drugs, hypothermia, hypoxic cerebral damage or other causes of coma may mask the response to naloxone. Diamorphine has a short elimination half-life of 2–4 hours, so prolonged treatment may not be needed. Some other drugs such as buprenorphine, dihydrocodeine, methadone and dextropropoxyphene have considerably longer elimination half-lives, so that a longer recovery period should be anticipated.

Once the diagnosis of opioid toxicity has been confirmed by administration of naloxone, the choice can be made between continuing reversal with an infusion of naloxone sufficient to maintain adequate spontaneous breathing, or controlled ventilation pending recovery while the drug is eliminated by metabolism. As a rough guide, twice the amount required to produce the initial response should be given over the first half hour, and subsequent doses should be titrated against the patient's condition. The patient who regains consciousness may insist on leaving, but should be restrained because the duration of action of naloxone (20–30 minutes) is shorter than the half-life of most opioid drugs, and the patient may relapse. Complications of heroin overdose include a chemical pneumonitis due to aspiration of vomit, non-cardiogenic pulmonary oedema and non-traumatic rhabdomyolysis. Infective complications of intravenous drug abuse include lung abscesses, septicaemia, right-sided endocarditis, hepatitis B and C and HIV infection.

Cocaine

Cocaine abuse produces marked hypertension, and patients may present with confusion, convulsions, myocardial infarction, acute heart failure or cerebral haemorrhage. There is no specific antidote, but control of blood pressure with intravenous nitrates is essential. Other management is supportive. Cocaine is known to lead to premature atheroma[45] and cocaine abuse should be suspected in any young patient with a myocardial infarction.

Amphetamine sulphate and methamphetamine

The amphetamines are frequently abused for their stimulant effect. Amphetamine sulphate may be injected, inhaled or taken orally. Methamphetamine ('ice') usually comes in the form of crystals which are smoked, and are about 20 times as potent as amphetamine sulphate, but its effects are similar. The main problems with amphetamine toxicity include convulsions, tachycardia and hypertension, hyperthermia, rhabdomyolysis and renal failure. Myocardial infarction and cerebral haemorrhage have been reported. There is no specific pharmacological antidote, and management is mainly supportive. Sedation and anticonvulsants will be needed for the patient who is agitated or convulsing. An acid urine markedly shortens the elimination half-life of amphetamine by increasing its renal excretion. However, if there is suspicion of rhabdomyolysis, for example after repeated seizures or if the plasma creatine phosphokinase (CPK) is raised, or if there is myoglobinuria, then alkalinization of the urine is more appropriate in order to prevent myoglobinuric renal failure.

Hallucinogenic amphetamines: MDMA ('Ecstasy'), MDA and MDEA

MDMA (3,4-methylenedioxymethamphetamine), also known as 'Ecstasy', 'XTC' or 'E', is an amphetamine derivative with properties unlike those of amphetamine. MDMA is metabolized to 3,4-methylenedioxyamphetamine (MDA, 'Adam') which is also an abused drug, and 3,4-methylenedioxyethamphetamine (MDEA, 'Eve'), a related drug. MDMA was first patented in 1914, and was briefly used as a mood modifying agent, but was banned in 1985. In the usual dose (up to 200 mg orally), it has few adverse effects in the majority of people, but sudden death from ventricular fibrillation has been reported.[46] Some users develop trismus and muscle pain and stiffness. When the drug is used as a 'dance drug' collapse and convulsions may occur and hyperthermia may lead to disseminated intravascular coagulation and collapse.[47] Initial investigation should include measurement of fibrin degradation products, plasma CPK and urine myoglobin. Initial management consists of restoring fluid volume to facilitate thermoregulation and the use of dantrolene in established heatstroke.[14]

LSD (lysergic acid diethylamide)

LSD is a synthetic hallucinogen, which may occasionally bring the patient to the intensive therapy unit because of hypertension or cardiac arrhythmias, or because of accidents or suicide attempts. If the patient is hallucinating or violent, chlorpromazine is the most suitable drug, but large doses may lower blood pressure, depressing cardiac function.

Volatile substance abuse

Volatile substance abuse (solvent abuse, 'glue sniffing') causes about 150 deaths per year in Britain. The main substances abused are toluene, a constituent of impact adhesives, and butane in lighter fuels. The main causes of death are either a direct effect on the heart, causing ventricular fibrillation, hypotension, respiratory depression or aspiration of vomit. Inhalation of butane may cause hypoxia due to an asphyxiant effect. The very low temperature may cause collapse due to a vagal effect. Management consists essentially of resuscitation ensuring an airway, and providing oxygen; the effects are usually short-lived. Since these agents sensitize the myocardium to noradrenaline, cardiac arrhythmias may respond to intravenous administration of a β-blocker such as atenolol.[48]

Drug couriers

Drug couriers ('body packers') smuggle drugs by bodily concealment. This can be by swallowing condoms, plastic bags or vials of drug (usually heroin, cocaine or amphetamine), or by rectal or vaginal concealment. Leakage results in severe toxicity, and the patient may be moribund when seen. There is usually a history of a recent plane journey. Abdominal x-ray may be revealing.[49] Heroin toxicity can be treated with large doses of naloxone. Cocaine toxicity has no specific antidote. Sodium nitroprusside is probably the most useful agent in a patient with a hypertensive crisis or neurological complications. Removal by purgation is contraindicated. Surgical removal may be the only way to prevent worsening toxicity.

REFERENCES

1. Olson KR, Pentel PR, Kelley MT. Physical assessment and differential diagnosis of the poisoned patient. *Medical Toxicology* 1987; **2**: 52–81.

2. Orr DA, Bramble MG. Tricyclic antidepressant poisoning and prolonged external cardiac massage during asystole. *British Medical Journal* 1981; **283**: 1107–8.

3 Vincent JL, Thijs L, Weil MH, Michaels S, Silverberg RA. Clinical and experimental studies on electro-mechanical dissociation. *Circulation* 1981; **64**: 18–27.

4. Henry JH, Cassidy SL. Membrane stabilising activity: a major cause of fatal poisoning. *Lancet* 1986; **1**: 1414–17.

5. Holt DW, Helliwell M, O'Keeffe B, Hayler AM, Marshall CB, Cook G. Successful management of serious disopyramide poisoning. *Postgraduate Medical Journal* 1980; **56**: 256–60.

6. Krantz T, Thisted B, Strom J, *et al.* Severe, acute propoxyphene overdose treated with dopamine. *Clinical Toxicology* 1985; **23**: 347–52.

7. Nozaki R, Ishibashi K, Adachi N, Nishihara S, Adashi S. Accidental profound hypothermia. *New England Journal of Medicine* 1986; **315**: 1680.

8. Maclean D. Emergency management of accidental hypothermia: a review. *Journal of the Royal Society of Medicine* 1986; **79**: 528–31.

9. Drenk NE, von Staffeldt H. Repeated deep accidental hypothermia. *Anaesthesia* 1986; **41**: 731–3.

10. Althaus U, Aeberhard P, Schupbach P, Nachbur B, Muhlemann W. Management of profound hypothermia with cardiorespiratory arrest. *Annals of Surgery* 1982; **195**: 492–5.

11. Osborne L, Kamal El-Din AS, Smith JE. Survival after prolonged cardiac arrest and accidental hypothermia. *British Medical Journal* 1984; **289**: 881–2.

12. Schaeffer M, Fischer M, Perret M. Hyperkalemia. *Journal of the American Medical Association* 1990; **264**: 1642–5.

13. Guze B, Baxter LR. Current concepts: neuroleptic malignant syndrome. *New England Journal of Medicine* 1985; **313**: 163–6.

14. Hall AH, Kulig KW, Rumack BW. Drug and chemical induced methaemoglobinaemia. Clinical features and management. *Medical Toxicology* 1986; **1**: 253–60.

15. Ward A, Chaffman MO, Sorkin EM. Dantrolene. A review of its pharmacodynamic and pharmacokinetic properties and therapeutic use in malignant hyperthermia, the neuroleptic malignant syndrome and an update of its use in muscle spasticity. *Drugs* 1986; **32**: 130–68.

16. Votey SR, Bosse GM, Bayer MJ, Hoffman JR. Flumazenil – a new benzodiazepine antagonist. *Annals of Emergency Medicine* 1991; **20**: 181–8.

17. Aarseth HP, Bredesen JE, Grynne B. Benzodiazepine-receptor antagonist, a clinical double blind study. *Clinical Toxicology* 1988; **26**: 283–92.

18. Lheureux P, Askenasi R. Specific treatment of benzodiazepine overdose. *Human Toxicology* 1988; **7**: 165–70.

19. Fluckiger A, Hartman D, Leishman B, Ziegler WH. Lack of effect of the benzodiazepine antagonist flumazenil (Ro 15–1788) on the performance of healthy subjects during experimentally induced ethanol intoxication. *European Journal of Clinical Pharmacology* 1988; **34**: 273–6.

20. Bowyer K, Glasser SP. Chloral hydrate overdose and cardiac arrhythmias. *Chest* 1980; **77**: 232–6.

21. Baud F, Barriot P, Tottis V, *et al.* Elevated blood cyanide concentrations in victims of smoke inhalation. *New England Journal of Medicine* 1991; **325**: 1761–6.

22. Senanayake N, Karalliedde L. Pattern of acute poisoning in a medical unit in central Sri Lanka. *Forensic Science International* 1988; **36**: 101–4.

23. Jacobsen D, McMartin KE. Methanol and ethylene glycol poisonings. Mechanism of toxicity, clinical course, diagnosis and treatment. *Medical Toxicology* 1986; **1**: 309–34.

24. Swartz RD, Millman RP, Billi JE, *et al.* Epidemic methanol poisoning: clinical and biochemical analysis of a recent episode. *Medicine* 1981; **60**: 373–82.

25. Crouzette J, Vicaut E, Palombo S, *et al.* Experimental assessment of the protective activity of diazepam on the acute toxicity of chloroquine. *Journal of Clinical Toxicology* 1983; **20**: 271–9.

26. Zucker AR, Lacina SJ, Das-Gupta DS, *et al.* Fab fragments of digoxin-specific antibodies used to reverse ventricular fibrillation induced by digoxin ingestion in a child. *Pediatrics* 1982; **70**: 468–71.

27. Smith TW, Butler Jr VP, Haber E, *et al.* Treatment of life-threatening digitalis intoxication with digoxin-specific Fab antibody fragments. Experience in 26 cases. *New England Journal of Medicine* 1982; **307**: 1357–62.

28. Gibb I, Adams PC, Parnham AJ, Jennings K. Plasma digoxin:

assay anomalies in Fab-treated patients. *British Journal of Clinical Pharmacology* 1983; **16**: 445–7.

29. Weinstein RS. Recognition and management of poisoning with beta adrenergic blocking agents. *Annals of Emergency Medicine* 1984; **13**: 1123–31.

30. Pentel PR, Benowitz NL. Tricyclic antidepressant poisoning: management of arrhythmias. *Medical Toxicology* 1986; **1**: 101–21.

31. Boehnert MT, Lovejoy FN. Value of the QRS duration versus the serum drug level in predicting seizures and ventricular arrhythmias after an acute overdose of tricyclic antidepressants. *New England Journal of Medicine* 1985; **313**: 474–9.

32. Swartz CM, Sherman AJ. The treatment of tricyclic antidepressant overdose with repeated charcoal. *Journal of Clinical Psychopharmacology* 1984; **4**: 336–40.

33. Brown TC, Barker GA, Dunlop MG. The use of sodium bicarbonate in the treatment of tricyclic antidepressant-induced arrythmias. *Anaesthesia and Intensive Care* 1973; **1**: 203–10.

34. Molloy DW, Penner SB, Rabson J, *et al.* Use of bicarbonate to treat tricyclic antidepressant-induced arrhythmias in a patient with alkalosis. *Canadian Medical Association Journal* 1984; **130**: 1452–9.

35. Pentel P, Peterson CD. Asystole complicating physostigmine treatment of tricyclic antidepressant overdose. *Annals of Emergency Medicine* 1980; **9**: 588–90.

36. Vance MA, Ross SM, Millington WR, *et al.* Potentiation of tricyclic antidepressant toxicity by physostigmine in mice. *Clinical Toxicology* 1977; **11**: 413–21.

37. Lippmann SB, Nash K. Monoamine oxidase inhibitor update: potential adverse food and drug interactions. *Drug Safety* 1990; **5**: 195–204.

38. Litovitz TL, Troutman WG. Amoxapine overdose – seizures and fatalities. *Journal of the American Medical Association* 1983; **250**: 1069–71.

39. Jennings AE, Levey AS, Harrington JT. Amoxapine-associated acute renal failure. *Archives of Internal Medicine* 1983; **143**: 1525–7.

40. Dyson EH, Simpson D, Prescott LF, Proudfoot AT. Self-poisoning and therapeutic intoxication with lithium. *Human Toxicology* 1987; **6**: 325–9.

41. Hansen HE, Amdisen A. Lithium intoxication (report of 23 cases and review of 100 cases from the literature). *Quarterly Journal of Medicine* 1978; **186**: 123–44.

42. Petersen CD. Seizure induced by acute loxapine overdose *American Journal of Psychiatry* 1981; **138**: 1089–91.

43. Mauro LS, Mauro VF, Brown DL, Somani P. Enhancement of phenytoin elimination by multiple-dose activated charcoal. *Annals of Emergency Medicine* 1987; **16**: 1132–5.

44. Dupuis RE, Lichtman SN, Pollack GM. Acute valproic acid overdose – clinical course and pharmacokinetic disposition of valproic acid and metabolites. *Drug Safety* 1990; **5**: 65–71.

45. Escobedo LG, Ruttenber AJ, Anda RF, Sweeney PA, Wetei CV. Coronary artery disease, left ventricular hypertrophy, and the risk of cocaine overdose death. *Coronary Artery Disease* 1992; **3**: 853–7.

46. Dowling GP, McDonough ET, Bost RO. 'Eve' and 'Ecstasy' – a report of five deaths associated with the use of MDEA and MDMA. *Journal of the American Medical Association* 1987; **257**: 1615–17.

47. Henry JA, Jeffreys, Dawling S. Toxicity and deaths from 3,4-methylenedioxymethamphetamine ('ecstasy'). *Lancet* 1992; **1**: 384–7.

48. Flanagan RJ, Ruprah M, Meredith TJ, Ramsey JD. An introduction to the clinical toxicology of volatile substances. *Drug Safety* 1990; **5**: 359–83.

49. Lancashire MJ, Legg PK, Lowe M, Davidson SM, Ellis BW. Surgical aspects of international drug smuggling. *British Medical Journal* 1988; **296**: 1035–7.

Section Four

Anaesthesia and Society

Chapter 73

Medicolegal Aspects: an American Perspective

Donald A. Kroll

The American legal system	Malpractice issues
The American health care delivery system	The elements of negligence
Non-malpractice medicolegal issues	Summary

The practice of medicine has many interfaces with the law. These occur because the State is responsible for the protection of its citizens and regulates the behaviour of citizens with respect to society. The degree to which the State regulates a person or group reflects the balance among the values of the people, the perceived need for regulation and the relative consequences of the acts of a person or group. Health care is highly valued by the people; there is, therefore, a perception that it is necessary to regulate the behaviour of physicians and hospitals because the consequences of a physician's actions are serious for the patient. It should not be surprising, therefore, that the State monitors a physician from the moment of matriculation to a medical school to the moment practice ceases. Such monitoring takes the form of: licensing, practice regulation by business and professional codes, required reports, National Practitioner Database maintenance, and determining the conditions for reimbursement of services to State-sponsored patients.

The State also makes provisions under its civil codes, which enable its citizens to resolve disputes about the extent and quality of services rendered. If the question concerns a civil wrongdoing, tort law applies. Negligence is one type of tort, and malpractice is one type of negligence. Increasingly complex forms of health care are leading to increased activity in the application of contract law to the practice of medicine. This in turn increases the interest of the State in protecting the rights of its citizens with respect to the business of medicine, and an increased application of antitrust laws to medical practice.

New technologies continue to emerge and to be applied in the practice of medicine. The application of a new technology may create new interfaces between medicine and jurisprudence. In some cases the interface may broadly overlap with ethics and economics, such as occurs with transplantation surgery and fetal interventions, whilst in others there is an overlap with business law, such as the use of computerized databases in health care delivery. The introduction of a new technology may change the standards of anaesthetic practice, as has been the case with pulse oximetry.

The practice of anaesthesia is neither insulated from nor immunized against the dynamic evolution of medical jurisprudence. Anaesthetists have been in the forefront of change in some areas, but are only indirectly involved in others. A detailed discussion of all aspects of the interface of law and medicine is beyond the scope of this chapter, but there are some recurring issues of interest to anaesthetists, and some dynamically evolving issues of major importance to the practice of anaesthesia. This chapter provides an overview of the medicolegal aspects of anaesthetic practice from an American perspective. It is important to discuss the American legal system and health care delivery system for readers who may not be familiar with these systems. The subsequent sections will deal first with non-malpractice issues, and then with malpractice in anaesthesia.

The American legal system

The issues presented in this chapter will be bewildering to the non-American reader unless there is at least a cursory description of the American legal system. Medical practice in the USA is governed not only by laws and statutes but also by a variety of rules and regulations which have the force of law. Executive orders and administrative regulations are not laws in the sense of being enacted by a legislative body, but have the same compliance requirements as laws because the agency promulgating them is empowered to do so by a legislative body or the Constitution. The Constitution's Tenth Amendment established broad latitude for the individual states to self-govern. This is why the specific laws governing the conduct of a physician vary from state to state. The derivation of laws will be addressed superficially in the appropriate sections and the reader is referred elsewhere for a detailed discussion of the sections of the US Constitution pertinent to medicine.[1] The day-to-day practice of anaesthesia is influenced far more heavily by rules and regulations derived from these sources than by malpractice law.

In addition to the statutory law discussed above, the American legal system also provides a body of case law which represents the legal principles derived from judicial decisions. Most American case law is derived from English common law principles. Common law refers to those unwritten laws grounded in custom and sanctioned by usage. Under this system, people have certain rights and obligations in their interactions with other people. When a person believes that someone has violated his or her rights, they may sue under the civil law system. Civil liability implies a wrongful act against a person's rights, whilst criminal liability implies a wrongful act against society. The term 'person' refers to a legally recognized entity, which may be a corporation or a group. In practice, the two areas of civil liability of importance for physicians are contract law and tort law.

Under the American system of jurisprudence, malpractice is an unintentional civil wrongdoing (tort), specifically the tort of negligence. Although other legal theories have been applied to medical malpractice, the tort of negligence is the basis for virtually all malpractice claims. Under negligence theories, the conduct of the defendant is measured against arbitrary standards of reasonableness and prudence, which in the case of physician conduct are determined by other physicians (expert witnesses). The rationale for the requirement of expert testimony is that the scope of medical knowledge is beyond the scope of knowledge of lay jurors, and hence requires interpretation.

In order to be successful in a malpractice action, the patient-plaintiff must prove each of the four elements of negligence:

1. *Duty* – the patient-plaintiff must prove that the anaesthetist owed him or her a particular duty or obligation.
2. *Breach of duty* – the patient-plaintiff must show that the anaesthetist failed to fulfil his or her obligation.
3. *Causation* – the patient-plaintiff must demonstrate that a reasonable and close causal relation exists between the anaesthetist's acts and the resultant injury.
4. *Damages* – the plaintiff must show that actual damage resulted because of the acts of the anaesthetist.

Proof of each of these elements is essential if the plaintiff is to prevail. Proof in civil law differs from proof in criminal law: in a civil procedure one need only prove that something is more likely to be true than not whereas criminal law requires certainty to beyond a shadow of doubt. Indeed, a 51 per cent probability is sufficient in civil trials. The American Society of Anesthetists has published a concise monograph on the topic of malpractice liability.[2]

The American health care delivery system

The delivery of anaesthetic care occurs in public and private hospitals and in free-standing surgery centres and offices. Anaesthetists may be reimbursed on a fee-for-service (FFS) basis or be paid a salary. Recent trends towards a new system of reimbursement have embroiled anaesthetists in legal battles to retain their hospital privileges and prevent replacement by new groups. It is therefore important to understand the development of managed care medicine.

Managed care medicine

There is only one reason for the change to managed care systems – cost containment. It has been said that the end for fee-for-service medicine was sounded when the automobile industry discovered that it was spending more for health care than for steel. In 1960 health care consumed 5.2 per cent of the gross national product (GNP). By 1980 it was 9.1 per cent of the GNP, and in 1986 had increased to 10.9 per cent of GNP. We are currently in the teens. Expressed in another way, the average per-capita costs of a worker's medical bills have increased from $142 in 1960 to $1054 in 1980, $1724 in 1985 and $3217 in 1990. It is projected to be $15 463 by the year 2000. Clearly, both the federal government and the private sector are strongly motivated to curb what they view to be runaway health care costs. Even though physician reimbursement accounts for only 20 per cent of all costs, physicians are regarded as the decision-makers in the process, and hence the controllers of costs. Thus, the central thrust of the movement towards cost-containment has been directed at modifying physician behaviour both through utilization review mechanisms (audit) and by changing the economic driving forces for physicians by capitated payment schemes.

Anaesthetists do not control the decision to operate, but are needed once the decision has been made by the surgeon. Therefore anaesthetists have no effective bargaining power in the process. This is why there was a major move by the Health Care Finance Administration (HCFA) in the mid-1980s to consider the 'hospital-based' physicians as part of hospital services and subject to a diagnosis-related-group (DRG) payment system, a form of capitated payment. This was coupled with an attempt to make the assignment of Medicare cases mandatory (see below).

The two extremes of health care reimbursement are represented by fully retrospective FFS on one hand, and fully prospective capitated payment on the other. These are functional descriptors of the process, and there are business forms of organization that have developed to follow function. The sole-proprietor private practitioner characterizes FFS, whilst the staff health maintenance organization (HMO) characterizes capitated payment. As third-party payers began attempting to extend the perceived cost advantages of the HMO toward the FFS sector, several new business forms appeared which could fulfill this function. These organizations – preferred provider organizations (PPOs), independent practice associations (IPAs), networks, along with the traditional HMO – comprise what has become known as 'managed care' systems. The functional distinction is that FFS provides health care services but not financing, whilst managed care systems provide both services and financing. The distinction between an HMO and other managed care systems is that an HMO requires the use of a plan member physician, whereas other forms such as a PPO provide incentives to use plan physicians and hospitals in the form of less patient cost. It is anticipated that within a very few years organizations that provide financing without services (the classic indemnity health insurance plans) will cease to exist. These changes in health care insurance have fundamentally altered the *business* of medicine by changing the physician's customer from the patient to the payer.

In managed care medicine the potential patient selects an insurance plan offered by an employer or may purchase one privately. Federal law requires (within certain limits) that both public and private employers, with 25 or more employees, must offer their employees the option of HMO membership. In 1990, 37 per cent of all employers offered a PPO option. This federal requirement is undoubtedly responsible for a large part of the rapid growth in HMO enrolment in the late 1980s, the majority of which is accounted for by IPA HMOs rather than group HMOs.

The selection of a plan by the patient is driven by cost unless quality is a significant issue. The incentive for patients is that, by giving up the autonomy of provider selection, the cost, to them, is less. The greater the loss of autonomy, the greater the personal savings. Many patients view PPOs or other managed care schemes as the best balance between provider selection and cost. This is why managed care plans advertise that they have the greatest number of highly skilled physicians and quality hospitals in their plan; they are trying to appeal to their customers (patients and businesses) by emphasizing that the loss of selection autonomy is minimal or of no importance.

Once enrolled in a managed care plan, a patient may receive a particular anaesthetist's services only if the anaesthetist is a member of the plan, or if the patient is willing to pay a substantial surcharge out of pocket. From a practical perspective, the only way a hospital or physician group may influence the patient is at the initial plan selection stage or by providing a service unavailable elsewhere.

Although the health care reimbursement industry has been able to adapt by creating new business organizational forms, physicians have been slow to respond. Perhaps this is a greater problem for anaesthetists because they tend to work alone and are unaccustomed to collaborative arrangements. The result is that unorganized individuals have little strength against a well-organized PPO. The PPO may control a large share of the local market whereas the individual doctor controls only a tiny fraction of the care delivery. The formation of IPAs is in a sense the physicians' answer to the power of large PPOs. By joining together, physicians hope to control a greater share of the delivery and thereby gain negotiating power.

Non-malpractice medicolegal issues

Contracts

A contract is a legally enforceable agreement between two or more persons in which there is an exchange of promises to either do or not do something. The term 'persons' may mean individuals or groups of individuals or corporate entities. Although there are many types of contracts that may be important to anaesthetists, contract issues are generally becoming major factors in the relation between the anaesthetist and the hospital.

The American Society of Anesthesiologists (ASA) regards the practice of anaesthesia as the practice of medicine and specifically excludes the idea that it may be considered as an institutional service. The ASA has taken the position that anaesthetists, in common with other physicians, must be free to enter whatever contractual relations they wish but that, as a general principle, the ability of an anaesthetist to practise in a hospital in which other clinical departments have 'open' staffs should not be arbitrarily restrained by requirements to join a group or enter into a financial arrangement with the hospital. Hospital administrators would prefer to consider the practice of anaesthesia as a hospital service and contract with anaesthetists for the provision of this service. Anaesthetists who are unwilling to enter into contracts with hospitals may face the loss of their privileges, although this has been difficult to enforce owing to 'due process' protection. The Fifth Amendment to the Constitution establishes that individuals may not be deprived of their rights without due process of law. The Fourteenth Amendment extends due process protection to the states. Because hospitals receive a great deal of federal funding, and are considered to be involved in interstate commerce, medical staff privileges could not be revoked without due process.

Mandatory assignment

The federal government is a large provider of health insurance through Medicare and Medicaid plans. The effect of reimbursement decisions by a federal or state provider carries the weight of any other administrative decree and is effectively a law. The penalties for not participating in the administrative regulation are severe because the government agency may link the availability of funds for hospitals to the agreement to provide services at the rate specified. Initially, only hospital costs were included in these plans, but there have been attempts at both federal and state level to link mandatory acceptance of Medicare assignments by physicians to the retention of their licensure. The net effect is that physician fees will be prospectively determined by the government agency and physicians will be forced to accept the fee offered. It seems fairly obvious that the fee offered will be substantially less than what the physician would otherwise charge. These attempts have been thwarted so far because the Constitution allows only a state to regulate business activities within the state. Thus, the granting of a licence to a physician is a state and not a federal function. In a sense, linking the licence to the acceptance of federally funded patients at predetermined fees is a generalized form of what individual hospitals have attempted by means of 'economic credentialling'.

Economic credentialling

In any business system there are external factors which must be considered and may serve to thwart even the best of plans. In anaesthetic practices the most potent of these external influences is the threat of economic credentialling. This term refers to attempts by hospitals or medical staffs to withdraw the privileges of anaesthetists who refuse to accept large discounts in reimbursement. In some cases, entire groups of anaesthetists have been replaced. The hospital management, of course, attempts to disguise this manoeuvre using as an excuse inadequate performance by the existing group, and in some cases has been able to present good evidence to establish the truth of this claim.

There have been several legal cases in recent years that have established the liability of hospitals for the negligence of physicians who have been granted practice privileges in the hospital. Hospitals are expected to maintain quality control and peer review mechanisms that monitor the quality of care given by physicians. Hospital administrators have reasoned that, if the hospital is to be held responsible for the acts of physicians, it should have control over the actions of those physicians and disciplinary authority over them. Most of the cases, in which a hospital has replaced an anaesthetic group, have been successful because the hospital was able to demonstrate a quality assurance basis for the manoeuvre.

Membership in an IPA or PPO may actually make the process of group displacement for economic reasons easier. The contractual arrangements with IPAs and PPOs depend on the physicians being members of an economic bargaining organization separate and distinct from the medical staff. There is therefore no longer any requirement for due process. The due process protection clauses that are almost universally required in hospital bylaws, as a precondition for the receipt by the hospital of state and federal funds, may not apply to anaesthetists as members of the PPO or IPA. Once involved in this type of arrangement, there is no protection for the anaesthetist against forced fee reductions in subsequent contracts, and there is no protection of medical staff privileges. This means that a physician may be effectively barred from practice without medical staff privileges ever being revoked. An anaesthetist might be perfectly free to practise in the hospital, except for the fact that no patients are referred to him because his membership in the economic unit has been terminated. A more direct approach is to include a clause in the contract stipulating that removal from the contracting unit automatically results in revocation of medical staff privileges.

Antitrust laws

Antitrust laws are becoming more important as a result of developments in exclusive service contracts, managed care medicine and the Health Care Quality Improvement Act. The significance for the anaesthetist in avoiding antitrust liability is that the damages awarded in antitrust cases may be triple the amount of the actual damages, and malpractice insurance does not cover this type of liability.

The foundation of antitrust law is the Sherman Antitrust Act of 1890, although several other statutes such as the Federal Trade Commission Act and the Clayton Act are significant for the practice of medicine. Section 1 of the Sherman Act considers unlawful any contract, combination, agreement or understanding which unreasonably restrains trade in interstate commerce. Section 2 deals with monopolies. Since virtually any contract or agreement between individuals or businesses is designed to 'restrain trade' in some fashion, there has never been a literal interpretation of the Sherman Act, and it is precisely the requirement for interpretation that caused the most confusion. Thus, antitrust liability is difficult to assess in advance and often appears capricious in the final determination because of the overwhelming importance of the specific facts of each case.

The 'learned professions' were initially exempt from the provisions of the Sherman Act because the interpretation of the words 'commerce or trade' did not include the provision of professional services. Ironically, we owe the application of antitrust law to medicine to a landmark case involving minimum fee setting in the legal profession, an abuse that eliminated the learned professions exemption. That medicine may be subject to federal interstate commerce regulation is obvious, one need only consider the patient's domicile, location of hospital supplies and insurance carriers or the involvement of hospitals themselves. This leaves two major areas for interpretation: what constitutes a concerted activity, and when does restraint become unreasonable.

Concerted activities

A single individual cannot act in restraint of trade because such activity can exist only with two or more legal entities which interact by means of a contract, combination or conspiracy. There have been many interpretations based upon individual cases, but membership in a professional association has been held to be sufficient to satisfy the requirements for liability. Although professional associations are entitled to exclude members for reasons of competence or ethics, forbidding association with a particular group may be viewed as a boycott, especially when applied to economic competitors. This has been the basis of antitrust cases involving disputes between anaesthetic groups and nurse anaesthetists (a section 2 violation), and the 1979 challenge to the ASA relative value scale (a section 1 violation). There is little doubt that membership in an economic unit such as a PPO or IPA would satisfy the concerted activities requirement, and there is ample precedent that organized group practices likewise qualify.

Unreasonable restraint

Not all restraint of trade is 'unreasonable'. There are two tests that may be applied. The 'per se' rule applies when the activity is so likely to restrict competition that all one must prove is that the activity existed. There is no need to prove that any actual restraint occurred, although mitigating

gains in efficiency may be considered. Price fixing and market allocations are considered to be 'per se' violations, but group boycotts and tying arrangements are now generally considered under the 'rule of reason'.

Under rule of reason analyses, the court will weigh the anticompetitive effect against the potentially legitimate purpose of the restraint. Application of the rule of reason analysis to medical practice has been restricted to cases where the discriminatory behaviour was based on the so-called 'patient care defence'.[3] As the application of this type of analysis to section 1 cases involving the health care industry increases, so does the importance of determining proof of market power, because actual harm to competition must be shown. It is difficult to demonstrate harm caused by the actions of a single group unless the group has clear market power.

Exclusive contracts

There have been several antitrust cases involving exclusive service contracts with hospitals. In a manoeuvre that may be related to economic credentialling (see above), the hospital first disallows practice by anaesthetists who are not members of the group with whom it has a contract. The excluded anaesthetist then alleges that the group has established an illegal monopoly by tying the provision of anaesthetic care within the hospital to membership of the group. The hospital responds that it is only by establishing an exclusive contract that quality patient care may be assured.

It must be proved that the hospital requires patients, who use its operating rooms, to purchase anaesthetic services solely from the group in order to establish a tying arrangement. No tying arrangement exists unless it limits the number of anaesthetists eligible to become members of the group.

Whether or not the hospital has market power is also important. If it can be shown that the hospital has sufficient market power to force patients to select only from their group, the 'per se' rule against tying arrangements might apply. It would not be a sufficient defence that the restraint was adopted to improve patient care (a rule of reason approach).

Economic bargaining units

Preferred provider organizations (PPOs) may be formed by brokers, purchasers or providers. Broker-type PPOs are not likely to have antitrust problems as long as they avoid appearing to become the agent of either the purchaser or the provider. Purchaser PPOs may encounter price-fixing or group boycott antitrust problems if more than one payer is involved in control of the PPO. Even a single purchaser PPO might encounter exclusive dealing, territorial restraint or tying arrangement problems, particularly if it has market power.

A provider-sponsored PPO is most at risk for antitrust problems if it is viewed as a horizontal arrangement among competing providers. If it is viewed as a joint venture, it becomes a single legal entity and no concerted activity can be shown.

Individual IPAs and other novel types of economic bargaining units will be analysed using the same principles as for a provider PPO. The IPA must be viewed as a bona fide joint venture that offers some new capabilities. The restraints in question must be reasonably related to the function of the joint venture, and the overall market power of the joint venture must not be so large as to be excessively anticompetitive.

Antitrust immunity

Since membership in a professional association may be sufficient evidence of concerted activity, and professional associations are increasingly involved in lobbying activities, it is significant that lobbying efforts are protected from antitrust litigation. The Noerr–Pennington Doctrine[3] exempts those seeking to influence government action or legislation even if their sole purpose is anticompetitive. Attempts to influence private associations are not exempt.

The second area of immunity is peer review activity under the State Action Doctrine.[4] Under the State Action Doctrine, hospitals have a common law duty to exercise care in selecting qualified medical staff. States regulate and supervise the process, and therefore are immune as long as they actively supervise the policies and the policies are clearly articulated and expressed. The Health Care Quality Improvement Act of 1986 (HCQIA) was a direct result of an antitrust action involving hospital peer review activities, and attempts to define the conditions under which peer review is exempt. The Act contains provisions that serve to broaden the applicability of the State Action Doctrine to include all medical peer review, and also to specify standards of fairness for peer review.

The HCQIA also, in an attempt to protect the public from incompetent physicians, established the National Practitioner Data Bank for Adverse Information on Physicians and Other Health Care Practitioners.

The National Practitioner Data Bank (NPDB)

Although the NPDB was established as part of the HCQIA in 1986, full promulgation into law did not occur until 1989,

and implementation was delayed even longer. In its final form the Act also included the requirements of the MediCare/Medicaid Patient and Program Protection Act of 1987.

The NPDB was established by the HCQIA as a nationwide information system because the licensure of physicians is a proper function of the states, and also because fleeing from one state to another was a recognized method of avoiding loss of medical livelihood. This theoretically should provide licensing boards and hospitals with improved access to adverse information about physicians. Simply moving into another state would no longer provide a haven for physicians.[5]

Mandatory data entry became effective only on 1 September 1990, so it is too early to know what impact the NPDB will have. Attempts to modify some of the provisions of the HCQIA continue, and it is therefore likely that the criteria for mandatory reporting will change.

Malpractice issues

The germination of a suit

All patients may be considered to be potentially litigious. There are several factors that will determine whether or not a patient will become a plaintiff, and several others that will determine whether the suit will be successful. The idea that there is a subset of patients who are more likely to sue than others is probably inaccurate. The so-called 'professional plaintiff' is very rare.

Perhaps the most direct evidence of a litigious patient is one who is currently involved in a malpractice lawsuit, or has been a plaintiff in a prior malpractice case. Unless a physician asks the question directly, or the patient volunteers the information, there is no way to determine the legal history of a patient.

A second category of patients who may be likely to sue are those who have had an adverse outcome from previous care or who have had improper or questionable care. This might include delayed or improper diagnosis, delayed treatment, improper therapy or an adverse reaction to treatment. Such patients may not yet have initiated a lawsuit but, depending upon the degree of injury or damages, they may be considered as highly likely to sue. Even though the current physicians may not have been directly involved in the initial injury, *all* physicians whose names appear in the records may be sued.

There are other clues which may indicate that a particular patient is more likely to sue. Any patient who appears hostile to a physician or the hospital may be considering a lawsuit. Patients who take copious notes or tape record an interview are suspect, as are those who exhibit 'doctor-shopping' behaviour or have numerous admissions to different hospitals for the same complaint. Unusually demanding patients and those who insist upon dictating treatment choices may be impossible to satisfy, and are therefore likely to sue.

The overriding factor that determines whether or not a patient sues a doctor is his or her satisfaction with the care provided. No patient comes to a doctor with the intent of suing him or her; they come with the expectation of receiving good medical care for their problems. Only unhappy patients sue their doctors. The satisfaction of the patient is influenced by many factors, and there are physician, patient, media, attorney and economic causes for the increasingly litigious patient population. Depersonalization of health care, unrealistic patient expectations resulting from lay medical literature and the media, and the pro-plaintiff legal system all contribute to the problem.

On the other hand, the degree of expected damages is probably the major factor in the likelihood of a lawsuit. The greater the claim for damages, the greater the probability of a suit. Stated simply, if no harm has come to the patient, the probability of a successful suit is small. Barring the occurrence of the million-dollar anaesthetic disaster, the anaesthetist may be able to influence directly the probability of a suit. It has been estimated that only one in eight negligent acts results in a suit. The prevention of a suit in the face of an adverse outcome is a function of risk management.

Risk management strategies[5]

Improve the doctor–patient relationship

Communication with the patient is essential. This requires to both talk and listen, with a willingness to listen being the more important. It is important to give patients the opportunity to ask questions and to make sure that their consent to proceed is as fully informed as possible. It is also important to be truthful in assessing risks and benefits in order to avoid offering guarantees of safety or efficacy or warranties of treatment.

Anaesthetists are more likely to be sued than primary care physicians, for several reasons. Perhaps the most significant is that the patient is (hopefully) unaware of us for most of the time that we are caring for them. The doctor–patient relationship is quickly established and even more quickly forgotten. There is very little, if any, involvement with the patient's family. The anaesthetist, because of this, is not always recognized as a dedicated and skilled physician. It is much easier to blame the anaesthetist for a less than perfect outcome than to blame the surgeon or the internist who spent hours at the patient's bedside and offered comfort and consolation to a worried or bereaved family. Although the obvious solution would be to extend our involvement with the patient and the patient's family, it is impractical in most cases to make daily visits to every patient while they are hospitalized or to increase the duration of preoperative visits sufficiently to include a discussion of every conceivable eventuality, every remotely possible complication and every alternative method or technique.

It is possible, however, for the anaesthetist to increase awareness of the problem, and to improve contact with patients who seem unusually anxious, ill or prone to a specific complication. Spouses and family members may be informed separately about the risks when patients request not to be informed. This is more of a 'public relations' technique than a legal safeguard, but it is effective in that it establishes good rapport with the family, and may have substantial medicolegal 'prophylactic' value. It must be remembered that, from a legal point of view, in informing the spouse or family member, the anaesthetist is not relieved from the burden of informing the patient unless the spouse or family member is the legal guardian of the patient.

Even if the anaesthetist does not personally conduct a full physical examination or take a complete history, the patient must be told that his or her condition is known about from other sources, and the medical record should include this information in the preoperative note. Active follow-up is essential and full explanations should be offered. The anaesthetist who gives the impression of being unconcerned or disinterested when a complication occurs, which may be related to anaesthesia, may suffer a lawsuit.

It is also important for the doctor to project a professional image to patients. Appearance and demeanour at the bedside are major contributors toward establishing this image. If the patient and his or her family have the initial impression that the doctor is sloppy, flippant, careless or poorly informed, they are not likely to be very forgiving of a less than perfect outcome. Likewise, it is important to guard against careless conversation in the hospital corridors and in the operating room.

If it becomes apparent, after the preoperative interview with the patient, that a reasonable doctor–patient relationship cannot be established, the anaesthetist should consider very carefully whether or not to provide the anaesthetic care. The option of refusing may not be available in an emergency. Nevertheless, there will be situations in which it would be unwise to become involved. One example is the patient who insists on dictating the type of anaesthetic, despite the anaesthetist's best judgement. A similar situation occurs when the patient's religious beliefs prohibit the use of blood products. The anaesthetist who cannot agree with this wish should not accept responsibility for the care of the patient. The important point is that the anaesthetist should avoid situations in which the provision of the best possible care is compromised.

Another valid reason for the anaesthetist refusing to provide care for a patient is lack of the necessary skill, experience, equipment or facilities to deliver adequate care.

Adhere to a 'standard of care'

One of the tests of negligence is whether or not the anaesthetist adhered to the standard of care in the treatment of the patient. The standard is often determined retrospectively, by review of the records. It may therefore not be feasible in all cases to know what the 'standard' is. In practical terms this requires that one keep current in knowledge and provide medical care consistent with this knowledge. This does not mean that the anaesthetist must be continually at the cutting edge of medical research, but it does require that one remains aware of, and conforms to, accepted guidelines for the provision of anaesthetic care. The American Society of Anesthesiologists has published general guidelines for the practice of anaesthesia, and the anaesthetist may be held accountable to these guidelines or to departmental guidelines or hospital procedures.

In terms of the medical and anaesthetic management of patients, adhering to the standard of care requires that the choice of agents and techniques is appropriate, and that the anaesthetist is competent in the use of these agents and techniques. It is necessary to know the contraindications for the use of drugs and anaesthetic agents. An anaesthetist who uses a barbiturate to induce anaesthesia in patients with acute porphyria or who gives suxamethonium (succinylcholine) to a burnt patient or to a quadriplegic, and who uses halothane in the presence of elevated hepatic enzymes, following the previous use of the agent, is likely to be sued successfully. If, on the other hand, one is prudent in the choice of agents, the likelihood of a successful legal action is decreased.

The appropriately chosen agents and techniques must be used appropriately. As a general rule, an anaesthetic record which shows that the vital signs were maintained within a reasonable range for the patient is evidence of the appropriate conduct of anaesthesia.

Keep good records

The importance of good record keeping cannot be overemphasized. It is of little benefit in legal terms to have delivered good anaesthetic care if it is impossible to identify what was done and when it was done. The record may have to be defended years later, long after the specifics

of the case have been forgotten. A general rule is that if it wasn't written, it wasn't done. When an incident occurs during the conduct of anaesthesia or during the care of the patient, the facts should be documented in the medical record. Avoid using terms such as 'inadvertently', which convey a message of guilt or negligence. Do not make written comments that admit wrongdoing, or are accusatory of others. Simply record the pertinent facts about the incident as accurately and completely as possible, including observations such as skin colour, colour of blood in the operative field, auscultatory findings or other information not typically charted on the anaesthetic record. If the problem occurred during the course of a regional or conduction technique, indicate the areas of anaesthesia or spinal level of somatic block. It is often helpful to include a differential diagnosis of the problem in a note on the chart, the laboratory tests or observations performed to make the diagnosis, and the treatments initiated. A frequent complaint is that there was no time to write notes while responding to an emergency situation. The solution is to write a 'perioperative' note as soon as possible, making reference to the approximate times of the events. Such records, written at the time of occurrence, are invaluable in distinguishing the difference between a bad result or known complication and actual negligence. It should be obvious to every physician that records should *never* be altered after the fact, but instances of record alteration continue to occur. These suspicious records cannot be defended in court, and may result in the loss of a defendable legal case or even a criminal prosecution for fraud. If an error is made in record keeping, a line should be drawn through the error, leaving it legible, and the correction initialled and timed.

Respond appropriately when an incident occurs

In addition to documenting the facts, there are other options that may help to prevent legal action. The most important is to ensure that optimal medical care is provided. Consultation should be obtained where appropriate to ensure that all diagnostic and therapeutic steps have been taken and that the continued care of the patient is provided by the most suitable specialists. One consultation frequently overlooked is that with another anaesthetist at the time when an adverse situation is occurring. When one is asked to help another anaesthetist in an emergency, and if it is believed that the patient may suffer harm because of the incident, a note should be made in the record verifying the events.

Many hospitals now use a system of risk management whereby specially trained personnel are available to intervene when a question of liability arises. These personnel are described by various titles such as 'patient–staff relations', 'patient advocates', 'ombudsmen' or 'risk management coordinators'. They may be under the administrative supervision of the hospital attorney or a hospital department such as administration. Such personnel should be notified immediately of any incidents that occur. They will help to collect information about the incident, offer support for the patient and the patient's family, and act as liaison between the family and the medical staff involved. They may be able to reduce the likelihood of a lawsuit or the damages eventually awarded.

The malpractice insurance carrier should be notified of any events that may lead to a lawsuit. Many companies require such notification as part of their insurance agreement. It is most unlikely that one would be adversely affected by giving notice of an incident to an insurance company.

The anaesthetist should continue to follow up the patient, during hospitalization, after any incident that might be related to anaesthesia. The anaesthetist's role is that of a consultant to the primary care physicians. Failure to do so might be construed as abandoning the patient, and at best indicates disinterest and disregard for the welfare of the patient. If the incident is clearly unrelated to the anaesthetic care, this should be made clear in the medical record with the reasons that support this view and that the services of the anaesthetist as a consultant is no longer required but will be available should the need arise.

Recognize malpractice 'prodromes'

Anaesthetists may first become suspicious about a pending lawsuit when they receive letters from former patients or requests for medical records from lawyers. Until a complaint has been served, no lawsuit has actually been filed, and the events are potentially reversible. The anaesthetist should, as a first step, notify his or her insurer, sending copies of the correspondence. The insurer may then be able to advise on how to proceed. Most of the 'angry letter' complaints from former patients are meritless claims, or claims for very low damages because claims with merit and potentially high financial awards will in all probability be in the hands of a plaintiff's lawyer. An example of a low cost claim is the request for payment of dental bills accrued a year after an uneventful anaesthetic. This type of nuisance claim is unlikely to result in a suit, but the anaesthetist may feel compelled to offer a settlement to avoid further interaction with the former patient. If it is decided to offer a settlement privately, this should only be after consultation with a lawyer who can ensure that the appropriate releases from further liability are obtained. If, after consultation with the insurer, it is decided to ignore the claim, a polite letter may be sent in answer to the patient's letter, explaining that no offer can be made in settlement for a meritless claim. The response should not be hostile or inflammatory. The goal is to explain why there is no liability for the claim and to reduce

the chance of a suit being filed. It is important not to appear unsympathetic or callous towards the patient, but direct correspondence with the patient should be kept to a minimum. It is generally best to work through the insurer or attorney.

A request by a plaintiff's lawyer for records may be a cause for anxiety, but should not be construed as meaning that a suit will be filed. The insurer should be notified. The insurer may wish to review the medical records, or may ask to be provided with the medical facts of the case. Other than following the instructions of your insurer, there is little else to do at this time. It is likely that nothing more will be heard about the case unless there is a genuine question of malpractice. It is necessary for the patient to sign an authorization for the release of medical records. Records must not be changed, nor should the records be added to by writing a covering letter. Since anaesthetists, with the exception of those working in pain clinics, usually keep no detailed records apart from those that are included as part of the patient's medical record, it is unlikely that there will be much to send in response to such a request. In general, requests for medical records do not include business records or billing statements.

Avoid vicarious liability

Instructions to ward nurses, circulating nurses or others may expose the anaesthetist to vicarious liability. Such situations occasionally arise as a result of the positioning of the patient, attaching electrocautery grounding pads or the use of various warming devices when nerve damage or burns may occur. The anaesthetist may be considered to be supervising and liable for the actions of anyone acting upon his or her instructions.

The anaesthetist should not agree to supervise anyone who is incompetent. This may seem obvious but the reality of the situation is that well-entrenched personnel may be very hard to remove because of institutional policies or politics. If it is believed that someone who must be supervised is incompetent, supervision must be very close, such as by double-checking preoperative findings, specifying all agents and techniques to be used and being present in the operating room during the entire case. Such extreme 'supervision' may preclude involvement in other cases but this must be weighed against the possibility of preventing injury to a patient with its attendant professional and emotional costs. The department and/or hospital administration should also be consulted in order that incompetence can be documented. No one, except in a teaching hospital, should be supervised who is not fully trained and certified. Anaesthetic 'technicians' or 'aides' are not qualified to provide direct patient care and should never be expected to function in that capacity. The American Society of Anesthesiologists specifies that the only personnel who may provide direct patient care under supervision are residents, student nurse anaesthetists, and certified registered nurse anaesthetists.

There is no agreement about the number of simultaneous cases that may be supervised by a single anaesthetist; however, if as a result of involvement with a patient the anaesthetist is unavailable during an emergency, or has not been supervising closely enough to prevent an adverse event, successful malpractice litigation may result. It may not matter that the result might have been the same even if the anaesthetist had been present for the entire case. Do not agree to supervise more simultaneous cases than can be handled safely. Issues of inadequate supervision and patient abandonment are hard to defend.

Notification of a lawsuit

An anaesthetist may receive a notice of intent to sue within a specified period of time, or may receive a summons and complaint which outlines the alleged wrongdoing. In either case, the first action is to notify the malpractice insurance carrier. There is a finite period of time for a response to the summons, and the assistance of a lawyer will be required. Failure to respond within the specified time will result in a directed decision for the plaintiff, and the malpractice insurer will not be responsible for payment if it has not been notified. Upon notification, the insurer will appoint an attorney to represent the physician.

The anaesthetist should refrain from discussing the case with anyone other than the attorney, as certain statements may be discoverable (see 'Discovery', below) and used by the plaintiff during the trial. Attorney–client discussions are protected as privileged communications.

Access to the patient's medical records for the purpose of reviewing the anaesthetic record or notes is permissible, but the temptation to add notes or improve the record by way of an additional explanation must be avoided. The plaintiff's legal advisers will already have obtained a copy of the medical record, and any alterations after notification of a suit will be taken as evidence of negligence.

It is uncommon to seek legal representation other than that provided by the insurer, but there are some circumstances in which it may be prudent. The physician may require private counsel when there is a conflict between the interests of the insurer and the physician. Such a situation may arise when the potential damages exceed the policy limits, thus exposing the physician's personal assets to risk, or when the physician's privileges or licensure are at risk as issues separate from the lawsuit. Finally, the physician may not have confidence in the attorney provided by the insurer and may ask for a substitute. The physician–attorney relationship will probably extend over one or two years, or even longer in complicated cases, during the process of discovery, so it is important that the relationship is sound.

Discovery

After the complaint is answered, both sides begin the process known as discovery. The purpose of discovery is to ascertain the 'facts' of the case in preparation for trial or settlement. It is important to consider briefly what constitutes a 'fact' in a malpractice trial. When two opposing sides arrive at radically different views about the events and their significance, there must be two sets of 'facts'. One set of facts will eventually be 'proved' to be true, and the other will be rejected. It is the responsibility of the jury to decide which facts to believe. Juries are, for this reason, called 'factfinders'. Judges may also act as factfinders before a case goes to trial. It is apparent that a 'fact' is 'proved' whenever it is determined that its truth is *more likely than not*. Both sides attempt to determine their strengths and weaknesses during discovery. A strength is a fact that is favourable, and a weakness is a fact that is unfavourable.

The medical record is the primary source of the facts. Observations and events that were written at the time of the event are the most credible source of information. Factfinders will believe what is written before they believe what is said. It is assumed that the record was contemporaneous and, therefore, the notes were completed before any adverse outcome was known. Because there is no logical reason to have lied, the records are generally trusted.

A second source of fact is the testimony of those who witnessed the event. The anaesthetist may testify that something was done or seen that was not recorded in the record. This is not as credible as the medical record because it relies upon a specific recall of events that may have happened in the remote past. Opposing lawyers will often follow up this type of recollection with a series of questions about the case that preceded or followed the matter in question in order to establish that the anaesthetist's memory is incomplete. Such recollections tend to be self-serving and are not trusted by the factfinder.

A third source of fact is the usual practice pattern of the anaesthetist. The anaesthetist may not have recorded an action and has no independent recall of the specific case, but asserts the action that would have been taken because it is part of a routine. This obviously does not have great credibility. The maxim that 'if it isn't written, it wasn't done' is difficult to overcome.

A fourth source of fact is expert witness testimony. Experts are necessary because the subject of medicine is held to be beyond the knowledge or understanding of lay jurors and therefore requires interpretation. Experts will review the entire medical record, not just the anaesthetic record, and may also review the depositions of the doctors involved. As excellent perioperative note that includes details of the event may certainly mitigate against a poor or incomplete anaesthetic record but does not substitute for a good anaesthetic record. It has been the author's experience that the order of credibility is:

1. The anaesthetic record and notes.
2. The expert's interpretation.
3. Specific recall.
4. Usual and customary practice.

Although a deposition or statement by the defendant anaesthetist will provide evidence from recall and usual practices that are less credible than the record or the expert's testimony, the statement is of critical importance in establishing the strength of the defence. In a system in which a 'fact' is 'true' once it is believed, the credibility of the doctor is of major importance.

Deposition testimony

The defendant anaesthetist's testimony is taken on behalf of the plaintiff, and is conducted by the plaintiff's attorney. The defence already knows what the doctor will say. The plaintiff's attorney will attempt to uncover facts favourable to the plaintiff to ascertain the defence position on the issues in question. Typically, all the defendants in the case will be represented by counsel at the deposition, and in cases in which the anaesthetist may wish to 'criticize' the hospital or the surgeon, counsel may have substantial interest in protecting their clients from adverse testimony.

The deposition is usually arranged to take place at a mutually convenient time and place. The time of the deposition should be arranged so that the defendant is well rested and there should be sufficient time allowed so that there is no need to rush. In general, it is better to have the deposition in the office of the defence attorney than in the doctor's office. An observant plaintiff's attorney may learn from the diplomas on the wall, the books on the shelf, or the neatness of the office, matters that he would otherwise not be entitled to know.

Prior to the deposition, the defence attorney should meet the anaesthetist to explain the conduct of the procedure, what to bring and what to wear.

Depositions usually begin with introductory comments by the plaintiff's attorney regarding the conduct of the proceeding. These are general statements about the need to speak slowly and clearly, to understand the questions and to wait until the question is fully asked before answering. There will then be a series of questions about the

credentials of the anaesthetist. It helps to have a current curriculum vitae and to be familiar with the dates of training, licence and certifications.

One of the purposes of the deposition is to give the plaintiff's attorney an opportunity to assess the defendant as a witness, so dress and demeanor are important. Conservative, professional attire creates a better impression than casual or flamboyant dress. A serious but relaxed and calm demeanour gives the impression of competence. Questions may be asked, or postures adopted, by the plaintiff's lawyer designed to intimidate or elicit anger. Showing anger or hostility will give the impression that the physician has poor self-control. The plaintiff's attorney may welcome the opportunity to demonstrate that the witness is pompous or arrogant or has inadequate knowledge or is evasive and unresponsive.

It is wise to pause for a moment before answering a question to allow for any interposed objections to be stated. Answers should be brief, and to the point. Do not embellish or expand upon the question asked. If a 'yes' or 'no' is sufficient, stop there. Do not anticipate the questions. If a question is vague or ambiguous, ask to have it rephrased. If a question is not understood, always ask for clarification.

Expert witnesses

The courts have recognized that certain topics are beyond the knowledge and experience of most jurors. They demand that witnesses be called so that the jury can benefit from the expertise and opinions of uninvolved parties for the interpretation of the facts. Witnesses who are allowed to give opinions are called expert witnesses. All other witnesses are restricted in testimony to what was actually observed or done.

An expert witness must first be qualified by the court. The qualifications of an expert are established by their training and experience. All that is required, in most jurisdictions, is that the medical expert is a licenced physician who professes familiarity with the standard of practice of anaesthetists. Occasionally, a doctor who is not a practising anaesthetist may be called to testify in a case involving anaesthesia. The rules vary from state to state and by discipline. For example, in California only an emergency medicine physician can testify against another emergency medicine physician. No other specialty receives this statutory protection. Academic appointments, published papers, memberships of societies, prestigious positions, etc. may be useful in establishing the power of the expert to influence a jury, but are not necessary to be qualified by a court as an expert.

The elements of negligence

Duty

In simple terms, establishing duty identifies that the anaesthetist had an obligation to the patient. This obligation is proved by the existence of a doctor–patient relationship. Once the anaesthetist has agreed to provide care for the patient, the doctor–patient relationship exists. Therefore, duty is usually not a major issue in malpractice. An exception occurs when the anaesthetist acts as a 'good samaritan' or rescuing physician in an emergency situation. In most states, 'good samaritans' have statutory immunity from malpractice suits.

Anaesthetists accept all the general responsibilities owed by any doctor to any patient, and a specific duty to adhere to the standards of practice of anaesthesia. General duties include an appropriate examination of the patient, obtaining consultations when necessary, obtaining informed consent, keeping good records and providing appropriate follow-up care. As the trend towards outpatient and same-day surgery has increased, so have the medicolegal problems of anaesthetists in fulfilling these general duties. It may be difficult to prove that an anaesthetist who first saw a patient on a stretcher outside the operating room conducted an appropriate evaluation. The institution of preoperative screening clinics is one solution to the problem, but has the drawback that the liability for injuries, which occur because of an improper evaluation by the anaesthetist who conducted the evaluation, is increased. The anaesthetist providing intraoperative care must rely on an evaluation by another anaesthetist. If no anaesthetist has evaluated the patient, the situation is more tenuous because an internist or surgeon may not be aware of the information required for the conduct of an anaesthetic.

Another problem arises from the general duty to provide appropriate follow-up care. If an anaesthetist orders a chest x-ray as part of a preoperative evaluation, he or she is responsible for following up any abnormalities detected. This has nothing to do with being an anaesthetist; it arises from being the ordering physician.

Patients have the right of self-determination regarding their bodies and what is done to them. In order for them to make such decisions, the law imposes upon physicians a duty to inform the patient of any material risks associated with the planned procedures. In order to be successful in a malpractice action, based on the duty to obtain informed consent, the patient/plaintiff must prove that:

1. the physician failed to disclose material risks and dangers *inherently* and *potentially* involved in the procedure;

2. the unrevealed risks actually materialized and were the proximate cause of the injuries sustained;
3. that a reasonable person would have decided against the treatment had the dangers been disclosed.

Three distinct tests are applied to determine the informed consent requirements: the reasonable doctor standard, the reasonable patient standard and the subjective patient standard. Most states have adopted the reasonable patient standard and, under this test, a risk is considered to be material when a reasonable person, in the circumstances of the patient, would be likely to attach significance to the risk(s) when deciding whether or not to forgo the treatment. The duty to disclose risks is not based on the doctor's practice (or upon expert witness testimony), but rather upon the patient's need for full disclosure of serious risks and the feasible alternatives in order to make an intelligent and informed choice.

Those risks that are inherently present would apply to all patients without regard to their individual condition, whilst those dangers that are potentially present depend to a great extent upon the individual patient. For example, the risk of dental damage is inherently present whenever an intubation is performed. The risks of a myocardial infarction or stroke are potentially present, with a greater degree of probability, in patients with vascular disease.

A special circumstance is imposed when there are non-medical risks associated with a procedure. The two issues with the highest public awareness in this regard are HIV testing and mandatory drug testing in the workplace. In either case the common theme is that the test results may affect the patient's ability to work, obtain insurance or be free from discriminatory practices. Additional standards for informed consent and confidentiality are imposed because of these special risks. Given the evolving nature of the law in these areas, it would be advisable to obtain the patient's consent for drug or HIV testing if the test is to be performed electively prior to surgery. On the other hand, if the tests are indicated for the proper diagnosis and treatment of an adverse reaction, the burden of obtaining special informed consent is reduced. Nevertheless, if the patient is able to give consent, it is always a good idea to get it.

The standard of care

The specific duty owed by an anaesthetist is to adhere to the standards of practice for the administration of anaesthetic care. Practice standards change over time, but there is no longer a locality rule for specialists. In almost all cases, adherence to the standard of care is established by expert witness testimony. Because it is impossible to delineate specific standards for all aspects of medical practice and all eventualities, the test that is applied is whether or not the conduct of the anaesthetist was reasonable and prudent. The conduct of the anaesthetist need be neither state-of-the-art nor the best possible care, and it is not necessary that it even be what the majority of anaesthetists would do. It only has to be reasonable and prudent behaviour in the specific circumstances. It is not sufficient, however, that the anaesthetist acted to the limits of his or her potential, acted in good faith or did what was considered normal in the hospital or the hospital down the street. On the other hand, to be considered a standard, the practice must be universally applicable. Although an anaesthetist or a department or a hospital may elect to hold themselves to a higher level of conduct or expertise than elsewhere, they cannot unilaterally hold anyone else to that standard. If an expert witness fails to understand that 'standard of care' implies the lowest level of acceptable practice, no matter how well intentioned or altruistic they may be, they may do great injustice to the defence of the case.

The standard of care may also be determined by the written policies of a department, a hospital, an accrediting organization or a professional society. It is imperative that written policies be unambiguous, enforceable and agreed to by all who practise under them. If an injury occurs because of failure to comply with a written standard, the occurrence of a breach of duty is obvious.

Breach of duty

If an anaesthetist either has done something that should not have been done or fails to do something that should have been done, there is a breach of duty, and the anaesthetist's conduct will have fallen below an acceptable standard of practice. The breach of duty is the exact act that was unreasonable or imprudent under the circumstances.

The fact that an injury occurred is not sufficient evidence for a breach of duty because this would imply a warranty or guarantee of a successful outcome. An honest mistake in diagnosis or treatment or an error in judgement 'will not form the basis for liability in the absence of additional affirmative evidence that the physician's conduct deviated from an acceptable standard of care'.[6]

It is also not sufficient to show that an alternative treatment was available or for the expert to assert that he or she would have done things differently. As long as the act of omission or commission was within a standard of care acceptable to even a respectable minority of anaesthetists, no breach will have occurred.

Causation

The breach in the standard of care must be tied to the injury. The link between the breach and the injury is known as the proximate cause of the injury. Proximate cause is not identical with the medical cause of the injury. It is the event that sets in motion a chain of events which leads to the injury. There may be a series of events rather than a simple relation between the breach and the injury. The law, therefore, allows for one of two tests to confirm the link. The method used varies from state to state.

The first test is the 'but for' test. If the injury would not have occurred 'but for' the acts of the anaesthetist, the test is true and proximate cause is established. Under this doctrine, if there were other factors that might have led to the same injury, the test is false and no proximate cause is established.

The second test is the 'substantial factor' test. Under this doctrine the act need not be the only possible cause of the injury. It must only be shown to have been a substantial factor in causing the injury. This test is less favourable to the medical defendant because recovery of damages is not prevented by the presence of other possible causes as is the case with the 'but for' test.

Damages

The term 'damages' as generally used applies to the injuries sustained by the patient, but is usually understood to refer to the translation of the injuries into a fiscal value. It is obvious that the legal system cannot restore a patient to his or her previous state of health, but it can compensate the patient with money. The money awarded is the damages, and the amount should be closely related to the degree of injury sustained.

There are two types of compensatory damages: general damages and special damages. *General damages* occur as a result of the injury but may be intangible. Pain and suffering, humiliation, anxiety and derivative claims such as loss of consortium are examples of general damages. *Special damages* have a measurable value and include such things as actual medical expenses, future expenses, loss of wages and earning capacity, and rehabilitation costs.

Much attention has been directed towards tort reforms that place a limit on general damages. Although largely applauded by physicians and decried by attorneys, such measures have relatively little effect on damages because the majority of damages are special damages, and the limit may not apply when wrongful death has resulted.

In cases of gross misconduct, the law also allows for exemplary or punitive damages. These damages are awarded in excess of the compensatory damages as a punishment, or in order to make an example of the case as a warning to other anaesthetists. Malpractice insurance typically does not cover punitive or exemplary damages.

Damages are decided by a jury after the amounts allowed have been described by the judge. There has been a trend towards increased awards in cases of medical malpractice even after adjusting for inflation, and there has been increased variability in the value of awards for similar cases. This has been attributed in part to the reduction in jury size in some states from twelve to six members.[7] Similar cases tried before juries with twelve members will result in more consistent decisions than when tried by six-member juries.

An additional factor has been the publicity that a multimillion dollar award gains for the plaintiff's attorney. What is often not appreciated is that these sums may represent the total estimated award over the lifetime of the patient – i.e. a structured settlement. The actual cost of purchasing the annuity is far less.

Closing a case: dismissal, settlement or trial verdict

Once a lawsuit has been filed with the courts, some form of court document is required to terminate the physician-defendant's liability. Approximately one in ten lawsuits is carried through to a jury trial. The reason for this low percentage is that negligence, if it occurred, usually becomes obvious during the discovery phase. If the physician was negligent, the insurer would much rather settle the case for a reasonable amount than risk a high jury award.

The plaintiff might refuse to settle, hoping for a very high jury award. If there had been no negligence, a small settlement might be offered because the cost of defence is higher than settlement, or the physician anaesthetist may be released with the promise not to attempt recovery of attorney's fees. It is very unprofitable for the plaintiff's attorney to pursue lawsuits that have little chance of winning because, in a contingency fee system, the attorney must bear all the costs.

If the plaintiff fails to establish sufficient facts in the case, the defence may file a motion 'with the court' for summary judgment by the judge in favour of the defendant physician anaesthetist. From the physician's point of view, a summary judgment is better than a settlement because it vindicates him or her of the charges. Because a summary judgment eliminates the possibility of a trial by jury on the issues of the case, judges are reluctant to grant these unless there is clear and irrefutable evidence that the case against the defendant has no merit. In order to block such a motion the plaintiff must find one physician anaesthetist willing to state that there may have been negligence.

The practical wisdom that 'settlements should be made whenever it is cheaper to settle' may alter because there is no minimum payment threshold for reporting in the National Practitioner Database. Physicians are more reluctant to approve settlements when they know that a mandatory report will follow. The fact that insurers must report to the National Practitioner Database all payments made on behalf of a physician may result in a decline in token settlements.

The physician is 'released' from liability if a settlement is made, as the patient then abandons the claim of malpractice. The patient executes a release document and is barred from making any further claims against the physician.

Even if a case proceeds to a trial, settlement offers may be made up to the time when a verdict is reached. Cases go to trial as a general rule only when both sides think they have a good chance of winning. A victory does not necessarily signify a favourable verdict; it only implies a lower monetary cost than would have been required to settle. If the case is complicated or emotionally charged, jury selection may be crucial to the outcome, and the lawyers may wait until after the jury is selected to estimate the odds of winning. Most cases that go all the way through the process to the point of a jury verdict are decided in favour of the defendant physician. This is not really surprising as the burden of proof rests with the plaintiff.

The adversarial legal system in malpractice: personalization of the process

The American judicial system is based on the belief that 'justice' is best served by having each side represented by lawyers whose sole responsibility is to their client. Thus, there are no ethical or moral restraints placed upon an attorney with respect to the opposing side; in fact, failure to do everything legally possible to discredit the opposing side would be considered unethical. This competition between opposing counsel, each side presenting their own version of the facts and the law, is supposed to enable the judge or jury to determine the truth and to arrive at a just verdict. Physicians who fail to understand the adversarial system are vulnerable to personalization of the process. The legal system therefore reinforces any pre-existent self-doubts, guilt feelings or simple regret over the bad outcome.

Bargaining in the malpractice process is an integral part of the discovery phase leading to a determination of whether to settle or go to trial. The critical issue for the attorneys and insurance companies is the limitation of the damages. It is simple economics and good business to minimize losses. At this point, the defendant physician is actually not a major player in the game. He or she has relatively little control over the events, even though they may feel that their concept of professional self-worth is being bargained away in the process. The defendant physician who believes he or she is innocent of any wrongdoing is particularly likely to want public vindication and exoneration. Such personalization of process may interfere with the expeditious settlement of the case, and, in the extreme case, refusal to cooperate with the insurer may result in withdrawal of coverage.

On the other hand, defendant anaesthetists who realize that they made an error will find the process of discovery agonizing in that every expert for the plaintiff will make the same accusations over and over again. They will be told on the record repeatedly and in great detail exactly how and when they made the mistake. This is part of the adversarial system, and may cause or exacerbate depression in the defendant.

It is not surprising, given the high stakes involved, that depression may occur at virtually any point in the process. Only the most callous and uncaring physician could fail to be saddened by the death or serious injury of their patient. When the charge is made that the physician was responsible for the outcome due to negligence, it becomes difficult to continue in the belief that the outcome was unavoidable or unforeseeable, or occurred despite the best that medicine had to offer.

Anaesthetists are at least as likely as other physicians to experience job-related stress. They may be at an increased risk, relative to other medical specialties, because of the 'solo practice' nature of anaesthesia and the resultant isolation from mainstream medicine. Anaesthetic-related injuries are probably more likely to be disabling or lethal than injuries that occur during care by other specialties, and the emotional trauma, guilt and self-doubts will therefore be more severe. The adversarial system of malpractice litigation, by its very nature, causes substantial additional stress. Anaesthetists involved in malpractice suits should become familiar with the system, learn what to expect and avoid taking the proceedings personally.[8]

Asset protection

A malpractice lawsuit may expose a physician's personal assets to risk. The most common means of protecting personal assets is the purchase of malpractice insurance.

There are three basic types of malpractice insurance: occurrence, claims made and umbrella. An occurrence policy requires only that the insurance was in effect at the time of the incident. This makes it difficult for the insurer to estimate how much risk they are exposed to at any given moment, as the lawsuit may not occur for several years after the incident. Most policies today are based on claims made. The difference is that the insurance must be in effect not only at the time of the occurrence but also when a claim is made. This has resulted in the need for 'nose' and 'tail' coverage, and mandates a clear definition of what

constitutes a claim. Tail coverage is commonly provided when a physician leaves practice or changes insurance carriers. It typically extends coverage with the company beyond the statute of limitations for filing a suit. Nose coverage is less common, but is offered by some companies to provide retroactive coverage when the physician purchases a new policy. Some companies allow notification of an incident to count as a claim. Physicians changing to another insurer may flood the insurer with reports of incidents, which might even remotely turn into a suit, in an attempt to avoid purchasing tail coverage. Some companies require notification of intent to sue, or the filing of a suit, before a claim can be made.

Umbrella coverage is used to insure against the possibility of exceeding the limits of coverage provided by the primary policy. This type of insurance is not commonly used, but may be of value if the anaesthetist practises in a high-risk specialty which may suffer multi-million dollar damages. In most such cases, there are multiple defendants who share in the damages and structured settlements. For practical purposes, policies providing $1 million in coverage for a single claim and $3 million in any year are adequate.

The malpractice insurance industry is complex,[9] and there are many factors to consider when purchasing a policy. As the primary purpose of insurance is the protection of personal assets, it is important that the physician has confidence in the company. The least expensive policy does not always provide adequate peace of mind and financial security. Although a detailed discussion of all variables cannot be provided in this chapter, some general guidelines may be useful.

The financial security of the company is of primary importance. The failure of some liability carriers has exposed physicians to substantial personal risk. As a general rule, newer companies are the least secure. New companies may not be able to select good-risk physicians, may not have acquired large capital reserves, or may not be well enough known to qualify for the lowest rates for reinsurance. These added uncertainties may be covered by making the policies assessable. In the event that the company suffers excessive losses, the physicians may be assessed a surcharge. The need to assess a surcharge depends on the reserves and net worth of the company. A rule of thumb for the structure of reinsurance relative to policy limits is that the retained risk on any one policy should not exceed 10 per cent of the surplus (net worth). A company's assets should be evaluated for adequacy, liquidity and the risk of the investments. If the total net retained annual premium income exceeds four times its surplus, a company may be too heavily leveraged. In other words, the company may be relying upon new premium funds to pay for losses incurred. Such arrangements are inherently unstable, as in any other pyramid scheme.

Once the financial security of an insurer is established, the physician anaesthetist should consider the quality of the services offered. This may be evaluated by references from insured physicians as well as by marketing brochures. An ongoing risk management and educational programme is desirable. Requiring higher standards of practice or advanced monitoring may indicate that the company has high underwriting standards. The company should retain high-quality legal counsel and be able to provide the names of the firms involved. Finally, the company should be responsive and accessible to the anaesthetist.

There are many methods other than insurance that may be used to shelter personal assets from risk.[10] As a general rule, any asset that is owned by the physician may be seized to satisfy a debt owed by the physician. The most obvious way to protect an asset is to transfer ownership to another party. The Uniform Fraudulent Conveyance Act effectively precludes any transfer of assets that occurs in the face of a debt or lawsuit. In practical terms, this requires preplanning and will make it very hard to avoid payment of any pending obligations. Assets may be transferred to entities such as trusts, partnerships or corporations, or to individuals such as family members. It is important to keep in mind that transferring ownership of an asset may also cause loss of control over that asset. In the case of a mortgaged asset such as your home or other property, ownership transfer may be deemed a sale and require repayment of any loans in full to the lender.

There are many types of trusts which are distinguished by legal differences that enable them to act in different ways in response to events, but all trusts require that the assets are owned by the trust rather than by individuals. Therefore, any income from the asset goes into the trust, and the beneficiaries of the trust may have legally enforceable rights of ownership.

Partners typically are liable for the acts of other partners that are carried out within the scope of the partnership business. Members of a corporation are typically not responsible for the debts of the corporation. Unfortunately, unless the rules for maintaining a corporation are strictly adhered to, the court may treat the corporation as a partnership and 'pierce the corporate veil' to reach the personal assets of the shareholders.

Transfers to individuals such as a spouse or children will result in loss of control of the asset by the physician anaesthetist. In the event of a divorce or other family difficulty, the physician will probably lose the asset altogether.

The major asset of many physician anaesthetists is their home. There are special provisions for protecting the home in some states, but there is great variability in the means by which this may be accomplished. The options available will vary from state to state.

There are ways for a physician to protect his or her assets, but there are advantages and disadvantages in each method, and state laws which will determine which methods are available. Competent individual legal advice is imperative.

Summary

This chapter has outlined the American system of medical jurisprudence in broad terms as it relates to the practice of anaesthesia. It is not intended to be, and should not be used as, a substitute for competent legal counsel. The intent has been to provide the reader with sufficient descriptive information to enable the physician anaesthetist to understand the important medicolegal issues that regulate or influence their practice. There are several textbooks devoted to medical jurisprudence which go into far greater detail for the specialty of anaesthesia[11,12] or for the laws of specific states.[6,13–15] They also provide reference material for many legal subjects such as vicarious liability, product liability and the doctrine of *res ipsa loquitur* that under special circumstances concern anaesthetists, in addition to discussions about the many ethical dilemmas facing anaesthetists.

Finally, it is important to recognize that, as multifaceted as the interfaces of law and medicine may be, the legal system addresses only a portion of a larger problem. If anaesthetists always and undeniably did the right things under all circumstances, there would never be any question of whether or not the care was reasonable or prudent. There are differences of opinion regarding the appropriateness of care that may be resolved on many levels with different intended outcomes. Peer review, quality assurance, risk management and utilization review look at different aspects of the larger question of whether or not the correct things were done. A malpractice suit is not a substitute for these activities and properly addresses only the issue of whether a plaintiff is entitled to receive monetary damages. Thus, a malpractice suit is inherently backward-looking and contingent upon the details of the specific case. The true value in understanding medicolegal principles lies in the recognition that the central issue is quality of care, and in the prospective application of these principles to influence decisions. Asking 'Can I adequately explain what I am about to do to a jury of lay people?' is one way of answering the broader question 'Is this the right thing to do for this patient under these circumstances?'

REFERENCES

1. Peters JD, Fineberg K, Kroll DA. *The law of medical practice in Michigan*. Ann Arbor MI: Health Administration Press, 1981: 21–5.
2. Kroll DA. *Professional liability and the anaesthetist*. Park Ridge IL: American Society of Anaesthesiologists, 1987.
3. Gilmore DA. The antitrust implications of boycotts by health care professionals: professional standards, professional ethics and the First Amendment. *American Journal of Law and Medicine* 1988; **14**, (2,3): 221–48.
4. McDowell TN, Rainer JM. The State Action Doctrine and the Local Government Antitrust Act: the restructured public hospital model. *American Journal of Law and Medicine* 1988; **14**, (2,3): 171–219.
5. Kroll DA. What to do when sued for malpractice. Part I. Prodromes of a lawsuit. *American Society of Anaesthesiologists Newsletter* 1986; **50**, (1): 7.
6. LeBlang TR, Basanta WE, Peters JD, Fineberg KS, Kroll DA. *The law of medical practice in Illinois*. Rochester NY: Lawyers Cooperative Publishing, 1986: 573.
7. Saks MJ. In search of the 'lawsuit crisis'. *Law, Medicine, and Healthcare* 1986; **14**, (2): 77–9.
8. Kroll DA. The trauma of malpractice suits. *Seminars in Anesthesia* 1989; **8**, (4): 347–52.
9. MacKenzie RA. Professional liability insurance. In: Dornette WHL, ed. *Legal issues in anesthesia practice*. Philadelphia PA: FA Davis, 1991: Ch 29.
10. Kroll DA. Litigation and anesthesia: why does it happen and what to do if involved. In: 40th Annual Refresher Course Lectures and Clinical Update Program 115. Park Ridge IL: American Society of Anesthesiologists, 1989.
11. Peters JD, Fineberg K, Kroll DA, Collins V. *Anaesthesia and the law*. Ann Arbor MI: Health Administration Press, 1983.
12. Dornette WHL. *Legal issues in anesthesia practice*. Philadelphia PA: FA Davis, 1991.
13. Peters JD, Fineberg K, Kroll DA. *The law of medical practice in Michigan*. Ann Arbor MI: Health Administration Press, 1981.
14. Post BL, Peters BM, Stahl SP, Peters JD, Fineberg KS, Kroll DA. *The law of medical practice in Pennsylvania and New Jersey*. Rochester NY: Lawyers Cooperative Publishing, 1984.
15. Woodside III FC, Lawson NA, Lyden DR, Peters JD, Fineberg KS, Kroll DA. *The law of medical practice in Ohio*. Rochester NY: Lawyers Cooperative Publishing, 1989.

Chapter 74

The Anaesthetist's Duty of Care: a British Perspective

C. J. Hull and Daniel Brennan QC

The Bolam test – *Bolam v Friern HMC*
A contract to treat?
A duty to provide
The duty to attend and assess
The duty to explain

The duty to ensure that the patient consents to treatment
The duty to provide safe anaesthesia
The duty to tell when things go wrong
Failure to fulfil the duty of care

In the UK, an anaesthetist owes a duty of care toward his or her patient. This is a legal concept, having nothing to do with clinical care. Indeed, a chartered accountant has exactly the same duty of care towards his client, who indeed may be the anaesthetist. Having accepted that duty, the professional person is obliged to fulfil it, while (in private practice) the client is equally bound to meet any reasonable charges. The duty encompasses the anaesthetist's obligation to attend the patient, to warn of likely hazards and, if it is in the patient's best interests, to apply such technical procedures as are appropriate. The standard of care to be provided is defined by a legal yardstick: the so-called Bolam test. This was established in *Bolam v Friern Hospital Management Committee* in 1957, and has since become the fundamental and guiding principle on the standard of care in medical negligence.[1]

The Bolam test – *Bolam v Friern HMC*

As well as being fundamental to the anaesthetist's duty of care, this case is of particular interest because it involves anaesthesia. In 1954 a mentally ill patient was 'advised' by a consultant psychiatrist that electroconvulsive therapy (ECT) was necessary. A standard consent form was completed, but this did not make mention of any risk of injury. Since ECT in 1954 was often conducted without the benefit of neuromuscular blockade (or, for that matter, anaesthesia of any kind), that omission might seem odd by today's standards. The patient underwent repeated ECT without benefit of anaesthesia or relaxant drugs, and the only precaution against injury was jaw support and a male nurse standing on each side, presumably to prevent him

being thrown off the couch by the force of his convulsions. During the second treatment the patient sustained a fracture, and subsequently sued the hospital claiming compensation for his injury. He claimed that muscle relaxants should have been used, but that if they were not he should have been manually restrained to minimize injury, having been warned beforehand that injury was a possible consequence. All these claims were denied. At trial it was stated that the risk of fracture and other injury was well known, and that anaesthetics with muscle relaxation were widely used as a protective measure (no suggestion, of course, that such techniques helped to mitigate the terror of repeated unmodified ECT). However, contrary evidence was heard to the effect that the risk of fracture was, in fact, quite small, and that the method used had been standard practice at that hospital. Furthermore, it was claimed that the risks associated with anaesthesia were just as great as (albeit different from) those of injury in the unmodified method. Only under special circumstances, therefore, was anaesthesia used, and Mr Bolam was not such a case. Similarly, it was stated that manual control of patients undergoing unmodified ECT was by no means a universal practice, and that warning patients of possible injury would be likely to dissuade them from undergoing treatment!

Faced with two seemingly irreconcilable views, Mr Justice McNair held that there were at least two quite distinct schools of thought, all perfectly respectable and applied by practitioners of equal standing. He could find no reason to prefer one such school of thought over the others, and on that basis enunciated the celebrated principle in his directions to the jury:

> The test is the standard of the ordinary skilled man exercising and professing to have that special skill . . . . A doctor is not negligent if he is acting in accordance with a practice accepted as proper by a responsible body of medical men skilled in that particular art, merely because there is a body of opinion that takes a contrary view.

Not surprisingly, the jury found for the defendant. By today's standards that might seem to be a perverse decision, as the risks from anaesthesia for ECT must be uncommonly small compared with those of physical injury. However, the case must be seen in the context of 1954 practice, whereby the anaesthetic might well have been administered by a doctor with little or no anaesthetic training.

In *Sidaway v Bethlem Royal Hospital Governors* (1985) Lord Scarman restated the test:

> A doctor is not negligent if he acts in accordance with the practice accepted at the time as proper by a responsible body of medical opinion, even though other doctors adopt a different practice.[2]

The test is applied to diagnosis,[3] to advice[2] and to teatment.[4] Liability is determined by the practice and medical thinking at the time of the alleged negligence rather than at the date of trial.[5]

Of course, the original concept leaves great scope for interpretation in individual cases, especially in deciding the precise meaning of 'proper' and 'responsible'. In *Gold v Haringey Health Authority* (1987), the Bolam principle was expressed in uncompromising terms, i.e. that the 'practice regarded as proper by a responsible body of medical men' would be accepted uncritically, regardless of context or clinical wisdom.[6] So long as the body of opinion was regarded as responsible, the court would not inquire further as to the reasoning behind that opinion.

Thus, an anaesthetist who, in the early 1970s, maintained anaesthesia during the pre-delivery phase of caesarean section using only 50 per cent nitrous oxide in oxygen but subsequently faced a claim that the patient had suffered pain and awareness as a consequence of his or her negligence, had a clear defence in showing that a very substantial number of consultant anaesthetists would have done likewise. The anaesthetist would not have to demonstrate the *wisdom* of so doing, but simply that the practice was considered reasonable by that body of anaesthetists.

There must, of course, be limits to doctors defending ludicrous clinical decisions on the sole ground that others could be found who would have done the same. *Gold* has not been slavishly followed. The opinion of the responsible body of doctors that is relied upon will usually be required to 'stand up to analysis'.

A contract to treat?

In private medical practice, the duty of care is based upon the contract (real or implied) between a professional person and the client. The patient–doctor relationship commences when the anaesthetist accepts any kind of clinical responsibility for the patient concerned. Of course, much depends upon the nature of the professional relationship between surgeon and anaesthetist. If they work together on a regular basis, such that by common consent Dr X anaesthetizes Mr Y's patients on a Wednesday afternoon, a list of names from Mr Y's secretary a few days earlier may be sufficient to establish his professional responsibility towards those patients. In the case of a more casual relationship, where Mr Y (or his secretary) telephones Dr X with a request to attend and anaesthetize Mrs Z, he has no obligations whatever until and unless he agrees to do so. Even then, his responsibility is to attend and assess the patient and to offer his professional advice. At that stage he has made no commitment to anaesthetize the patient, but simply to do that which is in the patient's best interests. If, for instance, the patient requires further preoperative investigation before safe anaesthesia can be contemplated (as might apply in the case of a young woman presenting for varicose vein surgery who has clinical signs of anaemia but no haemoglobin estimation), his duty toward that patient is clearly set: he must explain the need for additional tests to both patient and surgeon, and ensure that the former has a clear understanding of the risks involved.

If an anaesthetist in private practice is sufficiently unwise as to give an express guarantee as to outcome, that becomes part of the contract. Thus, if the anaesthetist expressly guarantees that the patient will not suffer pain and awareness during surgery, but in the event that patient does so suffer, it matters not whether the anaesthetist was negligent because he or she clearly is in breach of contract and may be obliged to compensate the patient accordingly.

A duty to provide

In hospital practice, the situation is more complicated, because the ultimate duty of care is borne by the clinical provider, be it health authority or NHS trust. The provider has a clear duty to ensure that its employed doctors are adequate in both quantity and quality. In addition, it is more than likely that the provider also makes express warranty as to the adequacy of clinical care in its contract with the budget-holding purchasers.

In appointing anaesthetists at any grade, employers must fulfil their duty of care (and perhaps contractual obligations too) by ensuring that the process is conducted properly in terms of the statutory instrument (i.e. the Whitley terms and conditions of service). Posts must be advertised and appointments advisory committees must be properly constituted if it is to be shown subsequently that all reasonable steps had been taken to ensure an appointee's suitability for the post.

In the case of short-term or locum posts, many hospitals have taken a far more relaxed view of their obligations. It has become commonplace for locum consultant posts to be offered to doctors who do not even hold the FRCA (or equivalent), let alone a certificate of accreditation or completion of specialist training. It has not been unknown for locum anaesthetists to be hired, even at consultant level, without seeking formal references and without any formal check of their medical qualifications or postgraduate diplomas. Such doctors then have been assigned clinical duties without any assessment of their clinical skills, and disasters have followed. The clinical director who instigates or even sanctions such a process in the interests of maintaining some kind of clinical service may be highly vulnerable in law. His or her employer is likely to find that any economies made by such appointments are likely to prove very expensive, in both financial and human terms.

The anaesthetists themselves have a duty towards their patients but there is no contract, real or implied. Thus, if a patient is scheduled to undergo panproctocolectomy on a scheduled operating list and is anaesthetized by a senior house officer with 6 months' anaesthetic experience, a mishap occurring as result of that doctor's inexperience might well be regarded as evidence of negligence on the part of the employer (i.e. the health authority or NHS trust). Generally, however, individual clinicians must expect to answer for their own clinical decisions and actions. It is particularly important to note that they are likely to be judged by the standards of the post in which they are employed. Thus the actions of senior house officers are judged by the standard of technical proficiency (and appreciation of their own limitations) and duty to consult their seniors where appropriate to be expected of an average senior house officer, and those of consultants by the standard to be expected of a consultant. This point should be well understood by anaesthetists with limited formal training who are appointed to locum consultant posts, and also by their employers. Although such doctors may have but 2 years' training and a Diploma in Anaesthetics (DA), in the event of error or mishap they will be judged as consultants. The likely cost of a major settlement should deter all but the most foolhardy clinical directors from making such appointments.

Equally, the law will probably require that trainees be judged by the same stardard as the more experienced at that level. Therefore, their work should be adequately supervised and they must be made aware of their duty to seek advice where appropriate.

The duty to attend and assess

It is quite obvious that an anaesthetist cannot make an informed judgement as to an individual patient's fitness for anaesthesia or to the type of anaesthetic procedure to be employed without careful preoperative assessment.

The assessment should establish the medical history, with particular emphasis on previous anaesthetic procedures. Records of such procedures may give valuable indications of special risks such as difficult intubation, adverse drug reactions, etc. Patients who have not previously been anaesthetized should be asked about their family history, which may reveal potential problems such as suxamethonium apnoea, sickle cell disease or malignant hyperpyrexia.

Physical examination should identify any abnormalities of physique that might influence the choice of technique, together with particular risk factors such as cardiovascular or respiratory disease. No matter what anaesthetic technique is planned, the assessment should always include a formal consideration of dentition and probable intubation difficulties. This should be self-evident, because the need to intubate may develop both suddenly and unexpectedly. Under such circumstances the discovery of a mouth full of loose and rotten teeth together with severe limitation of mandibular movement may transform an emergency into a disaster.

The assessment should include discovery of all essential documentation, including the results of essential preoperative assessment. In the event that such results are not available, it should be remembered that a measured but unreported preoperative serum potassium concentration of 7.5 mmol/l would be difficult to explain in the event of cardiac arrest occurring during the induction of anaesthesia.

The duty to explain

Having established all the relevant facts, the anaesthetist now has a clear duty to give the patient an explanation of what he or she intends to do, and to explain the nature of likely hazards. This cannot be accomplished satisfactorily at the doors of the operating theatre, and requires that the anaesthetist who is to administer the anaesthetic visits the patient some time before surgery. Obviously, there are limits to such a policy, as in the case of emergency surgery where the opportunities for preoperative discussion and explanation may be very limited. In day-case surgery, too, the very short period between admission and surgery may preclude an effective preoperative visit. Under such circumstances the value of a pre-admission anaesthetic clinic is self-evident.

Anaesthetists who intend to use spinal anaesthesia alone must tell the patients that they are to remain awake during surgery; but must the anaesthetist also inform patients of possible severe headache or even of major neurological complications? Here the lines are less clearly drawn, but the strength of duty is clearly related to the probability of these complications occurring.

This principle was clearly established in the celebrated case of *Sidaway v Bethlem Royal Hospital Governors*, wherein a patient suffered major neurological sequelae (of which she had not been warned) following a neurosurgical procedure.[2] In the House of Lords, Lord Bridge observed:

> ... the Judge might in certain circumstances come to the conclusion that disclosure of a particular risk was so obviously necessary to an informed choice on the part of the patient that no reasonably prudent medical man would fail to make it.

He went on to explain that a procedure known to carry a 10 per cent risk of some serious complication would be so regarded. In the same case in the Court of Appeal, Lord Donaldson had gone even further:

> ... a judge would be entitled to reject a *unanimous* medical view if he were satisfied that it was manifestly wrong and that the doctors must have been misdirecting themselves as to their duty in law.

He expressed a view that the key definition in the *Bolam* test should be re-worded as:

> The duty is fulfilled if the doctor acts in accordance with a practice *rightly* accepted as proper by a body of skilled and experienced medical men.

Although the House of Lords did not embrace this wording, it accords with justice and sense. It also would prevent doctors from playing God and setting the rules to suit themselves.

Since, in reasonably skilled hands, using modern needles and drugs, severe headache occurs in much less than 1 per cent of patients, the duty to inform would not seem to be strong. In the case of major neurological sequelae, it must be even weaker. However, by common consent the great majority of anaesthetists would respond positively to any patient making a direct request for information. Thus, if asked for the exact risk of major sequelae, they would tell the patient, to the best of their knowledge, the odds of such an event occurring.

The same obligation rests on the anaesthetist in the case of an anaesthetic procedure which results in some adverse outcome. There is a clear duty to inform the patient, so far

as possible, as to the cause of the accident. Indeed, a civil wrong may be committed in failing so to do.

However, it should be noted that the rigid Bolam interpretation set by *Gold v Haringey HA*[6] was reinforced by the Court of Appeal in *Blythe v Bloomsbury HA*, in which it was held that a patient's direct request for information regarding possible risks of drug treatment could, quite properly, be disregarded if it was the contemporary practice of a responsible body of doctors so to do.[7] Clearly, in this respect anaesthetists working in the UK might appear to have far greater discretion than do their colleagues in the USA, who have virtually none. In English law what should be said to the patient depends ultimately on the characteristics of the particular patient and a reasonable medical judgement of the extent of the information that should be given to that patient.

The duty to ensure that the patient consents to treatment

Having established that a patient has been properly informed of the planned surgical and anaesthetic procedures, it is essential that consent be obtained. Normally, this involves the patient signing a witnessed declaration that a proper explanation has been given and that he or she consents to the surgical procedure set out above. In the case of a child aged less than 16, the parent or guardian signs the form. However, minors may be considered competent to give their own consent if they have sufficient understanding and intelligence to make that decision.[8]

Most hospital consent forms are somewhat vague with regard to anaesthesia, and often refer to 'local or general anaesthesia'. From a legal point of view, the essential step is to establish that the patient, while of sound mind, does consent to the procedure. Thus verbal consent may be perfectly satisfactory so long as there are witnesses who may be relied upon to establish the fact at some later date. However, it should be borne in mind that, in the event of a legal claim three or more years after the event, finding those witnesses and establishing direct recollection of the consent procedure may be a formidable undertaking. Clearly, a retrievable document is to be preferred. But it is *not* satisfactory to rely upon a signature obtained from an elderly and heavily sedated patient, sitting on a trolley in the theatre reception area *sans* teeth or spectacles, when it is discovered that consent has not been obtained! Clearly, under such circumstances that patient is in no fit state to give informed consent, and there is no certainty that anyone has explained anything.

Occasionally, a patient refuses consent for a procedure

that is judged to be necessary or even life saving. Because such patients may, at some later date, give an entirely different account of events, it is essential that such refusal be documented and witnessed with just as much care as consent itself. A documented refusal should be preserved in the case records.

Consent for research procedures

Where research procedures are to be part of or associated with the conduct of anaesthesia, special considerations apply. It is essential that the patient has a full understanding:

1. of what is experimental and what is not;
2. as to whether any part of the planned procedure is subject to randomization, and, if so, what are the randomized alternatives;
3. that refusal to participate will not detract from the adequacy of the anaesthetic care;
4. that consent for research may be withdrawn at any time.

Most ethical committees quite properly insist on the use of patients' information sheets written in lay English, with consent forms that are quite separate from those used for the surgical procedure itself. It is good practice for a research protocol to be explained and consent obtained by one of the research team rather than the anaesthetist who is to administer the anaesthetic itself.

The duty to provide safe anaesthesia

Doctors who describe themselves as anaesthetists or are employed as such, should be capable of delivering safe, effective anaesthesia. In order to do so they must:

1. Use anaesthetic equipment that is fully functional, properly maintained and, above all, safe. Today, it might be difficult to justify the use of an anaesthetic machine capable of delivering oxygen-free gas mixtures or which has a carbon dioxide rotameter calibrated up to 2 litres per minute.
2. Use monitoring equipment that warns of unsafe gas mixtures, inadequate blood oxygen saturation, inappropriate pulmonary ventilation, cardiac arrhythmias and abnormalities in heart rate, blood pressure or temperature.
3. Check that all equipment is fully functional, with

particular attention paid to any items that have just returned from service or repair.
4. Be able to demonstrate adequate training for the post they hold. In the case of trainees this is self-evident, but even in the case of highly experienced clinicians, continuing education is essential if skills are to be maintained and developed through the course of a professional lifetime.
5. Be physically capable of delivering a high-quality service to the patient. It is inappropriate that patients should unknowingly trust their lives to the care of a doctor who is exhausted, sick, demented or under the influence of alcohol or other inebriants.
6. Use techniques that lie within the limits of currently accepted practice. Thus any anaesthetists who relied upon 50 per cent nitrous oxide to ensure their patient's insensibility might find themselves called upon to justify that decision in defence of an awareness claim.
7. Keep an adequate written record of the anaesthetic procedure and of physiological monitoring data. Any hard copy printouts from monitoring devices should be labelled with the patient's name and date, and then attached to the anaesthetic record. A vexatious claim may easily be discounted by reference to a good quality clinical record with a wealth of detail to demonstrate the anaesthetist's diligence. Conversely, a poor record may raise suspicions that the clinical care may also have been poor. This is particularly important in cases of patients complaining that they have been aware and in pain during surgery under 'general anaesthesia'. Such a claim may be impossible to defend unless the anaesthetist can produce a contemporaneous record showing what agents were used and in what concentrations or doses they were administered. Indeed, this difficulty prompted the Medical Protection Society to voice the following warning in its 1987 Annual Report:

> If members (whenever possible) ensure that they adhere to accepted techniques, and keep full anaesthetic notes (including details of preoperative assessment, gas flows, delivery volumes, physiological parameters, circuit used, and the concentration of volatile agents selected, together with timing and usage), then and only then, may it be possible to defend a claim that awareness was due to negligence on the part of the anaesthetist.

8. Ensure that their responsibilities are limited to caring for a single patient at any one time. Any anaesthetist who by intent was simultaneously caring for two anaesthetized patients would have no possible defence if misfortune were to befall either of them.
9. Ensure that their patients enjoy their undivided

attention at all times. At its most basic level this requires that they be present throughout the procedure. In the event of any hazardous incident occurring in an anaesthetist's absence, it might be necessary to prove that his or her actions were at all times in the patient's best interest. Since patient safety during the maintenance phase of anaesthesia may well depend on their vigilance, they should not permit any distractions. Clearly, this depends upon individual circumstances, but might well preclude inappropriate banter with or between other members of the theatre team, conversation with visitors or teaching activities unrelated to the immediate welfare of the patient concerned. The practice of playing music in the operating theatre may be a dangerous distraction.

10. Ensure that, if an inexperienced trainee is to administer the anaesthetic, he or she is closely supervised so that the patient enjoys exactly the same standard of care as would have been provided by the consultant anaesthetist. Inadequate supervision forms a major element of many negligence claims, and although not part of the direct chain of causation may influence a judge strongly in favour of the plaintiff.

11. Ensure that they maintain proper care of their patients until they may be handed safely over to the care of a nurse. This will depend upon local circumstances, as patients may be passed to the care of a fully trained recovery nurse at a far earlier stage than to a student nurse with no recovery training whatever.

12. Wherever possible, follow accepted clinical practice. However, an anaesthetist not following the accepted method of treatment (if indeed one exists) is not necessarily negligent.[9] In the Scots case of *Hunter v Hanley* the Court said:

> Even substantial deviation from normal practice may be warranted by the particular circumstances. To establish liability by a doctor when deviation from normal practice is alleged, three facts require to be established. First of all it must be proved that there is a usual and normal practice, secondly it must be proved that the defendant has not adopted that practice, and thirdly (and this is of crucial importance) it must be established that the course the doctor adopted is one which no professional man of ordinary skill would have taken if he had been acting with ordinary care.

13. When using new techniques, ensure that the patients have given informed consent to any new treatment that exposes them to a risk of which they should be aware. Patients should not be subjected to unacceptable risk. A new technique must be justified as to its reasonable use in any particular case. If it results in damage to a patient which ought not to have occurred with the exercise of reasonable care, the doctor will be liable in negligence, as occurred in *Landau v Werner*.[10]

The duty to tell when things go wrong

Just as anaesthetists have a clear duty to warn their patients of likely dangers and complications associated with the planned course of action, they also have a duty to inform the patients when those complications have occurred. Failure to do so, especially when some error has been made, may itself provide grounds for legal action.[11] Many patients discover that an error has been made only when they experience the result of that error. An obvious example is the patient who discovers that he has one less dental crown than he had on admission, but cannot get any kind of explanation as to what has happened to it. When such events occur it is always better to tell the patient immediately, before he has time to become angry. In the great majority of cases, patients become claimants only because no one will tell them what happened. In the case of intraoperative awareness, the need for absolute honesty is paramount. Any denial or evasion on the part of the anaesthetist is likely to intensify the mental trauma suffered by the patient, and may indeed contribute to the onset of a post-traumatic stress disorder.

Failure to fulfil the duty of care

If a patient suffers some accidental damage in the course of an anaesthetic procedure, he or she may claim that this occurred as a direct consequence of negligence on the part of the anaesthetist. Legal action may be brought against the anaesthetist and/or the employer in pursuit of compensation for the damage. Because (1) the introduction of NHS indemnity makes it no longer necessary to establish the liability of every doctor involved and (2) individual doctors may be difficult to identify and subsequently locate, it has become commonplace for health authorities and NHS trusts to be named as defendants, with the negligence alleged as committed by their servants or agents. Such civil actions can succeed only if a number of strictly defined critieria can be met:

1. The patient suffered damage in the course of clinical management.
2. The anaesthetist had a duty of care with respect to that patient.
3. The anaesthetist failed, in one or more respects, to fulfil his or her duty of care.
4. The failing(s) caused the damage.

The onus of proof lies with the plaintiff, who must demonstrate *on the balance of probability* that the above conditions have been met. Such actions, if they cannot be settled by negotiation, are heard by a judge sitting alone. Damages are assessed as financial compensation for pain and suffering, and for both past and future financial loss. Their intention, in as much as money can compensate for suffering, is to restore the plaintiff to the position he or she would have been in had the tort not occurred. There is no provision for punitive or exemplary damages in English or Scottish law, so the award must be seen as compensation, not punishment. Such cases often depend upon a judgment as to what standard of clinical care was to be expected of the anaesthetist concerned, and to this end the judge hears expert witnesses from both plaintiff and defence. There is an implicit assumption that in such an adversarial system the truth will emerge as somewhere between the opposing views.

Res ipsa loquitur

Occasionally, an event occurs which is so far outside the natural expectation of events that the plaintiff may claim that it could only have happened as a result of negligence. Thus, if it is agreed that an anaesthetized and intubated patient suffered an undetected accidental disconnection causing severe hypoxic cerebral damage, it is difficult to imagine any explanation that would not involve negligence. The plaintiff then may invoke the legal doctrine of *res ipsa loquitur*, meaning, quite literally, 'the thing speaks for itself'. Under such circumstances the onus of proof may pass to the defendant, who then must demonstrate, on the balance of probabilities, that the accident did *not* occur as a result of his or her negligence.

This applied in *Saunders v Leeds Western Health Authority*, in which a previously healthy 4-year-old child suffered a cardiac arrest in the course of an uncomplicated anaesthetic.[12] Mr Justice Mann held for the plaintiff on the grounds that healthy children do not, in the ordinary course of events, suffer such catastrophes without premonitory signs being evident from properly conducted physiological monitoring, and that the defendants had not adequately discharged their burden of proof to show a non-negligent

explanation. He held that, under those circumstances, the plaintiff did not need to demonstrate the specific cause of the cardiac arrest.

The chain of causation

Often it is assumed that if an anaesthetist is in some respect negligent *and* the patient sustains damage of some kind, that patient has grounds for making a claim. This is true only when it can be shown that the negligence actually *caused* or *materially contributed to* the damage. Thus a patient who suffers pain and awareness during caesarean section because her anaesthetist had relied upon an empty isoflurane vaporizer has an excellent prospect of success. However, if instead of pain and awareness during surgery the patient claims that she suffered postoperative jaundice as a result of his incorrect choice of anaesthetic agent, his undoubted negligence has no causative link with the supposed damage and her claim must fail.

Even if there is a feasible chain of causation, the plaintiff remains in severe difficulties unless he or she can show that, on the balance of probabilities, it was what actually happened. Thus, if there are alternative explanations, albeit less likely, that have not properly been excluded, causation has not been established. This principle is well illustrated by the case of *Wilsher v Essex Health Authority*, wherein a premature baby was accidentally exposed to high concentrations of oxygen and subsequently developed retinal fibrosis.[13] Since cause and effect appeared to be obvious at trial, judgment was given in favour of the plaintiff with £116 000 damages. On appeal the judgment was affirmed. However, in the House of Lords it was argued that there were five other pathological conditions which could, possibly, have resulted in similar retinal damage and had not been excluded.[13] Thus, despite the clear aetiological link between the hyperoxia and the damage, it was held that causation had not been established and the appeal was allowed. It was for the plaintiff to prove the *specific* causative link between the negligence and the injury.

Criminal charges

A tort is a civil wrong in which the state has no direct interest. Libel, slander, defamation, breach of contract and professional negligence are all torts. In certain circumstances, however, an anaesthetist's actions may go beyond that limit. If, for instance, an anaesthetist performs some procedure in direct contravention of the patient's wishes, which then results in physical harm, he or she may be accused of assault and battery.

If it appears that a patient's death was caused by an

anaesthetist's criminal neglect or reckless behaviour, an accusation of manslaughter may follow. An obvious scenario would be that in which a patient suffered hypoxic cardiac arrest following accidental circuit disconnection while the anaesthetist took an unauthorized coffee break. At an even more sinister level, if it can be proved that the anaesthetist actually intended to cause the patient's death (however noble the motivation, as may occur in the case of a terminally ill patient in uncontrolled agony) the accusation may be one of murder. Fortunately, such events are rare and likely never to trouble the typical anaesthetist.

REFERENCES

1. *Bolam v Friern Hospital Management Committee* [1957] 2 All England Law Reports 118; [1957] 1 Weekly Law Reports 582.
2. *Sidaway v Bethlem Royal Hospital Governors and others* [1985] 1 All England Law Reports 643 HL.
3. *Whitehouse v Jordan* [1981] 1 Weekly Law Reports 246.
4. *Maynard v West Midlands Region Health Authority* [1984] 1 Weekly Law Reports 643.
5. *Roe v Ministry of Health* [1954] 2 Queen's Bench 66.
6. *Gold v Haringey Health Authority* [1987] 2 All England Law Reports 888.
7. *Blythe v Bloomsbury Health Authority* 1987 The Times 5 February, CA.
8. *Gillick v West Norfolk & Wisbech Area Health Authority and the DHSS* [1985] 3 All England Law Reports 402, HL.
9 *Hunter v Hanley* [1955] Sessions (Scottish) Cases 200.
10. *Landau v Werner* [1961] 105 Solicitors Journal 1008.
11. *Naylor v Preston Health Authority* [1987] 2 All England Law Reports 643.
12. *Saunders v Leeds Western Health Authority and another* (1984) [1993] 4 Medical Law Reports.
13. *Wilsher v Essex Area Health Authority* [1986] 3 All England Law Reports 801.
14. *Wilsher v Essex Area Health Authority* [1988] 1 All England Law Reports 871, HL.

Chapter 75

Quality Assurance and Cost Management

Clifford B. Franklin

Quality assurance (QA) activities, whilst not new in medical practice, have assumed a higher profile in relation to clinical services in recent years throughout the developed world. The pressures leading to this tendency have in large measure come from a growing expectation on the part of the general public for accountability within the professions. In the case of medicine this has been fuelled in part by the increasing levels of litigation against doctors, and in part by purchasers of clinical services requiring assurances about the arrangments for maintaining standards of care. Of all contemporary themes, QA in respect of clinical care can provoke even the mildest of people to reactions ranging from fervent criticism to ardent support. Between these extremes, however, there are individuals whose responses include: What is this QA and how will it involve me? What is this medical audit that has been added to my contractual commitments? and How can QA help me in my work?

Definitions

In any discussion on the subject of QA it is important to prevent ambiguity or misunderstanding by defining key terms that will be employed, and a selection is set out below.

Quality

'Quality' is impossible to define precisely and this is reflected in the number of possible interpretations of the term. It is a subjective assessment of the attributes of services or products that reflect culture, personal experiences and individual expectations. Like beauty, it is something that largely exists in the eye of the beholder – one can recognize it when one sees it. It is assessed by making subjective comparisons with similar items or events and can be expressed simply only in terms of 'good quality' or 'bad quality', and in relation to an individual service or product there can only be 'better quality' or 'worse quality'. Also, over a period of time as standards and expectations change so do perceptions of quality.

Quality assurance

Quality assurance is in the nature of an agreement between the supplier of services and/or goods and consumers. Insofar as it is possible these agreements attempt to quantify the intention of a provider to fulfil the expectations of the customers normally in terms of agreed levels of service and/or standards. In a health care setting this requires expressions of intent, preferably in writing, to set out levels of activity, standards of care and any particular requirements relating to profiles of care in the management of specific clinical disorders agreed upon with prospective purchasers.

Quality control

'Quality control' is a term applied to the routine checking of processes, products or services on a sampling basis as a method of ensuring reliability and suitability for the intended purpose. It is in the nature of a reassurance to purchasers of goods and/or services of reliability. In commercial settings, items found by purchasers to fail to meet guaranteed standards are either repaired or replaced as determined either by the supplier or by an independent arbitrator. Such an approach would clearly not be applicable in relation to clinical care. Quality control should, however, be an integral part of QA programmes in the production area of a hospital pharmacy, hospital sterile supply departments, pathology laboratories and departments of radiology. The need for quality control of processes and the standards to be attained in such areas are normally determined by the relevant professional body and appropriate external organizations.

Medical audit

Medical audit – as defined in Working Paper 6 of the National Health Service (NHS) White Paper *Working for patients*[1] – is 'the systematic critical analysis of the quality of medical care, including the procedures used for diagnosis and treatment, the use of resources, and the resulting outcome and quality of life for the patient'. This activity is an aspect of QA undertaken by doctors to review their own medical practice against agreed standards of care, the main emphasis in this particular activity being upon educational benefits that might result from peer review of clinical work. An important byproduct of this activity should be the definition of condition-specific protocols of care. Such protocols would normally include criteria for admission to hospital, a normal profile of investigations, clinical management procedures, paramedical treatments and criteria for discharge. These protocols of clinical care are in effect the standards against which clinical practice can then be assessed, either retrospectively or prospectively, by professionally trained audit support staff who will be able to prepare objective reports on clinical practice for consideration by a peer group. This variant form of audit of practice, often referred to as criterion-based audit, can be employed to assess quality within the standard framework for assessing the quality of care defined by Donabedian[2] in which he identified the three elements of care to be reviewed as:

Structure → Process → Outcome

The Process element may be further subdivided into:

Input → Action → Output

Quality in each of its primary elements should be assessed against agreed standards/indices; whilst there is relatively little difficulty in defining what acceptable standards of structure and process might be, there are considerable problems when it comes to defining quantifiable indicators of outcome. This is partly because of the range of perceptions of outcome which will differ between the members of a clinical team, management and the patient. There is a particular problem in relation to quality of care as perceived by the patient, which will almost inevitably change as time elapses between the completion of treatment and any review of the outcome. Equally, perceptions of quality of outcome will vary between the individual professional groups involved in the clinical management process – it is, for example, unlikely that a surgeon and an anaesthetist will measure quality of outcome against the same indicators. Also, the professional manager concerned with the economic consequences will have a further set of indicators of quality of outcome which will focus upon the use of resources and maintaining the financial viability of the organization.

The key issues to be addressed in medical audit are:

- quality of medical care
- the use of resources
- the outcome of treatment

Clinical audit

Clinical audit extends the audit process to broader groups, to include nurses and professions allied to medicine (PAMs) (e.g. clinical therapists and pharmacists) – similar in many ways to those groups in the 'therapeutic teams' that have existed for some time in such disciplines as geriatrics, paediatrics and psychiatry. Audit in such professional groupings makes it possible to review quality of care against jointly agreed care profiles negotiated with internal customer groups.

Multidisciplinary audit

Multidisciplinary audit widens the range of participants involved in the audit process further to include individuals whose primary concern relates to management of resources, which inevitably extends the debate to include activity, quality of outcome, and cost as measured against previously agreed targets and standards.

Resource management (RM)

The Resource Management Initiative, which was formally announced in Health Notice HN(86)34 in November 1986, named six acute hospitals in the UK to pilot the development of innovative general management practice in the National Health Service. More specifically, the Resource Management Initiative was a programme designed to improve effectiveness in providing health care by involving doctors in management and assisting them with appropriate information and business management support.

Casemix management

Casemix management (CMM) systems were developed in the USA, employing modern electronic data-handling systems to monitor hospital activity and the use of resources – placing medical conditions with similar cost implications into groups (diagnostic-related groups or DRGs). These DRGs are used to determine payments made by health insurance organizations. The data contained in CMM systems can be aggregated, employing standard spreadsheet techniques, to provide information for scheduling work in order to optimize the use of resources – accommodation, equipment, consumables and staff. These CMM systems are used to predict the likelihood of meeting annual targets of activity within an agreed budget. They can also be used to test the effects of rescheduling activities – changing the case mix – on the use of resources, which can be an important contributing factor to the effectiveness of departmental budgetary control.

As CMM systems become more widely available in the UK they will enable heads of services to carry out quality assurance activities related to clinical activity and workload concurrently with their other business planning routines. Also, in the course of managing departmental activity, CMM systems should simplify the process of identifying issues relating to medical care that need to be addressed in the audit process, either in the form of a full case note review of individual patients or as some form of criterion-based audit on particular diagnostic groups. CMM systems will assist in the process of identifying profiles of clinical care but beyond this they have no direct function in the routine clinical management of patients on a day-to-day basis.

Collaborative care profiles

These combine care plans employed in the nursing process with medical condition specific protocols as a basis for better management of resources, improving clinical outcomes and meeting agreed targets for clinical activity. Such care profiles, when reviewed during the progress of treatment, become a real-time form of QA activity.

Total quality management

Total quality management (TQM) is a 'system for improvement', setting out in unambiguous terms what is necessary for improvement of quality to occur *throughout* an organization. It was taught largely by American experts sent out to Japan in the 1950s[3] to help in restoring manufacturing industries. Berwick has proposed eight principles[4] of such a system for improvement, as follows:

- Intention to improve
- Definition of quality
- Measurement of quality
- Understanding interdependence
- Understanding systems
- Investment in learning
- Reduction in costs
- Leadership commitment

The first and the last of these principles are crucial to the success or failure of any quality improvement programme which places an important responsibility upon those doctors who fulfil leadership roles because of their dual function as teachers and practitioners of quality improvement methods. Indeed, within a TQM culture 'everyone has two jobs – their job as they usually know it, and the job of making their own job better'.[4] Areas of particular concern that are likely to impede the involvement of doctors in TQM are time constraints, coming to terms with new concepts of mutual interdependency, the traditional role patterns of the profession and trust.

BS 5750

BS 5750 is the UK version of International Standards ISO 9000–9003, which sets out strict quality standards for total integration and control of all of the elements within a particular area of operation. This exacting standard, which is in many respects a minimum standard, is being looked upon as the 'gold standard' by private nursing homes and a range of other institutions.

Risk management

Risk management is a formal process for the avoidance of unplanned major expenditure arising from mishaps or inconvenience to patients, staff or visitors. A major contributor to this problem is the cost of compensation to patients for the consequences of human error or untimely failure of hospital equipment. An important activity in this connection is 'untoward incident reporting', being promoted in the UK by the Medical Defence Union Limited (MDU Ltd).[5]

Establishing a QA programme

The primary aim of QA is to improve the quality of care and services, provided both for patients and for internal customers by departments/directorates. This is achieved through three basic activities: formal meetings; monitoring using CMM systems and other data aggregation/reporting routines employed in the course of management of resources; and a selection of other review methods such as satisfaction surveys to identify changing consumer expectations.

In the USA, functioning departmental quality assurance committees are mandated by the Joint Commission on Accreditation of Healthcare Organizations (JCAHO). Their activities are similar in scope to quality assurance as conducted in the UK. For example, departments of anaesthesia are obliged to conduct continuous contemporaneous audits of monitoring standards, pre- and post-anaesthetic visits, completeness of anaesthetic records, evidence of anaesthetic machine 'check-out' procedures, use of regional anaesthesia, and major and minor complications of anaesthesia. Death reports must be filed in a timely fashion, whether or not events during anaesthesia were responsible. Problems such as chipped teeth, nerve damage, myalgias, intraoperative awareness and post-lumbar puncture headache would be subject to review. Case discussion conferences, often on a weekly basis, are an important part of quality assurance. In most jurisdictions in the USA, any items discussed at such conferences are considered legally privileged and may not be subpoenaed in the event of litigation. This provides the complete freedom necessary for conducting meaningful and honest discussions of all untoward incidents.

A useful starting point for discussions could be at a formal departmental or unit meeting focusing on quality issues, held at regular intervals, preferably during normal working hours, as a part of a structured QA programme. The frequency of such meetings will be determined by the need for such discussion. Those attending should be provided with adequate opportunities to raise issues of concern in problem areas and to suggest methods of improving the provision of care. Encouragement should also be given to individuals to become involved in the process by commissioning audits or undertaking topic review which could help to bring about the necessary changes. The focal points in discussions in clinical areas about patient care are, more often than not, primarily related to the management of individual patients, ranging from a review of progress to determine how to proceed or to highlight unusual features of particular interest or concern. The simple process of transferring the discussion into the setting of a formal meeting concerned with quality issues in relation to medical care should broaden the debate to include comparison of observed current practices with agreed standards either of care to a particular group of patients or of services provided to support the work of colleagues. Such meetings should be held with the intention of promoting changes in practice where problems exist, which could lead to either improved medical care or to more

efficient use of departmental assets – buildings, equipment time and staff – with a greater sense of job satisfaction for those contributing to these processes. Evidence of benefits can help in promoting interest in QA outside the immediate area involved.

An important but inescapable feature of setting up any sort of QA initiative is making realistic assessments of both the difficulties and the opportunities that can influence progress. This can be a particular problem when other activities that are essential to obtain the benefits from changes of practice are lagging behind the establishment of a QA programme. Whilst a considerable amount of progress can be made using existing sources of data employing manual data handling techniques, more rapid progress is often made possible by employing modern IT systems with automatic handling of data downloaded from feeder systems used in associated departments. Also, in any QA activity there is a need to be able to monitor the process, hence the widely voiced plea for an 'audit of the audit' – a predictable cry from doubters of the value of the medical audit, particularly when a sum in the region of £200 million had already been committed by the Department of Health to support this activity in the UK between 1989 and 1995, which in their opinion might have been spent more productively on direct improvements in patient care. Such individuals would appear to discount the time scale normally required to bring about a culture change, which is what medical audit is, and the inevitability of delays in such processes reaching full maturity.

In all endeavours of this type there is an inevitable plateau after the first flush of enthusiasm, normally associated with a slump in morale. At such times in any endeavour the greatest need is for support and reassurance before the group slides back to business as usual – an event all too acceptable to those who felt theatened by the process in the first place. Industrial experience suggests that this problem is inevitable in the development of QA initiatives, and that with sustained effort supported by commitment from management at the appropriate level or levels it can be surmounted. Within organizations providing health care it is essential to recognize that grafting QA activity onto already busy personal timetables, with the inevitable conflicts over diverting limited medical time away from direct patient care, could for some be the last straw.

For these reasons a 'control loop' to demonstrate the effectiveness or otherwise of such activities must be set in place at the outset if possible. In the case of medical audit this might be one of the functions of the infrastructure to support the activity together with regular stewardship reporting to funding bodies. The feedback from arrangements of this sort can be a vital ingredient which enables organizations to maintain their commitment. Self-sustaining QA takes time to mature and it can be difficult to maintain morale in the period between the beginnings of such activities and the point where the efforts/results ratio is sufficiently rewarding to remove the need for external support. Commercial experience suggests that this point usually coincides with the realization that improvement in quality 'is not simply about doing the same things better, but more about doing fundamentally different things in a fundamentally different way'.[6]

Review of the effectiveness of QA programmes can be addressed in a variety of ways, some of which are referred to later, namely: morbidity and mortality meetings, activity reporting, topic/criterion-based audit and untoward incident reporting. As the process of auditing activity matures, patterns of care should emerge that will lead to a better framework for prospective monitoring of disease-specific clinical care profiles, and as awareness of the expectation of customers becomes more complete it should become easier to determine what the important quality issues are. Practical points to be taken into account in preparing for monitoring quality issues include an awareness of:

1. The usefulness of existing sources of data, methods of data collection and aggregation, and the facilities for the preparation of presentation material
2. The time required to conduct reviews and to obtain access to experienced support staff able to provide effective contributions to QA studies
3. The possibility of collecting data on a prospective basis because of the well-known difficulties associated with record keeping
4. The aim of QA reviews, which is to detect remediable problems in patient care or services in order to correct them within a realistic time frame

A framework for QA in anaesthesia

With the relentless advance of the science and techniques of medical practice, it becomes increasingly important that credible arrangements are in place to demonstrate that improved technology, new pharmaceuticals and updated treatment protocols offer a greater likelihood of better outcomes following treatment. This becomes particularly important in the context of the cost of care because the majority of advances in care that lead to significant improvement in outcome consume more resources than their less effective predecessors. For the purpose of ensuring credibility of a QA programme it is imperative that all concerned:

1. Are aware that they can have a direct influence upon both the structure and the processes employed – that

is, their concerns will be promptly and competently attended to.
2. Feel free to criticize systems employed by the organization without fear of punishment or reprisal.
3. Are encouraged by the structure of systems or standards incorporated into service agreements with internal and external customers rather than in issues arising from the personal agendas of their immediate supervisors.

Continuous quality improvement, which has been a long-standing feature of medical practice, embraces a range of self-imposed disciplines, including keeping up to date with the medical literature, attendance at formal postgraduate training, and research, along with a whole range of other less formal professional activities. This in the past has ensured that all members of medical teams work from similar frames of reference, and share a similar outlook on quality of care. Effective QA programmes require that teams address two basic questions: Are we doing the right things (strategic quality)? and Are we doing them right (process quality)?[7] This demands an understanding not only of the requirements of the processes involved in providing a service but also of the needs and expectations of the customer on an individual basis – both of which are best understood by the staff providing the service. An important prerequisite for an effective QA programme is an explicit commitment from appropriate levels of management, which in the case of departments/directorates must include the clinical director. The reason for this is twofold: first, the need to support the educational aspects of QA and, second, to secure staff commitment to meeting the standards incorporated into service agreements entered into with customers.

To ensure the success of a QA programme, it is vital that a formal framework for action is defined to be sure that all concerned understand how it relates to the aims and objectives of the group, can identify the customer base and are able to recognize the constraints on providing quality care. A possible approach for a department of anaesthesia would be as set out below.

Formulate a 'mission statement'

The mission statement will acknowledge that quality is consistently meeting agreed standards of practice, which truly reflect customer expectations as determined by acceptable outcomes of clinical care, and through improvement projects that place a premium on the *prevention* of problems – the key objective being 'to get it right first time all of the time'.

Define the department's aims

Defining the aims of the department should include details of levels of activity to be incorporated into service agreements with internal customers to enable them to fulfil their own contracts with external purchasing bodies – set out as targets for numbers of general anaesthetics, regional blocks, acute pain relief service, support in high dependency areas, procedures for the relief of chronic pain and contributions to the resuscitation service.

Define standards of care

These will include reference to preoperative assessment and preparation; routine pre-anaesthesia equipment checks; monitoring during anaesthesia; record keeping during anaesthesia; and arrangements for postoperative care. There will of course also need to be agreement upon standards for the supervision and training of junior medical staff, and levels of assistance during the administration of anaesthesia.

Identify the customer base

The customer base comprises two groups:

1. *External customers* – patients and, in the case of children and the elderly, their parents/carers
2. *Internal customers* – clinical colleagues requiring anaesthesia services for their patients, nursing staff caring for patients who have undergone procedures requiring anaesthesia, and also paramedical staff involved with preparation and/or aftercare of patients undergoing anaesthesia procedures.

Establish a regular programme of meetings

These should embrace a range of topics, including the following.

Regular reports on departmental activity

Normally presented by the clinical director, these reports highlight any noteworthy variances or trends likely to have a bearing on the use of resources. This is an exercise that will become increasingly important in the UK as the National Health Service internal market moves away from its traditional 'block' contracts to a contracting process based on cost and volume and cost per individual case.

Topic- and criterion-based audit

This assesses current practice against agreed standards of quality on the three elements of care identified by Donabedian,[2] and subsequently reviews the effects of any resulting changes in practice on the outcome of clinical care. This discipline can be addressed within the framework of the 'audit cycle': identifying that a problem exists, setting standards, evaluating practice against those standards; negotiating change of practice; and, after a suitable interval, repeating the audit to find out whether the change has influenced outcomes. In drawing up a protocol for audit it is vital that the indicators of quality should be unambiguous; this is particularly relevant when defining indicators of quality of outcome. A pro forma for use in paediatric anaesthesia is given in Fig. 75.1.

Reports on untoward incidents

These have two purposes: to record the events that have caused injuries so that they can be investigated and documented; and to record 'near misses' so that steps can be taken to prevent possible harm to others in the future.

This activity can be carried out using either an in-house method or, alternatively, the criterion-based untoward incident reporting system developed by the MDU Limited, which comes complete with pads of report forms together with a list of specific criteria that identify the majority of incidents likely to lead to complaints or claims for compensation and a software package (MDUBASE) to provide an analysis of any potential contingent liabilities arising from untoward incidents.

It is important, in establishing arrangements for reporting untoward incidents, to include suitable safeguards for confidentiality – for the patient, for staff and for the hospital – along the lines set out by the MDU Limited.

Morbidity and mortality meetings

These should include the presentation of cases included in the National Confidential Enquiry into Perioperative Deaths (National CEPOD) in the UK.

Record decisions taken

Establish arrangements for the systematic recording of decisions taken at meetings to review quality issues to ensure that agreed aims and standards are achieved. For example, reports on criterion-based audit should include in summary form details of the subject, the methodology employed, the conclusion reached, any action to be taken and the date of a future review of the process to determine

Paediatric Anaesthesia Audit

PATIENT CODE: AGE: YEARS: MONTHS: WARD:

THEATRE: DATE: TIME: AM/PM

OPERATION PERFORMED:

GRADE OF PROCEDURE: MAJOR/MINOR/INTERMEDIATE

ANAESTHETIST   SURGEON:

Consultant code: Consultant code:
Trainee code: Trainee code:

WAS SURGERY CARRIED OUT ON THE DAY OF ADMISSION?   YES/NO

WAS DISCHARGE ON THE DAY OF ADMISSION?   YES/NO

DID AN ANAESTHETIST SEE THE PATIENT ON THE WARD PRE-OP?   YES/NO

  IF YES – WAS A PARENT/GUARDIAN PRESENT?   YES/NO

  State grade of anaesthetist – Consultant/Trainee

WAS A PARENT OR GUARDIAN GIVEN AN OPPORTUNITY TO BE PRESENT DURING THE
INDUCTION OF ANAESTHESIA?   YES/NO

WAS THE PARENT OR GUARDIAN PRESENT DURING THE INDUCTION OF ANAESTHESIA? YES/NO

  IF NO – WHY NOT? ....

WAS THE PARENT/GUARDIAN SATISFIED WITH THE INFORMATION GIVEN ABOUT THE
ANAESTHETIC MANAGEMENT OF THE CHILD?   YES/NO

  IF NO – WHY NOT? ....

DID A CONSULTANT SUPERVISE THE ANAESTHETIC?   YES/NO

WERE WRITTEN POSTOPERATIVE INSTRUCTIONS GIVEN BY THE ANAESTHETIST   YES/NO

  In relation to:

  a) Postoperative analgesia – YES/NO

  b) other – state nature ....

Fig. 75.1 Pro forma for criterion-based audit of paediatric anaesthesia practice.

whether any change of practice negotiated has actually influenced the outcome of care. An example is given in Fig. 75.2.

Review

The work of the directorate should be reviewed annually. The review should include objective assessment of success in achieving goals and meeting any targets or quality standards included in service agreements.

Identify factors that affect outcomes

In anaesthesia, factors that can influence the quality of care and outcome include training, equipment available, the technical quality of anaesthesia, local protocols, local arrangements for handling crash/emergency calls and unexpected untoward incidents.

Setting standards for QA in anaesthesia

Formulating standards for optimal care locally is the key to successful QA practices, the essence of which is that, whilst standards should be subject to local agreement, they must not represent lowest common denominators of quality – such standards should be both realistic and attainable. In many instances common sense will determine appropriate standards, and where there are national bodies that have issued guidance this should simplify the process. Any

---

AUDIT OF ANAESTHESIA RECORD

Objective - To assess the completeness of records of anaesthesia and their value as an account of clinical management of patients undergoing general anaesthesia.

Method - Criterion-based audit.

Key items of information relating to managing anaesthesia, including attention to detail in record keeping, to be included in a proforma for carrying out a criterion-based audit, were agreed. A review of clinical records relating to 60 anaesthetics administered in the Main Operating Theatre Suite at Wythenshawe Hospital in the period November 1990 to February 1991 was carried out.

Result - The overall standard of recording of agreed data during management of anaesthesia was of acceptable standard and all of the information likely to be required for planning future anaesthetic management was in a majority of instances available somewhere in the clinical notes.

Conclusion - The current method of collecting together data relating to the management of anaesthesia was considered to be inadequate to meet comtemporary needs, mainly because the present arrangement stores relevant items in separate locations in the clinical notes.

Recommendation - It was agreed to commission a revision of the form for recording of anaesthesia management.

Arrangements for further audit - A similar review employing new criteria will be carried out six months after the proposed new form has been introduced.

Fig. 75.2 Report on criterion-based audit of anaesthesia record.

known expectations of patients should be taken into account where appropriate.

The regulation of standards for the practice of anaesthesia is carried out by a number of agencies, including national and international bodies, which are principally concerned with reducing the incidence of anaesthetic morbidity and mortality by providing guidelines on the working environments, clinical practice and, more recently, on the outcome of care. This function in the UK rests primarily with the Royal College of Anaesthetists and the Association of Anaesthetists of Great Britain and Ireland, together with numerous other bodies that advise on issues related to patient safety during the course of anaesthesia. The need for such bodies is reflected in the statistics of the MDU Limited, which show mishaps associated with anaesthesia accounting for around 12 per cent of all medicolegal compensation claims in the UK and adverse occurrences related to surgical procedures (which include those for anaesthesia) accounting for about 60 per cent of the total.

In order to explore possible arrangements for QA in anaesthesia, it is once again helpful to fall back on the framework set out by Donabedian: structure, process and outcome. 'Structure' includes the environment and the resources, physical and human, required for the safe practice of anaesthesia. The other two elements are self-explanatory.

Structure and process

Both structure and process fall within the domain of the Royal College of Anaesthetists and the Association of Anaesthetists of Great Britain and Ireland, the two major professional bodies concerned with standards in the practice of anaesthesia in the UK. Also involved is the Department of Health as the major provider of health care services and the British Standards Institute which, often in collaboration with international standards organizations, deals with safety aspects of design, evaluation and approval of equipment used by anaesthetists.

The Royal College of Anaesthetists, as the professional body primarily responsible for both setting and maintaining standards for the practice of anaesthesia in the UK, carries out its functions under four main headings: examinations systems, recognition of hospital training schemes, education and liaison with other bodies including the national government. College representatives serve on all consultant and staff doctor appointments committees in the UK, and College regional educational advisers review and approve the content of these posts to ensure that they conform with College recommendations.

Through its examination system the College confers two Diplomas by a three-part examination system. The successful completion of the first part leads to the Diploma in Anaesthesia, which can be obtained after practising in a recognized hospital for 12 months. Fellowship of the College is gained after completing at least 2 years in recognized training posts and passing the remaining two parts of the examination. The recognition of hospital/regional training schemes during the basic/pre-fellowship period, the time spent in the grades of senior house officer and registrar, is carried out by the College. Higher professional training (HPT), the period that follows obtaining the diploma of Fellow of the Royal College of Anaesthetists (FRCA), is overseen by the Joint Committee for Higher Training in Anaesthesia (JCHTA). The JCHTA includes members from the Irish Faculty of Anaesthetists, the Association of Anaesthetists of Great Britain and Ireland, the Association of Professors in Anaesthesia, the Department of Health and a representative of Associates in Training. Successful completion of HPT is marked by the award of a certificate of accreditation, which serves as an indication to an advisory appointments committee of the eligibility of the holder to be considered for appointment to a post in the consultant grade. Further contributions of the College to maintaining standards are made in giving advice to the Department of Health, particularly on staffing levels, and through its Quality of Practice Committee which is promoting national and local quality initiatives.

In the USA, certification as a specialist anaesthetist is the resonsibility of the American Board of Anesthesiology (ABA). Certification is granted on the basis of a written and oral examination taken during and following completion of an approved residency. In contrast to practice in the UK, Board certification is not necessarily required for hospital practice in the USA. To a large extent, each hospital determines the criteria for accrediting physicians working within its walls. At present, a significant number of anaesthetics are administered by licensed physicians *not* certified by the ABA, as well as by non-physician certified registered nurse anesthetists (CRNAs). It is most probable that, in coming years, an increasing number of hospitals will require ABA certification before allowing a physician to practise anaesthesia in their institution, although it is unlikely that CRNA practice will disappear.

It is a truism that learning does not cease after specialist certification. The ABA has recently begun a voluntary programme for 'recertification' of its diplomates. From a less voluntary point of view in the USA, many states require physicians to demonstrate evidence of continuing medical education (CME) in order to maintain their medical licence. Attendance at approved lectures (e.g. 'Refresher Courses' conducted by the ASA) is the usual way in which CME may be documented, although some credit may be given for participation in case conferences within the physician's hospital. Each state has the authority to specify particular CME requirements for continued licensure.

In its commitment to improving standards of medical care the Royal College of Anaesthetists has lent its support to the inclusion of participation in medical audit activities in the job plans of all NHS consultants.

The Association of Anaesthetists of Great Britain and Ireland has, through studies carried out by ad hoc working parties on specific problem areas, provided guidelines on a number of topics related to the organization of departments of anaesthesia and health hazards, including HIV and hepatitis B, in relation to anaesthesia.The Association influences safety standards for anaesthetic equipment through members of its council serving on British Standards Institute working groups advising on these issues. The Association played a leading role in the introduction of the National CEPOD, which, in addition to conducting enquiries into both the anaesthetic and surgical management of all patients dying within 30 days of undergoing surgical procedures, carries out normative surveys of management in diagnostic/age group specific areas of surgical treatment.

Since the implementation of the National Health Service and Community Care Act in April 1991, the quality issues in relation to clinical care throughout the NHS are now being addressed in medical audit programmes held by all hospital departments/directorates, out of which numerous models of best practice – a balance between activity, cost and quality – are now beginning to emerge.

Cost management aspects of QA

The heightened interest in the role of doctors in management as espoused in the 1991 reforms in the NHS in the UK and the widespread introduction of medical directorates has placed responsibility for budgets high on the agenda of those consultants who have undertaken the role of clinical director.

In this context, monitoring progress towards agreed targets of activity and standards of clinical practice become key aspects of quality of care/service to customers, which include major purchasing bodies, fund-holding general medical practitioners and individual patients. Hence the importance of effective casemix management as a business tool for safeguarding organizational financial viability.

An important area in which QA/medical audit can impinge upon financial control is that of risk management – by taking action to reduce the likelihood of unplanned major expenditure. This has become an issue of particular concern in the NHS since health authorities and trusts became responsible for medical negligence claims, because these bodies are not empowered to insure themselves against negligence, and also in view of the increasing number of claims, with higher awards than previously in those that are successful. A further reason for the importance of risk management in departments/directorates of anaesthesia is to be found in the high incidence of negligence claims against anaesthetists published by the MDU Limited (see p. 1476). One of the purposes of untoward incident reporting must be to ensure that formal documentation is available which describes both the incident and what action was taken to reduce the likelihood of a recurrence of the same mishap. Also, the available evidence would suggest that aggrieved patients and their advocates in organizations such as Action for the Victims of Medical Accidents contend that the victims of medical mistakes do not necessarily want compensation but an admission that something went wrong, an apology and a reassurance that the circumstances which led to the mistake will be rectified. Whether the patient subsequently sues or not, this approach is correct. With this in mind, it is not possible to overemphasize the importance of keeping detailed records not only of all anaesthetics but also of those of any untoward incidents associated with anaesthesia, in order to safeguard the interests of all parties – the patient, the doctor and the employing authority. In this context suffice it to say that, in medical negligence cases involving any aspect of anaesthetic practice, either no record or an incomplete anaesthetic record has no case in law for the defence. It should also be remembered that, in the eyes of a lay person, the quality of an anaesthetic record is likely to be regarded as an index of the quality of the anaesthesia administered.

Lastly, but far from being least important, is the need to be circumspect in deciding upon the most appropriate way to determine priorities in medical care when limited resources place limitations upon choices in medical care. This dilemma has become sufficiently common for doctors for the General Medical Council in the UK to have included guidance to doctors on decisions about access to medical care in the section on contractual arrangements in health care of the 1992 edition of its handbook *Professional conduct and discipline: fitness to practise*:

> The Council endorses the principle that a doctor should always seek to give priority to the investigation and treatment of patients solely on the basis of clinical need. Acknowledging this, doctors have to work within resource constraints and, whatever the circumstances, they must make the best use of resources available for their patients, recognising the effects their decisions may have on the resources and choices available to others.

Consumerism and QA

The question, 'Who has right of access to information on QA within health care organizations?' can be a sensitive issue, particularly where it relates to individual patients and individual doctors because of the possible risk of breaches of confidentiality. However, there can be no logical objection to the results of QA expressed in general terms being made available in the public domain either in the form of an annual report or a report on the outcome of a particular QA activity together with any recommendations for future action.

In the particular case of medical audit, interested parties might include the clinicians whose practice has been the subject of audit, professional colleagues, nursing staff and other paramedical groups with whom the specialty concerned has entered into service agreements that include undertakings about standards of care or service. Other bodies with a legitimate interest in the general results of QA activities include purchasing bodies, major purchasers and fund-holding general medical practitioners, and patients or those with an interest in their welfare – in the case of children or the elderly this will include their immediate relatives or carers.

Potential benefits of QA programmes

Numerous claims are made for the potential benefits of QA but few of these are likely to be realized without ownership of departmental/directorate aims and objectives at operational levels endorsed by meaningful commitment from the chief executive officer, and in the case of individual departments/directorates the clinical directors. This commitment must be matched by action to enable those delivering care/services to provide the sort of effectiveness and efficiency from which benefits will accrue both to the recipients of the services provided by health care organizations and to individual providers.

An effective QA programme should enable the organization to:

1. Anticipate the requirements of consumers in the future and how it will fulfil these expectations
2. Ensure that individual members of the team have a clear understanding of their personal contribution to fulfilling corporate aims and how they can make an effective contribution to the smooth running of the organization
3. Reduce/eliminate inappropriate variation in patterns of care and, as a result, improve the quality of clinical outcomes which, together with measures for optimizing the use of assets (building, equipment and people) and hence eliminate waste, should release funds to develop new services.

Learning from litigation in the USA

An important aspect of quality assurance and audit in the USA is the Closed Claim Project developed by the ASA in the 1980s. Information was gathered by examination of insurance company files dealing with closed malpractice claims against anaesthetists in the state of Washington. This important educational and quality assurance activity has focused on potential causes of injury, a wide variety of liability concerns, and evaluation of how the anaesthetic community makes crucial decisions regarding standards of care.[8, 9]

The way forward

QA as a vehicle for formally declaring an intention to provide services and goods in quantities and at standards agreed with prospective purchasers has only recently become a part of the culture in health care organizations in the UK, and it has a long way to go before it reaches full maturity. A start has been made in such areas as medical audit but in many instances this has reached a plateau and will require persistence to realize its full potential in the management of clinical care and services. New momentum will almost certainly be stimulated from within the NHS as the contracting process matures, particularly when it is perceived that attention to quality issues can influence financial viability. The natural train of events as the majority of NHS hospitals acquire trust status is that quality assurance is likely to evolve into total quality management and all that this philosophy entails – including the concept of continuous quality improvement. The critical success factors likely to influence this change are not immediately apparent, but there could be an increased emphasis on corporate identity, greater degree of decision making at operational level, and promoting the concept that 'the patient not the supervisor is the primary customer' . . . not to mention the realities of true competition materializing in the NHS internal market.

A major defect in the arrangements for improving the quality of medical care in the UK is the lack of any formal need for doctors to take part in programmes of continuing professional education after the completion of HPT. This matter is being addressed by the General Medical Council and the Royal Colleges together with their Faculties and other bodies, including the Standing Committee on Postgraduate Medical Education, and it is likely that changes will result from this activity in the not too distant future.

REFERENCES

1. *Working for Patients*, Working Paper 6, Cmnd 555. London: HMSO, 1989.
2. Donabedian A. The quality of care: how can it be assessed? *Journal of the American Medical Association* 1988; **260**: 1743–8.
3. Deming WE. *Out of the crisis*. Cambridge MA: MIT Center for Advanced Engineering Study, 1986.
4. Berwick DM. Sounding board: continuous improvement as an ideal in health care. *New England Journal of Medicine* 1989: **320**: 53–6.
5. *Healthcare risk management: criterion based reporting of untoward incidents*. London: Medical Defence Union Limited, 1990.
6. James BC. *Quality management for health care delivery*. London: PA Consulting Group (in-house publication), 1989.
7. *Total quality experience: a guide for the continuing journey*. London: PA Consulting Group (in-house publication), 1991. [PA Consultant Group Publications are available from: PA Consulting Group, 38 Hans Crescent, London SW1X 0LZ]
8. Caplan RA, Ward RJ, Posner KL, Cheney FW. Unexpected cardiac arrest during spinal anaesthesia: a closed claims analysis of predisposing factors. *Anesthesiology* 1988; **68**: 5–11.
9. Caplan RA, Posner KL, Ward RJ, Cheney FW. Adverse respiratory events in anesthesia: a closed claims analysis. *Anesthesiology* 1990; **72**: 828–33.

FURTHER READING

Koch HCH. *Implementing and sustaining total quality management in health care*. Harlow: Longman, 1992.
Øvretveit J. *Health service quality – an introduction to quality methods for health services*. Oxford: Blackwell Scientific, 1992.
Pawsey M. *Quality assurance for health services – a practical approach*. Sydney: New South Wales Department of Health: 1990.
Taylor TH, Goldhill DR. *Standards of care in anaesthesia*. Oxford: Butterworth-Heinemann, 1992.

Chapter 76

Ethical Considerations in Anaesthesia

Cynthia B. Cohen

What does informed consent mean?
When is it ethically acceptable to stop treatment?

Should anaesthetists practise euthanasia?
What should anaesthetists do about chemically dependent colleagues?

No one enters medicine without becoming vividly aware of its ethical tradition. The science and art are imparted along with the ethical imperatives: seek the patient's benefit, do no harm, be respectful and compassionate, preserve confidences. These and other medical values are well known, and their vitality in the behaviour of many individual physicians is remarkable. It is clear that the current renewal of interest in medical ethics has not been stimulated by perceptions of immorality among physicians.

Instead, it has developed because the traditional values of medical ethics are being challenged by new medical capabilities and social forces. Novel medical technologies and surgical procedures are proliferating, health care systems are becoming more bureaucratic and financially complex, and the public has become increasingly sophisticated about medical possibilities and limitations. These and other changes are creating perplexing issues for physicians and patients, and are impelling us to search for new ways in which to respond.

Many question whether the ancient Hippocratic imperative, 'Do no harm', can be applied simplistically to today's patients. Is it harmful to fail to send the elderly patient suffering from chronic lung disease to the modern intensive therapy unit for one last attempt to keep him alive or, on the contrary, is it harmful to do so? How can the ideal, 'Preserve confidences', be maintained in an era when patients receive treatment from many different health care professionals and their records are reviewed by countless administrative staff?

When long-standing medical values no longer seem applicable in the face of modern medical capabilities and social forces, they must be interpreted in the light of new circumstances rather than abandoned. In the field of anaesthesia, the values of medical ethics must be applied to such issues as:

1. What role should the patient play in making medical treatment decisions in an era of increasingly complex medical knowledge?
2. When is it appropriate to end the use of technological machinery and procedures?
3. Is it ethically sound not to resuscitate the patient with a 'Do not resuscitate' order during surgery?
4. Should anaesthetists be made responsible for carrying out euthanasia?
5. At what point does the provision of drugs no longer constitute a form of palliation, but of euthanasia?
6. What should be done about the impaired colleague?

Although medical professionals have special obligations, these are meaningful only in the context of our general ethical framework. Because we live in a pluralistic society that tolerates varying beliefs and values, we sometimes forget that we share a core set of values. We agree that those who suffer should be treated with compassion, that confidences should be preserved and that we should deal with others honestly. These values are translated into the sphere of medicine through the principles of medical ethics. We need to apply such widely shared values to the novel concrete situations that arise in the provision of anaesthesia today.

What does informed consent mean?

A major shift in the understanding of the physician–patient relationship took place in the 1970s. There was a change from a traditional, almost singular, focus on the benefit of the patient to a new emphasis on patient self-determination. Medical paternalism, it was maintained, had to be corrected by giving greater weight to patient autonomy. To be autonomous is to choose one's actions intentionally with understanding and without being controlled by others.[1] An underlying justification for an emphasis on patient autonomy is the belief that patients are ordinarily in the best position to identify their best interests in the light of their own value system. Respect for patient autonomy requires that they be provided with information that is material to making well-formed decisions about their medical care.

In the decade of the 1970s, this meant that patients ought to have the opportunity to give an informed consent to their treatment. This new emphasis on informed consent and patient autonomy, while giving greater prominence to patients' values, created problems and contradictions for both patients and physicians. Medical practitioners believed that they could not convey to patients the full range of technical knowledge they would need to understand in order to give a truly informed consent. Yet they saw the burden of decision-making gradually shift to patients. Many physicians believed that patients were being pushed to make decisions about their own medical care although they did not necessarily appreciate the significance of these decisions. As a consequence, physicians increasingly came to view informed consent as a meaningless fiction that amounted to no more than a signature on a piece of paper.

In the 1980s, this shift toward informed consent was both reinforced and challenged. The concept of 'shared decision-making' was proposed as a way to correct the exaggerated individualism that had sometimes been produced by the emphasis on patient autonomy.[2] The physician was no longer to be viewed as an impartial bystander who gave information to patients and then acted upon their decisions, but instead was considered to be an active, mutual participant in decisions about treatment. Yet an appreciation of the importance of patient autonomy and a demand that it receive the safeguard of informed consent prevailed, even as 'shared decision-making' was adopted.

In the 1990s, there are good reasons for considering how to give practical application to a modified view of informed consent that takes into account 'shared decision-making'. This is particularly necessary in the context of anaesthetic practice where decisions must often be made in advance of treatment. In becomes important to develop new ways of involving patients in their medical planning before they receive anaesthesia so that, if circumstances change during the course of surgery, there is a mutual understanding of how the anaesthetist will proceed.

Patients must do more than express agreement with a proposed medical treatment when they give informed consent. They must actively authorize the proposed course.[3] Legal, medical, psychological and regulatory writings have enunciated the following elements of informed consent:[4] (1) competence, (2) disclosure, (3) understanding, (4) voluntariness and (5) consent or 'shared decision-making'. Patients give an informed consent to a medical intervention if they are competent to choose, receive a thorough disclosure about their options, understand what is disclosed to them, choose voluntarily, and consent to the intervention or reach a mutual decision about it with their physician.

Competence

Competence is more appropriately described as a precondition to informed consent than an element of it.[3] Patients must possess sufficient decision-making capacity to consent to or refuse a proposed treatment.[5] Physicians must begin with the presumption that adult patients are competent to make decisions, according to the law. The burden of proof falls on physicians to show that an individual patient is incompetent.

The law traditionally presumed that incompetence was global and that a person who was incompetent to make one decision was incompetent to make all others. Competence, however, can vary from context to context.[6] A person may be incompetent to drive an automobile but competent to make medical treatment decisions. By the same token, a person may be competent to make minor treatment decisions but incompetent to make major ones about the use of life-sustaining treatment. Courts are beginning to accept the notion that patient competence can be decision-specific. For practical and policy reasons, however, they need a dividing line on the continuum from competence to incompetence so that any person below a certain level of decisional capacity will be treated as incompetent. When physicians have difficulty in assessing whether a specific patient falls below this dividing line, it may be necessary for them to evaluate that patient over a period of time with regard to understanding, deliberative capacity and coherence before coming to a conclusion in the clinical context.

Disclosure

Disclosure is a primary condition of informed consent. In a leading American legal case, *Canterbury v Spence*,[7] the court held that patients should be told of (1) the nature of the procedure, (2) its probable benefits, (3) its probable risks or negative side effects and (4) the alternative kinds of treatment that could be provided and the consequences of these. Patients need not be told everything known to medical science about a procedure, but should be given information material to making a decision.

A debate about the standards that should govern the disclosure of information to patients has been taking place in the courts. Two major standards have emerged: the professional practice standard and the reasonable person standard.

The professional practice standard holds that physicians should disclose what it has been customary to disclose within the medical profession. If physicians do not generally tell their patients of certain risks, these need not be revealed. This rule creates several difficulties. It may encourage the perpetuation of negligence should the medical community tend to offer little information about a certain practice. Moreover, the standard can undermine the patient's right of autonomous choice, because decisions for or against treatment involve not only medical judgements but also value judgements that belong to the province of the patient. Because of such difficulties, the professional practice standard has been replaced by the reasonable person standard in some jurisdictions.

This second standard holds that the physician should disclose information that the reasonable person in the patient's situation would consider material. It requires that the physician consider what the reasonable patient would want to know, not what a particular patient would want to know. This means that all significant risks of a procedure and alternative procedures should be disclosed to patients. Even if the behaviour of physicians conforms to recognized and routine professional practice, on this standard, they may be guilty of negligent disclosure.

The law allows exceptions to the rule of informed consent in the doctrine of therapeutic privilege. In cases of emergency, incompetency, waiver and the like, physicians need not meet all the elements of informed consent if this would put the patient at great risk. They can intentionally and validly withhold information based on a medical judgement that to divulge the information would be potentially harmful to a depressed, emotionally drained or unstable patient.[8] The formulation of this therapeutic privilege varies across legal jurisdictions. Some allow physicians to withhold information if disclosure would cause any deterioration in the condition of the patient, but others permit this only if to impart the information would have serious consequences for the patient's health.

Understanding

The element of understanding is an important component of the process of obtaining an informed consent. Patients must have the ability to appreciate or understand the nature of the options for treatment available and their consequences. They must be able to communicate a preference.[5] When patients are sick, frightened or ignorant, what can physicians do to help them make good decisions based on sound understanding? Establishing an atmosphere that encourages patients to ask questions and eliciting the interests and concerns of patients are as important to their understanding as is providing them with medical information. Patients need not have a complete understanding of a course of treatment,

but they should have a substantial grasp of the central significant facts.

There are constraints on the amount of information that patients can meaningfully process and retain. Indeed, too much information may be an obstacle to understanding and may be as likely as too little information to produce uninformed decisions. Even though information must often be given to patients prior to anaesthesia in a compressed manner, it should be offered in digestible amounts and the central points should be reviewed with patients.

Patients often have difficulty in processing information about risks.[3] In one study, physicians, patients and graduate students were asked to make a hypothetical choice between two alternative therapies for lung cancer: surgery and radiation therapy.[9] The preferences of each group were affected by whether the information about outcomes was framed in terms of the probability of survival or the probability of dying. When the outcomes were framed in terms of survival rates, only 25 per cent chose radiation over surgery. However, when the outcomes were framed in terms of mortality rates, 42 per cent preferred radiation. The way in which the risk of immediate death from surgical complications, which has no analogue in radiation therapy, was presented appears to have made the difference. To avoid such effects, physicians need to provide patients with both the mortality and the survival information.

Voluntariness

By 'voluntariness' is meant that a person is free from manipulative and coercive influences exerted by others in order to control them. A consent manipulated by deliberate misrepresentation or by outright threats is no consent. When a physician threatens to abandon a patient unless that patient undergoes a certain procedure, this is coercion – the patient becomes controlled by the will of the physician. Similarly, when a physician indicates that the patient has no other choice than to undergo a particular treatment, this allows the patient no discretion and amounts to a form of coercion. Deception that involves such strategies as lying, withholding information, omitting a vital qualification or using a misleading exaggeration constitutes unacceptable manipulation. Those who solicit consent by such means fail to respect patient autonomy.

Consent or shared decision-making

The final stage in the act of giving informed consent is the authorization for treatment. Whilst this remains the responsibility of patients, physicians have an important role to play as collaborators in the process of making a treatment decision.[10] During this process, physicians need not remain value-neutral. In the course of disclosing information, it is appropriate for them to express their opinion about the optimal course, to give their reasons for this and to share their own values with patients. Physicians and patients reach a joint decision that does not ignore the views of physicians but that must ultimately be guided by the values and beliefs of patients. Shared decision-making requires physicians to engage in a delicate balancing. They must act as advocate for their patients' health and well-being, while also being prepared ultimately to respect their patients' self-determination.

When is it ethically acceptable to stop treatment?

At times, anaesthetists face the question whether there is an obligation to extend human life for as long as possible. In some instances, they are asked to provide anaesthesia for patients when it is doubtful that surgery will be of benefit to them. This can occur when surgery, contemplated for terminally ill patients, may only perpetuate a brief period of life in which they will experience additional pain and suffering. When the path of patients toward death is irreversible, there is no ethical duty to attempt to prolong their lives. Indeed, in such circumstances, it can be argued that there is an obligation *not* to do so. Patients who are terminally ill and near death are owed the only kind of care that can be offered at this stage: palliation and comfort.[11] Kass stated that 'For physicians to adhere to efforts at indefinite prolongation not only reduces them to slavish technicians without any intelligible goal, but also degrades and assaults the gravity and solemnity of a life in its close.'[12] Physicians who withhold treatment from such patients do not cause their death. They only affect the time of death. The death of these terminally ill patients is brought about by an underlying disease or lesion that cannot be reversed.

There are other situations in which patients are not terminally ill, but might survive for some period of time in poor condition. Are physicians required to make every attempt possible to prolong their lives? Medical treatment is not obligatory if its burdens outweigh its benefits to patients.[13] Life is a primary good and a peculiarly fundamental value, as it is the necessary condition for the realization of all other values. Yet life is not an absolute value that takes precedence over all other values. Life need not be sustained regardless of the burden to the patient. McCormick observed that '. . . life is not a value to be preserved in and for itself. . . . It is a value to be preserved

precisely as a condition for other values, and therefore insofar as these other values remain attainable.'[14] He went on to maintain that life is a value to be preserved only insofar as it contains some potentiality for human relationships. 'When in human judgment this potentiality is totally absent or would be, because of the condition of the individual, totally subordinated to the mere effort for survival, that life can be said to have achieved its potential.'[14] Respect for human life, for McCormick and others in the Judaeo–Christian tradition, does not require that life must be extended for as long as possible. The question that must be asked in such instances is whether patients will be able to relate to others after treatment.

Some object to the introduction of 'quality of life' considerations into decisions regarding patient care. They contend that these are akin to judgements made by the Nazis that some patients should not be kept alive because they are a burden to society.[15] Those who argue against the use of 'quality of life' considerations maintain that these require invidious comparisons between the worth of equally valuable human beings and that such comparisons have no place in medicine.

This is a serious concern, but it is based on a misunderstanding of the concept of 'quality of life'. The 'quality' of a patient's life refers to whether that life has certain properties related to his or her well-being. This must be considered in terms of that patient's own values and beliefs, not in terms of the interests of society or others.[16] The social worth of a patient in a certain condition is not relevant to this sort of decision. 'Quality of life' considerations require reflection about whether the good of a valuable human being can be achieved if treatment will leave him or her in a certain condition. They focus on the best interests of the patient as these are defined in terms of his or her own values and goals. As McCormick stated, 'It is neither inhuman nor unchristian to say that there comes a point where an individual's condition itself represents the negation of any truly human – i.e. relational – potential. When that point is reached is not the best treatment no treatment?'[14] In certain instances, the lives of patients have reached a point at which they bear no reasonable hope that their minimal well-being, as this is defined in terms of the patient's own values, can be achieved. In such instances, it is ethically sound not to treat.

The competent adult patient has the right to refuse treatment on 'quality of life' grounds, even if this leads to that patient's death. When a patient refuses aggressive treatment, however, this does not necessarily constitute a form of suicide. Suicide requires having the specific intent to die and deliberately setting in motion the agent that will cause death.[17] These conditions are not usually present in situations in which patients refuse further treatment.

Incompetent patients also have the right to refuse treatment, but this right must be exercised for them by a surrogate.[18,19] Some patients, when competent, may have developed an advance directive, a document that has legal force in the USA, indicating whom they wish to make treatment decisions for them should they become decisionally incapacitated. Others who have not done so or who have never been competent may have a family member or close friend who can act as surrogate for them. In some jurisdictions in the USA, families can serve as surrogates without court involvement for patients who have not executed advance directives. In others, however, it may be necessary to seek a court-appointed surrogate, especially if the situation is controversial or there are clear conflicts of interest.[20]

The responsibility of surrogates is to make a 'substituted judgement' for the patient.[18] That is, surrogates are to make the decision that the patient would have made had he or she been able to choose. They are to interpret what the patient would have wanted done, based on their knowledge of the patient's values, beliefs and goals. If this is not possible, surrogates are to make the decision on the basis of what would promote the patient's 'best interests'.[18] They are to consider what would be most conducive to the patient's welfare, since it is not possible to rest the decision on the value of self-determination.

Orders to withhold resuscitation in the perioperative setting

In the last two decades, orders to withhold resuscitation, 'Do not resuscitate' (DNR) orders, have been introduced into many institutional health care settings in the USA owing to ethical concerns about the inappropriate application of cardiopulmonary resuscitation (CPR). DNR orders are entered into the charts of patients because they and their physicians believe that CPR would be useless or disproportionately burdensome to them.[21,22] In Britain, in 1991 the Government's Chief Medical Officer sent a request to all hospital consultants, asking them to ensure that their policy on resuscitation and its withholding, whether written or not, is understood by all staff.[23] An editorial in the *Journal of the Royal College of Physicians of London* in 1992 maintained that it is 'indefensibly paternalistic' not to have written policy statements about the use of resuscitation.[24] Major ethical difficulties for anaesthetists have arisen in the USA, however, when patients with DNR orders written according to hospital policies have entered the operating rooms.

Many interventions that are considered emergency resuscitative interventions elsewhere in the hospital are routinely performed in the operating room setting. Resuscitative measures instituted on the wards or in the intensive care unit range from basic CPR and tracheal intubation to advanced cardiac life support with

defibrillation, positive pressure respiration and administration of vasoactive drugs. Most of these measures, however, are part of ordinary anaesthetic practice in the operating theatre to maintain patient stability and reverse the effects of the anaesthetic. If arrest were to occur in this setting, the only new measures to be introduced would be cardiac massage and countershock. Anaesthetists, consequently, are faced with the problem that a decision made in other health care settings to withhold resuscitation from certain patients could be medically and ethically inappropriate to carry out in the operating room setting.

In response to this difficulty, some hospitals in the USA have adopted a policy of automatically suspending DNR orders during anaesthesia and surgery.[25] Such policies, however, do not allow patients and physicians to weigh the distinctive benefits and burdens of resuscitation in the perioperative setting and then to decide whether or not to withhold resuscitation if the patient suffers cardiac arrest during surgery. An individualized policy of required reconsideration of DNR orders seems preferable.[26]

An initial issue to consider is whether it is appropriate to perform surgery on patients with DNR orders. There is considerable agreement that the entry of a DNR order does not require that all other interventions besides CPR must also be withheld or withdrawn.[18,21,27] Other interventions may have the potential for curing or ameliorating certain conditions in the patient, even though resuscitation does not. This means that surgery should not be denied to patients with DNR orders *solely* because they have such orders. Instead, the decision about surgery should be based on whether it will offer patients a reasonable probability of meeting their immediate and overall treatment goals.

When surgery on patients with DNR orders is warranted, plans must be made for the course of action to be taken if they suffer cardiac arrest during surgery or immediately after it. A distinction must be made between arrest related to the provision of anaesthesia and arrest related to the underlying disease process. When sudden cardiac arrest in the perioperative setting is in all probability due to the effects of the anaesthetic, it is appropriate to attempt to resuscitate affected patients and go on to meet the goals for which they had entered the operating room in the first place. To bring a patient to the operating room suite for a procedure in which there is a high probability that resuscitation will be needed because of the anaesthetic used, and to require the anaesthetist to withhold resuscitation when arrest occurs, is tantamount to requiring the anaesthetist to kill the patient.

However, cardiac arrest may also be due to the underlying disease. In that case, it can be ethically sound to continue the DNR order, as this is the situation to which the order was meant to apply when it was originally instituted. Although it is not always possible to determine the cause of an arrest, it is usually possible to predict whether resuscitation will contribute to the patient's surgical and overall treatment goals, and this prediction should guide the approach used.[28]

When surgery is considered appropriate for patients with DNR orders, that order should be discussed and re-evaluated by all involved physicians and patients or surrogates. When possible, patients should be asked what they mean by resuscitation and what kind of treatment or postoperative status they want to avoid through the use of a DNR order. When a decision is taken to suspend a DNR order, patients or surrogates and physicians should discuss the circumstances under which it will be reinstated. A shared individualized decision should be reached that is in accord with the values of patients and of physicians. Once patients understand that resuscitation during the peri-operative period involves interventions identical to those used in routine anaesthetic management and that a perioperative cardiac arrest has a substantially different outcome from that of arrests occurring at other times,[22,26,28–31] most will choose to suspend the DNR order during the perioperative period.

Discussions with patients or surrogates should be described in patients' medical records, and a summary of the final resuscitation plan and its rationale should be entered. If a DNR order is to be retained during surgery, the resuscitative measures to be withheld should be indicated. The duration of the DNR order should be discussed before surgery so that agreement is reached in advance about when it is to be reinstated. If agreement cannot be reached about the many issues associated with the use of the DNR order in the perioperative setting, the preferences of the informed patient or surrogate must prevail. If serious disagreement remains, it is appropriate for a responsible physician to withdraw from the case after ensuring continuity of care. An institutional ethics committee may be of assistance in such situations.[32,33]

In summary, the purpose of applying DNR orders is to protect patients from useless or disproportionately burdensome treatment. CPR, however, can offer benefits and burdens in the perioperative setting that differ from those it offers elsewhere in the health care institution. Although there are strong reasons for suspending DNR orders during surgery in most cases, the question of resuscitation should be reconsidered with each patient on an individual basis once the case for surgery has been made. When surgery is appropriate for patients with DNR orders and there is a reasonable chance of achieving their treatment objectives, surgery should be advised with a recommendation that DNR orders be suspended perioperatively. A recommendation should be made to reinstate the DNR order if cardiac arrest occurs during surgery and it is apparent that the cause of the arrest is irreversible underlying disease or a complication that cannot be corrected.

Should anaesthetists practise euthanasia?

The power of medical technology to sustain life can be a source of fear to patients, rather than comfort. The spectre of minimal, even unconscious, existence is frightening to many, as is the possibility of experiencing great pain and suffering if they remain conscious as they die.[34] Continued life under the control of machines is viewed as perilous, and an easy death appears desirable. Some have begun to ask, if it is right to stop treatment for certain patients, why not end their lives directly by killing them?[35]

In a 1988 article in the *Journal of the American Medical Association*,[36] an anonymous gynaecology resident wrote that in the middle of the night, he or she injected morphine into a 20-year-old woman known as 'Debbie' who had metastatic ovarian cancer. Publication of this admission of mercy killing, the first of its kind in an American medical journal, created a huge outcry against the actions of the physician. Physicians maintained that morphine, properly used, could have kept Debbie alive and comfortable.[37] The article initiated an international debate over euthanasia within the medical profession.

The debate accelerated when it became known that a Michigan physician, Jack Kevorkian, was openly assisting individuals with such ailments as Alzheimer's disease, multiple sclerosis, amyotrophic lateral sclerosis (Lou Gehrig's disease) and emphysema with congestive heart disease who were not his patients to commit suicide. At the time of writing, Dr Kevorkian has provided his 'services' to 15 people since 1990.[38]

In 1991, another physician, Timothy Quill, wrote an article in the *New England Journal of Medicine* in which he related how he had assisted a young woman who had been his patient for eight years to commit suicide.[39] She had been diagnosed with acute leukaemia and had rejected chemotherapy that would have given her a one-in-four chance of continuing to live. After extended discussions with her, Dr Quill prescribed the barbiturates that he knew she would use to end her life. He was generally applauded within the medical profession for this action,[40] and a grand jury decided against indicting him.

Recently it was suggested that if euthanasia becomes accepted in our society, a medical specialty in euthanasia should be developed that is limited to certified physicians.[41] The author proposed that anaesthetists, with their background in pain management, would be best qualified to practise this specialty. If the anaesthetist and the patient agreed that continued life would mean only 'the pointless prolongation of agony', the anaesthetist would 'perform the procedure'. If this proposal were adopted, anaesthetists would play a central role as practitioners of euthanasia.

Euthanasia literally means 'good death'. It connotes intentionally taking the life of a person who requests this in order to prevent or end suffering. Euthanasia does not include withdrawing or withholding treatment that is useless or disproportionately burdensome. Nor does it include administering drugs to relieve the pain of a dying patient, knowing that this may hasten death. It involves directly killing a person. Is the provision of active euthanasia a 'specialty' that anaesthetists should embrace?

There are three main reasons that have been given for involving physicians in euthanasia. The first is that physicians have an obligation of beneficence to their patients.[42] They have a specific obligation to relieve their patients' pain and suffering, particularly when physicians have inadvertently brought this about by their treatment. When they cannot do so, they have a special duty to assist patients to die quickly with a minimum of pain.[43] Although killing is usually an injury, it is not an injury for these patients, for it relieves their discomfort quickly and is more acceptable than allowing them to progress toward death slowly and miserably. Physicians are well qualified by their medical training to end the lives of their patients. Moreover, they have a virtual monopoly on the prescription of the medication necessary for doing so. Therefore, they must be involved in providing drugs to patients who wish to be killed.

A second argument for physician involvement in euthanasia is, ironically, a preventive one. It maintains that if euthanasia were an accepted medical practice, patients would be less likely to commit suicide prematurely. They would know that a physician was available who would be required to end their lives at a time of their choosing. Moreover, most physicians would feel such pressure to avoid euthanasia because of their faithfulness to the Hippocratic tradition that they would provide good palliative care. This would make it unnecessary for patients to request euthanasia.

Finally, some argue that it is preferable to legalize euthanasia and have it monitored by the government and the medical profession than to rely on the potentially arbitrary decisions of prosecutors and juries in 'mercy killing' cases. We need socially recognized procedures to ensure that patients who request aid in dying do so without being coerced and only after they have considered other possibilities. Physicians must be involved in such procedures, as only they have the necessary qualifications to assess the medical and mental condition of patients who ask for euthanasia.

There are distinctive arguments against euthanasia by physicians. The major one is that physicians would break a professional prohibition that is over 2000 years old against killing patients if they were to provide euthanasia. The Hippocratic Oath states that 'If any shall ask of me a drug to produce death, I will not give it nor will I suggest such counsel.'[44] The goal of the medical profession is to heal and protect life, not to end it.[45] Even when their intentions are merciful, physicians, just because they are physicians, should not terminate the lives of their patients. Further, opponents argue, if physicians were to be allowed to kill their patients, this would undermine trust, an essential element of the physician–patient relationship. What patient would want to be treated by a physician who might kill him? Patients would doubt their physician's commitment to their welfare if these same physicians at times served as 'euthanizers'.

A second argument given by opponents of euthanasia by physicians is a variant of what is known as the 'slippery slope' argument.[46] It begins by asking who would receive euthanasia from physicians. All terminally ill patients? Those diagnosed with Alzheimer's disease? Babies with AIDS? Those with chronic depression? Any patient who asks for it? Although euthanasia might begin with the compassionate, voluntary killing of competent dying people by physicians, it would gradually encompass the elderly, mentally disabled and uninsured.[47] Where would physicians draw the line and say that beyond this point they will not practise euthanasia? Once physicians started to kill for humane reasons, they would begin down a slippery slope that would lead them to accept killing in cases where it would amount to abuse of the vulnerable. The practice of euthanasia, on this argument, should be banned, as it is subject to the risk of great abuse.

This position is sometimes buttressed by recent information from the Netherlands.[48] Although euthanasia is officially a crime in that country, there have been no prosecutions of physicians who have followed certain guidelines for euthanasia that were established in 1985 by the State Commission on Euthanasia.[49] These require that the patient must be competent; must request euthanasia voluntarily, repeatedly and currently; and must suffer from unbearable physical or mental pain with no prospect for relief. Two physicians must concur with the decision for euthanasia, which must be carried out by a physician.[50]

Those who make the 'slippery slope' argument point out that in the Netherlands, euthanasia has spread to encompass patients who do not meet these conditions. A nationwide study sponsored by the Dutch government found that, in addition to the 2318 patients killed by physicians in conformity with the guidelines, there were 1030 patients who were killed in ways that violated them.[51,52] Indeed, some patients from whom life-sustaining treatment was withdrawn who had not requested euthanasia were given lethal doses of drugs.[53] These patients were not included in the euthanasia statistics because this practice is not considered euthanasia in that country.

Moreover, in at least three of eight neonatal centres in the Netherlands, infants born with impairments were killed immediately after birth in violation of the guidelines.[53] These reports suggest to some that, once euthanasia is accepted for certain groups of people, its scope will broaden rapidly and it will be performed on those whom it was not originally meant to encompass. The social stakes in the legitimization of killing, they maintain, are too high to endorse euthanasia.

A third argument against euthanasia is that physicians should not eliminate pain by eliminating life. The driving force behind the argument for euthanasia is that some patients experience extreme pain as they die and that this can be averted by the practice of euthanasia. In the USA this may occur because their physicians are not adequately trained in pain management.[54] Killing by physicians would not need to be contemplated if physicians used available knowledge about the alleviation of pain to eliminate the suffering of the dying, rather than eliminate the dying.

The balance of the reasons presented, in my judgement, weighs against accepting the practice of active euthanasia by physicians. Compassion alone cannot justify killing. Although it is surely an ethical motive, it does not make any action that it governs ethical. Compassion for the poor inspired Robin Hood, but a society that wishes to function and survive as a community must repudiate the private redistribution of property by compassionate Robin Hoods.

Yet compassion need not be abandoned by physicians. The fear of pain is the major reason that patients offer for requesting euthanasia.[55] Some physicians are reluctant to use adequate medication to relieve pain because they are concerned about inducing respiratory depression and hastening death.[56] There is widespread agreement among religious and secular ethicists that to provide pain-relieving medication to the dying, even though this may accelerate the process of dying, is ethically acceptable, because it is not done to kill but to alleviate pain. Indeed, it has been found that some dying patients who receive adequate medication for pain live longer than they would have without such medication.

There is a small percentage of patients with terminal malignancy, however, in whom pain cannot be alleviated by optimal use of systemic analgesics and adjuvants without allowing intolerable side effects.[57] Truog maintains that good relief can be provided in such cases by invasive methods of analgesia, including epidural and subarachnoid administration of opioids and local anaesthetics, neurolytic blockade, and neurosurgical procedures including antero-lateral cordotomy. He observes, 'Anesthesiologists' methods of pain management can play an important role in giving these patients viable alternatives to euthanasia or assisted suicide.'

Pieter Admiraal, an anaesthetist who is a leading Dutch practitioner of euthanasia, believes that the justification for euthanasia is not the elimination of pain, but of a kind of suffering at the end of life that leads to a complete loss of human dignity.[58] This suffering can be experienced by patients who may be free from pain, but whose life deteriorates as they are overtaken by incontinence, dyspnoea, immobility, nausea and paralysis. The dying process is not always gentle, free from suffering and dignified.

The loss of dignity that concerns Admiraal, however, is not inevitable. The hospice movement has shown that there are ways of caring for terminally ill patients that will allow them to die without undergoing suffering and loss of dignity.[59] Death need not represent a threat to personhood; it need not signify a loss of self and of meaning. Instead, it can be a significant part of life that is free from emptiness. Pressure to allow active euthanasia arises largely because we do not provide appropriate care for chronically ill and disabled persons who are slowly dying.[60] Many need better pain control, emotional support, spiritual counselling and human companionship while they are dying. Instead of leaving those who are suffering so bereft of hope that they prefer being killed to living, we owe it to them to develop effective ways of providing them with care and support through the process of dying.

What should anaesthetists do about chemically dependent colleagues?

Chemical dependence is a progressive illness that, if untreated, can lead to disability and death. Its appearance in physicians creates a problem that is compounded by misconceptions about the disease, misunderstandings by peers and a conspiracy of silence.[61] The failure to detect chemical dependence in its early stages can lead to the destruction of medical careers and death from suicide or overdose.[62] Physicians, who are called by their profession to heal, have a responsibility to provide assistance, rather than punishment, for addicted fellow-physicians. The impaired physicians' movement provides an effective and humane model for treating problems of physician addiction. Responsible administrators should be prepared to contact experienced professionals from recognized impaired physicians' programmes to provide early intervention, therapy and a re-entry plan for chemically dependent colleagues.

The scope of drug addiction and alcoholism among physicians was first discussed in the 1950s when the US Commissioner of Narcotics reported an addiction rate of 1 in 100 doctors, as compared with 1 in 3000 for the general population.[63] This estimate was subsequently challenged.[64] However, several later studies have confirmed that from 7 to 14 per cent of practising physicians have been chemically dependent at some point in their lives.[65–69]

Before organized medicine acknowledged the extent of the problem and its implications for professional performance, the typical fate of severely addicted or alcoholic physicians was loss of hospital privileges and/or revocation of licence.[70] In 1973, however, the American Medical Association (AMA) officially recognized the problem of chemical dependence and fundamentally changed the way in which the profession addressed it. In its report, which took account of studies in the UK, the AMA recommended that impaired physicians be treated and rehabilitated, rather than censured morally and discharged from their positions.[71] This is the view that governs treatment of the chemically dependent physician today in the USA and the UK.[72]

The disease concept and anaesthetists

Chemical dependence is a primary illness that encompasses the use of such drugs as alcohol, cocaine, opioids and a variety of depressants.[73–75] It is characterized by compulsive use of chemicals for which the addict develops abnormal tolerance, increasing physical and psychological dependence, and severe discomfort upon withdrawal. There may be an underlying biochemical and/or neurophysiological mechanism which, in conjunction with the use of chemicals, produces the syndrome.[76] These findings have added to the growing body of data which support the hypothesis that chemical dependence is a disease with a genetic basis.[74,77,78]

Chemical dependence is a recognized hazard of training and practice in anaesthesia.[79–88] Retrospective surveys suggest that the prevalence of the disease among anaesthetists is between 1 and 2 per cent.[79–81] There is conflicting evidence about whether the rate of chemical dependence among anaesthetists is significantly higher than that of physicians in other specialties. Although anaesthetists comprise 4 per cent of the physicians in the USA, 12–14 per cent of physicians treated in three programmes were anaesthetists.[82–84] Other treatment programmes, however, report a lower percentage of anaesthetists among their patients.[85,86] A survey sponsored by the American Society of Anesthesiologists (ASA) indicates that anaesthetists who develop chemical dependence tend to be in the top half of their medical school class; 28 per cent of them are members of Alpha Omega Alpha, the honorary medical school society.[89] One anaesthetist reported that he was president of his country's society of anaesthetists and chief of staff at his hospital, and was using nearly lethal doses of drugs.[90]

Diagnosis and intervention

Detection and diagnosis of chemical dependence among physicians is difficult.[91] The illness is often gradual in onset and is not readily recognized until its later stages when work performance is significantly impaired.[61,90,92] This is compounded by the fact that denial is a primary symptom of the disease.[89,92] In addition, physicians tend to view themselves as invulnerable healers who do not get sick. They generally avoid and resist treatment and believe that they can treat themselves successfully alone.[93] Colleagues may note that the chemically dependent physician has frequent unexplained illnesses or extended working hours, has withdrawn from family, friends and outside interests, and has undergone personality changes.[61,90,92] The Chemical Dependence Guidelines for Departments of Anesthesiology published by the ASA recognize the difficulties in developing a definitive diagnosis. They indicate that, 'without the advice of an expert on chemical dependence from outside the department, [the Chief of Anesthesia] . . . is unlikely to have the qualifications to make this judgment. Lacking previous experience, the Chief of Anesthesia may be able to obtain advice from the state medical society about recognized consultants on chemical dependence.'[75]

Intervention is a method of interrupting the symptoms of addiction. It involves a structured meeting in which the individual believed to be chemically dependent is presented with evidence that he or she has a disease and needs treatment. When practised appropriately, intervention is approached as advocacy rather than as punishment. The objectives are to present evidence that those involved are ill rather than evil, that they are not alone in their illness, that recovery is possible, and that sobriety will relieve the craving and pain of the disease.[94] Experts in the field of addiction advise that colleagues should not confront affected physicians direct, but should seek the help of specialists in the treatment of addictions with records of success.[90] These professionals have had considerable experience in managing denial, a major component of the disease. Physicians who initially decline treatment must be informed that, if the diagnosis is confirmed by qualified specialists and they do not begin treatment in a recognized programme, they risk losing their licence to practise. If it is determined that no addiction exists, the case is closed without prejudice to the involved physician.

The Association of Anaesthetists of Great Britain and Ireland (AAGBI) 'Sick Doctor' scheme provides, through a voluntary system, an opportunity for the doctor to seek independent, confidential advice and treatment.

Treatment and return to work

Many who specialize in addiction medicine recommend that physicians be treated at a facility experienced in the care of physicians.[80,81,83] An ASA-sponsored survey indicates that the vast majority of anaesthetists were treated in comprehensive, multidisciplinary programmes geared to physicians rather than in one-on-one treatment;[89] most received therapy in both inpatient and outpatient programmes that lasted for several months. During treatment, affected physicians are assisted in developing a strong relationship with peer support groups, such as Alcoholics Anonymous. In the last stage of treatment, they may become involved in evaluating new patients. The family is often provided treatment to complement that provided for the physician.

In the USA, committees affiliated with medical societies in various states have developed programmes to provide education, investigation, intervention, referral for treatment, return to practice and aftercare monitoring for the impaired physician in all 50 states.[79,81,83] They prepare an 'aftercare contract' with affected physicians that usually requires them to be followed by a monitoring physician, attend group meetings, submit to random drug screening tests, and follow a prescribed procedure in the event of a relapse. Physicians who refuse the assistance of the state medical society's committee will have their case referred to their hospital and the state's physician licensing agency. In the UK, physicians who do not voluntarily accept examination and treatment for addiction are referred to the Health Committee of the General Medical Council, the statutory body that regulates the medical profession.[95]

Once involved in treatment, physicians tend to have a better prognosis than do members of the general population.[96] Reports from recognized treatment programmes indicate that most recovering physicians are able to return to the practice of their calling.[83,86,92,96–98] The return to work can be a difficult step for physicians in recovery. Arnold notes that the way can be eased by compassionate colleagues who are willing to provide emotional support to recovering physicians.[92] He recommends that they re-enter the practice of anaesthesia in stages. They should return to meetings and journal discussions with colleagues, start observing in the operating rooms, begin to assist in the operating rooms and gradually return to practice. On-going education of the rest of the department is essential to destigmatize the returning colleague and to avoid reinforcing denial by colleagues.[90]

Legal issues

Institutional peer review committees in the USA that conduct reviews in good faith are immune from liability for damages under most federal laws.[99,100] This special legal protection creates a special responsibility on the part of these committees to develop an understanding of the nature of chemical dependency as a treatable disease. Otherwise, they risk a return to an earlier era in which valuable health care professionals were lost to their profession out of ignorance and misplaced self-protection. Administration and risk management sometimes view recovering physicians as a legal risk and urge peer review committees not to allow them to return to work. However, there is no evidence that there is a relation between addiction and the likelihood of being sued for malpractice.[101,102] It is neither ethical nor legal to deny employment to a recovering individual with a history of chemical dependency. The Americans with Disabilities Act (ADA), which took effect in 1992, includes alcoholism and previous, but not current, drug addiction as covered disabilities. This law requires an employer to demonstrate that a recovering anaesthetist poses a significant risk of substantial harm to the health or safety of others and that this risk cannot be eliminated or reduced by reasonable accommodation before re-employment can be denied.[103]

Many states in the USA have developed legislation that protects the licence and confidentiality of impaired physicians as long as they participate in a prescribed rehabilitation programme. However, confidentiality is not guaranteed in all states.[91] Concern has been raised about the adequacy of the safeguards to ensure confidentiality of information lodged in the National Practitioner Data Bank in the USA, which collects notifications of adverse actions taken against impaired practitioners.[104] In addition, the usual confidentiality afforded to communications between physician and patients often does not apply to communications between the impaired physician and the treating therapist.[91] As long as confidentiality cannot be assured, impaired physicians are likely to refrain from seeking assistance voluntarily.

Institutional responsibility

Appropriate departmental[105] and institutional[93] policies should be developed for the identification of affected physicians, intervention, referral for treatment and assistance with re-entry. Development of an effective programme preserves a valuable community resource and encourages physicians and their families to seek help, rather than remain silent for fear of punitive measures. Physicians, as members of a healing profession, have a special ethical obligation to assist their affected colleagues to receive adequate treatment and to return to their careers. One concerned anaesthetist maintains that 'The more we learn, the less likely we will shun addicted colleagues. Instead, we will treat them with the same compassion we offer to other patients. We will also be more willing to assist them in their return to the practice of medicine.'[90]

REFERENCES

1. Faden RR, Beauchamp TL. *A history and theory of informed consent.* New York: Oxford University Press, 1986.

2. Brock DW. The ideal of shared decision making between physicians and patients. *Kennedy Institute of Ethics Journal* 1991; **1**: 28–45.

3. Beauchamp TL, Childress JF. *Principles of biomedical ethics,* 3rd ed. New York: Oxford University Press, 1989: 67–119.

4. Meisel A, Roth L. What we do and do not know about informed consent. *Journal of the American Medical Association* 1981; **246**: 2473–7.

5. President's Commission for the Study of Ethical Problems in Medicine and Biomedical and Behavioral Research. *Making health care decisions,* vol. 1. Washington DC: US Government Printing Office, 1982: 55–62.

6. Buchanan A, Brock D. *Deciding for others: the ethics of surrogate decision making.* New York: Cambridge University Press, 1989: 18–23.

7. *Canterbury v Spence,* 464 F.2d 772, 785–87 (DC Cir. 1972).

8. Miller LJ. Informed consent: II. *Journal of the American Medical Association* 1980; **244**: 2348–9.

9. Eraker SE, Sox HC. Assessment of patients' preferences for therapeutic outcome. *Medical Decision Making* 1981; **1**: 29–39.

10. President's Commission for the Study of Ethical Problems in Medicine and Biomedical and Behavioral Research. *Deciding to forego life-sustaining treatment: a report on the ethical, medical and legal issues in treatment decisions.* Washington DC: US Government Printing Office, 1983: 43–90.

11. Ramsey P. *The patient as person.* New Haven: Yale University Press, 1970: 113–64.

12. Kass LR. Ethical dilemmas in the care of the ill. I. What is the patient's good? *Journal of the American Medical Association* 1980; **244**: 1946–9.

13. Beauchamp T, Childress J. *Principles of Biomedical Ethics,* 3rd ed. New York: Oxford University Press, 1989: 120–93.

14. McCormick R. To save or let die: the dilemma of modern medicine. *Journal of the American Medical Association* 1974; **229**: 175–6.

15. Cohen CB. The Nazi analogy in bioethics. *Hastings Center Report* 1988; **18**: 32–3.

16. Cohen CB. 'Quality of life' and the analogy with the Nazis. *Journal of Medicine and Philosophy* 1983; **8**: 113–35.

17. Byrn RM. Compulsory life-saving treatment for the competent adult. *Fordham Law Review* 1975; **44**: 16–24.

18. President's Commission for the Study of Ethical Problems in Medicine and Biomedical and Behavioral Research. *Deciding to forego life-sustaining treatment: a report on the ethical, medical and legal issues in treatment decisions.* Washington DC: Government Printing Office, 1983: 121–70.

19. British Medical Association. Statement on advance directives, May 1992. Available from the Medical Ethics Committee Secretariat, Professional and Scientific Division, BMA House, Tavistock Square, London WC1H 9JP.

20. MacDonald M, Meyer K, Essig B. *Health care law.* New York: Matthew Bender, 1991.

21. Hastings Center. *Guidelines on the termination of life-sustaining treatment and care of the dying.* Briarcliff Manor, NY: Hastings Center, 1987.

22. Cohen CB, Cohen PJ. Required consideration of 'Do-not-resuscitate' orders in the operating room and certain other treatment settings. *Law, Medicine and Health Care* 1992; **20**: 354–63.

23. Letter from the Chief Medical Officer to all consultants on resuscitation policy, dated 20.12.1991. Department of Health reference PL/CMO (91)22.

24. Saunders J. Who's for CPR? *Journal of the Royal College of Physicians of London* 1992; **26**: 254–7.

25. Truog RD. 'Do-not-resuscitate' orders during anesthesia and surgery. *Anesthesiology* 1991; **74**: 606–8.

26. Cohen CB, Cohen PJ. Do-not-resuscitate orders in the operating room. *New England Journal of Medicine* 1991; **325**: 1879–82.

27. Youngner S. Decisions not to resuscitate. In: Youngner SJ, ed. *Human values in critical care medicine.* New York: Praeger, 1986: 15–34.

28. Hovi-Viander M. Death associated with anaesthesia in Finland. *British Journal of Anaesthesia* 1980; **52**: 483–9.

29. Keenan RL, Boyan CP. Cardiac arrest due to anesthesia: a study of incidence and causes. *Journal of the American Medical Association* 1985; **253**: 2373–7.

30. Olsson GL, Hallen B. Cardiac arrest during anaesthesia: a computer-aided study in 250,543 anaesthetics. *Acta Anaesthesiologica Scandinavica* 1988; **32**: 653–64.

31. Mackey DC, Carpenter RL, Thompson GE, Brown DL, Bodily MN. Bradycardia and asystole during spinal anesthesia: a report of three cases without morbidity. *Anesthesiology* 1989; **70**: 866–8.

32. Cohen C. Interdisciplinary consultation on the care of the critically ill and dying: the role of one hospital ethics committee. *Critical Care Medicine* 1982; **10**: 776–84.

33. Cohen C. Avoiding 'Cloudcuckooland' in ethics committee case review: matching models to issues and concerns. *Law, Medicine and Health Care* 1992; **20**: 294–9.

34. Wanzer SH, Adelstein SJ, Cranford RE, *et al.* The physician's responsibility toward hopelessly ill patients. *New England Journal of Medicine* 1984; **310**: 955–9.

35. Blendon RJ, Szalay US, Knox RA. Should physicians aid their patients in dying? The public perspective. *Journal of the American Medical Association* 1992; **267**: 2658–62.

36. Anonymous. It's over, Debbie. *Journal of the American Medical Association* 1988; **259**: 272.

37. *Time* 15 February, 1988, p. 88.

38. *Newsweek* 8 March, 1993, pp. 46–9.

39. Quill TE. Death and dignity: case of individualized decision making. *New England Journal of Medicine* 1991; **324**: 691–4.

40. Cassel CK, Meir DE. Morals and moralism in the debate over euthanasia and assisted suicide. *New England Journal of Medicine* 1990; **323**: 750–2.

41. Benrubi GI. Euthanasia – the need for procedural safeguards. *New England Journal of Medicine* 1992; **326**: 197–9.

42. Kohl M. *Beneficent euthanasia.* Buffalo: Prometheus, 1975.

43. Fletcher J. *Morals and medicine.* Boston: Beacon, 1954.

44. Reiser SJ, Dyck A, Curran WJ, eds. *Ethics in medicine.*

Historical perspectives and contemporary concerns. Cambridge MA: MIT Press, 1977: 5.

45. Gaylin W, Kass LR, Pellegrino E, Siegler M. Doctors must not kill. *Journal of the American Medical Association* 1988; **259**: 2139–40.

46. Pellegrino ED. Compassion needs reason too. *Journal of the American Medical Association* 1993; **270**: 874–5.

47. Siegler M, Gomez CF. US consensus on euthanasia? *Lancet* 1992; **339**: 1164–5.

48. Capron AM. Euthanasia in the Netherlands. American observations. *Hastings Center Report* 1992; **22**: 30–3.

49. de Wachter MA. Active euthanasia in the Netherlands. *Journal of the American Medical Association* 1989; **262**: 3316–19.

50. Rigter H, Borst-Eilers E, Leenen HJJ. Euthanasia across the North Sea. *British Medical Journal* 1988; **297**: 1593–5.

51. Van der Maas PJ, Van Delden JJM, Pijnenborg L, Looman CWN. Euthanasia and other medical decisions concerning the end of life. *Lancet* 1991; **338**: 669–74.

52. Pijnenborg L, Van der Maas PJ, Van Delden JJM, Looman CWN. Life-terminating acts without explicit request of patient. *Lancet* 1993; **341**: 1196–9.

53. de Wachter MAM. Euthanasia in the Netherlands. *Hastings Center Report* 1992; **22**: 23–30.

54. Rhyme J. Hospice care in America. *Journal of the American Medical Association* 1990; **264**: 369–72.

55. Foley KM. The relationship of pain and symptom management to patient requests for physician-assisted suicide. *Journal of Pain and Symptom Management* 1991; **6**: 289–97.

56. Truog RD, Berde CB, Mitchell C, Grier HE. Barbiturates in the care of the terminally ill. *New England Journal of Medicine* 1992; **327**: 1678–81.

57. Truog RD, Berde CB. Pain, euthanasia, and anesthesiologists. *Anesthesiology* 1993; **78**: 353–60.

58. Parachini A. The Netherlands debates the legal limits of euthanasia. *Los Angeles Times* 5 July, 1987, part VI, pp. 1, 8–9.

59. Twycross RG. Assisted death: a reply. *Lancet* 1990; **336**: 796–8.

60. Lynn J. The health care professional's role when active euthanasia is sought. *Journal of Palliative Care* 1988; **4**: 100–2.

61. Summer GL. Intervention: how much evidence do I need? American Society of Anesthesiologists, Workshop on Substance Abuse Concerns for Anesthesiologists, 7–8 November, 1992.

62. Menk EJ, Baumgarten MC, Kingsley CP, Lulling RO, Middaugh R. Success of re-entry into anesthesiology training programs by residents with a history of substance abuse. *Journal of the American Medical Association* 1990; **263**: 3060–2.

63. Anslinger H, Chapman KW. Narcotic addiction: an interview. *Modern Medicine* 1957; **25**: 170–1.

64. Brewster JM. Prevalence of alcohol and other drug problems among physicians. *Journal of the American Medical Association* 1986; **255**: 1913–20.

65. Vaillant GI, Brighton JR, MacArthur C. Physicians' use of mood-altering drugs. *New England Journal of Medicine* 1970; **282**: 365–70.

66. Talbott CD, Benson R. Impaired physicians: the dilemma of identification. *Postgraduate Medicine* 1980; **68:** 56–63.

67. Moore RD, Mead L, Pearson TA. Youthful precursors of

alcohol abuse in physicians. *American Journal of Medicine* 1990; **88**: 332–6.

68. Regier DA, Farmer ME, Rae DS, *et al.* Comorbidity of mental disorders with alcohol and other drug abuse: results from the Epidemiologic Catchment Area study. *Journal of the American Medical Association* 1990; **264**: 2511–18.

69. Hughes PH, Brandenburg N, Baldwin DC, *et al.* Prevalence of substance use among US physicians. *Journal of the American Medical Association* 1992; **267**: 2333–9.

70. Morrow CK. Sick doctors: the social construction of professional deviance. *Social Problems* 1982; **30**: 92–108.

71. American Medical Association. The sick physician: impairment by psychiatric disorders, including alcoholism and drug dependence. Prepared by the Council on Mental Health of the American Medical Association. *Journal of the American Medical Association* 1973; **223**: 684–7.

72. Adshead F, Clare A. Doctors' double standards on alcohol. *British Medical Journal* 1986; **293**: 1590–1.

73. Jellinek EM. *The disease concept of alcoholism.* Highland Park NJ: Hellhouse Press, 1960.

74. Vaillant GE. *The natural history of alcoholism.* Cambridge MA: Harvard University Press, 1983.

75. *Chemical dependence guidelines for departments of anesthesiology.* Park Ridge IL: American Society of Anesthesiologists, 1991: 3.

76. Morse RM, Flavin DK, for the Joint Committee of the National Council on Alcoholism and Drug Dependence and the American Society of Addiction Medicine to Study the Definition and Criteria for the Diagnosis of Alcoholism. The definition of alcoholism. *Journal of the American Medical Association* 1992; **268**: 1012–14.

77. Cloninger CR. Neurogenetic adaptive mechanisms in alcoholism. *Science* 1987; **236**: 410–16.

78. Uhl GR, Persico AM, Smith SS. Current excitement with D2 dopamine receptor gene alleles in substance abuse. *Archives of General Psychiatry* 1992; **49**: 157–60.

79. Ward CF, Ward GC, Saidman LJ. Drug abuse in anesthesia training programs. *Journal of the American Medical Association* 1983; **250**: 922–5.

80. Gravenstein JS, Kory WP, Marks RG. Drug abuse by anesthesia personnel. *Anesthesia and Analgesia* 1983; **62**: 467–72.

81. Menk EJ, Baumgarten MC, Kingsley CP, Lulling RO, Middaugh R. Success of re-entry into anesthesiology training programs by residents with a history of substance abuse. *Journal of the American Medical Association* 1990; **263**: 3060–2.

82. Herrington RE, Benzer DG, Jacobson GR, Hawkins MK. Treating substance abuse disorders among physicians. *Journal of the American Medical Association* 1982; **247**: 2253–7.

83. Gualtieri AC, Cosentino JP, Becker JS. The California experience with a diversion program for impaired physicians. *Journal of the American Medical Association* 1982; **249**: 226–9.

84. Talbott GD, Gallegos KV, Wilson PO, Porter TL. The medical association of Georgia's impaired physicians program. *Journal of the American Medical Association* 1987; **257**: 2927–30.

85. McAuliffe WE, Rohman M, Santangelo S, *et al.* Psychoactive

drug use among practising physicians and medical students. *New England Journal of Medicine* 1986; **315**: 805–10.

86. Shore JH. The Oregon experience with impaired physicians on probation. An eight-year follow-up. *Journal of the American Medical Association* 1987; **257**: 2931–4.

87. Spiegelman WG, Saunders L, Mazze RI. Addiction and anesthesiologists. *Anesthesiology* 1984; **60**: 335–41.

88. Lutsky IV, Abram SE, Jacobson GR, Hopwood M, Kampine JP. Substance abuse by anesthesiology residents. *Academic Medicine* 1991; **66**: 164–6.

89. Arnold III WP. Substance abuse prospective study in progress. *American Society of Anesthesiologists Newsletter* 1992; **56**: 11–14.

90. Farley WJ, Arnold III WP. Chemical dependency in anesthesiology, a brochure to accompany videotape. American Society of Anesthesiologists, Workshop on Substance Abuse Concerns for Anesthesiologists, 7–8 November, 1992.

91. Walzer RS. Impaired physicians. An overview and update of the legal issues. *Journal of Legal Medicine* 1989; **11**: 131–98.

92. Arnold WP. Environmental safety including chemical dependency. In: Miller R, ed. *Anesthesia*, 3rd ed. New York: Churchill Livingstone, 1990: 2407–20.

93. Aach RD, Girard DE, Humphrey H, *et al.* Alcohol and other substance abuse and impairment among physicians in residency training. *Annals of Internal Medicine* 1992; **116**: 245–54.

94. Carroll JR, Rinella VJ. The impaired lawyer. In: Greene RM, ed. *The quality pursuit. Assuring standards in the practice of law.* Chicago: American Bar Association, 1989: 258–73.

95. General Medical Council of Great Britain. *Professional conduct and discipline: fitness to practise.* London: GMCGB, 1990: 13.

96. Morse RM, Martin MA, Swenson WM, Niven RG. Prognosis of physicians treated for alcoholism and drug dependence. *Journal of the American Medical Association* 1984; **251**: 743–8.

97. Vogtsberger KN. Treatment outcomes of substance-abusing physicians. *American Journal of Drug and Alcohol Abuse* 1984; **10**: 23–37.

98. Blevins JW, Bowers RNC, Gallegos KV, *et al.* Relapse and recovery: five to ten year followup study of chemically dependent physicians. *Maryland Medical Journal* 1992; **41**: 315–19.

99. Health Care Quality Improvement Act of 1986. Pub L No. 99–660, 100 Stat 3784; amended by the Public Health Service Amendments of 1987, Pub L No. 100–177, 101 Stat 986; further amended by amendments to the Public Health Service Amendments of 1987, Omnibus Budget Reconciliation Act of 1989, HR 3299, 101st Cong, 1st Sess.

100. Peer Review Immunity Task Group. *Immunity for peer review participants in hospitals: what is it? where does it come from? how do you protect it?* Chicago: American Academy of Hospital Attorneys of the American Hospital Association, 1989.

101. Robertson JJ, ed. The impaired physician: building well-being. *Proceedings of the fourth AMA conference on the impaired physician.* Chicago: American Medical Association, 1980.

102. Vaillant GE. Physician, cherish thyself. The hazards of self-prescribing. *Journal of the American Medical Association* 1992; **267**: 2373–4.

103. Semo JJ. Americans with disabilities act: shield or sword? *American Society of Anesthesiologists Newsletter* 1993; **57**: 10–12.

104. Mullan F, Politzer RM, Lewis CT, Bastacky S, Rodak J, Harmon RG. The National Practitioner Data Bank. Report from the first year. *Journal of the American Medical Association* 1992; **268**: 73–9.

105. Lecky JH, Aukberg SJ, Conahan III T, *et al.* A departmental policy addressing chemical substance abuse. *Anesthesiology* 1986; **65**: 414–17.

Index